Biological and Life Sciences

General Interest

Applied Calculus

FOR THE MANAGERIAL, LIFE, AND SOCIAL SCIENCES

S. T. Tan
Stonehill College

Petra Menz
Simon Fraser University

Dan Ashlock
University of Guelph

NELSON EDUCATION

Applied Calculus for the Managerial, Life, and Social Sciences, First Canadian Edition
by S. T. Tan, Petra Menz, and Dan Ashlock

Associate Vice President, Editorial Director:
Evelyn Veitch

Editor-in-Chief, Higher Education:
Anne Williams

Acquisitions Editor:
Shannon White

Marketing Manager:
Sean Chamberland

Developmental Editor:
My Editor Inc.

Photo Researcher/ Permissions Coordinator:
Indu Arora

Content Production Manager:
Imoinda Romain

Production Service:
Lachina Publishing Services

Copy Editor:
Shirley Corriveau

Proofreader:
Lachina Publishing Services

Indexer:
Lachina Publishing Services

Production Coordinator:
Ferial Suleman

Design Director:
Ken Phipps

Managing Designer:
Katherine Strain

Interior Design:
Lachina Publishing Services

Cover Design:
Dianna Little

Cover Image:
© Corel

Compositor:
Lachina Publishing Services

Printer:
RR Donnelley

Printed and bound in the United States of America
1 2 3 4 11 10 09 08

For more information contact Nelson Education Ltd., 1120 Birchmount Road, Toronto, Ontario, M1K 5G4. Or you can visit our Internet site at http://www.nelson.com

Additional Photo Credits: Page i: Thinkstock/Jupiter Images; Page iii: Dynamic Graphics/Jupiter Images; Page vii and page 901: Comstock/Jupiter Images. Group Discussion, Self-Check Exercises, Theorem, Chapter Summary, and Portfolio heads: Comstock/Jupiter Images; Exploring with Technology head: Ron Chapple/Jupiter Images.

Library and Archives Canada Cataloguing in Publication

Tan, Soo Tang

Applied calculus for the managerial, life, and social sciences / S.T. Tan, Petra Menz, Dan Ashlock. — 1st Canadian ed.

Includes index.
ISBN 978-0-17-644185-2

1. Calculus—Textbooks. I. Ashlock, Daniel II. Menz, Petra III. Title.

QA303.2.T35 2008
515 C2007-905196-0

ISBN-10: 0-17-644185-9
ISBN-13: 978-0-17-644185-2

CONTENTS

Chapter 5 Exponential and Logarithmic Functions

Chapter 6 Trigonometric Functions

Chapter 7 Applications of the Derivative

Chapter 8 Integration

Chapter 13 Probability and Calculus

Chapter 14 Taylor Polynomials and Infinite Series

PREFACE

Math is an integral part of our daily life. *Applied Calculus for the Managerial, Life, and Social Sciences, First Canadian Edition*, attempts to illustrate this point with its applied approach to mathematics. This text is appropriate for use in a two-semester or three-quarter introductory calculus course for students in the managerial, life, and social sciences. Our objective for this First Canadian Edition is twofold: (1) to write an applied text that motivates students and (2) to make the book a useful tool for instructors. We hope that with this present edition we have come closer to realizing our goal.

General Approach

Level Presentation Our approach is intuitive, and we state the results informally. However, we have taken special care to ensure that this approach does not compromise the mathematical content and accuracy.

Approach A problem-solving approach is stressed throughout the book. Numerous examples and applications are used to illustrate each new concept and result in order to help the students comprehend the material presented. An emphasis is placed on helping the students formulate, solve, and interpret the results of the problems involving applications. Very early on in the text, students are given practice in setting up word problems (Section 2.8) and developing modeling skills. As another example, when optimization problems are covered, the problems are presented in two sections. First students are asked to solve optimization problems in which the objective function to be optimized is given (Section 7.4) and then students are asked to solve problems where they have to formulate the optimization problems to be solved (Section 7.5).

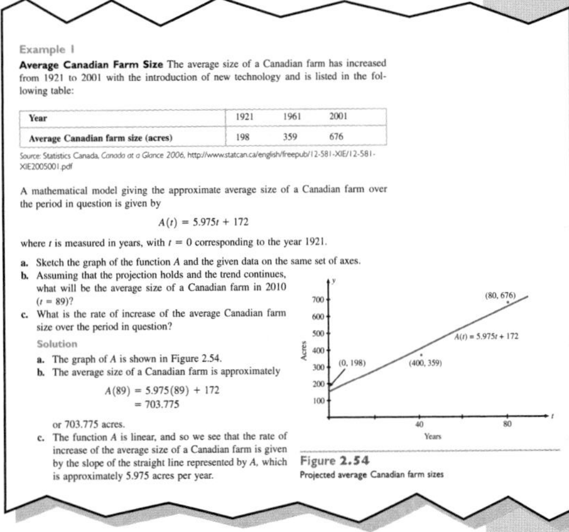

Example 1

Average Canadian Farm Size The average size of a Canadian farm has increased from 1921 to 2001 with the introduction of new technology and is listed in the following table:

Year	1921	1961	2001
Average Canadian farm size (acres)	198	359	676

Source: Statistics Canada, *Canada at a Glance 2006*, http://www.statcan.ca/english/freepub/12-581-XIE/12-581-XIE2005001.pdf

A mathematical model giving the approximate average size of a Canadian farm over the period in question is given by

$$A(t) = 5.975t + 172$$

where t is measured in years, with $t = 0$ corresponding to the year 1921.

a. Sketch the graph of the function A and the given data on the same set of axes.
b. Assuming that the projection holds and the trend continues, what will be the average size of a Canadian farm in 2010 ($t = 89$)?
c. What is the rate of increase of the average Canadian farm size over the period in question?

Solution

a. The graph of A is shown in Figure 2.54.
b. The average size of a Canadian farm is approximately

$$A(89) = 5.975(89) + 172 = 703.775$$

or 703.775 acres.

c. The function A is linear, and so we see that the rate of increase of the average size of a Canadian farm is given by the slope of the straight line represented by A, which is approximately 5.975 acres per year.

Figure 2.54
Projected average Canadian farm sizes

Intuitive Introduction to Concepts Mathematical concepts are introduced with concrete real-life examples, wherever appropriate. The goal is to capture students' interest and show the relevance of mathematics to their everyday lives. For example, differentiation (Section 5.3) is introduced with an analysis of the income distribution of Canadian families.

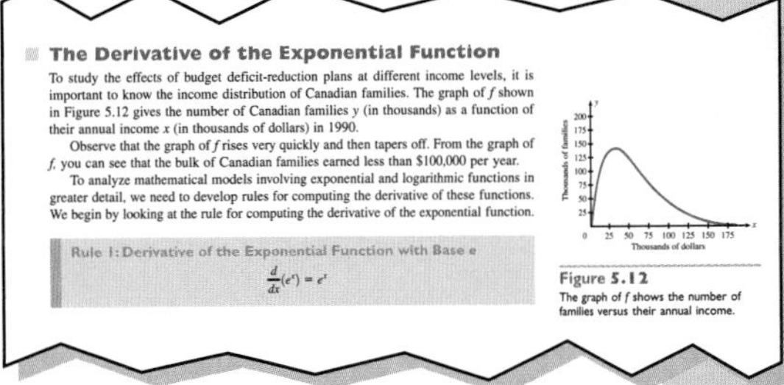

The Derivative of the Exponential Function

To study the effects of budget deficit-reduction plans at different income levels, it is important to know the income distribution of Canadian families. The graph of f shown in Figure 5.12 gives the number of Canadian families y (in thousands) as a function of their annual income x (in thousands of dollars) in 1990.

Observe that the graph of f rises very quickly and then tapers off. From the graph of f, you can see that the bulk of Canadian families earned less than $100,000 per year.

To analyze mathematical models involving exponential and logarithmic functions in greater detail, we need to develop rules for computing the derivative of these functions. We begin by looking at the rule for computing the derivative of the exponential function.

Rule 1: Derivative of the Exponential Function with Base e

$$\frac{d}{dx}(e^x) = e^x$$

Figure 5.12
The graph of f shows the number of families versus their annual income.

Motivation

Illustrating the practical value of mathematics in applied areas is an important objective of our approach. What follows are examples of how we have implemented this relevant approach throughout the text.

Example 5

Automobile Pollution An environmental impact study conducted for a certain city indicates that, under existing environmental protection laws, the level of carbon monoxide (CO) present in the air due to pollution from automobile exhaust will be $0.01x^{2/3}$ parts per million when the number of motor vehicles is x thousand. A separate study conducted by a governmental agency estimates that t years from now the number of motor vehicles in this city will be $0.2t^2 + 4t + 64$ thousand.

a. Find an expression for the concentration of CO in the air due to automobile exhaust t years from now.
b. What will be the level of concentration 5 years from now?

Solution

a. The level of CO present in the air due to pollution from automobile exhaust is described by the function $g(x) = 0.01x^{2/3}$, where x is the number (in thousands) of motor vehicles. But the number of motor vehicles x (in thousands) t years from now may be estimated by the rule $f(t) = 0.2t^2 + 4t + 64$. Therefore, the concentration of CO due to automobile exhaust t years from now is given by

$$C(t) = (g \circ f)(t) = g(f(t)) = 0.01(0.2t^2 + 4t + 64)^{2/3}$$

parts per million.

Real-life Applications Current and relevant examples and exercises are drawn from the fields of business, economics, social and behavioural sciences, life sciences, physical sciences, and other fields of interest.

Developing Modelling Skills We believe that one of the most important skills a student can learn is the ability to translate a real problem into a real model that can provide insight into the problem. Many of the applications are based on mathematical models (functions) that the author has constructed using data drawn from various sources, including current newspapers, magazines, and data obtained through the Internet Sources are given in the text for these applied problems. In Functions and Mathematical Models (Section 2.8), the modelling process is discussed and students are asked to use models (functions) constructed from real-life data to answer questions about the Average Canadian Farm Size and Email Usage.

81. **Bird Population** At Point Pelee National Park in the southernmost point in mainland Canada, an annual Christmas bird count is held. The total number of birds counted during the 1990s follows closely the line-of-best-fit described by the equation

$$y = 1981.4t + 25{,}928$$

where t is measured in years, with $t = 0$ corresponding to the beginning of 1990.

a. Sketch the line with the given equation.
b. What is the slope and the y-intercept of the line graphed in part (a)?
c. Give an interpretation of the slope and the y-intercept of the line found in part (a).
d. Suppose the number of birds counted continued to increase over the years. By what year can we expect to have 60,000 birds at Point Pelee National Park?

Source: Parks Canada, Point Pelee National Park

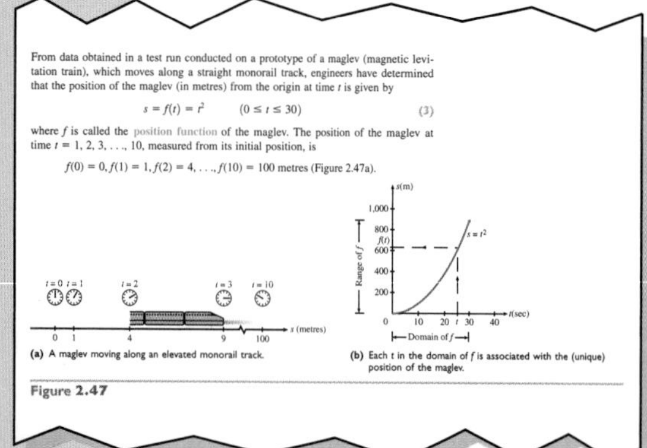
From data obtained in a test run conducted on a prototype of a maglev (magnetic levitation train), which moves along a straight monorail track, engineers have determined that the position of the maglev (in metres) from the origin at time t is given by

$$s = f(t) = t^2 \qquad (0 \le t \le 30) \qquad (3)$$

where f is called the position function of the maglev. The position of the maglev at time $t = 1, 2, 3, \ldots, 10$, measured from its initial position, is

$$f(0) = 0, f(1) = 1, f(2) = 4, \ldots, f(10) = 100 \text{ metres (Figure 2.47a).}$$

(a) A maglev moving along an elevated monorail track.

(b) Each t in the domain of f is associated with the (unique) position of the maglev.

Figure 2.47

Connections One example (the Maglev example) is used as a common thread throughout the development of calculus—from limits through integration. The goal here is to show students the connection between the concepts presented—limits, continuity, rates of change, the derivative, the definite integral, and so on.

Utilizing Tools Students Use

- **Technology** Technology is used to explore mathematical ideas and as a tool to solve problems throughout the text.
- **Exploring with Technology Questions** Here technology is used to explore mathematical concepts and to shed further light on examples in the text. These optional questions appear throughout the main body of the text and serve to enhance the student's understanding of the concepts and theory presented. Complete solutions to these exercises are given in the *Instructor's Solution Manual.*

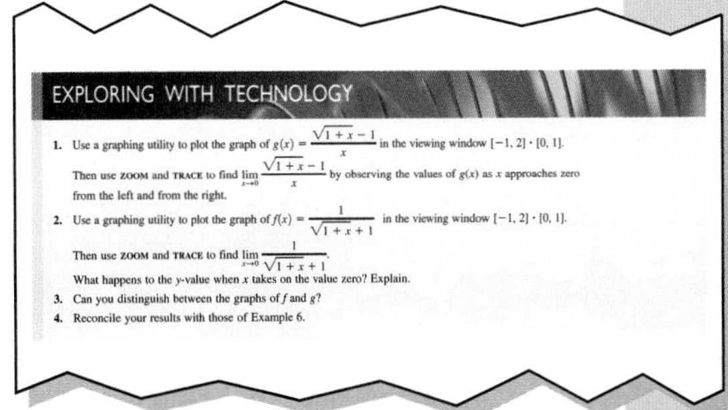

EXPLORING WITH TECHNOLOGY

1. Use a graphing utility to plot the graph of $g(x) = \frac{\sqrt{1+x}-1}{x}$ in the viewing window $[-1, 2] \cdot [0, 1]$.
   Then use ZOOM and TRACE to find $\lim_{x \to 0} \frac{\sqrt{1+x}-1}{x}$ by observing the values of $g(x)$ as x approaches zero from the left and from the right.
2. Use a graphing utility to plot the graph of $f(x) = \frac{1}{\sqrt{1+x}+1}$ in the viewing window $[-1, 2] \cdot [0, 1]$.
   Then use ZOOM and TRACE to find $\lim_{x \to 0} \frac{1}{\sqrt{1+x}+1}$.
   What happens to the y-value when x takes on the value zero? Explain.
3. Can you distinguish between the graphs of f and g?
4. Reconcile your results with those of Example 6.

Exercise Sets

The exercise sets are designed to help students understand and apply the concepts developed in each section. Two types of exercises are included in these sets.

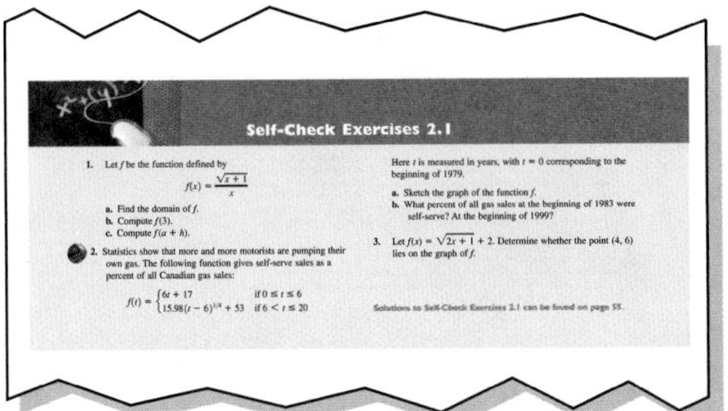

Self-Check Exercises 2.1

1. Let f be the function defined by
   $$f(x) = \frac{\sqrt{x+1}}{x}$$
   a. Find the domain of f.
   b. Compute $f(3)$.
   c. Compute $f(a+h)$.
2. Statistics show that more and more motorists are pumping their own gas. The following function gives self-serve sales as a percent of all Canadian gas sales:
   $$f(t) = \begin{cases} 6t + 17 & \text{if } 0 \le t \le 6 \\ 15.98(t-6)^{1/4} + 53 & \text{if } 6 < t \le 20 \end{cases}$$
   Here t is measured in years, with $t = 0$ corresponding to the beginning of 1979.
   a. Sketch the graph of the function f.
   b. What percent of all gas sales at the beginning of 1983 were self-serve? At the beginning of 1999?
3. Let $f(x) = \sqrt{2x+1} + 2$. Determine whether the point $(4, 6)$ lies on the graph of f.

Solutions to Self-Check Exercises 2.1 can be found on page 55.

- **Self-Check Exercises** offer students immediate feedback on key concepts with worked-out solutions following the section exercises.

- **Exercises** provide an ample set of problems of a routine computational nature that will help students master new techniques. The routine problems are followed by an extensive set of application-oriented problems that test students' mastery of the topics.

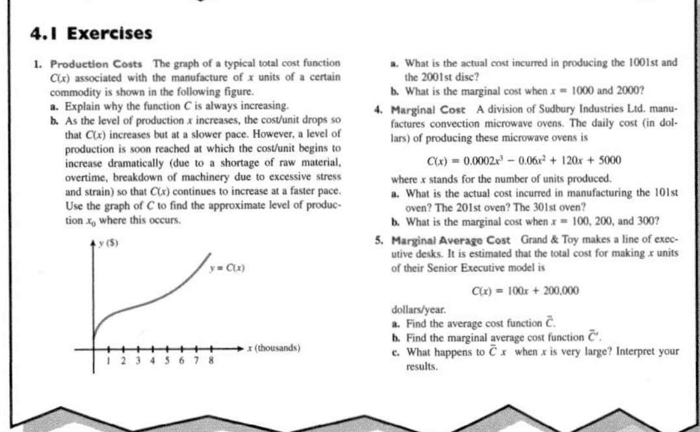

4.1 Exercises

1. **Production Costs** The graph of a typical total cost function $C(x)$ associated with the manufacture of x units of a certain commodity is shown in the following figure.
   a. Explain why the function C is always increasing.
   b. As the level of production x increases, the cost/unit drops so that $C(x)$ increases but at a slower pace. However, a level of production is soon reached at which the cost/unit begins to increase dramatically (due to a shortage of raw material, overtime, breakdown of machinery due to excessive stress and strain) so that $C(x)$ continues to increase at a faster pace. Use the graph of C to find the approximate level of production x_0 where this occurs.

   a. What is the actual cost incurred in producing the 1001st and the 2001st disc?
   b. What is the marginal cost when $x = 1000$ and 2000?
4. **Marginal Cost** A division of Sudbury Industries Ltd. manufactures convection microwave ovens. The daily cost (in dollars) of producing these microwave ovens is
   $$C(x) = 0.0002x^3 - 0.06x^2 + 120x + 5000$$
   where x stands for the number of units produced.
   a. What is the actual cost incurred in manufacturing the 101st oven? The 201st oven? The 301st oven?
   b. What is the marginal cost when $x = 100, 200$, and 300?
5. **Marginal Average Cost** Grand & Toy makes a line of executive desks. It is estimated that the total cost for making x units of their Senior Executive model is
   $$C(x) = 100x + 200{,}000$$
   dollars/year.
   a. Find the average cost function $\bar{C}$.
   b. Find the marginal average cost function $\bar{C}'$.
   c. What happens to $\bar{C}\ x$ when x is very large? Interpret your results.

Review Sections

These sections are designed to help students review the material in each section and assess their understanding of the basic concepts as well as problem-solving skills.

- **Summary of Principal Formulas and Terms** highlights important equations and terms with page numbers and terms with page numbers given for quick review.

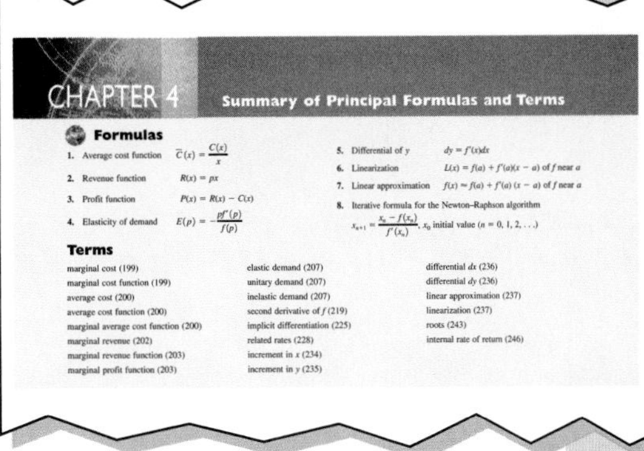

CHAPTER 4 Summary of Principal Formulas and Terms

Formulas

1. Average cost function $\bar{C}(x) = \frac{C(x)}{x}$
2. Revenue function $R(x) = px$
3. Profit function $P(x) = R(x) - C(x)$
4. Elasticity of demand $E(p) = -\frac{pf'(p)}{f(p)}$
5. Differential of y $dy = f'(x)dx$
6. Linearization $L(x) = f(a) + f'(a)(x-a)$ of f near a
7. Linear approximation $f(x) \approx f(a) + f'(a)(x-a)$ of f near a
8. Iterative formula for the Newton–Raphson algorithm
   $x_{n+1} = \frac{x_n - f(x_n)}{f'(x_n)}$, x_0 initial value ($n = 0, 1, 2, \ldots$)

Terms

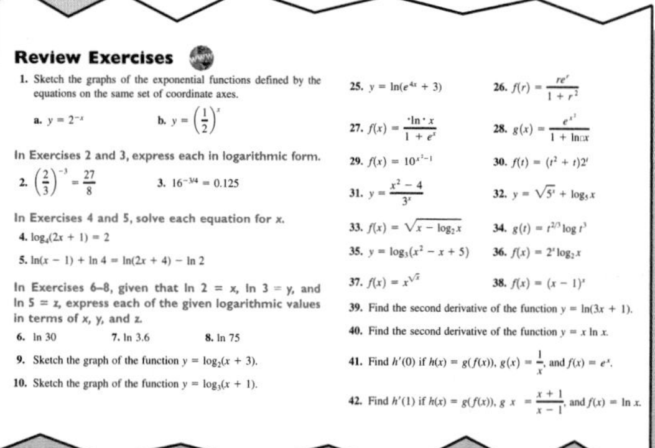

Review Exercises

1. Sketch the graphs of the exponential functions defined by the equations on the same set of coordinate axes.

   a. $y = 2^{-x}$ b. $y = \left(\frac{1}{2}\right)^x$

In Exercises 2 and 3, express each in logarithmic form.

2. $\left(\frac{2}{3}\right)^{-3} = \frac{27}{8}$ 3. $16^{-3/4} = 0.125$

In Exercises 4 and 5, solve each equation for x.

4. $\log_4(2x + 1) = 2$
5. $\ln(x - 1) + \ln 4 = \ln(2x + 4) - \ln 2$

In Exercises 6–8, given that $\ln 2 = x$, $\ln 3 = y$, and $\ln 5 = z$, express each of the given logarithmic values in terms of x, y, and z.

6. $\ln 30$ 7. $\ln 3.6$ 8. $\ln 75$

9. Sketch the graph of the function $y = \log_2(x + 3)$.
10. Sketch the graph of the function $y = \log_3(x + 1)$.

25. $y = \ln(e^{4x} + 3)$ 26. $f(r) = \frac{re^r}{1 + r^2}$
27. $f(x) = \frac{\cdot\ln\cdot x}{1 + e^x}$ 28. $g(x) = \frac{e^{x^2}}{1 + \ln x}$
29. $f(x) = 10^{x^2-1}$ 30. $f(t) = (t^2 + t)2^t$
31. $y = \frac{x^2 - 4}{3^x}$ 32. $y = \sqrt{5^x} + \log_5 x$
33. $f(x) = \sqrt{x - \log_2 x}$ 34. $g(t) = t^{2/3} \log t^3$
35. $y = \log_3(x^2 - x + 5)$ 36. $f(x) = 2^x \log_2 x$
37. $f(x) = x^{\sqrt{x}}$ 38. $f(x) = (x - 1)^x$
39. Find the second derivative of the function $y = \ln(3x + 1)$.
40. Find the second derivative of the function $y = x \ln x$.
41. Find $h'(0)$ if $h(x) = g(f(x))$, $g(x) = \frac{1}{x}$, and $f(x) = e^x$.
42. Find $h'(1)$ if $h(x) = g(f(x))$, $g\ x = \frac{x + 1}{x - 1}$, and $f(x) = \ln x$.

Review Exercises offer routine computational exercises followed by applied problems.

Portfolios The real-life experiences of a variety of professionals who use mathematics in the workplace are related in these interviews. Among those interviewed are a Product Manager who uses mathematical models and methods to determine and predict geostationary satellites' trajectories (Kerry Skipper at Telesat Canada) and a Director General who uses differential equations as well as linear and nonlinear programming in fuel cell research (Maja Veljkovic at National Research Council of Canada).

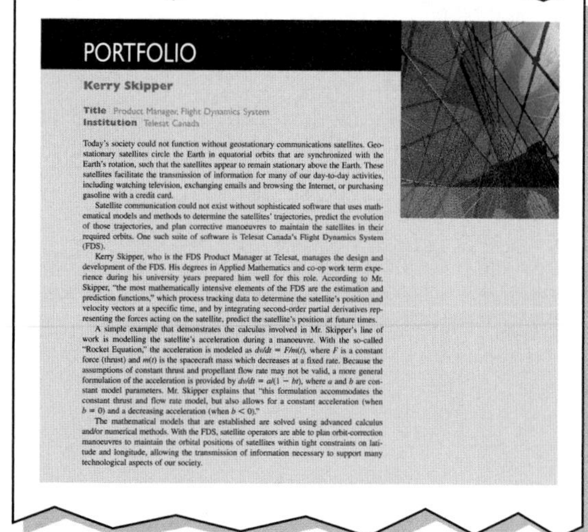

PORTFOLIO

Kerry Skipper

Title Product Manager, Flight Dynamics System
Institution Telesat Canada

Today's society could not function without geostationary communications satellites. Geostationary satellites circle the Earth in equatorial orbits that are synchronized with the Earth's rotation, such that the satellites appear to remain stationary above the Earth. These satellites facilitate the transmission of information for many of our day-to-day activities, including watching television, exchanging emails and browsing the Internet, or purchasing gasoline with a credit card.

Satellite communication could not exist without sophisticated software that uses mathematical models and methods to determine the satellites' trajectories, predict the evolution of those trajectories, and plan corrective manoeuvres to maintain the satellites in their required orbits. One such suite of software is Telesat Canada's Flight Dynamics System (FDS).

Kerry Skipper, who is the FDS Product Manager at Telesat, manages the design and development of the FDS. His degrees in Applied Mathematics and co-op work term experience during his university years prepared him well for this role. According to Mr. Skipper, "the most mathematically intensive elements of the FDS are the estimation and prediction functions," which process tracking data to determine the satellite's position and velocity vectors at a specific time, and by integrating second-order partial derivatives representing the forces acting on the satellite, predict the satellite's position at future times.

A simple example that demonstrates the calculus involved in Mr. Skipper's line of work is modelling the satellite's acceleration during a manoeuvre. With the so-called "Rocket Equation," the acceleration is modeled as $dv/dt = F/m(t)$, where F is a constant force (thrust) and $m(t)$ is the spacecraft mass which decreases at a fixed rate. Because the assumptions of constant thrust and propellant flow rate may not be valid, a more general formulation of the acceleration is provided by $dv/dt = a/(1 - bt)$, where a and b are constant model parameters. Mr. Skipper explains that "this formulation accommodates the constant thrust and flow rate model, but also allows for a constant acceleration (when $b = 0$) and a decreasing acceleration (when $b < 0$)."

The mathematical models that are established are solved using advanced calculus and/or numerical methods. With the FDS, satellite operators are able to plan orbit-correction manoeuvres to maintain the orbital positions of satellites within tight constraints on latitude and longitude, allowing the transmission of information necessary to support many technological aspects of our society.

GROUP DISCUSSION

Consider the graph of the function h whose graph is depicted in the following figure.

y
1
x
−1

It has the property that as x approaches zero from either the right or the left, the curve oscillates more and more frequently between the lines $y = -1$ and $y = 1$.

1. Explain why $\lim_{x \to 0} h(x)$ does not exist.
2. Compare this function with those in Example 3. More specifically, discuss the different ways the functions fail to have a limit at $x = 0$.

Group Discussion Questions These are optional questions, appearing throughout the main body of the text, that can be discussed in class or assigned as homework. These questions generally require more thought and effort than the usual exercises. They may also be used to add a writing component to the class. Complete solutions to these exercises are given in the *Instructor's Solutions Manual.*

New in the First Canadian Edition

- Examples, applications, and exercises are drawn from current Canadian statistical and financial information such as bird counts in Point Pelee National Park; average sizes of Canadian farms; Canadian email behaviour; provincial and territorial spending; revenue, deficit, and surplus; Canadian Air Travel Demand Elasticities; Guaranteed Investment Certificates; Canada Savings Bonds; and much more.
- All measurements, where relevant, are in metric units. Care has been taken to provide examples and applications that introduce new concepts predominantly in metric units.
- In Chapter 1, the concept of distance on the real number line has been added as an application of the absolute value function.
- In Chapter 2, extensive revisions have been made to enhance the mathematical explanations and descriptions of concepts. Section 2.2 is a new addition introducing transformation of functions visually and algebraically. The treatment of functions has been separated into classes of functions such as polynomial, rational, and power, and absolute value functions dealt with in Sections 2.3, 2.4, and 2.5, respectively. The inverse of a function was formerly handled as an appendix and has now been integrated into Chapter 2 as Section 2.7.
- The former Chapter 2, *Functions, Limits, and the Derivative,* and Chapter 3, *Differentiation,* have been split into three chapters dealing with functions first, then limits and the introduction of the derivative, and finally differentiation in general.
- In Chapter 3, the treatments of limits, continuity, and derivative have been grouped together. Continuity has been extended to introduce types of discontinuities along with visuals as well as more algebraically complex examples. Throughout this chapter, examples have been added that extend the level of difficulty of problem solving. The treatment of composite functions in Section 3.6 has been rewritten and a new approach taken that will enhance learning by the students.
- In Chapter 4, transformation of exponential and logarithmic functions has been added along with a general base b for exponential and logarithmic functions in terms of graphing and differentiation.
- The treatment of trigonometric functions along with revisions to derivatives has now been moved to appear earlier in the text in Chapter 6, immediately following the treatment of exponential and logarithmic functions.
- *Applications of Derivatives* in Chapter 7 has been moved to appear after the treatment of transcendental functions to include them as examples, applications, and exercises earlier on in the text, which extends this chapter dramatically.
- A new Chapter 12 has been added to introduce the mathematics of finance. Compound interest is no longer handled in the chapter on exponential functions, but rather as a sub-topic in financial mathematics. Special care has been taken to introduce decreasing annuity in general and then to explain how Canadian mortgages are calculated in a slightly different manner, which is a distinct feature of Canadian mathematics of finance.

Teaching Aids

- **Instructor's Resource CD** contains a complete manual of solutions to all exercises along with Microsoft PowerPoint slide presentations, the ExamView® Computerized Test Bank, and a printable test bank. ISBN 978-0-17-644193-7.
- **ExamView® Computerized Testing Suite** with algorithmic equations allows instructors to create, deliver, and customize tests and study guides (both print and online) in minutes with this easy-to-use assessment and tutorial system.

Learning Aids

- **Student Solutions Manual** contains complete solutions for all odd-numbered exercises in the text. ISBN 978-0-17-644177-7.
- **Student Resources on the Web** Students and instructors now have access to these additional materials at the Applied Calculus website: www.appliedcalculus.nelson.com.
- ***Succeeding in Applied Calculus: Algebra Essentials,*** by Warren Gordon, Baruch College—City College of New York, provides a clear and concise algebra review. This text is written so that students in need of an algebra refresher may have a convenient source for reference and review. This text may be especially useful before or while taking most college-level quantitative courses, including applied calculus or economics. ISBN 978-0-534-40122-1.

Acknowledgments

We wish to express our personal appreciation to each of the following reviewers, whose many suggestions have helped us revise the text for a Canadian audience:

Elena Devdariani
Carleton University
Brody Jozef
Concordia University
Matheus Grasselli
McMaster University
James Verner
Simon Fraser University
Howard Anderson
Trinity Western University
Peter Penner
University of Manitoba
Richard Blute
University of Ottawa
Steven Rosenfeld
Vanier College

We also wish to thank the My Editor Inc. team for their guidance and support during the development process of this Canadian edition. Our thanks and appreciation also extend to the Nelson staff, especially Imoinda Romain for the management of content throughout the production phase; Shirley Corriveau for ensuring the readability of the text; Jamie Mulholland for the technical check and ensuring the accuracy of the solutions; Lachina Publishing Services for the interior design revisions; and Dianna Little for the cover design. Simply put, the team we have been working with is outstanding, and we truly appreciate all their hard work and effort.

Petra Menz
Dan Ashlock

PRELIMINARIES

Photo: Thinkstock/Jupiter Images

The first two sections of this chapter contain a brief review of algebra. We then introduce the Cartesian coordinate system, which allows us to represent points in the plane in terms of ordered pairs of real numbers. This in turn enables us to compute the distance between two points algebraically. This chapter also covers straight lines. The slope of a straight line plays an important role in the study of calculus.

Photo: Thinkstock/Jupiter Images

What bird population can be predicted for next year? In Exercises 78 and 79 of Section 1.4, page 40, you will see how the line-of-best-fit can be used on bird counts from previous years to predict the bird count for next year.

Photo: © Don Johnston/Alamy

1.1 Precalculus Review I

Sections 1.1 and 1.2 review some of the basic concepts and techniques of algebra that are essential in the study of calculus. The material in this review will help you work through the examples and exercises in this book. You can read through this material now and do the exercises in areas where you feel a little "rusty," or you can review the material on an as-needed basis as you study the text. We begin our review with a discussion of real numbers.

The Real Number Line

The real number system is made up of the set of real numbers together with the usual operations of addition, subtraction, multiplication, and division.

Real numbers may be represented geometrically by points on a line. Such a line is called the **real number** or **coordinate line** and can be constructed as follows. Arbitrarily select a point on a straight line to represent the number 0. This point is called the **origin.** If the line is horizontal, then a point at a convenient distance to the right of the origin is chosen to represent the number 1. This determines the scale for the number line. Each positive real number lies at an appropriate distance to the right of the origin, and each negative real number lies at an appropriate distance to the left of the origin (Figure 1.1). The number 0 itself is neither positive nor negative.

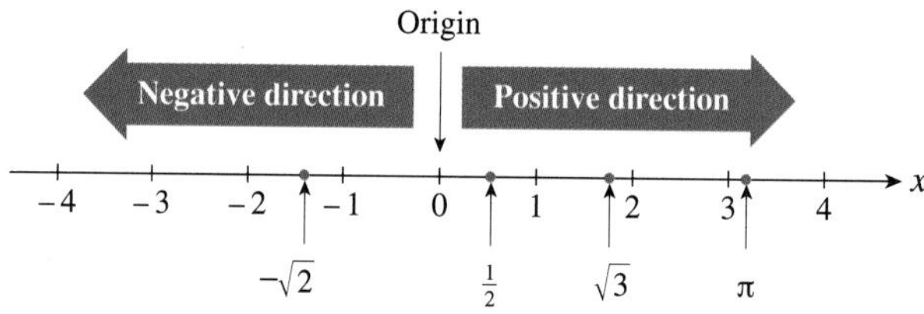

Figure 1.1
The real number line

A *one-to-one correspondence* is set up between the set of all real numbers and the set of points on the number line; that is, exactly one point on the line is associated with each real number. Conversely, exactly one real number is associated with each point on the line. The real number that is associated with a point on the real number line is called the **coordinate** of that point.

Intervals

Throughout this book, we will often restrict our attention to certain subsets of the set of real numbers. For example, if x denotes the number of cars rolling off a plant assembly line each day, then x must be nonnegative—that is, $x \geq 0$. Further, suppose management decides that the daily production must not exceed 200 cars. Then, x must satisfy the inequality $0 \le x \le 200$.

More generally, we will be interested in the following subsets of real numbers: open intervals, closed intervals, and half-open intervals. The set of all real numbers that lie *strictly* between two fixed numbers a and b is called an **open interval** (a, b). It consists of all real numbers x that satisfy the inequalities $a < x < b$, and it is called "open" because neither of its end points is included in the interval. A **closed interval** contains *both* of its end points. Thus, the set of all real numbers x that satisfy the inequalities $a \le x \le b$ is the closed interval $[a, b]$. Notice that square brackets are used to indicate that the end points are included in this interval. **Half-open intervals** contain only *one* of their end points. Thus, the interval $[a, b)$ is the set of all real numbers x that satisfy $a \le x < b$, whereas the interval $(a, b]$ is described by the inequalities $a < x \le b$. Examples of these **finite intervals** are illustrated in Table 1.1.

Table 1.1 Finite Intervals

Interval		Graph	Example	
Open	(a, b)	a b x	$(-2, 1)$	-3 -2 -1 0 1 2 3 x
Closed	$[a, b]$	a b x	$[-1, 2]$	-1 0 1 2 x
Half-open	$(a, b]$	a b x	$\left(\frac{1}{2}, 3\right]$	0 $\frac{1}{2}$ 1 2 3 x
Half-open	$[a, b)$	a b x	$\left[-\frac{1}{2}, 3\right)$	$-\frac{1}{2}$ 0 1 2 3 x

In addition to finite intervals, we will encounter **infinite intervals.** Examples of infinite intervals are the half lines (a, ∞), $[a, \infty)$, $(-\infty, a)$, and $(-\infty, a]$ defined by the set of all real numbers that satisfy $x > a$, $x \ge a$, $x < a$, and $x \le a$, respectively. The symbol ∞, called *infinity*, is not a real number. It is used here only for notational purposes in conjunction with the definition of infinite intervals. The notation $(-\infty, \infty)$ is used for the set of all real numbers x since, by definition, the inequalities $-\infty < x < \infty$ hold for any real number x. Infinite intervals are illustrated in Table 1.2.

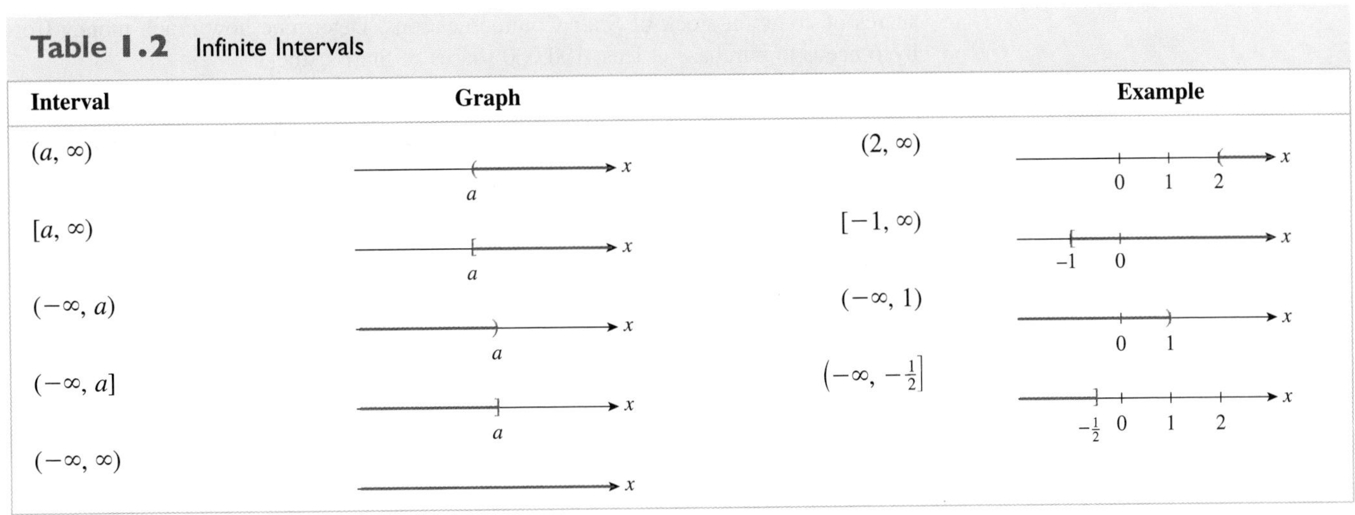

Table 1.2 Infinite Intervals

Interval	Graph	Example	
(a, ∞)	a x	$(2, \infty)$	0 1 2 x
$[a, \infty)$	a x	$[-1, \infty)$	-1 0 x
$(-\infty, a)$	a x	$(-\infty, 1)$	0 1 x
$(-\infty, a]$	a x	$\left(-\infty, -\frac{1}{2}\right]$	$-\frac{1}{2}$ 0 1 2 x
$(-\infty, \infty)$	x		

Properties of Inequalities

In practical applications, intervals are often found by solving one or more inequalities involving a variable. In such situations, the following properties may be used to advantage.

Properties of Inequalities

If a, b, and c are any real numbers, then

		Example
Property 1	If $a < b$ and $b < c$, then $a < c$.	$2 < 3$ and $3 < 8$, so $2 < 8$.
Property 2	If $a < b$, then $a + c < b + c$.	$-5 < -3$, so $-5 + 2 < -3 + 2$; that is, $-3 < -1$.
Property 3	If $a < b$ and $c > 0$, then $ac < bc$.	$-5 < -3$, and since $2 > 0$, we have $(-5)(2) < (-3)(2)$; that is, $-10 < -6$.
Property 4	If $a < b$ and $c < 0$, then $ac > bc$.	$-2 < 4$, and since $-3 < 0$, we have $(-2)(-3) > (4)(-3)$; that is, $6 > -12$.

Similar properties hold if each inequality sign, $<$, between a and b is replaced by $\geq$, $>$, or $\leq$.

A real number is a *solution of an inequality* involving a variable if a true statement is obtained when the variable is replaced by that number. The set of all real numbers satisfying the inequality is called the *solution set.*

Example 1

Find the set of real numbers that satisfy $-1 \leq 2x - 5 < 7$.

Solution

Add 5 to each member of the given double inequality, obtaining

$$4 \leq 2x < 12$$

Next, multiply each member of the resulting double inequality by $\frac{1}{2}$, yielding

$$2 \leq x < 6$$

Thus, the solution is the set of all values of x lying in the interval $[2, 6)$.

Example 2

Stock Purchase The management of Corbyco, a giant conglomerate, has estimated that x thousand dollars is needed to purchase

$$100{,}000(-1 + \sqrt{1 + 0.001x})$$

shares of common stock of Starr Communications. Determine how much money Corbyco needs to purchase at least 100,000 shares of Starr's stock.

Solution

The amount of cash Corbyco needs to purchase at least 100,000 shares is found by solving the inequality

$$100{,}000(-1 + \sqrt{1 + 0.001x}) \geq 100{,}000$$

Proceeding, we find

$$-1 + \sqrt{1 + 0.001x} \geq 1$$

$$\sqrt{1 + 0.001x} \geq 2$$

$$1 + 0.001x \geq 4 \quad \text{Square both sides.}$$

$$0.001x \geq 3$$

$$x \geq 3000$$

so Corbyco needs at least $3,000,000.

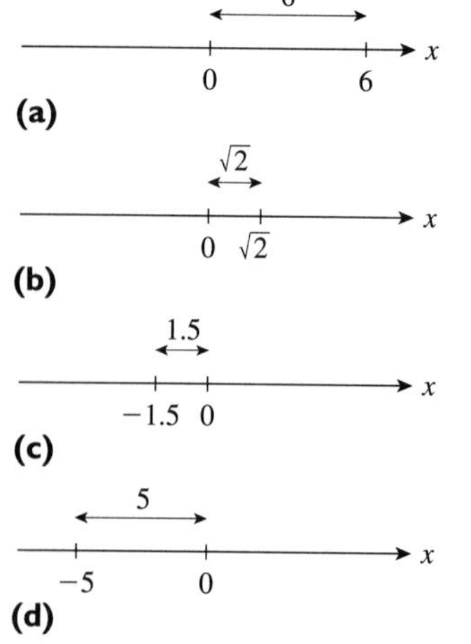

Figure 1.2
The distance of a number

Figure 1.2 depicts how this distance can be visualized. Notice that the answer to each question above is always the positive part of the number. We can now introduce the **absolute value,** which is a mathematical invention to find the distance of a real number a from 0.

Absolute Value

Another concept that can be visualized with the real number line is that of distance. Suppose we want to know how far away a real number a is from the origin. We calculate the distance between a and zero by subtracting the smaller number from the larger number. This process ensures that the distance is always positive.

Remark

The number zero is of course zero distance away from itself. We also want to remind you that the number zero does not have a sign, as it is neither positive nor negative. ◀

Example 3

Find the distance between 0 and the following numbers.

a. 6 **b.** $\sqrt{2}$ **c.** -5 **d.** -1.5

Solution

a. Since $6 > 0$, we subtract 0 from 6. Therefore, $6 - 0 = 6$. The distance between 6 and 0 is 6.

b. Since $\sqrt{2} > 0$, we subtract 0 from $\sqrt{2}$. Therefore, $\sqrt{2} - 0 = \sqrt{2}$. The distance between $\sqrt{2}$ and 0 is $\sqrt{2}$.

c. Since $0 > -5$, we subtract -5 from 0. Therefore, $0 - (-5) = 0 + 5 = 5$. The distance between -5 and 0 is 5.

d. Since $0 > -1.5$, we subtract -1.5 from 0. Therefore, $0 - (-1.5) = 0 + 1.5 = 1.5$. The distance between -1.5 and 0 is 1.5.

Absolute Value

The absolute value of a number a is denoted by $|a|$ and is defined by

$$|a| = \begin{cases} a & \text{if } a \geq 0 \\ -a & \text{if } a < 0 \end{cases}$$

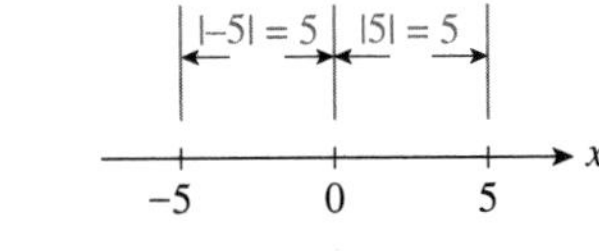

(a)

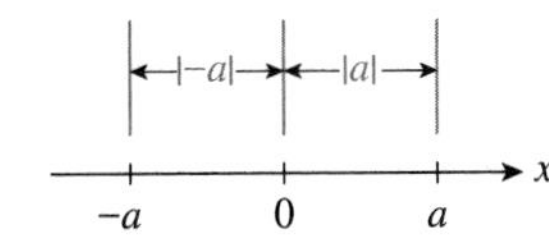

(b)

Figure 1.3
The absolute value of a number $a > 0$

Since $-a$ is a positive number when a is negative, it follows that the absolute value of a number is always nonnegative. For example, $|5| = 5$ and $|-5| = -(-5) = 5$. Geometrically, $|a|$ is the distance between the origin and the point on the number line that represents the number a, depicted for the case $a > 0$ (Figure 1.3).

Absolute Value Properties

If a and b are any real numbers, then

		Example
Property 5	$|-a| = |a|$	$|-3| = -(-3) = 3 = |3|$
Property 6	$|ab| = |a|\,|b|$	$|(2)(-3)| = |-6| = 6 = (2)(3) = |2|\,|-3|$
Property 7	$\left|\frac{a}{b}\right| = \frac{|a|}{|b|}$ $(b \neq 0)$	$\left|\frac{(-3)}{(-4)}\right| = \left|\frac{3}{4}\right| = \frac{3}{4} = \frac{|-3|}{|-4|}$
Property 8	$|a + b| \leq |a| + |b|$	$|8 + (-5)| = |3| = 3 \leq |8| + |-5| = 13$

Property 8 is called the triangle inequality.

Example 4

Evaluate each of the following expressions:

a. $|\pi - 5| + 3$ **b.** $|\sqrt{3} - 2| + |2 - \sqrt{3}|$

Solution

a. Since $\pi - 5 < 0$, we see that $|\pi - 5| = -(\pi - 5)$. Therefore, $|\pi - 5| + 3 = -(\pi - 5) + 3 = 8 - \pi$

b. Since $\sqrt{3} - 2 < 0$, we see that $|\sqrt{3} - 2| = -(\sqrt{3} - 2)$. Next, observe that $2 - \sqrt{3} > 0$, so $|2 - \sqrt{3}| = 2 - \sqrt{3}$. Therefore,

$$\begin{aligned} |\sqrt{3} - 2| + |2 - \sqrt{3}| &= -(\sqrt{3} - 2) + (2 - \sqrt{3}) \\ &= 4 - 2\sqrt{3} = 2(2 - \sqrt{3}) \end{aligned}$$

We can also find the distance between any two real numbers a and b on the number line. If $a \geq b$, then the distance between them is $a - b$, otherwise the distance between them is $b - a$. Since we could be in a situation where we do not know which of a and b is the larger number, we can simply express the distance using the absolute value, because $|a - b| = |b - a|$ as can be seen in Figure 1.4.

Distance Between Two Numbers

The distance between any two real numbers a and b is given by $|a - b|$.

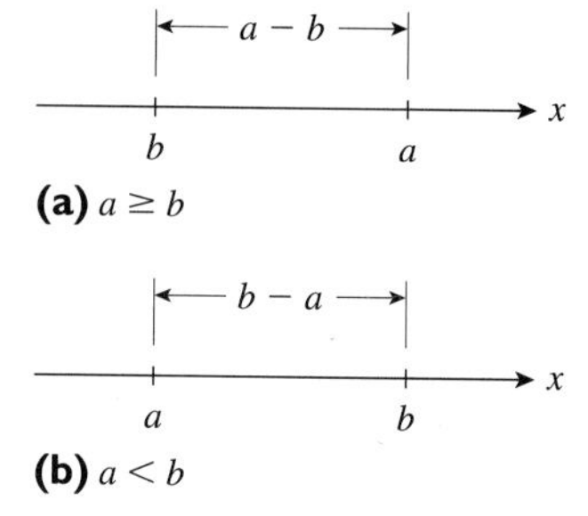

Figure 1.4

Exponents and Radicals

Recall that if b is any real number and n is a positive integer, then the expression b^n (read "b to the power n") is defined as the number

$$b^n = \underbrace{b \cdot b \cdot b \cdot \cdots \cdot b}_{n \text{ factors}}$$

The number b is called the base, and the superscript n is called the power of the exponential expression b^n. For example,

$$2^5 = 2 \cdot 2 \cdot 2 \cdot 2 \cdot 2 = 32 \quad \text{and} \quad \left(\frac{2}{3}\right)^3 = \left(\frac{2}{3}\right)\left(\frac{2}{3}\right)\left(\frac{2}{3}\right) = \frac{8}{27}$$

If $b \neq 0$, we define

$$b^0 = 1$$

For example, $2^0 = 1$ and $(-\pi)^0 = 1$, but the expression 0^0 is undefined.

Next, recall that if n is a positive integer, then the expression $b^{1/n}$ is defined to be the number that, when raised to the nth power, is equal to b. Thus,

$$(b^{1/n})^n = b$$

Such a number, if it exists, is called the nth root of b, also written $\sqrt[n]{b}$.

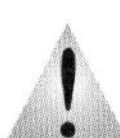

Observe that the nth root of a negative number is not defined when n is even. For example, the square root of -2 is not defined because there is no real number b such that $b^2 = -2$. Also, given a number b, more than one number might satisfy our definition of the nth root. For example, both 3 and -3 squared equal 9, and each is a square root of 9. So, to avoid ambiguity, we define $b^{1/n}$ to be the positive nth root of b whenever it exists. Thus, $\sqrt{9} = 9^{1/2} = 3$.

Next, recall that if p/q (p, q, positive integers with $q \neq 0$) is a rational number in lowest terms, then the expression $b^{p/q}$ is defined as the number $(b^{1/q})^p$ or, equivalently, $\sqrt[q]{b^p}$, whenever it exists. Expressions involving negative rational exponents are taken care of by the definition

$$b^{-p/q} = \frac{1}{b^{p/q}}$$

Table 1.3 Rules for Defining a^n

Definition of a^n $(a > 0)$	Example	Definition of a^n $(a > 0)$	Example
Integer exponent: If n is a positive integer, then $a^n = a \cdot a \cdot a \cdot \cdots \cdot a$ (n factors of a)	$2^5 = 2 \cdot 2 \cdot 2 \cdot 2 \cdot 2$ (5 factors) $= 32$	**Fractional exponent:** **a.** If n is a positive integer, then $a^{1/n}$ or $\sqrt[n]{a}$ denotes the nth root of a.	$16^{1/2} = \sqrt{16}$ $= 4$
Zero exponent: If n is equal to zero, then $a^0 = 1$ (0^0 is not defined.)	$7^0 = 1$	**b.** If m and n are positive integers, then $a^{m/n} = \sqrt[n]{a^m} = (\sqrt[n]{a})^m$	$8^{2/3} = (\sqrt[3]{8})^2$ $= 4$
Negative exponent: If n is a positive integer, then $a^{-n} = \frac{1}{a^n}$ $(a \neq 0)$	$6^{-2} = \frac{1}{6^2}$ $= \frac{1}{36}$	**c.** If m and n are positive integers, then $a^{-m/n} = \frac{1}{a^{m/n}}$ $(a \neq 0)$	$9^{-3/2} = \frac{1}{9^{3/2}}$ $= \frac{1}{27}$

Examples are

$$2^{3/2} = (2^{1/2})^3 \approx (1.4142)^3 \approx 2.8283$$

and

$$4^{-5/2} = \frac{1}{4^{5/2}} = \frac{1}{(4^{1/2})^5} = \frac{1}{2^5} = \frac{1}{32}$$

The rules defining the exponential expression a^n, where $a > 0$ for all rational values of n, are given in Table 1.3.

The first three definitions in Table 1.3 are also valid for negative values of a, whereas the fourth definition is valid only for negative values of a when n is odd. Thus,

$$(-8)^{1/3} = \sqrt[3]{-8} = -2 \qquad n \text{ is odd.}$$

$$(-8)^{1/2} \text{ has no real value} \qquad n \text{ is even.}$$

Finally, note that it can be shown that a^n has meaning for *all* real numbers n. For example, using a pocket calculator with a $\boxed{y^x}$ key, we see that $2^{\sqrt{2}} \approx 2.665144$.

The five laws of exponents are listed in Table 1.4.

Table 1.4 Laws of Exponents

Law	Example
1. $a^m \cdot a^n = a^{m+n}$	$x^2 \cdot x^3 = x^{2+3} = x^5$
2. $\frac{a^m}{a^n} = a^{m-n}$ $(a \neq 0)$	$\frac{x^7}{x^4} = x^{7-4} = x^3$
3. $(a^m)^n = a^{m \cdot n}$	$(x^4)^3 = x^{4 \cdot 3} = x^{12}$
4. $(ab)^n = a^n \cdot b^n$	$(2x)^4 = 2^4 \cdot x^4 = 16x^4$
5. $\left(\frac{a}{b}\right)^n = \frac{a^n}{b^n}$ $(b \neq 0)$	$\left(\frac{x}{2}\right)^3 = \frac{x^3}{2^3} = \frac{x^3}{8}$

These laws are valid for any real numbers a, b, m, and n whenever the quantities are defined.

Remember, $(x^2)^3 \neq x^5$. The correct equation is $(x^2)^3 = x^{2 \cdot 3} = x^6$.

The next several examples illustrate the use of the laws of exponents.

Example 5

Simplify the expressions:

a. $(3x^2)(4x^3)$ **b.** $\dfrac{16^{5/4}}{16^{1/2}}$ **c.** $(6^{2/3})^3$ **d.** $(x^3y^{-2})^{-2}$ **e.** $\left(\dfrac{y^{3/2}}{x^{1/4}}\right)^{-2}$

Solution

a. $(3x^2)(4x^3) = 12x^{2+3} = 12x^5$ Law 1

b. $\dfrac{16^{5/4}}{16^{1/2}} = 16^{5/4-1/2} = 16^{3/4} = (\sqrt[4]{16})^3 = 2^3 = 8$ Law 2

c. $(6^{2/3})^3 = 6^{(2/3)(3)} = 6^{6/3} = 6^2 = 36$ Law 3

d. $(x^3y^{-2})^{-2} = (x^3)^{-2}(y^{-2})^{-2} = x^{(3)(-2)}y^{(-2)(-2)} = x^{-6}y^4 = \dfrac{y^4}{x^6}$ Law 4

e. $\left(\dfrac{y^{3/2}}{x^{1/4}}\right)^{-2} = \dfrac{y^{(3/2)(-2)}}{x^{(1/4)(-2)}} = \dfrac{y^{-3}}{x^{-1/2}} = \dfrac{x^{1/2}}{y^3}$ Law 5

We can also use the laws of exponents to simplify expressions involving radicals, as illustrated in the next example.

Example 6

Simplify the expressions. (Assume x, y, and n are positive.)

a. $\sqrt[4]{16x^4y^8}$ **b.** $\sqrt{12m^3n} \cdot \sqrt{3m^5n}$ **c.** $\dfrac{\sqrt[3]{-27x^6}}{\sqrt[3]{8y^3}}$

Solution

a. $\sqrt[4]{16x^4y^8} = (16x^4y^8)^{1/4} = 16^{1/4} \cdot x^{4/4}y^{8/4} = 2xy^2$

b. $\sqrt{12m^3n} \cdot \sqrt{3m^5n} = \sqrt{36m^8n^2} = (36m^8n^2)^{1/2} = 36^{1/2} \cdot m^{8/2}n^{2/2} = 6m^4n$

c. $\dfrac{\sqrt[3]{-27x^6}}{\sqrt[3]{8y^3}} = \dfrac{(-27x^6)^{1/3}}{(8y^3)^{1/3}} = \dfrac{-27^{1/3}x^{6/3}}{8^{1/3}y^{3/3}} = -\dfrac{3x^2}{2y}$

When a radical appears in the numerator or denominator of an algebraic expression, we often try to simplify the expression by eliminating the radical from the numerator or denominator. This process, called **rationalization**, is illustrated in the next two examples.

Example 7

Rationalize the denominator of the expression $\dfrac{3x}{2\sqrt{x}}$.

Solution

$$\frac{3x}{2\sqrt{x}} = \frac{3x}{2\sqrt{x}} \cdot \frac{\sqrt{x}}{\sqrt{x}} = \frac{3x\sqrt{x}}{2\sqrt{x^2}} = \frac{3x\sqrt{x}}{2x} = \frac{3}{2}\sqrt{x}, x \neq 0$$

Example 8

Rationalize the numerator of the expression $\dfrac{3\sqrt{x}}{2x}$.

Solution

$$\frac{3\sqrt{x}}{2x} = \frac{3\sqrt{x}}{2x} \cdot \frac{\sqrt{x}}{\sqrt{x}} = \frac{3\sqrt{x^2}}{2x\sqrt{x}} = \frac{3x}{2x\sqrt{x}} = \frac{3}{2\sqrt{x}}$$

1.1 Exercises

In Exercises 1–4, determine whether the statement is true or false.

1. $-3 < -20$
2. $-5 \le -5$
3. $\frac{2}{3} > \frac{5}{6}$
4. $-\frac{5}{6} < -\frac{11}{12}$

In Exercises 5–10, show the given interval on a number line.

5. $(3, 6)$
6. $(-2, 5]$
7. $[-1, 4)$
8. $\left[-\frac{6}{5}, -\frac{1}{2}\right]$
9. $(0, \infty)$
10. $(-\infty, 5]$

In Exercises 11–20, find the values of *x* that satisfy the inequality (inequalities).

11. $2x + 4 < 8$
12. $-6 > 4 + 5x$
13. $-4x \ge 20$
14. $-12 \le -3x$
15. $-6 < x - 2 < 4$
16. $0 \le x + 1 \le 4$
17. $x + 1 > 4$ or $x + 2 < -1$
18. $x + 1 > 2$ or $x - 1 < -2$
19. $x + 3 > 1$ and $x - 2 < 1$
20. $x - 4 \le 1$ and $x + 3 > 2$

In Exercises 21–30, evaluate the expression.

21. $|-6 + 2|$
22. $4 + |-4|$
23. $\frac{|-12 + 4|}{|16 - 12|}$
24. $\left|\frac{0.2 - 1.4}{1.6 - 2.4}\right|$
25. $\sqrt{3}|-2| + 3|-\sqrt{3}|$
26. $|-1| + \sqrt{2}|-2|$
27. $|\pi - 1| + 2$
28. $|\pi - 6| - 3$
29. $|\sqrt{2} - 1| + |3 - \sqrt{2}|$
30. $|2\sqrt{3} - 3| - |\sqrt{3} - 4|$

In Exercises 31–36, suppose *a* and *b* are real numbers other than zero and that *a* > *b*. State whether the inequality is true or false.

31. $b - a > 0$
32. $\frac{a}{b} > 1$
33. $a^2 > b^2$
34. $\frac{1}{a} > \frac{1}{b}$
35. $a^3 > b^3$
36. $-a < -b$

In Exercises 37–42, determine whether the statement is true for all real numbers *a* and *b*.

37. $|-a| = a$
38. $|b^2| = b^2$
39. $|a - 4| = |4 - a|$
40. $|a + 1| = |a| + 1$
41. $|a + b| = |a| + |b|$
42. $|a - b| = |a| - |b|$

In Exercises 43–58, evaluate the expression.

43. $27^{2/3}$
44. $8^{-4/3}$
45. $\left(\frac{1}{\sqrt{3}}\right)^0$
46. $(7^{1/2})^4$
47. $\left[\left(\frac{1}{8}\right)^{1/3}\right]^{-2}$
48. $\left[\left(-\frac{1}{3}\right)^2\right]^{-3}$
49. $\left(\frac{7^{-5} \cdot 7^2}{7^{-2}}\right)^{-1}$
50. $\left(\frac{9}{16}\right)^{-1/2}$
51. $(125^{2/3})^{-1/2}$
52. $\sqrt[3]{2^6}$
53. $\frac{\sqrt{32}}{\sqrt{8}}$
54. $\sqrt[3]{\frac{-8}{27}}$
55. $\frac{16^{5/8}16^{1/2}}{16^{7/8}}$
56. $\left(\frac{9^{-3} \cdot 9^5}{9^{-2}}\right)^{-1/2}$
57. $16^{1/4} \cdot (8)^{-1/3}$
58. $\frac{6^{2.5} \cdot 6^{-1.9}}{6^{-1.4}}$

In Exercises 59–68, determine whether the statement is true or false. Give a reason for your choice.

59. $x^4 + 2x^4 = 3x^4$
60. $3^2 \cdot 2^2 = 6^2$
61. $x^3 \cdot 2x^2 = 2x^6$
62. $3^3 + 3 = 3^4$
63. $\frac{2^{4x}}{1^{3x}} = 2^{4x-3x}$
64. $(2^2 \cdot 3^2)^2 = 6^4$
65. $\frac{1}{4^{-3}} = \frac{1}{64}$
66. $\frac{4^{3/2}}{2^4} = \frac{1}{2}$
67. $(1.2^{1/2})^{-1/2} = 1$
68. $5^{2/3} \cdot (25)^{2/3} = 25$

In Exercises 69–74, rewrite the expression using positive exponents only.

69. $(xy)^{-2}$
70. $3s^{1/3} \cdot s^{-7/3}$
71. $\frac{x^{-1/3}}{x^{1/2}}$
72. $\sqrt{x^{-1}} \cdot \sqrt{9x^{-3}}$
73. $12^0(s + t)^{-3}$
74. $(x - y)(x^{-1} + y^{-1})$

In Exercises 75–90, simplify the expression. (Assume *x*, *y*, *r*, *s*, and *t* are positive.)

75. $\frac{x^{7/3}}{x^{-2}}$
76. $(49x^{-2})^{-1/2}$
77. $(x^2y^{-3})(x^{-5}y^3)$
78. $\frac{5x^6y^3}{2x^2y^7}$
79. $\frac{x^{3/4}}{x^{-1/4}}$
80. $\left(\frac{x^3y^2}{z^2}\right)^2$
81. $\left(\frac{x^3}{-27y^{-6}}\right)^{-2/3}$
82. $\left(\frac{e^x}{e^{x-2}}\right)^{-1/2}$
83. $\left(\frac{x^{-3}}{y^{-2}}\right)^2\left(\frac{y}{x}\right)^4$
84. $\frac{(r^n)^4}{r^{5-2n}}$
85. $\sqrt[3]{x^{-2}} \cdot \sqrt{4x^5}$
86. $\sqrt{81x^6y^{-4}}$

87. $-\sqrt[4]{16x^4y^8}$

88. $\sqrt[3]{x^{3a+b}}$

89. $\sqrt[6]{64x^8y^3}$

90. $\sqrt[3]{27r^6} \cdot \sqrt{s^2t^4}$

In Exercises 91–94, use the fact that the $2^{1/2} \approx 1.414$ and $3^{1/2} \approx 1.732$ to evaluate the expression without using a calculator.

91. $2^{3/2}$

92. $8^{1/2}$

93. $9^{3/4}$

94. $6^{1/2}$

In Exercises 95–98, use the fact that $10^{1/2} \approx 3.162$ and $10^{1/3} \approx 2.154$ to evaluate the expression without using a calculator.

95. $10^{3/2}$

96. $1000^{3/2}$

97. $10^{2.5}$

98. $(0.0001)^{-1/3}$

In Exercises 99–104, rationalize the denominator of the expression.

99. $\dfrac{3}{2\sqrt{x}}$

100. $\dfrac{3}{\sqrt{xy}}$

101. $\dfrac{2y}{\sqrt{3y}}$

102. $\dfrac{5x^2}{\sqrt{3x}}$

103. $\dfrac{1}{\sqrt[3]{x}}$

104. $\sqrt{\dfrac{2x}{y}}$

In Exercises 105–110, rationalize the numerator of the expression.

105. $\dfrac{2\sqrt{x}}{3}$

106. $\dfrac{\sqrt[3]{x}}{24}$

107. $\sqrt{\dfrac{2y}{x}}$

108. $\sqrt[3]{\dfrac{2x}{3y}}$

109. $\dfrac{\sqrt[3]{x^2z}}{y}$

110. $\dfrac{\sqrt[3]{x^2y}}{2x}$

Business and Economics Applications

111. **Finding Cost** Find the minimum cost C (in dollars), given that

$$5(C - 25) \geq 1.75 + 2.5C$$

112. **Finding Profit** Find the maximum profit P (in dollars) given that

$$6(P - 2500) \leq 4(P + 2400)$$

113. **Meeting Sales Targets** A salesman's monthly commission is 15% on all sales over \$12,000. If his goal is to make a commission of at least \$3000/month, what minimum monthly sales figures must he attain?

114. **Markup on a Car** The markup on a used car was at least 30% of its current wholesale price. If the car was sold for \$5600, what was the maximum wholesale price?

115. **Meeting Profit Goals** A manufacturer of a certain commodity has estimated that her profit in thousands of dollars is given by the expression

$$-6x^2 + 30x - 10$$

where x (in thousands) is the number of units produced. What production level will enable the manufacturer to realize a profit of at least \$14,000 on the commodity?

116. **Distribution of Incomes** The distribution of income in a certain city can be described by the exponential model $y = (2.8 \cdot 10^{11})(x)^{-1.5}$, where y is the number of families with an income of x or more dollars.
   a. How many families in this city have an income of \$30,000 or more?
   b. How many families have an income of \$60,000 or more?
   c. How many families have an income of \$150,000 or more?

Biological and Life Sciences Applications

117. **Driving Range of a Car** An advertisement for a certain car states that the fuel economy is 8 km/L city and 11 km/L highway and that the car's fuel-tank capacity is 72 L. Assuming ideal driving conditions, determine the driving range for the car from the foregoing data.

118. **Celsius and Fahrenheit Temperatures** The relationship between Celsius (°C) and Fahrenheit (°F) temperatures is given by the formula

$$C = \frac{5}{9}(F - 32)$$

   a. If the temperature range for Montreal during the month of January is $-15° < °C < -5°$, find the range in degrees Fahrenheit in Montreal for the same period.
   b. If the temperature range for New York City during the month of June is $63° < °F < 80°$, find the range in degrees Celsius in New York City for the same period.

119. **Quality Control** PAR Manufacturing manufactures steel rods. Suppose the rods ordered by a customer are manufactured to a specification of 0.5 cm and are acceptable only if they are within the *tolerance limits* of 0.49 cm and 0.51 cm. Letting x denote the diameter of a rod, write an inequality using absolute value to express a criterion involving x that must be satisfied in order for a rod to be acceptable.

120. **Quality Control** The diameter x (in centimetres) of a batch of ball bearings manufactured by PAR Manufacturing satisfies the inequality

$$|x - 1| \leq 0.1$$

What is the smallest diameter a ball bearing in the batch can have? The largest diameter?

In Exercises 121–124, determine whether the statement is true or false. If it is true, explain why it is true. If it is false, give an example to show why it is false.

121. If $a < b$, then $a - c > b - c$.

122. $|a - b| = |b - a|$

123. $|a - b| \leq |b| + |a|$

124. $\sqrt{a^2 - b^2} = |a| - |b|$

1.2 Precalculus Review II

Operations with Algebraic Expressions

In calculus we often work with algebraic expressions such as

$$2x^{4/3} - x^{1/3} + 1, \qquad 2x^2 - x - \frac{2}{\sqrt{x}}, \qquad \frac{3xy + 2}{x + 1}, \qquad 2x^3 + 2x + 1$$

An algebraic expression of the form ax^n, where the coefficient a is a real number and n is a nonnegative integer, is called a monomial, meaning it consists of one term. For example, $7x^2$ is a monomial. A polynomial is a monomial or the sum of two or more monomials. For example,

$$x^2 + 4x + 4, \qquad x^3 + 5, \qquad x^4 + 3x^2 + 3, \qquad x^2y + xy + y$$

are all polynomials.

Constant terms and terms containing the same variable factor are called like or similar terms. Like terms may be combined by adding or subtracting their numerical coefficients. For example,

$$3x + 7x = 10x \qquad \text{and} \qquad \frac{1}{2}xy + 3xy = \frac{7}{2}xy$$

The distributive property of the real number system,

$$ab + ac = a(b + c)$$

is used to justify this procedure.

To add or subtract two or more algebraic expressions, first remove the parentheses and then combine like terms. The resulting expression is written in order of decreasing degree from left to right.

R *Example 1

a. $(2x^4 + 3x^3 + 4x + 6) - (3x^4 + 9x^3 + 3x^2)$

$= 2x^4 + 3x^3 + 4x + 6 - 3x^4 - 9x^3 - 3x^2$ Remove parentheses.

$= 2x^4 - 3x^4 + 3x^3 - 9x^3 - 3x^2 + 4x + 6$

$= -x^4 - 6x^3 - 3x^2 + 4x + 6$ Combine like terms.

b. $2t^3 - \{t^2 - [t - (2t - 1)] + 4\}$

$= 2t^3 - \{t^2 - [t - 2t + 1] + 4\}$

$= 2t^3 - \{t^2 - [-t + 1] + 4\}$ Remove parentheses and combine like terms within brackets.

$= 2t^3 - \{t^2 + t - 1 + 4\}$ Remove brackets.

$= 2t^3 - \{t^2 + t + 3\}$ Combine like terms within braces.

$= 2t^3 - t^2 - t - 3$ Remove braces.

An algebraic expression is said to be simplified if none of its terms are similar. Observe that when the algebraic expression in Example 1b was simplified, the innermost grouping symbols were removed first; that is, the parentheses ( ) were removed first, the brackets [ ] second, and the braces { } third.

When algebraic expressions are multiplied, each term of one algebraic expression is multiplied by each term of the other. The resulting algebraic expression is then simplified.

R Example 2

Perform the indicated operations:

a. $(x^2 + 1)(3x^2 + 10x + 3)$ **b.** $(e^t + e^{-t})e^t - e^t(e^t - e^{-t})$

*The symbol R indicates that these examples were selected from the calculus portion of the text in order to help you review the algebraic computations you will *actually* be using in calculus.

Solution

a. $(x^2 + 1)(3x^2 + 10x + 3) = x^2(3x^2 + 10x + 3) + 1(3x^2 + 10x + 3)$
$= 3x^4 + 10x^3 + 3x^2 + 3x^2 + 10x + 3$
$= 3x^4 + 10x^3 + 6x^2 + 10x + 3$

b. $(e^t + e^{-t})e^t - e^t(e^t - e^{-t}) = e^{2t} + e^0 - e^{2t} + e^0$
$= e^{2t} - e^{2t} + e^0 + e^0$
$= 1 + 1$ Recall that $e^0 = 1$.
$= 2$

Certain product formulas that are frequently used in algebraic computations are given in Table 1.5.

Table 1.5 Product Formulas Frequently Used in Algebraic Computations

Formula	Example
$(a + b)^2 = a^2 + 2ab + b^2$	$(2x + 3y)^2 = (2x)^2 + 2(2x)(3y) + (3y)^2$ $= 4x^2 + 12xy + 9y^2$
$(a - b)^2 = a^2 - 2ab + b^2$	$(4x - 2y)^2 = (4x)^2 - 2(4x)(2y) + (2y)^2$ $= 16x^2 - 16xy + 4y^2$
$(a + b)(a - b) = a^2 - b^2$	$(2x + y)(2x - y) = (2x)^2 - (y)^2$ $= 4x^2 - y^2$

Factoring

Factoring is the process of expressing an algebraic expression as a product of other algebraic expressions. For example, by applying the distributive property, we may write

$$3x^2 - x = x(3x - 1)$$

The first step in factoring an algebraic expression is to determine whether it contains any common terms. If it does, the greatest common term is then factored out. For example, the common factor of the algebraic expression $2a^2x + 4ax + 6a$ is $2a$, because

$$2a^2x + 4ax + 6a = 2a \cdot ax + 2a \cdot 2x + 2a \cdot 3 = 2a(ax + 2x + 3)$$

R

Example 3

Factor out the greatest common factor in each of the following expressions:

a. $-0.3t^2 + 3t$ **b.** $2x^{3/2} - 3x^{1/2}$ **c.** $2ye^{xy^2} + 2xy^3e^{xy^2}$

d. $4x(x + 1)^{1/2} - 2x^2\left(\frac{1}{2}\right)(x + 1)^{-1/2}$

Solution

a. $-0.3t^2 + 3t = -0.3t(t - 10)$

b. $2x^{3/2} - 3x^{1/2} = x^{1/2}(2x - 3)$

c. $2ye^{xy^2} + 2xy^3e^{xy^2} = 2ye^{xy^2}(1 + xy^2)$

d. $4x(x + 1)^{1/2} - 2x^2\left(\frac{1}{2}\right)(x + 1)^{-1/2} = 4x(x + 1)^{1/2} - x^2(x + 1)^{-1/2}$
$= x(x + 1)^{-1/2}[4(x + 1)^{1/2}(x + 1)^{1/2} - x]$
$= x(x + 1)^{-1/2}[4(x + 1) - x]$
$= x(x + 1)^{-1/2}(4x + 4 - x) = x(x + 1)^{-1/2}(3x + 4)$

Here we select $(x + 1)^{-1/2}$ as the common factor because it is "contained" in each algebraic term. In particular, observe that

$$(x + 1)^{-1/2}(x + 1)^{1/2}(x + 1)^{1/2} = (x + 1)^{1/2}$$

Sometimes an algebraic expression may be factored by regrouping and rearranging its terms and then factoring out a common term. This technique is illustrated in Example 4.

Example 4

Factor:

a. $2ax + 2ay + bx + by$ **b.** $3x\sqrt{y} - 4 - 2\sqrt{y} + 6x$

Solution

a. First, factor the common term $2a$ from the first two terms and the common term b from the last two terms. Thus,

$$2ax + 2ay + bx + by = 2a(x + y) + b(x + y)$$

Since $(x + y)$ is common to both terms of the polynomial, we may factor it out. Hence,

$$2a(x + y) + b(x + y) = (2a + b)(x + y)$$

b.
$$\begin{aligned} 3x\sqrt{y} - 4 - 2\sqrt{y} + 6x &= 3x\sqrt{y} - 2\sqrt{y} + 6x - 4 \\ &= \sqrt{y}(3x - 2) + 2(3x - 2) \\ &= (3x - 2)(\sqrt{y} + 2) \end{aligned}$$

The first step in factoring a polynomial is to find the common factors. The next step is to express the polynomial as the product of a constant and/or one or more prime polynomials.

Certain product formulas that are useful in factoring binomials and trinomials are listed in Table 1.6.

Table 1.6 Product Formulas Used in Factoring

Formula	Example
Difference of two squares	
$x^2 - y^2 = (x + y)(x - y)$	$x^2 - 36 = (x + 6)(x - 6)$
	$8x^2 - 2y^2 = 2(4x^2 - y^2)$
	$= 2(2x + y)(2x - y)$
	$9 - a^6 = (3 + a^3)(3 - a^3)$
Perfect-square trinomial	
$x^2 + 2xy + y^2 = (x + y)^2$	$x^2 + 8x + 16 = (x + 4)^2$
$x^2 - 2xy + y^2 = (x - y)^2$	$4x^2 - 4xy + y^2 = (2x - y)^2$
Sum of two cubes	
$x^3 + y^3 = (x + y)(x^2 - xy + y^2)$	$z^3 + 27 = z^3 + (3)^3$
	$= (z + 3)(z^2 - 3z + 9)$
Difference of two cubes	
$x^3 - y^3 = (x - y)(x^2 + xy + y^2)$	$8x^3 - y^6 = (2x)^3 - (y^2)^3$
	$= (2x - y^2)(4x^2 + 2xy^2 + y^4)$

The factors of the second-degree polynomial with integral coefficients

$$px^2 + qx + r$$

are $(ax + b)(cx + d)$, where $ac = p$, $ad + bc = q$, and $bd = r$. Since only a limited number of choices are possible, we use a trial-and-error method to factor polynomials having this form.

For example, to factor $x^2 - 2x - 3$, we first observe that the only possible first-degree terms are

$$(x \quad)(x \quad) \qquad \text{Since the coefficient of } x^2 \text{ is 1}$$

Next, we observe that the product of the constant term is (-3). This gives us the following possible factors:

$$(x - 1)(x + 3)$$

$$(x + 1)(x - 3)$$

Looking once again at the polynomial $x^2 - 2x - 3$, we see that the coefficient of x is -2. Checking to see which set of factors yields -2 for the coefficient of x, we find that

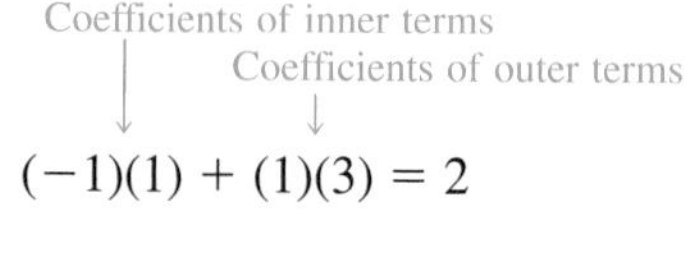

$$(-1)(1) + (1)(3) = 2$$

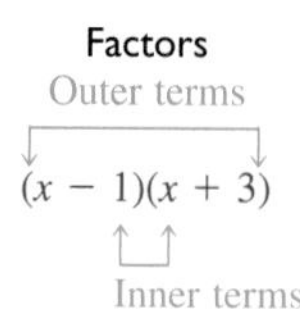

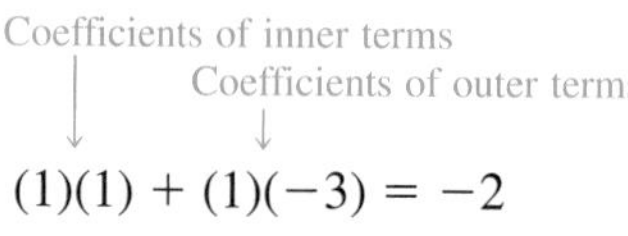

$$(1)(1) + (1)(-3) = -2$$

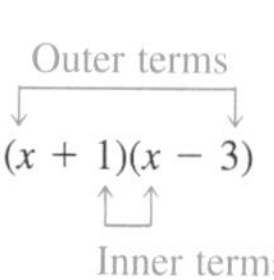

and we conclude that the correct factorization is

$$x^2 - 2x - 3 = (x + 1)(x - 3)$$

With practice, you will soon find that you can perform many of these steps mentally and the need to write out each step will be eliminated.

R **Example 5**

Factor:

a. $3x^2 + 4x - 4$ **b.** $3x^2 - 6x - 24$

Solution

a. Using trial and error, we find that the correct factorization is

$$3x^2 + 4x - 4 = (3x - 2)(x + 2)$$

b. Since each term has the common factor 3, we have

$$3x^2 - 6x - 24 = 3(x^2 - 2x - 8)$$

Using the trial-and-error method of factorization, we find that

$$x^2 - 2x - 8 = (x - 4)(x + 2)$$

Thus, we have

$$3x^2 - 6x - 24 = 3(x - 4)(x + 2)$$

Roots of Polynomial Equations

A polynomial equation of degree n in the variable x is an equation of the form

$$a_n x^n + a_{n-1} x^{n-1} + \cdots + a_0 = 0$$

where n is a nonnegative integer and $a_0, a_1, \ldots, a_n$ are real numbers with $a_n \neq 0$. For example, the equation

$$-2x^5 + 8x^3 - 6x^2 + 3x + 1 = 0$$

is a polynomial equation of degree 5 in x.

We will study and work with many polynomial equations throughout this textbook but particularly polynomial equations of degree 0, 1, and 2. The simplest polynomial equation is of degree 0 and given by $a = 0$. The only constant a for which this equation is true is when a is equal to 0. A polynomial equation of degree 1 is referred to as *linear* and often given as $mx + b = 0$, where $m \neq 0$. A polynomial equation of degree 2 is referred to as *quadratic* and often given as $ax^2 + bx + c = 0$, where $a \neq 0$.

The roots of a polynomial equation are precisely the values of x that satisfy the given equation.* One way of finding the roots of a polynomial equation is to first factor the polynomial and then solve the resulting equation. For example, the polynomial equation

$$x^3 - 3x^2 + 2x = 0$$

may be rewritten in the form

$$x(x^2 - 3x + 2) = 0 \qquad \text{or} \qquad x(x - 1)(x - 2) = 0$$

Since the product of two real numbers can be equal to zero if and only if one (or both) of the factors is equal to zero, we have

$$x = 0, \qquad x - 1 = 0, \qquad \text{or} \qquad x - 2 = 0$$

from which we see that the desired roots are $x = 0$, 1, and 2.

The roots of a linear equation $mx + b = 0$ are easy to find by solving for x.

$$\begin{aligned} mx + b &= 0 \\ mx &= -b \\ x &= \frac{-b}{m} \end{aligned}$$

The only root for the linear equation $mx + b = 0$ is given by $x = \dfrac{-b}{m}$.

The Quadratic Formula

In general, the problem of finding the roots of a polynomial equation is a difficult one. But the roots of a quadratic equation are easily found either by factoring or by using the following quadratic formula.

Quadratic Formula

The solutions of the equation $ax^2 + bx + c = 0$ $(a \neq 0)$ are given by

$$x = \frac{-b \pm \sqrt{b^2 - 4ac}}{2a}$$

Example 6

Solve each of the following quadratic equations:

a. $2x^2 + 5x - 12 = 0$ **b.** $x^2 = -3x + 8$

Solution

a. The equation is in standard form, with $a = 2$, $b = 5$, and $c = -12$. Using the quadratic formula, we find

$$\begin{aligned} x &= \frac{-b \pm \sqrt{b^2 - 4ac}}{2a} = \frac{-5 \pm \sqrt{5^2 - 4(2)(-12)}}{2(2)} \\ &= \frac{-5 \pm \sqrt{121}}{4} = \frac{-5 \pm 11}{4} \\ &= -4 \text{ or } \frac{3}{2} \end{aligned}$$

*In this book, we are interested only in the *real* roots of an equation.

This equation can also be solved by factoring. Thus,

$$2x^2 + 5x - 12 = (2x - 3)(x + 4) = 0$$

from which we see that the desired roots are $x = \frac{3}{2}$ or $x = -4$, as obtained earlier.

b. We first rewrite the given equation in the standard form $x^2 + 3x - 8 = 0$, from which we see that $a = 1$, $b = 3$, and $c = -8$. Using the quadratic formula, we find

$$x = \frac{-b \pm \sqrt{b^2 - 4ac}}{2a} = \frac{-3 \pm \sqrt{3^2 - 4(1)(-8)}}{2(1)}$$
$$= \frac{-3 \pm \sqrt{41}}{2}$$

That is, the solutions are

$$\frac{-3 + \sqrt{41}}{2} \approx 1.7 \quad \text{and} \quad \frac{-3 - \sqrt{41}}{2} \approx -4.7$$

In this case, the quadratic formula proves quite handy!

Rational Expressions

Quotients of polynomials are called **rational expressions.** Examples of rational expressions are

$$\frac{6x - 1}{2x + 3}, \qquad \frac{3x^2y^3 - 2xy}{4x}, \qquad \frac{2}{5ab}$$

Since rational expressions are quotients in which the variables represent real numbers, the properties of the real numbers apply to rational expressions as well, and operations with rational fractions are performed in the same manner as operations with arithmetic fractions. For example, using the properties of the real number system, we may write

$$\frac{ac}{bc} = \frac{a}{b} \cdot \frac{c}{c} = \frac{a}{b} \cdot 1 = \frac{a}{b}$$

where a, b, and c are any real numbers and b and c are not zero.

Similarly, using the same properties of real numbers, we may write

$$\frac{(x + 2)(x - 3)}{(x - 2)(x - 3)} = \frac{x + 2}{x - 2} \qquad (x \neq 2, 3)$$

after "canceling" the common factors.

An example of incorrect cancellation is

$$\frac{\cancel{3} + 4x}{\cancel{3}} \neq 1 + 4x$$

Instead, we need to write

$$\frac{3 + 4x}{3} = \frac{3}{3} + \frac{4x}{3} = 1 + \frac{4x}{3}$$

A rational expression is simplified, or in lowest terms, when the numerator and denominator have no common factors other than 1 and -1 and the expression contains no negative exponents.

Example 7

Simplify the following expressions:

a. $\dfrac{x^2 + 2x - 3}{x^2 + 4x + 3}$ **b.** $\dfrac{[(t^2 + 4)(2t - 4) - (t^2 - 4t + 4)(2t)]}{(t^2 + 4)^2}$

Solution

a. $\dfrac{x^2 + 2x - 3}{x^2 + 4x + 3} = \dfrac{(x + 3)(x - 1)}{(x + 3)(x + 1)} = \dfrac{x - 1}{x + 1}$

b. $\dfrac{[(t^2 + 4)(2t - 4) - (t^2 - 4t + 4)(2t)]}{(t^2 + 4)^2}$

$= \dfrac{2t^3 - 4t^2 + 8t - 16 - 2t^3 + 8t^2 - 8t}{(t^2 + 4)^2}$ Carry out the indicated multiplication.

$= \dfrac{4t^2 - 16}{(t^2 + 4)^2}$ Combine like terms.

$= \dfrac{4(t^2 - 4)}{(t^2 + 4)^2}$ Factor.

The operations of multiplication and division are performed with algebraic fractions in the same manner as with arithmetic fractions (Table 1.7).

Table 1.7 Rules of Multiplication and Division: Algebraic Fractions

Operation	Example
If P, Q, R, and S are polynomials, then	
Multiplication	
$\dfrac{P}{Q} \cdot \dfrac{R}{S} = \dfrac{PR}{QS}$ $(Q, S \neq 0)$	$\dfrac{2x}{y} \cdot \dfrac{(x + 1)}{(y - 1)} = \dfrac{2x(x + 1)}{y(y - 1)} = \dfrac{2x^2 + 2x}{y^2 - y}$
Division	
$\dfrac{P}{Q} \div \dfrac{R}{S} = \dfrac{P}{Q} \cdot \dfrac{S}{R} = \dfrac{PS}{QR}$ $(Q, R, S \neq 0)$	$\dfrac{x^2 + 3}{y} \div \dfrac{y^2 + 1}{x} = \dfrac{x^2 + 3}{y} \cdot \dfrac{x}{y^2 + 1} = \dfrac{x^3 + 3x}{y^3 + y}$

When rational expressions are multiplied and divided, the resulting expressions should be simplified.

Example 8

Perform the indicated operations and simplify:

$$\frac{2x - 8}{x + 2} \cdot \frac{x^2 + 4x + 4}{x^2 - 16}$$

Solution

$$\frac{2x - 8}{x + 2} \cdot \frac{x^2 + 4x + 4}{x^2 - 16} = \frac{2(x - 4)}{x + 2} \cdot \frac{(x + 2)^2}{(x + 4)(x - 4)}$$

$= \dfrac{2(x - 4)(x + 2)(x + 2)}{(x + 2)(x + 4)(x - 4)}$ Cancel the common factors $(x + 2)(x - 4)$.

$= \dfrac{2(x + 2)}{x + 4}$

The equality only holds for $x \neq -2$ and $x \neq 4$ due to cancellation of common factors.

For rational expressions, the operations of addition and subtraction are performed by finding a common denominator of the fractions and then adding or subtracting the fractions. Table 1.8 shows the rules for fractions with equal denominators.

Table 1.8 Rules of Addition and Subtraction: Fractions with Equal Denominators

Formula	Example
If P, Q, and R are polynomials, then	
Addition	
$\frac{P}{R} + \frac{Q}{R} = \frac{P + Q}{R} \quad (R \neq 0)$	$\frac{2x}{x + 2} + \frac{6x}{x + 2} = \frac{2x + 6x}{x + 2} = \frac{8x}{x + 2}$
Subtraction	
$\frac{P}{R} - \frac{Q}{R} = \frac{P - Q}{R} \quad (R \neq 0)$	$\frac{3y}{y - x} - \frac{y}{y - x} = \frac{3y - y}{y - x} = \frac{2y}{y - x}$

To add or subtract fractions that have different denominators, first find a common denominator, preferably the least common denominator (LCD). Then carry out the indicated operations following the procedure described in Table 1.8.

To find the LCD of two or more rational expressions:

1. *Find the prime factors of each denominator.*
2. *Form the product of the different prime factors that occur in the denominators.* Each prime factor in this product should be raised to the highest power of that factor appearing in the denominators.

$$\frac{x}{2 + y} \neq \frac{x}{2} + \frac{x}{y}$$

Example 9

Simplify:

a. $\frac{2x}{x^2 + 1} + \frac{6(3x^2)}{x^3 + 2}$ **b.** $\frac{1}{x + h} - \frac{1}{x}$

Solution

a. $$\frac{2x}{x^2 + 1} + \frac{6(3x^2)}{x^3 + 2} = \frac{2x(x^3 + 2) + 6(3x^2)(x^2 + 1)}{(x^2 + 1)(x^3 + 2)}$$ LCD $= (x^2 + 1)(x^3 + 2)$

$$= \frac{2x^4 + 4x + 18x^4 + 18x^2}{(x^2 + 1)(x^3 + 2)}$$

$$= \frac{20x^4 + 18x^2 + 4x}{(x^2 + 1)(x^3 + 2)}$$

$$= \frac{2x(10x^3 + 9x + 2)}{(x^2 + 1)(x^3 + 2)}$$

b. $$\frac{1}{x + h} - \frac{1}{x} = \frac{x - (x + h)}{x(x + h)}$$ LCD $= x(x + h)$

$$= \frac{x - x - h}{x(x + h)}$$

$$= \frac{-h}{x(x + h)}$$

Other Algebraic Fractions

The techniques used to simplify rational expressions may also be used to simplify algebraic fractions in which the numerator and denominator are not polynomials, as illustrated in Example 10.

Example 10

Simplify: **a.** $\dfrac{1+\dfrac{1}{x+1}}{x-\dfrac{4}{x}}$ **b.** $\dfrac{x^{-1}+y^{-1}}{x^{-2}-y^{-2}}$

Solution

a. $$\frac{1+\dfrac{1}{x+1}}{x-\dfrac{4}{x}}=\frac{\dfrac{x+1+1}{x+1}}{\dfrac{x^2-4}{x}}$$

$$=\frac{x+2}{x+1}\cdot\frac{x}{x^2-4}=\frac{x+2}{x+1}\cdot\frac{x}{(x+2)(x-2)}$$

$$=\frac{x}{(x+1)(x-2)}$$

b. $$\frac{x^{-1}+y^{-1}}{x^{-2}-y^{-2}}=\frac{\dfrac{1}{x}+\dfrac{1}{y}}{\dfrac{1}{x^2}-\dfrac{1}{y^2}}=\frac{\dfrac{y+x}{xy}}{\dfrac{y^2-x^2}{x^2y^2}}$$

$x^{-n}=\dfrac{1}{x^n}$

$$=\frac{y+x}{xy}\cdot\frac{x^2y^2}{y^2-x^2}=\frac{y+x}{xy}\cdot\frac{(xy)^2}{(y+x)(y-x)}$$

$$=\frac{xy}{y-x}$$

Example 11

Perform the given operations and simplify:

a. $\dfrac{x^2(2x^2+1)^{1/2}}{x-1}\cdot\dfrac{4x^3-6x^2+x-2}{x(x-1)(2x^2+1)}$ **b.** $\dfrac{12x^2}{\sqrt{2x^2+3}}+6\sqrt{2x^2+3}$

Solution

a. $$\frac{x^2(2x^2+1)^{1/2}}{x-1}\cdot\frac{4x^3-6x^2+x-2}{x(x-1)(2x^2+1)}=\frac{x(4x^3-6x^2+x-2)}{(x-1)^2(2x^2+1)^{1-1/2}}$$

$$=\frac{x(4x^3-6x^2+x-2)}{(x-1)^2(2x^2+1)^{1/2}}$$

b. $$\frac{12x^2}{\sqrt{2x^2+3}}+6\sqrt{2x^2+3}=\frac{12x^2}{(2x^2+3)^{1/2}}+6(2x^2+3)^{1/2}$$

$$=\frac{12x^2+6(2x^2+3)^{1/2}(2x^2+3)^{1/2}}{(2x^2+3)^{1/2}}$$

$$=\frac{12x^2+6(2x^2+3)}{(2x^2+3)^{1/2}}$$

$$=\frac{24x^2+18}{(2x^2+3)^{1/2}}=\frac{6(4x^2+3)}{\sqrt{2x^2+3}}$$

Rationalizing Algebraic Fractions

When the denominator of an algebraic fraction contains sums or differences involving radicals, we may **rationalize the denominator**—that is, transform the fraction into an equivalent one with a denominator that does not contain radicals. In doing so, we make use of the fact that

$$\begin{aligned}(\sqrt{a} + \sqrt{b})(\sqrt{a} - \sqrt{b}) &= (\sqrt{a})^2 - (\sqrt{b})^2 \\ &= a - b\end{aligned}$$

This procedure is illustrated in Example 12.

Example 12

Rationalize the denominator: $\dfrac{1}{1 + \sqrt{x}}$.

Solution

Upon multiplying the numerator and the denominator by $(1 - \sqrt{x})$, we obtain

$$\begin{aligned}\frac{1}{1 + \sqrt{x}} &= \frac{1}{1 + \sqrt{x}} \cdot \frac{1 - \sqrt{x}}{1 - \sqrt{x}} \\ &= \frac{1 - \sqrt{x}}{1 - (\sqrt{x})^2} \\ &= \frac{1 - \sqrt{x}}{1 - x}\end{aligned}$$

In other situations, it may be necessary to rationalize the numerator of an algebraic expression. In calculus, for example, one encounters the following problem.

Example 13

Rationalize the numerator: $\dfrac{\sqrt{1 + h} - 1}{h}$.

Solution

$$\begin{aligned}\frac{\sqrt{1 + h} - 1}{h} &= \frac{\sqrt{1 + h} - 1}{h} \cdot \frac{\sqrt{1 + h} + 1}{\sqrt{1 + h} + 1} \\ &= \frac{(\sqrt{1 + h})^2 - (1)^2}{h(\sqrt{1 + h} + 1)} \\ &= \frac{1 + h - 1}{h(\sqrt{1 + h} + 1)} \\ &= \frac{h}{h(\sqrt{1 + h} + 1)} \\ &= \frac{1}{\sqrt{1 + h} + 1}\end{aligned}$$

$$\begin{aligned}(\sqrt{1 + h})^2 &= \sqrt{1 + h} \cdot \sqrt{1 + h} \\ &= 1 + h\end{aligned}$$

1.2 Exercises

In Exercises 1–22, perform the indicated operations and simplify each expression.

1. $(7x^2 - 2x + 5) + (2x^2 + 5x - 4)$
2. $(3x^2 + 5xy + 2y) + (4 - 3xy - 2x^2)$
3. $(5y^2 - 2y + 1) - (y^2 - 3y - 7)$
4. $3(2a - b) - 4(b - 2a)$
5. $x - \{2x - [-x - (1 - x)]\}$
6. $3x^2 - \{x^2 + 1 - x[x - (2x - 1)]\} + 2$
7. $\left(\frac{1}{3} - 1 + e\right) - \left(-\frac{1}{3} - 1 + e^{-1}\right)$
8. $-\frac{3}{4}y - \frac{1}{4}x + 100 + \frac{1}{2}x + \frac{1}{4}y - 120$
9. $3\sqrt{8} + 8 - 2\sqrt{y} + \frac{1}{2}\sqrt{x} - \frac{3}{4}\sqrt{y}$
10. $\frac{8}{9}x^2 + \frac{2}{3}x + \frac{16}{3}x^2 - \frac{16}{3}x - 2x + 2$
11. $(x + 8)(x - 2)$
12. $(5x + 2)(3x - 4)$
13. $(a + 5)^2$
14. $(3a - 4b)^2$
15. $(x + 2y)^2$
16. $(6 - 3x)^2$
17. $(2x + y)(2x - y)$
18. $(3x + 2)(2 - 3x)$
19. $(x^2 - 1)(2x) - x^2(2x)$
20. $(x^{1/2} + 1)\left(\frac{1}{2}x^{-1/2}\right) - (x^{1/2} - 1)\left(\frac{1}{2}x^{-1/2}\right)$
21. $2(t + \sqrt{t})^2 - 2t^2$
22. $2x^2 + (-x + 1)^2$

In Exercises 23–30, factor out the greatest common factor from each expression.

23. $4x^5 - 12x^4 - 6x^3$
24. $4x^2y^2z - 2x^5y^2 + 6x^3y^2z^2$
25. $7a^4 - 42a^2b^2 + 49a^3b$
26. $3x^{2/3} - 2x^{1/3}$
27. $e^{-x} - xe^{-x}$
28. $2ye^{xy^2} + 2xy^3e^{xy^2}$
29. $2x^{-5/2} - \frac{3}{2}x^{-3/2}$
30. $\frac{1}{2}\left(\frac{2}{3}u^{3/2} - 2u^{1/2}\right)$

In Exercises 31–44, factor each expression.

31. $6ac + 3bc - 4ad - 2bd$
32. $3x^3 - x^2 + 3x - 1$
33. $4a^2 - b^2$
34. $12x^2 - 3y^2$
35. $10 - 14x - 12x^2$
36. $x^2 - 2x - 15$
37. $3x^2 - 6x - 24$
38. $3x^2 - 4x - 4$
39. $12x^2 - 2x - 30$
40. $(x + y)^2 - 1$
41. $9x^2 - 16y^2$
42. $8a^2 - 2ab - 6b^2$
43. $x^6 + 125$
44. $x^3 - 27$

In Exercises 45–52, perform the indicated operations and simplify each expression.

45. $(x^2 + y^2)x - xy(2y)$
46. $2kr(R - r) - kr^2$
47. $2(x - 1)(2x + 2)^3[4(x - 1) + (2x + 2)]$
48. $5x^2(3x^2 + 1)^4(6x) + (3x^2 + 1)^5(2x)$
49. $4(x - 1)^2(2x + 2)^3(2) + (2x + 2)^4(2)(x - 1)$
50. $(x^2 + 1)(4x^3 - 3x^2 + 2x) - (x^4 - x^3 + x^2)(2x)$
51. $(x^2 + 2)^2[5(x^2 + 2)^2 - 3](2x)$
52. $(x^2 - 4)(x^2 + 4)(2x + 8) - (x^2 + 8x - 4)(4x^3)$

In Exercises 53–58, find the real roots of each equation by factoring.

53. $x^2 + x - 12 = 0$
54. $3x^2 - x - 4 = 0$
55. $4t^2 + 2t - 2 = 0$
56. $-6x^2 + x + 12 = 0$
57. $\frac{1}{4}x^2 - x + 1 = 0$
58. $\frac{1}{2}a^2 + a - 12 = 0$

In Exercises 59–64, solve the equation by using the quadratic formula.

59. $4x^2 + 5x - 6 = 0$
60. $3x^2 - 4x + 1 = 0$
61. $8x^2 - 8x - 3 = 0$
62. $x^2 - 6x + 6 = 0$
63. $2x^2 + 4x - 3 = 0$
64. $2x^2 + 7x - 15 = 0$

In Exercises 65–70, simplify the expression.

65. $\dfrac{x^2 + x - 2}{x^2 - 4}$
66. $\dfrac{2a^2 - 3ab - 9b^2}{2ab^2 + 3b^3}$
67. $\dfrac{12t^2 + 12t + 3}{4t^2 - 1}$
68. $\dfrac{x^3 + 2x^2 - 3x}{-2x^2 - x + 3}$
69. $\dfrac{(4x - 1)(3) - (3x + 1)(4)}{(4x - 1)^2}$
70. $\dfrac{(1 + x^2)^2(2) - 2x(2)(1 + x^2)(2x)}{(1 + x^2)^4}$

In Exercises 71–88, perform the indicated operations and simplify each expression.

71. $\dfrac{2a^2 - 2b^2}{b - a} \cdot \dfrac{4a + 4b}{a^2 + 2ab + b^2}$
72. $\dfrac{x^2 - 6x + 9}{x^2 - x - 6} \cdot \dfrac{3x + 6}{2x^2 - 7x + 3}$
73. $\dfrac{3x^2 + 2x - 1}{2x + 6} \div \dfrac{x^2 - 1}{x^2 + 2x - 3}$
74. $\dfrac{3x^2 - 4xy - 4y^2}{x^2y} \div \dfrac{(2y - x)^2}{x^3y}$
75. $\dfrac{58}{3(3t + 2)} + \dfrac{1}{3}$
76. $\dfrac{a + 1}{3a} + \dfrac{b - 2}{5b}$
77. $\dfrac{2x}{2x - 1} - \dfrac{3x}{2x + 5}$
78. $\dfrac{-xe^x}{x + 1} + e^x$

79. $\dfrac{4}{x^2-9} - \dfrac{5}{x^2-6x+9}$

80. $\dfrac{x}{1-x} + \dfrac{2x+3}{x^2-1}$

81. $\dfrac{1+\dfrac{1}{x}}{1-\dfrac{1}{x}}$

82. $\dfrac{\dfrac{1}{x}+\dfrac{1}{y}}{1-\dfrac{1}{xy}}$

83. $\dfrac{4x^2}{2\sqrt{2x^2+7}} + \sqrt{2x^2+7}$

84. $6(2x+1)^2\sqrt{x^2+x} + \dfrac{(2x+1)^4}{2\sqrt{x^2+x}}$

85. $\dfrac{2x(x+1)^{-1/2} - (x+1)^{1/2}}{x^2}$

86. $\dfrac{(x^2+1)^{1/2} - 2x^2(x^2+1)^{-1/2}}{1-x^2}$

87. $\dfrac{(2x+1)^{1/2} - (x+2)(2x+1)^{-1/2}}{2x+1}$

88. $\dfrac{2(2x-3)^{1/3} - (x-1)(2x-3)^{-2/3}}{(2x-3)^{2/3}}$

In Exercises 89–94, rationalize the denominator of each expression.

89. $\dfrac{1}{\sqrt{3}-1}$

90. $\dfrac{1}{\sqrt{x}+5}$

91. $\dfrac{1}{\sqrt{x}-\sqrt{y}}$

92. $\dfrac{a}{1-\sqrt{a}}$

93. $\dfrac{\sqrt{a}+\sqrt{b}}{\sqrt{a}-\sqrt{b}}$

94. $\dfrac{2\sqrt{a}+\sqrt{b}}{2\sqrt{a}-\sqrt{b}}$

In Exercises 95–100, rationalize the numerator of each expression.

95. $\dfrac{\sqrt{x}}{3}$

96. $\dfrac{\sqrt[3]{y}}{x}$

97. $\dfrac{1-\sqrt{3}}{3}$

98. $\dfrac{\sqrt{x}-1}{x}$

99. $\dfrac{1+\sqrt{x+2}}{\sqrt{x+2}}$

100. $\dfrac{\sqrt{x+3}-\sqrt{x}}{3}$

In Exercises 101–104, determine whether the statement is true or false. If it is true, explain why it is true. If it is false, give an example to show why it is false.

101. If $b^2 - 4ac > 0$, then $ax^2 + bx + c = 0$ $(a \neq 0)$ has two real roots.

102. If $b^2 - 4ac < 0$, then $ax^2 + bx + c = 0$ $(a \neq 0)$ has no real roots.

103. $\dfrac{a}{b+c} = \left(\dfrac{a}{b} + \dfrac{a}{c}\right)$

104. $\sqrt{(a+b)(b-a)} = \sqrt{b^2-a^2}$ for all real numbers a and b.

1.3 The Cartesian Coordinate System

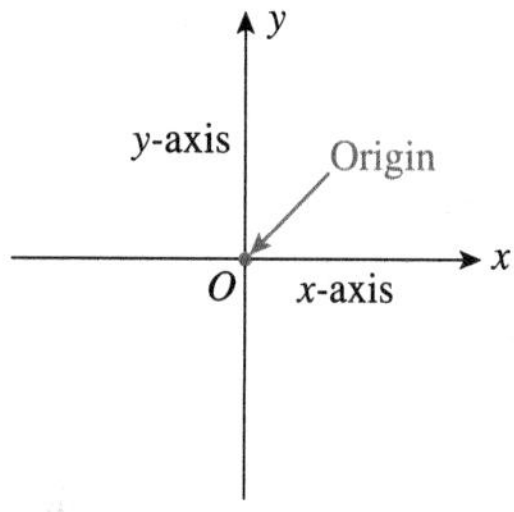

Figure 1.5
The Cartesian coordinate system

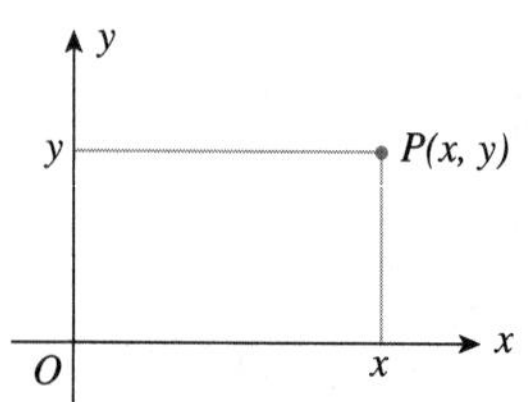

Figure 1.6
An ordered pair (x, y)

The Cartesian Coordinate System

In Section 1.1 we saw how a one-to-one correspondence between the set of real numbers and the points on a straight line leads to a coordinate system on a line (a one-dimensional space).

A similar representation for points in a plane (a two-dimensional space) is realized through the **Cartesian coordinate system,** which is constructed as follows: Take two perpendicular lines, one of which is normally chosen to be horizontal. These lines intersect at a point O, called the **origin** (Figure 1.5). The horizontal line is called the **x-axis,** and the vertical line is called the **y-axis.** A number scale is set up along the x-axis, with the positive numbers lying to the right of the origin and the negative numbers lying to the left of it. Similarly, a number scale is set up along the y-axis, with the positive numbers lying above the origin and the negative numbers lying below it.

The number scales on the two axes need not be the same. Indeed, in many applications different quantities are represented by x and y. For example, x may represent the number of typewriters sold and y the total revenue resulting from the sales. In such cases it is often desirable to choose different number scales to represent the different quantities. Note, however, that the zeros of both number scales coincide at the origin of the two-dimensional coordinate system.

A point in the plane can now be represented uniquely in this coordinate system by an **ordered pair** of numbers—that is, a pair (x, y), where x is the first number and y the second. To see this, let P be any point in the plane (Figure 1.6). Draw perpendiculars from P to the x-axis and y-axis, respectively. Then the number x is precisely the number that corresponds to the point on the x-axis at which the perpendicular through P hits the x-axis. Similarly, y is the number that corresponds to the point on the y-axis at which the perpendicular through P crosses the y-axis.

Conversely, given an ordered pair (x, y) with x as the first number and y the second, a point P in the plane is uniquely determined as follows: Locate the point on the x-axis represented by the number x and draw a line through that point parallel to the y-axis. Next, locate the point on the y-axis represented by the number y and draw a line through that point parallel to the x-axis. The point of intersection of these two lines is the point P (see Figure 1.6).

In the ordered pair (x, y), x is called the **abscissa,** or **x-coordinate,** y is called the **ordinate,** or **y-coordinate,** and x and y together are referred to as the **coordinates** of the point P.

Letting (a, b) denote the point P with x-coordinate a and y-coordinate b, the points $A = (2, 3)$, $B = (-2, 3)$, $C = (-2, -3)$, $D = (2, -3)$, $E = (3, 2)$, $F = (4, 0)$, and $G = (0, -5)$ are plotted in Figure 1.7. The fact that, in general, $(x, y) \neq (y, x)$ is clearly illustrated by points A and E.

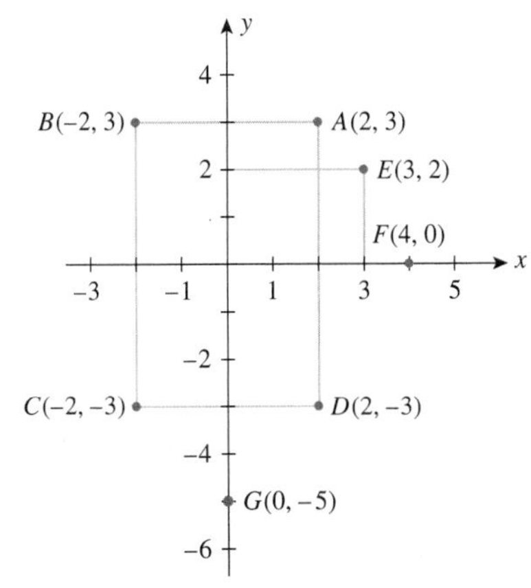

Figure 1.7
Several points in the Cartesian plane

The axes divide the plane into four quadrants. Quadrant I consists of the points (x, y) that satisfy $x > 0$ and $y > 0$ and is therefore involved $(+, +)$ in Figure 1.8; Quadrant II, the points (x, y), where $x < 0$ and $y > 0$, $(-, +)$; Quadrant III, the points (x, y), where $x < 0$ and $y < 0$, $(-, -)$; and Quadrant IV, the points (x, y), where $x > 0$ and $y < 0$, $(+, -)$ (Figure 1.8).

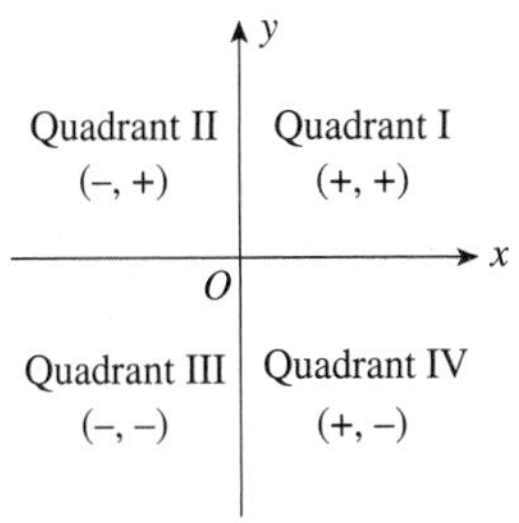

Figure 1.8
The four quadrants in the Cartesian plane

The Distance Formula

One immediate benefit that arises from using the Cartesian coordinate system is that the distance between any two points in the plane may be expressed solely in terms of their coordinates. Suppose, for example, (x_1, y_1) and (x_2, y_2) are any two points in the plane (Figure 1.9). Then the distance between these two points can be computed using the following formula.

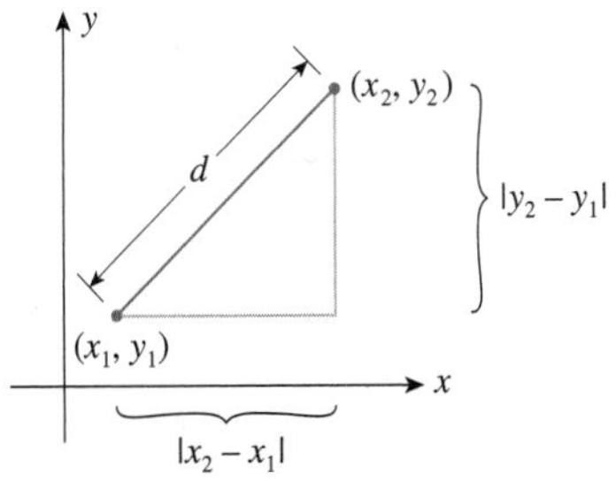

Figure 1.9
The distance d between the points (x_1, y_1) and (x_2, y_2)

Distance Formula

The distance d between two points $P_1(x_1, y_1)$ and $P_2(x_2, y_2)$ in the plane is given by

$$\begin{aligned} d &= \sqrt{(|x_2 - x_1|)^2 + (|y_2 - y_1|)^2} \\ &= \sqrt{(x_2 - x_1)^2 + (y_2 - y_1)^2} \end{aligned} \qquad (1)$$

We can drop the absolute value since squaring also ensures a positive answer.

For a proof of this result, see Exercise 37, page 27.

In what follows, we give several applications of the distance formula.

Example 1

Find the distance between the points $(-4, 3)$ and $(2, 6)$.

Solution

Let $P_1(-4, 3)$ and $P_2(2, 6)$ be points in the plane. Then, we have

$$x_1 = -4, \qquad y_1 = 3, \qquad x_2 = 2, \qquad y_2 = 6$$

Using Formula (1), we have

$$\begin{aligned} d &= \sqrt{[2 - (-4)]^2 + (6 - 3)^2} \\ &= \sqrt{6^2 + 3^2} \\ &= \sqrt{45} = 3\sqrt{5} \end{aligned}$$

GROUP DISCUSSION

Refer to Example 1. Suppose we label the point (2, 6) as P_1 and the point $(-4, 3)$ as P_2. (1) Show that the distance d between the two points is the same as that obtained earlier. (2) Prove that, in general, the distance d in Formula (1) is independent of the way we label the two points.

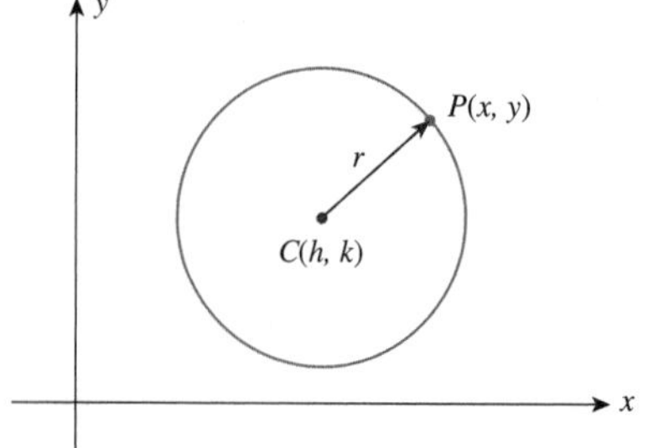

Figure 1.10
A circle with radius r and centre $C(h, k)$

Example 2

Let $P(x, y)$ denote a point lying on the circle with radius r and centre $C(h, k)$ (Figure 1.10). Find a relationship between x and y.

Solution

By the definition of a circle, the distance between $C(h, k)$ and $P(x, y)$ is r. Using Formula (1), we have

$$\sqrt{(x-h)^2+(y-k)^2}=r$$

which, upon squaring both sides, gives the equation

$$(x-h)^2+(y-k)^2=r^2$$

that must be satisfied by the variables x and y.

A summary of the result obtained in Example 2 follows.

Equation of a Circle

An equation of the circle with centre $C(h, k)$ and radius r is given by

$$(x-h)^2+(y-k)^2=r^2 \qquad (2)$$

Example 3

Find an equation of the circle with

a. Radius 2 and centre $(-1, 3)$.

b. Radius 3 and centre located at the origin.

Solution

a. We use Formula (2) with $r = 2$, $h = -1$, and $k = 3$, obtaining

$$[x-(-1)]^2+(y-3)^2=2^2 \qquad \text{or} \qquad (x+1)^2+(y-3)^2=4$$

(Figure 1.11a).

b. Using Formula (2) with $r = 3$ and $h = k = 0$, we obtain

$$x^2+y^2=3^2 \qquad \text{or} \qquad x^2+y^2=9$$

(Figure 1.11b).

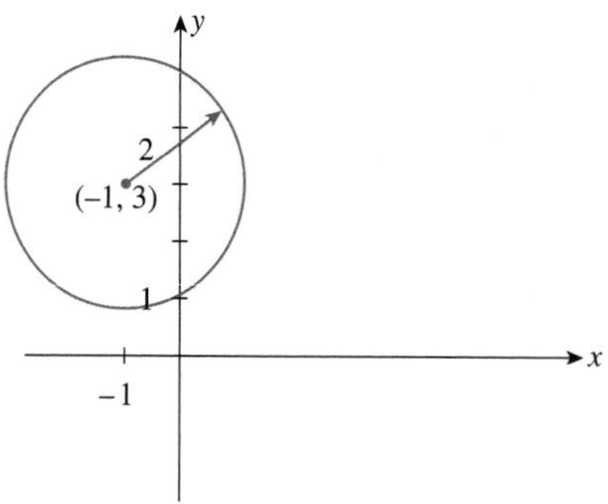

(a) The circle with radius 2 and centre $(-1, 3)$

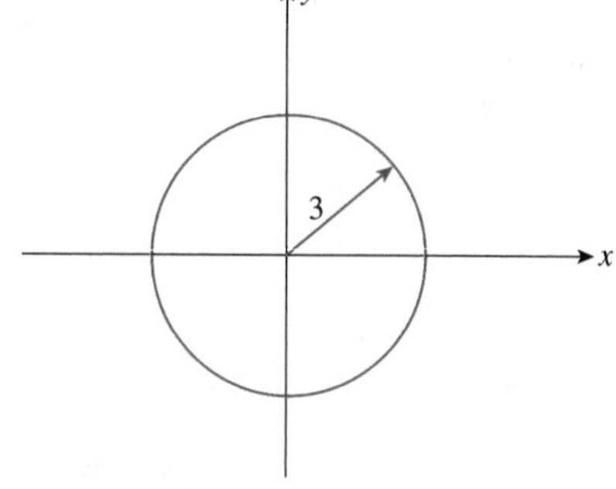

(b) The circle with radius 3 and centre (0, 0)

Figure 1.11

GROUP DISCUSSION

1. Use the distance formula to help you describe the set of points in the xy-plane satisfying each of the following inequalities.
   a. $(x - h)^2 + (y - k)^2 \leq r^2$ **b.** $(x - h)^2 + (y - k)^2 < r^2$
   c. $(x - h)^2 + (y - k)^2 \geq r^2$ **d.** $(x - h)^2 + (y - k)^2 > r^2$
2. Consider the equation $x^2 + y^2 = 4$.
   a. Show that $y = \pm\sqrt{4 - x^2}$.
   b. Describe the set of points (x, y) in the xy-plane satisfying the following equations:
   (i) $y = \sqrt{4 - x^2}$ **(ii)** $y = -\sqrt{4 - x^2}$

Application

Example 4

Cost of Laying Cable In Figure 1.12, S represents the position of a power relay station located on a straight coastal highway, and M shows the location of a marine biology experimental station on an island. A cable is to be laid connecting the relay station with the experimental station. If the cost of running the cable on land is \$1.50 per running metre and the cost of running the cable under water is \$2.50 per running metre, find the total cost for laying the cable.

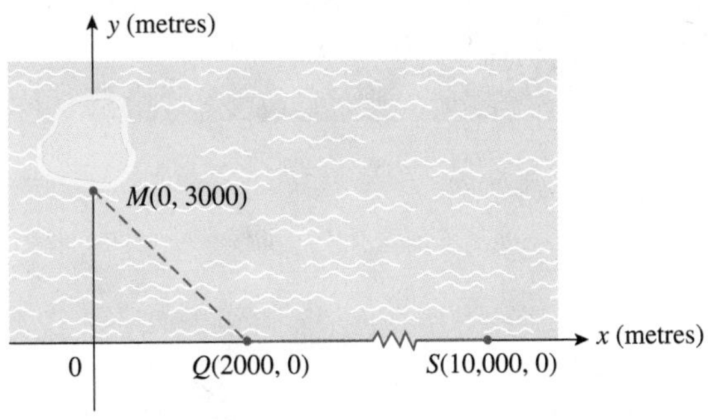

Figure 1.12
Cable connecting relay station S to experimental station M

Solution

The length of cable required on land is given by the distance from S to Q. This distance is (10,000 − 2000), or 8000 metres. Next, we see that the length of cable required underwater is given by the distance from M to Q. This distance is

$$\begin{aligned}\sqrt{(0 - 2000)^2 + (3000 - 0)^2} &= \sqrt{2000^2 + 3000^2}\\ &= \sqrt{13{,}000{,}000}\\ &\approx 3605.55\end{aligned}$$

or approximately 3605.55 metres. Therefore, the total cost for laying the cable is

$$1.5(8000) + 2.5(3605.55) = 21{,}013.875$$

or approximately \$21,014.

GROUP DISCUSSION

In the Cartesian coordinate system, the two axes are perpendicular to each other. Consider a coordinate system in which the x- and y-axes are noncollinear (that is, the axes do not lie along a straight line) and are not perpendicular to each other (see the accompanying figure).

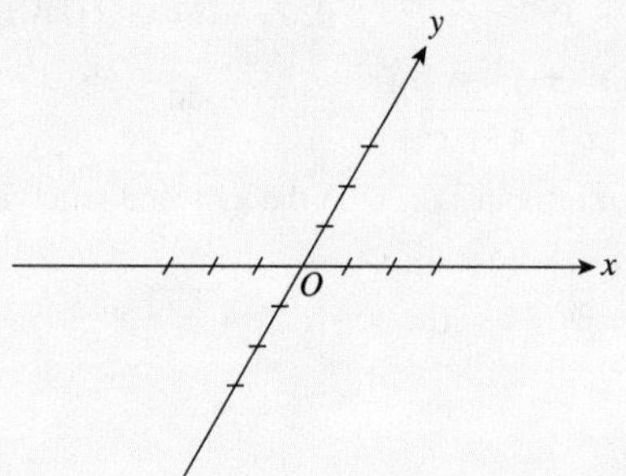

1. Describe how a point is represented in this coordinate system by an ordered pair (x, y) of real numbers. Conversely, show how an ordered pair (x, y) of real numbers uniquely determines a point in the plane.
2. Suppose you want to find a formula for the distance between two points $P_1(x_1, y_1)$ and $P_2(x_2, y_2)$ in the plane. What is the advantage that the Cartesian coordinate system has over the coordinate system under consideration? Comment on your answer.

Self-Check Exercises 1.3

1. **a.** Plot the points $A(4, -2)$, $B(2, 3)$, and $C(-3, 1)$.

   b. Find the distance between the points A and B; between B and C; between A and C.

   c. Use the Pythagorean theorem to show that the triangle with vertices A, B, and C is a right triangle.

2. The following figure shows the location of cities A, B, and C. Suppose a pilot wishes to fly from city A to city C but must make a mandatory stopover in city B. If the single-engine light plane has a range of 650 kilometres, can it make the trip without refueling in city B?

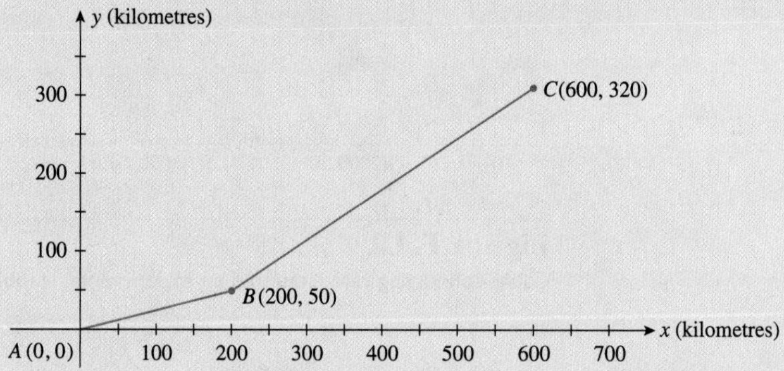

Solutions to Self-Check Exercises 1.3 can be found on page 29.

1.3 Exercises

In Exercises 1–6, refer to the following figure and determine the coordinates of each point and the quadrant in which it is located.

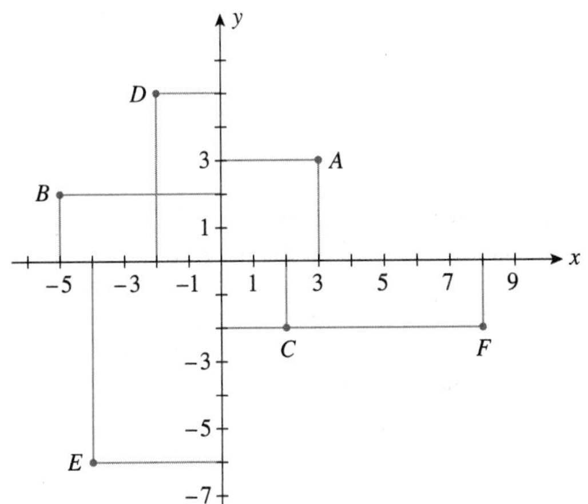

1. A	2. B	3. C
4. D	5. E	6. F

In Exercises 7–12, refer to the following figure.

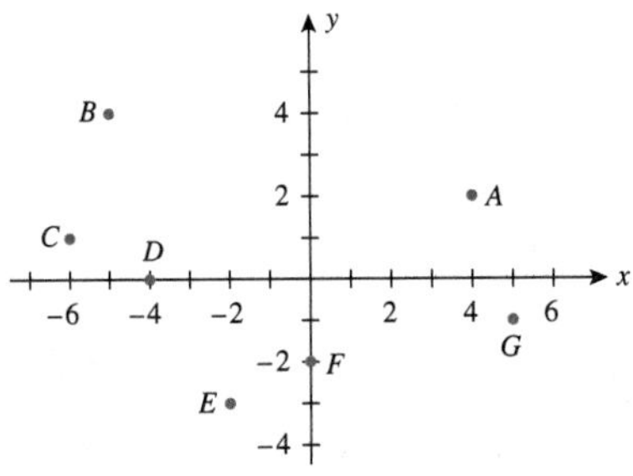

7. Which point has coordinates (4, 2)?
8. What are the coordinates of point B?
9. Which points have negative y-coordinates?
10. Which point has a negative x-coordinate and a negative y-coordinate?
11. Which point has an x-coordinate that is equal to zero?
12. Which point has a y-coordinate that is equal to zero?

In Exercises 13–20, sketch a set of coordinate axes and plot each point.

13. $(-2, 5)$	14. $(1, 3)$
15. $(3, -1)$	16. $(3, -4)$
17. $\left(8, -\frac{7}{2}\right)$	18. $\left(-\frac{5}{2}, \frac{3}{2}\right)$
19. $(4.5, -4.5)$	20. $(1.2, -3.4)$

In Exercises 21–24, find the distance between the given points.

21. $(1, 3)$ and $(4, 7)$	22. $(1, 0)$ and $(4, 4)$
23. $(-1, 3)$ and $(4, 9)$	24. $(-2, 1)$ and $(10, 6)$

25. Find the coordinates of the points that are 10 units away from the origin and have a y-coordinate equal to -6.
26. Find the coordinates of the points that are 5 units away from the origin and have an x-coordinate equal to 3.
27. Show that the points $(3, 4)$, $(-3, 7)$, $(-6, 1)$, and $(0, -2)$ form the vertices of a square.
28. Show that the triangle with vertices $(-5, 2)$, $(-2, 5)$, and $(5, -2)$ is a right triangle.

In Exercises 29–34, find an equation of the circle that satisfies the given conditions.

29. Radius 5 and centre $(2, -3)$
30. Radius 3 and centre $(-2, -4)$
31. Radius 5 and centre at the origin
32. Centre at the origin and passes through $(2, 3)$
33. Centre $(2, -3)$ and passes through $(5, 2)$
34. Centre $(-a, a)$ and radius $2a$
35. Two ships leave port at the same time. Ship A sails north at a speed of 20 km/h while ship B sails east at a speed of 30 km/h.
    a. Find an expression in terms of the time t (in hours) giving the distance between the two ships.
    b. Using the expression obtained in part (a), find the distance between the two ships 2 h after leaving port.
36. Ship A leaves port sailing north at a speed of 25 km/h. A half hour later, ship B leaves the same port sailing east at a speed of 20 km/h. Let t (in hours) denote the time ship B has been at sea.
    a. Find an expression in terms of t giving the distance between the two ships.
    b. Use the expression obtained in part (a) to find the distance between the two ships 2 h after ship A has left port.
37. Let (x_1, y_1) and (x_2, y_2) be two points lying in the xy-plane. Show that the distance between the two points is given by

$$d = \sqrt{(x_2 - x_1)^2 + (y_2 - y_1)^2}$$

Hint: Refer to the accompanying figure and use the Pythagorean theorem.

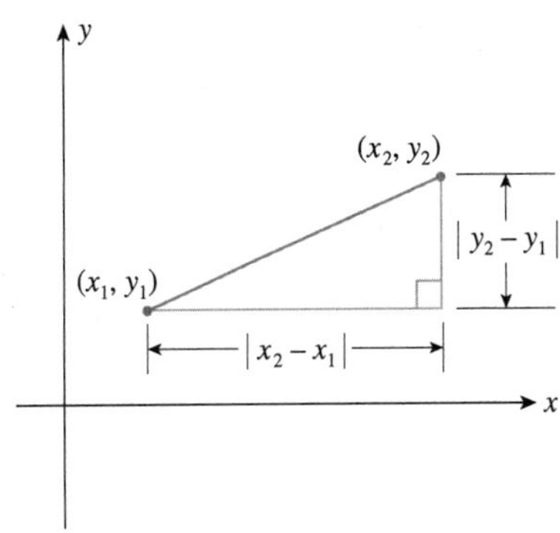

38. Show that an equation of a circle can be written in the form

$$x^2 + y^2 + Cx + Dy + E = 0$$

where C, D, and E are constants. This is called the general form of an equation of a circle.

39. **Distance Travelled** A grand tour of four cities begins at city A and makes successive stops at cities B, C, and D before returning to city A. If the cities are located as shown in the following figure, find the total distance covered on the tour.

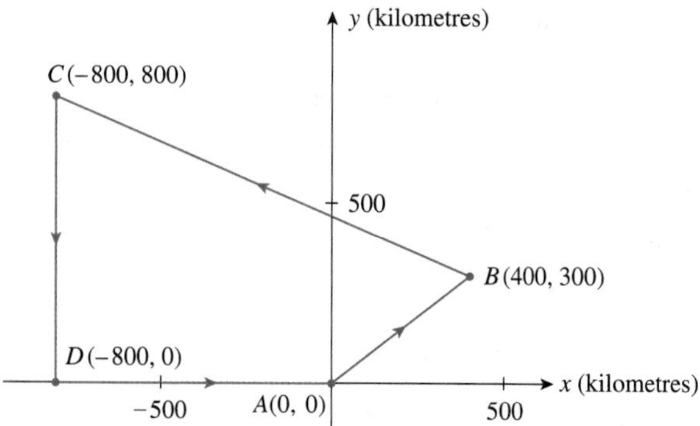

Business and Economics Applications

40. **Delivery Charges** A furniture store offers free setup and delivery services to all points within a 25-km radius of its warehouse distribution centre. If you live 20 km east and 14 km south of the warehouse, will you incur a delivery charge? Justify your answer.

41. **Travel Time** Towns A, B, C, and D are located as shown in the following figure. Two highways link town A to town D. Route 1 runs from town A to town D via town B, and Route 2 runs from town A to town D via town C. If a salesperson wishes to drive from town A to town D and traffic conditions are such that he could expect to average the same speed on either route, which highway should he take in order to arrive in the shortest time?

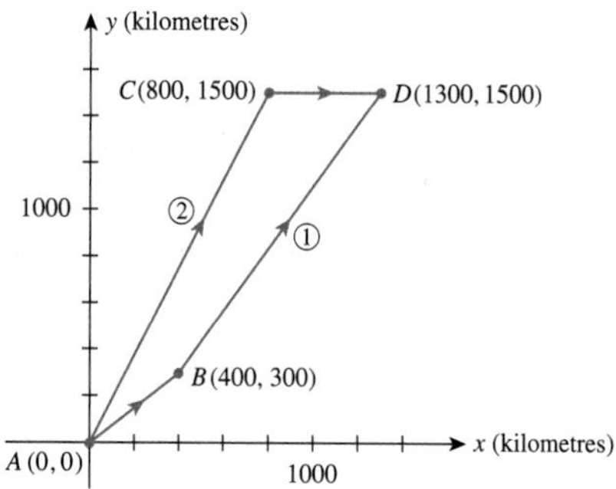

42. **Minimizing Shipping Costs** Refer to the figure for Exercise 41. Suppose a fleet of 100 automobiles are to be shipped from an assembly plant in town A to town D. They may be shipped either by freight train along Route 1 at a cost of 22¢/km per automobile or by truck along Route 2 at a cost of 21¢/km per automobile. Which means of transportation minimizes the shipping cost? What is the net savings?

43. **Consumer Decisions** Ivan wishes to determine which antenna he should purchase for his home. The TV store has supplied him with the following information:

Range in Kilometres			
VHF	UHF	Model	Price
30	20	A	\$40
45	35	B	\$50
60	40	C	\$60
75	55	D	\$70

Ivan wishes to receive Channel 17 (VHF), which is located 25 km east and 35 km north of his home, and Channel 38 (UHF), which is located 20 km south and 32 km west of his home. Which model will allow him to receive both channels at the least cost? (Assume that the terrain between Ivan's home and both broadcasting stations is flat.)

44. **Cost of Laying Cable** In the following diagram, S represents the position of a power relay station located on a straight coastal highway, and M shows the location of a marine biology experimental station on an island. A cable is to be laid connecting the relay station with the experimental station. If the cost of running the cable on land is \$1.50/running metre and the cost of running cable under water is \$2.50/running metre, find an expression in terms of x that gives the total cost for laying the cable. What is the total cost when $x = 2500$? when $x = 3000$?

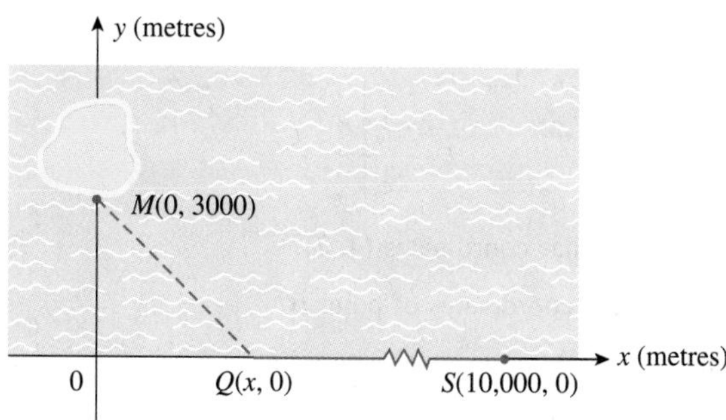

In Exercises 45–48, determine whether the statement is true or false. If it is true, explain why it is true. If it is false, give an example to show why it is false.

45. The point $(-a, b)$ is symmetric to the point (a, b) with respect to the y-axis.

46. The point $(-a, -b)$ is symmetric to the point (a, b) with respect to the origin.

47. If the distance between the points $P_1(a, b)$ and $P_2(c, d)$ is D, then the distance between the points $P_1(a, b)$ and $P_3(kc, kd)$, $(k \neq 0)$, is given by $|k|D$.

48. The circle with equation $kx^2 + ky^2 = a^2$ lies inside the circle with equation $x^2 + y^2 = a^2$, provided $k > 1$.

Solutions to Self-Check Exercises 1.3

1. a. The points are plotted in the following figure:

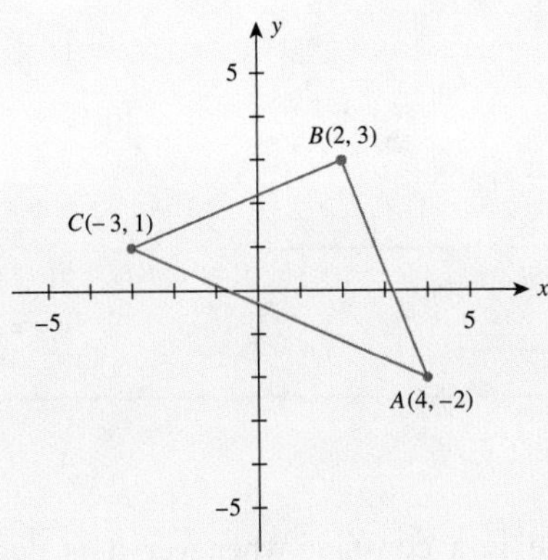

b. The distance between A and B is

$$d(A, B) = \sqrt{(2-4)^2 + [3-(-2)]^2}$$
$$= \sqrt{(-2)^2 + 5^2} = \sqrt{4+25} = \sqrt{29}$$

The distance between B and C is

$$d(B, C) = \sqrt{(-3-2)^2 + (1-3)^2}$$
$$= \sqrt{(-5)^2 + (-2)^2} = \sqrt{25+4} = \sqrt{29}$$

The distance between A and C is

$$d(A, C) = \sqrt{(-3-4)^2 + [1-(-2)]^2}$$
$$= \sqrt{(-7)^2 + 3^2} = \sqrt{49+9} = \sqrt{58}$$

c. We will show that

$$[d(A, C)]^2 = [d(A, B)]^2 + [d(B, C)]^2$$

From part (b), we see that $[d(A, B)]^2 = 29$, $[d(B, C)]^2 = 29$, and $[d(A, C)]^2 = 58$, and the desired result follows.

2. The distance between city A and city B is

$$d(A, B) = \sqrt{200^2 + 50^2} \approx 206$$

or 206 km. The distance between city B and city C is

$$d(B, C) = \sqrt{[600-200]^2 + [320-50]^2}$$
$$= \sqrt{400^2 + 270^2} \approx 483$$

or 483 km. Therefore, the total distance the pilot would have to cover is 689 km, so she must refuel in city B.

1.4 Straight Lines

In computing income tax, business firms are allowed by law to depreciate certain assets such as buildings, machines, furniture, automobile, and so on, over a period of time. Linear depreciation, or the straight-line method, is often used for this purpose. The graph of the straight line shown in Figure 1.13 describes the book value V of a computer that has an initial value of \$100,000 and that is being depreciated linearly over 5 years with a scrap value of \$30,000. Note that only the solid portion of the straight line is of interest here.

The book value of the computer at the end of year t, where t lies between 0 and 5, can be read directly from the graph. But there is one shortcoming in this approach: The result depends on how accurately you draw and read the graph. A better and more accurate method is based on finding an *algebraic* representation of the depreciation line.

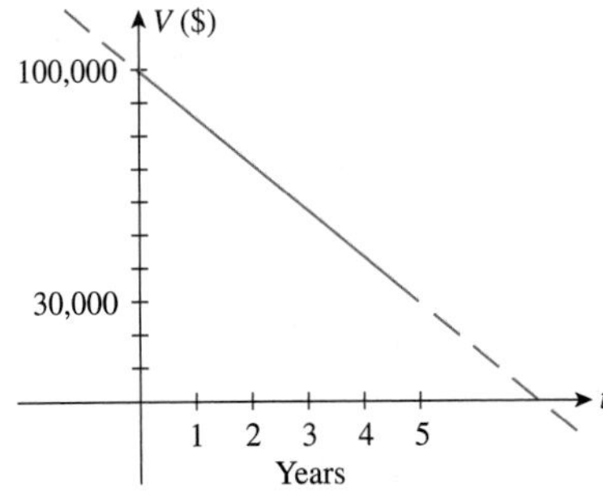

Figure 1.13
Linear depreciation of an asset

Slope of a Line

To see how a straight line in the xy-plane may be described algebraically, we need to first recall certain properties of straight lines. Let L denote the unique straight line that passes through the two distinct points (x_1, y_1) and (x_2, y_2). If $x_1 = x_2$, then L is a vertical line, and the slope is undefined (Figure 1.14).

If $x_1 \neq x_2$, we define the slope of L as follows:

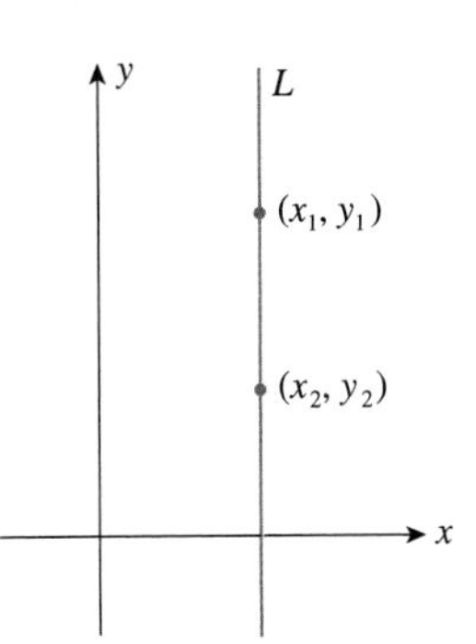

Figure 1.14
Slope m is undefined.

Slope of a Nonvertical Line

If (x_1, y_1) and (x_2, y_2) are any two distinct points on a nonvertical line L, then the slope m of L is given by

$$m = \frac{\Delta y}{\Delta x} = \frac{y_2 - y_1}{x_2 - x_1} \qquad (3)$$

See Figure 1.15.

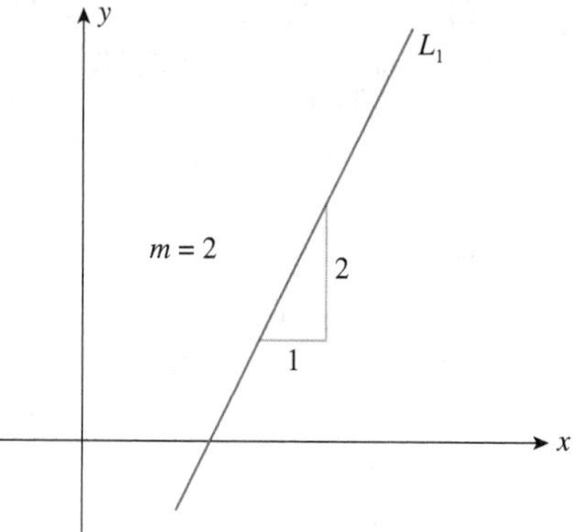

(a) The line rises ($m > 0$).

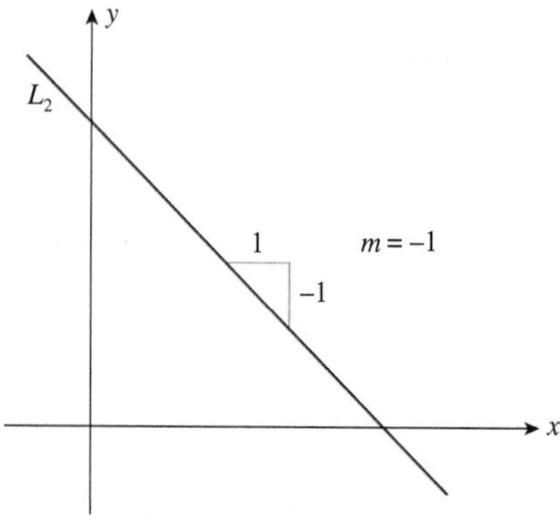

(b) The line falls ($m < 0$).

Figure 1.16

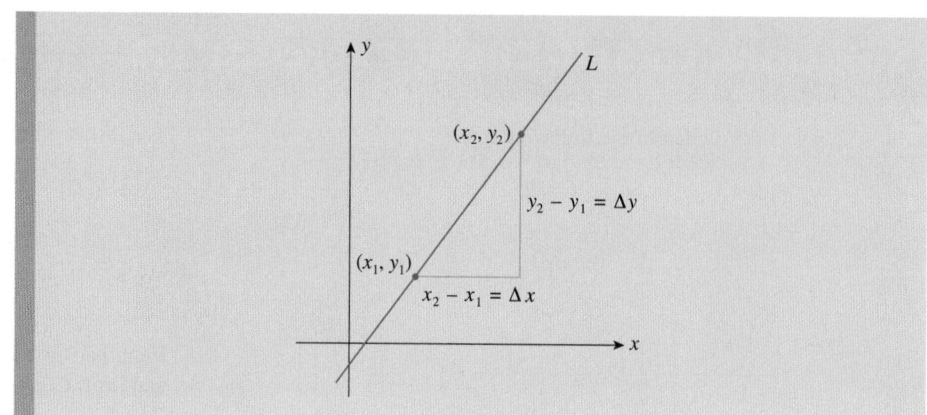

Figure 1.15

Observe that the slope of a straight line is a constant whenever it is defined. The number $\Delta y = y_2 - y_1$ (Δy is read "delta y") is a measure of the vertical change in y, and $\Delta x = x_2 - x_1$ is a measure of the horizontal change in x, as shown in Figure 1.15. From this figure we can see that the slope m of a straight line L is a measure of the *rate of change of y with respect to x*.

Figure 1.16a shows a straight line L_1 with slope 2. Observe that L_1 has the property that a 1-unit increase in x results in a 2-unit increase in y. To see this, let $\Delta x = 1$ in Formula (3) so that $m = \Delta y$. Since $m = 2$, we conclude that $\Delta y = 2$. Similarly, Figure 1.16b shows a line L_2 with slope -1. Observe that a straight line with positive slope slants upward from left to right (y increases as x increases), whereas a line with negative slope slants downward from left to right (y decreases as x increases). Finally, Figure 1.17 shows a family of straight lines passing through the origin with indicated slopes.

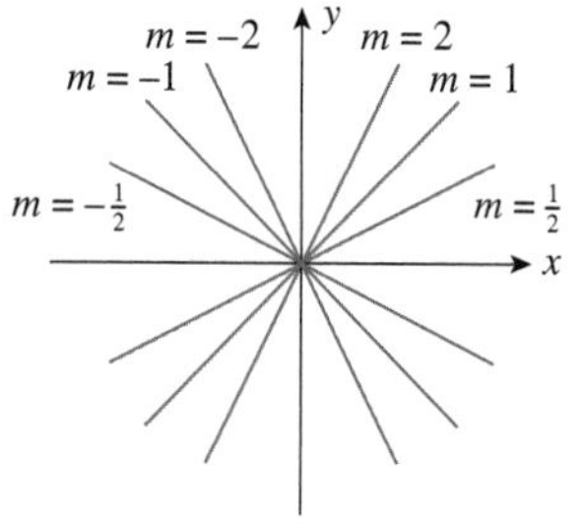

Figure 1.17
A family of straight lines with y-intercept 0

Example 1

Sketch the straight line that passes through the point $(-2, 5)$ and has slope $-\frac{4}{3}$.

Solution

First, plot the point $(-2, 5)$ (Figure 1.18). Next, recall that a slope of $-\frac{4}{3}$ indicates that an increase of 1 unit in the x-direction produces a *decrease* of $\frac{4}{3}$ units in the y-direction, or equivalently, a 3-unit increase in the x-direction produces a $3\left(\frac{4}{3}\right)$, or 4-unit, decrease in the y-direction. Using this information, we plot the point (1, 1) and draw the line through the two points.

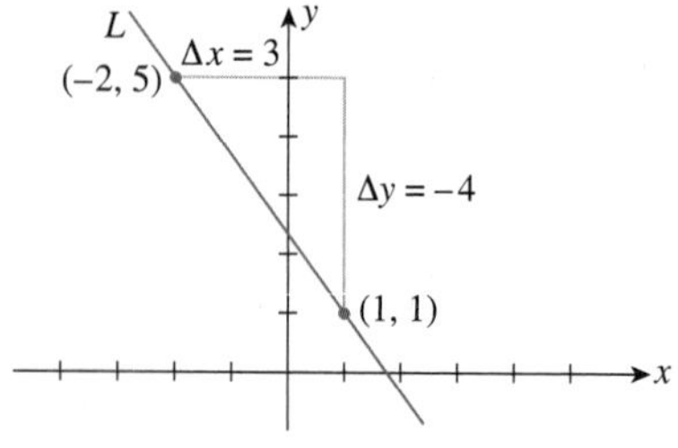

Figure 1.18
L has slope $-\frac{4}{3}$ and passes through $(-2, 5)$.

GROUP DISCUSSION

Show that the slope of a nonvertical line is independent of the two distinct points $P_1(x_1, y_1)$ and $P_2(x_2, y_2)$ used to compute it.

Hint: Suppose we pick two other distinct points, $P_3(x_3, y_3)$ and $P_4(x_4, y_4)$ lying on L. Draw a picture and use similar triangles to demonstrate that using P_3 and P_4 gives the same value as that obtained using P_1 and P_2.

Example 2

Find the slope m of the line that passes through the points $(-1, 1)$ and $(5, 3)$.

Solution

Choose (x_1, y_1) to be the point $(-1, 1)$ and (x_2, y_2) to be the point $(5, 3)$. Then, with $x_1 = -1$, $y_1 = 1$, $x_2 = 5$, and $y_2 = 3$, we find

$$m = \frac{y_2 - y_1}{x_2 - x_1} = \frac{3 - 1}{5 - (-1)} = \frac{1}{3} \qquad \text{Use Formula (3).}$$

(Figure 1.19). Try to verify that the result obtained would have been the same had we chosen the point $(-1, 1)$ to be (x_2, y_2) and the point $(5, 3)$ to be (x_1, y_1).

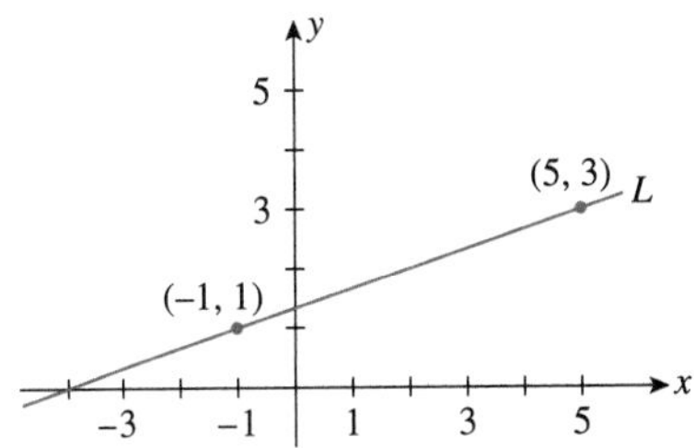

Figure 1.19
L passes through (5, 3) and (−1, 1).

Example 3

Find the slope of the line that passes through the points $(-2, 5)$ and $(3, 5)$.

Solution

The slope of the required line is given by

$$m = \frac{5 - 5}{3 - (-2)} = \frac{0}{5} = 0$$

(Figure 1.20).

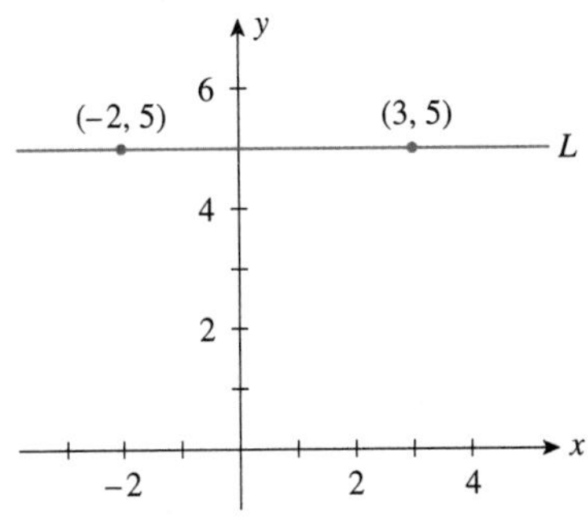

Figure 1.20
The slope of the horizontal line L is 0.

Remark

In general, the slope of a horizontal line is zero. ◀

We can use the slope of a straight line to determine whether a line is parallel to another line.

> **Parallel Lines**
>
> Two distinct lines are **parallel** if and only if their slopes are equal or their slopes are undefined.

Example 4

Let L_1 be a line that passes through the points $(-2, 9)$ and $(1, 3)$ and let L_2 be the line that passes through the points $(-4, 10)$ and $(3, -4)$. Determine whether L_1 and L_2 are parallel.

Solution

The slope m_1 of L_1 is given by

$$m_1 = \frac{3 - 9}{1 - (-2)} = -2$$

The slope m_2 of L_2 is given by

$$m_2 = \frac{-4 - 10}{3 - (-4)} = -2$$

Since $m_1 = m_2$, the lines L_1 and L_2 are in fact parallel (Figure 1.21).

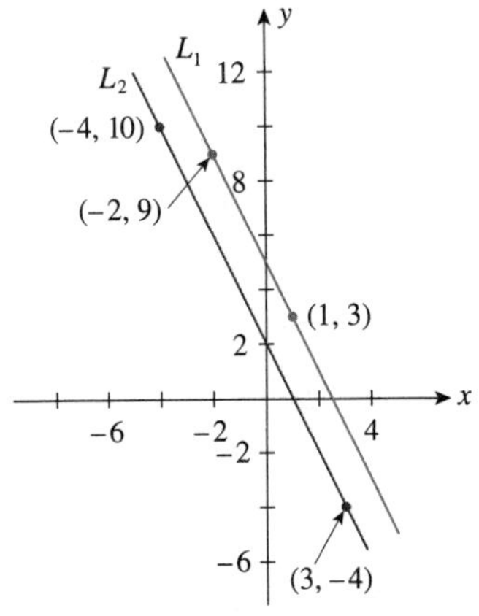

Figure 1.21
L_1 and L_2 have the same slope and hence are parallel.

Equations of Lines

We will now show that every straight line lying in the xy-plane may be represented by an equation involving the variables x and y. One immediate benefit of this is that problems involving straight lines may be solved algebraically.

Let L be a straight line parallel to the y-axis (perpendicular to the x-axis) (Figure 1.22). Then, L crosses the x-axis at some point $(a, 0)$ with the x-coordinate given by $x = a$, where a is some real number. Any other point on L has the form $(a, \bar{y})$, where $\bar{y}$ is an appropriate number. Therefore, the vertical line L is described by the sole condition

$$x = a$$

and this is, accordingly, the equation of L. For example, the equation $x = -2$ represents a vertical line 2 units to the left of the y-axis, and the equation $x = 3$ represents a vertical line 3 units to the right of the y-axis (Figure 1.23).

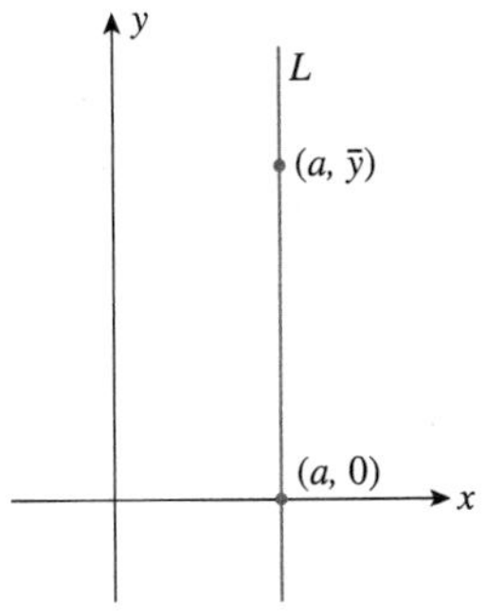

Figure 1.22
The vertical line $x = a$

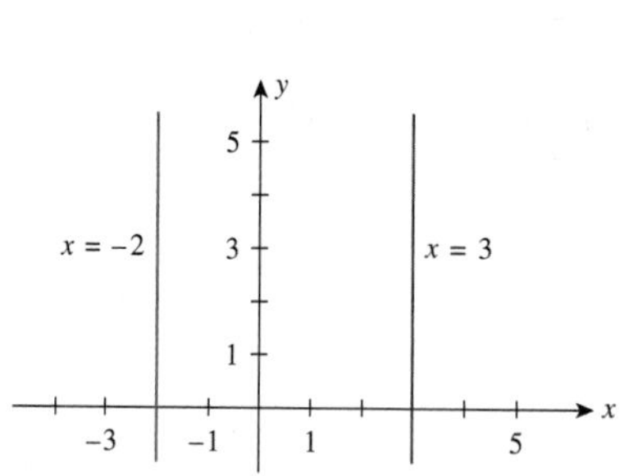

Figure 1.23
The vertical lines $x = -2$ and $x = 3$

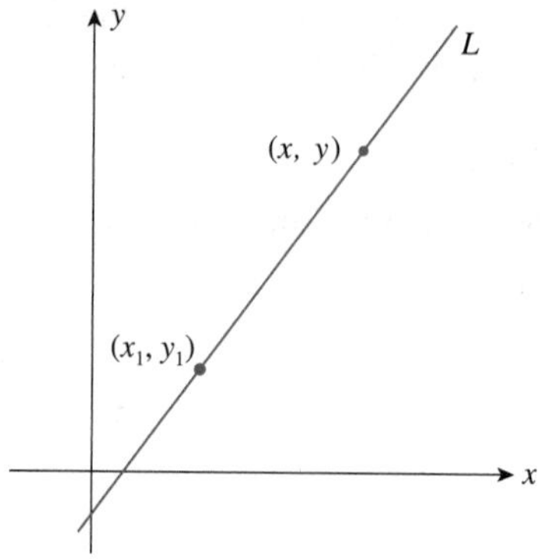

Figure 1.24
L passes through (x_1, y_1) and has slope m.

Next, suppose L is a nonvertical line so that it has a well-defined slope m. Suppose (x_1, y_1) is a fixed point lying on L and (x, y) is a variable point on L distinct from (x_1, y_1) (Figure 1.24).

Using Formula (3) with the point $(x_2, y_2) = (x, y)$, we find that the slope of L is given by

$$m = \frac{y - y_1}{x - x_1}$$

Upon multiplying both sides of the equation by $x - x_1$, we obtain Formula (4).

Point-Slope Form

An equation of the line that has slope m and passes through the point (x_1, y_1) is given by

$$y - y_1 = m(x - x_1) \qquad (4)$$

Equation (4) is called the **point-slope form of the equation of a line** since it utilizes a given point (x_1, y_1) on a line and the slope m of the line.

Remark

If the two distinct points on a line L are given by (x_1, b) and (x_2, b) with $x_1 \neq x_2$, then the slope is given by

$$m = \frac{b - b}{x_2 - x_1} = \frac{0}{x_2 - x_1} = 0$$

The equation describing the line L can be found by

$$\begin{aligned} y - y_1 &= m(x - x_1) \\ y - b &= 0(x - x_1) \\ y &= b \end{aligned}$$

Therefore $y = b$ describes a horizontal line with slope 0 (Figure 1.25). ◀

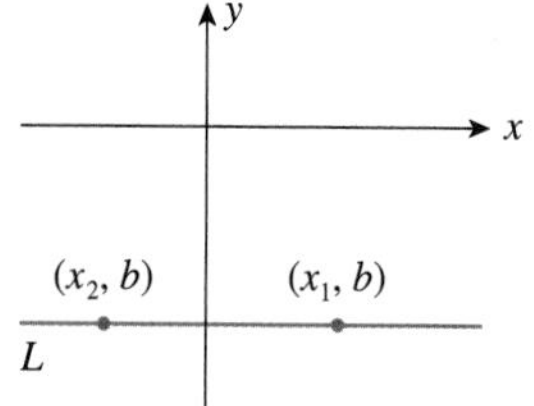

Figure 1.25
The horizontal line $y = b$

Example 5

Find an equation of the line that passes through the point (1, 3) and has slope 2.

Solution

Using the point-slope form of the equation of a line with the point (1, 3) and $m = 2$, we obtain

$$y - 3 = 2(x - 1) \qquad (y - y_1) = m(x - x_1)$$

which, when simplified, becomes

$$2x - y + 1 = 0$$

(Figure 1.26).

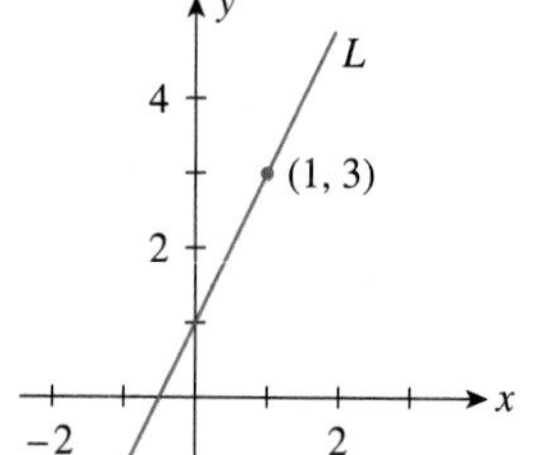

Figure 1.26
L passes through (1, 3) and has slope 2.

Example 6

Find an equation of the line that passes through the points $(-3, 2)$ and $(4, -1)$.

Solution

The slope of the line is given by

$$m = \frac{-1 - 2}{4 - (-3)} = -\frac{3}{7}$$

Using the point-slope form of an equation of a line with the point $(4, -1)$ and the slope $m = -\frac{3}{7}$, we have

$$\begin{aligned} y + 1 &= -\frac{3}{7}(x - 4) \qquad (y - y_1) = m(x - x_1) \\ 7y + 7 &= -3x + 12 \\ 3x + 7y - 5 &= 0 \end{aligned}$$

(Figure 1.27).

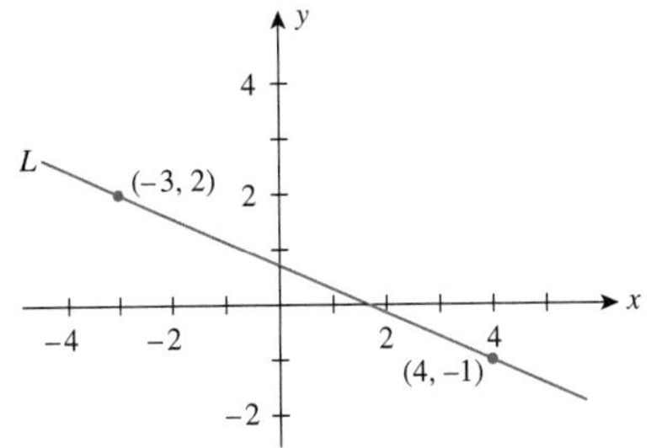

Figure 1.27
L passes through $(-3, 2)$ and $(4, -1)$

GROUP DISCUSSION

Consider the slope-intercept form of an equation of a straight line $y = mx + b$. Describe the family of straight lines obtained by keeping

1. The value of m fixed and allowing the value of b to vary.
2. The value of b fixed and allowing the value of m to vary.

We can use the slope of a straight line to determine whether a line is perpendicular to another line.

Perpendicular Lines

If L_1 and L_2 are two distinct nonvertical lines that have slopes m_1 and m_2, respectively, then L_1 is **perpendicular** to L_2 (written $L_1 \perp L_2$) if and only if

$$m_1 = -\frac{1}{m_2}$$

If the line L_1 is vertical (so that its slope is undefined), then L_1 is perpendicular to another line, L_2, if and only if L_2 is horizontal (so that its slope is zero). For a proof of these results, see Exercise 76, page 40.

Example 7

Find an equation of the line that passes through the point $(3, 1)$ and is perpendicular to the line of Example 5.

Solution

Since the slope of the line in Example 5 is 2, the slope of the required line is given by $m = -\frac{1}{2}$, the negative reciprocal of 2. Using the point-slope form of the equation of a line, we obtain

$$\begin{aligned} y - 1 &= -\frac{1}{2}(x - 3) \qquad (y - y_1) = m(x - x_1) \\ 2y - 2 &= -x + 3 \\ x + 2y - 5 &= 0 \end{aligned}$$

(See Figure 1.28).

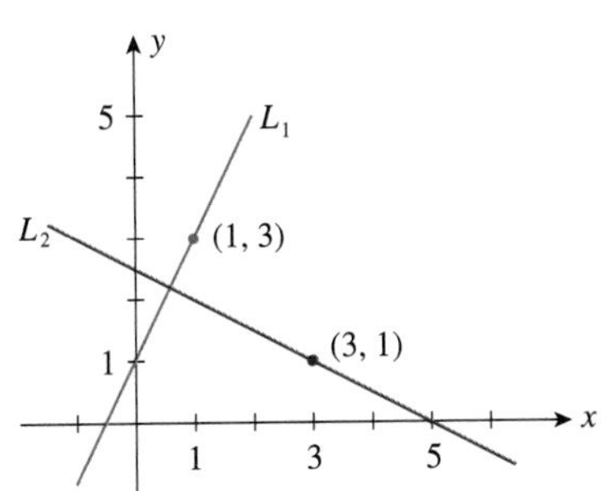

Figure 1.28
L_2 is perpendicular to L_1 and passes through $(3, 1)$.

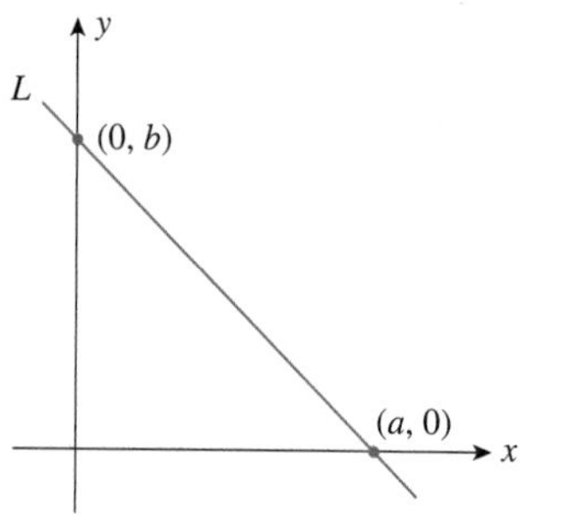

Figure 1.29
The line L has x-intercept a and y-intercept b.

A straight line L that is neither horizontal nor vertical cuts the x-axis and the y-axis at, say, points $(a, 0)$ and $(0, b)$, respectively (Figure 1.29). The numbers a and b are called the **x-intercept** and **y-intercept,** respectively, of L.

Now, let L be a line with slope m and y-intercept b. Using Formula (4), the point-slope form of the equation of a line, with the point $(0, b)$ and slope m, we have

$$y - b = m(x - 0)$$
$$y = mx + b$$

Slope-Intercept Form

An equation of the line that has slope m and intersects the y-axis at the point $(0, b)$ is given by

$$y = mx + b \qquad (5)$$

Example 8

Find an equation of the line that has slope 3 and y-intercept -4.

Solution

Using Equation (5) with $m = 3$ and $b = -4$, we obtain the required equation

$$y = 3x - 4$$

Example 9

Determine the slope and y-intercept of the line whose equation is $3x - 4y = 8$.

Solution

Rewrite the given equation in the slope-intercept form. Thus,

$$3x - 4y = 8$$
$$-4y = 8 - 3x$$
$$y = \frac{3}{4}x - 2$$

Comparing this result with Equation (5), we find $m = \frac{3}{4}$ and $b = -2$, and we conclude that the slope and y-intercept of the given line are $\frac{3}{4}$ and -2, respectively.

GROUP DISCUSSION

1. Graph the straight lines with equations $y = -2x + 3$, $y = -x + 3$, $y = x + 3$, and $y = 2.5x + 3$ on the same set of axes. What effect does changing the coefficient m of x in the equation $y = mx + b$ have on its graph?
2. Graph the straight lines with equations $y = 2x - 2$, $y = 2x - 1$, $y = 2x$, $y = 2x + 1$, and $y = 2x + 4$ on the same set of axes. What effect does changing the constant b in the equation $y = mx + b$ have on its graph?
3. Describe in words the effect of changing both m and b in the equation $y = mx + b$.

Applications

Example 10

Sales of a Sporting Goods Store The sales manager of a local sporting goods store plotted sales versus time for the last 5 years and found the points to lie approximately along a straight line (Figure 1.30). By using the points corresponding to the first and fifth years, find an equation of the trend line. What sales figure can be predicted for the sixth year?

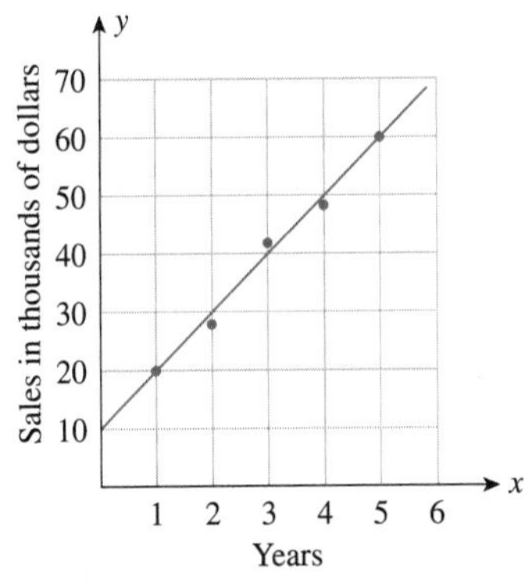

Figure 1.30
Sales of a sporting goods store

Solution

Using Formula (3) with the points (1, 20) and (5, 60), we find that the slope of the required line is given by

$$m = \frac{60 - 20}{5 - 1} = 10$$

Next, using the point-slope form of the equation of a line with the point (1, 20) and $m = 10$, we obtain

$$y - 20 = 10(x - 1) \qquad (y - y_1) = m(x - x_1)$$

$$y = 10x + 10$$

as the required equation.

The sales figure for the sixth year is obtained by letting $x = 6$ in the last equation, giving

$$y = 70$$

or $70,000.

Example 11

Appreciation in Value of an Art Object Suppose an art object purchased for $50,000 is expected to appreciate in value at a constant rate of $5000 per year for the next 5 years. Use Formula (5) to write an equation predicting the value of the art object in the next several years. What will be its value 3 years from the date of purchase?

Solution

Let x denote the time (in years) that has elapsed since the date the object was purchased and let y denote the object's value (in dollars). Then, $y = 50{,}000$ when $x = 0$. Furthermore, the slope of the required equation is given by $m = 5000$, since each unit increase in x (1 year) implies an increase of 5000 units (dollars) in y. Using (5) with $m = 5000$ and $b = 50{,}000$, we obtain

$$y = 5000x + 50{,}000 \qquad y = mx + b$$

Three years from the date of purchase, the value of the object will be given by

$$y = 5000(3) + 50{,}000$$

or $65,000.

GROUP DISCUSSION

Refer to Example 11. Can the equation predicting the value of the art object be used to predict long-term growth?

General Form of an Equation of a Line

We have considered several forms of an equation of a straight line in the plane. These different forms of the equation are equivalent to each other. In fact, each is a special case of the following equation.

General Form of a Linear Equation

The equation

$$Ax + By + C = 0 \qquad (6)$$

where A, B, and C are constants and A and B are not both zero, is called the general form of a linear equation in the variables x and y.

We will now state (without proof) an important result concerning the algebraic representation of straight lines in the plane.

THEOREM 1

An equation of a straight line is a linear equation; conversely, every linear equation represents a straight line.

This result justifies the use of the adjective *linear* describing Equation (6).

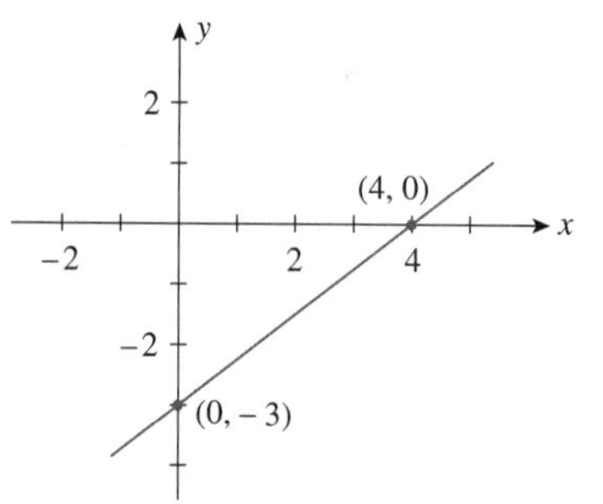

Figure 1.31
To sketch $3x - 4y - 12 = 0$, first find the x-intercept, 4, and the y-intercept, -3.

Example 12

Sketch the straight line represented by the equation

$$3x - 4y - 12 = 0$$

Solution

Since every straight line is uniquely determined by two distinct points, we need find only two such points through which the line passes in order to sketch it. For convenience, let's compute the x- and y-intercepts. Setting $y = 0$, we find $x = 4$; thus, the x-intercept is 4. Setting $x = 0$ gives $y = -3$, and the y-intercept is -3. A sketch of the line appears in Figure 1.31.

GROUP DISCUSSION

1. Graph the straight lines L_1 and L_2 with equations $2x + y - 5 = 0$ and $41x + 20y - 11 = 0$ on the same set of axes.
   a. Can you tell if the lines L_1 and L_2 are parallel to each other?
   b. Verify your observations by computing the slopes of L_1 and L_2 algebraically.
2. Graph the straight lines L_1 and L_2 with equations $x + 2y - 5 = 0$ and $5x - y + 5 = 0$ on the same set of axes.
   a. Can you tell if the lines L_1 and L_2 are perpendicular to each other?
   b. Verify your observation by computing the slopes of L_1 and L_2 algebraically.

Following is a summary of the common forms of the equations of straight lines discussed in this section.

Equations of Straight Lines

Vertical line:	$x = a$
Horizontal line:	$y = b$
Point-slope form:	$y - y_1 = m(x - x_1)$
Slope-intercept form:	$y = mx + b$
General form:	$Ax + By + C = 0$

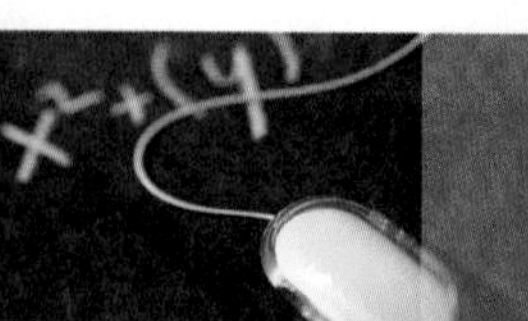

Self-Check Exercises 1.4

1. Determine the number a so that the line passing through the points $(a, 2)$ and $(3, 6)$ is parallel to a line with slope 4.
2. Find an equation of the line that passes through the point $(3, -1)$ and is perpendicular to a line with slope $-\frac{1}{2}$.
3. Does the point $(3, -3)$ lie on the line with equation $2x - 3y - 12 = 0$? Sketch the graph of the line.
4. The percent of wages for local government business enterprises in the public employment sector of Canada is summarized in the following table:

Year, x	2002	2003	2004	2005	2006
Percent of wages, y	1.94	1.90	1.96	1.97	2.03

Source: Statistics Canada, CANSIM, Table 183-0002

a. Plot the percent of wages for local government business enterprises (y) versus the year (x).

b. Draw the straight line L through the points (2002, 1.94) and (2006, 2.03).

c. Find an equation of the line L.

d. Assuming this trend continued, estimate the percent of wages for local government business enterprises by 2010.

Solutions to Self-Check Exercises 1.4 can be found on page 41.

1.4 Exercises

In Exercises 1–6, match the statement with one of the graphs (a)–(f).

1. The slope of the line is zero.
2. The slope of the line is undefined.
3. The slope of the line is positive, and its y-intercept is positive.
4. The slope of the line is positive, and its y-intercept is negative.
5. The slope of the line is negative, and its x-intercept is negative.
6. The slope of the line is negative, and its x-intercept is positive.

(a)

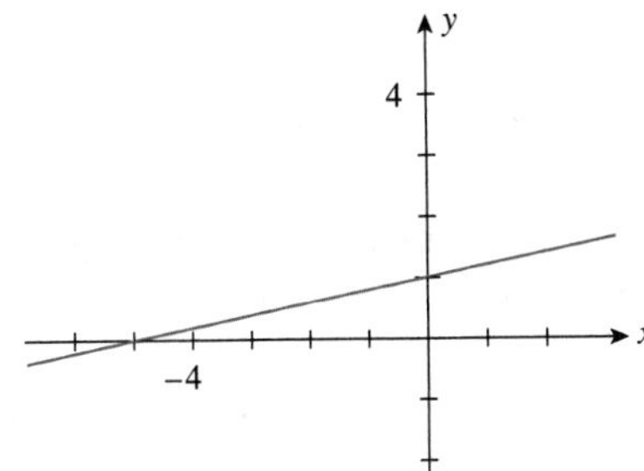

(b)

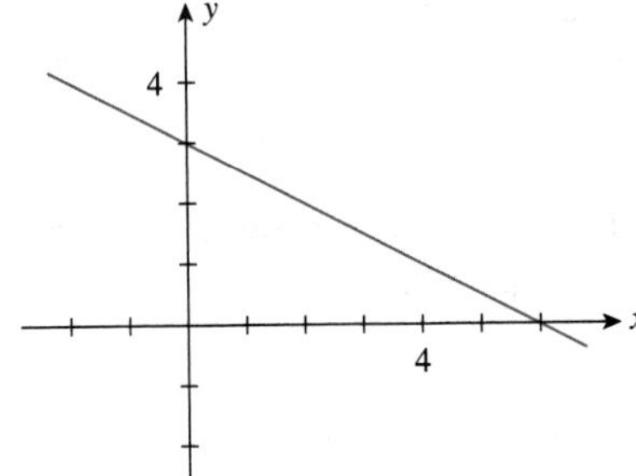

(c)

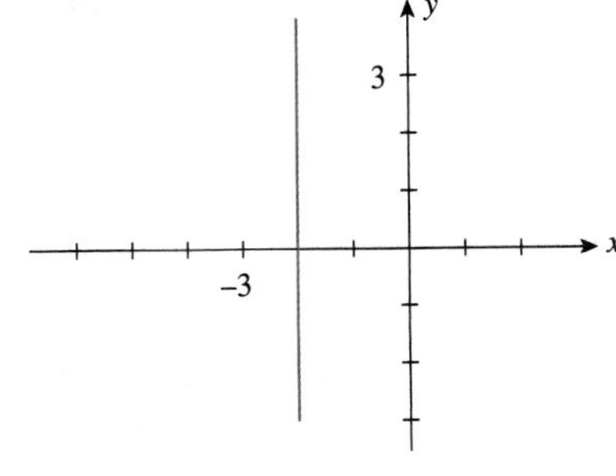

(d)

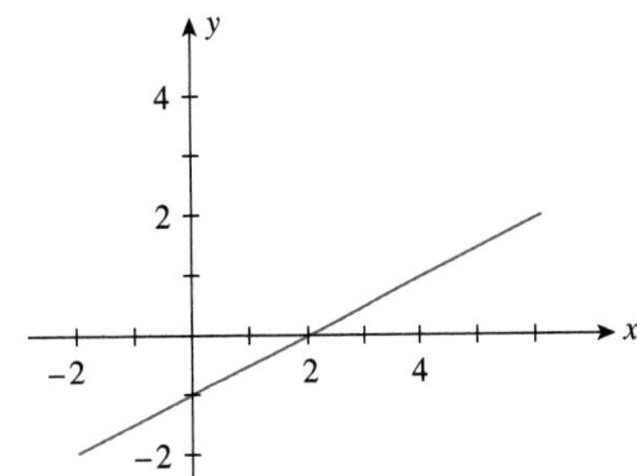

(e)

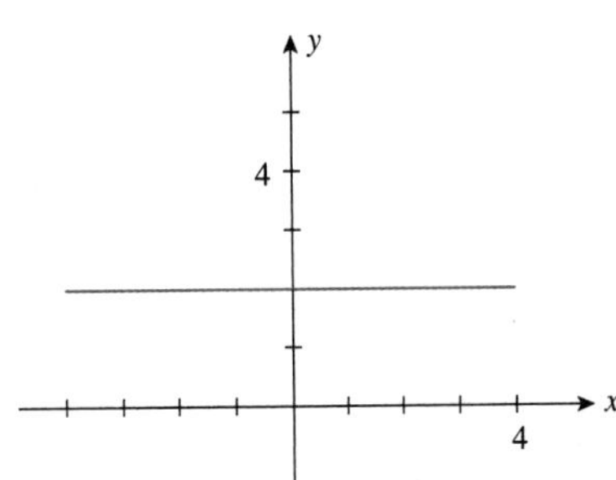

(f)

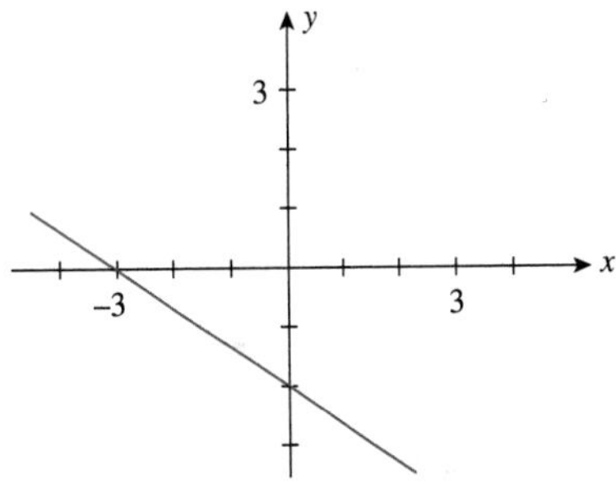

In Exercises 7–14, find the slope of the line shown in each figure.

7.

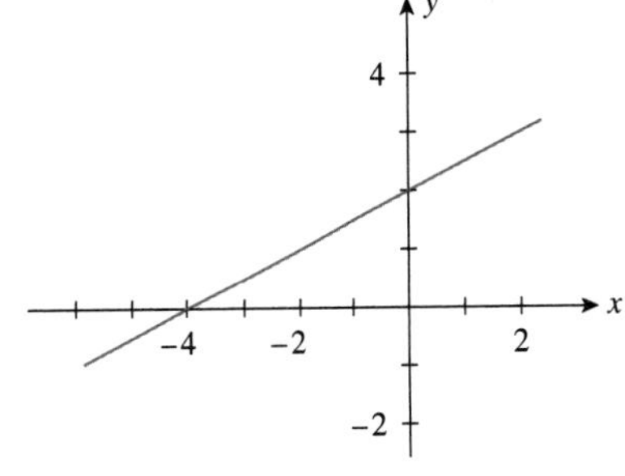

8.

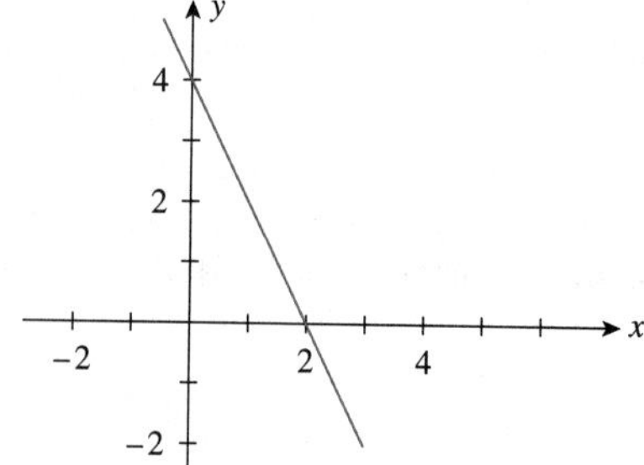

9.

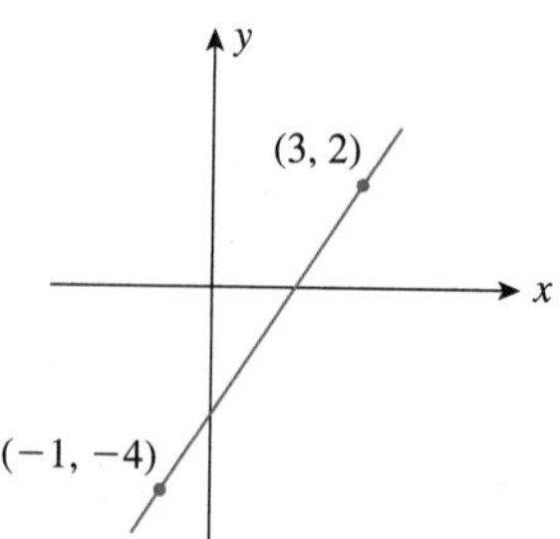

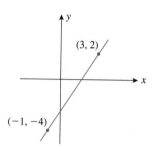

10.

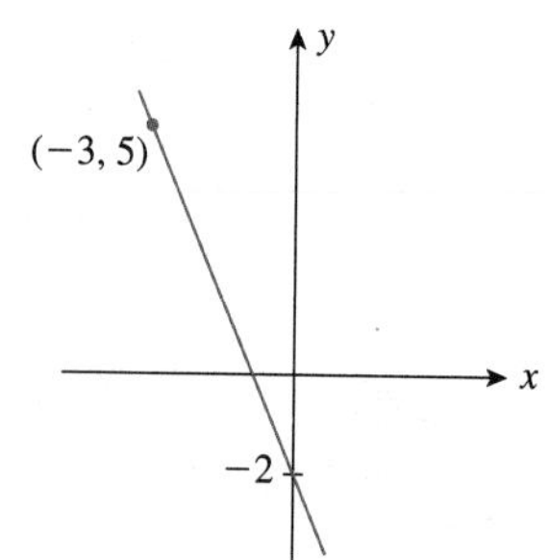

11.

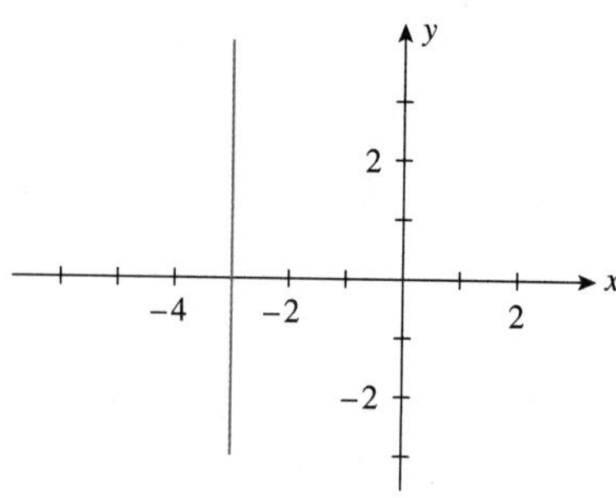

12.

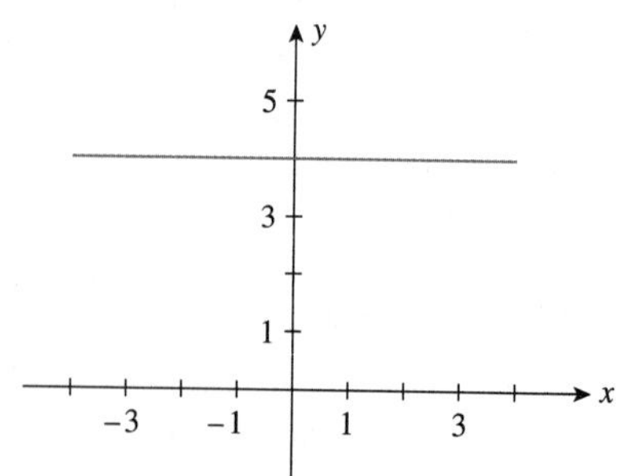

13.

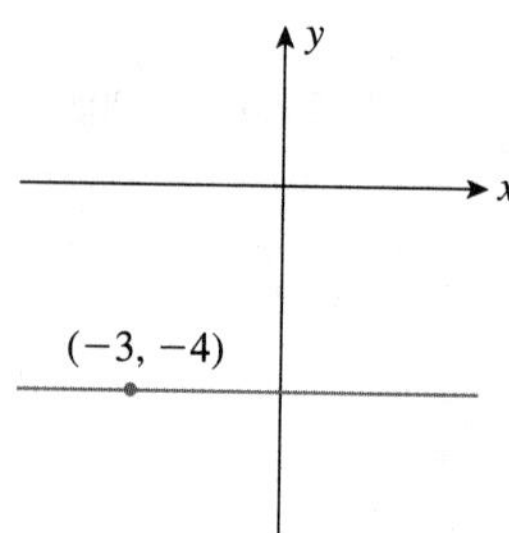

14.

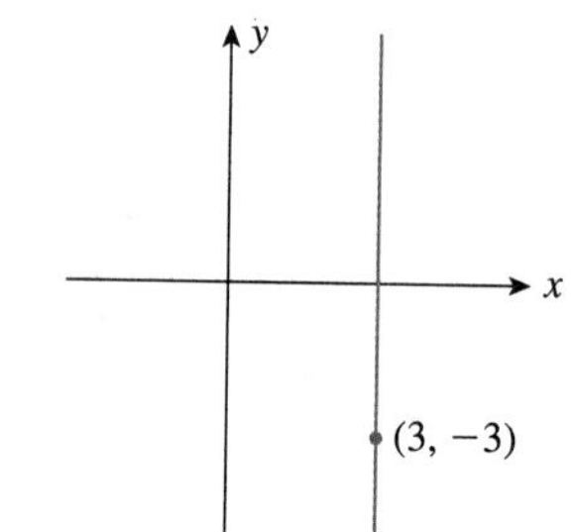

In Exercises 15–20, find the slope of the line that passes through each pair of points.

15. (4, 3) and (5, 8)

16. (4, 5) and (3, 8)

17. (−2, 3) and (4, 8)

18. (−2, −2) and (4, −4)

19. (a, b) and (c, d)

20. $(-a + 1, b - 1)$ and $(a + 1, -b)$

21. Given the equation $y = 4x - 3$, answer the following questions:
   a. If x increases by 1 unit, what is the corresponding change in y?
   b. If x decreases by 2 units, what is the corresponding change in y?

22. Given the equation $2x + 3y = 4$, answer the following questions:
   a. Is the slope of the line described by this equation positive or negative?
   b. As x increases in value, does y increase or decrease?
   c. If x decreases by 2 units, what is the corresponding change in y?

In Exercises 23 and 24, determine whether the line through each pair of points is parallel.

23. $A(1, -2)$, $B(-3, -10)$ and $C(1, 5)$, $D(-1, 1)$

24. $A(2, 3)$, $B(2, -2)$ and $C(-2, 4)$, $D(-2, 5)$

In Exercises 25 and 26, determine whether the line through each pair of points is perpendicular.

25. $A(-2, 5)$, $B(4, 2)$ and $C(-1, -2)$, $D(3, 6)$

26. $A(2, 0)$, $B(1, -2)$ and $C(4, 2)$, $D(-8, 4)$

27. If the line passing through the points $(1, a)$ and $(4, -2)$ is parallel to the line passing through the points $(2, 8)$ and $(-7, a + 4)$, what is the value of a?

28. If the line passing through the points $(a, 1)$ and $(5, 8)$ is parallel to the line passing through the points $(4, 9)$ and $(a + 2, 1)$, what is the value of a?

29. Find an equation of the horizontal line that passes through $(-4, -3)$.

30. Find an equation of the vertical line that passes through $(0, 5)$.

In Exercises 31–34, find an equation of the line that passes through the point and has the indicated slope *m*.

31. $(3, -4)$; $m = 2$

32. $(2, 4)$; $m = -1$

33. $(-3, 2)$; $m = 0$

34. $(1, 2)$; $m = -\frac{1}{2}$

In Exercises 35–38, find an equation of the line that passes through the given points.

35. $(2, 4)$ and $(3, 7)$

36. $(2, 1)$ and $(2, 5)$

37. $(1, 2)$ and $(-3, -2)$

38. $(-1, -2)$ and $(3, -4)$

In Exercises 39–42, find an equation of the line that has slope *m* and *y*-intercept *b*.

39. $m = 3$; $b = 4$

40. $m = -2$; $b = -1$

41. $m = 0$; $b = 5$

42. $m = -\frac{1}{2}$; $b = \frac{3}{4}$

In Exercises 43–48, write the equation in the slope-intercept form and then find the slope and *y*-intercept of the corresponding line.

43. $x - 2y = 0$

44. $y - 2 = 0$

45. $2x - 3y - 9 = 0$

46. $3x - 4y + 8 = 0$

47. $2x + 4y = 14$

48. $5x + 8y - 24 = 0$

49. Find an equation of the line that passes through the point $(-2, 2)$ and is parallel to the line $2x - 4y - 8 = 0$.

50. Find an equation of the line that passes through the point $(2, 4)$ and is perpendicular to the line $3x + 4y - 22 = 0$.

In Exercises 51–56, find an equation of the line that satisfies the given condition.

51. The line parallel to the *x*-axis and 6 units below it

52. The line passing through the origin and parallel to the line joining the points $(2, 4)$ and $(4, 7)$

53. The line passing through the point (a, b) with slope equal to zero

54. The line passing through $(-3, 4)$ and parallel to the *x*-axis

55. The line passing through $(-5, -4)$ and parallel to the line joining $(-3, 2)$ and $(6, 8)$

56. The line passing through (a, b) with undefined slope

57. Given that the point $P(-3, 5)$ lies on the line $kx + 3y + 9 = 0$, find *k*.

58. Given that the point $P(2, -3)$ lies on the line $-2x + ky + 10 = 0$, find *k*.

In Exercises 59–64, sketch the straight line defined by the given linear equation by finding the *x*- and *y*-intercepts.

Hint: See Example 12, page 36.

59. $3x - 2y + 6 = 0$

60. $2x - 5y + 10 = 0$

61. $x + 2y - 4 = 0$

62. $2x + 3y - 15 = 0$

63. $y + 5 = 0$

64. $-2x - 8y + 24 = 0$

65. Show that an equation of a line through the points $(a, 0)$ and $(0, b)$ with $a \neq 0$ and $b \neq 0$ can be written in the form

$$\frac{x}{a} + \frac{y}{b} = 1$$

(Recall that the numbers *a* and *b* are the *x*- and *y*-intercepts, respectively, of the line. This form of an equation of a line is called the *intercept form.*)

In Exercises 66–69, use the results of Exercise 65 to find an equation of a line with the given *x*- and *y*-intercepts.

66. *x*-intercept 3; *y*-intercept 4

67. *x*-intercept -2; *y*-intercept -4

68. *x*-intercept $-\frac{1}{2}$; *y*-intercept $\frac{3}{4}$

69. *x*-intercept 4; *y*-intercept $-\frac{1}{2}$

In Exercises 70 and 71, determine whether the given points lie on a straight line.

70. $A(-1, 7)$, $B(2, -2)$, and $C(5, -9)$

71. $A(-2, 1)$, $B(1, 7)$, and $C(4, 13)$

72. **Temperature Conversion** The relationship between the temperature in degrees Fahrenheit (°F) and the temperature in degrees Celsius (°C) is

$$F = \frac{9}{5}C + 32$$

a. Sketch the line with the given equation.
b. What is the slope of the line? What does it represent?
c. What is the *F*-intercept of the line? What does it represent?

73. **The Narrowing Gender Gap** When universities first opened their doors, there were only male instructors. A Canadian study of the wage differences between male and female university professors found that in 1970 ($t = 0$) the percentage of women teaching full-time at Canadian universities was 13%, and in 2001 ($t = 31$) it was 29%. If this percentage of female instructors continues to grow linearly, what percentage of female university instructors can we expect in 2010?

Source: Statistics Canada, *The Daily*, 08/12/06.

74. Is there a difference between the statements "The slope of a straight line is zero" and "The slope of a straight line does not exist (is not defined)"? Explain your answer.

75. Show that two distinct lines with equations $a_1x + b_1y + c_1 = 0$ and $a_2x + b_2y + c_2 = 0$, respectively, are parallel if and only if $a_1b_2 - b_1a_2 = 0$.

Hint: Write each equation in the slope-intercept form and compare.

76. Prove that if a line L_1 with slope m_1 is perpendicular to a line L_2 with slope m_2, then $m_1m_2 = -1$.

Hint: Refer to the following figure. Show that $m_1 = b$ and $m_2 = c$. Next, apply the Pythagorean theorem to triangles OAC, OCB, and OBA to show that $1 = -bc$.

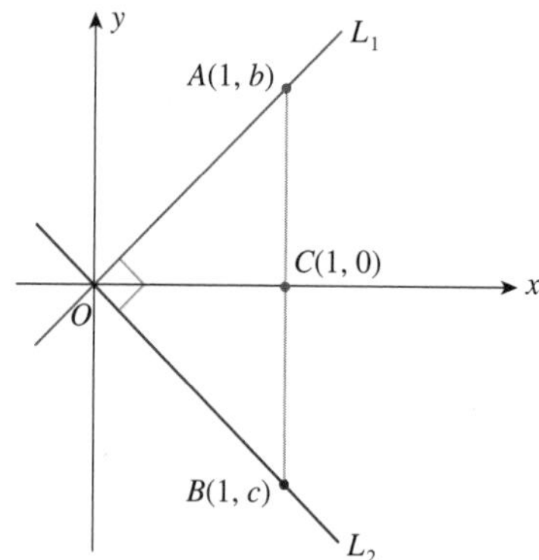

77. **College Admissions** Using data compiled by the office of analytical studies at a university in B.C., the office of the registrar estimates that 55% of the students who are offered admission to the freshman class at the university will actually enroll.
   a. Find an equation that expresses the relationship between the number of students who actually enroll (y) and the number of students who are offered admission to the university (x).
   b. If the desired freshman class size for the upcoming academic year is 1100 students, how many students should be admitted?

78. **Digital TV Services** The percent of homes with digital TV services, which stood at 5% at the beginning of 1999 ($t = 0$), is projected to grow linearly so that at the beginning of 2003 ($t = 4$) the percent of such homes is projected to be 25%.
   a. Derive an equation of the line passing through the points $A(0, 5)$ and $B(4, 25)$.
   b. Plot the line with the equation found in part (a).
   c. Using the equation found in part (a), find the percent of homes with digital TV services at the beginning of 2001.

Source: Paul Kagan Associates

Business and Economics Applications

79. **Sales Growth** Metro Department Store's annual sales (in millions of dollars) during the past 5 yr were

Annual Sales, y	5.8	6.2	7.2	8.4	9.0
Year, x	1	2	3	4	5

   a. Plot the annual sales (y) versus the year (x).
   b. Draw a straight line L through the points corresponding to the first and fifth years.
   c. Derive an equation of the line L.
   d. Using the equation found in part (c), estimate Metro's annual sales 4 yr from now ($x = 9$).

80. **Cost of a Commodity** A manufacturer obtained the following data relating the cost y (in dollars) to the number of units (x) of a commodity produced:

Units Produced, x	0	20	40	60	80	100
Cost, y	200	208	222	230	242	250

   a. Plot the cost (y) versus the quantity produced (x).
   b. Draw the straight line through the points (0, 200) and (100, 250).
   c. Derive an equation of the straight line of part (b).
   d. Taking this equation to be an approximation of the relationship between the cost and the level of production, estimate the cost of producing 54 units of the commodity.

Biological and Life Sciences Applications

81. **Bird Population** At Point Pelee National Park in the southernmost point in mainland Canada, an annual Christmas bird count is held. The total number of birds counted during the 1990s follows closely the line-of-best-fit described by the equation

$$y = 1981.4t + 25{,}928$$

where t is measured in years, with $t = 0$ corresponding to the beginning of 1990.
   a. Sketch the line with the given equation.
   b. What is the slope and the y-intercept of the line graphed in part (a)?
   c. Give an interpretation of the slope and the y-intercept of the line found in part (a).
   d. Suppose the number of birds counted continued to increase over the years. By what year can we expect to have 60,000 birds at Point Pelee National Park?

Source: Parks Canada, Point Pelee National Park

82. **Biodiversity of Birds** Refer to Exercise 81. The total number of different bird species counted during the nineties follows closely the line-of-best-fit described by the equation

$$y = -1.4242t + 100.73$$

where t is measured in years, with $t = 0$ corresponding to the beginning of 1990.
   a. Sketch the line with the given equation.
   b. What is the slope and the y-intercept of the line graphed in part (a)?
   c. Give an interpretation of the slope and the y-intercept of the line found in part (a).
   d. Suppose the number of different bird species counted continued to decrease over the years. By what year can we expect to have only 75 different bird species left at Point Pelee National Park?

Source: Parks Canada, Point Pelee National Park

83. **Weight of Whales** The equation $W = 3.51L - 192$, expressing the relationship between the length L (in metres) and the expected weight W (in metric tons) of adult blue whales, was adopted in the late 1960s by the International Whaling Commission.
   a. What is the expected weight of an 80-m blue whale?
   b. Sketch the straight line that represents the equation.

In Exercises 84–88, determine whether the statement is true or false. If it is true, explain why it is true. If it is false, give an example to show why it is false.

84. Suppose the slope of a line L is $-\frac{1}{2}$ and P is a given point on L. If Q is the point on L lying 4 units to the left of P, then Q is situated 2 units above P.

85. The line with equation $Ax + By + C = 0$, $(B \neq 0)$, and the line with equation $ax + by + c = 0$, $(b \neq 0)$, are parallel if $Ab - aB = 0$.

86. If the slope of the line L_1 is positive, then the slope of a line L_2 perpendicular to L_1 may be positive or negative.

87. The lines with equations $ax + by + c_1 = 0$ and $bx - ay + c_2 = 0$, where $a \neq 0$ and $b \neq 0$, are perpendicular to each other.

88. If L is the line with equation $Ax + By + C = 0$, where $A \neq 0$, then L crosses the x-axis at the point $(-C/A, 0)$.

Solutions to Self-Check Exercises 1.4

1. The slope of the line that passes through the points $(a, 2)$ and $(3, 6)$ is

$$m = \frac{6 - 2}{3 - a} = \frac{4}{3 - a}$$

Since this line is parallel to a line with slope 4, m must be equal to 4; that is,

$$\frac{4}{3 - a} = 4$$

or, upon multiplying both sides of the equation by $3 - a$,

$$\begin{aligned} 4 &= 4(3 - a) \\ 4 &= 12 - 4a \\ 4a &= 8 \\ a &= 2 \end{aligned}$$

2. Since the required line L is perpendicular to a line with slope $-1/2$, the slope of L is

$$\frac{-1}{-\frac{1}{2}} = 2$$

Next, using the point-slope form of the equation of a line, we have

$$\begin{aligned} y - (-1) &= 2(x - 3) \\ y + 1 &= 2x - 6 \\ y &= 2x - 7 \end{aligned}$$

3. Substituting $x = 3$ and $y = -3$ into the left-hand side of the given equation, we find

$$2(3) - 3(-3) - 12 = 3$$

which is not equal to zero (the right-hand side). Therefore, $(3, -3)$ does not lie on the line with equation $2x - 3y - 12 = 0$. (See the accompanying figure.)

Setting $x = 0$, we find $y = -4$, the y-intercept. Next, setting $y = 0$ gives $x = 6$, the x-intercept. We now draw the line passing through the points $(0, -4)$ and $(6, 0)$ as shown.

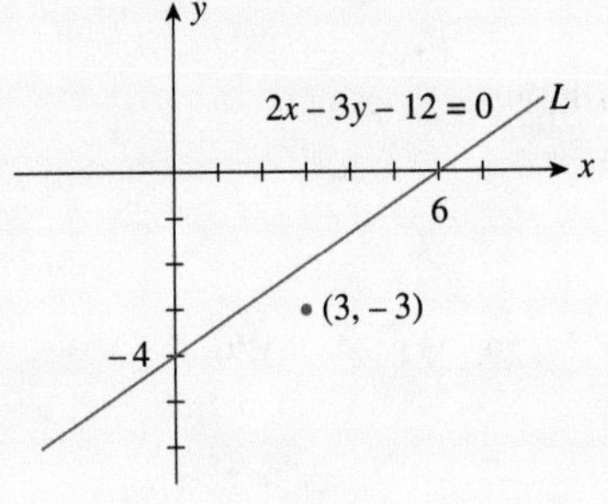

4. a. and **b.** See the accompanying figure.

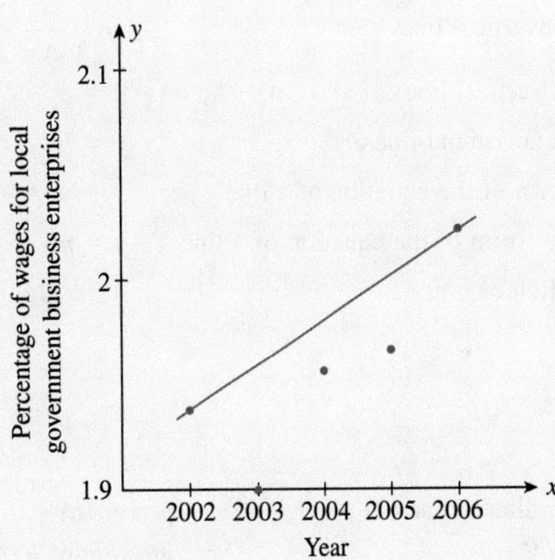

c. The slope of L is

$$m = \frac{2.03 - 1.94}{2006 - 2002} = \frac{0.09}{4} = \frac{9}{400}$$

Using the point-slope form of the equation of a line with the point (2002, 1.94), we find

$$y - 1.94 = \frac{9}{400}(x - 2002) = \frac{9}{400}x - \frac{9009}{200}$$

$$y = \frac{9}{400}x - \frac{9009}{200} + 1.94 = \frac{9}{400}x - \frac{9009}{200} + \frac{194}{100}$$

$$= \frac{9}{400}x - \frac{8621}{200}$$

d. To estimate the percent of wages for local government business enterprises by 2010, let $x = 2010$ in the equation obtained in part (c). Thus, the required estimate is

$$y = \frac{9}{400}(2010) - \frac{8621}{200} \approx 2.12$$

or approximately 2.12%.

CHAPTER 1 Summary of Principal Formulas and Terms

Formulas

1. Quadratic formula $x = \dfrac{-b \pm \sqrt{b^2 - 4ac}}{2a}$
2. Distance between two points $d = \sqrt{(x_2 - x_1)^2 + (y_2 - y_1)^2}$
3. Equation of a circle $(x - h)^2 + (y - k)^2 = r^2$
4. Slope of a nonvertical line $m = \dfrac{\Delta y}{\Delta x} \dfrac{y_2 - y_1}{x_2 - x_1}$
5. Equation of a vertical line $x = a$
6. Equation of a horizontal line $y = b$
7. Point-slope form of the equation of a line $y - y_1 = m(x - x_1)$
8. Slope-intercept form of the equation of a line $y = mx + b$
9. General equation of a line $Ax + By + C = 0$

Terms

real number (coordinate) line (2)
origin (2)
coordinate (2)
open interval (3)
closed interval (3)
half-open interval (3)
finite interval (3)
infinite interval (3)
absolute value (4)
distance (5)
triangle inequality (6)
base (6)
power (6)
*n*th root of *b* (6)
rationalization (8)
monomial (11)
polynomial (11)
like, or similar terms (11)
simplified (11)
factoring (12)
roots of a polynomial equation (15)
rational expressions (16)
rationalize the denominator (20)
Cartesian coordinate system (22)
origin (22)
x-axis (22)
y-axis (22)
ordered pair (22)
abscissa (*x*-coordinate) (23)
ordinate (*y*-coordinate) (23)
coordinate (23)
parallel lines (31)
point slope form of the equation of a line (32)
perpendicular lines (33)
x-intercept (34)
y-intercept (34)

Review Exercises

In Exercises 1–4, find the values of *x* that satisfy the inequality (inequalities).

1. $-x + 3 \le 2x + 9$
2. $-2 \le 3x + 1 \le 7$
3. $x - 3 > 2$ or $x + 3 < -1$
4. $2x^2 > 50$

In Exercises 5–8, evaluate the expression.

5. $|-5 + 7| + |-2|$
6. $\left|\dfrac{5 - 12}{-4 - 3}\right|$
7. $|2\pi - 6| - \pi$
8. $|\sqrt{3} - 4| + |4 - 2\sqrt{3}|$

In Exercises 9–14, evaluate the expression.

9. $\left(\dfrac{9}{4}\right)^{3/2}$
10. $\dfrac{5^6}{5^4}$
11. $(3 \cdot 4)^{-2}$
12. $(-8)^{5/3}$
13. $\dfrac{(3 \cdot 2^{-3})(4 \cdot 3^5)}{2 \cdot 9^3}$
14. $\dfrac{3\sqrt[3]{54}}{\sqrt[3]{18}}$

In Exercises 15–20, simplify the expression.

15. $\dfrac{4(x^2 + y)^3}{x^2 + y}$
16. $\dfrac{a^6b^{-5}}{(a^3b^{-2})^{-3}}$
17. $\dfrac{\sqrt[4]{16x^5yz}}{\sqrt[4]{81xyz^5}}$ (x, y, and z positive)
18. $(2x^3)(-3x^{-2})\left(\dfrac{1}{6}x^{-1/2}\right)$
19. $\left(\dfrac{3xy^2}{4x^3y}\right)^{-2}\left(\dfrac{3xy^3}{2xt^2}\right)^3$
20. $\sqrt[3]{81x^5y^{10}}\,\sqrt[3]{9xy^2}$

In Exercises 21–24, factor the expression.

21. $-2\pi^2r^3 + 100\pi r^2$

22. $2v^3w + 2vw^3 + 2u^2vw$

23. $16 - x^2$

24. $12t^3 - 6t^2 - 18t$

In Exercises 25–28, solve the equation by factoring.

25. $8x^2 + 2x - 3 = 0$

26. $-6x^2 - 10x + 4 = 0$

27. $-x^3 - 2x^2 + 3x = 0$

28. $2x^4 + x^2 = 1$

In Exercises 29 and 30, use the quadratic formula to solve the quadratic equation.

29. $x^2 - 2x - 5 = 0$

30. $2x^2 + 8x + 7 = 0$

In Exercises 31–34, perform the indicated operations and simplify the expression.

31. $\dfrac{(t + 6)(60) - (60t + 180)}{(t + 6)^2}$

32. $\dfrac{6x}{2(3x^2 + 2)} + \dfrac{1}{4(x + 2)}$

33. $\dfrac{2}{3}\left(\dfrac{4x}{2x^2 - 1}\right) + 3\left(\dfrac{3}{3x - 1}\right)$

34. $\dfrac{-2x}{\sqrt{x + 1}} + 4\sqrt{x + 1}$

35. Rationalize the numerator: $\dfrac{\sqrt{x} - 1}{x - 1}$.

36. Rationalize the denominator: $\dfrac{\sqrt{x} - 1}{2\sqrt{x}}$.

In Exercises 37 and 38, find the distance between the two points.

37. $(-2, -3)$ and $(1, -7)$

38. $\left(\dfrac{1}{2}, \sqrt{3}\right)$ and $\left(-\dfrac{1}{2}, 2\sqrt{3}\right)$

In Exercises 39–44, find an equation of the line L that passes through the point $(-2, 4)$ and satisfies the condition.

39. L is a vertical line.

40. L is a horizontal line.

41. L passes through the point $\left(3, \dfrac{7}{2}\right)$.

42. The x-intercept of L is 3.

43. L is parallel to the line $5x - 2y = 6$.

44. L is perpendicular to the line $4x + 3y = 6$.

45. Find an equation of the straight line that passes through the point (2, 3) and is parallel to the line with equation $3x + 4y - 8 = 0$.

46. Find an equation of the straight line that passes through the point $(-1, 3)$ and is parallel to the line passing through the points $(-3, 4)$ and $(2, 1)$.

47. Find an equation of the line that passes through the point $(-3, -2)$ and is parallel to the line passing through the points $(-2, -4)$ and $(1, 5)$.

48. Find an equation of the line that passes through the point $(-2, -4)$ and is perpendicular to the line with equation $2x - 3y - 24 = 0$.

49. Sketch the graph of the equation $3x - 4y = 24$.

50. Sketch the graph of the line that passes through the point (3, 2) and has slope $-2/3$.

51. Find the minimum cost C (in dollars) given that

$$2(1.5C + 80) \le 2(2.5C - 20)$$

52. Find the maximum revenue R (in dollars) given that

$$12(2R - 320) \le 4(3R + 240)$$

FUNCTIONS

Photo: Thinkstock/Jupiter Images

In this chapter we define a function, a special relationship between two variables. The concept of a function enables us to describe many relationships that exist in applications. To this end, we will introduce some of the essential functions, namely polynomial, rational, root, and absolute value functions, which mimic some of the behaviour of data from real-life applications. We will also need tools to manipulate and work with functions graphically and algebraically, and so we will take a look at transformations, algebraic operations, and finding the inverse. Lastly, we explore how functions are used in mathematical modelling.

Photo: Thinkstock/Jupiter Images

What percentage of Canadian households used the Internet for sending and receiving emails in 2006? In Example 3 on page 100 you will see how a function that models this percentage based on real data from previous years can help answer this question.

Photo: © Helene Rogers/Alamy

2.1 Functions and Their Graphs

Functions

A manufacturer would like to know how his company's profit is related to its production level; a biologist would like to know how the size of the population of a certain culture of bacteria will change over time; a psychologist would like to know the relationship between the learning time of an individual and the length of a vocabulary list; and a chemist would like to know how the initial speed of a chemical reaction is related to the amount of substrate used. In each instance we are concerned with the same question: How does one quantity depend upon another? The relationship between two quantities is conveniently described in mathematics by using the concept of a function.

> **Function**
>
> A **function** is a rule that assigns to each element in a set A one and only one element in a set B.

The set A is called the **domain** of the function. It is customary to denote a function by a letter of the alphabet, such as the letter f. If x is an element in the domain of a function f, then the element in B that f associates with x is written $f(x)$ (read "f of x") and is called the value of f at x. The set comprising all the values assumed by $y = f(x)$ as x takes on all possible values in its domain is called the **range** of the function f.

We can think of a function f as a machine. The domain is the set of inputs x (raw material) for the machine, the rule f describes how the input is to be processed, and the value(s) of the function are the outputs $y = f(x)$ of the machine (Figure 2.1).

We can also think of a function f as a mapping in which an element x in the domain of f is mapped onto a unique element $f(x)$ in B (Figure 2.2).

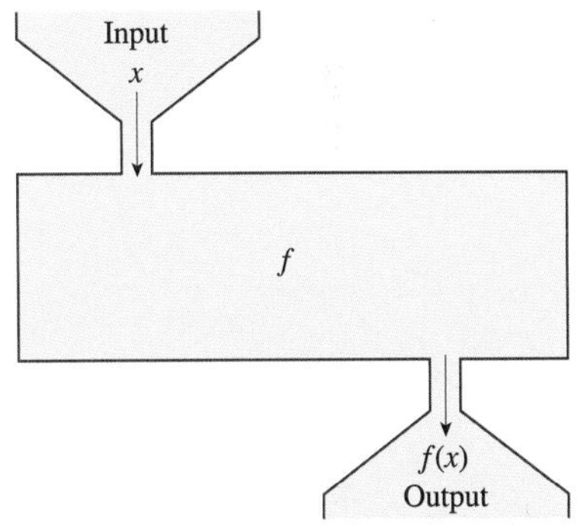

Figure 2.1
A function machine

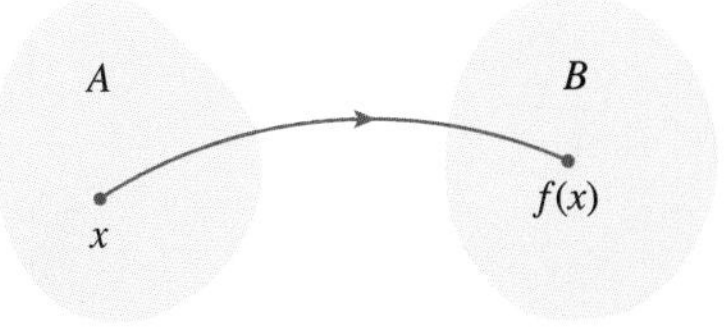

Figure 2.2
The function f viewed as a mapping

Remark

1. The output $f(x)$ associated with an input x is unique. To appreciate the importance of this uniqueness property, consider a rule that associates with each item x in a department store its selling price y. Then, each x must correspond to *one and only one* y. Notice, however, that different x's may be associated with the same y. In the context of the present example, this says that different items may have the same price.
2. Although the sets A and B that appear in the definition of a function may be quite arbitrary, in this book they will denote sets of real numbers. ◀

An example of a function may be taken from the familiar relationship between the area of a circle and its radius. Letting x and y denote the radius and area of a circle, respectively, we have, from elementary geometry,

$$y = \pi x^2 \qquad (1)$$

Equation (1) defines y as a function of x since for each admissible value of x (that is, for each nonnegative number representing the radius of a certain circle) there corresponds precisely one number $y = \pi x^2$ that gives the area of the circle. The rule defining this "area function" may be written as

$$f(x) = \pi x^2 \tag{2}$$

To compute the area of a circle of radius 5 inches, we simply replace x in Equation (2) with the number 5. Thus, the area of the circle is

$$f(5) = \pi 5^2 = 25\pi$$

or 25π square inches.

In general, to evaluate a function at a specific value of x, we replace x with that value, as illustrated in Examples 1 and 2.

Example 1

Let the function f be defined by the rule $f(x) = 2x^2 - x + 1$. Compute:

a. $f(1)$ **b.** $f(-2)$ **c.** $f(a)$ **d.** $f(a + h)$

Solution

a. $f(1) = 2(1)^2 - (1) + 1 = 2 - 1 + 1 = 2$
b. $f(-2) = 2(-2)^2 - (-2) + 1 = 8 + 2 + 1 = 11$
c. $f(a) = 2(a)^2 - (a) + 1 = 2a^2 - a + 1$
d. $f(a + h) = 2(a + h)^2 - (a + h) + 1 = 2a^2 + 4ah + 2h^2 - a - h + 1$

Example 2

Profit Functions ThermoMaster manufactures an indoor–outdoor thermometer at its Mexican subsidiary. Management estimates that the profit (in dollars) realizable by ThermoMaster in the manufacture and sale of x thermometers per week is

$$P(x) = -0.001x^2 + 8x - 5000$$

Find ThermoMaster's weekly profit if its level of production is (a) 1000 thermometers per week and (b) 2000 thermometers per week.

Solution

a. The weekly profit when the level of production is 1000 units per week is found by evaluating the profit function P at $x = 1000$. Thus,

$$P(1000) = -0.001(1000)^2 + 8(1000) - 5000 = 2000$$

or \$2000.

b. When the level of production is 2000 units per week, the weekly profit is given by

$$P(2000) = -0.001(2000)^2 + 8(2000) - 5000 = 7000$$

or \$7000.

Determining the Domain of a Function

Suppose we are given the function $y = f(x)$.* Then, the variable x is called the **independent variable.** The variable y, whose value depends on x, is called the **dependent variable.**

In determining the domain of a function, we need to find what restrictions, if any, are to be placed on the independent variable x. In many practical applications, the domain of a function is dictated by the nature of the problem, as illustrated in Example 3.

*It is customary to refer to a function f as $f(x)$ or by the equation $y = f(x)$ defining it.

Example 3

Packaging An open box is to be made from a rectangular piece of cardboard 16 centimetres long and 10 centimetres wide by cutting away identical squares (x by x centimetres) from each corner and folding up the resulting flaps (Figure 2.3). Find an expression that gives the volume V of the box as a function of x. What is the domain of the function?

Solution

The dimensions of the box are $(16 - 2x)$ centimetres long, $(10 - 2x)$ centimetres wide, and x inches high, so its volume (in cubic inches) is given by

$$\begin{aligned} V = f(x) &= (16 - 2x)(10 - 2x)x \qquad \text{Length} \cdot \text{width} \cdot \text{height} \\ &= (160 - 52x + 4x^2)x \\ &= 4x^3 - 52x^2 + 160x \end{aligned}$$

Since the length of each side of the box must be greater than or equal to zero, we see that

$$16 - 2x \geq 0, \qquad 10 - 2x \geq 0, \qquad x \geq 0$$

simultaneously; that is,

$$x \leq 8, \qquad x \leq 5, \qquad x \geq 0$$

All three inequalities are satisfied simultaneously provided that $0 \leq x \leq 5$. Thus, the domain of the function f is the interval $[0, 5]$.

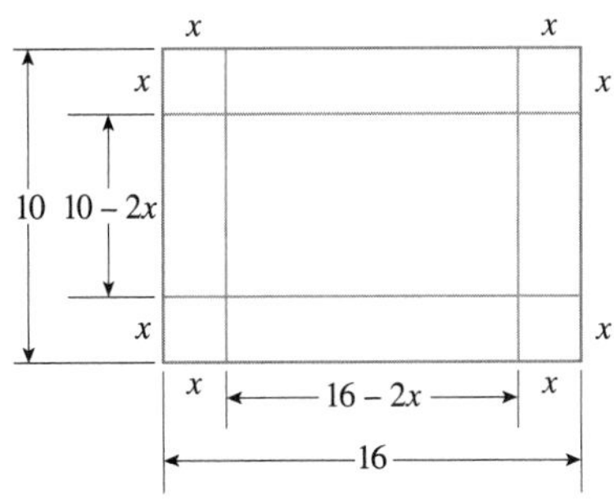

(a) The box is constructed by cutting x- by x-centimetre squares from each corner.

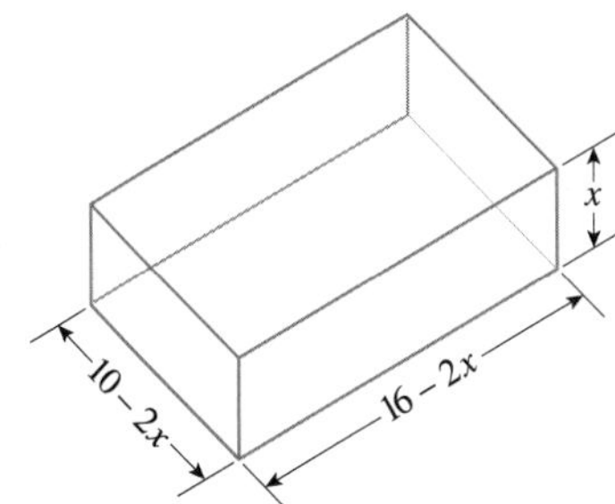

(b) The dimensions of the resulting box are $(10 - 2x)$ by $(16 - 2x)$ by x centimetres.

Figure 2.3

In general, if a function is defined by a rule relating x to $f(x)$ without specific mention of its domain, it is understood that the domain will consist of all values of x for which $f(x)$ is a real number. In this connection, you should keep in mind that (1) division by zero is not permitted and (2) the square root of a negative number is not defined.

Example 4

Find the domain of each of the functions defined by the following equations:

a. $f(x) = \sqrt{x - 1}$ **b.** $f(x) = \dfrac{1}{x^2 - 4}$ **c.** $f(x) = x^2 + 3$

Solution

a. Since the square root of a negative number is undefined, it is necessary that $x - 1 \geq 0$. The inequality is satisfied by the set of real numbers $x \geq 1$. Thus, the domain of f is the interval $[1, \infty)$.

b. The only restriction on x is that $x^2 - 4$ be different from zero since division by zero is not allowed. But $(x^2 - 4) = (x + 2)(x - 2) = 0$ if $x = -2$ or $x = 2$. Thus, the domain of f in this case consists of the intervals $(-\infty, -2)$, $(-2, 2)$, and $(2, \infty)$.

c. Here, any real number satisfies the equation, so the domain of f is the set of all real numbers.

Graphs of Functions

If f is a function with domain A, then corresponding to each real number x in A there is precisely one real number $f(x)$. We can also express this fact by using ordered pairs of real numbers. Write each number x in A as the first member of an ordered pair and each number $f(x)$ corresponding to x as the second member of the ordered pair. This gives exactly one ordered pair $(x, f(x))$ for each x in A. This observation leads to an alternative definition of a function f:

A function f with domain A is the set of all ordered pairs $(x, f(x))$ where x belongs to A.

Observe that the condition that there be one and only one number $f(x)$ corresponding to each number x in A translates into the requirement that *no two ordered pairs have the same first number.*

Since ordered pairs of real numbers correspond to points in the plane, we have found a way to exhibit a function graphically.

Graph of a Function of One Variable

The **graph of a function** f is the set of all points (x, y) in the xy-plane such that x is in the domain of f and $y = f(x)$.

Figure 2.4 shows the graph of a function f. Observe that the y-coordinate of the point (x, y) on the graph of f gives the height of that point (the distance above the x-axis), if $f(x)$ is positive. If $f(x)$ is negative, then $-f(x)$ gives the depth of the point (x, y) (the distance below the x-axis). Also, observe that the domain of f is a set of real numbers lying on the x-axis, whereas the range of f lies on the y-axis.

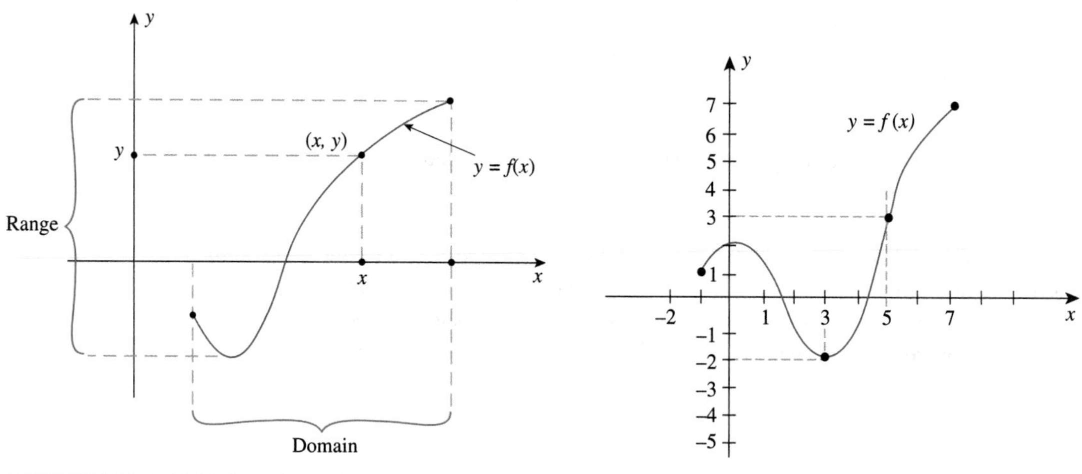

Figure 2.4
The graph of f

Figure 2.5

Example 5

The graph of a function f is shown in Figure 2.5.

a. What is the value of $f(3)$? The value of $f(5)$?
b. What is the height or depth of the point $(3, f(3))$ from the x-axis? The point $(5, f(5))$ from the x-axis?
c. What is the domain of f? The range of f?

Solution

a. From the graph of f, we see that $y = -2$ when $x = 3$ and conclude that $f(3) = -2$. Similarly, we see that $f(5) = 3$.

b. Since the point $(3, -2)$ lies below the x-axis, we see that the depth of the point $(3, f(3))$ is $-f(3) = -(-2) = 2$ units below the x-axis. The point $(5, f(5))$ lies above the x-axis and is located at a height of $f(5)$, or 3 units above the x-axis.

c. Observe that x may take on all values between $x = -1$ and $x = 7$, inclusive, and so the domain of f is $[-1, 7]$. Next, observe that as x takes on all values in the domain of f, $f(x)$ takes on all values between -2 and 7, inclusive. (You can easily see this by running your index finger along the x-axis from $x = -1$ to $x = 7$ and observing the corresponding values assumed by the y-coordinate of each point of the graph of f.) Therefore, the range of f is $[-2, 7]$.

Much information about the graph of a function can be gained by plotting a few points on its graph. Later on we will develop more systematic and sophisticated techniques for graphing functions.

Example 6

Sketch the graph of the function defined by the equation $y = x^2 + 1$. What is the range of f?

Solution

The domain of the function is the set of all real numbers. By assigning several values to the variable x and computing the corresponding values for y, we obtain the following solutions to the equation $y = x^2 + 1$:

x	−3	−2	−1	0	1	2	3
y	10	5	2	1	2	5	10

By plotting these points and then connecting them with a smooth curve, we obtain the graph of $y = f(x)$, which is a parabola (Figure 2.6). To determine the range of f, we observe that $x^2 \geq 0$, if x is any real number and so $x^2 + 1 \geq 1$ for all real numbers x. We conclude that the range of f is $[1, \infty)$. The graph of f confirms this result visually.

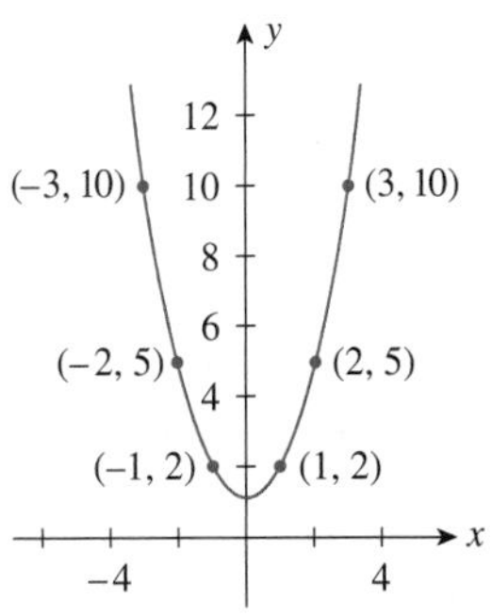

Figure 2.6
The graph of $y = x^2 + 1$ is a parabola.

EXPLORING WITH TECHNOLOGY

Let $f(x) = x^2$.

1. Plot the graphs of $F(x) = x^2 + c$ on the same set of axes for $c = -2, -1, -\frac{1}{2}, 0, \frac{1}{2}, 1, 2$.
2. Plot the graphs of $G(x) = (x + c)^2$ on the same set of axes for $c = -2, -1, -\frac{1}{2}, 0, \frac{1}{2}, 1, 2$.
3. Plot the graphs of $H(x) = cx^2$ on the same set of axes for $c = -2, -1, -\frac{1}{2}, -\frac{1}{4}, 0, \frac{1}{4}, \frac{1}{2}, 1, 2$.
4. Study the family of graphs in parts 1–3 and describe the relationship between the graph of a function f and the graphs of the functions defined by (a) $y = f(x) + c$, (b) $y = f(x + c)$, and (c) $y = cf(x)$, where c is a constant.

A function that is defined by more than one rule is called a **piecewise defined function.**

Example 7

Bank Deposits A finance company based in Montréal plans to open two branch offices 2 years from now in two separate locations: an industrial complex and a newly developed commercial centre in the city. As a result of these expansion plans, the company's total deposits during the next 5 years are expected to grow in accordance with the rule

$$f(x) = \begin{cases} \sqrt{2x} + 20 & \text{if } 0 \leq x \leq 2 \\ \frac{1}{2}x^2 + 20 & \text{if } 2 < x \leq 5 \end{cases}$$

where $y = f(x)$ gives the total amount of money (in millions of dollars) on deposit with the company in year x ($x = 0$ corresponds to the present). Sketch the graph of the function f.

Solution

The function f is defined in a piecewise fashion on the interval $[0, 5]$. In the subdomain $[0, 2]$, the rule for f is given by $f(x) = \sqrt{2x} + 20$. The values of $f(x)$ corresponding to $x = 0$, 1, and 2 may be tabulated as follows:

x	0	1	2
$f(x)$	20	21.4	22

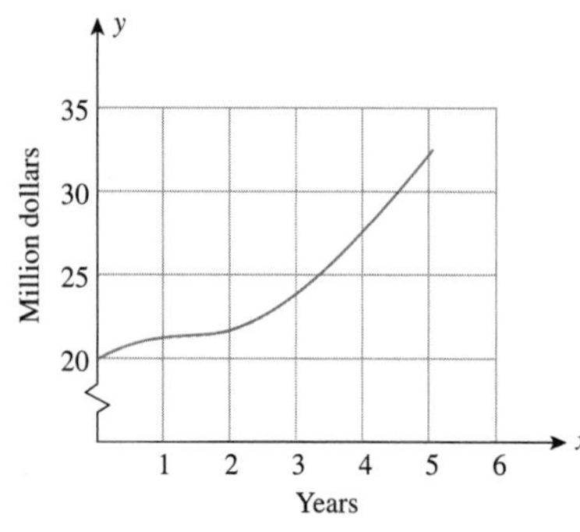

Figure 2.7
We obtain the graph of the function $y = f(x)$ by graphing $y = \sqrt{2x} + 20$ over $[0, 2]$ and $y = \frac{1}{2}x^2 + 20$ over $(2, 5]$.

Next, in the subdomain $(2, 5]$, the rule for f is given by $f(x) = \frac{1}{2}x^2 + 20$. The values of $f(x)$ corresponding to $x = 3$, 4, and 5 are shown in the following table:

x	3	4	5
$f(x)$	24.5	28	32.5

Using the values of $f(x)$ in this table, we sketch the graph of the function f as shown in Figure 2.7.

Example 8

Sketch the graph of the function f defined by

$$f(x) = \begin{cases} -x & \text{if } x < 0 \\ \sqrt{x} & \text{if } x \geq 0 \end{cases}$$

Solution

The function f is defined in a piecewise fashion on the set of all real numbers. In the subdomain $(-\infty, 0)$, the rule for f is given by $f(x) = -x$. The equation $y = -x$ is a linear equation in the slope-intercept form (with slope -1 and intercept 0). Therefore, the graph of f corresponding to the subdomain $(-\infty, 0)$ is the half line shown in Figure 2.8. Next, in the subdomain $[0, \infty)$, the rule for f is given by $f(x) = \sqrt{x}$. The values of $f(x)$ corresponding to $x = 0$, 1, 2, 3, 4, 9, and 16 are shown in the following table:

x	0	1	2	3	4	9	16
$f(x)$	0	1	$\sqrt{2}$	$\sqrt{3}$	2	3	4

Using these values, we sketch the graph of the function f as shown in Figure 2.8.

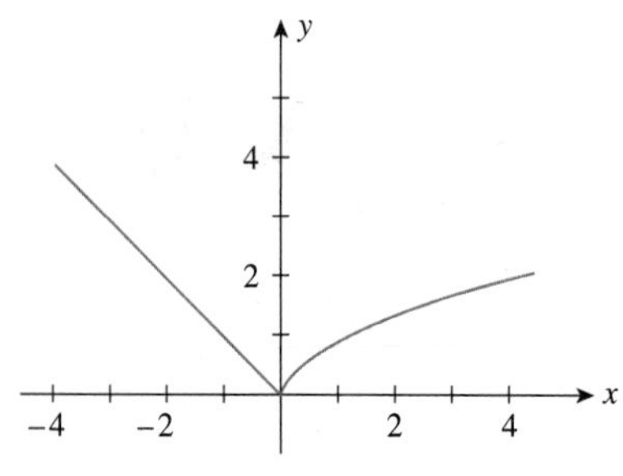

Figure 2.8
The graph of $y = f(x)$ is obtained by graphing $y = -x$ over $(-\infty, 0)$ and $y = \sqrt{x}$ over $[0, \infty)$.

The Vertical-Line Test

Although it is true that every function f of a variable x has a graph in the xy-plane, it is important to realize that not every curve in the xy-plane is the graph of a function. For example, consider the curve depicted in Figure 2.9. This is the graph of the equation $y^2 = x$. In general, the **graph of an equation** is the set of all ordered pairs (x, y) that satisfy the given equation. Observe that the points $(9, -3)$ and $(9, 3)$ both lie on the curve. This implies that the number $x = 9$ is associated with *two* numbers: $y = -3$ and $y = 3$. But this clearly violates the uniqueness property of a function. Thus, we conclude that the curve under consideration cannot be the graph of a function.

This example suggests the following **vertical-line test** for determining when a curve is the graph of a function.

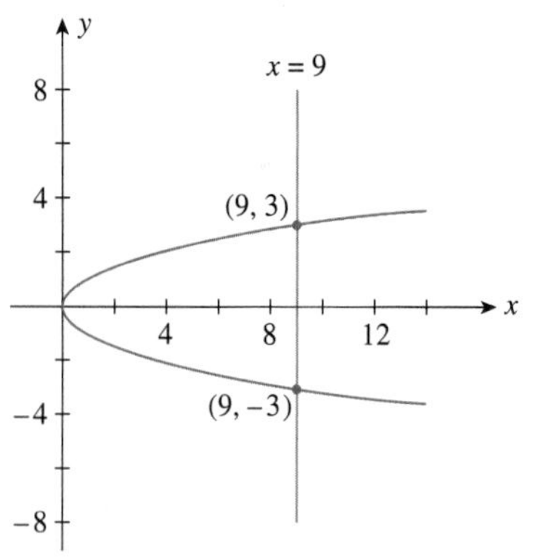

Figure 2.9
Since a vertical line passes through the curve at more than one point, we deduce that it is *not* the graph of a function.

> **Vertical-Line Test**
> A curve in the xy-plane is the graph of a function $y = f(x)$ if and only if each vertical line intersects it in at most one point.

Example 9

Determine which of the curves shown in Figure 2.10 are the graphs of functions of x.

Solution

The curves depicted in Figure 2.10a, c, and d are graphs of functions because each curve satisfies the requirement that each vertical line intersects the curve in at most one point. Note that the vertical line shown in Figure 2.10c does *not* intersect the graph because the point on the x-axis through which this line passes does not lie in the domain of the function. The curve depicted in Figure 2.10b is *not* the graph of a function because the vertical line shown there intersects the graph at three points.

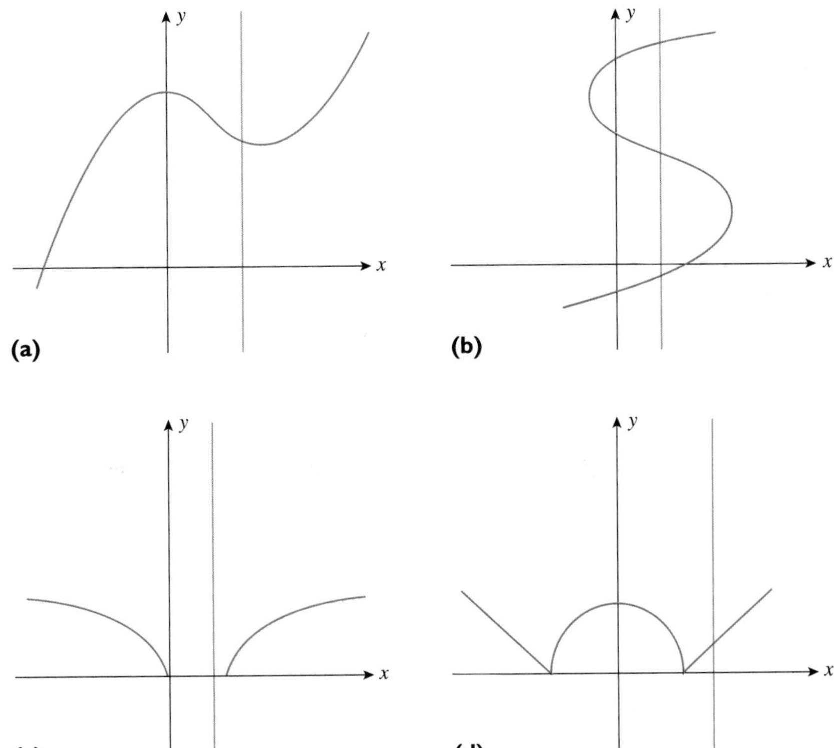

Figure 2.10
The vertical-line test can be used to determine which of these curves are graphs of functions.

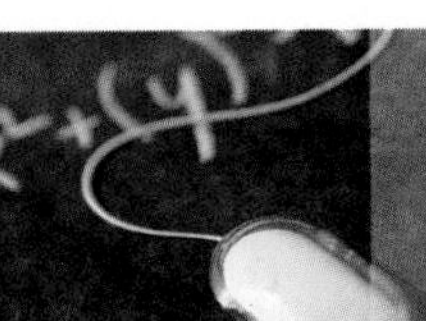

Self-Check Exercises 2.1

1. Let f be the function defined by

$$f(x) = \frac{\sqrt{x + 1}}{x}$$

 a. Find the domain of f.
 b. Compute $f(3)$.
 c. Compute $f(a + h)$.

2. Statistics show that more and more motorists are pumping their own gas. The following function gives self-serve sales as a percent of all Canadian gas sales:

$$f(t) = \begin{cases} 6t + 17 & \text{if } 0 \le t \le 6 \\ 15.98(t - 6)^{1/4} + 53 & \text{if } 6 < t \le 20 \end{cases}$$

Here t is measured in years, with $t = 0$ corresponding to the beginning of 1979.

 a. Sketch the graph of the function f.
 b. What percent of all gas sales at the beginning of 1983 were self-serve? At the beginning of 1999?

3. Let $f(x) = \sqrt{2x + 1} + 2$. Determine whether the point (4, 6) lies on the graph of f.

Solutions to Self-Check Exercises 2.1 can be found on page 55.

2.1 Exercises

1. Let f be the function defined by $f(x) = 5x + 6$. Find $f(3)$, $f(-3)$, $f(a)$, $f(-a)$, and $f(a + 3)$.

2. Let f be the function defined by $f(x) = 4x - 3$. Find $f(4)$, $f(\frac{1}{4})$, $f(0)$, $f(a)$, and $f(a + 1)$.

3. Let g be the function defined by $g(x) = 3x^2 - 6x - 3$. Find $g(0)$, $g(-1)$, $g(a)$, $g(-a)$, and $g(x + 1)$.

4. Let h be the function defined by $h(x) = x^3 - x^2 + x + 1$. Find $h(-5)$, $h(0)$, $h(a)$, and $h(-a)$.

5. Let f be the function defined by $f(x) = 2x + 5$. Find $f(a + h)$, $f(-a)$, $f(a^2)$, $f(a - 2h)$, and $f(2a - h)$.

6. Let g be the function defined by $g(x) = -x^2 + 2x$. Find $g(a + h)$, $g(-a)$, $g(\sqrt{a})$, $a + g(a)$, and $\dfrac{1}{g(a)}$.

7. Let s be the function defined by $s(t) = \dfrac{2t}{t^2 - 1}$. Find $s(4)$, $s(0)$, $s(a)$, $s(2 + a)$, and $s(t + 1)$.

8. Let g be the function defined by $g(u) = (3u - 2)^{3/2}$. Find $g(1)$, $g(6)$, $g(\frac{11}{3})$, and $g(u + 1)$.

9. Let f be the function defined by $f(t) = \dfrac{2t^2}{\sqrt{t - 1}}$. Find $f(2)$, $f(a)$, $f(x + 1)$, and $f(x - 1)$.

10. Let f be the function defined by $f(x) = 2 + 2\sqrt{5 - x}$. Find $f(-4)$, $f(1)$, $f(\frac{11}{4})$, and $f(x + 5)$.

11. Let f be the function defined by
$$f(x) = \begin{cases} x^2 + 1 & \text{if } x \le 0 \\ \sqrt{x} & \text{if } x > 0 \end{cases}$$
Find $f(-2)$, $f(0)$, and $f(1)$.

12. Let g be the function defined by
$$g(x) = \begin{cases} -\frac{1}{2}x + 1 & \text{if } x < 2 \\ \sqrt{x - 2} & \text{if } x \ge 2 \end{cases}$$
Find $g(-2)$, $g(0)$, $g(2)$, and $g(4)$.

13. Let f be the function defined by
$$f(x) = \begin{cases} -\frac{1}{2}x^2 + 3 & \text{if } x < 1 \\ 2x^2 + 1 & \text{if } x \ge 1 \end{cases}$$
Find $f(-1)$, $f(0)$, $f(1)$, and $f(2)$.

14. Let f be the function defined by
$$f(x) = \begin{cases} 2 + \sqrt{1 - x} & \text{if } x \le 1 \\ \dfrac{1}{1 - x} & \text{if } x > 1 \end{cases}$$
Find $f(0)$, $f(1)$, and $f(2)$.

15. Refer to the graph of the function f in the following figure.

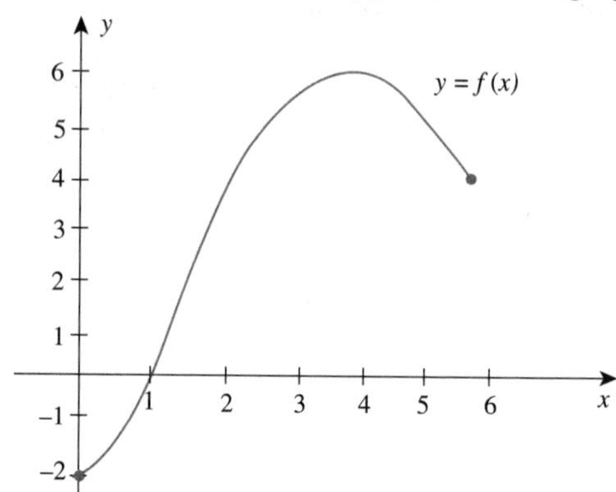

a. Find the value of $f(0)$.
b. Find the value of x for which (i) $f(x) = 3$ and (ii) $f(x) = 0$.
c. Find the domain of f.
d. Find the range of f.

16. Refer to the graph of the function f in the following figure.

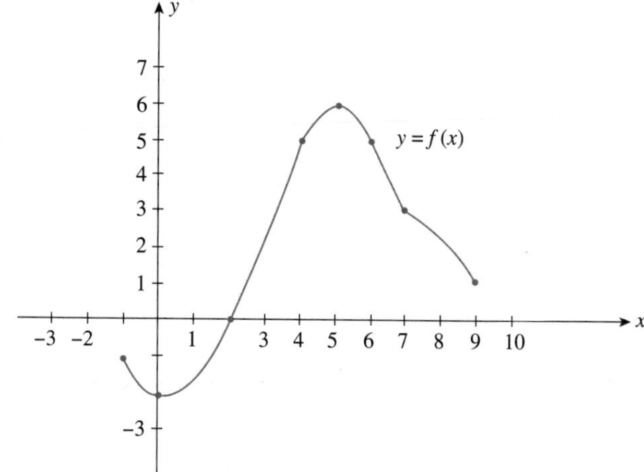

a. Find the value of $f(7)$.
b. Find the values of x corresponding to the point on the graph of f located at a height of 5 units from the x-axis.
c. Find the point on the x-axis at which the graph of f crosses it. What is the value of $f(x)$ at this point?
d. Find the domain and range of f.

In Exercises 17–20, determine whether the point lies on the graph of the function.

17. $(2, \sqrt{3})$; $g(x) = \sqrt{x^2 - 1}$

18. $(3, 3)$; $f(x) = \dfrac{x + 1}{\sqrt{x^2 + 7}} + 2$

19. $(-2, -3)$; $f(t) = \dfrac{|t - 1|}{t + 1}$

20. $\left(-3, -\dfrac{1}{13}\right)$; $h(t) = \dfrac{|t + 1|}{t^3 + 1}$

In Exercises 21–34, find the domain of the function.

21. $f(x) = x^2 + 3$

22. $f(x) = 7 - x^2$

23. $f(x) = \dfrac{3x + 1}{x^2}$

24. $g(x) = \dfrac{2x + 1}{x - 1}$

25. $f(x) = \sqrt{x^2 + 1}$

26. $f(x) = \sqrt{x - 5}$

27. $f(x) = \sqrt{5 - x}$

28. $g(x) = \sqrt{2x^2 + 3}$

29. $f(x) = \dfrac{x}{x^2 - 1}$

30. $f(x) = \dfrac{1}{x^2 + x - 2}$

31. $f(x) = (x + 3)^{3/2}$

32. $g(x) = 2(x - 1)^{5/2}$

33. $f(x) = \dfrac{\sqrt{1 - x}}{x^2 - 4}$

34. $f(x) = \dfrac{\sqrt{x - 1}}{(x + 2)(x - 3)}$

35. Let f be a function defined by the rule $f(x) = x^2 - x - 6$.
 a. Find the domain of f.
 b. Compute $f(x)$ for $x = -3, -2, -1, 0, \frac{1}{2}, 1, 2, 3$.
 c. Use the results obtained in parts (a) and (b) to sketch the graph of f.

36. Let f be a function defined by the rule $f(x) = 2x^2 + x - 3$.
 a. Find the domain of f.
 b. Compute $f(x)$ for $x = -3, -2, -1, -\frac{1}{2}, 0, 1, 2, 3$.
 c. Use the results obtained in parts (a) and (b) to sketch the graph of f.

In Exercises 37–48, sketch the graph of the function with the given rule. Find the domain and range of the function.

37. $f(x) = 2x^2 + 1$

38. $f(x) = 9 - x^2$

39. $f(x) = 2 + \sqrt{x}$

40. $g(x) = 4 - \sqrt{x}$

41. $f(x) = \sqrt{1 - x}$

42. $f(x) = \sqrt{x - 1}$

43. $f(x) = |x| - 1$

44. $f(x) = |x| + 1$

45. $f(x) = \begin{cases} x & \text{if } x < 0 \\ 2x + 1 & \text{if } x \geq 0 \end{cases}$

46. $f(x) = \begin{cases} 4 - x & \text{if } x < 2 \\ 2x - 2 & \text{if } x \geq 2 \end{cases}$

47. $f(x) = \begin{cases} -x + 1 & \text{if } x \leq 1 \\ x^2 - 1 & \text{if } x > 1 \end{cases}$

48. $f(x) = \begin{cases} -x - 1 & \text{if } x < -1 \\ 0 & \text{if } -1 \leq x \leq 1 \\ x + 1 & \text{if } x > 1 \end{cases}$

In Exercises 49–56, use the vertical-line test to determine whether the graph represents y as a function of x.

49.

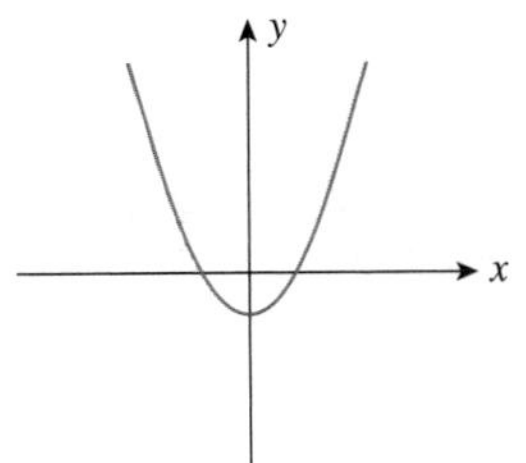

50.

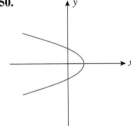

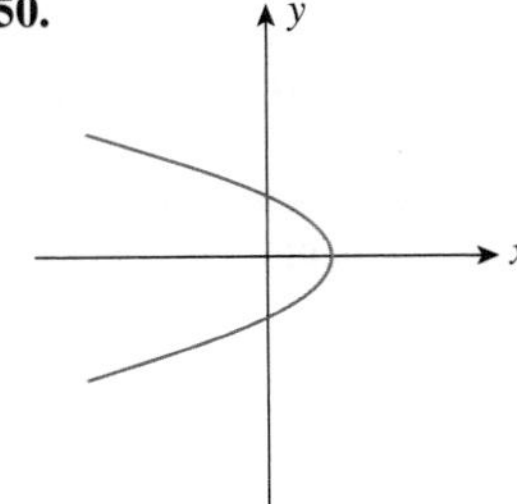

51.

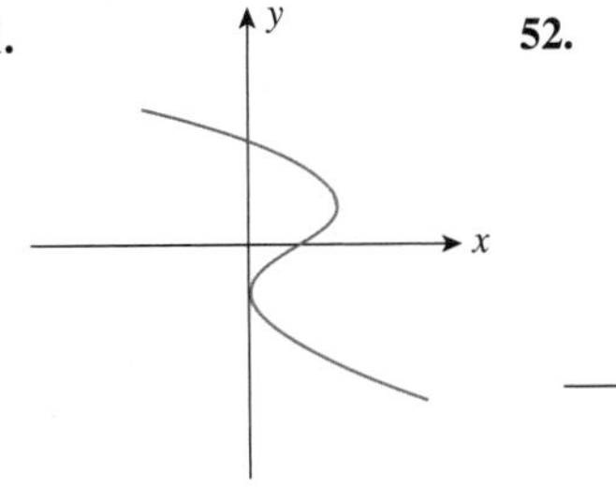

52.

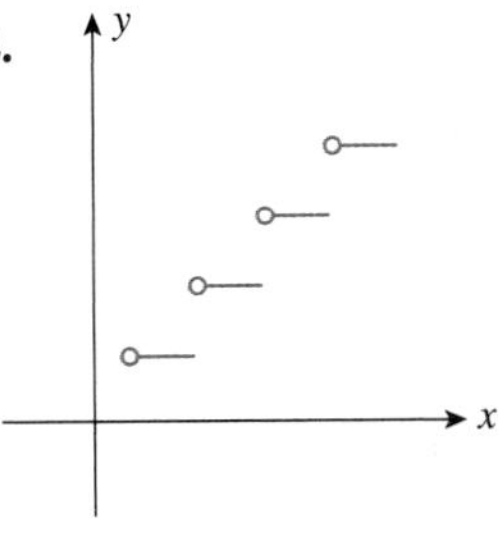

53. 54.

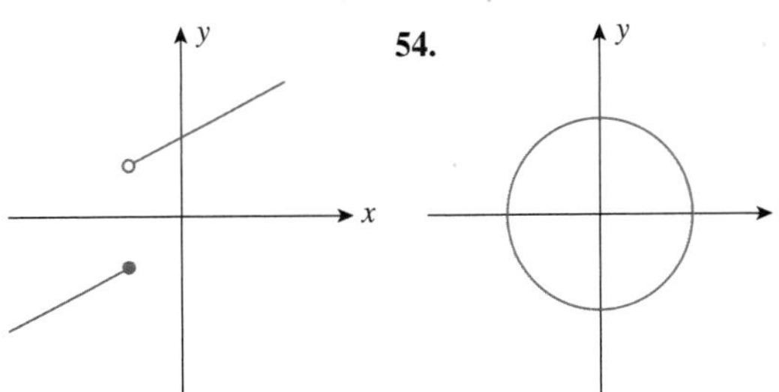

55. 56.

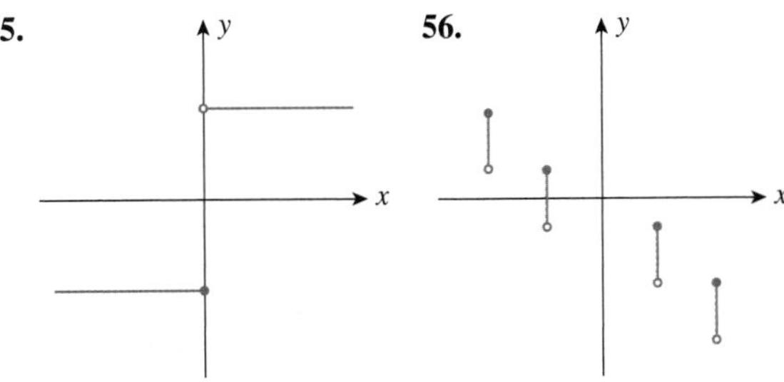

57. The circumference of a circle is given by $C(r) = 2\pi r$, where r is the radius of the circle. What is the circumference of a circle with a 5-cm. radius?

58. The volume of a sphere of radius r is given by $V(r) = \frac{4}{3}\pi r^3$. Compute $V(2.1)$ and $V(2)$. What does the quantity $V(2.1) - V(2)$ measure?

Business and Economics Applications

59. **Sales of Prerecorded Music** The following graphs show the sales y of prerecorded music (in millions of dollars) by format as a function of time t (in years), with $t = 0$ corresponding to 1985.

 a. In what years were the sales of prerecorded cassettes greater than those of prerecorded CDs?
 b. In what years were the sales of prerecorded CDs greater than those of prerecorded cassettes?
 c. In what year were the sales of prerecorded cassettes the same as those of prerecorded CDs? Estimate the level of sales in each format at that time.

60. **The Gender Gap** The following graph shows the ratio of women's earnings to men's from 1960 through 2000.

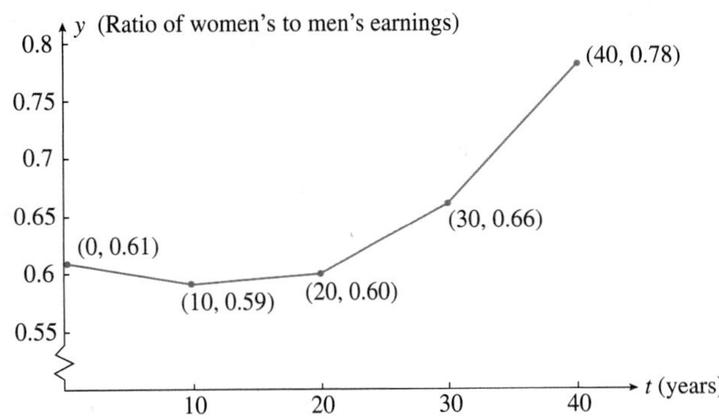

a. Write the rule for the function f giving the ratio of women's earnings to men's in year t, with $t = 0$ corresponding to 1960.

Hint: The function f is defined piecewise and is linear over each of four subintervals.

b. In what decade(s) was the gender gap expanding? Shrinking?

c. Refer to part (b). How fast was the gender gap (the ratio/year) expanding or shrinking in each of these decades?

61. Closing the Gender Gap in Education The following graph shows the ratio of bachelor's degrees earned by women to men from 1960 through 1990.

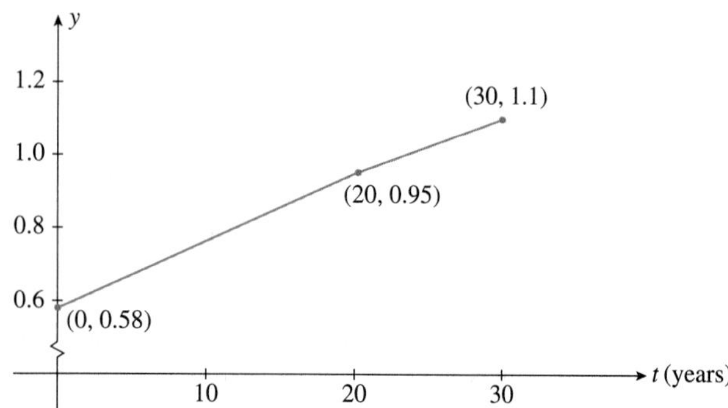

a. Write the rule for the function f giving the ratio of bachelor's degrees earned by women to men in year t, with $t = 0$ corresponding to 1960.

Hint: The function f is defined piecewise and is linear over each of two subintervals.

b. How fast was the ratio changing in the period from 1960 to 1980? From 1980 to 1990?

c. In what year (approximately) was the number of bachelor's degrees earned by women equal for the first time to that earned by men?

62. Consumption Function The consumption function in a certain economy is given by the equation

$$C(y) = 0.75y + 6$$

where $C(y)$ is the personal consumption expenditure, y is the disposable personal income, and both $C(y)$ and y are measured in billions of dollars. Find $C(0)$, $C(50)$, and $C(100)$.

63. Goods and Services Tax In Canada, the goods and services tax (GST) T on the amount of taxable goods is 6% of the value of the goods purchased (x), where both T and x are measured in dollars.

a. Express T as a function of x.

b. Find $T(200)$ and $T(5.65)$.

64. Income Adjustment Some employers offer retroactive automatic income adjustments every year based on the consumer price index (CPI), which allows for the calculation of the inflation rate for any particular year. The CPI for the year 1999 is 163.4 and the year 2000 is 166.6.

a. The inflation rate as a percentage gets calculated according to the formula 100 (year2 − year1)/(year1), where year 1 and year 2 are any consecutive years. Find the inflation rate for the year 2000.

b. Express the adjusted yearly income as a function of the yearly income.

c. If Barrington's annual income is $55,000, what will be his adjusted income?

65. Cost of Renting a Truck Ace Truck leases its 3 m box truck at $30/day and $.45/km, whereas Acme Truck leases a similar truck at $25/day and $.50/km.

a. Find the daily cost of leasing from each company as a function of the number of kilometres driven.

b. Sketch the graphs of the two functions on the same set of axes.

c. Which company should a customer rent a truck from for 1 day if she plans to drive at most 70 km and wishes to minimize her cost?

66. Linear Depreciation A new machine was purchased by National Textile for $120,000. For income tax purposes, the machine is depreciated linearly over 10 yr; that is, the book value of the machine decreases by a constant amount each year, so that at the end of 10 yr the book value is zero.

a. Express the book value of the machine (V) as a function of the age, in years, of the machine (n).

b. Sketch the graph of the function in part (a).

c. Find the book value of the machine at the end of the sixth year.

d. Find the rate at which the machine is being depreciated each year.

67. Linear Depreciation Refer to Exercise 66. An office building worth $1 million when completed in 1986 was depreciated linearly over 50 yr. What was the book value of the building in 2001? What will be the book value in 2005? In 2009? (Assume that the book value of the building will be zero at the end of the 50th year.)

68. Postal Regulations In 2006 the Canadian postage for domestic lettermail was $0.51 for the first 30 grams or a fraction thereof, $0.89 over 30 g and up to 50 g, $1.05 over 50 g and up to 100 g, $1.78 over 100 g and up to 200 g, and $2.49 over 200 g and up to 500 g. Any lettermail not exceeding 500 grams may be sent by domestic mail. Letting x denote the weight of a parcel in grams and $f(x)$ the postage in dollars, complete the following description of the "lettermail function" f:

$$f(x) = \begin{cases} 0.51 & 0 < x \le 30 \\ \vdots & \\ 2.49 & 200 < x \le 500 \end{cases}$$

a. What is the domain of f?

b. Sketch the graph of f.

Biological and Life Sciences Applications

69. Growth of a Cancerous Tumour The volume of a spherical cancerous tumour is given by the function

$$V(r) = \frac{4}{3}\pi r^3$$

where r is the radius of the tumour in centimetres. By what factor is the volume of the tumour increased if its radius is doubled?

70. Growth of a Cancerous Tumour The surface area of a spherical cancerous tumour is given by the function

$$S(r) = 4\pi r^2$$

where r is the radius of the tumour in centimetres. After extensive chemotherapy treatment, the surface area of the tumour is reduced by 75%. What is the radius of the tumour after treatment?

71. **Surface Area of a Single-Celled Organism** The surface area S of a single-celled organism may be found by multiplying 4π times the square of the radius r of the cell. Express S as a function of r.

72. **Friend's Rule** Friend's rule, a method for calculating pediatric drug dosages, is based on a child's age. If a denotes the adult dosage (in milligrams) and if t is the age of the child (in years), then the child's dosage is given by

$$D(t) = \frac{2}{25}ta$$

If the adult dose of a substance is 500 mg, how much should a 4-yr-old child receive?

73. **Boyle's Law** As a consequence of Boyle's law, the pressure P of a fixed sample of gas held at a constant temperature is related to the volume V of the gas by the rule

$$P = f(V) = \frac{k}{V}$$

where k is a constant. What is the domain of the function f? Sketch the graph of the function f.

74. **Poiseuille's Law** According to a law discovered by the nineteenth-century physician Poiseuille, the velocity (in centimetres/second) of blood r cm from the central axis of an artery is given by

$$v(r) = k(R^2 - r^2)$$

where k is a constant and R is the radius of the artery. Suppose that for a certain artery, $k = 1000$ and $R = 0.2$ so that $v(r) = 1000(0.04 - r^2)$.

a. What is the domain of the function $v(r)$?
b. Compute $v(0)$, $v(0.1)$, and $v(0.2)$ and interpret your results.
c. Sketch the graph of the function v on the interval $[0, 0.2]$.
d. What can you say about the velocity of blood as we move away from the central axis toward the artery wall?

75. **Population Growth** A study projected that the population of a town in the next 3 yr will grow according to the rule

$$P(x) = 50{,}000 + 30x^{3/2} + 20x$$

where $P(x)$ denotes the population x mo from now. By how much will the population increase during the next 9 mo? During the next 16 mo?

76. **Worker Efficiency** An efficiency study conducted for Elektra Electronics showed that the number of "Space Commander" walkie-talkies assembled by the average worker t hr after starting work at 8:00 A.M. is given by

$$N(t) = -t^3 + 6t^2 + 15t \qquad (0 \le t \le 4)$$

How many walkie-talkies can an average worker be expected to assemble between 8:00 and 9:00 A.M.? Between 9:00 and 10:00 A.M.?

77. **Politics** Political scientists have discovered the following empirical rule, known as the "cube rule," which gives the relationship between the proportion of seats in the House of Representatives won by Democratic candidates $s(x)$ and the proportion of popular votes x received by the Democratic presidential candidate:

$$s(x) = \frac{x^3}{x^3 + (1 - x)^3} \qquad (0 \le x \le 1)$$

Compute $s(0.6)$ and interpret your result.

78. **Rising Median Age** Increased longevity and the aging of the baby boom generation—those born between 1946 and 1965—are the primary reasons for a rising median age. The median age (in years) of the North American population from 1900 through 2000 is approximated by the function

$$f(t) = \begin{cases} 1.3t + 22.9 & 0 \le t \le 3 \\ -0.7t^2 + 7.2t + 11.5 & 3 < t \le 7 \\ 2.6t + 9.4 & 7 < t \le 10 \end{cases}$$

where t is measured in decades, with $t = 0$ corresponding to 1900.

a. What was the median age of the North American population at the beginning of 1900? At the beginning of 1950? At the beginning of 1990?
b. Sketch the graph of f.

In Exercises 79–82, determine whether the statement is true or false. If it is true, explain why it is true. If it is false, give an example to show why it is false.

79. If $a = b$, then $f(a) = f(b)$.

80. If $f(a) = f(b)$, then $a = b$.

81. If f is a function, then $f(a + b) = f(a) + f(b)$.

82. A vertical line must intersect the graph of $y = f(x)$ at exactly one point.

Solutions to Self-Check Exercises 2.1

1. a. The expression under the radical sign must be nonnegative, so $x + 1 \ge 0$ or $x \ge -1$. Also, $x \ne 0$ because division by zero is not permitted. Therefore, the domain of f is $[-1, 0) \cup (0, \infty)$.

b. $f(3) = \frac{\sqrt{3 + 1}}{3} = \frac{\sqrt{4}}{3} = \frac{2}{3}$

c. $f(a + h) = \frac{\sqrt{(a + h) + 1}}{a + h} = \frac{\sqrt{a + h + 1}}{a + h}$

2. a. For t in the subdomain $[0, 6]$, the rule for f is given by $f(t) = 6t + 17$. The equation $y = 6t + 17$ is a linear equation, so that portion of the graph of f is the line segment joining the points $(0, 17)$ and $(6, 53)$. Next, in the subdomain $(6, 20]$,

continued

the rule for f is given by $f(t) = 15.98\,(t - 6)^{1/4} + 53$. Using a calculator, we construct the following table of values of $f(t)$ for selected values of t.

t	6	8	10	12	14	16	18	20
$f(t)$	53	72	75.6	78	79.9	81.4	82.7	83.9

We have included $t = 6$ in the table, although it does not lie in the subdomain of the function under consideration, in order to help us obtain a better sketch of that portion of the graph of f in the subdomain (6, 20]. The graph of f is as follows:

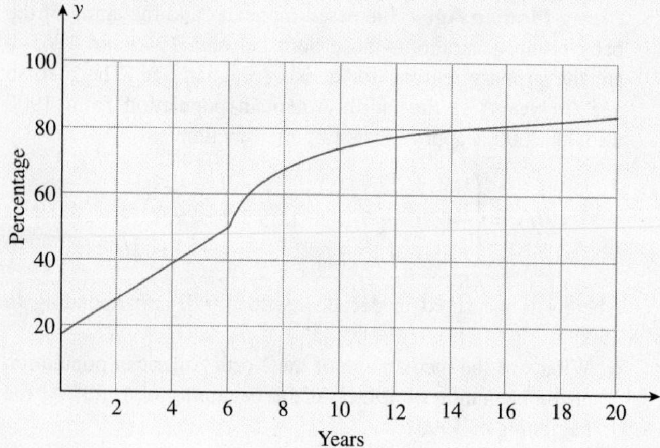

b. The percent of all self-serve gas sales at the beginning of 1983 is found by evaluating f at $t = 4$. Since this point lies in the interval [0, 6], we use the rule $f(t) = 6t + 17$ and find

$$f(4) = 6(4) + 17 = 41$$

giving 41% as the required figure. The percent of all self-serve gas sales at the beginning of 1999 is given by

$$f(20) = 15.98(20 - 6)^{1/4} + 53 = 83.9$$

or approximately 83.9%.

3. A point (x, y) lies on the graph of the function f if and only if the coordinates satisfy the equation $y = f(x)$. Now,

$$f(4) = \sqrt{2(4) + 1} + 2 = \sqrt{9} + 2 = 5 \neq 6$$

and we conclude that the given point does *not* lie on the graph of f.

2.2 Transformations of Functions

When graphing families of functions it is often beneficial to understand how a graph is derived from the graph of a more basic function which is readily graphed. We will therefore take a closer look at geometric tools such as reflection, translation, and stretching, which are grouped under the term **transformations** of functions.

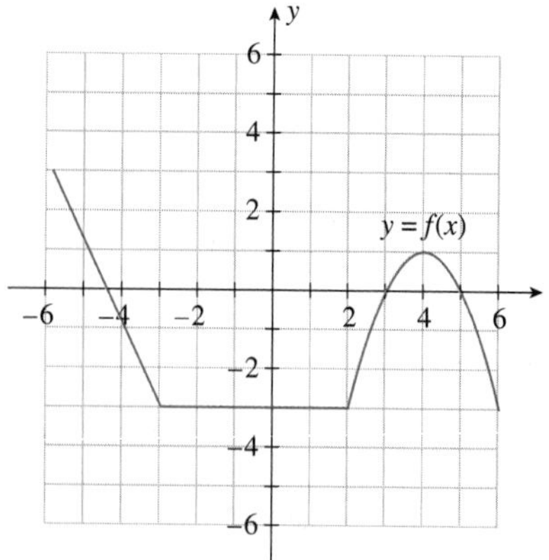

Figure 2.11
The graph of $y = f(x)$

Translation

Suppose a function $y = f(x)$ is given whose graph is shown in Figure 2.11. How does the graph of f change when we add some positive constant k to the function f? Since the operation $f(x) + k$ changes the y-value, we can imagine how every point (x, y) on the graph of $y = f(x)$ gets shifted upwards by k units when $k > 0$ to become the new point $(x, y + k)$, which generates the graph of $y = f(x) + k$. This is called a **vertical translation.** The graph of $y = f(x) + 4$ is shown in Figure 2.12 together with the original graph.

Similarly, the operation $f(x) - k$ changes the y-value, and we can imagine that every point (x, y) on the graph of $y = f(x)$ gets shifted downwards by k units when $k > 0$ to become the new point $(x, y - k)$, which generates the graph of $y = f(x) - k$. The graph of $y = f(x) - 2$ is shown in Figure 2.13 together with the original graph.

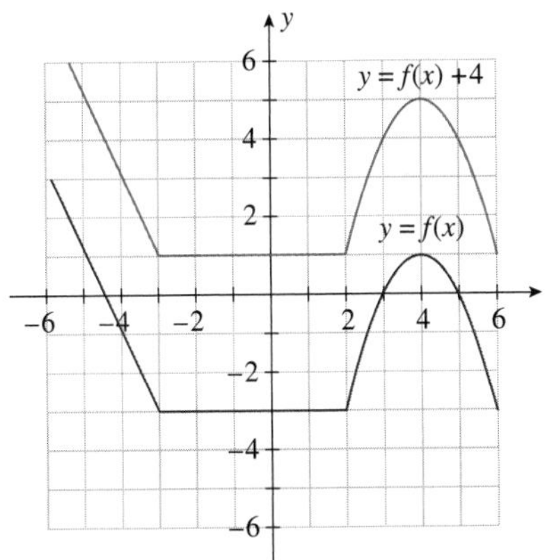

Figure 2.12

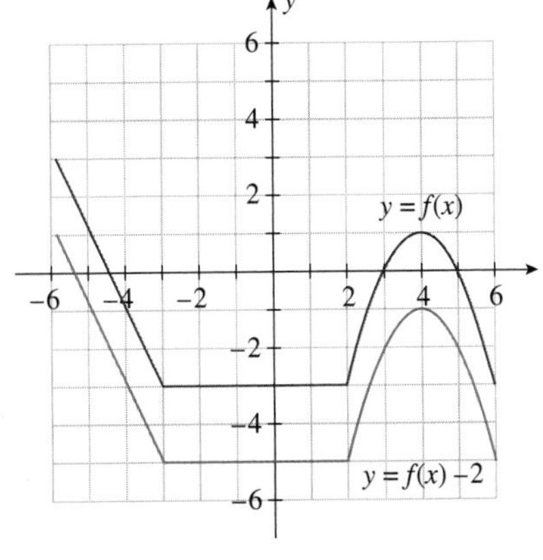

Figure 2.13

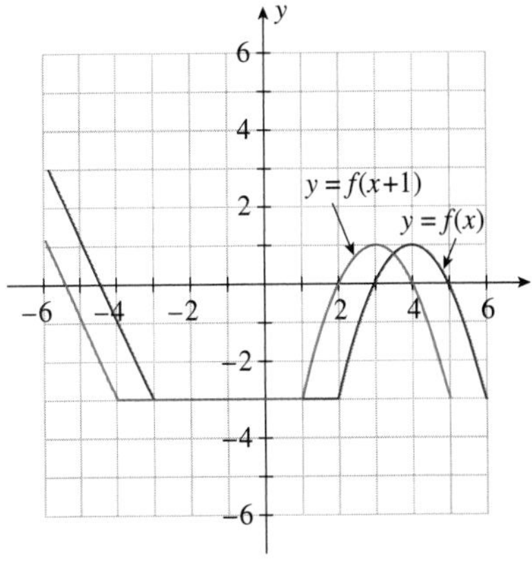

Figure 2.14

Vertical Translation, $k > 0$

The graph of $y = f(x) + k$ is the graph of $y = f(x)$ translated vertically up by k units.
The graph of $y = f(x) - k$ is the graph of $y = f(x)$ translated vertically down by k units.

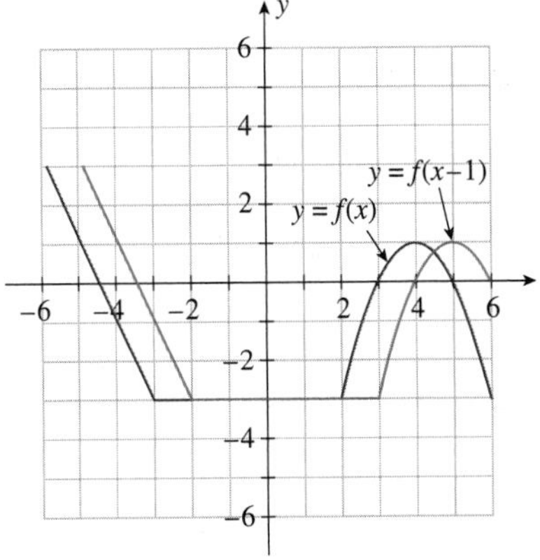

Figure 2.15

What happens to the graph of $y = f(x)$ if instead of adding to the function value (y-value) we add a positive constant h to the x-value? We can argue that similar to the vertical translation this addition must shift every point on the graph of $y = f(x)$ to the right by h units to generate the graph of $y = f(x + h)$ when $h > 0$. However, this argument is incorrect as we can observe from the graph of $y = f(x + 1)$ shown in Figure 2.14, which is clearly shifted to the left by one unit. What happened?

It will help our understanding of what happened by observing a few x- and y-values for $y = f(x)$ and $y = f(x + 1)$ given in the Table 2.1. Notice that the point (3, 0) transformed to the point (2, 0), and that the point (4, 1) transformed to the point (3, 1). Generally, every point (x, y) on the graph of $y = f(x)$ is transformed to the point $(x - 1, y)$ on the graph of $y = f(x + 1)$, which corresponds to a shift to the left when comparing the graphs of $y = f(x)$ and $y = f(x + 1)$. Likewise, if we subtract the positive constant h from the x-value of the graph of $y = f(x)$ then every point (x, y) on the graph of $y = f(x)$ gets shifted to the right by h units when $h > 0$, which create the graph of $y = f(x - h)$. This is referred to as **horizontal translation.** Table 2.1 also shows some of the x- and y-values of the function $y = f(x - 1)$, whose graph is shown in Figure 2.15.

Table 2.1

$y = f(x)$		$y = f(x + 1)$		$y = f(x - 1)$	
x	y	x	y	x	y
1	−3	1	−3	1	−3
2	−3	2	0	2	−3
3	0	3	1	3	−3
4	1	4	0	4	0
5	0	5	−3	5	1

Horizontal Translation, $h > 0$

The graph of $y = f(x + h)$ is the graph of $y = f(x)$ translated horizontally left by h units.
The graph of $y = f(x - h)$ is the graph of $y = f(x)$ translated horizontally right by h units.

Example 1

The graph of the function $y = g(x)$ is given below. Graph the function $y = g(x - 1) - 3$.

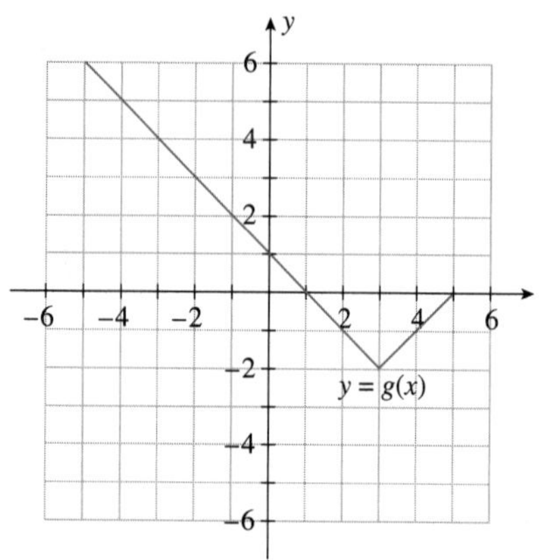

Solution

One approach is to look at special points that stand out on the graph of $y = g(x)$ such as $(-5, 6)$, $(3, -2)$, and $(5, 0)$. Using $y = g(x - 1) - 3$, the point $(-5, 6)$ gets translated to the point $(-5 + 1, 6 - 3) = (-4, 3)$, the point $(3, -2)$ gets translated to the point $(3 + 1, -2 - 3) = (4, -5)$, and the point $(5, 0)$ gets translated to the point $(5 + 1, -3) = (6, -3)$. Graph the new points and connect them in a similar fashion as the original graph. The graph of $y = g(x - 1) - 3$ is shown in Figure 2.16. Another approach is to use the whole graph of $y = g(x)$ and perform the required translations for $y = g(x - 1) - 3$, which means a horizontal translation to the right by 1 unit and a vertical translation down by 3 units. Use this information to translate the entire graph of $y = g(x)$ 1 unit right and 3 units down to yield the graph of $y = g(x - 1) - 3$, which is shown in Figure 2.16.

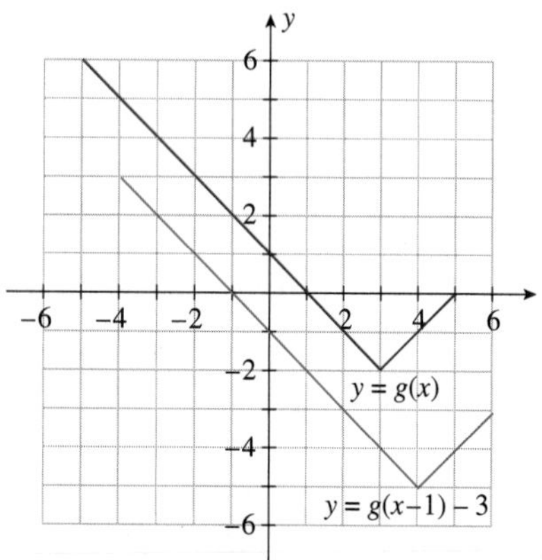

Figure 2.16
The graphs of $y = g(x)$ and $y = g(x - 1) - 3$

Example 2

The graph of the function $y = f(x)$ is given below together with its graph translated horizontally and vertically. Find h and k and label the new function.

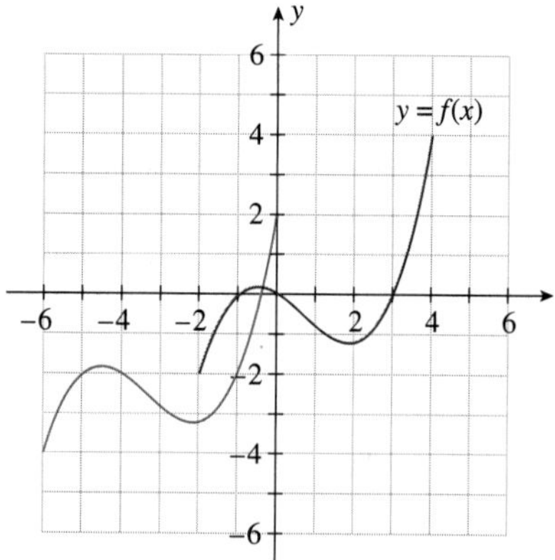

Solution

Notice that the translated graph is obtained from the graph of $y = f(x)$ by translating it 4 units to the left, so $h = 4$, and 2 units down, so $k = 2$. This corresponds to the function $y = f(x + 4) - 2$.

Reflection

Let us revisit our function $y = f(x)$ from the beginning of this section whose graph is shown in Figure 2.11. How does the graph of f change when we replace x with $-x$ in the function f? We observe that every point (x, y) on the graph of $y = f(x)$ gets transformed to the new point $(-x, y)$, which is like holding a mirror on the y-axis reflecting the graph to the opposing quadrant. This is called a **horizontal reflection.** The graph of $y = f(-x)$ is shown in Figure 2.17 together with the original graph.

A similar question is what happens to the graph of $y = f(x)$ if we do the same with y, i.e., what does the graph of $y = -f(x)$ look like? Here we observe that every point (x, y) on the graph of $y = f(x)$ gets transformed to the new point $(x, -y)$, which is like holding a mirror on the x-axis reflecting the graph to the opposing quadrant. This is called a **vertical reflection.** The graph of $y = -f(x)$ is shown in Figure 2.18 together with the original graph.

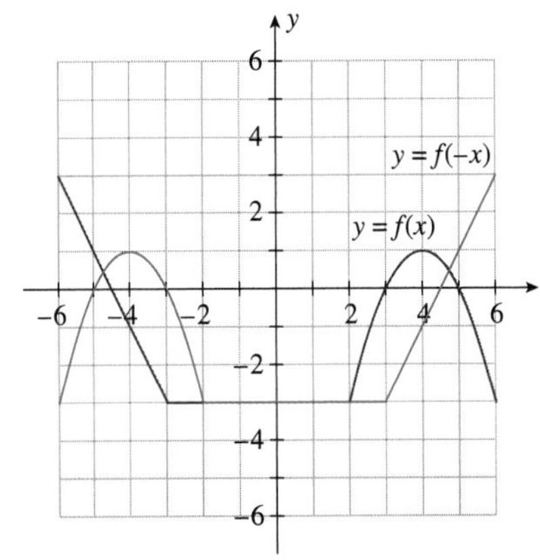

Figure 2.17

Reflection

The graph of $y = f(-x)$ is the graph of $y = f(x)$ reflected horizontally in the y-axis.
The graph of $y = -f(x)$ is the graph of $y = f(x)$ reflected vertically in the x-axis.

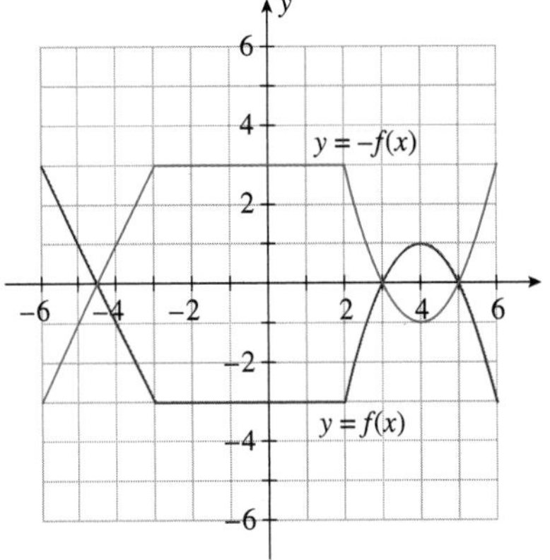

Figure 2.18

Now that we have looked at graphical changes when x is replaced with $-x$ or y is replaced with $-y$, we can introduce the definition for two special types of functions, namely an **even function** and an **odd function,** which classify a function based on symmetries it has with respect to the x- and y-axes.

Even Function

If $f(x) = f(-x)$ for all values of x in the domain of f, then the function is called *even.* The graph of an even function is symmetric about the y-axis.

Odd Function

If $-f(x) = f(-x)$ for all values of x in the domain of f, then the function is called *odd.* The graph of an odd function is symmetric about the origin, which corresponds to a reflection in both the x-axis and y-axis.

Example 3

The graph of the function $P(x) = \frac{1}{10}(x + 3)(x - 1)(x - 5)$ for $-4 \leq x \leq 6$ is shown below.

a. Show that the function is neither even nor odd.
b. Show that the function $y = P(x + 1)$ for $-5 \leq x \leq 5$ is odd.

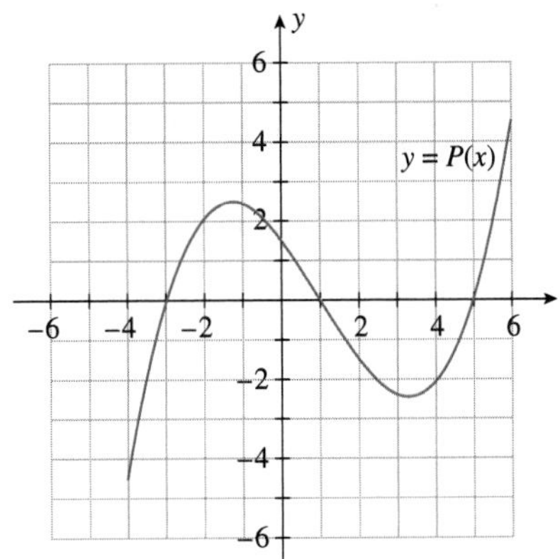

Solution

a. We observe from the graph that a reflection in the y-axis does not match with the graph of $y = P(x)$, so P is not even. Furthermore, a reflection in the x-axis and the y-axis also does not match the graph of $y = P(x)$, so P is not odd.

Alternatively, we could have shown this algebraically by finding $P(-x)$

$$\begin{aligned} P(-x) &= \frac{1}{10}(-x+3)(-x-1)(-x-5) \\ &= \frac{1}{10}(-1)(x-3)(-1)(x+1)(-1)(x+5) \\ &= -\frac{1}{10}(x-3)(x+1)(x+5) \end{aligned}$$

and comparing it to $P(x)$ and $-P(x)$. We observe that

$$\frac{1}{10}(x+3)(x-1)(x-5) \neq -\frac{1}{10}(x-3)(x+1)(x+5)$$
$$P(x) \neq P(-x)$$

so P is not even, and

$$-\frac{1}{10}(x+3)(x-1)(x-5) \neq -\frac{1}{10}(x-3)(x+1)(x+5)$$
$$-P(x) \neq P(-x)$$

so P is not odd.

b. We take an algebraic approach by finding $y = P(x+1)$. For ease of comparison we introduce a new name and let $A(x) = P(x+1)$.

$$\begin{aligned} A(x) &= P(x+1) \\ &= \frac{1}{10}((x+1)+3)((x+1)-1)((x+1)-5) \\ &= \frac{1}{10}(x+4)x(x-4) \end{aligned}$$

We now find $A(-x)$

$$\begin{aligned} A(-x) &= \frac{1}{10}((-x)+4)(-x)((-x)-4) \\ &= \frac{1}{10}(-x+4)(-x)(-x-4) \\ &= \frac{1}{10}(-1)(x-4)(-1)x(-1)(x+4) \\ &= -\frac{1}{10}(x-4)x(x+4) \end{aligned}$$

and observe that

$$A(-x) = -\frac{1}{10}(x-4)x(x+4) = -A(x)$$

We use the definition of an odd function to deduce that $A(x) = P(x+1)$ is indeed odd.

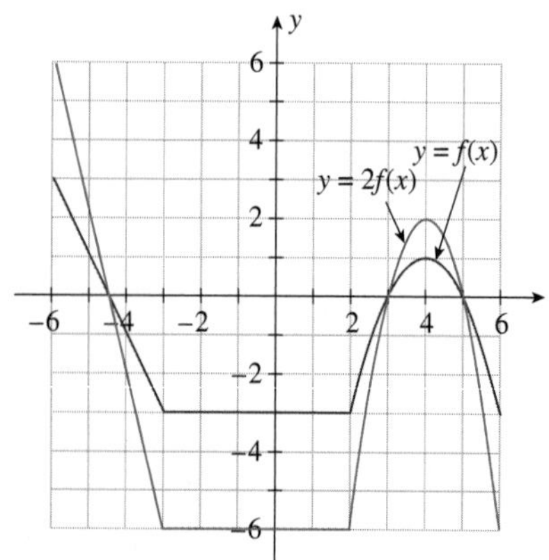

Figure 2.19

Stretching

Lastly, we want to observe how the graph of $y = f(x)$ changes when we multiply the function value or the x-value with a positive constant a. Let us first look at $y = af(x)$, for $a > 0$. Once again we notice that this means a change in the y-value and so the direction of y, which is vertical. Any point (x, y) on the graph of $y = f(x)$ gets transformed to the point (x, ay) on the graph of $y = af(x)$. If $a > 1$ then the graph gets expanded vertically by a factor of a as observed in Figure 2.19, which shows the graph of $y = 2f(x)$.

If $0 < a < 1$ then the graph gets compressed vertically by a factor of a as observed in Figure 2.20, which shows the graph of $y = \frac{1}{2}f(x)$. Nothing changes of course if $a = 1$.

Remark

If the constant a is negative, i.e. $a < 0$, then we can think of this as a vertical stretching by $|a|$ together with a vertical reflection across the x-axis. ◀

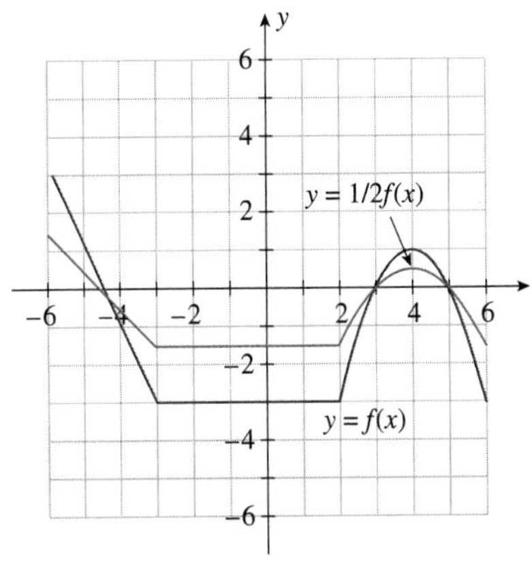

Figure 2.20

Next, let us look at $y = f(ax)$, for $a > 0$. This time it means a change in the x-value and so the direction of x, which is horizontal. Observe from Table 2.2 that any point (x, y) on the graph of $y = f(x)$ gets transformed to the point $(\frac{x}{2}, y)$ on the graph of $y = f(2x)$. Furthermore, any point (x, y) on the graph of $y = f(x)$ gets transformed to the point $(2x, y)$ on the graph of $y = f(\frac{1}{2}x)$. Looking at Figure 2.21 and Figure 2.22, we deduce that if $a > 1$ then the graph gets compressed horizontally by a factor of $\frac{1}{a}$ as is shown for the graph of $y = f(2x)$ in Figure 2.21, and if $0 < a < 1$ then the graph gets expanded horizontally by a factor of $\frac{1}{a}$ as is shown for the graph of $y = f(\frac{1}{2}x)$ in Figure 2.22.

Table 2.2

$y = f(x)$		$y = f(2x)$		$y = f(\frac{1}{2}x)$	
x	y	x	y	x	y
0	−3	0	−3	0	−3
2	−3	1	−3	4	−3
4	1	2	1	8	1

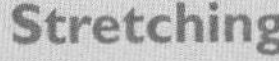
Stretching

For $a > 1$:
The graph of $y = af(x)$ is the graph of $y = f(x)$ expanded vertically by a factor of a.
The graph of $y = f(ax)$ is the graph of $y = f(x)$ compressed horizontally by a factor of $\frac{1}{a}$.

For $0 < a < 1$:
The graph of $y = af(x)$ is the graph of $y = f(x)$ compressed vertically by a factor of a.
The graph of $y = f(ax)$ is the graph of $y = f(x)$ expanded horizontally by a factor of $\frac{1}{a}$.

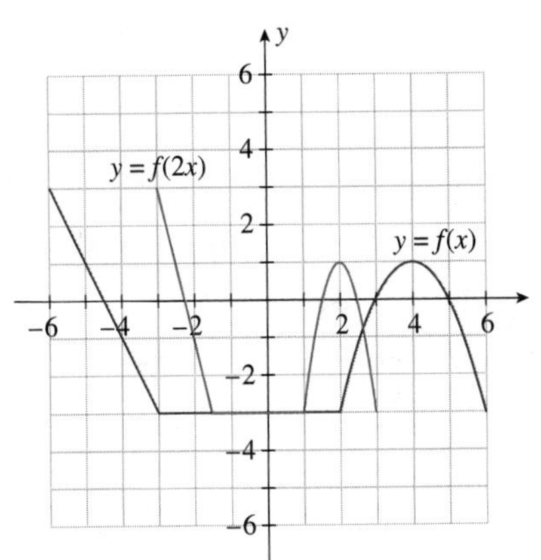

Figure 2.21

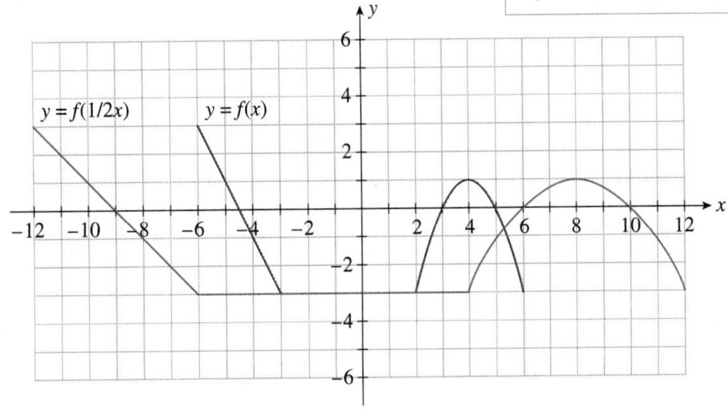

Figure 2.22

Example 4

The graph of $y = C(x)$ together with a transformed graph $y = aC(x + h) + k$ is shown below. Find the values of a, h and k.

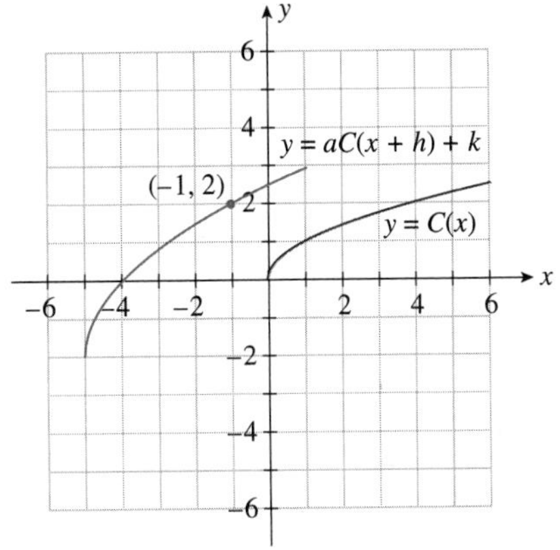

Solution

We observe that the transformed graph is vertically translated down by 2 units, resulting in $y = C(x) - 2$, which means that $k = -2$. We also notice that the transformed graph is horizontally translated left by 5 units, resulting in $y = C(x + 5) - 2$, which means that $h = 5$. Lastly, the transformed graph is also vertically expanded by a factor of a with $a > 0$, resulting in $y = aC(x + 5) - 2$. We now use the point $(-1, 2)$ on the transformed graph in the equation $y = aC(x + 5) - 2$ to find the value of a. We get

$$2 = aC(-1 + 5) - 2 = aC(4) - 2$$
$$4 = aC(4)$$
$$a = \frac{4}{C(4)}$$

From the graph we observe that $C(4) = 2$ and so

$$a = \frac{4}{C(4)} = \frac{4}{2} = 2$$

Remark

Notice that all the transformations we have looked at—translations, reflections, and stretching—preserve the property that the new graph obtained from $y = f(x)$ is still a function. ◀

Self-Check Exercises 2.2

1. The graphs of $y = f(x) = x^2$ and $y = g(x)$ are shown here. Name the sequence of transformations if the graph of g is obtained from the graph of f and describe g using f.

2. If the function $y = f(x)$ is given, name the sequence of transformations needed to get the graph of $y = 3f(-x + 2) + 6$.

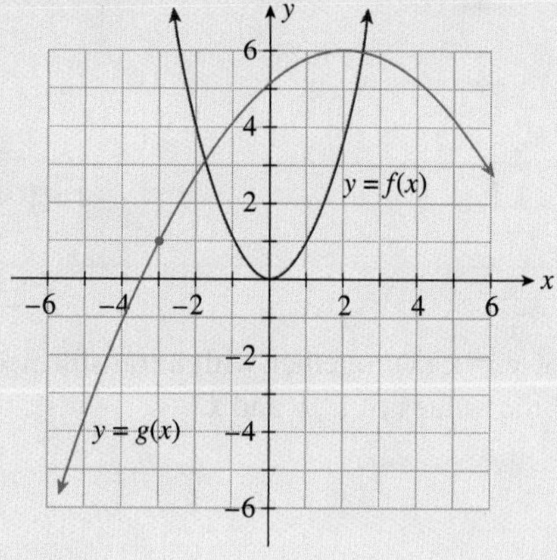

Solutions to Self-Check Exercises 2.2 can be found on page 63.

2.2 Exercises

In Exercises 1–6, use the graph of $y = f(x)$ below to graph the indicated transformed function.

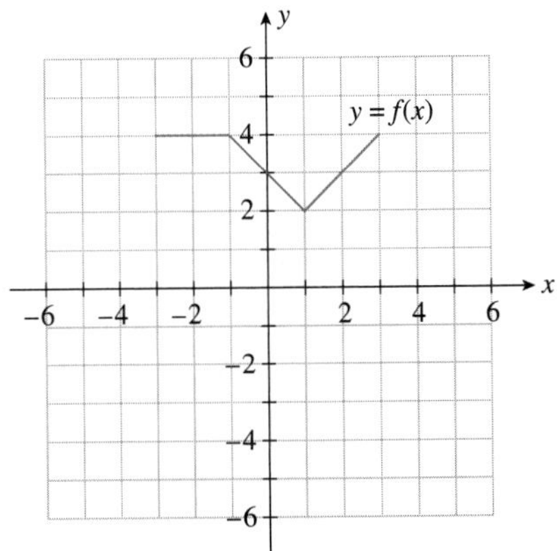

1. $y = f(x - 3)$
2. $y = f(x + 5)$
3. $y = f(x) - 2$
4. $y = f(x) + 4$
5. $y = f(-x)$
6. $y = 2f(x)$

In Exercises 7–14, use the graph of $y = g(x)$ below to graph the indicated transformed function.

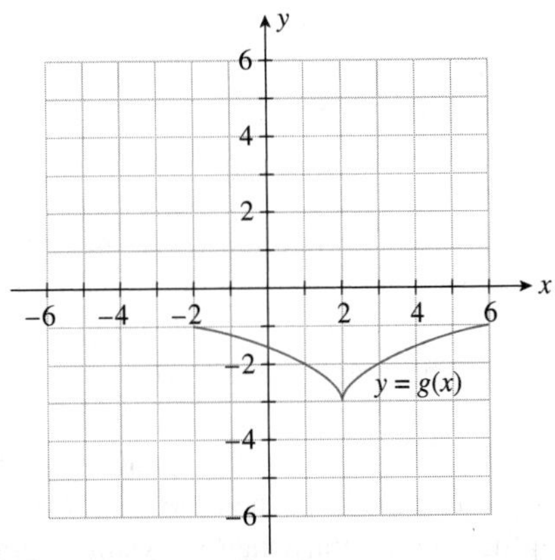

7. $y = g(x - 1) - 3$
8. $y = g(x + 2) - 2$
9. $y = g(x - 2) + 4$
10. $y = g(x + 4) + 5$
11. $y = -2g(x + 1)$
12. $y = \frac{1}{2}g(x + 3)$
13. $y = g(-x - 1)$
14. $y = g(-x + 2)$

In Exercises 15–20, name the sequence of transformations performed on the graph of $y = f(x)$ to obtain the graph of the indicated function.

15. $y = -f(x + 2) - 5$
16. $y = -f(x - 2) + 1$
17. $y = \frac{1}{3}f(x - 5) - 1$
18. $y = 4f(x + 3) + 2$
19. $y = 2f(-x) + 5$
20. $y = f(-3x) - 1$

In Exercises 21–26, find an equation of the function g, which is obtained when the indicated sequence of transformations are preformed on the graph of $y = f(x)$.

21. The graph of f is vertically translated down by 2 units and horizontally translated left by 5 units.
22. The graph of f is vertically translated up by 5 units and horizontally translated left by 3 units.
23. The graph of f is vertically translated up by 4 units, horizontally translated right by 2 units, and vertically expanded by a factor of 3.
24. The graph of f is vertically translated down by 6 units, horizontally translated right by 1 unit and vertically compressed by a factor of $\frac{2}{3}$.
25. The graph of f is vertically compressed by a factor of $\frac{3}{7}$ and reflected in the origin.
26. The graph of f is horizontally expanded by a factor of 3 and reflected in the y-axis.

Solutions to Self-Check Exercises 2.2

1. From the graphs we observe that in order to obtain the graph of g, the graph of f must be

   - horizontally translated right by 2 units, resulting in $y = f(x - 2)$
   - reflected in the x-axis, resulting in $y = -f(x - 2)$
   - vertically translated up by 6 units, resulting in $y = -f(x - 2) + 6$
   - and vertically compressed by a factor of a with $a > 0$, resulting in $y = -af(x - 2) + 6$

   To find a, we use the given point $(-3, 1)$ on the graph of g and substitute it in the equation $y = -af(x - 2) + 6$. We obtain

   $$1 = -af(-3 - 2) + 6 = -af(-5) + 6$$

   Now we use the fact that $f(x) = x^2$ and get

   $$1 = -a(-5)^2 + 6$$
   $$1 = -25a + 6$$
   $$-5 = -25a$$
   $$a = \frac{1}{5}$$

   Finally, $g(x) = -\frac{1}{5}f(x - 2) + 6$.

2. Studying $y = 3f(-x + 2) + 6$, we observe that the function value is multiplied with 3. This corresponds to a vertical expansion by a factor of 3. We also notice that 6 is added to the function value, which corresponds to a vertical translation up by 6 units. The expression $-x + 2$ contains two pieces of information, namely a horizontal translation left by 2 units and a reflection in the y-axis. Notice that the reflection comes last.

2.3 Polynomial Functions

In this section we introduce a special class of functions, namely the polynomial functions.

Polynomial Function

A **polynomial function** of *degree n* is a function of the form
$p(x) = a_nx^n + a_{n-1}x^{n-1} + \ldots + a_2x^2 + a_1x + a_0$
with $a_n \neq 0$, where $a_0, a_1, \ldots, a_n$ are constants and n is a nonnegative integer.

Notice that the degree of a polynomial function can be zero. In other words, $p(x) = a_0$ or $y = a_0$ are also polynomial functions. They are called *constant functions* and are the familiar horizontal lines when graphed.

The constants $a_0, a_1, \ldots, a_n$ in the above definition are referred to as **coefficients** and the coefficient a_n is called the **leading coefficient.** A polynomial function of degree 1 ($n = 1$)

$$p(x) = a_1x + a_0 \qquad a_1 \neq 0$$

is the equation of a straight line in the slope-intercept form with slope $m = a_1$ and y-intercept $b = a_0$ (see Section 1.4). For this reason, a polynomial function of degree 1 is called a **linear function.** A polynomial function of degree 2 is referred to as a **quadratic function.** A polynomial function of degree 3 is called a **cubic function,** and so on.

The *domain* of a polynomial function consists of all the real numbers, because there are no restrictions on the independent variable x in the equation $p(x) = a_nx^n + a_{n-1}x^{n-1} + \ldots + a_2x^2 + a_1x + a_0$. The *range* of a polynomial function depends on its degree. Let us investigate how the degree influences the range. The graphs of several polynomial functions are shown in Figure 2.23.

There are three behaviours we should notice. First of all, the number of *ups* and *downs* (read the graph from left to right) on the graph depends on the degree of the polynomial function. For example, the polynomial function q has degree 1 and only goes *down.* The polynomial function t has degree 3 and goes *up-down-up.* The polynomial function u has degree 4 and goes *down-up-down-up.* Secondly, the sign of the leading coefficient determines if the end of the graph goes up or down in quadrant I. Notice that when the leading coefficient is negative as it is for the functions q, s, and v, then the end of the graph goes down for large x (Quadrant I), and when the leading coefficient is positive as it is for the functions r, t, and u, then the end of the graph goes up for large x (Quadrant I). This is helpful information when we want to graph a polynomial function. Thirdly, when the degree is even such as for the functions r, s, and u, then either both ends of the graph go up or both ends go down. When the degree is odd such as for the functions q, t, and v, then one end of the graph goes up and the other end goes down. This means that the *range* of a polynomial function of odd degree consists of all the real numbers, while the range of a polynomial function of even degree will be of the form $(-\infty, k]$ (leading coefficient is negative) or $[k, \infty)$ (leading coefficient is positive) for some real number k.

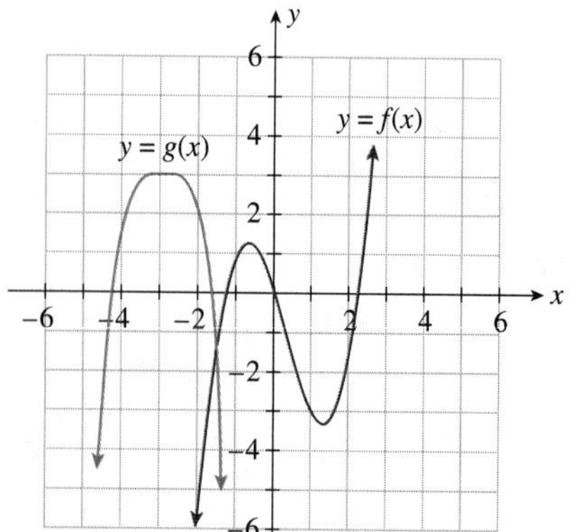

Example 1

The graphs of the two polynomial functions $f(x) = x^3 - x^2 - 3x$ and $g(x) = -(x + 3)^4 - 3$ are shown at left. Indicate the domain and range for both f and g.

Solution

Since both f and g are polynomials the domain for each function consists of all the real numbers, i.e., $D_f = (-\infty, \infty)$ and $D_g = (-\infty, \infty)$. From the graph we determine that the range for g is given by $(-\infty, 3]$ and the range for f is given by all the real numbers, i.e., $R_g = (-\infty, 3]$ and $R_f = (-\infty, \infty)$.

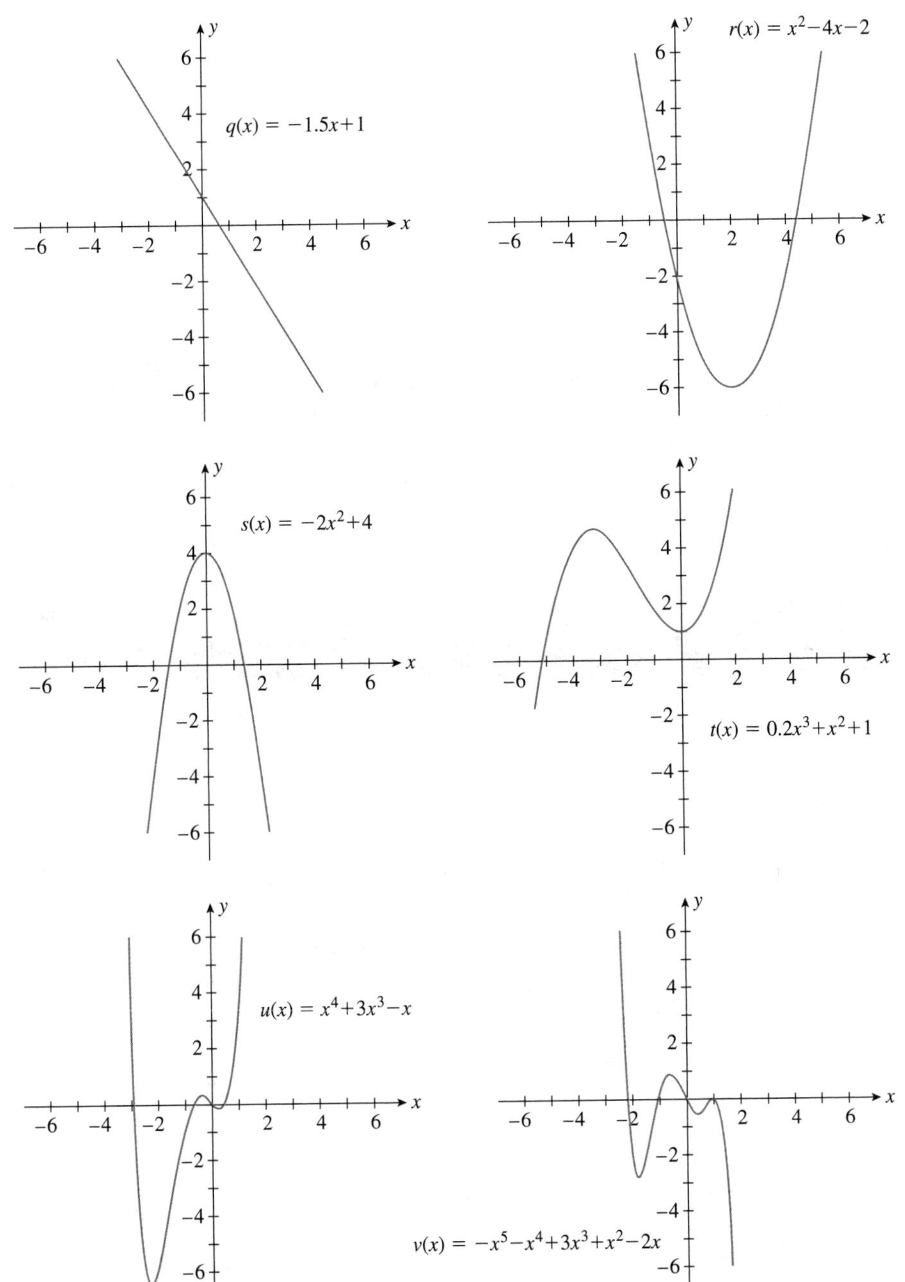

Figure 2.23

Quadratic Functions

Recall from the beginning of this section that quadratic functions are polynomial functions of degree 2. We are having a closer look at them here, since quadratic functions are often applied in optimization problems such as maximizing profit or minimizing cost (see Section 7.4).

Quadratic Function

A **quadratic function** is a polynomial function of degree 2 and has the *general* form

$$p(x) = ax^2 + bx + c$$

with $a \neq 0$, where a, b, c are constants.

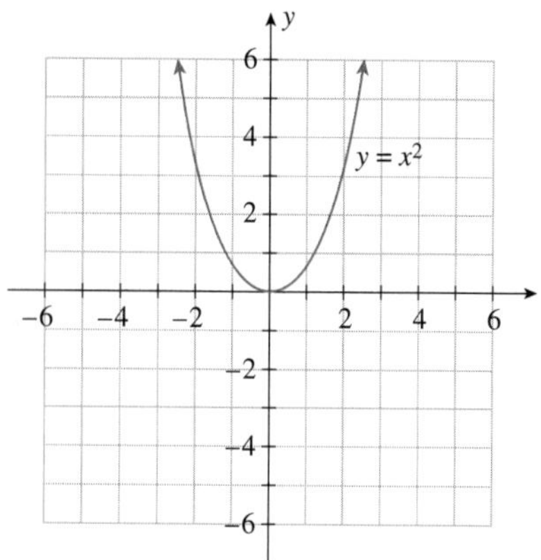

Figure 2.24

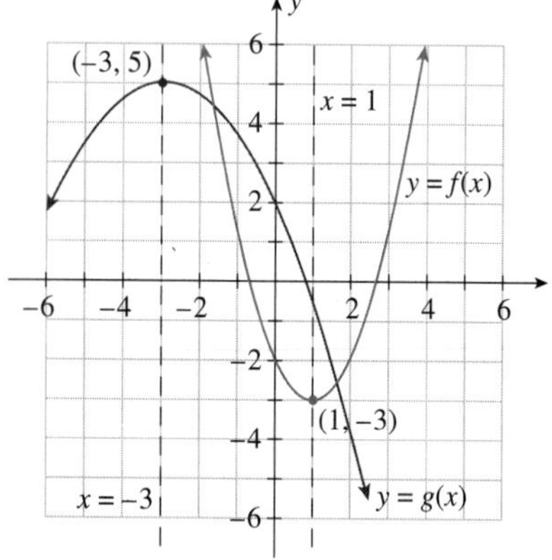

Figure 2.25
Graphs of $f(x) = x^2 - 6x + 7$ and $g(x) = -0.35x^2 - 2.1x + 1.85$

The *basic* quadratic function is $y = x^2$ and shown in Figure 2.24. The graph of a quadratic function is called a **parabola.** Depending on the sign of the leading coefficient a, the graph either opens *upward* ($a > 0$) as can be seen with the red graph of $f(x) = x^2 - 6x + 7$ in Figure 2.25 or *downward* ($a < 0$) as can be seen with the blue graph of $g(x) = -0.35x^2 - 2.1x + 1.85$ in Figure 2.25. If a parabola opens upward/downward, it has a lowest/highest point. This point is called the *vertex.* Furthermore, the vertical line through the vertex is called the *axis of symmetry.* Notice how the axis of symmetry acts like a mirror for any of the parabolas in Figure 2.25. For example, the vertex of the parabola for f is $(1, -3)$ and its axis of symmetry is given by $x = 1$, while the vertex of the parabola for g is $(-3, 5)$ and its axis of symmetry is given by $x = -3$.

The general form $p(x) = ax^2 + bx + c$ of a quadratic function does not lend itself well to graphing. By an algebraic method called *completing the square* we can manipulate the general form and get the *standard* form of a quadratic function $p(x) = a(x - h)^2 + k$. Recall from Section 2.2 that a, h, and k transform the graph of $y = x^2$ as is indicated in Table 2.3. Moreover, this form allows us to immediately read off the vertex as (h, k) as well as the axis of symmetry given by $x = h$, which aids in the graphing process.

GROUP DISCUSSION

Consider the quadratic function $f(x) = 3x^2 - 6x + 5$. Use the following sequence of steps to complete the square:

1. Factor out the leading coefficient of the two x-terms. There is nothing to do if the leading coefficient is 1.
2. Read off the coefficient of the linear x-term and call it $2B$.
3. Recall that $x^2 + 2Bx + B^2 = (x + B)^2$ and use it to write f similar to $x^2 + 2Bx = (x + B)^2 - B^2$. Collect like terms at the end.
4. Check that the standard form you obtained still represents $f(x) = 3x^2 - 6x + 5$ by expanding and collecting like terms. Now try to generalize the process with $p(x) = ax^2 + bx + c$ to obtain $p(x) = a(x - h)^2 + k$ and show that $\frac{b}{2a} = -h$ and $c - a(\frac{b}{2a})^2 = k$.

Table 2.3

Form of Equation	Description	Transformation
$y = x^2$	Basic parabola shown in Figure 2.24	
$y = ax^2$	If $a > 0$, then parabola opens upward. If $a < 0$, then parabola opens downward.	Reflection
	If $\lvert a\rvert > 1$, then the parabola is expanded vertically by a factor of a. If $0 < \lvert a\rvert < 1$, then the parabola is compressed vertically by a factor of a.	Stretching
$y = (x - h)^2$	If $h < 0$, then the parabola is translated horizontally left by h units. If $h > 0$, then the parabola is translated horizontally right by h units.	Horizontal Translation
$y = x^2 + k$	If $k > 0$, then the parabola is translated vertically up by k units. If $k < 0$, then the parabola is translated vertically down by k units.	Vertical Translation

Quadratic Function

A **quadratic function** is a polynomial function of degree 2 and has the *standard* form

$$p(x) = a(x - h)^2 + k$$

with $a \neq 0$, where a, h, k are constants, the *vertex* is (h, k) and the *axis of symmetry* is $x = h$.

Example 2

Graph $f(x) = \frac{1}{2}x^x - 2x - 1$. Then label its vertex and axis of symmetry. Does the vertex represent the highest point or the lowest point?

Solution

We must first write the quadratic function f in standard form by completing the square. We start by factoring out $\frac{1}{2}$ to get

$$f(x) = \frac{1}{2}(x^2 - 4x) - 1$$

Then we divide -4 by 2 to get -2 and use it to write

$$f(x) = \frac{1}{2}(x - 2)^2 - 1 - ?$$

where ? represents the third term in the expansion of $\frac{1}{2}(x - 2)^2$. This third term is obtained by calculating $[\frac{1}{2}(-2)^2]$. Therefore,

$$\begin{aligned} f(x) &= \frac{1}{2}(x - 2)^2 - 1 - \left[\frac{1}{2}(-2)^2\right] \\ &= \frac{1}{2}(x - 2)^2 - 1 - 2 \\ &= \frac{1}{2}(x - 2)^2 - 3 \end{aligned}$$

which is the standard form. We can now read of the sequence of transformations performed on the basic function $y = x^2$ to obtain the graph of f. Since the leading coefficient is $\frac{1}{2}$, the parabola opens upward and is vertically expanded by a factor of $\frac{1}{2}$. The vertex is $(2, -3)$ and represents the lowest point of f. The axis of symmetry is given by $x = 2$. Finally, we can either use the factor of $\frac{1}{2}$ to graph the parabola or find one other point on the parabola such as when $x = 4$

$$f(4) = \frac{1}{2}(4 - 2)^2 - 3 = \frac{1}{2}(2)^2 - 3 = 2 - 3 = -1$$

which gives the point $(4, -1)$. The parabola representing f is shown below.

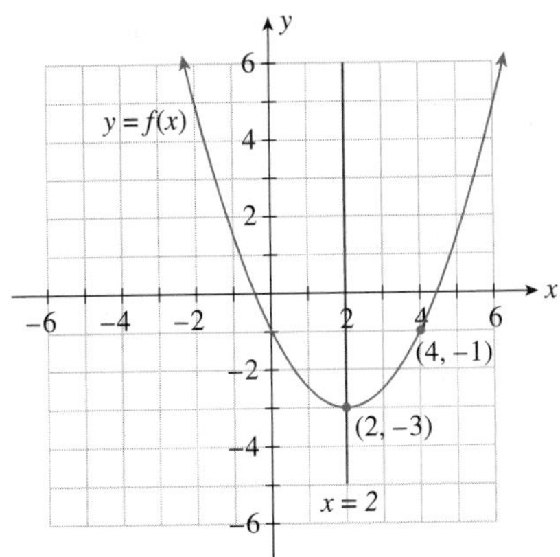

The further analysis of polynomial functions will be left to the tools of calculus, which are developed in later chapters.

Self-Check Exercises 2.3

1. Graph $f(x) = 3(x + 2)^2 + 1$.
2. Describe the sequence of transformations performed on the graph of $y = x^2$ to obtain the graph of $f(x) = -3(x + 9)^2 - 11$.

Solutions to Self-Check Exercises 2.3 can be found on page 69.

2.3 Exercises

In Exercises 1–6, match the quadratic functions described with the graphs labelled A through F shown below.

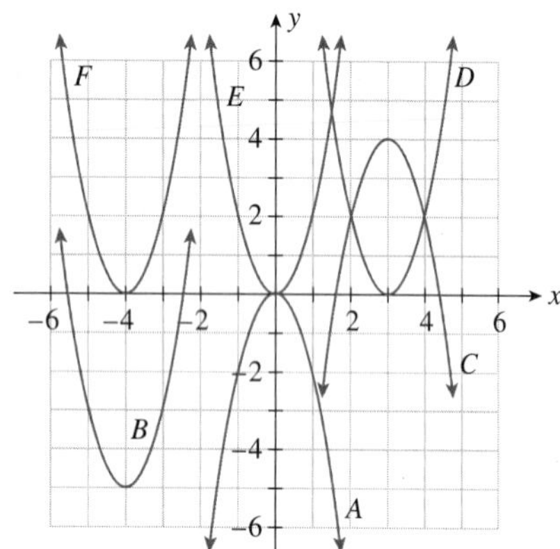

1. $a(x) = 2x^2$
2. $b(x) = -2x^2$
3. $c(x) = 2(x - 3)^2$
4. $d(x) = 2(x + 4)^2$
5. $e(x) = -2(x - 3)^2 + 4$
6. $f(x) = 2(x + 4)^2 - 5$

In Exercises 7–12, match the quadratic functions described with the graphs labelled A through F shown below.

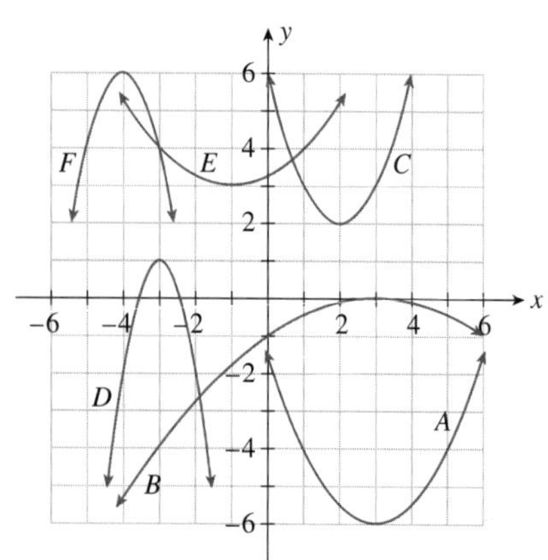

7. $a(x) = -2(x + 4)^2 + 2$
8. $b(x) = \frac{1}{4}(x + 1)^2 + 3$
9. $c(x) = \frac{1}{2}(x - 3)^2 - 6$
10. $d(x) = -3(x + 3)^2 + 1$
11. $e(x) = (x - 2)^2 + 2$
12. $f(x) = -\frac{1}{9}(x - 3)^2$

In Exercises 13–18, answer parts (a)–(d) using the indicated quadratic function.

a. Indicate if the parabola opens upward or downward.
b. Find the vertex and indicate if it is the highest or lowest point on the parabola.
c. Find the equation for the axis of symmetry.
d. State the range.

13. $f(x) = 3(x + 8)^2$
14. $f(x) = -(x + 1)^2$
15. $f(x) = x^2 - 3$
16. $f(x) = \frac{1}{3}x^2 - 2$
17. $f(x) = (x + 7)^2 + 2$
18. $f(x) = -5(x - 1)^2 + 6$

In Exercises 19–24, describe the sequence of transformations performed on the graph of $y = x^2$ to obtain the graph of the indicated function.

19. $f(x) = (x - 2)^2 - 3$
20. $f(x) = (x + 3)^2 + 7$
21. $f(x) = 3(x - 5)^2$
22. $f(x) = -\frac{1}{2}x^2 - 3$
23. $f(x) = 4(x - 3)^2 - 1$
24. $f(x) = -(x + 1)^2 - 2$

In Exercises 25–30, graph the indicated parabola.

25. $f(x) = x^2 + 2x + 5$
26. $f(x) = x^2 - 16x + 80$
27. $f(x) = 3x^2 - 24x + 50$
28. $f(x) = 2x^2 - 8x$
29. $f(x) = -x^2 + 4x + 3$
30. $f(x) = -2x^2 + 12x - 11$

Solutions to Self-Check Exercises 2.3

1. From $f(x) = 3(x + 2)^2 + 1$ we can read off the vertex, which is $(-2, 1)$. Since the leading coefficient is $3 > 0$, the parabola opens upward and is vertically expanded by a factor of 3. The parabola is shown below.

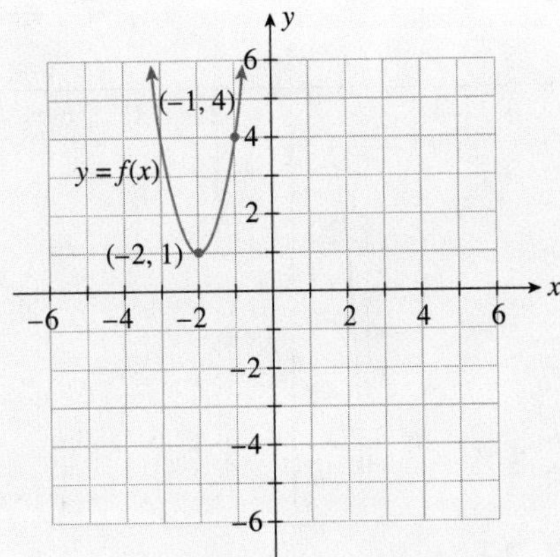

2. From $f(x) = -3(x + 9)^2 - 11$ we observe that in order to obtain the graph of f, the graph of $y = x^2$ must be
   - reflected in the y-axis, resulting in $y = -x^2$
   - horizontally translated left by 9 units, resulting in $y = -(x + 9)^2$
   - vertically translated down by 11 units, resulting in $y = -(x + 9)^2 - 11$
   - vertically expanded by a factor of 3, resulting in $y = -3(x + 9)^2 - 11$

2.4 Rational and Power Functions

In this section we introduce two special classes of functions, namely the rational and power functions.

Rational Functions

A rational function is simply the quotient of two polynomial functions. Examples of rational functions are

$$f(x) = \frac{x^2 + 1}{x^2 - 1}$$

$$g(x) = \frac{3x^3 + x^2 - x + 1}{x - 2}$$

Rational Function

A **rational function** has the form

$$r(x) = \frac{p(x)}{q(x)}$$

where $p(x)$ and $q(x)$ are polynomial functions with $q(x) \neq 0$ since division by zero is not allowed.

The domain of a rational function consists of all the real numbers excluding the roots of the polynomial function $q(x)$ in the denominator (recall Section 1.2), that is all the numbers x for which $q(x) = 0$.

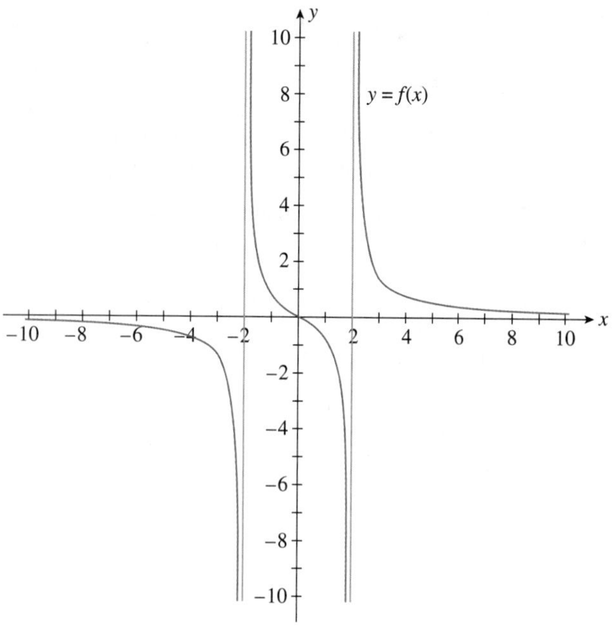

Figure 2.26

The graph of $f(x) = \dfrac{6x - 1}{3x^2 - 12}$

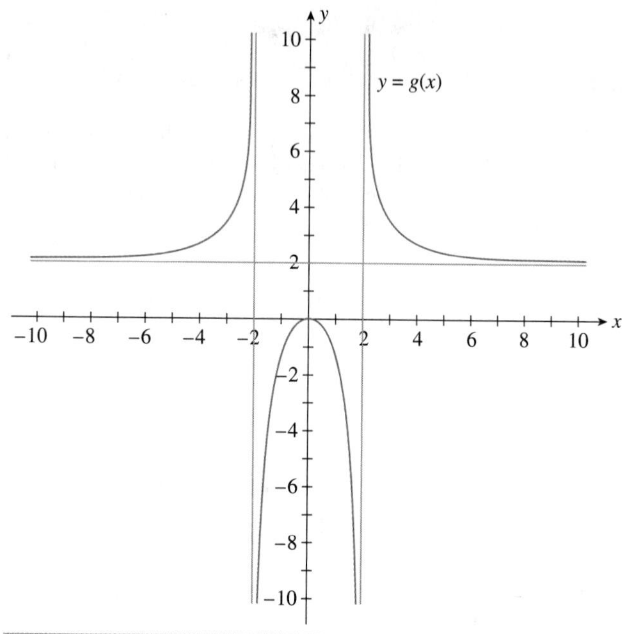

Figure 2.27

The graph of $g(x) = \dfrac{6x^2 - 1}{3x^2 - 12}$

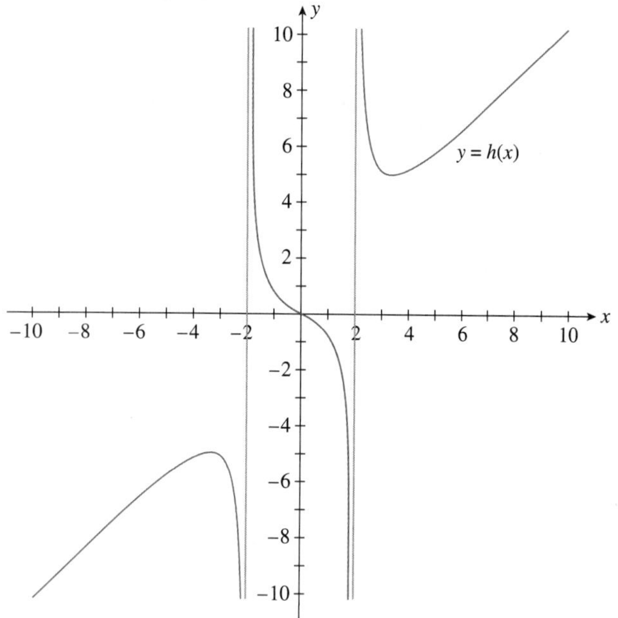

Figure 2.28

The graph of $h(x) = \dfrac{3x^3 - 1}{3x^2 - 12}$

The graph of a rational function $y = r(x)$ is characterized by one or more *asymptotes.* We will only consider asymptotic lines, which are lines that the graph of r approaches more and more closely. Asymptotic lines in general are discussed analytically in Section 7.3; here we are only concerned with finding asymptotes for rational functions. There are three types of asymptotes: *vertical, horizontal,* and *slant.* We will not concern ourselves with finding slant asymptotes. Vertical asymptotes are easily found by identifying the roots of the polynomial function in the denominator. If these roots are not also roots of the numerator, then they represent vertical asymptotes. Figures 2.26, 2.27, and 2.28 are the graphs of three different rational functions f, g, and h with the same denominator $3x^2 - 12$. The roots are easy to find by equating the denominator to zero and solving for x.

$$3x^2 - 12 = 0$$

$$3x^2 = 12$$

$$x^2 = \frac{12}{3} = 4$$

$$x = \pm 2$$

Now we have to check if $x = 2$ is also the root of the numerator $6x - 1$ of f. If it is, then the factor $(x - 2)$ can be cancelled from the numerator and denominator (recall Rational Expressions in Section 1.2). This would result in a hole in the graph of f at $x = 2$. However, $x = 2$ is not a root of the numerator $6x - 1$ of f, and so instead $x = 2$ is a vertical asymptote for the graph of f. In a similar fashion, we check that neither $x = 2$ nor $x = -2$ are roots of the numerators of f, g, and h, and so we know that the vertical asymptotes for f, g, and h are $x = 2$ and $x = -2$ as can be seen in Figures 2.26, 2.27, and 2.28.

Vertical asymptotes for the graph of any function can never be crossed, since the function is undefined there.

Horizontal asymptotes of rational functions can be found by considering the degree of the polynomials in the numerator and denominator. This will be done in more detail using calculus in Section 7.3.

Horizontal Asymptotes of a Rational Function $r(x) = \frac{p(x)}{q(x)}$

If degree $[p(x)] < $ degree $[q(x)]$, then the graph of r has **horizontal asymptote** $y = 0$ (x-axis).

If degree $[p(x)] = $ degree $[q(x)]$, then the graph of r has horizontal asymptote

$$y = \frac{\text{leading coefficient of } p}{\text{leading coefficient of } q}.$$

If degree $[p(x)] > $ degree $[q(x)]$, then the graph of r has no horizontal asymptote.

Notice that the graph of the rational function f has horizontal asymptote $y = 0$ ($x =$ axis) as is shown in Figure 2.26. The graph of the rational function g has horizontal asymptote $y = \frac{6}{3} = 2$ as is shown in Figure 2.27. Finally, the graph of the rational function h has no horizontal asymptote as is shown in Figure 2.28.

It is possible that the graph of a function crosses its horizontal asymptote, which we will encounter in Section 7.3.

Example 1

Consider the rational function f defined by $f(x) = \dfrac{-2(x^2 + 9)}{x^2 - 9}$.

a. State the domain of f.
b. Use the method outlined above to find the vertical asymptotes of f, if any.
c. Use the method outlined above to find the horizontal asymptotes of f, if any.
d. Find the y-intercept.
e. Develop a table of points chosen from either side of the vertical asymptotes of the graph of f.
f. Determine if f is even, odd, or neither.
g. Use the asymptotes, y-intercept, and table of points to provide a sketch of the graph of f.

Solution

a. We need to find the roots of the polynomial $x^2 - 9$ in the denominator

$$\begin{aligned} x^2 - 9 &= 0 \\ x^2 &= 9 \\ x &= \pm 3 \end{aligned}$$

So, the domain of f consists of all real numbers except ± 3, which we write as $D_f = (\infty, -3) \cup (-3, 3) \cup (3, \infty)$. The symbol $\cup$ means *union* and allows us to join two intervals.

b. We have to check if either $x = 3$ or $x = -3$ are roots of the polynomial $-2(x^2 + 9)$ in the numerator of f, which they are not. The roots of the polynomial in the denominator give us the vertical asymptotes of f, namely $x = 3$ and $x = -3$.

c. The degree of $-2(x^2 + 9)$ is 2 and the degree of $x^2 - 9$ is also 2. Therefore, the horizontal asymptote is given by

$$y = \frac{\text{leading coefficient of } -2(x^2 + 9)}{\text{leading coefficient of } x^2 + 9} = \frac{-2}{1} = -2$$

So, the horizontal asymptote is $y = -2$.

d. We find the y-intercept of the graph of f by setting $x = 0$ in $f(x) = \dfrac{-2(x^2 + 9)}{x^2 - 9}$. We calculate

$$f(0) = \frac{-2(0^2 + 9)}{0^2 - 9} = \frac{-18}{-9} = 2$$

which gives the point (0, 2) as the y-intercept.

e. We choose some x-values on either side of the two vertical asymptotes $x = 3$ and $x = -3$ such as $x = -5, -2, 2, 5$, and calculate the function value. These points are shown in Table 2.4.

Table 2.4

Points of $f(x) = \dfrac{-2(x^2 + 9)}{x^2 - 9}$ at $x = -5, -2, 2, 5$.

x	y
-5	$f(-5) = -4.25$
-2	$f(-2) = 5.2$
2	$f(2) = 5.2$
5	$f(5) = -4.25$

f. We need to calculate $f(-x)$ and compare it with $f(x)$ and $-f(x)$ to determine if f is an even or odd function.

$$f(-x) = \frac{-2((-x)^2 + 9)}{(-x)^2 - 9} = \frac{-2(x^2 + 9)}{x^2 - 9} = f(x)$$

Therefore, f is an even function, i.e., the graph of f is symmetric with respect to the y-axis.

g. Finally, the graph of the even function f with its two vertical asymptotes $x = 3$ and $x = -3$, horizontal asymptote $y = -2$, y-intercept (0, 2) as well as the points from Table 2.4 is shown in Figure 2.29.

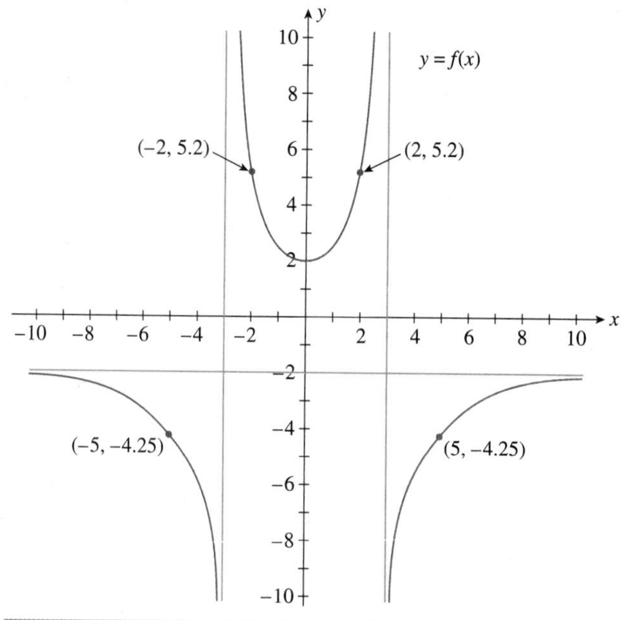

Figure 2.29

Graph of $f(x) = \dfrac{-2(x^2 + 9)}{x^2 - 9}$

Example 2

Consider again the function $g(x) = \dfrac{6x^2 - 1}{3x^2 - 12}$, whose graph is shown in Figure 2.27. Describe the sequence of transformations performed on the graph of g to obtain the graph of the function f defined by $f(x) = -g(x - 5) + 3$. Then provide the graph of f.

Solution

From $f(x) = -g(x - 5) + 3$ we observe that in order to obtain the graph of f, the graph of $g(x) = \dfrac{6x^2 - 1}{3x^2 - 12}$ must be

- reflected in the y-axis, resulting in $y = -\dfrac{6x^2 - 1}{3x^2 - 12}$
- horizontally translated right by 5 units, resulting in $y = -\dfrac{6(x - 5)^2 - 1}{3(x - 5)^2 - 12}$
- vertically translated up by 3 units, resulting in $y = -\dfrac{6(x - 5)^2 - 1}{3(x - 5)^2 - 12} + 3$

If we would not make use of our transformation knowledge, we would have to algebraically manipulate $y = -\dfrac{6(x - 5)^2 - 1}{3(x - 5)^2 - 12} + 3$ by expanding, combining like terms, and writing both fractions with a common denominator, which is a lot of work; and then start from scratch by finding its asymptotes, y-intercept, etc., to finally be able to graph g. Instead, we analyze what happens to the asymptotes under the transformations. Recall that $x = -2$ and $x = 2$ are the vertical asymptotes and $y = 2$ is the horizontal asymptote of f.

- Reflection in the y-axis does not change the vertical asymptotes, but the horizontal asymptote changes from $y = 2$ to $y = -2$.
- Horizontal translation right by 5 units does not change the horizontal asymptote, but moves the vertical asymptote $x = -2$ to $x - 5 = -2$, which simplifies to $x = -2 + 5 = 3$ and vertical asymptote $x = 2$ to $x - 5 = 2$, which simplifies to $x = 2 + 5 = 7$.
- Vertical translation up by 3 units does not change the vertical asymptotes, but moves the horizontal asymptote from $y = -2$ to $y = -2 + 3 = 1$.

Finally, we can graph $f(x) = -g(x - 5) + 3$ from $g(x) = \dfrac{6x^2 - 1}{3x^2 - 12}$ by considering the transformations and the new vertical asymptotes $x = 3$ and $x = 7$ and horizontal asymptote $y = 1$. The graph of f is shown in Figure 2.30.

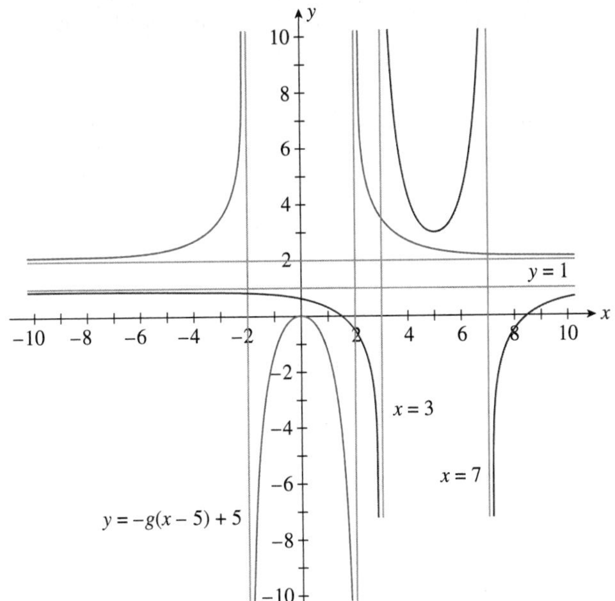

Figure 2.30
Graph of $f(x) = -g(x - 5) + 3$ from $g(x) = \dfrac{6x^2 - 1}{3x^2 - 12}$

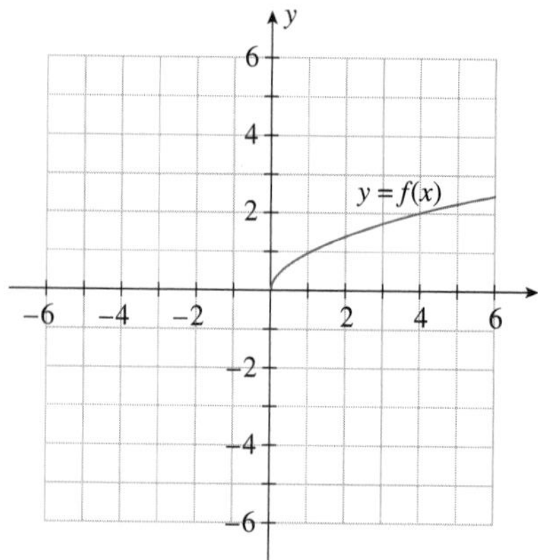

Figure 2.31

Graph of the square root function
$f(x) = \sqrt{x} = x^{1/2}$

GROUP DISCUSSION

Consider the following pairs of functions.

1. $f(x) = x^2 - 1$ and $g(x) = \dfrac{1}{x^2 - 1}$
2. $f(x) = x^2 - 4$ and $g(x) = \dfrac{1}{x^2 - 4}$
3. $f(x) = x^2 - 9$ and $g(x) = \dfrac{1}{x^2 - 9}$

Graph each pair of functions on the same coordinate system. Mark the roots of f and the vertical asymptotes of the graph of g. What do you notice? What else can you say about the graphs of f and g if you consider their symmetry?

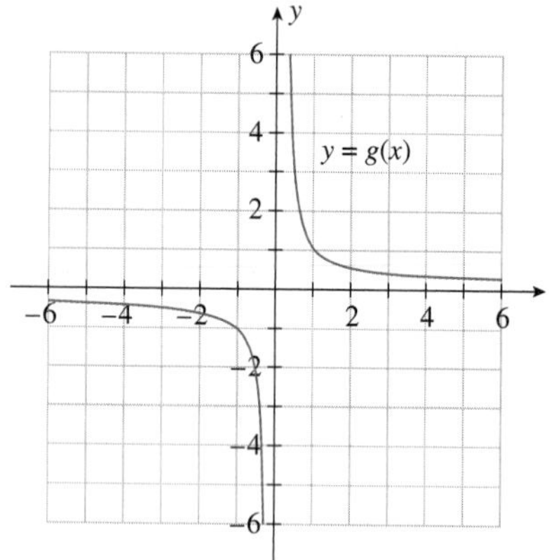

Figure 2.32

Graph of the rational function
$g(x) = \dfrac{1}{x} = x^{-1}$

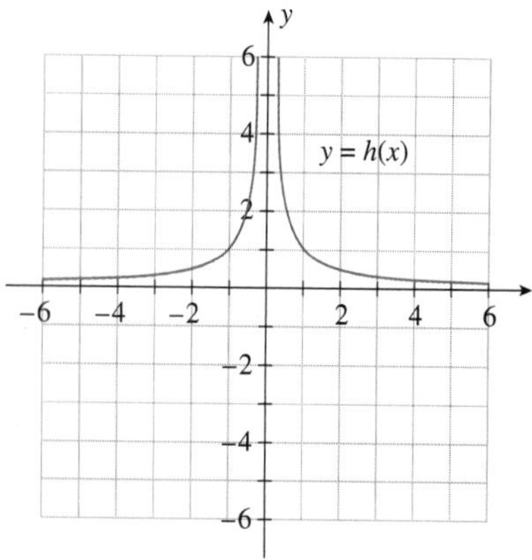

Figure 2.33

Graph of the rational function
$h(x) = \dfrac{1}{x^2} = x^{-2}$

Power Functions

A **power function** is any function of the form $f(x) = x^r$, where r is any real number. We encountered examples of power functions earlier in our work. Most importantly, any polynomial function is a combination of power functions with r any nonnegative integer. Other examples of power functions are **root functions** and some rational functions shown in Figures 2.31, 2.32, and 2.33.

Many of the functions we will encounter later will involve combinations of the functions introduced here. For example, the following functions may be viewed as suitable combinations of such functions:

$$f(x) = \sqrt{\frac{1 - x^2}{1 + x^2}}$$

$$g(x) = \sqrt{x^2 - 3x + 4}$$

$$h(x) = (1 + 2x)^{1/2} + \frac{1}{(x^2 + 2)^{3/2}}$$

As with polynomials of degree 3 or greater, analyzing the properties of these functions is facilitated by using the tools of calculus, to be developed later.

Example 3

The graph of the power function $f(x) = x^{2/3}$ is shown in Figure 2.34. By considering a sequence of transformations, graph the function $g(x) = -(x - 3)^{2/3} + 4$.

Solution

Notice that the graph of f has three nice grid points $(-8, 4)$, $(0, 0)$, and $(8, 4)$. By transforming these three points as indicated with $g(x) = -(x - 3)^{2/3} + 4$, we can graph g. The first transformation is a reflection in the x-axis. This moves the point $(-8, 4)$ to $(-8, -4)$, leaves $(0, 0)$ untouched, and moves the point $(8, 4)$ to $(8, -4)$. The next transformation is a horizontal translation right by 3 units. This moves the point $(-8, -4)$ to $(-5, -4)$, the point $(0, 0)$ to $(3, 0)$, and the point $(8, -4)$ to $(11, -4)$. The last transformation is a vertical translation up by 4 units. This moves the point $(-5, -4)$ to $(-5, 0)$, the point $(3, 0)$ to $(3, 4)$, and the point $(-5, -4)$ to $(-5, 0)$. The graph of $g(x) = -(x - 3)^{2/3} + 4$ is shown in Figure 2.35.

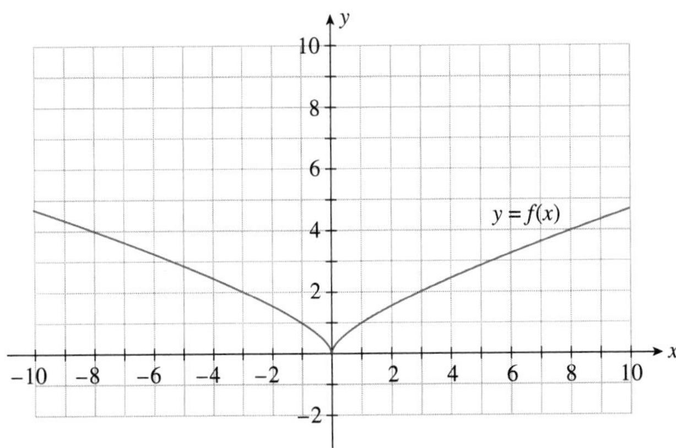

Figure 2.34
The graph of $f(x) = x^{2/3}$

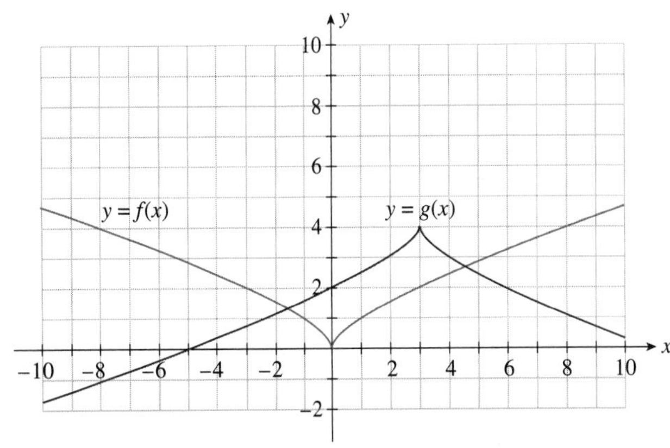

Figure 2.35
The graph of $g(x) = -(x - 3)^{2/3} + 4$, where $f(x) = x^{2/3}$

Self-Check Exercises 2.4

1. Label the following functions as either a linear, quadratic, polynomial, rational, root, or power function.

a. $y = 2(x - 4)^{-3} + 1$
b. $y = 2(x - 4)^2 + 1$
c. $y = 2(x - 4)^{2/3} + 1$
d. $y = 2(x - 4)^{1/3} + 1$
e. $y = 2(x - 4)^4 + 1$
f. $y = 2(x - 4) + 1$

2. Find the vertical and horizontal asymptotes of the following rational functions.

a. $f(x) = \dfrac{3(x + 4)^2}{x^2 + 3x - 4}$

b. $g(x) = \dfrac{6 - 7x^5}{(2x^2 - 3)^3}$

Solutions to Self-Check Exercises 2.4 can be found on page 77.

2.4 Exercises

In Exercises 1–6, answer parts (a)–(d) using the indicated rational function.

a. State the domain of f.

b. Find the vertical asymptotes of f, if any.

c. Find the horizontal asymptotes of f, if any.

d. Find the y-intercept.

1. $f(x) = \dfrac{9x - 1}{x^2 - 25}$

2. $f(x) = \dfrac{x + 15}{5x^2 - 20}$

3. $f(x) = \dfrac{8x^2 + 9}{2x^2 - 18}$

4. $f(x) = \dfrac{x^2 - 32}{4x^2 - 16}$

5. $f(x) = \dfrac{3(x^2 + 10)}{x^2 + 2x - 15}$

6. $f(x) = \dfrac{x^2 - 9}{2x^2 + 5x - 3}$

In Exercises 7–12, match the rational functions described with the graphs labelled A through F shown on page 76.

7. $a(x) = \dfrac{7}{x^2 + x - 12}$

8. $b(x) = \dfrac{x^2}{x^2 + x - 12}$

9. $c(x) = \dfrac{3x^2 - 5}{x^2 - 2x + 1}$

10. $d(x) = \dfrac{5}{x^2 - 2x - 8}$

11. $e(x) = \dfrac{x^2 - 25}{x^2 + 3x - 10}$

12. $f(x) = \dfrac{x^2 - 2x}{x^2 - 7 + 10}$

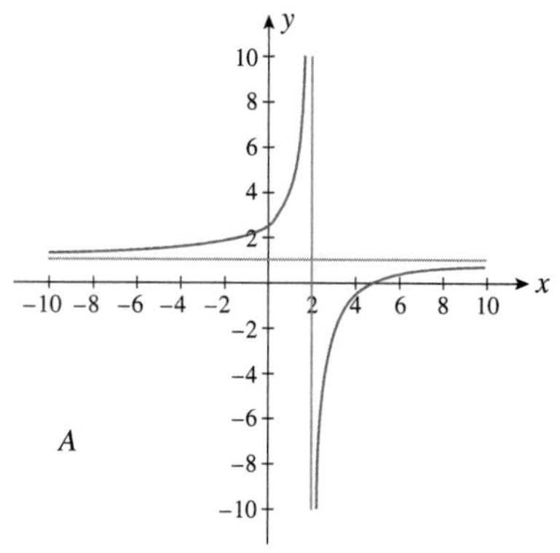

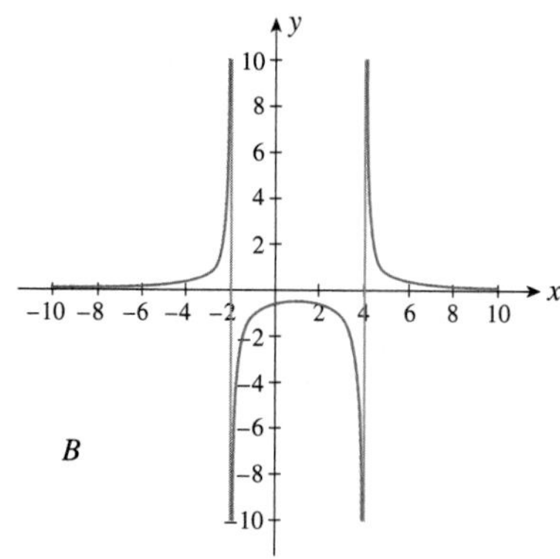

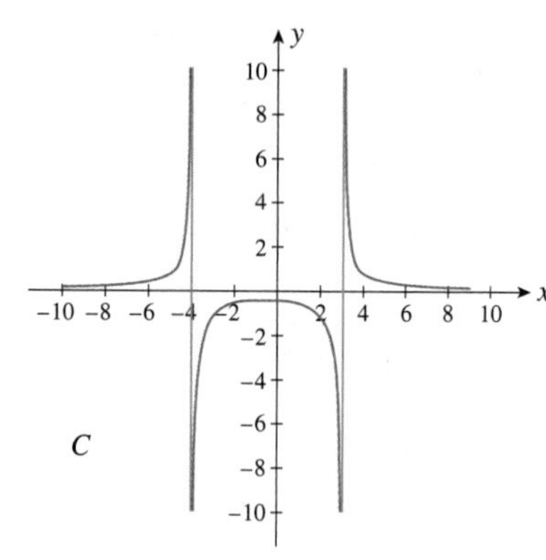

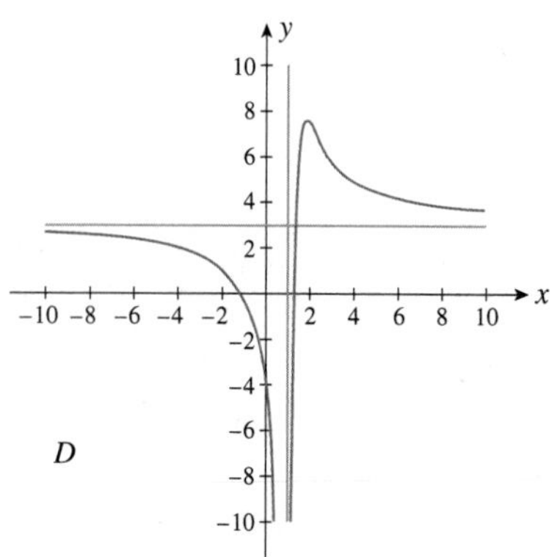

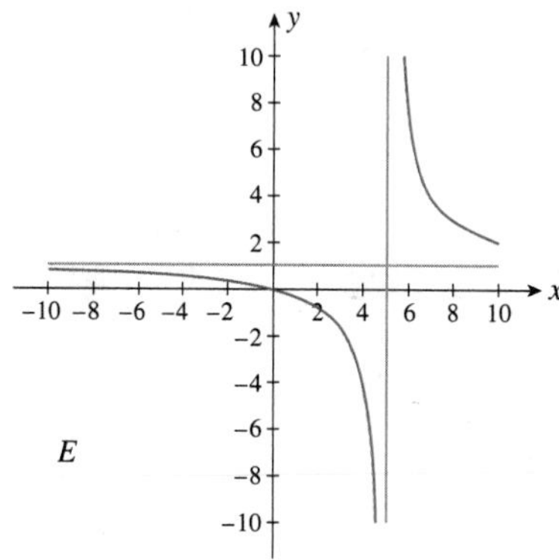

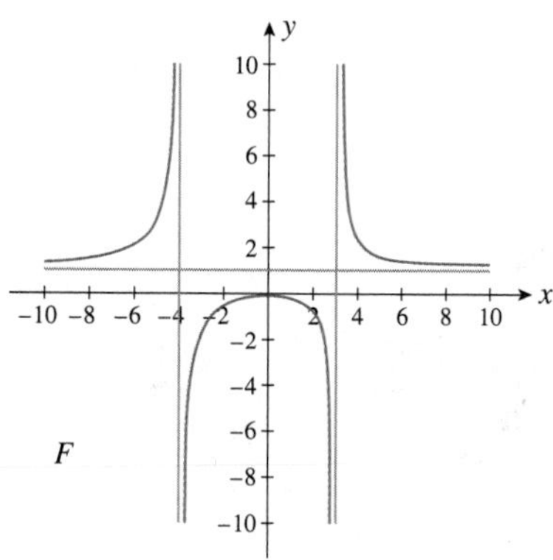

In Exercises 13–18, describe the sequence of transformations performed on the graph of $g(x) = \frac{3x^2}{x^2-4}$ shown below to obtain the graph of the indicated function including the changes to the vertical and horizontal asymptotes.

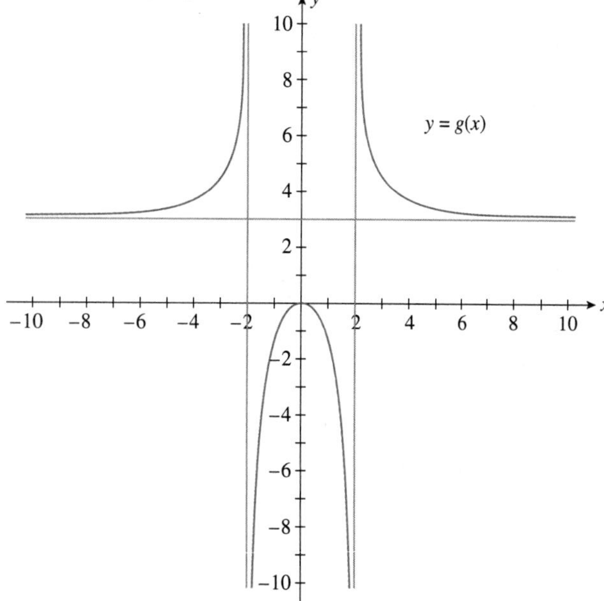

13. $f(x) = g(x + 2) - 4$

14. $f(x) = g(x - 3) - 5$

15. $f(x) = 2g(x + 5)$

16. $f(x) = \frac{1}{3}g(x) + 2$

17. $f(x) = -g(x) + 4$

18. $f(x) = -g(x - 1)$

In Exercises 19–22, sketch the graph of the indicated function using the sequence of transformations performed on the graph of $y = \sqrt{x}$ shown in Figure 2.31.

19. $f(x) = g(x + 2) + 3$

20. $f(x) = 2g(x - 3)$

21. $f(x) = \frac{1}{2}g(x + 5)$

22. $f(x) = -g(x) - 2$

In Exercises 23–26, sketch the graph of the indicated function using the sequence of transformations performed on the graph of $y = \frac{1}{x^2}$ shown in Figure 2.32.

23. $f(x) = g(x + 2) + 3$

24. $f(x) = 2g(x - 3)$

25. $f(x) = \frac{1}{2}g(x + 5)$

26. $f(x) = -g(x) - 2$

In Exercises 27–31, determine whether the given function is a polynomial function, a rational function, or some other function. State the degree of each polynomial function.

27. $f(x) = 3x^6 - 2x^2 + 1$

28. $f(x) = \frac{x^2 - 9}{x - 3}$

29. $G(x) = 2(x^2 - 3)^3$

30. $H(x) = 2x^{-3} + 5x^{-2} + 6$

31. $f(t) = 2t^2 + 3\sqrt{t}$

32. $f(r) = \frac{6r}{r^3 - 8}$

Solutions to Self-Check Exercises 2.4

1. The vertical expansion by a factor of 2, the horizontal translation right by 4, and the vertical translation up by 1 are just transformations. We have to pay attention to the value of the exponent to decide what type each given function represents.

   a. $y = 2(x - 4)^{-3} + 1$ rational function
   b. $y = 2(x - 4)^2 + 1$ quadratic function
   c. $y = 2(x - 4)^{2/3} + 1$ power function
   d. $y = 2(x - 4)^{1/3} + 1$ root function
   e. $y = 2(x - 4)^4 + 1$ polynomial function
   f. $y = 2(x - 4) + 1$ linear function

2. **a.** We find the roots of the denominator

$$x^2 + 3x - 4 = 0$$

$$(x + 4)(x - 1) = 0$$

$$x = -4, 1$$

and observe that $x = -4$ is also a root of the numerator of f. Therefore, f has only one vertical asymptote, namely $x = 1$. The degree of the numerator $3(x + 4)^2$ is 2 and the degree of the denominator $x^2 + 3x - 4$ is also 2. Therefore, the horizontal asymptote is

$$y = \frac{\text{leading coefficient of } 3(x + 4)^2}{\text{leading coefficient of } x^2 + 3x - 4} = \frac{3}{1} = 3$$

b. We find the roots of the denominator

$$(2x^2 - 3)^3 = 0$$
$$2x^2 - 3 = 0$$
$$2x^2 = 3$$
$$x^2 = \frac{3}{2}$$
$$x = \pm\sqrt{\frac{3}{2}}$$

and observe that neither $x = \sqrt{\frac{3}{2}}$ nor $x = -\sqrt{\frac{3}{2}}$ are roots of the numerator. So, the vertical asymptotes are $x = \sqrt{\frac{3}{2}}$ and $x = -\sqrt{\frac{3}{2}}$. The degree of the numerator $6 - 7x^5$ is 5, which is smaller than the degree of the denominator $(2x^2 - 3)^3$, which is 6. Therefore, the horizontal asymptote is $y = 0$ (x-axis).

2.5 Absolute Value Functions

We will take a closer look at a rather special function called absolute value function. Furthermore, now that we know a variety of functions, let us consider what happens to them if we apply the absolute value to them.

Absolute Value Function

Recall the definition of the absolute value from Section 1.1. We can now define an **absolute value function** $f(x) = |x|$ (read *f of x equals the absolute value of x*), which takes as input any real number and returns as output the input with its sign stripped off. This means that any positive number remains positive, zero stays zero, and any negative number becomes positive.

Let us start by rewriting the absolute value function $f(x) = |x|$ without the use of the absolute value sign. This can be done in a variety of ways. Let us first consider

$$f(x) = |x| = \sqrt{x^2}$$

This means that $\sqrt{x^2} \neq x$, which is best explained with a few examples. Consider $x = 3$ and $x = -3$. Then $\sqrt{3^2} = \sqrt{9} = 3$ but $\sqrt{(-3)^2} = \sqrt{9} = 3 \neq -3$. These examples also help illustrate why we can write $|x| = \sqrt{x^2}$, because the *squaring* of the input x ensures that the output is nonnegative and the *square rooting* undoes the *squaring* with the result that the input is returned as output but with the sign stripped off. This form of the absolute value function is useful in differentiation and integration, which are Calculus tools discussed in later chapters.

Notice that $\sqrt{x^2}$ is defined for all real numbers but that $(\sqrt{x})^2$ is only defined for nonnegative real numbers. Therefore, $\sqrt{x^2} \neq (\sqrt{x})^2$.

Another way of rewriting the absolute value function considers what happens to the sign of the input in a different manner. For example, if $x = 3$, then $|3| = 3$. This occurs if we ensure that the input x is zero or positive. If $x = -3$, then, $|-3| = 3$ which we can think of as multiplying -3 with a negative sign, i.e.,

$|-3| = -(-3) = 3$. This can be done every time we ensure that the input x is a negative number. This allows us to rewrite the absolute value function as the piecewise defined function

$$f(x) = |x| = \begin{cases} -x & \text{if } x < 0 \\ x & \text{if } x \geq 0 \end{cases}$$

This form of the absolute value function is also very useful in calculus, especially when considering limits, which are discussed in Chapter 3. In addition, this form is useful for graphing the absolute value function, which will be discussed next.

Absolute Value Function

An absolute value function is written as $f(x) = |x|$ and can have the form

$$f(x) = |x| = \sqrt{x^2} \text{ or}$$

$$f(x) = |x| = \begin{cases} -x & \text{if } x < 0 \\ x & \text{if } x \geq 0 \end{cases}$$

with domain $D_f = -\infty, \infty$ and range $R_f = [0, \infty)$.

Example 1

Complete the following table.

x	$f(x) = \lvert x \rvert$	$f(x) = \sqrt{x^2}$	$f(x) = \begin{cases} -x & \text{if } x < 0 \\ x & \text{if } x \geq 0 \end{cases}$
-5.3			
-2			
0			
4.5			
17			

Solution

x	$f(x) = \lvert x \rvert$	$f(x) = \sqrt{x^2}$	$f(x) = \begin{cases} -x & \text{if } x < 0 \\ x & \text{if } x \geq 0 \end{cases}$
-5.3	5.3	$\sqrt{(-5.3)^2} = \sqrt{28.09} = 5.3$	$-(-5.3) = 5.3$
-2	2	$\sqrt{(-2)^2} = \sqrt{4} = 2$	$-(-2) = 2$
0	0	$\sqrt{(0)^2} = \sqrt{0} = 0$	0
4.5	4.5	$\sqrt{(4.5)^2} = \sqrt{20.25} = 4.5$	4.5
17	17	$\sqrt{(17)^2} = \sqrt{289} = 17$	17

Example 2

Rewrite as a piecewise defined function.

Solution

We need to find out when $2x - 1$ is positive and negative. Let us start by finding when $2x - 1$ is negative.

$$2x - 1 < 0$$
$$2x < 1$$
$$x < \frac{1}{2}$$

Since the expression $2x - 1$ is linear, this means that $2x - 1$ is positive or zero for $x \geq \frac{1}{2}$ and negative for $x < \frac{1}{2}$. Now we can write $f(x) = 5|2x - 1| - 3$ as a piecewise defined function.

$$f(x) = 5|2x - 1| - 3 = \begin{cases} 5(-(2x - 1)) - 3 & \text{if } x < \dfrac{1}{2} \\ 5(2x - 1) - 3 & \text{if } x \geq \dfrac{1}{2} \end{cases}$$

After simplifying this becomes

$$f(x) = 5|2x - 1| - 3 = \begin{cases} -10x + 2 & \text{if } x < \dfrac{1}{2} \\ 10x - 8 & \text{if } x \geq \dfrac{1}{2} \end{cases}$$

Graphing the Absolute Value Function

If we consider the piecewise defined function form of the absolute value function, then we can see that its graph consists of two straight lines, namely $y = x$ (positive slope) for all $x \geq 0$ and $y = -x$ (negative slope) for all $x < 0$, which can be seen in Figure 2.36. The graph of the absolute value function is characterized by its V-shape and the *corner* where the function changes direction sharply. This corner is also referred to as *vertex* and is the point (0, 0).

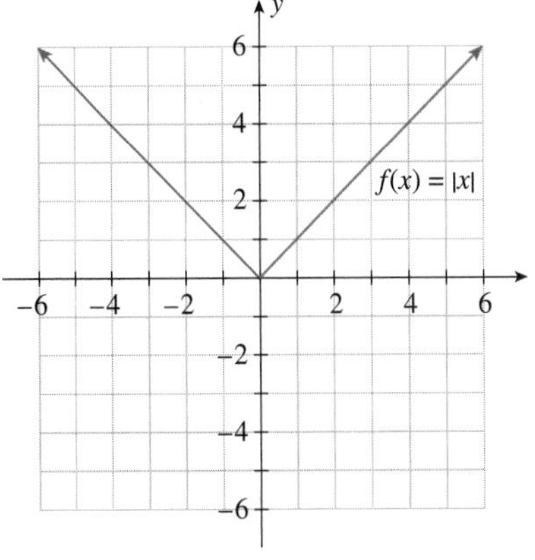

Figure 2.36

The graph of

$$f(x) = |x| = \begin{cases} -x & \text{if } x < 0 \\ x & \text{if } x \geq 0 \end{cases}$$

Example 3

Write the equation of the four graphs shown in Figure 2.37 labelled A–D, which are obtained by transforming $y = |x|$. Give two forms for each equation, the absolute value form and the piecewise defined function form.

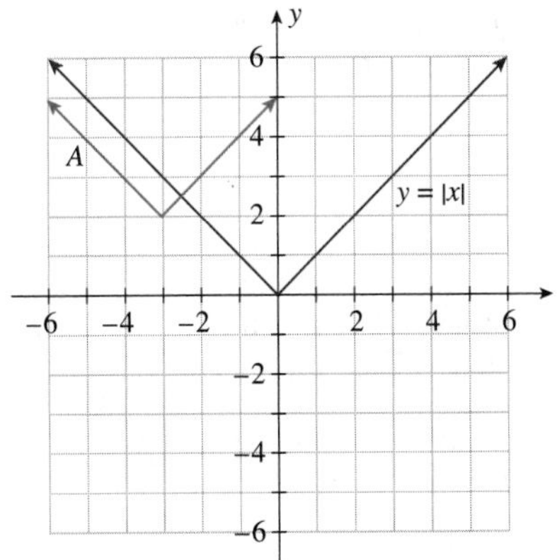

(a) Graph A

(b) Graph B

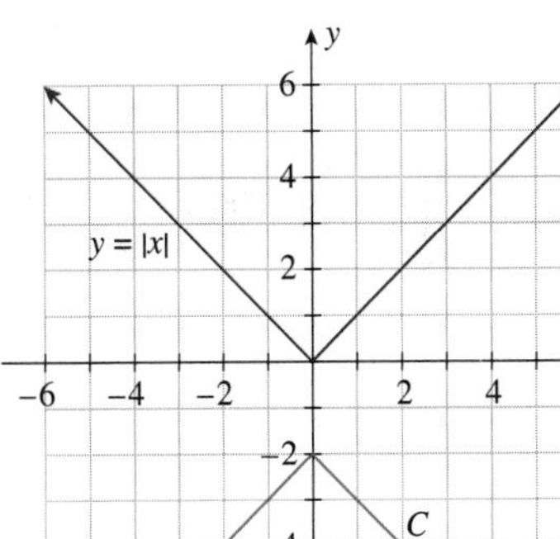

(c) Graph C

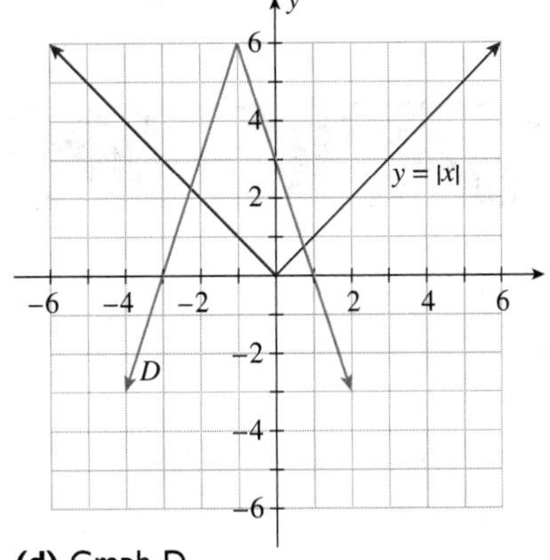

(d) Graph D

Figure 2.37

Solution

Graph A: The vertex is translated 3 units horizontally left and 2 units vertically up. There is no stretching and no reflection. Therefore, the equation is $y = |x + 3| + 2$ in absolute value form. To obtain the piecewise defined function form, notice that the root of $y = |x + 3|$ is -3. This means that $x + 3 \geq 0$ for all $x \geq -3$, i.e., $|x + 3| = x + 3$ for all $x \geq -3$, and $x + 3 < 0$ for all $x < -3$, i.e., $|x + 3| = x + 3$ for all $x < -3$. Therefore,

$$y = \begin{cases} -(x + 3) + 2 & \text{if } x < -3 \\ (x + 3) + 2 & \text{if } x \geq -3 \end{cases} \quad \text{or}$$

$$y = \begin{cases} -x - 1 & \text{if } x < -3 \\ x + 5 & \text{if } x \geq -3 \end{cases}$$

Graph B: The graph of $y = |x|$ is vertically compressed by a factor of $\frac{1}{2}$ and its vertex is translated 2 units horizontally right. There is no reflection and vertical translation. The equation is $y = \frac{1}{2}|x - 2|$ in absolute value form. The piecewise defined function form is

$$y = \begin{cases} -\dfrac{1}{2}(x - 2) & \text{if } x < 2 \\ \dfrac{1}{2}(x - 2) & \text{if } x \geq 2 \end{cases} \quad \text{or}$$

$$y = \begin{cases} -\dfrac{1}{2}x + 1 & \text{if } x < 2 \\ \dfrac{1}{2}x - 1 & \text{if } x \geq 2 \end{cases}$$

Graph C: The graph of $y = |x|$ is reflected in the x-axis and its vertex is translated 2 units vertically down. There is no stretching and horizontal translation. The equation is $y = -|x| - 2$ in absolute value form. The piecewise defined function form is

$$y = \begin{cases} -(-x) - 2 & \text{if } x < 0 \\ -(x) - 2 & \text{if } x \geq 0 \end{cases} \quad \text{or}$$

$$y = \begin{cases} x - 2 & \text{if } x < 0 \\ -x - 2 & \text{if } x \geq 0 \end{cases}$$

Graph D: The graph of $y = |x|$ is reflected in the x-axis, vertically expanded by a factor of 3, and its vertex is translated 1 unit horizontally left and 6 units vertically up. The equation is $y = -3|x + 1| + 6$ in absolute value form. The piecewise defined function form is

$$y = \begin{cases} -[-3(x + 1)] + 6 & \text{if } x < -1 \\ -3(x + 1) + 6 & \text{if } x \geq -1 \end{cases} \quad \text{or}$$

$$y = \begin{cases} 3x + 9 & \text{if } x < -1 \\ -3x + 3 & \text{if } x \geq -1 \end{cases}$$

GROUP DISCUSSION

Consider the following pairs of functions.

a. $f(x) = x - 3$ and $g(x) = |x - 3|$

b. $f(x) = x + 4$ and $g(x) = |x + 4|$

c. $f(x) = x^2 - 1$ and $g(x) = |x^2 - 1|$

Graph each pair of functions on the same coordinate system. Use your function knowledge to graph f and make a table of values for g, if necessary. Mark the roots of f and the corners of the graph of g. What do you notice? What else can you say about the graphs of f and g? How can the graph of g be obtained from the graph of f?

Absolute Value Function Applied to Other Functions

Consider graphing the line $y = x$ and then analyzing what happens when we apply the absolute value function to this line. When x is nonnegative, nothing happens to the line $y = x$, since the function value is positive. However, when x is negative, then $y = |x|$ has the function value $-x$. Recall from our function transformation in Section 2.2 that multiplying the function value with a negative sign means the graph is reflected vertically in the x-axis. In other words, the part of the line $y = x$ where $x < 0$ gets reflected in the x-axis to produce the graph for $y = |x|$, which is shown in Figure 2.38.

The above analysis of the absolute value function applied to the function $f(x) = x$ can be generalized to any function $y = f(x)$. Whenever the function value of f is nonnegative, the graph of $y = f(x)$ is the same as the graph of $y = |f(x)|$. However, when the function value of f is negative, then that part of the graph of f must be reflected in the x-axis to produce the graph of $y = |f(x)|$. Consider $y = f(x) = x^2 - 4$, which is the familiar parabola shifted 4 units vertically down, then the graph of $y = |f(x)| = |x^2 - 4|$ is constructed by reflecting the negative function values in the x-axis as is shown in Figure 2.39.

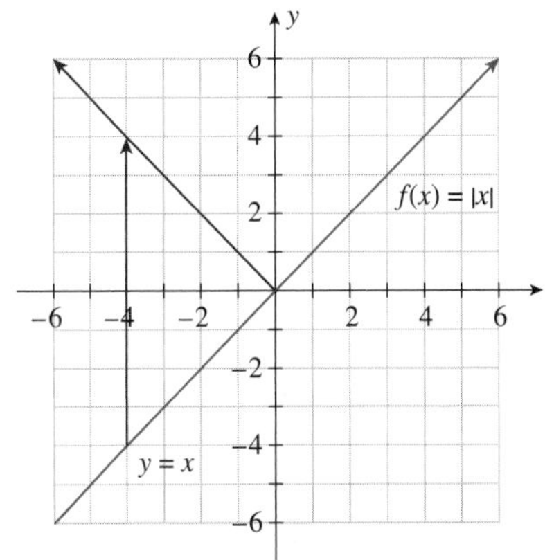

Figure 2.38
The graph of $y = |x|$ constructed from the graph of $y = x$

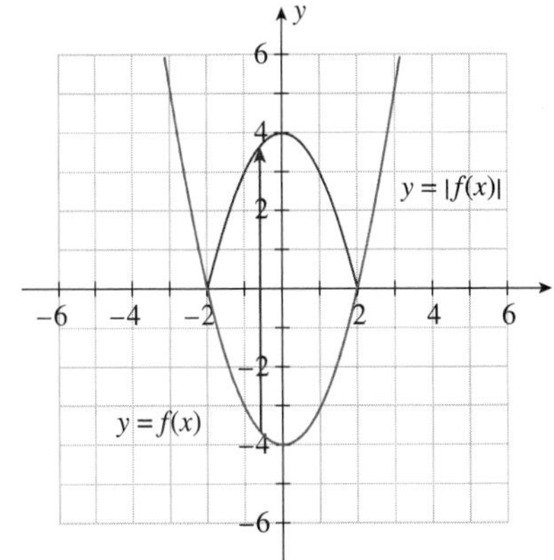

Figure 2.39
The graph of $y = |f(x)| = |x^2 - 4|$ constructed from the graph of $y = f(x) = x^2 - 4$.

Example 4

The graph of the function $y = f(x)$ is shown in Figure 2.40. Graph the functions $y = |f(x)|$ and $y = f(|x|)$. Should the graphs be the same? Why, or why not?

Solution

The graph of $y = |f(x)|$ is constructed from the graph of $y = f(x)$ by reflecting all negative y-values in the x-axis. The graph of $y = f(|x|)$ is constructed from the graph of $y = f(x)$ by ignoring the part of the graph on the left side of the y-axis and reflecting the part of the graph on the right side in the y-axis, because every negative x-value has as output the same function value as its positive counterpart. Therefore, the graphs of $y = |f(x)|$ (Figure 2.41a) and $y = f(|x|)$ (Figure 2.41b) cannot be the same.

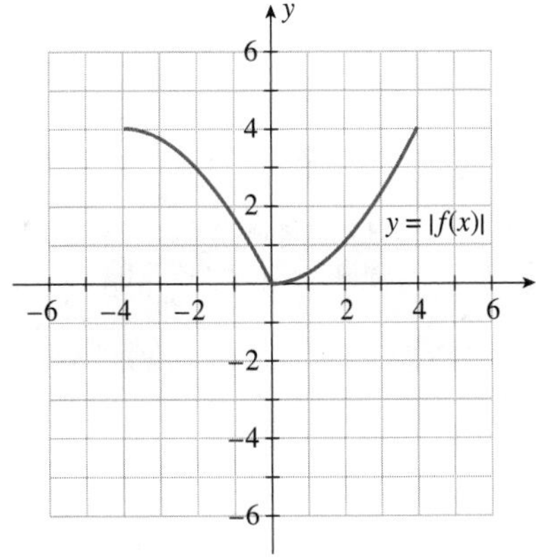

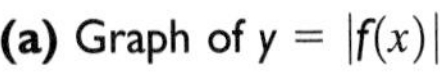

(a) Graph of $y = |f(x)|$

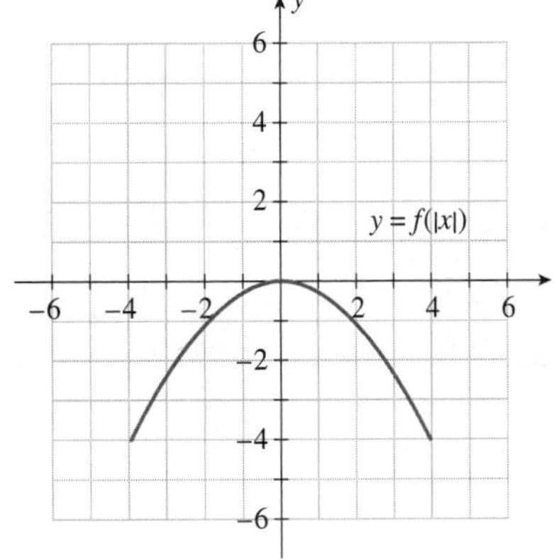

(b) Graph of $y = f(|x|)$

Figure 2.41

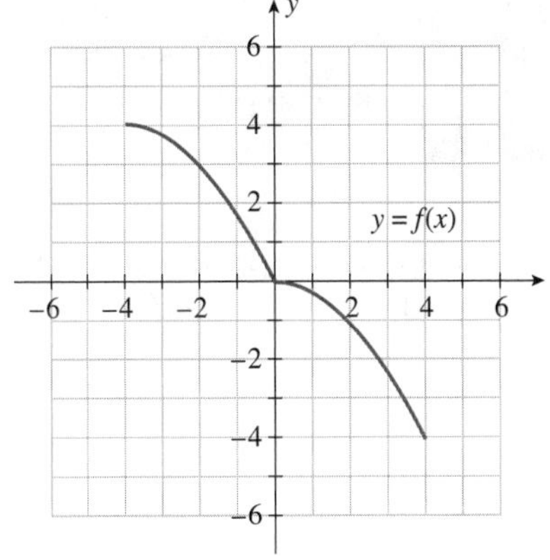

Figure 2.40
The graph of some function f.

Example 5

Graph $y = |-2\sqrt{x + 5} + 4|$.

Solution

We start by graphing $y = -2\sqrt{x + 5} + 4$ (Figure 2.42a), which is the square root function reflected vertically in the x-axis, expanded vertically by a factor of 2, translated horizontally left by 5 units, and translated vertically up by 4 units as is shown in the graph on page 82 to the left. We then proceed by reflecting the negative function values of $y = -2\sqrt{x+5} + 4$ in the x-axis. For this we need to find out for which x-values $y = -2\sqrt{x+5} + 4$ is negative.

$$-2\sqrt{x+5} + 4 < 0$$

$$-2\sqrt{x+5} < -4$$

$$\sqrt{x+5} > 2$$

$$(\sqrt{x+5})^2 > 2^2$$

$$x + 5 > 4$$

$$x > -1$$

For all $x > -1$, we need to reflect the y-values in the x-axis. The graph of $y = |-2\sqrt{x+5} + 4|$ is shown in Figure 2.42b.

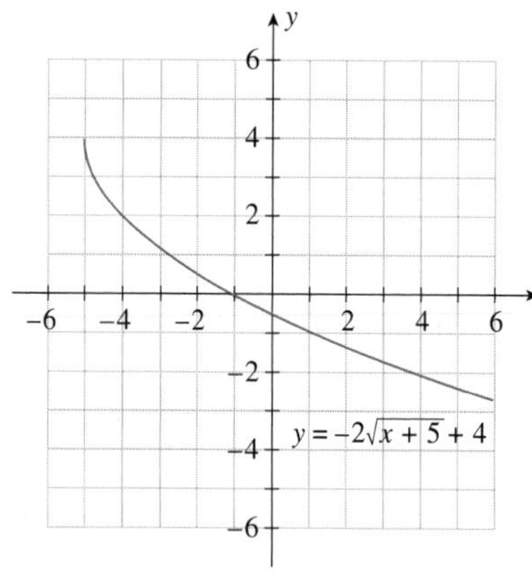

(a) Graph of $y = -2\sqrt{x+5} + 4$

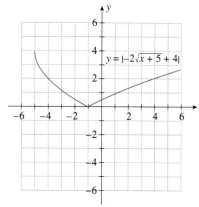

(b) Graph of $y = |-2\sqrt{x+5} + 4|$

Figure 2.42

Self-Check Exercises 2.5

1. Name the sequence of transformations and special reflections needed to get the graph of $y = f(x)$ from the graph of $y = g(x)$.

a. $f(x) = \left|\dfrac{1}{x-2} + 3\right|, g(x) = \dfrac{1}{x}$

b. $f(x) = \left|\dfrac{1}{2}x^2 - 5\right|, g(x) = x^2$

2. Consider the function $g(x) = |2x + 5| - 3$.

a. For which x-values is $2x + 5$ negative?
b. What are the roots of g, if any?
c. Graph g.

Solutions to Self-Check Exercises 2.5 can be found on page 84.

2.5 Exercises

In Exercises 1–10, rewrite the given function as a piecewise defined function.

1. $f(x) = |x + 3|$

2. $f(x) = |x - 5|$

3. $f(x) = |2x| + 4$

4. $f(x) = \left|\dfrac{1}{3}x\right| - 7$

5. $f(x) = -|x - 4|$

6. $f(x) = -|x + 2|$

7. $f(x) = |x - 1| + 3$

8. $f(x) = |x - 5| - 3$

9. $f(x) = \left|\frac{1}{2}x - 4\right| + 6$

10. $f(x) = |3x - 2| + 6$

In Exercises 11–16, sketch the graph of the indicated function using a sequence of transformations performed on the graph of y = |x| shown in Figure 2.36.

11. $f(x) = -|x - 2|$

12. $f(x) = -|x| - 5$

13. $f(x) = |x + 3| - 4$

14. $f(x) = |x - 4| - 1$

15. $f(x) = 3|x| + 2$

16. $f(x) = \frac{1}{4}|x + 1|$

In Exercises 17–22, match the functions described with the graphs labelled A through F shown below.

17. $a(x) = \left|\frac{x^2}{(x-3)(x+4)}\right|$

18. $b(x) = \left|\frac{x-5}{x-2}\right|$

19. $c(x) = \left|\frac{1}{12}(x+8)(x+2)(x-5)\right|$

20. $d(x) = \left|\frac{1}{2}(x-4)^2 - 6\right|$

21. $e(x) = |2\sqrt{8-x} - 4|$

22. $f(x) = |4\sqrt{x+9} - 8|$

In Exercises 23–26, describe the sequence of transformations performed on the graph of y = |x| shown in Figure 2.36.

23. $f(x) = 2|x - 5| - 1$

24. $f(x) = 5|x + 3| + 4$

25. $f(x) = \frac{2}{3}|x - 1| + 4$

26. $f(x) = \frac{1}{2}|x + 4| + 3$

In Exercises 27–34, do the following for each indicated function *f*:

a. Find the domain.

b. Find the roots.

c. Use the roots to divide the domain into intervals. Then choose an x-value from each interval, evaluate the function for each of those x-values, and use the result to decide over which intervals the indicated function is negative.

d. Graph the function $g(x) = |f(x)|$.

27. $f(x) = (x - 1)^2 - 4$

28. $f(x) = (x - 3)^2 - 9$

29. $f(x) = 2(x + 1)^2 - 8$

30. $f(x) = 3(x + 2)^2 - 27$

31. $f(x) = \sqrt{x + 5} - 2$

32. $f(x) = \sqrt{x - 1} - 4$

33. $f(x) = 4\sqrt{x - 3} - 8$

34. $f(x) = 5\sqrt{x + 1} - 10$

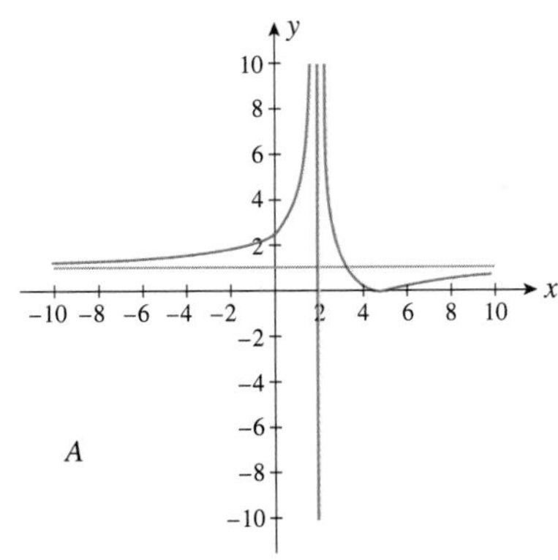

A

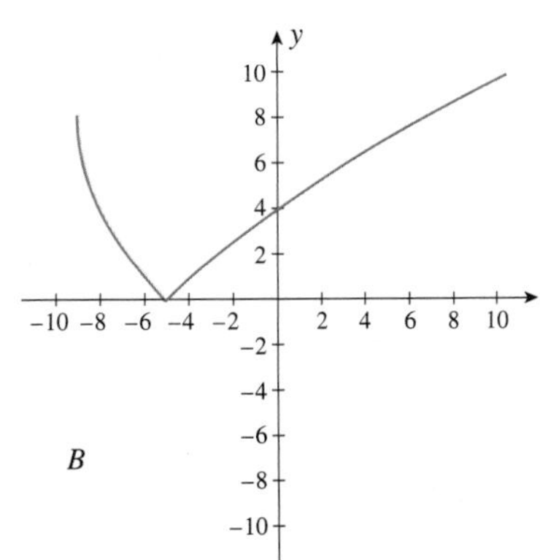

B

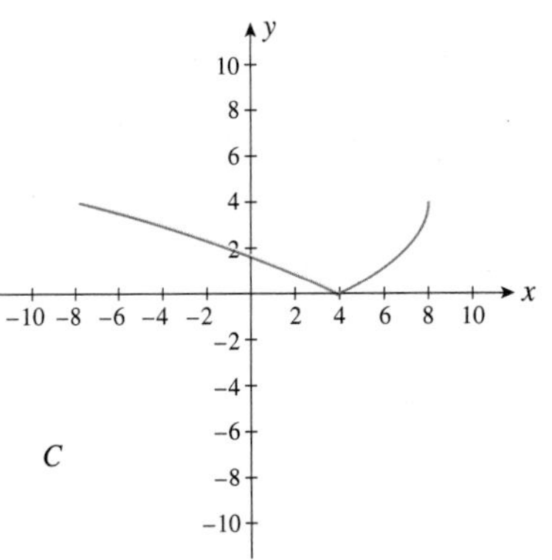

C

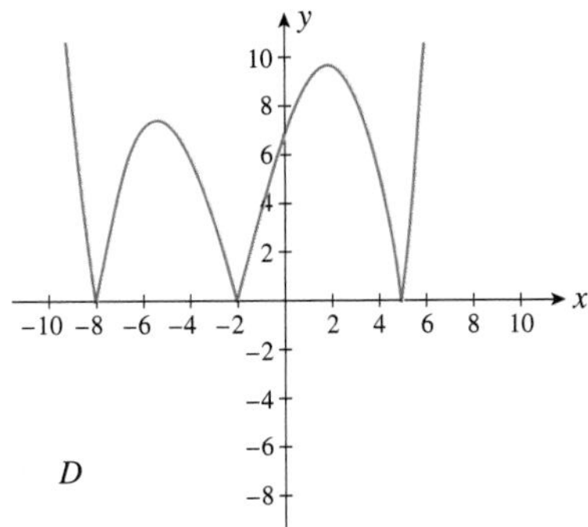

D

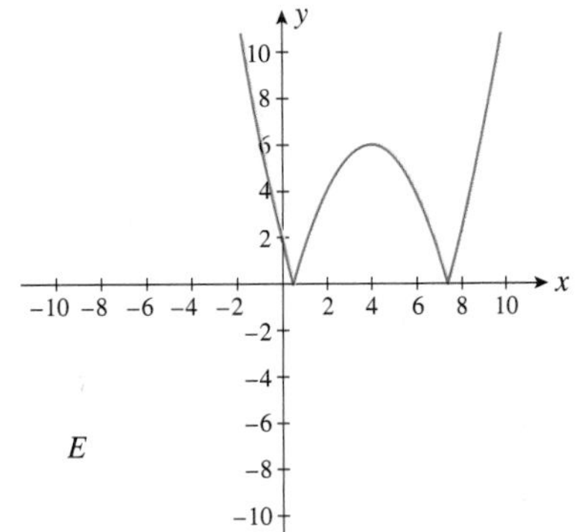

E

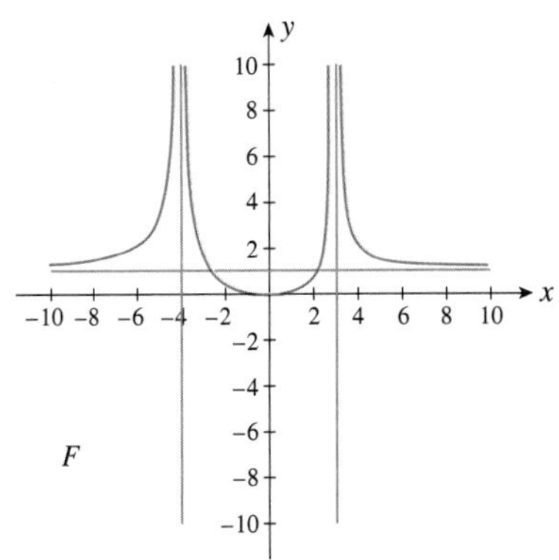

F

Solutions to Self-Check Exercises 2.5

1. a. From $f(x) = \left|\frac{1}{x-2} + 3\right|$ we observe that in order to obtain the graph of f, the graph of $g(x) = \frac{1}{x}$ must be

- horizontally translated right by 2 units, resulting in $y = \frac{1}{x-2}$
- vertically translated up by 3 units, resulting in $y = \frac{1}{x-2} + 3$
- and all negative function values reflected in the x-axis, resulting in $f(x) = \left|\frac{1}{x-2} + 3\right|$

b. From $f(x) = \left|\frac{1}{2}x^2 - 5\right|$ we observe that in order to obtain the graph of f, the graph of $g(x) = x^2$ must be

- vertically compressed by a factor of $\frac{1}{2}$, resulting in $y = \frac{1}{2}x^2$
- vertically translated down by 5 units, resulting in $y = \frac{1}{2}x^2 - 5$
- and all negative function values reflected in the x-axis, resulting in $f(x) = |\frac{1}{2}x^2 - 5|$

2. a.

$$2x + 5 < 0$$
$$2x < -5$$
$$x < -\frac{5}{2}$$

Therefore, $2x + 5$ is negative for all $x < -\frac{5}{2}$.

b. To find the roots of g, we solve $g(x) = 0$.

$$|2x + 5| - 3 = 0$$
$$|2x + 5| = 3$$

If $x < -\frac{5}{2}$, then

$$-(2x + 5) = 3$$
$$2x + 5 = -3$$
$$2x = -8$$
$$x = -4$$

Since $-4 < -\frac{5}{2}$ we have that $x = -4$ is a root of g.

If $x \geq -\frac{5}{2}$, then

$$2x + 5 = 3$$
$$2x = -2$$
$$x = -1$$

Since $-1 \geq -\frac{5}{2}$ we have that $x = -1$ is another root of g.

c. From part (b) we have two roots, namely $x = -4$ and $x = -1$. Furthermore, the vertex of $g(x) = |2x + 5| - 3$ is the point $(-\frac{5}{2}, -3)$. These three points allow us to graph g as shown below.

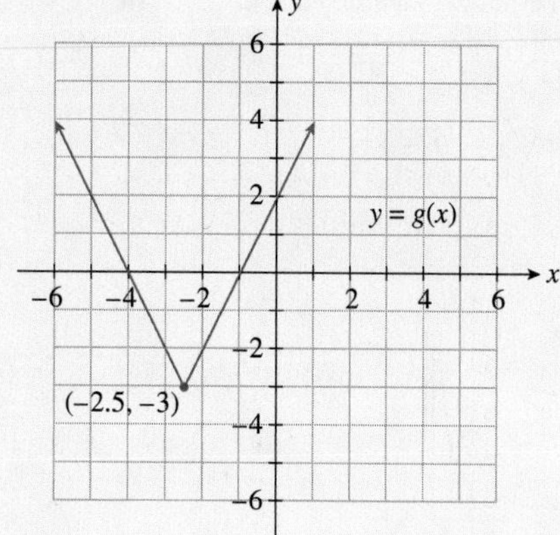

2.6 The Algebra of Functions

The Sum, Difference, Product, and Quotient of Functions

We will take a closer look at the Canadian federal and provincial-territorial budgets for the years 1981 to 1999, by examining the graphical information for spending and revenue, and then deducing information from the graphs regarding the provincial-territorial surplus/deficit. Let $S_f(t)$ and $R_f(t)$ denote, respectively, the federal spending and revenue at any time t, measured in billions of dollars. Similarly, let $S_p(t)$ and $R_p(t)$ denote, respectively, the provincial-territorial's spending and revenue at any time t, measured in billions of dollars. The graphs of these functions as a percentage of the gross domestic product (GDP) for the period between 1980 and 1999 are shown in Figure 2.43.

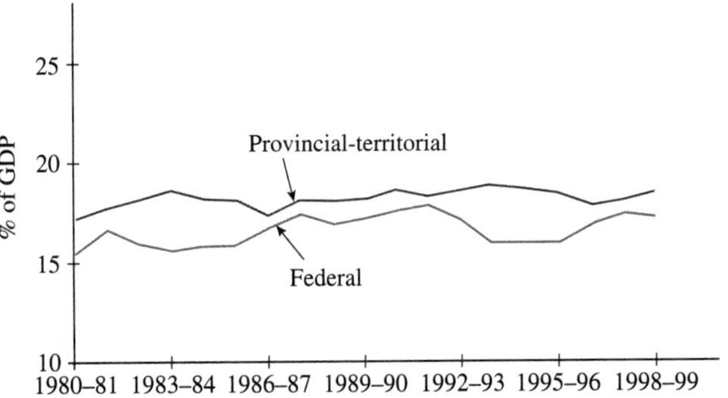

(a) Federal and provincial-territorial revenues

Source: From the Department of Finance Annex 2 of the November 1999 Economic and Fiscal Update charts: Federal and provincial-territorial revenues, Total federal and provincial-territorial surplus/deficit (−), Federal and provincial-territorial program spending. (http://www.fin.gc.ca/update99/annex_2e.html). Reproduced with the permission of the Minister of Public Works and Government Services, 2007.

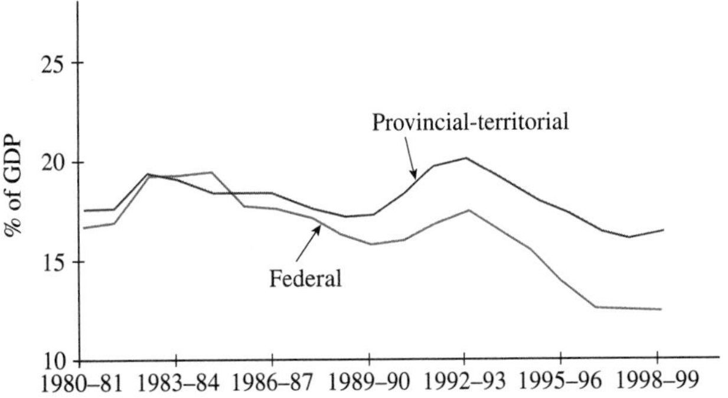

(b) Federal and provincial-territorial spending

Source: From the Department of Finance Annex 2 of the November 1999 Economic and Fiscal Update charts: Federal and provincial-territorial revenues, Total federal and provincial-territorial surplus/deficit (−), Federal and provincial-territorial program spending. (http://www.fin.gc.ca/update99/annex_2e.html). Reproduced with the permission of the Minister of Public Works and Government Services, 2007.

Figure 2.43

If we now define $R(t) = R_f(t) + R_p(t)$ and $S(t) = S_f(t) + S_p(t)$, then the difference $R(t) - S(t)$ gives the deficit (surplus) in billions of dollars at any time t if $R(t) - S(t)$ is negative (positive). This observation suggests that we can define a function D whose value at any time t is given by $R(t) - S(t)$. The function D, the *difference* of the two functions R and S, is written $D(t) = R(t) - S(t)$ and may be called the "deficit (surplus) function" since it gives the budget deficit or surplus at any time t. It has the same domain as the functions R and S and so is also measured in billions of dollars. The graph of the function D is shown in Figure 2.44.

Most functions are built up from other, generally simpler functions. For example, we may view the function $f(x) = 2x + 4$ as the sum of the two functions $g(x) = 2x$ and $h(x) = 4$. The function $g(x) = 2x$ may in turn be viewed as the product of the functions $p(x) = 2$ and $q(x) = x$.

In general, given the functions f and g, we define the sum $f + g$, the difference $f - g$, the product fg, and the quotient f/g of f and g as follows.

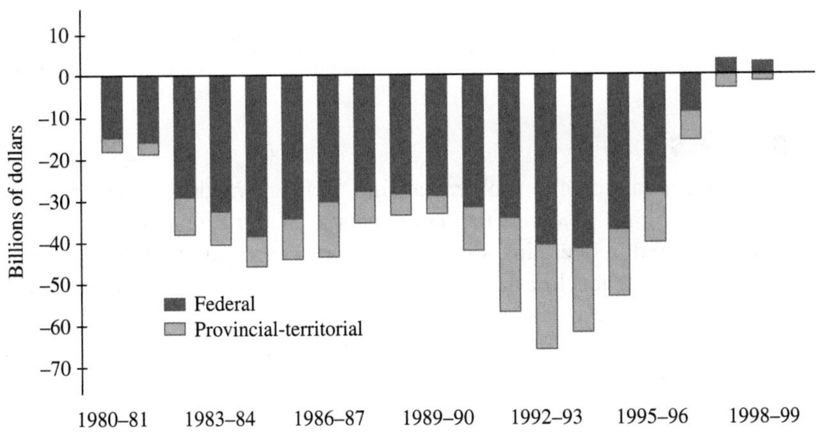

The bar graph of $D(t) = R(t) - S(t)$

Source: From the Department of Finance Annex 2 of the November 1999 Economic and Fiscal Update charts: Federal and provincial-territorial revenues, Total federal and provincial-territorial surplus/deficit (−), Federal and provincial-territorial program spending. (http://www.fin.gc.ca/update99/annex_2e.html). Reproduced with the permission of the Minister of Public Works and Government Services, 2007.

Figure 2.44

The Sum, Difference, Product, and Quotient of Functions

Let f and g be functions with domains A and B, respectively. Then the sum $f + g$, difference $f - g$, and product fg of f and g are functions with domain $A \cap B$* and rule given by

$$(f + g)(x) = f(x) + g(x) \qquad \text{Sum}$$
$$(f - g)(x) = f(x) - g(x) \qquad \text{Difference}$$
$$(fg)(x) = f(x)g(x) \qquad \text{Product}$$

The quotient f/g of f and g has domain $A \cap B$ excluding all points x such that $g(x) = 0$ and rule given by

$$\left(\frac{f}{g}\right)(x) = \frac{f(x)}{g(x)} \qquad \text{Quotient}$$

*$A \cap B$ is read "A intersected with B" and denotes the set of all points common to both A and B.

Example 1

Let $f(x) = \sqrt{x + 1}$ and $g(x) = 2x + 1$. Find the sum s, the difference d, the product p, and the quotient q of the functions f and g.

Solution

Since the domain of f is $A = [-1, \infty)$ and the domain of g is $B = (-\infty, \infty)$, we see that the domain of s, d, and p is $A \cap B = [-1, \infty)$. The rules follow.

$$s(x) = (f + g)(x) = f(x) + g(x) = \sqrt{x + 1} + 2x + 1$$
$$d(x) = (f - g)(x) = f(x) - g(x) = \sqrt{x + 1} - (2x + 1) = \sqrt{x + 1} - 2x - 1$$
$$p(x) = (fg)(x) = f(x)g(x) = \sqrt{x + 1}(2x + 1) = (2x + 1)\sqrt{x + 1}$$

The quotient function q has rule

$$q(x) = \left(\frac{f}{g}\right)(x) = \frac{f(x)}{g(x)} = \frac{\sqrt{x + 1}}{2x + 1}$$

Its domain is $[-1, \infty)$ together with the restriction $x \neq -\frac{1}{2}$. We denote this by $[-1, -\frac{1}{2}) \cup (-\frac{1}{2}, \infty)$.

Applications

The mathematical formulation of a problem arising from a practical situation often leads to an expression that involves the combination of functions. Consider, for example, the costs incurred in operating a business. Costs that remain more or less constant regardless of the firm's level of activity are called fixed costs. Examples of fixed costs are rental fees and executive salaries. On the other hand, costs that vary with production or sales are called variable costs. Examples of variable costs are wages and costs of raw materials. The total cost of operating a business is thus given by the *sum* of the variable costs and the fixed costs, as illustrated in the next example.

Example 2

Cost Functions Suppose Puritron, a manufacturer of water filters, has a monthly fixed cost of \$10,000 and a variable cost of

$$-0.0001x^2 + 10x \qquad (0 \leq x \leq 40{,}000)$$

dollars, where x denotes the number of filters manufactured per month. Find a function C that gives the total cost incurred by Puritron in the manufacture of x filters.

Solution

Puritron's monthly fixed cost is always \$10,000, regardless of the level of production, and it is described by the constant function $F(x) = 10{,}000$. Next, the variable cost is described by the function $V(x) = -0.0001x^2 + 10x$. Since the total cost incurred by Puritron at any level of production is the sum of the variable cost and the fixed cost, we see that the required total cost function is given by

$$\begin{aligned} C(x) &= V(x) + F(x) \\ &= -0.0001x^2 + 10x + 10{,}000 \qquad (0 \le x \le 40{,}000) \end{aligned}$$

Next, the **total profit** realized by a firm in operating a business is the *difference* between the total revenue realized and the total cost incurred; that is,

$$P(x) = R(x) - C(x)$$

Example 3

Profit Functions Refer to Example 2. Suppose the total revenue realized by Puritron from the sale of x water filters is given by the total revenue function

$$R(x) = -0.0005x^2 + 20x \qquad (0 \le x \le 40{,}000)$$

a. Find the total profit function—that is, the function that describes the total profit Puritron realizes in manufacturing and selling x water filters per month.

b. What is the profit when the level of production is 10,000 filters per month?

Solution

a. The total profit realized by Puritron in manufacturing and selling x water filters per month is the difference between the total revenue realized and the total cost incurred. Thus, the required total profit function is given by

$$\begin{aligned} P(x) &= R(x) - C(x) \\ &= (-0.0005x^2 + 20x) - (-0.0001x^2 + 10x + 10{,}000) \\ &= -0.0004x^2 + 10x - 10{,}000 \end{aligned}$$

b. The profit realized by Puritron when the level of production is 10,000 filters per month is

$$P(10{,}000) = -0.0004(10{,}000)^2 + 10(10{,}000) - 10{,}000 = 50{,}000$$

or \$50,000 per month.

Composition of Functions

Another way to build up a function from other functions is through a process known as the *composition of functions*. Consider, for example, the function h, whose rule is given by $h(x) = \sqrt{x^2 - 1}$. Let f and g be functions defined by the rules $f(x) = x^2 - 1$ and $g(x) = \sqrt{x}$. Evaluating the function g at the point $f(x)$ [remember that for each real number x in the domain of f, $f(x)$ is simply a real number], we find that

$$g(f(x)) = \sqrt{f(x)} = \sqrt{x^2 - 1}$$

which is just the rule defining the function h!

In general, the composition of a function g with a function f is defined as follows.

> **The Composition of Two Functions**
>
> Let f and g be functions. Then the composition of g and f is the function $g \circ f$ defined by
>
> $$(g \circ f)(x) = g(f(x))$$
>
> The domain of $g \circ f$ is the set of all x in the domain of f such that $f(x)$ lies in the domain of g.

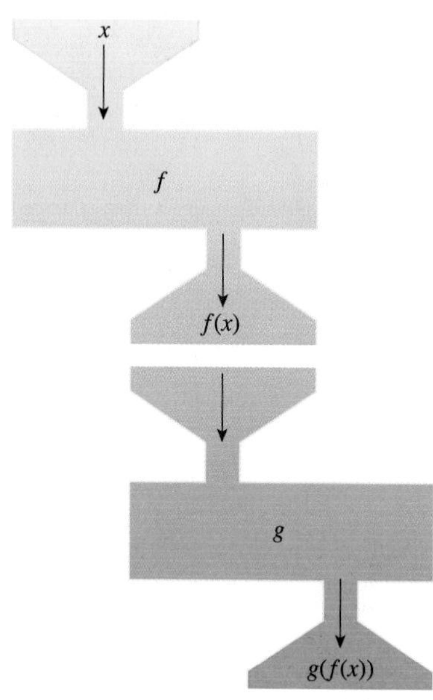

Figure 2.45
The composite function $h = g \circ f$ viewed as a machine

The function $g \circ f$ (read "g circle f") is also called a **composite function.** The interpretation of the function $h = g \circ f$ as a machine is illustrated in Figure 2.45, and its interpretation as a mapping is shown in Figure 2.46.

Example 4

Let $f(x) = x^2 - 1$ and $g(x) = \sqrt{x} + 1$. Use the rule for the composite function to compute:

a. $g \circ f$.
b. $f \circ g$.

Solution

a. To find the rule for the composite function $g \circ f$, evaluate the function g at $f(x)$. We obtain,

$$(g \circ f)(x) = g(f(x)) = \sqrt{f(x)} + 1 = \sqrt{x^2 - 1} + 1$$

b. To find the rule for the composite function $f \circ g$, evaluate the function f at $g(x)$. Thus,

$$\begin{aligned}(f \circ g)(x) &= f(g(x)) = (g(x))^2 - 1 = (\sqrt{x} + 1)^2 - 1 \\ &= x + 2\sqrt{x} + 1 - 1 = x + 2\sqrt{x}\end{aligned}$$

Example 4 reminds us that in general $g \circ f$ is different from $f \circ g$, so care must be taken when finding the rule for a composite function.

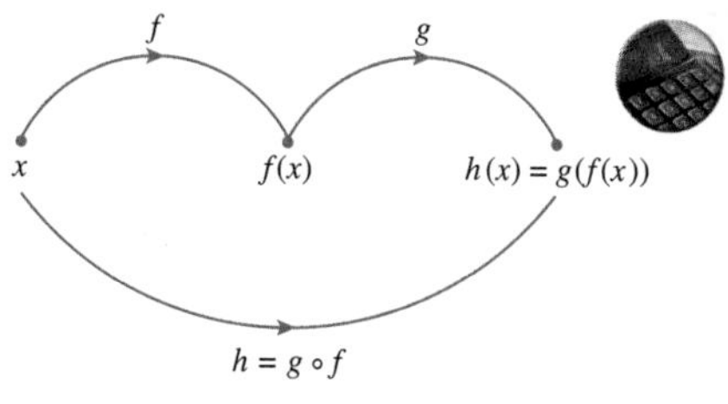

Figure 2.46
The function $h = g \circ f$ viewed as a mapping

GROUP DISCUSSION

Let $f(x) = \sqrt{x} + 1$ for $x \geq 0$ and let $g(x) = (x - 1)^2$ for $x \geq 1$.

1. Show that $(g \circ f)(x)$ and $(f \circ g)(x) = x$. (*Remark:* The function g is said to be the *inverse* of f and vice versa.)
2. Plot the graphs of f and g together with the straight line $y = x$. Describe the relationship between the graphs of f and g.

Example 5

Automobile Pollution An environmental impact study conducted for a certain city indicates that, under existing environmental protection laws, the level of carbon monoxide (CO) present in the air due to pollution from automobile exhaust will be $0.01x^{2/3}$ parts per million when the number of motor vehicles is x thousand. A separate study conducted by a governmental agency estimates that t years from now the number of motor vehicles in this city will be $0.2t^2 + 4t + 64$ thousand.

a. Find an expression for the concentration of CO in the air due to automobile exhaust t years from now.
b. What will be the level of concentration 5 years from now?

Solution

a. The level of CO present in the air due to pollution from automobile exhaust is described by the function $g(x) = 0.01x^{2/3}$, where x is the number (in thousands) of motor vehicles. But the number of motor vehicles x (in thousands) t years from now may be estimated by the rule $f(t) = 0.2t^2 + 4t + 64$. Therefore, the concentration of CO due to automobile exhaust t years from now is given by

$$C(t) = (g \circ f)(t) = g(f(t)) = 0.01(0.2t^2 + 4t + 64)^{2/3}$$

parts per million.

b. The level of concentration 5 years from now will be

$$C(5) = 0.01[0.2(5)^2 + 4(5) + 64]^{2/3}$$
$$= (0.01)89^{2/3} \approx 0.20$$

parts per million.

PORTFOLIO

Michael Marchlik

Title Project Manager
Institution Ebasco Services Incorporated

Calculus wasn't Mike Marchlik's favourite college subject. In fact, it was not until he started his first job that the "lights went on" and he realized how using calculus allowed him to solve real problems in his everyday work.

Marchlik emphasizes that he doesn't do "number crunching" himself. "I don't work out integrals, but in my work I use computer models that do that." The important issue is not computation but how the answers relate to client problems.

Marchlik's clients typically process highly toxic or explosive materials. Ebasco evaluates a client site, such as a chemical plant, to determine how safety systems might fail and what the probable consequences might be.

To avoid a major disaster like the one that occurred in Bhopal, India, in 1984, Marchlik and his team might be asked to determine how quickly a poisonous chemical would spread if a leak occurred. Or they might help avert a disaster like the one that rocked the Houston area in 1989. In that incident, hydrocarbon vapour exploded at a Phillips 66 chemical plant, shaking office buildings in Houston, 12 miles away. Several people were killed, and hundreds were injured. Property damage totalled $1.39 billion.

In assessing risks for a fuel-storage depot, Ebasco considers variables such as weather conditions, including probable wind speed, the flow properties of a gas, and possible ignition sources. With today's more powerful computers, the models can involve a system of very complex equations to project likely scenarios.

Mathematical models vary, however. One model might forecast how much gas will flow out of a hole and how quickly it will disperse. Another model might project where the gas will go, depending on local factors such as temperature and wind speed. Choosing the right model is essential. A model based on flat terrain when the client's storage depot is set among hills is going to produce the wrong answer.

Marchlik and his team run several models together to come up with their projections. Each model uses "equations that have to be integrated to come up with solutions." The bottom line? Marchlik stresses that "calculus is at the very heart" of Ebasco's risk-assessment work.

Ebasco is a diversified engineering and construction company. In addition to designing and building electric generating facilities, the company provides clients with studies. Their studies include analyses of potential hazards, recommendations for safe work practices, and evaluations of responses to emergencies in chemical plants, fuel-storage depots, nuclear facilities, and hazardous waste sites. As a project manager, Marchlik deals directly with clients to help them understand and implement Ebasco's recommendations to ensure a safer environment.

Self-Check Exercises 2.6

1. Let f and g be functions defined by the rules
$$f(x) = \sqrt{x} + 1 \quad \text{and} \quad g(x) = \frac{x}{1+x}$$
respectively. Find the rules for
   a. The sum s, the difference d, the product p, and the quotient q of f and g.
   b. The composite functions $f \circ g$ and $g \circ f$.

2. Suppose that health care spending per person by the private sector includes payments by individuals, corporations, and their insurance companies and is approximated by the function
$$f(t) = 2.48t^2 + 18.47t + 509 \qquad (0 \le t \le 6)$$
where $f(t)$ is measured in dollars and t is measured in years, with $t = 0$ corresponding to the beginning of 1994. The corresponding government spending is
$$g(t) = -1.12t^2 + 29.09t + 429 \qquad (0 \le t \le 6)$$
where t has the same meaning as before.
   a. Find a function that gives the difference between private and government health care spending per person at any time t.
   b. What was the difference between private and government expenditures per person at the beginning of 1995? At the beginning of 2000?

Solutions to Self-Check Exercises 2.6 can be found on page 92.

2.6 Exercises

In Exercises 1–8, let $f(x) = x^3 + 5$, $g(x) = x^2 - 2$, and $h(x) = 2x + 4$. Find the rule for each function.

1. $f + g$
2. $f - g$
3. fg
4. gf
5. $\frac{f}{g}$
6. $\frac{f-g}{h}$
7. $\frac{fg}{h}$
8. fgh

In Exercises 9–18, let $f(x) = x - 1$, $g(x) = \sqrt{x+1}$, and $h(x) = 2x^3 - 1$. Find the rule for each function.

9. $f + g$
10. $g - f$
11. fg
12. gf
13. $\frac{g}{h}$
14. $\frac{h}{g}$
15. $\frac{fg}{h}$
16. $\frac{fh}{g}$
17. $\frac{f-h}{g}$
18. $\frac{gh}{g-f}$

In Exercises 19–24, find the functions $f + g$, $f - g$, fg, and f/g.

19. $f(x) = x^2 + 5$; $g(x) = \sqrt{x} - 2$
20. $f(x) = \sqrt{x-1}$; $g(x) = x^3 + 1$
21. $f(x) = \sqrt{x+3}$; $g(x) = \frac{1}{x-1}$
22. $f(x) = \frac{1}{x^2+1}$; $g(x) = \frac{1}{x^2-1}$
23. $f(x) = \frac{x+1}{x-1}$; $g(x) = \frac{x+2}{x-2}$
24. $f(x) = x^2 + 1$; $g(x) = \sqrt{x+1}$

In Exercises 25–30, find the rules for the composite functions $f \circ g$ and $g \circ f$.

25. $f(x) = x^2 + x + 1$; $g(x) = x^2$
26. $f(x) = 3x^2 + 2x + 1$; $g(x) = x + 3$
27. $f(x) = \sqrt{x} + 1$; $g(x) = x^2 - 1$
28. $f(x) = 2\sqrt{x} + 3$; $g(x) = x^2 + 1$
29. $f(x) = \frac{x}{x^2+1}$; $g(x) = \frac{1}{x}$
30. $f(x) = \sqrt{x+1}$; $g(x) = \frac{1}{x-1}$

In Exercises 31–34, evaluate $h(2)$, where $h = g \circ f$.

31. $f(x) = x^2 + x + 1$; $g(x) = x^2$
32. $f(x) = \sqrt[3]{x^2 - 1}$; $g(x) = 3x^3 + 1$
33. $f(x) = \frac{1}{2x+1}$; $g(x) = \sqrt{x}$
34. $f(x) = \frac{1}{x-1}$; $g(x) = x^2 + 1$

In Exercises 35–42, find functions f and g such that $h = g \circ f$. (*Note:* The answer is *not* unique.)

35. $h(x) = (2x^3 + x^2 + 1)^5$
36. $h(x) = (3x^2 - 4)^{-3}$
37. $h(x) = \sqrt{x^2 - 1}$
38. $h(x) = (2x - 3)^{3/2}$
39. $h(x) = \frac{1}{x^2 - 1}$
40. $h(x) = \frac{1}{\sqrt{x^2 - 4}}$
41. $h(x) = \frac{1}{(3x^2 + 2)^{3/2}}$
42. $h(x) = \frac{1}{\sqrt{2x+1}} + \sqrt{2x+1}$

In Exercises 43–46, find $f(a + h) - f(a)$ for each function. Simplify your answer.

43. $f(x) = 3x + 4$

44. $f(x) = -\frac{1}{2}x + 3$

45. $f(x) = 4 - x^2$

46. $f(x) = x^2 - 2x + 1$

In Exercises 47–52, find and simplify

$$\frac{f(a + h) - f(a)}{h} \qquad (h \neq 0)$$

for each function.

47. $f(x) = x^2 + 1$

48. $f(x) = 2x^2 - x + 1$

49. $f(x) = x^3 - x$

50. $f(x) = 2x^3 - x^2 + 1$

51. $f(x) = \frac{1}{x}$

52. $f(x) = \sqrt{x}$

Business and Economics Applications

53. **Restaurant Revenue** Nicole owns and operates two restaurants. The revenue of the first restaurant at time t is $f(t)$ dollars, and the revenue of the second restaurant at time t is $g(t)$ dollars. What does the function $F(t) = f(t) + g(t)$ represent?

54. **Value of an Investment** The number of IBM shares that Nancy owns is given by $f(t)$. The price per share of the stock of IBM at time t is $g(t)$ dollars. What does the function $f(t)g(t)$ represent?

55. **Production Costs** The total cost incurred by time t in the production of a certain commodity is $f(t)$ dollars. The number of products produced by time t is $g(t)$ units. What does the function $f(t)/g(t)$ represent?

56. **Effect of Advertising on Revenue** The revenue of Leisure Travel is given by $f(x)$ dollars, where x is the dollar amount spent by the company on advertising. The amount spent by Leisure at time t on advertising is given by $g(t)$ dollars. What does the function $f \circ g$ represent?

57. **Manufacturing Costs** TMI, a manufacturer of blank audiocassette tapes, has a monthly fixed cost of \$12,100 and a variable cost of \$.60/tape. Find a function C that gives the total cost incurred by TMI in the manufacture of x tapes/month.

58. **Cost of Producing PDAs** Apollo manufactures PDAs at a variable cost of

$$V(x) = 0.000003x^3 - 0.03x^2 + 200x$$

dollars, where x denotes the number of units manufactured per month. The monthly fixed cost attributable to the division that produces these PDAs is \$100,000. Find a function C that gives the total cost incurred by the manufacture of x PDAs. What is the total cost incurred in producing 2000 units/month?

59. **Profit from Sale of PDAs** Refer to Exercise 58. Suppose the total revenue realized by Apollo from the sale of x PDAs is given by the total revenue function

$$R(x) = -0.1x^2 + 500x \qquad (0 \leq x \leq 5000)$$

where $R(x)$ is measured in dollars.

a. Find the total profit function.

b. What is the profit when 1500 units are produced and sold each month?

60. **Profit from Sale of Pagers** A division of Chapman Corporation manufactures a pager. The weekly fixed cost for the division is \$20,000, and the variable cost for producing x pagers/week is

$$V(x) = 0.000001x^3 - 0.01x^2 + 50x$$

dollars. The company realizes a revenue of

$$R(x) = -0.02x^2 + 150x \qquad (0 \leq x \leq 7500)$$

dollars from the sale of x pagers/week.

a. Find the total cost function.

b. Find the total profit function.

c. What is the profit for the company if 2000 units are produced and sold each week?

61. **Effect of Mortgage Rates on Housing Starts** A study prepared for the National Association of Realtors estimated that the number of housing starts per year over the next 5 yr will be

$$N(r) = \frac{7}{1 + 0.02r^2}$$

million units, where r (percent) is the mortgage rate. Suppose the mortgage rate over the next t mo is

$$r(t) = \frac{10t + 150}{t + 10} \qquad (0 \leq t \leq 24)$$

percent/year.

a. Find an expression for the number of housing starts per year as a function of t, t mo from now.

b. Using the result from part (a), determine the number of housing starts at present, 12 mo from now, and 18 mo from now.

62. **Hotel Occupancy Rate** The occupancy rate of the all-suite Wonderland Hotel, located near an amusement park, is given by the function

$$r(t) = \frac{10}{81}t^3 - \frac{10}{3}t^2 + \frac{200}{9}t + 55 \qquad (0 \leq t \leq 11)$$

where t is measured in months and $t = 0$ corresponds to the beginning of January. Management has estimated that the monthly revenue (in thousands of dollars) is approximated by the function

$$R(r) = -\frac{3}{5000}r^3 + \frac{9}{50}r^2 \qquad (0 \leq r \leq 100)$$

where r is the occupancy rate.

a. What is the hotel's occupancy rate at the beginning of January? At the beginning of June?

b. What is the hotel's monthly revenue at the beginning of January? At the beginning of June?
Hint: Compute $R(r(0))$ and $R(r(5))$.

63. **Housing Starts and Construction Jobs** The president of a major housing construction firm reports that the number of construction jobs (in millions) created is given by

$$N(x) = 1.42x$$

where x denotes the number of housing starts. Suppose the number of housing starts in the next t mo is expected to be

$$x(t) = \frac{7(t + 10)^2}{(t + 10)^2 + 2(t + 15)^2}$$

million units/year. Find an expression for the number of jobs created per month in the next t mo. How many jobs will have been created 6 mo and 12 mo from now?

Biological and Life Sciences Applications

64. **Birthrate of Endangered Species** The birthrate of an endangered species of whales in year t is $f(t)$ whales/year. This species of whales is dying at the rate of $g(t)$ whales/year in year t. What does the function $F(t) = f(t) - g(t)$ represent?

65. **Carbon Monoxide Pollution** The number of cars running in the business district of town at time t is given by $f(t)$. Carbon monoxide pollution coming from these cars is given by $g(x)$ parts per million, where x is the number of cars being operated in the district. What does the function $g \circ f$ represent?

66. **Biomass Production** Suppose that the biomass production by photosynthesis of a type grass is given as $f(t)$ grams per day and that sheep convert grass into body weight at a rate of $h(g)$ grams of sheep per gram of grass consumed. If the sheep keep the amount of grass in the field where they are pastured more or less constant, what function gives the biomass increase of the sheep?

67. If pelt growth in alpacas due to low temperature, measured in grams of new hair per day, is given by a function $f(t)$ while pelt growth due to nutrition is given by the function $g(t)$, then what function gives total pelt growth? Assume low temperature and nutrition are the only causes of pelt growth.

68. The total mass of clams that grow in a pond in kilograms per day is determined by the concentration of a critical nutrient so that the added mass of clams per day is

$$F(n) = 5 + 0.7n + 0.01n^2 - 0.002n^3$$

where n is the number of kilos of the nutrient present in the pond. If the concentration of the nutrient t days into the growing season is

$$n(t) = 2.6 + \frac{1.4}{\sqrt{t}}$$

write but do not simplify an expression for the added mass of clams as a function of time.

69. The efficiency of a biological survey is the fraction of animals actually in the survey area sighted by the volunteers performing the survey. Modelling the improvement of the volunteers with time, by testing them on a plot where the locations of all nests are known, it is found that the average efficiency of a survey for a type of flycatcher is

$$E(t) = \frac{t}{t + 3.14}$$

after t weeks. If the number of birds sighted by the volunteers per hectare as they move across Ontario is given by

$$f(t) = 26 + 0.74t$$

build a function that estimates the actual number of birds in week t.

In Exercises 70–73, determine whether the statement is true or false. If it is true, explain why it is true. If it is false, give an example to show why it is false.

70. If f and g are functions with domain D, then $f + g = g + f$.

71. If $g \circ f$ is defined at $x = a$, then $f \circ g$ must also be defined at $x = a$.

72. If f and g are functions, then $f \circ g = g \circ f$.

73. If f is a function, then $f \circ f = f^2$.

Solutions to Self-Check Exercises 2.6

1. a. $s(x) = f(x) + g(x) = \sqrt{x} + 1 + \dfrac{x}{1 + x}$

$$d(x) = f(x) - g(x) = \sqrt{x} + 1 - \frac{x}{1 + x}$$

$$p(x) = f(x)g(x) = (\sqrt{x} + 1)\cdot\frac{x}{1 + x} = \frac{x(\sqrt{x} + 1)}{1 + x}$$

$$q(x) = \frac{f(x)}{g(x)} = \frac{\sqrt{x} + 1}{\frac{x}{1 + x}} = \frac{(\sqrt{x} + 1)(1 + x)}{x}$$

b. $(f \circ g)(x) = f(g(x)) = \sqrt{\dfrac{x}{1 + x}} + 1$

$$(g \circ f)(x) = g(f(x)) = \frac{\sqrt{x} + 1}{1 + (\sqrt{x} + 1)} = \frac{\sqrt{x} + 1}{\sqrt{x} + 2}$$

2. a. The difference between private and government health care spending per person at any time t is given by the function d with the rule

$$\begin{aligned} d(t) = f(t) - g(t) &= (2.48t^2 + 18.47t + 509) \\ &\quad - (-1.12t^2 + 29.09t + 429) \\ &= 3.6t^2 - 10.62t + 80 \end{aligned}$$

b. The difference between private and government expenditures per person at the beginning of 1995 is given by

$$d(1) = 3.6(1)^2 - 10.62(1) + 80$$

or \$72.98/person.

The difference between private and government expenditures per person at the beginning of 2000 is given by

$$d(6) = 3.6(6)^2 - 10.62(6) + 80$$

or \$145.88/person.

2.7 The Inverse of a Function

From data obtained in a test run conducted on a prototype of a maglev (magnetic levitation train), which moves along a straight monorail track, engineers have determined that the position of the maglev (in metres) from the origin at time t is given by

$$s = f(t) = t^2 \qquad (0 \le t \le 30) \tag{3}$$

where f is called the position function of the maglev. The position of the maglev at time $t = 1, 2, 3, \ldots, 10$, measured from its initial position, is

$$f(0) = 0, f(1) = 1, f(2) = 4, \ldots, f(10) = 100 \text{ metres (Figure 2.47a).}$$

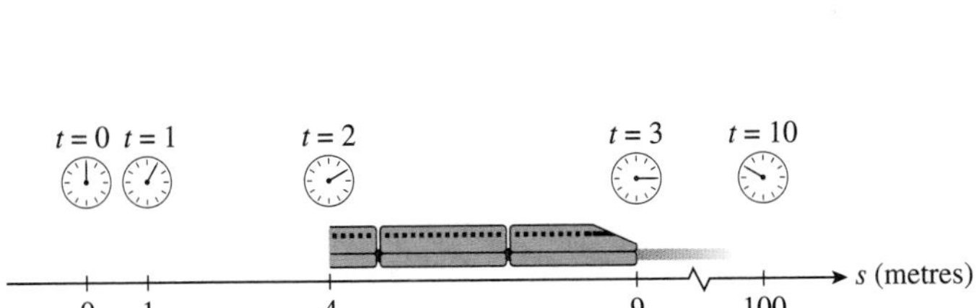

(a) A maglev moving along an elevated monorail track.

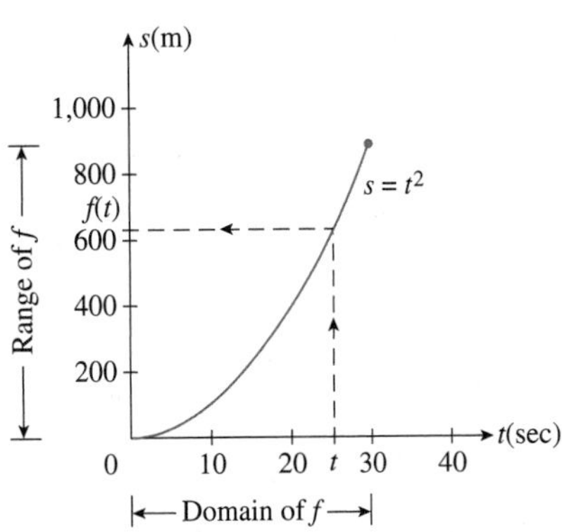

(b) Each t in the domain of f is associated with the (unique) position of the maglev.

Figure 2.47

Equation (3) enables us to compute algebraically the position of the maglev at any time t. Geometrically, we can find the position of the maglev at any given time t by following the path indicated in Figure 2.47b.

Now consider the reverse problem: Knowing the position function of the maglev, can we find some way of obtaining the time it takes for the maglev to reach a given position? Geometrically, this problem is easily solved: Locate the point on the s-axis corresponding to the given position. Follow the path considered earlier but traced in the *opposite* direction. This path associates the given position s with the desired time t.

Algebraically, we can obtain a formula for the time t it takes for the maglev to get to the position s by solving (1) for t in terms of s. Thus,

$$s = t^2$$

$$t^2 = s$$

$$t = \pm\sqrt{s}$$

We reject the negative root because t lies in [0, 30], and so

$$t = \sqrt{s}$$

Observe that the function g defined by

$$t = g(s) = \sqrt{s}$$

has domain [0, 900] (the range of f) and range [0, 30] (the domain of f) (Figure 2.48).

The functions f and g have the following properties:

1. The domain of g is the range of f and vice versa
2. $(g \circ f)(t) = g[f(t)] = \sqrt{f(t)} = \sqrt{t^2} = t$, since t is positive, and
3. $(f \circ g)(t) = f[g(t)] = [g(t)]^2 = [\sqrt{t}]^2 = t$

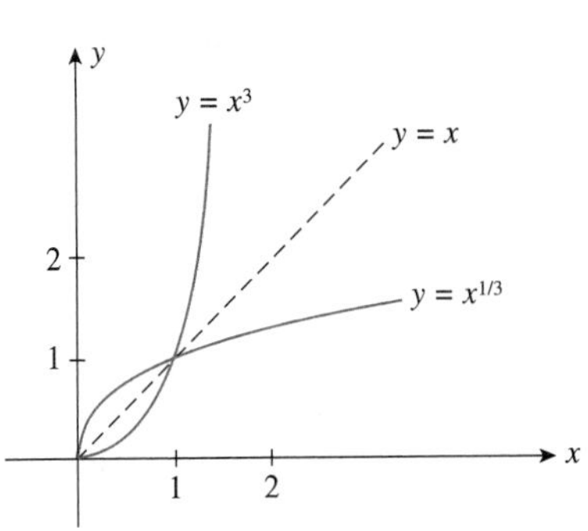

Figure 2.48
Each s in the domain of g is associated with the (unique) time $t = g(s)$.

In other words, one function undoes what the other function does. This is to be expected because f maps t onto and g maps back onto t.

The functions f and g are said to be *inverses* of each other. More generally, we have the following definition.

Inverse Functions

A function g is the inverse of the function f if

$$f[g(x)] = x \text{ for every } x \text{ in the domain of } g$$
$$g[f(x)] = x \text{ for every } x \text{ in the domain of } f$$

Equivalently, g is the inverse of f if the following condition is satisfied:

$$y = f(x) \quad \text{if and only if} \quad x = g(y)$$

for every x in the domain of f and for every y in its range.

Note: The inverse of f is normally denoted by f^{-1} (read "f inverse"), and we will use this notation throughout the text.

Do not confuse $f^{-1}(x)$ with $[f(x)]^{-1} = \frac{1}{f(x)}$!

Example 1

Show that the functions $f(x) = x^{1/3}$ and $g(x) = x^3$ are inverses of each other.

Solution

First, observe that the domain and range of both f and g are $(-\infty, \infty)$. Therefore, both composite functions $f \circ g$ and $g \circ f$ are defined. Next, we compute

$$(f \circ g)(x) = f[g(x)] = [g(x)]^{1/3} = [x^3]^{1/3} = x$$

and

$$(g \circ f)(x) = g[f(x)] = [f(x)]^3 = (x^{1/3})^3 = x$$

Since $f[g(x)] = g[f(x)] = x$, we conclude that f and g are inverses of each other. In short, $f^{-1}(x) = x^3$.

Interpreting Our Results We can view f as a cube root extracting machine and g as a "cubing" machine. In this light, it is easy to see that one function does undo what the other does. So f and g are indeed inverses of each other.

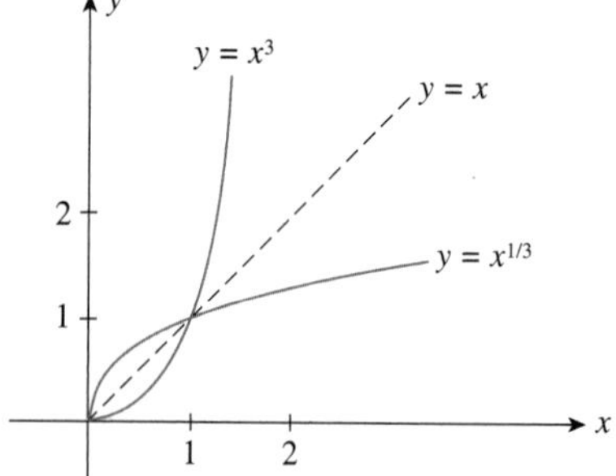

Figure 2.49
The functions $y = x^{1/3}$ and $f^{-1}(x) = x^3$ are inverses of each other. Their graphs are symmetric with respect to the line $y = x$.

The Graphs of Inverse Functions

The graphs of $f(x) = x^{1/3}$ and $f^{-1}(x) = x^3$ are shown in Figure 2.49.

They seem to suggest that the graphs of inverse functions are mirror images of each other with respect to the line $y = x$. This is true in general, a fact that we will not prove. Remember though that the function f maps x onto y, and the inverse function f^1 maps y onto x, as we have seen with the introductory example.

The Graphs of Inverse Functions

The graph of $y = f^{-1}(x)$ is the reflection of the graph of $y = f(x)$ with respect to the line $y = x$ and vice versa.

Functions That Have Inverses

Not every function has an inverse. Consider, for example, the function f defined by $y = x^2$ with domain $(-\infty, \infty)$ and range $[0, \infty)$. From the graph of f shown in Figure 2.50, you can see that each value of y in the range $[0, \infty)$ of f is associated with exactly *two* points $x = \pm\sqrt{y}$ (except for $y = 0$) in the domain $(-\infty, \infty)$ of f. This implies that f does not have an inverse because the uniqueness requirement of a function cannot be satisfied in this case. Observe that any horizontal line $y = c$ $(c > 0)$ intersects the graph of f at more than one point.

Next, consider the function g defined by the same rule as that of f, namely $y = x^2$, but with domain restricted to $[0, \infty)$. From the graph of g shown in Figure 2.51, you can see that each value of y in the range $[0, \infty)$ of g is mapped onto exactly *one* point $x = \sqrt{y}$ in the domain $[0, \infty)$ of g.

Thus, in this case, we can define the inverse function of g, from the range $[0, \infty)$ of g onto the domain $[0, \infty)$ of g. To find the rule for g^{-1}, we solve the equation $y = x^2$ for x in terms of y. Thus, $x = \sqrt{y}$ (because $x \geq 0$) and so $g^{-1}(y) = \sqrt{y}$, or, since y is a dummy variable, we can write $g^{-1}(x) = \sqrt{x}$. Also, observe that every horizontal line intersects the graph of g at no more than one point.

Why does g have an inverse but f does not? Observe that f takes on the same value twice; that is, there are two values of x that are mapped onto each value of y (except $y = 0$). On the other hand, g never takes on the same value more than once; that is, any two values of x have different images. The function g is said to be one-to-one.

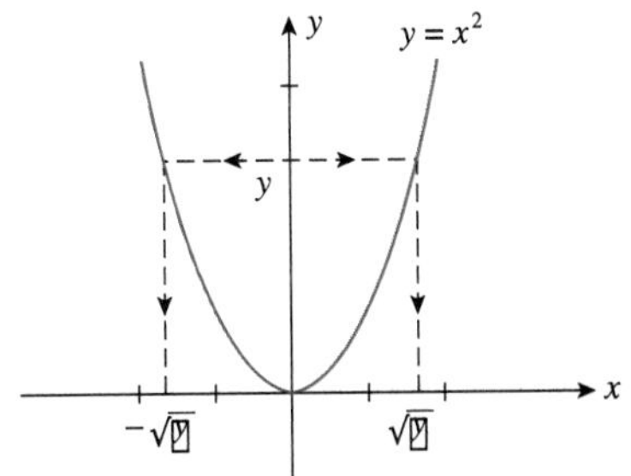

Figure 2.50
Each value of y is associated with two values of x.

One-to-One Function

A function f with domain D is one-to-one if no two points in D have the same image: that is, $f(x_1) \neq f(x_2)$ whenever $x_1 \neq x_2$.

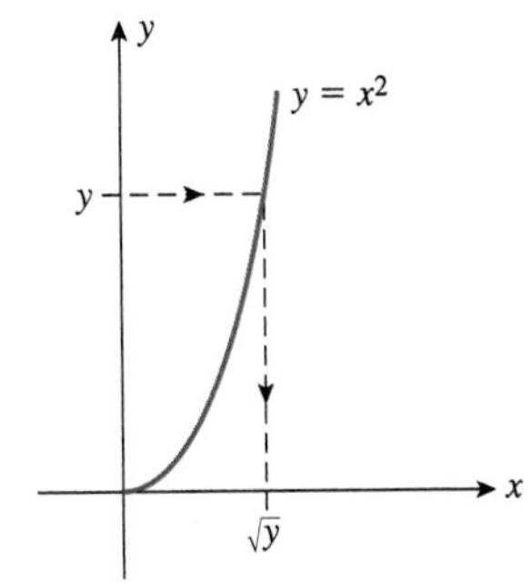

Figure 2.51
Each value of y is associated with exactly one value of x.

The example above suggests the following horizontal line test for determining geometrically when a function is one-to one.

Horizontal Line Test

A function is one-to-one if and only if every horizontal line intersects its graph at no more than one point.

The next theorem tells us when an inverse function exists.

The Existence of an Inverse Function

A function has an inverse if and only if it is one-to-one.

Finding the Inverse of a Function

Here is a summary of the steps for finding the inverse of a function (if it exists).

Guidelines for Finding the Inverse of a Function

1. Write $y = f(x)$.
2. Solve for x in terms of y (if possible).
3. Interchange x and y to obtain $y = f^{-1}(x)$.

Example 2

Find the inverse of the function defined by $f(x) = \dfrac{1}{\sqrt{2x - 3}}$ with domain $(\frac{3}{2}, \infty)$ and range $(0, \infty)$.

Solution

To find the rule for this inverse, write

$$y = \frac{1}{\sqrt{2x - 3}}$$

and then solve the equation for x:

$$y^2 = \frac{1}{2x - 3} \qquad \text{Square both sides.}$$

$$2x - 3 = \frac{1}{y^2} \qquad \text{Take reciprocals.}$$

$$2x = \frac{1}{y^2} + 3 = \frac{3y^2 + 1}{y^2}$$

$$x = \frac{3y^2 + 1}{2y^2}$$

Finally, interchanging x and y, we obtain

$$y = \frac{3x^2 + 1}{2x^2}$$

giving the rule for f^{-1} as

$$f^{-1}(x) = \frac{3x^2 + 1}{2x^2}$$

with domain $(0, \infty)$ and range $(\frac{3}{2}, \infty)$. Notice that the function $g(x) = \frac{3x^2 + 1}{2x^2}$ has domain $(-\infty, 0) \cup (0, \infty)$, but for the inverse function of f we had to satisfy our definition for inverse functions from page 94.

The graphs of both f and f^{-1} are shown in Figure 2.52.

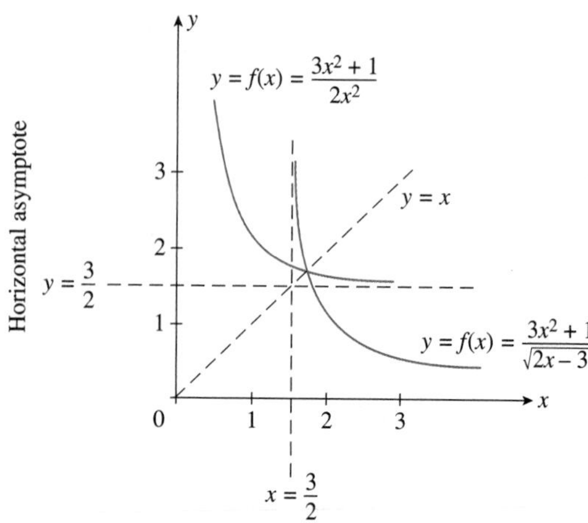

Figure 2.52
The graphs of f and f^{-1}. Notice that they are reflections of each other about the line $y = x$.

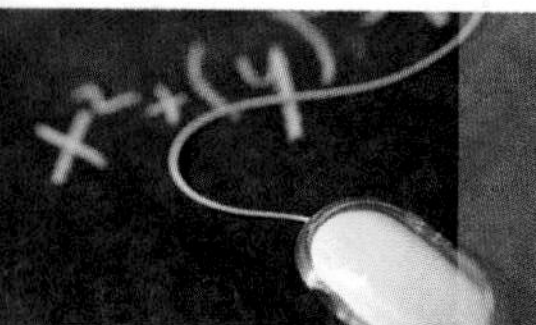

Self-Check Exercises 2.7

1. The function f is described as $f(x) = \frac{x - 2}{x + 3}$.
   a. Describe the domain of f.
   b. Describe the range of f by finding the domain of its inverse.

Solution to Self-Check Exercise 2.7 can be found on page 98.

2.7 Exercises

In Exercises 1–6, show that f and g are inverses of each other by showing that $f[g(x)]$ and $g[f(x)] = x$.

1. $f(x) = \frac{1}{3}x^3$; $g(x) = \sqrt[3]{3x}$ **2.** $f(x) = \frac{1}{x}$; $g(x) = \frac{1}{x}$

3. $f(x) = 2x + 3$; $g(x) = \frac{x - 3}{2}$

4. $f(x) = x^2 + 1 (x \leq 0)$; $g(x) = -\sqrt{x - 1}$

5. $f(x) = 4(x + 1)^{2/3}$ $(x \geq -1)$; $g(x) = \frac{1}{8}(x^{3/2} - 8)$ $(x \geq 0)$

6. $f(x) = \frac{1 + x}{1 - x}$; $g(x) = \frac{x - 1}{x + 1}$

In Exercises 7–12, you are given the graph of a function f. Determine whether f is one-to-one.

7. 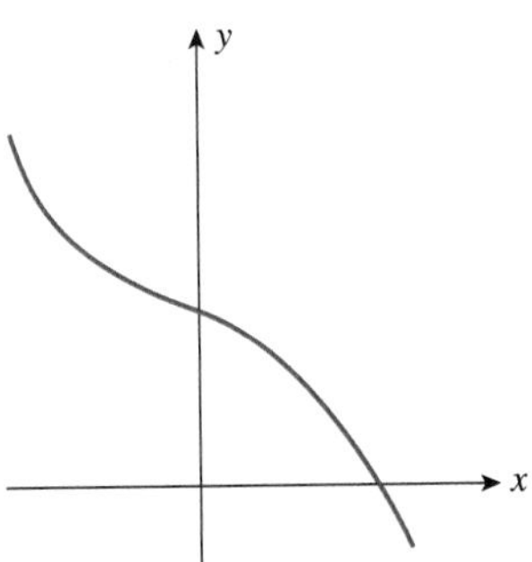

8. 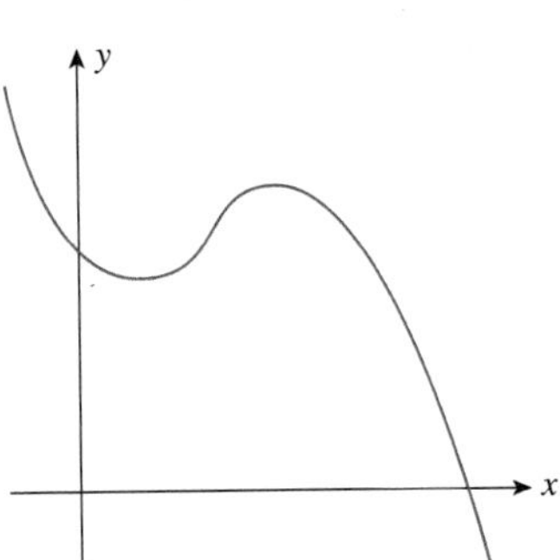

9. 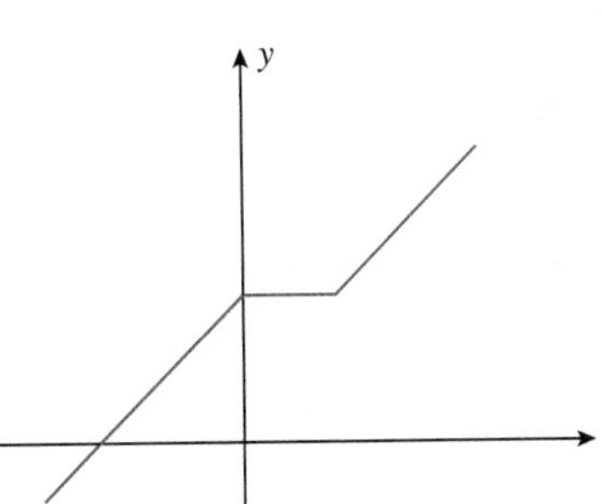

10. 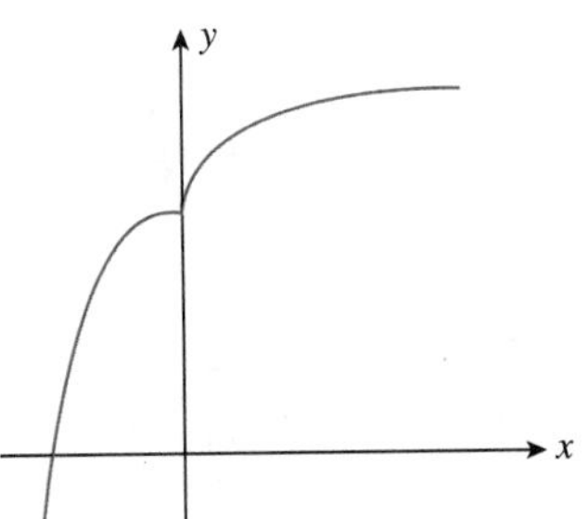

11. 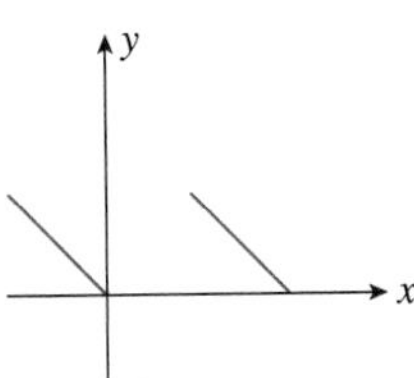

12. 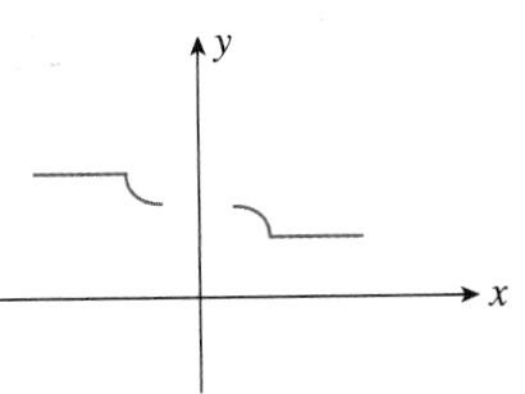

In Exercises 13–18, find the inverse of f. Then sketch the graphs of f and f^{-1} on the same set of axes.

13. $f(x) = 3x - 2$
14. $f(x) = x^2, x \le 0$
15. $f(x) = x^3 + 1$
16. $f(x) = 2\sqrt{x + 3}$
17. $f(x) = \sqrt{9 - x^2}, x \ge 0$
18. $f(x) = x^{3/5} + 1$
19. A hot air balloon rises vertically from the ground so that its height after t sec is $h = \frac{1}{2}t^2 + \frac{1}{2}t$ m $(0 \le t \le 60)$.
    a. Find the inverse of the function $f(t) = \frac{1}{2}t^2 + \frac{1}{2}t$ and explain what it represents.
    b. Use the result of part (a) to find the time when the balloon is between an altitude of 120 and 210 m.
20. **Aging Population** The population of Canadians age 55 and over as a percent of the total population is approximated by the function

$$f(t) = 10.72(0.9t + 10)^{0.3} \qquad (0 \le t \le 20)$$

where t is measured in years and $t = 0$ corresponds to the year 2000.
    a. Find the rule for f^{-1}.
    b. Evaluate $f^{-1}(25)$ and interpret your result.

Business and Economics Applications

21. Within a range of possible application levels, the yield increase of Canola, when given added fertilizer, is found to be

$$P(k) = 3.1\sqrt[3]{k/17.4}\% \text{ per kilogram/acre}$$

Find the inverse of this function and use it to estimate the number of kilograms of fertilizer per acre that cause a 10% increase in yield.

Biological and Life Sciences Applications

22. The response of a marker protein to supplemental hormone injections is found to be given by the formula

$$A(r) = \sqrt{1.7r - 3}$$

micrograms of marker per micrograms of supplementary hormone. On a particular day a patient chart is misplaced but the measurement of the marker protein returned by the lab finds 11.4 micrograms of the marker in the patient's blood. Invert the formula and estimate the amount of supplementary hormone injected.

23. In the spring, during the time the dandelions are coming up, the number of dandelions that sprout in a field after t hours is given by the function

$$F(t) = 216t^2$$

Find the inverse of this function and use it to compute how many hours it is since dandelions started sprouting, if 572 are counted in the field.

24. The amount of a standard sugar water solution spread in a test plot in a provincial park predicts the number of ants present after an hour according to the formula

$$A(L) = \frac{112L}{L + 4} \text{ litres.}$$

When a two-litre bottle of soda is spilled in the test spot, 470 ants are present after an hour. Find the inverse of the formula and use it to compute how much standard sugar solution two litres of the soda is equivalent to, from the perspective of attracting ants.

Solutions to Self-Check Exercise 2.7

1. **a.** Since f is not defined for $x = -3$, the domain of f is given by $D_f = \{x \neq -3\}$.

   b. We first need to find the inverse of f and write

$$y = \frac{x - 2}{x + 3}$$

and then solve the equation for x:

$$y(x + 3) = x - 2$$

$$xy + 3y = x - 2$$

$$3y + 2 = x - xy$$

$$3y + 2 = x(1 - y)$$

$$x = \frac{3y + 2}{1 - y}$$

Interchanging x and y, we obtain the inverse function

$$f^{-1}(x) = \frac{3x + 2}{1 - x}$$

The domain of f^{-1} is given by $D_{f^{-1}} = \{x \neq 1\}$ and so the range of f must be $R_f = \{y \neq 1\}$.

2.8 Functions and Mathematical Models

Mathematical Models

One of the fundamental goals in this book is to show how mathematics and, in particular, calculus can be used to solve real-world problems such as those arising from the world of business and the social, life, and physical sciences. You have already seen some of these problems earlier.

Regardless of the field from which the real-world problem is drawn, the problem is analyzed using a process called **mathematical modelling.** The four steps in this process, as illustrated in Figure 2.53, follow.

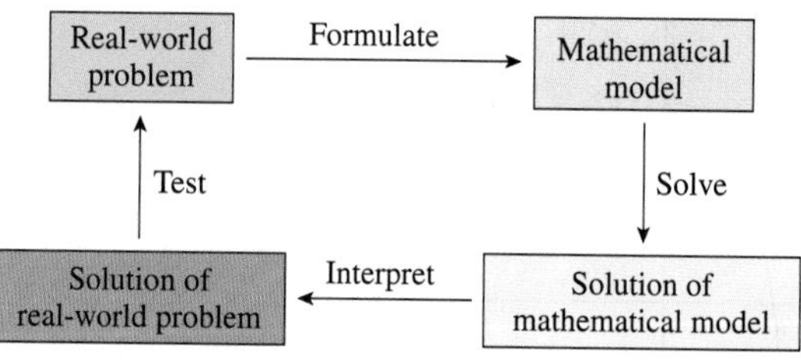

Figure 2.53

1. **Formulate** Given a real-world problem, our first task is to formulate the problem, using the language of mathematics. The many techniques used in constructing mathematical models range from theoretical consideration of the problem on the one extreme to an interpretation of data associated with the problem on the other. For example, the mathematical model giving the accumulated amount at any time when a certain sum of money is deposited in the bank can be derived theoretically (see Chapter 12). On the other hand, many of the mathematical models in this book are constructed by studying the data associated with the problem. In calculus, we are primarily concerned with how one (dependent) variable depends on one or more (independent) variables. Consequently, most of our mathematical models will involve functions of one or more variables or equations defining these functions (implicitly).

2. **Solve** Once a mathematical model has been constructed, we can use the appropriate mathematical techniques, which we will develop throughout the book, to solve the problem.
3. **Interpret** Bearing in mind that the solution obtained in step 2 is just the solution of the mathematical model, we need to interpret these results in the context of the original real-world problem.
4. **Test** Some mathematical models of real-world applications describe the situations with complete accuracy. For example, the model describing a deposit in a bank account gives the exact accumulated amount in the account at any time. But other mathematical models give, at best, an approximate description of the real-world problem. In this case we need to test the accuracy of the model by observing how well it describes the original real-world problem and how well it predicts past and/or future behaviour. If the results are unsatisfactory, then we may have to reconsider the assumptions made in the construction of the model or, in the worst case, return to step 1.

Before going on, let's look at two mathematical models. The first one is used to estimate the U.S. market for cholesterol-reducing drugs from 1999 through 2004, and the second is used to project the growth of the number of people enrolled in health maintenance organizations (HMOs). These models are derived from data using the least-squares technique. In Section 10.4 we will show how the model in Example 1 is actually constructed using the method of least squares.

Example 1

Average Canadian Farm Size The average size of a Canadian farm has increased from 1921 to 2001 with the introduction of new technology and is listed in the following table:

Year	1921	1961	2001
Average Canadian farm size (acres)	198	359	676

Source: Adapted from Statistics Canada website http://www40.statcan.ca/l01/cst01/comm09a.htm, table entitled Household Internet use at home by Internet activity (All households).

A mathematical model giving the approximate average size of a Canadian farm over the period in question is given by

$$A(t) = 5.975t + 172$$

where t is measured in years, with $t = 0$ corresponding to the year 1921.

a. Sketch the graph of the function A and the given data on the same set of axes.

b. Assuming that the projection holds and the trend continues, what will be the average size of a Canadian farm in 2010 ($t = 89$)?

c. What is the rate of increase of the average Canadian farm size over the period in question?

Solution

a. The graph of A is shown in Figure 2.54.

b. The average size of a Canadian farm is approximately

$$A(89) = 5.975(89) + 172$$
$$= 703.775$$

or 703.775 acres.

c. The function A is linear, and so we see that the rate of increase of the average size of a Canadian farm is given by the slope of the straight line represented by A, which is approximately 5.975 acres per year.

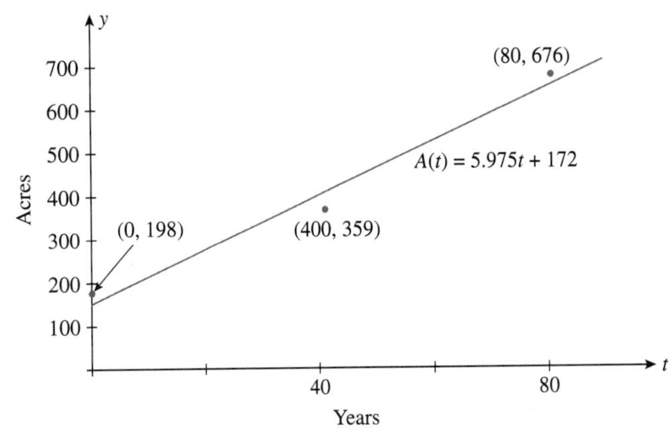

Figure 2.54
Projected average Canadian farm sizes

Example 2

HMO Membership In the United States, where medical care is not generally provided by a government plan, many people receive their health care from a corporation called a Health Maintenance Organization (HMO) through private insurance. The number of people (in millions) enrolled in HMOs from 1988 to 1998 is given in the following table:

Year	1988	1990	1992	1994	1996	1998
People	32.7	36.5	41.4	51.1	66.5	78.0

A mathematical model approximating the number of people, $N(t)$, enrolled in HMOs during this period is

$$N(t) = -0.0258t^3 + 0.7465t^2 - 0.3491t + 33.1444 \qquad (0 \leq t \leq 10)$$

where t is measured in years, with $t = 0$ corresponding to 1988.

a. Sketch the graph of the function N and the given data on the same set of axes.

b. Assume that this trend continues and use the model to predict how many people will be enrolled in HMOs at the beginning of 2003.

Source: Interstudy Publications

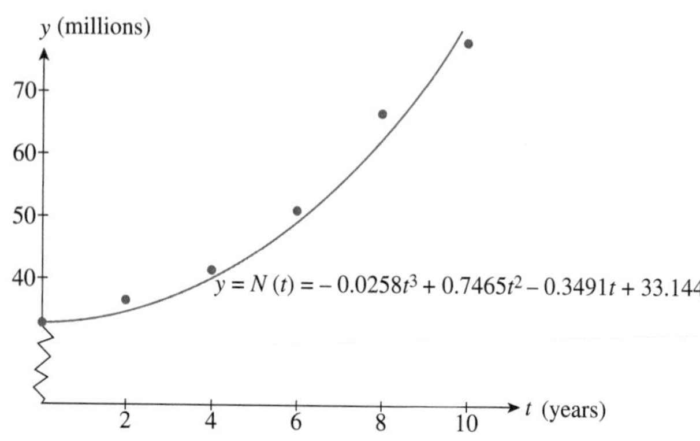

Figure 2.55
The graph of $y = N(t)$ approximates the number of people enrolled in HMOs from 1988 to 1998.

Solution

a. The graph of the function N is shown in Figure 2.55.

b. The number of people that will be enrolled in HMOs at the beginning of 2003 is given by

$$\begin{aligned} N(15) &= -0.0258(15)^3 + 0.7465(15)^2 - 0.3491(15) + 33.1444 \\ &= 108.795 \end{aligned}$$

or approximately 108.8 million people.

Example 3

Email The Internet is increasingly used by Canadian households to perform a variety of Internet activities such as financial banking, viewing the news, browsing, etc. The percentage of all Canadian households that use the Internet for sending and receiving email from 1999 to 2003 is listed in the following table:

Year	1999	2000	2001	2002	2003
Household Internet use at home by Email (% of all household)	26.3	37.4	46.1	48.9	52.1

Source: Adapted from Statistics Canada website http://www40.statcan.ca/l01/cst01/comm09a.htm, table entitled Household Internet use at home by Internet activity (All households).

A mathematical model giving the percentage of all households that use the Internet for email over the period in question is given by

$$E(t) = 11.1(t)^{0.65} + 26.3$$

where t is measured in years, with $t = 0$ corresponding to the year 1999.

a. What percentage of all households will be using the Internet for email in the year 2006? ($t = 7$)

b. In approximately which year will the percentage of households have tripled compared to 1999?

Solution

a. The percentage of Canadian households using the Internet for email is approximately

$$E(7) = 11.1(7)^{0.65} + 26.3$$
$$\approx 65.6$$

or 65.6%. The graph of E is shown in Figure 2.56.

b. We need to set up the equation and get

$$11.1(t)^{0.65} + 26.3 = 3 \times 26.3$$
$$11.1(t)^{0.65} + 26.3 = 78.9$$

Solving for t we get

$$11.1(t)^{0.65} = 78.9 - 26.3$$
$$11.1(t)^{0.65} = 52.6$$
$$(t)^{0.65} = \frac{52.6}{11.1}$$
$$t = \left(\frac{52.6}{11.1}\right)^{1/0.65}$$
$$t \approx 10.95$$

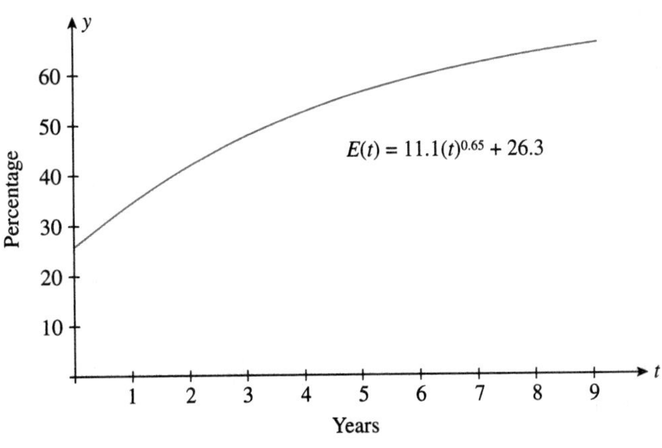

Figure 2.56
Percentages of all Canadian households that use the Internet for sending and receiving email.

Therefore, the percentage of Canadian households using the Internet for email will have tripled approximately by the year 2010.

Some Economic Models

In the remainder of this section, we look at some economic models.

In a free market economy, consumer demand for a particular commodity depends on the commodity's unit price. A demand equation expresses the relationship between the unit price and the quantity demanded. The graph of the demand equation is called a demand curve. In general, the quantity demanded of a commodity decreases as the commodity's unit price increases, and vice versa. Accordingly, a demand function defined by $p = f(x)$, where p measures the unit price and x measures the number of units of the commodity in question, is generally characterized as a decreasing function of x; that is, $p = f(x)$ decreases as x increases. Since both x and p assume only nonnegative values, the demand curve is that part of the graph of $f(x)$ that lies in the first quadrant (Figure 2.57).

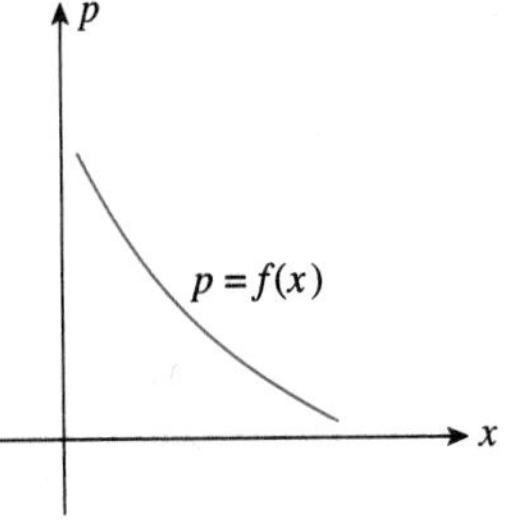

Figure 2.57
A demand curve

In a competitive market, a relationship also exists between the unit price of a commodity and the commodity's availability in the market. In general, an increase in the commodity's unit price induces the producer to increase the supply of the commodity. Conversely, a decrease in the unit price generally leads to a drop in the supply. The equation that expresses the relation between the unit price and the quantity supplied is called a supply equation, and its graph is called a supply curve. A supply function defined by $p = f(x)$ is generally characterized as an increasing function of x; that is, $p = f(x)$ increases as x increases. Since both x and p assume only nonnegative values, the supply curve is that part of the graph of $f(x)$ that lies in the first quadrant (Figure 2.58).

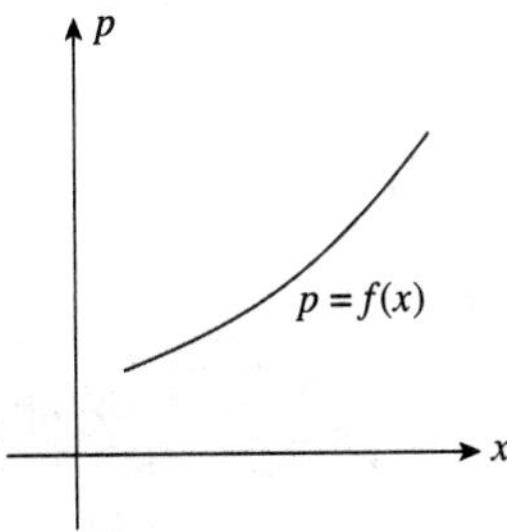

Figure 2.58
A supply curve

Under pure competition, the price of a commodity will eventually settle at a level dictated by the following condition: The supply of the commodity will be equal to the demand for it. If the price is too high, the consumer will not buy; if the price is too low, the supplier will not produce. Market equilibrium prevails when the quantity produced is equal to the quantity demanded. The quantity produced at market equilibrium is called the equilibrium quantity, and the corresponding price is called the equilibrium price.

Market equilibrium corresponds to the point at which the demand curve and the supply curve intersect. In Figure 2.59 x_0 represents the equilibrium quantity and p_0 the equilibrium price. The point (x_0, p_0) lies on the supply curve and therefore satisfies the supply equation. At the same time, it also lies on the demand curve and therefore satisfies the demand equation. Thus, to find the point (x_0, p_0), and hence the equilibrium quantity and price, we solve the demand and supply equations simultaneously for x and p. For meaningful solutions, x and p must both be positive.

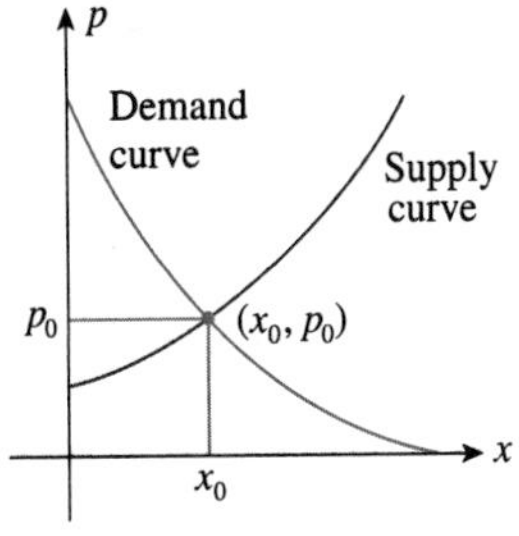

Figure 2.59
Market equilibrium corresponds to (x_0, p_0), the point at which the supply and demand curves intersect.

Example 4

Supply-Demand The demand function for a certain brand of DVD is given by

$$p = d(x) = -0.01x^2 - 0.2x + 8$$

and the corresponding supply function is given by

$$p = s(x) = 0.01x^2 + 0.1x + 3$$

where p is expressed in dollars and x is measured in units of a thousand. Find the equilibrium quantity and price.

Solution

We solve the following system of equations:

$$p = -0.01x^2 - 0.2x + 8$$
$$p = 0.01x^2 + 0.1x + 3$$

Substituting the first equation into the second yields

$$-0.01x^2 - 0.2x + 8 = 0.01x^2 + 0.1x + 3$$

which is equivalent to

$$0.02x^2 + 0.3x - 5 = 0$$
$$2x^2 + 30x - 500 = 0$$
$$x^2 + 15x - 250 = 0$$
$$(x + 25)(x - 10) = 0$$

Thus, $x = -25$ or $x = 10$. Since x must be nonnegative, the root $x = -25$ is rejected. Therefore, the equilibrium quantity is 10,000 DVDs. The equilibrium price is given by

$$p = 0.01(10)^2 + 0.1(10) + 3 = 5$$

or $5 per DVD (Figure 2.60).

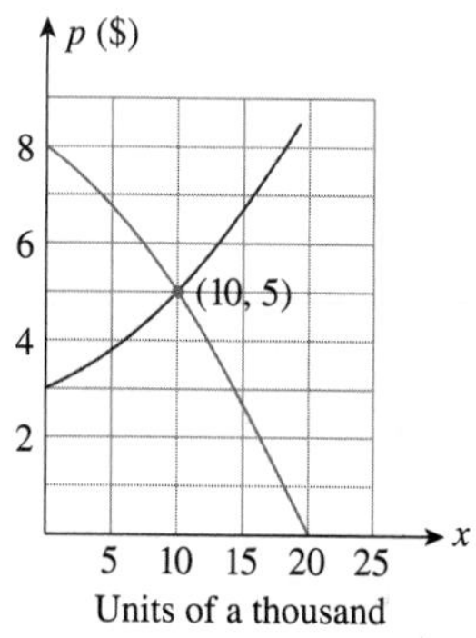

Figure 2.60
The supply curve and the demand curve intersect at the point (10, 5).

EXPLORING WITH TECHNOLOGY

1. a. Use a graphing utility to plot the straight lines L_1 and L_2 with equations $y = 2x - 1$ and $y = 2.1x + 3$, respectively, on the same set of axes, using the standard viewing window. Do the lines appear to intersect?
   b. Plot the straight lines L_1 and L_2, using the viewing window $[-100, 100] \cdot [-100, 100]$. Do the lines appear to intersect? Can you find the point of intersection using TRACE and ZOOM? Using the "intersection" function of your graphing utility?
   c. Find the point of intersection of L_1 and L_2 algebraically.
   d. Comment on the effectiveness of the methods of solutions in parts (b) and (c).
2. a. Use a graphing utility to plot the straight lines L_1 and L_2 with equations $y = 3x - 2$ and $y = -2x + 3$, respectively, on the same set of axes, using the standard viewing window. Then use TRACE and ZOOM to find the point of intersection of L_1 and L_2. Repeat using the "intersection" function of your graphing utility.
   b. Find the point of intersection of L_1 and L_2 algebraically.
   c. Comment on the effectiveness of the methods.

Constructing Mathematical Models

We close this section by showing how some mathematical models can be constructed using elementary geometric and algebraic arguments.

Example 5

Enclosing an Area The owner of a cattle ranch in the interior of B.C. has 3000 metres of fencing material with which to enclose a rectangular piece of grazing land along the straight portion of a river. Fencing is not required along the river. Letting x denote the width of the rectangle, find a function f in the variable x giving the area of the grazing land if she uses all of the fencing material (Figure 2.61).

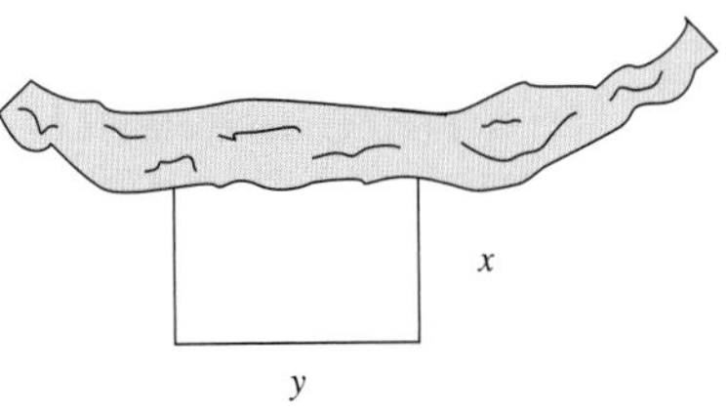

Figure 2.61
The rectangular grazing land has width x and length y.

Solution

The area of the rectangular grazing land is $A = xy$. Next, observe that the amount of fencing is $2x + y$ and this must be equal to 3000 since all the fencing material is used; that is,

$$2x + y = 3000$$

From the equation we see that $y = 3000 - 2x$. Substituting this value of y into the expression for A gives

$$A = xy = x(3000 - 2x) = 3000x - 2x^2$$

Finally, observe that both x and y must be nonnegative since they represent the width and length of a rectangle, respectively. Thus, $x \geq 0$ and $y \geq 0$. But the latter is equivalent to $3000 - 2x \geq 0$, or $x \leq 1500$. So the required function is $f(x) = 3000x - 2x^2$ with domain $0 \leq x \leq 1500$.

Remark

Observe that if we view the function $f(x) = 3000x - 2x^2$ strictly as a mathematical entity, then its domain is the set of all real numbers. But physical consideration dictates that its domain should be restricted to the interval [0, 1500]. ◀

Example 6

Charter-Flight Revenue If exactly 200 people sign up for a charter flight, Leisure World Travel Agency charges \$300 per person. However, if more than 200 people sign up for the flight (assume this is the case), then each fare is reduced by \$1 for each additional person. Letting x denote the number of passengers above 200, find a function giving the revenue realized by the company.

Solution

If there are x passengers above 200, then the number of passengers signing up for the flight is $200 + x$. Furthermore, the fare will be \$$(300 - x)$ per passenger. Therefore, the revenue will be

$$\begin{aligned} R &= (200 + x)(300 - x) \\ &= -x^2 + 100x + 60{,}000 \end{aligned}$$

Number of passengers × the fare per passenger

Clearly, x must be nonnegative, and $300 - x \geq 0$, or $x \leq 300$. So the required function is $f(x) = -x^2 + 100x + 60{,}000$ with domain [0, 300].

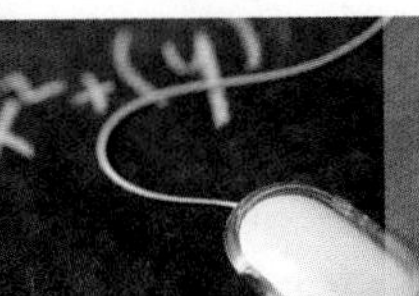

Self-Check Exercises 2.8

1. Thomas Young has suggested the following rule for calculating the dosage of medicine for children from ages 1 to 12 yr. If a denotes the adult dosage (in milligrams) and t is the age of the child (in years), then the child's dosage is given by

$$D(t) = \frac{at}{t + 12}$$

If the adult dose of a substance is 500 mg, how much should a 4-yr-old child receive?

2. The demand function for Mrs. Baker's cookies is given by

$$d(x) = -\frac{2}{15}x + 4$$

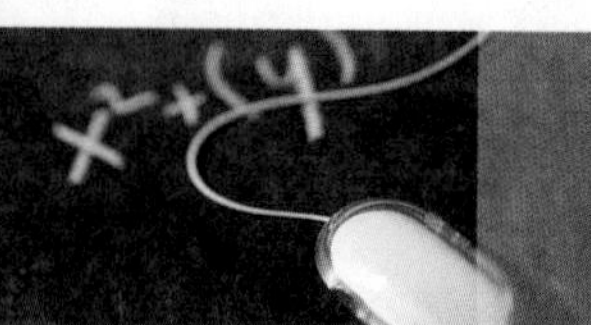

Self-Check Exercises 2.8, *continued*

where $d(x)$ is the wholesale price in dollars/kilogram and x is the quantity demanded each week, measured in thousands of kilograms. The supply function for the cookies is given by

$$s(x) = \frac{1}{75}x^2 + \frac{1}{10}x + \frac{3}{2}$$

where $s(x)$ is the wholesale price in dollars/kilogram and x is the quantity, in thousands of kilograms, that will be made available in the market each week by the supplier.

a. Sketch the graphs of the functions d and s.
b. Find the equilibrium quantity and price.

Solutions to Self-Check Exercises 2.8 can be found on page 108.

2.8 Exercises

In Exercises 1–8, determine whether the equation defines y as a linear function of x. If so, write it in the form y = mx + b.

1. $2x + 3y = 6$

2. $-2x + 4y = 7$

3. $x = 2y - 4$

4. $2x = 3y + 8$

5. $2x - 4y + 9 = 0$

6. $3x - 6y + 7 = 0$

7. $2x^2 - 8y + 4 = 0$

8. $3\sqrt{x} + 4y = 0$

9. Find the constants m and b in the linear function $f(x) = mx + b$ so that $f(0) = 2$ and $f(3) = -1$.

10. Find the constants m and b in the linear function $f(x) = mx + b$ so that $f(2) = 4$ and the straight line represented by f has slope -1.

For the demand equations in Exercises 11–14, where x represents the quantity demanded in units of a thousand and p is the unit price in dollars, (a) sketch the demand curve and (b) determine the quantity demanded when the unit price is set at \$p.

11. $p = -x^2 + 36; p = 11$

12. $p = -x^2 + 16; p = 7$

13. $p = \sqrt{9 - x^2}; p = 2$

14. $p = \sqrt{18 - x^2}; p = 3$

For the supply equations in Exercises 15–18, where x is the quantity supplied in units of a thousand and p is the unit price in dollars, (a) sketch the supply curve and (b) determine the price at which the supplier will make 2000 units of the commodity available in the market.

15. $p = 2x^2 + 18$

16. $p = x^2 + 16x + 40$

17. $p = x^3 + x + 10$

18. $p = x^3 + 2x + 3$

For each pair of supply and demand equations in Exercises 19–22, where x represents the quantity demanded in units of a thousand and p the unit price in dollars, find the equilibrium quantity and the equilibrium price.

19. $p = -2x^2 + 80$ and $p = 15x + 30$

20. $p = -x^2 - 2x + 100$ and $p = 8x + 25$

21. $11p + 3x - 66 = 0$ and $2p^2 + p - x = 10$

22. $p = 60 - 2x^2$ and $p = x^2 + 9x + 30$

23. Enclosing an Area Patricia wishes to have a rectangular-shaped garden on her acreage. She has 80 m of fencing material with which to enclose her garden. Letting x denote the width of the garden, find a function f in the variable x giving the area of the garden. What is its domain?

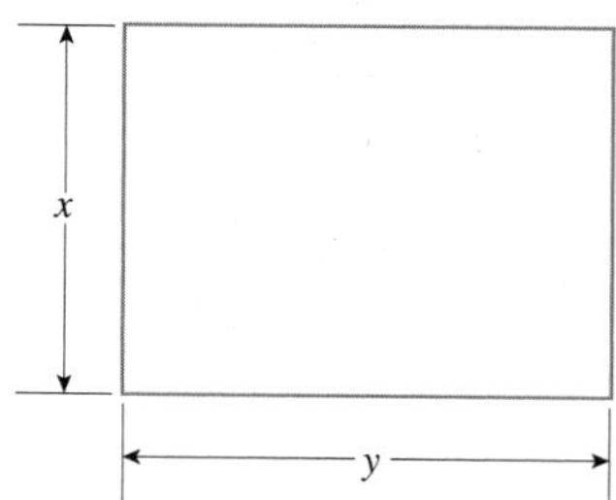

24. Enclosing an Area Patricia's neighbour, Juanita, also wishes to have a rectangular-shaped garden on her acreage. But Juanita wants her garden to have an area of 250 m². Letting x denote the width of the garden, find a function f in the variable x giving the length of the fencing material required to construct the garden. What is the domain of the function?
Hint: Refer to the figure for Exercise 23. The amount of fencing material required is equal to the perimeter of the rectangle, which is twice the width plus twice the length of the rectangle.

25. Packaging By cutting away identical squares from each corner of a rectangular piece of cardboard and folding up the resulting flaps, an open box may be made. If the cardboard is 15 cm long and 8 cm wide and the square cutaways have dimensions of x in. by x cm, find a function giving the volume of the resulting box.

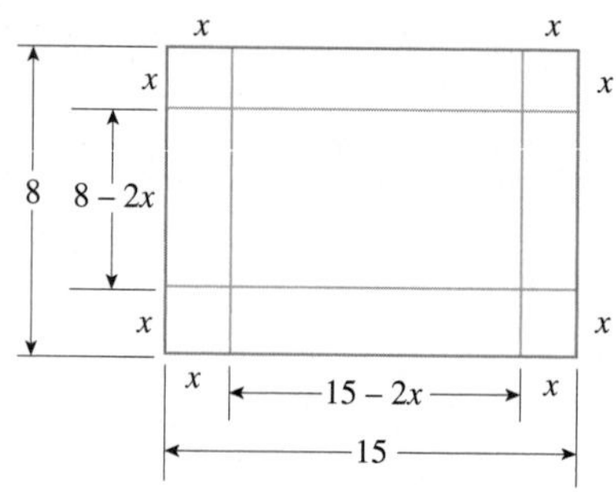

26. Construction Costs A rectangular container is to have a square base and a volume of 20 m^3. The material for the base costs 30¢/m^2, the material for the sides costs 10¢/m^2, and the material for the top costs 20¢/m^2. Letting x denote the length of one side of the base, find a function in the variable x giving the cost of constructing the container.

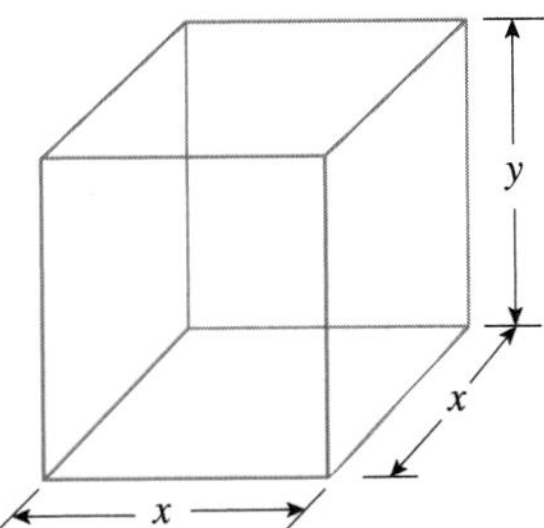

27. Area of a Norman Window A Norman window has the shape of a rectangle surmounted by a semicircle (see the accompanying figure). Suppose a Norman window for a church is to have a perimeter of 28 m; find a function in the variable x giving the area of the window.

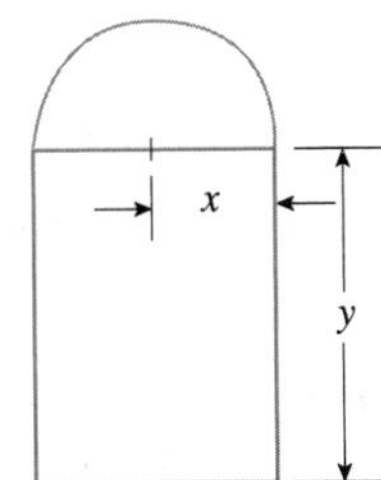

28. Yield of an Apple Orchard An apple orchard has an average yield of 36 Imperial bushels of apples/tree if tree density is 22 trees/acre. For each unit increase in tree density, the yield decreases by 2 Imperial bushels. Letting x denote the number of trees beyond 22/acre, find a function in x that gives the yield of apples.

29. Book Design A book designer has decided that the pages of a book should have 2.5-cm. margins at the top and bottom and 1.25-cm. margins on the sides. She further stipulated that each page should have an area of 125 cm.^{2} Find a function in the variable x, giving the area of the printed page. What is the domain of the function?

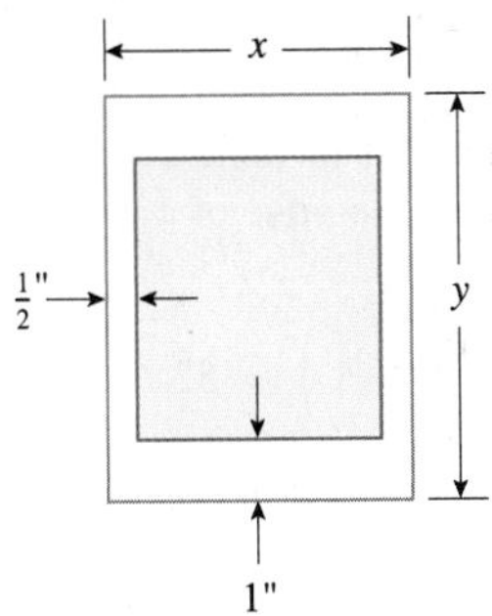

30. Profit of a Vineyard Phillip, the proprietor of a vineyard, estimates that if 10,000 bottles of wine were produced this season, then the profit would be \$5/bottle. But if more than 10,000 bottles were produced, then the profit/bottle would drop by \$0.0002 for each additional bottle sold. Assume at least 10,000 bottles of wine are produced and sold and let x denote the number of bottles produced and sold above 10,000.

a. Find a function P giving the profit in terms of x.
b. What is the profit Phillip can expect from the sale of 16,000 bottles of wine from his vineyard?

31. Charter Revenue The owner of a luxury motor yacht that sails among the 4000 Greek islands charges \$600/person/day if exactly 20 people sign up for the cruise. However, if more than 20 people sign up (up to the maximum capacity of 90) for the cruise, then each fare is reduced by \$4 for each additional passenger. Assume at least 20 people sign up for the cruise and let x denote the number of passengers above 20.

a. Find a function R giving the revenue/day realized from the charter.
b. What is the revenue/day if 60 people sign up for the cruise?
c. What is the revenue/day if 80 people sign up for the cruise?

32. Oil Spills The oil spilling from the ruptured hull of a grounded tanker spreads in all directions in calm waters. Suppose the area polluted is a circle of radius r and the radius is increasing at the rate of 2 m/s.

a. Find a function f giving the area polluted in terms of r.
b. Find a function g giving the radius of the polluted area in terms of t.
c. Find a function h giving the area polluted in terms of t.
d. What is the size of the polluted area 30 s after the hull was ruptured?

33. Inflating a Balloon A spherical balloon is being inflated at a rate of $\frac{9}{2}\pi$ m^3/min.

a. Find a function f giving the radius r of the balloon in terms of its volume V.
Hint: $V = \frac{4}{3}\pi r^3$.
b. Find a function g giving the volume of the balloon in terms of time t.
c. Find a function h giving the radius of the balloon in terms of time.
d. What is the radius of the balloon after 8 min?

Business and Economics Applications

34. Profit–Loss A manufacturer has a monthly fixed cost of \$40,000 and a production cost of \$8 for each unit produced. The product sells for \$12/unit.

a. What is the cost function?
b. What is the revenue function?
c. What is the profit function?
d. Compute the profit (loss) corresponding to production levels of 8000 and 12,000 units.

35. Profit–Loss A manufacturer has a monthly fixed cost of \$100,000 and a production cost of \$14 for each unit produced. The product sells for \$20/unit.

a. What is the cost function?
b. What is the revenue function?
c. What is the profit function?
d. Compute the profit (loss) corresponding to production levels of 12,000 and 20,000 units.

36. Disposable Income Economists define the *disposable annual income* for an individual by the equation $D = (1 - r)T$, where T is the individual's total income and r is the net rate at which he or she is taxed. What is the disposable income for an individual whose income is \$40,000 and whose net tax rate is 28%?

37. Revenue Functions The revenue (in dollars) realized by Apollo from the sale of its inkjet printers is given by

$$R(x) = -0.1x^2 + 500x$$

where x denotes the number of units manufactured each month. What is Apollo's revenue when 1000 units are produced?

38. Effect of Advertising on Sales The quarterly profit of Cunningham Realty depends on the amount of money x spent on advertising/quarter according to the rule

$$P(x) = -\frac{1}{8}x^2 + 7x + 30 \qquad (0 \le x \le 50)$$

where $P(x)$ and x are measured in thousands of dollars. What is Cunningham's profit when its quarterly advertising budget is \$28,000?

39. Document Management The size (measured in millions of dollars) of the document-management business is described by the function

$$f(t) = 0.22t^2 + 1.4t + 3.77 \qquad (0 \le t \le 6)$$

where t is measured in years, with $t = 0$ corresponding to the beginning of 1996.

a. Sketch the graph of f.

b. What was the size of the document-management business at the beginning of 2002?

Source: Sun Trust Equitable Securities

40. Forecasting Sales The annual sales of Crimson Drug Store are expected to be given by

$$S(t) = 2.3 + 0.4t$$

million dollars t yr from now, whereas the annual sales of Cambridge Drug Store are expected to be given by

$$S(t) = 1.2 + 0.6t$$

million dollars t yr from now. When will the annual sales of Cambridge first surpass the annual sales of Crimson?

41. Linear Depreciation In computing income tax, businesses are allowed by law to depreciate certain assets such as buildings, machines, furniture, automobiles, and so on, over a period of time. The linear depreciation, or straight-line method, is often used for this purpose. Suppose an asset has an initial value of \$$C$ and is to be depreciated linearly over n yr with a scrap value of \$$S$. Show that the book value of the asset at any time t $(0 \le t \le n)$ is given by the linear function

$$V(t) = C - \frac{(C - S)}{n}t$$

Hint: Find an equation of the straight line that passes through the points $(0, C)$ and (n, S). Then rewrite the equation in the slope-intercept form.

42. Linear Depreciation Using the linear depreciation model of Exercise 38, find the book value of a printing machine at the end of the second year if its initial value is \$100,000 and it is depreciated linearly over 5 yr with a scrap value of \$30,000.

43. Price of Ivory According to the World Wildlife Fund, a group in the forefront of the fight against illegal ivory trade, the price of ivory (in dollars/kilogram) compiled from a variety of legal and black market sources is approximated by the function

$$f(t) = \begin{cases} 8.37t + 7.44 & \text{if } 0 \le t \le 8 \\ 2.84t + 51.68 & \text{if } 8 < t \le 30 \end{cases}$$

where t is measured in years, with $t = 0$ corresponding to the beginning of 1970.

a. Sketch the graph of the function f.

b. What was the price of ivory at the beginning of 1970? At the beginning of 1990?

44. Sales of Digital TVs The number of Canadian homes with digital TVs is expected to grow according to the function

$$f(t) = 17.14t^2 + 66.57t + 71.43 \qquad (0 \le t \le 6)$$

where t is measured in years, with $t = 0$ corresponding to the beginning of 2000, and $f(t)$ is measured in thousands of homes.

a. How many homes had digital TVs at the beginning of 2000?

b. How many homes will have digital TVs at the beginning of 2005?

45. Credit Card Debt Following the introduction in 1950 of the nation's first credit card, the Diners Club Card, credit cards have proliferated over the years. More than 720 different cards are now used at more than 4 million locations in the United States. The average U.S. credit card debt (per household) in thousands of dollars is approximately given by

$$D(t) = \begin{cases} 4.77(1 + t)^{0.2676} & \text{if } 0 \le t \le 2 \\ 5.6423t^{0.1818} & \text{if } 2 < t \le 6 \end{cases}$$

where t is measured in years, with $t = 0$ corresponding to the beginning of 1994. What was the average U.S. credit card debt (per household) at the beginning of 1994? At the beginning of 1996? At the beginning of 1999?

Source: David Evans and Richard Schmalensee, *Paying with Plastic: The Digital Revolution in Buying and Borrowing*

46. Price of Automobile Parts For years, automobile manufacturers had a monopoly on the replacement-parts market, particularly for sheet metal parts such as fenders, doors, and hoods, the parts most often damaged in a crash. Beginning in the late 1970s, however, competition appeared on the scene. In a report conducted by an insurance company to study the effects of the competition, the price of an OEM (original equipment manufacturer) fender for a particular 1983 model car was found to be

$$f(t) = \frac{110}{\frac{1}{2}t + 1} \qquad (0 \le t \le 2)$$

where $f(t)$ is measured in dollars and t is in years. Over the same period of time, the price of a non-OEM fender for the car was found to be

$$g(t) = 26\left(\frac{1}{4}t^2 - 1\right)^2 + 52 \qquad (0 \le t \le 2)$$

where $g(t)$ is also measured in dollars. Find a function $h(t)$ that gives the difference in price between an OEM fender and a non-OEM fender. Compute $h(0)$, $h(1)$, and $h(2)$. What does the result of your computation seem to say about the price gap between OEM and non-OEM fenders over the 2 yr?

47. Demand for Clock Radios In the accompanying figure, L_1 is the demand curve for the model A clock radios manufactured by Ace Radio, and L_2 is the demand curve for their model B clock radios. Which line has the greater slope? Interpret your results.

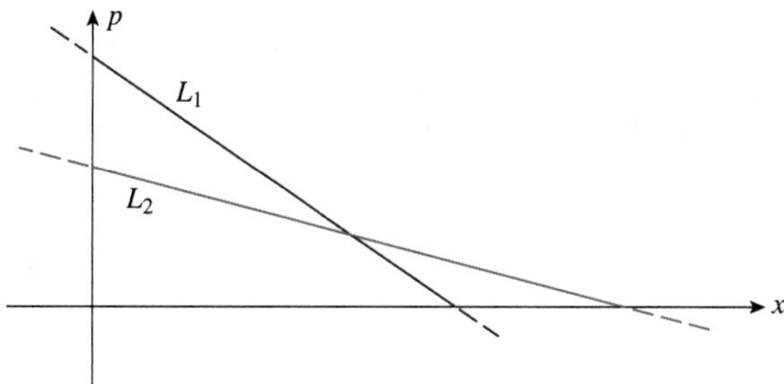

48. Supply of Clock Radios In the accompanying figure, L_1 is the supply curve for the model A clock radios manufactured by Ace Radio, and L_2 is the supply curve for their model B clock radios. Which line has the greater slope? Interpret your results.

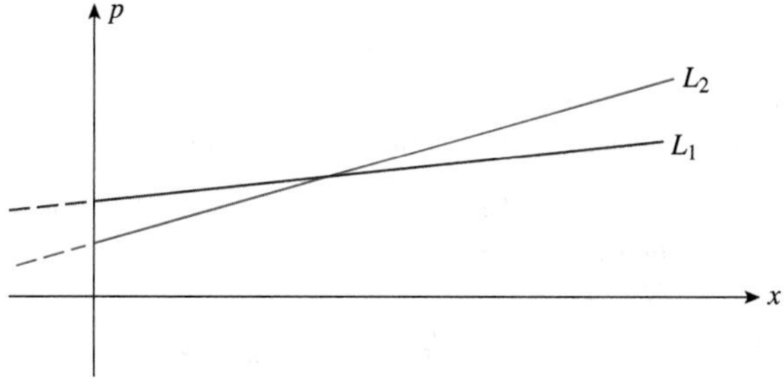

49. Demand for Smoke Alarms The demand function for the Sentinel smoke alarm is given by

$$p = \frac{30}{0.02x^2 + 1} \qquad (0 \le x \le 10)$$

where x (measured in units of a thousand) is the quantity demanded per week and p is the unit price in dollars. Sketch the graph of the demand function. What is the unit price that corresponds to a quantity demanded of 10,000 units?

50. Demand for Commodities Assume that the demand function for a certain commodity has the form

$$p = \sqrt{-ax^2 + b} \qquad (a \ge 0, b \le 0)$$

where x is the quantity demanded, measured in units of a thousand and p is the unit price in dollars. Suppose the quantity demanded is 6000 ($x = 6$) when the unit price is \$8.00 and 8000 ($x = 8$) when the unit price is \$6.00. Determine the demand equation. What is the quantity demanded when the unit price is set at \$7.50?

51. Supply Functions The supply function for the Luminar desk lamp is given by

$$p = 0.1x^2 + 0.5x + 15$$

where x is the quantity supplied (in thousands) and p is the unit price in dollars. Sketch the graph of the supply function. What unit price will induce the supplier to make 5000 lamps available in the marketplace?

52. Supply Functions Suppliers of transistor radios will market 10,000 units when the unit price is \$20 and 62,500 units when the unit price is \$35. Determine the supply function if it is known to have the form

$$p = a\sqrt{x} + b \qquad (a \ge 0, b \ge 0)$$

where x is the quantity supplied and p is the unit price in dollars. Sketch the graph of the supply function. What unit price will induce the supplier to make 40,000 transistor radios available in the marketplace?

53. Suppose the demand and supply equations for a certain commodity are given by $p = ax + b$ and $p = cx + d$, respectively, where $a < 0$, $c > 0$, and $b > d > 0$ (see the accompanying figure).

a. Find the equilibrium quantity and equilibrium price in terms of a, b, c, and d.

b. Use part (a) to determine what happens to the market equilibrium if c is increased, while a, b, and d remain fixed. Interpret your answer in economic terms.

c. Use part (a) to determine what happens to the market equilibrium if b is decreased while a, c, and d remain fixed. Interpret your answer in economic terms.

54. Market Equilibrium The weekly demand and supply functions for Sportsman 5 × 7 tents are given by

$$p = -0.1x^2 - x + 40$$

$$p = 0.1x^2 + 2x + 20$$

respectively, where p is measured in dollars and x is measured in units of a hundred. Find the equilibrium quantity and price.

55. Market Equilibrium The management of Titan Tire Company has determined that the weekly demand and supply functions for their Super Titan tires are given by

$$p = 144 - x^2$$

$$p = 48 + \frac{1}{2}x^2$$

respectively, where p is measured in dollars and x is measured in units of a thousand. Find the equilibrium quantity and price.

Biological and Life Sciences Applications

56. Drug Dosages A method sometimes used by pediatricians to calculate the dosage of medicine for children is based on the child's surface area. If a denotes the adult dosage (in milligrams) and S is the surface area of the child (in square metres), then the child's dosage is given by

$$D(S) = \frac{Sa}{1.7}$$

If the adult dose of a substance is 500 mg, how much should a child whose surface area is 0.4 m^2 receive?

57. Cowling's Rule Cowling's rule is a method for calculating pediatric drug dosages. If a denotes the adult dosage (in milligrams) and t is the age of the child (in years), then the child's dosage is given by

$$D(t) = \left(\frac{t+1}{24}\right)a$$

If the adult dose of a substance is 500 mg, how much should a 4-yr-old child receive?

58. Worker Efficiency An efficiency study showed that the average worker at Delphi Electronics assembled cordless telephones at the rate of

$$f(t) = -\frac{3}{2}t^2 + 6t + 10 \qquad (0 \le t \le 4)$$

phones/hour, t h after starting work during the morning shift. At what rate does the average worker assemble telephones 2 h after starting work?

59. Email Usage The number of international emailings originating from Canada per day (in millions) is approximated by the function

$$f(t) = 3.857t^2 - 2.429t + 7.914 \qquad (0 \leq t \leq 4)$$

where t is measured in years, with $t = 0$ corresponding to the beginning of 1998.

a. Sketch the graph of f.

b. How many international emailings/day were there at the beginning of 2002?

60. Reaction of a Frog to a Drug Experiments conducted by A. J. Clark suggest that the response $R(x)$ of a frog's heart muscle to the injection of x units of acetylcholine (as a percent of the maximum possible effect of the drug) may be approximated by the rational function

$$R(x) = \frac{100x}{b + x} \qquad (x \geq 0)$$

where b is a positive constant that depends on the particular frog.

a. If a concentration of 40 units of acetylcholine produces a response of 50% for a certain frog, find the "response function" for this frog.

b. Using the model found in part (a), find the response of the frog's heart muscle when 60 units of acetylcholine are administered.

61. Cricket Chirping and Temperature Entomologists have discovered that a linear relationship exists between the number of chirps of crickets of a certain species and the air temperature. When the temperature is 21°C, the crickets chirp at the rate of 120 times/minute, and when the temperature is 27°C, they chirp at the rate of 160 times/minute.

a. Find an equation giving the relationship between the air temperature T and the number of chirps/minute, N, of the crickets.

b. Find N as a function of T and use this formula to determine the rate at which the crickets chirp when the temperature is 39°C.

62. Patient's Temperature The response of the patient's temperature to a new febrifuge is found to be linear. If a dose of 100 milligrams lowers the patient's temperature 0.8°C, while a dose of 125 milligrams lowers the patient's temperature 1.1°C, find the line that models the response of the patient's temperature to the drug.

63. Mass of Fish The mass of mussels, clams, and other shellfish, measured in kilograms, in a part of the B.C. shore per square metre on pier pilings and sea walls, is found to depend on the depth d below high tide in metres, according to the rule

$$K(d) = \frac{1.6}{d^2 - 2d + 1.3}$$

Find the mass of shellfish per square metre at depths of 50 cm, 100 cm, and 200 cm below the high tide mark.

64. Number of Species The number of a species of rare bird nesting per hectare in a provincial park is found to be dependent on the noise level. If d is the average number of decibels of noise in an area, then the number of nests per hectare is found to be

$$N(d) + \frac{234}{2.8 + \sqrt{1.97d}}$$

Find the number of nests at noise levels of 5, 10, and 15 decibels and decide if the birds like or dislike noise.

65. Fish Population The size of a population of small fish accidentally introduced into the great lakes is found to be modelled by the rule

$$P(t) = 750\sqrt{t + 1.2} \text{ thousand fish}$$

during the spring months. Assume that $t = 0$ is February. Compute the number of fish present in March ($t = 1$), April ($t = 2$), and May ($t = 3$).

In Exercises 66–69, determine whether the statement is true or false. If it is true, explain why it is true. If it is false, give an example to show why it is false.

66. A polynomial function is a sum of constant multiples of power functions.

67. A polynomial function is a rational function, but the converse is false.

68. If $r > 0$, then the power function $f(x) = x^r$ is defined for all values of x.

69. The function $f(x) = 2^x$ is a power function.

Solutions to Self-Check Exercises 2.8

1. Since the adult dose of the substance is 500 mg, $a = 500$; thus, the rule in this case is

$$D(t) = \frac{500t}{t + 12}$$

A 4-yr-old should receive

$$D(4) = \frac{500(4)}{4 + 12}$$

or 125 mg of the substance.

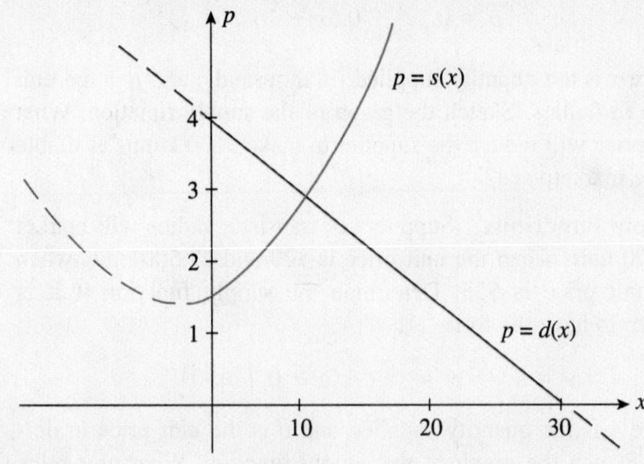

Solutions to Self-Check Exercises 2.8, *continued*

2. **a.** The graphs of the functions d and s are shown in the following figure:

b. Solve the following system of equations:

$$p = -\frac{2}{15}x + 4$$

$$p = \frac{1}{75}x^2 + \frac{1}{10}x + \frac{3}{2}$$

Substituting the first equation into the second yields

$$\frac{1}{75}x^2 + \frac{1}{10}x + \frac{3}{2} = -\frac{2}{15}x + 4$$

$$\frac{1}{75}x^2 + \left(\frac{1}{10} + \frac{2}{15}\right)x - \frac{5}{2} = 0$$

$$\frac{1}{75}x^2 + \frac{7}{30}x - \frac{5}{2} = 0$$

Multiplying both sides of the last equation by 150, we have

$$2x^2 + 35x - 375 = 0$$
$$(2x - 15)(x + 25) = 0$$

Thus, $x = -25$ or $x = 15/2 = 7.5$. Since x must be nonnegative, we take $x = 7.5$, and the equilibrium quantity is 7500 kg. The equilibrium price is given by

$$p = -\frac{2}{15}\left(\frac{15}{2}\right) + 4$$

or \$3/kg.

CHAPTER 2 Summary of Principal Formulas and Terms

Formulas

1. Polynomial function $p(x) = a_nx^n + a_{n-1}x^{n-1} + \dots + a_2x^2 + a_1x + a_0$, $a_n \neq 0$ and $a_0, a_1, \dots, a_n$ real number
2. Quadratic function in general form $p(x) = ax^2 + bx + c$, $a \neq 0$ and a, b, c real number
3. Quadratic function in standard form $p(x) = a(x - h)^2 + k$, $a \neq 0$ and a, h, k real number
4. Rational function $r(x) = \dfrac{p(x)}{q(x)}$, $q(x) \neq 0$ and $p(x), q(x)$ are polynomials
5. Power function $f(x) = x^r$, r real number
6. Absolute value function $f(x) = |x| = \begin{cases} -x & \text{if } x < 0 \\ x & \text{if } x \geq 0 \end{cases}$

Terms

function (45)
domain (45)
range (45)
independent variable (46)
dependent variable (46)
ordered pairs (47)
graph of a function (48)
piecewise defined function (49)
graph of an equation (50)
vertical-line test (50)
transformation (56)
vertical translation (56)
horizontal translation (57)
horizontal reflection (59)
vertical reflection (59)
even function (59)
odd function (59)
stretching (61)
polynomial function (64)
coefficients (64)
leading coefficient (64)
linear function (64)
quadratic function (64)
cubic function (64)
parabola (66)
rational function (69)
vertical asymptote for rational function (70)
horizontal asymptote for rational function (71)
power function (74)
root functions (74)
absolute value function (77)
sum (86)
difference (86)
product (86)
quotient (86)
fixed costs (86)
variable costs (86)
total cost (86)
total profit (87)
composite function (88)
position function (93)
inverse function (94)
graphs of inverse functions (94)
one-to-one (95)
horizontal line test (95)
mathematical modelling (98)

demand equation (101)
demand curve (101)
demand function (101)
supply equation (101)
supply curve (101)
supply function (101)
market equilibrium (101)
equilibrium quantity (101)
equilibrium price (101)

Review Exercises 

In Exercises 1–8, find the domain of each function.

1. $f(x) = x^2 - 4x - 5$
2. $f(x) = (x + 2)^2 + 1$
3. $f(x) = \sqrt{9 - x}$
4. $f(x) = \sqrt{x + 4} - 5$
5. $f(x) = \dfrac{x + 3}{2x^2 - x - 3}$
6. $f(x) = \dfrac{x - 1}{x^2 + 2x - 3}$
7. $f(x) = x^{2/3} - 4$
8. $f(x) = |2x + 5| + 3$

9. Let $f(x) = 3x^2 + 5x - 2$. Find:
   a. $f(-2)$
   b. $f(a + 2)$
   c. $f(2a)$
   d. $f(a + h)$

10. Let $y^2 = 2x + 1$.
   a. Sketch the graph of this equation.
   b. Is y a function of x? Why?
   c. Is x a function of y? Why?

In Exercises 11–24, sketch the graph of each function and include any vertex and any asymptotes, if they exist.

11. $f(x) = \begin{cases} 2 & \text{if } x < -3 \\ -\dfrac{2}{3}x - 4 & \text{if } x \geq -3 \end{cases}$

12. $f(x) = \begin{cases} x + 1 & \text{if } x < 1 \\ x^2 + 4x - 1 & \text{if } x \geq 1 \end{cases}$

13. $f(x) = \dfrac{1}{2}(x - 4)^2 + 1$
14. $f(x) = 3(x + 1)^2 - 2$
15. $f(x) = x^2 - 6x + 3$
16. $f(x) = x^2 + 2x + 5$
17. $f(x) = \dfrac{8x^2 + 3}{2x^2 - 18}$
18. $f(x) = \dfrac{x^2 - 1}{3(x^2 + 5x + 4)}$
19. $f(x) = \dfrac{2}{x - 3}$
20. $f(x) = \dfrac{1}{(x + 2)^2}$
21. $f(x) = \sqrt{x - 3} - 4$
22. $f(x) = 3\sqrt{x} + 2$
23. $f(x) = 2|x + 1| - 2$
24. $f(x) = -|x - 2| + 1$

In Exercises 25–28, write the equation for the graph of each function shown below based on the indicated basic function.

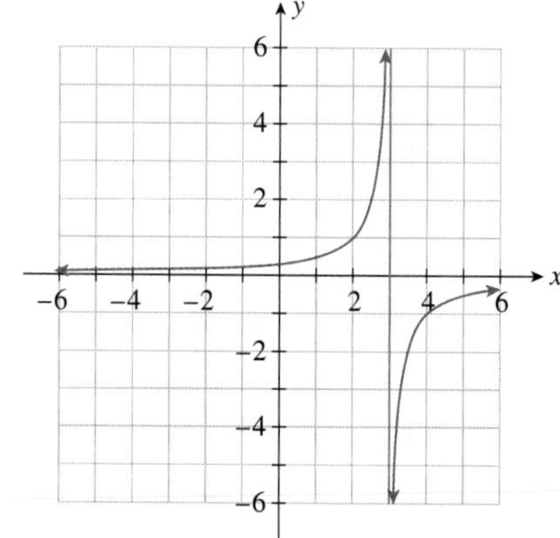

25. $f(x) = x^2$

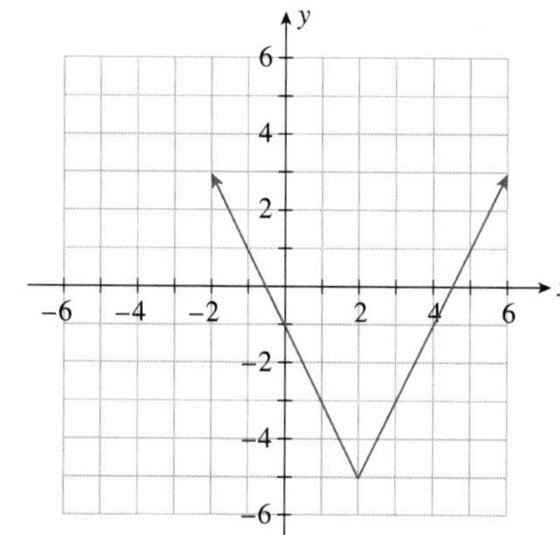

26. $f(x) = \sqrt{x}$

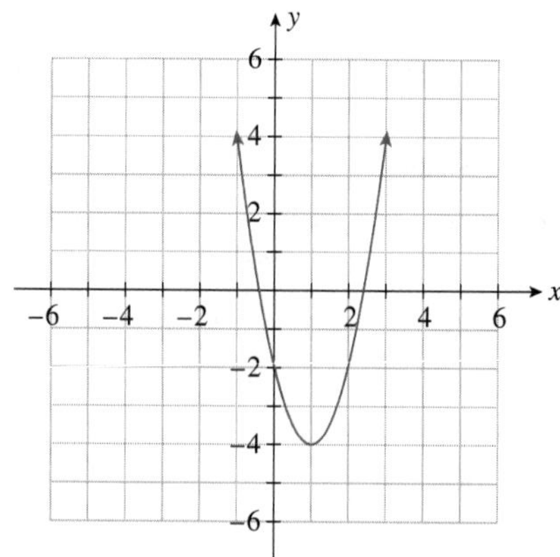

27. $f(x) = \dfrac{1}{x}$

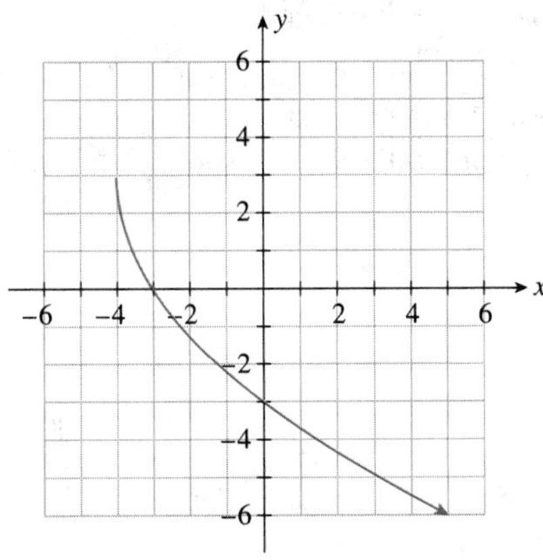

28. $f(x) = |x|$

In Exercises 29–32, let $f(x) = 1/x$ and $g(x) = 2x + 3$. Find:

29. $f(x)g(x)$

30. $f(x)/g(x)$

31. $f(g(x))$

32. $g(f(x))$

In Exercises 33–36, find the inverse for each function.

33. $f(x) = -\frac{2}{3}x + 5$

34. $f(x) = 2x^3 - 1$

35. $f(x) = \sqrt{x+2} + 3$

36. $f(x) = (x - 1)^{3/2}$

37. Find the point of intersection of the two straight lines having the equations $2x + 3y = 18$ and $x - 6y = -6$.

38. Find the point of intersection of the two straight lines having the equations $y = \frac{1}{2}x + 1$ and $y = -\frac{1}{2}x + 5$.

39. Find the point of intersection of the two straight lines having the equations $y = \frac{3}{4}x + 6$ and $3x - 2y + 3 = 0$.

Business and Economics Applications

40. **Sales of Clock Radios** Sales of a certain clock radio are approximated by the relationship $S(x) = 6000x + 30{,}000$ $(0 \leq x \leq 5)$, where $S(x)$ denotes the number of clock radios sold in year x ($x = 0$ corresponds to the year 2000). Find the number of clock radios expected to be sold in 2004.

41. **Sales of a Company** A company's total sales (in millions of dollars) are approximately linear as a function of time (in years). Sales in 1999 were \$2.4 million, whereas sales in 2004 amounted to \$7.4 million.
    a. Find an equation that gives the company's sales as a function of time.
    b. What were the sales in 2002?

42. **Profit Functions** A company has a fixed cost of \$30,000 and a production cost of \$6 for each unit it manufactures. A unit sells for \$10.
    a. What is the cost function?
    b. What is the revenue function?
    c. What is the profit function?
    d. Compute the profit (loss) corresponding to production levels of 6000, 8000, and 12,000 units, respectively.

43. The cost and revenue functions for a certain firm are given by $C(x) = 12x + 20{,}000$ and $R(x) = 20x$, respectively. Find the company's break-even point.

44. **Market Equilibrium** Given the demand equation $3x + p - 40 = 0$ and the supply equation $2x - p + 10 = 0$, where p is the unit price in dollars and x represents the quantity in units of a thousand, determine the equilibrium quantity and the equilibrium price.

45. **Revenue Functions** The monthly revenue R (in hundreds of dollars) realized in the sale of Royal electric shavers is related to the unit price p (in dollars) by the equation

$$R(p) = \frac{1}{2}p^2 + 30p$$

Find the revenue when an electric shaver is priced at \$30.

46. **Market Equilibrium** The monthly demand and supply functions for the Luminar desk lamp are given by

$$p = d(x) = -1.1x^2 + 1.5x + 40$$
$$p = s(x) = 0.1x^2 + 0.5x + 15$$

respectively, where p is measured in dollars and x in units of a thousand. Find the equilibrium quantity and price.

Biological and Life Sciences Applications

47. **Clark's Rule** Clark's rule is a method for calculating pediatric drug dosages based on a child's weight. If a denotes the adult dosage (in milligrams) and if w is the weight of the child (in pounds), then the child's dosage is given by

$$D(w) = \frac{aw}{150}$$

If the adult dose of a substance is 500 mg, how much should a child who weighs 35 lb receive?

48. **Health Club Membership** The membership of the newly opened Venus Health Club is approximated by the function

$$N(x) = 200(4 + x)^{1/2} \qquad (1 \leq x \leq 24)$$

where $N(x)$ denotes the number of members x mo after the club's grand opening. Find $N(0)$ and $N(12)$ and interpret your results.

49. **Thurstone Learning Curve** Psychologist L. L. Thurstone discovered the following model for the relationship between the learning time T and the length of a list n:

$$T = f(n) = An\sqrt{n - b}$$

where A and b are constants that depend on the person and the task. Suppose that, for a certain person and a certain task, $A = 4$ and $b = 4$. Compute $f(4), f(5), \ldots, f(12)$ and use this information to sketch the graph of the function f. Interpret your results.

50. **Blood Flow** Jean Louis Marie Poiseuille was a French physicist and mathematician who first experimentally derived and later formulated the so-called Poiseuille Law. This law describes in general the flow of a liquid through a cylindrical tube such as blood flow through an artery. The rate of flow is given by the equation $v(x) = a(r^2 - x^2)$, where a is some positive constant, r is the radius of the tube, and x is the distance from the centre of the tube with $0 \leq x \leq r$. Graph the rate of flow for $a = 100$ and $r = 1$. Where does the liquid flow the fastest?

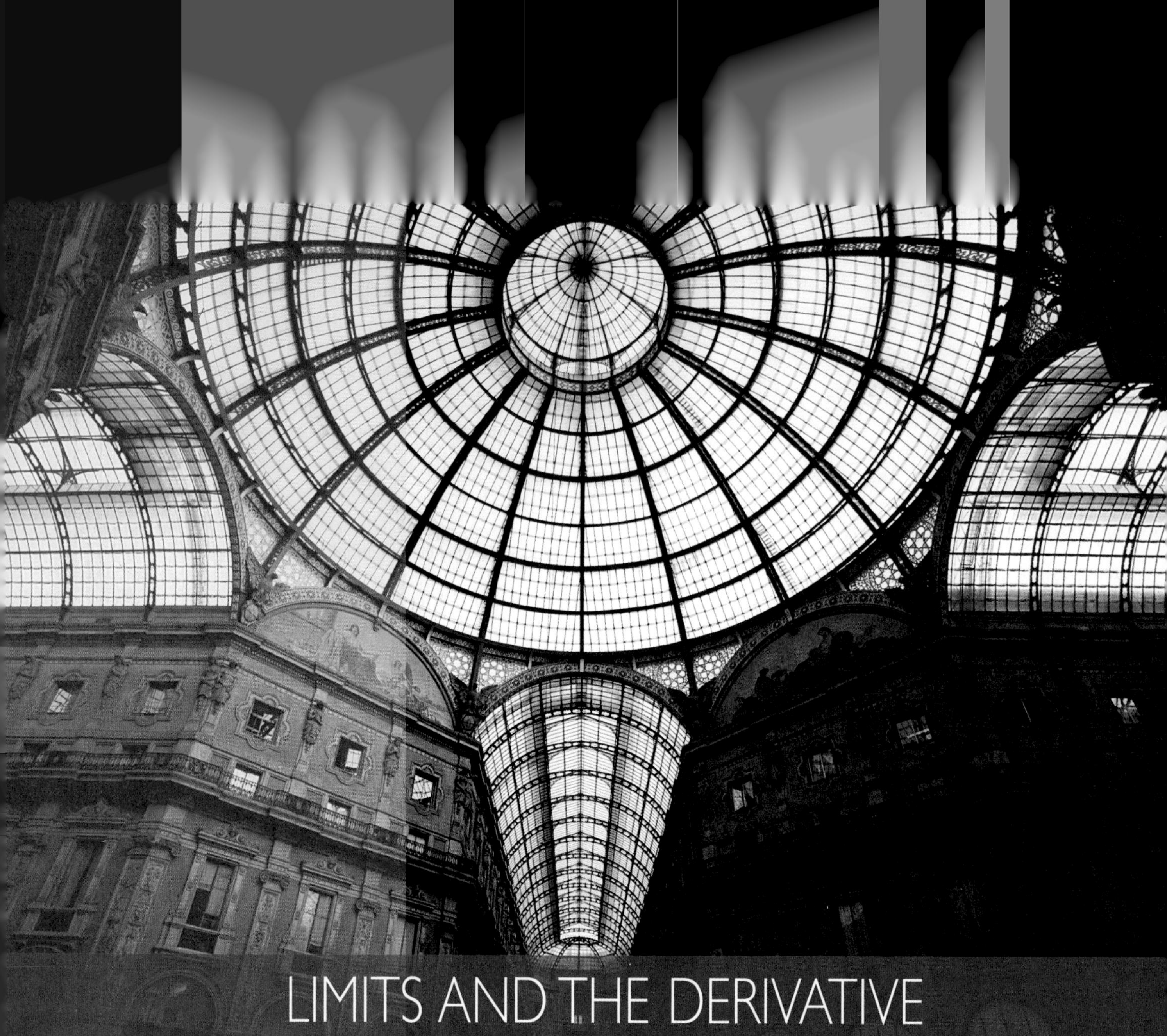

LIMITS AND THE DERIVATIVE

Photo: Thinkstock/Jupiter Images

In this chapter we begin the study of differential calculus. Historically, differential calculus was developed in response to the problem of finding the tangent line to an arbitrary curve. But it quickly became apparent that solving this problem provided mathematicians with a method for solving many practical problems involving the rate of change of one quantity with respect to another. The basic tool used in differential calculus is the *derivative* of a function. The concept of the derivative is based, in turn, on a more fundamental notion—that of the *limit* of a function. This chapter also gives several rules that will greatly simplify the task of finding the derivative of a function, thus enabling us to study how fast one quantity is changing with respect to another in many real-world situations.

Photo: Thinkstock/Jupiter Images

How does the change in the demand for a certain make of tires affect the unit price of the tires? The management of the Titan Tire Company has determined the demand function that relates the unit price of its Super Titan tires to the quantity demanded. In Example 8, page 153, you will see how this function can be used to compute the rate of change of the unit price of the Super Titan tires with respect to the quantity demanded.

Photo: © Rubens Abboud/Alamy

3.1 Limits

Introduction to Calculus

Historically, the development of calculus by Isaac Newton (1642–1727) and Gottfried Wilhelm Leibniz (1646–1716) resulted from the investigation of the following problems:

1. Finding the tangent line to a curve at a given point on the curve (Figure 3.1a)
2. Finding the area of a planar region bounded by an arbitrary curve (Figure 3.1b)

The tangent-line problem might appear to be unrelated to any practical applications of mathematics, but as you will see later, the problem of finding the *rate of change* of one quantity with respect to another is mathematically equivalent to the geometric problem of finding the slope of the *tangent line* to a curve at a given point on the curve. It is precisely the discovery of the relationship between these two problems that spurred the development of calculus in the seventeenth century and made it such an indispensable tool for solving practical problems. The following are a few examples of such problems:

- Finding the velocity of an object
- Finding the rate of change of a bacteria population with respect to time
- Finding the rate of change of a company's profit with respect to time
- Finding the rate of change of a travel agency's revenue with respect to the agency's expenditure for advertising

The study of the tangent-line problem led to the creation of *differential calculus,* which relies on the concept of the *derivative* of a function. The study of the area problem led to the creation of *integral calculus,* which relies on the concept of the *antiderivative,* or *integral,* of a function. (The derivative of a function and the integral of a function are intimately related, as you will see in Section 8.4.) Both the derivative of a function and the integral of a function are defined in terms of a more fundamental concept—the limit—our next topic.

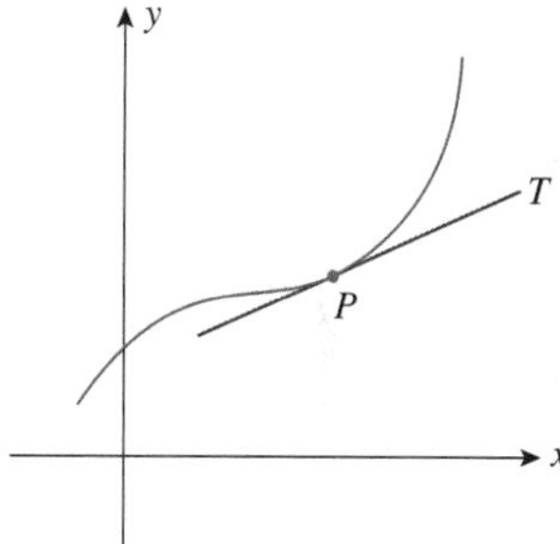

(a) What is the slope of the tangent line T at point P?

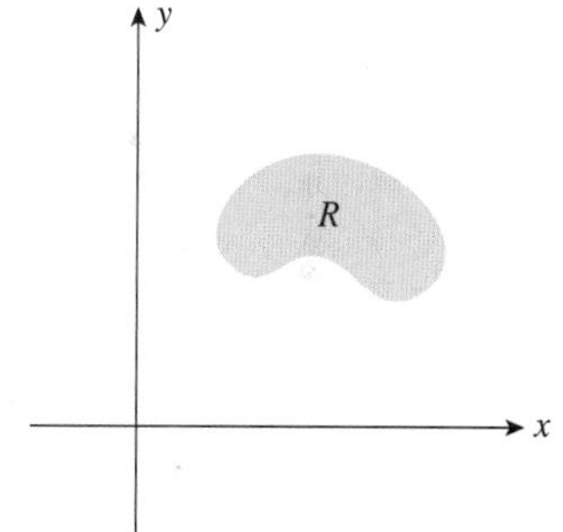

(b) What is the area of the region R?

Figure 3.1

A Real-Life Example

Recall the maglev problem from Section 2.7 about a magnetic levitation train, which moves along a straight monorail track and for which engineers have gathered data about certain positions (in metres) from the origin at time t (in seconds). The relationship between the position of the maglev and time is called the **position function** and given by

$$s = f(t) = t^2 \qquad (0 \le t \le 30) \qquad (1)$$

We can picture the maglev moving along its track (Figure 3.2) and determine its position at time $t = 1, 2, 3, \ldots, 10$, measured from its initial position, using the position function:

$$f(0) = 0, f(1) = 1, f(2) = 4, \ldots, f(10) = 100 \text{ metres}$$

Photo: Mirenska Olga/Shutterstock

Suppose we want to find the velocity of the maglev at $t = 2$. This is just the velocity of the maglev as shown on its speedometer at that precise instant of time. Offhand, calculating this quantity using only Equation (1) appears to be an impossible task; but

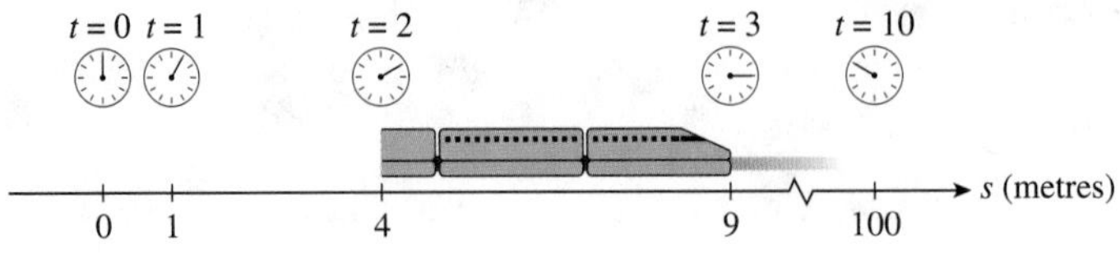

Figure 3.2
A maglev moving along an elevated monorail track

consider what quantities we *can* compute using this relationship. Obviously, we can compute the position of the maglev at any time t as we did earlier for some selected values of t. Using these values, we can then compute the *average velocity* of the maglev over an interval of time. For example, the average velocity of the train over the time interval [2, 4] is given by

$$\begin{aligned}\frac{\text{distance covered}}{\text{time elapsed}} &= \frac{f(4) - f(2)}{4 - 2} \\ &= \frac{(4)^2 - (2)^2}{4 - 2} \\ &= \frac{16 - 4}{4 - 2} \\ &= 6\end{aligned}$$

or 6 metres/second.

Although this is not quite the velocity of the maglev at $t = 2$, it does provide us with an approximation of its velocity at that time.

Can we do better? Intuitively, the smaller the time interval we pick (with $t = 2$ as the left end point), the better the average velocity over that time interval will approximate the actual velocity of the maglev at $t = 2$. In fact, any interval containing $t = 2$ will do.

Now, let's describe this process in general terms. Let $t > 2$. Then, the average velocity of the maglev over the time interval $[2, t]$ is given by

$$\frac{f(t) - f(2)}{t - 2} = \frac{t^2 - 2^2}{t - 2} = \frac{t^2 - 4}{t - 2} \tag{2}$$

By choosing the values of t closer and closer to 2, we obtain a sequence of numbers that give the average velocities of the maglev over smaller and smaller time intervals. As we observed earlier, this sequence of numbers should approach the *instantaneous velocity* of the train at $t = 2$.

Let's try some sample calculations. Using Equation (2) and taking the sequence $t = 2.5$, 2.1, 2.01, 2.001, and 2.0001, which approaches 2, we find the following:

The average velocity over [2, 2.5] is $\dfrac{2.5^2 - 4}{2.5 - 2} = 4.5$, or 4.5 metres/second;

The average velocity over [2, 2.1] is $\dfrac{2.1^2 - 4}{2.1 - 2} = 4.1$, or 4.1 metres/second; and so forth. These results are summarized in Table 3.1.

Table 3.1

	t approaches 2 from the right.				
t	2.5	2.1	2.01	2.001	2.0001
Average Velocity over [2, *t*]	4.5	4.1	4.01	4.001	4.0001
	Average velocity approaches 4 from the right.				

From Table 3.1, we see that the average velocity of the maglev seems to approach the number 4 as it is computed over smaller and smaller time intervals. These computations suggest that the instantaneous velocity of the train at $t = 2$ is 4 metres/second.

Remark

Notice that we cannot obtain the instantaneous velocity for the maglev at $t = 2$ by substituting $t = 2$ into Equation (2) because this value of t is not in the domain of the average velocity function. ◀

Intuitive Definition of a Limit

Consider the function g defined by

$$g(t) = \frac{t^2 - 4}{t - 2}$$

which gives the average velocity of the maglev [see Equation (2)]. Suppose we are required to determine the value that $g(t)$ approaches as t approaches the (fixed) number 2. If we take the sequence of values of t approaching 2 from the right-hand side, as we did earlier, we see that $g(t)$ approaches the number 4. Similarly, if we take a sequence of values of t approaching 2 from the left, such as $t = 1.5, 1.9, 1.99, 1.999$, and 1.9999, we obtain the results shown in Table 3.2.

Table 3.2

	t approaches 2 from the left.				
t	1.5	1.9	1.99	1.999	1.9999
$g(t)$	3.5	3.9	3.99	3.999	3.9999
	Average velocity approaches 4 from the left.				

Observe that $g(t)$ approaches the number 4 as t approaches 2—this time from the left-hand side. In other words, as t approaches 2 from *either* side of 2, $g(t)$ approaches 4. In this situation, we say the limit of $g(t)$ as t approaches 2 is 4, written

$$\lim_{t \to 2} g(t) = \lim_{t \to 2} \frac{t^2 - 4}{t - 2} = 4$$

The graph of the function g, shown in Figure 3.3, confirms this observation.

Observe that $t = 2$ is not in the domain of the function g [for this reason, the point (2, 4) is missing from the graph of g]. This, however, is inconsequential because the value, if any, of $g(t)$ at $t = 2$ plays no role in computing the limit.

This example leads to the following informal definition.

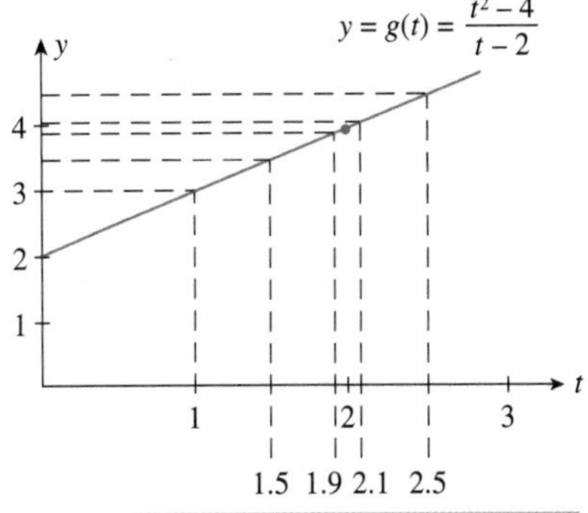

Figure 3.3
As t approaches $t = 2$ from either direction, $g(t)$ approaches $y = 4$.

Limit of a Function

The function f has the **limit** L as x approaches a, written

$$\lim_{x \to a} f(x) = L$$

if the value $f(x)$ can be made as close to the number L as we please by taking x sufficiently close to (but not equal to) a.

EXPLORING WITH TECHNOLOGY

1. Use a graphing utility to plot the graph of

$$g(x) = \frac{x^2 - 4}{x - 2}$$

in the viewing window [0, 3] · [0, 20].

2. Use **ZOOM** and **TRACE** to describe what happens to the values of $g(x)$ as x approaches 2, first from the right and then from the left.
3. What happens to the y-value when you try to evaluate $g(x)$ at $x = 2$? Explain.
4. Reconcile your results with those of the preceding example.

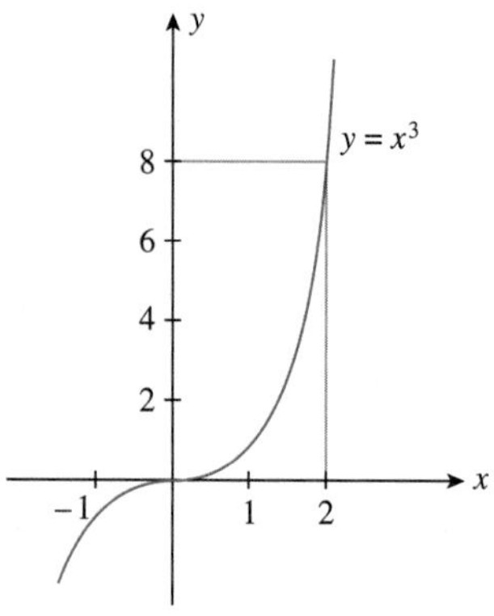

Figure 3.4
$f(x)$ is close to 8 whenever x is close to 2.

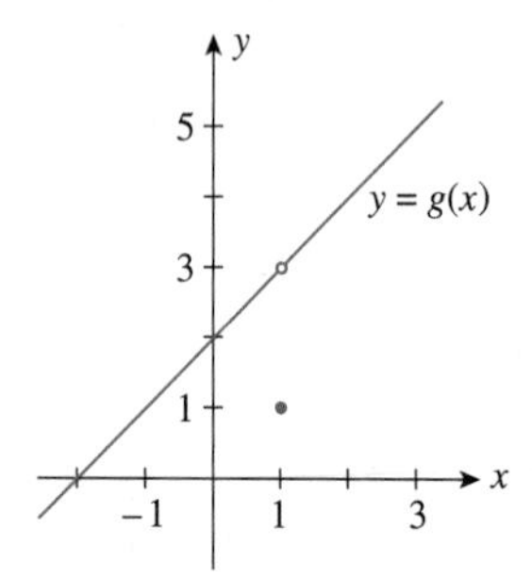

Figure 3.5
$\lim_{x \to 1} g(x) = 3$

Evaluating the Limit of a Function

Let's now consider some examples involving the computation of limits.

Example 1

Let $f(x) = x^3$ and evaluate $\lim_{x \to 2} f(x)$.

Solution

The graph of f is shown in Figure 3.4. You can see that $f(x)$ can be made as close to the number 8 as we please by taking x sufficiently close to 2. Therefore,

$$\lim_{x \to 2} x^3 = 8$$

Example 2

Let

$$g(x) = \begin{cases} x + 2 & \text{if } x \neq 1 \\ 1 & \text{if } x = 1 \end{cases}$$

Evaluate $\lim_{x \to 1} g(x)$.

Solution

The domain of g is the set of all real numbers. From the graph of g shown in Figure 3.5, we see that $g(x)$ can be made as close to 3 as we please by taking x sufficiently close to 1. Therefore,

$$\lim_{x \to 1} g(x) = 3$$

Observe that $g(1) = 1$, which is not equal to the limit of the function g as x approaches 1. [Once again, the value of $g(x)$ at $x = 1$ has no bearing on the existence or value of the limit of g as x approaches 1.]

Example 3

Evaluate the limit of the following functions as x approaches the indicated point.

a. $f(x) = \begin{cases} -1 & \text{if } x < 0 \\ 1 & \text{if } x \geq 0 \end{cases}$; $x = 0$ **b.** $g(x) = \frac{1}{x^2}$; $x = 0$

Solution

The graphs of the functions f and g are shown in Figure 3.6.

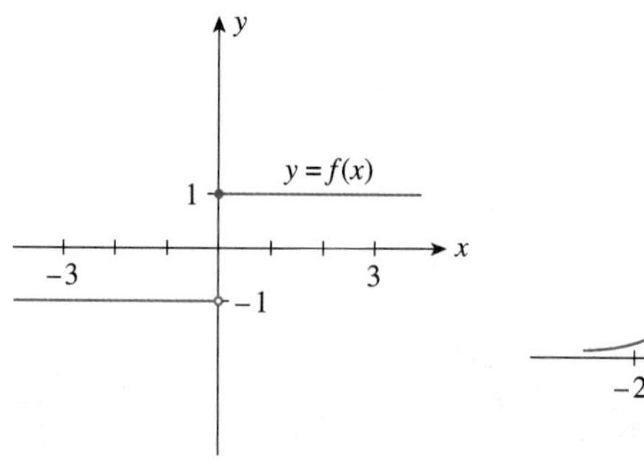

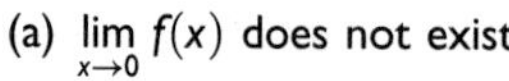

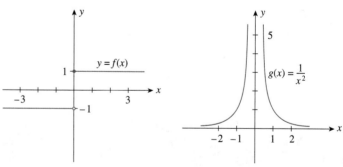

(a) $\lim_{x \to 0} f(x)$ does not exist. (b) $\lim_{x \to 0} g(x)$ does not exist.

Figure 3.6

a. Referring to Figure 3.6a, we see that no matter how close x is to $x = 0$, $f(x)$ takes on the values 1 or -1, depending on whether x is positive or negative. Thus, there is no *single* real number L that $f(x)$ approaches as x approaches zero. We conclude that the limit of $f(x)$ does *not* exist as x approaches zero.

b. Referring to Figure 3.6b, we see that as x approaches $x = 0$ (from either side), $g(x)$ increases without bound and thus does not approach any specific real number. We conclude, accordingly, that the limit of $g(x)$ does *not* exist as x approaches zero. The function g is an example of an **unbounded function.**

GROUP DISCUSSION

Consider the graph of the function h whose graph is depicted in the following figure.

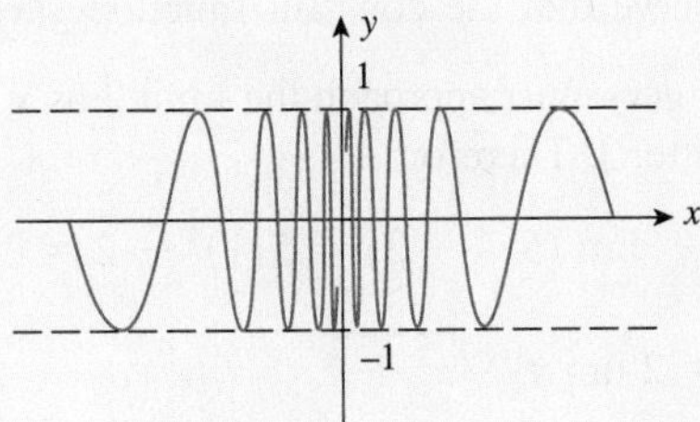

It has the property that as x approaches zero from either the right or the left, the curve oscillates more and more frequently between the lines $y = -1$ and $y = 1$.

1. Explain why $\lim_{x \to 0} h(x)$ does not exist.
2. Compare this function with those in Example 3. More specifically, discuss the different ways the functions fail to have a limit at $x = 0$.

Until now, we have relied on knowing the actual values of a function or the graph of a function near $x = a$ to help us evaluate the limit of the function $f(x)$ as x approaches a. The following properties of limits, which we list without proof, enable us to evaluate limits of functions algebraically.

THEOREM 1

Properties of Limits

Suppose

$$\lim_{x \to a} f(x) = L \quad \text{and} \quad \lim_{x \to a} g(x) = M$$

Then,

1. $\lim_{x \to a} [f(x)]^r = [\lim_{x \to a} f(x)]^r = L^r$ $\quad$ r, a real number
2. $\lim_{x \to a} cf(x) = c \lim_{x \to a} f(x) = cL$ $\quad$ c, a real number

3a. $\lim_{x \to a} [f(x) + g(x)] = \lim_{x \to a} f(x) + \lim_{x \to a} g(x) = L + M$

3b. $\lim_{x \to a} [f(x) - g(x)] = \lim_{x \to a} f(x) - \lim_{x \to a} g(x) = L - M$

4. $\lim_{x \to a} [f(x)g(x)] = [\lim_{x \to a} f(x)][\lim_{x \to a} g(x)] = LM$
5. $\lim_{x \to a} \dfrac{f(x)}{g(x)} = \dfrac{\lim_{x \to a} f(x)}{\lim_{x \to a} g(x)} = \dfrac{L}{M}$ $\quad$ Provided that $M \neq 0$

Example 4

Use Theorem 1 to evaluate the following limits.

a. $\lim_{x \to 2} x^3$ $\quad$ **b.** $\lim_{x \to 4} 5x^{3/2}$ $\quad$ **c.** $\lim_{x \to 1} (5x^4 - 2)$

d. $\lim_{x \to 3} 2x^3 \sqrt{x^2 + 7}$ $\quad$ **e.** $\lim_{x \to 2} \dfrac{2x^2 + 1}{x + 1}$

Solution

a. $\lim_{x\to 2} x^3 = [\lim_{x\to 2} x]^3$ Property 1

$= 2^3 = 8$ $\lim_{x\to 2} x = 2$

b. $\lim_{x\to 4} 5x^{3/2} = 5[\lim_{x\to 4} x^{3/2}]$ Property 2

$= 5(4)^{3/2} = 40$

c. $\lim_{x\to 1}(5x^4 - 2) = \lim_{x\to 1} 5x^4 - \lim_{x\to 1} 2$ Property 3

To evaluate $\lim_{x\to 1} 2$, observe that the constant function $g(x) = 2$ has value 2 for all values of x. Therefore, $g(x)$ must approach the limit 2 as x approaches $x = 1$ (or any other point for that matter!). Therefore,

$$\lim_{x\to 1} (5x^4 - 2) = 5(1)^4 - 2 = 3$$

d. $\lim_{x\to 3} 2x^3\sqrt{x^2 + 7} = 2 \lim_{x\to 3} x^3\sqrt{x^2 + 7}$ Property 2

$= 2 \lim_{x\to 3} x^3 \lim_{x\to 3} \sqrt{x^2 + 7}$ Property 4

$= 2(3)^3\sqrt{3^2 + 7}$ Property 1

$= 2(27)\sqrt{16} = 216$

e. $\lim_{x\to 2} \dfrac{2x^2 + 1}{x + 1} = \dfrac{\lim_{x\to 2} (2x^2 + 1)}{\lim_{x\to 2} (x + 1)}$ Property 5

$= \dfrac{2(2)^2 + 1}{2 + 1} = \dfrac{9}{3} = 3$

Indeterminate Forms

Let's emphasize once again that Property 5 of limits is valid only when the limit of the function that appears in the denominator is not equal to zero at the point in question.

If the numerator has a limit different from zero and the denominator has a limit equal to zero, then the limit of the quotient does not exist at the point in question. This is the case with the function $g(x) = 1/x^2$ in Example 3b. Here, as x approaches zero, the numerator approaches 1 but the denominator approaches zero, so the quotient becomes arbitrarily large. Thus, as observed earlier, the limit does not exist.

Next, consider

$$\lim_{x\to 2} \frac{4(x^2 - 4)}{x - 2}$$

which we evaluated earlier by looking at the values of the function for x near $x = 2$. If we attempt to evaluate this expression by applying Property 5 of limits, we see that both the numerator and denominator of the function

$$\frac{4(x^2 - 4)}{x - 2}$$

approach zero as x approaches 2; that is, we obtain an expression of the form 0/0. In this event, we say that the limit of the quotient $f(x)/g(x)$ as x approaches 2 has the **indeterminate form 0/0.**

We need to evaluate limits of this type when we discuss the derivative of a function, a fundamental concept in the study of calculus. As the name suggests, the meaningless expression 0/0 does not provide us with a solution to our problem. One strategy that can be used to solve this type of problem follows.

Strategy for Evaluating Indeterminate Forms

1. Replace the given function with an appropriate one that takes on the same values as the original function everywhere except at $x = a$.
2. Evaluate the limit of this function as x approaches a.

Examples 5 and 6 illustrate this strategy.

Example 5

Evaluate:

$$\lim_{x\to 2} \frac{4(x^2 - 4)}{x - 2}$$

Solution

Since both the numerator and the denominator of this expression approach zero as x approaches 2, we have the indeterminate form 0/0. We rewrite

$$\frac{4(x^2 - 4)}{x - 2} = \frac{4(x - 2)(x + 2)}{(x - 2)}$$

which, upon cancelling the common factors, is equivalent to $4(x + 2)$, provided $x \neq 2$. Next, we replace $4(x^2 - 4)/(x - 2)$ with $4(x + 2)$ and find that

$$\lim_{x\to 2} \frac{4(x^2 - 4)}{x - 2} = \lim_{x\to 2} 4(x + 2) = 16$$

The graphs of the functions

$$f(x) = \frac{4(x^2 - 4)}{x - 2} \qquad \text{and} \qquad g(x) = 4(x + 2)$$

are shown in Figure 3.7. Observe that the graphs are identical except when $x = 2$. The function g is defined for all values of x and, in particular, its value at $x = 2$ is $g(2) = 4(2 + 2) = 16$. Thus, the point (2, 16) is on the graph of g. However, the function f is not defined at $x = 2$. Since $f(x) = g(x)$ for all values of x except $x = 2$, it follows that the graph of f must look exactly like the graph of g, with the exception that the point (2, 16) is missing from the graph of f. This illustrates graphically why we can evaluate the limit of f by evaluating the limit of the "equivalent" function g.

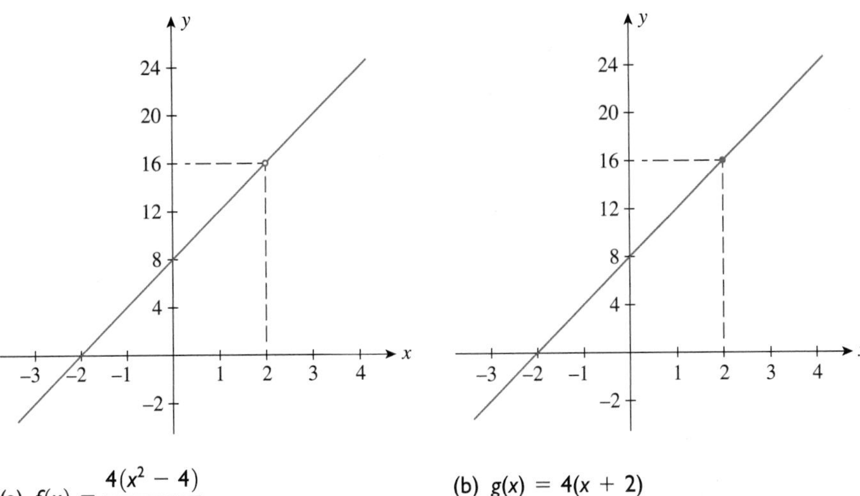

Figure 3.7
The graphs of $f(x)$ and $g(x)$ are identical except at the point (2, 16).

Be aware that

$$\frac{4(x-2)(x+2)}{x-2} \neq 4(x+2)$$

since the domain on the left side is the set of all real numbers *except 2,* while the domain on the right side is the set of all real numbers (see Section 1.2).

EXPLORING WITH TECHNOLOGY

1. Use a graphing utility to plot the graph of

$$f(x) = \frac{4(x^2-4)}{x-2}$$

in the viewing window [0, 3] · [0, 20]. Then use ZOOM and TRACE to find

$$\lim_{x\to 2} \frac{4(x^2-4)}{x-2}$$

2. Use a graphing utility to plot the graph of $g(x) = 4(x+2)$ in the viewing window [0, 3] · [0, 20]. Then use **ZOOM** and **TRACE** to find $\lim_{x\to 2} 4(x+2)$. What happens to the y-value when you try to evaluate $f(x)$ at $x = 2$? Explain.
3. Can you distinguish between the graphs of f and g?
4. Reconcile your results with those of Example 5.

Example 6

Evaluate:

$$\lim_{h\to 0} \frac{\sqrt{1+h}-1}{h}$$

Solution

Letting h approach zero, we obtain the indeterminate form 0/0. Next, we rationalize the numerator of the quotient (see page xx) by multiplying both the numerator and the denominator by the expression $(\sqrt{1+h}+1)$, obtaining

$$\frac{\sqrt{1+h}-1}{h} = \frac{(\sqrt{1+h}-1)(\sqrt{1+h}+1)}{h(\sqrt{1+h}+1)}$$

$$= \frac{1+h-1}{h(\sqrt{1+h}+1)} \qquad (\sqrt{a}-\sqrt{b})(\sqrt{a}+\sqrt{b}) = a-b$$

$$= \frac{h}{h(\sqrt{1-h}+1)}$$

$$= \frac{1}{\sqrt{1+h}+1}$$

Therefore,

$$\lim_{h\to 0} \frac{\sqrt{1+h}-1}{h} = \lim_{h\to 0} \frac{1}{\sqrt{1+h}+1}$$

$$= \frac{1}{\sqrt{1}+1} = \frac{1}{2}$$

EXPLORING WITH TECHNOLOGY

1. Use a graphing utility to plot the graph of $g(x) = \dfrac{\sqrt{1+x}-1}{x}$ in the viewing window [−1, 2] · [0, 1].

   Then use **ZOOM** and **TRACE** to find $\lim\limits_{x\to 0} \dfrac{\sqrt{1+x}-1}{x}$ by observing the values of $g(x)$ as x approaches zero from the left and from the right.

2. Use a graphing utility to plot the graph of $f(x) = \dfrac{1}{\sqrt{1+x}+1}$ in the viewing window [−1, 2] · [0, 1].

   Then use **ZOOM** and **TRACE** to find $\lim\limits_{x\to 0} \dfrac{1}{\sqrt{1+x}+1}$.

   What happens to the y-value when x takes on the value zero? Explain.

3. Can you distinguish between the graphs of f and g?

4. Reconcile your results with those of Example 6.

Limits at Infinity

Up to now we have studied the limit of a function as x approaches a (finite) number a. There are occasions, however, when we want to know whether $f(x)$ approaches a unique number as x increases without bound. Consider, for example, the function P, giving the number of fruit flies (*Drosophila*) in a container under controlled laboratory conditions, as a function of a time t. The graph of P is shown in Figure 3.8. You can see from the graph of P that, as t increases without bound (gets larger and larger), $P(t)$ approaches the number 400. This number, called the *carrying capacity* of the environment, is determined by the amount of living space and food available, as well as other environmental factors.

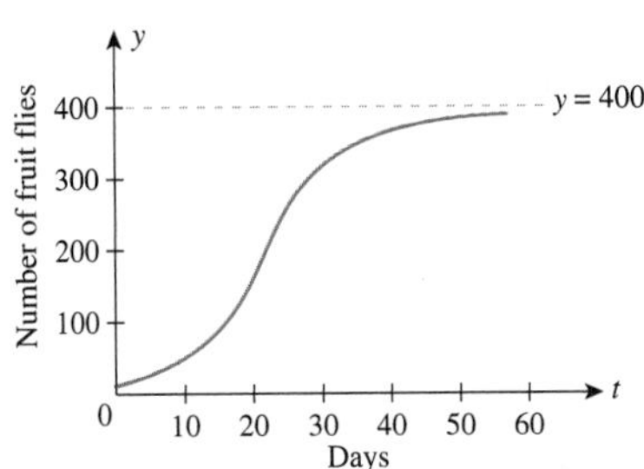

Figure 3.8
The graph of $P(t)$ gives the population of fruit flies in a laboratory experiment.

As another example, suppose we are given the function

$$f(x) = \frac{2x^2}{1+x^2}$$

and we want to determine what happens to $f(x)$ as x gets larger and larger. Picking the sequence of numbers 1, 2, 5, 10, 100, and 1000 and computing the corresponding values of $f(x)$, we obtain the following table of values:

x	1	2	5	10	100	1000
$f(x)$	1	1.6	1.92	1.98	1.9998	1.999998

From the table, we see that as x gets larger and larger, $f(x)$ gets closer and closer to 2. The graph of the function f shown in Figure 3.9 confirms this observation. We call the line $y = 2$ a *horizontal asymptote*.* In this situation we say that the limit of the function

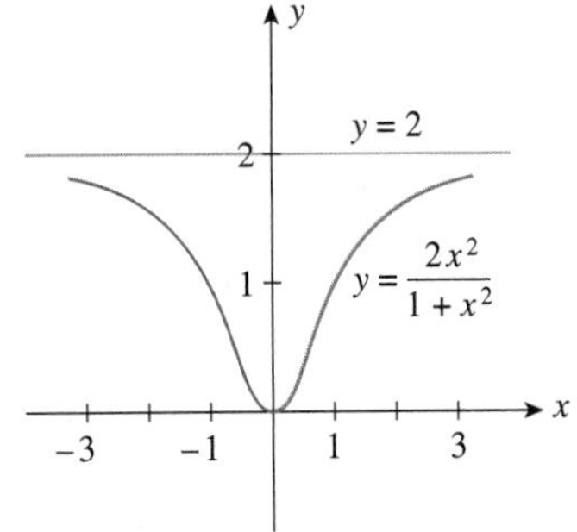

Figure 3.9
The graph of $y = \dfrac{2x^2}{1+x^2}$ has a horizontal asymptote at $y = 2$.

$$f(x) = \frac{2^2}{1+x^2}$$

as x increases without bound is 2, written

$$\lim_{x\to\infty} \frac{2x^2}{1+x^2} = 2$$

In the general case, the following definition for a **limit of a function at infinity** is applicable.

*We will discuss asymptotes in greater detail in Section 7.3.

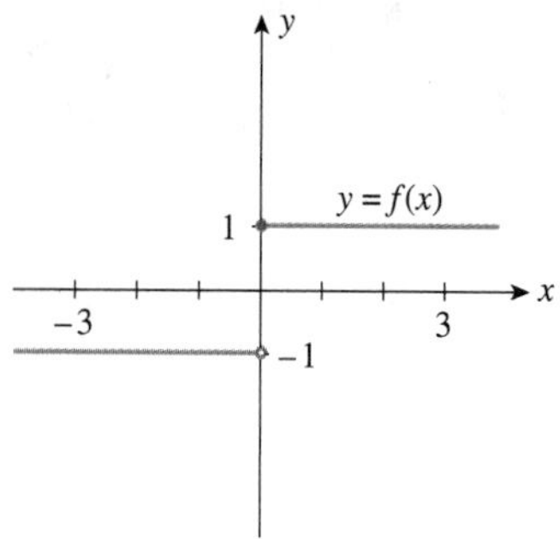

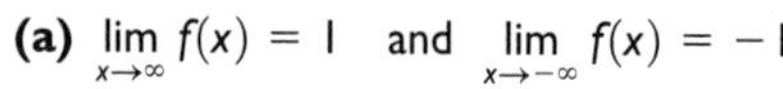

(a) $\lim_{x \to \infty} f(x) = 1$ and $\lim_{x \to -\infty} f(x) = -1$

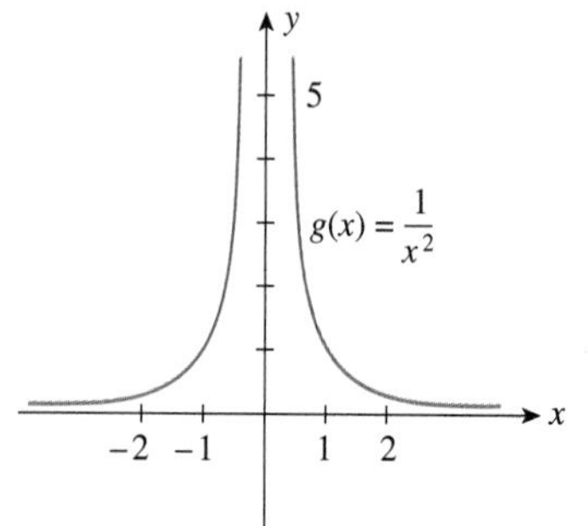

(b) $\lim_{x \to \infty} g(x) = 0$ and $\lim_{x \to -\infty} g(x) = 0$

Figure 3.10

Limit of a Function at Infinity

The function f has the limit L as x increases without bound (or, as x approaches infinity), written

$$\lim_{x \to \infty} f(x) = L$$

if $f(x)$ can be made arbitrarily close to L by taking x large enough.

Similarly, the function f has the limit M as x decreases without bound (or as x approaches negative infinity), written

$$\lim_{x \to -\infty} f(x) = M$$

if $f(x)$ can be made arbitrarily close to M by taking x to be negative and sufficiently large in absolute value.

Example 7

Let f and g be the functions

$$f(x) = \begin{cases} -1 & \text{if } x < 0 \\ 1 & \text{if } x \geq 0 \end{cases} \quad \text{and} \quad g(x) = \frac{1}{x^2}$$

Evaluate:

a. $\lim_{x \to \infty} f(x)$ and $\lim_{x \to -\infty} f(x)$ **b.** $\lim_{x \to \infty} g(x)$ and $\lim_{x \to -\infty} g(x)$

Solution

The graphs of $f(x)$ and $g(x)$ are shown in Figure 3.10. Referring to the graphs of the respective functions, we see that

a. $\lim_{x \to \infty} f(x) = 1$ and $\lim_{x \to -\infty} f(x) = -1$

b. $\lim_{x \to \infty} \frac{1}{x^2} = 0$ and $\lim_{x \to -\infty} \frac{1}{x^2} = 0$

All the properties of limits listed in Theorem 1 are valid when a is replaced by ∞ or $-\infty$. In addition, we have the following property for the limit at infinity.

THEOREM 2

For all $n > 0$,

$$\lim_{x \to \infty} \frac{1}{x^n} = 0 \quad \text{and} \quad \lim_{x \to -\infty} \frac{1}{x^n} = 0$$

provided that $\frac{1}{x^n}$ is defined.

EXPLORING WITH TECHNOLOGY

1. Use a graphing utility to plot the graphs of

$$y_1 = \frac{1}{x^{0.5}}, \quad y_2 = \frac{1}{x}, \quad y_3 = \frac{1}{x^{1.5}}$$

in the viewing window [0, 200] · [0, 0.5]. What can you say about $\lim_{x \to \infty} \frac{1}{x^n}$ if $n = 0.5$, $n = 1$, and $n = 1.5$? Are these results predicted by Theorem 2?

2. Use a graphing utility to plot the graphs of

$$y_1 = \frac{1}{x} \quad \text{and} \quad y_2 = \frac{1}{x^{5/3}}$$

in the viewing window $[-50, 0] \cdot [-0.5, 0]$. What can you say about $\lim_{x \to -\infty} \frac{1}{x^n}$ if $n = 1$ and $n = \frac{5}{3}$? Are these results predicted by Theorem 2?

Hint: To graph y_2, write it in the form $y_2 = 1/(x\text{^}(1/3))\text{^}5$.

In evaluating the limit at infinity of a rational function, the following technique is often used: *Divide the numerator and denominator of the expression by x^n, where n is the highest power present in the denominator of the expression.*

Example 8

Evaluate

$$\lim_{x \to \infty} \frac{x^2 - x + 3}{2x^3 + 1}$$

Solution

Since the limits of both the numerator and the denominator do not exist as x approaches infinity, the property pertaining to the limit of a quotient (Property 5) is not applicable. Let's divide the numerator and denominator of the rational expression by x^3, obtaining

$$\lim_{x \to \infty} \frac{x^2 - x + 3}{2x^3 + 1} = \lim_{x \to \infty} \frac{\frac{1}{x} - \frac{1}{x^2} + \frac{3}{x^3}}{2 + \frac{1}{x^3}}$$

$$= \frac{0 - 0 + 0}{2 + 0} = \frac{0}{2} \qquad \text{Use Theorem 2.}$$

$$= 0$$

Example 9

Let

$$f(x) = \frac{3x^2 + 8x - 4}{2x^2 + 4x - 5}$$

Compute $\lim_{x \to \infty} f(x)$ if it exists.

Solution

Again, we see that Property 5 is not applicable. Dividing the numerator and the denominator by x^2, we obtain

$$\lim_{x \to \infty} \frac{3x^2 + 8x - 4}{2x^2 + 4x - 5} = \lim_{x \to \infty} \frac{3 + \frac{8}{x} - \frac{4}{x^2}}{2 + \frac{4}{x} - \frac{5}{x^2}}$$

$$= \frac{\lim_{x \to \infty} 3 + 8 \lim_{x \to \infty} \frac{1}{x} - 4 \lim_{x \to \infty} \frac{1}{x^2}}{\lim_{x \to \infty} 2 + 4 \lim_{x \to \infty} \frac{1}{x} - 5 \lim_{x \to \infty} \frac{1}{x^2}}$$

$$= \frac{3 + 0 - 0}{2 + 0 - 0} \qquad \text{Use Theorem 2.}$$

$$= \frac{3}{2}$$

Example 10

Let $f(x) = \dfrac{2x^3 - 3x^2 + 1}{x^2 + 2x + 4}$ and evaluate:

a. $\lim_{x\to\infty} f(x)$

b. $\lim_{x\to-\infty} f(x)$

Solution

a. Dividing the numerator and the denominator of the rational expression by x^2, we obtain

$$\lim_{x\to\infty} \frac{2x^3 - 3x^2 + 1}{x^2 + 2x + 4} = \lim_{x\to\infty} \frac{2x - 3 + \frac{1}{x^2}}{1 + \frac{2}{x} + \frac{4}{x^2}}$$

Since the numerator becomes arbitrarily large whereas the denominator approaches 1 as x approaches infinity, we see that the quotient $f(x)$ gets larger and larger as x approaches infinity. In other words, the limit does not exist. We indicate this by writing

$$\lim_{x\to\infty} \frac{2x^3 - 3x^2 + 1}{x^2 + 2x + 4} = \infty$$

b. Once again, dividing both the numerator and the denominator by x^2, we obtain

$$\lim_{x\to-\infty} \frac{2x^3 - 3x^2 + 1}{x^2 + 2x + 4} = \lim_{x\to-\infty} \frac{2x - 3 + \frac{1}{x^2}}{1 + \frac{2}{x} + \frac{4}{x^2}}$$

In this case the numerator becomes arbitrarily large in magnitude but negative in sign, whereas the denominator approaches 1 as x approaches negative infinity. Therefore, the quotient $f(x)$ decreases without bound, and the limit does not exist. We indicate this by writing

$$\lim_{x\to-\infty} \frac{2x^3 - 3x^2 + 1}{x^2 + 2x + 4} = -\infty$$

Example 11 gives an application of the concept of the limit of a function at infinity.

Example 11

Average Cost Functions Custom Office makes a line of executive desks. It is estimated that the total cost of making x Senior Executive Model desks is $C(x) = 100x + 200{,}000$ dollars per year, so the average cost of making x desks is given by

$$\overline{C}(x) = \frac{C(x)}{x} = \frac{100x + 200{,}000}{x} = 100 + \frac{200{,}000}{x}$$

dollars per desk. Evaluate $\lim_{x\to\infty} \overline{C}(x)$ and interpret your results.

Solution

$$\begin{aligned}\lim_{x\to\infty} \overline{C}(x) &= \lim_{x\to\infty} \left(100 + \frac{200{,}000}{x}\right) \\ &= \lim_{x\to\infty} 100 + \lim_{x\to\infty} \frac{200{,}000}{x} = 100\end{aligned}$$

A sketch of the graph of the function $\overline{C}(x)$ appears in Figure 3.11. The result we obtained is fully expected if we consider its economic implications. Note that as the level of production increases, the fixed cost per desk produced, represented by the term (200,000/x), drops steadily. The average cost should approach a constant unit cost of production—$100 in this case.

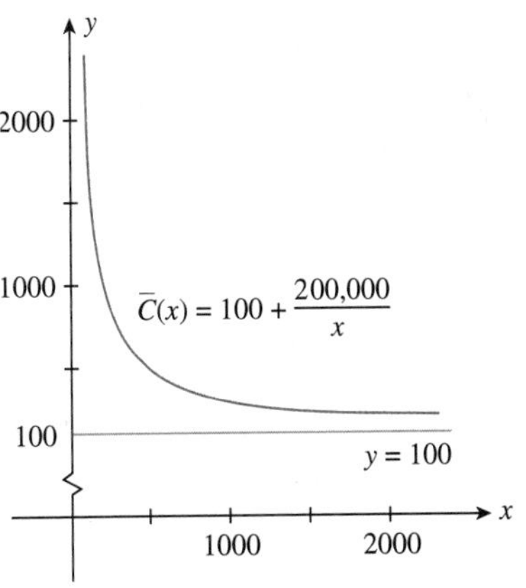

Figure 3.11

As the level of production increases, the average cost approaches $100 per desk.

Example 12

Oxygen Content of a Pond When organic waste is dumped into a pond, the oxidation process that takes place reduces the pond's oxygen content. However, given time, nature will restore the oxygen content to its natural level. Suppose the oxygen content t days after the organic waste has been dumped into the pond is given by

$$f(t) = 100\left(\frac{t^2 + 10t + 100}{t^2 + 20t + 100}\right)$$

percent of its normal level. The graph of f is shown in Figure 3.12.

a. What can you say about $f(t)$ when t is very large?

b. Verify your observation in part (a) by evaluating $\lim_{t\to\infty} f(t)$.

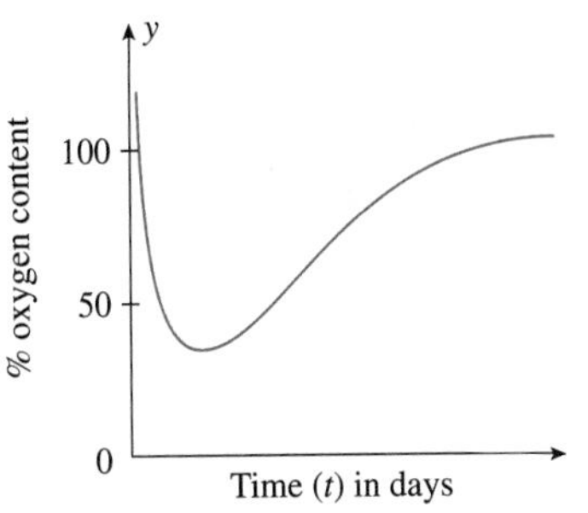

Figure 3.12

The graph of

$$f(t) = 100\left(\frac{t^2 + 10t + 100}{t^2 + 20t + 100}\right)$$

Solution

a. From the graph of f, it appears that $f(t)$ approaches 100 steadily as t gets larger and larger. This observation tells us that eventually the oxygen content of the pond will be restored to its natural level.

b. To verify the observation made in part (a), we compute

$$\begin{aligned}\lim_{t\to\infty} f(t) &= \lim_{t\to\infty} 100\left(\frac{t^2 + 10t + 100}{t^2 + 20t + 100}\right)\\ &= 100 \lim_{t\to\infty}\left(\frac{1 + \frac{10}{t} + \frac{100}{t^2}}{1 + \frac{20}{t} + \frac{100}{t^2}}\right) = 100\end{aligned}$$

GROUP DISCUSSION

Consider the graph of the function f depicted in the following figure:

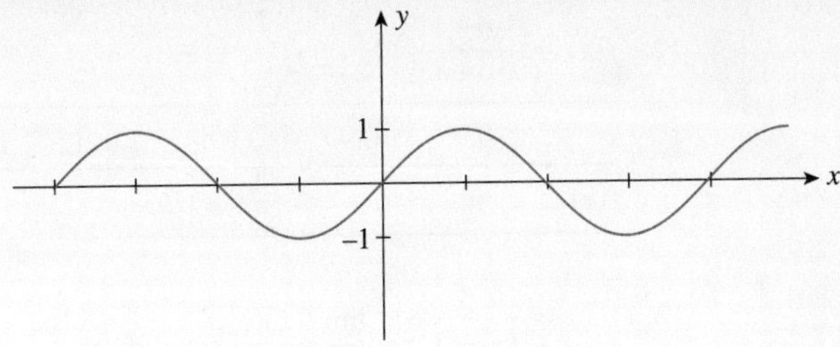

It has the property that the curve oscillates between $y = -1$ and $y = 1$ indefinitely in either direction.

1. Explain why $\lim_{x\to-\infty} f(x)$ and $\lim_{x\to\infty} f(x)$ do not exist.
2. Compare this function with those of Example 10. More specifically, discuss the different ways each function fails to have a limit at infinity or minus infinity.

Self-Check Exercises 3.1

1. Find the indicated limit if it exists.

 a. $\lim_{x\to3} \dfrac{\sqrt{x^2+7}+\sqrt{3x-5}}{x+2}$

 b. $\lim_{x\to-1} \dfrac{x^2-x-2}{2x^2-x-3}$

2. The average cost/disc (in dollars) incurred by Herald Records in pressing x compact discs (CDs) is given by the average cost function

$$\overline{C}(x) = 1.8 + \frac{3000}{x}$$

Evaluate $\lim_{x\to\infty} \overline{C}(x)$ and interpret your result.

Solutions to Self-Check Exercises 3.1 can be found on page 130.

3.1 Exercises

In Exercises 1–8, use the graph of the given function f to determine $\lim_{x\to a} f(x)$ at the indicated value of a, if it exists.

1. 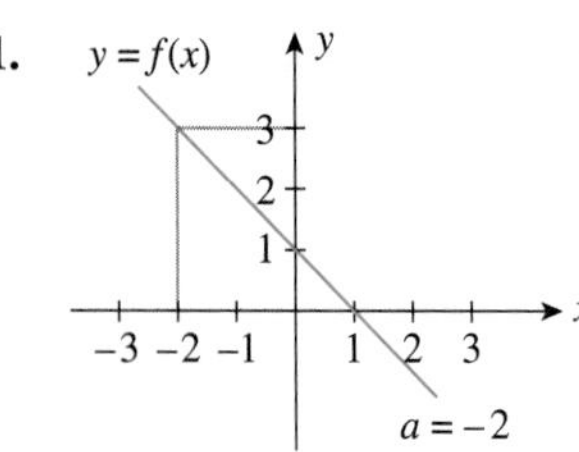

2. 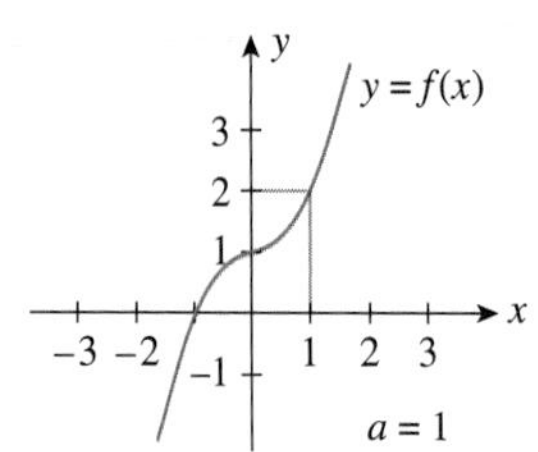

3. 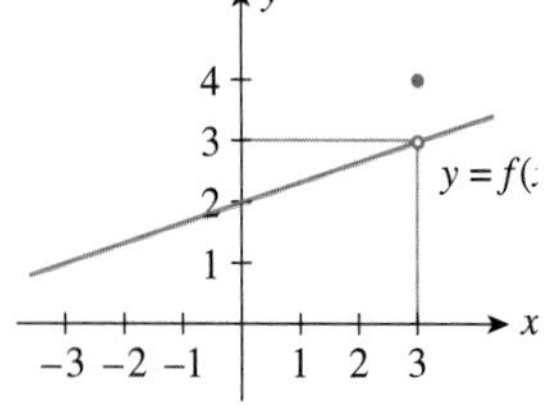

4. 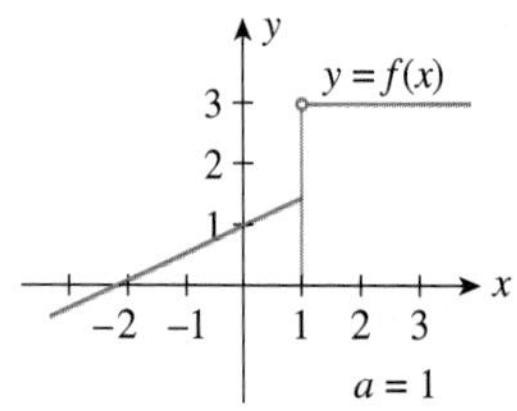

5. 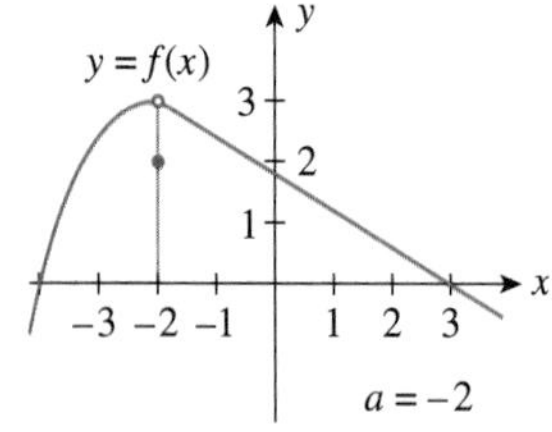

6. 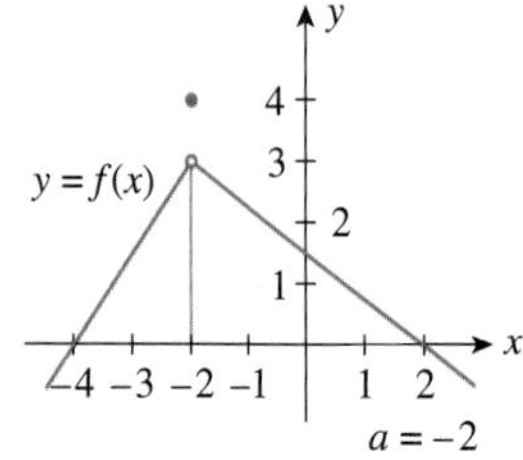

7. 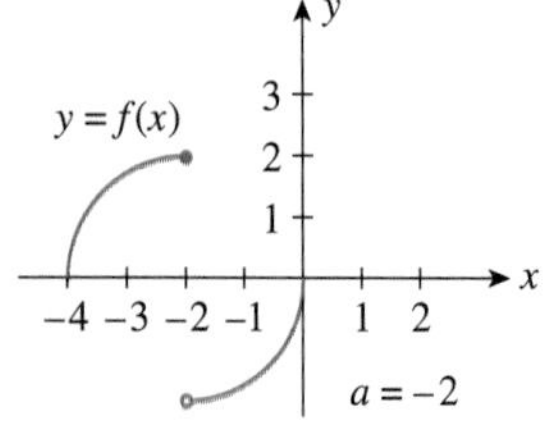

8. 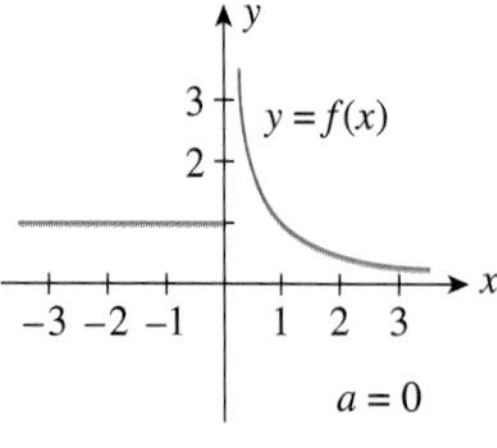

In Exercises 9–16, complete the table by computing $f(x)$ at the given values of x. Use these results to estimate the indicated limit (if it exists).

9. $f(x) = x^2 + 1$; $\lim_{x\to2} f(x)$

x	1.9	1.99	1.999	2.001	2.01	2.1
$f(x)$						

10. $f(x) = 2x^2 - 1$; $\lim_{x\to1} f(x)$

x	0.9	0.99	0.999	1.001	1.01	1.1
$f(x)$						

11. $f(x) = \dfrac{|x|}{x}$; $\lim_{x\to0} f(x)$

x	−0.1	−0.01	−0.001	0.001	0.01	0.1
$f(x)$						

12. $f(x) = \dfrac{|x-1|}{x-1}$; $\lim_{x\to1} f(x)$

x	0.9	0.99	0.999	1.001	1.01	1.1
$f(x)$						

13. $f(x) = \dfrac{1}{(x-1)^2}$; $\lim_{x\to1} f(x)$

x	0.9	0.99	0.999	1.001	1.01	1.1
$f(x)$						

14. $f(x) = \dfrac{1}{x-2}$; $\lim_{x\to2} f(x)$

x	1.9	1.99	1.999	2.001	2.01	2.1
$f(x)$						

15. $f(x) = \dfrac{x^2 + x - 2}{x - 1}$; $\lim_{x \to 1} f(x)$

x	0.9	0.99	0.999	1.001	1.01	1.1
$f(x)$						

16. $f(x) = \dfrac{x - 1}{x - 1}$; $\lim_{x \to 1} f(x)$

x	0.9	0.99	0.999	1.001	1.01	1.1
$f(x)$						

In Exercises 17–22, sketch the graph of the function f and evaluate $\lim_{x \to a} f(x)$, if it exists, for the given values of a.

17. $f(x) = \begin{cases} x - 1 & \text{if } x \le 0 \\ -1 & \text{if } x > 0 \end{cases}$ $\quad (a = 0)$

18. $f(x) = \begin{cases} x - 1 & \text{if } x \le 3 \\ -2x + 8 & \text{if } x > 3 \end{cases}$ $\quad (a = 3)$

19. $f(x) = \begin{cases} x & \text{if } x < 1 \\ 0 & \text{if } x = 1 \\ -x + 2 & \text{if } x > 1 \end{cases}$ $\quad (a = 1)$

20. $f(x) = \begin{cases} -2x + 4 & \text{if } x < 1 \\ 4 & \text{if } x = 1 \\ x^2 + 1 & \text{if } x > 1 \end{cases}$ $\quad (a = 1)$

21. $f(x) = \begin{cases} |x| & \text{if } x \ne 0 \\ 1 & \text{if } x = 0 \end{cases}$ $\quad (a = 0)$

22. $f(x) = \begin{cases} |x - 1| & \text{if } x \ne 1 \\ 0 & \text{if } x = 1 \end{cases}$ $\quad (a = 1)$

In Exercises 23–40, find the indicated limit.

23. $\lim_{x \to 2} 3$

24. $\lim_{x \to -2} -3$

25. $\lim_{x \to 3} x$

26. $\lim_{x \to -2} -3x$

27. $\lim_{x \to 1} (1 - 2x^2)$

28. $\lim_{t \to 3} (4t^2 - 2t + 1)$

29. $\lim_{x \to 1} (2x^3 - 3x^2 + x + 2)$

30. $\lim_{x \to 0} (4x^5 - 20x^2 + 2x + 1)$

31. $\lim_{s \to 0} (2s^2 - 1)(2s + 4)$

32. $\lim_{x \to 2} (x^2 + 1)(x^2 - 4)$

33. $\lim_{x \to 2} \dfrac{2x + 1}{x + 2}$

34. $\lim_{x \to 1} \dfrac{x^3 + 1}{2x^2 + 2}$

35. $\lim_{x \to 2} \sqrt{x + 2}$

36. $\lim_{x \to -2} \sqrt[3]{5x + 2}$

37. $\lim_{x \to -3} \sqrt{2x^4 + x^2}$

38. $\lim_{x \to 2} \sqrt{\dfrac{2x^3 + 4}{x^2 + 1}}$

39. $\lim_{x \to -1} \dfrac{\sqrt{x^2 + 8}}{2x + 4}$

40. $\lim_{x \to 3} \dfrac{x\sqrt{x^2 + 7}}{2x - \sqrt{2x + 3}}$

In Exercises 41–48, find the indicated limit given that $\lim_{x \to a} f(x) = 3$ and $\lim_{x \to a} g(x) = 4$.

41. $\lim_{x \to a} [f(x) - g(x)]$

42. $\lim_{x \to a} 2f(x)$

43. $\lim_{x \to a} [2f(x) - 3g(x)]$

44. $\lim_{x \to a} [f(x)g(x)]$

45. $\lim_{x \to a} \sqrt{g(x)}$

46. $\lim_{x \to a} \sqrt[3]{5f(x) + 3g(x)}$

47. $\lim_{x \to a} \dfrac{2f(x) - g(x)}{f(x)g(x)}$

48. $\lim_{x \to a} \dfrac{g(x) - f(x)}{f(x) + \sqrt{g(x)}}$

In Exercises 49–62, find the indicated limit, if it exists.

49. $\lim_{x \to 1} \dfrac{x^2 - 1}{x - 1}$

50. $\lim_{x \to -2} \dfrac{x^2 - 4}{x + 2}$

51. $\lim_{x \to 0} \dfrac{x^2 - x}{x}$

52. $\lim_{x \to 0} \dfrac{2x^2 - 3x}{x}$

53. $\lim_{x \to -5} \dfrac{x^2 - 25}{x + 5}$

54. $\lim_{b \to -3} \dfrac{b + 1}{b + 3}$

55. $\lim_{x \to 1} \dfrac{x}{x - 1}$

56. $\lim_{x \to 2} \dfrac{x + 2}{x - 2}$

57. $\lim_{x \to -2} \dfrac{x^2 - x - 6}{x^2 + x - 2}$

58. $\lim_{z \to 2} \dfrac{z^3 - 8}{z - 2}$

59. $\lim_{x \to 1} \dfrac{\sqrt{x} - 1}{x - 1}$

Hint: Multiply by $\dfrac{\sqrt{x} + 1}{\sqrt{x} + 1}$.

60. $\lim_{x \to 4} \dfrac{x - 4}{\sqrt{x} - 2}$

Hint: See Exercise 59.

61. $\lim_{x \to 1} \dfrac{x - 1}{x^3 + x^2 - 2x}$

62. $\lim_{x \to -2} \dfrac{4 - x^2}{2x^2 + x^3}$

In Exercises 63–68, use the graph of the function f to determine $\lim_{x \to \infty} f(x)$ and $\lim_{x \to -\infty} f(x)$, if they exist.

63.

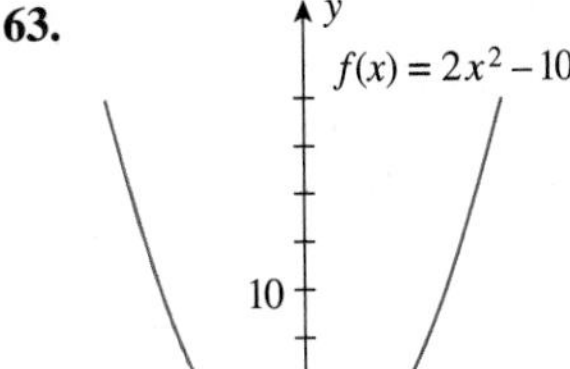

64.

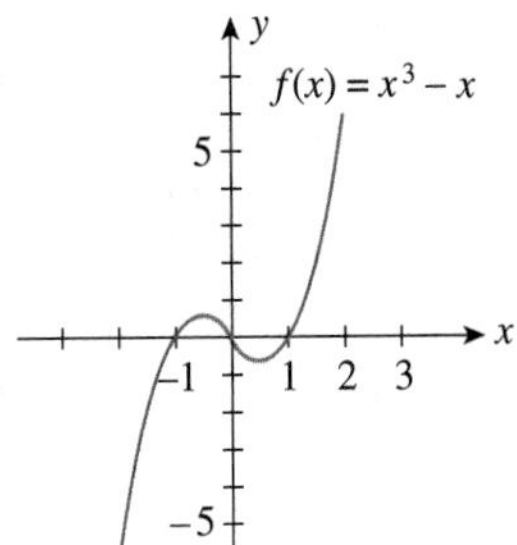

65.

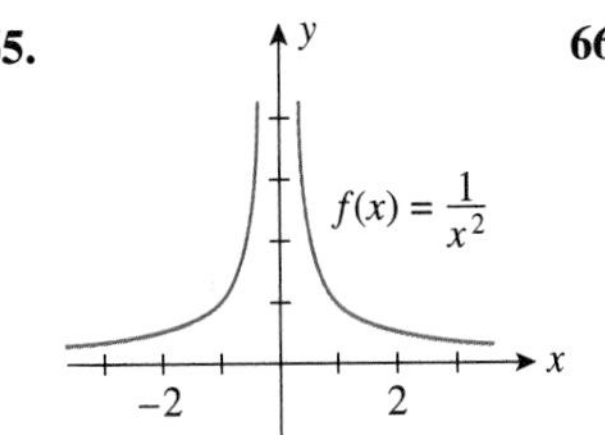

66.

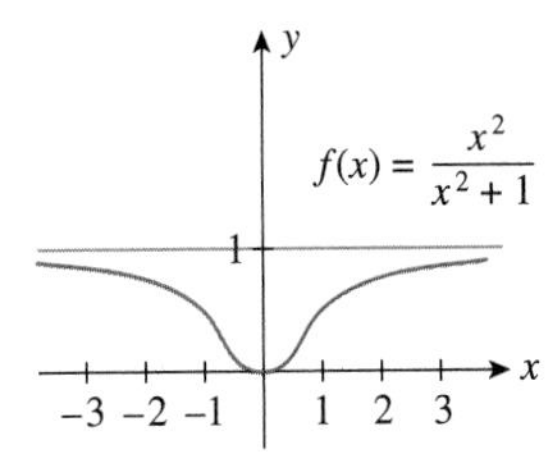

67.

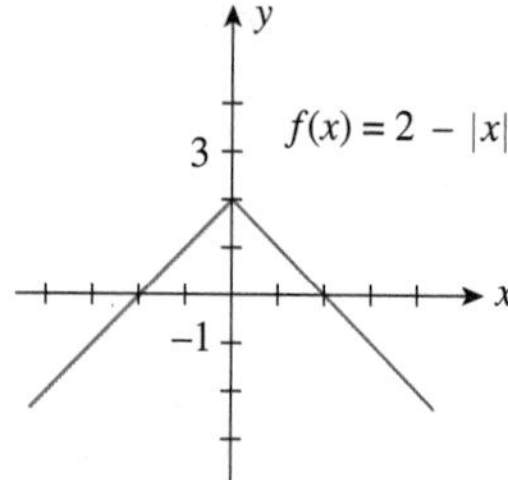

68.

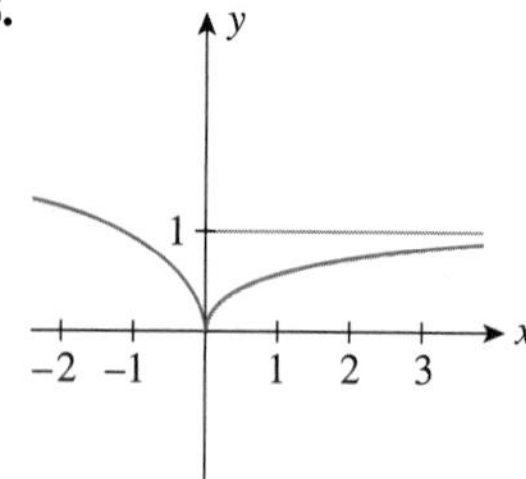

$$f(x) = \begin{cases} \sqrt{-x} & \text{if } x \le 0 \\ \dfrac{x}{x+1} & \text{if } x > 0 \end{cases}$$

In Exercises 69–72, complete the table by computing $f(x)$ at the given values of x. Use the results to guess at the indicated limits, if they exist.

69. $f(x) = \dfrac{1}{x^2+1}$; $\lim_{x\to\infty} f(x)$ and $\lim_{x\to-\infty} f(x)$

x	1	10	100	1000
$f(x)$				

x	-1	-10	-100	-1000
$f(x)$				

70. $f(x) = \dfrac{2x}{x+1}$; $\lim_{x\to\infty} f(x)$ and $\lim_{x\to-\infty} f(x)$

x	1	10	100	1000
$f(x)$				

x	-5	-10	-100	-1000
$f(x)$				

71. $f(x) = 3x^3 - x^2 + 10$; $\lim_{x\to\infty} f(x)$ and $\lim_{x\to-\infty} f(x)$

x	1	5	10	100	1000
$f(x)$					

x	-1	-5	-10	-100	-1000
$f(x)$					

72. $f(x) = \dfrac{|x|}{x}$; $\lim_{x\to\infty} f(x)$ and $\lim_{x\to-\infty} f(x)$

x	1	10	100	-1	-10	-100
$f(x)$						

In Exercises 73–80, find the indicated limits, if they exist.

73. $\lim_{x\to\infty} \dfrac{3x+2}{x-5}$

74. $\lim_{x\to-\infty} \dfrac{4x^2-1}{x+2}$

75. $\lim_{x\to-\infty} \dfrac{3x^3+x^2+1}{x^3+1}$

76. $\lim_{x\to\infty} \dfrac{2x^2+3x+1}{x^4-x^2}$

77. $\lim_{x\to-\infty} \dfrac{x^4+1}{x^3-1}$

78. $\lim_{x\to\infty} \dfrac{4x^4-3x^2+1}{2x^4+x^3+x^2+x+1}$

79. $\lim_{x\to\infty} \dfrac{x^5-x^3+x-1}{x^6+2x^2+1}$

80. $\lim_{x\to\infty} \dfrac{2x^2-1}{x^3+x^2+1}$

81. $\lim_{x\to1} \dfrac{2x^3-2x^2+3x-3}{x-1}$

82. $\lim_{x\to-2} \dfrac{2x^3+3x^2-x+2}{x+2}$

83. $\lim_{x\to-1} \dfrac{x^3+1}{x+1}$

84. $\lim_{x\to-1} \dfrac{x^4-1}{x-1}$

85. $\lim_{x\to1} \dfrac{x^3-x^2-x+1}{x^3-3x+2}$

86. $\lim_{x\to2} \dfrac{x^3+2x^2-16}{2x^3-x^2+2x-16}$

87. $\lim_{x\to0} \dfrac{\sqrt{x+1}-1}{x}$

88. Show by means of an example that $\lim_{x\to a} [f(x) + g(x)]$ may exist even though neither $\lim_{x\to a} f(x)$ nor $\lim_{x\to a} g(x)$ exists. Does this example contradict Theorem 1?

89. Show by means of an example that $\lim_{x\to a} [f(x)g(x)]$ may exist even though neither $\lim_{x\to a} f(x)$ nor $\lim_{x\to a} g(x)$ exists. Does this example contradict Theorem 1?

90. Show by means of an example that $\lim_{x\to a} f(x)/g(x)$ may exist even though neither $\lim_{x\to a} f(x)$ nor $\lim_{x\to a} g(x)$ exists. Does this example contradict Theorem 1?

Business and Economics Applications

91. **Toxic Waste** A city's main well was recently found to be contaminated with trichloroethylene, a cancer-causing chemical, as a result of an abandoned chemical dump leaching chemicals into the water. A proposal submitted to city council members indicates that the cost, measured in millions of dollars, of removing x% of the toxic pollutant is given by

$$C(x) = \frac{0.5x}{100-x} \qquad (0 < x < 100)$$

a. Find the cost of removing 50%, 60%, 70%, 80%, 90%, and 95% of the pollutant.

b. Evaluate

$$\lim_{x\to100} \frac{0.5x}{100-x}$$

and interpret your result.

92. **City Planning** A major developer is building a 5000-hectare complex of homes, offices, stores, schools, and churches in the rural community of Marlboro. As a result of this development, the planners have estimated that Marlboro's population (in thousands) t yr from now will be given by

$$P(t) = \frac{25t^2+125t+200}{t^2+5t+40}$$

What will be the population of Marlboro in the long run?

Hint: Find $\lim_{t \to \infty} P(t)$.

93. Amount of Rainfall During a heavy rainfall in the north of Vancouver Island, the total amount of rain (in millimetres) after t hr is given by

$$T(t) = \frac{128x}{x + 3.6}$$

What is the total amount of rain during this rainfall?

Hint: Find $\lim_{t \to \infty} T(t)$.

94. Average Cost The average cost/disc in dollars incurred by Herald Records in pressing x videodiscs is given by the average cost function

$$\overline{C}(x) = 2.2 + \frac{2500}{x}$$

Evaluate $\lim_{x \to \infty} \overline{C}(x)$ and interpret your result.

95. Box-Office Receipts The total worldwide box-office receipts for a long-running blockbuster movie are approximated by the function

$$T(x) = \frac{120x^2}{x^2 + 4}$$

where $T(x)$ is measured in millions of dollars and x is the number of months since the movie's release.

a. What are the total box-office receipts after the first month? The second month? The third month?

b. What will the movie gross in the long run?

96. Driving Costs A study of driving costs of 1992 model subcompact (four-cylinder) cars found that the average cost (car payments, gas, insurance, upkeep, and depreciation), measured in cents/kilometre, is approximated by the function

$$C(x) = \frac{2010}{x^{2.2}} + 17.80$$

where x denotes the number of kilometres (in thousands) the car is driven in a year.

a. What is the average cost of driving a subcompact car 5000 km/yr? 10,000 km/yr? 15,000 km/yr? 20,000 km/yr? 25,000 km/yr?

b. Use part (a) to sketch the graph of the function C.

c. What happens to the average cost as the number of kilometres driven increases without bound?

Biological and Life Sciences Applications

97. A Doomsday Situation The population of a certain breed of rabbits introduced into an isolated island is given by

$$P(t) = \frac{72}{9 - t} \qquad (0 \le t < 9)$$

where t is measured in months.

a. Find the number of rabbits present on the island initially (at $t = 0$).

b. Show that the population of rabbits is increasing without bound.

c. Sketch the graph of the function P.

(*Comment:* This phenomenon is referred to as a *doomsday situation.*)

98. Concentration of a Drug in the Bloodstream The concentration of a certain drug in a patient's bloodstream t hr after injection is given by

$$C(t) = \frac{0.2t}{t^2 + 1}$$

mg/cm^3. Evaluate $\lim_{t \to \infty} C(t)$ and interpret your result.

99. Population Growth A major corporation is building a 1700-hectare complex of homes, offices, stores, schools, and churches in the rural community of Tweed, Ontario. As a result of this development, the planners have estimated that Tweed's population (in thousands) t yr from now will be given by

$$P(t) = \frac{25t^2 + 125t + 200}{t^2 + 5t + 40}$$

a. What is the current population of Tweed?

b. What will be the population in the long run (when x is very large)?

100. Photosynthesis The rate of production R in photosynthesis is related to the light intensity I by the function

$$R(I) = \frac{aI}{b + I^2}$$

where a and b are positive constants.

a. Taking $a = b = 1$, compute $R(I)$ for $I = 0, 1, 2, 3, 4$, and 5.

b. Evaluate $\lim_{I \to \infty} R(I)$.

c. Use the results of parts (a) and (b) to sketch the graph of R. Interpret your results.

101. Speed of a Chemical Reaction Certain proteins, known as enzymes, serve as catalysts for chemical reactions in living things. In 1913 Leonor Michaelis and L. M. Menten discovered the following formula giving the initial speed V (in moles/litre/second) at which the reaction begins in terms of the amount of substrate x (the substance being acted upon, measured in moles/litres) present:

$$V = \frac{ax}{x + b}$$

where a and b are positive constants. Evaluate

$$\lim_{x \to \infty} \frac{ax}{x + b}$$

and interpret your result.

In Exercises 102–107, determine whether the statement is true or false. If it is true, explain why it is true. If it is false, give an example to show why it is false.

102. If $\lim_{x \to a} (x)$ exists, then f is defined at $x = a$.

103. If $\lim_{x \to 0} f(x) = 4$ and $\lim_{x \to 0} g(x) = 0$, then $\lim_{x \to 0} f(x)g(x) = 0$.

104. If $\lim_{x \to 2} f(x) = 3$ and $\lim_{x \to 2} g(x) = 0$, then $\lim_{x \to 2} [f(x)/g(x)]$ does not exist.

105. If $\lim_{x \to 3} f(x) = 0$ and $\lim_{x \to 3} g(x) = 0$, then $\lim_{x \to 3} [f(x)]/[g(x)]$ does not exist.

106. $\lim_{x \to 2} \left(\frac{x}{x+1} + \frac{3}{x-1} \right) = \lim_{x \to 2} \frac{x}{x+1} + \lim_{x \to 2} \frac{3}{x-1}$

107. $\lim_{x \to 1} \left(\frac{2x}{x-1} - \frac{2}{x-1} \right) = \lim_{x \to 1} \frac{2x}{x-1} - \lim_{x \to 1} \frac{2}{x-1}$

Solutions to Self-Check Exercises 3.1

1. a. $$\lim_{x\to 3}\frac{\sqrt{x^2+7}+\sqrt{3x-5}}{x+2}=\frac{\sqrt{9+7}+\sqrt{3(3)-5}}{3+2}$$
$$=\frac{\sqrt{16}+\sqrt{4}}{5}$$
$$=\frac{6}{5}$$

b. Letting x approach -1 leads to the indeterminate form 0/0. Thus, we proceed as follows:

$$\lim_{x\to -1}\frac{x^2-x-2}{2x^2-x-3}=\lim_{x\to -1}\frac{(x+1)(x-2)}{(x+1)(2x-3)}$$
$$=\lim_{x\to -1}\frac{x-2}{2x-3}$$ Cancel the common factors.
$$=\frac{-1-2}{2(-1)-3}$$
$$=\frac{3}{5}$$

2. $$\lim_{x\to\infty}\overline{C}(x)=\lim_{x\to\infty}\left(1.8+\frac{3000}{x}\right)$$
$$=\lim_{x\to\infty}1.8+\lim_{x\to\infty}\frac{3000}{x}$$
$$=1.8$$

Our computation reveals that, as the production of CDs increases "without bound," the average cost drops and approaches a unit cost of \$1.80/disc.

3.2 One-Sided Limits and Continuity

One-Sided Limits

Consider the function f defined by

$$f(x)=\begin{cases}x-1 & \text{if } x<0\\ x+1 & \text{if } x\ge 0\end{cases}$$

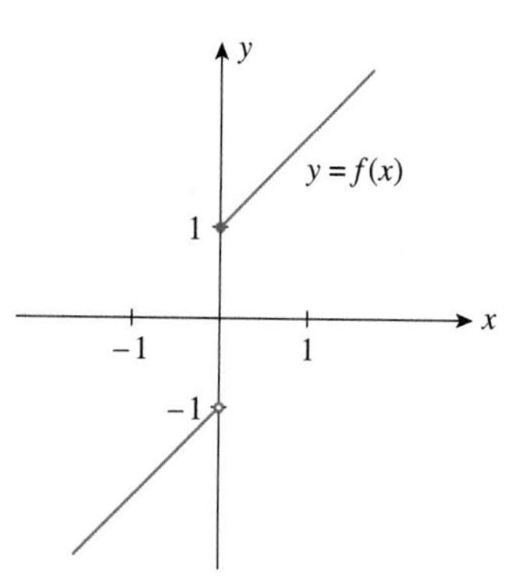

Figure 3.13
The function f does not have a limit as x approaches zero.

From the graph of f shown in Figure 3.13, we see that the function f does not have a limit as x approaches zero because, no matter how close x is to zero, $f(x)$ takes on values that are close to 1 if x is positive and values that are close to -1 if x is negative. Therefore, $f(x)$ cannot be close to a single number L—no matter how close x is to zero. Now, if we restrict x to be greater than zero (to the right of zero), then we see that $f(x)$ can be made as close to 1 as we please by taking x sufficiently close to zero. In this situation we say that the right-hand limit of f as x approaches zero (from the right) is 1, written

$$\lim_{x\to 0^+}f(x)=1$$

Similarly, we see that $f(x)$ can be made as close to -1 as we please by taking x sufficiently close to, but to the left of, zero. In this situation we say that the left-hand limit of f as x approaches zero (from the left) is -1, written

$$\lim_{x\to 0^-}f(x)=-1$$

These limits are called **one-sided limits.** More generally, we have the following definitions.

One-Sided Limits

The function f has the **right-hand limit** L as x approaches a from the right, written

$$\lim_{x \to a^+} f(x) = L$$

if the values $f(x)$ can be made as close to L as we please by taking x sufficiently close to (but not equal to) a and to the right of a.

Similarly, the function f has the **left-hand limit** M as x approaches a from the left, written

$$\lim_{x \to a^-} f(x) = M$$

if the values $f(x)$ can be made as close to M as we please by taking x sufficiently close to (but not equal to) a and to the left of a.

The connection between one-sided limits and the two-sided limit defined earlier is given by the following theorem.

THEOREM 3

Let f be a function that is defined for all values of x close to $x = a$ with the possible exception of a itself. Then

$$\lim_{x \to a} f(x) = L \quad \text{if and only if} \quad \lim_{x \to a^+} f(x) = \lim_{x \to a^-} f(x) = L$$

Thus, the two-sided limit exists if and only if the one-sided limits exist and are equal.

Example 1

Let

$$f(x) = \begin{cases} \sqrt{x} & \text{if } x > 0 \\ -x & \text{if } x \le 0 \end{cases} \quad \text{and} \quad g(x) = \begin{cases} -1 & \text{if } x < 0 \\ 1 & \text{if } x \ge 0 \end{cases}$$

a. Show that $\lim_{x \to 0} f(x)$ exists by studying the one-sided limits of f as x approaches $x = 0$.

b. Show that $\lim_{x \to 0} g(x)$ does not exist.

Solution

a. For $x > 0$, we find

$$\lim_{x \to 0^+} f(x) = \lim_{x \to 0^+} \sqrt{x} = 0$$

and for $x \le 0$

$$\lim_{x \to 0^-} f(x) = \lim_{x \to 0^-} (-x) = 0$$

Thus,

$$\lim_{x \to 0} f(x) = 0$$

(Figure 3.14a).

b. We have

$$\lim_{x \to 0^-} g(x) = -1 \quad \text{and} \quad \lim_{x \to 0^+} g(x) = 1$$

and since these one-sided limits are not equal, we conclude that $\lim_{x \to 0} g(x)$ does not exist (Figure 3.14b).

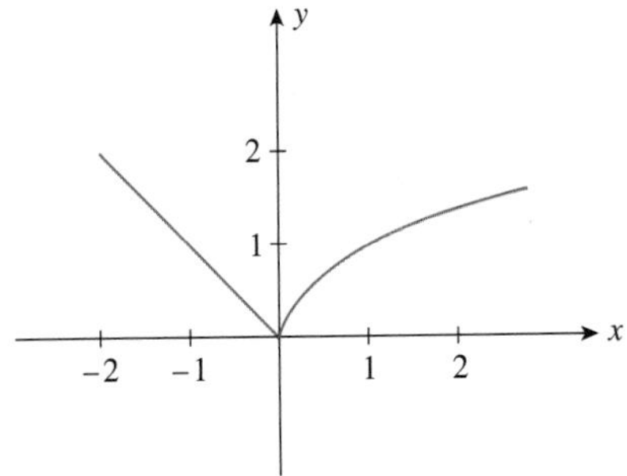

(a) $\lim_{x \to 0} f(x)$ exists.

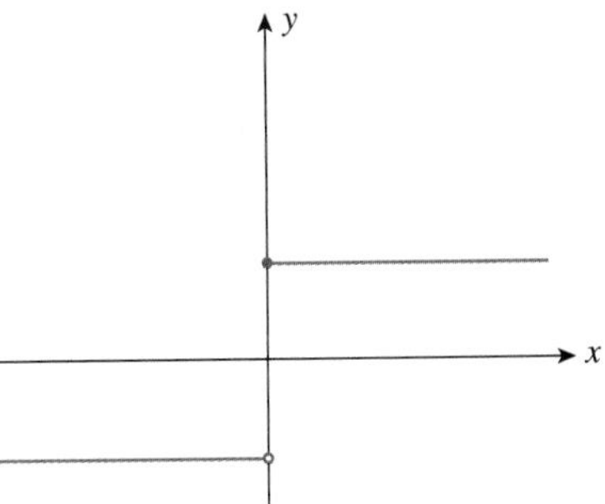

(b) $\lim_{x \to 0} g(x)$ does not exist.

Figure 3.14

Continuous Functions

Continuous functions will play an important role throughout most of our study of calculus. Loosely speaking, a function is continuous at a point if the graph of the function at that point is devoid of holes, gaps, jumps, or breaks. Consider, for example, the graph of the function f depicted in Figure 3.15.

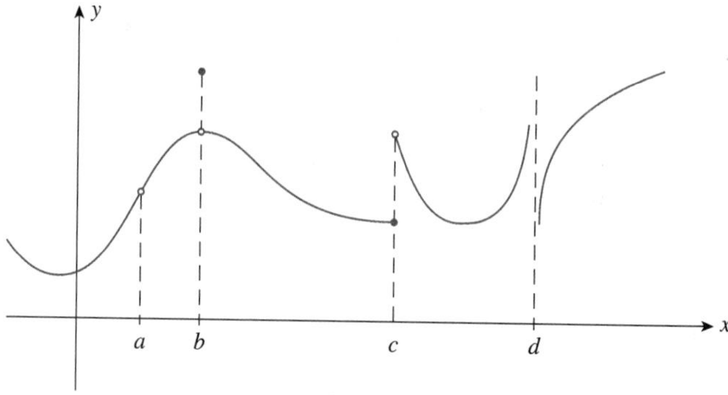

Figure 3.15
The graph of this function is not continuous at $x = a$, $x = b$, $x = c$, and $x = d$.

Let's take a closer look at the behaviour of f at or near each of the points $x = a$, $x = b$, $x = c$, and $x = d$. First, note that f is not defined at $x = a$; that is, the point $x = a$ is not in the domain of f, thereby resulting in a "hole" in the graph of f. Next, observe that the value of f at b, $f(b)$, is not equal to the limit of $f(x)$ as x approaches b, resulting in a "jump" in the graph of f at that point. The function f does not have a limit at $x = c$ since the left-hand and right-hand limits of $f(x)$ are not equal, also resulting in a jump in the graph of f at that point. Finally, the limit of f does not exist at $x = d$, resulting in a break in the graph of f. The function f is *discontinuous* at each of these points. It is *continuous* at all other points.

Continuity at a Point

A function f is **continuous at the point** $x = a$ if the following conditions are satisfied.
1. $f(a)$ is defined. **2.** $\lim_{x \to a} f(x)$ exists. **3.** $\lim_{x \to a} f(x) = f(a)$

Thus, a function f is continuous at the point $x = a$ if the limit of f at the point $x = a$ exists and has the value $f(a)$. Geometrically, f is continuous at the point $x = a$ if proximity of x to a implies the proximity of $f(x)$ to $f(a)$.

If f is not continuous at $x = a$, then f is said to be **discontinuous** at $x = a$. Also, f is **continuous on an interval** if f is continuous at every point in the interval.

Figure 3.16 depicts the graph of a continuous function on the interval (a, b). Notice that the graph of the function over the stated interval can be sketched without lifting one's pencil from the paper.

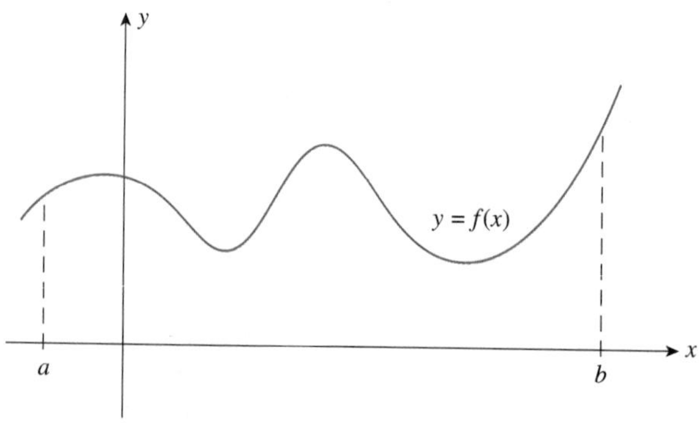

Figure 3.16
The graph of f is continuous on the interval (a, b).

Example 2

Find the values of x for which each of the following functions is continuous.

a. $f(x) = x + 2$ **b.** $g(x) = \dfrac{x^2 - 4}{x - 2}$ **c.** $h(x) = \begin{cases} x + 2 & \text{if } x \neq 2 \\ 1 & \text{if } x = 2 \end{cases}$

d. $F(x) = \begin{cases} -1 & \text{if } x < 0 \\ 1 & \text{if } x \geq 0 \end{cases}$ **e.** $G(x) = \begin{cases} \dfrac{1}{x} & \text{if } x > 0 \\ -1 & \text{if } x \leq 0 \end{cases}$ **f.** $H(x) = |x - 3|$

The graph of each function is shown in Figure 3.17.

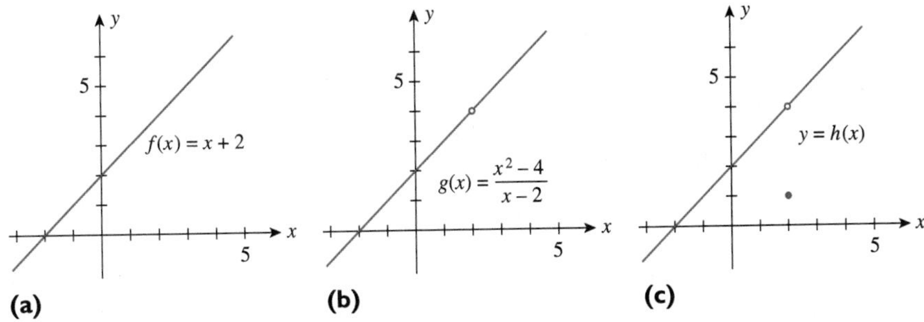

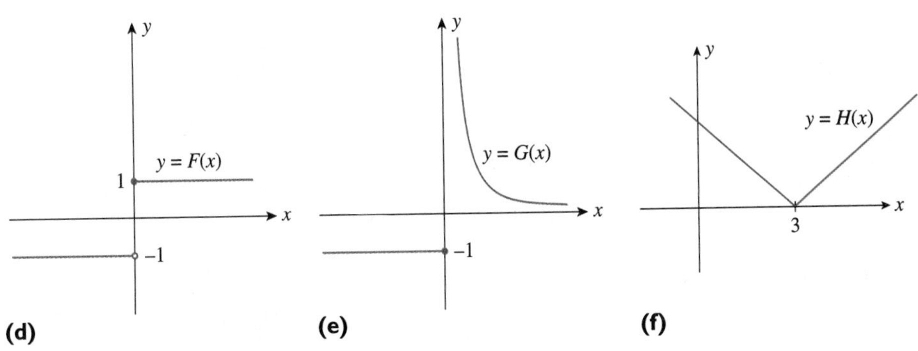

Figure 3.17

Solution

a. The function f is continuous everywhere because the three conditions for continuity are satisfied for all values of x.

b. The function g is discontinuous at the point $x = 2$ because g is not defined at that point. It is continuous everywhere else.

c. The function h is discontinuous at $x = 2$ because the third condition for continuity is violated; the limit of $h(x)$ as x approaches 2 exists and has the value 4, but this limit is not equal to $h(2) = 1$. It is continuous for all other values of x.

d. The function F is continuous everywhere except at the point $x = 0$, where the limit of $F(x)$ fails to exist as x approaches zero (see Example 3a, Section 3.1).

e. Since the limit of $G(x)$ does not exist as x approaches zero, we conclude that G fails to be continuous at $x = 0$. The function G is continuous at all other points.

f. The function H is continuous everywhere because the three conditions for continuity are satisfied for all values of x.

Example 2 shows that there are several types of discontinuities. It often helps to classify a discontinuity by type to get graphical and algebraic information about the function. Figures 3.17b and 3.17c show that the functions g and h respectively are discontinuous due to a *hole* in their graphs at $x = 2$ and $x = 2$ respectively. This type of discontinuity is referred to as a **removable discontinuity at $x = a$.** Figure 3.17d depicts a different type of discontinuity. Here, we need to *jump with our pen* at $x = 0$ to finish graphing the function F. This is called a **jump discontinuity at $x = a$**. Lastly, Figure 3.17e shows a graph that grows to infinity as we approach $x = 0$ from the right side. This type of discontinuity is named **infinite discontinuity at $x = a$**. Notice that any rational function with a vertical asymptote at $x = a$ has an infinite discontinuity there.

We now introduce the notion of *left* and *right* continuity for a function similar to left-hand and right-hand limits using the function $y = F(x)$ from Example 2. If we study the definition of *continuity at a point* on the previous page, then we see that we can define *left-continuity at a point* and *right-continuity at a point* by simply approaching the point from only the left side and right side respectively and ensuring that the existing limit agrees with the function value there. Graphically this means that we have left-continuity at a point if the function is defined there and the same value as the limit when x approaches a. Figure 3.17d shows that

$$\lim_{x\to 0^-} F(x) = -1 \neq F(0) = 1$$

Therefore the function F is not left-continuous at $x = 0$. However, when $x = 0$ is approached from the right side, then

$$\lim_{x\to 0^+} F(x) = 1 = F(0) = 1$$

as shown in Figure 3.17d. This means that the function F is defined at $x = 0$, that the limit exists as x approaches 0 from the right side, and that the limit value and function value agree, thereby satisfying all three conditions of continuity. We say that F is right-continuous at $x = 0$.

Left-Continuity at a Point

A function f is **left-continuous at the point $x = a$** if the following conditions are satisfied.

1. $f(a)$ is defined. **2.** $\lim_{x\to a^-} f(x)$ exists. **3.** $\lim_{x\to a^-} f(x) = f(a)$

Right-Continuity at a Point

A function f is **right-continuous at the point $x = a$** if the following conditions are satisfied.

1. $f(a)$ is defined. **2.** $\lim_{x\to a^+} f(x)$ exists. **3.** $\lim_{x\to a^+} f(x) = f(a)$

Example 3

Given the function f below, do the following.

$$f(x) = \begin{cases} \dfrac{x^2 + 5x + 6}{x + 3}, & -6 < x < -1 \\ \sqrt{x + 1} + 2, & -1 \le x \le 3 \\ (x - 3)^2, & 3 < x \le 6 \end{cases}$$

a. Sketch the graph of the function f.
b. List all points of discontinuities by type.
c. Consider $x = 4$. Is f left-continuous there? Is f right-continuous there?

Solution

a. The first branch $y = \dfrac{x^2 + 5x + 6}{x + 3}$ of f is a rational function with

$$y = \frac{x^2 + 5x + 6}{x + 3} = \frac{(x + 3)(x + 2)}{x + 3} = x + 2, x \neq -3$$

upon simplifying. This means that on the domain $(-6, -1)$ the graph of f is a straight line with slope 1 and y-intercept 2 as well as a hole at $x = -3$. The middle branch $y = \sqrt{x+1} + 2$ of f is a basic root function translated horizontally left by 1 unit and vertically up by 2 units. The last branch $(x - 3)^2$ of f is a basic parabola horizontally translated right by 3 units with vertex $(3, 0)$. The graph of f is shown in Figure 3.18.

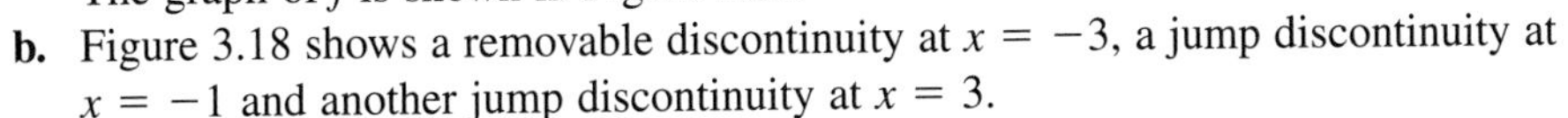

b. Figure 3.18 shows a removable discontinuity at $x = -3$, a jump discontinuity at $x = -1$ and another jump discontinuity at $x = 3$.

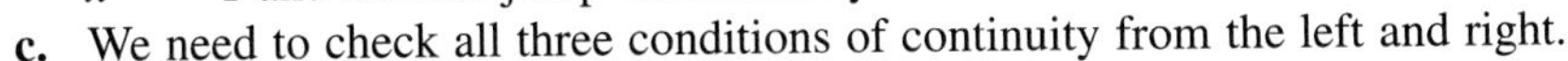

c. We need to check all three conditions of continuity from the left and right.

1. $f(3) = \sqrt{3+1} + 2 = 4$.

2. $\lim_{x\to 3^-} f(x) = \lim_{x\to 3^-} \sqrt{x+1} + 2 = 4$ and $\lim_{x\to 3^+} f(x) = \lim_{x\to 3^+} (x-3)^2 = 0$.

3. $\lim_{x\to 3^-} f(x) = 4 = f(3)$ but $\lim_{x\to 3^+} f(x) = 0 \neq 4 = f(3)$.

Therefore, f is left-continuous at $x = 3$ but not right-continuous there.

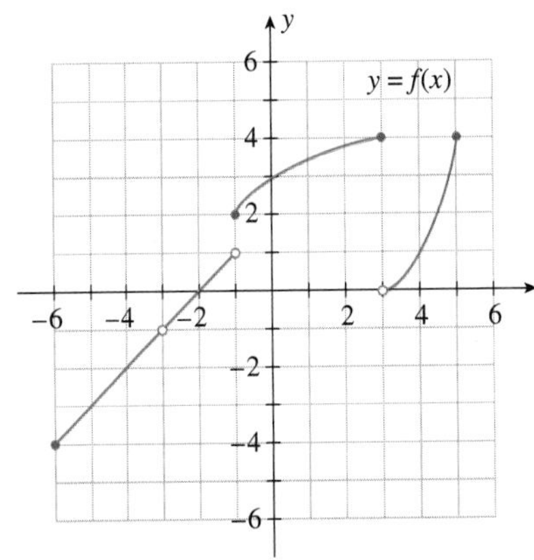

Figure 3.18
The graph of $y = f(x)$

Properties of Continuous Functions

The following properties of continuous functions follow directly from the definition of continuity and the corresponding properties of limits. They are stated without proof.

Properties of Continuous Functions

1. The constant function $f(x) = c$ is continuous everywhere.
2. The identity function $f(x) = x$ is continuous everywhere.

If f and g are continuous at $x = a$, then

3. $[f(x)]^n$, where n is a real number, is continuous at $x = a$ whenever it is defined at the point.
4. $f \pm g$ is continuous at $x = a$.
5. fg is continuous at $x = a$.
6. f/g is continuous at $x = a$ provided $g(a) \neq 0$.

Using these properties of continuous functions, we can prove the following results. (A proof is sketched in Exercise 90, page 142.)

Continuity of Polynomial and Rational Functions

1. A polynomial function $y = P(x)$ is continuous at every point x.
2. A rational function $R(x) = p(x)/q(x)$ is continuous at every point x where $q(x) \neq 0$.

Example 4

Find the values of x for which each of the following functions is continuous.

a. $f(x) = 3x^3 + 2x^2 - x + 10$ **b.** $g(x) = \dfrac{8x^{10} - 4x + 1}{x^2 + 1}$

c. $h(x) = \dfrac{4x^3 - 3x^2 + 1}{x^2 - 3x + 2}$

Solution

a. The function f is a polynomial function of degree 3, so $f(x)$ is continuous for all values of x.

b. The function g is a rational function. Observe that the denominator of g—namely, $x^2 + 1$—is never equal to zero. Therefore, we conclude that g is continuous for all values of x.

c. The function h is a rational function. In this case, however, the denominator of h is equal to zero at $x = 1$ and $x = 2$, which can be seen by factoring it. Thus,

$$x^2 - 3x + 2 = (x - 2)(x - 1)$$

We therefore conclude that h is continuous everywhere except at $x = 1$ and $x = 2$, where it is discontinuous.

Example 5

Find the value of k for which the function f below is continuous at $x = 2$.

$$f(x) = \begin{cases} 3x + k, & x < 2 \\ kx^2, & x \geq 2 \end{cases}$$

Solution

For the function f to be continuous at $x = 2$, it needs to satisfy all three conditions of continuity. We start by looking at the limit as x approaches 2 using Theorem 3, since f is a piecewise defined function.

$$\lim_{x\to 2^-} f(x) = \lim_{x\to 2^-} (3x + k) = 6 + k \quad \text{and}$$

$$\lim_{x\to 2^+} f(x) = \lim_{x\to 2^+} (kx^2) = 4k$$

By Theorem 3, the limit exists, if the right-hand limit and left-hand limit exist and are equal. We compute

$$\lim_{x\to 2^-} f(x) = \lim_{x\to 2^+} f(x)$$

$$6 + k = 4k$$

and solve for k to get

$$6 + k = 4k$$

$$6 = 3k$$

$$k = 2$$

We now have $\lim_{x\to 2} f(x) = \lim_{x\to 2^-} f(x) = \lim_{x\to 2^+} f(x) = 8$. This also means that the function f is defined as

$$f(x) = \begin{cases} 3x + 2, & x < 2 \\ 2x^2, & x \geq 2 \end{cases}$$

We check that $f(2) = 2(2)^2 = 8$ and observe that this is the same value as the limit value for f as x approaches 2. Therefore, f is continuous at $x = 2$ for $k = 2$.

Applications

Up to this point, most of the applications we have discussed involved functions that are continuous everywhere. In Example 6 we consider an application from the field of educational psychology that involves a discontinuous function.

Example 6

Learning Curves Figure 3.19 depicts the learning curve associated with a certain individual. Beginning with no knowledge of the subject being taught, the individual makes steady progress toward understanding it over the time interval $0 \leq t < t_1$. In this instance, the individual's progress slows as we approach time t_1 because he fails to grasp a particularly difficult concept. All of a sudden, a breakthrough occurs at time t_1, propelling his knowledge of the subject to a higher level. The curve is discontinuous at t_1.

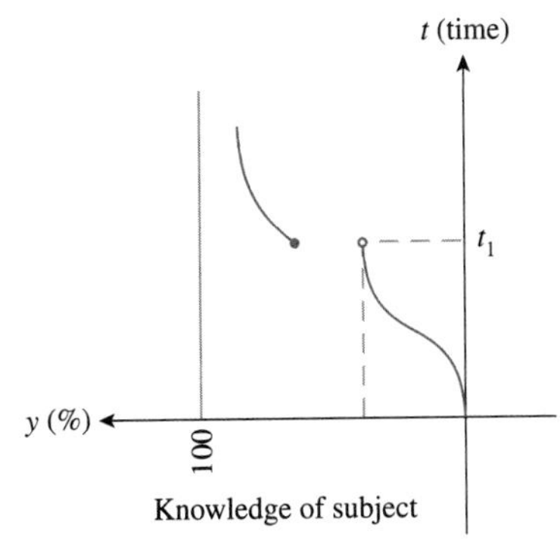

Figure 3.19
A learning curve that is discontinuous at $t = t_1$

Intermediate Value Theorem

Let's look again at our model of the motion of the maglev on a straight stretch of track. We know that the train cannot vanish at any instant of time and it cannot skip portions of the track and reappear someplace else. To put it another way, the train cannot occupy the positions s_1 and s_2 without at least, at some time, occupying an intermediate position (Figure 3.20).

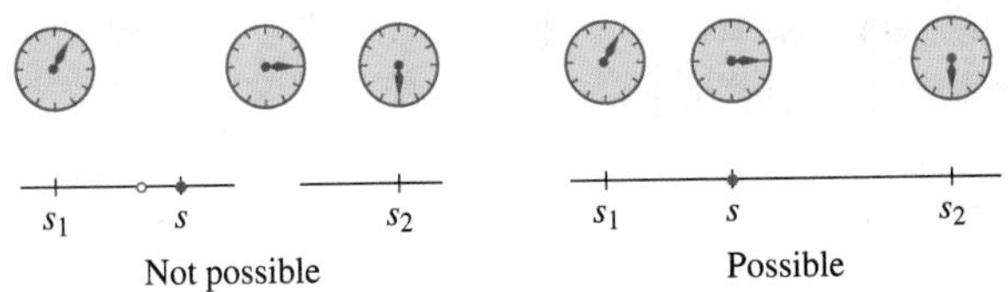

Figure 3.20
The position of the maglev

To state this fact mathematically, recall that the position of the maglev as a function of time is described by

$$f(t) = t^2$$

with restricted domain $0 \leq t \leq 10$. Suppose the position of the maglev is s_1 at some time t_1 and its position is s_2 at some time t_2 (Figure 3.21). Then, if s_3 is any number between s_1 and s_2 giving an intermediate position of the maglev, there must be at least one t_3 between t_1 and t_2 giving the time at which the train is at s_3—that is, $f(t_3) = s_3$.

This discussion carries the gist of the intermediate value theorem. The proof of this theorem can be found in most advanced calculus texts.

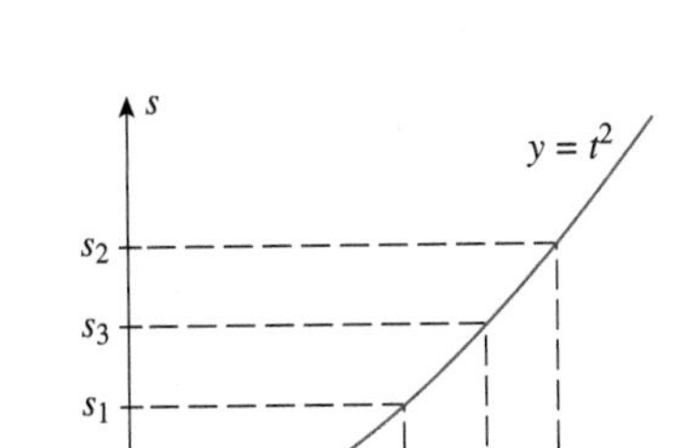

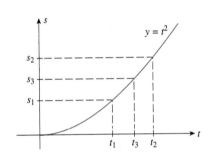

Figure 3.21
If $s_1 \leq s_3 \leq s_2$, then there must be at least one t_3 ($t_1 \leq t_3 \leq t_2$) such that $f(t_3) = s_3$.

The Intermediate Value Theorem

If f is a continuous function on a closed interval $[a, b]$ and M is any number between $f(a)$ and $f(b)$, then there is at least one number c in $[a, b]$ such that $f(c) = M$ (Figure 3.22).

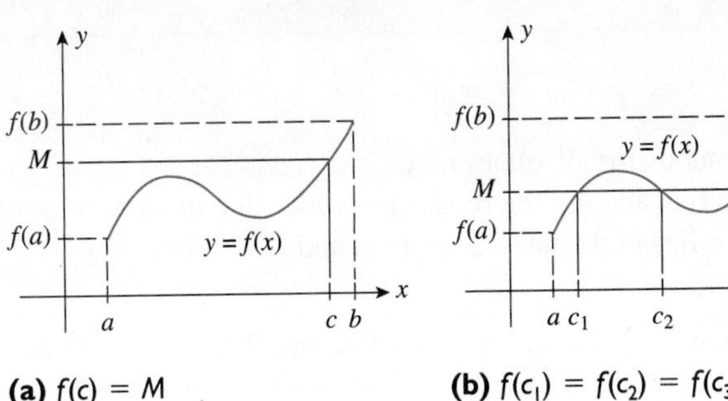

(a) $f(c) = M$

(b) $f(c_1) = f(c_2) = f(c_3) = M$

Figure 3.22

To illustrate the intermediate value theorem, let's look at the example involving the motion of the maglev again (see Figure 3.2, page 113). Notice that the initial position of the train is $f(0) = 0$ and the position at the end of its test run is $f(10) = 100$. Furthermore, the function f is continuous on $[0, 10]$. So, the intermediate value theorem guarantees that if we arbitrarily pick a number between 0 and 100—say, 40—giving the position of the maglev, there must be a $\bar{t}$ (read "t bar") between 0 and 10 at which time the train is at the position $s = 40$. To find the value of $\bar{t}$, we solve the equation $f(\bar{t}) = s$, or

$$\bar{t}^2 = 40$$

giving $\bar{t} = 2\sqrt{10}$ (t must lie between 0 and 10).

It is important to remember when we use Theorem 4 that the function f must be continuous. The conclusion of the intermediate value theorem may not hold if f is not continuous (see Exercise 91, page 143).

The next theorem is an immediate consequence of the intermediate value theorem. It not only tells us when a zero of a function f [root of the equation $f(x) = 0$] exists but also provides the basis for a method of approximating it.

THEOREM 5

Existence of Zeros of a Continuous Function

If f is a continuous function on a closed interval $[a, b]$, and if $f(a)$ and $f(b)$ have opposite signs, then there is at least one solution of the equation $f(x) = 0$ in the interval (a, b) (Figure 3.23).

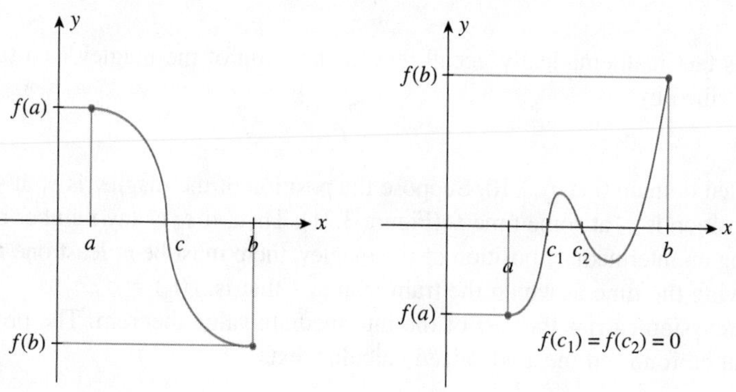

Figure 3.23
If $f(a)$ and $f(b)$ have opposite signs, there must be at least one number c ($a < c < b$) such that $f(c) = 0$.

Geometrically, this property states that if the graph of a continuous function goes from above the x-axis to below the x-axis, or vice versa, it must *cross* the x-axis. This is not necessarily true if the function is discontinuous (Figure 3.24).

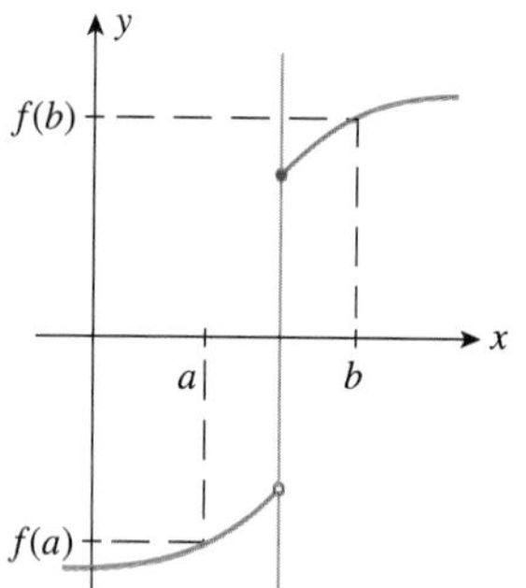

Figure 3.24
$f(a) < 0$ and $f(b) > 0$, but the graph of f does not cross the x-axis between a and b because f is discontinuous.

Example 7

Let $f(x) = x^3 + x + 1$.

a. Show that f is continuous for all values of x.

b. Compute $f(-1)$ and $f(1)$ and use the results to deduce that there must be at least one point $x = c$, where c lies in the interval $(-1, 1)$ and $f(c) = 0$.

Solution

a. The function f is a polynomial function of degree 3 and is therefore continuous everywhere.

b. $f(-1) = (-1)^3 + (-1) + 1 = -1$
$f(1) = 1^3 + 1 + 1 = 3$

Since $f(-1)$ and $f(1)$ have opposite signs, Theorem 5 tells us that there must be at least one point $x = c$ with $-1 < c < 1$ such that $f(c) = 0$.

The next example shows how the intermediate value theorem can be used to help us find the zero of a function.

Example 8

Let $f(x) = x^3 + x - 1$. Since f is a polynomial function, it is continuous everywhere. Observe that $f(0) = -1$ and $f(1) = 1$ so that Theorem 5 guarantees the existence of at least one root of the equation $f(x) = 0$ in $(0, 1)$.*

We can locate the root more precisely by using Theorem 5 once again as follows: Evaluate $f(x)$ at the midpoint of $[0, 1]$, obtaining

$$f(0.5) = -0.375$$

Because $f(0.5) < 0$ and $f(1) > 0$, Theorem 5 now tells us that the root must lie in $(0.5, 1)$.

Repeat the process: Evaluate $f(x)$ at the midpoint of $[0.5, 1]$, which is

$$\frac{0.5 + 1}{2} = 0.75$$

Thus,

$$f(0.75) = 0.1719$$

Because $f(0.5) < 0$ and $f(0.75) > 0$, Theorem 5 tells us that the root is in $(0.5, 0.75)$. This process can be continued. Table 3.3 summarizes the results of our computations through nine steps.

From Table 3.3 we see that the root is approximately 0.68, accurate to two decimal places. By continuing the process through a sufficient number of steps, we can obtain as accurate an approximation to the root as we please.

Table 3.3

Step	Root of $f(x) = 0$ Lies In
1	(0, 1)
2	(0.5, 1)
3	(0.5, 0.75)
4	(0.625, 0.75)
5	(0.625, 0.6875)
6	(0.65625, 0.6875)
7	(0.671875, 0.6875)
8	(0.6796875, 0.6875)
9	(0.6796875, 0.6835937)

Remark

The process of finding the root of $f(x) = 0$ used in Example 8 is called the **method of bisection.** It is crude but effective. ◀

Self-Check Exercises 3.2

1. Evaluate $\lim_{x \to -1^-} f(x)$ and $\lim_{x \to -1^+} f(x)$, where

$$f(x) = \begin{cases} 1 & \text{if } x < -1 \\ 1 + \sqrt{x + 1} & \text{if } x \geq -1 \end{cases}$$

Does $\lim_{x \to -1} f(x)$ exist?

2. Determine the values of x for which the given function is discontinuous. At each point of discontinuity, indicate which condition(s) for continuity are violated. Sketch the graph of the function.

a. $f(x) = \begin{cases} -x^2 + 1 & \text{if } x \leq 1 \\ x - 1 & \text{if } x > 1 \end{cases}$

b. $g(x) = \begin{cases} -x + 1 & \text{if } x < -1 \\ 2 & \text{if } -1 < x \leq 1 \\ -x + 3 & \text{if } x > 1 \end{cases}$

Solutions to Self-Check Exercises 3.2 can be found on page 144.

* It can be shown that f has precisely one zero in (0, 1) (see Exercise 114, Section 7.1).

3.2 Exercises

In Exercises 1–8, use the graph of the function f to find $\lim_{x \to a^-} f(x)$, $\lim_{x \to a^+} f(x)$, and $\lim_{x \to a} f(x)$ at the indicated value of a, if the limit exists.

1.

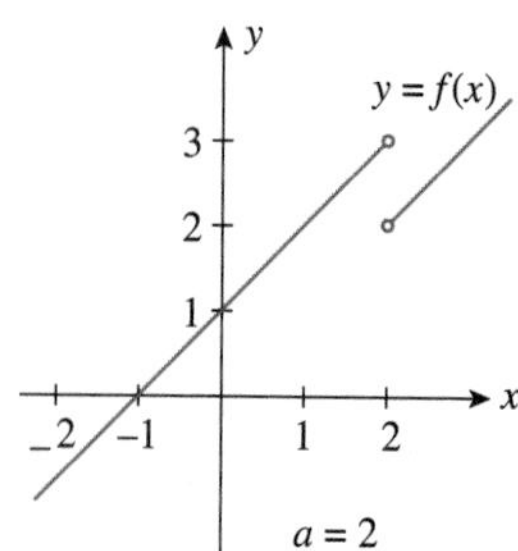

2.

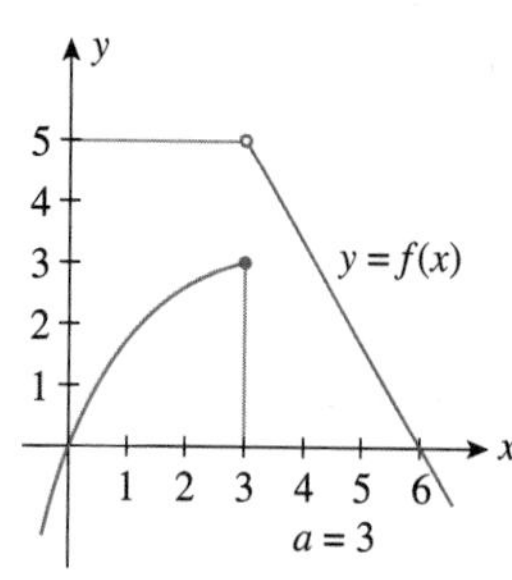

3.

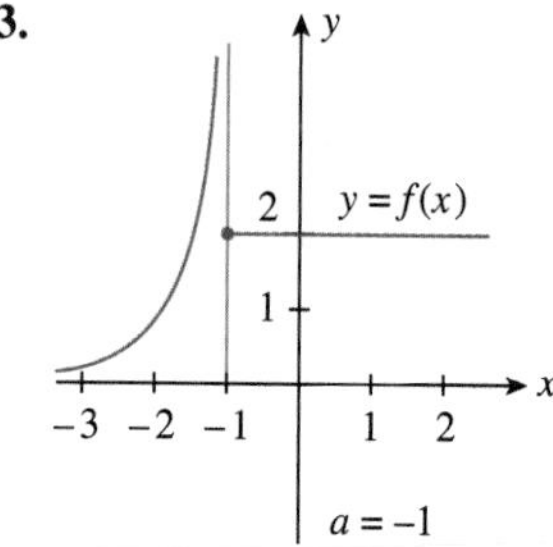

4.

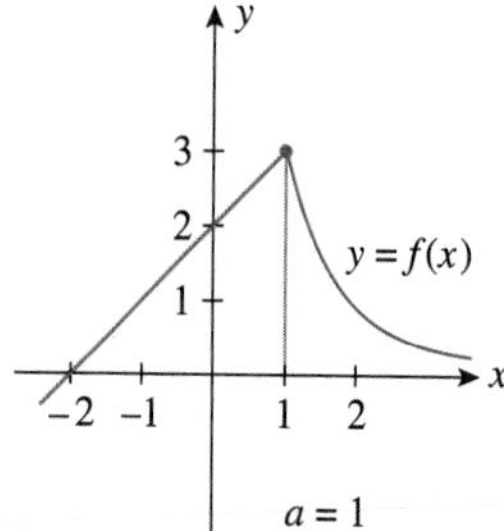

5.

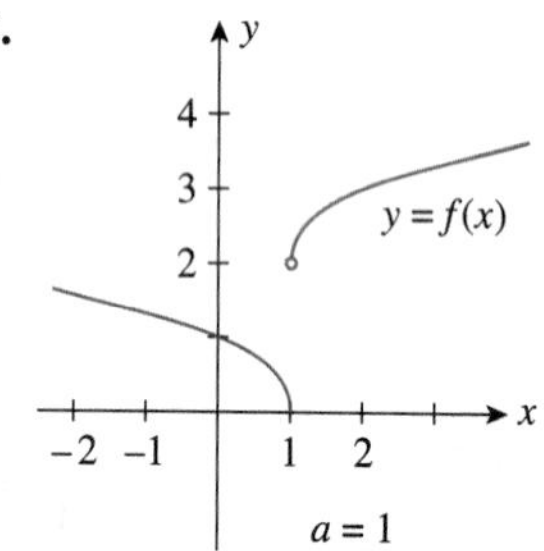

6.

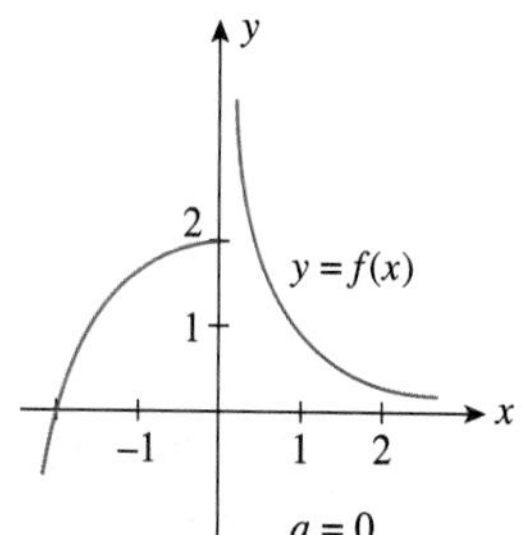

7.

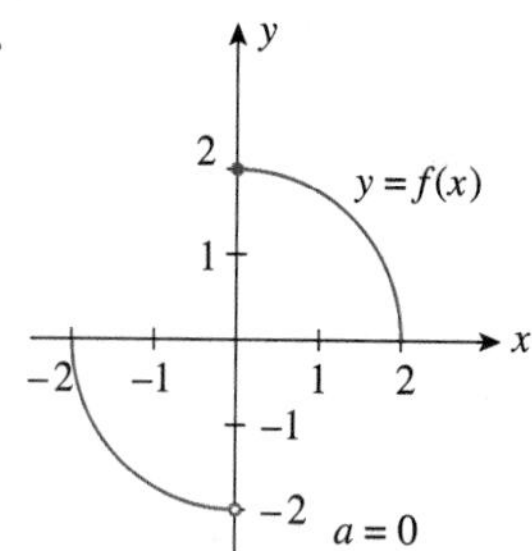

8.

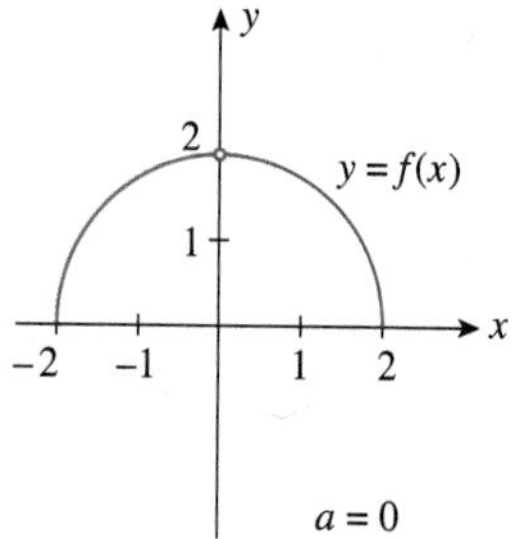

In Exercises 9–14, refer to the graph of the function f and determine whether each statement is true or false.

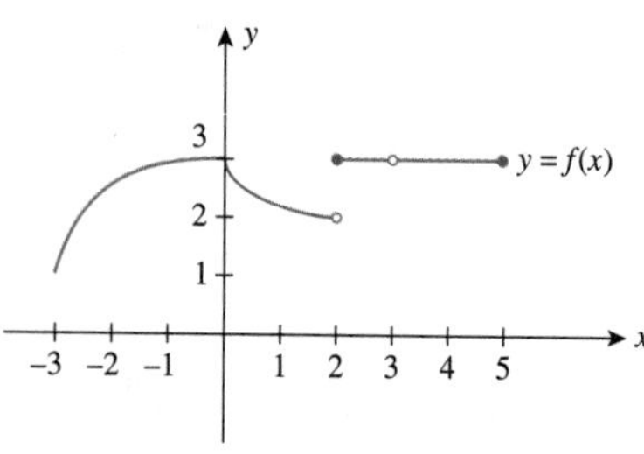

9. $\lim_{x \to -3^+} f(x) = 1$

10. $\lim_{x \to 0} f(x) = f(0)$

11. $\lim_{x \to 2^-} f(x) = 2$

12. $\lim_{x \to 2^+} f(x) = 3$

13. $\lim_{x \to 3} f(x)$ does not exist.

14. $\lim_{x \to 5^-} f(x) = 3$

In Exercises 15–20, refer to the graph of the function f and determine whether each statement is true or false.

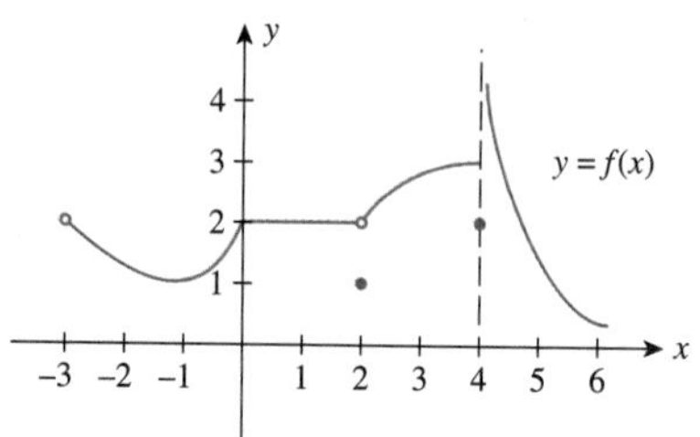

15. $\lim_{x \to -3^+} f(x) = 2$

16. $\lim_{x \to 0} f(x) = 2$

17. $\lim_{x \to 2} f(x) = 1$

18. $\lim_{x \to 4^-} f(x) = 3$

19. $\lim_{x \to 4^+} f(x)$ does not exist.

20. $\lim_{x \to 4} f(x) = 2$

In Exercises 21–42, find the indicated one-sided limit, if it exists.

21. $\lim_{x \to 1^+} (2x + 4)$

22. $\lim_{x \to 1^-} (3x - 4)$

23. $\lim_{x \to 2^-} \dfrac{x - 3}{x + 2}$

24. $\lim_{x \to 1^+} \dfrac{x + 2}{x + 1}$

25. $\lim_{x \to 0^+} \dfrac{1}{x}$

26. $\lim_{x \to 0^-} \dfrac{1}{x}$

27. $\lim_{x \to 0^+} \dfrac{x - 1}{x^2 + 1}$

28. $\lim_{x \to 2^+} \dfrac{x + 1}{x^2 - 2x + 3}$

29. $\lim_{x \to 0^+} \sqrt{x}$

30. $\lim_{x \to 2^+} \sqrt{x - 2}$

31. $\lim_{x \to -2^+} (2x + \sqrt{2 + x})$

32. $\lim_{x \to -5^+} x(1 + \sqrt{5 + x})$

33. $\lim_{x \to 1^-} \dfrac{1 + x}{1 - x}$

34. $\lim_{x \to 1^+} \dfrac{1 + x}{1 - x}$

35. $\lim_{x \to 2^-} \dfrac{x^2 - 4}{x - 2}$

36. $\lim_{x \to -3^+} \dfrac{\sqrt{x + 3}}{x^2 + 1}$

37. $\lim_{x \to 3^+} \dfrac{x^2 - 9}{x + 3}$

38. $\lim_{x \to -2^-} \dfrac{\sqrt[3]{x + 10}}{2x^2 + 1}$

39. $\lim_{x \to 0^+} f(x)$ and $\lim_{x \to 0^-} f(x)$, where

$$f(x) = \begin{cases} 2x & \text{if } x < 0 \\ x^2 & \text{if } x \geq 0 \end{cases}$$

40. $\lim_{x \to 0^+} f(x)$ and $\lim_{x \to 0^-} f(x)$, where

$$f(x) = \begin{cases} -x + 1 & \text{if } x \leq 0 \\ 2x + 3 & \text{if } x > 0 \end{cases}$$

41. $\lim_{x \to 1^+} f(x)$ and $\lim_{x \to 1^-} f(x)$, where

$$f(x) = \begin{cases} \sqrt{x+3} & \text{if } x \geq 1 \\ 2 + \sqrt{x} & \text{if } x < 1 \end{cases}$$

42. $\lim_{x \to 1^+} f(x)$ and $\lim_{x \to 1^-} f(x)$, where

$$f(x) = \begin{cases} x + 2\sqrt{x-1} & \text{if } x \geq 1 \\ 1 - \sqrt{1-x} & \text{if } x < 1 \end{cases}$$

In Exercises 43–50, determine the values of x, if any, at which each function is discontinuous. At each point of discontinuity, state the condition(s) for continuity that are violated.

43. $f(x) = \begin{cases} 2x - 4 & \text{if } x \leq 0 \\ 1 & \text{if } x > 0 \end{cases}$

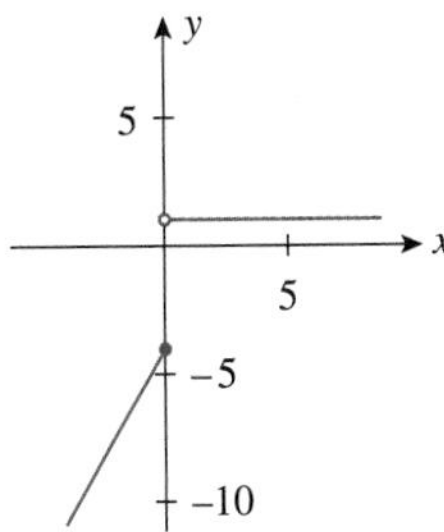

44. $f(x) = \begin{cases} x^2 + 1 & \text{if } x \neq 0 \\ 0 & \text{if } x = 0 \end{cases}$

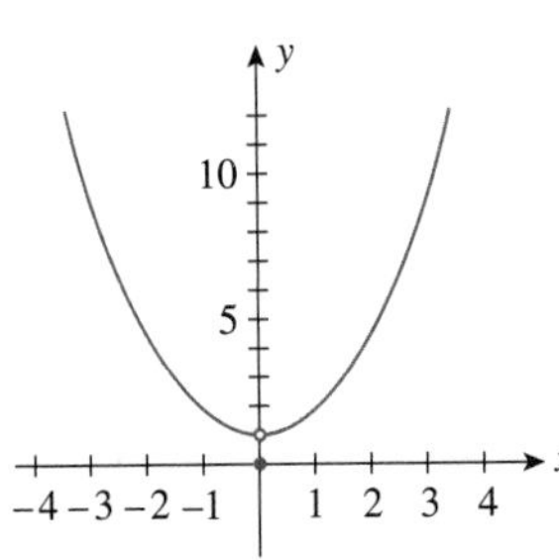

45. $f(x) = \begin{cases} x + 5 & \text{if } x \leq 0 \\ -x^2 + 5 & \text{if } x > 0 \end{cases}$

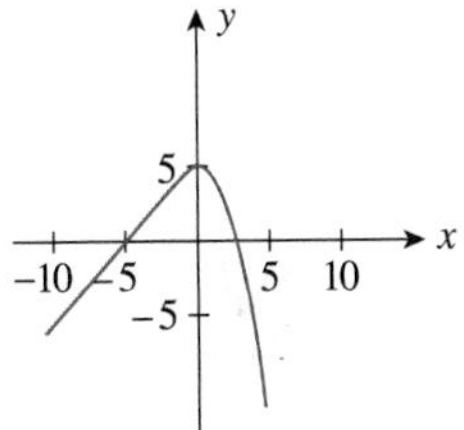

46. $f(x) = |x - 1|$

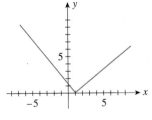

47. $f(x) = \begin{cases} x + 5 & \text{if } x < 0 \\ 2 & \text{if } x = 0 \\ -x^2 + 5 & \text{if } x > 0 \end{cases}$

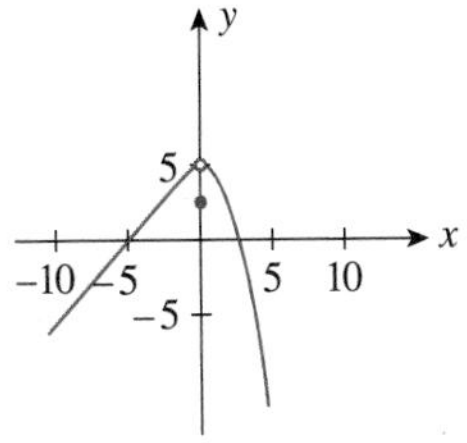

48. $f(x) = \begin{cases} \dfrac{x^2 - 1}{x + 1} & \text{if } x \neq -1 \\ 1 & \text{if } x = -1 \end{cases}$

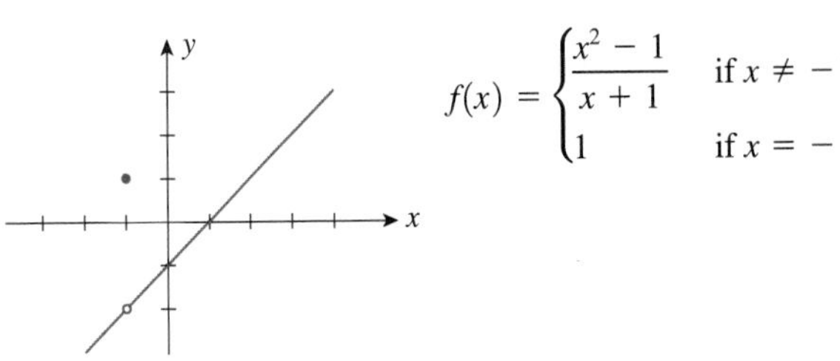

49. $f(x) = \begin{cases} -|x| + 1 & \text{if } x \neq 0 \\ 0 & \text{if } x = 0 \end{cases}$

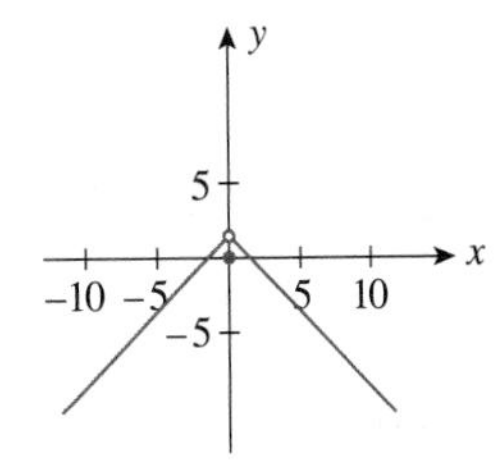

50. $f(x) = \begin{cases} \dfrac{1}{x^2} & \text{if } x \neq 0 \\ 1 & \text{if } x = 0 \end{cases}$

In Exercises 51–66, find the values of x for which each function is continuous.

51. $f(x) = 2x^2 + x - 1$

52. $f(x) = x^3 - 2x^2 + x - 1$

53. $f(x) = \dfrac{2}{x^2 + 1}$

54. $f(x) = \dfrac{x}{2x^2 + 1}$

55. $f(x) = \dfrac{2}{2x - 1}$

56. $f(x) = \dfrac{x + 1}{x - 1}$

57. $f(x) = \dfrac{2x + 1}{x^2 + x - 2}$

58. $f(x) = \dfrac{x - 1}{x^2 + 2x - 3}$

59. $f(x) = \begin{cases} x & \text{if } x \leq 1 \\ 2x - 1 & \text{if } x > 1 \end{cases}$

60. $f(x) = \begin{cases} -x + 1 & \text{if } x \leq -1 \\ x + 1 & \text{if } x > -1 \end{cases}$

61. $f(x) = \begin{cases} -2x + 1 & \text{if } x < 0 \\ x^2 + 1 & \text{if } x \geq 0 \end{cases}$

62. $f(x) = \begin{cases} x + 1 & \text{if } x \leq 1 \\ -x^2 + 1 & \text{if } x > 1 \end{cases}$

63. $f(x) = \begin{cases} \dfrac{x^2 - 1}{x - 1} & \text{if } x \neq 1 \\ 2 & \text{if } x = 1 \end{cases}$

64. $f(x) = \begin{cases} \dfrac{x^2 - 4}{x + 2} & \text{if } x \neq -2 \\ 1 & \text{if } x = -2 \end{cases}$

65. $f(x) = |x + 1|$

66. $f(x) = \dfrac{|x - 1|}{x - 1}$

In Exercises 67–70, determine all values of x at which the function is discontinuous.

67. $f(x) = \dfrac{2x}{x^2 - 1}$

68. $f(x) = \dfrac{1}{(x - 1)(x - 2)}$

69. $f(x) = \dfrac{x^2 - 2x}{x^2 - 3x + 2}$

70. $f(x) = \dfrac{x^2 - 3x + 2}{x^2 - 2x}$

In Exercises 71–74, (a) show that the function *f* is continuous for all values of *x* in the interval [*a*, *b*] and (b) prove that *f* must have at least one zero in the interval (a, b) by showing that *f*(*a*) and *f*(*b*) have opposite signs.

71. $f(x) = x^2 - 6x + 8$; $a = 1$, $b = 3$

72. $f(x) = 2x^3 - 3x^2 - 36x + 14$; $a = 0$, $b = 1$

73. $f(x) = x^3 - 2x^2 + 3x + 2$; $a = -1$, $b = 1$

74. $f(x) = 2x^{5/3} - 5x^{4/3}$; $a = 14$, $b = 16$

In Exercises 75 and 76, use the intermediate value theorem to find the value of *c* such that *f*(*c*) = *M*.

75. $f(x) = x^2 - 4x + 6$ on $[0, 3]$; $M = 4$

76. $f(x) = x^2 - x + 1$ on $[-1, 4]$; $M = 7$

77. Let

$$f(x) = \begin{cases} x + 2 & \text{if } x \leq 1 \\ kx^2 & \text{if } x > 1 \end{cases}$$

Find the value of k that will make f continuous on $(-\infty, \infty)$.

78. Let

$$f(x) = \begin{cases} \dfrac{x^2 - 4}{x + 2} & \text{if } x \neq -2 \\ k & \text{if } x = -2 \end{cases}$$

For what value of k will f be continuous on $(-\infty, \infty)$?

79. Let

$$f(x) = \begin{cases} a\sqrt{x + 7}, & x < 2 \\ 2x - a, & x \geq 2 \end{cases}$$

For what value(s) of k will f be continuous at $x = 2$?

80. Let

$$f(x) = \begin{cases} -\dfrac{x}{a}, & x < -3 \\ a(x + 1)^2 - 4, & x \geq -3 \end{cases}$$

For what value(s) of k will f be continuous at $x = -3$?

81. **a.** Suppose f is continuous at a and g is discontinuous at a. Is the sum $f + g$ discontinuous at a? Explain.
b. Suppose f and g are both discontinuous at a. Is the sum $f + g$ necessarily discontinuous at a? Explain.

82. **a.** Suppose f is continuous at a and g is discontinuous at a. Is the product fg necessarily discontinuous at a? Explain.
b. Suppose f and g are both discontinuous at a. Is the product fg necessarily discontinuous at a? Explain.

83. Use the method of bisection (see Example 8) to find the root of the equation $x^5 + 2x - 7 = 0$ accurate to two decimal places.

84. Use the method of bisection (see Example 8) to find the root of the equation $x^3 - x + 1 = 0$ accurate to two decimal places.

85. **Falling Object** Joan is looking straight out a window of a highrise at a height of 32 m from the ground. A boy shoots a ball with an air gun straight up by the side of the building where the window is located. Suppose the height of the ball (measured in metres) from the ground at time t is $h(t) = 4 + 64t - 16t^2$.
a. Show that $h(0) = 4$ and $h(2) = 68$.
b. Use the intermediate value theorem to conclude that the ball must cross Joan's line of sight at least once.
c. At what time(s) does the ball cross Joan's line of sight? Interpret your results.

86. **Oxygen Content of a Pond** The oxygen content t days after organic waste has been dumped into a pond is given by

$$f(t) = 100\left(\frac{t^2 + 10t + 100}{t^2 + 20t + 100}\right)$$

percent of its normal level.
a. Show that $f(0) = 100$ and $f(10) = 75$.
b. Use the intermediate value theorem to conclude that the oxygen content of the pond must have been at a level of 80% at some time.
c. At what time(s) is the oxygen content at the 80% level?
Hint: Use the quadratic formula.

87. Suppose f is continuous on $[a, b]$ and $f(a) < f(b)$. If M is a number that lies outside the interval $[f(a), f(b)]$, then there does not exist a number $a < c < b$ such that $f(c) = M$. Does this contradict the intermediate value theorem?

88. Let $f(x) = x - \sqrt{1 - x^2}$.
a. Show that f is continuous for all values of x in the interval $[-1, 1]$.
b. Show that f has at least one zero in $[-1, 1]$.
c. Find the zeros of f in $[-1, 1]$ by solving the equation $f(x) = 0$.

89. Let $f(x) = \dfrac{x^2}{x^2 + 1}$.
a. Show that f is continuous for all values of x.
b. Show that $f(x)$ is nonnegative for all values of x.
c. Show that f has a zero at $x = 0$. Does this contradict Theorem 5?

90. **a.** Prove that a polynomial function $y = P(x)$ is continuous at every point x. Follow these steps:
(1) Use Properties 2 and 3 of continuous functions to establish that the function $g(x) = x^n$, where n is a positive integer, is continuous everywhere.

(2) Use Properties 1 and 5 to show that $f(x) = cx^n$, where c is a constant and n is a positive integer, is continuous everywhere.

(3) Use Property 4 to complete the proof of the result.

b. Prove that a rational function $R(x) = p(x)/q(x)$ is continuous at every point x, where $q(x) \neq 0$.

Hint: Use the result of part (a) and Property 6.

91. Show that the conclusion of the intermediate value theorem does not hold if f is discontinuous on $[a, b]$.

Business and Economics Applications

92. The Postage Function The graph of the "postage function" for 2002,

$$f(x) = \begin{cases} 37 & \text{if } 0 < x \leq 1 \\ 60 & \text{if } 1 < x \leq 2 \\ \cdot & \\ \cdot & \\ \cdot & \\ 290 & \text{if } 11 < x \leq 12 \end{cases}$$

where x denotes the weight of a parcel in ounces and $f(x)$ the postage in cents, is shown in the accompanying figure. Determine the values of x for which f is discontinuous.

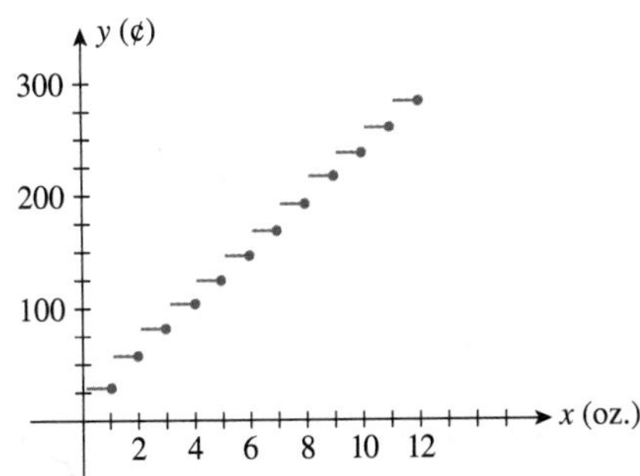

93. Inventory Control As part of an optimal inventory policy, the manager of an office supply company orders 500 reams of photocopy paper every 20 days. The accompanying graph shows the *actual* inventory level of paper in an office supply store during the first 60 business days of 2002. Determine the values of t for which the "inventory function" is discontinuous and give an interpretation of the graph.

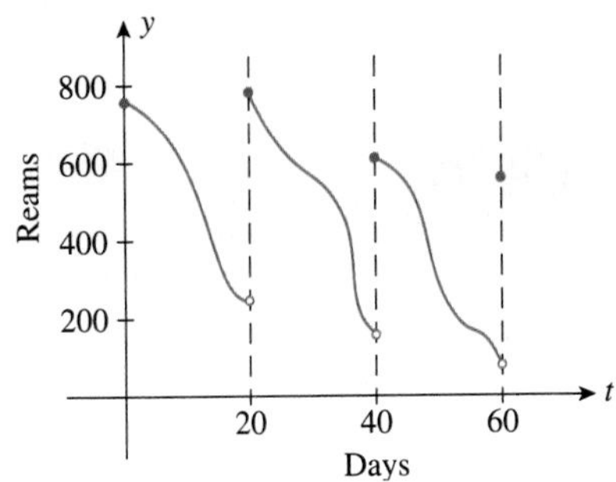

94. Ailing Financial Institutions Victoria Savings and Loan acquired two ailing financial institutions in 1992. One of them was acquired at time $t = T_1$, and the other was acquired at time $t = T_2$ ($t = 0$ corresponds to the beginning of 1992). The following graph shows the total amount of money on deposit with Victoria. Explain the significance of the discontinuities of the function at T_1 and T_2.

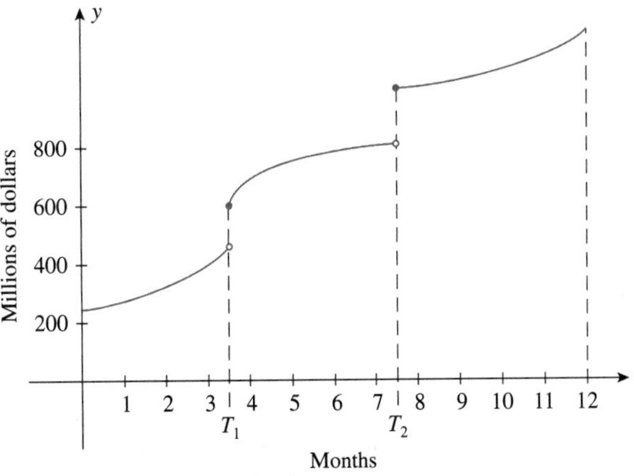

95. Prime Interest Rate The function P, whose graph follows, gives the prime rate (the interest rate banks charge their best corporate customers) as a function of time for the first 32 wk in 1989. Determine the values of t for which P is discontinuous and interpret your results.

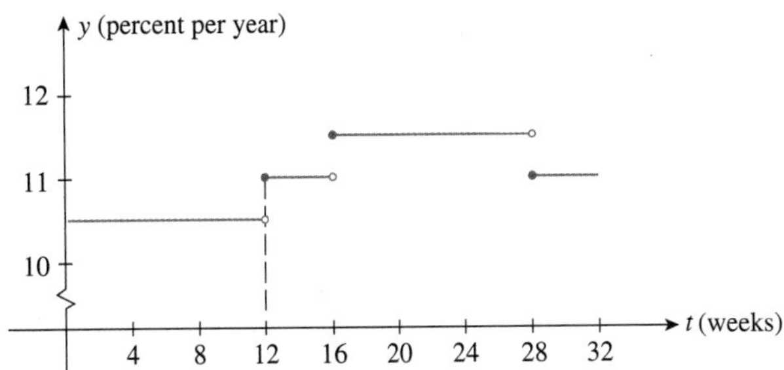

96. Commissions The base salary of a salesman working on commission is \$12,000. For each \$50,000 of sales beyond \$100,000, he is paid a \$1000 commission. Sketch a graph showing his earnings as a function of the level of his sales x. Determine the values of x for which the function f is discontinuous.

97. Parking Fees The fee charged per car in a downtown parking lot is \$1.00 for the first half hour and \$.50 for each additional half hour or part thereof, subject to a maximum of \$5.00. Derive a function f relating the parking fee to the length of time a car is left in the lot. Sketch the graph of f and determine the values of x for which the function f is discontinuous.

98. Commodity Prices The function that gives the cost of a certain commodity is defined by

$$C(x) = \begin{cases} 5x & \text{if } 0 < x < 10 \\ 4x & \text{if } 10 \leq x < 30 \\ 3.5x & \text{if } 30 \leq x < 60 \\ 3.25x & \text{if } x \geq 60 \end{cases}$$

where x is the number of kilograms of a certain commodity sold and $C(x)$ is measured in dollars. Sketch the graph of the function C and determine the value of x for which the function C is discontinuous.

Biological and Life Sciences Applications

99. Learning Curves The following graph describes the progress Michael made in solving a problem correctly during a mathematics quiz. Here, y denotes the percent of work completed, and x is measured in minutes. Give an interpretation of the graph.

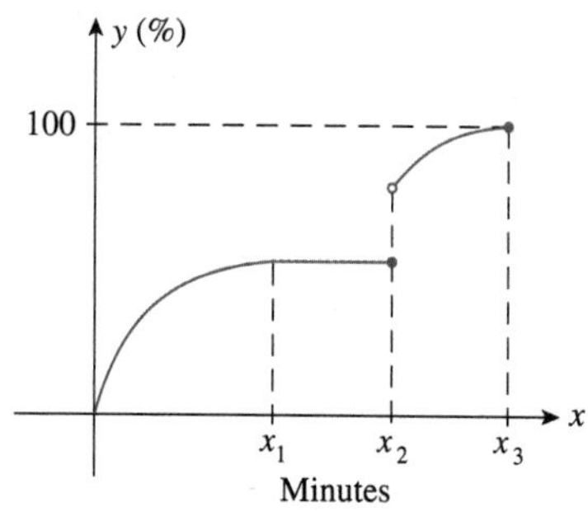

100. Energy Consumption Older homes in B.C. still have 200-gal oil tanks. The following graph shows the amount of home heating oil remaining in a 200-gal tank over a 120-day period ($t = 0$ corresponds to October 1). Explain why the function is discontinuous at $t = 40$, 70, 95, and 110.

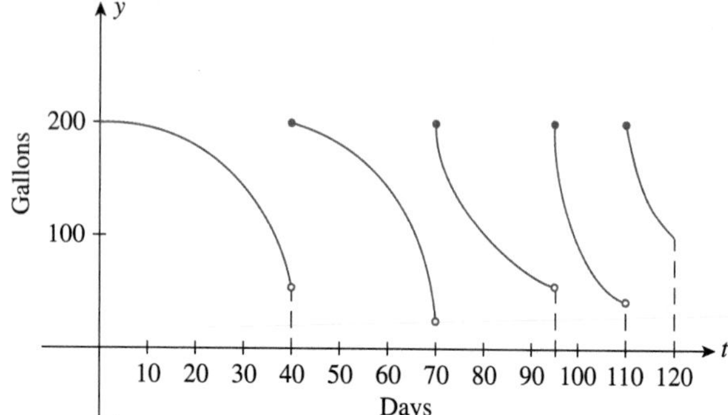

101. Administration of an Intravenous Solution A dextrose solution is being administered to a patient intravenously. The 1-litre (L) bottle holding the solution is removed and replaced by another as soon as the contents drop to approximately 5% of the initial (1-L) amount. The rate of discharge is constant, and it takes 6 h to discharge 95% of the contents of a full bottle. Draw a graph showing the amount of dextrose solution in a bottle in the IV system over a 24-h period, assuming that we started with a full bottle.

102. Weiss's Law According to Weiss's law of excitation of tissue, the strength S of an electric current is related to the time t the current takes to excite tissue by the formula

$$S(t) = \frac{a}{t} + b \qquad (t > 0)$$

where a and b are positive constants.

a. Evaluate $\lim_{t \to 0^+} S(t)$ and interpret your result.

b. Evaluate $\lim_{t \to \infty} S(t)$ and interpret your result.

(*Remark:* The limit in part (b) is called the threshold strength of the current. Why?)

103. Energy Expended by a Fish Suppose a fish swimming a distance of L m at a speed of v m/sec relative to the water and against a current flowing at the rate of u m/sec ($u < v$) expends a total energy given by

$$E(v) = \frac{aLv^3}{v - u}$$

where E is measured joules (J) and a is a constant.

a. Evaluate $\lim_{v \to u^+} E(v)$ and interpret your result.

b. Evaluate $\lim_{v \to \infty} E(v)$ and interpret your result.

In Exercises 104–112, determine whether the statement is true or false. If it is true, explain why it is true. If it is false, give an example to show why it is false.

104. If $f(2) = 4$, then $\lim_{x \to 2} f(x) = 4$.

105. If $\lim_{x \to 0} f(x) = 3$, then $f(0) = 3$.

106. Suppose the function f is defined on the interval $[a, b]$. If $f(a)$ and $f(b)$ have the same sign, then f has no zero in $[a, b]$.

107. If $\lim_{x \to a} f(x) = L$, then $\lim_{x \to a^+} f(x) - \lim_{x \to a^-} f(x) \neq 0$.

108. If $\lim_{x \to a^-} f(x) = L$ and $\lim_{x \to a^+} f(x) = L$, then $f(a) = L$.

109. If $\lim_{x \to a} f(x) = L$ and $g(a) = M$, then $\lim_{x \to a} f(x)g(x) = LM$.

110. If f is continuous for all $x \neq 0$ and $f(0) = 0$, then $\lim_{x \to 0} f(x) = 0$.

111. If f is continuous at $x = 5$ and $f(5) = 2$, then $\lim_{x \to 5^-} f(x) = 2$.

112. If f is continuous on $[-2, 3]$, $f(-2) = 3$, and $f(3) = 1$, then there exists at least one number c in $[-2, 3]$ such that $f(c) = 2$.

Solutions to Self-Check Exercises 3.2

1. For $x < -1$, $f(x) = 1$, and so

$$\lim_{x \to -1^-} f(x) = \lim_{x \to -1^-} 1 = 1$$

For $x \geq -1$, $f(x) = 1 + \sqrt{x + 1}$, and so

$$\lim_{x \to -1^+} f(x) = \lim_{x \to -1^+} (1 + \sqrt{x + 1}) = 1$$

Since the left-hand and right-hand limits of f exist as x approaches $x = -1$ and both are equal to 1, we conclude that

$$\lim_{x \to -1} f(x) = 1$$

continued

Solutions to Self-Check Exercises 3.2, *continued*

2\. **a.** The graph of f is as follows:

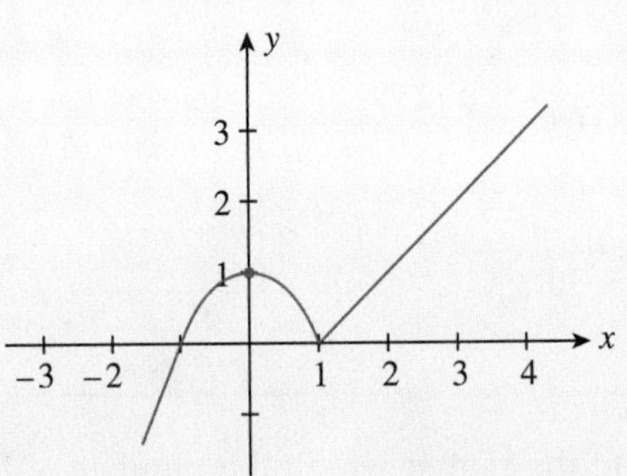

We see that f is continuous everywhere.

b. The graph of g is as follows:

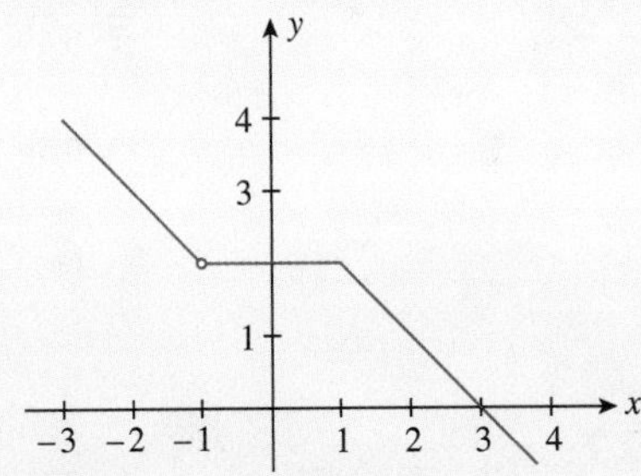

Since g is not defined at $x = -1$, it is discontinuous there. It is continuous everywhere else.

3.3 The Derivative

An Intuitive Example

We mentioned in Section 3.1 that the problem of finding the *rate of change* of one quantity with respect to another is mathematically equivalent to the problem of finding the *slope of the tangent line* to a curve at a given point on the curve. Before going on to establish this relationship, let's show its plausibility by looking at it from an intuitive point of view.

Consider the motion of the maglev discussed in Section 3.1. Recall that the position of the maglev at any time t is given by

$$s = f(t) = 4t^2 \qquad (0 \leq t \leq 30)$$

where s is measured in feet and t in seconds. The graph of the function f is sketched in Figure 3.25.

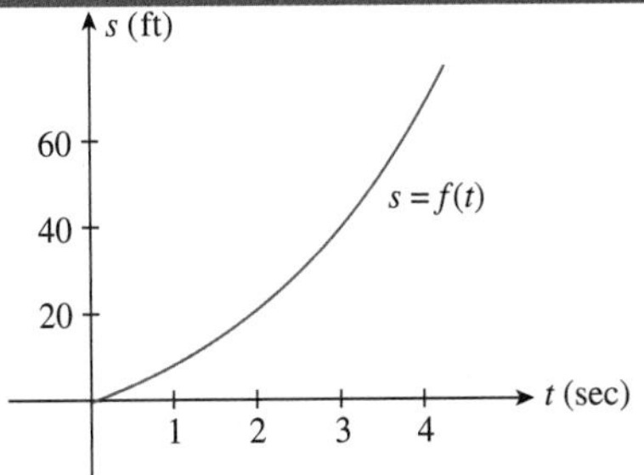

Figure 3.25
Graph showing the position s of a maglev at time t

Observe that the graph of f rises slowly at first but more rapidly as t increases, reflecting the fact that the speed of the maglev is increasing with time. This observation suggests a relationship between the speed of the maglev at any time t and the *steepness* of the curve at the point corresponding to this value of t. Thus, it would appear that we can solve the problem of finding the speed of the maglev at any time if we can find a way to measure the steepness of the curve at any point on the curve.

To discover a yardstick that will measure the steepness of a curve, consider the graph of a function f such as the one shown in Figure 3.26a. Think of the curve as representing a stretch of roller coaster track (Figure 3.26b). When the car is at the point P on the curve, a passenger sitting erect in the car and looking straight ahead will have a line of sight that is parallel to the line T, the tangent to the curve at P.

As Figure 3.26a suggests, the steepness of the curve—that is, the rate at which y is increasing or decreasing with respect to x—is given by the slope of the tangent line to the graph of f at the point $P(x, f(x))$. But for now we will show how this relationship can be used to estimate the rate of change of a function from its graph.

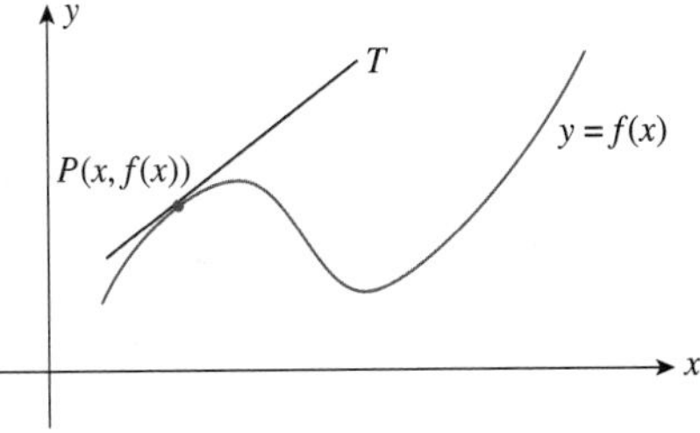

(a) T is the tangent line to the curve at P.

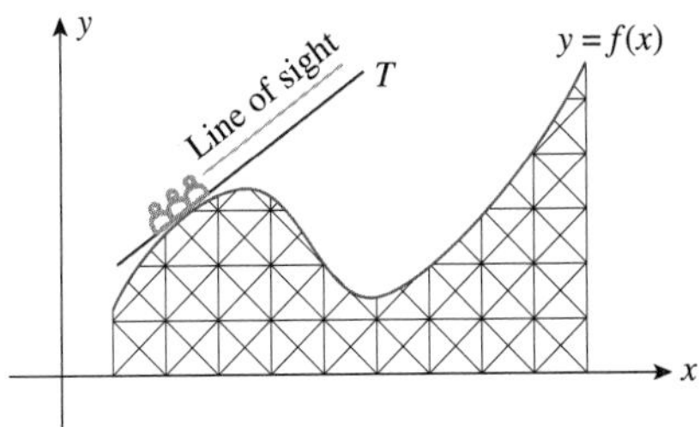

(b) T is parallel to the line of sight.

Figure 3.26

Example 1

Disability Beneficiaries The graph of the function $y = N(t)$, shown in Figure 3.27, gives the number of Canadians receiving disability benefits from the beginning of 1990 ($t = 0$) through the year 2045 ($t = 55$).

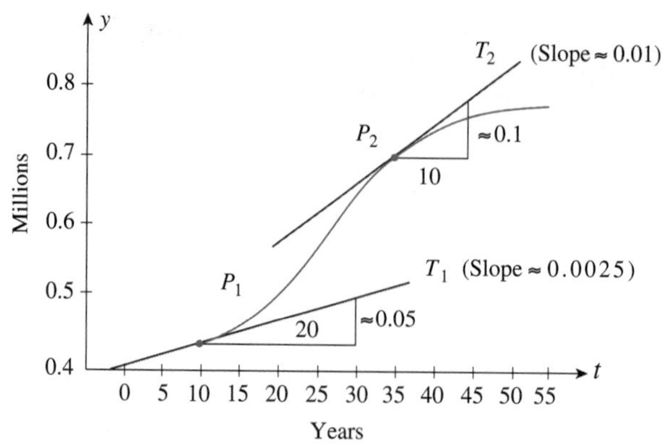

Figure 3.27

The number of Canadians receiving disability benefits from 1990 through 2045. We can use the slope of the tangent line at the indicated points to estimate the rate at which the number of beneficiaries will be changing.

Use the graph of $y = N(t)$ to estimate the rate at which the number of beneficiaries was growing at the beginning of the year 2000 ($t = 10$). How fast will the number be growing at the beginning of 2025 ($t = 35$)? [Assume that the rate of change of the function N at any value of t is given by the slope of the tangent line at the point $P(t, N(t))$.]

Solution

From the figure, we see that the slope of the tangent line T_1 to the graph of $y = N(t)$ at $P_1(10, 0.447)$ is approximately 0.0025. This tells us that the quantity y is increasing at the rate of 0.0025 unit per unit increase in t, when $t = 10$. In other words, at the beginning of the year 2000, the number of beneficiaries was increasing at the rate of approximately 0.0025 million, or 2500, per year.

The slope of the tangent line T_2 at $P_2(35, 0.719)$ is approximately 0.01. This tells us that at the beginning of 2025 the number of beneficiaries will be growing at the rate of approximately 0.01 million, or 10,000, per year.

Slope of a Tangent Line

In Example 1 we answered the questions raised by drawing the graph of the function N and estimating the position of the tangent lines. Ideally, however, we would like to solve a problem analytically whenever possible. To do this we need a precise definition of the slope of a tangent line to a curve.

To define the tangent line to a curve C at a point P on the curve, fix P and let Q be any point on C distinct from P (Figure 3.28). The straight line passing through P and Q is called a **secant line.**

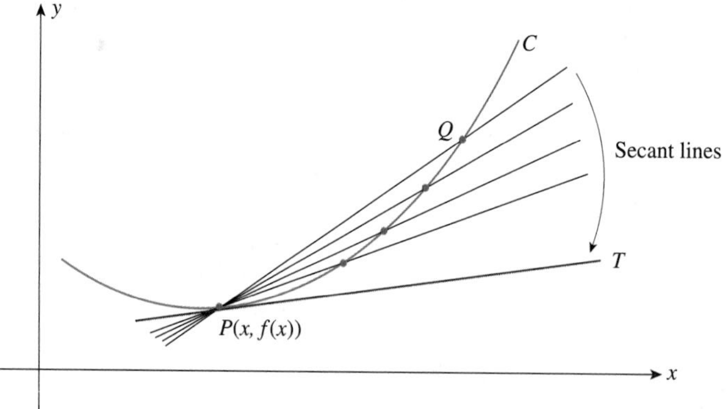

Figure 3.28

As Q approaches P along the curve C, the secant lines approach the tangent line T.

Now, as the point Q is allowed to move toward P along the curve, the secant line through P and Q rotates about the fixed point P and approaches a fixed line through P. This fixed line, which is the limiting position of the secant lines through P and Q as Q approaches P, is the **tangent line to the graph of f** at the point P.

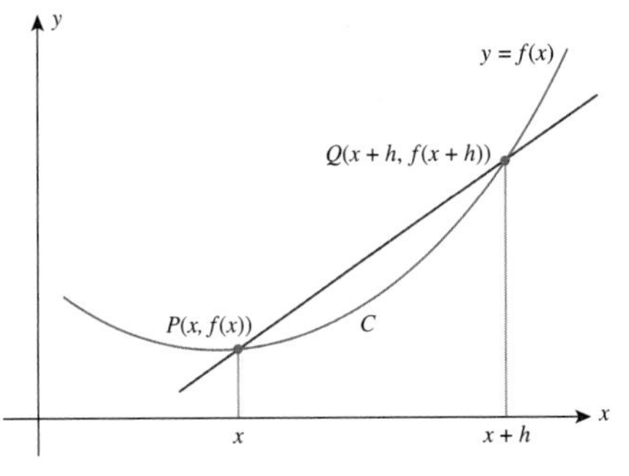

(a) The points $P(x, f(x))$ and $Q(x + h, f(x + h))$

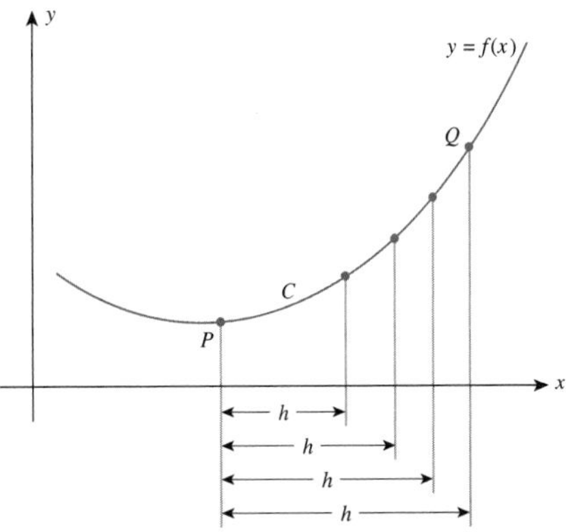

(b) As h approaches zero, Q approaches P.

Figure 3.29

We can describe the process more precisely as follows. Suppose the curve C is the graph of a function f defined by $y = f(x)$. Then the point P is described by $P(x, f(x))$ and the point Q by $Q(x + h, f(x + h))$, where h is some appropriate nonzero number (Figure 3.29a). Observe that we can make Q approach P along the curve C by letting h approach zero (Figure 3.29b).

Next, using the formula for the slope of a line, we can write the slope of the secant line passing through $P(x, f(x))$ and $Q(x + h, f(x + h))$ as

$$\frac{f(x+h) - f(x)}{(x+h) - x} = \frac{f(x+h) - f(x)}{h} \qquad (3)$$

As observed earlier, Q approaches P, and therefore the secant line through P and Q approaches the tangent line T as h approaches zero. Consequently, we might expect that the slope of the secant line would approach the slope of the tangent line T as h approaches zero. This leads to the following definition.

Slope of a Tangent Line

The slope of the tangent line to the graph f at the point $P(x, f(x))$ is given by

$$\lim_{h \to 0} \frac{f(x+h) - f(x)}{h} \qquad (4)$$

if it exists.

Rates of Change

We now show that the problem of finding the slope of the tangent line to the graph of a function f at the point $P(x, f(x))$ is mathematically equivalent to the problem of finding the rate of change of f at x. To see this, suppose we are given a function f that describes the relationship between the two quantities x and y—that is, $y = f(x)$. The number $f(x + h) - f(x)$ measures the change in y that corresponds to a change h in x (Figure 3.30).

Then, the **difference quotient**

$$\frac{f(x+h) - f(x)}{h} \qquad (5)$$

measures the **average rate of change of y with respect to x** over the interval $[x, x + h]$. For example, if y measures the position of a car at time x, then quotient (5) gives the average velocity of the car over the time interval $[x, x + h]$.

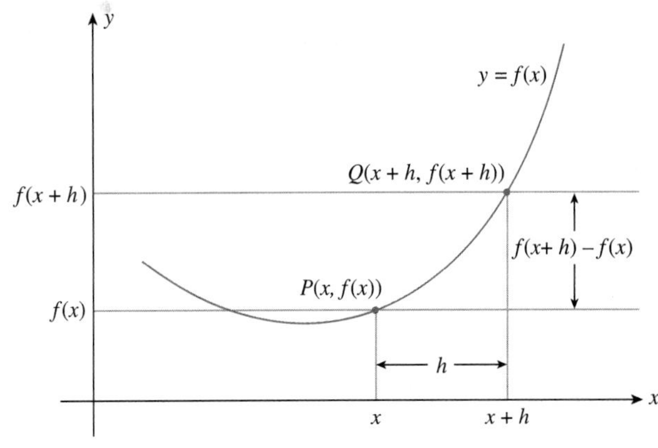

Figure 3.30
$f(x + h) - f(x)$ is the change in y that corresponds to a change h in x.

Observe that the difference quotient (5) is the same as (3). We conclude that the difference quotient (5) also measures the slope of the secant line that passes through the two points $P(x, f(x))$ and $Q(x + h, f(x + h))$ lying on the graph of $y = f(x)$. Next, by taking the limit of the difference quotient (5) as h goes to zero—that is, by evaluating

$$\lim_{h \to 0} \frac{f(x + h) - f(x)}{h} \qquad (6)$$

we obtain the **rate of change of f at x.** For example, if y measures the position of a car at time x, then the limit (6) gives the velocity of the car at time x. For emphasis, the rate of change of a function f at x is often called the **instantaneous rate of change of f at x.** This distinguishes it from the average rate of change of f, which is computed over an *interval* $[x, x + h]$ rather than at a *point* x.

Observe that the limit (6) is the same as (4). Therefore, the limit of the difference quotient also measures the slope of the tangent line to the graph of $y = f(x)$ at the point $(x, f(x))$. The following summarizes this discussion.

GROUP DISCUSSION

Explain the difference between the average rate of change of a function and the instantaneous rate of change of a function.

Average and Instantaneous Rates of Change

The **average rate of change** of f over the interval $[x, x + h]$ or **slope of the secant line** to the graph of f through the points $(x, f(x))$ and $(x + h, f(x + h))$ is

$$\frac{f(x + h) - f(x)}{h} \qquad (7)$$

The **instantaneous rate of change** of f at x or **slope of the tangent line** to the graph of f at $(x, f(x))$ is

$$\lim_{h \to 0} \frac{f(x + h) - f(x)}{h} \qquad (8)$$

The Derivative

The limit (4) or (8), which measures both the slope of the tangent line to the graph of $y = f(x)$ at the point $P(x, f(x))$ and the (instantaneous) rate of change of f at x, is given a special name: the **derivative of f at x.**

Derivative of a Function

The derivative of a function f with respect to x is the function f' (read "f prime"),

$$f'(x) = \lim_{h \to 0} \frac{f(x + h) = f(x)}{h} \qquad (9)$$

The domain of f' is the set of all x where the limit exists.

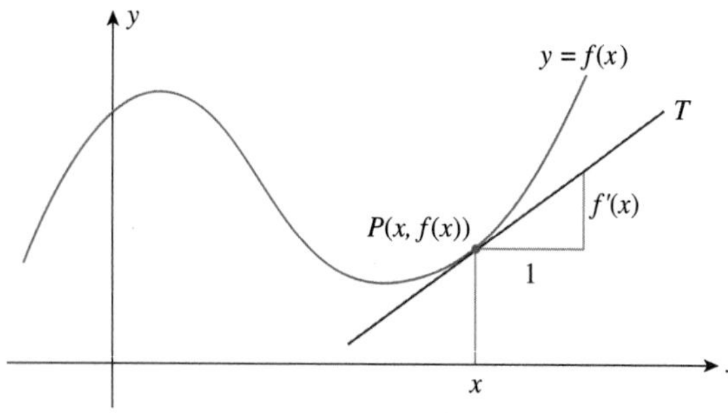

Figure 3.31
The slope of the tangent line at $P(x, f(x))$ is $f'(x)$; f changes at the rate of $f'(x)$ units per unit change in x at x.

Thus, the derivative of a function f is a function f' that gives the slope of the tangent line to the graph of f at *any* point $(x, f(x))$ and also the rate of change of f at x (Figure 3.31).

Other notations for the derivative of f include:

$D_x f(x)$	Read "d sub x of f of x"
$\frac{dy}{dx}$	Read "$d\ y\ d\ x$"
y'	Read "y prime"

The last two are used when the rule for f is written in the form $y = f(x)$.

The calculation of the derivative of f is facilitated using the following four-step process.

Four-Step Process for Finding $f'(x)$

1. Compute $f(x + h)$.
2. Form the difference $f(x + h) - f(x)$.
3. Form the quotient $\dfrac{f(x + h) - f(x)}{h}$.
4. Compute $f'(x) = \lim_{h \to 0} \dfrac{f(x + h) - f(x)}{h}$.

Example 2

Find the slope of the tangent line to the graph of $f(x) = 3x + 5$ at any point $(x, f(x))$.

Solution

The slope of the tangent line at any point on the graph of f is given by the derivative of f at x. To find the derivative, we use the four-step process:

Step 1 $f(x + h) = 3(x + h) + 5 = 3x + 3h + 5$

Step 2 $f(x + h) - f(x) = (3x + 3h + 5) - (3x + 5) = 3h$

Step 3 $\dfrac{f(x + h) - f(x)}{h} = \dfrac{3h}{h} = 3$

Step 4 $f'(x) = \lim_{h \to 0} \dfrac{f(x + h) - f(x)}{h} = \lim_{h \to 0} 3 = 3$

We expect this result since the tangent line to any point on a straight line must coincide with the line itself and therefore must have the same slope as the line. In this case the graph of f is a straight line with slope 3.

Example 3

Let $f(x) = x^2$.

a. Compute $f'(x)$.
b. Compute $f'(2)$ and interpret your result.

Solution

a. To find $f'(x)$, we use the four-step process:

Step 1 $f(x + h) = (x + h)^2 = x^2 + 2xh + h^2$

Step 2 $f(x + h) - f(x) = x^2 + 2xh + h^2 - x^2 = 2xh + h^2 = h(2x + h)$

Step 3 $\dfrac{f(x + h) - f(x)}{h} = \dfrac{h(2x + h)}{h} = 2x + h$

Step 4 $f'(x) = \lim_{h \to 0} \dfrac{f(x + h) - f(x)}{h} = \lim_{h \to 0} (2x + h) = 2x$

b. $f'(2) = 2(2) = 4$. This result tells us that the slope of the tangent line to the graph of f at the point $(2, 4)$ is 4. It also tells us that the function f is changing at the rate of 4 units per unit change in x at $x = 2$. The graph of f and the tangent line at $(2, 4)$ are shown in Figure 3.32.

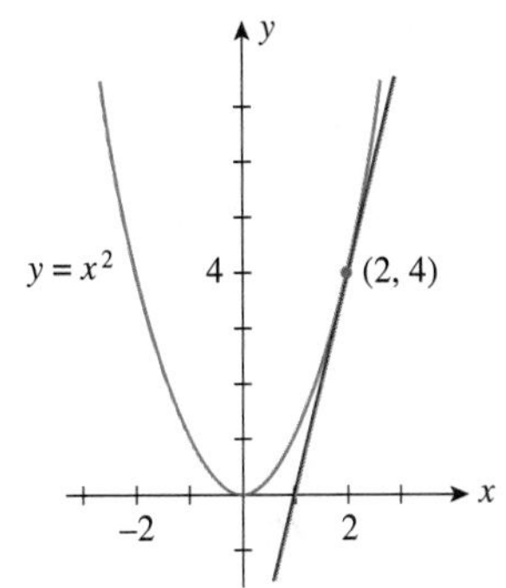

Figure 3.32
The tangent line to the graph of $f(x) = x^2$ at $(2, 4)$

EXPLORING WITH TECHNOLOGY

1. Consider the function $f(x) = x^2$ of Example 3. Suppose we want to compute $f'(2)$, using Equation (9). Thus,

$$f'(2) = \lim_{h\to 0} \frac{f(2+h) - f(2)}{h} = \lim_{h\to 0} \frac{(2+h)^2 - 2^2}{h}$$

Use a graphing utility to plot the graph of

$$g(x) = \frac{(2+x)^2 - 4}{x}$$

in the viewing window $[-3, 3] \cdot [-2, 6]$.

2. Use **ZOOM** and **TRACE** to find $\lim_{x\to 0} g(x)$.
3. Explain why the limit found in part 2 is $f'(2)$.

Example 4

Let $f(x) = x^2 - 4x$.

a. Compute $f'(x)$.
b. Find the point on the graph of f where the tangent line to the curve is horizontal.
c. Sketch the graph of f and the tangent line to the curve at the point found in part (b).
d. What is the rate of change of f at this point?

Solution

a. To find $f'(x)$, we use the four-step process:

Step 1 $f(x+h) = (x+h)^2 - 4(x+h) = x^2 + 2xh + h^2 - 4x - 4h$

Step 2 $f(x+h) - f(x) = x^2 + 2xh + h^2 - 4x - 4h - (x^2 - 4x) = 2xh + h^2 - 4h$
$= h(2x + h - 4)$

Step 3 $\dfrac{f(x+h) - f(x)}{h} = \dfrac{h(2x+h-4)}{h} = 2x + h - 4$

Step 4 $f'(x) = \lim_{h\to 0} \dfrac{f(x+h) - f(x)}{h} = \lim_{h\to 0} (2x + h - 4) = 2x - 4$

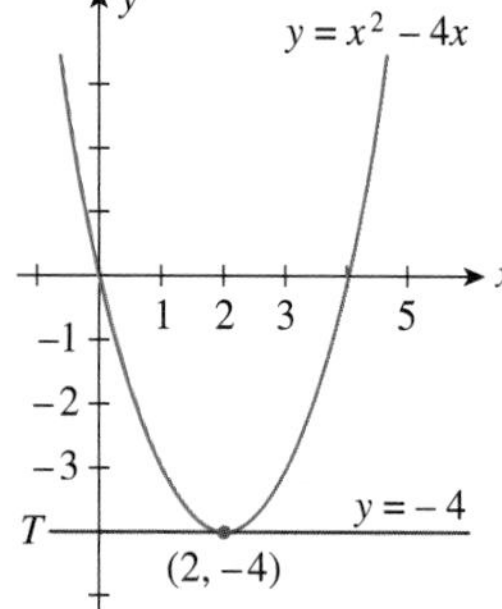

Figure 3.33
The tangent line to the graph of $y = x^2 - 4x$ at $(2, -4)$ is $y = -4$.

b. At a point on the graph of f where the tangent line to the curve is horizontal and hence has slope zero, the derivative f' of f is zero. Accordingly, to find such point(s) we set $f'(x) = 0$, which gives $2x - 4 = 0$, or $x = 2$. The corresponding value of y is given by $y = f(2) = 24$, and the required point is $(2, -4)$.
c. The graph of f and the tangent line are shown in Figure 3.33.
d. The rate of change of f at $x = 2$ is zero.

GROUP DISCUSSION

Can the tangent line to the graph of a function intersect the graph at more than one point? Explain your answer using illustrations.

Example 5

Let $f(x) = \dfrac{1}{x}$.

a. Compute $f'(x)$.
b. Find the slope of the tangent line T to the graph of f at the point where $x = 1$.
c. Find an equation of the tangent line T in part (b).

Solution

a. To find $f'(x)$, we use the four-step process:

Step 1 $f(x+h) = \dfrac{1}{x+h}$

Step 2 $f(x+h) - f(x) = \dfrac{1}{x+h} - \dfrac{1}{x} = \dfrac{x - (x+h)}{x(x+h)} = -\dfrac{h}{x(x+h)}$

Step 3 $\dfrac{f(x+h)-f(x)}{h} = -\dfrac{h}{x(x+h)}\cdot\dfrac{1}{h} = -\dfrac{1}{x(x+h)}$

Step 4 $f'(x) = \lim\limits_{h\to 0}\dfrac{f(x+h)-f(x)}{h} = \lim\limits_{h\to 0} -\dfrac{1}{x(x+h)} = -\dfrac{1}{x^2}$

b. The slope of the tangent line T to the graph of f where $x = 1$ is given by $f'(1) = -1$.

c. When $x = 1$, $y = f(1) = 1$ and T is tangent to the graph of f at the point $(1, 1)$. From part (b), we know that the slope of T is -1. Thus, an equation of T is

$$y - 1 = -1(x - 1)$$
$$y = -x + 2$$

(Figure 3.34).

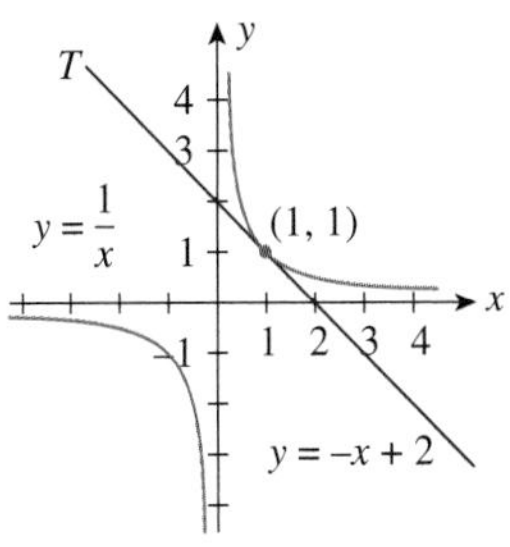

Figure 3.34
The tangent line to the graph of $f(x) = 1/x$ at $(1, 1)$

EXPLORING WITH TECHNOLOGY

1. Use the results of Example 5 to draw the graph of $f(x) = 1/x$ and its tangent line at the point (1, 1) by plotting the graphs of $y_1 = 1/x$ and $y_2 = -x + 2$ in the viewing window $[-4, 4] \cdot [-4, 4]$.
2. Some graphing utilities draw the tangent line to the graph of a function at a given point automatically—you need only specify the function and give the x-coordinate of the point of tangency. If your graphing utility has this feature, verify the result of part 1 without finding an equation of the tangent line.

Example 6

Let $f(x) = \sqrt{x+2}$ and compute $f'(2)$.

Solution

To find $f'(x)$, we use the four-step process:

Step 1 $f(x+h) = \sqrt{x+h+2}$

Step 2 $f(x+h) - f(x) = \sqrt{x+h+2} - \sqrt{x+2}$

Step 3 $\dfrac{f(x+h)-f(x)}{h} = \dfrac{\sqrt{x+h+2} - \sqrt{x+2}}{h}$

Step 4 $f'(x) = \lim\limits_{h\to 0}\dfrac{f(x+h)-f(x)}{h} = \lim\limits_{h\to 0}\dfrac{\sqrt{x+h+2} - \sqrt{x+2}}{h}$

Notice that this limit has the indeterminate form $\frac{0}{0}$ as h approaches zero. The technique to resolve this type of limit is multiplying the numerator and denominator with the conjugate of $\sqrt{x+h+2} - \sqrt{x+2}$, which yields

$$\begin{aligned} f'(x) &= \lim_{h\to 0}\frac{\sqrt{x+h+2} - \sqrt{x+2}}{h}\cdot\frac{\sqrt{x+h+2} + \sqrt{x+2}}{\sqrt{x+h+2} + \sqrt{x+2}} \\ &= \lim_{h\to 0}\frac{(x+h+2)-(x+2)}{h(\sqrt{x+h+2} + \sqrt{x+2})} \\ &= \lim_{h\to 0}\frac{\cancel{h}}{\cancel{h}(\sqrt{x+h+2} + \sqrt{x+2})} = \lim_{h\to 0}\frac{1}{\sqrt{x+h+2} + \sqrt{x+2}} \\ &= \frac{1}{\sqrt{x+2} + \sqrt{x+2}} = \frac{1}{2\sqrt{x+2}} \end{aligned}$$

Finally, we can compute $f'(2)$ and get

$$f'(2) = \frac{1}{2\sqrt{2+2}} = \frac{1}{2\sqrt{4}} = \frac{1}{4}$$

GROUP DISCUSSION

Consider the following alternative approach to the definition of the derivative of a function: Let h be a positive number and suppose $P(x - h, f(x - h))$ and $Q(x + h, f(x + h))$ are two points on the graph of f.

1. Give a geometric and a physical interpretation of the quotient

$$\frac{f(x + h) - f(x - h)}{2h}$$

Make a sketch to illustrate your answer.

2. Give a geometric and a physical interpretation of the limit

$$\lim_{h \to 0} \frac{f(x + h) - f(x - h)}{2h}$$

Make a sketch to illustrate your answer.

3. Explain why it makes sense to define

$$f'(x) = \lim_{h \to 0} \frac{f(x + h) - f(x - h)}{2h}$$

4. Using the definition given in part 3, formulate a four-step process for finding $f'(x)$ similar to that given on page 149 and use it to find the derivative of $f(x) = x^2$. Compare your answer with that obtained in Example 3 on page 149.

Applications

Example 7

Average Velocity of a Car Suppose the distance (in metres) covered by a car moving along a straight road t seconds

after starting from rest is given by the function $f(t) = \frac{t^2}{2}$ $(0 \leq t \leq 30)$.

a. Calculate the average velocity of the car over the time intervals [22, 23], [22, 22.1], and [22, 22.01].

b. Calculate the (instantaneous) velocity of the car when $t = 22$.

c. Compare the results obtained in part (a) with that obtained in part (b).

Solution

a. We first compute the average velocity (average rate of change of f) over the interval $[t, t + h]$ using Formula (7). We find

$$\frac{f(t + h) - f(t)}{h} = \frac{\frac{(t + h)^2}{2} - \frac{t^2}{2}}{h} = \frac{t^2 + 2ht + h^2 - t^2}{2h} = \frac{2ht + h^2}{2h} = \frac{2t + h}{2}$$

Next, using $t = 22$ and $h = 1$, we find that the average velocity of the car over the time interval [22, 23] is

$$\frac{2.22 + 1}{2}$$

or 22.5 metres per second. Similarly, using $t = 22$, $h = 0.1$, and $h = 0.01$, we find that its average velocities over the time intervals [22, 22.1] and [22, 22.01] are 22.05 and 22.005 metres per second, respectively.

b. Using the limit (8), we see that the instantaneous velocity of the car at any time t is given by

$$\lim_{h \to 0} \frac{f(t+h) - f(t)}{h} = \lim_{h \to 0} \frac{2t + h}{2} = \frac{2t}{2} = t$$

In particular, the velocity of the car 22 seconds from rest ($t = 22$) is given by

$$v = 22$$

or 22 metres per second.

c. The computations in part (a) show that, as the time intervals over which the average velocity of the car are computed become smaller and smaller, the average velocities over these intervals do approach 22 metres per second, the instantaneous velocity of the car at $t = 22$.

Example 8

Demand for Tires The management of Titan Tire Company has determined that the weekly demand function of their Super Titan tires is given by

$$p = f(x) = 144 - x^2$$

where p is measured in dollars and x is measured in units of a thousand (Figure 3.35).

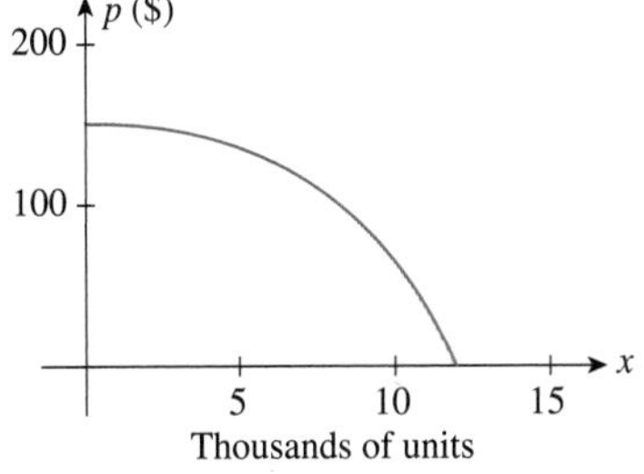

Figure 3.35
The graph of the demand function $p = 144 - x^2$

a. Find the average rate of change in the unit price of a tire if the quantity demanded is between 5000 and 6000 tires, between 5000 and 5100 tires, and between 5000 and 5010 tires.

b. What is the instantaneous rate of change of the unit price when the quantity demanded is 5000 units?

Solution

a. The average rate of change of the unit price of a tire if the quantity demanded is between x and $x + h$ is

$$\begin{aligned} \frac{f(x+h) - f(x)}{h} &= \frac{[144 - (x+h)^2] - (144 - x^2)}{h} \\ &= \frac{144 - x^2 - 2x(h) - h^2 - 144 + x^2}{h} \\ &= -2x - h \end{aligned}$$

To find the average rate of change of the unit price of a tire when the quantity demanded is between 5000 and 6000 tires (that is, over the interval [5, 6]), we take $x = 5$ and $h = 1$, obtaining

$$-2(5) - 1 = -11$$

or $-\$11$ per 1000 tires. (Remember, x is measured in units of a thousand.) Similarly, taking $h = 0.1$ and $h = 0.01$ with $x = 5$, we find that the average rates of change of the unit price when the quantities demanded are between 5000 and 5100 and between 5000 and 5010 are $-\$10.10$ and $-\$10.01$ per 1000 tires, respectively.

b. The instantaneous rate of change of the unit price of a tire when the quantity demanded is x units is given by

$$\begin{aligned} \lim_{h \to 0} \frac{f(x+h) - f(x)}{h} &= \lim_{h \to 0} (-2x - h) \qquad \text{Use the results from part (a).} \\ &= -2x \end{aligned}$$

In particular, the instantaneous rate of change of the unit price per tire when the quantity demanded is 5000 is given by $-2(5)$, or $-\$10$ per 1000 tires.

The derivative of a function provides us with a tool for measuring the rate of change of one quantity with respect to another. Table 3.4 lists several other applications involving this limit.

Table 3.4 Applications Involving Rate of Change

x Stands for	y Stands for	$\dfrac{f(a+h)-f(a)}{h}$ Measures	$\lim_{h\to 0}\dfrac{f(a+h)-f(a)}{h}$ Measures
Time	**Concentration of a drug** in the bloodstream at time x	Average rate of change in the concentration of the drug over the time interval $[a, a+h]$	Instantaneous rate of change in the concentration of the drug in the bloodstream at time $x = a$
Number of items sold	**Revenue** at a sales level of x units	Average rate of change in the revenue when the sales level is between $x = a$ and $x = a + h$	Instantaneous rate of change in the revenue when the sales level is a units
Time	**Volume of sales** at time x	Average rate of change in the volume of sales over the time interval $[a, a+h]$	Instantaneous rate of change in the volume of sales at time $x = a$
Time	**Population** of *Drosophila* (fruit flies) at time x	Average rate of growth of the fruit fly population over the time interval $[a, a+h]$	Instantaneous rate of change of the fruit fly population at time $x = a$
Temperature in a chemical reaction	**Amount of product formed in the chemical reaction** when the temperature is x degrees	Average rate of formation of chemical product over the temperature range $[a, a+h]$	Instantaneous rate of formation of chemical product when the temperature is a degrees

Differentiability and Continuity

In practical applications, one encounters continuous functions that fail to be **differentiable**—that is, do not have a derivative at certain values in the domain of the function f. It can be shown that a continuous function f fails to be differentiable at a point $x = a$ when the graph of f makes an abrupt change of direction at that point. We call such a point a "corner." A function also fails to be differentiable at a point where the tangent line is vertical since the slope of a vertical line is undefined. These cases are illustrated in Figure 3.36.

The next example illustrates a function that is not differentiable at a point.

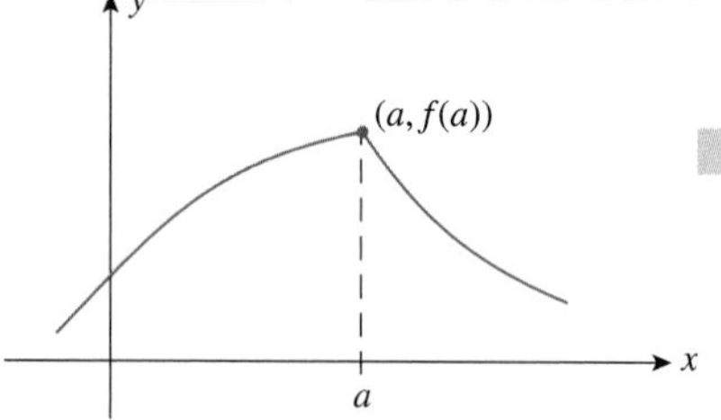

(a) The graph makes an abrupt change of direction at $x = a$.

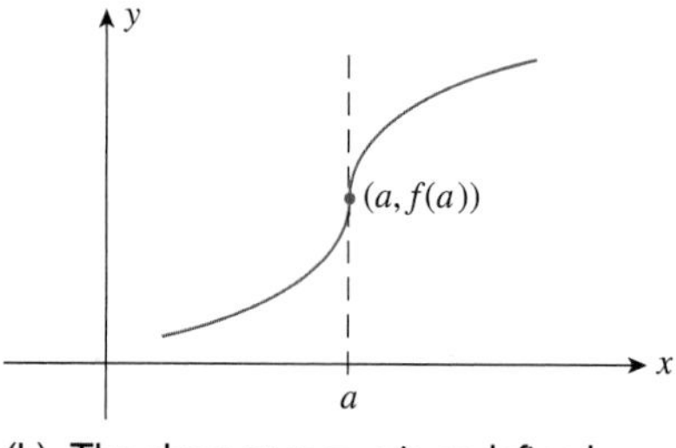

(b) The slope at $x = a$ is undefined.

Figure 3.36

Example 9

Wages Your mom, Mary, tells you that before your time she worked at a thrift store. On a weekday, she was paid \$6 an hour for the first 8 hours and \$9 an hour for overtime. The function

$$f(x) = \begin{cases} 6x & \text{if } 0 \le x \le 8 \\ 9x - 24 & \text{if } 8 < x \end{cases}$$

gives Mary's earnings on a weekday in which she worked x hours. Sketch the graph of the function f and explain why it is not differentiable at $x = 8$.

Solution

The graph of f is shown in Figure 3.37. Observe that the graph of f has a corner at $x = 8$ and consequently is not differentiable at $x = 8$.

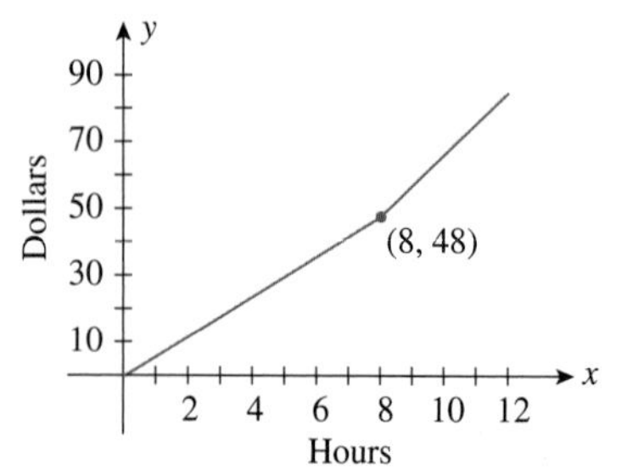

Figure 3.37
The function f is not differentiable at (8, 48).

We close this section by mentioning the connection between the continuity and the differentiability of a function at a given value $x = a$ in the domain of f. By reexamining the function of Example 9, it becomes clear that f is continuous everywhere and, in particular, when $x = 8$. This shows that in general the continuity of a function at a point $x = a$ does not necessarily imply the differentiability of the function at that point. The converse, however, is true: If a function f is differentiable at a point $x = a$, then it is continuous there.

Differentiability and Continuity

If a function is differentiable at $x = a$, then it is continuous at $x = a$.

For a proof of this result, see Exercise 49, page 158.

Remark

We can conclude from the above theorem that if a function f is discontinuous, then it cannot be differentiable, which is another useful statement to remember when working with differentiability of a function. It means that any function with a removable, jump, or infinite discontinuity at a point a is automatically not differentiable at the point a. ◀

EXPLORING WITH TECHNOLOGY

1. Use a graphing utility to plot the graph of $f(x) = x^{1/3}$ in the viewing window $[-2, 2] \cdot [-2, 2]$.
2. Use a graphing utility to draw the tangent line to the graph of f at the point (0, 0). Can you explain why the process breaks down?

Example 10

Figure 3.38 depicts a portion of the graph of a function. Explain why the function fails to be differentiable at each of the points $x = a, b, c, d, e, f$, and g.

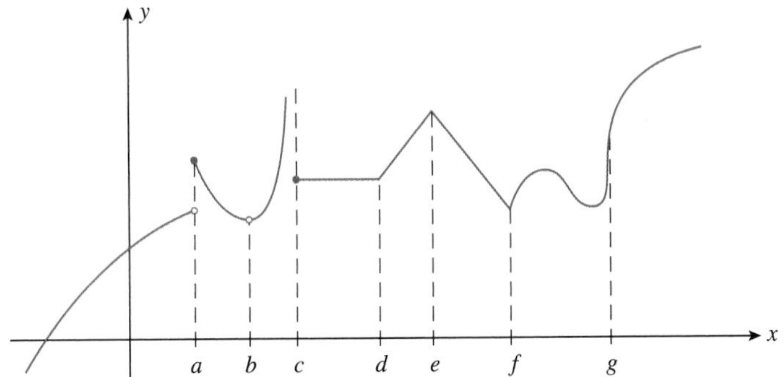

Figure 3.38
The graph of this function is not differentiable at the points a–g.

GROUP DISCUSSION

Suppose a function f is differentiable at $x = a$. Can there be two tangent lines to the graph of f at the point $(a, f(a))$? Explain your answer.

Solution

The function fails to be differentiable at the points $x = a, b$, and c because it is discontinuous at each of these points. The derivative of the function does not exist at $x = d$, e, and f because there is a corner on the graph at each of these points. Finally, the function is not differentiable at $x = g$ because the tangent line is vertical at that point.

Example 11

Show that $f(x) = |x - 3|$ is differentiable everywhere except at $x = 3$ by (a) graphing and (b) using calculus.

Solution

a. The graph of f is the basic absolute value function horizontally translated right by 3 units with vertex at (3, 0) shown in Figure 3.39.

We can see from the graph that f has a corner at the vertex (3, 0), and so f is not differentiable at $x = 3$. On the domain $x > 3$, the graph of f is a line and therefore differentiable everywhere. Similarly for the domain $x < 3$, the graph of f is also a line and therefore differentiable everywhere.

b. We rewrite f in piecewise defined form

$$f(x) = |x - 3| = \begin{cases} -(x - 3) & \text{if } x < 3 \\ x - 3 & \text{if } x \geq 3 \end{cases}$$

$$= \begin{cases} -x + 3 & \text{if } x < 3 \\ x - 3 & \text{if } x \geq 3 \end{cases}$$

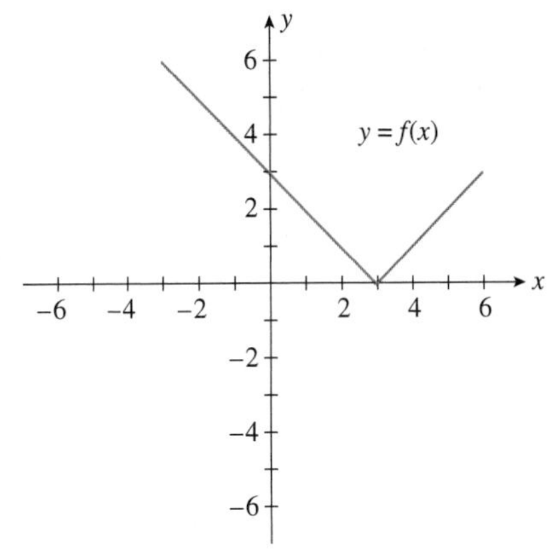

Figure 3.39
The graph of $f(x) = |x - 3|$

If $x > 3$, then

$$\begin{aligned}\lim_{h\to 0}\frac{f(x+h)-(f)x}{h} &= \lim_{h\to 0}\frac{|x+h-3|-|x-3|}{h}\\ &= \lim_{h\to 0}\frac{(x+h-3)-(x-3)}{h}\\ &= \lim_{h\to 0}\frac{h}{h}\\ &= 1\end{aligned}$$

and so f is differentiable for $x > 3$.

Similarly, if $x < 3$, then

$$\begin{aligned}\lim_{h\to 0}\frac{f(x+h)-f(x)}{h} &= \lim_{h\to 0}\frac{|x+h-3|-|x-3|}{h}\\ &= \lim_{h\to 0}\frac{-(x+h-3)-(-(x-3))}{h}\\ &= \lim_{h\to 0}\frac{-h}{h}\\ &= -1\end{aligned}$$

which means that f is differentiable for $x < 3$.

Finally, let us look at the case when $x = 3$ for which

$$\lim_{h\to 0}\frac{f(3+h)-f(3)}{h} = \lim_{h\to 0}\frac{|3+h-3|-|3-3|}{h} = \lim_{h\to 0}\frac{|h|}{h}$$

We use Theorem 3 to continue and resolve the limit. We first compute the left-hand limit.

$$\lim_{h\to 0^-}\frac{|h|}{h} = \lim_{h\to 0^-}\frac{-h}{h} = \lim_{h\to 0^-}(-1) = -1$$

Next, we compute the right-hand limit.

$$\lim_{h\to 0^+}\frac{|h|}{h} = \lim_{h\to 0^+}\frac{h}{h} = \lim_{h\to 0^+} 1 = 1$$

Since

$$\lim_{h\to 0^+}\frac{|h|}{h} = 1 \neq \lim_{h\to 0^-}\frac{|h|}{h} = -1$$

we conclude by Theorem 3 that $\lim_{h\to 0}\frac{|h|}{h}$ does not exist and therefore, f is not differentiable at $x = 3$.

Self-Check Exercises 3.3

1. Let $f(x) = -x^2 - 2x + 3$.
   a. Find the derivative f' of f, using the definition of the derivative.
   b. Find the slope of the tangent line to the graph of f at the point (0, 3).
   c. Find the rate of change of f when $x = 0$.
   d. Find an equation of the tangent line to the graph of f at the point (0, 3).
   e. Sketch the graph of f and the tangent line to the curve at the point (0, 3).

2. The losses (in millions of dollars) due to bad loans extended chiefly in agriculture, real estate, shipping, and energy by the Franklin Bank are estimated to be

$$A = f(t) = -t^2 + 10t + 30 \qquad (0 \le t \le 10)$$

where t is the time in years ($t = 0$ corresponds to the beginning of 1994). How fast were the losses mounting at the beginning of 1997? At the beginning of 1999? At the beginning of 2001?

Solutions to Self-Check Exercises 3.3 can be found on page 160.

3.3 Exercises

1. The position of car A and car B, starting out side by side and traveling along a straight road, is given by $s = f(t)$ and $s = g(t)$, respectively, where s is measured in metres and t is measured in seconds (see the accompanying figure).

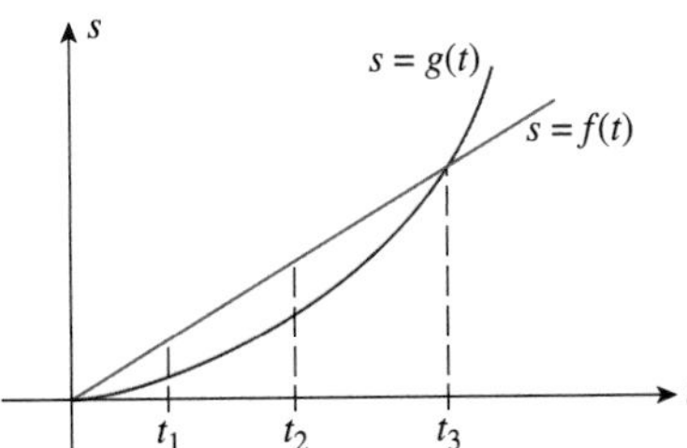

   a. Which car is travelling faster at t_1?
   b. What can you say about the speed of the cars at t_2?
   Hint: Compare tangent lines.
   c. Which car is travelling faster at t_3?
   d. What can you say about the positions of the cars at t_3?

2. The velocity of car A and car B, starting out side by side and travelling along a straight road, is given by $v = f(t)$ and $v = g(t)$, respectively, where v is measured in metres/second and t is measured in seconds (see the accompanying figure).

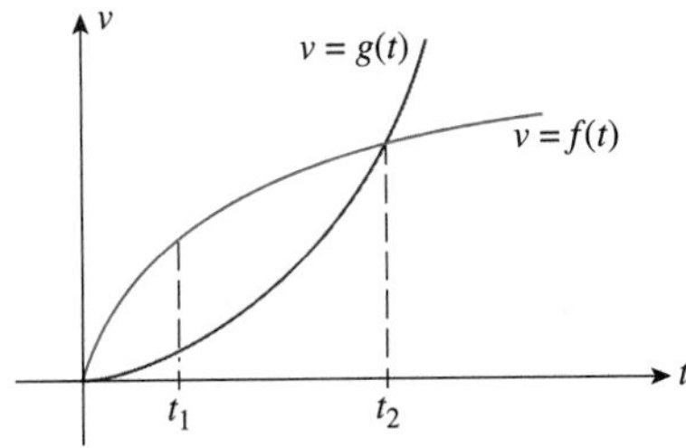

   a. What can you say about the velocity and acceleration of the two cars at t_1? (Acceleration is the rate of change of velocity.)
   b. What can you say about the velocity and acceleration of the two cars at t_2?

In Exercises 3–12, use the four-step process to find the slope of the tangent line to the graph of the given function at any point.

3. $f(x) = 13$
4. $f(x) = -6$
5. $f(x) = 2x + 7$
6. $f(x) = 8 - 4x$
7. $f(x) = 3x^2$
8. $f(x) = -\frac{1}{2}x^2$
9. $f(x) = -x^2 + 3x$
10. $f(x) = 2x^2 + 5x$
11. $f(x) = \sqrt{x - 5}$
12. $f(x) = \sqrt{3 - x}$

In Exercises 13–20, find the slope of the tangent line to the graph of each function at the given point and determine an equation of the tangent line.

13. $f(x) = 2x + 7$ at $(2, 11)$
14. $f(x) = -3x + 4$ at $(-1, 7)$
15. $f(x) = 3x^2$ at $(1, 3)$
16. $f(x) = 3x - x^2$ at $(-2, -10)$
17. $f(x) = -\frac{1}{x}$ at $(3, -\frac{1}{3})$
18. $f(x) = \frac{3}{2x}$ at $\left(1, \frac{3}{2}\right)$
19. $f(x) = 2\sqrt{x + 1}$ at $(3, 4)$
20. $f(x) = \sqrt{2x + 1}$ at $(4, 3)$

21. Let $f(x) = 2x^2 + 1$.
   a. Find the derivative f' of f.
   b. Find an equation of the tangent line to the curve at the point $(1, 3)$.
   c. Sketch the graph of f.

22. Let $f(x) = x^2 + 6x$.
   a. Find the derivative f' of f.
   b. Find the point on the graph of f where the tangent line to the curve is horizontal.
   Hint: Find the value of x for which $f'(x) = 0$.
   c. Sketch the graph of f and the tangent line to the curve at the point found in part (b).

23. Let $f(x) = x^2 - 2x + 1$.
   a. Find the derivative f' of f.
   b. Find the point on the graph of f where the tangent line to the curve is horizontal.
   c. Sketch the graph of f and the tangent line to the curve at the point found in part (b).
   d. What is the rate of change of f at this point?

24. Let $f(x) = \frac{1}{x - 1}$.
   a. Find the derivative f' of f.
   b. Find an equation of the tangent line to the curve at the point $(-1, -\frac{1}{2})$.
   c. Sketch the graph of f.

25. Let $y = f(x) = x^2 + x$.
   a. Find the average rate of change of y with respect to x in the interval from $x = 2$ to $x = 3$, from $x = 2$ to $x = 2.5$, and from $x = 2$ to $x = 2.1$.
   b. Find the (instantaneous) rate of change of y at $x = 2$.
   c. Compare the results obtained in part (a) with that of part (b).

26. Let $y = f(x) = x^2 - 4x$.
   a. Find the average rate of change of y with respect to x in the interval from $x = 3$ to $x = 4$, from $x = 3$ to $x = 3.5$, and from $x = 3$ to $x = 3.1$.
   b. Find the (instantaneous) rate of change of y at $x = 3$.
   c. Compare the results obtained in part (a) with that of part (b).

27. **Velocity of a Car** Suppose the distance s (in kilometres) covered by a car moving along a straight road after t hours is given by the function $f(t) = 2t^2 + 48t$.
   a. Calculate the average velocity of the car over the time intervals [20, 21], [20, 20.1], and [20, 20.01].
   b. Calculate the (instantaneous) velocity of the car when $t = 20$.
   c. Compare the results of part (a) with that of part (b).

28. **Velocity of a Ball Thrown into the Air** A ball is thrown straight up with an initial velocity of 128 m/sec, so that its height (in metres) after t sec is given by $s(t) = 32t - 4t^2$.
   a. What is the average velocity of the ball over the time intervals [2, 3], [2, 2.5], and [2, 2.1]?
   b. What is the instantaneous velocity at time $t = 2$?
   c. What is the instantaneous velocity at time $t = 5$? Is the ball rising or falling at this time?
   d. When will the ball hit the ground?

29. During the construction of a high-rise building, a worker accidentally dropped his portable electric screwdriver from a height of 400 ft. After t sec, the screwdriver had fallen a distance of $s = 16t^2$ ft.
   a. How long did it take the screwdriver to reach the ground?
   b. What was the average velocity of the screwdriver between the time it was dropped and the time it hit the ground?
   c. What was the velocity of the screwdriver at the time it hit the ground?

30. A hot-air balloon rises vertically from the ground so that its height after t sec is $h = \frac{1}{2}t^2 + \frac{1}{2}t$ ft $(0 \le t \le 60)$.
   a. What is the height of the balloon at the end of 40 sec?
   b. What is the average velocity of the balloon between $t = 0$ and $t = 40$?
   c. What is the velocity of the balloon at the end of 40 sec?

In Exercises 31–36, let x and $f(x)$ represent the given quantities. Fix $x = a$ and let h be a small positive number. Give an interpretation of the quantities

$$\frac{f(a+h) - f(a)}{h} \quad \text{and} \quad \lim_{h \to 0} \frac{f(a+h) - f(a)}{h}$$

31. x denotes time and $f(x)$ denotes the population of seals at time x.

32. x denotes time and $f(x)$ denotes the prime interest rate at time x.

33. x denotes time and $f(x)$ denotes a country's industrial production.

34. x denotes the level of production of a certain commodity, and $f(x)$ denotes the total cost incurred in producing x units of the commodity.

35. x denotes altitude and $f(x)$ denotes atmospheric pressure.

36. x denotes the speed of a car (in km/h), and $f(x)$ denotes the fuel economy of the car measured in kilometres per gallon (km/g).

In each of Exercises 37–42, the graph of a function is shown. For each function, state whether or not (a) $f(x)$ has a limit at $x = a$, (b) $f(x)$ is continuous at $x = a$, and (c) $f(x)$ is differentiable at $x = a$. Justify your answers.

37.
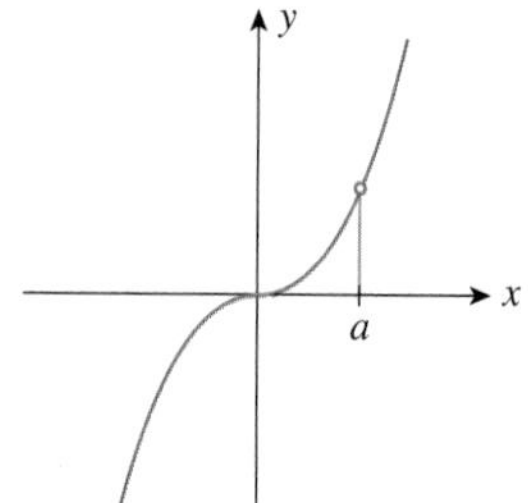

38.
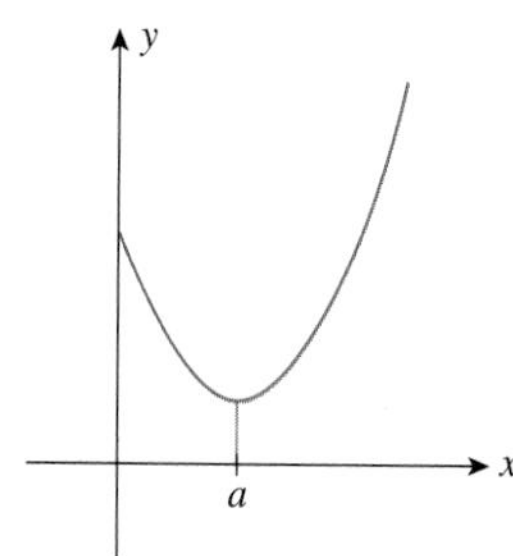

39.
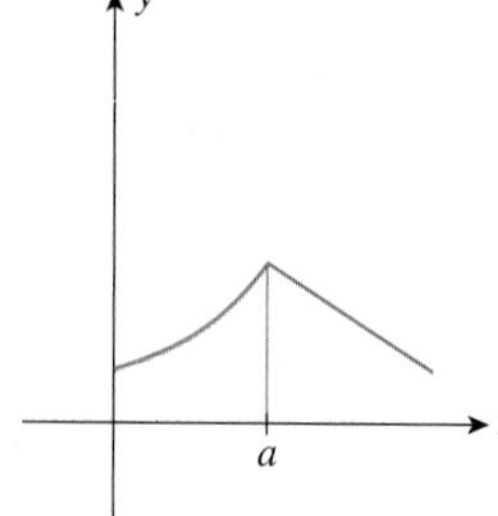

40.
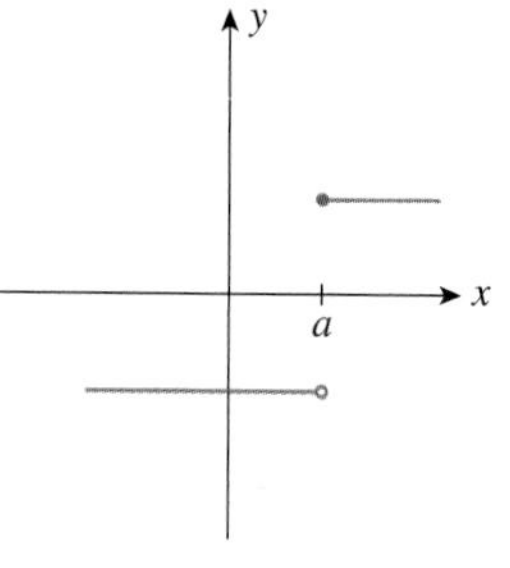

41.
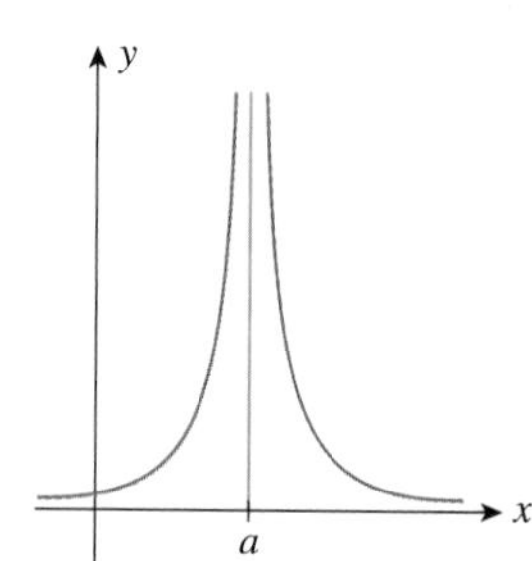

42.
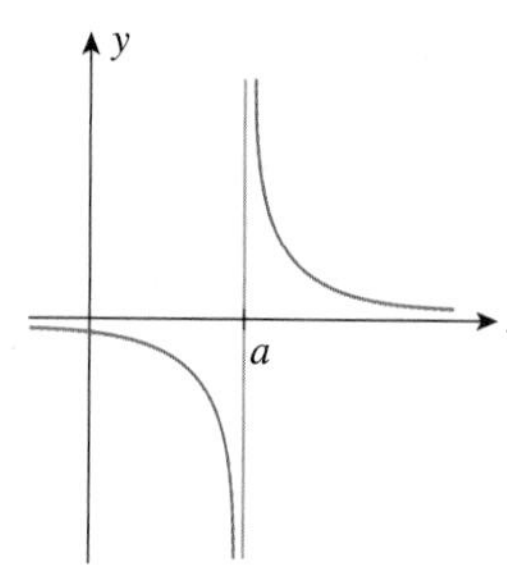

43. The distance s (in metres) covered by a glider traveling in a straight line and starting from rest in t sec is given by the function

$$s(t) = -0.1t^3 + 2t^2 + 24t$$

Calculate the glider's average velocity over the time interval [2, 2 + h] for $h = 1, 0.1, 0.01, 0.001, 0.0001$, and 0.00001 and use your results to guess at the glider's instantaneous velocity at $t = 2$.

44. Sketch the graph of the function $f(x) = |x + 1|$ and show that the function does not have a derivative at $x = -1$.

45. Sketch the graph of the function $f(x) = 1/(x - 1)$ and show that the function does not have a derivative at $x = 1$.

46. Let

$$f(x) = \begin{cases} x^2 & \text{if } x \le 1 \\ ax + b & \text{if } x > 1 \end{cases}$$

Find the values of a and b so that f is continuous and has a derivative at $x = 1$. Sketch the graph of f.

47. Sketch the graph of the function $f(x) = x^{2/3}$. Is the function continuous at $x = 0$? Does $f'(0)$ exist? Why or why not?

48. Show that the derivative of the function $f(x) = |x|$ for $x \ne 0$ is given by

$$f'(x) = \begin{cases} 1 & \text{if } x > 0 \\ -1 & \text{if } x < 0 \end{cases}$$

Hint: Rewrite the function as piecewise defined.

49. Show that if a function f is differentiable at a point $x = a$, then f must be continuous at that point.
Hint: Write

$$f(x) - f(a) = \left[\frac{f(x) - f(a)}{x - a}\right](x - a)$$

Use the product rule for limits and the definition of the derivative to show that

$$\lim_{x \to a} [f(x) - f(a)] = 0$$

Business and Economics Applications

50. **Market Share** The following figure shows the devastating effect the opening of a new discount department store had on an established department store in a small town. The revenue of the discount store at time t (in months) is given by $f(t)$ million dollars, whereas the revenue of the established department store at time t is given by $g(t)$ million dollars. Answer the following questions by giving the value of t at which the specified event took place.

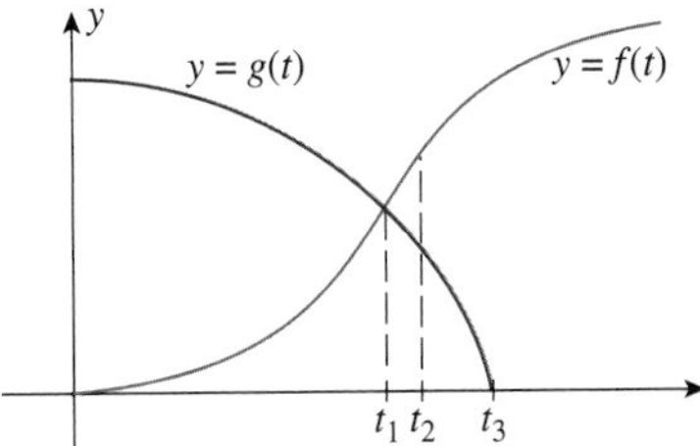

a. The revenue of the established department store is decreasing at the slowest rate.
b. The revenue of the established department store is decreasing at the fastest rate.
c. The revenue of the discount store first overtakes that of the established store.
d. The revenue of the discount store is increasing at the fastest rate.

51. **Cost of Producing Surfboards** The total cost $C(x)$ (in dollars) incurred by Aloha Company in manufacturing x surfboards a day is given by

$$C(x) = -10x^2 + 300x + 130 \qquad (0 \leq x \leq 15)$$

a. Find $C'(x)$.
b. What is the rate of change of the total cost when the level of production is ten surfboards a day?
c. What is the average cost Aloha incurs in manufacturing ten surfboards a day?

52. **Effect of Advertising on Profit** The quarterly profit (in thousands of dollars) of Cunningham Realty is given by

$$P(x) = -\tfrac{1}{3}x^2 + 7x + 30 \qquad (0 \leq x \leq 50)$$

where x (in thousands of dollars) is the amount of money Cunningham spends on advertising per quarter.

a. Find $P'(x)$.
b. What is the rate of change of Cunningham's quarterly profit if the amount it spends on advertising is \$10,000/quarter ($x = 10$) and \$30,000/quarter ($x = 30$)?

53. **Demand for Tents** The demand function for Sportsman 5 × 7 tents is given by

$$p = f(x) = -0.1x^2 - x + 40$$

where p is measured in dollars and x is measured in units of a thousand.

a. Find the average rate of change in the unit price of a tent if the quantity demanded is between 5000 and 5050 tents; between 5000 and 5010 tents.
b. What is the rate of change of the unit price if the quantity demanded is 5000?

54. **A Country's GDP** The gross domestic product (GDP) of a certain country is projected to be

$$N(t) = t^2 + 2t + 50 \qquad (0 \leq t \leq 5)$$

billion dollars t yr from now. What will be the rate of change of the country's GDP 2 yr and 4 yr from now?

55. The daily total cost $C(x)$ incurred by Trappee and Sons for producing x cases of TexaPep hot sauce is given by

$$C(x) = 0.000002x^3 + 5x + 400$$

Calculate

$$\frac{C(100 + h) - C(100)}{h}$$

for $h = 1, 0.1, 0.01, 0.001$, and 0.0001 and use your results to estimate the rate of change of the total cost function when the level of production is 100 cases/day.

Biological and Life Sciences Applications

56. **Average Weight of an Infant** The following graph shows the weight measurements of the average infant from the time of birth ($t = 0$) through age 2 ($t = 24$). By computing the slopes of the respective tangent lines, estimate the rate of change of the average infant's weight when $t = 3$ and when $t = 18$. What is the average rate of change in the average infant's weight over the first year of life?

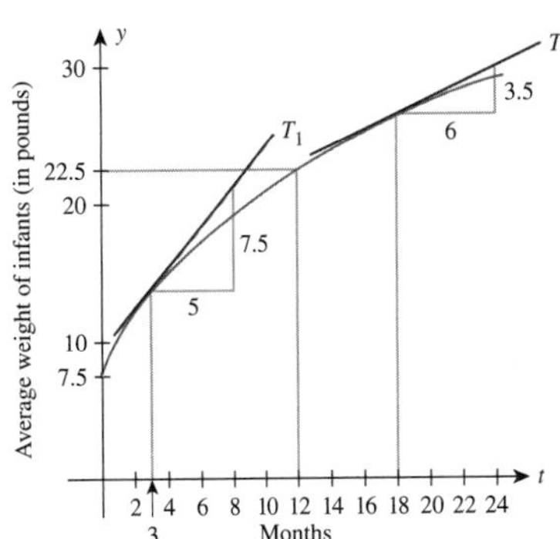

57. **Forestry** The following graph shows the volume of wood produced in a single-species forest. Here $f(t)$ is measured in cubic metres/hectare and t is measured in years. By computing the slopes of the respective tangent lines, estimate the rate at which the wood grown is changing at the beginning of year 10 and at the beginning of year 30.

Source: The Random House Encyclopedia

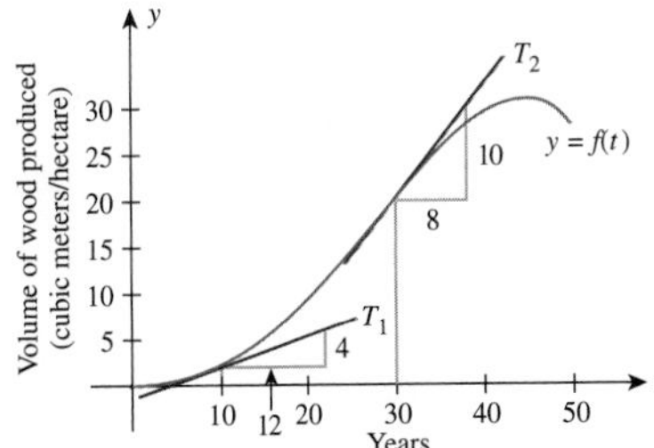

58. **TV-Viewing Patterns** The following graph shows the percent of North American households watching television during a 24-h period on a weekday ($t = 0$ corresponds to 6 A.M.). By computing the slopes of the respective tangent lines, estimate the rate of change of the percent of households watching television at 4 P.M. and 11 P.M.

Source: A. C. Nielsen Company

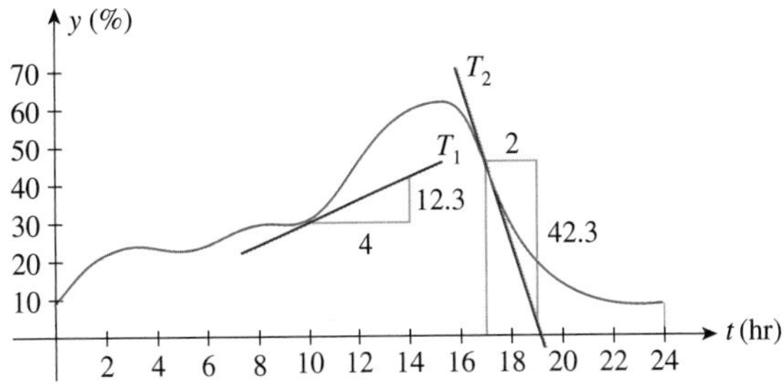

59. Crop Yield Productivity and yield of cultivated crops are often reduced by insect pests. The following graph shows the relationship between the yield of a certain crop, $f(x)$, as a function of the density of aphids x. (Aphids are small insects that suck plant juices.) Here, $f(x)$ is measured in kilograms/4000 square metres, and x is measured in hundreds of aphids/bean stem. By computing the slopes of the respective tangent lines, estimate the rate of change of the crop yield with respect to the density of aphids when that density is 200 aphids/bean stem and when it is 800 aphids/bean stem.

Source: The Random House Encyclopedia

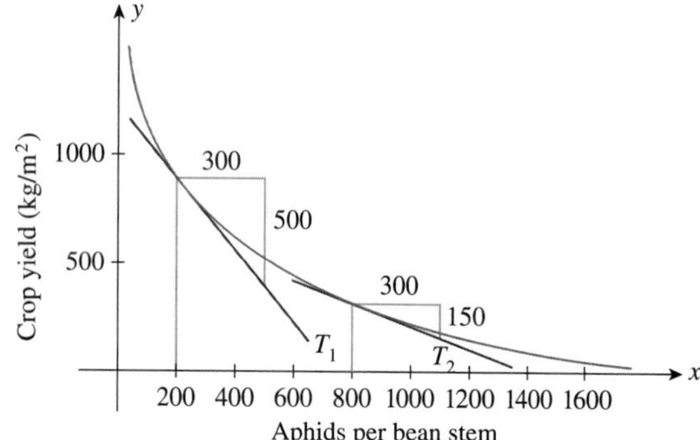

60. Effect of a Bactericide on Bacteria In the following figure, $f(t)$ gives the population P_1 of a certain bacteria culture at time t after a portion of bactericide A was introduced into the population at $t = 0$, and $g(t)$ gives the population P_2 of a similar bacteria culture at time t after a portion of bactericide B was introduced into the population at $t = 0$.

a. Which population is decreasing faster at t_1?

b. Which population is decreasing faster at t_2?

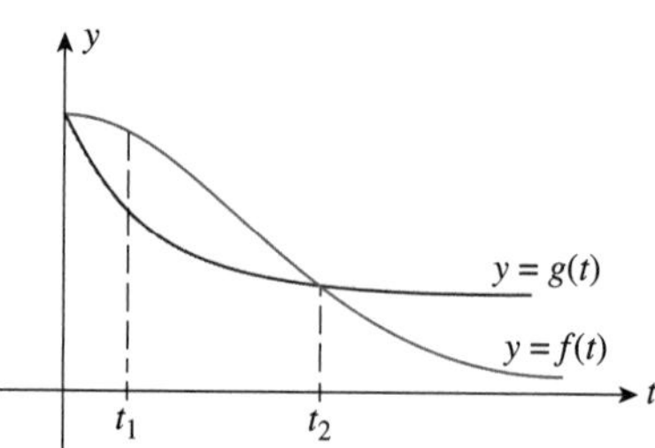

c. Which bactericide is more effective in reducing the population of bacteria in the short run? In the long run?

61. At a temperature of 20°C, the volume V (in litres) of 1.33 g of O_2 is related to its pressure p (in atmospheres) by the formula $V = 1/p$.

a. What is the average rate of change of V with respect to p as p increases from $p = 2$ to $p = 3$?

b. What is the rate of change of V with respect to p when $p = 2$?

62. Growth of Bacteria Under a set of controlled laboratory conditions, the size of the population of a certain bacteria culture at time t (in minutes) is described by the function

$$P = f(t) = 3t^2 + 2t + 1$$

Find the rate of population growth at $t = 10$ min.

In Exercises 63 and 64, determine whether the statement is true or false. If it is true, explain why it is true. If it is false, give an example to show why it is false.

63. If f is continuous at $x = a$, then f is differentiable at $x = a$.

64. If f is continuous at $x = a$ and g is differentiable at $x = a$, then $\lim_{x \to a} f(x)g(x) = f(a)g(a)$.

Solutions to Self-Check Exercises 3.3

1. a.

$$\begin{aligned} f'(x) &= \lim_{h \to 0} \frac{f(x+h) - f(x)}{h} \\ &= \lim_{h \to 0} \frac{[-(x+h)^2 - 2(x+h) + 3] - (-x^2 - 2x + 3)}{h} \\ &= \lim_{h \to 0} \frac{-x^2 - 2xh - h^2 - 2x - 2h + 3 + x^2 + 2x - 3}{h} \\ &= \lim_{h \to 0} \frac{h(-2x - h - 2)}{h} \\ &= \lim_{h \to 0} (-2x - h - 2) = -2x - 2 \end{aligned}$$

b. From the result of part (a), we see that the slope of the tangent line to the graph of f at any point $(x, f(x))$ is given by

$$f'(x) = -2x - 2$$

In particular, the slope of the tangent line to the graph of f at (0, 3) is

$$f'(0) = -2$$

c. The rate of change of f when $x = 0$ is given by $f'(0) = -2$, or -2 units/unit change in x.

d. Using the result from part (b), we see that an equation of the required tangent line is

$$\begin{aligned} y - 3 &= -2(x - 0) \\ y &= -2x + 3 \end{aligned}$$

e.

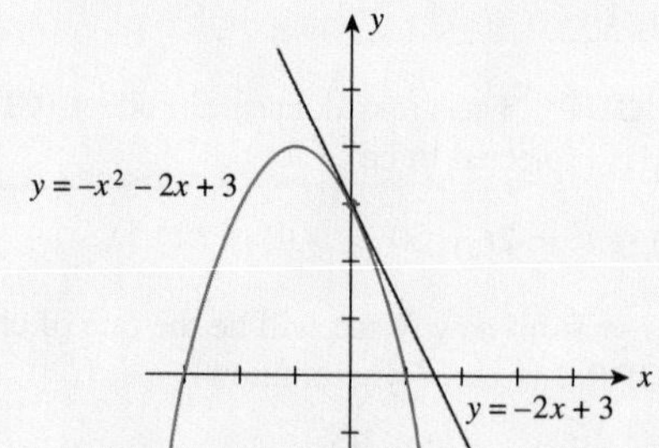

Solutions to Self-Check Exercises 3.3, *continued*

2. The rate of change of the losses at any time t is given by

$$
\begin{aligned}
f'(t) &= \lim_{h\to 0} \frac{f(t+h) - f(t)}{h} \\
&= \lim_{h\to 0} \frac{[-(t+h)^2 + 10(t+h) + 30] - (-t^2 + 10t + 30)}{h} \\
&= \lim_{h\to 0} \frac{-t^2 - 2th - h^2 + 10t + 10h + 30 + t^2 - 10t - 30}{h} \\
&= \lim_{h\to 0} \frac{h(-2t - h + 10)}{h} \\
&= \lim_{h\to 0} (-2t - h + 10) \\
&= -2t + 10
\end{aligned}
$$

Therefore, the rate of change of the losses suffered by the bank at the beginning of 1997 ($t = 3$) was

$$f'(3) = -2(3) + 10 = 4$$

That is, the losses were increasing at the rate of \$4 million/year. At the beginning of 1999 ($t = 5$),

$$f'(5) = -2(5) + 10 = 0$$

and we see that the growth in losses due to bad loans was zero at this point. At the beginning of 2001 ($t = 7$),

$$f'(7) = -2(7) + 10 = -4$$

and we conclude that the losses were decreasing at the rate of \$4 million/year.

3.4 Basic Rules of Differentiation

Four Basic Rules

The method used in Section 3.3 for computing the derivative of a function is based on a faithful interpretation of the definition of the derivative as the limit of a quotient. Thus, to find the rule for the derivative f' of a function f, we first computed the difference quotient

$$\frac{f(x+h) - f(x)}{h}$$

and then evaluated its limit as h approached zero. As you have probably observed, this method is tedious even for relatively simple functions.

The main purpose of this chapter is to derive certain rules that will simplify the process of finding the derivative of a function. Throughout this book we will use the notation

$$\frac{d}{dx}[f(x)] \qquad \text{Read "}d, d\,x \text{ of } f \text{ of } x\text{"}$$

to mean "the derivative of f with respect to x at x." In stating the rules of differentiation, we assume that the functions f and g are differentiable.

Rule 1: Derivative of a Constant

$$\frac{d}{dx}(c) = 0 \qquad (c, \text{ a constant})$$

The derivative of a constant function is equal to zero.

We can see this from a geometric viewpoint by recalling that the graph of a constant function is a straight line parallel to the x-axis (Figure 3.40). Since the tangent line to a straight line at any point on the line coincides with the straight line itself, its slope [as given by the derivative of $f(x) = c$] must be zero. We can also use the definition of the derivative to prove this result by computing

$$
\begin{aligned}
f'(x) &= \lim_{h\to 0} \frac{f(x+h) - f(x)}{h} \\
&= \lim_{h\to 0} \frac{c - c}{h} \\
&= \lim_{h\to 0} 0 = 0
\end{aligned}
$$

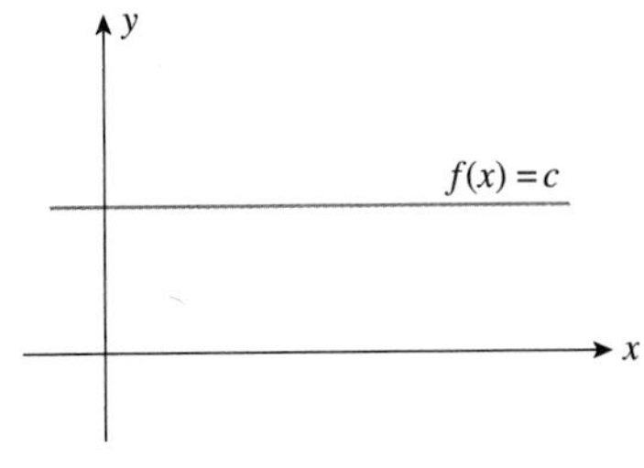

Figure 3.40
The slope of the tangent line to the graph of $f(x) = c$, where c is a constant, is zero.

Example 1

a. If $f(x) = 28$, then

$$f'(x) = \frac{d}{dx}(28) = 0$$

b. If $f(x) = -2$, then

$$f'(x) = \frac{d}{dx}(-2) = 0$$

Rule 2: The Power Rule

If n is any real number, then $\frac{d}{dx}(x^n) = nx^{n-1}$.

Let's verify the power rule for the special case $n = 2$. If $f(x) = x^2$, then

$$\begin{aligned} f'(x) = \frac{d}{dx}(x^2) &= \lim_{h \to 0} \frac{f(x+h) - f(x)}{h} \\ &= \lim_{h \to 0} \frac{(x+h)^2 - x^2}{h} \\ &= \lim_{h \to 0} \frac{x^2 + 2xh + h^2 - x^2}{h} \\ &= \lim_{h \to 0} \frac{2xh + h^2}{h} = \lim_{h \to 0} (2x + h) = 2x \end{aligned}$$

as we set out to show.

The proof of the power rule for the general case is not easy to prove and will be omitted. However, you will be asked to prove the rule for the special case $n = 3$ in Exercise 56, page 169.

Example 2

a. If $f(x) = x$, then

$$f'(x) = \frac{d}{dx}(x) = 1 \cdot x^{1-1} = x^0 = 1$$

b. If $f(x) = x^8$, then

$$f'(x) = \frac{d}{dx}(x^8) = 8x^7$$

c. If $f(x) = x^{5/2}$, then

$$f'(x) = \frac{d}{dx}(x^{5/2}) = \frac{5}{2}x^{3/2}$$

To differentiate a function whose rule involves a radical, we first rewrite the rule using fractional powers. The resulting expression can then be differentiated using the power rule.

Example 3

Find the derivative of the following functions:

a. $f(x) = \sqrt{x}$ **b.** $g(x) = \frac{1}{\sqrt[3]{x}}$

Solution

a. Rewriting $\sqrt{x}$ in the form $x^{1/2}$, we obtain

$$f'(x) = \frac{d}{dx}(x^{1/2})$$
$$= \frac{1}{2}x^{-1/2} = \frac{1}{2x^{1/2}} = \frac{1}{2\sqrt{x}}$$

b. Rewriting $\frac{1}{\sqrt[3]{x}}$ in the form $x^{-1/3}$, we obtain

$$g'(x) = \frac{d}{dx}(x^{-1/3})$$
$$= -\frac{1}{3}x^{-4/3} = -\frac{1}{3x^{4/3}}$$

Rule 3: Derivative of a Constant Multiple of a Function

$$\frac{d}{dx}[cf(x)] = c\frac{d}{dx}[f(x)] \qquad (c\text{, a constant})$$

The derivative of a constant times a differentiable function is equal to the constant times the derivative of the function.

This result follows from the following computations.

If $g(x) = cf(x)$, then

$$g'(x) = \lim_{h \to 0} \frac{g(x + h) - g(x)}{h} = \lim_{h \to 0} \frac{cf(x + h) - cf(x)}{h}$$
$$= c \lim_{h \to 0} \frac{f(x + h) - f(x)}{h}$$
$$= cf'(x)$$

Example 4

a. If $f(x) = 5x^3$, then

$$f'(x) = \frac{d}{dx}(5x^3) = 5\frac{d}{dx}(x^3)$$
$$= 5(3x^2) = 15x^2$$

b. If $f(x) = \frac{3}{\sqrt{x}}$, then

$$f'(x) = \frac{d}{dx}(3x^{-1/2})$$
$$= 3\left(-\frac{1}{2}x^{-3/2}\right) = -\frac{3}{2x^{3/2}}$$

Rule 4: The Sum Rule

$$\frac{d}{dx}[f(x) \pm g(x)] = \frac{d}{dx}[f(x)] \pm \frac{d}{dx}[g(x)]$$

The derivative of the sum (difference) of two differentiable functions is equal to the sum (difference) of their derivatives.

This result may be extended to the sum and difference of any finite number of differentiable functions. Let's verify the rule for a sum of two functions.

If $s(x) = f(x) + g(x)$, then

$$\begin{aligned} s'(x) &= \lim_{h \to 0} \frac{s(x+h) - s(x)}{h} \\ &= \lim_{h \to 0} \frac{[f(x+h) + g(x+h)] - [f(x) + g(x)]}{h} \\ &= \lim_{h \to 0} \frac{[f(x+h) - f(x)] + [g(x+h) - g(x)]}{h} \\ &= \lim_{h \to 0} \frac{f(x+h) - f(x)}{h} + \lim_{h \to 0} \frac{g(x+h) - g(x)}{h} \\ &= f'(x) + g'(x) \end{aligned}$$

Example 5

Find the derivatives of the following functions:

a. $f(x) = 4x^5 + 3x^4 - 8x^2 + x + 3$ **b.** $g(t) = \frac{t^2}{5} + \frac{5}{t^3}$

Solution

a. $$\begin{aligned} f'(x) &= \frac{d}{dx}(4x^5 + 3x^4 - 8x^2 + x + 3) \\ &= \frac{d}{dx}(4x^5) + \frac{d}{dx}(3x^4) - \frac{d}{dx}(8x^2) + \frac{d}{dx}(x) + \frac{d}{dx}(3) \\ &= 20x^4 + 12x^3 - 16x + 1 \end{aligned}$$

b. Here, the independent variable is t instead of x, so we differentiate with respect to t. Thus,

$$\begin{aligned} g'(t) &= \frac{d}{dt}\left(\frac{1}{5}t^2 + 5t^{-3}\right) && \text{Rewrite } \frac{1}{t^3} \text{ as } t^{-3}. \\ &= \frac{2}{5}t - 15t^{-4} \\ &= \frac{2t^5 - 75}{5t^4} && \text{Rewrite } t^{-4} \text{ as } \frac{1}{t^4} \text{ and simplify.} \end{aligned}$$

Example 6

Find the slope and an equation of the tangent line to the graph of $f(x) = 2x + 1/\sqrt{x}$ at the point (1, 3).

Solution

The slope of the tangent line at any point on the graph of f is given by

$$\begin{aligned} f'(x) &= \frac{d}{dx}\left(2x + \frac{1}{\sqrt{x}}\right) \\ &= \frac{d}{dx}(2x + x^{-1/2}) && \text{Rewrite } \frac{1}{\sqrt{x}} \text{ as } \frac{1}{x^{1/2}} = x^{-1/2} \\ &= 2 - \frac{1}{2}x^{-3/2} && \text{Use the sum rule.} \\ &= 2 - \frac{1}{2x^{3/2}} \end{aligned}$$

In particular, the slope of the tangent line to the graph of f at (1, 3) (where $x = 1$) is

$$f'(1) = 2 - \frac{1}{2(1^{3/2})} = 2 - \frac{1}{2} = \frac{3}{2}$$

Using the point-slope form of the equation of a line with slope $\frac{3}{2}$ and the point (1, 3), we see that an equation of the tangent line is

$$y - 3 = \frac{3}{2}(x - 1) \qquad (y - y_1) = m(x - x_1)$$

or, upon simplification,

$$y = \frac{3}{2}x + \frac{3}{2}$$

Applications

Example 7

The demand function for a certain product is given by $p(x) = \frac{\sqrt{x}}{2} - \frac{x}{40} + 2000$, where p is the price measured in dollars and the quantity x is measured in units.

a. Find the rate of change of price p per thousand products with respect to quantity x.

b. How fast is the price changing with respect to x when $x = 25$ and $x = 400$? Interpret your result.

Solution

a. The rate of change of the price with respect to quantity is given by

$$\begin{aligned} p'(x) &= \frac{d}{dx}\left(\frac{\sqrt{x}}{2} - \frac{x}{40} + 2000\right) \\ &= \frac{d}{dx}\left(\frac{1}{2}x^{1/2} - \frac{1}{40}x + 2000\right) \\ &= \frac{1}{2}\cdot\frac{1}{2}x^{-1/2} - \frac{1}{40} \\ &= \frac{1}{4}x^{-1/2} - \frac{1}{40} \end{aligned}$$

b. When $x = 25$, we have $p'(25) = \frac{1}{4}(25)^{-1/2} - \frac{1}{40} = \frac{1}{20} - \frac{1}{40} = \frac{1}{40} = 0.025$

This means that when 25 products are demanded, one additional product demanded by consumers increases the price by \$0.025.

When $x = 400$, we have

$$p'(400) = \frac{1}{4}(400)^{-1/2} - \frac{1}{40} = \frac{1}{80} - \frac{1}{40} = -\frac{1}{80} = -0.0125$$

This means that when 400 products are demanded, one additional product demanded by consumers decreases the price by \$0.0125. The graph of p is shown in Figure 3.41. Notice that although the price is decreasing for large numbers of products demanded, the decrease is minimal.

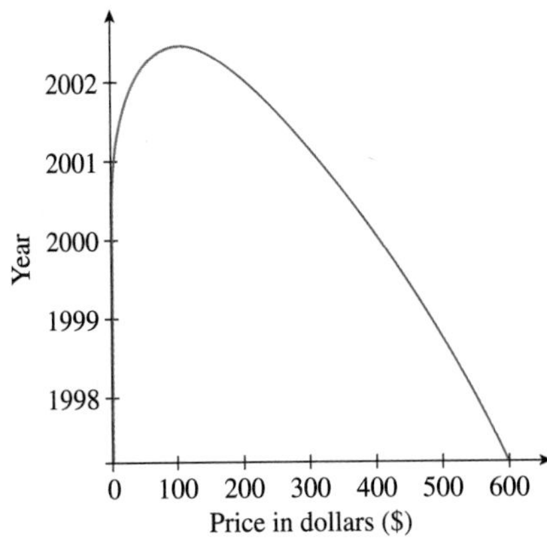

Figure 3.41

The graph of the demand function $p(x) = \frac{\sqrt{x}}{2} - \frac{x}{40} + 2000$

Example 8

Conservation of a Species A group of marine biologists at the Neptune Institute of Oceanography recommended that a series of conservation measures be carried out over the next decade to save a certain species of whale from extinction. After implementing the conservation measures, the population of this species is expected to be

$$N(t) = 3t^3 + 2t^2 - 10t + 600 \qquad (0 \le t \le 10)$$

where $N(t)$ denotes the population at the end of year t. Find the rate of growth of the whale population when $t = 2$ and $t = 6$. How large will the whale population be 8 years after implementing the conservation measures?

Solution

The rate of growth of the whale population at any time t is given by

$$N'(t) = 9t^2 + 4t - 10$$

In particular, when $t = 2$ and $t = 6$, we have

$$\begin{aligned} N'(2) &= 9(2)^2 + 4(2) - 10 \\ &= 34 \\ N'(6) &= 9(6)^2 + 4(6) - 10 \\ &= 338 \end{aligned}$$

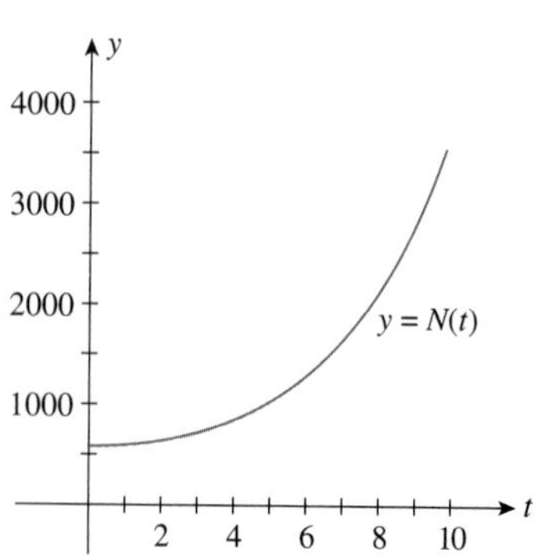

Figure 3.42
The whale population after year t is given by $N(t)$.

Thus, the whale population's rate of growth will be 34 whales per year after 2 years and 338 per year after 6 years.

The whale population at the end of the eighth year will be

$$\begin{aligned} N(8) &= 3(8)^3 + 2(8)^2 - 10(8) + 600 \\ &= 2184 \text{ whales} \end{aligned}$$

The graph of the function N appears in Figure 3.42. Note the rapid growth of the population in the later years, as the conservation measures begin to pay off, compared with the growth in the early years.

Example 9

Altitude of a Rocket The altitude of a rocket (in metres) t seconds into flight is given by

$$s = f(t) = -0.3t^3 + 28.8t^2 + 58.5t + 1.5$$

a. Find an expression v for the rocket's velocity at any time t.
b. Compute the rocket's velocity when $t = 0$, 30, 50, 65, and 70. Interpret your results.
c. Using the results from the solution to part (b) and the observation that at the highest point in its trajectory the rocket's velocity is zero, find the maximum altitude attained by the rocket.

Solution

a. The rocket's velocity at any time t is given by

$$v = f'(t) = -0.9t^2 + 57.6t + 58.5$$

b. The rocket's velocity when $t = 0$, 30, 50, 65, and 70 is given by

$$\begin{aligned} f'(0) &= -0.9(0)^2 + 57.6(0) + 58.5 = 58.5 \\ f'(30) &= -0.9(30)^2 + 57.6(30) + 58.5 = 976.5 \\ f'(50) &= -0.9(50)^2 + 57.6(50) + 58.5 = 688.5 \\ f'(65) &= -0.9(65)^2 + 57.6(65) + 58.5 = 0 \\ f'(70) &= -0.9(70)^2 + 57.6(70) + 58.5 = -319.5 \end{aligned}$$

or 58.5, 976.5, 688.5, 0, and -319.5 metres per second (m/s).

Thus, the rocket has an initial velocity of 58.5 m/s at $t = 0$ and accelerates to a velocity of 976.5 m/s at $t = 30$. Fifty seconds into the flight, the rocket's velocity is 688.5 m/s, which is less than the velocity at $t = 30$. This means that the rocket begins to decelerate after an initial period of acceleration. (Later on we will learn how to determine the rocket's maximum velocity.)

The deceleration continues: The velocity is 0 m/s at $t = 65$ and -319.5 m/s when $t = 70$. This result tells us that 70 seconds into the flight the rocket is heading back to Earth with a speed of 319.5 m/s.

c. The results of part (b) show that the rocket's velocity is zero when $t = 65$. At this instant, the rocket's maximum altitude is

$$s = f(65) = -0.3(65)^3 + 28.8(65)^2 + 58.5(65) + 1.5 = 43{,}096.5$$

or 43,096.5 metres. A sketch of the graph of f appears in Figure 3.43.

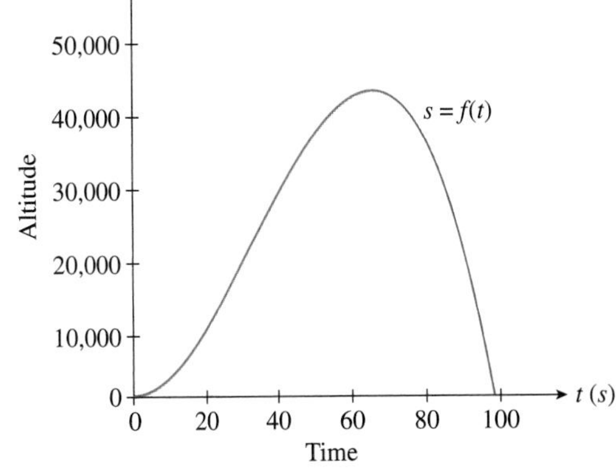

Figure 3.43
The rocket's altitude t seconds into flight is given by $s = f(t)$.

EXPLORING WITH TECHNOLOGY

Refer to Example 9.

1. Use a graphing utility to plot the graph of the velocity function
$$v = f'(t) = -3t^2 + 192t + 195$$
using the viewing window [0, 120] · [−5000, 5000]. Then, using **ZOOM** and **TRACE** or the root-finding capability of your graphing utility, verify that $f'(65) = 0$.
2. Plot the graph of the position function of the rocket
$$s = f(t) = -t^3 + 96t^2 + 195t + 5$$
using the viewing window [0, 120] · [0, 150,000]. Then, using **ZOOM** and **TRACE** repeatedly, verify that the maximum altitude of the rocket is 143,655 feet.
3. Use **ZOOM** and **TRACE** or the root-finding capability of your graphing utility to find when the rocket returns to Earth.

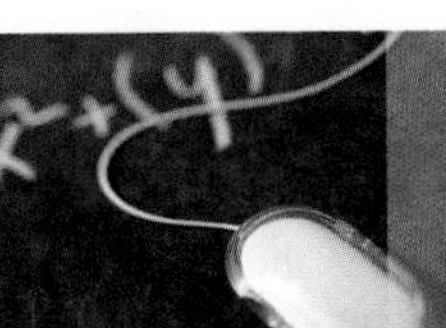

Self-Check Exercises 3.4

1. Find the derivative of each of the following functions using the rules of differentiation.
   a. $f(x) = 1.5x^2 + 2x^{1.5}$
   b. $g(x) = 2\sqrt{x} + \dfrac{3}{\sqrt{x}}$
2. Let $f(x) = 2x^3 - 3x^2 + 2x - 1$.
   a. Compute $f'(x)$.
   b. What is the slope of the tangent line to the graph of f when $x = 2$?
   c. What is the rate of change of the function f at $x = 2$?
3. A certain country's gross domestic product (GDP) (in millions of dollars) is described by the function
$$G(t) = -2t^3 + 45t^2 + 20t + 6000 \qquad (0 \le t \le 11)$$
where $t = 0$ corresponds to the beginning of 1994.
   a. At what rate was the GDP changing at the beginning of 1999? At the beginning of 2001? At the beginning of 2004?
   b. What was the average rate of growth of the GDP over the period 1999–2004?

Solutions to Self-Check Exercises 3.4 can be found on page 170.

3.4 Exercises

In Exercises 1–34, find the derivative of the function f by using the rules of differentiation.

1. $f(x) = -3$
2. $f(x) = 365$
3. $f(x) = x^5$
4. $f(x) = x^7$
5. $f(x) = x^{2.1}$
6. $f(x) = x^{0.8}$
7. $f(x) = 3x^2$
8. $f(x) = -2x^3$
9. $f(r) = \pi r^2$
10. $f(r) = \dfrac{4}{3}\pi r^3$
11. $f(x) = 9x^{1/3}$
12. $f(x) = \dfrac{5}{4}x^{4/5}$
13. $f(x) = 3\sqrt{x}$
14. $f(u) = \dfrac{2}{\sqrt{u}}$

15. $f(x) = 7x^{-12}$

16. $f(x) = 0.3x^{-1.2}$

17. $f(x) = 5x^2 - 3x + 7$

18. $f(x) = x^3 - 3x^2 + 1$

19. $f(x) = -x^3 + 2x^2 - 6$

20. $f(x) = x^4 - 2x^2 + 5$

21. $f(x) = 0.03x^2 - 0.4x + 10$

22. $f(x) = 0.002x^3 - 0.05x^2 + 0.1x - 20$

23. $f(x) = \dfrac{x^3 - 4x^2 + 3}{x}$

24. $f(x) = \dfrac{x^3 + 2x^2 + x - 1}{x}$

25. $f(x) = 4x^4 - 3x^{5/2} + 2$

26. $f(x) = 5x^{4/3} - \dfrac{2}{3}x^{3/2} + x^2 - 3x + 1$

27. $f(x) = 3x^{-1} + 4x^{-2}$

28. $f(x) = -\dfrac{1}{3}(x^{-3} - x^6)$

29. $f(t) = \dfrac{4}{t^4} - \dfrac{3}{t^3} + \dfrac{2}{t}$

30. $f(x) = \dfrac{5}{x^3} - \dfrac{2}{x^2} - \dfrac{1}{x} + 200$

31. $f(x) = 2x - 5\sqrt{x}$

32. $f(t) = 2t^2 + \sqrt{t^3}$

33. $f(x) = \dfrac{2}{x^2} - \dfrac{3}{x^{1/3}}$

34. $f(x) = \dfrac{3}{x^3} + \dfrac{4}{\sqrt{x}} + 1$

35. Let $f(x) = 2x^3 - 4x$. Find:
   a. $f'(-2)$
   b. $f'(0)$
   c. $f'(2)$

36. Let $f(x) = 4x^{5/4} + 2x^{3/2} + x$. Find:
   a. $f'(0)$
   b. $f'(16)$

In Exercises 37–40, find the given limit by evaluating the derivative of a suitable function at an appropriate point.

Hint: Look at the definition of the derivative.

37. $\displaystyle\lim_{h \to 0} \frac{(1 + h)^3 - 1}{h}$

38. $\displaystyle\lim_{x \to 1} \frac{x^5 - 1}{x - 1}$
   Hint: Let $h = x - 1$.

39. $\displaystyle\lim_{h \to 0} \frac{3(2 + h)^2 - (2 + h) - 10}{h}$

40. $\displaystyle\lim_{t \to 0} \frac{1 - (1 + t)^2}{t(1 + t)^2}$

In Exercises 41–44, find the slope and an equation of the tangent line to the graph of the function f at the specified point.

41. $f(x) = 2x^2 - 3x + 4$; $(2, 6)$

42. $f(x) = -\dfrac{5}{3}x^2 + 2x + 2$; $\left(-1, -\dfrac{5}{3}\right)$

43. $f(x) = x^4 - 3x^3 + 2x^2 - x + 1$; $(1, 0)$

44. $f(x) - \sqrt{x} + \dfrac{1}{\sqrt{x}}$; $\left(4, \dfrac{5}{2}\right)$

45. Let $f(x) = x^3$.
   a. Find the point on the graph of f where the tangent line is horizontal.
   b. Sketch the graph of f and draw the horizontal tangent line.

46. Let $f(x) = x^3 - 4x^2$. Find the point(s) on the graph of f where the tangent line is horizontal.

47. Let $f(x) = x^3 + 1$.
   a. Find the point(s) on the graph of f where the slope of the tangent line is equal to 12.
   b. Find the equation(s) of the tangent line(s) of part (a).
   c. Sketch the graph of f showing the tangent line(s).

48. Let $f(x) = \frac{2}{3}x^3 + x^2 - 12x + 6$. Find the values of x for which:
   a. $f'(x) = -12$
   b. $f'(x) = 0$
   c. $f'(x) = 12$

49. Let $f(x) = \frac{1}{4}x^4 - \frac{1}{3}x^3 - x^2$. Find the point(s) on the graph of f where the slope of the tangent line is equal to:
   a. $-2x$
   b. 0
   c. $10x$

50. A straight line perpendicular to and passing through the point of tangency of the tangent line is called the *normal* to the curve. Find an equation of the tangent line and the normal to the curve $y = x^3 - 3x + 1$ at the point $(2, 3)$.

51. **Effect of Stopping on Average Speed** According to data from a study, the average speed of your trip A (in km/h) is related to the number of stops/kilometre you make on the trip x by the equation

$$A = \frac{26.5}{x^{0.45}}$$

Compute dA/dx for $x = 0.25$ and $x = 2$. How is the rate of change of the average speed of your trip affected by the number of stops/kilometre?

52. **Flight of a Rocket** The altitude (in metres) of a rocket t sec into flight is given by

$$s = f(t) = s(t) = -0.6t^3 + 34.2t^2 + 144t + 0.3 \quad (t \geq 0)$$

   a. Find an expression v for the rocket's velocity at any time t.
   b. Compute the rocket's velocity when $t = 0$, 20, 40, and 60. Interpret your results.
   c. Using the results from the solution to part (b), find the maximum altitude attained by the rocket.
   Hint: At its highest point, the velocity of the rocket is zero.

53. **Stopping Distance of a Racing Car** During a test by the editors of an auto magazine, the stopping distance s (in metres) of the MacPherson X-2 racing car conformed to the rule

$$s = f(t) = 40t - 5t^2 \quad (t \geq 0)$$

where t was the time (in seconds) after the brakes were applied.
   a. Find an expression for the car's velocity v at any time t.
   b. What was the car's velocity when the brakes were first applied?
   c. What was the car's stopping distance for that particular test?
   Hint: The stopping time is found by setting $v = 0$.

54. **Average Speed of a Vehicle on a Highway** The average speed of a vehicle on a stretch of Route 134 between 6 A.M. and 10 A.M. on a typical weekday is approximated by the function

$$f(t) = 20t - 40\sqrt{t} + 50 \qquad (0 \le t \le 4)$$

where $f(t)$ is measured in km/h and t is measured in hours, with $t = 0$ corresponding to 6 A.M.

a. Compute $f'(t)$.
b. What is the average speed of a vehicle on that stretch of Route 134 at 6 A.M.? At 7 A.M.? At 8 A.M.?
c. How fast is the average speed of a vehicle on that stretch of Route 134 changing at 6:30 A.M.? At 7 A.M.? At 8 A.M.?

55. **Portable Phones** The percent of the North American population with portable phones is projected to be

$$P(t) = 24.4t^{0.34} \qquad (1 \le t \le 10)$$

where t is measured in years, with $t = 1$ corresponding to the beginning of 1998.

a. What percent of the North American population is expected to have portable phones by the beginning of 2006?
b. How fast is the percent of the North American population with portable phones expected to be changing at the beginning of 2006?

56. Prove the power rule (Rule 2) for the special case $n = 3$.

Hint: Compute $\lim_{h \to 0} \left[\dfrac{(x+h)^3 - x^3}{h} \right]$.

Business and Economics Applications

57. **Consumer Price Index** An economy's consumer price index (CPI) is described by the function

$$I(t) = -0.2t^3 + 3t^2 + 100 \qquad (0 \le t \le 10)$$

where $t = 0$ corresponds to 1994.

a. At what rate was the CPI changing in 1999? In 2001? In 2004?
b. What was the average rate of increase in the CPI over the period from 1999 to 2004?

58. **Effect of Advertising on Sales** The relationship between the amount of money x that Cannon Precision Instruments spends on advertising and the company's total sales $S(x)$ is given by the function

$$S(x) = -0.002x^3 + 0.6x^2 + x + 500 \qquad (0 \le x \le 200)$$

where x is measured in thousands of dollars. Find the rate of change of the sales with respect to the amount of money spent on advertising. Are Cannon's total sales increasing at a faster rate when the amount of money spent on advertising is (a) \$100,000 or (b) \$150,000?

59. **Demand Functions** The demand function for the Luminar desk lamp is given by

$$p = f(x) = -0.1x^2 - 0.4x + 35$$

where x is the quantity demanded (measured in thousands) and p is the unit price in dollars.

a. Find $f'(x)$.
b. What is the rate of change of the unit price when the quantity demanded is 10,000 units ($x = 10$)? What is the unit price at that level of demand?

60. **Sales of DSPs** The sales of digital signal processors (DSPs) from Canadian manufacturers in millions of dollars is projected to be

$$S(t) = 0.14t^2 + 0.68t + 3.1 \qquad (0 \le t \le 6)$$

where t is measured in years, with $t = 0$ corresponding to the beginning of 1997.

a. What were the sales of DSPs at the beginning of 1997? What were the sales at the beginning of 2002?
b. How fast was the level of sales increasing at the beginning of 1997? How fast were sales increasing at the beginning of 2002?

61. **Supply Functions** The supply function for a certain make of transistor radio is given by

$$p = f(x) = 0.0001x^{5/4} + 10$$

where x is the quantity supplied and p is the unit price in dollars.

a. Find $f'(x)$.
b. What is the rate of change of the unit price if the quantity supplied is 10,000 transistor radios?

62. **Online Shopping** Retail revenue each year from Internet shopping is approximated by the function

$$f(t) = 0.075t^3 + 0.025t^2 + 2.45t + 2.4 \qquad (0 \le t \le 4)$$

where $f(t)$ is measured in millions of dollars and t is measured in years, with $t = 0$ corresponding to the beginning of 1997.

a. Find an expression giving the rate of change of the retail revenue/year from Internet shopping at any time t.
b. How fast was the retail revenue/year from Internet shopping changing at the beginning of the year 2000?
c. What was the retail revenue/year from Internet shopping at the beginning of the year 2000?

63. **Health-Care Spending** Despite efforts at cost containment, the cost of medical care is increasing. Two major reasons for this increase are an aging population and extensive use by physicians of new technologies. Health-care spending through the year 2000 may be approximated by the function

$$S(t) = 0.02836t^3 - 0.05167t^2 + 9.60881t + 41.9 \qquad (0 \le t \le 35)$$

where $S(t)$ is the spending in millions of dollars and t is measured in years, with $t = 0$ corresponding to the beginning of 1965.

a. Find an expression for the rate of change of health-care spending at any time t.
b. How fast was health-care spending changing at the beginning of 1980? At the beginning of 2000?
c. What was the amount of health-care spending at the beginning of 1980? At the beginning of 2000?

Biological and Life Sciences Applications

64. **Growth of a Cancerous Tumour** The volume of a spherical cancerous tumour is given by the function

$$V(r) = \frac{f}{3}\pi r^3$$

where r is the radius of the tumour in centimetres. Find the rate of change in the volume of the tumour when

a. $r = \frac{2}{3}$ cm b. $r = \frac{5}{4}$ cm

65. **Velocity of Blood in an Artery** The velocity (in centimetres/second) of blood r cm from the central axis of an artery is given by

$$v(r) = k(R^2 - r^2)$$

where k is a constant and R is the radius of the artery (see the accompanying figure). Suppose $k = 1000$ and $R = 0.2$ cm. Find $v(0.1)$ and $v'(0.1)$ and interpret your results.

Blood vessel

66. **Worker Efficiency** An efficiency study conducted for Elektra Electronics showed that the number of Space Commander walkie-talkies assembled by the average worker t hr after starting work at 8 A.M. is given by

$$N(t) = -t^3 + 6t^2 + 15t$$

a. Find the rate at which the average worker will be assembling walkie-talkies t hr after starting work.
b. At what rate will the average worker be assembling walkie-talkies at 10 A.M.? At 11 A.M.?
c. How many walkie-talkies will the average worker assemble between 10 A.M. and 11 A.M.?

67. **Population Growth** A study prepared for a Sunbelt town's chamber of commerce projected that the town's population in the next 3 yr will grow according to the rule

$$P(t) = 50{,}000 + 30t^{3/2} + 20t$$

where $P(t)$ denotes the population t mo from now. How fast will the population be increasing 9 mo and 16 mo from now?

68. **Curbing Population Growth** Five years ago, the government of a Pacific Island state launched an extensive propaganda campaign toward curbing the country's population growth. According to the Census Department, the population (measured in thousands of people) for the following 4 yr was

$$P(t) = -\frac{1}{3}t^3 + 64t + 3000$$

where t is measured in years and $t = 0$ at the start of the campaign. Find the rate of change of the population at the end of years 1, 2, 3 and 4. Was the plan working?

69. **Conservation of Species** A certain species of turtle faces extinction because dealers collect truckloads of turtle eggs to be sold as aphrodisiacs. After severe conservation measures are implemented, it is hoped that the turtle population will grow according to the rule

$$N(t) = 2t^3 + 3t^2 - 4t + 1000 \quad (0 \le t \le 10)$$

where $N(t)$ denotes the population at the end of year t. Find the rate of growth of the turtle population when $t = 2$ and $t = 8$. What will be the population 10 yr after the conservation measures are implemented?

70. **Increase in Temporary Workers** The number of temporary workers in North America (in millions) is estimated to be

$$N(t) = 0.025t^2 + 0.255t + 1.505 \quad (0 \le t \le 5)$$

where t is measured in years, with $t = 0$ corresponding to 1991.
a. How many temporary workers were there at the beginning of 1994?
b. How fast was the number of temporary workers growing at the beginning of 1994?

71. **Fisheries** The total groundfish population off the shores of Nova Scotia between 1989 and 1999 is approximated by the function

$$f(t) = 5.303t^2 - 53.977t + 253.8 \quad (0 \le t \le 10)$$

where $f(t)$ is measured in thousands of metric tons and t is measured in years, with $t = 0$ corresponding to the beginning of 1989.
a. What was the rate of change of the groundfish population at the beginning of 1994? At the beginning of 1996?
b. Fishing restrictions were imposed on Dec. 7, 1994. Were the conservation measures effective?

In Exercises 72 and 73, determine whether the statement is true or false. If it is true, explain why it is true. If it is false, give an example to show why it is false.

72. If f and g are differentiable, then

$$\frac{d}{dx}[2f(x) - 5g(x)] = 2f'(x) - 5g'(x)$$

73. If $f(x) = \pi^x$, then $f'(x) = x\pi^{x-1}$.

Solutions to Self-Check Exercises 3.4

1. a. $f'(x) = \frac{d}{dx}(1.5x^2) + \frac{d}{dx}(2x^{1.5})$
$= (1.5)(2x) + (2)(1.5x^{0.5})$
$= 3x + 3\sqrt{x} = 3(x + \sqrt{x})$

b. $g'(x) = \frac{d}{dx}(2x^{1/2}) + \frac{d}{dx}(3x^{-1/2})$
$= (2)\left(\frac{1}{2}x^{-1/2}\right) + (3)\left(-\frac{1}{2}x^{-3/2}\right)$
$= x^{-1/2} - \frac{3}{2}x^{-3/2}$
$= \frac{1}{2}x^{-3/2}(2x - 3) = \frac{2x - 3}{2x^{3/2}}$

Solutions to Self-Check Exercises 3.4, *continued*

2. a. $f'(x) = \frac{d}{dx}(2x^3) - \frac{d}{dx}(3x^2) + \frac{d}{dx}(2x) - \frac{d}{dx}(1)$

$$= (2)(3x^2) - (3)(2x) + 2$$
$$= 6x^2 - 6x + 2$$

   b. The slope of the tangent line to the graph of f when $x = 2$ is given by

$$f'(2) = 6(2)^2 - 6(2) + 2 = 14$$

   c. The rate of change of f at $x = 2$ is given by $f'(2)$. Using the results of part (b), we see that the required rate of change is 14 units/unit change in x.

3. a. The rate at which the GDP was changing at any time t $(0 < t < 11)$ is given by

$$G'(t) = -6t^2 + 90t + 20$$

In particular, the rates of change of the GDP at the beginning of the years 1999 $(t = 5)$, 2001 $(t = 7)$, and 2004 $(t = 10)$ are given by

$$G'(5) = 320, \quad G'(7) = 356, \quad G'(10) = 320$$

respectively—that is, by \$320 million/year, \$356 million/year, and \$320 million/year, respectively.

   b. The average rate of growth of the GDP over the period from the beginning of 1999 $(t = 5)$ to the beginning of 2004 $(t = 10)$ is given by

$$\frac{G(10) - G(5)}{10 - 5} = \frac{[-2(10)^3 + 45(10)^2 + 20(10) + 6000]}{5} - \frac{[-2(5)^3 + 45(5)^2 + 20(5) + 6000]}{5}$$
$$= \frac{8700 - 6975}{5}$$

or \$345 million/year.

3.5 The Product and Quotient Rules

In this section we study two more rules of differentiation: the **product rule** and the **quotient rule.**

The Product Rule

The derivative of the product of two differentiable functions is given by the following rule:

Rule 5: The Product Rule

$$\frac{d}{dx}[f(x)g(x)] = f(x)g'(x) + g(x)f'(x)$$

The derivative of the product of two functions is the first function times the derivative of the second plus the second function times the derivative of the first.

The product rule may be extended to the case involving the product of any finite number of functions (see Exercise 50, p. 178). We prove the product rule at the end of this section.

The derivative of the product of two functions is *not* given by the product of the derivatives of the functions; that is, in general

$$\frac{d}{dx}[f(x)g(x)] \neq f'(x)g'(x)$$

Example 1

Find the derivative of the function

$$f(x) = (2x^2 - 1)(x^3 + 3)$$

Solution

By the product rule,

$$\begin{aligned} f'(x) &= (2x^2 - 1)\frac{d}{dx}(x^3 + 3) + (x^3 + 3)\frac{d}{dx}(2x^2 - 1) \\ &= (2x^2 - 1)(3x^2) + (x^3 + 3)(4x) \\ &= 6x^4 - 3x^2 + 4x^4 + 12x \\ &= 10x^4 - 3x^2 + 12x \\ &= x(10x^3 - 3x + 12) \end{aligned}$$

Example 2

Differentiate (that is, find the derivative of) the function

$$f(x) = x^3(\sqrt{x} + 1)$$

Solution

First, we express the function in exponential form, obtaining

$$f(x) = x^3(x^{1/2} + 1)$$

By the product rule,

$$\begin{aligned} f'(x) &= x^3\frac{d}{dx}(x^{1/2} + 1) + (x^{1/2} + 1)\frac{d}{dx}x^3 \\ &= x^3\left(\frac{1}{2}x^{-1/2}\right) + (x^{1/2} + 1)(3x^2) \\ &= \frac{1}{2}x^{5/2} + 3x^{5/2} + 3x^2 \\ &= \frac{7}{2}x^{5/2} + 3x^2 \end{aligned}$$

Remark

We can also solve the problem by first expanding the product before differentiating f. Examples for which this is not possible will be considered in Section 3.6, where the true value of the product rule will be appreciated. ◀

The Quotient Rule

The derivative of the quotient of two differentiable functions is given by the following rule:

Rule 6: The Quotient Rule

$$\frac{d}{dx}\left[\frac{f(x)}{g(x)}\right] = \frac{g(x)f'(x) - f(x)g'(x)}{[g(x)]^2} \qquad (g(x) \neq 0)$$

As an aid to remembering this expression, observe that it has the following form:

$$\frac{d}{dx}\left[\frac{f(x)}{g(x)}\right]$$

$$= \frac{(\text{Denominator})\begin{pmatrix}\text{Derivative of}\\ \text{numerator}\end{pmatrix} - (\text{Numerator})\begin{pmatrix}\text{Derivative of}\\ \text{denominator}\end{pmatrix}}{(\text{Square of denominator})}$$

For a proof of the quotient rule, see Exercise 51, page 176.

The derivative of a quotient is not equal to the quotient of the derivatives; that is,

$$\frac{d}{dx}\left[\frac{f(x)}{g(x)}\right] \neq \frac{f'(x)}{g'(x)}$$

For example, if $f(x) = x^3$ and $g(x) = x^2$, then

$$\frac{d}{dx}\left[\frac{f(x)}{g(x)}\right] = \frac{d}{dx}\left(\frac{x^3}{x^2}\right) = \frac{d}{dx}(x) = 1$$

which is *not* equal to

$$\frac{f'(x)}{g'(x)} = \frac{\frac{d}{dx}(x^3)}{\frac{d}{dx}(x^2)} = \frac{3x^2}{2x} = \frac{3}{2}x$$

Example 3

Find $f'(x)$ if $f(x) = \frac{x}{2x - 4}$.

Solution

Using the quotient rule, we obtain

$$\begin{aligned} f'(x) &= \frac{(2x - 4)\frac{d}{dx}(x) - x\frac{d}{dx}(2x - 4)}{(2x - 4)^2} \\ &= \frac{(2x - 4)(1) - x(2)}{(2x - 4)^2} \\ &= \frac{2x - 4 - 2x}{(2x - 4)^2} = -\frac{4}{(2x - 4)^2} \end{aligned}$$

Example 4

Find $f'(x)$ if $f(x) = \frac{x^2 + 1}{x^2 - 1}$.

Solution

By the quotient rule,

$$\begin{aligned} f'(x) &= \frac{(x^2 - 1)\frac{d}{dx}(x^2 + 1) - (x^2 + 1)\frac{d}{dx}(x^2 - 1)}{(x^2 - 1)^2} \\ &= \frac{(x^2 - 1)(2x) - (x^2 + 1)(2x)}{(x^2 - 1)^2} \\ &= \frac{2x^3 - 2x - 2x^3 - 2x}{(x^2 - 1)^2} \\ &= -\frac{4x}{(x^2 - 1)^2} \end{aligned}$$

Example 5

Find $h'(x)$ if $h(x) = \frac{\sqrt{x}}{x^2 + 1}$.

Solution

Rewrite $h(x)$ in the form $h(x) = \frac{x^{1/2}}{x^2 + 1}$. By the quotient rule, we find

$$h'(x) = \frac{(x^2+1)\frac{d}{dx}(x^{1/2}) - x^{1/2}\frac{d}{dx}(x^2+1)}{(x^2+1)^2}$$

$$= \frac{(x^2+1)\left(\frac{1}{2}x^{-1/2}\right) - x^{1/2}(2x)}{(x^2+1)^2}$$

$$= \frac{\frac{1}{2}x^{-1/2}(x^2+1-4x^2)}{(x^2+1)^2}$$

Factor out $\frac{1}{2}x^{-1/2}$ from the numerator.

$$= \frac{1-3x^2}{2\sqrt{x}(x^2+1)^2}$$

Applications

Example 6

Rate of Change of DVD Sales The sales (in millions of dollars) of a DVD recording of a hit movie t years from the date of release is given by

$$S(t) = \frac{5t}{t^2+1}$$

a. Find the rate at which the sales are changing at time t.

b. How fast are the sales changing at the time the DVDs are released ($t = 0$)? Two years from the date of release?

Solution

a. The rate at which the sales are changing at time t is given by $S'(t)$. Using the quotient rule, we obtain

$$S'(t) = \frac{d}{dt}\left[\frac{5t}{t^2+1}\right] = 5\frac{d}{dt}\left[\frac{t}{t^2+1}\right]$$

$$= 5\left[\frac{(t^2+1)(1) - t(2t)}{(t^2+1)^2}\right]$$

$$= 5\left[\frac{t^2+1-2t^2}{(t^2+1)^2}\right] = \frac{5(1-5^2)}{(t^2+1)^2}$$

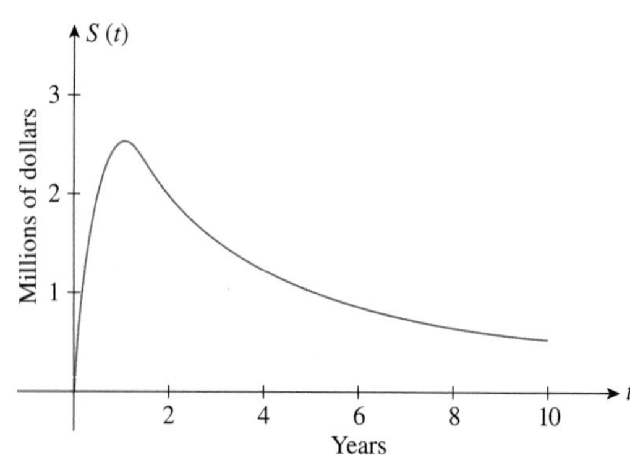

Figure 3.44
After a spectacular rise, the sales begin to taper off.

b. The rate at which the sales are changing at the time the DVDs are released is given by

$$S'(0) = \frac{5(1-0)}{(0+1)^2} = 5$$

That is, they are increasing at the rate of \$5 million per year.

Two years from the date of release, the sales are changing at the rate of

$$S'(2) = \frac{5(1-4)}{(4+1)^2} = -\frac{3}{5} = -0.6$$

That is, they are decreasing at the rate of \$600,000 per year.

The graph of the function S is shown in Figure 3.44.

EXPLORING WITH TECHNOLOGY

Refer to Example 6.

1. Use a graphing utility to plot the graph of the function S, using the viewing window $[0, 10] \cdot [0, 3]$.
2. Use TRACE and ZOOM to determine the coordinates of the highest point on the graph of S in the interval $[0, 10]$. Interpret your results.

GROUP DISCUSSION

Suppose the revenue of a company is given by $R(x) = xp(x)$, where x is the number of units of the product sold at a unit price of $p(x)$ dollars.

1. Compute $R'(x)$ and explain, in words, the relationship between $R'(x)$ and $p(x)$ and/or its derivative.
2. What can you say about $R'(x)$ if $p(x)$ is constant? Is this expected?

Example 7

Oxygen-Restoration Rate in a Pond When organic waste is dumped into a pond, the oxidation process that takes place reduces the pond's oxygen content. However, given time, nature will restore the oxygen content to its natural level. Suppose the oxygen content t days after organic waste has been dumped into the pond is given by

$$f(t) = 100\left[\frac{t^2 + 10t + 100}{t^2 + 20t + 100}\right] \qquad (0 < t < \infty)$$

percent of its normal level.

a. Derive a general expression that gives the rate of change of the pond's oxygen level at any time t.

b. How fast is the pond's oxygen content changing 1 day, 10 days, and 20 days after the organic waste has been dumped?

Solution

a. The rate of change of the pond's oxygen level at any time t is given by the derivative of the function f. Thus, the required expression is

$$\begin{aligned}
f'(t) &= 100\frac{d}{dt}\left[\frac{t^2 + 10t + 100}{t^2 + 20t + 100}\right] \\
&= 100\left[\frac{(t^2 + 20t + 100)\frac{d}{dt}(t^2 + 10t + 100) - (t^2 + 10t + 100)\frac{d}{dt}(t^2 + 20t + 100)}{(t^2 + 20t + 100)^2}\right] \\
&= 100\left[\frac{(t^2 + 20t + 100)(2t + 10) - (t^2 + 10t + 100)(2t + 20)}{(t^2 + 20t + 100)^2}\right] \\
&= 100\left[\frac{2t^3 + 10t^2 + 40t^2 + 200t + 200t + 1000 - 2t^3 - 20t^2 - 20t^2 - 200t - 2000}{(t^2 + 20t + 100)^2}\right] \\
&= 100\left[\frac{10t^2 - 1000}{(t^2 + 20t + 100)^2}\right]
\end{aligned}$$

b. The rate at which the pond's oxygen content is changing 1 day after the organic waste has been dumped is given by

$$f'(1) = 100\left[\frac{10 - 1000}{(1 + 20 + 100)^2}\right] = -6.76$$

That is, it is dropping at the rate of 6.8% per day. After 10 days the rate is

$$f'(10) = 100\left[\frac{10(10)^2 - 1000}{(100 + 200 + 100)^2}\right] = 0$$

That is, it is neither increasing nor decreasing. After 20 days the rate is

$$f'(20) = 100\left[\frac{10(20)^2 - 1000}{(400 + 400 + 100)^2}\right] = 0.37$$

That is, the oxygen content is increasing at the rate of 0.37% per day, and the restoration process has indeed begun.

GROUP DISCUSSION

Consider a particle moving along a straight line. Newton's second law of motion states that the external force F acting on the particle is equal to the rate of change of its momentum. Thus,

$$F = \frac{d}{dt}(mv)$$

where m, the mass of the particle, and v, its velocity, are both functions of time t.

1. Use the product rule to show that

$$F = m\frac{dv}{dt} + v\frac{dm}{dt}$$

and explain the expression on the right-hand side in words.

2. Use the results of part 1 to show that if the mass of a particle is constant, then $F = ma$, where a is the acceleration of the particle.

Verification of the Product Rule

We will now verify the product rule. If $p(x) = f(x)g(x)$, then

$$\begin{aligned} p'(x) &= \lim_{h\to 0}\frac{p(x+h) - p(x)}{h} \\ &= \lim_{h\to 0}\frac{f(x+h)g(x+h) - f(x)g(x)}{h} \end{aligned}$$

By adding $-f(x + h)g(x) + f(x + h)g(x)$ (which is zero!) to the numerator and factoring, we have

$$\begin{aligned} p'(x) &= \lim_{h\to 0}\frac{f(x+h)[g(x+h) - g(x)] + g(x)[f(x+h) - f(x)]}{h} \\ &= \lim_{h\to 0}\left\{f(x+h)\left[\frac{g(x+h) - g(x)}{h}\right] + g(x)\left[\frac{f(x+h) - f(x)}{h}\right]\right\} \\ &= \lim_{h\to 0} f(x+h)\left[\frac{g(x+h) - g(x)}{h}\right] \\ &\quad + \lim_{h\to 0} g(x)\left[\frac{f(x+h) - f(x)}{h}\right] \qquad \text{By Property 3 of limits} \end{aligned}$$

$$= \lim_{h\to 0} f(x+h) \cdot \lim_{h\to 0} \frac{g(x+h) - g(x)}{h}$$

$$+ \lim_{h\to 0} g(x) \cdot \lim_{h\to 0} \frac{f(x+h) - f(x)}{h} \qquad \text{By Property 4 of limits}$$

$$= f(x)g'(x) + g(x)f'(x)$$

Observe that in the second from the last link in the chain of equalities, we have used the fact that $\lim_{h\to 0} f(x+h) = f(x)$ because f is continuous at x.

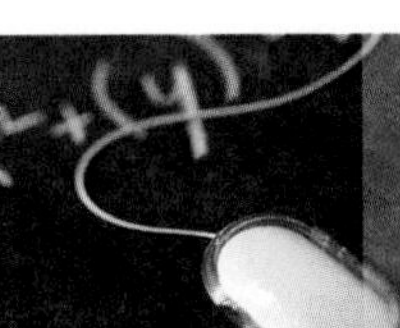

Self-Check Exercises 3.5

1. Find the derivative of $f(x) = \dfrac{2x+1}{x^2-1}$.

2. What is the slope of the tangent line to the graph of

$$f(x) = (x^2+1)(2x^3 - 3x^2 + 1)$$

at the point (2, 25)? How fast is the function f changing when $x = 2$?

3. The total sales of Security Products in its first 2 yr of operation are given by

$$S = f(t) = \frac{0.3t^3}{1 + 0.4t^2} \qquad (0 \le t \le 2)$$

where S is measured in millions of dollars and $t = 0$ corresponds to the date Security Products began operations. How fast were the sales increasing at the beginning of the company's second year of operation?

Solutions to Self-Check Exercises 3.5 can be found on page 180.

3.5 Exercises

In Exercises 1–30, find the derivative of the given function.

1. $f(x) = 2x(x^2+1)$
2. $f(x) = 3x^2(x-1)$
3. $f(t) = (t-1)(2t+1)$
4. $f(x) = (2x+3)(3x-4)$
5. $f(x) = (3x+1)(x^2-2)$
6. $f(x) = (x+1)(2x^2-3x+1)$
7. $f(x) = (x^3-1)(x+1)$
8. $f(x) = (x^3-12x)(3x^2+2x)$
9. $f(w) = (w^3 - w^2 + w - 1)(w^2+2)$
10. $f(x) = \frac{1}{5}x^5 + (x^2+1)(x^2-x-1) + 28$
11. $f(x) = (5x^2+1)(2\sqrt{x} - 1)$
12. $f(t) = (1+\sqrt{t})(2t^2-3)$
13. $f(x) = (x^2 - 5x + 2)\left(x - \frac{2}{x}\right)$
14. $f(x) = (x^3 + 2x + 1)\left(2 + \frac{1}{x^2}\right)$
15. $f(x) = \dfrac{1}{x-2}$
16. $g(x) = \dfrac{3}{2x+4}$
17. $f(x) = \dfrac{x-1}{2x+1}$
18. $f(t) = \dfrac{1-2t}{1+3t}$
19. $f(x) = \dfrac{1}{x^2+1}$
20. $f(u) = \dfrac{u}{u^2+1}$
21. $f(s) = \dfrac{s^2-4}{s+1}$
22. $f(x) = \dfrac{x^3-2}{x^2+1}$
23. $f(x) = \dfrac{\sqrt{x}}{x^2+1}$
24. $f(x) = \dfrac{x^2+1}{\sqrt{x}}$
25. $f(x) = \dfrac{x^2+2}{x^2+x+1}$
26. $f(x) = \dfrac{x+1}{2x^2+2x+3}$
27. $f(x) = \dfrac{(x+1)(x^2+1)}{x-2}$
28. $f(x) = (3x^2-1)\left(x^2 - \frac{1}{x}\right)$
29. $f(x) = \dfrac{x}{x^2-4} - \dfrac{x-1}{x^2+4}$
30. $f(x) = \dfrac{x + \sqrt{3x}}{3x-1}$

In Exercises 31–34, suppose f and g are functions that are differentiable at $x = 1$ and that $f(1) = 2$, $f'(1) = -1$, $g(1) = -2$, and $g'(1) = 3$. Find the value of $h'(1)$.

31. $h(x) = f(x)g(x)$
32. $h(x) = (x^2+1)g(x)$
33. $h(x) = \dfrac{xf(x)}{x + g(x)}$
34. $h(x) = \dfrac{f(x)g(x)}{f(x) - g(x)}$

In Exercises 35–38, find the derivative of each of the given functions and evaluate $f'(x)$ at the given value of x.

35. $f(x) = (2x - 1)(x^2 + 3);\ x = 1$

36. $f(x) = \dfrac{2x + 1}{2x - 1};\ x = 2$

37. $f(x) = \dfrac{x}{x^4 - 2x^2 - 1};\ x = -1$

38. $f(x) = (\sqrt{x} + 2x)(x^{3/2} - x);\ x - 4$

In Exercises 39–42, find the slope and an equation of the tangent line to the graph of the function f at the specified point.

39. $f(x) = (x^3 + 1)(x^2 - 2);\ (2, 18)$

40. $f(x) = \dfrac{x^2}{x + 1};\ \left(2, \dfrac{4}{3}\right)$

41. $f(x) = \dfrac{x + 1}{x^2 + 1};(1, 1)$

42. $f(x) = \dfrac{1 + 2x^{1/2}}{1 + x^{3/2}};\ \left(4, \dfrac{5}{9}\right)$

43. Find an equation of the tangent line to the graph of the function $f(x) = (x^3 + 1)(3x^2 - 4x + 2)$ at the point $(1, 2)$.

44. Find an equation of the tangent line to the graph of the function $f(x) = \dfrac{3x}{x^2 - 2}$ at the point $(2, 3)$.

45. Let $f(x) = (x^2 + 1)(2 - x)$. Find the point(s) on the graph of f where the tangent line is horizontal.

46. Let $f(x) = \dfrac{x}{x^2 + 1}$. Find the point(s) on the graph of f where the tangent line is horizontal.

47. Find the point(s) on the graph of the function $f(x) = (x^2 + 6)(x - 5)$ where the slope of the tangent line is equal to -2.

48. Find the point(s) on the graph of the function $f(x) = \dfrac{x + 1}{x - 1}$ where the slope of the tangent line is equal to $-\frac{1}{2}$.

49. A straight line perpendicular to and passing through the point of tangency of the tangent line is called the *normal* to the curve. Find the equation of the tangent line and the normal to the curve $y = \dfrac{1}{1 + x^2}$ at the point $(1, \frac{1}{2})$.

50. Extend the product rule for differentiation to the following case involving the product of three differentiable functions: Let $h(x) = u(x)v(x)w(x)$ and show that $h'(x) = u(x)v(x)w'(x) + u(x)v'(x)w(x) + u'(x)v(x)w(x)$.

Hint: Let $f(x) = u(x)v(x)$, $g(x) = w(x)$, and $h(x) = f(x)g(x)$ and apply the product rule to the function h.

51. Prove the quotient rule for differentiation (Rule 6).

Hint: Let $k(x) = f(x)/g(x)$ and verify the following steps:

a. $$\frac{k(x + h) - k(x)}{h} = \frac{f(x + h)g(x) - f(x)g(x + h)}{hg(x + h)g(x)}$$

b. By adding $[-f(x)g(x) + f(x)g(x)]$ to the numerator and simplifying, show that

$$\frac{k(x + h) - k(x)}{h} = \frac{1}{g(x + h)g(x)} \times \left[\frac{f(x + h) - f(x)}{h}\right]\cdot g(x) - \left[\frac{g(x + h) - g(x)}{h}\right]\cdot f(x)$$

c. $$k'(x) = \lim_{h \to 0} \frac{k(x + h) - k(x)}{h} = \frac{g(x)f'(x) - f(x)g'(x)}{[g(x)]^2}$$

Business and Economics Applications

52. Demand Function The demand function for digital cameras sold in Edmonton is given by

$$D(x) = (x^2 - 10x + 25)(8x - x^2) + 500$$

where x is the quantity demanded in thousands of units and $d(x)$ is measured in dollars.

a. Find $D'(x)$.

b. What is the rate of change of the demand when the quantity demanded is 7000 units? What is the unit price at that level of demand?

53. Revenue Function The total revenue in dollars for a video game is given by

$$R(x) = \frac{1}{100}(x + 2000)(1600 - x) - 36{,}000$$

where x is the number of units sold. What is the rate of change of revenue with respect to x when 600 units are sold? Interpret your result.

54. Sales of Outdoor Gear In British Columbia, the sales of specialized high-endurance, light-weight outdoor gear in millions of dollars is projected by

$$S(t) = \frac{4t}{t^2 + 2} \qquad 0 \le t \le 10$$

where t is measured in years, with $t = 0$ corresponding to the beginning of 2000. How fast were sales increasing at the beginning of 2000? How fast was the level of sales increasing at the beginning of 2005?

55. Cost of Removing Toxic Waste a city's main well was recently found to be contaminated with trichloroethylene, a cancer-causing chemical, as a result of an abandoned chemical dump leaching chemicals into the water. a proposal submitted to the city's council members indicates that the cost, measured in millions of dollars, of removing $x\%$ of the toxic pollutant is given by

$$C(x) = \frac{0.5x}{100 - x}$$

Find $C'(80)$, $C'(90)$, $C'(95)$, and $C'(99)$. What does your result tell you about the cost of removing *all* of the pollutant?

56. **Demand Functions** The demand function for the Sicard wristwatch is given by

$$d(x) = \frac{50}{0.01x^2 + 1} \qquad (0 \le x \le 20)$$

where x (measured in units of a thousand) is the quantity demanded per week and $d(x)$ is the unit price in dollars.
a. Find $d'(x)$.
b. Find $d'(5)$, $d'(10)$, and $d'(15)$ and interpret your results.

57. **Box-Office Receipts** The total worldwide box-office receipts for a long-running movie are approximated by the function

$$T(x) = \frac{120x^2}{x^2 + 4}$$

where $T(x)$ is measured in millions of dollars and x is the number of years since the movie's release. How fast are the total receipts changing 1 yr, 3 yr, and 5 yr after its release?

Biological and Life Sciences Applications

58. **Concentration of a Drug in the Bloodstream** The concentration of a certain drug in a patient's bloodstream t hr after injection is given by

$$C(t) = \frac{0.2t}{t^2 + 1}$$

a. Find the rate at which the concentration of the drug is changing with respect to time.
b. How fast is the concentration changing $\frac{1}{2}$ hr, 1 hr, and 2 hr after the injection?

59. **Drug Dosages** Thomas Young has suggested the following rule for calculating the dosage of medicine for children 1 to 12 yr old. If a denotes the adult dosage (in milligrams) and if t is the child's age (in years), then the child's dosage is given by

$$D(t) = \frac{at}{t + 12}$$

Suppose the adult dosage of a substance is 500 mg. Find an expression that gives the rate of change of a child's dosage with respect to the child's age. What is the rate of change of a child's dosage with respect to his or her age for a 6-yr-old child? A 10-yr-old child?

60. **Effect of Bactericide** The number of bacteria $N(t)$ in a certain culture t min after an experimental bactericide is introduced obeys the rule

$$N(t) = \frac{10{,}000}{1 + t^2} + 200$$

Find the rate of change of the number of bacteria in the culture 1 min and 2 min after the bactericide is introduced. What is the population of the bacteria in the culture 1 min and 2 min after the bactericide is introduced?

61. **Learning Curves** From experience, Emory Secretarial School knows that the average student taking Advanced Typing will progress according to the rule

$$N(t) = \frac{60t + 180}{t + 6} \qquad (t \ge 0)$$

where $N(t)$ measures the number of words/minute the student can type after t wk in the course.
a. Find an expression for $N'(t)$.
b. Compute $N'(t)$ for $t = 1, 3, 4,$ and 7 and interpret your results.
c. Sketch the graph of the function N. Does it confirm the results obtained in part (b)?
d. What will be the average student's typing speed at the end of the 12-wk course?

62. **Formaldehyde Levels** A study on formaldehyde levels in 900 homes indicates that emissions of various chemicals can decrease over time. The formaldehyde level (parts per million) in an average home in the study is given by

$$f(t) = \frac{0.055t + 0.26}{t + 2} \qquad (0 \le t \le 12)$$

where t is the age of the house in years. How fast is the formaldehyde level of the average house dropping when it is new? At the beginning of its fourth year?

Source: Bonneville Power Administration

63. **Population Growth** A major corporation is building a 4325-acre complex of homes, offices, stores, schools, and churches in the rural community of Glen Cove. As a result of this development, the planners have estimated that Glen Cove's population (in thousands) t yr from now will be given by

$$P(t) = \frac{25t^2 + 125t + 200}{t^2 + 5t + 40}$$

a. Find the rate at which Glen Cove's population is changing with respect to time.
b. What will be the population after 10 yr? At what rate will the population be increasing when $t = 10$?

In Exercises 64–67, determine whether the statement is true or false. If it is true, explain why it is true. If it is false, give an example to show why it is false.

64. If f and g are differentiable, then

$$\frac{d}{dx}[f(x)g(x)] = f'(x)g'(x)$$

65. If f is differentiable, then

$$\frac{d}{dx}[xf(x)] = f(x) + xf'(x)$$

66. If f is differentiable, then

$$\frac{d}{dx}\left[\frac{f(x)}{x^2}\right] = \frac{f'(x)}{2x}$$

67. If f, g, and h are differentiable, then

$$\frac{d}{dx}\left[\frac{f(x)g(x)}{h(x)}\right] = \frac{f'(x)g(x)h(x) + f(x)g'(x)h(x) - f(x)g(x)h'(x)}{[h(x)]^2}$$

Solutions to Self-Check Exercises 3.5

1. We use the quotient rule to obtain

$$f'(x) = \frac{(x^2 - 1)\dfrac{d}{dx}(2x + 1) - (2x + 1)\dfrac{d}{dx}(x^2 - 1)}{(x^2 - 1)^2}$$

$$= \frac{(x^2 - 1)(2) - (2x + 1)(2x)}{(x^2 - 1)^2}$$

$$= \frac{2x^2 - 2 - 4x^2 - 2x}{(x^2 - 1)^2}$$

$$= \frac{-2x^2 - 2x - 2}{(x^2 - 1)^2}$$

$$= \frac{-2(x^2 + x + 1)}{(x^2 - 1)^2}$$

2. The slope of the tangent line to the graph of f at any point is given by

$$f'(x) = (x^2 + 1)\frac{d}{dx}(2x^3 - 3x^2 + 1)$$
$$+ (2x^3 - 3x^2 + 1)\frac{d}{dx}(x^2 + 1)$$
$$= (x^2 + 1)(6x^2 - 6x) + (2x^3 - 3x^2 + 1)(2x)$$

In particular, the slope of the tangent line to the graph of f when $x = 2$ is

$$f'(2) = (2^2 + 1)[6(2^2) - 6(2)]$$
$$+ [2(2^3) - 3(2^2) + 1][2(2)]$$
$$= 60 + 20 = 80$$

Note that it is not necessary to simplify the expression for $f'(x)$ since we are required only to evaluate the expression at $x = 2$. We also conclude, from this result, that the function f is changing at the rate of 80 units/unit change in x when $x = 2$.

3. The rate at which the company's total sales are changing at any time t is given by

$$S'(t) = \frac{(1 + 0.4t^2)\dfrac{d}{dt}(0.3t^3) - (0.3t^3)\dfrac{d}{dt}(1 + 0.4t^2)}{(1 + 0.4t^2)^2}$$

$$= \frac{(1 + 0.4t^2)(0.9t^2) - (0.3t^3)(0.8t)}{(1 + 0.4t^2)^2}$$

Therefore, at the beginning of the second year of operation, Security Products' sales were increasing at the rate of

$$S'(1) = \frac{(1 + 0.4)(0.9) - (0.3)(0.8)}{(1 + 0.4)^2} = 0.520408$$

or \$520,408/year.

3.6 The Chain Rule

This section introduces another rule of differentiation called the **chain rule.** When used in conjunction with the rules of differentiation developed in the last two sections, the chain rule enables us to greatly enlarge the class of functions that we are able to differentiate.

The Chain Rule

Consider the function $h(x) = (x^2 + x + 1)^2$. If we were to compute $h'(x)$ using only the rules of differentiation from the previous sections, then our approach might be to expand $h(x)$. Thus,

$$h(x) = (x^2 + x + 1)^2 = (x^2 + x + 1)(x^2 + x + 1)$$
$$= x^4 + 2x^3 + 3x^2 + 2x + 1$$

from which we find

$$h'(x) = 4x^3 + 6x^2 + 6x + 2$$

But what about the function $H(x) = (x^2 + x + 1)^{100}$? The same technique may be used to find the derivative of the function H, but the amount of work involved in this case would be prodigious! Consider, also, the function $G(x) = \sqrt{x^2 + 1}$. For each of the two functions H and G, the rules of differentiation of the previous sections cannot be applied directly to compute the derivatives H' and G'.

Observe that both H and G are **composite functions;** that is, each is composed of, or built up from, simpler functions. For example, the function H is composed of the two simpler functions $f(x) = x^{100}$ and $g(x) = x^2 + x + 1$ as follows:

$$H(x) = f[g(x)] = f[x^2 + x + 1]$$
$$= (x^2 + x + 1)^{100}$$

In a similar manner, we see that the function G is composed of the two simpler functions $f(x) = \sqrt{x}$ and $g(x) = x^2 + 1$ as follows:

$$G(x) = f[g(x)] = f[x^2 + 1]$$
$$= \sqrt{x^2 + 1}$$

As a first step toward finding the derivative h' of a composite function $h = f \circ g$ defined by $h(x) = f(g(x))$, we write

$$u = g(x) \qquad \text{and} \qquad y = f(g(x)) = f(u)$$

The dependency of h on f and g is illustrated in Figure 3.45. Since u is a function of x, we may compute the derivative of u with respect to x, if g is a differentiable function, obtaining $du/dx = dg(x)/dx = g'(x)$. Next, if f is a differentiable function of u, we may compute the derivative of f with respect to u, obtaining $dy/du = df(u)/du = f'(u)$. Now, since the function h is composed of the function f and the function g, we might suspect that the rule $h'(x)$ for the derivative h' of h will be given by an expression that involves the rules for the derivatives of f and g. But how do we combine these derivatives to yield h'?

This question can be answered by interpreting the derivative of each function as giving the rate of change of that function. For example, suppose $u = g(x)$ changes three times as fast as x—that is,

$$g'(x) = \frac{dg(x)}{dx} = \frac{du}{dx} = 3$$

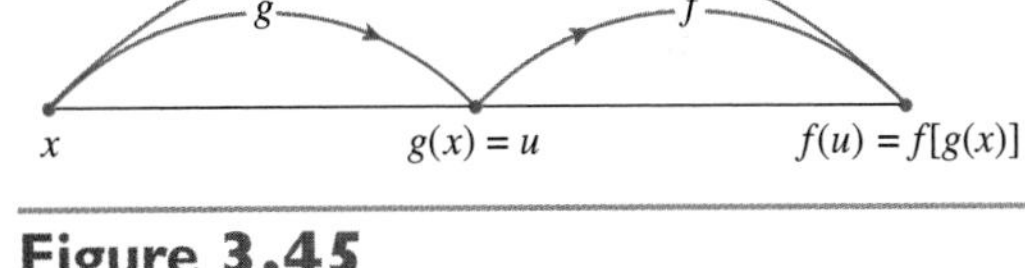

Figure 3.45
The composite function $h(x) = f[g(x)]$

And suppose $y = f(u)$ changes twice as fast as u—that is,

$$f'(u) = \frac{df(u)}{du} = \frac{dy}{du} = 2$$

Then, we would expect $y = h(x)$ to change six times as fast as x—that is,

$$h'(x) = f'(u)g'(x) = f'(g(x))g'(x) = (2)(3) = 6$$

or equivalently

$$h'(x) = \frac{dh(x)}{dx} = \frac{df(g(x))}{dx} = \frac{df(g(x))}{dg(x)} \cdot \frac{dg(x)}{dx} = (2)(3) = (6)$$

or equivalently

$$\frac{dy}{dx} = \frac{dy}{du} \cdot \frac{du}{dx} = (2)(3) = (6)$$

This observation suggests the following result, which we state without proof.

Rule 7: The Chain Rule

If $h(x) = f(g(x))$, then

$$h'(x) = f'(g(x))g'(x) \text{ or} \qquad (10)$$

$$\frac{dh(x)}{dx} = \frac{df(g(x))}{dx} = \frac{df(g(x))}{dg(x)} \cdot \frac{dg(x)}{dx} \qquad (11)$$

Equivalently, if we write $y = h(x) = f(u)$, where $u = g(x)$, then

$$\frac{dy}{dx} = \frac{dy}{du} \cdot \frac{du}{dx} \qquad (12)$$

Remarks

1. If we label the composite function h in the following manner

Inside function
↓
$$h(x) = f[g(x)]$$
↑
Outside function

then $h'(x)$ is just the *derivative* of the "outside function" *evaluated* at the "inside function" times the *derivative* of the "inside function."

2. Equation 12 can be remembered by observing that if we "cancel" the du's, then

$$\frac{dy}{dx} = \frac{dy}{\cancel{du}} \cdot \frac{\cancel{du}}{dx} = \frac{dy}{dx}$$ ◀

The Chain Rule for Powers of Functions

Many composite functions have the special form $h(x) = g(f(x))$, where g is defined by the rule $g(x) = x^n$ (n, a real number)—that is,

$$h(x) = [f(x)]^n$$

In other words, the function h is given by the power of a function f. The functions

$$h(x) = (x^2 + x + 1)^2, \qquad H = (x^2 + x + 1)^{100}, \qquad g = \sqrt{x^2 + 1}$$

discussed earlier are examples of this type of composite function. By using the following corollary of the chain rule, the general power rule, we can find the derivative of this type of function much more easily than by using the chain rule directly.

The General Power Rule

If the function f is differentiable and $h(x) = [f(x)]^n$ (n, a real number), then

$$h'(x) = \frac{d}{dx}[f(x)]^n = n[f(x)]^{n-1}f'(x) \qquad (13)$$

To see this, we observe that $h(x) = g(f(x))$, where $g(x) = x^n$, so that, by virtue of the chain rule, we have

$$\begin{aligned} h'(x) &= g'(f(x))f'(x) \\ &= n[f(x)]^{n-1}f'(x) \end{aligned}$$

since $g'(x) = nx^{n-1}$.

Example 1

Let $F(x) = (3x + 1)^2$.

a. Find $F'(x)$, using the general power rule.

b. Verify your result without the benefit of the general power rule.

Solution

a. Using the general power rule, we obtain

$$\begin{aligned} F'(x) &= 2(3x + 1)^1\frac{d}{dx}(3x + 1) \\ &= 2(3x + 1)(3) \\ &= 6(3x + 1) \end{aligned}$$

b. We first expand $F(x)$. Thus,

$$F(x) = (3x + 1)^2 = 9x^2 + 6x + 1$$

Next, differentiating, we have

$$\begin{aligned} F'(x) &= \frac{d}{dx}(9x^2 + 6x + 1) \\ &= 18 + 6 \\ &= 6(3x + 1) \end{aligned}$$

as before.

Example 2

Differentiate the function $G(x) = \sqrt{x^2 + 1}$.

Solution

We rewrite the function $G(x)$ as

$$G(x) = (x^2 + 1)^{1/2}$$

and apply the general power rule, obtaining

$$\begin{aligned} G'(x) &= \frac{1}{2}(x^2 + 1)^{-1/2}\frac{d}{dx}(x^2 + 1) \\ &= \frac{1}{2}(x^2 + 1)^{-1/2} \cdot 2x = \frac{x}{\sqrt{x^2 + 1}} \end{aligned}$$

Example 3

Differentiate the function $f(x) = x^2(2x + 3)^5$.

Solution

Applying the product rule followed by the general power rule, we obtain

$$\begin{aligned} f'(x) &= x^2\frac{d}{dx}(2x + 3)^5 + (2x + 3)^5\frac{d}{dx}(x^2) \\ &= (x^2)5(2x + 3)^4 \cdot \frac{d}{dx}(2x + 3) + (2x + 3)^5(2x) \\ &= 5x^2(2x + 3)^4(2) + 2x(2x + 3)^5 \\ &= 2x(2x + 3)^4(5x + 2x + 3) = 2x(7x + 3)(2x + 3)^4 \end{aligned}$$

Example 4

Find $f'(x)$ if $f(x) = (2x^2 + 3)^4(3x - 1)^5$.

Solution

Applying the product rule, we have

$$f'(x) = (2x^2 + 3)^4\frac{d}{dx}(3x - 1)^5 + (3x - 1)^5\frac{d}{dx}(2x^2 + 3)^4$$

Next, we apply the general power rule to each term, obtaining

$$\begin{aligned} f'(x) &= (2x^2 + 3)^4 \cdot 5(3x - 1)^4\frac{d}{dx}(3x - 1) + (3x - 1)^5 \cdot 4(2x^2 + 3)^3\frac{d}{dx}(2x^2 + 3) \\ &= 5(2x^2 + 3)^4 (3x - 1)^4 \cdot 3 + 4(3x - 1)^5(2x^2 + 3)^3(4x) \end{aligned}$$

Finally, observing that $(2x^2 + 3)^3(3x - 1)^4$ is common to both terms, we can factor and simplify as follows:

$$\begin{aligned} f'(x) &= (2x^2 + 3)^3(3x - 1)^4[15(2x^2 + 3) + 16x(3x - 1)] \\ &= (2x^2 + 3)^3(3x - 1)^4(30x^2 + 45 + 48x^2 - 16x) \\ &= (2x^2 + 3)^3(3x - 1)^4(78x^2 - 16x + 45) \end{aligned}$$

Example 5

Find $f'(x)$ if $f(x) = \dfrac{1}{(4x^2 - 7)^2}$.

Solution

Rewriting $f(x)$ and then applying the general power rule, we obtain

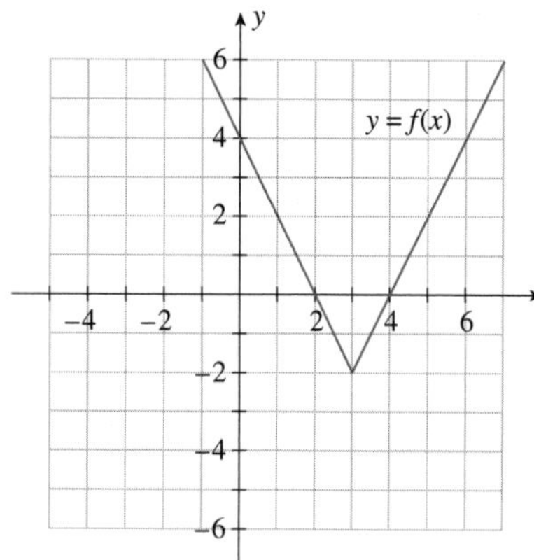

(a)

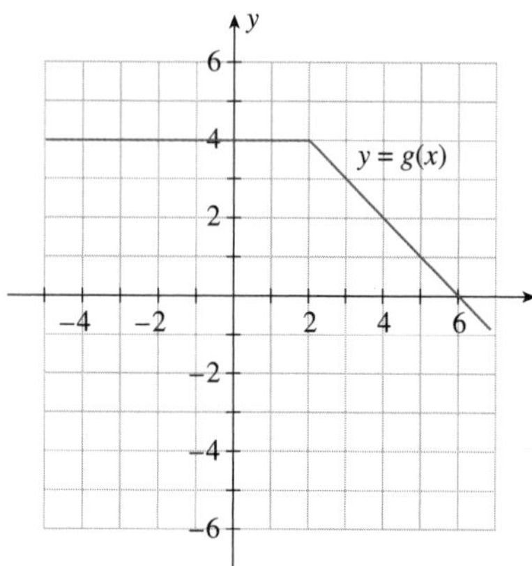

(b)

Figure 3.46

$$f'(x) = \frac{d}{dx}\left[\frac{1}{(4x^2 - 7)^2}\right] = \frac{d}{dx}(4x^2 - 7)^{-2}$$
$$= -2(4x^2 - 7)^{-3}\frac{d}{dx}(4x^2 - 7)$$
$$= -2(4x^2 - 7)^{-3}(8x) = -\frac{16x}{(4x^2 - 7)^3}$$

Example 6

The graphs of the functions $y = f(x)$ and $y = g(x)$ are given as shown in Figure 3.46a and Figure 3.46b respectively. Suppose $h(x) = f(g(x))$ and $k(x) = g(f(x))$.

a. Find $h(4)$ and $k(4)$. Are those values the same?
b. Find $h'(4)$ and $k'(4)$. Are those values the same?

Solution

a. Since $h(x) = f(g(x))$, we have

$$h(4) = f(g(4))$$

From the graph of g (Figure 3.46b) we read off that $g(4) = 2$, therefore

$$h(4) = f(g(4)) = f(2)$$

From the graph of f (Figure 3.46a) we read off that $f(2) = 0$, therefore

$$h(4) = f(2) = 0$$

Since $k(x) = g(f(x))$, we similarly read off from the graphs of f and g respectively to get

$$k(4) = g(f(4)) = g(0) = 4$$

Comparing values, we conclude that $h(4) \neq k(4)$.

b. By Equation (10) we have

$$h'(x) = f'(g(x))\,g'(x) \qquad \text{and}$$
$$k'(x) = g'(f(x))f'(x)$$

We use the derivative computed in (b) to find $h'(4)$.

$$h'(4) = f'(g(4))g'(4)$$

We already know from (a) that $g(4) = 2$, and so

$$h'(4) = f'(2)g'(4)$$

Now, the slope of the graph of f at $x = 2$ is read off as $-2/1 = -2$, and the slope of the graph of g at $x = 4$ is read off as $-1/1 = -1$. We compute

$$h'(4) = f'(2)g'(4) = (-2)(-1) = 2$$

Similarly, we compute $k'(4)$.

$$k'(4) = g'(f(4))f'(4) = g'(0)f'(4) = (0)(-2/1) = 0$$

Comparing values, we conclude that $h'(4) \neq k'(4)$.

In general,

$$f(g(x)) \neq g(f(x)) \qquad \text{and}$$
$$\frac{df(g(x))}{dx} \neq \frac{dg(f(x))}{dx}$$

as was shown in Example 6.

Example 7

Find the slope of the tangent line to the graph of the function

$$f(x) = \left(\frac{2x + 1}{3x + 2}\right)^3$$

at the point $\left(0, \frac{1}{8}\right)$.

Solution

The slope of the tangent line to the graph of f at any point is given by $f'(x)$. To compute $f'(x)$, we use the general power rule followed by the quotient rule, obtaining

$$\begin{aligned} f'(x) &= 3\left(\frac{2x + 1}{3x + 2}\right)^2 \frac{d}{dx}\left(\frac{2x + 1}{3x + 2}\right) \\ &= 3\left(\frac{2x + 1}{3x + 2}\right)^2 \left[\frac{(3x + 2)(2) - (2x + 1)(3)}{(3x + 2)^2}\right] \\ &= 3\left(\frac{2x + 1}{3x + 2}\right)^2 \left[\frac{6x + 4 - 6x - 3}{(3x + 2)^2}\right] \\ &= \frac{3(2x + 1)^2}{(3x + 2)^4} \end{aligned}$$

In particular, the slope of the tangent line to the graph of f at $\left(0, \frac{1}{8}\right)$ is given by

$$f'(0) = \frac{3(0 + 1)^2}{(0 + 2)^4} = \frac{3}{16}$$

Example 8

To see the full power of the chain rule, we shall consider more complex compositions of functions. Suppose we are given the functions $f(x) = \sqrt[5]{x}$, $g(x) = (x^2 + 3x)^{24}$, and $h(x) = \frac{1}{x + 2}$. Find the first derivative of the following compositions:

a. $(f \circ h \circ g)(x)$
b. $(g \circ f \circ h)(x)$

Solution

We will first make some general observations. The first derivative of $(f \circ g \circ h)(x)$ is simply $\frac{d}{dx}((f \circ g \circ h)(x)) = f'[g[h(x)]] \cdot g'[h(x)] \cdot h'(x)$, since

$$\begin{aligned} \frac{d}{dx}((f \circ g \circ h)(x)) &= \frac{d}{dx}(f[g[h(x)]]) \\ &= f'[g[h(x)]] \cdot \frac{d}{dx}(g[h(x)]) \quad \text{now apply the chain rule a second time} \\ &= f'[g[h(x)]] \cdot g'[h(x)] \cdot \frac{d}{dx}(h(x)) \\ &= f'[g[h(x)]] \cdot g'[g[h(x)] \cdot h'(x) \end{aligned}$$

The derivative of f is

$$f'(x) = \frac{d}{dx}(x)^{1/5} = \frac{1}{5}(x)^{-4/5}$$

The derivative of g is

$$\begin{aligned} g'(x) &= 24(x^2 + 3x)^{23} \cdot \frac{d}{dx}(x^2 + 3x) \\ &= 24(x^2 + 3x)^{23} \cdot (2x + 3) \\ &= 24(2x + 3)(x^2 + 3x)^{23} \end{aligned}$$

The derivative of h is

$$\begin{aligned} h'(x) &= \frac{d}{dx}\left(\frac{1}{x + 2}\right) = \frac{\frac{d}{dx}(1)(x + 2) - 1\frac{d}{dx}(x + 2)}{(x^3 + 2^2)} \\ &= \frac{0(x + 2) - 1(1)}{(x + 2)^2} = -\frac{1}{(x + 2)^2} \end{aligned}$$

a. The derivative of $(f \circ h \circ g)(x)$ is therefore given by

$$\frac{d}{dx}((f \circ h \circ g)(x)) = f'[h[g(x)]] \cdot h'[g(x)] \cdot g'(x)$$

We can now use the derivatives of f, g, and h we calculated above and evaluate them at the appropriate inner functions,

$$\begin{aligned} f'[h[g(x)]] &= f'[h[(x^2 + 3x)^{24}]] \\ &= f'\left[\frac{1}{(x^2 + 3x)^{24} + 2}\right] \\ &= \frac{1}{5}\left(\frac{1}{(x^2 + 3x)^{24} + 2^{-4/5}}\right) \\ &= \frac{1}{5}((x^2 + 3x)^{24} + 2)^{4/5} \end{aligned}$$

and

$$h'[g(x)] = h'[(x^2 + 3x)^{24}] = -\frac{1}{((x^2 + 3x)^{24} + 2)^2}$$

We finally can put together the derivative we are seeking,

$$\begin{aligned} \frac{d}{dx}((f \circ h \circ g)(x)) &= \frac{1}{5}((x^2 + 3x)^{24} + 2)^{4/5} \cdot \left[-\frac{1}{((x^2 + 3x)^{24} + 2)^2}\right] \cdot 24(2x + 3)(x^2 + 3x)^{23} \\ &= -\frac{24(2x + 3)(x^2 + 3x)^{23}((x^2 + 3x)^{24} + 2)^{4/5}}{5((x^2 + 3x)^{24} + 2)^2} \end{aligned}$$

b. In a similar fashion as in part (a), we can find the first derivative of $(g \circ f \circ h)(x)$ to be

$$\begin{aligned} \frac{d}{dx}((g \circ f \circ h)(x)) &= 24\left(2\sqrt[5]{\frac{1}{x + 2}} + 3\right)\left(\sqrt[5]{\frac{1}{x + 2}}^{\,2} + 3\sqrt[5]{\frac{1}{x + 2}}\right)^{23} \cdot \frac{1}{5}\left(\frac{1}{x + 2}\right)^{-4/5} \cdot \left[-\frac{1}{(x + 2)^2}\right] \\ &= 24\left(2\sqrt[5]{\frac{1}{x + 2}} + 3\right)\left(\sqrt[5]{\frac{1}{x + 2}}^{\,2} + 3\sqrt[5]{\frac{1}{x + 2}}\right)^{23} \cdot \frac{1}{5}(x + 2)^{4/5} \cdot \left[-\frac{1}{x + 2^2}\right] \\ &= -\frac{24\left(2\sqrt[5]{\frac{1}{x + 2}} + 3\right)\left(\sqrt[5]{\frac{1}{x + 2}}^{\,2} + 3\sqrt[5]{\frac{1}{x + 2}}\right)^{23}(x + 2)^{4/5}}{5(x + 2)^2} \end{aligned}$$

Example 9

Suppose the only data available to you are some points on the graphs of a function $y = f(x)$ and its derivative $f'(x)$.

$$f(2) = -1, f(-1) = 3, f'(2) = 4, f'(-1) = -5, \text{ and}$$
$$g(2) = 2,\ g(-1) = -2,\ g'(2) = 7,\ g'(-1) = 0$$

If possible, find the following derivatives:

a. $(f \circ g)'(2)$ **b.** $(f \circ f)'(2)$ **c.** $(g \circ f)'(-1)$

Solution

a. $(f \circ g)'(2) = f'[g(2)]g'(2) = f'[2] \cdot 7 = 4 \cdot 7 = 28$
b. $(f \circ f)'(2) = f'[f(2)]f'(2) = f'[-1] \cdot 4 = (-5) \cdot 4 = -20$
c. $(g \circ f)'(-1) = g'[f(-1)]\,f'(-1) = g'[3] \cdot (-5)$, which cannot be found since we do not know what $g'[3]$ evaluates to.

EXPLORING WITH TECHNOLOGY

Refer to Example 6.

1. Use a graphing utility to plot the graph of the function f, using the viewing window $[-2, 1] \cdot [-1, 2]$. Then draw the tangent line to the graph of f at the point $(0, \frac{1}{8})$.
2. For a better picture, repeat part 1 using the viewing window $[-1, 1] \cdot [-0.1, 0.3]$.
3. Use the numerical differentiation capability of the graphing utility to verify that the slope of the tangent line at $(0, \frac{1}{8})$ is $\frac{3}{16}$.

Applications

Example 10

Growth in a Health Club Membership The membership of The Fitness Centre, which opened a few years ago, is approximated by the function

$$N(t) = 100(64 + 4t)^{2/3} \qquad (0 \le t \le 52)$$

where $N(t)$ gives the number of members at the beginning of week t.

a. Find $N'(t)$.
b. How fast was the centre's membership increasing initially ($t = 0$)?
c. How fast was the membership increasing at the beginning of the 40th week?
d. What was the membership when the centre first opened? At the beginning of the 40th week?

Solution

a. Using the general power rule, we obtain

$$\begin{aligned} N'(t) &= \frac{d}{dt}[100(64 + 4t)^{2/3}] \\ &= 100\frac{d}{dt}(64 + 4t)^{2/3} \\ &= 100\left(\frac{2}{3}\right)(64 + 4t)^{-1/3}\frac{d}{dt}(64 + 4t) \\ &= \frac{200}{3}(64 + 4t)^{-1/3}(4) \\ &= \frac{800}{3(64 + 4t)^{1/3}} \end{aligned}$$

b. The rate at which the membership was increasing when the centre first opened is given by

$$N'(0) = \frac{800}{3(64)^{1/3}} \approx 66.7$$

or approximately 67 people per week.

c. The rate at which the membership was increasing at the beginning of the 40th week is given by

$$N'(40) = \frac{800}{3(64 + 160)^{1/3}} \approx 43.9$$

or approximately 44 people per week.

d. The membership when the centre first opened is given by

$$N(0) = 100(64)^{2/3} = 100(16)$$

or approximately 1600 people. The membership at the beginning of the 40th week is given by

$$N(40) = 100(64 + 160)^{2/3} \approx 3688.3$$

or approximately 3688 people.

GROUP DISCUSSION

The profit P of a one-product software manufacturer depends on the number of units of its products sold. The manufacturer estimates that it will sell x units of its product per week. Suppose $P = g(x)$ and $x = f(t)$, where g and f are differentiable functions.

1. Write an expression giving the rate of change of the profit with respect to the number of units sold.
2. Write an expression giving the rate of change of the number of units sold per week.
3. Write an expression giving the rate of change of the profit per week.

Example 11

Cost Function The total cost in thousands of dollars to operate a privately owned campground on the Sunshine Coast from the beginning of April to the end of September in 2006 was given by

$$C(t) = \frac{30t}{\sqrt{2 + t^2}} \qquad 0 \le t \le 5$$

where t is measured in months, with $t = 0$ corresponding to the beginning of April.

a. What was the total cost at the beginning of July?

b. Find $C'(t)$.

c. How fast was the total cost increasing at the beginning of July?

Solution

a. At the beginning of July we have $t = 3$, and so the total cost is

$$C(3) = \frac{30(3)}{\sqrt{2 + 3^2}} \approx 27.13602$$

or $27,136.02.

b. We rewrite the function $C(t)$ as

$$C(t) = \frac{30t}{(2 + t^2)^{1/2}}$$

Applying the quotient rule and simplifying, we have

$$C'(t) = \frac{(2 + t^2)^{1/2}\dfrac{d}{dx}(30t) - 30t\dfrac{d}{dx}((2 + t^2)^{1/2})}{[(2 + t^2)]^{1/2}]^2}$$

$$= \frac{30(2 + t^2)^{1/2} - 30t\dfrac{d}{dx}((2 + t^2)^{1/2})}{2 + t^2}$$

Using the general power rule and simplifying, we have

$$C'(t) = \frac{30(2 + t^2)^{1/2} - 30t\frac{1}{2}(2 + t^2)^{-1/2}2t}{2 + t^2}$$

$$= \frac{30(2 + t^2)^{1/2} - 30t^2(2 + t^2)^{-1/2}}{2 + t^2}$$

$$= \frac{30(2 + t^2) - 30t^2}{(2 + t^2)^{1/2}(2 + t^2)}$$

$$= \frac{60}{(2 + t^2)^{3/2}}$$

c. The rate at which the total operating cost was increasing at the beginning of July ($t = 3$) is given by

$$C'(3) = \frac{60}{(2 + 3^2)^{3/2}} \approx 1.64461$$

or approximately \$1644.61 per month.

Example 12

Arteriosclerosis Arteriosclerosis begins during childhood when plaque (soft masses of fatty material) forms in the arterial walls, blocking the flow of blood through the arteries and leading to heart attacks, strokes, and gangrene. Suppose the idealized cross-section of the aorta is circular with radius a cm and by year t the thickness of the plaque (assume it is uniform) is $h = f(t)$ cm (Figure 3.47). Then the area of the opening is given by $A = \pi(a - h)^2$ square centimetres (cm^2).

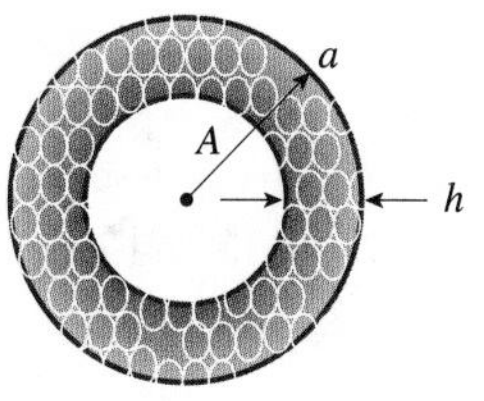

Figure 3.47
Cross-section of the aorta

Suppose the radius of an individual's artery is 1 cm ($a = 1$) and the thickness of the plaque in year t is given by

$$h = g(t) = 1 - 0.01(10{,}000 - t^2)^{1/2} \text{ cm}$$

Since the area of the arterial opening is given by

$$A = f(h) = \pi(1 - h)^2$$

the rate at which A is changing with respect to time is given by

$$\frac{dA}{dt} = \frac{dA}{dh} \cdot \frac{dh}{dt} = f'(h) \cdot g'(t) \qquad \text{By the chain rule}$$

$$= 2\pi(1 - h)(-1)\left[-0.01\left(\frac{1}{2}\right)(10{,}000 - t^2)^{-1/2}(-2t)\right] \qquad \text{Use the chain rule twice.}$$

$$= -2\pi(1 - h)\left[\frac{0.01t}{(10{,}000 - t^2)^{1/2}}\right]$$

$$= -\frac{0.02\pi(1 - h)t}{\sqrt{10{,}000 - t^2}}$$

For example, when $t = 50$,

$$h = g(50) = 1 - 0.01(10{,}000 - 2500)^{1/2} \approx 0.134$$

so that

$$\frac{dA}{dt} = -\frac{0.02\pi(1 - 0.134)50}{\sqrt{10{,}000 - 2500}} \approx -0.03$$

That is, the area of the arterial opening is decreasing at the rate of 0.03 cm^2 per year.

GROUP DISCUSSION

Suppose the population P of a certain bacteria culture is given by $P = f(T)$, where T is the temperature of the medium. Further, suppose the temperature T is a function of time t in seconds—that is, $T = g(t)$. Give an interpretation of each of the following quantities:

1. $\dfrac{dP}{dT}$ **2.** $\dfrac{dT}{dt}$ **3.** $\dfrac{dP}{dt}$ **4.** $(f \circ g)(t)$ **5.** $f'(g(t))g'(t)$

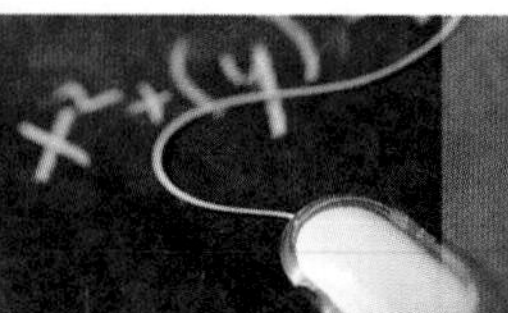

Self-Check Exercises 3.6

1. Find the derivative of

$$f(x) = -\frac{1}{\sqrt{2x^2 - 1}}$$

2. Suppose the life expectancy at birth (in years) of a female in a certain country is described by the function

$$g(t) = 50.02(1 + 1.09t)^{0.1} \qquad (0 \le t \le 150)$$

where t is measured in years, with $t = 0$ corresponding to the beginning of 1900.

a. What is the life expectancy at birth of a female born at the beginning of 1980? At the beginning of 2000?

b. How fast is the life expectancy at birth of a female born at any time t changing?

Solutions to Self-Check Exercises 3.6 can be found on page 194.

3.6 Exercises

In Exercises 1–48, find the derivative of the given function.

1. $f(x) = (2x - 1)^4$

2. $f(x) = (1 - x)^3$

3. $f(x) = (x^2 + 2)^5$

4. $f(t) = 2(t^3 - 1)^5$

5. $f(x) = (2x - x^2)^3$

6. $f(x) = 3(x^3 - x)^4$

7. $f(x) = (2x + 1)^{-2}$

8. $f(t) = \frac{1}{2}(2t^2 + t)^{-3}$

9. $f(x) = (x^2 - 4)^{3/2}$

10. $f(t) = (3t^2 - 2t + 1)^{3/2}$

11. $f(x) = \sqrt{3x - 2}$

12. $f(t) = \sqrt{3t^2 - t}$

13. $f(x) = \sqrt[3]{1 - x^2}$

14. $f(x) = \sqrt{2x^2 - 2x + 3}$

15. $f(x) = \dfrac{1}{(2x + 3)^2}$

16. $f(x) = \dfrac{2}{(x^2 - 1)^4}$

17. $f(t) = \dfrac{1}{\sqrt{2t - 3}}$

18. $f(x) = \dfrac{1}{\sqrt{2x^2 - 1}}$

19. $y = \dfrac{1}{(4x^4 + x)^{3/2}}$

20. $f(t) = \dfrac{4}{\sqrt[3]{2t^2 + t}}$

21. $f(x) = (3x^2 + 2x + 1)^{-2}$

22. $f(t) = (5t^3 + 2t^2 - t + 4)^{-3}$

23. $f(x) = (x^2 + 1)^3 - (x^3 + 1)^2$

24. $f(t) = (2t - 1)^4 + (2t + 1)^4$

25. $f(t) = (t^{-1} - t^{-2})^3$

26. $f(v) = (v^{-3} + 4v^{-2})^3$

27. $f(x) = \sqrt{x + 1} + \sqrt{x - 1}$

28. $f(u) = (2u + 1)^{3/2} + (u^2 - 1)^{-3/2}$

29. $f(x) = 2x^2(3 - 4x)^4$

30. $h(t) = t^2(3t + 4)^3$

31. $f(x) = (x - 1)^2(2x + 1)^4$

32. $g(u) = (1 + u^2)^5(1 - 2u^2)^8$

33. $f(x) = \left(\dfrac{x + 3}{x - 2}\right)^3$ 34. $f(x) = \left(\dfrac{x + 1}{x - 1}\right)^5$

35. $s(t) = \left(\dfrac{t}{2t + 1}\right)^{3/2}$ 36. $g(s) = \left(s^2 + \dfrac{1}{s}\right)^{3/2}$

37. $g(u) = \sqrt{\dfrac{u + 1}{3u + 2}}$ 38. $g(x) = \sqrt{\dfrac{2x + 1}{2x - 1}}$

39. $f(x) = \dfrac{x^2}{(x^2 - 1)^4}$ 40. $g(u) = \dfrac{2u^2}{(u^2 + u)^3}$

41. $h(x) = \dfrac{(3x^2 + 1)^3}{(x^2 - 1)^4}$ 42. $g(t) = \dfrac{(2t - 1)^2}{(3t + 2)^4}$

43. $f(x) = \dfrac{\sqrt{2x + 1}}{x^2 - 1}$ 44. $f(t) = \dfrac{4t^2}{\sqrt{2t^2 + 2t - 1}}$

45. $g(t) = \dfrac{\sqrt{t + 1}}{\sqrt{t^2 + 1}}$ 46. $f(x) = \dfrac{\sqrt{x^2 + 1}}{\sqrt{x^2 - 1}}$

47. $f(x) = (3x + 1)^4(x^2 - x + 1)^3$

48. $g(t) = (2t + 3)^2(3t^2 - 1)^{-3}$

In Exercises 49–52, find the first derivative of the stated composition given that $f(x) = \sqrt{x^2 - 1}$, $g(x) = (x^3 + 5)^{17}$, and $h(x) = \dfrac{2}{x}$.

49. $(f \circ g \circ h)(x)$ 50. $(f \circ h \circ g)(x)$

51. $(g \circ f \circ h)(x)$ 52. $(h \circ g \circ f)(x)$

In Exercises 53–58, use the graphical information of the functions f and g to find the derivative of the composition function at the given x-value.

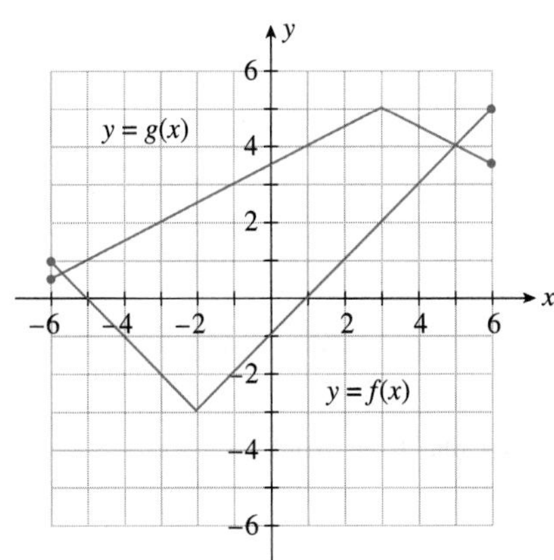

53. $(f \circ g)'(2)$ 54. $(f \circ f)'(1)$

55. $(g \circ f)'(3)$ 56. $(f \circ g)'(3)$

57. $(g \circ f)'(-2)$ 58. $(g \circ f)'(-4)$

In Exercises 59–64, find $\dfrac{dy}{du}$, $\dfrac{du}{dx}$, and $\dfrac{dy}{dx}$.

59. $y = u^{4/3}$ and $u = 3x^2 - 1$

60. $y = \sqrt{u}$ and $u = 7x - 2x^2$

61. $y = u^{-2/3}$ and $u = 2x^3 - x + 1$

62. $y = 2u^2 + 1$ and $u = x^2 + 1$

63. $y = \sqrt{u} + \dfrac{1}{\sqrt{u}}$ and $u = x^3 - x$

64. $y = \dfrac{1}{u}$ and $u = \sqrt{x} + 1$

65. Suppose $f(x) = g(f(x))$ and $f(2) = 3$, $f'(2) = -3$, $g(3) = 5$, and $g'(3) = 4$. Find $F'(2)$.

66. Suppose $h = f \circ g$. Find $h'(0)$ given that $f(0) = 6$, $f'(5) = -2$, $g(0) = 5$, and $g'(0) = 3$.

67. Suppose $F(x) = f(x^2 + 1)$. Find $F'(1)$ if $f'(2) = 3$.

68. Let $F(x) = f(f(x))$. Does it follow that $F'(x) = [f'(x)]^2$?
Hint: Let $f(x) = x^2$.

69. Suppose $h = g \circ f$. Does it follow that $h' = g' \circ f'$?
Hint: Let $f(x) = x$ and $g(x) = x^2$.

70. Suppose $h = f \circ g$. Show that $h' = (f' \circ g)g'$.

In Exercises 71–74, find an equation of the tangent line to the graph of the function at the given point.

71. $f(x) = (1 - x)(x^2 - 1)^2$; $(2, -9)$

72. $f(x) = \left(\dfrac{x + 1}{x - 1}\right)^2$; $(3, 4)$

73. $f(x) = x\sqrt{2x^2 + 7}$; $(3, 15)$

74. $f(x) = \dfrac{8}{\sqrt{x^2 + 6x}}$; $(2, 2)$

A calculator is recommended for Exercises 75–92.

75. In Section 3.4, we proved that

$$\frac{d}{dx}(x^n) = nx^{n-1}$$

for the special case when $n = 2$. Use the chain rule to show that

$$\frac{d}{dx}(x^{1/n}) = \frac{1}{n}x^{1/n-1}$$

for any nonzero integer n, assuming that $f(x) = x^{1/n}$ is differentiable.
Hint: Let $f(x) = x^{1/n}$ so that $[f(x)]^n = x$. Differentiate both sides with respect to x.

76. With the aid of Exercise 75, prove that

$$\frac{d}{dx}(x^r) = rx^{r-1}$$

for any rational number r.
Hint: Let $r = m/n$, where m and n are integers, with $n \neq 0$, and write $x^r = (x^m)^{1/n}$.

77. **Television Viewing** The number of viewers of a television series introduced several years ago is approximated by the function

$$N(x) = (60 + 2x)^{2/3} \qquad (1 \le x \le 26)$$

where $N(x)$ (measured in thousands) denotes the number of weekly viewers of the series in the xth week. Find the rate of increase of the weekly audience at the end of week 2 and at the end of week 12. How many viewers were there in week 2? In week 24?

78. **Working Mothers** The percent of mothers in North America who work outside the home and have children younger than age 6 yr is approximated by the function

$$P(t) = 32.4(t + 5)^{0.205} \qquad (0 \le t \le 21)$$

where t is measured in years, with $t = 0$ corresponding to the beginning of 1980. Compute $P'(t)$. At what rate was the percent of these mothers changing at the beginning of 2000? What was the percent of these mothers at the beginning of 2000?

79. **Continuing Education Enrollment** The registrar of a prominent city in Montréal estimates that the total student enrollment in the Continuing Education division will be given by

$$N(t) = -\frac{20{,}000}{\sqrt{1 + 0.2t}} + 21{,}000$$

where $N(t)$ denotes the number of students enrolled in the division t yr from now. Find an expression for $N'(t)$. How fast is the student enrollment increasing currently? How fast will it be increasing 5 yr from now?

80. **Traffic Flow** The Don Valley Parkway in Toronto was designed to move 75,000 vehicles a day. The number of vehicles moved per day is approximated by the function

$$x = f(t) = 6.25t^2 + 19.75t + 74.75 \qquad (0 \le t \le 5)$$

where x is measured in thousands and t in decades, with $t = 0$ corresponding to the beginning of 1959. Suppose the average speed of traffic flow in km/h is given by

$$S = g(x) = -0.00075x^2 + 67.5 \qquad (75 \le x \le 350)$$

where x has the same meaning as before. What was the rate of change of the average speed of traffic flow at the beginning of 1999? What was the average speed of traffic flow at that time?

81. **Effect of Housing Starts on Jobs** The president of a major housing construction firm claims that the number of construction jobs created is given by

$$N(x) = 1.42x$$

where x denotes the number of housing starts. Suppose the number of housing starts in the next t mo is expected to be

$$x(t) = \frac{7t^2 + 140t + 700}{3t^2 + 80t + 550}$$

million units/year. Find an expression that gives the rate at which the number of construction jobs will be created t mo from now. At what rate will construction jobs be created 1 yr from now?

Business and Economics Applications

82. **Investment Funds** The assets of a certain type of investment fund (millions of dollars) from 1991 through 2001 is given by

$$f(t) = 23.7(0.2t + 1)^{1.32} \qquad (0 \le t \le 11)$$

where $t = 0$ corresponds to the beginning of 1991.

a. Find the rate at which the assets were changing at the beginning of 2000.
b. What were the assets at the beginning of 2000?

83. **Effect of Luxury Tax on Consumption** Government economists of a developing country determined that the purchase of imported perfume is related to a proposed "luxury tax" by the formula

$$N(x) = \sqrt{10{,}000 - 40x - 0.02x^2} \qquad (0 \le x \le 200)$$

where $N(x)$ measures the percentage of normal consumption of perfume when a "luxury tax" of x% is imposed on it. Find the rate of change of $N(x)$ for taxes of 10%, 100%, and 150%.

84. **Hotel Occupancy Rates** The occupancy rate of the all-suite Wonderland Hotel, located near an amusement park, is given by the function

$$r(t) = \frac{10}{81}t^3 - \frac{10}{3}t^2 + \frac{200}{9}t + 60 \qquad (0 \le t \le 12)$$

where t is measured in months, with $t = 0$ corresponding to the beginning of January. Management has estimated that the monthly revenue (in thousands of dollars/month) is approximated by the function

$$R(r) = -\frac{3}{5000}r^3 + \frac{9}{50}r^2 \qquad (0 \le r \le 100)$$

where r is the occupancy rate.

a. Find an expression that gives the rate of change of Wonderland's occupancy rate with respect to time.
b. Find an expression that gives the rate of change of Wonderland's monthly revenue with respect to the occupancy rate.
c. What is the rate of change of Wonderland's monthly revenue with respect to time at the beginning of January? At the beginning of June?

Hint: Use the chain rule to find $R'(r(0))r'(0)$ and $R'(r(6))r'(6)$.

85. **Demand for PCs** The quantity demanded per month, x, of a certain make of personal computer (PC) is related to the average unit price, p (in dollars), of PCs by the equation

$$x = f(p) = \frac{100}{9}\sqrt{810{,}000 - p^2}$$

It is estimated that t mo from now, the average price of a PC will be given by

$$p(t) = \frac{400}{1 + \frac{1}{8}\sqrt{t}} + 200 \qquad (0 \le t \le 60)$$

dollars. Find the rate at which the quantity demanded per month of the PCs will be changing 16 mo from now.

86. **Demand for Watches** The demand equation for the Sicard wristwatch is given by

$$x = f(p) = 10\sqrt{\frac{50 - p}{p}} \qquad (0 < p \le 50)$$

where x (measured in units of a thousand) is the quantity demanded each week and p is the unit price in dollars. Find the rate of change of the quantity demanded of the wristwatches with respect to the unit price when the unit price is $25.

87. Cruise Ship Bookings The management of Cruise World, operators of Caribbean luxury cruises, expects that the percent of young adults booking passage on their cruises in the years ahead will rise dramatically. They have constructed the following model, which gives the percent of young adult passengers in year t:

$$p = f(t) = 50\left(\frac{t^2 + 2t + 4}{t^2 + 4t + 8}\right) \qquad (0 \le t \le 5)$$

Young adults normally pick shorter cruises and generally spend less on their passage. The following model gives an approximation of the average amount of money R (in dollars) spent per passenger on a cruise when the percent of young adults is p:

$$R(p) = 1000\left(\frac{p + 4}{p + 2}\right)$$

Find the rate at which the price of the average passage will be changing 2 yr from now.

Biological and Life Sciences Applications

88. Concentration of Carbon Monoxide (CO) in the Air According to a joint study conducted by Oxnard's Environmental Management Department and a government agency, the concentration of CO in the air due to automobile exhaust t yr from now is given by

$$C(t) = 0.01(0.2t^2 + 4t + 64)^{2/3}$$

parts per million.

a. Find the rate at which the level of CO is changing with respect to time.

b. Find the rate at which the level of CO will be changing 5 yr from now.

89. Aging Population The population of North Americans age 55 and over as a percent of the total population is approximated by the function

$$f(t) = 10.8(0.9) + 10)^{0.3} \qquad (0 \le t \le 20)$$

where t is measured in years, with $t = 0$ corresponding to the year 2000. At what rate was the percent of North Americans age 55 and over changing at the beginning of 2000? At what rate will the percent of North Americans age 55 and over be changing in 2010? What will be the percent of the population of North Americans age 55 and over in 2010?

90. Air Pollution According to an air quality testing, the level of nitrogen dioxide, a brown gas that impairs breathing, present in the atmosphere on a certain July day in the downtown area of a popular Canadian city is approximated by

$$A(t) = 0.03t^3(t - 7)^4 + 60.2 \qquad (0 \le t \le 7)$$

where $A(t)$ is measured in pollutant standard index and t is measured in hours, with $t = 0$ corresponding to 7 A.M.

a. Find $A'(t)$.

b. Find $A'(1)$, $A'(3)$, and $A'(4)$ and interpret your results.

91. Pulse Rate of an Athlete The pulse rate (the number of heartbeats/minute) of a long-distance runner t sec after leaving the starting line is given by

$$P(t) = \frac{300\sqrt{\frac{1}{2}t^2 + 2t + 25}}{t + 25} \qquad (t \ge 0)$$

Compute $P'(t)$. How fast is the athlete's pulse rate increasing 10 sec, 60 sec, and 2 min into the run? What is her pulse rate 2 min into the run?

92. Thurstone Learning Model Psychologist L. L. Thurstone suggested the following relationship between learning time t and the length of a list n:

$$T = f(n) - An\sqrt{n - b}$$

where A and b are constants that depend on the person and the task.

a. Compute dT/dn and interpret your result.

b. For a certain person and a certain task, suppose $A = 4$ and $b = 4$. Compute $f'(13)$ and $f'(29)$ and interpret your results.

93. Oil Spills In calm waters the oil spilling from the ruptured hull of a grounded tanker spreads in all directions. Assuming that the area polluted is a circle and that its radius is increasing at a rate of 2 m/sec, determine how fast the area is increasing when the radius of the circle is 40 m.

94. Arteriosclerosis Refer to Example 12, page 189. Suppose the radius of an individual's artery is 1 cm and the thickness of the plaque (in centimetres) t yr from now is given by

$$h = g(t) = \frac{0.5t^2}{t^2 + 10} \qquad (0 \le t \le 10)$$

How fast will the arterial opening be decreasing 5 yr from now?

In Exercises 95–98, determine whether the statement is true or false. If it is true, explain why it is true. If it is false, give an example to show why it is false.

95. If f and g are differentiable and $h = f \circ g$, then $h'(x) = f'[g(x)]g'(x)$.

96. If f is differentiable and c is a constant, then

$$\frac{d}{dx}[f(cx)] = cf'(cx).$$

97. If f is differentiable, then

$$\frac{d}{dx}\sqrt{f(x)} = \frac{f'(x)}{2\sqrt{f(x)}}$$

98. If f is differentiable, then

$$\frac{d}{dx}\left[f\left(\frac{1}{x}\right)\right] = f'\left(\frac{1}{x}\right)$$

Solutions to Self-Check Exercises 3.6

1. Rewriting, we have

$$f(x) = -(2x^2 - 1)^{-1/2}$$

Using the general power rule, we find

$$\begin{aligned} f'(x) &= -\frac{d}{dx}(2x^2 - 1)^{-1/2} \\ &= -\left(-\frac{1}{2}\right)(2x^2 - 1)^{-3/2}\frac{d}{dx}(2x^2 - 1) \\ &= -\left(\frac{1}{2}\right)(2x^2 - 1)^{-3/2}(4x) \\ &= \frac{2x}{(2x^2 - 1)^{3/2}} \end{aligned}$$

2. **a.** The life expectancy at birth of a female born at the beginning of 1980 is given by

$$g(80) = 50.02[1 + 1.09(80)]^{0.1} \approx 78.29$$

or approximately 78 yr. Similarly, the life expectancy at birth of a female born at the beginning of the year 2000 is given by

$$g(100) = 50.02[1 + 1.09(100)]^{0.1} \approx 80.04$$

or approximately 80 yr.

b. The rate of change of the life expectancy at birth of a female born at any time t is given by $g'(t)$. Using the general power rule, we have

$$\begin{aligned} g'(t) &= 50.02\frac{d}{dt}(1 + 1.09t)^{0.1} \\ &= (50.02)(0.1)(1 + 1.09t)^{-0.9}\frac{d}{dt}(1 + 1.09t) \\ &= (50.02)(0.1)(1.09)(1 + 1.09)^{-0.9} \\ &= 5.45218(1 + 1.09t)^{-0.9} \\ &= \frac{5.45218}{(1 + 1.09t)^{0.9}} \end{aligned}$$

CHAPTER 3 Summary of Principal Formulas and Terms

 Formulas

1. Average rate of change of f over $[x, x + h]$ or Slope of the secant line to the graph of f through $(x, f(x))$ and $(x + h, f(x + h))$ or Difference quotient — $\dfrac{f(x + h) - f(x)}{h}$

2. Instantaneous rate of change of f at $(x, f(x))$ or Slope of tangent line to the graph of f at $(x, f(x))$ at x or Derivative of f — $\displaystyle\lim_{h\to 0}\frac{f(x + h) - f(x)}{h}$

3. Derivative of a constant — $\dfrac{d}{dx}(c) = 0$ (c, a constant)

4. Power rule — $\dfrac{d}{dx}(x^n) = nx^{n-1}$

5. Constant multiple rule — $\dfrac{d}{dx}[cf(x)] = cf'(x)$

6. Sum rule — $\dfrac{d}{dx}[f(x) \pm g(x)] = f'(x) \pm g'(x)$

7. Product rule — $\dfrac{d}{dx}[f(x)g(x)] = f(x)g'(x) + g(x)f'(x)$

8. Quotient rule — $\dfrac{d}{dx}\left[\dfrac{f(x)}{g(x)}\right] = \dfrac{g(x)f'(x) - f(x)g'(x)}{[g(x)]^2}$

9. Chain rule — $\dfrac{d}{dx}g(f(x)) = g'(f(x))f'(x)$

10. General power rule — $\dfrac{d}{dx}[f(x)]^n = n[f(x)]^{n-1}f'(x)$

Terms

position function (113)
limit of a function (115)
unbounded function (117)
indeterminate form (118)
limit of a function at infinity (121)
one-sided limit (130)
right-hand limit of a function (131)
left-hand limit of a function (131)
continuity of a function at a point (132)
discontinuous (132)
continuous on an interval (132)
removable discontinuity at $x = a$ (133)
jump discontinuity at $x = a$ (134)
infinite discontinuity at $x = a$ (134)
left-continuity of a function at a point (134)
right-continuity of a function at a point (134)
method of bisection (139)
secant line (146)
tangent line to the graph of f (146)
difference quotient (147)
average rate of change of y with respect to x (147)
rate of change of f at x (148)
instantaneous rate of change of f at x (148)
average rate of change (148)
slope of the secant line (148)
instantaneous rate of change (148)
slope of the tangent line (148)
derivative of f at x (148)
differentiable function (154)
product rule (171)
quotient rule (171)
chain rule (180)
composite functions (180)

Review Exercises 

In Exercises 1–14, find the indicated limits, if they exist.

1. $\lim_{x\to 0}(5x - 3)$
2. $\lim_{x\to 1}(x^2 + 1)$
3. $\lim_{x\to -1}(3x^2 + 4)(2x - 1)$
4. $\lim_{x\to 3}\dfrac{x - 3}{x + 4}$
5. $\lim_{x\to 2}\dfrac{x + 3}{x^2 - 9}$
6. $\lim_{x\to -2}\dfrac{x^2 - 2x - 3}{x^2 + 5x + 6}$
7. $\lim_{x\to 3}\sqrt{2x^3 - 5}$
8. $\lim_{x\to 3}\dfrac{4x - 3}{\sqrt{x + 1}}$
9. $\lim_{x\to 1^+}\dfrac{x - 1}{x(x - 1)}$
10. $\lim_{x\to 1^-}\dfrac{\sqrt{x} - 1}{x - 1}$
11. $\lim_{x\to \infty}\dfrac{x^2}{x^2 - 1}$
12. $\lim_{x\to -\infty}\dfrac{x + 1}{x}$
13. $\lim_{x\to \infty}\dfrac{3x^2 + 2x + 4}{2x^2 - 3x + 1}$
14. $\lim_{x\to -\infty}\dfrac{x^2}{x + 1}$

15. Sketch the graph of the function

$$f(x) = \begin{cases} 2x - 3 & \text{if } x \le 2 \\ -x + 3 & \text{if } x > 2 \end{cases}$$

and evaluate $\lim_{x\to a^+} f(x)$, $\lim_{x\to a^-} f(x)$, and $\lim_{x\to a} f(x)$ at the point $a = 2$, if the limits exist.

16. Sketch the graph of the function

$$f(x) = \begin{cases} 4 - x & \text{if } x \le 2 \\ x + 2 & \text{if } x > 2 \end{cases}$$

and evaluate $\lim_{x\to a^+} f(x)$, $\lim_{x\to a^-} f(x)$, and $\lim_{x\to a} f(x)$ at the point $a = 2$, if the limits exist.

In Exercises 17–20, determine all values of x for which each function is discontinuous.

17. $g(x) = \begin{cases} x + 3 & \text{if } x \ne 2 \\ 0 & \text{if } x = 2 \end{cases}$

18. $f(x) = \dfrac{3x + 4}{4x^2 - 2x - 2}$

19. $f(x) = \begin{cases} \dfrac{1}{(x + 1)^2} & \text{if } x \ne -1 \\ 2 & \text{if } x = -1 \end{cases}$

20. $f(x) = \dfrac{|2x|}{x}$

21. Let $y = x^2 + 2$.
    a. Find the average rate of change of y with respect to x in the intervals [1, 2], [1, 1.5], and [1, 1.1].
    b. Find the (instantaneous) rate of change of y at $x = 1$.

22. Use the definition of the derivative to find the slope of the tangent line to the graph of the function $f(x) = 3x + 5$ at any point $P(x, f(x))$ on the graph.

23. Use the definition of the derivative to find the slope of the tangent line to the graph of the function $f(x) = -1/x$ at any point $P(x, f(x))$ on the graph.

24. Use the definition of the derivative to find the slope of the tangent line to the graph of the function $f(x) = \frac{3}{2}x + 5$ at the point $(-2, 2)$ and determine an equation of the tangent line.

25. Use the definition of the derivative to find the slope of the tangent line to the graph of the function $f(x) = -x^2$ at the point $(2, -4)$ and determine an equation of the tangent line.

26. The graph of the function f is shown in the accompanying figure.
    a. Is f continuous at $x = a$? Why?
    b. Is f differentiable at $x = a$? Justify your answers.

In Exercises 27–56, find the derivative of the given function.

27. $f(x) = 3x^5 - 2x^4 + 3x^2 - 2x + 1$

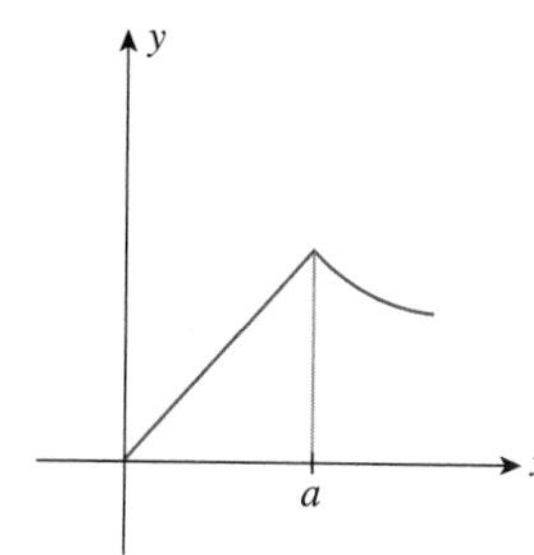

28. $f(x) = 4x^6 + 2x^4 + 3x^2 - 2$

29. $g(x) = -2x^{-3} + 3x^{-1} + 2$

30. $f(t) = 2t^2 - 3t^3 - t^{-1/2}$

31. $g(t) = 2t^{-1/2} + 4t^{-3/2} + 2$

32. $h(x) = x^2 + \dfrac{2}{x}$

33. $g(s) = 2s^2 - \dfrac{4}{s} + \dfrac{2}{\sqrt{s}}$

34. $f(x) = \dfrac{x + 1}{2x - 1}$

35. $h(t) = \dfrac{\sqrt{t}}{\sqrt{t} + 1}$

36. $f(t) = \dfrac{t}{2t^2 + 1}$

37. $f(t) = t + \dfrac{2}{t} + \dfrac{3}{t^2}$

38. $h(x) = x^2 - \dfrac{2}{x^{3/2}}$

39. $g(t) = \dfrac{t^2}{2t^2 + 1}$

40. $f(x) = \dfrac{\sqrt{x} - 1}{\sqrt{x} + 1}$

41. $f(x) = \dfrac{x^2(x^2 + 1)}{x^2 - 1}$

42. $f(x) = (2x^2 + x)^3$

43. $h(x) = (\sqrt{x} + 2)^5$

44. $g(t) = \sqrt[3]{1 - 2t^3}$

45. $f(x) = (2x^3 - 3x^2 + 1)^{-3/2}$

46. $h(x) = \left(x + \dfrac{1}{x}\right)^2$

47. $h(t) = (t^2 + t)^4(2t^2)$

48. $g(x) = \sqrt{x}(x^2 - 1)^3$

49. $h(x) = \dfrac{\sqrt{3x + 2}}{4x - 3}$

50. $f(x) = (3x^3 - 2)^8$

51. $f(t) = \sqrt{2t^2 + 1}$

52. $s(t) = (3t^2 - 2t + 5)^{-2}$

53. $h(x) = \dfrac{1 + x}{(2x^2 + 1)^2}$

54. $f(x) = (2x + 1)^3(x^2 + x)^2$

55. $f(x) = \dfrac{x}{\sqrt{x^3 + 2}}$

56. $f(t) = \dfrac{\sqrt{2t + 1}}{(t + 1)^3}$

57. Let $f(x) = 2x^3 - 3x^2 - 16x + 3$.
   a. Find the points on the graph of f at which the slope of the tangent line is equal to -4.
   b. Find the equation(s) of the tangent line(s) of part (a).

58. Let $f(x) = \frac{1}{3}x^3 + \frac{1}{2}x^2 - 4x + 1$.
   a. Find the points on the graph of f at which the slope of the tangent line is equal to -2.
   b. Find the equation(s) of the tangent line(s) of part (a).

59. Find an equation of the tangent line to the graph of $y = \sqrt{4 - x^2}$ at the point $(1, \sqrt{3})$.

60. Find an equation of the tangent line to the graph of $y = x \cdot (x + 1)^5$ at the point $(1, 32)$.

Business and Economics Applications

61. **Film Conversion Prices** PhotoMart transfers movie films to videocassettes. The fees charged for this service are shown in the following table. Find a function C relating the cost $C(x)$ to the number of feet x of film transferred. Sketch the graph of the function C and discuss its continuity.

Length of Film in Feet, x	Price ($) for Conversion
$1 \le x \le 100$	5.00
$100 < x \le 200$	9.00
$200 < x \le 300$	12.50
$300 < x \le 400$	15.00
$x > 400$	$7 + 0.02x$

62. **Average Price of a Commodity** The average cost (in dollars) of producing x units of a certain commodity is given by

$$\overline{C}(x) = 20 + \frac{400}{x}$$

Evaluate $\lim_{x \to \infty} \overline{C}(x)$ and interpret your results.

DIFFERENTIATION

Photo: Thinkstock/Jupiter Images

In this chapter we see how rules of differentiation facilitate the study of marginal analysis, which is the study of the rate of change of economic quantities, as well as how fast one quantity is changing with respect to another in many real-world situations such as how fast an economy's consumer price index (CPI) is changing at any time.

We visually differentiate by analyzing the graph of a function in terms of the slope of its tangent lines. We consider the derivative as a function itself, and find its derivative. We introduce the notion of the differential of a function. Using differentials is a relatively easy way of approximating the change in one quantity due to a small change in a related quantity. Linearization is introduced as an approximation technique. Finally, we also look at a method, called the Newton-Raphson method, for finding the zeros of a function. Such zeros are also the critical points of a function, which we will study in Chapter 7.

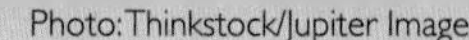
Photo: Thinkstock/Jupiter Images

What are the actual revenue and actual profit for selling loudspeaker systems? How is the demand for loudspeakers related to the revenue made from the sale of this product? In Examples 5 through 7, page 203 to page 207, we demonstrate how knowledge of the derivative and its interpretation can answer these questions.

Photo: PM/Reportage/Getty Images

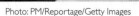

4.1 Marginal Functions in Economics

Marginal analysis is the study of the rate of change of economic quantities. For example, an economist is not merely concerned with the value of an economy's gross domestic product (GDP) at a given time but is equally concerned with the rate at which it is growing or declining. In the same vein, a manufacturer is not only interested in the total cost corresponding to a certain level of production of a commodity but also is interested in the rate of change of the total cost with respect to the level of production, and so on. Let's begin with an example to explain the meaning of the adjective *marginal,* as used by economists.

Cost Functions

Example 1

Rate of Change of Cost Functions Suppose the total cost in dollars incurred each week by Polaraire for manufacturing x refrigerators is given by the total cost function

$$C(x) = 8000 + 200x - 0.2x^2 \qquad (0 \le x \le 400)$$

a. What is the actual cost incurred for manufacturing the 251st refrigerator?
b. Find the rate of change of the total cost function with respect to x when $x = 250$.
c. Compare the results obtained in parts (a) and (b).

Solution

a. The actual cost incurred in producing the 251st refrigerator is the difference between the total cost incurred in producing the first 251 refrigerators and the total cost of producing the first 250 refrigerators:

$$\begin{aligned} C(251) - C(250) &= [8000 + 200(251) - 0.2(251)^2] \\ &\quad - [8000 + 200(250) - 0.2(250)^2] \\ &= 45{,}599.8 - 45{,}500 \\ &= 99.80 \end{aligned}$$

or \$99.80.

b. The rate of change of the total cost function C with respect to x is given by the derivative of C—that is, $C'(x) = 200 - 0.4x$. Thus, when the level of production is 250 refrigerators, the rate of change of the total cost with respect to x is given by

$$\begin{aligned} C'(250) &= 200 - 0.4(250) \\ &= 100 \end{aligned}$$

or \$100.

c. From the solution to part (a), we know that the actual cost for producing the 251st refrigerator is \$99.80. This answer is very closely approximated by the answer to part (b), \$100. To see why this is so, observe that the difference $C(251) - C(250)$ may be written in the form

$$\frac{C(251) - C(250)}{1} = \frac{C(250 + 1) - C(250)}{1} = \frac{C(250 + h) - C(250)}{h}$$

where $h = 1$. In other words, the difference $C(251) - C(250)$ is precisely the average rate of change of the total cost function C over the interval [250, 251], or, equivalently, the slope of the secant line through the points (250, 45,500) and (251, 45,599.8). However, the number $C'(250) = 100$ is the instantaneous rate of change of the total cost function C at $x = 250$, or, equivalently, the slope of the tangent line to the graph of C at $x = 250$.

Now when h is small, the average rate of change of the function C is a good approximation to the instantaneous rate of change of the function C, or, equivalently, the slope of the secant line through the points in question is a good approximation to the slope of the tangent line through the point in question. Thus, we may expect

$$\begin{aligned} C(251) - C(250) &= \frac{C(251) - C(250)}{1} = \frac{C(250 + h) - C(250)}{h} \\ &\approx \lim_{h \to 0} \frac{C(250 + h) - C(250)}{h} = C'(250) \end{aligned}$$

which is precisely the case in this example.

The actual cost incurred in producing an additional unit of a certain commodity given that a plant is already at a certain level of operation is called the **marginal cost.** Knowing this cost is very important to management in their decision-making processes. As we saw in Example 1, the marginal cost is approximated by the rate of change of the total cost function evaluated at the appropriate point. For this reason, economists have defined the **marginal cost function** to be the derivative of the corresponding total cost function. In other words, if C is a total cost function, then the marginal cost function is defined to be its derivative C'. Thus, the adjective *marginal* is synonymous with *derivative of.*

Example 2

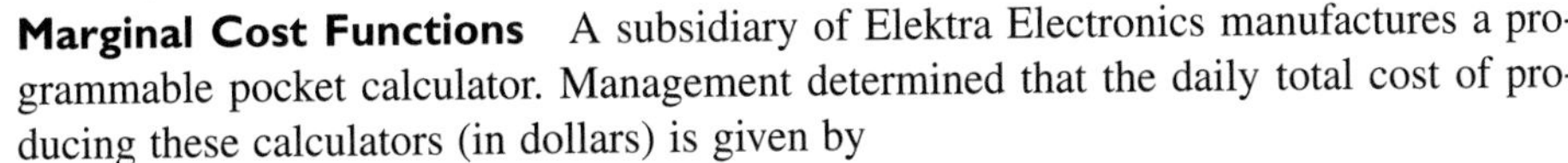

Marginal Cost Functions A subsidiary of Elektra Electronics manufactures a programmable pocket calculator. Management determined that the daily total cost of producing these calculators (in dollars) is given by

$$C(x) = 0.0001x^3 - 0.08x^2 + 40x + 5000$$

where x stands for the number of calculators produced.

a. Find the marginal cost function.

b. What is the marginal cost when $x = 200$, 300, 400, and 600?

c. Interpret your results.

Solution

a. The marginal cost function C' is given by the derivative of the total cost function C. Thus,

$$C'(x) = 0.0003x^2 - 0.16x + 40$$

b. The marginal cost when $x = 200, 300, 400$, and 600 is given by

$$C'(200) = 0.0003(200)^2 - 0.16(200) + 40 = 20$$

$$C'(300) = 0.0003(300)^2 - 0.16(300) + 40 = 19$$

$$C'(400) = 0.0003(400)^2 - 0.16(400) + 40 = 24$$

$$C'(600) = 0.0003(600)^2 - 0.16(600) + 40 = 52$$

or \$20, \$19, \$24, and \$52, respectively.

c. From the results of part (b), we see that Elektra's actual cost for producing the 201st calculator is approximately \$20. The actual cost incurred for producing one additional calculator when the level of production is already 300 calculators is approximately \$19, and so on. Observe that when the level of production is already 600 units, the actual cost of producing one additional unit is approximately \$52. The higher cost for producing this additional unit when the level of production is 600 units may be the result of several factors, among them excessive costs incurred because of overtime or higher maintenance, production breakdown caused by greater stress and strain on the equipment, and so on. The graph of the total cost function appears in Figure 4.1.

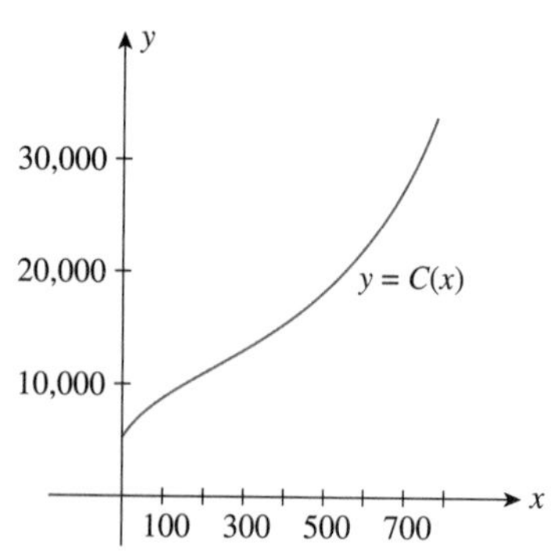

Figure 4.1
The cost of producing x calculators is given by $C(x)$.

Average Cost Functions

Let's now introduce another marginal concept closely related to the marginal cost. Let $C(x)$ denote the total cost incurred in producing x units of a certain commodity. Then the **average cost** of producing x units of the commodity is obtained by dividing the total production cost by the number of units produced. This leads to the following definition:

Average Cost Function

Suppose $C(x)$ is a total cost function. Then the **average cost function,** denoted by $\bar{C}(x)$ (read "C bar of x"), is

$$\bar{C}(x) = \frac{C(x)}{x} \qquad (1)$$

The derivative $\bar{C}'(x)$ of the average cost function, called the **marginal average cost function,** measures the rate of change of the average cost function with respect to the number of units produced.

Example 3

Marginal Average Cost Functions The total cost of producing x units of a certain commodity is given by

$$C(x) = 400 + 20x$$

dollars.

a. Find the average cost function $\bar{C}$.

b. Find the marginal average cost function $\bar{C}'$.

c. What are the economic implications of your results?

Solution

a. The average cost function is given by

$$\bar{C}(x) = \frac{C(x)}{x} = \frac{400 + 20x}{x}$$
$$= 20 + \frac{400}{x}$$

b. The marginal average cost function is

$$\bar{C}'(x) = -\frac{400}{x^2}$$

c. Since the marginal average cost function is negative for all admissible values of x, the rate of change of the average cost function is negative for all $x > 0$; that is, $\bar{C}(x)$ decreases as x increases. However, the graph of $\bar{C}$ always lies above the horizontal line $y = 20$, but it approaches the line since

$$\lim_{x\to\infty}\bar{C}(x) = \lim_{x\to\infty}\left(20 + \frac{400}{x}\right) = 20$$

A sketch of the graph of the function $\bar{C}(x)$ appears in Figure 4.2. This result is fully expected if we consider the economic implications. Note that as the level of production increases, the fixed cost per unit of production, represented by the term ($400/x$), drops steadily. The average cost approaches the constant unit of production, which is \$20 in this case.

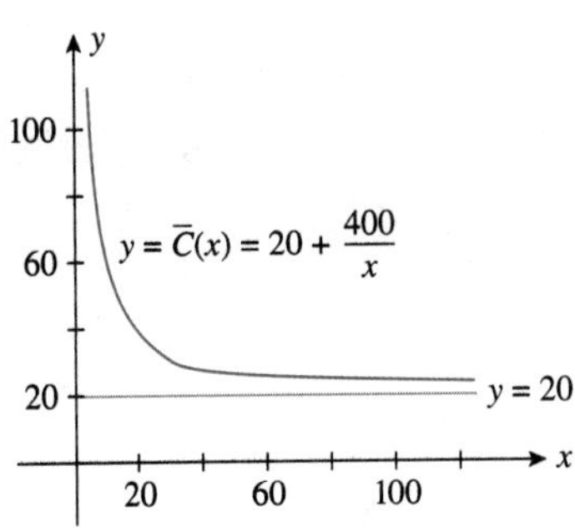

Figure 4.2
As the level of production increases, the average cost approaches \$20.

Example 4

Marginal Average Cost Functions Once again consider the subsidiary of Elektra Electronics. The daily total cost for producing its programmable calculators is given by

$$C(x) = 0.0001x^3 - 0.08x^2 + 40x + 5000$$

dollars, where x stands for the number of calculators produced (see Example 2).

a. Find the average cost function $\bar{C}$.

b. Find the marginal average cost function $\bar{C}'$. Compute $\bar{C}'(500)$.

c. Sketch the graph of the function $\bar{C}$ and interpret the results obtained in parts (a) and (b).

Solution

a. The average cost function is given by

$$\bar{C}(x) = \frac{C(x)}{x} = 0.0001x^2 - 0.08x + 40 + \frac{5000}{x}$$

b. The marginal average cost function is given by

$$\bar{C}'(x) = 0.0002x - 0.08 - \frac{5000}{x^2}$$

Also,

$$\bar{C}'(500) = 0.0002(500) - 0.08 - \frac{5000}{(500)^2} = 0$$

c. To sketch the graph of the function $\bar{C}$, observe that if x is a small positive number, then $\bar{C}(x) > 0$. Furthermore, $\bar{C}(x)$ becomes arbitrarily large as x approaches zero from the right, since the term ($5000/x$) becomes arbitrarily large as x approaches zero. Next, the result $\bar{C}'(500) = 0$ obtained in part (b)

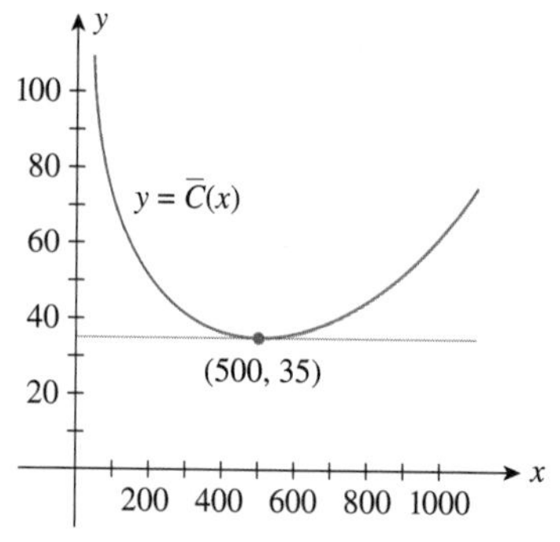

Figure 4.3
The average cost reaches a minimum of \$35 when 500 calculators are produced.

tells us that the tangent line to the graph of the function $\bar{C}$ is horizontal at the point (500, 35) on the graph. Finally, plotting the points on the graph corresponding to, say, $x = 100, 200, 300, \ldots, 900$, we obtain the sketch in Figure 4.3. As expected, the average cost drops as the level of production increases. But in this case, as opposed to the case in Example 3, the average cost reaches a minimum value of \$35, corresponding to a production level of 500, and *increases* thereafter.

This phenomenon is typical in situations where the marginal cost increases from some point on as production increases, as in Example 2. This situation is in contrast to that of Example 3, in which the marginal cost remains constant at any level of production.

EXPLORING WITH TECHNOLOGY

Refer to Example 4.

1. Use a graphing utility to plot the graph of the average cost function

$$\bar{C}(x) = 0.0001x^2 - 0.08x + 40 + \frac{5000}{x}$$

using the viewing window $[0, 1000] \cdot [0, 100]$. Then, using ZOOM and TRACE, show that the lowest point on the graph of $\bar{C}$ is (500, 35).

2. Draw the tangent line to the graph of $\bar{C}$ at (500, 35). What is its slope? Is this expected?
3. Plot the graph of the marginal average cost function

$$\bar{C}'(x) = 0.0002x - 0.08 - \frac{5000}{x^2}$$

using the viewing window $[0, 2000] \cdot [-1, 1]$. Then use ZOOM and TRACE to show that the zero of the function $\bar{C}'$ occurs at $x = 500$. Verify this result using the root-finding capability of your graphing utility. Is this result compatible with that obtained in part (2)? Explain your answer.

Revenue Functions

Recall that a revenue function $R(x)$ gives the revenue realized by a company from the sale of x units of a certain commodity. If the company charges p dollars per unit, then

$$R(x) = px \tag{2}$$

However, the price that a company can command for the product depends on the market in which it operates. If the company is one of many—none of which is able to dictate the price of the commodity—then in this competitive market environment the price is determined by market equilibrium (see Section 2.8). On the other hand, if the company is the sole supplier of the product, then under this monopolistic situation it can manipulate the price of the commodity by controlling the supply. The unit selling price p of the commodity is related to the quantity x of the commodity demanded. This relationship between p and x is called a *demand equation* (see Section 2.8). Solving the demand equation for p in terms of x, we obtain the unit price function f. Thus,

$$p = f(x)$$

and the revenue function R is given by

$$R(x) = px = xf(x)$$

The marginal revenue gives the actual revenue realized from the sale of an additional unit of the commodity given that sales are already at a certain level. Following an argument parallel to that applied to the cost function in Example 1, you can convince yourself that the marginal revenue is approximated by $R'(x)$. Thus, we define the marginal revenue function to be $R'(x)$, where R is the revenue function. The derivative R' of the function R measures the rate of change of the revenue function.

Example 5

Marginal Revenue Functions Suppose the relationship between the unit price p in dollars and the quantity demanded x of the Acrosonic model F loudspeaker system is given by the equation

$$p = -0.02x + 400 \qquad (0 \le x \le 20{,}000)$$

a. Find the revenue function R.
b. Find the marginal revenue function R'.
c. Compute $R'(2000)$ and interpret your result.

Solution

a. The revenue function R is given by

$$\begin{aligned} R(x) &= px \\ &= x(-0.02x + 400) \\ &= -0.02x^2 + 400x \qquad (0 \le x \le 20{,}000) \end{aligned}$$

b. The marginal revenue function R' is given by

$$R'(x) = -0.04x + 400$$

c.

$$R'(2000) = -0.04(2000) + 400 = 320$$

Thus, the actual revenue to be realized from the sale of the 2001st loudspeaker system is approximately \$320.

Profit Functions

Our final example of a marginal function involves the profit function. The profit function P is given by

$$P(x) = R(x) - C(x) \qquad (3)$$

where R and C are the revenue and cost functions and x is the number of units of a commodity produced and sold. The marginal profit function $P'(x)$ measures the rate of change of the profit function P and provides us with a good approximation of the actual profit or loss realized from the sale of the $(x + 1)$st unit of the commodity (assuming the xth unit has been sold).

Example 6

Marginal Profit Functions Refer to Example 5. Suppose the cost of producing x units of the Acrosonic model F loudspeaker is

$$C(x) = 100x + 200{,}000$$

dollars.

a. Find the profit function P.
b. Find the marginal profit function P'.
c. Compute $P'(2000)$ and interpret your result.
d. Sketch the graph of the profit function P.

Solution

a. From the solution to Example 5a, we have

$$R(x) = -0.02x^2 + 400x$$

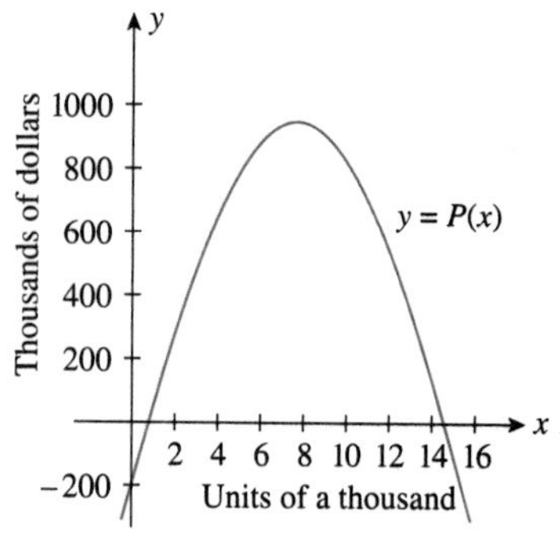

Figure 4.4
The total profit made when x loudspeakers are produced is given by $P(x)$.

Thus, the required profit function P is given by

$$P(x) = R(x) - C(x)$$
$$= (-0.02x^2 + 400x) - (100x + 200{,}000)$$
$$= -0.02x^2 + 300x - 200{,}000$$

b. The marginal profit function P' is given by

$$P'(x) = -0.04x + 300$$

c. $$P'(2000) = -0.04(2000) + 300 = 220$$

Thus, the actual profit realized from the sale of the 2001st loudspeaker system is approximately \$220.

d. The graph of the profit function P appears in Figure 4.4.

Elasticity of Demand

Finally, let's use the marginal concepts introduced in this section to derive an important criterion used by economists to analyze the demand function: elasticity of demand.

Let us first discuss a concrete situation. How does the demand for air travel influence the revenue on tickets sold? One would think that an increase in the cost for an airplane ticket would cause the revenue to decrease, while a decrease in the ticket cost would increase revenue. We will later find out that this is referred to as an elastic demand. Businesses with an elastic demand such as start-up airlines in Canada can collapse more readily under this condition. A look at a study on *Air Travel Demand Elasticities: Concepts, Issues and Measurement: 1* published by the Department of Finance in Canada (http://www.fin.gc.ca/consultresp/Airtravel/airtravStdy_e.html) shows that we must distinguish among six different types of air travel grouped in pairs, namely business and leisure, long-haul and short-haul, and international long-haul and North American long-haul air travel. Findings of the study confirm that the demand for leisure air travel is more elastic than that of business travel. This makes sense, because an expensive holiday trip can be postponed, while a business trip would more likely have to take place regardless of the cost. The study also finds that the demand for short-haul flights is more elastic than long-haul, and the demand for North American flights is more elastic than international. The further the destination, the less likely one will be able to find an alternative mode of transport as a substitute for an expensive flight.

We will now derive mathematically how the elasticity of demand can be determined when the demand function is given.

In what follows, it will be convenient to write the demand function f in the form $x = f(p)$; that is, we will think of the quantity demanded of a certain commodity as a function of its unit price. Since the quantity demanded of a commodity usually decreases as its unit price increases, the function f is typically a decreasing function of p (Figure 4.5a).

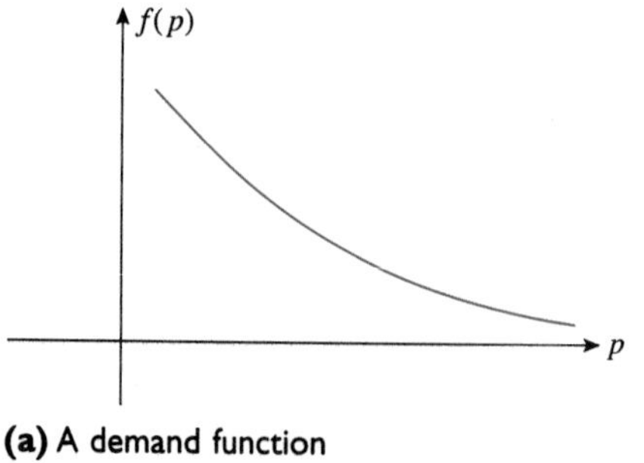

(a) A demand function

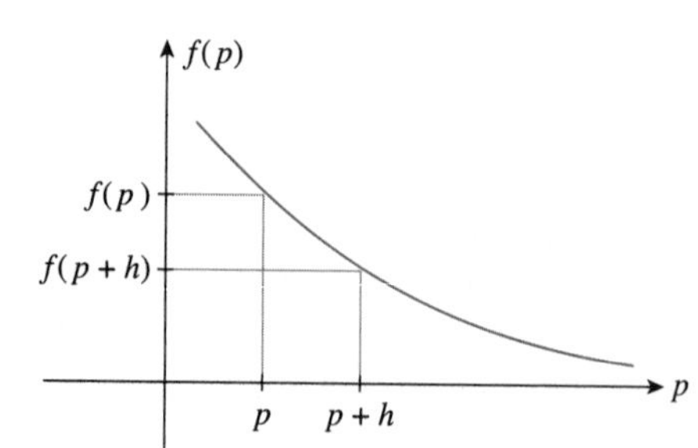

(b) $f(p + h)$ is the quantity demanded when the unit price increases from p to $p + h$ dollars.

Figure 4.5

Suppose the unit price of a commodity is increased by h dollars from p dollars to $(p + h)$ dollars (Figure 4.5b). Then the quantity demanded drops from $f(p)$ units to $f(p + h)$ units, a change of $[f(p + h) - f(p)]$ units. The percentage change in the unit price is

$$\frac{h}{p}(100) \qquad \frac{\text{change in unit price}}{\text{Price } p}(100)$$

and the corresponding percentage change in the quantity demanded is

$$100\left[\frac{f(p+h) - f(p)}{f(p)}\right] \qquad \frac{\text{Change in quantity demanded}}{\text{Quantity demanded at price } p}(100)$$

Now, one good way to measure the effect that a percentage change in price has on the percentage change in the quantity demanded is to look at the ratio of the latter to the former. We find

$$\frac{\text{Percentage change in the quantity demanded}}{\text{Percentage change in the unit price}} = \frac{100\left[\dfrac{f(p+h) - f(p)}{f(p)}\right]}{100\left(\dfrac{h}{p}\right)} = \frac{\dfrac{f(p+h) - f(p)}{h}}{\dfrac{f(p)}{p}}$$

If f is differentiable at p, then

$$\frac{f(p+h) - f(p)}{h} \approx f'(p)$$

when h is small. Therefore, if h is small, then the ratio is approximately equal to

$$\frac{f'(p)}{\dfrac{f(p)}{p}} = \frac{pf'(p)}{f(p)}$$

Economists call the negative of this quantity the elasticity of demand.

Elasticity of Demand

If f is a differentiable demand function defined by $x = f(p)$, then the **elasticity of demand** at price p is given by

$$E(p) = -\frac{pf'(p)}{f(p)} \tag{4}$$

Remark

It will be shown later (Section 7.1) that if f is decreasing on an interval, then $f'(p) < 0$ for p in that interval. In light of this, we see that since both p and $f(p)$ are positive, the quantity $\dfrac{pf'(p)}{f(p)}$ is negative. Because economists would rather work with a positive value, the elasticity of demand $E(p)$ is defined to be the negative of this quantity. ◀

PORTFOLIO

Rebecca Oulton

Title Financial Advisor—Senior Associate
Institution World Financial Group Insurance Agency of Canada Inc. & WFG Securities of Canada Inc.

When Rebecca Oulton graduated in 2005 with a master's degree in Mathematical Finance, she knew she wanted to work in her field of study taking a less traditional approach. She also knew she wanted to make a difference. Her current position at World Financial Group (WFG) enables her to do just that. As Oulton notes, "our company's focus is on educating families around the world about money and how it works."

Aside from acting as a financial services professional and providing support to some of the most well-known companies in the financial services and insurance industry, Oulton has taken on the role of an educator on financial services and segregated funds at WFG.

As Oulton explains, segregated funds "are instruments like mutual funds and equity linked contracts; however, they have guarantees involved." With marked uncertainty and a great deal of the baby-boomer generation looking for security, segregated funds have become the solution for many investors. The pricing of a segregated fund is calculated by one of two approaches, using either probability or stochastic calculus.

Aside from this specific financial mathematical aspect of her job, Oulton also finds that in her industry knowledge of compound interest, valuation of a person's insurance quote, pricing contracts, formulation of portfolios, risk calculations, etc.—and the ability to apply this knowledge—are of utmost importance to being successful.

When Oulton sits down with clients to discuss their financial situation, she needs to assess their current portfolio or lack thereof, provide education on financial terms and options, and offer sound solutions that will guide clients toward a healthy financial portfolio that leads them into retirement. Through WFG, Oulton represents banks, trust companies, and insurance companies, and has access to industry analysts and up-to-date information, which allows her to make knowledgeable decisions for her clients. If Oulton encounters a problem, her extensive study of mathematics has trained her well to be an effective problem solver. As Oulton points out, "sometimes it is as simple as trying another perspective or to liken the situation to one that you are already familiar with. The beauty is that there is so much out there, and yet still much more to be discovered."

Example 7

Elasticity of Demand Consider the demand equation

$$p = -0.02x + 400 \qquad (0 \le x \le 20{,}000)$$

which describes the relationship between the unit price in dollars and the quantity demanded x of the Acrosonic model F loudspeaker systems.

a. Find the elasticity of demand $E(p)$.
b. Compute $E(100)$ and interpret your result.
c. Compute $E(300)$ and interpret your result.

Solution

a. Solving the given demand equation for x in terms of p, we find

$$x = f(p) = -50p + 20{,}000$$

from which we see that

$$f'(p) = -50$$

Therefore,

$$E(p) = -\frac{pf'(p)}{f(p)} = -\frac{p(-50)}{-50p + 20{,}000}$$

$$= \frac{p}{400 - p}$$

b. $$E(100) = \frac{100}{400 - 100} = \frac{1}{3}$$

which is the elasticity of demand when $p = 100$. To interpret this result, recall that $E(100)$ is the negative of the ratio of the percentage change in the quantity demanded to the percentage change in the unit price when $p = 100$. Therefore, our result tells us that when the unit price p is set at \$100 per speaker, an increase of 1% in the unit price will cause a decrease of approximately 0.33% in the quantity demanded.

c. $$E(300) = \frac{300}{400 - 300} = 3$$

which is the elasticity of demand when $p = 300$. It tells us that when the unit price is set at \$300 per speaker, an increase of 1% in the unit price will cause a decrease of approximately 3% in the quantity demanded.

Economists often use the following terminology to describe demand in terms of elasticity.

Elasticity of Demand

The demand is said to be **elastic** if $E(p) > 1$.
The demand is said to be **unitary** if $E(p) = 1$.
The demand is said to be **inelastic** if $E(p) < 1$.

As an illustration, our computations in Example 7 revealed that demand for Acrosonic loudspeakers is elastic when $p = 300$ but inelastic when $p = 100$. These computations confirm that when demand is elastic, a small percentage change in the unit price will result in a greater percentage change in the quantity demanded; and when demand is inelastic, a small percentage change in the unit price will cause a smaller percentage change in the quantity demanded. Finally, when demand is unitary, a small percentage change in the unit price will result in the same percentage change in the quantity demanded.

We can describe the way revenue responds to changes in the unit price using the notion of elasticity. If the quantity demanded of a certain commodity is related to its unit price by the equation $x = f(p)$, then the revenue realized through the sale of x units of the commodity at a price of p dollars each is

$$R(p) = px = pf(p)$$

The rate of change of the revenue with respect to the unit price p is given by

$$R'(p) = f(p) + pf'(p)$$

$$= f(p)\left[1 + \frac{pf'(p)}{f(p)}\right]$$

$$= f(p)[1 - E(p)]$$

Now, suppose demand is elastic when the unit price is set at a dollars. Then $E(a) > 1$, and so $1 - E(a) < 0$. Since $f(p)$ is positive for all values of p, we see that

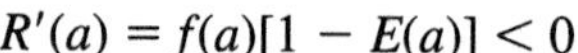

$$R'(a) = f(a)[1 - E(a)] < 0$$

and so $R(p)$ is decreasing at $p = a$. This implies that a small increase in the unit price when $p = a$ results in a decrease in the revenue, whereas a small decrease in the unit price will result in an increase in the revenue. Similarly, you can show that if the demand is inelastic when the unit price is set at a dollars, then a small increase in the unit price will cause the revenue to increase, and a small decrease in the unit price will cause the revenue to decrease. Finally, if the demand is unitary when the unit price is set at a dollars, then $E(a) = 1$ and $R'(a) = 0$. This implies that a small increase or decrease in the unit price will not result in a change in the revenue. The following statements summarize this discussion.

Figure 4.6

The revenue is increasing on an interval where the demand is inelastic, decreasing on an interval where the demand is elastic, and stationary at the point where the demand is unitary.

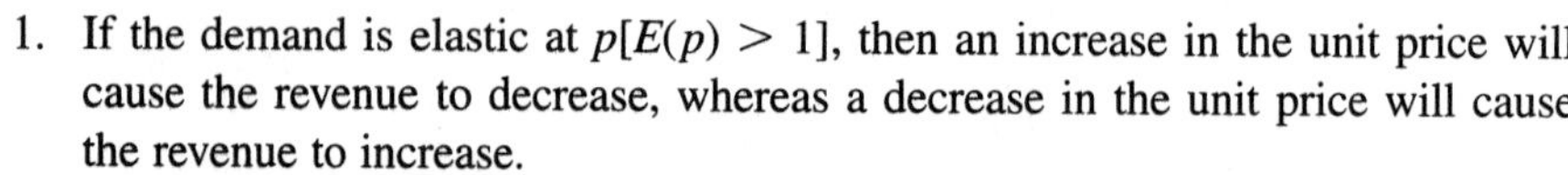

1. If the demand is elastic at $p[E(p) > 1]$, then an increase in the unit price will cause the revenue to decrease, whereas a decrease in the unit price will cause the revenue to increase.
2. If the demand is inelastic at $p[E(p) < 1]$, then an increase in the unit price will cause the revenue to increase, and a decrease in the unit price will cause the revenue to decrease.
3. If the demand is unitary at $p[E(p) = 1]$, then an increase in the unit price will cause the revenue to stay about the same.

These results are illustrated in Figure 4.6.

Remark

As an aid to remembering this, note the following:

1. If demand is elastic, then the change in revenue and the change in the unit price move in opposite directions.
2. If demand is inelastic, then they move in the same direction. ◀

Example 8

Elasticity of Demand Refer to Example 7.

a. Is demand elastic, unitary, or inelastic when $p = 100$? When $p = 300$?

b. If the price is \$100, will raising the unit price slightly cause the revenue to increase or decrease?

Solution

a. From the results of Example 7, we see that $E(100) = \frac{1}{3} < 1$ and $E(300) = 3 > 1$. We conclude accordingly that demand is inelastic when $p = 100$ and elastic when $p = 300$.

b. Since demand is inelastic when $p = 100$, raising the unit price slightly will cause the revenue to increase.

Self-Check Exercises 4.1

1. The weekly demand for Pulsar VCRs is given by the demand equation

$$p = -0.02x + 300 \qquad (0 \le x \le 15{,}000)$$

where p denotes the wholesale unit price in dollars and x denotes the quantity demanded. The weekly total cost function associated with manufacturing these VCRs is

$$C(x) = 0.000003x^3 - 0.04x^2 + 200x + 70{,}000 \text{ dollars}$$

a. Find the revenue function R and the profit function P.

b. Find the marginal cost function C', the marginal revenue function R', and the marginal profit function P'.

c. Find the marginal average cost function $\bar{C}'$.

d. Compute $C'(3000)$, $R'(3000)$, and $P'(3000)$ and interpret your results.

2. Refer to the preceding exercise. Determine whether the demand is elastic, unitary, or inelastic when $p = 100$ and when $p = 200$.

Solutions to Self-Check Exercises 4.1 can be found on page 211.

4.1 Exercises

1. **Production Costs** The graph of a typical total cost function $C(x)$ associated with the manufacture of x units of a certain commodity is shown in the following figure.
   a. Explain why the function C is always increasing.
   b. As the level of production x increases, the cost/unit drops so that $C(x)$ increases but at a slower pace. However, a level of production is soon reached at which the cost/unit begins to increase dramatically (due to a shortage of raw material, overtime, breakdown of machinery due to excessive stress and strain) so that $C(x)$ continues to increase at a faster pace. Use the graph of C to find the approximate level of production x_0 where this occurs.

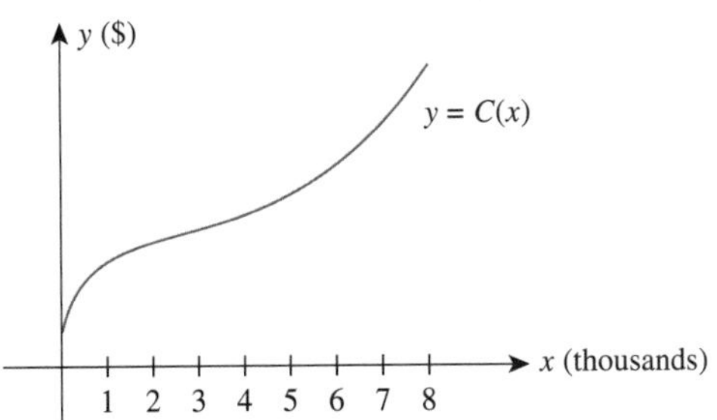

2. **Production Costs** The graph of a typical average cost function $A(x) = C(x)/x$, where $C(x)$ is a total cost function associated with the manufacture of x units of a certain commodity, is shown in the following figure.
   a. Explain in economic terms why $A(x)$ is large if x is small and why $A(x)$ is large if x is large.
   b. What is the significance of the numbers x_0 and y_0, the x- and y-coordinates of the lowest point on the graph of the function A?

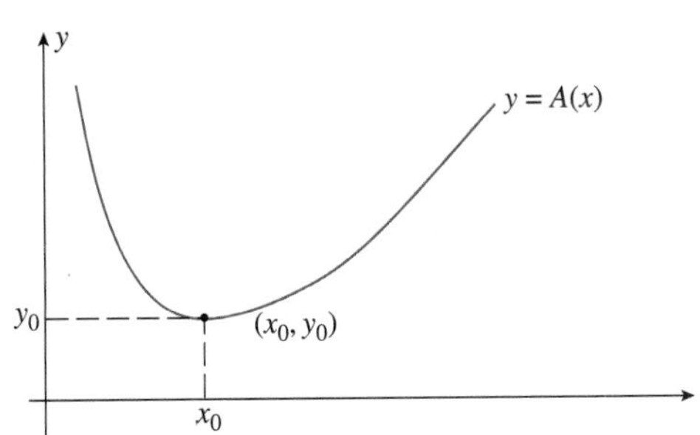

A calculator is recommended for Exercises 3–34.

Business and Economics Applications

3. **Marginal Cost** The total weekly cost (in dollars) incurred by Markham Records in pressing x compact discs is

$$C(x) = 2000 + 2x - 0.0001x^2 \qquad (0 \leq x \leq 6000)$$

   a. What is the actual cost incurred in producing the 1001st and the 2001st disc?
   b. What is the marginal cost when $x = 1000$ and 2000?

4. **Marginal Cost** A division of Sudbury Industries Ltd. manufactures convection microwave ovens. The daily cost (in dollars) of producing these microwave ovens is

$$C(x) = 0.0002x^3 - 0.06x^2 + 120x + 5000$$

where x stands for the number of units produced.
   a. What is the actual cost incurred in manufacturing the 101st oven? The 201st oven? The 301st oven?
   b. What is the marginal cost when $x = 100$, 200, and 300?

5. **Marginal Average Cost** Grand & Toy makes a line of executive desks. It is estimated that the total cost for making x units of their Senior Executive model is

$$C(x) = 100x + 200{,}000$$

dollars/year.
   a. Find the average cost function $\bar{C}$.
   b. Find the marginal average cost function $\bar{C}'$.
   c. What happens to $\bar{C}(x)$ when x is very large? Interpret your results.

6. **Marginal Average Cost** The management of ThermoMaster Company, whose Mexican subsidiary manufactures an indoor–outdoor thermometer, has estimated that the total weekly cost (in dollars) for producing x thermometers is

$$C(x) = 5000 + 2x$$

   a. Find the average cost function $\bar{C}$.
   b. Find the marginal average cost function $\bar{C}'$.
   c. Interpret your results.

7. Find the average cost function $\bar{C}$ and the marginal average cost function $\bar{C}'$ associated with the total cost function C of Exercise 3.

8. Find the average cost function $\bar{C}$ and the marginal average cost function $\bar{C}'$ associated with the total cost function C of Exercise 4.

9. **Marginal Revenue** Calgary Commuter Air Service realizes a monthly revenue of

$$R(x) = 8000x - 100x^2$$

dollars when the price charged per passenger is x dollars.
   a. Find the marginal revenue R'.
   b. Compute $R'(39)$, $R'(40)$, and $R'(41)$.
   c. Based on the results of part (b), what price should the airline charge in order to maximize their revenue?

10. **Marginal Revenue** The management of Sudbury Industries Ltd. plans to market the ElectroStat, an electrostatic speaker system. The marketing department has determined that the

demand for these speakers is

$$p = -0.04x + 800 \qquad (0 \le x \le 20{,}000)$$

where p denotes the speaker's unit price (in dollars) and x denotes the quantity demanded.

a. Find the revenue function R.
b. Find the marginal revenue function R'.
c. Compute $R'(5000)$ and interpret your results.

11. Marginal Profit Refer to Exercise 10. Sudbury Industries Ltd.'s production department estimates that the total cost (in dollars) incurred in manufacturing x ElectroStat speaker systems in the first year of production will be

$$C(x) = 200x + 300{,}000$$

a. Find the profit function P.
b. Find the marginal profit function P'.
c. Compute $P'(5000)$ and $P'(8000)$.
d. Sketch the graph of the profit function and interpret your results.

12. Marginal Profit Don Valley West, an apartment complex, has 100 two-bedroom units. The monthly profit (in dollars) realized from renting x apartments is

$$P(x) = -10x^2 + 1760x - 50{,}000$$

a. What is the actual profit realized from renting the 51st unit, assuming that 50 units have already been rented?
b. Compute the marginal profit when $x = 50$ and compare your results with that obtained in part (a).

13. Marginal Cost, Revenue, and Profit The weekly demand for the Pulsar 25 colour console television is

$$p = 600 - 0.05x \qquad (0 \le x \le 12{,}000)$$

where p denotes the wholesale unit price in dollars and x denotes the quantity demanded. The weekly total cost function associated with manufacturing the Pulsar 25 is given by

$$C(x) = 0.000002x^3 - 0.03x^2 + 400x + 80{,}000$$

where $\bar{C}(x)$ denotes the total cost incurred in producing x sets.

a. Find the revenue function R and the profit function P.
b. Find the marginal cost function C', the marginal revenue function R', and the marginal profit function P'.
c. Compute $C'(2000)$, $R'(2000)$, and $P'(2000)$ and interpret your results.
d. Sketch the graphs of the functions C, R, and P and interpret parts (b) and (c), using the graphs obtained.

14. Marginal Cost, Revenue, and Profit Pulsar also manufactures a series of 19-in. colour television sets. The quantity x of these sets demanded each week is related to the wholesale unit price p by the equation

$$p = -0.006x + 180$$

The weekly total cost incurred by Pulsar for producing x sets is

$$C(x) = 0.000002x^3 - 0.02x^2 + 120x + 60{,}000$$

dollars. Answer the questions in Exercise 13 for these data.

15. Marginal Average Cost Find the average cost function $\bar{C}$ associated with the total cost function C of Exercise 13.

a. What is the marginal average cost function $\bar{C}'$?
b. Compute $\bar{C}'(5000)$ and $\bar{C}'(10{,}000)$ and interpret your results.
c. Sketch the graph of $\bar{C}$.

16. Marginal Average Cost Find the average cost function $\bar{C}$ associated with the total cost function C of Exercise 14.

a. What is the marginal average cost function $\bar{C}'$?
b. Compute $\bar{C}'(5000)$ and $\bar{C}'(10{,}000)$ and interpret your results.

17. Marginal Revenue The quantity of Sicard wristwatches demanded each month is related to the unit price by the equation

$$p = \frac{50}{0.01x^2 + 1} \qquad (0 \le x \le 20)$$

where p is measured in dollars and x in units of a thousand.

a. Find the revenue function R.
b. Find the marginal revenue function R'.
c. Compute $R'(2)$ and interpret your result.

18. Marginal Propensity to Consume The consumption function of a certain economy for the 1930s is

$$C(x) = 0.712x + 95.05$$

where $C(x)$ is the personal consumption expenditure and x is the personal income, both measured in billions of dollars. Find the rate of change of consumption with respect to income, dC/dx. This quantity is called the *marginal propensity to consume.*

19. Marginal Propensity to Consume Refer to Exercise 18. Suppose a certain economy's consumption function is

$$C(x) = 0.873x^{1.1} + 20.34$$

where $C(x)$ and x are measured in billions of dollars. Find the marginal propensity to consume when $x = 10$.

20. Marginal Propensity to Save Suppose $C(x)$ measures an economy's personal consumption expenditure and x the personal income, both in billions of dollars. Then,

$$S(x) = x - C(x) \qquad \text{Income minus consumption}$$

measures the economy's savings corresponding to an income of x billion dollars. Show that

$$\frac{dS}{dx} = 1 - \frac{dC}{dx}$$

The quantity dS/dx is called the *marginal propensity to save.*

21. Refer to Exercise 20. For the consumption function of Exercise 18, find the marginal propensity to save.

22. Refer to Exercise 20. For the consumption function of Exercise 19, find the marginal propensity to save when $x = 10$.

For each demand equation in Exercises 23–28, compute the elasticity of demand and determine whether the demand is elastic, unitary, or inelastic at the indicated price.

23. $x = -\frac{5}{4}p + 20;\ p = 10$

24. $x = -\frac{3}{2}p + 9;\ p = 2$

25. $x + \frac{1}{3}p - 20 = 0;\ p = 30$

26. $0.4x + p - 20 = 0; p = 10$

27. $p = 169 - x^2; p = 29$ **28.** $p = 144 - x^2; p = 96$

29. Elasticity of Demand The demand equation for the Roland portable hair dryer is given by

$$x = \frac{1}{5}(225 - p^2) \qquad (0 \leq f \leq 15)$$

where x (measured in units of a hundred) is the quantity demanded per week and p is the unit price in dollars.

a. Is the demand elastic or inelastic when $p = 8$ and when $p = 10$?

b. When is the demand unitary?

Hint: Solve $E(p) = 1$ for p.

c. If the unit price is lowered slightly from \$10, will the revenue increase or decrease?

d. If the unit price is increased slightly from \$8, will the revenue increase or decrease?

30. Elasticity of Demand The management of Canadian Tire has determined that the quantity demanded x of their Super Grip tires per week is related to the unit price p by the equation

$$x = \sqrt{144 - p} \qquad (0 \leq p \leq 12)$$

where p is measured in dollars and x in units of a thousand.

a. Compute the elasticity of demand when $p = 63$, 96, and 108.

b. Interpret the results obtained in part (a).

c. Is the demand elastic, unitary, or inelastic when $p = 63$, 96, and 108?

31. Elasticity of Demand The proprietor of the Showplace, a video club, has estimated that the rental price p (in dollars) of prerecorded videocassette tapes is related to the quantity x rented/week by the demand equation

$$x = \frac{2}{3}\sqrt{36 - p^2} \qquad (0 \leq p \leq 6)$$

Currently, the rental price is \$2/tape.

a. Is the demand elastic or inelastic at this rental price?

b. If the rental price is increased, will the revenue increase or decrease?

32. Elasticity of Demand The quantity demanded each week x (in units of a hundred) of the Mikado miniature camera is related to the unit price p (in dollars) by the demand equation

$$x = \sqrt{400 - 5p} \qquad (0 \leq p \leq 80)$$

a. Is the demand elastic or inelastic when $p = 40$? When $p = 60$?

b. When is the demand unitary?

c. If the unit price is lowered slightly from \$60, will the revenue increase or decrease?

d. If the unit price is increased slightly from \$40, will the revenue increase or decrease?

33. Elasticity of Demand The demand function for a certain make of exercise bicycle sold exclusively through cable television is

$$p = \sqrt{9 - 0.02x} \qquad (0 \leq x \leq 450)$$

where p is the unit price in hundreds of dollars and x is the quantity demanded/week. Compute the elasticity of demand and determine the range of prices corresponding to inelastic, unitary, and elastic demand.

Hint: Solve the equation $E(p) = 1$.

34. Elasticity of Demand The demand equation for the Sicard wristwatch is given by

$$x = 10\sqrt{\frac{50 - p}{p}} \qquad (0 < p \leq 50)$$

where x (measured in units of a thousand) is the quantity demanded/week and p is the unit price in dollars. Compute the elasticity of demand and determine the range of prices corresponding to inelastic, unitary, and elastic demand.

In Exercises 35 and 36, determine whether the statement is true or false. If it is true, explain why it is true. If it is false, give an example to show why it is false.

35. If C is a differentiable total cost function, then the marginal average cost function is

$$\bar{C}'(x) = \frac{xC'(x) - C(x)}{x^2}$$

36. If the marginal profit function is positive at $x = a$, then it makes sense to decrease the level of production.

Solutions to Self-Check Exercises 4.1

1. a.
$$\begin{aligned} R(x) &= px \\ &= x(-0.02x + 300) \\ &= -0.02x^2 + 300x \qquad (0 \leq x \leq 15{,}000) \\ P(x) &= R(x) - C(x) \\ &= -0.02x^2 + 300x \\ &\quad -(0.000003x^3 - 0.04x^2 + 200x + 70{,}000) \\ &= -0.000003x^3 + 0.02x^2 + 100x - 70{,}000 \end{aligned}$$

b.
$$\begin{aligned} C'(x) &= 0.000009x^2 - 0.08x + 200 \\ R'(x) &= -0.04x + 300 \\ P'(x) &= -0.000009x^2 + 0.04x + 100 \end{aligned}$$

c. The average cost function is

$$\begin{aligned} \bar{C}(x) &= \frac{C(x)}{x} \\ &= \frac{0.000003x^3 - 0.04x^2 + 200x + 70{,}000}{x} \\ &= 0.000003x^2 - 0.04x + 200 + \frac{70{,}000}{x} \end{aligned}$$

Solutions to Self-Check Exercises 4.1, *continued*

Therefore, the marginal average cost function is

$$\bar{C}'(x) = 0.000006x - 0.04 - \frac{70{,}000}{x^2}$$

d. Using the results from part (b), we find

$$C'(3000) = 0.000009(3000)^2 - 0.08(3000) + 200$$
$$= 41$$

That is, when the level of production is already 3000 VCRs, the actual cost of producing one additional VCR is approximately $41. Next,

$$R'(3000) = -0.04(3000) + 300 = 180$$

That is, the actual revenue to be realized from selling the 3001st VCR is approximately $180. Finally,

$$P'(3000) = -0.000009(3000)^2 + 0.04(3000) + 100$$
$$= 139$$

That is, the actual profit realized from selling the 3001st VCR is approximately $139.

2. We first solve the given demand equation for x in terms of p, obtaining

$$x = f(p) = -50p + 15{,}000$$
$$f'(p) = -50$$

Therefore,

$$E(p) = -\frac{pf'(p)}{f(p)} = -\frac{p}{-50p + 15{,}000}(-50)$$
$$= \frac{p}{300 - p} \qquad (0 \le p < 300)$$

Next, we compute

$$E(100) = \frac{100}{300 - 100} = \frac{1}{2} < 1$$

and we conclude that demand is inelastic when $p = 100$. Also,

$$E(200) = \frac{200}{300 - 200} = 2 > 1$$

and we see that demand is elastic when $p = 200$.

4.2 Visual Differentiation

Visual Differentiation

In Example 5 of the previous section, we found the marginal revenue for selling 2000 loudspeaker systems by first calculating the marginal revenue from the revenue function $R(x) = -0.02x^2 + 400x$ with domain $0 \le x \le 20{,}000$, and then evaluating the marginal revenue function at $x = 2000$. The process of calculating the derivative was straightforward, but what if the information for this problem had been given to us in graphical form as is often the case when analyzing data? That is, we would have only had the graph of $y = R(x)$ as shown in Figure 4.7. Then recall from Section 3.3 that the derivative represents the slope of the tangent line at a point on the graph of the function, so we can estimate the slope of the tangent line at $x = 2000$ by sketching a tangent line and then calculating its slope with rise and run as shown in Figure 4.8. We see that the marginal revenue at $x = 2000$ is given by

$$\frac{500{,}000}{1500} \approx 333.33$$

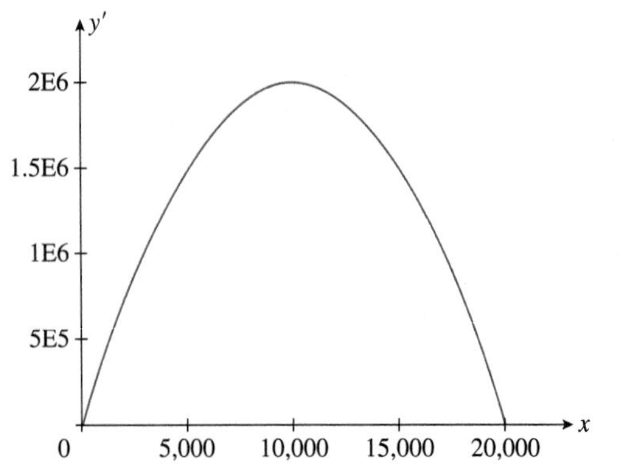

Figure 4.7
The graph of $y = R(x)$

Comparing this to the value of $320 from Example 5, we observe that $333.33 is a reasonable approximation for the marginal revenue at $x =$ 2000.

Ultimately, we want to be able to find the marginal revenue for any value of x, and so we should really be graphing the marginal revenue function $y = R'(x)$. Observe from the graph of $y = R(x)$ that there are three important features of the graph. First of all, at $x = 10{,}000$ the graph of $y = R(x)$ has a horizontal tangent line, i.e., the slope and hence its derivative there are zero. This means that the graph of $y = R'(x)$ intercepts the x-axis at $x = 10{,}000$. Secondly, the slope of any tangent line to the left of $x = 10{,}000$ is positive. This means that for values of x less than 10,000 the marginal revenue is positive. Moreover, as x increases from 0 to 10,000, we observe that the slope is decreasing from a certain positive slope size to zero. Thirdly, the slope of any tangent

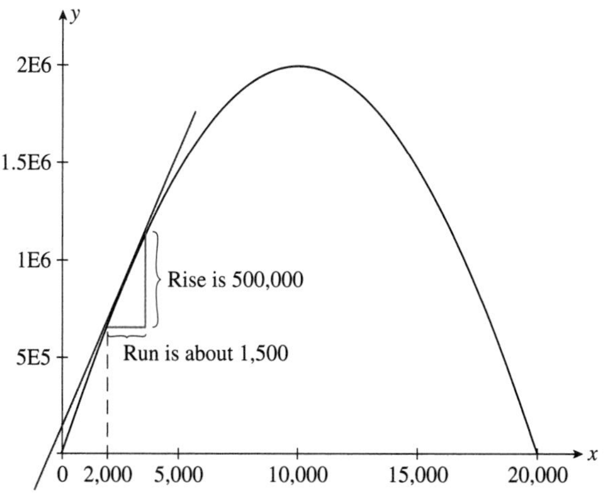

Figure 4.8
The graph of $y = R(x)$ and the tangent line with its slope at $x = 2000$

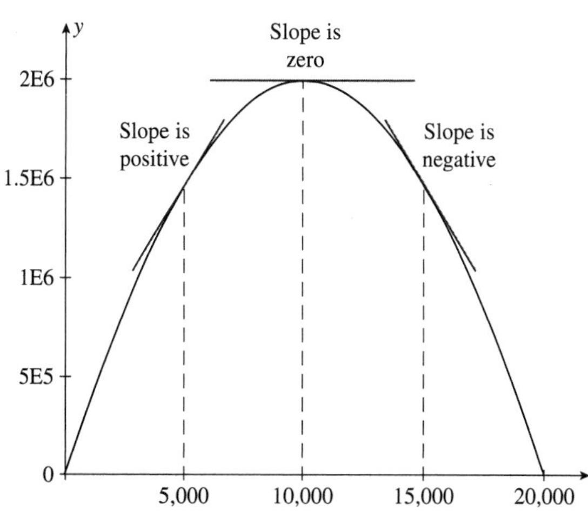

Figure 4.9
The graph of $y = R(x)$ and an analysis of its slope at various x-values

line to the right of $x = 10{,}000$ is negative. This means that for values of x larger than 10,000 the marginal revenue is negative. Here, we observe that the slope continues to decrease from zero to a certain negative slope value as x is increasing from 10,000 to 20,000. Next, we note that the graph looks symmetric about the y-axis. Remember that without the function description we cannot be certain if the function is even or odd. Nonetheless, we make use of the apparent symmetry we observe about the graph, i.e., the slopes at an equal distance from $x = 10{,}000$ to either side must have the same magnitude. All this information is indicated in Figure 4.9. Now we estimate a few slopes in a similar fashion as for Figure 4.8 such as near $x = 0$ $\left(\dfrac{1{,}000{,}000}{2500} = 400\right)$, $x = 5000$ $\left(\dfrac{300{,}000}{1500} = 200\right)$, $x = 15{,}000\left(\dfrac{300{,}000}{-1500} = -200\right)$, and $x = 20{,}000$ $\left(\dfrac{1{,}000{,}000}{-2500} = -400\right)$, and sketch the graph of $y = R'(x)$ (Figure 4.10).

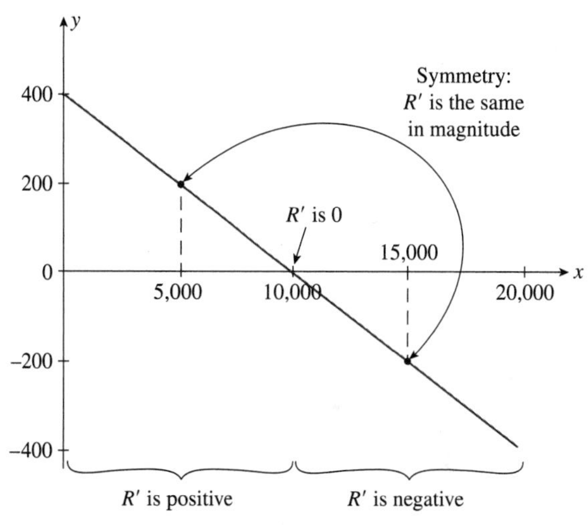

Figure 4.10
The graph of $y = R'(x)$

Remark 1

Finding the derivative graphically from the graph of any function f assumes that slopes may need to be estimated. The consequence is that the graph of the derivative may not be exact in its values, but the general shape of the graph is identified using this method. ◀

Remark 2

When we develop the graph of the derivative from the graph of any function f, it helps to do the following: Hold a pen between your index finger and thumb, and trace the graph of f from left to right by pretending that the place where you hold the pen is the point $(x, f(x))$, and the pen represents the tangent line at that point $(x, f(x))$. Looking at the direction the pen slopes, we can deduce whether the slope is positive, negative, or zero, and estimate its size if it is not zero. ◀

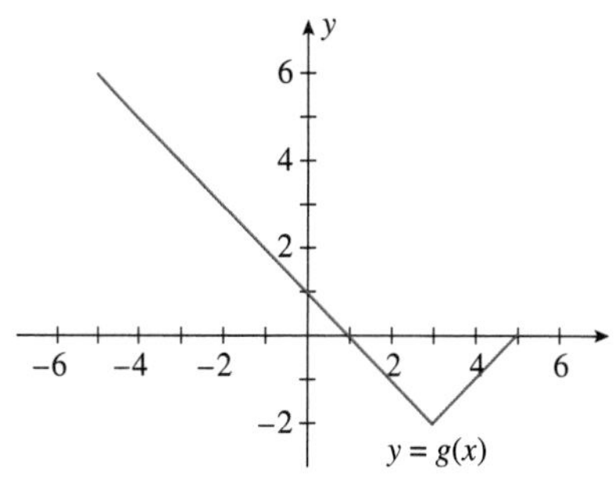

Figure 4.11
The graph of $y = g(x)$

Example 1

The graph of the function $y = g(x)$ is shown in Figure 4.11. Sketch the graph of its derivative.

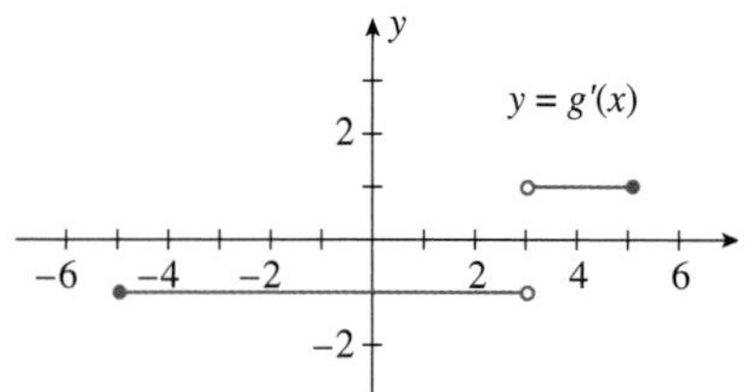

Figure 4.12
The graph of $y = g'(x)$

Solution

For $-5 < x < 3$ the slope is constant and can be read off the graph as having value -1. For $3 < x < 5$ the slope is constant and can be read off the graph as having value 1. Observe that the derivative does not exist at $x = 3$. The graph of the derivative $y = g'(x)$ is shown in Figure 4.12.

Example 2

The graph of the function $y = f(x)$ is shown in Figure 4.13. Sketch the graph of its derivative.

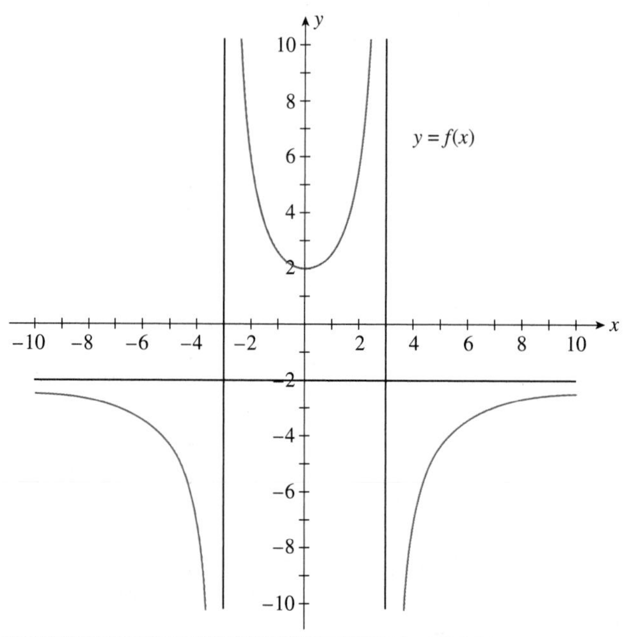

Figure 4.13
The graph of $y = f(x)$

Solution

For $x < -3$ the slope is negative. When we trace the curve from left to right in this section while holding a pen between our index finger and thumb, we observe that the slope decreases from a value close to zero to negative infinity as x increases to -3. Notice that the derivative is not defined for $x = -3$. For $-3 < x < 3$ we can see from the graph that the tangent line is horizontal at $x = 0$, and therefore has slope zero there. When we trace the curve from left to right in this section pretending that our pen is the tangent line, we observe that the slope increases from negative infinity to zero. As x increases to 3, we observe that the slope continues to increase to infinity. Notice that the derivative is not defined for $x = 3$. For $x > 3$ the slope is positive. When we trace the curve with our tangent line pen from left to right in this section, we observe that the slope is decreasing from infinity to a value close to zero. Furthermore, we observe that the graph of f seems to be symmetric about the y-axis. We use this information to mean that the slopes of tangent lines placed at an equal distance from $x = 0$ are the same size in magnitude, where the left tangent line has negative slope and the right one has positive slope as already stated above. In other words, the derivative should be symmetric with respect to the origin. Lastly, we estimate a few slopes in between to give a better approximation of how the graph of the derivative curves. At $(-5, f(-5))$ the slope is about -1.5, and at $(-2, f(-2))$ the slope is about -5. Using the symmetry we observed the slope at $(5, f(5))$ must be about 1.5, and the slope at $(2, f(2))$ must be about 5. The graph of the derivative $y = f'(x)$ is shown in Figure 4.14.

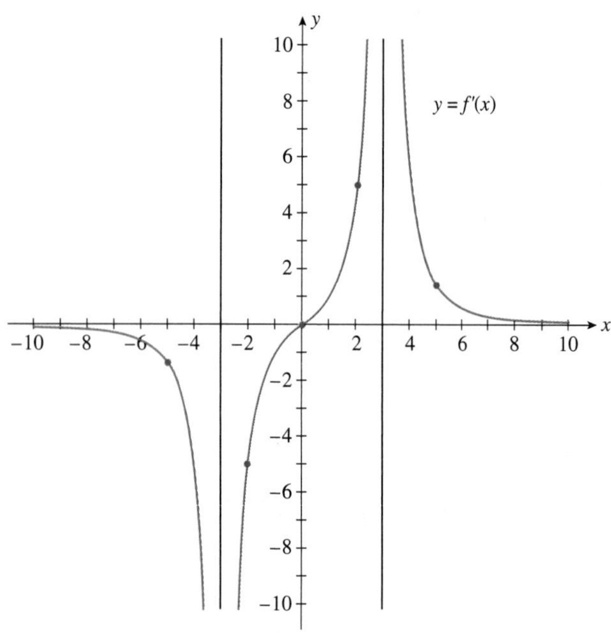

Figure 4.14
The graph of $y = f'(x)$

Example 3

Determine which of the two graphs in Figure 4.15 represents the function f and its derivative f'.

Solution

We start by checking each graph for which values of x the tangent line is horizontal. The derivative must have a value of zero at these x-values. However, we observe that the blue graph has horizontal tangent lines at $x = -6, -2, 2, 6$, which agrees with the red graph having zero value there, but likewise we observe that the red graph has horizontal tangent lines at $x = -4, 0, 4$, which agrees with the blue graph having zero value there. Next, we need to look at the sign of the slope between those zero slopes we have pointed out already. For $-6 < x < -2$ the slope on the blue graph is negative, which agrees with the function values of the red graph. For $-4 < x < 0$ the slope on the red graph is positive, which does not agree with the function values of the blue graph. Therefore, the blue graph represents the function $y = f(x)$, and the red graph represents its derivative $y = f'(x)$.

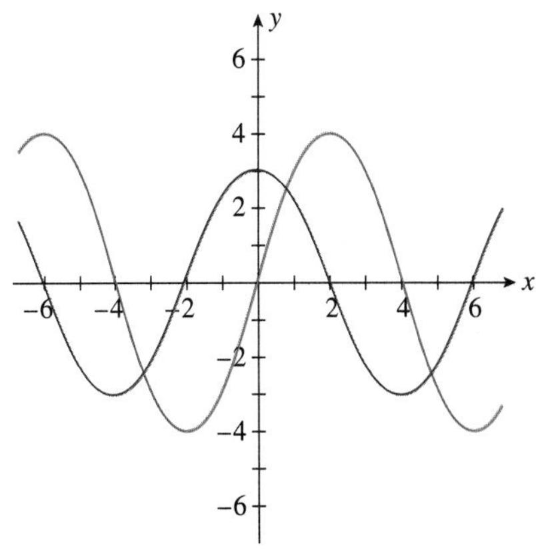

Figure 4.15
The graphs of $y = f(x)$ and $y = f'(x)$

Self-Check Exercise 4.2

1. The graph of the function $y = f(x)$ is shown at right. Sketch the graph of its derivative.

Solution to Self-Check Exercise 4.2 can be found on page 218.

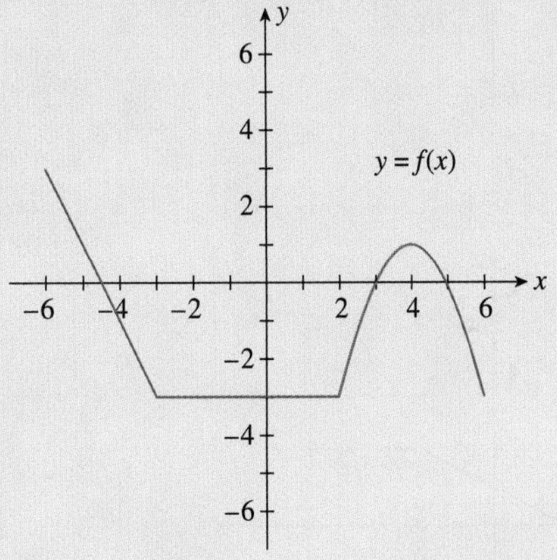

4.2 Exercises

In Exercises 1–8, determine which of the two graphs represents the function $y = f(x)$ and its derivative $y = f'(x)$.

1.

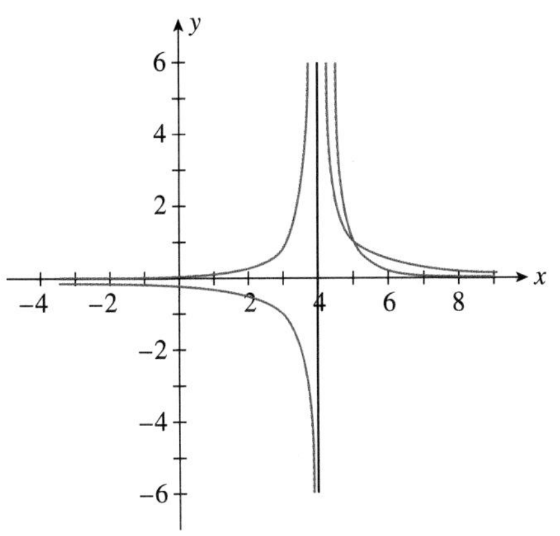

2.

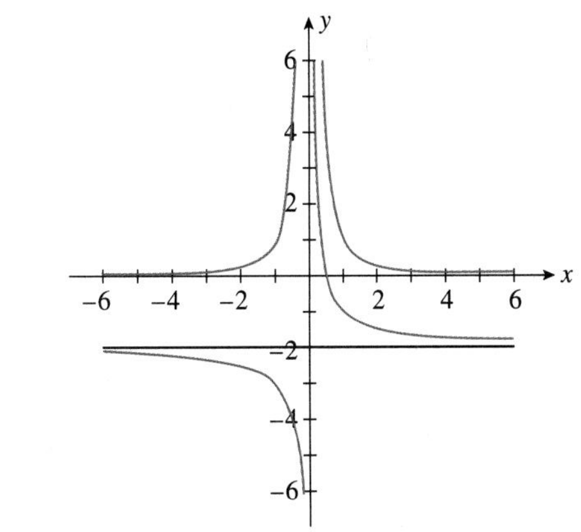

3.

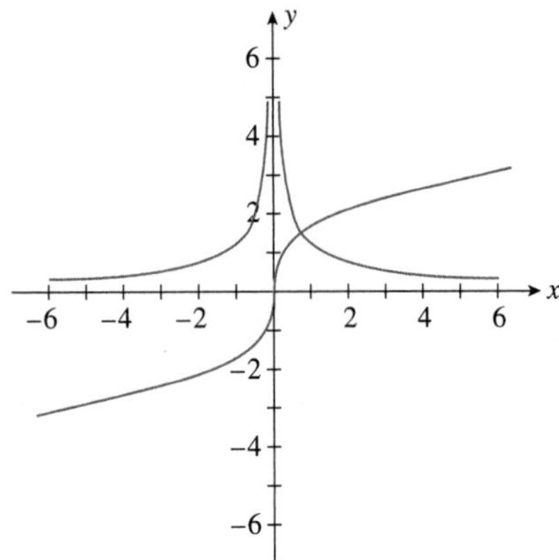

4.

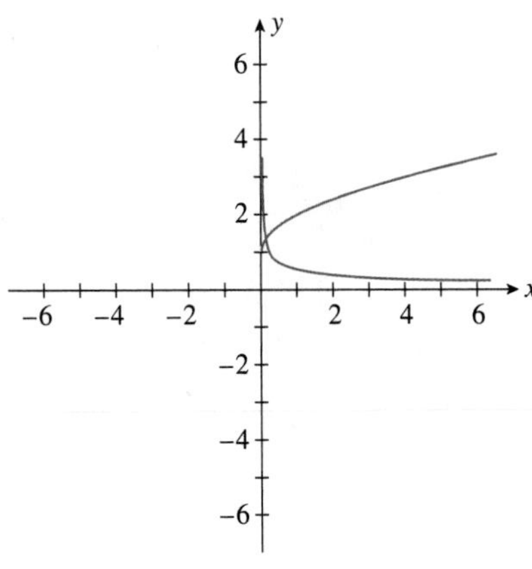

5.

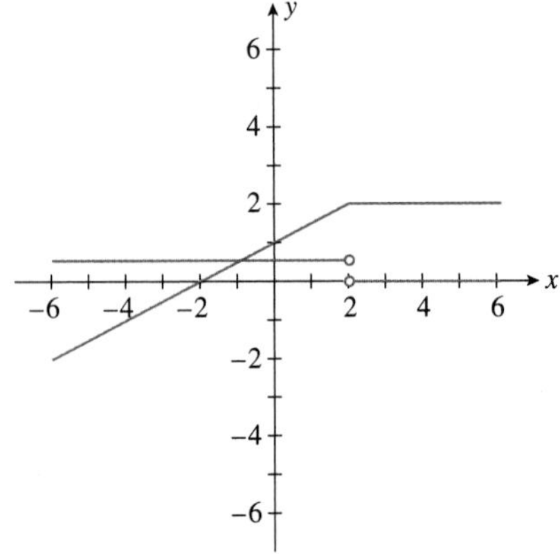

6.

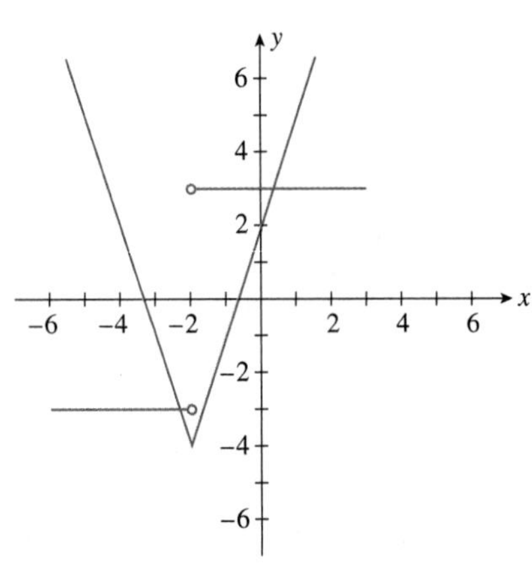

7.

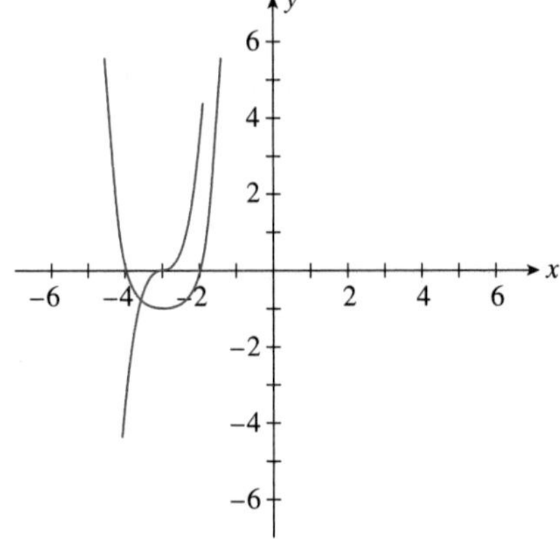

8.

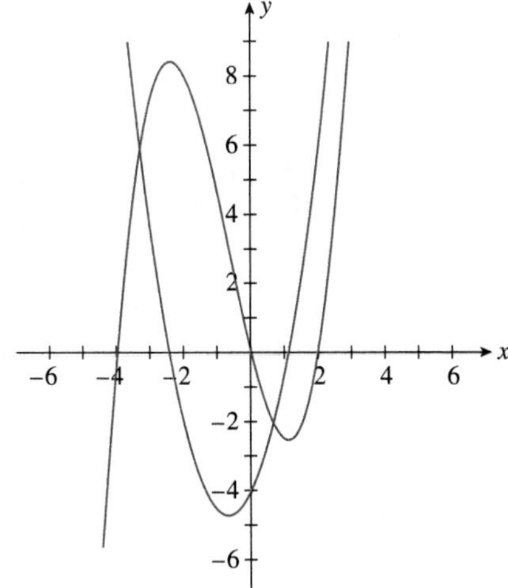

In Exercises 9–16, use the graph of $y = f(x)$ below to graph its derivative.

9.

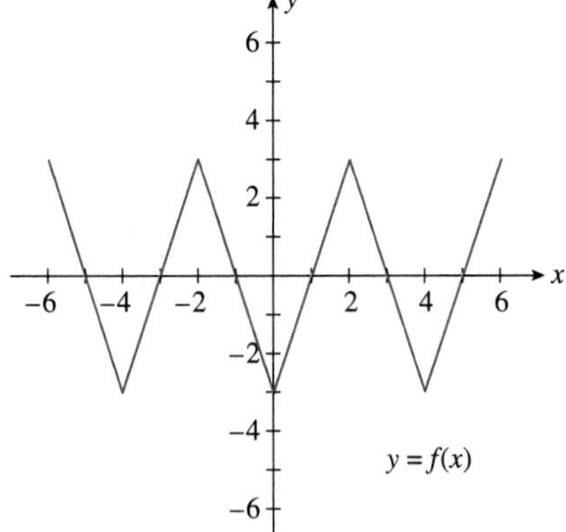

10.

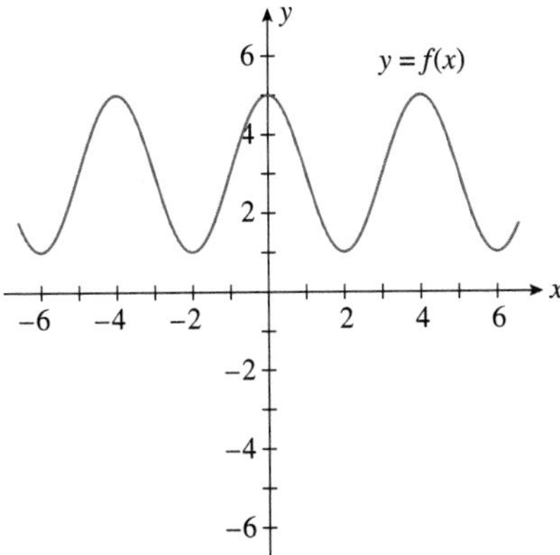

11.

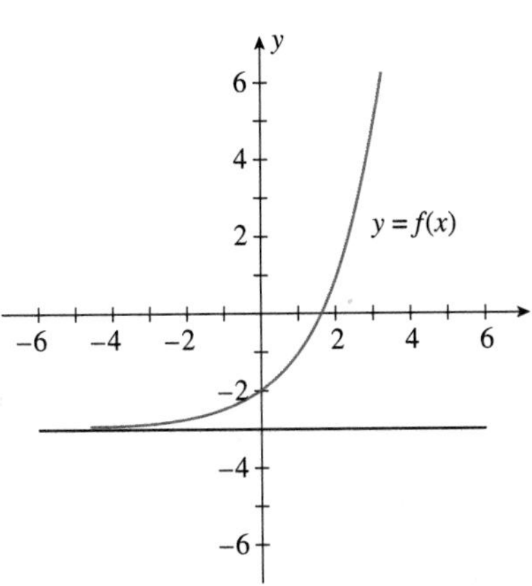

12.

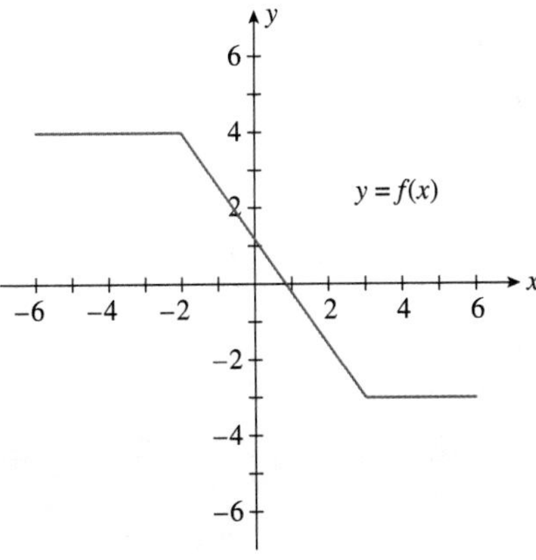

13.

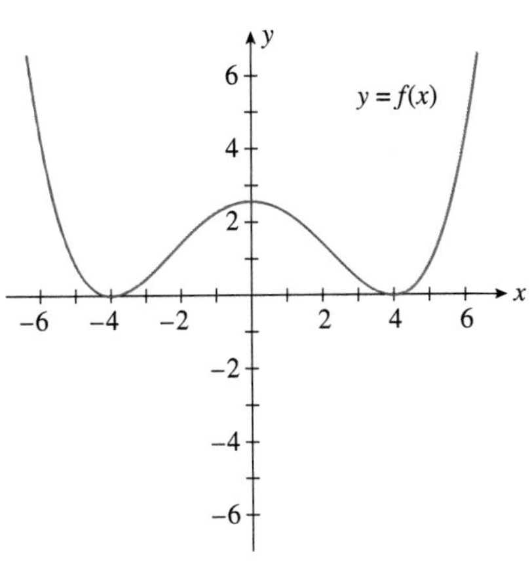

14.

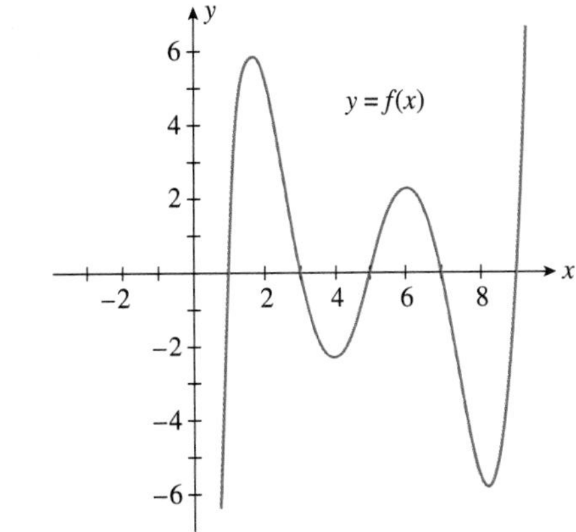

15.

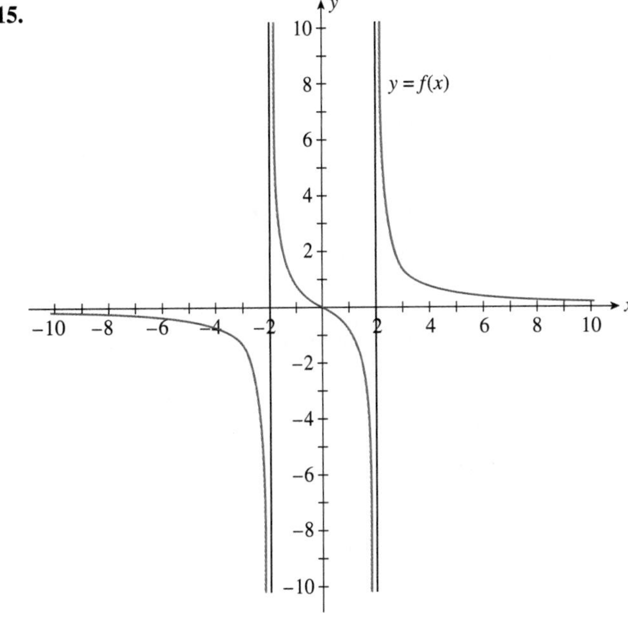

16.

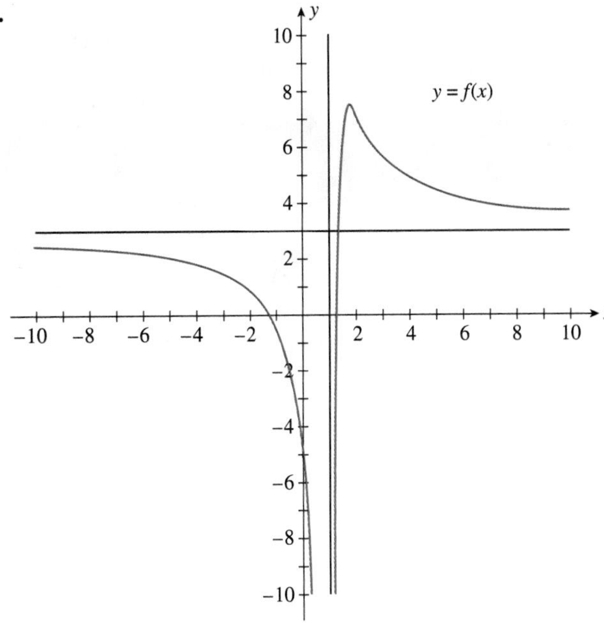

17. **Marginal Revenue** The graph below displays the monthly revenue $y = R(x)$ in dollars of a bus company servicing the commute between two nearby cities in Ontario when the price charged per passenger is x dollars.

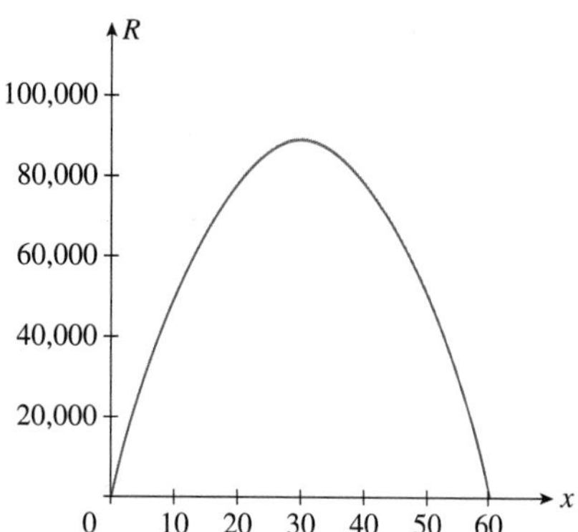

a. Estimate $R'(10)$, $R'(20)$, $R'(30)$, $R'(40)$, and $R'(50)$.
b. Graph the marginal revenue function $y = R'(x)$.
c. Based on the graph in part (b), what price should the bus company charge per person in order to maximize their revenue?

Business and Economics Applications

18. **Marginal Revenue** The graph below displays the seasonal revenue $y = R(x)$ in dollars from season passes sold at an amusement park outside a large city when the price charged per person per season pass is x dollars.

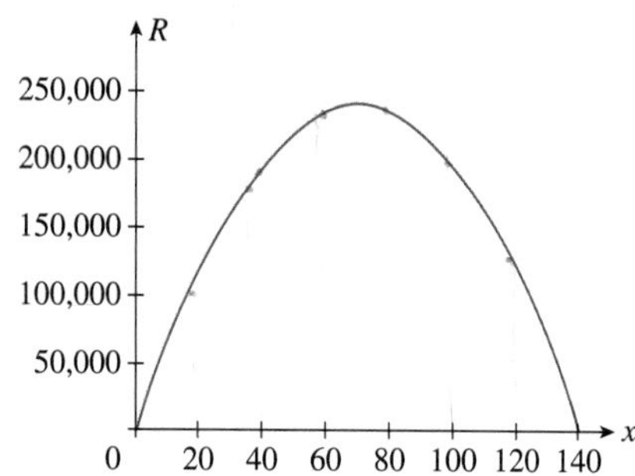

a. Estimate $R'(20)$, $R'(40)$, $R'(70)$, $R'(100)$, and $R'(120)$.
b. Graph the marginal revenue function $y = R'(x)$.
c. Based on the graph in part (b), what price should the amusement park charge per person per season pass in order to maximize their revenue?

Biological and Life Sciences Applications

19. **Concentration of a Drug in the Bloodstream** The graph below displays the concentration of a certain drug in a patient's bloodstream $y = C(t)$ when t is time in hours.

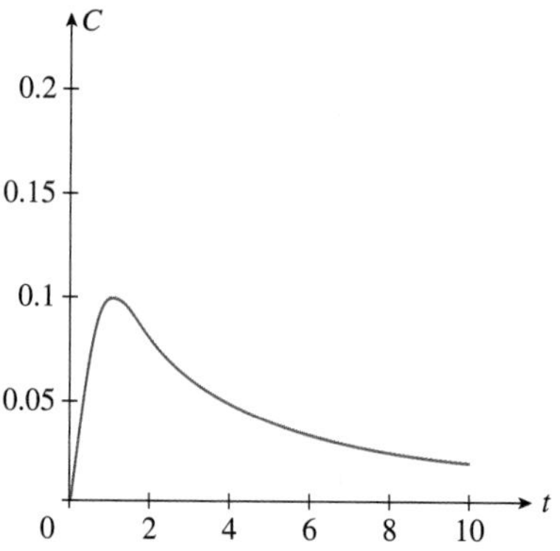

a. Estimate how fast the concentration is changing $\frac{1}{2}$ hr, 1 hr, and 2 hr after the injection.
b. Graph $y = C'(t)$.

20. **Population Growth** The graph below displays the projected growth of a town's population $y = P(t)$ over the next 3 years when t is time in months.

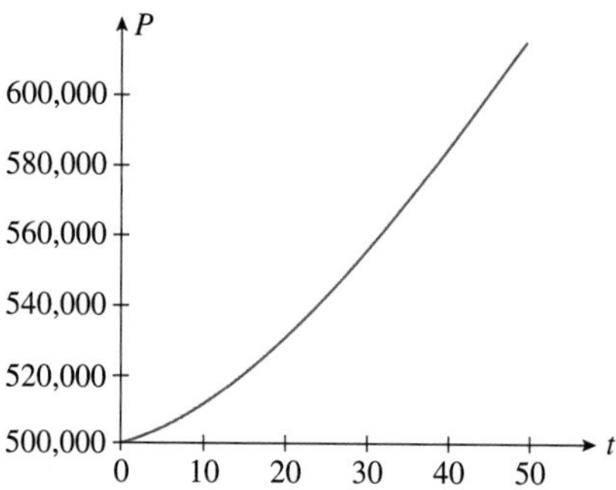

a. Estimate how fast the population is increasing 5 months and 20 months from now.
b. Graph $y = P'(t)$.

Solution to Self-Check Exercise 4.2

1. For $-6 < x < -3$ the slope is constant and can be read off the graph as having value -2. Notice that the derivative does not exist at $x = -3$. For $-3 < x < 2$ the slope is constant and can be read off the graph as having value 0. Notice that the derivative does not exist at $x = 2$. At $x = 4$ the tangent line is horizontal, so the slope is zero. For $2 < x < 4$ the slope is positive and decreasing, while for $4 < x < 6$ the slope is negative and decreasing. The graph of the derivative $y = f'(x)$ is shown at right.

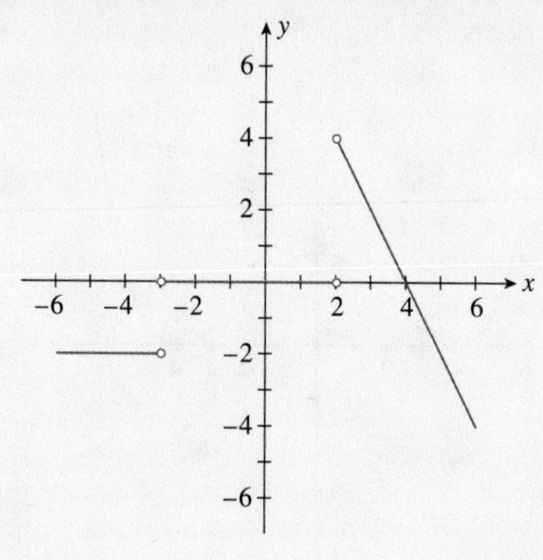

Higher-Order Derivatives

The derivative f' of a function f is also a function. As such, the differentiability of f' may be considered. Thus, the function f' has a derivative f'' at a point x in the domain of f' if the limit of the quotient

$$\frac{f'(x+h) - f'(x)}{h}$$

exists as h approaches zero. In other words, it is the derivative of the first derivative.

The function f'' obtained in this manner is called the **second derivative of the function f**, just as the derivative f' of f is often called the first derivative of f. Continuing in this fashion, we are led to considering the third, fourth, and higher-order derivatives of f whenever they exist. Notations for the first, second, third, and, in general, nth derivatives of a function f at a point x are

$$f'(x), f''(x), f'''(x), \ldots, f^{(n)}(x)$$

or

$$D^1f(x), D^2f(x), D^3f(x), \ldots, D^nf(x)$$

If f is written in the form $y = f(x)$, then the notations for its derivatives are

$$y', y'', y''', \ldots, y^{(n)}$$

$$\frac{dy}{dx}, \frac{d^2y}{dx^2}, \frac{d^3y}{dx^3}, \ldots, \frac{d^ny}{dx^n}$$

or

$$D^1y, D^2y, D^3y, \ldots, D^ny$$

respectively.

Example 1

Find the derivatives of all orders of the polynomial function

$$f(x) = x^5 - 3x^4 + 4x^3 - 2x^2 + x - 8$$

Solution

We have

$$f'(x) = 5x^4 - 12x^3 + 12x^2 - 4x + 1$$

$$f''(x) = \frac{d}{dx}f'(x) = 20x^3 - 36x^2 + 24x - 4$$

$$f'''(x) = \frac{d}{dx}f''(x) = 60x^2 - 72x + 24$$

$$f^{(4)}x = \frac{d}{dx}f'''(x) = 120x - 72$$

$$f^{(5)}x = \frac{d}{dx}f^{(4)}(x) = 120$$

and, in general,

$$f^{(n)}(x) = 0 \qquad (\text{for } n > 5)$$

Example 2

Find the third derivative of the function f defined by $y = x^{2/3}$. What is its domain?

Solution

We have

$$y' = \frac{2}{3}x^{-1/3}$$

$$y'' = \left(\frac{2}{3}\right)\left(-\frac{1}{3}\right)x^{-4/3} = -\frac{2}{9}x^{-4/3}$$

so the required derivative is

$$y''' = \left(-\frac{2}{9}\right)\left(-\frac{4}{3}\right)x^{-7/3} = \frac{8}{27}x^{-7/3} = \frac{8}{27x^{7/3}}$$

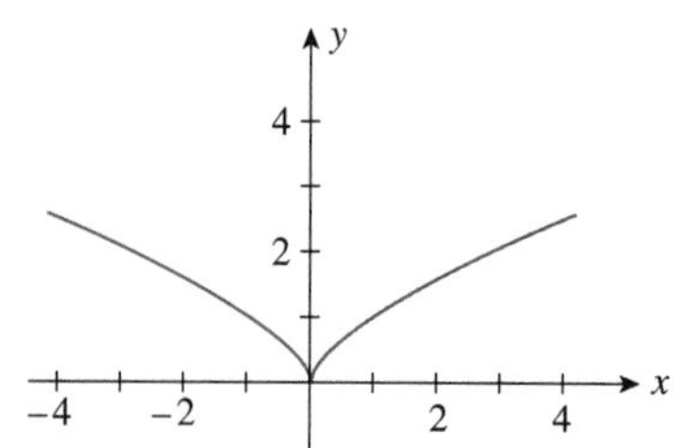

Figure 4.16
The graph of the function $y = x^{2/3}$

The common domain of the functions f', f'', and f''' is the set of all real numbers except $x = 0$. The domain of $y = x^{2/3}$ is the set of all real numbers. The graph of the function $y = x^{2/3}$ appears in Figure 4.16.

Remark

Always simplify an expression before differentiating it to obtain the next order derivative. ◀

Example 3

Find the second derivative of the function $y = (2x^2 + 3)^{3/2}$.

Solution

We have, using the general power rule,

$$y' = \frac{3}{2}(2x^2 + 3)^{1/2}(4x) = 6x(2x^2 + 3)^{1/2}$$

Next, using the product rule and then the chain rule, we find

$$\begin{aligned} y'' &= (6x)\cdot\frac{d}{dx}(2x^2 + 3)^{1/2} + \left[\frac{d}{dx}(6x)\right](2x^2 + 3)^{1/2} \\ &= (6x)\left(\frac{1}{2}\right)(2x^2 + 3)^{-1/2}(4x) + 6(2x^2 + 3)^{1/2} \\ &= 12x^2(2x^2 + 3)^{-1/2} + 6(2x^2 + 3)^{1/2} \\ &= 6(2x^2 + 3)^{-1/2}[2x^2 + (2x^2 + 3)] \\ &= \frac{6(4x^2 + 3)}{\sqrt{2x^2 + 3}} \end{aligned}$$

Applications

Just as the derivative of a function f at a point x measures the rate of change of the function f at that point, the second derivative of f (the derivative of f') measures the rate of change of the derivative f' of the function f. The third derivative of the function f, f''', measures the rate of change of f'', and so on.

In Chapter 7 we will discuss applications involving the geometric interpretation of the second derivative of a function. The following example gives an interpretation of the second derivative in a familiar role.

Example 4

Acceleration of a Maglev Refer to the example on page 113. The distance s (in metres) covered by a maglev moving along a straight track t seconds after starting from rest is given by the function $s = t^2$ $(0 \leq t \leq 30)$. What is the maglev's acceleration at the end of 30 seconds?

Solution

The velocity of the maglev t seconds from rest is given by

$$v = \frac{ds}{dt} = \frac{d}{dt}(t^2) = 2t$$

The acceleration of the maglev t seconds from rest is given by the rate of change of the velocity of t—that is,

$$a = \frac{d}{dt}v = \frac{d}{dt}\left(\frac{ds}{dt}\right) = \frac{d^2s}{dt^2} = \frac{d}{dt}2t = 2$$

or 2 metres per second per second, normally abbreviated 8 m/sec^2.

Example 5

Acceleration and Velocity of a Falling Object A ball is thrown straight up into the air from the roof of a building. The height of the ball as measured from the ground is given by

$$s(t) = -4.9t^2 + 7.3t + 36.6$$

where s is measured in metres and t in seconds. Find the velocity and acceleration of the ball 3 seconds after it is thrown into the air.

Solution

The velocity v and acceleration a of the ball at any time t are given by

$$v = \frac{ds}{dt} = \frac{d}{dt}(-4.9t^2 + 7.3t + 36.6) = -9.8t + 7.3$$

and

$$a = \frac{d^2s}{dt^2} = \frac{d}{dt}\left(\frac{ds}{dt}\right) = \frac{d}{dt}(-9.8t + 7.3) = -9.8$$

Therefore, the velocity of the ball 3 seconds after it is thrown into the air is

$$v(3) = -9.8(3) + 7.3 = -22.1$$

That is, the ball is falling downward at a speed of 22.1 m/s. The acceleration of the ball is 9.8 m/s^2 downward at any time during the motion.

Another interpretation of the second derivative of a function—this time from the field of economics—follows. Suppose the consumer price index (CPI) of an economy between the years a and b is described by the function $I(t)$ $(a \leq t \leq b)$ (Figure 4.17). Then the first derivative of I at $t = c$, $I'(c)$, where $a < c < b$, gives the rate of change of I at c. The quantity

$$\frac{I'(c)}{I(c)}$$

called the *relative rate of change of* $I(t)$ with respect to t at $t = c$, measures the *inflation rate* of the economy at $t = c$. The second derivative of I at $t = c$, $I''(c)$, gives the rate of change of I' at $t = c$. Now, it is possible for $I'(t)$ to be positive and $I''(t)$ to be negative at $t = c$ (see Example 6). This tells us that at $t = c$ the economy is experiencing inflation

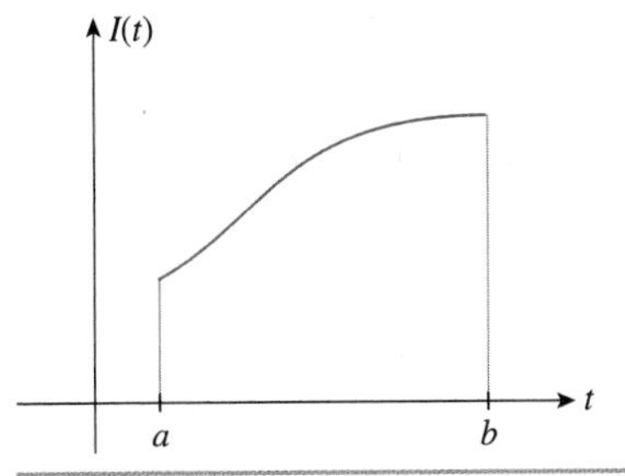

Figure 4.17
The CPI of a certain economy from year a to year b is given by $I(t)$.

(the CPI is increasing) but the rate at which inflation is growing is in fact decreasing. This is precisely the situation described by an economist or a politician when she claims that "inflation is slowing." One may not jump to the conclusion from the aforementioned quote that prices of goods and services are about to drop!

Example 6

Inflation Rate of an Economy The function

$$f(t) = -0.2t^3 + 3t^2 + 100 \qquad (0 \le t \le 9)$$

gives the CPI of an economy, where $t = 0$ corresponds to the beginning of 1995.

a. Find the inflation rate at the beginning of 2001 ($t = 6$).

b. Show that inflation was moderating at that time.

Solution

a. We find $I'(t) = -0.6t^2 + 6t$. Next, we compute

$$I'(6) = -0.6(6)^2 + 6(6) = 14.4 \quad \text{and} \quad I(6) = -0.2(6)^3 + 3(6)^2 + 100 = 164.8$$

from which we see that the inflation rate is

$$\frac{I'(6)}{I(6)} = \frac{14.4}{164.8} \approx 0.0874$$

or approximately 8.7%.

b. We find

$$I''(t) = \frac{d}{dt}(-0.6t^2 + 6t) = -1.2t + 6$$

Since

$$I''(6) = -1.2(6) + 6 = -1.2$$

we see that I' is indeed decreasing at $t = 6$ and conclude that inflation was moderating at that time (Figure 4.18).

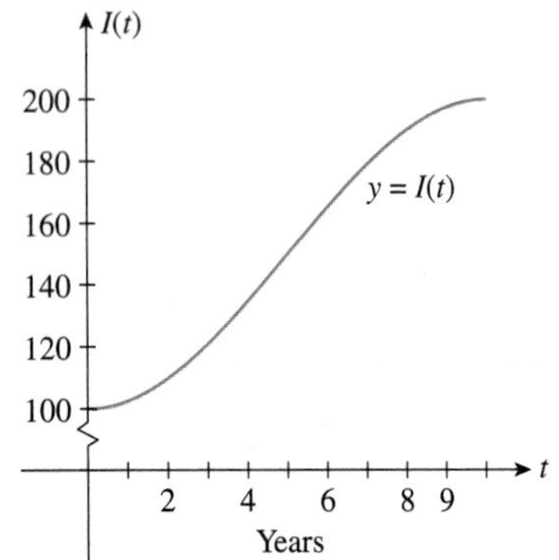

Figure 4.18
The CPI of an economy is given by $I(t)$.

Self-Check Exercises 4.3

1. Find the third derivative of
$$f(x) = 2x^5 - 3x^3 + x^2 - 6x + 10$$

2. Let
$$f(x) = \frac{1}{1 + x}$$
Find $f'(x)$, $f''(x)$, and $f'''(x)$.

3. A certain species of turtle faces extinction because dealers collect truckloads of turtle eggs to be sold as aphrodisiacs. After severe conservation measures are implemented, it is hoped that the turtle population will grow according to the rule
$$N(t) = 2t^3 + 3t^2 - 4t + 1000 \qquad (0 \le t \le 10)$$
where $N(t)$ denotes the population at the end of year t. Compute $N''(2)$ and $N''(8)$. What do your results tell you about the effectiveness of the program?

Solutions to Self-Check Exercises 4.3 can be found on page 224.

4.3 Exercises

In Exercises 1–20, find the first and second derivatives of the given function.

1. $f(x) = 4x^2 - 2x + 1$

2. $f(x) = -0.2x^2 + 0.3x + 4$

3. $f(x) = 2x^3 - 3x^2 + 1$

4. $g(x) = -3x^3 + 24x^2 + 6x - 64$

5. $h(t) = t^4 - 2t^3 + 6t^2 - 3t + 10$

6. $f(x) = x^5 - x^4 + x^3 - x^2 + x - 1$

7. $f(x) = (x^2 + 2)^5$

8. $g(t) = t^2(3t + 1)^4$

9. $g(t) = (2t^2 - 1)^2(3t^2)$

10. $h(x) = (x^2 + 1)^2(x - 1)$

11. $f(x) = (2x^2 + 2)^{7/2}$

12. $h(w) = (w^2 + 2w + 4)^{5/2}$

13. $f(x) = x(x^2 + 1)^2$

14. $g(u) = u(2u - 1)^3$

15. $f(x) = \dfrac{x}{2x + 1}$

16. $g(t) = \dfrac{t^2}{t - 1}$

17. $f(s) = \dfrac{s - 1}{s + 1}$

18. $f(u) = \dfrac{u}{u^2 + 1}$

19. $f(u) = \sqrt{4 - 3u}$

20. $f(x) = \sqrt{2x - 1}$

In Exercises 21–28, find the third derivative of the given function.

21. $f(x) = 3x^4 - 4x^3$

22. $f(x) = 3x^5 - 6x^4 + 2x^2 - 8x + 12$

23. $f(x) = \dfrac{1}{x}$

24. $f(x) = \dfrac{2}{x^2}$

25. $g(s) = \sqrt{3s - 2}$

26. $g(t) = \sqrt{2t + 3}$

27. $f(x) = (2x - 3)^4$

28. $g(t) = \left(\dfrac{1}{2}t^2 - 1\right)^5$

29. Let f be the function defined by the rule $f(x) = x^{7/3}$. Show that f has first- and second-order derivatives at all points x, and in particular at $x = 0$. Show also that the third derivative of f does *not* exist at $x = 0$.

30. Construct a function f that has derivatives of order up through and including n at a point a but fails to have the $(n + 1)$st derivative there.
Hint: See Exercise 29.

31. Show that a polynomial function has derivatives of all orders.
Hint: Let $P(x) = a_0x^n + a_1x^{n-1} + a_2x^{n-2} + \cdots + a_n$ be a polynomial of degree n, where n is a positive integer and $a_0, a_1, \ldots, a_n$ are constants with $a_0 \neq 0$. Compute $P'(x), P''(x), \ldots$.

32. **Acceleration of a Falling Object** During the construction of an office building, a hammer is accidentally dropped from a height of 78 m. The distance the hammer falls in t sec is $s(t) = 4.875t^2$. What is the hammer's velocity when it strikes the ground? What is its acceleration?

33. **Acceleration of a Car** The distance s (in metres) covered by a car after t sec is given by

$$s(t) = -0.3t^3 + 2.4t^2 + 6.1 \qquad (0 \le t \le 6)$$

Find a general expression for the car's acceleration at any time t $(0 \le t \le 6)$. Show that the car is decelerating after $2\frac{2}{3}$ sec.

34. **Test Flight of a VTOL** In a test flight of the McCord Terrier, McCord Aviation's experimental VTOL (vertical takeoff and landing) aircraft, it was determined that t sec after liftoff, when the craft was operated in the vertical takeoff mode, its altitude (in metres) was

$$h(t) = \frac{1}{48}t^4 - \frac{1}{3}t^3 + \frac{4}{3}t^2 \qquad (0 \le t \le 8)$$

a. Find an expression for the craft's velocity at time t.
b. Find the craft's velocity when $t = 0$ (the initial velocity), $t = 4$, and $t = 8$.
c. Find an expression for the craft's acceleration at time t.
d. Find the craft's acceleration when $t = 0$, 4, and 8.
e. Find the craft's height when $t = 0$, 4, and 8.

Business and Economics Applications

35. **GDP of a Developing Country** A developing country's gross domestic product (GDP) from 1992 to 2000 is approximated by the function

$$G(t) = -0.2t^3 + 2.4t^2 + 60 \qquad (0 \le t \le 8)$$

where $G(t)$ is measured in billions of dollars, with $t = 0$ corresponding to 1992.
a. Compute $G'(0), G'(1), \ldots, G'(8)$.
b. Compute $G''(0), G''(1), \ldots, G''(8)$.
c. Using the results obtained in parts (a) and (b), show that after a spectacular growth rate in the early years, the growth of the GDP cooled off.

36. **Disability Benefits** The number of persons aged 18–64 receiving disability benefits through Insurance in Eastern Canada from 1990 through 2000 is approximated by the function

$$N(t) = 0.00037t^3 - 0.0242t^2 + 0.52t + 5.3 \qquad (0 \le t \le 10)$$

where $N(t)$ is measured in units of a thousand and t is measured in years, with $t = 0$ corresponding to the beginning of 1990. Compute $N(8)$, $N'(8)$, and $N''(8)$ and interpret your results.

37. **Crime Rates** The number of major crimes committed in York between 1988 and 1995 is approximated by the function

$$N(t) = -0.1t^3 + 1.5t^2 + 100 \qquad (0 \le t \le 7)$$

where $N(t)$ denotes the number of crimes committed in year t, with $t = 0$ corresponding to 1988. Enraged by the dramatic increase in the crime rate, York's citizens, with the help of the local police, organized "Neighborhood Crime Watch" groups in early 1992 to combat this menace.
a. Verify that the crime rate was increasing from 1988 through 1995.
Hint: Compute $N'(0), N'(1), \ldots, N'(7)$.
b. Show that the Neighborhood Crime Watch program was working by computing $N''(4)$, $N''(5)$, $N''(6)$, and $N''(7)$.

38. **Canadian Population** The number of Canadians aged 45 to 54 can be approximated by

$$N(x) = -0.233t^4 + 0.633t^3 - 5.417t^2 + 134.67t + 2500$$

thousand people in year t, with $t = 0$ corresponding to the beginning of 1990. Compute $N'(10)$ and $N''(10)$ and interpret your results.

39. **Air Purification** During testing of a certain brand of air purifier, the amount of smoke remaining t min after the start of the test was

$$A(t) = -0.00006t^5 + 0.00468t^4 - 0.1316t^3 + 1.915t^2 - 17.63t + 100$$

percent of the original amount. Compute $A'(10)$ and $A''(10)$ and interpret your results.

Source: Consumer Reports

40. **Aging Population** The population of Canadians age 55 yr and over as a percent of the total population is approximated by the function

$$f(t) = 10.72(0.9t + 10)^{0.3} \qquad (0 \le t \le 20)$$

where t is measured in years, with $t = 0$ corresponding to 2000. Compute $f''(10)$ and interpret your result.

41. **Working Mothers** The percent of mothers who work outside the home and have children younger than age 6 yr is approximated by the function

$$P(t) = 33.55(t + 5)^{0.205} \qquad (0 \le t \le 21)$$

where t is measured in years, with $t = 0$ corresponding to the beginning of 1980. Compute $P''(20)$ and interpret your result.

In Exercises 42–46, determine whether the statement is true or false. If it is true, explain why it is true. If it is false, give an example to show why it is false.

42. If the second derivative of f exists at $x = a$, then $f''(a) = [f'(a)]^2$.

43. If $h = fg$ where f and g have second-order derivatives, then

$$h''(x) = f''(x)g(x) + 2f'(x)g'(x) + f(x)g''(x)$$

44. If $f(x)$ is a polynomial function of degree n, then $f^{(n+1)}(x) = 0$.

45. Suppose $P(t)$ represents the population of bacteria at time t and suppose $P'(t) > 0$ and $P''(t) < 0$; then the population is increasing at time t but at a decreasing rate.

46. If $h(x) = f(2x)$, then $h''(x) = 4f''(2x)$.

Solutions to Self-Check Exercises 4.3

1. $f'(x) = 10x^4 - 9x^2 + 2x - 6$

$f''(x) = 40x^3 - 18x + 2$

$f'''(x) = 120x^2 - 18$

2. We write $f(x) = (1 + x)^{-1}$ and use the general power rule, obtaining

$$f'(x) = (-1)(1+x)^{-2}\frac{d}{dx}(1+x) = -(1+x)^{-2}(1)$$
$$= -(1+x)^{-2} = -\frac{1}{(1+x)^2}$$

Continuing, we find

$$f''(x) = -(-2)(1 + x)^{-3}$$
$$= 2(1+x)^{-3} = \frac{2}{(1+x)^3}$$

$$f'''(x) = 2(-3)(1 + x)^{-4}$$
$$= -6(1 + x)^{-4}$$
$$= -\frac{6}{(1+x)^4}$$

3. $N'(t) = 6t^2 + 6t - 4$ and $N''(t) = 12t + 6 = 6(2t + 1)$

Therefore, $N''(2) = 30$ and $N''(8) = 102$. The results of our computations reveal that at the end of year 2, the *rate* of growth of the turtle population is increasing at the rate of 30 turtles/year/year. At the end of year 8, the rate is increasing at the rate of 102 turtles/year/year. Clearly, the conservation measures are paying off handsomely.

4.4 Implicit Differentiation and Related Rates

Differentiating Implicitly

Up to now we have dealt with functions expressed in the form $y = f(x)$; that is, the dependent variable y is expressed *explicitly* in terms of the independent variable x. However, not all functions are expressed in this form. Consider, for example, the equation

$$x^2y + y - x^2 + 1 = 0 \qquad \textbf{(5)}$$

This equation does express y *implicitly* as a function of x. In fact, solving (5) for y in terms of x, we obtain

$$(x^2 + 1)y = x^2 - 1 \qquad \text{Implicit equation}$$

$$y = f(x) = \frac{x^2 - 1}{x^2 + 1} \qquad \text{Explicit equation}$$

which gives an explicit representation of f.

Next, consider the equation

$$y^4 - y^3 - y + 2x^3 - x = 8$$

When certain restrictions are placed on x and y, this equation defines y as a function of x. But in this instance, we would be hard pressed to find y explicitly in terms of x. The following question arises naturally: How does one go about computing dy/dx in this case?

As it turns out, thanks to the chain rule, a method *does* exist for computing the derivative of a function directly from the implicit equation defining the function. This method is called **implicit differentiation** and is demonstrated in the next several examples.

Example 1

Given the equation $y^2 = x$, find $\frac{dy}{dx}$.

Solution

Differentiating both sides of the equation with respect to x, we obtain

$$\frac{d}{dx}(y^2) = \frac{d}{dx}(x)$$

To carry out the differentiation of the term $\frac{d}{dx}y^2$, we note that y is a function of x. Writing $y = f(x)$ to remind us of this fact, we find that

$$\begin{aligned} \frac{d}{dx}(y^2) &= \frac{d}{dx}[f(x)]^2 && \text{Write } y = f(x). \\ &= 2f(x)f'(x) && \text{Use the chain rule.} \\ &= 2y\frac{dy}{dx} && \text{Return to using } y \text{ instead of } f(x). \end{aligned}$$

Therefore, the equation

$$\frac{d}{dx}(y^2) = \frac{d}{dx}(x)$$

is equivalent to

$$2y\frac{dy}{dx} = 1$$

Solving for $\frac{dy}{dx}$ yields

$$\frac{dy}{dx} = \frac{1}{2y}$$

Before considering other examples, let's summarize the important steps involved in implicit differentiation. (Here we assume that dy/dx exists.)

Finding $\frac{dy}{dx}$ by Implicit Differentiation

1. Differentiate both sides of the equation *with respect to x*. (Make sure that the derivative of any term involving y includes the factor dy/dx.)
2. Solve the resulting equation for dy/dx in terms of x and y.

Example 2

Find $\frac{dy}{dx}$ given the equation

$$y^3 - y + 2x^3 - x = 8$$

Solution

Differentiating both sides of the given equation with respect to x, we obtain

$$\frac{d}{dx}(y^3 - y + 2x^3 - x) = \frac{d}{dx}(8)$$

$$\frac{d}{dx}(y^3) - \frac{d}{dx}(y) + \frac{d}{dx}(2x^3) - \frac{d}{dx}(x) = 0$$

Now, recalling that y is a function of x, we apply the chain rule to the first two terms on the left. Thus,

$$3y^2\frac{dy}{dx} - \frac{dy}{dx} + 6x^2 - 1 = 0$$

$$(3y^2 - 1)\frac{dy}{dx} = 1 - 6x^2$$

$$\frac{dy}{dx} = \frac{1 - 6x^2}{3y^2 - 1}$$

GROUP DISCUSSION

Refer to Example 2. Suppose we think of the equation $y^3 - y + 2x^3 - x = 8$ as defining x implicitly as a function of y. Find dx/dy and justify your method of solution.

Example 3

Consider the equation $x^2 + y^2 = 4$.

a. Find dy/dx by implicit differentiation.

b. Find the slope of the tangent line to the graph of the function $y = f(x)$ at the point $(1, \sqrt{3})$.

c. Find an equation of the tangent line of part (b).

Solution

a. Differentiating both sides of the equation with respect to x, we obtain

$$\frac{d}{dx}(x^2 + y^2) = \frac{d}{dx}(4)$$

$$\frac{d}{dx}(x^2) + \frac{d}{dx}(y^2) = 0$$

$$2x + 2y\frac{dy}{dx} = 0$$

$$\frac{dy}{dx} = -\frac{x}{y} \qquad y \neq 0$$

b. The slope of the tangent line to the graph of the function at the point $(1, \sqrt{3})$ is given by

$$\left.\frac{dy}{dx}\right|_{(1,\sqrt{3})} = -\left.\frac{x}{y}\right|_{(1,\sqrt{3})} = -\frac{1}{\sqrt{3}}$$

(*Note:* This notation is read "dy/dx evaluated at the point $(1, \sqrt{3})$.")

c. An equation of the tangent line in question is found by using the point-slope form of the equation of a line with the slope $m = -1\sqrt{3}$ and the point $(1, \sqrt{3})$. Thus,

$$y - \sqrt{3} = -\frac{1}{\sqrt{3}}(x - 1)$$

$$\sqrt{3}y - 3 = -x + 1$$

$$x + \sqrt{3}y - 4 = 0$$

A sketch of this tangent line is shown in Figure 4.19.

We can also solve the equation $x^2 + y^2 = 4$ explicitly for y in terms of x. If we do this, we obtain

$$y = \pm\sqrt{4 - x^2}$$

From this, we see that the equation $x^2 + y^2 = 4$ defines the two functions

$$y = f(x) = \sqrt{4 - x^2}$$

$$y = g(x) = -\sqrt{4 - x^2}$$

Since the point $(1, \sqrt{3})$ does not lie on the graph of $y = g(x)$, we conclude that

$$y = f(x) = \sqrt{4 - x^2}$$

is the required function. The graph of f is the upper semicircle shown in Figure 4.19.

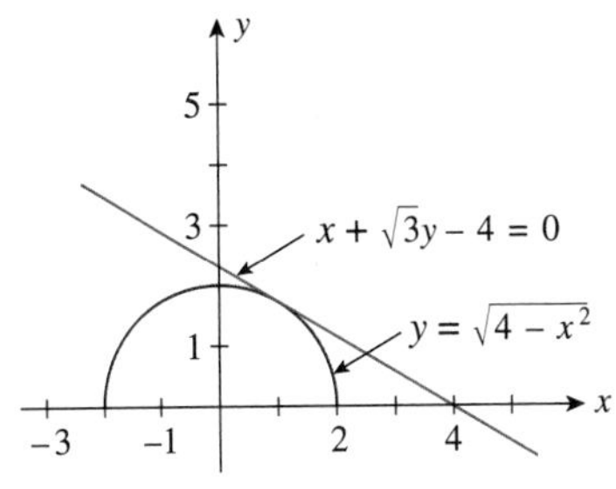

Figure 4.19
The line $x + \sqrt{3}y - 4 = 0$ is tangent to the graph of the function $y = f(x)$.

Remark
The notation

$$\left.\frac{dy}{dx}\right|_{(a,b)}$$

is used to denote the value of dy/dx at the point (a, b). ◀

GROUP DISCUSSION

Refer to Example 3. Yet another function defined implicitly by the equation $x^2 + y^2 = 4$ is the function

$$y = h(x) = \begin{cases} \sqrt{4 - x^2} & \text{if } -2 \le x < 0 \\ -\sqrt{4 - x^2} & \text{if } 0 \le x \le 2 \end{cases}$$

1. Sketch the graph of h.
2. Show that $h'(x) = -x/y$.
3. Find an equation of the tangent line to the graph of h at the point $(1, -\sqrt{3})$.

To find dy/dx at a *specific point* (a, b), differentiate the given equation implicitly with respect to x and then replace x and y by a and b, respectively, *before* solving the equation for dy/dx. This often simplifies the amount of algebra involved.

Example 4

Find $\frac{dy}{dx}$ given that x and y are related by the equation

$$x^2y^3 + 6x^2 = y + 12$$

and that $y = 2$ when $x = 1$.

Solution

Differentiating both sides of the given equation with respect to x, we obtain

$$\frac{d}{dx}(x^2y^3) + \frac{d}{dx}(6x^2) = \frac{d}{dx}(y) + \frac{d}{dx}(12)$$

$$x^2 \cdot \frac{d}{dx}(y^3) + y^3 \cdot \frac{d}{dx}(x^2) + 12x = \frac{dy}{dx} \qquad \text{Use the product rule on } \frac{d}{dx}(x^2y^3).$$

$$3x^2y^2\frac{dy}{dx} + 2xy^3 + 12x = \frac{dy}{dx}$$

Substituting $x = 1$ and $y = 2$ into this equation gives

$$3(1)^2(2)^2\frac{dy}{dx} + 2(1)(2)^3 + 12(1) = \frac{dy}{dx}$$

$$12\frac{dy}{dx} + 16 + 12 = \frac{dy}{dx}$$

and, solving for $\frac{dy}{dx}$,

$$\frac{dy}{dx} = -\frac{28}{11}$$

Note that it is not necessary to find an explicit expression for dy/dx.

Remark

In Examples 3 and 4, you can verify that the points at which we evaluated dy/dx actually lie on the curve in question by showing that the coordinates of the points satisfy the given equations. ◀

Example 5

Find $\frac{dy}{dx}$ given that x and y are related by the equation

$$\sqrt{x^2 + y^2} - x^2 = 5$$

Solution

Differentiating both sides of the given equation with respect to x, we obtain

$$\frac{d}{dx}(x^2 + y^2)^{1/2} - \frac{d}{dx}(x^2) = \frac{d}{dx}(5)$$

Write $\sqrt{x^2 + y^2} = (x^2 + y^2)^{1/2}$.

$$\frac{1}{2}(x^2 + y^2)^{-1/2}\frac{d}{dx}(x^2 + y^2) - 2x = 0$$

Use the general power rule on the first term.

$$\frac{1}{2}(x^2 + y^2)^{-1/2}\left(2x + 2y\frac{dy}{dx}\right) - 2x = 0$$

$$2x + 2y\frac{dy}{dx} = 4x(x^2 + y^2)^{1/2}$$

Transpose $2x$ and multiply both sides by $2(x^2 + y^2)^{1/2}$.

$$2y\frac{dy}{dx} = 4x(x^2 + y^2)^{1/2} - 2x$$

$$\frac{dy}{dx} = \frac{2x\sqrt{x^2 + y^2} - x}{y}$$

Related Rates

Implicit differentiation is a useful technique for solving a class of problems known as **related rates** problems. For example, suppose x and y are each functions of a third variable t. Here, x might denote the mortgage rate and y the number of single-family homes sold at any time t. Further, suppose we have an equation that gives the relationship between x and y (the number of houses sold y is related to the mortgage rate x). Differentiating both sides of this equation implicitly with respect to t, we obtain an equation that gives a relationship between dx/dt and dy/dt. In the context of our example, this equation gives us a relationship between the rate of change of the mortgage rate and the rate of change of the number of houses sold, as a function of time. Thus, knowing

$$\frac{dx}{dt}$$

How fast the mortgage rate x is changing at time t.

we can determine

$$\frac{dy}{dt}$$ How fast the sale of houses y is changing at that instant of time.

Example 6

Rate of Change of Housing Starts It is estimated that the number of housing starts, $N(t)$ (in units of a million), over the next 5 years is related to the mortgage rate $r(t)$ (percent per year) by the equation

$$9N^2 + r = 36$$

What is the rate of change of the number of housing starts with respect to time when the mortgage rate is 6% per year and is increasing at the rate of 0.25% per year?

Solution

We are given that

$$r = 6 \quad \text{and} \quad \frac{dr}{dt} = 0.25$$

at a certain instant of time, and we are required to find dN/dt. First, by substituting $r = 6$ into the given equation, we find

$$9N^2 + 6 = 36$$

$$N^2 = \frac{30}{9}$$ How fast the sale of houses is changing at that instant of time.

or $N = 10/3$ (we reject the negative root). Next, differentiating the given equation implicitly on both sides with respect to t, we obtain

$$\frac{d}{dt}(9N^2) + \frac{d}{dt}(r) = \frac{d}{dt}36$$

$$18N\frac{dN}{dt} + \frac{dr}{dt} = 0$$ Use the chain rule on the first term.

Then, substituting $N = 10/3$ and $dr/dt = 0.25$ into this equation gives

$$18\left(\frac{10}{3}\right)\frac{dN}{dt} + 0.25 = 0$$

Solving this equation for dN/dt then gives

$$\frac{dN}{dt} = -\frac{0.25}{6\sqrt{30}} = -\frac{1}{24\sqrt{30}}$$

Thus, at the instant of time under consideration, the number of housing starts is decreasing at the rate of 7607 units per year.

Example 7

Supply-Demand A major audiotape manufacturer is willing to make x thousand ten-packs of metal alloy audiocassette tapes available in the marketplace each week when the wholesale price is \$$p$ per ten-pack. It is known that the relationship between x and p is governed by the supply equation

$$x^2 - 3xp + p^2 = 5$$

How fast is the supply of tapes changing when the price per ten-pack is \$11, the quantity supplied is 4000 ten-packs, and the wholesale price per ten-pack is increasing at the rate of \$.10 per ten-pack each week?

Solution

We are given that

$$p = 11, \qquad x = 4, \qquad \frac{dp}{dt} = 0.1$$

at a certain instant of time, and we are required to find dx/dt. Differentiating the given equation on both sides with respect to t, we obtain

$$\frac{d}{dt}(x^2) - \frac{d}{dt}(3xp) + \frac{d}{dt}(p^2) = \frac{d}{dt}(5)$$

$$2x\frac{dx}{dt} - 3\left(p\frac{dx}{dt} + x\frac{dp}{dt}\right) + 2p\frac{dp}{dt} = 0$$

Use the product rule on the second term.

Substituting the given values of p, x, and dp/dt into the last equation, we have

$$2(4)\frac{dx}{dt} - 3\left[(11)\frac{dx}{dt} + 4(0.1)\right] + 2(11)(0.1) = 0$$

$$8\frac{dx}{dt} - 33\frac{dx}{dt} - 1.2 + 2.2 = 0$$

$$25\frac{dx}{dt} = 1$$

$$\frac{dx}{dt} = 0.04$$

Thus, at the instant of time under consideration the supply of ten-pack audiocassettes is increasing at the rate of (0.04)(1000), or 40, ten-packs per week.

In certain related rates problems, we need to formulate the problem mathematically before analyzing it. The following guidelines can be used to help solve problems of this type.

Solving Related Rates Problems

1. Assign a variable to each quantity. Draw a diagram if needed.
2. Write the *given* values of the variables and their rates of change with respect to t.
3. Find an equation giving the relationship between the variables.
4. Differentiate both sides of this equation implicitly with respect to t.
5. Replace the variables and their derivatives by the numerical data found in step 2 and solve the equation for the required rate of change.

Example 8

Watching a Rocket Launch At a distance of 1200 metres from the launch site, a spectator is observing a rocket being launched. If the rocket lifts off vertically and is rising at a speed of 180 metres/second when it is at an altitude of 900 metres, how fast is the distance between the rocket and the spectator changing at that instant?

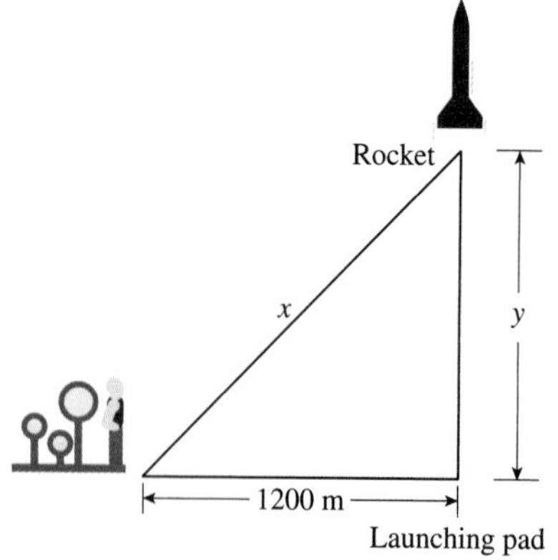

Figure 4.20
The rate at which x is changing with respect to time is related to the rate of change of y with respect to time.

Solution

Step 1 Let

y = the altitude of the rocket
x = the distance between the rocket and the spectator

at any time t (Figure 4.20).

Step 2 We are given that at a certain instant of time

$$y = 900 \quad \text{and} \quad \frac{dy}{dt} = 180$$

and are asked to find dx/dt at that instant.

Step 3 Applying the Pythagorean theorem to the right triangle in Figure 4.20, we find that

$$x^2 = y^2 + 1200^2$$

Therefore, when $y = 900$,

$$x = \sqrt{900^2 + 1200^2} = 1500$$

Step 4 Next, we differentiate the equation $x^2 = y^2 + 1200^2$ with respect to t, obtaining

$$2x\frac{dx}{dt} = 2y\frac{dy}{dt}$$

(Remember, both x and y are functions of t.)

Step 5 Substituting $x = 1500$, $y = 900$, and $dy/dt = 180$, we find

$$2(1500)\frac{dx}{dt} = 2(900)(180)$$

$$\frac{dx}{dt} = 108$$

Therefore, the distance between the rocket and the spectator is changing at a rate of 108 metres/second.

Be sure that you do not replace the variables in the equation found in step 3 by their numerical values before differentiating the equation.

Self-Check Exercises 4.4

1. Given the equation $x^3 + 3xy + y^3 = 4$, find dy/dx by implicit differentiation.

2. Find an equation of the tangent line to the graph of $16x^2 + 9y^2 = 144$ at the point

$$\left(2, -\frac{4\sqrt{5}}{3}\right).$$

Solutions to Self-Check Exercises 4.4 can be found on page 234.

4.4 Exercises

In Exercises 1–8, find the derivative dy/dx (a) by solving each of the given implicit equations for y explicitly in terms of x and (b) by differentiating each of the given equations implicitly. Show that, in each case, the results are equivalent.

1. $x + 2y = 5$
2. $3x + 4y = 6$
3. $xy = 1$
4. $xy - y - 1 = 0$
5. $x^3 - x^2 - xy = 4$
6. $x^2y - x^2 + y - 1 = 0$
7. $\frac{x}{y} - x^2 = 1$
8. $\frac{y}{x} - 2x^3 = 4$

In Exercises 9–30, find dy/dx by implicit differentiation.

9. $x^2 + y^2 = 16$
10. $2x^2 + y^2 = 16$
11. $x^2 - 2y^2 = 16$
12. $x^3 + y^3 + y - 4 = 0$
13. $x^2 - 2xy = 6$
14. $x^2 + 5xy + y^2 = 10$
15. $x^2y^2 - xy = 8$
16. $x^2y^3 - 2xy^2 = 5$
17. $x^{1/2} + y^{1/2} = 1$
18. $x^{1/3} + y^{1/3} = 1$
19. $\sqrt{x + y} = x$
20. $(2x + 3y)^{1/3} = x^2$
21. $\frac{1}{x^2} + \frac{1}{y^2} = 1$
22. $\frac{1}{x^3} + \frac{1}{y^3} = 5$
23. $\sqrt{xy} = x + y$
24. $\sqrt{xy} = 2x + y^2$

25. $\dfrac{x + y}{x - y} = 3x$

26. $\dfrac{x - y}{2x + 3y} = 2x$

27. $xy^{3/2} = x^2 + y^2$

28. $x^2y^{1/2} = x + 2y^3$

29. $(x + y)^3 + x^3 + y^3 = 0$

30. $(x + y^2)^{10} = x^2 + 25$

In Exercises 31–34, find an equation of the tangent line to the graph of the function f defined by the given equation at the indicated point.

31. $4x^2 + 9y^2 = 36$; $(0, 2)$

32. $y^2 - x^2 = 16$; $(2, 2\sqrt{5})$

33. $x^2y^3 - y^2 + xy - 1 = 0$; $(1, 1)$

34. $(x - y - 1)^3 = x$; $(1, -1)$

In Exercises 35–38, find the second derivative d^2y/dx^2 of each of the functions defined implicitly by the given equation.

35. $xy = 1$

36. $x^3 + y^3 = 28$

37. $y^2 - xy = 8$

38. $x^{1/3} + y^{1/3} = 1$

39. The volume of a right circular cylinder of radius r and height h is $V = \pi r^2 h$. Suppose the radius and height of the cylinder are changing with respect to time t.
   a. Find a relationship between dV/dt, dr/dt, and dh/dt.
   b. At a certain instant of time, the radius and height of the cylinder are 2 and 6 cm and are increasing at the rate of 0.1 and 0.3 cm/sec, respectively. How fast is the volume of the cylinder increasing?

40. A car leaves an intersection travelling west. Its position 4 sec later is 20 m from the intersection. At the same time, another car leaves the same intersection heading north so that its position 4 sec later is 28 m from the intersection. If the speed of the cars at that instant of time is 9 m/sec and 11 m/sec, respectively, find the rate at which the distance between the two cars is changing.

41. The volume V of a cube with sides of length x cm is changing with respect to time. At a certain instant of time, the sides of the cube are 5 cm long and increasing at the rate of 0.1 cm/sec. How fast is the volume of the cube changing at that instant of time?

42. Two ships leave the same port at noon. Ship A sails north at 15 km/h, and ship B sails east at 12 km/h. How fast is the distance between them changing at 1 P.M.?

43. A car leaves an intersection travelling east. Its position t sec later is given by $x = t^2 + t$ m. At the same time, another car leaves the same intersection heading north, travelling $y = t^2 + 3t$ m in t sec. Find the rate at which the distance between the two cars will be changing 5 sec later.

44. At a distance of 15 m from the pad, a man observes a helicopter taking off from a heliport. If the helicopter lifts off vertically and is rising at a speed of 13 m/s when it is at an altitude of 36 m, how fast is the distance between the helicopter and the man changing at that instant?

45. A spectator watches a rowing race from the edge of a river bank. The lead boat is moving in a straight line that is 36 m from the river bank. If the boat is moving at a constant speed of 6 m/s, how fast is the boat moving away from the spectator when it is 15 m past her?

46. A boat is pulled toward a dock by means of a rope which is wound on a drum that is located 4 m above the bow of the boat. If the rope is being pulled in at the rate of 3 m/sec, how fast is the boat approaching the dock when it is 25 m from the dock?

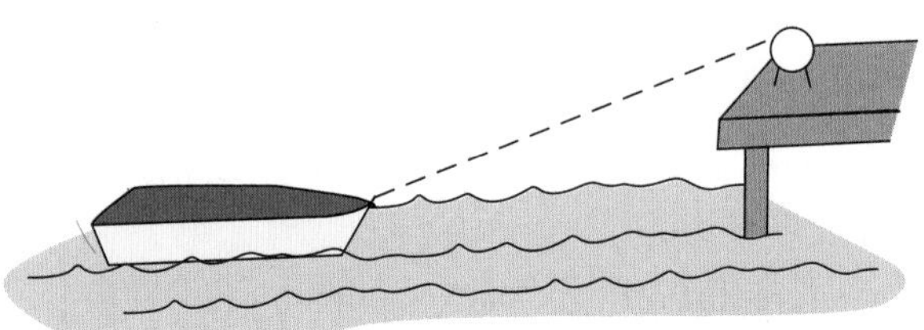

47. Assume that a snowball is in the shape of a sphere. If the snowball melts at a rate that is proportional to its surface area, show that its radius decreases at a constant rate.
Hint: Its volume is $V = (4/3)\pi r^3$, and its surface area is $S = 4\pi r^2$.

48. A coffee pot in the form of a circular cylinder of radius 10 cm is being filled with water flowing at a constant rate. If the water level is rising at the rate of 1 cm/sec, what is the rate at which water is flowing into the coffee pot?

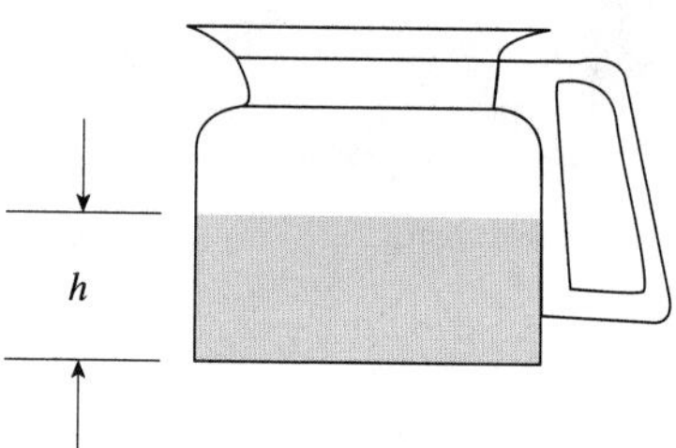

49. A 1.8-m-tall man is walking away from a street light 5.4 m high at a speed of 1.5 m/s. How fast is the tip of his shadow moving along the ground?

50. A 20-m ladder leaning against a wall begins to slide. How fast is the top of the ladder sliding down the wall at the instant of time when the bottom of the ladder is 12 m from the wall and sliding away from the wall at the rate of 5 m/sec?
Hint: Refer to the adjacent figure. By the Pythagorean theorem, $x^2 + y^2 = 400$. Find dy/dt when $x = 12$ and $dx/dt = 5$.

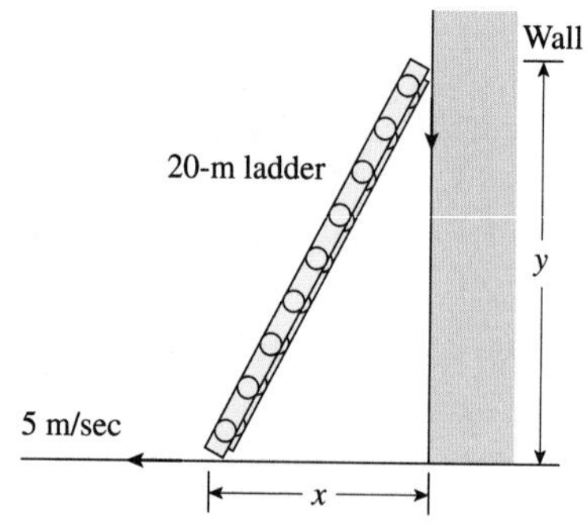

51. The base of a 13-m ladder leaning against a wall begins to slide away from the wall. At the instant of time when the base is 12 m from the wall, the base is moving at the rate of 8 m/sec. How fast is the top of the ladder sliding down the wall at that instant of time?
Hint: Refer to the hint in Exercise 50.

52. **Coast Guard Patrol Search Mission** The pilot of a Coast Guard patrol aircraft on a search mission had just spotted a disabled fishing trawler and decided to go in for a closer look. Flying at a constant altitude of 300 m and at a steady speed of 80 m/s, the aircraft passed directly over the trawler. How fast was the aircraft receding from the trawler when it was 450 m from it?

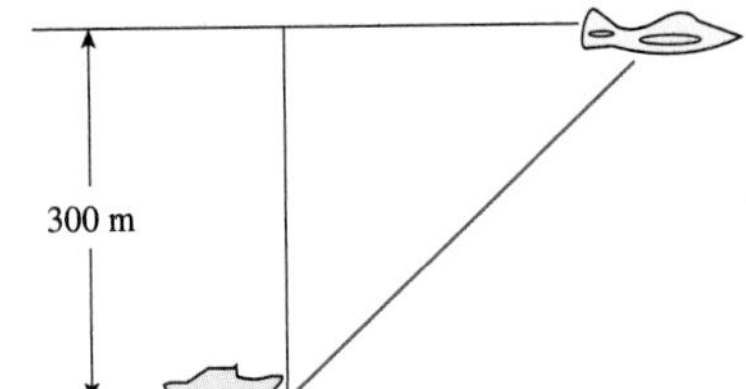

Business and Economics Applications

53. **Price-Demand** Suppose the quantity demanded weekly of Super Grip radial tires is related to its unit price by the equation

$$p + x^2 = 144$$

where p is measured in dollars and x is measured in units of a thousand. How fast is the quantity demanded changing when $x = 9$, $p = 63$, and the price/tire is increasing at the rate of \$2/week?

54. **Price-Supply** Suppose the quantity x of Super Grip radial tires made available each week in the marketplace is related to the unit-selling price by the equation

$$p - \frac{1}{2}x^2 = 48$$

where x is measured in units of a thousand and p is in dollars. How fast is the weekly supply of Super Grip radial tires being introduced into the marketplace when $x = 6$, $p = 66$, and the price/tire is decreasing at the rate of \$3/week?

55. **Price-Demand** The demand equation for a certain brand of metal alloy audiocassette tape is

$$100x^2 + 9p^2 = 3600$$

where x represents the number (in thousands) of ten-packs demanded each week when the unit price is \$$p$. How fast is the quantity demanded increasing when the unit price/ten-pack is \$14 and the selling price is dropping at the rate of \$.15/ten-pack/week?
Hint: To find the value of x when $p = 14$, solve the equation

$$100x^2 + 9p^2 = 3600 \text{ for } x \text{ when } p = 14.$$

56. **Effect of Price on Supply** Suppose the wholesale price of a certain brand of medium-size eggs p (in dollars/carton) is related to the weekly supply x (in thousands of cartons) by the equation

$$625p^2 - x^2 = 100$$

If 25,000 cartons of eggs are available at the beginning of a certain week and the price is falling at the rate of 2¢/carton/week, at what rate is the supply falling?
Hint: To find the value of p when $x = 25$, solve the supply equation for p when $x = 25$.

57. **Supply-Demand** Refer to Exercise 56. If 25,000 cartons of eggs are available at the beginning of a certain week and the supply is falling at the rate of 1000 cartons/week, at what rate is the wholesale price changing?

58. **Elasticity of Demand** The demand function for a certain make of ink-jet cartridge is

$$p = -0.01x^2 - 0.1x + 6$$

where p is the unit price in dollars and x is the quantity demanded each week, measured in units of a thousand. Compute the elasticity of demand and determine whether the demand is inelastic, unitary, or elastic when $x = 10$.

59. **Elasticity of Demand** The demand function for a certain brand of compact disc is

$$p = -0.01x^2 - 0.2x + 8$$

where p is the wholesale unit price in dollars and x is the quantity demanded each week, measured in units of a thousand. Compute the elasticity of demand and determine whether the demand is inelastic, unitary, or elastic when $x = 15$.

Biological and Life Sciences Applications

60. **Oil Spills** In calm waters oil spilling from the ruptured hull of a grounded tanker spreads in all directions. If the area polluted is a circle and its radius is increasing at a rate of 2 m/sec, determine how fast the area is increasing when the radius of the circle is 40 m.

61. **Blowing Soap Bubbles** Carlos is blowing air into a soap bubble at the rate of 8 cm^3/sec. Assuming that the bubble is spherical, how fast is its radius changing at the instant of time when the radius is 10 cm? How fast is the surface area of the bubble changing at that instant of time?

In Exercises 62 and 63, determine whether the statement is true or false. If it is true, explain why it is true. If it is false, give an example to show why it is false.

62. If f and g are differentiable and $f(x)g(y) = 0$, then

$$\frac{dy}{dx} = -\frac{f'(x)g(y)}{f(x)g'(y)} \qquad (f(x) \neq 0 \text{ and } g'(y) \neq 0)$$

63. If f and g are differentiable and $f(x) + g(y) = 0$, then

$$\frac{dy}{dx} = -\frac{f'(x)}{g'(y)}$$

Solutions to Self-Check Exercises 4.4

1. Differentiating both sides of the equation with respect to x, we have

$$3x^2 + 3y + 3xy' + 3y^2y' = 0$$
$$(x^2 + y) + (x + y^2)y' = 0$$
$$y' = -\frac{x^2 + y}{x + y^2}$$

2. To find the slope of the tangent line to the graph of the function at any point, we differentiate the equation implicitly with respect to x, obtaining

$$32x + 18yy' = 0$$
$$y' = -\frac{16x}{9y}$$

In particular, the slope of the tangent line at $\left(2, -\frac{4\sqrt{5}}{3}\right)$ is

$$m = -\frac{16(2)}{9\left(-\frac{4\sqrt{5}}{3}\right)} = \frac{8}{3\sqrt{5}}$$

Using the point-slope form of the equation of a line, we find

$$y - \left(-\frac{4\sqrt{5}}{3}\right) = \frac{8}{3\sqrt{5}}(x - 2)$$
$$y = \frac{8\sqrt{5}}{15}x - \frac{36\sqrt{5}}{15} = \frac{8\sqrt{5}}{15}x - \frac{12\sqrt{5}}{5}$$

4.5 Differentials and Linear Approximation

The Millers are planning to buy a house in the near future and estimate that they will need a 30-year fixed-rate mortgage of \$120,000. If the interest rate increases from the present rate of 6% per year to 6.25% per year between now and the time the Millers decide to secure the loan, approximately how much more per month will their mortgage be? (See Section 14.4.)

Questions such as this, in which one wishes to *estimate* the change in the dependent variable (monthly mortgage payment) corresponding to a small change in the independent variable (interest rate per year), occur in many real-life applications. For example:

- An economist would like to know how a small increase in a country's capital expenditure will affect the country's gross domestic output.
- A sociologist would like to know how a small increase in the amount of capital investment in a housing project will affect the crime rate.
- A businesswoman would like to know how raising a product's unit price by a small amount will affect her profit.
- A bacteriologist would like to know how a small increase in the amount of a bactericide will affect a population of bacteria.

To calculate these changes and estimate their effects, we use the *differential* of a function, a concept that will be introduced shortly.

Increments

Let x denote a variable quantity and suppose x changes from x_1 to x_2. This change in x is called the **increment in x** and is denoted by the symbol Δx (read "delta x"). Thus,

$$\Delta x = x_2 - x_1 \quad \text{Final value} - \text{initial value} \tag{6}$$

Example 1

Find the increment in x as x changes (a) from 3 to 3.2 and (b) from 3 to 2.7.

Solution

a. Here, $x_1 = 3$ and $x_2 = 3.2$, so

$$\Delta x = x_2 - x_1 = 3.2 - 3 = 0.2$$

b. Here, $x_1 = 3$ and $x_2 = 2.7$. Therefore,

$$\Delta x = x_2 - x_1 = 2.7 - 3 = -0.3$$

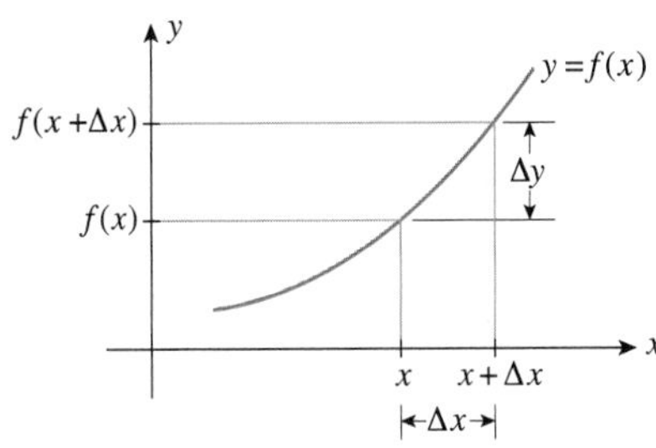

Figure 4.21
An increment of Δx in x induces an increment of $\Delta y = f(x + \Delta x) - f(x)$ in y.

Observe that Δx plays the same role that h played in Section 3.1.

Now, suppose two quantities, x and y, are related by an equation $y = f(x)$, where f is a function. If x changes from x to $x + \Delta x$, then the corresponding change in y is called the increment in y. It is denoted by Δy and is defined in Figure 4.21 by

$$\Delta y = f(x + \Delta x) - f(x) \qquad (7)$$

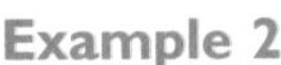

Example 2

Let $y = x^3$. Find Δx and Δy when x changes (a) from 2 to 2.01 and (b) from 2 to 1.98.

Solution

Let $f(x) = x^3$.

a. Here, $\Delta x = 2.01 - 2 = 0.01$. Next,

$$\Delta y = f(x + \Delta x) - f(x) = f(2.01) - f(2)$$

$$= (2.01)^3 - 2^3 = 8.120601 - 8 = 0.120601$$

b. Here, $\Delta x = 1.98 - 2 = -0.02$. Next,

$$\Delta y = f(x + \Delta x) - f(x) = f(1.98) - f(2)$$

$$= (1.98)^3 - 2^3 = 7.762392 - 8 = -0.237608$$

Differentials

We can obtain a relatively quick and simple way of approximating Δy, the change in y due to a small change Δx, by examining the graph of the function f shown in Figure 4.22.

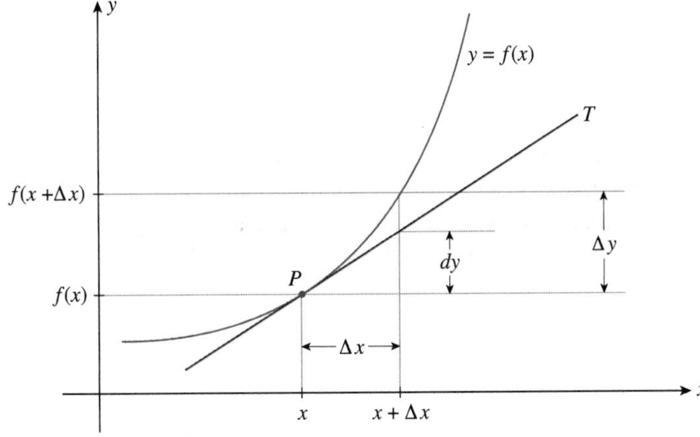

Figure 4.22
If Δx is small, dy is a good approximation of Δy.

Observe that near the point of tangency P, the tangent line T is close to the graph of f. Therefore, if Δx is small, then dy is a good approximation of Δy. We can find an expression for dy as follows: Notice that the slope of T is given by

$$\frac{dy}{\Delta x} \qquad \text{Rise} \div \text{run}$$

However, the slope of T is given by $f'(x)$. Therefore, we have

$$\frac{dy}{\Delta x} = f'(x)$$

or $dy = f'(x)\Delta x$. Thus, we have the approximation

$$\Delta y \approx dy = f'(x)\Delta x \tag{8}$$

in terms of the derivative of f at x. The quantity dy is called the *differential of y*.

The Differential

Let $y = f(x)$ define a differentiable function of x. Then,

1. The **differential** dx of the independent variable x is $dx = \Delta x$.
2. The **differential** dy of the dependent variable y is

$$dy = f'(x)\Delta x = f'(x)dx \tag{9}$$

Remarks

1. For the independent variable x: There is no difference between Δx and dx—both measure the change in x from x to $x + \Delta x$.
2. For the dependent variable y: Δy measures the *actual* change in y as x changes from x to $x + \Delta x$, whereas dy measures the *approximate* change in y corresponding to the same change in x.
3. The differential dy depends on both x and dx, but for fixed x, dy is a linear function of dx. ◀

As noted in the introduction to this section, differentials are used to estimate the change in the dependent variable corresponding to a small change in the independent variable. The following is a summary of how differentials are used in marginal analysis of cost, revenue, and profit functions.

Differentials in Marginal Analysis

Provided that the cost function $y = C(x)$, the revenue function $y = R(x)$, and the profit function $y = P(x)$ are differentiable, then

$$dC = C'(x)dx \qquad dR = R'(x)dx \qquad dP = P'(x)dx$$
$$\Delta C \approx C'(x)\Delta x \qquad \Delta R \approx R'(x)\Delta x \qquad \Delta P \approx P'(x)\Delta x$$

Example 3

Let $y = x^3$.

a. Find the differential dy of y.
b. Use dy to approximate Δy when x changes from 2 to 2.01.
c. Use dy to approximate Δy when x changes from 2 to 1.98.
d. Compare the results of part (b) with those of Example 2.

Solution

a. Let $f(x) = x^3$. Then,

$$dy = f'(x)\,dx = 3x^2dx$$

b. Here, $x = 2$ and $dx = 2.01 - 2 = 0.01$. Therefore,

$$dy = 3x^2dx = 3(2)^2(0.01) = 0.12$$

c. Here, $x = 2$ and $dx = 1.98 - 2 = -0.02$. Therefore,

$$dy = 3x^2dx = 3(2)^2(-0.02) = -0.24$$

d. As you can see, both approximations 0.12 and -0.24 are quite close to the actual changes of Δy obtained in Example 2: 0.120601 and -0.237608.

Observe how much easier it is to find an approximation to the exact change in a function with the help of the differential, rather than calculating the exact change in the function itself. In the following examples, we take advantage of this fact.

Example 4

Approximate the value of $\sqrt{26.5}$ using differentials. Verify your result using the $\boxed{\sqrt{\ }}$ key on your calculator.

Solution

Since we want to compute the square root of a number, let's consider the function $y = f(x) = \sqrt{x}$. Since 25 is the number nearest 26.5 whose square root is readily recognized, let's take $x = 25$. We want to know the change in y, Δy, as x changes from $x = 25$ to $x = 26.5$, an increase of $\Delta x = 1.5$ units. Using Equation (9), we find

$$\begin{aligned}\Delta y \approx dy &= f'(x)\Delta x \\ &= \left(\frac{1}{2\sqrt{x}}\bigg|_{x=25}\right) \cdot (1.5) = \left(\frac{1}{10}\right)(1.5) = 0.15\end{aligned}$$

Therefore,

$$\begin{aligned}\sqrt{26.5} - \sqrt{25} &= \Delta y \approx 0.15 \\ \sqrt{26.5} &\approx \sqrt{25} + 0.15 = 5.15\end{aligned}$$

The exact value of $\sqrt{26.5}$, rounded off to five decimal places, is 5.14782. Thus, the error incurred in the approximation is 0.00218.

Notice that in Example 4, we are using Formula (8) in a particular way. We used the fact that

$$\Delta y \approx dy = f'(x)\Delta x \quad \text{near } x \text{, and} \quad \Delta y = f(x + \Delta x) - f(x)$$

and wrote

$$f(x + \Delta x) - f(x) \approx f'(x)\Delta x$$

From this it follows that

$$f(x + \Delta x) \approx f(x) + f'(x)\Delta x$$

This is often written as

$$f(x) \approx f(a) + f'(a)(x - a) \tag{10}$$

where $(a, f(a))$ (formerly $(x, f(x))$) is the point of tangency referred to in Figure 4.22 and $f(x)$ (formerly $f(x + \Delta x)$) is the function value we want to approximate. Formula (10) basically says that we can approximate $f(x)$ with $f(a) + f'(a)(x - a)$ as long as the x-value we choose is near a. Notice also, as mentioned earlier, that the expression $f(a) + f'(a)(x - a)$ is linear in x. Therefore, Formula (10) is called the **linear approximation** of f at a, and the function defined as $L(x) = f(a) + f'(a)(x - a)$ is called the **linearization** of f at a.

Linear Approximation

The linearization of f at $x = a$ is given by

$$L(x) = f(a) + f'(a)(x - a) \tag{11}$$

and the linear approximation of f at $x = a$ is given by

$$f(x) \approx L(x) = f(a) + f'(a)(x - a) \tag{12}$$

provided that f is differentiable at $x = a$.

Example 5

Find the linear approximation of $f(x) = \sqrt{x + 1}$ at $a = 0$. Would you use it to approximate $x = \frac{3}{2}$ or $x = \frac{2}{3}$? Why?

Solution

We use Formula (11). For this, we need to find the derivative of f at $a = 0$.

$$f'(x) = \frac{1}{2\sqrt{x+1}} \qquad \text{and so}$$

$$f'(0) = \frac{1}{2\sqrt{0+1}} = \frac{1}{2}$$

Next, we evaluate the function f at $a = 0$.

$$f(0) = \sqrt{0+1} = 1$$

Finally,

$$L(x) = f(0) + f'(0)(x-0) = 1 + \frac{1}{2}x$$

By Formula (12) we have

$$f(x) \approx L(x) = 1 + \frac{1}{2}x$$

Since $x = \frac{2}{3}$ is closer to $a = 0$ than $x = \frac{3}{2}$, we should approximate $f(\frac{2}{3})$. Therefore,

$$f\left(\frac{2}{3}\right) \approx 1 + \frac{1}{2}\cdot\frac{2}{3} = \frac{4}{3} = 1.\bar{3}$$

Compare this to the first few digits of the actual value of $f(\frac{2}{3}) = 1.2909\ldots$ and we notice that the approximation is only accurate to the unit digit.

Applications

Example 6

The Effect of Speed on Vehicular Operating Cost The total cost incurred in operating a certain type of truck on a 500-kilometre trip, travelling at an average speed of v km/h, is estimated to be

$$C(v) = 125 + v + \frac{7200}{v}$$

dollars. Find the approximate change in the total operating cost when the average speed is increased from 80 km/h to 85 km/h.

Solution

With $v = 80$ and $\Delta v = dv = 5$, we find

$$\Delta C \approx dC = C'(v)dv = \left(1 - \frac{7200}{v^2}\right)\Bigg|_{v=80} \cdot (5)$$

$$= \left(1 - \frac{7200}{6400}\right)(5) \approx -0.63$$

so the total operating cost is found to decrease by $0.63. This might explain why so many independent truckers often exceed the 80 kmh speed limit.

Example 7

The Effect of Advertising on Sales The relationship between the amount of money x spent by Cannon Precision Instruments on advertising and Cannon's total sales $S(x)$ is given by the function

$$S(x) = -0.002x^3 + 0.6x^2 + x + 500 \qquad (0 \le x \le 200)$$

where x is measured in thousands of dollars. Use differentials to estimate the change in Cannon's total sales if advertising expenditures are increased from \$100,000 ($x = 100$) to \$105,000 ($x = 105$).

Solution

The required change in sales is given by

$$\begin{aligned} \Delta S \approx dS &= S'(100)dx \\ &= -0.006x^2 + 1.2x + 1|_{x=100} \cdot (5) \qquad dx = 105 - 100 = 5 \\ &= (-60 + 120 + 1)(5) = 305 \end{aligned}$$

—that is, an increase of \$305,000.

Example 8

The Rings of Neptune

a. A ring has an inner radius of r units and an outer radius of R units, where $(R - r)$ is small in comparison to r (Figure 4.23a). Use differentials to estimate the area of the ring.

b. Recent observations, including those of *Voyager I* and *II,* showed that Neptune's ring system is considerably more complex than had been believed. For one thing, it is made up of a large number of distinguishable rings rather than one continuous great ring as previously thought (Figure 4.23b). The outermost ring, 1989N1R, has an inner radius of approximately 62,900 kilometres (measured from the centre of the planet), and a radial width of approximately 50 kilometres. Using these data, estimate the area of the ring.

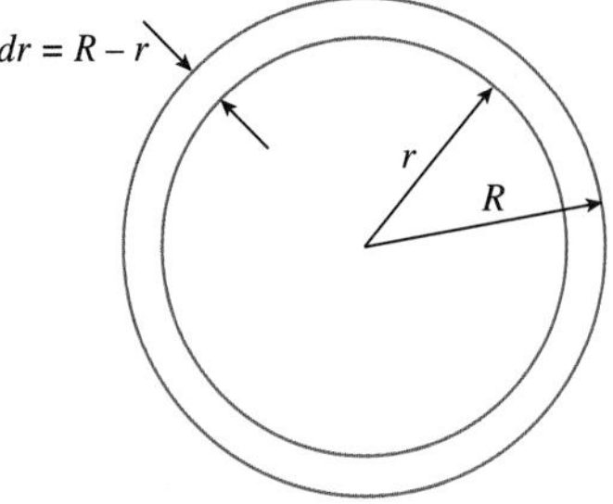

(a) The area of the ring is the circumference of the inner circle times the thickness.

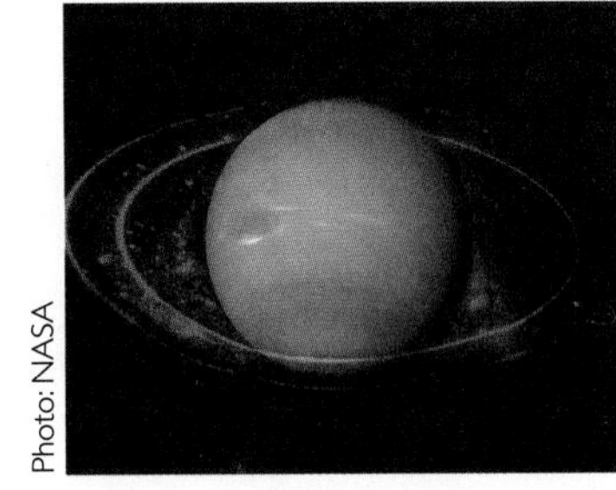

(b) Neptune and its rings

Figure 4.23

Solution

a. Using the fact that the area of a circle of radius x is $A = f(x) = px^2$, we find

$$\begin{aligned} \pi R^2 - \pi r^2 &= f(R) - f(r) \\ &= \Delta A \qquad \text{Remember, } \Delta A = \text{change in } f \text{ when } x \text{ changes from } x = r \text{ to } x = R. \\ &\approx dA \\ &= f'(r)dr \end{aligned}$$

where $dr = R - r$. So, we see that the area of the ring is approximately $2\pi r(R - r)$ square units. In words, the area of the ring is approximately equal to

Circumference of the inner circle $\times$ Thickness of the ring

b. Applying the results of part (a) with $r = 62{,}900$ and $dr = 50$, we find that the area of the ring is approximately $2\pi(62{,}900)(50)$, or 19,760,618 square kilometres, which is roughly 4% of Earth's surface.

Before looking at the next example, we need to familiarize ourselves with some terminology regarding how error is measured when working with differentials (Table 4.1). Suppose we are given a function f, then f has exact value $f(a)$ for $x = a$. If f at $x = a$ is measured or calculated with an error of Δf, then the quantity $\Delta f/f(a)$ is called the *relative error* in the measurement or calculation of $f(a)$. If the quantity $\Delta f/f(a)$ is expressed as a percentage, it then called the *percentage error*. Because Δf is approximated by df, we normally approximate the relative error $\Delta f/f(a)$ by $df/f(a)$.

Table 4.1

Error	True Value	Approximate Value
Absolute	Δf	df
Relative	$\dfrac{\Delta f}{f(a)}$	$\dfrac{df}{f(a)}$
Percentage	$\dfrac{\Delta f}{f(a)} \cdot 100\%$	$\dfrac{df}{f(a)} \cdot 100\%$

Example 9

Estimating Errors in Measurement Suppose the radius of a ball-bearing is measured to be 0.5 cm, with a maximum error of ± 0.0002 cm. Then, the relative error in r is

$$\frac{dr}{r} = \frac{\pm 0.0002}{0.5} = \pm 0.0004$$

and the percentage error is $\pm 0.04\%$.

Example 10

Estimating Errors in Measurement Suppose the side of a cube is measured with a maximum percentage error of 2%. Use differentials to estimate the maximum percentage error in the calculated volume of the cube.

Solution

Suppose the side of the cube is x, so its volume is

$$V = x^3$$

We are given that $\left|\frac{dx}{x}\right| \le 0.02$. Now,

$$dV = 3x^2 dx$$

and so

$$\frac{dV}{V} = \frac{3x^2 dx}{x^3} = 3\frac{dx}{x}$$

Therefore,

$$\left|\frac{dV}{V}\right| = 3\left|\frac{dx}{x}\right| \le 3(0.02) = 0.06$$

and we see that the maximum percentage error in the measurement of the volume of the cube is 6%.

Finally, we want to point out that if at some point in reading this section you have a sense of déjà vu, do not be surprised, because the notion of the differential was first used in Section 4.1 (see Example 1). There we took $\Delta x = 1$ since we were interested in finding the marginal cost when the level of production was increased from $x = 250$ to $x = 251$. If we had used differentials, we would have found

$$C(251) - C(250) \approx C'(250)dx$$

so that taking $dx = \Delta x = 1$, we have $C(251) - C(250) \approx C'(250)$, which agrees with the result obtained in Example 1. Thus, in Section 4.1, we touched upon the notion of the differential, albeit in the special case in which $dx = 1$.

Self-Check Exercises 4.5

1. Find the differential of $f(x) = \sqrt{x} + 1$.
2. A certain country's government economists have determined that the demand equation for corn in that country is given by

$$p = f(x) = \frac{125}{x^2 + 1}$$

where p is expressed in dollars/bushel and x, the quantity demanded each year, is measured in billions of bushels. The economists are forecasting a harvest of 6 billion bushels for the year. If the actual production of corn were 6.2 billion bushels for the year instead, what would be the approximate drop in the predicted price of corn/bushel?

Solutions to Self-Check Exercises 4.5 can be found on page 243.

4.5 Exercises

In Exercises 1–14, find the differential of the given function.

1. $f(x) = 2x^2$
2. $f(x) = 3x^2 + 1$
3. $f(x) = x^3 - x$
4. $f(x) = 2x^3 + x$
5. $f(x) = \sqrt{x + 1}$
6. $f(x) = \dfrac{3}{\sqrt{x}}$
7. $f(x) = 2x^{3/2} + x^{1/2}$
8. $f(x) = 3x^{5/6} + 7x^{2/3}$
9. $f(x) = x + \dfrac{2}{x}$
10. $f(x) = \dfrac{3}{x - 1}$
11. $f(x) = \dfrac{x - 1}{x^2 + 1}$
12. $f(x) = \dfrac{2x^2 + 1}{x + 1}$
13. $f(x) = \sqrt{3x^2 - x}$
14. $f(x) = (2x^2 + 3)^{1/3}$

In Exercises 15–20, find the linear approximation $L(x)$ of the given function at a, and use it to approximate the function value at the given x-value.

15. $f(x) = \sqrt{x}$, $a = 9$, $x = 10$
16. $f(x) = \sqrt[3]{x}$, $a = -27$, $x = -28$
17. $f(x) = \dfrac{1}{2}$, $a = 3$, $x = 3.2$
18. $f(x) = \dfrac{1}{x^2}$, $a = 2$, $x = 2.05$
19. $f(x) = x^2 + 3$, $a = 5$, $x = 5.65$
20. $f(x) = (x - 2)^3$, $a = 4$, $x = 3.95$
21. Let f be the function defined by

$$y = f(x) = x^2 - 1$$

   a. Find the differential of f.
   b. Use your result from part (a) to find the approximate change in y if x changes from 1 to 1.02.
   c. Find the actual change in y if x changes from 1 to 1.02 and compare your result with that obtained in part (b).

22. Let f be the function defined by

$$y = f(x) = 3x^2 - 2x + 6$$

   a. Find the differential of f.
   b. Use your result from part (a) to find the approximate change in y if x changes from 2 to 1.97.
   c. Find the actual change in y if x changes from 2 to 1.97 and compare your result with that obtained in part (b).

23. Let f be the function defined by

$$y = f(x) = \frac{1}{x}$$

   a. Find the differential of f.
   b. Use your result from part (a) to find the approximate change in y if x changes from -1 to -0.95.
   c. Find the actual change in y if x changes from -1 to -0.95 and compare your result with that obtained in part (b).

24. Let f be the function defined by

$$y = f(x) = \sqrt{2x + 1}$$

   a. Find the differential of f.
   b. Use your result from part (a) to find the approximate change in y if x changes from 4 to 4.1.
   c. Find the actual change in y if x changes from 4 to 4.1 and compare your result with that obtained in part (b).

In Exercises 25–32, use differentials to approximate the given quantity.

25. $\sqrt{10}$
26. $\sqrt{17}$
27. $\sqrt{49.5}$
28. $\sqrt{99.7}$
29. $\sqrt[3]{7.8}$
30. $\sqrt[4]{81.6}$
31. $\sqrt{0.089}$
32. $\sqrt[3]{0.00096}$
33. Use a differential to approximate $\sqrt{4.02} + \dfrac{1}{\sqrt{4.02}}$.

   Hint: Let $f(x) = \sqrt{x} + \dfrac{1}{\sqrt{x}}$ and compute dy with $x = 4$ and $dx = 0.02$.

34. Use a differential to approximate $\dfrac{2(4.98)}{(4.98)^2 + 1}$.

   Hint: Study the hint for Exercise 33.

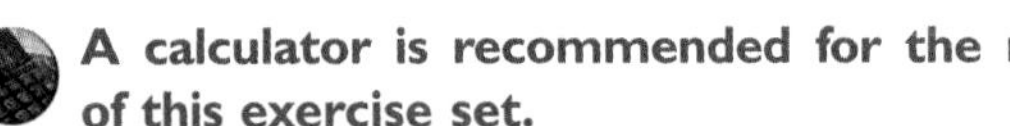

A calculator is recommended for the remainder of this exercise set.

35. **Error Estimation** The length of each edge of a cube is 12 cm, with a possible error in measurement of 0.02 cm. Use differentials to estimate the error that might occur when the volume of the cube is calculated.
36. **Estimating the Amount of Paint Required** A coat of paint of thickness 0.05 cm is to be applied uniformly to the faces of a cube of edge 30 cm. Use differentials to find the approximate amount of paint required for the job.
37. **Error Estimation** A hemisphere-shaped dome of radius 18 m is to be coated with a layer of rust-proofer before painting. Use differentials to estimate the amount of rust-proofer needed if the coat is to be 0.03 cm thick.

   Hint: The volume of a hemisphere of radius r is $V = \frac{2}{3}\pi r^3$.

Business and Economics Applications

38. **Gross Domestic Product** An economist has determined that a certain country's gross domestic product (GDP) is approximated by the function $f(x) = 640x^{1/5}$, where $f(x)$ is measured in

billions of dollars and x is the capital outlay in billions of dollars. Use differentials to estimate the change in the country's GDP if the country's capital expenditure changes from $243 billion to $248 billion.

39. **Effect of Advertising on Profits** The relationship between Cunningham Realty's quarterly profits, $P(x)$, and the amount of money x spent on advertising per quarter is described by the function

$$P(x) = -\frac{1}{8}x^2 + 7x + 30 \qquad (0 \le x \le 50)$$

where both $P(x)$ and x are measured in thousands of dollars. Use differentials to estimate the increase in profits when advertising expenditure each quarter is increased from $24,000 to $26,000.

40. **Effect of Mortgage Rates on Housing Starts** A study prepared for the National Association of Realtors estimates that the number of housing starts per year over the next 5 yr will be

$$N(r) = \frac{7}{1 + 0.02r^2}$$

million units, where r (percent) is the mortgage rate. Use differentials to estimate the decrease in the number of housing starts when the mortgage rate is increased from 12% to 12.5%.

41. **Supply-Price** The supply equation for a certain brand of transistor radio is given by

$$p = s(x) = 0.3\sqrt{x} + 10$$

where x is the quantity supplied and p is the unit price in dollars. Use differentials to approximate the change in price when the quantity supplied is increased from 10,000 units to 10,500 units.

42. **Demand-Price** The demand function for the Sentinel smoke alarm is given by

$$p = d(x) = \frac{30}{0.02x^2 + 1}$$

where x is the quantity demanded (in units of a thousand) and p is the unit price in dollars. Use differentials to estimate the change in the price p when the quantity demanded changes from 5000 to 5500 units/week.

43. **Forecasting Profits** The management of Trappee and Sons forecast that they will sell 200,000 cases of their TexaPep hot sauce next year. Their annual profit is described by

$$P(x) = -0.000032x^3 + 6x - 100$$

thousand dollars, where x is measured in thousands of cases. If the maximum error in the forecast is 15%, determine the corresponding error in Trappee's profits.

44. **Forecasting Commodity Prices** A certain country's government economists have determined that the demand equation for soybeans in that country is given by

$$p = f(x) = \frac{55}{2x^2 + 1}$$

where p is expressed in dollars/bushel and x, the quantity demanded each year, is measured in millions of bushels. The economists are forecasting a harvest of 1.8 million bushels for the year, with a maximum error of 15% in their forecast. Determine the corresponding maximum error in the predicted price per bushel of soybeans.

45. **Financing a Home** The Millers are planning to buy a home in the near future and estimate that they will need a 30-yr fixed-rate mortgage for $120,000. Their monthly payment P (in dollars) can be computed using the formula

$$P = \frac{10{,}000r}{1 - \left(1 + \frac{r}{12}\right)^{-360}}$$

where r is the interest rate per year.

a. Find the differential of P.

b. If the interest rate increases from the present rate of 9%/year to 9.2%/year between now and the time the Millers decide to secure the loan, approximately how much more will their monthly mortgage payment be? How much more will it be if the interest rate increases to 9.3%/year? To 9.4%/year? To 9.5%/year?

46. **Investments** Lupé deposits a sum of $10,000 into an account that pays interest at the rate of r/year compounded monthly. Her investment at the end of 10 yr is given by

$$A = 10{,}000\left(1 + \frac{r}{12}\right)^{120}$$

a. Find the differential of A.

b. Approximately how much more would Lupé's account be worth at the end of the term if her account paid 8.1%/year instead of 8%/year? 8.2%/year instead of 8%/year? 8.3%/year instead of 8%/year?

47. **Keogh Accounts** Ian, who is self-employed, contributes $2000 a month into a Keogh account earning interest at the rate of r/year compounded monthly. At the end of 25 yr, his account will be worth

$$S = \frac{24{,}000\left[\left(1 + \frac{r}{12}\right)^{300} - 1\right]}{r}$$

dollars.

a. Find the differential of S.

b. Approximately how much more would Ian's account be worth at the end of 25 yr if his account earned 9.1%/year instead of 9%/year? 9.2%/year instead of 9%/year? 9.3%/year instead of 9%/year?

Biological and Life Sciences Applications

48. **Growth of a Cancerous Tumour** The volume of a spherical cancerous tumour is given by

$$V(r) = \frac{4}{3}\pi r^3$$

If the radius of a tumour is estimated at 1.1 cm, with a maximum error in measurement of 0.005 cm, determine the error that might occur when the volume of the tumour is calculated.

49. **Unclogging Arteries** Research done in the 1930s by the French physiologist Jean Poiseuille showed that the resistance R of a blood vessel of length l and radius r is $R = kl/r^4$, where k is a constant. Suppose a dose of the drug TPA increases r by 10%. How will this affect the resistance R? Assume that l is constant.

50. **Learning Curves** The length of time (in seconds) a certain individual takes to learn a list of n items is approximated by

$$f(n) = 4n\sqrt{n - 4}$$

Use differentials to approximate the additional time it takes the individual to learn the items on a list when n is increased from 85 to 90 items.

51. **Surface Area of an Animal** Animal physiologists use the formula

$$S = kW^{2/3}$$

to calculate an animal's surface area (in square metres) from its weight W (in kilograms), where k is a constant that depends on the animal under consideration. Suppose a physiologist calculates the surface area of a horse ($k = 0.1$). If the horse's weight is estimated at 300 kg, with a maximum error in measurement of 0.6 kg, determine the percentage error in the calculation of the horse's surface area.

52. **Crime Studies** A sociologist has found that the number of serious crimes in a certain city each year is described by the function

$$N(x) = \frac{500(400 + 20x)^{1/2}}{5 + 0.2x^2}$$

where x (in cents/dollar deposited) is the level of reinvestment in the area in conventional mortgages by the city's ten largest banks. Use differentials to estimate the change in the number of crimes if the level of reinvestment changes from 20¢/dollar deposited to 22¢/dollar deposited.

In Exercises 53 and 54, determine whether the statement is true or false. If it is true, explain why it is true. If it is false, give an example to show why it is false.

53. If $y = ax + b$ where a and b are constants, then $\Delta y = dy$.

54. If $A = f(x)$, then the percentage change in A is

$$\frac{100f'(x)}{f(x)}dx$$

Solutions to Self-Check Exercises 4.5

1. We find

$$f'(x) = \frac{1}{2}x^{-1/2} = \frac{1}{2\sqrt{x}}$$

Therefore, the required differential of f is

$$dy = \frac{1}{2\sqrt{x}}dx$$

2. We first compute the differential

$$dp = -\frac{250x}{(x^2 + 1)^2}dx$$

Next, using Equation (8) with $x = 6$ and $dx = 0.2$, we find

$$\Delta p \approx dp = -\frac{250(6)}{(36 + 1)^2}(0.2) = -0.22$$

or a drop in price of 22¢/bushel.

4.6 The Newton–Raphson Method

The Newton–Raphson Method (Optional)

Previously we have had occasion to find the zeros of a function f or, equivalently, the **roots** of the equation $f(x) = 0$. For example, the x-intercepts of a function f are precisely the values of x when $f(x) = 0$; the critical points of f include the roots of the equation $f'(x) = 0$; and the candidates for the inflection points of f include the roots of the equation $f''(x) = 0$.

If $f(x)$ is a linear or quadratic function or $f(x)$ is a polynomial that is easily factored, the roots of f are readily found. In practice, however, we often encounter functions with zeros that cannot be found as readily. For example, the function

$$f(t) = -0.3t^3 + 28.8t^2 + 58.5t + 1.5$$

of Example 9, page 240, that gives the altitude (in metres) of a rocket t seconds into flight, is not easily factored, so its zeros cannot be found by elementary algebraic methods. (We will find the zeros of this function in Example 2.) Another example of a function with roots that are not easily found is

$$g(x) = e^{2x} - 3x - 2$$

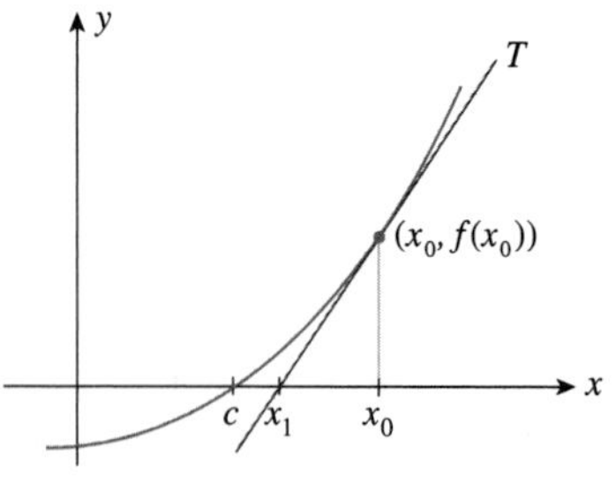

Figure 4.24

x_0 is our first guess in estimating the value of c, the zero of $f(x)$.

This type of function involves an exponential function, which we will discuss in Chapter 5.

In this section we develop an algorithm, based on the approximation of a function $f(x)$ by a first-degree Taylor polynomial, to approximate the value of a zero of $f(x)$ to any desired degree of accuracy.

Suppose the function f has a zero at $x = c$ (Figure 4.24). Let x_0 be an initial estimate of the actual zero c of $f(x)$. Now, the first-degree Taylor polynomial of $f(x)$ at $x = x_0$ is

$$P_1(x) = f(x_0) + f'(x_0)(x - x_0)$$

and, as observed earlier, is just the equation of the tangent line to the graph of $y = f(x)$ at the point $(x_0, f(x_0))$. Since the tangent line T is an approximation of the graph of $y = f(x)$ near $x = x_0$ (and x_0 is assumed to be close to c!), it may be expected that the zero of $P_1(x)$ is close to the zero $x = c$ of $f(x)$. But the zero of $P_1(x)$, a linear function, is found by setting $P_1(x) = 0$. Thus,

$$f(x_0) + f'(x_0)(x - x_0) = 0$$

$$f'(x_0)(x - x_0) = -f(x_0)$$

$$x - x_0 = -\frac{f(x_0)}{f'(x_0)}$$

$$x = x_0 - \frac{f(x_0)}{f'(x_0)}$$

This number provides us with another estimate of the zero of $f(x)$:

$$x_1 = x_0 - \frac{f(x_0)}{f'(x_0)}$$

In general, the estimate x_1 is better than the initial estimate x_0.

This process may be repeated with the initial estimate x_0 replaced by the recent estimate x_1. This leads to yet another estimate,

$$x_2 = x_1 - \frac{f(x_1)}{f'(x_1)}$$

which is usually better than x_1. In this manner, we generate a sequence of approximations $x_0, x_1, x_2, \ldots, x_n, x_{n+1}, \ldots$, with

$$x_{n+1} = x_n - \frac{f(x_n)}{f'(x_n)}$$

which, in most instances, approaches the zero $x = c$ of $f(x)$.

A summary of this algorithm follows.

Newton–Raphson Algorithm

1. Pick an initial estimate x_0 of the root c.
2. Find a new estimate using the iterative formula

$$x_{n+1} = x_n - \frac{f(x_n)}{f'(x_n)} \qquad (n = 0, 1, 2, \ldots) \qquad \textbf{(13)}$$

3. Compute $|x_n - x_{n+1}|$. If this number is less than a prescribed positive number, stop. The required approximation to the root $x = c$ is $x = x_{n+1}$.

Solving Equations Using the Newton–Raphson Method

Example 1

Use the Newton–Raphson algorithm to approximate the zero of $f(x) = x^2 - 2$. Start the iteration with initial guess $x_0 = 1$ and terminate the process when two successive approximations differ by less than 0.00001.

Solution

We have

$$f(x) = x^2 - 2f'(x) = 2x$$

so that, by (13), the required iterative formula is

$$x_{n+1} = x_n - \frac{x_n^2 - 2}{2x_n} = \frac{x_n^2 + 2}{2x_n}$$

With $x_0 = 1$, we find

$$x_1 = \frac{1^2 + 2}{2(1)} = 1.5$$

$$x_2 = \frac{(1.5)^2 + 2}{2(1.5)} \approx 1.416667$$

$$x_3 = \frac{(1.416667)^2 + 2}{2(1.416667)} \approx 1.414216$$

$$x_4 = \frac{(1.414216)^2 + 2}{2(1.414216)} \approx 1.414214$$

Since $x_3 - x_4 = 0.000002 < 0.00001$, we terminate the process. The sequence generated converges to $\sqrt{2}$, which is one of the two roots of the equation $x^2 - 2 = 0$. Note that, to six places, $\sqrt{2} = 1.414214$!

Example 2

Flight of a Rocket Refer to Example 9, page 240. The altitude in feet of a rocket t seconds into flight is given by

$$s = f(t) = -0.3t^3 + 28.8t^2 + 58.5t + 1.5 \qquad (t \geq 0)$$

Find the time T when the rocket hits Earth.

Solution

The rocket hits Earth when the altitude is equal to zero. So we are required to solve the equation

$$s = f(t) = -0.3t^3 + 28.8t^2 + 58.5t + 1.5 = 0$$

Let's use the Newton–Raphson algorithm with initial guess $t_0 = 100$ (Figure 4.25). Here

$$f'(t) = -0.9t^2 + 57.6t + 58.5$$

so the required iterative formula takes the form

$$t_{n+1} = t_n - \frac{-0.3t_n^3 + 28.8t_n^2 + 58.5t_n + 1.5}{-0.9t_n^2 + 57.6t_n + 58.5}$$

$$= t_n + \frac{-0.3t_n^3 + 28.8t_n^2 + 58.5t_n + 1.5}{0.9t_n^2 - 57.6t_n - 58.5}$$

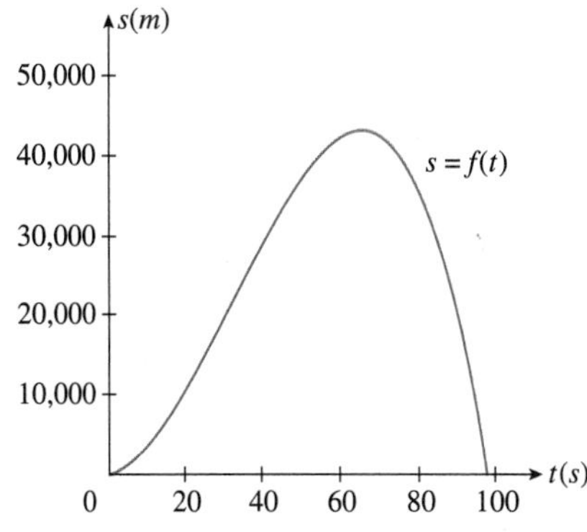

Figure 4.25
The rocket's altitude t seconds into flight is given by $f(t)$.

$$= \frac{0.9t_n^3 - 57.6t_n^2 - 58.5t_n - 0.3t_n^3 + 28.8t_n^2 + 58.5t_n + 1.5}{0.9t_n^2 - 57.6t_n - 58.5}$$

$$= \frac{0.6t_n^3 - 28.8t_n^2 + 1.5}{0.9t_n^t - 57.6t_n - 58.5}$$

We have

$$t_0 = 100$$

$$t_1 = \frac{0.6(100)^3 - 28.8(100)^2 + 1.5}{0.9(100)^2 - 57.6(100) - 58.5} \approx 98.067421$$

$$t_2 = \frac{0.6(98.067421)^3 - 28.8(98.067421)^2 + 1.5}{0.9(98.067421)^2 - 57.6(98.067421) - 58.5} \approx 97.990629$$

$$t_3 = \frac{0.6(97.9990629)^3 - 28.8(97.990629)^2 + 1.5}{0.9(97.990629)^2 - 57.6(97.990629) - 58.5} \approx 97.990509$$

Thus, we see that $T \approx 97.990509$, so that rocket hits Earth approximately 98 seconds after liftoff.

The Internal Rate of Return on an Investment

Yet another use of the Newton–Raphson algorithm is in finding a quantity called the internal rate of return on an investment. Suppose a company has an initial outlay of C dollars in an investment that yields returns of $R_1, R_2, \ldots, R_n$ dollars at the end of the first, second, . . . , nth periods, respectively. Then the *net present value* of the investment is

$$\frac{R_1}{1+r} + \frac{R_2}{(1+r)^2} + \frac{R_3}{(1+r)^3} + \cdots + \frac{R_n}{(1+r)^n} - C$$

where r denotes the interest rate per period earned on the investment. Now the **internal rate of return** on the investment is defined as the rate of return for which the net present value of the investment is equal to zero; that is, it is the value of r that satisfies the equation

$$\frac{R_1}{1+r} + \frac{R_2}{(1+r)^2} + \frac{R_3}{(1+r)^3} + \cdots + \frac{R_n}{(1+r)^n} - C = 0$$

or, equivalently, the equation

$$C(1+r)^n - R_1(1+r)^{n-1} - R_2(1+r)^{n-2} - R_3(1+r)^{n-3} - \cdots - R_n = 0$$

obtained by multiplying both sides of the former by $(1+r)^n$. The internal rate of return is used by management to decide on the worthiness or profitability of an investment.

Example 3

Internal Rate of Return The management of A-1 Rental—a tool and equipment rental service for industry, contractors, and homeowners—is contemplating purchasing new equipment. The initial outlay for the equipment, which has a useful life of 4 years, is $45,000. It is expected that the investment will yield returns of $15,000 at the end of the first year, $18,000 at the end of the second year, $14,000 at the end of the third year, and $10,000 at the end of the fourth year. Find the internal rate of return on this investment.

Solution

Here $n = 4$, $C = 45{,}000$, $R_1 = 15{,}000$, $R_2 = 18{,}000$, $R_3 = 14{,}000$, and $R_4 = 10{,}000$. So we are required to solve the equation

$$45{,}000(1 + r)^4 - 15{,}000(1 + r)^3 - 18{,}000(1 + r)^2 - 14{,}000(1 + r) - 10{,}000 = 0$$

for r. The equation may be written more simply by letting $x = 1 + r$. Thus,

$$f(x) = 45{,}000x^4 - 15{,}000x^3 - 18{,}000x^2 - 14{,}000x - 10{,}000 = 0$$

To solve the equation $f(x) = 0$, using the Newton–Raphson algorithm, we first compute

$$f'(x) = 180{,}000x^3 - 45{,}000x^2 - 36{,}000x - 14{,}000$$

The required iterative formula is

$$x_{n+1} = x_n - \frac{45{,}000x_n^4 - 15{,}000x_n^4 - 18{,}000x_n^2 - 14{,}000(x)_n - 10{,}000}{180{,}000x_n^3 - 45{,}000x_n^2 - 36{,}000x_n - 14{,}000}$$

Starting with the initial estimate $x_0 = 1.1$, we find

$$x_1 = 1.1 - \frac{45{,}000(1.1)^4 - 15{,}000(1.1)^3 - 18{,}000(1.1)^2 - 14{,}000x_n - 10{,}000}{180{,}000(1.1)^3 - 45{,}000(1.1)^2 - 36{,}000(1.1) - 14{,}000}$$

$$\approx 1.10958$$

$$x_2 \approx 1.10941$$

$$x_3 \approx 1.10941$$

Therefore, we may take $x \approx 1.1094$, in which case we see that $r \approx 0.1094$. Thus, the rate of return on the investment is approximately 10.94% per year.

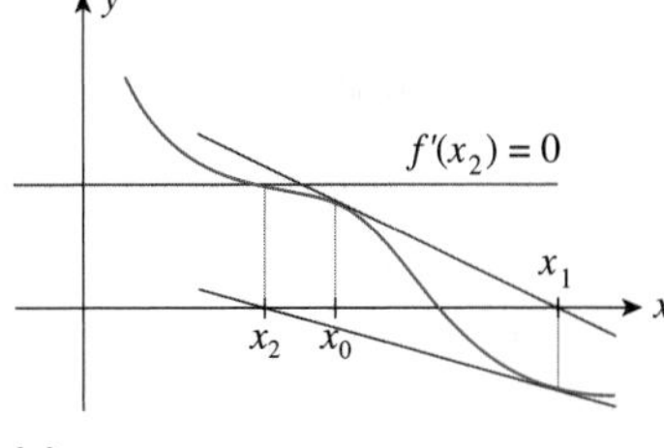

(a)

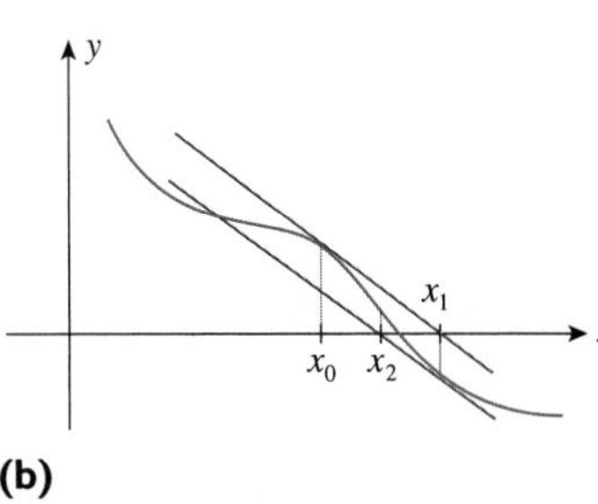

(b)

Figure 4.26

In **(a)**, the Newton–Raphson method fails to work because $f'(x_2) = 0$, but this situation is remedied in **(b)** by selecting a different initial estimate x_0.

Having seen how effective the Newton–Raphson method can be in finding the zeros of a function, we want to close this section by reminding you that there are situations in which the method fails and that care must be exercised in applying it. Figure 4.26a illustrates a situation where $f'(x_n) = 0$ for some n (in this case, $n = 2$). Since the iterative Formula (13) involves division by $f'(x_n)$, it should be clear why the method fails to work in this case. However, if you choose a different initial estimate x_0, the situation may yet be salvaged (Figure 4.26b).

GROUP DISCUSSION

For a concrete example of a situation similar to that depicted in Figure 4.26, consider the function

$$f(x) = x^3 - 1.5x^2 - 6x + 2$$

1. Show that the Newton–Raphson method fails to work if we choose $x_0 = -1$ or $x_0 = 2$ for an initial estimate.
2. Using the initial estimates $x_0 = -2.5$, $x_0 = 1$, and $x_0 = 2.5$, show that the three roots of $f(x) = 0$ are -2, 0.313859, and 3.186141, respectively.
3. Using a graphing utility, plot the graph of f in the viewing window $[-3, 4] \cdot [-10, 7]$. Verify the results of part 2, using **TRACE** and **ZOOM** or the root-finding function of your calculator.

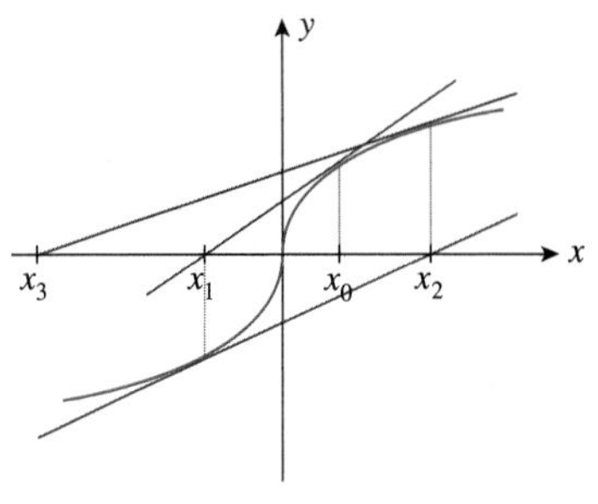

Figure 4.27

The Newton–Raphson method fails here because the sequence of estimates diverges.

The next situation, shown in Figure 4.27, is more serious, and the method will not work for any choice of the initial estimate x_0 other than the actual zero of the function $f(x)$ because the sequence, $x_1, x_2, \ldots x_n$ diverges.

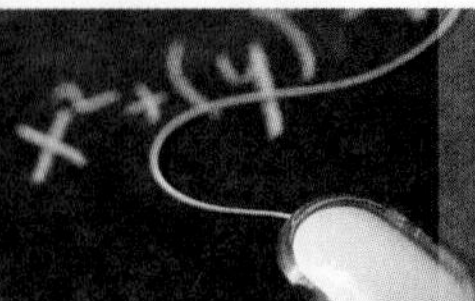

Self-Check Exercises 4.6

1. Use three iterations of the Newton–Raphson algorithm on an appropriate function $f(x)$ and initial guess x_0 to obtain an estimate of $\sqrt[3]{10}$.

2. A study projects that a certain town's population in the next 3 yr will grow according to the rule

$$P(t) = 50{,}000 + 30t^{3/2} + 20t$$

where $P(t)$ denotes the population t mo from now. Using this model, estimate the time when the population will reach 55,000.

Solutions to Self-Check Exercises 4.6 can be found on page 249.

4.6 Exercises

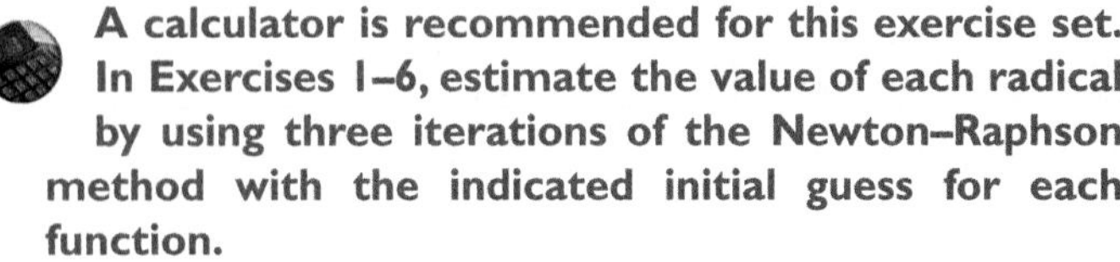

A calculator is recommended for this exercise set. In Exercises 1–6, estimate the value of each radical by using three iterations of the Newton–Raphson method with the indicated initial guess for each function.

1. $\sqrt{3}; f(x) = x^2 - 3; x_0 = 1.5$

2. $\sqrt{5}; f(x) = x^2 - 5; x_0 = 2$

3. $\sqrt{7}; f(x) = x^2 - 7; x_0 = 2.5$

4. $\sqrt[3]{6}; f(x) = x^3 - 6; x_0 = 2$

5. $\sqrt[3]{14}; f(x) = x^3 - 14; x_0 = 2.5$

6. $\sqrt[4]{50}; f(x) = x^4 - 50; x_0 = 2.5$

In Exercises 7–11, use the Newton–Raphson method to approximate the indicated zero of each function. Continue with the iteration until two successive approximations differ by less than 0.0001.

7. The zero of $f(x) = x^2 - x - 3$ between $x = 2$ and $x = 3$

8. The zero of $f(x) = x^3 + x - 3$ between $x = 1$ and $x = 2$

9. The zero of $f(x) = x^3 + 2x^2 + x - 5$ between $x = 1$ and $x = 2$

10. The zero of $f(x) = x^5 + x - 1$ between $x = 0$ and $x = 1$

11. The zero of $f(x) = \sqrt{x + 1} - x$ between $x = 1$ and $x = 2$

12. Let $f(x) = 2x^3 - 9x^2 + 12x - 2$.
    a. Show that $f(x) = 0$ has a root between $x = 0$ and $x = 1$.
    Hint: Compute $f(0)$ and $f(1)$ and use the fact that f is continuous.
    b. Use the Newton–Raphson method to find the zero of f in the interval (0, 1).

13. Let $f(x) = x^3 - x - 1$.
    a. Show that $f(x) = 0$ has a root between $x = 1$ and $x = 2$.
    Hint: See Exercise 12.
    b. Use the Newton–Raphson method to find the zero of f in the interval (1, 2).

14. Let $f(x) = x^3 - 3x - 1$.
    a. Show that f has a zero between $x = 1$ and $x = 2$.
    Hint: See Exercise 12.
    b. Use the Newton–Raphson method to find the zero of f.

15. Let $f(x) = x^4 - 4x^3 + 10$.
    a. Show that $f(x) = 0$ has a root between $x = 1$ and $x = 2$.
    b. Use the Newton–Raphson method to find the zero of f.

16. Apply the Newton–Raphson method to an appropriate function $f(x)$ to obtain an estimate of $\sqrt[3]{12}$.

17. Use the Newton–Raphson method to find the root of the equation $x^3 + x^2 - 1 = 0$ between $x = 0$ and $x = 1$.

In Exercises 18–20, make a rough sketch of the graphs of each of the given pairs of functions. Use your sketch to approximate the point(s) of intersection of the two graphs and then apply the Newton–Raphson method to refine the approximation of the x-coordinate of the point of intersection.

18. $f(x) = 2\sqrt{x + 3}; g(x) = 2x - 1$

19. $f(x) = \sqrt{x}; g(x) = e^{-x}$

20. $f(x) = 2 - x^2; g(x) = \ln x$

21. a. Show that using the Newton–Raphson method for finding the nth root of a real number a leads to the iterative formula

$$x_{i+1} = \left(\frac{n-1}{n}\right)x_i + \frac{a}{nx_i^{n-1}}$$

    b. Use the result of part (a) to find $\sqrt[4]{42}$, accurate to three decimal places.

Business and Economics Applications

22. **Minimizing Average Cost** A division of Ditton Industries manufactures the "Futura" model microwave oven. Given that the daily cost of producing these microwave ovens (in dollars) obeys the rule

$$C(x) = 0.0002x^3 - 0.06x^2 + 120x + 5000$$

where x stands for the number of units produced, find the level of production that minimizes the daily average cost per unit.

23. **Internal Rate of Return on an Investment** The proprietor of Qwik Film Lab recently purchased $12,000 of new film-processing equipment. She expects that this investment, which has a useful life of 4 yr, will yield returns of $4000 at the end of the first year, $5000 at the end of the second year, $4000 at the end of the third year, and $3000 at the end of the fourth year. Find the internal rate of return on the investment.

24. **Internal Rate of Return on an Investment** Executive Limousine Service recently acquired limousines worth \$120,000. The projected returns over the next 3 yr, the time period the limousines will be in service, are \$80,000 at the end of the first year, \$60,000 at the end of the second year, and \$40,000 at the end of the third year. Find the internal rate of return on the investment.

25. **Internal Rate of Return on an Investment** Suppose an initial outlay of \$$C$ in an investment yields returns of \$$R$ at the end of each period over N periods.
   a. Show that the internal rate of return on the investment, r, may be obtained by solving the equation

$$Cr + R[(1 + r)^{-N} - 1] = 0$$

Hint: $1 + x + x^2 + \cdots + x^{N-1} = \dfrac{1 - x^{N+1}}{1 - x}$

   b. Show that r can be found by performing the iteration

$$r_{n+1} = r_n - \frac{Cr_n + R[(1 + r_n)^{-N} - 1]}{C - NR(1 + r_n)^{-N-1}}$$

[r_0 (positive), an initial guess]

Hint: Apply the Newton–Raphson method to the function $f(r) = Cr + R[(1 + r)^{-N} - 1]$.

26. **Home Mortgages** Refer to Exercise 23. The Flemings secured a loan of \$100,000 from a bank to finance the purchase of a house. They have agreed to repay the loan in equal monthly installments of \$1053 over 25 yr. The bank charges interest at the rate of $12r$/year on the unpaid balance, and interest computations are made at the end of each month. Find r.

27. **Home Mortgages** The Blakelys borrowed a sum of \$80,000 from a bank to help finance the purchase of a house. The bank charges interest at the rate of $12r$/year on the unpaid balance, with interest being computed at the end of each month. The Blakelys have agreed to repay the loan in equal monthly installments of \$643.70 over 30 yr. What is the true rate of interest charged by the bank?
Hint: See Exercise 23.

28. **Car Loans** The price of a certain new car is \$8000. Suppose an individual makes a down payment of 25% toward the purchase of the car and secures financing for the balance over 4 yr. If the monthly payment is \$152.18, what is the true rate of interest charged by the finance company?
Hint: See Exercise 23.

29. **Real Estate Investment Groups** Refer to Exercise 23. A group of private investors purchased a condominium complex for \$2 million. They made an initial down payment of 10% and have obtained financing for the balance. If the loan is amortized over 15 yr with quarterly repayments of \$65,039, determine the interest rate charged by the bank. Assume that interest is calculated at the end of each quarter and is based on the unpaid balance.

30. **Demand for Wristwatches** The quantity of "Sicard" wristwatches demanded per month is related to the unit price by the equation

$$p = d(x) = \frac{50}{0.01x^2 + 1} \qquad (1 \le x \le 20)$$

where p is measured in dollars and x is measured in units of a thousand. The supplier is willing to make x thousand wristwatches available per month when the price per watch is given by $p = s(x) = 0.1x + 20$ dollars. Find the equilibrium quantity and price.

31. **Demand for TV Sets** The weekly demand for the Pulsar 25-in. colour console television is given by the demand equation

$$p = -0.05x + 600 \qquad (0 \le x \le 6000)$$

where p denotes the wholesale unit price in dollars and x denotes the quantity demanded. The weekly total cost function associated with the manufacture of these sets is given by

$$C = 0.000002x^3 - 0.03x^2 + 400x + 80{,}000$$

where $C(x)$ denotes the total cost incurred in the production of x sets. Find the break-even level(s) of operation for the company.
Hint: Solve the equation $P(x) = 0$, where P is the total profit function.

Biological and Life Sciences Applications

32. **Altitude of a Rocket** The altitude (in metres) of a rocket t sec into flight is given by

$$s = f(t) = -2t^3 + 114t^2 + 480t + 1 \qquad (t \ge 0)$$

Find the time T when the rocket hits Earth.

33. **Temperature** The temperature at 6 A.M. on a certain December day was measured at 15.6°C. In the next t hr, the temperature was given by the function

$$T = -0.05t^3 + 0.4t^2 + 3.8t \qquad (0 \le t \le 15)$$

where T is measured in degrees Celsius. At what time was the temperature 0°C?

Solutions to Self-Check Exercises 4.6

1. Since $\sqrt[3]{10}$ is one of the cube roots of the equation $x^3 - 10 = 0$, let's take $f(x) = x^3 - 10$. We have

$$f'(x) = 3x^2$$

and so the required iteration formula is

$$x_{n+1} = x_n - \frac{f(x_n)}{f'(x_n)} = x_n - \frac{x_n^3 - 10}{3x_n^2} = \frac{2x_n^3 + 10}{3x_n^2}$$

Solutions to Self-Check Exercises 4.6, *continued*

Taking $x_0 = 2$, we find

$$x_1 = \frac{2(2^3) + 10}{3(2^2)} \approx 2.166667$$

$$x_2 = \frac{2(2.166667)^3 + 10}{3(2.166667)^2} \approx 2.154504$$

$$x_3 = \frac{2(2.154504)^3 + 10}{3(2.154504)^2} \approx 2.154435$$

Therefore, $\sqrt[3]{10} \approx 2.1544$.

2. We need to solve the equation $P(t) = 55{,}000$ or

$$50{,}000 + 30t^{3/2} + 20t = 55{,}000$$

Rewriting, we have

$$30t^{3/2} + 20t - 5000 = 0$$
$$3t^{3/2} + 2t - 500 = 0$$

Thus, the problem reduces to that of finding the zero of the function

$$f(t) = 3t^{3/2} + 2t - 500$$

Using the Newton–Raphson method, we have

$$f'(t) = \frac{9}{2}t^{1/2} + 2 = \frac{9t^{1/2} + 4}{2}$$

So the required iteration formula is

$$\begin{aligned} t_{n+1} &= t_n - \frac{3t_n^{3/2} + 2t_n - 500}{\frac{9t_n^{1/2} + 4}{2}} \\ &= t_n - \frac{6t_n^{3/2} + 4t_n - 1000}{9t_n^{1/2} + 4} \\ &= \frac{9t_n^{3/2} + 4t_n - (6t_n^{3/2} + 4t_n - 1000)}{9t_n^{1/2} + 4} \\ &= \frac{3t_n^{3/2} + 1000}{9t_n^{1/2} + 4} \end{aligned}$$

Using the initial guess $t_0 = 24$, we find

$$t_1 = \frac{3(24)^{3/2} + 1000}{9(24)^{1/2} + 4} \approx 28.128584$$

$$t_2 = \frac{3(28.128584)^{3/2} + 1000}{9(28.128584)^{1/2} + 4} \approx 27.981338$$

$$t_3 = \frac{3(27.981338)^{3/2} + 1000}{9(27.981338)^{1/2} + 4} \approx 27.981160$$

So we may take $t \approx 27.98$. Thus, the population will reach 55,000 approximately 28 mo from now.

CHAPTER 4 Summary of Principal Formulas and Terms

 Formulas

1. Average cost function $\quad \overline{C}(x) = \dfrac{C(x)}{x}$
2. Revenue function $\quad R(x) = px$
3. Profit function $\quad P(x) = R(x) - C(x)$
4. Elasticity of demand $\quad E(p) = -\dfrac{pf'(p)}{f(p)}$
5. Differential of y $\quad dy = f'(x)dx$
6. Linearization $\quad L(x) = f(a) + f'(a)(x - a)$ of f near a
7. Linear approximation $\quad f(x) \approx f(a) + f'(a)(x - a)$ of f near a
8. Iterative formula for the Newton–Raphson algorithm

$$x_{n+1} = \frac{x_n - f(x_n)}{f'(x_n)}, \; x_0 \text{ initial value } (n = 0, 1, 2, \ldots)$$

Terms

marginal cost (199)
marginal cost function (199)
average cost (200)
average cost function (200)
marginal average cost function (200)
marginal revenue (202)
marginal revenue function (203)
marginal profit function (203)
elastic demand (207)
unitary demand (207)
inelastic demand (207)
second derivative of f (219)
implicit differentiation (225)
related rates (228)
increment in x (234)
increment in y (235)
differential dx (236)
differential dy (236)
linear approximation (237)
linearization (237)
roots (243)
internal rate of return (246)

Review Exercises 

In Exercises 1–6, match the graph of the function with the graph of its derivative below.

a.
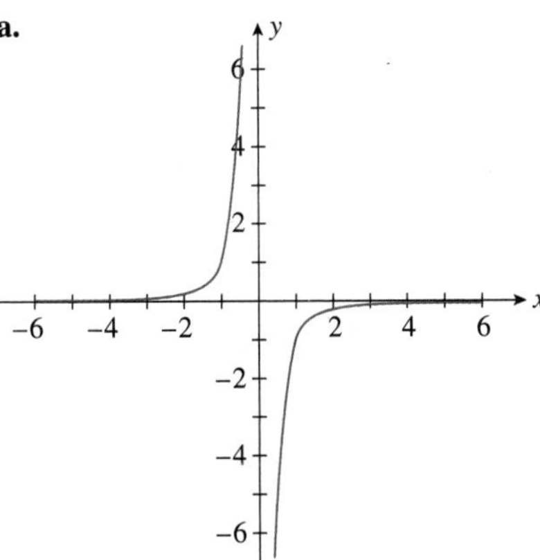

b.
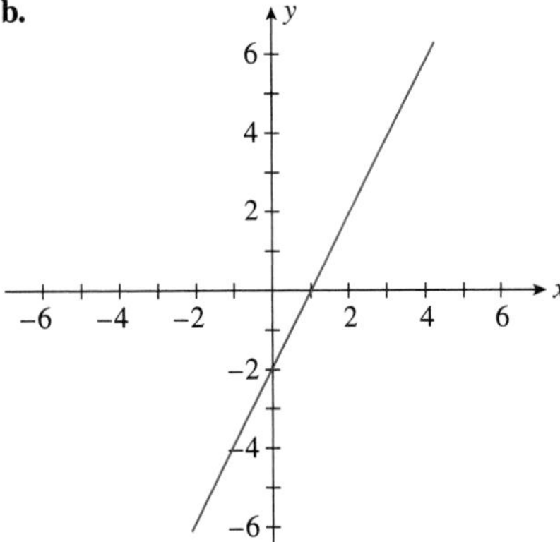

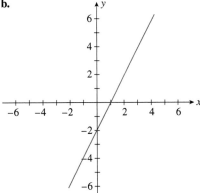

c.
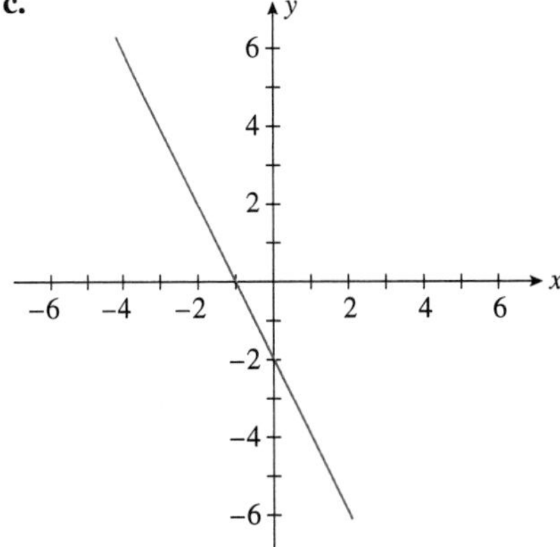

d.
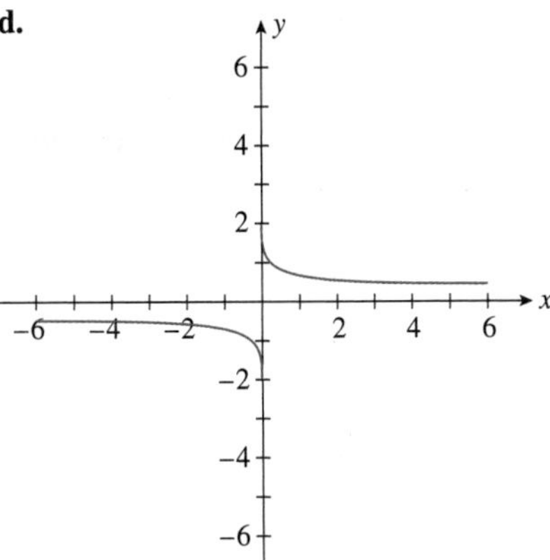

e.
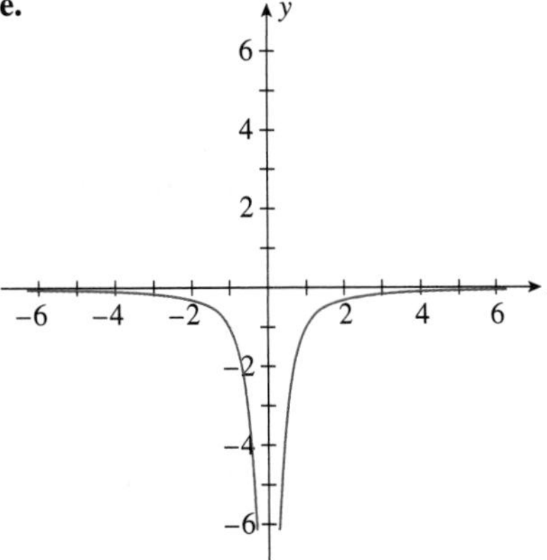

f.
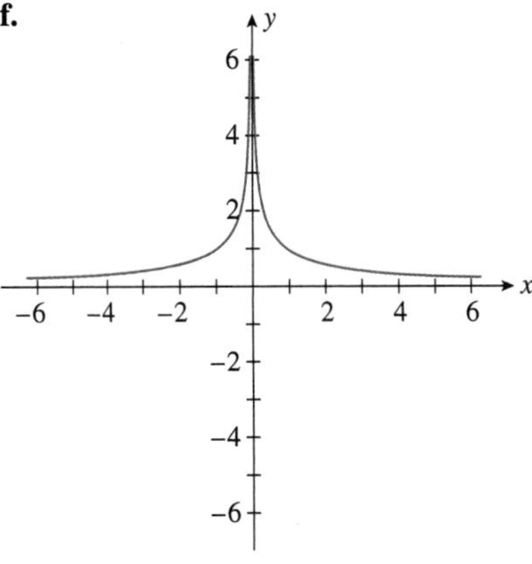

1.
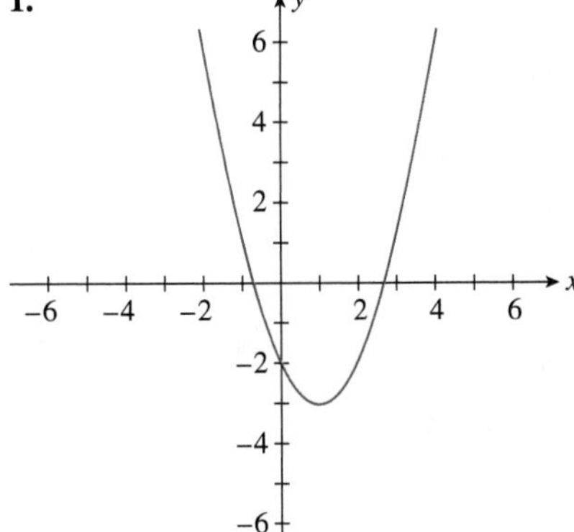

2.
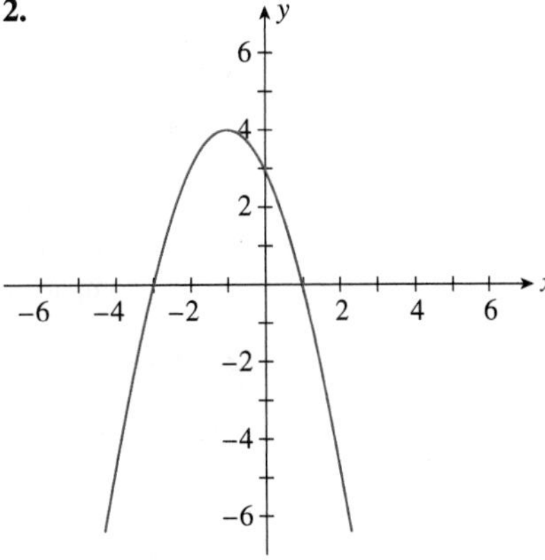

3.
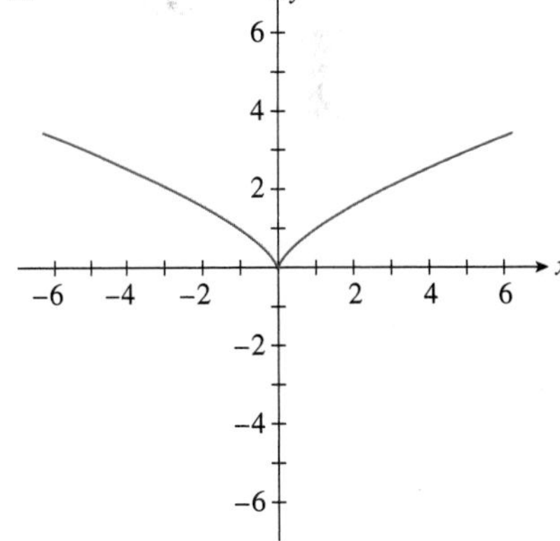

4.
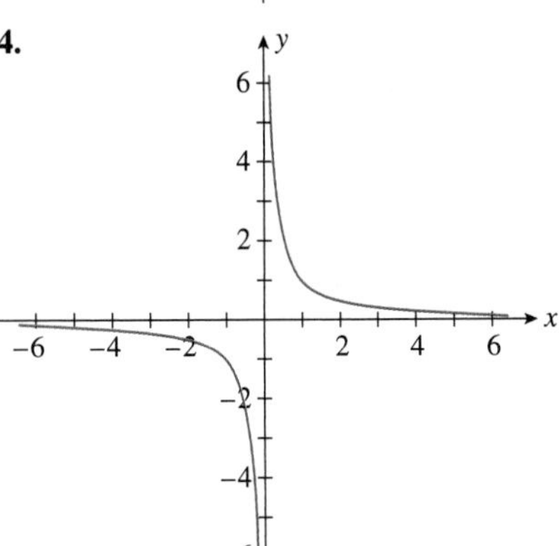

5.
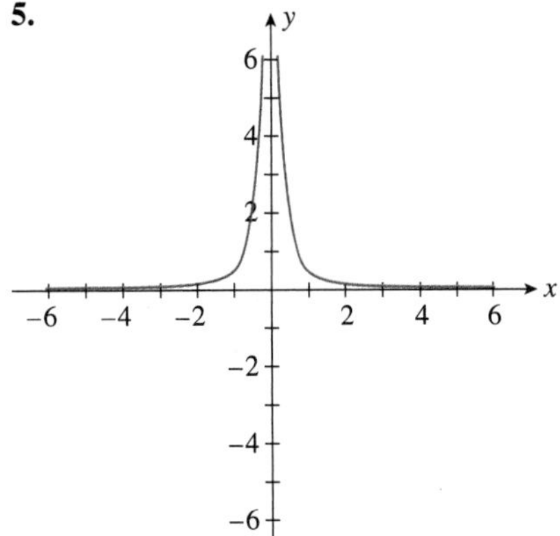

6.
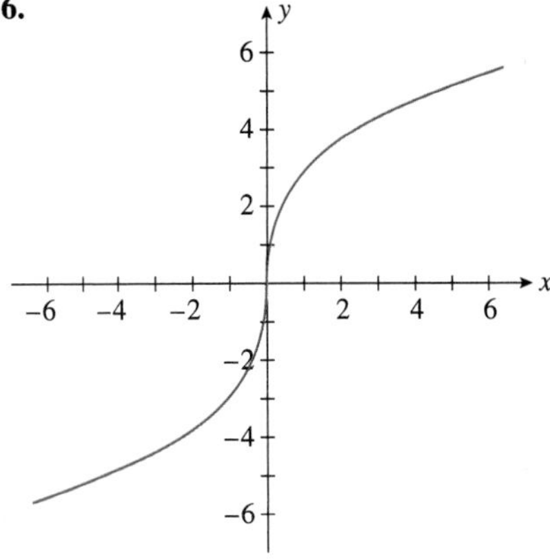

In Exercises 7–10, use the graph of $y = f(x)$ below to graph its derivative.

7.

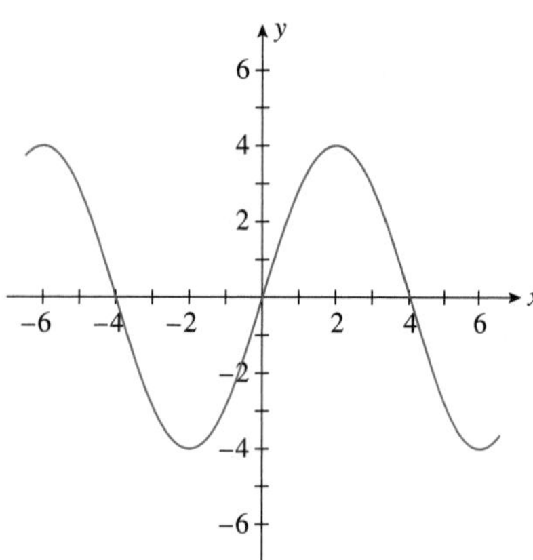

8.

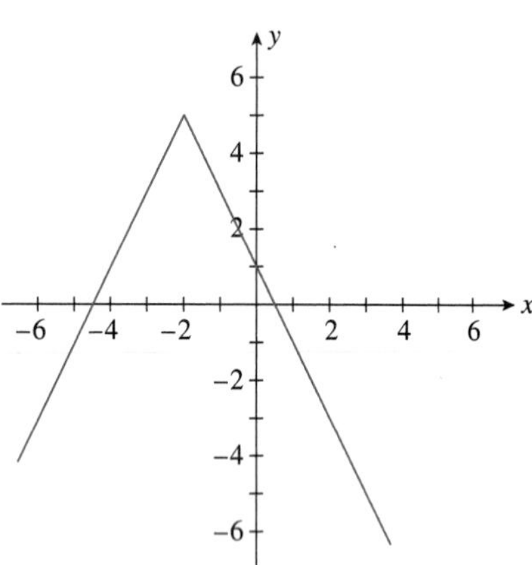

9.

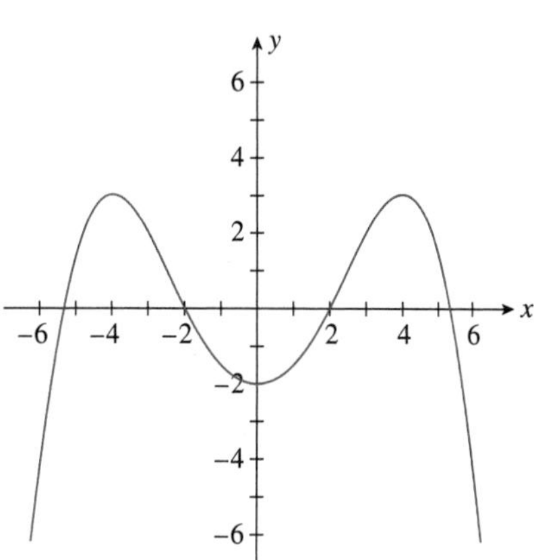

10.

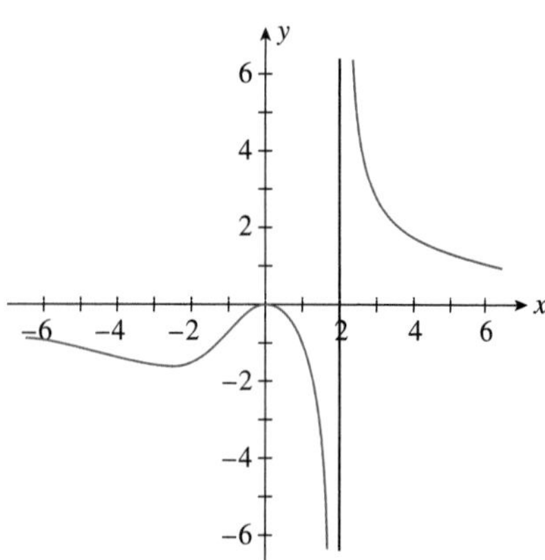

In Exercises 11–16, find the second derivative of the given function.

11. $f(x) = 2x^4 - 3x^3 + 2x^2 + x + 4$

12. $g(x) = \sqrt{x} + \dfrac{1}{\sqrt{x}}$

13. $h(t) = \dfrac{t}{t^2 + 4}$

14. $f(x) = (x^3 + x + 1)^2$

15. $f(x) = \sqrt{2x^2 + 1}$

16. $f(t) = t(t^2 + 1)^3$

In Exercises 17–20, find the linear approximation $L(x)$ of the given function at a, and use it to approximate the function value at the given x-value.

17. $f(x) = x^{2/3}$, $a = 27$, $x = 26$

18. $f(x) = \sqrt{x} - x$, $a = 16$, $x = 15$

19. $f(x) = \frac{4}{x - 1}$, $a = 3$, $x = 2.9$

20. $f(x) = x^2 + 3x - 1$, $a = 5$, $x = 5.2$

In Exercises 21–26, find dy/dx by implicit differentiation.

21. $6x^2 - 3y^2 = 9$

22. $2x^3 - 3xy = 4$

23. $y^3 + 3x^2 = 3y$

24. $x^2 + 2x^2y^2 + y^2 = 10$

25. $x^2 - 4xy - y^2 = 12$

26. $3x^2y - 4xy + x - 2y = 6$

27. Find the differential of $f(x) = x^2 + \dfrac{1}{x^2}$.

28. Find the differential of $f(x) = \dfrac{1}{\sqrt{x^3 + 1}}$.

29. Let f be the function defined by $f(x) = \sqrt{2x^2 + 4}$.

a. Find the differential of f.

b. Use your result from part (a) to find the approximate change in $y = f(x)$ if x changes from 4 to 4.1.

c. Find the actual change in y if x changes from 4 to 4.1 and compare your result with that obtained in part (b).

30. Use a differential to approximate $\sqrt[3]{26.8}$.

31. Find the third derivative of the function

$$f(x) = \frac{1}{2x - 1}$$

What is its domain?

Business and Economics Applications

32. The demand equation for a certain product is $2x + 5p - 60 = 0$, where p is the unit price and x is the quantity demanded of the product. Find the elasticity of demand and determine whether the demand is elastic or inelastic, at the indicated prices.

a. $p = 3$ **b.** $p = 6$ **c.** $p = 9$

33. The demand equation for a certain product is

$$x = \frac{25}{\sqrt{p}} - 1$$

where p is the unit price and x is the quantity demanded for the product. Compute the elasticity of demand and determine the range of prices corresponding to inelastic, unitary, and elastic demand.

34. The demand equation for a certain product is $x = 100 - 0.01p^2$.
a. Is the demand elastic, unitary, or inelastic when $p = 40$?
b. If the price is \$40, will raising the price slightly cause the revenue to increase or decrease?

35. The demand equation for a certain product is $p = 9\sqrt[3]{1000 - x}$.
a. Is the demand elastic, unitary, or inelastic when $p = 60$?
b. If the price is \$60, will raising the price slightly cause the revenue to increase or decrease?

36. Cost of Wireless Phone Calls As cellular phone usage continues to soar, the airtime costs have dropped. The average price per minute of use (in cents) can be projected to be

$$f(t) = 31.88(1 + t)^{-0.45} \qquad (0 \le t \le 6)$$

where t is measured in years and $t = 0$ corresponds to the beginning of 1998. Compute $f'(t)$. How fast was the average price/minute of use changing at the beginning of 2000? What was the average price/minute of use at the beginning of 2000?

37. Cost of Pressing DVDs The total weekly cost in dollars incurred by Herald Records in pressing x DVDs is given by the total cost function

$$C(x) = 2500 + 2.2x \qquad (0 \le x \le 8000)$$

a. What is the marginal cost when $x = 1000$ and 2000?
b. Find the average cost function $\overline{C}$ and the marginal average cost function $\overline{C}'$.
c. Using the results from part (b), show that the average cost incurred by Herald in pressing a DVD approaches \$2.20/ disc when the level of production is high enough.

38. Demand for Cordless Phones The marketing department of a province's telephone company has determined that the demand for their cordless phones obeys the relationship

$$p = -0.02x + 600 \qquad (0 \le x \le 30{,}000)$$

where p denotes the phone's unit price (in dollars) and x denotes the quantity demanded.
a. Find the revenue function R.
b. Find the marginal revenue function R'.
c. Compute $R'(10{,}000)$ and interpret your result.

39. Demand for Photocopying Machines The weekly demand for the LectroCopy photocopying machine is given by the demand equation

$$p = 2000 - 0.04x \qquad (0 \le x \le 10{,}000)$$

where p denotes the wholesale unit price in dollars and x denotes the quantity demanded. The weekly total cost function for manufacturing these copiers is given by

$$C(x) = 0.000002x^3 - 0.02x^2 + 1000x + 120{,}000$$

where $C(x)$ denotes the total cost incurred in producing x units.
a. Find the revenue function R, the profit function P, and the average cost function $\overline{C}$.
b. Find the marginal cost function C', the marginal revenue function R', the marginal profit function P', and the marginal average cost function $\overline{C}'$.
c. Compute $C'(3000)$, $R'(3000)$, and $P'(3000)$.
d. Compute $\overline{C}'(5000)$ and $\overline{C}'(8000)$ and interpret your results.

Biological and Life Sciences Applications

40. Worldwide Networked PCs The number of worldwide networked PCs (in millions) is given by

$$N(t) = 3.136t^2 + 3.954t + 116.468 \qquad (0 \le t \le 9)$$

where t is measured in years, with $t = 0$ corresponding to the beginning of 1991.
a. How many worldwide networked PCs were there at the beginning of 1997?
b. How fast was the number of worldwide networked PCs changing at the beginning of 1997?

41. Cable TV Subscribers The number of subscribers to a cable television company in the town of Squamish is approximated by the function

$$N(x) = 1000(1 + 2x)^{1/2} \qquad (1 \le x \le 30)$$

where $N(x)$ denotes the number of subscribers to the service in the xth week. Find the rate of increase in the number of subscribers at the end of the 12th week.

42. Male Life Expectancy Suppose the life expectancy of a male at birth in a certain country is described by the function

$$f(t) = 46.9(1 + 1.09t)^{0.1} \qquad (0 \le t \le 150)$$

where t is measured in years, with $t = 0$ corresponding to the beginning of 1900. How long can a male born at the beginning of 2000 in that country expect to live? What is the rate of change of the life expectancy of a male born in that country at the beginning of 2000?

EXPONENTIAL AND LOGARITHMIC FUNCTIONS

Photo: Thinkstock/Jupiter Images

The exponential function is, without doubt, the most important function in mathematics and its applications. After a brief introduction to the exponential function and its *inverse,* the logarithmic function, we learn how to differentiate such functions. This lays the foundation for exploring the many applications involving exponential functions. For example, we look at the role played by exponential functions in computing earned interest on a bank account and in studying the growth of a bacteria population in the laboratory, the way radioactive matter decays, the rate at which a factory worker learns a certain process, and the rate at which a communicable disease is spread over time.

Photo: Thinkstock/Jupiter Images

How many bacteria will there be in a culture at the end of a certain period of time? How fast will the bacteria population be growing at the end of that time? Example 1, page 290, answers these questions.

Photo: Geoff Tompkinson/Photo Researchers, Inc.

5.1 Exponential Functions

Exponential Functions and Their Graphs

Suppose you deposit \$1000 in an account earning interest at the rate of 10% per year *compounded continuously* (the way most financial institutions compute interest). Then, the accumulated amount at the end of t years $(0 \leq t \leq 20)$ is described by the function f, whose graph appears in Figure 5.1.* Such a function is called an *exponential function.* Observe that the graph of f rises rather slowly at first but very rapidly as time goes by. For purposes of comparison, we have also shown the graph of the function $y = g(t) = 1000\ (1 + 0.10t)$, giving the accumulated amount for the same principal (\$1000) but earning *simple* interest at the rate of 10% per year. The moral of the story: It is never too early to save.

Exponential functions play an important role in many real-world applications, as you will see throughout this chapter.

Observe that whenever b is a positive number and n is any real number, the expression b^n is a real number. This enables us to define an exponential function as follows:

Figure 5.1
Under continuous compounding, a sum of money grows exponentially.

> **Exponential Function**
> The function defined by
> $$f(x) = b^x \qquad (b > 0, b \neq 1)$$
> is called an **exponential function with base b and exponent x.** The domain of f is the set of all real numbers.

For example, the exponential function with base 2 is the function

$$f(x) = 2^x$$

with domain $(-\infty, \infty)$. The values of $f(x)$ for selected values of x follow:

$$f(3) = 2^3 = 8, \qquad f\left(\frac{3}{2}\right) = 2^{3/2} = 2 \cdot 2^{1/2} = 2\sqrt{2}, \qquad f(0) = 2^0 = 1$$

$$f(-1) = 2^{-1} = \frac{1}{2}, \qquad f\left(-\frac{2}{3}\right) = 2^{-2/3} = \frac{1}{2^{2/3}} = \frac{1}{\sqrt[3]{4}}$$

*We will derive the rule for f in Section 12.2.

Computations involving exponentials are facilitated by the laws of exponents. These laws were stated in Section 1.1, and you might want to review the material there. For convenience, however, we will restate these laws.

Laws of Exponents

Let a and b be positive numbers and let x and y be real numbers. Then,

1. $b^x \cdot b^y = b^{x+y}$

2. $\dfrac{b^x}{b^y} = b^{x-y}$

3. $(b^x)^y = b^{xy}$

4. $(ab)^x = a^x b^x$

5. $\left(\dfrac{a}{b}\right)^x = \dfrac{a^x}{b^x}$

The use of the laws of exponents is illustrated in the next example.

Example 1

a. $16^{7/4} \cdot 16^{-1/2} = 16^{7/4-1/2} = 16^{5/4} = 2^5 = 32$ Law 1

b. $\dfrac{8^{5/3}}{8^{-1/3}} = 8^{5/3-(-1/3)} = 8^2 = 64$ Law 2

c. $(64^{4/3})^{-1/2} = 64^{(4/3)(-1/2)} = 64^{-2/3}$

$$= \frac{1}{64^{2/3}} = \frac{1}{(64^{1/3})^2} = \frac{1}{4^2} = \frac{1}{16} \quad \text{Law 3}$$

d. $(16 \cdot 81)^{-1/4} = 16^{-1/4} \cdot 81^{-1/4} = \dfrac{1}{16^{1/4}} \cdot \dfrac{1}{81^{1/4}} = \dfrac{1}{2} \cdot \dfrac{1}{3} = \dfrac{1}{6}$ Law 4

e. $\left(\dfrac{3^{1/2}}{2^{1/3}}\right)^4 = \dfrac{3^{4/2}}{2^{4/3}} = \dfrac{9}{2^{4/3}}$ Law 5

Example 2

Let $f(x) = 2^{2x-1}$. Find the value of x for which $f(x) = 16$.

Solution

We want to solve the equation

$$2^{2x-1} = 16 = 2^4$$

But this equation holds if and only if

$$2x - 1 = 4 \qquad b^m = b^n \Leftrightarrow m = n$$

giving $x = \dfrac{5}{2}$.

Exponential functions play an important role in mathematical analysis. Because of their special characteristics, they are some of the most useful functions and are found in virtually every field where mathematics is applied. To mention a few examples: Under ideal conditions the number of bacteria present at any time t in a culture may be described by an exponential function of t; radioactive substances decay over time in accordance with an "exponential" law of decay; money left on fixed deposit and earning compound interest grows exponentially; and some of the most important distribution functions encountered in statistics are exponential.

Let's begin our investigation into the properties of exponential functions by studying their graphs.

Example 3

Sketch the graph of the exponential function $y = 2^x$.

Solution

First, as discussed earlier, the domain of the exponential function $y = f(x) = 2^x$ is the set of real numbers. Next, putting $x = 0$ gives $y = 2^0 = 1$, the y-intercept of f. There is no x-intercept since there is no value of x for which $y = 0$. To find the range of f, consider the following table of values:

x	−5	−4	−3	−2	−1	0	1	2	3	4	5
y	1/32	1/16	1/8	1/4	1/2	1	2	4	8	16	32

We see from these computations that 2^x decreases and approaches zero as x decreases without bound and that 2^x increases without bound as x increases without bound. Thus, the range of f is the interval $(0, \infty)$—that is, the set of positive real numbers. Finally, we sketch the graph of $y = f(x) = 2^x$ in Figure 5.2.

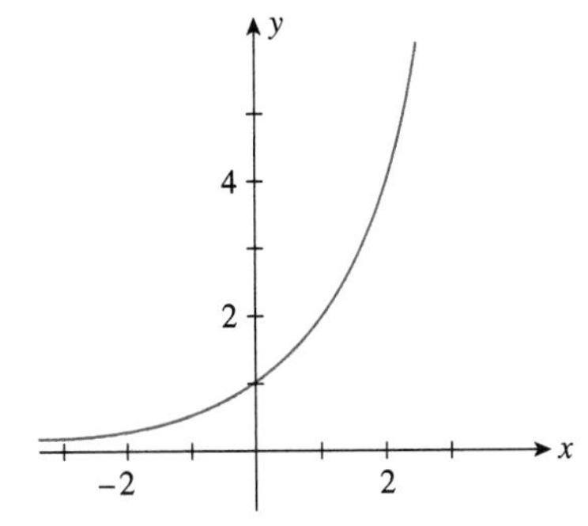

Figure 5.2
The graph of $y = 2^x$

Example 4

Sketch the graph of the exponential function $y = (1/2)^x$.

Solution

The domain of the exponential function $y = (1/2)^x$ is the set of all real numbers. The y-intercept is $(1/2)^0 = 1$; there is no x-intercept since there is no value of x for which $y = 0$. From the following table of values

x	−5	−4	−3	−2	−1	0	1	2	3	4	5
y	32	16	8	4	2	1	1/2	1/4	1/8	1/16	1/32

we deduce that $(1/2)^x = 1/2^x$ increases without bound as x decreases without bound and that $(1/2)^x$ decreases and approaches zero as x increases without bound. Thus, the range of f is the interval $(0, \infty)$. The graph of $y = f(x) = (1/2)^x$ is sketched in Figure 5.3.

The functions $y = 2^x$ and $y = (1/2)^x$, whose graphs you studied in Examples 3 and 4, are special cases of the exponential function $y = f(x) = b^x$, obtained by setting $b = 2$ and $b = 1/2$, respectively. In general, the exponential function $y = b^x$ with $b > 1$ has a graph similar to $y = 2^x$, whereas the graph of $y = b^x$ for $0 < b < 1$ is similar to that of $y = (1/2)^x$ (Exercises 27 and 28 on page 261). When $b = 1$, the function $y = b^x$ reduces to the constant function $y = 1$. For comparison, the graphs of all three functions are sketched in Figure 5.4.

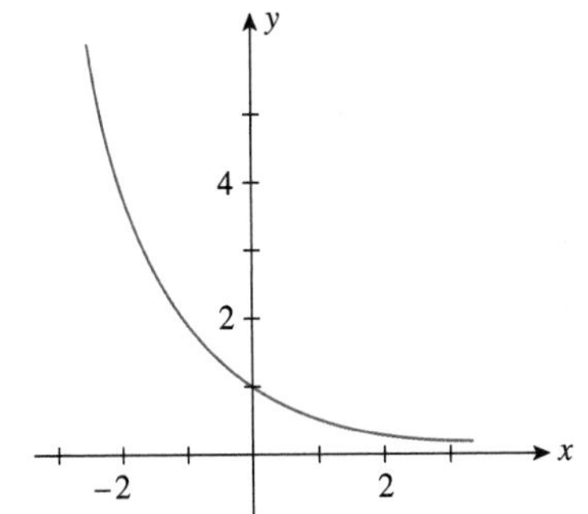

Figure 5.3
The graph of $y = \left(\frac{1}{2}\right)^x$

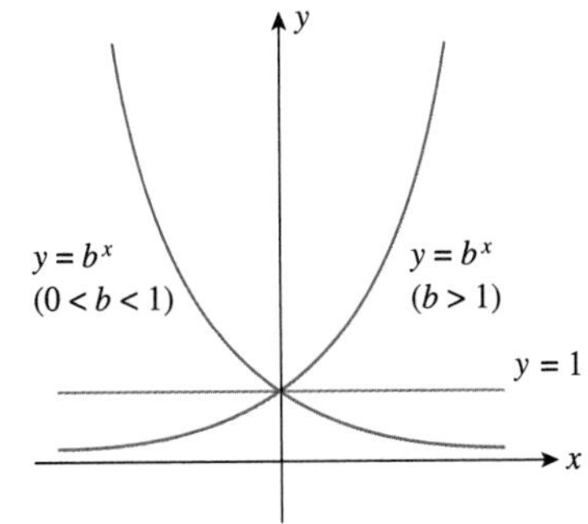

Figure 5.4
$y = b^x$ is an increasing function of x if $b > 1$, a constant function if $b = 1$, and a decreasing function if $0 < b < 1$.

Remark

If $b > 1$, then

$$\lim_{x\to\infty} b^x = \infty \text{ and } \lim_{x\to-\infty} b^x = \lim_{x\to\infty} b^{-x} = \lim_{x\to\infty} \frac{1}{b^x} = \frac{1}{\infty} = 0$$

This means that the asymptotic behaviour only occurs for very small x-values, as was observed for the graph of $y = 2^x$ in Figure 5.2. If $0 < b < 1$, we can write $b = \frac{1}{a}$ for $a > 0$. Then

$$\lim_{x\to\infty} b^x = \lim_{x\to\infty} \left(\frac{1}{a}\right)^x = \lim_{x\to\infty} \frac{1}{a^x} = \frac{1}{\infty} = 0 \quad \text{and}$$

$$\lim_{x\to-\infty} b^x = \lim_{x\to-\infty} \left(\frac{1}{a}\right)^x = \lim_{x\to-\infty} \frac{1}{a^x} = \lim_{x\to\infty} \frac{1}{a^{-x}} = \lim_{x\to\infty} a^x = \infty$$

This means that the asymptotic behaviour only occurs for very large x-values, as was observed for the graph of $y = (\frac{1}{2})^x$ in Figure 5.3. ◀

Properties of the Exponential Function

The exponential function $y = b^x$ ($b > 0$, $b \neq 1$) has the following properties:

1. Its domain is $(-\infty, \infty)$.

2. Its range is $(0, \infty)$.

3. Its graph passes through the point $(0, 1)$.

4. It is continuous on $(-\infty, \infty)$.

5. It is increasing on $(-\infty, \infty)$ if $b > 1$ and decreasing on $(-\infty, \infty)$ if $b < 1$.

6. It has horizontal asymptote $y = 0$ (the x-axis).

Transformation of the Exponential Function

The transformations from Section 2.2 were presented for functions in general, and thus also apply to exponential functions. We simply summarize the different translations, stretchings, and reflections below.

Transformation of the Exponential Function

Suppose we are given the exponential function

$$y = b^x (b > 0), \qquad b \neq 1$$

Vertical Translation, $k > 0$

The graph of $y = b^x + k$ is the graph of $y = b^x$ translated vertically up by k units.
The graph of $y = b^x - k$ is the graph of $y = b^x$ translated vertically down by k units.

Horizontal Translation, $h > 0$

The graph of $y = b^{x+h}$ is the graph of $y = b^x$ translated horizontally left by h units.
The graph of $y = b^{x-h}$ is the graph of $y = b^x$ translated horizontally right by h units.

Reflection

The graph of $y = b^{-x}$ is the graph of $y = b^x$ reflected horizontally in the y-axis.
The graph of $y = -b^x$ is the graph of $y = b^x$ reflected vertically in the x-axis.

Stretching

For $a > 1$:
The graph of $y = ab^x$ is the graph of $y = b^x$ expanded vertically by a factor of a.
The graph of $y = b^{ax}$ is the graph of $y = b^x$ compressed horizontally by a factor of $\frac{1}{a}$.
For $0 < a < 1$:
The graph of $y = ab^x$ is the graph of $y = b^x$ compressed vertically by a factor of a.
The graph of $y = b^{ax}$ is the graph of $y = b^x$ expanded horizontally by a factor of $\frac{1}{a}$.

Notice that $y = b^{ax} = b^a b^x = cb^x$, for $c = b^a$. Therefore, there is a natural flow between vertical stretching ($y = b^{ax}$) and horizontal stretching ($y = ab^x$), and it is up to us to decide which way we want to think about it.

Remark

When graphing transformed exponential functions using the methods described in Section 2.2, it helps to begin by translating the horizontal axis, followed by transforming the point $(0, 1)$. ◀

The Base e

We now want to take a closer look at Property (3) (p. 258) of the exponential function. For this, we consider the slope of the tangent line at the point $(0, b^0) = (0, 1)$ on the graph of the function $y = b^x$, which is denoted by

$$\left.\frac{d}{dx}(b^x)\right|_{x=0}$$

Figure 5.5a shows the graphs of $y = 2^x$, $y = 3^x$, and $y = 5^x$, while Figure 5.5b shows the corresponding tangent lines at the point (0, 1). According to the point-slope form for a line, the general equation of the tangent line is

$$y - 1 = m(x - 0)$$

$$y = mx + 1$$

where $m = \left.\frac{d}{dx}(b^x)\right|_{x=0}$. For example, the tangent line equation for $y = 2^x$ at (0, 1) is

$$y = \left(\left.\frac{d}{dx}(2^x)\right|_{x=0}\right)x + 1$$

In Section 5.3 we will learn how to differentiate an exponential function, but it is not necessary to do so now. Figure 5.5c shows the three introduced tangent lines together with the graph for the line $y = x + 1$. Notice that this line lies between the tangent lines for the graphs of $y = 2^x$ and $y = 3^x$. The question we want to raise is, what is the value of the base b, for which the tangent line equation is given by $y = x + 1$, i.e., the slope has value one? All we know so far is that $2 < b < 3$.

We need to solve the equation

$$1 = \left.\frac{d}{dx}(b^x)\right|_{x=0} \tag{1}$$

Recall the definition for the slope of a function f at $x = a$

$$f'(a) = \lim_{h\to 0}\frac{f(a + h) - f(a)}{h}$$

Let us write $f(x) = b^x$ and $a = 0$, then Equation (1) becomes

$$1 = \lim_{h\to 0}\frac{b^{0+h} - b^0}{h}$$

which can be simplified to

$$1 = \lim_{h\to 0}\frac{b^h - 1}{h}$$

We will not prove it here, but the special base b that makes this equation true is an irrational number called e, whose value is 2.7182818 . . . (continuing infinitely without repetition). This number is also known as Euler's number after the German mathematician Leonard Euler, who studied logarithmic functions, which are the inverse functions of the exponential functions (see Section 5.2). For completion, we present this famous limit

$$1 = \lim_{h\to 0}\frac{e^h - 1}{h}$$

Exponential functions to the base e play an important role in both theoretical and applied problems. It can be shown, although we will not do so here, that

$$e = \lim_{m\to\infty}\left(1 + \frac{1}{m}\right)^m \tag{2}$$

However, you may convince yourself of the plausibility of this definition of the number e by examining Table 5.1, which may be constructed with the help of a calculator.

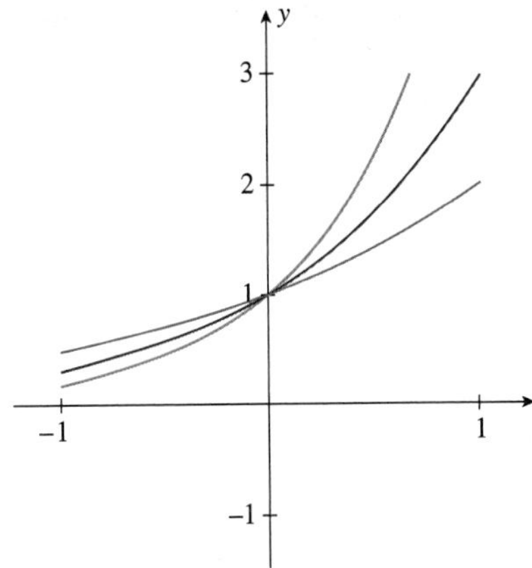

(a) The graphs of $y = 2^x$, $y = 3^x$, and $y = 5^x$

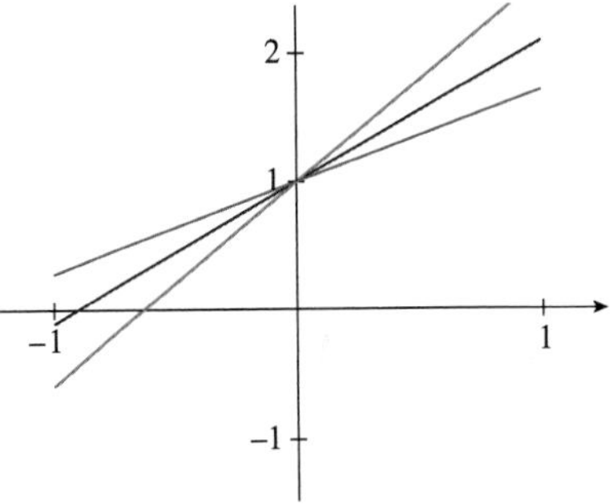

(b) The corresponding tangent lines

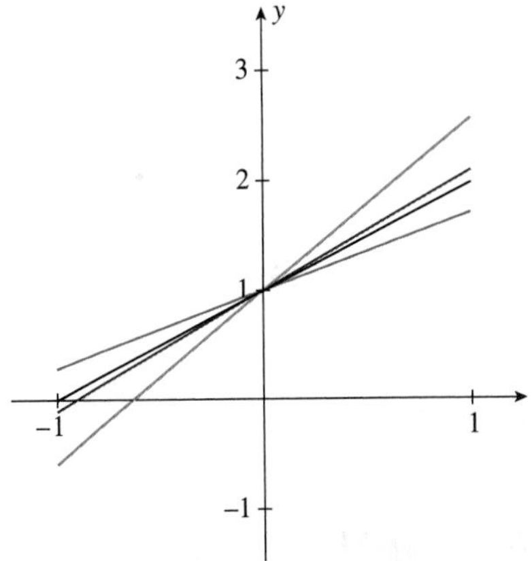

(c) The line $y = x + 1$ added

Figure 5.5

Table 5.1

m	$\left(1 + \frac{1}{m}\right)^m$
10	2.59374
100	2.70481
1000	2.71692
10,000	2.71815
100,000	2.71827
1,000,000	2.71828

EXPLORING WITH TECHNOLOGY

To obtain a visual confirmation of the fact that the expression $(1 + 1/m)^m$ approaches the number $e = 2.71828...$ as m increases without bound, plot the graph of $f(x) = (1 + 1/x)^x$ in a suitable viewing window and observe that $f(x)$ approaches 2.71828... as x increases without bound. Use **ZOOM** and **TRACE** to find the value of $f(x)$ for large values of x.

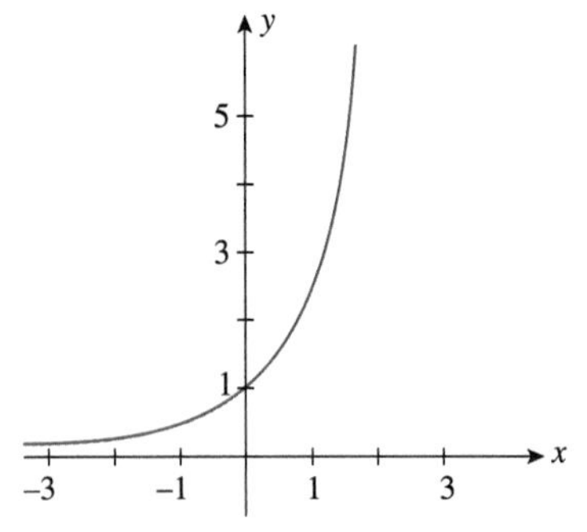

Figure 5.6
The graph of $y = e^x$

Example 5

Sketch the graph of the function $y = e^x$.

Solution

Since $e > 1$, it follows from our previous discussion that the graph of $y = e^x$ is similar to the graph of $y = 2^x$ (see Figure 5.2). With the aid of a calculator, we obtain the following table the following approximate values for y:

x	−3	−2	−1	0	1	2	3
y	0.05	0.14	0.37	1	2.72	7.39	20.09

The graph of $y = e^x$ is sketched in Figure 5.6.

Next, we consider another exponential function to the base e that is closely related to the previous function and is particularly useful in constructing models that describe "exponential decay."

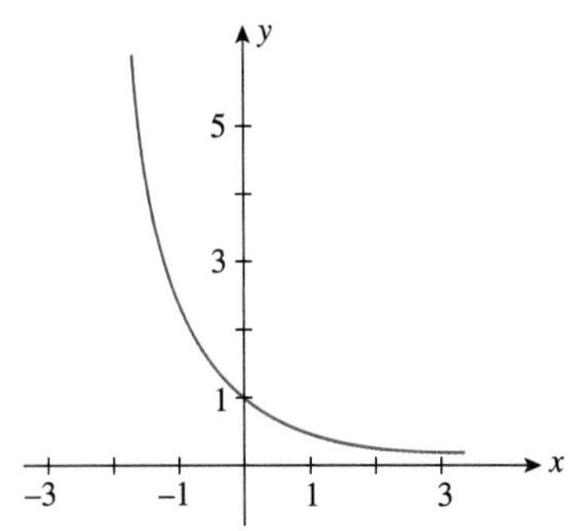

Figure 5.7
The graph of $y = e^{-x}$

Example 6

Sketch the graph of the function $y = e^{-x}$.

Solution

Since $e > 1$, it follows that $0 < 1/e < 1$, so $f(x) = e^{-x} = 1/e^x = (1/e)^x$ is an exponential function with base less than 1. Therefore, it has a graph similar to that of the exponential function $y = (1/2)^x$. As before, we construct the following table of approximate values of $y = e^{-x}$ for selected values of x:

x	−3	−2	−1	0	1	2	3
y	20.09	7.39	2.72	1	0.37	0.14	0.05

Using this table, we sketch the graph of $y = e^{-x}$ in Figure 5.7.

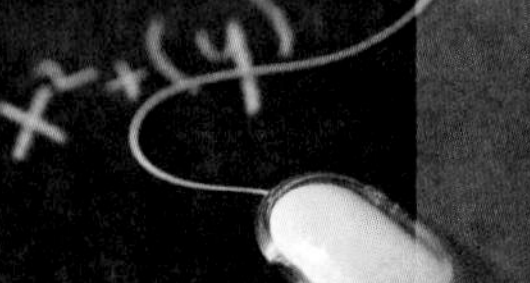

Self-Check Exercises 5.1

1. Solve the equation $2^{2x+1} \cdot 2^{-3} = 2^{x-1}$.
2. Sketch the graph of $y = e^{0.4x}$.

Solutions to Self-Check Exercises 5.1 can be found on page 262.

5.1 Exercises

In Exercises 1–8, evaluate the expression.

1. a. $4^{-3} \cdot 4^5$ b. $3^{-3} \cdot 3^6$

2. a. $(2^{-1})^3$ b. $(3^{-2})^3$

3. a. $9(9)^{-1/2}$ b. $5(5)^{-1/2}$

4. a. $\left[\left(-\frac{1}{2}\right)^3\right]^{-2}$ b. $\left[\left(-\frac{1}{3}\right)^2\right]^{-3}$

5. a. $\dfrac{(-3)^4(-3)^5}{(-3^8)}$ b. $\dfrac{(2^{-4})(2^6)}{2^{-1}}$

6. a. $3^{1/4} \cdot 9^{-5/8}$ b. $2^{3/4} \cdot 4^{-3/2}$

7. a. $\dfrac{5^{3.3} \cdot 5^{-1.6}}{5^{-0.3}}$ b. $\dfrac{4^{2.7} \cdot 4^{-1.3}}{4^{-0.4}}$

8. a. $\left(\frac{1}{16}\right)^{-1/4}\left(\frac{27}{64}\right)^{-1/3}$ b. $\left(\frac{8}{27}\right)^{-1/3}\left(\frac{8}{256}\right)^{-1/4}$

In Exercises 9–16, simplify the expression.

9. a. $(64x^9)^{1/3}$ b. $(25x^3y^4)^{1/2}$

10. a. $(2x^3)(-4x^{-2})$ b. $(4x^{-2})(-3x^5)$

11. a. $\dfrac{6a^{-5}}{3a^{-3}}$ b. $\dfrac{4b^{-4}}{12b^{-6}}$

12. a. $y^{-3/2}\,y^{5/3}$ b. $x^{-3/5}\,x^{8/3}$

13. a. $(2x^3y^2)^3$ b. $(4x^2y^2z^3)^2$

14. a. $(x^{r/s})^{s/r}$ b. $(x^{-b/a})^{-a/b}$

15. a. $\dfrac{5^0}{(2^{-3}x^{-3}y^2)^2}$ b. $\dfrac{(x+y)(x-y)}{(x-y)^0}$

16. a. $\dfrac{(a^m \cdot a^{-n})^{-2}}{(a^{m+n})^2}$ b. $\left(\dfrac{x^{2n-2}y^{2n}}{x^{5n+1}y^{-n}}\right)^{1/3}$

In Exercises 17–26, solve the equation for x.

17. $6^{2x} = 6^4$ 18. $5^{-x} = 5^3$

19. $3^{3x-4} = 3^5$ 20. $10^{2x-1} = 10^{x+3}$

21. $(2.1)^{x+2} = (2.1)^5$ 22. $(-1.3)^{x-2} = (-1.3)^{2x+1}$

23. $8^x = \left(\frac{1}{32}\right)^{x-2}$ 24. $3^{x-x^2} = \dfrac{1}{9^x}$

25. $3^{2x} - 12 \cdot 3^x + 27 = 0$ 26. $2^{2x} - 4 \cdot 2^x + 4 = 0$

In Exercises 27–36, sketch the graphs of the given functions on the same axes. A calculator is recommended for these exercises.

27. $y = 2^x$, $y = 3^x$, and $y = 4^x$

28. $y = \left(\frac{1}{2}\right)^x$, $y = \left(\frac{1}{3}\right)^x$, and $y = \left(\frac{1}{4}\right)^x$

29. $y = 2^{-x}$, $y = 3^{-x}$, and $y = 4^{-x}$

30. $y = 4^{0.5x}$ and $y = 4^{-0.5x}$

31. $y = 4^{0.5x}$, $y = 4^x$, and $y = 4^{2x}$

32. $y = e^x$, $y = 2e^x$, and $y = 3e^x$

33. $y = e^{0.5x}$, $y = e^x$, and $y = e^{1.5x}$

34. $y = e^{-0.5x}$, $y = e^{-x}$, and $y = e^{-1.5x}$

35. $y = 0.5e^{-x}$, $y = e^{-x}$, and $y = 2e^{-x}$

36. $y = 1 - e^{-x}$ and $y = 1 - e^{-0.5x}$

A calculator is recommended for Exercises 37–42.

37. **Growth of Web Sites** In a study conducted in 2000, the projected number of Web addresses (in billions) is approximated by the function

$$N(t) = 0.45e^{0.5696t} \qquad (0 \le t \le 5)$$

where t is measured in years, with $t = 0$ corresponding to 1997.

a. Complete the following table by finding the number of Web addresses in each year:

Year	0	1	2	3	4	5
Number of Web Addresses (billions)						

b. Sketch the graph of N.

Biological and Life Sciences Applications

38. **Infant Mortality Rate** The infant mortality rate (per 1000 live births) is approximated by the function

$$N(t) = 12.5e^{-0.03t} \qquad (0 \le t \le 21)$$

where t is measured in years, with $t = 0$ corresponding to 1980.

a. What was the mortality rate in 1980? In 1990? In 2000?

b. Sketch the graph of N.

39. **Absorption of Drugs** The concentration of a drug in an organ at any time t (in seconds) is given by

$$C(t) = \begin{cases} 0.3t - 18(1 - e^{-t/60}) & \text{if } 0 \le t \le 20 \\ 18e^{-t/60} - 12e^{-(t-20)/60} & \text{if } t > 20 \end{cases}$$

where $C(t)$ is measured in grams/cubic centimetre (g/cm^3).

a. What is the initial concentration of the drug in the organ?

b. What is the concentration of the drug in the organ after 10 s?

c. What is the concentration of the drug in the organ after 30 s?

d. What will be the concentration of the drug in the long run?

40. **Absorption of Drugs** The concentration of a drug in an organ at any time t (in seconds) is given by

$$x(t) = 0.08(1 - e^{-0.02t})$$

where $x(t)$ is measured in grams/cubic centimetre (g/cm^3).

a. What is the initial concentration of the drug in the organ?

b. What is the concentration of the drug in the organ after 30 s?

c. What will be the concentration of the drug in the organ in the long run?

d. Sketch the graph of x.

41. **Absorption of Drugs** The concentration of a drug in an organ at any time t (in seconds) is given by

$$x(t) = 0.08 + 0.12(1 - e^{-0.02t})$$

where $x(t)$ is measured in grams/cubic centimetre (g/cm^3).

a. What is the initial concentration of the drug in the organ?
b. What is the concentration of the drug in the organ after 20 s?
c. What will be the concentration of the drug in the organ in the long run?
d. Sketch the graph of x.

42. **Absorption of Drugs** Jane took 100 mg of a drug in the morning and another 100 mg of the same drug at the same time the following morning. The amount of the drug in her body t days after the first dosage was taken is given by

$$A(t) = \begin{cases} 100e^{-1.4t} & \text{if } 0 \le t < 1 \\ 100(1 + e^{1.4})e^{-1.4t} & \text{if } t \ge 1 \end{cases}$$

a. What was the amount of drug in Jane's body immediately after taking the second dose? After 2 days? In the long run?
b. Sketch the graph of A.

In Exercises 43–46, determine whether the statement is true or false. If it is true, explain why it is true. If it is false, give an example to show why it is false.

43. $(x^2 + 1)^3 = x^6 + 1$

44. $e^{xy} = e^x e^y$

45. If $x < y$, then $e^x < e^y$.

46. If $0 < b < 1$ and $x < y$, then $b^x > b^y$.

Solutions to Self-Check Exercises 5.1

1. $2^{2x+1} \cdot 2^{-3} = 2^{x-1}$

$$\frac{2^{2x+1}}{2^{x-1}} \cdot 2^{-3} = 1 \qquad \text{Divide both sides by } 2^{x-1}.$$

$$2^{(2x+1)-(x-1)-3} = 1$$

$$2^{x-1} = 1$$

This is true if and only if $x - 1 = 0$ or $x = 1$.

2. We first construct the following table of values:

x	−3	−2	−1	0	1	2	3	4
$y = e^{0.4x}$	0.3	0.4	0.7	1	1.5	2.2	3.3	5

Next, we plot these points and join them by a smooth curve to obtain the graph of f shown in the accompanying figure.

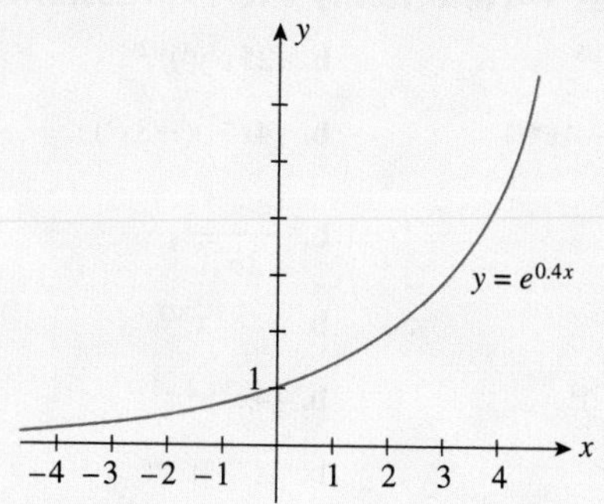

5.2 Logarithmic Functions

Logarithms

You are already familiar with exponential equations of the form

$$b^y = x \qquad (b > 0, b \neq 1)$$

where the variable x is expressed in terms of a real number b and a variable y. But what about solving this same equation for y? You may recall from your study of algebra that the number y is called the **logarithm of x to the base b** and is denoted by **$\log_b x$**. It is the power to which the base b must be raised in order to obtain the number x.

Logarithm of x to the Base b

$$y = \log_b x \quad \text{if and only if} \quad x = b^y \qquad (b > 0, b \neq 1, x > 0)$$

Observe that the logarithm $\log_b x$ is defined only for positive values of x.

Example 1

a. $\log_{10} 100 = 2$ since $100 = 10^2$
b. $\log_5 125 = 3$ since $125 = 5^3$
c. $\log_3 \frac{1}{27} = -3$ since $\frac{1}{27} = \frac{1}{3^3} = 3^{-3}$
d. $\log_{20} 20 = 1$ since $20 = 20^1$

Example 2

Solve each of the following equations for x.

a. $\log_3 x = 4$ **b.** $\log_{16} 4 = x$ **c.** $\log_x 8 = 3$

Solution

a. By definition, $\log_3 x = 4$ implies $x = 3^4 = 81$.

b. $\log_{16} 4 = x$ is equivalent to $4 = 16^x = (4^2)^x = 4^{2x}$, or $4^1 = 4^{2x}$, from which we deduce that

$$2x = 1 \qquad b^m = b^n \Leftrightarrow m = n$$

$$x = \frac{1}{2}$$

c. Referring once again to the definition, we see that the equation $\log_x 8 = 3$ is equivalent to

$$8 = 2^3 = x^3$$

$$x = 2 \qquad b^m = b^n \Leftrightarrow m = n$$

The two widely used systems of logarithms are the system of **common logarithms,** which uses the number 10 as the base, and the system of **natural logarithms,** which uses the irrational number $e = 2.71828\ldots$ as the base. Also, it is standard practice to write **log** for $\log_{10}$ and **ln** for $\log_e$.

Logarithmic Notation

$\log x = \log_{10} x$ Common logarithm

$\ln x = \log_e x$ Natural logarithm

The system of natural logarithms is widely used in theoretical work. Using natural logarithms rather than logarithms to other bases often leads to simpler expressions.

Laws of Logarithms

Computations involving logarithms are facilitated by the following **laws of logarithms.**

Laws of Logarithms

If m and n are positive numbers $b > 0$, $b \neq 1$, then

1. $\log_b mn = \log_b m + \log_b n$
2. $\log_b \frac{m}{n} = \log_b m - \log_b n$
3. $\log_b m^n = n \log_b m$
4. $\log_b 1 = 0$
5. $\log_b b = 1$
6. $\log_b m = \frac{\log_a m}{\log_a b}$, in particular

$$\log_b m = \frac{\ln m}{\ln b}$$

Do not confuse the expression $\log m/n$ (Law 2) with the expression $\log m/\log n$. For example,

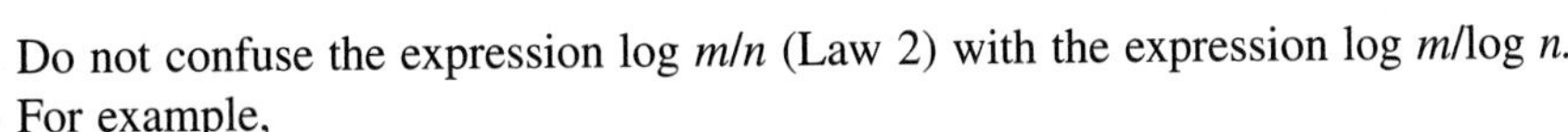

$$\log \frac{100}{10} = \log 100 - \log 10 = 2 - 1 = 1 \neq \frac{\log 100}{\log 10} = \frac{2}{1} = 2$$

Remark

The Logarithmic Law 6 is useful when we need to evaluate $\log_b m$ and b is not a base provided on the calculator. ◀

You will be asked to prove some of these laws in Exercises 60–62 on page 269. Their derivations are based on the definition of a logarithm and the corresponding laws of exponents. The following examples illustrate the properties of logarithms.

Example 3

a. $\log(2 \cdot 3) = \log 2 + \log 3$ **b.** $\ln \frac{5}{3} = \ln 5 - \ln 3$

c. $\log\sqrt{7} = \log 7^{1/2} = \frac{1}{2}\log 7$ **d.** $\log_5 1 = 0$

e. $\log_{45} 45 = 1$

Example 4

Given that $\log 2 \approx 0.3010$, $\log 3 \approx 0.4771$, and $\log 5 \approx 0.6990$, use the laws of logarithms to find

a. $\log 15$ **b.** $\log 7.5$ **c.** $\log 81$ **d.** $\log 50$ **e.** $\log_3 5$

Solution

a. Note that $15 = 3 \cdot 5$, so by Law 1 for logarithms,

$$\begin{aligned}\log 15 &= \log 3 \cdot 5\\ &= \log 3 + \log 5\\ &\approx 0.4771 + 0.6990\\ &= 1.1761\end{aligned}$$

b. Observing that $7.5 = 15/2 = (3 \cdot 5)/2$, we apply Laws 1 and 2, obtaining

$$\begin{aligned}\log 7.5 &= \log\frac{(3)(5)}{2}\\ &= \log 3 + \log 5 - \log 2\\ &\approx 0.4771 + 0.6990 - 0.3010\\ &= 0.8751\end{aligned}$$

c. Since $81 = 3^4$, we apply Law 3 to obtain

$$\begin{aligned}\log 81 &= \log 3^4\\ &= 4 \log 3\\ &\approx 4(0.4771)\\ &= 1.9084\end{aligned}$$

d. We write $50 = 5 \cdot 10$ and find

$$\begin{aligned}\log 50 &= \log(5)(10)\\ &= \log 5 + \log 10\\ &\approx 0.6990 + 1 && \text{Use Law 5}\\ &= 1.6990\end{aligned}$$

e. We use Law 6 to rewrite

$$\begin{aligned}\log_3 5 &= \frac{\log 5}{\log 3}\\ &\approx \frac{0.6990}{0.4771}\\ &\approx 1.4651\end{aligned}$$

Example 5

Expand and simplify the following expressions:

a. $\log_3 x^2y^3$ **b.** $\log_2 \frac{x^2 + 1}{2^x}$ **c.** $\ln \frac{x^2\sqrt{x^2 - 1}}{e^x}$

Solution

a.
$$\begin{aligned}\log_3 x^2y^3 &= \log_3 x^2 + \log_3 y^3 && \text{Law 1}\\ &= 2 \log_3 x + 3 \log_3 y && \text{Law 3}\end{aligned}$$

b. $\log_2 \dfrac{x^2 + 1}{2^x} = \log_2(x^2 + 1) - \log_2 2^x$ Law 2

$= \log_2(x^2 + 1) - x \log_2 2$ Law 3

$= \log_2(x^2 + 1) - x$ Law 5

c. $\ln \dfrac{x^2\sqrt{x^2 - 1}}{e^x} = \ln \dfrac{x^2(x^2 - 1)^{1/2}}{e^x}$ Rewrite

$= \ln x^2 + \ln(x^2 - 1)^{1/2} - \ln e^x$ Laws 1 and 2

$= 2 \ln x + \dfrac{1}{2} \ln(x^2 - 1) - x \ln e$ Law 3

$= 2 \ln x + \dfrac{1}{2} \ln(x^2 - 1) - x$ Law 5

Logarithmic Functions and Their Graphs

The definition of a logarithm implies that if b and n are positive numbers and b is different from 1, then the expression $\log_b n$ is a real number. This enables us to define a logarithmic function as follows:

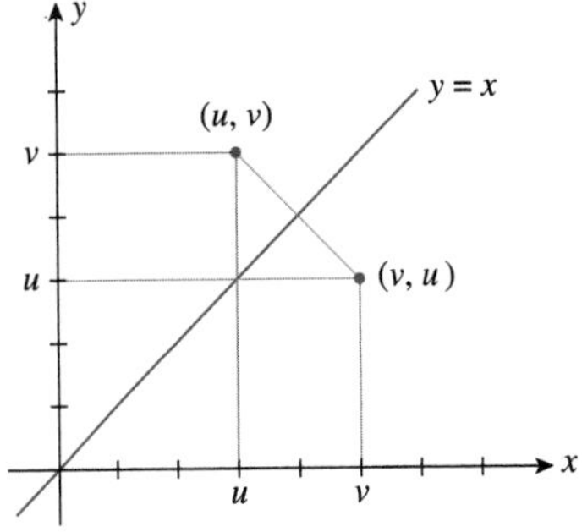

Figure 5.8
The points (u, v) and (v, u) are mirror reflections of each other.

Logarithmic Function

The function defined by

$$f(x) = \log_b x \qquad (b > 0, b \neq 1)$$

is called the **logarithmic function with base b**. The domain of f is the set of all positive numbers.

One easy way to obtain the graph of the logarithmic function $y = \log_b x$ is to construct a table of values of the logarithm (base b). However, another method—and a more instructive one—is based on exploiting the intimate relationship between logarithmic and exponential functions.

If a point (u, v) lies on the graph of $y = \log_b x$, then

$$v = \log_b u$$

But we can also write this equation in exponential form as

$$u = b^v$$

So the point (v, u) also lies on the graph of the function $y = b^x$. Let's look at the relationship between the points (u, v) and (v, u) and the line $y = x$ (Figure 5.8). If we think of the line $y = x$ as a mirror, then the point (v, u) is the mirror reflection of the point (u, v). Similarly, the point (u, v) is a mirror reflection of the point (v, u). Recall from Section 2.7 that one-to-one functions, and $f(x) = b^x$ certainly is one, have an inverse function f^{-1}, whose graph is the reflection of the graph of f with respect to the line $y = x$. We can take advantage of this relationship to help us draw the graph of logarithmic functions. For example, if we wish to draw the graph of $y = \log_b x$, where $b > 1$, then we need only draw the mirror reflection of the graph of $y = b^x$ with respect to the line $y = x$ (Figure 5.9).

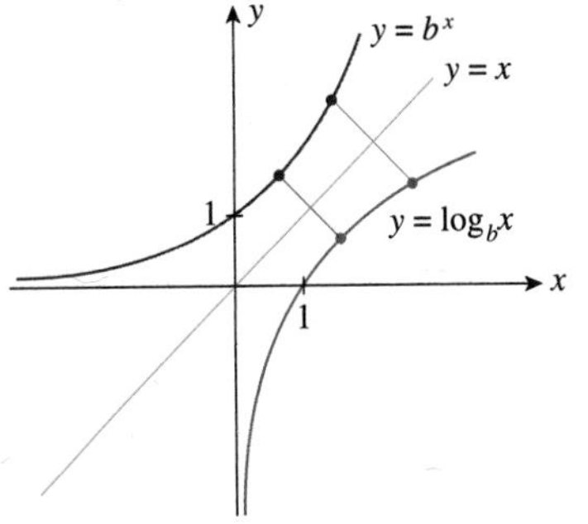

Figure 5.9
The graphs of $y = b^x$ and $y = \log_b x$ are mirror reflections of each other.

You may discover the following properties of the logarithmic function by taking the reflection of the graph of an appropriate exponential function (Exercises 35 and 36 on page 269).

Properties of the Logarithmic Function

The logarithmic function $y = \log_b x$ $(b > 0, b \neq 1)$ has the following properties:

1. Its domain is $(0, \infty)$.
2. Its range is $(-\infty, \infty)$.
3. Its graph passes through the point $(1, 0)$.
4. It is continuous on $(0, \infty)$.
5. It is increasing on $(0, \infty)$ if $b > 1$ and decreasing on $(0, \infty)$ if $b < 1$.
6. It has vertical asymptote $x = 0$ (the y-axis).

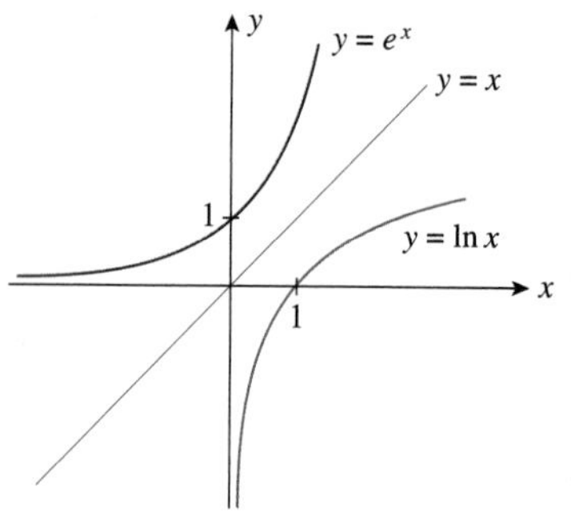

Figure 5.10
The graph of $y = \ln x$ is the mirror reflection of the graph of $y = e^x$.

Example 6

Sketch the graph of the function $y = f(x) = \ln x$.

Solution

We first sketch the graph of $y = e^x$. Then, the required graph is obtained by tracing the mirror reflection of the graph of $y = e^x$ with respect to the line $y = x$ (Figure 5.10).

Transformation of the Logarithmic Function

Once again, we summarize the transformations that apply to the graph of the function $y = \log_b x$, as it was done for the exponential function in the previous section.

Transformation of the Logarithmic Function

Suppose we are given the logarithmic function

$$y = \log_b x \qquad (b > 0, b \neq 1)$$

Vertical Translation, $k > 0$

The graph of $y = \log_b x + k$ is the graph of $y = \log_b x$ translated vertically up by k units.
The graph of $y = \log_b x - k$ is the graph of $y = \log_b x$ translated vertically down by k units.

Horizontal Translation, $h > 0$

The graph of $y = \log_b (x + h)$ is the graph of $y = \log_b x$ translated horizontally left by h units.

The graph of $y = \log_b (x - h)$ is the graph of $y = \log_b x$ translated horizontally right by h units.

Reflection

The graph of $y = \log_b (-x)$ is the graph of $y = \log_b x$ reflected horizontally in the y-axis.
The graph of $y = -\log_b x$ is the graph of $y = \log_b x$ reflected vertically in the x-axis.

Stretching

For $a > 1$:
The graph of $y = a \log_b x$ is the graph of $y = \log_b x$ expanded vertically by a factor of a.
The graph of $y = \log_b (ax)$ is the graph of $y = \log_b x$ compressed horizontally by a factor of $\frac{1}{a}$.

For $0 < a < 1$:
The graph of $y = a \log_b x$ is the graph of $y = \log_b x$ compressed vertically by a factor of a.
The graph of $y = \log_b (ax)$ is the graph of $y = \log_b x$ expanded horizontally by a factor of $\frac{1}{a}$.

Remark

When graphing transformed logarithmic functions using the methods described in Section 2.2, it helps to begin by translating the vertical axis, followed by transforming the point (1, 0). ◀

Example 7

Refer to Example 6, sketch the graph of $y = 2f(x + 4) + 1$.

Solution

As can be observed from Figure 5.10, but also by Properties 3 and 6, the graph of $y = \ln x$ passes through the point (1, 0) and has vertical asymptote $x = 0$. We begin by performing the required transformations on this point and line. Take note that stretching does not affect these two graphical items. $y = 2f(x + 4) + 1$ indicates that the graph of the function $y = \ln x$ is translated horizontally to the left by 4 units, and translated vertically up by 1 unit. Therefore, the point (1, 0) moves to the point $(-3, 1)$, while the vertical asymptote $x = 0$ becomes $x = -4$. Finally, we vertically stretch the graph of $y = \ln x$ by a factor of 2, and move it to its required position. For this we can simply choose one other point on the graph of $y = \ln x$, say $(e, \ln e) = (e, 1)$, and after applying all three transformations it becomes $(e - 4, 2 \cdot 1 + 1) = (e - 4, 3)$, see Figure 5.11.

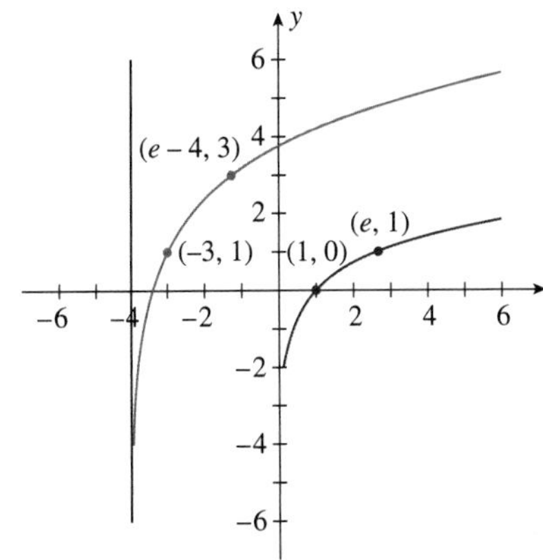

Figure 5.11
The graph of $y = 2\ln(x + 4) + 1$

Properties Relating the Exponential and Logarithmic Functions

We made use of the relationship that exists between the exponential function $f(x) = e^x$ and the logarithmic function $g(x) = \ln x$ when we sketched the graph of g in Example 6. This relationship is further described by the following properties, which are an immediate consequence of the definition of the logarithm of a number.

Properties Relating e^x and $\ln x$

$$e^{\ln x} = x \qquad (x > 0) \qquad (3)$$

$$\ln e^x = x \qquad (\text{for any real number } x) \qquad (4)$$

(Try to verify these properties.)

The relationships expressed in Equations (3) and (4) are useful in solving equations that involve exponentials and logarithms.

EXPLORING WITH TECHNOLOGY

You can demonstrate the validity of Properties 2 and 3, which state that the exponential function $f(x) = e^x$ and the logarithmic function $g(x) = \ln x$ are inverses of each other as follows:

1. Sketch the graph of $(f \circ g)(x) = e^{\ln x}$, using the viewing window $[0, 10] \cdot [0, 10]$. Interpret the result.
2. Sketch the graph of $(g \circ f)(x) = \ln e^x$, using the standard viewing window. Interpret the result.

Example 8

Solve the equation $2e^{x+2} = 5$.

Solution

Note that $x \in 1R$. We first divide both sides of the equation by 2 to obtain

$$e^{x+2} = \frac{5}{2} = 2.5$$

Next, taking the natural logarithm of each side of the equation and using Equation (3), we have

$$\ln e^{x+2} = \ln 2.5$$

$$x + 2 = \ln 2.5$$

$$x = -2 + \ln 2.5$$

$$\approx -1.08$$

GROUP DISCUSSION

Consider the equation $y = y_0 b^{kx}$, where y_0 and k are positive constants and $b > 0$, $b \neq 1$. Suppose we want to express y in the form $y = y_0 e^{px}$. Use the laws of logarithms to show that $p = k \ln b$ and hence that $y = y_0 e^{(k \ln b)x}$ is an alternative form of $y = y_0 b^{kx}$ using the base e.

Example 9

Solve the equation $5 \ln x + 3 = 0$.

Solution

Note that $x > 0$, which means that we must check the solution against this restriction. Adding -3 to both sides of the equation leads to

$$5 \ln x = -3$$
$$\ln x = -\frac{3}{5} = -0.6$$

and so

$$e^{\ln x} = e^{-0.6}$$

Using Equation (2), we conclude that

$$x = e^{-0.6}$$
$$\approx 0.55$$

and $e^{-0.6} > 0$, so it is a solution for our equation.

Example 10

Solve the equation $\log_5(x + 1) + \log_5(x - 3) = 1$.

Solution

Note that the restrictions are $x > -1$ and $x > 3$, which simplifies to $x > 3$. Applying Law 1, we get

$$\log_5[(x + 1)(x - 3)] = 1$$

which we rewrite using the definition of the logarithm and simplify

$$5^{\log_5[(x+1)(x-3)]} = 5^1$$
$$(x + 1)(x - 3) = 5$$
$$x^2 - 2x - 3 = 5$$
$$x^2 - 2x - 8 = 0$$
$$(x - 4)(x + 2) = 0$$

Possible solutions are $x = 4$ and $x = -2$. However, only $x = 4$ passes the restriction that $x > 3$, and so it is the only solution to the above equation.

Self-Check Exercises 5.2

1. Sketch the graph of $y = 3^x$ and $y = \log_3 x$ on the same set of axes.
2. Solve the equation $3e^{x+1} - 2 = 4$.

Solutions to Self-Check Exercises 5.2 can be found on page 271.

5.2 Exercises

In Exercises 1–10, express the given equation in logarithmic form.

1. $2^6 = 64$
2. $3^5 = 243$
3. $3^{-2} = \frac{1}{9}$
4. $5^{-3} = \frac{1}{125}$
5. $\left(\frac{1}{3}\right)^1 = \frac{1}{3}$
6. $\left(\frac{1}{2}\right)^{-4} = 16$
7. $32^{3/5} = 8$
8. $81^{3/4} = 27$
9. $10^{-3} = 0.001$
10. $16^{-1/4} = 0.5$

In Exercises 11–20, use the facts that log 3 = 0.4771 and log 4 = 0.6021 to find the value of the given logarithm.

11. $\log 12$
12. $\log \frac{3}{4}$
13. $\log 16$
14. $\log \sqrt{3}$
15. $\log 48$
16. $\log \frac{1}{300}$
17. $\log 64$
18. $\log_4 36$
19. $\log_{12} 3$
20. $\log_9 \sqrt{4}$

In Exercises 21–24, write the given expression as the logarithm of a single quantity.

21. $2 \ln a + 3 \ln b$
22. $\frac{1}{2} \ln x + 2 \ln y - 3 \ln z$
23. $\ln 3 + \frac{1}{2} \ln x + \ln y - \frac{1}{3} \ln z$
24. $\ln 2 + \frac{1}{2} \ln(x + 1) - 2 \ln(1 + \sqrt{x})$

In Exercises 25–32, use the laws of logarithms to simplify the given expression.

25. $\log x(x + 1)^4$
26. $\log x(x^2 + 1)^{-1/2}$
27. $\log \frac{\sqrt{x + 1}}{x^2 + 1}$
28. $\ln \frac{e^x}{1 + e^x}$
29. $\ln xe^{-x^2}$
30. $\ln x(x + 1)(x + 2)$
31. $\ln \frac{x^{1/2}}{x^2\sqrt{1 + x^2}}$
32. $\ln \frac{x^2}{\sqrt{x}(1 + x)^2}$

In Exercises 33–40, sketch the graph of the given equation.

33. $y = \log_3 x$
34. $y = \log_{1/3} x$
35. $y = \ln 2x$
36. $y = \ln \frac{1}{2} x$
37. $y = -3 \ln x + 2$
38. $y = \frac{1}{2} \ln(x - 3)$
39. $y = \ln (x + 5) - 2$
40. $y = -2 \ln (x - 4) - 3$

In Exercises 41 and 42, sketch the graphs of the given equations on the same coordinate axes.

41. $y = 2^x$ and $y = \log_2 x$
42. $y = e^{3x}$ and $y = \ln 3x$

In Exercises 43–58, use properties and definitions of exponents and logarithms to solve the given equation for t.

43. $e^{0.4t} = 8$
44. $\frac{1}{3} e^{-3t} = 0.9$
45. $5e^{-2t} = 6$
46. $4e^{t-1} = 4$
47. $2e^{-0.2t} - 4 = 6$
48. $12 - e^{0.4t} = 3$
49. $\frac{50}{1 + 4e^{0.2t}} = 20$
50. $\frac{200}{1 + 3e^{-0.3t}} = 100$
51. $A = Be^{-t/2}$
52. $\frac{A}{1 + Be^{t/2}} = C$
53. $\log_3 (t + 4) + \log_3 (t - 4) = 2$
54. $\log_2 t + \log_2 (t - 2) = 3$
55. $\ln t^2 - \ln (2t - 1) = 0$
56. $\ln t + \ln (1 - e^2 - t) = 2$
57. $1 - \log^{16} t^2 = \log^{16} (t - 3)^2$
58. $\log_2 2 (2 - t) + \log_2 (1 - t) = 5$
59. **a.** Given that $2^x = e^{kx}$, find k.
    b. Show that, in general, if b is a nonnegative real number, then any equation of the form $y = b^x$ may be written in the form $y = e^{kx}$, for some real number k.
60. Use the definition of a logarithm to prove
    a. $\log_b mn = \log_b m + \log_b n$
    b. $\log_b \frac{m}{n} = \log_b m - \log_b n$
    Hint: Let $\log_b m = p$ and $\log_b n = q$. Then, $b^p = m$ and $b^q = n$.
61. Use the definition of a logarithm to prove $\log_b m^n = n \log_b m$
62. Use the definition of a logarithm to prove
    a. $\log_b 1 = 0$
    b. $\log_b b = 1$

Business and Economics Applications

63. **Drug Sales** Annual sales in Canada of a particular cancer-fighting drug can be modelled by

$$S(t) = \frac{160}{1 + 64{,}000e^{-2t}}, \qquad 0 \le t \le 10$$

measured in millions of dollars, where $t = 0$ corresponds to the year 1995. In what year did sales reach 80 million dollars?

64. **Bike Sales** Monthly sales from April to August of 2002 of a locally manufactured specialty bike in Vancouver is described by the function

$$S(t) = 14.6 \ln (1.8t - 3.3)^4$$

where $t = 3$ corresponds to the beginning of April in 2001, and S is measured in thousands of dollars. In what month did sales reach \$100,000?

65. Manufacturing Cost The cost of manufacturing clogs made of recycled plastic and sold in Western Canada projected over the period of one year can be modelled by

$$C(t) = \ln \frac{1}{(11.5t + 7.1)^{30}} + 185$$

C is measured in thousands of dollars and t is measured in months. Compare the costs for the months of February and August.

66. Art Value The value of a collection of paintings owned by a well-known philanthropist living in Montreal is modelled by the function

$$S(t) = \frac{0.8}{0.5 + 130e^{-2t}}$$

over time t in years ($t =$ 1corresponds to the beginning of 2001 when the collection was acquired), where S is measured in million of dollars. By how much have the paintings increased in value from 2001 to 2003?

Biological and Life Sciences Applications

67. Blood Pressure A normal child's systolic blood pressure may be approximated by the function

$$p(x) = m(\ln x) + b$$

where $p(x)$ is measured in millimetres of mercury, x is measured in kilograms, and m and b are constants. Given that $m = 19.4$ and $b = 33.3$, determine the systolic blood pressure of a child who weighs 42 kg.

68. Magnitude of Earthquakes On the Richter scale, the magnitude R of an earthquake is given by the formula

$$R = \log \frac{I}{I_0}$$

where I is the intensity of the earthquake being measured and I_0 is the standard reference intensity.

a. Express the intensity I of an earthquake of magnitude $R = 5$ in terms of the standard intensity I_0.

b. Express the intensity I of an earthquake of magnitude $R = 8$ in terms of the standard intensity I_0. How many times greater is the intensity of an earthquake of magnitude 8 than one of magnitude 5?

c. In modern times the greatest loss of life attributable to an earthquake occurred in eastern China in 1976. Known as the Tangshan earthquake, it registered 8.2 on the Richter scale. How does the intensity of this earthquake compare with the intensity of an earthquake of magnitude $R = 5$?

69. Sound Intensity The relative loudness of a sound D of intensity I is measured in decibels (db), where

$$D = 10 \log \frac{I}{I_0}$$

and I_0 is the standard threshold of audibility.

a. Express the intensity I of a 30-db sound (the sound level of normal conversation) in terms of I_0.

b. Determine how many times greater the intensity of an 80-db sound (rock music) is than that of a 30-db sound.

c. Prolonged noise above 150 db causes immediate and permanent deafness. How does the intensity of a 150-db sound compare with the intensity of an 80-db sound?

70. Barometric Pressure Halley's law states that the barometric pressure (in millimetres of mercury) at an altitude of x km above sea level is approximated by the equation

$$p(x) = 760e^{-0.124x} \qquad (x \geq 0)$$

If the barometric pressure as measured by a hot-air balloonist is 500 mm of mercury, what is the balloonist's altitude?

71. Height of Trees The height (in metres) of a certain kind of tree is approximated by

$$h(t) = \frac{48}{1 + 240e^{-0.2t}}$$

where t is the age of the tree in years. Estimate the age of an 24-m-tall tree.

72. Newton's Law of Cooling The temperature of a cup of coffee t min after it is poured is given by

$$T = 21 + 55e^{-0.0446t}$$

where T is measured in degrees Celsius.

a. What is the temperature of the coffee when it was poured?

b. When will the coffee be cool enough to drink (say, 60°C)?

73. Lengths of Fish The length (in centimetres) of a typical Pacific halibut t yr old is approximately

$$f(t) = 200(1 - 0.956e^{-0.18t})$$

Suppose a Pacific halibut caught by Mike measures 140 cm. What is its approximate age?

74. Absorption of Drugs The concentration of a drug in an organ at any time t (in seconds) is given by

$$x(t) = 0.08(1 - e^{-0.02t})$$

where $x(t)$ is measured in grams/cubic centimetre (g/cm^3).

a. How long would it take for the concentration of the drug in the organ to reach 0.02 g/cm^3?

b. How long would it take for the concentration of the drug in the organ to reach 0.04 g/cm^3?

75. Absorption of Drugs The concentration of a drug in an organ at any time t (in seconds) is given by

$$x(t) = 0.08 + 0.12e^{-0.02t}$$

where $x(t)$ is measured in grams/cubic centimetre (g/cm^3).

a. How long would it take for the concentration of the drug in the organ to reach 0.18 g/cm^3?

b. How long would it take for the concentration of the drug in the organ to reach 0.16 g/cm^3?

76. Forensic Science Forensic scientists use the following law to determine the time of death of accident or murder victims. If T denotes the temperature of a body t hr after death, then

$$T = T_0 + (T_1 - T_0)(0.97)^t$$

where T_0 is the air temperature and T_1 is the normal body temperature. John Doe was found murdered at midnight in his house, when the room temperature was 21°C and his body temperature was 26.6°C. When was he killed? Assume that the normal body temperature is 37°C.

In Exercises 77–80, determine whether the statement is true or false. If it is true, explain why it is true. If it is false, given an example to show why it is false.

77. $(\ln x)^3 = 3 \ln x$ for all x in $(0, \infty)$.

78. $\ln a - \ln b = \ln(a - b)$ for all positive real numbers a and b.

79. The function $f(x) = \dfrac{1}{\ln{}^\circ x}$ is continuous on $(1, \infty)$.

80. The function $f(x) = \ln|x|$ is continuous for all $x \neq 0$.

Solutions to Self-Check Exercises 5.2

1. First, sketch the graph of $y = 3^x$ with the help of the following table of values:

x	-3	-2	-1	0	1	2	3
$y = 3^x$	1/27	1/9	1/3	0	3	9	27

Next, take the mirror reflection of this graph with respect to the line $y = x$ to obtain the graph of $y = \log_3 x$.

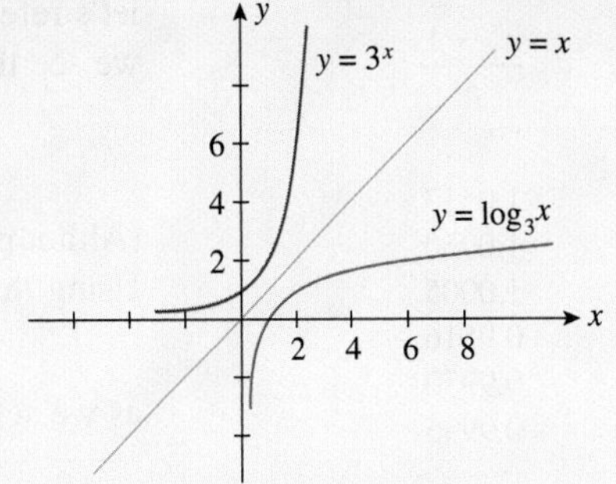

2.

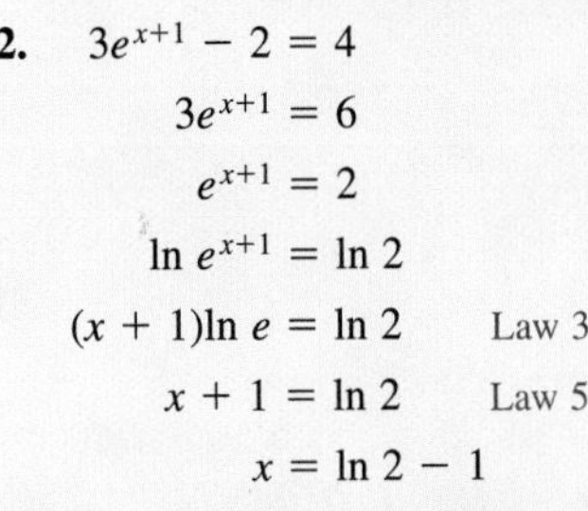

$$
\begin{aligned}
3e^{x+1} - 2 &= 4 \\
3e^{x+1} &= 6 \\
e^{x+1} &= 2 \\
\ln e^{x+1} &= \ln 2 \\
(x + 1)\ln e &= \ln 2 \qquad \text{Law 3} \\
x + 1 &= \ln 2 \qquad \text{Law 5} \\
x &= \ln 2 - 1 \\
&\approx -0.3069
\end{aligned}
$$

5.3 Differentiation of Exponential Functions

The Derivative of the Exponential Function

To study the effects of budget deficit-reduction plans at different income levels, it is important to know the income distribution of Canadian families. The graph of f shown in Figure 5.12 gives the number of Canadian families y (in thousands) as a function of their annual income x (in thousands of dollars) in 1990.

Observe that the graph of f rises very quickly and then tapers off. From the graph of f, you can see that the bulk of Canadian families earned less than $100,000 per year.

To analyze mathematical models involving exponential and logarithmic functions in greater detail, we need to develop rules for computing the derivative of these functions. We begin by looking at the rule for computing the derivative of the exponential function.

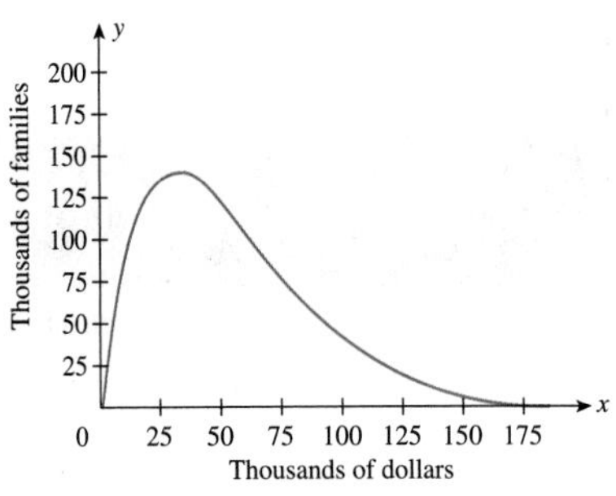

Figure 5.12
The graph of f shows the number of families versus their annual income.

Rule 1: Derivative of the Exponential Function with Base e

$$\frac{d}{dx}(e^x) = e^x$$

Thus, the derivative of the exponential function with base e is equal to the function itself. To demonstrate the validity of this rule, we compute

$$
\begin{aligned}
f'(x) &= \lim_{h \to 0} \frac{f(x + h) - f(x)}{h} \\
&= \lim_{h \to 0} \frac{e^{x+h} - e^x}{h}
\end{aligned}
$$

$$= \lim_{h \to 0} \frac{e^x(e^h - 1)}{h} \quad \text{Write } e^{x+h} = e^x e^h \text{ and factor.}$$

$$= e^x \lim_{h \to 0} \frac{e^h - 1}{h} \quad \text{Why?}$$

We have already encountered this limit in our discussion about e in Section 5.1. To evaluate

$$\lim_{h \to 0} \frac{e^h - 1}{h}$$

let's refer to Table 5.2, which is constructed with the aid of a calculator. From the table, we see that

$$\lim_{h \to 0} \frac{e^h - 1}{h} = 1$$

(Although a rigorous proof of this fact is possible, it is beyond the scope of this book.) Using this result, we conclude that

$$f'(x) = e^x \cdot 1 = e^x$$

as we set out to show.

Table 5.2

h	$\frac{e^h - 1}{h}$
0.1	1.0517
0.01	1.0050
0.001	1.0005
−0.1	0.9516
−0.01	0.9950
−0.001	0.9995

Example 1

Compute the derivative of each of the following functions:

a. $f(x) = x^2e^x$ **b.** $g(t) = (e^t + 2)^{3/2}$

Solution

a. The product rule gives

$$f'(x) = \frac{d}{dx}(x^2e^x) = x^2\frac{d}{dx}(e^x) + e^x\frac{d}{dx}(x^2)$$
$$= x^2e^x + e^x(2x) = xe^x(x + 2)$$

b. Using the general power rule, we find

$$g'(t) = \frac{3}{2}(e^t + 2)^{1/2}\frac{d}{dt}(e^t + 2) = \frac{3}{2}(e^t + 2)^{1/2}e^t = \frac{3}{2}e^t(e^t + 2)^{1/2}$$

EXPLORING WITH TECHNOLOGY

Consider the exponential function $f(x) = b^x$ ($b > 0$, $b \neq 1$).

1. Use the definition of the derivative of a function to show that
$$f'(x) = b^x \cdot \lim_{h \to 0} \frac{b^h - 1}{h}$$
2. Use the result of part 1 to show that
$$\frac{d}{dx}(2^x) = 2^x \cdot \lim_{h \to 0} \frac{2^h - 1}{h}$$
$$\frac{d}{dx}(3^x) = 3^x \cdot \lim_{h \to 0} \frac{3^h - 1}{h}$$
3. Use graphs to show that (to two decimal places)
$$\lim_{h \to 0} \frac{2^h - 1}{h} = 0.69 \quad \text{and} \quad \lim_{h \to 0} \frac{3^h - 1}{h} = 1.10$$
4. Conclude from the results of parts 2 and 3 that
$$\frac{d}{dx}(2^x) \approx (0.69)2^x \quad \text{and} \quad \frac{d}{dx}(3^x) \approx (1.10)3^x$$
Thus,
$$\frac{d}{dx}(b^x) = k \cdot b^x$$
where k is an appropriate constant.
5. The results of part 4 suggest that, for convenience, we pick the base b, where $2 < b < 3$, so that $k = 1$. This value of b is $e \approx 2.718281828. \ldots$ Thus,
$$\frac{d}{dx}(e^x) = e^x$$
This is why we prefer to work with the exponential function $f(x) = e^x$.

Applying the Chain Rule to Exponential Functions

To enlarge the class of exponential functions to be differentiated, we appeal to the chain rule to obtain the following rule for differentiating composite functions of the form $h(x) = e^{f(x)}$. An example of such a function is $h(x) = e^{x^2-2x}$. Here, $f(x) = x^2 - 2x$.

Rule 2: Chain Rule for Exponential Functions with Base e

If $f(x)$ is a differentiable function, then

$$\frac{d}{dx}(e^{f(x)}) = e^{f(x)}f'(x)$$

To see this, observe that if $h(x) = g[f(x)]$, where $g(x) = e^x$, then by virtue of the chain rule,

$$h'(x) = g'(f(x))f'(x) = e^{f(x)}f'(x)$$

since $g'(x) = e^x$.

As an aid to remembering the chain rule for exponential functions, observe that it has the following form:

$$\frac{d}{dx}(e^{f(x)}) = e^{f(x)} \cdot \text{derivative of exponent}$$

↑ Same ↑

Example 2

Find the derivative of each of the following functions:

a. $f(x) = e^{2x}$ **b.** $y = e^{-3x}$ **c.** $g(t) = e^{2t^2+t}$

Solution

a. $f'(x) = e^{2x}\frac{d}{dx}(2x) = e^{2x} \cdot 2 = 2e^{2x}$

b. $\frac{dy}{dx} = e^{-3x}\frac{d}{dx}(-3x) = -3e^{-3x}$

c. $g'(t) = e^{2t^2+t} \cdot \frac{d}{dt}(2t^2 + t) = (4t + 1)e^{2t^2+t}$

Example 3

Differentiate the function $y = xe^{-2x}$.

Solution

Using the product rule, followed by the chain rule, we find

$$\begin{aligned}\frac{dy}{dx} &= x\frac{d}{dx}e^{-2x} + e^{-2x}\frac{d}{dx}(x) \\ &= xe^{-2x}\frac{d}{dx}(-2x) + e^{-2x} \quad \text{Use the chain rule on the first term.} \\ &= -2xe^{-2x} + e^{-2x} \\ &= e^{-2x}(1 - 2x)\end{aligned}$$

Example 4

Differentiate the function $g(t) = \frac{e^t}{e^t + e^{-t}}$.

Solution

Using the quotient rule, followed by the chain rule, we find

$$g'(t) = \frac{(e^t + e^{-t})\frac{d}{dt}(e^t) - e^t\frac{d}{dt}(e^t + e^{-t})}{(e^t + e^{-t})^2}$$

$$= \frac{(e^t + e^{-t})e^t - e^t(e^t - e^{-t})}{(e^t + e^{-t})^2}$$

$$= \frac{e^{2t} + 1 - e^{2t} + 1}{(e^t + e^{-t})^2} \qquad e^0 = 1$$

$$= \frac{2}{(e^t + e^{-t})^2}$$

Example 5

In Section 5.5 we will discuss some practical applications of the exponential function

$$Q(t) = Q_0 e^{kt}$$

where Q_0 and k are positive constants and $t \in [0, \infty)$. A quantity $Q(t)$ growing according to this Law experiences exponential growth. Show that for a quantity $Q(t)$ experiencing exponential growth, the rate of growth of the quantity $Q'(t)$ at any time t is directly proportional to the amount of the quantity present.

Solution

Using the chain rule for exponential functions, we compute the derivative Q' of the function Q. Thus,

$$Q'(t) = q_0 e^{kt}\frac{d}{dt}(kt)$$

$$= Q_0 e^{kt}(k)$$

$$= kQ_0 e^{kt}$$

$$= kQ(t) \qquad Q(t) = Q_0 e^{kt}$$

which is the desired conclusion.

The Derivative of the Exponential Function with Base *b*

We will now develop the rule for differentiating an exponential function with any base b. For this we need to rewrite $y = b^x$ using the properties relating e^x and $\ln x$, and Logarithm Law 3 as follows

$$y = b^x = e^{\ln b^x} = e^{x \ln b}$$

Applying the chain rule for differentiating exponential functions with base e, we get

$$\frac{dy}{dx} = \frac{d}{dx}(b^x)$$

$$= \frac{d}{dx}(e^{x \ln b})$$

$$= e^{x \ln b}\frac{d}{dx}(x \ln b)$$

$$= e^{x \ln b} \ln b$$

which can be simplified to

$$\frac{dy}{dx} = (\ln b)b^x$$

Rule 3: Derivative of the Exponential Function with Base b

$$\frac{d}{dx}(b^x) = (\ln b)b^x$$

Rule 4: Chain Rule for Exponential Function with Base b

If $f(x)$ is a differentiable function, then

$$\frac{d}{dx}(b^{f(x)}) = (\ln b)b^{f(x)}f'(x)$$

Rule 4 is simply an extension of Rule 2, which should be readily observed.

Example 6

Find the derivative of each of the following functions:

a. $f(x) = 3^{5-x^2}$ **b.** $R(q) = \dfrac{q}{10^q + 5}$ **c.** $y = \sqrt{x}\,2^x$

Solution

a. Using Rule 4, we get

$$f'(x) = (\ln 3)3^{5-x^2}\frac{d}{dx}(5 - x^2) = (\ln 3)3^{5-x^2}(-2x) = -2(\ln 3)x \cdot 3^{5-x^2}$$

b. Using the quotient rule with Rule 3, we calculate

$$\begin{aligned} R'(q) &= \frac{\frac{d}{dx}(q)\cdot(10^q + 5) - q\cdot\frac{d}{dx}(10^q + 5)}{(10^q + 5)^2} \\ &= \frac{(1)(10^q + 5) - q(\ln 10)10^q}{(10^q + 5)^2} \\ &= \frac{5 + 10^q - (\ln 10)q10^q}{(10^q + 5)^2} \end{aligned}$$

c. Using the product rule followed by Rule 3, we find

$$\begin{aligned} y' &= \frac{d}{dx}(\sqrt{x})2^x + \sqrt{x}\frac{d}{dx}(2^x) = \frac{2^x}{2\sqrt{x}} + \sqrt{x}(\ln 2)2^x \\ &= \frac{2^{x-1}}{\sqrt{x}} + (\ln 2)\sqrt{x}\,2^x \end{aligned}$$

Example 7

Find the points of inflection of the function $f(x) = e^{-x^2}$.

Solution

The first derivative of f is

$$f'(x) = 2e^{-x^2}$$

Differentiating $f'(x)$ with respect to x yields

$$\begin{aligned} f''(x) &= (-2x)(-2xe^{-x^2}) - 2e^{-x^2} \\ &= 2e^{-x^2}(2x^2 - 1) \end{aligned}$$

Setting $f''(x) = 0$ gives

$$2e^{-x^2}(2x^2 - 1) = 0$$

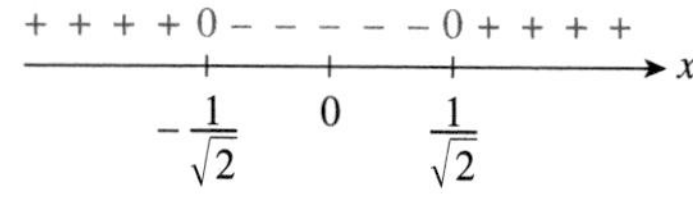

Figure 5.13
Sign diagram for f''

Since e^{-x^2} never equals zero for any real value of x, we see that $x = \pm 1/\sqrt{2}$ are the only candidates for inflection points of f. The sign diagram of f'', shown in Figure 5.13, tells us that both $x = -1/\sqrt{2}$ and $x = 1/\sqrt{2}$ give rise to inflection points of f.

Next,

$$f\left(-\frac{1}{\sqrt{2}}\right) = f\left(\frac{1}{\sqrt{2}}\right) = e^{-1/2}$$

and the inflection points of f are $(-1/\sqrt{2}, e^{-1/2})$ and $(1/\sqrt{2}, e^{-1/2})$. The graph of f appears in Figure 5.14.

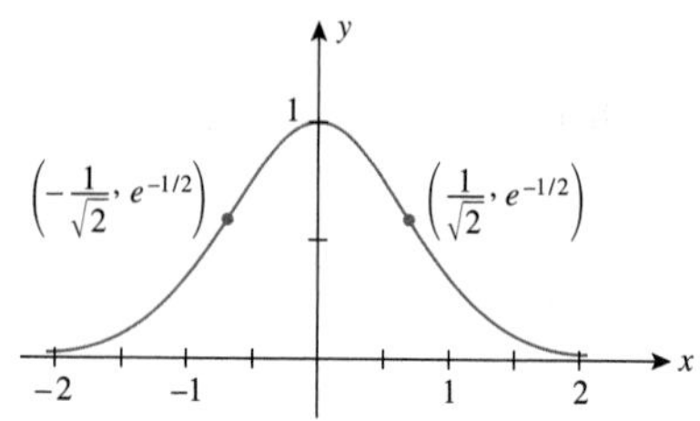

Figure 5.14
The graph of $y = e^{-x^2}$ has two inflection points.

Application

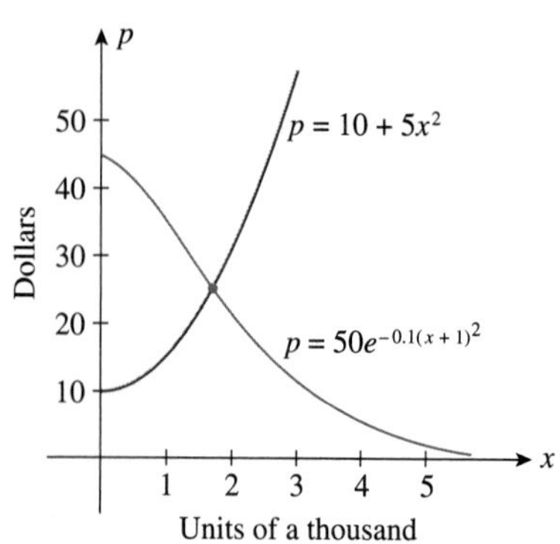

Figure 5.15
The equilibrium point is the point of intersection of the demand curve and the supply curve.

Example 8

Market Equilibrium The demand equation for the "Tempus" quartz wristwatch is given by

$$p = 50e^{-0.1(x+1)^2}$$

where x (measured in units of a thousand) is the quantity demanded per week and p is the unit wholesale price in dollars. National Importers, the supplier of the watches, will make x units (in thousands) available in the market if the unit wholesale price is

$$p = 10 + 5x^2$$

dollars. Find the equilibrium quantity and price.

Solution

We determine the equilibrium point by finding the point of intersection of the demand curve and the supply curve (Figure 5.15). To solve the system of equations

$$p = 50e^{-0.1(x+1)^2}$$
$$p = 10 + 5x^2$$

we substitute the second equation into the first, obtaining

$$10 + 5x^2 = 50e^{-0.1(x+1)^2} = 0$$
$$10 + 5x^2 - 50e^{-0.1(x+1)^2} = 0$$

To solve the last equation, we use the Newton–Raphson method with

$$\begin{aligned} f(x) &= 10 + 5x^2 - 50e^{-0.1(x+1)^2} \\ &= 5[2 + x^2 - 10e^{-0.1(x+1)^2}] \\ f'(x) &= 10x - 50e^{-0.1(x+1)^2} \cdot [-0.2(x + 1)] \\ &= 10[x + (x + 1)e^{-0.1(x+1)^2}] \end{aligned}$$

leading to the iterative formula

$$x_{n+1} = x_n - \frac{2 + x_n^2 - 10e^{-0.1(x_n+1)^2}}{2[x_n + (x_n + 1)e^{-0.1(x_2+1)^2}]}$$

Referring to Figure 5.15, we see that a reasonable initial estimate is $x_0 = 2$. We have

$$x_1 = 2 - \frac{2 + 2^2 - 10e^{-0.1(3)^2}}{2[2 + 3e^{-0.1(3)^2}]} \approx 1.69962$$

$$x_2 = 1.69962 - \frac{2 + (1.69962)^2 - 10e^{-0.1(2.69962)^2}}{2[1.69962 + 2.69962e^{-0.1(2.6692)^2}]} \approx 1.68899$$

$$x_3 = 1.68899 - \frac{2 + (1.68899)^2 - 10e^{-0.1(2.68899)^2}}{2[1.68899 + 2.68899e^{-0.1(2.68899)^2}]} \approx 1.68898$$

Thus, the equilibrium quantity is approximately 1.689 units, and the equilibrium wholesale price is given by

$$p = 10 + 5(1.68898)^2 \approx 24.263$$

or approximately $24.26 per watch.

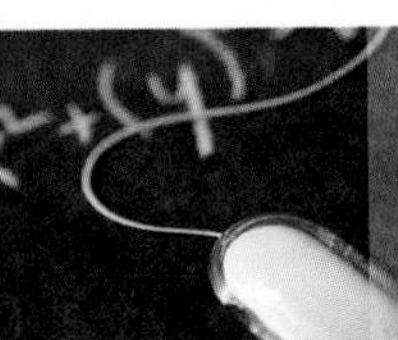

Self-Check Exercises 5.3

1. Let $f(x) = xe^{-x}$.
   a. Find the first and second derivatives of f.
   b. Find the relative extrema of f.
   c. Find the inflection points of f.

2. An industrial asset is being depreciated at a rate so that its book value t yr from now will be

$$V(t) = 50{,}000e^{-0.4t}$$

dollars. How fast will the book value of the asset be changing 3 yr from now?

Solutions to Self-Check Exercises 5.3 can be found on page 280.

5.3 Exercises

In Exercises 1–39, find the derivatives of the function.

1. $f(x) = e^{3x}$

2. $f(x) = 3e^x$

3. $g(t) = e^{-t}$

4. $f(x) = e^{-2x}$

5. $f(x) = e^x + x$

6. $f(x) = 2e^x - x^2$

7. $f(x) = x^3e^x$

8. $f(u) = u^2e^{-u}$

9. $f(x) = \frac{2e^x x}{x}$

10. $f(x) = \frac{x}{e^x}$

11. $f(x) = 3(e^x + e^{-x})$

12. $f(x) = \frac{e^x + e^{-x}}{2}$

13. $f(w) = \frac{e^w + 1}{e^w}$

14. $f(x) = \frac{e^x}{e^x + 1}$

15. $f(x) = 2e^{3x-1}$

16. $f(t) = 4e^{3t+2}$

17. $h(x) =$

18. $f(x) =$

19. $f(x) = 3e^{-1/x}$

20. $f(x) = e^{1/(2x)}$

21. $f(x) = (e^x + 1)^{25}$

22. $f(x) = (4 - e^{-3x})^3$

23. $f(x) = e^{\sqrt{x}}$

24. $f(t) = -e^{-\sqrt{2t}}$

25. $f(x) = (x - 1)e^{3x+2}$

26. $f(s) = (s^2 + 1)e^{-s^2}$

27. $f(x) = \frac{e^x - 1}{e^x + 1}$

28. $g(t) = \frac{e^{-t}}{1 + t^2}$

29. $f(x) = \frac{3^x}{5x}$

30. $f(x) = 7^x + 3x^4$

31. $f(x) = \sqrt{2^x - 4}$

32. $y = 10^{\sqrt{x}}$

33. $f(x) = x(1 - 4^x)$

34. $f(x) = 3^{1/x}$

35. $f(t) = 10^t\sqrt{t - 5}$

36. $f(t) = -2^{3t^5}$

37. $f(t) = 3^t 5^t$

38. $g(t) = \dfrac{5^{2-t} + 3}{t}$

39. $g(t) = \dfrac{2^t - 3}{2^t + t}$

In Exercises 40–47, find the second derivative of the function.

40. $f(x) = e^{-4x} + 2e^{3x}$

41. $f(t) = 3e^{-2t} - 5e^{-t}$

42. $f(x) = 2xe^{3x}$

43. $f(t) = t^2 e^{-2t}$

44. $f(x) = 2^{x^2}$

45. $f(t) = (t - 5)3^t$

46. $y = \dfrac{x}{5^x}$

47. $y = 6^{\sqrt{x}}$

48. Find an equation of the tangent line to the graph of $y = e^{2x-3}$ at the point $(\frac{3}{2}, 1)$.

49. Find an equation of the tangent line to the graph of $y = e^{-x^2}$ at the point $(1, 1/e)$.

In Exercises 50–52, use the Newton–Raphson method to approximate the indicated zero of each function. Continue with the iteration until two successive approximations differ by less than 0.0001.

50. The zero of $f(x) = e^{-x} - x$ between $x = 0$ and $x = 1$

51. The zero of $f(x) = e^x - (1/x)$ between $x = 0$ and $x = 1$

52. The zero of $f(x) = \ln x^2 - 0.7x + 1$ between $x = 6$ and $x = 7$

In Exercises 53–55, make a rough sketch of the graphs of each of the given pairs of functions. Use your sketch to approximate the point(s) of intersection of the two graphs and then apply the Newton–Raphson method to refine the approximation of the x-coordinate of the point of intersection.

53. $f(x) = e^{-x^2}$; $g(x) = x^2$

54. $f(x) = e^{-x}$; $g(x) = x - 1$

55. $f(x) = \ln x$; $g(x) = 2 - x$

A calculator is recommended for Exercises 56–70.

Business and Economics Applications

56. **Sales Promotion** The Lady Bug, a women's clothing chain store, found that t days after the end of a sales promotion the volume of sales was given by

$$S(t) = 20{,}000(1 + e^{-0.5t}) \qquad (0 \le t \le 5)$$

dollars.

a. Find the rate of change of The Lady Bug's sales volume when $t = 1$, $t = 2$, $t = 3$, and $t = 4$.

b. After how many days will the sales volume drop below \$27,400?

57. **Marginal Revenue** The unit selling price p (in dollars) and the quantity x demanded (in pairs) of a certain brand of women's gloves is given by the demand equation

$$p = 100e^{-0.0001x} \qquad (0 \le x \le 20{,}000)$$

a. Find the revenue function R.

Hint: $R(x) = px$.

b. Find the marginal revenue function R'.

c. What is the marginal revenue when $x = 10$?

58. **Price of Perfume** The monthly demand for a certain brand of perfume is given by the demand equation

$$p = 100e^{-0.0002x} + 150$$

where p denotes the retail unit price (in dollars) and x denotes the quantity (in 30 mL bottles) demanded.

a. Find the rate of change of the price per bottle when $x = 1000$ and when $x = 2000$.

b. What is the price per bottle when $x = 1000$? When $x = 2000$?

59. **Price of Wine** The monthly demand for a certain brand of table wine is given by the demand equation

$$p = 240\left(1 - \frac{3}{3 + e^{-0.0005x}}\right)$$

where p denotes the wholesale price per case (in dollars) and x denotes the number of cases demanded.

a. Find the rate of change of the price per case when $x = 1000$.

b. What is the price per case when $x = 1000$?

60. **Price of a Commodity** The price of a certain commodity in dollars per unit at time t (measured in weeks) is given by $p = 18 - 3e^{-2t} - 6e^{-t/3}$.

a. What is the price of the commodity at $t = 0$?

b. How fast is the price of the commodity changing at $t = 0$?

c. Find the equilibrium price of the commodity.

Hint: It is given by $\lim_{t \to \infty} p$.

61. **Price of a Commodity** The price of a certain commodity in dollars per unit at time t (measured in weeks) is given by $p = 8 + 4e^{-2t} + te^{-2t}$.

a. What is the price of the commodity at $t = 0$?

b. How fast is the price of the commodity changing at $t = 0$?

c. Find the equilibrium price of the commodity.

Hint: It's given by $\lim_{t \to \infty} p$. Also, use the fact that $\lim_{t \to \infty} te^{-2t} = 0$.

62. **Percentage of Population Relocating** Based on data obtained from Statistics Canada, the manager of Moving Van Lines estimates that the percent of the total population relocating in year t ($t = 0$ corresponds to the year 1960) may be approximated by the formula

$$P(t) = 20.6e^{-0.009t} \qquad (0 \le t \le 35)$$

Compute $P'(10)$, $P'(20)$, and $P'(30)$ and interpret your results.

63. **Online Banking** The percent of households using online banking may be approximated by the formula

$$f(t) = 1.5e^{0.78t} \qquad (0 \le t \le 4)$$

where t is measured in years, with $t = 0$ corresponding to the beginning of 2000.

a. What is the projected percent of households using online banking at the beginning of 2003?

b. How fast will the projected percent of households using online banking be changing at the beginning of 2003?

c. How fast will the rate of the projected percent of households using online banking be changing at the beginning of 2003?
Hint: We want $f''(3)$. Why?

64. **Air Travel** Air travel has been rising dramatically in the past 30 yr. A recent study by an independent travel organization predicts further exponential growth for air travel through 2010. The function

$$f(t) = 666e^{0.0413t} \qquad (0 \le t \le 11)$$

gives the number of passengers (in thousands) in year t, with $t = 0$ corresponding to 2000.

a. How many air passengers were there in 2000? What is the projected number of air passengers for 2005?

b. What is the rate of change of the number of air passengers in 2005?

65. **Energy Consumption of Appliances** The average energy consumption of the typical refrigerator/freezer manufactured by York Industries is approximately

$$C(t) = 1486e^{-0.073t} + 500 \qquad (0 \le t \le 20)$$

kilowatt-hours (kWh) per year, where t is measured in years, with $t = 0$ corresponding to 1972.

a. What was the average energy consumption of the York refrigerator/freezer at the beginning of 1972?

b. What is the rate of change of the average energy consumption?

c. All refrigerator/freezers manufactured as of January 1, 1990, must meet a 950-kWh/year maximum energy-consumption standard. Show that the York refrigerator/freezer satisfies this requirement.

Biological and Life Sciences Applications

66. **Blood Alcohol Level** The percentage of alcohol in a person's bloodstream t hr after drinking 240 mL of whiskey is given by

$$A(t) = 0.23te^{-0.4t} \qquad (0 \le t \le 12)$$

a. What is the percentage of alcohol in a person's bloodstream after $\frac{1}{2}$ hr? After 8 hr?

b. How fast is the percentage of alcohol in a person's bloodstream changing after $\frac{1}{2}$ hr? After 8 hr?
Source: Encyclopedia Britannica

67. **Polio Immunization** Polio, a once-feared killer, declined markedly in North America in the 1950s after Jonas Salk developed the inactivated polio vaccine and mass immunization of children took place. The number of polio cases from the beginning of 1959 to the beginning of 1963 is approximated by the function

$$N(t) = 5.3e^{0.0095t^2-0.85t-0.85t} \qquad (0 \le t \le 4)$$

where $N(t)$ gives the number of polio cases (in thousands) and t is measured in years with $t = 0$ corresponding to the beginning of 1959.

a. Show that the function N is decreasing over the time interval under consideration.

b. How fast was the number of polio cases decreasing at the beginning of 1959? At the beginning of 1962? (*Comment:* Following the introduction of the oral vaccine developed by Dr. Albert B. Sabin in 1963, polio in North America has, for all practical purposes, been eliminated.)

68. **Spread of an Epidemic** During a flu epidemic, the total number of students at a Canadian college who had contracted influenza by the xth day was given by

$$N(x) = \frac{3000}{1 + 99e^{-x}} \qquad (x \ge 0)$$

a. How many students had influenza initially?

b. Derive an expression for the rate at which the disease was being spread.

c. Sketch the graph of N. What was the total number of students who contracted influenza during that particular epidemic?

69. **Weights of Children** The Ehrenberg equation

$$W = 2.4e^{1.84h}$$

gives the relationship between the height (in metres) and the average weight W (in kilograms) for children between 5 and 13 yr of age.

a. What is the average weight of a 10-yr-old child who stands at 1.6 m tall?

b. Use differentials to estimate the change in the average weight of a 10-yr-old child whose height increases from 1.6 m to 1.65 m.

70. **Oil Used to Fuel Productivity** A study on worldwide oil use was prepared for a major oil company. The study predicted that the amount of oil used to fuel productivity in a certain country is given by

$$f(t) = 1.5 + 1.8te^{-1.2t} \qquad (0 \le t \le 4)$$

where $f(t)$ denotes the number of barrels per \$1000 of economic output and t is measured in decades ($t = 0$ corresponds to 1965). Compute $f'(0)$, $f'(1)$, $f'(2)$, and $f'(3)$ and interpret your results.

In Exercises 71–74, determine whether the statement is true or false. If it is true, explain why it is true. If it is false, give an example to show why it is false.

71. If $f(x) = 3^x$, then $f'(x) = x \cdot 3^{x-1}$.

72. If $f(x) = e^{\pi}$, then $f'(x) = e^{\pi}$.

73. If $f(x) = \pi^x$, then $f'(x) = \pi^x$.

74. If $x^2 + e^y = 10$, then .

Solutions to Self-Check Exercises 5.3

1. a. Using the product rule, we obtain

$$f'(x) = x\frac{d}{dx}e^{-x} = e^{-x}\frac{d}{dx}x$$

$$= -xe^{-x} + e^{-x} = (1-x)e^{-x}$$

Using the product rule once again, we obtain

$$f''(x) = (1-x)\frac{d}{dx}e^{-x} + e^{-x}\frac{d}{dx}(1-x)$$

$$= (1-x)(-e^{-x}) + e^{-x}(-1)$$

$$= -e^{-x} + xe^{-x} - e^{-x} = (x-2)e^{-x}$$

b. Setting $f'(x) = 0$ gives

$$(1-x)e^{-x} = 0$$

Since $e^{-x} \neq 0$, we see that $1 - x = 0$, and this gives $x = 1$ as the only critical point of f. The sign diagram of f' shown in the accompanying figure tells us that the point $(1, e^{-1})$ is a relative maximum of f.

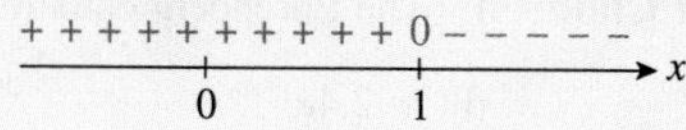

c. Setting $f''(x) = 0$ gives $x - 2 = 0$, so $x = 2$ is a candidate for an inflection point of f. The sign diagram of f'' (see the accompanying figure) shows that $(2, 2e^{-2})$ is an inflection point of f.

- - - - - - - - - - - 0 + + + + +
x
0   2

2. The rate change of the book value of the asset t yr from now is

$$V'(t) = 50{,}000\frac{d}{dt}e^{-0.4t}$$

$$= 50{,}000(-0.4)e^{-0.4t} = -20{,}000e^{-0.4t}$$

Therefore, 3 yr from now the book value of the asset will be changing at the rate of

$$V'(3) = -20{,}000e^{-0.4(3)} = -20{,}000e^{-1.2} \approx -6023.88$$

—that is, decreasing at the rate of approximately \$6024/year.

5.4 Differentiation of Logarithmic Functions

The Derivative of ln x

Let's now turn our attention to the differentiation of logarithmic functions.

Rule 5: Derivative of ln x

$$\frac{d}{dx}\ln|x| = \frac{1}{x} \qquad (x \neq 0)$$

To derive Rule 5, suppose $x > 0$ and write $f(x) = \ln x$ in the equivalent form

$$x = e^{f(x)}$$

Differentiating both sides of the equation with respect to x, we find, using the chain rule,

$$1 = e^{f(x)} \cdot f'(x)$$

from which we see that

$$f'(x) = \frac{1}{e^{f(x)}}$$

or, since $e^{f(x)} = x$,

$$f'(x) = \frac{1}{x}$$

as we set out to show. You are asked to prove the rule for the case $x < 0$ in Exercise 72, page 289.

Example 1

Compute the derivative of each of the following functions:

a. $f(x) = x\ln x$ b. $g(x) = \dfrac{\ln x}{x}$

Solution

a. Using the product rule, we obtain

$$f'(x) = \frac{d}{dx}(x \ln x) = x\frac{d}{dx}(\ln x) + (\ln x)\frac{d}{dx}(x)$$

$$= x\left(\frac{1}{x}\right) + \ln x = 1 + \ln x$$

b. Using the quotient rule, we obtain

$$g'(x) = \frac{x\frac{d}{dx}(\ln x) - (\ln x)\frac{d}{dx}(x)}{x^2} = \frac{x\left(\frac{1}{x}\right) - \ln x}{x^2} = \frac{1 - \ln x}{x^2}$$

GROUP DISCUSSION

You can derive the formula for the derivative of $f(x) = \ln x$ directly from the definition of the derivative, as follows.

1. Show that

$$f'(x) = \lim_{h \to 0} \frac{f(x + h) - f(x)}{h} = \lim_{h \to 0} \ln\left(1 + \frac{h}{x}\right)^{1/h}$$

2. Put $m = x/h$ and note that $m \to \infty$ as $h \to 0$. Then, $f'(x)$ can be written in the form

$$f'(x) = \lim_{m \to \infty} \ln\left(1 + \frac{1}{m}\right)^{m/x}$$

3. Finally, use both the fact that the natural logarithmic function is continuous and the definition of the number e to show that

$$f'(x) = \frac{1}{x}\ln\left[\lim_{m \to \infty}\left(1 + \frac{1}{m}\right)^m\right] = \frac{1}{x}$$

The Chain Rule for Logarithmic Functions

To enlarge the class of logarithmic functions to be differentiated, we appeal once more to the chain rule to obtain the following rule for differentiating composite functions of the form $h(x) = \ln f(x)$, where $f(x)$ is assumed to be a positive differentiable function.

Rule 6: Chain Rule for Logarithmic Functions with Base e

If $f(x)$ is a differentiable function, then

$$\frac{d}{dx}[\ln f(x)] = \frac{f'(x)}{f(x)} \qquad [f(x) > 0]$$

To see this, observe that $h(x) = g[f(x)]$, where $g(x) = \ln x$ $(x > 0)$. Since $g'(x) = 1/x$, we have, using the chain rule,

$$h'(x) = g'(f(x))f'(x)$$

$$= \frac{1}{f(x)}f'(x) = \frac{f'(x)}{f(x)}$$

Observe that in the special case $f(x) = x$, $h(x) = \ln x$, so the derivative of h is, by Rule 5, given by $h'(x) = 1/x$.

Example 2

Find the derivative of the function $f(x) = \ln(x^2 + 1)$.

Solution

Using Rule 6, we see immediately that

$$f'(x) = \frac{\frac{d}{dx}(x^2 + 1)}{x^2 + 1} = \frac{2x}{x^2 + 1}$$

The Derivative of the Logarithmic Function with Base *b*

We will now generalize our findings from the previous sections to the derivative of the logarithmic function with any base b by considering

$$y = \log_b x$$

We apply the definition of the logarithm to rewrite this equation as

$$b^y = x$$

and differentiate implicitly. Using Rule 4 of Section 5.3 we obtain

$$\frac{d}{dx}(b^y) = \frac{d}{dx}(x)$$

$$(\ln b)b^y \frac{dy}{dx} = 1$$

$$\frac{dy}{dx} = \frac{1}{(\ln b)b^y}$$

This can be expressed in terms of x as

$$\frac{dy}{dx} = \frac{1}{(\ln b)x}$$

Observe that the domain of $y = \log_b x$ is given by $D = \{x > 0\}$. Since there are no further restrictions for the derivative, this domain also carries over to the derivative. Notice that we could have obtained the derivative without using implicit differentiation by applying Logarithmic Law 6 to $y = \log_b x$ and using Rule 3 from Basic Rules of Differentiation (Section 3.4).

GROUP DISCUSSION

Let $y = \log_b f(x)$. Assume that f is differentiable and $f(x) > 0$. Show that Rule 8 is correct by first using the logarithmic definition to rewrite the given function, and then differentiating implicitly.

Rule 7: Derivative of the Logarithmic Function with Base b, ($b > 0, b \neq 1$)

$$\frac{d}{dx}(\log_b x) = \frac{1}{(\ln b)x}, \; x > 0$$

Rule 8: Chain Rule for Logarithmic Functions with Base b, ($b > 0, b \neq 1$)

If $f(x)$ is a differentiable function, then

$$\frac{d}{dx}(\log_b f(x)) = \frac{f'(x)}{(\ln b)f(x)}, \; f(x) > 0$$

Example 3

Find the derivative of each of the following functions:

a. $C(x) = \frac{\log x}{x + 5}$ **b.** $y = \log_2(\sqrt{x} - 5)$ **c.** $f(t) = (\log_5(4t))^{2/3}$

Solution

a. Using the quotient rule with Rule 7, we get

$$C'(x) = \frac{\frac{d}{dx}(\log x) \cdot (x + 5) - \log x \cdot \frac{d}{dx}(x + 5)}{(x + 5)^2}$$

$$= \frac{\frac{1}{(\ln 10)x}(x + 5) - \log x \cdot (1)}{(x + 5)^2}$$

$$= \frac{\frac{x + 5}{(\ln 10)x} - \log x}{(x + 5)^2}$$

which can be simplified to

$$C'(x) = \frac{x + 5 - (\ln 10)x \log x}{(\ln 10)x(x + 5)^2}$$

b. By Rule 8, we obtain

$$y' = \frac{1}{(\ln 2)\sqrt{x - 5}} \frac{d}{dx}(\sqrt{x - 5})$$

$$= \frac{1}{(\ln 2)\sqrt{x - 5}} \frac{1}{2\sqrt{x - 5}}$$

$$= \frac{1}{2(\ln 2)(\sqrt{x - 5})^2}$$

Note that we must have $x > 5$. Therefore, we need to be careful when we want to simplify the expression on the right side of the equation, and state the restriction as well:

$$y' = \frac{1}{2(\ln 2)(x - 5)}, x > 5$$

c. Using the chain rule followed by Rule 8, we get

$$f'(t) = \frac{2}{3}(\log_5(4t))^{-1/3} \frac{d}{dx}(\log_5(4t))$$

$$= \frac{2}{3}(\log_5(4t))^{-1/3} \frac{1}{(\ln 5)\log_5(4t)} \frac{d}{dx}(4t)$$

$$= \frac{2}{3}(\log_5(4t))^{-1/3} \frac{1}{(\ln 5)\log_5(4t)} 4$$

which becomes

$$f'(t) = \frac{8}{3(\ln 5)(\log_5(4t))^{4/3}}$$

after we simplify.

When differentiating functions involving logarithms, the rules of logarithms may be used to advantage, as shown in Examples 4 and 5.

Example 4

Differentiate the function $y = \ln[(x^2 + 1)(x^3 + 2)^6]$.

Solution

We first rewrite the given function using the properties of logarithms:

$$y = \ln[(x^2 + 1)(x^3 + 2)^6]$$

$$= \ln(x^2 + 1) + \ln(x^3 + 2)^6 \qquad \ln mn = \ln m + \ln n$$

$$= \ln(x^2 + 1) + 6\ln(x^3 + 2) \qquad \ln m^n = n \ln m$$

Differentiating and using Rule 6, we obtain

$$y' = \frac{\frac{d}{dx}(x^2 + 1)}{x^2 + 1} + \frac{6\frac{d}{dx}(x^3 + 2)}{x^3 + 2}$$

$$= \frac{2x}{x^2 + 1} + \frac{6(3x^2)}{x^3 + 2} = \frac{2x}{x^2 + 1} + \frac{18x^2}{x^3 + 2}$$

EXPLORING WITH TECHNOLOGY

Use a graphing utility to plot the graphs of $f(x) = \ln x$; its first derivative function, $f'(x) = 1/x$; and its second derivative function $f''(x) = -1/x^2$, using the same viewing window $[0, 4] \cdot [-3, 3]$.

1. Describe the properties of the graph of f revealed by studying the graph of $f'(x)$. What can you say about the rate of increase of f for large values of x?
2. Describe the properties of the graph of f revealed by studying the graph of $f''(x)$. What can you say about the concavity of f for large values of x?

Example 5

Find the derivative of the function $g(t) = \ln(t^2 e^{-t^2})$.

Solution

Here again, to save a lot of work, we first simplify the given expression using the properties of logarithms. We have

$$g(t) = \ln(t^2 e^{-t^2})$$

$$= \ln t^2 + \ln e^{-t^2} \qquad \ln mn = \ln m + \ln n$$

$$= 2\ln t - t^2 \qquad \ln m^n = n \ln m \quad \text{and} \quad \ln e = 1$$

Therefore,

$$g'(t) = \frac{2}{t} - 2t = \frac{2(1 - t^2)}{t}$$

Logarithmic Differentiation

As we saw in the last two examples, the task of finding the derivative of a given function can be made easier by first applying the laws of logarithms to simplify the function. We now illustrate a process called logarithmic differentiation, which not only simplifies the calculation of the derivatives of certain functions but also enables us to compute the derivatives of functions we could not otherwise differentiate using the techniques developed thus far.

Example 6

Differentiate $y = x(x + 1)(x^2 + 1)$, using logarithmic differentiation.

Solution

First, we take the natural logarithm on both sides of the given equation, obtaining

$$\ln y = \ln x(x + 1)(x^2 + 1)$$

Next, we use the properties of logarithms to rewrite the right-hand side of this equation, obtaining

$$\ln y = \ln x + \ln(x + 1) + \ln(x^2 + 1)$$

If we differentiate both sides of this equation, we have

$$\frac{d}{dx}\ln y = \frac{d}{dx}[\ln x + \ln(x + 1) + \ln(x^2 + 1)]$$

$$= \frac{1}{x} + \frac{1}{x + 1} + \frac{2x}{x^2 + 1} \qquad \text{Use Rule 4.}$$

To evaluate the expression on the left-hand side, note that y is a function of x. Therefore, writing $y = f(x)$ to remind us of this fact, we have

$$\frac{d}{dx}\ln y = \frac{d}{dx}\ln[f(x)] \qquad \text{Write } y = f(x).$$

$$= \frac{f'(x)}{f(x)} \qquad \text{Use Rule 6.}$$

$$= \frac{y'}{y} \qquad \text{Return to using } y \text{ instead of } f(x).$$

Therefore, we have

$$\frac{y'}{y} = \frac{1}{x} + \frac{1}{x + 1} + \frac{2x}{x^2 + 1}$$

Finally, solving for y', we have

$$y' = y\left(\frac{1}{x} + \frac{1}{x + 1} + \frac{2x}{x^2 + 1}\right)$$

$$= x(x + 1)(x^2 + 1)\left(\frac{1}{x} + \frac{1}{x + 1} + \frac{2x}{x^2 + 1}\right)$$

Before considering other examples, let's summarize the important steps involved in logarithmic differentiation.

Finding $\frac{dy}{dx}$ by Logarithmic Differentiation

1. Take the natural logarithm on both sides of the equation and simplify the resulting equation using the properties of logarithms.
2. Differentiate both sides of the equation with respect to x.
3. Solve the resulting equation for $\frac{dy}{dx}$.

Example 7

Differentiate $y = x^2(x - 1)(x^2 + 4)^3$.

Solution

Taking the natural logarithm on both sides of the given equation and using the laws of logarithms, we obtain

$$\ln y = \ln x^2(x - 1)(x^2 + 4)^3$$

$$= \ln x^2 + \ln(x - 1) + \ln(x^2 + 4)^3$$

$$= 2\ln x + \ln(x - 1) + 3\ln(x^2 + 4)$$

Differentiating both sides of the equation with respect to x, we have

$$\frac{d}{dx}\ln y = \frac{y'}{y} = \frac{2}{x} + \frac{1}{x-1} + 3\cdot\frac{2x}{x^2+4}$$

Finally, solving for y', we have

$$\begin{aligned} y' &= y\left(\frac{2}{x} + \frac{1}{x-1} + \frac{6x}{x^2+4}\right) \\ &= x^2(x-1)(x^2+4)^3\left(\frac{2}{x} + \frac{1}{x-1} + \frac{6x}{x^2+4}\right) \end{aligned}$$

Example 8

Differentiate $f(x) = \dfrac{3\sqrt{x-2}(x^3+5)^2e^{2x}}{x^2-4x}$.

Solution

We first rewrite f as

$$y = \frac{3(x-2)^{1/2}(x^3+5)^2e^{2x}}{x(x-4)}$$

Next, we take the natural logarithm on both sides of the given equation and use the laws of logarithm to obtain

$$\begin{aligned} \ln y &= \ln\left(\frac{3(x-2)^{1/2}(x^3+5)^2e^{2x}}{x(x-4)}\right) \\ &= \ln(3(x-2)^{1/2}(x^3+5)^2e^{2x}) - \ln(x(x-4)) \\ &= \ln 3 + \ln(x-2)^{1/2} + \ln(x^3+5)^2 + \ln e^{2x} - \ln(x) - \ln(x-4) \\ &= \ln 3 + \frac{1}{2}\ln(x-2) + 2\ln(x^3+5) + 2x - \ln x - \ln(x-4) \end{aligned}$$

Differentiating both sides of the equation with respect to x, we have

$$\begin{aligned} \frac{d}{dx}\ln y &= \frac{d}{dx}\left(\ln 3 + \frac{1}{2}\ln(x-2) + 2\ln(x^3+5) + 2x - \ln x - \ln(x-4)\right) \\ \frac{1}{y}\frac{dy}{dx} &= 0 + \frac{1}{2}\frac{1}{x-2} + 2\frac{3x^2}{x^3+5} + 2 - \frac{1}{x} - \frac{1}{x-4} \\ &= 2 + \frac{1}{2(x-2)} + \frac{2}{x^3+5} - \frac{1}{x} - \frac{1}{x-4} \end{aligned}$$

Solving for $\dfrac{dy}{dx}$, we finally get

$$\begin{aligned} \frac{dy}{dx} &= y\left(2 + \frac{1}{2(x-2)} + \frac{2}{x^3+5} - \frac{1}{x} - \frac{1}{x-4}\right) \\ &= \frac{3(x-2)^{1/2}(x^3+5)^2e^{2x}}{x(x-4)}\left(2 + \frac{1}{2(x-2)} + \frac{2}{x^3+5} - \frac{1}{x} - \frac{1}{x-4}\right) \end{aligned}$$

Example 9

Find the derivative of $f(x) = x^x$ $(x > 0)$.

Solution

A word of caution! This function is neither a power function nor an exponential function. Taking the natural logarithm on both sides of the equation gives

$$\ln f(x) = \ln x^x = x \ln x$$

Differentiating both sides of the equation with respect to x, we obtain

$$\begin{aligned}\frac{f'(x)}{f(x)} &= x\frac{d}{dx}\ln x + (\ln x)\frac{d}{dx}x \\ &= x\left(\frac{1}{x}\right) + \ln x \\ &= 1 + \ln x\end{aligned}$$

Therefore,

$$f'(x) = f(x)(1 + \ln x) = x^x(1 + \ln x)$$

EXPLORING WITH TECHNOLOGY

Refer to Example 9.

1. Use a graphing utility to plot the graph of $f(x) = x^x$, using the viewing window $[0, 2] \cdot [0, 2]$. Then use ZOOM and TRACE to show that
$$\lim_{x \to 0^+} f(x) = 1$$
2. Use the results of part 1 and Example 9 to show that $\lim_{x \to 0^+} f'(x) = -\infty$. Justify your answer.

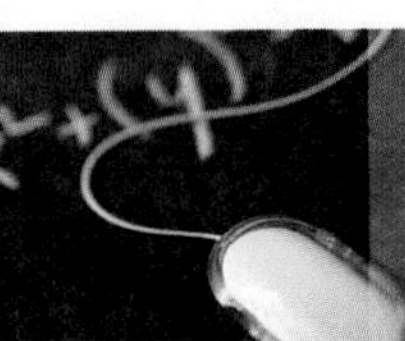

Self-Check Exercises 5.4

1. Find an equation of the tangent line to the graph of $f(x) = x \ln(2x + 3)$ at the point $(-1, 0)$.
2. Use logarithmic differentiation to compute y', given $y = (2x + 1)^3(3x + 4)^5$.

Solutions to Self-Check Exercises 5.4 can be found on page 289.

5.4 Exercises

In Exercises 1–43, find the derivative of the function.

1. $f(x) = 5 \ln x$
2. $f(x) = \ln 5x$
3. $f(x) = \ln(x + 1)$
4. $g(x) = \ln(2x + 1)$
5. $f(x) = \ln x^8$
6. $h(t) = 2 \ln t^5$
7. $f(x) = \ln\sqrt{x}$
8. $f(x) = \ln(\sqrt{x} + 1)$
9. $f(x) = \ln\dfrac{1}{x^2}$
10. $f(x) = \ln\dfrac{1}{2x^3}$
11. $f(x) = \ln(4x^2 - 6x + 3)$
12. $f(x) = \ln(3x^2 - 2x + 1)$
13. $f(x) = \ln\dfrac{2x}{x + 1}$
14. $f(x) = \ln\dfrac{x + 1}{x - 1}$
15. $f(x) = x^2 \ln x$
16. $f(x) = 3x^2 \ln 2x$
17. $f(x) = \dfrac{2 \ln x}{x}$
18. $f(x) = \dfrac{3 \ln x}{x^2}$
19. $f(u) = \ln(u - 2)^3$
20. $f(x) = \ln(x^3 - 3)^4$
21. $f(x) = \sqrt{\ln x}$
22. $f(x) = \sqrt{\ln x + x}$

23. $f(x) = (\ln x)^3$

24. $f(x) = 2(\ln x)^{3/2}$

25. $f(x) = \ln(x^3 + 1)$

26. $f(x) = \ln\sqrt{x^2 - 4}$

27. $f(x) = e^x \ln x$

28. $f(x) = e^x \ln\sqrt{x + 3}$

29. $f(t) = e^{2t} \ln(t + 1)$

30. $g(t) = t^2 \ln(e^{2t} + 1)$

31. $f(x) = \dfrac{\ln x}{x}$

32. $g(t) = \dfrac{t}{\ln t}$

33. $f(x) = \log_3 x(1 - x)$

34. $f(x) = (\log_3 x)^{1/3}$

35. $f(t) = \log_5(6 - \sqrt{t})$

36. $f(t) = 2^{\log_5 t}$

37. $f(t) = t \log_2 t$

38. $g(t) = \dfrac{t + 3}{\log_2 t}$

39. $g(t) = \dfrac{\log(3t)}{1 - t}$

40. $f(x) = \dfrac{\log x}{x}$

41. $f(x) = \sqrt{\log_6 x}$

42. $f(x) = \sqrt{x} \log_2 x$

43. $y = \dfrac{\log_7 x}{x^2 + 3}$

In Exercises 44–51, find the second derivative of the function.

44. $f(x) = \ln 2x$

45. $f(x) = \ln(x + 5)$

46. $f(x) = \ln(x^2 + 2)$

47. $f(x) = (\ln x)^2$

48. $f(x) = (\log_2 x)^5$

49. $f(t) = \log_5 t^3$

50. $y = \log\sqrt{x}$

51. $y = \dfrac{\log_2 x}{x^2}$

In Exercises 52–61, use logarithmic differentiation to find the derivative of the function.

52. $y = (x + 1)^2(x + 2)^3$

53. $y = (3x + 2)^4(5x - 1)^2$

54. $y = (x - 1)^2(x + 1)^3(x + 3)^4$

55. $y = \sqrt{3x + 5}(2x - 3)^4$

56. $y = \dfrac{(2x^2 - 1)^5}{\sqrt{x + 1}}$

57. $y = \dfrac{\sqrt{4 + 3x^2}}{\sqrt[3]{x^2 + 1}}$

58. $y = 3^x$

59. $y = x^{x+2}$

60. $y = (x^2 + 1)^x$

61. $y = x^{\ln x}$

62. Find an equation of the tangent line to the graph of $y = x \ln x$ at the point (1, 0).

63. Find an equation of the tangent line to the graph of $y = \ln x^2$ at the point (2, ln 4).

Biological and Life Sciences Applications

64. **Absorption of Light** When light passes through a window glass, some of it is absorbed. It can be shown that if $r\%$ of the light is absorbed by a glass of thickness w, then the percent of light absorbed by a piece of glass of thickness nw is

$$A(n) = 100\left[1 - \left(1 - \frac{r}{100}\right)^n\right] \qquad (0 \leq r \leq 100)$$

a. Show that A is an increasing function of n on $(0, \infty)$ if $0 < r < 100$.
Hint: Use logarithmic differentiation.
b. Sketch the graph of A for the special case where $r = 10$.
c. Evaluate and interpret your result.

65. **Lambert's Law of Absorption** Lambert's law of absorption states that the light intensity $I(x)$ (in calories/square centimetre/second) at a depth of x m as measured from the surface of a material is given by $I = I_0 a^x$, where I_0 and a are positive constants.
a. Find the rate of change of the light intensity with respect to x at a depth of x m from the surface of the material.
b. Using the result of part (a), conclude that the rate of change $I'(x)$ at a depth of x m is proportional to $I(x)$. What is the constant of proportion?

66. **Magnitude of Earthquakes** On the Richter scale, the magnitude R of an earthquake is given by the formula

$$R = \log \frac{I}{I_0}$$

where I is the intensity of the earthquake being measured and I_0 is the standard reference intensity.
a. What is the magnitude of an earthquake that has intensity 1 million times that of I_0?
b. Suppose an earthquake is measured with a magnitude of 6 on the Richter scale with an error of at most 2%. Use differentials to find the error in the intensity of the earthquake.
Hint: Observe that $I = I_0 10^R$ and use logarithmic differentiation.

67. **Strain on Vertebrae** The strain (percent of compression) on the lumbar vertebral disks in an adult human as a function of the load x (in kilograms) is given by

$$f(x) = 7.2956 \ln(0.0645012x^{0.95} + 1)$$

What is the rate of change of the strain with respect to the load when the load is 100 kg? When the load is 500 kg?

Source: Benedek and Villars, *Physics with Illustrative Examples from Medicine and Biology*

68. **Heights of Children** For children between the ages of 5 and 13 years old, the Ehrenberg equation

$$\ln W = \ln 2.4 + 1.84h$$

gives the relationship between the weight W (in kilograms) and the height h (in metres) of a child. Use differentials to estimate the change in the weight of a child who grows from 1 m to 1.1 m.

69. **Weber–Fechner Law** The Weber–Fechner law

$$R = k \ln \frac{S}{S_0}$$

where k is a positive constant, describes the relationship between a stimulus S and the resulting response R. Here, S_0, a positive constant, is the threshold level.
a. Show that $R = 0$ if the stimulus is at the threshold level S_0.
b. The derivative dR/dS is the *sensitivity* corresponding to the stimulus level S and measures the capability to detect small changes in the stimulus level. Show that dR/dS is inversely proportional to S and interpret your result.

In Exercises 70 and 71, determine whether the statement is true or false. If it is true, explain why it is true. If it is false, give an example to show why it is false.

70. If $f(x) = \ln 5$, then $f'(x) = \frac{1}{5}$.

71. If $f(x) = \ln a^x$, then $f'(x) = \ln a$.

72. Prove that $\frac{d}{dx} \ln|x| = \frac{1}{x} (x \neq 0)$ for the case $x < 0$.

73. Use the definition of the derivative to show that

$$\lim_{x \to 0} \frac{\ln(x+1)}{x} = 1$$

Solutions to Self-Check Exercises 5.4

1. The slope of the tangent line to the graph of f at any point $(x, f(x))$ lying on the graph of f is given by $f'(x)$. Using the product rule, we find

$$f'(x) = \frac{d}{dx}[x \ln(2x+3)]$$
$$= x \frac{d}{dx} \ln(2x+3) + \ln(2x+3) \cdot \frac{d}{dx}(x)$$
$$= x\left(\frac{2}{2x+3}\right) + \ln(2x+3) \cdot 1$$
$$= \frac{2x}{2x+3} + \ln(2x+3)$$

In particular, the slope of the tangent line to the graph of f at the point $(-1, 0)$ is

$$f'(-1) = \frac{-2}{-2+3} + \ln 1 = -2$$

Therefore, using the point-slope form of the equation of a line, we see that a required equation is

$$y - 0 = -2(x+1)$$
$$y = -2x - 2$$

2. Taking the logarithm on both sides of the equation gives

$$\ln y = \ln(2x+1)^3(3x+4)^5$$
$$= \ln(2x+1)^3 + \ln(3x+4)^5$$
$$= 3\ln(2x+1) + 5\ln(3x+4)$$

Differentiating both sides of the equation with respect to x, keeping in mind that y is a function of x, we obtain

$$\frac{d}{dx}(\ln y) = \frac{y'}{y} = 3 \cdot \frac{2}{2x+1} + 5 \cdot \frac{3}{3x+4}$$
$$3\left(\frac{2}{2x+1} + \frac{5}{3x+4}\right)$$

and

$$y' = 3(2x+1)^3(3x+4)^5\left(\frac{2}{2x+1} + \frac{5}{3x+4}\right)$$

5.5 Exponential Functions as Mathematical Models

Exponential Growth

Many problems arising from practical situations can be described mathematically in terms of exponential functions or functions closely related to the exponential function. In this section we look at some applications involving exponential functions from the fields of the life and social sciences.

In Section 5.1 we saw that the exponential function $f(x) = b^x$ is an increasing function when $b > 1$. In particular, the function $f(x) = e^x$ shares this property. From this result one may deduce that the function $Q(t) = Q_0 e^{kt}$, where Q_0 and k are positive constants, has the following properties:

1. $Q(0) = Q_0$
2. $Q(t)$ increases "rapidly" without bound as t increases without bound (Figure 5.16).

Property 1 follows from the computation

$$Q(0) = Q_0 e^0 = Q_0$$

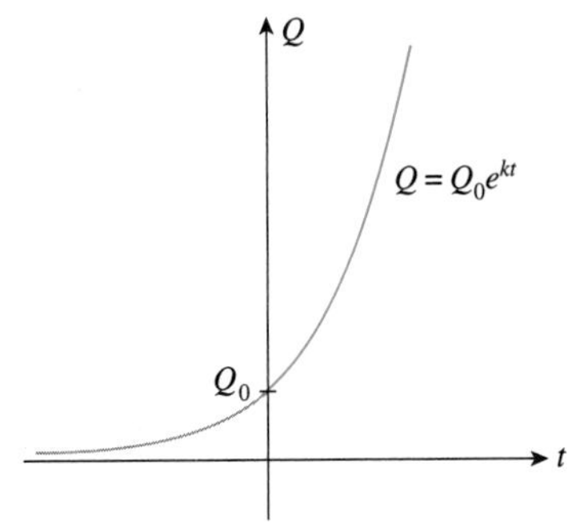

Figure 5.16
Exponential growth

Next, to study the rate of change of the function $Q(t)$, we differentiate it with respect to t, obtaining

$$\begin{aligned} Q'(t) &= \frac{d}{dt}(Q_0 e^{kt}) \\ &= Q_0 \frac{d}{dt}(e^{kt}) \\ &= kQ_0 e^{kt} \\ &= kQ(t) \end{aligned} \tag{5}$$

Since $Q(t) > 0$ (because Q_0 is assumed to be positive) and $k > 0$, we see that $Q'(t) > 0$ and so $Q(t)$ is an increasing function of t. Our computation has in fact shed more light on an important property of the function $Q(t)$. Equation (5) says that the rate of increase of the function $Q(t)$ is proportional to the amount $Q(t)$ of the quantity present at time t. The implication is that as $Q(t)$ increases, so does the *rate of increase* of $Q(t)$, resulting in a very rapid increase in $Q(t)$ as t increases without bound.

Thus, the exponential function

$$Q(t) = Q_0 e^{kt} \; (0 \leq t < \infty) \tag{6}$$

provides us with a mathematical model of a quantity $Q(t)$ that is initially present in the amount of $Q(0) = Q_0$ and whose rate of growth at any time t is directly proportional to the amount of the quantity present at time t. Such a quantity is said to exhibit **exponential growth,** and the constant k is called the **growth constant.** Interest earned on a fixed deposit when compounded continuously exhibits exponential growth. Other examples of exponential growth follow.

Example 1

Growth of Bacteria Under ideal laboratory conditions, the number of bacteria in a culture grows in accordance with the law $Q(t) = Q_0 e^{kt}$, where Q_0 denotes the number of bacteria initially present in the culture, k is some constant determined by the strain of bacteria under consideration, and t is the elapsed time measured in hours. Suppose 10,000 bacteria are present initially in the culture and 60,000 present 2 hours later.

a. How many bacteria will there be in the culture at the end of 4 hours?
b. What is the rate of growth of the population after 4 hours?

Solution

a. We are given that $Q(0) = Q_0 = 10{,}000$, so $Q(t) = 10{,}000e^{kt}$. Next, the fact that 60,000 bacteria are present 2 hours later translates into $Q(2) = 60{,}000$. Thus,

$$\begin{aligned} 60{,}000 &= 10{,}000e^{2k} \\ e^{2k} &= 6 \end{aligned}$$

Taking the natural logarithm on both sides of the equation, we obtain

$$\begin{aligned} \ln e^{2k} &= \ln 6 \\ 2k \ln e &= \ln 6 \\ 2k &= \ln 6 \qquad \text{Since } \ln e = 1 \\ k &= \frac{1}{2} \ln 6 \\ (k &\approx 0.8959) \end{aligned}$$

Thus, the number of bacteria present at any time t is given by

$$\begin{aligned} Q(t) &= 10{,}000e^{t \cdot \frac{1}{2} \ln 6} \\ &= 10{,}000e^{\frac{t}{2} \ln 6} \end{aligned}$$

$$= 10{,}000e^{\ln(6)^{t/2}}$$

$$= 10{,}000(6)^{t/2}$$

Or we could have approximated Q with $Q(t) = 10{,}000e^{0.8959t}$, which is less accurate and does not give as much information. Notice that the factor $(6)^{t/2}$ means that every two hours the initial quantity multiplies by a factor of 6.

In particular, the number of bacteria present in the culture at the end of 4 hours is given by

$$Q(4) = 10{,}000(6)^{4/2} = 10{,}000(6)^2 = 360{,}000$$

Compare this to the approximation $Q(4) = 10{,}000e^{0.8959 \cdot 4} \approx 360{,}029$.

b. The rate of growth of the bacteria population at any time t is given by

$$Q'(t) = kQ(t)$$

Thus, using the result from part (a), we find that the rate at which the population is growing at the end of 4 hours is

$$\begin{aligned} Q'(4) &= kQ(4) \\ &= \left(\frac{1}{2}\ln 6\right)360{,}000 \\ &\approx 322{,}517 \end{aligned}$$

or approximately 322,517 bacteria per hour. Once again the approximation would yield a less accurate result with $Q'(4) = kQ(4) \approx 0.8959 \cdot 360{,}029 \approx 322{,}550$ bacteria per hour.

Exponential Decay

In contrast to exponential growth, a quantity exhibits **exponential decay** if it decreases at a rate that is directly proportional to its size. Such a quantity may be described by the exponential function

$$Q(t) = Q_0e^{-kt} \qquad t \in [0, \infty) \tag{7}$$

where the positive constant Q_0 measures the amount present initially ($t = 0$) and k is some suitable positive number, called the **decay constant.** The choice of this number is determined by the nature of the substance under consideration. The graph of this function is sketched in Figure 5.17.

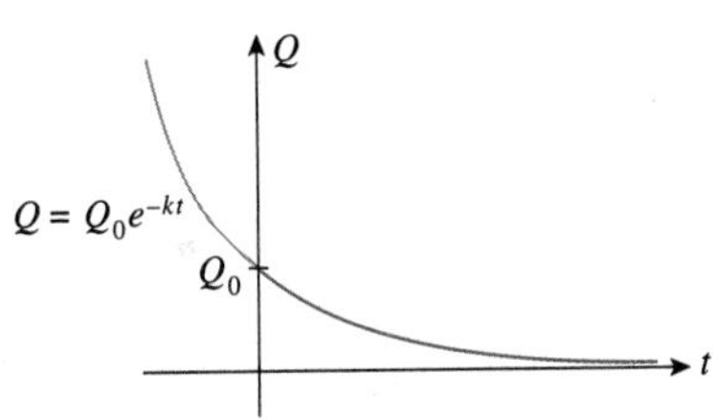

Figure 5.17
Exponential decay

To verify the properties ascribed to the function $Q(t)$, we simply compute

$$\begin{aligned} Q(0) &= Q_0e^0 = Q_0 \\ Q'(t) &= \frac{d}{dt}(Q_0e^{-kt}) \\ &= Q_0\frac{d}{dt}(e^{-kt}) \\ &= -kQ_0e^{-kt} = -kQ(t) \end{aligned}$$

Example 2

Radioactive Decay Radioactive substances decay exponentially. For example, the amount of radium present at any time t obeys the law $Q(t) = Q_0e^{-kt}$, where Q_0 is the initial amount present and k is a suitable positive constant. The **half-life of a radioactive element** is the time required for a given amount to be reduced by one-half. Now, it is known that the half-life of radium is approximately 1600 years. Suppose initially there are 200 milligrams of pure radium. Find the amount left after t years. What is the amount left after 800 years?

Solution

The initial amount of radium present is 200 milligrams, so $Q(0) = Q_0 = 200$. Thus, $Q(t) = 200e^{-kt}$. Next, the datum concerning the half-life of radium implies that $Q(1600) = 100$, and this gives

$$100 = 200e^{-1600k}$$

$$e^{-1600k} = \frac{1}{2}$$

Taking the natural logarithm on both sides of this equation yields

$$-1600k \ln e = \ln \frac{1}{2}$$

$$-1600k = \ln \frac{1}{2} \qquad \ln e = 1$$

$$k = -\frac{1}{1600} \ln \left(\frac{1}{2}\right) (\approx 0.0004322)$$

Therefore, the amount of radium left after t years is

$$\begin{aligned} Q(t) &= 200e^{-(-1/1600) \ln (1/2) \cdot t} \\ &= 200e^{(t/1600) \ln (1/2)} \\ &= 200e^{\ln (1/2)^{t/1600}} \\ &= 200\left(\frac{1}{2}\right)^{t/1600} \end{aligned}$$

We want to point out again that the factor $(\frac{1}{2})^{t/1600}$ means that every 1600 years the initial amount halves. For a less accurate model, we could have used $Q(t) = 200e^{-0.0004334t}$. We compute the amount of radium left after 800 years to be

$$\begin{aligned} Q(800) &= 200\left(\frac{1}{2}\right)^{800/1600} \\ &= 200\left(\frac{1}{2}\right)^{1/2} \\ &\approx 141 \end{aligned}$$

or approximately 141 milligrams.

Example 3

Carbon 14, a radioactive isotope of carbon, has a half-life of 5770 years. What is its decay constant?

Solution

We have $Q(t) = Q_0e^{-kt}$. Since the half-life of the element is 5770 years, half of the substance is left at the end of that period; that is,

$$Q(5770) = Q_0e^{-5770k} = \frac{1}{2}Q_0$$

$$e^{-5770k} = \frac{1}{2}$$

Taking the natural logarithm on both sides of this equation, we have

$$\ln e^{-5770k} = \ln \frac{1}{2}$$

$$-5770k \ln e = \ln \frac{1}{2}$$

$$-5770k = \ln \frac{1}{2}$$

$$k = -\frac{\ln(1/2)}{5770}$$

$$(k \approx 0.00012)$$

Carbon-14 dating is a well-known method used by anthropologists to establish the age of animal and plant fossils. This method assumes that the proportion of carbon 14 (C-14) present in the atmosphere has remained constant over the past 50,000 years. Professor Willard Libby, recipient of the Nobel Prize in chemistry in 1960, proposed this theory.

The amount of C-14 in the tissues of a living plant or animal is constant. However, when an organism dies, it stops absorbing new quantities of C-14, and the amount of C-14 in the remains diminishes because of the natural decay of the radioactive substance. Thus, the approximate age of a plant or animal fossil can be determined by measuring the amount of C-14 present in the remains.

Example 4

Carbon-14 Dating A skull from an archeological site has one-tenth the amount of C-14 that it originally contained. Determine the approximate age of the skull.

Solution

Here,

$$\begin{aligned} Q(t) &= Q_0 e^{-kt} \\ &= Q_0 e^{-(-1/5770) \ln (1/2) \cdot t} \\ &= Q_0 e^{\ln (1/2) t/5770} \\ &= Q_0 \left(\frac{1}{2}\right)^{t/5770} \end{aligned}$$

where Q_0 is the amount of C-14 present originally and k, the decay constant, is equal to $(-1/5770) \ln (1/2)$ (see Example 3). Since $Q(t) = (1/10)Q_0$, we have

$$\frac{1}{10} Q_0 = Q_0 \left(\frac{1}{2}\right)^{t/5770}$$

$$\ln \frac{1}{10} = \ln \left(\frac{1}{2}\right)^{t/5770} \quad \text{Take the natural logarithm on both sides}$$

$$\ln 10 = \ln (2)^{t/5770} \quad \text{Simplify the logarithms}$$

$$\ln 10 = (t/5770) \ln 2$$

$$t = \frac{5770 \ln 10}{\ln 2}$$

$$\approx 19{,}168$$

Learning Curves

The next example shows how the exponential function may be applied to describe certain types of learning processes. Consider the function

$$Q(t) = C - Ae^{-kt}$$

where C, A, and k are positive constants. To sketch the graph of the function Q, observe that its y-intercept is given by $Q(0) = C - A$. Next, we compute

$$Q'(t) = kAe^{-kt}$$

Since both k and A are positive, we see that $Q'(t) > 0$ for all values of t. Thus, $Q(t)$ is an increasing function of t. Also,

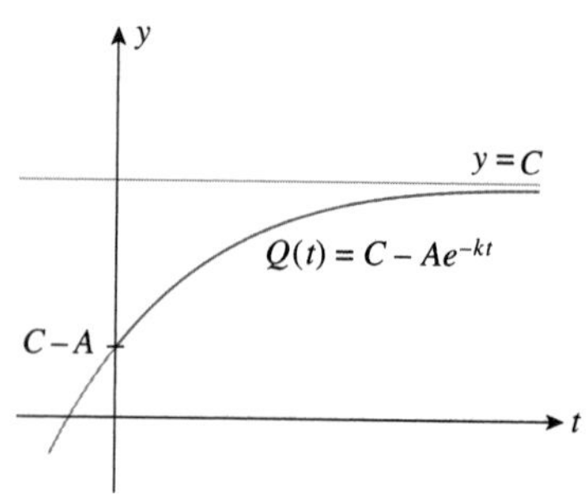

Figure 5.18
A learning curve

$$\begin{aligned} \lim_{t\to\infty} Q(t) &= \lim_{t\to\infty}(C - A^{-kt}) \\ &= \lim_{t\to 0} C - \lim_{t\to\infty} Ae^{-kt} \\ &= C \end{aligned}$$

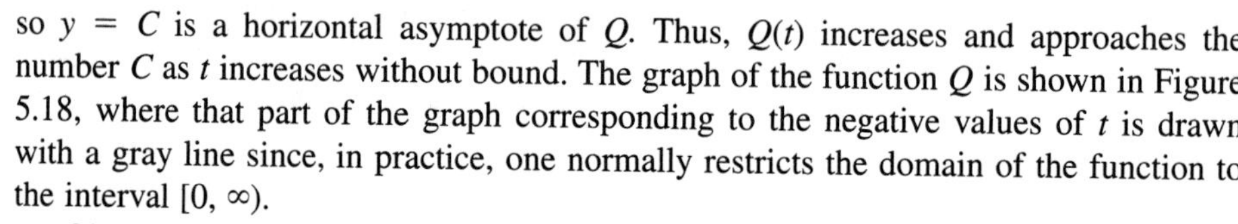

so $y = C$ is a horizontal asymptote of Q. Thus, $Q(t)$ increases and approaches the number C as t increases without bound. The graph of the function Q is shown in Figure 5.18, where that part of the graph corresponding to the negative values of t is drawn with a gray line since, in practice, one normally restricts the domain of the function to the interval $[0, \infty)$.

Observe that $Q(t)$ $(t > 0)$ increases rather rapidly initially but that the rate of increase slows down considerably after a while. To see this, we compute

$$\lim_{t\to\infty} Q'(t) = \lim_{t\to\infty} kAe^{-kt} = 0$$

This behaviour of the graph of the function Q closely resembles the learning pattern experienced by workers engaged in highly repetitive work. For example, the productivity of an assembly-line worker increases very rapidly in the early stages of the training period. This productivity increase is a direct result of the worker's training and accumulated experience. But the rate of increase of productivity slows as time goes by, and the worker's productivity level approaches some fixed level due to the limitations of the worker and the machine. Because of this characteristic, the graph of the function $Q(t) = C - Ae^{-kt}$ is often called a **learning curve.**

Example 5

Assembly Time The Camera Division of Eastman Optical produces a 35-mm single-lens reflex camera. Eastman's training department determines that after completing the basic training program, a new, previously inexperienced employee will be able to assemble

$$Q(t) = 50 - 30e^{-0.5t}$$

model F cameras per day, t months after the employee starts work on the assembly line.

a. How many model F cameras can a new employee assemble per day after basic training?

b. How many model F cameras can an employee with 1 month of experience assemble per day? An employee with 2 months of experience? An employee with 6 months of experience?

c. How many model F cameras can the average experienced employee assemble per day?

Solution

a. The number of model F cameras a new employee can assemble is given by

$$Q(0) = 50 - 30 = 20$$

b. The number of model F cameras that an employee with 1 month of experience, 2 months of experience, and 6 months of experience can assemble per day is given by

$$Q(1) = 50 - 30e^{-0.5} \approx 31.80$$

$$Q(2) = 50 - 30e^{-1} \approx 38.96$$

$$Q(6) = 50 - 30e^{-3} \approx 48.51$$

or approximately 32, 39, and 49, respectively.

c. As t increases without bound, $Q(t)$ approaches 50. Hence, the average experienced employee can ultimately be expected to assemble 50 model F cameras per day.

Other applications of the learning curve are found in models that describe the dissemination of information about a product or the velocity of an object dropped into a viscous medium.

Logistic Growth Functions

Our last example of an application of exponential functions to the description of natural phenomena involves the **logistic** (also called the **S-shaped,** or **sigmoidal) curve,** which is the graph of the function

$$Q(t) = \frac{A}{1 + Be^{-kt}}$$

where A, B, and k are positive constants. The function Q is called a **logistic growth function,** and the graph of the function Q is sketched in Figure 5.19.

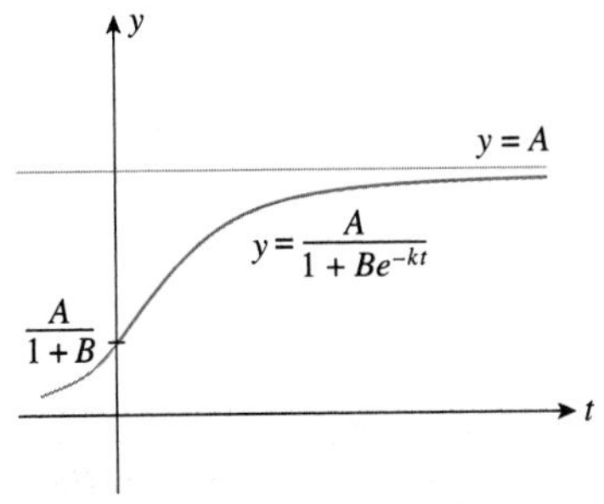

Figure 5.19
A logistic curve

Observe that $Q(t)$ increases rather rapidly for small values of t. In fact, for small values of t, the logistic curve resembles an exponential growth curve. However, the *rate of growth* of $Q(t)$ decreases quite rapidly as t increases and $Q(t)$ approaches the number A as t increases without bound.

Thus, the logistic curve exhibits both the property of rapid growth of the exponential growth curve and the "saturation" property of the learning curve. Because of these characteristics, the logistic curve serves as a suitable mathematical model for describing many natural phenomena. For example, if a small number of rabbits were introduced to a tiny island in the South Pacific, the rabbit population might be expected to grow very rapidly at first, but the growth rate would decrease quickly as overcrowding, scarcity of food, and other environmental factors affected it. The population would eventually stabilize at a level compatible with the life-support capacity of the environment. Models describing the spread of rumours and epidemics are other examples of the application of the logistic curve.

Example 6

Spread of Flu The number of people in a small town in the Yukon who contracted influenza after t days during a flu epidemic is approximated by the exponential model

$$Q(t) = \frac{5000}{1 + 1249e^{-kt}}$$

If 40 people contracted the flu by day 7, find how many people contracted the flu by day 15.

Solution

The given information implies that

$$Q(7) = 40 \quad \text{and} \quad Q(7) = \frac{5000}{1 + 1249e^{-7k}} = 40$$

Thus,

$$40(1 + 1249e^{-7k}) = 5000$$

$$1 + 1249e^{-7k} = \frac{5000}{40} = 125$$

$$e^{-7k} = \frac{124}{1249}$$

$$-7k = \ln \frac{124}{1249}$$

$$k = -\frac{\ln \frac{124}{1249}}{7} \approx 0.33$$

Therefore, the number of people who contracted the flu after t days is given by

$$Q(t) = \frac{5000}{1 + 1249e^{-0.33t}}$$

In particular, the number of people who contracted the flu by day 15 is given by

$$Q(15) = \frac{5000}{1 + 1249e^{-15(0.33)}}$$

$$\approx 508$$

or approximately 508 people.

EXPLORING WITH TECHNOLOGY

Refer to Example 6.

1. Use a graphing utility to plot the graph of the function Q, using the viewing window [0, 40] · [0, 5000].
2. Find how long it takes for the first 1000 people to contract the flu.

Hint: Plot the graphs of $y_1 = Q(t)$ and $y_2 = 1000$ and find the point of intersection of the two graphs.

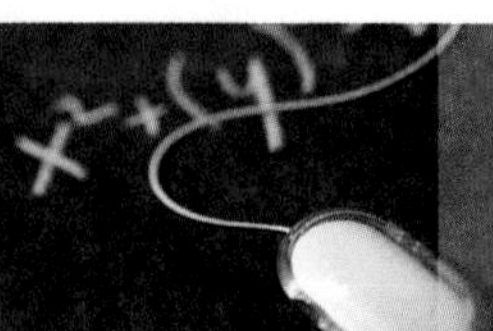

Self-Check Exercise 5.5

Suppose the population (in millions) of a country at any time t grows in accordance with the rule

$$P = \left(P_0 + \frac{I}{k}\right)e^{kt} - \frac{I}{k}$$

where P denotes the population at any time t, k is a constant reflecting the natural growth rate of the population, I is a constant giving the (constant) rate of immigration into the country, and P_0 is the total population of the country at time $t = 0$. The population of Canada in 1980 ($t = 0$) was 24.5 million. If the natural growth rate is 0.9% annually ($k = 0.007$) and net immigration is allowed at the rate of 0.23 million people per year ($I = 0.23$) until the end of the century, what is the population of Canada expected to be in 2010?

Source: Statistics Canada

Solutions to Self-Check Exercise 5.5 can be found on page 299.

5.5 Exercises

 **A calculator is recommended for this exercise set.**

1. **Exponential Growth** Given that a quantity $Q(t)$ is described by the exponential growth function

$$Q(t) = 400e^{0.05t}$$

where t is measured in minutes, answer the following questions:

a. What is the growth constant?

b. What quantity is present initially?

c. Complete the following table of values:

t	0	10	20	100	1000
Q					

2. **Exponential Decay** Given that a quantity $Q(t)$ exhibiting exponential decay is described by the function

$$Q(t) = 2000e^{-0.06t}$$

where t is measured in years, answer the following questions:

a. What is the decay constant?

b. What quantity is present initially?

c. Complete the following table of values:

t	0	5	10	20	100
Q					

Business and Economics Applications

3. **Reliability of Computer Chips** The percent of a certain brand of computer chips that will fail after t yr of use is estimated to be

$$P(t) = 100(1 - e^{-0.1t})$$

a. What percent of this brand of computer chips are expected to be usable after 3 yr?

b. Evaluate . Did you expect this result?

4. **Lay Teachers at Roman Catholic Schools** The change from religious to lay teachers at Roman Catholic schools has been partly attributed to the decline in the number of women and men entering religious orders. The percent of teachers who are lay teachers is given by

$$f(t) = \frac{98}{1 + 2.77e^{-t}} \qquad (0 \le t \le 4)$$

where t is measured in decades, with $t = 0$ corresponding to the beginning of 1960.

a. What percent of teachers were lay teachers at the beginning of 1990?

b. How fast was the percent of lay teachers changing at the beginning of 1990?

5. **Percent of Households with VCRs** According to estimates by Paul Kroger Associates, the percent of households that own VCRs is given by

$$P(t) = \frac{68}{1 + 21.67e^{-0.62t}} \qquad (0 \le t \le 12)$$

where t is measured in years, with $t = 0$ corresponding to the beginning of 1985. What percent of households owned VCRs at the beginning of 1985? At the beginning of 1995?

6. **Radioactive Decay** A radioactive substance decays according to the formula

$$Q(t) = Q_0e^{-kt}$$

where $Q(t)$ denotes the amount of the substance present at time t (measured in years), Q_0 denotes the amount of the substance present initially, and k (a positive constant) is the decay constant.

a. Show that half-life of the substance is $\bar{t} = \ln 2/k$.

b. Suppose a radioactive substance decays according to the formula

$$Q(t) = 20e^{-0.0001238\,\bar{t}}$$

How long will it take for the substance to decay to half the original amount?

7. **Resale Value** A certain piece of machinery was purchased 3 yr ago by Garland Mills for \$500,000. Its present resale value is \$320,000. Assuming that the machine's resale value decreases exponentially, what will it be 4 yr from now?

8. **Effect of Advertising on Sales** Metro Department Store found that t wk after the end of a sales promotion the volume of sales was given by a function of the form

$$S(t) = B + Ae^{-kt} \qquad (0 \le t \le 4)$$

where $B = 50{,}000$ and is equal to the average weekly volume of sales before the promotion. The sales volumes at the end of the first and third weeks were \$83,515 and \$65,055, respectively. Assume that the sales volume is decreasing exponentially.

a. Find the decay constant k.

b. Find the sales volume at the end of the fourth week.

c. How fast is the sales volume dropping at the end of the fourth week?

9. **Demand for Computers** Universal Instruments found that the monthly demand for its new line of Galaxy Home Computers t mo after placing the line on the market was given by

$$D(t) = 2000 - 1500e^{-0.05t} \qquad (t > 0)$$

Graph this function and answer the following questions:

a. What is the demand after 1 mo? After 1 yr? After 2 yr? After 5 yr?

b. At what level is the demand expected to stabilize?

c. Find the rate of growth of the demand after the tenth month.

10. **Price of a Commodity** The unit price of a certain commodity is given by

$$p = f(t) = 6 + 4e^{-2t}$$

where p is measured in dollars and t is measured in months.

a. Find the rate of change of the unit price.

b. Evaluate $\lim_{t \to \infty} f(t)$. (*Remark:* This value is called the *equilibrium price* of the commodity, and in this case, we have *price stability.*)

Biological and Life Sciences Applications

11. **Growth of Bacteria** The growth rate of the bacterium *Escherichia coli,* a common bacterium found in the human intestine, is proportional to its size. Under ideal laboratory conditions, when this bacterium is grown in a nutrient broth medium, the number of cells in a culture doubles approximately every 20 min.

a. If the initial cell population is 100, determine the function $Q(t)$ that expresses the exponential growth of the number of cells of this bacterium as a function of time t (in minutes).

b. How long will it take for a colony of 100 cells to increase to a population of 1 million?

c. If the initial cell population were 1000, how would this alter our model?

12. **World Population** The world population at the beginning of 1990 was 5.3 billion. Assume that the population continues to grow at its present rate of approximately 2%/year and find the function $Q(t)$ that expresses the world population (in billions) as a function of time t (in years), with $t = 0$ corresponding to the beginning of 1990.

a. Using this function, complete the following table of values and sketch the graph of the function Q.

Year	1990	1995	2000	2005
World Population				

Year	2010	2015	2020	2025
World Population				

b. Find the estimated rate of growth in 2000.

13. **World Population** Refer to Exercise 12.
a. If the world population continues to grow at its present rate of approximately 2%/year, find the length of time t_0 required for the world population to triple in size.
b. Using the time t_0 found in part (a), what would be the world population if the growth rate were reduced to 1.8%/yr?

14. **Atmospheric Pressure** If the temperature is constant, then the atmospheric pressure P (in kilopascals) varies with the altitude above sea level h in accordance with the law

$$P = p_0 e^{-kh}$$

where p_0 is the atmospheric pressure at sea level and k is a constant. If the atmospheric pressure is 100 kPa at sea level and 85 kPa at 1200 m, find the atmospheric pressure at an altitude of 3600 m. How fast is the atmospheric pressure changing with respect to altitude at an altitude of 3600 m?

15. **Radioactive Decay** The radioactive element polonium decays according to the law

$$Q(t) = Q_0 \cdot 2^{-(t/140)}$$

where Q_0 is the initial amount and the time t is measured in days. If the amount of polonium left after 280 days is 20 mg, what was the initial amount present?

16. **Radioactive Decay** Phosphorus 32 has a half-life of 14.2 days. If 100 g of this substance is present initially, find the amount present after t days. What amount will be left after 7.1 days? How fast is the phosphorus 32 decaying when $t = 7.1$?

17. **Nuclear Fallout** Strontium 90, a radioactive isotope of strontium, is present in the fallout resulting from nuclear explosions. It is especially hazardous to animal life, including humans, because, upon ingestion of contaminated food, it is absorbed into the bone structure. Its half-life is 27 yr. If the amount of strontium 90 in a certain area is found to be four times the "safe" level, find how much time must elapse before an "acceptable level" is reached.

18. **Carbon-14 Dating** Wood deposits recovered from an archeological site contain 20% of the carbon 14 they originally contained. How long ago did the tree from which the wood was obtained die?

19. **Carbon-14 Dating** Skeletal remains of the so-called "Pittsburgh Man," unearthed in Pennsylvania, had lost 82% of the carbon 14 they originally contained. Determine the approximate age of the bones.

20. **Learning Curves** The Canadian Association of Court Reporters finds that the average student taking Advanced Machine Shorthand, an intensive 20-wk course, progresses according to the function

$$Q(t) = 120(1 - e^{-0.05t}) + 60 \qquad (0 \le t \le 20)$$

where $Q(t)$ measures the number of words (per minute) of dictation that the student can take in machine shorthand after t wk in the course. Sketch the graph of the function Q and answer the following questions:
a. What is the beginning shorthand speed for the average student in this course?
b. What shorthand speed does the average student attain halfway through the course?
c. How many words per minute can the average student take after completing this course?

21. **Lengths of Fish** The length (in centimetres) of a typical Pacific halibut t yr old is approximately

$$f(t) = 200(1 - 0.956e^{-0.18t})$$

a. What is the length of a typical 5-yr-old Pacific halibut?
b. How fast is the length of a typical 5-yr-old Pacific halibut increasing?
c. What is the maximum length a typical Pacific halibut can attain?

22. **Spread of an Epidemic** During a flu epidemic, the number of children in the Woodbridge Community School System who contracted influenza after t days was given by

$$Q(t) = \frac{1000}{1 + 199e^{-0.8t}}$$

a. How many children were stricken by the flu after the first day?
b. How many children had the flu after 10 days?
c. How many children eventually contracted the disease?

23. **Growth of a Fruit Fly Population** On the basis of data collected during an experiment, a biologist found that the growth of the fruit fly (*Drosophila*) with a limited food supply could be approximated by the exponential model

$$N(t) = \frac{400}{1 + 39e^{-0.16t}}$$

where t denotes the number of days since the beginning of the experiment.
a. What was the initial fruit fly population in the experiment?
b. What was the maximum fruit fly population that could be expected under this laboratory condition?
c. What was the population of the fruit fly colony on the 20th day?
d. How fast was the population changing on the 20th day?

24. **Population Growth in the Twenty-First Century** The North American population can be approximated by the function

$$P(t) = \frac{616.5}{1 + 4.02e^{-0.5t}}$$

where $P(t)$ is measured in millions of people and t is measured in 30-yr intervals, with $t = 0$ corresponding to 1930. What is the expected population of North America in 2020 ($t = 3$)?

25. **Dissemination of Information** Three hundred students attended the dedication ceremony of a new building on a college campus. The president of the traditionally female college

announced a new expansion program, which included plans to make the college coeducational. The number of students who learned of the new program t hr later is given by the function

$$f(t) = \frac{3000}{1 + Be^{-kt}}$$

If 600 students on campus had heard about the new program 2 hr after the ceremony, how many students had heard about the policy after 4 hr? How fast was the news spreading 4 hr after the ceremony?

26. **Chemical Mixtures** Two chemicals react to form another chemical. Suppose the amount of the chemical formed in time t (in hours) is given by

$$x(t) = \frac{15[1 - (\frac{2}{3})^{3t}]}{1 - \frac{1}{4}(\frac{2}{3})^{3t}}$$

where $x(t)$ is measured in pounds. How many pounds of the chemical are formed eventually?
Hint: You need to evaluate $\lim_{t\to\infty} x(t)$.

27. **Von-Bertalanffy Growth Function** The length (in centimetres) of a common commercial fish is approximated by the von-Bertalanffy growth function

$$f(t) = a(1 - be^{-kt})$$

where a, b, and k are positive constants.
a. Find the rate of change of the growth function.
b. Show that $\lim_{t\to\infty} f(t) = a$.

Solutions to Self-Check Exercise 5.5

We are given that $P_0 = 24.5$, $k = 0.007$, and $I = 0.23$. So

$$P(t) = \left(24.5 + \frac{0.23}{0.007}\right)e^{0.007t} - \frac{0.23}{0.007}$$
$$\approx 57.357e^{0.007t} - 32.857$$

Therefore, the population in 2010 will be given by

$$P(30) \approx 57.357e^{0.007 \cdot 30} - 32.857 \approx 37.903$$

or approximately 38 million.

CHAPTER 5 Summary of Principal Formulas and Terms

Formulas

1. Exponential function with base b	$y = b^x$
	$1 = \lim_{n\to 0} \frac{e^n - 1}{n}$
2. The number e	$e = \lim_{m\to\infty}\left(1 + \frac{1}{m}\right)^m = 2.71828\ldots$
3. Exponential function with base e	$y = e^x$
4. Logarithmic function with base b	$y = \log_b x$
5. Logarithmic function with base e	$y = \ln x$
6. Inverse properties of $\ln x$ and e	$\ln e^x = x, x \in 1R$ and $e^{\ln x} = x, x > 0$
7. Derivative of the exponential function base b	$\frac{d}{dx}(e^x) = e^x$
8. Chain rule for exponential functions base b	$\frac{d}{dx}\left(e^u = e^u \frac{du}{dx}\right)$
9. Derivative of the exponential function base b	$\frac{d}{dx}(b^x) = (\ln b)b^x$
10. Chain rule for exponential function base b	$\frac{d}{dx}(b^u) = (\ln b)b^u \frac{dn}{dx}$
11. Derivative of the logarithmic function base b	$\frac{d}{dx}\ln\lvert x\rvert = \frac{1}{x}$

16. Chain rule for logarithmic functions base b $\qquad \frac{d}{dx}(\ln u) = \frac{1}{u}\frac{du}{dx}$

17. Derivative of the logarithmic function base b $\qquad \frac{d}{dx}(\log_b |x|) = \frac{1}{(\ln b)x}$

18. Chain rule for logarithmic function base b $\qquad \frac{d}{dx}(\log_b u) = \frac{1}{(\ln b)u}\frac{du}{dx}$

Terms

exponential function with base b and exponent x (255)
logarithm of x to the base b ($\log_b x$) (262)
common logarithm (263)
natural logarithm (283)
laws of logarithms (263)
log (263)
ln (263)
logarithmic function with base b (265)
logarithmic differentiation (284)
exponential growth (290)
growth constant (290)
exponential decay (291)
decay constant (291)
half-life of a radioactive element (291)
learning curve (294)
logistic (S-shaped, or sigmoidal) curve (295)
logistic growth function (295)

Review Exercises 

1. Sketch the graphs of the exponential functions defined by the equations on the same set of coordinate axes.

a. $y = 2^{-x}$ $\qquad$ b. $y = \left(\frac{1}{2}\right)^x$

In Exercises 2 and 3, express each in logarithmic form.

2. $\left(\frac{2}{3}\right)^{-3} = \frac{27}{8}$ $\qquad$ 3. $16^{-3/4} = 0.125$

In Exercises 4 and 5, solve each equation for x.

4. $\log_4(2x + 1) = 2$

5. $\ln(x - 1) + \ln 4 = \ln(2x + 4) - \ln 2$

In Exercises 6–8, given that ln 2 = x, ln 3 = y, and ln 5 = z, express each of the given logarithmic values in terms of x, y, and z.

6. ln 30 $\qquad$ 7. ln 3.6 $\qquad$ 8. ln 75

9. Sketch the graph of the function $y = \log_2(x + 3)$.

10. Sketch the graph of the function $y = \log_3(x + 1)$.

In Exercises 11–38, find the derivative of the function.

11. $f(x) = xe^{2x}$ $\qquad$ 12. $f(t) = \sqrt{te^t + t}$

13. $g(t) = \sqrt{t}e^{-2t}$ $\qquad$ 14. $g(x) = e^x\sqrt{1 + x^2}$

15. $y = \frac{e^{2x}}{1 + e^{-2x}}$ $\qquad$ 16. $f(x) = e^{2x^2-1}$

17. $f(x) = xe^{-x^2}$ $\qquad$ 18. $g(x) = (1 + e^{2x})^{3/2}$

19. $f(x) = x^2e^x + e^x$ $\qquad$ 20. $g(t) = t \ln t$

21. $f(x) = \ln(e^{x^2} + 1)$ $\qquad$ 22. $f(x) = \frac{x}{\ln x}$

23. $f(x) = \frac{\ln x}{x + 1}$ $\qquad$ 24. $y = (x + 1)e^x$

25. $y = \ln(e^{4x} + 3)$ $\qquad$ 26. $f(r) = \frac{re^r}{1 + r^2}$

27. $f(x) = \frac{\ln x}{1 + e^x}$ $\qquad$ 28. $g(x) = \frac{e^{x^2}}{1 + \ln x}$

29. $f(x) = 10^{x^2-1}$ $\qquad$ 30. $f(t) = (t^2 + t)2^t$

31. $y = \frac{x^2 - 4}{3^x}$ $\qquad$ 32. $y = \sqrt{5^x} + \log_5 x$

33. $f(x) = \sqrt{x - \log_2 x}$ $\qquad$ 34. $g(t) = t^{2/3} \log t^3$

35. $y = \log_3(x^2 - x + 5)$ $\qquad$ 36. $f(x) = 2^x \log_2 x$

37. $f(x) = x^{\sqrt{x}}$ $\qquad$ 38. $f(x) = (x - 1)^x$

39. Find the second derivative of the function $y = \ln(3x + 1)$.

40. Find the second derivative of the function $y = x \ln x$.

41. Find $h'(0)$ if $h(x) = g(f(x))$, $g(x) = \frac{1}{x}$, and $f(x) = e^x$.

42. Find $h'(1)$ if $h(x) = g(f(x))$, $g(x) = \frac{x + 1}{x - 1}$, and $f(x) = \ln x$.

43. Use logarithmic differentiation to find the derivative of $f(x) = (2x^3 + 1)(x^2 + 2)^3$.

44. Use logarithmic differentiation to find the derivative of

$$f(x) = \frac{x(x^2 - 2)^2}{(x - 1)}.$$

45. Find an equation of the tangent line to the graph of $y = e^{-2x}$ at the point $(1, e^{-2})$.

46. Find an equation of the tangent line to the graph of $y = xe^{-x}$ at the point $(1, e^{-1})$.

47. Use the Newton–Raphson method to find the point of intersection of the graphs of the functions $y = f(x) = 2x$ and $y = g(x) = e^{-x}$.

48. A suitcase released from rest at the top of a plane metal slide moves a distance of

$$x = f(t) = 27(t + 3e^{-t/3} - 3)$$

metres in t sec. If the metal slide is 24 m long, how long does it take the suitcase to reach the bottom?
Hint: Use the Newton–Raphson method.

Business and Economics Applications

49. **Demand for DVD Players** VCA Television found that the monthly demand for its new line of DVD players t mo after placing the players on the market is given by:

$$D(t) = 4000 - 3000e^{-0.06t} \qquad (t \geq 0)$$

Answer the following questions:
a. What was the demand after 1 mo? After 1 yr? After 2 yr?
b. At what level is the demand expected to stabilize?

50. **Drug Sales** Annual sales in Canada of a particular cancer-fighting drug can be modelled by

$$S(t) = \frac{16}{1 + 64{,}000e^{-2}}, \qquad 0 \leq t \leq 10$$

measured in millions of dollars, where $t = 0$ corresponds to the year 1995. Find $S'(3)$ and interpret your result.

51. **Bike Sales** Monthly sales from April to August of 2002 of a locally manufactured specialty bike in Vancouver is described by the function

$$S(t) = 14.6 \ln(1.8t - 3.3)^4$$

where $t = 3$ corresponds to the beginning of April in 2001, and S is measured in thousands of dollars. Find the rate of change of sales for the beginning of July.

Biological and Life Sciences Applications

52. **Growth of Bacteria** A culture of bacteria that initially contained 2000 bacteria has a count of 18,000 bacteria after 2 hr.
a. Determine the function $Q(t)$ that expresses the exponential growth of the number of cells of this bacterium as a function of time t (in minutes).
b. Find the number of bacteria present after 4 hr.

53. **Radioactive Decay** The radioactive element radium has a half-life of 1600 yr. What is its decay constant?

54. **Flu Epidemic** During a flu epidemic, the number of students at a certain university who contracted influenza after t days could be approximated by the exponential model

$$Q(t) = \frac{3000}{1 + 499e^{-kt}}$$

If 90 students contracted the flu by day 10, how many students contracted the flu by day 20?

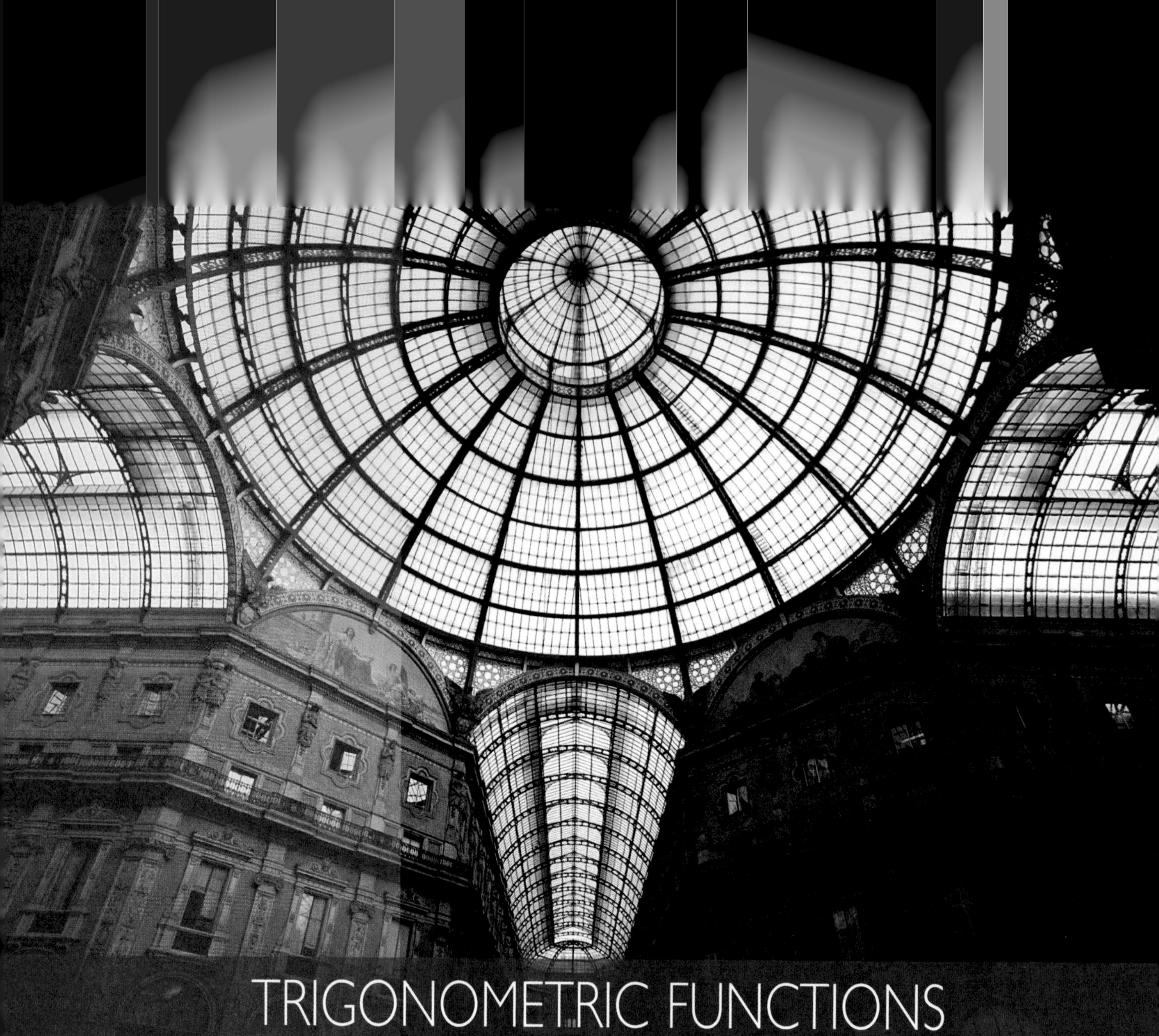

TRIGONOMETRIC FUNCTIONS

Photo: Thinkstock/Jupiter Images

During the golden period of Greek civilization, Apollonius (262–200 B.C.) developed the trigonometric techniques necessary for calculating the radii of various circles. This was done in an effort to describe planetary motion, a physical process that exhibits a cyclical, or periodic, mode of behaviour. Many other real-life phenomena, such as business cycles, earthquake vibrations, respiratory cycles, sales trends, and sound waves, exhibit cyclical behaviour patterns.

In this chapter we extend our study of calculus to an important class of functions called the trigonometric functions. These functions are periodic and hence lend themselves readily to describing many natural phenomena.

Photo: Thinkstock/Jupiter Images

The revenue of McMenamy's Fish Shanty follows a cyclical pattern. When is the revenue of the restaurant increasing most rapidly? In Example 6, page 334, we attempt to answer this question.

Photo: © Gunter Marx Photography/CORBIS

6.1 Measurement of Angles

Angles

An angle consists of two rays that intersect at a common end point. If we rotate the ray l_1 in a *counterclockwise* direction about the point O, the angle generated is *positive* (Figure 6.1a). On the other hand, if we rotate the ray l_1 in a *clockwise* direction about the point O, the angle generated is *negative* (Figure 6.1b). We refer to the ray l_1 as the **initial ray,** the ray l_2 as the **terminal ray,** and the end point O as the **vertex** of the angle. If A and B are points on l_2 and l_1, respectively, then we refer to the angle as angle AOB (Figure 6.1c).

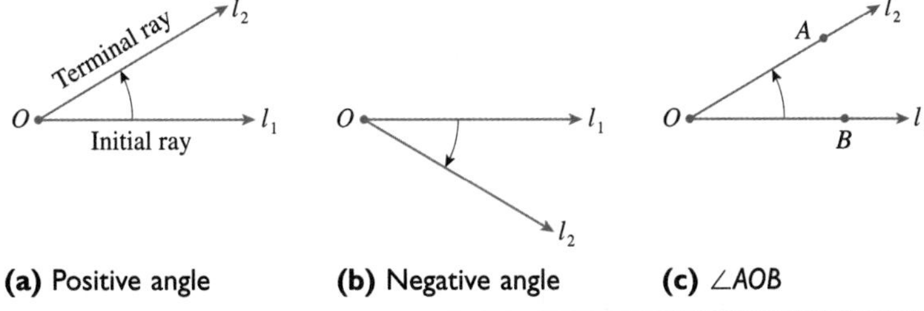

(a) Positive angle **(b)** Negative angle **(c)** $\angle AOB$

Figure 6.1

We can represent an angle in the rectangular coordinate system. An angle is in **standard position** if the vertex of the angle is centred at the origin and its initial side coincides with the positive x-axis (Figures 6.2a and b).

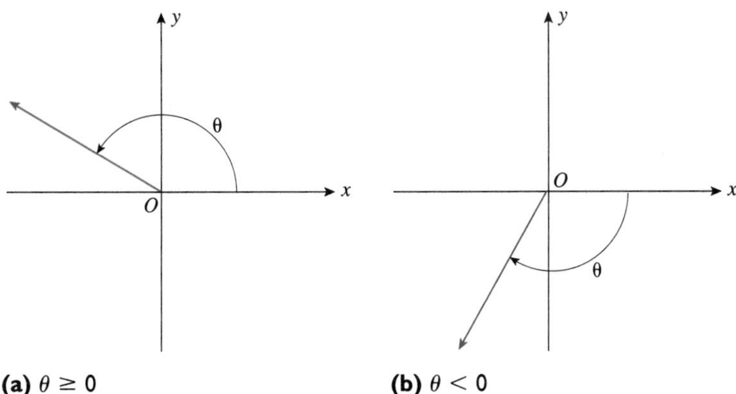

(a) $\theta \geq 0$ **(b)** $\theta < 0$

Figure 6.2

When we say that an angle lies in a certain quadrant, we are referring to the quadrant in which the terminal ray lies. The angle shown in Figure 6.3a lies in Quadrant II, and the angles shown in Figures 6.3b and c lie in Quadrant IV.

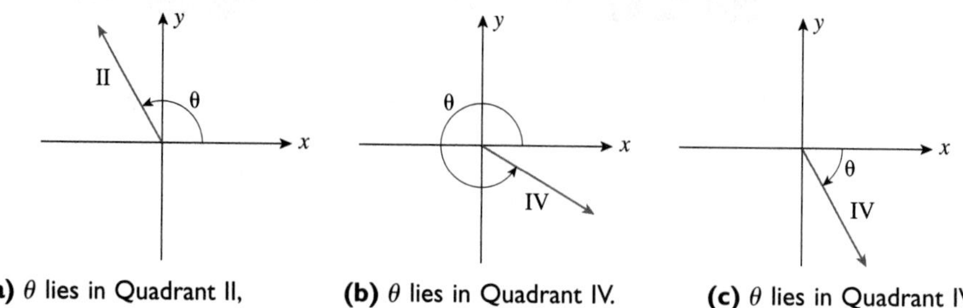

(a) θ lies in Quadrant II, **(b)** θ lies in Quadrant IV. **(c)** θ lies in Quadrant IV.

Figure 6.3

Degree and Radian Measure

An angle may be measured in either degrees or radians. A **degree** is the measure of the angle formed by $\frac{1}{360}$ of one complete revolution. If we rotate an initial ray in standard position through one complete revolution, we obtain an angle of 360°.

A **radian** is the measure of the central angle subtended by an arc equal in length to the radius of the circle. In Figure 6.4, if s is the length of the arc subtended by a central angle θ in a circle of radius r, then

$$\theta = \frac{s}{r} \text{ radians} \qquad (1)$$

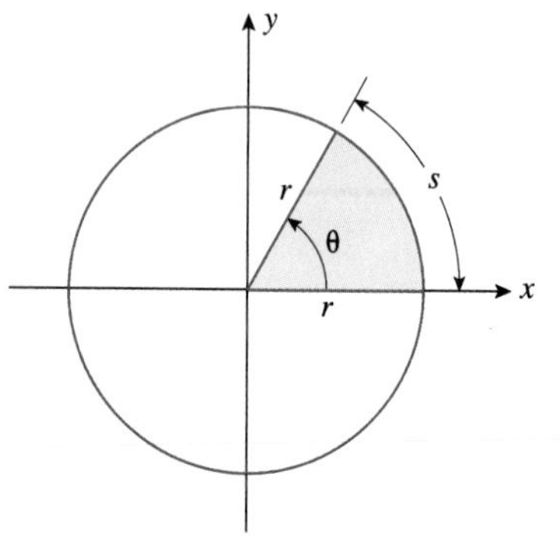

Figure 6.4
$\theta = \frac{s}{r}$ radians

For convenience, we consider a circle of radius 1 centred at the origin. We refer to this circle as the **unit circle.** Then an angle of 1 radian is subtended by an arc of length 1 (Figure 6.5). The circumference of the unit circle is 2π, and, consequently, the angle subtended by one complete revolution is 2π radians. In terms of degrees, the angle subtended by one complete revolution is 360°. Therefore,

$$360 \text{ degrees} = 2\pi \text{ radians}$$

so that

$$1 \text{ degree} = \frac{\pi}{180} \text{ radians}$$

and

$$x \text{ degrees} = \frac{\pi}{180} x \text{ radians} \qquad (2)$$

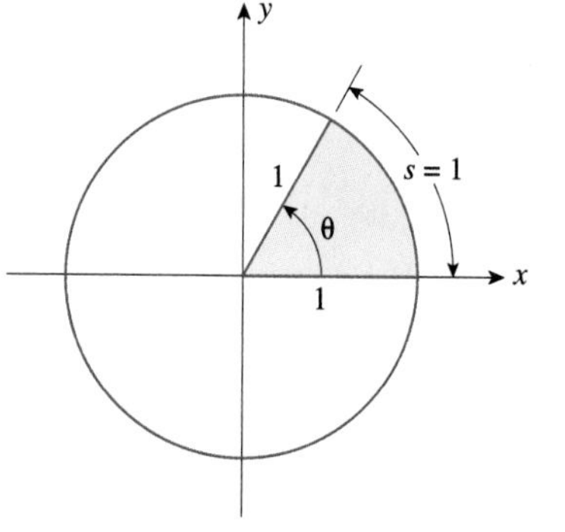

Figure 6.5
$\theta = \frac{s}{r} = 1$ radian

Since this relationship is linear, we may specify the rule for converting degree measure to radian measure in terms of the following linear function.

> **Converting Degrees to Radians**
>
> $$f(x) = \frac{\pi}{180} x \text{ radians} \qquad (3)$$
>
> where x is the number of degrees and $f(x)$ is the number of radians.

Example 1

Convert each angle to radian measure:

a. 30° **b.** 45° **c.** 300° **d.** 450° **e.** −240°

Solution

Using Formula (3), we have

a. $f(30) = \frac{\pi}{180}(30)$ radians, or $\frac{\pi}{6}$ radians

b. $f(45) = \frac{\pi}{180}(45)$ radians, or $\frac{\pi}{4}$ radians

c. $f(300) = \frac{\pi}{180}(300)$ radians, or $\frac{5\pi}{3}$ radians

d. $f(450) = \frac{\pi}{180}(450)$ radians, or $\frac{5\pi}{2}$ radians

e. $f(-240) = \frac{\pi}{180}(-240)$ radians, or $\frac{4\pi}{3}$ radians

By multiplying both sides of Equation (2) by $\frac{180}{\pi}$, we have

$$x \text{ radians} = \frac{180}{\pi}x \text{ degrees}$$

Thus, we have the following rule for converting radian measure to degree measure.

Converting Radians to Degrees

$$g(x) = \frac{180}{\pi}x \text{ degrees} \qquad (4)$$

where x is the number of radians and $g(x)$ is the number of degrees.

Example 2

Convert each angle to degree measure:

a. $\frac{\pi}{2}$ radians **b.** $\frac{5\pi}{4}$ radians **c.** $-\frac{3\pi}{4}$ radians **d.** $\frac{7\pi}{2}$ radians

Solution

Using Formula (4), we have

a. $g\left(\frac{\pi}{2}\right) = \frac{180}{\pi}\left(\frac{\pi}{2}\right)$ degrees, or 90 degrees

b. $g\left(\frac{5\pi}{4}\right) = \frac{180}{\pi}\left(\frac{5\pi}{4}\right)$ degrees, or 225 degrees

c. $g\left(-\frac{3\pi}{4}\right) = \frac{180}{\pi}\left(-\frac{3\pi}{4}\right)$ degrees, or -135 degrees

d. $g\left(\frac{7\pi}{2}\right) = \frac{180}{\pi}\left(\frac{7\pi}{2}\right)$ degrees, or 630 degrees

Take time to familiarize yourself with the radian and degree measures of the common angles given in Table 6.1.

Table 6.1

Degrees	0°	30°	45°	60°	90°	120°	135°	150°	180°	270°	360°
Radians	0	$\frac{\pi}{6}$	$\frac{\pi}{4}$	$\frac{\pi}{3}$	$\frac{\pi}{2}$	$\frac{2\pi}{3}$	$\frac{3\pi}{4}$	$\frac{5\pi}{6}$	π	$\frac{3\pi}{2}$	2π

You may have observed by now that more than one angle may be described by the same initial and terminal rays. The two angles in Figure 6.6a and b illustrate this case. The angle $\theta = 5\pi/4$ radians is generated by rotating the initial ray in a counterclockwise direction, and the angle $\theta = -3\pi/4$ radians is generated by rotating the initial ray in a clockwise direction. We refer to such angles as **coterminal angles.**

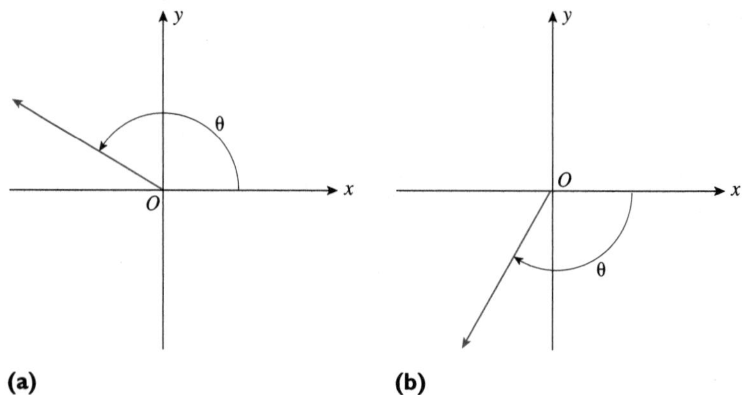

Figure 6.6
Coterminal angles

You may also have observed from the preceding examples that an angle may be greater than 2π radians. For example, an angle of 3π radians is generated by rotating a ray through 1.5 revolutions. As illustrated in Figure 6.7, an angle of 3π radians has its initial and terminal rays in the same position as an angle of π radians. A similar statement is true for the negative angles $-\pi$ and -3π radians. Thus, in identifying an angle, we must be careful to specify the *direction* of the rotation and the *number of revolutions* through which the angle has gone.

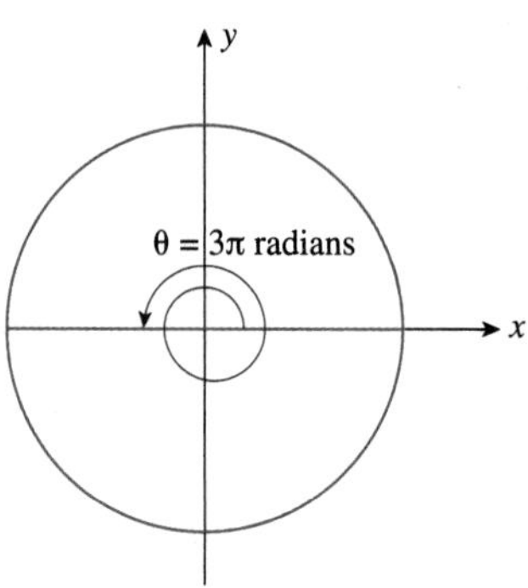

Figure 6.7
$\theta = 3\pi$ radians has its initial and terminal rays in the same position as $\theta = \pi$ radians.

Self-Check Exercises 6.1

1. **a.** Convert 315° to radian measure.
   b. Convert $-\dfrac{5\pi}{4}$ radians to degree measure.
2. Make a sketch of the angle $-\dfrac{2\pi}{3}$ radians.

Solutions to Self-Check Exercises 6.1 can be found on page 308.

6.1 Exercises

In Exercises 1–4, express the angle shown in the figure in radian measure.

1.

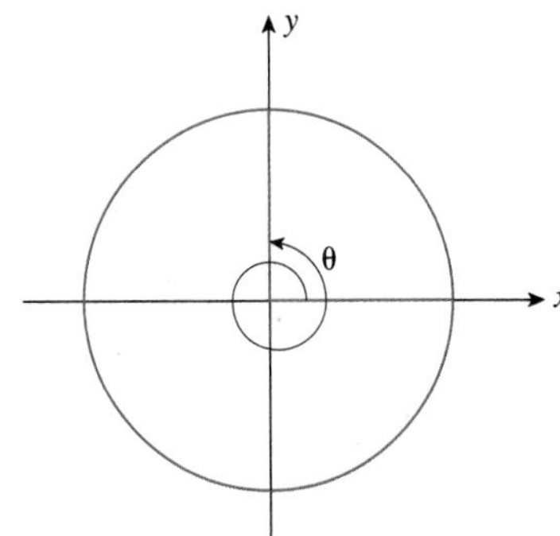

2.

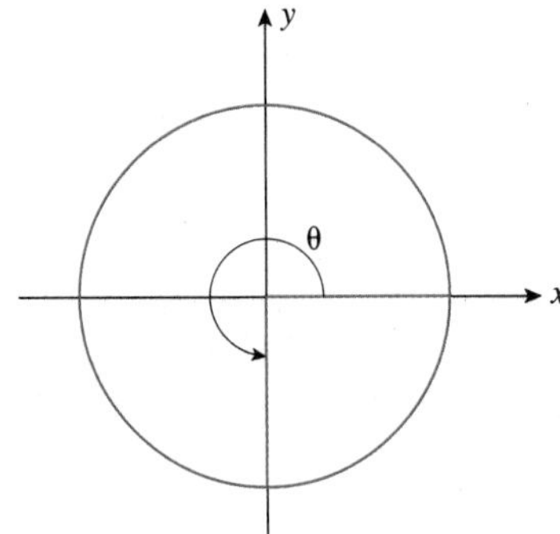

3.

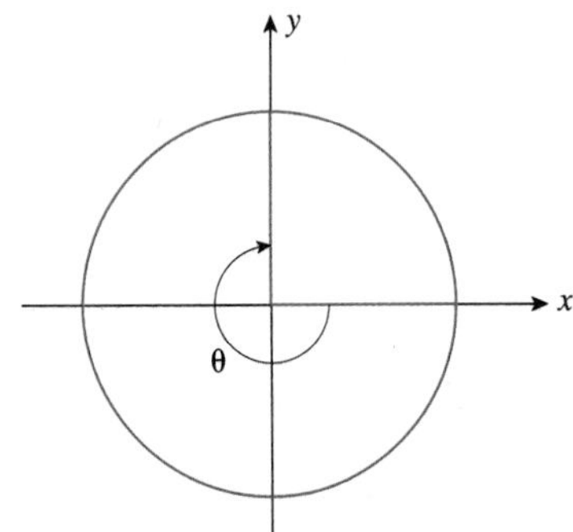

4.

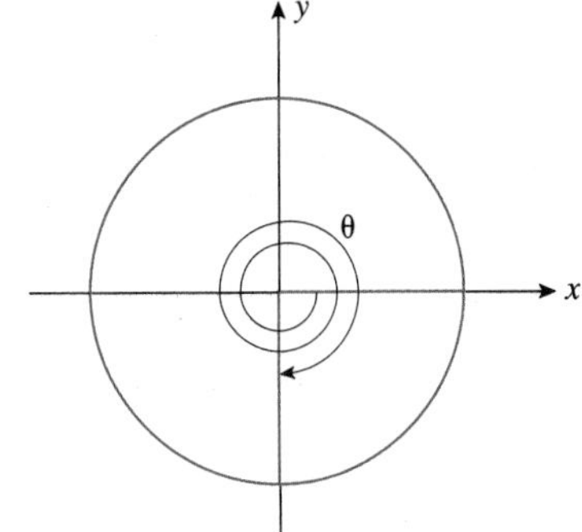

5. Identify the quadrant in which the angle lies.
   a. 220° **b.** −110°
   c. 460° **d.** −310°

6. Identify the quadrant in which the angle lies.
   a. $\frac{13}{6}\pi$ radians **b.** $-\frac{11}{4}\pi$ radians
   c. $\frac{17}{3}\pi$ radians **d.** $-\frac{25}{12}\pi$ radians

In Exercises 7–12, convert the angle to radian measure.

7. 75° 8. 330° 9. 160°

10. −210° 11. 630° 12. −420°

In Exercises 13–18, convert the angle to degree measure.

13. $\frac{2}{3}\pi$ radians 14. $\frac{7}{6}\pi$ radians 15. $-\frac{3}{2}\pi$ radians

16. $-\frac{13}{12}\pi$ radians 17. $\frac{22}{18}\pi$ radians 18. $-\frac{21}{6}\pi$ radians

In Exercises 19–22, make a sketch of the angle on a unit circle centred at the origin.

19. 225° 20. −120°

21. $\frac{7}{3}\pi$ radians 22. $-\frac{13}{6}\pi$ radians

In Exercises 23–26, determine a positive angle and a negative angle that are coterminal angles of the angle θ. Use degree measure.

23.

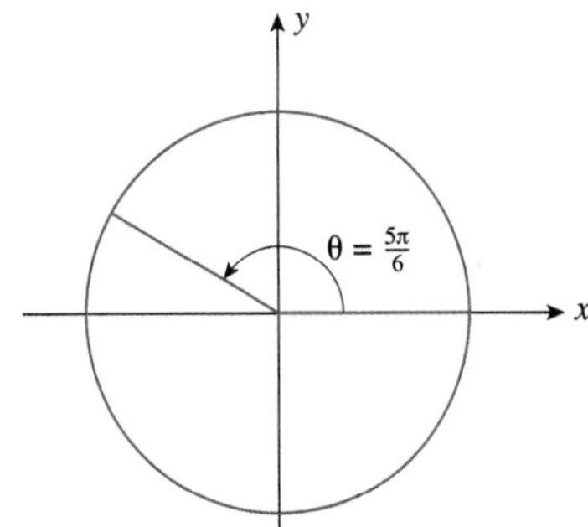

24.

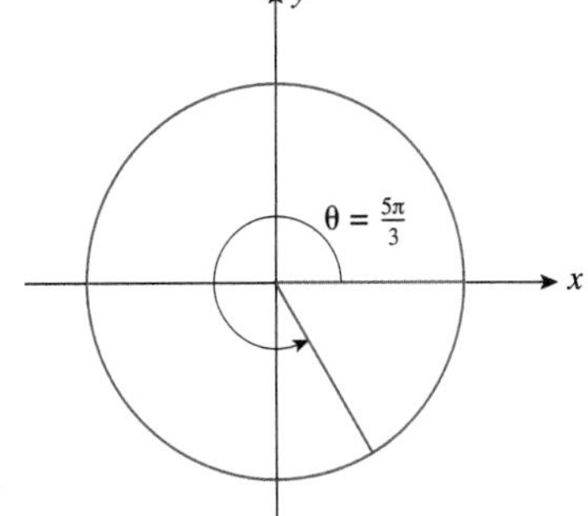

25.

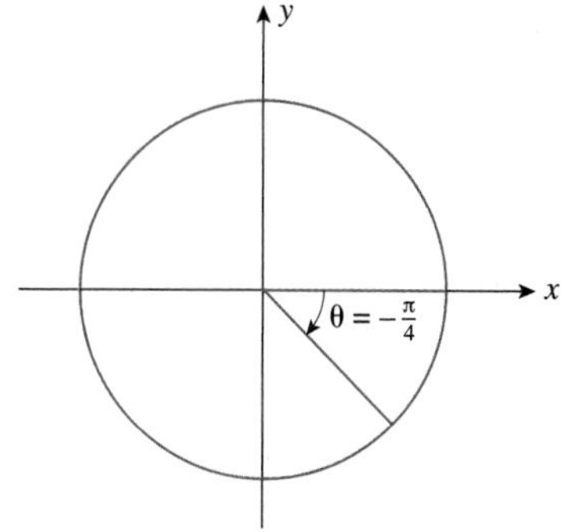

26.

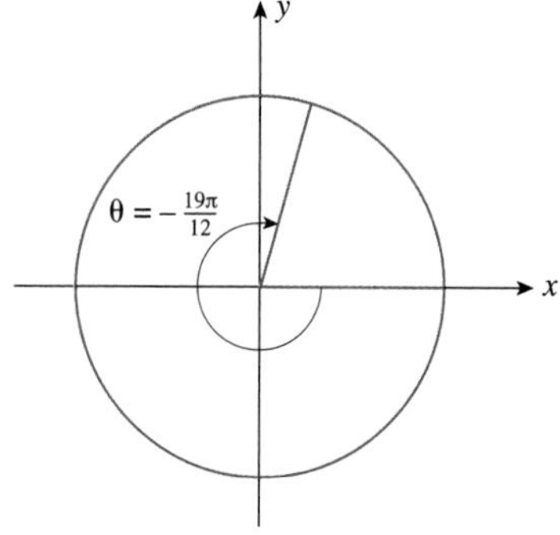

In Exercises 27–30, determine whether the statement is true or false. If it is true, explain why it is true. If it is false, give an example to show why it is false.

27. The angle 3630° lies in the first quadrant.

28. The angle $\dfrac{103\pi}{6}$ radians lies in the second quadrant.

29. If θ is any angle, then $\theta + n(360)$, where n is a nonzero integer, is coterminal with θ (all angles measured in degrees).

30. If x, y, and z are measured in degrees, then $(x + y + z)$ degrees is equal to $\dfrac{\pi}{180}(x + y + z)$ radians.

Solutions to Self-Check Exercises 6.1

1. a. Using Formula (3), we find that the required radian measure is

$$f(315) = \frac{\pi}{180}(315) = \frac{7\pi}{4}, \qquad \text{or} \qquad \frac{7\pi}{4} \text{ radians}$$

b. Using Formula (4), we find that the required degree measure is

$$g\left(-\frac{5\pi}{4}\right) = \frac{180}{\pi}\left(-\frac{5\pi}{4}\right) = -225, \qquad \text{or} \qquad -225°$$

2.

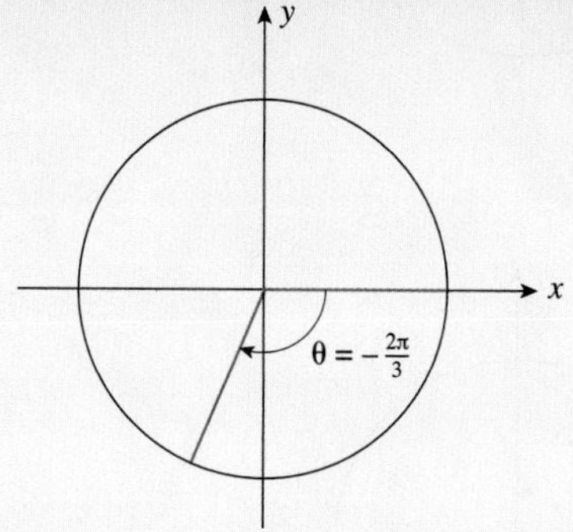

6.2 The Trigonometric Functions

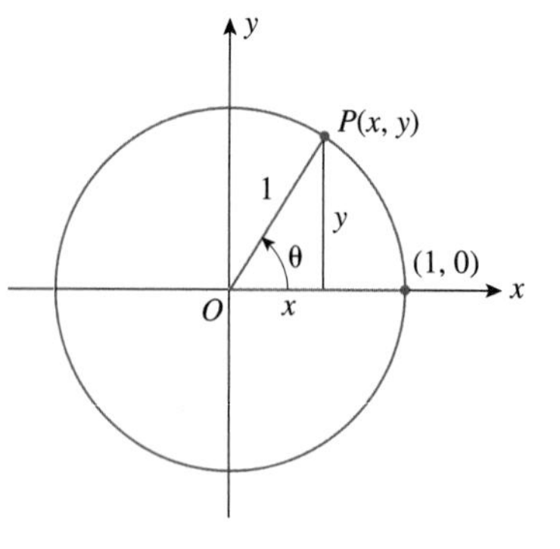

Figure 6.8
P is a point on the unit circle with coordinates $x = \cos\theta$ and $y = \sin\theta$.

Before we begin this section, we need to point out that in calculus all angles need to be given in radian measurement. The proof for this is technical, therefore omitted.

Remember to set your calculators to radian mode.

The Trigonometric Functions and Their Graphs

Let $P(x, y)$ be a point on the unit circle so that the radius OP forms an angle of θ radians $(0 \le \theta < 2\pi)$ with respect to the positive x-axis (see Figure 6.8).

We define the sine of the angle θ, written $\sin\theta$, to be the y-coordinate of P. Similarly, the cosine of the angle θ, written $\cos\theta$, is defined to be the x-coordinate of P. The other trigonometric functions, tangent, cosecant, secant, and cotangent of θ—written $\tan\theta$, $\csc\theta$, $\sec\theta$, and $\cot\theta$, respectively—are defined in terms of the sine and cosine functions.

Trigonometric Functions

If P is a point on the unit circle and the coordinates of P are (x, y), then

$$\cos\theta = x \quad \text{and} \quad \sin\theta = y$$

$$\tan\theta = \frac{y}{x} = \frac{\sin\theta}{\cos\theta} \quad (x \neq 0)$$

$$\csc\theta = \frac{1}{y} = \frac{1}{\sin\theta} \quad (y \neq 0)$$

$$\sec\theta = \frac{1}{x} = \frac{1}{\cos\theta} \quad (x \neq 0)$$

$$\cot\theta = \frac{x}{y} = \frac{\cos\theta}{\sin\theta} \quad (y \neq 0)$$

As you work with trigonometric functions, it is useful to remember the values of the sine, cosine, and tangent of some important angles, such as $\theta = 0, \pi/6, \pi/4, \pi/3, \pi/2$, and so on. These values may be found using elementary trigonometry. For example, if $\theta = 0$, then the point P has coordinates $(1, 0)$ (see Figure 6.9a), and we see that

$$\sin 0 = y = 0 \qquad \cos 0 = x = 1 \qquad \tan 0 = \frac{y}{x} = 0$$

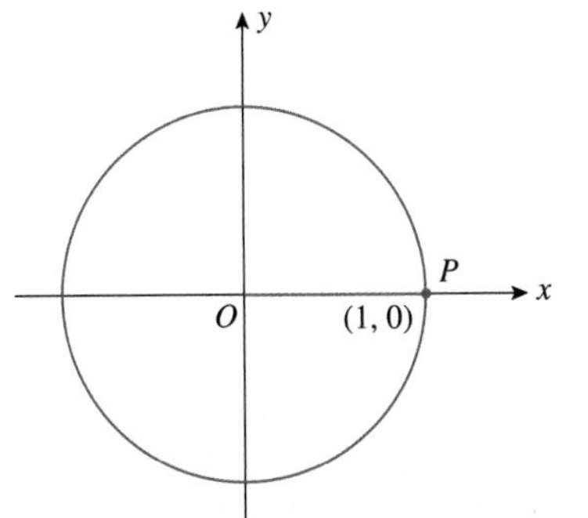

(a) $P = (1, 0), \theta = 0$, and $\cos 0 = 1, \sin 0 = 0$

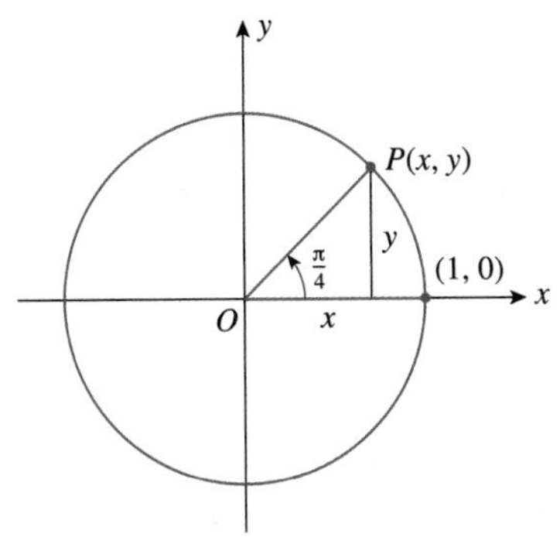

(b) $\theta = \pi/4, x = y$, and $\cos\frac{\pi}{4} = \sin\frac{\pi}{4} = \sqrt{2}/2$

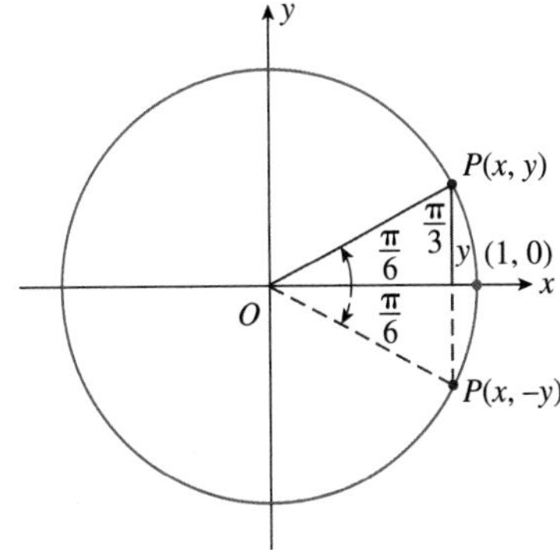

(c) $\theta = \frac{\pi}{6}, y = \frac{1}{2}, x = \frac{\sqrt{3}}{2}, \cos\frac{\pi}{6} = \sin\frac{\pi}{3} = \frac{\sqrt{3}}{2}$, and $\cos\frac{\pi}{3} = \sin\frac{\pi}{6} = \frac{1}{2}$

Figure 6.9

As another example, the diagram for $\theta = \pi/4$ (Figure 6.9b) suggests that $x = y$, so that, by the Pythagorean theorem, we have

$$x^2 + y^2 = 2x^2 = 1$$

and $x = y = \frac{\sqrt{2}}{2}$. Therefore,

$$\sin\frac{\pi}{4} = y = \frac{\sqrt{2}}{2} \qquad \cos\frac{\pi}{4} = x = \frac{\sqrt{2}}{2} \qquad \tan\frac{\pi}{4} = \frac{y}{x} = 1$$

As a last example, the diagram for $\theta = \pi/6$ (Figure 6.9c) suggests that $2y = 1$, which implies that $y = 1/2$. Once again using the Pythagorean theorem, we have

$$x^2 + y^2 = 1$$

$$x^2 + \left(\frac{1}{2}\right)^2 = 1$$

$$x^2 = \frac{3}{4}$$

and $x = \frac{\sqrt{3}}{2}$. Therefore,

$$\sin\frac{\pi}{6} = y = \frac{1}{2} \qquad \cos\frac{\pi}{6} = x = \frac{\sqrt{3}}{2} \qquad \tan\frac{\pi}{6} = \frac{y}{x} = \frac{1}{\sqrt{3}}$$

and furthermore,

$$\sin\frac{\pi}{3} = x = \frac{\sqrt{3}}{2} \qquad \cos\frac{\pi}{3} = y = \frac{1}{2} \qquad \tan\frac{\pi}{3} = \frac{x}{y} = \sqrt{3}$$

GROUP DISCUSSION

Evaluate each of the three primary trigonometric functions (sine, cosine, and tangent) for each of the four angles $\theta_1 = \frac{\pi}{3}$, $\theta_2 = \frac{2\pi}{3}$, $\theta_3 = \frac{4\pi}{3}$, and $\theta_4 = \frac{5\pi}{3}$ by drawing diagrams similar to the ones in Figure 6.9. Be careful which angle is your reference angle in the right triangle you set up. Notice that an angle from each of the four quadrants has been chosen. Observe the sign of the x- and y-values in each of the diagrams. What conclusion can you draw about the sign of each of the primary trigonometric functions for each quadrant?

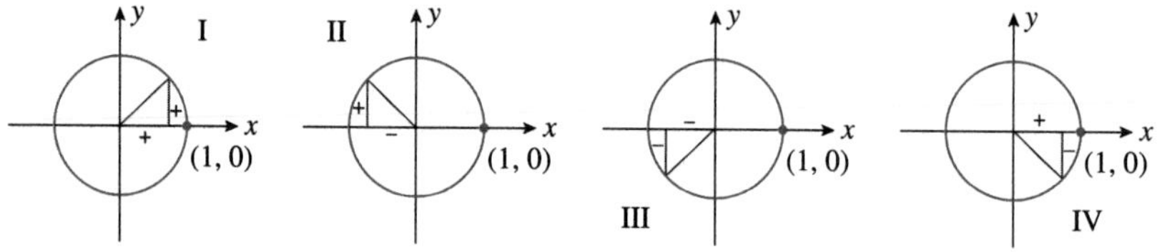

Figure 6.10

Now observe the sign of x and y as an angle moves through the four quadrants as shown in Figure 6.10. We have $\sin\theta = y$, so the sign of the sine function is only dependent on the sign of the y-value. Therefore, sine is positive only in Quadrants I and II. Since $\cos\theta = x$, the sign of the cosine function is only dependent on the sign of the x-value. Therefore, cosine is positive only in Quadrants I and IV. Lastly, $\tan\theta = \frac{y}{x}$, and so the tangent is positive only in Quadrants I and III, when either both x and y are positive or negative. These results are summarized in Figure 6.11, where **A** stands for "all primary trigonometric functions are positive in Quadrant I," **S** stands for "only the sine function is positive in Quadrant II," **T** stands for "only the tangent function is positive in Quadrant III," and **C** stands for "only the cosine function is positive in Quadrant IV." From now on, we will refer to this result as the **ASTC Rule,** which can be remembered as the infamous mnemonic *All Students Take Calculus.*

S	*A*
T	*C*

Figure 6.11
The graph of $y = x^2 + 1$ is a parabola.

Using the results of Figure 6.11 as well as the special angles discussed in Figure 6.9, we can develop the trigonometric ratios of all special angles. Table 6.2 summarizes the values of the sine, cosine, and tangent of some of these angles.

Table 6.2

θ in radians	0	$\frac{\pi}{6}$	$\frac{\pi}{4}$	$\frac{\pi}{3}$	$\frac{\pi}{2}$	$\frac{2}{3}\pi$	$\frac{3}{4}\pi$	$\frac{5}{6}\pi$	π	$\frac{7}{6}\pi$	$\frac{5}{4}\pi$	$\frac{4}{3}\pi$	$\frac{3}{2}\pi$	$\frac{5}{3}\pi$	$\frac{7}{4}\pi$	$\frac{11}{6}\pi$	2π
sin θ	0	$\frac{1}{2}$	$\frac{\sqrt{2}}{2}$	$\frac{\sqrt{3}}{2}$	1	$\frac{\sqrt{3}}{2}$	$\frac{\sqrt{2}}{2}$	$\frac{1}{2}$	0	$-\frac{1}{2}$	$-\frac{\sqrt{2}}{2}$	$-\frac{\sqrt{3}}{2}$	-1	$-\frac{\sqrt{3}}{2}$	$-\frac{\sqrt{2}}{2}$	$-\frac{1}{2}$	0
cos θ	1	$\frac{\sqrt{3}}{2}$	$\frac{\sqrt{2}}{2}$	$\frac{1}{2}$	0	$-\frac{1}{2}$	$-\frac{\sqrt{2}}{2}$	$-\frac{\sqrt{3}}{2}$	-1	$-\frac{\sqrt{3}}{2}$	$-\frac{\sqrt{2}}{2}$	$-\frac{1}{2}$	0	$\frac{1}{2}$	$\frac{\sqrt{2}}{2}$	$\frac{\sqrt{3}}{2}$	1
tan θ	0	$\frac{1}{\sqrt{3}}$	1	$\sqrt{3}$	∞	$-\sqrt{3}$	-1	$-\frac{1}{\sqrt{3}}$	0	$\frac{1}{\sqrt{3}}$	1	$\sqrt{3}$	$-\infty$	$-\sqrt{3}$	-1	$-\frac{1}{\sqrt{3}}$	0

Example 1

Evaluate $\cos\frac{8\pi}{3}$ without using Table 6.2.

Solution

The angle is $\theta = \frac{8\pi}{3} = 2\pi + \frac{2\pi}{3}$, and so coterminal with the angle $\frac{2\pi}{3}$, which resides in Quadrant II (Figure 6.12a). The reference angle α is then given by $\alpha = \pi - \frac{2\pi}{3} = \frac{\pi}{3}$ (Figure 6.12a), which is a special angle. The trigonometric ratio associated with this angle can be read of the special triangle with angles $\frac{\pi}{6}, \frac{\pi}{3}$, and $\frac{\pi}{2}$ shown in Figure 6.12b, and so $\cos\frac{\pi}{3} = \frac{1}{2}$. Lastly, we need to consider the sign of the cosine function. We already mentioned that θ terminates in Quadrant II, which makes the cosine negative there according to the ASTC Rule shown in Figure 6.12a. Therefore,

$$\cos\frac{8\pi}{3} = -\cos\frac{\pi}{3} = -\frac{1}{2}$$

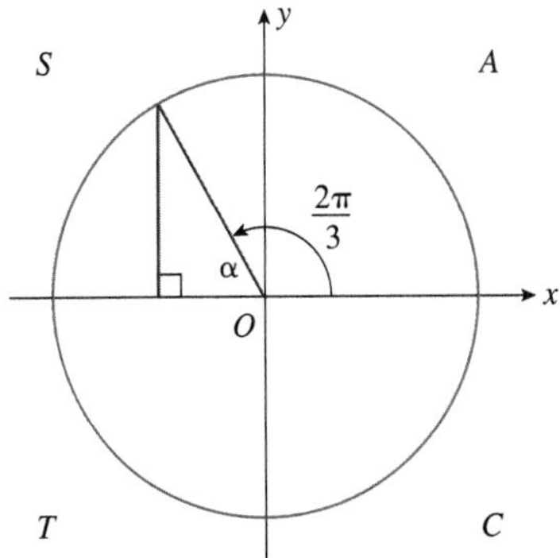

(a) $\theta = \frac{8\pi}{3}$ is coterminal with angle $\frac{2\pi}{3}$

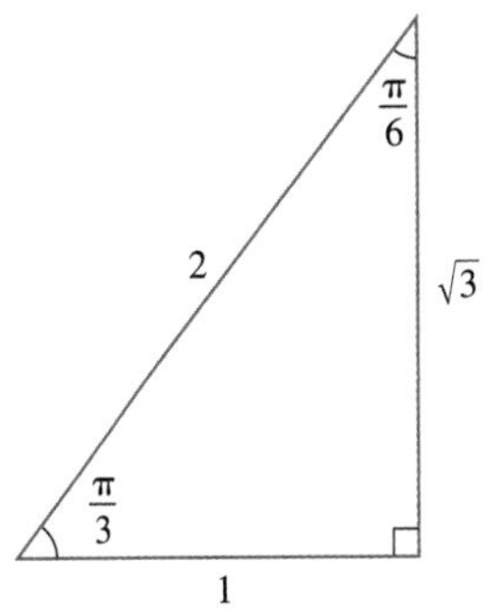

(b) Special triangle with angles $\frac{\pi}{6}, \frac{\pi}{3}$, *and* $\frac{\pi}{2}$

Figure 6.12

Example 2

Without using Table 6.2, find all values of the angle θ, for which $\sin\theta = -\frac{1}{\sqrt{2}}$ and $0 \le \theta \le 2\pi$ (in other words we only consider one complete revolution, and do not take coterminal angles into account).

Solution

Notice that the trigonometric ratio is negative. According to the ASTC Rule sine is negative in Quadrants III and IV. We deduce that there are two solutions for the angle θ as indicated in Figure 6.13a. To find the reference angle α, we analyze the ratio $\frac{1}{\sqrt{2}}$, and note that it is a special ratio given by the special triangle with angles $\frac{\pi}{4}, \frac{\pi}{4}$, and $\frac{\pi}{2}$ shown in Figure 6.13b. We now have that $\alpha = \frac{\pi}{4}$. Therefore, the two angles we are looking for are

$$\theta = \pi + \frac{\pi}{4} = \frac{5\pi}{4} \text{ in Quadrant III (Figure 6.13a)}$$

and

$$\theta = 2\pi - \frac{\pi}{4} = \frac{7\pi}{4} \text{ in Quadrant IV (Figure 6.13a)}$$

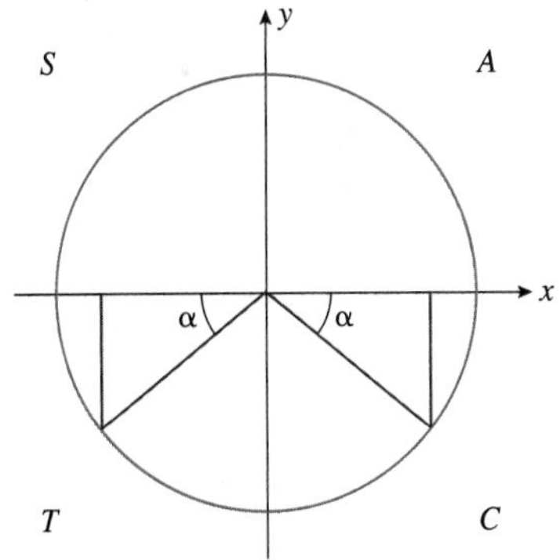

(a) Reference angle α

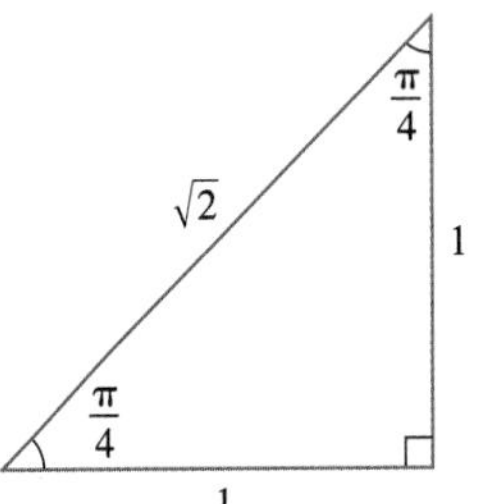

(b) Special triangle with angles $\frac{\pi}{4}, \frac{\pi}{4}$, *and* $\frac{\pi}{2}$

Figure 6.13

Since the rotation of the radius OP by 2π radians leaves it in its original configuration (see Figure 6.14), we see that

$$\sin(\theta + 2\pi) = \sin\theta \quad \text{and} \quad \cos(\theta + 2\pi) = \cos\theta$$

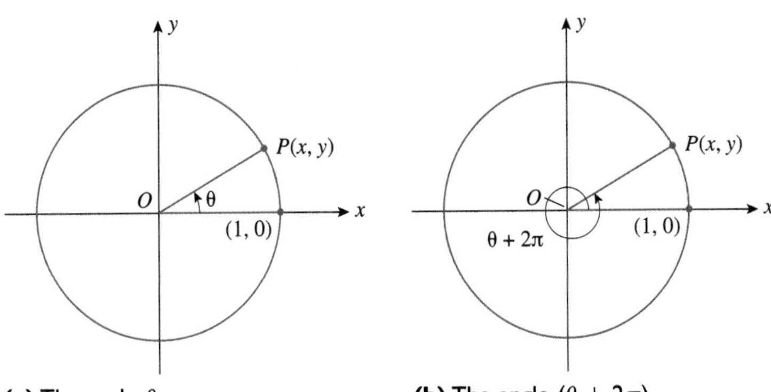

(a) The angle θ **(b)** The angle $(\theta + 2\pi)$

Figure 6.14

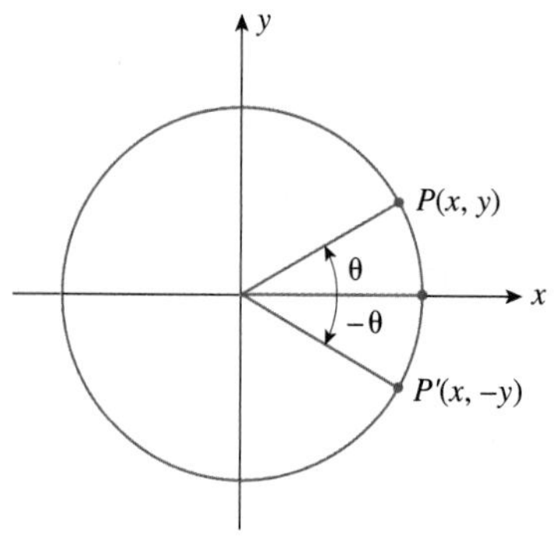

Figure 6.15
$\sin(-\theta) = -\sin\theta$ and $\cos(-\theta) = \cos\theta$.

It can be shown that 2π is the smallest positive number for which the preceding equations hold true. That is, the sine and cosine function are periodic with **period 2π**. It follows that

$$\sin(\theta + 2n\pi) = \sin\theta \quad \text{and} \quad \cos(\theta + 2n\pi) = \cos\theta \tag{5}$$

whenever n is an integer. Also, from Figure 6.15, we see that

$$\sin(-\theta) = -\sin\theta \quad \text{and} \quad \cos(-\theta) = \cos\theta \tag{6}$$

Example 3

Evaluate

a. $\sin\dfrac{7\pi}{2}$ **b.** $\cos 5\pi$ **c.** $\sin\left(-\dfrac{5\pi}{2}\right)$ **d.** $\cos\left(-\dfrac{11\pi}{4}\right)$

Solution

a. Using (5) and Table 6.2, we find

$$\sin\left(\frac{7\pi}{2}\right) = \sin\left(2\pi + \frac{3\pi}{2}\right) = \sin\frac{3\pi}{2} = -1$$

b. Using (5) and Table 6.2, we have

$$\cos 5\pi = \cos(4\pi + \pi) = \cos\pi = -1$$

c. Using (5), (6), and Table 6.2, we have

$$\sin\left(-\frac{5\pi}{2}\right) = -\sin\left(\frac{5\pi}{2}\right) = -\sin\left(2\pi + \frac{\pi}{2}\right) = -\sin\frac{\pi}{2} = -1$$

d. Using (5), (6), and Table 6.2, we have

$$\cos\left(-\frac{11\pi}{4}\right) = \cos\frac{11\pi}{4} = \cos\left(2\pi + \frac{3\pi}{4}\right) = \cos\frac{3\pi}{4} = -\frac{\sqrt{2}}{2}$$

To draw the graph of the function $y = f(x) = \sin x$, we first note that $\sin x$ is defined for every real number x so that the domain of the sine function is $(-\infty, \infty)$. Next, since the sine function is periodic with period 2π, it suffices to concentrate on sketching that part of the graph of $y = \sin x$ on the interval $[0, 2\pi]$ and repeating it as necessary. With the help of Table 6.2, we sketch Figure 6.16, the graph of $y = \sin x$.

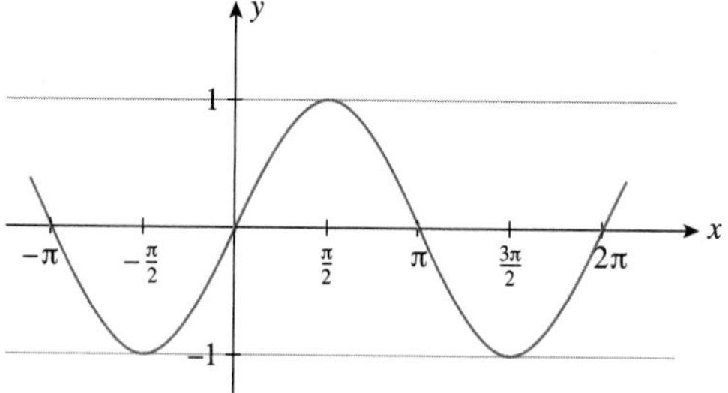

Figure 6.16
The graph of $y = \sin x$

In a similar manner, we can sketch the graph of $y = \cos x$ (Figure 6.17).

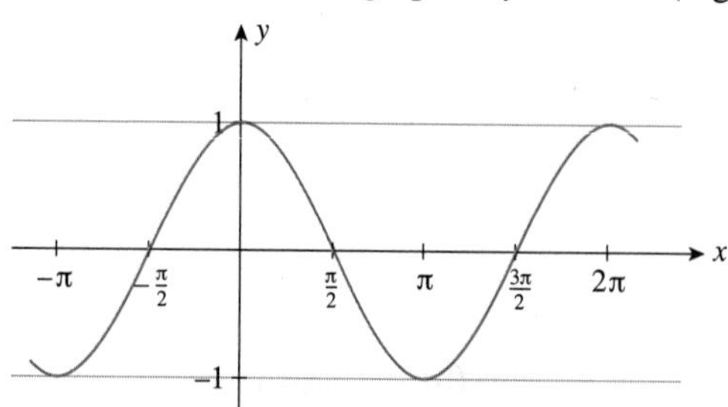

Figure 6.17
The graph of $y = \cos x$

EXPLORING WITH TECHNOLOGY

Plot the graphs of $f(x) = \sin x$ and $g(x) = \cos x$ in the viewing window $[-10, 10] \cdot [-1.1, 1.1]$.

1. What do the graphs suggest about the relationship between the two trigonometric functions?
2. Confirm your observation by plotting the graphs of $h(x) = \sin\left(x + \frac{\pi}{2}\right)$ and $g(x) = \cos x$ in the viewing window $[-10, 10] \cdot [-1.1, 1.1]$.

PORTFOLIO

Kerry Skipper

Title Product Manager, Flight Dynamics System
Institution Telesat Canada

Today's society could not function without geostationary communications satellites. Geostationary satellites circle the Earth in equatorial orbits that are synchronized with the Earth's rotation, such that the satellites appear to remain stationary above the Earth. These satellites facilitate the transmission of information for many of our day-to-day activities, including watching television, exchanging emails and browsing the Internet, or purchasing gasoline with a credit card.

Satellite communication could not exist without sophisticated software that uses mathematical models and methods to determine the satellites' trajectories, predict the evolution of those trajectories, and plan corrective manoeuvres to maintain the satellites in their required orbits. One such suite of software is Telesat Canada's Flight Dynamics System (FDS).

Kerry Skipper, who is the FDS Product Manager at Telesat, manages the design and development of the FDS. His degrees in Applied Mathematics and co-op work term experience during his university years prepared him well for this role. According to Mr. Skipper, "the most mathematically intensive elements of the FDS are the estimation and prediction functions," which process tracking data to determine the satellite's position and velocity vectors at a specific time, and by integrating second-order partial derivatives representing the forces acting on the satellite, predict the satellite's position at future times.

A simple example that demonstrates the calculus involved in Mr. Skipper's line of work is modelling the satellite's acceleration during a manoeuvre. With the so-called "Rocket Equation," the acceleration is modeled as $dv/dt = F/m(t)$, where F is a constant force (thrust) and $m(t)$ is the spacecraft mass which decreases at a fixed rate. Because the assumptions of constant thrust and propellant flow rate may not be valid, a more general formulation of the acceleration is provided by $dv/dt = a/(1 - bt)$, where a and b are constant model parameters. Mr. Skipper explains that "this formulation accommodates the constant thrust and flow rate model, but also allows for a constant acceleration (when $b = 0$) and a decreasing acceleration (when $b < 0$)."

The mathematical models that are established are solved using advanced calculus and/or numerical methods. With the FDS, satellite operators are able to plan orbit-correction manoeuvres to maintain the orbital positions of satellites within tight constraints on latitude and longitude, allowing the transmission of information necessary to support many technological aspects of our society.

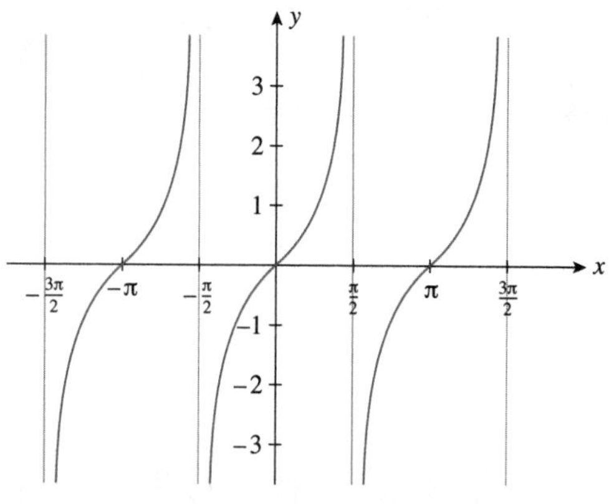

Figure 6.18
The graph of $y = \tan x$

To sketch the graph of $y = \tan x$, recall that $\tan x = \sin x/\cos x$, and so $\tan x$ is not defined when $\cos x = 0$—that is, when $x = (\pi/2) \pm n\pi$ $(n = 0, 1, 2, 3, \ldots)$. The function is defined at all other points so that the domain of the tangent function is the set of all real numbers with the exception of the points just noted. Next, we can show that the vertical lines with equation $x = (\pi/2) \pm n\pi (n = 0, 1, 2, 3, \ldots)$ are vertical asymptotes of the function $f(x) = \tan x$. For example, since

$$\lim_{x \to \frac{\pi}{2}^-} \tan x = \infty \qquad \text{and} \qquad \lim_{x \to \frac{\pi}{2}^+} \tan x = -\infty$$

which we readily verify with the help of a calculator, we conclude that $x = \pi/2$ is a vertical asymptote of $y = \tan x$. Finally, using Table 6.2, we can sketch the graph of $y = \tan x$ (Figure 6.18).

Observe that the tangent function is periodic with period π. It follows that

$$\tan(x + n\pi) = \tan x \tag{7}$$

whenever n is an integer.

The graphs of $y = \sec x$, $y = \csc x$, and $y = \cot x$ shown in Figure 6.19 may be sketched in a similar manner (see Exercises 23–25).

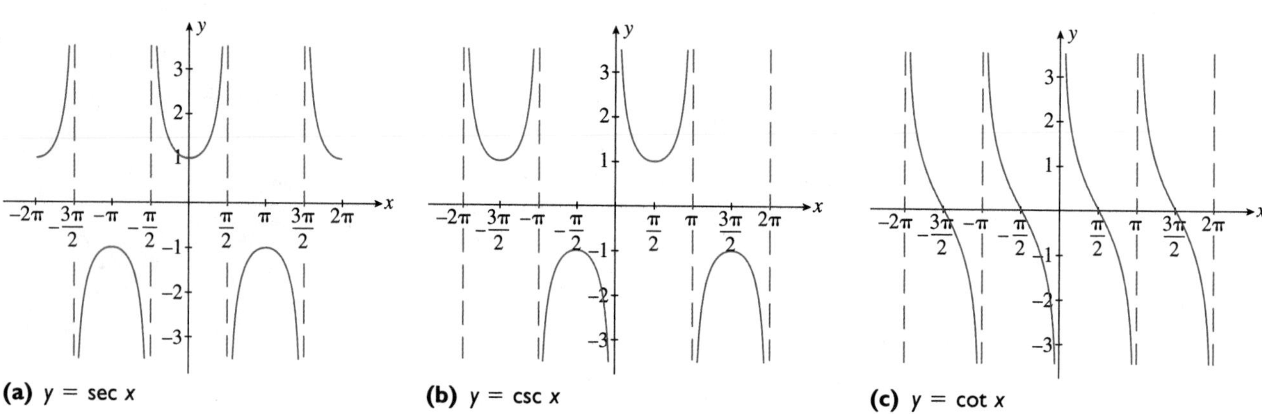

(a) $y = \sec x$ **(b)** $y = \csc x$ **(c)** $y = \cot x$

Figure 6.19

The Predator–Prey Population Model

We will now look at a specific mathematical model of a phenomenon exhibiting cyclical behaviour—the so-called **predator–prey population model.**

Photo: © Corel

Example 4

The population of owls (predators) in a certain region over a 2-year period is estimated to be

$$P(t) = 1000 + 100 \sin\left(\frac{\pi t}{12}\right)$$

in month t, and the population of mice (prey) in the same area at time t is given by

$$p(t) = 20{,}000 + 4000 \cos\left(\frac{\pi t}{12}\right)$$

Sketch the graphs of these two functions and explain the relationship between the sizes of the two populations.

Solution

We first observe that both of the given functions are periodic with period 24 (months). To see this, recall that both the sine and cosine functions are periodic with period 2π. Now the smallest value of $t > 0$ such that $\sin(\pi t/12) = 0$ is obtained by solving the equation

$$t = \frac{2\pi}{\pi/12}$$

giving $t = 24$ as the period of $\sin(\pi t/12)$. Since $P(t + 24) = P(t)$, we see that the function P is periodic with period 24. Similarly, one verifies that the function p is also periodic, with period 24, as asserted. Next, recall that both the sine and cosine functions oscillate between -1 and $+1$ so that $P(t)$ is seen to oscillate between $[1000 + 100(-1)]$, or 900, and $[1000 + 100(1)]$, or 1100, while $p(t)$ oscillates between $[20{,}000 + 4000(-1)]$, or 16,000, and $[20{,}000 + 4000(1)]$, or 24,000. Finally, plotting a few points on each graph for—say, $t = 0, 2, 3$, and so on—we obtain the graphs of the functions P and p as shown in Figure 6.20.

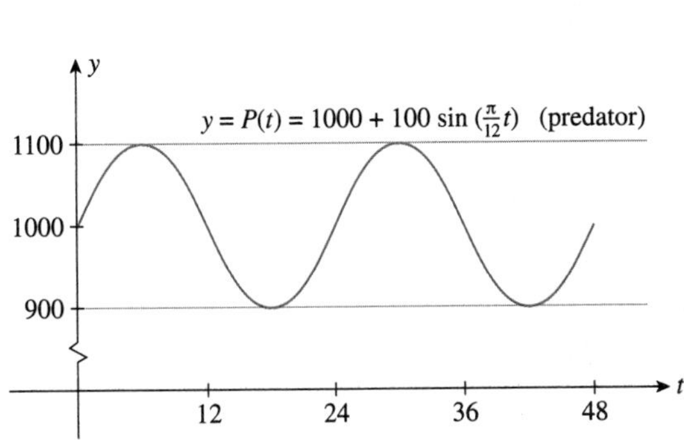

(a) The graph of the predator function $P(t)$

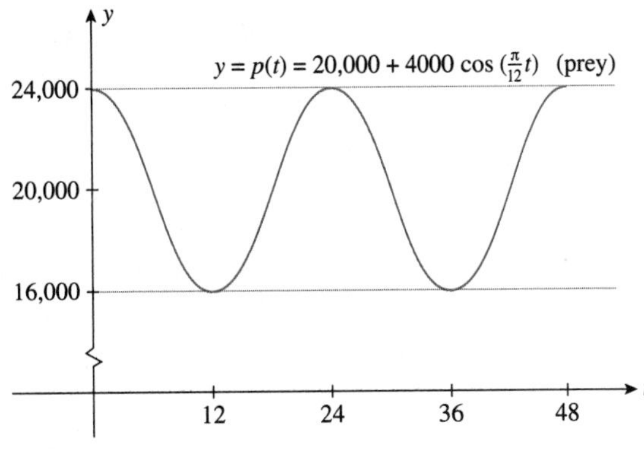

(b) The graph of the prey function $p(t)$

Figure 6.20

From the graphs, we see that at time $t = 0$ the predator population stands at 1000 owls. As it increases, the prey population decreases from 24,000 mice at that instant of time. Eventually, this decrease in the food supply causes the predator population to decrease, which in turn causes an increase in the prey population. But as the prey population increases, resulting in an increase in food supply, the predator population once again increases. The cycle is complete and starts all over again.

Transformation of the Sine and Cosine Functions

For the solution of Example 4, we described how to plot a sine and cosine functions, when they have been transformed by some translations, reflections, and stretchings in the horizontal and vertical directions. Recall from Section 2.2 that this has already been described for functions in general. However, since trigonometric functions are special in that they have periodic behaviour, we will summarize transformations of the sine and cosine functions. This information will help when graphing these functions in general, if we know the graphs of the basic functions $y = \sin x$ and $y = \cos x$ as shown in Figures 6.16 and 6.17 respectively. The tangent function can be dealt with in a similar manner; just remember that it has period π. This will be done in the Group Discussion at the end of this section.

Transformation of the Sine and Cosine Functions

Suppose we are given the sine and cosine functions

$$y = A\sin(B(x - C)) + D \quad \text{and} \quad y = A\cos(B(x - C)) + D$$

Vertical Translation

The graphs of the basic sine function and basic cosine function are translated vertically up by D units if $D > 0$, and translated vertically down by $|D|$ units if $D < 0$.

Horizontal Translation

The graphs of the basic sine function and basic cosine function are translated horizontally right by C units if $C > 0$, and translated horizontally left by $|C|$ units if $|C| < 0$. C is referred to as the phase shift.

Vertical Stretching and Reflecting

Vertical stretching is given by $|A|$, and stretches the graphs of the basic sine function and basic cosine function vertically by a factor of $|A|$ units. $|A|$ is referred to as the amplitude. If $A < 0$, then the graphs of the basic sine function and basic cosine function reflect in the x-axis.

Horizontal Stretching and Reflecting

Horizontal Stretching is given by $\frac{1}{|B|}$, $B \neq 0$ and stretches the graphs of the basic sine function and basic cosine function horizontally by a factor of $\frac{1}{|B|}$ units. If $B < 0$, then the graphs of the basic sine function and basic cosine function reflect in the y-axis. $1\pi \cdot \frac{1}{|B|} = \frac{2\pi}{|B|}$ is referred to as the period or wavelength.

Remember the order in which to perform the translations to produce the correct graph: first reflections and stretchings, then translations.

Example 5

Graph the function defined by $f(x) = 3\cos(2x) + 4$.

Solution

The graph of the basic cosine function $y = \cos x$ is translated vertically up by 4 units as indicated in Figure 6.21a resulting in $y = \cos x + 4$. The amplitude of this cosine function is 3, which means we now stretch the graph of the cosine function

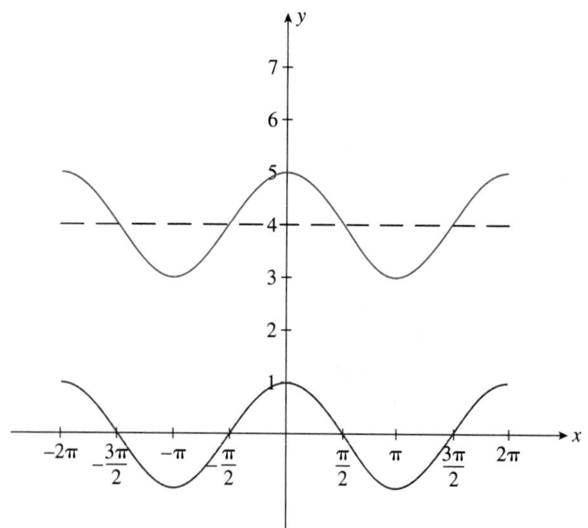

(a) Graphing the vertical translation $y = \cos x + 4$

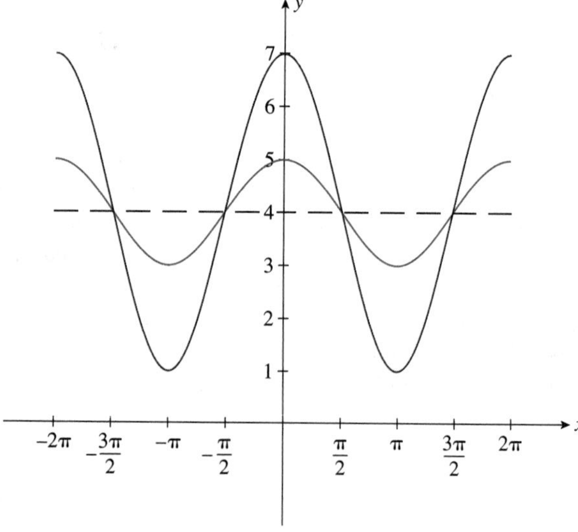

(b) Graphing the amplitude $y = 3\cos x + 4$

Figure 6.21

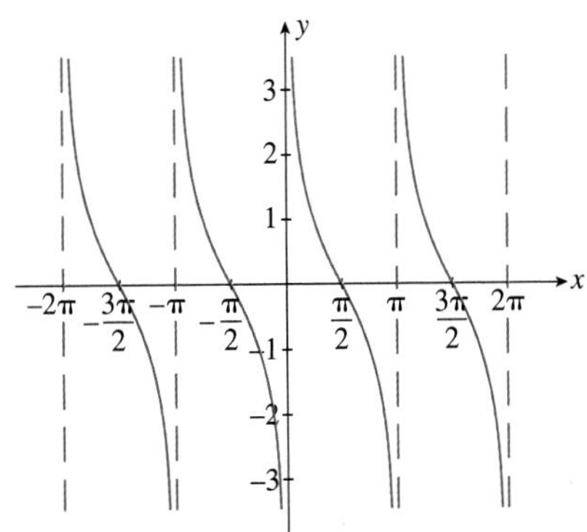

(c) Graphing the period $f(x) = 3\cos(2x) + 4$

vertically by 3 units as indicated in Figure 6.21b resulting in $y\ 3\cos x + 4$. Lastly, the period is $\frac{2\pi}{2} = \pi$, which means the graph of the cosine function is compressed by a factor of $\frac{1}{2}$ as indicated in Figure 6.21c, resulting finally in the graph of f as desired. There are no horizontal or vertical reflections, since both $3 > 0$ and $2 > 0$.

Suppose we are given the tangent function

$$y = A\tan(B(x - C)) + D$$

Trigonometric Identities

Equations expressing the relationships between trigonometric functions, such as

$$\sin(-\theta) = -\sin\theta \quad \text{and} \quad \cos(-\theta) = \cos\theta$$

are called **trigonometric identities.** Some other important trigonometric identities are listed in Table 6.3. Each identity holds true for every value of θ in the domain of the specified function. The proofs of these identities may be found in any elementary trigonometry book.

Table 6.3 Trigonometric Identities

Pythagorean Identities	Half-Angle Formulas	Sum and Difference Formulas
$\sin^2\theta + \cos^2\theta = 1$	$\cos^2\theta = \frac{1}{2}(1 + \cos 2\theta)$	$\sin(A \pm B) = \sin A\cos B \pm \cos A\sin B$
$\tan^2\theta + 1 = \sec^2\theta$	$\sin^2\theta = \frac{1}{2}(1 - \cos 2\theta)$	$\cos(A \pm B) = \cos A\cos B \mp \sin A\sin B$
$\cot^2\theta + 1 = \csc^2\theta$		
Double-Angle Formulas	**Cofunctions of Complementary Angles**	
$\sin 2A = 2\sin A\cos A$	$\sin\theta = \cos\left(\frac{\pi}{2} - \theta\right)$	
$\cos 2A = \cos^2 A - \sin^2 A$	$\cos\theta = \sin\left(\frac{\pi}{2} - 0\right)$	

As we will see later, these identities are useful in simplifying trigonometric expressions and equations and in deriving other trigonometric relationships. They are also used to verify other trigonometric identities, as illustrated in the next example.

Example 6

Verify the identity

$$\sin \theta(\csc \theta - \sin \theta) = \cos^2 \theta$$

Solution

We verify this identity by showing that the expression on the left side of the equation can be transformed into the expression on the right side. Thus,

$$\begin{aligned}\sin \theta(\csc \theta - \sin \theta) &= \sin \theta \csc \theta - \sin^2 \theta \\ &= \sin \theta \frac{1}{\sin \theta} - \sin^2 \theta \\ &= 1 - \sin^2 \theta \\ &= \cos^2 \theta\end{aligned}$$

EXPLORING WITH TECHNOLOGY

1. You can confirm some of the trigonometric identities graphically in many ways. For example, explain why an identity given in the form $f(\theta) = g(\theta)$ over some domain D is equivalent to the following statements:
   a. The graphs of f and g coincide when viewed in any appropriate viewing window.
   b. The graphs of f and $-g$ are reflections of each other with respect to the θ-axis when viewed in any appropriate viewing window.
   c. The graph of $h = f - g$ is the graph of the zero function $h(\theta) = 0$ in the domain of f (and g).
2. Use the observations made in part 1 to verify graphically the identity $\sin^2 \theta + \cos^2 \theta = 1$ by taking $f(x) = \sin^2 x + \cos^2 x$ and $g(x) = 1$.

   Hint: Enter $f(x)$ as $y_1 = (\sin x)\text{^}2 + (\cos x)\text{^}2$ and use the viewing window $[-10, 10] \cdot [-1.1, 1.1]$.
3. Verify graphically the half-angle formula $\cos^2 \theta = \frac{1}{2}(1 + \cos 2\theta)$.
4. Verify graphically the formula $\sin \theta = \cos\left(\frac{\pi}{2} - \theta\right)$.

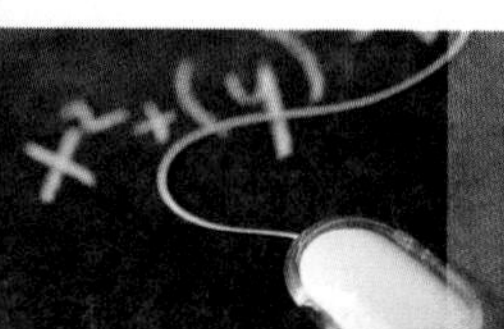

Self-Check Exercises 6.2

1. Evaluate $\cos\left(-\frac{13\pi}{3}\right)$.
2. Solve the equation $\cos \theta = -\frac{\sqrt{2}}{2}$ for $0 \le \theta \le 2\pi$.
3. Sketch the graph of $y = 2 \cos x$.

Solutions to Self-Check Exercises 6.2 can be found on page 320.

6.2 Exercises

In Exercises 1–10, evaluate the trigonometric function.

1. $\sin 3\pi$
2. $\cos\left(-\frac{3}{2}\pi\right)$
3. $\sin \frac{9}{2}\pi$
4. $\cos \frac{13}{6}\pi$
5. $\sin\left(-\frac{4}{3}\pi\right)$
6. $\cos\left(-\frac{5}{4}\pi\right)$
7. $\tan \frac{\pi}{6}$
8. $\cot\left(-\frac{\pi}{3}\right)$
9. $\sec\left(-\frac{5}{8}\pi\right)$
10. $\csc \frac{9}{4}\pi$

In Exercises 11–14, find the six trigonometric functions of the angle.

11. $\frac{\pi}{2}$
12. $-\frac{\pi}{6}$
13. $\frac{5}{3}\pi$
14. $-\frac{3}{4}\pi$

In Exercises 15–22, find all values of θ that satisfy the equation over the interval $[0, 2\pi]$.

15. $\sin \theta = -\frac{1}{2}$
16. $\tan \theta = 1$
17. $\cot \theta = -\sqrt{3}$
18. $\csc \theta = \sqrt{2}$
19. $\sec \theta = -1$
20. $\cos \theta = \sin \theta$
21. $\sin \theta = \sin\left(-\frac{4}{3}\pi\right)$
22. $\cos \theta = \cos\left(-\frac{\pi}{6}\right)$

In Exercises 23–26, sketch the graph of the function over the interval $[0, 2\pi]$.

23. $y = \csc x$
24. $y = \sec x$
25. $y = \cot x$
26. $y = \tan 2x$

In Exercises 27–34, sketch the graph of the functions over the interval $[0, 2\pi]$ by comparing it to the graph of $y = \sin x$.

27. $y = \sin 2x$
28. $y = 2 \sin x$
29. $y = -\sin x$
30. $y = -2 \sin 2x$
31. $y = 3 \sin (4x) - 5$
32. $y = \sin\left(2x - \frac{\pi}{2}\right)$
33. $y = -\sin (x + \pi)$
34. $y = 2 \sin\left(x - \frac{\pi}{2}\right) + 1$

In Exercises 35–38, sketch the graph of the functions over the interval $[-2\pi, 2\pi]$ by comparing it to the graph of $y = \tan x$.

35. $y = \tan\left(\frac{x}{3}\right)$
36. $y = -\tan x + 2$
37. $y = \tan\left(x - \frac{\pi}{2}\right)$
38. $y = \tan\left(\frac{1}{2}\left(x + \frac{\pi}{4}\right)\right)$

In Exercises 39–46, verify each identity.

39. $\cos^2 \theta - \sin^2 \theta = 2 \cos^2 \theta - 1$
40. $1 - 2 \sin^2 \theta = 2 \cos^2 \theta - 1$
41. $(\sec \theta + \tan \theta)(1 - \sin \theta) = \cos \theta$
42. $\dfrac{\sin^2 \theta}{1 + \cos^2 \theta} = \dfrac{1 - \cos^2 \theta}{2 - \sin^2 \theta}$
43. $(1 + \cot^2 \theta)\tan^2 \theta = \sec^2 \theta$
44. $\dfrac{\sec \theta - \cos \theta}{\tan \theta} = \sin \theta$
45. $\dfrac{\csc \theta}{\tan \theta + \cot \theta} = \cos \theta$
46. $\tan(A + B) = \dfrac{\tan A + \tan B}{1 - \tan A \tan B}$
47. The accompanying figure shows a right triangle ABC superimposed over a unit circle in the xy-coordinate system. By considering similar triangles, show that

$$\sin \theta = \frac{BC}{AC} = \frac{\text{Opposite side}}{\text{Hypotenuse}} \qquad \csc \theta = \frac{AC}{BC} = \frac{\text{Hypotenuse}}{\text{Opposite side}}$$

$$\cos \theta = \frac{AB}{AC} = \frac{\text{Adjacent side}}{\text{Hypotenuse}} \qquad \sec \theta = \frac{AC}{AB} = \frac{\text{Hypotenuse}}{\text{Adjacent side}}$$

$$\tan \theta = \frac{BC}{AB} = \frac{\text{Opposite side}}{\text{Adjacent side}} \qquad \cot \theta = \frac{AB}{BC} = \frac{\text{Adjacent side}}{\text{Opposite side}}$$

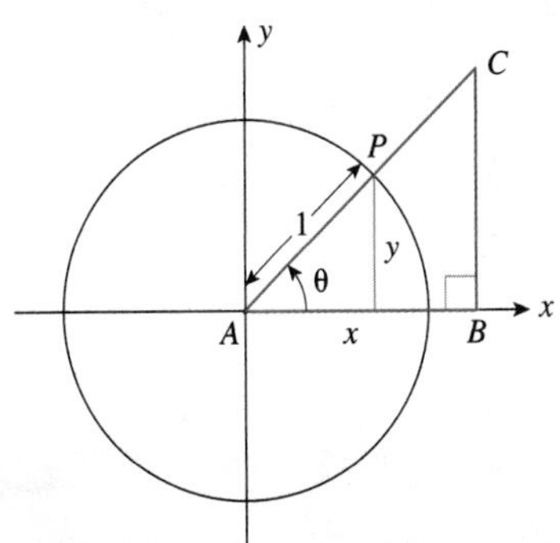

48. Refer to Exercise 47. Find the values of the six trigonometric functions of θ, where θ is the angle shown in the right triangle in the accompanying figure.

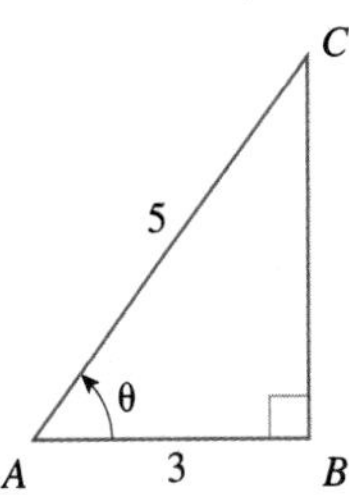

49. Refer to Exercise 47. Find the values of the six trigonometric functions of θ, where θ is the angle shown in the right triangle in the accompanying figure.

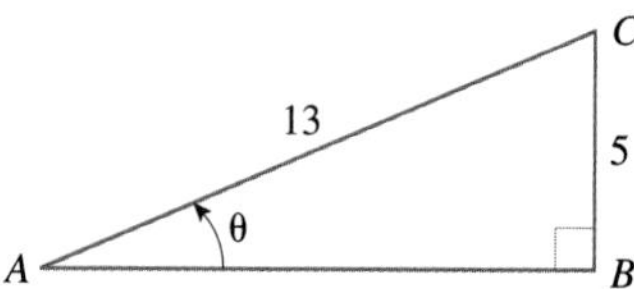

Business and Economics Applications

50. **Marginal Revenue** The revenue received from the sale of electronic fans is seasonal, with the maximum revenue in the summer. Let the revenue in thousands of dollars received from the sales of fans be approximated by

$$R(x) = -100\cos(2\pi x) + 120$$

where x is time in years, measured from January 1.

a. Sketch the graph of the revenue function for one year.

b. What is the maximum revenue for the sale of electronic fans?

c. Find the values of x when the revenues are highest during one year.

51. **Marginal Revenue** Sales of large kitchen appliances such as ovens and fridges are usually subject to seasonal fluctuations. Everything Kitchen's sales of fridge models from the beginning of 2001 to the end of 2002 can be approximated by

$$S(x) = \frac{1}{10}\sin\left(\frac{\pi}{2}(x+1)\right) + \frac{1}{2}$$

where x is time in quarters, $x = 1$ represents the end of the first quarter of 2001, and S is measured in millions of dollars.

a. What are the maximum and minimum quarterly revenues for fridges?

b. Find the values of x when the quarterly revenues are highest and lowest.

Biological and Life Sciences Applications

52. **Predator–Prey Population** The population of foxes (predators) in a certain region over a 2-year period is estimated to be

$$P(t) = 400 + 50\sin\left(\frac{\pi t}{12}\right)$$

in month t, and the population of rabbits (prey) in the same region at time t is given by

$$P(t) = 3000 + 500\cos\left(\frac{\pi t}{12}\right)$$

Sketch the graphs of each of these two functions and explain the relationship between the sizes of the two populations.

53. **Blood Pressure** The arterial blood pressure of an individual in a state of relaxation is given by

$$P(t) = 100 + 20\sin 6t$$

where $P(t)$ is measured in mm of mercury (Hg) and t is the time in seconds.

a. Show that the individual's systolic pressure (maximum blood pressure) is 120 and his diastolic pressure (minimum blood pressure) is 80.

b. Find the values of t when the individual's blood pressure is highest and lowest.

In Exercises 54–58, determine whether the statement is true or false. If it is true, explain why it is true. If it is false, give an example to show why it is false.

54. $\sin(a - b) = -\sin(b - a)$

55. If $\sin\theta = -\frac{\sqrt{3}}{2}$ and $0 \le \theta \le 2\pi$, then $\theta = \frac{5\pi}{3}$.

56. If $\tan\theta + \frac{1}{3}$, then $\sin\theta = \frac{\sqrt{10}}{10}$.

57. $\cos 2\theta = 2\cos^2\theta - 1$

58. $\sin 3\theta = 3\sin\theta + 4\sin^3\theta$

Solutions to Self-Check Exercises 6.2

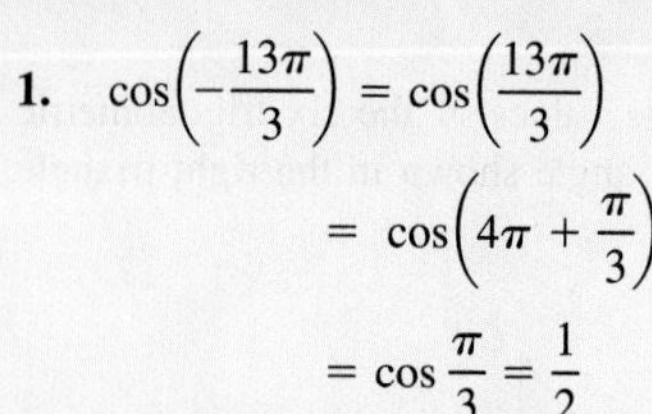

1. $\cos\left(-\frac{13\pi}{3}\right) = \cos\left(\frac{13\pi}{3}\right)$
$= \cos\left(4\pi + \frac{\pi}{3}\right)$
$= \cos\frac{\pi}{3} = \frac{1}{2}$

2. Cos θ is negative for θ in Quadrants II and III. From Table 6.2, we see that the required values of θ are $3\pi/4$ and $5\pi/4$.

3. By comparison with the graph of $y = \cos x$, we obtain the accompanying graph of $y = 2\cos x$.

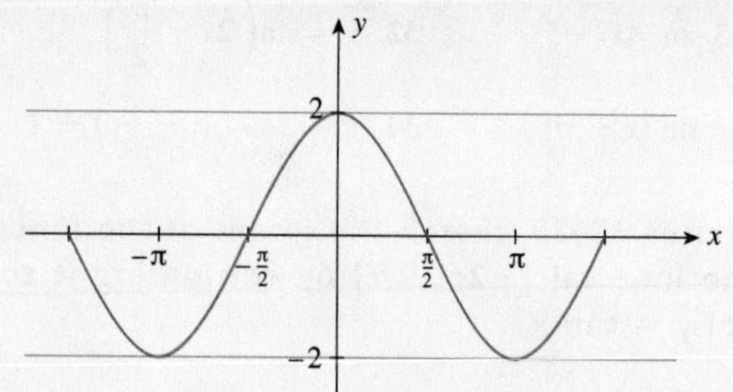

6.3 Inverse Trigonometric Functions

In the previous section we introduced the six trigonometric functions sine, cosine, tangent, secant, cosecant, and cotangent. Recall from Section 2.7, Theorem 1 (page 95) that functions have inverse functions, if they are one-to-one. By their very nature of being periodic functions, trigonometric functions are not one-to-one. However, we can restrict the domain so that the trigonometric function is one-to-one, and then we can find its inverse function. We will now develop the inverse function for the primary trigonometric function sine.

Inverse Sine

Figure 6.22a shows the sine function, and Figure 6.22b shows the sine function reflected in the line $y = x$, which recalling Section 2.7 is a necessary graphical relationship between a function and its inverse function. However, the curve in Figure 6.22 does not pass the vertical line test, and is therefore not a function, which we already knew since the sine function is not one-to-one. Clearly, we need to restrict the domain of the sine function. There is a choice of subinterval on which we define a restricted trigonometric function so that its inverse will be a function, but conventionally, we choose the interval $\left[-\frac{\pi}{2}, \frac{\pi}{2}\right]$ as shown in Figure 6.23a. Now we can reflect again in the line $y = x$, which gives us the inverse function for sine called **inverse sine** and denoted by $\sin^{-1} x$ shown in Figure 6.23b. The inverse sine has domain $[-1, 1]$ and range $\left[-\frac{\pi}{2}, \frac{\pi}{2}\right]$ (see Table 6.4 at the end of this section). We want to emphasize here that the sine function takes as a function argument an angle θ and returns as a function value the sine ratio $\sin \theta$. This means that the inverse sine function takes as a function argument a ratio represented by x and returns as a function value an angle θ denoted by $\sin^{-1} x$.

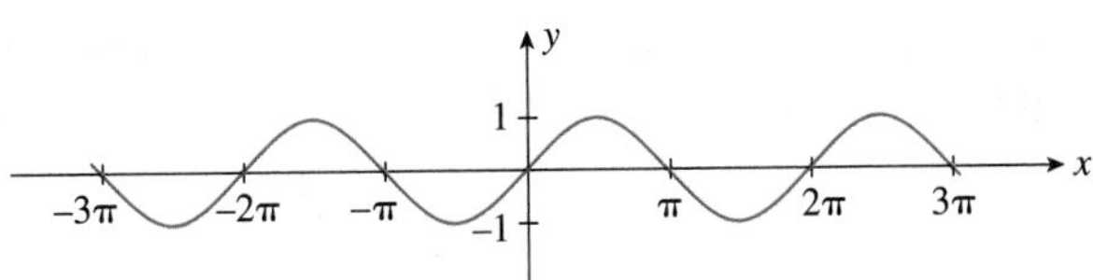

(a) $y = \sin x$ with unrestricted domain

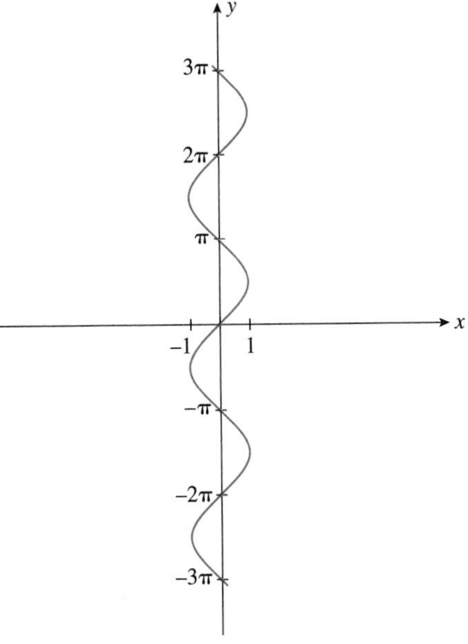

(b) Sine reflected in the line $y = x$

Figure 6.22

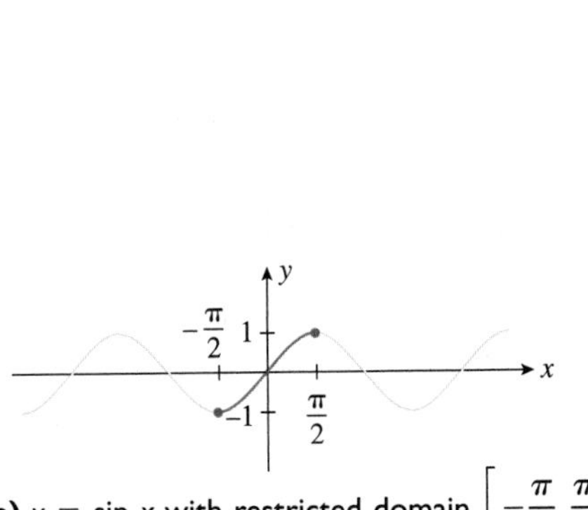

(a) $y = \sin x$ with restricted domain $\left[-\frac{\pi}{2}, \frac{\pi}{2}\right]$

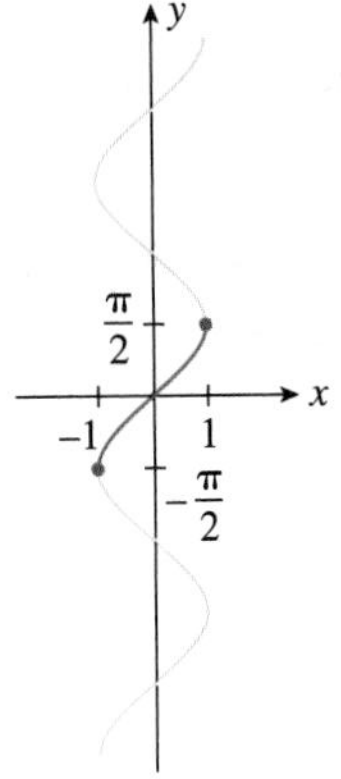

(b) $y = \sin^{-1} x$

Figure 6.23

Remark

The inverse sine is sometimes also called the arcsine and denoted by arcsin x. ◀

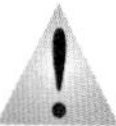

Do not confuse $\sin^{-1} x$ with $(\sin x)^{-1}$! The inverse sine is $\sin^{-1} x$, while the reciprocal of sine is the cosecant given by $(\sin x)^{-1} = \frac{1}{\sin x} = \csc x$.

GROUP DISCUSSION

We have seen that restricting the domain of the sine function to $\left[-\frac{\pi}{2}, \frac{\pi}{2}\right]$ allows us to define the inverse sine function. Hypothetically speaking, what other intervals could you have chosen to get an inverse for sine? What would its graph look like? Why do you think the interval $\left[-\frac{\pi}{2}, \frac{\pi}{2}\right]$ was chosen to develop the inverse sine?

Example 1

Evaluate $\sin^{-1}\left(-\frac{\sqrt{2}}{2}\right)$, if defined.

Solution

Since $-\frac{\sqrt{2}}{2}$ is in the domain of inverse sine, the angle $\sin^{-1}\left(-\frac{\sqrt{2}}{2}\right)$ is defined. We are looking for the angle $\theta \in \left[-\frac{\pi}{2}, \frac{\pi}{2}\right]$ such that $\sin\theta = -\frac{\sqrt{2}}{2}$. We begin by writing $\frac{\sqrt{2}}{2} = \frac{1}{\sqrt{2}}$, and recognizing this special ratio from the special triangle with angles $\frac{\pi}{4}, \frac{\pi}{4}$, and $\frac{\pi}{2}$ (see Figure 6.13b). Therefore, the reference angle is $\alpha = \frac{\pi}{4}$. Since the ratio $-\frac{\sqrt{2}}{2}$ is negative, we know by the ASTC Rule that sine is only negative in Quadrants III and IV. However, we also have domain $\left[-\frac{\pi}{2}, \frac{\pi}{2}\right]$, and therefore θ must be in Quadrant IV. Putting all the information together, we get

$$\theta = -\alpha = -\frac{\pi}{4}$$

Remark

Notice that we simply measured clockwise in the above example to get θ. We could have also continued the following way: Since θ must be in Quadrant IV, we calculate

$$2\pi - \alpha = 2\pi - \frac{\pi}{4} = \frac{7\pi}{4}$$

However, $\frac{7\pi}{4} \notin \left[-\frac{\pi}{2}, \frac{\pi}{2}\right]$. Our angle θ is coterminal with $\frac{7\pi}{4}$, and so we keep adding/subtracting 2π until we get a coterminal angle for θ which resides in the domain $\left[-\frac{\pi}{2}, \frac{\pi}{2}\right]$. This yields

$$\theta = \frac{7\pi}{4} - 2\pi = -\frac{\pi}{4}$$

We see that these calculations are much more cumbersome than to simply measure counterclockwise. ◀

The Inverse Function for the Remaining Trigonometric Functions

The inverse function for each of the remaining five trigonometric functions can be developed in a similar fashion to the inverse sine above. Care has to be taken to restrict the domain of the trigonometric function whose inverse we are looking for, so that over the domain the function is one-to-one, and therefore has an inverse. Table 6.4 lists the inverses of all six trigonometric functions together with their domain, range, and graph (Figures 6-24(a)–(f)).

Do not mix up the notation for an inverse trigonometric function with the reciprocal of a trigonometric function.

Remark

The inverse trigonometric functions can also be denoted with the prefix arc: arcsin x, arccos x, arctan x, arcsec x, arccsc x, and arccot x. ◀

Table 6.4

Trigonometric Inverse Function	Domain	Range	Graph
$y = \sin^{-1} x$	$[-1, 1]$	$\left[-\frac{\pi}{2}, \frac{\pi}{2}\right]$	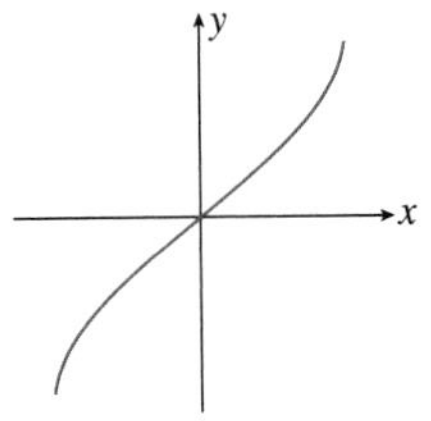 Figure 6.24(a)
$y = \cos^{-1} x$	$[-1, 1]$	$[0, \pi]$	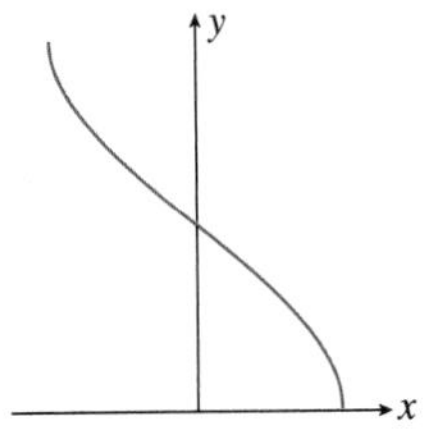 Figure 6.24(b)
$y = \tan^{-1} x$	$(-\infty, \infty)$	$\left(-\frac{\pi}{2}, \frac{\pi}{2}\right)$	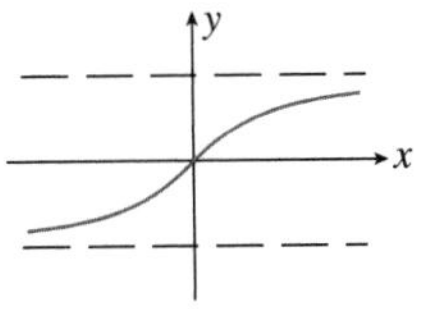 Figure 6.24(c)
$y = \cot^{-1} x$	$(-\infty, \infty)$	$[0, \pi]$	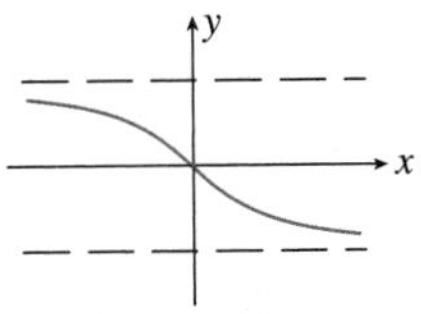 Figure 6.24(d)
$y = \sec^{-1} x$	$(-\infty, -1] \cup [1,\infty)$	$\left[0, \frac{\pi}{2}\right) \cup \left(\frac{\pi}{2}, \pi\right]$	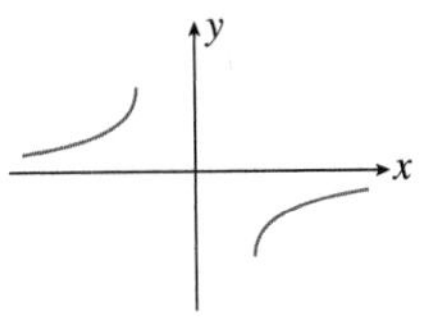 Figure 6.24(e)
$y = \csc^{-1} x$	$(-\infty, -1] \cup [1, \infty)$	$\left[-\frac{\pi}{2}, 0\right) \cup \left(0, \frac{\pi}{2}\right]$	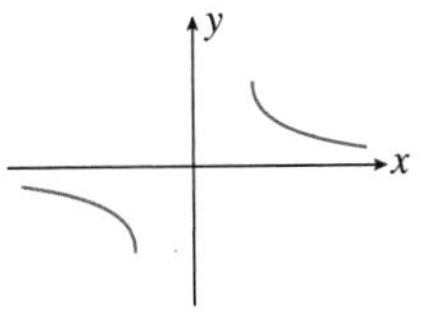 Figure 6.24(f)

Example 2

Calculate $\cot\left(\cos^{-1}\left(\frac{1}{2}\right)\right)$.

Solution

We begin by letting $\theta = \cos^{-1}\left(\frac{1}{2}\right)$, where the angle $\theta \in [0, \pi]$. Now,

$$\theta = \cos^{-1}\left(\frac{1}{2}\right) \Leftrightarrow \cos\theta = \frac{1}{2}$$

We recognize that $\frac{1}{2}$ is a special ratio from the special triangle with angles $\frac{\pi}{6}, \frac{\pi}{3}$, and $\frac{\pi}{2}$ (see Figure 6.12b). Therefore, the reference angle is $\alpha = \frac{\pi}{3}$. The ASTC Rule indicates that we need to consider Quadrants I and IV, since the ratio $\frac{1}{2}$ is positive. However, we also have domain $[0, \pi]$, which represents Quadrants I and II. So θ must be in Quadrant I. Putting all the information together, we get

$$\theta = \alpha = \frac{\pi}{3}$$

This implies that

$$\cot\left(\cos^{-1}\left(\frac{1}{2}\right)\right) = \cot\left(\frac{\pi}{3}\right)$$

Using sine and cosine, we can write

$$\cot\left(\frac{\pi}{3}\right) = \frac{\cos\left(\frac{\pi}{3}\right)}{\sin\left(\frac{\pi}{3}\right)}$$

which we can solve using Table 6.2 or by the method of Example 1 in Section 6.2 to get

$$\cot\left(\frac{\pi}{3}\right) = \frac{\cos\left(\frac{\pi}{3}\right)}{\sin\left(\frac{\pi}{3}\right)} = \frac{\frac{1}{2}}{\frac{\sqrt{3}}{2}} = \frac{1}{\sqrt{3}}$$

Example 3

Evaluate $\sec^{-1}(-2.5)$ to two decimal places, if defined.

Solution

Since -2.5 is in the domain of inverse secant, the angle $\sec^{-1}(-2.5)$ is defined. We are looking for the angle $\theta \in \left[0, \frac{\pi}{2}\right)$ or $\theta \in \left(\frac{\pi}{2}, \pi\right]$ such that $\sec\theta = -2.5$. Notice that the ratio -2.5 is not special, which means we need a calculator to answer this question. Most calculators do not have a secant function, so we need to express the secant in terms of cosine

$$\sec\theta = -2.5$$

$$\Leftrightarrow \frac{1}{\cos\theta} = -2.5$$

$$\Leftrightarrow \cos\theta = \frac{1}{-2.5} = -0.4$$

Finally, we can calculate the angle by using the inverse cosine function on our calculator and setting the mode to *radian* to get

$$\theta = \cos^{-1}(-0.4) \approx 1.98$$

Notice that $1.98 \in \left(\frac{\pi}{2}, \pi\right]$.

Example 4

Find $\sin\theta$ and $\cos\theta$ if $\theta = \tan^{-1}\left(\frac{3}{5}\right)$.

Solution

Since

$$\theta = \tan^{-1}\left(\frac{3}{5}\right) \Leftrightarrow \tan\theta = \frac{3}{5}$$

we start by drawing a right triangle with the following information

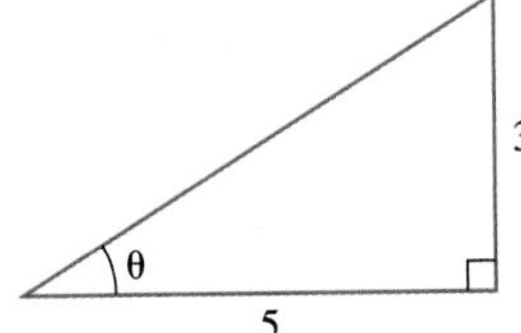

Let us call the missing side of the triangle x. By the Pythagorean theorem, we can calculate the length of x to be

$$x = \sqrt{3^2 + 5^2} = \sqrt{34}$$

Rescaling the triangle to fit into the unit circle we get

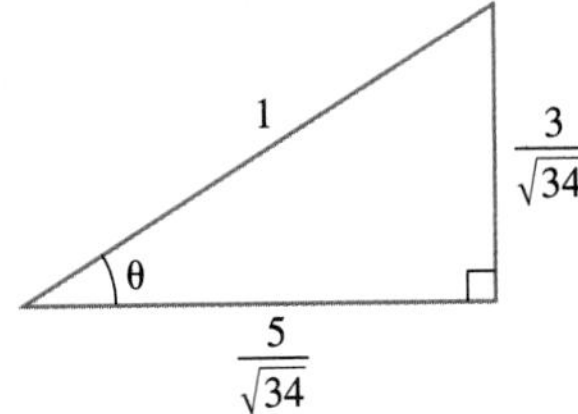

Therefore,

$$\sin\theta = 3/\sqrt{34} \text{ and } \cos\theta = 5/\sqrt{34}$$

Self-Check Exercises 6.3

1. Evaluate the following angles, if defined.
   a. $\cos^{-1}(-1)$
   b. $\tan^{-1}(0)$
   c. $\csc^{-1}(0)$
2. Is $\tan^{-1}\left(\tan\left(\frac{5\pi}{4}\right)\right) = \frac{5\pi}{4}$ true or false? Why?

Solutions to Self-Check Exercises 6.3 can be found on page 326.

6.3 Exercises

In Exercises 1–12, evaluate the given angles using the special triangle with angles $\frac{\pi}{4}, \frac{\pi}{4}$, and $\frac{\pi}{2}$ or the special triangle with angles $\frac{\pi}{6}, \frac{\pi}{3}$, and $\frac{\pi}{2}$.

1. $\sin^{-1}\left(\frac{\sqrt{3}}{2}\right)$
2. $\sin^{-1}\left(-\frac{1}{\sqrt{2}}\right)$
3. $\cos^{-1}\left(-\frac{1}{2}\right)$
4. $\cos^{-1}\left(\frac{1}{2}\right)$
5. $\tan^{-1}\left(\frac{1}{\sqrt{3}}\right)$
6. $\tan^{-1}(-\sqrt{3})$
7. $\sec^{-1}(-\sqrt{2})$
8. $\sec^{-1}\left(\frac{2}{\sqrt{3}}\right)$
9. $\csc^{-1}(2)$
10. $\csc^{-1}\left(-\frac{2}{\sqrt{3}}\right)$
11. $\cot^{-1}(\sqrt{3})$
12. $\cot^{-1}\left(-\frac{1}{\sqrt{3}}\right)$

In Exercises 13–18, use a calculator to evaluate the given angles to three decimal places.

13. $\cos^{-1}(0.8)$
14. $\sin^{-1}(-0.25)$
15. $\sin^{-1}(0.92)$
16. $\cos^{-1}(-0.87)$
17. $\tan^{-1}(-5.3)$
18. $\tan^{-1}(2.1)$

In Exercises 19–22, evaluate the given expressions.

19. $\cos\left(\tan^{-1}\left(\frac{1}{\sqrt{3}}\right)\right)$
20. $\csc\left(\sin^{-1}\left(\frac{1}{\sqrt{2}}\right)\right)$
21. $\sec\left(\cos^{-1}\left(-\frac{1}{\sqrt{2}}\right)\right)$
22. $\cos\left(\sin^{-1}\left(-\frac{1}{2}\right)\right)$

In Exercises 23–26, given θ find $\sin\theta$, $\cos\theta$, $\tan\theta$, $\sec\theta$, $\csc\theta$, and $\cot\theta$.

23. $\theta = \sec^{-1}\left(\frac{\sqrt{5}}{2}\right)$
24. $\theta = \tan^{-1}\left(\frac{1}{4}\right)$
25. $\theta = \cos^{-1}\left(\frac{3}{7}\right)$
26. $\theta = \sin^{-1}\left(\frac{2}{\sqrt{6}}\right)$

In Exercises 27–32, determine whether the statement is true or false. If it is true, explain why it is true. If it is false, give an example to show why it is false.

27. If $\sin^{-1}(x) = \frac{9\pi}{4}$, then $x = \frac{1}{\sqrt{2}}$.
28. If $\cos^{-1}(x) = \frac{5\pi}{3}$, then $x = \frac{1}{2}$.
29. If $\sin^{-1}(2) = \theta$, then $\sin\theta = 2$.
30. If $\sec^{-1}\left(-\frac{1}{2}\right) = \theta$, then $\sec\theta = -\frac{1}{2}$.
31. $\sin^{-1}\left(\sin\left(-\frac{\pi}{3}\right)\right) = -\frac{\pi}{3}$
32. $\cos^{-1}\left(\cos\left(-\frac{\pi}{4}\right)\right) = -\frac{\pi}{4}$

Solutions to Self-Check Exercises 6.3

1. Evaluate the following angles, if defined.

   a. $\cos^{-1}(-1)$
   Since $-1 \in [-1, 1]$, the angle $\cos^{-1}(-1)$ is defined. We read off the graph of the inverse cosine (Figure 6.24b) that $\cos^{-1}(-1) = \pi$.

   b. $\tan^{-1}(0)$
   Since $0 \in (-\infty, \infty)$, the angle $\tan^{-1}(0)$ is defined. We read off the graph of the inverse tangent (Figure 6.24c) that $\tan^{-1}(0) = 0$.

   c. $\csc^{-1}(0)$
   Since $0 \notin (-\infty, -1] \cup [1, \infty)$, the angle $\csc^{-1}(0)$ is not defined.

2. The statement is false, because $\frac{5\pi}{4}$ does not belong in the range of the inverse tangent, which is $\left(-\frac{\pi}{2}, \frac{\pi}{2}\right)$. The correct answer is

$$\tan^{-1}\left(\tan\left(\frac{5\pi}{4}\right)\right) = \frac{5\pi}{4} - \pi = \frac{\pi}{4} \in \left(-\frac{\pi}{2}, \frac{\pi}{2}\right)$$

6.4 Differentiation Rules

In this section we develop rules for differentiating trigonometric functions and their inverses. Knowledge of the derivatives of the trigonometric functions helps us to study the properties of functions involving the trigonometric functions in much the same way we analyzed the algebraic, exponential, and logarithmic functions in our earlier work.

Derivatives of the Sine and Cosine Functions

First we develop the rule for differentiating the sine and cosine functions. Then, using these rules and the rules of differentiation, we derive the rules for differentiating the other trigonometric functions.

To derive the rule for differentiating the sine function, we need the following two results:

$$\lim_{h\to 0}\frac{\sin h}{h} = 1 \quad \text{and} \quad \lim_{h\to 0}\frac{\cos h - 1}{h} = 0 \tag{8}$$

The plausibility of the first limit can be seen by examining Table 6.5, which is constructed with the aid of a calculator.

Table 6.5

Values of $\frac{\sin h}{h}$ for Selected Values of h [in Radians] Approaching Zero

h	± 0.5	± 0.1	± 0.01	± 0.001
$\frac{\sin h}{h}$	0.9588511	0.9983342	0.9999833	0.9999998

The second limit can be derived from the first using the appropriate trigonometric identity and the properties of limits (see Exercise 54).

EXPLORING WITH TECHNOLOGY

You can obtain a visual confirmation of the limits

$$\lim_{h\to 0}\frac{\sin h}{h} = 1 \quad \text{and} \quad \lim_{h\to 0}\frac{\cos h - 1}{h} = 0$$

given in Equation (8) as follows:

1. Plot the graph of $f(x) = (\sin x)/x$, using the viewing window $[-1, 1] \cdot [0, 2]$. Then use ZOOM and TRACE to find the values of $f(x)$ for values of x close to $x = 0$. What happens when you try to evaluate $f(0)$? Explain your answer.
2. Plot the graph of $g(x) = (\cos x - 1)/x$, using the viewing window $[-1, 1] \cdot [-0.5, 0.5]$. Then use ZOOM and TRACE to show that $g(x)$ is close to zero if x is close to zero.

Caution: Convincing as these arguments are, they do not establish the stated results. As noted earlier, the proofs are given at the end of the section.

Now suppose $f(x) = \sin x$. Then

$$\begin{aligned} f'(x) &= \lim_{h\to 0}\frac{f(x+h) - f(x)}{h} \\ &= \lim_{h\to 0}\frac{\sin(x+h) - \sin x}{h} \end{aligned}$$

$$= \lim_{h \to 0} \frac{\sin x \cos h + \cos x \sin h - \sin x}{h}$$ Use the sine rule for sums.

$$= \lim_{h \to 0} \frac{\sin x(\cos h - 1) + \cos x \sin h}{h}$$

$$= \lim_{h \to 0} \sin x\left(\frac{\cos h - 1}{h}\right) + \lim_{h \to 0} \cos x\left(\frac{\sin h}{h}\right)$$

$$= \sin x \lim_{h \to 0}\left(\frac{\cos h - 1}{h}\right) + \cos x \lim_{h \to 0}\left(\frac{\sin h}{h}\right)$$

$$= (\sin x)(0) + (\cos x)(1)$$ Use Equation (8).

$$= \cos x$$

With the help of the general power rule, this result can be generalized as follows: Suppose

$$y = h(x) = \sin f(x)$$

where $f(x)$ is a differentiable function of x. Then putting

$$u = f(x)$$

so that $y = \sin u$, we find, upon applying the chain rule,

$$h'(x) = \frac{dy}{du} \cdot \frac{du}{dx} = (\cos u)\frac{du}{dx} = [\cos f(x)]f'(x)$$

Derivatives of the Sine and Generalized Sine Functions

Rule 1a: $\frac{d}{dx}(\sin x) = \cos x$

Rule lb: $\frac{d}{dx}[\sin f(x)] = [\cos f(x)]f'(x)$

Observe that Rule 1b reduces to Rule 1a when $f(x) = x$, as expected.

Example I

Differentiate each of the following functions:

a. $f(x) = x^2 \sin x$ **b.** $g(x) = \sin(2x + 1)$

c. $h(x) = (x + \sin x^2)^{10}$ **d.** $k(x) = \ln(\sin x) + \sin(2^x)$

Solution

a. Using the product rule followed by Rule 1a, we find

$$f'(x) = 2x \sin x + x^2 \frac{d}{dx}(\sin x)$$

$$= 2x \sin x + x^2 \cos x = x(2 \sin x + x \cos x)$$

b. Using Rule 1b, we find

$$g'(x) = [\cos(2x + 1)]\frac{d}{dx}(2x + 1) = 2\cos(2x + 1)$$

c. We first use the general power rule followed by Rule 1b. We obtain

$$h'(x) = 10(x + \sin x^2)^9 \frac{d}{dx}(x + \sin x^2)$$

$$= 10(x + \sin x^2)^9\left[1 + \cos x^2 \frac{d}{dx}(x^2)\right]$$

$$= 10(x + \sin x^2)^9(1 + 2x \cos x^2)$$

d. Using Rule 1a and 1b, we find

$$k'(x) = \frac{1}{\sin x}\frac{d}{dx}(\sin x) + \cos(2^x)\frac{d}{dx}(2^x)$$

$$= \frac{\cos^x}{\sin^x} + \cos(2^x)2^x(\ln 2)$$

$$= \frac{\cos x}{\sin x} + (\ln 2)\, 2^x \cos(2^x)$$

To derive the rule for differentiating the cosine function, we make use of the following relationships between the cosine function and the sine function:

$$\cos x = \sin\left(\frac{\pi}{2} - x\right) \quad \text{and} \quad \sin x = \cos\left(\frac{\pi}{2} - x\right)$$

(see Table 6.3). Now, with the help of Rule 1b, we see that if $f(x) = \cos x$, then

$$f'(x) = \frac{d}{dx}\cos x$$

$$= \frac{d}{dx}\sin\left(\frac{\pi}{2} - x\right)$$

$$= \cos\left(\frac{\pi}{2} - x\right)\frac{d}{dx}\left(\frac{\pi}{2} - x\right)$$

$$= (\sin x)(-1)$$

$$= -\sin x$$

This result may be generalized immediately using the general power rule. In fact, if

$$y = h(x) = \cos f(x)$$

where $f(x)$ is a differentiable function of x, then with $u = f(x)$, we have $y = \cos u$ so that

$$\frac{dy}{dx} = \frac{dy}{du}\frac{du}{dx} = (-\sin u)\frac{du}{dx} = -[\sin f(x)]f'(x)$$

Derivatives of the Cosine and Generalized Cosine Functions

Rule 2a: $\frac{d}{dx}(\cos x) = -\sin x$

Rule 2b: $\frac{d}{dx}[\cos f(x)] = -[\sin f(x)]f'(x)$

Observe that Rule 2b reduces to Rule 2a when $f(x) = x$, as expected.

GROUP DISCUSSION

Derive Rule 2a,

$$\frac{d}{dx}(\cos x) = -\sin x,$$

directly from the definition of the derivative and the relationships given in Equation (8).

Example 2

Find the derivative of each function:

a. $f(x) = \cos(2x^2 - 1)$ **b.** $g(x) = \sqrt{\cos 2x}$

c. $h(x) = e^{\sin 2x + \cos 3x}$ **d.** $k(x) = \sin(\cos^2 x)$

Solution

a. Using Rule 2b, we find

$$f'(x) = -\sin(2x^2 - 1)\frac{d}{dx}(2x^2 - 1)$$

$$= -[\sin(2x^2 - 1)]4x$$

$$= -4x\sin(2x^2 - 1)$$

b. We first rewrite $g(x)$ as $g(x) = (\cos 2x)^{1/2}$. Using the general power rule followed by Rule 2b, we find

$$\begin{aligned} g'(x) &= \frac{1}{2}(\cos 2x)^{-1/2}\frac{d}{dx}(\cos 2x) \\ &= \frac{1}{2}(\cos 2x)^{-1/2}(-\sin 2x)\frac{d}{dx}(2x) \\ &= \frac{1}{2}(\cos 2x)^{-1/2}(-\sin 2x)(2) \\ &= -\frac{\sin 2x}{\sqrt{\cos 2x}} \end{aligned}$$

c. Using the corollary to the chain rule for exponential functions, we find

$$\begin{aligned} h'(x) &= e^{\sin 2x + \cos 3x} \cdot \frac{d}{dx}(\sin 2x + \cos 3x) \\ &= e^{\sin 2x + \cos 3x}\left[(\cos 2x)\frac{d}{dx}(2x) - (\sin 3x)\frac{d}{dx}(3x)\right] \\ &= (2 \cos 2x - 3 \sin 3x)e^{\sin 2x + \cos 3x} \end{aligned}$$

d. We first rewrite $k(x)$ as $k(x) = \sin((\cos x)^2)$. Using Rule 1b with the general power rule and Rule 2a, we find

$$\begin{aligned} k'(x) &= \cos(\cos^2 x)\frac{d}{dx}((\cos x)^2) \\ &= \cos(\cos^2 x)\, 2\cos x \frac{d}{dx}(\cos x) \\ &= \cos(\cos^2 x)\, 2\cos x\,(-\sin x) \\ &= -2 \sin x \cos x \cos(\cos^2 x) \end{aligned}$$

Derivatives of the Other Trigonometric Functions

We are now in a position to find the rule for differentiating the tangent function. In fact, using the quotient rule, we see that if $f(x) = \tan x$, then

$$\begin{aligned} f'(x) &= \frac{d}{dx}\tan x = \frac{d}{dx}\left(\frac{\sin x}{\cos x}\right) \\ &= \frac{(\cos x)\frac{d}{dx}\sin x - (\sin x)\frac{d}{dx}\cos x}{\cos^2 x} \\ &= \frac{(\cos x)(\cos x) - (\sin x)(-\sin x)}{\cos^2 x} \\ &= \frac{\cos^2 x + \sin^2 x}{\cos^2 x} = \frac{1}{\cos^2 x} = \sec^2 x \end{aligned}$$

Once again, this result may be generalized with the help of the general power rule, and we are led to the next two rules.

Derivatives of the Tangent and the Generalized Tangent Functions

Rule 3a: $\frac{d}{dx}(\tan x) = \sec^2 x$

Rule 3b: $\frac{d}{dx}[\tan f(x)] = [\sec^2 f(x)]f'(x)$

The rules for differentiating the remaining three trigonometric functions are derived in a similar manner. We state the generalized versions of these rules. Here $f(x)$ is a differentiable function of x.

Derivatives of the Cosecant, Secant, and Cotangent Functions

Rule 4: $\frac{d}{dx}[\csc f(x)] = -[\csc f(x)][\cot f(x)]f'(x)$

Rule 5: $\frac{d}{dx}[\sec f(x)] = [\sec f(x)][\tan f(x)]f'(x)$

Rule 6: $\frac{d}{dx}[\cot f(x)] = -[\csc^2 f(x)]f'(x)$

Remark

As an aid to remembering the signs of the derivatives of the trigonometric functions, observe that those functions beginning with a "c" [$\cos x$, $\csc x$, $\cot x$, $\cos f(x)$, $\csc f(x)$, and $\cot f(x)$] have a minus sign attached to their derivatives. ◀

Example 3

Find an equation of the tangent line to the graph of the function $f(x) = \tan 2x$ at the point $(\pi/8, 1)$.

Solution

The slope of the tangent line at any point on the graph of f is given by

$$f'(x) = 2 \sec^2 2x$$

In particular, the slope of the tangent line at the point $(\pi/8, 1)$ is given by

$$f'\left(\frac{\pi}{8}\right) = 2 \sec^2 \frac{\pi}{4}$$

$$= 2\left(\frac{2}{\sqrt{2}}\right)^2 = 4$$

Therefore, a required equation is given by

$$y - 1 = 4\left(x - \frac{\pi}{8}\right)$$

$$y = 4x + \left(1 - \frac{\pi}{2}\right)$$

Derivative of the Inverse Sine Function

We show how to derive the derivative for the inverse sine function using implicit method (Section 4.4), which is a method that can also be applied to finding the derivative of the other five trigonometric inverse functions.

To find the derivative of $y = \sin^{-1} x$, we first write an equivalent equation using the sine function

$$y = \sin^{-1} x \Leftrightarrow \sin y = x \tag{9}$$

and understand that we are looking for $\frac{dy}{dx}$. Now we differentiate $\sin y = x$ implicitly

$$\frac{d}{dx}(\sin y) = \frac{d}{dx}(x)$$

$$\cos y \frac{dy}{dx} = 1$$

We need to ensure that $\cos y \neq 0$, otherwise we cannot divide by it to solve for $\frac{dy}{dx}$. Since $\cos\left(-\frac{\pi}{2}\right) = 0$ and $\cos\left(\frac{\pi}{2}\right) = 0$, we restrict the domain for the derivative of $y = \sin^{-1} x$ to the interval $(-1, 1)$, which gives us the necessary corresponding interval $\left(-\frac{\pi}{2}, \frac{\pi}{2}\right)$ for the range. Upon division by $\cos y$ we get

$$\frac{dy}{dx} = \frac{1}{\cos y}$$

Even though we have already found $\frac{dy}{dx}$, we want to express this derivative in terms of x. For this, we construct a right triangle based on Equation (9)

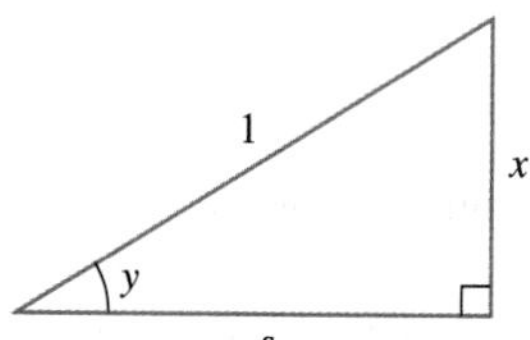

and label the missing side s. We calculate the length of s using the Pythagorean theorem

$$s = \sqrt{1^2 - x^2} = \sqrt{1 - x^2}$$

Finally, we can read off the triangle the value of $\cos y$. Therefore

$$\frac{dy}{dx} = \frac{1}{\cos y} = \frac{1}{\sqrt{1 - x^2}}$$

We can generalize this result. If $u = f(x)$ is a differentiable function of x with range $-1 < f(x) < 1$ (or $|f(x)| < 1$), then we apply the chain rule to

$$y = h(x) = \sin^{-1} f(x) = \sin^{-1} u$$

and get

$$\frac{dy}{dx} = \frac{dy}{du}\frac{du}{dx} = \frac{1}{\sqrt{1 - u^2}}\frac{du}{dx} = \frac{f'(x)}{\sqrt{1 - [f(x)]^2}}$$

Derivative of the Inverse Sine and Generalized Inverse Sine Function

Rule 7a: $\frac{d}{dx}[\sin^{-1} x] = \frac{1}{\sqrt{1 - x^2}}, \quad |x| < 1$

Rule 7b: $\frac{d}{dx}[\sin^{-1} f(x)] = \frac{f'(x)}{\sqrt{1 - [f(x)]^2}}, \quad |f(x)| < 1$

Example 4

Find the derivative of $f(x) = \log_2 (\sin^{-1} (x^2 - 3x))$.

Solution

Using the derivative of the general logarithmic function followed by Rule 7b, we find

$$f'(x) = \frac{1}{(\ln 2)\sin^{-1}(x^2 - 3x)} \frac{d}{dx}(\sin^{-1}(x^2 - 3x))$$

$$= \frac{1}{(\ln 2)\sin^{-1}(x^2 - 3x)}\left(\frac{1}{\sqrt{1 - (x^2 - 3x)^2}}\frac{d}{dx}(x^2 - 3x)\right)$$

$$= \frac{1}{(\ln 2)\sin^{-1}(x^2 - 3x)} \frac{1}{\sqrt{1 - (x^2 - 3x)^2}}(2x - 3)$$

$$= \frac{2x - 3}{(\ln 2)\sin^{-1}(x^2 - 3x)\sqrt{1 - (x^2 - 3x)^2}}$$

Derivatives of the Other Inverse Trigonometric Functions

The rules for differentiating the remaining five trigonometric functions are derived in a similar manner. We state the generalized versions of these rules. Here $f(x)$ is a differentiable function of x. Restrictions for the range of f are indicated when necessary.

Derivatives of the Inverse Cosine, Inverse Tangent, Inverse Secant, Inverse Cosecant, and Inverse Cotangent Functions

Rule 8: $\frac{d}{dx}[\cos^{-1} f(x)] = -\frac{f'(x)}{\sqrt{1 - [f(x)]^2}}, \quad |f(x)| < 1$

Rule 9: $\frac{d}{dx}[\tan^{-1} f(x)] = \frac{f'(x)}{1 + [f(x)]^2}$

Rule 10: $\frac{d}{dx}[\cot^{-1} f(x)] = -\frac{f'(x)}{1 + [f(x)]^2}$

Rule 11: $\frac{d}{dx}[\sec^{-1} f(x)] = \frac{f'(x)}{|f(x)|\sqrt{[f(x)]^2 - 1}}, \quad |f(x)| > 1$

Rule 12: $\frac{d}{dx}[\csc^{-1} f(x)] = -\frac{f'(x)}{|f(x)|\sqrt{[f(x)]^2 - 1}}, \quad |f(x)| > 1$

Applications

The techniques developed in Chapter 7 may be used to study the properties of functions involving trigonometric functions, as the next examples show.

Example 5

Predator–Prey Population The owl population in a certain area is estimated to be

$$P(t) = 1000 + 100 \sin\left(\frac{\pi t}{12}\right)$$

in month t, and the mouse population in the same area at time t is given by

$$p(t) = 20{,}000 + 4000 \cos\left(\frac{\pi t}{12}\right)$$

Find the rate of change of the population when $t = 2$.

Solution

The rate of change of the owl population at any time t is given by

$$P'(t) = 100\left[\cos\left(\frac{\pi t}{12}\right)\right]\left(\frac{\pi}{12}\right)$$

$$= \frac{25\pi}{3} \cos\left(\frac{\pi t}{12}\right)$$

and the rate of change of the mouse population at any time t is given by

$$p'(t) = 4000\left[-\sin\left(\frac{\pi t}{12}\right)\right]\left(\frac{\pi}{12}\right)$$
$$= -\frac{1000\pi}{3}\sin\left(\frac{\pi t}{12}\right)$$

In particular, when $t = 2$ the rate of change of the owl population is

$$P'(2) = \frac{25\pi}{3}\cos\frac{\pi}{6}$$
$$= \frac{25\pi}{3}\left(\frac{\sqrt{3}}{2}\right) \approx 22.7$$

That is, the predator population is increasing at the rate of approximately 22.7 owls per month, and the rate of change of the mouse population is

$$p'(2) = -\frac{1000\pi}{3}\sin\frac{\pi}{6}$$
$$= -\frac{1000\pi}{3}\left(\frac{1}{2}\right) \approx -523.6$$

That is, the prey population is decreasing at the rate of approximately 523.6 mice per month.

Example 6

Restaurant Revenue The revenue of McMenamy's Fish Shanty, located at a popular summer resort, is approximately

$$R(t) = 2\left(5 - 4\cos\frac{\pi}{6}t\right) \qquad (0 \le t \le 12)$$

during the tth week ($t = 1$ corresponds to the first week of June), where R is measured in thousands of dollars. The graph of R is shown in Figure 6.25.

a. Find the marginal revenue function R'.
b. Compute $R'(3)$ and $R'(9)$ and interpret your results.

Solution

The revenue function R is increasing at the rate of

$$R'(t) = -8\left(-\sin\frac{\pi t}{6}\right)\left(\frac{\pi}{6}\right)$$
$$= \frac{4\pi}{3}\sin\frac{\pi t}{6}$$

thousand dollars per week. We compute

$$R'(3) = \frac{4\pi}{3}\sin\frac{3\pi}{6} = \frac{4\pi}{3}\sin\frac{\pi}{2} = \frac{4\pi}{3}(1)\frac{4\pi}{3} \approx 4.18879$$

and

$$R'(9) = \frac{4\pi}{3}\sin\frac{9\pi}{6} = \frac{4\pi}{3}\sin\frac{3\pi}{2} = \frac{4\pi}{3}(-1) = -\frac{4\pi}{3} \approx -4.18879$$

Thus, at time $t = 3$, which corresponds to the third week of June, the revenue is increasing by \$4188.79 per week, while at time $t = 9$, which corresponds to the first week of August, the revenue is decreasing by \$4188.79 per week. This can also be observed from the graph of R (Figure 6.25), which shows that revenue is increasing for $t = 3$ and decreasing for $t = 9$.

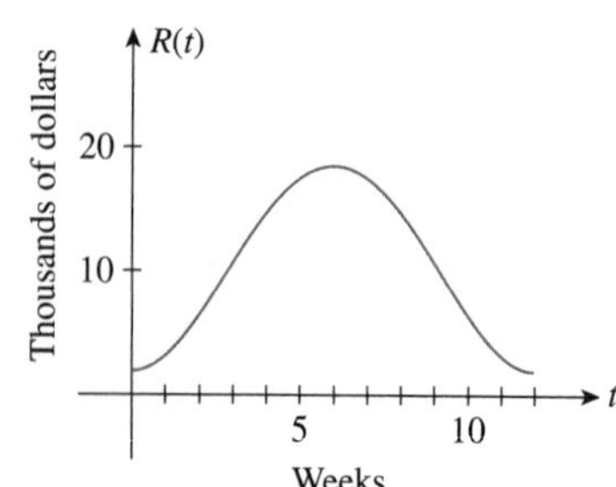

Figure 6.25

The graph of
$R(t) = 2\left[5 - 4\cos\left(\frac{\pi}{6}t\right)\right]$

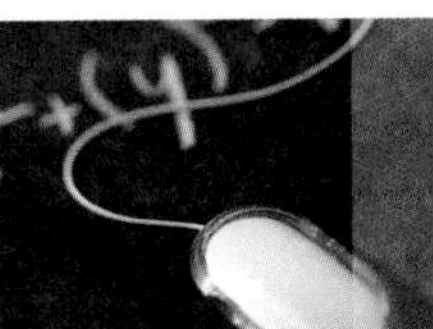

Self-Check Exercises 6.4

1. Find the derivative of $f(x) = x \cos 2x$.

2. If $f(x) = x$ 1 $\tan x$, find $f'(\pi/4)$.

3. Find $\frac{dy}{dx}$ for $\cos^{-1}(xy) = x^2$ by implicit differentiation.

Solutions to Self-Check Exercises 6.4 can be found on page 336.

6.4 Exercises

In Exercises 1–42, find the derivative of the function.

1. $f(x) = \cos 3x$
2. $f(x) = \sin 5x$
3. $f(x) = 2 \cos \pi x$
4. $f(x) = \pi \sin 2x$
5. $f(x) = \sin(x^2 + 1)$
6. $f(x) = \cos \pi x^2$
7. $f(x) = \tan 2x^2$
8. $f(x) = \cot \sqrt{x}$
9. $f(x) = x \sin x$
10. $f(x) = x^2 \cos x$
11. $f(x) = 2 \sin 3x + 3 \cos 2x$
12. $f(x) = 2 \cot 2x + \sec 3x$
13. $f(x) = x^2 \cos 2x$
14. $f(x) = \sqrt{x} \sin \pi x$
15. $f(x) = \sin \sqrt{x^2 - 1}$
16. $f(x) = \csc(x^2 + 1)$
17. $f(x) = e^x \sec x$
18. $f(x) = e^{-x} \csc x$
19. $f(x) = x \cos \frac{1}{x}$
20. $f(x) = x^2 \sin \frac{1}{x}$
21. $f(x) = \frac{x - \sin x}{1 + \cos x}$
22. $f(x) = \frac{\sin 2x}{1 + \cos 3x}$
23. $f(x) = \sqrt{\tan x}$
24. $f(x) = \sqrt{\cos x + \sin x}$
25. $f(x) + \frac{\sin x}{x}$
26. $f(x) = \frac{\cos x}{x^2 + 1}$
27. $f(x) = \tan^2 x$
28. $f(x) = \sqrt{\cot x}$
29. $f(x) = e^{\cot x}$
30. $f(x) = e^{\tan x + \sec x}$
31. $f(x) = (3x + 1)\sin x$
32. $f(x) = (x^2 - 5x) \tan x$
33. $f(x) = \sec(\log_{10}(x^3))$
34. $f(x) = \ln(\cot x)$
35. $f(x) = \csc^{-1}(5x^2 + 1)$
36. $f(x) = (\tan^{-1}(3x))^3$
37. $f(x) = \sqrt{e^{\cos^{-1} x}}$
38. $f(x) = \ln(\sin^{-1} x)$
39. $f(x) = \sec^{-1} x^{3/2}$
40. $f(x) = \cos^{-1}(\log_2 x)$
41. $f(x) = (\cot^{-1} x)^{1/3}$
42. $f(x) = \sin^{-1}(3^x)$

In Exercises 43–48, find dy/dx by implicit differentiation.

43. $\cos(xy) - \sin x = 1$
44. $x \sec(y) = \ln(\sin x)$
45. $\sin(x + y) - \sin^{-1} y = 0$
46. $(x^2 - y^2)\tan(y) = \sqrt{y}$
47. $\sin^{-1}(xy) + xy = x$
48. $\tan^{-1}(x - y) = xy$

49. Find the equation of the tangent line to the graph of the function $f(x) = \cot 2x$ at the point $(\pi/4, 0)$.

50. Find an equation of the tangent line to the graph of the function $f(x) = e^{\sec x}$ at the point $(\pi/4, e^{\sqrt{2}})$.

51. Prove Rule 4:

    If $h(x) = \csc f(x)$

    then $h'(x) = -[\csc f(x)][\cot f(x)]f'(x)$

    Hint: $\csc x = 1/\sin x$

52. Prove Rule 5:

    If $h(x) = \sec f(x)$

    then $h'(x) = [\sec f(x)][\tan f(x)]f'(x)$

    Hint: $\sec x = 1/\cos x$

53. Prove Rule 6:

    If $h(x) = \cot f(x)$

    then $h'(x) = -[\csc^2 f(x)]f'(x)$

    Hint: $\cot x = 1/\tan x = (\cos x)/(\sin x)$

54. Prove $\lim_{h \to 0} \frac{\cos h - 1}{h} = 0$

    Hint: Multiply by $\frac{\cos h + 1}{\cos h + 1}$.

Business and Economics Applications

55. **Stock Prices** The closing price (in dollars) per share of stock of Tempco Electronics on the tth day it was traded is approximated by

$$P(t) = 20 + 12 \sin \frac{\pi t}{30} - 6 \sin \frac{\pi t}{15} + 4 \sin \frac{\pi t}{10} - 3 \sin \frac{2\pi t}{15} \quad (0 \le t \le 20)$$

where $t = 0$ corresponds to the time the stock was first listed in a major stock exchange. What was the rate of change of the stock's price at the close of the 15th day of trading? What was the closing price on that day?

56. Restaurant Revenue The revenue of McMenamy's Fish Shanty located at a popular summer resort is approximately

$$R(t) = 2\left(5 - 4\cos\frac{\pi}{6}t\right) \qquad (0 \le t \le 12)$$

√during the tth week ($t = 1$ corresponds to the first week of June), where R is measured in thousands of dollars. On what week is the marginal revenue zero?

57. Marginal Revenue The revenue in thousands of dollars received from the sale of electronic fans is seasonal, with the maximum revenue in the summer. Let the revenue received from the sales of fans be approximated by

$$R(x) = -100\cos(2\pi x) + 120$$

where x is time in years, measured from January 1. Calculate the marginal revenue for September 1. (Take that 1 month represents 1/12 of a year.)

58. Marginal Sales Sales of large kitchen appliances such as ovens and fridges are usually subject to seasonal fluctuations. Everything Kitchen's sales of fridge models from the beginning of 2001 to the end of 2002 can be approximated by

$$S(x) = \frac{1}{10}\sin\left(\frac{\pi}{2}(x+1)\right) + \frac{1}{2}$$

where x is time in quarters, $x = 1$ represents the end of the first quarter of 2001, and S is measured in millions of dollars. Find the rate of change of sales for the end of the third quarter of 2002.

Biological and Life Sciences Applications

59. Predator–Prey Population The wolf population in a certain northern region is estimated to be

$$P(t) = 8000 + 1000\sin\left(\frac{\pi t}{24}\right)$$

in month t, and the caribou population in the same region is given by

$$p(t) = 40{,}000 + 12{,}00\cos\left(\frac{\pi t}{24}\right)$$

Find the rate of change of each population when $t = 12$.

60. Number of Hours of Daylight The number of hours of daylight on a particular day of the year in Fredericton is approximated by the function

$$f(t) = 3\sin\frac{2\pi}{365}(t - 79) + 12$$

where $t = 0$ corresponds to January 1. Compute $f'(79)$ and interpret your result.

61. Water Level in a Harbour The water level (in metres) off the coast of Newfoundland during a certain 24-hr period is approximated by the formula

$$H = 4.8\sin\frac{\pi}{6}(t - 10) + 7.6 \qquad (0 \le t \le 24)$$

where $t = 0$ corresponds to 12 A.M. Find the rate of change of the water level at 8 A.M. What was the water level at that time?

62. Average Daily Temperature The average daily temperature (in degrees Fahrenheit) at a tourist resort in Cameron Highlands is approximated by

$$T = 62 - 18\cos\frac{2\pi(t - 23)}{365}$$

on the t-th day ($t = 1$ corresponds to January 1). On what day is the rate of change of temperature zero?

Solutions to Self-Check Exercises 6.4

1. $f'(x) = x\frac{d}{dx}\cos 2x + (\cos 2x)\frac{d}{dx}(x)$

$$= x(-\sin 2x)(2) + (\cos 2x)(1)$$
$$= \cos 2x - 2x\sin 2x$$

2. $f'(x) = 1 + \sec^2 x$. Therefore,

$$f'\left(\frac{\pi}{4}\right) = 1 + \sec^2\left(\frac{\pi}{4}\right) = 1 + (\sqrt{2})^2 = 3$$

3. $\frac{d}{dx}(\cos^{-1}(xy)) = \frac{d}{dx}(x^2)$

$$-\frac{1}{\sqrt{1-(xy)^2}}\frac{d}{dx}(xy) = 2x$$

$$-\frac{(1)y + x\frac{dy}{dx}}{\sqrt{1-(xy)^2}} = 2x$$

$$-\frac{x}{\sqrt{1-(xy)^2}}\frac{dy}{dx} = 2x + \frac{y}{\sqrt{1-(xy)^2}}$$

$$\frac{dy}{dx} = -\frac{\sqrt{1-(xy)^2}}{x}\left(2x + \frac{y}{\sqrt{1-(xy)^2}}\right)$$

CHAPTER 6 Summary of Principal Formulas and Terms

Formulas

1. **Degree-radian conversion** 360 degrees $= 2\pi$ radians
2. **Derivatives of the trigonometric functions:**

Sine	$\frac{d}{dx}(\sin u) = \cos u \frac{du}{dx}$
Cosine	$\frac{d}{dx}(\cos u) = -\sin u \frac{du}{dx}$
Tangent	$\frac{d}{dx}(\tan u) = \sec^2 u \frac{du}{dx}$
Cosecant	$\frac{d}{dx}(\csc u) = -\csc u \cot u \frac{du}{dx}$
Secant	$\frac{d}{dx}(\sec u) = \sec u \tan u \frac{du}{dx}$
Cotangent	$\frac{d}{dx}(\cot u) = -\csc^2 u \frac{du}{dx}$

3. **Derivatives of the inverse trigonometric functions:**

Arcsine	$\frac{d}{dx}[\sin^{-1} u] = \frac{1}{\sqrt{1-u^2}} \frac{du}{dx}$,	$\lvert u \rvert < 1$
Arccosine	$\frac{d}{dx}[\cos^{-1} u] = -\frac{1}{\sqrt{1-u^2}} \frac{du}{dx}$,	$\lvert u \rvert < 1$
Arctangent	$\frac{d}{dx}[\tan^{-1} u] = \frac{1}{1+u^2} \frac{du}{dx}$	
Arcsecant	$\frac{d}{dx}[\cot^{-1} u] = -\frac{1}{1+u^2} \frac{du}{dx}$	
Arccosecant	$\frac{d}{dx}[\sec^{-1} u] = \frac{1}{\lvert u \rvert \sqrt{u^2-1}} \frac{du}{dx}$,	$\lvert u \rvert > 1$
Arccotangent	$\frac{d}{dx}[\csc^{-1} u] = -\frac{1}{\lvert u \rvert \sqrt{u^2-1}} \frac{du}{dx}$,	$\lvert u \rvert > 1$

Terms

angle (303)
initial ray (303)
terminal ray (303)
vertex (303)
standard position (303)
degree (304)
radian (304)
unit circle (304)
coterminal angles (305)
sine (308)
cosine (308)
ASTC Rule (310)
special triangle with angles $\frac{\pi}{4}, \frac{\pi}{4}$ and $\frac{\pi}{2}$ (311)
special triangle with angles $\frac{\pi}{6}, \frac{\pi}{3}$ and $\frac{\pi}{2}$ (311)
periodic (312)
predator–prey population model (314)
phase shift (316)
amplitude (316)
period or wave length (316)
trigonometric identity (317)
inverse sine (321)
arcsine (322)
arc (322)

Review Exercises 

In Exercises 1–3, convert the angle to radian measure.

1. 120° **2.** 450° **3.** −225°

In Exercises 4–6, convert the angle to degree measure.

4. $\frac{11}{6}\pi$ radians **5.** $-\frac{5}{2}\pi$ radians **6.** $-\frac{7}{4}\pi$ radians

In Exercises 7 and 8, find all values of θ that satisfy the equation over the interval $[0, 2\pi]$.

7. $\cos\theta = \frac{1}{2}$ **8.** $\cot\theta = \sqrt{3}$

In Exercises 9 and 10, sketch the graph of the functions over the interval $[-2\pi, 2\pi]$ by comparing it to the graph of $y = \cos x$.

9. $y = 2\cos\left(x - \frac{\pi}{3}\right) - 4$ **10.** $y = -\cos\left(\frac{x}{2}\right)$

In Exercises 11–14, evaluate the given angles using the special triangle with angles $\frac{\pi}{4}, \frac{\pi}{4}$, and $\frac{\pi}{2}$ or the special triangle with angles $\frac{\pi}{6}, \frac{\pi}{3}$, and $\frac{\pi}{2}$.

11. $\cos^{-1}\left(\frac{\sqrt{3}}{2}\right)$ **12.** $\sin^{-1}\left(-\frac{1}{2}\right)$

13. $\sec^{-1}\left(-\frac{2}{\sqrt{3}}\right)$ **14.** $\tan^{-1}(\sqrt{3})$

In Exercises 15 and 16, given θ find $\sin\theta$, $\cos\theta$, $\tan\theta$, $\sec\theta$, $\csc\theta$, and $\cot\theta$.

15. $\theta = \sin^{-1}\left(\frac{3}{5}\right)$ **16.** $\theta = \cos^{-1}\left(\frac{2}{\sqrt{7}}\right)$

In Exercises 17–34, find the derivative of the function f.

17. $f(x) = \sin 3x$ **18.** $f(x) = 2\cos\frac{x}{2}$

19. $f(x) = 2\sin x - 3\cos 2x$

20. $f(x) = \sec^2\sqrt{x}$ **21.** $f(x) = e^{-x}\tan 3x$

22. $f(x) = (1 - \csc 2x)^2$ **23.** $f(x) = 4\sin x\cos x$

24. $f(x) = \frac{\cos x}{1 - \cos x}$ **25.** $f(x) = \frac{1 - \tan x}{1 - \cot x}$

26. $f(x) = \ln(\cos^2 x)$ **27.** $f(x) = \sin(\sin x)$

28. $f(x) = e^{\sin x}\cos x$ **29.** $f(x) = \cos(3^x + \log_3 x)$

30. $f(x) = \tan(\ln(\sin x))$ **31.** $f(x) = \cos^{-1}(x^3)$

32. $f(x) = \log_2(\sin^{-1} x)$ **33.** $f(x) = \ln x \tan^{-1}(e^x)$

34. $f(x) = \frac{\sec^{-1} x}{\cos x}$

35. Find the equation of the tangent line to the graph of the function $f(x) = \tan^2 x$ at the point $(\pi/4, 1)$.

Business and Economics Applications

36. Hotel Occupancy Rate The occupancy rate of a large hotel in the Okanagan in month t is described by the function

$$R(t) = 60 + 37\sin^2\left(\frac{\pi t}{12}\right) \qquad (0 \le t \le 12)$$

where $t = 0$ corresponds to the beginning of January. What is the marginal occupancy rate at the beginning of August?

37. Marginal Profit The monthly profit from the sales of ski and snowboard equipment in a town in Eastern Canada is approximately

$$P(t) = 23\cos\left(\frac{\pi}{6}(t - 1)\right) + 32 \qquad (0 \le t \le 12)$$

thousands of dollars, where $t = 0$ corresponds to the beginning of January. What is the marginal profit at the beginning of November?

Biological and Life Sciences Applications

38. Average Daily Temperature The average daily temperature (in degrees Celsius) at a tourist resort in Whistler is approximated by

$$T(t) = 15 - 16\cos\left(\frac{2\pi}{365}(t - 23)\right) \qquad 1 \le t \le 365$$

on the t-th day ($t = 1$ corresponds to January 1). At what rate is the temperature changing on July 1? What is the temperature that day?

39. Predator–Prey Population The population of foxes in a certain region over a 2-yr period is estimated to be

$$P(t) = 400 + 50\sin\left(\frac{\pi t}{12}\right)$$

in month t, and the population of rabbits in the same region at time t is given by

$$p(t) = 3000 + 500\cos\left(\frac{\pi t}{12}\right)$$

Find the rate of change of the populations when $t = 4$.

7

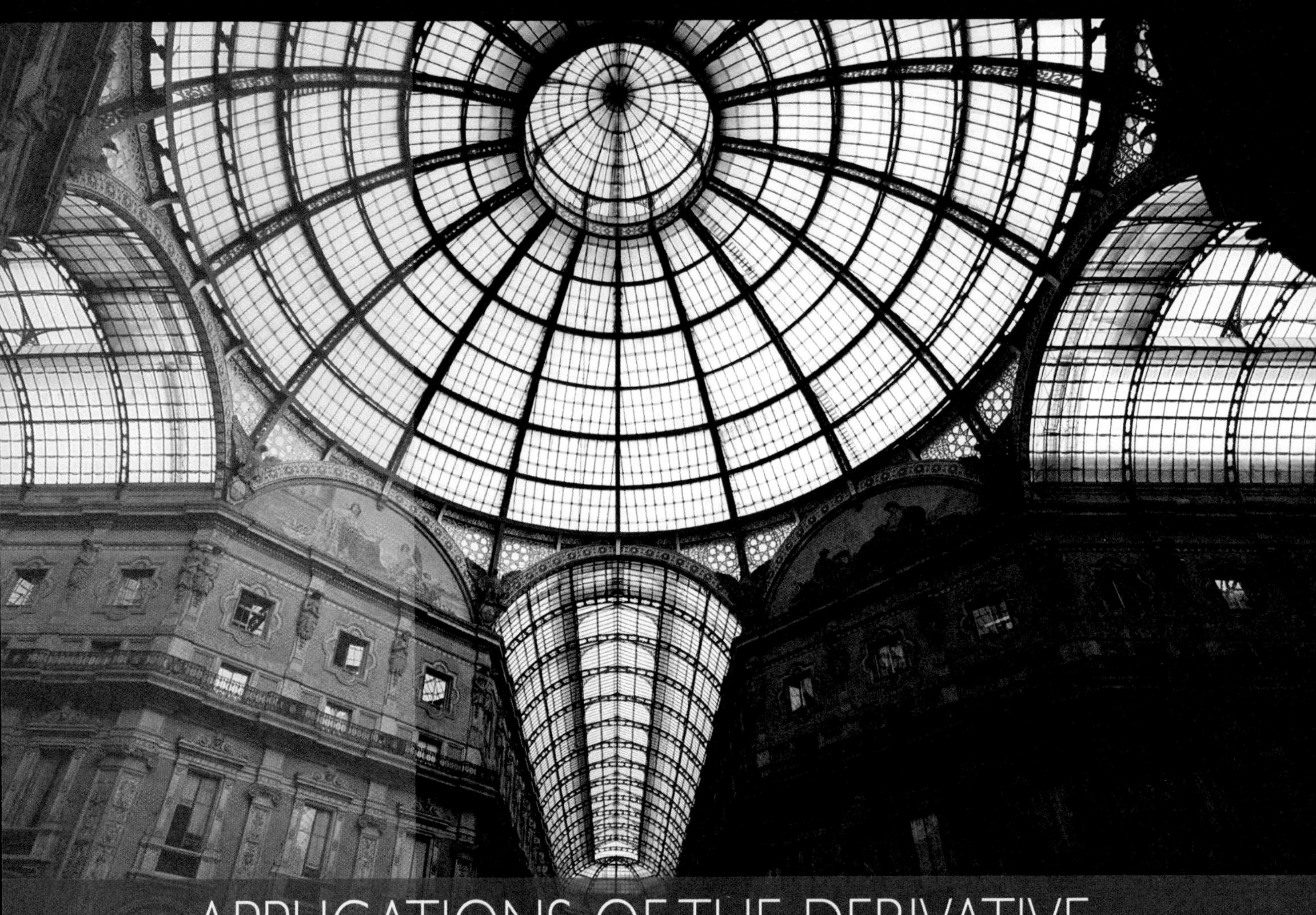

APPLICATIONS OF THE DERIVATIVE

Photo: Thinkstock/Jupiter Images

This chapter further explores the power of the derivative, which we use to help analyze the properties of functions. The information obtained can then be used to accurately sketch graphs of functions. We also see how the derivative is used in solving a large class of optimization problems, including finding what level of production will yield a maximum profit for a company, finding what level of production will result in minimal cost to a company finding the maximum height attained by a rocket, finding the maximum velocity at which air is expelled when a person coughs, and a host of other problems. We will also present a technique that uses the derivative to make difficult limits easier.

Photo: Thinkstock/Jupiter Images

Photo: Eurockot Launch Services, eurockot@astrium-space.com

What is the maximum altitude and the maximum velocity attained by the rocket? In Example 8, page 391, you will see how the techniques of calculus can be used to help answer these questions.

7.1 Applications of the First Derivative

Determining the Intervals Where a Function Is Increasing or Decreasing

According to a study by the U.S. Department of Energy and the Shell Development Company, a typical car's fuel economy as a function of its speed is described by the graph shown in Figure 7.1. Observe that the fuel economy $f(x)$ in kilometres per litre (km/L) improves as x, the vehicle's speed in kilometres per hour (km/h), increases from 0 to 42, and then drops as the speed increases beyond 42 kilometres per hour (km/h). We use the terms *increasing* and *decreasing* to describe the behaviour of a function as we move from left to right along its graph.

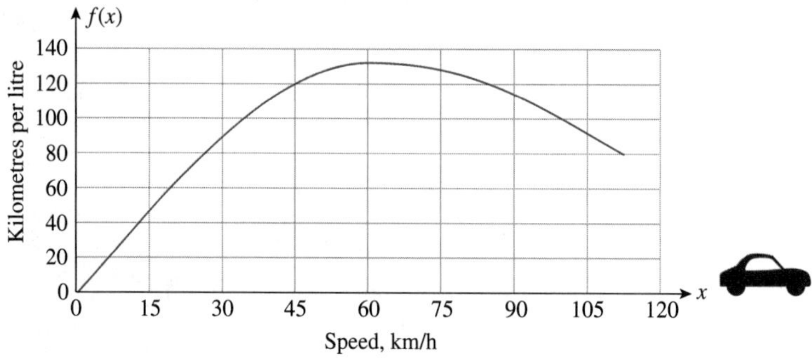

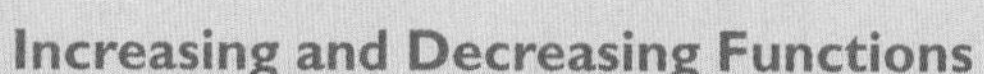

Figure 7.1
A typical car's fuel economy improves as the speed at which it is driven increases from 0 km/h to 42 km/h and drops at speeds greater than 42 km/h.

More precisely, we have the following definitions.

Increasing and Decreasing Functions

A function f is increasing on an interval (a, b) if for any two numbers x_1 and x_2 in (a, b), $f(x_1) < f(x_2)$ whenever $x_1 < x_2$ (Figure 7.2a).

A function f is decreasing on an interval (a, b) if for any two numbers x_1 and x_2 in (a, b), $f(x_1) > f(x_2)$ whenever $x_1 < x_2$ (Figure 7.2b).

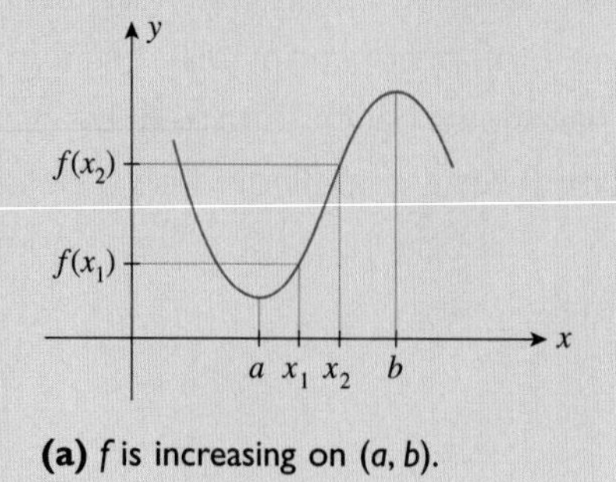

(a) f is increasing on (a, b).

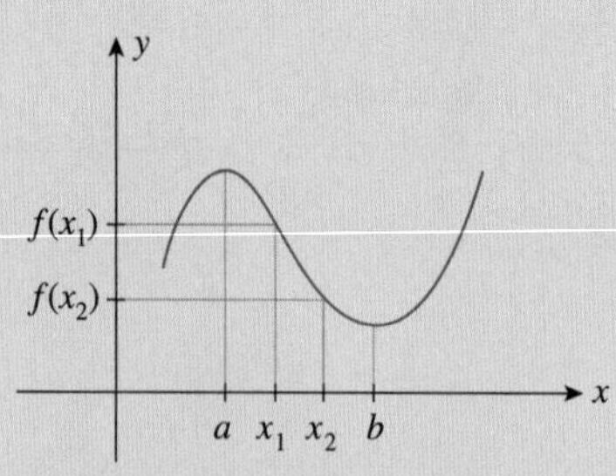

(b) f is decreasing on (a, b).

Figure 7.2

We say that f is *increasing at a point c* if there exists an interval (a, b) containing c such that f is increasing on (a, b). Similarly, we say that f is *decreasing at a point c* if there exists an interval (a, b) containing c such that f is decreasing on (a, b). Increasing functions slope upward from left to right while decreasing functions slope downward from left to right, as can be observed in Figures 7.2 and 7.3 respectively.

Since the rate of change of a function at a point $x = c$ is given by the derivative of the function at that point, the derivative lends itself naturally to being a tool for determining the intervals where a differentiable function is increasing or decreasing. Indeed, as we saw in Chapter 3, the derivative of a function at a point measures both the slope of the tangent line to the graph of the function at that point and the rate of change of the function at the same point. In fact, at a point where the derivative is positive, the slope of the tangent line to the graph is positive, and the function is increasing. At a point where the derivative is negative, the slope of the tangent line to the graph is negative, and the function is decreasing (Figure 7.3).

These observations lead to the following important theorem, which we state without proof.

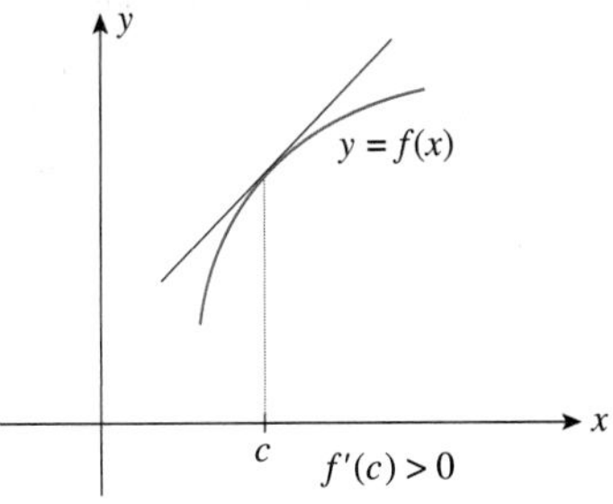

(a) f is increasing at $x = c$.

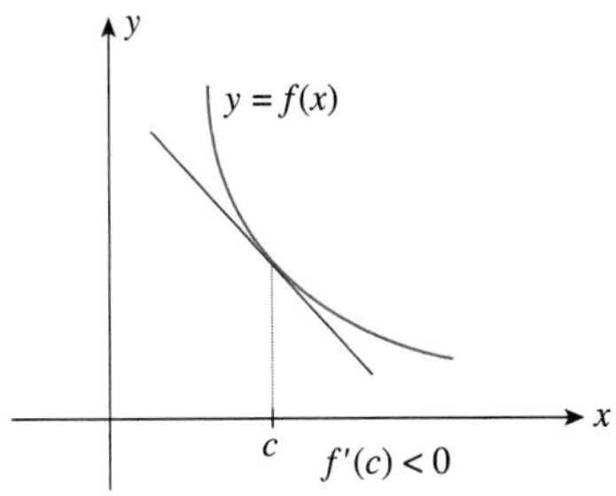

(b) f is decreasing at $x = c$.

Figure 7.3

THEOREM 1

a. If $f'(x) > 0$ for each value of x in an interval (a, b), then f is increasing on (a, b).

b. If $f'(x) < 0$ for each value of x in an interval (a, b), then f is decreasing on (a, b).

c. If $f'(x) = 0$ for each value of x in an interval (a, b), then f is constant on (a, b).

Example 1

Find the interval where the function $f(x) = x^2$ is increasing and the interval where it is decreasing.

Solution

The derivative of $f(x) = x^2$ is $f'(x) = 2x$. Since

$$f'(x) = 2x > 0 \quad \text{if } x > 0 \qquad \text{and} \qquad f'(x) = 2x < 0 \quad \text{if } x < 0$$

f is increasing on the interval $(0, \infty)$ and decreasing on the interval $(-\infty, 0)$ (Figure 7.4).

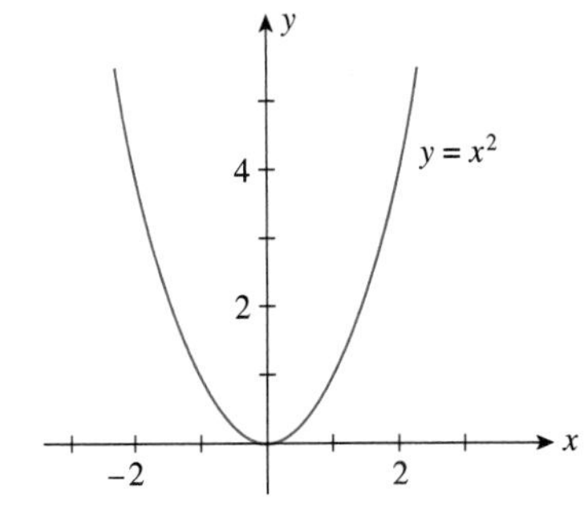

Figure 7.4
The graph of f falls on $(-\infty, 0)$ where $f'(x) < 0$ and rises on $(0, \infty)$ where $f'(x) > 0$.

Recall that the graph of a continuous function cannot have any breaks. As a consequence, a continuous function cannot change sign unless it equals zero for some value of x. (See Theorem 5, page 138.) This observation suggests the following procedure for determining the sign of the derivative f' of a function f, and hence the intervals where the function f is increasing and where it is decreasing.

Determining the Intervals Where a Function Is Increasing or Decreasing

1. Find all values of x for which $f'(x) = 0$ or f' is discontinuous and identify the open intervals determined by these points.
2. Select a test point c in each interval found in step 1 and determine the sign of $f'(c)$ in that interval.
   a. If $f'(c) > 0$, f is increasing on that interval.
   b. If $f'(c) < 0$, f is decreasing on that interval.

GROUP DISCUSSION

True or false? If f is continuous at c and f is increasing at c, then $f'(c) \neq 0$. Explain your answer.

Hint: Consider $f(x) = x^3$ and $c = 0$.

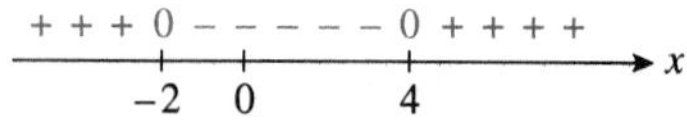

Figure 7.5
Sign diagram for f'

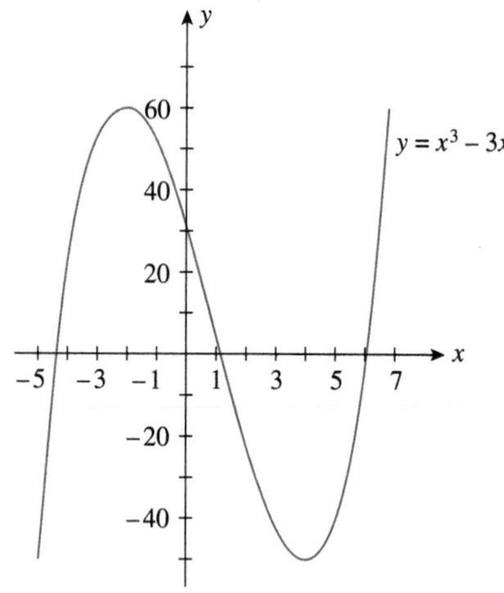

Figure 7.6
The graph of f rises on $(-\infty, -2)$, falls on $(-2, 4)$, and rises again on $(4, \infty)$.

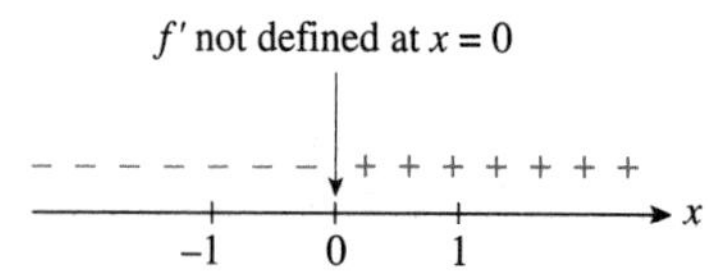

Figure 7.7
Sign diagram for f'

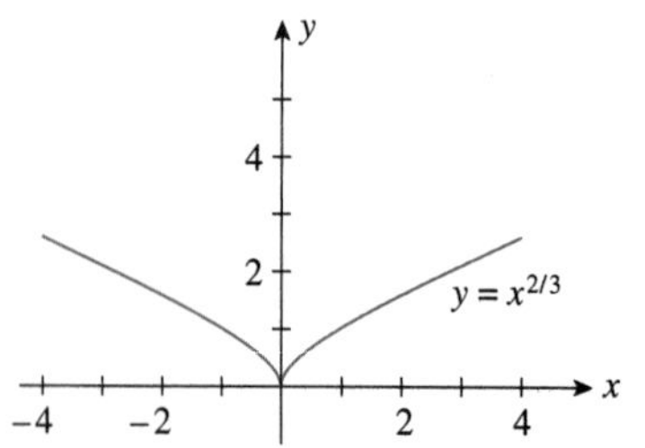

Figure 7.8
f decreases on $(-\infty, 0)$ and increases on $(0, \infty)$.

Example 2

Determine the intervals where the function $f(x) = x^3 - 3x^2 - 24x + 32$ is increasing and where it is decreasing.

Solution

1. The derivative of f is

$$f'(x) = 3x^2 - 6x - 24 = 3(x + 2)(x - 4)$$

and it is continuous everywhere. The zeros of $f'(x)$ are $x = -2$ and $x = 4$, and these points divide the real line into the intervals $(-\infty, -2)$, $(-2, 4)$, and $(4, \infty)$.

2. To determine the sign of $f'(x)$ in the intervals $(-\infty, -2)$, $(-2, 4)$, and $(4, \infty)$, compute $f'(x)$ at a convenient test point in each interval. The results are shown in the following table.

Interval	Test Point c	$f'(c)$	Sign of $f'(x)$
$(-\infty, -2)$	-3	21	+
$(-2, 4)$	0	-24	−
$(4, \infty)$	5	21	+

Using these results, we obtain the sign diagram shown in Figure 7.5. We conclude that f is increasing on the intervals $(-\infty, -2)$ and $(4, \infty)$ and is decreasing on the interval $(-2, 4)$. Figure 7.6 shows the graph of f.

Example 3

Find the interval where the function $f(x) = x^{2/3}$ is increasing and the interval where it is decreasing.

Solution

1. The derivative of f is

$$f'(x) = \frac{2}{3}x^{-1/3} = \frac{2}{3x^{1/3}}$$

The function f' is not defined at $x = 0$, so f' is discontinuous there. It is continuous everywhere else. Furthermore, f' is not equal to zero anywhere. The point $x = 0$ divides the real line (the domain of f) into the intervals $(-\infty, 0)$ and $(0, \infty)$.

2. Pick a test point (say, $x = -1$) in the interval $(-\infty, 0)$ and compute

$$f'(-1) = -\frac{2}{3}$$

Since $f'(-1) < 0$, we see that $f'(x) < 0$ on $(-\infty, 0)$. Next, we pick a test point (say, $x = 1$) in the interval $(0, \infty)$ and compute

$$f'(1) = \frac{2}{3}$$

Since $f'(1) > 0$, we see that $f'(x) > 0$ on $(0, \infty)$. Figure 7.7 shows these results in the form of a sign diagram.

We conclude that f is decreasing on the interval $(-\infty, 0)$ and increasing on the interval $(0, \infty)$. The graph of f, shown in Figure 7.8, confirms these results.

Remark

Do not be concerned with how the graphs in this section are obtained. We will learn how to sketch these graphs later. However, if you are familiar with the use of a graphing utility, you may go ahead and verify each graph. ◀

GROUP DISCUSSION

If a farm pond is stocked with tilapia, a type of herbivorous fish, the population grows until it approaches the *carrying capacity* of the pond. The carrying capacity of an environment is the maximum population the environment can support. The first graph below shows how the population grows with time if no fish are harvested. The second graph shows the derivative of the first, the rate at which new fish are entering the population. Both graphs give numbers as the fraction of carrying capacity rather than as the number of fish.

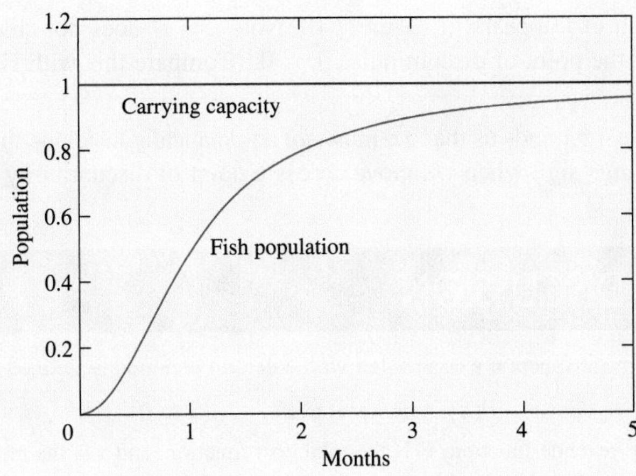

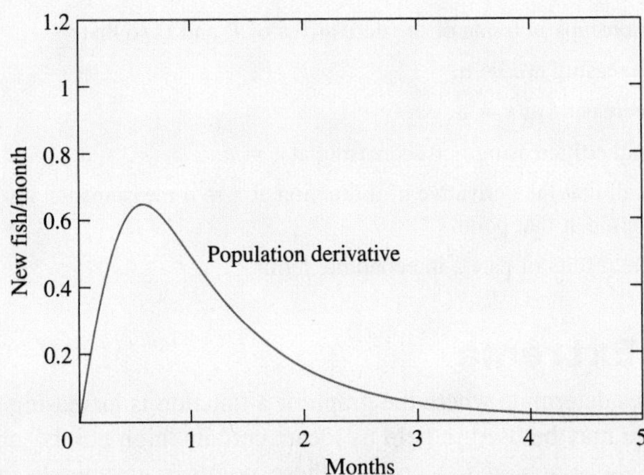

Suppose that you are in charge of setting harvesting policy for farm ponds of this type. What is the best time to start harvesting adult fish?

Example 4

Find the intervals where the function $f(x) = x + \dfrac{1}{x}$ is increasing and where it is decreasing.

Solution

1. The derivative of f is

$$f'(x) = 1 - \frac{1}{x^2} = \frac{x^2 - 1}{x^2}$$

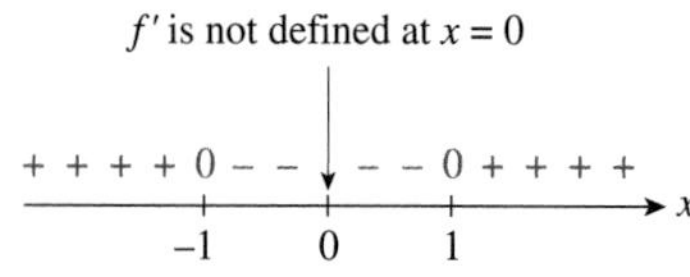

Figure 7.9
f' does not change sign as we move across $x = 0$.

Since f' is not defined at $x = 0$, it is discontinuous there. Furthermore, $f'(x)$ is equal to zero when $x^2 - 1 = 0$ or $x = \pm 1$. These values of x partition the domain of f' into the open intervals $(-\infty, -1)$, $(-1, 0)$, $(0, 1)$, and $(1, \infty)$, where the sign of f' is different from zero.

2. To determine the sign of f' in each of these intervals, we compute $f'(x)$ at the test points $x = -2, -\frac{1}{2}, \frac{1}{2}$, and 2, respectively, obtaining $f'(-2) = \frac{3}{4}$, $f'(-\frac{1}{2}) = -3$, $f'(\frac{1}{2}) = -3$, and $f'(2) = \frac{3}{4}$. From the sign diagram for f' (Figure 7.9), we conclude that f is increasing on $(-\infty, -1)$ and $(1, \infty)$ and decreasing on $(-1, 0)$ and $(0, 1)$.

The graph of f appears in Figure 7.10. Note that f' does not change sign as we move across the point of discontinuity, $x = 0$. (Compare this with Example 3.)

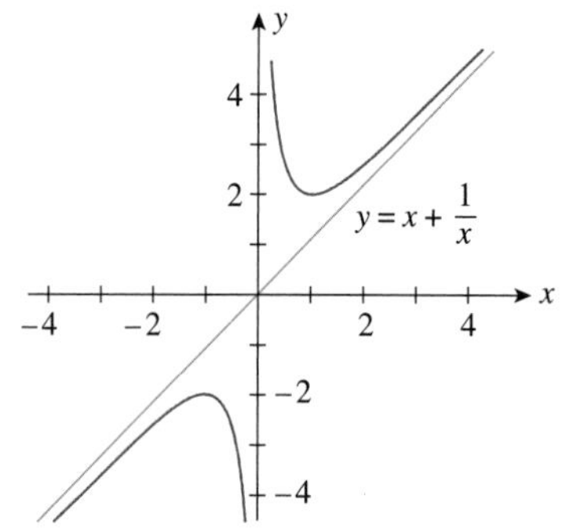

Figure 7.10
The graph of f rises on $(-\infty, -1)$, falls on $(-1, 0)$, and $(0, 1)$, and rises again on $(1, \infty)$.

Example 4 reminds us that we must *not* automatically conclude that the derivative f' must change sign when we move across a point of discontinuity or a zero of f'.

GROUP DISCUSSION

Consider the profit function P associated with a certain commodity defined by

$$P(x) = R(x) - C(x) \qquad (x \geq 0)$$

where R is the revenue function, C is the total cost function, and x is the number of units of the product produced and sold.

1. Find an expression for $P'(x)$.
2. Find relationships in terms of the derivatives of R and C so that
   a. P is increasing at $x = a$.
   b. P is decreasing at $x = a$.
   c. P is neither increasing nor decreasing at $x = a$.

   Hint: Recall that the derivative of a function at $x = a$ measures the rate of change of the function at that point.
3. Explain the results of part 2 in economic terms.

Relative Extrema

Besides helping us determine where the graph of a function is increasing and decreasing, the first derivative may be used to help us locate certain "high points" and "low points" on the graph of f. Knowing these points is invaluable in sketching the graphs of functions and solving optimization problems. These "high points" and "low points" correspond to the *relative (local) maxima* and *relative minima* of a function. They are so called because they are the highest or the lowest points when compared with points nearby.

The graph shown in Figure 7.11 gives the number of mourning doves spotted during the breeding bird survey in Ontario from 1966 to 1986. The relative maxima and minima are indicated on the graph.

More generally, we have the following definition:

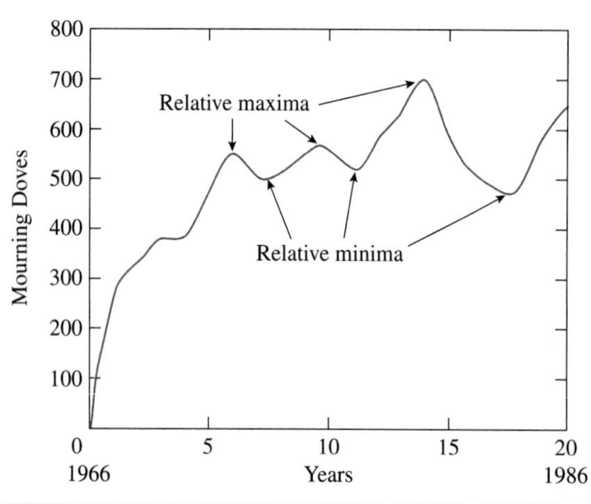

Figure 7.11
Ontario breeding mourning doves pairs, 1966–1986
Source: Breeding bird survey.

Relative Maximum

A function f has a **relative maximum** at $x = c$ if there exists an open interval (a, b) containing c such that $f(x) \leq f(c)$ for all x in (a, b).

Geometrically, this means that there is *some* interval containing $x = c$ such that no point on the graph of f with its x-coordinate in that interval can lie above the point $(c, f(c))$; that is, $f(c)$ is the largest value of $f(x)$ in some interval around $x = c$. Figure 7.12 depicts the graph of a function f that has a relative maximum at $x = x_1$ and another at $x = x_3$.

Observe that all the points on the graph of f with x-coordinates in the interval I_1 containing x (shown in blue) lie on or below the point $(x_1, f(x_1))$. This is also true for the point $(x_3, f(x_3))$ and the interval I_3. Thus, even though there are points on the graph of f that are "higher" than the points $(x_1, f(x_1))$ and $(x_3, f(x_3))$, the latter points are "highest" relative to points in their respective neighbourhoods (intervals). Points on the graph of a function f that are "highest" and "lowest" with respect to *all* points in the domain of f will be studied in Section 7.4.

The definition of the relative minimum of a function parallels that of the relative maximum of a function.

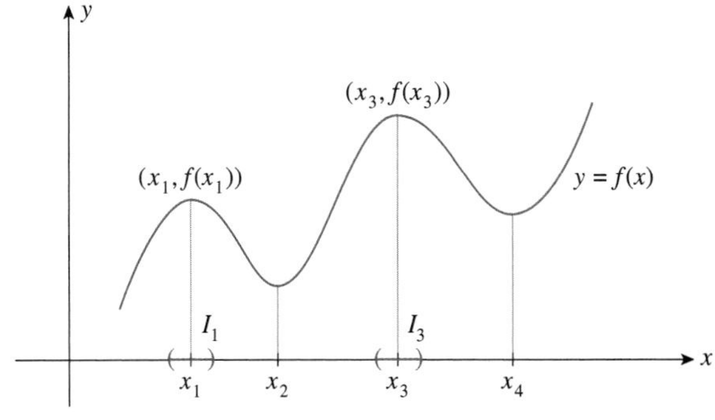

Figure 7.12
f has a relative maximum at $x = x_1$ and at $x = x_3$.

Relative Minimum

A function f has a **relative minimum** at $x = c$ if there exists an open interval (a, b) containing c such that $f(x) \geq f(c)$ for all x in (a, b).

The graph of the function f, depicted in Figure 7.12, has a relative minimum at $x = x_2$ and another at $x = x_4$.

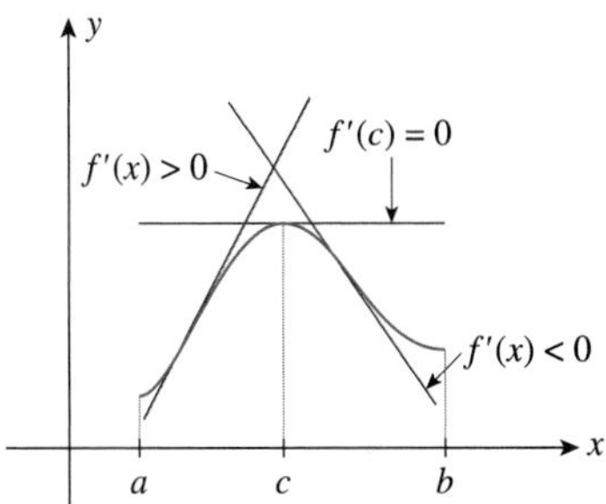

(a) f has a relative maximum at $x = c$.

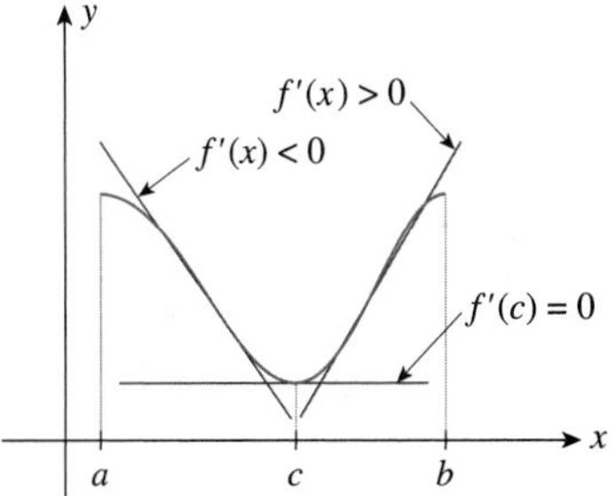

(b) f has a relative maximum at $x = c$.

Figure 7.13

Finding the Relative Extrema

We refer to the relative maximum and relative minimum of a function as the **relative extrema** of that function. As a first step in our quest to find the relative extrema of a function, we consider functions that have derivatives at such points. Suppose that f is a function that is differentiable in some interval (a, b) that contains a point $x = c$ and that f has a relative maximum at $x = c$ (Figure 7.13a).

Observe that the slope of the tangent line to the graph of f must change from positive to negative as we move across the point $x = c$ from left to right. Therefore, the tangent line to the graph of f at the point $(c, f(c))$ must be horizontal; that is, $f'(c) = 0$ (Figure 7.13a).

Using a similar argument, it may be shown that the derivative f' of a differentiable function f must also be equal to zero at a point $x = c$, where f has a relative minimum (Figure 7.13b).

! This analysis reveals an important characteristic of the relative extrema of a differentiable function f: *At any point c where f has a relative extremum,* and $f'(c)$ exists, $f'(c) = 0$. Before we develop a procedure for finding such points, a few words of caution are in order. First, this result tells us that if a differentiable function f has a relative extremum at a point $x = c$, then $f'(c) = 0$. The converse of this statement—if $f'(c) = 0$ at some point $x = c$, then f must have a relative extremum at that point—is *not* true. Consider, for example, the function $f(x) = x^3$. Here, $f'(x) = 3x^2$, so $f'(0) = 0$. Yet, f has neither a relative maximum nor a relative minimum at $x = 0$ (Figure 7.14).

Second, our result assumes that the function is differentiable and thus has a derivative at a point that gives rise to a relative extremum. The functions $f(x) = |x|$ and $g(x) = x^{2/3}$ demonstrate that a relative extremum of a function may exist at a point at which the derivative does not exist. Both these functions fail to be differentiable at $x = 0$, but each has a relative minimum there. Figure 7.15 shows the graphs of these functions. Note that the slopes of the tangent lines change from negative to positive as we move across $x = 0$, just as in the case of a function that is differentiable at a value of x that gives rise to a relative minimum.

We refer to a point in the domain of f that *may* give rise to a relative extremum as a critical point.

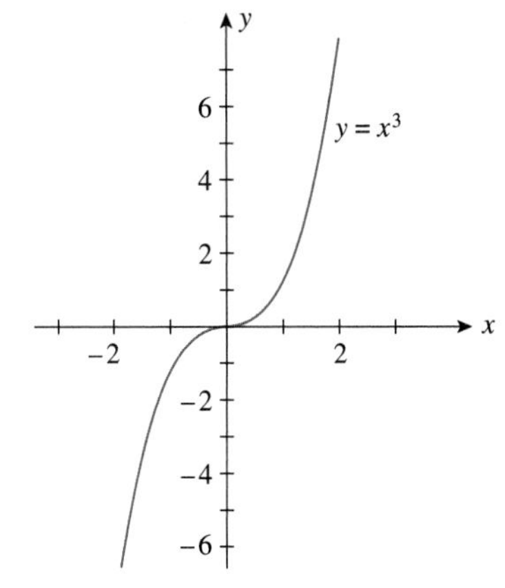

Figure 7.14
$f'(0) = 0$, but f does not have a relative extremum at $(0, 0)$.

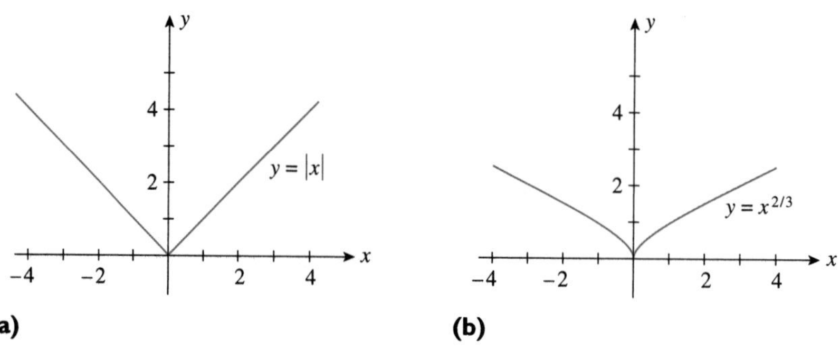

Figure 7.15
Each of these functions has a relative extremum at (0, 0), but the derivative does not exist there.

Critical Point of f

A **critical point** of a function f is any point x in the domain of f such that $f'(x) = 0$ or $f'(x)$ does not exist.

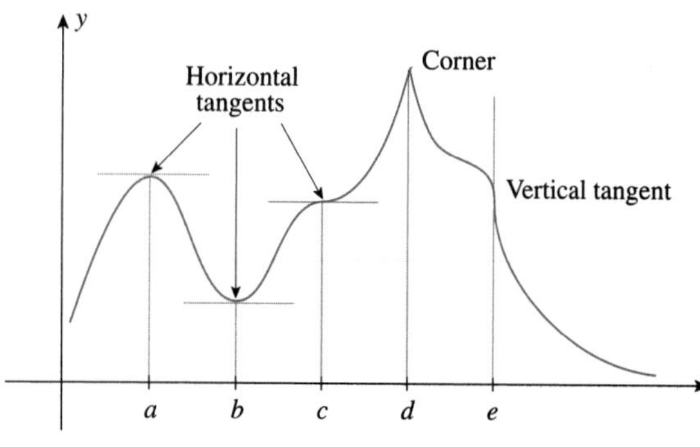

Figure 7.16
Critical points of f

Figure 7.16 depicts the graph of a function that has critical points at $x = a, b, c, d$, and e. Observe that $f'(x) = 0$ at $x = a, b$, and c. Next, since there is a corner at $x = d$, $f'(x)$ does not exist there. Finally, $f'(x)$ does not exist at $x = e$ because the tangent line there is vertical. Also, observe that the critical points $x = a, b$, and d give rise to relative extrema of f, whereas the critical points $x = c$ and $x = e$ do not.

Having defined what a critical point is, we can now state a formal procedure for finding the relative extrema of a continuous function that is differentiable everywhere except at isolated values of x. Incorporated into the procedure is the so-called **first derivative test,** which helps us determine whether a point gives rise to a relative maximum or a relative minimum of the function f.

The First Derivative Test

Procedure for Finding Relative Extrema of a Continuous Function f

1. Determine the critical points of f.
2. Determine the sign of $f'(x)$ to the left and right of each critical point.
   a. If $f'(x)$ changes sign from *positive* to *negative* as we move across a critical point $x = c$, then $f(c)$ is a relative maximum.
   b. If $f'(x)$ changes sign from *negative* to *positive* as we move across a critical point $x = c$, then $f(c)$ is a relative minimum.
   c. If $f'(x)$ does not change sign as we move across a critical point $x = c$, then $f(c)$ is not a relative extremum.

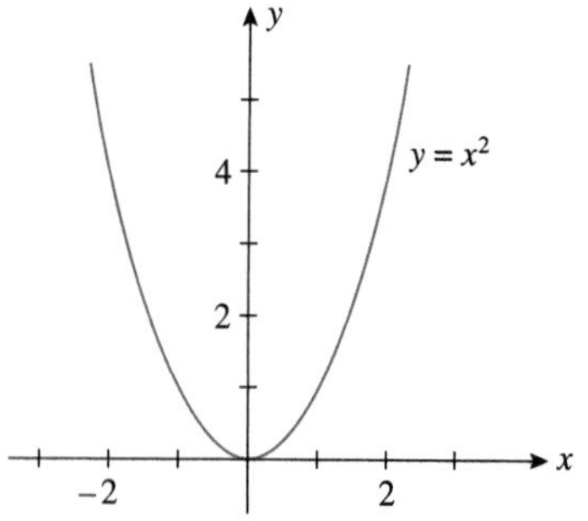

Figure 7.17
f has a relative minimum at $x = 0$.

Example 5

Find the relative maxima and relative minima of the function $f(x) = x^2$.

Solution

The derivative of $f(x) = x^2$ is given by $f'(x) = 2x$. Setting $f'(x) = 0$ yields $x = 0$ as the only critical point of f. Since

$$f'(x) < 0 \quad \text{if } x < 0 \qquad \text{and} \qquad f'(x) > 0 \quad \text{if } x > 0$$

we see that $f'(x)$ changes sign from negative to positive as we move across the critical point $x = 0$. Thus, we conclude that $f(0) = 0$ is a relative minimum of f (Figure 7.17).

Example 6

Find the relative maxima and relative minima of the function $f(x) = x^{2/3}$ (see Example 3).

Solution

The derivative of f is $f'(x) = \frac{2}{3}x^{-1/3}$. As noted in Example 3, f' is not defined at $x = 0$, is continuous everywhere else, and is not equal to zero in its domain. Thus, $x = 0$ is the only critical point of the function f.

The sign diagram obtained in Example 3 is reproduced in Figure 7.18. We can see that the sign of $f'(x)$ changes from negative to positive as we move across $x = 0$ from left to right. Thus, an application of the first derivative test tells us that $f(0) = 0$ is a relative minimum of f (Figure 7.19).

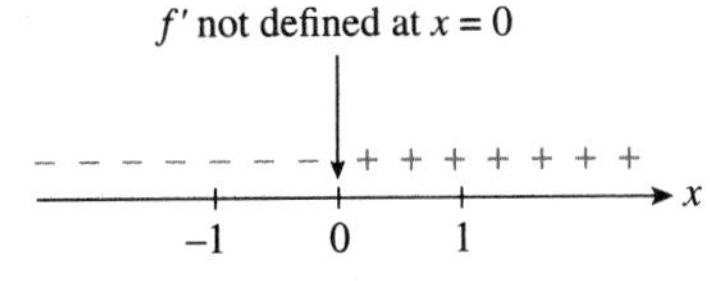

Figure 7.18
Sign diagram for f'

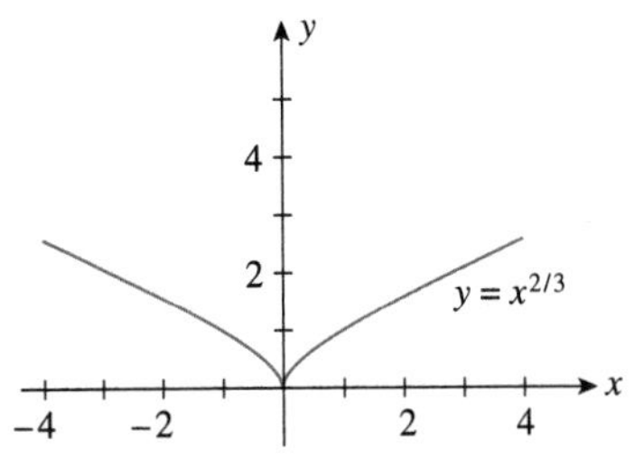

Figure 7.19
f has a relative minimum at $x = 0$.

GROUP DISCUSSION

Recall that the average cost function $\overline{C}$ is defined by

$$\overline{C} = \frac{C(x)}{x}$$

where $C(x)$ is the total cost function and x is the number of units of a commodity manufactured (see Section 7.1).

1. Show that

$$\overline{C}'(x) = \frac{C'(x) - \overline{C}(x)}{x} \qquad (x > 0)$$

2. Use the result of part 1 to conclude that $\overline{C}$ is decreasing for values of x at which $C'(x) < \overline{C}(x)$.Find similar conditions for which $\overline{C}$ is increasing and for which $\overline{C}$ is constant.
3. Explain the results of part 2 in economic terms.

Figure 7.20
Sign diagram for f'

Example 7

Find the relative maxima and relative minima of the function

$$f(x) = x^3 - 3x^2 - 24x + 32$$

Solution

The derivative of f is

$$f'(x) = 3x^2 - 6x - 24 = 3(x + 2)(x - 4)$$

and it is continuous everywhere. The zeros of $f'(x)$, $x = -2$ and $x = 4$, are the only critical points of the function f. The sign diagram for f' is shown in Figure 7.20. Examine the two critical points $x = -2$ and $x = 4$ for a relative extremum using the first derivative test and the sign diagram for f':

1. *The critical point* $x = -2$: Since the function $f'(x)$ changes sign from positive to negative as we move across $x = -2$ from left to right, we conclude that a relative maximum of f occurs at $x = -2$. The value of $f(x)$ when $x = -2$ is

$$f(-2) = (-2)^3 - 3(-2)^2 - 24(-2) + 32 = 60$$

2. *The critical point* $x = 4$: $f'(x)$ changes sign from negative to positive as we move across $x = 4$ from left to right, so $f(4) = -48$ is a relative minimum of f. The graph of f appears in Figure 7.21.

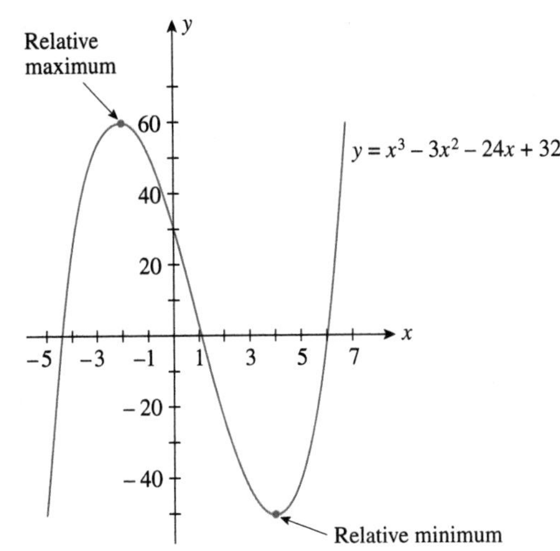

Figure 7.21
f has a relative maximum at $x = -2$ and a relative minimum at $x = 4$.

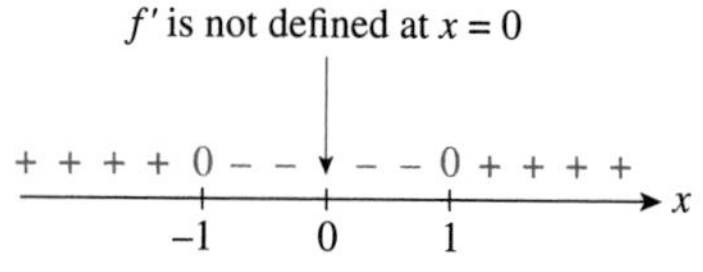

Figure 7.22
$x = 0$ is not a critical point because f is not defined at $x = 0$.

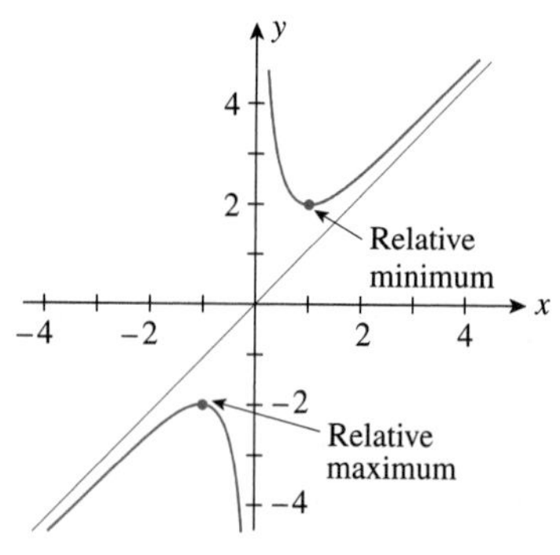

Figure 7.23
$f(x) = x + \frac{1}{x}$

Example 8

Find the relative maxima and the relative minima of the function

$$f(x) = x + \frac{1}{x}$$

Solution

The derivative of f is

$$f'(x) = 1 - \frac{1}{x^2} = \frac{x^2 - 1}{x^2} = \frac{(x+1)(x-1)}{x^2}$$

Since f' is equal to zero at $x = -1$ and $x = 1$, these are critical points for the function f. Next, observe that f' is discontinuous at $x = 0$. However, because f *is not defined at that point,* the point $x = 0$ does not qualify as a critical point of f (it is not in the domain of f). Figure 7.22 shows the sign diagram for f'.

Since $f'(x)$ changes sign from positive to negative as we move across $x = -1$ from left to right, the first derivative test implies that $f(-1) = -2$ is a relative maximum of the function f. Next, $f'(x)$ changes sign from negative to positive as we move across $x = 1$ from left to right, so $f(1) = 2$ is a relative minimum of the function f. The graph of f appears in Figure 7.23. Note that this function has a relative maximum that lies below its relative minimum.

Applications

Example 9

Optimizing Bacterial Nutrients A *bacterial reactor* is a tank containing a population of bacteria with inlets for nutrients and outlets for harvesting useful biochemicals. The *productivity* of a reactor is the quantity of useful biochemicals recovered. Too great a quantity of nutrients can poison a reactor, so the amount of nutrients added must be controlled carefully. Experimentally the productivity of a reactor in grams per hour is found to be

$$P(x) = 0.52x - 2.61x^3 \quad \text{for} \quad 0 \le x \le 0.446,$$

where x is the kilograms of nutrients added per hour. Find the optimum nutrient level x for the reactor.

Solution

Compute the derivative

$$P'(x) = 0.52 - 7.83x^2.$$

Find the critical points.

$$0.52 - 7.83x^2 = 0$$
$$0.52 = 7.83x^2$$
$$x^2 = \frac{0.52}{7.83}$$
$$x = \pm\sqrt{\frac{0.52}{7.83}}$$

The negative root is inappropriate and so we obtain

$$x = \sqrt{\frac{0.52}{7.83}} = 0.258 \text{ kg/h}$$

Notice that $P'(x)$ is increasing on (0, 0.258) and it is decreasing on (0.258, 0.446) so we see that the critical point is a maximum (Figure 7.24).

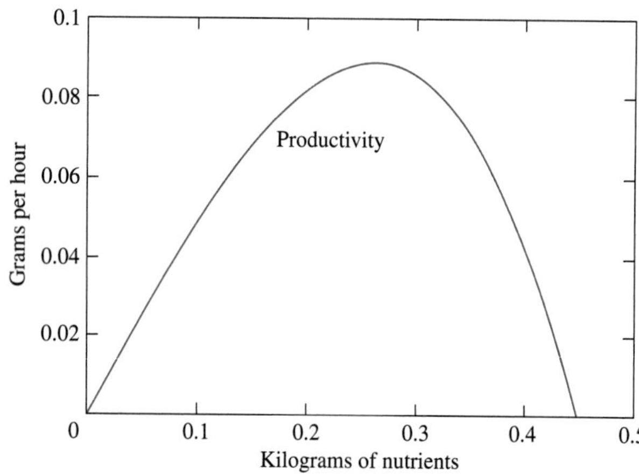

Figure 7.24
Productivity as a function of nutrients

Example 10

Profit Functions The profit function of Acrosonic Company is given by

$$P(x) = -0.02x^2 + 300x - 200{,}000$$

dollars, where x is the number of Acrosonic model F loudspeaker systems produced. Find where the function P is increasing and where it is decreasing.

Solution

The derivative P' of the function P is

$$P'(x) = -0.04x + 300 = -0.04(x - 7500)$$

Thus, $P'(x) = 0$ when $x = 7500$. Furthermore, $P'(x) > 0$ for x in the interval $(0, 7500)$, and $P'(x) < 0$ for x in the interval $(7500, \infty)$. This means that the profit function P is increasing on $(0, 7500)$ and decreasing on $(7500, \infty)$ (Figure 7.25).

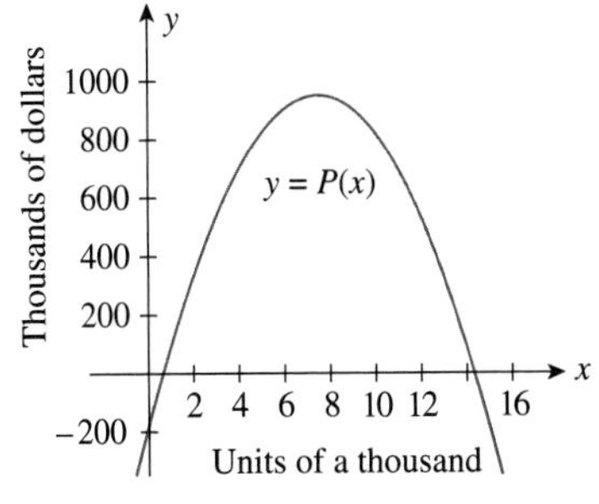

Figure 7.25
The profit function is increasing on (0, 7500) and increasing on (7500, ∞).

Example 11

Crime Rates The number of major crimes committed in the city of York from 1993 to 2000 is approximated by the function

$$N(t) = -0.1t^3 + 1.5t^2 + 100 \qquad (0 \le t \le 7)$$

where $N(t)$ denotes the number of crimes committed in year t, with $t = 0$ corresponding to the beginning of 1993. Find where the function N is increasing and where it is decreasing.

Solution

The derivative N' of the function N is

$$N'(t) = -0.3t^2 + 3t = -0.3t(t - 10)$$

Since $N'(t) > 0$ for t in the interval $(0, 7)$, the function N is increasing throughout that interval (Figure 7.26).

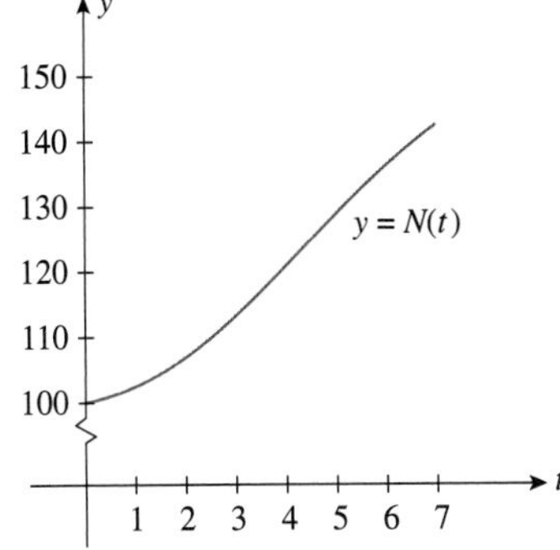

Figure 7.26
The number of crimes, $N(t)$, is increasing over the 7-year interval.

Self-Check Exercises 7.1

1. Find the intervals where the function $f(x) = \frac{2}{3}x^3 - x^2 - 12x + 3$ is increasing and the intervals where it is decreasing.

2. Find the relative extrema of $f(x) = \dfrac{x^2}{1 - x^2}$.

Solutions to Self-Check Exercises 7.1 can be found on page 355.

7.1 Exercises

In Exercises 1–8, you are given the graph of a function f. Determine the intervals where f is increasing, constant, or decreasing.

1.

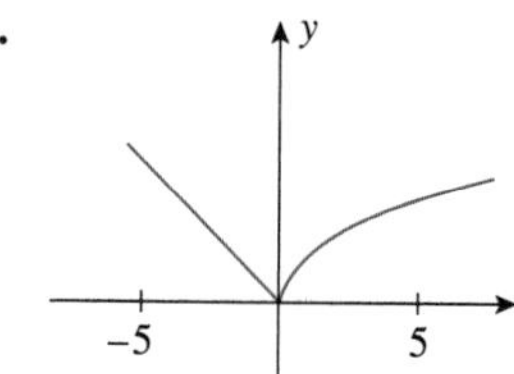

2.

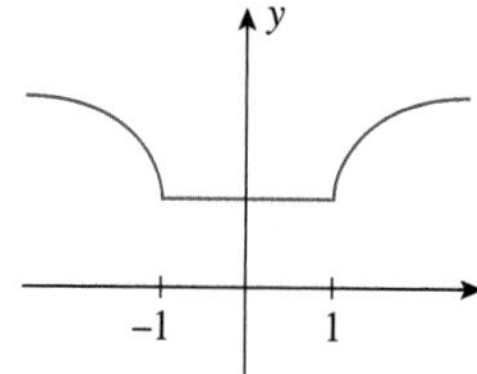

3.

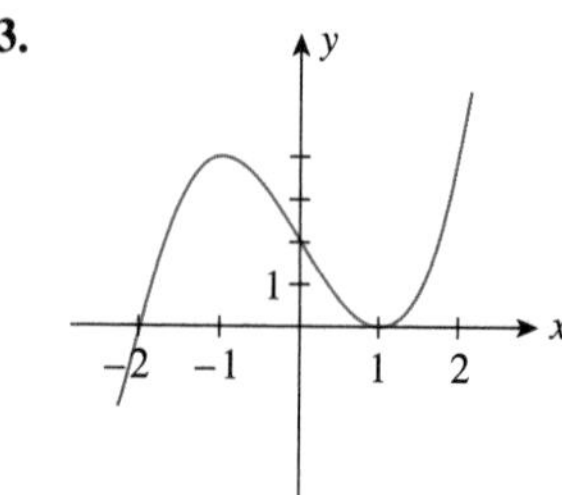

4.

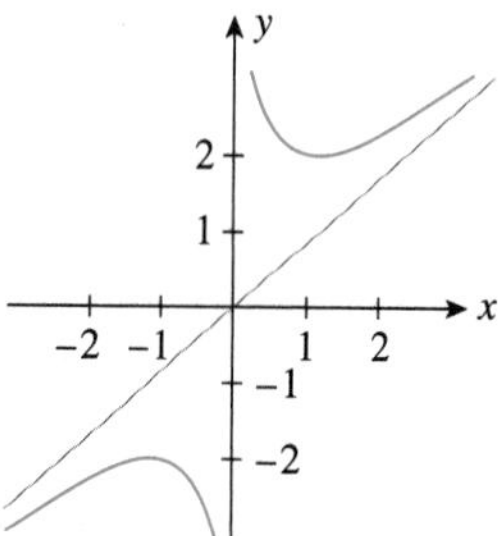

5.

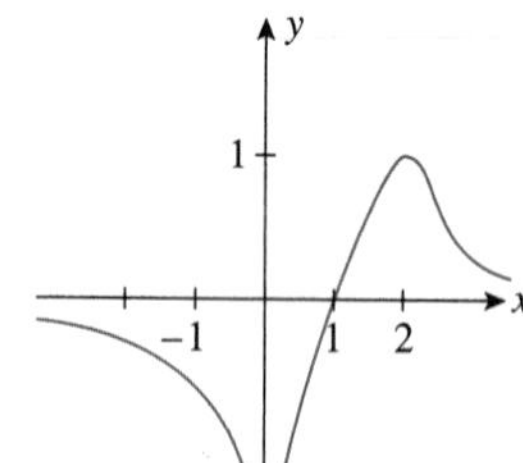

6.

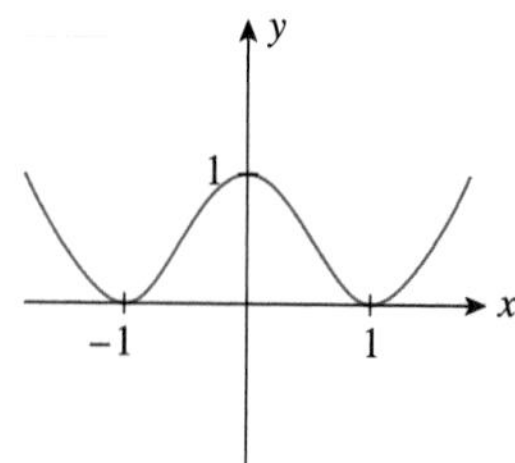

7.

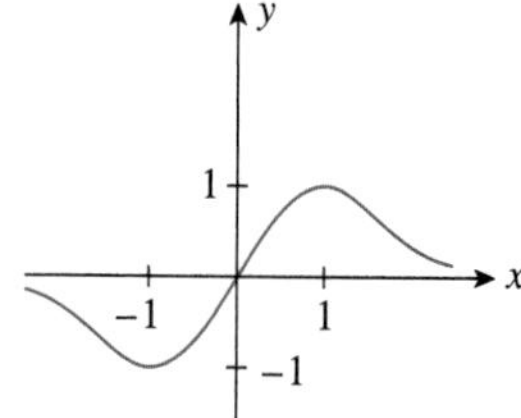

8.

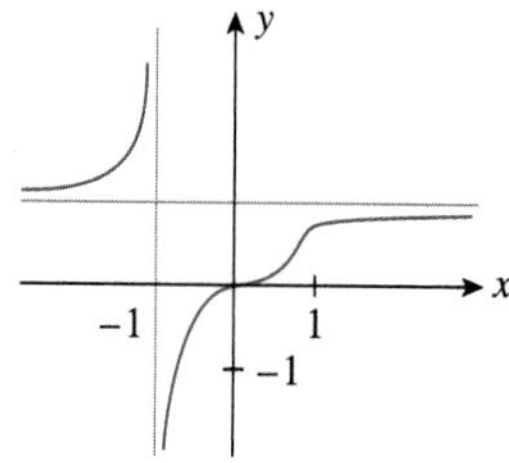

9. The graph of the function f shown in the accompanying figure gives the elevation of that part of the Boston Marathon course that includes the notorious Heartbreak Hill. Determine the intervals (stretches of the course) where the function f is increasing (the runner is labouring), where it is constant (the runner is taking a breather), and where it is decreasing (the runner is coasting).

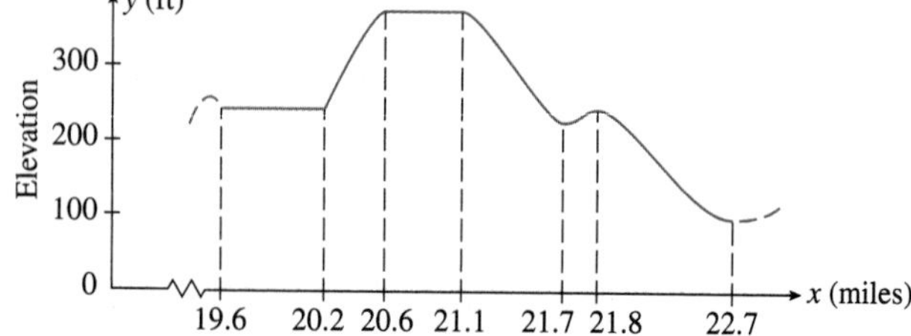

10. Among the important factors in determining the structural integrity of an aircraft is its age. Advancing age makes planes more likely to crack. The graph of the function f, shown in the accompanying figure, is referred to as a "bathtub curve" in the airline industry. It gives the fleet damage rate (damage due to corrosion, accident, and metal fatigue) of a typical fleet of commercial aircraft as a function of the number of years of service.

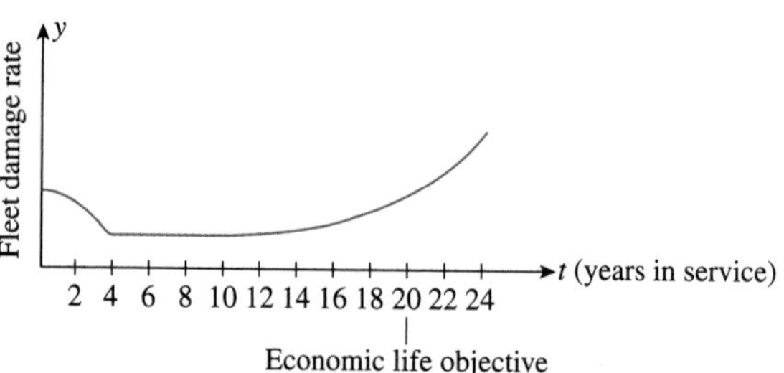

a. Determine the interval where f is decreasing. This corresponds to the time period when the fleet damage rate is dropping as problems are found and corrected during the initial "shakedown" period.

b. Determine the interval where f is constant. After the initial shakedown period, planes have few structural problems, and this is reflected by the fact that the function is constant on this interval.

c. Determine the interval where f is increasing. Beyond the time period mentioned in part (b), the function is increasing—reflecting an increase in structural defects due mainly to metal fatigue.

In Exercises 11–54, find the interval(s) where each function is increasing and the interval(s) where it is decreasing.

11. $f(x) = 3x + 5$

12. $f(x) = 4 - 5x$

13. $f(x) = x^2 - 3x$

14. $f(x) = 2x^2 + x + 1$

15. $g(x) = x - x^3$

16. $f(x) = x^3 - 3x^2$

17. $g(x) = x^3 + 3x^2 + 1$

18. $f(x) = x^3 - 3x + 4$

19. $f(x) = \frac{1}{3}x^3 - 3x^2 + 9x + 20$

20. $f(x) = \frac{2}{3}x^3 - 2x^2 - 6x - 2$

21. $h(x) = x^4 - 4x^3 + 10$

22. $g(x) = x^4 - 2x^2 + 4$

23. $f(x) = \frac{1}{x - 2}$

24. $h(x) = \frac{1}{2x + 3}$

25. $h(t) = \frac{t}{t - 1}$

26. $g(t) = \frac{2t}{t^2 + 1}$

27. $f(x) = x^{3/5}$

28. $f(x) = x^{2/3} + 5$

29. $f(x) = \sqrt{x + 1}$

30. $f(x) = (x - 5)^{2/3}$

31. $f(x) = \sqrt{16 - x^2}$

32. $g(x) = x\sqrt{x + 1}$

33. $f(x) = \dfrac{x^2 - 1}{x}$

34. $h(x) = \dfrac{x^2}{x - 1}$

35. $f(x) = \dfrac{1}{(x - 1)^2}$

36. $g(x) = \dfrac{x}{(x + 1)^2}$

37. $f(x) = e^{2x}$

38. $g(x) = e^{-x}$

39. $h(x) = e^{-x^2}$

40. $g(x) = x \cdot e^{-x}$

41. $g(x) = x \cdot e^{-x/3}$

42. $f(x) = x^2 \cdot e^{-x}$

43. $g(x) = \ln(x^2 + 1)$

44. $f(x) = \ln(x^2 + 3x + 1)$

45. $g(x) = \ln(10 - x^2)$

46. $h(x) = \ln\left(\dfrac{x}{x + 1}\right)$

47. $f(x) = \sin(x)$ for $0 \le x \le 2\pi$

48. $f(x) = \cos(x)$ for $0 \le x \le 2\pi$

49. $h(x) = \sin(2x)$ for $0 \le x \le 2\pi$

50. $g(x) = \sin\left(x + \dfrac{\pi}{2}\right)$ for $0 \le x \le 2\pi$

51. $g(x) = \arctan(x)$

52. $h(x) = \arctan(1 - x)$

53. $f(x) = \sin\left(\dfrac{1}{x^2 + 1}\right)$

54. $h(x) = \sin^2(x)$

In Exercises 55–62, you are given the graph of a function *f*. Determine the relative maxima and relative minima, if any.

55.

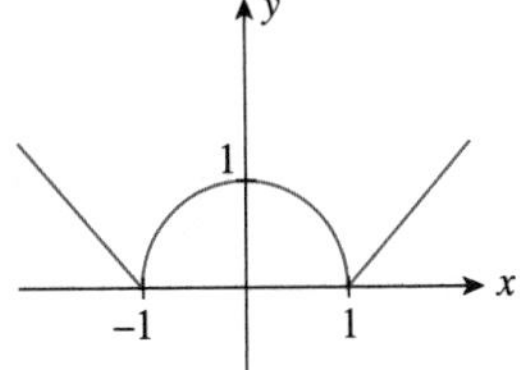

56.

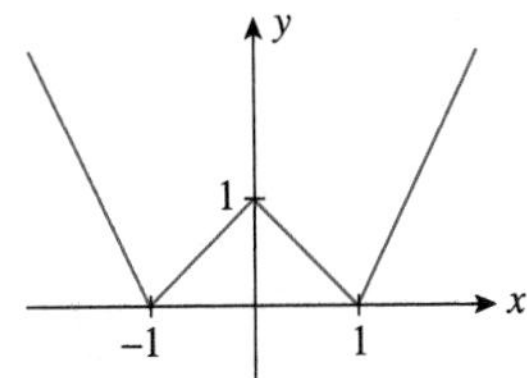

57.

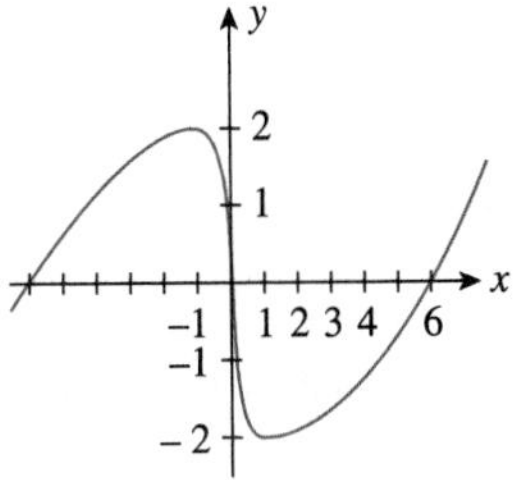

58.

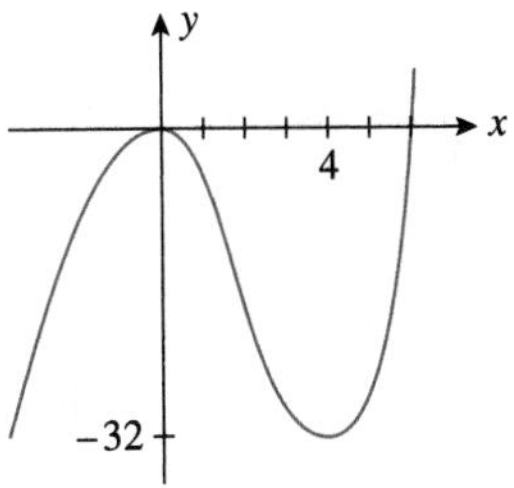

59.

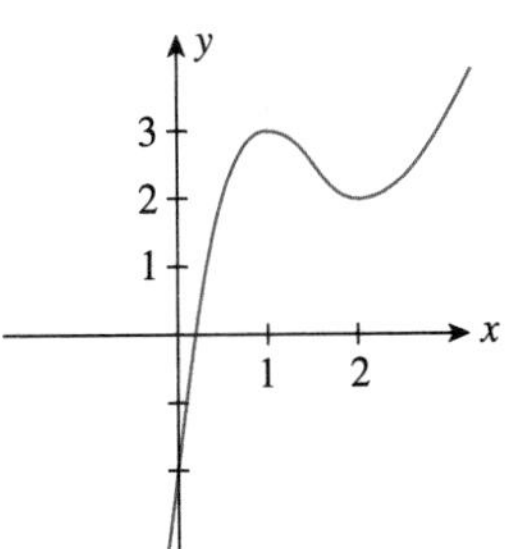

60.

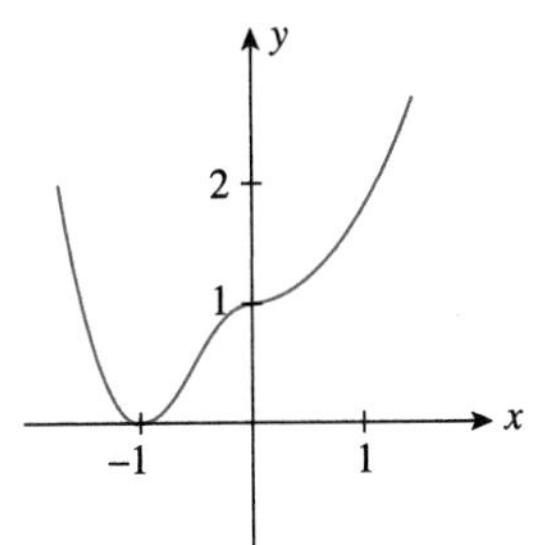

61.

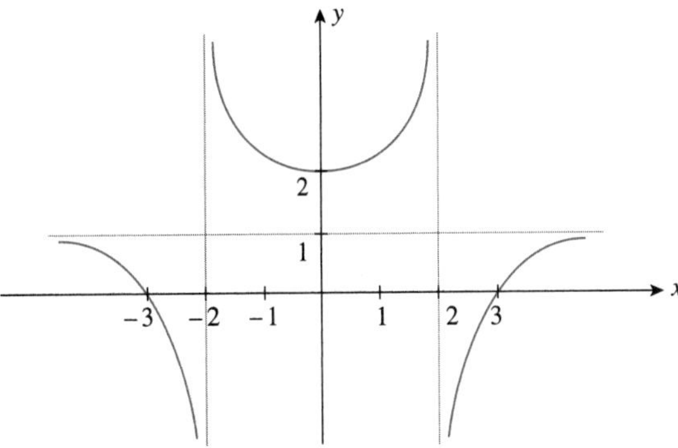

62.

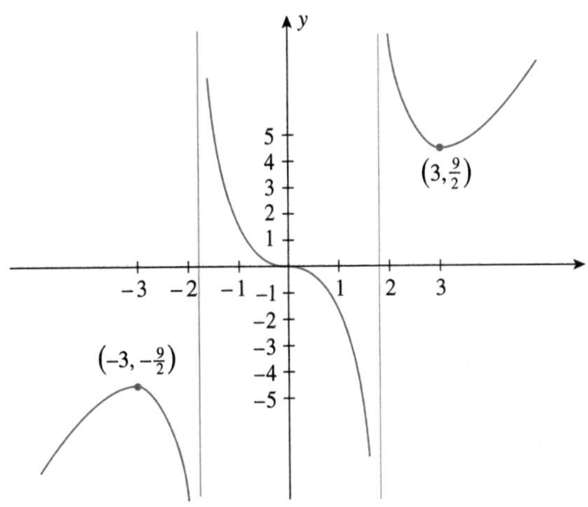

In Exercises 63–66, match the graph of the function with the graph of its derivative in (a)–(d).

63.

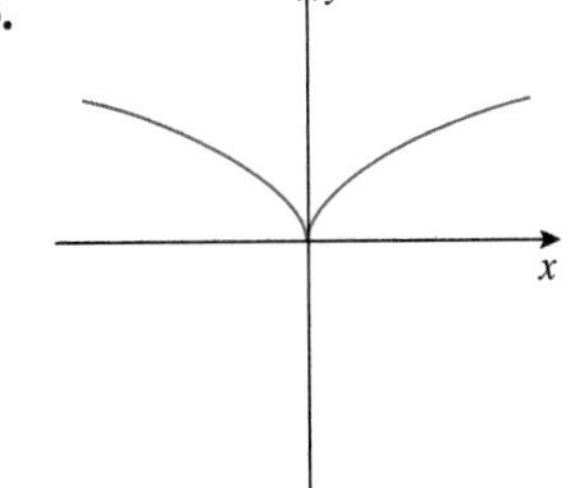

64.

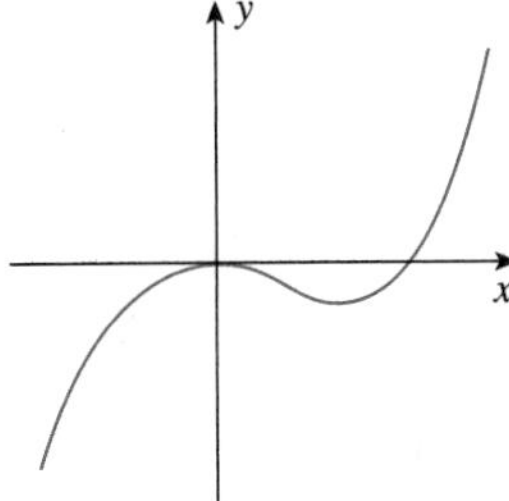

65.

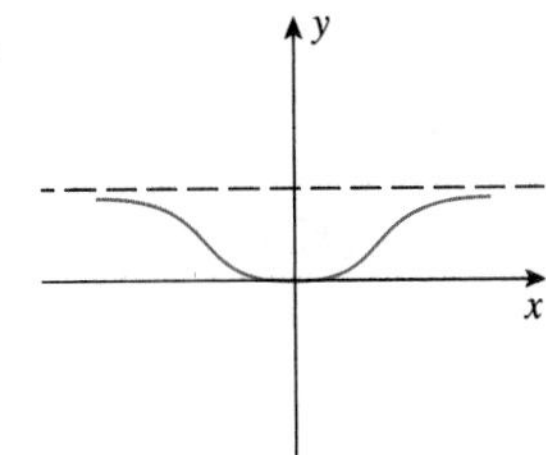

66.

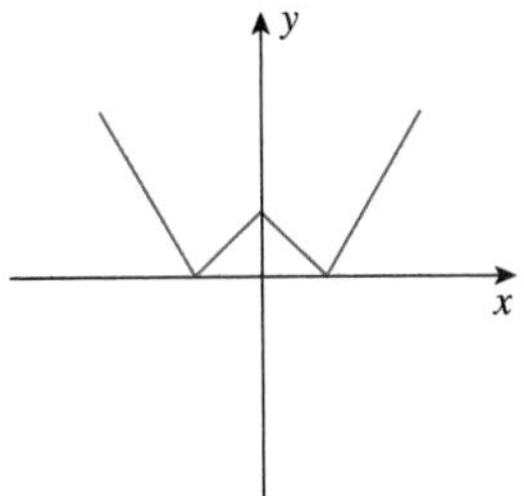

a.

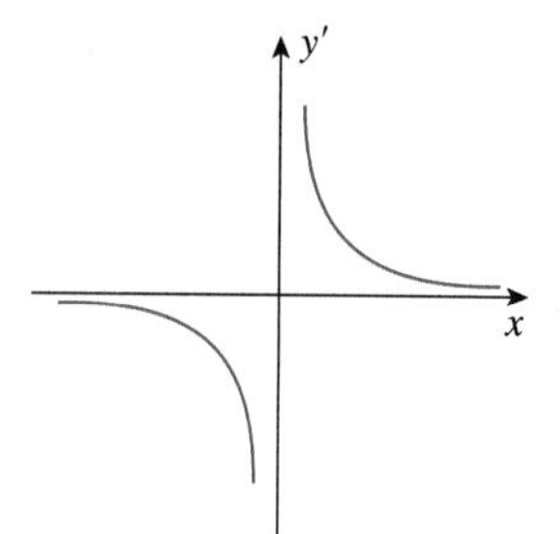

b.

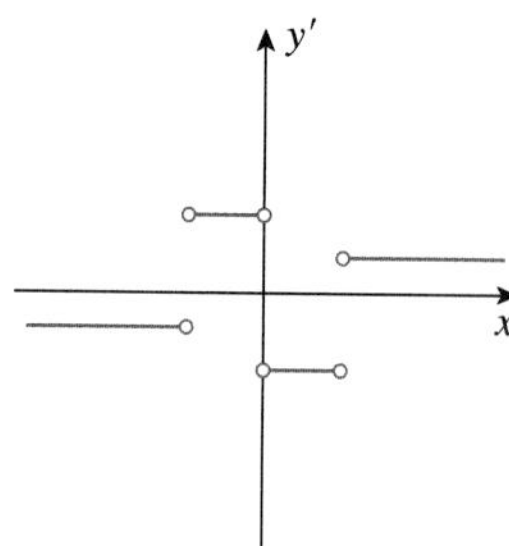

c.

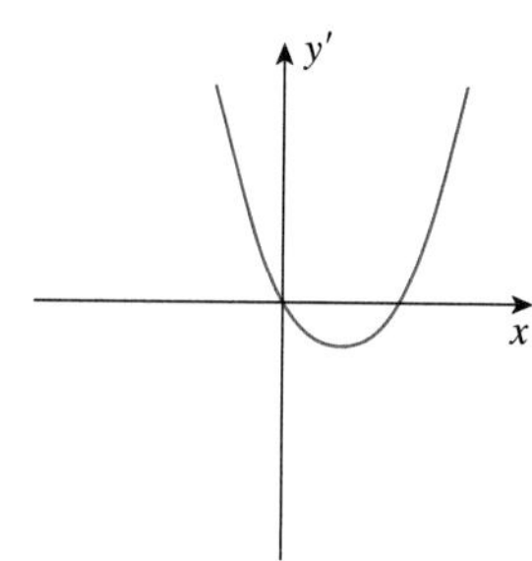

d.

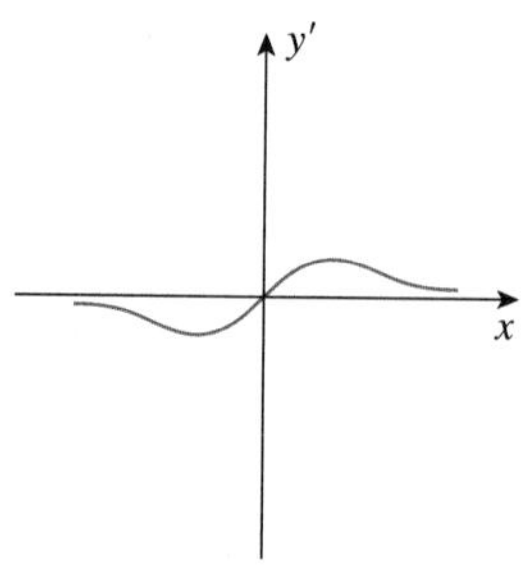

In Exercises 67–102, find the relative maxima and relative minima, if any, of each function.

67. $f(x) = x^2 - 4x$

68. $g(x) = x^2 + 3x + 8$

69. $h(t) = -t^2 + 6t + 6$

70. $f(x) = \frac{1}{2}x^2 - 2x + 4$

71. $f(x) = x^{5/3}$

72. $f(x) = x^{2/3} + 2$

73. $g(x) = x^3 - 3x^2 + 4$

74. $f(x) = x^3 - 3x + 6$

75. $f(x) = \frac{1}{2}x^4 - x^2$

76. $h(x) = \frac{1}{2}x^4 - 3x^2 + 4x - 8$

77. $F(x) = \frac{1}{3}x^3 - x^2 - 3x + 4$

78. $F(t) = 3t^5 - 20t^3 + 20$

79. $g(x) = x^4 - 4x^3 + 8$

80. $f(x) = 3x^4 - 2x^3 + 4$

81. $g(x) = \frac{x + 1}{x}$

82. $h(x) = \frac{x}{x + 1}$

83. $f(x) = x + \frac{9}{x} + 2$

84. $g(x) = 2x^2 + \frac{4000}{x} + 10$

85. $f(x) = \frac{x}{1 + x^2}$

86. $g(x) = \frac{x}{x^2 - 1}$

87. $f(x) = \frac{x^2}{x^2 - 4}$

88. $g(t) = \frac{t^2}{1 + t^2}$

89. $f(x) = x \cdot e^{-x}$ for $x \geq 0$

90. $f(x) = x \cdot e^{-2x}$ for $x \geq 0$

91. $f(x) = x^2 \cdot e^{-x}$

92. $f(x) = \frac{1}{2}(e^x + e^{-x})$

93. $f(x) = \ln(x^2 + 1)$

94. $f(x) = \ln(x^2 - 2x + 3)$

95. $f(x) = \ln(x^3 - 3x^2 + 5)$

96. $f(x) = \ln(x^3 - 6x^2 + 9x + 7)$

97. $f(x) = \sin(x) + \sin(2x)/2$

98. $f(x) = \sin(x) + \cos(x)$

99. $f(x) = \sin\left(\frac{1}{x^2 + 1}\right)$

100. $f(x) = x \cdot \tan(x)$ for $-\frac{\pi}{2} < x < \frac{\pi}{2}$

101. $f(x) = (x - 1)^{2/3}$

102. $g(x) = x\sqrt{x - 4}$

103. A stone is thrown straight up from the roof of an 80-m building. The distance (in metres) of the stone from the ground at any time t (in seconds) is given by

$$h(t) = -16t^2 + 64t + 80$$

When is the stone rising, and when is it falling? If the stone were to miss the building, when would it hit the ground? Sketch the graph of h.

Hint: The stone is on the ground when $h(t) = 0$.

104. Flight of a Rocket The height (in metres) attained by a rocket t sec into flight is given by the function

$$-\frac{1}{9}t^3 + 5t^2 + 33t + 3$$

When is the rocket rising, and when is it descending?

105. Average Speed of a Highway Vehicle The average speed of a vehicle on a stretch of Highway 401 between 6 A.M. and 10 A.M. on a typical weekday is approximated by the function

$$f(t) = 20t - 40\sqrt{t} + 50 \qquad (0 \leq t \leq 4)$$

where $f(t)$ is measured in kilometres per hour and t is measured in hours, with $t = 0$ corresponding to 6 A.M. Find the interval where f is increasing and the interval where f is decreasing and interpret your results.

106. Prison Overcrowding The 1980s saw a trend toward old-fashioned punitive deterrence as opposed to the more liberal penal policies and community-based corrections popular in the 1960s and early 1970s. As a result, prisons became more crowded, and the gap between the number of people in prison and the prison capacity widened. The number of prisoners (in thousands) in federal and provincial prisons is approximated by the function

$$N(t) = 3.5t^2 + 26.7t + 436.2 \qquad (0 \leq t \leq 10)$$

where t is measured in years, with $t = 0$ corresponding to 1984. The number of inmates for which prisons were designed is given by

$$C(t) = 24.3t + 365 \qquad (0 \leq t \leq 10)$$

where $C(t)$ is measured in thousands and t has the same meaning as before. Show that the gap between the number of prisoners and the number for which the prisons were designed has been widening at any time t.

Hint: First, write a function G that gives the gap between the number of prisoners and the number for which the prisons were designed at any time t. Then show that $G'(t) > 0$ for all values of t in the interval $(0, 10)$.

107. Using Theorem 1, verify that the linear function $f(x) = mx + b$ is (a) increasing everywhere if $m > 0$, (b) decreasing everywhere if $m < 0$, and (c) constant if $m = 0$.

108. In what interval is the quadratic function

$$f(x) = ax^2 + bx + c \qquad (a \neq 0)$$

increasing? In what interval is f decreasing?

109. Show that the function $f(x) = x^3 + x + 1$ has no relative extrema on $(-\infty, \infty)$.

110. Let

$$f(x) = \begin{cases} -3x & \text{if } x < 0 \\ 2x + 4 & \text{if } x \geq 0 \end{cases}$$

a. Compute $f'(x)$ and show that it changes sign from negative to positive as we move across $x = 0$.
b. Show that f does not have a relative minimum at $x = 0$. Does this contradict the first derivative test? Explain your answer.

111. Let

$$f(x) = \begin{cases} -x^2 + 3 & \text{if } x \neq 0 \\ 2 & \text{if } x = 0 \end{cases}$$

a. Compute $f'(x)$ and show that it changes sign from positive to negative as we move across $x = 0$.
b. Show that f does not have a relative maximum at $x = 0$. Does this contradict the first derivative test? Explain your answer.

112. Let

$$f(x) = \begin{cases} \dfrac{1}{x^2} & \text{if } x > 0 \\ x^2 & \text{if } x \leq 0 \end{cases}$$

a. Compute $f'(x)$ and show that it does not change sign as we move across $x = 0$.
b. Show that f has a relative minimum at $x = 0$. Does this contradict the first derivative test? Explain your answer.

113. Show that the quadratic function

$$f(x) = ax^2 + bx + c \qquad (a \neq 0)$$

has a relative extremum when $x = -b/2a$. Also, show that the relative extremum is a relative maximum if $a < 0$ and a relative minimum if $a > 0$.

114. Show that the cubic function

$$f(x) = ax^3 + bx^2 + cx + d \qquad (a \neq 0)$$

has no relative extremum if and only if $b^2 - 3ac \leq 0$.

115. Consider the function $2x^2 - 1$.
a. Show that f is increasing on the interval $(0, 1)$.
b. Show that $f(0) = -1$ and $f(1) = 1$ and use the result of part (a) together with the intermediate value theorem to conclude that there is exactly one root of $f(x) = 0$ in $(0, 1)$.

116. Show that the function

$$f(x) = \frac{ax + b}{cx + d}$$

does not have a relative extremum if $ad - bc \neq 0$. What can you say about f if $ad - bc = 0$?

Business and Economics Applications

117. **Profit Functions** A subsidiary of ThermoMaster manufactures an indoor–outdoor thermometer. Management estimates that the profit (in dollars) realizable by the company for the manufacture and sale of x units of thermometers each week is

$$P(x) = -0.001x^2 + 8x - 5000$$

Find the intervals where the profit function P is increasing and the intervals where P is decreasing.

118. **Average Cost** The average cost (in dollars) incurred by Lincoln Records each week in pressing x compact discs is given by

$$\overline{C}(x) = -0.0001x + 2 + \frac{2000}{x} \qquad (0 < x \leq 6000)$$

Show that $\overline{C}(x)$ is always decreasing over the interval $(0, 6000)$.

119. **Projected Retirement Funds** Based on data from the Central Provident Fund of a certain country (a government agency similar to the Canada Pension Plan), the estimated cash in the fund in 1995 is given by

$$A(t) = -96.6t^4 + 403.6t^3 + 660.9t^2 + 250 \qquad (0 \leq t \leq 5)$$

where $A(t)$ is measured in billions of dollars and t is measured in decades, with $t = 0$ corresponding to 1995. Find the interval where A is increasing and the interval where A is decreasing and interpret your results.
Hint: Use the quadratic formula.

120. **Growth of Managed Services** Almost half of companies let other firms manage some of their Web operations—a practice called Web hosting. Managed services—monitoring a customer's technology services—is the fastest growing part of Web hosting. Managed services sales are expected to grow in accordance with the function

$$f(t) = 0.469t^2 + 0.758t + 0.44 \qquad (0 \leq t \leq 6)$$

where $f(t)$ is measured in billions of dollars and t is measured in years, with $t = 0$ corresponding to 1999.
a. Find the interval where f is increasing and the interval where f is decreasing.
b. What does your result tell you about sales in managed services from 1999 through 2005?

Source: International Data Corp.

121. **Web Hosting** Refer to Exercise 119. Sales in the Web-hosting industry are projected to grow in accordance with the function

$$f(t) = -0.05t^3 + 0.56t^2 + 5.47t + 7.5 \qquad (0 \leq t \leq 6)$$

where $f(t)$ is measured in billions of dollars and t is measured in years, with $t = 0$ corresponding to 1999.

a. Find the interval where f is increasing and the interval where f is decreasing.
Hint: Use the quadratic formula.

b. What does your result tell you about sales in the Web-hosting industry from 1999 through 2005?
Source: International Data Corp.

122. Cellular Phone Revenue According to a study conducted in 1997, the revenue (in millions of dollars) in the Canadian cellular market in the next 6 years is approximated by the function

$$R(t) = 0.03056t^3 - 0.45357t^2 + 4.81111t + 31.7 \qquad (0 \le t \le 6)$$

where t is measured in years, with $t = 0$ corresponding to 1997.

a. Find the interval where R is increasing and the interval where R is decreasing.

b. What does your result tell you about the revenue in the Canadian cellular market in the years under consideration?
Hint: Use the quadratic formula.

123. Cellular Phone Subscription According to a study conducted in 1997, the number of subscribers (in thousands) in the Canadian cellular market in the next 6 years is approximated by the function

$$N(t) = 0.09444t^3 - 1.44167t^2 + 10.65695t + 52 \qquad (0 \le t \le 6)$$

where t is measured in years, with $t = 0$ corresponding to 1997.

a. Find the interval where N is increasing and the interval where N is decreasing.

b. What does your result tell you about the number of subscribers in the Canadian cellular market in the years under consideration?
Hint: Use the quadratic formula.

124. Sales of Functional Food Products The sales of functional food products—those that promise benefits beyond basic nutrition—have risen sharply in recent years. The sales (in billions of dollars) of foods and beverages with herbal and other additives is approximated by the function

$$S(t) = 0.46t^3 - 2.22t^2 + 6.21t + 17.25 \qquad (0 \le t \le 4)$$

where t is measured in years, with $t = 0$ corresponding to the beginning of 1997. Show that S is increasing on the interval $[0, 4]$.
Hint: Use the quadratic formula.
Source: Frost & Sullivan

Biological and Life Sciences Applications

125. Prevalence of Alzheimer's Patients Based on a study conducted in 1997, the percent of the Canadian population by age afflicted with Alzheimer's disease is given by the function

$$P(x) = 0.0726x^2 + 0.7902x + 4.9623 \qquad (0 \le x \le 25)$$

where x is measured in years, with $x = 0$ corresponding to age 65. Show that P is an increasing function of x on the interval (0, 25). What does your result tell you about the relationship between Alzheimer's disease and age for the population that is age 65 and over?
Source: Alzheimer's Association

126. Environment of Forests Following the lead of the National Wildlife Federation, the Department of the Interior of a South American country began to record an index of environmental quality that measured progress and decline in the environmental quality of its forests. The index for the years 1984 through 1994 is approximated by the function

$$I(t) = \frac{1}{3}t^3 - \frac{5}{2}t^2 + 80 \qquad (0 \le t \le 10)$$

where $t = 0$ corresponds to 1984. Find the intervals where the function I is increasing and the intervals where it is decreasing. Interpret your results.
Source: World Almanac

127. Air Pollution According to the Ministry of Natural Resources, the level of nitrogen dioxide, a brown gas that impairs breathing, present in the atmosphere on a certain August day in downtown Toronto is approximated by

$$A(t) = 0.03t^3(t - 7)^4 + 60.2 \qquad (0 \le t \le 7)$$

where $A(t)$ is measured in pollutant standard index (PSI) and t is measured in hours, with $t = 0$ corresponding to 7 A.M. At what time of day is the air pollution increasing, and at what time is it decreasing?

128. Drug Concentration in the Blood The concentration (in milligrams/cubic centimetre) of a certain drug in a patient's body t hr after injection is given by

$$C(t) = \frac{t^2}{2t^3 + 1} \qquad (0 \le t \le 4)$$

When is the concentration of the drug increasing, and when is it decreasing?

129. Age of Drivers in Crash Fatalities The number of crash fatalities per 100,000 vehicle kilometres of travel (based on 1994 data) is approximated by the model

$$f(x) = \frac{15}{0.08333x^2 + 1.91667x + 1} \qquad (0 \le x \le 11)$$

where x is the age of the driver in years, with $x = 0$ corresponding to age 16. Show that f is decreasing on (0, 11) and interpret your result.
Source: Ministry of Transportation

130. Air Pollution The amount of nitrogen dioxide, a brown gas that impairs breathing, present in the atmosphere on a certain May day in the city of Long Beach is approximated by

$$A(t) = \frac{136}{1 + 0.25(t - 4.5)^2} + 28 \qquad (0 \le t \le 11)$$

where $A(t)$ is measured in pollutant standard index (PSI) and t is measured in hours, with $t = 0$ corresponding to 7 A.M. Find the intervals where A is increasing and where A is decreasing and interpret your results.

In Exercises 131–136, determine whether the statement is true or false. If it is true, explain why it is true. If it is false, give an example to show why it is false.

131. If f is decreasing on (a, b), then $f'(x) < 0$ for each x in (a, b).

132. If f and g are both increasing on (a, b), then $f + g$ is increasing on (a, b).

133. If f and g are both decreasing on (a, b), then $f - g$ is decreasing on (a, b).

134. If $f(x)$ and $g(x)$ are positive on (a, b) and both f and g are increasing on (a, b), then fg is increasing on (a, b).

135. If $f'(c) = 0$, then f has a relative maximum or a relative minimum at $x = c$.

136. If f has a relative minimum at $x = c$, then $f'(c) = 0$.

Solutions to Self-Check Exercises 7.1

1. The derivative of f is

$$f'(x) = 2x^2 - 2x - 12 = 2(x + 2)(x - 3)$$

and it is continuous everywhere. The zeros of $f'(x)$ are $x = -2$ and $x = 3$. The sign diagram of f' is shown in the accompanying figure. We conclude that f is increasing on the intervals $(-\infty, -2)$ and $(3, \infty)$ and decreasing on the interval $(-2, 3)$.

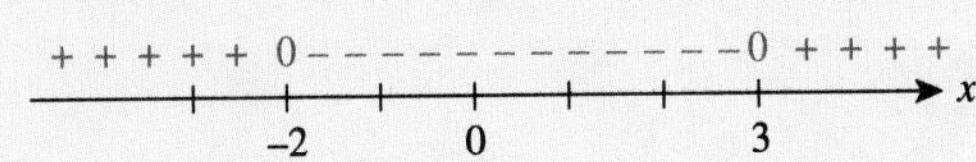

2. The derivative of f is

$$f'(x) = \frac{(1 - x^2)\frac{d}{dx}(x^2) - x^2\frac{d}{dx}(1 - x^2)}{(1 - x^2)^2}$$

$$= \frac{(1 - x^2)(2x) - x^2(-2x)}{(1 - x^2)^2} = \frac{2x}{(1 - x^2)^2}$$

and it is continuous everywhere except at $x = \pm 1$. Since $f'(x)$ is equal to zero at $x = 0$, $x = 0$ is a critical point of f. Next, observe that $f'(x)$ is discontinuous at $x = \pm 1$, but since these points are not in the domain of f, they do not qualify as critical points of f. Finally, from the sign diagram of f' shown in the accompanying figure, we conclude that $f(0) = 0$ is a relative minimum of f.

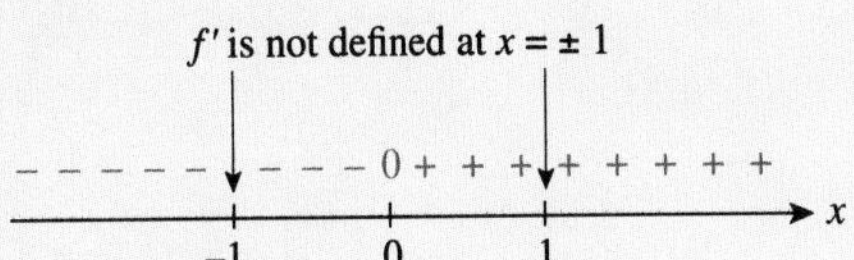

7.2 Applications of the Second Derivative

Determining the Intervals of Concavity

Consider the graphs shown in Figure 7.27, which give the estimated population of the world and of Canada through the year 2000. Both graphs are rising, indicating that both the Canadian population and the world population continued to increase through the year 2000. But observe that the graph in Figure 7.27a opens upward, whereas the graph in Figure 7.27b opens downward. What is the significance of this? To answer this question, let's look at the slopes of the tangent lines to various points on each graph (Figure 7.28).

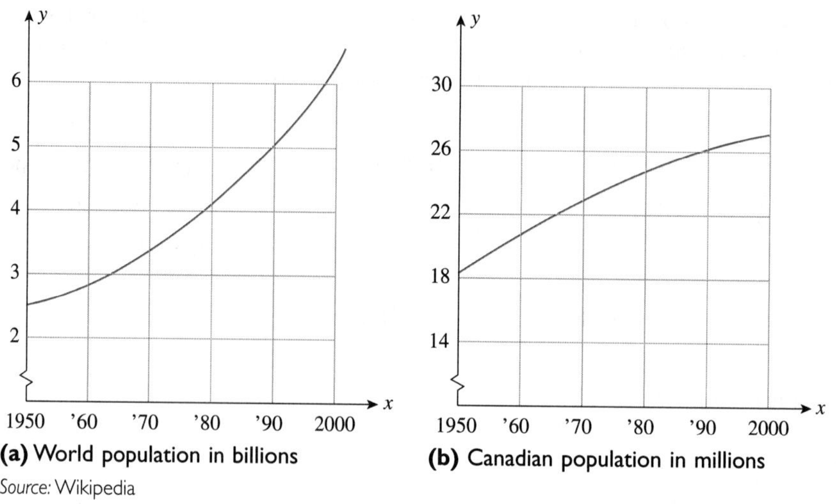

(a) World population in billions

Source: Wikipedia

(b) Canadian population in millions

Figure 7.27

In Figure 7.28a we see that the slopes of the tangent lines to the graph are increasing as we move from left to right. Since the slope of the tangent line to the graph at a point on the graph measures the rate of change of the function at that point, we conclude that the world population is not only increasing through the year 2000 but is increasing at an *increasing* pace. A similar analysis of Figure 7.28b reveals that the Canadian population is increasing, but at a *decreasing* pace.

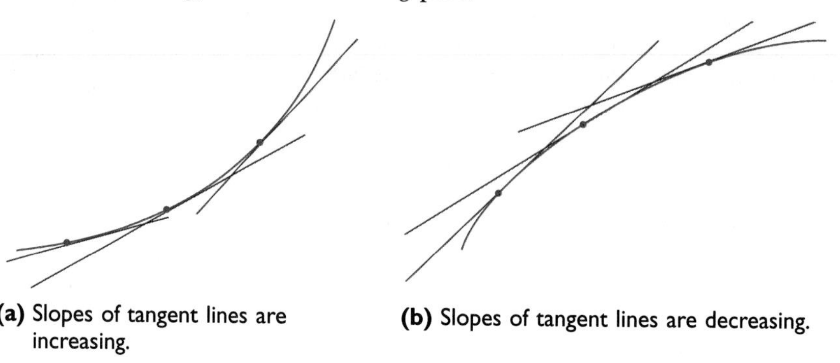

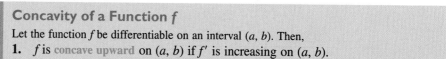

(a) Slopes of tangent lines are increasing.

(b) Slopes of tangent lines are decreasing.

Figure 7.28

The shape of a curve can be described using the notion of concavity.

Concavity of a Function f

Let the function f be differentiable on an interval (a, b). Then,

1. f is **concave upward** on (a, b) if f' is increasing on (a, b).
2. f is **concave downward** on (a, b) if f' is decreasing on (a, b).

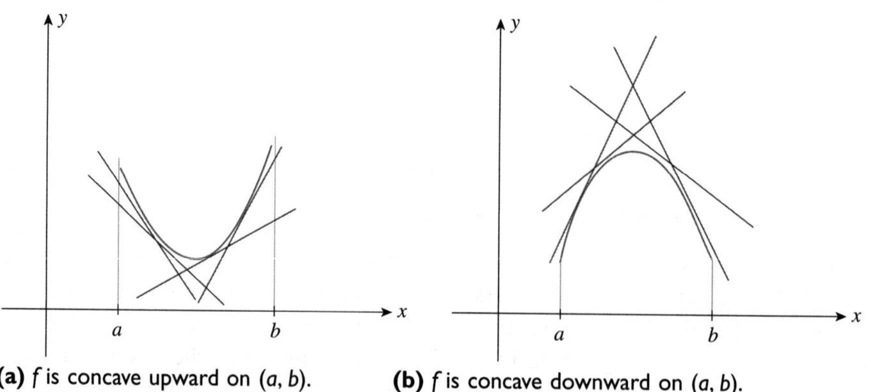

(a) f is concave upward on (a, b).

(b) f is concave downward on (a, b).

Figure 7.29

Geometrically, a curve is concave upward if it lies above its tangent lines (Figure 7.29a). Similarly, a curve is concave downward if it lies below its tangent lines (Figure 7.29b).

We also say that f is *concave upward at a point* $x = c$ if there exists an interval (a, b) containing c in which f is concave upward. Similarly, we say that f is *concave downward at a point* $x = c$ if there exists an interval (a, b) containing c in which f is concave downward.

If a function f has a second derivative f'', we can use f'' to determine the intervals of concavity of the function. Recall that $f''(x)$ measures the rate of change of the slope $f'(x)$ of the tangent line to the graph of f at the point $(x, f(x))$. Thus, if $f''(x) > 0$ on an interval (a, b), then the slopes of the tangent lines to the graph of f are increasing on (a, b), and so f is concave upward on (a, b). Similarly, if $f''(x) < 0$ on (a, b), then f is concave downward on (a, b). These observations suggest the following theorem.

THEOREM 2

a. If $f''(x) > 0$ for each value of x in (a, b), then f is concave upward on (a, b).

b. If $f''(x) < 0$ for each value of x in (a, b), then f is concave downward on (a, b).

The following procedure, based on the conclusions of Theorem 2, may be used to determine the intervals of concavity of a function.

Determining the Intervals of Concavity of f

1. Determine the values of x for which f'' is zero or where f'' is not defined, and identify the open intervals determined by these points.
2. Determine the sign of f'' in each interval found in step 1. To do this, compute $f''(c)$, where c is any conveniently chosen test point in the interval.
   a. If $f''(c) > 0$, f is concave upward on that interval.
   b. If $f''(c) < 0$, f is concave downward on that interval.

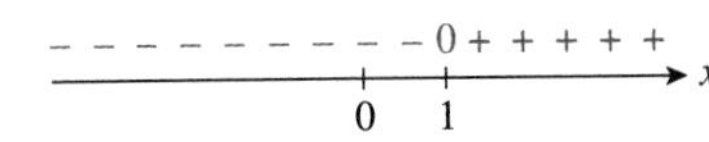

Figure 7.30
Sign diagram for f''

Example 1

Determine where the function $f(x) = x^3 - 3x^2 - 24x + 32$ is concave upward and where it is concave downward.

Solution

Here,

$$f'(x) = 3x^2 - 6x - 24$$
$$f''(x) = 6x - 6 = 6(x - 1)$$

and f'' is defined everywhere. Setting $f''(x) = 0$ gives $x = 1$. The sign diagram of f'' appears in Figure 7.30. We conclude that f is concave downward on the interval $(-\infty, 1)$ and is concave upward on the interval $(1, \infty)$. Figure 7.31 shows the graph of f.

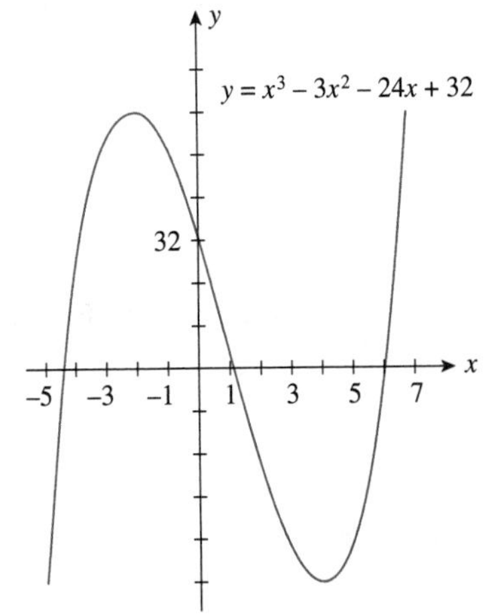

Figure 7.31
f is concave downward on $(-\infty, 1)$ and concave upward on $(1, \infty)$.

Example 2

Determine the intervals where the function $f(x) = x + \frac{1}{x}$ is concave upward and where it is concave downward.

Solution

We have

$$f'(x) = 1 - \frac{1}{x^2}$$

$$f''(x) = \frac{2}{x^3}$$

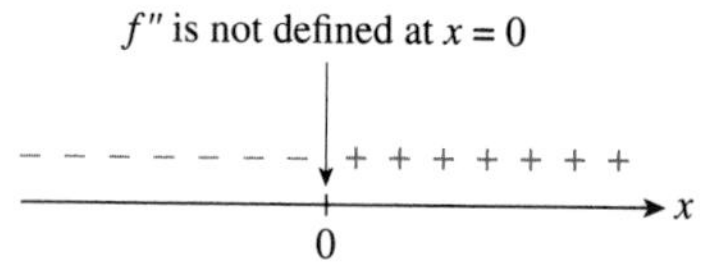

Figure 7.32
The sign diagram for f''

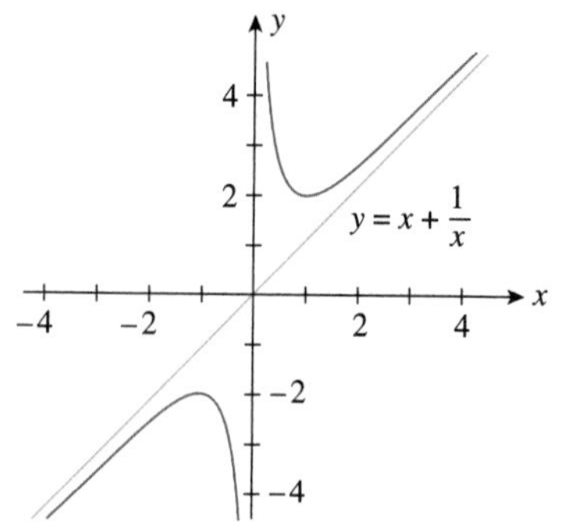

Figure 7.33
f is concave downward on $(-\infty, 0)$ and concave upward on $(0, \infty)$.

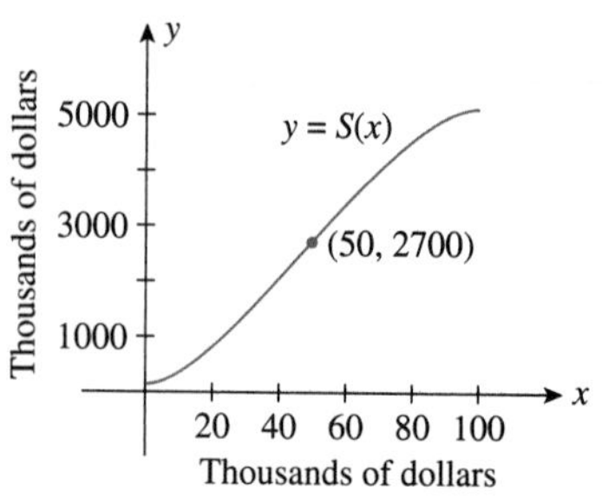

Figure 7.34
The graph of S has a point of inflection at (50, 2700).

We deduce from the sign diagram for f'' (Figure 7.32) that the function f is concave downward on the interval $(-\infty, 0)$ and concave upward on the interval $(0, \infty)$. The graph of f is sketched in Figure 7.33.

Inflection Points

Figure 7.34 shows the total sales S of a manufacturer of automobile air conditioners versus the amount of money x that the company spends on advertising its product. Notice that the graph of the continuous function $y = S(x)$ changes concavity—from upward to downward—at the point (50, 2700). This point is called an inflection point of S. To understand the significance of this inflection point, observe that the total sales increase rather slowly at first, but as more money is spent on advertising, the total sales increase rapidly. This rapid increase reflects the effectiveness of the company's ads. However, a point is soon reached after which any additional advertising expenditure results in increased sales but at a slower rate of increase. This point, commonly known as the *point of diminishing returns,* is the point of inflection of the function S. We will return to this example later.

Let's now state formally the definition of an inflection point.

Inflection Point

A point on the graph of a function f where the tangent line exists and where the concavity changes is called an **inflection point.**

Observe that the graph of a function crosses its tangent line at a point of inflection (Figure 7.35).

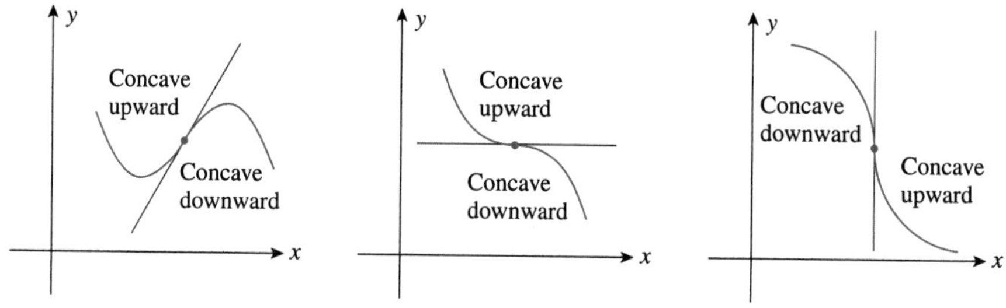

Figure 7.35
At each point of inflection, the graph of a function crosses its tangent line.

The following procedure may be used to find inflection points.

Finding Inflection Points

1. Compute $f''(x)$.
2. Determine the points in the domain of f for which $f''(x) = 0$ or $f''(x)$ does not exist.
3. Determine the sign of $f''(x)$ to the left and right of each point $x = c$ found in step 2. If there is a change in the sign of $f''(x)$ as we move across the point $x = c$, then $(c, f(c))$ is an inflection point of f.

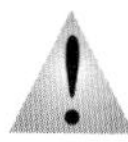

The points determined in step 2 are only *candidates* for the inflection points of f. For example, you can easily verify that $f''(0) = 0$ if $f(x) = x^4$, but a sketch of the graph of f will show that (0, 0) is *not* an inflection point of f.

Example 3

Find the points of inflection of the function $f(x) = x^3$.

Solution

$$f'(x) = 3x^2$$

$$f''(x) = 6x$$

Observe that f'' is continuous everywhere and is zero if $x = 0$. The sign diagram of f'' is shown in Figure 7.36. From this diagram, we see that $f''(x)$ changes sign as we move across $x = 0$. Thus, the point (0, 0) is an inflection point of the function f (Figure 7.37).

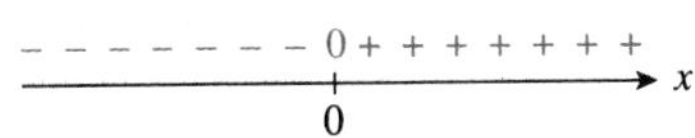

Figure 7.36
Sign diagram for f''

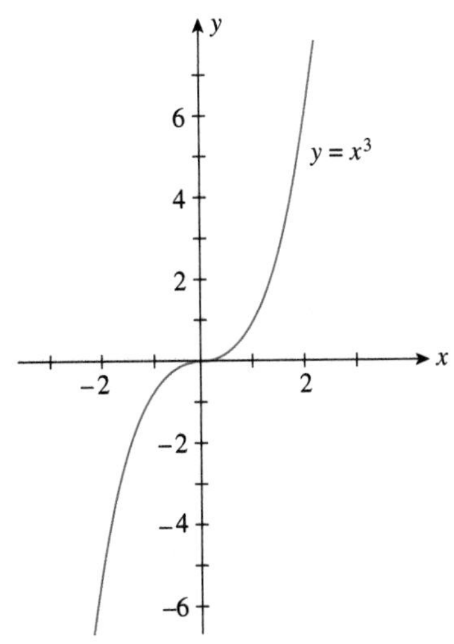

Figure 7.37
f has an inflection point at (0, 0).

Example 4

Determine the intervals where the function $f(x) = (x - 1)^{5/3}$ is concave upward and where it is concave downward and find the inflection points of f.

Solution

The first derivative of f is

$$f'(x) = \frac{5}{3}(x - 1)^{2/3}$$

and the second derivative of f is

$$f''(x) = \frac{10}{9}(x - 1)^{-1/3} = \frac{10}{9(x - 1)^{1/3}}$$

We see that f'' is not defined at $x = 1$. Furthermore, $f''(x)$ is not equal to zero anywhere. The sign diagram of f'' is shown in Figure 7.38. From the sign diagram, we see that f is concave downward on $(-\infty, 1)$ and concave upward on $(1, \infty)$. Next, since $x = 1$ does lie in the domain of f, our computations also reveal that the point (1, 0) is an inflection point of f (Figure 7.39).

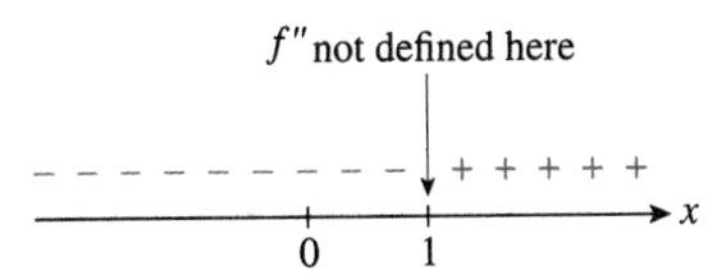

Figure 7.38
Sign diagram for f''

Example 5

Determine the intervals where the function

$$f(x) = \frac{1}{x^2 + 1}$$

is concave upward and where it is concave downward and find the inflection points of f.

Solution

The first derivative of f is

$$f'(x) = \frac{d}{dx}(x^2 + 1)^{-1} = -2x(x^2 + 1)^{-2} \qquad \text{Use the general power rule.}$$

$$= -\frac{2x}{(x^2 + 1)^2}$$

Next, using the quotient rule, we find

$$f''(x) = \frac{(x^2 + 1)^2(-2) + (2x)2(x^2 + 1)(2x)}{(x^2 + 1)^4}$$

$$= \frac{(x^2 + 1)[-2(x^2 + 1) + 8x^2]}{(x^2 + 1)^4} = \frac{(x^2 + 1)(6x^2 - 2)}{(x^2 + 1)^4}$$

$$= \frac{2(3x^2 - 1)}{(x^2 + 1)^3} \qquad \text{Cancel the common factors.}$$

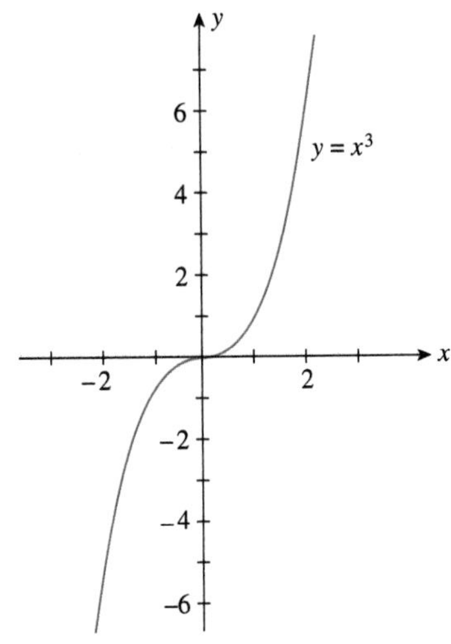

Figure 7.39
f has an inflection point at (1, 0).

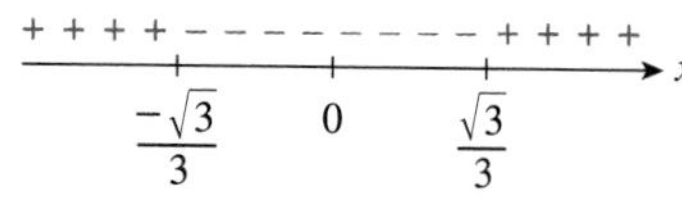

Figure 7.40
Sign diagram for f''

Observe that f'' is continuous everywhere and is zero if

$$3x^2 - 1 = 0$$

$$x^2 = \frac{1}{3}$$

or $x = \pm\sqrt{3}/3$. The sign diagram for f'' is shown in Figure 7.40. From the sign diagram for f'', we see that f is concave upward on $(-\infty, \sqrt{3}/3) \cup (\sqrt{3}/3, \infty)$ and concave downward on $(\sqrt{3}/3, \sqrt{3}/3)$. Also, observe that $f''(x)$ changes sign as we move across the points $x = -\sqrt{3}/3$ and $x = \sqrt{3}/3$. Since

$$f\left(-\frac{\sqrt{3}}{3}\right) = \frac{1}{\frac{1}{3}+1} = \frac{3}{4} \quad \text{and} \quad f\left(\frac{\sqrt{3}}{3}\right) = \frac{3}{4}$$

we see that the points $(-\sqrt{3}/3, 3/4)$ and $(-\sqrt{3}/3, 3/4)$ are inflection points of f. The graph of f is shown in Figure 7.41.

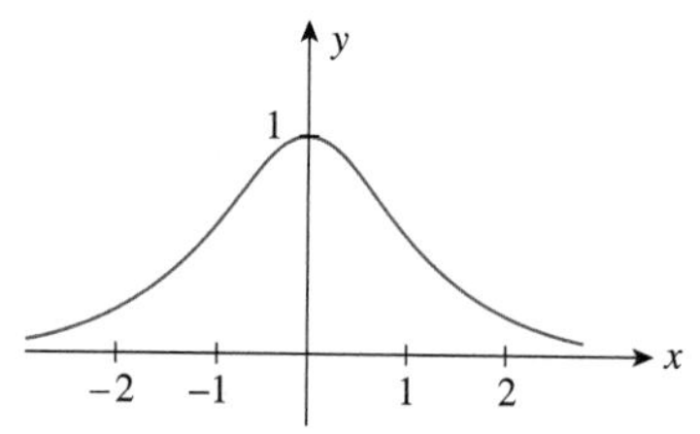

Figure 7.41
The graph of $f(x) = \dfrac{1}{x^2+1}$ is concave upward on $(-\infty, -\sqrt{3}/3) \cup (\sqrt{3}/3, \infty)$ and concave downward on $(-\sqrt{3}/3, \sqrt{3}/3)$.

GROUP DISCUSSION

1. Suppose $(c, f(c))$ is an inflection point of f. Can you conclude that f has no relative extremum at $x = c$? Explain your answer.
2. True or false: A polynomial function of degree 3 has exactly one inflection point. **Hint:** Study the function $f(x) = ax^3 + bx^2 + cx + d$ $(a \neq 0)$.

Applications

Examples 6 and 7 illustrate familiar interpretations of the significance of the inflection point of a function.

Example 6

Effect of Advertising on Sales The total sales S (in thousands of dollars) of Arctic Air Corporation, a manufacturer of automobile air conditioners, is related to the amount of money x (in thousands of dollars) the company spends on advertising its products by the formula

$$S = -0.01x^3 + 1.5x^2 + 200 \qquad (0 \leq x \leq 100)$$

Find the inflection point of the function S.

Solution

The first two derivatives of S are given by

$$S' = -0.03x^2 + 3x$$

$$S'' = -0.006x + 3$$

Setting $S'' = 0$ gives $x = 50$ as the only candidate for an inflection point of S. Moreover, since

$$S'' > 0 \quad \text{for} \quad x < 50$$

and

$$S'' < 0 \quad \text{for} \quad x > 50$$

the point (50, 2700) is an inflection point of the function S. The graph of S appears in Figure 7.42. Notice that this is the graph of the function we discussed earlier.

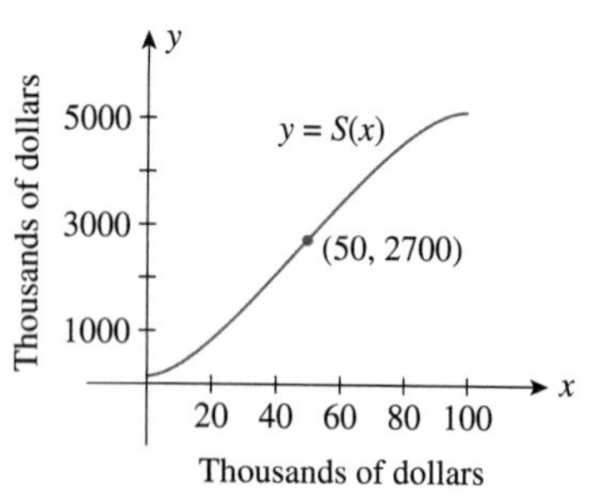

Figure 7.42
The graph of $S(x)$ has a point of inflection at (50, 2700).

Example 7

Consumer Price Index An economy's consumer price index (CPI) is described by the function

$$I(t) = -0.2t^3 + 3t^2 + 100 \qquad (0 \le t \le 9)$$

where $t = 0$ corresponds to the year 1995. Find the point of inflection of the function I and discuss its significance.

Solution

The first two derivatives of I are given by

$$I'(t) = -0.6t^2 + 6t$$
$$I''(t) = -1.2t + 6 = -1.2(t - 5)$$

Setting $I''(t) = 0$ gives $t = 5$ as the only candidate for an inflection point of I. Next, we observe that

$$I'' > 0 \quad \text{for} \quad t < 5$$
$$I'' < 0 \quad \text{for} \quad t > 5$$

so the point (5, 150) is an inflection point of I. The graph of I is sketched in Figure 7.43.

Since the second derivative of I measures the rate of change of the inflation rate, our computations reveal that the rate of inflation had in fact peaked at $t = 5$. Thus, relief actually began at the beginning of 2000.

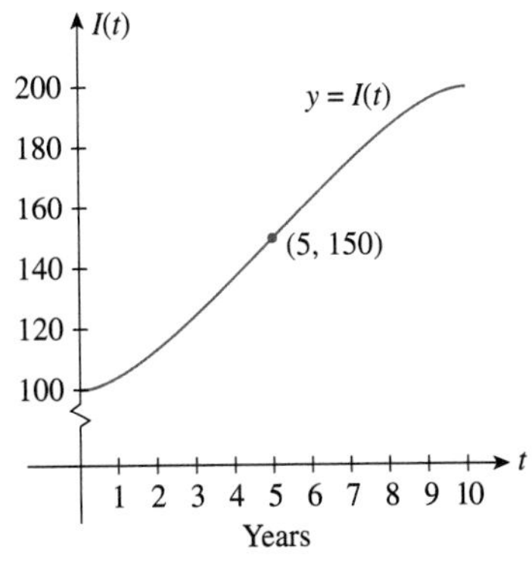

Figure 7.43
The graph of $I(t)$ has a point of inflection at (5, 150)

The Second Derivative Test

We now show how the second derivative f'' of a function f can be used to help us determine whether a critical point of f is a relative extremum of f. Figure 7.44a shows the graph of a function that has a relative maximum at $x = c$. Observe that f is concave downward at that point. Similarly, Figure 7.44b shows that at a relative minimum of f the graph is concave upward. But from our previous work we know that f is concave downward at $x = c$ if $f''(c) < 0$ and f is concave upward at $x = c$ if $f''(c) > 0$. These observations suggest the following alternative procedure for determining whether a critical point of f gives rise to a relative extremum of f. This result is called the second derivative test and is applicable when f'' exists.

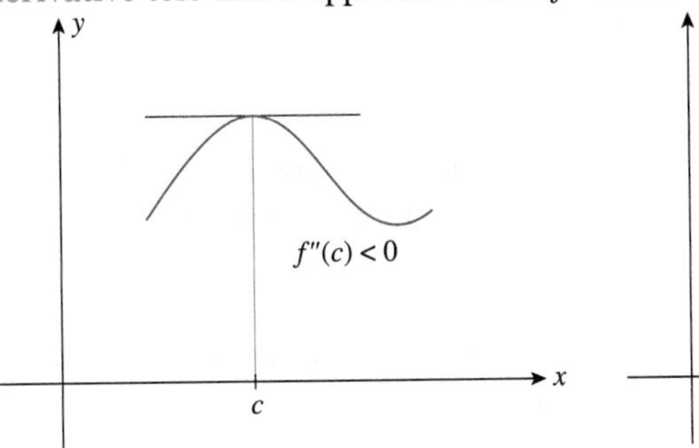

(a) f has a relative maximum at $x = c$.

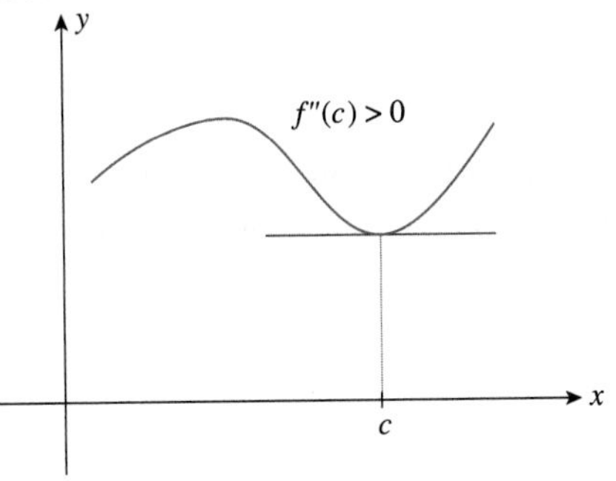

(b) f has a relative minimum at $x = c$.

Figure 7.44

The Second Derivative Test

1. Compute $f'(x)$ and $f''(x)$.
2. Find all the critical points of f at which $f'(x) = 0$.
3. Compute $f''(c)$ for each such critical point c.
   a. If $f''(c) < 0$, then f has a relative maximum at c.
   b. If $f''(c) > 0$, then f has a relative minimum at c.
   c. If $f''(c) = 0$, the test tells us nothing; c must be checked some other way.

Remark

As stated in step 3c, the second derivative test does not yield a conclusion if $f''(c) = 0$ or if $f''(c)$ does not exist. In other words, $x = c$ may give rise to a relative extremum or an inflection point (see Exercise 92, page 368). In such cases, you should revert to the first derivative test. ◀

Example 8

Determine the relative extrema of the function

$$f(x) = x^3 - 3x^2 - 24x + 32$$

using the second derivative test. (See Example 7, Section 7.1.)

Solution

We have

$$f'(x) = 3x^2 - 6x - 24 = 3(x + 2)(x - 4)$$

so $f'(x) = 0$ gives $x = -2$ and $x = 4$, the critical points of f, as in Example 7. Next, we compute

$$f''(x) = 6x - 6 = 6(x - 1)$$

Since

$$f''(-2) = 6(-2 - 1) = -18 < 0$$

the second derivative test implies that $f(-2) = 60$ is a relative maximum of f. Also,

$$f''(4) = 6(4 - 1) = 18 > 0$$

and the second derivative test implies that $f(4) = -48$ is a relative minimum of f, which confirms the results obtained earlier.

When $P(t)$ gives the size of a population at time t, assuming no individuals will be harvested from the population, then the best time to start harvesting is when the slope of the graph of $P(t)$ is at a maximum. This is because new individuals are entering the population at the greatest rate at that point. This means that maximizing $P'(t)$ yields the optimal time to begin harvesting a population. In order to maximize $P'(t)$ we need to find critical points of $P'(t)$ by solving the equation

$$P''(t) = 0$$

In this kind of problem we use the second derivative to maximize the first derivative. Optimal harvesting times of a population model $P(t)$ are at inflection points of $P(t)$.

Example 9

Optimal Harvesting Time Suppose that the population, in thousands of fish, t months after stocking a breeding pond is given by

$$P(t) = \frac{13t^2 + 24}{t^2 + 3}$$

for $t \geq 0$. Find the optimal time to begin harvesting by maximizing $P'(t)$.

Solution

Begin by computing $P'(t)$ with the quotient rule.

$$P'(t) = \frac{(t^2 + 3) \cdot 26t - (13t^2 + 24) \cdot 2t}{(t^2 + 3)^2}$$

$$= \frac{26t^3 + 78t - 26t^3 - 48t}{(t^2 + 3)^2}$$

$$= \frac{30t}{(t^2 + 3)^2}$$

Next, we need the derivative of $P'(t)$.

$$P''(t) = \frac{(t^2 + 3)^2 \cdot 30 - 30t \cdot 2(t^2 + 3) \cdot (2t)}{(t^2 + 3)^4}$$

$$= \frac{(t^2 + 3)^{\cancel{2}} \cdot 30 - 30t \cdot 2\cancel{(t^2 + 3)} \cdot (2t)}{(t^2 + 3)^{\cancel{4}3}}$$

$$= \frac{(t^2 + 3) \cdot 30 - 120t^2}{(t^2 + 3)^3}$$

$$= \frac{90 - 90t^2}{(t^2 + 3)^3}$$

We now need the critical points of $P'(t)$ so we solve $P''(t) = 0$. The denominator of $P''(t)$ is $(t^2 + 3)^3$; notice $t^2 + 3$ is strictly positive and so $P''(t) = 0$ only when

$$90 - 90t^2 = 0.$$

Solve.

$$\begin{aligned} 90 - 90t^2 &= 0 \\ 90(1 - t^2) &= 0 \\ 90(1 - t)(1 + t) &= 0 \\ t &= \pm 1 \end{aligned}$$

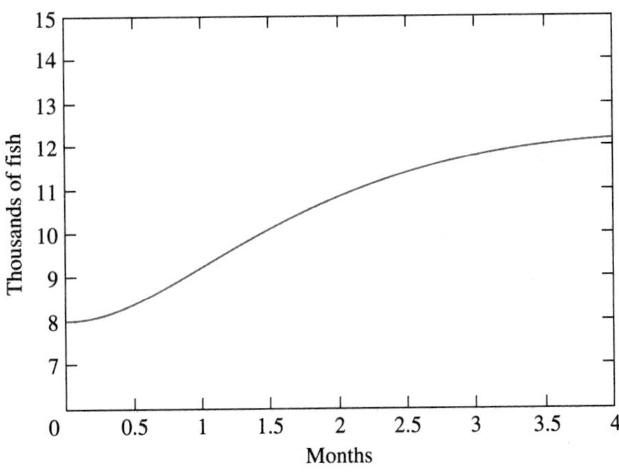

Figure 7.45
Thousands of fish after t months.

The problem states the model is only good for $t \geq 0$ and so $t = 1$ month is our sole candidate point. Examining the graph of $P(t)$ (Figure 7.45) we see that this is a point where the slope is steepest. We conclude that harvesting should commence one month after the breeding pond is started.

Finding Optimal Harvesting Times

Suppose that $P(t)$ is a function that gives a population at a time $t \geq 0$ if no individuals are harvested. It is assumed that $P(t)$ is an increasing function (otherwise it is not a population that anyone should be harvesting from). To find the best time to harvest we want the maximum rate of increase—the point where $P'(t)$ is largest. To find it we follow these steps.

1. Compute $P'(t)$ and $P''(t)$.
2. Solve for those t so that $P''(t) = 0$.
3. For each candidate time found in step (ii), plug the time into $P''(t)$ and take the time that yields the largest increase. This is the optimal time to begin harvesting.

It is important to note that the population model used to find an optimal harvesting time assumes that no harvesting will take place. It gives the behaviour of an undisturbed population. Harvesting is a large disturbance. The amount of harvesting that takes place is such that the number of creatures in the population stays near the value that yields the maximum rate of $P'(t)$. New creatures come from those you already have by biological reproduction. When a population is small there are few creatures breeding. When a population is large there is competition for food, space, or other resources that reduces the rate at which the population can grow. This means that at some point in between a small and a large population, these exists a best population size for producing new creatures—and this is the optimal harvesting point. Since population models are stated as functions of time, we find an optimal harvesting time, but we could convert this into an optimal population size for harvesting by simply computing $P(t)$ for the optimal harvesting time t.

GROUP DISCUSSION

Suppose a function f has the following properties:

1. $f''(x) > 0$ for all x in an interval (a, b).
2. There is a point c between a and b such that $f'(c) = 0$.

What special property can you ascribe to the point $(c, f(c))$? Answer the question if Property 1 is replaced by the property that $f''(x) < 0$ for all x in (a, b).

Comparing the First and Second Derivative Tests

Notice that both the first derivative test and the second derivative test are used to classify the critical points of f. What are the pros and cons of the two tests? Since the second derivative test is applicable only when f'' exists, it is less versatile than the first derivative test. For example, it cannot be used to locate the relative minimum $f(0) = 0$ of the function $f(x) = x^{2/3}$.

Furthermore, the second derivative test is inconclusive when f'' is equal to zero at a critical point of f, whereas the first derivative test always yields positive conclusions. The second derivative test is also inconvenient to use when f'' is difficult to compute. On the plus side, if f'' is computed easily, then we use the second derivative test since it involves just the evaluation of f'' at the critical point(s) of f. Also, the conclusions of the second derivative test are important in theoretical work.

We close this section by summarizing the different roles played by the first derivative f' and the second derivative f'' of a function f in determining the properties of the graph of f. The first derivative f' tells us where f is increasing and where f is decreasing, whereas the second derivative f'' tells us where f is concave upward and where f is concave downward. These different properties of f are reflected by the signs of f' and f'' in the interval of interest. The following table shows the general characteristics of the function f for various possible combinations of the signs of f' and f'' in the interval (a, b).

Signs of f' and f''	Properties of the Graph of f	General Shape of the Graph of f
$f'(x) > 0$ $f''(x) > 0$	f increasing f concave upward	
$f'(x) > 0$ $f''(x) < 0$	f increasing f concave downward	
$f'(x) < 0$ $f''(x) > 0$	f decreasing f concave upward	
$f'(x) < 0$ $f''(x) < 0$	f decreasing f concave downward	

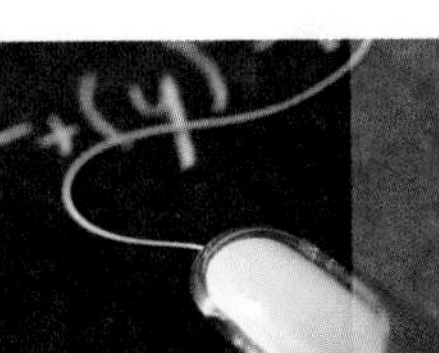

Self-Check Exercises 7.2

1. Determine where the function $f(x) = 4x^3 - 3x^2 + 6$ is concave upward and where it is concave downward.

2. Using the second derivative test, if applicable, find the relative extrema of the function $f(x) = 2x^3 - \frac{1}{2}x^2 - 12x - 10$.

3. A certain country's gross domestic product (GDP) (in millions of dollars) in year t is described by the function

$$G(t) = -2t^3 + 45t^2 + 20t + 6000 \qquad (0 \le t \le 11)$$

where $t = 0$ corresponds to the beginning of 1992. Find the inflection point of the function G and discuss its significance.

Solutions to Self-Check Exercises 7.2 can be found on page 371.

7.2 Exercises

In Exercises 1–8, you are given the graph of a function f. Determine the intervals where f is concave upward and where it is concave downward. Also, find all inflection points of f, if any.

1.

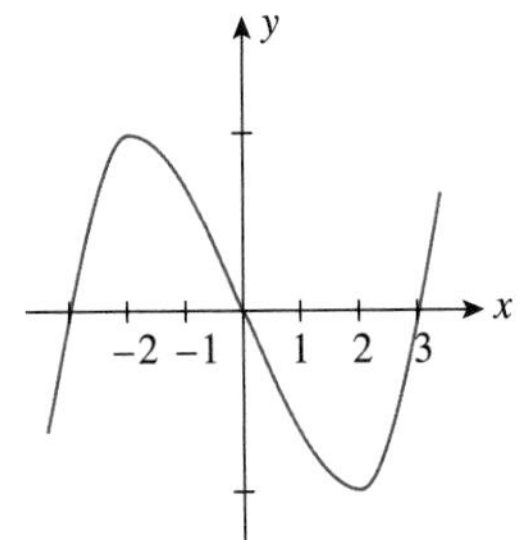

2.

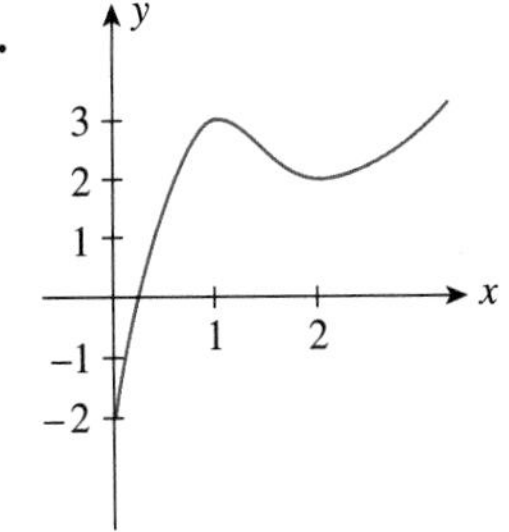

3.

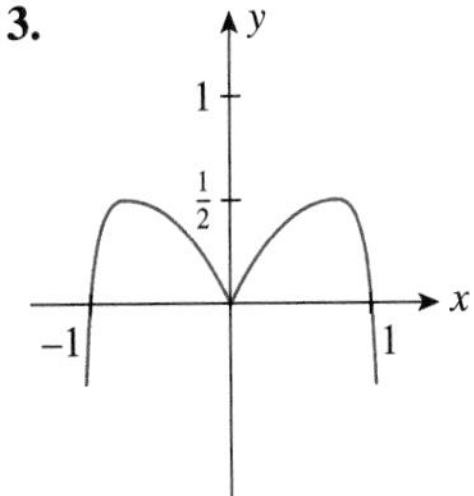

4.

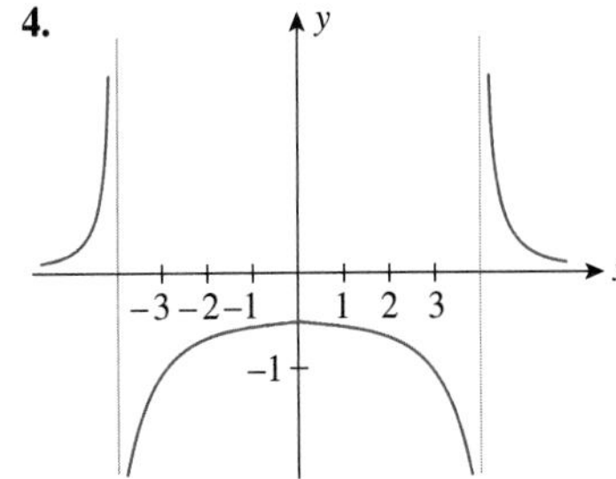

5.

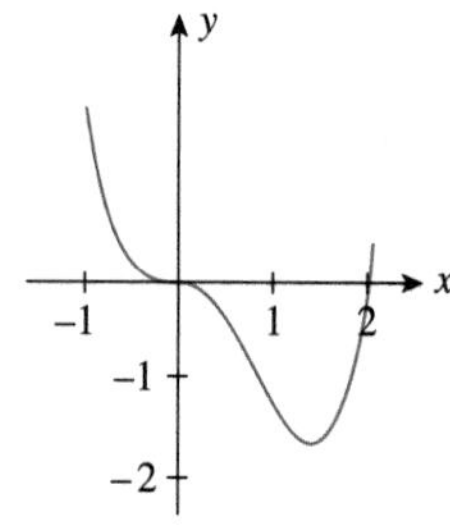

6.

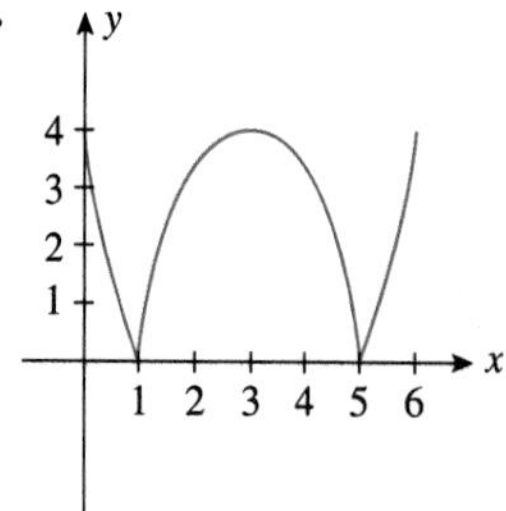

7.

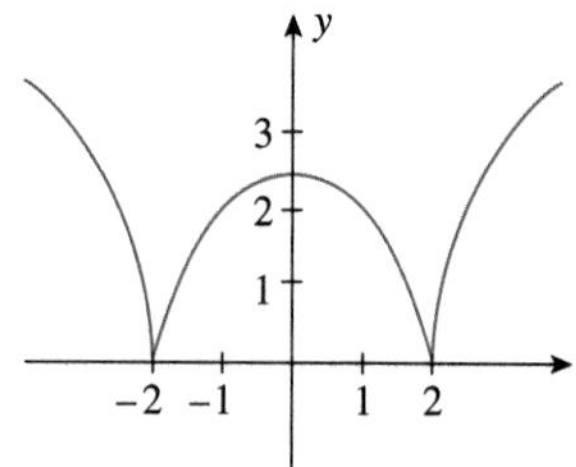

8.

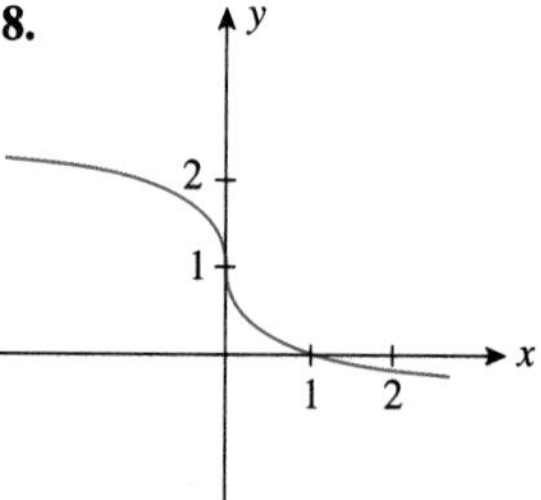

In Exercises 9–12, determine which graph—(a), (b), or (c)—is the graph of the function f with the specified properties. Explain.

9. $f(2) = 1, f'(2) > 0$, and $f''(2) < 0$

a.

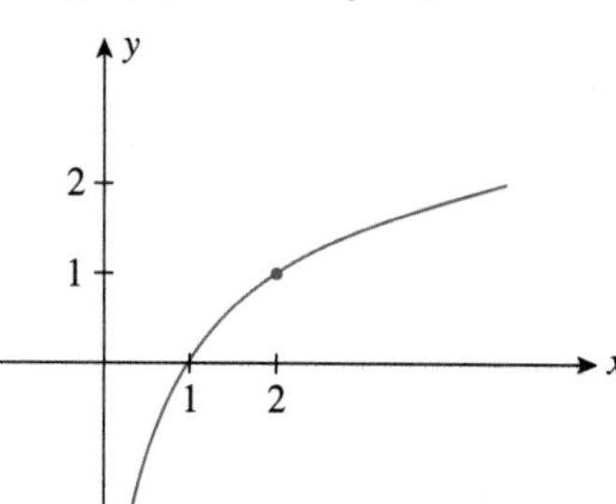

b.

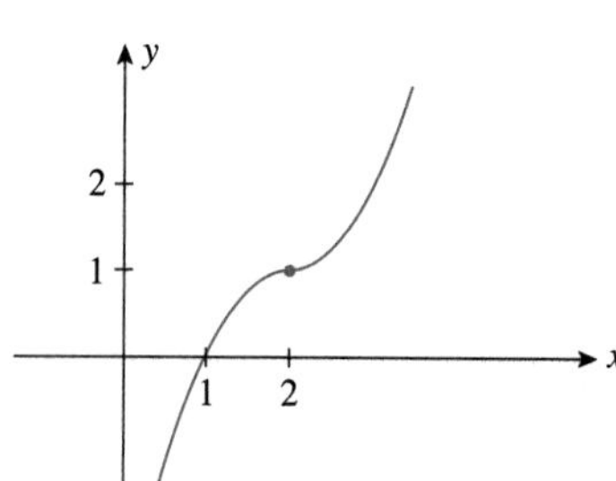

c.

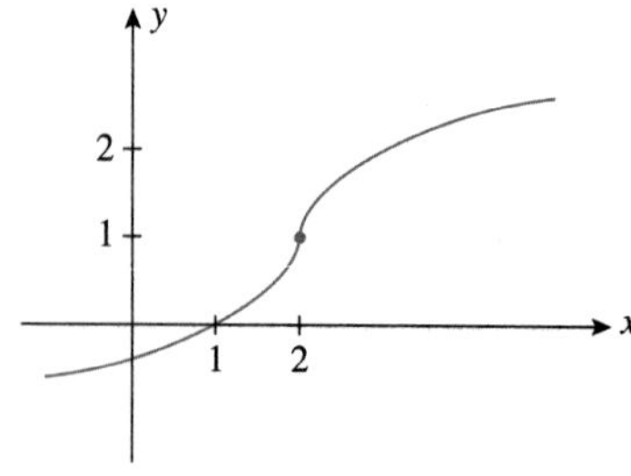

10. $f(1) = 2, f'(x) > 0$ on $(-\infty, 1) \cup (1, \infty)$, and $f''(1) = 0$

a.

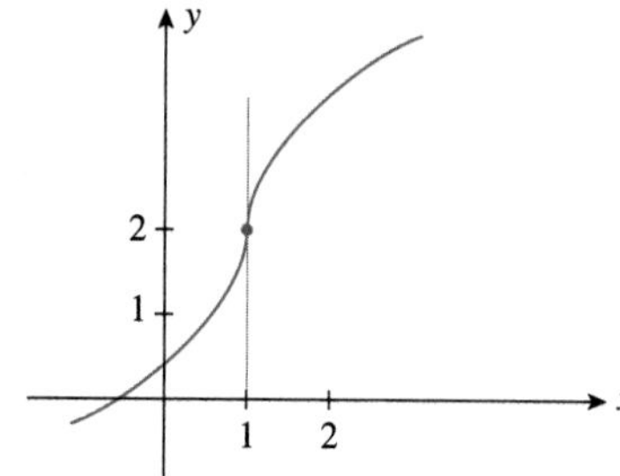

b.

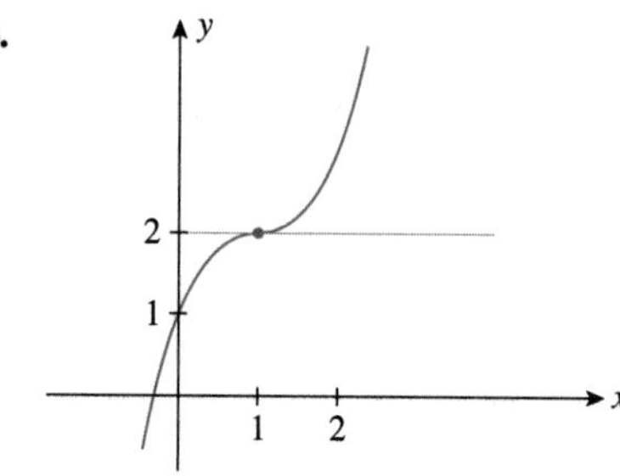

c.

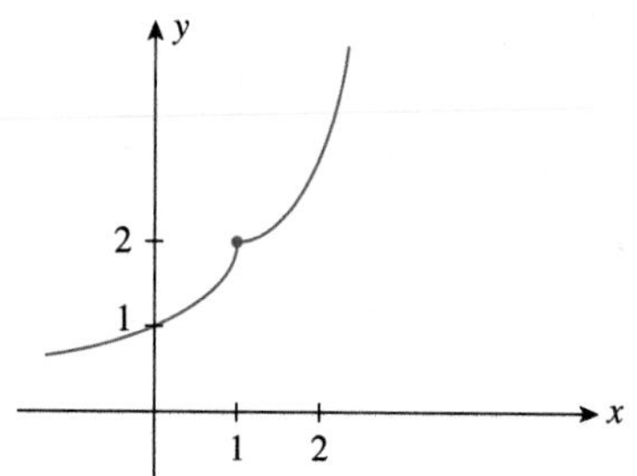

11. $f'(0)$ is undefined, f is decreasing on $(-\infty, 0)$, f is concave downward on $(0, 3)$, and f has an inflection point at $x = 3$.

a.

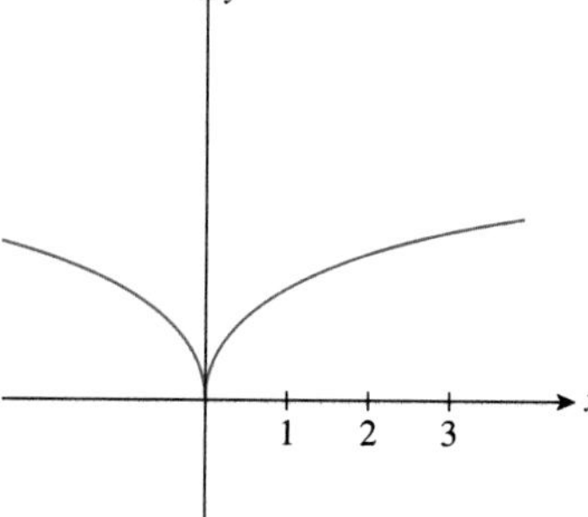

b.

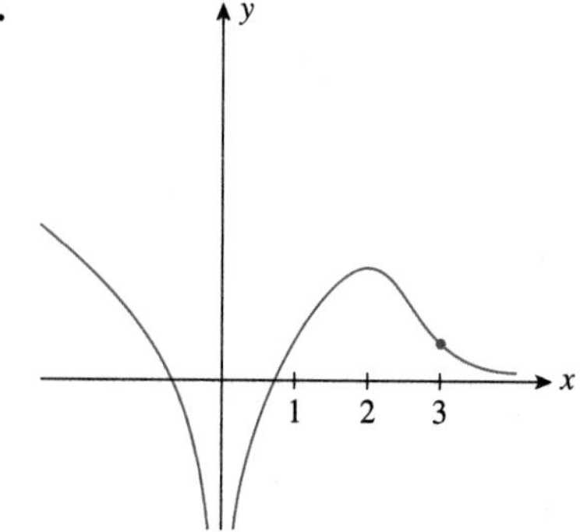

c.

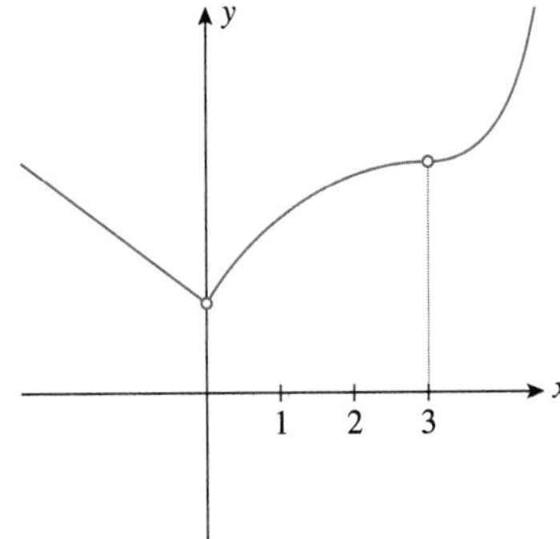

12. f is decreasing on $(-\infty, 2)$ and increasing on $(2, \infty)$, f is concave upward on $(1, \infty)$, and f has inflection points at $x = 0$ and $x = 1$.

a.

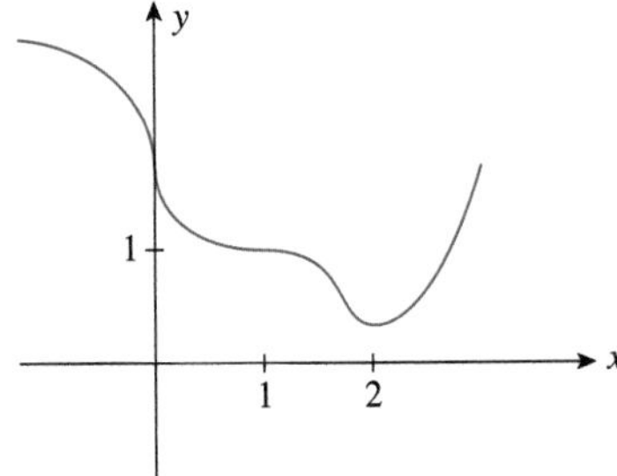

b.

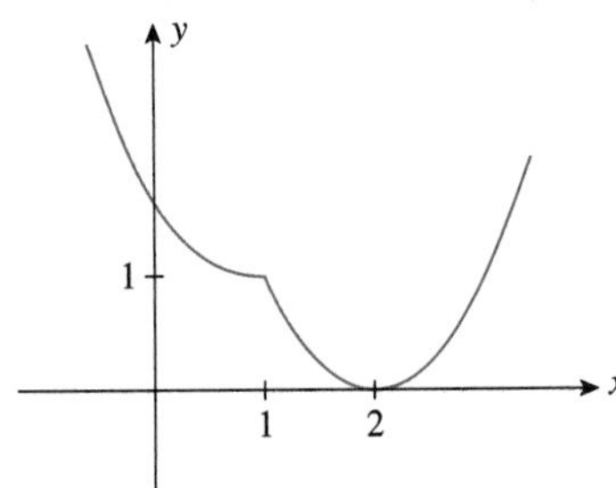

c.

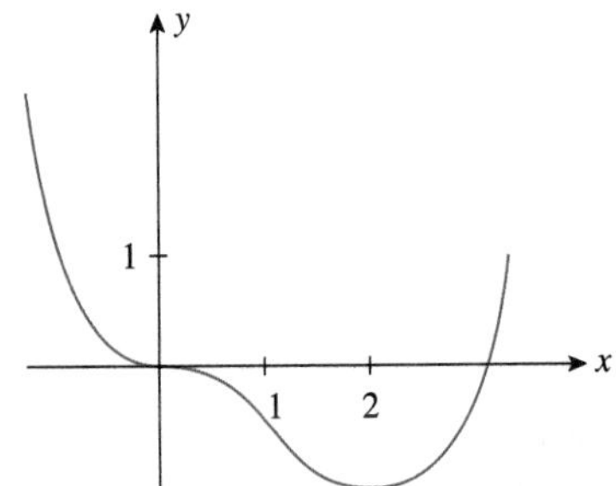

In Exercises 13–18, show that the function is concave upward wherever it is defined.

13. $f(x) = 4x^2 - 12x + 7$

14. $g(x) = x^4 + \frac{1}{2}x^2 + 6x + 10$

15. $f(x) = \frac{1}{x^4}$

16. $g(x) = -\sqrt{4 - x^2}$

17. $f(x) = e^x$

18. $g(x) = e^{x+2}$

In Exercises 19–46, determine where the function is concave upward and where it is concave downward.

19. $f(x) = 2x^2 - 3x + 4$

20. $g(x) = -x^2 + 3x + 4$

21. $f(x) = x^3 - 1$

22. $g(x) = x^3 - x$

23. $f(x) = x^4 - 6x^3 + 2x + 8$

24. $f(x) = 3x^4 - 6x^3 + x - 8$

25. $f(x) = x^{4/7}$

26. $f(x) = \sqrt[3]{x}$

27. $f(x) = \sqrt{4 - x}$

28. $g(x) = \sqrt{x - 2}$

29. $f(x) = \frac{1}{x - 2}$

30. $g(x) = \frac{x}{x + 1}$

31. $f(x) = \frac{1}{2 + x^2}$

32. $g(x) = \frac{x}{1 + x^2}$

33. $h(t) = \frac{t^2}{t - 1}$

34. $f(x) = \frac{x + 1}{x - 1}$

35. $g(x) = x + \frac{1}{x^2}$

36. $h(r) = -\frac{1}{(r - 2)^2}$

37. $g(t) = (2t - 4)^{1/3}$

38. $f(x) = (x - 2)^{2/3}$

39. $f(s) = e^{-s}$

40. $f(s) = e^{s^2 - 9}$

41. $g(x) = \ln(x)$

42. $f(x) = \ln(x^2 + 1)$

43. $f(x) = \sin(x)$

44. $f(x) = \cos(x)$

45. $f(x) = \sin(2x)$

46. $h(x) = \arctan(x)$

In Exercises 47–62, find the inflection points, if any, of each function.

47. $f(x) = x^3 - 2$

48. $g(x) = x^3 - 6x$

49. $f(x) = 6x^3 - 18x^2 + 12x - 15$

50. $g(x) = 2x^3 - 3x^2 + 18x - 8$

51. $f(x) = 3x^4 - 4x^3 + 1$

52. $f(x) = x^4 - 2x^3 + 6$

53. $g(t) = \sqrt[3]{t}$

54. $f(x) = \sqrt[5]{x}$

55. $f(x) = (x - 1)^3 + 2$

56. $f(x) = (x - 2)^{4/3}$

57. $f(x) = \frac{2}{1 + x^2}$

58. $f(x) = 2 + \frac{3}{x}$

59. $f(t) = t \cdot e^{-4t}$

60. $f(t) = x^2 \cdot e^t$

61. $f(a) = \cos(3a)$

62. $f(x) = \sin(x) - \cos(x)$

In Exercises 63–82, find the relative extrema, if any, of each function. Use the second derivative test, if applicable.

63. $f(x) = -x^2 + 2x + 4$

64. $g(x) = 2x^2 + 3x + 7$

65. $f(x) = 2x^3 + 1$

66. $g(x) = x^3 - 6x$

67. $f(x) = \frac{1}{3}x^3 - 2x^2 - 5x - 10$

68. $f(x) = 2x^3 + 3x^2 - 12x - 4$

69. $g(t) = t + \frac{9}{t}$

70. $f(t) = 2t + \frac{3}{t}$

71. $f(x) = \frac{x}{1 - x}$

72. $f(x) = \frac{2x}{x^2 + 1}$

73. $f(t) = t^2 - \frac{16}{t}$

74. $g(x) = x^2 + \frac{2}{x}$

75. $g(s) = \frac{s}{1 + s^2}$

76. $g(x) = \frac{1}{1 + x^2}$

77. $f(x) = \frac{x^4}{x - 1}$

78. $f(x) = \frac{x^2}{x^2 + 1}$

79. $h(s) = s \cdot e^{-s/4}$

80. $f(x) = x^2 \cdot e^{1-x}$

81. $g(t) = 2 - \sin(2t)$

82. $f(x) = \cos\left(x + \frac{\pi}{3}\right)$

In Exercises 83–88, sketch the graph of a function having the given properties.

83. $f(2) = 4, f'(2) = 0, f''(x) < 0$ on $(-\infty, \infty)$

84. $f(2) = 2, f'(2) = 0, f'(x) > 0$ on $(-\infty, 2), f'(x) > 0$ on $(2, \infty)$, $f''(x) < 0$ on $(-\infty, 2), f''(x) > 0$ on $(2, \infty)$

85. $f(-2) = 4, f(3) = -2, f'(-2) = 0, f'(3) = 0, f'(x) > 0$ on $(-\infty, -2) \cup (3, -\infty), f'(x) < 0$ on $(-2, 3)$, inflection point at $(1, 1)$

86. $f(0) = 0, f'(0)$ does not exist, $f''(x) < 0$ if $x \neq 0$

87. $f(0) = 1, f'(0) = 0, f(x) > 0$

on $(-\infty, \infty), f''(x) < 0$ on $(-\sqrt{2}/2, \sqrt{2}/2), f''(x) > 0$ on $(-\infty, -\sqrt{2}/2) \cup (\sqrt{2}/2, \infty)$

88. f has domain

$[-1, 1], f(-1) = -1, f(-\frac{1}{2}) = -2, f'(-\frac{1}{2})$

$= 0, f''(x) > 0$ on $(-1, 1)$

89. In the following figure, water is poured into the vase at a constant rate (in appropriate units), and the water level rises to a height of $f(t)$ units at time t as measured from the base of the vase. The graph of f follows. Explain the shape of the curve in terms of its concavity. What is the significance of the inflection point?

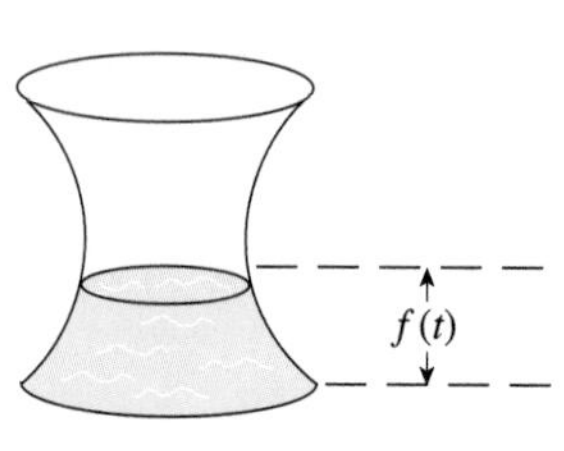

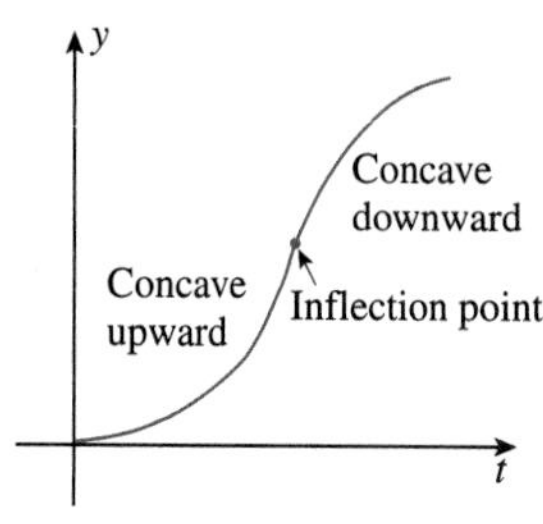

90. In the following figure, water is poured into an urn at a constant rate (in appropriate units), and the water level rises to a height of $f(t)$ units at time t as measured from the base of the urn. Sketch the graph of f and explain its shape, indicating where it is concave upward and concave downward. Indicate the inflection point on the graph and explain its significance.
Hint: Study Exercise 88.

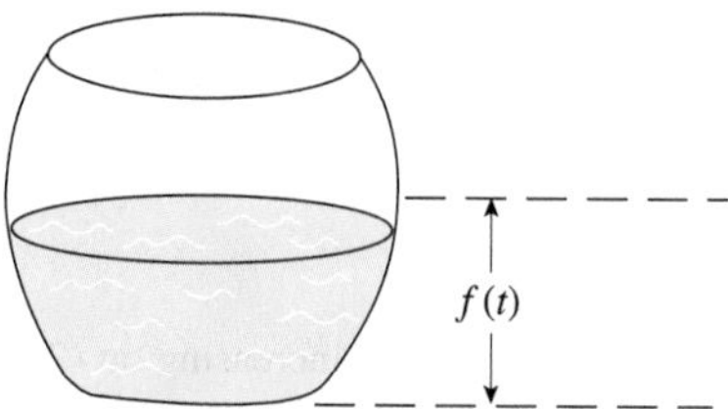

91. Show that the quadratic function

$$f(x) = ax^2 + bx + c \qquad (a \neq 0)$$

is concave upward if $a > 0$ and concave downward if $a < 0$. Thus, by examining the sign of the coefficient of x^2, one can tell immediately whether the parabola opens upward or downward.

92. Consider the functions $f(x) = x^3$, $g(x) = x^4$, and $h(x) = -x^4$.
  a. Show that $x = 0$ is a critical point of each of the functions f, g, and h.
  b. Show that the second derivative of each of the functions f, g, and h equals zero at $x = 0$.
  c. Show that f has neither a relative maximum nor a relative minimum at $x = 0$, that g has a relative minimum at $x = 0$, and that h has a relative maximum at $x = 0$.

93. **Spread of a Rumour** Initially, a handful of students heard a rumour on campus. The rumour spread and, after t hours, the number had grown to $N(t)$. The graph of the function N is shown in the following figure:

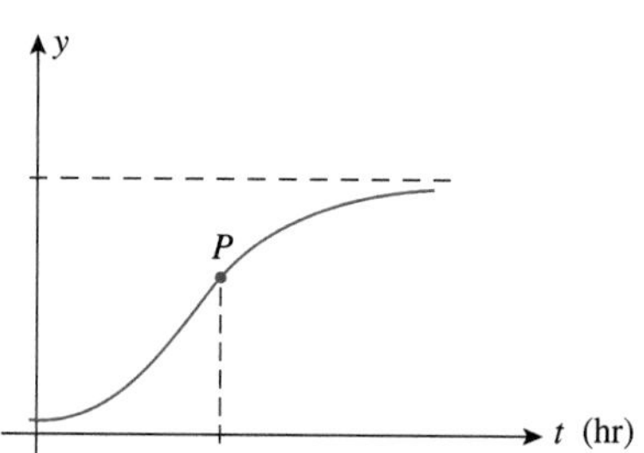

Describe the spread of the rumour in terms of the speed at which it was spread. In particular, explain the significance of the inflection point P of the graph of N.

94. **Demand for RNs** The following graph gives the total number of help-wanted ads for RNs (registered nurses) in 22 cities over the last 12 months as a function of time t (t measured in months).
  a. Explain why $N'(t)$ is positive on the interval $(0, 12)$.
  b. Determine the signs of $N''(t)$ on the interval $(0, 6)$ and the interval $(6, 12)$.
  c. Interpret the results of part (b).

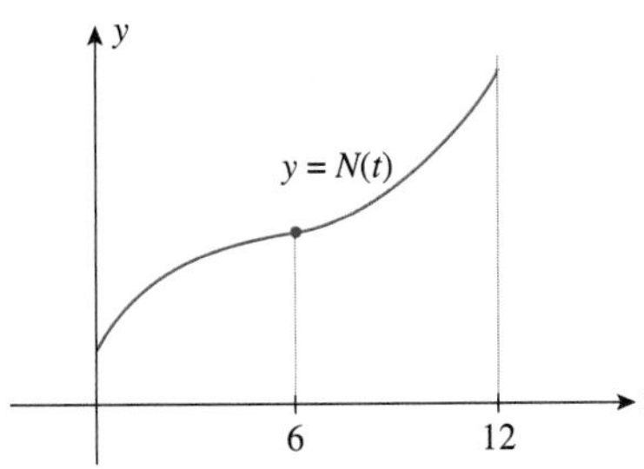

Business and Economics Applications

95. Effect of Advertising on Bank Deposits The following graphs were used by the CEO of the Madison Savings Bank to illustrate what effect a projected promotional campaign would have on its deposits over the next year. The functions D_1 and D_2 give the projected amount of money on deposit with the bank over the next 12 months with and without the proposed promotional campaign, respectively.

a. Determine the signs of $D_1'(t)$, $D_2'(t)$, $D_1''(t)$, and $D_2''(t)$ on the interval (0, 12).

b. What can you conclude about the rate of change of the growth rate of the money on deposit with the bank with and without the proposed promotional campaign?

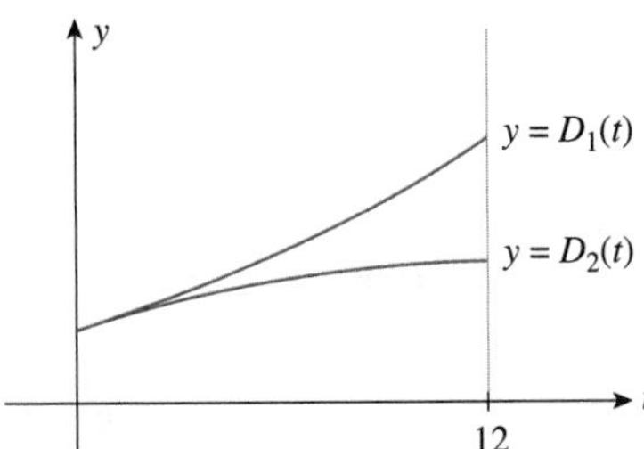

96. Effect of Budget Cuts on Drug-Related Crimes The graphs below were used by a police commissioner to illustrate what effect a budget cut would have on crime in the city. The number $N_1(t)$ gives the projected number of drug-related crimes in the next 12 months. The number $N_2(t)$ gives the projected number of drug-related crimes in the same time frame if next year's budget is cut.

a. Explain why $N_1'(t)$ and $N_2'(t)$ are both positive on the interval (0, 12).

b. What are the signs of $N_1''(t)$ and $N_2''(t)$ on the interval (0, 12)?

c. Interpret the results of part (b).

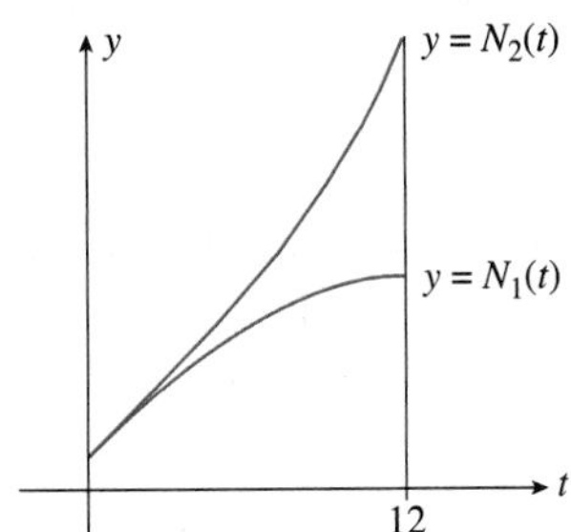

A calculator is recommended for Exercises 97–104.

97. Effect of Advertising on Hotel Revenue The total annual revenue R of the Miramar Resorts Hotel is related to the amount of money x the hotel spends on advertising its services by the function

$$R(x) = -0.003x^3 + 1.35x^2 + 2x + 8000 \qquad (0 \le x \le 400)$$

where both R and x are measured in thousands of dollars.

a. Find the interval where the graph of R is concave upward and the interval where the graph of R is concave downward. What is the inflection point of R?

b. Would it be more beneficial for the hotel to increase its advertising budget slightly when the budget is \$140,000 or when it is \$160,000?

98. Effect of Advertising on Sales The total sales S of Cannon Precision Instruments is related to the amount of money x that Cannon spends on advertising its products by the function

$$S(x) = -0.002x^3 + 0.6x^2 + x + 500 \qquad (0 < x \le 200)$$

where S and x are measured in thousands of dollars. Find the inflection point of the function S and discuss its significance.

99. Forecasting Profits As a result of increasing energy costs, the growth rate of the profit of the 4-year old Venice Glassblowing Company has begun to decline. Venice's management, after consulting with energy experts, decides to implement certain energy-conservation measures aimed at cutting energy bills. The general manager reports that, according to his calculations, the growth rate of Venice's profit should be on the increase again within 4 yr. If Venice's profit (in hundreds of dollars) x yr from now is given by the function

$$P(x) = x^3 - 9x^2 + 40x + 50 \qquad (0 \le x \le 8)$$

determine whether the general manager's forecast will be accurate.

Hint: Find the inflection point of the function P and study the concavity of P.

100. Worker Efficiency An efficiency study conducted for Elektra Electronics showed that the number of Space Commander walkie-talkies assembled by the average worker t hr after starting work at 8 A.M. is given by

$$N(t) = -t^3 + 6t^2 + 15t \qquad (0 \le t \le 4)$$

At what time during the morning shift is the average worker performing at peak efficiency?

101. Cost of Producing Calculators A subsidiary of Elektra Electronics manufactures programmable calculators. Management determines that the daily cost $C(x)$ (in dollars) of producing these calculators is

$$C(x) = 0.0001x^3 - 0.08x^2 + 40x + 5000$$

where x is the number of calculators produced. Find the inflection point of the function C and interpret your result.

102. Assembly Time of a Worker In the following graph, $N(t)$ gives the number of transistor radios assembled by the average worker by the tth hr, where $t = 0$ corresponds to 8 A.M. and $0 \le t \le 4$. The point P is an inflection point of N.

a. What can you say about the rate of change of the number of transistor radios assembled by the average worker between 8 A.M. and 10 A.M.? Between 10 A.M. and 12 A.M.?

b. At what time is the rate at which the transistor radios are being assembled by the average worker greatest?

103. **Flight of a Rocket** The altitude (in metres) of a rocket t seconds into flight is given by

$$s = f(t) = -\frac{1}{3}t^3 + 18t^2 + 160t + 2$$

Find the point of inflection of the function f and interpret your result. What is the maximum velocity attained by the rocket?

Biological and Life Sciences Applications

104. **Water Pollution** When organic waste is dumped into a pond, the oxidation process that takes place reduces the pond's oxygen content. However, given time, nature will restore the oxygen content to its natural level. In the following graph, $P(t)$ gives the oxygen content (as a percent of its normal level) t days after organic waste has been dumped into the pond. Explain the significance of the inflection point Q.

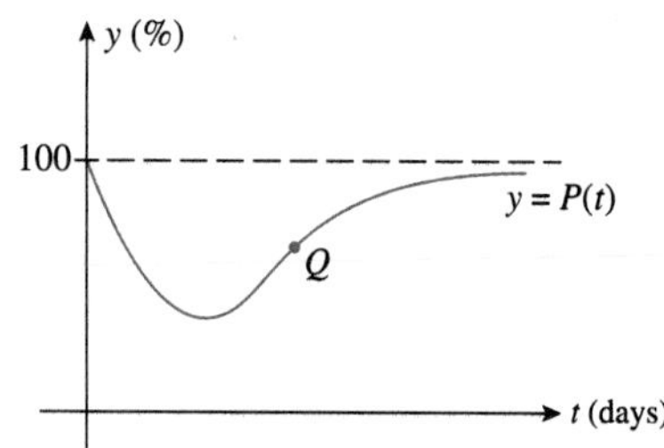

105. **Population Growth in Brampton-Springdale Riding** Brampton-Springdale Riding, an administrative district in Ontario, has experienced rapid growth recently. This growth is approximated by the function

$$P(t) = 4456t^3 - 8939t^2 + 23463t + 27328 \qquad (0 \leq t \leq 4)$$

where t is measured in decades, with $t = 0$ corresponding to the beginning of 1970.

a. Show that the population of Brampton-Springdale riding was always increasing over the time period in question.
Hint: Show that $P'(t) > 0$ for all t in the interval (0, 4).

b. Show that the population of Brampton-Springdale riding was increasing at the slowest pace some time toward the middle of August 1976.
Hint: Find the inflection point of P in the interval (0, 4).

106. **Air Pollution** The level of ozone, an invisible gas that irritates and impairs breathing, present in the atmosphere on a certain August day in the city of North York was approximated by

$$A(t) = 1.0974t^3 - 0.0915t^4 \qquad (0 \leq t \leq 11)$$

where $A(t)$ is measured in pollutant standard index (PSI) and t is measured in hours, with $t = 0$ corresponding to 7 A.M. Use the second derivative test to show that the function A has a relative maximum at approximately $t = 9$. Interpret your results.

107. **Rocket Takeoff** Suppose that a rocket one kilometre from a camera takes off vertically so that its height at a time t is 100 t^2 metres. Assume the camera tracks the rocket. Find the following

a. A formula for the rate of rotation of the camera in radians/second.

b. The time after takeoff at which the camera is spinning fastest.

c. The maximum speed of the camera in radians per second.

Problems 108–111 are optimal harvesting problems.

108. **Yeast Production** Suppose that a type of yeast used to produce medical compounds, if permitted to grow indefinitely in a growth tank, has a total weight of

$$W(t) = \frac{265t^2}{t^2 + 2}$$

after t weeks if left undisturbed. At what time t is it best to start removing yeast from the tank?

109. **Pike Production** If the number of pike in a lake t years after they are introduced is found to be modelled by the equation

$$P(t) = 6500\frac{e^t}{2 + e^t}$$

then in what year is it best to permit fishing for the pike to start if we are maximizing the production of pike by the lake ecosystem?

110. **Grass Harvesting** Suppose that a field of grass is being used to provide cellulose for a biofuel plant. If the mass of grass per square metre of the field is found to be

$$M(t) = \frac{6t^2 + 2}{t^2 + 1}$$

after t weeks of undisturbed growth, at what time should grass be harvested to maximize the production of biomass?

111. **Deer Population** A native deer species, hunted to extinction in South Carolina, is reintroduced from Canadian reserves of the species. The population's size t years after reintroduction is found to be close to

$$P(t) = \frac{1200t^2 + 4}{t^2 + 1}$$

Lacking a natural enemy, the deer population must be controlled by hunting. If we wish to maximize the number of deer that can be hunted while keeping the deer population relatively steady, at what population level of deer would hunting commence?
Hint: Find the optimal time first, then use the population formula to find what the population is at that time.

In Exercises 112–115, determine whether the statement is true or false. If it is true, explain why it is true. If it is false, give an example to show why it is false.

112. If the graph of f is concave upward on (a, b), then the graph of $-f$ is concave downward on (a, b).

113. If the graph of f is concave upward on (a, c) and concave downward on (c, b), where $a < c < b$, then f has an inflection point at $x = c$.

114. If $x = c$ is a critical point of f where $a < c < b$ and $f''(x) < 0$ on (a, b), then f has a relative maximum at $x = c$.

115. A polynomial function of degree n $(n \geq 3)$ can have at most $(n - 2)$ inflection points.

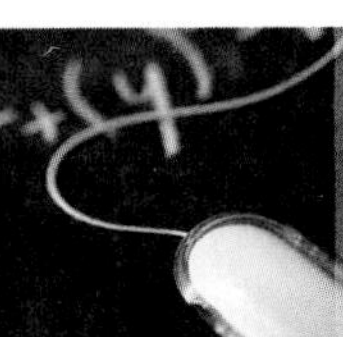

Solutions to Self-Check Exercises 7.2

1. We first compute

$$f'(x) = 12x^2 - 6x$$

$$f''(x) = 24x - 6 = 6(4x - 1)$$

Observe that f'' is continuous everywhere and has a zero at $x = \frac{1}{4}$. The sign diagram of f'' is shown in the accompanying figure.

- - - - - - 0 + + + + + +

0 $\frac{1}{4}$ x

From the sign diagram for f'', we see that f is concave upward on $(\frac{1}{4}, \infty)$ and concave downward on $(-\infty, \frac{1}{4})$.

2. First, we find the critical points of f by solving the equation

$$f'(x) = 6x^2 - x - 12 = 0$$

That is,

$$(3x + 4)(2x - 3) = 0$$

giving $x = -\frac{4}{3}$ and $x = \frac{3}{2}$. Next, we compute

$$f''(x) = 12x - 1$$

Since

$$f''\left(-\frac{4}{3}\right) = 12\left(-\frac{4}{3}\right) - 1 = -17 < 0$$

the second derivative test implies that $f(-\frac{4}{3}) = \frac{10}{27}$ is a relative maximum of f. Also,

$$f''\left(\frac{3}{2}\right) = 12\left(\frac{3}{2}\right) - 1 = 17 > 0$$

and we see that $f(\frac{3}{2}) = -\frac{179}{8}$ is a relative minimum.

3. We compute the second derivative of G. Thus,

$$G'(t) = -6t^2 + 90t + 20$$

$$G''(t) = -12t + 90$$

Now, G'' is continuous everywhere, and $G''(t) = 0$, where $t = \frac{15}{2}$, giving $t = \frac{15}{2}$ as the only candidate for an inflection point of G. Since $G''(t) > 0$ for $t < \frac{15}{2}$ and $G''(t) < 0$ for $t > \frac{15}{2}$, we see that the point $(\frac{15}{2}, \frac{15{,}675}{2})$ is an inflection point of G. The results of our computations tell us that the country's GDP was increasing most rapidly at the beginning of July 1999.

7.3 Curve Sketching

A Real-Life Example

As we have seen on numerous occasions, the graph of a function is a useful aid for visualizing the function's properties. From a practical point of view, the graph of a function also gives, at one glance, a complete summary of all the information captured by the function.

Consider, for example, the graph of the function giving the Toronto Stock Exchange Index (TSX) on Black Monday, October 19, 1987 (Figure 7.46). Here, $t = 0$ corresponds to 8:30 A.M., when the market was open for business, and $t = 7.5$ corresponds to 4 P.M., the closing time. The following information may be gleaned from studying the graph.

The graph is *decreasing* rapidly from $t = 0$ to $t = 1$, reflecting the sharp drop in the index in the first hour of trading. The point (1, 2047) is a *relative minimum* point of the function, and this turning point coincides with the start of an aborted recovery. The short-lived rally, represented by the portion of the graph that is *increasing* on the interval (1, 2), quickly fizzled out at $t = 2$ (10:30 A.M.). The *relative maximum* point (2, 2150) marks the highest point of the recovery. The function is decreasing in the rest of the interval. The point (4, 2006) is an *inflection point* of the function; it shows that there was a temporary respite at $t = 4$ (12:30 P.M.). However, selling pressure continued unabated, and the TSX continued to fall until the closing bell. Finally, the graph also shows that the index opened at the high of the day [$f(0) = 2247$ is the *absolute maximum* of the function] and closed at the low of the day [$f(\frac{15}{2}) = 1739$ is the *absolute minimum* of the function], a drop of 508 points!*

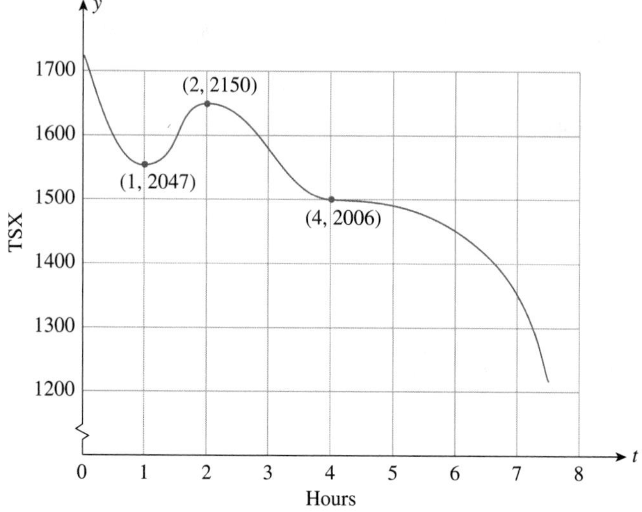

Figure 7.46
The Toronto Stock Exchange Index on Black Monday

Before we turn our attention to the actual task of sketching the graph of a function, let's look at some properties of graphs that will be helpful in this connection.

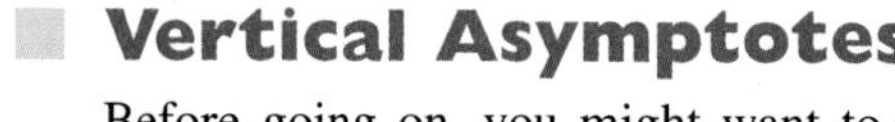

Vertical Asymptotes

Before going on, you might want to review the material on one-sided limits and the limit at infinity of a function (Sections 3.1 and 3.2).

Consider the graph of the function

$$f(x) = \frac{x + 1}{x - 1}$$

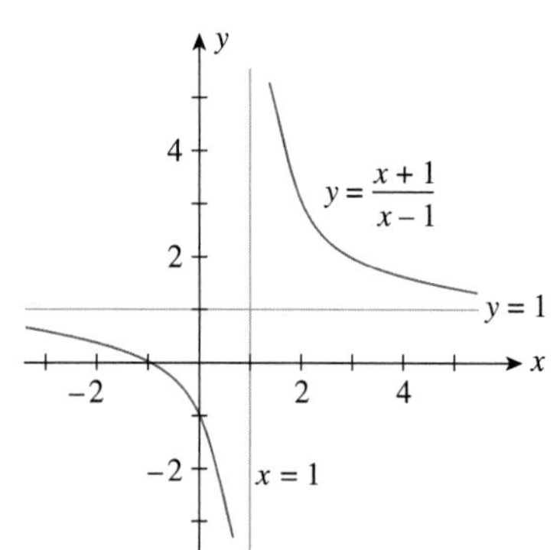

Figure 7.47
The graph of f has a vertical asymptote at $x = 1$.

shown in Figure 7.47. Observe that $f(x)$ increases without bound (tends to infinity) as x approaches $x = 1$ from the right; that is,

$$\lim_{x \to 1^+} \frac{x + 1}{x - 1} = \infty$$

You can verify this by taking a sequence of values of x approaching $x = 1$ from the right and looking at the corresponding values of $f(x)$.

Here is another way of looking at the situation: Observe that if x is a number that is a little larger than 1, then both $(x + 1)$ and $(x - 1)$ are positive, so $(x + 1)/(x - 1)$ is also positive. As x approaches $x = 1$, the numerator $(x + 1)$ approaches the number 2, but the denominator $(x - 1)$ approaches zero, so the quotient $(x + 1)/(x - 1)$ approaches infinity, as observed earlier. The line $x = 1$ is called a vertical asymptote of the graph of f.

For the function $f(x) = (x + 1)/(x - 1)$, you can show that

$$\lim_{x \to 1^-} \frac{x + 1}{x - 1} = -\infty$$

and this tells us how $f(x)$ approaches the asymptote $x = 1$ from the left.

More generally, we have the following definition:

*Absolute maxima and absolute minima of functions are covered in Section 7.4.

Vertical Asymptote

The line $x = a$ is a **vertical asymptote** of the graph of a function f if either

$$\lim_{x \to a^+} f(x) = \infty \quad \text{or} \quad -\infty$$

or

$$\lim_{x \to a^-} f(x) = \infty \quad \text{or} \quad -\infty$$

Remark

Although a vertical asymptote of a graph is not part of the graph, it serves as a useful aid for sketching the graph. ◀

For rational functions

$$f(x) = \frac{P(x)}{Q(x)}$$

there is a simple criterion for determining whether the graph of f has any vertical asymptotes.

Finding Vertical Asymptotes of Rational Functions

Suppose f is a rational function

$$f(x) = \frac{P(x)}{Q(x)}$$

where P and Q are polynomial functions. Then, the line $x = a$ is a vertical asymptote of the graph of f if $Q(a) = 0$ but $P(a) \neq 0$.

For the function

$$f(x) = \frac{x + 1}{x - 1}$$

considered earlier, $P(x) = x + 1$ and $Q(x) = x - 1$. Observe that $Q(1) = 0$ but $P(1) = 2 \neq 0$, so $x = 1$ is a vertical asymptote of the graph of f.

Example 1

Find the vertical asymptotes of the graph of the function

$$f(x) = \frac{x^2}{4 - x^2}$$

Solution

The function f is a rational function with $P(x) = x^2$ and $Q(x) = 4 - x^2$. The zeros of Q are found by solving

$$4 - x^2 = 0$$

—that is,

$$(2 - x)(2 + x) = 0$$

giving $x = -2$ and $x = 2$. These are candidates for the vertical asymptotes of the graph of f. Examining $x = -2$, we compute $P(-2) = (-2)^2 = 4 \neq 0$, and we see that $x = -2$ is indeed a vertical asymptote of the graph of f. Similarly, we find $P(2) = 2^2 = 4 \neq 0$, and so $x = 2$ is also a vertical asymptote of the graph of f. The graph of f sketched in Figure 7.48 confirms these results.

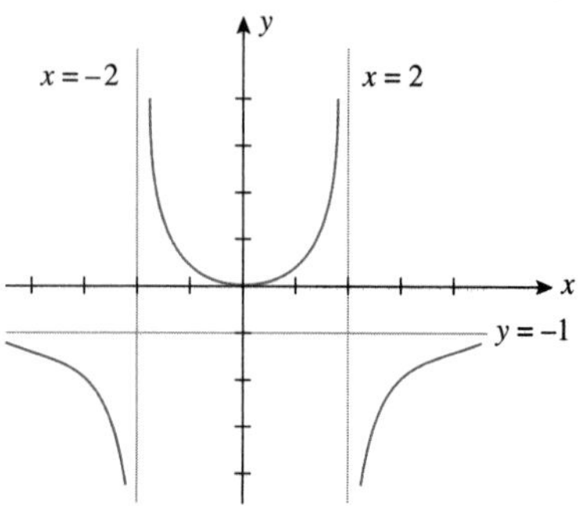

Figure 7.48
$x = -2$ and $x = 2$ are vertical asymptotes of the graph of f.

Recall that in order for the line $x = a$ to be a vertical asymptote of the graph of a rational function f, *only* the denominator of $f(x)$ must be equal to zero at $x = a$. If *both* $P(a)$ and $Q(a)$ are equal to zero, then $x = a$ need *not* be a vertical asymptote. For example, look at the function

$$f(x) = \frac{4(x^2 - 4)}{x - 2}$$

whose graph appears in Figure 3.5 on page 116.

Horizontal Asymptotes

Let's return to the function f defined by

$$f(x) = \frac{x + 1}{x - 1}$$

(Figure 7.49).

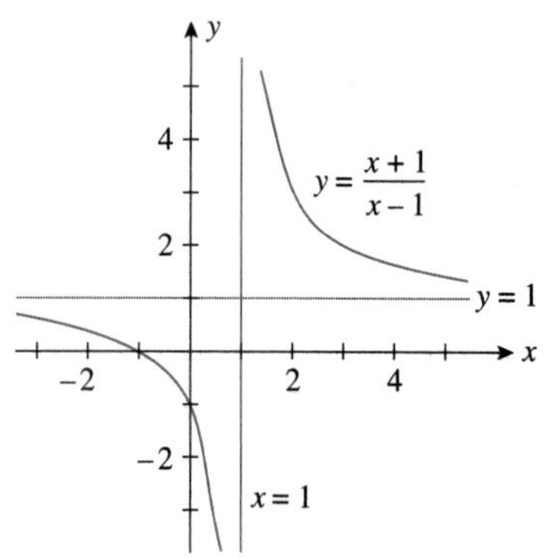

Figure 7.49
The graph of f has a horizontal asymptote at $y = 1$.

Observe that $f(x)$ approaches the horizontal line $y = 1$ as x approaches infinity, and, in this case, $f(x)$ approaches $y = 1$ as x approaches minus infinity as well. The line $y = 1$ is called a horizontal asymptote of the graph of f. More generally, we have the following definition:

Horizontal Asymptote

The line $y = b$ is a **horizontal asymptote** of the graph of a function f if either

$$\lim_{x \to \infty} f(x) = b \quad \text{or} \quad \lim_{x \to -\infty} f(x) = b$$

For the function

$$f(x) = \frac{x + 1}{x - 1}$$

we see that

$$\lim_{x \to \infty} \frac{x + 1}{x - 1} = \lim_{x \to \infty} \frac{1 + \frac{1}{x}}{1 - \frac{1}{x}} \qquad \text{Divide numerator and denominator by } x.$$

$$= 1$$

Also,

$$\lim_{x \to -\infty} \frac{x + 1}{x - 1} = \lim_{x \to -\infty} \frac{1 + \frac{1}{x}}{1 - \frac{1}{x}}$$

$$= 1$$

In either case, we conclude that $y = 1$ is a horizontal asymptote of the graph of f, as observed earlier.

Example 2

Find the horizontal asymptotes of the graph of the function

$$f(x) = \frac{x^2}{4 - x^2}$$

Solution

We compute

$$\lim_{x\to\infty}\frac{x^2}{4-x^2} = \lim_{x\to\infty}\frac{1}{\frac{4}{x^2}-1} \quad \text{Divide numerator and denominator by } x^2.$$

$$= -1$$

and so $y = -1$ is a horizontal asymptote, as before. The graph of f sketched in Figure 7.50 confirms this result.

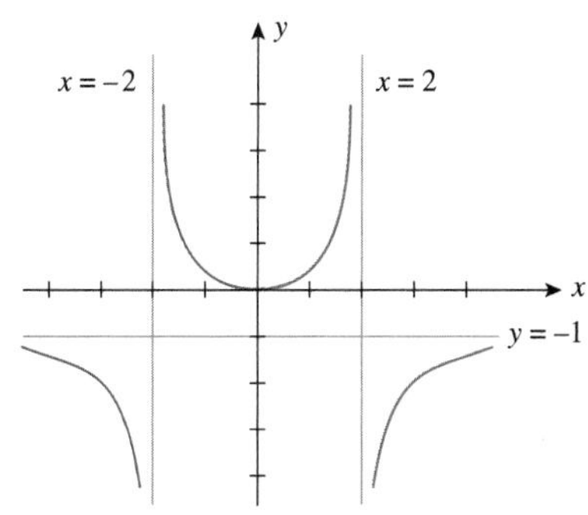

Figure 7.50
The graph of f has a horizontal asymptote at $y = -1$.

We next state an important property of polynomial functions.

> A polynomial function has no vertical or horizontal asymptotes.

To see this, note that a polynomial function $P(x)$ can be written as a rational function with denominator equal to 1. Thus,

$$P(x) = \frac{P(x)}{1}$$

Since the denominator is never equal to zero, P has no vertical asymptotes. Next, if P is a polynomial of degree greater than or equal to 1, then

$$\lim_{x\to\infty} P(x) \qquad \text{and} \qquad \lim_{x\to-\infty} P(x)$$

are either infinity or minus infinity; that is, they do not exist. Therefore, P has no horizontal asymptotes.

In the last two sections, we saw how the first and second derivatives of a function are used to reveal various properties of the graph of a function f. We now show how this information can be used to help us sketch the graph of f. We begin by giving a general procedure for curve sketching.

A Guide to Curve Sketching

1. Determine the domain of f.
2. Find the x- and y-intercepts of f.*
3. Determine the behaviour of f for large absolute values of x.
4. Find all horizontal and vertical asymptotes of f.
5. Determine the intervals where f is increasing and where f is decreasing.
6. Find the relative extrema of f.
7. Determine the concavity of f.
8. Find the inflection points of f.
9. Plot a few additional points to help further identify the shape of the graph of f and sketch the graph.

*The equation $f(x) = 0$ may be difficult to solve, in which case one may decide against finding the x-intercepts or to use technology, if available, for assistance.

We now illustrate the techniques of curve sketching with several examples.

Two Step-by-Step Examples

Example 3

Sketch the graph of the function

$$y = f(x) = x^3 - 6x^2 + 9x + 2$$

Solution

Obtain the following information on the graph of f.

1. The domain of f is the interval $(-\infty, -\infty)$.
2. By setting $x = 0$, we find that the y-intercept is 2. The x-intercept is found by setting $y = 0$, which in this case leads to a cubic equation. Since the solution is not readily found, we will not use this information.
3. Since

$$\lim_{x\to-\infty} f(x) = \lim_{x\to-\infty} (x^3 - 6x^2 + 9x + 2) = -\infty$$

$$\lim_{x\to\infty} f(x) = \lim_{x\to\infty} (x^3 - 6x^2 + 9x + 2) = \infty$$

we see that f decreases without bound as x decreases without bound and that f increases without bound as x increases without bound.

4. Since f is a polynomial function, there are no asymptotes.
5.

$$f'(x) = 3x^2 - 12x + 9 = 3(x^2 - 4x + 3) = 3(x - 3)(x - 1)$$

Setting $f'(x) = 0$ gives $x = 1$ or $x = 3$. The sign diagram for f' shows that f is increasing on the intervals $(-\infty, 1)$ and $(3, \infty)$ and decreasing on the interval $(1, 3)$ (Figure 7.51).

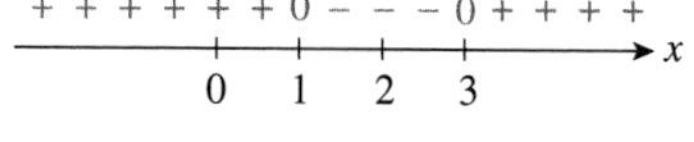

Figure 7.51
Sign diagram for f'

6. From the results of step 5, we see that $x = 1$ and $x = 3$ are critical points of f. Furthermore, f' changes sign from positive to negative as we move across $x = 1$, so a relative maximum of f occurs at $x = 1$. Similarly, we see that a relative minimum of f occurs at $x = 3$. Now,

$$f(1) = 1 - 6 + 9 + 2 = 6$$

$$f(3) = 3^3 - 6(3)^2 + 9(3) + 2 = 2$$

so $f(1) = 6$ is a relative maximum of f and $f(3) = 2$ is a relative minimum of f.

7. $f''(x) = 6x - 12 = 6(x - 2)$

which is equal to zero when $x = 2$. The sign diagram of f'' shows that f is concave downward on the interval $(-\infty, 2)$ and concave upward on the interval $(2, \infty)$ (Figure 7.52).

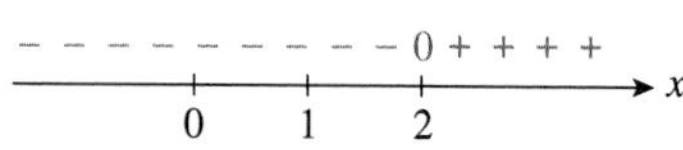

Figure 7.52
Sign diagram for f''

8. From the results of step 7, we see that f'' changes sign as we move across the point $x = 2$. Next,

$$f(2) = 2^3 - 6(2)^2 + 9(2) + 2 = 4$$

and so the required inflection point of f is $(2, 4)$.

Summarizing, we have the following:

> Domain: $(-\infty, \infty)$
> Intercept: $(0, 2)$
> $\lim_{x\to-\infty} f(x)$; $\lim_{x\to\infty} f(x)$: $-\infty$; ∞
> Asymptotes: None
> Intervals where f is ↗ or ↘: ↗ on $(-\infty, 1) \cup (3, \infty)$; ↘ on $(1, 3)$
> Relative extrema: Relative maximum at $(1, 6)$; relative minimum at $(3, 2)$
> Concavity: Downward on $(-\infty, 2)$; upward on $(2, \infty)$

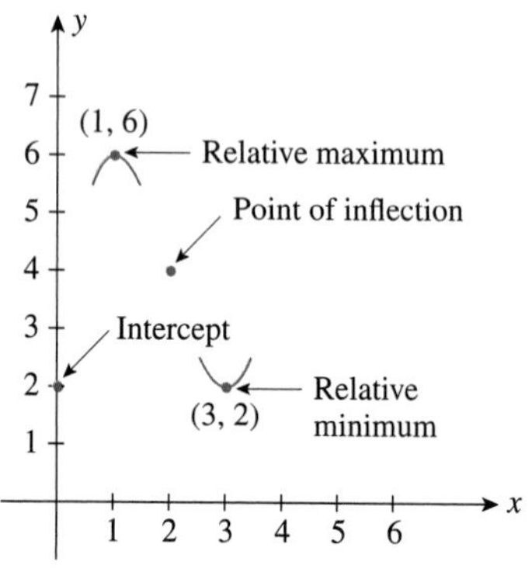

Figure 7.53
We first plot the intercept, the relative extrema, and the inflection point.

In general, it is a good idea to start graphing by plotting the intercept, relative extrema, and inflection point (Figure 7.53). Then, using the rest of the information, we complete the graph of f, as sketched in Figure 7.54.

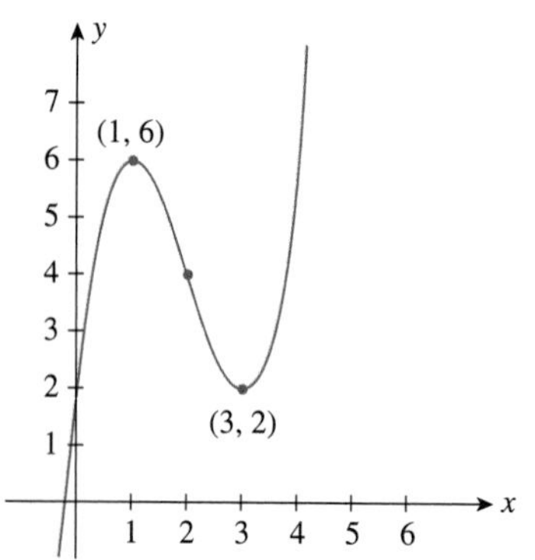

Figure 7.54
The graph of $y = x^3 - 6x^2 + 9x + 2$

GROUP DISCUSSION

The average price of gasoline at the pump over a 3-month period, during which there was a temporary shortage of oil, is described by the function f defined on the interval [0, 3]. During the first month, the price was increasing at an increasing rate. Starting with the second month, the good news was that the rate of increase was slowing down, although the price of gas was still increasing. This pattern continued until the end of the second month. The price of gas peaked at the end of $t = 2$ and began to fall at an increasing rate until $t = 3$.

1. Describe the signs of $f'(t)$ and $f''(t)$ over each of the intervals (0, 1), (1, 2), and (2, 3).
2. Make a sketch showing a plausible graph of f over [0, 3].

Example 4

Sketch the graph of the function

$$y = f(x) = \frac{x+1}{x-1}$$

Solution

Obtain the following information:

1. f is undefined when $x = 1$, so the domain of f is the set of all real numbers other than $x = 1$.
2. Setting $y = 0$ gives -1, the x-intercept of f. Next, setting $x = 0$ gives -1 as the y-intercept of f.
3. Earlier we found that

$$\lim_{x\to\infty}\frac{x+1}{x-1} = 1 \qquad \text{and} \qquad \lim_{x\to-\infty}\frac{x+1}{x-1} = 1$$

(see p. 374). Consequently, we see that $f(x)$ approaches the line $y = 1$ as $|x|$ becomes arbitrarily large. For $x > 1, f(x) > 1$ and $f(x)$ approaches the line $y = 1$ from above. For $x < 1, f(x) < 1$, so $f(x)$ approaches the line $y = 1$ from below.

4. The straight line $x = 1$ is a vertical asymptote of the graph of f. Also, from the results of step 3, we conclude that $y = 1$ is a horizontal asymptote of the graph of f.
5. $$f'(x) = \frac{(x-1)(1) - (x+1)(1)}{(x-1)^2} = -\frac{2}{(x-1)^2}$$

and is discontinuous at $x = 1$. The sign diagram of f' shows that $f'(x) < 0$ whenever it is defined. Thus, f is decreasing on the intervals $(-\infty, 1)$ and $(1, \infty)$ (Figure 7.55).

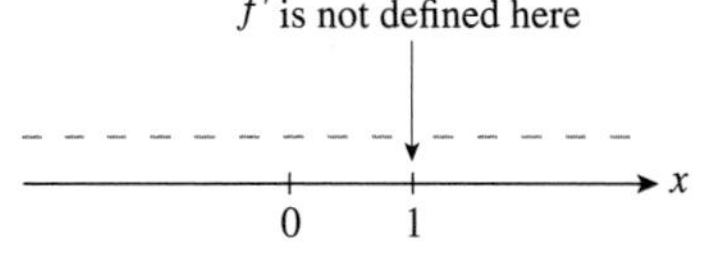

Figure 7.55
The sign diagram for f'

6. From the results of step 5, we see that there are no critical points of f since $f'(x)$ is never equal to zero for any value of x in the domain of f.
7. $$f''(x) = \frac{d}{dx}[-2(x-1)^{-2}] = 4(x-1)^{-3} = \frac{4}{(x-1)^3}$$

The sign diagram of f'' shows immediately that f is concave downward on the interval $(-\infty, 1)$ and concave upward on the interval $(1, \infty)$ (Figure 7.56).

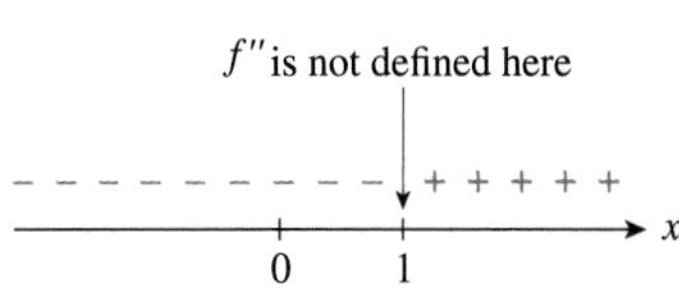

Figure 7.56
The sign diagram for f''

8. From the results of step 7, we see that there are no candidates for inflection points of f since $f''(x)$ is never equal to zero for any value of x in the domain of f. Hence, f has no inflection points.

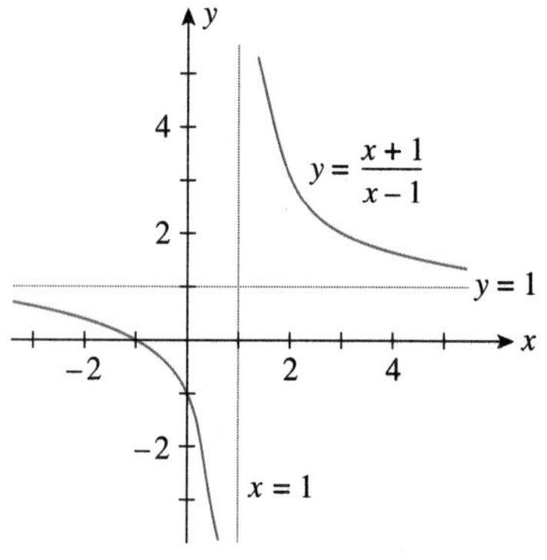

Figure 7.57
The graph of f has a horizontal asymptote at $y = 1$ and a vertical asymptote at $x = 1$.

Summarizing, we have the following:

> Domain: $(-\infty, 1) \cup (1, \infty)$
> Intercepts: $(0\ -1); (-1, 0)$
> $\lim_{x\to-\infty} f(x); \lim_{x\to\infty} f(x)$: 1;1
> Asymptotes: $x = 1$ is a vertical asymptote
> $y = 1$ is a horizontal asymptote
> Intervals where f is ↗ or ↘: ↘ on $(-\infty, 1) \cup (1, \infty)$
> Relative extrema: None
> Concavity: Downward on $(-\infty, 1)$; upward on $(1, \infty)$
> Points of inflection: None

The graph of f is sketched in Figure 7.57.

When a biological population is introduced into a new area it often grows rapidly and then settles into a final population level. When we have a model of a biological population, the level of the population after it finishes adapting to its new environment can be estimated by finding the horizontal asymptote of the population model.

Example 5

Eventual Level of a Biological Population Experimentally it is found that a population of fish introduced into a system of lakes has a population of

$$y(t) = \frac{6.05e^{0.51t}}{e^{0.51t} + 2.1}$$

thousands of fish after t years. Sketch the graph of the function and report the level the population of fish will settle down to in the long term by computing the horizontal asymptote. Assume the fish are introduced at time zero.

Solution

1. Because the fish are introduced at time zero the domain of the function is the set of nonnegative real numbers.
2. Setting $y(t) = 0$ we see that there is no t-intercept for the population model (the population is always positive). Setting $t = 0$ gives 1.95 thousand fish for the value of the y-intercept.
3. Compute the horizontal asymptote in the positive direction (the asymptote in the negative direction is not required because the negative t-axis is not part of the domain of the function).

$$\lim_{x\to\infty} \frac{6.05e^{0.51t}}{e^{0.51t} + 2.1} = \lim_{x\to\infty} \frac{6.05}{1 + \frac{2.1}{e^{0.51t}}}$$
$$= \frac{6.05}{1 + 0}$$
$$= 6.05 \text{ thousand fish}$$

So the population approaches a horizontal line representing 6.05 thousand fish as the time becomes arbitrarily large.

4. The denominator of the population model is a positive constant plus a function that is always larger than zero and so there are no vertical asymptotes.
5. Rewrite $y(t) = 6.05 \cdot \frac{e^{0.51t}}{e^{0.51t} + 2.1}$ and compute the derivative.

$$y'(t) = 6.05 \cdot \frac{(2.1 + e^{0.51t}) \cdot e^{0.51t} \cdot 0.51 - (0 + e^{0.51t} \cdot 0.51)e^{0.51t}}{(2.1 + e^{0.51t})^2}$$
$$= 6.05 \cdot \frac{0.51e^{0.51t}(2.1 + e^{0.51t} - e^{0.51t})}{(2.1 + e^{0.51t})^2}$$

$$= 3.0855\frac{2.1e^{0.51t}}{(2.1 + e^{0.51t})^2}$$

$$= \frac{6.47955e^{0.51t}}{(2.1 + e^{0.51t})^2}$$

This function is always positive and so the graph is increasing on its entire domain.

6. Remember that $e^x > 0$ for any value of x. Knowing that, the results of step 5 show that $y'(t)$ is never zero and so there are no critical points. This means there are no candidate points for local extrema.

7. Rewrite $6.47955\frac{e^{0.51t}}{(2.1 + e^{0.51t})^2}$ and compute the second derivative.

$$y''(t) = 6.47955 \cdot \frac{(2.1 + e^{0.51t})^2 e^{0.51t} \cdot 0.51 - e^{0.51t} \cdot 2(2.1 + e^{0.51t}) \cdot e^{0.51t} 0.51}{(2.1 + e^{0.51t})^4}$$

$$= 6.47955 \cdot \frac{(2.1 + e^{0.51t})^{\cancel{2}} e^{0.51t} \cdot 0.51 - e^{0.51t} \cdot 2(\cancel{2.1 + e^{0.51t}}) \cdot e^{0.51t} 0.51}{(2.1 + e^{0.51t})^{\cancel{4}3}}$$

$$= 6.47955 \cdot \frac{0.51e^{0.51t} \cdot (2.1 + e^{0.51t} - 2e^{0.51t})}{(2.1 + e^{0.51t})^3}$$

$$= 6.47955 \cdot \frac{0.51e^{0.51t} \cdot (2.1 - e^{0.51t})}{(2.1 + e^{0.51t})^3}$$

The only way for this expression to equal zero is if the term $(2.1 - e^{0.51t})$ is zero and so we solve this expression set equal to zero.

$$2.1 - e^{0.51t} = 0$$

$$2.1 = e^{0.51}$$

$$\ln(2.1) = \ln(e^{0.51t})$$

$$\ln(2.1) = 0.51t$$

$$t = \frac{\ln(2.1)}{0.51}$$

$$t = 1.45 \text{ years}$$

There is a single zero at 1.45 years. The sign of the second derivative is positive before this and negative after it, and so the population model is concave up before 1.45 years and concave down after 1.45 years.

8. There is a single inflection point, found in step 7, at 1.45 years with a population of $y(1.45) = 3.021$ thousand fish.

Summarizing, we have the following:

Domain: $[0, \infty)$
Intercept: $(0, 3.01)$
$\lim_{t \to \infty} y(t) = 6.05$
Asymptotes: $y = 6.05$ is a horizontal asymptote
The function is increasing everywhere.
Relative extrema: None
Concavity: Upward on $[0, 1.45)$; downward on $(1.45, \infty)$.
Points of inflection: $(1.45, 3.021)$

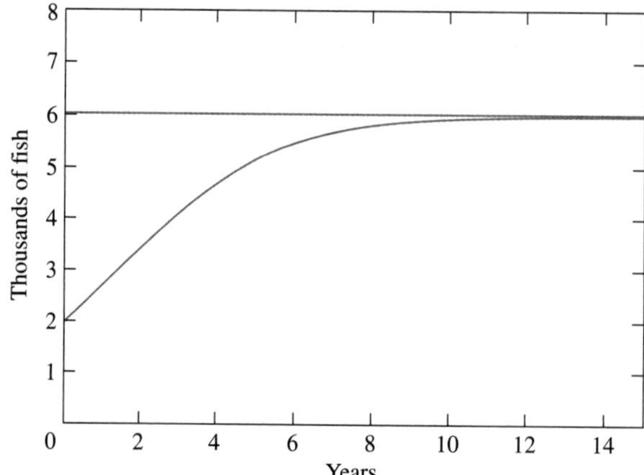

Figure 7.58
Thousands of fish after t years.

The sketch of the population model is shown in Figure 7.58. The value of the horizontal asymptote tells us that the population of fish settles down to a value of just over 6000 fish in the long term.

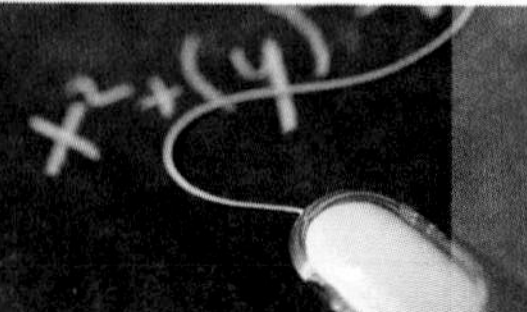

Self-Check Exercises 7.3

1. Find the horizontal and vertical asymptotes of the graph of the function

$$f(x) = \frac{2x^2}{x^2 - 1}$$

2. Sketch the graph of the function

$$f(x) = \frac{2}{3}x^3 - 2x^2 - 6x + 4$$

Solutions to Self-Check Exercises 7.3 can be found on page 384.

7.3 Exercises

In Exercises 1–10, find the horizontal and vertical asymptotes of the graph.

1.

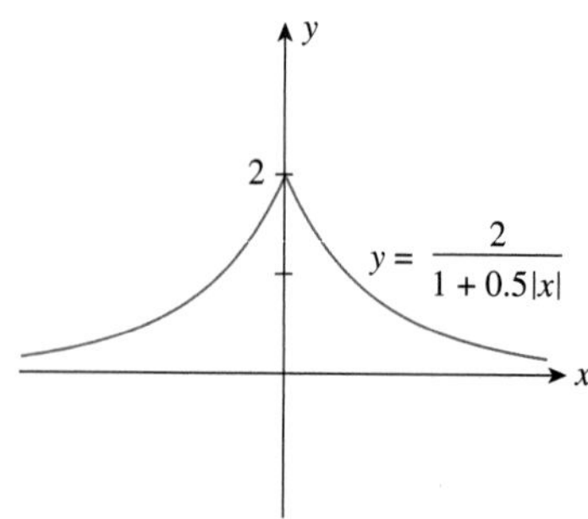

2.

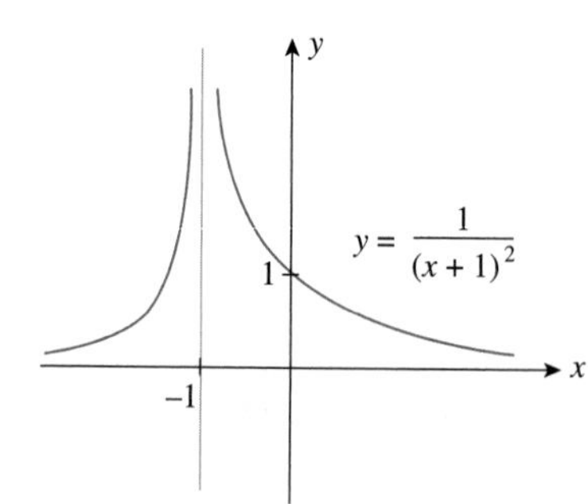

3.

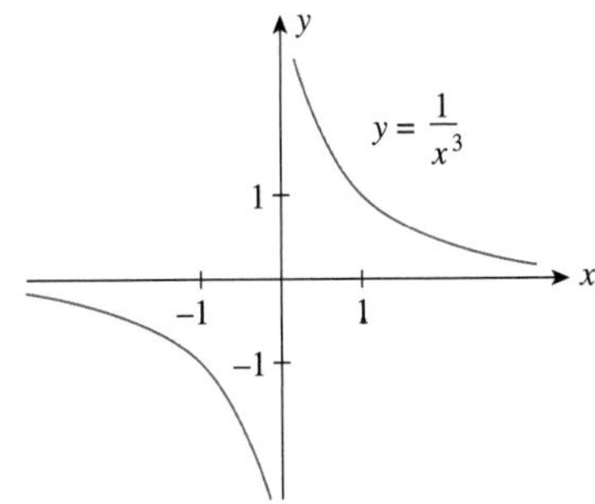

4.

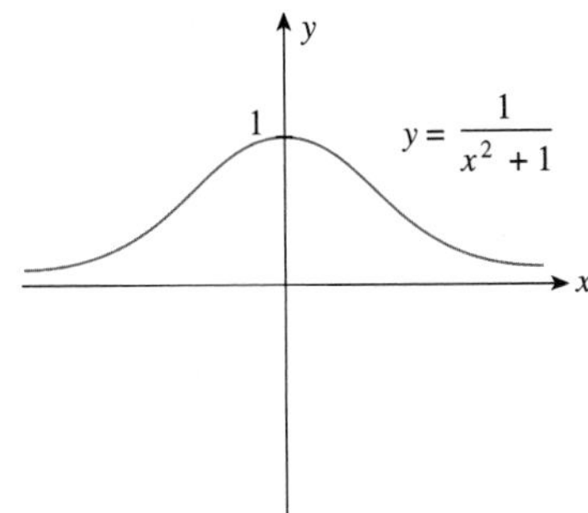

5.

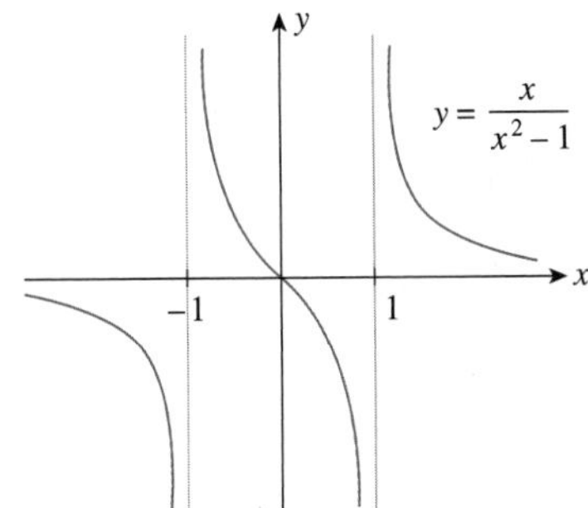

6.

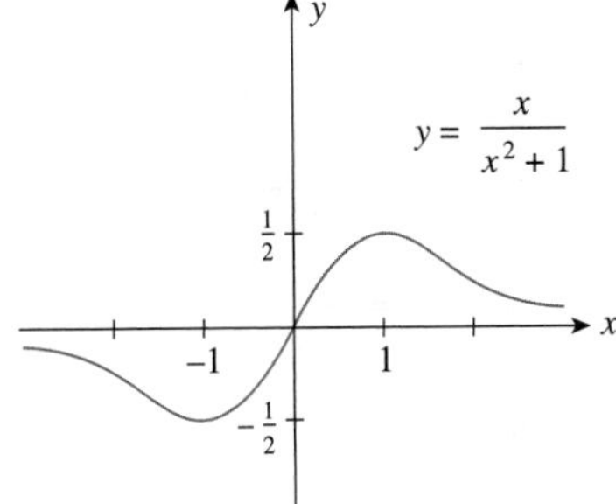

7.

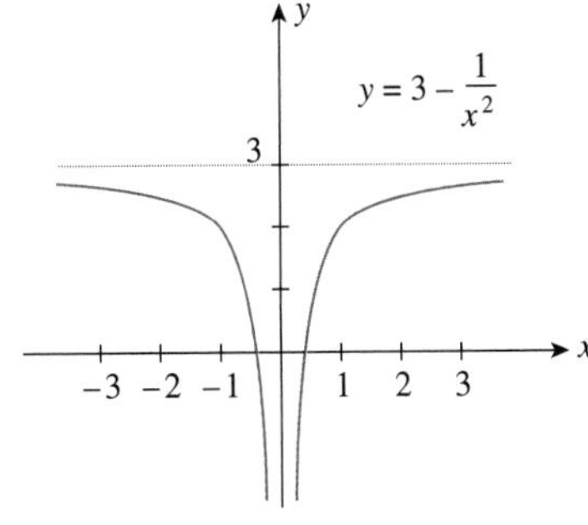

8.

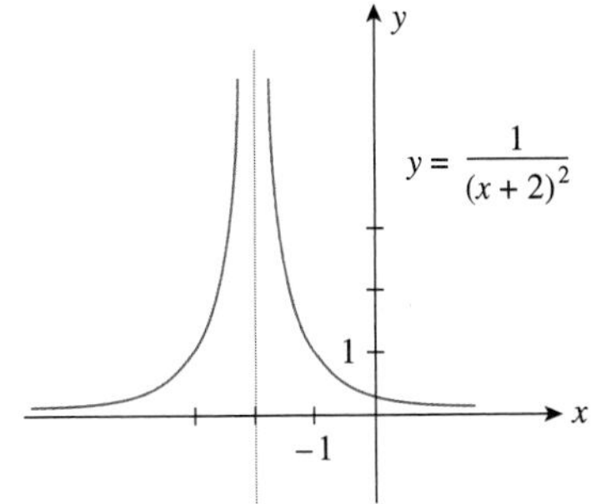

9.

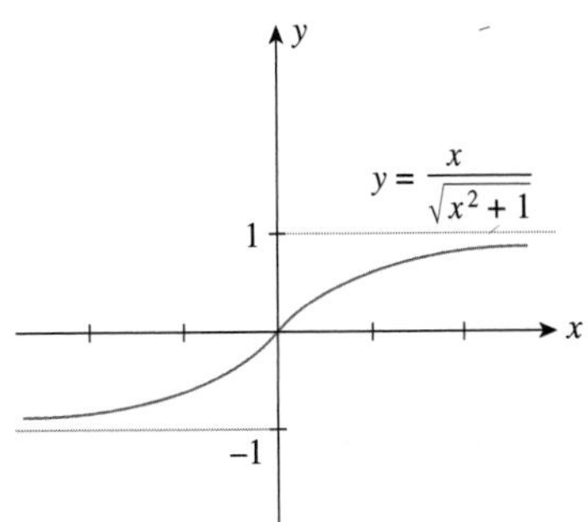

10.

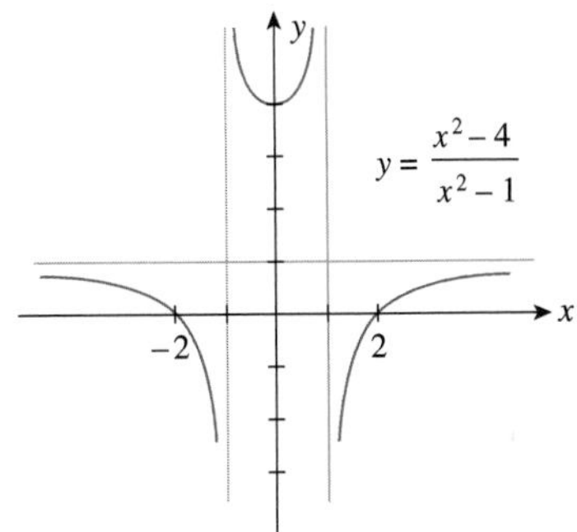

In Exercises 11–28, find the horizontal and vertical asymptotes of the graph of the function. (You need not sketch the graph.)

11. $f(x) = \frac{1}{x}$

12. $f(x) = \frac{1}{x + 2}$

13. $f(x) = -\frac{2}{x^2}$

14. $g(x) = \frac{1}{1 + 2x^2}$

15. $f(x) = \frac{x - 1}{x + 1}$

16. $g(t) = \frac{t + 1}{2t - 1}$

17. $h(x) = x^3 - 3x^2 + x + 1$

18. $g(x) = 2x^3 + x^2 + 1$

19. $f(t) = \frac{t^2}{t^2 - 9}$

20. $g(x) = \frac{x^3}{x^2 - 4}$

21. $f(x) = \frac{3x}{x^2 - x - 6}$

22. $g(x) = \frac{2x}{x^2 + x - 2}$

23. $g(t) = 2 + \frac{5}{(t - 2)^2}$

24. $f(x) = 1 + \frac{2}{x - 3}$

25. $f(x) = \dfrac{x^2 - 2}{x^2 - 4}$

26. $h(x) = \dfrac{2 - x^2}{x^2 + x}$

27. $g(x) = \dfrac{x^3 - x}{x(x + 1)}$

28. $f(x) = \dfrac{x^4 - x^2}{x(x - 1)(x + 2)}$

In Exercises 29 and 30, you are given the graphs of two functions f and g. One function is the derivative function of the other. Identify each of them.

29.

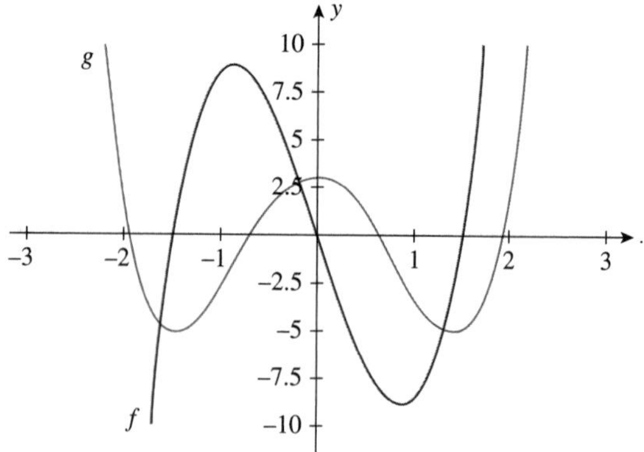

30.

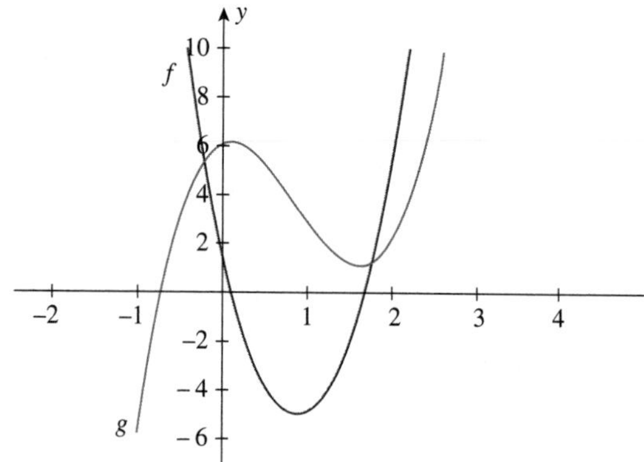

In Exercises 31–34, use the information summarized in the table to sketch the graph of f.

31. $f(x) = x^3 - 3x^2 + 1$

Domain: $(-\infty, \infty)$
Intercept: y-intercept: 1
Asymptotes: None
Intervals where f is ↗ and ↘: ↗ on $(-\infty, 0) \cup (2, \infty)$;
↘ on $(0, 2)$
Relative extrema: Rel. max. at $(0, 1)$; rel. min. at $(2, -3)$
Concavity: Downward on $(-\infty, 1)$; upward on $(1, \infty)$
Point of inflection: $(1, -1)$

32. $f(x) = \dfrac{1}{9}(x^4 - 4x^3)$

Domain: $(-\infty, \infty)$
Intercepts: x-intercepts: 0, 4; y-intercept: 0
Asymptotes: None
Intervals where f is ↗ and ↘: ↗ on $(3, -\infty)$;
↘ on $(-\infty, 0) \cup (0, 3)$
Relative extrema: Rel. min. at $(3, -3)$
Concavity: Downward on $(0, 2)$; upward on $(-\infty, 0) \cup (2, \infty)$
Points of inflection: $(0, 0)$ and $(2, -\frac{16}{9})$

33. $f(x) = \dfrac{4x - 4}{x^2}$

Domain: $(-\infty, 0) \cup (0, \infty)$
Intercept: x-intercept: 1
Asymptotes: x-axis and y-axis
Intervals where f is ↗ and ↘: ↗ on $(0, 2)$;
↘ on $(-\infty, 0) \cup (2, \infty)$
Relative extrema: Rel. max. at $(2, 1)$
Concavity: Downward on $(-\infty, 0) \cup (0, 3)$; upward on $(3, \infty)$
Point of inflection: $(3, \frac{8}{9})$

34. $f(x) = x - 3x^{1/3}$

Domain: $(-\infty, \infty)$
Intercepts: x-intercepts: $\pm 3\sqrt{3}, 0$
Asymptotes: None
Intervals where f is ↗ and ↘: ↗ on $(-\infty, -1) \cup (1, \infty)$;
↘ on $(-1, 1)$
Relative extrema: Rel. max. at $(-1, 2)$; rel. min. at $(1, -2)$
Concavity: Downward on $(-\infty, 0)$; upward on $(0, \infty)$
Point of inflection: $(0, 0)$

In Exercises 35–66, sketch the graph of the function, using the curve-sketching guide of this section.

35. $g(x) = 4 - 3x - 2x^3$

36. $f(x) = x^2 - 2x + 3$

37. $h(x) = x^3 - 3x + 1$

38. $f(x) = 2x^3 + 1$

39. $f(x) = -2x^3 + 3x^2 + 12x + 2$

40. $f(t) = 2t^3 - 15t^2 + 36t - 20$

41. $h(x) = \dfrac{3}{2}x^4 - 2x^3 - 6x^2 + 8$

42. $f(t) = 3t^4 + 4t^3$

43. $f(t) = \sqrt{t^2 - 4}$

44. $f(x) = \sqrt{x^2 + 5}$

45. $g(x) = \dfrac{1}{2}x - \sqrt{x}$

46. $f(x) = \sqrt[3]{x^2}$

47. $g(x) = \dfrac{2}{x - 1}$

48. $f(x) = \dfrac{1}{x + 1}$

49. $h(x) = \dfrac{x + 2}{x - 2}$

50. $g(x) = \dfrac{x}{x - 1}$

51. $f(t) = \dfrac{t^2}{1 + t^2}$

52. $g(x) = \dfrac{x}{x^2 - 4}$

53. $g(t) = -\dfrac{t^2 - 2}{t - 1}$

54. $f(x) = \dfrac{x^2 - 9}{x^2 - 4}$

55. $g(t) = \dfrac{t + 1}{t^2 - 2t - 1}$

56. $h(x) = \dfrac{1}{x^2 - x - 2}$

57. $h(x) = (x - 1)^{2/3} + 1$

58. $g(x) = (x + 2)^{3/2} + 1$

59. $f(x) = x \cdot e^x$

60. $f(x) = x^2 \cdot e^x$

61. $f(x) = x \cdot e^{-x}$

62. $f(x) = \ln(x^2 + 1)$

63. $g(x) = \ln(x^2 - 4x + 5)$

64. $g(x) = \arctan(x)$

65. $g(x) = \arctan(x^2)$

66. $h(x) = \cos^2(\arctan(x))$

67. Terminal Velocity A skydiver leaps from the gondola of a hot-air balloon. As she free-falls, air resistance, which is proportional to her velocity, builds up to a point where it balances the force due to gravity. The resulting motion may be described in terms of her velocity as follows: Starting at rest (zero velocity), her velocity increases and approaches a constant velocity, called the *terminal velocity*. Sketch a graph of her velocity v versus time t.

Business and Economics Applications

68. Average Cost of Producing Videodiscs The average cost per disc (in dollars) incurred by Herald Records in pressing x videodiscs is given by the average cost function

$$\overline{C}(x) = 2.2 + \frac{2500}{x}$$

a. Find the horizontal asymptote of $\overline{C}(x)$.

b. What is the limiting value of the average cost?

69. GDP of a Developing Country A developing country's gross domestic product (GDP) from 1992 to 2000 is approximated by the function

$$G(t) = -0.2t^3 + 2.4r^2 + 60 \qquad (0 \le t \le 8)$$

where G(t) is measured in billions of dollars, with $t = 0$ corresponding to 1992. Sketch the graph of the function G and interpret your results.

70. Worker Efficiency An efficiency study showed that the total number of cordless telephones assembled by an average worker at Delphi Electronics t hr after starting work at 8 A.M. is given by

$$N(t) = -\frac{1}{2}t^3 + 3t^2 + 10t \qquad (0 \le t \le 4)$$

Sketch the graph of the function N and interpret your results.

71. Box-Office Receipts The total worldwide box-office receipts for a long-running movie are approximated by the function

$$T(x) = \frac{120x^2}{x^2 + 4}$$

where $T(x)$ is measured in millions of dollars and x is the number of years since the movie's release. Sketch the graph of the function T and interpret your results.

Biological and Life Sciences Applications

72. Spread of a Flu Epidemic Initially, 10 students at a junior high school contracted influenza. The flu spread over time, and the total number of students who eventually contracted the flu approached but never exceeded 200. Let $P(t)$ denote the number of students who had contracted the flu after t days, where P is an appropriate function.

a. Make a sketch of the graph of P. (Your answer will *not* be unique.)

b. Where is the function increasing?

c. Does P have a horizontal asymptote? If so, what is it?

d. Discuss the concavity of P. Explain its significance.

e. Is there an inflection point on the graph of P? If so, explain its significance.

73. Cost of Removing Toxic Pollutants A city's main well was recently found to be contaminated with trichloroethylene (a cancer-causing chemical) as a result of an abandoned chemical dump leaching chemicals into the water. A proposal submitted to the city council indicated that the cost, measured in millions of dollars, of removing x% of the toxic pollutants is given by

$$C(x) = \frac{0.5x}{100 - x}$$

a. Find the vertical asymptote of $C(x)$.

b. Is it possible to remove 100% of the toxic pollutant from the water?

74. Concentration of a Drug in the Bloodstream The concentration (in milligrams/cubic centimetre) of a certain drug in a patient's bloodstream t hr after injection is given by

$$C(t) = \frac{0.2t}{t^2 + 1}$$

a. Find the horizontal asymptote of $C(t)$.

b. Interpret your result.

75. Effect of Enzymes on Chemical Reactions Certain proteins, known as enzymes, serve as catalysts for chemical reactions in living things. In 1913 Leonor Michaelis and L. M. Menten discovered the following formula giving the initial speed V (in moles/litre/second) at which the reaction begins in terms of the amount of substrate x (the substance that is being acted upon, measured in moles/litre):

$$V = \frac{ax}{x + b}$$

where a and b are positive constants.

a. Find the horizontal asymptote of V.

b. What does the result of part (a) tell you about the initial speed at which the reaction begins, if the amount of substrate is very large?

76. Concentration of a Drug in the Bloodstream The concentration (in millimetres/cubic centimetre) of a certain drug in a patient's bloodstream t hr after injection is given by

$$C(t) = \frac{0.2t}{t^2 + 1}$$

Sketch the graph of the function C and interpret your results.

77. Oxygen Content of a Pond When organic waste is dumped into a pond, the oxidation process that takes place reduces the pond's oxygen content. However, given time, nature will restore the oxygen content to its natural level. Suppose the oxygen content t days after organic waste has been dumped into the pond is given by

$$f(t) = 100\left(\frac{t^2 - 4t + 4}{t^2 + 4}\right) \qquad (0 \le t < \infty)$$

percent of its normal level. Sketch the graph of the function f and interpret your results.

78. Cost of Removing Toxic Pollutants A city's main well was recently found to be contaminated with trichloroethylene, a cancer-causing chemical, as a result of an abandoned chemical dump leaching chemicals into the water. A proposal submitted to the city council indicates that the cost, measured in millions of dollars, of removing x% of the toxic pollutant is given by

$$C(x) = \frac{0.5x}{100 - x}$$

Sketch the graph of the function C and interpret your results.

Solutions to Self-Check Exercises 7.3

1. Since

$$\lim_{x\to\infty}\frac{2x^2}{x^2-1}=\lim_{x\to\infty}\frac{2}{1-\dfrac{1}{x^2}}\qquad\text{Divide the numerator and denominator by } x^2.$$

$$=2$$

we see that $y = 2$ is a horizontal asymptote. Next, since

$$x^2 - 1 = (x+1)(x-1) = 0$$

implies $x = -1$ or $x = 1$, these are candidates for the vertical asymptotes of f. Since the numerator of f is not equal to zero for $x = -1$ or $x = 1$, we conclude that $x = -1$ and $x = 1$ are vertical asymptotes of the graph of f.

2. We obtain the following information on the graph of f.

(1) The domain of f is the interval $(-\infty, \infty)$.

(2) By setting $x = 0$, we find the y-intercept is 4.

(3) Since

$$\lim_{x\to-\infty} f(x) = \lim_{x\to-\infty}\left(\frac{2}{3}x^3 - 2x^2 - 6x + 4\right) = -\infty$$

$$\lim_{x\to\infty} f(x) = \lim_{x\to\infty}\left(\frac{2}{3}x^3 - 2x^2 - 6x + 4\right) = \infty$$

we see that $f(x)$ decreases without bound as x decreases without bound and that $f(x)$ increases without bound as x increases without bound.

(4) Since f is a polynomial function, there are no asymptotes.

$$f'(x) = 2x^2 - 4x - 6 = 2(x^2 - 2x - 3)$$

$$= 2(x+1)(x-3)$$

(5) Setting $f'(x) = 0$ gives $x = -1$ or $x = 3$. The accompanying sign diagram for f' shows that f is increasing on the intervals $(-\infty, -1)$ and $(3, \infty)$ and decreasing on $(-1, 3)$.

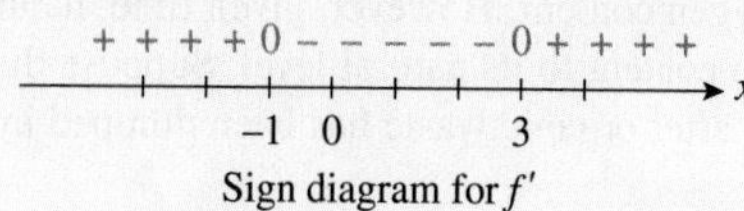

Sign diagram for f'

(6) From the results of step 5, we see that $x = -1$ and $x = 3$ are critical points of f. Furthermore, the sign diagram of f' tells us that $x = -1$ gives rise to a relative maximum of f and $x = 3$ gives rise to a relative minimum of f. Now,

$$f(-1) = \frac{2}{3}(-1)^3 - 2(-1)^2 - 6(-1) + 4 = \frac{22}{3}$$

$$f(3) = \frac{2}{3}(3)^3 - 2(3)^2 - 6(3) + 4 = -14$$

so $f(-1) = \frac{22}{3}$ is a relative maximum of f and $f(3) = -14$ is a relative minimum of f.

(7) $$f''(x) = 4x - 4 = 4(x-1)$$

Which is equal to zero when $x = 1$. The accompanying sign diagram of f'' shows that f is concave downward on the interval $(-\infty, 1)$ and concave upward on the interval $(1, -\infty)$,

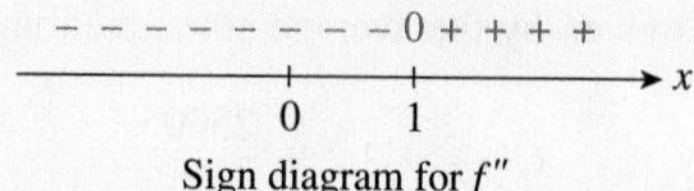

Sign diagram for f''

(8) From the results of step 7, we see that $x = 1$ is the only candidate for an inflection point of f. Since $f''(x)$ changes sign as we move across the point $x = 1$ and

$$f(1) = \frac{2}{3}(1)^3 - 2(1)^2 - 6(1) + 4 = -\frac{10}{3}$$

we see that the required inflection point is $(1, -\frac{10}{3})$.

(9) Summarizing this information, we have the following:

Domain: $(-\infty, \infty)$
Intercept: $(0, 4)$
$\lim_{x\to-\infty} f(x)$; $\lim_{x\to\infty} f(x)$: $-\infty$; ∞
Asymptotes: None
Intervals where f is ↗ or ↘: ↗ on $(-\infty, -1) \cup (3, \infty)$
↘ on $(-1, 3)$
Relative extrema: Rel. max. at $(-1, \frac{22}{3})$; rel. min. at $(3, -14)$
Concavity: Downward on $(-\infty, 1)$; upward on $(1, \infty)$
Points of inflection: $(1, -\frac{10}{3})$

The graph of f is sketched in the accompanying figure.

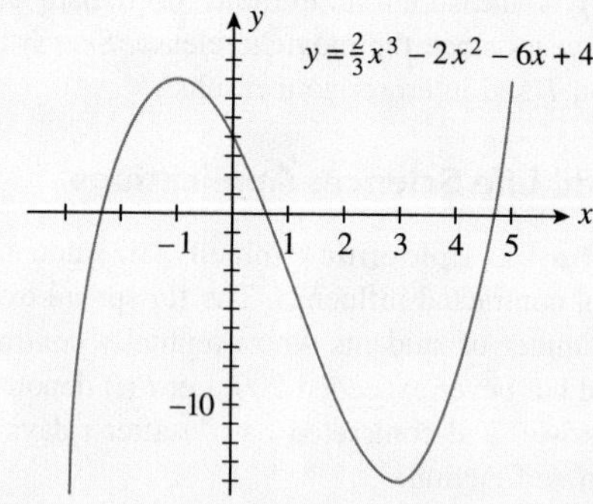

Absolute Extrema

The graph of the function f in Figure 7.59 shows the average age of cars in use in North America from the beginning of 1946 ($t = 0$) to the beginning of 1990 ($t = 44$). Observe that the highest average age of cars in use during this period is 9 years, whereas the lowest average age of cars in use during the same period is $5\frac{1}{2}$ years. The number 9, the largest value of $f(t)$ for all values of t in the interval $[0, 44]$ (the domain of f), is called the *absolute maximum value of f* on that interval. The number $5\frac{1}{2}$, the smallest value of $f(t)$ for all values of t in $[0, 44]$, is called the *absolute minimum value of f* on that interval. Notice, too, that the absolute maximum value of f is attained at the end point $t = 0$ of the interval, whereas the absolute minimum value of f is attained at the two interior points $t = 12$ (corresponding to 1958) and $t = 23$ (corresponding to 1969).

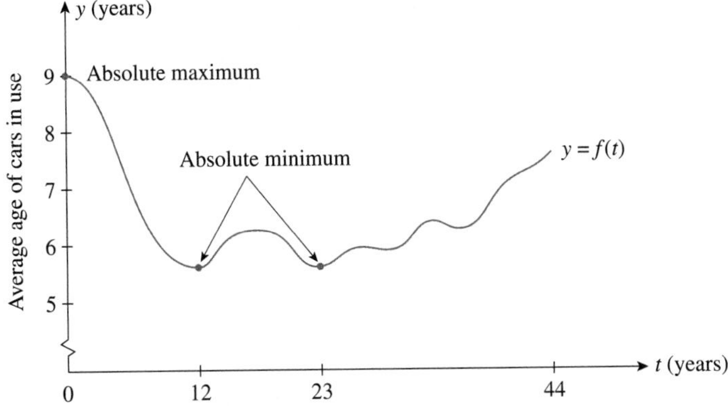

Source: Canadian Automobile Association

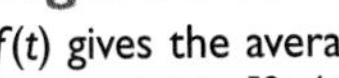

Figure 7.59
$f(t)$ gives the average age of cars in use in year t, t in $[0, 44]$.

A precise definition of the absolute extrema (absolute maximum or absolute minimum) of a function follows.

> **The Absolute Extrema of a Function f**
>
> If $f(x) \le f(c)$ for all x in the domain of f, then $f(c)$ is called the absolute maximum value of f.
>
> If $f(x) \ge f(c)$ for all x in the domain of f, then $f(c)$ is called the absolute minimum value of f.

Figure 7.60 shows the graphs of several functions and gives the absolute maximum and absolute minimum of each function, if they exist.

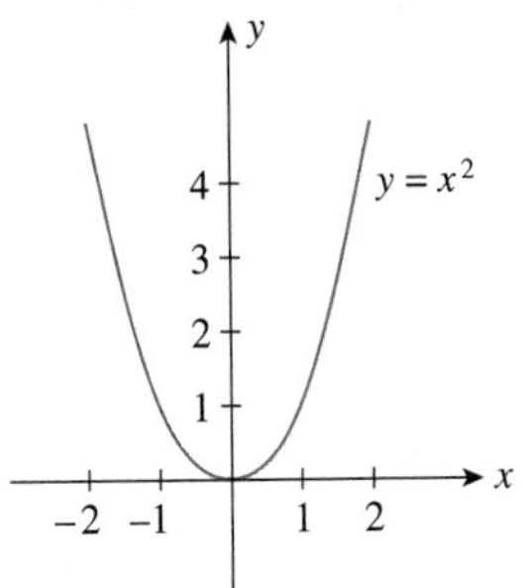

(a) $f(0) = 0$ is the absolute minimum of f; f has no absolute maximum.

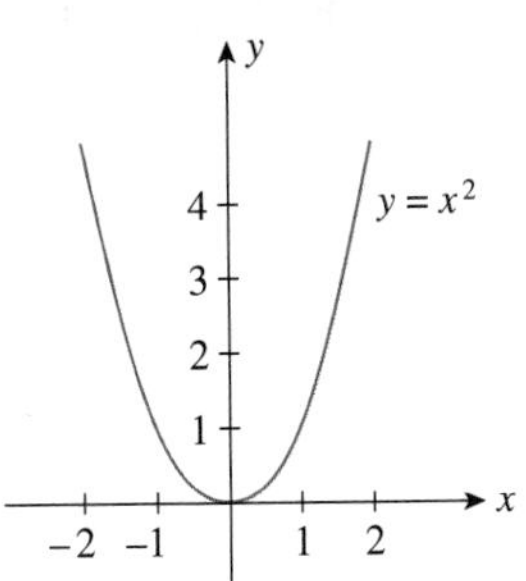

(b) $f(0) = 4$ is the absolute maximum of f; f has no absolute minimum.

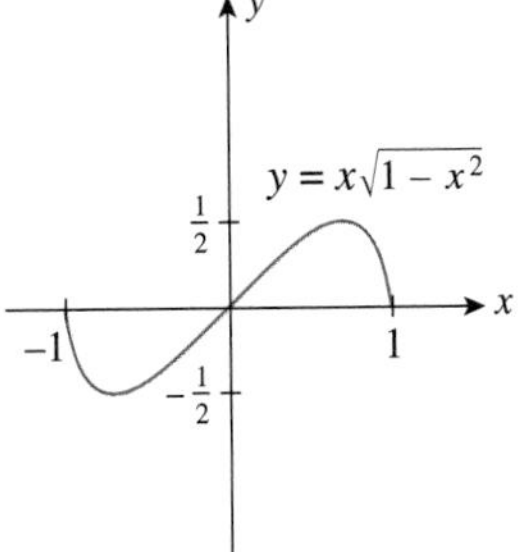

(c) $f(\sqrt{2}/2) = 1/2$ is the absolute maximum of f. $f(-\sqrt{2}/2) = -1/2$ is the absolute minimum of f.

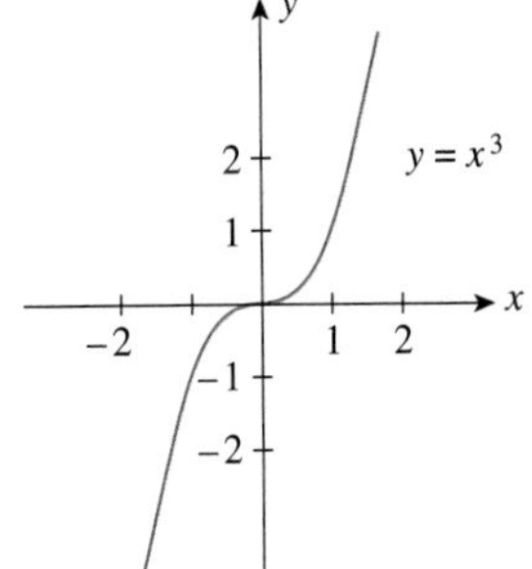

(d) f has no absolute extrema.

Figure 7.60

Absolute Extrema on a Closed Interval

As the preceding examples show, a continuous function defined on an arbitrary interval does not always have an absolute maximum or an absolute minimum. But an important case arises often in practical applications in which both the absolute maximum and the absolute minimum of a function are guaranteed to exist. This occurs where a continuous function is defined on a *closed* interval. Let's state this important result in the form of a theorem, whose proof we will omit.

If a function f is continuous on a closed interval $[a, b]$, then f has both an absolute maximum value and an absolute minimum value on $[a, b]$.

Observe that if an absolute extremum of a continuous function f occurs at a point in an open interval (a, b), then it must be a relative extremum of f and hence its x-coordinate must be a critical point of f. Otherwise, the absolute extremum of f must occur at one or both of the end points of the interval $[a, b]$. A typical situation is illustrated in Figure 7.61.

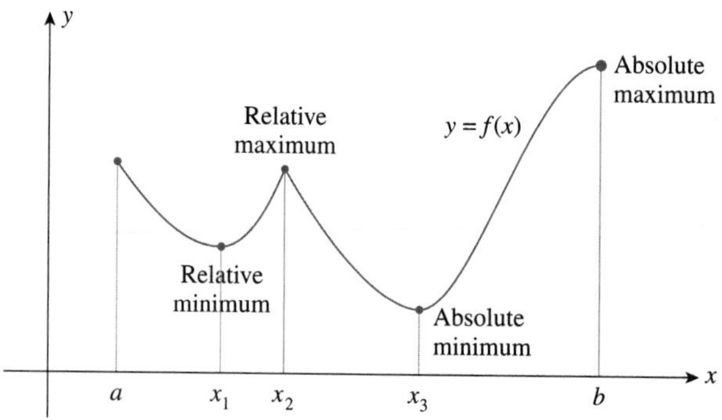

Figure 7.61
The relative minimum of f at x_3 is the absolute minimum of f. The right end point b gives rise to the absolute maximum value $f(b)$ of f.

Here x_1, x_2, and x_3 are critical points of f. The absolute minimum of f occurs at x_3, which lies in the open interval (a, b) and is a critical point of f. The absolute maximum of f occurs at b, an end point. This observation suggests the following procedure for finding the absolute extrema of a continuous function on a closed interval.

Finding the Absolute Extrema of a Continuous f on a Closed Interval

1. Find the critical points of f that lie in (a, b).
2. Compute the value of f at each critical point found in step 1 and compute $f(a)$ and $f(b)$.
3. The absolute maximum value and absolute minimum value of f will correspond to the largest and smallest numbers, respectively, found in step 2.

Example 1

Find the absolute extrema of the function $F(x) = x^2$ defined on the interval $[-1, 2]$.

Solution

The function F is continuous on the closed interval $[-1, 2]$ and differentiable on the open interval $(-1, 2)$. The derivative of F is

$$F'(x) = 2x$$

so $x = 0$ is the only critical point of F. Next, evaluate $F(x)$ at $x = -1$, $x = 0$, and $x = 2$. Thus,

$$F(-1) = 1, \qquad F(0) = 0, \qquad F(2) = 4$$

It follows that 0 is the absolute minimum value of F and 4 is the absolute maximum value of F. The graph of F, in Figure 7.62, confirms our results.

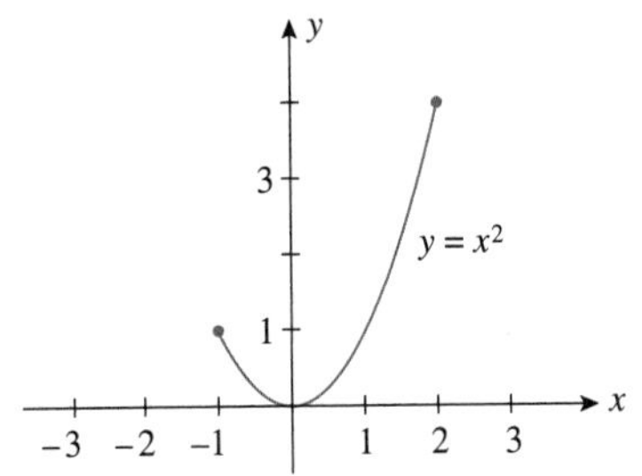

Figure 7.62

F has an absolute minimum value of 0 and an absolute maximum value of 4.

Example 2

Find the absolute extrema of the function

$$f(x) = x^3 - 2x^2 - 4x + 4$$

defined on the interval $[0, 3]$.

Solution

The function f is continuous on the closed interval $[0, 3]$ and differentiable on the open interval $(0, 3)$. The derivative of f is

$$f'(x) = 3x^2 - 4x - 4 = (3x + 2)(x - 2)$$

and it is equal to zero when $x = -\frac{2}{3}$ and $x = 2$. Since the point $x = -\frac{2}{3}$ lies outside the interval $[0, 3]$, it is dropped from further consideration, and $x = 2$ is seen to be the sole critical point of f. Next, we evaluate $f(x)$ at the critical point of f as well as the end points of f, obtaining

$$f(0) = 4, \qquad f(2) = -4, \qquad f(3) = 1$$

From these results, we conclude that -4 is the absolute minimum value of f and 4 is the absolute maximum value of f. The graph of f, which appears in Figure 7.63, confirms our results. Observe that the absolute maximum of f occurs at the end point $x = 0$ of the interval $[0, 3]$, while the absolute minimum of f occurs at $x = 2$, which is a point in the interval $(0, 3)$.

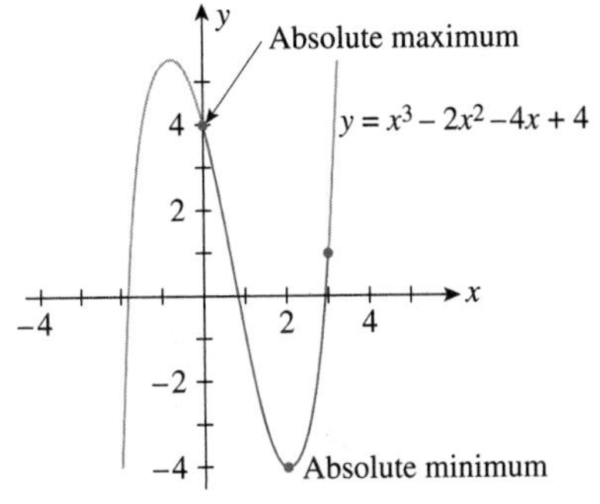

Figure 7.63

f has an absolute maximum value of 4 and an absolute minimum value of -4.

Example 3

Find the absolute maximum and absolute minimum values of the function $f(x) = x^{2/3}$ on the interval $[-1, 8]$.

Solution

The derivative of f is

$$f'(x) = \frac{2}{3}x^{-1/3} = \frac{2}{3x^{1/3}}$$

Note that f' is not defined at $x = 0$, is continuous everywhere else, and does not equal zero for all x. Therefore, $x = 0$ is the only critical point of f. Evaluating $f(x)$ at $x = -1$, 0, and 8, we obtain

$$f(-1) = 1, \qquad f(0) = 0, \qquad f(8) = 4$$

We conclude that the absolute minimum value of f is 0, attained at $x = 0$, and the absolute maximum value of f is 4, attained at $x = 8$ (Figure 7.64).

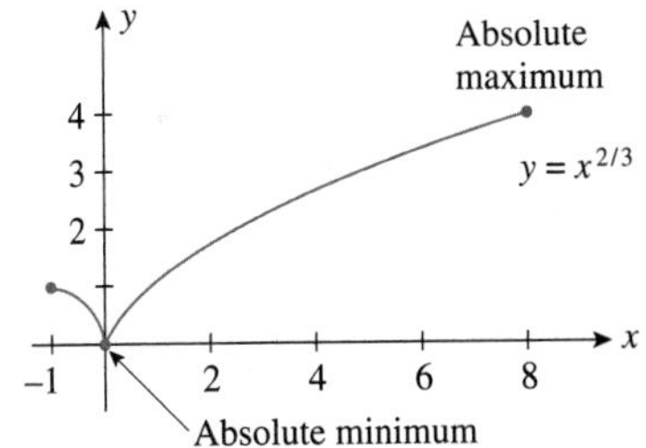

Figure 7.64

f has an absolute minimum value of $f(0) = 0$ and an absolute maximum value of $f(8) = 4$.

Applications

Many real-world applications call for finding the absolute maximum value or the absolute minimum value of a given function. For example, management is interested in finding what level of production will yield the maximum profit for a company; a farmer is interested in finding the right amount of fertilizer to maximize crop yield; a doctor is interested in finding the maximum concentration of a drug in a patient's body and the time at which it occurs; and an engineer is interested in finding the dimension of a container with a specified shape and volume that can be constructed at a minimum cost.

Example 4

Maximizing Profits Acrosonic's total profit (in dollars) from manufacturing and selling x units of their model F loudspeaker system is given by

$$P(x) = -0.02x^2 + 300x - 200{,}000 \qquad (0 \le x \le 20{,}000)$$

How many units of the loudspeaker system must Acrosonic produce to maximize its profits?

Solution

To find the absolute maximum of P on [0, 20,000], first find the critical points of P on the interval (0, 20,000). To do this, compute

$$P'(x) = -0.04x + 300$$

Solving the equation $P'(x) = 0$ gives $x = 7500$. Next, evaluate $P(x)$ at $x = 7500$ as well as the end points $x = 0$ and $x = 20{,}000$ of the interval [0, 20,000], obtaining

$$P(0) = -200{,}000$$

$$P(7500) = 925{,}000$$

$$P(20{,}000) = -2{,}200{,}000$$

From these computations we see that the absolute maximum value of the function P is 925,000. Thus, by producing 7500 units, Acrosonic will realize a maximum profit of $925,000. The graph of P is sketched in Figure 7.65.

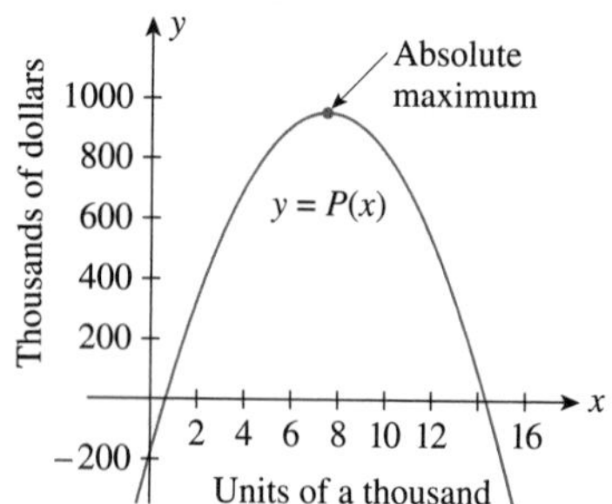

Figure 7.65
P has an absolute maximum at (7500, 925,000).

GROUP DISCUSSION

Recall that the total profit function P is defined as $P(x) = R(x) - C(x)$, where R is the total revenue function, C is the total cost function, and x is the number of units of a product produced and sold. (Assume all derivatives exist.)

1. Show that at the level of production x_0 that yields the maximum profit for the company, the following two conditions are satisfied:

$$R'(x_0) = C'(x_0) \quad \text{and} \quad R''(x_0) < C''(x_0)$$

2. Interpret the two conditions in part 1 in economic terms and explain why they make sense.

Example 5

Maximum of a Biological Population Suppose that a population of mice introduced into Alberta is modelled by the equation

$$P(t) = 5000t \cdot e^{-t/2}$$

After how many years is the population at a maximum and what is the maximum number of mice? The graph of the population of mice is shown in Figure 7.66.

Solution

To find critical points we first take the derivative of $P(t)$. Taking the derivative of $P(t)$ requires the product rule and, for $e^{t/2}$, the chain rule.

$$P'(t) = 5000\left(t \cdot e^{-t/2} \cdot \frac{-1}{2} + 1 \cdot e^{-t/2}\right)$$

$$= 5000\left(-\frac{1}{2}t \cdot e^{-t/2} + e^{-t/2}\right)$$

$$= 5000(-t/2 + 1)e^{-t/2}$$

We now solve for the critical points. Recall that the exponential function is never zero and so if $5000(-t/2 + 1)e^{t/2} = 0$ the only possibility is that

$$5000(-t/2 + 1) = 0$$

and so

$$5000(-t/2 + 1) = 0$$

$$-2500t + 5000 = 0$$

$$-2500t = -5000$$

$$t = 2 \text{ years}$$

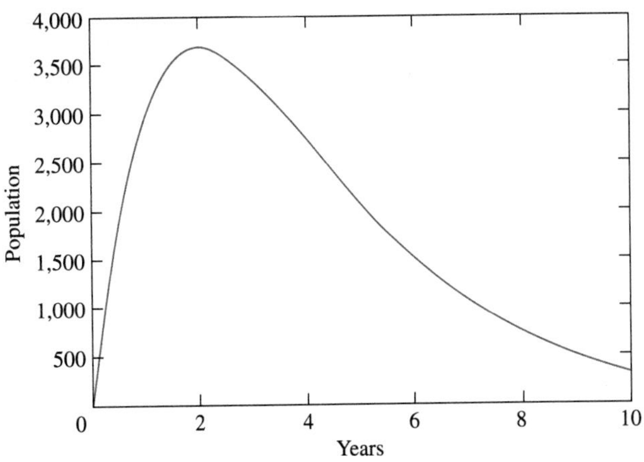

Figure 7.66
The population of a type of introduced mouse

There is a single critical point at two years. Examining the graph of the function shown in Figure 7.66 we see that it is a maximum. To find the number of mice at 2 years we compute $P(2) = 10{,}000e^{-2/2} = \frac{10{,}000}{e} = 3679$ mice, rounded to the nearest whole mouse.

Example 6

Trachea Contraction during a Cough When a person coughs, the trachea (windpipe) contracts, allowing air to be expelled at a maximum velocity. It can be shown that during a cough the velocity v of airflow is given by the function

$$v = f(r) = kr^2(R - r)$$

where r is the trachea's radius (in centimetres) during a cough, R is the trachea's normal radius (in centimeters), and k is a positive constant that depends on the length of the trachea. Find the radius r for which the velocity of airflow is greatest.

Solution

To find the absolute maximum of f on $[0, R]$, first find the critical points of f on the interval $(0, R)$. We compute

$$f'(r) = 2kr(R - r) - kr^2 \quad \text{Use the product rule.}$$

$$= -3kr^2 + 2kRr = kr(-3r + 2R)$$

Setting $f'(r) = 0$ gives $r = 0$ or $r = \frac{2}{3}R$, and so $r = \frac{2}{3}R$ is the sole critical point of f ($r = 0$ is an end point). Evaluating $f(r)$ at $r = \frac{2}{3}R$, as well as at the end points $r = 0$ and $r = R$, we obtain

$$f(0) = 0$$

$$f\left(\frac{2}{3}R\right) = \frac{4k}{27}R^3$$

$$f(R) = 0$$

from which we deduce that the velocity of airflow is greatest when the radius of the contracted trachea is $\frac{2}{3}R$—that is, when the radius is contracted by approximately 33%. The graph of the function f is shown in Figure 7.67.

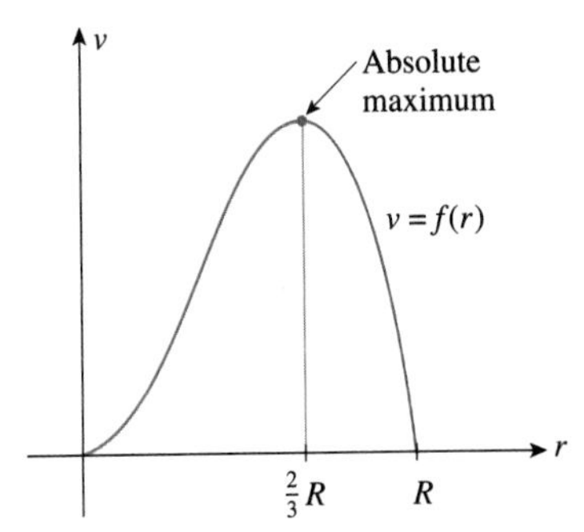

Figure 7.67
The velocity of airflow is greatest when the radius of the contracted trachea is $\frac{2}{3}R$.

GROUP DISCUSSION

Prove that if a cost function $C(x)$ is concave upward $[C''(x) > 0]$, then the level of production that will result in the smallest average production cost occurs when

$$\overline{C}(x) = C'(x)$$

—that is, when the average cost $\overline{C}(x)$ is equal to the marginal cost $C'(x)$.

Hints:

1. Show that

$$\overline{C}'(x) = \frac{xC'(x) - C(x)}{x^2}$$

so that the critical point of the function $\overline{C}$ occurs when

$$xC'(x) - C(x) = 0$$

2. Show that at a critical point of $\overline{C}$

$$\overline{C}''(x) = \frac{C''(x)}{x}$$

Use the second derivative test to reach the desired conclusion.

Example 7

Minimizing Average Cost The daily average cost function (in dollars per unit) of Elektra Electronics is given by

$$\overline{C}(x) = 0.0001x^2 - 0.08x + 40 + \frac{5000}{x} \qquad (x > 0)$$

where x stands for the number of programmable calculators that Elektra produces. Show that a production level of 500 units per day results in a minimum average cost for the company.

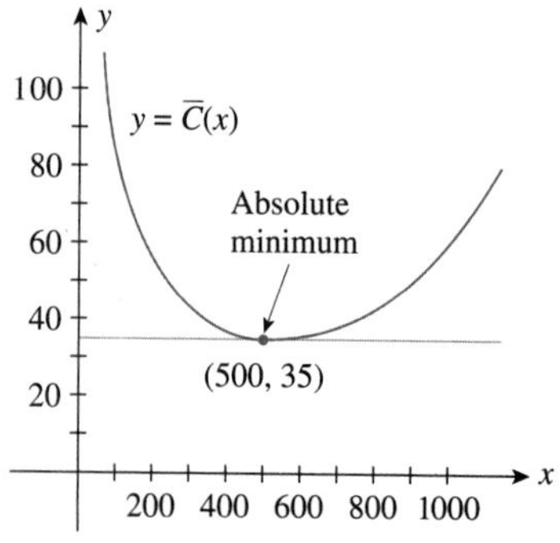

Figure 7.68
The minimum average cost is $35 per unit.

Solution

The domain of the function $\overline{C}$ is the interval $(0, \infty)$, which is not closed. To solve the problem, we resort to the graphic method. Using the techniques of graphing from the last section, we sketch the graph of $\overline{C}$ (Figure 7.68).

Now,

$$\overline{C}'(x) = 0.0002x - 0.08 - \frac{5000}{x^2}$$

Substituting the given value of x, 500, into $\overline{C}'(x)$ gives $\overline{C}'(500) = 0$, so $x = 500$ is a critical point of C. Next,

$$\overline{C}''(x) = 0.0002 + \frac{10{,}000}{x^3}$$

Thus,

$$\overline{C}''(500) = 0.0002 + \frac{10{,}000}{(500)^3} > 0$$

and by the second derivative test, a relative minimum of the function $\overline{C}$ occurs at the point $x = 500$. Furthermore, $\overline{C}''(x) > 0$ for $x > 0$, which implies that the graph of $\overline{C}$ is concave upward everywhere, so the relative minimum of $\overline{C}$ must be the absolute minimum of $\overline{C}$. The minimum average cost is given by

$$\begin{aligned}\overline{C}(500) &= 0.0001(500)^2 - 0.08(500) + 40 + \frac{5000}{500}\\ &= 35\end{aligned}$$

or \$35 per unit.

Example 8

Flight of a Rocket The altitude (in metres) of a rocket t seconds into flight is given by

$$s = f(t) = -\frac{1}{3}t^3 + 32t^2 + 65t + 2 \qquad (t \geq 0)$$

a. Find the maximum altitude attained by the rocket.
b. Find the maximum velocity attained by the rocket.

Solution

a. The maximum altitude attained by the rocket is given by the largest value of the function f in the closed interval $[0, T]$, where T denotes the time the rocket impacts Earth. We know that such a number exists because the dominant term in the expression for the continuous function f is $-\frac{1}{3}t^3$. So for t large enough, the value of $f(t)$ must change from positive to negative and, in particular, it must attain the value 0 for some T.

To find the absolute maximum of f, compute

$$\begin{aligned}f'(t) &= -t^2 + 64t + 65\\ &= -(t - 65)(t + 1)\end{aligned}$$

and solve the equation $f'(t) = 0$, obtaining $t = -1$ and $t = 65$. Ignore $t = -1$ since it lies outside the interval $[0, T]$. This leaves the critical point $t = 65$ of f. Continuing, we compute

$$f(0) = 2, \qquad f(65) = 47{,}885\frac{1}{3} \qquad f(T) = 0$$

and conclude, accordingly, that the absolute maximum value of f is $47{,}885\frac{1}{3}$. Thus, the maximum altitude of the rocket is $47{,}885\frac{1}{3}$ metres, attained 65 seconds into flight. The graph of f is sketched in Figure 7.69.

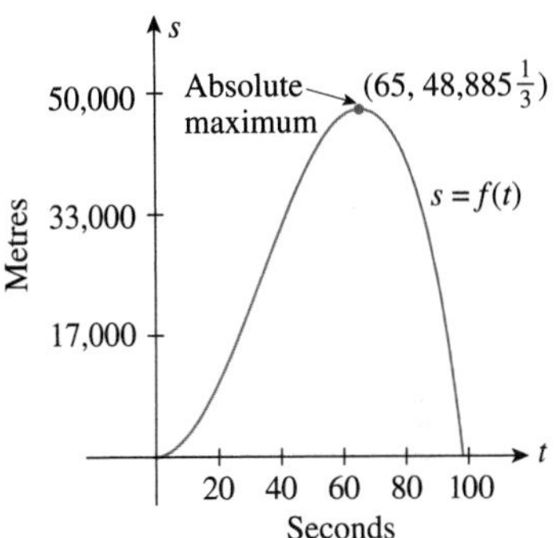

Figure 7.69
The maximum altitude of the rocket is $47{,}885\frac{1}{3}$ metres.

b. To find the maximum velocity attained by the rocket, find the largest value of the function that describes the rocket's velocity at any time t—namely,

$$v = f'(t) = -t^2 + 64t + 65 \qquad (t \geq 0)$$

We find the critical point of v by setting $v' = 0$. But

$$v' = -2t + 64$$

and the critical point of v is $t = 32$. Since

$$v'' = -2 < 0$$

the second derivative test implies that a relative maximum of v occurs at $t = 32$. Our computation has in fact clarified the property of the "velocity curve." Since $v'' < 0$ everywhere, the velocity curve is concave downward everywhere. With this observation, we assert that the relative maximum must in fact be the

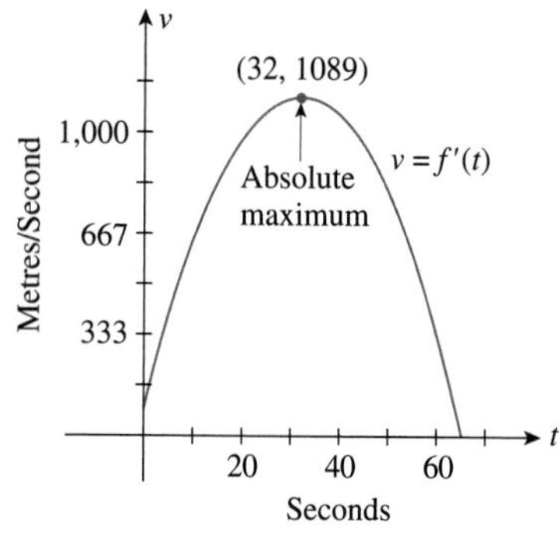

Figure 7.70
The maximum velocity of the rocket is 1089 metres per second.

absolute maximum of v. The maximum velocity of the rocket is given by evaluating v at $t = 32$,

$$f'(32) = -(32)^2 + 64(32) + 65$$

or 1089 metres per second. The graph of the velocity function v is sketched in Figure 7.70.

Monotone Functions

We say that a function $f(x)$ is *monotone increasing* if for any a and b with $a < b$ we have $f(a) < f(b)$. One way to tell if a function is monotone increasing is to check that the derivative is always positive. Monotone increasing functions go uphill from left to right.

A function is *monotone decreasing* if for any a and b with $a < b$ we have $f(a) > f(b)$. A monotone decreasing function has a first derivative that is always negative. Functions of this type go downhill from left to right.

We call a function that is either monotone increasing or monotone decreasing a **monotone function.**

The next example will show why knowledge of monotone functions can make optimization simpler.

Example 9

Monotone Functions Suppose that we want to find the maximum value of

$$g(x) = e^{2x-x^2} \quad \text{for} \quad 0 \leq x \leq 2$$

We also want to use the properties of monotone functions to make the calculus problem simpler.

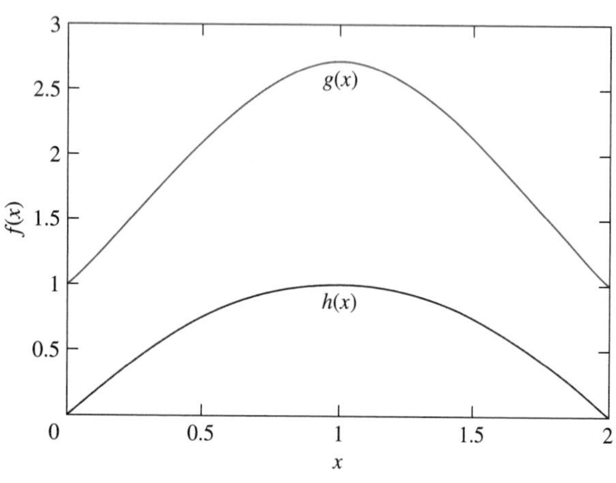

Figure 7.71
The graph shows $g(x) = e^{2x-x^2}$ and $h(x) = 2x - x^2$.

Solution

Remember that if $f(x) = e^x$ then $f'(x) = e^x$ and that for all x we know that $e^x > 0$. This means $f(x)$ is a monotone increasing function. Let us give the exponent in $g(x)$ a name by setting $h(x) = 2x - x^2$. This makes $g(x) = e^{h(x)}$. The fact that e^x is monotone increasing means that $g(x)$ is largest when $h(x)$ is largest, so we can just find the value of x that makes $h(x)$ largest and that will also be the correct value to make $g(x)$ largest. Find the zeros of the derivative of $h(x)$.

$$h'(x) = 0$$
$$2 - 2x = 0$$
$$2 = 2x$$
$$x = 1$$

There is a single candidate value, $x = 1$. Examining Figure 7.71, which shows the graphs of both $g(x)$ and $h(x)$, we see this is in fact a maximum. Notice that the graphs of $g(x)$ and $h(x)$ go up and downhill in the same way even though they take on different values. Monotone increasing functions, like e^x, preserve relative height in this fashion.

Monotone Functions and Optimization

Suppose that $f(x)$ is a monotone function and that we want to find relative maxima and minima of $f(g(x))$. As long as $f(x)$ has a domain that includes the range of $g(x)$, then the relative maxima and minima of $f(g(x))$ occur at the same values of x as the relative maxima and minima of $g(x)$.

If $f(x)$ is a monotone increasing function then the relative maxima of $f(g(x))$ appear at the same values of x as the relative maxima of $g(x)$. Likewise the relative minima of $f(g(x))$ happen at the same values of x as the relative minima of $g(x)$.

If $f(x)$ is a monotone decreasing function then the relative maxima of $f(g(x))$ appear at the same values of x as the relative minima of $g(x)$. The relative minima of $f(g(x))$ happen at the same values of x as the relative maxima of $g(x)$. Monotone decreasing functions reverse the maxima and minima of functions they are composed with.

Examples of monotone functions

Increasing

- $f(x) = e^x$
- $g(x) = \ln(x)$
- $h(x) = \arctan(x)$
- $a(x) = \sqrt{x}$

Decreasing

- $b(x) = -e^x$
- $c(x) = -\ln(x)$
- $d(x) = -\arctan(x)$
- $k(x) = \frac{1}{x}$

Example 10

Monotone Functions Suppose that we want to find the minimum value of

$$g(x) = \ln(x^2 - 4x + 5)$$

We also want to use the properties of monotone functions to make the calculus problem simpler.

Solution

The function $\ln(x)$ is one of the examples of a monotone increasing function. That means that it will have its minimum value at the minimum of the function $h(x) = x^2 - 4x + 5$. This function is a quadratic that opens upward and so has a single minimum value. We find the single critical point by solving the derivative equal to zero.

$$h'(x) = 2x - 4$$
$$2x - 4 = 0$$
$$2x = 4$$
$$x = 2$$

The minimum value of $h(x)$ and $g(x)$ occur at $x = 2$. The minimum value we want is $g(2) = \ln(2^2 - 4 \cdot 2 + 5) = \ln(1) = 0$.

When using monotone function techniques we solve for the minima and maxima of a simpler function to get the x values where the minima and maxima of the original function occur. It is important to plug these x values into the *original* function, not the simpler one we used to solve for the x values.

When a new species is introduced to an area it sometimes grows rapidly for a time and then dies back as the local environment adapts to it. Populations of this type can be modelled as a polynomial times an exponential with a negative exponent. Finding the extreme values of such a model lets us compute the maximum population that was present at any time.

Self-Check Exercises 7.4

1. Let $f(x) = x - 2\sqrt{x}$.
   a. Find the absolute extrema of f on the interval $[0, 9]$.
   b. Find the absolute extrema of f.
2. Find the absolute extrema of $f(x) = 3x^4 + 4x^3 + 1$ on $[-2, 1]$.
3. The operating rate (expressed as a percent) of factories, mines, and utilities in a certain region of the country on the tth day of 2000 is given by the function

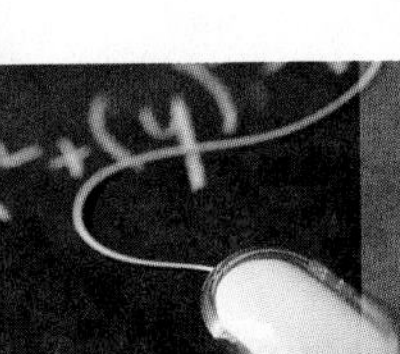

$$f(t) = 80 + \frac{1200t}{t^2 + 40{,}000} \qquad (0 \le t \le 250)$$

On which day of the first 250 days of 2000 was the manufacturing capacity operating rate highest?

Solutions to Self-Check Exercises 7.4 can be found on page 398.

7.4 Exercises

In Exercises 1–8, you are given the graph of some function f defined on the indicated interval. Find the absolute maximum and the absolute minimum of f, if they exist.

1.

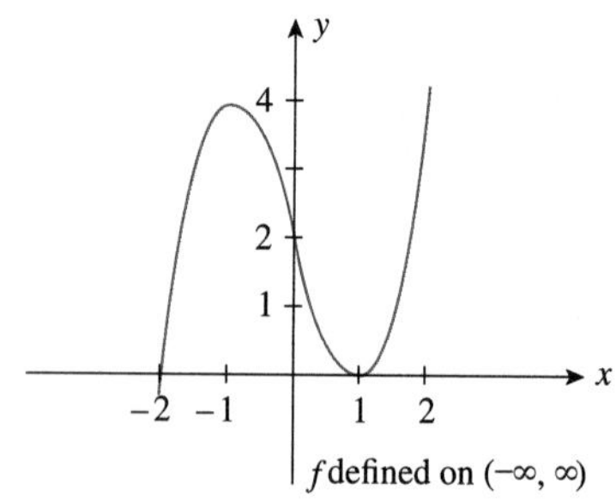

2.

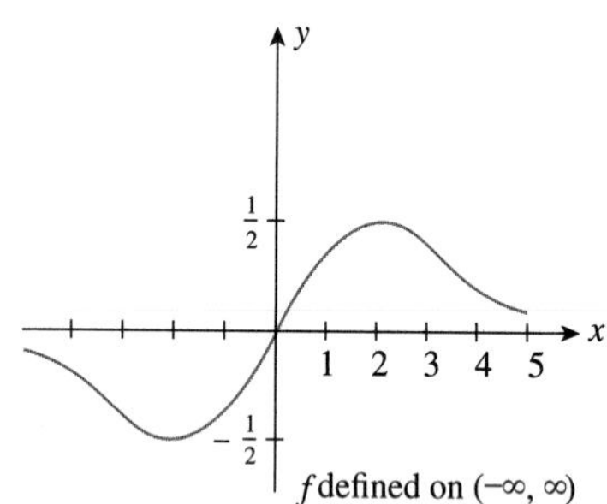

3.

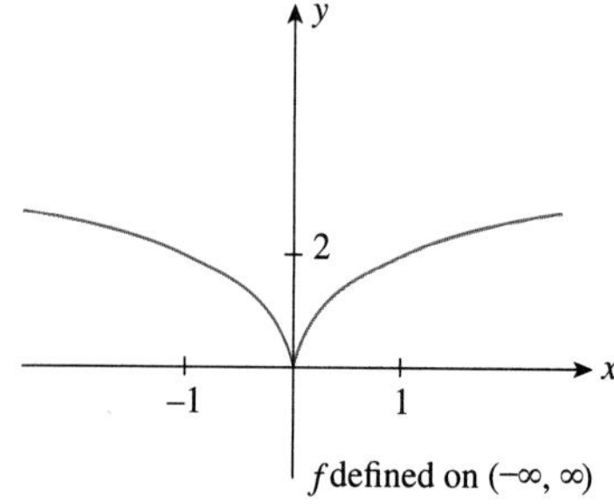

4.

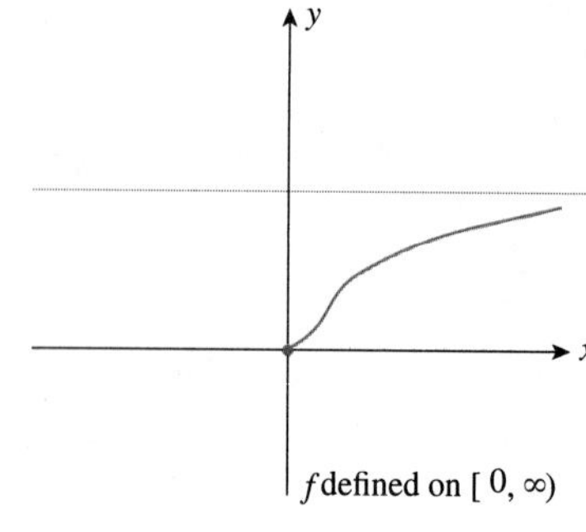

5.

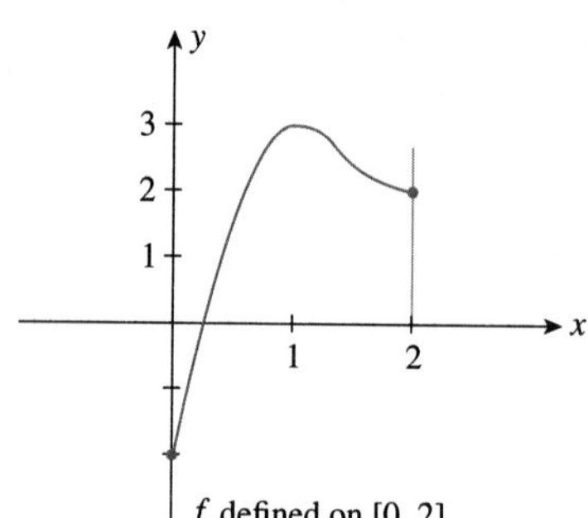

6.

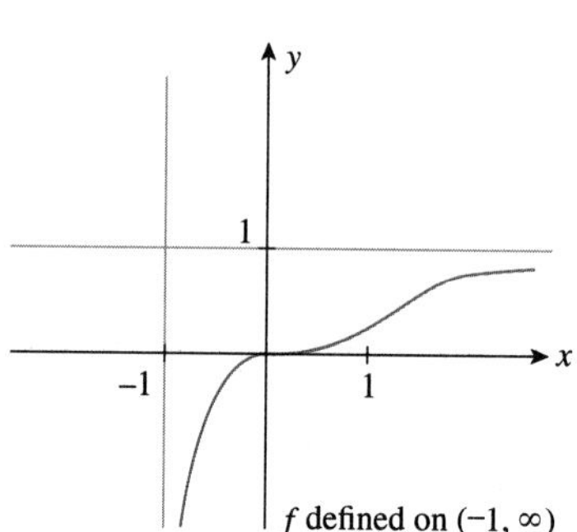

7.

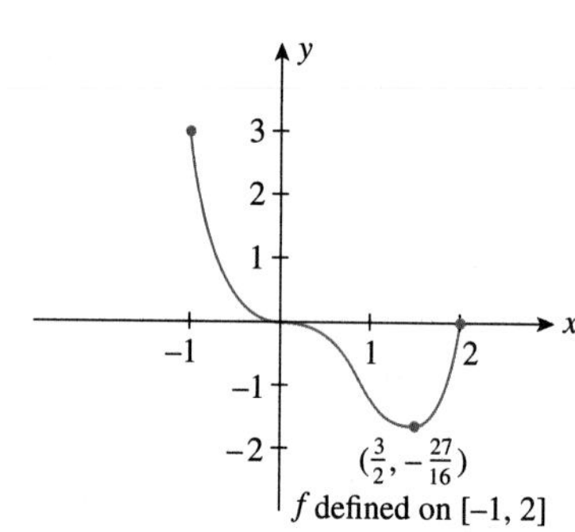

8.

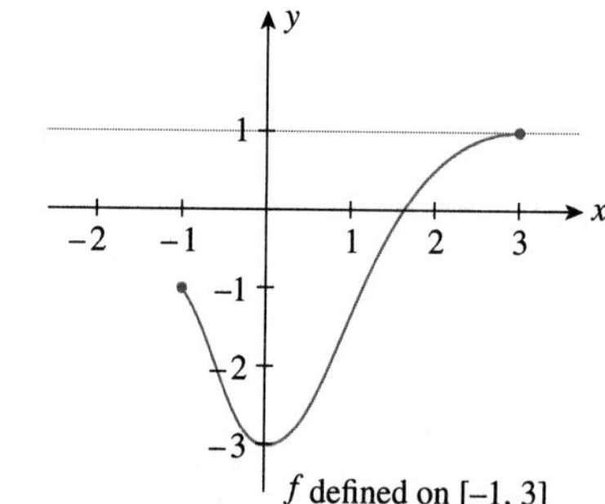

In Exercises 9–53, find the absolute maximum value and the absolute minimum value, if any, of the given function.

9. $f(x) = 2x^2 + 3x - 4$

10. $g(x) = -x^2 + 4x + 3$

11. $h(x) = x^{1/3}$

12. $f(x) = x^{2/3}$

13. $f(x) = \dfrac{1}{1 + x^2}$

14. $f(x) = \dfrac{x}{1 + x^2}$

15. $f(x) = x^2 - 2x - 3$ on $[-2, 3]$

16. $g(x) = x^2 - 2x - 3$ on $[0, 4]$

17. $f(x) = -x^2 + 4x + 6$ on $[0, 5]$

18. $f(x) = -x^2 + 4x + 6$ on $[3, 6]$

19. $f(x) = x^3 + 3x^2 - 1$ on $[-3, 2]$

20. $g(x) = x^3 + 3x^2 - 1$ on $[-3, 1]$

21. $g(x) = 3x^4 + 4x^3$ on $[-2, 1]$

22. $f(x) = \frac{1}{2}x^4 - \frac{2}{3}x^3 - 2x^2 + 3$ on $[-2, 3]$

23. $f(x) = \frac{x+1}{x-1}$ on $[2, 4]$

24. $g(t) = \frac{t}{t-1}$ on $[2, 4]$

25. $f(x) = 4x + \frac{1}{x}$ on $[1, 3]$

26. $f(x) = 9x - \frac{1}{x}$ on $[1, 3]$

27. $f(x) = \frac{1}{2}x^2 - 2\sqrt{x}$ on $[0, 3]$

28. $g(x) = \frac{1}{8}x^2 - 4\sqrt{x}$ on $[0, 9]$

29. $f(x) = \frac{1}{x}$ on $(0, \infty)$

30. $g(x) = \frac{1}{x+1}$ on $(0, \infty)$

31. $f(x) = 3x^{2/3} - 2x$ on $[0, 3]$

32. $g(x) = x^2 + 2x^{2/3}$ on $[-2, 2]$

33. $f(x) = x^{2/3}(x^2 - 4)$ on $[-1, 2]$

34. $f(x) = x^{2/3}(x^2 - 4)$ on $[-1, 3]$

35. $f(x) = \frac{x}{x^2 + 2}$ on $[-1, 2]$

36. $f(x) = \frac{1}{x^2 + 2x + 5}$ on $[-2, 1]$

37. $f(x) = \frac{x}{\sqrt{x^2 + 1}}$ on $[-1, 1]$

38. $g(x) = x\sqrt{4 - x^2}$ on $[0, 2]$

39. $f(x) = e^{-x}$ on $[0, 20]$

40. $g(x) = x \cdot e^{-x}$ on $[0, 5]$

41. $f(x) = (x + 1) \cdot e^{-x}$ on $[0, 3]$

42. $g(x) = (x + 1) \cdot e^{-x/2}$ on $[0, 10]$

43. $h(x) = x^2 \cdot e^{-x/3}$ on $[0, 12]$

44. $f(x) = (x^2 + 2x + 2) \cdot e^{-x}$ on $[0, 10]$

45. $f(x) = (x^2 + 2x + 1) \cdot e^{-x/2}$ on $[0, 4]$

46. $g(x) = \ln(x^2 + 1)$ on $[-3, 3]$

47. $f(x) = x \cdot \ln(x)$ on $(0, 2]$

48. $f(x) = \cos(x)$ on $\left[\frac{\pi}{4}, \frac{\pi}{3}\right]$

49. $f(x) = \sin(x)$ on $[0, 11\pi]$

50. $f(x) = \sin\left(\frac{1}{x}\right)$ on $[1, 20]$

51. $f(x) = \cos\left(\frac{1}{x^2 + 1}\right)$ on $[-5, 5]$

52. $f(x) = \sin(x) - \cos(x)$ on $[0, \pi]$

53. $f(x) = \arctan\left(\frac{1}{x^2 + 1}\right)$ on $[1, 20]$

In Exercises 54–59, use the properties of monotone functions to help find the absolute minimum and absolute maximum values, if any, of the function.

54. $f(x) = e^{2x^2+3x-4}$

55. $f(x) = e^{x/\sqrt{x^2+1}}$ on $[-1, 1]$

56. $f(x) = e^{1/x^2+2x+5}$ on $[-2, 1]$

57. $f(x) = \ln\left(\frac{1}{1 + x^2}\right)$

58. $f(x) = \ln((x + 1) \cdot e^{-x})$ on $[0, 3]$

59. $g(x) = \arctan\left(\frac{1}{2}x^2 - 2\sqrt{x}\right)$ on $[0, 3]$

60. Let f be a constant function—that is, let $f(x) = c$, where c is some real number. Show that every point $x = a$ is an absolute maximum and, at the same time, an absolute minimum of f.

61. Show that a polynomial function defined on the interval $(-\infty, \infty)$ cannot have both an absolute maximum and an absolute minimum unless it is a constant function.

62. One condition that must be satisfied before Theorem 3 (p. 386) is applicable is that the function f must be continuous on the closed interval $[a, b]$. Define a function f on the closed interval $[-1, 1]$ by

$$f(x) = \begin{cases} \frac{1}{x} & \text{if } x \in [-1, 1] \quad (x \neq 0) \\ 0 & \text{if } x = 0 \end{cases}$$

a. Show that f is not continuous at $x = 0$.
b. Show that $f(x)$ does not attain an absolute maximum or an absolute minimum on the interval $[-1, 1]$.
c. Confirm your results by sketching the function f.

63. One condition that must be satisfied before Theorem 3 (page 386) is applicable is that the interval on which f is defined must be a closed interval $[a, b]$. Define a function f on the *open* interval $(-1, 1)$ by $f(x) = x$. Show that f does not attain an absolute maximum or an absolute minimum on the interval $(-1, 1)$.
Hint: What happens to $f(x)$ if x is close to but not equal to $x = -1$? If x is close to but not equal to $x = 1$?

A calculator is recommended for Exercises 64–90.

64. A stone is thrown straight up from the roof of a 20-metre building. The height (in metres) of the stone at any time t (in seconds), measured from the ground, is given by

$$h(t) = -5t^2 + 20t + 20$$

What is the maximum height the stone reaches?

65. **Flight of a Rocket** The altitude (in metres) attained by a model rocket t sec into flight is given by the function

$$h(t) = -\frac{1}{3}t^3 + t^2 + 8t + 2$$

Find the maximum altitude attained by the rocket.

66. **Average Speed of a Vehicle on a Highway** The average speed of a vehicle on a stretch of Highway 401 between 6 A.M. and 10 A.M. on a typical weekday is approximated by the function

$$f(t) = 32t - 64\sqrt{t} + 80 \qquad (0 \le t \le 4)$$

where $f(t)$ is measured in kilometres per hour and t is measured in hours, with $t = 0$ corresponding to 6 A.M. At what time of the morning commute is the traffic moving at the slowest rate? What is the average speed of a vehicle at that time?

Business and Economics Applications

67. **Maximizing Profits** Lynbrook West, an apartment complex, has 100 two-bedroom units. The monthly profit (in dollars) realized from renting out x apartments is given by

$$P(x) = -10x^2 + 1760x - 50{,}000$$

To maximize the monthly rental profit, how many units should be rented out? What is the maximum monthly profit realizable?

68. **Maximizing Profits** The estimated monthly profit (in dollars) realizable by Cannon Precision Instruments for manufacturing and selling x units of its model M1 camera is

$$P(x) = -0.04x^2 + 240x - 10{,}000$$

To maximize its profits, how many cameras should Cannon produce each month?

69. **Maximizing Profits** The management of Trappee and Sons, producers of the famous TexaPep hot sauce, estimate that their profit (in dollars) from the daily production and sale of x cases (each case consisting of 24 bottles) of the hot sauce is given by

$$P(x) = -0.000002x^3 + 6x - 400$$

What is the largest possible profit Trappee can make in 1 day?

70. **Maximizing Profits** The quantity demanded each month of the Walter Serkin recording of Beethoven's Moonlight Sonata, manufactured by Phonola Record Industries, is related to the price/compact disc. The equation

$$p = -0.00042x + 6 \qquad (0 \le x \le 12{,}000)$$

where p denotes the unit price in dollars and x is the number of discs demanded, relates the demand to the price. The total monthly cost (in dollars) for pressing and packaging x copies of this classical recording is given by

$$C(x) = 600 + 2x - 0.00002x^2 \qquad (0 \le x \le 20{,}000)$$

To maximize its profits, how many copies should Phonola produce each month?

Hint: The revenue is $R(x) = px$, and the profit is $P(x) = R(x) - C(x)$.

71. **Maximizing Profit** A manufacturer of tennis rackets finds that the total cost $C(x)$ (in dollars) of manufacturing x rackets/day is given by $C(x) = 400 + 4x + 0.0001x^2$. Each racket can be sold at a price of p dollars, where p is related to x by the demand equation $p = 10 - 0.0004x$. If all rackets that are manufactured can be sold, find the daily level of production that will yield a maximum profit for the manufacturer.

72. **Maximizing Profit** The weekly demand for the Pulsar 25-inch colour console television is given by the demand equation

$$p = -0.05x + 600 \qquad (0 \le x \le 12{,}000)$$

where p denotes the wholesale unit price in dollars and x denotes the quantity demanded. The weekly total cost function associated with manufacturing these sets is given by

$$C(x) = 0.000002x^3 - 0.03x^2 + 400x + 80{,}000$$

where $C(x)$ denotes the total cost incurred in producing x sets. Find the level of production that will yield a maximum profit for the manufacturer.

Hint: Use the quadratic formula.

73. **Maximizing Profit** A division of Chapman Corporation manufactures a pager. The weekly fixed cost for the division is \$20,000, and the variable cost for producing x pagers/week is

$$V(x) = 0.000001x^3 - 0.01x^2 + 50x$$

dollars. The company realizes a revenue of

$$R(x) = -0.02x^2 + 150x \qquad (0 \le x \le 7500)$$

dollars from the sale of x pagers/week. Find the level of production that will yield a maximum profit for the manufacturer.

Hint: Use the quadratic formula.

74. **Minimizing Average Cost** Suppose the total cost function for manufacturing a certain product is $C(x) = 0.2(0.01x^2 + 120)$ dollars, where x represents the number of units produced. Find the level of production that will minimize the average cost.

75. **Minimizing Production Costs** The total monthly cost (in dollars) incurred by Cannon Precision Instruments for manufacturing x units of the model M1 camera is given by the function

$$C(x) = 0.0025x^2 + 80x + 10{,}000$$

a. Find the average cost function $\overline{C}$.
b. Find the level of production that results in the smallest average production cost.
c. Find the level of production for which the average cost is equal to the marginal cost.
d. Compare the result of part (c) with that of part (b).

76. **Minimizing Production Costs** The daily total cost (in dollars) incurred by Trappee and Sons for producing x cases of TexaPep hot sauce is given by the function

$$C(x) = 0.000002x^3 + 5x + 400$$

Using this function, answer the questions posed in Exercise 76.

77. **Maximizing Revenue** Suppose the quantity demanded per week of a certain dress is related to the unit price p by the demand equation $p = \sqrt{800 - x}$, where p is in dollars and x is the number of dresses made. To maximize the revenue, how many dresses should be made and sold each week?
Hint: $R(x) = px$.

78. **Maximizing Revenue** The quantity demanded each month of the Sicard wristwatch is related to the unit price by the equation

$$p = \frac{50}{0.01x^2 + 1} \qquad (0 \leq x \leq 20)$$

where p is measured in dollars and x is measured in units of a thousand. To yield a maximum revenue, how many watches must be sold?

79. **Maximizing Revenue** The average revenue is defined as the function

$$\overline{R}(x) = \frac{R(x)}{x} \qquad (x > 0)$$

Prove that if a revenue function $R(x)$ is concave downward $[R''(x) < 0]$, then the level of sales that will result in the largest average revenue occurs when $\overline{R}(x) = R'(x)$.

80. **GDP of a Developing Country** A developing country's gross domestic product (GDP) from 1993 to 2001 is approximated by the function

$$G(t) = -0.2t^3 + 2.4t^2 + 60 \qquad (0 \leq t \leq 8)$$

where $G(t)$ is measured in billions of dollars and $t = 0$ corresponds to 1993. Show that the growth rate of the country's GDP was maximal in 1997.

81. **Canada Pension Plan Surplus** Data show that the estimated cash in the Canada Pension Plan trust fund may be approximated by

$$f(t) = -0.0129t^4 + 0.3087t^3 + 2.1760t^2 + 62.8466t + 506.2955$$

$$(0 \leq t \leq 35)$$

where $f(t)$ is measured in billions of dollars and t is measured in years, with $t = 0$ corresponding to 1995. Show that the pension will be at its highest level at approximately the middle of the year 2018.
Hint: Show that $t = 23.6811$ is an approximate critical point of $f'(t)$.

82. **Female Self-Employed Workforce** Data show that the number of nonfarm, full-time, self-employed women can be approximated by

$$N(t) = 0.81t - 1.14\sqrt{t} + 1.53 \qquad (0 \leq t \leq 6)$$

where $N(t)$ is measured in millions and t is measured in 5-yr intervals, with $t = 0$ corresponding to the beginning of 1963. Determine the absolute extrema of the function N on the interval $[0, 6]$. Interpret your results.

Biological and Life Sciences Applications

83. **Oxygen Content of a Pond** When organic waste is dumped into a pond, the oxidation process that takes place reduces the pond's oxygen content. However, given time, nature will restore the oxygen content to its natural level. Suppose the oxygen content t days after organic waste has been dumped into the pond is given by

$$f(t) = 100\left[\frac{t^2 - 4t + 4}{t^2 + 4}\right] \qquad (0 \leq t \leq \infty)$$

percent of its normal level.
a. When is the level of oxygen content lowest?
b. When is the rate of oxygen regeneration greatest?

84. **Air Pollution** The amount of nitrogen dioxide, a brown gas that impairs breathing, present in the atmosphere on a certain August day in the city of Etobicoke is approximated by

$$A(t) = \frac{136}{1 + 0.25(t - 4.5)^2} + 28 \qquad (0 \leq t \leq 11)$$

where $A(t)$ is measured in pollutant standard index (PSI) and t is measured in hours, with $t = 0$ corresponding to 7 A.M. Determine the time of day when the pollution is at its highest level.

85. **Velocity of Blood** According to a law discovered by the nineteenth-century physician Jean Louis Marie Poiseuille, the velocity (in centimetres/second) of blood r cm from the central axis of an artery is given by

$$v(r) = k(R^2 - r^2)$$

where k is a constant and R is the radius of the artery. Show that the velocity of blood is greatest along the central axis.

86. **Crime Rates** The number of major crimes committed in the city of York between 1987 and 1994 is approximated by the function

$$N(t) = -0.1t^3 + 1.5t^2 + 100 \qquad (0 \leq t \leq 7)$$

where $N(t)$ denotes the number of crimes committed in year t ($t = 0$ corresponds to 1987). Enraged by the dramatic increase in the crime rate, the citizens of York with the help of the local police, organized "Neighborhood Crime Watch" groups in early 1991 to combat this menace. Show that the growth in the crime rate was maximal in 1992, giving credence to the claim that the Neighborhood Crime Watch program was working.

87. **Energy Expended by a Fish** It has been conjectured that a fish swimming a distance of L m at a speed of v m/sec relative to the water and against a current flowing at the rate of u m/sec ($u < v$) expends a total energy given by

$$E(v) = \frac{2aLv^3}{3(v - u)}$$

where E is measured in calorie and a is a constant. Find the speed v at which the fish must swim in order to minimize the total energy expended. (*Note:* This result has been verified by biologists.)

88. **Reaction to a Drug** The strength of a human body's reaction R to a dosage D of a certain drug is given by

$$R = D^2\left(\frac{k}{2} - \frac{D}{3}\right)$$

where k is a positive constant. Show that the maximum reaction is achieved if the dosage is k units.

89. Refer to Exercise 89. Show that the rate of change in the reaction R with respect to the dosage D is maximal if $D = k/2$.

90. **Population Density** An accidentally introduced species of fly in Nova Scotia is found, in a laboratory test plot, to have a population density in wetlands of about

$$P(t) = (7.6t + 2.5)e^{-t}$$

flies per square metre. At what time is the density the highest and what is that density?

91. **Maximum Density** Some weeds have seeds that lie dormant until the ground is turned over, at which point the seeds sprout rapidly. The weeds quickly make and disperse seeds and then die back. The number of such weeds in a test plot is found to be

$$P(t) = 11t^2e^{-t/3}$$

t weeks after the ground in the plot was turned over. When are the weeds at maximum density and what is that density?

92. **Maximum Number of Fish** A type of Siberian catfish is released from the ballast water of a freighter in a stream in British Columbia. After t months the number of individuals is estimated to be

$$Q(t) = t^3e^{-1.3t}$$

Find the absolute maximum number of these catfish in the stream. Does the model predict that the catfish will eventually die out?

93. **Number of Plants** After introducing a beetle that eats purple loosestrife it is found that the population of loosestrife plants in a test plot is given by

$$f(t) = 120 + 20\cos(t) - 20\sin(t)$$

What is the absolute maximum and minimum number of plants after the beetle has been introduced?

94. **Maximum Number of Beetles** A population of aphid-eating ladybird beetles is introduced into a soybean field to test the stability of the beetle population. The number of beetles in the population is modelled as

$$B(t) = e\frac{2t^2 + 12}{t^2 + 2}$$

after t months. If the experiment is run for three months what is the absolute maximum number of beetles in the test plot? Does the model suggest that the beetles can persist indefinitely?

95. **Number of Hectares** Suppose that the number of hectares covered by a type of grass after a small number of seeds are placed in a test area is found to be

$$A(t) = \ln\left(\frac{27x}{x^2 + 2}\right)$$

after t years. When is the number of acres covered at a maximum and what is the number of acres covered at that time?

In Exercises 96–99, determine whether the statement is true or false. If it is true, explain why it is true. If it is false, give an example to show why it is false.

96. If f is defined on a closed interval $[a, b]$, then f has an absolute maximum value.

97. If f is continuous on an open interval (a, b), then f does not have an absolute minimum value.

98. If f is not continuous on the closed interval $[a, b]$, then f cannot have an absolute maximum value.

99. If $f''(x) < 0$ on (a, b) and $f'(c) = 0$ where $a < c < b$, then $f(c)$ is the absolute maximum value of f on $[a, b]$.

Application

Our final example involves finding the absolute maximum of an exponential function.

Example 11

Optimal Market Price The present value of the market price of the Blakely Office Building is given by

$$P(t) = 300{,}000e^{-0.09t - \sqrt{t}/2} \qquad (0 \le t \le 0)$$

where t is measured in years and P in dollars. Find the optimal present value of the building's market price.

Solution

To find the maximum value of P over [0, 10], we compute

$$P'(t) = 300{,}000e^{-0.09t-\sqrt{t}/2}\frac{d}{dt}\left(-0.09t + \frac{1}{2}t^{1/2}\right)$$

$$= 300{,}000e^{-0.09t+\sqrt{t}/2}\left(-0.09 + \frac{1}{4}t^{-1/2}\right)$$

Setting $P'(t) = 0$ gives

$$-0.09 + \frac{1}{4t^{1/2}} = 0$$

since $e^{-0.09t+\sqrt{t}/2}$ is never zero for any value of t. Solving this equation, we find.

$$\frac{1}{4t^{1/2}} = 0.09$$

$$t^{1/2} = \frac{1}{4(0.09)}$$

$$= \frac{1}{0.36}$$

$$t \approx 7.72$$

the sole critical point of the function P. Finally, evaluating $P(t)$ at the critical point as well as at the end points of [0, 10], we have

t	0	7.72	10
$P(t)$	300,000	600,779	592,838

We conclude, accordingly, that the optimal present value of the property's market price is \$600,779 and that this will occur 7.72 years from now.

Solutions to Self-Check Exercises 7.4

1. **a.** The function f is continuous in its domain and differentiable in the interval (0, 9). The derivative of f is

$$f'(x) = 1 - x^{-1/2} = \frac{x^{1/2} - 1}{x^{1/2}}$$

and it is equal to zero when $x = 1$. Evaluating $f(x)$ at the end points $x = 0$ and $x = 9$ and at the critical point $x = 1$ of f, we have

$$f(0) = 0, \quad f(1) = -1, \quad f(9) = 3$$

From these results, we see that -1 is the absolute minimum value of f and 3 is the absolute maximum value of f.

b. In this case, the domain of f is the interval $[0, \infty)$, which is not closed. Therefore, we resort to the graphic method.

Using the techniques of graphing, we sketch in the accompanying figure the graph of f.

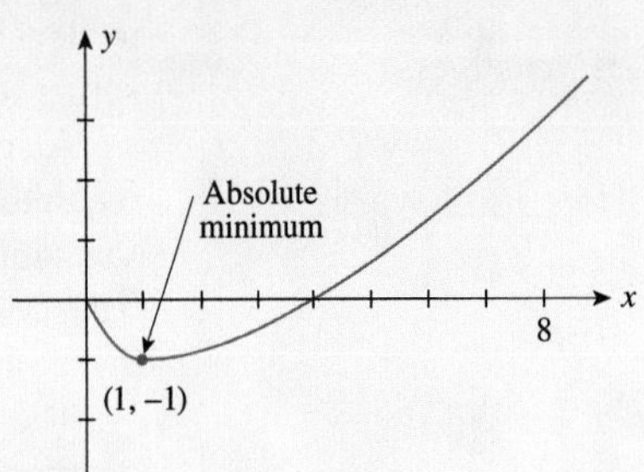

The graph of f shows that -1 is the absolute minimum value of f, but f has no absolute maximum since $f(x)$ increases without bound as x increases without bound.

Solutions to Self-Check Exercises 7.4, *continued*

2. The function f is continuous on the interval $[-2, 1]$. It is also differentiable on the open interval $(-2, 1)$. The derivative of f is

$$f'(x) = 12x^3 + 12x^2 = 12x^2(x + 1)$$

and it is continuous on $(-2, 1)$. Setting $f'(x) = 0$ gives $x = -1$ and $x = 0$ as critical points of f. Evaluating $f(x)$ at these critical points of f as well as at the end points of the interval $[-2, 1]$, we obtain

$$f(-2) = 17, \quad f(-1) = 0, \quad f(0) = 1, \quad f(1) = 8$$

From these results, we see that 0 is the absolute minimum value of f and 17 is the absolute maximum value of f.

3. The problem is solved by finding the absolute maximum of the function f on $[0, 250]$. Differentiating $f(t)$, we obtain

$$f'(t) = \frac{(t^2 + 40{,}000)(1200) - 1200t(2t)}{(t^2 + 40{,}000)^2}$$

$$= \frac{-1200(t^2 - 40{,}000)}{(t^2 + 40{,}000)^2}$$

Upon setting $f'(t) = 0$ and solving the resulting equation, we obtain $t = -200$ or 200. Since -200 lies outside the interval $[0, 250]$, we are interested only in the critical point $t = 200$ of f. Evaluating $f(t)$ at $t = 0$, $t = 200$, and $t = 250$, we find

$$f(0) = 80, \quad f(200) = 83, \quad f(250) = 82.93$$

We conclude that the manufacturing capacity operating rate was the highest on the 200th day of 2000—that is, a little past the middle of July 2000.

Buying women's clothes for a chain of 50 discount clothing stores is a demanding job, requiring Meiselman to be part fashion arbiter and part accountant. She must balance her fashion choices against a bottom line that has to show a profit. To make her job even more difficult,

PORTFOLIO

Lilli Meiselman

Title Buyer

Lilli Meiselman's job is challenging. She visits New York at least twice a month to buy clothes from a number of different manufacturers. She works with the store's in-house advertising agency, planning advertising as well as approving copy. Meiselman notes that the vast majority of the ads appear in local newspapers, the store's principal advertising medium.

For all the demands on her time, Meiselman enjoys her work. "It's gratifying to see things I bought for the stores being sold," she says. Although Meiselman decides how much to spend on particular items and which styles, sizes, and colours to select, her decisions are the end result of detailed plans that guide her buying decisions.

Based on the previous year's sales, Meiselman will "work with a departmental planner" to develop a seasonal merchandise plan (in retail there are only two seasons, fall and spring, each covering a 6-month time span).

As part of that merchandising plan, Meiselman must determine how much inventory is required per month. If she needs \$600,000 worth of inventory at the beginning of June, she subtracts projected sales of \$400,000 from the May inventory of \$800,000, leaving a balance of \$400,000. To meet her June inventory goal of \$600,000, she has to increase her inventory by an additional \$200,000.

The finances have to be coordinated with actual quantities of suits, dresses, blouses, outerwear, and so on. Using a sales-to-stock ratio, Meiselman estimates that perhaps half of a particular item will sell in any given month. To meet her financial goals, she needs sufficient quantity on hand to sell. For example, Meiselman would need 4000 raincoats in April retailing at \$99 each to reach a \$200,000 sales goal if the stores' sales were expected to reach a 50% sales-to-stock ratio.

Sales volume is the key to success in any discount business, whether clothing or home appliances. Meiselman's merchandise is marked up or down depending on that volume. Her goal is a 45% markup on all merchandise. If a line of suits doesn't sell, she may mark it down, achieving only a 35% markup over the wholesale price. To balance out the loss, she marks up another line of items so that her *average* markup hits 45%.

With its high overhead, the chain has to generate sufficient sales volume and profits to stay in business. Meiselman's fashion choices have helped make that goal a continuing reality.

Meiselman's selections are judged by women from northern New England to the midwestern states. What sells in one area may not sell in another. An incorrect choice can be a costly mistake.

7.5 Optimization II

Section 7.4 outlined how to find the solution to certain optimization problems in which the objective function is given. In this section we consider problems in which we are required to first find the appropriate function to be optimized. The following guidelines will be useful for solving these problems.

Guidelines for Solving Optimization Problems

1. Assign a letter to each variable mentioned in the problem. If appropriate, draw and label a figure.
2. Find an expression for the quantity to be optimized.
3. Use the conditions given in the problem to write the quantity to be optimized as a function f of *one* variable. Note any restrictions to be placed on the domain of f from physical considerations of the problem.
4. Optimize the function f over its domain using the methods of Section 7.4.

Remark

In carrying out step 4, remember that if the function f to be optimized is continuous on a closed interval, then the absolute maximum and absolute minimum of f are, respectively, the largest and smallest values of $f(x)$ on the set composed of the critical points of f and the end points of the interval. If the domain of f is not a closed interval, then we resort to the graphic method. ◀

Maximization Problems

Example 1

Fencing a Garden A man wishes to have a rectangular-shaped garden in his backyard. He has 50 metres of fencing material with which to enclose his garden. Find the dimensions for the largest garden he can have if he uses all of the fencing material.

Solution

Step 1 Let x and y denote the dimensions (in metres) of two adjacent sides of the garden (Figure 7.72) and let A denote its area.

Step 2 The area of the garden

$$A = xy \tag{1}$$

is the quantity to be maximized.

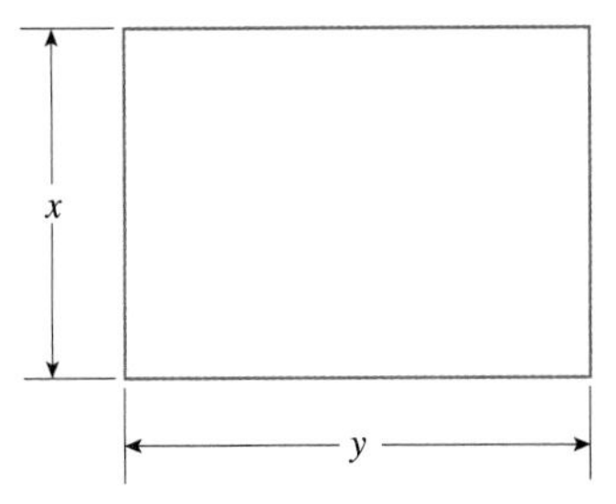

Figure 7.72
What is the maximum rectangular area that can be enclosed with 50 metres of fencing?

Step 3 The perimeter of the rectangle, $(2x + 2y)$ feet, must equal 50 metres. Therefore, we have the equation

$$2x + 2y = 50$$

Next, solving this equation for y in terms of x yields

$$y = 25 - x \tag{2}$$

which, when substituted into Equation (1), gives

$$\begin{aligned} A &= x(25 - x) \\ &= -x^2 + 25x \end{aligned}$$

(Remember, the function to be optimized must involve just one variable.) Since the sides of the rectangle must be nonnegative, we must have $x \geq 0$ and $y = 25 - x \geq 0$; that is, we must have $0 \leq x \leq 25$. Thus, the problem is reduced to that of finding the absolute maximum of $A = f(x) = -x^2 + 25x$ on the closed interval $[0, 25]$.

Step 4 Observe that f is continuous on $[0, 25]$, so the absolute maximum value of f must occur at the end point(s) or at the critical point(s) of f. The derivative of the function A is given by

$$A' = f'(x) = -2x + 25$$

Setting $A' = 0$ gives

$$-2x + 25 = 0$$

or $x = 12.5$, as the critical point of A. Next, we evaluate the function $A = f(x)$ at $x = 12.5$ and at the end points $x = 0$ and $x = 25$ of the interval $[0, 25]$, obtaining

$$f(0) = 0, \qquad f(12.5) = 156.25, \qquad f(25) = 0$$

We see that the absolute maximum value of the function f is 156.25. From Equation (2) we see that $y = 12.5$ when $x = 12.5$. Thus, the garden of maximum area (156.25 square metres) is a square with sides of length 12.5 metres.

Example 2

Packaging By cutting away identical squares from each corner of a rectangular piece of cardboard and folding up the resulting flaps, the cardboard may be turned into an open box. If the cardboard is 16 centimetres long and 10 centimetres wide, find the dimensions of the box that will yield the maximum volume.

Solution

Step 1 Let x denote the length (in centimetres) of one side of each of the identical squares to be cut out of the cardboard (Figure 7.73) and let V denote the volume of the resulting box.

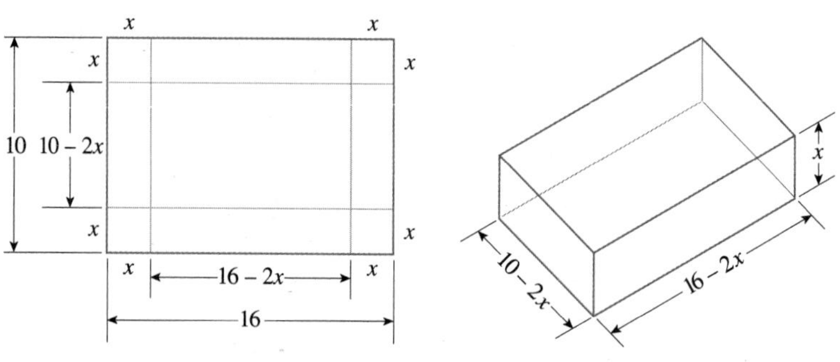

Figure 7.73

The dimensions of the open box are $(16 - 2x)$ by $(10 - 2x)$ by x centimetres.

Step 2 The dimensions of the box are $(16 - 2x)$ centimetres long, $(10 - 2x)$ centimetres wide, and x centimetres high. Therefore, its volume (in cubic centimetres),

$$V = (16 - 2x)(10 - 2x)x$$
$$= 4(x^3 - 13x^2 + 40x) \quad \text{Expand the expression.}$$

is the quantity to be maximized.

Step 3 Since each side of the box must be nonnegative, x must satisfy the inequalities $x \geq 0$, $16 - 2x \geq 0$, and $10 - 2x \geq 0$. This set of inequalities is satisfied if $0 \leq x \leq 5$. Thus, the problem at hand is equivalent to that of finding the absolute maximum of

$$V = f(x) = 4(x^3 - 13x^2 + 40x)$$

on the closed interval $[0, 5]$.

Step 4 Observe that f is continuous on $[0, 5]$, so the absolute maximum value of f must be attained at the end point(s) or at the critical point(s) of f.

Differentiating $f(x)$, we obtain

$$f'(x) = 4(3x^2 - 26x + 40)$$
$$= 4(3x - 20)(x - 2)$$

Upon setting $f'(x) = 0$ and solving the resulting equation for x, we obtain $x = \frac{20}{3}$ or $x = 2$. Since $\frac{20}{3}$ lies outside the interval $[0, 5]$, it is no longer considered, and we are interested only in the critical point $x = 2$ of f. Next, evaluating $f(x)$ at $x = 0$, $x = 5$ (the end points of the interval $[0, 5]$), and $x = 2$, we obtain

$$f(0) = 0, \qquad f(2) = 144, \qquad f(5) = 0$$

Thus, the volume of the box is maximized by taking $x = 2$. The dimensions of the box are $12 \times 6 \times 2$ centimetres, and the volume is 144 cubic centimetres.

Example 3

Optimal Subway Fare A city's Metropolitan Transit Authority (MTA) operates a subway line for commuters from a certain suburb to the downtown metropolitan area. Currently, an average of 6000 passengers a day take the trains, paying a fare of \$3.00 per ride. The board of the MTA, contemplating raising the fare to \$3.50 per ride in order to generate a larger revenue, engages the services of a consulting firm. The firm's study reveals that for each \$.50 increase in fare, the ridership will be reduced by an average of 1000 passengers a day. Thus, the consulting firm recommends that MTA stick to the current fare of \$3.00 per ride, which already yields a maximum revenue. Show that the consultants are correct.

Solution

Step 1 Let x denote the number of passengers per day, p denote the fare per ride, and R be MTA's revenue.

Step 2 To find a relationship between x and p, observe that the given data imply that when $x = 6000$, $p = 3$, and when $x = 5000$, $p = 3.50$. Therefore, the points $(6000, 3)$ and $(5000, 3.50)$ lie on a straight line. (Why?) To find the linear relationship between p and x, use the point-slope form of the equation of a straight line. Now, the slope of the line is

$$m = \frac{3.50 - 3}{5000 - 6000} = -0.0005$$

Therefore, the required equation is

$$p - 3 = -0.0005(x - 6000)$$
$$= -0.0005x + 3$$
$$p = -0.0005x + 6$$

Therefore, the revenue

$$R = f(x) = xp = -0.0005x^2 + 6x \qquad \text{Number of riders} \times \text{unit fare}$$

is the quantity to be maximized.

Step 3 Since both p and x must be nonnegative, we see that $0 \le x \le 12{,}000$, and the problem is that of finding the absolute maximum of the function f on the closed interval [0, 12,000].

Step 4 Observe that f is continuous on [0, 12,000]. To find the critical point of R, we compute

$$f'(x) = -0.001x + 6$$

and set it equal to zero, giving $x = 6000$. Evaluating the function f at $x = 6000$, as well as at the end points $x = 0$ and $x = 12{,}000$, yields

$$f(0) = 0$$
$$f(6000) = 18{,}000$$
$$f(12{,}000) = 0$$

We conclude that a maximum revenue of $18,000 per day is realized when the ridership is 6000 per day. The optimum price of the fare per ride is therefore $3.00, as recommended by the consultants. The graph of the revenue function R is shown in Figure 7.74.

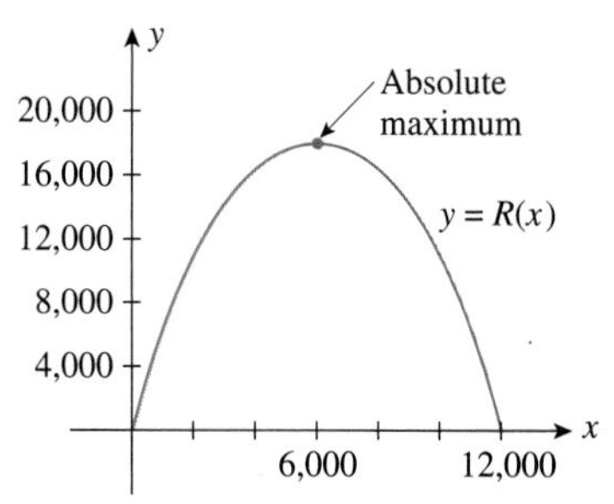

Figure 7.74
f has an absolute maximum of 18,000 when $x = 6000$.

Minimization Problems

Example 4

Packaging Betty Moore Company requires that its tuna containers have a capacity of 54 cubic centimetres, have the shape of right circular cylinders, and be made of aluminum. Determine the radius and height of the container that requires the least amount of metal.

Solution

Step 1 Let the radius and height of the container be r and h inches, respectively, and let S denote the surface area of the container (Figure 7.75).

Step 2 The amount of aluminum used to construct the container is given by the total surface area of the cylinder. Now, the area of the base and the top of the cylinder are each πr^2 square centimetres and the area of the side is $2\pi rh$ square centimetres. Therefore,

$$S = 2\pi r^2 + 2\pi rh \tag{3}$$

is the quantity to be minimized.

Step 3 The requirement that the volume of a container be 54 cubic centimetres implies that

$$\pi r^2 h = 54 \tag{4}$$

Solving Equation (4) for h, we obtain

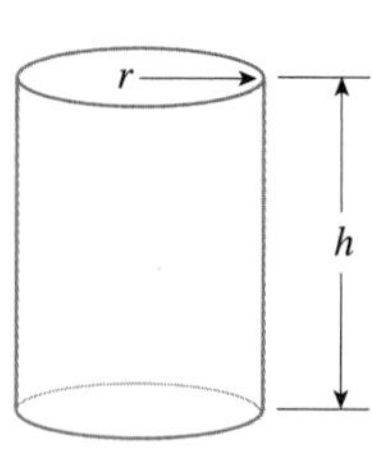

Figure 7.75
We want to minimize the amount of material used to construct the container.

$$h = \frac{54}{\pi r^2} \tag{5}$$

which, when substituted into (3), yields

$$S = 2\pi r^2 + 2\pi r\left(\frac{54}{\pi r^2}\right)$$

$$= 2\pi r^2 + \frac{108}{r}$$

Clearly, the radius r of the container must satisfy the inequality $r > 0$. The problem now is reduced to finding the absolute minimum of the function $S = f(r)$ on the interval $(0, \infty)$.

Step 4 Using the curve-sketching techniques of Section 7.3, we obtain the graph of f in Figure 7.76.

To find the critical point of f, we compute

$$S' = 4\pi r - \frac{108}{r^2}$$

and solve the equation $S' = 0$ for r:

$$4\pi r - \frac{108}{r^2} = 0$$

$$4\pi r^3 - 108 = 0$$

$$r^3 = \frac{27}{\pi}$$

$$r = \frac{3}{\sqrt[3]{\pi}} \approx 2 \tag{6}$$

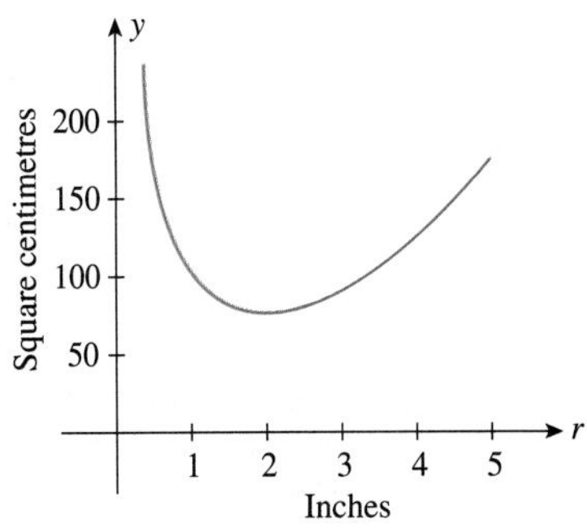

Figure 7.76
The total surface area of the right cylindrical container is graphed as a function of r.

Next, let's show that this value of r gives rise to the absolute minimum of f. To show this, we first compute

$$S'' = 4\pi + \frac{216}{r^3}$$

Since $S'' > 0$ for $r = 3/\sqrt[3]{\pi}$, the second derivative test implies that the value of r in Equation (6) gives rise to a relative minimum of f. Finally, this relative minimum of f is also the absolute minimum of f since f is always concave upward ($S'' > 0$ for all $r > 0$). To find the height of the given container, we substitute the value of r given in (6) into (5). Thus,

$$h = \frac{54}{\pi r^2} = \frac{54}{\pi\left(\dfrac{3}{\pi^{1/3}}\right)^2}$$

$$= \frac{54\pi^{2/3}}{(\pi)9}$$

$$= \frac{6}{\pi^{1/3}} = \frac{6}{\sqrt[3]{\pi}}$$

$$\approx 4.1$$

We conclude that the required container has a radius of approximately 2 centimetres and a height of approximately 4 centimetres, or twice the size of the radius.

An Inventory Problem

One problem faced by many companies is that of controlling the inventory of goods carried. Ideally, the manager must ensure that the company has sufficient stock to meet customer demand at all times. At the same time, she must make sure that this is accomplished without overstocking (incurring unnecessary storage costs) and also without having to place orders too frequently (incurring reordering costs).

Example 5

Inventory Control and Planning Dixie Import-Export is the sole agent for the Excalibur 250-cc motorcycle. Management estimates that the demand for these motorcycles is 10,000 per year and that they will sell at a uniform rate throughout the year. The cost incurred in ordering each shipment of motorcycles is \$10,000, and the cost per year of storing each motorcycle is \$200.

Dixie's management faces the following problem: Ordering too many motorcycles at one time ties up valuable storage space and increases the storage cost. On the other hand, placing orders too frequently increases the ordering costs. How large should each order be, and how often should orders be placed, to minimize ordering and storage costs?

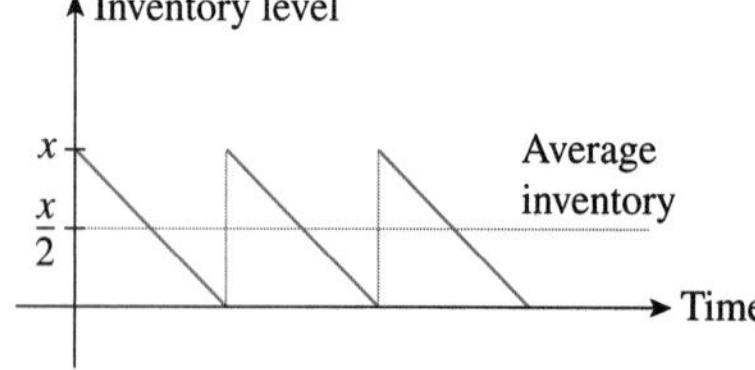

Figure 7.77
As each lot is depleted, the new lot arrives. The average inventory level is $x/2$ if x is the lot size.

Solution

Let x denote the number of motorcycles in each order (the lot size). Then, assuming that each shipment arrives just as the previous shipment has been sold, the average number of motorcycles in storage during the year is $x/2$. You can see that this is the case by examining Figure 7.77. Thus, Dixie's storage cost for the year is given by $200(x/2)$, or $100x$ dollars.

Next, since the company requires 10,000 motorcycles for the year and since each order is for x motorcycles, the number of orders required is

$$\frac{10{,}000}{x}$$

This gives an ordering cost of

$$10{,}000\left(\frac{10{,}000}{x}\right) = \frac{100{,}000{,}000}{x}$$

dollars for the year. Thus, the total yearly cost incurred by Dixie, which includes the ordering and storage costs attributed to the sale of these motorcycles, is given by

$$C(x) = 100x + \frac{100{,}000{,}000}{x}$$

The problem is reduced to finding the absolute minimum of the function C in the interval $(0, 10{,}000]$. To accomplish this, we compute

$$C'(x) = 100 - \frac{100{,}000{,}000}{x^2}$$

Setting $C'(x) = 0$ and solving the resulting equation, we obtain $x = \pm 1000$. Since the number -1000 is outside the domain of the function C, it is rejected, leaving $x = 1000$ as the only critical point of C. Next, we find

$$C''(x) = \frac{200{,}000{,}000}{x^3}$$

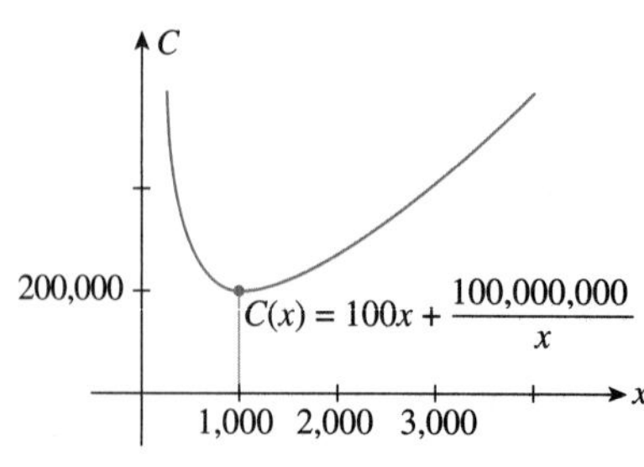

Figure 7.78
C has an absolute minimum at (1000, 200,000).

Since $C''(1000) > 0$, the second derivative test implies that the critical point $x = 1000$ is a relative minimum of the function C (Figure 7.78). Also, since $C''(x) > 0$ for all x in $(0, 10{,}000]$, the function C is concave upward everywhere so that the point $x = 1000$ also gives the absolute minimum of C. Thus, to minimize the ordering and storage costs, Dixie should place 10,000/1000, or 10, orders a year, each for a shipment of 1000 motorcycles.

Self-Check Exercises 7.5

1. A man wishes to have an enclosed vegetable garden in his backyard. If the garden is to be a rectangular area of 300 square metres, find the dimensions of the garden that will minimize the amount of fencing material needed.

2. The demand for the Super Titan tires is 1,000,000/year. The setup cost for each production run is \$4000, and the manufacturing cost is \$20/tire. The cost of storing each tire over the year is \$2. Assuming uniformity of demand throughout the year and instantaneous production, determine how many tires should be manufactured per production run in order to keep the production cost to a minimum.

Solutions to Self-Check Exercises 7.5 can be found on page 410.

7.5 Exercises

1. **Enclosing the Largest Area** The owner of the Champion Ranch has 3000 m of fencing material with which to enclose a rectangular piece of grazing land along the straight portion of a river. If fencing is not required along the river, what are the dimensions of the largest area that he can enclose? What is this area?

2. **Enclosing the Largest Area** Refer to Exercise 1. As an alternative plan, the owner of the Champion Ranch might use the 3000 m of fencing material to enclose the rectangular piece of grazing land along the straight portion of the river and then subdivide it by means of a fence running parallel to the sides. Again, no fencing is required along the river. What are the dimensions of the largest area that can be enclosed? What is this area? (See the accompanying figure.)

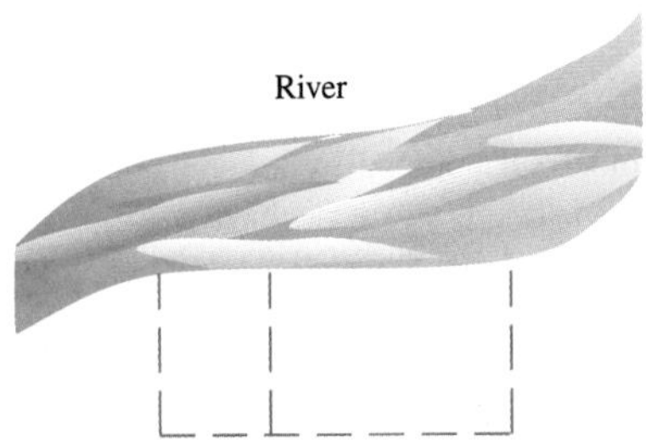

3. **Minimizing Construction Costs** The management of the UNICO department store has decided to enclose an 800-m^2 area outside the building for displaying potted plants and flowers. One side will be formed by the external wall of the store, two sides will be constructed of pine boards, and the fourth side will be made of galvanized steel fencing material. If the pine board fencing costs \$18/running metre and the steel fencing costs \$9/running metre, determine the dimensions of the enclosure that can be erected at minimum cost.

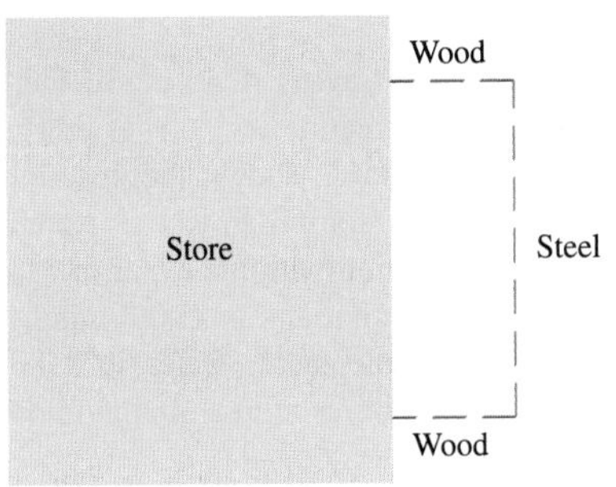

4. **Packaging** By cutting away identical squares from each corner of a rectangular piece of cardboard and folding up the resulting flaps, an open box may be made. If the cardboard is 15 cm long and 8 cm wide, find the dimensions of the box that will yield the maximum volume.

5. **Metal Fabrication** If an open box is made from a tin sheet 8 cm square by cutting out identical squares from each corner and bending up the resulting flaps, determine the dimensions of the largest box that can be made.

6. **Minimizing Packaging Costs** If an open box has a square base and a volume of 108 cm^3 and is constructed from a tin sheet, find the dimensions of the box, assuming a minimum amount of material is used in its construction.

7. **Minimizing Packaging Costs** What are the dimensions of a closed rectangular box that has a square cross section, has a capacity of 128 cm^3, and is constructed using the least amount of material?

8. **Minimizing Packaging Costs** A rectangular box is to have a square base and a volume of 2 m^2. If the material for the base costs 90¢/square metre, the material for the sides costs 30¢/square metre, and the material for the top costs 60¢/square metre, determine the dimensions of the box that can be constructed at minimum cost.

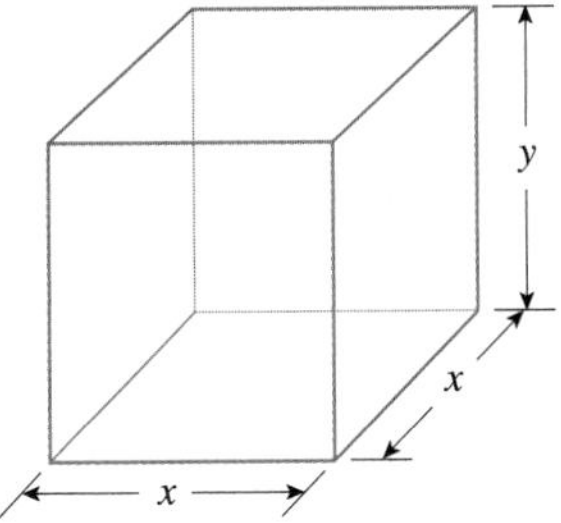

9. Parcel Post Regulations Postal regulations specify that a parcel sent by parcel post may have a combined length and girth of no more than 252 cm. Find the dimensions of a rectangular package that has a square cross section and the largest volume that may be sent through the mail. The girth of a cylinder is its circumference. What is the volume of such a package?

Hint: The length plus the girth is $4x + h$ (see the accompanying figure).

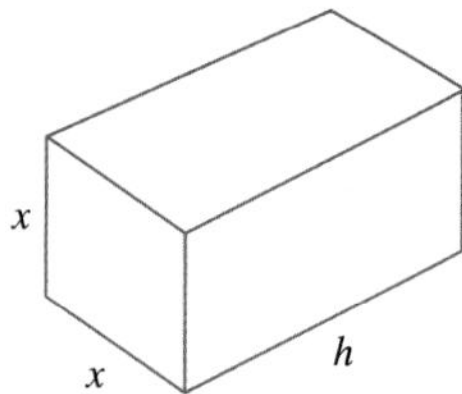

10. Book Design A book designer has decided that the pages of a book should have 2-cm margins at the top and bottom and 1-cm margins on the sides. She further stipulated that each page should have an area of 300 cm^2 (see the accompanying figure). Determine the page dimensions that will result in the maximum printed area on the page.

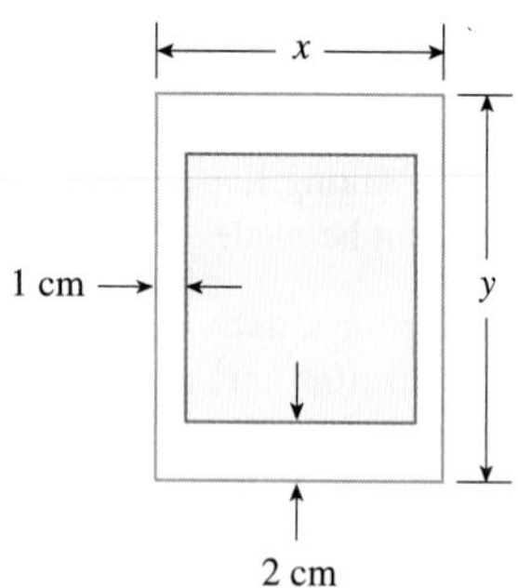

11. Parcel Post Regulations Postal regulations specify that a parcel sent by parcel post may have a combined length and girth of no more than 252 cm. Find the dimensions of the cylindrical package of greatest volume that may be sent through the mail. What is the volume of such a package? Compare with Exercise 9.

Hint: The length plus the girth is $2\pi r + l$.

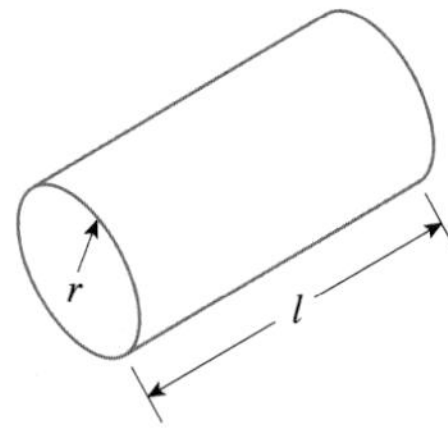

12. Minimizing Costs For its beef stew, Betty Moore Company uses aluminum containers that have the form of right circular cylinders. Find the radius and height of a container if it has a capacity of 216 cm^3 and is constructed using the least amount of metal.

13. Product Design The cabinet that will enclose the Acrosonic model D loudspeaker system will be rectangular and will have an internal volume of 1.0 m^3. For aesthetic reasons, it has been decided that the height of the cabinet is to be 1.5 times its width. If the top, bottom, and sides of the cabinet are constructed of veneer costing \$1.20/square metre and the front (ignore the cutouts in the baffle) and rear are constructed of particle board costing 60¢/square metre, what are the dimensions of the enclosure that can be constructed at a minimum cost?

14. Designing a Norman Window A Norman window has the shape of a rectangle surmounted by a semicircle (see the accompanying figure). If a Norman window is to have a perimeter of 9 m, what should its dimensions be in order to allow the maximum amount of light through the window?

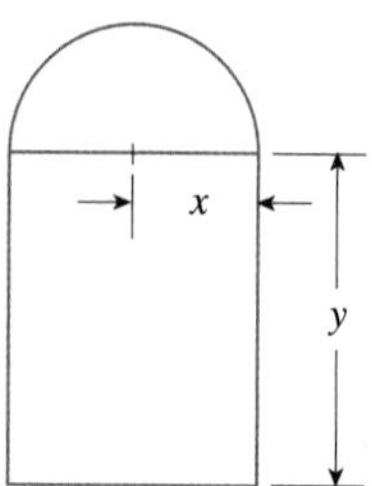

15. Strength of a Beam A wooden beam has a rectangular cross section of height h cm and width w cm (see the accompanying figure). The strength S of the beam is directly proportional to its width and the square of its height. What are the dimensions of the cross section of the strongest beam that can be cut from a round log of diameter 24 cm?

Hint: $S = kh^2w$, where k is a constant of proportionality.

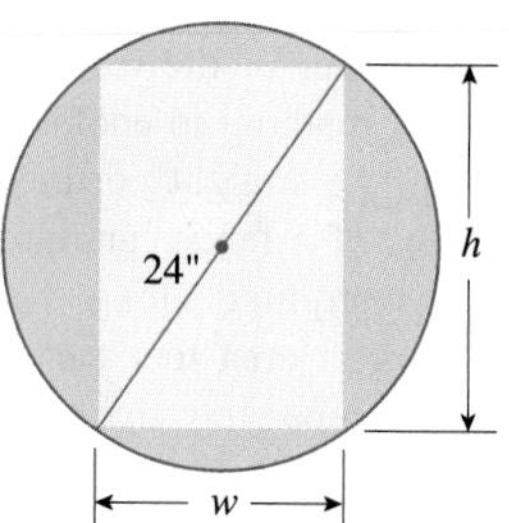

16. Designing a Grain Silo A grain silo has the shape of a right circular cylinder surmounted by a hemisphere (see the accompanying figure). If the silo is to have a capacity of 168π m^3, find the radius and height of the silo that requires the least amount of material to construct.

Hint: The volume of the silo is $\pi r^2h + \frac{2}{3}\pi r^3$, and the surface area (including the floor) is $\pi(3r^2 + 2rh)$.

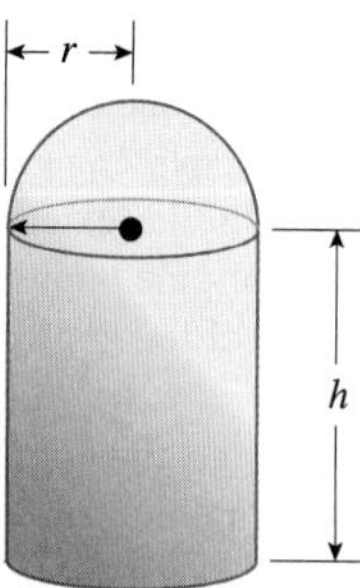

17. Minimizing Cost of Laying Cable In the following diagram, S represents the position of a power relay station located on a straight coast, and E shows the location of a marine biology experimental station on an island. A cable is to be laid connecting the relay station with the experimental station. If the cost of running the cable on land is \$5.00/running metre and the cost of running the cable under water is \$18/running metre, locate the point P that will result in a minimum cost (solve for x).

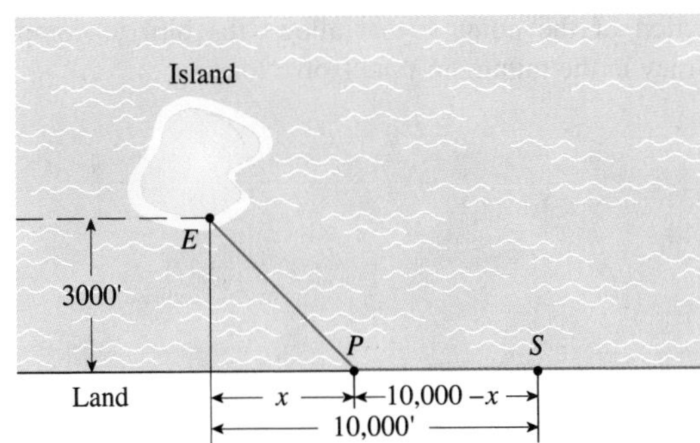

18. **Storing Radioactive Waste** A cylindrical container for storing radioactive waste is to be constructed from lead and have a thickness of 15 cm (see the accompanying figure). If the volume of the outside cylinder is to be π m^3, find the radius and the height of the inside cylinder that will result in a container of maximum storage capacity.
Hint: Show that the storage capacity (inside volume) is given by

$$V(r) = \pi r^2\left[\frac{1{,}000{,}000}{(r + 15)^2} - 30\right] \qquad (0 \le r \le 167.6 \text{ cm})$$

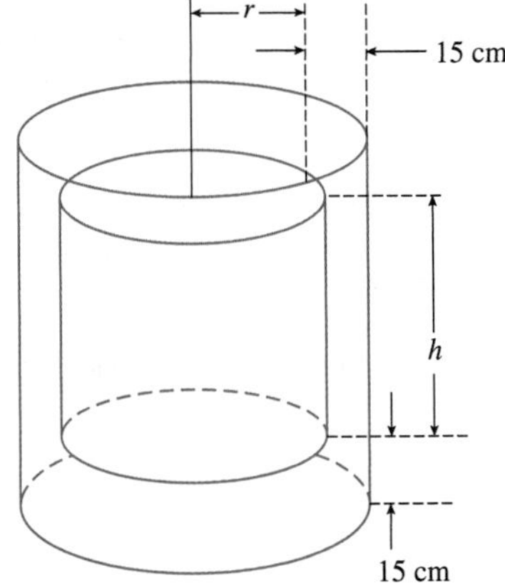

19. **Optimal Speed of a Truck** A truck gets $1200/x$ km/l when driven at a constant speed of x mph (between 80 and 110 km/h). If the price of fuel is $1/litre and the driver is paid $8/hour, at what speed between 80 and 110 km/h is it most economical to drive?

20. **Racetrack Design** The accompanying figure depicts a racetrack with ends that are semicircular in shape. The length of the track is 600 metres. Find l and r so that the area enclosed by the rectangular region of the racetrack is as large as possible. What is the area enclosed by the track in this case?

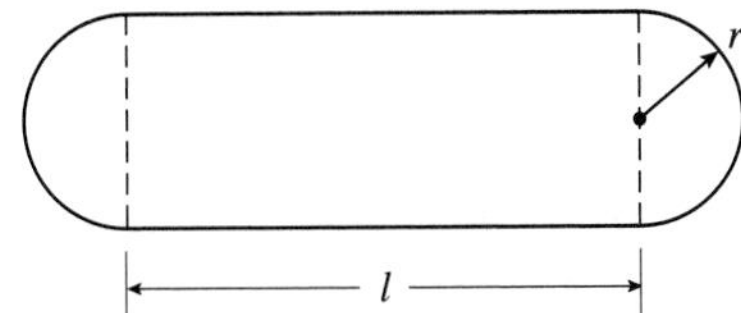

Business and Economics Applications

21. **Optimal Charter-Flight Fare** If exactly 200 people sign up for a charter flight, Leisure World Travel Agency charges $300/person. However, if more than 200 people sign up for the flight (assume this is the case), then each fare is reduced by $1 for each additional person. Determine how many passengers will result in a maximum revenue for the travel agency. What is the maximum revenue? What would be the fare per passenger in this case?
Hint: Let x denote the number of passengers above 200. Show that the revenue function R is given by $R(x) = (200 + x)(300 - x)$.

22. **Charter Revenue** The owner of a luxury motor yacht that sails among the 4000 Greek islands charges $600/person/day if exactly 20 people sign up for the cruise. However, if more than 20 people sign up (up to the maximum capacity of 90) for the cruise, then each fare is reduced by $4 for each additional passenger. Assuming at least 20 people sign up for the cruise, determine how many passengers will result in the maximum revenue for the owner of the yacht. What is the maximum revenue? What would be the fare/passenger in this case?

23. **Profit of a Vineyard** Phillip, the proprietor of a vineyard, estimates that the first 10,000 bottles of wine produced this season will fetch a profit of $5/bottle. However, the profit from each bottle beyond 10,000 drops by $0.0002 for each additional bottle sold. Assuming at least 10,000 bottles of wine are produced and sold, what is the maximum profit? What would be the price/bottle in this case?

24. **Inventory Control and Planning** The demand for motorcycle tires imported by Dixie Import-Export is 40,000/year and may be assumed to be uniform throughout the year. The cost of ordering a shipment of tires is $400, and the cost of storing each tire for a year is $2. Determine how many tires should be in each shipment if the ordering and storage costs are to be minimized. (Assume that each shipment arrives just as the previous one has been sold.)

25. **Inventory Control and Planning** McDuff Preserves expects to bottle and sell 2,000,000 500-ml jars of jam. The company orders its containers from Consolidated Bottle Company. The cost of ordering a shipment of bottles is $200, and the cost of storing each empty bottle for a year is $.40. How many orders should McDuff place per year and how many bottles should be in each shipment if the ordering and storage costs are to be minimized? (Assume that each shipment of bottles is used up before the next shipment arrives.)

26. **Inventory Control and Planning** Neilsen Cookie Company sells its assorted butter cookies in containers that have a net content of 500 g. The estimated demand for the cookies is 1,000,000 500-g containers. The setup cost for each production run is $500, and the manufacturing cost is $.50 for each container of cookies. The cost of storing each container of cookies over the year is $.40. Assuming uniformity of demand throughout the year and instantaneous production, how many containers of cookies should Neilsen produce per production run in order to minimize the production cost?
Hint: Following the method of Example 5, show that the total production cost is given by the function

$$C(x) = \frac{500{,}000{,}000}{x} + 0.2x + 500{,}000$$

Then minimize the function C on the interval $(0, 1{,}000{,}000)$.

Biological and Life Sciences Applications

27. **Maximizing Yield** An apple orchard has an average yield of 36 bushels of apples/tree if tree density is 22 trees/acre. For each unit increase in tree density, the yield decreases by 2 bushels. How many trees should be planted in order to maximize the yield?

28. **Flights of Birds** During daylight hours, some birds fly more slowly over water than over land because some of their energy is expended in overcoming the downdrafts of air over open bodies of water. Suppose a bird that flies at a constant speed of 8 km/h over water and 12 km/h over land starts its journey at the point E on an island and ends at its nest N on the shore of the mainland, as shown in the accompanying figure. Find the location of the point P that allows the bird to complete its journey in the minimum time (solve for x).

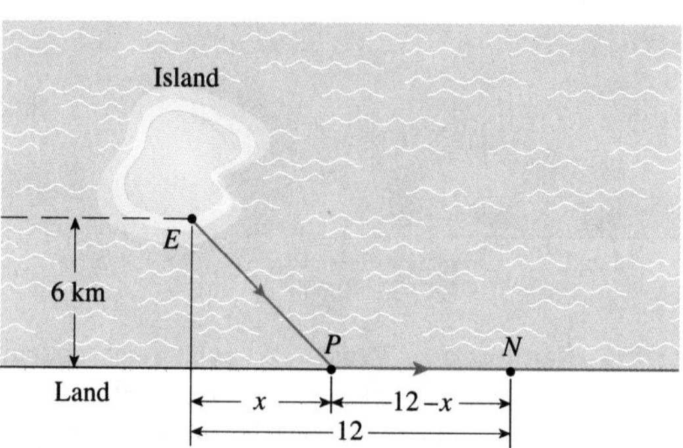

Solutions to Self-Check Exercises 7.5

1. Let x and y (measured in metres) denote the length and width of the rectangular garden.

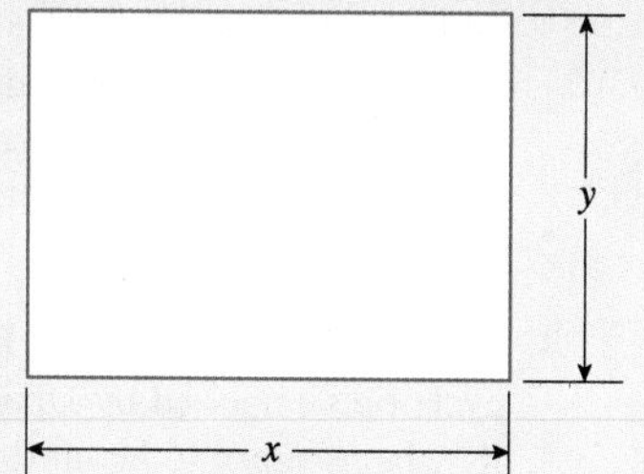

Since the area is to be 300 square metres, we have

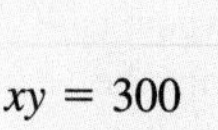

$$xy = 300$$

Next, the amount of fencing to be used is given by the perimeter, and this quantity is to be minimized. Thus, we want to minimize

$$2x + 2y$$

or, since $y = 300/x$ (obtained by solving for y in the first equation), we see that the expression to be minimized is

$$f(x) = 2x + 2\left(\frac{300}{x}\right)$$
$$= 2x + \frac{600}{x}$$

for positive values of x. Now

$$f'(x) = 2 - \frac{600}{x^2}$$

Setting $f'(x) = 0$ yields $x = -\sqrt{300}$ or $x = \sqrt{300}$. We consider only the critical point $x = \sqrt{300}$ since $-\sqrt{300}$ lies outside the interval $(0, -\infty)$. We then compute

$$f''(x) = \frac{1200}{x^3}$$

Since

$$f''(300) > 0$$

the second derivative test implies that a relative minimum of f occurs at $x = \sqrt{300}$. In fact, since $f''(x) > 0$ for all x in $(0, \infty)$, we conclude that $x = \sqrt{300}$ gives rise to the absolute minimum of f. The corresponding value of y, obtained by substituting this value of x into the equation $xy = 300$, is $y = \sqrt{300}$. Therefore, the required dimensions of the vegetable garden are approximately 17.3 metres × 17.3 metres.

2. Let x denote the number of tires in each production run. Then, the average number of tires in storage is $x/2$, so the storage cost incurred by the company is $2(x/2)$, or x dollars. Next, since the company needs to manufacture 1,000,000 tires for the year in order to meet the demand, the number of production runs is $1{,}000{,}000/x$. This gives setup costs amounting to

$$4000\left(\frac{1{,}000{,}000}{x}\right) = \frac{4{,}000{,}000{,}000}{x}$$

dollars for the year. The total manufacturing cost is \$20,000,000. Thus, the total yearly cost incurred by the company is given by

$$C(x) = x + \frac{4{,}000{,}000{,}000}{x} + 20{,}000{,}000$$

Differentiating $C(x)$, we find

$$C'(x) = 1 - \frac{4{,}000{,}000{,}000}{x^2}$$

Setting $C'(x) = 0$ gives $x = 63{,}246$ as the critical point in the interval $(0, 1{,}000{,}000)$. Next, we find

$$C''(x) = \frac{8{,}000{,}000{,}000}{x^3}$$

Since $C''(x) > 0$ for all $x > 0$, we see that C is concave upward for all $x > 0$. Furthermore, $C''(63{,}246) > 0$ implies that $x = 63{,}246$ gives rise to a relative minimum of C (by the second derivative test). Since C is always concave upward for $x > 0$, $x = 63{,}246$ gives the absolute minimum of C. Therefore, the company should manufacture 63,246 tires in each production run.

7.6 L'Hôpital's Rule: Indeterminate Forms

In Chapter 2 we studied how to take limits even when those limits had an inconvenient form such as the ratio of two functions that were both going to zero. An example of this type of computation is shown at the right.

$$\lim_{x\to 2}\frac{x^2-4}{x-2}=\lim_{x\to 2}\frac{(x-2)(x+2)}{(x-2)}$$
$$=\lim_{x\to 2}\frac{\cancel{(x-2)}(x+2)}{\cancel{(x-2)}}$$
$$=\lim_{x\to 2}(x+2)$$
$$=4$$

Resolving such limits required that we apply algebra to the expression we were taking the limit of until the inconvenient character of the expression changed. One or both of the zeros went away. In this section we learn a more powerful and often easier way to resolve this kind of limit, **L'Hôpital's rule.** This rule only works when the limit is in the form of a fraction where the numerator and denominator are both going to zero or are both going to infinity. These forms $\frac{0}{0}$ and $\frac{\infty}{\infty}$ are called **indeterminate forms.** As with other limits, it is sometimes necessary to use algebra to put the problem into the appropriate form. The problems to which L'Hôpital's rule can be applied are called indeterminate forms.

An indeterminate form is a limit where the quantities involved are both going to zero or infinity in a fashion that makes the value of the limit hard to determine. If we look at the ratio of a quantity going to zero divided by a quantity going to infinity, such as

$$\lim_{x\to 0}\frac{\sin(x)}{\frac{1}{x^2}}$$

then the limit can be resolved by common sense. The numerator is going to zero while the denominator is going to positive infinity. The ratio must be going to zero (and our usual techniques of algebraic manipulation could quickly show this was so). If, however we look at a limit like

$$\lim_{x\to 0}\frac{\sin(x)}{x}$$

then both numerator and denominator are going to zero and the value of the limit is not at all clear. It is ambiguous cases like the second limit above that L'Hôpital's rule is useful for.

L'Hôpital's Rule for the Indeterminate Form $\frac{0}{0}$

Suppose that

$$\lim_{x\to a}f(x)=0 \quad\text{and}\quad \lim_{x\to a}g(x)=0$$

Then

$$\lim_{x\to a}\frac{f(x)}{g(x)}=\lim_{x\to a}\frac{f'(x)}{g'(x)}$$

Remark

Notice that the derivatives of the numerator and denominator are taken *separately,* not together. Resolving limits of ratios with L'Hôpital's rule does *not* use the chain rule. ◀

Example 1

Compute the limit

$$\lim_{x\to 2}\frac{x^2-4}{x-2}$$

using L'Hôpital's rule.

Solution

Step 1 First notice that $\lim_{x\to 2}(x^2-4)=0$ and that $\lim_{x\to 2}(x-2)=0$, meaning that we have the indeterminate form $\frac{0}{0}$, for which L'Hôpital's rule is appropriate.

Step 2 The derivative of the numerator $(x^2 - 4)$ is $2x$ and the derivative of the denominator $x - 2$ is 1, so if we apply L'Hôpital's rule we get

$$\lim_{x\to 2}\frac{x^2-4}{x-2} = \lim_{x\to 2}\frac{2x}{1}$$
$$= 4$$

Compare the answer to the limit taken in Figure 3.7 on page 119. Is the result the same? There are many limits that could be done without using L'Hôpital's rule but the problems are often easier to do, because the concentration required to find the correct algebraic transformation to resolve the limit is replaced with the simple computation of derivatives.

Example 2

Compute

$$\lim_{x\to 0}\frac{\sin(x)}{x}$$

using L'Hôpital's rule.

Solution

Step 1 First notice that $\lim_{x\to 0}\sin(x) = 0$ and that $\lim_{x\to 0} x = 0$ as well, meaning that L'Hôpital's rule is appropriate.

Step 2 Compute

$$\lim_{x\to 0}\frac{\sin(x)}{x} = \lim_{x\to 0}\frac{(\sin(x))'}{(x)'}$$
$$= \lim_{x\to 0}\frac{\cos(x)}{1}$$
$$= \frac{1}{1}$$
$$= 1$$

In this example we could not have used the technique of factoring and cancelling as in Figure 3.7.

GROUP DISCUSSION

Suppose we are computing

$$\lim_{x\to 3}\frac{x^3-2x^2-4x+3}{x^3-7x^2+16x-12}$$

It turns out that both the numerator and denominator of the fraction are approaching zero as x approaches 3. Using the techniques of Chapter 2 we could factor $(x - 3)$ out of the top and bottom of the fraction, cancel them, and then try to compute the limit again. The problem is also appropriate for L'Hôpital's rule. To use L'Hôpital's rule we compute the derivative of the numerator and denominator, individually, and then try to compute the limit of their ratio again. Discuss which technique is easier to use and which one is less likely to lead to a student making errors while working the problem.

The second form of L'Hôpital's rule is very similar to the first, except that the functions in the numerator and denominator are not going to zero; instead they are diverging toward plus or minus infinity. The application of the rule does not change.

The derivatives of both numerator and denominator are both computed and then the limit of the ratio is attempted again using the new expressions.

L'Hôpital's Rule for the Indeterminate Form $\frac{\infty}{\infty}$

Suppose that

$$\lim_{x\to a} f(x) = \pm\infty \quad \text{and} \quad \lim_{x\to a} g(x) = \pm\infty$$

Then

$$\lim_{x\to a} \frac{f(x)}{g(x)} = \lim_{x\to a} \frac{f'(x)}{g'(x)}$$

Example 3

Compute

$$\lim_{x\to\infty} \frac{x+1}{(\sqrt{x}+1)^2}$$

using L'Hôpital's rule.

Solution

Step 1 Both the numerator and the denominator of the fraction are going to infinity and so it is appropriate to apply L'Hôpital's rule.

Step 2 Take the derivative of the numerator and denominator and try the limit again (the denominator requires the chain rule).

$$\lim_{x\to\infty} \frac{x+1}{(\sqrt{x}+1)^2} = \lim_{x\to\infty} \frac{1}{2(\sqrt{x}+1)\cdot\frac{1}{2\sqrt{x}}}$$

Next, we need to simplify the expression before applying the limit to get

$$\lim_{x\to\infty} \frac{1}{1+\frac{1}{\sqrt{x}}}$$

$$= \frac{1}{1+0}$$

$$= 1$$

Example 3 is a problem which, if we multiplied out the denominator and used fractional powers, could have been done with the methods we learned in Chapter 2. When you have multiple methods available, it is important to ask yourself which method will be easier. The answer to the question "which method is easier" may be individual to each student on each problem.

Using L'Hôpital's rule does not guarantee an answer, as we will see in Example 4. If, after you apply L'Hôpital's rule, the new limit clearly does not exist, then the original limit did not exist either. All that L'Hôpital's rule does is transform a limit into a different limit that you *may* be able to resolve. The new limit is still a limit problem and may or may not have a numerical solution.

Example 4

Compute

$$\lim_{x\to 0^+} \frac{\frac{1}{x^2}}{\ln(x)}$$

using L'Hôpital's rule.

Solution

Step 1 As x approaches zero from the right the numerator approaches $+\infty$ while the denominator approaches $-\infty$. This is an appropriate situation for L'Hôpital's rule.

Step 2 Compute:

$$\lim_{x\to 0^+} \frac{\frac{1}{x^2}}{\ln(x)} = \lim_{x\to 0^+} \frac{\frac{-2}{x^3}}{\frac{1}{x}}$$

$$= \lim_{x\to 0^+} \frac{-2}{x^2}$$

$$= -\infty$$

The limit is undefined, because the numerator of the fraction is minus 2 while the limit of the denominator is approaching zero from above. In this case L'Hôpital's rule permits us to see that the limit does not exist.

We have already seen, in the previous example, that L'Hôpital's rule may turn an indeterminate form into a form in which the limit clearly does not exist. It can also turn an indeterminate form into another indeterminate form of the correct type for L'Hôpital's rule. One option, in this case, is to apply L'Hôpital's rule again. It is possible to find examples that require any number of applications of L'Hôpital's rule; the following example requires two.

Example 5

Multiple Applications of L'Hôpital's Rule Compute

$$\lim_{x\to 1} \frac{x^3 + x^2 - 5x + 3}{x^4 - x^3 - x + 1}$$

Solution

Step 1 Notice that the numerator and denominator of the fraction are zero when $x = 1$. This means we can use L'Hôpital's rule.

Step 2 Apply L'Hôpital's rule.

$$\lim_{x\to 1} \frac{x^3 + x^2 - 5x + 3}{x^4 - x^3 - x + 1} = \lim_{x\to 1} \frac{3x^2 + 2x - 5}{4x^3 - 3x^2 - 1}$$

The numerator and denominator of the new expression are also both zero when $x = 1$, and so we may apply L'Hôpital's rule again.

Step 3 Apply L'Hôpital's rule a second time.

$$\lim_{x\to 1} \frac{3x^2 + 2x - 5}{4x^3 - 3x^2 - 1} = \lim_{x\to 1} \frac{6x + 2}{12x^2 - 6x}$$

$$= \frac{8}{6}$$

$$= \frac{4}{3}$$

The reason that the preceding example required two applications of L'Hôpital's rule to resolve is that the polynomial $(x - 1)^2$ divided both the numerator and the denominator of the ratio. If we had noticed this we could have factored $(x - 1)^2$ out of the numerator and denominator, cancelled, and then substituted in 1 into the simplified numerator and denominator to obtain $\frac{4}{3}$. If the numerator and denominator are polynomials with common factors raised to powers, as with $(x - 1)^2$ in the example, then each application of L'Hôpital's rule will eliminate one power of the common factor. If you find taking derivatives of polynomials easier than factoring them, then L'Hôpital's rule is an easier way to resolve this kind of limit than factoring.

GROUP DISCUSSION

There is no version of L'Hôpital's rule for the forms $\frac{0}{\infty}$ and $\frac{\infty}{0}$. If the rule were applied to these forms the results might be incorrect and the computations would certainly be inappropriate. Discuss what the limits would be for these indeterminate forms.

It is also sometimes possible to create a situation where L'Hôpital's rule applies even though we must use some algebra to set up the situation. So far, both the indeterminate forms we have for L'Hôpital's rule are ratios. If we have the limit of a quantity going to zero times another quantity that is going to infinity, then algebraic rearrangement of the expression can turn it into a ratio of the sort for which we can apply L'Hôpital's rule.

There are two more indeterminate forms $0 \cdot \infty$ and $\infty - \infty$ that L'Hôpital's rule is helpful for, but they require that the problems be algebraically transformed into one of the two indeterminate forms $\frac{0}{0}$ or $\frac{\infty}{\infty}$. The next two examples show one example of each of these situations.

Example 6

Indeterminates of the Form $0 \cdot \infty$ Compute

$$\lim_{x \to 0^+} x \cdot \ln(x)$$

Solution

Step 1 This problem is not yet an indeterminate form of the sort we know can be used for L'Hôpital's rule. As x approaches zero from above, $\ln(x)$ approaches negative infinity, meaning the form is $0 \cdot (-\infty)$. If a quantity is going to zero, its reciprocal is going to one of $\pm\infty$, and so we can rewrite the expression in an indeterminate form in which both numerator and denominator are going to $\pm\infty$.

$$\lim_{x \to 0^+} x \cdot \ln(x) = \lim_{x \to 0^+} \frac{\ln(x)}{\frac{1}{x}}$$

Step 2 Since we now have one of the indeterminate forms that is appropriate for L'Hôpital's rule, we apply the rule.

$$\begin{aligned} \lim_{x \to 0^+} \frac{\ln(x)}{\frac{1}{x}} &= \lim_{x \to 0^+} \frac{\frac{1}{x}}{\frac{-1}{x^2}} \\ &= \lim_{x \to 0^+} \frac{1x^2}{x \cdot (-1)} \\ &= \lim_{x \to 0^+} (-x) \\ &= 0 \end{aligned}$$

We see that $x \cdot \ln(x)$ goes to zero as x goes to zero.

In the example, the reciprocal of the quantity going to zero was put in the denominator, yielding the form $\frac{\infty}{\infty}$. It would also have been possible to take the reciprocal of the quantity going to infinity and put it in the denominator. This would have given us the form $\frac{0}{0}$. Either algebraic transformation yields a form for which we can apply L'Hôpital's rule. You should choose whichever of these two techniques yields an easier problem.

Dealing with the Indeterminate Form $0 \cdot \infty$

If

$$\lim_{x \to a} f(x) = 0$$

and

$$\lim_{x \to a} g(x) = \infty$$

then

$$\lim_{x \to a} f(x) \cdot g(x) = \lim_{x \to a} \frac{g(x)}{\frac{1}{f(x)}}$$

which is an indeterminate of the form $\frac{\infty}{\infty}$ because the reciprocal of a quantity going to zero is a quantity going to infinity. Once the limit is put in this form we may apply L'Hôpital's rule.

Alternatively you may use

$$\lim_{x \to a} f(x) \cdot g(x) = \lim_{x \to a} \frac{f(x)}{\frac{1}{g(x)}}$$

which is an indeterminate of the form $\frac{0}{0}$ because the reciprocal of a quantity going to infinity is a quantity going to zero.

The second indeterminate form that L'Hôpital's rule can be used to resolve is one where we have the limit of the difference of two quantities that are both going to infinity. While it may have other forms, a difference of functions diverging to infinity is often in the form of the difference of ratios of functions. If we find a common denominator and combine the two ratios, the resulting function *may* be in the correct form for L'Hôpital's rule.

Example 7

Indeterminates of the Form $\infty - \infty$ Compute

$$\lim_{x \to 0^+} \frac{1}{\sin(x)} - \frac{1}{x}$$

Solution

Step 1 This problem is not yet an indeterminate form of the sort we know can be used for L'Hôpital's rule. Both parts of the difference are going to infinity as x approaches zero from above. If we combine the fractions, the problem becomes

$$\lim_{x \to 0^+} \frac{x - \sin(x)}{x \cdot \sin(x)}$$

which is an indeterminate of the proper form.

Step 2 Apply L'Hôpital's rule (the denominator requires a product rule).

$$\lim_{x \to 0^+} \frac{x - \sin(x)}{x \cdot \sin(x)} = \lim_{x \to 0^+} \frac{1 - \cos(x)}{\sin(x) + x \cdot \cos(x)}$$

This is still of the form $\frac{0}{0}$, and so we apply L'Hôpital's rule again.

Step 3

$$\lim_{x\to 0^+} \frac{1 - \cos(x)}{\sin(x) + x \cdot \cos(x)} = \lim_{x\to 0^+} \frac{\sin(x)}{2\cos(x) - x \cdot \sin(x)}$$

$$= \frac{0}{2 - 0}$$

$$= 0$$

Remark

One application of limits is to see which of two functions fundamentally grows faster. When we compute the limit of the ratio of two functions, the limit goes to ∞ if the top function grows faster, the ratio goes to zero if the bottom function grows faster, and the ratio goes to a constant other than zero if the functions have similar growth rates (one is not fundamentally faster than the other). L'Hôpital's rule makes it easy to compute this type of limit and so is a useful tool for finding out if one of two functions grows fundamentally faster.

The ideal of *fundamentally faster* growth requires some additional explanation. The function $f(x) = 2x^2$ does grow faster than $g(x) = x^2$, exactly twice as fast in fact. Since the ratio

$$\frac{f(x)}{g(x)} = \frac{2x^2}{x^2} = 2$$

is a constant, then two functions have fundamentally the same growth rate. Suppose a chemical reaction is increasing the amount of two chemicals in the reaction vessel. If the rate at which the concentrations grow is not fundamentally different, then both chemicals will be present in a roughly constant ratio at different times during the reaction. If, however, one chemical grows fundamentally faster, then after the reaction has run, its concentration should be much higher. ◄

Remark

An example of this type of reaction is the *polymerase chain reaction* used to amplify DNA samples for forensics or biotechnology. A quantity of two small DNA molecules, called primers, are added to a sample containing DNA with an enzyme called *DNA polymerase.* The sample is heated to make the double-stranded DNA fall apart and then cooled. If a primer matches the DNA in the reaction, it sticks to it while the reaction cools. The enzyme takes DNA bases from solution and adds them at one end of the primer to complete a single strand of DNA into a new double strand (DNA has an intrinsic direction, which is opposite on each strand, and which the enzyme recognizes). The reason that two primers are used is that one is designed to match a location on each strand of the DNA that the forensic technician or researcher wants to amplify. Between the primers on the DNA strand each primer sticks to one or the other strand of the DNA and builds a new strand. This means the DNA between the primers is (roughly) doubled each time the reaction is heated and cooled. Other DNA is ignored or, if it is on the strand for one primer but past the point where the other primer anneals to the opposite strand, it is amplified only linearly. Since the function $f(x) = 2^x$ (amplification of DNA between the primers) does grow fundamentally faster than $g(x) = x$ (amplification of DNA just beyond the primers), the DNA present in the reaction ends up being almost entirely the target DNA once the reaction has been heated and cooled 20–40 times.

The fundamentally faster growth rate of the DNA between the primers permits an investigation to take a tiny sample and test it for the presence of specific DNA. Even if the DNA of hundreds of people are in a sample, the polymerase chain reaction is highly specific for spotting the DNA of one person for whom the two primers were designed. The technique can also be applied to so-called *gene fishing*, where primers designed for a gene found on one organism can be used to amplify that gene, if it is present, in another. This is a standard method for checking for the presence of highly similar genes between two species when the genetic code of only one is known. ◄

Finding the Relative Growth Rate of Two Functions

Suppose that $f(x)$ and $g(x)$ are functions that are defined for all positive real numbers. Compute the limit

$$\lim_{x\to\infty}\frac{f(x)}{g(x)}$$

Three outcomes are possible:

$$\lim_{x\to\infty}\frac{f(x)}{g(x)} = 0 \qquad g(x) \text{ grows faster}$$

$$\lim_{x\to\infty}\frac{f(x)}{g(x)} = C > 0 \qquad f(x) \text{ and } g(x) \text{ grow at similar rates}$$

$$\lim_{x\to\infty}\frac{f(x)}{g(x)} = \infty \qquad f(x) \text{ grows faster}$$

When two functions have similar graphs, it is possible to get an idea which graph grows faster by examining the graphs. A careful demonstration that one function actually grows faster requires computing the limit of the ratios, as in the following example.

Example 8

Comparing Growth Rates of Functions Examine the graphs of $f(x)$ and $g(x)$ below. The graphs have similar but different shapes. Find out which function grows faster or if they grow at similar rates.

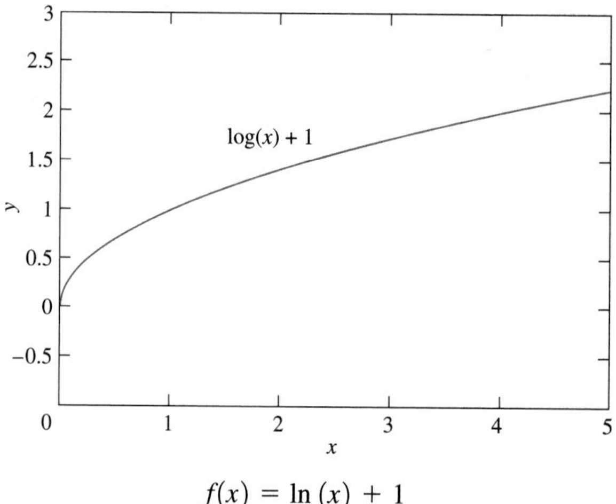

$f(x) = \ln(x) + 1$

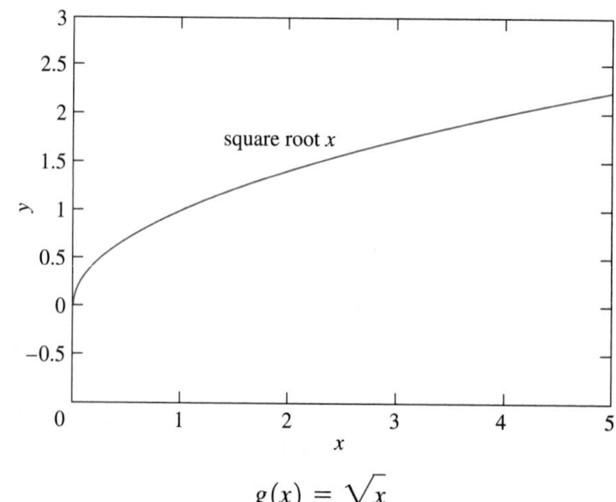

$g(x) = \sqrt{x}$

Solution

Step 1 Both functions go to ∞ as x goes to infinity, so

$$\lim_{x\to\infty}\frac{f(x)}{g(x)}$$

is an indeterminate form of the kind appropriate for L'Hôpital's rule.

Step 2 Compute

$$\lim_{x\to\infty}\frac{\ln^{\circ}(x)+1}{\sqrt{x}} = \lim_{x\to\infty}\frac{\frac{1}{x}}{\frac{1}{2\sqrt{x}}}$$

$$= \lim_{x\to\infty}\frac{2\sqrt{x}}{x}$$

$$= \lim_{x\to\infty}\frac{2}{\sqrt{x}}$$

$$= 0$$

So we see that $f(x) = \ln(x) + 1$ grows fundamentally more slowly than $g(x) = \sqrt{x}$. When two curves have similar growth rates, that doesn't mean they are the same. It means they grow in a similar fashion. In the next example we will see that two functions that are both logarithms of linear functions grow tin the same fashion.

Example 9

Comparing Growth Rates of Functions Compare the growth rates of

$$f(x) = \ln(3x + 5) \text{ and } g(x) = \ln(x + 4).$$

Solution

Step 1 Both functions go to ∞ as x goes to infinity, so

$$\lim_{x\to\infty}\frac{f(x)}{g(x)}$$

is an indeterminate from of the kind appropriate for L'Hôpital's rule.

Step 2 Compute

$$\lim_{x\to\infty}\frac{\ln(3x+5)}{\ln(x+4)} = \lim_{x\to\infty}\frac{\frac{3}{3x+5}}{\frac{1}{x+4}}$$

$$= \lim_{x\to\infty}\frac{3(x+4)}{3x+5}$$

$$= \lim_{x\to\infty}\frac{3x+12}{3x+5}$$

$$= \lim_{x\to\infty}\frac{\frac{1}{x}(3x+12)}{\frac{1}{x}(3x+5)}$$

$$= \lim_{x\to\infty}\frac{3+\frac{12}{x}}{3+\frac{5}{x}}$$

$$= \frac{3}{3}$$

$$= 1$$

So even though the two functions in the example are different, they both grow in a similar fashion. We say both functions grow logarithmically.

Relative rates of growth have applications in biology. The following example takes rates of growth approximated in a laboratory and uses L'Hôpital's rule to see if one strain of bacteria will displace another in a setting in which both are present.

Example 10

Comparing Growth Rates of Bacterial Populations An application of relative growth rates of populations occurs in biology when we want to know if one type of bacteria will displace another. There is a bacteria that grows on strawberry plants that has an outer protein coat that helps ice crystals form. When this bacteria is present, frost forms on strawberry plants at a slightly higher temperature. Getting rid of this bacteria will decrease the number of strawberries lost to frost. In order to displace the bacteria, a closely related bacteria without the frost-nucleating protein is located. The size of the population of the frost-forming bacteria, in millions of bacteria per kilogram of strawberry plants after introduction, is found experimentally to be close to

$$f(t) = 2.75 + 3.4e^{0.16t}$$

while the size of the population of the bacteria without the frost-forming capability is found experimentally to be close to

$$g(t) = 1.46 + 1.4e^{0.18t}$$

The variable t is the number of weeks since the bacteria were introduced. Using L'Hôpital's rule, find which sort of bacteria grow faster and so will eventually dominate a strawberry field where both are present.

Solution

Step 1 As always we must check that the functions are both going to 0 or $\pm\infty$. In this case both functions grow without limit and thus are going to ∞, so L'Hôpital's rule is appropriate.

Step 2 Compute

$$\begin{aligned}
\lim_{t\to\infty} \frac{2.75 + 3.4e^{0.16t}}{1.46 + 1.4e^{0.18t}} &= \lim_{t\to\infty} \frac{0.16 \cdot 3.4e^{0.16t}}{0.18 \cdot 1.4e^{0.18t}} \\
&= \lim_{t\to\infty} \frac{0.544 \cdot e^{0.16t}}{0.2628 \cdot e^{0.18t}} \\
&= \lim_{t\to\infty} 2.07 \cdot e^{(0.16-0.18)t} \\
&= \lim_{t\to\infty} 2.07 \cdot e^{-0.02t} \\
&= 2.07 \cdot 0 \\
&= 0
\end{aligned}$$

So the bacteria that do not help frost crystals form grow faster and using them to treat strawberry fields will provide a biological control for frost, at least at temperatures near zero degrees Celsius.

Since we are not going to have a strawberry field that grows until an infinite amount of time has passed, it may seem odd to talk about the limit of a ratio at infinity. Strawberries are harvested within a year, not after infinite time. The reason techniques like the one in the last problem are used is because the relative growth rate, computed at infinity, is informative of the short-term behaviour. If one population is growing in a different manner, i.e., much faster, then it will quickly come to dominate a system.

GROUP DISCUSSION

Suppose that two populations of bacteria $P(t)$ and $Q(t)$ are found to grow in a similar fashion. Does this mean that in the short term both populations will coexist in nearly equal numbers or that both will persist though one may be more common than the other? Try to find examples of functions with similar growth rates and see if they stay equal or if they simply don't grow apart from one another too quickly. Does the starting population of each type of bacteria matter?

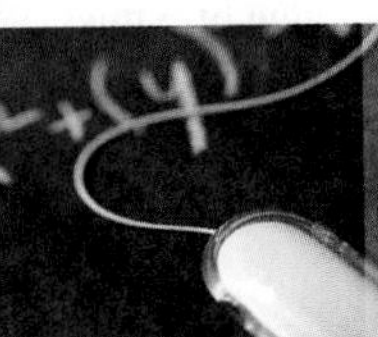

Self-Check Exercises 7.6

For each of the following limit problems, check that the limit is an indeterminate form that is appropriate for L'Hôpital's rule. If it is, then compute the limit using L'Hôpital's rule; if not, then use algebra to change it into a form where L'Hôpital's rule can be applied and then compute the limit or use a different technique.

1. $\lim\limits_{x\to 2} \dfrac{x^2 + x - 6}{x^2 - 4}$

2. $\lim\limits_{x\to 0} \dfrac{x}{1 - \cos(x)}$

3. $\lim\limits_{x\to 1} + (x - 1)\ln(x - 1)$

Solutions to Self-Check Exercises 7.6 can be found on page 432.

7.6 Exercises

In Exercises 1–10, determine if the limit in question is an indeterminate form that is appropriate for L'Hôpital's rule.

1. $\lim\limits_{x\to 5} \dfrac{x^2 - 25}{x^2 - 10x + 25}$

2. $\lim\limits_{x\to 3} \dfrac{2x^2 - 8x + 6}{x^2 + x + 1}$

3. $\lim\limits_{x\to 0} \dfrac{1 - \cos(x)}{x^2}$

4. $\lim\limits_{x\to 3} \dfrac{2x^2 - 8x + 6}{x^2 - 4x + 3}$

5. $\lim\limits_{x\to \frac{\pi}{2}} \dfrac{\cos(x)}{x - \dfrac{\pi}{2}}$

6. $\lim\limits_{x\to 1} \dfrac{x^3 + 3x^2 - x - 3}{x^3 - 6x^2 + 11x - 6}$

7. $\lim\limits_{x\to \infty} \dfrac{1 - e^x}{1 + e^x}$

8. $\lim\limits_{x\to \infty} \dfrac{\ln(x)}{\sqrt[3]{x} + 1}$

9. $\lim\limits_{x\to 0^+} \dfrac{\sin(3x)}{\sin(2x)}$

10. $\lim\limits_{x\to 0^-} \dfrac{\sin(4x)}{\sin(5x)}$

In Exercises 11–20, use L'Hôpital's rule to compute the limit or state that the limit is not an appropriate one for L'Hôpital's rule if it is not.

11. $\lim\limits_{x\to 1} \dfrac{x^3 - x^2 + 2x - 2}{x^3 + x^2 + x - 3}$

12. $\lim\limits_{x\to 4} \dfrac{x^2 - 8x + 16}{x^2 - 16}$

13. $\lim\limits_{x\to 4} \dfrac{x^2 - 16}{x^2 - 8x + 16}$

14. $\lim\limits_{x\to -2} \dfrac{x^3 + 2x^2 + 3x + 2}{x^3 + 4x^2 + 7x + 6}$

15. $\lim\limits_{x\to 0} \dfrac{1 - \cos(x)}{x}$

16. $\lim\limits_{x\to 0} \dfrac{\sin(2x)}{x}$

17. $\lim\limits_{x\to \infty} \dfrac{\ln\cdot(5x)}{\sqrt[4]{x}}$

18. $\lim\limits_{x\to \infty} \dfrac{4 + e^x}{1 + e^x}$

19. $\lim\limits_{x\to 0^+} \dfrac{\sin(x)}{\sin(3x)}$

20. $\lim\limits_{x\to 0^-} \dfrac{\sin(3x)}{\sin(7x)}$

In Exercises 21–26, the indeterminate form is $0 \cdot \infty$. Compute the limit by using algebra to put the limit in the correct form for L'Hôpital's rule and then applying the rule. Some of these exercises will require multiple applications of L'Hôpital's rule.

21. $\lim\limits_{x\to 0^+} \sqrt{x} \cdot \ln(x)$

22. $\lim\limits_{x\to 1^+} (x - 1) \cdot \ln(x^2 - 1)$

23. $\lim\limits_{x\to 0^+} \sec(2x) \cdot \sin(5x)$

24. $\lim\limits_{x\to 0^+} x^{0.25} \cdot \ln(x)$

25. $\lim\limits_{x\to 0^+} x \cdot \cot(x)$

26. $\lim\limits_{x\to \frac{\pi}{2}^-} x \cdot \tan(x)$

In Exercises 27–32, the indeterminate form is $\infty - \infty$. Compute the limit by using algebra to put the limit in the correct form for L'Hôpital's rule and then applying the rule. Some of these exercises will require multiple applications of L'Hôpital's rule.

27. $\lim_{x\to 0^-} \frac{1}{2x} - \frac{1}{\sin(2x)}$

28. $\lim_{x\to 1^+} \frac{1}{\ln(x)} - \frac{1}{1-x}$

29. $\lim_{x\to\infty} \frac{x^3+x^2-x+1}{x^2-9} - x$

30. $\lim_{x\to\infty} \sec(x) - \frac{1}{x - \frac{\pi}{2}}$

31. $\lim_{x\to\infty} \frac{1}{x} - \frac{1}{1-e^x}$

32. $\lim_{x\to\infty} \frac{x+1}{x} - \frac{1}{\sin(x)}$

In Exercises 33–37, compute the given limit using as many applications of L'Hôpital's rule as are required.

33. $\lim_{x\to -1} \frac{2x^3+5x^2+4x+1}{x^3+9x^2+15x+7}$

34. $\lim_{x\to 2} \frac{x^4-4x^3+5x^2-4x+4}{x^3-x^2-8x+12}$

35. $\lim_{x\to\infty} \frac{e^x+3x+5}{e^x+2x-7}$

36. $\lim_{x\to 0} \frac{1-\cos(2x)}{1-\cos(3x)}$

37. $\lim_{x\to 0} \frac{x+e^{2x}}{1-e^{3x}}$

In Exercises 38–43, determine which of the two given functions grows faster or say if they grow at similar rates. You may need to use L'Hôpital's rule more than once for some of these exercises.

38. $f(x) = 2x + 1, g(x) = 3x - 4$

39. $f(x) = x^2, g(x) = (x-2)\cdot(x+1)$

40. $f(x) = \sqrt[5]{2x}, g(x) = \ln(x)$

41. $f(x) = e^x, g(x) = 2^x$

42. $f(x) = e^{1.3x} + 7, g(x) = e^{3.1x} + 5$

43. $f(x) = e^{1.1x} + 6x + 2, g(x) = e^{1.2x} - 2x + 7$

44. Suppose that $r > 0$. Show that the limit

$$\lim_{x\to\infty} x^r \cdot \ln(x)$$

is always zero. From this we can conclude that the logarithmic function grows more slowly than any positive power of x.

45. Consider functions of the type $f(x) = A \cdot x^b$ where both A and b are positive real numbers.

a. If we increase A does that make the function grow faster or leave it growing in a similar fashion in the sense of Example 8?

b. What if we increase b?

46. Consider the limit

$$\lim_{x\to\infty} \frac{e^{ax}}{x^n}$$

for a positive whole number n. Depending on the value of n this limit may take many applications of L'Hôpital's rule to resolve. Do examples for $n = 1, 2, 3$ and then try to generalize from these examples. Does e^x grow faster than any whole number power of n?

47. Displacement In a part of the Amazon basin a type of mouse lives that harbours a bacteria that causes an asymptomatic disease in the mice but can cause a severe respiratory infection if humans breathe particles of mouse droppings. A closely related mouse species that is immune to the disease-causing bacteria lives in an area a few hundred miles away. Raising both sorts of mice in a laboratory habitat arranged to simulate the jungle, it is found that the bacteria-harbouring mice have a population function, in total mice after t months, of close to

$$P(t) = 250 + 7.6e^{0.41t}$$

while the bacteria-resistant mice have a population function in the same units close to

$$Q(t) = 250 + 8.5e^{0.37t}$$

Based on these approximate population functions, will the bacteria-resistant mice displace the bacteria-carrying mice if they are introduced?

48. Invasive Species An *invasive species* is a species introduced to a new area that does well in the new area. An example of an invasive species is purple loosestrife, a plant with vivid purple flowers. It was brought to North America from Europe as an ornamental plant. When introduced into North American wetlands this plant spreads rapidly, meaning its population curve grows faster than those of most native plants. The population of an aquatic grass native to Manitoba is found to have a population of plants per square metre in a test plot of

$$P(t) = 100 + 2.3x^{2.57}$$

after t months. The corresponding equation for purple loosestrife is found to be

$$Q(t) = 100 + 1.6x^{3.14}$$

Will the purple loosestrife simply outgrow the aquatic grass?

49. Comparing Growth Curves Read Exercise 48. Searching in Europe, the home of purple loosestrife, a root-mining weevil is found that feeds on purple loosestrife. The experiment to find the population growth of purple loosestrife is performed but with the addition of a population of the weevils. The new growth curve for the purple loosestrife is found to be

$$Q(t) = 100 + 1.1x^{2.61}$$

Is this weevil likely to be an effective bio-control? Decide by comparing the new growth curve with that for the aquatic grass. Assume the weevil is tested and found not to attack the aquatic grass.

50. Comparing Growth Curves Read Exercise 48. Unsatisfied with the results obtained by seeding plots of purple loosestrife with root-mining weevils, researchers continue their search for natural enemies. Eventually, a flower-feeding weevil is found that feeds on purple loosestrife. The experiment to find the population growth of purple loosestrife is performed but with the addition of a population of the flower-feeding weevils. The new growth curve for the purple loosestrife is found to be

$$Q(t) = 100 + 1.8x^{2.52}$$

Is the flower-feeding weevil likely to be an effective bio-control? Decide by comparing the new growth curve with that for the aquatic grass. Assume the weevil is tested and found not to attack the aquatic grass.

Solutions to Self-Check Exercises 7.6

1. The limit is an indeterminate form correct for L'Hôpital's rule; the numerator and denominator are both going to zero. Applying L'Hôpital's rule we obtain

$$\lim_{x\to 2}\frac{x^2+x-6}{x^2-4} = \lim_{x\to 2}\frac{2x+1}{2x} = \frac{5}{4}$$

2. The limit is an indeterminate form correct for L'Hôpital's rule; the numerator and denominator are both going to zero. Applying L'Hôpital's rule we obtain

$$\lim_{x\to 0}\frac{x}{1-\cos(x)} = \lim_{x\to 0}\frac{1}{\sin(x)}$$

The left-hand side is I divided by a quantity going to zero, so the limit fails to exist.

3. The limit $\lim_{x\to 1^+}(x-1)\ln(x-1)$ is not one of the indeterminate forms for which L'Hôpital's rule applies because it is a quantity going to zero times one going to negative infinity. By algebraic manipulation we can transform it into an indeterminate form of the correct type.

$$\lim_{x\to 1^+}\frac{\ln(x-1)}{\frac{1}{x-1}}$$

Applying L'Hôpital's rule,

$$\lim_{x\to 1^+}\frac{\ln(2x)}{\frac{1}{x-1}} = \lim_{x\to 1^+}\frac{\frac{1}{x-1}}{\frac{-1}{(x-1)^2}}$$

$$= \lim_{x\to 1^+} -(x-1)$$

$$= 0$$

CHAPTER 7 Summary of Principal Terms

Terms

increasing function (340)
decreasing function (340)
relative maximum (344)
relative minimum (345)
relative extrema (345)
critical point (346)
first derivative test (346)
concave upward (356)
concave downward (356)
inflection point (358)
second derivative test (361)
vertical asymptote (373)
horizontal asymptote (374)
absolute extrema (385)
absolute maximum value (385)
absolute minimum value (385)
monotone function (392)
L'Hôpital's rule (411)
indeterminate form (411)

Review Exercises 

In Exercises 1–14, (a) find the intervals where the given function f is increasing and where it is decreasing, (b) find the relative extrema of f, (c) find the intervals where f is concave upward and where it is concave downward, and (d) find the inflection points, if any, of f.

1. $f(x) = \frac{1}{3}x^3 - x^2 + x - 6$

2. $f(x) = (x - 2)^3$

3. $f(x) = x^4 - 2x^2$

4. $f(x) = x + \frac{4}{x}$

5. $f(x) = \frac{x^2}{x - 1}$

6. $f(x) = \sqrt{x - 1}$

7. $f(x) = (1 - x)^{1/3}$

8. $f(x) = x\sqrt{x - 1}$

9. $f(x) = \frac{2x}{x + 1}$

10. $f(x) = \frac{-1}{1 + x^2}$

11. $f(x) = x^3 e^{-x/2}$

12. $f(x) = \ln(x^2 + 4)$

13. $f(x) = \frac{e^{2x}}{1 + e^{2x}}$

14. $g(x) = \sin(2x) - \cos(2x)$

15. **Absorption of Drugs** The concentration or a drug in grams/cubic centimetre (g/cm^3) t min after it has been injected into the bloodstream is given by

$$C(t) = \frac{k}{b - a}(e^{-at} - e^{-bt})$$

where a, b, and k are positive constants, with $b > a$.
   a. At what time is the concentration of the drug the greatest?
   b. What will be the concentration of the drug in the long run?

16. **Concentration of Glucose in the Bloodstream** A glucose solution is administered intravenously into the bloodstream at a constant rate of r mg/hr. As the glucose is being administered, it is converted into other substances and removed from the bloodstream. Suppose the concentration of the glucose solution at time t is given by

$$C(t) = \frac{r}{k} - \left[\left(\frac{r}{k}\right) - C_0\right]e^{-kt}$$

where C_0 is the concentration at time $t = 0$ and k is a positive constant. Assuming that $C_0 < r/k$, evaluate $\lim_{t\to\infty} C(t)$.
   a. What does your result say about the concentration of the glucose solution in the long run?
   b. Show that the function C is increasing on $(0, \infty)$
   c. Show that the graph of C is concave downward on $(0, \infty)$.
   d. Sketch the graph of the function C.

17. **Gompertz Growth Curve** Consider the function

$$Q(t) = Ce^{-Ae^{-kt}}$$

where $Q(t)$ is the size of a quantity at time t and A, C, and k are positive constants. The graph of this function, called the *Gompertz growth curve*, is used by biologists to describe restricted population growth.
   a. Show that the function Q is always increasing.
   b. Find the time t at which the growth rate $Q'(t)$ is increasing most rapidly.
   Hint: Find the inflection point of Q.
   c. Show that $\lim_{x\to\infty} Q(t) = C$ and interpret your result.

In Exercises 18–29, obtain as much information as possible on each of the given functions. Then use this information to sketch the graph of the function.

18. $f(x) = x^2 - 5x + 5$

19. $f(x) = -2x^2 - x + 1$

20. $g(x) = 2x^3 - 6x^2 + 6x + 1$

21. $g(x) = \frac{1}{3}x^3 - x^2 + x - 3$

22. $h(x) = x\sqrt{x - 2}$

23. $h(x) = \frac{2x}{1 + x^2}$

24. $f(x) = \frac{x - 2}{x + 2}$

25. $f(x) = x - \frac{1}{x}$

26. $f(x) = (x + 3)e^{-x/4}$

27. $g(x) = \ln(x^2 + 4x + 6)$

28. $f(x) = 3\sin(x) + 3\cos(x)$

29. $g(x) = \arctan\left(\frac{x}{\pi}\right)$

In Exercises 30–35, find the horizontal and vertical asymptotes of the graphs of the given functions. Do not sketch the graphs.

30. $f(x) = \frac{1}{2x + 3}$

31. $f(x) = \frac{2x}{x + 1}$

32. $f(x) = \frac{5x}{x^2 - 2x - 8}$

33. $f(x) = \frac{x^2 + x}{x(x - 1)}$

34. $f(x) = \frac{6e^{2x}}{e^{2x} + 5}$

35. $g(x) = e^{\frac{x}{x-2}}$

In Exercises 36–51, find the absolute maximum value and the absolute minimum value, if any, of the given function.

36. $f(x) = 2x^2 + 3x - 2$

37. $g(x) = x^{2/3}$

38. $g(t) = \sqrt{25 - t^2}$

39. $f(x) = \frac{1}{3}x^3 - x^2 + x + 1$ on $[0, 2]$

40. $h(t) = t^3 - 6t^2$ on $[2, 5]$

41. $g(x) = \frac{x}{x^2 + 1}$ on $[0, 5]$

42. $f(x) = x - \frac{1}{x}$ on $[1, 3]$

43. $h(t) = 8t - \frac{1}{t^2}$ on $[1, 3]$

44. $f(s) = s\sqrt{1 - s^2}$ on $[-1, 1]$

45. $f(x) = \frac{x^2}{x - 1}$ on $[-1, 3]$

46. $f(x) = \ln(9 - x^2)$ on $[-3, 3]$

47. $f(x) = \ln\left(\frac{x^2 + 1}{x^2 + 3}\right)$ on $[-1, 5]$

48. $f(x) = (x + 2)e^{-3x/2}$ on $[0, 10]$

49. $f(x) = (x + 1)e^{-x/2}$ on $[0, 10]$

50. $f(x) = \cos(x/4)$ on $[-\pi, \pi]$

51. $f(x) = \frac{1}{2}\cos(2x) + \sin(x)$ on $[0, 2\pi]$

In Exercises 52–57, resolve the limit with L'Hôpital's rule. Use multiple applications of the rule if it is required.

52. $\lim_{x \to 0} \frac{\sin(2x)}{3x}$

53. $\lim_{x \to 2} \frac{x^2 - 4}{x^2 + x - 6}$

54. $\lim_{x \to 3} \frac{x^3 - 2x^2 - 2x + 3}{x^3 - 5x^2 + 9x - 9}$

55. $\lim_{x \to 1} \frac{x^3 + x^2 - 5x + 3}{x^3 + 3x^2 + 4}$

56. $\lim_{x \to 4} \frac{x^3 + 7x^2 + 8x - 16}{x^3 + 6x^2 - 32}$

57. $\lim_{x \to \infty} \frac{5 + e^{2x}}{7 - e^{2x}}$

In Exercises 58 and 59, determine which of the two functions given grows more quickly or if they grow at similar rates.

58. $f(x) = x^2 + 2$, $g(x) = x^{5/2} + 1$

59. $f(x) = 3.6e^{2x}$, $g(x) = 1.7r^{2x+1}$

60. Determine the intervals where the function $f(x) = e^{-x^2/2}$ is increasing and where it is decreasing.

61. Determine the intervals where the function $f(x) = x^2e^{-x}$ is increasing and where it is decreasing.

62. Determine the intervals of concavity for the function

$$f(x) = \frac{e^x - e^{-x}}{2}.$$

63. Determine the intervals of concavity for the function $f(x) = xe^x$.

64. Find the inflection point of the function $f(x) = xe^{-2x}$.

65. Find the inflection point(s) of the function $f(x) = 2e^{-x^2}$.

In Exercises 66–69, find the absolute extrema of the function.

66. $f(x) = e^{-x^2}$ on $[-1, 1]$

67. $h(x) = e^{x^2-4}$ on $[-2, 2]$

68. $g(x) = (2x - 1)e^{-x}$ on $[0, \infty)$

69. $f(x) = xe^{-x^2}$ on $[0, 2]$

In Exercises 70–73, use the curve-sketching guidelines on page 375 to sketch the graph of the function.

70. $f(t) = e^t - t$

71. $h(x) = \frac{e^x + e^{-x}}{2}$

72. $f(x) = 2 - e^{-x}$

73. $f(x) = \frac{3}{1 + e^{-x}}$

74. Determine the intervals where the function $f(x) = \ln x^2$ is increasing and where it is decreasing.

75. Determine the intervals where the function $f(x) = \frac{\ln x}{x}$ is increasing and where it is decreasing.

76. Determine the intervals of concavity for the function $f(x) = x^2 + \ln x^2$.

77. Determine the intervals of concavity for the function $f(x) = \frac{\ln x}{x}$.

78. Find the inflection points of the function $f(x) = \ln(x^2 + 1)$.

79. Find the inflection points of the function $f(x) = x^2 \ln x$.

80. Find the absolute extrema of the function $f(x) = x - \ln x$ on $[\frac{1}{2}, 3]$.

81. Find the absolute extrema of the function $g(x) = \frac{x}{\ln x}$ on $[2, 5]$.

82. **Learning Curves** The average worker at Wakefield Avionics can assemble

$$N(t) = -2t^3 + 12t^2 + 2t \qquad (0 \le t \le 4)$$

ready-to-fly radio-controlled model airplanes t hr into the 8 A.M. to 12 noon morning shift. At what time during this shift is the average worker performing at peak efficiency?

83. **Maximizing the Volume of a Box** A box with an open top is to be constructed from a square piece of cardboard, 10 cm wide, by cutting out a square from each of the four corners and bending up the sides. What is the maximum volume of such a box?

84. **Minimizing Construction Costs** A man wishes to construct a cylindrical barrel with a capacity of 32π m^3. The cost/square metre of the material for the side of the barrel is half that of the cost/square metre for the top and bottom. Help him find the dimensions of the barrel that can be constructed at a minimum cost in terms of material used.

85. **Packaging** You wish to construct a closed rectangular box that has a volume of 1 m^3. The length of the base of the box will be twice as long as its width. The material for the top and

bottom of the box costs $1.50/m^2. The material for the sides of the box costs $1.00/m^2. Find the dimensions of the least expensive box that can be constructed.

86. Let

$$f(x) = \begin{cases} x^3 + 1 & \text{if } x \neq 0 \\ 2 & \text{if } x = 0 \end{cases}$$

a. Compute $f'(x)$ and show that it does not change sign as we move across $x = 0$.
b. Show that f has a relative maximum at $x = 0$. Does this contradict the first derivative test? Explain your answer.

Business and Economics Applications

87. Maximizing Profits Odyssey Travel Agency's monthly profit (in thousands of dollars) depends on the amount of money x (in thousands of dollars) spent on advertising each month according to the rule

$$P(x) = -x^2 + 8x + 20$$

To maximize its monthly profits, what should be Odyssey's monthly advertising budget?

88. Online Hotel Reservations The online lodging industry is expected to grow dramatically. In a study conducted in 1999, analysts projected the online travel spending for lodging to be approximately

$$f(t) = 0.157t^2 + 1.175t + 2.03 \qquad (0 \leq t \leq 6)$$

billion dollars, where t is measured in years, with $t = 0$ corresponding to 1999.
a. Show that f is increasing on the interval (0, 6).
b. Show that the graph of f is concave upward on (0, 6).
c. What do your results from parts (a) and (b) tell you about the growth of online travel spending over the years in question?

89. Maximizing Profits The weekly demand for DVDs manufactured by Herald Records is given by

$$p = -0.0005x^2 + 60$$

where p denotes the unit price in dollars and x denotes the quantity demanded. The weekly total cost function associated with producing these discs is given by

$$C(x) = -0.001x^2 + 18x + 4000$$

where $C(x)$ denotes the total cost incurred in pressing x discs. Find the production level that will yield a maximum profit for the manufacturer.
Hint: Use the quadratic formula.

90. Minimizing Costs The total monthly cost (in dollars) incurred by Carlota Music in manufacturing x units of its Professional Series guitars is given by the function

$$C(x) = 0.001x^2 + 100x + 4000$$

a. Find the average cost function $\overline{C}$.
b. Determine the production level that will result in the smallest average production cost.

91. Inventory Control and Planning Lehen Vinters imports a certain brand of beer. The demand, which may be assumed to be uniform, is 800,000 cases/year. The cost of ordering a shipment of beer is $500, and the cost of storing each case of beer for a year is $2. Determine how many cases of beer should be in each shipment if the ordering and storage costs are to be kept at a minimum. (Assume that each shipment of beer arrives just as the previous one has been sold.)

Biological and Life Sciences Applications

92. Index of Environmental Quality The Department of the Interior of an African country began to record an index of environmental quality to measure progress or decline in the environmental quality of its wildlife. The index for the years 1984 through 1994 is approximated by the function

$$I(t) = \frac{50t^2 + 600}{t^2 + 10} \qquad (0 \leq t \leq 10)$$

a. Compute $I'(t)$ and show that $I(t)$ is decreasing on the interval (0, 10).
b. Compute $I''(t)$. Study the concavity of the graph of I.
c. Sketch the graph of I.
d. Interpret your results.

93. Spread of a Contagious Disease The incidence (number of new cases/day) of a contagious disease spreading in a population of M people is given by

$$R(x) = kx(M - x)$$

where k is a positive constant and x denotes the number of people already infected. Show that the incidence R is greatest when half the population is infected.

94. Absorption of Drugs A liquid carries a drug into an organ of volume V cm^3 at the rate of a cm^3/sec and leaves at the same rate. The concentration of the drug in the entering liquid is c g/cm^3. Letting $x(t)$ denote the concentration of the drug in the organ at any time t, we have $x(t) = c(1 - e^{-at}/V)$.
a. Show that x is an increasing function on $(0, \infty)$.
b. Sketch the graph of x.

95. Absorption of Drugs Refer to Exercise 94. Suppose the maximum concentration of the drug in the organ must *not* exceed m g/cm^3, where $m < c$. Show that the liquid must not be allowed to enter the organ for a time longer than

$$T = \left(\frac{V}{a}\right) \cdot \ln\left(\frac{c}{c - m}\right)$$

minutes.

96. Concentration of a Drug in the Bloodstream The concentration of a drug in the bloodstream t sec after injection into a muscle is given by

$$y = c(e^{-bt} - e^{-at}) \qquad (\text{for } t \geq 0)$$

where a, b, and c are positive constants, with $a > b$.

a. Find the time at which the concentration is maximal.

b. Find the time at which the concentration of the drug in the bloodstream is decreasing most rapidly.

97. Absorption of Drugs The concentration of a drug in an organ at any time t (in seconds) is given by

$$C(t) = \begin{cases} 0.3t - 18(1 - e^{-t/60}) & \text{if } 0 \le t \le 20 \\ 18e^{-t/60} - 12e^{-(t-20)/60} & \text{if } t > 20 \end{cases}$$

where $C(t)$ is measured in grams/cubic centimetre (g/cm^3).

a. How fast is the concentration of the drug in the organ changing after 10 sec?

b. How fast is the concentration of the drug in the organ changing after 30 sec?

c. When will the concentration of the drug in the organ reach a maximum?

d. What is the maximum drug concentration in the organ?

98. Absorption of Drugs Jane took 100 mg of a drug in the morning and another 100 mg of the same drug at the same time the following morning. The amount of the drug in her body t days after the first dosage was taken is given by

$$A(t) = \begin{cases} 100e^{-1.4t} & \text{if } 0 \le t < 1 \\ 100(1 + e^{1.4})e^{-1.4t} & \text{if } t \ge 1 \end{cases}$$

a. How fast was the amount of drug in Jane's body changing after 12 hr $t = \frac{1}{2}$? After 2 days?

b. When was the amount of drug in Jane's body a maximum?

c. What was the maximum amount of drug in Jane's body?

99. Additive Test A company is testing the ability of an additive to permit concrete to prevent moss from growing on it. Suppose that the number of square metres of concrete covered by a type of moss after a small patch is transplanted on top of a concrete surface is found to be close to

$$M(t) = \frac{30t + 1}{t^2 + 4}$$

after t weeks. What is the maximum area covered by moss and does the moss die off eventually?

100. Sequestering Toxic Metals Suppose that a type of bacteria that sequesters toxic metals is used in a treatment lagoon. If the bacteria are permitted to grow indefinitely, they have a density of

$$D(t) = \frac{2400t^2}{t^2 + 3}$$

bacteria/cc after t hours. At what time t is it best to start removing bacteria from the the lagoon to maximize the sequestration of toxic metals?

Hint: This is an optimal harvesting problem.

INTEGRATION

Photo: Thinkstock/Jupiter Images

Differential calculus is concerned with the problem of finding the rate of change of one quantity with respect to another. In this chapter we begin the study of the other branch of calculus, known as integral calculus. Here we are interested in precisely the opposite problem: If we know the rate of change of one quantity with respect to another, can we find the relationship between the two quantities? The principal tool used in the study of integral calculus is the *antiderivative* of a function, and we develop rules for antidifferentiation, or *integration*, as the process of finding the antiderivative is called. We also show that a link is established between differential and integral calculus—via the fundamental theorem of calculus.

Photo: Thinkstock/Jupiter Images

How much will the solar cell panels cost? The head of Soloron Corporation's research and development department has projected that the cost of producing solar cell panels will drop at a certain rate in the next several years. In Example 10, page 453, you will see how this information can be used to predict the cost of solar cell panels in the coming years.

Photo: Toronto Star Archives

8.1 Antiderivatives and the Rules of Integration

Antiderivatives

Let's return, once again, to the example involving the motion of the maglev (Figure 8.1).

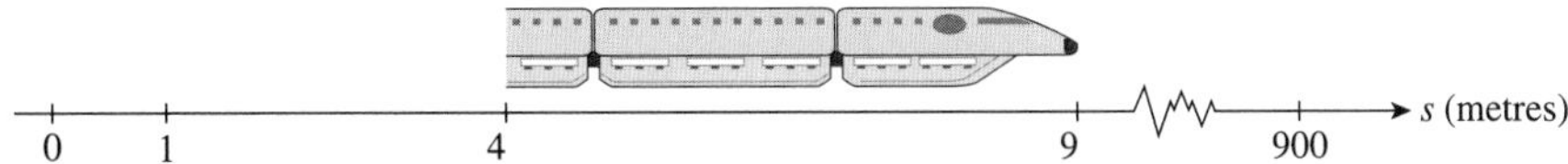

Figure 8.1
A maglev moving along an elevated monorail track

In Chapter 3, we discussed the following problem:

If we know the position of the maglev at any time t, can we find its velocity at time t?

As it turns out, if the position of the maglev is described by the position function f, then its velocity at any time t is given by $f'(t)$. Here f'—the velocity function of the maglev—is just the derivative of f.

Now, in Chapters 8 and 9, we will consider precisely the opposite problem:

If we know the velocity of the maglev at any time t, can we find its position at time t?

Stated another way, if we know the velocity function f' of the maglev, can we find its position function f?

To solve this problem, we need the concept of an antiderivative of a function.

Antiderivative

A function F is an **antiderivative** of f on an interval I if $F'(x) = f(x)$ for all x in I.

Thus, an antiderivative of a function f is a function F whose derivative is f. For example, $F(x) = x^2$ is an antiderivative of $f(x) = 2x$ because

$$F'(x) = \frac{d}{dx}(x^2) = 2x = f(x)$$

and $F(x) = x^3 + 2x + 1$ is an antiderivative of $f(x) = 3x^2 + 2$ because

$$F'(x) = \frac{d}{dx}(x^3 + 2x + 1) = 3x^2 + 2 = f(x)$$

Example 1

Let $F(x) = \frac{1}{3}x^3 - 2x^2 + x - 1$. Show that F is an antiderivative of $f(x) = x^2 - 4x + 1$.

Solution

Differentiating the function F, we obtain

$$F'(x) = x^2 - 4x + 1 = f(x)$$

and the desired result follows.

Example 2

Let $F(x) = x$, $G(x) = x + 2$, and $H(x) = x + C$, where C is a constant. Show that F, G, and H are all antiderivatives of the function f defined by $f(x) = 1$.

Solution

Since

$$F'(x) = \frac{d}{dx}(x) = 1 = f(x)$$

$$G'(x) = \frac{d}{dx}(x + 2) = 1 = f(x)$$

$$H'(x) = \frac{d}{dx}(x + C) = 1 = f(x)$$

we see that F, G, and H are indeed antiderivatives of f.

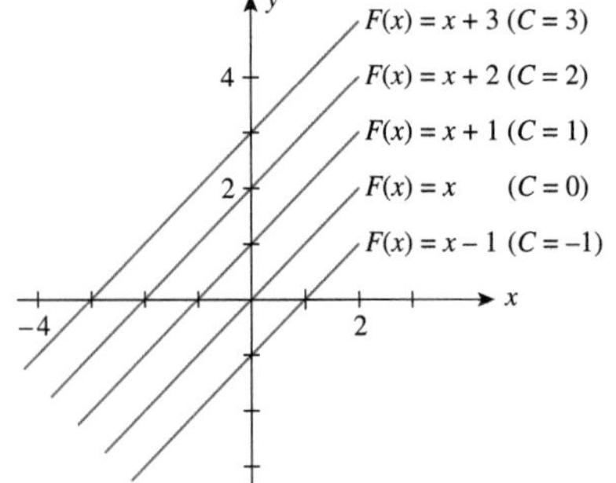

Figure 8.2
The graphs of some antiderivatives of $f(x) = 1$

Example 2 shows that once an antiderivative G of a function f is known, then another antiderivative of f may be found by adding an arbitrary constant to the function G. The following theorem states that no function other than one obtained in this manner can be an antiderivative of f. (We omit the proof.)

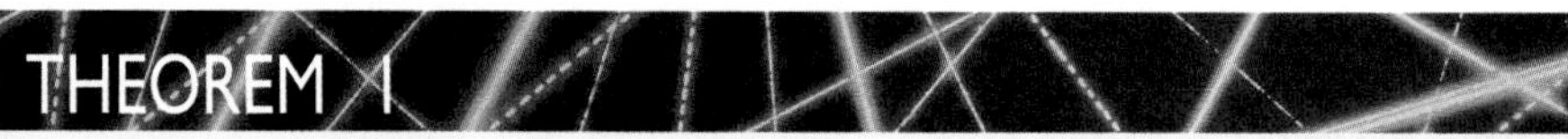

Let G be an antiderivative of a function f. Then, every antiderivative F of f must be of the form $F(x) = G(x) + C$, where C is a constant.

Returning to Example 2, we see that there are infinitely many antiderivatives of the function $f(x) = 1$. We obtain each one by specifying the constant C in the function $F(x) = x + C$. Figure 8.2 shows the graphs of some of these antiderivatives for selected values of C. These graphs constitute part of a family of infinitely many parallel straight lines, each having a slope equal to 1. This result is expected since there are infinitely many curves (straight lines) with a given slope equal to 1. The antiderivatives $F(x) = x + C$ (C, a constant) are precisely the functions representing this family of straight lines.

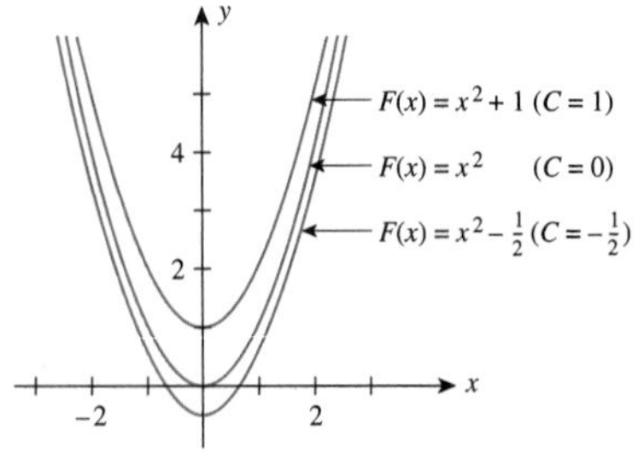

Figure 8.3
The graphs of some antiderivatives of $f(x) = 2x$

Example 3

Prove that the function $G(x) = x^2$ is an antiderivative of the function $f(x) = 2x$. Write a general expression for the antiderivatives of f.

Solution

Since $G'(x) = 2x = f(x)$, we have shown that $G(x) = x^2$ is an antiderivative of $f(x) = 2x$. By Theorem 1, every antiderivative of the function $f(x) = 2x$ has the form $F(x) = x^2 + C$, where C is some constant. The graphs of a few of the antiderivatives of f are shown in Figure 8.3.

EXPLORING WITH TECHNOLOGY

Let $f(x) = x^2 - 1$.

1. Show that $F(x) = \frac{1}{3}x^3 - x + C$, where C is an arbitrary constant, is an antiderivative of f.
2. Use a graphing utility to plot the graphs of the antiderivatives of f corresponding to $C = -2$, $C = -1$, $C = 0$, $C = 1$, and $C = 2$ on the same set of axes, using the viewing window $[-4, 4] \cdot [-4, 4]$.
3. If your graphing utility has the capability, draw the tangent line to each of the graphs in part 2 at the point whose x-coordinate is 2. What can you say about this family of tangent lines?
4. What is the slope of a tangent line in this family? Explain how you obtained your answer.

The Indefinite Integral

The process of finding all antiderivatives of a function is called **antidifferentiation**, or **integration**. We use the symbol $\int$, called an **integral sign**, to indicate that the operation of integration is to be performed on some function f. Thus,

$$\int f(x)\,dx = F(x) + C$$

[read "the indefinite integral of $f(x)$ with respect to x equals $F(x)$ plus C"] tells us that the **indefinite integral** of f is the family of functions given by $F(x) + C$, where $F'(x) = f(x)$. The function f to be integrated is called the **integrand**, and the constant C is called a **constant of integration**. The expression dx following the integrand $f(x)$ reminds us that the operation is performed with respect to x. If the independent variable is t, we write $\int f(t)\,dt$ instead. In this sense both t and x are "dummy variables."

Using this notation, we can write the results of Examples 2 and 3 as

$$\int 1\,dx = x + C \qquad \text{and} \qquad \int 2x\,dx = x^2 + K$$

where C and K are arbitrary constants.

Basic Integration Rules

Our next task is to develop some rules for finding the indefinite integral of a given function f. Because integration and differentiation are reverse operations, we discover many of the rules of integration by first making an "educated guess" at the antiderivative F of the function f to be integrated. Then this result is verified by demonstrating that $F' = f$.

Rule 1: The Indefinite Integral of a Constant

$$\int k\,dx = kx + C \qquad (k\text{, a constant})$$

To prove this result, observe that

$$F'(x) = \frac{d}{dx}(kx + C) = k$$

Example 4

Find each of the following indefinite integrals:

a. $\int 2\,dx$ **b.** $\int \pi^2\,dx$

Solution

Each of the integrands has the form $f(x) = k$, where k is a constant. Applying Rule 1 in each case yields

a. $\int 2\,dx = 2x + C$ **b.** $\int \pi^2\,dx = \pi^2 x + C$

Next, from the rule of differentiation,

$$\frac{d}{dx}x^n = nx^{n-1}$$

we obtain the following rule of integration.

Rule 2: The Power Rule

$$\int x^n\,dx = \frac{1}{n+1}x^{n+1} + C \qquad (n \neq -1)$$

An antiderivative of a power function is another power function obtained from the integrand by increasing its power by 1 and dividing the resulting expression by the new power.

To prove this result, observe that

$$F'(x) = \frac{d}{dx}\left[\frac{1}{n+1}x^{n+1} + C\right]$$

$$= \frac{n+1}{n+1}x^n$$

$$= x^n$$

$$= f(x)$$

Example 5

Find each of the following indefinite integrals:

a. $\int x^3\,dx$ **b.** $\int x^{3/2}\,dx$ **c.** $\int \frac{1}{x^{3/2}}\,dx$

Solution

Each integrand is a power function with exponent $n \neq -1$. Applying Rule 2 in each case yields the following results:

a. $\int x^3\,dx = \frac{1}{4}x^4 + C$

b. $\int x^{3/2}\,dx = \frac{1}{\frac{5}{2}}x^{5/2} + C = \frac{2}{5}x^{5/2} + C$

c. $\int \frac{1}{x^{3/2}}\,dx = \int x^{-3/2}\,dx = \frac{1}{-\frac{1}{2}}x^{-1/2} + C = -2x^{-1/2} + C = -\frac{2}{x^{1/2}} + C$

These results may be verified by differentiating each of the antiderivatives and showing that the result is equal to the corresponding integrand.

The next rule tells us that a constant factor may be moved through an integral sign.

Rule 3: The Indefinite Integral of a Constant Multiple of a Function

$$\int cf(x)\,dx = c\int f(x)\,dx \qquad (c\text{, a constant})$$

The indefinite integral of a constant multiple of a function is equal to the constant multiple of the indefinite integral of the function.

This result follows from the corresponding rule of differentiation (see Rule 3, Section 3.4).

Only a constant can be "moved out" of an integral sign. For example, it would be incorrect to write

$$\int x^2\,dx = x^2\int 1\,dx$$

In fact, $\int x^2\,dx = \frac{1}{3}x^3 + C$, whereas $x^2\int 1\,dx = x^2(x + C) = x^3 + Cx^2$.

Example 6

Find each of the following indefinite integrals:

a. $\displaystyle\int 2t^3\,dt$ **b.** $\displaystyle\int -3x^{-2}\,dx$

Solution

Each integrand has the form $cf(x)$, where c is a constant. Applying Rule 3, we obtain:

a. $\displaystyle\int 2t^3\,dt = 2\int t^3\,dt = 2\left[\frac{1}{4}t^4 + K\right] = \frac{1}{2}t^4 + 2K = \frac{1}{2}t^4 + C$

where $C = 2K$. From now on, we will write the constant of integration as C, since any nonzero multiple of an arbitrary constant is an arbitrary constant.

b. $\displaystyle\int -3x^{-2}\,dx = -3\int x^{-2}\,dx = (-3)(-1)x^{-1} + C = \frac{3}{x} + C$

Rule 4: The Sum Rule

$$\int [f(x) + g(x)]\,dx = \int f(x)\,dx + \int g(x)\,dx$$

$$\int [f(x) - g(x)]\,dx = \int f(x)\,dx - \int g(x)\,dx$$

The indefinite integral of a sum (difference) of two integrable functions is equal to the sum (difference) of their indefinite integrals.

This result is easily extended to the case involving the sum and difference of any finite number of functions. As in Rule 3, the proof of Rule 4 follows from the corresponding rule of differentiation (see Rule 4, Section 3.4).

Example 7

Find the indefinite integral

$$\int (3x^5 + 4x^{3/2} - 2x^{-1/2})\,dx$$

Solution

Applying the extended version of Rule 4, we find that

$$\int (3x^5 + 4x^{3/2} - 2x^{-1/2})\, dx$$

$$= \int 3x^5\, dx + \int 4x^{3/2}\, dx - \int 2x^{-1/2}\, dx$$

$$= 3\int x^5\, dx + 4\int x^{3/2}\, dx - 2\int x^{-1/2}\, dx \qquad \text{Rule 3}$$

$$= (3)\left(\frac{1}{6}\right)x^6 + (4)\left(\frac{2}{5}\right)x^{5/2} - (2)(2)x^{1/2} + C \qquad \text{Rule 2}$$

$$= \frac{1}{2}x^6 + \frac{8}{5}x^{5/2} - 4x^{1/2} + C$$

Observe that we have combined the three constants of integration, which arise from evaluating the three indefinite integrals, to obtain one constant C. After all, the sum of three arbitrary constants is also an arbitrary constant.

Rule 5: The Indefinite Integral of Exponential Functions

$$\int a^x\, dx = \frac{a^x}{\ln a} + C$$

where $a > 0$ and $a \neq 1$.

The indefinite integral of the exponential function with base e is equal to the function itself (except, of course, for the constant of integration).

Example 8

Find the indefinite integral

$$\int (2e^x + 3^x - x^3)\, dx$$

Solution

We have

$$\int (2e^x + 3^x - x^3)\, dx = \int 2e^x\, dx + \int 3^x\, dx - \int x^3\, dx$$

$$= 2\int e^x\, dx + \int 3^x\, dx - \int x^3\, dx$$

$$= 2e^x + \frac{3^x}{\ln 3} - \frac{1}{4}x^4 + C$$

The next rule of integration in this section covers the integration of the function $f(x) = x^{-1}$. Remember that this function constituted the only exceptional case in the integration of the power function $f(x) = x^n$ (see Rule 2).

Rule 6: The Indefinite Integral of the Function $f(x) = x^{-1}$

$$\int x^{-1}\, dx = \int \frac{1}{x}\, dx = \ln|x| + C \qquad (x \neq 0)$$

To prove Rule 6, observe that

$$\frac{d}{dx}\ln|x| = \frac{1}{x}$$ See Rule 5 Section 5.4.

Example 9

Find the indefinite integral

$$\int\left(2x + \frac{3}{x} + \frac{4}{x^2}\right)dx$$

Solution

$$\begin{aligned}\int\left(2x + \frac{3}{x} + \frac{4}{x^2}\right)dx &= \int 2x\,dx + \int\frac{3}{x}dx + \int\frac{4}{x^2}dx\\ &= 2\int x\,dx + 3\int\frac{1}{x}dx + 4\int x^{-2}\,dx\\ &= 2\left(\frac{1}{2}\right)x^2 + 3\ln|x| + 4(-1)x^{-1} + C\\ &= x^2 + 3\ln|x| - \frac{4}{x} + C\end{aligned}$$

Each rule for derivatives we have learned so far has a corresponding rule for integration. We are now ready to harvest the integration rules associated with derivatives of trigonometric functions.

Rule 7: The Indefinite Integral of Sine

$$\int\sin(x)\,dx = -\cos(x) + C$$

Rule 8: The Indefinite Integral of Cosine

$$\int\cos(x)\,dx = -\sin(x) + C$$

Example 10

Find the indefinite integral

$$\int(\sin(x) - \cos(x))dx$$

Solution

$$\begin{aligned}\int(\cos(x) - \sin(x))dx &= \int\cos(x)\,dx - \int\sin(x)dx\\ &= \sin(x) - (-\cos(x)) + C\\ &= \sin(x) + \cos(x) + C\end{aligned}$$

Let's review the derivative formulas for tan (x) and sec (x). In Chapter 6 we saw that if

$$f(x) = \tan(x)$$

then

$$f'(x) = \sec^2(x)$$

Likewise, if

$$g(x) = \sec(x)$$

then

$$g'(x) = \sec(x)\tan(x)$$

This means that the integration rules we get from these derivatives are not integrals of a single function.

Rule 9: The Indefinite Integral of $\sec^2(x)$

$$\int \sec^2(x)\,dx = \tan(x) + C$$

Rule 10: The Indefinite Integral of $\sec(x)\tan(x)$

$$\int \sec(x)\tan(x)\,dx = \sec(x) + C$$

Example 11

Find the indefinite integral

$$\int \left(3\sec^2(x) + \frac{\sec(x)\tan(x)}{2}\right)dx$$

Solution

$$\begin{aligned}\int \left(3\sec^2(x) + \frac{\sec(x)\tan(x)}{2}\right)dx &= 3\int \sec^2(x)dx + \frac{1}{2}\int \sec(x)\tan(x)dx \\ &= 3\tan(x) + \frac{1}{2}\sec(x) + C\end{aligned}$$

There are two more trigonometric functions that we know the derivative of, the cosecant and the cotangent. These derivative rules yield integration rules that are similar to those for the derivatives of the tangent and secant.

Rule 11: The Indefinite Integral of $\csc^2(x)$

$$\int \csc^2(x)dx = -\cot(x) + C$$

Rule 12: The Indefinite Integral of $\csc(x)\cot(x)$

$$\int \csc(x)\cot(x)dx = -\csc(x) + C$$

Example 12

Find the indefinite integral

$$\int (\csc^2(x) - 2\csc(x)\cot(x))dx$$

Solution

$$\begin{aligned}\int (\csc^2(x) - 2\csc(x)\cot(x))dx \\ &= \int \csc^2(x)dx - 2\int \csc(x)\cot(x)dx \\ &= -\cot(x) - 2(-\csc(x)) + C \\ &= -\cot(x) + 2\csc(x) + C\end{aligned}$$

The inverse trigonometric functions also have derivatives that yield integration rules. We saw in Chapter 6 that if

$$f(x) = \arctan(x)$$

then

$$f'(x) = \frac{1}{1 + x^2}$$

and that if

$$g(x) = \arcsin(x)$$

then

$$g'(x) = \frac{1}{\sqrt{1 - x^2}}$$

These two derivatives give us the following integration formulas.

Rule 13: The Indefinite Integral of $\frac{1}{1 + x^2}$

$$\int \frac{dx}{1 + x^2} = \arctan(x) + C$$

Rule 14: The Indefinite Integral of $\frac{1}{\sqrt{1 - x^2}}$

$$\int \frac{dx}{\sqrt{1 - x^2}} = \arcsin(x) + C$$

Example 13

Find the indefinite integral

$$\int \left(4 + \frac{1}{1 + x^2}\right)dx$$

Solution

$$\int\left(4+\frac{1}{1+x^2}\right)dx$$

$$= \int 4\,dx + \int \frac{1}{1+x^2}dx$$

$$= 4x + \arctan(x) + C$$

Example 14

Find the indefinite integral

$$\int\left(\frac{1}{\sqrt{1+x^2}} + 2x + 1\right)dx$$

Solution

$$\int\left(\frac{1}{\sqrt{1+x^2}} + 2x + 1\right)dx = \int \frac{1}{\sqrt{1+x^2}}dx + 2\int x dx + \int dx$$

$$= \arcsin(x) + 2\cdot\frac{1}{2}x^2 + x + C$$

$$= \arcsin(x) + x^2 + x + C$$

Differential Equations

Let's return to the problem posed at the beginning of the section: *Given the derivative of a function f', can we find the function f?* As an example, suppose we are given the function

$$f'(x) = 2x - 1 \qquad (1)$$

and we wish to find $f(x)$. From what we now know, we can find f by integrating Equation (1). Thus,

$$f(x) = \int f'(x)dx = \int (2x-1)dx = x^2 - x + C \qquad (2)$$

where C is an arbitrary constant. Thus, infinitely many functions have the derivative f', each differing from the other by a constant.

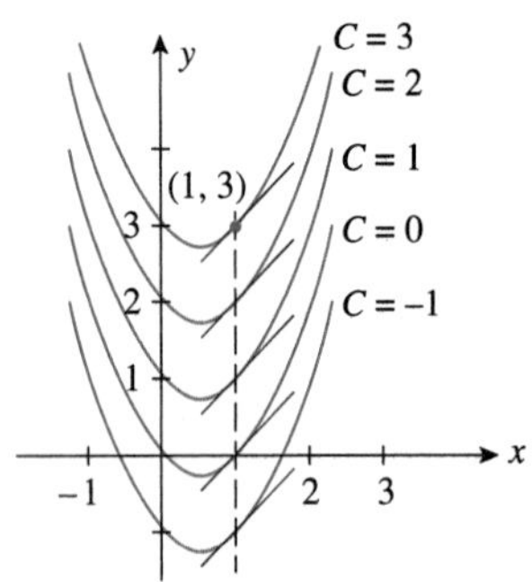

Figure 8.4

The graphs of some of the functions having the derivative $f'(x) = 2x - 1$. Observe that the slopes of the tangent lines to the graphs are the same for a fixed value of x.

Equation (1) is called a differential equation. In general, a **differential equation** is an equation that involves the derivative or differential of an unknown function. [In the case of Equation (1), the unknown function is f.] A **solution** of a differential equation is any function that satisfies the differential equation. Thus, Equation (2) gives *all* the solutions of the differential Equation (1), and it is, accordingly, called the **general solution** of the differential equation $f'(x) = 2x - 1$.

The graphs of $f(x) = x^2 - x + C$ for selected values of C are shown in Figure 8.4. These graphs have one property in common: For any fixed value of x, the tangent lines to these graphs have the same slope. This follows because any member of the family $f(x) = x^2 - x + C$ must have the same slope at x—namely, $2x - 1$!

Although there are infinitely many solutions to the differential equation $f'(x) = 2x - 1$, we can obtain a **particular solution** by specifying the value the function must assume at a certain value of x. For example, suppose we stipulate that the function f under consideration must satisfy the condition $f(1) = 3$ or, equivalently, the graph of f must pass through the point (1, 3). Then, using the condition on the general solution

$f(x) = x^2 - x + C$, we find that

$$f(1) = 1 - 1 + C = 3$$

and $C = 3$. Thus, the particular solution is $f(x) = x^2 - x + 3$ (see Figure 8.4).

The condition $f(1) = 3$ is an example of an initial condition. More generally, an **initial condition** is a condition imposed on the value of f at a point $x = a$.

Initial Value Problems

An **initial value problem** is one in which we are required to find a function satisfying (1) a differential equation and (2) one or more initial conditions. The following are examples of initial value problems.

Example 15

Find the function f if it is known that

$$f'(x) = 3x^2 - 4x + 8 \qquad \text{and} \qquad f(1) = 9$$

Solution

We are required to solve the initial value problem

$$\begin{aligned} f'(x) &= 3x^2 - 4x + 8 \\ f(1) &= 9 \end{aligned}$$

Integrating the function f', we find

$$\begin{aligned} f(x) &= \int f'(x)\,dx \\ &= \int (3x^2 - 4x + 8)\,dx \\ &= x^3 - 2x^2 + 8x + C \end{aligned}$$

Using the condition $f(1) = 9$, we have

$$9 = f(1) = 1^3 - 2(1)^2 + 8(1) + C = 7 + C$$

or $\quad C = 2$

Therefore, the required function f is given by $f(x) = x^3 - 2x^2 + 8x + 2$.

Applications

Example 16

Velocity of a Maglev In a test run of a maglev along a straight elevated monorail track, data obtained from reading its speedometer indicate that the velocity of the maglev at time t can be described by the velocity function

$$v(t) = 2t \qquad (0 \le t \le 30)$$

Find the position function of the maglev. Assume that initially the maglev is located at the origin of a coordinate line.

Solution

Let $s(t)$ denote the position of the maglev at any time t $(0 \leq t \leq 30)$. Then, $s'(t) = v(t)$. So, we have the initial value problem

$$\left.\begin{aligned} s'(t) &= 2t \\ s(0) &= 0 \end{aligned}\right\}$$

Integrating both sides of the differential equation $s'(t) = 8t$, we obtain

$$s(t) = \int s'(t)\, dt = \int 2t\, dt = t^2 + C$$

where C is an arbitrary constant. To evaluate C, we use the initial condition $s(0) = 0$ to write

$$s(0) = (0) + C = 0 \qquad \text{or} \qquad C = 0$$

Therefore, the required position function is $s(t) = t^2$ $(0 \leq t \leq 30)$.

Example 17

Magazine Circulation The current circulation of the *Investor's Digest* is 3000 copies per week. The managing editor of the weekly projects a growth rate of

$$4 + 5t^{2/3}$$

copies per week, t weeks from now, for the next 3 years. Based on her projection, what will the circulation of the digest be 125 weeks from now?

Solution

Let $S(t)$ denote the circulation of the digest t weeks from now. Then $S'(t)$ is the rate of change in the circulation in the tth week and is given by

$$S'(t) = 4 + 5t^{2/3}$$

Furthermore, the current circulation of 3000 copies per week translates into the initial condition $S(0) = 3000$. Integrating the differential equation with respect to t gives

$$\begin{aligned} S(t) &= \int S'(t)\, dt = \int (4 + 5t^{2/3})\, dt \\ &= 4t + 5\left(\frac{t^{5/3}}{\frac{5}{3}}\right) + C = 4t + 3t^{5/3} + C \end{aligned}$$

To determine the value of C, we use the condition $S(0) = 3000$ to write

$$S(0) = 4(0) + 3(0) + C = 3000$$

which gives $C = 3000$. Therefore, the circulation of the digest t weeks from now will be

$$S(t) = 4t + 3t^{5/3} + 3000$$

In particular, the circulation 125 weeks from now will be

$$s(125) = 4(125) + 3(125)^{5/3} + 3000 = 12{,}875$$

copies per week.

Example 18

Loss from a Leak A leak is found in a tank at a chemical plant. Because of cyclic variations in the use of the equipment of the plant the total amount of fluid leaking has a volume of

$$f(t) = 3.4 + 2.7\sin(t) \text{ litres/hour}$$

at a time of t hours after midnight on the day that measurement is started. Compute the total amount of fluid that leaks in the next week.

Solution

The leakage is per hour. There are $7 \cdot 24 = 168$ hours in a week. Since we are measuring the total leakage during a week we start with an initial volume of $V(0) = 0$. We now need to find a formula for the total leakage by integrating the amount of fluid leaking per hour.

$$\begin{aligned}\int (3.4 + 2.7\sin(t))\,dt &= \int 3.4\,dt + \int 2.7\sin(t)\,dt \\ &= \int 3.4\,dt + 2.7\int \sin(t)\,dt \\ &= 3.4t + 2.7(-\cos(t)) + C \\ &= 3.4t - 2.7\cos(t) + C\end{aligned}$$

where C is an arbitrary constant. We use the condition $V(0) = 0$ to solve for C.

$$\begin{aligned}V(0) &= 3.40 - 2.7\cos(0) + C \\ 0 &= 3.4 - 2.7 \cdot 1 + C \\ -C &= 3.4 - 2.7 \\ C &= -0.7\end{aligned}$$

The amount of fluid that leaks t hours after midnight the day that measurement started is

$$F(t) = 3.4t - 2.7\cos(t) - 0.7.$$

Evaluating this formula at 168 hours gives us

$$\begin{aligned}F(168) = 3.4 \cdot 168 - 2.7\cos(168) - 0.7 &= 571.2 - 2.7 \cdot (-0.075) - 0.7 \\ &= 571.2 + 0.2 - 0.7 \\ &= 570.7\end{aligned}$$

Therefore the total amount of fluid that leaks in the next week is 570.7 litres.

Self-Check Exercises 8.1

1. Evaluate $\displaystyle\int \left(\frac{1}{\sqrt{x}} - \frac{2}{x} + 3e^x\right) dx$.

2. Find the rule for the function f given that (1) the slope of the tangent line to the graph of f at any point $P(x, f(x))$ is given by the expression $3x^2 - 6x + 3$ and (2) the graph of f passes through the point (2, 9).

3. Suppose United Motors' share of the new cars sold in a certain country is changing at the rate of

$$f(t) = -0.01875t^2 + 0.15t - 1.2 \qquad (0 \le t \le 12)$$

percent at year t ($t = 0$ corresponds to the beginning of 1988). The company's market share at the beginning of 1988 was 48.4%. What was United Motors' market share at the beginning of 2000?

Solutions to Self-Check Exercises 8.1 can be found on page 446.

8.1 Exercises

In Exercises 1–6, verify directly that *F* is an antiderivative of *f*.

1. $F(x) = \frac{1}{3}x^3 + 2x^2 - x + 2; f(x) = x^2 + 4x - 1$

2. $F(x) = xe^x + \pi; f(x) = e^x(1 + x)$

3. $F(x) = \sqrt{2x^2 - 1}; f(x) = \frac{2x}{\sqrt{2x^2 - 1}}$

4. $F(x) = x \ln x - x; f(x) = \ln x$

5. $F(x) = \frac{1}{2}\cos(2x) + 4; f(x) = \sin(2x)$

6. $F(x) = x \cdot \arctan(x); f(x) = \arctan(x) + \frac{x}{x^2 + 1}$

In Exercises 7–12, (a) verify that *G* is an antiderivative of *f*, (b) find all antiderivatives of *f*, and (c) sketch the graphs of a few of the family of antiderivatives found in part (b).

7. $G(x) = 2x; f(x) = 2$

8. $G(x) = 2x^2; f(x) = 4x$

9. $G(x) = \frac{1}{3}x^3; f(x) = x^2$

10. $G(x) = e^x; f(x) = e^x$

11. $G(x) = \cos(2x); f(x) = 2\sin(2x)$

12. $G(x) = 4\arctan(x); f(x) = \frac{4}{1 + x^2}$

In Exercises 13–60, find the indefinite integral.

13. $\int 6\,dx$

14. $\int \sqrt{2}\,dx$

15. $\int x^3\,dx$

16. $\int 2x^5\,dx$

17. $\int x^{-4}\,dx$

18. $3t^{-7}\,dt$

19. $\int x^{2/3}\,dx$

20. $\int 2u^{3/4}\,du$

21. $\int x^{-5/4}\,dx$

22. $\int 3x^{-2/3}\,dx$

23. $\int \frac{2}{x^2}\,dx$

24. $\int \frac{1}{3x^5}\,dx$

25. $\int \pi\sqrt{t}\,dt$

26. $\int \frac{3}{\sqrt{t}}\,dt$

27. $\int (3 - 2x)\,dx$

28. $\int (1 + u + u^2)\,du$

29. $\int (x^2 + x + x^{-3})\,dx$

30. $\int (0.3t^2 + 0.02t + 2)\,dt$

31. $\int 4e^x\,dx$

32. $\int (1 + e^x)\,dx$

33. $\int (1 + x + e^x)\,dx$

34. $\int (2 + x + 2x^2 + e^x)\,dx$

35. $\int \left(4x^3 - \frac{2}{x^2} - 1\right) dx$

36. $\int \left(6x^3 + \frac{3}{x^2} - x\right) dx$

37. $\int (x^{5/2} + 2x^{3/2} - x)\,dx$

38. $\int (t^{3/2} + 2t^{1/2} - 4t^{-1/2})\,dt$

39. $\int \left(\sqrt{x} + \frac{3}{\sqrt{x}}\right) dx$

40. $\int \left(\sqrt[3]{x^2} - \frac{1}{x^2}\right) dx$

41. $\int \left(\frac{u^3 + 2u^2 - u}{3u}\right) du$

Hint: $\frac{u^3 + 2u^2 - u}{3u} = \frac{1}{3}u^2 + \frac{2}{3}u - \frac{1}{3}$

42. $\int \frac{x^4 - 1}{x^2}\,dx$

Hint: $\frac{x^4 - 1}{x^2} = x^2 - x^{-2}$

43. $\int (2t + 1)(t - 2)\,dt$

44. $\int u^{-2}(1 - u^2 + u^4)\,du$

45. $\int \frac{1}{x^2}(x^4 - 2x^2 + 1)\,dx$

46. $\int \sqrt{t}(t^2 + t - 1)\,dt$

47. $\int \frac{ds}{(s + 1)^{-2}}$

48. $\int \left(\sqrt{x} + \frac{3}{x} - 2e^x\right) dx$

49. $\int (e^t + t^e)\,dt$

50. $\int \left(\frac{1}{x^2} - \frac{1}{\sqrt[3]{x^2}} + \frac{1}{\sqrt{x}}\right) dx$

51. $\int \left(\frac{x^3 + x^2 - x + 1}{x^2}\right) dx$

Hint: Simplify the integrand first.

52. $\int \frac{t^3 + \sqrt[3]{t}}{t^2}\,dt$

Hint: Simplify the integrand first.

53. $\int \frac{(\sqrt{x} - 1)^2}{x^2}\,dx$

Hint: Simplify the integrand first.

54. $\int (x + 1)^2\left(1 - \frac{1}{x}\right) dx$

Hint: Simplify the integrand first.

55. $\int (2\cos(x) + 3\sin(x))\,dx$

56. $\int (t^2 + t + \cos(t))\,dt$

57. $\int \left(s + \frac{3}{1 + s^2}\right) ds$

58. $\int (\sin(x)(\sin(x) - 1) + \cos^2(x))\,dx$

Hint: Simplify the integrand first using a trig identity.

59. $\int (s - \sec^2(s))\,ds$

60. $\int (\sec(t)\tan(t) - \csc^2(t))\,dt$

In Exercises 61–70, find $f(x)$ by solving the initial value problem.

61. $f'(x) = 2x + 1; f(1) = 3$

62. $f'(x) = 3x^2 - 6x; f(2) = 4$

63. $f'(x) = 3x^2 + 4x - 1; f(2) = 9$

64. $f'(x) = \dfrac{1}{\sqrt{x}}; f(4) = 2$

65. $f'(x) = 1 + \dfrac{1}{x^2}; f(1) = 2$

66. $f'(x) = e^x - 2x; f(0) = 2$

67. $f'(x) = \dfrac{x+1}{x}; f(1) = 1$

68. $f'(x) = 1 + e^x + \dfrac{1}{x}; f(1) = 3 + e$

69. $f'(x) = \dfrac{2}{1+x^2}; f(1) = \dfrac{\pi}{2}$

70. $f'(x) = \cos(x); f\left(\dfrac{\pi}{6}\right) = 0$

In Exercises 71–74, find the function f given that the slope of the tangent line to the graph of f at any point $(x, f(x))$ is $f'(x)$ and that the graph of f passes through the given point.

71. $f'(x) = \dfrac{1}{2}x^{-1/2}; (2, \sqrt{2})$

72. $f'(t) = t^2 - 2t + 3; (1, 2)$

73. $f'(x) = e^x + x; (0, 3)$

74. $f'(x) = \dfrac{2}{x} + 1; (1, 2)$

75. **Velocity of a Car** The velocity of a car (in metres/second) t sec after starting from rest is given by the function

$$f(t) = \frac{2}{3}\sqrt{t} \qquad (0 \le t \le 30)$$

Find the car's position at any time t.

76. **Velocity of a Maglev** The velocity (in metres/second) of a maglev is

$$v(t) = 0.05t + 1 \qquad (0 \le t \le 120)$$

At $t = 0$, it is at the station. Find the function giving the position of the maglev at time t, assuming that the motion takes place along a straight stretch of track.

77. **Measuring Temperature** The temperature on a certain day as measured at the airport of a city is changing at the rate of

$$T'(t) = 0.11t^2 - 1.8t + 2.2 \qquad (0 \le t \le 4)$$

°C/hr, where t is measured in hours, with $t = 0$ corresponding to 6 A.M. The temperature at 6 A.M. was 4°C.

a. Find an expression giving the temperature T at the airport at any time between 6 A.M. and 10 A.M.
b. What was the temperature at 10 A.M.?

78. **Ballast Dropped from a Balloon** A ballast is dropped from a stationary hot-air balloon that is hovering at an altitude of 120 metres. Its velocity after t sec is -9.8 m/sec.
a. Find the height $h(t)$ of the ballast from the ground at time t.
Hint: $h'(t) = -9.8t$ and $h(0) = 120$.
b. When will the ballast strike the ground?
c. Find the velocity of the ballast when it hits the ground.

79. **Flight of a Rocket** The velocity, in metres/second, of a rocket t sec into vertical flight is given by

$$v(t) = -t^2 + 64t + 40$$

Find an expression $h(t)$ that gives the rocket's altitude, in metres, t sec after liftoff. What is the altitude of the rocket 30 sec after liftoff?
Hint: $h'(t) = v(t)$; $h(0) = 0$.

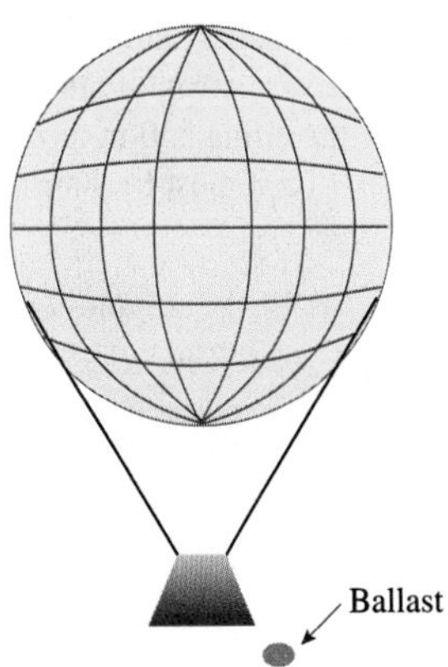

80. **Acceleration of a Car** A car travelling along a straight road at 21 m/sec accelerated to a speed of 43 m/sec over a distance of 120 m. What was the acceleration of the car, assuming it was constant?

81. **Deceleration of a Car** What constant deceleration would a car moving along a straight road have to be subjected to if it were brought to rest from a speed of 29 m/sec in 9 sec? What would be the stopping distance?

82. A tank has a constant cross-sectional area of 8 m^2 and an orifice of constant cross-sectional area of $\frac{1}{5}$m^2 located at the bottom of the tank (see the accompanying figure).

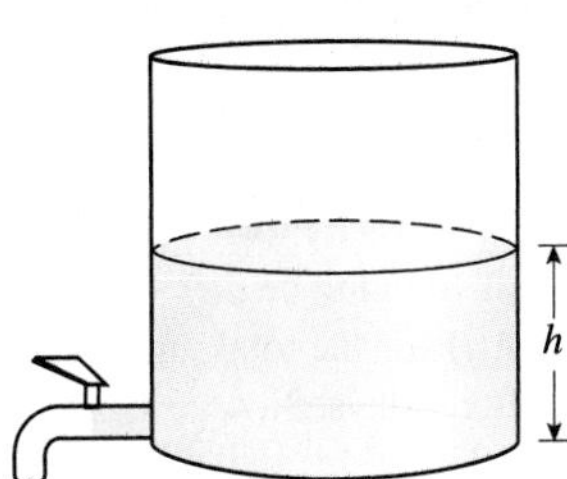

If the tank is filled with water to a height of h ft and allowed to drain, then the height of the water decreases at a rate that is described by the equation

$$\frac{dh}{dt} = -\frac{1}{25}\left(\sqrt{20} - \frac{t}{8}\right) \qquad (0 \leq t \leq 8\sqrt{20})$$

Find an expression for the height of the water at any time t if its height initially is 3 m.

83. **Amount of Rainfall** During a thunderstorm, rain was falling at the rate of

$$\frac{20}{(t+4)^2} \qquad (0 \leq t \leq 2)$$

cm/h.

a. Find an expression giving the total amount of rainfall after t hours.

Hint: The total amount of rainfall at $t = 0$ is zero.

b. How much rain had fallen after 1 h? After 2 h?

84. **Launching a Fighter Aircraft** A fighter aircraft is launched from the deck of a Nimitz-class aircraft carrier with the help of a steam catapult. If the aircraft is to attain a takeoff speed of at least 240 m/sec after travelling 250 m along the flight deck, find the minimum acceleration it must be subjected to, assuming it is constant.

85. **Velocity of a Car** Two cars, side by side, start from rest and travel along a straight road. The velocity of car A is given by $v = f(t)$, and the velocity of car B is given by $v = g(t)$. The graphs of f and g are shown in the following figure. Are the cars still side by side after T sec? If not, which car is ahead of the other? Justify your answer.

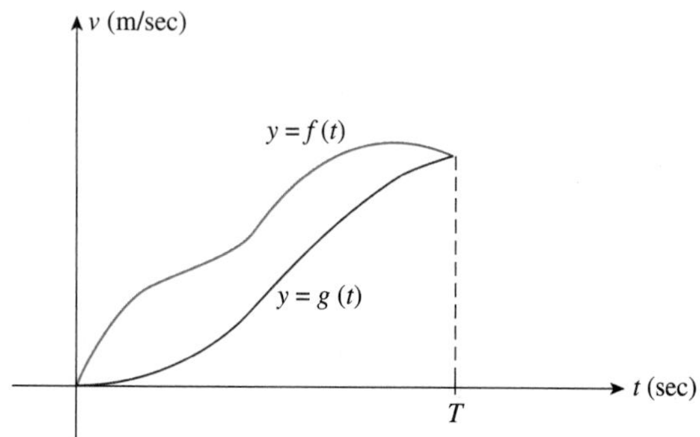

86. **Total Leakage** A leakage of gasoline from underground storage tanks into the soil is found to have a rate of $1.2 + \sqrt{t}$ mL/h. If the total amount of gas missing from the tank at time $t = 0$ is found to be 127 mL find an expression for the total amount of gas lost into the soil at any time $t > 0$.

87. **Periodically Varying Water Flow** Suppose that daily flow of water from a glacier is found to be $a(t) = 2300 + 300 \sin(t)$ L/d. Find an expression $A(t)$ for the total amount of water that flows out of the glacier after t days. Now assume that the amount of flow is just $b(t) = 2300$ L/d and find an expression for the amount $B(t)$ of water that flows out of the glacier after t days. Graph and compare both functions and discuss briefly how well they agree.

88. **Total Capture of Periodically Varying Energy** If the energy harvested from the sun is $f(t) = 200 + 20 \cos(t)$ J/h t hours after sunrise, then what is the total energy harvested in the 8 h after sunrise?

Business and Economics Applications

89. **Cost of Producing Clocks** Lorimar Watch Company manufactures travel clocks. The daily marginal cost function associated with producing these clocks is

$$C'(x) = 0.000009x^2 - 0.009x + 8$$

where $C'(x)$ is measured in dollars/unit and x denotes the number of units produced. Management has determined that the daily fixed cost incurred in producing these clocks is \$120. Find the total cost incurred by Lorimar in producing the first 500 travel clocks/day.

90. **Revenue Functions** The management of Lorimar Watch Company has determined that the daily marginal revenue function associated with producing and selling their travel clocks is given by

$$R'(x) = -0.009x + 12$$

where x denotes the number of units produced and sold and $R'(x)$ is measured in dollars/unit.

a. Determine the revenue function $R(x)$ associated with producing and selling these clocks.

b. What is the demand equation that relates the wholesale unit price with the quantity of travel clocks demanded?

91. **Profit Functions** Cannon Precision Instruments makes an automatic electronic flash with Thyrister circuitry. The estimated marginal profit associated with producing and selling these electronic flashes is

$$-0.004x + 20$$

dollars/unit/month when the production level is x units per month. Cannon's fixed cost for producing and selling these electronic flashes is \$16,000/month. At what level of production does Cannon realize a maximum profit? What is the maximum monthly profit?

92. **Cost of Producing Guitars** Carlota Music Company estimates that the marginal cost of manufacturing its Professional Series guitars is

$$C'(x) = 0.002x + 100$$

dollars/month when the level of production is x guitars/month. The fixed costs incurred by Carlota are \$4000/month. Find the total monthly cost incurred by Carlota in manufacturing x guitars/month.

93. **Quality Control** As part of a quality-control program, the chess sets manufactured by Jones Brothers are subjected to a final inspection before packing. The rate of increase in the number of sets checked per hour by an inspector t hr into the 8 A.M. to 12 noon morning shift is approximately

$$N'(t) = -3t^2 + 12t + 45 \qquad (0 \leq t \leq 4)$$

a. Find an expression $N(t)$ that approximates the number of sets inspected at the end of t hours.
Hint: $N(0) = 0$.
b. How many sets does the average inspector check during a morning shift?

94. Health-Care Costs The average out-of-pocket costs for people enrolled in Health Canada–sponsored programs (including premiums, cost sharing, and prescription drugs not covered) is projected to grow at the rate of

$$C'(t) = 12.288t^2 - 150.5594t + 695.23$$

dollars/year, where t is measured in 5-yr intervals, with $t = 0$ corresponding to 2000. The out-of-pocket costs for beneficiaries in 2000 were \$3142.
a. Find an expression giving the average out-of-pocket costs for beneficiaries in year t.
b. What is the projected average out-of-pocket costs for beneficiaries in 2010?

A calculator is recommended for Exercises 95–99.

95. Cable TV Subscribers A study conducted by TeleCable estimates that the number of cable TV subscribers will grow at the rate of

$$100 + 210t^{3/4}$$

new subscribers/month t mo from the start date of the service. If 5000 subscribers signed up for the service before the starting date, how many subscribers will there be 16 mo from that date?

96. Online Ad Sales In a study conducted in 2000, the share of online advertisement, worldwide, as a percent of the total ad market is expected to grow at the rate of

$$R(t) = -0.033t^2 + 0.3428t + 0.07 \qquad (0 \le t \le 6)$$

percent/year at time t (in years), with $t = 0$ corresponding to the beginning of 2000. The online ad market at the beginning of 2000 was 2.9% of the total ad market.
a. What is the projected online ad market share at any time t?
b. What is the projected online ad market share at the beginning of 2005?
Source: Jupiter Media Metrix, Inc.

97. Bank Deposits Madison Finance opened two branches on September 1 ($t = 0$). Branch A is located in an established industrial park, and branch B is located in a fast-growing new development. The net rate at which money was deposited into branch A and branch B in the first 180 business days is given by the graphs of f and g, respectively (see the figure). Which branch has a larger amount on deposit at the end of 180 business days? Justify your answer.

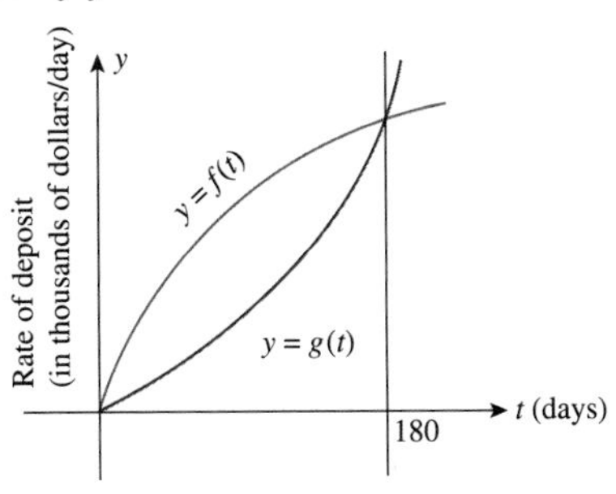

Biological and Life Sciences Applications

98. Air Pollution On an average summer day, the level of carbon monoxide (CO) in a city's air is 2 parts per million (ppm). An environmental protection agency's study predicts that, unless more stringent measures are taken to protect the city's atmosphere, the CO concentration present in the air will increase at the rate of

$$0.003t^2 + 0.06t + 0.1$$

ppm/year t yr from now. If no further pollution-control efforts are made, what will be the CO concentration on an average summer day 5 yr from now?

99. Population Growth The development of AstroWorld ("The Amusement Park of the Future") on the outskirts of a city will increase the city's population at the rate of

$$4500\sqrt{t} + 1000$$

people/year t yr from the start of construction. The population before construction is 30,000. Determine the projected population 9 yr after construction of the park has begun.

100. Ozone Pollution The rate of change of the level of ozone, an invisible gas that is an irritant and impairs breathing, present in the atmosphere on a certain August day in the city of North York is given by

$$R(t) = 3.2922t^2 - 0.366t^3 \qquad (0 < t < 11)$$

(measured in pollutant standard index/hour). Here, t is measured in hours, with $t = 0$ corresponding to 7 A.M. Find the ozone level $A(t)$ at any time t, assuming that at 7 A.M. it is zero.
Hint: $A'(t) = R(t)$ and $A(0) = 0$.

101. Surface Area of a Human Empirical data suggest that the surface area of a 180-cm-tall human body changes at the rate of

$$S'(W) = 0.131773W^{-0.575}$$

square metres/kilogram, where W is the weight of the body in kilograms. If the surface area of a 180-cm-tall human body weighing 70 kg is 1.886277 m^2, what is the surface area of a human body of the same height weighing 75 kg?

102. Heights of Children According to the Jenss model for predicting the height of preschool children, the rate of growth of a typical preschool child is

$$R(t) = 25.8931e^{-0.993t} + 6.39 \qquad \left(\frac{1}{4} \le t \le 6\right)$$

cm/yr, where t is measured in years. The height of a typical 3-mo-old preschool child is 60.2952 cm.
a. Find a model for predicting the height of a typical preschool child at age t.
b. Use the result of part (a) to estimate the height of a typical 1-yr-old child.

103. Blood Flow in an Artery Nineteenth-century physician Jean Louis Marie Poiseuille discovered that the rate of change of the velocity of blood r cm from the central axis of an artery (in centimetres/second/centimetre) is given by

$$a(r) = -kr$$

where k is a constant. If the radius of an artery is R cm, find an expression for the velocity of blood as a function of r (see the accompanying figure).
Hint: $v'(r) = a(r)$ and $v(R) = 0$. (Why?)

Blood vessel

104. Integral Rule Remember that

$$\sin^2(\theta) = \frac{1 - \cos(2\theta)}{2}$$

Using this trigonometric identity, compute an integral rule for

$$\int \sin^2(x)\,dx$$

105. Integral Rule Remember that

$$\cos^2(\theta) = \frac{1 + \cos(2\theta)}{2}$$

Using this trigonometric identity, compute an integral rule for

$$\int \cos^2(x)\,dx$$

106. Accumulation of Shells The population of fresh-water crustaceans in a lake is seasonal with a population that casts off shells at a rate of $P(t) = 400 + 300 \sin(t)$ shells/month. Each shell contains about 0.16 grams of calcium carbonate. Assume that the lake bottom was dredged to remove all existing sediment. How much calcium carbonate is deposited in the lake bottom sediment per year?

107. Production of a Biochemical The production of a biochemical in a bacterial reactor drops off as the concentration increases because the biochemical interferes with the bacteria's ability to produce it. If the amount of the biochemical present in the reactor grows at a rate of $c(t) = \dfrac{2.3}{1 + t^2}$ g/h and starts with none of the biochemical present at time $t = 0$ find an expression for the total amount of the biochemical $C(t)$ after t hours. Will the amount of the biochemical in the reactor ever exceed a kilogram?

In Exercises 108–111, determine whether the statement is true or false. If it is true, explain why it is true. If it is false, give an example to show why it is false.

108. If F and G are antiderivatives of f on an interval I, then $F(x) = G(x) + C$ on I.

109. If F is an antiderivative of f on an interval I, then $\int f(x)\,dx = F(x)$.

110. If f and g are integrable, then $\int [2f(x) - 3g(x)]\,dx = 2\int f(x)\,dx - 3\int g(x)\,dx$.

111. If f and g are integrable, then $\int f(x)g(x)\,dx = [\int f(x)\,dx][\int g(x)\,dx]$.

Solutions to Self-Check Exercises 8.1

1. $$\int \left(\frac{1}{\sqrt{x}} - \frac{2}{x} + 3e^x\right) dx = \int \left(x^{-1/2} - \frac{2}{x} + 3e^x\right) dx$$
$$= \int x^{-1/2}\,dx - 2\int \frac{1}{x}\,dx + 3\int e^x\,dx$$
$$= 2x^{1/2} - 2\ln|x| + 3e^x + C$$
$$= 2\sqrt{x} - 2\ln|x| + 3e^x + C$$

Solutions to Self-Check Exercises 8.1, *continued*

2. The slope of the tangent line to the graph of the function f at any point $P(x, f(x))$ is given by the derivative f' of f. Thus, the first condition implies that

$$f'(x) = 3x^2 - 6x + 3$$

which, upon integration, yields

$$f(x) = \int (3x^2 - 6x + 3)\, dx$$

$$= x^3 - 3x^2 + 3x + k$$

where k is the constant of integration.

To evaluate k, we use the initial condition (2), which implies that $f(2) = 9$, or

$$9 = f(2) = 2^3 - 3(2)^2 + 3(2) + k$$

or $k = 7$. Hence, the required rule of definition of the function f is

$$f(x) = x^3 - 3x^2 + 3x + 7$$

3. Let $M(t)$ denote United Motors' market share at year t. Then,

$$M(t) = \int f(t)\, dt$$

$$= \int (-0.01875t^2 + 0.15t - 1.2)\, dt$$

$$= -0.00625t^3 + 0.075t^2 - 1.2t + C$$

To determine the value of C, we use the initial condition $M(0) = 48.4$, obtaining $C = 48.4$. Therefore,

$$M(t) = -0.00625t^3 + 0.075t^2 - 1.2t + 48.4$$

In particular, United Motors' market share of new cars at the beginning of 2000 is given by

$$M(12) = -0.00625(12)^3 + 0.075(12)^2$$

$$-1.2(12) + 48.4 = 34$$

or 34%.

8.2 Integration by Substitution

In Section 8.1 we developed certain rules of integration that are closely related to the corresponding rules of differentiation in Chapters 3 and 5. In this section we introduce a method of integration called the **method of substitution,** which is related to the chain rule for differentiating functions. When used in conjunction with the rules of integration developed earlier, the method of substitution is a powerful tool for integrating a large class of functions.

How the Method of Substitution Works

Consider the indefinite integral

$$\int 2(2x + 4)^5\, dx \qquad (3)$$

One way of evaluating this integral is to expand the expression $(2x + 4)^5$ and then integrate the resulting integrand term by term. As an alternative approach, let's see if we can simplify the integral by making a change of variable. Write

$$u = 2x + 4$$

with differential

$$du = 2\, dx$$

If we formally substitute these quantities into Equation (3), we obtain

$$\int 2(2x + 4)^5\, dx = \int (2x + 4)^5(2\, dx) = \int u^5\, du$$

↑ Rewrite

↑ $\begin{bmatrix} u = 2x + 4 \\ du = 2\, dx \end{bmatrix}$

Now, the last integral involves a power function and is easily evaluated using Rule 2 of Section 8.1. Thus,

$$\int u^5\, du = \frac{1}{6}u^6 + C$$

Therefore, using this result and replacing u by $u = 2x + 4$, we obtain

$$\int 2(2x + 4)^5\, dx = \frac{1}{6}(2x + 4)^6 + C$$

We can verify that the foregoing result is indeed correct by computing

$$\frac{d}{dx}\left[\frac{1}{6}(2x + 4)^6\right] = \frac{1}{6}\cdot 6(2x + 4)^5(2) \qquad \text{Use the chain rule.}$$

$$= 2(2x + 4)^5$$

and observing that the last expression is just the integrand of (3).

The Method of Integration by Substitution

To see why the approach used in evaluating the integral in (3) is successful, write

$$f(x) = x^5 \qquad \text{and} \qquad g(x) = 2x + 4$$

Then, $g'(x) = 2\, dx$. Furthermore, the integrand of (3) is just the composition of f and g. Thus,

$$(f \circ g)(x) = f(g(x))$$

$$= [g(x)]^5 = (2x + 4)^5$$

Therefore, (3) can be written as

$$\int f(g(x))g'(x)\, dx \qquad (4)$$

Next, let's show that an integral having the form (4) can always be written as

$$\int f(u)\, du \qquad (5)$$

Suppose F is an antiderivative of f. By the chain rule, we have

$$\frac{d}{dx}[F(g(x))] = F'(g(x))g'(x)$$

Therefore,

$$\int F'(g(x))g'(x)\, dx = F(g(x)) + C$$

Letting $F' = f$ and making the substitution $u = g(x)$, we have

$$\int f(g(x))g'(x)\, dx = F(u) + C = \int F'(u)\, du = \int f(u)\, du$$

as we wished to show. Thus, if the transformed integral is readily evaluated, as is the case with the integral (3), then the method of substitution will prove successful.

Before we look at more examples, let's summarize the steps involved in integration by substitution.

Integration by Substitution

Step 1 Let $u = g(x)$, where $g(x)$ is part of the integrand, usually the "inside function" of the composite function $f(g(x))$.

Step 2 Compute $du = g'(x)\,dx$.

Step 3 Use the substitution $u = g(x)$ and $du = g'(x)\,dx$ to convert the *entire* integral into one involving *only* u.

Step 4 Evaluate the resulting integral.

Step 5 Replace u by $g(x)$ to obtain the final solution as a function of x.

Remark Sometimes we need to consider different choices of g for the substitution $u = g(x)$ in order to carry out Step 3 and/or Step 4. ◀

Example 1

Find $\int 2x(x^2 + 3)^4\,dx$.

Solution

Step 1 Observe that the integrand involves the composite function $(x^2 + 3)^4$ with "inside function" $g(x) = x^2 + 3$. So, we choose $u = x^2 + 3$.

Step 2 Compute $du = 2x\,dx$.

Step 3 Making the substitution $u = x^2 + 3$ and $du = 2x\,dx$, we obtain

$$\int 2x(x^2 + 3)^4\,dx = \underset{\text{Rewrite}}{\underset{\uparrow}{=}} \int (x^2 + 3)^4(2x\,dx) = \int u^4\,du$$

an integral involving only the variable u.

Step 4 Evaluate

$$\int u^4\,du = \frac{1}{5}u^5 + C$$

Step 5 Replacing u by $x^2 + 3$, we obtain

$$\int 2x(x^2 + 3)^4\,dx = \frac{1}{5}(x^2 + 3)^5 + C$$

Example 2

Find $\int 3\sqrt{3x + 1}\,dx$.

Solution

Step 1 The integrand involves the composite function $\sqrt{3x + 1}$ with "inside function" $g(x) = 3x + 1$. So, let $u = 3x + 1$.

Step 2 Compute $du = 3\,dx$.

Step 3 Making the substitution $u = 3x + 1$ and $du = 3\,dx$, we obtain

$$\int 3\sqrt{3x + 1}\,dx = \int \sqrt{3x + 1}\,(3\,dx) = \int \sqrt{u}\,du$$

an integral involving only the variable u.

Step 4 Evaluate

$$\int \sqrt{u}\,du = \int u^{1/2}\,du = \frac{2}{3}u^{3/2} + C$$

Step 5 Replacing u by $3x + 1$, we obtain

$$\int 3\sqrt{3x + 1}\, dx = \frac{2}{3}(3x + 1)^{3/2} + C$$

Example 3

Find $\int x^2 (x^3 + 1)^{3/2}\, dx$.

Solution

Step 1 The integrand contains the composite function $(x^3 + 1)^{3/2}$ with "inside function" $g(x) = x^3 + 1$. So, let $u = x^3 + 1$.

Step 2 Compute $du = 3x^2\, dx$.

Step 3 Making the substitution $u = x^3 + 1$ and $du = 3x^2\, dx$, or $x^2\, dx = \frac{1}{3}\, du$, we obtain

$$\int x^2 (x^3 + 1)^{3/2}\, dx = \int (x^3 + 1)^{3/2} (x^2\, dx)$$

$$= \int u^{3/2}\left(\frac{1}{3}\, du\right) = \frac{1}{3}\int u^{3/2}\, du$$

an integral involving only the variable u.

Step 4 We evaluate

$$\frac{1}{3}\int u^{3/2}\, du = \frac{1}{3}\cdot\frac{2}{5}u^{5/2} + C = \frac{2}{15}u^{5/2} + C$$

Step 5 Replacing u by $x^3 + 1$, we obtain

$$\int x^2(x^3 + 1)^{3/2}\, dx = \frac{2}{15}(x^3 + 1)^{5/2} + C$$

GROUP DISCUSSION

Let $f(x) = x^2(x^3 + 1)^{3/2}$. Using the result of Example 3, we see that an antiderivative of f is $F(x) = \frac{2}{15}(x^3 + 1)^{5/2}$. However, in terms of u (where $u = x^3 + 1$), an antiderivative of f is $G(u) = \frac{2}{15}u^{5/2}$. Compute $F(2)$. Next, suppose we want to compute $F(2)$ using the function G instead. At what value of u should you evaluate $G(u)$ in order to obtain the desired result? Explain your answer.

In the remaining examples, we drop the practice of labelling the steps involved in evaluating each integral.

Example 4

Find $\int e^{-3x}\, dx$.

Solution

Let $u = -3x$ so that $du = -3\, dx$, or $dx = -\frac{1}{3}du$. Then,

$$\int e^{-3x}\, dx = \int e^u\left(-\frac{1}{3}\, du\right) = -\frac{1}{3}\int e^u\, du$$

$$= -\frac{1}{3}e^u + C = -\frac{1}{3}e^{-3x} + C$$

Example 5

Find $\int \frac{x}{3x^2 + 1}\,dx$.

Solution

Let $u = 3x^2 + 1$. Then, $du = 6x\,dx$, or $x\,dx = \frac{1}{6}du$. Making the appropriate substitutions, we have

$$\begin{aligned}\int \frac{x}{3x^2 + 1}\,dx &= \int \frac{\frac{1}{6}}{u}\,du \\ &= \frac{1}{6}\int \frac{1}{u}\,du \\ &= \frac{1}{6}\ln|u| + C \\ &= \frac{1}{6}\ln(3x^2 + 1) + C \qquad \text{Since } 3x^2 + 1 > 0\end{aligned}$$

Example 6

Find $\int \frac{(\ln x)^2}{2x}\,dx$.

Solution

Let $u = \ln x$. Then,

$$du = \frac{d}{dx}(\ln x)\,dx = \frac{1}{x}\,dx$$

$$\begin{aligned}\int \frac{(\ln x)^2}{2x}\,dx &= \frac{1}{2}\int \frac{(\ln x)^2}{x}\,dx \\ &= \frac{1}{2}\int u^2\,du \\ &= \frac{1}{6}u^3 + C \\ &= \frac{1}{6}(\ln x)^3 + C\end{aligned}$$

Example 7

Find

$$\int \left(\cos\left(\frac{x}{2\pi}\right)\right)dx$$

Solution

Let $u = \frac{x}{2\pi}$ then $du = \frac{dx}{2\pi}$ and so $2\pi du = dx$.

$$\begin{aligned}\int \left(\cos\left(\frac{x}{2\pi}\right)\right)dx &= \int \cos(u)2\pi du \\ &= 2\pi\int \cos(u)\,ds \\ &= 2\pi\sin(u) + C \\ &= 2\pi\sin\left(\frac{x}{2\pi}\right) + C\end{aligned}$$

Example 8

Find

$$\int (x \sec^2(x^2))\,dx$$

Solution

Let $u = x^2$ then $du = 2x\,dx$ and so $\frac{1}{2}du = x\,dx$.

$$\begin{aligned}\int (x \sec^2(x^2))\,dx &= \int (\sec^2(x^2))\,x\,dx \\ &= \int (\sec^2(u))\,\frac{1}{2}\,du \\ &= \frac{1}{2}\int \sec^2(u)\,du \\ &= \frac{1}{2}\tan(u) + C \\ &= \frac{1}{2}\tan(x^2) + C\end{aligned}$$

In the next example the choice of u is not as simple. It is sometimes necessary to try different possible substitutions before one is found that works properly.

Example 9

Find

$$\int \frac{dx}{x^2 + 2x + 2}\,dx$$

Solution

First recall that

$$\int \frac{dx}{1 + x^2} = \arctan(x) + C$$

This is similar to the problem we are trying to solve. What is needed is a choice of u so that $u^2 + 1 = x^2 + 2x + 2$. If we let $u = x + 1$ then

$$\begin{aligned}u^2 + 1 &= (x + 1)^2 + 1 \\ &= x^2 + 2x + 1 + 1 \\ &= x^2 + 2x + 2\end{aligned}$$

Which means this substitution will transform the problem into a known integral. If $u = x + 1$ then $du = dx$.

$$\begin{aligned}\int \frac{dx}{x^2 + 2x + 2}\,dx &= \int \frac{dx}{(x + 1)^2 + 1} \\ &= \int \frac{du}{u^2 + 1} \\ &= \arctan(u) + C \\ &= \arctan(x + 1) + C\end{aligned}$$

GROUP DISCUSSION

Suppose $\int f(u)\,du = F(u) + C$.

1. Show that $\int f(ax + b)\,dx = \frac{1}{a}F(ax + b) + C$.
2. How can you use this result to facilitate the evaluation of integrals such as $\int (2x + 3)^5\,dx$ and $\int e^{3x-2}\,dx$? Explain your answer.

Applications

Examples 10 through 12 show how the method of substitution can be used in practical situations.

Example 10

Cost of Producing Solar Cell Panels In 1990 the head of the research and development department of Soloron Corporation claimed that the cost of producing solar cell panels would drop at the rate of

$$\frac{58}{(3t+2)^2} \qquad (0 \le t \le 10)$$

dollars per peak watt for the next t years, with $t = 0$ corresponding to the beginning of 1990. (A peak watt is the power produced at noon on a sunny day.) In 1990 the panels, which are used for photovoltaic power systems, cost \$10 per peak watt. Find an expression giving the cost per peak watt of producing solar cell panels at the beginning of year t. What was the cost at the beginning of 2000?

Solution

Let $C(t)$ denote the cost per peak watt for producing solar cell panels at the beginning of year t. Then,

$$C'(t) = -\frac{58}{(3t+2)^2}$$

Integrating, we find that

$$\begin{aligned} C(t) &= \int \frac{-58}{(3t+2)^2}\,dt \\ &= -58\int (3t+2)^{-2}\,dt \end{aligned}$$

Let $u = 3t + 2$ so that

$$du = 3\,dt \qquad \text{or} \qquad dt = \frac{1}{3}du$$

Then,

$$\begin{aligned} C(t) &= -58\left(\frac{1}{3}\right)\int u^{-2}\,du \\ &= -\frac{58}{3}(-1)u^{-1} + k \\ &= \frac{58}{3(3t+2)} + k \end{aligned}$$

where k is an arbitrary constant. To determine the value of k, note that the cost per peak watt of producing solar cell panels at the beginning of 1990 ($t = 0$) was 10, or $C(0) = 10$. This gives

$$C(0) = \frac{58}{3(2)} + k = 10$$

or $k = \frac{1}{3}$. Therefore, the required expression is given by

$$\begin{aligned} C(t) &= \frac{58}{3(3t + 2)} + \frac{1}{3} \\ &= \frac{58 + (3t + 2)}{3(3t + 2)} = \frac{3t + 60}{3(3t + 2)} \\ &= \frac{t + 20}{3t + 2} \end{aligned}$$

The cost per peak watt for producing solar cell panels at the beginning of 2000 is given by

$$C(10) = \frac{10 + 20}{3(10) + 2} \approx 0.94$$

or approximately $.94 per peak watt.

EXPLORING WITH TECHNOLOGY

Refer to Example 10.

1. Use a graphing utility to plot the graph of

$$C(t) = \frac{t + 20}{3t + 2}$$

using the viewing window $[0, 10] \cdot [0, 5]$. Then, use the numerical differentiation capability of the graphing utility to compute $C'(10)$.

2. Plot the graph of

$$C'(t) = -\frac{58}{(3t + 2)^2}$$

using the viewing window $[0, 10] \cdot [-10, 0]$. Then, use the evaluation capability of the graphing utility to find $C'(10)$. Is this value of $C'(10)$ the same as that obtained in part 1? Explain your answer.

Example 11

Computer Sales Projections A study prepared by the marketing department of Universal Instruments forecasts that, after its new line of Galaxy Home Computers is introduced into the market, sales will grow at the rate of

$$2000 - 1500e^{-0.05t} \qquad (0 \le t \le 60)$$

units per month. Find an expression that gives the total number of computers that will sell t months after they become available on the market. How many computers will Universal sell in the first year they are on the market?

Solution

Let $N(t)$ denote the total number of computers that may be expected to be sold t months after their introduction in the market. Then, the rate of growth of sales is given by $N'(t)$ units per month. Thus,

$$N'(t) = 2000 - 1500e^{-0.05t}$$

so that

$$N(t) = \int (2000 - 1500e^{-0.05t})\, dt$$

$$= \int 2000\, dt - 1500 \int e^{-0.05t}\, dt$$

Upon integrating the second integral by the method of substitution, we obtain

$$N(t) = 2000t + \frac{1500}{0.05} e^{-0.05t} + C$$

Let $u = -0.05t$, then $du = -0.05\, dt$.

$$= 2000t + 30{,}000e^{-0.05t} + C$$

To determine the value of C, note that the number of computers sold at the end of month 0 is nil, so $N(0) = 0$. This gives

$$N(0) = 30{,}000 + C = 0$$

Since $e^0 = 1$

or $C = -30{,}000$. Therefore, the required expression is given by

$$N(t) = 2000t + 30{,}000e^{-0.05t} - 30{,}000$$

$$= 2000t + 30{,}000(e^{-0.05t} - 1)$$

The number of computers that Universal can expect to sell in the first year is given by

$$N(12) = 2000(12) + 30{,}000(e^{-0.05(12)} - 1)$$

$$= 10{,}464 \text{ units}$$

Example 12

In earlier examples that used periodic functions, the fact we have not yet learned to do integration by substitution meant we had to live with the natural period of the function, $[0, 2\pi]$ for the sine and cosine functions. In order to adjust the period we multiply or divide the argument of the sine or cosine by a constant. This example uses this capability.

During a period of snow melt the amount of water flowing into a stream in Vancouver is found to be

$$6000 + 250 \sin\left(\frac{12}{\pi}t\right)$$

litres of water per hour if $t = 0$ is taken to be sunrise. Compute the amount of water that flows in the four hours following sunrise and from hours 4 to 8 after sunrise. Which period has the greatest total water flow?

Solution

Because we are measuring flow since sunrise we can assume that the total water flow starts at zero. We treat this problem as two initial value problems, one from 0 to 4 hours and one from 0 to 8 hours. The flow during hours 4 to 8 is then the flow for hours 0 to 8 minus the flow from hours 0 to 4. Both initial value problems have the same formula $F(t)$ as their general solution.

Let $u = \frac{12}{\pi} t$ then $du = \frac{12}{\pi} dt$ and so $\frac{\pi}{12} du = dt$.

$$\begin{aligned} F(t) &= \int 6000 + 250 \sin\left(\frac{12}{\pi} t\right) dt \\ &= \int 6000\, dt + 250 \int \sin\left(\frac{12}{\pi} t\right) dt \\ &= \int 6000\, dt + 250 \int \sin(u) \frac{\pi}{12}\, du \\ &= \int 6000\, dt + 250 \frac{\pi}{12} \int \sin(u)\, du \\ &= \int 6000\, dt + \frac{250\pi}{12} \int \sin(u)\, du \\ &= 6000t + \frac{250\pi}{12}(-\cos(u)) + C \\ &= 6000t - \frac{250\pi}{12} \cos\left(\frac{12}{\pi} t\right) + C \end{aligned}$$

We now need to use the fact that $F(0) = 0$ to find C.

$$\begin{aligned} F(0) &= 6000t - \frac{250\pi}{12}\left(\cos\left(\frac{\pi}{12}\right) \cdot 0\right) + C \\ 0 &= 60{,}000 - \frac{250\pi}{12} \cos(0) + C \\ -C &= -\frac{250\pi}{12} \cdot 1 \\ C &= \frac{250\pi}{12} \end{aligned}$$

And so the flow at t hours after sunrise is

$$\begin{aligned} F(t) &= 6000t - \frac{250\pi}{12} \cos\left(\frac{12}{\pi} t\right) + \frac{250\pi}{12} \\ &= 6000t + \frac{250\pi}{12}\left[-\cos\left(\frac{12}{\pi} t\right) + 1\right] \\ &= 6000t + \frac{250\pi}{12}\left[1 - \cos\left(\frac{12}{\pi} t\right)\right] \end{aligned}$$

After 4 hours the total amount of water is

$$\begin{aligned} F(4) &= 6000.4 + \frac{250\pi}{12}[1 - \cos(4 \times 12/\pi) \\ &= 24{,}000 + \frac{250\pi}{12}(1 - \cos(48/\pi)) \\ &= 24{,}000 + \frac{250\pi}{12}(1 - (-0.909)) \\ &= 24{,}000 + 124.9 \text{ litres} \\ &= 24{,}124.9 \text{ litres} \end{aligned}$$

After the first 8 hours the flow is

$$\begin{aligned} F(8) &= 6000.8 + \frac{250\pi}{12}(1 - \cos(8 \times 12/\pi)) \\ &= 48{,}000 + \frac{250\pi}{12}(1 - \cos(96/\pi)) \\ &= 48{,}000 + \frac{250\pi}{12}(1 - 0.654) \\ &= 48{,}000 + \frac{250\pi}{12}(0.346) \\ &= 48{,}000 + 22.6 \text{ litres} \\ &= 48{,}022.6 \text{ litres} \end{aligned}$$

Subtracting to obtain the flow in hours 4–8 we get

$$48022.6 - 24124.9 = 23897.7$$

And we see that the water flow is great during the first 4-hour period after sunrise.

Finding the indefinite integrals of the remaining trigonometric functions. Integration by substitution will also permit us to integrate the four trigonometric functions we do not yet have integration formulas for. The integrals for the tangent and cotangent functions are simple substitutions while the substitutions for the secant and cosecant functions require an inspired algebraic step.

Example 13

Find the indefinite integral of $f(x) = \tan(x)$.

Solution

Recall that $\tan(x) = \dfrac{\sin(x)}{\cos(x)}$. This means we can solve the problem by finding

$$\int \frac{\sin(x)}{\cos(x)}\,dx$$

Choose $u = \cos(x)$ then $du = -\sin(x)\,dx$ so $-du = \sin(x)\,dx$. Perform the integral:

$$\begin{aligned} \int \frac{\sin(x)}{\cos(x)}\,dx &= \int \frac{\sin(x)\,dx}{\cos(x)} \\ &= \int \frac{-du}{u} \\ &= -\int \frac{du}{u} \\ &= -\ln(|u|) + C \\ &= -\ln(|\cos(x)|) + C \end{aligned}$$

We usually simplify this expression a little more to get a form without a minus sign by using one of the algebraic rules for logarithms. Recall that $a\ln(b) = \ln(b^a)$. Then:

$$\begin{aligned} -\ln(|\cos(x)|) + C &= \ln(|\cos(x)|^{-1}) + C \\ &= \ln\left(\frac{1}{|\cos(x)|}\right) + C \\ &= \ln(|\sec(x)|) + C \end{aligned}$$

Rule 15: The Indefinite Integral of tan(x)

$$\int \tan(x)\,dx = \ln(|\sec(x)|) + C$$

Rule 16: The Indefinite Integral of cot(x)

$$\int \cot(x)\,dx = \ln(|\sin(x)|) + C$$

The next example gives us the indefinite integral of the secant function by making one algebraic step that is difficult to motivate except by noticing that it works.

Example 14

Find the indefinite integral of $f(x) = \sec(x)$.

Solution

$$\int \sec(x)\,dx = \int \sec(x)\frac{\sec(x) + \tan(x)}{\sec(x) + \tan(x)}\,dx$$

$$= \int \frac{\sec^2(x) + \sec(x)\tan(x)}{\sec(x) + \tan(x)}\,dx$$

Remember that the derivative of $\tan(x)$ is $\sec^2(x)$ and the derivative of $\sec(x)$ is $\sec(x)\tan(x)$. This means that the numerator of the expression above is the derivative of the denominator. That is the correct form for a substitution integral that yields a logarithm function as the result. Set $u = \tan(x) + \sec(x)$, then $du = \sec^2(x) + \sec(x)\tan(x)$.

$$\int \frac{\sec^2(x) + \sec(x)\tan(x)}{\sec(x) + \tan(x)}\,dx = \int \frac{(\sec^2(x) + \sec(x)\tan(x)\,dx}{\tan(x) + \sec(x)}$$

$$= \int \frac{du}{u}$$

$$= \ln(|u|) + C$$

$$= \ln(|\tan(x) + \sec(x)|) + C$$

Rule 17: The Indefinite Integral of sec(x)

$$\int \sec(x)\,dx = \ln(|\tan(x) + \sec(x)|) + C$$

Rule 18: The Indefinite Integral of csc(x)

$$\int \csc(x)\,dx = -\ln(|\cot(x) + \csc(x)|) + C$$

GROUP DISCUSSION

Read Example 14 carefully. Suppose that you were asked to find the integral $\int \sec(x)\,dx$ *without* being told to multiply and divide by $\sec(x) + \tan(x)$. What methods could be used to discover the technique? Is there any pattern that points to the leap of algebra that permits us to perform the integral in Example 14?

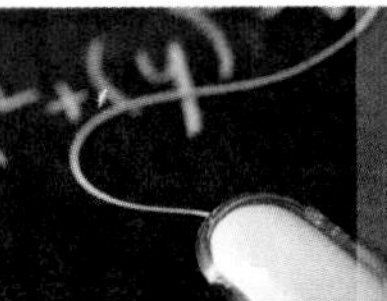

Self-Check Exercises 8.2

1. Evaluate $\int \sqrt{2x+5}\,dx$.

2. Evaluate $\int \frac{x^2}{(2x^3+1)^{3/2}}\,dx$.

3. Evaluate $\int xe^{2x^2-1}\,dx$.

4. According to a joint study conducted by a university Environmental Management Department and a provincial government agency, the concentration of carbon monoxide (CO) in the air due to automobile exhaust is increasing at the rate given by

$$f(t) = \frac{8(0.1t+1)}{300(0.2t^2+4t+64)^{1/3}}$$

parts per million (ppm) per year t. Currently, the CO concentration due to automobile exhaust is 0.16 ppm. Find an expression giving the CO concentration t yr from now.

Solutions to Self-Check Exercises 8.2 can be found on page 462.

8.2 Exercises

In Exercises 1–70, find the indefinite integral.

1. $\int 4(4x+3)^4\,dx$

2. $\int 4x(2x^2+1)^7\,dx$

3. $\int (x^3-2x)^2(3x^2-2)\,dx$

4. $\int (3x^2-2x+1)(x^3-x^2+x)^4\,dx$

5. $\int \frac{4x}{(2x^2+3)^3}\,dx$

6. $\int \frac{3x^2+2}{(x^3+2x)^2}\,dx$

7. $\int 3t^2\sqrt{t^3+2}\,dt$

8. $\int 3t^2(t^3+2)^{3/2}\,dt$

9. $\int (x^2-1)^9 x\,dx$

10. $\int x^2(2x^3+3)^4\,dx$

11. $\int \frac{x^4}{1-x^5}\,dx$

12. $\int \frac{x^2}{\sqrt{x^3-1}}\,dx$

13. $\int \frac{2}{x-2}\,dx$

14. $\int \frac{x^2}{x^3-3}\,dx$

15. $\int \frac{0.3x-0.2}{0.3x^2-0.4x+2}\,dx$

16. $\int \frac{2x^2+1}{0.2x^3+0.3x}\,dx$

17. $\int \frac{x}{3x^2-1}\,dx$

18. $\int \frac{x^2-1}{x^3-3x+1}\,dx$

19. $\int e^{-2x}\,dx$

20. $\int e^{-0.02x}\,dx$

21. $\int e^{2-x}\,dx$

22. $\int e^{2t+3}\,dt$

23. $\int xe^{-x^2}\,dx$

24. $\int x^2e^{x^3-1}\,dx$

25. $\int (e^x-e^{-x})\,dx$

26. $\int (e^{2x}+e^{-3x})\,dx$

27. $\int \frac{e^x}{1+e^x}\,dx$

28. $\int \frac{e^{2x}}{1+e^{2x}}\,dx$

29. $\int \frac{e^{\sqrt{x}}}{\sqrt{x}}\,dx$

30. $\int \frac{e^{-1/x}}{x^2}\,dx$

31. $\int \frac{e^{3x}+x^2}{(e^{3x}+x^3)^3}\,dx$

32. $\int \frac{e^x-e^{-x}}{(e^x+e^{-x})^{3/2}}\,dx$

33. $\int e^{2x}(e^{2x}+1)^3\,dx$

34. $\int e^{-x}(1+e^{-x})\,dx$

35. $\int \frac{\ln 5x}{x}\,dx$

36. $\int \frac{(\ln u)^3}{u}\,du$

37. $\int \frac{1}{x\ln x}\,dx$

38. $\int \frac{1}{x(\ln x)^2}\,dx$

39. $\int \sqrt{\frac{\ln x}{x}}\,dx$

40. $\int \frac{(\ln x^{7/2})}{x}\,dx$

41. $\int \left(xe^{x^2}-\frac{x}{x^2+2}\right)dx$

42. $\int \left(xe^{-x^2}+\frac{e^x}{e^x+3}\right)dx$

43. $\int \frac{x+1}{\sqrt{x}-1}\,dx$

Hint: Let $u=\sqrt{x}-1$.

44. $\int \frac{e^{-u}-1}{e^{-u}+u}\,du$

Hint: Let $v=e^{-u}+u$.

45. $\int x(x-1)^5\,dx$

Hint: $u=x-1$ implies $x=u+1$.

46. $\int \frac{t}{t+1}\,dt$

Hint: $\int \frac{t}{t+1} = 1 - \frac{1}{t+1}$.

47. $\int \frac{1-\sqrt{x}}{1+\sqrt{x}}\,dx$

Hint: Let $u = 1 + \sqrt{x}$.

48. $\int \frac{1+\sqrt{x}}{1-\sqrt{x}}\,dx$

Hint: Let $u = 1 - \sqrt{x}$.

49. $\int v^2(1-v)^6\,dv$

Hint: Let $u = 1 - v$.

50. $\int x^3(x^2+1)^{3/2}\,dx$

Hint: Let $u = x^2 + 1$.

51. $\int f(x) = \sin(2x)\,dx$

52. $\int t^2\cos(t^3)\,dr$

53. $\int \sec^2(3x+1)\,dx$

54. $\int x\sec(x^2)\tan(x^2)\,dx$

55. $\int \frac{\cos(\ln(t))}{t}\,dt$

Hint: What is the derivative of ln (t)?

56. $\int \frac{du}{u^2+4u+5}$

57. $\int \tan(5x+4)\,dx$

58. $\int \frac{\cos(x)}{\sin(x)}\,dx$

59. $\int \cos(x)\sec^2(\sin(x))\,dx$

60. $\int x^4\csc(x^5+1)\,dx$

61. $\int x\cot(x^2+4)\,dx$

62. $\int \sin(t)e^{-\cos(t)}\,dt$

63. $\int e^x\cos(e^x)dx$

64. $\int \frac{du}{u^2-2u+2}$

65. $\int (2x+1)\sin(x^2+x+1)\,dx$

66. $\int \frac{e^x}{\sqrt{1-e^{2x}}}\,dx$

Hint: $(e^x)^2 = e^{2x}$

67. $\int \frac{\cos(\frac{1}{x})}{x^2}\,dx$

68. $\int \csc(2x+1)\,dx$

69. $\int \sin(u)\sec(u)du$

Hint: Rearrange the integrand with trig identities.

70. $\int x\sec(1-x^2)\,dx$

In Exercises 71–74, find the function f given that the slope of the tangent line to the graph of f at any point $(x, f(x))$ is $f'(x)$ and that the graph of f passes through the given point.

71. $f'(x) = 5(2x-1)^4$; $(1, 3)$

72. $f'(x) = \frac{3x^2}{2\sqrt{x^3-1}}$; $(1, 1)$

73. $f'(x) = -2xe^{-x^2+1}$; $(1, 0)$

74. $f'(x) = 1 - \frac{2x}{x^2+1}$; $(0, 2)$

75. Using techniques similar to (but not identical to) those used in Example 13, show that Rule 16 is correct.

76. Using techniques similar to (but not identical to) those used in Example 14, show that Rule 18 is correct.

77. Examine the problem:

$$\int \frac{dx}{x^2+x+1}$$

Use completing the square in the denominator and other algebra to force this problem to match Rule 13.

78. Remember that

$$\sin^2(\theta) + \cos^2(\theta) = 1$$

Using this trigonometric identity, find an integral rule for

$$\int \sin^3(x)dx$$

You will need to use both algebra and integration by substitution.

79. Remember that

$$\sin^2(\theta) + \cos^2(\theta) = 1$$

Using this trigonometric identity, compute an integral rule for

$$\int \cos^3(x)\,dx$$

You will need to use both algebra and integration by substitution.

80. **Student Enrollment** The registrar of Laurentian University estimates that the total student enrollment in the Continuing Education division will grow at the rate of

$$N'(t) = 2000(1 + 0.2t)^{-3/2}$$

students/year t yr from now. If the current student enrollment is 1000, find an expression giving the total student enrollment t yr from now. What will be the student enrollment 5 yr from now?

81. **TV Viewers: Newsmagazine Shows** The number of viewers of a weekly TV newsmagazine show, introduced in the 1998 season, has been increasing at the rate of

$$3\left(2 + \frac{1}{2}t\right)^{-1/3} \qquad (1 \le t \le 6)$$

million viewers/year in its tth year on the air. The number of viewers of the program during its first year on the air is given by $9(5/2)^{2/3}$ million. Find how many viewers were expected in the 2003 season.

Business and Economics Applications

82. **Learning Curves** The average student enrolled in the 20-wk Court Reporting I course at the Institute of Court Reporting progresses according to the rule

$$N'(t) = 6e^{-0.05t} \qquad (0 \le t \le 20)$$

where $N'(t)$ measures the rate of change in the number of words/minute of dictation the student takes in machine shorthand after t wk in the course. Assuming that the average student enrolled in this course begins with a dictation speed of 60 words/minute, find an expression $N(t)$ that gives the dictation speed of the student after t wk in the course.

83. **Demand: Women's Boots** The rate of change of the unit price p (in dollars) of Apex women's boots is given by

$$p'(x) = \frac{-250x}{(16 + x^2)^{3/2}}$$

where x is the quantity demanded daily in units of a hundred. Find the demand function for these boots if the quantity demanded daily is 300 pairs ($x = 3$) when the unit price is \$50/pair.

A calculator is recommended for the remaining exercises.

84. **Supply: Women's Boots** The rate of change of the unit price p (in dollars) of Apex women's boots is given by

$$p'(x) = \frac{240x}{(5 - x)^2}$$

where x is the number of pairs that the supplier will make available in the market daily when the unit price is \$$p$/pair. Find the supply equation for these boots if the quantity the supplier is willing to make available is 200 pairs daily ($x = 2$) when the unit price is \$50/pair.

85. **Sales: Loudspeakers** In the first year they appeared in the market, 2000 pairs of Acrosonic model F loudspeaker systems were sold. Since then, sales of these loudspeaker systems have been growing at the rate of

$$f'(t) = 2000(3 - 2e^{-t})$$

units/yr, where t denotes the number of years these systems have been on the market. Determine the number of systems that were sold in the first 5 yr after their introduction.

Biological and Life Sciences Applications

86. **Oil Spill** In calm waters the oil spilling from the ruptured hull of a grounded tanker forms an oil slick that is circular in shape. If the radius r of the circle is increasing at the rate of

$$r'(t) = \frac{10}{\sqrt{2t + 4}}$$

metres/minute t min after the rupture occurs, find an expression for the radius at any time t. How large is the polluted area 16 min after the rupture occurred?
Hint: $r(0) = 0$.

87. **Life Expectancy of a Female** Suppose in a certain country the life expectancy at birth of a female is changing at the rate of

$$g'(t) = \frac{5.45218}{(1 + 1.09t)^{0.9}}$$

years/year. Here, t is measured in years, with $t = 0$ corresponding to the beginning of 1900. Find an expression $g(t)$ giving the life expectancy at birth (in years) of a female in that country if the life expectancy at the beginning of 1900 is 50.02 yr. What is the life expectancy at birth of a female born in the year 2000 in that country?

88. **Population Growth** The population of a certain city is projected to grow at the rate of

$$r(t) = 400\left(1 + \frac{2t}{24 + t^2}\right) \qquad (0 \leq t \leq 5)$$

people/year, t years from now. The current population is 60,000. What will be the population 5 yr from now?

89. **Average Birth Height of Boys** Using data collected at Kaiser Hospital, pediatricians estimate that the average height of male children changes at the rate of

$$h'(t) = \frac{52.8706e^{-0.3277t}}{(1 + 2.449e^{-0.3277t})^2}$$

in./yr, where the child's height $h(t)$ is measured in inches and t, the child's age, is measured in years, with $t = 0$ corresponding to the age at birth. Find an expression $h(t)$ for the average height of a boy at age t if the height at birth of an average child is 19.4 in. What is the height of an average 8-yr-old boy?

90. **Amount of Glucose in the Bloodstream** Suppose a patient is given a continuous intravenous infusion of glucose at a constant rate of r mg/min. Then, the rate at which the amount of glucose in the bloodstream is changing at time t due to this infusion is given by

$$A'(t) = re^{-at}$$

mg/min, where a is a positive constant associated with the rate at which excess glucose is eliminated from the bloodstream and is dependent on the patient's metabolism rate. Derive an expression for the amount of glucose in the bloodstream at time t.
Hint: $A(0) = 0$.

91. **Concentration of a Drug in an Organ** A drug is carried into an organ of volume V cm^3 by a liquid that enters the organ at the rate of a cm^3/sec and leaves it at the rate of b cm^3/sec. The concentration of the drug in the liquid entering the organ is c g/cm^3. If the concentration of the drug in the organ at time t is increasing at the rate of

$$x'(t) = \frac{1}{V}(ac - bx_0)e^{-bt/v}$$

g/cm^3/sec, and the concentration of the drug in the organ initially is x_0 g/cm^3, show that the concentration of the drug in the organ at time t is given by

$$x(t) = \frac{ac}{b} + \left(x_0 - \frac{ac}{b}\right)e^{-bt/V}$$

92. **Alcohol Production in Wine-Making** Because of self-poisoning the total amount of alcohol being produced in wine by bacteria while it ferments drops off with time. If the alcohol production is

$$c(t) = \frac{4.5}{t^2 - 6t + 10} \text{ mL/d}$$

and the initial amount of alcohol in the cask is zero, find an expression for the total amount of alcohol in the cask after t days.

93. **Hormone Production** A hormone is produced at a rate of

$$g(t) = 2.6 + 0.45\cos\left(\frac{t}{8.6}\right) \text{ mg/d}$$

on day t of a woman's menstrual cycle. In some women with a genetic disease the hormone is not processed but stays in the bloodstream. A protein that degrades the hormone into safe substances has its dose set by the amount of hormone it must process. How much of the hormone builds up in the bloodstream after 28 days?

Solutions to Self-Check Exercises 8.2

1. Let $u = 2x + 5$. Then, $du = 2\,dx$, or $dx = \frac{1}{2}\,du$. Making the appropriate substitutions, we have

$$\begin{aligned}\int \sqrt{2x+5}\,dx &= \int \sqrt{u}\left(\frac{1}{2}du\right) = \frac{1}{2}\int u^{1/2}\,du\\ &= \frac{1}{2}\left(\frac{2}{3}\right)u^{3/2} + C\\ &= \frac{1}{3}(2x+5)^{3/2} + C\end{aligned}$$

2. Let $u = 2x^3 + 1$, so that $du = 6x^2\,dx$, or $x^2dx = \frac{1}{6}\,du$.

$$\begin{aligned}\int \frac{x^2}{(2x^3+1)^{3/2}}\,dx &= \int \frac{\left(\frac{1}{6}\right)du}{u^{3/2}} = \frac{1}{6}\int u^{-3/2}\,du\\ &= \left(\frac{1}{6}\right)(-2)u^{-1/2} + C\\ &= -\frac{1}{3}(2x^3+1)^{-1/2} + C\\ &= -\frac{1}{3\sqrt{2x^3+1}} + C\end{aligned}$$

3. Let $u = 2x^2 - 1$, so that $du = 4x\,dx$, or $x\,dx = \frac{1}{4}du$. Then,

$$\begin{aligned}\int xe^{2x^2-1}\,dx &= \frac{1}{4}\int e^u\,du\\ &= \frac{1}{4}e^u + C\\ &= \frac{1}{4}e^{2x^2-1} + C\end{aligned}$$

4. Let $C(t)$ denote the CO concentration in the air due to automobile exhaust t yr from now. Then.

$$\begin{aligned}C'(t) = f(t) &= \frac{8(0.1t+1)}{300(0.2t^2+4t+64)^{1/3}}\\ &= \frac{8}{300}(0.1t+1)(0.2t^2+4t+64)^{-1/3}\end{aligned}$$

Integrating, we find

$$\begin{aligned}C(t) &= \int \frac{8}{300}(0.1t+1)(0.2t^2+4t+64)^{-1/3}\,dt\\ &= \frac{8}{300}\int (0.1t+1)(0.2t^2+4t+64)^{-1/3}\,dt\end{aligned}$$

Let $u = 0.2t^2 + 4t + 64$, so that $du = (0.4t + 4)\,dt = 4(0.1t + 1)\,dt$, or

$$(0.1t+1)\,dt = \frac{1}{4}\,du$$

Then,

$$\begin{aligned}C(t) &= \frac{8}{300}\left(\frac{1}{4}\right)\int u^{-1/3}\,du\\ &= \frac{1}{150}\left(\frac{3}{2}u^{2/3}\right) + k\\ &= 0.01(0.2t^2+4t+64)^{2/3} + k\end{aligned}$$

Where k is an arbitrary constant. To determine the value of k, we use the condition $C(0) = 0.16$, obtaining

$$\begin{aligned}C(0) = 0.16 &= 0.01(64)^{2/3} + k\\ 0.16 &= 0.16 + k\\ k &= 0\end{aligned}$$

Therefore,

$$C(t) = 0.01(0.2t^2+4t+64)^{2/3}$$

8.3 Area and the Definite Integral

An Intuitive Look

Suppose a certain province's annual rate of petroleum consumption over a 4-year period is constant and is given by the function

$$f(t) = 1.2 \qquad (0 \le t \le 4)$$

where t is measured in years and $f(t)$ in millions of barrels per year. Then, the province's total petroleum consumption over the period of time in question is

$$(1.2)(4 - 0) \qquad \text{Rate of consumption} \times \text{Time elapsed}$$

or 4.8 million barrels. If you examine the graph of f shown in Figure 8.5, you will see that this total is just the area of the rectangular region bounded above by the graph of f, below by the t-axis, and to the left and right by the vertical lines $t = 0$ (the y-axis) and $t = 4$, respectively.

Figure 8.6 shows the actual petroleum consumption of a certain maritime province over a 4-year period from 1990 ($t = 0$) to 1994 ($t = 4$). Observe that the rate of consumption is not constant; that is, the function f is not a constant function. What is the state's total petroleum consumption over this 4-year period? It seems reasonable to conjecture that it is given by the "area" of the region bounded above by the graph of f, below by the t-axis, and to the left and right by the vertical lines $t = 0$ and $t = 4$, respectively.

This example raises two questions:

1. What is the "area" of the region shown in Figure 8.6?
2. How do we compute this area?

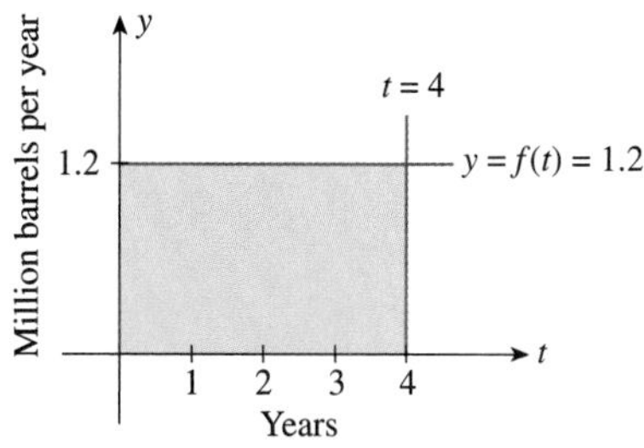

Figure 8.5
The total petroleum consumption is given by the area of the rectangular region.

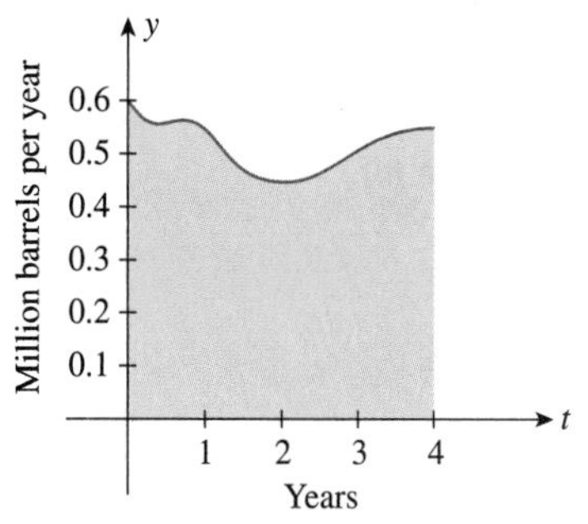

Figure 8.6
The daily petroleum consumption is given by the "area" of the shaded region.

The Area Problem

The preceding example touches on the second fundamental problem in calculus: Calculate the area of the region bounded by the graph of a nonnegative function f, the x-axis, and the vertical lines $x = a$ and $x = b$ (Figure 8.7). This area is called the **area under the graph of f** on the interval $[a, b]$, or from a to b.

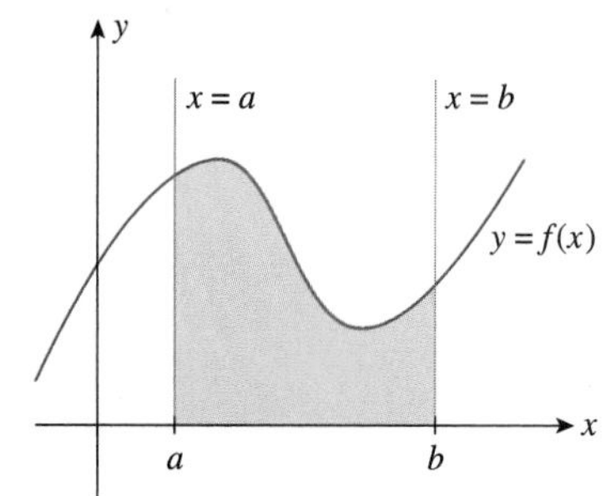

Figure 8.7
The area under the graph of f on $[a, b]$

Defining Area—Two Examples

Just as we used the slopes of secant lines (quantities that we could compute) to help us define the slope of the tangent line to a point on the graph of a function, we now adopt a parallel approach and use the areas of rectangles (quantities that we can compute) to help us define the area under the graph of a function. We begin by looking at a specific example.

Example 1

Let $f(x) = x^2$ and consider the region R under the graph of f on the interval $[0, 1]$ (Figure 8.8a). To obtain an approximation of the area of R, let's construct four nonoverlapping rectangles as follows: Divide the interval $[0, 1]$ into four subintervals

$$\left[0, \frac{1}{4}\right], \left[\frac{1}{4}, \frac{1}{2}\right], \left[\frac{1}{2}, \frac{3}{4}\right], \left[\frac{3}{4}, 1\right]$$

of equal length $\frac{1}{4}$. Next, construct four rectangles with these subintervals as bases and with heights given by the values of the function at the midpoints

$$\frac{1}{8}, \frac{3}{8}, \frac{5}{8}, \frac{7}{8}$$

of each subinterval. Then, each of these rectangles has width $\frac{1}{4}$ and height

$$f\left(\frac{1}{8}\right), f\left(\frac{3}{8}\right), f\left(\frac{5}{8}\right), f\left(\frac{7}{8}\right)$$

respectively (Figure 8.8b).

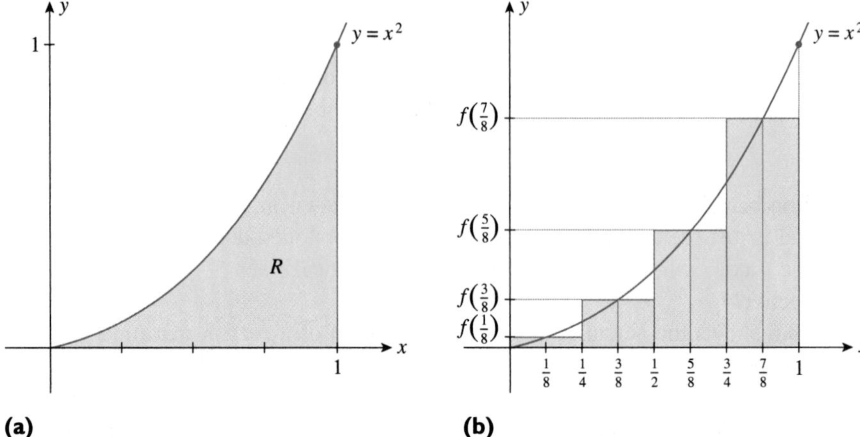

Figure 8.8
The area of the region under the graph of f on [0, 1] in (a) is approximated by the sum of the areas of the four rectangles in (b).

If we approximate the area A of R by the sum of the areas of the four rectangles, we obtain

$$\begin{aligned} A &\approx \frac{1}{4}f\left(\frac{1}{8}\right) + \frac{1}{4}f\left(\frac{3}{8}\right) + \frac{1}{4}f\left(\frac{5}{8}\right) + \frac{1}{4}f\left(\frac{7}{8}\right) \\ &= \frac{1}{4}\left[f\left(\frac{1}{8}\right) + f\left(\frac{3}{8}\right) + f\left(\frac{5}{8}\right) + f\left(\frac{7}{8}\right)\right] \\ &= \frac{1}{4}\left[\left(\frac{1}{8}\right)^2 + \left(\frac{3}{8}\right)^2 + \left(\frac{5}{8}\right)^2 + \left(\frac{7}{8}\right)^2\right] \\ &= \frac{1}{4}\left(\frac{1}{64} + \frac{9}{64} + \frac{25}{64} + \frac{49}{64}\right) = \frac{21}{64} \end{aligned}$$

Recall that $f(x) = 16 - x^2$.

or approximately 0.328125 square unit.

Following the procedure of Example 1, we can obtain approximations of the area of the region R using any number n of rectangles ($n = 4$ in Example 1). Figure 8.9a shows the approximation of the area A of R using 8 rectangles ($n = 8$), and Figure 8.9b shows the approximation of the area A of R using 16 rectangles.

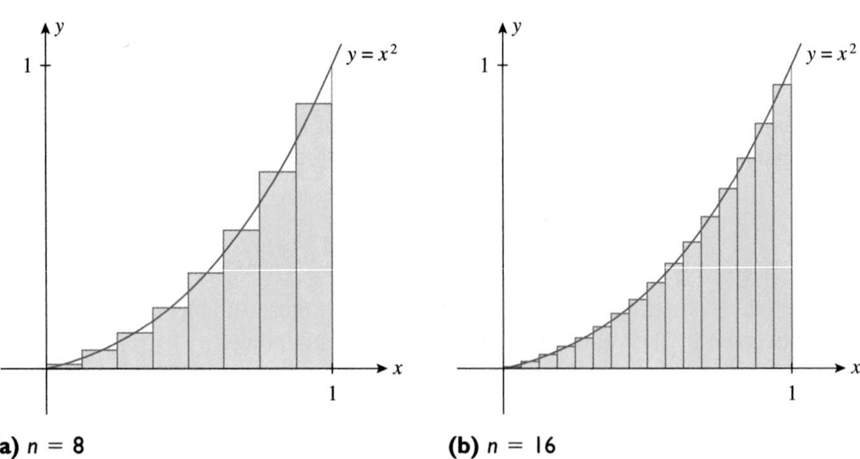

Figure 8.9
As n increases, the number of rectangles increases, and the approximation improves.

These figures suggest that the approximations seem to get better as n increases. This is borne out by the results given in Table 8.1, which were obtained using a computer.

Table 8.1

Number of Rectangles, n	4	8	16	32	64	100	200
Approximation of A	0.328125	0.332031	0.333008	0.333252	0.333313	0.333325	0.333331

Our computations seem to suggest that the approximations approach the number $\frac{1}{3}$ as n gets larger and larger. This result suggests that we *define* the area of the region under the graph of $f(x) = x^2$ on the interval $[0, 1]$ to be $\frac{1}{3}$ square unit.

In Example 1 we chose the *midpoint* of each subinterval as the point at which to evaluate $f(x)$ to obtain the height of the approximating rectangle. Let's consider another example, this time choosing the *left end point* of each subinterval.

Example 2

Let R be the region under the graph of $f(x) = 16 - x^2$ on the interval $[1, 3]$. Find an approximation of the area A of R using four subintervals of $[1, 3]$ of equal length and picking the left end point of each subinterval to evaluate $f(x)$ to obtain the height of the approximating rectangle.

Solution

The graph of f is sketched in Figure 8.10a. Since the length of $[1, 3]$ is 2, we see that the length of each subinterval is $\frac{2}{4}$, or $\frac{1}{2}$. Therefore, the four subintervals are

$$\left[1, \frac{3}{2}\right], \left[\frac{3}{2}, 2\right], \left[2, \frac{5}{2}\right], \left[\frac{5}{2}, 3\right]$$

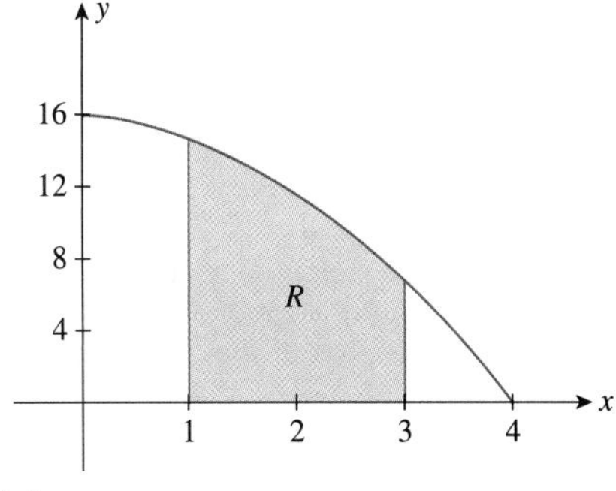

(a)

The left end points of these subintervals are $1, \frac{3}{2}, 2$, and $\frac{5}{2}$, respectively, so the heights of the approximating rectangles area $f(1)$, $f(\frac{3}{2})$, $f(2)$,and $f(\frac{5}{2})$, respectively (Figure 8.10b). Therefore, the required approximation is

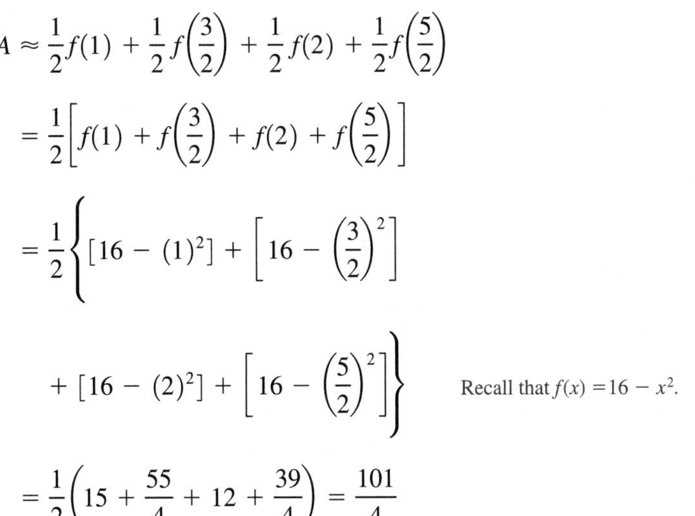

$$A \approx \frac{1}{2}f(1) + \frac{1}{2}f\left(\frac{3}{2}\right) + \frac{1}{2}f(2) + \frac{1}{2}f\left(\frac{5}{2}\right)$$

$$= \frac{1}{2}\left[f(1) + f\left(\frac{3}{2}\right) + f(2) + f\left(\frac{5}{2}\right)\right]$$

$$= \frac{1}{2}\left\{[16 - (1)^2] + \left[16 - \left(\frac{3}{2}\right)^2\right]\right.$$

$$\left. + [16 - (2)^2] + \left[16 - \left(\frac{5}{2}\right)^2\right]\right\}$$

Recall that $f(x) = 16 - x^2$.

$$= \frac{1}{2}\left(15 + \frac{55}{4} + 12 + \frac{39}{4}\right) = \frac{101}{4}$$

or approximately 25.25 square units.

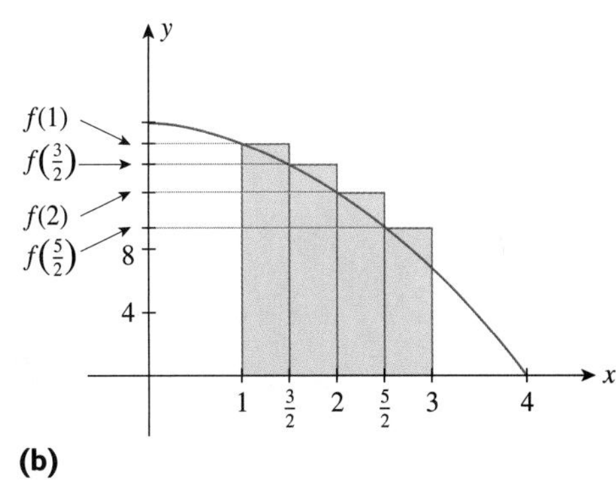

(b)

Figure 8.10
The area of R in (a) is approximated by the sum of the areas of the four rectangles in (b).

Table 8.2 shows the approximations of the area A of the region R of Example 2 when n rectangles are used for the approximation and the heights of the approximating rectangles are found by evaluating $f(x)$ at the left end points.

Table 8.2

Number of Rectangles, n	4	10	100	1,000	10,000	50,000	100,000
Approximation of A	25.2500	24.1200	23.4132	23.3413	23.3341	23.3335	23.3334

Once again, we see that the approximations seem to approach a unique number as n gets larger and larger—this time the number is $23\frac{1}{3}$. This result suggests that we *define* the area of the region under the graph of $f(x) = 16 - x^2$ on the interval $[1, 3]$ to be $23\frac{1}{3}$ square units.

Defining Area—The General Case

Examples 1 and 2 point the way to defining the area A under the graph of an arbitrary but continuous and nonnegative function f on an interval $[a, b]$ (Figure 8.11a).

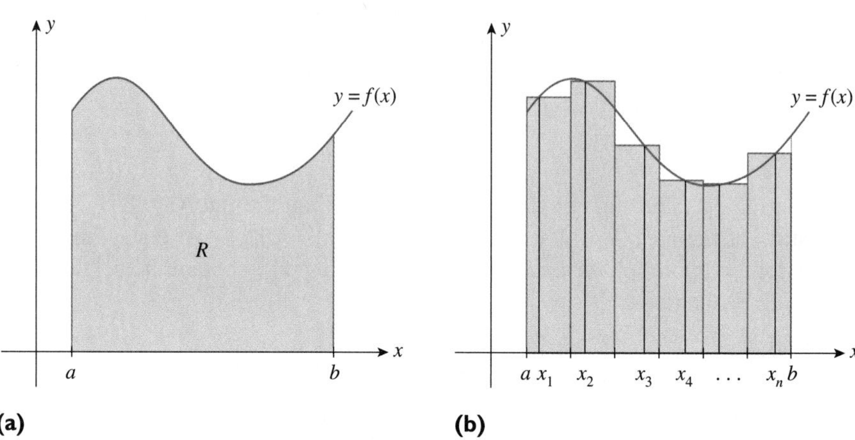

Figure 8.11
The area of the region under the graph of f on $[a, b]$ in (a) is approximated by the sum of the areas of the n rectangles shown in (b).

Divide the interval $[a, b]$ into n subintervals of equal length $\Delta x = (b - a)/n$. Next, pick n arbitrary points $x_1, x_2, \ldots, x_n$, called *representative points*, from the first, second, . . . , and nth subintervals, respectively (Figure 8.11b). Then, approximating the area A of the region R by the n rectangles of width Δx and heights $f(x_1), f(x_2), \ldots, f(x_n)$, so that the areas of the rectangles are $f(x_1)\Delta x, f(x_2)\Delta x, \ldots, f(x_n)\Delta x$, we have

$$A \approx f(x_1)\Delta x + f(x_2)\Delta x + \cdots + f(x_n)\Delta x$$

The sum on the right-hand side of this expression is called a **Riemann sum** in honour of the German mathematician Bernhard Riemann (1826–1866). Now, as the earlier examples seem to suggest, the Riemann sum will approach a unique number as n becomes arbitrarily large.* We define this number to be the area A of the region R.

The Area under the Graph of a Function

Let f be a nonnegative continuous function on $[a, b]$. Then, the area of the region under the graph of f is

$$A = \lim_{n\to\infty}[f(x_1) + f(x_2) + \cdots + f(x_n)]\Delta x \qquad (6)$$

where $x_1, x_2, \ldots, x_n$ are arbitrary points in the n subintervals of $[a, b]$ of equal width $\Delta x = (b - a)/n$.

The Definite Integral

As we have just seen, the area under the graph of a continuous *nonnegative* function f on an interval $[a, b]$ is defined by the limit of the Riemann sum

$$\lim_{n\to\infty}[f(x_1)\Delta x + f(x_2)\Delta x + \cdots + f(x_n)\Delta x]$$

*Even though we chose the representative points to be the midpoints of the subintervals in Example 1 and the left end points in Example 2, it can be shown that each of the respective sums will always approach a unique number as n approaches infinity.

We now turn our attention to the study of limits of Riemann sums involving functions that are not necessarily nonnegative. Such limits arise in many applications of calculus.

For example the calculation of the distance covered by a body travelling along a straight line involves evaluating a limit of this form. The computation of the total revenue realized by a company over a certain time period, the calculation of the total amount of electricity consumed in a typical home over a 24-hour period, the average concentration of a drug in a body over a certain interval of time, and the volume of a solid—all involve limits of this type.

We begin with the following definition.

The Definite Integral

Let f be a continuous function defined on $[a, b]$. If

$$\lim_{n\to\infty}[f(x_1)\Delta x + f(x_2)\Delta x + \cdots + f(x_n)\Delta x]$$

exists for all choices of representative points $x_1, x_2, \ldots, x_n$ in the n subintervals of $[a, b]$ of equal width $\Delta x = (b - a)/n$, then this limit is called the **definite integral** of f from a to b and is denoted by $\int_a^b f(x)dx$. Thus,

$$\int_a^b f(x)\,dx = \lim_{n\to\infty}[f(x_1)\Delta x + f(x_2)\Delta x + \cdots + f(x_n)\Delta x] \qquad (7)$$

The number a is the **lower limit of integration**, and the number b is the **upper limit of integration.**

Remark

1. If f is nonnegative, then the limit in (7) is the same as the limit in (6); therefore, the definite integral gives the area under the graph of f on $[a, b]$.
2. The limit in (7) is denoted by the integral sign $\int$ because, as we will see later, the definite integral and the antiderivative of a function f are related.
3. It is important to realize that the definite integral $\int_a^b f(x)\,dx$ is a number, whereas the indefinite integral $\int f(x)\,dx$ represents a family of functions (the antiderivatives of f).
4. If the limit in (7) exists, we say that f is **integrable** on the interval $[a, b]$. ◀

When Is a Function Integrable?

The following theorem, which we state without proof, guarantees that a continuous function is integrable.

Integrability of a Function

Let f be continuous on $[a, b]$. Then, f is integrable on $[a, b]$; that is, the definite integral $\int_a^b f(x)\,dx$ exists.

Geometric Interpretation of the Definite Integral

If f is nonnegative and integrable on $[a, b]$, then we have the following geometric interpretation of the definite integral $\int_a^b f(x)\,dx$.

Geometric Interpretation of $\int_a^b f(x)\,dx$ for $f(x) \geq 0$ on $[a, b]$

If f is nonnegative and continuous on $[a, b]$, then

$$\int_a^b f(x)\,dx \tag{8}$$

is equal to the area of the region under the graph of f on $[a, b]$ (Figure 8.12).

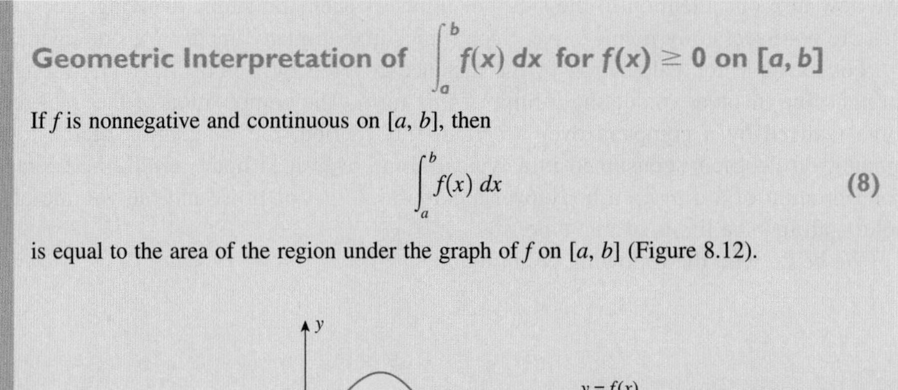

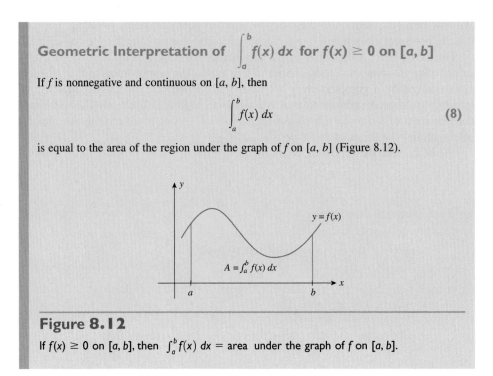

Figure 8.12
If $f(x) \geq 0$ on $[a, b]$, then $\int_a^b f(x)\,dx$ = area under the graph of f on $[a, b]$.

GROUP DISCUSSION

Suppose f is nonpositive [that is, $f(x) \leq 0$] and continuous on $[a, b]$. Explain why the area of the region below the x-axis and above the graph of f is given by $-\int_a^b f(x)\,dx$.

Next, let's extend our geometric interpretation of the definite integral to include the case where f assumes both positive as well as negative values on $[a, b]$. Consider a typical Riemann sum of the function f,

$$f(x_1)\Delta x + f(x_2)\Delta x + \cdots + f(x_n)\Delta x$$

corresponding to a partition of $[a, b]$ into n subintervals of equal width $(b - a)/n$, where $x_1, x_2, \ldots, x_n$ are representative points in the subintervals. The sum consists of n terms in which a positive term corresponds to the area of a rectangle of height $f(x_k)$ (for some positive integer k) lying above the x-axis and a negative term corresponds to the area of a rectangle of height $-f(x_k)$ lying below the x-axis. (See Figure 8.13, which depicts a situation with $n = 6$.)

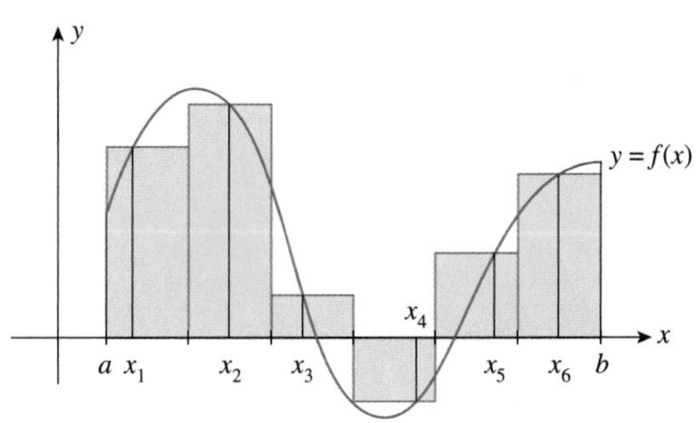

Figure 8.13
The positive terms in the Riemann sum are associated with the areas of the rectangles that lie above the x-axis, and the negative terms are associated with the areas of those that lie below the x-axis.

As n gets larger and larger, the sums of the areas of the rectangles lying above the x-axis seem to give a better and better approximation of the area of the region lying above the x-axis (Figure 8.14). Similarly, the sums of the areas of those rectangles lying below the x-axis seem to give a better and better approximation of the area of the region lying below the x-axis.

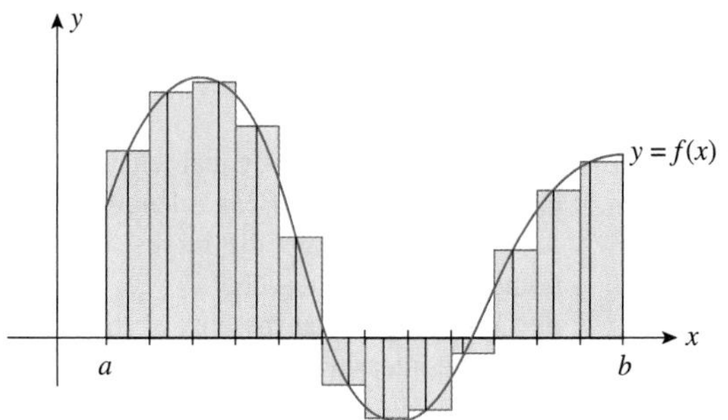

Figure 8.14

As n gets larger, the approximations get better. Here, $n = 12$ and we are approximating with twice as many rectangles as in Figure 6.13.

These observations suggest the following geometric interpretation of the definite integral for an arbitrary continuous function on an interval $[a, b]$.

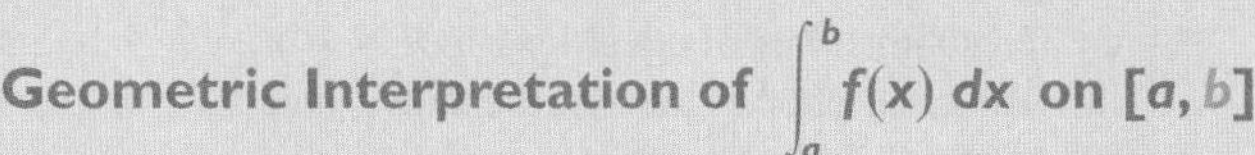

Geometric Interpretation of $\int_a^b f(x)\,dx$ on $[a, b]$

If f is continuous on $[a, b]$, then

$$\int_a^b f(x)\,dx$$

is equal to the area of the region above $[a, b]$ minus the area of the region below $[a, b]$ (Figure 8.15).

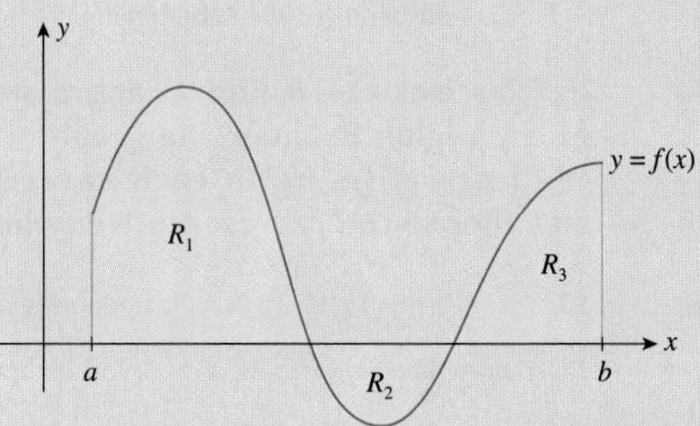

Figure 8.15

$\int_a^b f(x)\,dx$ = Area of R_1 − Area of R_2 + Area of R_3

Self-Check Exercise 8.3

Find an approximation of the area of the region R under the graph of $f(x) = 2x^2 + 1$ on the interval $[0, 3]$, using four subintervals of $[0, 3]$ of equal length and picking the midpoint of each subinterval as a representative point.

The Solution to Self-Check Exercise 8.3 can be found on page 471.

8.3 Exercises

In Exercises 1 and 2, find an approximation of the area of the region R under the graph of f by computing the Riemann sum of f corresponding to the partition of the interval into the subintervals shown in the accompanying figures. In each case, use the midpoints of the subintervals as the representative points.

1.

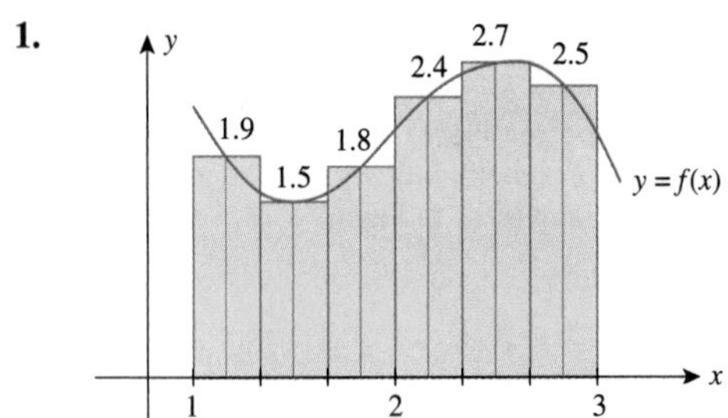

2.

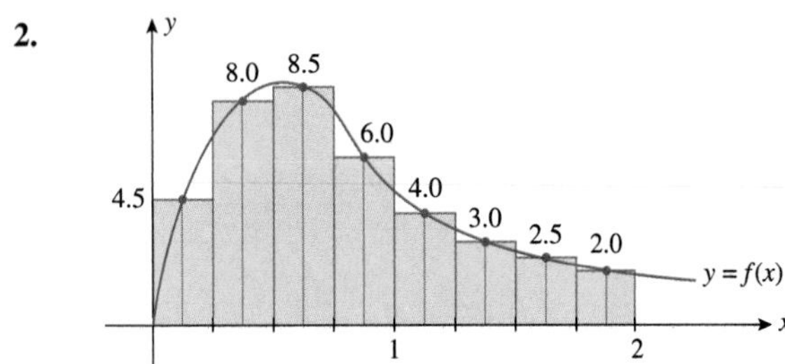

3. Let $f(x) = 3x$.
   a. Sketch the region R under the graph of f on the interval $[0, 2]$ and find its exact area using geometry.
   b. Use a Riemann sum with four subintervals of equal length ($n = 4$) to approximate the area of R. Choose the representative points to be the left end points of the subintervals.
   c. Repeat part (b) with eight subintervals of equal length ($n = 8$).
   d. Compare the approximations obtained in parts (b) and (c) with the exact area found in part (a). Do the approximations improve with larger n?

4. Repeat Exercise 3, choosing the representative points to be the right end points of the subintervals.

5. Let $f(x) = 4 - 2x$.
   a. Sketch the region R under the graph of f on the interval $[0, 2]$ and find its exact area using geometry.
   b. Use a Riemann sum with five subintervals of equal length ($n = 5$) to approximate the area of R. Choose the representative points to be the left end points of the subintervals.
   c. Repeat part (b) with ten subintervals of equal length ($n = 10$).
   d. Compare the approximations obtained in parts (b) and (c) with the exact area found in part (a). Do the approximations improve with larger n?

6. Repeat Exercise 5, choosing the representative points to be the right end points of the subintervals.

7. Let $f(x) = x^2$ and compute the Riemann sum of f over the interval $[2, 4]$, using
   a. Two subintervals of equal length ($n = 2$).
   b. Five subintervals of equal length ($n = 5$).
   c. Ten subintervals of equal length ($n = 10$).

   In each case, choose the representative points to be the midpoints of the subintervals.
   d. Can you guess at the area of the region under the graph of f on the interval $[2, 4]$?

8. Repeat Exercise 7, choosing the representative points to be the left end points of the subintervals.

9. Repeat Exercise 7, choosing the representative points to be the right end points of the subintervals.

10. Let $f(x) = x^3$ and compute the Riemann sum of f over the interval $[0, 1]$, using
    a. Two subintervals of equal length ($n = 2$).
    b. Five subintervals of equal length ($n = 5$).
    c. Ten subintervals of equal length ($n = 10$).

    In each case, choose the representative points to be the midpoints of the subintervals.
    d. Can you guess at the area of the region under the graph of f on the interval $[0, 1]$?

11. Repeat Exercise 10, choosing the representative points to be the left end points of the subintervals.

12. Repeat Exercise 10, choosing the representative points to be the right end points of the subintervals.

In Exercises 13–18, find an approximation of the area of the region R under the graph of the function f on the interval $[a, b]$. In each case, use n subintervals and choose the representative points as indicated.

13. $f(x) = x^2 + 1$; $[0, 2]$; $n = 5$; midpoints

14. $f(x) = 4 - x^2$; $[-1, 2]$; $n = 6$; left end points

15. $f(x) = \dfrac{1}{x}$; $[1, 3]$; $n = 4$; right end points

16. $f(x) = e^x$; $[0, 3]$; $n = 5$; midpoints

17. $ʃf(t) = \sin(t)$; $[0, \pi]$; $n = 8$; left end points (remember to use radians).

18. $ʃf(x) = \cos(x)$; $[0, 4]$; $n = 6$; midpoints (remember to use radians).

19. **Real Estate** Figure (a) shows a vacant lot with a 100-m frontage in a development. To estimate its area, we introduce a coordinate system so that the x-axis coincides with the edge of the straight road forming the lower boundary of the property, as shown in Figure (b). Then, thinking of the upper boundary of the property as the graph of a continuous function f over the interval $[0, 100]$, we see that the problem is mathematically equivalent to that of finding the area under the graph of f on $[0, 100]$. To estimate the area of the lot using a Riemann sum,

we divide the interval [0, 100] into five equal subintervals of length 20 m. Then, using surveyor's equipment, we measure the distance from the midpoint of each of these subintervals to the upper boundary of the property. These measurements give the values of $f(x)$ at $x = 10, 30, 50, 70$, and 90. What is the approximate area of the lot?

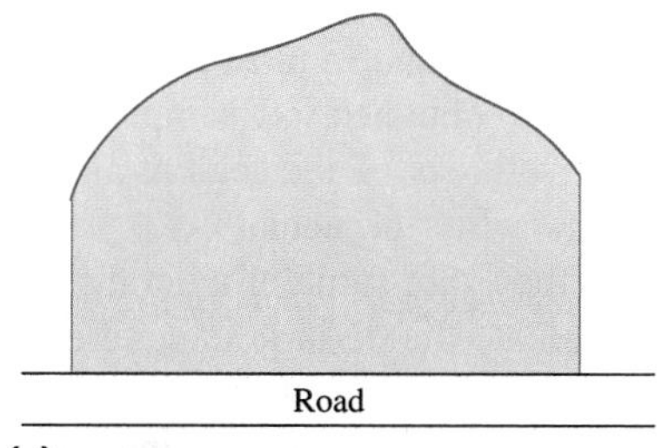

(a)

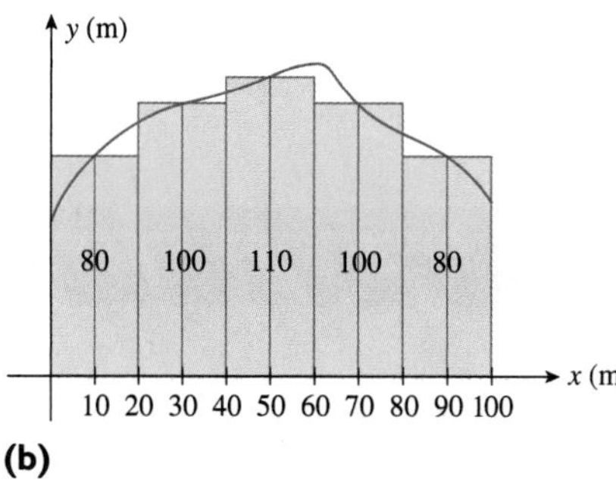

(b)

20. **Real Estate** Use the technique of Exercise 19 to obtain an estimate of the area of the vacant lot shown in the accompanying figures.

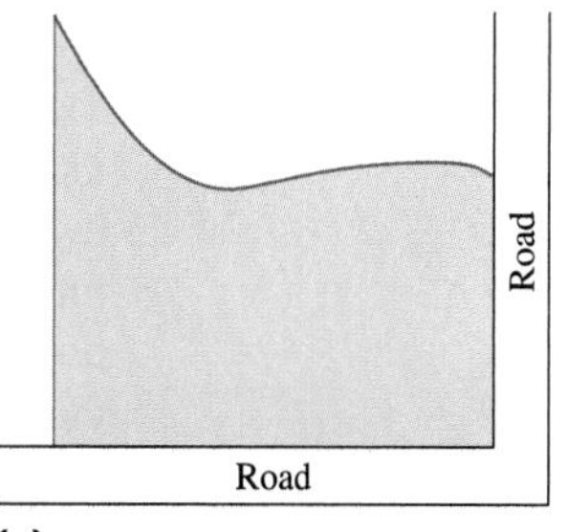

(a)

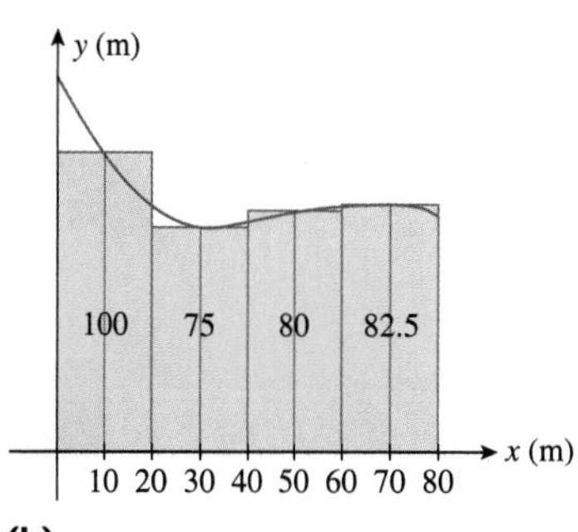

(b)

Solution to Self-Check Exercise 8.3

The length of each subinterval is $\frac{3}{4}$. Therefore, the four subintervals are

$$\left[0, \frac{3}{4}\right], \left[\frac{3}{4}, \frac{3}{2}\right], \left[\frac{3}{2}, \frac{9}{4}\right], \left[\frac{9}{4}, 3\right]$$

The representative points are $\frac{3}{8}$, $\frac{9}{8}$, $\frac{15}{8}$, and $\frac{21}{8}$, respectively. Therefore, the required approximation is

$$\begin{aligned} A &= \frac{3}{4}f\left(\frac{3}{8}\right) + \frac{3}{4}f\left(\frac{9}{8}\right) + \frac{3}{4}f\left(\frac{15}{8}\right) + \frac{3}{4}f\left(\frac{21}{8}\right) \\ &= \frac{3}{4}\left[f\left(\frac{3}{8}\right) + f\left(\frac{9}{8}\right) + f\left(\frac{15}{8}\right) + f\left(\frac{21}{8}\right)\right] \\ &= \frac{3}{4}\left\{\left[2\left(\frac{3}{8}\right)^2 + 1\right] + \left[2\left(\frac{9}{8}\right)^2 + 1\right] + \left[2\left(\frac{15}{8}\right)^2 + 1\right] + \left[2\left(\frac{21}{8}\right)^2 + 1\right]\right\} \\ &= \frac{3}{4}\left(\frac{41}{32} + \frac{113}{32} + \frac{257}{32} + \frac{473}{32} = \frac{663}{32}\right) \end{aligned}$$

or approximately 20.72 square units.

The Fundamental Theorem of Calculus

In Section 8.3 we defined the definite integral of an arbitrary continuous function on an interval $[a, b]$ as a limit of Riemann sums. Calculating the value of a definite integral by actually taking the limit of such sums is tedious and in most cases impractical. It is important to realize that the numerical results we obtained in Examples 1 and 2 of Section 8.3 were *approximations* of the respective areas of the regions in question, even though these results enabled us to *conjecture* what the actual areas might be. Fortunately, there is a much better way of finding the exact value of a definite integral.

The following theorem shows how to evaluate the definite integral of a continuous function provided we can find an antiderivative of that function. Because of its importance in establishing the relationship between differentiation and integration, this theorem—discovered independently by Sir Isaac Newton (1642–1727) in England and Gottfried Wilhelm Leibniz (1646–1716) in Germany—is called the **fundamental theorem of calculus.**

THEOREM 2

The Fundamental Theorem of Calculus

Let f be continuous on $[a, b]$. Then,

$$\int_a^b f(x)\,dx = F(b) - F(a) \tag{9}$$

where F is any antiderivative of f; that is, $F'(x) = f(x)$.

We will explain why this theorem is true at the end of this section.

When applying the fundamental theorem of calculus, it is convenient to use the notation

$$F(x)\Big|_a^b = F(b) - F(a)$$

For example, using this notation, Equation (9) is written

$$\int_a^b f(x)\,dx = F(x)\Big|_a^b = F(b) - F(a)$$

Example 1

Let R be the region under the graph of $f(x) = x$ on the interval $[1, 3]$. Use the fundamental theorem of calculus to find the area A of R and verify your result by elementary means.

Solution

The region R is shown in Figure 8.16a. Since f is nonnegative on $[1, 3]$, the area of R is given by the definite integral of f from 1 to 3; that is,

$$A = \int_1^3 x\,dx$$

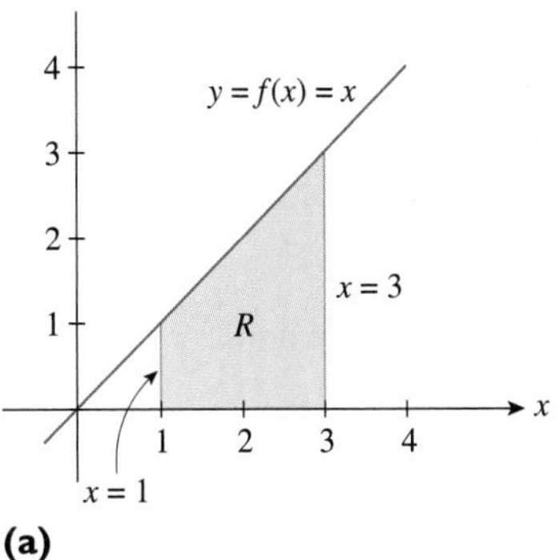

(a)

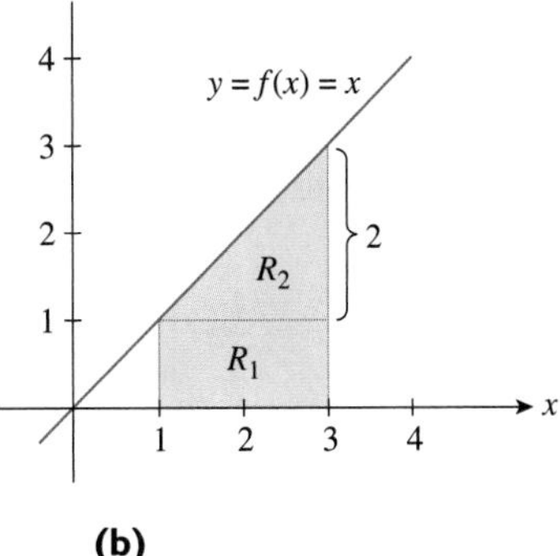

(b)

Figure 8.16
The area of R can be computed in two different ways.

To evaluate the definite integral, observe that an antiderivative of $f(x) = x$ is $F(x) = \frac{1}{2}x^2 + C$, where C is an arbitrary constant. Therefore, by the fundamental theorem of calculus, we have

$$A = \int_1^3 x\,dx = \frac{1}{2}x^2 + C\Big|_1^3$$

$$= \left(\frac{9}{2} + C\right) - \left(\frac{1}{2} + C\right) = 4 \text{ square units}$$

To verify this result by elementary means, observe that the area A is the area of the rectangle R_1 (width $\times$ height) plus the area of the triangle R_2 ($\frac{1}{2}$ base $\times$ height) (see Figure 8.16b); that is,

$$2(1) + \frac{1}{2}(2)(2) = 2 + 2 = 4$$

which agrees with the result obtained earlier.

Observe that in evaluating the definite integral in Example 1, the constant of integration "dropped out." This is true in general, for if $F(x) + C$ denotes an antiderivative of some function f, then

$$F(x) + C\Big|_a^b = [F(b) + C] - [F(a) + C]$$

$$= F(b) + C - F(a) - C$$

$$= F(b) - F(a)$$

With this fact in mind, we may, in all future computations involving the evaluations of a definite integral, drop the constant of integration from our calculations.

Finding the Area under a Curve

Having seen how effective the fundamental theorem of calculus is in helping us find the area of simple regions, we now use it to find the area of more complicated regions.

Example 2

In Section 8.3 we conjectured that the area of the region R under the graph of $f(x) = x^2$ on the interval $[0, 1]$ was $\frac{1}{3}$ square unit. Use the fundamental theorem of calculus to verify this conjecture.

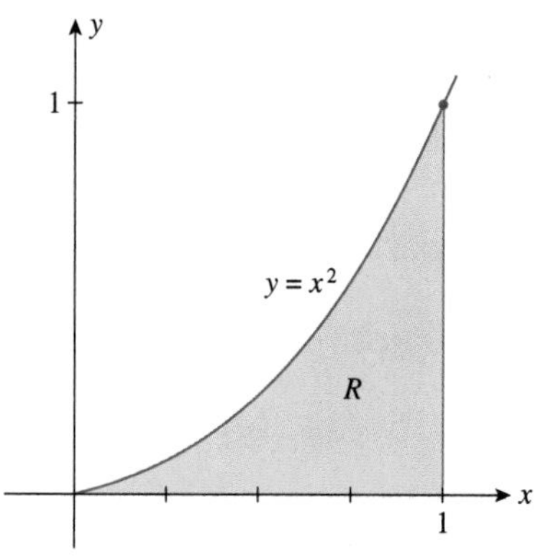

Figure 8.17

The area of R is $\int_0^1 x^2\,dx = \frac{1}{3}$.

Solution

The region R is reproduced in Figure 8.17. Observe that f is nonnegative on [0, 1], so the area of R is given by $A = \int_0^1 x^2\,dx$. Since an antiderivative of $f(x) = x^2$ is $F(x) = \frac{1}{3}x^3$, we see, using the fundamental theorem of calculus, that

$$A = \int_0^1 x^2\,dx = \frac{1}{3}x^3\Big|_0^1 = \frac{1}{3}(1) - \frac{1}{3}(0) = \frac{1}{3} \text{ square unit}$$

as we wished to show.

Remark

It is important to realize that the value, $\frac{1}{3}$, is by definition the exact value of the area of R. ◀

Example 3

Find the area of the region R under the graph of $y = x^2 + 1$ from $x = -1$ to $x = 2$.

Solution

The region R under consideration is shown in Figure 8.18. Using the fundamental theorem of calculus, we find that the required area is

$$\int_{-1}^{2} (x^2 + 1)\,dx = \left(\frac{1}{3}x^3 + x\right)\Big|_{-1}^{2}$$

$$= \left[\frac{1}{3}(8) + 2\right] - \left[\frac{1}{3}(-1)^3 + (-1)\right] = 6$$

or 6 square units.

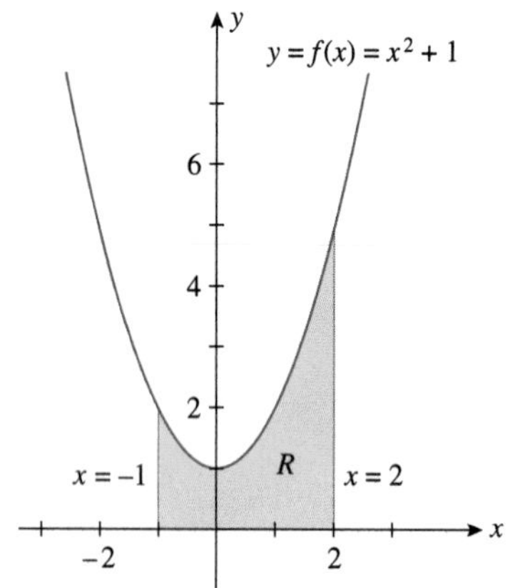

Figure 8.18

The area of R is $\int_{-1}^{2} (x^2 + 1)\,dx$.

Evaluating Definite Integrals

In Examples 4 and 5 we use the rules of integration of Section 8.1 to help us evaluate the definite integrals.

Example 4

Evaluate $\displaystyle\int_1^3 (3x^2 + e^x)\,dx$.

Solution

$$\int_1^3 (3x^2 + e^x)\,dx = x^3 + e^x\Big|_1^3$$

$$= (27 + e^3) - (1 + e) = 26 + e^3 - e$$

Example 5

Evaluate $\displaystyle\int_1^2 \left(\frac{1}{x} - \frac{1}{x^2}\right) dx$.

Solution

$$\int_1^2 \left(\frac{1}{x} - \frac{1}{x^2}\right) dx = \int_1^2 \left(\frac{1}{x} - x^{-2}\right) dx$$

$$= \ln|x| + \frac{1}{x}\Big|_1^2$$

$$= \left(\ln 2 + \frac{1}{2}\right) - (\ln 1 + 1)$$

$$= \ln 2 - \frac{1}{2} \qquad \text{Recall, } \ln 1 = 0.$$

GROUP DISCUSSION

Consider the definite integral $\int_{-1}^{1} \frac{1}{x^2}\,dx$.

1. Show that a formal application of Equation (9) leads to

$$\int_{-1}^{1} \frac{1}{x^2}\,dx = -\frac{1}{x}\Big|_{-1}^{1} = -1 - 1 = -2$$

2. Observe that $f(x) = 1/x^2$ is positive at each value of x in $[-1, 1]$ where it is defined. Therefore, one might expect that the definite integral with integrand f has a positive value, if it exists.
3. Explain this apparent contradiction in the result (1) and the observation (2).

Applications

Example 6

Production Costs The management of Staedtler Office Equipment has determined that the daily marginal cost function associated with producing battery-operated pencil sharpeners is given by

$$C'(x) = 0.000006x^2 - 0.006x + 4$$

where $C'(x)$ is measured in dollars per unit and x denotes the number of units produced. Management has also determined that the daily fixed cost incurred in producing these pencil sharpeners is \$100. Find Staedtler's daily total cost for producing (a) the first 500 units and (b) the 201st through 400th units.

Solution

a. Since $C'(x)$ is the marginal cost function, its antiderivative $C(x)$ is the total cost function. The daily fixed cost incurred in producing the pencil sharpeners is $C(0)$ dollars. Since the daily fixed cost is given as \$100, we have $C(0) = 100$. We are required to find $C(500)$. Let's compute $C(500) - C(0)$, the net change in the total cost function $C(x)$ over the interval $[0, 500]$. Using the fundamental theorem of calculus, we find

$$C(500) - C(0) = \int_0^{500} C'(x)\,dx$$

$$= \int_0^{500} (0.000006x^2 - 0.006x + 4)\,dx$$

$$= 0.000002x^3 - 0.003x^2 + 4x\Big|_0^{500}$$

$$= [0.000002(500)^3 - 0.003(500)^2 + 4(500)] - [0.000002(0)^3 - 0.003(0)^2 + 4(0)]$$

$$= 1500$$

Therefore, $C(500) = 1500 + C(0) = 1500 + 100 = 1600$, so the total cost incurred daily by Staedtler in producing 500 pencil sharpeners is \$1600.

b. The daily total cost incurred by Staedtler in producing the 201st through 400th units of battery-operated pencil sharpeners is given by

$$C(400) - C(200) = \int_{200}^{400} C'(x)\,dx$$

$$= \int_{200}^{400} (0.000006x^2 - 0.006x + 4)\,dx$$

$$= 0.000002x^3 - 0.003x^2 + 4x \Big|_{200}^{400}$$

$$= 552$$

or \$552.

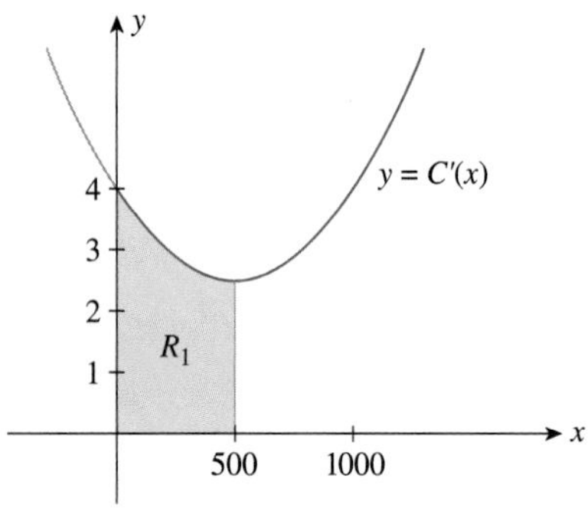

(a) Area of $R_1 = \int_1^{500} C'(x)\,dx$

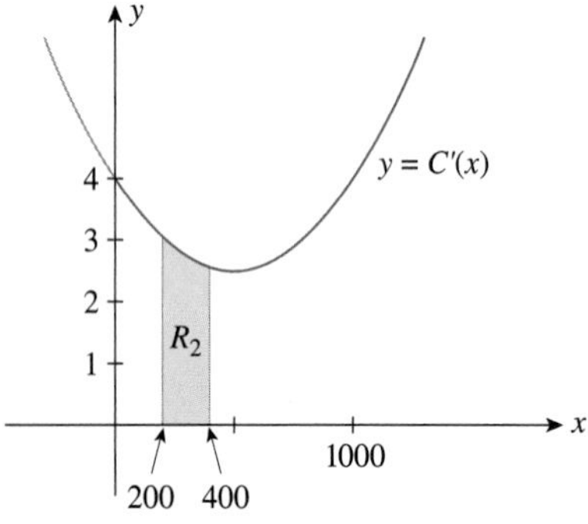

(b) Area of $R_2 = \int_{200}^{400} C'(x)\,dx$

Figure 8.19

Since $C'(x)$ is nonnegative for x in the interval $(0, \infty)$, we have the following geometric interpretation of the two definite integrals in Example 6: $\int_0^{500} C'(x)\,dx$ is the area of the region under the graph of the function C' from $x = 0$ to $x = 500$, shown in Figure 8.19a, and $\int_{200}^{400} C'(x)\,dx$ is the area of the region from $x = 200$ to $x = 400$, shown in Figure 8.19b.

Example 7

Assembly Time of Workers An efficiency study conducted for Elektra Electronics showed that the rate at which Space Commander walkie-talkies are assembled by the average worker t hours after starting work at 8 A.M. is given by the function

$$f(t) = -3t^2 + 12t + 15 \qquad (0 \le t \le 4)$$

Determine how many walkie-talkies can be assembled by the average worker in the first hour of the morning shift.

Solution

Let $N(t)$ denote the number of walkie-talkies assembled by the average worker t hours after starting work in the morning shift. Then, we have

$$N'(t) = f(t) = -3t^2 + 12t + 15$$

Therefore, the number of units assembled by the average worker in the first hour of the morning shift is

$$N(1) - N(0) = \int_0^1 N'(t)\,dt = \int_0^1 (-3t^2 + 12t + 15)\,dt$$

$$= -t^3 + 6t^2 + 15t \Big|_0^1 = -1 + 6 + 15$$

$$= 20$$

or 20 units.

EXPLORING WITH TECHNOLOGY

You can demonstrate graphically that $\int_0^x t\,dt = \frac{1}{2}x^2$ as follows:

1. Plot the graphs of $y1 = \text{fnInt}(t, t, 0, x) = \int_0^x t\,dt$ and $y2 = \frac{1}{2}x^2$ on the same set of axes, using the viewing window $[-5, 5] \cdot [0, 10]$.
2. Compare the graphs of $y1$ and $y2$ and draw the desired conclusion.

Example 8

Projected Demand for Electricity A certain city's rate of electricity consumption is expected to grow exponentially with a growth constant of $k = 0.04$. If the present rate of consumption is 40 million kilowatt-hours (kWh) per year, what should be the total production of electricity over the next 3 years in order to meet the projected demand?

Solution

If $R(t)$ denotes the expected rate of consumption of electricity t years from now, then

$$R(t) = 40e^{0.04t}$$

million kWh per year. Next, if $C(t)$ denotes the expected total consumption of electricity over a period of t years, then

$$C'(t) = R(t)$$

Therefore, the total consumption of electricity expected over the next 3 years is given by

$$\begin{aligned}\int_0^3 C'(t)\,dt &= \int_0^3 40e^{0.04t}\,dt \\ &= \frac{40}{0.04}e^{0.04t}\Big|_0^3 \\ &= 1000(e^{0.12} - 1) \\ &= 127.5\end{aligned}$$

or 127.5 million kWh, the amount that must be produced over the next 3 years in order to meet the demand.

GROUP DISCUSSION

The definite integral $\int_{-3}^{3}\sqrt{9 - x^2}\,dx$ cannot be evaluated by the means available up to now to find an antiderivative of the integrand. But the integral can be evaluated by interpreting it as the area of a certain plane region. What is the region? And what is the value of the integral?

Example 9

Suppose that the flow of waste water though a pipe is given by the expression

$$a(t) = 60 + 20\sin\left(\frac{t}{4\pi}\right) + 36\cos\left(\frac{t}{12\pi}\right) \text{ L/min}$$

at t minutes after midnight. If the total discharge from midnight to 4 A.M. is not permitted to be more than 15,000 litres, is this waste water flow consistent with regulations?

Solution

First we must compute an antiderivative $A(t)$ for the flow rate to get the total amount of water. In a group discussion in Section 8.2 on page 453 we found that if $F(x)$ is an antiderivative for $f(x)$ then $\frac{1}{a}\,F(ax + b)$ is an antiderivative for $f(ax + b)$. This rule, with $b = 0$, makes it easy to compute the antiderivative.

$$\begin{aligned} A(t) &= \int \left(60 + 20\sin\left(\frac{t}{4\pi}\right) + 36\cos\left(\frac{t}{12\pi}\right)\right) dt \\ &= 60t + 204\pi\left(-\cos\left(\frac{t}{4\pi}\right)\right) + 3612\pi\sin\left(\frac{t}{12\pi}\right) + C \\ &= 60t - 80\pi\cos\left(\frac{t}{4\pi}\right) + 432\pi\sin\left(\frac{t}{12\pi}\right) + C \end{aligned}$$

Between midnight and 4 A.M. is 240 minutes so the total flow is

$$\begin{aligned} A(t)\Big|_0^{240} &= [60240 - 80\pi\cos(240/4\pi) + 432\pi\sin(240/12\pi)] \\ &\quad [-600 - 80\pi\cos(0/4\pi) + 432\pi\sin(0/12\pi)] \\ &\approx 14400 - 251.30.9688 + 1357 \times 0.087 - 0 + 251.3 \times 0 - 1357 \times 1 \\ &\approx 1219.5 \text{ litres} \end{aligned}$$

We see that the total water flow from midnight to 4 A.M. is well within the amount allowed by regulations.

The Definite Integral of Sine or Cosine over a Full Period

Look at Figure 8.20, which shows a graph of the sine function from 0 to 2π. The period of the sine function is 2π and so the graph covers exactly one period. The graph shows the same amount of area above and below the x-axis. What does that say about the definite integral

$$\int_0^{2\pi} \sin(x)\,dx$$

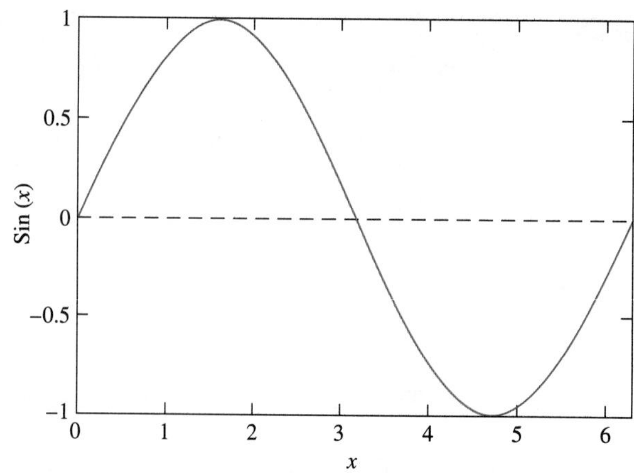

Figure 8.20
The area of $f(x) = \sin(x)$ above and below the axis over one period.

Let's compute the integral and find out.

$$\begin{aligned}\int_0^{2\pi} \sin(x)\,dx &= -\cos(x)\Big|_0^{2\pi} \\ &= -\cos(2\pi) - (-\cos(0)) \\ &= -1 - (-1) \\ &= -1 + 1 \\ &= 0\end{aligned}$$

This phenomenon works for both the sine and cosine function over a full period.

The Definite Integral of sin(x), cos(x) over a Full Period

Suppose that $f(x) = \sin(ax + b)$ or that $g(x) = \cos(ax + b)$. In either case the function is periodic with period equal to $\frac{2\pi}{a}$. In this case if n is an integer we have

$$\int_u^{u+n\frac{2\pi}{a}} f(x)\,dx = 0$$

and

$$\int_u^{u+n\frac{2\pi}{a}} g(x)\,dx = 0$$

Validity of the Fundamental Theorem of Calculus

To demonstrate the plausibility of the fundamental theorem of calculus for the case where f is nonnegative on an interval $[a, b]$, let's define an "area function" A as follows. Let $A(t)$ denote the area of the region R under the graph of $y = f(x)$ from $x = a$ to $x = t$, where $a \leq t \leq b$ (Figure 8.21).

If h is a small positive number, then $A(t + h)$ is the area of the region under the graph of $y = f(x)$ from $x = a$ to $x = t + h$. Therefore, the difference

$$A(t + h) - A(t)$$

is the area under the graph of $y = f(x)$ from $x = t$ to $x = t + h$ (Figure 8.22).

Now, the area of this last region can be approximated by the area of the rectangle of width h and height $f(t)$—that is, by the expression $h \cdot f(t)$ (Figure 8.23). Thus,

$$A(t + h) - A(t) \approx h \cdot f(t)$$

where the approximations improve as h is taken to be smaller and smaller.

Dividing both sides of the foregoing relationship by h, we obtain

$$\frac{A(t + h) - A(t)}{h} \approx f(t)$$

Taking the limit as h approaches zero, we find, by the definition of the derivative, that the left-hand side is

$$\lim_{h \to 0} \frac{A(t + h) - A(t)}{h} = A'(t)$$

The right-hand side, which is independent of h, remains constant throughout the limiting process. Because the approximation becomes exact as h approaches zero, we find that

$$A'(t) = f(t)$$

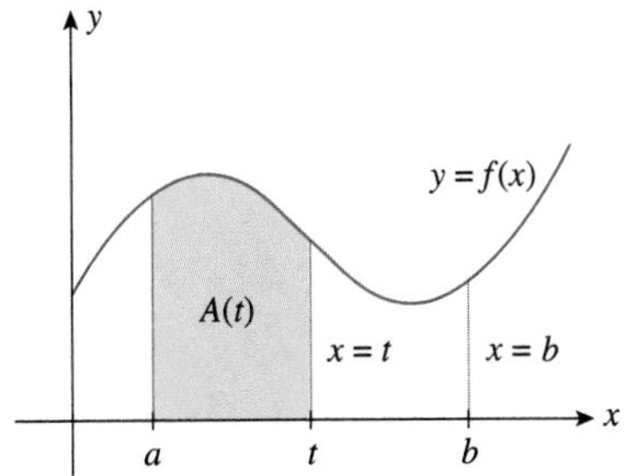

Figure 8.21
$A(t)$ = area under the graph of f from $x = a$ to $x = t$

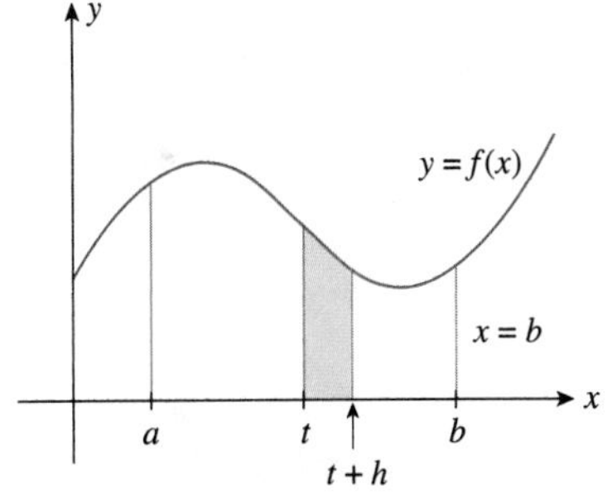

Figure 8.22
$A(t + h) - A(t)$ = area under the graph of f from $x = t$ to $x = t + h$

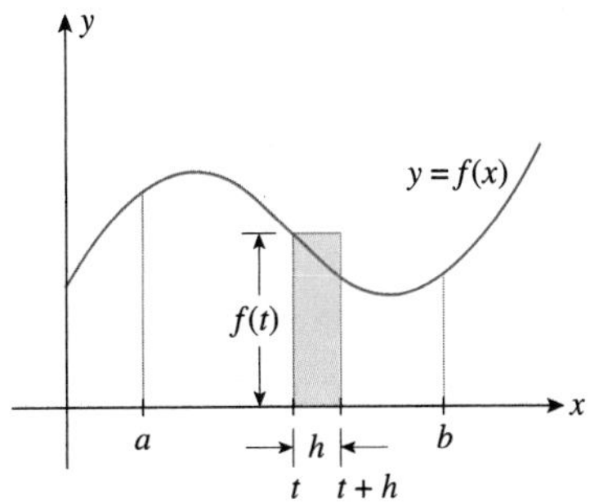

Figure 8.23
The area of the rectangle is $h \cdot f(t)$.

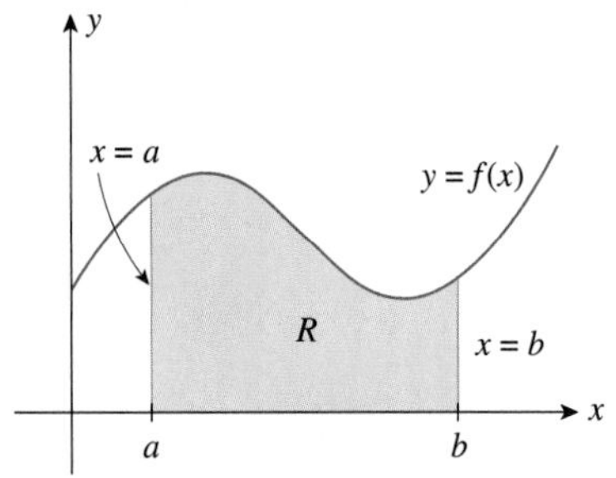

Figure 8.24
The area of R is given by $A(b)$.

Since the foregoing equation holds for all values of t in the interval $[a, b]$, we have shown that the *area function* A is an antiderivative of the function $f(x)$. By Theorem 1 of Section 6.1, we conclude that $A(x)$ must have the form

$$A(x) = F(x) + C$$

where F is any antiderivative of f and C is a constant. To determine the value of C, observe that $A(a) = 0$. This condition implies that

$$A(a) = F(a) + C = 0$$

or $C = -F(a)$. Next, since the area of the region R is $A(b)$ (Figure 8.24), we see that the required area is

$$\begin{aligned} A(b) &= F(b) + C \\ &= F(b) - F(a) \end{aligned}$$

Since the area of the region R is

$$\int_a^b f(x)\, dx$$

we have

$$\int_a^b f(x)\, dx = F(b) - F(a)$$

as we set out to show.

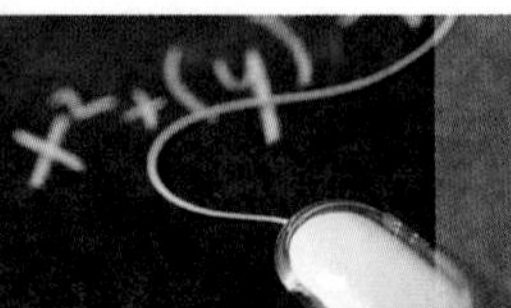

Self-Check Exercises 8.4

1. Evaluate $\int_0^2 (x + e^x)\, dx$.

2. The daily marginal profit function associated with producing and selling TexaPep hot sauce is

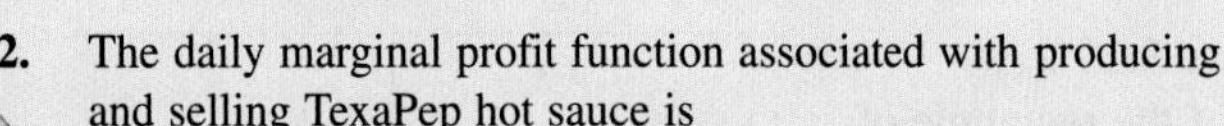

$$P'(x) = -0.000006x^2 + 6$$

where x denotes the number of cases (each case contains 24 bottles) produced and sold daily and $P'(x)$ is measured in dollars/units. The fixed cost is \$400.

a. What is the total profit realizable from producing and selling 1000 cases of TexaPep per day?
b. What is the additional profit realizable if the production and sale of TexaPep is increased from 1000 to 1200 cases/day?

Solutions to Self-Check Exercises 8.4 can be found on page 483.

8.4 Exercises

In Exercises 1–4, find the area of the region under the graph of the function f on the interval $[a, b]$, using the fundamental theorem of calculus. Then verify your result using geometry.

1. $f(x) = 2$; $[1, 4]$
2. $f(x) = 4$; $[-1, 2]$
3. $f(x) = 2x$; $[1, 3]$
4. $f(x) = -\frac{1}{4}x + 1$; $[1, 4]$

In Exercises 5–20, find the area of the region under the graph of the function f on the interval $[a, b]$.

5. $f(x) = 2x + 3$; $[-1, 2]$
6. $f(x) = 4x - 1$; $[2, 4]$
7. $f(x) = -x^2 + 4$; $[-1, 2]$
8. $f(x) = 4x - x^2$; $[0, 4]$
9. $f(x) = \frac{1}{x}$; $[1, 2]$
10. $f(x) = \frac{1}{x^2}$; $[2, 4]$
11. $f(x) = \sqrt{x}$; $[1, 9]$
12. $f(x) = x^3$; $[1, 3]$

13. $f(x) = 1 - \sqrt[3]{x}; [-8, -1]$

14. $f(x) = \dfrac{1}{\sqrt{x}}; [1, 9]$

15. $f(x) = e^x; [0, 2]$

16. $f(x) = e^x - x; [1, 2]$

17. $f(x) = \sin(x); [0, \pi]$

18. $f(x) = \dfrac{1}{1 + x^2}; [-1, 1]$

19. $f(x) = \cos(x); \left[0, \dfrac{\pi}{3}\right]$

20. $g(x) = x + \sin(x); \left[0, \dfrac{\pi}{4}\right]$

In Exercises 21–52, evaluate the definite integral.

21. $\displaystyle\int_2^4 3\,dx$

22. $\displaystyle\int_{-1}^2 -2\,dx$

23. $\displaystyle\int_1^3 (2x + 3)\,dx$

24. $\displaystyle\int_{-1}^0 (4 - x)\,dx$

25. $\displaystyle\int_{-1}^3 2x^2\,dx$

26. $\displaystyle\int_0^2 8x^3\,dx$

27. $\displaystyle\int_{-2}^2 (x^2 - 1)\,dx$

28. $\displaystyle\int_1^4 \sqrt{u}\,du$

29. $\displaystyle\int_1^8 4x^{1/3}\,dx$

30. $\displaystyle\int_1^4 2x^{-3/2}\,dx$

31. $\displaystyle\int_0^1 (x^3 - 2x^2 + 1)\,dx$

32. $\displaystyle\int_1^2 (t^5 - t^3 + 1)\,dt$

33. $\displaystyle\int_2^4 \frac{1}{x}\,dx$

34. $\displaystyle\int_1^3 \frac{2}{x}\,dx$

35. $\displaystyle\int_0^4 x(x^2 - 1)\,dx$

36. $\displaystyle\int_0^2 (x - 4)(x - 1)\,dx$

37. $\displaystyle\int_1^3 (t^2 - t)^2\,dt$

38. $\displaystyle\int_{-1}^1 (x^2 - 1)^2\,dx$

39. $\displaystyle\int_{-3}^{-1} \frac{1}{x^2}\,dx$

40. $\displaystyle\int_1^2 \frac{2}{x^3}\,dx$

41. $\displaystyle\int_1^4 \left(\sqrt{x} - \frac{1}{\sqrt{x}}\right) dx$

42. $\displaystyle\int_0^1 \sqrt{2x}(\sqrt{x} + \sqrt{2})\,dx$

43. $\displaystyle\int_1^4 \frac{3x^3 - 2x^2 + 4}{x^2}\,dx$

44. $\displaystyle\int_1^2 \left(1 + \frac{1}{u} + \frac{1}{u}\right) du$

45. $\displaystyle\int_0^{4\pi} \cos(x)\,dx$

46. $\displaystyle\int_0^{\sqrt{3}} \frac{dx}{x^2 + 1}$

47. $\displaystyle\int_2^{2+2\pi} \sin(x)\,dx$

48. $\displaystyle\int_0^{\pi/4} \sec^2(x)\,dx$

49. $\displaystyle\int_0^{\pi/3} \sec(x)\tan(x)\,dx$

50. $\displaystyle\int_0^5 (\sin^2(x) + \cos^2(x))\,dx$

51. $\displaystyle\int_0^{\pi} (2\sin(x))\,dx$

52. $\displaystyle\int_{\pi/4}^{\pi/3} (\cos(x) - \sin(x))\,dx$

A calculator is recommended for Exercises 53–65.

53. **Speedboat Racing** In a recent pretrial run for the world water speed record, the velocity of the *Sea Falcon II* t sec after firing the booster rocket was given by

$$v(t) = -\frac{2}{3}t^2 + 6t + 75 \qquad (0 \le t \le 20)$$

m/sec. Find the distance covered by the boat over the 20-sec period after the booster rocket was activated.

Hint: The distance is given by $\displaystyle\int_0^{20} v(t)\,dt$.

Business and Economics Applications

54. **Marginal Cost** A division of Ditton Industries manufactures a deluxe toaster oven. Management has determined that the daily marginal cost function associated with producing these toaster ovens is given by

$$C'(x) = 0.0003x^2 - 0.12x + 20$$

where $C'(x)$ is measured in dollars/unit and x denotes the number of units produced. Management has also determined that the daily fixed cost incurred in the production is $800.

a. Find the total cost incurred by Ditton in producing the first 300 units of these toaster ovens per day.

b. What is the total cost incurred by Ditton in producing the 201st through 300th units/day?

55. **Marginal Revenue** The management of Ditton Industries has determined that the daily marginal revenue function associated with selling x units of their deluxe toaster ovens is given by

$$R'(x) = -0.1x + 40$$

where $R'(x)$ is measured in dollars/unit.

a. Find the daily total revenue realized from the sale of 200 units of the toaster oven.

b. Find the additional revenue realized when the production (and sales) level is increased from 200 to 300 units.

56. **Marginal Profit** Refer to Exercise 54. The daily marginal profit function associated with the production and sales of the deluxe toaster ovens is known to be

$$P'(x) = -0.0003x^2 + 0.02x + 20$$

where x denotes the number of units manufactured and sold daily and $P'(x)$ is measured in dollars/unit.

a. Find the total profit realizable from the manufacture and sale of 200 units of the toaster ovens per day.

Hint: $P(200) - P(0) = \displaystyle\int_0^{200} P'(x)\,dx$, $P(0) = -800$.

b. What is the additional daily profit realizable if the production and sale of the toaster ovens are increased from 200 to 220 units/day?

57. **Efficiency Studies** Tempco Electronics, a division of Tempco Toys, manufactures an electronic football game. An efficiency study showed that the rate at which the games are assembled by

the average worker t hr after starting work at 8 A.M. is

$$-\frac{3}{2}t^2 + 6t + 20 \qquad (0 \leq t \leq 4)$$

units/hour.

a. Find the total number of games the average worker can be expected to assemble in the 4-h morning shift.

b. How many units can the average worker be expected to assemble in the first hour of the morning shift? In the second hour of the morning shift?

58. Hand-Held Computers Annual sales (in millions of units) of hand-held computers are expected to grow in accordance with the function

$$f(t) = 0.18t^2 + 0.16t + 2.64 \qquad (0 \leq t \leq 6)$$

where t is measured in years, with $t = 0$ corresponding to 1997. How many hand-held computers will be sold over the 6-yr period between the beginning of 1997 and the end of 2002?

Source: Dataquest, Inc.

Biological and Life Sciences Applications

59. Air Purification To test air purifiers, engineers ran a purifier in a smoke-filled 3×5-m room. While conducting a test for a certain brand of air purifier, it was determined that the amount of smoke in the room was decreasing at the rate of

$$R(t) = 0.00032t^4 - 0.01872t^3 + 0.3948t^2 - 3.83t + 17.63 \qquad (0 \leq t \leq 20)$$

percent of the (original) amount of the smoke per minute, t min after the start of the test. How much smoke was left in the room 5 min after the start of the test? Ten minutes after the start of the test?

Source: Consumer Reports

60. Senior Citizens The population aged 65 years old and older (in millions) from 2000 to 2050 is projected to be

$$f(t) = \frac{9}{1 + 1.859e^{-0.66t}} \qquad (0 \leq t \leq 5)$$

where t is measured in decades, with $t = 0$ corresponding to 2000. What will be the average population aged 65 years and older over the years from 2000 to 2030?

Hint: The average population is given by $(1/3)\int_0^3 f(t)\,dt$. Multiply the integrand by $e^{0.66t}/e^{0.66t}$ and then use the method of substitution.

61. Blood Flow Consider an artery of length L cm and radius R cm. Using Poiseuille's law (Problem 90, page 493), it can be shown that the rate at which blood flows through the artery (measured in cubic centimetres/second) is given by

$$V = \int_0^R \frac{k}{L}x(R^2 - x^2)\,dx$$

where k is a constant. Find an expression for V that does *not* involve an integral.

Hint: Use the substitution $u = R^2 - x^2$.

62. Total Production Due to accumulation of metabolic wastes the total amount of thiotimoline being produced in a bacterial reactor drops off with time. If the thiotimoline production is

$$c(t) = \frac{6.2}{t^2 - 4t + 5}\ \text{g/h}$$

and the amount present when the reactor is started is zero, find the total production from hour 0 to 2, from hour 1 to 3, and from hour 2 to 4. Which of these is the largest?

63. Total Amount of Hormone Suppose that production of a hormone varies periodically according to the function

$$q(t) = 212 + 47\sin\left(\frac{\pi t}{8}\right)\ \text{mg/h}$$

If $t = 0$ at midnight, compute the total amount of hormone from 2 A.M. to 4 P.M. and also during the entire day.

64. Total Production This problem refers to the same situation as Problem 63. Does the total production of the hormone during a day depend on when we start measuring that day? In other words, is the production from midnight to the following midnight different from the production from 7 A.M. to 7 P.M. the following day, or any other starting point?

65. Total Number of Rabbits The density of rabbits along a strip stretching one mile on either side of a 100-km stretch of the Trans-Canada highway is found to be well approximated by the function

$$23 + 12x\cos\left(\frac{x^2}{500}\right)\ \text{rabbits/km}$$

Using integration, estimate the total number of rabbits along that stretch of highway.

In Exercises 66–69, determine whether the statement is true or false. If it is true, explain why it is true. If it is false, give an example to show why it is false.

66. $\displaystyle\int_{-1}^{1} \frac{1}{x^3}\,dx = -\frac{1}{2x^2}\Big|_{-1}^{1} = -\frac{1}{2} - \left(-\frac{1}{2}\right) = 0$

67. $\displaystyle\int_{-1}^{1} \frac{1}{x}\,dx = \ln|x|\Big|_{-1}^{1} = \ln|1| - \ln|-1| = \ln 1 - \ln 1 = 0$

68. $\int_0^2 (1 - x)\,dx$ gives the area of the region under the graph of $f(x) = 1 - x$ on the interval [0, 2].

69. The total revenue realized in selling the first 5000 units of a product is given by

$$\int_0^{500} R'(x)\,dx = R(500) - R(0)$$

where $R(x)$ is the total revenue.

Solutions to Self-Check Exercises 8.4

1. $$\int_0^2 (x + e^x)\,dx = \frac{1}{2}x^2 + e^x \Big|_0^2$$
$$= \left[\frac{1}{2}(2)^2 + e^2\right] - \left[\frac{1}{2}(0) + e^0\right]$$
$$= 2 + e^2 - 1$$
$$= e^2 + 1$$

2. **a.** We want $P(1000)$, but

$$P(1000) - P(0) = \int_0^{1000} P'(x)\,dx = \int_0^{1000} (-0.000006x^2 + 6)\,dx$$
$$= -0.000002x^3 + 6x \Big|_0^{1000}$$
$$= -0.000002(1000)^3 + 6(1000)$$
$$= 4000$$

So, $P(1000) = 4000 + P(0) = 4000 - 400$, or \$3600/day $[P(0) = -C(0)]$.

b. The additional profit realizable is given by

$$\int_{1000}^{1200} P'(x)\,dx = -0.000002x^3 + 6x \Big|_{1000}^{1200}$$
$$= [-0.000002(1200)^3 + 6(1200)]$$
$$- [-0.000002(1000)^3 + 6(1000)]$$
$$= 3744 - 4000$$
$$= -256$$

That is, the company sustains a loss of \$256/day if production is increased to 1200 cases/day.

8.5 Evaluating Definite Integrals

This section continues our discussion of the applications of the fundamental theorem of calculus.

Properties of the Definite Integral

Before going on, we list the following useful properties of the definite integral, some of which parallel the rules of integration of Section 8.1.

Properties of the Definite Integral

Let f and g be integrable functions; then,

1. $\displaystyle\int_a^a f(x)\,dx = 0$
2. $\displaystyle\int_a^b f(x)\,dx = -\int_b^a f(x)\,dx$
3. $\displaystyle\int_a^b cf(x)\,dx = c\int_a^b f(x)\,dx \qquad (c, \text{ a constant})$
4. $\displaystyle\int_a^b [f(x) \pm g(x)]\,dx = \int_a^b f(x)\,dx \pm \int_a^b g(x)\,dx$
5. $\displaystyle\int_a^b f(x)\,dx = \int_a^c f(x)\,dx + \int_c^b f(x)\,dx \qquad (a < c < b)$

Property 5 states that if c is a number lying between a and b so that the interval $[a, b]$ is divided into the intervals $[a, c]$ and $[c, b]$, then the integral of f over the interval $[a, b]$ may be expressed as the sum of the integral of f over the interval $[a, c]$ and the integral of f over the interval $[c, b]$.

Property 5 has the following geometric interpretation when f is nonnegative. By definition

$$\int_a^b f(x)\,dx$$

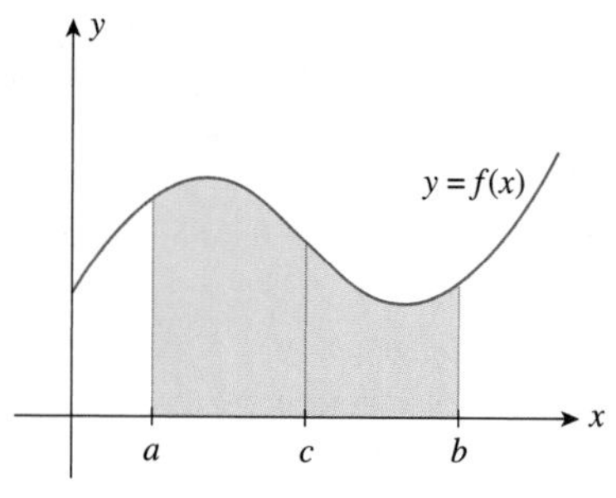

Figure 8.25

$\int_a^b f(x)\,dx = \int_a^c f(x)\,dx + \int_c^b f(x)\,dx$

is the area of the region under the graph of $y = f(x)$ from $x = a$ to $x = b$ (Figure 8.25). Similarly, we interpret the definite integrals

$$\int_a^c f(x)\,dx \qquad \text{and} \qquad \int_c^b f(x)\,dx$$

as the areas of the regions under the graph of $y = f(x)$ from $x = a$ to $x = c$ and from $x = c$ to $x = b$, respectively. Since the two regions do not overlap, we see that

$$\int_a^b f(x)\,dx = \int_a^c f(x)\,dx + \int_c^b f(x)\,dx$$

The Method of Substitution for Definite Integrals

Our first example shows two approaches generally used when evaluating a definite integral using the method of substitution.

Example 1

Evaluate $\int_0^4 x\sqrt{9 + x^2}\,dx.$

Solution

Method 1 We first find the corresponding indefinite integral:

$$I = \int x\sqrt{9 + x^2}\,dx$$

Make the substitution $u = 9 + x^2$ so that

$$\begin{aligned} du &= \frac{d}{dx}(9 + x^2)\,dx \\ &= 2x\,dx \\ x\,dx &= \frac{1}{2}\,du \qquad \text{Divide both sides by 2.} \end{aligned}$$

Then,

$$\begin{aligned} I &= \int \frac{1}{2}\sqrt{u}\,du = \frac{1}{2}\int u^{1/2}\,du \\ &= \frac{1}{3}u^{3/2} + C = \frac{1}{3}(9 + x^2)^{3/2} + C \qquad \text{Substitute } 9 + x^2 \text{ for } u. \end{aligned}$$

Using this result, we now evaluate the given definite integral:

$$\begin{aligned} \int_0^4 x\sqrt{9 + x^2}\,dx &= \frac{1}{3}(9 + x^2)^{3/2}\Big|_0^4 \\ &= \frac{1}{3}[(9 + 16)^{3/2} - 9^{3/2}] \\ &= \frac{1}{3}(125 - 27) = \frac{98}{3} = 32\frac{2}{3} \end{aligned}$$

Method 2 *Changing the Limits of Integration.* As before, we make the substitution

$$u = 9 + x^2 \qquad \textbf{(10)}$$

so that

$$du = 2x\,dx$$
$$x\,dx = \frac{1}{2}\,du$$

Next, observe that the given definite integral is evaluated *with respect to x* with the range of integration given by the interval [0, 4]. If we perform the integration *with respect to u* via the substitution (10), then we must adjust the range of integration to reflect the fact that the integration is being performed with respect to the new variable u. To determine the proper range of integration, note that when $x = 0$, Equation (10) implies that

$$u = 9 + 0^2 = 9$$

which gives the required lower limit of integration with respect to u. Similarly, when $x = 4$,

$$u = 9 + 16 = 25$$

is the required upper limit of integration with respect to u. Thus, the range of integration when the integration is performed with respect to u is given by the interval [9, 25]. Therefore, we have

$$\begin{aligned}\int_0^4 x\sqrt{9 + x^2}\,dx &= \int_9^{25} \frac{1}{2}\sqrt{u}\,du = \frac{1}{2}\int_9^{25} u^{1/2}\,du\\ &= \frac{1}{3}u^{3/2}\Big|_9^{25} = \frac{1}{3}(25^{3/2} - 9^{3/2})\\ &= \frac{1}{3}(125 - 27) = \frac{98}{3} = 32\frac{2}{3}\end{aligned}$$

which agrees with the result obtained using Method 1.

When you use the method of substitution, make sure you adjust the limits of integration to reflect integrating with respect to the new variable u.

EXPLORING WITH TECHNOLOGY

Refer to Example 1. You can confirm the results obtained there by using a graphing utility as follows:

1. Use the numerical integration operation of the graphing utility to evaluate

$$\int_0^4 x\sqrt{9 + x^2}\,dx$$

2. Evaluate $\frac{1}{2}\int_9^{25}\sqrt{u}\,du$.

3. Conclude that $\int_0^4 x\sqrt{9 + x^2}\,dx = \frac{1}{2}\int_9^{25}\sqrt{u}\,du$.

Example 2

Evaluate $\int_0^2 xe^{2x^2}\,dx$.

Solution

Let $u = 2x^2$ so that $du = 4x\,dx$, or $x\,dx = \frac{1}{4}\,du$. When $x = 0$, $u = 0$, and when $x = 2$, $u = 8$. This gives the lower and upper limits of integration with respect to u.

Making the indicated substitutions, we find

$$\int_0^2 xe^{2x^2}\,dx = \int_0^8 \frac{1}{4}e^u\,du = \frac{1}{4}e^u\Big|_0^8 = \frac{1}{4}(e^8 - 1)$$

Example 3

Evaluate $\displaystyle\int_0^1 \frac{x^2}{x^3 + 1}\,dx.$

Solution

Let $u = x^3 + 1$ so that $du = 3x^2\,dx$, or $x^2\,dx = \frac{1}{3}\,du$. When $x = 0$, $u = 1$, and when $x = 1$, $u = 2$. This gives the lower and upper limits of integration with respect to u. Making the indicated substitutions, we find

$$\int_0^1 \frac{x^2}{x^3 + 1}\,dx = \frac{1}{3}\int_1^2 \frac{du}{u} = \frac{1}{3}\ln|u|\Big|_1^2$$

$$= \frac{1}{3}(\ln 2 - \ln 1) = \frac{1}{3}\ln 2$$

Finding the Area under a Curve

Example 4

Find the area of the region R under the graph of $f(x) = e^{(1/2)x}$ from $x = -1$ to $x = 1$.

Solution

The region R is shown in Figure 8.26. Its area is given by

$$A = \int_{-1}^1 e^{(1/2)x}\,dx$$

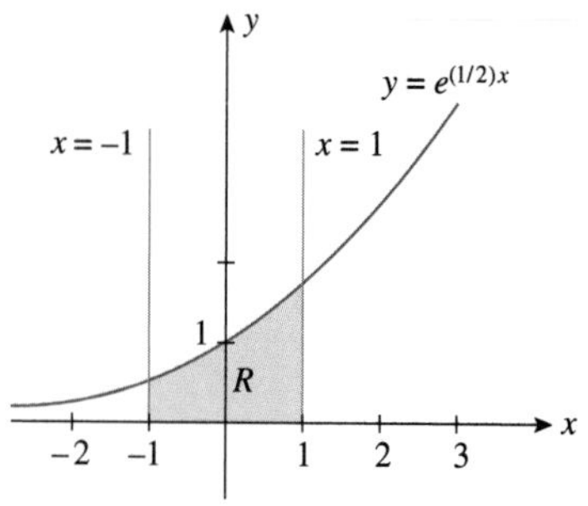

To evaluate this integral, we make the substitution

$$u = \frac{1}{2}x$$

so that

$$du = \frac{1}{2}\,dx$$
$$dx = 2\,du$$

When $x = -1$, $u = -\frac{1}{2}$, and when $x = 1$, $u = \frac{1}{2}$. Making the indicated substitutions, we obtain

$$A = \int_{-1}^1 e^{(1/2)x}\,dx = 2\int_{-1/2}^{1/2} e^u\,du$$

$$= 2e^u\Big|_{-1/2}^{1/2} = 2(e^{1/2} - e^{-1/2})$$

or approximately 2.08 square units.

Figure 8.26

Area of $R = \displaystyle\int_{-1}^{1} e^{(1/2)x}\,dx$

Example 5

Evaluate $\displaystyle\int_1^2 \frac{dx}{x^2 - 2x + 2}.$

Solution

Let $u = x - 1$ so that $x^2 - 2x + 2 = (x + 1)^2 + 1 = u^2 + 1$ and $du = dx$. When $x = 1$, $u = 0$ and when $x = 2$, $u = 1$. Making the indicated substitution we find that

$$\begin{aligned}\int_1^2 \frac{dx}{x^2 - 2x + 2} &= \int_0^1 \frac{du}{u^2 + 1} \\ &= \arctan(u)\Big|_0^1 \\ &= \arctan(1) - \arctan(0) \\ &= \frac{\pi}{4} - 0 \\ &= \frac{\pi}{4}\end{aligned}$$

GROUP DISCUSSION

Let f be a function defined piecewise by the rule

$$f(x) = \begin{cases} \sqrt{x} & \text{if } 0 \le x \le 1 \\ \dfrac{1}{x} & \text{if } 1 < x \le 2 \end{cases}$$

How would you use Property 5 of definite integrals to find the area of the region under the graph of f on $[0, 2]$? What is the area?

Average Value of a Function

The *average value* of a function over an interval provides us with an application of the definite integral. Recall that the average value of a set of n numbers is the number

$$\frac{y_1 + y_2 + \cdots + y_n}{n}$$

Now, suppose f is a continuous function defined on $[a, b]$. Let's divide the interval $[a, b]$ into n subintervals of equal length $(b - a)/n$. Choose points $x_1, x_2, \ldots, x_n$ in the first, second, ... , and nth subintervals, respectively. Then, the average value of the numbers $f(x_1), f(x_2), \ldots, f(x_n)$, given by

$$\frac{f(x_1) + f(x_2) + \cdots + f(x_n)}{n}$$

is an approximation of the average of all the values of $f(x)$ on the interval $[a, b]$. This expression can be written in the form

$$\begin{aligned}&\frac{(b - a)}{(b - a)}\left[f(x_1)\cdot\frac{1}{n} + f(x_2)\cdot\frac{1}{n} + \cdots + f(x_n)\cdot\frac{1}{n}\right] \\ &= \frac{1}{b - a}\left[f(x_1)\cdot\frac{b - a}{n} + f(x_2)\cdot\frac{b - a}{n} + \cdots + f(x_n)\cdot\frac{b - a}{n}\right] \\ &= \frac{1}{b - a}[f(x_1)\Delta x + f(x_2)\Delta x + \cdots + f(x_n)\Delta x] \qquad (11)\end{aligned}$$

As n gets larger and larger, the expression (11) approximates the average value of $f(x)$ over $[a, b]$ with increasing accuracy. But the sum inside the brackets in (11) is a Riemann sum of the function f over $[a, b]$. In view of this, we have

$$\lim_{n\to\infty}\left[\frac{f(x_1) + f(x_2) + \cdots + f(x_n)}{n}\right]$$

$$= \frac{1}{b-a}\lim_{n\to\infty}[f(x_1)\Delta x + f(x_2)\Delta x + \cdots + f(x_n)\Delta x]$$

$$= \frac{1}{b-a}\int_a^b f(x)dx$$

This discussion motivates the following definition.

The Average Value of a Function

Suppose f is integrable on $[a, b]$. Then the average value of f over $[a, b]$ is

$$\frac{1}{b-a}\int_a^b f(x)\,dx$$

Example 6

Find the average value of the function $f(x) = \sqrt{x}$ over the interval [0, 4].

Solution

The required average value is given by

$$\frac{1}{4-0}\int_0^4 \sqrt{x}\,dx = \frac{1}{4}\int_0^4 x^{1/2}\,dx$$

$$= \frac{1}{6}x^{3/2}\Big|_0^4 = \frac{1}{6}(4^{3/2})$$

$$= \frac{4}{3}$$

Applications

Example 7

Automobile Financing The interest rates charged by Madison Finance on auto loans for used cars over a certain 6-month period in 2000 are approximated by the function

$$r(t) = -\frac{1}{12}t^3 + \frac{7}{8}t^2 - 3t + 12 \qquad (0 + \leq t \leq 6)$$

where t is measured in months and $r(t)$ is the annual percentage rate. What is the average rate on auto loans extended by Madison over the 6-month period?

Solution

The average rate over the 6-month period in question is given by

$$\frac{1}{6-0}\int_0^6\left(-\frac{1}{12}t^3 + \frac{7}{8}t^2 - 3t + 12\right)dt$$

$$= \frac{1}{6}\left(-\frac{1}{48}t^4 + \frac{7}{24}t^3 - \frac{3}{2}t^2 + 12t\right)\Bigg|_0^6$$

$$= \frac{1}{6}\left[-\frac{1}{48}(6^4) + \frac{7}{24}(6^3) - \frac{3}{2}(6^2) + 12(6)\right]$$
$$= 9$$

or 9% per year.

Example 8

Drug Concentration in a Body The amount of a certain drug in a patient's body t days after it has been administered is

$$C(t) = 5e^{-0.2t}$$

units. Determine the average amount of the drug present in the patient's body for the first 4 days after the drug has been administered.

Solution

The average amount of the drug present in the patient's body for the first 4 days after it has been administered is given by

$$\begin{aligned}
\frac{1}{4-0}\int_0^4 5e^{-0.2t}\,dt &= \frac{5}{4}\int_0^4 e^{-0.2t}\,dt \\
&= \frac{5}{4}\left[\left(-\frac{1}{0.2}\right)e^{-0.2t}\Big|_0^4\right] \\
&= \frac{5}{4}(-5e^{-0.8} + 5) \\
&\approx 3.44
\end{aligned}$$

or approximately 3.44 units.

We now give a geometric interpretation of the average value of a function f over an interval $[a, b]$. Suppose $f(x)$ is nonnegative so that the definite integral

$$\int_a^b f(x)\,dx$$

gives the area under the graph of f from $x = a$ to $x = b$ (Figure 8.27). Observe that, in general, the "height" $f(x)$ varies from point to point. Can we replace $f(x)$ by a constant function $g(x) = k$ (which has constant height) such that the areas under each of the two functions f and g are the same? If so, since the area under the graph of g from $x = a$ to $x = b$ is $k(b - a)$, we have

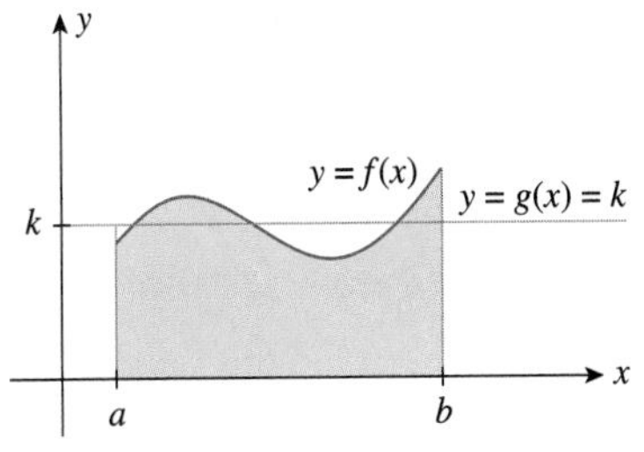

Figure 8.27
The average value of f over $[a, b]$ is k.

$$k(b-a) = \int_a^b f(x)\,dx$$

$$k = \frac{1}{b-a}\int_a^b f(x)\,dx$$

so that k is the average value of f over $[a, b]$. Thus, the average value of a function f over an interval $[a, b]$ is the height of a rectangle with base of length $(b - a)$ that has the same area as that of the region under the graph of f from $x = a$ to $x = b$.

Example 9

Stock Prices The weekly closing price of HAL Corporation stock in week t is approximated by the rule

$$f(t) = 30 + t\sin\frac{\pi}{6} \qquad (0 \le t \le 15)$$

where $f(t)$ is the price (in dollars) per share. Find the average weekly closing price of the stock over the 15-week period.

Solution

The average weekly closing price of the stock over the 15-week period in question is given by

$$A = \frac{1}{15 - 0} \int_0^{15} \left(30 + t \sin\frac{\pi}{6}t\right)$$
$$= \frac{1}{15}\int_0^{15} 30\, dt + \frac{1}{15}\int_0^{15} t \sin\frac{\pi}{6}t\, dt$$
$$= 30 + \frac{1}{15}\int_0^{15} t \sin\frac{\pi}{t}t\, dt$$

Integrating by parts with

$$u = t \quad \text{and} \quad dv = \sin\frac{\pi}{6}t\, dt$$

so that

$$du = dt \quad \text{and} \quad v = -\frac{6}{\pi}\cos\frac{\pi}{6}t$$

we have

$$A = 30 + \frac{1}{15}\left(-\frac{6}{\pi}t\cos\frac{\pi}{6}t\Big|_0^{15} + \frac{6}{\pi}\int_0^{15}\cos\frac{\pi}{6}t\, dt\right)$$
$$= 30 + \frac{1}{15}\left[\left(\frac{6}{\pi}\right)(15)\cos\frac{15\pi}{6} + \frac{6}{\pi}(0)\cos 0 + \left(\frac{6}{\pi}\right)^2 \sin\frac{\pi}{6}t\Big|_0^{15}\right.$$
$$= 30 + \frac{1}{15}\left[-\left(\frac{6}{\pi}\right)(15)(0) + \left(\frac{6}{\pi}\right)^2 \sin\frac{15\pi}{6} - \left(\frac{6}{\pi}\right)^2 \sin 0\right]$$
$$= 30 + \frac{1}{15}\left(\frac{6}{\pi}\right)(1) \approx 30.24$$

or approximately \$30.24 per share.

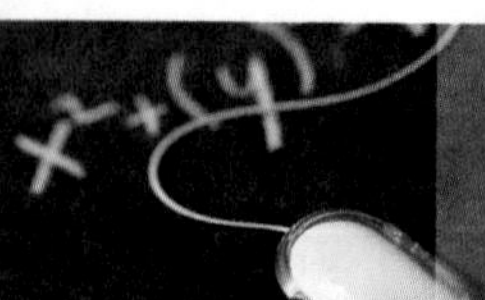

Self-Check Exercises 8.5

1. Evaluate $\int_0^2 \sqrt{2x + 5}\, dx$.

2. Find the average value of the function $f(x) = 1 - x^2$ over the interval $[-1, 2]$.

3. The median price of a house in a western province between January 1, 1995, and January 1, 2000, is approximated by the function

$$f(t) = t^3 - 7t^2 + 17t + 190 \qquad (0 \le t \le 5)$$

where $f(t)$ is measured in thousands of dollars and t is expressed in years, with $t = 0$ corresponding to the beginning of 1995. Determine the average median price of a house over that time interval.

Solutions to Self-Check Exercises 8.5 can be found on page 494.

8.5 Exercises

In Exercises 1–38, evaluate the given definite integral.

1. $\int_0^2 x(x^2 - 1)^3\, dx$
2. $\int_0^1 x^2(2x^3 - 1)^4\, dx$
3. $\int_0^1 x\sqrt{5x^2 + 4}\, dx$
4. $\int_1^3 x\sqrt{3x^2 - 2}\, dx$
5. $\int_0^2 x^2(x^3 + 1)^{3/2}\, dx$
6. $\int_1^5 (2x - 1)^{5/2}\, dx$
7. $\int_0^1 \frac{1}{\sqrt{2x + 1}}\, dx$
8. $\int_0^2 \frac{x}{\sqrt{x^2 + 5}}\, dx$
9. $\int_1^2 (2x - 1)^4\, dx$
10. $\int_1^2 (2x + 4)(x^2 + 4x - 8)^3\, dx$
11. $\int_{-1}^1 x^2(x^3 + 1)^4\, dx$
12. $\int_1^2 \left(x^3 + \frac{3}{4}\right)(x^4 + 3x)^{-2}\, dx$
13. $\int_1^5 x\sqrt{x - 1}\, dx$
14. $\int_1^4 x\sqrt{x + 1}\, dx$

    Hint: Let $u = x + 1$.
15. $\int_0^2 xe^{x^2}\, dx$
16. $\int_0^1 e^{-x}\, dx$
17. $\int_0^1 (e^{2x} + x^2 + 1)\, dx$
18. $\int_0^2 (e^t - e^{-t})\, dt$
19. $\int_{-1}^1 xe^{x^2+1}\, dx$
20. $\int_0^4 \frac{e^{\sqrt{x}}}{\sqrt{x}}\, dx$
21. $\int_3^6 \frac{2}{x - 2}\, dx$
22. $\int_0^1 \frac{x}{1 + 2x^2}\, dx$
23. $\int_1^2 \frac{x^2 + 2x}{x^3 + 3x^2 - 1}\, dx$
24. $\int_0^1 \frac{e^x}{1 + e^x}\, dx$
25. $\int_1^2 \left(4e^{2u} - \frac{1}{u}\right) du$
26. $\int_1^2 \left(1 + \frac{1}{x} + e^x\right) dx$
27. $\int_1^2 \left(2e^{-4x} - \frac{1}{x^2}\right) dx$
28. $\int_1^2 \frac{\ln x}{x}\, dx$
29. $\int_1^{\sqrt{3}} \frac{1}{x^2 + 1}\, dx$
30. $\int_{-\pi/2}^{\pi/3} \cos(2x)\, dx$
31. $\int_{-3\pi/2}^{3\pi/2} \sin(t)\, dt$
32. $\int_0^{\pi/3} \sec^2(u)\, du$
33. $\int_0^1 \frac{4z}{z^4 + 1}\, dz$
34. $\int_{-2}^2 x\cos(2x^2)\, dx$
35. $\int_0^1 \sec(x/2)\, dx$
36. $\int_0^{\pi} \frac{\sin(2x)}{\sin(x)}\, dx$

    Hint: Simplify with a trig identity before integrating.
37. $\int_{-\pi/4}^{\pi/3} \tan(u)\, du$
38. $\int_0^{\pi^2/4} \frac{\sin(\sqrt{x})}{\sqrt{x}}\, dx$

In Exercises 39–54, find the average value of the given function f over the indicated interval $[a, b]$.

39. $f(x) = 2x + 3$; $[0, 2]$
40. $f(x) = 8 - x$; $[1, 4]$
41. $f(x) = 2x^2 - 3$; $[1, 3]$
42. $f(x) = 4 - x^2$; $[-2, 3]$
43. $f(x) = x^2 + 2x - 3$; $[-1, 2]$
44. $f(x) = x^3$; $[-1, 1]$
45. $f(x) = \sqrt{2x + 1}$; $[0, 4]$
46. $f(x) = e^{-x}$; $[0, 4]$
47. $f(x) = xe^{x^2}$; $[0, 2]$
48. $f(x) = \frac{1}{x + 1}$; $[0, 2]$
49. $f(x) = \sin(x)$; $[0, \pi]$
50. $f(x) = \tan(x)$; $[0, \pi/3]$
51. $f(x) = x\sin(1 - x^2)$; $[0, 3]$
52. $f(x) = \cos(4x - 5)$; $[-\pi/3, 2\pi/3]$
53. $f(x) = \sec^2(2x)$; $[-\pi/4, \pi/4]$
54. $f(x) = 2\sin(x)\cos(x)$; $[\pi/2, \pi]$

55. The position of a charged particle in a magnetic field is found to be

$$p(t) = 0.05t + 2\cos(1.7t)$$

left of a measuring mark after t seconds. Find the average position of the particle from $t = 0$ to $t = 4$.

56. The number of tons of leaves collected by the city of Barrie in the fall is found to be

$$\frac{175}{w^2 - 5w + 7.25} \text{ metric tons/week}$$

Find the average amount of leaves collected in the second week ($w = 1$ to $w = 2$), the third week ($w = 2$ to $w = 3$), and the fourth week ($w = 3$ to $w = 4$). Which week has the largest number of leaves?

57. Prove Property 1 of the definite integral.

    Hint: Let F be an antiderivative of f and use the definition of the definite integral.

58. Prove Property 2 of the definite integral.

    Hint: See Exercise 57.

59. Verify by direct computation that

$$\int_1^3 x^2 dx = -\int_3^1 x^2 dx$$

60. Prove Property 3 of the definite integral.

    Hint: See Exercise 57.

61. Verify by direct computation that

$$\int_1^9 2\sqrt{x}\, dx = 2\int_1^9 \sqrt{x}\, dx$$

62. Verify by direct computation that

$$\int_0^1 (1 + x - e^x)\,dx = \int_0^1 dx + \int_0^1 x\,dx - \int_0^1 e^x\,dx$$

What properties of the definite integral are demonstrated in this exercise?

63. Verify by direct computation that

$$\int_0^3 (1 + x^3)dx = \int_0^1 (1 + x^3)dx + \int_1^3 (1 + x^3)\,dx$$

What property of the definite integral is demonstrated here?

64. Verify by direct computation that

$$\int_0^3 (1 + x^3)\,dx$$

$$= \int_0^1 (1 + x^3)dx + \int_1^2 (1 + x^3)dx + \int_2^3 (1 + x^3)dx$$

hence showing that Property 5 may be extended.

65. Evaluate $\int_3^3 (1 + \sqrt{x})e^{-x}\,dx$.

66. Evaluate $\int_3^0 f(x)\,dx$, given that $\int_0^3 f(x)\,dx = 4$.

67. Evaluate $\int_0^3 4f(x)\,dx$, given that $\int_0^3 f(x)\,dx = -1$.

68. Given that $\int_{-1}^2 f(x)\,dx = -2$ and $\int_{-1}^2 g(x)\,dx = 3$, evaluate

a. $\int_{-1}^2 [2f(x) + g(x)]\,dx$

b. $\int_{-1}^2 [g(x) - f(x)]\,dx$

c. $\int_{-1}^2 [2f(x) - 3g(x)]\,dx$

69. Given that $\int_{-1}^2 f(x)\,dx = 2$ and $\int_0^2 f(x)\,dx = 3$, evaluate

a. $\int_{-1}^0 f(x)\,dx$

b. $\int_0^2 f(x)\,dx - \int_{-1}^0 f(x)\,dx$

70. **Newton's Law of Cooling** A bottle of white wine at room temperature (20°C) is placed in a refrigerator at 4 P.M. Its temperature after t hr is changing at the rate of

$$-18e^{-0.6t}$$

°C/h. By how many degrees will the temperature of the wine have dropped by 7 P.M.? What will the temperature of the wine be at 7 P.M.?

71. **Velocity of a Car** A car moves along a straight road in such a way that its velocity (in metres/second) at any time t (in seconds) is given by

$$v(t) = t\sqrt{16 - t^2} \qquad (0 \le t \le 4)$$

Find the distance traveled by the car in the 4 sec from $t = 0$ to $t = 4$.

72. **Average Velocity of a Truck** A truck travelling along a straight road has a velocity (in metres/second) at time t (in seconds) given by

$$v(t) = \frac{1}{36}t^2 + \frac{2}{3}t + 15 \qquad (0 \le t \le 5)$$

What is the average velocity of the truck over the time interval from $t = 0$ to $t = 5$?

73. **Average Temperature** The temperature (in °C) in Toronto over a 12-hr period on a certain December day was given by

$$T = -0.05t^3 + 0.4t^2 + 3.8t - 16.4 \qquad (0 \le t \le 12)$$

where t is measured in hours, with $t = 0$ corresponding to 6 A.M. Determine the average temperature on that day over the 12-hr period from 6 A.M. to 6 P.M.

74. **Velocity of a Car** Refer to Exercise 71. Find the average velocity of the car over the time interval [0, 4].

Business and Economics Applications

75. **Stock Prices** The weekly closing price of TMA Corporation stock in week t is approximated by the rule

$$f(t) = 80 + 3t\cos\frac{\pi t}{6} \qquad (0 \le t \le 15)$$

where $f(t)$ is the price (in dollars) per share. Find the average weekly closing price of the stock over the 15-week period.

76. **Restaurant Revenue** The revenue of McMenamy's Fish Shanty, located at a popular summer resort, is approximately

$$R(t) = 2\left(5 - 4\cos\frac{\pi}{6}t\right) \qquad (0 \le t \le 12)$$

during the tth week ($t = 1$ corresponds to the first week of June), where R is measured in thousands of dollars. What is the total revenue realized by the restaurant over the 12-week period starting June 1?

77. **Stock Prices** Refer to Exercise 60, Section 6.4. What was the average price per share of Tempco Electronics stock over the first 20 days that it was traded on the stock exchange?

78. **World Production of Coal** A study proposed in 1980 by researchers from the major producers and consumers of the world's coal concluded that coal could and must play an important role in fueling global economic growth over the next 20 yr. The world production of coal in 1980 was 3.5 billion metric tons. If output increased at the rate of $3.5e^{0.05t}$ billion metric tons/year in year t ($t = 0$ corresponding to 1980), determine how much coal was produced worldwide between 1980 and the end of the twentieth century.

79. **Net Investment Flow** The net investment flow (rate of capital formation) of the giant conglomerate LTF incorporated is projected to be

$$t\sqrt{\frac{1}{2}t^2 + 1}$$

million dollars/year in year t. Find the accruement on the company's capital stock in the second year.
Hint: The amount is given by

$$\int_1^2 t\sqrt{\frac{1}{2}t^2 + 1}\,dt$$

80. **Average Price of a Commodity** The price of a certain commodity in dollars/unit at time t (measured in weeks) is given by

$$p = 18 - 3e^{-2t} - 6e^{-t/3}$$

What is the average price of the commodity over the 5-wk period from $t = 0$ to $t = 5$?

81. Oil Production Based on a preliminary report by a geological survey team, it is estimated that a newly discovered oil field can be expected to produce oil at the rate of

$$R(t) = \frac{600t^2}{t^3 + 32} + 5 \qquad (0 \le t \le 20)$$

thousand barrels/year, t yr after production begins. Find the amount of oil that the field can be expected to yield during the first 5 yr of production, assuming that the projection holds true.

82. Depreciation: Double Declining-Balance Method Suppose a tractor purchased at a price of \$60,000 is to be depreciated by the *double declining-balance method* over a 10-yr period. It can be shown that the rate at which the book value will be decreasing is given by

$$R(t) = 13388.61e^{-0.22314t} \qquad (0 \le t \le 10)$$

dollars/year at year t. Find the amount by which the book value of the tractor will depreciate over the first 5 yr of its life.

83. Cable TV Subscribers The manager of TeleStar Cable Service estimates that the total number of subscribers to the service in a certain city t yr from now will be

$$N(t) = -\frac{40{,}000}{\sqrt{1 + 0.2t}} + 50{,}000$$

Find the average number of cable television subscribers over the next 5 yr if this prediction holds true.

84. Average Yearly Sales The sales of Universal Instruments in the first t yr of its operation are approximated by the function

$$S(t) = t\sqrt{0.2t^2 + 4}$$

where $S(t)$ is measured in millions of dollars. What were Universal's average yearly sales over its first 5 yr of operation?

Biological and Life Sciences Applications

85. Whale Population A group of marine biologists estimates that if certain conservation measures are implemented, the population of an endangered species of whale will be

$$N(t) = 3t^3 + 2t^2 - 10t + 600 \qquad (0 \le t \le 10)$$

where $N(t)$ denotes the population at the end of year t. Find the average population of the whales over the next 10 yr.

86. Average Daily Temperature The average daily temperature in degrees Fahrenheit at a tourist resort in the Cameron Highlands is approximately

$$T = 62 - 18\cos\frac{2\pi(t - 23)}{365}$$

on the tth day ($t = 1$ corresponds to January 1). What is the average temperature at the resort in the month of January?
Hint: Find the average value of T over the interval [0, 31].

87. Volume of Air Inhaled During Respiration Suppose that the rate of air flow in and out of a person's lungs during respiration is

$$R(t) = 0.6\sin\frac{\pi t}{2}$$

liters per second, where t is the time in seconds. Find an expression for the volume of air in the person's lungs at any time t.

88. Average Volume of Inhaled Air During Respiration Refer to Exercise 87. What is the average volume of inspired air in the person's lungs during one respiratory cycle?
Hint: Find the average value of the function giving the volume of inhaled air (obtained in the solution to Exercise 75) over the interval [0, 4]

89. Concentration of a Drug in the Bloodstream The concentration of a certain drug in a patient's bloodstream t hr after injection is

$$C(t) = \frac{0.2t}{t^2 + 1}$$

mg/cm^3. Determine the average concentration of the drug in the patient's bloodstream over the first 4 hr after the drug is injected.

90. Flow of Blood in an Artery According to a law discovered by nineteenth-century physician Jean Louis Marie Poiseuille, the velocity of blood (in centimetres/second) r cm from the central axis of an artery is given by

$$v(r) = k(R^2 - r^2)$$

where k is a constant and R is the radius of the artery. Find the average velocity of blood along a radius of the artery (see the accompanying figure).

Hint: Evaluate $\frac{1}{R}\int_0^R v(r)\,dr$.

Blood vessel

91. Cancer Treatment For a cancer treatment to be effective, the level of a hormone in the bloodstream must have an average level of less than 36 milligrams while the treatment is being applied. The level in the blood is found to be close to

$$21 + 10\cos(1.23t) + 7\sin(0.76t)\text{mg}$$

at a time t hours after the patient eats breakfast together with a preparatory medication. Is it safe to have the treatment during the 4 hours after breakfast?

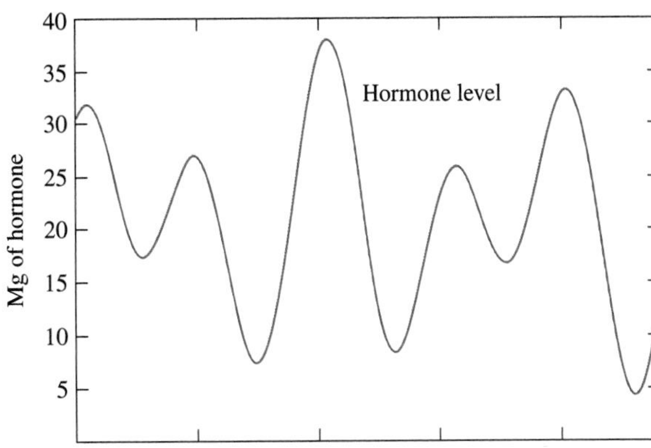

92. Average Hormone Level This problem refers to Exercise 91. A graph of the level of the hormone in the blood for a 24-hour period starting at $t = 0$ is shown above. Using the graph estimate which 4-hour period has the lowest average hormone level and compute the average level during that period.

In Exercises 93–102, determine whether the statement is true or false. If it is true, explain why it is true. If it is false, give an example to show why it is false.

93. $\displaystyle\int_2^2 \frac{e^x}{\sqrt{1+x}}\,dx = 0$

94. $\displaystyle\int_1^3 \frac{dx}{x-2} = -\int_3^1 \frac{dx}{x-2}$

95. $\displaystyle\int_0^1 x\sqrt{x+1}\,dx = \sqrt{x+1}\int_0^1 x\,dx = \frac{1}{2}x^2\sqrt{x+1}\Big|_0^1 = \frac{\sqrt{2}}{2}$

96. If f' is continuous on [0, 2], then $\displaystyle\int_0^2 f'(x)\,dx = f(2) - f(0)$.

97. If f and g are continuous on $[a, b]$ and k is a constant, then

$$\int_a^b [kf(x) + g(x)]\,dx = k\int_a^b f(x)\,dx + \int_a^b g(x)\,dx$$

98. If f is continuous on $[a, b]$ and $a < c < b$, then

$$\int_b^c f(x)\,dx = \int_a^c f(x)\,dx - \int_a^b f(x)\,dx$$

99. $\displaystyle\int_a^b \sin x\,dx = \int_{a+2\pi}^{b+2\pi} \sin x\,dx$

100. $\displaystyle\int_a^b \cos x\,dx = \int_a^{b+2\pi} \cos x\,dx$

101. $\displaystyle\int_{-\pi/2}^{\pi/2} \sqrt[3]{x}\sin 2x\,dx > 0$

102. $\displaystyle\int_{-\pi/2}^{\pi/2} |\sin x|\,dx = \int_{-\pi/2}^{\pi/2} |\cos x|\,dx$

Solutions to Self-Check Exercises 8.5

1. Let $u = 2x + 5$. Then, $du = 2\,dx$, or $dx = \frac{1}{2}\,du$. Also, when $x = 0$, $u = 5$, and when $x = 2$, $u = 9$. Therefore,

$$\begin{aligned}\int_0^2 \sqrt{2x+5}\,dx &= \int_0^2 (2x+5)^{1/2}\,dx\\ &= \frac{1}{2}\int_5^9 u^{1/2}\,du\\ &= \left(\frac{1}{2}\right)\left(\frac{2}{3}u^{3/2}\right)\Big|_5^9\\ &= \frac{1}{3}[9^{3/2} - 5^{3/2}]\\ &= \frac{1}{3}(27 - 5\sqrt{5})\end{aligned}$$

2. The required average value is given by

$$\begin{aligned}\frac{1}{2-(-1)}\int_{-1}^2 (1-x^2)\,dx &= \frac{1}{3}\int_{-1}^2 (1-x^2)\,dx\\ &= \frac{1}{3}\left(x - \frac{1}{3}x^3\right)\Big|_{-1}^2\\ &= \frac{1}{3}\left[\left(2 - \frac{8}{3}\right) - \left(-1 + \frac{1}{3}\right)\right] = 0\end{aligned}$$

3. The average median price of a house over the stated time interval is given by

$$\begin{aligned}&\frac{1}{5-0}\int_0^5 (t^3 - 7t^2 + 17t + 190)\,dt\\ &= \frac{1}{5}\left(\frac{1}{4}t^4 - \frac{7}{3}t^3 + \frac{17}{2}t^2 + 190t\right)\Big|_0^5\\ &= \frac{1}{5}\left[\frac{1}{4}(5)^4 - \frac{7}{3}(5)^3 + \frac{17}{2}(5)^2 + 190(5)\right]\\ &= 205.417\end{aligned}$$

or \$205,417.

8.6 Area between Two Curves

Suppose a certain country's petroleum consumption is expected to grow at the rate of $f(t)$ million barrels per year, t years from now, for the next 5 years. Then, the country's total petroleum consumption over the period of time in question is given by the area under the graph of f on the interval $[0, 5]$ (Figure 8.28).

Next, suppose that because of the implementation of certain energy-conservation measures, the rate of growth of petroleum consumption is expected to be $g(t)$ million barrels per year instead. Then, the country's projected total petroleum consumption over the 5-year period is given by the area under the graph of g on the interval $[0, 5]$ (Figure 8.29).

Therefore, the area of the shaded region S lying between the graphs of f and g on the interval $[0, 5]$ (Figure 8.30) gives the amount of petroleum that would be saved over the 5-year period because of the conservation measures.

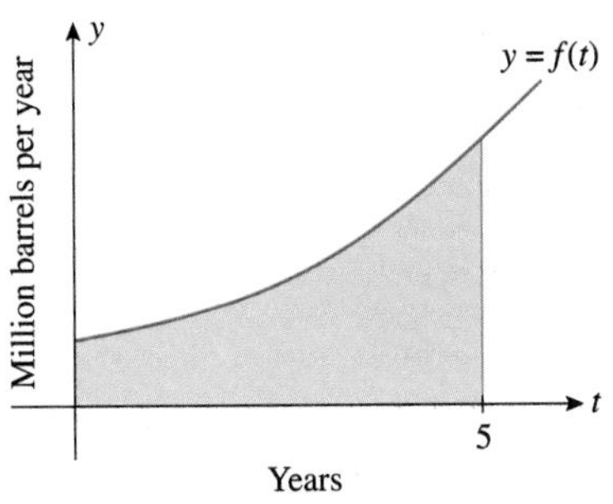

Figure 8.28
At a rate of consumption $f(t)$ million barrels per year, the total petroleum consumption is given by the area of the region under the graph of f.

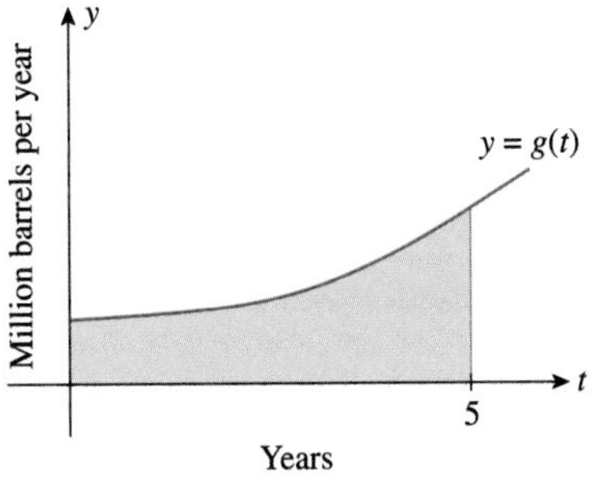

Figure 8.29
At a rate of consumption of $g(t)$ million barrels per year, the total petroleum consumption is given by the area of the region under the graph of g.

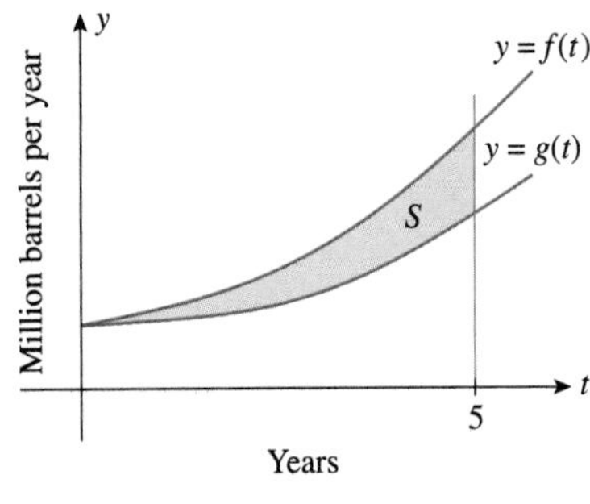

Figure 8.30
The area of S gives the amount of petroleum that would be saved over the 5-year period.

But the area of S is given by

Area under the graph of f on $[a, b]$ − Area under the graph of g on $[a, b]$

$$= \int_0^5 f(t)\,dt - \int_0^5 g(t)\,dt$$

$$= \int_0^5 [f(t) - g(t)]\,dt \qquad \text{By Property 4, Section 8.5}$$

This example shows that some practical problems can be solved by finding the area of a region between two curves, which in turn can be found by evaluating an appropriate definite integral.

Finding the Area between Two Curves

We now turn our attention to the general problem of finding the area of a plane region bounded both above and below by the graphs of functions. First, consider the situation in which the graph of one function lies above that of another. More specifically, let R be the region in the xy-plane (Figure 8.31) that is bounded above by the graph of a continuous function f, below by a continuous function g where $f(x) \geq g(x)$ on $[a, b]$ and to the left and right by the vertical lines $x = a$ and $x = b$, respectively. From the figure, we see that

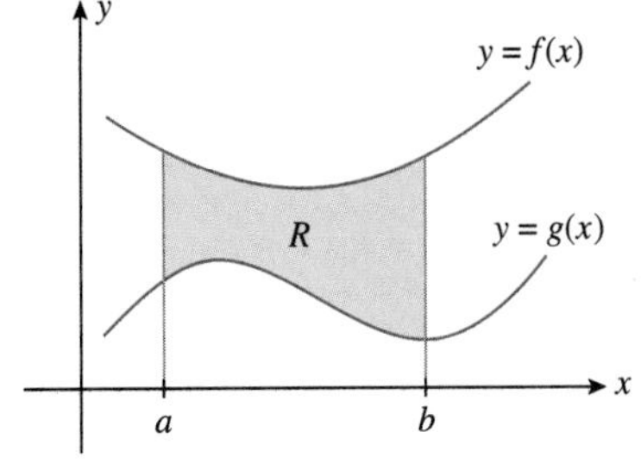

Figure 8.31
Area of $R = \int_a^b [f(x) - g(x)]\,dx$

$$\text{Area of } R = \text{Area under } f(x) - \text{Area under } g(x)$$

$$= \int_a^b f(x)\,dx - \int_a^b g(x)\,dx$$

$$= \int_a^b [f(x) - g(x)]\, dx$$

upon using Property 4 of the definite integral.

The Area between Two Curves

Let f and g be continuous functions such that $f(x) \geq g(x)$ on the interval $[a, b]$. Then, the area of the region bounded above by $y = f(x)$ and below by $y = g(x)$ on $[a, b]$ is given by

$$\int_a^b [f(x) - g(x)]\, dx \qquad (12)$$

Even though we assumed that both f and g were nonnegative in the derivation of (12), it may be shown that this equation is valid if f and g are not nonnegative (see Exercise 51). Also, observe that if $g(x)$ is 0 for all x—that is, when the lower boundary of the region R is the x-axis—Equation (12) gives the area of the region under the curve $y = f(x)$ from $x = a$ to $x = b$, as we would expect.

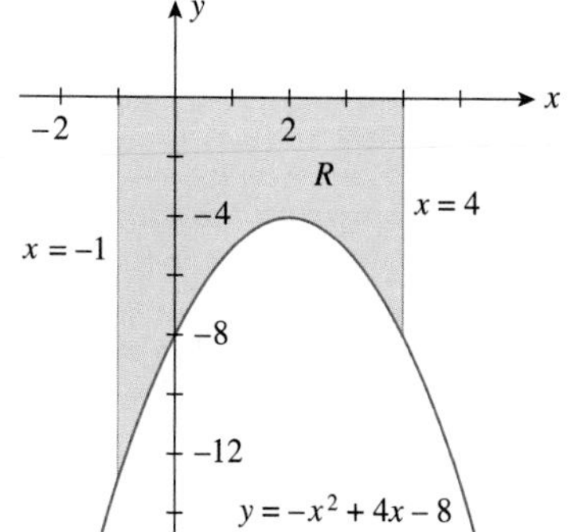

Figure 8.32
Area of $R = -\int_{-1}^{4} g(x)\, dx$

Example 1

Find the area of the region bounded by the x-axis, the graph of $y = -x^2 + 4x - 8$, and the lines $x = -1$ and $x = 4$.

Solution

The region R under consideration is shown in Figure 8.32. We can view R as the region bounded above by the graph of $f(x) = 0$ (the x-axis) and below by the graph of $g(x) = -x^2 + 4x - 8$ on $[-1, 4]$. Therefore, the area of R is given by

$$\int_a^b [f(x) - g(x)]\, dx = \int_{-1}^{4} [0 - (-x^2 + 4x - 8)]\, dx$$

$$= \int_{-1}^{4} (x^2 - 4x + 8)\, dx$$

$$= \frac{1}{3}x^3 - 2x^2 + 8x \Big|_{-1}^{4}$$

$$= \left[\frac{1}{3}(64) - 2(16) + 8(4)\right] - \left[\frac{1}{3}(-1) - 2(1) + 8(-1)\right]$$

$$= 31\tfrac{2}{3}$$

or $31\frac{2}{3}$ square units.

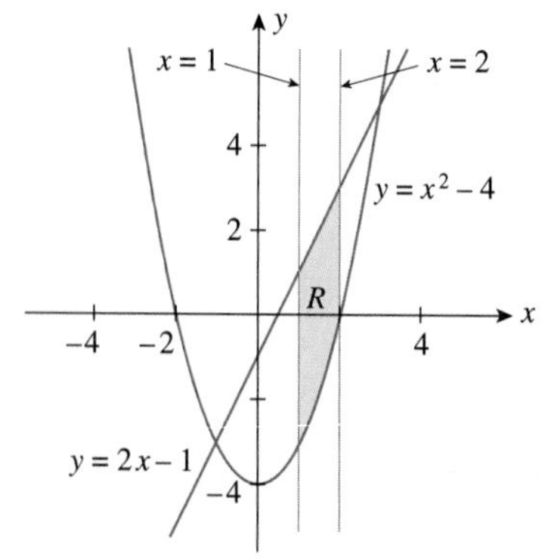

Figure 8.33
Area of $R = \int_1^2 [f(x) - g(x)]\, dx$

Example 2

Find the area of the region R bounded by the graphs of

$$f(x) = 2x - 1 \qquad \text{and} \qquad g(x) = x^2 - 4$$

and the vertical lines $x = 1$ and $x = 2$.

Solution

We first sketch the graphs of the functions $f(x) = 2x - 1$ and $g(x) = x^2 - 4$ and the vertical lines $x = 1$ and $x = 2$, and then we identify the region R whose area is to be calculated (Figure 8.33).

Since the graph of f always lies above that of g for x in the interval [1, 2], we see by Equation (12) that the required area is given by

$$\begin{aligned}\int_1^2 [f(x) - g(x)]\,dx &= \int_1^2 [(2x-1) - (x^2-4)]\,dx \\ &= \int_1^2 (-x^2 + 2x + 3)\,dx \\ &= -\frac{1}{3}x^3 + x^2 + 3x\Big|_1^2 \\ &= \left(-\frac{8}{3} + 4 + 6\right) - \left(-\frac{1}{3} + 1 + 3\right) = \frac{11}{3}\end{aligned}$$

or $\frac{11}{3}$ square units.

Example 3

Find the area of the region R that is completely enclosed by the graphs of the functions

$$f(x) = 2x - 1 \qquad \text{and} \qquad g(x) = x^2 - 4$$

Solution

The region R is shown in Figure 8.34. First, we find the points of intersection of the two curves. To do this, we solve the system that consists of the two equations $y = 2x - 1$ and $y = x^2 - 4$. Equating the two values of y gives

$$\begin{aligned}x^2 - 4 &= 2x - 1 \\ x^2 - 2x - 3 &= 0 \\ (x+1)(x-3) &= 0\end{aligned}$$

so $x = -1$ or $x = 3$. That is, the two curves intersect when $x = -1$ and $x = 3$.

Observe that we could also view the region R as the region bounded above by the graph of the function $f(x) = 2x - 1$, below by the graph of the function $g(x) = x^2 - 4$, and to the left and right by the vertical lines $x = -1$ and $x = 3$, respectively.

Next, since the graph of the function f always lies above that of the function g on $[-1, 3]$, we can use (12) to compute the desired area:

$$\begin{aligned}\int_a^b [f(x) - g(x)]\,dx &= \int_{-1}^3 [(2x-1) - (x^2-4)]\,dx \\ &= \int_{-1}^3 (-x^2 + 2x + 3)\,dx \\ &= -\frac{1}{3}x^3 + x^2 + 3x\Big|_{-1}^3 \\ &= (-9 + 9 + 9) - \left(\frac{1}{3} + 1 - 3\right) = \frac{32}{3} \\ &= 10\tfrac{2}{3}\end{aligned}$$

or $10\frac{2}{3}$ square units.

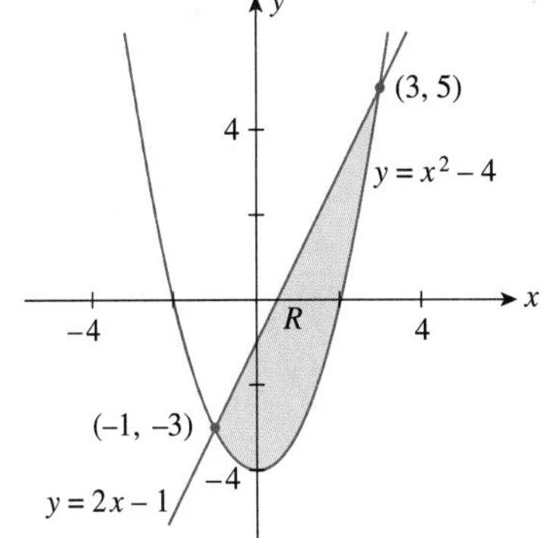

Figure 8.34
Area of $R = \int_{-1}^{3}[f(x) - g(x)]\,dx$

Example 4

Find the area of the region R bounded by the graphs of the functions

$$f(x) = x^2 - 2x - 1 \qquad \text{and} \qquad g(x) = -e^x - 1$$

and the vertical lines $x = -1$ and $x = 1$.

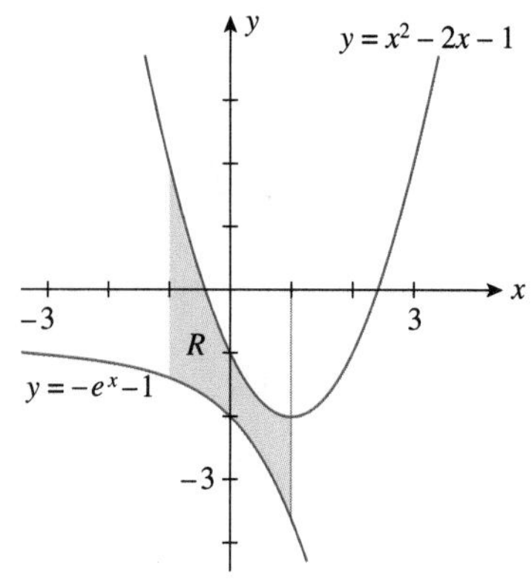

Figure 8.35
Area of R = $\int_{-1}^{1}[f(x) - g(x)]\,dx$

Solution

The region R is shown in Figure 8.35. Since the graph of the function f always lies above that of the function g, the area of the region R is given by

$$\begin{aligned}\int_a^b [f(x) - g(x)]\,dx &= \int_{-1}^{1} [(x^2 - 2x - 1) - (-e^x - 1)]\,dx \\ &= \int_{-1}^{1} (x^2 - 2x + e^x)\,dx \\ &= \frac{1}{3}x^3 - x^2 + e^x \Big|_{-1}^{1} \\ &= \left(\frac{1}{3} - 1 + e\right) - \left(-\frac{1}{3} - 1 + e^{-1}\right) \\ &= \frac{2}{3} + e - \frac{1}{e}, \text{ or } 3.02 \text{ square units}\end{aligned}$$

Equation (12), which gives the area of the region between the curves $y = f(x)$ and $y = g(x)$ for $a \le x \le b$, is valid when the graph of the function f lies above that of the function g over the interval $[a, b]$. Example 5 shows how to use (12) to find the area of a region when the latter condition does not hold.

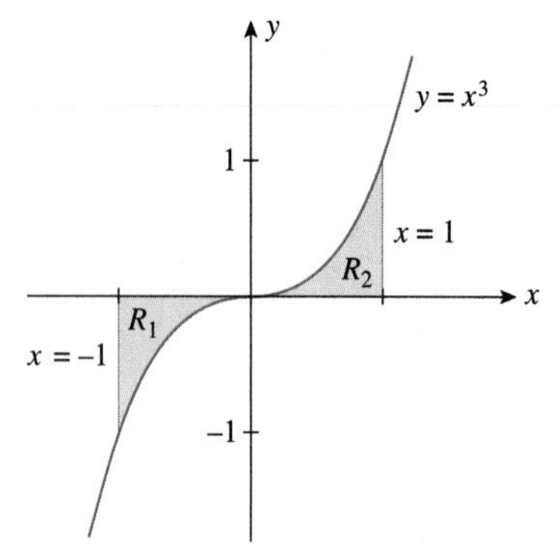

Figure 8.36
Area of R_1 = Area of R_2

Example 5

Find the area of the region bounded by the graph of the function $f(x) = x^3$, the x-axis, and the lines $x = -1$ and $x = 1$.

Solution

The region R under consideration can be thought of as composed of the two subregions R_1 and R_2, as shown in Figure 8.36.

Recall that the x-axis is represented by the function $g(x) = 0$. Since $g(x) \ge f(x)$ on $[-1, 0]$, we see that the area of R_1 is given by

$$\begin{aligned}\int_a^b [g(x) - f(x)]\,dx &= \int_{-1}^{0} (0 - x^3)\,dx = -\int_{-1}^{0} x^3\,dx \\ &= -\frac{1}{4}x^4\Big|_{-1}^{0} = 0 - \left(-\frac{1}{4}\right) = \frac{1}{4}\end{aligned}$$

To find the area of R_2, we observe that $f(x) \ge g(x)$ on $[0, 1]$, so it is given by

$$\begin{aligned}\int_a^b [f(x) - g(x)]\,dx &= \int_0^1 (x^3 - 0)\,dx = \int_0^1 x^3\,dx \\ &= \frac{1}{4}x^4\Big|_0^1 = \left(\frac{1}{4}\right) - 0 = \frac{1}{4}\end{aligned}$$

Therefore, the area of R is $\frac{1}{4} + \frac{1}{4}$, or $\frac{1}{2}$, square units.

By making use of symmetry, we could have obtained the same result by computing

$$-2\int_{-1}^{0} x^3\,dx \qquad \text{or} \qquad 2\int_0^1 x^3\,dx$$

as you may verify.

GROUP DISCUSSION

Recall from Section 2.2 that a function is *even* if it satisfies the condition $f(-x) = f(x)$, and it is *odd* if it satisfies the condition $f(-x) = -f(x)$. Show that the graph of an even function is symmetric with respect to the y-axis while the graph of an odd function is symmetric with respect to the origin. Explain why

$$\int_{-a}^{a} f(x)\,dx = 2\int_{0}^{a} f(x)\,dx \quad \text{if } f \text{ is even}$$

$$\int_{-a}^{a} f(x)\,dx = 0 \qquad \text{if } f \text{ is odd}$$

Example 6

Find the area of the region completely enclosed by the graphs of the functions

$$f(x) = x^3 - 3x + 3 \qquad \text{and} \qquad g(x) = x + 3$$

Solution

First, sketch the graphs of $y = x^3 - 3x + 3$ and $y = x + 3$ and then identify the required region R. We can view the region R as being composed of the two subregions R_1 and R_2, as shown in Figure 8.37. By solving the equations $y = x + 3$ and $y = x^3 - 3x + 3$ simultaneously, we find the points of intersection of the two curves. Equating the two values of y, we have

$$\begin{aligned} x^3 - 3x + 3 &= x + 3 \\ x^3 - 4x &= 0 \\ x(x^2 - 4) &= 0 \\ x(x + 2)(x - 2) &= 0 \\ x &= 0, -2, 2 \end{aligned}$$

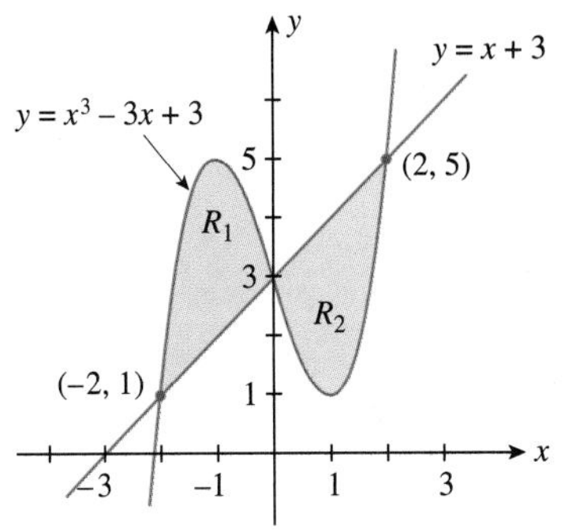

Figure 8.37
Area of R_1 + Area of R_2

$$= \int_{-2}^{0} [f(x) - g(x)]\,dx + \int_{0}^{2} [g(x) - f(x)]\,dx$$

Hence, the points of intersection of the two curves are $(-2, 1)$, $(0, 3)$, and $(2, 5)$.

For $-2 \leq x \leq 0$, we see that the graph of the function f lies above that of the function g, so the area of the region R_1 is, by virtue of (12),

$$\begin{aligned} \int_{-2}^{0} [(x^3 - 3x + 3) - (x + 3)]\,dx &= \int_{-2}^{0} (x^3 - 4x)\,dx \\ &= \frac{1}{4}x^4 - 2x^2 \Big|_{-2}^{0} \\ &= -(4 - 8) \\ &= 4 \end{aligned}$$

or 4 square units. For $0 \leq x \leq 2$, the graph of the function g lies above that of the function f, and the area of R_2 is given by

$$\begin{aligned} \int_{0}^{2} [(x + 3) - (x^3 - 3x + 3)]\,dx &= \int_{0}^{2} (-x^3 + 4x)\,dx \\ &= -\frac{1}{4}x^4 + 2x^2 \Big|_{0}^{2} \\ &= -4 + 8 \\ &= 4 \end{aligned}$$

or 4 square units. Therefore, the required area is the sum of the area of the two regions $R_1 + R_2$—that is, $4 + 4$, or 8 square units.

Application

Example 7

Conservation of Oil In a 1994 study for a developing country's Economic Development Board, government economists and energy experts concluded that if the Energy Conservation Bill were implemented in 1995, the country's oil consumption for the next 5 years would be expected to grow in accordance with the model

$$R(t) = 20e^{0.05t}$$

where t is measured in years ($t = 0$ corresponding to the year 1995) and $R(t)$ in millions of barrels per year. Without the government-imposed conservation measures, however, the expected rate of growth of oil consumption would be given by

$$R_1(t) = 20e^{0.08t}$$

millions of barrels per year. Using these models, determine how much oil would have been saved from 1995 through 2000 if the bill had been implemented.

Solution

Under the Energy Conservation Bill, the total amount of oil that would have been consumed between 1995 and 2000 is given by

$$\int_0^5 R(t)\,dt = \int_0^5 20e^{0.05t}\,dt \tag{13}$$

Without the bill, the total amount of oil that would have been consumed between 1995 and 2000 is given by

$$\int_0^5 R_1(t)\,dt = \int_0^5 20e^{0.08t}\,dt \tag{14}$$

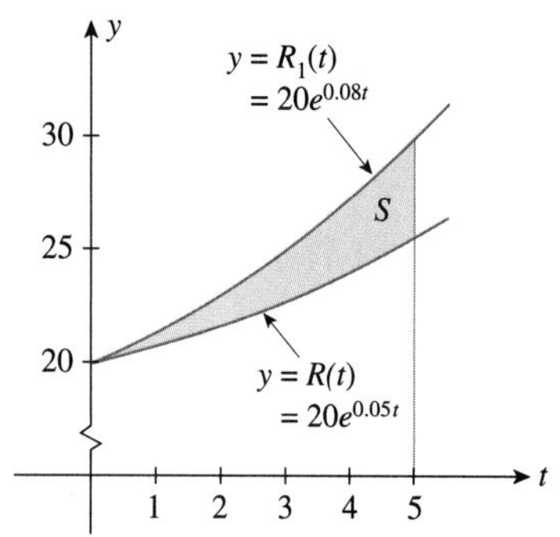

Figure 8.38
Area of $S = \int_0^5 [R_1(t) - R(t)]\,dt$

Equation (13) may be interpreted as the area of the region under the curve $y = R(t)$ from $t = 0$ to $t = 5$. Similarly, we interpret (14) as the area of the region under the curve $y = R_1(t)$ from $t = 0$ to $t = 5$. Furthermore, note that the graph of $y = R_1(t) = 20e^{0.08t}$ always lies on or above the graph of $y = R(t) = 20e^{0.05t}$ ($t \geq 0$). Thus, the area of the shaded region S in Figure 8.38 shows the amount of oil that would have been saved from 1995 to 2000 if the Energy Conservation Bill had been implemented. But the area of the region S is given by

$$\begin{aligned}\int_0^5 [R_1(t) - R(t)]\,dt &= \int_0^5 [20e^{0.08t} - 20^{0.05t}]\,dt \\ &= 20\int_0^5 (e^{0.08t} - e^{0.05t})\,dt \\ &= 20\left(\frac{e^{0.08t}}{0.08} - \frac{e^{0.05t}}{0.05}\right)\Bigg|_0^5 \\ &= 20\left[\left(\frac{e^{0.4}}{0.08} - \frac{e^{0.25}}{0.05}\right) - \left(\frac{1}{0.08} - \frac{1}{0.05}\right)\right] \\ &\approx 9.3\end{aligned}$$

or approximately 9.3 square units. Thus, the amount of oil that would have been saved is 9.3 million barrels.

EXPLORING WITH TECHNOLOGY

Refer to Example 7. Suppose we want to construct a mathematical model giving the amount of oil saved from 1995 through the year 1995 + x, where $x \geq 0$. For example, in Example 7, $x = 5$.

1. Show that this model is given by

$$F(x) = \int_0^x [R_1(t) - R(t)]\,dt$$
$$= 250e^{0.08x} - 400e^{0.05x} + 150$$

Hint: You may find it helpful to use some of the results of Example 7.

2. Use a graphing utility to plot the graph of F, using the viewing window [0, 10] · [0, 50].
3. Find $F(5)$ and thus confirm the result of Example 7.
4. What is the main advantage of this model?

Example 8

Intense sunlight, concentrated by reflectors, can decompose dioxins. The measured inflow of dioxin into a solar reactor intended to inactivate it is $I(t) = 212e^{-0.49t}$ g/h. The outflow of harmless compounds is $J(t) = 198e^{-0.51t}$ g/h. Material remaining in the reactor may be incompletely processed. Compute how many grams of material are still in the reactor after 20 hours.

Solution

The inflow per hour of dioxin to the reactor for the first 20 hours is

$$\int_0^{20} 212e^{-0.49t}\,dt$$

grams while the amount of treated compounds that left during the first 20 hours is

$$\int_0^{20} 198e^{-0.51t}\,dt$$

grams. The amount of compounds remaining in the reactor is given by the difference of $I(t)$ and $J(t)$, the area between these two curves shown in Figure 8.39. The area is given by

$$\begin{aligned}
\int_0^{20} (I(t) - J(t))\,dt &= \int_0^{20} (212e^{-0.49t} - 198e^{-0.51t})\,dt \\
&= \frac{212}{-0.49}e^{-0.49t} - \frac{198}{-0.51}e^{-0.51t}\Big|_0^{20} \\
&= -432.7e^{-0.49t} + 388.2e^{-0.51t}\Big|_0^{20} \\
&= -432.7e^{-9.8} + 388.2e^{-10.2} - (-432.7e^{0} + 388.2e^{0}) \\
&\approx -0.024 + 0.014 + 432.7 - 388.2 \\
&\approx 44.5
\end{aligned}$$

So there are approximately 44.5 grams of material remaining in the reactor after 20 hours.

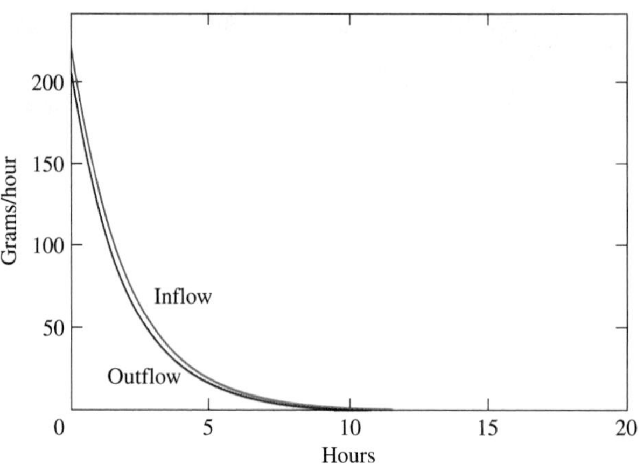

Figure 8.39

The difference of the inflow to the reactor and the outflow is the amount of compounds remaining in the reactor.

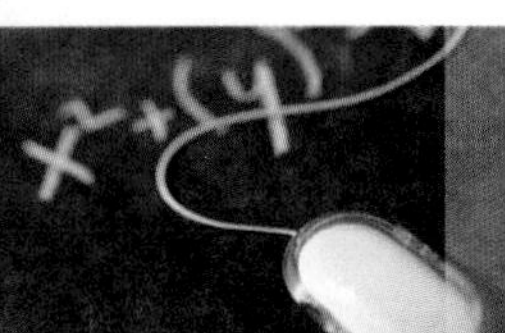

Self-Check Exercises 8.6

1. Find the area of the region bounded by the graphs of $f(x) = x^2 + 2$ and $g(x) = 1 - x$ and the vertical lines $x = 0$ and $x = 1$.
2. Find the area of the region completely enclosed by the graphs of $f(x) = -x^2 + 6x + 5$ and $g(x) = x^2 + 5$.
3. The management of Kane Corporation, which operates a chain of hotels, expects its profits to grow at the rate of $1 + t^{2/3}$ million dollars/year t yr from now. However, with renovations and improvements of existing hotels and proposed acquisitions of new hotels, Kane's profits are expected to grow at the rate of $t - 2\sqrt{t} + 4$ million dollars/year in the next decade. What additional profits are expected over the next 10 yr if the group implements the proposed plans?

Solutions to Self-Check Exercises 8.6 can be found on page 506.

8.6 Exercises

In Exercises 1–8, find the area of the shaded region.

1.

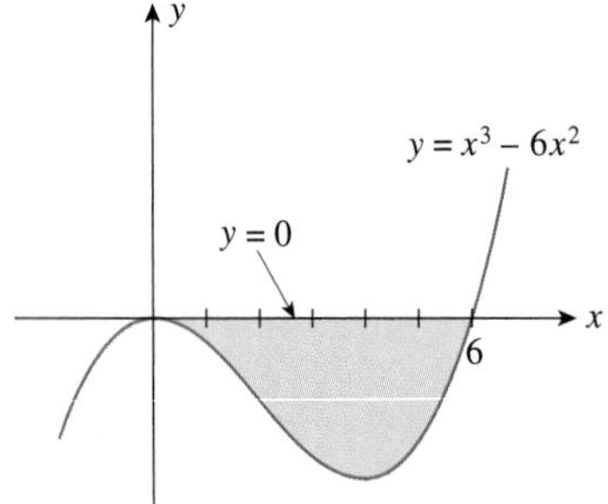

2.

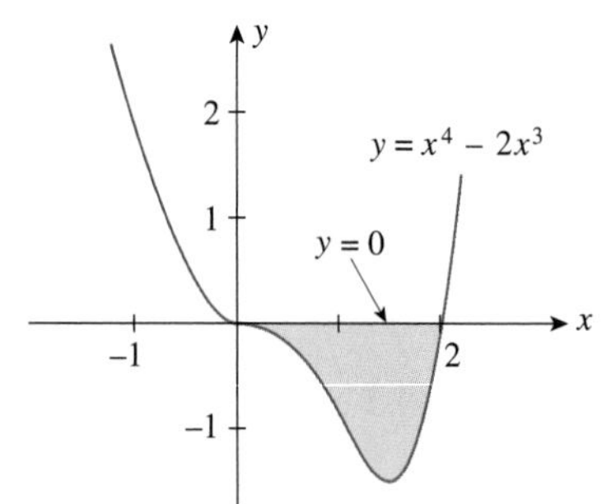

3.

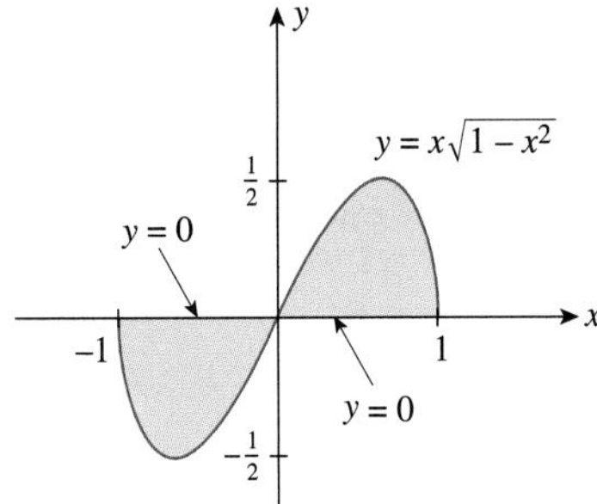

4.

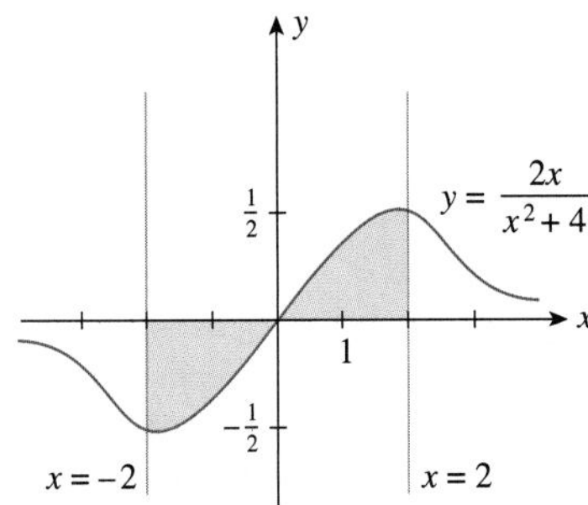

5.

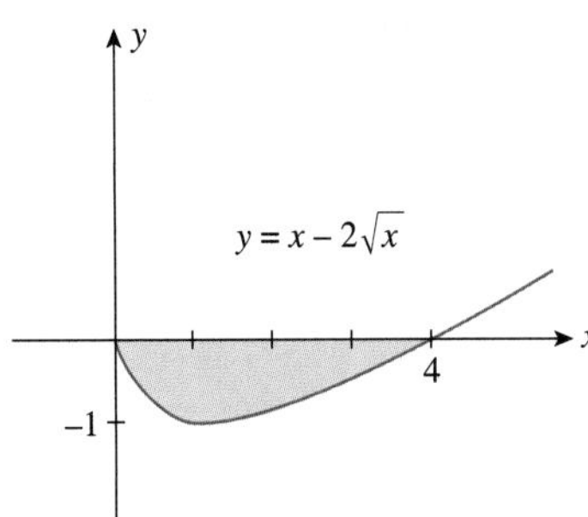

6.

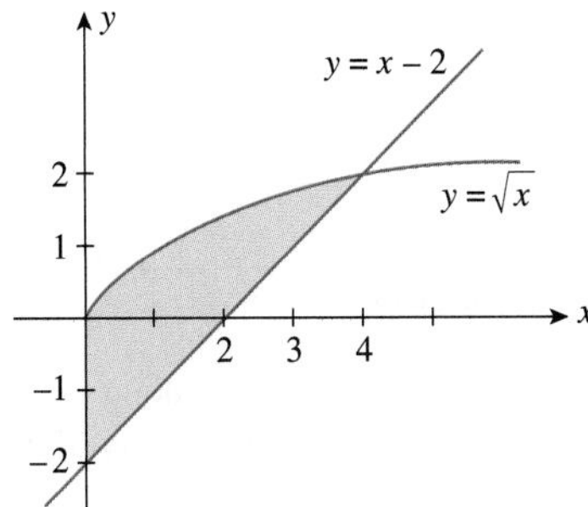

7.

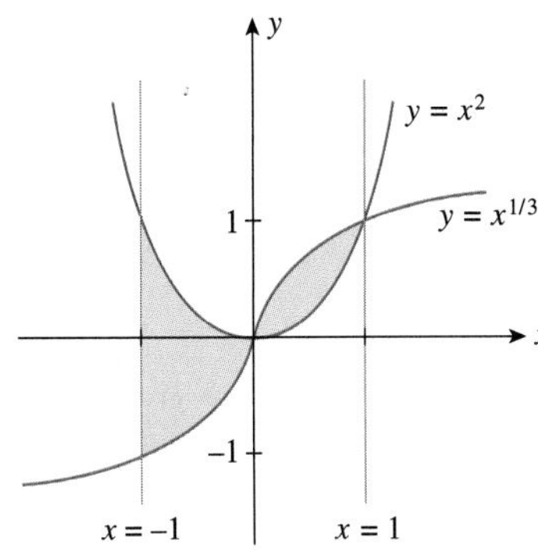

8.

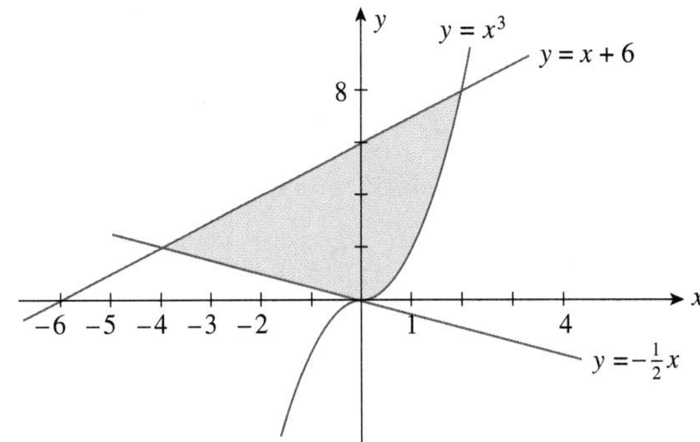

In Exercises 9–18, sketch the graph and find the area of the region bounded below by the graph of each of the given functions and above by the *x*-axis from $x = a$ to $x = b$.

9. $f(x) = -x^2$; $a = -1$, $b = 2$

10. $f(x) = x^2 - 4$; $a = -2$, $b = 2$

11. $f(x) = x^2 - 5x + 4$; $a = 1$, $b = 3$

12. $f(x) = x^3$; $a = -1$, $b = 0$

13. $f(x) = -1 - \sqrt{x}$; $a = 0$, $b = 9$

14. $f(x) = \frac{1}{2}x - \sqrt{x}$; $a = 0$, $b = 4$

15. $f(x) = -e^{(1/2)x}$; $a = -2$, $b = 4$

16. $f(x) = -xe^{-x^2}$; $a = 0$, $b = 1$

17. $f(x) = \sin(x)$; $a = \pi$, $b = \frac{3\pi}{2}$

18. $f(x) = \frac{-1}{1 + x^2}$; $a = 0$, $b = \sqrt{3}$

In Exercises 19–32, sketch the graphs of the functions *f* and *g* and find the area of the region enclosed by these graphs and the vertical lines $x = a$ and $x = b$.

19. $f(x) = x^2 + 3$, $g(x) = 1$; $a = 1$, $b = 3$

20. $f(x) = x + 2$, $g(x) = x^2 - 4$; $a = -1$, $b = 2$

21. $f(x) = -x^2 + 2x + 3$, $g(x) = -x + 3$; $a = 0$, $b = 2$

22. $f(x) = 9 - x^2$, $g(x) = 2x + 3$; $a = -1$, $b = 1$

23. $f(x) = x^2 + 1$, $g(x) = \frac{1}{3}x^3$; -1, $b = 2$

24. $f(x) = \sqrt{x}$, $g(x) = -\frac{1}{2}x - 1$; $a = 1$, $b = 4$

25. $f(x) = \frac{1}{x}$, $g(x) = 2x - 1$; $a = 1$, $b = 4$

26. $f(x) = x^2$, $g(x) = \frac{1}{x^2}$; $a = 1$, $b = 3$

27. $f(x) = e^x, g(x) = \dfrac{1}{x}; a = 1, b = 2$

28. $f(x) = x, g(x) = e^{2x}; a = 1, b = 3$

29. $f(x) = \cos(x), g(x) = \sin(x); a = 0, b = \pi$

30. $f(x) = \dfrac{1}{\sqrt{1 - x^2}}; g(x) = x; a = 0, b = 0.5$

31. $f(x) = x, g(x) = \sin(x); a = 0, b = \dfrac{\pi}{2}$

32. $f(x) = \cos(x), g(x) = x; a = -1, b = 0$

In Exercises 33–40, sketch the graph and find the area of the region bounded by the graph of the function f and the lines $y = 0$, $x = a$, and $x = b$.

33. $f(x) = x; a = -1, b = 2$

34. $f(x) = x^2 - 2x; a = -1, b = 1$

35. $f(x) = -x^2 + 4x - 3; a = -1, b = 2$

36. $f(x) = x^3 - x^2; a = -1, b = 1$

37. $f(x) = x^3 - 4x^2 + 3x; a = 0, b = 2$

38. $f(x) = 4x^{1/3} + x^{4/3}; a = -1, b = 8$

39. $f(x) = e^x - 1; a = -1, b = 3$

40. $f(x) = xe^{x^2}; a = 0, b = 2$

In Exercises 41–50, sketch the graph and find the area of the region completely enclosed by the graphs of the given functions f and g.

41. $f(x) = x + 2$ and $g(x) = x^2 - 4$

42. $f(x) = -x^2 + 4x$ and $g(x) = 2x - 3$

43. $f(x) = x^2$ and $g(x) = x^3$

44. $f(x) = x^3 + 2x^2 - 3x$ and $g(x) = 0$

45. $f(x) = x^3 - 6x^2 + 9x$ and $g(x) = x^2 - 3x$

46. $f(x) = \sqrt{x}$ and $g(x) = x^2$

47. $f(x) = x\sqrt{9 - x^2}$ and $g(x) = 0$

48. $f(x) = 2x$ and $g(x) = x\sqrt{x + 1}$

49. $f(x) = \cos(x)$ and $g(x) = \sin(x)$ between their first two intersections for $x > 0$.

50. $f(x) = \dfrac{1}{1 + x^2}$ and $g(x) = 0.5$.

51. Show that the area of a region R bounded above by the graph of a function f and below by the graph of a function g from $x = a$ to $x = b$ is given by

$$\int_a^b [f(x) - g(x)]\, dx$$

Hint: The validity of the formula was verified earlier for the case when both f and g were nonnegative. Now, let f and g be two functions such that $f(x) \geq g(x)$ for $a \leq x \leq b$. Then, there exists some nonnegative constant c such that the curves $y = f(x) + c$ and $y = g(x) + c$ are translated in the y-direction in such a way that the region R' has the same area as the region R (see the accompanying figures). Show that the area of R' is given by

$$\int_a^b \{[f(x) + c] - [g(x) + c]\}\, dx = \int_a^b [f(x) - g(x)]\, dx$$

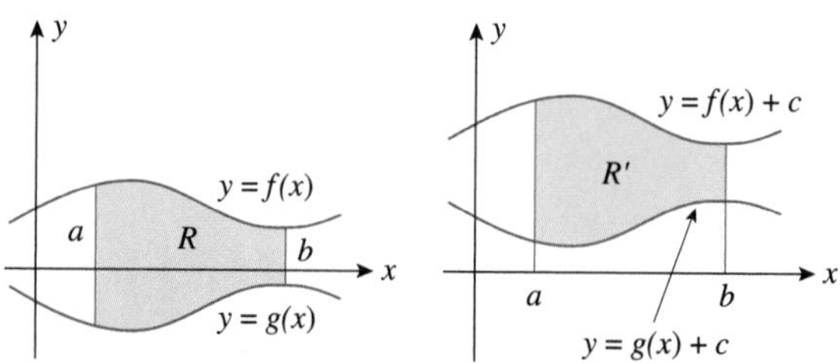

52. **Oil Production Shortfall** Energy experts disagree about when global oil production will begin to decline. In the following figure, the function f gives the annual world oil production in billions of barrels from 1980 to 2050, according to the Ministry of Natural Resources projection. The function g gives the world oil production in billions of barrels per year over the same period, according to longtime petroleum geologist Colin Campbell. Find an expression in terms of the definite integrals involving f and g, giving the shortfall in the total oil production over the period in question heeding Campbell's dire warnings.

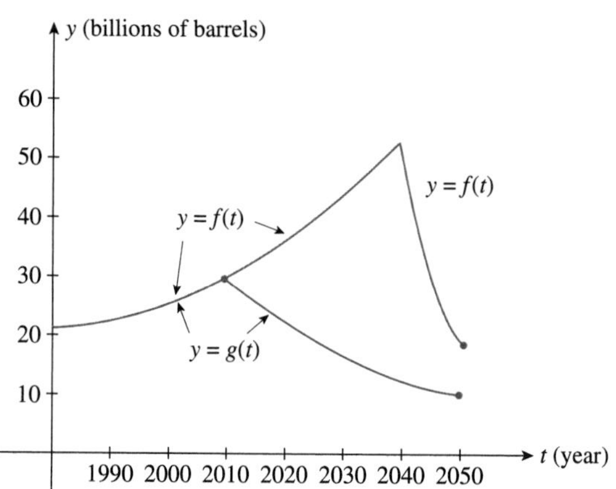

53. Two cars start out side by side and travel along a straight road. The velocity of car 1 is $f(t)$ m/sec, the velocity of car 2 is $g(t)$ m/sec over the interval $[0, T]$, and $0 < T_1 < T$. Furthermore, suppose the graphs of f and g are as depicted in the accompanying figure. Denote the area of region I by A_1 and the area of region II by A_2.

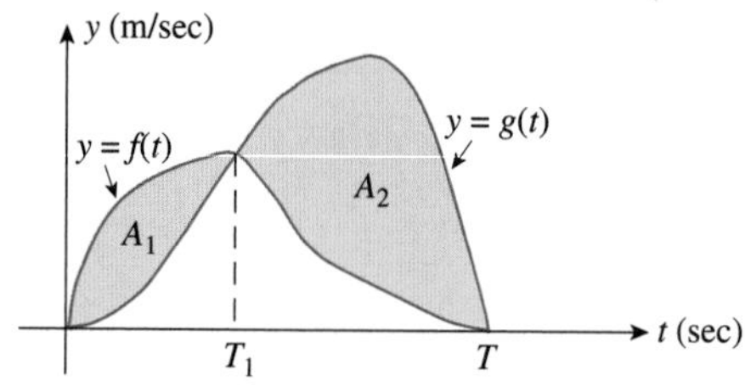

a. Write the number

$$\int_{T_1}^{T} [g(t) - f(t)]\,dt - \int_{0}^{T_1} [f(t) - g(t)]\,dt$$

in terms of A_1 and A_2.

b. What does the number obtained in part (a) represent?

54. Turbo-Charged Engine vs. Standard Engine In tests conducted by *Auto Test Magazine* on two identical models of the Phoenix Elite—one equipped with a standard engine and the other with a turbo-charger—it was found that the acceleration of the former is given by

$$a = f(t) = 1 + 0.5t \qquad (0 \le t \le 12)$$

m/sec/sec, t sec after starting from rest at full throttle, whereas the acceleration of the latter is given by

$$a = g(t) = 1 + 0.4t + 0.01t^2 \qquad (0 \le t \le 12)$$

m/sec/sec. How much faster is the turbo-charged model moving than the model with the standard engine at the end of a 10-sec test run at full throttle?

Business and Economics Applications

55. Effect of Advertising on Revenue In the accompanying figure, the function f gives the rate of change of Odyssey Travel's revenue with respect to the amount x it spends on advertising with their current advertising agency. By engaging the services of a different advertising agency, it is expected that Odyssey's revenue will grow at the rate given by the function g. Give an interpretation of the area A of the region S and find an expression for A in terms of a definite integral involving f and g.

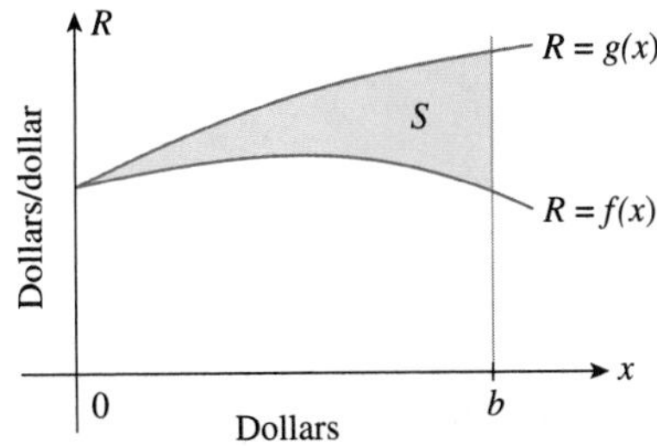

56. The rate of change of the revenue of company A over the (time) interval $[0, T]$ is $f(t)$ dollars/week, whereas the rate of change of the revenue of company B over the same period is $g(t)$ dollars/week. Suppose the graphs of f and g are as depicted in the accompanying figure. Find an expression in terms of definite integrals involving f and g giving the additional revenue that company B will have over company A in the period $[0, T]$.

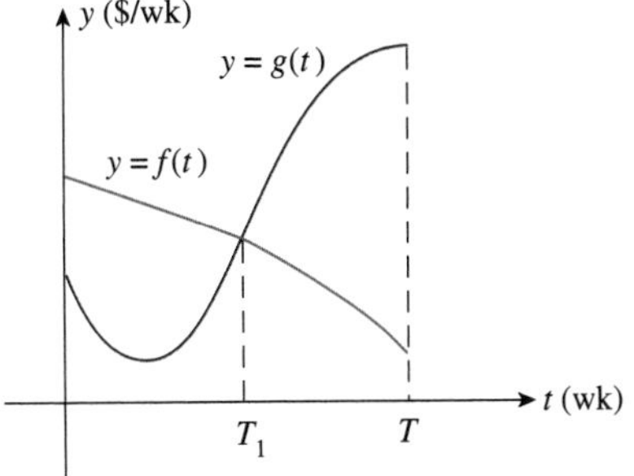

57. Effect of TV Advertising on Car Sales Carl Williams, the proprietor of Carl Williams Auto Sales, estimates that with extensive television advertising, car sales over the next several years could be increasing at the rate of

$$5e^{0.3t}$$

thousand cars/year t yr from now, instead of at the current rate of

$$(5 + 0.5t^{3/2})$$

thousand cars/year t yr from now. Find how many more cars Carl expects to sell over the next 5 yr by implementing his advertising plans.

58. Population Growth In an endeavor to curb population growth in a Southeast Asian island state, the government has decided to launch an extensive propaganda campaign. Without curbs, the government expects the rate of population growth to have been

$$60e^{0.02t}$$

thousand people/year t yr from now, over the next 5 yr. However, successful implementation of the proposed campaign is expected to result in a population growth rate of

$$-t^2 + 60$$

thousand people/year t yr from now, over the next 5 yr. Assuming that the campaign is mounted, how many fewer people will there be in that country 5 yr from now then there would have been if no curbs had been imposed?

Biological and Life Sciences Applications

59. Pulse Rate During Exercise In the accompanying figure, the function f gives the rate of increase of an individual's pulse rate when he walked a prescribed course on a treadmill 6 mo ago. The function g gives the rate of increase of his pulse rate when he recently walked the same prescribed course. Give an interpretation of the area A of the region S and find an expression for A in terms of a definite integral involving f and g.

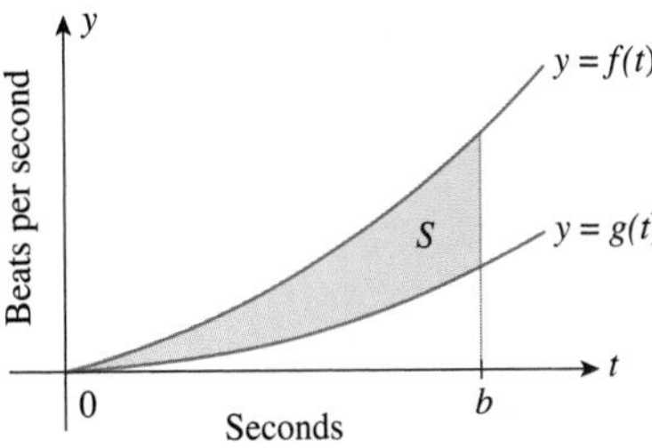

60. Air Purification To study the effectiveness of air purifiers in removing smoke, engineers ran each purifier in a smoke-filled 3- by 8-m room. In the accompanying figure, the function f gives the rate of change of the smoke level/minute, t min after the start of the test, when a brand A purifier is used. The function g gives the rate of change of the smoke level/minute when a brand B purifier is used.

a. Give an interpretation of the area of the region S.

b. Find an expression for the area of S in terms of a definite integral involving f and g.

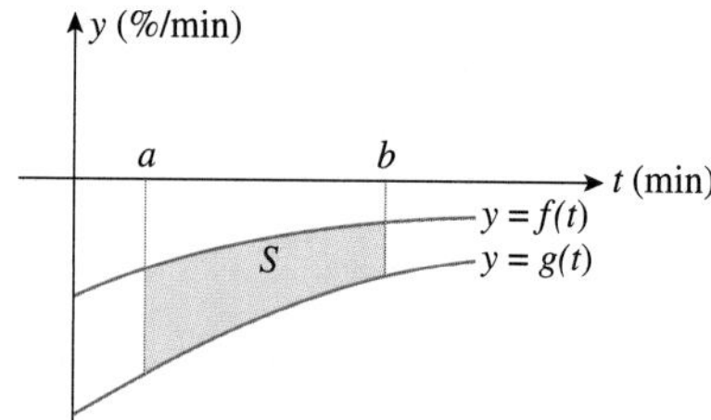

61. Alternative Energy Sources Because of the increasingly important role played by coal as a viable alternative energy source, the production of coal has been growing at the rate of

$$3.5e^{0.05t}$$

billion metric tons/year t yr from 1980 (which corresponds to $t = 0$). Had it not been for the energy crisis, the rate of production of coal since 1980 might have been only

$$3.5e^{0.05t}$$

billion metric tons/year t yr from 1980. Determine how much additional coal was produced between 1980 and the end of the century as an alternate energy source.

In Exercises 62 and 63, determine whether the statement is true or false. If it is true, explain why it is true. If it is false, give an example to show why it is false.

62. If f and g are continuous on $[a, b]$ and either $f(x) \geq g(x)$ for all x in $[a, b]$ or $f(x) \leq g(x)$ for all x in $[a, b]$, then the area of the region bounded by the graphs of f and g and the vertical lines $x = a$ and $x = b$ is given by $\int_a^b |f(x) - g(x)|\, dx$.

63. The area of the region bounded by the graphs of $f(x) = 2 - x$ and $g(x) = 4 - x^2$ and the vertical lines $x = 0$ and $x = 2$ is given by $\int_0^2 [f(x) - g(x)]\, dx$.

Solutions to Self-Check Exercises 8.6

1. The region in question is shown in the accompanying figure. Since the graph of the function f lies above that of the function g for $0 \leq x \leq 1$, we see that the required area is given by

$$\int_0^1 [(x^2 + 2) - (1 - x)]dx = \int_0^1 (x^2 + x + 1)\, dx$$

$$= \frac{1}{3}x^3 + \frac{1}{2}x^2 + x \Big|_0^1$$

$$= \frac{1}{3} + \frac{1}{2} + 1$$

$$= \frac{11}{6}$$

or $\frac{11}{6}$ square units.

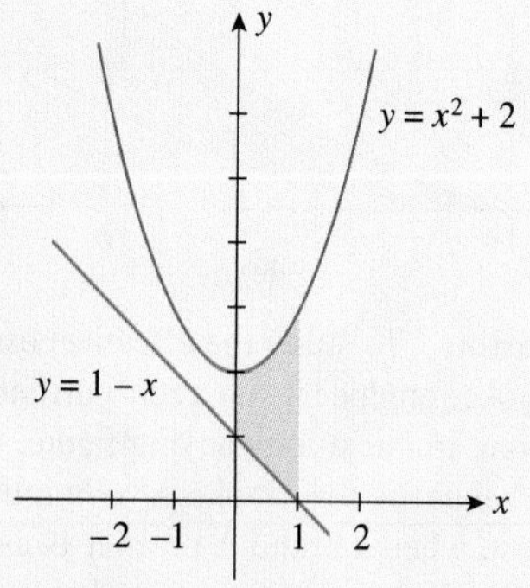

2. The region in question is shown in the accompanying figure. To find the points of intersection of the two curves, we solve the equations

$$-x^2 + 6x + 5 = x^2 + 5$$

$$2x^2 - 6x = 0$$

$$2x(x - 3) = 0$$

giving $x = 0$ or $x = 3$. Therefore, the points of intersection are (0, 5) and (3, 14).

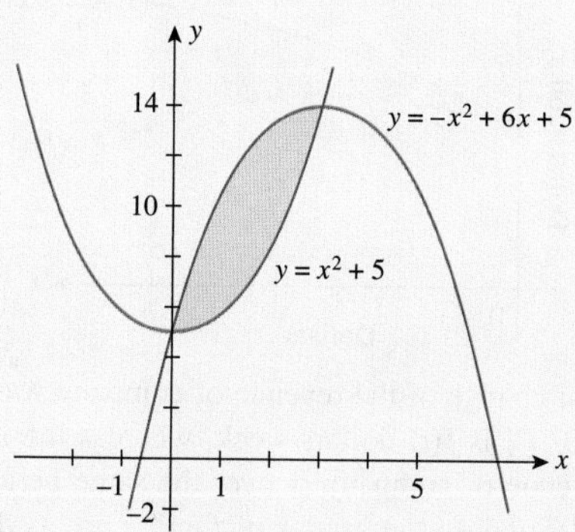

Since the graph of f always lies above that of g for $0 \leq x \leq 3$, we see that the required area is given by

$$\int_0^3 [(-x^2 + 6x + 5) - (x^2 + 5)]\, dx = \int_0^3 (-2x^2 + 6x)\, dx$$

$$= -\frac{2}{3}x^3 + 3x^2 \Big|_0^3$$

$$= -18 + 27$$

$$= 9$$

or 9 square units.

Solutions to Self-Check Exercises 8.6, *continued*

3. The additional profits realizable over the next 10 yr are given by

$$\int_0^{10} [(t - 2\sqrt{t} + 4) - (1 + t^{2/3})]\,dt$$

$$= \int_0^{10} (t - 2t^{1/2} + 3 - t^{2/3})\,dt$$

$$= \frac{1}{2}t^2 - \frac{4}{3}t^{3/2} + 3t - \frac{3}{5}t^{5/3}\Big|_0^{10}$$

$$= \frac{1}{2}(10)^2 - \frac{4}{3}(10)^{3/2} + 3(10) - \frac{3}{5}(10)^{5/3}$$

$$\approx 9.99$$

or approximately $10 million.

8.7 Applications of the Definite Integral to Business and Economics (Optional)

In this section we consider several applications of the definite integral in the fields of business and economics.

Consumers' and Producers' Surplus

We begin by deriving a formula for computing the consumers' surplus. Suppose $p = D(x)$ is the demand function that relates the unit price p of a commodity to the quantity x demanded of it. Furthermore, suppose a fixed unit market price $\bar{p}$ has been established for the commodity and corresponding to this unit price the quantity demanded is $\bar{x}$ units (Figure 8.40). Then, those consumers who would be willing to pay a unit price higher than $\bar{p}$ for the commodity would in effect experience a savings. This difference between what the consumers *would* be willing to pay for $\bar{x}$ units of the commodity and what they *actually* pay for them is called the **consumers' surplus.**

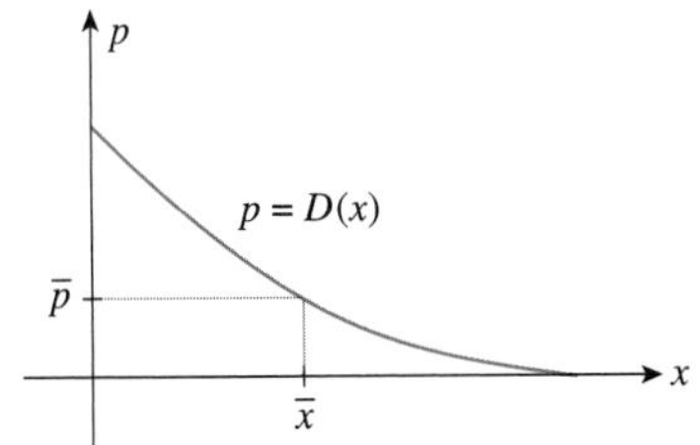

Figure 8.40
$D(x)$ is a demand function.

To derive a formula for computing the consumers' surplus, divide the interval $[0, \bar{x}]$ into n subintervals, each of length $\Delta x = \bar{x}/n$, and denote the right end points of these subintervals by $x_1, x_2, \ldots, x_n = \bar{x}$ (Figure 8.41).

We observe in Figure 8.41 that there are consumers who would pay a unit price of at least $D(x_1)$ dollars for the first Δx units of the commodity instead of the market price of $\bar{p}$ dollars per unit. The savings to these consumers is approximated by

$$D(x_1)\Delta x - \bar{p}\Delta x = [D(x_1) - \bar{p}]\Delta x$$

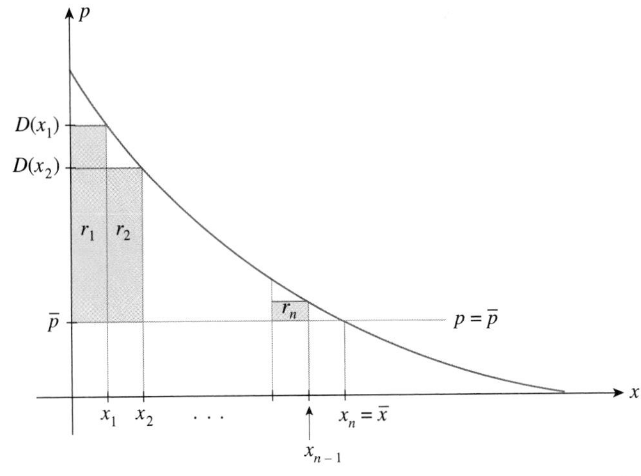

Figure 8.41
Approximating consumers' surplus by the sum of the rectangles $r_1, r_2, \ldots, r_n$

which is the area of the rectangle r_1. Pursuing the same line of reasoning, we find that the savings to the consumers who would be willing to pay a unit price of at least $D(x_2)$ dollars for the next Δx units (from x_1 through x_2) of the commodity, instead of the market price of $\bar{p}$ dollars per unit, is approximated by

$$D(x_2)\Delta x - \bar{p}\Delta x = [D(x_2) - \bar{p}]\Delta x$$

Continuing, we approximate the total savings to the consumers in purchasing $\bar{x}$ units of the commodity by the sum

$$\begin{aligned}
&[D(x_1) - \bar{p}]\Delta x + [D(x_2) - \bar{p}]\Delta x + \cdots + [D(x_n) - \bar{p}]\Delta x \\
&\quad = [D(x_1) + D(x_2) + \cdots + D(x_n)]\Delta x - \underbrace{[\bar{p}\Delta x + \bar{p}\Delta x + \cdots + \bar{p}\Delta x]}_{n \text{ terms}} \\
&\quad = [D(x_1) + D(x_2) + \cdots + D(x_n)]\Delta x - n\bar{p}\Delta x \\
&\quad = [D(x_1) + D(x_2) + \cdots + D(x_n)]\Delta x - \bar{p}\,\bar{x}
\end{aligned}$$

Now, the first term in the last expression is the Riemann sum of the demand function $p = D(x)$ over the interval $[0, \bar{x}]$ with representative points $x_1, x_2, \ldots, x_n$. Letting n approach infinity, we obtain the following formula for the consumers' surplus CS.

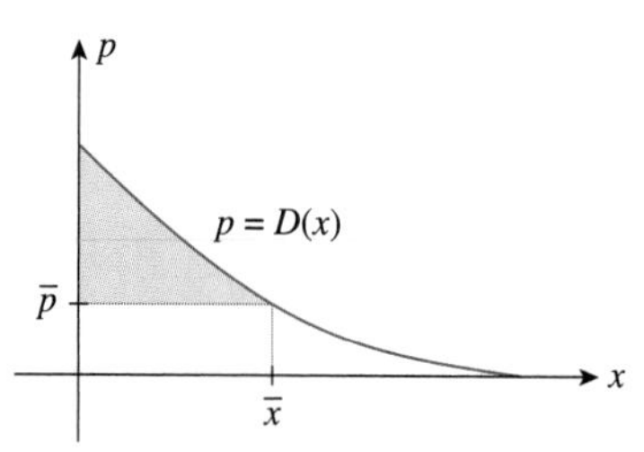

Figure 8.42
Consumers' surplus

Consumers' Surplus

The consumers' surplus is given by

$$CS = \int_0^{\bar{x}} D(x)\,dx - \bar{p}\,\bar{x} \qquad (15)$$

where D is the demand function, $\bar{p}$ is the unit market price, and $\bar{x}$ is the quantity sold.

The consumer's surplus is given by the area of the region bounded above by the demand curve $p = D(x)$ and below by the straight line $p = \bar{p}$ from $x = 0$ to $x = \bar{x}$ (Figure 8.42). We can also see this if we rewrite Equation (15) in the form

$$\int_0^{\bar{x}} [D(x) - \bar{p}]\,dx$$

and interpret the result geometrically.

Analogously, we can derive a formula for computing the producers' surplus. Suppose $P = S(x)$ is the supply equation that relates the unit price p of a certain commodity to the quantity x that the supplier will make available in the market at that price.

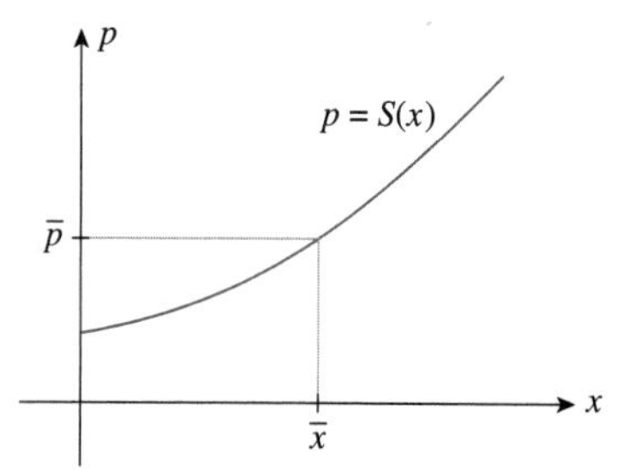

Figure 8.43
$S(x)$ is a supply function.

Again, suppose a fixed market price $\bar{p}$ has been established for the commodity and, corresponding to this unit price, a quantity of $\bar{x}$ units will be made available in the market by the supplier (Figure 8.43). Then, the suppliers who would be willing to make the commodity available at a lower price stand to gain from the fact that the market price is set as such. The difference between what the suppliers actually receive and what they would be willing to receive is called the **producers' surplus.** Proceeding in a manner similar to the derivation of the equation for computing the consumers' surplus, we find that the producers' surplus PS is defined as follows:

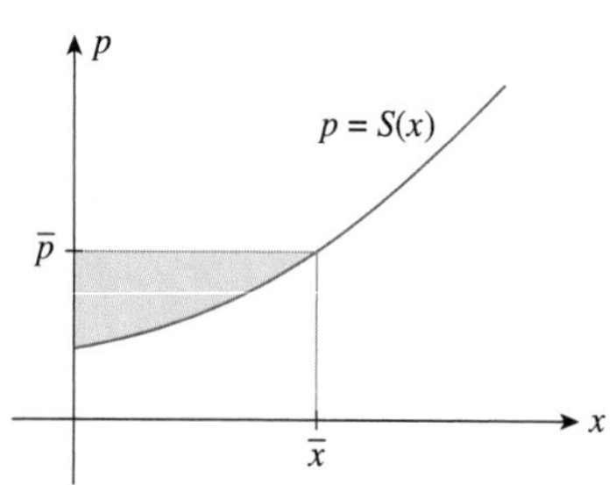

Figure 8.44
Producers' surplus

Producers' Surplus

The producers' surplus is given by

$$PS = \bar{p}\bar{x} - \int_0^{\bar{x}} S(x)\,dx \qquad (16)$$

where $S(x)$ is the supply function, $\bar{p}$ is the unit market price, and $\bar{x}$ is the quantity supplied.

Geometrically, the producers' surplus is given by the area of the region bounded above by the straight line $p = \bar{p}$ and below by the supply curve $p = S(x)$ from $x = 0$ to $x = \bar{x}$ (Figure 8.44).

We can also show that the last statement is true by converting Equation (16) to the form

$$\int_0^{\bar{x}} [\bar{p} - S(x)]\, dx$$

and interpreting the definite integral geometrically.

Example 1

The demand function for a certain make of 10-speed bicycle is given by

$$p = D(x) = -0.001x^2 + 250$$

where p is the unit price in dollars and x is the quantity demanded in units of a thousand. The supply function for these bicycles is given by

$$p = S(x) = 0.0006x^2 + 0.02x + 100$$

where p stands for the unit price in dollars and x stands for the number of bicycles that the supplier will put on the market, in units of a thousand. Determine the consumers' surplus and the producers' surplus if the market price of a bicycle is set at the equilibrium price.

Solution

Recall that the equilibrium price is the unit price of the commodity when market equilibrium occurs. We determine the equilibrium price by solving for the point of intersection of the demand curve and the supply curve (Figure 8.45). To solve the system of equations

$$p = -0.001x^2 \qquad + 250$$

$$p = 0.0006x^2 + 0.02x + 100$$

we simply substitute the first equation into the second, obtaining

$$0.0006x^2 + 0.02x + 100 = -0.001x^2 + 250$$

$$0.0016x^2 + 0.02x - 150 = 0$$

$$16x^2 + 200x - 1{,}500{,}000 = 0$$

$$2x^2 + 25x - 187{,}500 = 0$$

Factoring this last equation, we obtain

$$(2x + 625)(x - 300) = 0$$

Thus, $x = -625/2$ or $x = 300$. The first number lies outside the interval of interest, so we are left with the solution $x = 300$, with a corresponding value of

$$p = -0.001(300)^2 + 250 = 160$$

Thus, the equilibrium point is (300, 160); that is, the equilibrium quantity is 300,000, and the equilibrium price is \$160. Setting the market price at \$160 per unit and using Formula (15) with $\bar{p} = 160$ and $\bar{x} = 300$, we find that the consumers' surplus is given by

$$CS = \int_0^{300} (-0.001x^2 + 250)\, dx - (160)(300)$$

$$= \left(-\frac{1}{3000}x^3 + 250x\right)\Bigg|_0^{300} - 48{,}000$$

$$= -\frac{300^3}{3000} + (250)(300) - 48{,}000$$

$$= 18{,}000$$

or \$18,000,000.

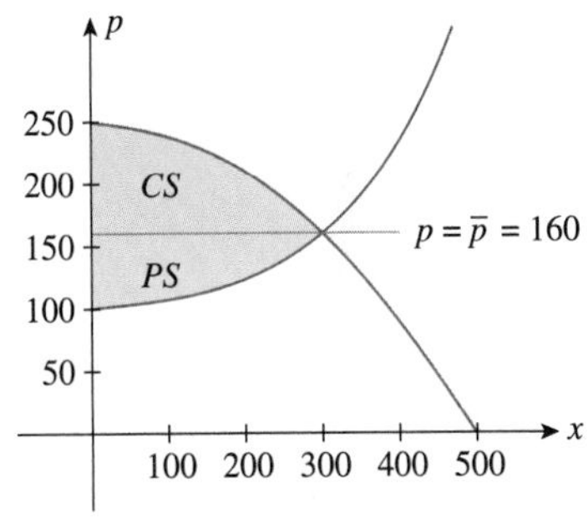

Figure 8.45
Consumers' surplus and producers' surplus when market price = equilibrium price

(Recall that x is measured in units of a thousand.) Next, using (16), we find that the producers' surplus is given by

$$\begin{aligned} PS &= (160)(300) - \int_0^{300} (0.0006x^2 + 0.02x + 100)\,dx \\ &= 48{,}000 - (0.0002x^3 + 0.01x^2 + 100x)\Big|_0^{300} \\ &= 48{,}000 - [(0.0002)(300)^3 + (0.01)(300)^2 + 100(300)] \\ &= 11{,}700 \end{aligned}$$

or \$11,700,000.

The Future and Present Value of an Income Stream

Suppose a firm generates a stream of income over a period of time—for example, the revenue generated by a large chain of retail stores over a 5-year period. As the income is realized, it is reinvested and earns interest at a fixed rate. The **accumulated future income stream** over the 5-year period is the amount of money the firm ends up with at the end of that period.

The definite integral can be used to determine this accumulated, or total, future income stream over a period of time. The total future value of an income stream gives us a way to measure the value of such a stream. To find the **total future value of an income stream,** suppose

$R(t)$ = Rate of income generation at any time t — Dollars per year

r = Interest rate compounded continuously

T = Term — In years

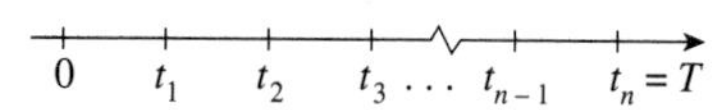

Figure 8.46
The time interval $[0, T]$ is partitioned into n subintervals.

Let's divide the time interval $[0, T]$ into n subintervals of equal length $\Delta t = T/n$ and denote the right end points of these intervals by $t_1, t_2, \ldots, t_n = T$, as shown in Figure 8.46.

If R is a continuous function on $[0, T]$, then $R(t)$ will not differ by much from $R(t_1)$ in the subinterval $[0, t_1]$ provided that the subinterval is small (which is true if n is large). Therefore, the income generated over the time interval $[0, t_1]$ is approximately

$$R(t_1)\Delta t \qquad \text{Constant rate of income} \times \text{Length of time}$$

dollars. The future value of this amount, T years from now, calculated as if it were earned at time t_1, is

$$[R(t_1)\Delta t]e^{r(T-t_1)} \qquad \text{Equation (10), Section 12.1}$$

dollars. Similarly, the income generated over the time interval $[t_1, t_2]$ is approximately $P(t_2)\Delta t$ dollars and has a future value, T years from now, of approximately

$$[R(t_2)\Delta t]e^{r(T-t_2)}$$

dollars. Therefore, the sum of the future values of the income stream generated over the time interval $[0, T]$ is approximately

$$\begin{aligned} &R(t_1)e^{r(T-t_1)}\Delta t + R(t_2)e^{r(T-t_2)}\Delta t + \cdots + R(t_n)e^{r(T-t_n)}\Delta t \\ &= e^{rT}[R(t_1)e^{-rt_1}\Delta t + R(t_2)e^{-rt_2}\Delta t + \cdots + R(t_n)e^{-rt_n}\Delta t] \end{aligned}$$

dollars. But this sum is just the Riemann sum of the function $e^{rT}R(t)e^{-rt}$ over the interval $[0, T]$ with representative points $t_1, t_2, \ldots, t_n$. Letting n approach infinity, we obtain the following result.

Accumulated or Total Future Value of an Income Stream

The accumulated, or total, future value after T years of an income stream of $R(t)$ dollars per year, earning interest at the rate of r per year compounded continuously, is given by

$$A = e^{rT}\int_0^T R(t)e^{-rt}\,dt \tag{17}$$

Example 2

Income Stream Crystal Car Wash recently bought an automatic car-washing machine that is expected to generate \$40,000 in revenue per year, t years from now, for the next 5 years. If the income is reinvested in a business earning interest at the rate of 12% per year compounded continuously, find the total accumulated value of this income stream at the end of 5 years.

Solution

We are required to find the total future value of the given income stream after 5 years. Using Equation (17) with $R(t) = 40{,}000$, $r = 0.12$, and $T = 5$, we see that the required value is given by

$$e^{0.12(5)}\int_0^5 40{,}000e^{-0.12t}\,dt$$

$$= e^{0.6}\left[-\frac{40{,}000}{0.12}e^{-0.12t}\right]\Bigg|_0^5 \qquad \text{Integrate using the substitution } u = -0.12t.$$

$$= -\frac{40{,}000e^{0.6}}{0.12}(e^{-0.6} - 1) \approx 274{,}039.60$$

or approximately \$274,040.

Another way of measuring the value of an income stream is by considering its present value. The **present value of an income stream** of $R(t)$ dollars per year over a term of T years, earning interest at the rate of r per year compounded continuously, is the principal P that will yield the same accumulated value as the income stream itself when P is invested today for a period of T years at the same rate of interest. In other words

$$Pe^{rT} = e^{rT}\int_0^T R(t)e^{-rt}\,dt$$

Dividing both sides of the equation by e^{rT} gives the following result.

Present Value of an Income Stream

The present value of an income stream of $R(t)$ dollars per year, earning interest at the rate of r per year compounded continuously, is given by

$$PV = \int_0^T R(t)e^{-rt}\,dt \tag{18}$$

Example 3

Investment Analysis The owner of a local cinema is considering two alternative plans for renovating and improving the theatre. Plan A calls for an immediate cash outlay of \$250,000, whereas plan B requires an immediate cash outlay of \$180,000. It has been estimated that adopting plan A would result in a net income stream generated at the rate of

$$f(t) = 630{,}000$$

dollars per year, whereas adopting plan B would result in a net income stream generated at the rate of

$$g(t) = 580{,}000$$

dollars per year for the next 3 years. If the prevailing interest rate for the next 5 years is 10% per year, which plan will generate a higher net income by the end of 3 years?

Solution

Since the initial outlay is \$250,000, we find—using Equation (18) with $R(t) = 630{,}000$, $r = 0.1$, and $T = 3$—that the present value of the net income under plan A is given by

$$\begin{aligned}\int_0^3 630{,}000e^{-0.1t}\,dt - 250{,}000 &= \frac{630{,}000}{-0.1}e^{-0.1t}\Big|_0^3 - 250{,}000 && \text{Integrate using the substitution } u = -0.1t. \\ &= -6{,}300{,}000e^{-0.3} + 6{,}300{,}000 - 250{,}000 \\ &\approx 1{,}382{,}845\end{aligned}$$

or approximately \$1,382,845.

To find the present value of the net income under plan B, we use (18) with $R(t) = 580{,}000$, $r = 0.1$, and $T = 3$, obtaining

$$\int_0^3 580{,}000e^{-0.1t}\,dt - 180{,}000$$

dollars. Proceeding as in the previous computation, we see that the required value is \$1,323,254 (see Exercise 8, page 517).

Comparing the present value of each plan, we conclude that plan A would generate a higher net income by the end of 3 years.

Remark

The function R in Example 3 is a constant function. If R is not a constant function, then we may need more sophisticated techniques of integration to evaluate the integral in (18). Exercises 9.2 and 9.3 contain problems of this type. ◀

The Amount and Present Value of an Annuity

An annuity is a sequence of payments made at regular time intervals. The time period in which these payments are made is called the *term* of the annuity. Although the payments need not be equal in size, they are equal in many important applications, and we will assume that they are equal in our discussion. Examples of annuities are regular deposits to a savings account, monthly home mortgage payments, and monthly insurance payments.

The amount of an annuity is the sum of the payments plus the interest earned. A formula for computing the amount of an annuity A can be derived with the help of (17). Let

P = Size of each payment in the annuity

r = Interest rate compounded continuously

T = Term of the annuity (in years)

m = Number of payments per year

The payments into the annuity constitute a constant income stream of $R(t) = mP$ dollars per year. With this value of $R(t)$, (17) yields

$$A = e^{rT}\int_0^T R(t)e^{-rt}\,dt = e^{rT}\int_0^T mPe^{-rt}\,dt$$

$$= mPe^{rT}\left[-\frac{e^{-rt}}{r}\right]\Bigg|_0^T$$

$$= mPe^{rT}\left[-\frac{e^{-rT}}{r}+\frac{1}{r}\right] = \frac{mP}{r}(e^{rT}-1)$$

This leads us to the following formula.

Amount of an Annuity

The amount of an annuity is

$$A = \frac{mP}{r}(e^{rT}-1) \qquad (19)$$

where P, r, T, and m are as defined earlier.

Example 4

RRSPs On January 1, 1990, Marcus Chapman deposited \$2000 into a Registered Retirement Savings Plan (RRSP) paying interest at the rate of 10% per year compounded continuously. Assuming that he deposits \$2000 annually into the account, how much will he have in his IRA at the beginning of the year 2006?

Solution

We use (19), with $P = 2000$, $r = 0.1$, $T = 16$, and $m = 1$, obtaining

$$A = \frac{2000}{0.1}(e^{1.6}-1)$$
$$\approx 79{,}060.65$$

Thus, Marcus will have approximately \$79,061 in his account at the beginning of the year 2006.

EXPLORING WITH TECHNOLOGY

Refer to Example 4. Suppose Marcus wishes to know how much he will have in his IRA at any time in the future, not just at the beginning of 2006, as you were asked to compute in the example.

1. Using Formula (17) and the relevant data from Example 4, show that the required amount at any time x (x measured in years, $x > 0$) is given by
$$A = f(x) = 20{,}000(e^{0.1x}-1)$$
2. Use a graphing utility to plot the graph of f, using the viewing window $[0, 30] \cdot [2000, 400{,}000]$.
3. Using ZOOM and TRACE, or using the function evaluation capability of your graphing utility, use the result of part 2 to verify the result obtained in Example 4. Comment on the advantage of the mathematical model found in part 1.

Using (18), we can derive the following formula for the present value of an annuity.

Present Value of an Annuity

The present value of an annuity is given by

$$PV = \frac{mP}{r}(1-e^{-rT}) \qquad (20)$$

where P, r, T, and m are as defined earlier.

Example 5

Sinking Funds Tomas Perez, the proprietor of a hardware store, wants to establish a fund from which he will withdraw \$1000 per month for the next 10 years. If the fund earns interest at the rate of 9% per year compounded continuously, how much money does he need to establish the fund?

Solution

We want to find the present value of an annuity with $P = 1000$, $r = 0.09$, $T = 10$, and $m = 12$. Using Equation (20), we find

$$PV = \frac{12{,}000}{0.09}(1 - e^{-(0.09)(10)})$$

$$\approx 79{,}124.04$$

Thus, Tomas needs approximately \$79,124 to establish the fund.

In Chapter 12 we will develop annuities from a different perspective.

Lorentz Curves and Income Distributions

One method used by economists to study the distribution of income in a society is based on the Lorentz curve, named after American statistician M. D. Lorentz. To describe the Lorentz curve, let $f(x)$ denote the proportion of the total income received by the poorest $100x\%$ of the population for $0 \leq x \leq 1$. Using this terminology, $f(0.3) = 0.1$ simply states that the lowest 30% of the income recipients receive 10% of the total income.

The function f has the following properties:

1. The domain of f is $[0, 1]$.
2. The range of f is $[0, 1]$.
3. $f(0) = 0$ and $f(1) = 1$.
4. $f(x) \leq x$ for every x in $[0, 1]$.
5. f is increasing on $[0, 1]$.

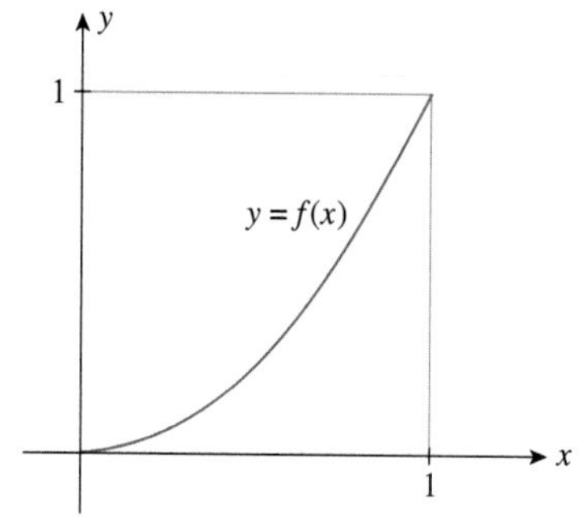

Figure 8.47
A Lorentz curve

The first two properties follow from the fact that both x and $f(x)$ are fractions of a whole. Property 3 is a statement that 0% of the income recipients receive 0% of the total income and 100% of the income recipients receive 100% of the total income. Property 4 follows from the fact that the lowest $100x\%$ of the income recipients cannot receive more than $100x\%$ of the total income. A typical Lorentz curve is shown in Figure 8.47.

Example 6

Lorentz Curves A developing country's income distribution is described by the function

$$f(x) = \frac{19}{20}x^2 + \frac{1}{20}x$$

a. Sketch the Lorentz curve for the given function.
b. Compute $f(0.2)$ and $f(0.8)$ and interpret your results.

Solution

a. The Lorentz curve is shown in Figure 8.48.

b. $f(0.2) = \frac{19}{20}(0.2)^2 + \frac{1}{20}(0.2) = 0.048$

Thus, the lowest 20% of the people receive 4.8% of the total income.

$$f(0.8) = \frac{19}{20}(0.8)^2 + \frac{1}{20}(0.8) = 0.648$$

Thus, the lowest 80% of the people receive 64.8% of the total income.

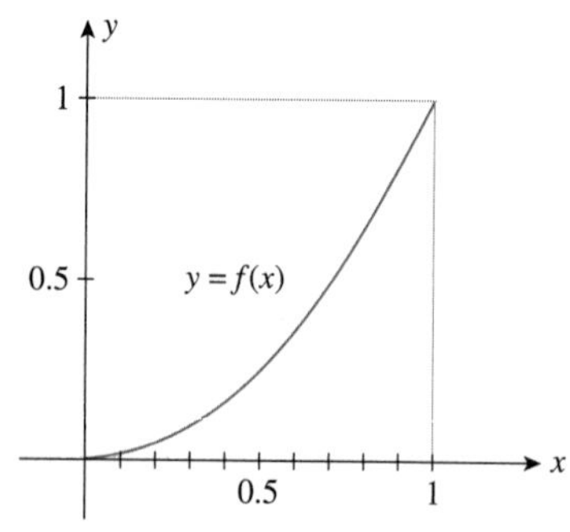

Figure 8.48
The Lorentz curve $f(x) = \frac{19}{20}x^2 + \frac{1}{20}x$

Next, let's consider the Lorentz curve described by the function $y = f(x) = x$. Since exactly $100x\%$ of the total income is received by the lowest $100x\%$ of income recipients, the line $y = x$ is called the **line of complete equality.** For example, 10% of the total income is received by the lowest 10% of income recipients, 20% of the total income is received by the lowest 20% of income recipients, and so on. Now, it is evident that the closer a Lorentz curve is to this line, the more equitable the income distribution is among the income recipients. But the proximity of a Lorentz curve to the line of complete equality is reflected by the area between the Lorentz curve and the line $y = x$ (Figure 8.49). The closer the curve is to the line, the smaller the enclosed area.

This observation suggests that we may define a number, called the coefficient of inequality of a Lorentz curve, as the ratio of the area between the line of complete equality and the Lorentz curve to the area under the line of complete equality. Since the area under the line of complete equality is $\frac{1}{2}$, we see that the coefficient of inequality is given by the following formula.

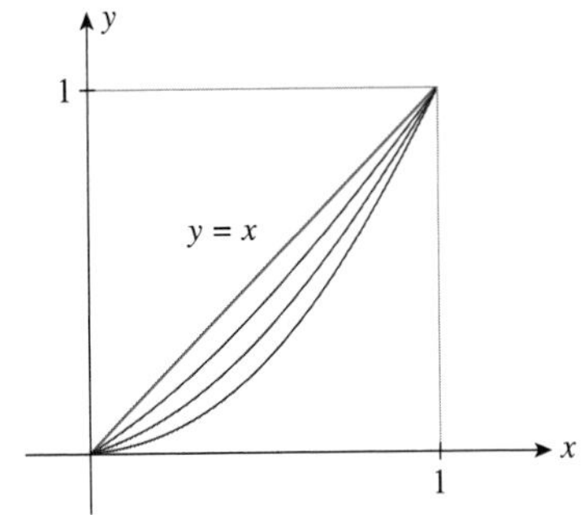

Figure 8.49
The closer the Lorentz curve is to the line, the more equitable the income distribution.

Coefficient of Inequality of a Lorentz Curve

The coefficient of inequality, or **Gini index,** of a Lorentz curve is

$$L = 2\int_0^1 [x - f(x)]\,dx \qquad (21)$$

The coefficient of inequality is a number between 0 and 1. For example, a coefficient of zero implies that the income distribution is perfectly uniform.

Example 7

Income Distributions In a study conducted by a certain country's Economic Development Board with regard to the income distribution of certain segments of the country's workforce, it was found that the Lorentz curves for the distribution of income of medical doctors and of movie actors are described by the functions

$$f(x) = \frac{14}{15}x^2 + \frac{1}{15}x \qquad \text{and} \qquad g(x) = \frac{5}{8}x^4 + \frac{3}{8}x$$

respectively. Compute the coefficient of inequality for each Lorentz curve. Which profession has a more equitable income distribution?

Solution

The required coefficients of inequality are, respectively,

$$\begin{aligned} L_1 &= 2\int_0^1 \left[x - \left(\frac{14}{15}x^2 + \frac{1}{15}x\right)\right] dx = 2\int_0^1 \left(\frac{14}{15}x - \frac{14}{15}x^2\right) dx \\ &= \frac{28}{15}\int_0^1 (x - x^2)\,dx = \frac{28}{15}\left(\frac{1}{2}x^2 - \frac{1}{3}x^3\right)\Bigg|_0^1 \\ &= \frac{14}{45} \approx 0.311 \end{aligned}$$

$$\begin{aligned} L_2 &= 2\int_0^1 \left[x - \left(\frac{5}{8}x^4 + \frac{3}{8}x\right)\right] dx = 2\int_0^1 \left(\frac{5}{8}x - \frac{5}{8}x^4\right) dx \\ &= \frac{5}{4}\int_0^1 (x - x^4)\,dx = \frac{5}{4}\left(\frac{1}{2}x^2 - \frac{1}{5}x^5\right)\Bigg|_0^1 \\ &= \frac{15}{40} \approx 0.375 \end{aligned}$$

We conclude that in this country the incomes of medical doctors are more evenly distributed than the incomes of movie actors.

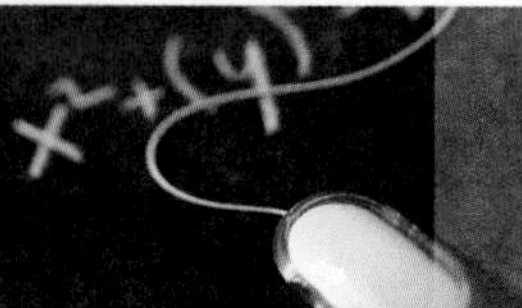

Self-Check Exercise 8.7

The demand function for a certain make of exercise bicycle that is sold exclusively through cable television is

$$p = d(x) = \sqrt{9 - 0.02x}$$

where p is the unit price in hundreds of dollars and x is the quantity demanded/week. The corresponding supply function is given by

$$p = s(x) = \sqrt{1 + 0.02x}$$

where p has the same meaning as before and x is the number of exercise bicycles the supplier will make available at price p. Determine the consumers' surplus and the producers' surplus if the unit price is set at the equilibrium price.

The solution to Self-Check Exercise 8.7 can be found on page 518.

8.7 Exercises

 **A calculator is recommended for this exercise set.**

Business and Economics Applications

1. **Consumers' Surplus** The demand function for a certain make of replacement cartridges for a water purifier is given by

$$p = -0.01x^2 - 0.1x + 6$$

where p is the unit price in dollars and x is the quantity demanded each week, measured in units of a thousand. Determine the consumers' surplus if the market price is set at $4/cartridge.

2. **Consumers' Surplus** The demand function for a certain brand of compact disc is given by

$$p = -0.01x^2 - 0.2x + 8$$

where p is the wholesale unit price in dollars and x is the quantity demanded each week, measured in units of a thousand. Determine the consumers' surplus if the wholesale market price is set at $5/disc.

3. **Consumers' Surplus** It is known that the quantity demanded of a certain make of portable hair dryer is x hundred units/week and the corresponding wholesale unit price is

$$p = \sqrt{225 - 5x}$$

dollars. Determine the consumers' surplus if the wholesale market price is set at $10/unit.

4. **Producers' Surplus** The supplier of the portable hair dryers in Exercise 3 will make x hundred units of hair dryers available in the market when the wholesale unit price is

$$p = \sqrt{36 + 1.8x}$$

dollars. Determine the producers' surplus if the wholesale market price is set at $9/unit.

5. **Producers' Surplus** The supply function for the compact discs of Exercise 2 is given by

$$p = 0.01x^2 + 0.1x + 3$$

where p is the unit wholesale price in dollars and x stands for the quantity that will be made available in the market by the supplier, measured in units of a thousand. Determine the producers' surplus if the wholesale market price is set at the equilibrium price.

6. **Consumers' and Producers' Surplus** The management of the Titan Tire Company has determined that the quantity demanded x of their Super Titan tires/week is related to the unit price p by the relation

$$p = 144 - x^2$$

where p is measured in dollars and x is measured in units of a thousand. Titan will make x units of the tires available in the market if the unit price is

$$p = 48 + \frac{1}{2}x^2$$

dollars. Determine the consumers' surplus and the producers' surplus when the market unit price is set at the equilibrium price.

7. **Consumers' and Producers' Surplus** The quantity demanded x (in units of a hundred) of the Mikado miniature cameras/week is related to the unit price p (in dollars) by

$$p = -0.2x^2 + 80$$

and the quantity x (in units of a hundred) that the supplier is willing to make available in the market is related to the unit price p (in dollars) by

$$p = 0.1x^2 + x + 40$$

If the market price is set at the equilibrium price, find the consumers' surplus and the producers' surplus.

8. Refer to Example 3, page 511. Verify that

$$\int_0^3 580{,}000e^{-0.1t}\,dt - 180{,}000 \approx 1{,}323{,}254$$

9. **Present Value of an Investment** Suppose an investment is expected to generate income at the rate of

$$R(t) = 200{,}000$$

dollars/year for the next 5 yr. Find the present value of this investment if the prevailing interest rate is 8%/year compounded continuously.

10. **Franchises** Camille purchased a 15-yr franchise for a computer outlet store that is expected to generate income at the rate of

$$R(t) = 400{,}000$$

dollars/year. If the prevailing interest rate is 10%/year compounded continuously, find the present value of the franchise.

11. **The Amount of an Annuity** Find the amount of an annuity if $250/month is paid into it for a period of 20 yr, earning interest at the rate of 8%/year compounded continuously.

12. **The Amount of an Annuity** Find the amount of an annuity if $400/month is paid into it for a period of 20 yr, earning interest at the rate of 8%/year compounded continuously.

13. **The Amount of an Annuity** Aiso deposits $150/month in a savings account paying 8%/year compounded continuously. Estimate the amount that will be in his account after 15 yr.

14. **Custodial Accounts** The Armstrongs wish to establish a custodial account to finance their children's education. If they deposit $200 monthly for 10 yr in a savings account paying 9%/year compounded continuously, how much will their savings account be worth at the end of this period?

15. **IRA Accounts** Refer to Example 4, page 513. Suppose Marcus makes his RRSP payment on April 1, 1990, and annually thereafter. If interest is paid at the same initial rate, approximately how much will Marcus have in his account at the beginning of the year 2006?

16. **Present Value of an Annuity** Estimate the present value of an annuity if payments are $800 monthly for 12 yr and the account earns interest at the rate of 10%/year compounded continuously.

17. **Present Value of an Annuity** Estimate the present value of an annuity if payments are $1200 monthly for 15 yr and the account earns interest at the rate of 10%/year compounded continuously.

18. **Lottery Payments** A lottery commission pays the winner of the "Million Dollar" lottery 20 annual installments of $50,000 each. If the prevailing interest rate is 8%/year compounded continuously, find the present value of the winning ticket.

19. **Reverse Annuity Mortgages** Sinclair wishes to supplement his retirement income by $300/month for the next 10 yr. He plans to obtain a reverse annuity mortgage (RAM) on his home to meet this need. Estimate the amount of the mortgage he will require if the prevailing interest rate is 12%/year compounded continuously.

20. **Reverse Annuity Mortgage** Refer to Exercise 19. Leah wishes to supplement her retirement income by $400/month for the next 15 yr by obtaining a RAM. Estimate the amount of the mortgage she will require if the prevailing interest rate is 9%/year compounded continuously.

21. **Lorentz Curves** A certain country's income distribution is described by the function

$$f(x) = \frac{15}{16}x^2 + \frac{1}{16}x$$

a. Sketch the Lorentz curve for this function.
b. Compute $f(0.4)$ and $f(0.9)$ and interpret your results.

22. **Lorentz Curves** In a study conducted by a certain country's Economic Development Board, it was found that the Lorentz curve for the distribution of income of college teachers was described by the function

$$f(x) = \frac{13}{14}x^2 + \frac{1}{14}x$$

and that of lawyers by the function

$$g(x) = \frac{9}{11}x^4 + \frac{2}{11}x$$

a. Compute the coefficient of inequality for each Lorentz curve.
b. Which profession has a more equitable income distribution?

23. **Lorentz Curves** A certain country's income distribution is described by the function

$$f(x) = \frac{14}{15}x^2 + \frac{1}{15}x$$

a. Sketch the Lorentz curve for this function.
b. Compute $f(0.3)$ and $f(0.7)$.

24. **Lorentz Curves** In a study conducted by a certain country's Economic Development Board, it was found that the Lorentz curve for the distribution of income of stockbrokers was described by the function

$$f(x) = \frac{11}{12}x^2 + \frac{1}{12}x$$

and that of high school teachers by the function

$$g(x) = \frac{5}{6}x^2 + \frac{1}{6}x$$

a. Compute the coefficient of inequality for each Lorentz curve.
b. Which profession has a more equitable income distribution?

Solution to Self-Check Exercise 8.7

We find the equilibrium price and equilibrium quantity by solving the system of equations

$$p = \sqrt{9 - 0.02x}$$

$$p = \sqrt{1 + 0.002x}$$

simultaneously. Substituting the first equation into the second, we have

$$\sqrt{9 - 0.02x} = \sqrt{1 + 0.02x}$$

Squaring both sides of the equation then leads to

$$9 - 0.02x = 1 + 0.02x$$

$$x = 200$$

Therefore,

$$p = \sqrt{9 - 0.02(200)}$$

$$= \sqrt{5} \approx 2.24$$

The equilibrium price is $224, and the equilibrium quantity is 200. The consumers' surplus is given by

$$CS = \int_0^{200} \sqrt{9 - 0.02x}\,dx - (2.24)(200)$$

$$= \int_0^{200} (9 - 0.02x)^{1/2}\,dx - 448$$

$$= -\frac{1}{0.02}\left(\frac{2}{3}\right)(9 - 0.02x)^{3/2}\Big|_0^{200} - 448 \quad \text{Integrate by substitution.}$$

$$= -\frac{1}{0.03}(5^{3/2} - 9^{3/2}) - 448$$

$$\approx 79.32$$

or approximately $7932.

Next, the producers' surplus is given by

$$PS = (2.24)(200) - \int_0^{200} \sqrt{1 + 0.02x}\,dx$$

$$= 448 - \int_0^{200} (1 + 0.02x)^{1/2}\,dx$$

$$= 448 - \frac{1}{0.02}\left(\frac{2}{3}\right)(1 + 0.02x)^{3/2}\Big|_0^{200}$$

$$= 448 - \frac{1}{0.03}(5^{3/2} - 1)$$

$$\approx 108.66$$

or approximately $10,866.

8.8 Volumes of Solids of Revolution (Optional)

Finding the Volume of a Solid of Revolution

In this section we use a Riemann sum to find a formula for computing the volume of a solid that results when a region is revolved about an axis. The solid is called a **solid of revolution.** To find the volume of a solid of revolution, suppose the plane region under the curve defined by a nonnegative continuous function $y = f(x)$ between $x = a$ and $x = b$ is revolved about the x-axis (Figure 8.50).

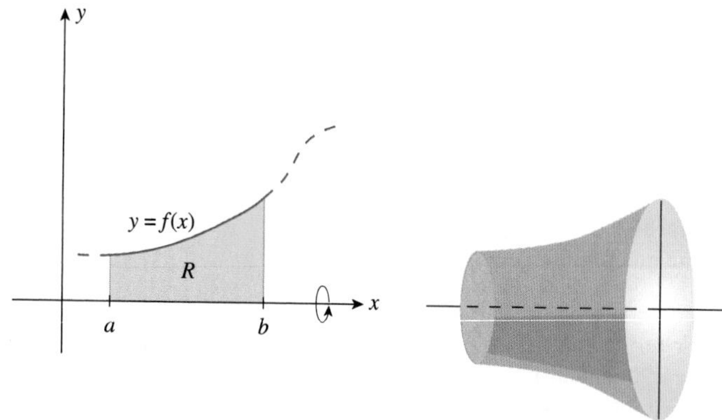

(a) Region R under the curve

(b) Solid obtained by revolving R about the x-axis

Figure 8.50

To derive a formula for finding the volume of the solid of revolution, we divide the interval $[a, b]$ into n subintervals, each of equal length, by means of the points $x_0 = a$, $x_1, x_2, \ldots, x_n = b$, so that the length of each subinterval is given by $\Delta x = (b - a)/n$. Also, let $p_1, p_2, \ldots, p_n$ be points in the subintervals $[x_0, x_1], [x_1, x_2], \ldots, [x_{n-1}, x_n]$, respectively (Figure 8.51).

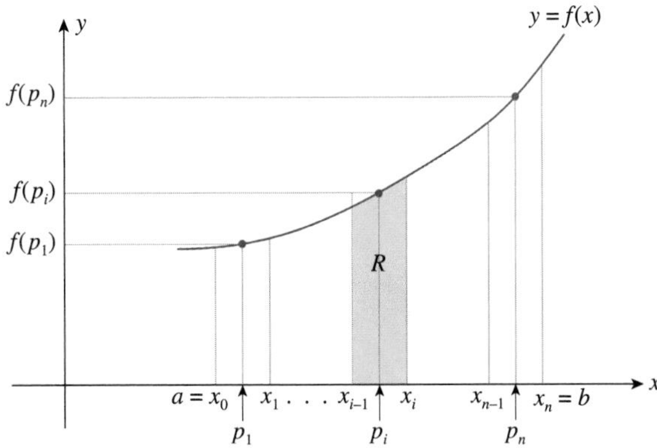

Figure 8.51
The region R is revolved about the x-axis.

Let's concentrate on that part of the solid of revolution swept out by revolving the region under the curve between $x = x_{i-1}$ and $x = x_i$ about the x-axis. The volume ΔV_i of this object, shown in Figure 8.52a, may be approximated by the volume of the disk shown in Figure 8.52b, of radius $f(p_i)$ and width Δx, obtained by revolving the rectangular region of height $f(p_i)$ and width Δx about the x-axis (Figure 8.52c).

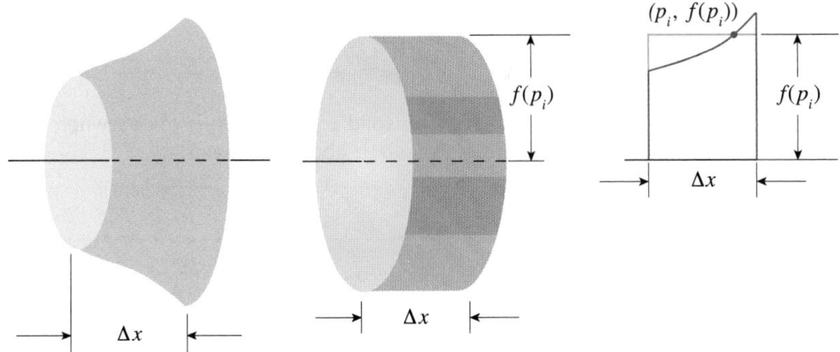

(a) Volume ΔV_i of object **(b)** Volume of disk **(c)** Approximating rectangle

Figure 8.52

Since the volume of a disk (cylinder) is equal to $\pi r^2 h$, we see that

$$\Delta V_i \approx \pi[f(p_i)]^2 \Delta x$$

This analysis suggests that the volume of the solid of revolution may be approximated by the sum of the volumes of n suitable disks—namely,

$$\pi[f(p_1)]^2 \Delta x + \pi[f(p_2)]^2 \Delta x + \cdots + \pi[f(p_n)]^2 \Delta x$$

(shown in Figure 8.53). This last expression is a Riemann sum of the function $g(x) = \pi[f(x)]^2$ over the interval $[a, b]$. Letting n approach infinity, we obtain the following formula.

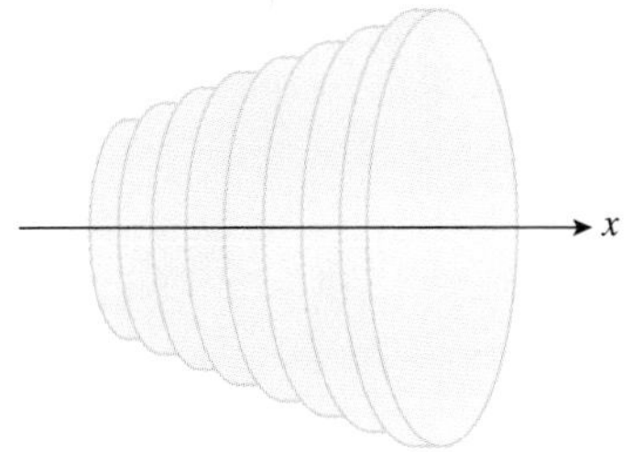

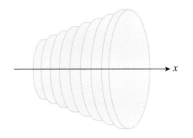

Figure 8.53
The solid of revolution is approximated by n disks.

Volume of a Solid of Revolution

The volume V of the solid of revolution obtained by revolving the region below the graph of $y = f(x)$ from $x = a$ to $x = b$ about the x-axis is

$$V = \pi \int_a^b [f(x)]^2\, dx \qquad (22)$$

Example 1

Find the volume of the solid of revolution obtained by revolving the region under the curve $y = f(x) = e^{-x}$ from $x = 0$ to $x = 1$ about the x-axis.

Solution

The region under the curve and the resulting solid of revolution are shown in Figure 8.54. Using Formula (22), we find that the required volume is given by

$$\pi \int_0^1 (e^{-x})^2 \, dx = \pi \int_0^1 e^{-2x} \, dx$$

$$= -\frac{\pi}{2} e^{-2x} \Big|_0^1 = \frac{\pi}{2}(1 - e^{-2})$$

cubic units.

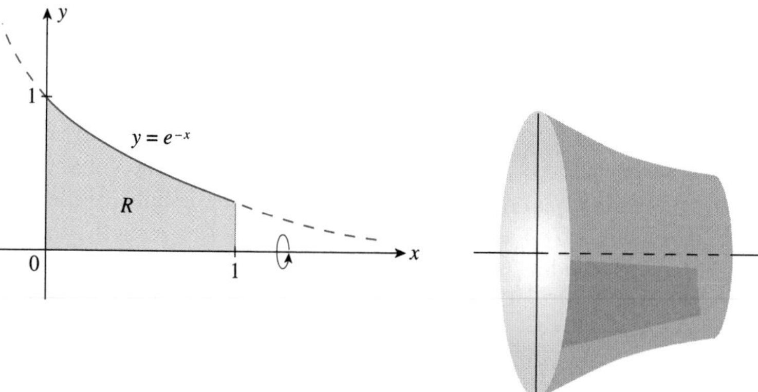

(a) The region under the curve $y = e^{-x}$ from $x = 0$ to $x = 1$

(b) The solid of revolution obtained when R is revolved about the x-axis

Figure 8.54

Example 2

Derive a formula for the volume of a right circular cone of radius r and height h.

Solution

The cone is shown in Figure 8.55a. It is the solid of revolution obtained by revolving the region R of Figure 8.55b about the x-axis. The region R is the region under the straight line $y = f(x) = (r/h)x$ between $x = 0$ and $x = h$.

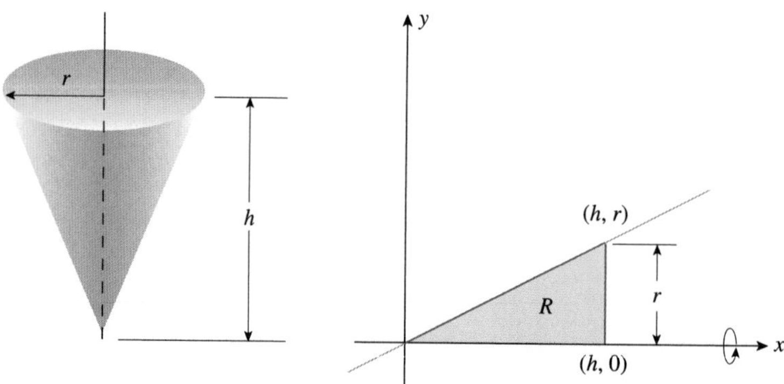

(a) A right circular cone of radius r and height h

(b) R is the region under the straight line $y = \frac{r}{h}x$ from $x = 0$ to $x = h$.

Figure 8.55

Using Formula (22), we see that the required volume is given by

$$V = \pi \int_0^h \left[\left(\frac{r}{h} \right) x \right]^2 dx = \frac{\pi r^2}{h^2} \cdot \int_0^h x^2\, dx$$

$$= \left(\frac{\pi r^2}{3h^2} \right) x^3 \Big|_0^h = \frac{1}{3} \pi r^2 h$$

cubic units.

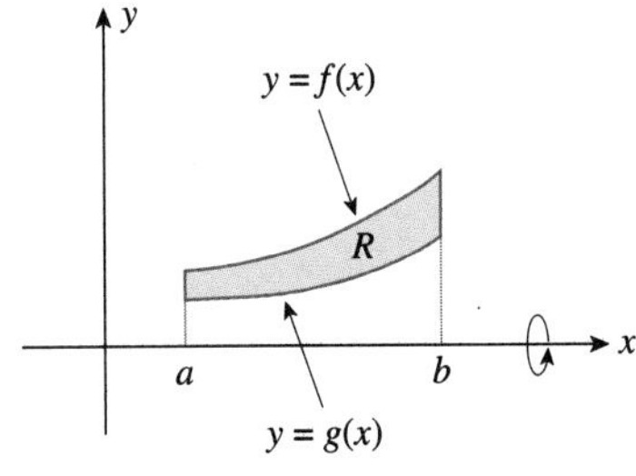

(a) R is the region bounded by the curves $y = f(x)$ and $y = g(x)$ from $x = a$ to $x = b$.

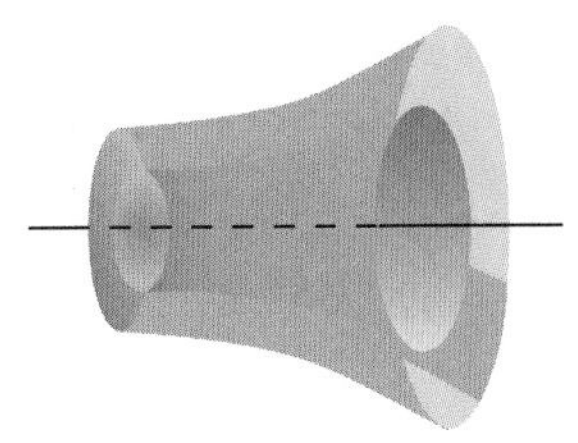

(b) The solid of revolution obtained by revolving R about the x-axis

Figure 8.56

Next, let's consider the solid of revolution obtained by revolving the region R bounded above by the graph of the nonnegative function $f(x)$ and below by the graph of the nonnegative function $g(x)$, from $x = a$ to $x = b$, about the x-axis (Figure 8.56).

To derive a formula for computing the volume of this solid of revolution, observe that the required volume is the volume of the solid of revolution obtained by revolving the region under the curve $y = f(x)$ from $x = a$ to $x = b$ about the x-axis *minus* the volume of the solid of revolution obtained by revolving the region under the curve $y = g(x)$ from $x = a$ to $x = b$ about the x-axis. Thus, the required volume is given by

$$V = \pi \int_a^b [f(x)]^2 dx - \pi \int_a^b [g(x)]^2\, dx \qquad (23)$$

or

$$V = \pi \int_a^b \{[f(x)]^2 - [g(x)]^2\}\, dx$$

Example 3

Find the volume of the solid of revolution obtained by revolving the region bounded by the curves $y = e^x$ and $y = x^2 + \frac{1}{2}$ from $x = 0$ to $x = 1$ about the x-axis (Figure 8.57).

Solution

Using Formula (23) with $f(x) = e^x$ and $g(x) = x^2 + \frac{1}{2}$, we see that the required volume is given by

$$V = \pi \int_0^1 \left\{ [e^x]^2 - \left[x^2 + \frac{1}{2} \right]^2 \right\} dx$$

$$= \pi \int_0^1 \left(e^{2x} - x^4 - x^2 - \frac{1}{4} \right) dx$$

$$= \pi \left(\frac{1}{2} e^{2x} - \frac{1}{5} x^5 - \frac{1}{3} x^3 - \frac{1}{4} x \right) \Big|_0^1$$

$$= \pi \left[\left(\frac{1}{2} e^2 - \frac{1}{5} - \frac{1}{3} - \frac{1}{4} \right) - \frac{1}{2} \right]$$

$$= \frac{\pi}{60} (30e^2 - 77)$$

or approximately 7.57 cubic units.

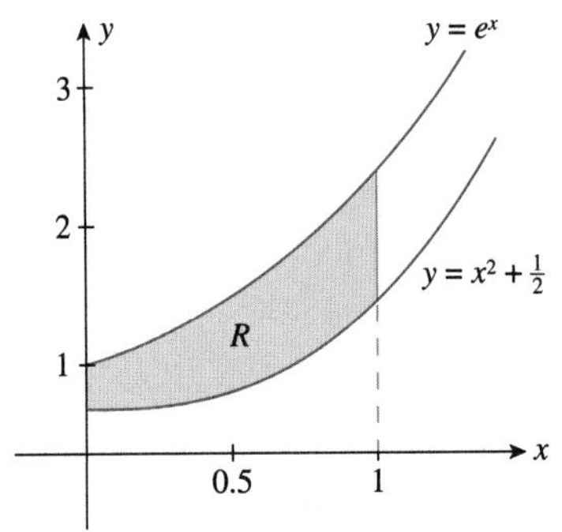

Figure 8.57

R is the region bounded by the curves $y = e^x$ and $y = x^2 + \frac{1}{2}$ from $x = 0$ to $x = 1$.

Example 4

Find the solid of revolution obtained by revolving the region bounded by the curves $y = f(x) = x^2$ and $y = g(x) = x^3$ about the x-axis.

Solution

We first find the point(s) of intersection of the two curves by solving the system of equations

$$y = x^2$$
$$y = x^3$$

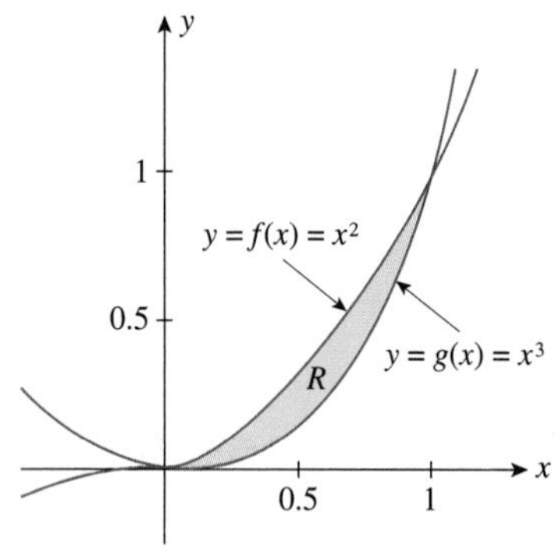

Figure 8.58
R is the region bounded by the curves $y = x^2$ and $y = x^3$.

We have

$$x^3 = x^2$$
$$x^3 - x^2 = 0$$
$$x^2(x - 1) = 0$$

so that $x = 0$ or $x = 1$, and the points of intersection are (0, 0) and (1, 1) (Figure 8.58).

Next, observe that $g(x) \leq f(x)$ for all values of x between $x = 0$ and $x = 1$ so that, using Formula (23), we find the required volume to be

$$\begin{aligned} V &= \pi \int_0^1 [(x^2)^2 - (x^3)^2]\, dx \\ &= \pi \int_0^1 (x^4 - x^6)\, dx \\ &= \pi\left(\frac{1}{5}x^5 - \frac{1}{7}x^7\right)\Bigg|_0^1 \\ &= \pi\left(\frac{1}{5} - \frac{1}{7}\right) \\ &= \frac{2\pi}{35} \end{aligned}$$

cubic units.

Application

Example 5

The external fuel tank for a space shuttle has a shape that may be obtained by revolving the region under the curve

$$f(x) = \begin{cases} 4\sqrt{10} & \text{if } -120 \leq x \leq 10 \\ \frac{1}{5}\sqrt{x(30 - x)} & \text{if } 10 < x \leq 30 \end{cases}$$

from $x = -120$ to $x = 30$ about the x-axis (Figure 8.59) where all measurements are given in feet. The tank carries liquid hydrogen for fueling the shuttle's three main engines. Estimate the capacity of the tank (231 cubic inches = 1 gallon).

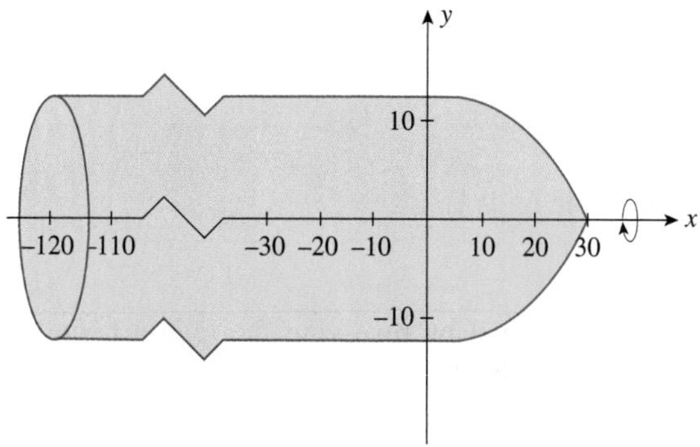

Figure 8.59
The solid of revolution obtained by revolving the region under the curve $y = f(x)$ about the x-axis.

Solution

The volume of the tank is given by

$$\begin{aligned}
V &= \pi\int_{-120}^{30} [f(x)]^2\,dx \\
&= \pi\int_{-120}^{10} (4\sqrt{10})^2\,dx + \pi\int_{10}^{30}\left[\frac{1}{5}\sqrt{x}(30-x)\right]^2 dx \\
&= 160\pi\int_{-120}^{10} dx + \frac{\pi}{25}\int_{10}^{30} x(30-x)^2\,dx \\
&= 160\pi x\Big|_{-120}^{10} + \frac{\pi}{25}\int_{10}^{30}(900x - 60x^2 + x^3)\,dx \\
&= 160\pi(130) + \frac{\pi}{25}\left(450x^2 - 20x^3 + \frac{1}{4}x^4\right)\Big|_{10}^{30} \\
&= 20{,}800\pi + \frac{\pi}{25}\left\{\left[450(30)^2 - 20(30)^3 + \frac{1}{4}(30)^4\right]\right. \\
&\quad \left. -\left[450(10)^2 - 20(10)^3 + \frac{1}{4}(10^4)\right]\right\} \\
&= 20{,}800\pi + 1600\pi = 22{,}400\pi
\end{aligned}$$

or approximately 70,372 cubic feet. Therefore, its capacity is approximately $(70{,}372)(12^3)/231$, or approximately 526,419 gallons.

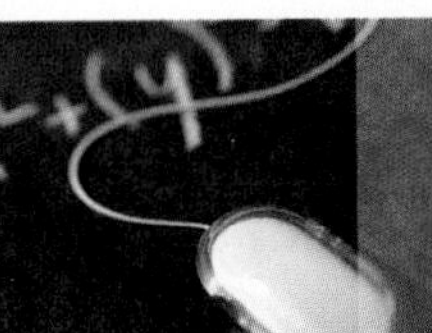

Self-Check Exercise 8.8

Find the volume of the solid of revolution obtained by revolving the region bounded by the graphs of $f(x) = 5 - x^2$ and $g(x) = 1$ about the x-axis.

The solution to Self-Check Exercise 8.8 can be found on page 525.

8.8 Exercises

In Exercises 1–14, find the volume of the solid of revolution obtained by revolving the region under the curve $y = f(x)$ from $x = a$ to $x = b$ about the x-axis.

1. $y = 3x$; $a = 0$, $b = 1$
2. $y = x + 1$; $a = 0$, $b = 2$
3. $y = \sqrt{x}$; $a = 1$, $b = 4$
4. $y = \sqrt{x+1}$; $a = 0$, $b = 3$
5. $y = \sqrt{1+x^2}$; $a = 0$, $b = 1$
6. $y = \sqrt{4-x^2}$; $a = -2$, $b = 2$
7. $y = 1 - x^2$; $a = -1$, $b = 1$
8. $y = 2x - x^2$; $a = 0$, $b = 2$
9. $y = e^x$; $a = 0$, $b = 1$
10. $y = e^{-3x}$; $a = 0$, $b = 1$
11. $y = \sin(x)$; $x = 0$, $x = \pi$
12. $y = \dfrac{1}{\sqrt{x^2+1}}$; $x = 0$, $x = \sqrt{3}$

13. $y = \cos(2x)$; $x = 0$, $x = \frac{\pi}{2}$

14. $y = \tan(x)$; $x = 0$, $x = \frac{\pi}{3}$

In Exercises 15–26, find the volume of the solid of revolution obtained by revolving the region bounded above by the curve $y = f(x)$ and below by the curve $y = g(x)$ from $x = a$ to $x = b$ about the x-axis.

15. $f(x) = x$ and $g(x) = x^2$; $a = 0$, $b = 1$

16. $f(x) = 1$ and $g(x) = \sqrt{x}$; $a = 0$, $b = 1$

17. $f(x) = 4 - x^2$ and $g(x) = 3$; $a = -1$, $b = 1$

18. $f(x) = 8$ and $g(x) = x^{3/2}$; $a = 0$, $b = 4$

19. $f(x) = \sqrt{16 - x^2}$ and $g(x) = x$; $a = 0$, $b = 2\sqrt{2}$

20. $f(x) = e^x$ and $g(x) = \frac{1}{x}$; $a = 1$, $b = 2$

21. $f(x) = e^x$ and $g(x) = e^{-x}$; $a = 0$, $b = 1$

22. $f(x) = e^x$ and $g(x) = \sqrt{x}$; $a = 1$, $b = 2$

23. $f(x) = 2\sin(x)$, $g(x) = x$; $a = 0$, $b = 1.9$

24. $f(x) = \sin(x)$, $g(x) = \cos(x)$; $a = \pi/4$, $b = \frac{\pi}{2}$

25. $f(x) = 4$, $g(x) = \cos^2(x)$; $a = 0$, $b = \pi$

26. $f(x) = \frac{1}{x}$, $g(x) = \sin(2x)$; $a = 1$, $b = 3$

In Exercises 27–36, find the volume of the solid of revolution obtained by revolving the region bounded by the graphs of the given functions about the x-axis.

27. $y = x$ and $y = \sqrt{x}$

28. $y = x^{1/3}$ and $y = x^2$

29. $y = \frac{1}{2}x + 3$ and $y = x^2$

30. $y = 2x$ and $y = x\sqrt{x + 1}$

31. $y = x^2$ and $y = 4 - x^2$

32. $y = x^2$ and $y = x(4 - x)$

33. $y = \frac{1}{x}$, $y = x$, and $y = 2x$

Hint: You will need to evaluate two integrals.

34. $y = \sqrt{x - 1}$ and $y = (x - 1)^2$

35. $f(x) = \sin(x)$ and $g(x) = -\sin(x)$ for any one of the enclosed areas.

36. $f(x) = \frac{1}{\sqrt{x^2 + 1}}$ and $g(x) = 0.5$

37. By computing the volume of the solid of revolution obtained by revolving the region under the semicircle $y = \sqrt{r^2 - x^2}$ from $x = -r$ to $x = r$ about the x-axis, show that the volume of a sphere of radius r is $\frac{4}{3}\pi r^3$ cubic units.

38. **Volume of a Football** Find the volume of the prolate spheroid (a solid of revolution in the shape of a football) obtained by revolving the region under the graph of the function $y = \frac{3}{5}\sqrt{25 - x^2}$ from $x = -5$ to $x = 5$ about the x-axis.

39. **Capacity of a Man-Made Lake** A man-made lake is approximately circular and has a cross section that is a region bounded above by the x-axis and below by the graph of

$$y = 10\left[\left(\frac{x}{100}\right)^2 - 1\right]$$

(see the accompanying figure). Find the approximate capacity of the lake.

Hint: Find the volume of the solid of revolution obtained by revolving the shaded region about the y-axis. Thus, $V = \pi\int_{-10}^{0}[f(y)]^2\,dy$, where $f(y)$ is obtained by solving the given equation for x in terms of y.

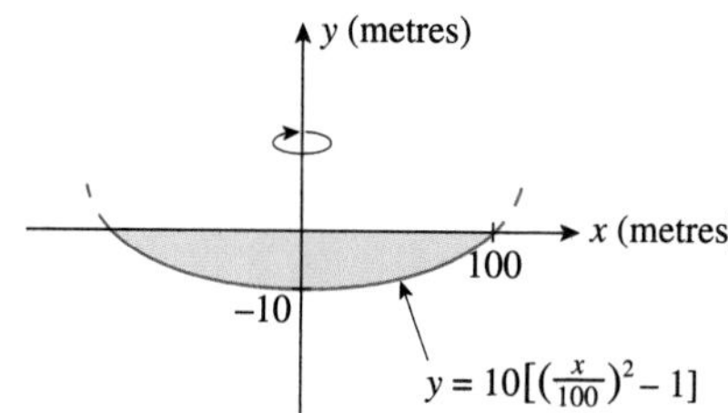

Solution to Self-Check Exercise 8.8

To find the points of intersection of the two curves, we solve the equation $5 - x^2 = 1$, giving $x = \pm 2$ (see the figure). Using Formula (23), we find the required volume:

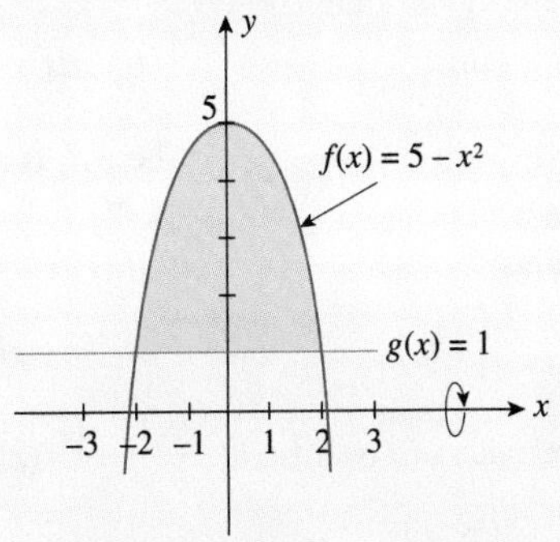

$$V = \pi \int_{-2}^{2} [(5 - x^2)^2 - 1^2]\, dx$$

$$= 2\pi \int_{0}^{2} (25 - 10x^2 + x^4 - 1)\, dx \qquad \text{By symmetry}$$

$$= 2\pi \int_{0}^{2} (x^4 - 10x^2 + 24)\, dx$$

$$= 2\pi \left(\frac{1}{5}x^5 - \frac{10}{3}x^3 + 24x \right) \Bigg|_{0}^{2}$$

$$= 2\pi \left(\frac{32}{5} - \frac{80}{3} + 48 \right)$$

$$= \frac{832\pi}{15} \text{ cubic units}$$

CHAPTER 8 Summary of Principal Formulas and Terms

Formulas

1. Indefinite integral of a constant — $\int k\, du = ku + C$
2. Power rule — $\int u^n\, du = \dfrac{u^{n+1}}{n+1} + C$
3. Constant multiple rule — $\int kf(u)\, du = k \int f(u)\, du$ (k, a constant)
4. Sum rule — $\int [f(u) \pm g(u)]\, du = \int f(u)\, du \pm \int g(u)\, du$
5. Indefinite integral of the exponential function — $\int e^u\, du = e^u + C$
6. Indefinite integral of $f(u) = \dfrac{1}{u}$ — $\int \dfrac{du}{u} = \ln |u| + C$

Indefinite integrals of trigonometric functions:

7. Sine. $\int \sin(x)\, dx = -\cos(x) + C$
8. Cosine. $\int \cos(x)\, dx = \sin(x) + C$
9. $\sec^2(x)$. $\int \sec^2(x)\, dx = \tan(x) + C$
10. $\sec(x)\tan(x)$. $\int \sec(x)\tan(x)\, dx = \sec(x) + C$
11. $\csc^2(x)$. $\int \csc(x)\cot(x)\, dx = -\csc(x) + C$
12. $\csc(x)\cos(x)$. $\int \csc(x)\cot(x)\, dx = -\csc(x) + C$
13. $\dfrac{1}{1 + x^2}$. $\int\int \dfrac{dx}{1 + x^2} = \arctan(x) + C$
14. $\dfrac{1}{\sqrt{1 - x^2}}$. $\int \dfrac{dx}{\sqrt{1 - x^2}} = \arcsin(x) + C$
15. Tangent. $\int \tan(x)\, dx = \ln(|\sec(x)|) + C$
16. Cotangent. $\int \cot(x)\, dx = \ln(|\sin(x)|) + C$
17. Secant. $\int \sec(x)\, dx = \ln(|\tan(x) + \sec(x)|) + C$
18. Cosecant. $\int \csc(x)\, dx = -\ln(|\cot(x) + \csc(x)|) + C$

19. Method of substitution $\int f'(g(x))g'(x)\,dx = \int f'(u)\,du$

20. Definite integral as the limit of a sum $\int_a^b f(x)\,dx = \lim_{n\to\infty} S_n$, where S_n is a Riemann sum

21. Fundamental theorem of calculus $\int_a^b f(x)\,dx = F(b) - F(a),\ F'(x) = F(x)$

22. Average value of f over $[a, b]$ $\frac{1}{b-a}\int_a^b f(x)\,dx$

23. Area between two curves $\int_a^b [f(x) - g(x)]\,dx,\ f(x) \geq g(x)$

24. Consumers' surplus $CS = \int_0^{\bar{x}} D(x)\,dx - \bar{p}\bar{x}$

25. Producers' surplus $PS = \bar{p}\bar{x} - \int_0^{\bar{x}} S(x)\,dx$

26. Accumulated (future) value of an income stream $A = e^{rT}\int_0^T R(t)e^{-rt}\,dt$

27. Present value of an income stream $PV = \int_0^T R(t)e^{-rt}\,dt$

28. Amount of an annuity $A = \frac{mP}{r}(e^{rT} - 1)$

29. Present value of an annuity $PV = \frac{mP}{r}(1 - e^{-rT})$

30. Coefficient of inequality of a Lorentz curve $L = 2\int_0^1 [x - f(x)]\,dx$

31. Volume of a solid of revolution $V = \pi\int_a^b [f(x)]^2\,dx$

Terms

antiderivative (429)
antidifferentiation (431)
integration (431)
$\int$, integral sign (431)
indefinite integral (431)
integrand (431)
constant of integration (431)
differential equation (438)
solution (438)
general solution (438)
particular solution (438)
initial condition (439)
initial value problem (439)
method of substitution (447)
area under the graph of f (463)
Riemann sum (466)
definite integral (467)
lower limit of integration (467)
upper limit of integration (467)
integrable (467)
fundamental theorem of calculus (472)
average value (488)
consumers' surplus (507)
producers' surplus (508)
accumulated future income stream (510)
total future value of an income stream (510)
present value of an income stream (511)
amount of an annuity (512)
Lorentz curve (514)
line of complete equality (515)
Gini index (515)
solid of revolution (518)

Review Exercises 

In Exercises 1–42, find each indefinite integral.

1. $\int (x^3 + 2x^2 - x)\,dx$
2. $\int \left(\frac{1}{3}x^3 - 2x^2 + 8\right) dx$
3. $\int \left(x^4 - 2x^3 + \frac{1}{x^2}\right) dx$
4. $\int (x^{1/3} - \sqrt{x} + 4)\,dx$
5. $\int x(2x^2 + x^{1/2})\,dx$
6. $\int (x^2 + 1)(\sqrt{x} - 1)\,dx$
7. $\int \left(x^2 - x + \frac{2}{x} + 5\right) dx$
8. $\int \sqrt{2x + 1}\,dx$
9. $\int (3x - 1)(3x^2 - 2x + 1)^{1/3}\,dx$
10. $\int x^2(x^3 + 2)^{10}\,dx$
11. $\int \frac{x - 1}{x^2 - 2x + 5}\,dx$
12. $\int 2e^{-2x}\,dx$
13. $\int \left(x + \frac{1}{2}\right) e^{x^2+x+1}\,dx$
14. $\int \frac{e^{-x} - 1}{(e^{-x} + x)^2}\,dx$
15. $\int \frac{(\ln x)^5}{x}\,dx$
16. $\int \frac{\ln x^2}{x}\,dx$
17. $\int x^3(x^2 + 1)^{10}\,dx$
18. $\int x\sqrt{x + 1}\,dx$
19. $\int \frac{x}{\sqrt{x - 2}\,dx}$
20. $\int \frac{3x}{\sqrt{x + 1}}\,dx$
21. $\int \sin 3x\,dx$
22. $\int \cos(x + \pi)\,dx$
23. $\int (3\sin x + 4\cos x)\,dx$
24. $\int (x^2 - \cos 2x)\,dx$
25. $\int \sec^2 2x\,dx$
26. $\int \csc^2 3x\,dx$
27. $\int x\cos x^2\,dx$

28. $\int x \sec x^2 \tan x^2\, dx$

29. $\int \csc \pi x \cot \pi x\, dx$

30. $\int \sec 2x \tan 2x\, dx$

31. $\int \sin^3 x \cos x\, dx$

32. $\int \cos^2 x \sin x\, dx$

33. $\int \sec \pi x\, dx$

34. $\int \csc(1 = x)\, dx$

35. $\int \sqrt{\cos x} \sin x\, dx$

36. $\int (\sin x)^{1/3} \cos x\, dx$

37. $\int \cos 3x \sqrt{1 - 2 \sin 3x}\, dx$

38. $\int \frac{\cos x\, dx}{\sqrt{1 + \sin x}}$

39. $\int \tan^3 x \sec^2 x\, dx$

40. $\int \sec^5 x \tan x\, dx$

41. $\int \csc^2 x(\cot x - 1)^3\, dx$

42. $\int \sec^2 x \sqrt{1 + \tan x}\, dx$

In Exercises 43–62, evaluate each definite integral.

43. $\int_0^1 (2x^3 - 3x^2 + 1)\, dx$

44. $\int_0^2 (4x^3 - 9x^2 + 2x - 1)\, dx$

45. $\int_1^4 (\sqrt{x} + x^{-3/2})\, dx$

46. $\int_0^1 20x(2x^2 + 1)^4\, dx$

47. $\int_{-1}^0 12(x^2 - 2x)(x^3 - 3x^2 + 1)^3\, dx$

48. $\int_4^7 x\sqrt{x - 3}\, dx$

49. $\int_0^2 \frac{x}{x^2 + 1}\, dx$

50. $\int_0^1 \frac{dx}{(5 - 2x)^2}$

51. $\int_0^2 \frac{4x}{\sqrt{1 + 2x^2}}\, dx$

52. $\int_0^2 xe^{(-1/2)x^2}\, dx$

53. $\int_{-1}^0 \frac{e^{-x}}{(1 + e^{-x})^2}\, dx$

54. $\int_1^e \frac{\ln x}{x}\, dx$

55. $\int_{-\pi/2}^{\pi/2} (\sin x + \cos x)\, dx$

56. $\int_0^{\pi} (2 \cos x + 3 \sin x)\, dx$

57. $\int_0^{\pi/6} \tan 2x\, dx$

58. $\int_{\pi/8}^{\pi/4} \cot 2x\, dx$

59. $\int_0^{\pi/12} \sec 3x\, dx$

60. $\int_0^{\pi/8} \tan 2x\, dx$

61. $\int_1^{e^{\pi}} \frac{(\sin(\ln x))}{x}\, dx$

62. $\int_0^{\pi/2} \frac{\cos x}{1 + \sin x}\, dx$

In Exercises 63–66, find the function f given that the slope of the tangent line to the graph at any point $(x, f(x))$ is $f'(x)$ and that the graph of f passes through the given point.

63. $f'(x) = 3x^2 - 4x + 1$; $(1, 1)$

64. $f'(x) = \frac{x}{\sqrt{x^2 + 1}}$; $(0, 1)$

65. $f'(x) = 1 - e^{-x}$; $(0, 2)$

66. $f'(x) = \frac{\ln x}{x}$; $(1, -2)$

67. Let $f(x) = -2x^2 + 1$ and compute the Riemann sum of f over the interval $[1, 2]$ by partitioning the interval into five subintervals of the same length ($n = 5$), where the points p_i ($1 \le i \le 5$) are taken to be the *right* end points of the respective subintervals.

In Exercises 68–71, find the area of the region under the graph of the function f from $x = a$ to $x = b$.

68. $f(x) = \cos \frac{x}{4}$; $a = 0, b = \pi$

69. $f(x) = x + \sin x$; $a = 0, b = \frac{\pi}{4}$

70. $f(x) = \tan x$; $a = 0, b = \frac{\pi}{4}$

71. $f(x) = \cot x + \csc x$; $a = \frac{\pi}{4}, b = \frac{\pi}{2}$

72. Find the area of the region bounded above by the graph of $y = x$ and below by the graph of $y = \sin x$ from $x = 0$ to $x = \pi$.

73. Find the area of the region bounded above by the graph of $y = \cos x$ and below by the graph of $y = x$ from $x = 0$ to $x = \pi/4$. Make a rough sketch of the region.

74. Find the area of the region under the curve $y = 3x^2 + 2x + 1$ from $x = -1$ to $x = 2$.

75. Find the area of the region under the curve $y = e^{2x}$ from $x = 0$ to $x = 2$.

76. Find the area of the region bounded by the graph of the function $y = 1/x^2$, the x-axis, and the lines $x = 1$ and $x = 3$.

77. Find the area of the region bounded by the curve $y = -x^2 - x + 2$ and the x-axis.

78. Find the area of the region bounded by the graphs of the functions $f(x) = e^x$ and $g(x) = x$ and the vertical lines $x = 0$ and $x = 2$.

79. Find the area of the region that is completely enclosed by the graphs of $f(x) = x^4$ and $g(x) = x$.

80. Find the area of the region between the curve $y = x(x - 1)(x - 2)$ and the x-axis.

81. Find the average value of the function

$$f(x) = \frac{x}{\sqrt{x^2 + 16}}$$

over the interval $[0, 3]$.

82. Find the volume of the solid of revolution obtained by revolving the region under the curve $f(x) = 1/x$ from $x = 1$ to $x = 3$ about the x-axis.

83. Find the volume of the solid of revolution obtained by revolving the region bounded above by the curve $f(x) = \sqrt{x}$ and below by the curve $g(x) = x^2$ about the x-axis.

84. Commuter Trends Due to the increasing cost of fuel, the manager of the City Transit Authority estimates that the number of commuters using the city subway system will increase at the rate of

$$3000(1 + 0.4t)^{-1/2} \qquad (0 \leq t \leq 36)$$

per month t mo from now. If 100,000 commuters are currently using the system, find an expression giving the total number of commuters who will be using the subway t mo from now. How many commuters will be using the subway 6 mo from now?

85. World Coal Production In 1980 the world produced 3.5 billion metric tons of coal. If output increased at the rate of

$$3.5e^{0.04t}$$

billion metric tons/year in year t ($t = 0$ corresponds to 1980), determine how much coal was produced worldwide between 1980 and the end of 1985.

86. Oil Production Based on current production techniques, the rate of oil production from a certain oil well t yr from now is estimated to be

$$R_1(t) = 100e^{0.05t}$$

thousand barrels/year. Based on a new production technique, however, it is estimated that the rate of oil production from that oil well t yr from now will be

$$R_2(t) = 100e^{0.08t}$$

thousand barrels/year. Determine how much additional oil will be produced over the next 10 yr if the new technique is adopted.

87. Shrimp Chitin Buildup The population of shrimp shed in a seasonally varying pattern that casts off shells at a rate of $P(t) = 2000 + 1200 \cos(\pi t/12)$ shells/month. An average shell contains about 0.24 grams of material. Assume that the lake bottom was dredged to remove all existing sediment. What total mass of shrimp shells is deposited in the lake bottom sediment per year?

88. Self-Sealing Tank Performance Added compounds in an underground fuel storage tank cause it to be self-sealing. Once a leak forms, the sealants interact with the soil to block the leak. A test leak causes a loss from the tank of

$$A(t) = 27.4e^{-0.86t} \text{ g/h}$$

What is the total loss in the day after the leak starts? In the week after the leak starts? Does the difference of these two quantities justify the term "self-sealing"?

89. Self-Sealing Tank Improvement Refer to Problem 88. In a test without the self-sealing additives the tank was found to leak at a rate of

$$A(t) = \frac{27.4}{x + 1} \text{ g/h}$$

How much leakage was prevented in the first day by using the self-sealant compounds?

Business and Economics Applications

90. Marginal Cost Functions The management of National Electric has determined that the daily marginal cost function associated with producing their automatic coffeemakers is given by

$$C'(x) = 0.00003x^2 - 0.03x + 20$$

where $C'(x)$ is measured in dollars/unit and x denotes the number of units produced. Management has also determined that the daily fixed cost incurred in producing these coffeemakers is \$500. What is the total cost incurred by National in producing the first 400 coffeemakers/day?

91. Marginal Revenue Functions Refer to Exercise 90. Management has also determined that the daily marginal revenue function associated with producing and selling their coffeemakers is given by

$$R'(x) = -0.03x + 60$$

where x denotes the number of units produced and sold and $R'(x)$ is measured in dollars/unit.

a. Determine the revenue function $R(x)$ associated with producing and selling these coffeemakers.

b. What is the demand equation relating the wholesale unit price to the quantity of coffeemakers demanded?

92. Computer Resale Value Franklin National Life Insurance Company purchased new computers for \$200,000. If the rate at which the computers' resale value changes is given by the function

$$V'(t) = 3800(t - 10)$$

where t is the length of time since the purchase date and $V'(t)$ is measured in dollars/year, find an expression $V(t)$ that gives the resale value of the computers after t yr. How much would the computers cost after 6 yr?

93. Projection TV Sales The marketing department of Vista Vision forecasts that sales of their new line of projection television systems will grow at the rate of

$$3000 - 2000e^{0.04t} \qquad (0 \leq t \leq 24)$$

units/month once they are introduced into the market. Find an expression giving the total number of the projection television systems that Vista may expect to sell t mo from the time they are put on the market. How many units of the television systems can Vista expect to sell during the first year?

94. Marginal Cost Functions The management of a division of Ditton Industries has determined that the daily marginal cost function associated with producing their hot-air corn poppers is given by

$$C'(x) = 0.00003x^2 - 0.03x + 10$$

where $C'(x)$ is measured in dollars/unit and x denotes the number of units manufactured. Management has also determined that the daily fixed cost incurred in producing these corn poppers is \$600. Find the total cost incurred by Ditton in producing the first 500 corn poppers.

95. **Demand for Digital Camcorder Tapes** The demand function for a brand of blank digital camcorder tapes is given by

$$p = -0.01x^2 - 0.2x + 23$$

where p is the wholesale unit price in dollars and x is the quantity demanded each week, measured in units of a thousand. Determine the consumers' surplus if the wholesale unit price is \$8/tape.

96. **Consumers' and Producers' Surplus** The quantity demanded x (in units of a hundred) of the Sportsman 5 × 7 tents, per week, is related to the unit price p (in dollars) by the relation

$$p = -0.1x^2 - x + 40$$

The quantity x (in units of a hundred) that the supplier is willing to make available in the market is related to the unit price by the relation

$$p = 0.1x^2 + 2x + 20$$

If the market price is set at the equilibrium price, find the consumers' surplus and the producers' surplus.

97. **Retirement Account Savings** Chi-Tai plans to deposit \$4000/year in his Keogh Retirement Account. If interest is compounded continuously at the rate of 8%/year, how much will he have in his retirement account after 20 yr?

98. **Installment Contracts** Glenda sold her house under an installment contract whereby the buyer gave her a down payment of \$9000 and agreed to make monthly payments of \$925/month for 30 yr. If the prevailing interest rate is 12%/year compounded continuously, find the present value of the purchase price of the house.

99. **Present Value of a Franchise** Alicia purchased a 10-yr franchise for a health spa that is expected to generate income at the rate of

$$P(t) = 80{,}000$$

dollars/year. If the prevailing interest rate is 10%/year compounded continuously, find the present value of the franchise.

100. **Income Distribution of a Country** A certain country's income distribution is described by the function

$$f(x) = \frac{17}{18}x^2 + \frac{1}{18}x$$

a. Sketch the Lorentz curve for this function.
b. Compute $f(0.3)$ and $f(0.6)$ and interpret your results.
c. Compute the coefficient of inequality for this Lorentz curve.

Biological and Life Sciences Applications

101. **Population Growth** The population of a certain west coast city, currently 80,000, is expected to grow exponentially in the next 5 yr with a growth constant of 0.05. If the prediction comes true, what will be the average population of the city over the next 5 yr?

ADDITIONAL TOPICS IN INTEGRATION

Photo: Thinkstock/Jupiter Images

Besides the basic rules of integration developed in Chapter 8, there are more sophisticated techniques for finding the antiderivatives of functions. We begin this chapter by looking at the method of integration by parts. We then look at a technique of integration that involves using tables of integrals that have been compiled for this purpose. We also look at numerical methods of integration, which enable us to obtain approximate solutions to definite integrals, especially those whose exact value cannot be found otherwise. More specifically, we study the trapezoidal rule and Simpson's rule. Numerical integration methods are especially useful when the integrand is known only at discrete points. Finally, we learn how to evaluate integrals in which the intervals of integration are unbounded. Such integrals, called *improper integrals*, play an important role in the study of probability.

Photo: Thinkstock/Jupiter Images

What is the area of the oil spill caused by a grounded tanker? In Example 5, page 564, you will see how to determine the area of the oil spill.

Photo: © Barrett & MacKay

9.1 Integration by Parts

The Method of Integration by Parts

Integration by parts is another technique of integration that, like the method of substitution discussed in Chapter 8, is based on a corresponding rule of differentiation. In this case, the rule of differentiation is the product rule, which asserts that if f and g are differentiable functions, then

$$\frac{d}{dx}[f(x)g(x)] = f(x)g'(x) + g(x)f'(x) \qquad (1)$$

If we integrate both sides of Equation (1) with respect to x, we obtain

$$\int \frac{d}{dx} f(x)g(x)\,dx = \int f(x)g'(x)\,dx + \int g(x)f'(x)\,dx$$

$$f(x)g(x) = \int f(x)g'(x)\,dx + \int g(x)f'(x)\,dx$$

This last equation, which may be written in the form

$$\int f(x)g'(x)\,dx = f(x)g(x) - \int g(x)f'(x)\,dx \qquad (2)$$

is called the formula for **integration by parts.** This formula is useful since it enables us to express one indefinite integral in terms of another that may be easier to evaluate. Formula (2) may be simplified by letting

$$u = f(x) \qquad dv = g'(x)\,dx$$
$$du = f'(x)\,dx \qquad v = g(x)$$

giving the following version of the formula for integration by parts.

Integration by Parts Formula

$$\int u\,dv = uv - \int v\,du \qquad (3)$$

Example 1

Evaluate $\int xe^x\,dx$.

Solution

No method of integration developed thus far enables us to evaluate the given indefinite integral in its present form. Therefore, we attempt to write it in terms of an indefinite integral that will be easier to evaluate. Let's use the integration by parts Formula (3) by letting

$$u = x \qquad \text{and} \qquad dv = e^x\,dx$$

so that

$$du = dx \qquad \text{and} \qquad v = e^x$$

Therefore,

$$\begin{aligned}\int xe^x\,dx &= \int u\,dv \\ &= uv - \int v\,du \\ &= xe^x - \int e^x\,dx \\ &= xe^x - e^x + C \\ &= (x-1)e^x + C\end{aligned}$$

The success of the method of integration by parts depends on the proper choice of u and dv. For example, if we had chosen

$$u = e^x \qquad \text{and} \qquad dv = x\,dx$$

in the last example, then

$$du = e^x\,dx \qquad \text{and} \qquad v = \frac{1}{2}x^2$$

Thus, (3) would have yielded

$$\begin{aligned}\int xe^x\,dx &= \int u\,dv \\ &= uv - \int v\,du \\ &= \frac{1}{2}x^2e^x - \int \frac{1}{2}x^2e^x\,dx\end{aligned}$$

Since the indefinite integral on the right-hand side of this equation is not readily evaluated (it is in fact more complicated than the original integral!), choosing u and dv as shown has not helped us evaluate the given indefinite integral.

In general, we can use the following guidelines.

Guidelines for Choosing u and dv

Choose u and dv so that

1. du is simpler than u.
2. dv is easy to integrate.

Example 2

Evaluate $\int x \ln x\, dx$.

Solution

Letting

$$u = \ln x \qquad \text{and} \qquad dv = x\, dx$$

we have

$$du = \frac{1}{x}\, dx \qquad \text{and} \qquad v = \frac{1}{2}x^2$$

Therefore,

$$\begin{aligned}\int x \ln x\, dx &= \int u\, dv = uv - \int v\, du \\ &= \frac{1}{2}x^2 \ln x - \int \frac{1}{2}x^2 \cdot \left(\frac{1}{x}\right) dx \\ &= \frac{1}{2}x^2 \ln x - \frac{1}{2}\int x\, dx \\ &= \frac{1}{2}x^2 \ln x - \frac{1}{4}x^2 + C \\ &= \frac{1}{4}x^2(2 \ln x - 1) + C\end{aligned}$$

In Example 1 we chose $u = x$ and it was a good choice because the differentiation of u removed the x from the integral, leaving only e^x to integrate. It would seem that when a single x is present that $u = x$ might be a good choice. In Example 2, however, this choice is not made and would be a bad choice. This is because we do not have an antiderivative for $v = \ln(x)$ which is what v would be. The choice of u and v is critical for integration by parts but there is no simple rule for the choice. Instead you must look at what will happen next and try to see which choice will give an easier integral, after applying the formula for integration by parts.

Example 3

Evaluate $\int \frac{xe^x}{(x + 1)^2}\, dx$.

Solution

Let

$$u = xe^x \qquad \text{and} \qquad dv = \frac{1}{(x + 1)^2}\, dx$$

Then,

$$du = (xe^x + e^x)\, dx = e^x(x + 1)\, dx \qquad \text{and} \qquad v = -\frac{1}{x + 1}$$

Therefore,

$$\begin{aligned}\int \frac{xe^x}{(x + 1)^2}\, dx &= \int u\, dv = uv - \int v\, du \\ &= xe^x\left(\frac{-1}{x + 1}\right) - \int \left(-\frac{1}{x + 1}\right) e^x(x + 1)\, dx\end{aligned}$$

$$= -\frac{xe^x}{x+1} + \int e^x\,dx$$
$$= -\frac{xe^x}{x+1} + e^x + C$$
$$= \frac{e^x}{x+1} + C$$

The next example shows that repeated applications of the technique of integration by parts is sometimes required to evaluate an integral.

Example 4

Find $\int x^2e^x\,dx$.

Solution

Let

$$u = x^2 \qquad \text{and} \qquad dv = e^x\,dx$$

so that

$$du = 2x\,dx \qquad \text{and} \qquad v = e^x$$

Therefore,

$$\int x^2e^x\,dx = \int u\,dv = uv - \int v\,du$$
$$= x^2e^x - \int e^x(2x)\,dx = x^2e^x - 2\int xe^x\,dx$$

To complete the solution of the problem, we need to evaluate the integral

$$\int xe^x\,dx$$

But this integral may be found using integration by parts. In fact, you will recognize that this integral is precisely that of Example 1. Using the results obtained there, we now find

$$\int x^2e^x\,dx = x^2e^x - 2[(x-1)e^x] + C = e^x(x^2 - 2x + 2) + C$$

GROUP DISCUSSION

1. Use the method of integration by parts to derive the formula
$$\int x^ne^{ax}\,dx = \frac{1}{a}x^ne^{ax} - \frac{n}{a}\int x^{n-1}e^{ax}\,dx$$
where n is a positive integer and a is a real number.
2. Use the formula of part 1 to evaluate
$$\int x^3e^x\,dx.$$

Hint: You may find the results of Example 4 helpful.

Example 5

Find $\int x \cdot \cos(x)\,dx$.

Solution

Let

$$u = x \qquad \text{and} \qquad dv = \cos(x)\,dx$$

so that

$$du = dx \qquad \text{and} \qquad v = \sin(x)$$

Therefore

$$\int x \cdot \cos(x)\,dx = \int u\,dv = uv - \int v\,du$$

$$= x\sin(x) - \int \sin(x)\,dx = x\sin(x) - (-\cos(x)) + C$$

$$= x\sin(x) + \cos(x) + C$$

Application

Example 6

Oil Production The estimated rate at which oil will be produced from a certain oil well t years after production has begun is given by

$$R(t) = 100te^{-0.1t}$$

thousand barrels per year. Find an expression that describes the total production of oil at the end of year t.

Solution

Let $T(t)$ denote the total production of oil from the well at the end of year $t (t \geq 0)$. Then, the rate of oil production will be given by $T'(t)$ thousand barrels per year. Thus,

$$T'(t) = R(t) = 100te^{-0.1t}$$

so

$$T(t) = \int 100te^{-0.1t}\,dt$$

$$= 100\int te^{-0.1t}\,dt$$

We use the technique of integration by parts to evaluate this integral. Let

$$u = t \qquad \text{and} \qquad dv = e^{-0.1t}dt$$

so that

$$du = dt \qquad \text{and} \qquad v = -\frac{1}{0.1}e^{-0.1t} = -10e^{-0.1t}$$

Therefore,

$$T(t) = 100\left[-10te^{-0.1t} + 10\int e^{-0.1t}dt\right]$$

$$= 100[-10te^{-0.1t} - 100e^{-0.1t}] + C$$

$$= -1000e^{-0.1t}(t + 10) + C$$

To determine the value of C, note that the total quantity of oil produced at the end of year 0 is nil, so $T(0) = 0$. This gives

$$T(0) = -1000(10) + C = 0$$

$$C = 10{,}000$$

Thus, the required production function is given by

$$T(t) = -1000e^{-0.1t}(t + 10) + 10{,}000$$

EXPLORING WITH TECHNOLOGY

Refer to Example 6.

1. Use a graphing utility to plot the graph of

$$T(t) = -1000e^{-0.1t}(t + 10) + 10{,}000$$

using the viewing window $[0, 100] \cdot [0, 12{,}000]$. Use **TRACE** to see what happens when t is very large.

2. Verify the result of part 1 by evaluating $\lim_{t\to\infty} T(t)$. Interpret your results.

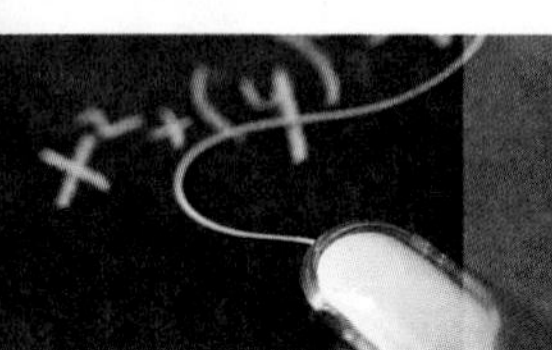

Self-Check Exercises 9.1

1. Evaluate $\int x^2 \ln x\, dx$.

2. Since the inauguration of Ryan's Express at the beginning of 2000, the number of passengers (in millions) flying on this commuter airline has been growing at the rate of

$$R(t) = 0.1 + 0.2te^{-0.4t}$$

passengers/year ($t = 0$ corresponds to the beginning of 2000). Assuming that this trend continues through 2004, determine how many passengers will have flown on Ryan's Express by that time.

Solutions to Self-Check Exercises 9.1 can be found on page 539.

9.1 Exercises

In Exercises 1–32, find each indefinite integral.

1. $\int xe^{2x}\, dx$
2. $\int xe^{-x}\, dx$
3. $\int xe^{x/4}\, dx$
4. $\int 6xe^{3x}\, dx$
5. $\int (e^x - x)^2\, dx$
6. $\int (e^{-x} + x)^2\, dx$
7. $\int (x + 1)e^x\, dx$
8. $\int (x - 3)e^{3x}\, dx$
9. $\int x(x + 1)^{-3/2}\, dx$
10. $\int x(x + 4)^{-2}\, dx$
11. $\int x\sqrt{x - 5}\, dx$
12. $\int \frac{x}{\sqrt{2x + 3}}\, dx$
13. $\int x \ln 2x\, dx$
14. $\int x^2 \ln 2x\, dx$
15. $\int x^3 \ln x\, dx$
16. $\int \sqrt{x} \ln x\, dx$
17. $\int \sqrt{x} \ln \sqrt{x}\, dx$
18. $\int \frac{\ln x}{\sqrt{x}}\, dx$
19. $\int \frac{\ln x}{x^2}\, dx$
20. $\int \frac{\ln x}{x^3}\, dx$
21. $\int \ln x\, dx$

    Hint: Let $u = \ln x$ and $dv = dx$.
22. $\int \ln(x + 1)\, dx$
23. $\int x^2e^{-x}\, dx$

    Hint: Integrate by parts twice.
24. $\int e^{-\sqrt{x}}\, dx$

    Hint: First, make the substitution $u = \sqrt{x}$; then, integrate by parts.

25. $\int x(\ln x)^2\,dx$

Hint: Integrate by parts twice.

26. $\int x\ln(x+1)\,dx$

Hint: First, make the substitution $u = x + 1$; then, integrate by parts.

27. $\int x\sin(x)\,dx$

28. $\int \arctan(x)\,dx$

Hint: Let $u = \arctan(x)$ and $dv = dx$.

29. $\int x\cos(3x)\,dx$

30. $\int x^2\cos(x)\,dx$

31. $\int x\sin(2x)\,dx$

32. $\int \frac{\frac{1}{x}\sin\left(\frac{1}{x}\right)}{x}\,dx$

In Exercises 33–43, determine if substitution or integration by parts is the correct method, and then find the indefinite integral. In some cases both methods may be required.

33. $\int \frac{2x}{x^2+1}\,dx$

34. $\int (3x+1)e^{2x}\,dx$

35. $\int x\sin(x)\cos(x)\,dx$

36. $\int x^2e^{1-x^3}\,dx$

37. $\int x\cos(2x^2)\,dx$

38. $\int x^2e^{2x}\,dx$

39. $\int 2x\ln(5x)\,dx$

40. $\int x\sec^2(x)\,dx$

41. $\int \frac{dx}{4x^2+4x+2}$

42. $\int (x+2)e^{2x}\,dx$

43. $\int xe^{2x^2}\,dx$

In Exercises 44–52, evaluate each definite integral by using the method of integration by parts.

44. $\int_0^{\ln 2} xe^x\,dx$

45. $\int_0^2 xe^{-x}\,dx$

46. $\int_1^4 \ln x\,dx$

47. $\int_1^2 x\ln x\,dx$

48. $\int_0^2 xe^{2x}\,dx$

49. $\int_0^1 x^2e^{-x}\,dx$

50. $\int_0^{\pi/2} x\cos(x)\,dx$

51. $\int_0^2 \arctan(x)\,dx$

52. $\int_0^{\pi/3} x^2\sin(x)\,dx$

53. Find the function f given that the slope or the tangent line to the graph of f at any point $(x, f(x))$ is xe^{-2x} and that the graph passes through the point (0, 3).

54. Find the function f given that the slope of the tangent line to the graph of f at any point $(x, f(x))$ is $x\sqrt{x+1}$ and that the graph passes through the point (3, 6).

55. Find the area of the region under the graph of $f(x) = \ln x$ from $x = 1$ to $x = 5$.

56. Find the area of the region under the graph of $f(x) = xe^{-x}$ from $x = 0$ to $x = 3$.

57. Using integration by parts twice, integrate

$$\int \sin(x)e^x\,dx$$

After the second integration by parts you will still have an integral. Finish the problem with algebra, not integration.

58. We saw in Chapter 8 that the integral of a sine or cosine function is zero if we integrate over whole periods of the function. Compute the integral

$$\int_0^{2\pi} x\sin(x)\,dx$$

The result should not be zero. Explain why the rule does not work when we multiply $\sin(x)$ by x.

59. **Velocity of a Dragster** The velocity of a dragster t sec after leaving the starting line is

$$100te^{-0.2t}$$

ft/sec. What is the distance covered by the dragster in the first 10 sec of its run?

60. **Production of Steam Coal** In keeping with the projected increase in worldwide demand for steam coal, the boiler-firing fuel used for generating electricity, the management of Consolidated Mining has decided to step up its mining operations. Plans call for increasing the yearly production of steam coal by

$$2te^{-0.05t}$$

million metric tons/year for the next 20 yr. The current yearly production is 20 million metric tons. Find a function that describes Consolidated's total production of steam coal at the end of t yr. How much coal will Consolidated have produced over the next 20 yr if this plan is carried out?

Business and Economics Applications

61. **Alcohol-Related Traffic Accidents** As a result of increasingly stiff laws aimed at reducing the number of alcohol-related

traffic accidents, preliminary data indicate that the number of such accidents has been changing at the rate of

$$R(t) = -10 - te^{0.1t}$$

accidents/month t mo after the laws took effect. There were 982 alcohol-related accidents for the year before the enactment of the laws. Determine how many alcohol-related accidents were expected during the first year the laws were in effect.

62. **Growth of HMOs** The membership of the Cambridge Community Health Plan (a health maintenance organization) is projected to grow at the rate of $9\sqrt{t+1}\ln\sqrt{t+1}$ thousand people/year, t yr from now. If the HMO's current membership is 50,000, what will be the membership 5 yr from now?

63. **A Mixture Problem** Two tanks are connected in tandem as shown in the following figure. Each tank contains 240 litres of water. Starting at time $t = 0$, brine containing 1 kg/L of salt flows into tank 1 at the rate of 8 L/min. The mixture then enters and leaves tank 2 at the same rate. The mixtures in both tanks are stirred uniformly. It can be shown that the amount of salt in tank 2 after t min is given by

$$A(t) = 90(1 - e^{-t/30}) - 6te^{-t/30}$$

where $A(t)$ is measured in kilograms.

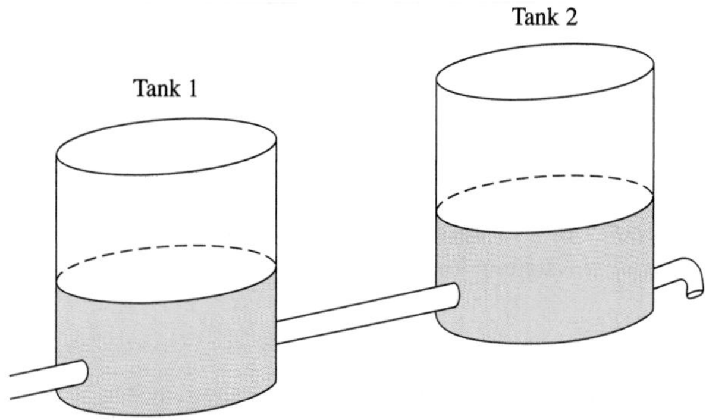

a. What is the initial amount of salt in tank 2?
b. What is the amount of salt in tank 2 after 3 h (180 min)?
c. What is the average amount of salt in tank 2 over the first 3 h?

64. **Compact Disc Sales** Sales of the latest recording by Brittania, a British rock group, are currently $2te^{-0.1t}$ units/week (each unit representing 10,000 discs), where t denotes the number of weeks since the recording's release. Find an expression that gives the total number of discs sold as a function of t.

65. **Average Price of a Commodity** The price of a certain commodity in dollars/unit at time t (measured in weeks) is given by

$$p = 8 + 4e^{-2t} + te^{-2t}$$

What is the average price of the commodity over the 4-wk period from $t = 0$ to $t = 4$?

66. **Rate of Return on an Investment** Suppose an investment is expected to generate income at the rate of

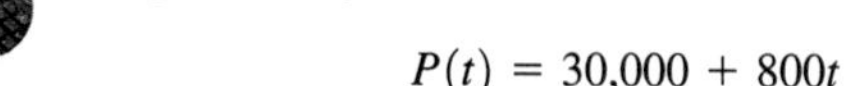

$$P(t) = 30{,}000 + 800t$$

dollars/year for the next 5 yr. Find the present value of this investment if the prevailing interest rate is 8%/year compounded continuously.

Hint: Use Formula (18), Section 8.7 (page 511).

67. **Present Value of a Franchise** Tracy purchased a 15-yr franchise for a computer outlet store that is expected to generate income at the rate of

$$P(t) = 50{,}000 + 3000t$$

dollars/year. If the prevailing interest rate is 10%/year compounded continuously, find the present value of the franchise.

Hint: Use Formula (18), Section 8.7 (page 511).

Biological and Life Sciences Applications

68. **Concentration of a Drug in the Bloodstream** The concentration (in milligrams/millilitre) of a certain drug in a patient's bloodstream t hr after it has been administered is given by $C(t) = 3te^{-t/3}$ mg/mL. Find the average concentration of the drug in the patient's bloodstream over the first 12 hr after administration.

69. **Average Load** A population of infectious bacteria is found to number $2.1te^{-1.4t}$ individuals t hours after an infection starts. Compute the average bacterial load in the first six hours.

70. **Hormone Produced** The production of hormones in a patient's bloodstream t hours after an infection is found to be well modelled by the equation

$$h(t) = 2.47\arctan(t)\text{ mg/h}$$

Compute the total number of milligrams of hormone produced in the first 12 hours.

In Exercises 71 and 72, determine whether the statement is true or false. If it is true, explain why it is true. If it is false, give an example to show why it is false.

71. $\int u\,dv + \int v\,du = uv$

72. $\int e^x g'(x)\,dx = e^x g(x) - \int e^x g(x)\,dx$

Solutions to Self-Check Exercises 9.1

1. Let $u = \ln x$ and $dv = x^2\,dx$ so that $du = \frac{1}{x}\,dx$ and $v = \frac{1}{3}x^3$.

Therefore,

$$\begin{aligned}\int x^2 \ln x\,dx &= \int u\,dv = uv - \int v\,du \\ &= \frac{1}{3}x^3 \ln x - \int \frac{1}{3}x^2\,dx \\ &= \frac{1}{3}x^3 \ln x - \frac{1}{9}x^3 + C \\ &= \frac{1}{9}x^3(3\ln x - 1) + C\end{aligned}$$

2. If $N(t)$ denotes the total number of passengers who will have flown on Ryan's Express by the end of year t, then $N'(t) = R(t)$, so that

$$\begin{aligned}N(t) &= \int R(t)\,dt \\ &= \int (0.1 + 0.2te^{-0.4t})\,dt \\ &= \int 0.1\,dt + 0.2\int te^{-0.4t}\,dt\end{aligned}$$

We now use the technique of integration by parts on the second integral. Letting $u = t$ and $dv = e^{-0.4t}\,dt$, we have

$$du = dt \qquad \text{and} \qquad v = -\frac{1}{0.4}e^{-0.4t} = -2.5e^{-0.4t}$$

Therefore,

$$\begin{aligned}N(t) &= 0.1t + 0.2\left[-2.5te^{-0.4t} + 2.5\int e^{-0.4t}\,dt\right] \\ &= 0.1t - 0.5te^{-0.4t} - \frac{0.5}{0.4}e^{-0.4t} + C \\ &= 0.1t - 0.5(t + 2.5)e^{-0.4t} + C\end{aligned}$$

To determine the value of C, note that $N(0) = 0$, which gives

$$N(0) = -0.5(2.5) + C = 0$$
$$C = 1.25$$

Therefore,

$$N(t) = 0.1t - 0.5(t + 2.5)e^{-0.4t} + 1.25$$

The number of passengers who will have flown on Ryan's Express by the end of 2004 is given by

$$\begin{aligned}N(5) &= 0.1(5) - 0.5(5 + 2.5)e^{-0.4(5)} + 1.25 \\ &= 1.242493\end{aligned}$$

—that is, 1,242,493 passengers.

9.2 Integration by Partial Fractions

The Method of Partial Fractions

Unlike many other methods of integration such as substitution from Chapter 8 or integration by parts in Section 9.1, partial fractions is *not* a neat reversal of a rule for differentiation. Rather it relies on an algebraic technique. The basis of this is the following fact:

Coefficients of Equal Polynomials

If two polynomials are equal then the coefficients of like terms are also equal.

Example 1

Suppose that $ax^2 + bx + c = x^2 + 3x + 5$. Find the value of a, b, and c.

Solution

Set the terms of the equal polynomials with the same powers equal to one another.

$$\begin{aligned}ax^2 &= x^2 \\ bx &= 3x \\ c &= 5\end{aligned}$$

From which we deduce that $a = 1$ and $b = 3$ as well as the obvious fact that $c = 5$.

This fact is an algebra fact, not a calculus fact, but it permits us to integrate a type of function we could not integrate before.

Example 2

Compute the indefinite integral $\displaystyle\int \frac{dx}{(x-1)(x-2)}$.

Solution

The first step in partial fractions is to assume that we can separate the fractions in the following fashion. We assume that there are some number A and B so that

$$\frac{1}{(x-1)(x-2)} = \frac{A}{x-1} + \frac{B}{x-2}$$

We then combine the fraction on the right and get

$$\frac{1}{(x-1)(x-2)} = \frac{A(x-2) + B(x-1)}{(x-1)(x-2)}$$

Since the denominators of the fractions are equal, we solve for the equality of the numerators and get

$$\begin{aligned} 1 &= A(x-2) + B(x-1) \\ 1 &= Ax - 2A + Bx - B \\ 1 &= (A+B)x + (-2A - B) \end{aligned}$$

This is a pair of equal polynomials, so we can apply the rule that says equal polynomials have equal coefficients. To make this clearer we rewrite the polynomial on the left-hand side of the equality to get

$$0x + 1 = (A+B)x + (-2A - B)$$

This gives us two equations in two unknowns,

$$\begin{cases} 0 = A + B \\ 1 = -2A - B \end{cases}$$

The first equation tells us that $A = -B$. Putting this information into the second equation gives us $1 = 2B - B$ or $B = 1$. This also tells us $A = -1$ and so in the end we have

$$\frac{1}{(x-1)(x-2)} = \frac{1}{x-2} - \frac{1}{x-1}$$

When working your own problems it is a good idea to combine the fractions you have computed to double-check the result. Let's recombine the fractions and make sure we did the algebra correctly.

$$\begin{aligned} \frac{1}{x-2} - \frac{1}{x-1} &= \frac{x-1}{(x-1)(x-2)} - \frac{(x-2)}{(x-1)(x-2)} \\ &= \frac{(x-1) - (x-2)}{(x-1)(x-2)} \\ &= \frac{x - x - 1 - (-2)}{(x-1)(x-2)} \\ &= \frac{1}{(x-1)(x-2)} \end{aligned}$$

So we see that our algebraic decomposition of the problem was correct.

Now that we have broken the original fraction into two simpler fractions we are ready to perform the integration.

$$\int \frac{dx}{(x-1)(x-2)} = \int \left(\frac{1}{x-2} - \frac{1}{x-1}\right) dx$$
$$= \int \frac{dx}{x-2} - \int \frac{dx}{x-1}$$

The two integrals can both be done by substitution to give us an answer of

$$\int \frac{dx}{(x-1)(x-2)} = \ln|x-2| - \ln|x-1| + C$$

Any integral with a quadratic denominator of the sort we had in Example 2, one with two different linear factors, is a candidate for integration by partial fractions. If the factors were the same, then the integral would be a type of substitution integral that we learned in Chapter 8. It is also possible to have a quadratic with no roots, which we will deal with in the next section.

Example 3

Compute the indefinite integral $\int \frac{2x+3}{(x+2)(x-3)}\,dx$.

Solution

As in Example 2 we use partial fractions to break up our integrand into simpler fractions using two unknown numbers, A and B.

$$\frac{2x+3}{(x+2)(x-3)} = \frac{A}{x+2} + \frac{B}{x-3}$$
$$= \frac{A(x-3)+B(x+2)}{(x+2)(x-3)}$$
$$= \frac{(A+B)x+(2B-3A)}{(x+2)(x-3)}$$

Applying the rule that equal polynomials have equal corresponding terms to the numerators now that the denominators are the same, we get two equations.

$$\begin{cases} 2 = A + B \\ 3 = 2B - 3A \end{cases}$$

Solving the first equation for B we obtain $B = 2 - A$. Substituting this into the second equation we get

$$3 = 2(2-A) - 3A$$
$$3 = 4 - 2A - 3A$$
$$-1 = -5A$$
$$\frac{1}{5} = A$$

Substituting $A = \frac{1}{5}$ into $B = 2 - A$ we get $B = \frac{10}{5} - \frac{1}{5} = \frac{9}{5}$. This yields the following simplification of our integrand:

$$\frac{2x+3}{(x+2)(x-3)} = \frac{1/5}{x+2} + \frac{9/5}{x-3}$$

With this simplification of the integrand we are now ready to perform the integration.

$$\int \frac{2x+3}{(x+2)(x-3)}\,dx = \int \left(\frac{1/5}{x+2} + \frac{9/5}{x-3}\right) dx$$

$$= \frac{1}{5}\int \frac{dx}{x+2} + \frac{9}{5}\int \frac{dx}{x-3}$$

$$= \frac{1}{5}\ln|x+2| + \frac{9}{5}\ln|x-3| + C$$

Notice that the two integrals produced by our simplification are, once again, simple substitution integrals.

In Examples 2 and 3 we found ourselves needing to solve two equations in two unknowns as a step in using partial fractions. While solving simultaneous equations is not too difficult, it is possible to find the unknown numbers an easier way as we demonstrate in the next example. We need another useful fact to make the simpler method work.

Polynomial Equality for Specific Values

If two polynomials $f(x)$ and $g(x)$ are equal, $f(x) = g(x)$, then for any real number a, $f(a) = g(a)$.

Example 4

Compute the indefinite integral $\int \frac{x+2}{(x-3)(x-4)}\,dx$.

Solution

Step one is to set up the partial fractions equations with unknowns A and B for this integrand.

$$\frac{x+2}{(x-3)(x-4)} = \frac{A}{x-3} + \frac{B}{x-4}$$

$$= \frac{A(x-4) + B(x-3)}{(x-3)(x-4)}$$

At this point, however, we extract the equal polynomials without simplifying the numerator on the right-hand side of the equation. This gives us

$$x + 2 = A(x-4) + B(x-3)$$

If we first look at what happens when $x = 4$, we get

$$4 + 2 = A(4-4) + B(4-3)$$

$$6 = 0A + 1B$$

$$6 = B$$

If, on the other hand, we plug in $x = 3$ we get

$$3 + 2 = A(3-4) + B(3-3)$$

$$5 = A(-1) + B(0)$$

$$-5 = A$$

So we have $A = -5$ and $B = 6$. With the unknown number from the partial fractions computed we are ready to perform the integral.

$$\int \frac{x+2}{(x-3)(x-4)}\,dx = \int \left(\frac{-5}{x-3} + \frac{6}{x-4}\right) dx$$

$$= 6\int \frac{dx}{x-4} - 5\int \frac{dx}{x-3}$$
$$= 6\ln|x-4| - 5\ln|x-3| + C$$

The key for using the new method is to plug in a value for x that makes the coefficient of the unknown numbers zero.

All of the examples we have done so far have had an integrand where the numerator is of a lower degree than the denominator. It turns out that this is necessary, but also that we can always achieve this state with algebra.

Example 5

Compute the indefinite integral $\int \frac{x^2}{(x-3)(x-4)}\,dx$.

Solution

Observe that the degrees of the numerator and denominator are both 2. First we divide the two polynomials to get a quotient and remainder.

$$\frac{x^2}{(x-3)(x-4)} = \frac{x^2}{x^2-7x+12}$$
$$= \frac{x^2-7x+12+7x-12}{x^2-7x+12}$$
$$= \frac{x^2-7x+12}{x^2-7x+12} + \frac{7x-12}{x^2-7x+12}$$
$$= 1 + \frac{7x-12}{x^2-7x+12}$$
$$= 1 + \frac{7x-12}{(x-3)(x-4)}$$

The division transforms the original problem to

$$\int \frac{x^2}{(x-3)(x-4)}\,dx = \int \left(1 + \frac{7x-12}{(x-3)(x-4)}\right)dx$$

We can now apply the partial fractions technique,

$$\frac{7x-12}{(x-3)(x-4)} = \frac{A}{x-3} + \frac{B}{x-4}$$
$$= \frac{A(x-4)+B(x-3)}{(x-3)(x-4)}$$

so that

$$7x - 12 = A(x-4) + B(x-3)$$

If $x = 4$, we get

$$7 \cdot 4 - 12 = 0 + B(1)$$
$$28 - 12 = B$$
$$B = 16$$

If we set $x = 3$, then we obtain

$$7 \cdot 3 - 12 = A(-1) + 0$$
$$21 - 12 = -A$$
$$-A = 9$$
$$A = -9$$

Once A and B are known we have finished transforming the original integral and are ready to integrate.

$$\int \frac{x^2}{(x-3)(x-4)}\,dx = \int \left(1 - \frac{9}{x-3} + \frac{16}{x-4}\right) dx$$
$$= \int dx - 9\int \frac{dx}{x-3} + 16\int \frac{dx}{x-4}$$
$$= x - 9\ln|x-3| + 16\ln|x-4| + C$$

It may be necessary to factor the denominator before continuing with partial fractions.

Example 6

Compute the indefinite integral $\int \frac{dx}{x^2 - 3x + 2}$.

Solution

We factor $x^2 - 3x + 2 = (x-1)(x-2)$ and so

$$\int \frac{dx}{x^2 - 3x + 2} = \int \frac{dx}{(x-1)(x-2)}$$

which is the problem solved in Example 2.

Steps for Partial Fractions

1. If the denominator has not yet been factored, verify that it factors into two distinct first-degree terms.
2. If the numerator is not lower degree than the denominator, then divide the numerator into the denominator to obtain a quotient and a remainder. The quotient is a polynomial and can be integrated by the methods from Chapter 8. The remainder is integrated with partial fractions.
3. Set up the partial fractions equation
$$\frac{ax+b}{(x-u)(x-v)} = \frac{A}{x-u} + \frac{B}{x-v}$$
4. Solve for the unknown values A and B.
5. Substitute A and B in and perform the resulting integrals by substitution.

GROUP DISCUSSION

Look at the indefinite integral

$$\int \frac{x}{x^2 - 4}\,dx = \int \frac{x}{(x-2)(x+2)}\,dx$$

The left-hand form could be done by substitution while the right-hand form is ready for partial fractions. Will either method work? If you perform the integrals, do the results look the same? Are they in fact equal?

Application

Example 7

Computing the Growth of a Bacterial Population Suppose that a population $P(t)$ of bacteria is limited by available food so that the population's growth rate is $P'(t) = \frac{2t+1}{(t+3)(t-4)}$ millions of bacteria per day. If there are $P(0) = 500{,}000$ bacteria at the beginning of the experiment, find a formula for $P(t)$.

Solution

First we perform the indefinite integral

$$\begin{aligned} P(t) &= \int P'(t)\,dt \\ &= \int \frac{2t+1}{(t+3)(t-4)}\,dt \end{aligned}$$

This integral is in the correct form for partial fractions,

$$\begin{aligned} \frac{2t+1}{(t+3)(t-4)} &= \frac{A}{t+3} + \frac{B}{t-4} \\ &= \frac{A(t-4) + B(t+3)}{(t+3)(t-4)} \end{aligned}$$

So

$$2t + 1 = A(t-4) + B(t+3)$$

If we set $t = 4$ we get

$$\begin{aligned} 2 \cdot 4 + 1 &= A(0) + B(4+3) \\ 9 &= 7B \\ B &= 9/7 \end{aligned}$$

If we set $t = -3$ the result is

$$\begin{aligned} 2(-3) + 1 &= A(-3-4) + B(0) \\ -6 + 1 &= -7A \\ -5 &= -7A \\ A &= 5/7 \end{aligned}$$

With A and B known we are ready to integrate.

$$\begin{aligned} \int \frac{2t+1}{(t+3)(t-4)}\,dt &= \int \left(\frac{9/7}{t-4} + \frac{5/7}{t+3} \right) dt \\ &= \frac{9}{7} \int \frac{dt}{t-4} + \frac{5}{7} \int \frac{dt}{t+3} \\ &= \frac{9}{7} \ln|t-4| + \frac{5}{7} \ln|t+3| + C \\ &= P(t) \end{aligned}$$

We now use the fact that $P(0) = 500{,}000$ to compute the value of the unknown constant C. Since 500,000 is one-half of one million and the growth rate was given in millions of bacteria per hour, we have that $P(0) = 0.5$ million.

$$\begin{aligned} 0.5 &= P(0) \\ 0.5 &= \frac{9}{7} \ln|0-4| + \frac{5}{7} \ln|0+3| + C \\ 0.5 &= \frac{9}{7} \ln|4| + \frac{5}{7} \ln|3| + C \\ C &= 0.5 - \frac{9}{7} \ln(4) - \frac{5}{7} \ln(3) \\ C &\approx -2.067 \end{aligned}$$

This yields the following formula for the bacterial population in millions after t hours.

$$P(t) = \frac{9}{7} \ln|t-4| + \frac{5}{7} \ln|t+3| - 2.067$$

More complicated versions of partial fractions can be used to integrate functions with a larger number of factors in the denominator, but they are beyond the scope of this text.

Self-Check Exercises 9.2

1. Suppose that $2x^2 + 5x - 3 = ax^2 + bx + c$. Compute the values of a, b, and c.
2. Perform the indefinite integral $\int \frac{2x + 1}{(x - 2)(x - 4)}\, dx$.

Solutions to Self-Check Exercises 9.2 can be found on page 548.

9.2 Exercises

In Exercises 1–6, find the unknown coefficients in the polynomial equalities. It may be necessary to simplify the polynomial expressions first.

1. $3x^2 + 2x + 7 = ax^2 + bx + c$
2. $(x - 2)(x + 4) = ax^2 + bx + c$
3. $(x + 2)^2 + 3(x + 1) + 2 = ax^2 + bx + c$
4. $ax^2 + 3x + 1 = 2x^2 + bx + c$
5. $(ax + b)(x + 1) = 2x^2 + 4x + 2$
6. $3x + 5 = ax^2 + bx + c$

In Exercises 7–16, use the method of partial fractions to compute the indefinite integral.

7. $\int \frac{1}{(x - 1)(x - 3)}\, dx$
8. $\int \frac{1}{(x + 4)(x - 4)}\, dx$
9. $\int \frac{1}{(x + 2)(x + 6)}\, dx$
10. $\int \frac{1}{x^2 + 9x + 20}\, dx$
11. $\int \frac{x + 3}{(x + 1)(x - 2)}\, dx$
12. $\int \frac{3x + 5}{(x + 4)(x - 3)}\, dx$
13. $\int \frac{2x + 1}{x^2 + 3x + 2}\, dx$
14. $\int \frac{x^2}{(x - 1)(x - 3)}\, dx$

    Hint: Remember to make the denominator lower degree than the numerator by dividing.

15. $\int \frac{x - 3}{x^2 + 4x}\, dx$
16. $\int \frac{x^2 + x + 1}{x^2 + 5x}\, dx$

    Hint: Remember to make the denominator lower degree than the numerator by dividing.

In Exercises 17–20, perform the specified definite integral.

17. $\int_2^4 \frac{1}{(x + 1)(x + 4)}\, dx$
18. $\int_0^1 \frac{1}{(x - 2)(x - 3)}\, dx$
19. $\int_1^3 \frac{x + 1}{x^2 + x}\, dx$
20. $\int_{-1}^1 \frac{2x + 3}{x^2 + 6x + 8}\, dx$

In Exercises 21–30, use both substitution and the method of partial fractions to compute the indefinite integral.

21. $\int \frac{2x}{(x^2 + 2)(x^2 - 2)}\, dx$
22. $\int \frac{e^x}{(e^x - 2)(e^x - 3)}\, dx$
23. $\int \frac{\ln(x) + 1}{x(\ln(x) - 1)(\ln(x) - 3)}\, dx$
24. $\int \frac{e^x}{e^{2x} + 3e^x}\, dx$
25. $\int \frac{\cos(x)}{\sin^2(x) - 3\sin(x) + 2}\, dx$
26. $\int \frac{\sec^2(x)}{\tan^2(x) - 5\tan(x) + 4}\, dx$

27. $\displaystyle\int \frac{x^2}{x^6 + 2x^3}\,dx$

28. $\displaystyle\int \frac{e^{2x} + e^x}{e^{2x} - 9}\,dx$

29. $\displaystyle\int \frac{\sin(x)}{\cos^2(x) + 5\cos(x) + 6}\,dx$

30. $\displaystyle\int \frac{2x}{x^4 - 7x^2 + 10}\,dx$

In Exercises 31–36, perform the specified indefinite integral. It may require substitution, integration by parts, or integration by partial fractions. State the method you use.

31. $\displaystyle\int \frac{x}{x^4 + 1}\,dx$

32. $\displaystyle\int \frac{\ln(x) + 1}{x}\,dx$

33. $\displaystyle\int \frac{x}{x^4 + 2x^2}\,dx$

34. $\displaystyle\int xe^{x^2+1}\,dx$

35. $\displaystyle\int \frac{\cos(x)}{\sin^2(x) + 6\sin(x) + 8}\,dx$

36. $\displaystyle\int \frac{e^x}{e^{2x} + 1}\,dx$

37. After an initial failure, leakage of hot water from a heating system is found to have a rate of

$$L(t) = \frac{2t + 1}{t^2 + 4t + 8}$$

g/min. Find the total amount of water that leaks in the first hour after the leak starts by integrating $L(t)$ from $t = 0$ to $t = 60$ minutes.

38. Water escaping from a crack in a dam is found to have a flow rate of

$$L(t) = \frac{190t + 65}{t^2 + 6t + 5}$$

L/min. Find the total water lost in the first 4 hours.

39. Thiotimoline is produced in a chemical reactor at a rate of

$$\frac{e^x}{e^{2x} + 3e^x + 2}$$

μg/d. Compute the total amount of thiotimoline produced in the first five days of operation.

40. Suppose that

$$F(x) = \int_0^x \frac{dt}{t^2 - 16}$$

Find the domain of $F(x)$.

Hint: Consider the graph of $f(t) = \dfrac{1}{t^2 - 16}$; which areas under this graph define $F(x)$?

41. The method of partial fractions is not needed for the indefinite integral

$$\int \frac{dx}{x^2 - 4x + 4}$$

Instead it is performed by substitution. The problem does have a quadratic in the denominator: What is different about it that makes it inappropriate for partial fractions? Once you have answered the question, perform the integration by substitution.

42. Suppose that

$$\frac{1}{(x-1)(x-2)(x-3)} = \frac{A}{x-1} + \frac{B}{x-2} + \frac{C}{x-3}$$

Can you use the technique of plugging in well-chosen values of x to find A, B, and C? If not, explain why; if yes, use the information to compute the indefinite integral

$$\int \frac{dx}{(x-1)(x-2)(x-3)}.$$

Business and Economics Applications

43. **Total Losses** As news spreads among hackers about a flaw in ATM software that permits small, unregistered withdrawals, the money lost is found to be modelled by

$$\frac{200t + 800}{t^2 + 8t + 15}$$

dollars/hour. If the bank notices the problem and takes the system offline at the end of 24 hours, what are the total losses from hour 0 to hour 24?

44. **Total Savings** The rate at which total savings increase as members join a purchasing network is found to be

$$S'(x) = \frac{300}{x^2 + 5x + 4}$$

cents/new member. Find the total savings of the network as the number of members increases from 10 to 20.

Biological and Life Sciences Applications

45. **Amount of Hormone Present** A common source of metabolic disease is the failure of the liver to remove a hormone from the bloodstream after it has done its job. Blood tests show that a patient has an increase in concentration of

$$\frac{0.04x + 0.2}{x^2 + 7x + 12}$$

μg/h per litre of blood. How many micrograms of the hormone are present after 96 hours?

46. **Number of Bacteria** Bacteria produce poisons that limit their own growth in a closed environment. A population of bacteria being grown in a Petri dish adds numbers (new bacteria minus those that die) equal to

$$P'(t) = \frac{2.1}{t^2 + 5t + 6}$$

million/hour. If the dish started with $P(0) = 1.4$ million bacteria, compute the number in the dish after three hours.

47. Amount of Calcium The population of a type of diatom in a test tank is found to be

$$P(t) = \frac{3.7t^2}{t^2 + 4x + 3}$$

million after t days. Diatoms take calcium from the water and fix it in their shells. Assuming that these diatoms fix 0.45 g of calcium per day per million diatoms, how much calcium is fixed in the first week? Remember that the method of partial fractions cannot be used until the numerator is a lower degree than the denominator and so long division may be required.

48. Alcohol Produced Alcohol is produced in a fermentation vat at a rate of

$$\frac{4x + 7}{6x^2 + 5x + 1}$$

L/h. Compute the total amount produced in the first day of operation.

Solutions to Self-Check Exercises 9.2

1. Equating like terms, we see that $a = 2$, $b = 5$, and $c = -3$.

2. The partial fractions equation for this problem is:

$$\frac{2x+1}{(x-2)(x-4)} = \frac{A}{x-2} + \frac{B}{x-4}$$

$$= \frac{A(x-4) + B(x-2)}{(x-2)(x-4)}$$

$$2x + 1 = A(x-4) + B(x-2)$$

If we set $x = 4$ then

$$2 \cdot 4 + 1 = A(0) + B(2)$$

$$9 = 2B$$

$$B = 9/2$$

If we set $x = 2$ then

$$2 \cdot 2 + 1 = A(-2) + B(0)$$

$$5 = -2A$$

$$A = -5/2$$

This gives us the information we need to transform and finish the integral:

$$\int \frac{2x+1}{(x-2)(x-4)}\,dx = \int \left(\frac{-5/2}{x-2} + \frac{9/5}{x-4}\right) dx$$

$$= -\frac{5}{2}\int \frac{dx}{x-2} + \frac{9}{5}\int \frac{dx}{x-4}$$

$$= -\frac{5}{2}\ln|x-2| + \frac{9}{5}\ln|x-4| + C$$

9.3 Integration Using Tables of Integrals (Optional)

A Table of Integrals

We have studied several techniques for finding an antiderivative of a function. However, useful as they are, these techniques are not always applicable. There are of course numerous other methods for finding an antiderivative of a function. Extensive lists of integration formulas have been compiled based on these methods.

A small sample of the integration formulas that can be found in many mathematical handbooks is given in the following table of integrals. The formulas are grouped according to the basic form of the integrand. Note that it may be necessary to modify the integrand of the integral to be evaluated in order to use one of these formulas.

Table of Integrals

Forms involving $a + bu$

1. $\int \frac{u\,du}{a + bu} = \frac{1}{b^2}[a + bu - a\ln|a + bu|] + C$

2. $\int \frac{u^2du}{a + bu} = \frac{1}{2b^3}[(a + bu)^2 - 4a(a + bu) + 2a^2\ln|a + bu|] + C$

3. $\int \frac{u\,du}{(a + bu)^2} = \frac{1}{b^2}\left[\frac{a}{a + bu} + \ln|a + bu|\right] + C$

4. $\int u\sqrt{a + bu}\,du = \frac{2}{15b^2}(3bu - 2a)(a + bu)^{3/2} + C$

5. $\int \frac{u\,du}{\sqrt{a + bu}} = \frac{2}{3b^2}(bu - 2a)\sqrt{a + bu} + C$

6. $\int \frac{du}{u\sqrt{a + bu}} = \frac{1}{\sqrt{a}}\ln\left|\frac{\sqrt{a + bu} - \sqrt{a}}{\sqrt{a + bu} + \sqrt{a}}\right| + C \qquad (\text{if } a > 0)$

7. $\int \frac{dx}{(ax + b)^2 + d^2} = \frac{1}{ad}\arctan\left(\frac{ax + b}{d}\right) + C$

Forms involving $\sqrt{a^2 + u^2}$

8. $\int \sqrt{a^2 + u^2}\,du = \frac{u}{2}\sqrt{a^2 + u^2} + \frac{a^2}{2}\ln|u + \sqrt{a^2 + u^2}| + C$

9. $\int u^2\sqrt{a^2 + u^2}\,du = \frac{u}{8}(a^2 + 2u^2)\sqrt{a^2 + u^2} - \frac{a^4}{8}\ln|u + \sqrt{a^2 + u^2}| + C$

10. $\int \frac{du}{\sqrt{a^2 + u^2}} = \ln|u + \sqrt{a^2 + u^2}| + C$

11. $\int \frac{du}{u\sqrt{a^2 + u^2}} = -\frac{1}{a}\ln\left|\frac{\sqrt{a^2 + u^2} + a}{u}\right| + C$

12. $\int \frac{du}{u^2\sqrt{a^2 + u^2}} = -\frac{\sqrt{a^2 + u^2}}{a^2u} + C$

13. $\int \frac{du}{(a^2 + u^2)^{3/2}} = \frac{u}{a^2\sqrt{a^2 + u^2}} + C$

Forms Involving $\sqrt{u^2 - a^2}$

14. $\int \sqrt{u^2 - a^2}\,du = \frac{u}{2}\sqrt{u^2 - a^2} - \frac{a^2}{2}\ln|u + \sqrt{u^2 - a^2}| + C$

15. $\int u^2\sqrt{u^2 - a^2}\,du = \frac{u}{8}(2u^2 - a^2)\sqrt{u^2 - a^2} - \frac{a^4}{8}\ln|u + \sqrt{u^2 - a^2}| + C$

16. $\int \frac{\sqrt{u^2 - a^2}}{u^2}\,du = -\frac{\sqrt{u^2 - a^2}}{u} + \ln|u + \sqrt{u^2 - a^2}| + C$

17. $\int \frac{du}{\sqrt{u^2 - a^2}} = \ln|u + \sqrt{u^2 - a^2}| + C$

18. $\displaystyle\int \frac{du}{u^2\sqrt{u^2 - a^2}} = \frac{\sqrt{u^2 - a^2}}{a^2u} + C$

19. $\displaystyle\int \frac{du}{(u^2 - a^2)^{3/2}} = -\frac{u}{a^2\sqrt{u^2 - a^2}} + C$

Forms Involving $\sqrt{a^2 - u^2}$

20. $\displaystyle\int \frac{\sqrt{a^2 - u^2}}{u}\, du = \sqrt{a^2 - u^2} - a \ln\left|\frac{a + \sqrt{a^2 - u^2}}{u}\right| + C$

21. $\displaystyle\int \frac{du}{u\sqrt{a^2 - u^2}} = -\frac{1}{a} \ln\left|\frac{a + \sqrt{a^2 - u^2}}{u}\right| + C$

22. $\displaystyle\int \frac{du}{u^2\sqrt{a^2 - u^2}} = -\frac{\sqrt{a^2 - u^2}}{a^2u} + C$

23. $\displaystyle\int \frac{du}{(a^2 - u^2)^{3/2}} = \frac{u}{a^2\sqrt{a^2 - u^2}} + C$

Forms Involving e^{au} and $\ln u$

24. $\displaystyle\int ue^{au}\, du = \frac{1}{a^2}(au - 1)e^{au} + C$

25. $\displaystyle\int u^n e^{au}\, du = \frac{1}{a} u^n e^{au} - \frac{n}{a} \int u^{n-1} e^{au}\, du$

26. $\displaystyle\int \frac{du}{1 + be^{au}} = u - \frac{1}{a} \ln(1 + be^{au}) + C$

27. $\displaystyle\int \ln u\, du = u \ln u - u + C$

28. $\displaystyle\int u^n \ln u\, du = \frac{u^{n+1}}{(n + 1)^2}[(n + 1) \ln u - 1] + C \qquad (n \neq -1)$

29. $\displaystyle\int \frac{du}{u \ln u} = \ln|\ln u| + C$

30. $\displaystyle\int (\ln u)^n\, du = u(\ln u)^n - n \int (\ln u)^{n-1}\, du$

Using a Table of Integrals

We now consider several examples that illustrate how the table of integrals can be used to evaluate an integral.

Example 1

Use the table of integrals to find $\displaystyle\int \frac{2x\, dx}{\sqrt{3 + x}}$.

Solution

We first write

$$\int \frac{2x\, dx}{\sqrt{3 + x}} = 2\int \frac{x\, dx}{\sqrt{3 + x}}$$

Since $\sqrt{3 + x}$ is of the form $\sqrt{a + bu}$, with $a = 3$, $b = 1$, and $u = x$, we use formula 5,

$$\int \frac{u\,du}{\sqrt{a + bu}} = \frac{2}{3b^2}(bu - 2a)\sqrt{a + bu} + C$$

obtaining

$$2\int \frac{x}{\sqrt{3 + x}}\,dx = 2\left[\frac{2}{3(1)}(x - 6)\sqrt{3 + x}\right] + C$$
$$= \frac{4}{3}(x - 6)\sqrt{3 + x} + C$$

GROUP DISCUSSION

All formulas given in the table of integrals can be verified by direct computation. Describe a method you would use and apply it to verify a formula of your choice.

Example 2

Use the table of integrals to find $\int x^2\sqrt{3 + x^2}\,dx$.

Solution

Observe that if we write 3 as $(\sqrt{3})^2$, then $3 + x^2$ has the form $\sqrt{a^2 + u^2}$ with $a = \sqrt{3}$ and $u = x$. Using Formula 9,

$$\int u^2\sqrt{a^2 + u^2}\,du = \frac{u}{8}(a^2 + 2u^2)\sqrt{a^2 + u^2} - \frac{a^4}{8}\ln|u + \sqrt{a^2 + u^2}| + C$$

we obtain

$$\int x^2\sqrt{3 + x^2}\,dx = \frac{x}{8}(3 + 2x^2)\sqrt{3 + x^2} - \frac{9}{8}\ln|x + \sqrt{3 + x^2}| + C$$

Example 3

Use the table of integrals to evaluate

$$\int_3^4 \frac{dx}{x^2\sqrt{50 - 2x^2}}$$

Solution

We first find the indefinite integral

$$I = \int \frac{dx}{x^2\sqrt{50 - 2x^2}}$$

Observe that $\sqrt{50 - 2x^2} = \sqrt{2(25 - x^2)} = \sqrt{2}\sqrt{25 - x^2}$, so we can write I as

$$I = \frac{1}{\sqrt{2}}\int \frac{dx}{x^2\sqrt{25 - x^2}} = \frac{\sqrt{2}}{2}\int \frac{dx}{x^2\sqrt{25 - x^2}} \qquad \frac{1}{\sqrt{2}} = \frac{\sqrt{2}}{2}$$

Next, using Formula 22,

$$\int \frac{du}{u^2\sqrt{a^2 - u^2}} = -\frac{\sqrt{a^2 - u^2}}{a^2u} + C$$

with $a = 5$ and $u = x$, we find

$$I = \frac{\sqrt{2}}{2}\left[-\frac{\sqrt{25 - x^2}}{25x}\right]$$
$$= -\left(\frac{\sqrt{2}}{50}\right)\frac{\sqrt{25 - x^2}}{x}$$

Finally, using this result, we obtain

$$\int_3^4 \frac{dx}{x^2\sqrt{50-2x^2}} = -\frac{\sqrt{2}}{50}\frac{\sqrt{25-x^2}}{x}\Bigg|_3^4$$

$$= -\frac{\sqrt{2}}{50}\frac{\sqrt{25-16}}{4} - \left(-\frac{\sqrt{2}}{50}\frac{\sqrt{25-9}}{3}\right)$$

$$= -\frac{3\sqrt{2}}{200} + \frac{2\sqrt{2}}{75} = \frac{7\sqrt{2}}{600}$$

Example 4

Use the table of integrals to find $\int e^{2x}\sqrt{5+2e^x}\,dx$.

Solution

Let $u = e^x$. Then $du = e^x\,dx$. Therefore, the given integral can be written

$$\int e^x\sqrt{5+2e^x}\,(e^x dx) = \int u\sqrt{5+2u}\,du$$

Using Formula 4,

$$\int u\sqrt{a+bu}\,du = \frac{2}{15b^2}(3bu-2a)(a+bu)^{3/2} + C$$

with $a = 5$ and $b = 2$, we see that

$$\int u\sqrt{5+2u}\,du = \frac{2}{15(4)}(6u-10)(5+2u)^{3/2} + C$$

$$= \frac{1}{15}(3u-5)(5+2u)^{3/2} + C$$

Finally, recalling the substitution $u = e^x$, we find

$$\int e^{2x}\sqrt{5+2e^x}\,dx = \frac{1}{15}(3e^x-5)(5+2e^x)^{3/2} + C$$

GROUP DISCUSSION

The formulas given in the table of integrals were derived using various techniques, including the method of substitution and the method of integration by parts studied earlier. For example, Formula 1,

$$\int \frac{u\,du}{a+bu} = \frac{1}{b^2}[a+bu-a\ln|a+bu|] + C$$

can be derived using the method of substitution. Show how this is done.

As illustrated in the next example, we may need to apply a formula more than once in order to evaluate an integral.

Example 5

Use the table of integrals to find $\int x^2 e^{(-1/2)x}\,dx$.

Solution

Scanning the table of integrals for a formula involving e^{ax} in the integrand, we are led to Formula 25,

$$\int u^n e^{au}\,du = \frac{1}{a} u^n e^{au} - \frac{n}{a}\int u^{n-1}e^{au}\,du$$

With $n = 2$, $a = -\frac{1}{2}$, and $u = x$, we have

$$\begin{aligned}\int x^2e^{(-1/2)x}\,dx &= \left(\frac{1}{-\frac{1}{2}}\right)x^2e^{(-1/2)x} - \frac{2}{(-\frac{1}{2})}\int xe^{(-1/2)x}\,dx \\ &= -2x^2e^{(-1/2)x} + 4\int xe^{(-1/2)x}\,dx\end{aligned}$$

If we use Formula 25 once again, with $n = 1$, $a = -\frac{1}{2}$, and $u = x$, to evaluate the integral on the right, we obtain

$$\begin{aligned}\int x^2e^{(-1/2)x}\,dx &= -2x^2e^{(-1/2)x} + 4\left[\left(\frac{1}{-\frac{1}{2}}\right)xe^{(-1/2)x} - \frac{1}{(-\frac{1}{2})}\int e^{(-1/2)x}\,dx\right] \\ &= -2x^2e^{(-1/2)x} + 4\left[-2xe^{(-1/2)x} + 2\cdot\frac{1}{(-\frac{1}{2})}e^{(-1/2)x}\right] + C \\ &= -2e^{(-1/2)x}(x^2 + 4x + 8) + C\end{aligned}$$

Example 6

Mortgage Rates A study prepared for the National Association of Realtors estimated that the mortgage rate over the next t months will be

$$r(t) = \frac{8t + 100}{t + 10} \qquad (0 \le t \le 24)$$

percent per year. If the prediction holds true, what will be the average mortgage rate over the next 12 months?

Solution

The average mortgage rate over the next 12 months will be given by

$$\begin{aligned}A &= \frac{1}{12 - 0}\int_0^{12}\frac{8t + 100}{t + 10}\,dt = \frac{1}{12}\left[\int_0^{12}\frac{8t}{t + 10}\,dt + \int_0^{12}\frac{100}{t + 10}\,dt\right] \\ &= \frac{8}{12}\int_0^{12}\frac{t}{t + 10}\,dt + \frac{100}{12}\int_0^{12}\frac{1}{t + 10}\,dt\end{aligned}$$

Using Formula 1,

$$\int\frac{u\,du}{a + bu} = \frac{1}{b^2}[a + bu - a\ln|a + bu|] + C \qquad a = 10, b = 1, u = t$$

to evaluate the first integral, we have

$$\begin{aligned}A &= \left(\frac{2}{3}\right)\left[10 + t - 10\ln(10 + t)\right]\Big|_0^{12} + \left(\frac{25}{3}\right)\ln(10 + t)\Big|_0^{12} \\ &= \left(\frac{2}{3}\right)[(22 - 10\ln 22) - (10 - 10\ln 10)] + \left(\frac{25}{3}\right)[\ln 22 - \ln 10] \\ &\approx 9.31\end{aligned}$$

or approximately 9.31% per year.

In Chapter 8 we learned to integrate a function of the form $f(x) = \frac{1}{(ax + b)^2}$ by substitution. This is an integral in which the denominator is a quadratic with a single root. In Section 9.2 we learned to use partial fractions to integrate functions of the form $g(x) = \frac{1}{(ax + b)(dx + e)}$ which have two distinct roots. There is a third type of quadratic, that with no real roots at all. In Chapter 8 we learned to integrate functions like $\int \frac{1}{x^2 + 1} = \arctan(x) + C$. Rule 7 makes this type of integral more general. In fact this type of integration can also be performed by *completing the square.* In the next two examples we will solve a problem using Rule 7 by completing the square in the denominator.

Example 7

Find the indefinite integral

$$\int \frac{dx}{x^2 + 4x + 5}$$

Solution

In order to complete the square we must turn $x^2 + 4x + 5$ into a perfect square plus a remainder. Since $(x + 2)^2 = x^2 + 4x + 4$ we see this form is $x^2 + 4x + 5 = (x + 2)^2 + 1$ and the integral becomes

$$\int \frac{dx}{(x + 2)^2 + 1}$$

In order to use Rule 7 we notice that $a = 1$, $b = 2$ and $d = 1$. This gives us the answer that

$$\begin{aligned}\int \frac{dx}{x^2 + 4x + 5} &= \frac{1}{1 \cdot 1} \arctan\left(\frac{x + 2}{1}\right) + C \\ &= \arctan(x + 2) + C.\end{aligned}$$

Example 8

Find the indefinite integral

$$\int \frac{dx}{x^2 + 6x + 13}$$

Solution

In order to complete the square we must turn $x^2 + 6x + 13$ into a perfect square plus a remainder. Since $(x + 3)^2 = x^2 + 6x + 9$ we see this form is $x^2 + 6x + 13 = (x + 3)^2 + 4$ and the integral becomes

$$\int \frac{dx}{(x + 3)^2 + 4} = \int \frac{dx}{(x + 3)^2 + 2^2}$$

In order to use Rule 7 we notice that $a = 1$, $b = 3$ and $d = 2$. This gives us the answer that

$$\begin{aligned}\int \frac{dx}{x^2 + 4x + 5} &= \frac{1}{1 \cdot 2} \arctan\left(\frac{x + 3}{2}\right) + C \\ &= \frac{1}{2} \arctan\left(\frac{x + 3}{2}\right) + C\end{aligned}$$

Completing the Square

When completing the square for a quadratic $ax^2 + b + c$ we first factor out a to get

$$a\left(x^2 + \frac{b}{a}x + \frac{c}{a}\right)$$

In this form the completed square will be the square of x plus half the coefficient $\frac{b}{a}$ of x. We use algebra to finish completing the square.

$$\begin{aligned} a\left(x^2 + \frac{b}{a}x + \frac{c}{a}\right) &= a\left(x^2 + 2\frac{b}{2a}x + \left(\frac{b}{2a}\right)^2 - \left(\frac{b}{2a}\right)^2 + \frac{c}{a}\right) \\ &= a\left(x^2 + 2\frac{b}{2a}x + \left(\frac{b}{2a}\right)^2\right) - a\left(\frac{b}{2a}\right)^2 + a\frac{c}{a} \\ &= a\left(x + \frac{b}{2a}\right)^2 - \frac{ab^2}{4a^2} + c \\ &= a\left(x + \frac{b}{2a}\right)^2 - \frac{b^2}{4a} + \frac{4ac}{4a} \\ &= a\left(x + \frac{b}{2a}\right)^2 + \frac{4ac - b^2}{4a} \end{aligned}$$

Remember that this form of the completed square is only useful in Formula 7 if the original quadratic equation has no roots.

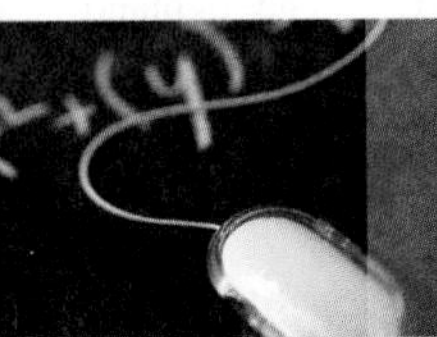

Self-Check Exercises 9.3

1. Use the table of integrals to evaluate

$$\int_0^2 \frac{dx}{(5 - x^2)^{3/2}}$$

2. During a flu epidemic, the number of children in Easton Middle School who contracted influenza t days after the outbreak began was given by

$$N(t) = \frac{200}{1 + 9e^{-0.8t}}$$

Determine the average number of children who contracted the flu in the first 10 days of the epidemic.

Solutions to Self-Check Exercises 9.3 can be found on page 557.

9.3 Exercises

In Exercises 1–32, use the table of integrals in this section to find each integral.

1. $\int \frac{2x}{2 + 3x}\,dx$

2. $\int \frac{x}{(1 + 2x)^2}\,dx$

3. $\int \frac{3x^2}{2 + 4x}\,dx$

4. $\int \frac{x^2}{3 + x}\,dx$

5. $\int x^2\sqrt{9 + 4x^2}\,dx$

6. $\int x^2\sqrt{4 + x^2}\,dx$

7. $\int \frac{dx}{x\sqrt{1 + 4x}}$

8. $\int_0^2 \frac{x + 1}{\sqrt{2 + 3x}}\,dx$

9. $\int_0^2 \frac{dx}{\sqrt{9 + 4x^2}}$

10. $\int \frac{dx}{x\sqrt{4 + 8x^2}}$

11. $\int \frac{dx}{(9 - x^2)^{3/2}}$

12. $\int \frac{dx}{(2 - x^2)^{3/2}}$

13. $\int x^2\sqrt{x^2 - 4}\,dx$

14. $\int_3^5 \frac{dx}{x^2\sqrt{x^2 - 9}}$

15. $\displaystyle\int \frac{\sqrt{4 - x^2}}{x}\, dx$ 16. $\displaystyle\int_0^1 \frac{dx}{(4 - x^2)^{3/2}}$

17. $\displaystyle\int xe^{2x}\, dx$ 18. $\displaystyle\int \frac{dx}{1 + e^{-x}}$

19. $\displaystyle\int \frac{dx}{(x + 1)\ln(1 + x)}$

Hint: First use the substitution $u = x + 1$.

20. $\displaystyle\int \frac{x}{(x^2 + 1)\ln(x^2 + 1)}\, dx$

Hint: First use the substitution $u = x^2 + 1$.

21. $\displaystyle\int \frac{e^{2x}}{(1 + 3e^x)^2}\, dx$ 22. $\displaystyle\int \frac{e^{2x}}{\sqrt{1 + 3e^x}}\, dx$

23. $\displaystyle\int \frac{3e^x}{1 + e^{(1/2)x}}\, dx$ 24. $\displaystyle\int \frac{dx}{1 - 2e^{-x}}$

25. $\displaystyle\int \frac{\ln x}{x(2 + 3\ln x)}\, dx$ 26. $\displaystyle\int_1^e (\ln x)^2\, dx$

27. $\displaystyle\int_0^1 x^2 e^x\, dx$ 28. $\displaystyle\int x^3 e^{2x}\, dx$

29. $\displaystyle\int x^2 \ln x\, dx$ 30. $\displaystyle\int x^3 \ln x\, dx$

31. $\displaystyle\int (\ln x)^3\, dx$ 32. $\displaystyle\int (\ln x)^4\, dx$

In Exercises 33–38, use the method of partial fractions to find each integral.

33. $\displaystyle\int \frac{dx}{x^2 + 6x + 10}$ 34. $\displaystyle\int \frac{ds}{4s^2 + 4s + 2}$

35. $\displaystyle\int \frac{dt}{t^2 + 4t + 13}$ 36. $\displaystyle\int \frac{dv}{v^2 + 8v + 20}$

37. $\displaystyle\int \frac{dx}{x^2 + x + 1}$ 38. $\displaystyle\int \frac{dx}{x^2 + 3x + 7}$

39. **Amusement Park Attendance** The management of AstroWorld ("The Amusement Park of the Future") estimates that the number of visitors (in thousands) entering the amusement park t hr after opening time at 9 A.M. is given by

$$R(t) = \frac{60}{(2 + t^2)^{3/2}}$$

per hour. Determine the number of visitors admitted by noon.

40. **Voter Registration** The number of voters in a certain district of a city is expected to grow at the rate of

$$R(t) = \frac{3000}{\sqrt{4 + t^2}}$$

people/year t yr from now. If the number of voters at present is 20,000, how many voters will be in the district 5 yr from now?

41. **VCR Ownership** The percent of households that own VCRs is given by

$$P(t) = \frac{68}{1 + 21.67\,e^{-0.62t}} \qquad (0 \le t \le 12)$$

where t is measured in years, with $t = 0$ corresponding to the beginning of 1981. Find the average percent of households owning VCRs from the beginning of 1981 to the beginning of 1993.

Source: Paul Kroger Associates

42. **Recycling Programs** The commissioner of the City of Newton Department of Public Works estimates that the number of people in the city who have been recycling their magazines in year t following the introduction of the recycling program at the beginning of 1990 is

$$N(t) = \frac{100{,}000}{2 + 3e^{-0.2t}}$$

Find the average number of people who will have recycled their magazines during the first 5 yr since the program was introduced.

Business and Economics Applications

43. **Consumers' Surplus** Refer to Section 6.7. The demand function for Apex women's boots is

$$p = \frac{250}{\sqrt{16 + x^2}}$$

where p is the wholesale unit price in dollars and x is the quantity demanded daily, in units of a hundred. Find the consumers' surplus if the wholesale price is set at \$50/pair.

44. **Producers' Surplus** Refer to Section 8.7. The supplier of Apex women's boots will make x hundred pairs of the boots available in the market daily when the wholesale unit price is

$$p = \frac{30x}{5 - x}$$

dollars. Find the producers' surplus if the wholesale price is set at \$50/pair.

45. **Franchises** Elaine purchased a 10-yr franchise for a fast-food restaurant that is expected to generate income at the rate of $R(t) = 250{,}000 + 2000t^2$ dollars/year, t yr from now. If the prevailing interest rate is 10%/year compounded continuously, find the present value of the franchise.

Hint: Use Formula (18), Section 8.7.

46. **Accumulated Value of an Income Stream** The revenue of Virtual Reality, a video-game arcade, is generated at the rate of $R(t) = 20{,}000t$ dollars. If the revenue is invested t yr from now in a business earning interest at the rate of 15%/year compounded continuously, find the accumulated value of this stream of income at the end of 5 yr.

Hint: Use Formula (18), Section 8.7.

47. **Lorentz Curves** In a study conducted by a certain country's Economic Development Board regarding the income distribution of certain segments of the country's workforce, it was

found that the Lorentz curve for the distribution of income of college professors is described by the function

$$g(x) = \frac{1}{3}x\sqrt{1 + 8x}$$

Compute the coefficient of inequality of the Lorentz curve.
Hint: Use Formula (21), Section 8.7.

Biological and Life Sciences Applications

48. Growth of Fruit Flies Based on data collected during an experiment, a biologist found that the number of fruit flies (*Drosophila*) with a limited food supply could be approximated by the exponential model

$$N(t) = \frac{1000}{1 + 24e^{-0.02t}}$$

where t denotes the number of days since the beginning of the experiment. Find the average number of fruit flies in the colony in the first 10 days of the experiment and in the first 20 days.

49. Amount of Bacteria A population of bacteria is found to have a population increase of

$$P'(t) = \frac{4.3}{t\sqrt{25 - t}}$$

million/hour for times $0 \leq t \leq 4$ hours. If the initial number of bacteria is $P(0) = 0.72$ million bacteria, how many are there by the end of the first hour?

50. Total Seedlings Suppose that the number of new larch seedlings found in a reforestation project is found empirically to be

$$S(t) = \sqrt{400 + 16t^2}$$

per week. Find the total number of seedlings after five weeks.

Solutions to Self-Check Exercises 9.3

1. Using Formula 23, page 550, with $a^2 = 5$ and $u = x$, we see that

$$\int_0^2 \frac{dx}{(5 - x^2)^{3/2}} = \left.\frac{x}{5\sqrt{5 - x^2}}\right|_0^2 = \frac{2}{5\sqrt{5 - 4}} = \frac{2}{5}$$

2. The average number of children who contracted the flu in the first 10 days of the epidemic is given by

$$A = \frac{1}{10}\int_0^{10} \frac{200}{1 + 9e^{-0.8t}}\,dt = 20\int_0^{10} \frac{dt}{1 + 9e^{-0.8t}}$$

$$= 20\left[t + \frac{1}{0.8}\ln\left(1 + 9e^{-0.8t}\right)\right]\Bigg|_0^{10} \qquad \text{Formula 25, } a = -0.8,\ b = 9,\ u = t$$

$$= 20\left[10 + \frac{1}{0.8}\ln\left(1 + 9e^{-8}\right)\right] - 20\left(\frac{1}{0.8}\right)\ln 10$$

$$\approx 200.07537 - 57.56463$$

$$\approx 143$$

or 143 students.

9.4 Numerical Integration

Approximating Definite Integrals

One method of measuring cardiac output is to inject 5 to 10 milligrams (mg) of a dye into a vein leading to the heart. After making its way through the lungs, the dye returns to the heart and is pumped into the aorta, where its concentration is measured at equal time intervals. The graph of the function c in Figure 9.1 shows the concentration of dye in a person's aorta, measured at 2-second intervals after 5 mg of dye have been injected. The person's cardiac output, measured in litres per minute (L/min), is computed using the formula

$$R = \frac{60\,D}{\int_0^{28} c(t)\,dt} \tag{4}$$

where D is the quantity of dye injected (see Exercise 40, on page 568).

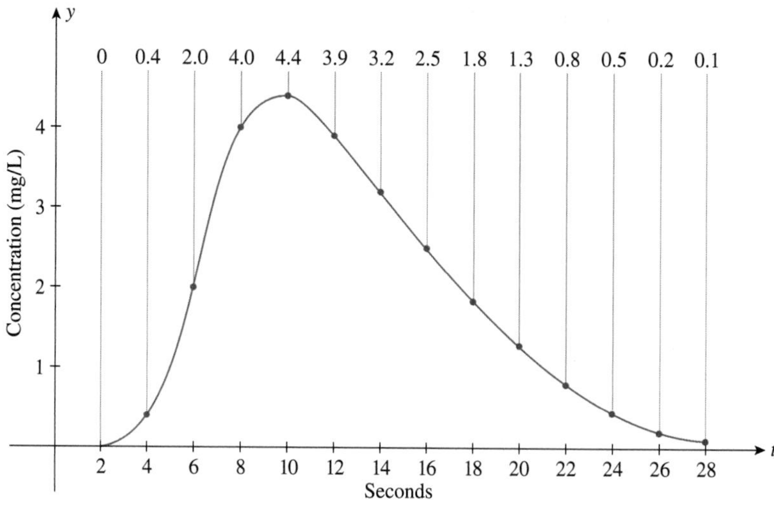

Figure 9.1

The function c gives the concentration of a dye measured as the aorta. The graph is constructed by drawing a smooth curve through a set of discrete points.

Now, to use Formula (4), we need to evaluate the definite integral

$$\int_0^{28} c(t)\, dt$$

But we do not have the algebraic rule defining the integrand c for all values of t in $[0, 28]$. In fact, we are given its values only at a set of discrete points in that interval. In situations such as this, the fundamental theorem of calculus proves useless because we cannot find an antiderivative of c. (We will complete the solution to this problem in Example 4.)

Other situations also arise in which an integrable function has an antiderivative that cannot be found in terms of elementary functions (functions that can be expressed as a finite combination of algebraic, exponential, logarithmic, and trigonometric functions). Examples of such functions are

$$f(x) = e^{x^2}, \qquad g(x) = x^{-1/2}e^x, \qquad h(x) = \frac{1}{\ln x}$$

Riemann sums provide us with a good approximation of a definite integral, provided the number of subintervals in the partitions is large enough. But there are better techniques and formulas, called *quadrature formulas,* that give a more efficient way of computing approximate values of definite integrals. In this section we look at two rather simple but effective ways of approximating definite integrals.

The Trapezoidal Rule

We assume that $f(x) \geq 0$ on $[a, b]$ in order to simplify the derivation of the trapezoidal rule, but the result is valid without this restriction. We begin by subdividing the interval $[a, b]$ into n subintervals of equal length Δx, by means of the $(n + 1)$ points $x_0 = a, x_1, x_2, \ldots, x_n = b$ where n is a positive integer (Figure 9.2).

Then, the length of each subinterval is given by

$$\Delta x = \frac{b - a}{n}$$

Furthermore, as we saw earlier, we may view the definite integral

$$\int_a^b f(x)\, dx$$

as the area of the region R under the curve $y = f(x)$ between $x = a$ and $x = b$. This area is given by the sum of the areas of the n nonoverlapping subregions $R_1, R_2, \ldots, R_n$,

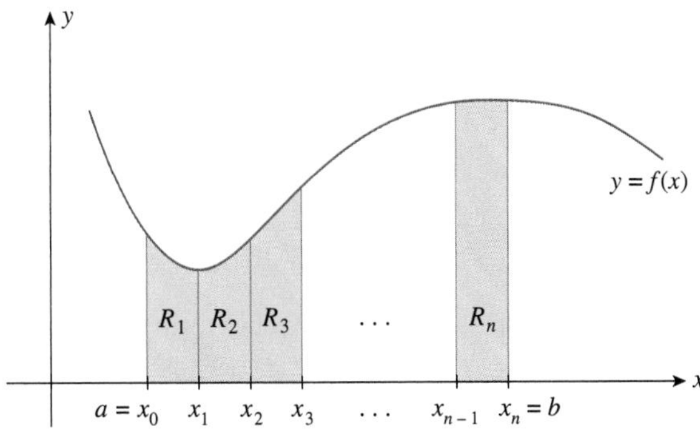

Figure 9.2

The area under the curve is equal to the sum of the n subregions $R_1, R_2, \ldots, R_n$.

such that R_1 represents the region under the curve $y = f(x)$ from $x = x_0$ to $x = x_1$, and so on.

The basis for the trapezoidal rule lies in the approximation of each of the regions $R_1, R_2, \ldots, R_n$ by a suitable trapezoid. This often leads to a much better approximation than one obtained by means of rectangles (a Riemann sum).

Let's consider the subregion R_1, shown magnified for the sake of clarity in Figure 9.3. Observe that the area of the region R_1 may be approximated by the trapezoid of width Δx whose parallel sides are of lengths $f(x_0)$ and $f(x_1)$. The area of the trapezoid is given by

$$\left[\frac{f(x_0) + f(x_1)}{2}\right]\Delta x$$

Average of the lengths of the parallel sides × Width

square units. Similarly, the area of the region R_2 may be approximated by the trapezoid of width Δx and sides of lengths $f(x_1)$ and $f(x_2)$. The area of the trapezoid is given by

$$\left[\frac{f(x_1) + f(x_2)}{2}\right]\Delta x$$

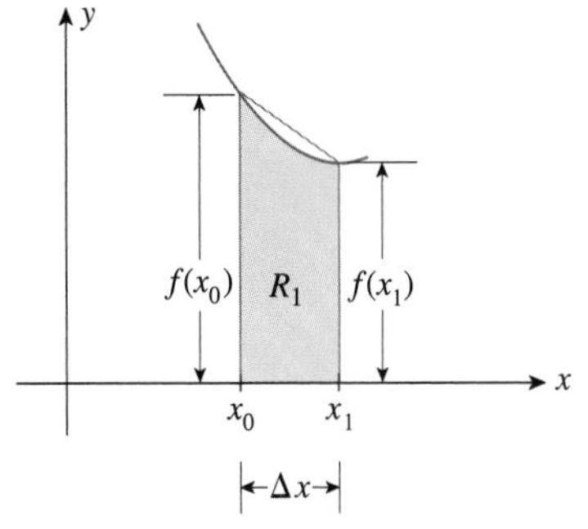

Figure 9.3

The area of R_1 is approximated by the area of the trapezoid.

Similarly, we see that the area of the last (nth) approximating trapezoid is given by

$$\left[\frac{f(x_{n-1}) + f(x_n)}{2}\right]\Delta x$$

Then, the area of the region R is approximated by the sum of the areas of the n trapezoids—that is,

$$\left[\frac{f(x_0) + f(x_1)}{2}\right]\Delta x + \left[\frac{f(x_1) + f(x_2)}{2}\right]\Delta x + \cdots + \left[\frac{f(x_{n-1}) + f(x_n)}{2}\right]\Delta x$$

$$= \frac{\Delta x}{2}[f(x_0) + f(x_1) + f(x_2) + f(x_2) + \cdots + f(x_{n-1}) + f(x_n)]$$

$$= \frac{\Delta x}{2}[f(x_0) + 2f(x_1) + 2f(x_2) + \cdots + 2f(x_{n-1}) + f(x_n)]$$

Since the area of the region R is given by the value of the definite integral we wished to approximate, we are led to the following approximation formula, which is called the **trapezoidal rule.**

Trapezoidal Rule

$$\int_a^b f(x)\,dx \approx \frac{\Delta x}{2}[f(x_0) + 2f(x_1) + 2f(x_2) + \cdots + 2f(x_{n-1}) + f(x_n)] \quad (5)$$

where $\Delta x = \dfrac{b-a}{n}$.

The approximation generally improves with larger values of n.

Example 1

Approximate the value of

$$\int_1^2 \frac{1}{x}\,dx$$

using the trapezoidal rule with $n = 10$. Compare this result with the exact value of the integral.

Solution

Here, $a = 1$, $b = 2$, and $n = 10$, so

$$\Delta x = \frac{b-a}{n} = \frac{1}{10} = 0.1$$

and

$$x_0 = 1, \quad x_1 = 1.1, \quad x_2 = 1.2, \quad x_3 = 1.3, \ldots, x_9 = 1.9, \quad x_{10} = 2$$

The trapezoidal rule yields

$$\begin{aligned}\int_1^2 \frac{1}{x}\,dx &\approx \frac{0.1}{2}\left[1 + 2\left(\frac{1}{1.1}\right) + 2\left(\frac{1}{1.2}\right) + 2\left(\frac{1}{1.3}\right) + \cdots + 2\left(\frac{1}{1.9}\right) + \frac{1}{2}\right]\\ &\approx 0.693771\end{aligned}$$

In this case we can easily compute the actual value of the definite integral under consideration. In fact,

$$\begin{aligned}\int_1^2 \frac{1}{x}\,dx = \ln x\Big|_1^2 &= \ln 2 - \ln 1 = \ln 2\\ &\approx 0.693147\end{aligned}$$

Thus, the trapezoidal rule with $n = 10$ yields a result with an error of 0.000624 to six decimal places.

Example 2

Consumer's Surplus The demand function for a certain brand of perfume is given by

$$p = D(x) = \sqrt{10{,}000 - 0.01x^2}$$

where p is the unit price in dollars and x is the quantity demanded each week, measured in ounces. Find the consumers' surplus if the market price is set at \$60 per ounce.

Solution

When $p = 60$, we have

$$\begin{aligned}\sqrt{10{,}000 - 0.01x^2} &= 60\\ 10{,}000 - 0.01x^2 &= 3{,}600\\ x^2 &= 640{,}000\end{aligned}$$

or $x = 800$ since x must be nonnegative. Next, using the consumers' surplus formula (page 508) with $\bar{p} = 60$ and $\bar{x} = 800$, we see that the consumers' surplus is given by

$$CS = \int_0^{800} \sqrt{10{,}000 - 0.01x^2}\,dx - (60)(800)$$

It is not easy to evaluate this definite integral by finding an antiderivative of the integrand. Instead, let's use the trapezoidal rule with $n = 10$.

With $a = 0$ and $b = 800$, we find that

$$\Delta x = \frac{b-a}{n} = \frac{800}{10} = 80$$

and

$$x_0 = 0, \quad x_1 = 80, \quad x_2 = 160, \quad x_3 = 240, \ldots, x_9 = 720, \quad x_{10} = 800$$

so

$$\begin{aligned}
&\int_0^{800} \sqrt{10{,}000 - 0.01x^2}\, dx \\
&\approx \frac{80}{2}\big[100 + 2\sqrt{10{,}000 - (0.01)(80)^2} \\
&\quad + 2\sqrt{10{,}000 - (0.01)(160)^2} + \cdots + 2\sqrt{10{,}000 - (0.01)(720)^2} \\
&\quad + \sqrt{10{,}000 - (0.01)(800)^2}\big] \\
&= 40[100 + 199.3590 + 197.4234 + 194.1546 + 189.4835 \\
&\quad + 183.3030 + 175.4537 + 165.6985 \\
&\quad + 153.6750 + 138.7948 + 60] \\
&\approx 70{,}293.82
\end{aligned}$$

Therefore, the consumers' surplus is approximately 70,294 − 48,000, or \$22,294.

GROUP DISCUSSION

Explain how you would approximate the value of $\int_0^2 f(x)\, dx$ using the trapezoidal rule with $n = 10$, where

$$f(x) = \begin{cases} \sqrt{1+x^2} & \text{if } 0 \le x \le 1 \\ \dfrac{2}{\sqrt{1+x^2}} & \text{if } 1 < x \le 2 \end{cases}$$

and find the value.

Simpson's Rule

Before stating Simpson's rule, let's review the two rules we have used in approximating a definite integral. Let f be a continuous nonnegative function defined on the interval $[a, b]$. Suppose the interval $[a, b]$ is partitioned by means of the $n + 1$ equally spaced points, $x_0 = a, x_1, x_2 \ldots, x_n = b$, where n is a positive integer, so that the length of each subinterval is $\Delta x = (b - a)/n$ (Figure 9.4).

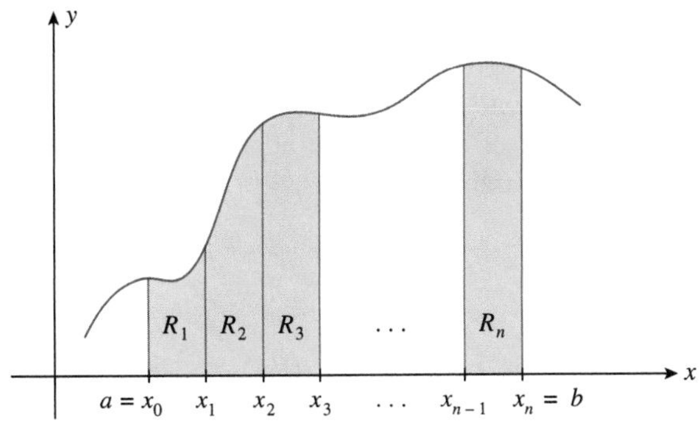

Figure 9.4
The area under the curve is equal to the sum of the n subregions $R_1, R_2, \ldots, R_n$.

Let's concentrate on the portion of the graph of $y = f(x)$ defined on the interval $[x_0, x_2]$. In using a Riemann sum to approximate the definite integral, we are in effect approximating the function $f(x)$ on $[x_0, x_1]$ by the *constant* function $y = f(p_1)$, where p_1 is chosen to be a point in $[x_0, x_1]$; the function $f(x)$ on $[x_1, x_2]$ by the constant function $y = f(p_2)$, where p_2 lies in $[x_1, x_2]$; and so on. Using a Riemann sum, we see that the area of the region under the curve $y = f(x)$ between $x = a$ and $x = b$ is approximated by the area under the approximating "step" function (Figure 9.5a).

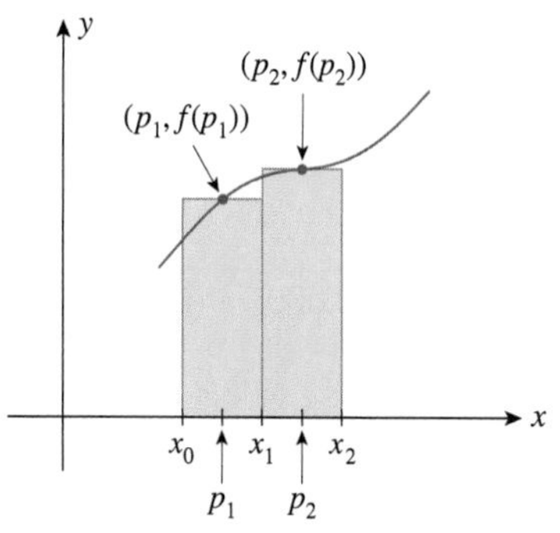

(a) The area under the curve is approximated by the area of the rectangles.

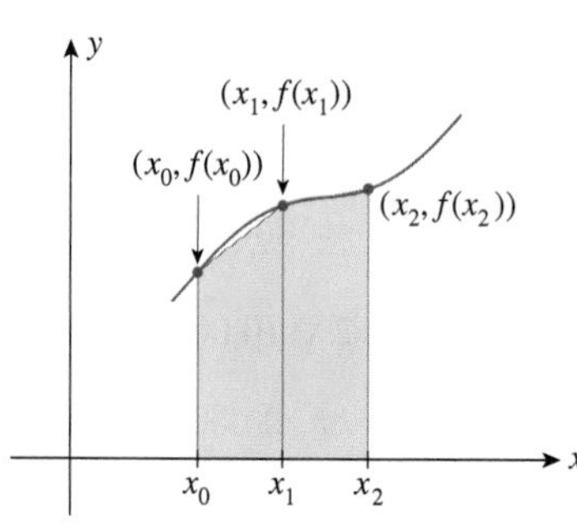

(b) The area under the curve is approximated by the area of the trapezoids.

Figure 9.5

When we use the trapezoidal rule, we are in effect approximating the function $f(x)$ on the interval $[x_0, x_1]$ by a *linear* function through the two points $(x_0, f(x_0))$ and $(x_1, f(x_1))$; the function $f(x)$ on $[x_1, x_2]$ by a *linear* function through the two points $(x_1, f(x_1))$ and $(x_2, f(x_2))$; and so on. Thus, the trapezoidal rule simply approximates the actual area of the region under the curve $y = f(x)$ from $x = a$ to $x = b$ by the area under the approximating polygonal curve (Figure 9.5b).

A natural extension of the preceding idea is to approximate portions of the graph of $y = f(x)$ by means of portions of the graphs of second-degree polynomials (parts of parabolas). It can be shown that given any three noncollinear points there is a unique parabola that passes through the given points. Choose the points $(x_0, f(x_0))$, $(x_1, f(x_1))$, and $(x_2, f(x_2))$ corresponding to the first three points of the partition. Then, we can approximate the function $f(x)$ on $[x_0, x_2]$ by means of a quadratic function whose graph contains these three points (Figure 9.6).

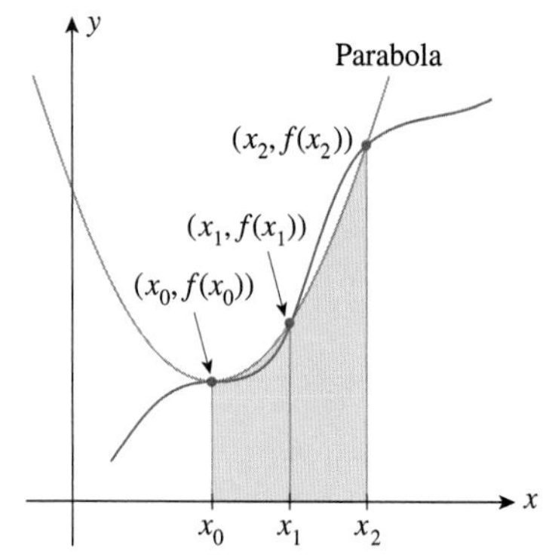

Figure 9.6

Simpson's rule approximates the area under the curve by the area under the parabola.

Although we will not do so here, it can be shown that the area under the parabola between $x = x_0$ and $x = x_2$ is given by

$$\frac{\Delta x}{3}\left[f(x_0) + 4f(x_1) + f(x_2)\right]$$

square units. Repeating this argument on the interval $[x_2, x_4]$, we see that the area under the curve between $x = x_2$ and $x = x_4$ is approximated by the area under the parabola between x_2 and x_4—that is, by

$$\frac{\Delta x}{3}\left[f(x_2) + 4f(x_3) + f(x_4)\right]$$

square units. Proceeding, we conclude that if n is even (Why?), then the area under the curve $y = f(x)$ from $x = a$ to $x = b$ may be approximated by the sum of the areas under the $n/2$ approximating parabolas—that is,

$$\frac{\Delta x}{3}\left[f(x_0) + 4f(x_1) + f(x_2)\right] + \frac{\Delta x}{3}\left[f(x_2) + 4f(x_3) + f(x_4)\right] + \cdots$$
$$+ \frac{\Delta x}{3}\left[f(x_{n-2}) + 4f(x_{n-1}) + f(x_n)\right]$$

$$= \frac{\Delta x}{3}[f(x_0) + 4f(x_1) + f(x_2) + f(x_2) + 4f(x_3) + f(x_4) + \cdots$$
$$+ f(x_{n-2}) + 4f(x_{n-1}) + f(x_n)]$$
$$= \frac{\Delta x}{3}[f(x_0) + 4f(x_1) + 2f(x_2) + 4f(x_3)$$
$$+ 2f(x_4) + \cdots + 4f(x_{n-1}) + f(x_n)]$$

The preceding is the derivation of the approximation formula known as **Simpson's rule.**

Simpson's Rule

$$\int_a^b f(x)\,dx \approx \frac{\Delta x}{3}[f(x_0) + 4f(x_1) + 2f(x_2) + 4f(x_3) + 2f(x_4)$$
$$+ \cdots + 4f(x_{n-1}) + f(x_n)] \qquad (6)$$

where $\Delta x = \dfrac{b-a}{n}$ and n is even.

In using this rule, remember that n must be even.

Example 3

Find an approximation of

$$\int_1^2 \frac{1}{x}\,dx$$

using Simpson's rule with $n = 10$. Compare this result with that of Example 1 and also with the exact value of the integral.

Solution

We have $a = 1$, $b = 2$, $f(x) = \dfrac{1}{x}$, and $n = 10$, so

$$\Delta x = \frac{b-a}{n} = \frac{1}{10} = 0.1$$

and

$$x_0 = 1, \quad x_1 = 1.1, \quad x_2 = 1.2, \quad x_3 = 1.3, \ldots, x_9 = 1.9, \quad x_{10} = 2$$

Simpson's rule yields

$$\int_1^2 \frac{1}{x}\,dx \approx \frac{0.1}{3}[f(1) + 4f(1.1) + 2f(1.2) + \cdots + 4f(1.9) + f(2)]$$
$$= \frac{0.1}{3}\left[1 + 4\left(\frac{1}{1.1}\right) + 2\left(\frac{1}{1.2}\right) + 4\left(\frac{1}{1.3}\right) + 2\left(\frac{1}{1.4}\right) + 4\left(\frac{1}{1.5}\right)\right.$$
$$\left. + 2\left(\frac{1}{1.6}\right) + 4\left(\frac{1}{1.7}\right) + 2\left(\frac{1}{1.8}\right) + 4\left(\frac{1}{1.9}\right) + \frac{1}{2}\right]$$
$$\approx 0.693150$$

The trapezoidal rule with $n = 10$ yielded an approximation of 0.693771, which is 0.000624 off the value of $\ln 2 \approx 0.693147$ to six decimal places. Simpson's rule yields an approximation with an error of 0.000003, a definite improvement over the trapezoidal rule.

Example 4

Cardiac Output Solve the problem posed at the beginning of this section. Recall that we wished to find a person's cardiac output by using the formula

$$R = \frac{60\,D}{\int_0^{28} c(t)\,dt}$$

where D (the quantity of dye injected) is equal to 5 milligrams and the function c has the graph shown in Figure 9.7. Use Simpson's rule with $n = 14$ to estimate the value of the integral.

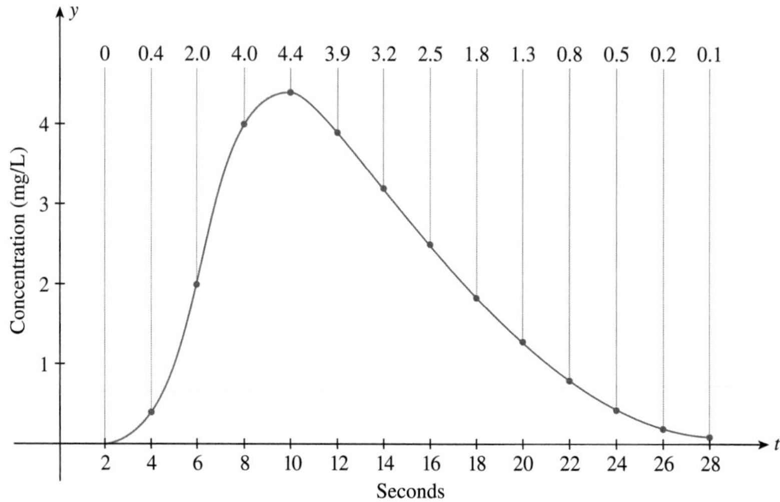

Figure 9.7
The function c gives the concentration of a dye measured at the aorta. The graph is constructed by drawing a smooth curve through a set of discrete points.

Solution

Using Simpson's rule with $n = 14$ and $\Delta t = 2$ so that

$$t_0 = 0, \quad t_1 = 2, \quad t_2 = 4, \quad t_3 = 6, \ldots, t_{14} = 28$$

we obtain

$$\begin{aligned}
\int_0^{28} c(t)\,dt &\approx \frac{2}{3}[c(0) + 4c(2) + 2c(4) + 4c(6) + \cdots \\
&\quad + 4c(26) + c(28)] \\
&= \frac{2}{3}[0 + 4(0) + 2(0.4) + 4(2.0) + 2(4.0) \\
&\quad + 4(4.4) + 2(3.9) + 4(3.2) + 2(2.5) + 4(1.8) \\
&\quad + 2(1.3) + 4(0.8) + 2(0.5) + 4(0.2) + 0.1] \\
&\approx 49.9
\end{aligned}$$

Therefore, the person's cardiac output is

$$R \approx \frac{6(5)}{49.9} \approx 6.0$$

or 6.0 L/min.

Example 5

Oil Spill An oil spill off the coastline was caused by a ruptured tank in a grounded oil tanker. Using aerial photographs, the Coast Guard was able to obtain the dimensions of the oil spill (Figure 9.8). Using Simpson's rule with $n = 10$, estimate the area of the oil spill.

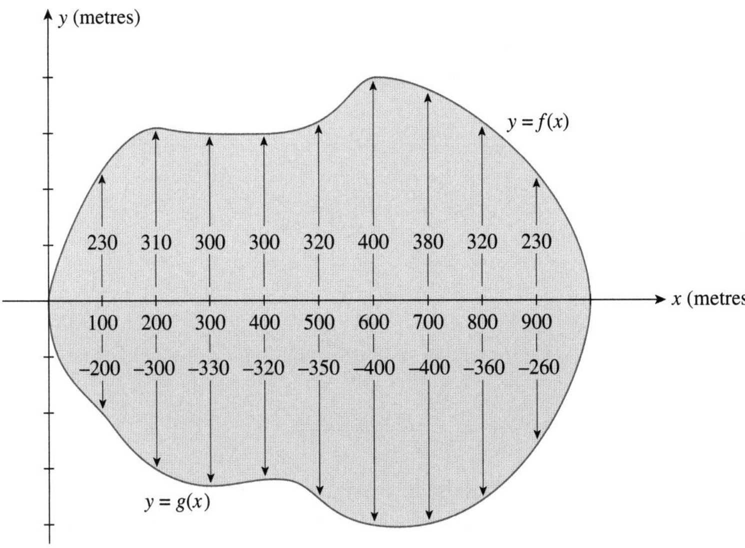

Figure 9.8

Simpson's rule can be used to calculate the area of the oil spill.

Solution

We may think of the area affected by the oil spill as the area of the plane region bounded above by the graph of the function $f(x)$ and below by the graph of the function $g(x)$ between $x = 0$ and $x = 1000$ (Figure 9.8). Then, the required area is given by

$$A = \int_0^{1000} [f(x) - g(x)]\,dx$$

Using Simpson's rule with $n = 10$ and $\Delta x = 100$ so that

$$x_0 = 0, \qquad x_1 = 1000, \qquad x_2 = 200, \ldots, x_{10} = 1000$$

we have

$$\begin{aligned}
A &= \int_0^{1000} [f(x) - g(x)]\,dx \\
&\approx \frac{\Delta x}{3}\{[f(x_0) - g(x_0)] + 4[f(x_1) - g(x_1)] + 2[f(x_2) - g(x_2)] \\
&\quad + \cdots + 4[f(x_9) - g(x_9)] + [f(x_{10}) - g(x_{10})]\} \\
&= \frac{100}{3}\{[0 - 0] + 4[230 - (-200)] + 2[310 - (-300)] \\
&\quad + 4[300 - (-330)] + 2[300 - (-320)] + 4[320 - (-350)] \\
&\quad + 2[400 - (-400)] + 4[380 - (-400)] + 2[320 - (-360)] \\
&\quad + 4[230 - (-260)] + [0 - 0]\} \\
&= \frac{100}{3}[0 + 4(430) + 2(610) + 4(630) + 2(620) + 4(670) \\
&\quad + 2(800) + 4(780) + 2(680) + 4(490) + 0] \\
&= \frac{100}{3}(17{,}420) \\
&\approx 580.667
\end{aligned}$$

or approximately 580,667 square metres.

GROUP DISCUSSION

Explain how you would approximate the value of $\int_0^2 f(x)\,dx$ using Simpson's rule with $n = 10$, where

$$f(x) = \begin{cases} \sqrt{1+x^2} & \text{if } 0 \le x \le 1 \\ \dfrac{2}{\sqrt{1+x^2}} & \text{if } 1 < x < 2 \end{cases}$$

and find the value.

Error Analysis

The following results give the bounds on the errors incurred when the trapezoidal rule and Simpson's rule are used to approximate a definite integral (proof omitted).

Errors in the Trapezoidal and Simpson Approximations

Suppose the definite integral

$$\int_a^b f(x)\,dx$$

is approximated with n subintervals.

1. The *maximum* error incurred in using the trapezoidal rule is

$$\frac{M(b-a)^3}{12n^2} \tag{7}$$

where M is a number such that $|f''(x)| \le M$ for all x in $[a, b]$.

2. The *maximum* error incurred in using Simpson's rule is

$$\frac{M(b-a)^5}{180n^4} \tag{8}$$

where M is a number such that $|f^{(4)}(x)| \le M$ for all x in $[a, b]$.

Remark

In many instances, the actual error is less than the upper error bounds given. ◀

Example 6

Find bounds on the errors incurred when

$$\int_1^2 \frac{1}{x}\,dx$$

is approximated using (a) the trapezoidal rule and (b) Simpson's rule with $n = 10$. Compare these with the actual errors found in Examples 1 and 3.

Solution

a. Here, $a = 1$, $b = 2$, and $f(x) = 1/x$. Next, to find a value for M, we compute

$$f'(x) = -\frac{1}{x^2} \quad \text{and} \quad f''(x) = \frac{2}{x^3}$$

Since $f''(x)$ is positive and decreasing on $(1, 2)$ (Why?), it attains its maximum value of 2 at $x = 1$, the left end point of the interval. Therefore, if we take $M = 2$, then $|f''(x)| \le 2$. Using (7), we see that the maximum error incurred is

$$\frac{2(2-1)^3}{12(10)^2} = \frac{2}{1200} = 0.0016667$$

The actual error found in Example 1, 0.000624, is much less than the upper bound just found.

b. We compute

$$f'''(x) = \frac{-6}{x^4} \quad \text{and} \quad f^{(4)}(x) = \frac{24}{x^5}$$

Since $f^{(4)}(x)$ is positive and decreasing on (1, 2) (just look at $f^{(5)}$ to verify this fact), it attains its maximum at the left end point of [1, 2]. Now,

$$f^{(4)}(1) = 24$$

and so we may take $M = 24$. Using (8), we obtain the maximum error of

$$\frac{24(2-1)^5}{180(10)^4} = 0.0000133$$

The actual error is 0.000003 (see Example 3).

GROUP DISCUSSION

Refer to the Group Discussion on pages 561 and 566. Explain how you would find the maximum error incurred in using (1) the trapezoidal rule and (2) Simpson's rule with $n = 10$ to approximate $\int_0^2 f(x)\,dx$.

Self-Check Exercises 9.4

1. Use the trapezoidal rule and Simpson's rule with $n = 8$ to approximate the value of the definite integral

$$\int_0^2 \frac{1}{\sqrt{1+x^2}}\,dx$$

2. The graph in the accompanying figure shows the consumption of petroleum in North America in quadrillion BTU, from 1976 to 1990. Using Simpson's rule with $n = 14$, estimate the average consumption during the 14-yr period.

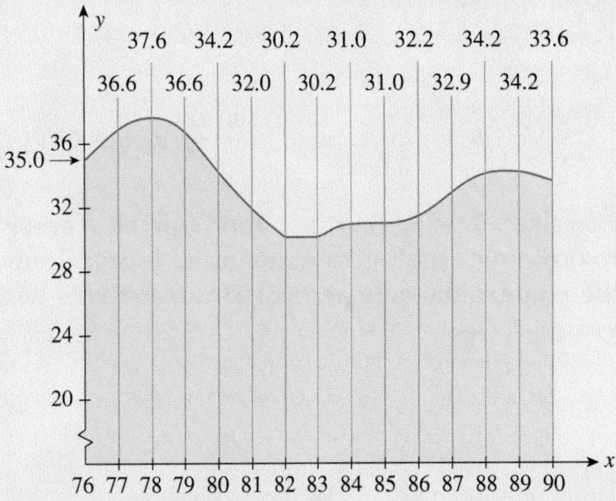

Source: The World Almanac

Solutions to Self-Check Exercises 9.4 can be found on page 571.

9.4 Exercises

A calculator is recommended for this exercise set.

In Exercises 1–18, use the trapezoidal rule and Simpson's rule to approximate the value of each definite integral. Compare your result with the exact value of the integral.

1. $\int_0^2 x^2\,dx;\ n = 6$

2. $\int_1^3 (x^2 - 1)\,dx;\ n = 4$

3. $\int_0^1 x^3\,dx;\ n = 4$

4. $\int_1^2 x^3\,dx;\ n = 6$

5. $\int_1^2 \frac{1}{x}\,dx;\ n = 4$

6. $\int_1^2 \frac{1}{x}\,dx;\ n = 8$

7. $\int_1^2 \frac{1}{x^2}\,dx;\ n = 4$

8. $\int_0^1 \frac{1}{1+x}\,dx;\ n = 4$

9. $\int_0^4 \sqrt{x}\,dx;\ n = 8$

10. $\int_0^2 x\sqrt{2x^2+1}\,dx;\ n = 6$

11. $\int_0^1 e^{-x}\,dx;\ n = 6$

12. $\int_0^1 xe^{-x^2}\,dx;\ n = 6$

13. $\int_1^2 \ln x\,dx;\ n = 4$

14. $\int_0^1 x\ln(x^2+1)\,dx;\ n = 8$

15. $\int_0^\pi \sin(x)\,dx;\ n = 6$

16. $\int_0^{\pi/2} \cos(x)\,dx;\ n = 4$

17. $\int_0^4 \frac{dx}{x^2+1};\ n = 8$

18. $\int_0^{0.5} \frac{dx}{\sqrt{1-x^2}};\ n = 4$

In Exercises 19–30, use the trapezoidal rule and Simpson's rule to approximate the value of each definite integral.

19. $\int_0^1 \sqrt{1+x^3}\,dx;\ n = 4$

20. $\int_0^2 x\sqrt{1+x^3}\,dx;\ n = 4$

21. $\int_0^2 \frac{1}{\sqrt{x^3+1}}\,dx;\ n = 4$

22. $\int_0^1 \sqrt{1-x^2}\,dx;\ n = 4$

23. $\int_0^2 e^{-x^2}\,dx;\ n = 4$

24. $\int_0^1 e^{x^2}\,dx;\ n = 6$

25. $\int_1^2 x^{-1/2}e^x\,dx;\ n = 4$

26. $\int_2^4 \frac{dx}{\ln x};\ n = 6$

27. $\int_0^2 \sin(x^2)\,dx;\ n = 4$

28. $\int_0^1 e^{\sin(3x)}\,dx;\ n = 4$

29. $\int_0^{\pi/2} \frac{\cos(x)}{x^2+1}\,dx;\ n = 6$

30. $\int_{-1}^1 \frac{\sin(\pi x)}{x+2}\,dx;\ n = 6$

In Exercises 31–38, find a bound on the error in approximating the given definite integral using (a) the trapezoidal rule and (b) Simpson's rule with *n* intervals.

31. $\int_{-1}^2 x^5\,dx;\ n = 10$

32. $\int_0^1 e^{-x}\,dx;\ n = 8$

33. $\int_1^3 \frac{1}{x}\,dx;\ n = 10$

34. $\int_1^3 \frac{1}{x^2}\,dx;\ n = 8$

35. $\int_0^2 \frac{1}{\sqrt{1+x}}\,dx;\ n = 8$

36. $\int_1^3 \ln x\,dx;\ n = 10$

37. $\int_0^\pi \sin(x)\,dx;\ n = 10$

38. $\int_0^{\pi/3} \cos(x)\,dx;\ n = 8$

39. **Trial Run of an Attack Submarine** In a submerged trial run of an attack submarine, a reading of the sub's velocity was made every quarter hour, as shown in the accompanying table. Use the trapezoidal rule to estimate the distance travelled by the submarine during the 2-h period.

Time, *t* (h)	0	$\frac{1}{4}$	$\frac{1}{2}$	$\frac{3}{4}$	
Velocity, *V*(*t*) (km/h)	31.5	36.4	41.2	54.6	
Time, *t* (h)	1	$\frac{5}{4}$	$\frac{3}{2}$	$\frac{7}{4}$	2
Velocity, *V*(*t*) (km/h)	44.1	38.5	27.1	23.5	12.1

40. Derive the formula

$$R = \frac{60\,D}{\int_0^T C(t)\,dt}$$

for calculating the cardiac output of a person in L/min. Here, $c(t)$ is the concentration of dye in the aorta (in mg/L) at time t (in seconds) for t in $[0, T]$, and D is the amount of dye (in mg) injected into a vein leading to the heart.

Hint: Partition the interval $[0, T]$ into n subintervals of equal length Δt. The amount of dye that flows past the measuring point in the aorta during the time interval $[0, \Delta t.]$ is approximately $c(t_i)(R\Delta t)/60$ (concentration times volume). Therefore, the total amount of dye measured at the aorta is

$$\frac{[c(t_1)R\Delta t + c(t_2)R\Delta t + \cdots + c(t_n)R\Delta t]}{60} = D$$

Take the limit of the Riemann sum to obtain

$$R = \frac{60\,D}{\int_0^T c(t)\,dt}$$

41. **Fuel Consumption of Domestic Cars** Thanks to smaller and more fuel-efficient models, carmakers have doubled their average fuel economy over a 13-yr period, from 1974 to 1987. The graph depicted in the figure below gives the average fuel consumption in kilometres per litre (km/L) of domestic-built cars over the period under consideration ($t = 0$ corresponds to the beginning of 1974). Use the trapezoidal rule to estimate the average fuel consumption of the domestic car built during this period.

Hint: Approximate the integral $\frac{1}{13}\int_0^{13} f(t)\,dt$.

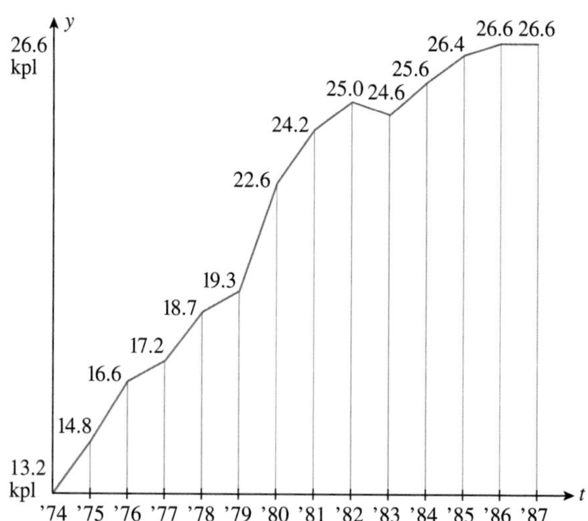

42. **Average Temperature** The graph depicted in the accompanying figure shows the daily mean temperatures recorded during one September in Cameron Highlands. Using (a) the trapezoidal rule and (b) Simpson's rule with $n = 10$, estimate the average temperature during that month.

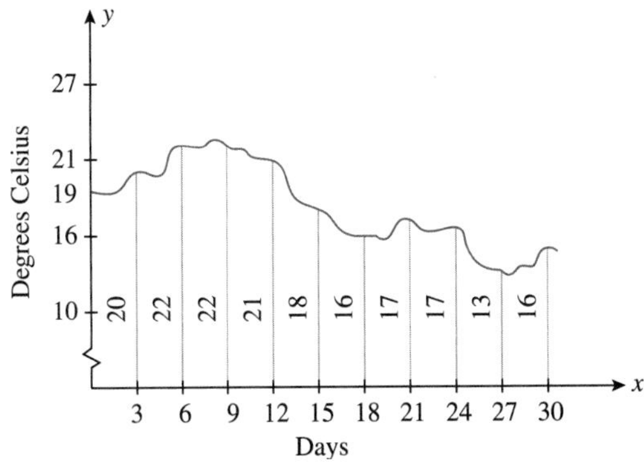

43. Surface Area of a Lake A manmade lake located in Lake View Condominiums has the shape depicted in the following figure. The measurements shown were taken at 15-metre intervals. Using Simpson's rule with $n = 10$, estimate the surface area of the lake.

Hint: Study Example 5.

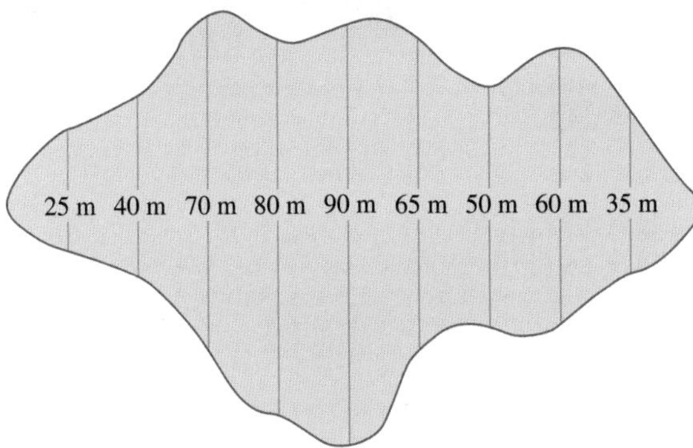

44. Air Pollution The amount of nitrogen dioxide, a brown gas that impairs breathing, present in the atmosphere on a certain May day in the city of Vancouver has been approximated by

$$A(t) = \frac{136}{1 + 0.25(t - 4.5)^2} + 28 \qquad (0 \le t \le 11)$$

where $A(t)$ is measured in pollutant standard index (PSI), and t is measured in hours, with $t = 0$ corresponding to 7 A.M. Use the trapezoidal rule with $n = 10$ to estimate the average PSI between 7 a.m. and noon.

Hint: $\frac{1}{5}\int_0^5 A(t)\,dt$.

45. Tread Lives of Tires Under normal driving conditions the percent of Super Titan radial tires expected to have a useful tread life of between 30,000 and 40,000 km is given by

$$P = 100\int_{30,000}^{40,000} \frac{1}{2000\sqrt{2\pi}} e^{-(1/2)[(x-40,000)/2000]^2}\,dx$$

Use Simpson's rule with $n = 10$ to estimate P.

Business and Economics Applications

46. Real Estate Cooper Realty is considering development of a time-sharing condominium resort complex along the oceanfront property illustrated in the accompanying graph. To obtain an estimate of the area of this property, measurements of the distances from the edge of a straight road, which defines one boundary of the property, to the corresponding points on the shoreline are made at 100-m intervals. Using Simpson's rule with $n = 10$, estimate the area of the oceanfront property.

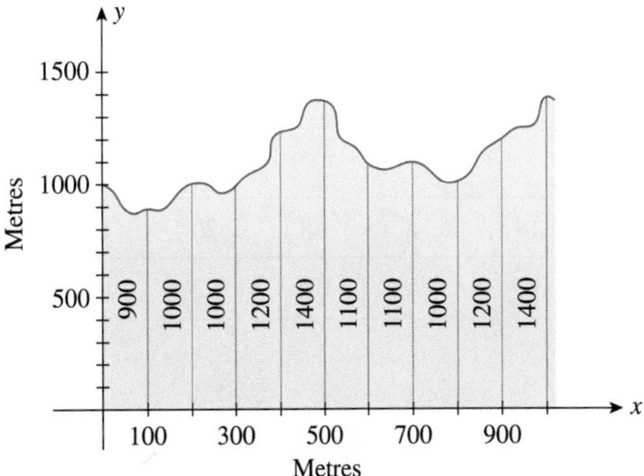

47. Consumers' Surplus Refer to Section 8.7. The demand equation for the Sicard wristwatch is given by

$$p = S(x) = \frac{50}{0.01x^2} + 1 \qquad (0 \le x \le 20)$$

where x (measured in units of a thousand) is the quantity demanded per week and p is the unit price in dollars. Use (a) the trapezoidal rule and (b) Simpson's rule (take $n = 8$) to estimate the consumers' surplus if the market price is \$25/watch.

48. Producers' Surplus Refer to Section 8.7. The supply function for the audio compact disc manufactured by Herald Records is given by

$$p = S(x) = \sqrt{0.01x^2 + 0.11x + 38}$$

where p is the unit wholesale price in dollars and x stands for the quantity that will be made available in the market by the supplier, measured in units of a thousand. Use (a) the trapezoidal rule and (b) Simpson's rule (take $n = 8$) to estimate the producers' surplus if the wholesale price is \$8/disc.

49. Growth of Service Industries It has been estimated that service industries, which currently make up 30% of the nonfarm workforce in a certain country, will continue to grow at the rate of

$$R(t) = 5e^{1/(t+1)}$$

percent/decade t decades from now. Estimate the percent of the nonfarm workforce in the service industries one decade from now.

Hint: (a) Show that the answer is given by $30 + \int_0^1 5e^{1/(t+1)}\,dt$ and (b) use Simpson's rule with $n = 10$ to approximate the definite integral.

Biological and Life Sciences Applications

50. Length of Infants at Birth Medical records of infants delivered at Kaiser Memorial Hospital show that the percent of infants whose length at birth is between 48 and 53 cm is given by

$$P = 100\int_{48}^{53} \frac{1}{2.6\sqrt{2\pi}} e^{-(1/2)[(x-20)/2.6]^2}\,dx$$

Use Simpson's rule with $n = 10$ to estimate P.

51. Measuring Cardiac Output Eight milligrams of a dye are injected into a vein leading to an individual's heart. The concentration of the dye in the aorta measured at 2-sec intervals is shown in the accompanying table. Use Simpson's rule and the formula of Example 4 to estimate the person's cardiac output.

t	0	2	4	6	8	10	12
$C(t)$	0	0	2.8	6.1	9.7	7.6	4.8

t	14	16	18	20	22	24
$C(t)$	3.7	1.9	0.8	0.3	0.1	0

In Exercises 52–55, determine whether the statement is true or false. It is true, explain why it is true. If it is false, give an example to show why it is false.

52. In using the trapezoidal rule, the number of subintervals n must be even.

53. In using Simpson's rule, the number of subintervals n may be chosen to be odd or even.

54. Simpson's rule is more accurate than the trapezoidal rule.

55. If f is a polynomial function of degree less than or equal to 3, then the approximation $\int_a^b f(x)\,dx$ using Simpson's rule is exact.

PORTFOLIO

James H. Chesebro, M.D.

Title Professor of Medicine
Institution Harvard Medical School, Massachusetts General Hospital

In 1992 Chesebro was appointed professor of medicine at Harvard Medical School, after having conducted cardiovascular research at the Mayo Clinic for a number of years. In addition, he was made Associate Director for Research in the Cardiac Unit at Massachusetts General Hospital, one of Harvard's teaching affiliates.

For over 20 years, James Chesebro has worked as an investigative cardiologist, diagnosing and treating patients suffering from heart and blood vessel diseases. He specializes in researching ways to prevent blood clots from narrowing the heart's arteries. Even when patients have coronary-bypass operations to correct problems, stresses Chesebro, "the piece of vein used to detour around blocked arteries is prone to plug up with clots."

Throughout his career, Chesebro has investigated the body's clotting mechanisms, as well as substances that may prevent, or even dissolve, clots. A study conducted in 1982 showed that bypass patients taking dipyridamole and aspirin were nearly three times less likely to develop clots in vein grafts than patients given placebos.

Recently, Chesebro has been working with a substance called hirudin derived from leech saliva. Hirudin has proved quite beneficial in blocking arterial blood-clot formations.

Determining the correct medication and dosage involves intense quantitative research. Colleagues from other branches of science such as nuclear medicine, molecular biology, and biochemistry play key roles in the research process. Chesebro and his colleagues have to understand how the heart's physiology and pharmacological variables such as distribution rates, concentration levels, elimination rates, and biological effects of particular substances dictate the choice and dosage of medication.

In considering whether a particular medication will bind effectively with specific body cells, researchers must determine bond tightness, speed, and length of time to reach optimal concentration. Researchers rely on complicated equations such as "integrating the area under a curve to find the amount of medication in the body at a given time," notes Chesebro. Without calculus, these equations can't be solved.

Whether physicians rely on medication or surgical procedures such as balloon angioplasty to open clogged passages, other factors play a major role in the outcome. Cholesterol and blood sugar levels, patient age, blood pressure, the geometry of the blockage, and the velocity and turbulence of blood flow all warrant consideration. According to Chesebro, "linear modelling can be used to predict the contribution of patient variables and local blood vessel variables in the outcome of opening blocked arteries."

The bottom line? Calculus has been a key contributor to Chesebro's success in preventing disabling arterial blood clots.

Solutions to Self-Check Exercises 9.4

1. We have $x = 0$, $b = 2$, and $n = 8$, so

$$\Delta x = \frac{b-a}{n} = \frac{2}{8} = 0.25$$

and $x_0 = 0$, $x_1 = 0.25$, $x_2 = 0.50$, $x_3 = 0.75, \ldots, x_7 = 1.75$, and $x_8 = 2$. The trapezoidal rule gives

$$\int_0^2 \frac{1}{\sqrt{1+x^2}}\,dx \approx \frac{0.25}{2}\left[1 + \frac{2}{\sqrt{1+(0.25)^2}} + \frac{2}{\sqrt{1+(0.5)^2}} + \cdots + \frac{2}{\sqrt{1+(1.75)^2}} + \frac{1}{\sqrt{5}}\right]$$

$$\approx 0.125(1 + 1.9403 + 1.7889 + 1.6000 + 1.4142 + 1.2494 + 1.1094 + 0.9923 + 0.4472)$$

$$\approx 1.4427$$

Using Simpson's rule with $n = 8$ gives

$$\int_0^2 \frac{1}{\sqrt{1+x^2}}\,dx \approx \frac{0.25}{3}\left[1 + \frac{4}{\sqrt{1+(0.25)^2}} + \frac{2}{\sqrt{1+(0.5)^2}} + \frac{4}{\sqrt{1+(0.75)^2}} + \cdots + \frac{4}{\sqrt{1+(1.75)^2}} + \frac{1}{\sqrt{5}}\right]$$

$$\approx \frac{0.25}{3}(1 + 3.8806 + 1.7889 + 3.2000 + 1.4142 + 2.4988 + 1.1094 + 1.9846 + 0.4472)$$

$$\approx 1.4436$$

2. The average consumption of petroleum during the 14-yr period is given by

$$\frac{1}{14}\int_0^{14} f(x)\,dx$$

where f is the function describing the given graph. Using Simpson's rule with $a = 0$, $b = 14$, and $n = 14$ so that $\Delta x = 1$ and

$$x_0 = 0, \qquad x_1 = 1, \qquad x_2 = 2, \ldots, x_{14} = 14$$

we have

$$\frac{1}{14}\int_0^{14} f(x)\,dx$$

$$\approx \left(\frac{1}{14}\right)\left(\frac{1}{3}\right)[f(x_0) + 4f(x_1) + 2f(x_2) + 4f(x_3) + \cdots + 4f(x_{13}) + f(x_{14})]$$

$$= \frac{1}{42}[35 + 4(36.6) + 2(37.6) + 4(36.6) + 2(34.2) + 4(32.0) + 2(30.2) + 4(30.2) + 2(31.0) + 4(31.0) + 2(32.2) + 4(32.9) + 2(34.2) + 4(34.2) + 33.6]$$

$$\approx 33.4$$

or approximately 33.4 quadrillion BTU/year.

9.5 Improper Integrals

Improper Integrals

All the definite integrals we have encountered have had finite intervals of integration. In many applications, however, we are concerned with integrals that have unbounded intervals of integration. Such integrals are called **improper integrals.**

To lead us to the definition of an improper integral of a function f over an infinite interval, consider the problem of finding the area of the region R under the curve $y = f(x) = 1/x^2$ and to the right of the vertical line $x = 1$, as shown in Figure 9.9. Because the interval over which the integration must be performed is unbounded, the

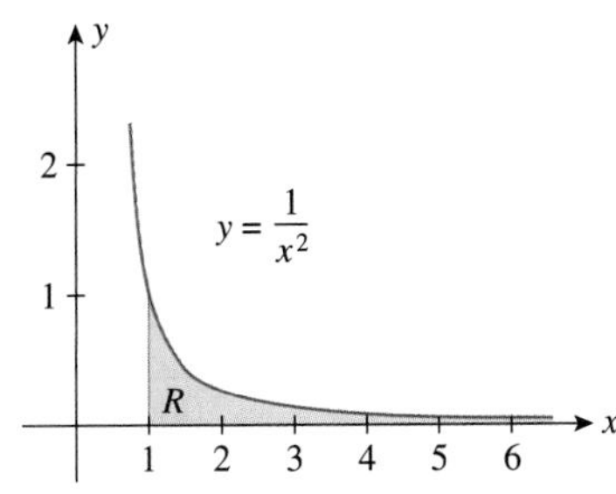

Figure 9.9
The area of the unbounded region R can be approximated by a definite integral.

method of integration presented previously cannot be applied directly in solving this problem. However, we can approximate the region R by the definite integral

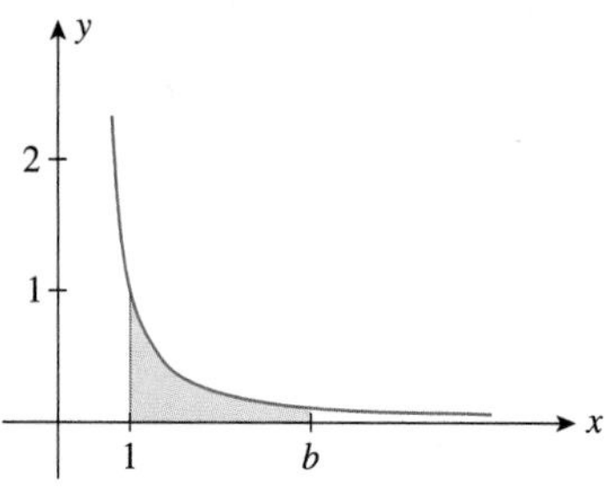

Figure 9.10

Area of shaded region $= \int_1^b \frac{1}{x^2}\,dx$

$$\int_1^b \frac{1}{x^2}\,dx \tag{9}$$

which gives the area of the region under the curve $y = f(x) = 1/x^2$ from $x = 1$ to $x = b$ (Figure 9.10). You can see that the approximation of the region R by the definite integral (9) improves as the upper limit of integration, b, becomes larger and larger. Figure 9.11 illustrates the situation for $b = 2$, 3, and 4, respectively.

This observation suggests that if we define a function $I(b)$ by

$$I(b) = \int_1^b \frac{1}{x^2}\,dx \tag{10}$$

then we can find the area of the required region R by evaluating the limit of $I(b)$ as b tends to infinity; that is, the area of R is given by

$$\lim_{b\to\infty} I(b) = \lim_{b\to\infty} \int_1^b \frac{1}{x^2}\,dx \tag{11}$$

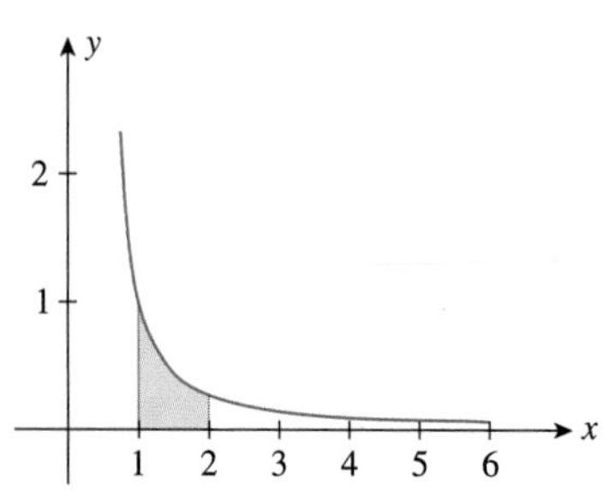

(a) Area of region under the graph of f on $[1, 2]$

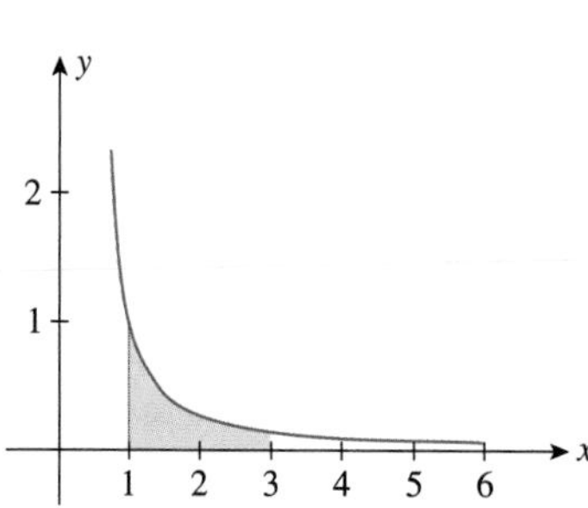

(b) Area of region under the graph of f on $[1, 3]$

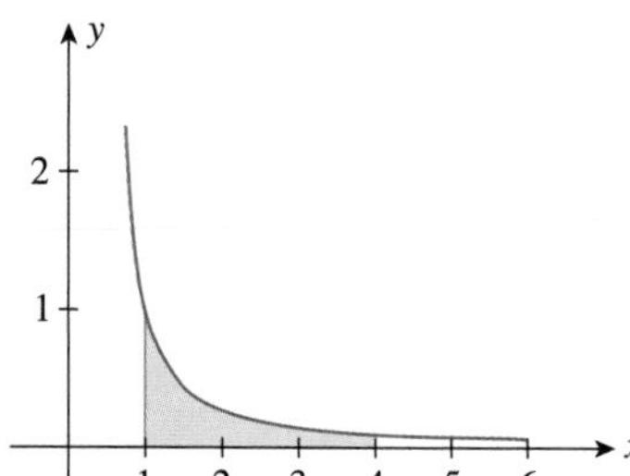

(c) Area of region under the graph of f on $[1, 4]$

Figure 9.11

As b increases, the approximation of R by the definite integral improves.

Example 1

a. Evaluate the definite integral $I(b)$ in Equation (10).
b. Compute $I(b)$ for $b = 10$, 100, 1000, 10,000.
c. Evaluate the limit in Equation (11).
d. Interpret the results of parts (b) and (c).

Solution

a. $I(b) = \int_1^b \frac{1}{x^2}\,dx = -\frac{1}{x}\Big|_1^b = -\frac{1}{b} + 1$

b. From the result of part (a),

$$I(b) = 1 - \frac{1}{b}$$

Therefore,

$$I(10) = 1 - \frac{1}{10} = 0.9$$
$$I(100) = 1 - \frac{1}{100} = 0.99$$
$$I(1000) = 1 - \frac{1}{1000} = 0.999$$
$$I(10{,}000) = 1 - \frac{1}{10{,}000} = 0.9999$$

c. Once again, using the result of part (a), we find

$$\begin{aligned}\lim_{b\to\infty} I(b) &= \lim_{b\to\infty}\int_1^b \frac{1}{x^2}\,dx\\ &= \lim_{b\to\infty}\left(1-\frac{1}{b}\right)\\ &= 1\end{aligned}$$

d. The result of part (c) tells us that the area of the region R is 1 square unit. The results of the computations performed in part (b) reinforce our expectation that $I(b)$ should approach 1, the area of the region R, as b approaches infinity.

The preceding discussion and the results of Example 1 suggest that we define the improper integral of a continuous function f over the unbounded interval $[a, \infty)$ as follows.

Improper Integral of f over $[a, \infty)$

Let f be a continuous function on the unbounded interval $[a, \infty)$. Then, the improper integral of f over $[a, \infty)$ is defined by

$$\int_a^\infty f(x)\,dx = \lim_{b\to\infty}\int_a^b f(x)\,dx \tag{12}$$

if the limit exists.

If the limit exists, the improper integral is said to be **convergent.** An improper integral for which the limit in Equation (12) fails to exist is said to be **divergent.**

Example 2

Evaluate $\displaystyle\int_2^\infty \frac{1}{x}\,dx$ if it converges.

Solution

$$\begin{aligned}\int_2^\infty \frac{1}{x}\,dx &= \lim_{b\to\infty}\int_2^b \frac{1}{x}\,dx\\ &= \lim_{b\to\infty} \ln x\,\Big|_2^b\\ &= \lim_{b\to\infty}(\ln b - \ln 2)\end{aligned}$$

Since $\ln b \to \infty$ as $b \to \infty$, the limit does not exist, and we conclude that the given improper integral is divergent.

GROUP DISCUSSION

1. Suppose f is continuous and nonnegative on $[0, \infty)$, Furthermore, suppose $\lim_{x\to\infty} f(x) = L$, where L is a positive number. What can you say about the convergence of the improper integral $\int_0^\infty f(x)\,dx$? Explain your answer and illustrate with an example.

2. Suppose f is continuous and nonnegative on $[0, \infty)$ and satisfies the condition $\lim_{x\to\infty} f(x) = 0$. What can you say about $\int_0^\infty f(x)\,dx$? Explain and illustrate your answer with examples.

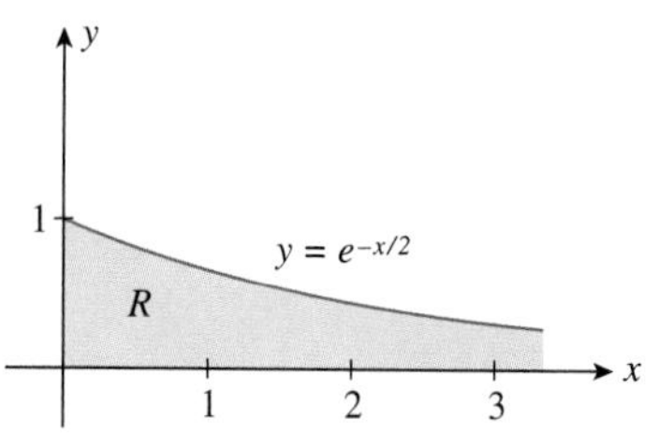

Figure 9.12

Area of $R = \int_0^{\infty} e^{-x/2}\, dx$

Example 3

Find the area of the region R under the curve $y = e^{-x/2}$ for $x \geq 0$.

Solution

The region R is shown in Figure 9.12. Taking $b > 0$, we compute the area of the region under the curve $y = e^{-x/2}$ from $x = 0$ to $x = b$—namely,

$$I(b) = \int_0^b e^{-x/2}\, dx = -2e^{-x/2}\Big|_0^b = -2e^{-b/2} + 2$$

Then, the area of the region R is given by

$$\lim_{b\to\infty} I(b) = \lim_{b\to\infty} (2 - 2e^{-b/2}) = 2 - 2\lim_{b\to\infty} \frac{1}{e^{b/2}}$$

$$= 2$$

or 2 square units.

EXPLORING WITH TECHNOLOGY

You can see how fast the improper integral in Example 3 converges, as follows:

1. Use a graphing utility to plot the graph of $I(b) = 2-2e^{-b/2}$, using the viewing window [0, 50] · [0, 3].
2. Use **TRACE** to follow the values of y for increasing values of x, starting at the origin.

The improper integral defined in Equation (12) has an interval of integration that is unbounded on the right. Improper integrals with intervals of integration that are unbounded on the left also arise in practice and are defined in a similar manner.

Improper Integral of f over $(-\infty, b]$

Let f be a continuous function on the unbounded interval $(-\infty, b]$. Then, the improper integral of f over $(-\infty, b]$ is defined by

$$\int_{-\infty}^{b} f(x)\, dx = \lim_{a\to-\infty} \int_a^b f(x)\, dx \qquad (13)$$

if the limit exists.

In this case, the improper integral is said to be convergent. Otherwise, the improper integral is said to be divergent.

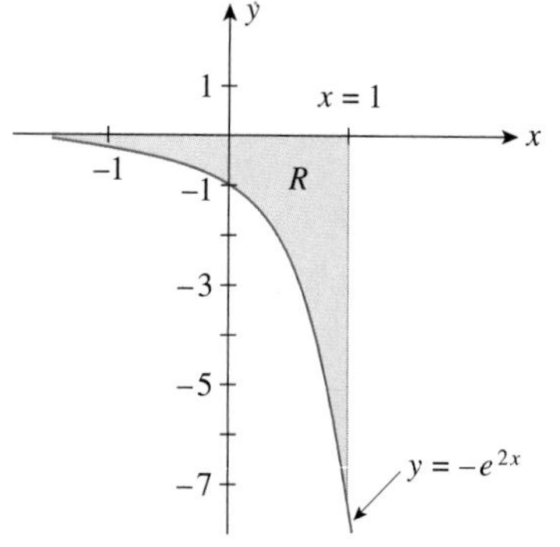

Figure 9.13

Area of $R = -\int_{-\infty}^{1} -e^{2x}\, dx$

Example 4

Find the area of the region R bounded above by the x-axis, below by the curve $y = -e^{2x}$, and on the right by the vertical line $x = 1$.

Solution

The region R is shown in Figure 9.13. Taking $a < 1$, compute the area of the region bounded above by the x-axis ($y = 0$), and below by the curve $y = -e^{2x}$ from $x = a$ to $x = 1$—namely,

$$I(a) = \int_a^1 [0 - (-e^{2x})]\, dx = \int_a^1 e^{2x}\, dx$$

$$= \frac{1}{2}e^{2x}\Big|_a^1 = \frac{1}{2}e^2 - \frac{1}{2}e^{2a}$$

Then, the area of the required region is given by

$$\begin{aligned}\lim_{a\to-\infty} I(a) &= \lim_{a\to-\infty}\left(\frac{1}{2}e^2 - \frac{1}{2}e^{2a}\right)\\ &= \frac{1}{2}e^2 - \frac{1}{2}\lim_{a\to-\infty} e^{2a}\\ &= \frac{1}{2}e^2\end{aligned}$$

Another improper integral found in practical applications involves the integration of a function f over the unbounded interval $(-\infty, \infty)$.

Improper Integral of f over $(-\infty, \infty)$

Let f be a continuous function over the unbounded interval $(-\infty, \infty)$. Let c be any real number and suppose both the improper integrals

$$\int_{-\infty}^{c} f(x)\,dx \quad \text{and} \quad \int_{c}^{\infty} f(x)\,dx$$

are convergent. Then, the improper integral of f over $(-\infty, \infty)$ is defined by

$$\int_{-\infty}^{\infty} f(x)\,dx = \int_{-\infty}^{c} f(x)\,dx + \int_{c}^{\infty} f(x)\,dx \tag{14}$$

In this case we say that the improper integral on the left in Equation (14) is convergent. If either one of the two improper integrals on the right in (14) is divergent, then the improper integral on the left is not defined.

Remark Usually, we choose $c = 0$. ◀

Example 5

Evaluate the improper integral

$$\int_{-\infty}^{\infty} xe^{-x^2}\,dx$$

and give a geometric interpretation of the results.

Solution

Take the point c in Equation (14) to be $c = 0$. Let's first evaluate

$$\begin{aligned}\int_{-\infty}^{0} xe^{-x^2}\,dx &= \lim_{a\to-\infty}\int_{a}^{0} xe^{-x^2}\,dx\\ &= \lim_{a\to-\infty} -\frac{1}{2}e^{-x^2}\Big|_a^0\\ &= \lim_{a\to-\infty} -\left[-\frac{1}{2} + \frac{1}{2}e^{-a^2}\right] = -\frac{1}{2}\end{aligned}$$

Next, we evaluate

$$\begin{aligned}\int_{0}^{\infty} xe^{-x^2}\,dx &= \lim_{b\to\infty}\int_{0}^{b} xe^{-x^2}\,dx\\ &= \lim_{b\to\infty} -\frac{1}{2}e^{-x^2}\Big|_0^b\\ &= \lim_{b\to\infty}\left[-\frac{1}{2}e^{-b^2} + \frac{1}{2}\right] = \frac{1}{2}\end{aligned}$$

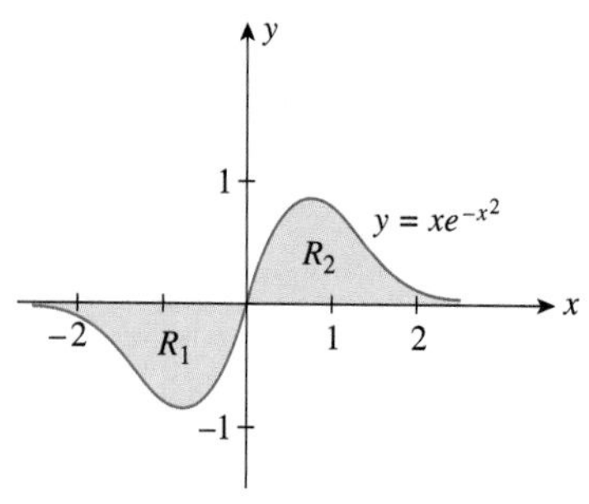

Figure 9.14

$$\int_{-\infty}^{\infty} xe^{-x^2}\,dx$$

$$= \int_{-\infty}^{0} xe^{-x^2}\,dx + \int_{0}^{\infty} xe^{-x^2}\,dx$$

Therefore,

$$\begin{aligned}\int_{-\infty}^{\infty} xe^{-x^2}\,dx &= \int_{-\infty}^{0} xe^{-x^2}\,dx + \int_{0}^{\infty} xe^{-x^2}\,dx\\ &= -\frac{1}{2} + \frac{1}{2}\\ &= 0\end{aligned}$$

The graph of $y = xe^{-x^2}$ is sketched in Figure 9.14. A glance at the figure tells us that the improper integral

$$\int_{\infty}^{0} xe^{-x^2}\,dx$$

gives the negative of the area of the region R_1, bounded above by the x-axis, below by the curve $y = xe^{-x^2}$, and on the right by the y-axis ($x = 0$).

However, the improper integral

$$\int_{0}^{\infty} xe^{-x^2}\,dx$$

gives the area of the region R_2 under the curve $y = xe^{-x^2}$ for $x \geq 0$. Since the graph of f is symmetric with respect to the origin, the area of R_1 is equal to the area of R_2. In other words,

$$\int_{-\infty}^{0} xe^{-x^2}\,dx = -\int_{0}^{\infty} xe^{-x^2}\,dx$$

Therefore,

$$\begin{aligned}\int_{-\infty}^{\infty} xe^{-x^2}\,dx &= \int_{-\infty}^{0} xe^{-x^2}\,dx + \int_{0}^{\infty} xe^{-x^2}\,dx\\ &= -\int_{0}^{\infty} xe^{-x^2}\,dx + \int_{0}^{\infty} xe^{-x^2}\,dx\\ &= 0\end{aligned}$$

as was shown earlier.

Perpetuities

Recall from Section 8.7 that the present value of an annuity is given by

$$PV \approx mP \int_{0}^{T} e^{-rt}\,dt = \frac{mP}{r}\left(1 - e^{-rT}\right) \tag{15}$$

Now, if the payments of an annuity are allowed to continue indefinitely, we have what is called a perpetuity. The present value of a perpetuity may be approximated by the improper integral

$$PV \approx mP \int_{0}^{\infty} e^{-rt}\,dt$$

obtained from Formula (15) by allowing the term of the annuity, T, to approach infinity. Thus,

$$mP \int_{0}^{\infty} e^{-rt}\,dt = \lim_{b\to\infty} mP \int_{0}^{b} e^{-rt}\,dt$$

$$= mP \lim_{b \to \infty} \int_0^b e^{-rt}\, dt$$

$$= mP \lim_{b \to \infty} \left[-\frac{1}{r} e^{-rt} \Big|_0^b \right]$$

$$= mP \lim_{b \to \infty} \left(-\frac{1}{r} e^{-rb} + \frac{1}{r} \right) = \frac{mP}{r}$$

The Present Value of a Perpetuity

The present value PV of a perpetuity is given by

$$PV = \frac{mP}{r} \qquad (16)$$

where m is the number of payments per year, P is the size of each payment, and r is the interest rate (compounded continuously).

Example 6

Endowments The Robinson family wishes to create a scholarship fund at a college. If a scholarship in the amount of \$5000 is awarded annually beginning 1 year from now, find the amount of the endowment they are required to make now. Assume that this fund will earn interest at a rate of 8% per year compounded continuously.

Solution

The amount of the endowment, A, is given by the present value of a perpetuity, with $m = 1$, $P = 5000$, and $r = 0.08$. Using Formula (16), we find

$$A = \frac{(1)(5000)}{0.08}$$

$$= 62{,}500$$

or \$62,500.

The improper integral also plays an important role in the study of probability theory.

It is sometimes necessary to use L'Hôpital's rule to compute the limits that arise during improper integration. The next two examples give improper integrals of this type.

Example 7

Improper Integrals That Use L'Hôpital's Rule Compute

$$\int_0^\infty xe^{-x}\, dx$$

Solution

Notice that this integral requires integration by parts. Let

$$u = x \quad \text{and} \quad dv = e^{-x}\, dx$$

Then

$$du = dx \quad \text{and} \quad v = -e^{-x}$$

Computing the integral:

$$\int_0^\infty xe^{-x}\,dx = \lim_{b\to\infty}\int_0^b xe^{-x}\,dx$$

$$= \lim_{b\to\infty} x(-e^{-x}) - \int_0^b - e^{-x}\,dx$$

$$= \lim_{b\to\infty} -xe^{-x} + \int_0^b e^{-x}\,dx$$

$$= \lim_{b\to\infty} (-xe^{-x} - e^{-x})\Big|_0^b$$

$$= \lim_{b\to\infty}\left(\frac{-(x+1)}{e^x}\right)\Big|_0^b$$

$$= \lim_{b\to\infty}\left(\frac{-(b+1)}{e^b}\right) - \frac{-1}{1}$$

$$= 1 - \lim_{b\to\infty}\frac{b+1}{e^b}$$

The remaining limit requires L'Hôpital's rule.

$$\lim_{b\to\infty}\frac{b+1}{e^b} = \lim_{b\to\infty}\frac{1}{e^b}$$
$$= 0$$

And so we see that the integral exists and is equal to $1 - 0$ or 1.

Example 8

Improper Integrals That Use L'Hôpital's Rule Compute

$$\int_1^\infty \frac{\ln(x)}{x^3}\,dx$$

Solution

This integral requires integration by parts. Let

$$u = \ln(x) \qquad \text{and} \qquad dv = \frac{dx}{x^3}$$

Then

$$du = \frac{dx}{x} \qquad \text{and} \qquad v = \frac{-1}{2x^2}$$

Compute the integral, using L'Hôpital's rule when required to resolve limits.

$$\int_1^\infty \frac{\ln(x)}{x^3}\,dx = \lim_{b\to\infty}\int_1^b \frac{\ln(x)}{x^3}\,dx$$

$$= \lim_{b\to\infty} \ln(x)\cdot\frac{-1}{2x^2} - \int_1^b \frac{-1}{2x^2}\cdot\frac{dx}{x}$$

$$= \lim_{b\to\infty}\frac{-\ln(x)}{2x^2} + \int_1^b \frac{dx}{2x^3}$$

$$= \lim_{b\to\infty}\frac{-\ln(x)}{2x^2} - \frac{1}{2\cdot 2x^2}\Big|_1^b$$

$$
\begin{aligned}
&= \lim_{b\to\infty} \left. \frac{-\ln(x)}{2x^2} - \frac{1}{4x^2} \right|_1^b \\
&= \lim_{b\to\infty} \frac{-\ln(b)}{2b^2} - \frac{1}{4b^2} - \left(-0 - \frac{1}{4}\right) \\
&= \frac{1}{4} - \lim_{b\to\infty} \frac{\ln(b)}{2b^2} + \frac{1}{4b^2} \\
&= \frac{1}{4} - \lim_{b\to\infty} \frac{\frac{1}{b}}{4b} + 0 \\
&= \frac{1}{4} - \lim_{b\to\infty} \frac{1}{4b^2} \\
&= \frac{1}{4} - 0 \\
&= \frac{1}{4}
\end{aligned}
$$

So the integral exists and has a value of $\frac{1}{4}$.

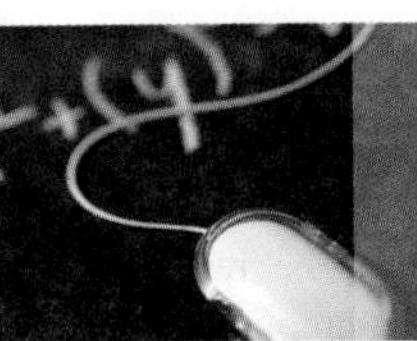

Self-Check Exercises 9.5

1. Evaluate $\displaystyle\int_{-\infty}^{\infty} \frac{x^3}{(1 + x^4)^{3/2}}\, dx.$

2. Suppose an income stream is expected to continue indefinitely. Then, the present value of such a stream can be calculated from the formula for the present value of an income stream by letting T approach infinity. Thus, the required present value is given by

$$PV = \int_0^{\infty} P(t)e^{-rt}\, dt$$

Suppose Marcia has an oil well in her backyard that generates a stream of income given by

$$P(t) = 20e^{-0.02t}$$

where $P(t)$ is expressed in thousands of dollars per year and t is the time in years from the present. Assuming that the prevailing interest rate in the foreseeable future is 10%/year compounded continuously, what is the present value of the income stream?

Solutions to Self-Check Exercises 9.5 can be found on page 581.

9.5 Exercises

In Exercises 1–10, find the area of the region under the given curve $y = f(x)$ over the indicated interval.

1. $f(x) = \dfrac{2}{x^2}; x \geq 3$

2. $f(x) = \dfrac{2}{x^3}; x \geq 2$

3. $f(x) = \dfrac{1}{(x-2)^2}; x \geq 3$

4. $f(x) = \dfrac{2}{(x+1)^3}; x \geq 0$

5. $f(x) = \dfrac{1}{x^{3/2}}; x \geq 1$

6. $f(x) = \dfrac{3}{x^{5/2}}; x \geq 4$

7. $f(x) = \dfrac{1}{(x+1)^{5/2}}; x \geq 0$

8. $f(x) = \dfrac{1}{(1-x)^{3/2}}; x \leq 0$

9. $f(x) = e^{2x}; x \leq 2$

10. $f(x) = xe^{-x^2}; x \geq 0$

11. Find the area of the region bounded by the x-axis and the graph of the function

$$f(x) = \frac{x}{(1 + x^2)^2}$$

12. Find the area of the region bounded by the x-axis and the graph of the function

$$f(x) = \frac{e^x}{(1 + e^x)^2}$$

13. Consider the improper integral

$$\int_0^{\infty} \sqrt{x}\, dx$$

a. Evaluate $I(b) = \int_0^b \sqrt{x}\, dx$.

b. Show that

$$\lim_{b \to \infty} I(b) = \infty$$

thus proving that the given improper integral is divergent.

14. Consider the improper integral

$$\int_1^{\infty} x^{-2/3}\, dx$$

a. Evaluate $I(b) = \int_1^b x^{-2/3}\, dx$.

b. Show that

$$\lim_{b \to \infty} I(b) = \infty$$

thus proving that the given improper integral is divergent.

In Exercises 15–42, evaluate each improper integral whenever it is convergent.

15. $\displaystyle\int_1^{\infty} \frac{3}{x^4}\, dx$

16. $\displaystyle\int_1^{\infty} \frac{1}{x^3}\, dx$

17. $\displaystyle\int_4^{\infty} \frac{2}{x^{3/2}}\, dx$

18. $\displaystyle\int_1^{\infty} \frac{1}{\sqrt{x}}\, dx$

19. $\displaystyle\int_1^{\infty} \frac{4}{x}\, dx$

20. $\displaystyle\int_2^{\infty} \frac{3}{x}\, dx$

21. $\displaystyle\int_{-\infty}^{0} \frac{1}{(x-2)^3}\, dx$

22. $\displaystyle\int_2^{\infty} \frac{1}{(x+1)^2}\, dx$

23. $\displaystyle\int_1^{\infty} \frac{1}{(2x-1)^{3/2}}\, dx$

24. $\displaystyle\int_{-\infty}^{0} \frac{1}{(4-x)^{3/2}}\, dx$

25. $\displaystyle\int_0^{\infty} e^{-x}\, dx$

26. $\displaystyle\int_0^{\infty} e^{-x/2}\, dx$

27. $\displaystyle\int_{-\infty}^{0} e^{2x}\, dx$

28. $\displaystyle\int_{-\infty}^{0} e^{3x}\, dx$

29. $\displaystyle\int_1^{\infty} \frac{e^{\sqrt{x}}}{\sqrt{x}}\, dx$

30. $\displaystyle\int_1^{\infty} \frac{e^{-\sqrt{x}}}{\sqrt{x}}\, dx$

31. $\displaystyle\int_{-\infty}^{0} xe^{x}\, dx$

32. $\displaystyle\int_0^{\infty} xe^{-2x}\, dx$

33. $\displaystyle\int_{-\infty}^{\infty} x\, dx$

34. $\displaystyle\int_{-\infty}^{\infty} x^3\, dx$

35. $\displaystyle\int_{-\infty}^{\infty} x^3(1+x^4)^{-2}\, dx$

36. $\displaystyle\int_{-\infty}^{\infty} x(x^2+4)^{-3/2}\, dx$

37. $\displaystyle\int_{-\infty}^{\infty} xe^{1-x^2}\, dx$

38. $\displaystyle\int_{-\infty}^{\infty} \left(x - \frac{1}{2}\right)e^{-x^2+x-1}\, dx$

39. $\displaystyle\int_{-\infty}^{\infty} \frac{e^{-x}}{1+e^{-x}}\, dx$

40. $\displaystyle\int_{-\infty}^{\infty} \frac{xe^{-x^2}}{1+e^{-x^2}}\, dx$

41. $\displaystyle\int_e^{\infty} \frac{1}{x \ln^3 x}\, dx$

42. $\displaystyle\int_{e^2}^{\infty} \frac{1}{x \ln x}\, dx$

In Exercises 43–48, use L'Hôpital's rule to evaluate the integral on the specified interval.

43. $\displaystyle\int (x+2)e^{-x}\, dx;\ x \geq 2$

44. $\displaystyle\int xe^{x}\, dx;\ x \leq 1$

45. $\displaystyle\int \frac{\ln(x)}{x^4}\, dx;\ x \geq 1$

46. $\displaystyle\int x^2e^{-x}\, dx;\ x \geq 3$

47. $\displaystyle\int \frac{\ln(x)+1}{x^3}\, dx;\ x \geq 2$

48. $\displaystyle\int (1-x)e^{x}\, dx;\ x \leq 4$

49. Show that an integral of the form $\int_a^{\infty} e^{-px}\, dx$ is convergent if $p > 0$ and divergent if $p < 0$.

50. Show that an integral of the form $\int_{-\infty}^{b} e^{px}\, dx$ is convergent if $p > 0$ and divergent if $p < 0$.

51. **Capital Value** The capital value (present sale value) CV of property that can be rented on a perpetual basis for R dollars annually is approximated by the formula

$$CV = \int_0^{\infty} Re^{-it}\, dt$$

where i is the prevailing continuous interest rate.

a. Show that $CV = R/i$.

b. Find the capital value of property that can be rented at \$10,000 annually when the prevailing continuous interest rate is 12%/year.

Business and Economics Applications

52. **The Amount of an Endowment** A university alumni group wishes to provide an annual scholarship in the amount of \$1500 beginning next year. If the scholarship fund will earn an interest rate of 8%/year compounded continuously, find the amount of the endowment the alumni are required to make now.

53. **The Amount of an Endowment** Mel Thompson wishes to establish a fund to provide a university medical centre with an annual research grant of \$50,000 beginning next year. If the fund will earn an interest rate of 9%/year compounded continuously, find the amount of the endowment he is required to make now.

54. **Perpetual Net Income Streams** The present value of a perpetual stream of income that flows continually at the rate of $P(t)$ dollars/year is given by the formula

$$PV = \int_0^{\infty} P(t)e^{-rt}\, dt$$

where r is the interest rate compounded continuously. Using this formula, find the present value of a perpetual net income stream that is generated at the rate of

$$P(t) = 10{,}000 + 4000t$$

dollars/year.

Hint: $\displaystyle\lim_{b \to \infty} \frac{b}{e^{rb}} = 0$.

55. **Establishing a Trust Fund** Becky Wilkinson wants to establish a trust fund that will provide her children and heirs with a perpetual annuity in the amount of

$$P(t) = 20 + t$$

thousand dollars/year beginning next year. If the trust fund will earn an interest rate of 10%/year compounded continuously, find the amount that she must place in the trust fund now.

Hint: Use the formula given in Exercise 54.

Biological and Life Sciences Applications

56. **Total Uptake** The uptake of toxic selenium by a grass used in bioremediation of toxic soils is found to be

$$\frac{5.6}{t^{1.7}}$$

g/d for each square metre planted with the grass under normal growing conditions. Use an improper integral to estimate the total uptake of the grass by integrating the daily uptake for all times $t \geq 0$.

57. **Reactor Production** Production of a valuable biochemical in a type of commercial bacterial reactor is modelled well as

$$\frac{17.2}{t^2 + 4t + 5}$$

g/h after the reactor is set running. Compute the production in the first day and also estimate the production if the reactor is left running for a long time by using an improper integral for all times $t \geq 0$. What fraction of the production in a given reactor run happens after the first day?

58. **Total Number of Bacteria** A population of subterranean bacteria that convert sulfites into sulfates are found to add members according to the rule

$$P'(t) = 1.4te^{-2t}$$

billion/day. Using an improper integral, estimate the total number of bacteria that grow in the sulfite-bearing strata for $t \geq 0$ by computing the integral of $P'(t)$ for all $t \geq 0$.

59. **Total Amount of Hormone** Production of a hormone that regulates development during the second week of fetal growth is modelled by the formula

$$(3.6t + 1.7)e^{-1.1t}$$

mg/d. Using an improper integral, estimate the total amount of hormone produced by integrating the production for all times $t \geq 0$.

60. **Total Toxin Released** A shipping accident releases industrial waste into the Gulf of St. Lawrence. While the waste is not dangerous, the action of salt water on it generates a toxic compound at the rate of

$$\frac{0.76\ln(t)}{t^2}$$

tons/d. Using an improper integral, estimate the total amount of the toxin released into the environment by integrating over all times $t \geq 0$.

In Exercises 61–63, determine whether the statement is true or false. If it is true, explain why it is true. If it is false, give an example to show why it is false.

61. If $\int_a^\infty f(x)\,dx$ exists, then $\int_a^\infty f(x)\,dx$ exists for every real number $b > a$.

62. If $\int_a^\infty f(x)\,dx$ exists, then $\int_{-\infty}^{-a} f(x)\,dx$ exists, and $\int_a^\infty f(x)\,dx = -\int_{-\infty}^{-a} f(x)\,dx$ exists.

63. If $\int_{-\infty}^\infty f(x)\,dx$ exists, then $\int_0^\infty f(x)\,dx$ exists, and $\int_{-\infty}^\infty f(x)\,dx = 2\int_0^\infty f(x)\,dx$.

Solutions to Self-Check Exercises 9.4

1. Write

$$\int_{-\infty}^{\infty} \frac{x^3}{(1+x^4)^{3/2}}\,dx = \int_{-\infty}^{0} \frac{x^3}{(1+x^4)^{3/2}}\,dx + \int_{0}^{\infty} \frac{x^3}{(1+x^4)^{3/2}}\,dx$$

Now,

$$\begin{aligned}\int_{-\infty}^{0} \frac{x^3}{(1+x^4)^{3/2}}\,dx &= \lim_{a\to-\infty}\int_a^0 x^3(1+x^4)^{-3/2}\,dx \\ &= \lim_{a\to-\infty}\frac{1}{4}(-2)(1+x^4)^{-1/2}\Big|_a^0 \quad \text{Integrate by substitution.} \\ &= -\frac{1}{2}\lim_{a\to-\infty}\left[1-\frac{1}{(1+a^4)^{1/2}}\right] \\ &= -\frac{1}{2}\end{aligned}$$

Similarly, you can show that

$$\int_0^\infty \frac{x^3}{(1+x^4)^{3/2}}\,dx = \frac{1}{2}$$

Therefore,

$$\int_{-\infty}^\infty \frac{x^3}{(1+x^4)^{3/2}}\,dx = -\frac{1}{2}+\frac{1}{2} = 0$$

2. The required present value is given by

$$\begin{aligned}PV &= \int_0^\infty 20e^{-0.02t}e^{-0.10t}\,dt \\ &= 20\int_0^\infty e^{-0.12t}\,dt \\ &= 20\lim_{b\to\infty}\int_0^b e^{-0.12t}\,dt \\ &= -\frac{20}{0.12}\lim_{b\to\infty}e^{-0.12t}\Big|_0^b \\ &= -\frac{500}{3}\lim_{b\to\infty}(e^{-0.12b}-1) \\ &= \frac{500}{3}\end{aligned}$$

or approximately $166,667.

CHAPTER 9 Summary of Principal Formulas and Terms

Formulas

1. Integration by parts $\int u\,dv = uv - \int v\,du$

2. Trapezoidal rule $\int_a^b f(x)\,dx \approx \frac{\Delta x}{2}[f(x_0) + 2f(x_1) + 2f(x_2) + \cdots + 2f(x_{n-1}) + f(x_n)]$

   where $\Delta x = \frac{b-a}{n}$

3. Simpson's rule $\int_a^b f(x)\,dx \approx \frac{\Delta x}{3}[f(x_0) + 4f(x_1) + 2f(x_2) + 4f(x_3) + 2f(x_4) + \cdots + 4f(x_{n-1}) + f(x_n)]$

   where $\Delta x = \frac{b-a}{n}$

4. Maximum error for trapezoidal rule $\frac{M(b-a)^3}{12n^2}$, where $|f''(x)| \leq M$ $(a \leq x \leq b)$

4. Maximum error for Simpson's rule $\frac{M(b-a)^5}{180n^4}$, where $|f^{(4)}(x)| \leq M$ $(a \leq x \leq b)$

6. Improper integral of f over $[a, \infty)$ $\int_a^\infty f(x)\,dx = \lim_{b\to\infty}\int_a^b f(x)\,dx$

7. Improper integral of f over $(-\infty, b]$ $\int_{-\infty}^b f(x)\,dx = \lim_{a\to-\infty}\int_a^b f(x)\,dx$

8. Improper integral of f over $(-\infty, \infty)$ $\int_{-\infty}^\infty f(x)\,dx = \int_{-\infty}^c f(x)\,dx + \int_c^\infty f(x)\,dx$

9. Present value of a perpetuity $PV = \frac{mP}{r}$

Terms

integration by parts (531)
partial fractions (539)
trapezoidal rule (559)
Simpson's rule (563)
improper integral (571)
convergent integral (573)
divergent integral (573
perpetuity (576)
present value PV of a perpetuity (577)

Review Exercises

In Exercises 1–6, evaluate the integral.

1. $\int 2xe^{-x}\,dx$

2. $\int xe^{4x}\,dx$

3. $\int \ln 5x\,dx$

4. $\int_1^4 \ln 2x\,dx$

5. $\int_0^1 xe^{-2x}\,dx$

6. $\int_0^2 xe^{2x}\,dx$

7. Find the function f given that the slope of the tangent line to the graph of f at any point $(x, f(x))$ is

$$f'(x) = \frac{\ln x}{\sqrt{x}}$$

and that the graph of f passes through the point $(1, -2)$.

8. Find the function f given that the slope of the tangent line to the graph of f at any point $(x, f(x))$ is

$$f'(x) = xe^{-3x}$$

and that the graph of f passes through the point $(0, 0)$.

In Exercises 9–14, use the method of partial fractions to perform the given integral.

9. $\int \frac{dx}{x^2 - x - 2}$

10. $\int \frac{x-1}{x^2-9}\,dx$

11. $\int \frac{2x+1}{x^2 - 3x + 2}\,dx$

12. $\int \frac{x^2}{x^2 + 4x + 3}\,dx$

Hint: Remember to use algebra to make the denominator lower degree.

13. $\int_0^5 \frac{dx}{x^2 + 7x + 12}$

14. $\int_0^1 \frac{dx}{x^2 - 5x + 6}$

In Exercises 15–20, use the table of integrals in Section 9.3 to evaluate the integral.

15. $\int \frac{x^2\,dx}{(3 + 2x)^2}$

16. $\int \frac{2x}{\sqrt{2x + 3}}\,dx$

17. $\int x^2 e^{4x}\,dx$

18. $\int \frac{dx}{(x^2 - 25)^{3/2}}$

19. $\int \frac{dx}{x^2\sqrt{x^2 - 4}}$

20. $\int 8x^3 \ln 2x\,dx$

In Exercises 21–28, evaluate each improper integral whenever it is convergent.

21. $\int_0^\infty e^{-2x}\,dx$

22. $\int_{-\infty}^0 e^{3x}\,dx$

23. $\int_3^\infty \frac{2}{x}\,dx$

24. $\int_2^\infty \frac{1}{(x + 2)^{3/2}}\,dx$

25. $\int_2^\infty \frac{dx}{(1 + 2x)^2}$

26. $\int_1^\infty 3e^{1-x}\,dx$

27. $\int_1^\infty \frac{\ln(x)}{x^5}\,dx$

28. $\int_0^\infty (x^2 + x + 1)e^{-x}\,dx$

In Exercises 29–32, use the trapezoidal rule and Simpson's rule to approximate the value of the definite integral.

29. $\int_1^3 \frac{dx}{1 + \sqrt{x}};\ n = 4$

30. $\int_0^1 e^{x^2}\,dx;\ n = 4$

31. $\int_{-1}^1 \sqrt{1 + x^4}\,dx;\ n = 4$

32. $\int_1^3 \frac{e^x}{x}\,dx;\ n = 4$

33. **Producer's Surplus** The supply equation for the GTC SlimPhone is given by

$$p = 2\sqrt{25 + x^2}$$

where p is the unit price in dollars and x is the quantity demanded per month in units of 10,000. Find the producers' surplus if the market price is \$26. Use the table of integrals in Section 9.3 to evaluate the definite integral.

Business and Economics Applications

34. **Oil Spills** Using aerial photographs, the Coast Guard was able to determine the dimensions of an oil spill along an embankment on a coastline, as shown in the accompanying figure. Using (a) the trapezoidal rule and (b) Simpson's rule with $n = 10$, estimate the area of the oil spill.

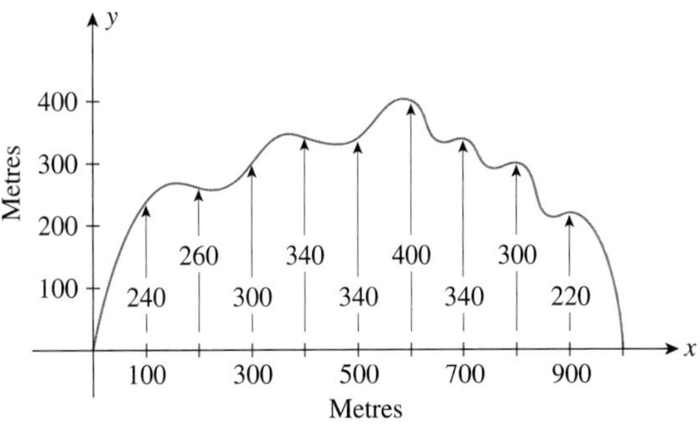

35. **Computer Game Sales** The sales of Starr Communication's newest computer game, Laser Beams, are currently

$$te^{-0.05t}$$

units/month (each unit representing 1000 games), where t denotes the number of months since the release of the game. Find an expression that gives the total number of games sold as a function of t. How many games will be sold by the end of the first year?

36. **Demand for Computer Software** The demand equation for a computer software program is given by

$$p = 2\sqrt{325 - x^2}$$

where p is the unit price in dollars and x is the quantity demanded each month in units of a thousand. Find the consumers' surplus if the market price is \$30. Evaluate the definite integral using Simpson's rule with $n = 10$.

37. **Perpetuities** Lindsey wishes to establish a memorial fund at Newtown Hospital in the amount of \$10,000/year beginning next year. If the fund earns interest at a rate of 9%/year compounded continuously, find the amount of endowment that he is required to make now.

Biological and Life Sciences Applications

38. **Total Amount of Enzyme** Production of a potentially dangerous enzyme by gut bacteria after they are stimulated by accidental ingestion of a incorrect medication is modelled by the formula

$$(22t + 6)e^{-2.3t}$$

mg/h. Using an improper integral, estimate the total amount of enzyme produced by integrating the production for all times $t \geq 0$.

39. **Total Uptake** The uptake of plutonium by Jimsonweed used to clean up old reactor sites is found to be

$$\frac{44.1}{t^{1.06}}$$

μg/d for each acre planted with Jimsonweed under normal growing conditions. Use an improper integral to estimate to total uptake of plutonium by the grass by integrating the daily uptake for all times $t \geq 0$.

Source: Los Alamos National Labs

10

CALCULUS OF SEVERAL VARIABLES

Photo: Thinkstock/Jupiter Images

Up to now we have dealt with functions involving one variable. In many real-life situations, however, we encounter quantities that depend on two or more quantities. For example, the Consumer Price Index (CPI) compiled every month by Statistics Canada depends on the price of more than 95,000 consumer items from gas to groceries. To study such relationships, we need the notion of a function of several variables, the first topic in this chapter. Next, generalizing the concept of the derivative of a function of one variable, we study the *partial derivatives* of a function of two or more variables. Partial derivatives enable us to study the rate of change of a function with respect to one variable while holding all other variables constant. We then learn how to find the extremum values of a function of several variables. As an application of optimization theory, we learn how to find an equation of the straight line that "best" fits a set of data points scattered about a straight line. Finally, we generalize the notion of the integral to the case involving a function of two variables.

Photo: Thinkstock/Jupiter Images

What should the dimensions of the swimming pool be? The operators of the *Viking Princess,* a luxury cruise ship, are thinking about adding another swimming pool to the *Princess.* The chief engineer has suggested that an area in the form of an ellipse, located in the rear of the promenade deck, would be suitable for this purpose. Subject to this constraint, what are the dimensions of the largest pool that can be built? See Example 5, page 635, to see how to solve this problem.

Photo: © Annie Griffiths Belt/CORBIS

10.1 Functions of Several Variables

Up to now, our study of calculus has been restricted to functions of one variable. In many practical situations, however, the formulation of a problem results in a mathematical model that involves a function of two or more variables. For example, suppose Ace Novelty determines that the profits are \$6, \$5, and \$4 for three types of souvenirs it produces. Let x, y, and z denote the number of type-A, type-B, and type-C souvenirs to be made; then the company's profit is given by

$$P = 6x + 5y + 4z$$

and P is a function of the three variables, x, y, and z.

Functions of Two Variables

Although this chapter deals with real-valued functions of several variables, most of our definitions and results are stated in terms of a function of two variables. One reason for adopting this approach, as you will soon see, is that there is a geometric interpretation for this special case, which serves as an important visual aid. We can then draw upon the experience gained from studying the two-variable case to help us understand the concepts and results connected with the more general case, which, by and large, is just a simple extension of the lower-dimensional case.

A Function of Two Variables

A real-valued **function of two variables,** f, consists of

1. A set A of ordered pairs of real numbers (x, y) called the **domain** of the function.
2. A rule that associates with each ordered pair in the domain of f one and only one real number, denoted by $z = f(x, y)$.

The variables x and y are called independent variables, and the variable z, which is dependent on the values of x and y, is referred to as a **dependent variable.**

As in the case of a real-valued function of one real variable, the number $z = f(x, y)$ is called the **value of f** at the point (x, y). And, unless specified, the domain of the function f will be taken to be the largest possible set for which the rule defining f is meaningful.

Example 1

Let f be the function defined by

$$f(x, y) = x + xy + y^2 + 2$$

Compute $f(0, 0)$, $f(1, 2)$, and $f(2, 1)$.

Solution

We have

$$f(0, 0) = 0 + (0)(0) + 0^2 + 2 = 2$$
$$f(1, 2) = 1 + (1)(2) + 2^2 + 2 = 9$$
$$f(2, 1) = 2 + (2)(1) + 1^2 + 2 = 7$$

The domain of a function of two variables $f(x, y)$ is a set of ordered pairs of real numbers and may therefore be viewed as a subset of the xy-plane.

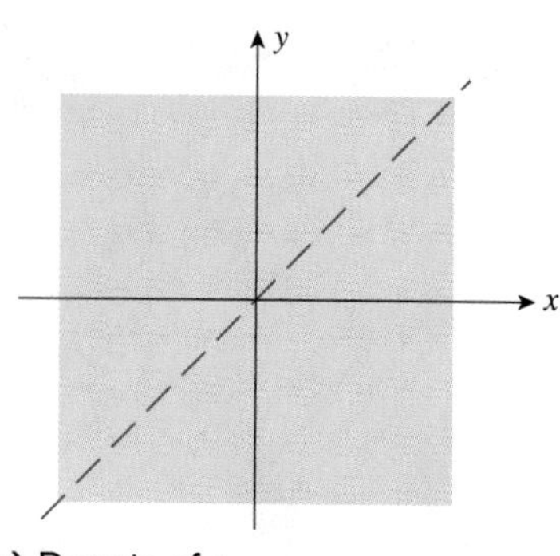

(a) Domain of g

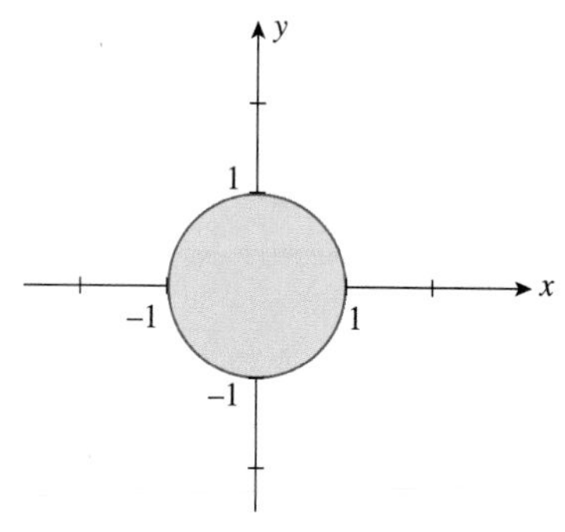

(b) Domain of h

Figure 10.1

Example 2

Find the domain of each of the following functions.

a. $f(x, y) = x^2 + y^2$

b. $g(x, y) = \dfrac{2}{x - y}$

c. $h(x, y) = \sqrt{1 - x^2 - y^2}$

Solution

a. $f(x, y)$ is defined for all real values of x and y, so the domain of the function f is the set of all points (x, y) in the xy-plane.

b. $g(x, y)$ is defined for all $x \neq y$, so the domain of the function g is the set of all points in the xy-plane except those lying on the line $y = x$ (Figure 10.1a).

c. We require that $1 - x^2 - y^2 \geq 0$ or $x^2 + y^2 \leq 1$, which is just the set of all points (x, y) lying on and inside the circle of radius 1 with centre at the origin (Figure 10.1b).

Applications

Example 3

Revenue Functions Acrosonic manufactures a bookshelf loudspeaker system that may be bought fully assembled or in a kit. The demand equations that relate the unit prices, p and q, to the quantities demanded weekly, x and y, of the assembled and kit versions of the loudspeaker systems are given by

$$p = 300 - \frac{1}{4}x - \frac{1}{8}y \qquad \text{and} \qquad q = 240 - \frac{1}{8}x - \frac{3}{8}y$$

a. What is the weekly total revenue function $R(x, y)$?

b. What is the domain of the function R?

Solution

a. The weekly revenue realizable from the sale of x units of the assembled speaker systems at p dollars per unit is given by xp dollars. Similarly, the weekly revenue realizable from the sale of y units of the kits at q dollars per unit is given by yq dollars. Therefore, the weekly total revenue function R is given by

$$\begin{aligned} R(x, y) &= xp + yq \\ &= x\left(300 - \frac{1}{4}x - \frac{1}{8}y\right) + y\left(240 - \frac{1}{8}x - \frac{3}{8}y\right) \\ &= -\frac{1}{4}x^2 - \frac{3}{8}y^2 - \frac{1}{4}xy + 300x + 240y \end{aligned}$$

b. To find the domain of the function R, let's observe that the quantities x, y, p, and q must be nonnegative. This observation leads to the following system of linear inequalities:

$$300 - \frac{1}{4}x - \frac{1}{8}y \geq 0$$
$$240 - \frac{1}{8}x - \frac{3}{8}y \geq 0$$
$$x \geq 0$$
$$y \geq 0$$

The domain of the function R is sketched in Figure 10.2.

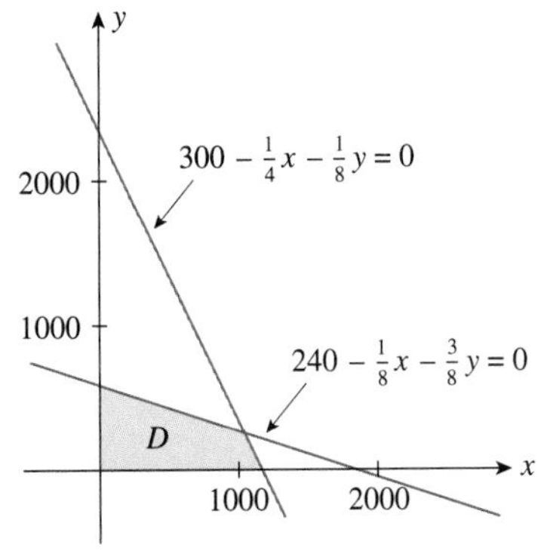

Figure 10.2
The domain of $R(x, y)$

GROUP DISCUSSION

Suppose the total profit of a two-product company is given by $P(x, y)$, where x denotes the number of units of the first product produced and sold and y denotes the number of units of the second product produced and sold. Fix $x = a$, where a is a positive number so that (a, y) is in the domain of P. Describe and give an economic interpretation of the function $f(y) = P(a, y)$. Next, fix $y = b$, where b is a positive number so that (x, b) is in the domain of P. Describe and give an economic interpretation of the function $g(x) = P(x, b)$.

Function of Several Variables

Functions of several variables are similar to functions of two variables. The only difference is the number of independent variables.

A Function of Multiple Variables

A real-valued function of multiple variables f consists of

1. A set A of ordered lists of n real numbers $(x_1, x_2, \ldots, x_n)$.
2. A rule that associates with each ordered list a single real value denoted by $f(x_1, x_2, \ldots, x_n)$.

Example 4

Home Mortgage Payments The monthly payment that amortizes a loan of A dollars in t years when the interest rate is r per year is given by

$$P = f(A, r, t) = \frac{Ar}{12\left[1 - \left(1 + \frac{r}{12}\right)^{-12t}\right]}$$

Find the monthly payment for a home mortgage of \$90,000 to be amortized over 30 years when the interest rate is 10% per year.

Solution

Letting $A = 90{,}000$, $r = 0.1$, and $t = 30$, we find the required monthly payment to be

$$P = f(90{,}000, 0.1, 30) = \frac{90{,}000(0.1)}{12\left[1 - \left(1 + \frac{0.1}{12}\right)^{-360}\right]}$$
$$\approx 789.81$$

or approximately \$789.81.

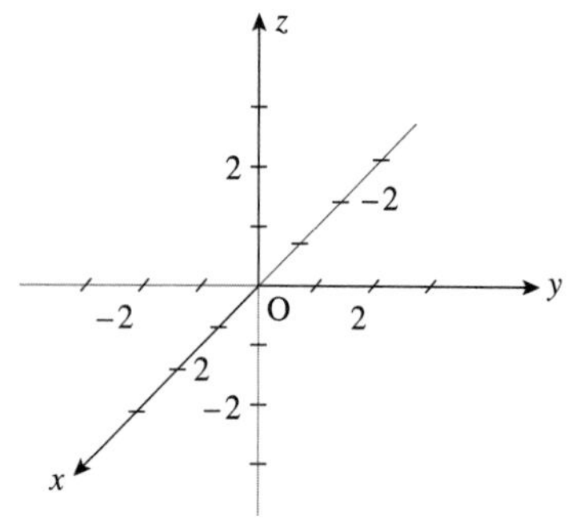

Figure 10.3
The three-dimensional Cartesian coordinate system

Graphs of Functions of Two Variables

To graph a function of two variables, we need a three-dimensional coordinate system. This is readily constructed by adding a third axis to the plane Cartesian coordinate system in such a way that the three resulting axes are mutually perpendicular and intersect at O. Observe that, by construction, the zeros of the three number scales coincide at the origin of the **three-dimensional Cartesian coordinate system** (Figure 10.3).

A point in three-dimensional space can now be represented uniquely in this coordinate system by an **ordered triple** of numbers (x, y, z), and, conversely, every ordered triple of real numbers (x, y, z) represents a point in three-dimensional space (Figure 10.4a). For example, the points $A(2, 3, 4)$, $B(1, -2, -2)$, $C(2, 4, 0)$, and $D(0, 0, 4)$ are shown in Figure 10.4b.

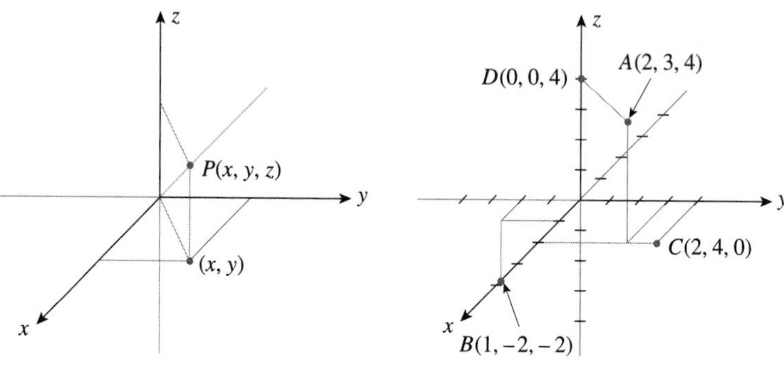

(a) A point in three-dimensional space **(b)** Some sample points in three-dimensional space

Figure 10.4

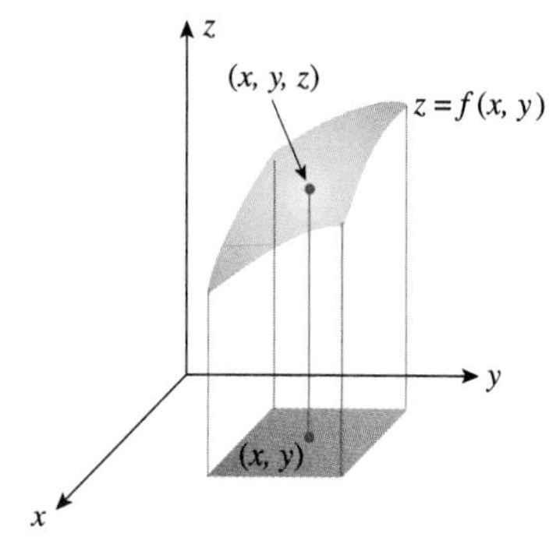

Figure 10.5
The graph of a function in three-dimensional space

Now, if $f(x, y)$ is a function of two variables x and y, the domain of f is a subset of the xy-plane. Let $z = f(x, y)$ so that there is one and only one point $(x, y, z) \equiv (x, y, f(x, y))$ associated with each point (x, y) in the domain of f. The totality of all such points makes up the **graph** of the function f and is, except for certain degenerate cases, a surface in three-dimensional space (Figure 10.5).

In interpreting the graph of a function $f(x, y)$, one often thinks of the value $z = f(x, y)$ of the function at the point (x, y) as the "height" of the point (x, y, z) on the graph of f. If $f(x, y) > 0$, then the point (x, y, z) is $f(x, y)$ units above the xy-plane; if $f(x, y) < 0$, then the point (x, y, z) is $|f(x, y)|$ units below the xy-plane.

In general, it is quite difficult to draw the graph of a function of two variables. But techniques have been developed that enable us to generate such graphs with minimum effort, using a computer. Figure 10.6 shows the computer-generated graphs of two functions.

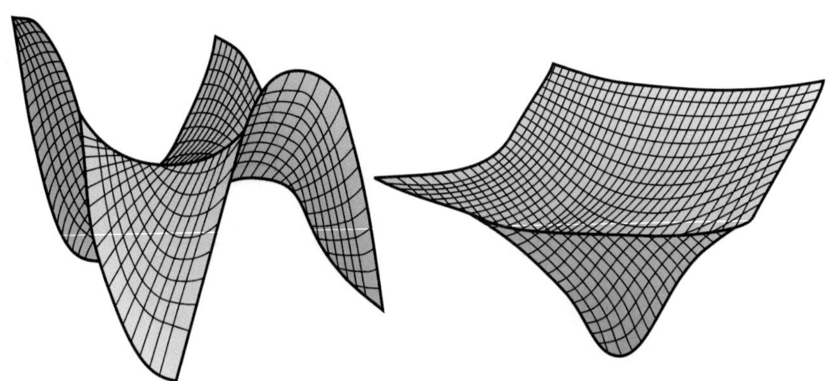

(a) $f(x, y) = x^3 - 3y^2x$ **(b)** $f(x, y) = \ln(x^2 + 2y^2 + 1)$

Figure 10.6
Two computer-generated graphs of functions of two variables

Level Curves

As mentioned earlier, the graph of a function of two variables is often difficult to sketch, and we will not develop a systematic procedure for sketching it. Instead, we describe a method that is used in constructing topographic maps. This method is relatively easy to apply and conveys sufficient information to enable one to obtain a feel for the graph of the function.

Suppose that $f(x, y)$ is a function of two variables x and y, with a graph as shown in Figure 10.7. If c is some value of the function f, then the equation $f(x, y) = c$ describes a curve lying on the plane $z = c$ called the **trace** of the graph of f in the plane $z = c$. If this trace is projected onto the xy-plane, the resulting curve in the xy-plane is called a **level curve.** By drawing the level curves corresponding to several admissible values of c, we obtain a **contour map.** Observe that, by construction, every point on a particular level curve corresponds to a point on the surface $z = f(x, y)$ that is a certain fixed distance from the xy-plane. Thus, by elevating or depressing the level curves that make up the contour map in one's mind, it is possible to obtain a feel for the general shape of the surface represented by the function f. Figure 10.8a shows a part of a mountain range with one peak; Figure 10.8b is the associated contour map.

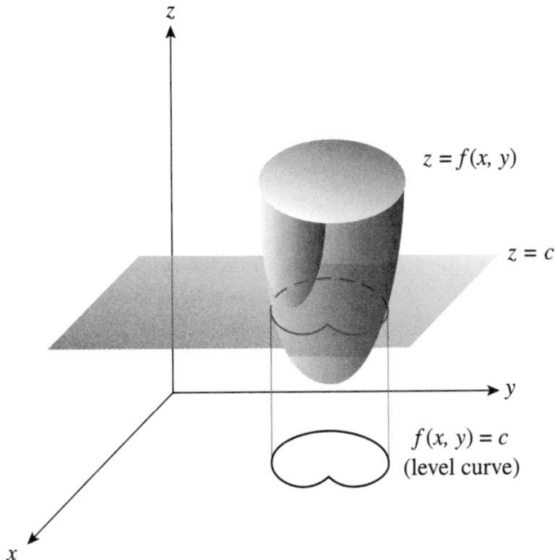

Figure 10.7
The graph of the function $z = f(x, y)$ and its intersection with the plane $z = c$

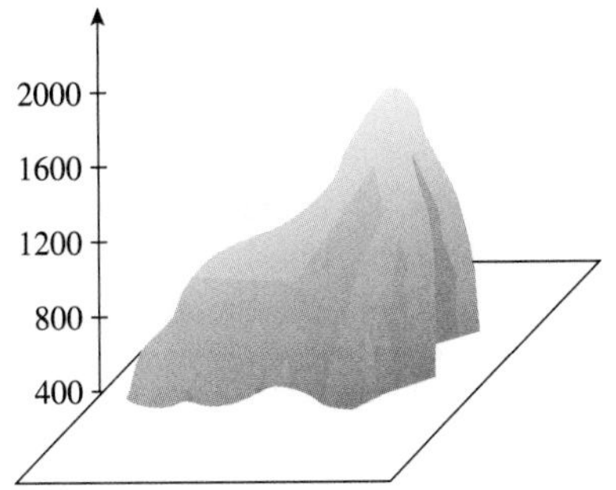

(a) A peak on a mountain range

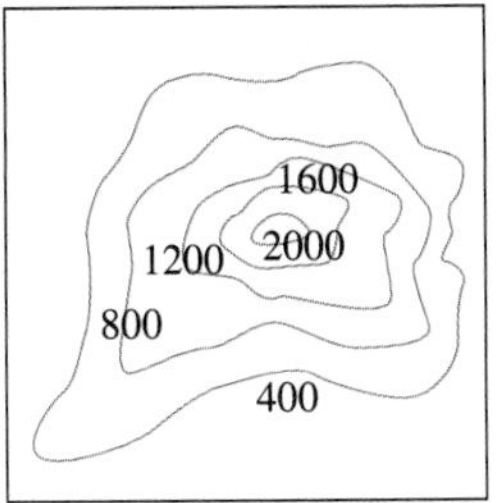

(b) A contour map for the mountain peak

Figure 10.8

Example 5

Sketch a contour map for the function $f(x, y) = x^2 + y^2$.

Solution

The level curves are the graphs of the equation $x^2 + y^2 = c$ for nonnegative numbers c. Taking $c = 0, 1, 4, 9,$ and 16, for example, we obtain

$$c = 0: x^2 + y^2 = 0$$

$$c = 1: x^2 + y^2 = 1$$

$$c = 4: x^2 + y^2 = 4 = 2^2$$

$$c = 9: x^2 + y^2 = 9 = 3^2$$

$$c = 16: x^2 + y^2 = 16 = 4^2$$

The five level curves are concentric circles with centre at the origin and radius given by $r = 0, 1, 2, 3$, and 4, respectively (Figure 10.9a). A sketch of the graph of $f(x, y) = x^2 + y^2$ is included for your reference in Figure 10.9b.

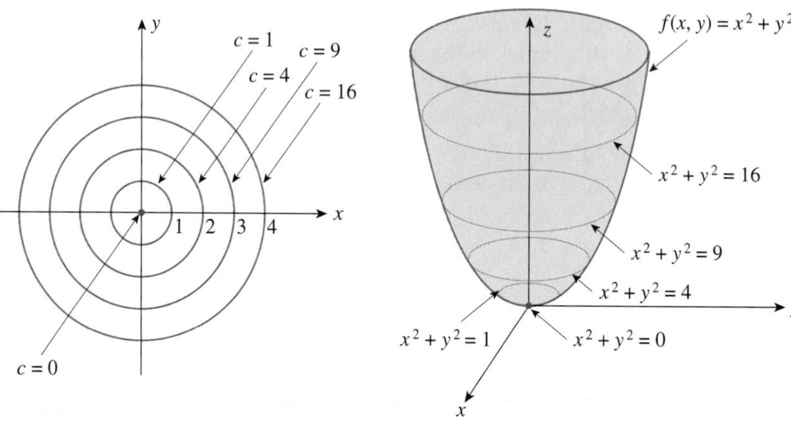

(a) Level curves of $f(x, y) = x^2 + y^2$ **(b)** The graph of $f(x, y) = x^2 + y^2$

Figure 10.9

Example 6

Sketch the level curves for the function $f(x, y) = 2x^2 - y$ corresponding to $z = -2$, -1, 0, 1, and 2.

Solution

The level curves are the graphs of the equation $2x^2 - y = k$ or $y = 2x^2 - k$ for $k = -2, -1, 0, 1$, and 2. The required level curves are shown in Figure 10.10.

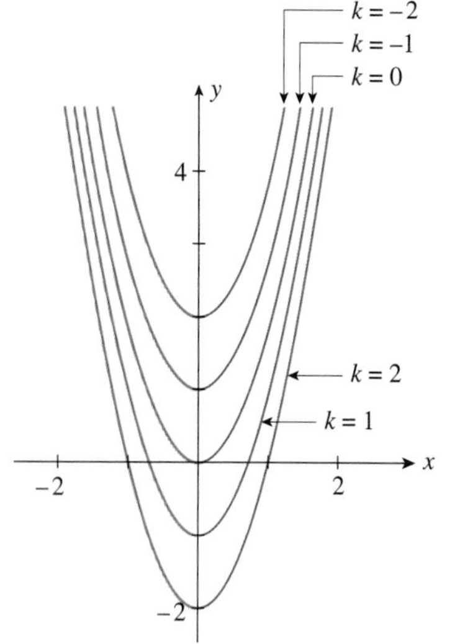

Figure 10.10
Level curves for $f(x, y) = 2x^2 - y$

Level curves of functions of two variables are found in many practical applications. For example, if $f(x, y)$ denotes the temperature at a location within North America with longitude x and latitude y at a certain time of day, then the temperature at the point (x, y) is given by the "height" of the surface, represented by $z = f(x, y)$. In this situation the level curve $f(x, y) = k$ is a curve superimposed on a map of North America, connecting points having the same temperature at a given time (Figure 10.11). These level curves are called **isotherms.**

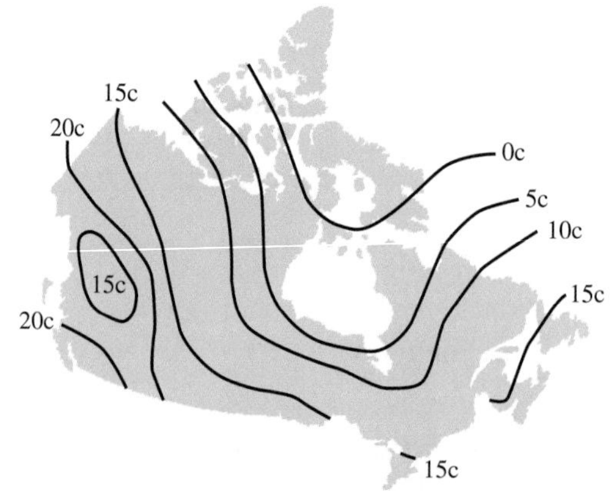

Figure 10.11
Isotherms: curves connecting points that have the same temperature

Similarly, if $f(x, y)$ gives the barometric pressure at the location (x, y), then the level curves of the function f are called isobars, lines connecting points having the same barometric pressure at a given time.

As a final example, suppose $P(x, y, z)$ is a function of three variables x, y, and z giving the profit realized when x, y, and z units of three products. A, B, and C, respectively, are produced and sold. Then, the equation $P(x, y, z) = k$, where k is a constant, represents a surface in three-dimensional space called a level surface of P. In this situation, the level surface represented by $P(x, y, z) = k$ represents the product mix that results in a profit of exactly k dollars. Such a level surface is called an isoprofit surface.

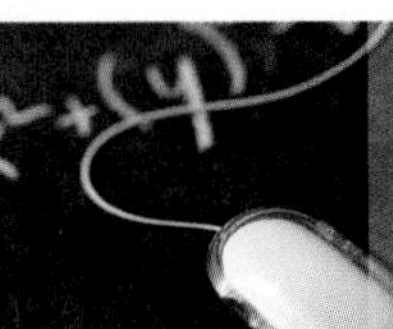

Self-Check Exercises 10.1

1. Let $f(x, y) = x^2 - 3xy + \sqrt{x + y}$. Compute $f(1, 3)$ and $f(-1, 1)$. Is the point $(-1, 0)$ in the domain of f?
2. Find the domain $f(x, y) = \frac{1}{x} + \frac{1}{x - y} - e^{x+y}$.
3. Odyssey Travel Agency has a monthly advertising budget of \$20,000. Odyssey's management estimates that if they spend x dollars on newspaper advertising and y dollars on television advertising, then the monthly revenue will be

$$f(x, y) = 30x^{1/4}y^{3/4}$$

dollars. What will be the monthly revenue if Odyssey spends \$5000/month on newspaper ads and \$15,000/month on television ads? If Odyssey spends \$4000/month on newspaper ads and \$16,000/month on television ads?

Solutions to Self-Check Exercises 10.1 can be found on page 594.

10.1 Exercises

1. Let $f(x, y) = 2x + 3y - 4$. Compute $f(0, 0)$, $f(1, 0)$, $f(0, 1)$, $f(1, 2)$, and $f(2, -1)$.
2. Let $g(x, y) = 2x^2 - y^2$. Compute $g(1, 2)$, $g(2, 1)$, $g(1, 1)$, $g(-1, 1)$, and $g(2, -1)$.
3. Let $f(x, y) = x^2 + 2xy - x + 3$. Compute $f(1, 2)$, $f(2, 1)$, $f(-1, 2)$, and $f(2, -1)$.
4. Let $h(x, y) = (x + y)/(x - y)$. Compute $h(0, 1)$, $h(-1, 1)$, $h(2, 1)$, and $h(\pi, -\pi)$.
5. Let $g(s, t) = 3s\sqrt{t} + t\sqrt{s} + 2$. Compute $g(1, 2)$, $g(2, 1)$, $g(0, 4)$, and $g(4, 9)$.
6. Let $f(x, y) = xye^{x^2+y^2}$. Compute $f(0, 0)$, $f(0, 1)$, $f(1, 1)$, and $f(-1, -1)$.
7. Let $h(s, t) = s \ln t - t \ln s$. Compute $h(1, e)$, $h(e, 1)$, and $h(e, e)$.
8. Let $f(u, v) = (u^2 + v^2)e^{uv^2}$. Compute $f(0, 1)$, $f(-1, -1)$, $f(a, b)$, and $f(b, a)$.
9. Let $g(r, s, t) = re^{s/t}$. Compute $g(1, 1, 1)$, $g(1, 0, 1)$, and $g(-1, -1, -1)$.
10. Let $g(u, v, w) = (ue^{vw} + ve^{uw} + we^{uv})/(u^2 + v^2 + w^2)$. Compute $g(1, 2, 3)$ and $g(3, 2, 1)$.
11. Let $f(x, y) = \sin(x) + \cos(y)$. Compute $f(0, \pi/2)$, $f(\pi/3, \pi/4)$, and $f(0, -2\pi/3)$.
12. Let $g(\theta, \phi) = \cos(\theta\pi) - \sin(\phi) - \sin(\pi)$. Compute $g(\pi/3, 2\pi/3)$ and $g(\pi/4, 5\pi/4)$.
13. Let $f(u, v) = u\cos(v)$. Compute $f(1,0)$, $f(\sqrt{3}, \pi/3)$, and $f(-1, -\pi/4)$.
14. Let $h(x, y) = \frac{\arctan(x)}{y^2 + 1}$. Compute $h(1, 1)$, $h(\sqrt{3}, \sqrt{3})$, and $h(-1, 0)$.

In Exercises 15–22, find the domain of the function.

15. $f(x, y) = 2x + 3y$
16. $g(x, y, z) = x^2 + y^2 + z^2$

17. $h(u, v) = \dfrac{uv}{u - v}$

18. $f(s, t) = \sqrt{s^2 + t^2}$

19. $g(r, s) = \sqrt{rs}$

20. $f(x, y) = e^{-xy}$

21. $h(x, y) = \ln(x + y - 5)$

22. $h(u, v) = \sqrt{4 - u^2 - v^2}$

In Exercises 23–28, sketch the level curves of the function corresponding to the given values of z.

23. $f(x, y) = 2x + 3y;\ z = -2, -1, 0, 1, 2$

24. $f(x, y) = -x^2 + y;\ z = -2, -1, 0, 1, 2$

25. $f(x, y) = 2x^2 + y;\ z = -2, -1, 0, 1, 2$

26. $f(x, y) = xy;\ z = -4, -2, 2, 4$

27. $f(x, y) = \sqrt{16 - x^2 - y^2};\ z = 0, 1, 2, 3, 4$

28. $f(x, y) = e^x - y;\ z = -2, -1, 0, 1, 2$

29. The volume of a cylindrical tank of radius r and height h is given by

$$V = f(r, h) = \pi r^2 h$$

Find the volume of a cylindrical tank of radius 1.5 m and height 4 m.

30. **IQs** The IQ (intelligence quotient) of a person whose mental age is m yr and whose chronological age is c yr is defined as

$$f(m, c) = \frac{100m}{c}$$

What is the IQ of a 9-yr-old child who has a mental age of 13.5 yr?

31. **Volume of a Gas** The volume of a certain mass of gas is related to its pressure and temperature by the formula

$$V = \frac{30.9T}{P}$$

where the volume V is measured in litres, the temperature T is measured in degrees Kelvin (obtained by adding 273° to the Celsius temperature), and the pressure P is measured in millimetres of mercury pressure.

a. Find the domain of the function V.

b. Calculate the volume of the gas at standard temperature and pressure—that is, when $T = 273$ K and $P = 760$ mm of mercury.

32. **Force Generated by a Centrifuge** A centrifuge is a machine designed for the specific purpose of subjecting materials to a sustained centrifugal force. The actual amount of centrifugal force, F, expressed in dynes (1 gram of force = 980 dynes) is given by

$$F = f(M, S, R) = \frac{\pi^2 S^2 MR}{900}$$

where S is in revolutions per minute (rpm), M is in grams, and R is in centimetres. Show that an object revolving at the rate of 600 rpm in a circle with radius of 10 cm generates a centrifugal force that is approximately 40 times gravity.

Business and Economics Applications

33. **Revenue Functions** Country Workshop manufactures both finished and unfinished furniture for the home. The estimated quantities demanded each week of its rolltop desks in the finished and unfinished versions are x and y units when the corresponding unit prices are

$$p = 200 - \frac{1}{5}x - \frac{1}{10}y$$

$$q = 160 - \frac{1}{10}x - \frac{1}{4}y$$

dollars, respectively.

a. What is the weekly total revenue function $R(x, y)$?

b. Find the domain of the function R.

34. For the total revenue function $R(x, y)$ of Exercise 33, compute $R(100, 60)$ and $R(60, 100)$. Interpret your results.

35. **Revenue Functions** Weston Publishing publishes a deluxe edition and a standard edition of its English language dictionary. Weston's management estimates that the number of deluxe editions demanded is x copies/day and the number of standard editions demanded is y copies/day when the unit prices are

$$p = 20 - 0.005x - 0.001y$$
$$q = 15 - 0.001x - 0.003y$$

dollars, respectively.

a. Find the daily total revenue function $R(x, y)$.

b. Find the domain of the function R.

36. For the total revenue function $R(x, y)$ of Exercise 35, compute $R(300, 200)$ and $R(200, 300)$. Interpret your results.

37. **Arson for Profit** A study of arson for profit was conducted by a team of paid civilian experts and police detectives appointed by the mayor of a large city. It was found that the number of suspicious fires in that city in 1992 was very closely related to the concentration of tenants in the city's public housing and to the level of reinvestment in the area in conventional mortgages by the ten largest banks. In fact, the number of fires was closely approximated by the formula

$$N(x, y) = \frac{100(1000 + 0.03x^2y)^{1/2}}{(5 + 0.2y)^2} \qquad (0 \le x \le 150;\ 5 \le y \le 35)$$

where x denotes the number of persons/census tract and y denotes the level of reinvestment in the area in cents/dollar deposited. Using this formula, estimate the total number of suspicious fires in the districts of the city where the concentration of public housing tenants was 100/census tract and the level of reinvestment was 20 cents/dollar deposited.

38. **Continuously Compounded Interest** If a principal of P dollars is deposited in an account earning interest at the rate of

r/year compounded continuously, then the accumulated amount at the end of t yr is given by

$$A = f(P, r, t) = Pe^{rt}$$

dollars. Find the accumulated amount at the end of 3 yr if a sum of $10,000 is deposited in an account earning interest at the rate of 10%/year.

39. **Home Mortgages** The monthly payment that amortizes a loan of A dollars in t yr when the interest rate is r per year is given by

$$P = f(A, r, t) = \frac{Ar}{12\left[1 - \left(1 + \frac{r}{12}\right)^{-12t}\right]}$$

a. What is the monthly payment for a home mortgage of $100,000 that will be amortized over 30 yr with an interest rate of 8%/year? An interest rate of 10%/year?

b. Find the monthly payment for a home mortgage of $100,000 that will be amortized over 20 yr with an interest rate of 8%/year.

40. **Home Mortgages** Suppose a home buyer secures a bank loan of A dollars to purchase a house. If the interest rate charged is r/year and the loan is to be amortized in t yr, then the principal repayment at the end of i mo is given by

$$B = f(A, r, t, i) = A\left[\frac{\left(1 + \frac{r}{12}\right)^{i} - 1}{\left(1 + \frac{r}{12}\right)^{12t} - 1}\right] \quad (0 \le i \le 12t)$$

Suppose the Blakelys borrow a sum of $80,000 from a bank to help finance the purchase of a house and the bank charges interest at a rate of 9%/year. If the Blakelys agree to repay the loan in equal installments over 30 yr, how much will they owe the bank after the 60th payment (5 yr)? The 240th payment (20 yr)?

41. **Wilson Lot-Size Formula** The Wilson lot-size formula in economics states that the optimal quantity Q of goods for a store to order is given by

$$Q = f(C, N, h) = \sqrt{\frac{2\,CN}{h}}$$

where C is the cost of placing an order, N is the number of items the store sells per week, and h is the weekly holding cost for each item. Find the most economical quantity of 10-speed bicycles to order if it costs the store $20 to place an order, $5 to hold a bicycle for a week, and the store expects to sell 40 bicycles a week.

Biological and Life Sciences Applications

42. **Body Mass** The body mass index (BMI) is used to identify, evaluate, and treat overweight and obese adults. The BMI value for an adult of weight w (in kilograms) and height h (in metres) is defined to be

$$M = f(w, h) = \frac{w}{h^2}$$

According to federal guidelines, an adult is overweight if he or she has a BMI value between 25 and 29.9 and is "obese" if the value is greater than or equal to 30.

a. What is the BMI of an adult who weighs in at 80 kg and stands 1.8 m tall?

b. What is the maximum weight for an adult of height 1.8 m, who is not classified as overweight or obese?

43. **Poiseuille's Law** Poiseuille's law states that the resistance R, measured in dynes, of blood flowing in a blood vessel of length l and radius r (both in centimetres) is given by

$$R = f(l, r) = \frac{kl}{r^4}$$

where k is the viscosity of blood (in dyne-sec/cm^2). What is the resistance, in terms of k, of blood flowing through an arteriole 4 cm long and of radius 0.1 cm?

44. **Surface Area of a Human Body** An empirical formula by E. F. Dubois relates the surface area S of a human body (in square metres) to its weight W (in kilograms) and its height H (in centimetres). The formula, given by

$$S = 0.007184W^{0.425}H^{0.725}$$

is used by physiologists in metabolism studies.

a. Find the domain of the function S.

b. What is the surface area of a human body that weighs 70 kg and has a height of 178 cm?

In Exercises 45–48, determine whether the statement is true or false. If it is true, explain why it is true. If it is false, give an example to show why it is false.

45. If h is a function of x and y, then there are functions f and g of one variable such that

$$h(x, y) = f(x) + g(y)$$

46. If f is a function of x and y and a is a real number, then

$$f(ax, ay) = af(x, y)$$

47. The domain of $f(x, y) = 1/(x^2 - y^2)$ is $\{(x, y) \mid y \neq x\}$.

48. Every point on the level curve $f(x, y) = c$ corresponds to a point on the graph of f that is c units above the xy-plane if $c > 0$ and $|c|$ units below the xy-plane if $c < 0$.

Solutions to Self-Check Exercises 10.1

1. $f(1, 3) = 1^2 - 3(1)(3) + \sqrt{1 + 3} = -6$

$f(-1, 1) = (-1)^2 - 3(-1)(1) + \sqrt{-1 + 1} = 4$

The point $(-1, 0)$ is not in the domain of f because the term $\sqrt{x + y}$ is not defined when $x = -1$ and $y = 0$. In fact, the domain of f consists of all real values of x and y that satisfy the inequality $x + y \geq 0$, the shaded half-plane shown in the accompanying figure.

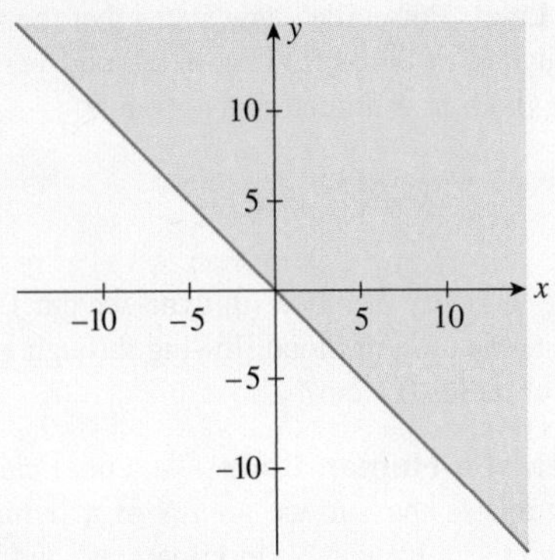

2. Since division by zero is not permitted, we see that $x \neq 0$ and $x - y \neq 0$. Therefore, the domain of f is the set of all points in the xy-plane not containing the y-axis ($x = 0$) and the straight line $x = y$.

3. If Odyssey spends \$5000/month on newspaper ads ($x = 5000$) and \$15,000/month on television ads ($y = 15{,}000$), then its monthly revenue will be given by

$$\begin{aligned} f(5000, 15{,}000) &= 30(5000)^{1/4}(15{,}000)^{3/4} \\ &\approx 341{,}926.06 \end{aligned}$$

or approximately \$341,926. If the agency spends \$4000/month on newspaper ads and \$16,000/month on television ads, then its monthly revenue will be given by

$$\begin{aligned} f(4000, 16{,}000) &= 30(4000)^{1/4}(16{,}000)^{3/4} \\ &\approx 339{,}411.26 \end{aligned}$$

or approximately \$339,411.

10.2 Partial Derivatives

Partial Derivatives

For a function $f(x)$ of one variable x, there is no ambiguity when we speak about the rate of change of $f(x)$ with respect to x since x must be constrained to move along the x-axis. The situation becomes more complicated, however, when we study the rate of change of a function of two or more variables. For example, the domain D of a function of two variables $f(x, y)$ is a subset of the plane (Figure 10.12), so if $P(a, b)$ is any point in the domain of f, there are infinitely many directions from which one can approach the point P. We may therefore ask for the rate of change of f at P along any of these directions.

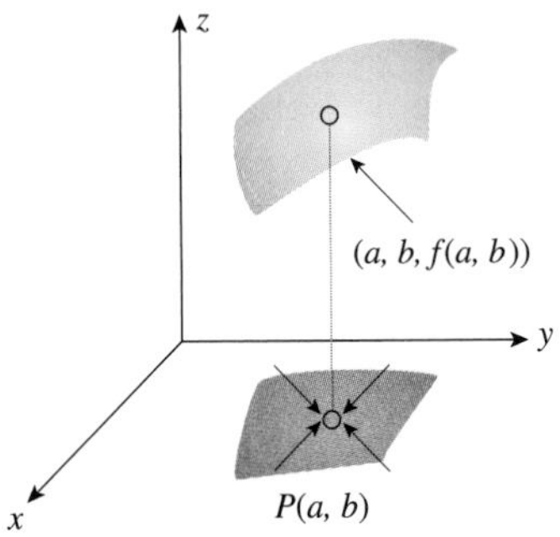

Figure 10.12
We can approach a point in the plane from infinitely many directions.

However, we will not deal with this general problem. Instead, we will restrict ourselves to studying the rate of change of the function $f(x, y)$ at a point $P(a, b)$ in each of two *preferred directions*—namely, the direction parallel to the x-axis and the direction parallel to the y-axis. Let $y = b$, where b is a constant, so that $f(x, b)$ is a function of the one variable x. Since the equation $z = f(x, y)$ is the equation of a surface, the equation $z = f(x, b)$ is the equation of the curve C on the surface formed by the intersection of the surface and the plane $y = b$ (Figure 10.13).

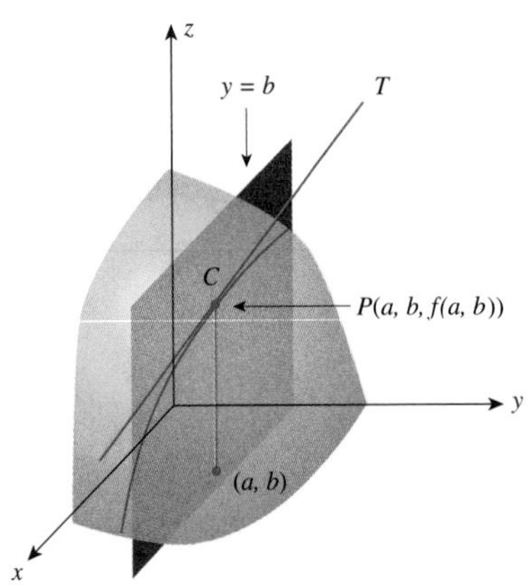

Figure 10.13
The curve C is formed by the intersection of the plane $y = b$ with the surface $z = f(x, y)$.

Because $f(x, b)$ is a function of one variable x, we may compute the derivative of f with respect to x at $x = a$. This derivative, obtained by keeping the variable y fixed and differentiating the resulting function $f(x, b)$ with respect to x, is called the **first partial derivative of f with respect to x** at (a, b), written

$$\frac{\partial z}{\partial x}(a, b) \quad \text{or} \quad \frac{\partial f}{\partial x}(a, b) \quad \text{or} \quad f_x(a, b)$$

Thus,

$$\frac{\partial z}{\partial x}(a, b) = \frac{\partial f}{\partial x}(a, b) = f_x(a, b) = \lim_{h \to 0} \frac{f(a + h, b) - f(a, b)}{h}$$

provided that the limit exists. The first partial derivative of f with respect to x at (a, b) measures both the slope of the tangent line T to the curve C and the rate of change of the function f in the x-direction when $x = a$ and $y = b$. We also write

$$\left.\frac{\partial f}{\partial x}\right|_{(a, b)} \equiv f_x(a, b)$$

Similarly, we define the **first partial derivative of f with respect to y** at (a, b), written

$$\frac{\partial z}{\partial y}(a, b) \quad \text{or} \quad \frac{\partial f}{\partial y}(a, b) \quad \text{or} \quad f_y(a, b)$$

as the derivative obtained by keeping the variable x fixed and differentiating the resulting function $f(a, y)$ with respect to y. That is,

$$\frac{\partial z}{\partial y}(a, b) = \frac{\partial f}{\partial y}(a, b) = f_y(a, b)$$

$$= \lim_{k \to 0} \frac{f(a, b + k) - f(a, b)}{k}$$

if the limit exists. The first partial derivative of f with respect to y at (a, b) measures both the slope of the tangent line T to the curve C, obtained by holding x constant (Figure 10.14), and the rate of change of the function f in the y-direction when $x = a$ and $y = b$. We write

$$\left.\frac{\partial f}{\partial y}\right|_{(a, b)} \equiv f_y(a, b)$$

Before looking at some examples, let's summarize these definitions.

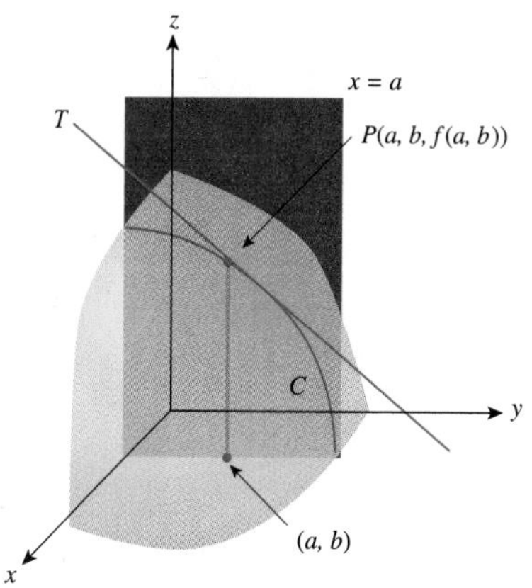

Figure 10.14
The first partial derivative of f with respect to y at (a, b) measures the slope of the tangent line T to the curve C with x held constant.

First Partial Derivatives of $f(x, y)$

Suppose $f(x, y)$ is a function of the two variables x and y. Then, the **first partial derivative of f with respect to x** at the point (x, y) is

$$\frac{\partial f}{\partial x} = \lim_{h \to 0} \frac{f(x + h, y) - f(x, y)}{h}$$

provided the limit exists. The first partial derivative of f with respect to y at the point (x, y) is

$$\frac{\partial f}{\partial y} = \lim_{k \to 0} \frac{f(x, y + k) - f(x, y)}{k}$$

provided the limit exists.

Example 1

Find the partial derivatives $\partial f/\partial x$ and $\partial f/\partial y$ of the function

$$f(x, y) = x^2 - xy^2 + y^3$$

What is the rate of change of the function f in the x-direction at the point $(1, 2)$? What is the rate of change of the function f in the y-direction at the point $(1, 2)$?

Solution

To compute $\partial f/\partial x$, think of the variable y as a constant and differentiate the resulting function of x with respect to x. Let's write

$$f(x, y) = x^2 - xy^2 + y^3$$

where the variable y to be treated as a constant is shown in colour. Then,

$$\frac{\partial f}{\partial x} = 2x - y^2$$

To compute $\partial f/\partial y$, think of the variable x as being fixed—that is, as a constant—and differentiate the resulting function of y with respect to y. In this case,

$$f(x, y) = x^2 - xy^2 + y^3$$

so that

$$\frac{\partial f}{\partial y} = -2xy + 3y^2$$

The rate of change of the function f in the x-direction at the point (1, 2) is given by

$$f_x(1, 2) = \left.\frac{\partial f}{\partial x}\right|_{(1,2)} = 2(1) - 2^2 = -2$$

That is, f decreases 2 units for each unit increase in the x-direction, y being kept constant ($y = 2$). The rate of change of the function f in the y-direction at the point (1, 2) is given by

$$f_y(1, 2) = \left.\frac{\partial f}{\partial y}\right|_{(1,2)} = -2(1)(2) + 3(2)^2 = 8$$

That is, f increases 8 units for each unit increase in the y-direction, x being kept constant ($x = 1$).

GROUP DISCUSSION

Refer to the Group Discussion on page 587. Suppose management has decided that the projected sales of the first product is a units. Describe how you might help management decide how many units of the second product the company should produce and sell in order to maximize the company's total profit. Justify your method to management. Suppose, however, management feels that b units of the second product can be manufactured and sold. How would you help management decide how many units of the first product to manufacture in order to maximize the company's total profit?

Example 2

Compute the first partial derivatives of each of the following functions.

a. $f(x, y) = \dfrac{xy}{x^2 + y^2}$

b. $g(s, t) = (s^2 - st + t^2)^5$

c. $h(u, v) = e^{u^2 - v^2}$

d. $f(x, y) = \ln(x^2 + 2y^2)$

Solution

a. To compute $\partial f/\partial x$, think of the variable y as a constant. Thus,

$$f(x, y) = \frac{xy}{x^2 + y^2}$$

so that, upon using the quotient rule, we have

$$\frac{\partial f}{\partial x} = \frac{(x^2 + y^2)y - xy(2x)}{(x^2 + y^2)^2}$$

$$= \frac{y(y^2 - x^2)}{(x^2 + y^2)^2}$$

upon simplification and factorization. To compute $\partial f/\partial y$, think of the variable x as a constant. Thus,

$$f(x, y) = \frac{xy}{x^2 + y^2}$$

so that, upon using the quotient rule once again, we obtain

$$\frac{\partial f}{\partial y} = \frac{(x^2 + y^2)x - xy(2y)}{(x^2 + y^2)^2}$$

$$= \frac{x(x^2 - y^2)}{(x^2 + y^2)^2}$$

b. To compute $\partial g/\partial s$, we treat the variable t as if it were a constant. Thus,

$$g(s, t) = (s^2 - st + t^2)^5$$

Using the general power rule, we find

$$\frac{\partial g}{\partial s} = 5(s^2 - st + t^2)^4 \cdot (2s - t)$$

$$= 5(2s - t)(s^2 - st + t^2)^4$$

To compute $\partial g/\partial t$, we treat the variable s as if it were a constant. Thus,

$$g(s, t) = (s^2 - st + t^2)^5$$

$$\frac{\partial g}{\partial t} = 5(s^2 - st + t^2)^4(-s + 2t)$$

$$= 5(2t - s)(s^2 - st + t^2)^4$$

c. To compute $\partial h/\partial u$, think of the variable v as a constant. Thus,

$$h(u, v) = e^{u^2 - v^2}$$

Using the chain rule for exponential functions, we have

$$\frac{\partial h}{\partial u} = e^{u^2 - v^2} \cdot 2u$$

$$= 2ue^{u^2 - v^2}$$

Next, we treat the variable u as if it were a constant,

$$h(u, v) = e^{u^2 - v^2}$$

and we obtain

$$\frac{\partial h}{\partial v} = e^{u^2-v^2} \cdot (-2v)$$
$$= -2ve^{u^2-v^2}$$

d. To compute $\partial f/\partial x$, think of the variable y as a constant. Thus,

$$f(x, y) = \ln(x^2 + 2y^2)$$

so that the chain rule for logarithmic functions gives

$$\frac{\partial f}{\partial x} = \frac{2x}{x^2 + 2y^2}$$

Next, treating the variable x as if it were a constant, we find

$$f(x, y) = \ln(x^2 + 2y^2)$$
$$\frac{\partial f}{\partial y} = \frac{4y}{x^2 + 2y^2}$$

To compute the partial derivative of a function of several variables with respect to one variable—say, x—we think of the other variables as if they were constants and differentiate the resulting function with respect to x.

GROUP DISCUSSION

1. Let (a, b) be a point in the domain of $f(x, y)$. Put $g(x) = f(x, b)$ and suppose g is differentiable at $x = a$. Explain why you can find $f_x(a, b)$ by computing $g'(a)$. How would you go about calculating $f_y(a, b)$ using a similar technique? Give a geometric interpretation of these processes.
2. Let $f(x, y) = x^2y^3 - 3x^2y + 2$. Use the method of Problem 1 to find $f_x(1, 2)$ and $f_y(1, 2)$.

Example 3

Compute the first partial derivatives of the function

$$w = f(x, y, z) = xyz - xe^{yz} + x \ln y$$

Solution

Here we have a function of three variables, x, y, and z, and we are required to compute

$$\frac{\partial f}{\partial x}, \frac{\partial f}{\partial y}, \frac{\partial f}{\partial z}$$

To compute f_x, we think of the other two variables, y and z, as fixed, and we differentiate the resulting function of x with respect to x, thereby obtaining

$$f_x = yz - e^{yz} + \ln y$$

To compute f_y, we think of the other two variables, x and z, as constants, and we differentiate the resulting function of y with respect to y. We then obtain

$$f_y = xz - xze^{yz} + \frac{x}{y}$$

Finally, to compute f_z, we treat the variables x and y as constants and differentiate the function f with respect to z, obtaining

$$f_z = xy - xye^{yz}$$

EXPLORING WITH TECHNOLOGY

Refer to the Group Discussion on page 598. Let

$$f(x, y) = \frac{e^{\sqrt{xy}}}{(1 + xy^2)^{3/2}}$$

1. Compute $g(x) = f(x, 1)$ and use a graphing utility to plot the graph of g in the viewing window $[0, 2] \cdot [0, 2]$.
2. Use the differentiation operation of your graphing utility to find $g'(1)$ and hence $f_x(1, 1)$.
3. Compute $h(y) = f(1, y)$ and use a graphing utility to plot the graph of g in the viewing window $[0, 2] \cdot [0, 2]$.
4. Use the differentiation operation of your graphing utility to find $h'(1)$ and hence $f_y(1, 1)$.

The Cobb–Douglas Production Function

For an economic interpretation of the first partial derivatives of a function of two variables, let's turn our attention to the function

$$f(x, y) = ax^b y^{1-b} \tag{1}$$

where a and b are positive constants with $0 < b < 1$. This function is called the **Cobb–Douglas production function.** Here, x stands for the amount of money expended for labour, y stands for the cost of capital equipment (buildings, machinery, and other tools of production), and the function f measures the output of the finished product (in suitable units) and is called, accordingly, the production function.

The partial derivative f_x is called the **marginal productivity of labour.** It measures the rate of change of production with respect to the amount of money expended for labour, with the level of capital expenditure held constant. Similarly, the partial derivative f_y, called the **marginal productivity of capital,** measures the rate of change of production with respect to the amount expended on capital, with the level of labour expenditure held fixed.

Example 4

Marginal Productivity A certain country's production in the early years following World War II is described by the function

$$f(x, y) = 30x^{2/3}y^{1/3}$$

units, when x units of labour and y units of capital were used.

a. Compute f_x and f_y.
b. What is the marginal productivity of labour and the marginal productivity of capital when the amounts expended on labour and capital are 125 units and 27 units, respectively?
c. Should the government have encouraged capital investment rather than increasing expenditure on labour to increase the country's productivity?

Solution

a. $$f_x = 30 \cdot \frac{2}{3} x^{-1/3} y^{1/3} = 20\left(\frac{y}{x}\right)^{1/3}$$

$$f_y = 30x^{2/3} \cdot \frac{1}{3} y^{-2/3} = 10\left(\frac{x}{y}\right)^{2/3}$$

b. The required marginal productivity of labour is given by

$$f_x(125, 27) = 20\left(\frac{27}{125}\right)^{1/3} = 20\left(\frac{3}{5}\right)$$

or 12 units per unit increase in labour expenditure (capital expenditure is held constant at 27 units). The required marginal productivity of capital is given by

$$f_y(125, 27) = 10\left(\frac{125}{27}\right)^{2/3} = 10\left(\frac{25}{9}\right)$$

or $27\frac{7}{9}$ units per unit increase in capital expenditure (labour outlay is held constant at 125 units).

c. From the results of part (b), we see that a unit increase in capital expenditure resulted in a much faster increase in productivity than a unit increase in labour expenditure would have. Therefore, the government should have encouraged increased spending on capital rather than on labour during the early years of reconstruction.

Substitute and Complementary Commodities

For another application of the first partial derivatives of a function of two variables in the field of economics, let's consider the relative demands of two commodities. We say that the two commodities are **substitute** (competitive) **commodities** if a decrease in the demand for one results in an increase in the demand for the other. Examples of competitive commodities are coffee and tea. Conversely, two commodities are referred to as **complementary commodities** if a decrease in the demand for one results in a decrease in the demand for the other as well. Examples of complementary commodities are automobiles and tires.

We now derive a criterion for determining whether two commodities A and B are substitute or complementary. Suppose the demand equations that relate the quantities demanded, x and y, to the unit prices, p and q, of the two commodities are given by

$$x = f(p, q) \qquad \text{and} \qquad y = g(p, q)$$

Let's consider the partial derivative $\partial f/\partial p$. Since f is the demand function for commodity A, we see that, for fixed q, f is typically a decreasing function of p—that is, $\partial f/\partial p < 0$. Now, if the two commodities were substitute commodities, then the quantity demanded of commodity B would increase with respect to p—that is, $\partial g/\partial p > 0$. A similar argument with p fixed shows that if A and B are substitute commodities, then $\partial f/\partial q > 0$. Thus, the two commodities A and B are substitute commodities if

$$\frac{\partial f}{\partial q} > 0 \qquad \text{and} \qquad \frac{\partial g}{\partial p} > 0$$

Similarly, A and B are complementary commodities if

$$\frac{\partial f}{\partial q} < 0 \qquad \text{and} \qquad \frac{\partial g}{\partial p} < 0$$

Substitute and Complementary Commodities

Two commodities A and B are substitute commodities if

$$\frac{\partial f}{\partial q} > 0 \qquad \text{and} \qquad \frac{\partial g}{\partial p} > 0 \tag{2}$$

Two commodities A and B are complementary commodities if

$$\frac{\partial f}{\partial q} < 0 \qquad \text{and} \qquad \frac{\partial g}{\partial p} < 0 \tag{3}$$

Example 5

Substitute and Complementary Commodities Suppose that the daily demand for butter is given by

$$x = f(p, q) = \frac{3q}{1 + p^2}$$

and the daily demand for margarine is given by

$$y = g(p, d) = \frac{2p}{1 + \sqrt{q}} \qquad (p > 0, q > 0)$$

where p and q denote the prices per pound (in dollars) of butter and margarine, respectively, and x and y are measured in millions of pounds. Determine whether these two commodities are substitute, complementary, or neither.

Solution

We compute

$$\frac{\partial f}{\partial q} = \frac{3}{1 + p^2} \quad \text{and} \quad \frac{\partial g}{\partial p} = \frac{2}{1 + \sqrt{q}}$$

Since

$$\frac{\partial f}{\partial q} > 0 \quad \text{and} \quad \frac{\partial g}{\partial p} > 0$$

for all values of $p > 0$ and $q > 0$, we conclude that butter and margarine are substitute commodities.

Example 6

Comparing Pest Control Strategies A population of mice that eat grain in Alberta can be controlled by deploying predators (cats) or by putting out poisoned bait. Suppose that the steady-state population of mice on a test farm is modelled by the equation

$$P(c, b) = \frac{5200}{1 + 0.5c + 0.2b}$$

where c is the number of cats and b is the number of packages of poisoned bait put out per week. If we have 10 cats and are deploying 30 packs of poisoned bait per week, which should be increased to have the greatest effect?

Solution

The effect of the control strategies, cats and poison, can be modelled by computing the partial derivatives at the current values of 10 cats and 30 packs of poison bait per week. Rewriting the model in the easily differentiated form

$$P(c,b) = 5200(1.0 + 0.5c + 0.2b)^{-1}$$

$$P_c(c, b) = -5200(1.0 + 0.5c + 0.2b)^{-2}(0.5)$$

$$= \frac{-2600}{(1.0 + 0.5b + 0.2b)^2}$$

$$P_b(c, b) = -5200(1.0 + 0.5c + 0.2b)^{-2}(0.2)$$

$$= \frac{-1040}{(1.0 + 0.5b + 0.2b)^2}$$

Substituting in the current values $c = 10$ and $b = 30$ we obtain

$$P_c(10, 30) = \frac{-2600}{(1.0 + 5.0 + 6.0)^2} = \frac{-2600}{144} = -18.1$$

$$P_b(10, 30) = \frac{-1040}{144} = -7.2$$

So we see that each additional cat reduces the load of mice by about 18 while each additional weekly poison bait packet kills about 7 mice. If cost is not a consideration, the cats are the more effective choice.

Second-Order Partial Derivatives

The first partial derivatives $f_x(x, y)$ and $f_y(x, y)$ of a function $f(x, y)$ of the two variables x and y are also functions of x and y. As such, we may differentiate each of the functions f_x and f_y to obtain the **second-order partial derivatives of f** (Figure 10.15). Thus, differentiating the function f_x with respect to x leads to the second partial derivative

$$f_{xx} \equiv \frac{\partial^2 f}{\partial x^2} = \frac{\partial}{\partial x}(f_x)$$

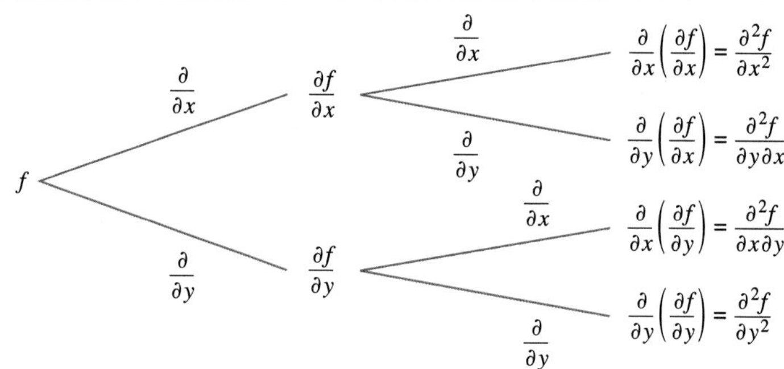

Figure 10.15
A schematic showing the four second-order partial derivatives of f

However, differentiation of f_x with respect to y leads to the second partial derivative

$$f_{xy} \equiv \frac{\partial^2 f}{\partial y \partial x} = \frac{\partial}{\partial y}(f_x)$$

Similarly, differentiation of the function f_y with respect to x and with respect to y leads to

$$f_{yx} \equiv \frac{\partial^2 f}{\partial x \partial y} = \frac{\partial}{\partial x}(f_y)$$

$$f_{yy} \equiv \frac{\partial^2 f}{\partial y^2} = \frac{\partial}{\partial y}(f_y)$$

respectively. Note that, in general, it is not true that $f_{xy} = f_{yx}$, but they are equal if both f_{xy} and f_{yx} are continuous. We might add that this is the case in most practical applications.

Example 7

Find the second-order partial derivatives of the function

$$f(x, y) = x^3 - 3x^2y + 3xy^2 + y^2$$

Solution

The first partial derivatives of f are

$$f_x = \frac{\partial}{\partial x}(x^3 - 3x^2y + 3xy^2 + y^2)$$
$$= 3x^2 - 6xy + 3y^2$$
$$f_y = \frac{\partial}{\partial y}(x^3 - 3x^2y + 3xy^2 + y^2)$$
$$= -3x^2 + 6xy + 2y$$

Therefore,

$$f_{xx} = \frac{\partial}{\partial x}(f_x) = \frac{\partial}{\partial x}(3x^2 - 6xy + 3y^2)$$
$$= 6x - 6y = 6(x - y)$$
$$f_{xy} = \frac{\partial}{\partial y}(f_x) = \frac{\partial}{\partial y}(3x^2 - 6xy + 3y^2)$$
$$= -6x + 6y = 6(y - x)$$
$$f_{yx} = \frac{\partial}{\partial x}(f_y) = \frac{\partial}{\partial x}(-3x^2 + 6xy + 2y)$$
$$= -6x + 6y = 6(y - x)$$
$$f_{yy} = \frac{\partial}{\partial y}(f_y) = \frac{\partial}{\partial y}(-3x^2 + 6xy + 2y)$$
$$= 6x + 2$$

Example 8

Find the second-order partial derivatives of the function

$$f(x, y) = e^{xy^2}$$

Solution

We have

$$f_x = \frac{\partial}{\partial x}(e^{xy^2})$$
$$= y^2e^{xy^2}$$
$$f_y = \frac{\partial}{\partial y}(e^{xy^2})$$
$$= 2xye^{xy^2}$$

so the required second-order partial derivatives of f are

$$f_{xx} = \frac{\partial}{\partial x}(f_x) = \frac{\partial}{\partial x}(y^2e^{xy^2})$$
$$= y^4e^{xy^2}$$
$$f_{xy} = \frac{\partial}{\partial y}(f_x) = \frac{\partial}{\partial y}(y^2e^{xy^2})$$
$$= 2ye^{xy^2} + 2xy^3e^{xy^2}$$
$$= 2ye^{xy^2}(1 + xy^2)$$
$$f_{yx} = \frac{\partial}{\partial x}(f_y) = \frac{\partial}{\partial x}(2xye^{xy^2})$$
$$= 2ye^{xy^2} + 2xy^3e^{xy^2}$$
$$= 2ye^{xy^2}(1 + xy^2)$$

$$f_{yy} = \frac{\partial}{\partial y}(f_y) = \frac{\partial}{\partial y}(2xye^{xy^2})$$
$$= 2xe^{xy^2} + (2xy)(2xy)e^{xy^2}$$
$$= 2xe^{xy^2}(1 + 2xy^2)$$

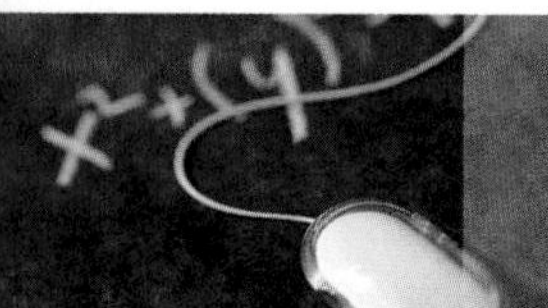

Self-Check Exercises 10.2

1. Compute the first partial derivatives of $f(x, y) = x^3 - 2xy^2 + y^2 - 8$.
2. Find the first partial derivatives of $f(x, y) = x \ln y + ye^x - x^2$ at (0, 1) and interpret your results.
3. Find the second-order partial derivatives of the function of Self-Check Exercise 1.
4. A certain country's production is described by the function

$$f(x, y) = 60x^{1/3}y^{2/3}$$

when x units of labour and y units of capital are used.

a. What is the marginal productivity of labour and the marginal productivity of capital when the amounts expended on labour and capital are 125 units and 8 units, respectively?
b. Should the government encourage capital investment rather than increased expenditure on labour at this time in order to increase the country's productivity?

Solutions to Self-Check Exercises 10.2 can be found on page 607.

10.2 Exercises

In Exercises 1–30, find the first partial derivatives of the function.

1. $f(x, y) = 2x + 3y + 5$
2. $f(x, y) = 2xy$
3. $g(x, y) = 2x^2 + 4y + 1$
4. $f(x, y) = 1 + x^2 + y^2$
5. $f(x, y) = \dfrac{2y}{x^2}$
6. $f(x, y) = \dfrac{x}{1 + y}$
7. $g(u, v) = \dfrac{u - v}{u + v}$
8. $f(x, y) = \dfrac{x^2 - y^2}{x^2 + y^2}$
9. $f(s, t) = (s^2 - st + t^2)^3$
10. $g(s, t) = s^2t + st^{-3}$
11. $f(x, y) = (x^2 + y^2)^{2/3}$
12. $f(x, y) = x\sqrt{1 + y^2}$
13. $f(x, y) = e^{xy+1}$
14. $f(x, y) = (e^x + e^y)^5$
15. $f(x, y) = x \ln y + y \ln x$
16. $f(x, y) = x^2e^{y^2}$
17. $g(u, v) = e^u \ln v$
18. $f(x, y) = \dfrac{e^{xy}}{x + y}$
19. $f(x, y, z) = xyz + xy^2 + yz^2 + zx^2$
20. $g(u, v, w) = \dfrac{2uvw}{u^2 + v^2 + w^2}$
21. $h(r, s, t) = e^{rst}$
22. $f(x, y, z) = xe^{y/z}$
23. $f(x, y) = x \sin(y)$
24. $g(u, v) = \cos(xy)$
25. $f(x, y) = xe^{\sin(y)}$
26. $h(a, b) = \ln(a \cos(b))$
27. $f(x, y) = \arctan\left(\dfrac{x}{y}\right)$
28. $f(x, y) = \tan(1 + x + y + xy)$

29. $h(s, t) = \dfrac{s\sin(t)}{t\cos(s)}$

30. $f(x, y) = \sin^2(x) + \cos^2(y) + 1$

In Exercises 31–46, evaluate the first partial derivatives of the function at the given point.

31. $f(x, y) = x^2y + xy^2$; $(1, 2)$

32. $f(x, y) = x^2 + xy + y^2 + 2x - y$; $(-1, 2)$

33. $f(x, y) = x\sqrt{y} + y^2$; $(2, 1)$

34. $g(x, y) = \sqrt{x^2 + y^2}$; $(3, 4)$

35. $f(x, y) = \dfrac{x}{y}$; $(1, 2)$

36. $f(x, y) = \dfrac{x + y}{x - y}$; $(1, -2)$

37. $f(x, y) = e^{xy}$; $(1, 1)$

38. $f(x, y) = e^x \ln y$; $(0, e)$

39. $f(x, y, z) = x^2yz^3$; $(1, 0, 2)$

40. $f(x, y, z) = x^2y^2 + z^2$; $(1, 1, 2)$

41. $f(x, y) = \sin(xy)$; $(\pi/3, \pi/3)$

42. $g(u, v) = (u - \cos(v))^2$; $(2, \pi/4)$

43. $h(a, b) = e^{\sin(ab)}$; $(\pi/4, 5)$

44. $f(s, t) = \tan(te^s)$; $(\pi/6, \ln(2))$

45. $g(x, y) = \dfrac{\sin(x)\cos(y)}{x + y}$; $(\pi/2, 3\pi/2)$

46. $h(u, v) = (3\tan(2u) + 2\cot(3v))^2$; $(\pi/4, \pi/6)$

In Exercises 47–60, find the second-order partial derivatives of the given function. In each case, show that the mixed partial derivatives f_{xy} and f_{yx} are equal.

47. $f(x, y) = x^2y + xy^3$

48. $f(x, y) = x^3 + x^2y + x + 4$

49. $f(x, y) = x^2 - 2xy + 2y^2 + x - 2y$

50. $f(x, y) = x^3 + x^2y^2 + y^3 + x + y$

51. $f(x, y) = \sqrt{x^2 + y^2}$

52. $f(x, y) = x\sqrt{y} + y\sqrt{x}$

53. $f(x, y) = e^{-x/y}$

54. $f(x, y) = \ln(1 + x^2y^2)$

55. $f(x, y) = \cos(xy)$

56. $f(x, y) = \cos(x + y)$

57. $f(x, y) = x\sin(y)$

58. $f(x, y) = \tan(x + y)$

59. $f(x, y) = x\sec(y)$

60. $f(x, y) = xe^{\tan(y)}$

61. **Wind Chill Factor** A formula used by meteorologists to calculate the wind chill temperature (the temperature that you feel in still air that is the same as the actual temperature when the presence of wind is taken into consideration) is

$$T = f(t, s) = 2.1 + 0.345t - 1.91s^{0.16} + 0.2189ts^{0.16} \quad (s \geq 1)$$

where t is the actual air temperature in °C and s is the wind speed in km/h.

a. What is the wind chill temperature when the actual air temperature is 0°C and the wind speed is 12 km/h?

b. If the temperature is 0°C, by how much approximately will the wind chill temperature change if the wind speed increases from 12 to 13 km/h?

62. **Engine Efficiency** The efficiency of an internal combustion engine is given by

$$E = \left(1 - \frac{v}{V}\right)^{0.4}$$

where V and v are the respective maximum and minimum volumes of air in each cylinder.

a. Show that $\partial E/\partial V > 0$ and interpret your result.

b. Show that $\partial E/\partial v < 0$ and interpret your result.

63. **Volume of a Gas** The volume V (in litres) of a certain mass of gas is related to its pressure P (in millimetres of mercury) and its temperature T (in degrees Kelvin) by the law

$$V = \frac{30.9T}{9}$$

Compute $\partial V/\partial T$ and $\partial V/\partial P$ when $T = 300$ and $P = 800$. Interpret your results.

64. According to the *ideal gas law*, the volume V (in litres) of an ideal gas is related to its pressure P (in pascals) and temperature T (in degrees Kelvin) by the formula

$$V = \frac{kT}{P}$$

where k is a constant. Show that

$$\frac{\partial V}{\partial T} \cdot \frac{\partial T}{\partial P} \cdot \frac{\partial P}{\partial V} = -1$$

65. **Kinetic Energy of a Body** The kinetic energy K of a body of mass m and velocity v is given by

$$K = \frac{1}{2}mv^2$$

Show that $(\partial K/\partial m)(\partial^2 K/\partial v^2) = K$.

Business and Economics Applications

66. **Productivity of a Country** The productivity of a South American country is given by the function

$$f(x, y) = 20x^{3/4}y^{1/4}$$

when x units of labour and y units of capital are used.

a. What is the marginal productivity of labour and the marginal productivity of capital when the amounts expended on labour and capital are 256 units and 16 units, respectively?

b. Should the government encourage capital investment rather than increased expenditure on labour at this time in order to increase the country's productivity?

67. **Productivity of a Country** The productivity of a country in Western Europe is given by the function

$$f(x, y) = 40x^{4/5}y^{1/5}$$

when x units of labour and y units of capital are used.

a. What is the marginal productivity of labour and the marginal productivity of capital when the amounts expended on labour and capital are 32 units and 243 units, respectively?

b. Should the government encourage capital investment rather than increased expenditure on labour at this time in order to increase the country's productivity?

68. **Land Prices** The rectangular region R shown in the accompanying figure represents a city's financial district. The price of land within the district is approximated by the function

$$p(x, y) = 200 - 10\left(x - \frac{1}{2}\right)^2 - 15(y - 1)^2$$

where $p(x, y)$ is the price of land at the point (x, y) in dollars per square metre and x and y are measured in kilometres. Compute

$$\frac{\partial p}{\partial x}(0, 1) \quad \text{and} \quad \frac{\partial p}{\partial y}(0, 1)$$

and interpret your results.

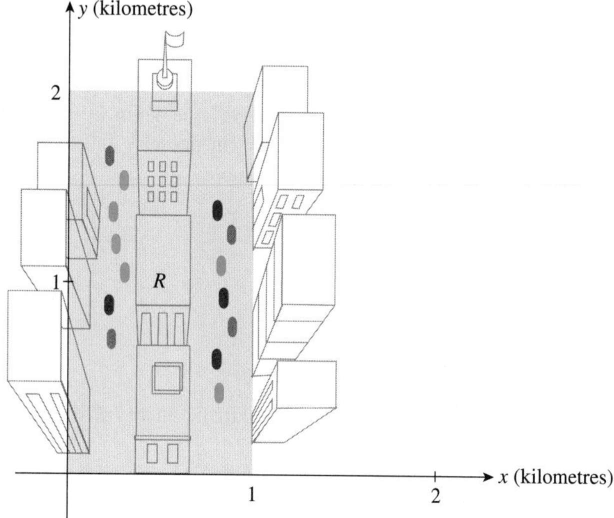

69. **Complementary and Substitute Commodities** In a survey conducted by *Home Entertainment* magazine, it was determined that the demand equation for VCRs is given by

$$x = f(p, q) = 10{,}000 - 10p + 0.2q^2$$

and the demand equation for DVD players is given by

$$y = g(p, q) = 5000 + 0.8p^2 - 20q$$

where p and q denote the unit prices (in dollars) for the VCRs and DVD players, respectively, and x and y denote the number of VCRs and DVD players demanded per week. Determine whether these two products are substitute, complementary, or neither.

70. **Complementary and Substitute Commodities** In a survey it was determined that the demand equation for VCRs is given by

$$x = f(p, q) = 10{,}000 - 10p - e^{0.5q}$$

The demand equation for blank VCR tapes is given by

$$y = g(p, q) = 50{,}000 - 4000q - 10p$$

where p and q denote the unit prices, respectively, and x and y denote the number of VCRs and the number of blank VCR tapes demanded each week. Determine whether these two products are substitute, complementary, or neither.

71. **Complementary and Substitute Commodities** Refer to Exercise 33, Exercises 10.1. Show that the finished and unfinished home furniture manufactured by Country Workshop are substitute commodities.
Hint: Solve the system of equations for x and y in terms of p and q.

72. **Revenue Functions** The total weekly revenue (in dollars) of Country Workshop associated with manufacturing and selling their rolltop desks is given by the function

$$R(x, y) = -0.2x^2 - 0.25y^2 - 0.2xy + 200x + 160y$$

where x denotes the number of finished units and y denotes the number of unfinished units manufactured and sold each week. Compute $\partial R/\partial x$ and $\partial R/\partial y$ when $x = 300$ and $y = 250$. Interpret your results.

73. **Profit Functions** The monthly profit (in dollars) of Bond and Barker Department Store depends on the level of inventory x (in thousands of dollars) and the floor space y (in thousands of square metres) available for display of the merchandise, as given by the equation

$$P(x, y) = -0.02x^2 - 15y^2 + xy + 39x + 25y - 20{,}000$$

Compute $\partial P/\partial x$ and $\partial P/\partial y$ when $x = 4000$ and $y = 150$. Interpret your results. Repeat with $x = 5000$ and $y = 150$.

Biological and Life Sciences Applications

74. **Surface Area of a Human Body** The formula

$$S = 0.007184W^{0.425}H^{0.725}$$

gives the surface area S of a human body (in square metres) in terms of its weight W (in kilograms) and its height H (in centimetres). Compute $\partial S/\partial W$ and $\partial S/\partial H$ when $W = 70$ kg and $H = 180$ cm. Interpret your results.

75. **Comparing Pest Control Strategies** A western province faced with an infestation of aphids has a choice to invest in biological controls in the form of ladybird beetles or to increase the amount of pesticide used. In order to compare the two methods of pest control a test is performed. The experiment is designed so that beetles are deployed in units b that cost as much as one complete spraying s of a test field. If the effect of deploying beetles and spraying on the population P of aphids is modelled by

$$P(b, s) = \frac{6400}{1.2 + 0.77b + 0.65s}$$

compare the impact of additional units of beetles with that of additional spraying at the points $b = 2, s = 0$ and $b = 0, s = 2$. You may assume that the ladybird beetles are not affected by the aphid spray.

76. **Comparing Intervention Strategies** Restoring fish stocks is a substantial concern of the Canadian government. A Maritime province is studying two different intervention strategies.

Either it can spend money to support people who normally make their living by fishing by finding alternate employment, subsidies for living expenses, and so on. This removes the pressure of harvesting on the fish stocks so that they can recover. The second intervention strategy is to seed nutrients offshore that increase phytoplankton and hence food available to the marine food chain. This also helps fish stocks recover. The recovery of the fish population is modelled by

$$R(p, n) = 12{,}300 + 1.2p + 0.66pn - 0.05p^2$$

where p is the number of millions of dollars spent supporting citizens of the province to avoid fishing and n is the number of millions spent on nutrient seeding. If the province's current spending is $p = 10.2$, $n = 1.0$, then increasing which sort of spending will have the largest effect?

77. **Controlling Fungus** The rate at which fungus grows in a storage facility is affected by temperature t and relative humidity h. If the rate of growth is proportional to

$$2.1 + 0.75t + 2.75h + 0.5th - 0.02t^2$$

then which factor has a larger impact at 20°C and relative humidity of 78% ($h = 0.78$)?

In Exercises 78–81, determine whether the statement is true or false. If it is true, explain why it is true. If it is false, give an example to show why it is false.

78. If $f_x(x, y)$ is defined at (a, b), then $f_y(x, y)$ must also be defined at (a, b).

79. If $f_x(a, b) < 0$, then f is decreasing with respect to x near (a, b).

80. If $f_{xy}(x, y)$ and $f_{yx}(x, y)$ are both continuous for all values of x and y, then $f_{xy} = f_{yx}$ for all values of x and y.

81. If both f_{xy} and f_{yx} are defined at (a, b), then f_{xx} and f_{yy} must be defined at (a, b).

Solutions to Self-Check Exercises 10.2

1. $$f_x = \frac{\partial f}{\partial x} = 3x^2 - 2y^2$$

$$f_y = \frac{\partial f}{\partial y} = -2x(2y) + 2y$$

$$= 2y(1 - 2x)$$

2. $$f_x = \ln y + ye^x - 2x; f_y = \frac{x}{y} + e^x$$

In particular,

$$f_x(0, 1) = \ln 1 + 1e^0 - 2(0) = 1$$

$$f_y(0, 1) = \frac{0}{1} + e^0 = 1$$

The results tell us that at the point (0, 1), $f(x, y)$ increases 1 unit for each unit increase in the x-direction, y being kept constant; $f(x, y)$ also increases 1 unit for each unit increase in the y-direction, x being kept constant.

3. From the results of Self-Check Exercise 1,

$$f_x = 3x^2 - 2y^2$$

Therefore,

$$f_{xx} = \frac{\partial}{\partial x}(3x^2 - 2y^2) = 6x$$

$$f_{xy} = \frac{\partial}{\partial y}(3x^2 - 2y^2) = -4y$$

Also, from the results of Self-Check Exercise 1,

$$f_y = 2y(1 - 2x)$$

Thus,

$$f_{yx} = \frac{\partial}{\partial x}[2y(1 - 2x)] = -4y$$

$$f_{yy} = \frac{\partial}{\partial y}[2y(1 - 2x)] = 2(1 - 2x)$$

4. a. The marginal productivity of labour when the amounts expended on labour and capital are x and y units, respectively, is given by

$$f_x(x, y) = 60\left(\frac{1}{3}x^{-2/3}\right)y^{2/3} = 20\left(\frac{y}{x}\right)^{2/3}$$

In particular, the required marginal productivity of labour is given by

$$f_x(125, 8) = 20\left(\frac{8}{125}\right)^{2/3} = 20\left(\frac{4}{25}\right)$$

or 3.2 units/unit increase in labour expenditure, capital expenditure being held constant at 8 units. Next, we compute

$$f_y(x, y) = 60x^{1/3}\left(\frac{2}{3}y^{-1/3}\right) = 40\left(\frac{x}{y}\right)^{1/3}$$

and deduce that the required marginal productivity of capital is given by

$$f_y(125, 8) = 40\left(\frac{125}{8}\right)^{1/3} = 40\left(\frac{5}{2}\right)$$

or 100 units/unit increase in capital expenditure, labour expenditure being held constant at 125 units.

b. The results of part (a) tell us that the government should encourage increased spending on capital rather than on labour.

10.3 Maxima and Minima of Functions of Several Variables

Maxima and Minima

In Chapter 7 we saw that the solution of a problem often reduces to finding the extreme values of a function of one variable. In practice, however, situations also arise in which a problem is solved by finding the absolute maximum or absolute minimum value of a function of two or more variables.

For example, suppose Scandi Company manufactures computer desks in both assembled and unassembled versions. Its profit P is therefore a function of the number of assembled units, x, and the number of unassembled units, y, manufactured and sold per week; that is, $P = f(x, y)$. A question of paramount importance to the manufacturer is, How many assembled and unassembled desks should the company manufacture per week in order to maximize its weekly profit? Mathematically, the problem is solved by finding the values of x and y that will make $f(x, y)$ a maximum.

In this section we will focus our attention on finding the extrema of a function of two variables. As in the case of a function of one variable, we distinguish between the relative (or local) extrema and the absolute extrema of a function of two variables.

Relative Extrema of a Function of Two Variables

Let f be a function defined on a region R containing the point (a, b). Then, f has a **relative maximum** at (a, b) if $f(x, y) \leq f(a, b)$ for all points (x, y) that are sufficiently close to (a, b). The number $f(a, b)$ is called a **relative maximum value.** Similarly, f has a **relative minimum** at (a, b), with **relative minimum value** $f(a, b)$ if $f(x, y) \geq f(a, b)$ for all points (x, y) that are sufficiently close to (a, b).

Loosely speaking, f has a relative maximum at (a, b) if the point $(a, b, f(a, b))$ is the highest point on the graph of f when compared with all nearby points. A similar interpretation holds for a relative minimum.

If the inequalities in this last definition hold for *all* points (x, y) in the domain of f, then f has an **absolute maximum** (or **absolute minimum**) at (a, b) with **absolute maximum value** (or **absolute minimum value**) $f(a, b)$. Figure 10.16 shows the graph of a function with relative maxima at (a, b) and (e, f) and a relative minimum at (c, d). The absolute maximum of f occurs at (e, f) and the absolute minimum of f occurs at (g, h).

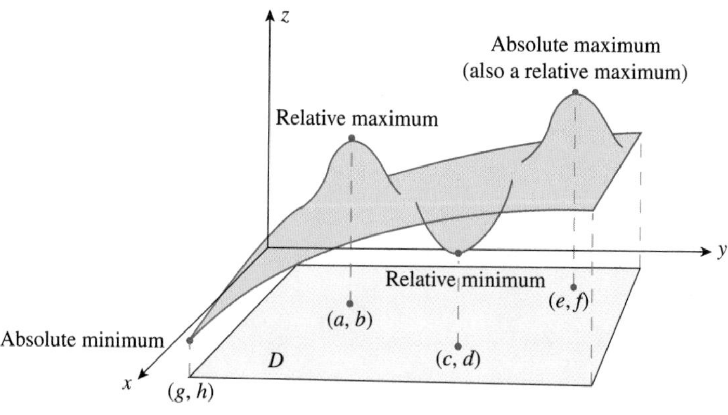

Figure 10.16

Observe that just as in the case of a function of one variable, a relative extremum (relative maximum or relative minimum) may or may not be an absolute extremum.

Now let's turn our attention to the study of relative extrema of a function. Suppose that a differentiable function $f(x, y)$ of two variables has a relative maximum (relative minimum) at a point (a, b) in the domain of f. From Figure 10.17 it is clear that at the point (a, b) the slope of the "tangent lines" to the surface in any direction must be zero. In particular, this implies that both

$$\frac{\partial f}{\partial x}(a, b) \quad \text{and} \quad \frac{\partial f}{\partial y}(a, b)$$

must be zero.

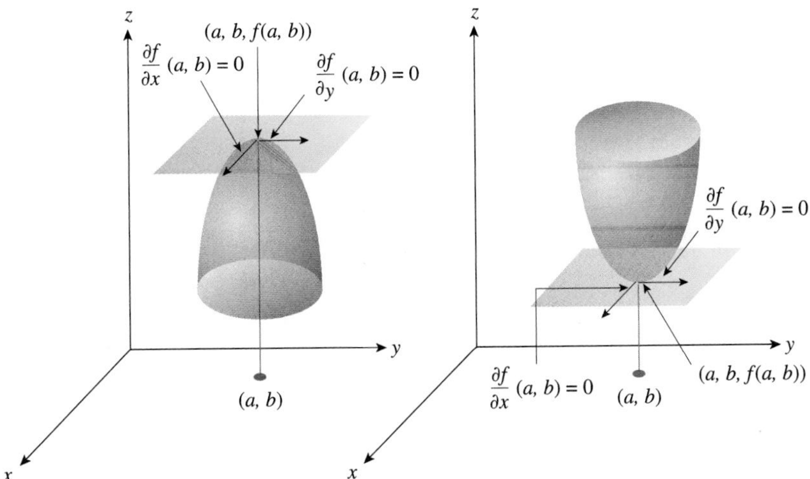

(a) f has a relative maximum at (a, b). **(b)** f has a relative minimum at (a, b).

Figure 10.17

Lest we are tempted to jump to the conclusion that a differentiable function f satisfying both the conditions

$$\frac{\partial f}{\partial x}(a, b) = 0 \quad \text{and} \quad \frac{\partial f}{\partial y}(a, b) = 0$$

at a point (a, b) must have a relative extremum at the point (a, b), let's examine the graph of the function f depicted in Figure 10.18. Here both

$$\frac{\partial f}{\partial x}(a, b) = 0 \quad \text{and} \quad \frac{\partial f}{\partial y}(a, b) = 0$$

but f has neither a relative maximum nor a relative minimum at the point (a, b) because some nearby points are higher and some are lower than the point $(a, b, f(a, b))$. The point $(a, b, f(a, b))$ is called a **saddle point.**

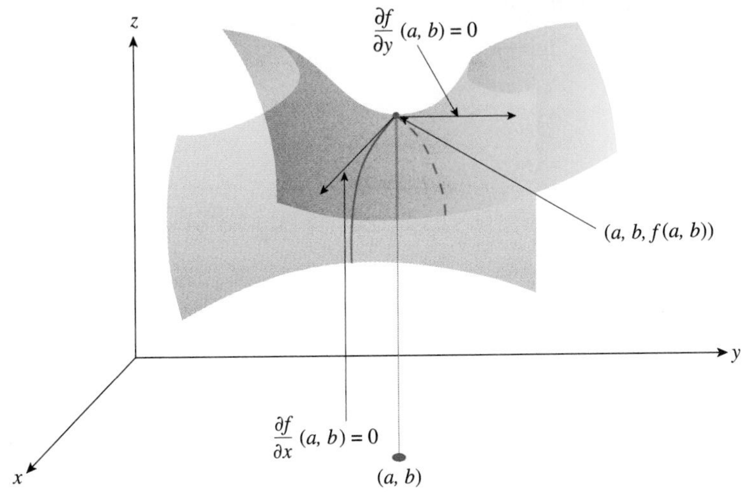

Figure 10.18
The point $(a, b, f(a, b))$ is called a saddle point.

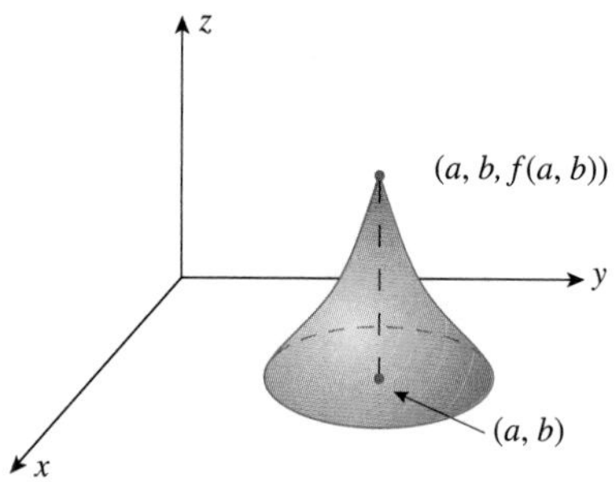

Figure 10.19

f has a relative maximum at (a, b), but neither $\partial f/\partial x$ nor $\partial f/\partial y$ exist at (a, b).

Finally, an examination of the graph of the function f depicted in Figure 10.19 should convince you that f has a relative maximum at the point (a, b). But both $\partial f/\partial x$ and $\partial f/\partial y$ fail to be defined at (a, b).

To summarize, a function f of two variables can only have a relative extremum at a point (a, b) in its domain where either $\partial f/\partial x$ or $\partial f/\partial y$ does not exist, or, in the case they both exist, their values must be zero at (a, b). As in the case of one variable, we refer to a point in the domain of f that *may* give rise to a relative extremum as a critical point. The precise definition follows.

Critical Point of f

A critical point of f is a point (a, b) in the domain of f such that both

$$\frac{\partial f}{\partial x}(a, b) = 0 \qquad \text{and} \qquad \frac{\partial f}{\partial y}(a, b) = 0$$

or at least one of the partial derivatives does not exist.

To determine the nature of a critical point of a function $f(x, y)$ of two variables, we use the second partial derivatives of f. The resulting test, which helps us classify these points, is called the second derivative test and is incorporated in the following procedure for finding and classifying the relative extrema of f.

Determining Relative Extrema

1. Find the critical points of $f(x, y)$ by solving the system of simultaneous equations
$$f_x = 0$$
$$f_y = 0$$
2. The second derivative test: Let
$$D(x, y) = f_{xx}f_{yy} - f_{xy}^2$$
Then,
   a. $D(a, b) > 0$ and $f_{xx}(a, b) < 0$ implies that $f(x, y)$ has a **relative maximum** at the point (a, b).
   b. $D(a, b) > 0$ and $f_{xx}(a, b) > 0$ implies that $f(x, y)$ has a **relative minimum** at the point (a, b).
   c. $D(a, b) < 0$ implies that $f(x, y)$ has neither a relative maximum nor a relative minimum at the point (a, b).
   d. $D(a, b) = 0$ implies that the test is inconclusive, so some other technique must be used to solve the problem.

Example 1

Find the relative extrema of the function

$$f(x, y) = x^2 + y^2$$

Solution

We have

$$f_x = 2x$$
$$f_y = 2y$$

To find the critical point(s) of f, we set $f_x = 0$ and $f_y = 0$ and solve the resulting system of simultaneous equations

$$2x = 0$$
$$2y = 0$$

obtaining $x = 0$, $y = 0$, or $(0, 0)$, as the sole critical point of f. Next, we apply the second derivative test to determine the nature of the critical point $(0, 0)$. We compute

$$f_{xx} = 2, \qquad f_{xy} = 0, \qquad f_{yy} = 2$$

and

$$D(x, y) = f_{xx}f_{yy} - f_{xy}^2 = (2)(2) - 0 = 4$$

In particular, $D(0, 0) = 4$. Since $D(0, 0) > 0$ and $f_{xx}(0, 0) = 2 > 0$, we conclude that $f(x, y)$ has a relative minimum at the point $(0, 0)$. The relative minimum value, 0, also happens to be the absolute minimum of f. The graph of the function f, shown in Figure 10.20, confirms these results.

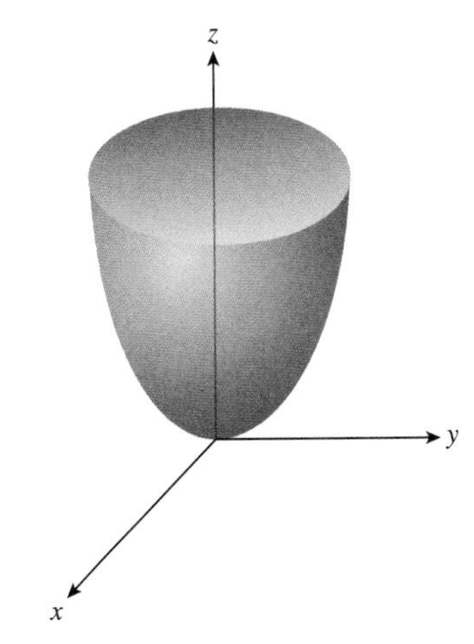

Figure 10.20
The graph of $f(x, y) = x^2 + y^2$.

Example 2

Find the relative extrema of the function

$$f(x, y) = 3x^2 - 4xy + 4y^2 - 4x + 8y + 4$$

Solution

We have

$$\begin{aligned} f_x &= 6x - 4y - 4 \\ f_y &= -4x + 8y + 8 \end{aligned}$$

To find the critical points of f, we set $f_x = 0$ and $f_y = 0$ and solve the resulting system of simultaneous equations

$$\begin{aligned} 6x - 4y &= 4 \\ -4x + 8y &= -8 \end{aligned}$$

Multiplying the first equation by 2 and the second equation by 3, we obtain the equivalent system

$$\begin{aligned} 12x - 8y &= 8 \\ -12x + 24y &= -24 \end{aligned}$$

Adding the two equations gives $16y = -16$, or $y = -1$. We substitute this value for y into either equation in the system to get $x = 0$. Thus, the only critical point of f is the point $(0, -1)$. Next, we apply the second derivative test to determine whether the point $(0, -1)$ gives rise to a relative extremum of f. We compute

$$f_{xx} = 6, \qquad f_{xy} = -4, \qquad f_{yy} = 8$$

and

$$D(x, y) = f_{xx}f_{yy} - f_{xy}^2 = (6)(8) - (-4)^2 = 32$$

Since $D(0, -1) = 32 > 0$ and $f_{xx}(0, -1) = 6 > 0$, we conclude that $f(x, y)$ has a relative minimum at the point $(0, -1)$. The value of $f(x, y)$ at the point $(0, -1)$ is given by

$$f(0, -1) = 3(0)^2 - 4(0)(-1) + 4(-1)^2 - 4(0) + 8(-1)4 = 0$$

GROUP DISCUSSION

Suppose $f(x, y)$ has a relative extremum (relative maximum or relative minimum) at a point (a, b). Let $g(x) = f(x, b)$ and $h(y) = f(a, y)$. Assuming that f and g are differentiable, explain why $g'(a) = 0$ and $h'(b) = 0$. Explain why these results are equivalent to the conditions $f_x(a, b) = 0$ and $f_y(a, b) = 0$.

Example 3

Find the relative extrema of the function

$$f(x, y) = 4y^3 + x^2 - 12y^2 - 36y + 2$$

Solution

To find the critical points of f, we set $f_x = 0$ and $f_y = 0$ simultaneously, obtaining

$$\begin{aligned} f_x &= 2x = 0 \\ f_y &= 12y^2 - 24y - 36 = 0 \end{aligned}$$

The first equation implies that $x = 0$. The second equation implies that

$$\begin{aligned} y^2 - 2y - 3 &= 0 \\ (y + 1)(y - 3) &= 0 \end{aligned}$$

—that is, $y = -1$ or 3. Therefore, there are two critical points of the function f—namely, $(0, -1)$ and $(0, 3)$.

Next, we apply the second derivative test to determine the nature of each of the two critical points. We compute

$$f_{xx} = 2, \qquad f_{xy} = 0, \qquad f_{yy} = 24y - 24 = 24(y - 1)$$

Therefore,

$$D(x, y) = f_{xx}f_{yy} - f_{xy}^2 = 48(y - 1)$$

For the point $(0, -1)$,

$$D(0, -1) = 48(-1 - 1) = -96 < 0$$

Since $D(0, -1) < 0$, we conclude that the point $(0, -1)$ gives a saddle point of f. For the point $(0, 3)$,

$$D(0, 3) = 48(3 - 1) = 96 > 0$$

Since $D(0, 3) > 0$ and $f_{xx}(0, 3) > 0$, we conclude that the function f has a relative minimum at the point $(0, 3)$. Furthermore, since

$$\begin{aligned} f(0, 3) &= 4(3)^3 + (0)^2 - 12(3)^2 - 36(3) + 2 \\ &= -106 \end{aligned}$$

we see that the relative minimum value of f is -106.

GROUP DISCUSSION

1. Refer to the second derivative test. Can the condition $f_{xx}(a, b) < 0$ in part 2a be replaced by the condition $f_{yy}(a, b) > 0$ in part 2b? Explain your answer. How about the condition $f_{xx}(a, b) > 0$ in part 2b?
2. Let $f(x, y) = x^4 + y^4$.
   a. Show that $(0, 0)$ is a critical point of f and that $D(0, 0) = 0$.
   b. Explain why f has a relative (in fact, an absolute minimum at $(0, 0)$. Does this contradict the second derivative test? Explain your answer.

Applications

As in the case of a practical optimization problem involving a function of one variable, the solution to an optimization problem involving a function of several variables calls for finding the *absolute* extremum of the function. Determining the absolute extremum of a function of several variables is more difficult than merely finding the relative extrema of the function. However, in many situations, the absolute extremum of a function actually coincides with the largest relative extremum of the function that occurs in the interior of its domain. We assume that the problems considered here belong to this category. Furthermore, the existence of the absolute extremum (solution) of a practical problem is often deduced from the geometric or physical nature of the problem.

Example 4

Maximizing Profits The total weekly revenue (in dollars) that Acrosonic realizes in producing and selling its bookshelf loudspeaker systems is given by

$$R(x, y) = -\frac{1}{4}x^2 - \frac{3}{8}y^2 - \frac{1}{4}xy + 300x + 240y$$

where x denotes the number of fully assembled units and y denotes the number of kits produced and sold each week. The total weekly cost attributable to the production of these loudspeakers is

$$C(x, y) = 180x + 140y + 5000$$

dollars, where x and y have the same meaning as before. Determine how many assembled units and how many kits Acrosonic should produce per week to maximize its profit.

Solution

The contribution to Acrosonic's weekly profit stemming from the production and sale of the bookshelf loudspeaker systems is given by

$$\begin{aligned} P(x, y) &= R(x, y) - C(x, y) \\ &= \left(-\frac{1}{4}x^2 - \frac{3}{8}y^2 - \frac{1}{4}xy + 300x + 240y\right) - (180x + 140y + 5000) \\ &= -\frac{1}{4}x^2 - \frac{3}{8}y^2 - \frac{1}{4}xy + 120x + 100y - 5000 \end{aligned}$$

To find the relative maximum of the profit function $P(x, y)$, we first locate the critical point(s) of P. Setting $P_x(x, y)$ and $P_y(x, y)$ equal to zero, we obtain

$$\begin{aligned} P_x &= -\frac{1}{2}x - \frac{1}{4}y + 120 = 0 \\ P_y &= -\frac{3}{4}y - \frac{1}{4}x + 100 = 0 \end{aligned}$$

Solving the first of these equations for y yields

$$y = -2x + 480$$

which, upon substitution into the second equation, yields

$$\begin{aligned} -\frac{3}{4}(-2x + 480) - \frac{1}{4}x + 100 &= 0 \\ 6x - 1440 - x + 400 &= 0 \\ x &= 208 \end{aligned}$$

We substitute this value of x into the equation $y = -2x + 480$ to get

$$y = 64$$

Therefore, the function P has the sole critical point (208, 64). To show that the point (208, 64) is a solution to our problem, we use the second derivative test. We compute

$$P_{xx} = -\frac{1}{2}, \qquad P_{xy} = -\frac{1}{4}, \qquad P_{yy} = -\frac{3}{4}$$

So,

$$D(x, y) = \left(-\frac{1}{2}\right)\left(-\frac{3}{4}\right) - \left(-\frac{1}{4}\right)^2 = \frac{3}{8} - \frac{1}{16} = \frac{5}{16}$$

In particular, $D(208, 64) = 5/16 > 0$.

Since $D(208, 64) > 0$ and $P_{xx}(208, 64) < 0$, the point (208, 64) yields a relative maximum of P. This relative maximum is also the absolute maximum of P. We conclude that Acrosonic can maximize its weekly profit by manufacturing 208 assembled units and 64 kits of their bookshelf loudspeaker systems. The maximum weekly profit realizable from the production and sale of these loudspeaker systems is given by

$$\begin{aligned} P(208, 64) &= -\frac{1}{4}(208)^2 - \frac{3}{8}(64)^2 - \frac{1}{4}(208)(64) \\ &\quad + 120(208) + 100(64) - 5000 \\ &= 10{,}680 \end{aligned}$$

or $10,680.

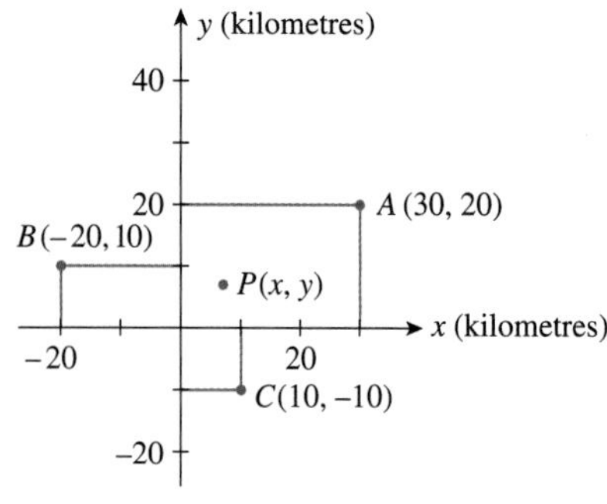

Figure 10.21
Locating a site for a television relay station

Example 5

Locating a Television Relay Station Site A television relay station will serve towns A, B, and C, whose relative locations are shown in Figure 10.21. Determine a site for the location of the station if the sum of the squares of the distances from each town to the site is minimized.

Solution

Suppose the required site is located at the point $P(x, y)$. With the aid of the distance formula, we find that the square of the distance from town A to the site is

$$(x - 30)^2 + (y - 20)^2$$

The respective distances from towns B and C to the site are found in a similar manner, so the sum of the squares of the distances from each town to the site is given by

$$\begin{aligned} f(x, y) &= (x - 30)^2 + (y - 20)^2 + (x + 20)^2 \\ &\quad + (y - 10)^2 + (x - 10)^2 + (y + 10)^2 \end{aligned}$$

To find the relative minimum of $f(x, y)$, we first find the critical point(s) of f. Using the chain rule to find $f_x(x, y)$ and $f_y(x, y)$ and setting each equal to zero, we obtain

$$f_x = 2(x - 30) + 2(x + 20) + 2(x - 10) = 6x - 40 = 0$$
$$f_y = 2(y - 20) + 2(y - 10) + 2(y + 10) = 6y - 40 - 0$$

from which we deduce that (20/3, 20/3) is the sole critical point of f. Since

$$f_{xx} = 6, \qquad f_{xy} = 0, \qquad f_{yy} = 6$$

we have

$$D(x, y) = f_{xx}f_{yy} - f_{xy}^2 = (6)(6) - 0 = 36$$

Since $D(20/3, 20/3) > 0$ and $f_{xx}(20/3, 20/3) > 0$, we conclude that the point (20/3, 20/3) yields a relative minimum of f. Thus, the required site has coordinates $x = 20/3$ and $y = 20/3$.

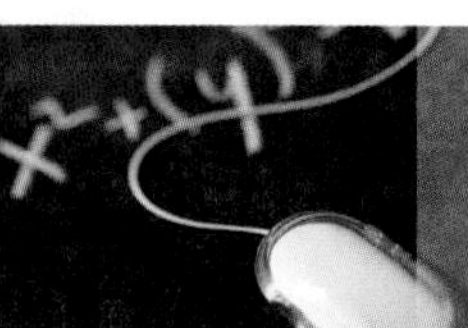

Self-Check Exercises 10.3

1. Let $f(x, y) = 2x^2 + 3y^2 - 4xy + 4x - 2y + 3$.
   a. Find the critical point of f.
   b. Use the second derivative test to classify the nature of the critical point.
   c. Find the relative extremum of f, if it exists.
2. Robertson Controls manufactures two basic models of setback thermostats: a standard mechanical thermostat and a deluxe electronic thermostat. Robertson's monthly revenue (in hundreds of dollars) is

$$R(x, y) = -\frac{1}{8}x^2 - \frac{1}{2}y^2 - \frac{1}{4}xy + 20x + 60y$$

where x (in units of a hundred) denotes the number of mechanical thermostats manufactured and y (in units of a hundred) denotes the number of electronic thermostats manufactured each month. The total monthly cost incurred in producing these thermostats is

$$C(x, y) = 7x + 20y + 280$$

hundred dollars. Find how many thermostats of each model Robertson should manufacture each month in order to maximize its profits. What is the maximum profit?

Solutions to Self-Check Exercises 10.3 can be found on page 617

10.3 Exercises

In Exercises 1–25, find the critical point(s) of the function. Then use the second derivative test to classify the nature of each point, if possible. Finally, determine the relative extrema of the function.

1. $f(x, y) = 1 - 2x^2 - 3y^2$
2. $f(x, y) = x^2 - xy + y^2 + 1$
3. $f(x, y) = x^2 - y^2 - 2x + 4y + 1$
4. $f(x, y) = 2x^2 + y^2 - 4x + 6y + 3$
5. $f(x, y) = x^2 + 2xy + 2y^2 - 4x + 8y - 1$
6. $f(x, y) = x^2 - 4xy + 2y^2 + 4x + 8y - 1$
7. $f(x, y) = 2x^3 + y^2 - 9x^2 - 4y + 12x - 2$
8. $f(x, y) = 2x^3 + y^2 - 6x^2 - 4y + 12x - 2$
9. $f(x, y) = x^3 + y^2 - 2xy + 7x - 8y + 4$
10. $f(x, y) = 2y^3 - 3y^2 - 12y + 2x^2 - 6x + 2$
11. $f(x, y) = x^3 - 3xy + y^3 - 2$
12. $f(x, y) = x^3 - 2xy + y^2 + 5$
13. $f(x, y) = xy + \frac{4}{x} + \frac{2}{y}$
14. $f(x, y) = \frac{x}{y^2} + xy$
15. $f(x, y) = x^2 - e^{y^2}$
16. $f(x, y) = e^{x^2 - y^2}$
17. $f(x, y) = e^{x^2 + y^2}$
18. $f(x, y) = e^{xy}$
19. $f(x, y) = \ln(1 + x^2 + y^2)$
20. $f(x, y) = xy + \ln x + 2y^2$
21. $f(x, y) = \cos\left(\frac{1}{x^2 + y^2}\right)$
22. $f(x, y) = \sin(x)\cos(y); 0 \le x, y \le 2\pi$
23. $f(x, y) = e^{x^2 + \sin(y)}; 0 \le y \le 2\pi$
24. $f(x, y) = \sin(xy); 0 \le x, y \le 2\pi$
25. $f(x, y) = \arctan(xy + 1)$

26. **Determining the Optimal Site** An auxiliary electric power station will serve three communities, A, B, and C, whose relative locations are shown in the accompanying figure. Determine where the power station should be located if the sum of the squares of the distances from each community to the site is minimized.

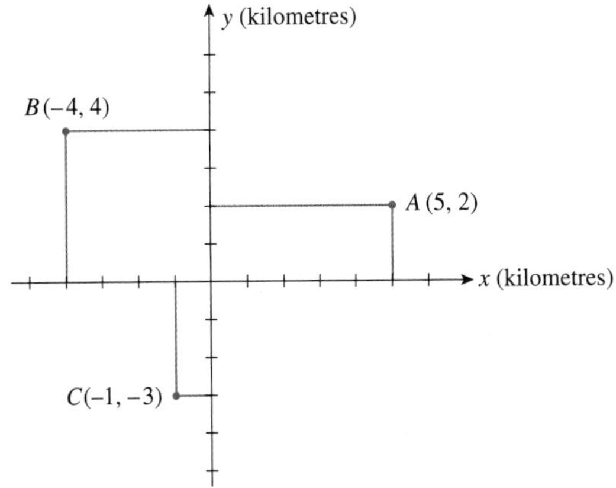

27. **Packaging** An open rectangular box having a volume of 108 cm^3 is to be constructed from a tin sheet. Find the dimensions of such a box if the amount of material used in its construction is to be minimal.
Hint: Let the dimensions of the box be x cm by y cm by z cm. Then, $xyz = 108$ and the amount of material used is given by $S = xy + 2yz + 2xz$. Show that
$$S = f(x, y) = xy + \frac{216}{x} + \frac{216}{y}$$
Minimize $f(x, y)$.

28. **Packaging** An open rectangular box having a surface area of 300 cm^2 is to be constructed from a tin sheet. Find the dimensions of the box if the volume of the box is to be as large as possible. What is the maximum volume?
Hint: Let the dimensions of the box be $x \times y \times z$ (see the figure that follows). Then the surface area is $xy + 2xz + 2yz$, and its volume is xyz.

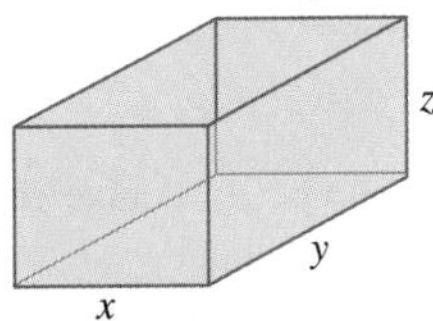

29. **Packaging** Postal regulations specify that the combined length and girth of a parcel sent by parcel post may not exceed 108 cm. Find the dimensions of the rectangular package that would have the greatest possible volume under these regulations.
Hint: Let the dimensions of the box be x cm by y cm by z cm (see the figure below). Then, $2x + 2z + y = 108$, and the volume $V = xyz$. Show that
$$V = f(x, y) = 108xz - 2x^2z - 2xz^2$$
Maximize $f(x, z)$.

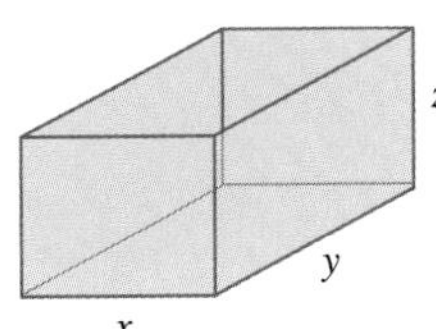

30. Packaging An open box having a volume of 48 cm^3 is to be constructed. If the box is to include a partition that is parallel to a side of the box, as shown in the figure, and the amount of material used is to be minimal, what should be the dimensions of the box?

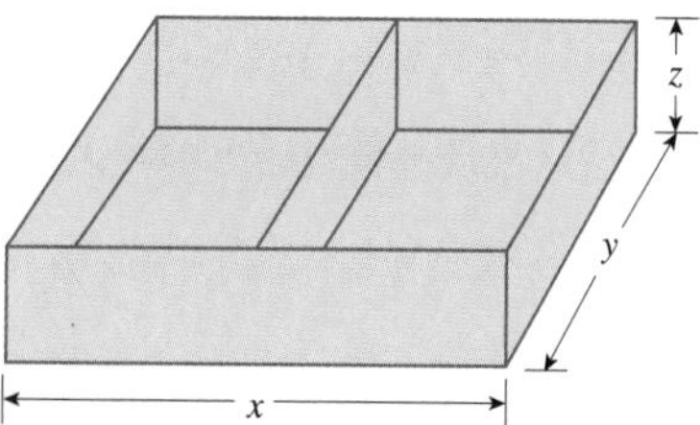

Business and Economics Applications

31. Maximizing Profit The total weekly revenue (in dollars) of the Country Workshop realized in manufacturing and selling its rolltop desks is given by

$$R(x, y) = -0.2x^2 - 0.25y^2 - 0.2xy + 200x + 160y$$

where x denotes the number of finished units and y denotes the number of unfinished units manufactured and sold each week. The total weekly cost attributable to the manufacture of these desks is given by

$$C(x, y) = 100x + 70y + 4000$$

dollars. Determine how many finished units and how many unfinished units the company should manufacture each week in order to maximize its profit. What is the maximum profit realizable?

32. Maximizing Profit The total daily revenue (in dollars) that Weston Publishing realizes in publishing and selling its English-language dictionaries is given by

$$R(x, y) = -0.005x^2 - 0.003y^2 - 0.002xy + 20x + 15y$$

where x denotes the number of deluxe copies and y denotes the number of standard copies published and sold daily. The total daily cost of publishing these dictionaries is given by

$$C(x, y) = 6x + 3y + 200$$

dollars. Determine how many deluxe copies and how many standard copies Weston should publish each day to maximize its profits. What is the maximum profit realizable?

33. Maximum Price The rectangular region R shown in the accompanying figure represents the financial district of a city. The price of land within the district is approximated by the function

$$p(x, y) = 200 - 10\left(x - \frac{1}{2}\right)^2 - 15(y - 1)^2$$

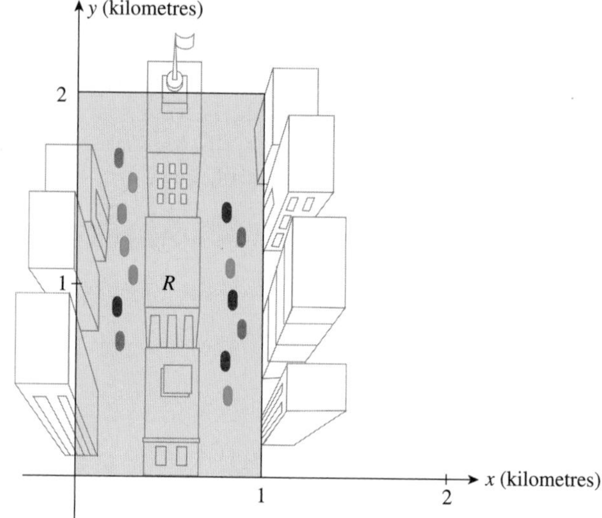

where $p(x, y)$ is the price of land at the point (x, y) in dollars/square metre and x and y are measured in kilometres. At what point within the financial district is the price of land highest?

34. Maximizing Profit C&G Imports imports two brands of white wine, one from Germany and the other from Italy. The German wine costs \$4/bottle, and the Italian wine costs \$3/bottle. It has been estimated that if the German wine retails at p dollars/bottle and the Italian wine is sold for q dollars/bottle, then

$$2000 - 150p + 100q$$

bottles of the German wine and

$$1000 + 80p - 120q$$

bottles of the Italian wine will be sold each week. Determine the unit price for each brand that will allow C&G to realize the largest possible weekly profit.

35. Minimizing Heating and Cooling Costs A building in the shape of a rectangular box is to have a volume of 12,000 m^3 (see the figure). It is estimated that the annual heating and cooling costs will be \$2/square metre for the top, \$4/square metre for the front and back, and \$3/square metre for the sides. Find the dimensions of the building that will result in a minimal annual heating and cooling cost. What is the minimal annual heating and cooling cost?

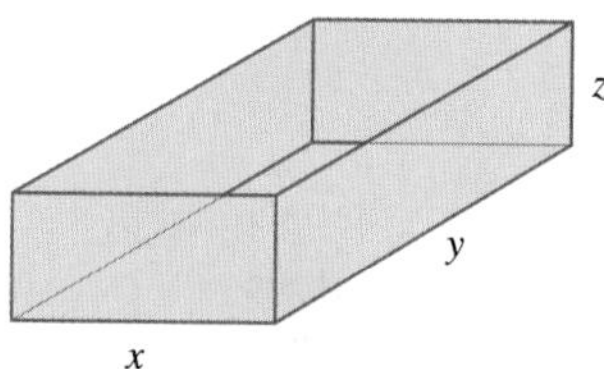

Biological and Life Sciences Applications

36. Optimizing Protein Production A bioreactor with a mixture of yeast and algae is to be used to produce protein feed stocks for animals on a proposed lunar base. Protein production at harvest depends on the initial amount of yeast x and algae y added, in kilograms. If the production is modelled by

$$Q(x, y) = \frac{3.47}{x^2 + y^2 - 4.2x - 3.8y + 41.7}$$

then what initial quantity of yeast and algae yield the maximum production of protein?

37. **Optimizing Weight Gain** Hogs being grown on a farm require a particular mixture of protein and fat to gain weight quickly. The growth rate as a function of the number of kilograms of fat f and protein p in the feed mixture per hog is found to be modelled well by

$$G(f, p) = \frac{3.77}{f^2 + p^2 - 4.6f - 6.1p + 114}$$

Find the values of f and p that optimize the hogs' weight gain.

38. **Maximizing Protein** The number of micrograms per litre of blood of a marker protein that is used as an effectiveness indicator for a two-drug cancer therapy depends on the dose in micrograms of each of the two drugs. If u is the dose of the first drug and v is the dose of the second drug in micrograms, then the amount of the marker protein is modelled by the formula

$$A(u, v) = 31.7 + 5.4u + 9.2v - u^2 - v^2 \mu\text{g/L}$$

whenever the quantity $A(u, v) > 0$. Find the amount of the two drugs that maximizes the amount of the marker protein.

In Exercises 39 and 40, determine whether the statement is true or false. If it is true, explain why it is true. If it is false, give an example to show why it is false.

39. If $f_x(a, b) = 0$ and $f_y(a, b) = 0$, then f must have a relative extremum at (a, b).

40. If (a, b) is a critical point of f and both the conditions $f_{xx}(a, b) < 0$ and $f_{yy}(a, b) < 0$ hold, then f has a relative maximum at (a, b).

Solutions to Self-Check Exercises 10.3

1. **a.** To find the critical point(s) of f, we solve the system of equations

$$f_x = 4x - 4y + 4 = 0$$
$$f_y = -4x + 6y - 2 = 0$$

obtaining $x = -2$ and $y = -1$. Thus, the only critical point of f is the point $(-2, -1)$.

b. We have $f_{xx} = 4, f_{xy} = -4$, and $f_{yy} = 6$, so

$$\begin{aligned} D(x, y) &= f_{xx} f_{yy} - f_{xy}^2 \\ &= (4)(6) - (-4)^2 = 8 \end{aligned}$$

Since $D(-2, -1) > 0$ and $f_{xx}(-2, -1) > 0$, we conclude that f has a relative minimum at the point $(-2, -1)$.

c. The relative minimum value of $f(x. y)$ at the point $(-2, -1)$ is

$$\begin{aligned} f(-2, -1) &= 2(-2)^2 + 3(-1)^2 - 4(-2)(-1) \\ &\quad + 4(-2) - 2(-1) + 3 \\ &= 0 \end{aligned}$$

2. Robertson's monthly profit is

$$\begin{aligned} P(x, y) &= R(x, y) - C(x, y) \\ &= \left(-\frac{1}{8}x^2 - \frac{1}{2}y^2 - \frac{1}{4}xy + 20x + 60y\right) - (7x + 20y + 280) \\ &= -\frac{1}{8}x^2 - \frac{1}{2}y^2 - \frac{1}{4}xy + 13x + 40y - 280 \end{aligned}$$

The critical point of P is found by solving the system

$$P_x = -\frac{1}{4}x - \frac{1}{4}y + 13 = 0$$
$$P_y = -\frac{1}{4}x - \quad y + 40 = 0$$

Solutions to Self-Check Exercises 10.3, *continued*

giving $x = 16$ and $y = 36$. Thus, (16, 36) is the critical point of P. Next,

$$P_{xx} = -\frac{1}{4}, \quad P_{xy} = -\frac{1}{4}, \quad P_{yy} = -1$$

and

$$D(x, y) = f_{xx}f_{yy} - f_{xy}^2$$
$$= \left(-\frac{1}{4}\right)(-1) - \left(-\frac{1}{4}\right)^2 = \frac{3}{16}$$

Since $D(16, 36) > 0$ and $P_{xx}(16,36) < 0$, the point (16, 36) yields a relative maximum of P. We conclude that the monthly profit is maximized by manufacturing 1600 mechanical and 3600 electronic setback thermostats each month. The maximum monthly profit realizable is

$$P(16, 36) = -\frac{1}{8}(16)^2 - \frac{1}{2}(36)^2 - \frac{1}{4}(16)(36)$$
$$+ 13(16) + 40(36) - 280$$
$$= 544$$

or \$54,400.

10.4 The Method of Least Squares

The Method of Least Squares

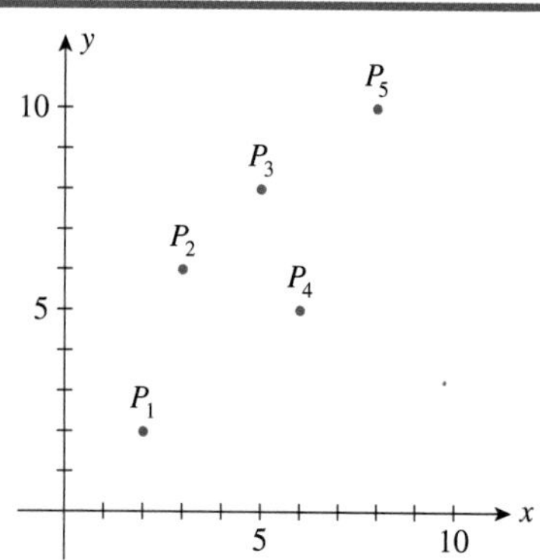

Figure 10.22
A scatter diagram

In Section 1.4, Example 10, we saw how a linear equation can be used to approximate the sales trend for a local sporting goods store. As we saw there, one use of a trend line is to predict a store's future sales. Recall that we obtained the line by requiring that it pass through two data points, the rationale being that such a line seems to *fit* the data reasonably well.

In this section we describe a general method, known as the method of least squares, for determining a straight line that, in some sense, *best* fits a set of data points when the points are scattered about a straight line. To illustrate the principle behind the method of least squares, suppose, for simplicity, that we are given five data points,

$$P_1(x_1, y_1), \quad P_2(x_2, y_2), \quad P_3(x_3, y_3) \quad P_4(x_4, y_4), \quad P_5(x_5, y_5)$$

that describe the relationship between the two variables x and y. By plotting these data points, we obtain a graph called a scatter diagram (Figure 10.22).

If we try to fit a straight line to these data points, the line will miss the first, second, third, fourth, and fifth data points by the amounts d_1, d_2, d_3, d_4, and d_5, respectively (Figure 10.23).

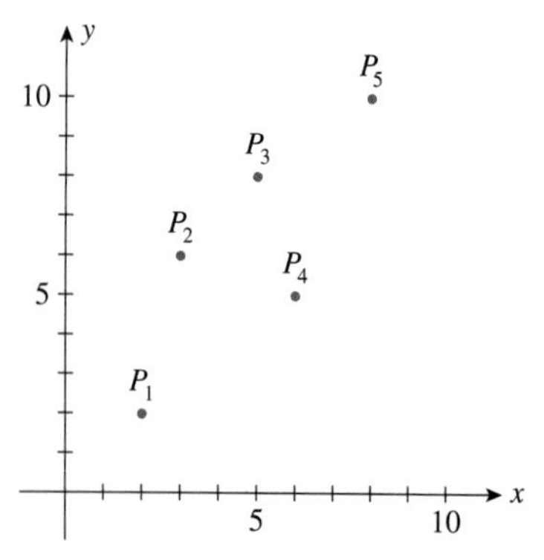

Figure 10.23
The approximating line misses each point by the amounts $d_1, d_2, \ldots, d_5$.

The principle of least squares states that the straight line L that fits the data points best is the one chosen by requiring that the sum of the squares of $d_1, d_2, \ldots, d_5$—that is,

$$d_1^2 + d_2^2 + d_3^2 + d_4^2 + d_5^2$$

be made as small as possible. If we think of the amount d_1 as the error made when the value y_1 is approximated by the corresponding value of y lying on the straight line L, and d_2 as the error made when the value y_2 is approximated by the corresponding value of y, and so on, then it can be seen that the least-squares criterion calls for minimizing the sum of the squares of the errors. The line L obtained in this manner is called the least-squares line, or regression line.

To find a method for computing the regression line L, suppose L has representation $y = f(x) = mx + b$, where m and b are to be determined. Observe that

$$\begin{aligned} d_1^2 &+ d_2^2 + d_3^2 + d_4^2 + d_5^2 \\ &= [f(x_1) - y_1]^2 + [f(x_2) - y_2]^2 + [f(x_3) - y_3]^2 \\ &\quad + [f(x_4) - y_4]^2 + [f(x_5) - y_5]^2 \\ &= (mx_1 + b - y_1)^2 + (mx_2 + b - y_2)^2 + (mx_3 + b - y_3)^2 \\ &\quad + (mx_4 + b - y_4)^2 + (mx_5 + b - y_5)^2 \end{aligned}$$

and may be viewed as a function of the two variables m and b. Thus, the least-squares criterion is equivalent to minimizing the function

$$\begin{aligned} f(m, b) &= (mx_1 + b - y_1)^2 + (mx_2 + b - y_2)^2 + (mx_3 + b - y_3)^2 \\ &\quad + (mx_4 + b - y_4)^2 + (mx_5 + b - y_5)^2 \end{aligned}$$

with respect to m and b. Using the chain rule, we compute

$$\begin{aligned} \frac{\partial f}{\partial m} &= 2(mx_1 + b - y_1)x_1 + 2(mx_2 + b - y_2)x_2 + 2(mx_3 + b - y_3)x_3 \\ &\quad + 2(mx_4 + b - y_4)x_4 + 2(mx_5 + b - y_5)x_5 \\ &= 2[mx_1^2 + bx_1 - x_1y_1 + mx_2^2 + bx_2 - x_2y_2 + mx_3^2 + bx_3 - x_3y_3 \\ &\quad + mx_4^2 + bx_4 - x_4y_4 + mx_5^2 + bx_5 - x_5y_5] \\ &= 2[(x_1^2 + x_2^2 + x_3^2 + x_4^2 + x_5^2)m + (x_1 + x_2 + x_3 + x_4 + x_5)b \\ &\quad - (x_1y_1 + x_2y_2 + x_3y_3 + x_4y_4 + x_5y_5)] \end{aligned}$$

and

$$\begin{aligned} \frac{\partial f}{\partial b} &= 2(mx_1 + b - y_1) + 2(mx_2 + b - y_2) + 2(mx_3 + b - y_3) \\ &\quad + 2(mx_4 + b - y_4) + 2(mx_5 + b - y_5) \\ &= 2[(x_1 + x_2 + x_3 + x_4 + x_5)m + 5b - (y_1 + y_2 + y_3 + y_4 + y_5)] \end{aligned}$$

Setting

$$\frac{\partial f}{\partial m} = 0 \quad \text{and} \quad \frac{\partial f}{\partial b} = 0$$

gives

$$\begin{aligned} (x_1^2 &+ x_2^2 + x_3^2 + x_4^2 + x_5^2)m + (x_1 + x_2 + x_3 + x_4 + x_5)b \\ &= x_1y_1 + x_2y_2 + x_3y_3 + x_4y_4 + x_5y_5 \end{aligned}$$

and

$$(x_1 + x_2 + x_3 + x_4 + x_5)m + 5b = y_1 + y_2 + y_3 + y_4 + y_5$$

Solving these two simultaneous equations for m and b then leads to an equation $y = mx + b$ of a straight line.

Before looking at an example, we state a more general result whose derivation is identical to the special case involving the five data points just discussed.

The Method of Least Squares

Suppose we are given n data points:

$$P_1(x_1, y_1),\quad P_2(x_2, y_2),\quad P_3(x_3, y_3), \ldots, P_n(x_n, y_n)$$

Then, the least-squares (regression) line for the data is given by the linear equation

$$y = f(x) = mx + b$$

where the constants m and b satisfy the equations

$$(x_1^2 + x_2^2 + x_3^2 + \cdots + x_n^2)m + (x_1 + x_2 + x_3 + \cdots + x_n)b$$
$$= x_1y_1 + x_2y_2 + x_3y_3 + \cdots + x_ny_n \qquad (4)$$

and

$$(x_1 + x_2 + x_3 + \cdots + x_n)m + nb$$
$$= y_1 + y_2 + y_3 + \cdots + y_n \qquad (5)$$

simultaneously. Equations (4) and (5) are called **normal equations.**

Example 1

Find an equation of the least-squares line for the data

$$P_1(1, 1),\quad P_2(2, 3),\quad P_3(3, 4),\quad P_4(4, 3),\quad P_5(5, 6)$$

Solution

Here, we have $n = 5$ and

$$x_1 = 1,\quad x_2 = 2,\quad x_3 = 3,\quad x_4 = 4,\quad x_5 = 5$$
$$y_1 = 1,\quad y_2 = 3,\quad y_3 = 4,\quad y_4 = 3,\quad y_5 = 6$$

so Equation (4) becomes

$$(1 + 4 + 9 + 16 + 25)m + (1 + 2 + 3 + 4 + 5)b = 1 + 6 + 12 + 12 + 30$$

or

$$55m + 15b = 61 \qquad (6)$$

and (5) becomes

$$(1 + 2 + 3 + 4 + 5)m + 5b = 1 + 3 + 4 + 3 + 6$$

or

$$15m + 5b = 17 \qquad (7)$$

Solving Equation (7) for b gives

$$b = -3m + \frac{17}{5} \qquad (8)$$

which, upon substitution into (6), gives

$$15\left(-3m + \frac{17}{5}\right) + 55m = 61$$
$$-45m + 51 + 55m = 61$$
$$10m = 10$$
$$m = 1$$

Substituting this value of m into (8) gives

$$b = -3 + \frac{17}{5} = \frac{2}{5} = 0.4$$

Therefore, the required least-squares line is

$$y = x + 0.4$$

The scatter diagram and the regression line are shown in Figure 10.24.

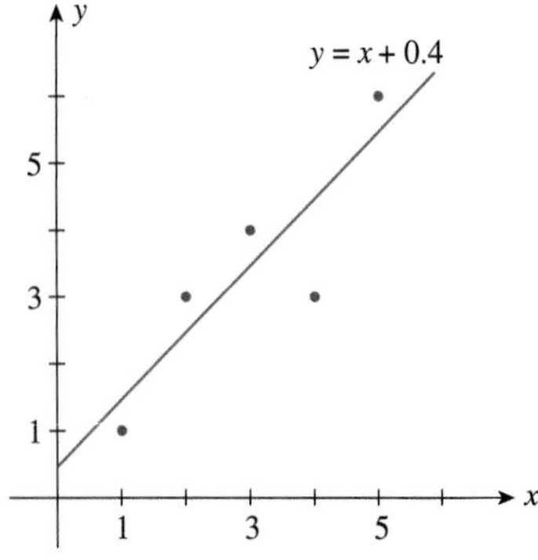

Figure 10.24
The scatter diagram and the least-squares line $y = x + 0.4$

Example 2

Advertising Expense and a Firm's Profit The proprietor of Leisure Travel Service compiled the following data relating the firm's annual profit to its annual advertising expenditure (both measured in thousands of dollars).

Annual Advertising Expenditure, *x*	12	14	17	21	26	30
Annual Profit, *y*	60	70	90	100	100	120

a. Determine an equation of the least-squares line for these data.
b. Draw a scatter diagram and the least-squares line for these data.
c. Use the result obtained in part (a) to predict Leisure Travel's annual profit if the annual advertising budget is $20,000.

Solution

a. The calculations required for obtaining the normal equations may be summarized as follows:

	x	y	x^2	xy
	12	60	144	720
	14	70	196	980
	17	90	289	1,530
	21	100	441	2,100
	26	100	676	2,600
	30	120	900	3,600
Sum	120	540	2,646	11,530

The normal equations are

$$6b + 120m = 540 \quad (9)$$
$$120b + 2646m = 11{,}530 \quad (10)$$

Solving Equation (9) for b gives

$$b = -20m + 90 \quad (11)$$

which, upon substitution into Equation (10), gives

$$\begin{aligned} 120(-20m + 90) + 2646m &= 11{,}530 \\ -2400m + 10{,}800 + 2646m &= 11{,}530 \\ 246m &= 730 \\ m &\approx 2.97 \end{aligned}$$

Substituting this value of m into Equation (11) gives

$$b = -20(2.97) + 90 = 30.6$$

Therefore, the required least-squares line is given by

$$y = f(x) = 2.97x + 30.6$$

b. The scatter diagram and the least-squares line are shown in Figure 10.25.
c. Leisure Travel's predicted annual profit corresponding to an annual budget of $20,000 is given by

$$\begin{aligned} f(20) &= 2.97(20) + 30.6 \\ &= 90 \end{aligned}$$

or $90,000.

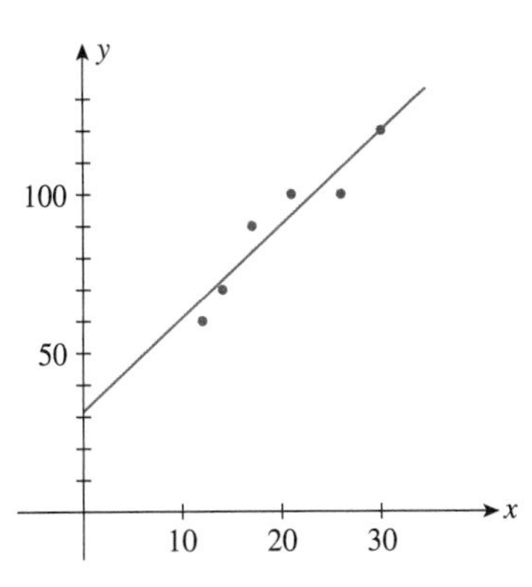

Figure 10.25
The scatter diagram and the least-squares line $y = 2.97x + 30.6$

Example 3

Maximizing Profit A market research study conducted for Century Communications provided the following data based on the projected monthly sales x (in thousands) of Century's DVD version of a box-office hit adventure movie with a proposed wholesale unit price of p dollars.

p	38	36	34.5	30	28.5
x	2.2	5.4	7.0	11.5	14.6

a. Find the demand equation if the demand curve is the least-squares line for these data.

b. The total monthly cost function associated with producing and distributing the DVD movies is given by

$$C(x) = 4x + 25$$

where x denotes the number of discs (in thousands) produced and sold and $C(x)$ is in thousands of dollars. Determine the unit wholesale price that will maximize Century's monthly profit.

Solution

a. The calculations required for obtaining the normal equations may be summarized as follows:

	x	p	x^2	xp
	2.2	38	4.84	83.6
	5.4	36	29.16	194.4
	7.0	34.5	49	241.5
	11.5	30	132.25	345
	14.6	28.5	213.16	416.1
Sum	40.7	167	428.41	1280.6

The normal equations are

$$5b + 40.7m = 167$$

$$40.7b + 428.41m = 1280.6$$

Solving this system of linear equations simultaneously, we find that

$$m \approx -0.81 \quad \text{and} \quad b \approx 39.99$$

Therefore, the required least-squares line is given by

$$p = f(x) = -0.81x + 39.99$$

which is the required demand equation, provided $0 \leq x \leq 49.37$.

b. The total revenue function in this case is given by

$$R(x) = xp = -0.81x^2 + 39.99x$$

and since the total cost function is

$$C(x) = 4x + 25$$

we see that the profit function is

$$\begin{aligned} P(x) &= -0.81x^2 + 39.99x - (4x + 25) \\ &= -0.81x^2 + 35.99x - 25 \end{aligned}$$

To find the absolute maximum of $P(x)$ over the closed interval [0, 49.37], we compute

$$P'(x) = -1.62x + 35.99$$

Since $P'(x) = 0$, we find $x \approx 22.22$ as the only critical point of P. Finally, from the table

x	0	22.22	49.37
$P(x)$	−25	374.78	−222.47

we see that the optimal wholesale price is \$22.22 per disc.

Example 4

Estimating Pesticide Impact Different amounts of pesticide are applied to test fields and then the fields are checked for the number of moth caterpillars per hectare. The results are given below.

Kilograms of Pesticide/Hectare	5	10	15	20	25	30
Thousands of Caterpillars/Hectare	12	10	9	7.5	6	4.5

Assuming that there is a linear relationship between pesticide amount and caterpillar mortality within the range of pesticide application in the experiment, use the method of least squares to find the relationship. Use this relationship to estimate (a) the amount of pesticide required to eliminate the caterpillars and (b) the number of caterpillars present if no pesticide is used.

Solution

We let k be kilograms of pesticide per hectare and c be the number of thousands of caterpillars per hectare. The following table gives the computations required for finding the normal equations.

	k	c	k^2	kp
	5	12	25	60
	10	10	100	100
	15	9	225	135
	20	7.5	400	150
	25	6	625	150
	30	4.5	900	120
Sum	105	49	2275	715

The normal equations are

$$\begin{aligned} 6b + 105m &= 49 \\ 105b + 2275m &= 715 \end{aligned}$$

Solving these equations simultaneously we find that

$$m = -0.326 \qquad b = 13.9$$

Thus the number of thousands of caterpillars as a function of pesticide application is modelled by the line $c(k) = 13.9 - 0.326k$. The scatter plot for the data versus the line is shown in Figure 10.26.

a. To estimate the amount of pesticide needed to eliminate the caterpillars we solve

$$\begin{aligned} 0 &= c(k) \\ 0 &= 13.9 - 0.326k \\ 0.326k &= 13.9 \\ k &= \frac{13.9}{0.326} \\ k &= 42.6 \end{aligned}$$

b. The number of caterpillars when no pesticide is applied ($k = 0$) is $c(0) = 13.9$ thousand per hectare. Compare this result with Figure 10.26.

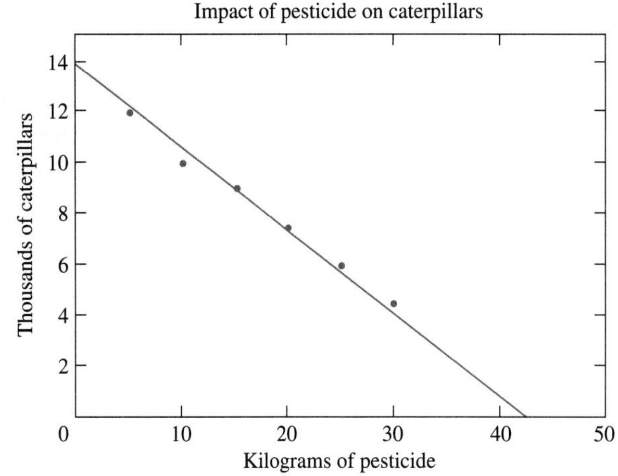

Figure 10.26
The least-squares line and data points for the effect of pesticides on caterpillars

Applying Least Squares to Exponential Models

The greatest weakness of the least-squares method of fitting a line is that a line is often not the appropriate model of the data to fit. It is possible to use the technique of least squares to derive other models, but it is also possible to use a technique similar to the change of variables we learned in Chapter 8 to change the data so that the least squares line is an appropriate type of model. One very common model for populations is the exponential model

$$P(t) = Ae^{ct}$$

The number A is the population at time $t = 0$ and the number c is a constant that dictates how fast the population grows. If we take the log of both sides of the model, then we can use algebra to change it into a linear function.

$$\begin{aligned} P &= Ae^{ct} \\ \ln(P) &= \ln(Ae^{ct}) \\ \ln(P) &= \ln(A) + \ln(e^{ct}) \\ \ln(P) &= \ln(A) + ct \\ \ln(P) &= ct + \ln(A) \end{aligned}$$

Where before the line was $y = mx + b$, it is now $\ln(P) = ct + \ln(A)$. The slope of the line is the value c and the intercept is the natural log of the initial population: $\ln(A)$. Notice also that if we have a set of points giving times and populations at those times, we must take the log of the population values (the dependent variables) before we can fit the line. The next example shows how to do this.

Example 5

Fitting Exponential Models with Least Squares The number of bacteria in a growth medium is estimated at four different times with an optical densitometer, yielding the data below.

Time in Hours	1	3	5	7
Millions of Bacteria	0.5	1.6	4.7	15.5

Take the log of the number of bacteria and then use the formula for a least-squares line to find a model for the log of the population of bacteria as a function of time. After finding the logarithmic model, use the exponential function to find a model for the number of bacteria after t hours. Use this model to find the population A at time zero and also the population after 10 hours.

Solution

First we build a new data table by taking the natural log of the number of bacteria (but not of the time!).

Time in Hours	1	3	5	7
ln (Millions of Bacteria)	−0.693	0.470	1.55	2.74

We now fit a line to these data in the usual manner. The following table gives the computations required for finding the normal equations.

	t	$\ln(P)$	t^2	$t \ln(P)$
	1	−0.693	1	−0.693
	3	0.470	9	1.41
	5	1.55	25	7.74
	7	2.74	49	19.2
Sum	16	4.07	84	27.7

The normal equations are

$$4 \ln (A) + 16c = 4.07$$

$$16 \ln (A) + 84c = 27.7$$

Remember to treat ln (*A*) as if it is just a variable. Simultaneously solving these equations gives us

$$c = 0.571 \qquad \ln (A) = -1.25$$

This gives a line that models the log of the population as

$$\ln (P) = 0.571t - 1.25$$

To get a model for the population, rather than its log, we take *e* to the power of both sides of the linear model.

$$e^{\ln (P)} = e^{0.571t-1.25}$$

$$P = e^{0.571t} \cdot e^{-1.25}$$

$$P(t) = e^{0.581t} \cdot 0.287$$

$$P(t) = 0.287e^{0.571t}$$

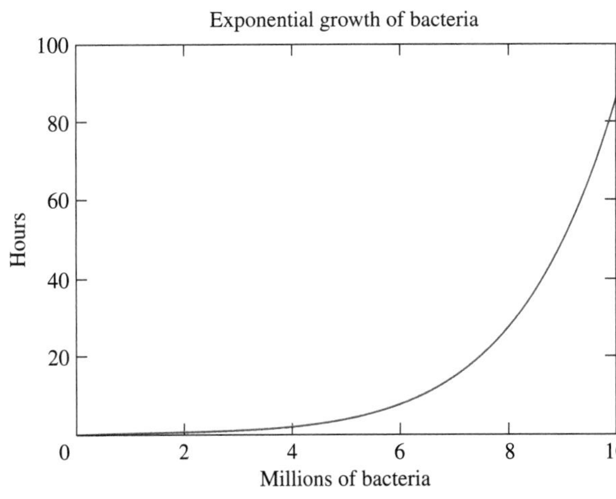

Figure 10.27
The exponential model fit with least-squares to the bacterial population model

The scatter plot for this model and the data are shown in Figure 10.27. With this model is it easy to see that the original population at $t = 0$ is $A = 0.287e^{0.571 \cdot 0} = 0.287$ million bacteria or about 287,000 bacteria. At time $t = 10$ hours the number of bacteria is $P(10) = 0.287e^{0.571 \cdot 10} = 86.6$ million bacteria.

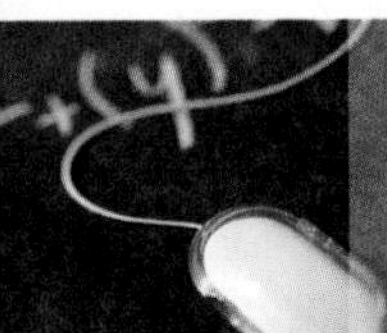

Self-Check Exercises 10.4

1. Find an equation of the least-squares line for the data

$P_1(0, 3)$, $P_2(2, 6.5)$, $P_3(4, 10)$ $P_4(6, 16)$, $P_5(7, 16.5)$

2. The following data give the percent of people over age 65 who have high school diplomas.

Year, *x*	0	6	11	16	22	26
Percentage with Diplomas, *y*	19	25	30	35	44	48

Here, $x = 0$ corresponds to the beginning of the year 1967.

a. Find an equation of the least-squares line for the given data.

b. Assuming that this trend continues, what percent of people over age 65 will have high school diplomas at the beginning of the year 2011 ($x = 44$)?

Solutions to Self-Check Exercises 10.4 can be found on page 628.

10.4 Exercises

A calculator is recommended for this exercise set.

In Exercises 1–6, (a) find the equation of the least-squares line for the given data and (b) draw a scatter diagram for the given data and graph the least-squares line.

1.

x	1	2	3	4
y	4	6	8	11

2.

x	1	3	5	7	9
y	9	8	6	3	2

3.

x	1	2	3	4	4	6
y	4.5	5	3	2	3.5	1

4.

x	1	1	2	3	4	4	5
y	2	3	3	3.5	3.5	4	5

5. $P_1(1, 3)$, $P_2(2, 5)$, $P_3(3, 5)$, $P_4(4, 7)$, $P_5(5, 8)$

6. $P_1(1, 8)$, $P_2(2, 6)$, $P_3(5, 6)$, $P_4(7, 4)$, $P_5(10, 1)$

7. **College Admissions** The following data, compiled by the admissions office at Faber College during the past 5 yr, relate the number of college brochures and follow-up letters (x) sent to a preselected list of high school juniors who had taken the Preliminary Admissions Scholastic Test and the number of completed applications (y) received from these students (both measured in units of 1000):

x	4	4.5	5	5.5	6
y	0.5	0.6	0.8	0.9	1.2

a. Determine the equation of the least-squares line for these data.
b. Draw a scatter diagram and the least-squares line for these data.
c. Use the result obtained in part (a) to predict the number of completed applications that might be expected if 6400 brochures and follow-up letters are sent out during the next year.

8. **Scholastic Admissions Test Verbal Scores** The following data, compiled by the superintendent of schools in a large metropolitan area, shows the average Scholastic Admissions Test verbal scores of high school seniors during the 5 yr since the district implemented the "back-to-basics" program:

Year, x	1	2	3	4	5
Average Score, y	436	438	428	430	426

a. Determine the equation of the least-squares line for these data.
b. Draw a scatter diagram and the least-squares line for these data.
c. Use the result obtained in part (a) to predict the average Scholastic Admissions Test verbal score of high school seniors 2 yr from now ($x = 7$).

9. **Size of Average Farm** The size of the average farm has been growing steadily over the years. The following data gives the size of the average farm y (in acres) from 1940 through 1991 ($x = 0$ corresponds to the beginning of the year 1940):

Year, x	0	10	20	30	40	57
Size of Farm, y	168	213	297	374	427	471

a. Find the equation of the least-squares line for these data.
b. Use the result of part (a) to estimate the size of the average farm in the year 1985.

10. **Production of All-Aluminum Cans** Steel has been playing a decreasing role in the manufacture of beverage cans in North America. The use of bimetallic cans has been dwindling, whereas the use of all-aluminum cans has been growing steadily. The accompanying table gives the production of all-aluminum cans (in billions) over the period from 1975 through 1989:

Year	1975	1977	1979	1981
Cans, y	16.7	26	33.3	48.3

Year	1983	1985	1987	1989
Cans, y	57	65.8	74.2	83.3

a. Find an equation of the least-squares line for these data. (Let $x = 1$ represent 1975.)
b. Use the result of part (a) to estimate the number of cans produced in 1993, assuming the trend continued.

Source: Can Manufacturing Institute

11. **Net-Connected Computers in Europe** The estimated number of computers (in millions) connected to the Internet in Europe from 2003 through 2007 is summarized in the accompanying table. (Here, $x = 0$ corresponds to the beginning of 2003.)

Year, x	0	1	2	3	4
Net-Connected Computers, y	21.7	32.1	45.0	58.3	69.6

a. Find an equation of the least-squares line for these data.
b. Use the result of part (a) to estimate the projected number of computers connected to the Internet in Europe at the beginning of 2008, assuming the trend continues.

12. **Authentication Technology** With computer security always a hot-button issue, demand is growing for technology that authenticates and authorizes computer users. The following table gives the authentication software sales (in billions of dollars), including projections, from 1999 through 2006 ($x = 0$ represents 1999):

Year, x	0	1	2	3	4	5
Sales, y	2.4	2.9	3.7	4.5	5.2	6.1

a. Find an equation of the least-squares line for these data.
b. Use the result of part (a) to estimate the sales for 2007, assuming the projection is accurate.

Source: International Data Corporation

13. **Portable-Phone Services** The projected number of wireless subscribers y (in millions) from the year 2000 through 2006 is summarized in the accompanying table. (Here, $x = 0$ corresponds to the beginning of the year 2000.)

Year, x	0	1	2	3	4	5	6
Subscribers, y	90.4	100.0	110.4	120.4	130.8	140.4	150.0

a. Find an equation of the least-squares line for these data.
b. Use the result of part (a) to estimate the projected number of wireless subscribers at the beginning of 2006. How does this result compare with the given data for that year?

Source: BancAmerica Robertson Stephens

14. **Health-Care Spending** The following data, compiled by the Organization for Economic Cooperation and Development (OECD) in 1990, gives the per capita Gross Domestic Product

(GDP) (in thousands of dollars) and the corresponding per capita spending on health care (in dollars) for selected countries.

Country	Turkey	Spain	Netherlands
GDP	4.25	10	14
Health-Care Spending	178	667	1194

Country	Sweden	Switzerland	Canada
GDP	15.5	17.8	19.5
Health-Care Spending	1500	1388	1640

a. Letting x denote a country's GDP (in thousands of dollars per capita) and y denote the per capita health-care spending (in dollars), find an equation of the least-squares line for these data giving the typical relationship between GDP and health-care spending for the selected countries.
b. The per capita GDP of the United States is $20,000. If the health-care spending of the United States were in line with that of these sample OECD countries, what would it be? (*Note:* The actual per capita health-care spending of the United States in 1990 was $2444.)

Source: Organization for Economic Cooperation and Development

Business and Economics Applications

15. **Starbucks' Store Count** According to company reports, the number of Starbucks stores in North America between 1996 and 2000 are as follows ($x = 0$ corresponds to 1996):

Year, x	0	1	2	3	4
Stores, y	929	1270	1622	2038	2446

a. Find an equation of the least-squares line for these data.
b. Use the result of part (a) to estimate the rate at which new stores were opened annually in North America for the period in question.

Source: Company reports

16. **Net Sales** The management of Kaldor, a manufacturer of electric motors, submitted the following data in the annual report to its shareholders. The table shows the net sales (in millions of dollars) during the 5 yr that have elapsed since the new management team took over. (The first year the firm operated under the new management corresponds to the time period $x = 1$, and the four subsequent years correspond to $x = 2, 3, 4, 5$.)

Year, x	1	2	3	4	5
Net Sales, y	426	437	460	473	477

a. Determine the equation of the least-squares line for these data.
b. Draw a scatter diagram and the least-squares line for these data.
c. Use the result obtained in part (a) to predict the net sales for the upcoming year.

17. **Mass Transit Subsidies** The following table gives the projected state subsidies (in millions of dollars) to the Massachusetts Bay Transit Authority (MBTA) over a 5-yr period:

Year, x	1	2	3	4	5
Subsidy, y	20	24	26	28	32

a. Find an equation of the least-squares line for these data.
b. Use the result of part (a) to estimate the state subsidy to the MBTA for the eighth year ($x = 8$).

Source: Massachusetts Bay Transit Authority

18. **Canadian Pension Plan Wage Base** The Canadian Pension Plan wage base (in thousands of dollars) from 2001 to 2006 is given in the following table:

Year	2001	2002	2003	2004	2005	2006
Wage Base, y	72.6	76.2	80.4	84.9	87.0	87.9

a. Find an equation of the least-squares line for these data. (Let $x = 1$ represent the year 2001.)
b. Use your result of part (a) to estimate the wage base in the year 2009.

19. **Online Banking** According to industry sources, online banking is expected to take off in the near future. The projected number of households (in millions) using this service is given in the following table. (Here, $x = 0$ corresponds to the beginning of 2002.)

Year, x	0	1	2	3	4	5
Households, y	4.5	7.5	10.0	13.0	15.6	18.0

a. Find an equation of the least-squares line for these data.
b. Use the result of part (a) to estimate the number of households using online banking at the beginning of 2008, assuming the projection is accurate.

20. **Online Travel** More and more travellers are purchasing their tickets online. According to industry projections, the North American online travel revenue (in billions of dollars) from 2003 through 2007 is given in the following table (here, $t = 0$ corresponds to 2001):

Year, t	0	1	2	3	4
Revenue, y	16.3	21.0	25.0	28.8	32.7

a. Find an equation of the least-squares line for these data.
b. Use the result of part (a) to estimate the online travel revenue for 2006.

21. **Market for Drugs** Because of new, lower standards, experts in a study conducted in early 2000 projected a rise in the market for cholesterol-reducing drugs. The market (in billions of dollars) for such drugs from 2002 through 2007 is given in the following table ($x = 0$ represents 2002):

Year	2002	2003	2004	2005	2006	2007
Market, y	12.07	14.07	16.21	18.28	20	21.72

a. Find an equation of the least-squares line for these data.
b. Use the result of part (a) to estimate the market for cholesterol-reducing drugs in 2008, assuming the trend continues.

Biological and Life Sciences Applications

22. **Male Life Expectancy at 65** The projections of male life expectancy at age 65 in North America are summarized in the following table ($x = 0$ corresponds to 2000):

Year, x	0	10	20	30	40	50
Years beyond 65, y	15.9	16.8	17.6	18.5	19.3	20.3

a. Find an equation of the least-squares line for these data.
b. Use the result of (a) to estimate the life expectancy at 65 of a male in 2040. How does this result compare with the given data for that year?
c. Use the result of (a) to estimate the life expectancy at 65 of a male in 2030.

23. **Female Life Expectancy at 65** The projections of female life expectancy at age 65 in North America are summarized in the following table ($x = 0$ corresponds to 2000):

Year, x	0	10	20	30	40	50
Years beyond 65, y	19.5	20.0	20.6	21.2	21.8	22.4

a. Find an equation of the least-squares line for these data.
b. Use the result of (a) to estimate life expectancy at 65 of a female in 2040. How does this result compare with the given data for that year?
c. Use the result of (a) to estimate the life expectancy at 65 of a female in 2030.

24. **Herbicide Level** An herbicide is tested to determine the effective level of application. The number of weeds found in a test plot for a given amount of herbicide applied are given in the table below.

Grams of Herbicide/m^2	4	8	12	16	20
Weeds/m^2	112	86	62	41	25

Use the method of least squares to fit a line to the data and estimate the minimum level of herbicide application that will kill all the weeds. Also graph the data and your least-squares line.

25. **Wolf Population Change** In an attempt to help wolf populations rebound in a provincial park, four test areas are set aside and the number of births minus deaths of wolves for the year are recorded. In different test areas, different numbers of dead domestic elk are left out for the wolves as supplementary food. Using the data in the table below, find a line that models the wolves' population change as a function of the number of elk left for the wolves as food. Give an estimate of the number of wolves gained per elk (this will be the slope of your line).

Elk	0	10	30	50
Wolves (born minus died)	−6	2	11	17

26. **Estimating Mouse Population** Reported observations are used to estimate the population of a kind of mice accidentally released from a pet store on an island in British Columbia. The population estimates at various times are given in the table below. Assuming that the mice are growing according to an exponential model

$$P(t) = Ae^{ct}$$

(at least initially before they grow too numerous), estimate the initial population of mice released A and the growth rate c. The store owner thinks four mice escaped initially: Does the model agree? How many mice will there be after 52 weeks?

Weeks	2	3	5	8	10
Number of Mice	41	58	110	310	616

27. **Doubling a Population** The mass of yeast in a bioreactor is estimated by measuring their metabolic wastes. The number of grams of yeast in the bioreactor is recorded at various times and is given in the table below. Use the method of least squares for fitting exponential models to estimate the growth rate c of the yeast. From this, compute the time it takes the yeast to double its initial population.

Hours	1	2	4	8
Grams of Yeast	6.0	7.2	10.4	21.5

In Exercises 28–31, determine whether the statement is true or false. If it is true, explain why it is true. If it is false, give an example to show why it is false.

28. The least-squares line must pass through at least one of the data points.

29. The sum of the squares of the errors incurred in approximating n data points using the least-squares linear function is zero if and only if the n data points lie along a straight line.

30. If the data consist of two distinct points, then the least-squares line is just the line that passes through the two points.

31. A data point lies on the least-squares line if and only if the vertical distance between the point and the line is equal to zero.

Solutions to Self-Check Exercises 10.4

1. We first construct the table:

	x	y	x^2	xy
	0	3	0	0
	2	6.5	4	13
	4	10	16	40
	6	16	36	96
	7	16.5	49	115.5
Sum	19	52	105	264.5

The normal equations are

$$5b + 19m = 52$$
$$19b + 105m = 264.5$$

Solving the first equation for b gives

$$b = -3.8m + 10.4$$

which, upon substitution into the second equation, gives

$$19(-3.8m + 10.4) + 105m = 264.5$$

Solutions to Self-Check Exercises 10.4, *continued*

$$-72.2m + 197.6 + 105m = 264.5$$
$$32.8m = 66.9$$
$$m \approx 2.04$$

Substituting this value of m into the expression for b found earlier gives

$$b = -3.8(2.04) + 10.4 \approx 2.65$$

Therefore, the required least-squares line has the equation given by

$$y = 2.04x + 2.65$$

2. a. The calculations required for obtaining the normal equations may be summarized as follows:

	x	y	x^2	xy
	0	19	0	0
	6	25	36	150
	11	30	121	330
	16	35	256	560
	22	44	484	968
	26	48	676	1248
Sum	81	201	1573	3256

The normal equations are

$$6b + 81m = 201$$
$$81b + 1573m = 3256$$

Solving this system of linear equations simultaneously, we find

$$m \approx 1.13 \qquad \text{and} \qquad b \approx 18.23$$

Therefore, the required least-squares line has the equation given by

$$y = f(x) = 1.13x + 18.23$$

b. The percent of people over the age of 65 who will have high school diplomas at the beginning of the year 2011 is given by

$$f(44) = 1.13(44) + 18.23$$
$$= 67.95$$

or approximately 68%.

10.5 Constrained Maxima and Minima and the Method of Lagrange Multipliers

Constrained Relative Extrema

In Section 10.3 we studied the problem of determining the relative extremum of a function $f(x, y)$ without placing any restrictions on the independent variables x and y—except, of course, that the point (x, y) lies in the domain of f. Such a relative extremum of a function f is referred to as an **unconstrained relative extremum** of f. However, in many practical optimization problems, we must maximize or minimize a function in which the independent variables are subjected to certain further constraints.

In this section we discuss a powerful method for determining the relative extrema of a function $f(x, y)$ whose independent variables x and y are required to satisfy one or more constraints of the form $g(x, y) = 0$. Such a relative extremum of a function f is called a **constrained relative extremum** of f. We can see the difference between an unconstrained extremum of a function $f(x, y)$ of two variables and a constrained extremum of f, where the independent variables x and y are subjected to a constraint of the form $g(x, y) = 0$, by considering the geometry of the two cases. Figure 10.28a depicts the graph of a function $f(x, y)$ that has an unconstrained relative minimum at the point $(0, 0)$. However, when the independent variables x and y are subjected to an equality constraint of the form $g(x, y) = 0$, the points (x, y, z) that satisfy both $z = f(x, y)$ and the constraint equation $g(x, y) = 0$ lie on a curve C. Therefore, the constrained relative minimum of f must also lie on C (Figure 10.28b).

Our first example involves an equality constraint $g(x, y) = 0$ in which we solve for the variable y explicitly in terms of x. In this case we may apply the technique used in Chapter 7 to find the relative extrema of a function of one variable.

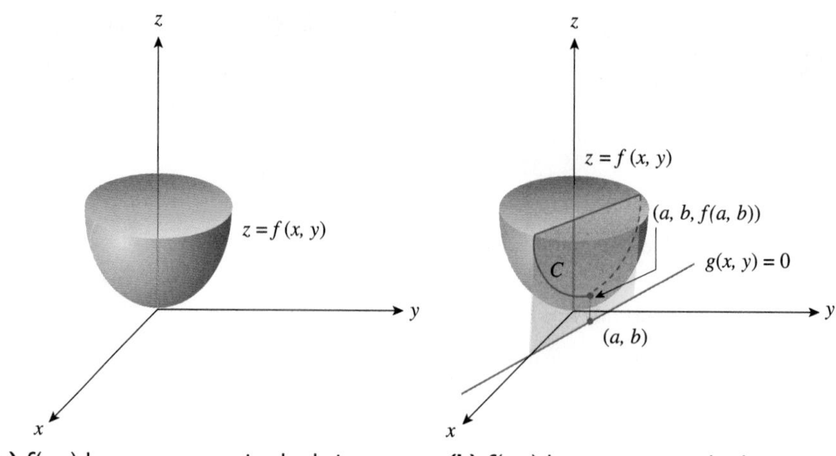

(a) $f(x, y)$ has an unconstrained relative extremum at (0, 0).

(b) $f(x, y)$ has a constrained relative extremum at $(a, b, f(a, b))$.

Figure 10.28

Example 1

Find the relative minimum of the function

$$f(x, y) = 2x^2 + y^2$$

subject to the constraint $g(x, y) = x + y - 1 = 0$.

Solution

Solving the constraint equation for y explicitly in terms of x, we obtain $y = -x + 1$. Substituting this value of y into the function $f(x, y) = 2x^2 + y^2$ results in a function of x,

$$h(x) = 2x^2 + (-x + 1)^2 = 3x^2 - 2x + 1$$

The function h describes the curve C lying on the graph of f on which the constrained relative minimum of f occurs. To find this point, use the technique developed in Chapter 7 to determine the relative extrema of a function of one variable:

$$h'(x) = 6x - 2 = 2(3x - 1)$$

Setting $h'(x) = 0$ gives $x = \frac{1}{3}$ as the sole critical point of the function h. Next, we find

$$h''(x) = 6$$

and, in particular,

$$h''\left(\frac{1}{3}\right) = 6 > 0$$

Therefore, by the second derivative test, the point $x = \frac{1}{3}$ gives rise to a relative minimum of h. Substitute this value of x into the constraint equation $x + y - 1 = 0$ to get $y = \frac{2}{3}$. Thus, the point $(\frac{1}{3}, \frac{2}{3})$ gives rise to the required constrained relative minimum of f. Since

$$f\left(\frac{1}{3}, \frac{2}{3}\right) = 2\left(\frac{1}{3}\right)^2 + \left(\frac{2}{3}\right)^2 = \frac{2}{3}$$

the required constrained relative minimum value of f is $\frac{2}{3}$ at the point $(\frac{1}{3}, \frac{2}{3})$. It may be shown that $\frac{2}{3}$ is in fact a constrained absolute minimum value of f (Figure 10.29).

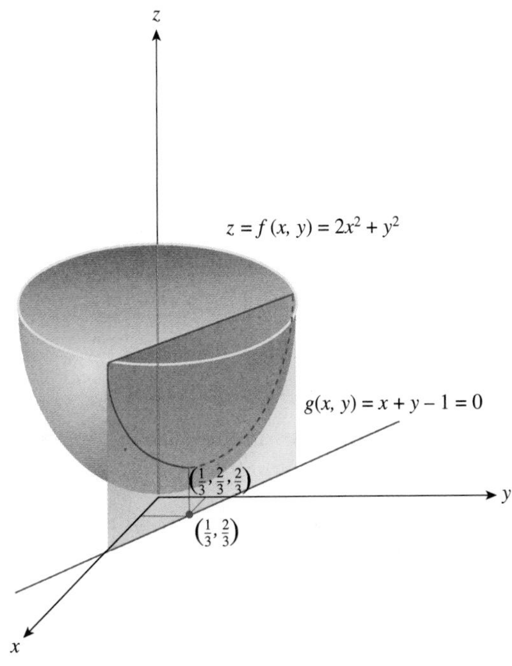

Figure 10.29
f has a constrained absolute minimum of $\frac{2}{3}$ at $(\frac{1}{3}, \frac{2}{3})$.

The Method of Lagrange Multipliers

The major drawback of the technique used in Example 1 is that it relies on our ability to solve the constraint equation $g(x, y) = 0$ for y explicitly in terms of x. This is not always an easy task. Moreover, even when we can solve the constraint equation $g(x, y) = 0$ for y explicitly in terms of x, the resulting function of one variable that is to be optimized may turn out to be unnecessarily complicated. Fortunately, an easier method exists. This method, called the **method of Lagrange multipliers** (Joseph Lagrange, 1736–1813), is as follows:

The Method of Lagrange Multipliers

To find the relative extremum of the function $f(x, y)$ subject to the constraint $g(x, y) = 0$ (assuming that these extreme values exist),

1. Form an auxiliary function

$$F(x, y, \lambda) = f(x, y) + \lambda g(x, y)$$

called the Lagrangian function (the variable λ is called the Lagrange multiplier).

2. Solve the system that consists of the equations

$$F_x = 0, \qquad F_y = 0, \qquad F_\lambda = 0$$

for all values of x, y, and λ.

3. Evaluate f at each of the points (x, y) found in step 2. The largest (smallest) of these values is the maximum (minimum) value of f.

Let's re-solve Example 1 using the method of Lagrange multipliers.

Example 2

Using the method of Lagrange multipliers, find the relative minimum of the function

$$f(x, y) = 2x^2 + y^2$$

subject to the constraint $x + y = 1$.

Solution

Write the constraint equation $x + y = 1$ in the form $g(x, y) = x + y - 1 = 0$. Then, form the Lagrangian function

$$\begin{aligned} F(x, y, \lambda) &= f(x, y) + \lambda g(x, y) \\ &= 2x^2 + y^2 + \lambda(x + y - 1) \end{aligned}$$

To find the critical point(s) of the function F, solve the system composed of the equations

$$\begin{aligned} F_x &= 4x + \lambda = 0 \\ F_y &= 2y + \lambda = 0 \\ F_\lambda &= x + y - 1 = 0 \end{aligned}$$

Solving the first and second equations in this system for x and y in terms of λ, we obtain

$$x = -\frac{1}{4}\lambda \quad \text{and} \quad y = -\frac{1}{2}\lambda$$

which, upon substitution into the third equation, yields

$$-\frac{1}{4}\lambda - \frac{1}{2}\lambda - 1 = 0 \quad \text{or} \quad \lambda = -\frac{4}{3}$$

Therefore, $x = \frac{1}{3}$ and $y = \frac{2}{3}$, and $(\frac{1}{3}, \frac{2}{3})$ affords a constrained minimum of the function f, in agreement with the result obtained earlier.

Remark

A disadvantage of the method of Lagrange multipliers is that there is no test analogous to the second derivative test mentioned in Section 10.3 for determining whether a critical point of a function of two or more variables leads to a relative maximum or relative minimum (and thus the absolute extrema) of the function. Here we have to rely on the geometric or physical nature of the problem to help us draw the necessary conclusions (see Example 2). ◀

The method of Lagrange multipliers may be used to solve a problem involving a function of three or more variables, as illustrated in the next example.

Example 3

Use the method of Lagrange multipliers to find the minimum of the function

$$f(x, y, z) = 2xy + 6yz + 8xz$$

subject to the constraint

$$xyz = 12{,}000$$

(*Note:* The existence of the minimum is suggested by the geometry of the problem.)

Solution

Write the constraint equation $xyz = 12{,}000$ in the form $g(x, y, z) = xyz - 12{,}000$. Then, the Lagrangian function is

$$\begin{aligned} F(x, y, z, \lambda) &= f(x, y, z) + \lambda g(x, y, z) \\ &= 2xy + 6yz + 8xz + \lambda(xyz - 12{,}000) \end{aligned}$$

To find the critical point(s) of the function F, we solve the system composed of the equations

$$\begin{aligned} F_x &= 2y + 8z + \lambda yz = 0 \\ F_y &= 2x + 6z + \lambda xz = 0 \end{aligned}$$

$$F_z = 6y + 8x + \lambda xy = 0$$
$$F_\lambda = xyz - 12{,}000 = 0$$

Solving the first three equations of the system for λ in terms of x, y, and z, we have

$$\lambda = -\frac{2y + 8z}{yz}$$
$$\lambda = -\frac{2x + 6z}{xz}$$
$$\lambda = -\frac{6y + 8x}{xy}$$

Equating the first two expressions for λ leads to

$$\frac{2y + 8z}{yz} = \frac{2x + 6z}{xz}$$
$$2xy + 8xz = 2xy + 6yz$$
$$x = \frac{3}{4}y$$

Next, equating the second and third expressions for l in the same system yields

$$\frac{2x + 6z}{xz} = \frac{6y + 8x}{xy}$$
$$2xy + 6yz = 6yz + 8xz$$
$$z = \frac{1}{4}y$$

Finally, substituting these values of x and z into the equation $xyz - 12{,}000 = 0$, the fourth equation of the first system of equations, we have

$$\left(\frac{3}{4}y\right)(y)\left(\frac{1}{4}y\right) - 12{,}000 = 0$$
$$y^3 = \frac{(12{,}000)(4)(4)}{3} = 64{,}000$$
$$y = 40$$

The corresponding values of x and z are given by $x = \frac{3}{4}(40) = 30$ and $z = \frac{1}{4}(40) = 10$. Therefore, we see that the point (30, 40, 10) gives the constrained minimum of f. The minimum value is

$$f(30, 40, 10) = 2(30)(40) + 6(40)(10) + 8(30)(10) = 7200$$

Applications

Example 4

Maximizing Profit Refer to Example 3, Section 10.1. The total weekly profit (in dollars) that Acrosonic realized in producing and selling its bookshelf loudspeaker systems is given by the profit function

$$P(x, y) = -\frac{1}{4}x^2 - \frac{3}{8}y^2 - \frac{1}{4}xy + 120x + 100y - 5000$$

where x denotes the number of fully assembled units and y denotes the number of kits produced and sold per week. Acrosonic's management decides that production of these loudspeaker systems should be restricted to a total of exactly 230 units each week. Under this condition, how many fully assembled units and how many kits should be produced each week to maximize Acrosonic's weekly profit?

Solution

The problem is equivalent to the problem of maximizing the function

$$P(x, y) = -\frac{1}{4}x^2 - \frac{3}{8}y^2 - \frac{1}{4}xy + 120x + 100y - 5000$$

subject to the constraint

$$g(x, y) = x + y - 230 = 0$$

The Lagrangian function is

$$\begin{aligned} F(x, y, \lambda) &= P(x, y) + \lambda g(x, y) \\ &= \frac{1}{4}x^2 - \frac{3}{8}y^2 - \frac{1}{4}xy + 120x + 100y \\ &\quad -5000 + \lambda(x + y - 230) \end{aligned}$$

To find the critical point(s) of F, solve the following system of equations:

$$\begin{aligned} F_x &= -\frac{1}{2}x - \frac{1}{4}y + 120 + \lambda = 0 \\ F_y &= -\frac{3}{4}y - \frac{1}{4}x + 100 + \lambda = 0 \\ F_\lambda &= x + y - 230 = 0 \end{aligned}$$

Solving the first equation of this system for λ, we obtain

$$\lambda = \frac{1}{2}x + \frac{1}{4}y - 120$$

which, upon substitution into the second equation, yields

$$\begin{aligned} -\frac{3}{4}y - \frac{1}{4}x + 100 + \frac{1}{2}x + \frac{1}{4}y - 120 &= 0 \\ -\frac{1}{2}y + \frac{1}{4}x - 20 &= 0 \end{aligned}$$

Solving the last equation for y gives

$$y = \frac{1}{2}x - 40$$

When we substitute this value of y into the third equation of the system, we have

$$\begin{aligned} x + \frac{1}{2}x - 40 - 230 &= 0 \\ x &= 180 \end{aligned}$$

The corresponding value of y is $\frac{1}{2}(180) - 40$, or 50. Thus, the required constrained relative maximum of P occurs at the point (180, 50). Again, we can show that the point (180, 50) in fact yields a constrained absolute maximum for P. Thus, Acrosonic's profit is maximized by producing 180 assembled and 50 kit versions of their bookshelf loudspeaker systems. The maximum weekly profit realizable is given by

$$\begin{aligned} P(180, 50) &= -\frac{1}{4}(180)^2 - \frac{3}{8}(50)^2 - \frac{1}{4}(180)(50) \\ &\quad + 120(180) + 100(50) - 5000 \\ &= 10{,}312.5 \end{aligned}$$

or \$10,312.50.

Example 5

Designing a Cruise-Ship Pool The operators of the *Viking Princess*, a luxury cruise liner, are contemplating the addition of another swimming pool to the ship. The chief engineer has suggested that an area in the form of an ellipse located in the rear of the promenade deck would be suitable for this purpose. This location would provide a poolside area with sufficient space for passenger movement and placement of deck chairs (Figure 10.30). It has been determined that the shape of the ellipse may be described by the equation $x^2 + 4y^2 = 400$, where x and y are measured in metres. *Viking*'s operators would like to know the dimensions of the rectangular pool with the largest possible area that would meet these requirements.

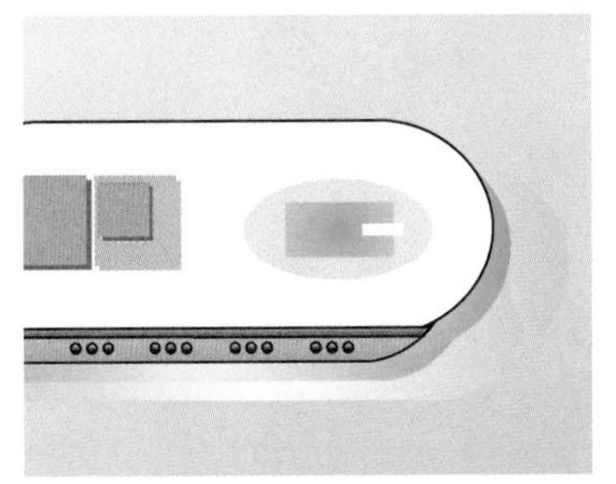

Figure 10.30
A rectangular-shaped pool will be built in the elliptical-shaped poolside area.

Solution

To solve this problem, we need to find the rectangle inscribed in the ellipse with equation $x^2 + 4y^2 = 400$ and having the largest area. Letting the sides of the rectangle be $2x$ and $2y$ metres, we see that the area of the rectangle is $A = 4xy$ (Figure 10.31). Furthermore, the point (x, y) must be constrained to lie on the ellipse so that it satisfies the equation $x^2 + 4y^2 = 400$. Thus, the problem is equivalent to the problem of maximizing the function

$$f(x, y) = 4xy$$

subject to the constraint $g(x, y) = x^2 + 4y^2 - 400 = 0$. The Lagrangian function is

$$\begin{aligned} F(x, y, \lambda) &= f(x, y) + \lambda g(x, y) \\ &= 4xy + \lambda(x^2 + 4y^2 - 400) \end{aligned}$$

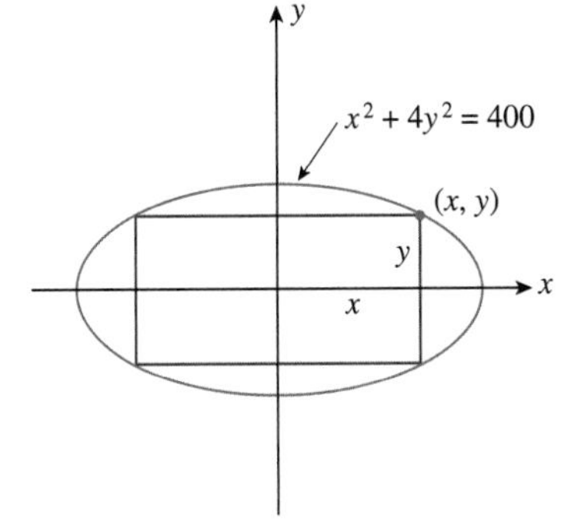

Figure 10.31
We want to find the largest rectangle that can be inscribed in the ellipse described by $x^2 + 4y^2 = 400$.

To find the critical point(s) of F, we solve the following system of equations:

$$\begin{aligned} F_x &= 4y + 2\lambda x = 0 \\ F_y &= 4x + 8\lambda y = 0 \\ F_\lambda &= x^2 + 4y^2 - 400 = 0 \end{aligned}$$

Solving the first equation of this system for λ, we obtain

$$\lambda = -\frac{2y}{x}$$

which, upon substitution into the second equation, yields

$$4x + 8\left(-\frac{2y}{x}\right)y = 0 \qquad \text{or} \qquad x^2 - 4y^2 = 0$$

—that is, $x = \pm 2y$. Substituting these values of x into the third equation of the system, we have

$$4y^2 + 4y^2 - 400 = 0$$

or, upon solving $y = \pm\sqrt{50} = \pm 5\sqrt{2}$. The corresponding values of x are $\pm 10\sqrt{2}$. Because both x and y must be nonnegative, we have $x = 10\sqrt{2}$ and $y = 5\sqrt{2}$. Thus, the dimensions of the pool with maximum area are $10\sqrt{2}$ metres $\times$ $10\sqrt{2}$ metres, or approximately 7×14 metres.

Example 6

Cobb–Douglas Production Function Suppose x units of labour and y units of capital are required to produce

$$f(x, y) = 100x^{3/4}y^{1/4}$$

units of a certain product (recall that this is a Cobb–Douglas production function). If each unit of labour costs \$200 and each unit of capital costs \$300 and a total of \$60,000 is available for production, determine how many units of labour and how many units of capital should be used in order to maximize production.

Solution

The total cost of x units of labour at \$200 per unit and y units of capital at \$300 per unit is equal to $200x + 300y$ dollars. But \$60,000 is budgeted for production, so $200x + 300y = 60{,}000$, which we rewrite as

$$g(x, y) = 200x + 300y - 60{,}000 = 0$$

To maximize $f(x, y) = 100x^{3/4}y^{1/4}$ subject to the constraint $g(x, y) = 0$, we form the Lagrangian function

$$\begin{aligned} F(x, y, \lambda) &= f(x, y) + \lambda g(x, y) \\ &= 100x^{3/4}y^{1/4} + \lambda(200x + 300y - 60{,}000) \end{aligned}$$

To find the critical point(s) of F, we solve the following system of equations:

$$\begin{aligned} F_x &= 75x^{-1/4}y^{1/4} + 200\lambda = 0 \\ F_y &= 25x^{3/4}y^{-3/4} + 300\lambda = 0 \\ F_\lambda &= 200x + 300y - 60{,}000 = 0 \end{aligned}$$

Solving the first equation for λ, we have

$$\lambda = -\frac{75x^{-1/4}y^{1/4}}{200} = -\frac{3}{8}\left(\frac{y}{x}\right)^{1/4}$$

which, when substituted into the second equation, yields

$$25\left(\frac{x}{y}\right)^{3/4} + 300\left(-\frac{3}{8}\right)\left(\frac{y}{x}\right)^{1/4} = 0$$

Multiplying the last equation by $\left(\frac{x}{y}\right)^{1/4}$ then gives

$$25\left(\frac{x}{y}\right) - \frac{900}{8} = 0$$

$$x = \left(\frac{900}{8}\right)\left(\frac{1}{25}\right)y = \frac{9}{2}y$$

Substituting this value of x into the third equation of the first system of equations, we have

$$200\left(\frac{9}{2}y\right) + 300y - 60{,}000 = 0$$

from which we deduce that $y = 50$. Hence, $x = 225$. Thus, maximum production is achieved when 225 units of labour and 50 units of capital are used.

When used in the context of Example 6, the negative of the Lagrange multiplier λ is called the **marginal productivity of money.** That is, if one additional dollar is available for production, then approximately $-\lambda$ units of a product can be produced. Here,

$$\lambda = -\frac{3}{8}\left(\frac{y}{x}\right)^{1/4} = -\frac{3}{8}\left(\frac{50}{225}\right)^{1/4} \approx -0.257$$

so, in this case, the marginal productivity of money is 0.257. For example, if \$65,000 is available for production instead of the originally budgeted figure of \$60,000, then the maximum production may be boosted from the original

$$f(225, 50) = 100(225)^{3/4}(50)^{1/4}$$

or 15,448 units, to

$$15{,}448 + 5000(0.257)$$

or 16,733 units.

Self-Check Exercises 10.5

1. Use the method of Lagrange multipliers to find the relative maximum of the function

$$f(x, y) = -2x^2 - y^2$$

subject to the constraint $3x + 4y = 12$.

2. The total monthly profit of Robertson Controls in manufacturing and selling x hundred of its standard mechanical setback thermostats and y hundred of its deluxe electronic setback thermostats each month is given by the total profit function

$$P(x, y) = -\frac{1}{8}x^2 - \frac{1}{2}y^2 - \frac{1}{4}xy + 13x + 40y - 280$$

where P is in hundreds of dollars. If the production of setback thermostats is to be restricted to a total of exactly 4000/month, how many of each model should Robertson manufacture in order to maximize its monthly profits? What is the maximum monthly profit?

Solutions to Self-Check Exercises 10.5 can be found on page 638.

10.5 Exercises

In Exercises 1–16, use the method of Lagrange multipliers to optimize the given function subject to the given constraint.

1. Minimize the function $f(x, y) = x^2 + 3y^2$ subject to the constraint $x + y - 1 = 0$.

2. Minimize the function $f(x, y) = x^2 + y^2 - xy$ subject to the constraint $x + 2y - 14 = 0$.

3. Maximize the function $f(x, y) = 2x + 3y - x^2 - y^2$ subject to the constraint $x + 2y = 9$.

4. Maximize the function $f(x, y) = 16 - x^2 - y^2$ subject to the constraint $x + y - 6 = 0$.

5. Minimize the function $f(x, y) = x^2 + 4y^2$ subject to the constraint $xy = 1$.

6. Minimize the function $f(x, y) = xy$ subject to the constraint $x^2 + 4y^2 = 4$.

7. Maximize the function $f(x, y) = x + 5y - 2xy - x^2 - 2y^2$ subject to the constraint $2x + y = 4$.

8. Maximize the function $f(x, y) = xy$ subject to the constraint $2x + 3y - 6 = 0$.

9. Maximize the function $f(x, y) = xy^2$ subject to the constraint $9x^2 + y^2 = 9$.

10. Minimize the function $f(x, y) = \sqrt{y^2 - x^2}$ subject to the constraint $x + 2y - 5 = 0$.

11. Find the maximum and minimum values of the function $f(x, y) = xy$ subject to the constraint $x^2 + y^2 = 16$.

12. Find the maximum and minimum values of the function $f(x, y) = e^{xy}$ subject to the constraint $x^2 + y^2 = 8$.

13. Find the maximum and minimum values of the function $f(x, y) = xy^2$ subject to the constraint $x^2 + y^2 = 1$.

14. Maximize the function $f(x, y, z) = xyz$ subject to the constraint $2x + 2y + z = 84$.

15. Minimize the function $f(x, y, z) = x^2 + y^2 + z^2$ subject to the constraint $3x + 2y + z = 6$.

16. Find the maximum value of the function $f(x, y, z) = x + 2y - 3z$ subject to the constraint $z = 4x^2 + y^2$.

17. Use the method of Lagrange multipliers to solve Exercise 35, Section 10.3.

18. **Minimizing Construction Costs** The management of UNICO Department Store decides to enclose an 100-m^2 area outside their building to display potted plants. The enclosed area will be a rectangle, one side of which is provided by the external walls of the store. Two sides of the enclosure will be made of pine board, and the fourth side will be made of galvanized steel fencing material. If the pine board fencing costs \$20/running metre and the steel fencing costs \$10/running metre, determine the dimensions of the enclosure that will cost the least to erect.

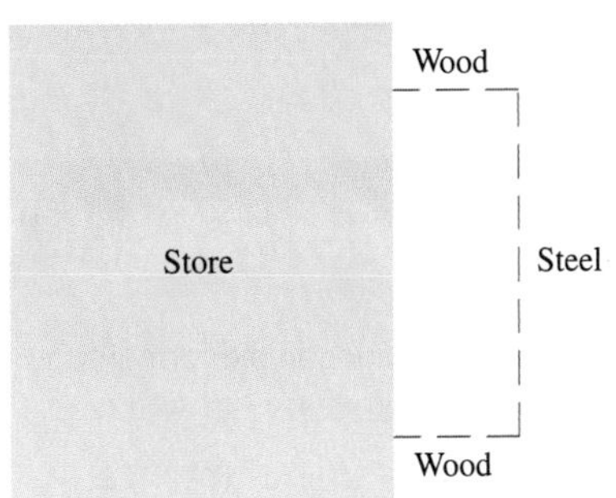

19. **Parcel Post Regulations** Postal regulations specify that a parcel sent by parcel post may have a combined length and girth of no more than 240 cm. Find the dimensions of the cylindrical package of greatest volume that may be sent through the mail. What is the volume of such a package?
Hint: The length plus the girth is $2\pi r + l$, and the volume is $\pi r^2 l$.

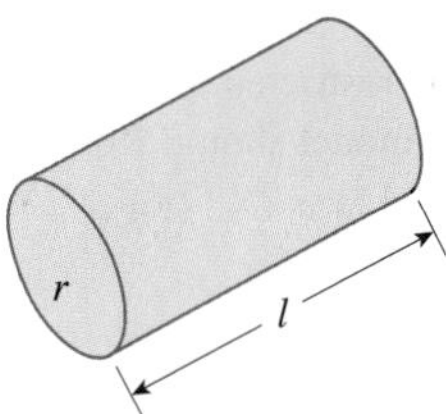

20. **Minimizing Container Costs** The Betty Moore Company requires that its corned beef hash containers have a capacity of 216 cm^3, be right circular cylinders, and be made of a tin alloy. Find the radius and height of the least expensive container that can be made if the metal for the side and bottom costs 1¢/cm^2 and the metal for the pull-off lid costs 0.5¢/cm^2.
Hint: Let the radius and height of the container be r and h cm, respectively. Then, the volume of the container is $\pi r^2 h = 216$, and the cost is given by $C(r, h) = 2\pi rh + 1.5\pi r^2$.

21. **Minimizing Construction Costs** An open rectangular box is to be constructed from material that costs \$10/m^2 for the bottom and \$3/m^2 for its sides. Find the dimensions of the box of greatest volume that can be constructed for \$36.

22. **Minimizing Construction Costs** A closed rectangular box having a volume of 1 m^3 is to be constructed. If the material for the sides costs \$8/m^2 and the material for the top and bottom costs \$6/m^2, find the dimensions of the box that can be constructed with minimum cost.

Business and Economics Applications

23. **Maximizing Profit** The total weekly profit (in dollars) realized by Country Workshop in manufacturing and selling its rolltop desks is given by the profit function

$$P(x, y) = -0.2x^2 - 0.25y^2 - 0.2xy + 100x + 90y - 4000$$

where x stands for the number of finished units and y denotes the number of unfinished units manufactured and sold each week. The company's management decides to restrict the manufacture of these desks to a total of exactly 200 units/week. How many finished and how many unfinished units should be manufactured each week to maximize the company's weekly profit?

24. **Maximizing Profit** The total daily profit (in dollars) realized by Weston Publishing in publishing and selling its dictionaries is given by the profit function

$$P(x, y) = -0.005x^2 - 0.003y^2 - 0.002xy + 14x + 12y - 200$$

where x stands for the number of deluxe editions and y denotes the number of standard editions sold daily. Weston's management decides that publication of these dictionaries should be restricted to a total of exactly 400 copies/day. How many deluxe copies and how many standard copies should be published each day to maximize Weston's daily profit?

25. **Maximizing Sales** The Ross–Simons Company has a monthly advertising budget of \$60,000. Their marketing department estimates that if they spend x dollars on newspaper advertising and y dollars on television advertising, then the monthly sales will be given by

$$z = f(x, y) = 90x^{1/4}y^{3/4}$$

dollars. Determine how much money Ross–Simons should spend on newspaper ads and on television ads each month to maximize its monthly sales.

26. **Maximizing Production** John Mills, the proprietor of Mills Engine Company, a manufacturer of model airplane engines, finds that it takes x units of labour and y units of capital to produce

$$f(x, y) = 100x^{3/4}y^{1/4}$$

units of the product. If a unit of labour costs \$100, a unit of capital costs \$200, and \$200,000 is budgeted for production, determine how many units should be expended on labour and how many units should be expended on capital in order to maximize production.

In Exercises 27 and 28, determine whether the statement is true or false. If it is true, explain why it is true. If it is false, give an example to show why it is false.

27. If (a, b) gives rise to a (constrained) relative extremum of f subject to the constraint $g(x, y) = 0$, then (a, b) also gives rise to the unconstrained relative extremum of f.

28. If (a, b) gives rise to a (constrained) relative extremum of f subject to the constraint $g(x, y) = 0$, then $f_x(a, b) = 0$ and $f_y(a, b) = 0$, simultaneously.

Solutions to Self-Check Exercises 10.5

1. Write the constraint equation in the form $g(x, y) = 3x + 4y - 12 = 0$. Then, the Lagrangian function is

$$F(x, y, \lambda) = -2x^2 - y^2 + \lambda(3x + 4y - 12)$$

To find the critical point(s) of F, we solve the system

$$F_x = -4x + 3\lambda = 0$$
$$F_y = -2y + 4\lambda = 0$$
$$F_\lambda = 3x + 4y - 12 = 0$$

Solutions to Self-Check Exercises 10.5, *continued*

Solving the first two equations for x and y in terms of λ, we find $x = \frac{3}{4}\lambda$ and $y = 2\lambda$. Substituting these values of x and y into the third equation of the system yields

$$3\left(\frac{3}{4}\lambda\right) + 4(2\lambda) - 12 = 0$$

or $\lambda = \frac{48}{41}$. Therefore, $x = \left(\frac{3}{4}\right)\left(\frac{48}{41}\right) = \frac{36}{41}$ and $y = 2\left(\frac{48}{41}\right) = \frac{96}{41}$, and we see that the point $\left(\frac{36}{41}, \frac{96}{41}\right)$ gives the constrained maximum of f. The maximum value is

$$\begin{aligned} f\left(\frac{36}{41}, \frac{96}{41}\right) &= -2\left(\frac{36}{41}\right)^2 - \left(\frac{96}{41}\right)^2 \\ &= -\frac{11{,}808}{1681} = -\frac{288}{41} \end{aligned}$$

2. We want to maximize

$$P(x, y) = -\frac{1}{8}x^2 - \frac{1}{2}y^2 - \frac{1}{4}xy + 13x + 40y - 280$$

subject to the constraint

$$g(x, y) = x + y - 40 = 0$$

The Lagrangian function is

$$\begin{aligned} F(x, y, \lambda) &= P(x, y) + \lambda g(x, y) \\ &= -\frac{1}{8}x^2 - \frac{1}{2}y^2 - \frac{1}{4}xy + 13x \\ &\quad + 40y - 280 + \lambda(x + y - 40) \end{aligned}$$

To find the critical points of F, solve the following system of equations:

$$\begin{aligned} F_x &= -\frac{1}{4}x - \frac{1}{4}y + 13 + \lambda = 0 \\ F_y &= -\frac{1}{4}x - y + 40 + \lambda = 0 \\ F_\lambda &= x + y - 40 = 0 \end{aligned}$$

Subtracting the first equation from the second gives

$$-\frac{3}{4}y + 27 = 0 \quad \text{or} \quad y = 36$$

Substituting this value of y into the third equation yields $x = 4$. Therefore, to maximize its monthly profits, Robertson should manufacture 400 standard and 3600 deluxe thermostats. The maximum monthly profit is given by

$$\begin{aligned} P(4, 36) &= -\frac{1}{8}(4)^2 - \frac{1}{2}(36)^2 - \frac{1}{4}(4)(36) \\ &\quad + 13(4) + 40(36) - 280 \\ &= 526 \end{aligned}$$

or \$52,600.

10.6 Total Differentials (Optional)

Differentials of Two or More Variables

In Section 4.5 we defined the differential of a function of one variable $y = f(x)$ to be

$$dy = f'(x)\,dx$$

In particular, we showed that if the actual change $dx = \Delta x$ in the independent variable x is small, then the differential dy provides us with an approximation of the actual change Δy in the dependent y; that is,

$$\begin{aligned} \Delta y &= f(x + \Delta x) - f(x) \\ &\approx f'(x)\,dx \end{aligned}$$

The concept of the differential extends readily to a function of two or more variables.

Total Differential

Let $z = f(x, y)$ define a differentiable function of x and y.

1. The differentials of the independent variables x and y are $dx = \Delta x$ and $dy = \Delta y$.
2. The differential of the dependent variable z is

$$dz = \frac{\partial f}{\partial x}\,dx + \frac{\partial f}{\partial y}\,dy \qquad (12)$$

Thus, analogous to the one-variable case, the total differential of z is a linear function of dx and dy. Furthermore, it provides us with an approximation of the exact change in z,

$$\Delta z = f(x + \Delta x, y + \Delta y) - f(x, y)$$

corresponding to a net change Δx in x from x to $x + \Delta x$ and a net change Δy in y from y to $y + \Delta y$; that is,

$$\Delta z \approx dz = \frac{\partial f}{\partial x}(x, y)\,dx + \frac{\partial f}{\partial y}(x, y)\,dy \qquad (13)$$

provided $\Delta x = dx$ and $\Delta y = dy$ are sufficiently small.

Example 1

Let $z = 2x^2y + y^3$.

a. Find the differential dz of z.

b. Find the approximate change in z when x changes from $x = 1$ to $x = 1.01$ and y changes from $y = 2$ to $y = 1.98$.

c. Find the actual change in z when x changes from $x = 1$ to $x = 1.01$ and y changes from $y = 2$ to $y = 1.98$. Compare the result with that obtained in part (b).

Solution

a. Let $f(x, y) = 2x^2y + y^3$. Then the required differential is

$$dz = \frac{\partial f}{\partial x}\,dx + \frac{\partial f}{\partial y}\,dy = 4xy\,dx + (2x^2 + 3y^2)\,dy$$

b. Here $x = 1$, $y = 2$, and $dx = 1.01 - 1 = 0.01$ and $dy = 1.98 - 2 = -0.02$. Therefore,

$$\Delta z \approx dz = 4(1)(2)(0.01) + [2(1) + 3(4)](-0.02) = -0.20$$

c. The actual change in z is given by

$$\begin{aligned}\Delta z &= f(1.01, 1.98) - f(1, 2)\\ &= [2(1.01)^2(1.98) + (1.98)^3] - [2(1)^2(2) + (2)^3]\\ &\approx 11.801988 - 12\\ &= -0.1980\end{aligned}$$

We see that $\Delta z \approx dz$, as expected.

Applications

Example 2

Approximating Changes in Revenue The weekly total revenue of Acrosonic Company resulting from the production and sales of x fully assembled bookshelf loudspeaker systems and y kit versions of the same loudspeaker system is

$$R(x, y) = -\frac{1}{4}x^2 - \frac{3}{8}y^2 - \frac{1}{4}xy + 300x + 240y$$

dollars. Determine the approximate change in Acrosonic's weekly total revenue when the level of production is increased from 200 assembled units and 60 kits per week to 206 assembled units and 64 kits per week.

Solution

The approximate change in the weekly total revenue is given by the total differential R at $x = 200$ and $y = 60$, $dx = 206 - 200 = 6$ and $dy = 64 - 60 = 4$; that is, by

$$
\begin{aligned}
dR &= \left.\frac{\partial R}{\partial x}\,dx + \frac{\partial R}{\partial y}\,dy\right|_{\substack{x=200,\ y=60\\ dx=6,\ dy=4}} \\
&= \left.\left(-\frac{1}{2}x - \frac{1}{4}y + 300\right)\right|_{(200,\ 60)} \cdot (6) \\
&\quad + \left.\left(-\frac{3}{4}y - \frac{1}{4}y + 300\right)\right|_{(200,\ 60)} \cdot (4) \\
&= (-100 - 15 + 300)6 + (-45 - 50 + 240)4 \\
&= 1690
\end{aligned}
$$

or \$1690.

Example 3

Cobb–Douglas Production Function The production for a certain country in the early years following World War II is described by the function

$$f(x, y) = 30x^{2/3}y^{1/3}$$

units, when x units of labour and y units of capital were utilized. Find the approximate change in output if the amount expended on labour had been decreased from 125 units to 123 units and the amount expended on capital had been increased from 27 to 29 units. Is your result as expected given the result of Example 4c, Section 10.2?

Solution

The approximate change in output is given by the total differential of f at $x = 125$, $y = 27$, $dx = 123 - 125 = -2$, and $dy = 29 - 27 = 2$; that is, by

$$
\begin{aligned}
df &= \left.\frac{\partial f}{\partial x}\,dx + \frac{\partial f}{\partial y}\,dy\right|_{\substack{x=125,\ y=27\\ dx=-2,\ dy=2}} \\
&= \left.20x^{-1/3}y^{1/3}\right|_{(125,\ 27)} \cdot (-2) + \left.10x^{2/3}y^{-2/3}\right|_{(125,\ 27)} \cdot (2) \\
&= 20\left(\frac{27}{125}\right)^{1/3}(-2) + 10\left(\frac{125}{27}\right)^{2/3}(2) \\
&= -20\left(\frac{3}{5}\right)(2) + 10\left(\frac{25}{9}\right)(2) = \frac{284}{9}
\end{aligned}
$$

or $31\frac{5}{9}$ units. This result is fully compatible with the result of Example 4, where the recommendation was to encourage increased spending on capital rather than on labour.

If f is a function of the three variables x, y, and z, then the total differential of $w = f(x, y, z)$ is defined to be

$$dw = \frac{\partial f}{\partial x}\,dx + \frac{\partial f}{\partial y}\,dy + \frac{\partial f}{\partial z}\,dz$$

where $dx = \Delta x$, $dy = \Delta y$, and $dz = \Delta z$ are the actual changes in the independent variables x, y, and z as x changes from $x = a$ to $x = a + \Delta x$, y changes from $y = b$ to $y = b + \Delta y$, and z changes from $z = c$ to $z = c + \Delta z$, respectively.

Example 4

Error Analysis Find the maximum percentage error in calculating the volume of a rectangular box if an error of at most 1% is made in measuring the length, width, and height of the box.

Solution

Let x, y, and z denote the length, width, and height, respectively, of the rectangular box. Then the volume of the box is given by $V = f(x, y, z) = xyz$ cubic units. Now suppose the true dimensions of the rectangular box are a, b, and c units, respectively. Since the error committed in measuring the length, width, and height of the box is at most 1%, we have

$$|\Delta x| = |x - a| \leq 0.01a$$
$$|\Delta y| = |y - b| \leq 0.01b$$
$$|\Delta z| = |z - c| \leq 0.01c$$

Therefore, the maximum error in calculating the volume of the box is

$$\begin{aligned}|\Delta V| \approx |dV| &= \left|\frac{\partial f}{\partial x}\,dx + \frac{\partial f}{\partial y}\,dy + \frac{\partial f}{\partial z}\,dz\right|\Bigg|_{x=a,\, y=b,\, z=c} \\ &= |yz\,dx + xz\,dy + xy\,dz|\Bigg|_{x=a,\, y=b,\, z=c} \\ &= |bc\,dx + ac\,dy + ab\,dz| \\ &\leq bc|dx| + ac|dy| + ab|dz| \\ &\leq bc(0.01a) + ac(0.01b) + ab(0.01c) \\ &= (0.03)abc\end{aligned}$$

Since the actual volume of the box is abc cubic units, we see that the maximum percentage error in calculating its volume is

$$\frac{|\Delta V|}{V\Big|_{(a,\, b,\, c)}} \approx \frac{(0.03)abc}{abc} = 0.03$$

—that is, approximately 3%.

GROUP DISCUSSION

Refer to Example 4, where we found the maximum percentage error in calculating the volume of the rectangular box to be *approximately* 3%. What is the precise maximum percentage error?

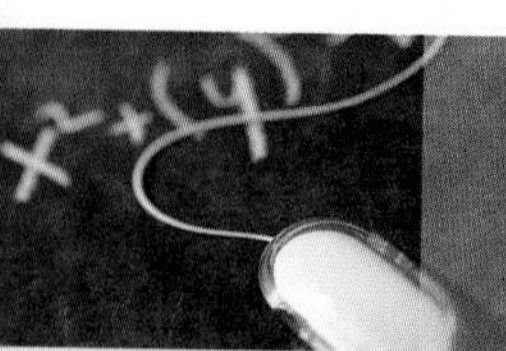

Self-Check Exercise 10.6

Let f be a function defined by $z = f(x, y) = 3xy^2 - 4y$. Find the total differential of f at $(-1, 3)$. Then find the approximate change in z when x changes from $x = -1$ to $x = -0.98$ and y changes from $y = 3$ to $y = 3.01$.

The solution to Self-Check Exercise 10.6 can be found on page 644.

10.6 Exercises

In Exercises 1–18, find the total differential of the function.

1. $f(x, y) = x^2 + 2y$
2. $f(x, y) = 2x^2 + 3y^2$
3. $f(x, y) = 2x^2 - 3xy + 4x$
4. $f(x, y) = xy^3 - x^2y^2$
5. $f(x, y) = \sqrt{x^2 + y^2}$
6. $f(x, y) = (x + 3y^2)^{1/3}$
7. $f(x, y) = \dfrac{5y}{x - y}$
8. $f(x, y) = \dfrac{x + y}{x - y}$
9. $f(x, y) = 2x^5 - ye^{-3x}$
10. $f(x, y) = xye^{x+y}$
11. $f(x, y) = x^2e^y + y \ln x$
12. $f(x, y) = \ln (x^2 + y^2)$
13. $f(x, y, z) = xy^2z^3$
14. $f(x, y, z) = x\sqrt{y} + y\sqrt{z}$
15. $f(x, y, z) = \dfrac{x}{y + z}$
16. $f(x, y, z) = \dfrac{x + y}{y + z}$
17. $f(x, y, z) = xyz + xe^{yz}$
18. $f(x, y, z) = \sqrt{e^x + e^y + ze^{xy}}$

In Exercises 19–30, find the approximate change in z when the point (x, y) changes from (x_0, y_0) to (x_1, y_1).

19. $f(x, y) = 4x^2 - xy$; from (1, 2) to (1.01, 2.02)
20. $f(x, y) = 2x^2 - 2x^3y^2 - y^3$; from (−1, 2) to (−0.98, 2.01)
21. $f(x, y) = x^{2/3}y^{1/2}$; from (8, 9) to (7.97, 9.03)
22. $f(x, y) = \sqrt{x^2 + y^2}$; from (1, 3) to (1.03, 3.03)
23. $f(x, y) = \dfrac{x}{x - y}$; from (−3, −2) to (−3.02, −1.98)
24. $f(x, y) = \dfrac{x - y}{x + y}$; from (−3, −2) to (−3.02, −1.98)
25. $f(x, y) = 2xe^{-y}$; from (4, 0) to (4.03, 0.03)
26. $f(x, y) = \sqrt{xe^y}$; from (1, 1) to (1.01, 0.98)
27. $f(x, y) = xe^{xy} - y^2$; from (−1, 0) to (−0.97, 0.03)
28. $f(x, y) = xe^{-y} + ye^{-x}$; from (1, 1) to (1.01, 0.90)
29. $f(x, y) = x \ln x + y \ln x$; from (2, 3) to (1.98, 2.89)
30. $f(x, y) = \ln(xy)^{1/2}$; from (5, 10) to (5.05, 9.95)
31. **Error in Calculating the Volume of a Cylinder** The radius and height of a right circular cylinder are measured with a maximum error of 0.1 cm in each measurement. Approximate the maximum error in calculating the volume of the cylinder if the measured dimensions $r = 8$ cm and $h = 20$ cm are used.
32. **Error in Calculating Total Resistance** The total resistance R of three resistors with resistance R_1, R_2, and R_3, connected in parallel, is given by the relationship

$$\frac{1}{R} = \frac{1}{R_1} + \frac{1}{R_2} + \frac{1}{R_3}$$

If R_1, R_2, and R_3 are measured at 100, 200, and 300 ohms, respectively, with a maximum error of 1% in each measurement, find the approximate maximum error in the calculated value of R.

Business and Economics Applications

33. **Effect of Inventory and Floor Space on Profit** The monthly profit (in dollars) of a department store depends on the level of inventory x (in thousands of dollars) and the floor space y (in thousands of square metres) available for display of the merchandise, as given by the equation

$$P(x, y) = -0.02x^2 - 15y^2 + xy + 39x + 25y - 20{,}000$$

Currently, the level of inventory is \$4,000,000 ($x = 4000$), and the floor space is 150,000 square metres ($y = 150$). Find the anticipated change in monthly profit if management increases the level of inventory by \$500,000 and decreases the floor space for display of merchandise by 10,000 square metres.

34. **Effect of Production on Profit** The Country Workshop's total weekly profit (in dollars) realized in manufacturing and selling its rolltop desks is given by

$$P(x, y) = -0.2x^2 - 0.25y^2 - 0.2xy + 100x + 90y - 4000$$

where x stands for the number of finished units and y denotes the number of unfinished units manufactured and sold per week. Currently, the weekly output is 190 finished and 105 unfinished units. Determine the approximate change in the total weekly profit if the sole proprietor of the Country Workshop decides to increase the number of finished units to 200 per week and decrease the number of unfinished units to 100 per week.

35. **Revenue of a Travel Agency** The Odyssey Travel Agency's monthly revenue (in thousands of dollars) depends on the amount of money x (in thousands) spent on advertising per month and the number of agents y in its employ in accordance with the rule

$$R(x, y) = -x^2 - 0.5y^2 + xy + 8x + 3y + 20$$

Currently, the amount of money spent on advertising is \$10,000 per month, and there are 15 agents in the agency's employ. Estimate the change in revenue resulting from an increase of \$1000 per month in advertising expenditure and a decrease of 1 agent.

36. **Effect of Capital and Labour on Productivity** The production of a South American country is given by the function

$$f(x, y) = 20x^{3/4}y^{1/4}$$

when x units of labour and y units of capital are utilized. Find the approximate change in output if the amount expended on labour is decreased from 256 to 254 units and the amount expended on capital is increased from 16 to 18 units.

37. **Price–Earnings Ratio** The price–earnings ratio (PE ratio) of a stock is given by

$$R(x, y) = \frac{x}{y}$$

where x denotes the price per share of the stock and y denotes the earnings per share. Estimate the change in the PE ratio of a stock if its price increases from \$60/share to \$62/share while its earnings decrease from \$4/share to \$3.80/share.

38. Effect of Capital and Labour on Productivity The production of a certain company is given by the function

$$f(x, y) = 50x^{1/3}y^{2/3}$$

when x units of labour and y units of capital are utilized. Find the approximate percentage change in the production of the company if labour is increased by 2% and capital is increased by 1%.

Biological and Life Sciences Applications

39. Error in Calculating the Surface Area of a Human The formula

$$S = 0.007184W^{0.425}H^{0.725}$$

gives the surface area S of a human body (in square metres) in terms of its weight W in kilograms and its height H in centimetres. If an error of 1% is made in measuring the weight of a person and an error of 2% is made in measuring the height, what is the percentage error in the measurement of the person's surface area?

40. Error in Measuring Arterial Blood Flow The flow of blood through an arteriole in cubic centimetres per second is given by

$$V = \frac{\pi p r^4}{8kl}$$

where l (in cm) is the length of the arteriole, r (in cm) is its radius, p (in dyne/cm²) is the difference in pressure between the two ends of the arteriole, and k is the viscosity of blood (in dyne-sec/cm²). Find the approximate percentage change in the flow of blood if an error of 2% is made in measuring the length of the arteriole and an error of 1% is made in measuring its radius. Assume that p and k are constant.

Solution to Self-Check Exercise 10.6

We find

$$\frac{\partial f}{\partial x} = 3y^2 \quad \text{and} \quad \frac{\partial f}{\partial y} = 6xy - 4$$

so that

$$\frac{\partial f}{\partial x}(-1, 3) = 3(3)^2 = 27$$

and

$$\frac{\partial f}{\partial y}(-1, 3) = 6(-1)(3) - 4 = -22$$

Therefore, the total differential is

$$dz = \frac{\partial f}{\partial x}(-1, 3)\,dx + \frac{\partial f}{\partial y}(-1, 3)dy$$

$$= 27\,dx - 22\,dy$$

Now $dx = -0.98 - (-1) = 0.02$ and $dy = 3.01 - 3 = 0.01$, so the approximate change in z is

$$dz = 27(0.02) - 22(0.01) = 0.32$$

10.7 Double Integrals (Optional)

A Geometric Interpretation of the Double Integral

To introduce the notion of the integral of a function of two variables, let's first recall the definition of the definite integral of a continuous function of one variable $y = f(x)$ over the interval $[a, b]$. We first divide the interval $[a, b]$ into n subintervals, each of equal length, by the points $x_0 = a < x_1 < x_2 < \cdots < x_n = b$ and define the **Riemann sum** by

$$S_n = f(p_1)h + f(p_2)h + \cdots + f(p_n)h$$

where $h = (b - a)/n$ and p_i is an arbitrary point in the interval $[x_{i-1}, x_i]$. The definite integral of f over $[a, b]$ is defined as the limit of the Riemann sum S_n as n tends to infinity, whenever it exists. Furthermore, recall that when f is a nonnegative continuous function on $[a, b]$, then the ith term of the Riemann sum, $f(p_i)h$, is an approximation (by the area of a rectangle) of the area under that part of the graph of $y = f(x)$ between

$x = x_{i-1}$ and $x = x_i$, so that the Riemann sum S_n provides us with an approximation of the area under the curve $y = f(x)$ from $x = a$ to $x = b$. The integral

$$\int_a^b f(x)\,dx = \lim_{n\to\infty} S_n$$

gives the *actual* area under the curve from $x = a$ to $x = b$.

Now suppose $f(x, y)$ is a continuous function of two variables defined over a region R. For simplicity, we assume for the moment that R is a rectangular region in the plane (Figure 10.32). Let's construct a Riemann sum for this function over the rectangle R by following a procedure that parallels the case for a function of one variable over an interval I. We begin by observing that the analogue of a *partition* in the two-dimensional case is a rectangular grid composed of mn rectangles, each of length h and width k, as a result of partitioning the side of the rectangle R of length $(b - a)$ into m segments and the side of length $(d - c)$ into n segments.

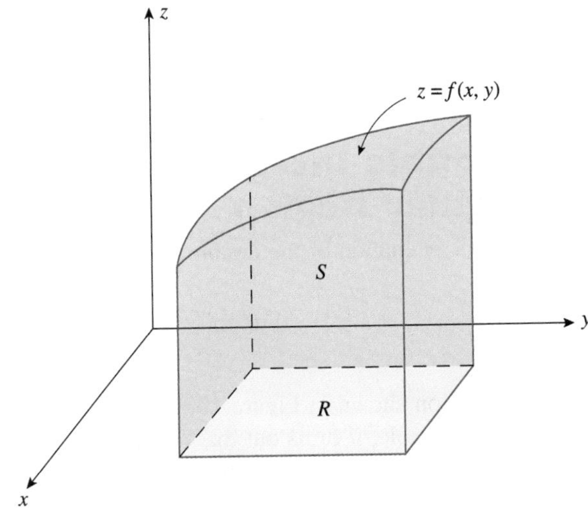

Figure 10.32
$f(x, y)$ is a function defined over a rectangular region R.

By construction

$$h = \frac{b - a}{m} \quad \text{and} \quad k = \frac{d - c}{n}$$

A sample grid with $m = 5$ and $n = 4$ is shown in Figure 10.33.

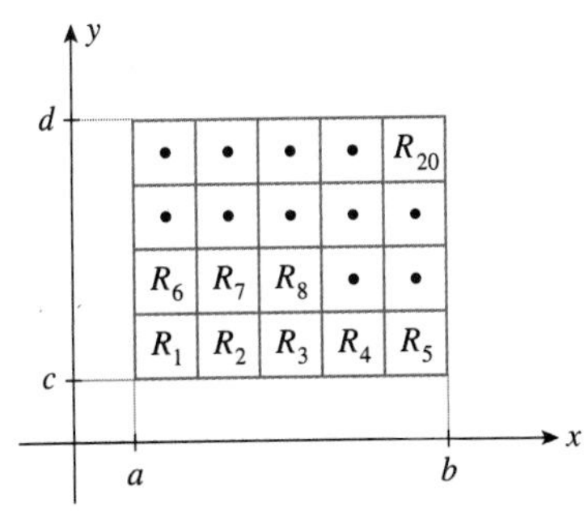

Figure 10.33
Grid with $m = 5$ and $n = 4$

Let's label the rectangles $R_1, R_2, R_3, \ldots, R_{mn}$. If (x_i, y_i) is *any* point in $R_i (1 \le i \le mn)$, then the **Riemann sum of $f(x, y)$ over the region R** is defined as

$$S(m, n) = f(x_1, y_1)hk + f(x_2, y_2)hk + \cdots + f(x_{mn}, y_{mn})hk$$

If the limit of $S(m, n)$ exists as both m and n tend to infinity, we call this limit the value of the **double integral of $f(x, y)$ over the region R** and denote it by

$$\iint_R f(x, y)\,dA$$

If $f(x, y)$ is a nonnegative function, then it defines a solid S bounded above by the graph of f and below by the rectangular region R. Furthermore, the solid S is the union of the mn solids bounded above by the graph of f and below by the mn rectangular regions corresponding to the partition of R (Figure 10.34). The volume of a typical solid S_i can be approximated by a parallelepiped with base R_i and height $f(x_i, y_i)$ (Figure 10.35).

GROUP DISCUSSION

Using a geometric interpretation, evaluate

$$\iint_R \sqrt{4 - x^2 - y^2}\,dA$$

where $R = \{(x, y) | x^2 + y^2 \le 4\}$.

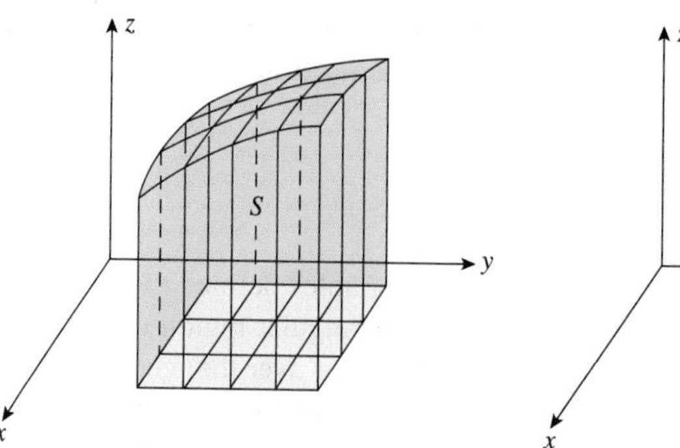

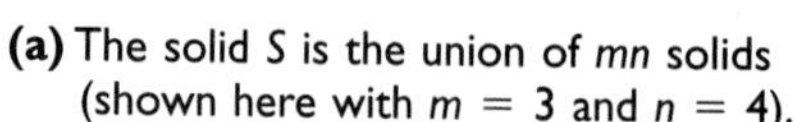
(a) The solid S is the union of mn solids (shown here with $m = 3$ and $n = 4$).

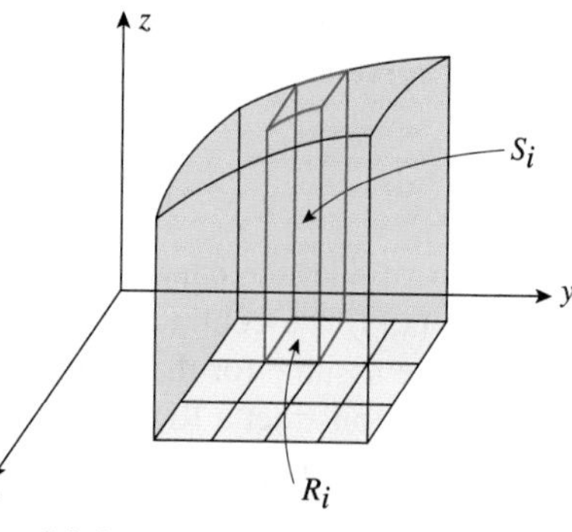

(b) A typical solid S_i is bounded above by the graph of f and lies above R_i.

Figure 10.34

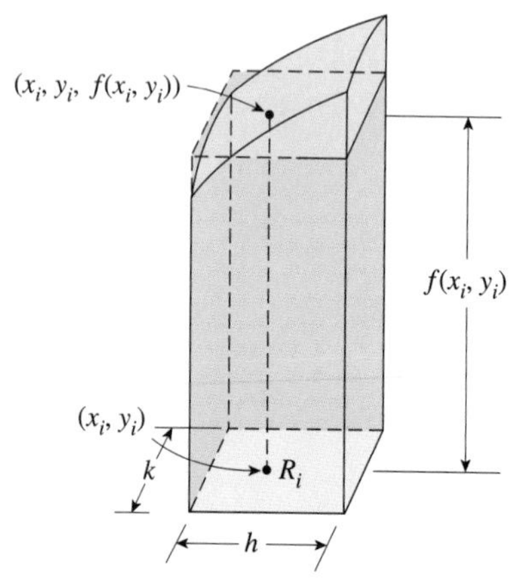

Figure 10.35
The volume of S_i is approximated by the parallelepiped with base R_i and height $f(x_i, y_i)$.

Therefore, the Riemann sum $S(m, n)$ gives us an approximation of the volume of the solid bounded above by the surface $z = f(x, y)$ and below by the plane region R. As both m and n tend to infinity, the Riemann sum $S(m, n)$ approaches the *actual* volume under the solid.

Evaluating a Double Integral over a Rectangular Region

Let's turn our attention to the evaluation of the double integral

$$\iint_R f(x, y)\, dA$$

where R is the rectangular region shown in Figure 10.32. As in the case of the definite integral of a function of one variable, it turns out that the double integral can be evaluated without our having to first find an appropriate Riemann sum and then take the limit of that sum. Instead, as we will now see, the technique calls for evaluating two single integrals—the so-called **iterated integrals**—in succession, using a process that might be called "antipartial differentiation." The technique is described in the following result, which we state without proof.

Let R be the rectangle defined by the inequalities $a \le x \le b$ and $c \le y \le d$ (see Figure 10.33). Then,

$$\iint_R f(x, y)\, dA = \int_c^d \left[\int_a^b f(x, y) dx \right] dy \tag{14}$$

where the iterated integrals on the right-hand side are evaluated as follows. We first compute the integral

$$\int_a^b f(x, y)\, dx$$

by treating y as if it were a constant and integrating the resulting function of x with respect to x (dx reminds us that we are integrating with respect to x). In this manner we obtain a value for the integral that may contain the variable y. Thus,

$$\int_a^b f(x, y)\, dx = g(y)$$

for some function g. Substituting this value into Equation (14) gives

$$\int_c^d g(y)\, dy$$

which may be integrated in the usual manner.

Example 1

Evaluate $\iint\limits_R f(x, y)\,dA$, where $f(x, y) = x + 2y$ and R is the rectangle defined by $1 \le x \le 4$ and $1 \le y \le 2$.

Solution

Using Equation (14), we find

$$\iint\limits_R f(x, y)\, dA = \int_1^2 \left[\int_1^4 (x + 2y)\, dx \right] dy$$

To compute

$$\int_1^4 (x + 2y)\, dx$$

we treat y as if it were a constant (remember that dx reminds us that we are integrating with respect to x). We obtain

$$\begin{aligned}\int_1^4 (x + 2y)\, dx &= \frac{1}{2}x^2 + 2xy \Big|_{x=1}^{x=4} \\ &= \left[\frac{1}{2}(16) + 2(4)y\right] - \left[\frac{1}{2}(1) + 2(1)y\right] \\ &= \frac{15}{2} + 6y\end{aligned}$$

Thus,

$$\begin{aligned}\iint\limits_R f(x, y)\, dA &= \int_1^2 \left(\frac{15}{2} + 6y\right) dy = \left(\frac{15}{2}y + 3y^2\right)\Big|_1^2 \\ &= (15 + 12) - \left(\frac{15}{2} + 3\right) = 16\tfrac{1}{2}\end{aligned}$$

Evaluating a Double Integral over a Plane Region

Up to now we have assumed that the region over which a double integral is to be evaluated is rectangular. In fact, however, it is possible to compute the double integral of functions over rather arbitrary regions. The next theorem, which we state without proof, expands the types of regions over which we may integrate.

THEOREM 1

a. Suppose $g_1(x)$ and $g_2(x)$ are continuous functions on $[a, b]$ and the region R is defined by $R = \{(x, y) | g_1(x) \le y \le g_2(x);\ a \le x \le b\}$. Then,

$$\iint\limits_R f(x, y)\, dA = \int_a^b \left[\int_{g_1(x)}^{g_2(x)} f(x, y)\, dy \right] dx \qquad (15)$$

(Figure 10.36a).

b. Suppose $h_1(y)$ and $h_2(y)$ are continuous functions on $[c, d]$ and the region R is defined by $R = \{(x, y)|h_1(y) \le x \le h_2(y); c \le y \le d\}$. Then,

$$\iint_R f(x, y)\, dA = \int_c^d \left[\int_{h_1(y)}^{h_2(y)} f(x, y) dx \right] dy \qquad \textbf{(16)}$$

(Figure 10.36b).

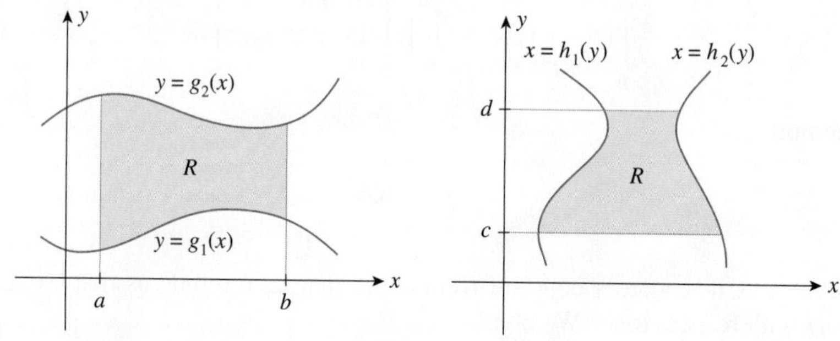

Figure 10.36

Remarks

1. Observe that in (15) the lower and upper limits of integration with respect to y are given by $y = g_1(x)$ and $y = g_2(x)$. This is to be expected since, for a fixed value of x lying between $x = a$ and $x = b$, y runs between the lower curve defined by $y = g_1(x)$ and the upper curve defined by $y = g_2(x)$ (see Figure 10.36a). Observe, too, that in the special case when $g_1(x) = c$ and $g_2(x) = d$, the region R is rectangular, and (15) reduces to (14).
2. For a fixed value of y, x runs between $x = h_1(y)$ and $x = h_2(y)$, giving the indicated limits of integration with respect to x in (16) (see Figure 10.36b).
3. Note that the two curves in Figure 10.36b are not graphs of functions of x (use the vertical-line test), but they are graphs of functions of y. It is this observation that justifies the approach leading to (16). ◀

We now look at several examples.

Example 2

Evaluate $\iint_R f(x, y)\, dA$ given that $f(x, y) = x^2 + y^2$ and R is the region bounded by the graphs of $g_1(x) = x$ and $g_2(x) = 2x$ for $0 \le x \le 2$.

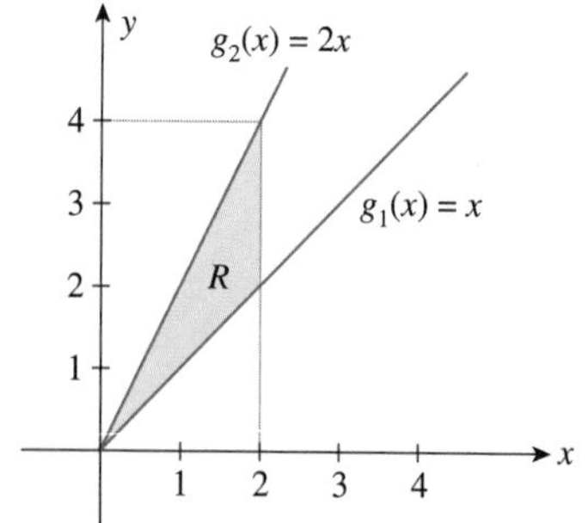

Figure 10.37
R is the region bounded by $g_1(x) = x$ and $g_2(x) = 2x$ for $0 \le x \le 2$.

Solution

The region under consideration is shown in Figure 10.37. Using Equation (15), we find

$$\iint_R f(x, y)\, dA = \int_0^2 \left[\int_x^{2x} (x^2 + y^2) dy \right] dx$$

$$= \int_0^2 \left[\left(x^2 y + \frac{1}{3} y^3 \right) \Big|_x^{2x} \right] dx$$

$$= \int_0^2 \left[\left(2x^3 + \frac{8}{3} x^3 \right) - \left(x^3 + \frac{1}{3} x^3 \right) \right] dx$$

$$= \int_0^2 \frac{10}{3} x^3\, dx = \frac{5}{6} x^4 \Big|_0^2 = 13\frac{1}{3}$$

Example 3

Evaluate $\iint_R f(x, y)\, dA$, where $f(x, y) = xe^y$ and R is the plane region bounded by the graphs of $y = x^2$ and $y = x$.

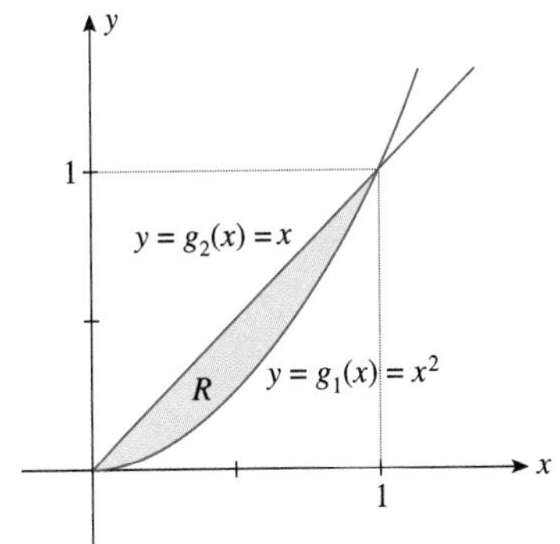

Figure 10.38
R is the region bounded by $y = x^2$ and $y = x$.

Solution

The region in question is shown in Figure 10.38. The point of intersection of the two curves is found by solving the equation $x^2 = x$, giving $x = 0$ and $x = 1$. Using Equation (15), we find

$$\iint_R f(x, y)\, dA = \int_0^1 \left[\int_{x^2}^{x} xe^y\, dy\right] dx = \int_0^1 \left[xe^y\Big|_{x^2}^{x}\right] dx$$

$$= \int_0^1 (xe^x - xe^{x^2})dx = \int_0^1 xe^x dx - \int_0^1 xe^{x^2}\, dx$$

and integrating the first integral on the right-hand side by parts,

$$= \left[(x - 1)e^x - \frac{1}{2}e^{x^2}\right]\Bigg|_0^1$$

$$= -\frac{1}{2}e - \left(-1 - \frac{1}{2}\right) = \frac{1}{2}(3 - e)$$

GROUP DISCUSSION

Refer to Example 3.

1. You can also view the region R as an example of the region shown in Figure 10.34b. Doing so, find the functions h_1 and h_2 and the numbers c and d.
2. Find an expression for $\iint_R f(x, y)\, dA$ in terms of iterated integrals using Formula (16).
3. Evaluate the iterated integrals of part 2 and hence verify the result of Example 3. **Hint:** Integrate by parts twice.
4. Does viewing the region R in two different ways make a difference?

The next example not only illustrates the use of Equation (16) but also shows that it may be the only viable way to evaluate the given double integral.

Example 4

Evaluate

$$\iint_R xe^{y^2}\, dA$$

where R is the plane region bounded by the y-axis, $x = 0$, the horizontal line $y = 4$, and the graph of $y = x^2$.

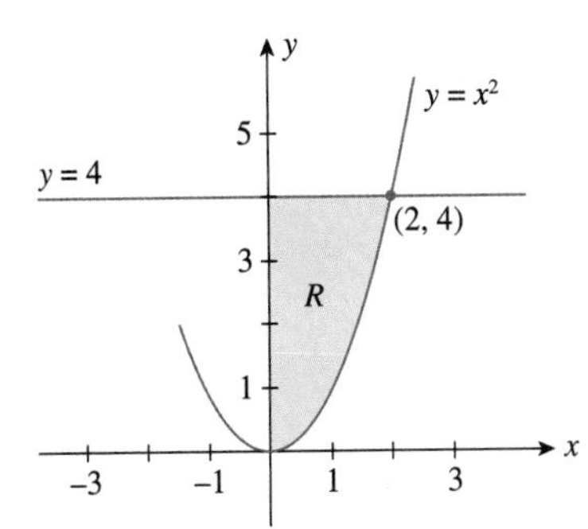

Figure 10.39
R is the region bounded by the y-axis, $x = 0$, $y = 4$, and $y = x^2$.

Solution

The region R is shown in Figure 10.39. The point of intersection of the line $y = 4$ and the graph of $y = x^2$ is found by solving the equation $x^2 = 4$, giving $x = 2$ and the required point (2, 4). Using Equation (15) with $y = g_1(x) = x^2$ and $y = g_2(x) = 4$ leads to

$$\iint_R xe^{y^2}\, dA = \int_0^2 \left[\int_{x^2}^{4} xe^{y^2}\, dy\right] dx$$

Now evaluation of the integral

$$\int_{x^2}^{4} xe^{y^2} dy = x\int_{x^2}^{4} e^{y^2}\, dy$$

calls for finding the antiderivative of the integrand e^{y^2} in terms of elementary functions, a task that, as was pointed out in Section 9.4, cannot be done. Let's begin afresh and attempt to make use of Equation (16).

Since the equation $y = x^2$ is equivalent to the equation $x = \sqrt{y}$, which clearly expresses x as a function of y, we may write, with $x = h_1(y) = 0$ and $h_2(y) = \sqrt{y}$,

$$\iint_R xe^{y^2} dA = \int_0^4 \left[\int_0^{\sqrt{y}} xe^{y^2} dx \right] dy = \int_0^4 \left[\frac{1}{2}x^2 e^{y^2} \Big|_0^{\sqrt{2}} \right] dy$$

$$= \int_0^4 \frac{1}{2} ye^{y^2} dy = \frac{1}{4} e^{y^2} \Big|_0^4 = \frac{1}{4}(e^{16} - 1)$$

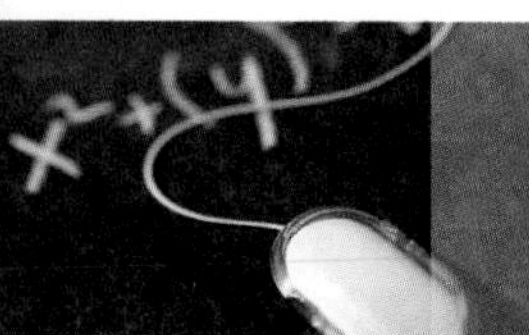

Self-Check Exercise 10.7

Evaluate $\iint_R (x + y)\, dA$, where R is the region bounded by the graphs of $g_1(x) = x$ and $g_2(x) = x^{1/3}$.

The solution to Self-Check Exercise 10.7 can be found on page 651.

10.7 Exercises

In Exercises 1–25, evaluate the double integral $\iint_R f(x, y)\, dA$ for the given function $f(x, y)$ and the region R.

1. $f(x, y) = y + 2x$; R is the rectangle defined by $1 \le x \le 2$ and $0 \le y \le 1$.

2. $f(x, y) = x + 2y$; R is the rectangle defined by $-1 \le x \le 2$ and $0 \le y \le 2$.

3. $f(x, y) = xy^2$; R is the rectangle defined by $-1 \le x \le 1$ and $0 \le y \le 1$.

4. $f(x, y) = 12xy^2 + 8y^3$; R is the rectangle defined by $0 \le x \le 1$ and $0 \le y \le 2$.

5. $f(x, y) = \dfrac{x}{y}$; R is the rectangle defined by $-1 \le x \le 2$ and $1 \le y \le e^3$.

6. $f(x, y) = \dfrac{xy}{1 + y^2}$; R is the rectangle defined by $-2 \le x \le 2$ and $0 \le y \le 1$.

7. $f(x, y) = 4xe^{2x^2+y}$; R is the rectangle defined by $0 \le x \le 1$ and $-2 \le y \le 0$.

8. $f(x, y) = \dfrac{y}{x^2} e^{y/x}$; R is the rectangle defined by $1 \le x \le 2$ and $0 \le y \le 1$.

9. $f(x, y) = \ln y$; R is the rectangle defined by $0 \le x \le 1$ and $1 \le y \le e$.

10. $f(x, y) = \dfrac{\ln y}{x}$; R is the rectangle defined by $1 \le x \le e^2$ and $1 \le y \le e$.

11. $f(x, y) = x + 2y$; R is bounded by $x = 0$, $x = 1$, $y = 0$, and $y = x$.

12. $f(x, y) = xy$; R is bounded by $x = 0$, $x = 1$, $y = 0$, and $y = x$.

13. $f(x, y) = 2x + 4y$; R is bounded by $x = 1$, $x = 3$, $y = 0$, and $y = x + 1$.

14. $f(x, y) = 2 - y$; R is bounded by $x = -1$, $x = 1 - y$, $y = 0$, and $y = 2$.

15. $f(x, y) = x + y$; R is bounded by $x = 0$, $x = \sqrt{y}$, $y = 0$, and $y = 4$.

16. $f(x, y) = x^2y^2$; R is bounded by $x = 0$, $x = 1$, $y = x^2$, and $y = x^3$.

17. $f(x, y) = y$; R is bounded by $x = 0$, $x = \sqrt{4 - y^2}$, $y = 0$, and $y = 2$.

18. $f(x, y) = \dfrac{y}{x^3 + 2}$; R is bounded by $x = 0$, $x = 1$, $y = 0$, and $y = x$.

19. $f(x, y) = 2xe^y$; R is bounded by $x = 0$, $x = 1$, $y = 0$, and $y = x$.

20. $f(x, y) = 2x$; R is bounded by $x = e^{2y}$, $x = y$, $y = 0$, and $y = 1$.

21. $f(x, y) = ye^x$; R is bounded by $y = \sqrt{x}$ and $y = x$.

22. $f(x, y) = xe^{-y^2}$; R is bounded by $x = 0$, $y = x^2$, and $y = 4$.

23. $f(x, y) = e^{y^2}$; R is bounded by $x = 0$, $x = 1$, $y = 2x$, and $y = 2$.

24. $f(x, y) = y$; R is bounded by $x = 1$, $x = e$, $y = 0$, and $y = \ln x$.

25. $f(x, y) = ye^{x^3}$; R is bounded by $x = \dfrac{y}{2}$, $x = 1$, $y = 0$, and $y = 2$.

In Exercises 26 and 27, determine whether the statement is true or false. If it is true, explain why it is true. If it is false, give an example to show why it is false.

26. If $h(x, y) = f(x)g(y)$, where f is continuous on $[a, b]$ and g is continuous on $[c, d]$, then

$$\iint_R h(x, y)\, dA = \left[\int_a^b f(x)\, dx\right]\left[\int_c^d g(y)\, dy\right]$$

where $R = \{(x, y) | a \le x \le b; c \le y \le d\}$.

27. If $\iint_{R_1} f(x, y)\, dA$ exists, where $c \le y \le d\}$, then $\iint_{R_2} f(x, y)\, dA$ exists, where $R_2 = \{(x, y) | c \le x \le d; a \le y \le b\}$.

Solution to Self-Check Exercise 10.7

The region R is shown in the accompanying figure. The points of intersection of the two curves are found by solving the equation $x = x^{1/3}$, giving $x = 0$ and $x = 1$. Using Equation (15), we find

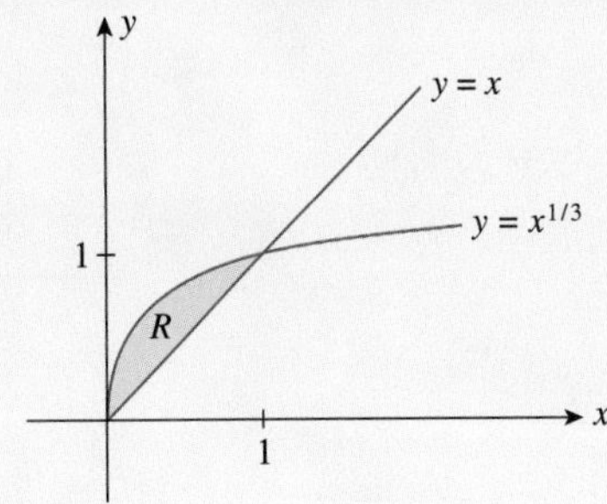

$$\iint_R (x + y)\, dA = \int_0^1 \left[\int_x^{x^{1/3}} (x + y)\, dy\right] dx$$

$$= \int_0^1 \left[xy + \frac{1}{2}y^2\Big|_x^{x^{1/3}}\right] dx$$

$$= \int_0^1 \left[\left(x^{4/3} + \frac{1}{2}x^{2/3}\right) - \left(x^2 + \frac{1}{2}x^2\right)\right] dx$$

$$= \int_0^1 \left(x^{4/3} + \frac{1}{2}x^{2/3} - \frac{3}{2}x^2\right) dx$$

$$= \frac{3}{7}x^{7/3} + \frac{3}{10}x^{5/3} - \frac{1}{2}x^3\Big|_0^1$$

$$= \frac{3}{7} + \frac{3}{10} - \frac{1}{2} = \frac{8}{35}$$

10.8 Applications of Double Integrals (Optional)

In this section we will give some sample applications involving the double integral.

Finding the Volume of a Solid by Double Integrals

As we saw earlier, the double integral

$$\iint_R f(x, y)\, dA$$

gives the volume of the solid bounded by the graph of $f(x, y)$ over the region R.

The Volume of a Solid under a Surface

Let R be a region in the xy-plane and let f be continuous and nonnegative on R. Then, the volume of the solid bounded above by the surface $z = f(x, y)$ and below by R is given by

$$V = \iint_R f(x, y)\, dA$$

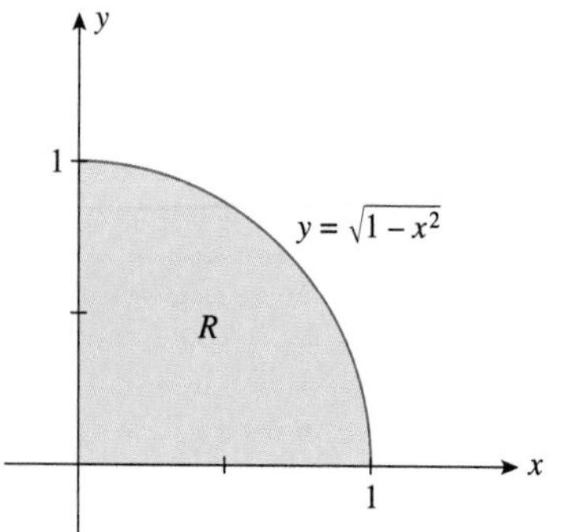

Figure 10.40
The plane region R defined by $y = \sqrt{1 - x^2}$ $(0 \le x \le 1)$

Example 1

Find the volume of the solid bounded above by the plane $z = f(x, y) = y$ and below by the plane region R defined by $y = \sqrt{1 - x^2}$ $(0 \le x \le 1)$.

Solution

The region R is sketched in Figure 10.40. Observe that $f(x, y) = y \ge 0$ for $(x, y) \in R$. Therefore, the required volume is given by

$$\iint_R y\, dA = \int_0^1 \left[\int_0^{\sqrt{1-x^2}} y\, dy \right] dx = \int_0^1 \left[\frac{1}{2} y^2 \Big|_0^{\sqrt{1-x^2}} \right] dx$$

$$= \int_0^1 \frac{1}{2}(1 - x^2)\, dx = \frac{1}{2}\left(x - \frac{1}{3}x^3\right)\Big|_0^1 = \frac{1}{3}$$

or $\frac{1}{3}$ cubic unit. The solid is shown in Figure 10.41. Note that it is not necessary to make a sketch of the solid in order to compute its volume.

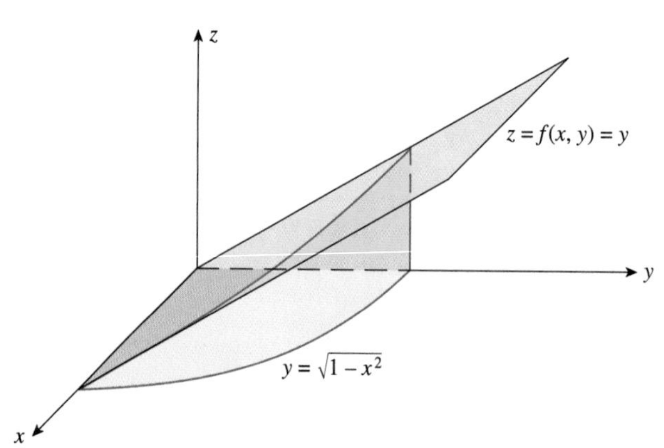

Figure 10.41
The solid bounded above by the plane $z = y$ and below by the plane region defined by $y = \sqrt{1 - x^2}$ $(0 \le x \le 1)$

Population of a City

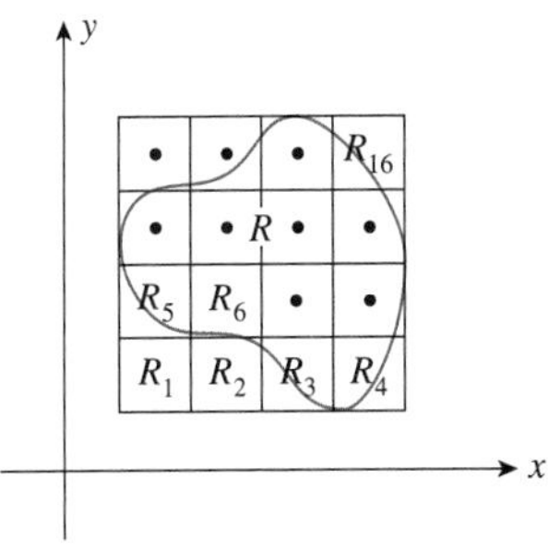

Figure 10.42

The rectangular region R representing a certain district of a city is enclosed by a rectangular grid.

Suppose the plane region R represents a certain district of a city and $f(x, y)$ gives the population density (the number of people per square kilometre) at any point (x, y) in R. Enclose the set R by a rectangle and construct a grid for it in the usual manner. In any rectangular region of the grid that has no point in common with R, set $f(x_i, y_i)hk = 0$ (Figure 10.42). Then, corresponding to any grid covering the set R, the general term of the Riemann sum $f(x_i, y_i)hk$ (population density times area) gives the number of people living in that part of the city corresponding to the rectangular region R_i. Therefore, the Riemann sum gives an approximation of the number of people living in the district represented by R and, in the limit, the double integral

$$\iint_R f(x, y)\, dA$$

gives the actual number of people living in the district under consideration.

Example 2

Population Density The population density of a certain city is described by the function

$$f(x, y) = 10{,}000e^{-0.2|x|-0.1|y|}$$

where the origin (0, 0) gives the location of the city hall. What is the population inside the rectangular area described by

$$R = \{(x, y) | -10 \le x \le 10; -5 \le y \le 5\}$$

where x and y are measured in kilometres? (See Figure 10.43.)

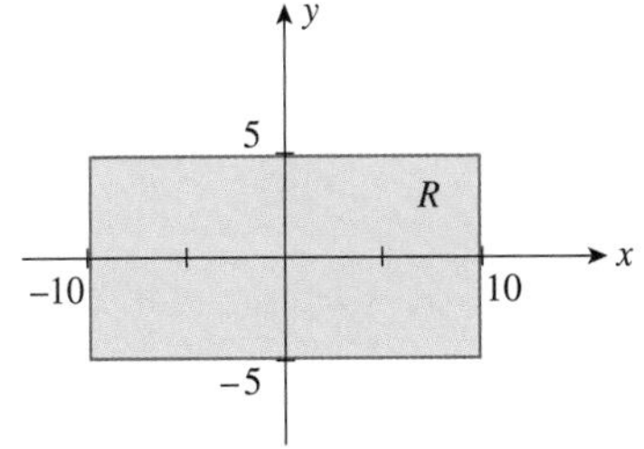

Figure 10.43

The rectangular region R represents a certain district of a city.

Solution

By symmetry, it suffices to compute the population in the first quadrant. (Why?) Then, upon observing that in this quadrant

$$f(x, y) = 10{,}000e^{-0.2x-0.1y} = 10{,}000e^{-0.2x}\, e^{-0.1y}$$

we see that the population in R is given by

$$\begin{aligned}\iint_R f(x, y)\, dA &= 4\int_0^{10}\left[\int_0^5 10{,}000e^{-0.2x}\, e^{-0.1y}\, dy\right] dx \\ &= 4\int_0^{10}\left[-100{,}000e^{-0.2x}\, e^{-0.1y}\Big|_0^5\right] dx \\ &= 400{,}000(1 - e^{-0.5})\int_0^{10} e^{-0.2x}\, dx \\ &= 2{,}000{,}000(1 - e^{-0.5})(1 - e^{-2})\end{aligned}$$

or approximately 680,438 people.

GROUP DISCUSSION

1. Consider the improper double integral $\iint_D f(x, y)\, dA$ of the continuous function f of two variables defined over the plane region

$$D = \{(x, y) | 0 \le x < \infty; 0 \le y < \infty\}$$

Using the definition of improper integrals of functions of one variable (Section 9.5), explain why it makes sense to define

$$\iint_D f(x, y)\, dA = \lim_{N\to\infty} \int_0^N \left[\lim_{M\to\infty} \int_0^M f(x, y)\, dx \right] dy$$

$$= \lim_{M\to\infty} \int_0^M \left[\lim_{N\to\infty} \int_0^N f(x, y)\, dy \right] dx$$

provided the limits exist.

2. Refer to Example 2. Assuming that the population density of the city is described by

$$f(x, y) = 10{,}000e^{-0.2|x|-0.1|y|}$$

for $-\infty < x < \infty$ and $-\infty < y < \infty$, show that the population outside the rectangular region

$$R = \{(x, y) | -10 < x < 10; -5 < y \leq 5\}$$

of Example 2 is given by

$$4 \iint_D f(x, y)\, dx\, dy - 680{,}438$$

(recall that 680,438 is the approximate population inside R).

3. Use the results of parts 1 and 2 to determine the population of the city outside the rectangular area R.

Average Value of a Function

In Section 8.5 we showed that the average value of a continuous function $f(x)$ over an interval $[a, b]$ is given by

$$\frac{1}{b-a} \int_a^b f(x)\, dx$$

That is, the average value of a function over $[a, b]$ is the integral of f over $[a, b]$ divided by the length of the interval. An analogous result holds for a function of two variables $f(x, y)$ over a plane region R. To see this, we enclose R by a rectangle and construct a rectangular grid. Let (x_i, y_i) be any point in the rectangle R_i of area hk. Now, the average value of the mn numbers $f(x_1, y_1), f(x_2, y_2), \ldots, f(x_{mn}, y_{mn})$ is given by

$$\frac{f(x_1, y_1) + f(x_2, y_2) + \cdots + f(x_{mn}, y_{mn})}{mn}$$

which can also be written as

$$\frac{hk}{hk}\left[\frac{f(x_1, y_1) + f(x_2, y_2) + \cdots + f(x_{mn}, y_{mn})}{mn}\right]$$

$$= \frac{1}{(mn)hk}[f(x_1, y_1) + f(x_2, y_2) + \cdots + f(x_{mn}, y_{mn})]hk$$

Now the area of R is approximated by the sum of the mn rectangles (*omitting* those having no points in common with R), each of area hk. Note that this is the denominator of the previous expression. Therefore, taking the limit as m and n both tend to infinity, we obtain the following formula for the *average value of $f(x, y)$ over R.*

Average Value of $f(x, y)$ over the Region R

If f is integrable over the plane region R, then its average value over R is given by

$$\frac{\displaystyle\iint_R f(x, y)\, dA}{\text{Area of } R} \quad \text{or} \quad \frac{\displaystyle\iint_R f(x, y)\, dA}{\displaystyle\iint_R dA} \tag{17}$$

Remark

If we let $f(x, y) = 1$ for all (x, y) in R, then

$$\iint_R f(x, y)\, dA = \iint_R dA = \text{Area of } R \quad ◀$$

Example 3

Find the average value of the function $f(x, y) = xy$ over the plane region defined by $y = e^x$ $(0 \leq x \leq 1)$.

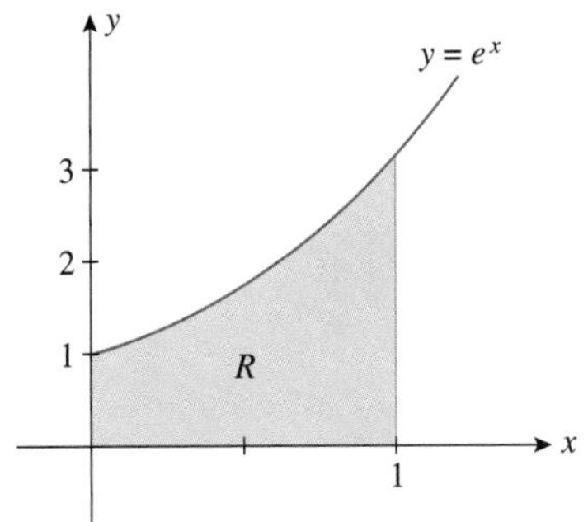

Figure 10.44

The plane region R defined by $y = e^x$ $(0 \leq x \leq 1)$

Solution

The region R is shown in Figure 10.44. The area of the region R is given by

$$\begin{aligned}\int_0^1 \left[\int_0^{e^x} dy\right] dx &= \int_0^1 \left[y\Big|_0^{e^x}\right] dx \\ &= \int_0^1 e^x\, dx \\ &= e^x\Big|_0^1 \\ &= e - 1\end{aligned}$$

square units. We would obtain the same result had we viewed the area of this region as the area of the region under the curve $y = e^x$ from $x = 0$ to $x = 1$. Next, we compute

$$\begin{aligned}\iint_R f(x, y)\, dA &= \int_0^1 \left[\int_0^{e^x} xy\, dy\right] dx \\ &= \int_0^1 \left[\frac{1}{2} xy^2\Big|_0^{e^x}\right] dx \\ &= \int_0^1 \frac{1}{2} xe^{2x}\, dx \\ &= \frac{1}{4} xe^{2x} - \frac{1}{8} e^{2x}\Big|_0^1 \qquad \text{Integrate by parts.} \\ &= \left(\frac{1}{4} e^2 - \frac{1}{8} e^2\right) + \frac{1}{8} \\ &= \frac{1}{8} (e^2 + 1)\end{aligned}$$

square units. Therefore, the required average value is given by

$$\frac{\iint_R f(x, y)\, dA}{\iint_R dA} = \frac{\frac{1}{8}(e^2 + 1)}{e - 1} = \frac{e^2 + 1}{8(e - 1)}$$

Example 4

Population Density (Refer to Example 2.) The population density of a certain city (number of people per square kilometre) is described by the function

$$f(x, y) = 10{,}000e^{-0.2|x|-0.1|y|}$$

where the origin gives the location of the city hall. What is the average population density inside the rectangular area described by

$$R = \{(x, y)| - 10 \le x \le 10; -5 \le y \le 5\}$$

where x and y are measured in kilometres?

Solution

From the results of Example 2, we know that

$$\iint_R f(x, y)\, dA \approx 680{,}438$$

From Figure 10.43, we see that the area of the plane rectangular region R is (20)(10), or 200, square kilometres. Therefore, the average population inside R is

$$\frac{\iint_R f(x, y)\, dA}{\iint_R dA} = \frac{680{,}438}{200} = 3402.19$$

or approximately 3402 people per square kilometre.

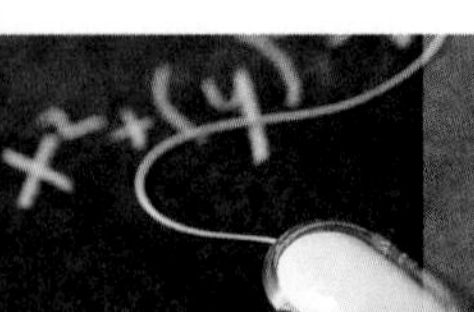

Self-Check Exercise 10.8

The population density of a coastal area located on an island is described by the function

$$f(x, y) = \frac{5000xe^y}{1 + 2x^2} \qquad (0 \le x \le 4; -2 \le y \le 0)$$

where x and y are measured in kilometres (see the accompanying figure).

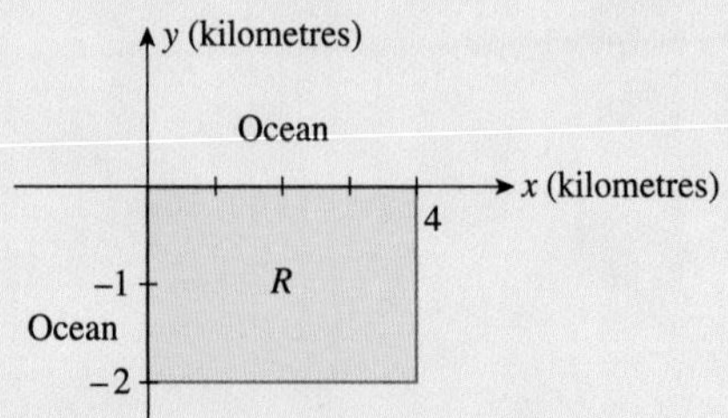

What is the population inside the rectangular area defined by $R = \{(x, y)|0 \le x \le 4; -2 \le y \le 0\}$? What is the average population density in the area?

The solution to Self-Check Exercise 10.8 can be found on page 659.

10.8 Exercises

In Exercises 1–8, use a double integral to find the volume of the solid shown in the figure.

1.

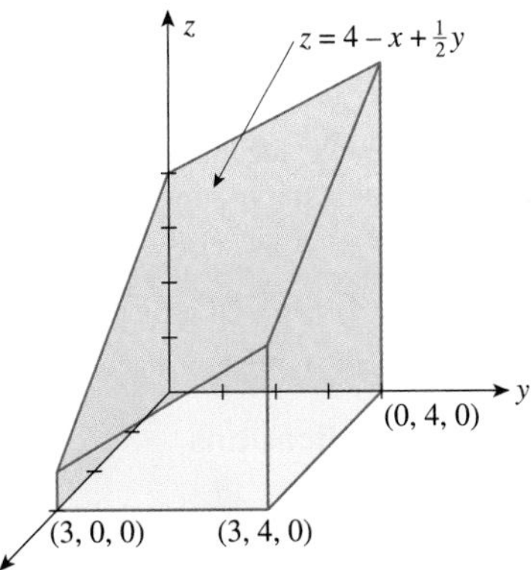

2.

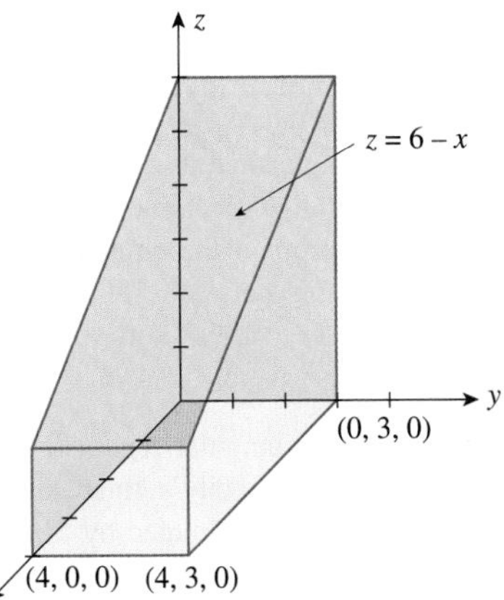

3.

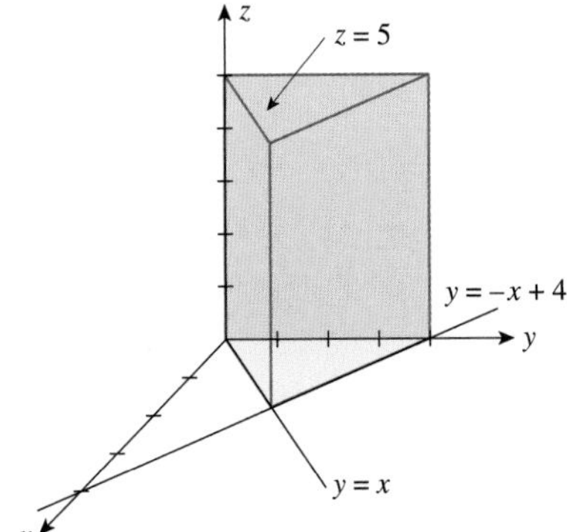

4.

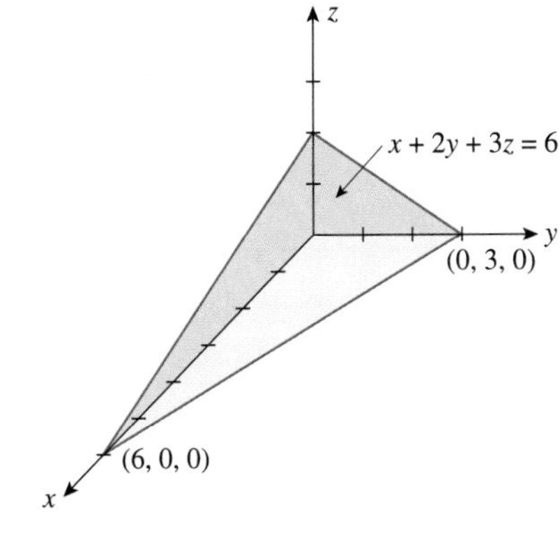

5.

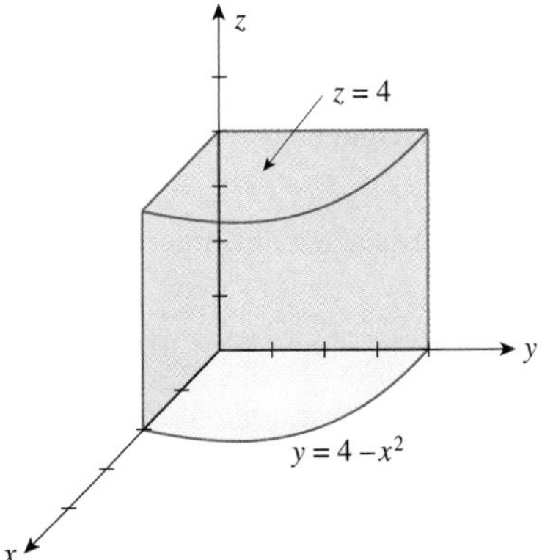

6.

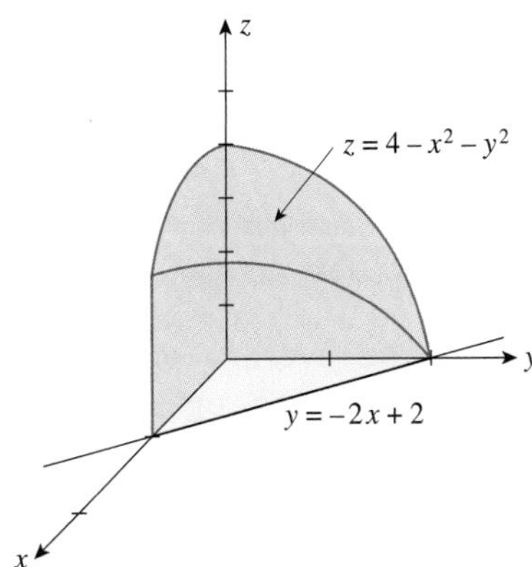

7.

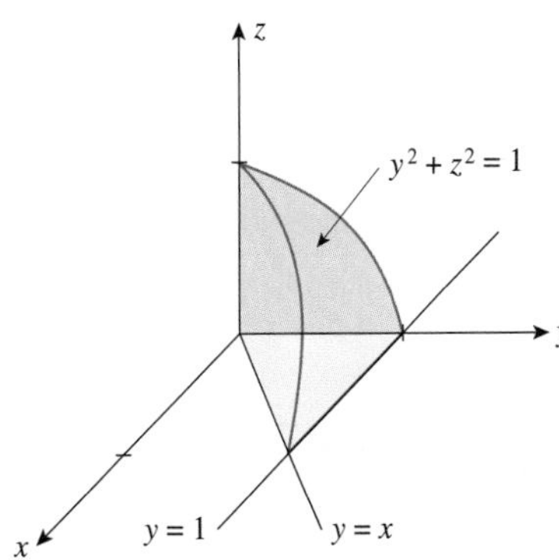

8.

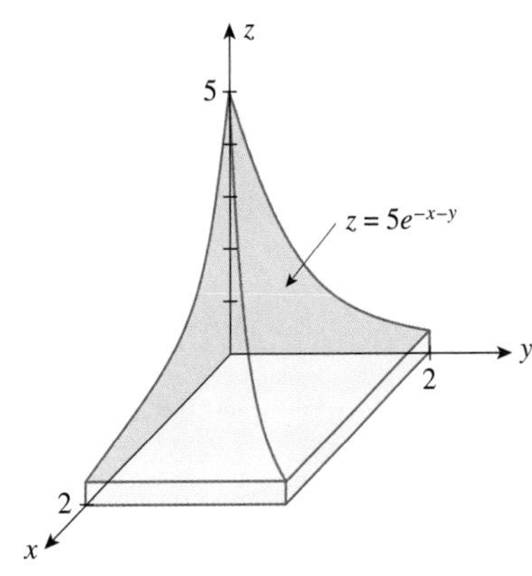

In Exercises 9–16, find the volume of the solid bounded above by the surface $z = f(x, y)$ and below by the plane region R.

9. $f(x, y) = 4 - 2x - y$; $R = \{(x, y) | 0 \le x \le 1; 0 \le y \le 2\}$

10. $f(x, y) = 2x + y$; R is the triangle bounded by $y = 2x$, $y = 0$, and $x = 2$.

11. $f(x, y) = x^2 + y^2$; R is the rectangle with vertices (0, 0), (1, 0), (1, 2), and (0, 2).

12. $f(x, y) = e^{x+2y}$; R is the triangle with vertices (0, 0), (1, 0), and (0, 1).

13. $f(x, y) = 2xe^y$; R is the triangle bounded by $y = x$, $y = 2$, and $x = 0$.

14. $f(x, y) = \dfrac{2y}{1 + x^2}$; R is the region bounded by $y = \sqrt{x}$, $y = 0$, and $x = 4$.

15. $f(x, y) = 2x^2y$; R is the region bounded by the graphs of $y = x$ and $y = x^2$.

16. $f(x, y) = x$; R is the region in the first quadrant bounded by the semicircle $y = \sqrt{16 - x^2}$, the x-axis, and the y-axis.

In Exercises 17–22, find the average value of the given function $f(x, y)$ over the plane region R.

17. $f(x, y) = 6x^2y^3$; $R = \{(x, y) | 0 \le x \le 2; 0 \le y \le 3\}$

18. $f(x, y) = x + 2y$; R is the triangle with vertices (0, 0), (1, 0), and (1, 1).

19. $f(x, y) = xy$; R is the triangle bounded by $y = x$, $y = 2 - x$, and $y = 0$.

20. $f(x, y) = e^{-x^2}$; R is the triangle with vertices (0, 0), (1, 0), and (1, 1).

21. $f(x, y) = xe^y$; R is the triangle with vertices (0, 0), (1, 0), and (1, 1).

22. $f(x, y) = \ln x$; R is the region bounded by the graphs of $y = 2x$ and $y = 0$ from $x = 1$ to $x = 3$.
Hint: Use integration by parts.

23. Population Density The population density of a coastal town is described by the function

$$f(x, y) = \frac{10{,}000e^y}{1 + 0.5|x|} \quad (-10 \le x \le 10; -4 \le y \le 0)$$

where x and y are measured in kilometres (see the accompanying figure). Find the population inside the rectangular area described by

$$R = \{(x, y) | -5 \le x \le 5; -2 \le y \le 0\}$$

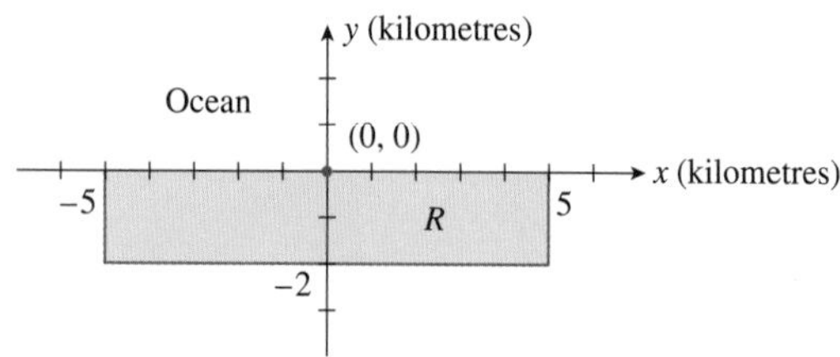

24 Average Population Density Refer to Exercise 23. Find the average population density inside the rectangular area R.

25. Population Density The population density of a certain city is given by the function

$$f(x, y) = \frac{50{,}000|xy|}{(x^2 + 20)(y^2 + 36)}$$

where the origin (0, 0) gives the location of the government centre. Find the population inside the rectangular area described by

$$R = \{(x, y) | -15 \le x \le 15; -20 \le y \le 20\}$$

Business and Economics Applications

26. Average Profit The Country Workshop's total weekly profit (in dollars) realized in manufacturing and selling its rolltop desks is given by the profit function

$$P(x, y) = -0.2x^2 - 0.25y^2 - 0.2xy + 100x + 90y - 4000$$

where x stands for the number of finished units and y stands for the number of unfinished units manufactured and sold each week. Find the average weekly profit if the number of finished units manufactured and sold varies between 180 and 200 and the number of unfinished units varies between 100 and 120/week.

27. Average Price of Land The rectangular region R shown in the accompanying figure represents a city's financial district. The price of land in the district is approximated by the function

$$p(x, y) = 200 - 10\left(x - \frac{1}{2}\right)^2 - 15(y - 1)^2$$

where $p(x, y)$ is the price of land at the point (x, y) in dollars/square metre and x and y are measured in kilometres. What is the average price of land per square metre in the district?

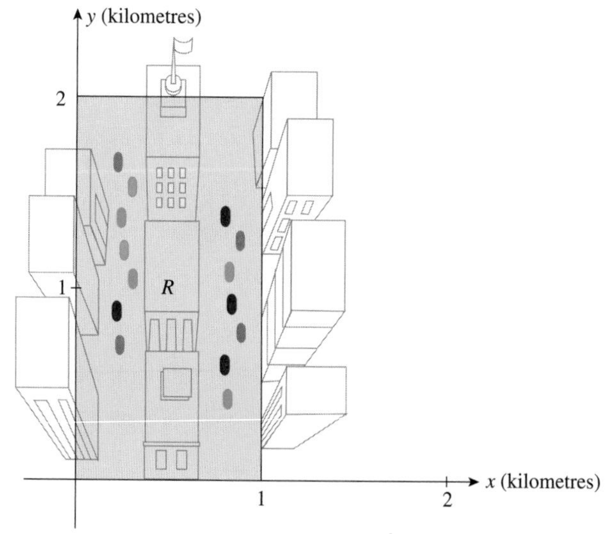

In Exercises 28–29, determine whether the statement is true or false. If it is true, explain why it is true. If it is false, give an example to show why it is false.

28. Let R be a region in the xy-plane and let f and g be continuous functions on R that satisfy the condition $f(x, y) \le g(x, y)$ for all (x, y) in R. Then, $\iint_R [g(x, y) - f(x, y)]\, dA$ gives the volume of the solid bounded above by the surface $z = g(x, y)$ and below by the surface $z = f(x, y)$.

29. Suppose f is nonnegative and integrable over the plane region R. Then, the average value of f over R can be thought of as the (constant) height of the cylinder with base R and volume that is exactly equal to the volume of the solid under the graph of $z = f(x, y)$. (*Note:* The cylinder referred to here has sides perpendicular to R.)

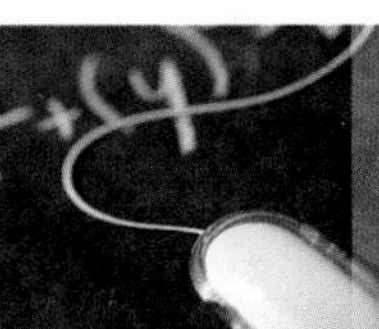

Solution to Self-Check Exercise 10.8

The population in R is given by

$$\iint_R f(x, y)\, dA = \int_0^4 \left[\int_{-2}^0 \frac{5000xe^{y}}{1 + 2^{2}}\, dy \right] dx$$

$$= \int_0^4 \left[\frac{5000xe^y}{1 + 2x^2} \Bigg|_{-2}^{0} \right] dx$$

$$= 5000(1 - e^{-2}) \int_0^4 \frac{x}{1 + 2x^2}\, dx$$

$$= 5000(1 - e^{-2}) \left[\frac{1}{4} \ln\, (1 + 2x^2) \Bigg|_0^4 \right]$$

$$= 5000(1 - e^{-2}) \left(\frac{1}{4} \right) \ln 33$$

or approximately 3779 people. The average population density inside R is

$$\frac{\iint_R f(x, y)\, dA}{\iint_R dA} = \frac{3779}{(2)(4)}$$

or approximately 472 people/square kilometre.

CHAPTER 10 Summary of Principal Terms

Terms

function of two variables (585)
domain (585)
independent variable (585)
dependent variable (585)
value of f (585)
three-dimensional Cartesian coordinate system (588)
ordered triple (588)
graph (588)
trace (589)
level curve (589)
contour map (589)
isotherm (590)
isobar (591)
level surface (591)
isoprofit surface (591)
first partial derivative of f with respect to x (594)
first partial derivative of f with respect to y (595)
first partial derivatives of f with respect to x (595)

Cobb–Douglas production function (599)
marginal productivity of labour (599)
marginal productivity of capital (599)
substitute commodities (600)
complementary commodities (600)
second-order partial derivative of f (602)
relative maximum (608)
relative maximum value (608)
relative minimum (608)
relative minimum value (608)
absolute maximum (608)
absolute minimum (608)
absolute maximum value (608)
absolute minimum value (608)
saddle point (609)
critical point (610)
second derivative test (610)
trend line (618)
method of least squares (618)
scatter diagram (618)
principle of least squares (618)
least-squares line (regression line) (618)
normal equation (620)
least squares method of fitting a line (624)
unconstrained relative extremum (629)
constrained relative extremum (629)
method of Lagrange multipliers (631)
marginal productivity of money (636)
total differential (640)
Riemann sum (644)
double integral (645)
iterated integrals (646)
volume of a solid under a surface (652)

Review Exercises 

1. Let $f(x, y) = \dfrac{xy}{x^2 + y^2}$. Compute $f(0, 1)$, $f(1, 0)$, and $f(1, 1)$. Does $f(0, 0)$ exist?

2. Let $f(x, y) = \dfrac{xe^y}{1 + \ln xy}$. Compute $f(1, 1)$, $f(1, 2)$, and $f(2, 1)$. Does $f(1, 0)$ exist?

3. Let $h(x, y, z) = xye^z + \dfrac{x}{y}$. Compute $h(1, 1, 0)$, $h(-1, 1, 1)$, and $h(1, -1, 1)$.

4. Find the domain of the function $f(u, v) = \dfrac{\sqrt{u}}{u - v}$.

5. Find the domain of the function $f(x, y) = \dfrac{x - y}{x + y}$.

6. Find the domain of the function $f(x, y) = x\sqrt{y} + y\sqrt{1 - x}$.

7. Find the domain of the function

$$f(x, y, z) = \frac{xy\sqrt{z}}{(1 - x)(1 - y)(1 - z)}$$

In Exercises 8–11, sketch the level curves of the function corresponding to the given values of x.

8. $z = f(x, y) = 2x + 3y$; $z = -2, -1, 0, 1, 2$

9. $z = f(x, y) = y - x^2$; $z = -2, -1, 0, 1, 2$

10. $z = f(x, y) = \sqrt{x^2 + y^2}$; $z = 0, 1, 2, 3, 4$

11. $z = f(x, y) = e^{xy}$; $z = 1, 2, 3$

In Exercises 12–21, compute the first partial derivatives of the function.

12. $f(x, y) = x^2y^3 + 3xy^2 + \dfrac{x}{y}$

13. $f(x, y) = x\sqrt{y} + y\sqrt{x}$

14. $f(u, v) = \sqrt{uv^2 - 2u}$

15. $f(x, y) = \dfrac{x - y}{y + 2x}$

16. $g(x, y) = \dfrac{xy}{x^2 + y^2}$

17. $h(x, y) = (2xy + 3y^2)^5$

18. $f(x, y) = (xe^y + 1)^{1/2}$

19. $f(x, y) = (x^2 + y^2)e^{x^2+y^2}$

20. $f(x, y) = \ln(1 + 2x^2 + 4_y{}^4)$

21. $f(x, y) = \ln\left(1 + \dfrac{x^2}{y^2}\right)$

In Exercises 22–27, compute the second-order partial derivatives of the function.

22. $f(x, y) = x^3 - 2x^2y + y^2 + x - 2y$

23. $f(x, y) = x^4 + 2x^2y^2 - y^4$

24. $f(x, y) = (2x^2 + 3y^2)^3$

25. $(x, y) = \dfrac{x}{x + y^2}$

26. $g(x, y) = e^{x^2+y^2}$

27. $h(s, t) = \ln\left(\dfrac{s}{t}\right)$

28. Let $f(x, y, z) = x^3y^2z + xy^2z + 3xy - 4z$. Compute $f_x(1, 1, 0)$, $f_y(1, 1, 0)$, and $f_z(1, 1, 0)$ and interpret your results.

In Exercises 29–34, find the critical point(s) of the functions. Then use the second derivative test to classify the nature of each of these points, if possible. Finally, determine the relative extrema of each function.

29. $f(x, y) = 2x^2 + y^2 - 8x - 6y + 4$

30. $f(x, y) = x^2 + 3xy + y^2 - 10x - 20y + 12$

31. $f(x, y) = x^3 - 3xy + y^2$

32. $f(x, y) = x^3 + y^2 - 4xy + 17x - 10y + 8$

33. $f(x, y) = e^{2x^2+y^2}$

34. $f(x, y) = \ln(x^2 + y^2 - 2x - 2y + 4)$

In Exercises 35–38, use the method of Lagrange multipliers to optimize the function subject to the given constraints.

35. Maximize the function $f(x, y) = -3x^2 - y^2 + 2xy$ subject to the constraint $2x + y = 4$.

36. Minimize the function $f(x, y) = 2x^2 + 3y^2 - 6xy + 4x - 9y + 10$ subject to the constraint $x + y = 1$.

37. Find the maximum and minimum values of the function $f(x, y) = 2x - 3y + 1$ subject to the constraint $2x^2 + 3y^2 - 125 = 0$.

38. Find the maximum and minimum values of the function $f(x, y) = e^{x-y}$ subject to the constraint $x^2 + y^2 = 1$.

In Exercises 39 and 40, find the total differential of each function at the given point.

39. $f(x, y) = (x^2 + y^4)^{3/2}$, (3, 2)

40. $f(x, y) = xe^{x-y} + x \ln y$; (1, 1)

In Exercises 41 and 42, find the approximate change in z when the point (x, y) changes from (x_0, y_0) to (x_1, y_1).

41. $f(x, y) = 2x^2y^3 + 3y^2x^2 - 2xy$; from (1, −1) to (1.02, −0.98)

42. $f(x, y) = 4x^{3/4}y^{1/4}$; from (16, 81) to (17, 80)

In Exercises 43–46, evaluate the double integrals.

43. $f(x, y) = 3x - 2y$; R is the rectangle defined by $2 \le x \le 4$ and $-1 \le y \le 2$.

44. $f(x, y) = e^{-x-2y}$; R is the rectangle defined by $0 \le x \le 2$ and $0 \le y \le 1$.

45. $f(x, y) = 2x^2y$; R is bounded by $x = 0$, $x = 1$, $y = x^2$, and $y = x^3$.

46. $f(x, y) = \dfrac{y}{x}$; R is bounded by $x = 1$, $x = 2$, $y = 1$, and $y = x$.

In Exercises 47 and 48, find the volume of the solid bounded above by the surface $z = f(x, y)$ and below by the plane region R.

47. $f(x, y) = 4x^2 + y^2$; $R = \{0 \le x \le 2; 0 \le y \le 1\}$

48. $f(x, y) = x + y$; R is the region bounded by $y = x^2$, $y = 4x$, and $y = 4$.

49. Find the average value of the function

$$f(x, y) = xy + 1$$

over the plane region R bounded by $y = x^2$ and $y = 2x$.

50. **Average Daily TV-Viewing Time** The following data were compiled by the Bureau of Television Advertising in a large metropolitan area, giving the average daily TV-viewing time per household in that area over the years 1999 to 2007.

Year	1999	2001	2003	2005	2007
Daily Viewing Time, y	6 hr 9 min	6 hr 30 min	6 hr 36 min	7 hr	7 hr 16 min

a. Find the least-squares line for these data. (Let $x = 1$ represent the year 1999.)
b. Estimate the average daily TV-viewing time per household in the year 2009.

Business and Economics Applications

51. **Revenue Functions** A division of Ditton Industries makes a 16-speed and a 10-speed electric blender. The company's management estimates that x units of the 16-speed model and y units of the 10-speed model are demanded daily when the unit prices are

$$p = 80 - 0.02x - 0.1y$$

$$q = 60 - 0.1x - 0.05y$$

dollars, respectively.
a. Find the daily total revenue function $R(x, y)$.
b. Find the domain of the function R.
c. Compute $R(100, 300)$ and interpret your result.

52. **Demand for CD Players** In a survey conducted by *Home Entertainment* magazine, it was determined that the demand equation for CD players is given by

$$x = f(p, q) = 900 - 9p - e^{0.4q}$$

whereas the demand equation for audio CDs is given by

$$y = g(p, q) = 20{,}000 - 3000q - 4p$$

where p and q denote the unit prices (in dollars) for the CD players and audio CDs, respectively, and x and y denote the number of CD players and audio CDs demanded per week. Determine whether these two products are substitute, complementary, or neither.

53. **Estimating Changes in Profit** The total daily profit function (in dollars) of Weston Publishing Company realized in publishing and selling its English language dictionaries is given by

$$P(x, y) = -0.0005x^2 - 0.003y^2 - 0.002xy + 14x + 12y - 200$$

where x denotes the number of deluxe copies and y denotes the number of standard copies published and sold daily. Currently the number of deluxe and standard copies of the dictionaries published and sold daily are 1000 and 1700, respectively. Determine the approximate daily change in the total daily profit if the number of deluxe copies is increased to 1050 and the number of standard copies is decreased to 1650 per day.

54. **Maximizing Revenue** Odyssey Travel Agency's monthly revenue depends on the amount of money x (in thousands of dollars) spent on advertising per month and the number of agents y in its employ in accordance with the rule

$$R(x, y) = -x^2 - 0.5y^2 + xy + 8x + 3y + 20$$

Determine the amount of money the agency should spend per month and the number of agents it should employ in order to maximize its monthly revenue.

55. **Minimizing Fencing Costs** The owner of the True North Ranch wants to enclose a rectangular piece of grazing land along the straight portion of a river and then subdivide it using a fence running parallel to the sides. No fencing is required along the river. If the material for the sides costs \$3/running metre and the material for the divider costs \$2/running metre, what will be the dimensions of a 303,750-metre pasture if the cost of fencing material is kept to a minimum?

56. **Cobb–Douglas Production Functions** The production of Q units of a commodity is related to the amount of labour x and the amount of capital y (in suitable units) expended by the equation

$$Q = f(x, y) = x^{3/4}y^{1/4}$$

If an expenditure of 100 units is available for production, how should it be apportioned between labour and capital so that Q is maximized?

Hint: Use the method of Lagrange multipliers to maximize the function Q subject to the constraint $x + y = 100$.

11

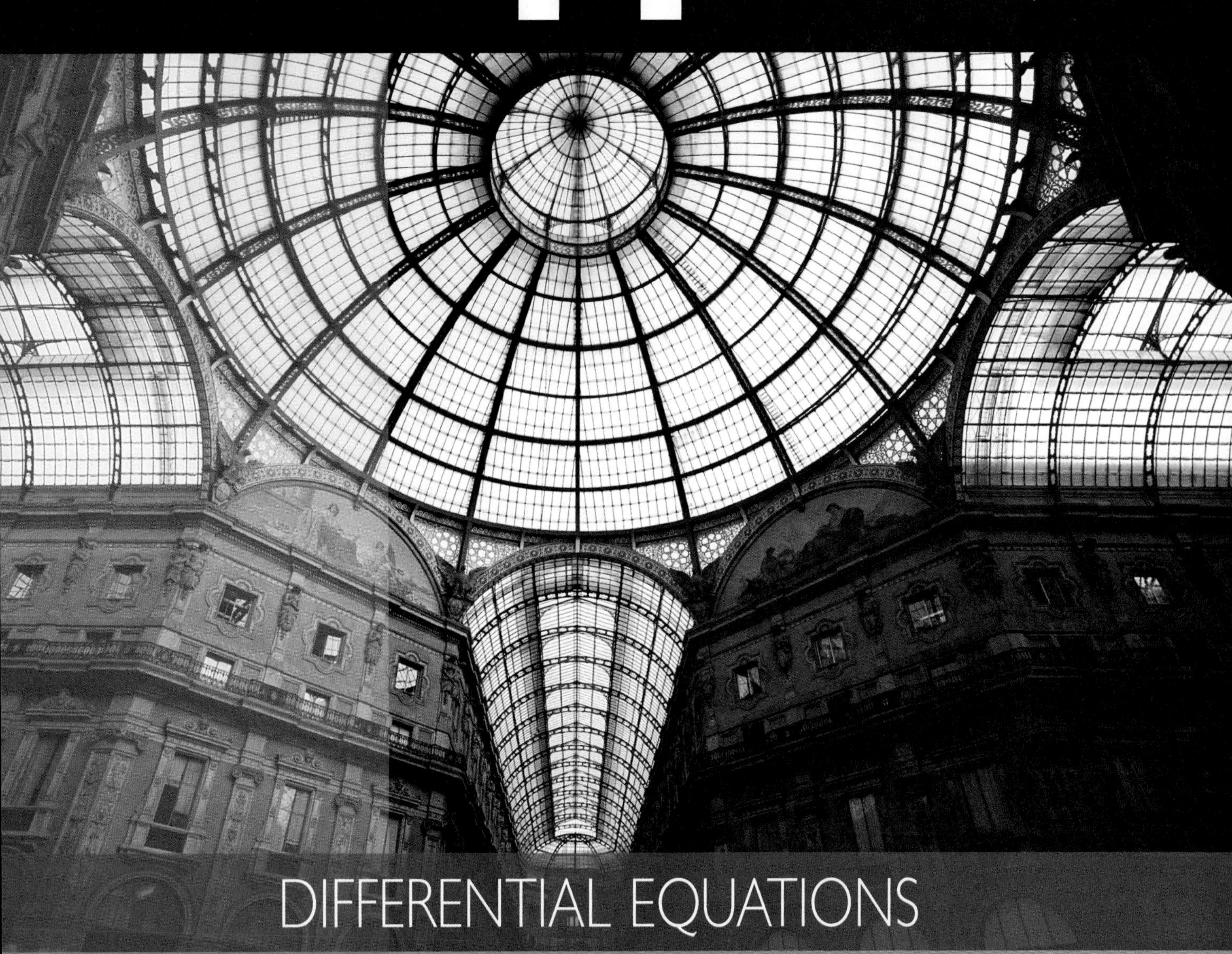

DIFFERENTIAL EQUATIONS

Photo: Thinkstock/Jupiter Images

An equation involving the derivative, or differential, of an unknown function is called a *differential equation.* In this chapter we show how differential equations are used to solve problems involving the growth of an amount of money earning interest compounded continuously, the growth of a population of bacteria, the decay of radioactive material, and the rate at which a person learns a new subject, to name just a few.

Photo: Thinkstock/Jupiter Images

If the amount of fertilizer used to cultivate land on a wheat farm is increased, will the crop yield increase substantially? Researchers at a midwestern university found that a new experimental fertilizer increased the wheat yield of the land at the university's experimental field station. In Example 2, page 678, you will see how they used a differential equation to estimate the crop yield.

Photo: © James Marshall/CORBIS

11.1 Differential Equations

Models Involving Differential Equations

We first encountered differential equations in Section 8.1. Recall that a **differential equation** is an equation that involves an unknown function and its derivative(s). Here are some examples of differential equations:

$$\frac{dy}{dx} = xe^x \qquad \frac{dy}{dx} + 2y = x^2 \qquad \frac{d^2y}{dt^2} + \left(\frac{dy}{dt}\right)^3 + ty - 8 = 0$$

Differential equations appear in practically every branch of applied mathematics, and the study of these equations remains one of the most active areas of research in mathematics. As you will see in the next few examples, models involving differential equations often arise from the mathematical formulation of practical problems.

Unrestricted Growth Models The unrestricted growth model was first discussed in Chapter 5. There we saw that the size of a population at any time t, $Q(t)$, increases at a rate that is proportional to $Q(t)$ itself. Thus,

$$\frac{dQ}{dt} = kQ \tag{1}$$

where k is a constant of proportionality. This is a differential equation involving the unknown function Q and its derivative Q'.

Restricted Growth Models In many applications the quantity $Q(t)$ does not exhibit unrestricted growth but approaches some definite upper bound. The learning curves and logistic functions we discussed in Chapter 5 are examples of restricted growth models. Let's derive the mathematical models that lead to these functions.

Suppose $Q(t)$ does not exceed some number C, called the *carrying capacity of the environment.* Furthermore, suppose the rate of growth of this quantity is *proportional* to the difference between its upper bound and its current size. The resulting differential equation is

$$\frac{dQ}{dt} = k(C - Q) \tag{2}$$

where k is a constant of proportionality. Observe that if the initial population is small relative to C, then the rate of growth of Q is relatively large. But as $Q(t)$ approaches C, the difference $C - Q(t)$ approaches zero, as does the rate of growth of Q. In Section 11.3 you will see that the solution of the differential Equation (2) is a function that describes a learning curve (Figure 11.1).

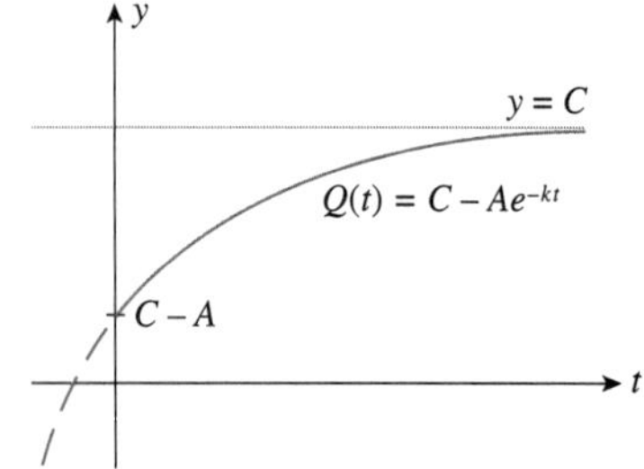

Figure 11.1
$Q(t)$, the solution of a differential equation, describes a learning curve.

Next, let's consider a restricted growth model in which the rate of growth of a quantity $Q(t)$ is *jointly proportional* to its current size and the difference between its upper bound and its current size; that is,

$$\frac{dQ}{dt} = kQ(C - Q) \tag{3}$$

where k is a constant of proportionality. Observe that when $Q(t)$ is small relative to C, the rate of growth of Q is approximately proportional to Q. But as $Q(t)$ approaches C, the growth rate slows down to zero. Next, if $Q > C$, then $dQ/dt < 0$ and the quantity is decreasing with time, with the decay rate slowing down as Q approaches C. We will show later that the solution of the differential Equation (3) is just the logistic function we discussed in Chapter 5. Its graph is shown in Figure 11.2.

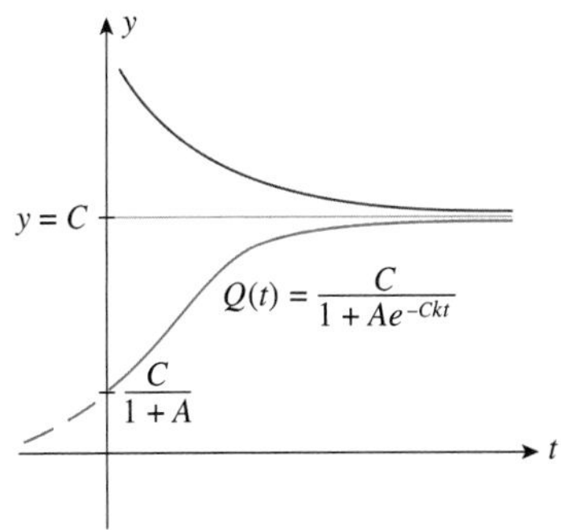

Figure 11.2
Two possible solutions of a differential equation, each of which describes a logistic function

Stimulus Response In the quantitative theory of psychology, one model that describes the relationship between a stimulus S and the resulting response R is the Weber–Fechner law. This law asserts that the rate of change of a reaction R is inversely proportional to the stimulus S. Mathematically, this law may be expressed as

$$\frac{dR}{dS} = \frac{k}{S} \tag{4}$$

where k is a constant of proportionality. Furthermore, suppose that the threshold level, the lowest level of stimulation at which sensation is detected, is S_0. Then we have the condition $R = 0$ when $S = S_0$; that is, $R(S_0) = 0$. The graph of R versus S is shown in Figure 11.3.

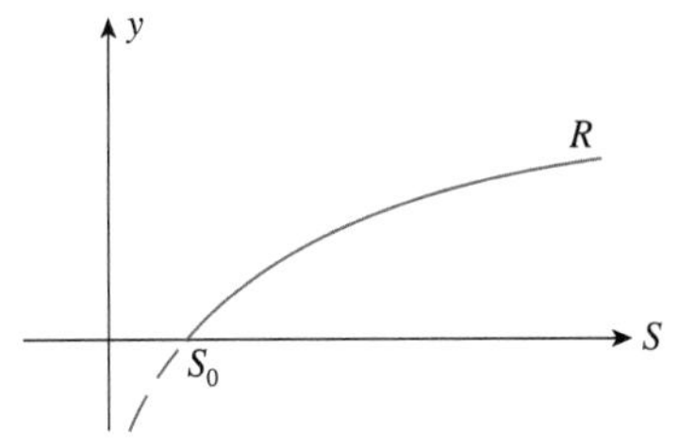

Figure 11.3
R, the solution to a differential equation, describes the response to a stimulus.

Mixture Problems Our next example is a typical mixture problem. Suppose a tank initially contains 40 litres of pure water. Brine containing 1 kilogram of salt per litre flows into the tank at a rate of 8 litres per minute, and the well-stirred mixture flows out of the tank at the same rate. How much salt is in the tank at any given time?

Let's formulate this problem mathematically. Suppose $A(t)$ denotes the amount of salt in the tank at any time t. Then the derivative dA/dt, the rate of change of the amount of salt at any time t, must satisfy the condition

$$\frac{dA}{dt} = (\text{rate of salt flowing in}) - (\text{rate of salt flowing out})$$

(Figure 11.4). But the rate at which salt flows into the tank is given by

$$(8 \text{ L/min})(1 \text{ kg/L}) \qquad \text{(rate of flow)} \times \text{(concentration)}$$

or 8 kilograms per minute. Since the rate at which the solution leaves the tank is the same as the rate at which the brine is poured into it, that tank contains 40 litres of the mixture at any time t. Since the salt content at any time t is A kilograms, the concentration of the mixture is $(A/10)$ kilograms per litre. Therefore, the rate at which salt flows out of the tank is given by

$$(8 \text{ L/min})\left(\frac{A}{40} \text{ kg/L}\right)$$

or $(A/5)$ kilograms per minute. Therefore, we are led to the differential equation

$$\frac{dA}{dt} = 8 - \frac{A}{5} \tag{5}$$

An additional condition arises from the fact that initially there is no salt in the solution. This condition may be expressed mathematically as $A = 0$ when $t = 0$ or, more concisely, $A(0) = 0$.

We will solve each of the differential equations we have introduced here in Section 11.3.

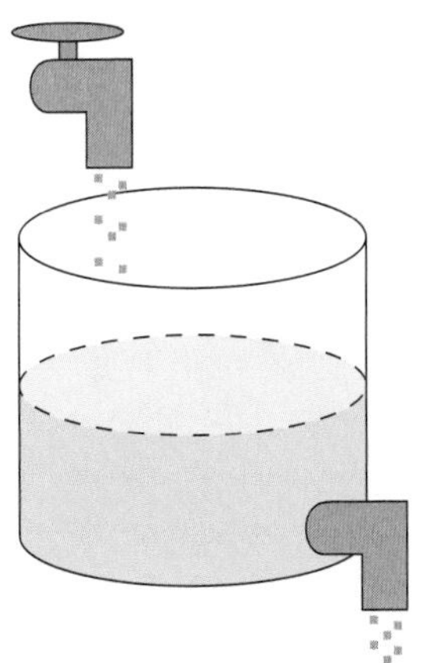

Figure 11.4
The rate of change of the amount of salt at time t = (rate of salt flowing in) − (rate of salt flowing out)

Solutions of Differential Equations

Suppose we are given a differential equation involving the derivative(s) of a function y. Recall that a solution to a differential equation is any function $f(x)$ that satisfies the differential equation. Thus, $y = f(x)$ is a solution of the differential equation, provided that the replacement of y and its derivative(s) by the function $f(x)$ and its corresponding derivatives reduces the given differential equation to an identity for all values of x.

Example 1

Show that the function $f(x) = e^{-x} + x - 1$ is a solution of the differential equation

$$y' + y = x$$

Solution

Let

$$y = f(x) = e^{-x} + x - 1$$

so that

$$y' = f'(x) = -e^{-x} + 1$$

Substituting this last equation into the left side of the given equation yields

$$\overbrace{(-e^{-x} + 1)}^{y'} + \overbrace{(e^{-x} + x - 1)}^{y} = -e^{-x} + 1 + e^{-x} + x - 1 = x$$

which is equal to the right side of the given equation for all values of x. Therefore, $f(x) = e^{-x} + x - 1$ is a solution of the given differential equation.

In Example 1, we verified that $y = e^{-x} + x - 1$ is a solution of the differential equation $y' + y = x$. This is by no means the only solution of the differential equation, as the next example shows.

Example 2

Show that any function of the form $f(x) = ce^{-x} + x - 1$, where c is a constant, is a solution of the differential equation

$$y' + y = x$$

Solution

Let

$$y = f(x) = ce^{-x} + x - 1$$

so that

$$y' = f'(x) = -ce^{-x} + 1$$

Substituting the last equation into the left side of the given differential equation yields

$$\overbrace{-ce^{-x} + 1}^{y'} + \overbrace{ce^{-x} + x - 1}^{y} = x$$

and we have verified the assertion.

It can be shown that *every* solution of the differential equation $y' + y = x$ must have the form $y = ce^{-x} + x - 1$, where c is a constant; therefore, this is the general solution of the differential equation $y' + y = x$. Figure 11.5 shows a family of solutions of this differential equation for selected values of c.

Recall that a solution obtained by assigning a specific value to the constant c is called a particular solution of the differential equation. For example, the particular solution $y = e^{-x} + x - 1$ of Example 1 is obtained from the general solution by taking $c = 1$. In practice, a particular solution of a differential equation is obtained from the general solution of the differential equation by requiring that the solution and/or its derivative(s) satisfy certain conditions at one or more values of x.

Example 3

Use the results of Example 2 to find the particular solution of the equation $y' + y = x$ that satisfies the condition $y(0) = 0$; that is, $f(0) = 0$, where f denotes the solution.

Solution

From the results of Example 2, we see that the general solution of the given differential equation is given by

$$y = f(x) = ce^{-x} + x - 1$$

Using the given condition, we see that

$$f(0) = ce^{0} + 0 - 1 = c - 1 = 0 \qquad \text{or} \qquad c = 1$$

Therefore, the required particular solution is $y = e^{-x} + x - 1$.

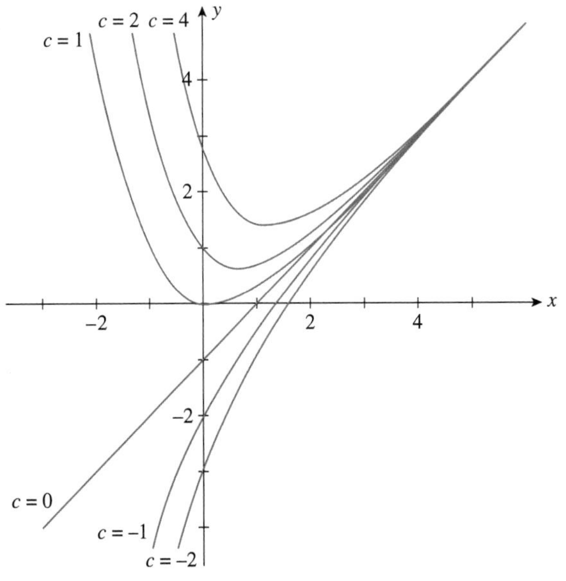

Figure 11.5
Some solutions of $y' + y = x$

GROUP DISCUSSION

Consider the differential equation $dy/dx = F(x, y)$ and suppose $y = f(x)$ is a solution of the differential equation.

1. If (a, b) is a point in the domain of F, explain why $F(a, b)$ gives the slope of f at $x = a$.
2. For the differential equation $dy/dx = x/y$, compute $F(x, y)$ for selected integral values of x and y. (For example, try $x = 0, \pm 1, \pm 2$ and $y = \pm 1, \pm 2, \pm 3$.) Verify that if you draw a lineal element (a tiny line segment) having slope $F(x, y)$ through each point (x, y), you obtain a *direction field* similar to the one shown in the figure:

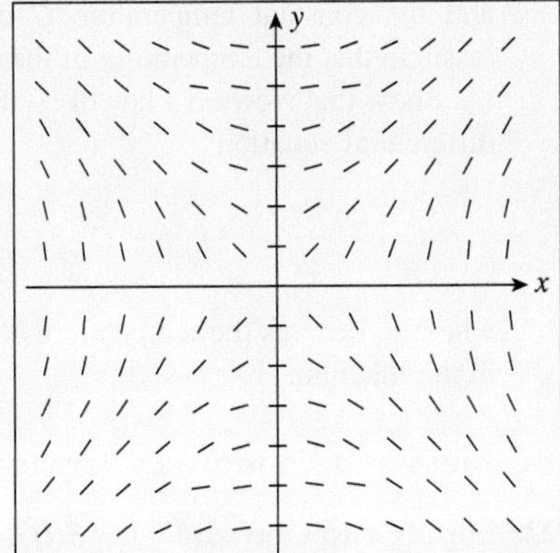

3. The direction field associated with the differential equation hints at the solution curves for the differential equation. Sketch a few solution curves for the differential equation. You will be asked to verify your answer to part 3 in the next section.

Self-Check Exercises 11.1

1. Consider the differential equation

$$xy' + 2y = 4x^2$$

   a. Show that $y = x^2 + (c/x^2)$ is the general solution of the differential equation.
   b. Find the particular solution of the differential equation that satisfies $y(1) = 4$.

2. The population of a certain species grows at a rate directly proportional to the square root of its size. If the initial population is N_0, find the population at any time t. Formulate but do not solve the problem.

Solutions to Self-Check Exercises 11.1 can be found on page 669.

11.1 Exercises

In Exercises 1–12, verify that y is a solution of the differential equation.

1. $y = x^2;\ xy' + y = 3x^2$
2. $y = e^x;\ y' - y = 0$
3. $y = \dfrac{1}{2} + ce^{-x^2}$, c any constant; $y' + 2xy = x$
4. $y = Ce^{kx}$, C any constant; $\dfrac{dy}{dx} = ky$
5. $y = e^{-2x};\ y'' + y' - 2y = 0$
6. $y = C_1e^x + C_2e^{2x};\ y'' - 3y' + 2y = 0$
7. $y = C_1e^{-2x} + C_2xe^{-2x};\ y'' + 4y' + 4y = 0$
8. $y = C_1 + C_2x^{1/3};\ 3xy'' + 2y' = 0$
9. $y = \dfrac{C_1}{x} + C_2\dfrac{\ln x}{x};\ x^2y'' + 3xy' + y = 0$
10. $y = C_1e^x + C_2xe^x + C_3x^2e^x;\ y''' - 3y'' + 3y' - y = 0$
11. $y = C - Ae^{-kt}$, A and C constants; $\dfrac{dy}{dt} = k(C - y)$
12. $y = \dfrac{C}{1 + Ae^{-Ckt}}$, A and C constants; $\dfrac{dy}{dt} = ky(C - y)$

In Exercises 13–18, verify that y is a general solution of the differential equation. Then find a particular solution of the differential equation that satisfies the side condition.

13. $y = Cx^2 - 2x;\ y' - 2\left(\dfrac{y}{x}\right) = 2;\ y(1) = 10$
14. $y = Ce^{-x^2};\ y' = -2xy;\ y(0) = y_0$
15. $y = \dfrac{C}{x};\ y' + \left(\dfrac{1}{x}\right)y = 0;\ y(1) = 1$
16. $y = Ce^{2x} - 2x - 1;\ y' - 2y - 4x = 0;\ y(0) = 3$
17. $y = \dfrac{Ce^x}{x} + \dfrac{1}{2}\ xe^x;\ y' + \left(\dfrac{1 - x}{x}\right)y = e^x;\ y(1) = -\dfrac{1}{2}\ e$
18. $y = C_1x^3 + C_2x^2;\ x^2y'' - 4xy' + 6y = 0;\ y(2) = 0$ and $y'(2) = 4$
19. **Radioactive Decay** A radioactive substance decays at a rate directly proportional to the amount present. If the substance is present in the amount of Q_0 g initially ($t = 0$), find the amount present at any time t. Formulate the problem in terms of a differential equation with a side condition. Do not solve it.
20. **Lambert's Law of Absorption** Lambert's law of absorption states that the percentage of incident light L, absorbed in passing through a thin layer of material x, is proportional to the thickness of the material. If, for a certain material, x_0 cm of the material reduces the light to half its intensity, how much additional material is needed to reduce the intensity to a quarter of its initial value? Formulate but do not solve the problem in terms of a differential equation with a side condition.
21. **Newton's Law of Cooling** Newton's law of cooling states that the temperature of a body drops at a rate that is proportional to the difference between the temperature y of the body and the constant temperature C of the surrounding medium (assume that the temperature of the body is initially greater than C). Show that Newton's law of cooling may be expressed as the differential equation

$$\frac{dy}{dt} = -k(y - C) \qquad [y(0) = y_0]$$

where y_0 denotes the temperature of the body before immersion in the medium.

Business and Economics Applications

22. **Supply and Demand** Let $S(t)$ denote the supply of a certain commodity as a function of time t. Suppose the rate of change of the supply is proportional to the difference between the demand $D(t)$ and the supply. Find a differential equation that describes this situation.
23. **Net Investment** The management of a company has decided that the level of investment should not exceed C dollars. Fur-

thermore, management has decided that the rate of net investment (the rate of change of the total capital invested) should be proportional to the difference between C and the total capital invested. Formulate but do not solve the problem in terms of a differential equation.

Biological and Life Sciences Applications

24. **Concentration of a Drug in the Bloodstream** The rate at which the concentration of a drug in the bloodstream decreases is proportional to the concentration at any time t. Initially, the concentration of the drug in the bloodstream is C_0 g/mL. What is the concentration of the drug in the bloodstream at any time t? Formulate but do not solve the problem in terms of a differential equation with a side condition.

25. **Amount of Glucose in the Bloodstream** Suppose glucose is infused into the bloodstream at a constant rate of C g/min and, at the same time, the glucose is converted and removed from the bloodstream at a rate proportional to the amount of glucose present. Show that the amount of glucose $A(t)$ present in the bloodstream at any time t is governed by the differential equation.

$$A' = C - kA$$

where k is a constant.

26. **Fisk's Law** Suppose a cell of volume V cc is surrounded by a homogeneous chemical solution of concentration C g/cc. Let y denote the concentration of the solute inside the cell at any time t and suppose that, initially, the concentration is y_0. Fisk's law, named after the German physiologist Adolf Fisk (1829–1901), states that the rate of change of the concentration of solute inside the cell at any time t is proportional to the difference between the concentration of the solute outside the cell and the concentration inside the cell and inversely proportional to the volume of the cell. Show that Fisk's law may be expressed as the differential equation

$$\frac{dy}{dt} = \frac{k}{V}(C - y) \qquad [y(0) = y_0]$$

where k is a constant. (*Note:* The constant of proportionality k depends on the area and permeability of the cell membrane.)

27. **Allometric Laws** Suppose $x(t)$ denotes the weight of an animal's organ at time t and $g(t)$ denotes the size of another organ in the same animal at the same time t. An allometric law (allometry is the study of the relative growth of a part in relation to an entire organism) states that the relative growth rate of one organ, $(dx/dt)/x$, is proportional to the relative growth rate of the other, $(dy/dt)/y$. Show that this allometric law may be stated in terms of the differential equation

$$\frac{1}{x}\frac{dx}{dt} = k\frac{1}{y}\frac{dy}{dt}$$

where k is a constant.

28. **Gompertz Growth Curve** Suppose a quantity $Q(t)$ does not exceed some number C; that is, $Q(t) \leq C$ for all t. Suppose further that the rate of growth of $Q(t)$ is jointly proportional to its current size and the difference between its upper bound and the natural logarithm of its current size. What is the size of the quantity $Q(t)$ at any time t? Show that the mathematical formulation of this problem leads to the differential equation

$$\frac{dQ}{dt} = kQ(C - \ln Q) \qquad [Q(0) = Q_0]$$

where Q_0 denotes the size of the quantity present initially. The graph of $Q(t)$ is called the *Gompertz growth curve.* This model, like the ones leading to the learning curve and the logistic curve, describes restricted growth.

In Exercises 29–34, determine whether the statement is true or false. If it is true, explain why it is true. If it is false, give an example to show why it is false.

29. The function $f(x) = x^2 + 2x + \frac{1}{x}$ is a solution of the differential equation $xy' + y = 3x^2 + 4x$.

30. The function $f(x) = \frac{1}{4}e^{3x} + ce^{-x}$ is a solution of the differential equation $y' + y = e^{3x}$.

31. The function $f(x) = 2 + ce^{-x^3}$ is a solution of the differential equation $y' + 3x^2y = x^2$.

32. The function $f(x) = 1 + cx^{-2}$ is a solution of the differential equation $xy' + 2y = 3$.

33. If $y = f(x)$ is a solution of a first-order differential equation, then $y = Cf(x)$ is also a solution.

34. If $y = f(x)$ is a solution of a first-order differential equation, then $y = f(x) + C$ is also a solution.

Solutions to Self-Check Exercises 11.1

1. **a.** We compute

$$y' = 2x - \frac{2c}{x^3}$$

Substituting this into the left side of the given differential equation gives

$$x\left(2x - \frac{2c}{x^3}\right) + 2\left(x^2 + \frac{c}{x^2}\right) = 2x^2 - \frac{2c}{x^2} + 2x^2 + \frac{2c}{x^2} = 4x^2$$

Solutions to Self-Check Exercises 11.1, *continued*

which equals the expression on the right side of the differential equation, and this verifies the assertion.

b. Using the given condition, we have

$$4 = 1^2 + \frac{c}{1^2} \quad \text{or} \quad c = 3$$

and the required particular solution is

$$y = x^2 + \frac{3}{x^2}$$

2. Let N denote the size of the population at any time t. Then the required differential equation is

$$\frac{dN}{dt} = kN^{1/2}$$

and the initial condition is $N(0) = N_0$.

11.2 Separation of Variables

The Method of Separation of Variables

Differential equations are classified according to their basic form. A compelling reason for this categorization is that different methods are used to solve different types of equations.

A differential equation may be classified by the order of its derivative. The **order** of a differential equation is the order of the highest derivative of the unknown function appearing in the equation. For example, the differential equations

$$y' = xe^x \quad \text{and} \quad y' + 2y = x^2$$

are **first-order equations,** whereas the differential equation

$$\frac{d^2y}{dt^2} + \left(\frac{dy}{dt}\right)^3 + ty - 8 = 0$$

is a **second-order equation.** For the remainder of this chapter, except for Section 11.4, we restrict our study to first-order differential equations.

In this section we describe a method for solving an important class of first-order differential equations that can be written in the form

$$\frac{dy}{dx} = f(x)g(y)$$

where $f(x)$ is a function of x only and $g(y)$ is a function of y only. Such differential equations are said to be **separable** because the variables can be separated. Equations (1) through (5) are first-order separable differential equations. As another example, the equation

$$\frac{dQ}{dt} = kQ(C - Q)$$

has the form $dQ/dt = f(t)g(Q)$, where $f(t) = k$ and $g(Q) = Q(C - Q)$, and so is separable. On the other hand, the differential equation

$$\frac{dy}{dx} = xy^2 + 2$$

is *not* separable.

Separable first-order equations can be solved using the **method of separation of variables.**

Method of Separation of Variables

Suppose we are given a first-order separable differential equation in the form

$$\frac{dy}{dx} = f(x)g(y) \qquad (6)$$

Step 1 Write Equation (6) in the form

$$\frac{dy}{g(y)} = f(x)dx \qquad (7)$$

When written in this form, the variables in (7) are said to be *separated.*

Step 2 Integrate each side of Equation (7) with respect to the appropriate variable.

We will justify this method at the end of this section.

Solving Separable Differential Equations

Example 1

Find the general solution of the first-order differential equation

$$y' = \frac{xy}{x^2 + 1}$$

Solution

Step 1 Observe that the given differential equation has the form

$$\frac{dy}{dx} = \left(\frac{x}{x^2 + 1}\right)y = f(x)g(y)$$

where $f(x) = x/(x^2 + 1)$ and $g(y) = y$, and is therefore separable. Separating the variables, we obtain

$$\frac{dy}{y} = \left(\frac{x}{x^2 + 1}\right) dx$$

Step 2 Integrating each side of the last equation with respect to the appropriate variable, we have

$$\int \frac{dy}{y} = \int \frac{x}{x^2 + 1}\, dx$$

or

$$\ln|y| + C_1 = \frac{1}{2}\ln(x^2 + 1) + C_2$$

$$\ln|y| = \frac{1}{2}\ln(x^2 + 1) + C_2 - C_1$$

where C_1 and C_2 are arbitrary constants of integration. Letting C denote the constant such that $C_2 - C_1 = \ln|C|$, we have

$$\begin{aligned} \ln|y| &= \frac{1}{2}\ln(x^2 + 1) + \ln|C| \\ &= \ln\sqrt{x^2 + 1} + \ln|C| \\ &= \ln|C\sqrt{x^2 + 1}| \qquad \ln A + \ln B = \ln AB \end{aligned}$$

so the general solution is

$$y = C\sqrt{x^2 + 1}$$

EXPLORING WITH TECHNOLOGY

Refer to Example 1, where it was shown that the general solution of the given differential equation is $y = C\sqrt{x^2 + 1}$. Use a graphing utility to plot the graphs of the members of this family of solutions corresponding to $C = -3, -2, -1, 0, 1, 2$, and 3. Use the standard viewing window.

Example 2

Find the particular solution of the differential equation

$$ye^x + (y^2 - 1)y' = 0$$

that satisfies the condition $y(0) = 1$.

Solution

Step 1 Writing the given differential equation in the form

$$ye^x + (y^2 - 1)\frac{dy}{dx} = 0 \qquad \text{or} \qquad (y^2 - 1)\frac{dy}{dx} = -ye^x$$

and separating the variables, we obtain

$$\frac{y^2 - 1}{y}\,dy = -e^x\,dx$$

Step 2 Integrating each side of this equation with respect to the appropriate variable, we have

$$\int \frac{y^2 - 1}{y}\,dy = -\int e^x\,dx$$

$$\int \left(y - \frac{1}{y}\right)dy = -\int e^x\,dx$$

$$\frac{1}{2}y^2 - \ln|y| = -e^x + C_1$$

$$y^2 - \ln y^2 = -2e^x + C \qquad (C = 2C_1)$$

Using the condition $y(0) = 1$, we have

$$1 - \ln 1 = -2 + C \qquad \text{or} \qquad C = 3$$

Therefore, the required solution is

$$y^2 - \ln y^2 = -2e^x + 3$$

Example 2 is an initial value problem. In general, an **initial value problem** consists of a differential equation with one or more side conditions specified at a point. Also observe that the solution of Example 2 appeared as an implicit equation involving x and y. This often happens when we solve separable differential equations.

Example 3

Find an equation describing f given that (1) the slope of the tangent line to the graph of f at any point $P(x, y)$ is given by the expression $-x/(2y)$ and (2) the graph of f passes through the point $P(1, 2)$.

Solution

The slope of the tangent line to the graph of f at any point $P(x, y)$ is given by the derivative

$$y' = \frac{dy}{dx} = -\frac{x}{2y}$$

which is a separable first-order differential equation. Separating the variables, we obtain

$$2y\,dy = -x\,dx$$

which, upon integration, yields

$$y^2 = -\frac{1}{2}x^2 + C_1$$

or

$$x^2 + 2y^2 = C \qquad (C = 2C_1)$$

where C is an arbitrary constant.

To evaluate C, we use the second condition, which implies that when $x = 1$, $y = 2$. This gives

$$1^2 + 2(2^2) = C \qquad \text{or} \qquad C = 9$$

Hence, the required equation is

$$x^2 + 2y^2 = 9$$

The graph of the equation f appears in Figure 11.6.

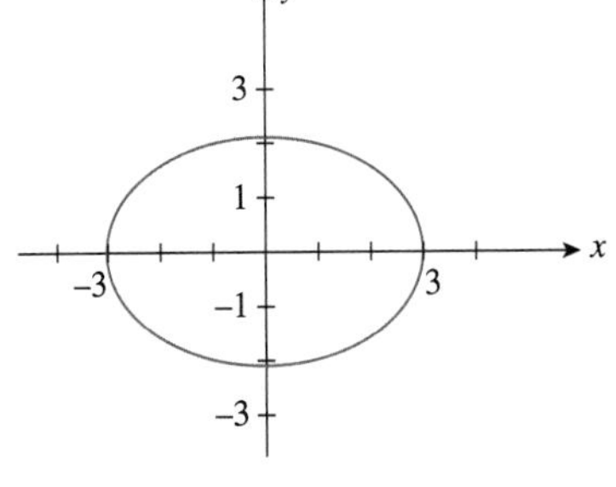

Figure 11.6
The graph of $x^2 + 2y^2 = 9$

We have already seen in this chapter that differential equations can have many different solutions. Graphing several solutions for different values of the initial conditions can give a sense of the effect that the initial conditions have. The next example shows how to use the graphing of differential equations to see how the initial conditions affect them.

Example 4

Solve the differential equation

$$y' = \frac{dy}{dx} = \frac{x}{3y}$$

and plot the solution for the initial conditions $y(x = 0) = 1$, $y(x = 0) = 2$, and $y(x = 0) = 4$.

Solution

This equation is separable and so we obtain

$$3y\,dy = x\,dx$$

which, after integration, yields

$$\frac{3}{2}y^2 = \frac{1}{2}x^2 + C$$

or

$$3y^2 - x^2 = 2C$$

The first initial condition $y(x = 0) = 1$ gives us $3 \cdot 1 - 0 = 2C$ or $C = 3/2$. Similarly the second set of initial conditions $y(x = 0) = 2$ gives us $C = 6$ while the third set of initial conditions $y(x = 0) = 4$ gives us $C = 24$. The three solutions to the differential equation are thus

$$3y^2 - x^2 = 3$$

$$3y^2 - x^2 = 12$$

$$3y^2 - x^2 = 48$$

The graph of these three solutions is shown in Figure 11.7.

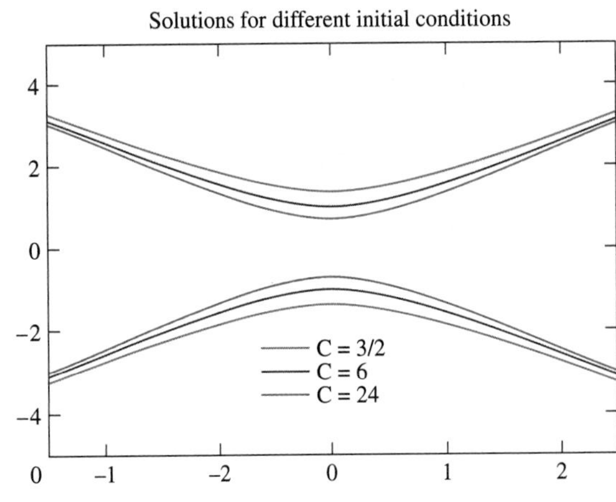

Figure 11.7
Graphs of three solutions to the same differential equation for different initial values

Justification of the Method of Separation of Variables

GROUP DISCUSSION

Refer to the Group Discussion Problem on page 667. Use the method of separation of variables to solve the differential equation $dy/dx = x/y$ and thus verify that your solution to part 3 in the problem is indeed correct.

To justify the method of separation of variables, let's consider the separable Equation (6) in its general form:

$$\frac{dy}{dx} = f(x)g(y)$$

If $g(y) \neq 0$, we may rewrite the equation in the form

$$\frac{1}{g(y)}\frac{dy}{dx} - f(x) = 0$$

Now, suppose that G is an antiderivative of $1/g$ and F is an antiderivative of f. Using the chain rule, we see that

$$\frac{d}{dx}[G(y) - F(x)] = G'(y)\frac{dy}{dx} - F'(x) = \frac{1}{g(y)}\frac{dy}{dx} - f(x)$$

Therefore,

$$\frac{d}{dx}[G(y) - F(x)] = 0$$

and so

$$G(y) - F(x) = C \qquad \text{C, a constant}$$

But the last equation is equivalent to

$$G(y) = F(x) + C \qquad \text{or} \qquad \int \frac{dy}{g(y)} = \int f(x)\,dx$$

which is precisely the result of step 2 in the method of separation of variables.

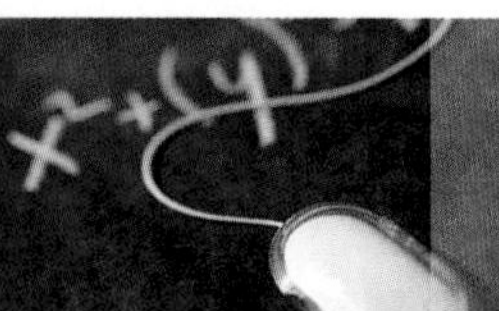

Self-Check Exercise 11.2

Find the solution of the differential equation $y' = 2x^2y + 2x^2$ that satisfies the equation $y(0) = 0$.

The solution to Self-Check Exercise 11.2 can be found on page 676.

11.2 Exercises

In Exercises 1–22, find the general solution of the first-order differential equation by separating variables.

1. $y' = \dfrac{x+1}{y^2}$
2. $y' = \dfrac{x^2}{y}$
3. $y' = \dfrac{e^x}{y^2}$
4. $y' = -\dfrac{x}{y}$
5. $y' = 2y$
6. $y' = 2(y - 1)$
7. $y' = xy^2$
8. $y' = \dfrac{2y}{x+1}$
9. $y' = -2(3y + 4)$
10. $y' = \dfrac{2y+3}{x^2}$
11. $y' = \dfrac{x^2+1}{3y^2}$
12. $y' = \dfrac{xe^x}{2y}$
13. $y' = \sqrt{\dfrac{y}{x}}$
14. $y' = \dfrac{xy^2}{\sqrt{1+x^2}}$
15. $y' = \dfrac{y \ln x}{x}$
16. $y' = \dfrac{(x-4)y^4}{x^3(y^2-3)}$
17. $y' = \sqrt{y}\sin(x)$
18. $y' = x(1 + y^2)$
19. $y' = y\cos(x)$

20. $y' = \dfrac{y}{\sqrt{1 - x^2}}$

21. $y' = x\sqrt{1 - y^2}$

22. $y' = y \sin(x)$

In Exercises 23–38, find the solution of the initial value problem.

23. $y' = \dfrac{2x}{y}$; $y(1) = -2$

24. $y' = xe^{-y}$; $y(0) = 1$

25. $y' = 2 - y$; $y(0) = 3$

26. $y' = \dfrac{y}{x}$; $y(1) = 1$

27. $y' = 3xy - 2x$; $y(0) = 1$

28. $y' = xe^{x^2}y$; $y(0) = 1$

29. $y' = \dfrac{xy}{x^2 + 1}$; $y(0) = 1$

30. $y' = x^2y^{-1/2}$; $y(1) = 1$

31. $y' = xye^{x}$; $y(1) = 1$

32. $y' = 2xe^{-y}$; $y(0) = 1$

33. $y' = 3x^2e^{-y}$; $y(0) = 1$

34. $y' = \dfrac{y^2}{x - 2}$; $y(3) = 1$

35. $y' = \sqrt{y}\cos(x)$; $y(0) = 4$

36. $y' = y^2 \cos(x)$; $y(0) = -\dfrac{1}{2}$

37. $y' = \sin(x)\cos^2(y)$; $y(0) = -\dfrac{\pi}{4}$

38. $y' = x^2 \cos^2(y)\sec(y)$; $y(0) = -\dfrac{\pi}{3}$

In Exercises 39–41, solve the given differential equation by the method of separation of variables and then graph the solution on the same axes for the different initial values given. A calculator is recommended for these exercises.

39. $y' = -3xy$; $y(0) = 2$, $y(0) = 4$, and $y(0) = 6$

40. $y' = 2xy$; $y(0) = 1$, $y(0) = 4$, and $y(0) = 7$

41. $y' = \dfrac{y}{x^2 + 1}$; $y(0) = 1$, $y(0) = 2$, and $y(0) = 3$

42. Find an equation defining a function f given that (1) the slope of the tangent line to the graph of f at any point $P(x, y)$ is given by the expression $dy/dx = (3x^2)/(2y)$ and (2) the graph of f passes through the point (1, 3).

43. Find a function f given that (1) the slope of the tangent line to the graph of f at any point $P(x, y)$ is given by the expression $dy/dx = 3xy$ and (2) the graph of f passes through the point (0, 2).

44. **Exponential Decay** Use separation of variables to solve the differential equation

$$\frac{dQ}{dt} = -kQ \qquad [Q(0) = Q_0]$$

where k and Q_0 are positive constants, describing exponential decay.

Business and Economics Applications

45. **Supply and Demand** Assume that the rate of change of the supply of a commodity is proportional to the difference between the demand and the supply so that

$$\frac{dS}{dt} = k(D - S)$$

where k is a constant of proportionality. Suppose that D is constant and $S(0) = S_0$. Find a formula for $S(t)$.

46. **Supply and Demand** Assume that the rate of change of the unit price of a commodity is proportional to the difference between the demand and the supply so that

$$\frac{dp}{dt} = k(D - S)$$

where k is a constant of proportionality. Suppose that $D = 50 - 2p$, $S = 5 + 3p$, and $p(0) = 4$. Find a formula for $p(t)$.

Biological and Life Sciences Applications

47. **Fisk's Law** Refer to Exercise 26, Section 11.1. Use separation of variables to solve the differential equation

$$\frac{dy}{dt} = \frac{k}{V}(C - y) \quad [y(0) = y_0]$$

where k, V, C, and y_0 are constants with $C - y > 0$. Find $\lim_{t\to\infty} y$ and interpret your result.

48. **Concentration of Glucose in the Bloodstream** Refer to Exercise 25, Section 11.1. Use separation of variables to solve the differential equation $A' = C - kA$, where C and k are positive constants.
Hint: Rewrite the given differential equation in the form

$$\frac{dA}{dt} = k\left(\frac{C}{k} - A\right)$$

49. **Allometric Laws** Refer to Exercise 27, Section 11.1. Use separation of variables to solve the differential equation

$$\frac{1}{x}\frac{dx}{dt} = k\frac{1}{y}\frac{dy}{dt}$$

where k is a constant.

In Exercises 50–55, determine whether the statement is true or false. If it is true, explain why it is true. If it is false, give an example to show why it is false.

50. The differential equation $y' = xy + 2x - y - 2$ is separable.

51. The differential equation $y' = x^2 - y^2$ is separable.

52. If the differentiable function $M(x, y)\,dx + N(x, y)\,dy = 0$ can be written so that $M(x, y) = f(x)g(y)$ and $N(x, y) = F(x)G(y)$ for functions f, g, F, and G, then it is separable.

53. The differential equation $(x^2 + 2)\,dx + (2x - 4xy)\,dy = 0$ is separable.

54. The differential equation $y\,dx - (y - xy^2)\,dy = 0$ is separable.

55. The differential equation $\dfrac{dy}{dx} = \dfrac{f(x)\,g(y)}{F(x) + G(y)}$ is separable.

Solution to Self-Check Exercise 11.2

Writing the differential equation in the form

$$\frac{dy}{dx} = 2x^2(y + 1)$$

and separating variables, we obtain

$$\frac{dy}{y + 1} = 2x^2\,dx$$

Integrating each side of the last equation with respect to the appropriate variable, we have

$$\int \frac{dy}{y + 1} = \int 2x^2\,dx \quad \text{or} \quad \ln|y + 1| = \frac{2}{3}x^3 + C$$

Using the initial condition $y(0) = 0$, we have

$$\ln 1 = C \quad \text{or} \quad C = 0$$

Therefore,

$$\ln|y + 1| = \frac{2}{3}x^3$$

$$y + 1 = e^{(2/3)x^3}$$

$$y = e^{(2/3)x^3} - 1$$

11.3 Applications of Separable Differential Equations

In this section we look at some applications of first-order separable differential equations. We begin by reexamining some of the applications discussed in Section 11.1.

Unrestricted Growth Models

The differential equation describing an unrestricted growth model is given by

$$\frac{dQ}{dt} = kQ$$

where $Q(t)$ represents the size of a certain population at any time t and k is a positive constant. Separating the variables in this differential equation, we have

$$\frac{dQ}{Q} = k\,dt$$

which, upon integration, yields

$$\int \frac{dQ}{Q} = \int k\,dt$$

$$\ln|Q| = kt + C_1$$

$$Q = e^{kt+C_1} = Ce^{kt}$$

where $C = e^{C_1}$ is an arbitrary positive constant. Thus, we may write the solution as

$$Q(t) = Ce^{kt}$$

Observe that if the quantity present initially is denoted by Q_0, then $Q(0) = Q_0$. Applying this condition yields the equation

$$Ce^0 = Q_0 \quad \text{or} \quad C = Q_0$$

Therefore, the model for unrestricted exponential growth with initial population Q_0 is given by

$$Q(t) = Q_0e^{kt} \tag{8}$$

(Figure 11.8).

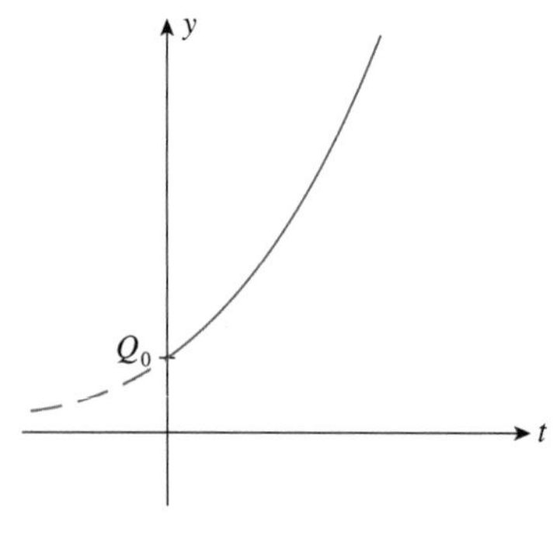

Figure 11.8
An unrestricted growth model

Example 1

Growth of Bacteria Under ideal laboratory conditions, the rate of growth of bacteria in a culture is proportional to the size of the culture at any time t. Suppose that 10,000 bacteria are present initially in a culture and 60,000 are present 2 hours later. How many bacteria will there be in the culture at the end of 4 hours?

Solution

Let $Q(t)$ denote the number of bacteria present in the culture at time t. Then

$$\frac{dQ}{dt} = kQ$$

where k is a constant of proportionality. Solving this separable first-order differential equation, we obtain

$$Q(t) = Q_0 e^{kt} \qquad \text{Equation (8)}$$

where Q_0 denotes the initial bacteria population. Since $Q_0 = 10{,}000$, we have

$$Q(t) = 10{,}000e^{kt}$$

Next, the condition that 60,000 bacteria are present 2 hours later translates into $Q(2) = 60{,}000$, or

$$\begin{aligned} 60{,}000 &= 10{,}000e^{2k} \\ e^{2k} &= 6 \\ e^k &= 6^{1/2} \end{aligned}$$

Thus, the number of bacteria present at any time t is given by

$$\begin{aligned} Q(t) &= 10{,}000e^{kt} = 10{,}000(e^k)^t \\ &= (10{,}000)6^{(t/2)} \end{aligned}$$

In particular, the number of bacteria present in the culture at the end of 4 hours is given by

$$\begin{aligned} Q(4) &= 10{,}000(6^{4/2}) \\ &= 360{,}000 \end{aligned}$$

Restricted Growth Models

From Section 11.1 we see that a differential equation describing a restricted growth model is given by

$$\frac{dQ}{dt} = k(C - Q) \tag{9}$$

where both k and C are positive constants. To solve this separable first-order differential equation, we first separate the variables, obtaining

$$\frac{dQ}{C - Q} = k\,dt$$

Integrating each side with respect to the appropriate variable yields

$$\int \frac{dQ}{C - Q} = \int k\,dt$$

$$-\ln|C - Q| = kt + d \qquad d\text{, an arbitrary constant}$$

$$\ln|C - Q| = -kt - d \tag{10}$$

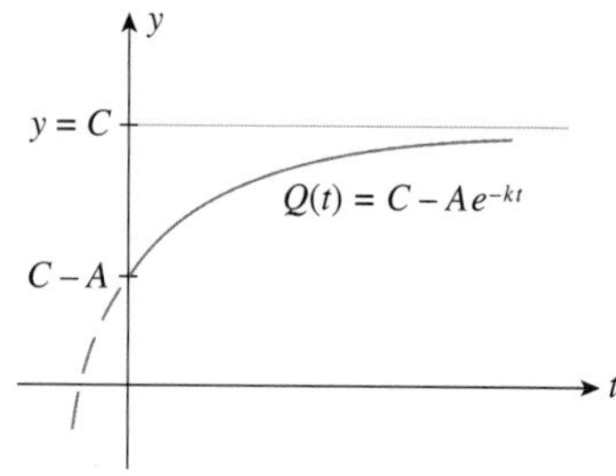

Figure 11.9
A restricted exponential growth model

$$C - Q = e^{-kt-d} = e^{-kt}\,e^{-d}$$

$$Q(t) = C - Ae^{-kt}$$

where we have denoted the constant e^{-d} by A. This is the equation of the learning curve (Figure 11.9) studied in Chapter 5.

Example 2

Yield of a Wheat Field In an experiment conducted by researchers of the Agriculture Department of a prairie university, it was found that the maximum yield of wheat in the university's experimental field station was 150 bushels per hectare. Furthermore, the researchers discovered that the rate at which the yield of wheat increased was governed by the differential equation

$$\frac{dQ}{dx} = k(150 - Q)$$

where $Q(x)$ denotes the yield in bushels per hectare and x is the amount in kilograms of an experimental fertilizer used per hectare of land. Data obtained in the experiment indicated that 10 kilograms of fertilizer per hectare of land would result in a yield of 80 bushels of wheat per hectare, whereas 20 kilograms of fertilizer per hectare of land would result in a yield of 120 bushels of wheat per hectare. Determine the yield if 30 pounds of fertilizer were used per hectare.

Solution

The given differential equation has the same form as Equation (9) with $C = 150$. Solving it directly or using the result obtained in the solution of Equation (9), we see that the yield per hectare is given by

$$Q(x) = 150 - Ae^{-kx}$$

The first of the given conditions implies that $Q(10) = 80$; that is,

$$150 - Ae^{-10k} = 80$$

or $A = 70e^{10k}$. Therefore,

$$\begin{aligned} Q(x) &= 150 - 70e^{10k}\,e^{-kx} \\ &= 150 - 70e^{-k(x-10)} \end{aligned}$$

The second of the given conditions implies that $Q(20) = 120$, or

$$\begin{aligned} 150 - 70e^{-k(20-10)} &= 120 \\ 70e^{-10k} &= 30 \\ e^{-10k} &= \frac{3}{7} \end{aligned}$$

Taking the logarithm of each side of the equation, we find

$$\begin{aligned} \ln e^{-10k} &= \ln\left(\frac{3}{7}\right) \\ -10k &= \ln 3 - \ln 7 \approx -0.8473 \\ k &\approx 0.085 \end{aligned}$$

Therefore,

$$Q(x) = 150 - 70e^{-0.085(x-10)}$$

In particular, when $x = 30$, we have

$$\begin{aligned} Q(30) &= 150 - 70e^{-0.085(20)} \\ &= 150 - 70e^{-1.7} \\ &\approx 137 \end{aligned}$$

So the yield would be 137 bushels per hectare if 30 kilograms of fertilizer were used per hectare. The graph of $Q(x)$ is shown in Figure 11.10. Note that it was not necessary to find the specific value of k in order to solve this problem. (Why?)

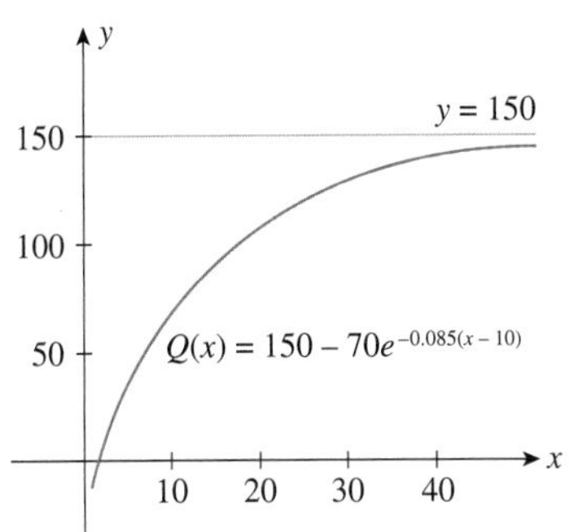

Figure 11.10
$Q(x)$ is a function relating crop yield to the amount of fertilizer used.

Next, let's consider a differential equation describing another type of restricted growth:

$$\frac{dQ}{dt} = kQ(C - Q)$$

where k and C are positive constants. Separating variables leads to

$$\frac{dQ}{Q(C - Q)} = k\,dt$$

Integrating each side of this equation with respect to the appropriate variable, we have

$$\int \frac{1}{Q(C - Q)}\,dQ = \int k\,dt$$

As it stands, the integrand on the left side of this equation is not in a form that can be easily integrated. However, observe that

$$\frac{1}{Q(C - Q)} = \frac{1}{C}\left[\frac{1}{Q} + \frac{1}{C - Q}\right]$$

as you may verify by adding the terms between the brackets on the right-hand side. Making use of this identity, we have

$$\int \frac{1}{C}\left[\frac{1}{Q} + \frac{1}{C - Q}\right] dQ = \int k\,dt$$

$$\int \frac{dQ}{Q} + \int \frac{dQ}{C - Q} = Ck\int dt$$

$$\ln|Q| - \ln|C - Q| = Ckt + b \qquad b\text{, an arbitrary constant}$$

$$\ln\left|\frac{Q}{C - Q}\right| = Ckt + b$$

$$\frac{Q}{C - Q} = e^{Ckt+b} = e^b e^{Ckt} = Be^{Ckt} \qquad (B = e^b)$$

$$Q = CBe^{Ckt} - QBe^{Ckt}$$

$$(1 + Be^{Ckt})Q = CBe^{Ckt}$$

and

$$Q = \frac{CBe^{Ckt}}{1 + Be^{Ckt}}$$

or

$$Q(t) = \frac{C}{1 + Ae^{-Ckt}} \qquad \left(A = \frac{1}{B}\right) \qquad (11)$$

(see Figure 11.2, page 665). In its final form, this function is equivalent to the logistic function encountered in Chapter 5.

Example 3

Spread of a Flu Epidemic During a flu epidemic, 5% of the 5000 army personnel stationed at a Canadian Forces base had contracted influenza at time $t = 0$.

Furthermore, the rate at which they were contracting influenza was jointly proportional to the number of personnel who had already contracted the disease and the noninfected population. If 20% of the personnel had contracted the flu by the 10th day, find the number of personnel who had contracted the flu by the 13th day.

Solution

Let $Q(t)$ denote the number of army personnel who had contracted the flu after t days. Then

$$\frac{dQ}{dt} = kQ(5000 - Q)$$

We may solve this separable differential equation directly, or we may appeal to the result for the more general problem obtained earlier. Opting for the latter and noting that $C = 5000$, we find that, using Equation (11),

$$Q(t) = \frac{5000}{1 + Ae^{-5000kt}}$$

The condition that 5% of the population had contracted influenza at time $t = 0$ implies that

$$Q(0) = \frac{5000}{1 + A} = 250$$

from which we see that $A = 19$. Therefore,

$$Q(t) = \frac{5000}{1 + 19e^{-5000kt}}$$

Next, the condition that 20% of the population had contracted influenza by the tenth day implies that

$$Q(10) = \frac{5000}{1 + 19e^{-50,000k}} = 1000$$

or

$$1 + 19e^{-50,000k} = 5$$

$$e^{-50,000k} = \frac{4}{19}$$

$$-50,000k = \ln 4 - \ln 19$$

and

$$k = -\frac{1}{50,000}(\ln 4 - \ln 19)$$

$$\approx 0.0000312$$

Therefore,

$$Q(t) = \frac{5000}{1 + 19e^{-0.156t}}$$

In particular, the number of army personnel who had contracted the flu by the 13th day is given by

$$Q(13) = \frac{5000}{1 + 19e^{-0.156(13)}} = \frac{5000}{1 + 19e^{-2.028}} \approx 1428$$

or approximately 1428, or 29%. The graph of $Q(t)$ is shown in Figure 11.11.

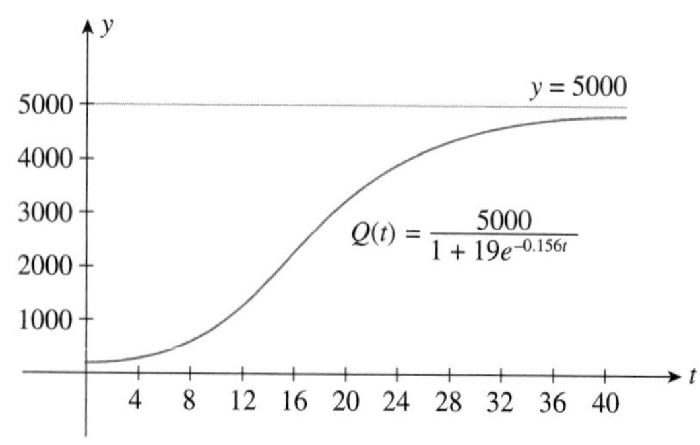

Figure 11.11
An epidemic model

GROUP DISCUSSION

Consider the model for restricted growth described by the differential equation $dQ/dt = kQ(C - Q)$ with the solution given in Equation (11).

1. Show that the rate of growth of Q is greatest at $t = (\ln A)/kC$.
   Hint: Use both the differential equation and Equation (11).
2. Refer to Example 3. At what time is the number of influenza cases increasing at the greatest rate?

Example 4

Weber–Fechner Law Derive the Weber–Fechner law describing the relationship between a stimulus S and the resulting response R by solving the differential Equation (4) subject to the condition $R = 0$ when $S = S_0$, where S_0 is the threshold level.

Solution

The differential equation under consideration is

$$\frac{dR}{dS} = \frac{k}{S}$$

and is separable. Separating the variables, we have

$$dR = k\,\frac{dS}{S}$$

which, upon integration, yields

$$\int dR = k\int \frac{dS}{S}$$

$$R = k \ln S + C$$

where C is an arbitrary constant. Using the condition $R = 0$ when $S = S_0$ gives

$$0 = k \ln S_0 + C$$

$$C = -k \ln S_0$$

Substituting this value of C in the expression for R leads to

$$R = k \ln S - k \ln S_0$$

$$= k \ln \frac{S}{S_0}$$

the required relationship between R and S. The graph of R is shown in Figure 11.12.

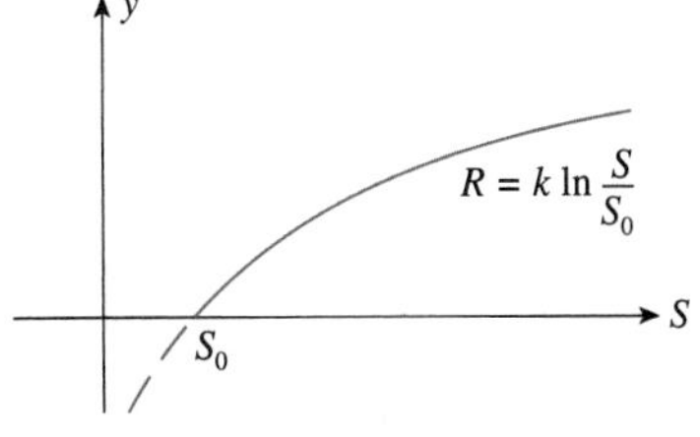

Figure 11.12
The Weber–Fechner law

Example 5

A Mixture Problem A tank initially contains 40 litres of pure water. Brine containing 1 kilogram of salt per litre flows into the tank at a rate of 8 litres per minute, and the well-stirred mixture flows out of the tank at the same rate. How much salt is present at the end of 10 minutes? How much salt is present in the long run?

Solution

The problem was formulated mathematically on page 665, and we were led to the differential equation

$$\frac{dA}{dt} = 8 - \frac{A}{5}$$

subject to the condition $A(0) = 0$. The differential equation

$$\frac{dA}{dt} = 8 - \frac{A}{5} = \frac{40 - A}{5}$$

is separable. Separating the variables and integrating, we obtain

$$\int \frac{dA}{40 - A} = \int \frac{1}{5}\,dt$$

$$-\ln|40 - A| = \frac{1}{5}t + b \qquad b\text{, a constant}$$

$$\ln|40 - A| = -\frac{1}{5}t - b$$

$$40 - A = e^{-b}e^{-t/5}$$

$$A = 40 - Ce^{-t/5} \qquad (C = e^{-b})$$

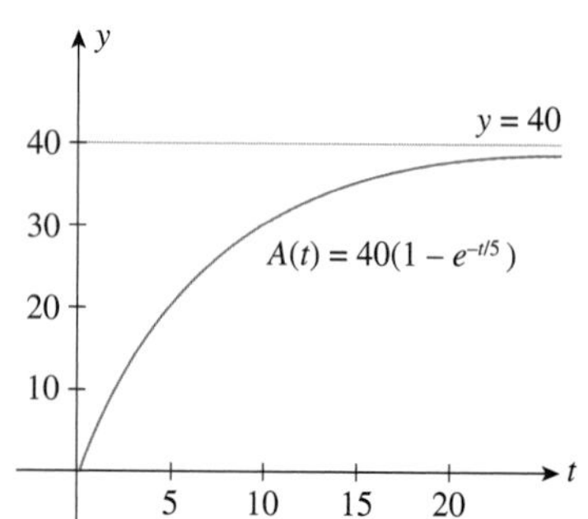

Figure 11.13

The solution of the differential equation $\frac{dA}{dt} = 8 - \frac{A}{5}$

The condition $A(0) = 0$ implies that

$$0 = 40 - C$$

giving $C = 40$, and so on

$$A(t) = 40(1 - e^{-t/5})$$

The amount of salt present after 10 minutes is given by

$$A(10) = 40(1 - e^{-2}) \approx 34.59$$

or 34.59 kilograms. The amount of salt present in the long run is given by

$$\lim_{t\to\infty} A(t) = \lim_{t\to\infty} 40(1 - e^{-t/5}) = 40$$

or 40 kilograms (Figure 11.13).

A More Complex Restricted Growth Model

In Section 9.2 we learned how to perform integration by partial fractions. This method of integration lets us solve a type of separable differential equation that describes a more complex type of limited growth. The model is also based on a growth constant k and a carrying capacity C in the differential equation describing the population at a time t by

$$P' = kP(C - P)$$

The first step is to separate the variables in this equation.

$$\frac{dP}{dt} = kP(C - P)$$

$$\frac{dP}{P(C - P)} = k\,dt$$

$$\int \frac{dP}{P(C - P)} = \int k\,dt$$

The integral on the left resulting from the separation of variables is one that can be solved with partial fractions. The next step is to perform the partial fractions steps to set up the integral.

$$\frac{1}{P(C - P)} = \frac{A}{P} + \frac{B}{C - P}$$

$$\frac{1}{P(C-P)} = \frac{A(C-P) + BP}{P(C-P)}$$

$$1 = A\,(C-P) + BP$$

If we set $P = 0$ we get $1 = AC$ or $A = 1/C$. If, on the other hand, we set $P = C$ then we get $1 = BP$ or $B = 1/C$ so both constants have the same value. Now that we know the value of A and B we can return to the integral.

$$\int \frac{dP}{P(C-P)} = \int k\,dt$$

$$\int \left(\frac{1/C}{P} + \frac{1/C}{C-P}\right) dP = \int k\,dt$$

$$\frac{1}{C}\int \frac{dP}{P} + \frac{1}{C}\int \frac{dP}{C-P} = \int k\,dt$$

$$\frac{\ln|P|}{C} - \frac{\ln|C-P|}{C} = kt + D$$

Notice that the constant of integration is named D, because C is already in use for the carrying capacity. We now use the algebraic properties of logarithms to solve for P.

$$\frac{\ln|P|}{C} - \frac{\ln|C-P|}{C} = kt + D$$

$$\ln|P| - \ln|C-P| = Ckt + CD$$

$$\ln\left|\frac{P}{C-P}\right| = Ckt + CD$$

$$e^{\ln\left|\frac{P}{C-P}\right|} = e^{\mathrm{Ckt+CD}}$$

$$\frac{P}{C-P} = e^{Ckt+CD}$$

$$P = (C-P)e^{Ckt+CD}$$

$$P = Ce^{Ckt+CD} - Pe^{Ckt+CD}$$

$$P + Pe^{Ckt+CD} = Ce^{Ckt+CD}$$

$$P(1 + e^{Ckt+CD}) = Ce^{Ckt+CD}$$

$$P = \frac{Ce^{Ckt+CD}}{1 + e^{Ckt+CD}}$$

If we rename the constants $r = Ck$ and $R = e^{CD}$ then we see that the general form of a solution to the new limited growth model is

$$P(t) = C\,\frac{Re^{rt}}{1 + Re^{rt}}$$

A graph of this solution is shown in Figure 11.14.

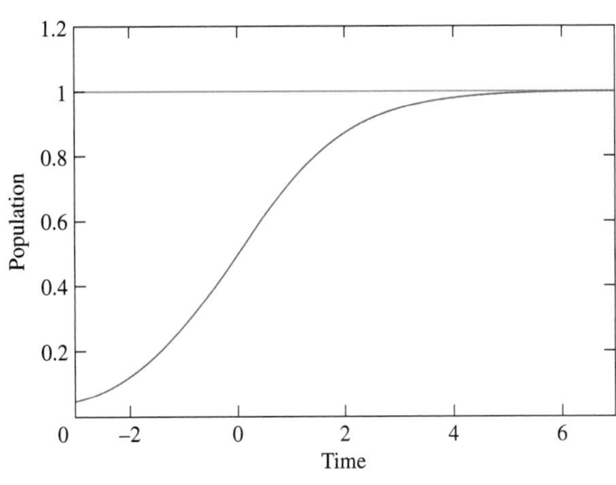

Figure 11.14
A graph of a solution of the limited growth model with $C = 1$, $k = 1$, and $y(0) = 0.5$. Notice that as time increases, the solution has an asymptote at $y = C$. Any solution to a limited growth model of this form has an asymptote at $y = C = 1$.

Example 6

Solve the limited growth model

$$P'(t) = 2P(20 - P)$$

if $y(0) = 5$.

Solution

First we separate the variables and obtain

$$\int \frac{dP}{P(20 - P)} = 2\,dt$$

Remember that we already know the values of the constants that arise during partial fractions; they are both equal to the reciprocal of the carrying capacity $C = 20$. This means that the integral is

$$\frac{1}{20}\int \frac{dP}{P} + \frac{1}{20}\int \frac{dP}{20 - P} = \int 2\,dt$$

Integrating, we obtain

$$\frac{1}{20}\ln|P| - \frac{1}{20}\ln|20 - P| = 2t + D$$

At this point we can use the information $P(0) = 5$ to get

$$D = \frac{1}{20}\ln(5) - \frac{1}{20}\ln(15)$$

$$D = \frac{1}{20}[\ln(5) - \ln(15)]$$

$$D \approx -0.549$$

We now solve for P.

$$\frac{1}{20}\ln|P| - \frac{1}{20}\ln|20 - P| = 2t - 0.549$$

$$\frac{1}{20}\ln\left|\frac{P}{20 - P}\right| = 2t - 0.549$$

$$\ln\left|\frac{P}{20 - P}\right| = 40t - 1.10$$

$$e^{\ln\left|\frac{P}{20-P}\right|} = e^{40t - 1.1}$$

$$\frac{P}{20 - P} = e^{40t} \cdot e^{-1.1}$$

$$\frac{P}{20 - P} = e^{40t} \cdot e^{-1.1}$$

$$\frac{P}{20 - P} = 0.333e^{40t}$$

$$P = (20 - P)\,0.333e^{40t}$$

$$P = 6.66e^{40t} - P0.333e^{40t}$$

$$P + P0.333e^{40t} = 6.66e^{40t}$$

$$P(1 + 0.333e^{40t}) = 6.66e^{40t}$$

$$P(t) = \frac{6.66e^{40t}}{1 + 0.333e^{40t}}$$

This type of limited growth equation, like the one we first encountered in Chapter 5, is also called a *logistic* equation. The term is used for both sorts of differential equations and it is necessary to be careful to tell from context which of the two models is being used. The type of limited growth described by the more complex model includes situations where adults of a species compete with (e.g., eat) their own young. Anyone who has tried to raise guppies knows that a separate tank (or a lot of weeds) are needed for the babies.

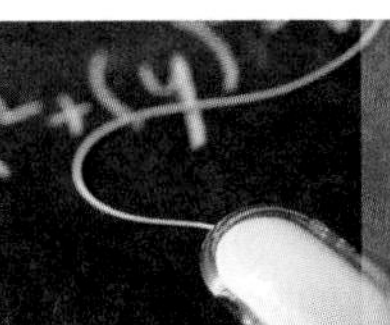

Self-Check Exercise 11.3

Newton's law of cooling states that the temperature of an object drops at a rate that is proportional to the difference in the temperature between the object and that of the surrounding medium. Suppose that an apple pie is taken out of the oven at a temperature of 200°F and placed on the counter in a room where the temperature is 70°F. If the temperature of the apple pie is 150°F after 5 min, find its temperature $y(t)$ as a function of time t.

The solution to Self-Check Exercise 11.3 can be found on page 687.

11.3 Exercises

A calculator is recommended for these exercises.

1. **Lambert's Law of Absorption** According to Lambert's law of absorption, the percentage of incident light L, absorbed in passing through a thin layer of material x, is proportional to the thickness of the material. If $\frac{1}{2}$ cm. of a certain material reduces the light to half of its intensity, how much additional material is needed to reduce the intensity to one-fourth of its initial value?

2. **Newton's Law of Cooling** Newton's law of cooling states that the rate at which the temperature of an object changes is directly proportional to the difference in temperature between the object and that of the surrounding medium. A horseshoe heated to a temperature of 100°C is immersed in a large tank of water at a (constant) temperature at 30°C at time $t = 0$. Three minutes later the temperature of the horseshoe is reduced to 70°C. Derive an expression that gives the temperature of the horseshoe at any time t. What is the temperature of the horseshoe 5 min after it has been immersed in the water?

3. **Newton's Law of Cooling** Newton's law of cooling states that the rate at which the temperature of an object changes is directly proportional to the difference in temperature between the object and that of the surrounding medium. A cup of coffee is prepared with boiling water (212°F) and left to cool on the counter in a room where the temperature is 72°F. If the temperature of the coffee is 140°F after 2 min, determine when the coffee will be cool enough to drink (say, 110°F).

4. **Training Personnel** The personnel manager of Gibraltar Insurance Company estimates that the number of insurance claims an experienced clerk can process in a day is 40. Furthermore, the rate at which a clerk can process insurance claims during the tth wk of training is proportional to the difference between the maximum number possible (40) and the number he or she can process in the tth wk. If the number of claims the average trainee can process after 2 wk on the job is 10/day, determine how many claims the average trainee can process after 6 wk on the job.

5. **Spread of a Rumour** The rate at which a rumour spreads through an Alpine village of 400 residents is jointly proportional to the number of residents who have heard it and the number who have not. Initially, 10 residents heard the rumour, but 2 days later this number increased to 80. Find the number of people who will have heard the rumour after 1 wk.

6. **Gompertz Growth Curves** Refer to Exercise 28, Section 11.1. Consider the differential equation

$$\frac{dQ}{dt} = kQ(C - \ln Q)$$

with the side condition $Q(0) = Q_0$. The solution $Q(t)$ describes restricted growth and has a graph known as the Gompertz curve. Using separation of variables, solve this differential equation.

7. **Mixture Problems** A tank initially contains 80 litres of pure water. Brine containing $\frac{1}{4}$ kg of salt per litre flows into the tank at a rate of 12 L/min, and the well-stirred mixture flows out of the tank at the same rate. How much salt is present in the tank at any time t? How much salt is present at the end of 20 min? How much salt is present in the long run?

8. **Mixture Problems** A tank initially contains 50 litres of brine, in which 10 kg of salt is dissolved. Brine containing 2 kg of dissolved salt per litre flows into the tank at the rate of 2 L/min, and the well-stirred mixture flows out of the tank at the same rate. How much salt is present in the tank at the end of 10 min?

Business and Economics Applications

9. **Savings Accounts** An amount of money deposited in a savings account grows at a rate proportional to the amount present. (It can be shown that an amount of money grows in this manner if it earns interest compounded continuously.) Suppose $10,000 is deposited in a fixed account earning interest at the rate of 10%/year compounded continuously.
   a. What is the accumulated amount after 5 yr?
   b. How long does it take for the original deposit to double in value?

10. **Sinking Funds** The proprietor of Carson Hardware Store has decided to set up a sinking fund for the purpose of purchasing a computer 2 yr from now. It is expected that the purchase will involve a sum of $30,000. The fund grows at the rate of

$$\frac{dA}{dt} = rA + P$$

where A denotes the size of the fund at any time t, r is the annual interest rate earned by the fund compounded continuously, and P is the amount (in dollars) paid into the fund by the proprietor per year (assume this is done on a frequent basis in small deposits over the year so that it is essentially continuous). If the fund earns 10% interest per year compounded continuously, determine the size of the yearly investment the proprietor should pay into the fund.

Biological and Life Sciences Applications

11. **Chemical Decomposition** The rate of decomposition of a certain chemical substance is directly proportional to the amount present at any time t. If y_0 g of the chemical are present at time $t = 0$, find an expression for the amount present at any time t.

12. **Growth of Bacteria** Under ideal laboratory conditions, the rate of growth of bacteria in a culture is proportional to the size of the culture at any time t. Suppose that 2000 bacteria are present initially in the culture and 5000 are present 1 hr later. How many bacteria will be in the culture at the end of 2 hr?

13. **World Population Growth** The world population at the beginning of 1980 was 4.5 billion. Assuming that the population continues to grow at its present rate of approximately 2%/year, find a function $Q(t)$ that expresses the world population (in billions) as a function of time t (in years). What will be the world population at the beginning of 2008?

14. **Population Growth** The population of a certain community is increasing at a rate directly proportional to the population at any time t. In the last 3 yr, the population has doubled. How long will it take for the population to triple?

15. **Chemical Reactions** In a certain chemical reaction, a substance is converted into another substance at a rate proportional to the square of the amount of the first substance present at any time t. Initially ($t = 0$) 50 g of the first substance was present; 1 hr later, only 10 g of it remained. Find an expression that gives the amount of the first substance present at any time t. What is the amount present after 2 hr?

16. **Learning Curves** The American Stenographic Institute finds that the average student taking Elementary Shorthand will progress at a rate given by

$$\frac{dQ}{dt} = k(80 - Q)$$

in a 20-wk course, where $Q(t)$ measures the number of words of dictation a student can take per minute after t wk in the course. If the average student can take 50 words of dictation per minute after 10 wk in the course, how many words per minute can the average student take after completing the course?

17. **Effect of Immigration on Population Growth** Suppose a country's population at any time t grows in accordance with the rule

$$\frac{dP}{dt} = kP + I$$

where P denotes the population at any time t, k is a positive constant reflecting the natural growth rate of the population, and I is a constant giving the (constant) rate of immigration into the country. If the total population of the country at time $t = 0$ is P_0, find an expression for the population at any time t.

18. **Effect of Immigration on Population Growth** Refer to Exercise 17. The population of Canada in the year 1990 ($t = 0$) was 22 million. Suppose the natural growth rate is 0.8% annually ($k = 0.008$) and net immigration is allowed at the rate of 0.2 million people/year ($I = 0.2$) until the end of the century. What will be the Canadian population in 2018?

19. **Growth of a Fruit Fly Colony** A biologist has determined that the maximum number of fruit flies that can be sustained in a carefully controlled environment (with a limited supply of space and food) is 400. Suppose that the rate at which the population of the colony increases obeys the rule

$$\frac{dQ}{dt} = kQ(C - Q)$$

where C is the carrying capacity (400) and Q denotes the number of fruit flies in the colony at any time t. If the initial population of fruit flies in the experiment is 10 and it grows to 45 after 10 days, determine the population of the colony of fruit flies on the 20th day.

20. **Chemical Reaction Rates** Two chemical solutions, one containing N molecules of chemical A and another containing M molecules of chemical B, are mixed together at time $t = 0$. The molecules from the two chemicals combine to form another chemical solution containing $y(AB)$ molecules. The rate at which the AB molecules are formed, dy/dt, is called the *reaction rate* and is jointly proportional to $(N - y)$ and $(M - y)$. Thus,

$$\frac{dy}{dt} = k(N - y)(M - y)$$

where k is a constant. (We assume the temperature of the chemical mixture remains constant during the interaction.) Solve this differential equation with the side condition $y(0) = 0$ assuming that $N - y > 0$ and $M - y > 0$.

Hint: Use the identity

$$\frac{1}{(N - y)(M - y)} = \frac{1}{M - N}\left(\frac{1}{N - y} - \frac{1}{M - y}\right)$$

21. **Calculating Growth** In a species of tetra, a small colourful fish often sold in pet stores, the parents will hunt for eggs right after laying them. This self-predation causes the fish to obey a growth law of the form

$$y' = 1.2y(200 - y)$$

fish/month in a pond where they are being raised for later sale. If the pond is started with $y(0) = 4$ fish, compute how many are present after six months and after one year.

22. **Calculating Growth** Competition for grazing space causes a species of antelope to have a growth rate of

$$P' = 2P(400 - P)$$

antelope/year. Assuming that the number of antelope is $P(0) = 50$ in year zero, find how many antelope there are after 5, 10, and 20 years.

Solution to Self-Check Exercise 11.3

We are required to solve the initial value problem

$$\frac{dy}{dt} = -k(y - 70) \qquad k\text{, a constant of proportionality}$$

$$y(0) = 200$$

To solve the differential equation, we separate variables and integrate, obtaining

$$\int \frac{dy}{y - 70} = \int -k\,dt$$

$$\ln|y - 70| = -kt + d \qquad d\text{, an arbitrary constant}$$

$$y - 70 = e^{-kt+d} = Ae^{-kt} \qquad (A = e^d)$$

$$y = 70 + Ae^{-kt}$$

Using the initial condition $y(0) = 200$, we find

$$200 = 70 + A$$

$$A = 130$$

Therefore,

$$y = 70 + 130e^{-kt}$$

To determine the value of k, we use the fact that $y(5) = 150$, obtaining

$$150 = 70 + 130e^{-5k}$$

$$130e^{-5k} = 80$$

$$e^{-5k} = \frac{80}{130}$$

$$-5k = \ln\frac{80}{130}$$

$$k = -\frac{1}{5}\ln\frac{80}{130}$$

$$= 0.097$$

Therefore, the required expression is

$$y(t) = 70 + 130e^{-0.097t}$$

11.4 Linear Differential Equations

A **linear differential equation** has the form

$$y' + uy = f(x)$$

if it is first-order and

$$y'' + uy' + vy = f(x)$$

if it is second-order. It is also possible to have higher-order linear differential equations—they consist of a sum of constant multiples of derivatives of y equal to a function of x—but here we will restrict our attention to second-order and lower equations. Linear equations describe a great many natural phenomena, such as radioactive decay or the unbounded growth model in Section 11.2. If the function $f(x) = 0$ then we call the differential equation **homogeneous.** We will start by learning to solve the homogeneous type of linear differential equation. The first-order homogeneous equation is separable.

Example 1

Solving a First-Order Homogeneous Equation Solve the equation

$$y' + uy = 0$$

where u is a real number.

Solution

$$y' + uy = 0$$

$$\frac{dy}{dt} + uy = 0$$

$$\frac{dy}{dt} = -uy$$

$$\frac{dy}{y} = -u\,dt$$

$$\int \frac{dy}{y} = \int -u\,dt$$

$$\ln|y| = -ut + C$$

$$e^{\ln|y|} = e^{-ut+C}$$

$$y = e^{-ut+C}$$

$$y = e^{-ut} \cdot e^{C}$$

$$y = Ae^{-ut}$$

Notice that we renamed the constant $e^C = A$ to get a more compact form for the solution. This solution is familiar—it is an exponential growth or decay function (depending on the sign of u).

Suppose that we have a second-order homogeneous equation. Since the equation is second-order it is not separable. If we assume that it has a solution of the same form as the first-order equation, it is possible to plug that solution in, as if we were verifying it, and so find a solution. Assume that the equation

$$y'' + uy' + vy = 0$$

has a solution $y = Ae^{ct}$. We start by computing the derivatives of the tentative solution.

$$y = Ae^{ct}$$

$$y' = Ace^{ct}$$

$$y'' = Ac^2e^{ct}$$

We then plug these derivatives into the differential equation and solve.

$$y'' + uy' + vy = 0$$

$$Ac^2e^{ct} + uAce^{ct} + vAe^{ct} = 0$$

$$Ae^{ct}(c^2 + uc + v) = 0$$

This last equation is equal to zero. When the product of two quantities equals zero then one or the other of those quantities must be zero. We know from studying the exponential function in Chapter 5 that $Ae^{ct} \neq 0$ and so we conclude that $c^2 + uc + v + 0$. Since this is a quadratic equation we can solve it and get

$$c = \frac{-u \pm \sqrt{u^2 - 4c}}{2}$$

It is a bit of a surprise to obtain two values for c, but it turns out that second-order homogeneous linear equations have two different fundamental solutions, both of the form Ae^{ct} but for different values of c. We will call the two values of c by the names c_1 and c_2. The general solution to a second-order homogeneous solution is of the form

$$y = Ae^{c_1 t} + Be^{c_2 t}$$

There are two different unknown constants A and B, analogous to the constant of integration we get when solving separable equations, and so differential equations of this type require two initial conditions.

Example 2

Solving a Second-Order Homogeneous Equation with Initial Conditions
Solve the equation

$$y'' - 3y' + 2y = 0$$

if $y(0) = 2$ and $y'(0) = 3$.

Solution

If we set $y = Ae^{ct}$ then, following the derivation for the form of a solution, we get the expression

$$Ae^{ct}(c^2 - 3c + 2) = 0$$

The quadratic part of the expression factors, giving us

$$(c - 2)(c - 1) = 0$$

and so $c = 1$ or $c = 2$. With the two possible values of c known we can find the general solution

$$y = Ae^{c_1 t} + Be^{c_2 t} = Ae^{t} + Be^{2t}$$

Since one of the two initial conditions is given in terms of y' we will need to compute the derivative of the solution. This derivative is

$$y' = Ae^{t} + 2Be^{2t}$$

Plugging in the initial conditions we obtain

$$y(0) = 2 = Ae^{0} + Be^{2 \cdot 0} = A + B$$

and

$$y'(0) = 3 = Ae^{0} + 2Be^{2 \cdot 0} = A + 2B$$

which yields a system of linear equations

$$2 = A + B \text{ and } 3 = A + 2B$$

Solving these equations simultaneously we find the values $A = 1$ and $B = 1$, so the solution to the second-order initial value problem is

$$y = e^{t} + e^{2t}$$

Homogeneous linear equations can be solved easily by extracting the polynomial that describes the constant c in the solution form Ae^{ct}. This polynomial has a name; it is called the **characteristic polynomial** of the differential equation.

Example 3

Finding the Characteristic Polynomial of a Differential Equation Find the characteristic equation of the differential equation

$$y'' - 5y' + 4y = 0$$

Solution

Following the solution form we get $Ae^{ct}(c^2 - 5c + 4) = 0$. The characteristic polynomial is the polynomial portion of the expression:

$$c^2 - 5c + 4$$

The characteristic polynomial can be read off the differential equation. The coefficient of y'' in the differential equation is the coefficient of c^2 in the characteristic polynomial; the coefficient of y' is the coefficient of c, and the coefficient of y is the constant coefficient of the characteristic polynomial.

Example 4

Solving a Homogeneous Equation with its Characteristic Polynomial Find the general solution of the differential equation

$$y'' - 5y' + 4y = 0$$

Solution

We already know the characteristic equation of this differential equation from Example 3; the characteristic polynomial factors, so

$$c^2 - 5c + 4 = (c - 4)(c - 1) = 0$$

giving $c = 1$ or $c = 4$. Applying our general solution form we find the general solution

$$y = Ae^t + Be^{4t}$$

We already know quadratic equations can have two solutions, one solution, or no solutions over the real numbers. It is possible for a second-order homogeneous linear differential equation to have any of these three possible types of quadratics as its characteristic polynomial. The examples we have done so far have all had characteristic polynomials with two distinct roots. The following solution rule gives the form of the solution in all three cases.

Solutions of Second-Order Homogeneous Differential Equations

1. If the characteristic polynomial has two distinct real roots c_1 and c_2 then

$$y = Ae^{c_1t} + Bte^{c_2t}$$

2. If the characteristic polynomial has a single, repeated root c then

$$y = Ae^{ct} + Bte^{ct}$$

3. If the characteristic polynomial has no real roots then it is of the form $c^2 + uc + v$ with $u^2 - 4v < 0$ (because the portion of the quadratic equation under the square root is negative). Set $h = -u/2$ and $k = \sqrt{4v - u^2}/2$. Then the solution has the form

$$y = e^{ht}[A \sin(kt) + B \cos(kt)]$$

The third part of the solution rule relies on properties of the *complex numbers*. Although complex numbers are beyond the scope of this text, one of their uses is taking the square root of negative numbers—that is why they are useful for solving differential equations whose characteristic polynomial has a solution that involves the square roots of a negative number. While we are not going to derive the second or third parts of the solution rule, we will do an example that helps explain why sine and cosine functions appear as solutions to second-order equations.

Example 5

A Second-Order Equation with Trigonometric Solutions Verify that $y = \sin(t)$ and $y = \cos(t)$ are both solutions to the second-order equation

$$y'' + y = 0$$

Compare this with the result obtained by applying the rule for solving second-order homogeneous equations.

Solution

First we compute the derivatives.

y	$\sin(t)$	$\cos(t)$
y'	$\cos(t)$	$-\sin(t)$
y'	$\sin(t)$	$-\cos(t)$

Notice that the second derivative of both $y = \sin(t)$ and $y = \cos(t)$ are their own negatives. Thus for both of these functions, $y'' + y = 0$. This verifies that both $y = \sin(t)$ and $y = \cos(t)$ are solutions to $y'' + y = 0$.

To apply the solution rule we must find the characteristic equation. This is $c^2 + 1 = 0$. This equation has no solution so, using the third part of the solution rule, we compute $h = -0/2 = 0$ and $k = \sqrt{4 \cdot 1 - 0/2} = 1$. The solution is therefore

$$y = e^{0t}[A\sin(t) + B\cos(t)] = A\sin(t) + B\cos(t)$$

The two solutions, $\sin(t)$ and $\cos(t)$, that we verified are the two fundamental solutions to the equation. The general solution is their sum with unknown coefficients A and B.

Example 6

Solve the initial value problem

$$y'' - 4y' + 4$$

if $y(0) = 3$ and $y'(0) = 11$.

Solution

The characteristic polynomial of this equation is $c^2 - 4c + 4 = (c - 2)^2$, and so we use part 2 of the solution rule. This tells us that the form of the solution is

$$y = Ae^{2t} + Bte^{2t}$$

We will need y' to plug in the second of the two initial values.

$$\begin{aligned} y' &= 2Ae^{2t} + Be^{2t} + 2Bte^{2t} \\ &= (2A + B + 2Bt)e^{2t} \end{aligned}$$

Plugging in the first initial value we get the equation $y(0) = 3 = A + B \cdot 0 = A$ and so $A = 3$. The second initial value gives us $y'(0) = 11 = 2A + B + 2b \cdot 0 = 2A + B$ or $11 = 2A + B$. Since $A = 3$ we see $B = 5$, and the solution is

$$y = 3e^{2t} + 5te^{2t}$$

Solution of Non-Homogeneous Equations

A particular solution to a differential equation is any function $y = h(x)$ that solves the equation. If we have a **non-homogeneous equation**

$$y' + uy = f(x)$$

or

$$y'' + uy' + vy = f(x)$$

then it turns out that finding one particular solution is enough, together with our techniques for solving homogeneous equations, to find a general solution to a nonhomogeneous equation.

Solutions of Non-Homogenous Differential Equations

Suppose that $h(x)$ is a particular solution to either of the non-homogenous equations

$$y' + uy = f(x)$$

or

$$y'' + uy' + vy = f(x)$$

and that $Y(t)$ is the general solution to the corresponding homogeneous equation

$$y' + uy = 0$$

or

$$y'' + uy' + vy = 0$$

Then the general solution to the non-homogenous differential equation is

$$Y(t) + h(x)$$

Finding a particular solution can be done by assuming a general form for the particular solution, plugging that form into the differential equation, and then solving the resulting system of equations. The next example does this when the right-hand side of the equation is a polynomial.

Example 7

Find a particular solution to the non-homogeneous equation

$$y'' + 5y' + 6y = x$$

Solution

The function on the right-hand side of the equation is a linear equation. If we assume the particular solution is a linear equation $f(x) = ax + b$ then $f'(x) = a$ and $f''(x) = 0$. If we plug these three derivatives into the differential equation then we get

$$0 - 5(a) + 6(ax + b) = x$$

$$6ax + (6b - 5a) = 1x + 0$$

Using the polynomial equivalence principle we learned in Chapter 10 we get two linear equations

$$6b - 5a = 0 \text{ and } 6a = 1$$

so we see that $a = \frac{1}{6}$. Knowing this we can solve for b, obtaining $b = \frac{5}{36}$. The particular solution to the differential equation is thus

$$f(x) = \frac{1}{6}x + \frac{5}{36}$$

Once we know a particular solution, it is not difficult to find the general solution to a non-homogeneous equation.

Example 8

Find the general solution to the non-homogeneous equation

$$y'' + 5y' + 6y = x$$

Solution

The rule for solving non-homogeneous equations says we need a particular solution, which we found for this equation in Example 7, and the general solution to the corresponding homogeneous equation. In this case the homogeneous equation is

$$y'' + 5y' + 6y = 0$$

The characteristic polynomial for this differential equation is $c^2 + 5c + 6$. This equation factors into $(c + 2)(c + 3)$, so $c = -2$ or $c = -3$. This means that the general solution to the homogeneous equation is

$$y = Ae^{-2t} + Be^{-3t}$$

Applying the rule for solving non-homogeneous equations, we add the general solution for the homogeneous equation to the particular solution we found, and obtain the general solution

$$y = Ae^{-2t} + Be^{-3t} + \frac{1}{6}x + \frac{5}{36}$$

A differential equation usually has an infinite number of different particular solutions. For the equation solved in Example 8 we can get a different particular solution by setting A and B to any value. Fortunately, it does not matter which particular solution you find—any one will do to construct a general solution to a non-homogeneous equation. This means that the particular solution that requires the least work to find is the best one.

The technique that we used to find a particular solution in Example 7 is called the **method of undetermined coefficients.** It will work whenever the right-hand side of a non-homogeneous equation is a polynomial.

Undetermined Coefficients for Polynomial Particular Solutions

In order to find a particular solution to the non-homogeneous equation

$$y' + uy = p(t)$$

or

$$y'' + uy'vy = p(t)$$

where $p(t)$ is a polynomial, use the following steps:

1. Pick $y = q(t)$ to be a polynomial of the same degree as $p(t)$ with unknown coefficients.
2. Compute the derivatives of $q(t)$ that appear in the equation, then plug those derivatives into the differential equation.
3. Use the fact that equal polynomials have equal coefficients to solve for the coefficients of $q(t)$.

Example 9

Find a particular solution to the non-homogeneous equation

$$y' + 2y = 4t^2$$

Solution

First set $q(t) = at^2 + bt + c$. This is a quadratic with unknown coefficients a, b, and c—and hence a polynomial the same degree as $4t^2$, the right-hand side of the differential equation. Next, we compute the derivatives of $q(t)$.

$$\begin{aligned} q(t) &= at^2 + bt + c \\ q'(t) &= 2at + b \end{aligned}$$

Plugging these into the differential equations yields

$$\begin{aligned} y' + 2y &= 4t^2 \\ 2at + b + 2(at^2 + bt + c) &= 4t^2 \\ 2at + b + 2at^2 + 2bt + 2c &= 4t^2 \\ (2a)t^2 + 2(a + b)t + (b + 2c) &= 4t^2 + 0t + 0 \end{aligned}$$

Setting equivalent coefficients equal yields the system of equations

$$\begin{cases} 2a = 4 \\ 2(a + b) = 0 \\ b + 2c = 0 \end{cases}$$

These equations are especially easy to solve. The first equation tells us that $a = 2$. Knowing this, the second equation simplifies to $b = -a = -2$. Plugging this into the third equation, we compute $-2 + 2c = 0$ or $c = 1$. The particular solution is thus

$$q(t) = 2t^2 - 2t + 1$$

Example 10

Find the general solution to the non-homogeneous equation

$$y' + 2y = 4t^2$$

Solution

From Example 9, we already have the particular solution $q(t) = 2t^2 - 2t + 1$. This means that all we require is the solution to the corresponding homogeneous equation

$$y' + 2y = 0$$

The characteristic equation is $c - 2 = 0$ (notice this is the first time we have used the characteristic polynomial with a first-degree differential equation). This means that $c = 2$, so the general solution is $y = Ae^{2t}$. Using the rule for solving non-homogeneous equations we add this general solution for the homogeneous equation to the particular solution, and obtain the general solution to the non-homogeneous equation:

$$y = Ae^{2t} + 2t^2 - 2t + 1$$

Applications

Example 11

A population of bacteria are found to grow according to the model

$$y'' + 6y' + 8y = 0$$

where $y(t)$ is the number of bacteria, in millions, after t hours. If the number of bacteria at $t = 0$ is 1.1 million $[y(0) = 1.1]$ and the initial rate at which the population is growing is 0.2 million bacteria per hour $[y'(0) = 0.2]$, solve to find $y(t)$.

Solution

This is a homogeneous equation. The characteristic polynomial is $c^2 + 6c + 8 = 0$, so $(c + 2)(c + 4) = 0$, and we have $c = -2$ or $c = -4$. This gives us the general solution

$$y = Ae^{-2t} + Be^{-4t}$$

In order to plug in the initial conditions, we also require

$$y' = -2Ae^{-2t} - 4Be^{-4t}$$

The first initial condition tells us that

$$y(0) = Ae^{-2\cdot 0} + Be^{-4\cdot 0}$$
$$1.1 = A + B$$

The second initial condition tells us that

$$y'(0) = -2Ae^{-2\cdot 0} - 4Be^{-4\cdot 0}$$
$$0.2 = -2A - 4B$$

Solving these equations simultaneously we get $A = 2.3$ and $B = -1.2$, and so the solution is

$$y(t) = 2.3e^{-2t} - 1.2e^{-4t}$$

Example 12

When first introduced to an island, a population of rabbits grows rapidly. As local predators learn to hunt rabbits the population drops off. The number of rabbits in a given month is modelled by the differential equation

$$y'' + 2y' + y = 0$$

The initial population is $y(0) = 20$ rabbits, while the growth rate when the rabbits were introduced is $y'(0) = 30$ rabbits/month. Solve the differential equation to find a formula for the number of rabbits after t months. Graph the solution for $t = 0$ to $t = 6$. Do the rabbits survive indefinitely or die out?

Solution

This equation is homogeneous with characteristic polynomial $c^2 - 2c + 1 = (c - 1)^2 = 0$, so $c = 1$. The rule for solving second-order homogeneous equations tells us that the general solution is

$$y = Ae^{-t} + Bte^{-t}$$

As before, we need the first derivative of the general solution to plug in the initial conditions. It is

$$y' = -Ae^{-t} + Be^{-t} - Bte^{-t} = (B - A - Bt)e^{-t}$$

Plugging in the first initial condition tells us that

$$y(0) = 20 = Ae^0 + B \cdot 0e^0 = A$$

and so $A = 20$. The second initial condition then becomes

$$y'(0) = 30 = (B - 20 - B \cdot 0)e^0 = (B - 20)$$

If $30 = B - 20$, then $B = 50$, and the solution to the differential equation is

$$y = 20e^{-t} + 50te^{-t} = (20 + 50t)e^{-t}$$

Figure 11.15 suggests that the rabbits die out.

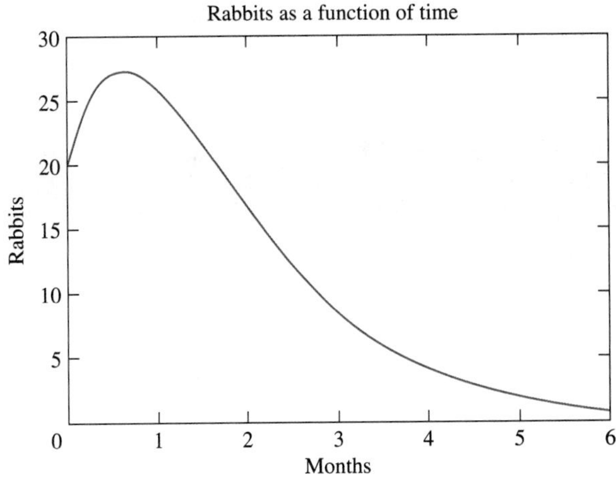

Figure 11.15
The population of rabbits as a function of time

Example 13

The demand for a new type of game console, in thousands of units per month, is found to obey the differential equation

$$y' + y = 397 - 3t$$

If the initial sales are $y(0) = 600$, solve for $y(t)$.

Solution

This is a non-homogeneous first-order equation. We start by using the method of undetermined coefficients to find a particular solution. The right-hand side of the differential equation is linear so we choose the particular solution $q(t) = at + b$. The derivative of q is $q'(t) = a$. Plugging it into the differential equation yields the expression

$$a + at + b = 397 - 3t$$

which gives us the system of equations

$$a + b = 397 \qquad \text{and} \qquad a = -3$$

Solving these simultaneously yields $a = -3$ and $b = 400$, so the particular solution is

$$q(t) = -3t + 400$$

To solve this equation we also need a solution to the homogeneous equation

$$y' + y = 0$$

which has characteristic equation $c + 1 = 0$ so $c = -1$. The general solution to the homogeneous equation is $y = Ae^{-t}$. Thus, the general solution to the original equation is

$$y = Ae^{-t} + 400 - 3t$$

Now, we plug in the initial condition to find the value of A.

$$\begin{aligned} y(0) &= 600 \\ Ae^{-0} + 400 - 3 \cdot 0 &= 600 \\ A + 400 &= 600 \\ A &= 200 \end{aligned}$$

This means that the demand function is

$$y(t) = 200e^{-t} + 400 - 3t \text{ consoles/month}$$

GROUP DISCUSSION

Suppose that a population of birds is modelled by the differential equation

$$y'' - y = 0$$

Find the general solution to this differential equation and answer the following questions.

1. Find a set of initial conditions in which there is a positive population of birds at some point but they later die off.
2. Does the type of population modelled by this differential equation usually die off?

Self-Check Exercises 11.4

1. Find the characteristic polynomial for $y'' + 3y' + 5y = 0$.
2. Compute the general solution to $y'' + 3y' + 2y = 0$.
3. Find a particular solution to $y' - 2y = -4t - 4$.

Solutions to Self-check Exercises 11.4 can be found on page 698.

11.4 Exercises

In Exercises 1–8, find the characteristic equation of the given homogeneous differential equation.

1. $y' + 4y = 0$
2. $y' - 2y = 0$
3. $y'' + 3y' - 7y = 0$
4. $y'' + y' + y = 0$
5. $y'' + 4y' + 2y = 0$
6. $y'' + 5y' + 16y = 0$
7. $y'' + 4y' + 4y = 0$
8. $3y'' + 17y' + 8y = 0$

In Exercises 9–18, find the general solution to the given homogeneous linear differential equation.

9. $y' + 2y = 0$
10. $y' - 5y = 0$
11. $y'' - 4y' + 3y = 0$
12. $y'' + 7y' + 12y = 0$
13. $y'' - 8y' + 15y = 0$
14. $y'' - y' - y = 0$
    Hint: Solve the characteristic equation using the quadratic formula.
15. $y'' - 6y' + 9y = 0$
16. $y'' - 2y' + y = 0$
17. $y'' + 4y = 0$
    Hint: The characteristic polynomial has no roots.
18. $y'' + 4y' + 5y = 0$
    Hint: The characteristic polynomial has no roots.

In Exercises 19–22, find a particular solution for the non-homogeneous differential equation and use it to find a general solution.

19. $y' + 2y = x$
20. $y' - y = x^2$
21. $y'' + 3y' + 2y = 2x + 1$
22. $y'' - 5y' + 6y = 12 - 5x$

In Exercises 23–28, find the solution to the given initial value problem. Some of the differential equations are homogeneous and some are not.

23. $y' - 5y = 0$; $y(0) = 12$
24. $y' + 2y = x + 1$; $y(0) = 4$
25. $y'' + 2y' + y = 0$; $y(0) = 4$, $y'(0) = 7$
26. $y'' - 6y' + 5y = 0$; $y(0) = 5$, $y'(0) = 13$
27. $y'' - 3y' + 2y = x$; $y(0) = 3$, $y'(0) = 5$
28. $y'' + 4y' + 4y = 2x + 1$; $y(0) = 5$, $y'(0) = 16$
29. The number of members $n(d)$ of a web site on a given day d after the site starts keeping statistics is found to follow the differential equation

$$n'' + 2n' + 2n = 40d + 140$$

If the number of members the first day statistics are taken is $n(0) = 60$ and the rate at which people are joining the site is $n'(0) = 24$, solve the initial value problem to find $n(d)$. How many members does the web site have after 150 days? After 300?

 **A calculator is recommended for Exercises 30–36.**

Business and Economics Applications

30. **Demand for Sunglasses** The demand for a type of designer sunglasses follows the differential equation

$$y'' + 4y' + 4y = 0$$

where $y(t)$ is millions of pairs sold per month. If sales the first month are $y(0) = 2.3$ million pairs and the rate at which sales are changing initially is $y'(0) = 0.6$ million additional pairs per month, then solve the initial value problem to find the demand function $y(t)$.

Biological and Life Sciences Applications

31. **Bacterial Growth** A population of bacteria is modelled by the differential equation $y' - 2y = 0$ where $y(t)$ is the number of bacteria, in millions, present after t weeks. If the initial number of bacteria is $y(0) = 6.8$ million, solve the differential equation to find $y(t)$.

32. **Grams of Yeast Producing Alcohol** The number of grams of live yeast in a bioreactor that makes alcohol after t days is modelled by the differential equation $y' + 0.04y = 0$. If the reactor is started with $y(0) = 800$ grams of yeast, solve the differential equation to find $y(t)$.

33. **Weed Growth** The growth of a type of weed imported from Europe is found to grow so that the number $w(t)$ of weeds in a five-hectare survey area follows the differential equation

$$6w'' - 5w' + w = 0$$

If the initial number of weeds in the test plot is $w(0) = 200$ and the initial growth rate of the weeds is $w'(0) = 40$, solve the initial problem to find $w(t)$.

34. **Bird Extinction Cause** A population of birds that were hunted for their red-and-yellow wing feathers is reintroduced to an island where they were hunted out. The population of birds present after t years is modelled by the differential equation

$$y' + y = -8x + 196$$

A population of $y(0) = 200$ birds is placed on the island at the beginning of the attempt at reintroduction. The population of birds is found to decline. Two theories are advanced: (i) the population can no longer live on the island and (ii) the birds are being poached. Solve the initial value problem, and graph the solution. Does the model suggest that the birds will die out even if there are no poachers?

35. **Mycoplasm Population** A population of mycoplasm has a population $P(t)$ that is modelled by the differential equation

$$P'' + P' + 0.25P = 0$$

after they are injected into a mouse. Find the general form of the solution to this differential equation and decide, based on that solution, if the mycoplasm will eventually die out or persist indefinitely.

36. **Accidental Release of Mice** A population of mice accidentally released on Prince Edward Island is modelled by the differential equation

$$y'' + y' - 2y = 0$$

where $y(t)$ is the number of mice present after t periods equal to 3 months. If the number of mice released originally is $y(0) = 40$ and the population of mice grows initially so that $y'(0) = 10$, solve the initial value problem to find $y(t)$. Will the population of mice die out on their own or are control measures needed?
Hint: Look at the fundamental solutions. Only one of them really matters.

Solution to Self-Check Exercises 11.4

1. $c^2 + 3c + 5$

2. The characteristic polynomial of this differential equation is $c^2 + 3c + 2 = (c + 1)(c + 2)$. Solving $(c + 1)(c + 2) = 0$, we obtain $c = -1$ or $c = -2$ which makes the general solution

$$y = Ae^{-t} + Be^{-2t}$$

3. The right-hand side of the equation is linear so we apply the method of undetermined coefficient to $q(t) = at + b$. Note that $q'(t) = a$ and plug this information into the differential equation. The resulting equation is

$$a - 2(at + b) = -2at + a - 2b = -4t - 4$$

This polynomial equality yields the system of equations

$$-2a = -4 \quad \text{and} \quad a - 2b = -4$$

Solving these simultaneously we get $a = 2$ and $b = 3$. The particular solution is thus $y = 2t + 3$.

11.5 Approximate Solutions of Differential Equations

Euler's Method

As in the case of definite integrals, there are many differential equations whose exact solutions cannot be found using any of the available methods. In such cases, we must once again resort to approximate solutions.

Many numerical methods have been developed for efficient computation of approximate solutions to differential equations. One of these, the Runge-Kutta method, is more accurate than Euler's method, but it is so computation intensive that it is typically done only with a computer. In this section we will look at a method for solving the problem.

$$\frac{dy}{dx} = F(x, y) \qquad [y(x_0) = y_0] \tag{12}$$

Euler's method, named after Leonhard Euler (1707–1783), illustrates the idea behind the method for finding the approximate solution of Equation (12). Basically, the technique calls for approximating the actual solution $y = f(x)$ at certain selected values of x. The values of f between two adjacent values of x are then found by linear interpolation. This situation is depicted geometrically in Figure 11.16. Thus, in Euler's method, the actual solution curve of the differential equation is approximated by a suitable polygonal curve.

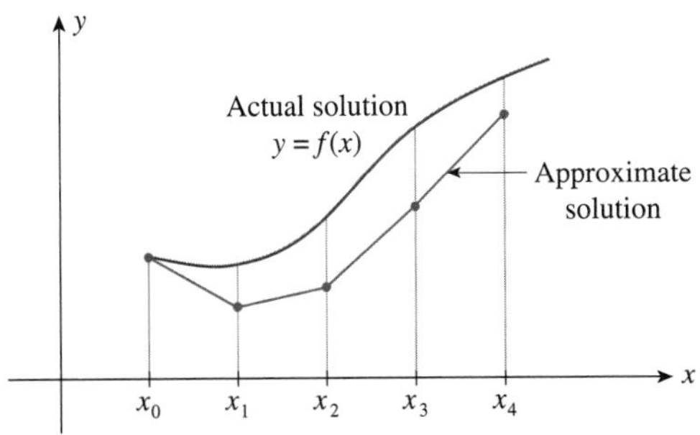

Figure 11.16
Using Euler's method, the actual solution curve of the differential equation is approximated by a polygonal curve.

We now describe the method in greater detail. Let h be a small positive number and let $x_n = x_0 + nh$ ($n = 0, 1, 2, 3, \ldots$). Then,

$$x_1 = x_0 + h \qquad x_2 = x_0 + 2h \qquad x_3 = x_0 + 3h \quad \ldots$$

Thus, the points $x_0, x_1, x_2, x_3, \ldots$ are spaced evenly apart, and the distance between any two adjacent points is h units.

We begin by finding an approximation $y = \tilde{f}_1(x)$ (read "f tilde sub-one of x") to the actual solution $y = f(x)$ on the subinterval $[x_0, x_1]$. Observe that the *initial* condition $y(x_0) = y_0$ of (12) tells us that the point (x_0, y_0) lies on the solution curve. Now, Euler's method calls for approximating the part of the graph of f on the interval $[x_0, x_1]$ by a straight-line segment that is tangent to the graph of f at the point (x_0, y_0). To find this approximating linear function $y = \tilde{f}_1(x)$, recall that the differential equation $y' = F(x, y)$ gives the slope of the tangent line to the graph of $y = f(x)$ at any point (x, y) lying on the graph. In particular, the slope of the required straight-line segment is equal to $F(x_0, y_0)$. Therefore, using the point-slope form of the equation of a line, we see that the equation of the straight line segment is

$$y - y_0 = F(x_0, y_0)(x - x_0)$$

$$y = y_0 + F(x_0, y_0)(x - x_0)$$

Thus, the required approximation to the actual solution $y = f(x)$ on the interval $[x_0, x_1]$ is given by the linear function

$$\tilde{f}_1(x) = y_0 + F(x_0, y_0)(x - x_0) \qquad (13)$$

This situation is depicted graphically in Figure 11.17.

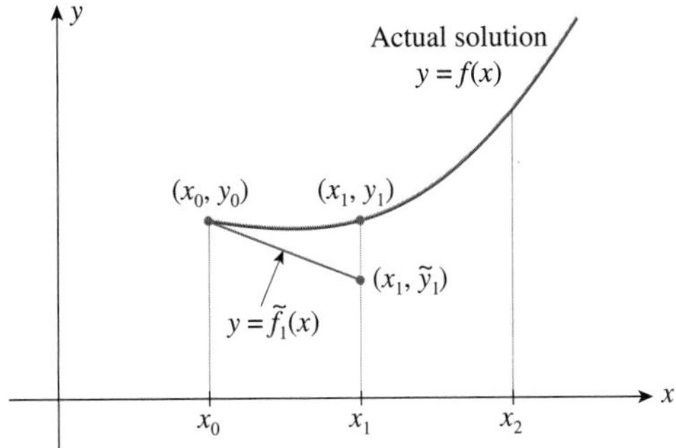

Figure 11.17
$\tilde{f}_1(x) = y_0 + F(x_0, y_0)(x - x_0)$ is the equation of the straight line used to approximate $f(x)$ on $[x_0, x_1]$.

Next, to find an approximation $y = \widetilde{f}_2(x)$ to the actual solution $y = f(x)$ on the subinterval $[x_1, x_2]$, observe that at $x = x_1$ the true value $y_1 = f(x)$ is approximated by the number

$$\widetilde{y}_1 = \widetilde{f}_1(x_1) = y_0 + F(x_0, y_0)(x_1 - x_0)$$
$$= y_0 + F(x_0, y_0 h) \qquad \text{Since } x_1 - x_0 = h$$

According to Euler's method, the approximating function in this interval has a graph that is just a line segment beginning at the point $(x_1, \widetilde{y}_1)$ and having an appropriate slope. Ideally, this slope should be the slope of the tangent line to the graph of $y = f(x)$ at the point (x_1, y_1). But since this point is supposedly unknown to us, a practical alternative is to use the point $(x_1, \widetilde{y}_1)$ in lieu of the point (x_1, y_1) to compute an approximation of the desired slope—namely, $F(x_1, \widetilde{y}_1)$. Once again, using the point-slope form of the equation of a line, we find that the equation of the required tangent line is

$$y - \widetilde{y}_1 = F(x_1, \widetilde{y}_1)(x - x_1)$$
$$y = \widetilde{y}_1 + F(x_1, \widetilde{y}_1)(x - x_1)$$

so the required approximating linear function on the interval $[x_1, x_2]$ is given by

$$\widetilde{f}_2(x) = \widetilde{y}_1 + F(x_1, \widetilde{y}_1)(x - x_1) \tag{14}$$

(Figure 11.18).

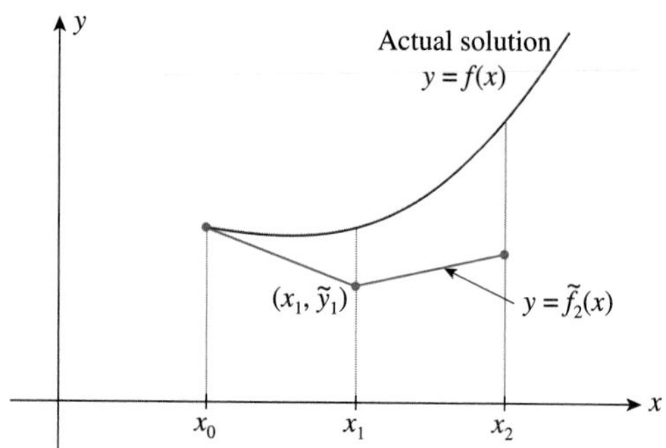

Figure 11.18
$\widetilde{f}_2(x) = \widetilde{y}_1 + F(x_1, \widetilde{y}_1)(x - x_1)$ is the equation of the straight line used to approximate $f(x)$ on $[x_1, x_2]$.

Proceeding, we find that the actual value $y_2 = f(x_2)$ is approximated by the number

$$\widetilde{y}_2 = \widetilde{f}_2(x_2) = \widetilde{y}_1 + F(x_1, \widetilde{y}_1)(x_2 - x_1)$$
$$= \widetilde{y}_1 + F(x_1, \widetilde{y}_1)h$$

and the approximating function on the interval $[x_2, x_3]$ is given by

$$\widetilde{f}_3(x) = \widetilde{y}_2 + F(x_2, \widetilde{y}_2)(x - x_2)$$

and so on.

In many practical applications, we are interested only in finding an approximation of the solution of the differential Equation (12) at some specific value of x—say, $x = b$. In such cases, it is not important to compute the approximating functions $\widetilde{f}_1, \widetilde{f}_2, \widetilde{f}_3, \ldots$. It suffices to compute the numbers $\widetilde{y}_1, \widetilde{y}_2, \widetilde{y}_3, \ldots, \widetilde{y}_n$ at the points $x_1, x_2, x_3, \ldots, x_n$ with h chosen to be

$$h = \frac{b - x_0}{n}$$

when n is a sufficiently large positive integer. In this case, we see that

$$x_1 = x_0 + h \qquad x_2 = x_0 + 2h, \ldots \qquad x_n = x_0 + nh = x_0 + (b - x_0) = b$$

so that $\widetilde{y}_n = \widetilde{f}_n(x_n) = \widetilde{f}_n(b)$ does in fact yield the required approximation.

We now summarize this procedure.

Euler's Method

Suppose we are given the differential equation

$$\frac{dy}{dx} = F(x, y)$$

subject to the initial condition $y(x_0) = y_0$ and we wish to find an approximation of $y(b)$, where b is a number greater than x_0. Compute

$$h = \frac{b - x_0}{n}$$

$$x_1 = x_0 + h \qquad x_2 = x_0 + 2h \qquad x_3 = x_0 + 3h, \ldots \qquad x_n = x_0 + nh = b$$

and

$$\begin{aligned} \tilde{y}_0 &= y_0 \\ \tilde{y}_1 &= \tilde{y}_0 + hF(x_0, \tilde{y}_0) \\ \tilde{y}_2 &= \tilde{y}_1 + hF(x_1, \tilde{y}_1) \\ &\vdots \\ \tilde{y}_n &= \tilde{y}_{n-1} + hF(x_{n-1}, \tilde{y}_{n-1}) \end{aligned}$$

Then, $\tilde{y}_n$ gives an approximation of the true value $y(b)$ of the solution to the initial value problem at $x = b$.

Solving Differential Equations with Euler's Method

Example 1

Use Euler's method with $n = 8$ to obtain an approximation of the initial value problem

$$y' = x - y \quad [y(0) = 1]$$

when $x = 2$.

Solution

Here, $x_0 = 0$ and $b = 2$ so that, taking $n = 8$, we find

$$h = \frac{2 - 0}{8} = \frac{1}{4}$$

and $\quad x_0 = 0 \quad x_1 = \frac{1}{4} \quad x_2 = \frac{1}{2} \quad x_3 = \frac{3}{4} \quad x_4 = 1$

$$x_5 = \frac{5}{4} \quad x_6 = \frac{3}{2} \quad x_7 = \frac{7}{4} \quad x_8 = b = 2$$

Also $\quad F(x, y) = x - y \quad$ and $\quad y_0 = y(0) = 1$

Therefore, the approximations of the actual solution at the points $x_0, x_1, x_2, \ldots, x_n = b$ are

$$\begin{aligned} \tilde{y}_0 &= y_0 = 1 \\ \tilde{y}_1 &= \tilde{y}_0 + hF(x_0, \tilde{y}_0) = 1 + \frac{1}{4}(0 - 1) = \frac{3}{4} \\ \tilde{y}_2 &= \tilde{y}_1 + hF(x_1, \tilde{y}_1) = \frac{3}{4} + \frac{1}{4}\left(\frac{1}{4} - \frac{3}{4}\right) = \frac{5}{8} \\ \tilde{y}_3 &= \tilde{y}_2 + hF(x_2, \tilde{y}_2) = \frac{5}{8} + \frac{1}{4}\left(\frac{1}{2} - \frac{5}{8}\right) = \frac{19}{32} \\ \tilde{y}_4 &= \tilde{y}_3 + hF(x_3, \tilde{y}_3) = \frac{19}{32} + \frac{1}{4}\left(\frac{3}{4} - \frac{19}{32}\right) = \frac{81}{128} \end{aligned}$$

$$\widetilde{y}_5 = \widetilde{y}_4 + hF(x_4, \widetilde{y}_4) = \frac{81}{128} + \frac{1}{4}\left(1 - \frac{81}{128}\right) = \frac{371}{512}$$

$$\widetilde{y}_6 = \widetilde{y}_5 + hF(x_5, \widetilde{y}_5) = \frac{371}{512} + \frac{1}{4}\left(\frac{5}{4} - \frac{371}{512}\right) = \frac{1753}{2048}$$

$$\widetilde{y}_7 = \widetilde{y}_6 + hF(x_6, \widetilde{y}_6) = \frac{1753}{2048} + \frac{1}{4}\left(\frac{3}{2} - \frac{1753}{2048}\right) = \frac{8331}{8192}$$

$$\widetilde{y}_8 = \widetilde{y}_7 + hF(x_7, \widetilde{y}_7) = \frac{8331}{8192} + \frac{1}{4}\left(\frac{7}{4} - \frac{8331}{8192}\right) = \frac{39{,}329}{32{,}768}$$

Thus, the approximate value of $y(2)$ is

$$\frac{39{,}329}{32{,}768} \approx 1.2002$$

PORTFOLIO

Maja Veljkovic

Title Director General
Institution National Research Council of Canada
Institute for Fuel Cell Innovation

The search for new energy solutions has become a foremost global concern. In this realm, fuel cells are considered as the primary enabling technology of the widely anticipated hydrogen economy. The National Research Council of Canada Institute for Fuel Cell Innovation (NRC-IFCI) is a worldwide leading institution for the support and development of fuel cell research. When Maja Veljkovic was appointed Director General of NRC-IFCI in June 2001, her immense experience as Senior Engineer and Research and Development (R&D) Manager at Syncrude Canada helped her set realistic visions for this institution.

Veljkovic has long been fascinated with the science of fuel cells, and as she notes, "obsessed with finding a clean energy solution." What makes fuel cells so interesting to her is their complex interdependency of various processes. Each process needs to be understood and fine-tuned to design an efficiently working source of energy.

Veljkovic explains, "a fuel cell is like a living organism: malfunctioning of one organ or unhealthy diet is likely to destroy the balance of the whole body in an incurable way." The fuel cell must be designed as a whole, and mathematical modelling plays an indispensable role in this undertaking.

The mathematics involved in fuel cell modelling mainly comprises ordinary and partial differential equations, numerical methods, and linear and nonlinear programming. For example, notes Veljkovic, "to investigate how the operating conditions (such as oxygen concentration) and design parameters (such as the thickness of electrode and catalyst loading) affect the performance of the cathode catalyst layer, a 1D, single phase, homogeneous mathematical model for the cathode catalyst layer is employed." This model is composed of three ordinary differential equations (DE) for the mass transport of each species in the layer. Generally speaking, these DEs do not have a set of analytical solutions. In order to solve the DEs, numerical methods are usually employed. Through modelling study, we can optimize the design of the cathode catalyst layer in order to reduce the cost and improve the performance.

For Veljkovic, the study of fuel cells is more than the excitement behind solving a mathematical problem; it is also a way to develop environmentally friendly energy during a time when modern society encounters many energy, health, and environmental challenges.

Example 2

Use Euler's method with (a) $n = 5$ and (b) $n = 10$ to approximate the solution of the initial value problem

$$y' = -2xy^2 \qquad [y(0) = 1]$$

on the interval [0, 0.5]. Find the actual solution of the initial value problem. Finally, sketch the graphs of the approximate solutions and the actual solution for $0 \le x \le 0.5$ on the same set of axes.

Solution

a. Here, $x_0 = 0$ and $b = 0.5$. Taking $n = 5$, we find

$$h = \frac{0.5 - 0}{5} = 0.1$$

and $x_0 = 0$, $x_1 = 0.1$, $x_2 = 0.2$, $x_3 = 0.3$, $x_4 = 0.4$, and $x_5 = b = 0.5$. Also,

$$F(x, y) = -2xy^2 \quad \text{and} \quad y_0 = y(0) = 1$$

Therefore,

$$\widetilde{y}_0 = y_0 = 1$$

$$\widetilde{y}_1 = \widetilde{y}_0 + hF(x_0, \widetilde{y}_0) = 1 + 0.1(-2)(0)(1)^2 = 1$$

$$\widetilde{y}_2 = \widetilde{y}_1 + hF(x_1, \widetilde{y}_1) = 1 + 0.1(-2)(0.1)(1)^2 = 0.98$$

$$\widetilde{y}_3 = \widetilde{y}_2 + hF(x_2, \widetilde{y}_2) = 0.98 + 0.1(-2)(0.2)(0.98)^2 = 0.9416$$

$$\widetilde{y}_4 = \widetilde{y}_3 + hF(x_3, \widetilde{y}_3) = 0.9416 + 0.1(-2)(0.3)(0.9416)^2 = 0.8884$$

$$\widetilde{y}_5 = \widetilde{y}_4 + hF(x_4, \widetilde{y}_4) = 0.8884 + 0.1(-2)(0.4)(0.8884)^2 = 0.8253$$

b. Here $x_0 = 0$ and $b = 0.5$. Taking $n = 10$, we find

$$h = \frac{0.5 - 0}{10}$$

and $x_0 = 0$, $x_1 = 0.05$, $x_2 = 0.10$, . . . , $x_9 = 0.45$, and $x_{10} = 0.5 = b$. Proceeding as in part (a), we obtain the approximate solutions listed in the following table:

x	0.00	0.05	0.10	0.15	0.20	0.25
$\widetilde{y}_n$	1.0000	1.0000	0.9950	0.9851	0.9705	0.9517

x	0.30	0.35	0.40	0.45	0.50
$\widetilde{y}_n$	0.9291	0.9032	0.8746	0.8440	0.8119

To obtain the actual solution of the differential equation, we separate variables, obtaining

$$\frac{dy}{y^2} = -2x\,dx$$

Integrating each side of the last equation with respect to the appropriate variables, we have

$$\int \frac{dy}{y^2} = -\int 2x\,dx$$

or

$$-\frac{1}{y} + C_1 = -x^2 + C_2$$

$$\frac{1}{y} = x^2 + C \qquad C = C_1 - C_2$$

$$y = \frac{1}{x^2 + C}$$

Using the condition $y(0) = 1$, we have

$$1 = \frac{1}{0 + C} \qquad \text{or} \qquad C = 1$$

Therefore, the required solution is given by

$$y = \frac{1}{x^2 + 1}$$

The graphs of the approximate solutions and the actual solution are sketched in Figure 11.19.

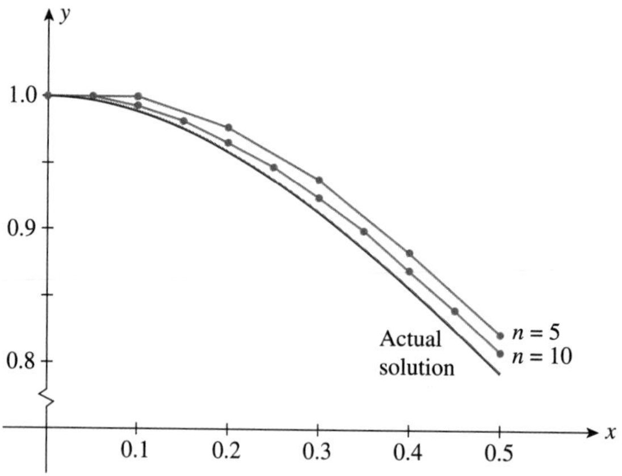

Figure 11.19
The approximate solutions and the actual solution to an initial value problem

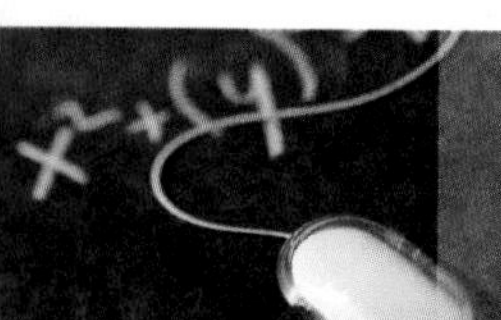

Self-Check Exercise 11.5

Use Euler's method with $n = 5$ to obtain an approximation of the initial value problem

$$y' = 2x + y \qquad [y(0) = 1]$$

when $x = 1$.

The solution to Self-Check Exercise 11.5 can be found on page 705.

11.5 Exercises

A calculator is recommended for this exercise set. In Exercises 1–10, use Euler's method with (a) $n = 4$ and (b) $n = 6$ to obtain an approximation of the initial value problem when $x = b$.

1. $y' = x + y$, $y(0) = 1$; $b = 1$
2. $y' = x - 2y$, $y(0) = 1$; $b = 2$
3. $y' = 2x - y + 1$, $y(0) = 2$; $b = 2$
4. $y' = 2xy$, $y(0) = 1$; $b = 0.5$
5. $y' = -2xy^2$, $y(0) = 1$; $b = 0.5$
6. $y' = x^2 + y^2$, $y(0) = 1$; $b = 1.5$
7. $y' = \sqrt{x + y}$, $y(1) = 1$; $b = 1.5$
8. $y' = (1 + x^2)^{-1}$, $y(0) = 0$; $b = 1$
9. $y' = \dfrac{x}{y}$, $y(0) = 1$; $b = 1$
10. $y' = xy^{1/3}$, $y(0) = 1$; $b = 1$

In Exercises 11–15, use Euler's method with $n = 5$ to obtain an approximate solution to the initial value problem over the indicated interval.

11. $y' = \dfrac{1}{2}xy$, $y(0) = 1$; $0 \le x \le 1$
12. $y' = x^2y$, $y(0) = 2$; $0 \le x \le 0.6$
13. $y' = 2x - y + 1$, $y(0) = 2$; $0 \le x \le 1$
14. $y' = x + y^2$, $y(0) = 0$; $0 \le x \le 0.5$
15. $y' = x^2 + y$, $y(0) = 1$; $0 \le x \le 0.5$

Business and Economics Applications

16. **Growth of Service Industries** It has been estimated that service industries, which currently make up 30% of the nonfarm workforce in a certain country, will continue to grow at the rate of

$$R(t) = 5e^{1/(t+1)}$$

percent per decade t decades from now. Estimate the percentage of the nonfarm workforce in the service industries one decade from now.

Hint: (a) Show that the desired answer is $P(1)$, where P is the solution of the initial value problem

$$P' = 5e^{1/(t+1)} \qquad [P(0) = 30]$$

(b) Use Euler's method with $n = 10$ to approximate the solution.

Solution to Self-Check Exercise 11.5

Here, $x_0 = 0$ and $b = 1$ so that, taking $n = 5$, we find

$$h = \frac{1 - 0}{5} = \frac{1}{5}$$

and

$$x_0 = 0 \quad x_1 = \frac{1}{5} \quad x_2 = \frac{1}{5}$$

$$x_3 = \frac{3}{5} \quad x_4 = \frac{4}{5} \quad x_5 = b = 1$$

Also,

$$F(x, y) = 2x + y \quad \text{and} \quad y_0 = y(0) = 1$$

Therefore, the approximations of the actual solution at the points $x_0, x_1, x_2, \ldots, x_5 = 1$ are

$$\tilde{y}_0 = y_0 = 1$$

$$\tilde{y}_1 = \tilde{y}_0 + hF(x_0, \tilde{y}_0) = 1 + \frac{1}{5}(0 + 1) = \frac{6}{5}$$

$$\tilde{y}_2 = \tilde{y}_1 + hF(x, \tilde{y}_1) = \frac{6}{5} + \frac{1}{5}\left(\frac{2}{5} + \frac{6}{5}\right) = \frac{38}{25}$$

$$\tilde{y}_3 = \tilde{y}_2 + hF(x_2, \tilde{y}_2) = \frac{38}{25} + \frac{1}{5}\left(\frac{4}{5} + \frac{38}{25}\right) = \frac{248}{125}$$

$$\tilde{y}_4 = \tilde{y}_3 + hF(x_3, \tilde{y}_3) = \frac{248}{125} + \frac{1}{5}\left(\frac{6}{5} + \frac{248}{125}\right) = \frac{1638}{625}$$

$$\tilde{y}_5 = \tilde{y}_4 + hF(x_4, \tilde{y}_4) = \frac{1638}{625} + \frac{1}{5}\left(\frac{8}{5} + \frac{1638}{625}\right) = \frac{10{,}828}{3125}$$

Thus, the approximate value of $y(1)$ is

$$\frac{10{,}828}{3125} \approx 3.4650$$

CHAPTER 11 **Summary of Principal Terms**

Terms

differential equation (664)
solution to a differential equation (666)
general solution of a differential equation (666)
particular solution of a differential equation (666)
order (670)
first-order differential equation (670)
second-order differential equation (670)
separable differential equation (670)
method of separation of variables (670)
initial value problem (672)
linear differential equation (687)
homogeneous equation (687)
fundamental solution (688)
characteristic polynomial (689)
non-homogeneous equation (691)
method of undermined coefficients (693)
Euler's method (698)

Review Exercises 

In Exercises 1–3, verify that y is a solution of the differential equation.

1. $y = C_1e^{2x} + C_2e^{-3x}$; $y'' + y' - 6y = 0$

2. $y = 2e^{2x} + 3x - 2$; $y'' - y' - 2y = -6x + 1$

3. $y = Cx^{-4/3}$; $4xy^3\,dx + 3x^2y^2\,dy = 0$

In Exercises 4 and 5, verify that y is a general solution of the differential equation and find a particular solution of the differential equation satisfying the side condition.

4. $y = \dfrac{1}{x^2 - C}$; $\dfrac{dy}{dx} = -2xy^2$; $y(0) = 1$

5. $y = (9x + C)^{-1/3}$; $\dfrac{dy}{dx} = -3y^4$; $y(0) = \dfrac{1}{2}$

In Exercises 6–14, solve the differential equation.

6. $y' = \dfrac{x^3 + 1}{y^2}$

7. $\dfrac{dy}{dt} = 2(4 - y)$

8. $y' = \dfrac{y \ln x}{x}$

9. $y' = 3x^2y^2 + y^2$; $y(0) = -2$

10. $y' = x^2(1 - y)$; $y(0) = -2$

11. $\dfrac{dy}{dx} = -\dfrac{3}{2}x^2y$; $y(0)$

12. $y'' - 4y' + 3y = 0$; $y(0) = 2$, $y'(0) = 4$

13. $y'' - 2y' + y = 0$; $y(0) = 4$, $y'(0) = 6$

14. $y' + 3y = x + 3$; $y(0) = 12$

15. Find a function f given that (1) the slope of the tangent line to the graph of f at any point $P(x, y)$ is given by the expression

$$y' = \frac{4xy}{x^2 + 1}$$

and (2) the graph of f passes through the point $(1, 1)$.

In Exercises 16–19, use Euler's method with (a) $n = 4$ and (b) $n = 6$ to obtain an approximation of the initial value problem when $x = b$.

16. $y' = x + y^2$, $y(0) = 0$; $b = 1$

17. $y' = x^2 + 2y^2$, $y(0) = 0$; $b = 1$

18. $y' = 1 + 2xy^2$, $y(0) = 0$; $b = 1$

19. $y' = e^x + y^2$, $y(0) = 0$; $b = 1$

In Exercises 20 and 21, use Euler's method with $n = 5$ to obtain an approximate solution to the initial value problem over the indicated interval.

20. $y' = 2xy$, $y(0) = 1$; $0 \le x \le 1$

21. $y' = x^2 + y^2$, $y(0) = 1$; $0 \le x \le 1$

22. Barbara placed a 5-kilogram rib roast, which had been standing at room temperature (68°F), into a 350°F oven at 4 P.M. At 6 P.M. the temperature of the roast was 118°F. At what time would the temperature of the roast have been 150°F (medium rare)?
Hint: Use Newton's law of cooling (or heating).

23. **Spread of a Rumour** A rumour to the effect that a rental increase was imminent was first heard by four residents of the Chatham West Condominium Complex. The rumour spread through the complex of 200 single-family dwellings at a rate jointly proportional to the number of families who had heard it and the number who had not. Two days later, the number of

families who had heard the rumour had increased to 40. Find how many families had heard the rumour after 5 days.

24. **A Mixture Problem** A tank initially contains 40 litres of pure water. Brine containing 3 kg of salt per litre flows into the tank at a rate of 4 L/min, and the well-stirred mixture flows out of the tank at the same rate. How much salt is in the tank at any time t? How much salt is in the tank in the long run?

Business and Economics Applications

25. **Resale Value of a Machine** The resale value of a certain machine decreases at a rate proportional to the machine's purchase price. The machine was purchased at \$50,000 and 2 yr later was worth \$32,000.
    a. Find an expression for the resale value of the machine at any time t.
    b. Find the value of the machine after 5 yr.

26. **Continuous Compound Interest** HAL Corporation invests P dollars/year (assume this is done on a frequent basis in small deposits over the year so that it is essentially continuous) into a fund earning interest at the rate of r%/year compounded continuously. Then the size of the fund A grows at a rate given by

$$\frac{dA}{dt} = rA + P$$

Suppose $A = 0$ when $t = 0$. Determine the size of the fund after t yr. What is the size of the fund after 5 yr if $P = \$50,000$ and $r = 12\%$/year?

27. **Demand for a Commodity** Suppose the demand D for a certain commodity is constant and the rate of change in the supply S over time t is proportional to the difference between the demand and the supply so that

$$\frac{dS}{dt} = k(D - S)$$

Find an expression for the supply at any time t if the supply at $t = 0$ is S_0.

28. **Demand for Cinnamon Rolls** The demand for a new type of cinnamon roll, in thousands of rolls per day, is found to be modelled by the differential equation

$$y' + 0.01y = 1 - 0.02t$$

If $y(0) = 800$ rolls are sold the first day they are available, solve the initial value problem to find the demand function $y(t)$.

Biological and Life Sciences Applications

29. **Brentano–Stevens Law** The Brentano–Stevens law, which describes the rate of change of a response R to a stimulus S, is given by

$$\frac{dR}{dS} = k \cdot \frac{R}{S}$$

where k is a positive constant. Solve this differential equation.

30. **A Learning Curve** The Court Reporting Institute finds that the average student taking the Advanced Stenotype course will progress at a rate given by

$$\frac{dQ}{dt} = k(120 - Q)$$

in a 20-wk course, where $Q(t)$ measures the number of words of dictation the student can take per minute after t wk in the course. [Assume $Q(0) = 60$.] If the average student can take 90 words of dictation per minute after 10 wk in the course, how many words per minute can the average student take after completing the course?

31. **Bacterial Growth** A population of bacteria is modelled by the differential equation $y' - 1.7y = 0$ where $y(t)$ is the number of millions of bacteria present after t days. If the initial number of bacteria is $y(0) = 0.76$ million, solve the differential equation to find $y(t)$.

32. **Ginseng Survival** A type of prized wild ginseng was harvested to extinction in a district where it once grew. A group of $y(0) = 1000$ plants are reintroduced in various locations about the district and these plants have a growth rate of $y'(0) = 200$ plants/month initially. If the plants are modeled by the differential equation

$$2y'' + 3y' + 2y = 0$$

solve the initial value problem to get a formula for $y(t)$. Graph the solution. Does model suggest that the population of plants will increase or die off?

33. **Accumulation of Protien** The number of grams of protein in a growth medium after t days is modelled by the differential equation $y' + 0.04y = 0$. If the growth medium initially contains $y(0) = 2$ grams of protein and the initial increase in protein is $y'(0) = 4$ g/d, then find the amount of protein at the end of two weeks.

12

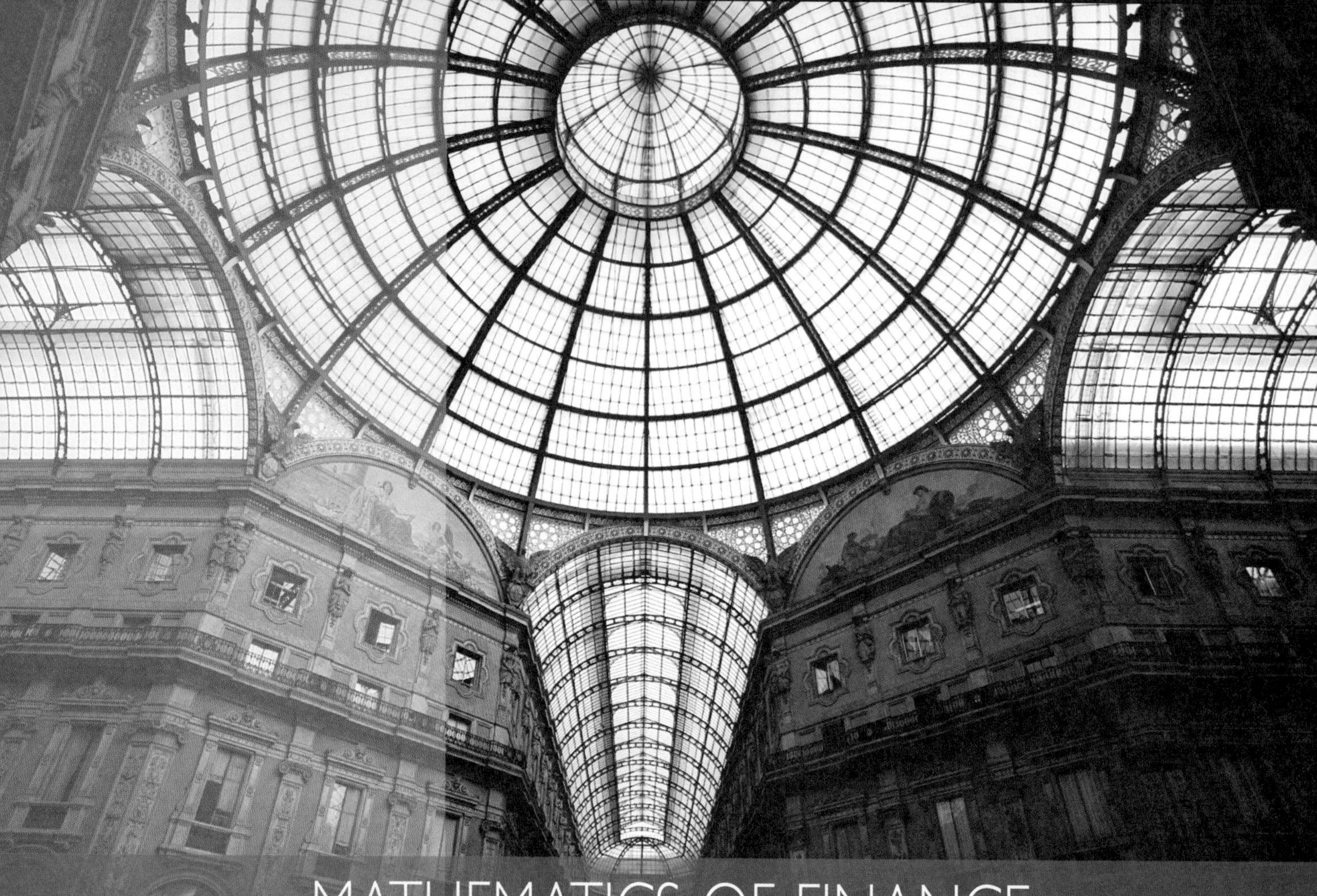

MATHEMATICS OF FINANCE

Photo: Thinkstock/Jupiter Images

This chapter describes how to model the value of money over time and under various conditions. The key concept is interest, which is the fee we are expected to pay when borrowing money. Mortgages, loans, investments, etc. are all based on interest and consumers benefit from knowing these financial terms as well as having the money skills to work with them. Several financial formulas will be introduced with an aim to gaining a better understanding of them and why they apply to a certain financial situation.

Photo: Thinkstock/Jupiter Images

Photo: © BananaStock/Alamy

When money is loaned for a certain purchase and paid back in regular installments that include compound interest, an amortization table should be developed, as it allows the purchaser to visualize the amortization process and to understand how the loan is paid off. The discussion on pages 739–741 establishes an amortization table based on a car purchase through a loan, which is worked out in Example 1 in Section 12.4, page 738.

12.1 Simple Interest

Interest

Interest is the amount of money that is paid for borrowing money for a certain amount of time and usually denoted by I. Borrowed money can be money that an institution loans a customer, in which case the customer pays the interest, or the money that a customer deposits into an account, in which case the institution pays the interest. Either form of borrowed money is called **principal,** denoted by P. The interest rate r is typically stated as an **annual percentage rate,** also referred to as **nominal annual rate,** but calculated in **decimal form.** Time t is usually measured in years or a fraction thereof unless specified otherwise. If the year is measured in 360 days, then the interest based on this year is referred to as **ordinary interest.** If the year is measured in 365 days, then the interest based on this year is referred to as **exact interest.**

Simple Interest

One application of interest is **simple interest,** which is the interest calculated only on a principal over time. If a principal P accumulates simple interest at an annual interest rate r over time t, then the interest I is calculated as

$$I = Prt$$

In other words, the interest is principal times rate times time. Unless stated otherwise, we will assume that simple interest is ordinary for the remainder of this chapter, i.e., the year is measured in 360 days.

Simple Interest

$$I = Prt \qquad (1)$$

where

I is the total interest paid
P is the principal in dollars
r is the annual interest rate expressed as a decimal
t is the time in years

Example 1

Lucas borrows \$5000 from his bank to purchase a mountain bike and gear for cross-country racing. The bank charges simple interest at an interest rate of 8% per annum. Lucas decides to pay back the amount he owes the bank after 7 months because he has won some prize money. How much interest is he charged?

Solution

We use Formula (1), where I is the simple interest we are looking for, $P = 5000$, $r = 8\% = 0.08$, and $t = 7/12$. We compute

$$I = 5000(0.08)(7/12) = 233.33$$

Lucas is charged with \$233.33 in simple interest.

Example 2

Ann has deposited \$14,000 into a savings account, which earns simple interest at an annual interest rate of 0.5%. After some time she clears her account and takes all her money out, which by now is worth \$14,350. How much interest has she earned and how long did she leave the money in her account?

Solution

\$14,350 − \$14,000 = \$350. So the simple interest earned is \$350.

To decide how long the money stayed in the account, we use Formula (1), where $I = 350$, $P = 14{,}000$, $r = 0.5\% = 0.005$, and t is the time in years we are looking for. Solving the formula for time

$$t = \frac{I}{Pr},$$

we can compute the time

$$t = \frac{350}{14{,}000(0.005)} = 5$$

Ann left her money deposited for 5 years.

Future Value for Simple Interest

When the simple interest earned is added to the principal we get the **maturity value** at time t. The maturity value is the value of the money in the future and so also referred to as the **future value,** often denoted by A. The principal is the value of the money in the present and so also referred to as the **present value.** Table 12.1 models how the future value of A dollars changes when a principal of P dollars accumulates simple interest I at a nominal interest rate r per year for a **term** of t years.

As can be seen from Table 12.1, the maturity value can be simplified to

$$A = P + I = P + Prt = P(1 + rt)$$

Table 12.1

Time	Principal	Interest in Year t	Accumulated Interest I	Future Value A
1	P	rP	rP	$P + rP = P(1 + r)$
2	P	rP	$rP + rP$	$P + rP + rP = P + 2rP = P(1 + 2r)$
3	P	rP	$rP + rP + rP$	$P + rP + rP + rP = P + 3rP = P(1 + 3r)$
...	...	...	...	...
t	P	rP	$\underbrace{rP + \ldots + rP}_{t \text{ terms}}$	$\underbrace{P + rP + \ldots + rP}_{t \text{ terms}} = P + rtP = P(1 + rt)$

Future Value for Simple Interest

$$A = P(1 + rt) \quad \text{and} \qquad (2)$$

Present Value for Simple Interest

$$P = \frac{A}{(1 + rt)} \qquad (3)$$

where
A is the future value in dollars
P is the principal in dollars
r is the annual interest rate expressed as a decimal
t is the time in years

Remark

Notice that the accumulated interest is given as the difference between the future value and the principal

$$I = A - P$$

◀

Example 3

What is the maturity value for a loan of \$22,500 at simple interest to be repaid in 4 years when the interest rate is 6.5% annually?

Solution

We use Formula (2), where A is the maturity value in dollars we are looking for, $P = 22{,}500$, $r = 6.5\% = 0.065$, and $t = 4$. We compute

$$A = 22{,}500(1 + (0.065)(4)) = 28{,}350$$

The maturity value of the loan is \$28,350.

GROUP DISCUSSION

Suppose that the principal of a loan is \$100. After one day the loan and simple interest on it get paid back. The interest rate is 1% per annum. How much is the interest if it is ordinary? How much is the interest if it is exact? If the loan gets paid back after only 360 days, discuss which interest—ordinary or exact—is higher and give your rationale for it.

Example 4

How much money was invested at the start of a 10-year period if the future value is \$60,000 based on simple interest at an interest rate of 4.3% per year?

Solution

We use Formula (3), where $A = 60{,}000$, P is the present value in dollars we are looking for, $r = 4.3\% = 0.043$, and $t = 10$. We compute

$$P = \frac{60{,}000}{(1 + (0.043)(10))} \approx 41{,}958.04$$

\$41,958.04 were invested over the 10-year period.

Remark

Notice that rounding has been done to the nearest cent in all examples given so far. This is the accepted method for mathematics of finance and we will continue to do it for the remainder of this chapter. ◀

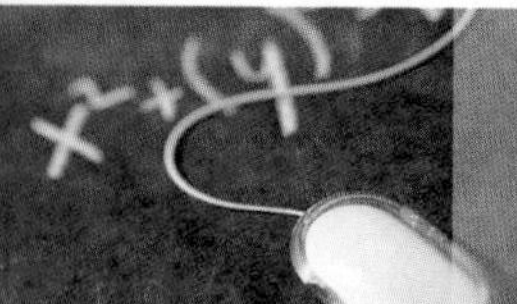

Self-Check Exercises 12.1

1. An amount of \$3000 is deposited into an account based on simple interest. The annual interest rate is 1.2%. Assume a 360-day year.
   a. What is the interest after 30 days?
   b. What is the future value after 30 days?

2. Repeat Exercise 1 assuming a 365-day year.

3. A loan of \$100,000 at simple interest was repaid after 13 years with a maturity value of \$158,500. What was the annual interest rate?

Solutions to Self-Check Exercises 12.1 can be found on page 713.

12.1 Exercises

A calculator is recommended for these exercises. Assume that a year has 360 days. In exercises asking for an interest rate, express the rate as a percentage accurate to two decimal places.

In Exercises 1–6, find the simple interest I if the principal P is invested at a simple interest rate of r per year for time t.

1. $P = 12{,}000$, $r = 7.2\%$, $t = 240$ days
2. $P = 85{,}500$, $r = 6.5\%$, $t = 175$ days
3. $P = 5{,}000$, $r = 4.6\%$, $t = 15$ months
4. $P = 7{,}000$, $r = 8.14\%$, $t = 21$ months
5. $P = 3{,}450$, $r = 0.85\%$, $t = 7$ years
6. $P = 2{,}575$, $r = 1.4\%$, $t = 12$ years

In Exercises 7–12, find the maturity value A if the principal P is invested at a simple interest rate of r per year for time t.

7. $P = 12{,}000$, $r = 7.2\%$, $t = 240$ days
8. $P = 85{,}500$, $r = 6.52\%$, $t = 175$ days
9. $P = 5{,}000$, $r = 4.68\%$, $t = 15$ months
10. $P = 7{,}000$, $r = 8.1\%$, $t = 21$ months
11. $P = 3{,}450$, $r = 0.8\%$, $t = 7$ years
12. $P = 2{,}575$, $r = 1.4\%$, $t = 12$ years

In Exercises 13–18, find the principal P if the maturity value A is returned at a simple interest rate of r per year for time t.

13. $A = 12{,}000$, $r = 7.2\%$, $t = 240$ days
14. $A = 85{,}500$, $r = 6.52\%$, $t = 175$ days
15. $A = 5{,}000$, $r = 4.68\%$, $t = 15$ months
16. $A = 7{,}000$, $r = 8.1\%$, $t = 21$ months
17. $A = 3{,}450$, $r = 0.8\%$, $t = 7$ years
18. $A = 2{,}575$, $r = 1.4\%$, $t = 12$ years

In Exercises 19–22, find the simple interest rate r per annum accurate to two decimal places if the principal P is invested for time t with a maturity value A.

19. $P = 14{,}500$, $A = 15{,}000$, $t = 335$ days
20. $P = 5{,}500$, $A = 5{,}700$, $t = 280$ days
21. $P = 10{,}000$, $A = 20{,}000$, $t = 10$ years
22. $P = 10{,}000$, $A = 20{,}000$, $t = 20$ years

In Exercises 23–26, find the time t in years accurate to one decimal place if the principal P is invested at a simple interest rate of r per year with a maturity value A.

23. $P = 15{,}000$, $A = 25{,}000$, $r = 3\frac{1}{2}\%$
24. $P = 85{,}500$, $A = 100{,}000$, $r = 5.2\%$
25. $P = 5{,}000$, $A = 10{,}000$, $r = 9.65\%$
26. $P = 7{,}000$, $A = 9{,}500$, $r = 2.75\%$
27. Find the future value after 6 yrs if \$4000 is loaned at a simple interest rate of 3.8% per year.
28. Find the future value after 3 yrs if \$15,000 is loaned at a simple interest rate of 2.33% per year.
29. Find the simple interest rate needed for a loan of \$14,000 to grow to an amount of \$30,000 in 10 yrs. Round the interest rate to three decimal places.
30. Find the simple interest rate needed for a loan of \$8,500 to grow to an amount of \$25,000 in 8 yrs. Round the interest rate to three decimal places.
31. Find the simple interest rate needed for an investment of \$5000 to double in 4 yrs.

32. Find the simple interest rate needed for an investment of $2000 to triple in 6 yrs.

33. How many days did it take for a loan of $5000 to accumulate a simple interest of $100 at an interest rate of 6.45% per year?

34. How many days did it take for a loan of $3000 to accumulate a simple interest of $100 at an interest rate of 4.73% per year?

35. How many years did it take for a loan of $2500 to accumulate a simple interest of $500 at an annual interest rate of 2.6%?

36. How many years did it take for a loan of $3500 to accumulate a simple interest of $500 at an annual interest rate of 3.1%?

Business and Economics Applications

Assume that a year has 365 days.

37. **Canada Savings Bonds** Anna purchases Canada Savings Bond (CSB) valued at $5000 earning simple interest at an annual interest rate of 2.75%. Assuming that the interest rate does not change for the next 3.5 yrs, how much have Anna's CSB earned in interest during that time?

38. **Canada Savings Bonds** Sandeep purchases Canada Savings Bond (CSB) valued at $2500 earning simple interest at an annual interest rate of 1.65%. Assuming that the interest rate does not change for the next 1.5 yrs, how much have Sandeep's CSB earned in interest during that time?

39. **Canada Premium Bonds** In 2006 Joshua purchased Canada Premium Bond (CPB) valued at $4000 earning simple interest at an annual interest rate of 3.00% the first year, 3.25% the second year and 4.00% the third year.
    a. How much have Joshua's CPB earned in interest in the third year?
    b. What is the future value of Joshua's CPB at the end of 3 years?

40. **Canada Premium Bonds** In 2004 Emma purchased Canada Premium Bond (CPB) valued at $10,000 earning simple interest at an annual interest rate of 1.85% the first year, 2.45% the second year and 3.40% the third year.
    a. How much have Emma's CPB earned in interest in the second year?
    b. What is the future value of Emma's CPB at the end of 3 years?

41. **Guaranteed Investment Certificate** Mike decides to invest in a Guaranteed Investment Certificate (GIC) short term for 180 days earning simple interest at an annual interest rate of 2.45%. How much money must Mike invest if he hopes to have a maturity value of $100,000?

42. **Guaranteed Investment Certificate** Ching decides to invest in a Guaranteed Investment Certificate (GIC) short term for 60 days earning simple interest at an annual interest rate of 2.25%. How much money must Ching invest if he hopes to have a maturity value of $20,000?

43. **Guaranteed Investment Certificate** A bank offers Guaranteed Investment Certificate (GIC) long term for 4 years earning simple interest at an annual interest rate of 4.45%. What is the maturity value of the GIC if the principal is $1500?

44. **Guaranteed Investment Certificate** A bank offers Guaranteed Investment Certificate (GIC) long term for 2 yrs earning simple interest at an annual interest rate of 4.20%. What is the maturity value of the GIC if the principal is $3000?

Solutions To Self-Check Exercises 12.1

1. In this exercise, $P = 3000$, A is the maturity value in dollars we are looking for, $r = 1.2\% = 0.012$, and $t = 30/360$.

    a. We use Formula (1) and compute the interest

$$I = 3000(0.012)(30/360) = 3$$

After 30 days the interest is $3.

    b. The future value is the addition of the present value and the simple interest earned:

$$A = 3000 + 3 = 3003$$

After 30 days the future value is $3003.

2. In this exercise, $P = 3000$, A is the maturity value in dollars we are looking for, $r = 1.2\% = 0.012$, and $t = 30/365$.

    a. We use Formula (1) and compute the interest

$$I = 3000(0.012)(30/365) = 2.96$$

After 30 days the interest is $2.96.

    b. The future value is the addition of the present value and the simple interest earned:

$$A = 3000 + 2.96 = 3002.96$$

After 30 days the future value is $3002.96.

3. We use Formula (2), where $P = 100{,}000$, $A = 158{,}500$, r is the interest rate we are looking for, and $t = 13$. We compute the interest rate

$$158{,}500 = 100{,}000(1 + 13r)$$

$$\frac{158{,}500}{100{,}000} = 1 + 13r$$

$$\frac{158{,}500}{100{,}000} - 1 = 13r$$

$$r = \left(\frac{158{,}500}{100{,}000} - 1\right)/13 = 0.045 = 4.5\%$$

Compound Interest

Compound interest is a natural application of the exponential function in the business world. Frequently, interest earned is periodically added to the principal and thereafter earns interest itself at the same interest rate. This is called compound interest. To find a formula for the accumulated amount, let's consider a numerical example. Suppose \$1000 (the principal) is deposited in a bank for a term of 3 years, earning interest at the rate of 8% per year compounded annually. Then, using Formula (2) with $P = 1000$, $r = 0.08$, and $t = 1$, we see that the accumulated amount at the end of the first year is

$$\begin{aligned} A_1 &= P(1 + rt) \\ &= 1000[1 + 0.08(1)] = 1000(1.08) = 1080 \end{aligned}$$

or \$1080.

To find the accumulated amount A_2 at the end of the second year, we use (2) once again, this time with $P = A_1$. (Remember, the principal *and* interest now earn interest over the second year.) We obtain

$$\begin{aligned} A_2 &= P(1 + rt) = A_1(1 + rt) \\ &= 1000[1 + 0.08(1)][1 + 0.08(1)] \\ &= 1000(1 + 0.08)^2 = 1000(1.08)^2 \approx 1166.40 \end{aligned}$$

or approximately \$1166.40.

Finally, the accumulated amount A_3 at the end of the third year is found using (4) with $P = A_2$, giving

$$\begin{aligned} A_3 &= P(1 + rt) = A_2(1 + rt) \\ &= 1000[1 + 0.08(1)]^2[1 + 0.08(1)] \\ &= 1000(1 + 0.08)^3 = 1000(1.08)^3 \approx 1259.71 \end{aligned}$$

or approximately \$1259.71.

If you reexamine our calculations in this example, you will see that the accumulated amounts at the end of each year have the following form:

$$\begin{aligned} &\text{First year:} & A_1 &= 1000(1 + 0.08) & &\text{or} & A_1 &= P(1 + r) \\ &\text{Second year:} & A_2 &= 1000(1 + 0.08)^2 & &\text{or} & A_2 &= P(1 + r)^2 \\ &\text{Third year:} & A_3 &= 1000(1 + 0.08)^3 & &\text{or} & A_3 &= P(1 + r)^3 \end{aligned}$$

These observations suggest the following general result: If P dollars are invested over a term of t years earning interest at the rate of r per year compounded annually, then the accumulated amount is

$$A = P(1 + r)^t \tag{4}$$

Formula (4) was derived under the assumption that interest was compounded *annually.* In practice, however, interest is usually compounded more than once a year. The interval of time between successive interest calculations is called compounding period. The number of compounding periods in a year is denoted by m. Table 12.2 lists common compounding periods along with their description, and value of m.

Table 12.2

Length of the Compounding Period	Description of Interest Rate Compounded ...	m, Number of Compounding Periods Per Year
1 year	Annually	1
6 months	Semiannually	2
3 months	Quarterly	4
1 month	Monthly	12
2 weeks	Biweekly	26
1 week	Weekly	52
1 day	Daily	365

If interest at a nominal rate of r per year is compounded m times a year on a principal of P dollars, then the simple interest rate per conversion period is

$$i = \frac{r}{m} \qquad \frac{\text{Annual interest rate}}{\text{Periods per year}}$$

For example, if the nominal interest rate is 8% per year ($r = 0.08$) and interest is compounded quarterly ($m = 4$), then

$$i = \frac{r}{m} = \frac{0.08}{4} = 0.02$$

or 2% per period.

To find a general formula we proceed as before with simple interest and use Table 12.3 to model how the maturity value of A dollars changes when a principal of P dollars accumulates compound interest I at a nominal interest rate r per year for a term of t years. Notice that there are $n = mt$ compounding periods in t years.

Table 12.3

Time in Compounding Periods	Principal	Interest per Compounding Period	Future Value A (is the sum of the principal and the interest)
1	P	$P\frac{r}{m}$	$P + P\frac{r}{m} = P\left(1 + \frac{r}{m}\right)$
2	$P\left(1 + \frac{r}{m}\right)$	$P\left(1 + \frac{r}{m}\right)\frac{r}{m}$	$P\left(1 + \frac{r}{m}\right) + P\left(1 + \frac{r}{m}\right)\frac{r}{m}$ $= \left[P\left(1 + \frac{r}{m}\right)\right]\left(1 + \frac{r}{m}\right) = P\left(1 + \frac{r}{m}\right)^2$
3	$P\left(1 + \frac{r}{m}\right)^2$	$P\left(1 + \frac{r}{m}\right)^2\frac{r}{m}$	$P\left(1 + \frac{r}{m}\right)^2 + P\left(1 + \frac{r}{m}\right)^2\frac{r}{m}$ $= \left[P\left(1 + \frac{r}{m}\right)^2\right]\left(1 + \frac{r}{m}\right) = P\left(1 + \frac{r}{m}\right)^3$
...	...	...	...
n	$P\left(1 + \frac{r}{m}\right)^{n-1}$	$P\left(1 + \frac{r}{m}\right)^{n-1}\frac{r}{m}$	$P\left(1 + \frac{r}{m}\right)^{n-1} + P\left(1 + \frac{r}{m}\right)^{n-1}\frac{r}{m}$ $= \left[P\left(1 + \frac{r}{m}\right)^{n-1}\right]\left(1 + \frac{r}{m}\right) = P\left(1 + \frac{r}{m}\right)^n$

As can be seen from Table 12.3, the future value can be simplified to

$$A = P\left(1 + \frac{r}{m}\right)^n$$

Future Value for Compound Interest

$$A = P\left(1 + \frac{r}{m}\right)^n \quad \text{and} \qquad (5)$$

Present Value for Compound Interest

$$P = \frac{A}{\left(1 + \frac{r}{m}\right)^n} = A\left(1 + \frac{r}{m}\right)^{-n} \qquad (6)$$

where
A is the future value in dollars
P is the principal in dollars
r is the annual interest rate expressed as a decimal
m is the number of compounding periods per year
t is the time in years
n is the total number of compounding periods and $n = mt$

Remark

Notice that the accumulated interest is given as the difference between the future value and the principal

$$I = A - P$$

◀

Example 1

Find the accumulated amount after 3 years if \$1000 is invested at 8% per year compounded (a) annually, (b) semiannually, (c) quarterly, (d) monthly, and (e) daily.

Solution

a. Here, $P = 1000$, $r = 0.08$, $m = 1$, and $t = 3$, so Formula (6) gives

$$A = 1000(1 + 0.08)^3$$
$$= 1259.71$$

or \$1259.71

b. Here, $P = 1000$, $r = 0.08$, $m = 2$, and $t = 3$, so (6) gives

$$A = 1000\left(1 + \frac{0.08}{2}\right)^{(2)(3)}$$
$$= 1265.32$$

or \$1265.32.

c. In this case, $P = 1000$, $r = 0.08$, $m = 4$, and $t = 3$, so (6) gives

$$A = 1000\left(1 + \frac{0.08}{4}\right)^{(4)(3)}$$
$$= 1268.24$$

or \$1268.24

d. Here, $P = 1000$, $r = 0.08$, $m = 12$, and $t = 3$, so (6) gives

$$A = 1000\left(1 + \frac{0.08}{12}\right)^{(12)(3)}$$
$$= 1270.24$$

or \$1270.24.

e. Here, $P = 1000$, $r = 0.08$, $m = 365$, and $t = 3$, so (6) gives

$$\begin{aligned} A &= 1000\left(1 + \frac{0.08}{365}\right)^{(365)(3)} \\ &= 1271.22 \end{aligned}$$

or \$1271.22. These results are summarized in Table 12.4.

Table 12.4

Nominal Rate, r	Description of Compounding Period	m	Term in Years	Principal	Maturity Value A
8%	Annually	1	3	\$1000	\$1259.71
8	Semiannually	2	3	1000	1265.32
8	Quarterly	4	3	1000	1268.24
8	Monthly	12	3	1000	1270.24
8	Daily	365	3	1000	1271.22

Example 2

Find how much money should be deposited in a bank paying interest at the rate of 6% per year compounded monthly so that at the end of 3 years the accumulated amount will be \$20,000.

Solution

Here, $A = 20{,}000$, $r = 0.06$, $m = 12$, and $t = 3$. Using Formula (6), we obtain

$$\begin{aligned} P &= 20{,}000\left(1 + \frac{0.06}{12}\right)^{-(12)(3)} \\ &\approx 16{,}713 \end{aligned}$$

or \$16,713.

Example 3

Find the present value of \$49,158.60 due in 5 years at an interest rate of 10% per year compounded quarterly.

Solution

Using Formula (6) with $A = 49{,}158.60$, $r = 0.1$, $m = 4$, and $t = 5$, we obtain

$$P = (49{,}158.6)\left(1 + \frac{0.1}{4}\right)^{-(4)(5)} \approx 30{,}000$$

or \$30,000.

The next two examples show how logarithms can be used to solve problems involving compound interest.

Example 4

How long will it take \$10,000 to grow to \$15,000 if the investment earns an interest rate of 12% per year compounded quarterly?

Solution

Using Formula (5) with $A = 15{,}000$, $P = 10{,}000$, $r = 0.12$, and $m = 4$, we obtain

$$15{,}000 = 10{,}000\left(1 + \frac{0.12}{4}\right)^{4t}$$

$$(1.03)^{4t} = \frac{15{,}000}{10{,}000} = 1.5$$

Taking the logarithm on each side of the equation gives

$$\ln(1.03)^{4t} = \ln 1.5$$

$$4t \ln 1.03 = \ln 1.5 \qquad \log_b m^n = n \log_b m$$

$$4t = \frac{\ln 1.5}{\ln 1.03}$$

$$t = \frac{\ln 1.5}{4 \ln 1.03} \approx 3.43$$

So it will take approximately 3.4 years for the investment to grow from \$10,000 to \$15,000.

Example 5

Find the interest rate needed for an investment of \$10,000 to grow to an amount of \$18,000 in 5 years if the interest is compounded monthly.

Solution

Using Formula (5) with $A = 18{,}000$, $P = 10{,}000$, $m = 12$, and $t = 5$, we obtain

$$18{,}000 = 10{,}000\left(1 + \frac{r}{12}\right)^{12(5)}$$

Dividing both sides of the equation by 10,000 gives

$$\frac{18{,}000}{10{,}000} = \left(1 + \frac{r}{12}\right)^{60}$$

or, upon simplification,

$$\left(1 + \frac{r}{12}\right)^{60} = 1.8$$

Now, we take the logarithm on each side of the equation, obtaining

$$\ln\left(1 + \frac{r}{12}\right)^{60} = \ln 1.8$$

$$60 \ln\left(1 + \frac{r}{12}\right) = \ln 1.8$$

$$\ln\left(1 + \frac{r}{12}\right) = \frac{\ln 1.8}{60} = 0.009796$$

$$\left(1 + \frac{r}{12}\right) = e^{0.009796} \qquad \text{Logarithm Definition Section 5.2}$$

$$= 1.009844$$

and

$$\frac{r}{12} = 1.009844 - 1$$

$$r = 0.1181$$

or 11.81% per year.

GROUP DISCUSSION

Suppose that the principal of an investment is \$1000. Suppose that the nominal interest rate is fixed at 1% per annum and interest only accrues for a period of one year. Show that more frequent compounding produces more interest by finding the accumulated interest for each number of compounding periods m stated in Table 12.2.

Remark

Often, the formula for compound interest is given with the compounded rate of interest i rather than the annual rate of interest r. Since $i = \frac{r}{m}$ we have

$$A = P\left(1 + \frac{r}{m}\right)^n = P(1 + i)^n \tag{7}$$

◀

Effective Rate of Interest

In the last example we saw that the interest actually earned on an investment depends on the frequency with which the interest is compounded. Thus, the stated, or nominal, rate of 8% per year does not reflect the actual rate at which interest is earned. This suggests that we need to find a common basis for comparing interest rates. One such way of comparing interest rates is provided by using the effective rate. The **effective rate** is the *simple* interest rate that would produce the same accumulated amount in 1 year as the nominal rate compounded m times a year. The effective rate is also called the **true rate.**

To derive a relation between the nominal interest rate, r per year compounded m times, and its corresponding effective rate, r_{eff} per year, let's assume an initial investment of P dollars. Then, the accumulated amount after 1 year at a simple interest rate of r_{eff} per year is

$$A = P(1 + r_{\text{eff}})$$

Also, the accumulated amount after 1 year at an interest rate of r per year compounded m times a year is

$$A = P\left(1 + \frac{r}{m}\right)^m \qquad \text{Since } t = 1$$

Equating the two expressions gives

$$P(1 + r_{\text{eff}}) = P\left(1 + \frac{r}{m}\right)^m$$

$$1 + r_{\text{eff}} = \left(1 + \frac{r}{m}\right)^m \qquad \text{Divide both sides by } P.$$

or, upon solving for r_{eff}, we obtain the formula for computing the effective rate of interest:

Effective Rate of Interest

$$r_{\text{eff}} = \left(1 + \frac{r}{m}\right)^m - 1 \tag{8}$$

where

r_{eff} is the effective rate of interest
r is the annual interest rate expressed as a decimal
m is the number of compounding periods per year

Example 6

Find the effective rate of interest corresponding to a nominal rate of 8% per year compounded (a) annually, (b) semiannually, (c) quarterly, (d) monthly, and (e) daily.

Solution

a. The effective rate of interest corresponding to a nominal rate of 8% per year compounded annually is of course given by 8% per year. This result is also confirmed by using Formula (8) with $r = 0.08$ and $m = 1$. Thus,

$$r_{\text{eff}} = (1 + 0.08) - 1 = 0.08$$

b. Let $r = 0.08$ and $m = 2$. Then, (7) yields

$$\begin{aligned} r_{\text{eff}} &= \left(1 + \frac{0.08}{2}\right)^2 - 1 \\ &= 0.0816 \end{aligned}$$

so the required effective rate is 8.16% per year.

c. Let $r = 0.08$ and $m = 4$. Then, (7) yields

$$\begin{aligned} r_{\text{eff}} &= \left(1 + \frac{0.08}{4}\right)^4 - 1 \\ &= 0.08243 \end{aligned}$$

so the corresponding effective rate in this case is 8.243% per year.

d. Let $r = 0.08$ and $m = 12$. Then, (7) yields

$$\begin{aligned} r_{\text{eff}} &= \left(1 + \frac{0.08}{12}\right)^{12} - 1 \\ &= 0.08300 \end{aligned}$$

so the corresponding effective rate in this case is 8.3% per year.

e. Let $r = 0.08$ and $m = 365$. Then, (7) yields

$$\begin{aligned} r_{\text{eff}} &= \left(1 + \frac{0.08}{365}\right)^{365} - 1 \\ &= 0.08328 \end{aligned}$$

so the corresponding effective rate in this case is 8.328% per year.

Now, if the effective rate of interest r_{eff} is known, then the accumulated amount after t years on an investment of P dollars may be more readily computed by using the formula

$$A = P(1 + r_{\text{eff}})^t$$

Thus, if the effective rates of interest found in Example 6 were known, the accumulated values of Example 1, shown in Table 12.5, could have been readily found.

Table 12.5

Nominal Rate, r	Frequency of Interest Payment	Effective Rate	Initial Investment	Maturity Value after 3 Years
8%	Annually	8%	\$1000	$1000(1 + 0.08)^3 = \$1259.71$
8	Semiannually	8.16	1000	$1000(1 + 0.0816)^3 = 1265.32$
8	Quarterly	8.243	1000	$1000(1 + 0.08243)^3 = 1268.23$
8	Monthly	8.300	1000	$1000(1 + 0.08300)^3 = 1270.24$
8	Daily	8.328	1000	$1000(1 + 0.08328)^3 = 1271.22$

Continuous Compounding of Interest

One question that arises naturally in the study of compound interest is: What happens to the accumulated amount over a fixed period of time if the interest is computed more and more frequently?

Intuition suggests that the more often interest is compounded, the larger the accumulated amount will be. This is confirmed by the results of Example 1, where we found that the accumulated amounts did in fact increase when we increased the number of conversion periods per year.

This leads us to another question: Does the accumulated amount approach a limit when the interest is computed more and more frequently over a fixed period of time?

To answer this question, let's look again at the compound interest formula:

$$A = P\left(1 + \frac{r}{m}\right)^{mt} \qquad (9)$$

Recall that m is the number of conversion periods per year. So to find an answer to our problem, we should let m get larger and larger (approach infinity) in (9). But first we will rewrite this equation in the form

$$A = P\left[\left(1 + \frac{r}{m}\right)^{m}\right]^{t} \qquad \text{By Exponent Law 3}$$

Now, letting $m \to \infty$, we find that

$$\lim_{m\to\infty}\left[P\left(1 + \frac{r}{m}\right)^{m}\right]^{t} = P\left[\lim_{m\to\infty}\left(1 + \frac{r}{m}\right)^{m}\right]^{t} \qquad \text{Why?}$$

Next, upon making the substitution $u = m/r$ and observing that $u \to \infty$ as $m \to \infty$, the foregoing expression reduces to

$$P\left[\lim_{u\to\infty}\left(1 + \frac{1}{u}\right)^{ur}\right]^{t} = P\left[\lim_{u\to\infty}\left(1 + \frac{1}{u}\right)^{u}\right]^{rt}$$

But

$$\lim_{u\to\infty}\left(1 + \frac{1}{u}\right)^{u} = e \qquad \text{Use (2) Section 5.1}$$

so

$$\lim_{m\to\infty} P\left[\left(1 + \frac{r}{m}\right)^{m}\right]^{t} = Pe^{rt}$$

Our computations tell us that as the frequency with which interest is compounded increases without bound, the accumulated amount approaches Pe^{rt}. In this situation, we say that interest is *compounded continuously* and it is called continuous compound interest. Let's summarize this important result.

Continuous Compound Interest

$$A = Pe^{rt} \qquad (10)$$

where
- P is the principal in dollars
- r is the annual interest rate compounded continuously expressed as a decimal
- t is the time in years
- A is the future value in dollars at the end of t years

Example 7

Find the accumulated amount after 3 years if \$1000 is invested at 8% per year compounded (a) daily (assume a 365-day year) and (b) continuously.

Solution

a. Using Formula (6) with $P = 1000$, $r = 0.08$, $m = 365$, and $t = 3$, we find

$$A = 1000\left(1 + \frac{0.08}{365}\right)^{(365)(3)} \approx 1271.22$$

or \$1271.22.

b. Here we use Formula (10) with $P = 1000$, $r = 0.08$, and $t = 3$, obtaining

$$A = 1000e^{(0.08)(3)}$$
$$\approx 1271.25$$

or \$1271.25.

Observe that the accumulated amounts corresponding to interest compounded daily and interest compounded continuously differ by very little. The continuous compound interest formula is a very important tool in theoretical work in financial analysis.

If we solve Formula (10) for P, we obtain

$$P = Ae^{-rt} \tag{11}$$

which gives the present value in terms of the future (accumulated) value for the case of continuous compounding.

Example 8

Real Estate Investment Blakely Investment Company owns an office building located in the commercial district of a city. As a result of the continued success of an urban renewal program, local business is enjoying a miniboom. The market value of Blakely's property is

$$V(t) = 300{,}000e^{\sqrt{t/2}}$$

where $V(t)$ is measured in dollars and t is the time in years from the present. If the expected rate of inflation is 9% compounded continuously for the next 10 years, find an expression for the present value $P(t)$ of the market price of the property valid for the next 10 years. Compute $P(7)$, $P(8)$, and $P(9)$, and interpret your results.

Solution

Using Formula (11) with $A = V(t)$ and $r = 0.09$, we find that the present value of the market price of the property t years from now is

$$P(t) = V(t)e^{-0.09t}$$
$$= 300{,}000e^{-0.09t+\sqrt{t/2}} \qquad (0 \le t \le 10)$$

Letting $t = 7$, 8, and 9, respectively, we find that

$$P(7) = 300{,}000e^{-0.09(7)+\sqrt{7/2}} \approx 599{,}837, \quad \text{or} \quad \$599{,}837$$

$$P(8) = 300{,}000e^{-0.09(8)+\sqrt{8/2}} \approx 600{,}640, \quad \text{or} \quad \$600{,}640$$

$$P(9) = 300{,}000e^{-0.09(9)+\sqrt{9/2}} \approx 598{,}115, \quad \text{or} \quad \$598{,}115$$

From the results of these computations, we see that the present value of the property's market price seems to decrease after a certain period of growth. This suggests that there is an optimal time for the owners to sell. Later we will show that the highest present value of the property's market price is \$600,779, which occurs at time $t = 7.72$ years.

EXPLORING WITH TECHNOLOGY

The effective rate of interest is given by

$$r_{\text{eff}} = \left(1 + \frac{r}{m}\right)^m - 1$$

where the number of conversion periods per year is m. In Exercise 41 you will be asked to show that the effective rate of interest r_{eff} corresponding to a nominal interest rate r per year compounded continuously is given by

$$\hat{r}_{\text{eff}} = e^r - 1$$

To obtain a visual confirmation of this result, consider the special case where $r = 0.1$ (10% per year).

1. Use a graphing utility to plot the graph of both

$$y_1 = \left(1 + \frac{0.1}{x}\right)^x - 1 \quad \text{and} \quad y_2 = e^{0.1} - 1$$

in the viewing window $[0, 3] \cdot [0, 0.12]$.

2. Does your result seem to imply that

$$\left(1 + \frac{r}{m}\right)^m - 1$$

approaches

$$\hat{r}_{\text{eff}} = e^r - 1$$

as m increases without bound for the special case $r = 0.1$?

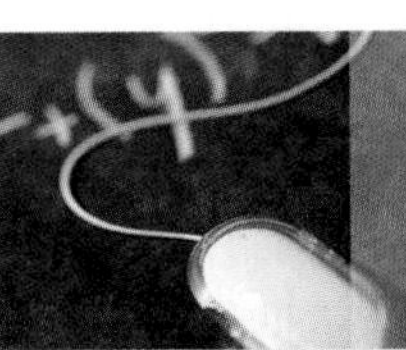

Self-Check Exercises 12.2

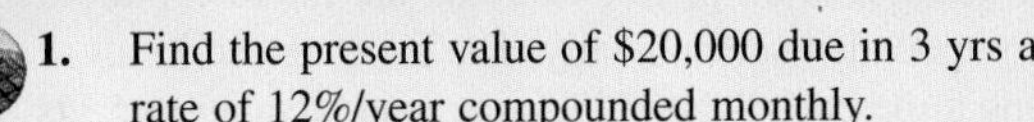

1. Find the present value of $20,000 due in 3 yrs at an interest rate of 12%/year compounded monthly.

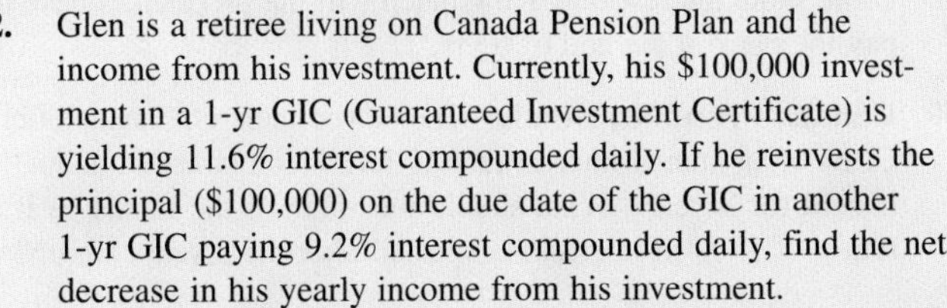

2. Glen is a retiree living on Canada Pension Plan and the income from his investment. Currently, his $100,000 investment in a 1-yr GIC (Guaranteed Investment Certificate) is yielding 11.6% interest compounded daily. If he reinvests the principal ($100,000) on the due date of the GIC in another 1-yr GIC paying 9.2% interest compounded daily, find the net decrease in his yearly income from his investment.

3. **a.** What is the accumulated amount after 5 yrs if $10,000 is invested at 10%/year compounded continuously?
   b. Find the present value of $10,000 due in 5 yrs at an interest rate of 10%/year compounded continuously.

Solutions to Self-Check Exercises 12.2 can be found on page 727.

12.2 Exercises

A calculator is recommended for these exercises.

In Exercises 1–4, find the future value A if the principal P is invested at an interest rate of r per year for t years.

1. $P = \$2500$, $r = 7\%$, $t = 10$, compounded semiannually
2. $P = \$12{,}000$, $r = 8\%$, $t = 10$, compounded quarterly
3. $P = \$150{,}000$, $r = 10\%$, $t = 4$, compounded monthly
4. $P = \$150{,}000$, $r = 9\%$, $t = 3$, compounded daily

In Exercises 5 and 6, find the effective rate corresponding to the given nominal rate.

5. **a.** 10%/year compounded semiannually
   b. 9%/year compounded quarterly
6. **a.** 8%/year compounded monthly
   b. 8%/year compounded daily

In Exercises 7 and 8, find the present value of $40,000 due in 4 yrs at the given rate of interest.

7. **a.** 8%/year compounded semiannually
   b. 8%/year compounded quarterly

8. **a.** 7%/year compounded monthly
   b. 9%/year compounded daily

9. Find the accumulated amount after 4 yrs if $5000 is invested at 8%/year compounded continuously.

10. An amount of $25,000 is deposited in a bank that pays interest at the rate of 7%/year, compounded annually. What is the total amount on deposit at the end of 6 yrs, assuming there are no deposits or withdrawals during those 6 yrs? What is the interest earned in that period of time?

11. Find the interest rate needed for an investment of $5000 to grow to an amount of $7500 in 3 yrs if interest is compounded monthly.

12. Find the interest rate needed for an investment of $5000 to grow to an amount of $7500 in 3 yrs if interest is compounded quarterly.

13. Find the interest rate needed for an investment of $5000 to grow to an amount of $8000 in 4 yrs if interest is compounded semiannually.

14. Find the interest rate needed for an investment of $5000 to grow to an amount of $5500 in 6 mo if interest is compounded monthly.

15. Find the interest rate needed for an investment of $2000 to double in 5 yrs if interest is compounded annually.

16. Find the interest rate needed for an investment of $2000 to triple in 5 yrs if interest is compounded monthly.

17. How long will it take $5000 to grow to $6500 if the investment earns interest at the rate of 12%/year compounded monthly?

18. How long will it take $12,000 to grow to $15,000 if the investment earns interest at the rate of 8%/year compounded monthly?

19. How long will it take an investment of $2000 to double if the investment earns interest at the rate of 9%/year compounded monthly?

20. How long will it take an investment of $5000 to triple if the investment earns interest at the rate of 8%/year compounded daily?

21. Find the interest rate needed for an investment of $5000 to grow to an amount of $6000 in 3 yrs if interest is compounded continuously.

22. Find the interest rate needed for an investment of $4000 to double in 5 yrs if interest is compounded continuously.

23. How long will it take an investment of $6000 to grow to $7000 if the investment earns interest at the rate of $7\frac{1}{2}$% compounded continuously?

24. How long will it take an investment of $8000 to double if the investment earns interest at the rate of 8% compounded continuously?

Business and Economics Applications

25. **Housing Prices** The Estradas are planning to buy a house 4 yrs from now. Housing experts in their area have estimated that the cost of a home will increase at a rate of 9%/year during that 4-yr period. If this economic prediction holds true, how much can they expect to pay for a house that currently costs $80,000?

26. **Energy Consumption** A utility company in a city in Ontario expects the consumption of electricity to increase by 8%/year during the next decade, due mainly to the expected population increase. If consumption does increase at this rate, find the amount by which the utility company will have to increase its generating capacity in order to meet the area's needs at the end of the decade.

27. **Guaranteed Investment Certificate** A Canadian bank offers some specially priced GICs at an annual interest rate of 3.50% compounded monthly for a 5-yr term. The same bank also offers long-term GICs at an annual interest rate of 3.55% compounded annually for a 5-yr term. Which GIC should you choose to make the most in interest when a principal of $10,000 is available?

28. **Savings Accounts** Bernie invested a sum of money 5 yrs ago in a savings account, which has since paid interest at the rate of 8%/yr compounded quarterly. His investment is now worth $22,289.22. How much did he originally invest?

29. **Loan Consolidation** The proprietors of the Coachmen Inn secured two loans from a bank, one for $8000 due in 3 yrs and one for $15,000 due in 6 yrs, both at an interest rate of 10%/year compounded semiannually. The bank agreed to allow the two loans to be consolidated into one loan payable in 5 yrs at the same interest rate. How much will the proprietors have to pay the bank at the end of 5 yrs?

30. **Canada Premium Bonds** The Government of Canada offers CPBs at an annual interest rate of 3.41% compounded annually if held for 3 years. What should the value of the bonds be if a return of $500 in interest is expected after the 3 years are up?

31. **Consumer Price Index** At an annual inflation rate of 7.5%, how long will it take the Consumer Price Index (CPI) to double?

32. **Investment Returns** Zoe purchased a house in 1996 for $160,000. In 2002 she sold the house and made a net profit of $56,000. Find the effective annual rate of return on her investment over the 6-yr period.

33. **Investment Returns** Julio purchased 1000 shares of a certain stock for $25,250 (including commissions). He sold the shares 2 yrs later and received $32,100 after deducting commissions. Find the effective annual rate of return on his investment over the 2-yr period.

34. **Investment Options** Investment A offers a 10% return compounded semiannually, and investment B offers a 9.75% return

compounded continuously. Which investment has a higher rate of return over a 4-yr period?

35. **Present Value** Find the present value of \$59,673 due in 5 yrs at an interest rate of 8%/year compounded continuously.

36. **Real Estate Investments** A condominium complex was purchased by a group of private investors for \$1.4 million and sold 6 yrs later for \$3.6 million. Find the annual rate of return (compounded continuously) on their investment.

37. **Saving for College** Having received a large inheritance, a child's parents wish to establish a trust for the child's college education. If 7 yrs from now they need an estimated \$70,000, how much should they set aside in trust now, if they invest the money at 10.5% compounded (a) quarterly? (b) continuously?

38. **Effect of Inflation on Salaries** Omar's current annual salary is \$35,000. How much will he need to earn 10 yrs from now in order to retain his present purchasing power if the rate of inflation over that period is 6%/year? Assume that inflation is continuously compounded.

39. **Pensions** Eleni, who is now 50 years old, is employed by a firm that guarantees her a pension of \$40,000/year at age 65. What is the present value of her first year's pension if inflation over the next 15 yrs is (a) 6%? (b) 8%? (c) 12%? Assume that inflation is continuously compounded.

40. **Real Estate Investments** An investor purchased a piece of waterfront property. Because of the development of a marina in the vicinity, the market value of the property is expected to increase according to the rule

$$V(t) = 80{,}000e^{\sqrt{t}/2}$$

where $V(t)$ is measured in dollars and t is the time in years from the present. If the inflation rate is expected to be 9% compounded continuously for the next 8 yrs, find an expression for the present value $P(t)$ of the property's market price valid for the next 8 yrs. What is $P(t)$ expected to be in 4 yrs?

41. **Effective Interest Rate** Show that the effective interest rate $\hat{r}_{\text{eff}}$ that corresponds to a nominal interest rate r per year compounded continuously is given by

$$\hat{r}_{\text{eff}} = e^r - 1$$

Hint: From Formula (7) we see that the effective rate $\hat{r}_{\text{eff}}$ corresponding to a nominal interest rate r per year compounded m times a year is given by

$$\hat{r}_{\text{eff}} = \left(1 + \frac{r}{m}\right)^m - 1$$

Let m tend to infinity in this expression.

42. **Effective Interest Rate** Refer to Exercise 41. Find the effective interest rate that corresponds to a nominal rate of 10%/year compounded (a) quarterly, (b) monthly, and (c) continuously.

43. **Investment Analysis** Refer to Exercise 41. Bank A pays interest on deposits at a 7% annual rate compounded quarterly, and Bank B pays interest on deposits at a $7\frac{1}{8}$% annual rate compounded continuously. Which bank has the higher effective rate of interest?

44. **Investment Analysis** Find the nominal interest rate that, when compounded monthly, yields an effective interest rate of 10%/year.
Hint: Use Equation (7).

45. **Investment Analysis** Find the nominal interest rate that, when compounded continuously, yields an effective interest rate of 10%/year.
Hint: See Exercise 41.

46. **Annuities** An annuity is a sequence of payments made at regular time intervals. The future value of an annuity of n payments of R dollars each paid at the end of each investment period into an account that earns an interest rate of i/period is

$$S = R\left[\frac{(1+i)^n - 1}{i}\right]$$

Determine

$$\lim_{i \to 0} R\left[\frac{(1+i)^n - 1}{i}\right]$$

and interpret your result.
Hint: Use the definition of the derivative.

PORTFOLIO

Misato Nakazaki

Title Assistant Vice President
Institution A large investment corporation

With an MBA from New York University, Nakazaki typifies the younger generation of Japanese women who have chosen to succeed in the business world. Since earning her degree, she has sold bonds for a global securities firm in New York City.

Nakazaki's client list reads like a *who's who* of the leading Japanese banks, insurance companies, mutual funds, and corporations. As institutional buyers, her clients purchase large blocks of American corporate bonds and mortgage-backed securities such as Ginnie Maes.

In the securities industry, buying and selling stocks and bonds has always required a mastery of concepts and formulas that outsiders find confusing. As a bond seller, Misato Nakazaki routinely uses terms such as *issue, maturity, current yield, callable* and *convertible bonds,* and so on.

These terms, however, are easily defined. When corporations issue bonds, they are borrowing money at a fixed rate of interest. The bonds are scheduled to mature—to be paid back—on a specific date as much as 30 years into the future. Callable bonds allow the issuer to pay off the loans prior to their expected maturity, reducing overall interest payments. In its simplest terms, current yield is the price of a bond multiplied by the interest rate at which the bond is issued. For example, a bond with a face value of \$1000 and an interest rate of 10% yields \$100 per year in interest payments. When that same bond is resold at a premium on the secondary market for \$1200, its current yield nets only an 8.3% rate of return based on the higher purchase price.

Bonds attract investors for many reasons. A key variable is the sensitivity of the bond's price to future changes in interest rates. If investors get locked into a low-paying bond when future bonds pay higher yields, they lose money. Nakazaki stresses that "no one knows for sure what rates will be over time." Employing differentials allows her to calculate interest-rate sensitivity for clients as they ponder purchase decisions.

Computerized formulas, "whose basis is calculus," says Nakazaki, help her factor the endless stream of numbers flowing across her desk.

On a typical day, Nakazaki might be given a bid on "10 million, GMAC, 8.5%, January 2011." Translation: Her customer wants her to buy General Motors Acceptance Corporation bonds with a face value of \$10 million and an interest rate of 8.5%, maturing in January 2011.

After she calls her firm's trader to find out the yield on the bond in question, Nakazaki enters the price and other variables, such as the interest rate and date of maturity, and the computer prints out the answers. Nakazaki can then relay to her client the bond's current yield, accrued interest, and so on. In Nakazaki's rapid-fire work environment, such speed is essential. Nakazaki cautions that "computer users have to understand what's behind the formulas." The software "relies on the basics of calculus. If people don't understand the formula, it's useless for them to use the calculations."

Solutions to Self-Check Exercises 12.2

1. Using Formula (6) with $A = 20{,}000$, $r = 0.12$, $m = 12$, and $t = 3$, we find the required present value to be

$$P = 20{,}000\left(1 + \frac{0.12}{12}\right)^{-(12)(3)}$$

$$= 13{,}978.50$$

or \$13,978.50.

2. The accumulated amount of Glen's current investment is found by using Formula (5) with $P = 100{,}000$, $r = 0.116$, and $m = 365$. Thus, the required accumulated amount is

$$A = 100{,}000\left(1 + \frac{0.166}{365}\right)^{365}$$

$$= 112{,}297.52$$

or \$112,297.52. Next, we compute the accumulated amount of Glen's reinvestment. Once again, using (6) with $P = 100{,}000$, $r = 0.092$, and $m = 365$, we find the required accumulated amount in this case to be

$$\bar{A} = 100{,}000\left(1 + \frac{0.092}{365}\right)^{365}$$

or \$109,635.21. Therefore, Glen can expect to experience a net decrease in yearly income of

$$112{,}297.52 - 109{,}635.21$$

or \$2,662.31.

3. a. Using Formula (10) with $P = 10{,}000$, $r = 0.1$, and $t = 5$, we find that the required accumulated amount is given by

$$A = 10{,}000e^{(0.1)(5)}$$

$$= 16{,}487.21$$

or \$16,487.21.

b. Using Formula (11) with $A = 10{,}000$, $r = 0.1$, and $t = 5$, we see that the required present value is given by

$$P = 10{,}000e^{-(0.1)(5)}$$

$$= 6065.31$$

or \$6065.31.

12.3 Increasing Annuity

Annuity

Let us recall from Section 8.7 that an **annuity** is a sequence of payments made at certain time intervals over a specified length of time. We will assume in this chapter that these time intervals, also called **periods,** are all equal and will match the compounding period. The total length of time in which these payments are made is called the **term** of the annuity. The individual payments are often referred to as **rent** and denoted by R. Although the payments need not be equal in size, they are equal in many important applications, and we will assume that they are equal in our discussion.

All annuities are investments that accumulate interest. However, they differ between each other depending on whether regular rents are deposited or withdrawn. Quarterly deposits to a savings account build a fund for the future. This type of annuity is called an **increasing annuity** and will be discussed in this section. A monthly home mortgage pays off a loan charged with interest for a property. This type of annuity is called a **decreasing annuity** and will be discussed in Section 12.4.

Lastly, we address when during a period a payment is made. Equal payments that are made at the beginning of the payment period refer to an **annuity due.** Equal payments that are made at the end of the payment period refer to an **ordinary annuity.** Throughout this chapter we will assume that an annuity is ordinary unless specified otherwise.

Increasing Annuity

Let us consider a numerical example to illustrate how increasing annuities are calculated. Suppose \$1000 is deposited into a bank at the end of every year for a term of

4 years, earning interest at the rate of 8% per year compounded annually. The problem is classified as an increasing annuity because interest is compounded and regular deposits are made at equal time intervals. The deposits are referred to as payments and denoted by R. To find the future value A of the annuity, let us consider each payment at a time. We will use subscripts to indicate the number of years that compound interest was earned in order to differentiate between the different future values we are calculating. The compound interest rate of our problem is

$$i = \frac{r}{m} = \frac{0.08}{1} = 0.08$$

When the first payment is made, there are three years left before the completion of the term. This means that the first payment earns compound interest for three years and its future value is

$$A_3 = 1000(1 + 0.08)^3 = 1000(1.08)^3$$

When the second payment is made, there are two years left before the completion of the term. This means that the second payment earns compound interest for two years and its future value is

$$A_2 = 1000(1 + 0.08)^2 = 1000(1.08)^2$$

When the third payment is made, there is only one year left before the completion of the term. This means that the third payment earns compound interest for only one year and its future value is

$$A_1 = 1000(1 + 0.08)^1 = 1000(1.08)^1$$

The fourth and final payment does not earn any interest but it contributes to the annuity, so let us denote it by A_0 and

$$A_0 = 1000$$

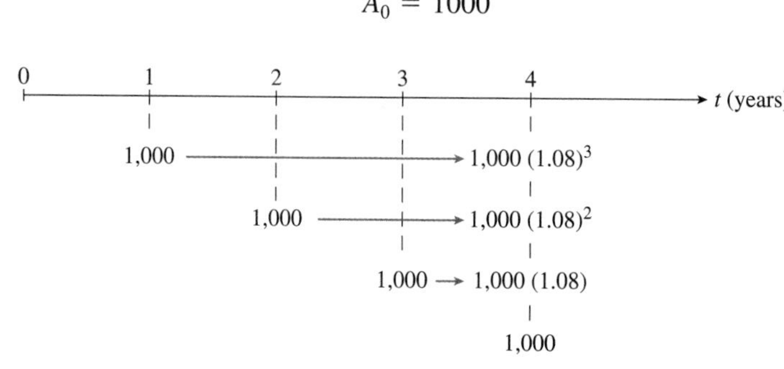

Figure 12.1

Figure 12.1 gives a visual representation of these calculations. The future value is the sum of the above four future values

$$\begin{aligned} A &= A_3 + A_2 + A_1 + A_0 \\ &= 1000(1.08)^3 + 1000(1.08)^2 + 1000(1.08)^1 + 1000 \\ &= 1259.71 + 1166.40 + 1080 + 1000 \\ &= \$4506.11 \end{aligned}$$

This was a really cumbersome calculation. Imagine the term of the annuity would be 20 years long. Then we would have to calculate 20 individual terms similar to the above to find the future value. Let us find a more efficient way of computing the future value and stop at

$$\begin{aligned} A &= A_3 + A_2 + A_1 + A_0 \\ &= 1000(1.08)^3 + 1000(1.08)^2 + 1000(1.08)^1 + 1000 \end{aligned}$$

Notice that the sum represents a partial geometric series. As long as $r \neq 1$ the sum of a partial geometric series with initial term a, common ratio r and $n + 1$ terms is given by

$$S_n = \frac{a(1 - r^{n+1})}{1 - r} = \frac{a(r^{n+1} - 1)}{r - 1} \qquad (12)$$

The first formula is more convenient if $|r| < 1$ and the second formula when $|r| > 1$. In our problem, the initial term a is given by 1000, the common ratio r given by 1.08 and we have four terms in total. Therefore, the future value calculation simplifies to

$$\begin{aligned} A &= 1000[(1.08)^3 + (1.08)^2 + (1.08)^1 + 1] \\ &= 1000\left[\frac{1((1.08)^{3+1} - 1)}{1.08 - 1}\right] \\ &= 1000\left[\frac{((1.08)^4 - 1)}{0.08}\right] \\ &\approx 4506.11 \end{aligned}$$

We can guess that this formula can be generalized to

$$A = R\left[\frac{((1 + i)^n - 1)}{i}\right]$$

However, let us be convinced of this fact. To find a general formula we proceed in our usual way and use Table 12.6 to model an increasing annuity with a nominal annual interest rate r compounded m times per year for a term of t years and equal payments R made at the end of the same compound period as for the nominal rate. Notice that there are $n = mt$ compounding periods in t years.

The future value at the end of a certain period gets calculated as the previous future value plus interest on the previous future value plus next payment. Keep this in mind when you study the column for the future value in Table 12.6.

As can be seen from Table 12.6, the future value is given by

$$A = R(1 + i)^{n-1} + \ldots + R(1 + i) + R$$

Table 12.6

Payment Number	Payment Made at the End of the Period	Interest per Compounding Period	Future Value A (is the sum of the previous future value, interest, and next payment)
1	R	0	R
2	R	$[R]i = Ri$	$R + Ri + R$ $= R(1 + i) + R$
3	R	$[R(1 + i) + R]i$ $= R(1 + i)i + Ri$	$R(1 + i) + R + R(1 + i)i + Ri + R$ $= R(1 + i) + R(1 + i)i + R + Ri + R$ $= R(1 + i)(1 + i) + R(1 + i) + R$ $= R(1 + i)^2 + R(1 + i) + R$
...	...	...	...
n	R	$[R(1 + i)^{n-2} + \ldots + R(1 + i) + R]i$ $= R(1 + i)^{n-2}i + \ldots + R(1 + i)i + Ri$	$= R(1 + i)^{n-2} + \ldots + R(1 + i) + R$ $+ R(1 + i)^{n-2}i + \ldots + R(1 + i)i + Ri$ $+ R$ $= R(1 + i)^{n-1} + \ldots + R(1 + i) + R$

Using Formula (12) with R as the initial value and $(1 + i)$ as the common ratio we get

$$A = R\left[\frac{(1 + i)^n - 1}{i}\right] \tag{13}$$

Remark

The expression $\left[\frac{(1 + i)^n - 1}{i}\right]$ in Formula (13) is commonly denoted by $s_{\overline{n}|i}$ in this chapter and is read *s sub n angle i*. ◀

Future Value of an Increasing Annuity

$$A = R\left[\frac{(1 + i)^n - 1}{i}\right] = R \cdot s_{\overline{n}|i} \quad \text{and} \tag{14}$$

Payment of an Increasing Annuity

$$R = \frac{A}{\left[\frac{(1 + i)^n - 1}{i}\right]} = A\left[\frac{(1 + i)^n - 1}{i}\right]^{-1} = \frac{A}{s_{\overline{n}|i}} \tag{15}$$

where

A is the future value of the annuity in dollars
R is the amount of each equal payment in dollars
m is the number of compounding periods per year
r is the annual interest rate expressed as a decimal
i is the interest per compounding period as a decimal and $i = \frac{r}{m}$
t is the time in years
n is the total number of compounding periods and $n = mt$

Remark

Notice that since future value contains the interest earned, the accumulated interest is given as the difference between the future value of the annuity and the total number of payments

$$I = A - n \cdot R$$ ◀

Example 1

Find the future value if \$400 is deposited at the end of every 3 months for 10 years into an account paying 6% compounded quarterly.

Solution

Notice that the interest is compounded quarterly and regular payments are made quarterly, so this is an increasing annuity. We use Formula (14), where A is the future value in dollars we are looking for, $R = 400$, $m = 4$, $r = 6\% = 0.06$, $i = \frac{r}{m} = \frac{0.06}{4} = 0.015$, $t = 10$ and $n = mt = 4 \cdot 10 = 40$. We compute

$$A = 400\left[\frac{(1 + 0.015)^{40} - 1}{0.015}\right] \approx 21{,}707.16$$

The future value of the annuity is \$21,707.16.

Example 2

Suppose a trust fund is set up into which regular payments of \$1500 are made at the end of every month. The trust earns 3.2% interest per annum compounded monthly. How many years does it take to accumulate \$1,000,000 in the trust?

Solution

Notice that the interest is compounded monthly and regular payments are made monthly, so this is an increasing annuity. We use Formula (14), where $A = 1{,}000{,}000$, $R = 1500$, $m = 12$, $r = 3.2\% = 0.032$, $i = \frac{0.032}{12} = \frac{0.008}{3}$, t is the number of years we are looking for and $n = mt = 12t$. We compute

$$1{,}000{,}000 = 1500\left[\frac{\left(1 + \frac{0.008}{3}\right)^{12t} - 1}{\frac{0.008}{3}}\right]$$

This is an equation that we need to solve for t.

$$\frac{1{,}000{,}000}{1500} = \frac{\left(1 + \frac{0.008}{3}\right)^{12t} - 1}{\frac{0.008}{3}}$$

$$\frac{1{,}000{,}000}{1500} \cdot \frac{0.008}{3} = \left(1 + \frac{0.008}{3}\right)^{12t} - 1$$

$$\frac{1{,}000{,}000}{1500} \cdot \frac{0.008}{3} + 1 = \left(1 + \frac{0.008}{3}\right)^{12t}$$

Now, we take the logarithm on each side of the equation, obtaining

$$\ln\left(\frac{1{,}000{,}000}{1500} \cdot \frac{0.008}{3} + 1\right) = \ln\left(1 + \frac{0.008}{3}\right)^{12t}$$

$$\ln\left(\frac{1{,}000{,}000}{1500} \cdot \frac{0.008}{3} + 1\right) = 12t\ln\left(1 + \frac{0.008}{3}\right) \qquad \text{Logarithm Law 3}$$

$$\frac{\ln\left(\frac{1{,}000{,}000}{1500} \cdot \frac{0.008}{3} + 1\right)}{\ln\left(1 + \frac{0.008}{3}\right)} = 12t$$

and

$$t = \frac{\ln\left(\frac{1{,}000{,}000}{1500} \cdot \frac{0.008}{3} + 1\right)}{12\ln\left(1 + \frac{0.008}{3}\right)} \approx 31.96915219$$

It takes approximately 32 years for the trust to accumulate to $1,000,000.

GROUP DISCUSSION

Return to the numerical example at the beginning of our discussion about increasing annuity that says $1000 is deposited into a bank at the end of every year for a term of 4 years, earning interest at the rate of 8% per year compounded annually. Suppose that instead of the *end* of the year it says *beginning* of the year. Recalculate the future value for this problem. Then derive a general formula by constructing a table similar to Table 12.6. The formula that you get for this future value describes an increasing annuity due.

Sinking Fund

A sinking fund is a fund into which regular deposits are made at equal time intervals in order to accumulate enough money to pay off a future expenditure.

Example 3

A hospital is expecting renovation costs in excess of \$500,000 in 5 years. How much should the executive committee plan to deposit at the end of each month into an account paying 5.75% interest per year compounded monthly, so that there is enough money for the upgrading?

Solution

Notice that the interest is compounded monthly and regular payments are made monthly, so this is an increasing annuity. We use Formula (15), where $A = 500{,}000$, R is the amount of each deposit in dollars we are looking for, $m = 12$, $r = 5.75\% = 0.0575$, $i = \frac{r}{m} = \frac{0.0585}{12} = 0.004875$, $t = 5$ and $n = mt = 12 \cdot 5 = 60$. We compute

$$R = 500{,}000\left[\frac{(1 + 0.004875)^{60} - 1}{0.004875}\right]^{-1} \approx 7194.07$$

The executive committee needs to plan for a deposit of \$7194.07 every month for 6 years.

Example 4

Nancy would like to have \$3500 to plan a trip to Asia in 3 years. How much should she deposit at the end of each week into an account paying 2.6% interest per year compounded weekly, so that she has enough money for the trip?

Solution

Notice that the interest is compounded weekly and regular payments are made weekly, so this is an increasing annuity. We use Formula (15), where $A = 3500$, R is the amount of each deposit in dollars we are looking for, $m = 52$, $r = 2.6\% = 0.026$, $i = \frac{r}{m} = \frac{0.026}{52} = 0.0005$, $t = 3$ and $n = mt = 52 \cdot 3 = 156$. We compute

$$R = 3500\left[\frac{(1 + 0.0005)^{156} - 1}{0.0005}\right]^{-1} \approx 21.58$$

Nancy needs to make a deposit of \$21.58 every week for 3 years.

The following example requires both compound interest and increasing annuity calculation.

Example 5

Tom deposited \$80 monthly into an account at his local bank for 15 years. The bank paid 3% interest per year compounded monthly for the first 6 years. During the remaining 9 years the interest per year compounded monthly was raised to 6%. What is the value of the account after the 15-year term?

Solution

Let us start by considering the first 6 years. Notice that the interest is compounded monthly and regular payments are made monthly, so this is an increasing annuity. We use Formula (14), where A_1 is the future value in dollars we are looking for, $R = 80$, $m = 12$, $r = 3\% = 0.03$, $i = \frac{r}{m} = \frac{0.03}{12} = 0.0025$, $t = 6$ and $n = mt = 12 \cdot 6 = 72$. We compute

$$A_1 = 80\left[\frac{(1 + 0.0025)^{72} - 1}{0.0025}\right] \approx 6302.35$$

Now, let us consider the next 9 years. We first observe that i, t, and n change to $i = \frac{r}{m} = \frac{0.06}{12} = 0.005$, $t = 9$ and $n = mt = 12 \cdot 9 = 108$. A_1 continues to stay in the account collecting compound interest for this time interval $t = 9$. In addition, monthly payments to the account continue, and so we have a second increasing annuity to calculate. First, we use Formula (7) to calculate the future value for the principal plus compound interest, where A_2 is the future value in dollars we are looking for and $P = A_1$. We compute

$$A_2 = 6302.35(1 + 0.005)^{108} \approx 10,800.33$$

Next, we use Formula (14) to calculate the second annuity, where A_3 is the future value in dollars we are looking for, and we still have $R = 80$. We compute

$$A_3 = 80\left[\frac{(1 + 0.005)^{108} - 1}{0.005}\right] \approx 11,419.19$$

Finally, the future value A we are looking for is the sum of A_2 and A_3

$$A = A_2 + A_3 = 10,800.33 + 11,419.19 = 22,219.52$$

The value of the account is \$22,219.52 after 15 years.

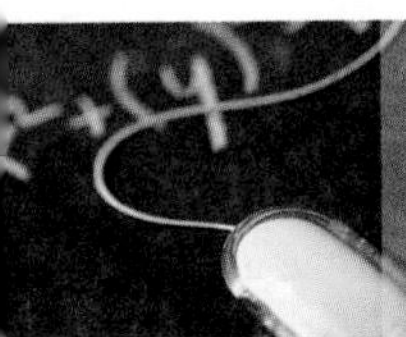

Self-Check Exercises 12.3

1. Every time Sally receives her biweekly paycheque she sets aside \$30 and deposits the amount into an account paying 5.2% annually compounded biweekly. Sally continues with this habit for 10 yrs.
   a. How much money will have accumulated into the account?
   b. What is the amount of interest that Sally will have earned?

2. Tom and Jerry inherit \$10,000 after taxes are deducted. Tom decides to deposit the whole amount into an account earning 3.75% per year compounded monthly. Jerry decides to invest his money into an increasing annuity earning 5.25% yearly compounded monthly with regular payments of \$100 at the end of every month.
   a. What is the value of Tom's investment after 5 yrs?
   b. What is the value of Jerry's investment after 5 yrs?
   c. By how much should Jerry increase his payments in order to make the same amount of money as Tom in 5 yrs? Does Jerry eventually need to contribute to the payments from his paycheque?

Solutions to Self-Check Exercises 12.3 can be found on page 735.

12.3 Exercises

A calculator is recommended for these exercises. Assume that the increasing annuity is ordinary unless otherwise clearly stated.

In Exercises 1–6, find the future value A of an increasing annuity with a nominal annual interest rate r compounded m times per year for a term of t years and regular equal payment R.

1. $R = 1800$, $r = 7.2\%$ compounded monthly, $t = 10$ years
2. $R = 2500$, $r = 6.5\%$ compounded quarterly, $t = 15$ years
3. $R = 75$, $r = 4.6\%$ compounded weekly, $t = 5$ years
4. $R = 45$, $r = 8.14\%$ compounded monthly, $t = 20$ years
5. $R = 18,000$, $r = 0.85\%$ compounded semiannually, $t = 7$ years
6. $R = 15,000$, $r = 1.4\%$ compounded biweekly, $t = 12$ years

In Exercises 7–12, find the regular equal payment R if the future value A is given of an increasing annuity with a nominal annual interest rate r compounded m times per year for a term of t years.

7. $A = 18,000$, $r = 7.2\%$ compounded weekly, $t = 10$ years
8. $A = 25,000$, $r = 6.5\%$ compounded quarterly, $t = 15$ years
9. $A = 7500$, $r = 4.6\%$ compounded quarterly, $t = 5$ years
10. $A = 4500$, $r = 8.14\%$ compounded monthly, $t = 20$ years

11. $A = 150{,}000$, $r = 0.85\%$ compounded biweekly, $t = 7$ years

12. $A = 500{,}000$, $r = 1.4\%$ compounded semiannually, $t = 12$ years

In Exercises 13–18, find the term t in years accurate to one decimal place for an increasing annuity with a nominal annual interest rate *r* compounded *m* times per year and regular equal payments *R* that accumulates to a future value of *A*.

13. $A = 17{,}500$, $R = 50$, $r = 3\frac{1}{2}\%$ compounded quarterly

14. $A = 31{,}000$, $R = 100$, $r = 2.75\%$ compounded monthly

15. $A = 250{,}000$, $R = 540$, $r = 3.6\%$ compounded monthly

16. $A = 375{,}000$, $R = 350$, $r = 2.6\%$ compounded biweekly

17. $A = 1{,}000{,}000$, $R = 2000$, $r = 1.96\%$ compounded weekly

18. $A = 1{,}500{,}000$, $R = 40{,}000$, $r = 4.44\%$ compounded quarterly

Business and Economics Applications

19. **Renovation Cost** The planning committee of a metropolitan hospital predicts renovation costs of $2,000,000 in 4 yrs. How much should be deposited monthly into a sinking fund earning 6.8% compounded monthly to cover the renovation costs when needed?

20. **Renovation Cost** The successful owner of a reclusive resort is planning to open another resort in 3 yrs. He has bought the rights to an island including all the buildings on it. How much should be deposited weekly into a sinking fund earning 5.6% compounded weekly to cover the predicted renovation costs of $1,500,000?

21. **Car Purchase** Daniela can afford to set aside 25% from every $1800 biweekly paycheque. She wants to move to the country in 4 yrs, where she needs a car. How much money will she have available to purchase a car if she deposits her savings into a sinking fund earning 5.2% compounded biweekly?

22. **Trip Purchase** Sunita receives $120 every month in spending money from her parents while attending a 3-yr program at a college. However, instead of spending it all, she decides to place 75% of the spending money into a sinking fund earning 3.9% compounded monthly to purchase a trip around the world at the end of her program. How much money will she have available for her trip?

23. **Kayak Purchase** Goran deposits $350 every three months into an account that pays 4.5% compounded quarterly. How long will it be until he can have $8000 to purchase a kayak and its accessories?

24. **Upgrade Cost** Howard wants to upgrade his computer equipment and will need $4000. He can afford to deposit $100 every month into an account that pays 2.5% compounded monthly. How long will it be until he can upgrade his computer?

25. **Savings** Mary loves chocolate and spends $5 every week on it. Suppose she would instead invest that amount each week in a savings account earning 2.5% interest per year compounded weekly. What amount has accumulated in the account after 40 yrs?

26. **Savings** A Lotto 6/49 player spends $10 every week. Suppose the player would instead invest that amount each week in a savings account earning 2.8% interest per year compounded weekly. What amount has accumulated in the account after 30 yrs? How does this compare to a payout if 5/6 numbers are correct, where the odds of winning are approximately 1 in 55,492? Suppose that the pool has $2,000,000 and the win is 4.75% of the pool's fund with 100 winners across the country.

27. **RSP** Francois contributes $5000 every year to a Retirement Savings Plan (RSP) offered at his local bank. Assume that the average rate of return is 8% compounded yearly. How much money will have accumulated in his RSP after 25 yrs when Francois retires?

28. **RSP** Mira contributes $300 every month to a Retirement Savings Plan (RSP) offered at her local bank. Assume that the average rate of return is 8% compounded monthly. How much money will have accumulated in her RSP after 28 yrs when Mira retires?

29. **Investment** Amarjeet deposits $1000 every three months into a savings account earning 6% interest per year compounded quarterly. After 3 yrs he stops making payments but leaves the accumulated amount in the account.
    a. How much interest has been earned after an additional 2 yrs have passed?
    b. When should Amarjeet stop making payments, if he wants to have $100,000 in his savings account?

30. **Investment** Ahmed deposits $100 every month into a savings account earning 5% interest per year compounded monthly. After 5 yrs he stops making payments but leaves the accumulated amount in the account.
    a. How much money has accumulated in the savings account after an additional 5 yrs have passed?
    b. For how many more months after the initial 5 yrs must the money earn interest to accumulate to $10,000?

31. **Investment** Kirsten sets up an investment plan based on variable interest rates with her bank. She contributes $50 every month for 10 yrs. Interest rates are compounded monthly. During the first 4 yrs the bank pays 4% interest per annum, while for the remaining years the annual interest rate increases to 4.5%. How much has accumulated in Kirsten's investment plan after the 10 yrs?

32. **Investment** Lucas deposits $80 every week for 15 yrs into a savings account based on variable interest rates. Interest rates are compounded weekly. During the first 5 yrs the bank pays 3.5% interest per annum, while for the remaining years the annual interest rate drops to 2.8%. This prompts Lucas to withdraw $20,000 after 5 yrs. How much has accumulated in his account at the end of the 15-yr period?

33. **Buying Collectibles** Colleen buys a set of rare coins for $3500 as a birthday gift for her father's coin collection. The seller agrees to receive the money in 4 yrs but wants 10% simple interest monthly on the price.
    a. What is Colleen's monthly payment to the seller?
    b. Colleen sets up a sinking fund to pay off the price of the coins. She finds a bank that offers 6.5% annual interest compounded weekly. What is Colleen's weekly payment to the bank?
    c. How much is Colleen paying beyond the actual cost of the coins?

34. **Buying Collectibles** Iris collects teddy bears and finds a rare 1921 Farnell bear for a mere $5000. The seller agrees to receive the money in 5 yrs but wants 7.5% simple interest quarterly on the price.
   a. What is Iris' quarterly payment to the seller?
   b. Iris sets up a sinking fund to pay off the price of the coins. She finds a bank that offers 5.85% annual interest compounded monthly. What is Iris' monthly payment to the bank?
   c. How much is Iris paying beyond the actual cost of the teddy bear?

Solutions To Self-Check Exercises 12.3

1. Regular deposits are made so this is an increasing annuity, where $R = 30$, $m = 26$, $r = 5.2\% = 0.052$, $i = \frac{r}{m} = \frac{0.052}{26} = 0.002$, $t = 10$ and $n = 26 \cdot 10 = 260$.

   a. We use Formula (14) to find the future value A. We compute

$$A = 30\left[\frac{(1 + 0.002)^{260} - 1}{0.002}\right] \approx 10{,}217.32$$

   After 10 years the account contains $10,217.32.

   b. The interest is the difference between the future value and the total payments that Sally deposited. First, we compute the total payments

$$R_{total} = R \cdot n = 30 \cdot 260 = 7800$$

   Next, we subtract the total payment amount from the future value calculated in 1a.

$$I = A - R_{total} = 10{,}217.32 - 7800 = 2417.32$$

   Sally will have earned $2417.32 in the 10-year term.

2. **a.** This is a compound interest problem, where A is the future value in dollars we are looking for, $P = 10{,}000$, $m = 12$, $r = 1.95\% = 0.0195$, $i = \frac{r}{m} = \frac{0.0195}{12} = 0.001625$, $t = 5$ and $n = 12 \cdot 5 = 60$. We use Formula (7) and compute

$$A = 10{,}000(1 + 0.001625)^{60} \approx 11{,}023.24$$

   Tom's investment is worth $11,023.24 after 5 years.

   b. As stated, this is an increasing annuity problem, where A is the future value in dollars we are looking for, $R = 100$, $m = 12$, $r = 5.25\% = 0.0525$, $i = \frac{r}{m} = \frac{0.0525}{12} = 0.004375$, $t = 5$ and $n = 12 \cdot 5 = 60$. We use Formula (14) and compute 0.004375

$$A = 100\left[\frac{(1 + 0.004375)^{60} - 1}{0.004375}\right] \approx 6844.17$$

   Jerry's investment is worth $6844.17 after 5 years.

   c. From 2a., the future value Jerry needs to achieve is $A = 11{,}023.24$. We use Formula (15) to find the rent R in dollars. All other values are the same as in 2b. We compute

$$R = 11{,}023.24\left[\frac{(1 + 0.004375)^{60} - 1}{0.004375}\right]^{-1} \approx 161.06$$

   The new payments are $161.06, so Jerry needs to make an increase of

$$\$161.06 - \$100 = \$61.06$$

   There are 60 payments in the 5-year term. The total payments that Jerry has to make is

$$161.06 \cdot 60 = 9663.60$$

   This amount is still less than the inheritance of $10,000 and so Jerry does not need to contribute to the annuity from his paycheques.

12.4 Decreasing Annuity and Amortization

Decreasing Annuity

In the previous section, we talked in general terms about annuities and in particular about increasing annuities, where regular payments are deposited to accumulate a future fund. The reverse of this would be the deposit of a present value of money into an account that pays compound interest, so that periodic payments can be withdrawn over a certain length of time. This financial situation is called a decreasing annuity and not only applies to setting up retirement payments, but also to a sequence of regular payments to pay off a loan and its interest charges. As in the previous section, we assume throughout this section that the annuity is ordinary, that is, the payments are made at the end of the payment period.

Let us consider a numerical example to illustrate how decreasing annuities are calculated. Suppose we want to retire at age 50 and be able to withdraw \$50,000 at the end of each year for 30 years. Surely we should not need the total value of $30 \cdot \$50{,}000 = \$1{,}500{,}000$; we should let the money work for us by collecting interest. Let us assume that the bank pays 6% per annum compounded annually. The question is, what is the present value of money that needs to be deposited to support these regular withdrawals? The problem is classified as a decreasing annuity because interest is compounded and regular withdrawals are made at equal time intervals. The withdrawals are referred to as payments and denoted by R. To find the present value P of the annuity, let us consider the following. Suppose the payments are withdrawn but immediately deposited into another account earning the same interest rate. These deposits will accumulate to a future value F. But this is an increasing annuity described by

$$A = R\left[\frac{(1+i)^n - 1}{i}\right], \tag{16}$$

where $R = 50{,}000$, $m = 1$, $i = \dfrac{0.06}{1} = 0.06$, and $n = 1 \cdot 30 = 30$. So,

$$A = 50{,}000\left[\frac{(1.06)^{30} - 1}{0.06}\right] \approx 3{,}952{,}909.31$$

This means that the future value is \$3,952,909.31. Now, our question can be phrased as, what is the principal that needs to be deposited earning the same compound interest to grow to this future value of \$3,952,909.31? But this is just a compound interest problem and can be described by

$$A = P(1+i)^n, \tag{17}$$

where $A = 50{,}000\left[\dfrac{(1.06)^{30} - 1}{0.06}\right] \approx \$3{,}952{,}909.31$ and i and n are as described above. Solving for P, we get

$$\begin{aligned} P &= 50{,}000\left[\frac{(1.06)^{30} - 1}{0.06}\right](1.06)^{-30} \\ &\approx 3{,}952{,}909.31(1.06)^{-30} \\ &\approx 688{,}241.56 \end{aligned}$$

This answers our question, i.e., we need to deposit \$688,241.56 at age 50 in order to be able to withdraw annual payments of \$50,000 for a 30-year term.

Let us make use of our description above to find a general formula for a decreasing annuity. The key understanding is that Formulas (16) and (17) above are calculating the same future value A with the same compounded interest rate i and the same number of compounding periods n. Setting the two formulas equal to each other we get

$$P(1+i)^n = R\left[\frac{(1+i)^n - 1}{i}\right]$$

Now we solve for P,

$$P = R\left[\frac{(1+i)^n - 1}{i(1+i)^n}\right]$$

which gives us a relationship between the payments and the present value of a decreasing annuity. This equation can be simplified by dividing through by the term $(1+i)^n$ to get

$$P = R\left[\frac{1 - (1+i)^{-n}}{i}\right] \tag{18}$$

We now have a formula for a decreasing annuity. It helps to think of the n in Formula (18) as the number of compounding periods or payment periods that are left in order to get P as the present value.

Remark

Notice that Formula (18) for a decreasing annuity is very similar to Formula (13) for an increasing annuity. However, the numerator for an increasing annuity is $(1 + i)^n - 1$, while the numerator for a decreasing annuity is $1 - (1 + i)^{-n}$. ◀

We want to emphasize here, that the typical application of a decreasing annuity is to loans, where money is borrowed from an institution and paid back with interest charges through a sequence of regular payments. Such loans can be for the purchase of a car, appliances, etc., as well as the purchase of a house, in which case the loan is referred to as a **mortgage.** Since mortgage calculations are handled somewhat differently in Canada as regular loans, we will defer looking at Canadian mortgages until the end of this section. When a purchase is made, the buyer may pay a **down payment** on the purchase price to reduce the loan amount that needs to be financed through a decreasing annuity. Also, at times an annuity is concluded with a final payment before the term is up. This is referred to as a **lump sum payment.**

Remark

The expression $\left[\frac{1 - (1 + i)^{-n}}{i}\right]$ in Formula (18) is commonly denoted by $a_{\overline{n}|i}$ in this chapter and is read *a sub n angle i.* ◀

GROUP DISCUSSION

Return to the numerical example at the beginning of our discussion about a decreasing annuity that is asking for the present value needed to be deposited at age 50 so that \$50,000 can be withdrawn annually for a 30-year term if interest is compounded annually at the rate of 0.06% per annum. Consider a different way of calculating P and thereby discovering the decreasing annuity formula by finding the present value at time $t = 0$ of each withdrawal over the 30-year period. Figure 12.2 assists you with a visual presentation of these withdrawals and their present values.

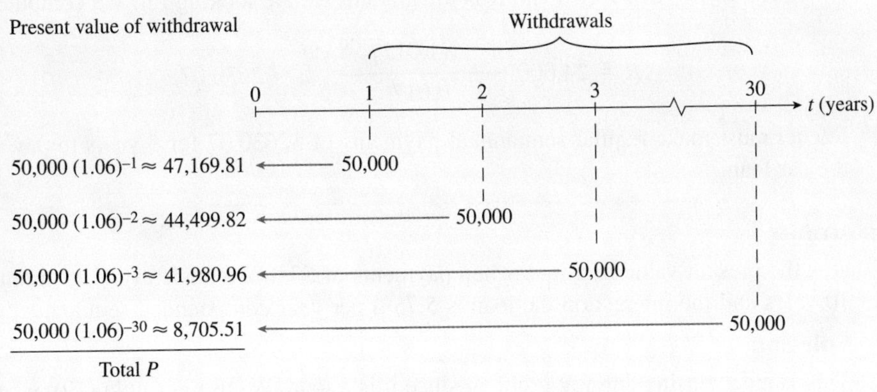

Figure 12.2

Study the figure and sum up all the present values to get P. Then derive the general formula for a decreasing annuity by describing each present value in general terms and summing a geometric series much like in Section 12.3 to get P.

Present Value of a Decreasing Annuity

$$P = R\left[\frac{1 - (1 + i)^{-n}}{i}\right] = R \cdot a_{\overline{n}|i} \quad \text{and} \tag{19}$$

Payment of a Decreasing Annuity

$$P = R\left[\frac{1 - (1 + i)^{-n}}{i}\right] = \frac{P}{a_{\overline{n}|i}} \tag{20}$$

where

P is the present value of the annuity in dollars

R is the amount of each equal payment in dollars

m is the number of compounding periods per year

r is the annual interest rate expressed as a decimal

i is the interest per compounding period as a decimal and $i = \frac{r}{m}$

t is the time in years

n is the total number of compounding periods and $n = mt$

Remark

Notice that since the payments include the interest earned, the accumulated interest is given as the difference between the total of all payments and the present value of the annuity

$$I = n \cdot R - P$$

◄

Example 1

Rachel is buying a family-sized car for \$34,000. She has a down payment of \$10,000 and is financing the remaining amount at 3.4% compounded semiannually over a 5-year period. What payments must Rachel make every half year?

Solution

The down payment reduces the amount that needs to be financed so that the total loan is

$$34{,}000 - 10{,}000 = 24{,}000$$

Notice that interest is compounded semiannually and regular payments must be made semiannually in this loan problem, so this is a decreasing annuity. We use Formula (20), where $P = 24{,}000$, $m = 2$, $r = 3.4\% = 0.034$, $i = \frac{0.034}{2} = 0.017$, $t = 5$, $n = mt = 2 \cdot 5 = 10$ and R is the amount we are looking for. We compute

$$R = 24{,}000\left[\frac{1 - (1.017)^{-10}}{0.017}\right]^{-1} \approx 2630.07$$

Rachel must make regular semiannual payments of \$2630.07 for 5 years to pay off the car loan.

Example 2

What is the present value of a loan when payments of \$350 are made every six months for 10 years and the interest on the loan is 5.75% per year compounded semiannually?

Solution

This is a decreasing annuity problem since it is a loan. We use Formula (19), where $R = 350$, $m = 2$, $r = 5.75\% = 0.0575$, $i = \frac{0.0575}{2} = 0.02875$, $t = 10$, $n = m \cdot t = 2 \cdot 10 = 20$ and P is the amount we are looking for. We compute

$$P = 350\left[\frac{1 - (1.02875)^{-20}}{0.02875}\right] \approx 5267.81$$

The present value of the loan is \$5267.81.

Example 3

You receive an inheritance of \$100,000 after taxes have been deducted and immediately place it into an account paying 6% per year compounded monthly. You would like to make monthly withdrawals to supplement your income for 20 years.

a. What is the amount of each withdrawal?

b. What is the total amount of interest earned over the 20-year period?

c. After 12 years you decide to stop making the withdrawals since you need a down payment for a house you want to purchase. What is the value of the account at this time?

Solution

a. Notice that the interest is compounded monthly and regular withdrawals are made monthly, so this is a decreasing annuity. We use Formula (20), where

$P = 100{,}000, \quad m = 12, \quad r = 6\% = 0.06, \quad i = \frac{0.06}{12} = 0.005, \quad t = 20,$

$n = mt = 12 \cdot 20 = 240$ and R is the amount we are looking for. We compute

$$R = 100{,}000\left[\frac{1 - (1.005)^{-240}}{0.005}\right]^{-1} \approx 716.43$$

You can withdraw \$716.43 at the end of every month for 20 years.

b. To find the total amount of interest that accumulates over 20 years, we first need to calculate the total amount of all withdrawals. We multiply the withdrawal R by the total number of compounding periods n to get

$$n \cdot R = 240 \cdot 716.43 = 171{,}943.20$$

The total amount of all withdrawals is \$171,943.20. If we now subtract the principal amount from this number, we get

$$171{,}943.20 - 100{,}000 = 71{,}943.20$$

and so the accumulated interest earned is \$71,943.20.

c. After 12 years, there are only 8 years left to make withdrawals, so $t = 8$. We use Formula (19), where $R = 716.43$, $m = 12$, $r = 6\% = 0.06$, $i = \frac{0.06}{12} = 0.005$, $t = 8$, $n = m \cdot t = 12 \cdot 8 = 96$ and P is the amount we are looking for. We compute

$$P = 716.43\left[\frac{1 - (1.005)^{-96}}{0.005}\right] \approx 54{,}516.90$$

After 12 years, the remaining amount in the savings account is \$54,516.90.

Amortization

Example 1 above about financing a car purchase is a special application of a decreasing annuity. When a loan—the principal—and its interest charges are paid back in a sequence of regular payments, then we say that the loan is **amortized.** The Latin root *mors* means *death.* In other words, when we say that we are amortizing a debt, we really mean that the debt is being killed by making all the required payments. Notice that the word mortgage also comes from the root *mors,* which is befitting, as we mean to pay off the debt for a house purchase. In this context, Formula (20) really gives the **amortization payments.**

Let us have a closer look at what it means to make these payments. With each payment, part of the principal and part of the interest incurred are paid off. As the principal grows smaller over time, the portion of the payment that goes towards the principal grows larger while the portion that goes towards the interest charges grows smaller.

This is best observed with an amortization table, which is a visual representation of the amortization process and helps us to understand how the loan is paid off. An amortization table lists the payment number, the amortization payment, the interest portion, the portion applied to the principal, and the principal at the end of each period, which basically is the unpaid balance. With the information given in an amortization table important decisions can be made such as when to stop payments and pay off the loan with a lump sum payment, or make a prepayment during the term of the annuity, which reduces the value of the principal.

Let us now revisit Example 1 above and prepare an amortization table. Recall that we used Formula (20) with $P = 24{,}000$, $m = 2$, $r = 3.4\% = 0.034$, $i = \dfrac{0.034}{2} = 0.017$, $t = 5$, $n = mt = 2 \cdot 5 = 10$ to calculate the payment R as

$$R = 24{,}000\left[\frac{1 - (1.017)^{-10}}{0.017}\right]^{-1} \approx 2630.07$$

Suppose Rachel is making her first payment of \$2630.07. How much of that payment goes towards the principal and the interest charges? We first calculate the interest that incurs at the end of the first payment period (which is the same as the first compound period) by multiplying the principal with the compounded interest rate

$$I = i \cdot P = 0.017 \cdot 24{,}000 \approx 408.00$$

We now subtract this amount from the payment amount and get

$$R - i \cdot P = 2630.07 - 408.00 = 2222.07$$

Therefore, \$408.00 is the portion of the first payment of \$2630.07 that goes towards interest charges and \$2222.07 is the portion that goes towards the principal. The unpaid balance remaining is calculated as

$$\begin{aligned} P_{new} &= P_{old} - (R - i \cdot P_{old}) \\ &= 24{,}000 - (2630.07 - 408.00) \\ &= 24{,}000 - 2222.07 \\ &= 21{,}777.93 \end{aligned}$$

So, the new principal at the end of the first payment period is \$21,777.93. We now have all the information to fill the first two lines of our amortization table shown in Table 12.7. To calculate the next line we repeat the process. First we find the interest charged at the end of the second payment period by multiplying the new principal with the compounded interest rate

$$I = i \cdot P = 0.017 \cdot 21{,}777.93 \approx 370.22$$

We now subtract this amount from the payment amount and get

$$R - i \cdot P = 2630.07 - 370.22 = 2259.85$$

Therefore, \$370.22 is the portion of the second payment of \$2630.07 that goes towards interest charges and \$2259.85 is the portion that goes towards the principal. The unpaid balance remaining is calculated as

$$\begin{aligned} P_{new} &= P_{old} - (R - i \cdot P_{old}) \\ &= 21{,}777.93 - (2630.07 - 370.22) \\ &= 21{,}777.93 - 2259.85 \\ &= 19{,}518.08 \end{aligned}$$

Table 12.7

Payment Number	Payment R	Interest Portion $I = i \cdot P$	Portion Applied to Principal $R - I = R - i \cdot P$	Principal (Unpaid Balance) $P_{new} = P_{old} - (R - i \cdot P_{old})$
0				\$24,000.00
1	\$2630.07	\$408.00	\$2222.07	21,777.93
2	2630.07	370.22	2259.85	19,518.08
3	2630.07	331.81	2298.26	17,219.82
4	2630.07	292.74	2337.33	14,882.48
5	2630.07	253.00	2377.07	12,505.41
6	2630.07	212.59	2417.48	10,087.93
7	2630.07	171.49	2458.58	7629.36
8	2630.07	129.70	2500.37	5128.99
9	2630.07	87.19	2542.88	2586.11
10	2630.07	43.96	2586.11	0

So, the principal at the end of the second payment period is \$23,660.80. We now have all the information to fill the next line of our amortization table shown in Table 12.7. This process is repeated to get the complete amortization table for this car loan in Table 12.7.

As you have just experienced, the calculations to produce an amortization table are tedious and time-consuming. You may also notice that while properly rounded amounts are shown in the table, the calculations are processed using the exact amounts calculated. Luckily, there are many Web-based mortgage calculators available, many of which are provided by your local bank.

GROUP DISCUSSION

Extend Table 12.7 by adding two columns to include the accumulating amount of interest as well as the accumulating amount applied to principal. Use the last line to read off the total amount of interest paid. Use Formula (19) to calculate the principal remaining after eight payments have been made. Compare your answer to that provided in the table. How can you explain the discrepancy?

Canadian Mortgage Calculation

Canadian mortgages are compounded semiannually; however, mostly the payments are made monthly. The monthly rate is calculated by equating the effective annual rate for monthly compounding with the effective annual rate for semiannual compounding:

$$\left(1 + \frac{r_{monthly}}{12}\right)^{12} - 1 = \left(1 + \frac{r_{semiannually}}{2}\right)^{2} - 1$$

$$\left(1 + \frac{r_{monthly}}{12}\right)^{12} = \left(1 + \frac{r_{semiannually}}{2}\right)^{2}$$

This we will now solve for the monthly rate $r_{monthly}$, and get

$$1 + \frac{r_{monthly}}{12} = \left(1 + \frac{r_{semiannually}}{2}\right)^{\frac{2}{12}}$$

$$\frac{r_{monthly}}{12} = \left(1 + \frac{r_{semiannually}}{2}\right)^{\frac{2}{12}} - 1$$

$$r_{monthly} = 12\left[\left(1 + \frac{r_{semiannually}}{2}\right)^{\frac{2}{12}} - 1\right]$$

Monthly Rate for Semicompounded Mortgage

$$r_{monthly} = 12\left[\left(1 + \frac{r_{semiannually}}{2}\right)\frac{2}{12} - 1\right] \qquad (21)$$

where

$r_{semiannually}$ is the semiannual interest rate expressed as a decimal.

Example 4

The Wong family purchases a house for \$228,000 in Quebec City amortized at 5.75% compounded semiannually over a 25-year period with monthly payments. How much are the interest charges?

Solution

We start by calculating the monthly rate using Formula (21):

$$r_{monthly} = 12\left[\left(1 + \frac{0.0757}{2}\right)\frac{2}{12} - 1\right] \approx 0.0745330$$

Next, we calculate the monthly payments based on this monthly rate by using Formula (20), where $P = 198{,}000$, $m = 12$, $r = 0.0745330$, $i = \dfrac{0.0745330}{12} \approx 0.0062111$, $t = 25$, and $n = mt = 12 \cdot 25 = 300$. We compute

$$R = 228{,}000\left[\frac{1 - (1.0062111)^{-300}}{0.0062111}\right]^{-1} \approx 1677.98$$

or each monthly payment is \$1677.98. The total amount paid for the house is therefore

$$\$1677.98 \times 300 = \$503{,}394.00$$

from which we subtract the purchase price for the house to yield the total interest charges:

$$\$503{,}394 - \$228{,}000 = \$275{,}394$$

It is often of meaningful to find the equity of a property you own to make further financial decisions. Equity is defined as the difference between the market value of the property and its unpaid loan balance. The following example illustrates how to calculate the equity.

Example 5

A family purchased a home for \$240,000 in 1995 by paying a down payment of \$60,000 and amortizing the rest with equal monthly payments over 20 years at 7.05% per year compounded semiannually. In 2006 the net worth of the house was \$330,000. How much equity did the family have in the house?

Solution

Since equity is the difference between the market value and the unpaid loan balance at that time, we need to find the loan amount and the monthly payments in order to calculate the unpaid loan balance. The down payment reduces the amount that needs to be financed

$$240{,}000 - 60{,}000 = 180{,}000$$

so that the total loan is \$180,000. This is clearly a mortgage problem and so we first have to calculate the correct monthly rate using Formula (21)

$$r_{monthly} = 12\left[\left(1 + \frac{0.0705}{2}\right)^{\frac{2}{12}} - 1\right] \approx 0.0694846$$

Now, we use Formula (20) to calculate the payment amount R, where $P = 180{,}000$, $m = 12$, $r = 0.0694846$, $i = \dfrac{0.0694863}{12} \approx 0.0057905$, $t = 20$, and $n = mt = 12 \cdot 20 = 240$. We compute

$$R = 180{,}000\left[\frac{1 - (1.0057905)^{-240}}{0.0057905}\right]^{-1} \approx 1389.99$$

The family made regular monthly payments of \$1389.99. Next, we calculate the unpaid loan balance in 2006, when only 9 years are left to amortize the house. We use Formula (19), where $R = 1389.99$, $m = 12$, $r = 0.0694846$, $i = \dfrac{0.0694863}{12} \approx 0.0057905$, $t = 9$, $n = m \cdot t = 12 \cdot 9 = 108$ and P is the amount we are looking for. We compute

$$P = 1389.99\left[\frac{1 - (1.0057905)^{-108}}{0.0057905}\right] \approx 155{,}139.96$$

In 2006, the principal is \$155,139.96. Finally, we can calculate the equity

$$330{,}000 - 155{,}139.96 = 174{,}860.04$$

Therefore, the equity of the family's house was \$174,860.04 in 2006.

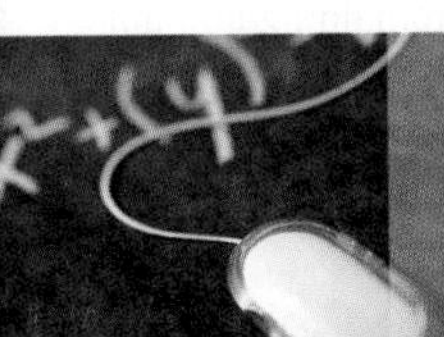

Self-Check Exercises 12.4

1. A kitchen costs \$8000. After a down payment of \$2500, the remaining amount is amortized with 36 equal monthly payments at 4.95% per annum compounded monthly.
   a. Find the amount of each payment.
   b. Find the total amount paid for the kitchen.
   c. Find the total amount of interest paid.
   d. Find the amount that is still owed after the 24th and 25th payments.
   e. Find the interest portion paid with the 25th payment. Can you find two ways to calculate this interest?

Solutions to Self-Check Exercises 12.4 can be found on page 745.

12.4 Exercises

A calculator is recommended for these exercises. Assume that the decreasing annuity is ordinary unless otherwise clearly stated.

In Exercises 1–6, find the present value P of a decreasing annuity with a nominal annual interest rate r compounded m times per year for a term of t years and regular equal payment R.

1. $R = 1800$, $r = 7.2\%$ compounded monthly, $t = 10$ years
2. $R = 2500$, $r = 6.5\%$ compounded quarterly, $t = 15$ years
3. $R = 75$, $r = 4.6\%$ compounded weekly, $t = 5$ years
4. $R = 45$, $r = 8.14\%$ compounded monthly, $t = 20$ years
5. $R = 18{,}000$, $r = 0.85\%$ compounded semiannually, $t = 7$ years
6. $R = 15{,}000$, $r = 1.4\%$ compounded biweekly, $t = 12$ years

In Exercises 7–12, find the regular equal payment R if the present value P is given of a decreasing annuity with a nominal annual interest rate r compounded m times per year for a term of t years.

7. $P = 18{,}000$, $r = 7.2\%$ compounded weekly, $t = 10$ years
8. $P = 25{,}000$, $r = 6.5\%$ compounded quarterly, $t = 15$ years

9. $P = 7500$, $r = 4.6\%$ compounded quarterly, $t = 5$ years

10. $P = 4500$, $r = 8.14\%$ compounded monthly, $t = 20$ years

11. $P = 150{,}000$, $r = 0.85\%$ compounded biweekly, $t = 7$ years

12. $P = 500{,}000$, $r = 1.4\%$ compounded semiannually, $t = 12$ years

In Exercises 13–18, find the term t in years accurate to one decimal place for a decreasing annuity with a nominal annual interest rate *r* compounded *m* times per year and regular equal payments *R* with a present value of *P*.

13. $P = 17{,}500$, $R = 200$, $r = 3\frac{1}{2}\%$ compounded quarterly

14. $P = 31{,}000$, $R = 500$, $r = 2.75\%$ compounded monthly

15. $P = 250{,}000$, $R = 1400$, $r = 3.6\%$ compounded monthly

16. $P = 375{,}000$, $R = 800$, $r = 2.6\%$ compounded biweekly

17. $P = 1{,}000{,}000$, $R = 2000$, $r = 1.96\%$ compounded weekly

18. $P = 1{,}500{,}000$, $R = 40{,}000$, $r = 4.44\%$ compounded quarterly

In Exercises 19–24, calculate the first six lines of the amortization table when a principal value of *P* is amortized over t years at a nominal annual interest rate *r* compounded *m* times per year and regular equal payments *R* whose payment period coincides with the compounding period.

19. $P = 77{,}500$, $t = 10$, $r = 4.25\%$ compounded quarterly

20. $P = 65{,}000$, $t = 15$, $r = 4.73\%$ compounded quarterly

21. $P = 520{,}000$, $t = 20$, $r = 3.9\%$ compounded biweekly

22. $P = 405{,}000$, $t = 25$, $r = 5.4\%$ compounded biweekly

23. $P = 1{,}000{,}000$, $t = 25$, $r = 1.96\%$ compounded monthly

24. $P = 150{,}000$, $t = 20$, $r = 1.25\%$ compounded monthly

Business and Economics Applications

25. **Car Purchase** Kristian is buying a car valued at \$45,000. He pays 10% of the purchase price as a down payment and finances the remaining amount at 2.6% compounded biweekly over a 5-yr period. What are the biweekly payments that Kristian has to deduct from his paycheque?

26. **Canoe Purchase** Aiko is buying a canoe for \$6000. He pays half of the purchase price up front and finances the remaining amount at 4.44% compounded quarterly over a 3-yr period. How much is each payment?

27. **Lottery** Your grandmother wins \$18,000 at a local fundraising lottery for sick children and immediately places the amount into a savings account earning 5.75% compounded monthly. How much money can she withdraw at the end of every month, if she wants the funds to last for 5 yrs?

28. **Inheritance** Liam inherits \$20,000 after taxes. He decides to invest the money into a Guaranteed Investment Account (GIA) paying 6.25% compounded monthly. How much money can he withdraw at the end of every month over a 10-yr term?

29. **Land Purchase** A developer purchased several hectares of land outside Fredericton to build a senior care complex. His purchase is completely amortized at an annual interest rate of 2.87% compounded monthly for 10 yrs. His monthly payments are \$2650.
    a. What is the purchase price for the land?
    b. How much did he pay in interest?

30. **Lottery** John ecstatically tells his parents over the phone that he won the lottery and is receiving \$1000 every week for 20 yrs. In his excitement, he hangs up before his parents find out how much the present value of the lottery win is. If the annual interest rate is 5% compounded weekly, what is the present value of the lottery win?

31. **House Purchase** The Banga family purchases a house for \$225,000 in Saskatoon. To avoid a mortgage loan insurance by CMHC (Canada Mortgage and Housing Corporation), they make a 30% down payment and amortize the remaining amount at 6.24% compounded monthly over a 30-yr period. After 200 payments, the family decides to pay off the mortgage with a lump sum payment. How much have they saved in interest charges?

32. **House Purchase** The Mulholland family purchases a house for \$185,000 in Regina. To avoid a mortgage loan insurance by CMHC (Canada Mortgage and Housing Corporation), they make a 25% down payment and amortize the remaining amount at 4.67% compounded monthly over a 15-yr period. After 70 payments, the family decides to pay off the mortgage with a lump sum payment. How much have they saved in interest charges?

33. **Estate Purchase** A CEO of a business placed in Toronto buys an estate for \$2,500,000 in Markham. She pays a down payment of \$1,000,000 and amortizes the remaining amount at 4.95% compounded semiannually over a 25-yr period with monthly payments.
    a. What is the amortization payment?
    b. What is the total amount of interest the CEO is paying for the house?
    c. After 10 yrs, the CEO makes a prepayment of \$100,000 due to a profitable investment. What are the new amortization payments?
    d. How much money is needed to pay off the house loan with a lump sum payment 5 yrs before the term is up?

34. **Estate Purchase** A couple buys an estate for \$1,900,000 in North Vancouver making a \$650,000 down payment. They amortize the remaining amount at 5.65% compounded semiannually over a 20-yr period with monthly payments.
    a. What is the amortization payment?
    b. What is the total amount of interest the couple is paying for the house?
    c. After 5 yrs, the couple is able to make a prepayment of \$200,000 due to an inheritance. What are the new amortization payments?
    d. Four years before the term is up, the couple decides to pay off the house loan with a lump sum payment. How much must they have saved up to do this?

35. **Decreasing Annuity** Find the present value of a decreasing annuity at an annual interest rate of 3% compounded monthly, if $2000 is due at the end of each month for 4 yrs and $3000 is due at the end of each month for 2 yrs.

36. **Decreasing Annuity** Find the present value of a decreasing annuity at an annual interest rate of 2.55% compounded biweekly, if $500 is due at the end of every two weeks for 20 weeks and $200 is due at the end of every two weeks for 30 weeks.

37. **Retirement** Maryam plans to retire in 30 yrs and starts to make monthly contributions to her retirement account. At the end of 30 yrs, she will begin to withdraw $4000 per month for 20 yrs. Assuming that the annual interest rate is 8.25% compounded monthly, how large are the contributions?

38. **Retirement** Stefan plans to retire in 25 yrs and starts to make biweekly contributions to his retirement account. At the end of 25 yrs, he will begin to withdraw $4500 per month for 20 yrs. Assuming that the annual interest rate is 7.85% compounded monthly, how large are the contributions?

39. **Equity** A Toronto business man purchased a company for $500,000 in 2001. He paid a down payment of $125,000 and amortized the rest with 240 equal monthly payments at 4.62% per year compounded monthly. Over a period of 4 yrs, he refurbishes his company spending an additional $100,000. In 2005 the net worth of the company was $640,000. How much equity did he have in the company?

40. **Equity** Sergio bought a yacht for $300,000. He paid $85,000 up front and financed the rest with 260 equal biweekly payments at 5.13% compounded biweekly. Five years after he bought the yacht there is a vandalism incidence and the net market value of the yacht is estimated at $275,000. How much equity does Sergio have in the yacht?

Solutions To Self-Check Exercises 12.4

1. Notice that the interest is compounded monthly and regular withdrawals are made monthly, so this is a decreasing annuity, where $P = 8000 - 2500 = 5500$, $m = 12$,

$r = 4.95\% = 0.0495$, $i = \dfrac{0.0495}{12} = 0.004125$, and $n = 36$.

a. We use Formula (20) to find the payment amount R. We compute

$$R = 5500\left[\frac{1 - (1.004125)^{-36}}{0.004125}\right]^{-1} \approx 164.72$$

The monthly payments are $164.72.

b. The total amount paid for the kitchen is comprised of the down payment and the 36 monthly payments of $164.72

$$2500 + 36 \cdot 164.72 = 8429.92$$

Therefore, $8429.92 were paid for the kitchen.

c. The total interest paid is the difference between the total paid for the kitchen and the present value of the kitchen at time of the purchase

$$8429.92 - 8000 = 429.92$$

which means that $429.92 was paid in interest.

d. After the 24th payment, there are still 12 payments to make. We use Formula (19), where $R = 164.72$ and $n = 36 - 24 = 12$. We compute the present value P

$$P = 164.72\left[\frac{1 - (1.004125)^{-12}}{0.004125}\right] \approx 1924.65$$

After the 24th payment the amount still owed is $1924.65. We repeat this process to find the amount owed after the 25th payment with $n = 36 - 25 = 11$

$$P = 164.72\left[\frac{1 - (1.004125)^{-11}}{0.004125}\right] \approx 1767.87$$

After the 25th payment the amount still owed is $1767.87.

e. There are two ways to calculate the interest amount that was paid with the 25th payment. In the first way, we think of the interest paid as the difference between the payment and the portion that is applied to the principal. However, the portion applied to the principal must be the difference between the present value after the 24th payment and after the 25th payment, since this is the amount that the principal was reduced by. Using the values computed in part 1d we get

$$1924.65 - 1767.87 = 156.78$$

So, the portion applied to the principal with the 25th payment is $156.78. Next, we subtract this amount from the payment and get

$$\$164.72 - \$156.78 = \$7.94$$

Therefore, the interest portion of the 25th payment is $7.94. In the second way, we recall how interest is calculated in the amortization table, where the previous principal is multiplied by the compounded interest rate to yield the interest portion of the next payment. This means that our calculations are based on the principal after the 24th payment and we get

$$0.004125 \cdot 1924.65 = 7.94$$

which is the same amount we have already calculated for the interest portion of the 25th payment.

CHAPTER 12 Summary of Principal Formulas and Terms

Formulas

1. Simple interest $I = Prt$

2. Future value for simple interest $A = P(1 + rt)$

3. Present value for simple interest $P = \dfrac{A}{(1 + rt)}$

4. Future value for compound interest $A = P\left(1 + \dfrac{r}{m}\right)^n$

5. Present value for compound interest $P = \dfrac{A}{\left(1 + \dfrac{r}{m}\right)^n} = A\left(1 + \dfrac{r}{m}\right)^{-n}$

6. Effective rate of interest $r_{eff} = \left(1 + \dfrac{r}{m}\right)^m - 1$

7. Continuous compound interest $A = Pe^{rt}$

8. Future value of an increasing annuity $A = R\left[\dfrac{(1 + i)^n - 1}{i}\right] = R \cdot s_{\overline{n}|i}$

9. Payment of an increasing annuity $R = \dfrac{A}{\left[\dfrac{(1 + i)^n - 1}{i}\right]} = A\left[\dfrac{(1 + i)^n - 1}{i}\right]^{-1} = \dfrac{A}{s_{\overline{n}|i}}$

10. Present value of a decreasing annuity $P = R\left[\dfrac{1 - (1 + i)^{-n}}{i}\right] = R \cdot a_{\overline{n}|i}$

11. Payment of a decreasing annuity $R = P\left[\dfrac{1 - (1 + i)^{-n}}{i}\right]^{-1} = \dfrac{P}{a_{\overline{n}|i}}$

12. Monthly rate for a semiannually compounded mortgage $r_{monthly} = 12\left[\left(1 + \dfrac{r_{semiannually}}{2}\right)^{\frac{2}{12}} - 1\right]$

Terms

interest (709)
principal (709)
ordinary interest (709)
exact interest (709)
annual percentage rate (709)
nominal annual rate (709)
decimal form (709)
simple interest (709)
future value or maturity value (710)
present value or principal (710)
term (710)
compound interest (714)
compounding period (714)
number of compounding periods (714)
effective rate of interest (true rate) (719)
continuous compound interest (721)
annuity (727)
periods (727)
rent or payments (727)
increasing annuity (727)
decreasing annuity (727)
annuity due (727)
ordinary annuity (727)
sinking fund (732)
loan (735)
mortgage (737)
down payment (737)
lump sum payment (737)
amortized (739)
amortization payments (739)
amortization table (740)
prepayment (740)
equity (742)

Review Exercises 

 **A calculator is recommended for these exercises.**

Assume that a year has 360 days. Assume that the annuity is ordinary unless otherwise clearly stated.

In Exercises 1–26, find the amount indicated.

1. $P = 5000$, $r = 3.4\%$, $t = 120$ days, simple interest
2. $P = 13{,}500$, $r = 4.75\%$, $t = 300$ days, simple interest
3. $P = 12{,}000$, $r = 7.1\%$, $t = 5$ years, future value for simple interest
4. $P = 7500$, $r = 5.2\%$, $t = 8$ years, future value for simple interest
5. $A = 2500$, $r = 7.2\%$, $t = 10$ years, present value for simple interest
6. $A = 18{,}000$, $r = 6.15\%$, $t = 7$ years, present value for simple interest
7. $P = 8500$, $A = 17{,}000$, $r = 3.85\%$ simple interest, term
8. $P = 10{,}000$, $A = 12{,}000$, $r = 5.44\%$ simple interest, term
9. $P = 12{,}000$, $r = 7.1\%$ compounded weekly, $t = 5$ years, future value for compound interest
10. $P = 7500$, $r = 5.2\%$ compounded monthly, $t = 8$ years, future value for compound interest
11. $A = 2500$, $r = 7.2\%$ compounded quarterly, $t = 10$ years, present value for compound interest
12. $A = 18{,}000$, $r = 6.15\%$ compounded biweekly, $t = 7$ years, present value for compound interest
13. $P = 8500$, $A = 17{,}000$, $r = 3.85\%$ compounded monthly, term
14. $P = 10{,}000$, $A = 12{,}000$, $r = 5.44\%$ compounded weekly, term
15. $R = 500$, $r = 2.5\%$ compounded quarterly, $t = 12$ years, future value of an increasing annuity
16. $R = 4000$, $r = 4.21\%$ compounded monthly, $t = 8$ years, future value of an increasing annuity
17. $A = 25{,}000$, $r = 7.25\%$ compounded weekly, $t = 15$ years, payment of an increasing annuity
18. $A = 340{,}000$, $r = 3.65\%$ compounded quarterly, $t = 18$ years, payment of an increasing annuity
19. $R = 300$, $A = 15{,}000$, $r = 1.94\%$ compounded quarterly, term of an increasing annuity
20. $R = 7000$, $A = 120{,}000$, $r = 4.83\%$ compounded monthly, term of an increasing annuity
21. $R = 500$, $r = 2.5\%$ compounded quarterly, $t = 12$ years, present value of a decreasing annuity
22. $R = 4000$, $r = 4.21\%$ compounded monthly, $t = 8$ years, present value of a decreasing annuity
23. $P = 25{,}000$, $r = 7.25\%$ compounded weekly, $t = 15$ years, payment of a decreasing annuity
24. $P = 340{,}000$, $r = 3.65\%$ compounded quarterly, $t = 18$ years, payment of a decreasing annuity
25. $R = 300$, $P = 15{,}000$, $r = 1.94\%$ compounded quarterly, term of a decreasing annuity
26. $R = 7000$, $P = 120{,}000$, $r = 4.83\%$ compounded monthly, term of a decreasing annuity

Business and Economics Applications

27. **Canada Savings Bonds** Karen purchases Canada Savings Bond (CSB) valued at \$2500 earning simple interest at an annual interest rate of 1.75%. Assuming that the interest rate does not change for the next 2 yrs, how much have Karen's CSB earned in interest during that time?
28. **Guaranteed Investment Certificate** Jonas invests \$5000 in a long term GIC offered by his local bank. The GIC earns simple interest at an annual interest rate of 3.65%. How much is Jonas receiving in interest after 5 yrs?
29. **Investment Options** Investment A offers a 4.05% return compounded semiannually, and investment B offers a 4% return compounded daily. Which investment has a higher return over a 10-yr period?
30. **Savings Accounts** A bank offers two savings accounts. Account A earns 3.75% annual interest compounded weekly, and account B earns 3.78% annual interest compounded monthly. Which account earns more interest on a deposit of \$1500 over a 5-yr period?
31. **Investment Analysis** Find the nominal interest rate correct to four decimal places that, when compounded quarterly, yields an effective interest rate of 5% per year.
32. **Investment Analysis** Find the effective rate correct to four decimal places of interest if an investment earns 4.5% interest per annum compounded monthly.
33. **Renovation Cost** The finance committee of a private school sets up a sinking fund with a local bank to accumulate money for future renovations. If \$5000 is deposited monthly earning 6.05% compounded monthly, what will be the value of the sinking fund in 5 yrs? How much interest was earned?
34. **Upgrade Cost** A dental office needs to secure money to be able to upgrade their equipment in the future. The dentist sets up a sinking fund with her local bank and deposits \$2000 every month into an account that pays 3.15% compounded monthly. How much money will have accumulated over a 6-yr term? How much interest was earned?
35. **Retirement Plans** How much money should you deposit every week into an account earning 5.45% compounded weekly, if you would like to have \$100,000 available for your retirement after a 15-yr period of deposits?
36. **Car Purchase** How much money should you deposit every three months into an account earning 3.95% compounded quarterly, if you would like to have \$60,000 available to purchase an SUV after a 5-yr period of deposits?

37. **Annuity Due** An increasing annuity due is an annuity where payments are made at the beginning of a payment period and are calculated using the following formula:

$$A = R\left[\frac{(1+i)^n - 1}{i}\right] - R = R(s_{\overline{n}|i} - 1).$$

If payments of \$400 are made at the beginning of each quarter for 10 yrs into an account paying 2.89% compounded quarterly, find the future value and interest earned.

38. **Annuity Due** Refer to the previous exercise. If payments of \$1500 are made at the beginning of each month for 15 yrs into an account paying 4.2% compounded monthly, find the future value and interest earned.

39. **Retirement** Eli plans to retire 22 yrs from now and starts to make monthly contributions to a GIA paying an annual interest rate of 7.94% compounded monthly. At the end of 22 yrs, he withdraws the whole amount and places it into a retirement account paying an annual interest rate of 8.5% compounded semiannually. In addition, he will begin to withdraw \$10,000 every six months in travel money for 8 yrs. How large are the contributions?

40. **Retirement** Manuela plans to retire 24 yrs from now and starts to make biweekly contributions to a GIA paying an annual interest rate of 6.47% compounded biweekly. At the end of 24 yrs, she withdraws the whole amount and places it into a retirement account paying an annual interest rate of 7.83% compounded monthly. In addition, she will begin to withdraw \$2000 every month in spending money for 15 yrs. How large are the contributions?

41. **Amortization Table** A loan for \$500,000 is amortized at 5.5% interest per year compounded biweekly over a 13-yr term. Prepare the first six lines of the amortization table.

42. **Amortization Table** A loan for \$600,000 is amortized at 2.85% interest per year compounded weekly over a 16-yr term. Prepare the first six lines of the amortization table.

43. **House Purchase** A couple buys a house for \$385,000 in Halifax. Since they are only making a 15% down payment, they require a mortgage loan insurance. The company the couple decides to get insurance from is asking for a 1.75% premium payable in the form of a lump sum when the deal is closed. They amortize the remaining amount at 4.75% compounded semiannually over a 25-yr period with monthly payments.
   a. How much is the couple paying up front for the house?
   b. What is the amortization payment?
   c. What is the total amount the couple is paying for the house?
   d. After 36 payments, the couple is able to make a prepayment of \$20,000. What are the new amortization payments?
   e. What amount is needed 5 yrs before the term is up to pay off the house loan with a lump sum payment?

44. **House Purchase** A single woman buys a house for \$240,000 in Winnipeg. She is only making a 10% down payment, and therefore requires a mortgage loan insurance. Her insurance company is asking for a 2% premium payable in the form of a lump sum when the deal is closed. The remaining amount is amortized at 5.64% compounded semiannually over a 20-yr period with monthly payments.
   a. How much is the woman paying up front for the house?
   b. What is the amortization payment?
   c. What is the total amount the woman is paying for the house?
   d. After 60 payments, the woman is able to make a prepayment of \$55,000. What are the new amortization payments?
   e. What amount is needed 5 yrs before the term is up to pay off the house loan with a lump sum payment?

45. **Equity** Refer to Exercise 43. After 20 yrs from the purchase date, the net worth of the house doubled from its original purchase price. How much equity did the couple have in the house?

46. **Equity** Refer to Exercise 44. After 15 yrs from the purchase date, the net worth of the house had increased by 35% from its original purchase price. How much equity did the woman have in the house?

13

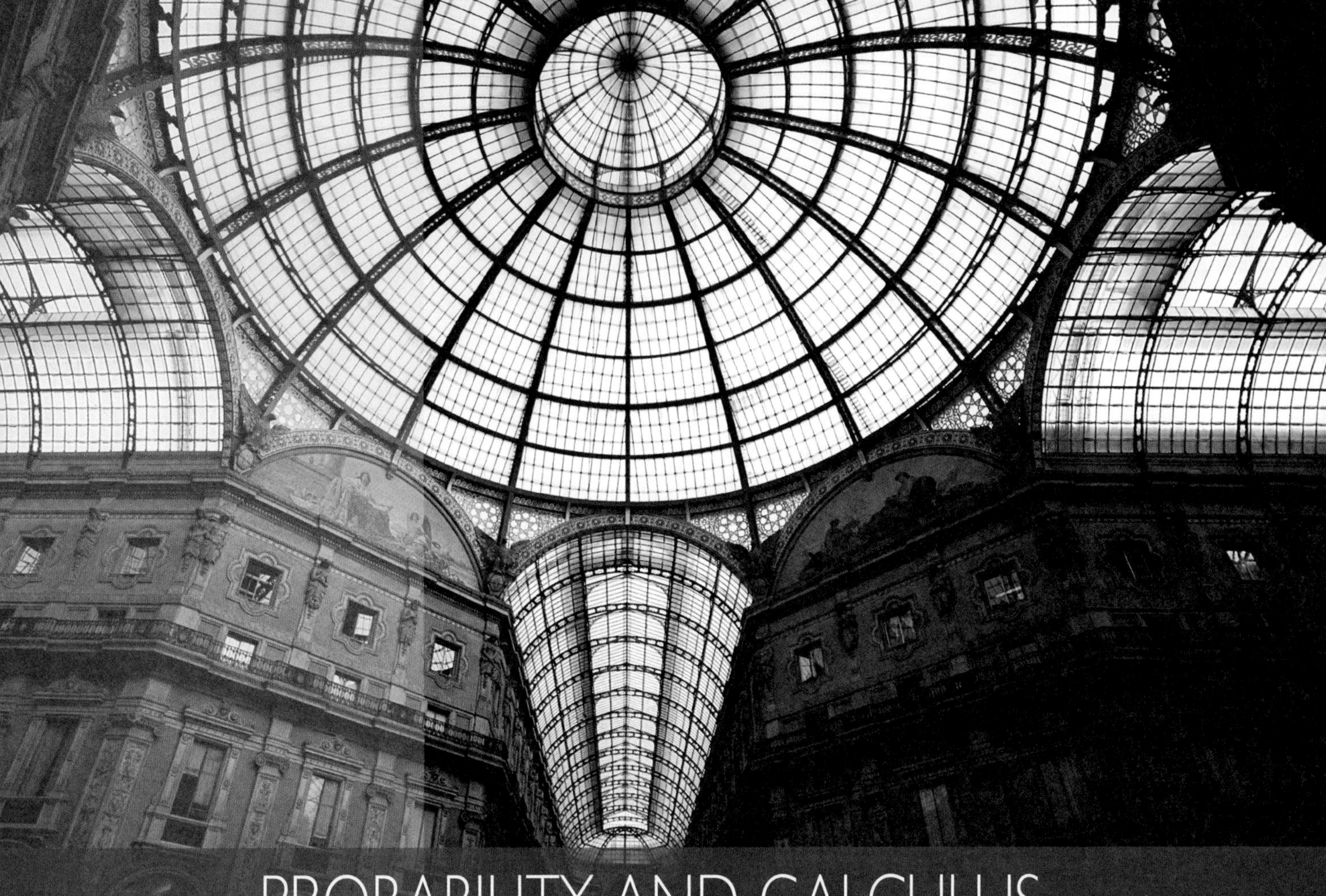

PROBABILITY AND CALCULUS

Photo: Thinkstock/Jupiter Images

The systematic study of probability began in the seventeenth century, when certain aristocrats wanted to discover superior strategies to use in the gaming rooms of Europe. Some of the best mathematicians of the period were engaged in this pursuit. Since then, applications of probability have evolved in virtually every sphere of human endeavour that contains an element of uncertainty.

In this chapter we take a look at the role of calculus in the study of *probability* involving a *continuous random variable*. We see how probability can be used to find the average life span of a certain brand of colour television tube, the average waiting time for patients in a health clinic, and the percentage of a current Mediterranean population who have serum cholesterol levels between 160 and 180 mg/dL—to name but a few applications.

Photo: Thinkstock/Jupiter Images

Where do we go from here? Some of the top 10% of this senior class will further their education at one of the nearby universities. In Example 6, page 779, we determine the minimum grade point average a senior needs to be eligible for admission to one of the nearby universities.

Photo: © Toronto Star Archives

13.1 Probability Distributions of Random Variables

Probability

We begin by mentioning some elementary notations important in the study of probability. For our purpose, an **experiment** is an activity with observable results called **outcomes,** or **sample points.** The totality of all outcomes is the **sample space** of the experiment. A subset of the sample space is called an **event** of the experiment.

Now, given an event associated with an experiment, our primary objective is to determine the likelihood that this event will occur. This likelihood, or **probability of the event,** is a number between 0 and 1 and may be viewed as the proportionate number of times that the event will occur if the experiment associated with the event is repeated indefinitely under independent and similar conditions. By way of example, let's consider the simple experiment of tossing an unbiased coin and observing whether it lands "heads" (H) or "tails" (T). Since the coin is *unbiased,* we see that the probability of each outcome is $\frac{1}{2}$, abbreviated

$$P(\text{H}) = \frac{1}{2} \qquad \text{and} \qquad P(\text{T}) = \frac{1}{2}$$

Discrete Random Variables

In many situations it is desirable to assign numerical values to the outcomes of an experiment. For example, suppose an experiment consists of casting a die and observing the face that lands up. If we let X denote the outcome of the experiment, then X assumes one of the values 1, 2, 3, 4, 5, or 6. Because the values assumed by X depend on the outcomes of a chance experiment, the outcome X is referred to as a **random variable.** In this case the random variable X is also said to be **finite discrete** since it can assume only a finite number of integer values.

The function P that associates with each value of a random variable its probability of occurrence is called a *probability function.* It is *discrete* since its domain consists of a finite set. In general, a discrete probability function is defined as follows.

Discrete Probability Function

A **discrete probability function** P with domain $[x_1, x_2, \ldots, x_n]$ satisfies these conditions:

1. $0 \le P(x_i) \le 1$

2. $P(x_1) + P(x_2) + \cdots + P(x_n) = 1$

Remark

These conditions imply that the probability assigned to an outcome must be nonnegative and less than or equal to 1 and that the sum of all probabilities must be 1. ◀

Histograms

A discrete probability function or distribution may be exhibited graphically by means of a histogram. To construct a histogram of a probability distribution, first locate the values of the random variable on a number line. Then, above each such number, erect a rectangle with width 1 and height equal to the probability associated with that value of the random variable.

For example, consider the data in Table 13.1, the number of cars observed waiting in line at 2-minute intervals between 3 and 5 P.M. on a certain Friday at the drive-in ABM of the Westwood Savings Bank and the corresponding frequency of occurrence. If we divide each number on the right of the table by 60 (the sum of these numbers), then we obtain the respective probabilities associated with the random variable X, when X assumes the values 0, 1, 2, . . . , 8. For example,

$$P(X = 0) = \frac{2}{60} \approx .03$$

$$P(X = 1) = \frac{9}{60} = .15, \ldots$$

The resulting probability distribution is shown in Table 13.2.

The histogram associated with this probability distribution is shown in Figure 13.1 on this page.

Observe that the area of a rectangle in a histogram is associated with a value of a random variable X and that this area gives precisely the probability associated with that value of X. This follows since each rectangle, by construction, has width 1 and height corresponding to the probability associated with the value of the random variable.

Another consequence arising from the method of construction of a histogram is that the probability associated with more than one value of the random variable X is given by the sum of the areas of the rectangles associated with those values of X. For example, the probability that three or four cars are in line is given by

$$P(X = 3) + P(X = 4)$$

which may be obtained from the histogram by adding the areas of the rectangles associated with the values 3 and 4 of the random variable X. Thus, the required probability is

$$P(X = 3) + P(X = 4) = .20 + .13 = .33$$

Table 13.1

Number of Cars	Frequency of Occurrence
0	2
1	9
2	16
3	12
4	8
5	6
6	4
7	2
8	1

Table 13.2
Probability Distribution for the Random Variable X

x	$P(X = x)$
0	.03
1	.15
2	.27
3	.20
4	.13
5	.10
6	.07
7	.03
8	.02

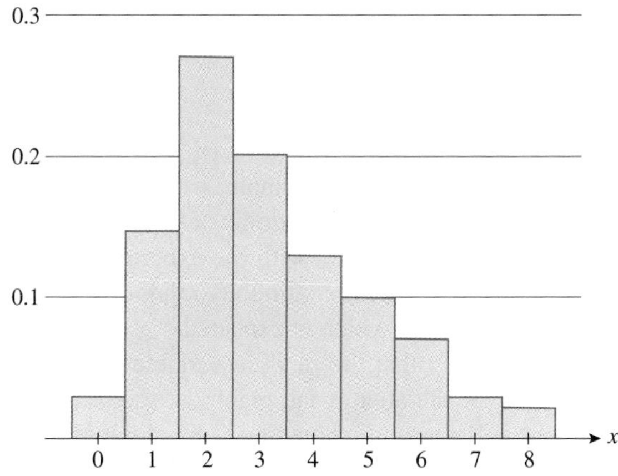

Figure 13.1
Probability distribution of the number of cars waiting in line

Continuous Random Variables

A random variable x that can assume any value in an interval is called a **continuous random variable.** Examples of continuous random variables are the life span of a light bulb, the length of a telephone call, the length of an infant at birth, the daily amount of rainfall in Halifax, and the life span of a certain plant species. For the remainder of this chapter, we will be interested primarily in continuous random variables.

Consider an experiment in which the associated random variable x has the interval $[a, b]$ as its sample space. Then an event of the experiment is any subset of $[a, b]$. For example, if x denotes the life span of a light bulb, then the sample space associated with the experiment is $[0, \infty)$, and the event that a light bulb selected at random has a life span between 500 and 600 hours inclusive is described by the interval $[500, 600]$ or, equivalently, by the inequality $500 \leq x \leq 600$. The probability that the light bulb will have a life span of between 500 and 600 hours is denoted by $P(500 \leq x \leq 600)$.

In general, we will be interested in computing $P(a \leq x \leq b)$, the probability that a random variable x assumes a value in the interval $a \leq x \leq b$. This computation is based on the notion of a probability density function, which we now introduce.

Probability Density Function

A **probability density function** of a random variable x is a nonnegative function f having the following properties:

1. The total area of the region under the graph of f is equal to 1 (see Figure 13.2a).
2. The probability that an observed value of the random variable x lies in the interval $[a, b]$ is given by

$$P(a \leq x \leq b) = \int_a^b f(x)dx$$

(see Figure 13.2b).

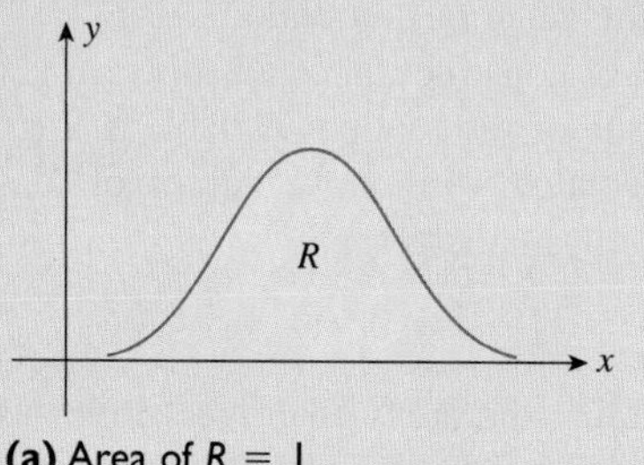

(a) Area of $R = 1$

(b) $P(a \leq x \leq b)$ is the probability that an outcome of an experiment will lie between a and b.

Figure 13.2

Remarks

1. A probability density function of a random variable x may be constructed using methods that range from theoretical considerations of the problem on the one extreme to an interpretation of data associated with the experiment on the other.
2. Property 1 states that the probability that a continuous random variable takes on a value lying in its range is 1, a certainty, which is expected.
3. Property 2 states that the probability that the random variable x assumes a value in an interval $a \leq x \leq b$ is given by the area of the region between the graph of f and the x-axis from $x = a$ to $x = b$. Because the area under one point of the graph of f is equal to zero, we see immediately that $P(a \leq x \leq b) = P(a < x \leq b) = P(a \leq x < b) = P(a < x < b)$. ◀

Example 1

Show that each of the following functions satisfies the nonnegativity condition and Property 1 of probability density functions.

a. $f(x) = \frac{2}{27}x(x - 1) \qquad (1 \le x \le 4)$

b. $f(x) = \frac{1}{3}e^{(-1/3)x} \qquad (0 \le x < \infty)$

Solution

a. Since the factors x and $(x - 1)$ are both nonnegative, we see that $f(x) \ge 0$ on $[1, 4]$. Next, we compute

$$\begin{aligned}\int_1^4 \frac{2}{27}x(x-1)\,dx &= \frac{2}{27}\int_1^4 (x^2 - x)\,dx \\ &= \frac{2}{27}\left(\frac{1}{3}x^3 - \frac{1}{2}x^2\right)\Big|_1^4 \\ &= \frac{2}{27}\left[\left(\frac{64}{3} - 8\right) - \left(\frac{1}{3} - \frac{1}{2}\right)\right] \\ &= \frac{2}{27}\left(\frac{27}{2}\right) \\ &= 1\end{aligned}$$

showing that Property 1 of probability density functions holds as well.

b. First, $f(x) = \frac{1}{3}e^{(-1/3)x} \ge 0$ for all values of x in $[0, \infty)$. Next,

$$\begin{aligned}\int_0^\infty \frac{1}{3}e^{(-1/3)x}dx &= \lim_{b\to\infty}\int_0^b \frac{1}{3}e^{(-1/3)x}dx \\ &= \lim_{b\to\infty} - e^{(-1/3)x}\Big|_0^b \\ &= \lim_{b\to\infty}(-e^{(-1/3)b} + 1) \\ &= 1\end{aligned}$$

so the area under the graph of $f(x) = \frac{1}{3}e^{(-1/3)x}$ is equal to 1, as we set out to show.

Example 2

a. Determine the value of the constant k such that the function $f(x) = kx^2$ is a probability density function on the interval $[0, 5]$.

b. If x is a continuous random variable with the probability density function given in part (a), compute the probability that x will assume a value between $x = 1$ and $x = 2$.

c. Find the probability that x will assume a value at $x = 3$.

Solution

a. We compute

$$\begin{aligned}\int_0^5 kx^2\,dx &= k\int_0^5 x^2\,dx \\ &= \frac{k}{3}x^3\Big|_0^5 \\ &= \frac{125}{3}k\end{aligned}$$

Since this value must be equal to 1, we find that $k = \frac{3}{125}$.

b. The required probability is given by

$$\begin{aligned} P(1 \le x \le 2) &= \int_1^2 f(x)\,dx \\ &= \int_1^2 \frac{3}{125}x^2\,dx \\ &= \frac{1}{125}x^3 \Big|_1^2 \\ &= \frac{1}{125}(8 - 1) \\ &= \frac{7}{125} \end{aligned}$$

The graph of the probability density function f and the area corresponding to the probability $P(1 \le x \le 2)$ are shown in Figure 13.3.

c. The required probability is given by

$$\begin{aligned} P(x = 3) &= \int_3^3 f(x)\,dx \\ &= \int_3^3 \frac{3}{125}x^2\,dx = 0 \end{aligned}$$

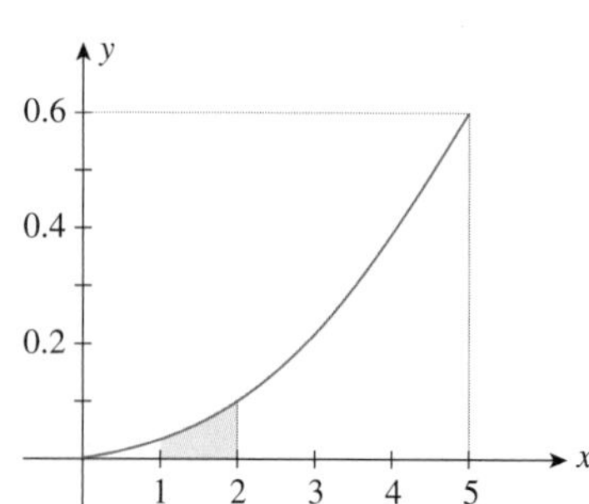

Figure 13.3

$P(1 \le x \le 2)$ for the probability density function $y = \frac{3}{125}x^2$

Remark

Observe that, in general, the probability that x will assume a value at a point $x = a$ is zero since $P(x = a) = \int_a^a f(x)\,dx = 0$ by Property 1 of the definite integral (page 483). ◀

Example 3

Life Span of Light Bulbs TKK Products manufactures a 200-watt electric light bulb. Laboratory tests show that the life spans of these light bulbs have a distribution described by the probability density function

$$f(x) = 0.001e^{-0.001x}$$

Determine the probability that a light bulb will have each of these life spans:

a. 500 hours or less

b. More than 500 hours

c. More than 1000 hours but less than 1500 hours

Solution

Let x denote the life span of a light bulb.

a. The probability that a light bulb will have a life span of 500 hours or less is given by

$$\begin{aligned} P(0 \le x \le 500) &= \int_0^{500} 0.001e^{-0.001x}\,dx \\ &= -e^{-0.001x} \Big|_0^{500} \\ &= -e^{-0.5} + 1 \\ &\approx .3935 \end{aligned}$$

b. The probability that a light bulb will have a life span of more than 500 hours is given by

$$
\begin{aligned}
P(x > 500) &= \int_{500}^{\infty} 0.001e^{-0.001x}\,dx \\
&= \lim_{b\to\infty} \int_{500}^{b} 0.001e^{-0.001x}\,dx \\
&= \lim_{b\to\infty} -e^{-0.001x}\Big|_{500}^{b} \\
&= \lim_{b\to\infty} (-e^{-0.001b} + e^{-0.5}) \\
&= e^{-0.5} \approx .6065
\end{aligned}
$$

This result may also be obtained by observing that

$$
\begin{aligned}
P(x > 500) &= 1 - P(x) \le 500) \\
&= 1 - .3935 \qquad \text{Using the result from part (a)} \\
&\approx .6065
\end{aligned}
$$

c. The probability that a light bulb will have a life span of more than 1000 hours but less than 1500 hours is given by

$$
\begin{aligned}
P(1000 < x < 1500) &= \int_{1000}^{1500} 0.001e^{-0.001x}dx \\
&= -e^{-0.001x}\Big|_{1000}^{1500} \\
&= -e^{-1.5} + e^{-1} \\
&\approx -.2231 + 0.3679 \\
&= .1448
\end{aligned}
$$

The probability density function of Example 3 has the form

$$f(x) = ke^{-kx}$$

where $x \ge 0$ and k is a positive constant. Its graph is shown in Figure 13.4. This probability function is called an **exponential density function,** and the random variable associated with it is said to be **exponentially distributed.** Exponential random variables are used to represent the life span of electronic components, the duration of telephone calls, the waiting time in a doctor's office, and the time between successive flight arrivals and departures in an airport, to mention but a few applications.

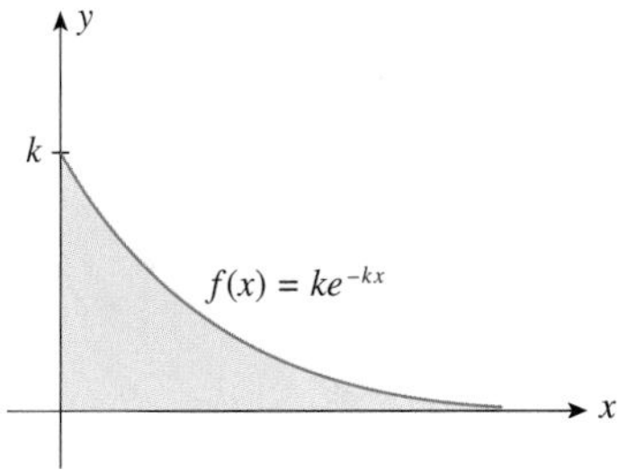

y
k
$f(x) = ke^{-kx}$
x

Figure 13.4
The area under the graph of the exponential density function is equal to 1.

Joint Probability Density Functions

Sometimes the outcomes of an experiment are associated with more than one random variable. For example, we might be interested in the relationship between the weight and height of newborn infants or the price and tread life of automobile tires. To study such problems, we need to extend the concept of a probability density function of a random variable to functions of more than one variable. In the case involving two random variables, we have the following definition.

Joint Probability Density Function

A joint probability density function of the variables x and y on D is a nonnegative function (x, y) having the following properties:

1. $\iint_D f(x, y)\,dA = 1$

Thus, the volume of the solid under the graph of f is equal to 1.

2. The probability that the observed values of the random variables x and y lie in a region $R \subset D$ given by

$$P[(x, y) \text{ in } R] = \iint_R f(x, y)\, dA$$

Example 4

Show that the function $f(x, y) = xy$ is a joint probability density function on $D = \{(x, y) | 0 \le x \le 1; 0 \le y \le 2\}$.

Solution

First, observe that $f(x, y) = xy$ is nonnegative on D. Next, we compute

$$\iint_D f(x, y)\, dA = \int_0^2 \int_0^1 xy\, dx\, dy$$

$$= \int_0^2 \left(\frac{1}{2}x^2y \Big|_0^1 \right) dy$$

$$\int_0^2 \frac{1}{2}y\, dy = \frac{1}{4}y^2 \Big|_0^2 = 1$$

Therefore, f is a joint probability density function on D.

Example 5

Let $f(x, y) = xy$ be a joint probability density function on $D = \{(x, y) | 0 \le x \le 1; 0 \le y \le 2\}$. Find (a) $P(0 \le x \le \frac{1}{2}; 1 \le y \le 2)$ and (b) $P(\{(x, y) | x + y \le 1\})$.

Solution

a. The required probability is given by

$$P\left(0 \le x \le \frac{1}{2}; 1 \le y \le 2\right) = \int_1^2 \int_0^{1/2} xy\, dx\, dy$$

$$= \int_1^2 \left(\frac{1}{2}x^2y \Big|_0^{1/2} \right) dy$$

$$= \int_1^2 \frac{1}{8}y\, dy = \frac{1}{16}y^2 \Big|_1^2 = \frac{3}{16}$$

b. The region $R = \{(x, y) | x + y \le 1\}$ is shown in Figure 13.5. The required probability is

$$P(\{(x, y) | x + y \le 1\}) = \int_0^1 \int_0^{1-x} xy\, dy\, dx$$

$$= \int_0^1 \left(\frac{1}{2}xy^2 \Big|_0^{1-x} \right) dx$$

$$= \int_0^1 \frac{1}{2}x(1 - x)^2 dx$$

$$= \frac{1}{2} \int_0^1 (x - 2x^2 + x^3) dx$$

$$= \frac{1}{2}\left(\frac{1}{2}x^2 - \frac{2}{3}x^3 + \frac{1}{4}x^4\right)\Big|_0^1$$

$$= \frac{1}{2}\left(\frac{1}{2} - \frac{2}{3} + \frac{1}{4}\right) = \frac{1}{24}$$

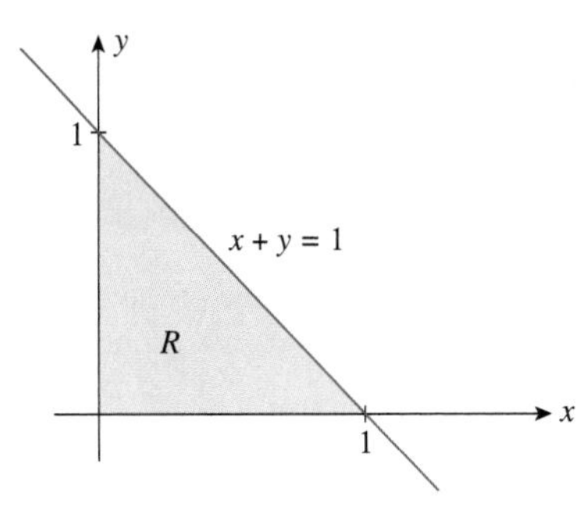

Figure 13.5

Example 6

Let $f(x, y) = 2e^{-x-2y}$ be a joint probability density function on $D = \{(x, y) | x \geq 0$ and $y \geq 0\}$.

a. Find the probability that x will assume a value between 0 and 1 and that y will assume a value between 1 and 2.

b. Find the probability that x will assume a value greater than 1 and that y will assume a value less than 2.

Solution

a. The required probability is given by

$$P(0 \leq x \leq 1; 1 \leq y \leq 2) = \int_1^2 \int_0^1 2e^{-x-2y} dx\, dy$$

$$= \int_1^2 \left(-2e^{-x-2y}\Big|_0^1\right) dy$$

$$= \int_1^2 (-2e^{-1-2y} + 2e^{-2y})\, dy$$

$$= \int_1^2 2(1 - e^{-1})e^{-2y} dy \qquad \text{Simplify the integrand.}$$

$$= 2(1 - e^{-1})\left(-\frac{1}{2}e^{-2y}\right)\Big|_1^2$$

$$= -(1 - e^{-1})e^{-2y}\Big|_1^2$$

$$= -(1 - e^{-1})(e^{-4} - e^{-2}) \approx .0740$$

b. The required probability is given by

$$P(x \geq 1; 0 \leq y \leq 2) = \int_0^2 \int_1^\infty 2e^{-x-2y}\, dx\, dy$$

$$= \int_0^2 \left(\lim_{b\to\infty} \int_1^b 2e^{-x-2y}\, dx\right) dy$$

$$= \int_0^2 \left(\lim_{b\to\infty} -2e^{-x-2y}\Big|_1^b\right) dy$$

$$= \int_0^2 [\lim_{b\to\infty} (-2e^{-b-2y} + 2e^{-1-2y})]\, dy$$

$$= \int_0^2 2e^{-1-2y}\, dy$$

$$= -e^{-1-2y}\Big|_0^2 = -e^{-5} + e^{-1} \approx .3611$$

Self-Check Exercises 13.1

1. Determine the value of the constant k such that the function $f(x) = k(4x - x^2)$ is a probability density function on the interval $[0, 4]$.

2. Suppose that x is a continuous random variable with the probability density function of Self-Check Exercise 1. Find the probability that x will assume a value between $x = 1$ and $x = 3$.

Solutions to Self-Check Exercises 13.1 can be found on page 760.

13.1 Exercises

In Exercises 1–12, show that the function is a probability density function on the indicated interval.

1. $f(x) = \frac{1}{3}$; $(3 \le x \le 6)$

2. $f(x) = \frac{1}{\pi}$; $(\pi \le x \le 2\pi)$

3. $f(x) = \frac{2}{32}x$; $(2 \le x \le 6)$

4. $f(x) = \frac{3}{8}x^2$; $(0 \le x \le 2)$

5. $f(x) = \frac{2}{9}(3x - x^2)$; $(0 \le x \le 3)$

6. $f(x) = \frac{3}{32}(x - 1)(5 - x)$; $(1 \le x \le 5)$

7. $f(x) = \frac{12 - x}{72}$; $(0 \le x \le 12)$

8. $f(x) = 20(x^3 - x^4)$; $(0 \le x \le 1)$

9. $f(x) = \frac{8}{7x^2}$; $(1 \le x \le 8)$

10. $f(x) = \frac{3}{14}\sqrt{x}$; $(1 \le x \le 4)$

11. $f(x) = \frac{x}{(x^2 + 1)^{3/2}}$; $(0 \le x \le \infty)$

12. $f(x) = 4xe^{-2x^2}$; $(0 \le x \le \infty)$

In Exercises 13–20, find the value of the constant k so that the function is a probability density function on the indicated interval.

13. $f(x) = k$; $[1, 4]$

14. $f(x) = kx$; $[0, 4]$

15. $f(x) = k(4 - x)$; $[0, 4]$

16. $f(x) = kx^3$; $[0, 1]$

17. $f(x) = k\sqrt{x}$; $[0, 4]$

18. $f(x) = \frac{k}{x}$; $[1, 5]$

19. $f(x) = \frac{k}{x^3}$; $[1, \infty)$

20. $f(x) = ke^{-x/2}$; $[0, \infty)$

In Exercises 21–30, f is a probability density function defined on the given interval. Find the indicated probabilities.

21. $f(x) = \frac{1}{12}x$; $[1, 5]$

a. $P(2 \le x \le 4)$ **b.** $P(1 \le x \le 4)$
c. $P(x \ge 2)$ **d.** $P(x = 2)$

22. $f(x) = \frac{1}{9}x^2$; $[0, 3]$

a. $P(1 \le x \le 2)$ **b.** $P(1 < x \le 3)$
c. $P(x \le 2)$ **d.** $P(x = 1)$

23. $f(x) = \frac{3}{32}(4 - x^2)$; $[-2, 2]$

a. $P(-1 \le x \le 1)$ **b.** $P(x \le 0)$
c. $P(x > -1)$ **d.** $P(x = 0)$

24. $f(x) = \frac{3}{16}\sqrt{x}$; $[0, 4]$

a. $P(1 < x < 3)$ **b.** $P(x \le 2)$
c. $P(x = 2)$ **d.** $P(x \ge 1)$

25. $f(x) = \frac{1}{4\sqrt{x}}$; $[1, 9]$

a. $P(x \ge 4)$ **b.** $P(1 \le x < 8)$
c. $P(x = 3)$ **d.** $P(x \le 4)$

26. $f(x) = \frac{1}{2}e^{-x/2}$; $[0, \infty)$

a. $P(x \le 4)$ **b.** $P(1 < x < 2)$
c. $P(x = 50)$ **d.** $P(x \ge 2)$

27. $f(x) = 4xe^{-2x^2}$; $[0, \infty)$

a. $P(0 \le x \le 4)$ **b.** $P(x \ge 1)$

28. $f(x) = \frac{1}{9}xe^{-x/3}$; $[0, \infty)$

a. $P(0 \le x \le 3)$ **b.** $P(x \ge 1)$

29. $f(x) = \begin{cases} x & \text{if } 0 \le x \le 1 \\ 2 - x & \text{if } 1 \le x \le 2 \end{cases}$; $[0, 2]$

a. $P(\frac{1}{2} \le x \le 1)$ **b.** $P(\frac{1}{2} \le x \le \frac{3}{2})$
c. $P(x \ge 1)$ **d.** $P(x \le \frac{3}{2})$

30. $f(x) = \begin{cases} \dfrac{3}{40}\sqrt{x} & \text{if } 0 \le x \le 4 \\ \dfrac{12}{5x^2} & \text{if } 4 < x < \infty \end{cases}; [0, \infty)$

a. $P(1 \le x \le 4)$ b. $P(0 \le x \le 5)$

In Exercises 31–34, show that the function is a joint probability density function on *D*.

31. $f(x, y) = \frac{1}{4}; D = \{0 \le x \le 2; 1 \le y \le 3\}$

32. $f(x, y) = \frac{1}{4}(x + 2y); D = \{0 \le x \le 2; 0 \le y \le 1\}$

33. $f(x, y) = \frac{1}{3}xy; D = \{0 \le x \le 2; 1 \le y \le 2\}$

34. $f(x, y) = 4(1 - x)(2 - y); D = \{0 \le x \le 1; 1 \le y \le 2\}$

In Exercises 35–38, find the value of the constant *k* so that the function is a joint probability density function on *D*.

35. $f(x, y) = kx^2y; D = \{0 \le x \le 1; 1 \le y \le 2\}$

36. $f(x, y) = k\sqrt{x}(2 - y); D = \{1 \le x \le 2; 0 \le y \le 2\}$

37. $f(x, y) = k(x - x^2)e^{-2y}; D = \{0 \le x \le 1; 1 \le y < \infty\}$

38. $f(x, y) = kxye^{-(x^2+y^2)}; D = \{0 < x < \infty; 0 < y < \infty\}$

In Exercises 39–42, *f* is a joint probability density function on *D*. Find the indicated probabilities.

39. $f(x, y) = xy; D = \{(x, y) | 0 \le x \le 1; 0 \le y \le 2\}$
   a. $P(0 \le x \le 1; 0 \le y \le 1)$ b. $P(\{(x, y) | x + 2y \le 1)$

40. $f(x, y) = \frac{1}{12}(x + y); D = \{(x, y) | 0 \le x \le 2; 1 \le y \le 3\}$
   a. $P(0 \le x \le 1; 1 \le y \le 2)$
   b. $P(\{(x, y) | y \ge 1 \text{ and } x + y \le 3\})$

41. $f(x, y) = \frac{9}{224}\sqrt{xy}; D = \{(x, y) | 1 \le x \le 4; 0 \le y \le 4\}$
   a. $P(1 \le x \le 2; 0 \le y \le 1)$
   b. $P(\{(x, y) | 1 \le x \le 4; 0 \le y \le \sqrt{x}\})$

42. $f(x, y) = 3\sqrt{x}e^{-2y}; D = \{(x, y) | 0 \le x \le 1; 0 \le y < \infty\}$
   a. $P(0 \le x \le 1; 1 \le y < \infty)$
   b. $P(\{(x, y) | \frac{1}{2} \le x \le 1; 0 \le y < \infty\})$

43. **Waiting Time at a Health Clinic** The average waiting time in minutes for patients arriving at the Newtown Health Clinic between 1 and 4 P.M. on a weekday is an exponentially distributed random variable x with associated probability density function $f(x) = \frac{1}{15}e^{-(1/15)x}$.
   a. What is the probability that a patient arriving at the clinic between 1 and 4 P.M. will have to wait longer than 15 min?
   b. What is the probability that a patient arriving at the clinic between 1 and 4 P.M. will have to wait between 10 and 12 min?

44. **Waiting Time at an Expressway Tollbooth** Suppose the time intervals in seconds between arrivals of successive cars at an expressway tollbooth during rush hour are exponentially distributed with associated probability density function $f(t) = \frac{1}{8}e^{-(1/8)t}$. Find the probability that the average time interval between arrivals of successive cars is more than 8 sec.

45. **Reliability of a Computer** The computers manufactured by United Motor Works, which are used in automobiles to regulate fuel consumption, are guaranteed against defects for 30,000 km of use. Tests conducted in the laboratory under simulated driving conditions reveal that the distances driven in kilometers before the computers break down are exponentially distributed with probability density function $f(x) = 0.00001e^{-0.00001t}$. What is the probability that a computer selected at random will fail during the warranty period?

46. **Probability of Snowfall** The amount of snowfall (in meters) in a remote region of the Yukon territory in the month of January is a continuous random variable with probability density function $f(x) = 6x(1 - x); 0 \le x \le 1$. Find the probability that the amount of snowfall will be between $\frac{1}{3}$ m and $\frac{2}{3}$ m; more than $\frac{1}{3}$ m.

Business and Economics Applications

47. **Reliability of Robots** National Welding Company uses industrial robots in some of its assembly-line operations. Management has determined that the lengths of time in hours between breakdowns are exponentially distributed with probability density function $f(t) = 0.001e^{-0.001t}$.
   a. What is the probability that a robot selected at random will break down after between 600 and 800 hr of use?
   b. What is the probability that a robot will break down after 1200 hr of use?

48. **Mail-Order Phone Calls** A study conducted by Uni-Mart, a mail-order department store, reveals that the time intervals in minutes between incoming telephone calls on its toll-free 800 line between 10 A.M. and 2 P.M. are exponentially distributed with probability density function $f(t) = \frac{1}{30}e^{-t/30}$. What is the probability that the time interval between successive calls is more than 2 min?

49. **Number of Chips in Chocolate Chip Cookies** The number of chocolate chips in each cookie of a certain brand has a distribution described by the probability density function

$$f(x) = \frac{1}{36}(6x - x^2) \qquad (0 \le x \le 6)$$

Find the probability that the number of chocolate chips in a randomly chosen cookie is fewer than two.

50. **Life Expectancy of Colour Television Tubes** The life expectancy (in years) of a certain brand of colour television tube is a continuous random variable with probability density function

$$f(t) = 9(9 + t^2)^{-3/2} \qquad (0 \le t < \infty)$$

Find the probability that a randomly chosen television tube will last more than 4 yr.
Hint: Integrate using a table of integrals.

51. **Product Reliability** The tread life (in thousands of kilometres) of a certain make of tire is a continuous random variable with a probability density function

$$f(x) = 0.02e^{-0.02x} \qquad (0 \le x < \infty)$$

a. Find the probability that a randomly selected tire of this make will have a tread life of at most 30,000 km.
b. Find the probability that a randomly selected tire of this make will have a tread life between 40,000 and 60,000 km.
c. Find the probability that a randomly selected tire of this make will have a tread life of at least 70,000 km.

52. **Cable TV Subscribers** The management of Rogers Cable estimates that the number of new subscribers (in thousands) to their service next year in North York, x, and the number of new subscribers (in thousands) next year in Scarborough, y, have a distribution given by the joint probability density function

$$f(x, y) = \frac{9}{4000}xy\sqrt{25 - x^2}(4 - y);$$

$$D = \{(x, y) | 0 \le x \le 5; 0 \le y \le 4\}$$

Find the probability that the number of new subscribers next year in North York will be between 2000 and 2500 and the number of new subscribers in Scarborough will be between 1000 and 2000.

53. **Cable TV Subscribers** Refer to Exercise 52. What is the probability that the total number of new subscribers next year in North York will be fewer than 3000 and the number of new subscribers next year in Scarborough will be fewer than 2000?

Biological and Life Sciences Applications

54. **Life Span of a Plant** The life span of a certain plant species (in days) is described by the probability density function

$$f(x) = \frac{1}{100}e^{-x/100}$$

a. Find the probability that a plant of this species will live for 100 days or less.
b. Find the probability that a plant of this species will live longer than 120 days.
c. Find the probability that a plant of this species will live longer than 60 days but less than 140 days.

In Exercises 55 and 56, determine whether the statement is true or false. If it is true, explain why it is true. If it is false, give an example to show why it is false.

55. If $\int_a^b f(x)dx = 1$, then f is a probability density function on $[a, b]$.

56. If f is a probability function on an interval $[a, b]$, then f is a probability function on $[c, d]$ for any real numbers c and d satisfying $a < c < d < b$.

Solutions to Self-Check Exercises 13.1

1. We compute

$$\int_0^4 k(4x - x^2)dx = k\left(2x^2 - \frac{1}{3}x^3\right)\Bigg|_0^4$$

$$= k\left(32 - \frac{64}{3}\right)$$

$$= \frac{32}{3}k$$

Since this value must be equal to 1, we have $k = \frac{3}{32}$.

2. The required probability is given by

$$P(1 \le x \le 3) = \int_1^3 f(x)dx$$

$$= \int_1^3 \frac{3}{32}(4x - x^2)dx$$

$$= \frac{3}{32}\left(2x^2 - \frac{1}{3}x^3\right)\Bigg|_1^3$$

$$= \frac{3}{32}\left[(18 - 9) - \left(2 - \frac{1}{3}\right)\right] = \frac{11}{16}$$

13.2 Expected Value and Standard Deviation

Expected Value

The average value of a set of numbers is a familiar notion to most people. For example, to compute the average of the four numbers 12, 16, 23, and 37, we simply add these numbers and divide the resulting sum by 4, giving the average as

$$\frac{12 + 16 + 23 + 37}{4} = \frac{88}{4}$$

or 22. In general, we have the following definition.

Average, or Mean

The **average,** or **mean,** of the n numbers $x_1, x_2, \ldots, x_n$ is $\bar{x}$ (read "x bar"), where

$$\bar{x} = \frac{x_1 + x_2 + \cdots + x_n}{n}$$

Example 1

The number of cars observed waiting in line at the beginning of each 2-minute interval between 3 and 5 P.M. on a certain Friday at the drive-in ABM of Westwood Savings Bank and the corresponding frequency of occurrence are shown in Table 13.3. Find the average number of cars observed waiting in line at the beginning of each 2-minute interval during the 2-hour period.

Table 13.3

Cars	0	1	2	3	4	5	6	7	8
Frequency of Occurrence	2	9	16	12	8	6	4	2	1

Solution

Observe from Table 13.3 that the number 0 (of cars) occurs twice, the number 1 occurs nine times, and so on. There are altogether

$$2 + 9 + 16 + 12 + 8 + 6 + 4 + 2 + 1 = 60$$

numbers to be averaged. Therefore, the required average is given by

$$\frac{0\cdot 2 + 1\cdot 9 + 2\cdot 16 + 3\cdot 12 + 4\cdot 8 + 5\cdot 6 + 6\cdot 4 + 7\cdot 2 + 8\cdot 1}{60} \approx 3.1 \quad (1)$$

or approximately 3.1 cars.

Let's reconsider the expression in Equation (1) that gives the average of the frequency distribution shown in Table 13.3. Dividing each term by the denominator, we can rewrite the expression in the form

$$0\cdot\left(\frac{2}{60}\right) + 1\cdot\left(\frac{9}{60}\right) + 2\cdot\left(\frac{16}{60}\right) + 3\cdot\left(\frac{12}{60}\right) + 4\cdot\left(\frac{8}{60}\right) + 5\cdot\left(\frac{6}{60}\right)$$
$$+ 6\cdot\left(\frac{4}{60}\right) + 7\cdot\left(\frac{2}{60}\right) + 8\cdot\left(\frac{1}{60}\right)$$

Observe that each term in the sum is a product of two factors. The first factor is the value assumed by the random variable X, where X denotes the number of cars observed waiting in line, and the second factor is just the probability associated with that value of the random variable. This observation suggests the following general method for calculating the expected value (that is, the average, or mean) of a random variable X that assumes a finite number of values from the knowledge of its probability distribution.

Expected Value of a Discrete Random Variable X

Let X denote a random variable that assumes the values $x_1, x_2, \ldots, x_n$ with associated probabilities $p_1, p_2, \ldots, p_n$, respectively. Then the **expected value** of X, $E(X)$, is given by

$$E(X) = x_1p_1 + x_2p_2 + \cdots + x_np_n \quad (2)$$

Remark

The numbers $x_1, x_2, \ldots, x_n$ may be positive, zero, or negative. For example, such a number will be positive if it represents a profit and negative if it represents a loss. ◀

Table 13.4
Probability Distribution for the Random Variable X

x	$P(X = x)$
0	.03
1	.15
2	.27
3	.20
4	.13
5	.10
6	.07
7	.03
8	.02

Example 2

Refer to Example 1. Let the random variable X denote the number of cars observed waiting in line. Use the probability distribution for the random variable X (Table 13.4) to solve Example 1 again.

Solution

Let X denote the number of cars observed waiting in line. Then, the average number of cars observed waiting in line is given by the expected value of X—that is, by

$$\begin{aligned} E(X) &= 0 \cdot (.03) + 1 \cdot (.15) + 2 \cdot (.27) + 3 \cdot (.20) + 4 \cdot (.13) \\ &\quad + 5 \cdot (.10) + 6 \cdot (.07) + 7 \cdot (.03) + 8 \cdot (.02) \\ &= 3.1 \end{aligned}$$

or 3.1 cars, which agrees with the earlier results.

The expected value of a random variable X is a measure of the central tendency of the probability distribution associated with X. In repeated trials of an experiment with a random variable X, the average of the observed values of X gets closer and closer to the expected value of X as the number of trials gets larger and larger. Geometrically, the expected value of a random variable X has the following simple interpretation: If a laminate is made of the histogram of a probability distribution associated with a discrete random variable X, then the expected value of X corresponds to the point on the base of the laminate at which the latter will balance perfectly when the point is directly over a fulcrum (Figure 13.6).

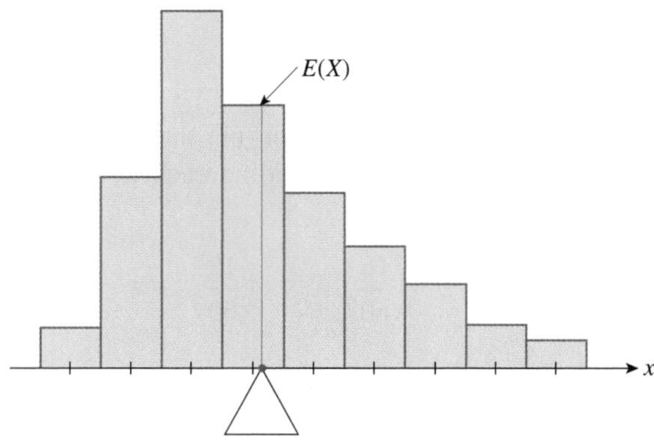

Figure 13.6
Expected value of a random variable X

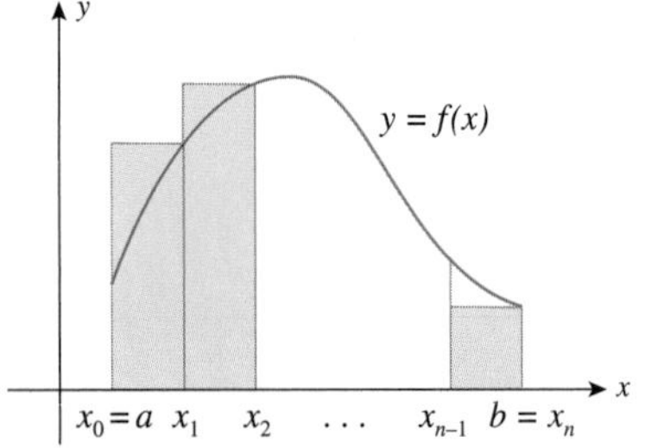

Figure 13.7
Approximating the expected value of a random variable x on $[a, b]$ by a Riemann sum

Now suppose x is a continuous random variable and f is the probability density function associated with it. For simplicity, let's first assume that $a \leq x \leq b$. Divide the interval $[a, b]$ into n subintervals of equal length $\Delta x = (b - a)/n$ by means of the $(n + 1)$ points $x_0 = a, x_1, x_2, \ldots, x_n = b$ (Figure 13.7).

To find an approximation of the average value, or expected value, of x on the interval $[a, b]$, let's treat x as if it were a discrete random variable that takes on the values $x_1, x_2, \ldots, x_n$ with probabilities $p_1, p_2, \ldots, p_n$. Then

$$E(x) \approx x_1 p_1 + x_2 p_2 + \cdots + x_n p_n$$

But p_1 is the probability that x is in the interval $[x_0, x_1]$, and this is just the area under the graph of f from $x = x_0$ to $x = x_1$, which may be approximated by $f(x_1)\,\Delta x$. The probabilities $p_2, \ldots, p_n$ may be approximated in a similar manner. Thus,

$$E(x) \approx x_1 f(x_1)\Delta x + x_2 f(x_2)\Delta x + \cdots + x_n f(x_n)\Delta x$$

which is seen to be the Riemann sum of the function $g(x) = xf(x)$ over the interval $[a, b]$. Letting n approach infinity, we obtain the following formula:

Expected Value of a Continuous Random Variable

Suppose the function f defined on the interval $[a, b]$ is the probability density function associated with a continuous random variable x. Thus, the expected value of x is

$$E(x) = \int_a^b xf(x)\,dx \qquad (3)$$

If either $a = -\infty$ or $b = \infty$, then the integral in (3) becomes an improper integral.

The expected value of a random variable is used in many practical applications. For example, if x represents the life span of a certain brand of electronic component, then the expected value of x gives the average life span of these components. If x measures the waiting time in a doctor's office, then $E(x)$ gives the average waiting time, and so on.

Example 3

Average Life Span of a Light Bulb Show that if a continuous random variable x is exponentially distributed with the probability density function

$$f(x) = ke^{-kx}$$

then the expected value of x, $E(x)$, is equal to $1/k$. Using this result, determine the average life span of a 200-watt electric light bulb manufactured by TKK Products of Example 3, page 754.

Solution

We compute

$$\begin{aligned} E(x) &= \int_0^\infty xf(x)\,dx \\ &= \int_0^\infty kxe^{-kx}dx \\ &= k\lim_{b\to\infty}\int_0^b xe^{-kx}dx \end{aligned}$$

Integrating by parts with

$$u = x \qquad \text{and} \qquad dv = e^{-kx}\,dx$$

so that

$$du = dx \qquad \text{and} \qquad v = -\frac{1}{k}e^{-kx}$$

we have

$$\begin{aligned} E(x) &= k\lim_{b\to\infty}\left(-\frac{1}{k}xe^{-kx}\Big|_0^b + \frac{1}{k}\int_0^b e^{-kx}dx\right) \\ &= k\lim_{b\to\infty}\left[-\left(\frac{1}{k}\right)be^{-kb} - \frac{1}{k^2}e^{-kx}\Big|_0^b\right] \end{aligned}$$

$$= k \lim_{b\to\infty}\left[-\left(\frac{1}{k}\right)be^{-kb} - \frac{1}{k^2}e^{-kb} + \frac{1}{k^2}\right]$$

$$= -\lim_{b\to\infty}\frac{b}{e^{kb}} - \frac{1}{k}\lim_{b\to\infty}\frac{1}{e^{kb}} + \frac{1}{k}\lim_{b\to\infty} 1$$

Now, by taking a sequence of values of b that approaches infinity, for example, $b = 10, 100, 1000, 10{,}000, \ldots$, we see that, for a fixed k,

$$\lim_{b\to\infty}\frac{b}{e^{kb}} = 0$$

Therefore,

$$E(x) = \frac{1}{k}$$

as we set out to show. From our previous work, we know that $k = 0.001$. So the average life span of the TKK light bulb is $1/(0.001) = 1000$ hours.

Before considering another example, let's summarize the result obtained in Example 3.

Average Value of an Exponential Density Function

If a continuous random variable x is exponentially distributed with probability density function

$$f(x) = ke^{-kx}$$

then the expected value of x is given by

$$E = \frac{1}{k}$$

EXPLORING WITH TECHNOLOGY

Refer to Example 3. Plot the graphs of $f(x) = x/e^{kx}$ for different positive values of k, using the appropriate viewing windows to demonstrate that $\lim_{b\to\infty} b/e^{kb} = 0$ for a fixed $k > 0$. Repeat for the function $f(x) = 1/e^{kx}$ to demonstrate that $\lim_{b\to\infty} 1/e^{kb} = 0$.

Example 4

Airplane Arrival Times On a typical Monday morning, the time between successive arrivals of planes at Edmonton International Airport is an exponentially distributed random variable x with expected value of 10 (minutes).

a. Find the probability density function associated with x.

b. What is the probability that between 6 and 8 minutes will elapse between successive arrivals of planes?

c. What is the probability that the time between successive arrivals will be more than 15 minutes?

Solution

a. Since x is exponentially distributed, the associated probability density function has the form $f(x) = ke^{-kx}$. Next, since the expected value of x is 10, we see that

$$E(x) = \frac{1}{k} = 10$$

$$k = \frac{1}{10} = 0.1$$

so the required probability density function is

$$f(x) = 0.1e^{-0.1x}$$

b. The probability that between 6 and 8 minutes will elapse between successive arrivals is given by

$$\begin{aligned} P(6 \leq x \leq 8) &= \int_6^8 0.1e^{-0.1x}dx = -e^{-0.1x}\Big|_6^8 \\ &= -e^{-0.8} + e^{-0.6} \\ &\approx .100 \end{aligned}$$

c. The probability that the time between successive arrivals will be more than 15 minutes is given by

$$\begin{aligned} P(x > 15) &= \int_{15}^{\infty} 0.1e^{-0.1x}dx \\ &= \lim_{b\to\infty} \int_{15}^{b} 0.1e^{-0.1x}dx \\ &= \lim_{b\to\infty} \left[-e^{-0.1x}\Big|_{15}^{b} \right] \\ &= \lim_{b\to\infty} (-e^{-0.1b} + e^{-1.5}) = e^{-1.5} \\ &\approx .22 \end{aligned}$$

Variance

The mean, or expected value, of a random variable enables us to express an important property of the probability distribution associated with the random variable in terms of a single number. But knowing the location, or central tendency, of a probability distribution alone is usually not enough to give a reasonably accurate picture of the probability distribution. Consider, for example, the two probability distributions whose histograms appear in Figure 13.8. Both distributions have the same expected value, or mean, $\mu = 4$ (the Greek letter μ is read "mu"). Note that the probability distribution with the histogram shown in Figure 13.8a is closely concentrated about its mean μ, whereas the one with the histogram shown in Figure 13.8b is widely dispersed, or spread, about its mean.

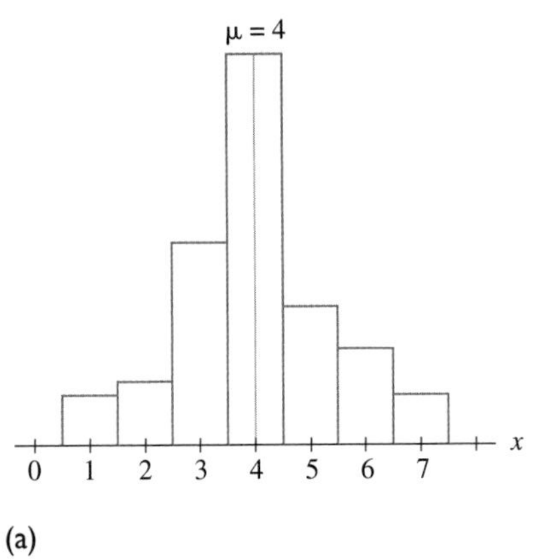

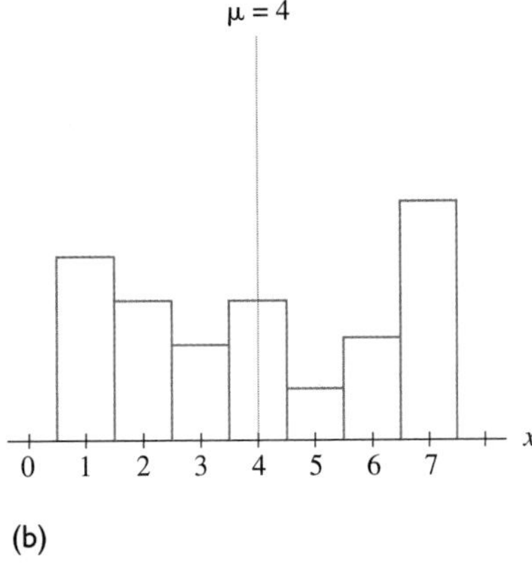

Figure 13.8
The histograms of two probability distributions

Table 13.5

Brand A	Brand B
16.1	16.3
16	15.7
15.8	15.8
16	16.2
15.9	15.9
16.1	16.1
15.9	15.7
16	16.2
16	16
16.2	16.1

As another example, suppose David Horowitz, host of the television show *The Consumer Advocate,* decides to demonstrate the accuracy of the weights of two popular brands of potato chips. Ten packages of potato chips of each brand are selected at random and weighed carefully. The weights in ounces are listed in Table 13.5.

We will verify in Example 7 that the mean weights for each of the two brands is 16 ounces. However, a cursory examination of the data now shows that the weights of the brand B packages exhibit much greater dispersion about the mean than those of brand A.

One measure of the degree of dispersion, or spread, of a probability distribution about its mean is the variance of the random variable associated with the probability distribution. A probability distribution with a small spread about its mean will have a small variance, whereas one with a larger spread will have a larger variance. Thus, the variance of the random variable associated with the probability distribution whose histogram appears in Figure 13.8a is smaller than the variance of the random variable associated with the probability distribution whose histogram is shown in Figure 13.8b. Also, as we will see in Example 7, the variance of the random variable associated with the weights of the brand A potato chips is smaller than that of the random variable associated with the weights of the brand B potato chips.

Variance of a Discrete Random Variable X

Suppose a random variable has the probability distribution

x	x_1	x_2	x_3	. . .	x_n
$P(X = x)$	p_1	p_2	p_3	. . .	p_n

and expected value

$$E(X) = \mu$$

Then the **variance of a random variable** X is

$$\text{Var}\ (X) = p_1(x_1 - \mu)^2 + p_2(x_2 - \mu)^2 + \cdots + p_n(x_n - \mu)^2 \tag{4}$$

Let's look a little closer at Formula (4). First, note that the numbers

$$x_1 - \mu, x_2 - \mu, \ldots, x_n - \mu \tag{5}$$

measure the **deviations** of $x_1, x_2, \ldots, x_n$ from μ, respectively. Thus, the numbers

$$(x_1 - \mu)^2, (x_2 - \mu)^2, \ldots, (x_n - \mu)^2 \tag{6}$$

measure the squares of the deviations of $x_1, x_2, \ldots, x_n$ from μ, respectively. Next, by multiplying each of the numbers in (6) by the probability associated with each value of the random variable X, we weight the numbers accordingly so that their sum is a measure of the variance of X about its mean. An attempt to define the variance of a random variable about its mean in a similar manner using the deviations rather than their squares would not be fruitful. Some of the deviations may be positive whereas others may be negative, and (because of cancellations) the sum will not give a satisfactory measure of the variance of the random variable.

Table 13.6

x	$P(X = x)$	y	$P(Y = y)$
1	.05	1	.20
2	.075	2	.15
3	.2	3	.1
4	.375	4	.15
5	.15	5	.05
6	.1	6	.1
7	.05	7	.25

Example 5

Find the variance of the random variable X and of the random variable Y whose probability distributions are given in Table 13.6. These are the probability distributions associated with the histograms shown in Figure 13.8a and b.

Solution

The mean of the random variable X is given by

$$\begin{aligned}\mu_X &= (1)(.05) + (2)(.075) + (3)(.2) + (4)(.375) + (5)(.15)\\ &\quad + (6)(.1) + (7)(.05)\\ &= 4\end{aligned}$$

Therefore, using Formula (4) and the data from the probability distribution of X, we find that the variance of X is given by

$$\begin{aligned}\text{Var}(X) &= (.05)(1-4)^2 + (.075)(2-4)^2 + (.2)(3-4)^2 \\ &\quad + (.375)(4-4)^2 + (.15)(5-4)^2 \\ &\quad + (.1)(6-4)^2 + (.05)(7-4)^2 \\ &= 1.95\end{aligned}$$

Next, we find that the mean of the random variable Y is given by

$$\begin{aligned}\mu_Y &= (1)(.2) + (2)(.15) + (3)(.1) + (4)(.15) + (5)(.05) \\ &\quad + (6)(.1) + (7)(.25) \\ &= 4\end{aligned}$$

and so the variance of Y is given by

$$\begin{aligned}\text{Var}(Y) &= (.2)(1-4)^2 + (.15)(2-4)^2 + (.1)(3-4)^2 \\ &\quad + (.15)(4-4)^2 + (.05)(5-4)^2 + (.1)(6-4)^2 \\ &\quad + (.25)(7-4)^2 \\ &= 5.2\end{aligned}$$

Note that Var (X) is smaller than Var (Y), which confirms the earlier observations about the spread, or dispersion, of the probability distribution of X and Y, respectively.

Standard Deviation

Because Formula (4), which gives the variance of the random variable X, involves the squares of the deviations, the unit of measurement of Var (X) is the square of the unit of measurement of the values of X. For example, if the values assumed by the random variable X are measured in units of a gram, then Var (X) will be measured in units involving the *square* of a gram. To remedy this situation, one normally works with the square root of Var (X) rather than Var (X) itself. The former is called the standard deviation of X.

Standard Deviation of a Discrete Random Variable X

The standard deviation of a random variable X, σ (pronounced "sigma"), is defined by

$$\begin{aligned}\sigma &= \sqrt{\text{Var}(X)} \\ &= \sqrt{p_1(x_1-\mu)^2 + p_2(x_2-\mu)^2 + \cdots + p_n(x_n-\mu)^2} \qquad (7)\end{aligned}$$

where $x_1, x_2, \ldots, x_n$ denote the values assumed by the random variable X and $p_1 = P(X = x_1), p_2 = P(X = x_2), \ldots, p_n = P(X = x_n)$.

Example 6

Find the standard deviations of the random variables X and Y of Example 5.

Solution

From the results of Example 5, we have Var (X) = 1.95 and Var (Y) = 5.2. Taking their respective square roots, we have

$$\begin{aligned}\sigma_X &= \sqrt{1.95} \\ &\approx 1.40 \\ \sigma_Y &= \sqrt{5.2} \\ &\approx 2.28\end{aligned}$$

Example 7

Let X and Y denote the random variables whose values are the weights of the brand A and brand B potato chips, respectively (see Table 13.5). Compute the means and standard deviations of X and Y and interpret your results.

Solution

The probability distributions of X and Y may be computed from the data in Table 13.7.

Table 13.7

x	Relative Frequency of Occurrence	$P(X = x)$	y	Relative Frequency of Occurrence	$P(Y = y)$
15.8	1	.1	15.7	2	.2
15.9	2	.2	15.8	1	.1
16.0	4	.4	15.9	1	.1
16.1	2	.2	16.0	1	.1
16.2	1	.1	16.1	2	.2
			16.2	2	.2
			16.3	1	.1

The means of X and Y are given by

$$\begin{aligned}\mu_X &= (.1)(15.8) + (.2)(15.9) + (.4)(16.0) + (.2)(16.1)\\ &\quad + (.1)(16.2)\\ &= 16\\ \mu_Y &= (.2)(15.7) + (.1)(15.8) + (.1)(15.9) + (.1)(16.0)\\ &\quad + (.2)(16.1) + (.2)(16.2) + (.1)(16.3)\\ &= 16\end{aligned}$$

Therefore,

$$\begin{aligned}\text{Var }(X) &= (.1)(15.8 - 16)^2 + (.2)(15.9 - 16)^2 + (.4)(16 - 16)^2\\ &\quad + (.2)(16.1 - 16)^2 + (.1)(16.2 - 16)^2\\ &= 0.012\\ \text{Var }(Y) &= (.2)(15.7 - 16)^2 + (.1)(15.8 - 16)^2 + (.1)(15.9 - 16)^2\\ &\quad + (.1)(16 - 16)^2 + (.2)(16.1 - 16)^2 + (.2)(16.2 - 16)^2\\ &\quad + (.1)(16.3 - 16)^2\\ &= 0.042\end{aligned}$$

Thus, the required standard deviations are

$$\begin{aligned}\sigma_X &= \sqrt{\text{Var }(X)}\\ &= \sqrt{0.012}\\ &\approx 0.11\\ \sigma_Y &= \sqrt{\text{Var }(Y)}\\ &= \sqrt{0.042}\\ &\approx 0.20\end{aligned}$$

The means of both X and Y are equal to 16. Therefore, the average weight of a package of potato chips of either brand is 16 ounces. However, the standard deviation of Y is greater than that of X. This tells us that the weights of the packages of brand B potato chips are more widely dispersed about the common mean of 16 than are those of brand A.

GROUP DISCUSSION

A useful alternative formula for the variance is

$$\sigma^2 = E(X^2) - \mu^2$$

where $E(X)$ is the expected value of X.

1. Establish the validity of the formula.
2. Use the formula to verify the calculations in Example 7.

Now suppose that x is a *continuous* random variable. Using an argument similar to that used earlier when we extended the result for the expected value from the discrete to the continuous case, we have the following definitions:

Variance and Standard Deviation of a Continuous Random Variable

Let x be a continuous random variable with probability density function $f(x)$ on $[a, b]$. Then the variance of x is

$$\text{Var}(x) = \int_a^b (x - \mu)^2 f(x)\, dx \tag{8}$$

and the standard deviation of x is

$$\sigma = \sqrt{\text{Var}(x)} \tag{9}$$

Example 8

Find the expected value, variance, and standard deviation of the random variable x associated with the probability density function

$$f(x) = \frac{32}{15x^3}$$

on $[1, 4]$.

Solution

Using (3), we find the mean of x:

$$\begin{aligned}
\mu &= \int_a^b xf(x)\, dx = \int_1^4 x \cdot \frac{32}{15x^3}\, dx \\
&= \frac{32}{15}\int_1^4 x^{-2}\, dx = \frac{32}{15}\left(-\frac{1}{x}\right)\Big|_1^4 \\
&= \frac{32}{15}\left(-\frac{1}{4} + 1\right) = \frac{8}{5}
\end{aligned}$$

Next, using (8), we find

$$\begin{aligned}
\text{Var}(x) &= \int_a^b (x - \mu)^2 f(x)\, dx = \int_1^4 \left(x - \frac{8}{5}\right)^2 \cdot \frac{32}{15x^3}\, dx \\
&= \frac{32}{15}\int_1^4 \left(x^2 - \frac{16}{5}x + \frac{64}{25}\right)\frac{1}{x^3}\, dx \\
&= \frac{32}{15}\int_1^4 \left(\frac{1}{x} - \frac{16}{5}x^{-2} + \frac{64}{25}x^{-3}\right) dx
\end{aligned}$$

$$= \frac{32}{15}\left[\ln x + \frac{16}{5x} - \frac{32}{25x^2}\right]\Bigg|_1^4$$

$$= \frac{32}{15}\left[\left(\ln 4 + \frac{16}{5 \cdot 4} - \frac{32}{25 \cdot 16}\right) - \left(\ln 1 + \frac{16}{5} - \frac{32}{25}\right)\right]$$

$$= \frac{32}{15}\left(\ln 4 + \frac{4}{5} - \frac{2}{25} - \frac{16}{5} + \frac{32}{25}\right)$$

$$= \frac{32}{15}\left(\ln 4 - \frac{6}{5}\right) \approx 0.40$$

Finally, using (9), we find the required standard deviation to be

$$\sigma = \sqrt{\text{Var }(x)} \approx 0.63$$

Alternative Formula for Variance Using (8) to calculate the variance of a continuous random variable can be rather tedious. The following formula often makes this task easier:

$$\text{Var }(x) = \int_a^b x^2 f(x)\,dx - \mu^2 \qquad \textbf{(10)}$$

This equation follows from these computations:

$$\text{Var }(x) = \int_a^b (x - \mu)^2 f(x)\,dx$$

$$= \int_a^b (x^2 - 2\mu x + \mu^2) f(x)\,dx$$

$$= \int_a^b x^2 f(x)\,dx - 2\mu \int_a^b x f(x)\,dx + \mu^2 \int_a^b f(x)\,dx$$

$$= \int_a^b x^2 f(x)\,dx - 2\mu \cdot \mu + \mu^2 \qquad \text{Since } \int_a^b x f(x)\,dx = \mu \text{ and } \int_a^b f(x)\,dx = 1$$

$$= \int_a^b x^2 f(x)\,dx - \mu^2$$

Example 9

Use Formula (10) to calculate the variance of the random variable of Example 8.

Solution

Using (10), we have

$$\text{Var }(x) = \int_a^b x^2 f(x)\,dx - \mu^2 = \int_1^4 x^2 \cdot \frac{32}{15x^3}\,dx - \left(\frac{8}{5}\right)^2$$

$$= \frac{32}{15}\int_1^4 \frac{1}{x}\,dx - \frac{64}{25} = \frac{32}{15}\ln x\Big|_1^4 - \frac{64}{25}$$

$$= \frac{32}{15}\ln 4 - \frac{64}{25} = \frac{32}{15}\left(\ln 4 - \frac{6}{5}\right)$$

as obtained earlier.

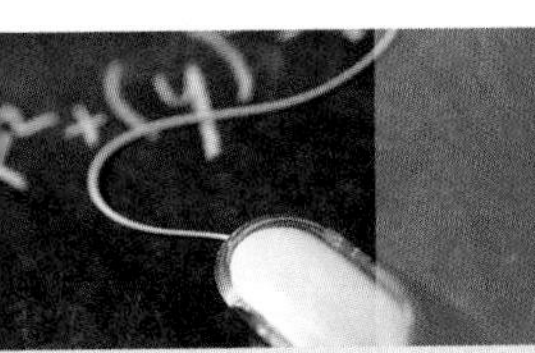

Self-Check Exercise 13.2

Find the expected value, variance, and standard deviation of the random variable x associated with the probability density function $f(x) = 6(x - x^2)$ on $[0, 1]$.

The solution to Self-Check Exercise 13.2 can be found on page 772.

13.2 Exercises

In Exercises 1–14, find the mean, variance, and standard deviation of the random variable x associated with the probability density function over the indicated interval.

1. $f(x) = \frac{1}{3}; [3, 6]$

2. $f(x) = \frac{1}{4}; [2, 6]$

3. $f(x) = \frac{3}{125}x^2; [0, 5]$

4. $f(x) = \frac{3}{8}x^2; [0, 2]$

5. $f(x) = \frac{3}{32}(x - 1)(5 - x); [1, 5]$

6. $f(x) = 20(x^3 - x^4); [0, 1]$

7. $f(x) = \frac{8}{7x^2}; [1, 8]$

8. $f(x) = \frac{4}{3x^2}; [1, 4]$

9. $f(x) = \frac{3}{14}\sqrt{x}; [1, 4]$

10. $f(x) = \frac{5}{2}x^{3/2}; [0, 1]$

11. $f(x) = \frac{3}{x^4}; [1, \infty)$

12. $f(x) = 3.5x^{-4.5}; [1, \infty)$

13. $f(x) = \frac{1}{4}e^{-x/4}; [0, \infty)$

Hint: $\lim_{x\to\infty} x^n e^{kx} = 0, k < 0$

14. $f(x) = \frac{1}{9}xe^{-x/3}; [0, \infty)$

Hint: $\lim_{x\to\infty} x^n e^{kx} = 0, k < 0$

In Exercises 15–20, find the median of the random variable x with the probability density function defined on the indicated interval I. The median of x is defined to be the number m such that Observe that half of the x-values lie below m and the other half lie above m.

15. $f(x) = \frac{1}{6}; [2, 8]$

16. $f(x) = \frac{2}{15}x; [1, 4]$

17. $f(x) = \frac{3}{16}\sqrt{x}; [0, 4]$

18. $f(x) = \frac{1}{6\sqrt{x}}; [1, 16]$

19. $f(x) = \frac{3}{x^2}; [1, \infty)$

20. $f(x) = \frac{1}{2}e^{-x/2}; [0, \infty)$

21. Number of Chips in Chocolate Chip Cookies The number of chocolate chips in a certain brand of cookies has a distribution described by the probability density function

$$f(x) = \frac{1}{36}(6x - x^2) \qquad (0 \le x \le 6)$$

Find the expected number of chips in a cookie selected at random.

22. Reaction Time of a Motorist The amount of time t (in seconds) it takes a motorist to react to a road emergency is a continuous random variable with probability density function

$$f(t) = \frac{6}{4t^3} \qquad (1 \le t \le 3)$$

What is the expected reaction time for a motorist chosen at random?

23. Expected Snowfall The amount of snowfall in metres in a remote region of the Yukon territory in the month of January is a continuous random variable with probability density function

$$f(x) = 6x(1 - x) \qquad (0 \le x \le 1)$$

Find the amount of snowfall one can expect in any given month of January in the region.

24. Life Expectancy of Colour Television Tubes The life expectancy (in years) of a certain brand of colour television tube is a continuous random variable with probability density function

$$f(t) = 9(9 + t^2)^{-3/2} \qquad (0 \le t < \infty)$$

How long is one of these colour television tubes expected to last?

Business and Economics Applications

25. Shopping Habits The amount of time t (in minutes) a shopper spends browsing in the magazine section of a supermarket is a continuous random variable with probability density function

$$f(t) = \frac{2}{25}t \qquad (0 \le t \le 5)$$

How much time is a shopper chosen at random expected to spend in the magazine section?

26. Gas Station Sales The amount of gas (in thousands of litres) Al's Gas Station sells on a typical Monday is a continuous random variable with probability density function

$$f(x) = 4(x-2)^3 \qquad (2 \le x \le 3)$$

How much gas can the gas station expect to sell each Monday?

27. Demand for Butter The quantity demanded x (in thousands of kilograms) of a certain brand of butter per week is a continuous random variable with probability density function

$$f(x) = \frac{6}{125}x(5-x) \qquad (0 \le x \le 5)$$

What is the expected demand for this brand of butter per week?

Biological and Life Sciences Applications

28. Life Span of a Plant The life span of a certain plant species (in days) is described by the probability density function

$$f(x) = \frac{1}{100}e^{-x/100}$$

If a plant of this species is selected at random, how long can the plant be expected to live?

In Exercises 29 and 30, determine whether the statement is true or false. If it is true, explain why it is true. If it is false, give an example to show why it is false.

29. If f is a probability density function of a continuous random variable x in the interval $[a, b]$, then the expected value of x is given by $\int_a^b x^2 f(x)\,dx$.

30. If f is a probability density function of a continuous random variable x in the interval $[a, b]$, then

$$\text{Var}(x) = \int_a^b x^2 f(x)\,dx - \left[\int_a^b xf(x)\,dx\right]^2$$

Solution to Self-Check Exercise 13.2

Using (3), we find

$$\mu = \int_0^1 xf(x) = 6\int_0^1 (x^2 - x^3)\,dx$$
$$= 6\left[\frac{1}{3}x^3 - \frac{1}{4}x^4\right]\Big|_0^1$$
$$= 6\left(\frac{1}{3} - \frac{1}{4}\right) = \frac{1}{2}$$

Using (8), we find

$$\text{Var}(x) = \int_0^1 (x-\mu)^2 f(x)\,dx$$
$$= \int_0^1 \left(x - \frac{1}{2}\right)^2 6(x - x^2)\,dx$$
$$= 6\int_0^1 \left(x^2 - x + \frac{1}{4}\right)(x - x^2)\,dx$$
$$= 6\int_0^1 \left(-x^4 + 2x^3 - \frac{5}{4}x^2 + \frac{1}{4}x\right)dx$$
$$= 6\left(-\frac{1}{5}x^5 + \frac{1}{2}x^4 - \frac{5}{12}x^3 + \frac{1}{8}x^2\right)\Big|_0^1$$
$$= 6\left(-\frac{1}{5} + \frac{1}{2} - \frac{5}{12} + \frac{1}{8}\right) = 0.05$$

Using (9), we see that

$$\sigma = \sqrt{\text{Var}(x)} \approx 0.224$$

13.3 Normal Distributions

Normal Distributions

In Section 13.2 we saw the useful role played by exponential density functions in many applications. In this section we look at yet another class of continuous probability distributions known as **normal distributions.** The normal distribution is without doubt the most important of all the probability distributions. Many phenomena, such as the heights of people in a given population, the weights of newborn infants, the IQs of college students, and the actual weights of 500-gram packages of cereals, have probability distributions that are normal. The normal distribution also provides us with an accurate approximation to the distributions of many random variables associated with random sampling problems.

The general **normal probability density function** with mean μ and standard deviation σ is defined to be

$$f(x) = \frac{e^{-(1/2)[(x-\mu)/\sigma]^2}}{\sigma\sqrt{2\pi}} \qquad (-\infty < x < \infty)$$

The graph of f, which is bell shaped, is called a **normal curve** (Figure 13.9).

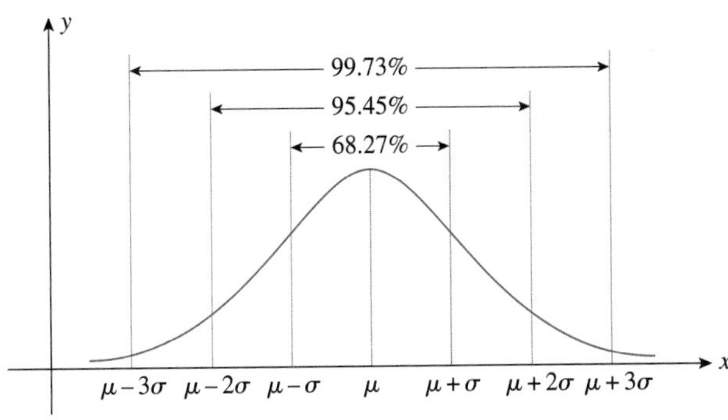

Figure 13.9
A normal curve

The normal curve (and therefore the corresponding normal distribution) is completely determined by its mean μ and standard deviation σ. In fact, the normal curve has the following characteristics, described in terms of these two parameters:

1. The curve has a peak at $x = \mu$.
2. The curve is symmetrical with respect to the vertical line $x = \mu$.
3. The curve always lies above the x-axis but approaches the x-axis as x extends indefinitely in either direction.
4. The area under the curve is 1.
5. For any normal curve, 68.27% of the area under the curve lies within 1 standard deviation of the mean (that is, between $\mu - \sigma$ and $\mu + \sigma$), 95.45% of the area lies within 2 standard deviations of the mean, and 99.73% of the area lies within 3 standard deviations of the mean.

Figure 13.10 shows two normal curves with different means μ_1 and μ_2 but the same deviation. Figure 13.11 shows two normal curves with the same mean but different standard deviations σ_1 and σ_2. (Which number is smaller?)

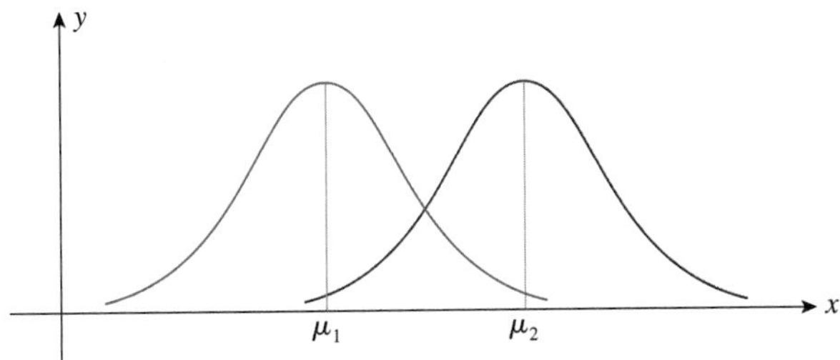

Figure 13.10
Two normal curves that have the same standard deviation but different means

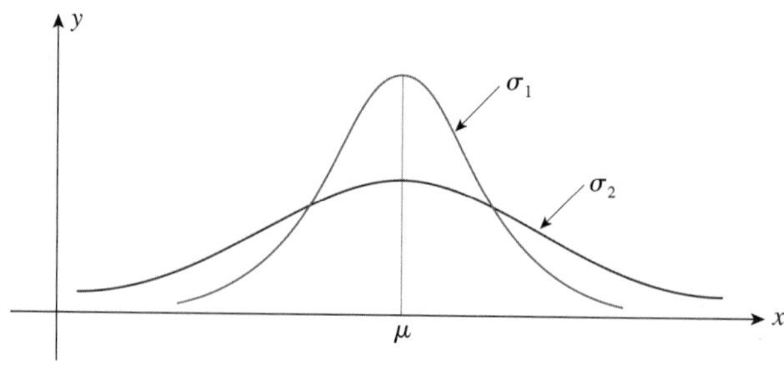

Figure 13.11
Two normal curves that have the same mean but different standard deviations

In general, the mean μ of a normal distribution determines where the centre of the curve is located, whereas the standard deviation σ of a normal distribution determines the sharpness (or flatness) of the curve.

As this discussion reveals, infinitely many normal curves correspond to different choices of the parameters μ and σ, which characterize such curves. Fortunately, any normal curve may be transformed into any other normal curve (as we see later on), so in the study of normal curves it suffices to single out one such particular curve for special attention. The normal curve with mean $\mu = 0$ and standard deviation $\sigma = 1$ is called the **standard normal curve.** The corresponding distribution is called the **standard normal distribution.** The random variable itself is called the **standard normal variable** and is commonly denoted by Z.

EXPLORING WITH TECHNOLOGY

Consider the probability density function

$$f(x) = \frac{1}{\sqrt{2\pi}} e^{-x^2/2}$$

which is the formula given on page 773 with $\mu = 0$ and $\sigma = 1$.

1. Use a graphing utility to plot the graph of f using the viewing window $[-4, 4] \cdot [0, 0.5]$.
2. Use the numerical integration function of a graphing utility to find the area of the region under the graph of f on the intervals $[-1, 1]$, $[-2, 2]$, and $[-3, 3]$ and thus verify Property 5 of normal distributions for the special case where $\mu = 0$ and $\sigma = 1$.

Computations of Probabilities Associated with Normal Distributions

Areas under the standard normal curve have been extensively computed and tabulated. The "Standard Normal Distribution Table" at the end of this chapter gives the areas of the regions under the standard normal curve to the left of the number z; these areas correspond, of course, to probabilities of the form $P(Z < z)$ or $P(Z \leq z)$. The next several examples illustrate the use of this table in computations involving the probabilities associated with the standard normal variable.

Example 1

Let Z be the standard normal variable. By first making a sketch of the appropriate region under the standard normal curve, find these values:

a. $P(Z < 1.24)$ **b.** $P(Z > 0.5)$ **c.** $P(0.24 < Z < 1.48)$
d. $P(-1.65 < Z < 2.02)$

Solution

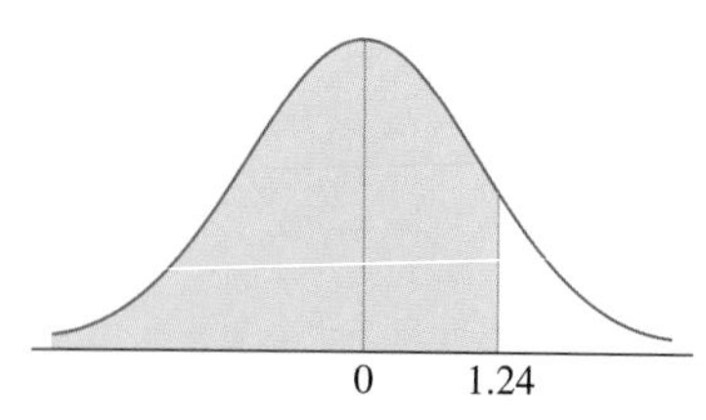

Figure 13.12
$P(Z < 1.24)$

a. The region under the standard normal curve associated with the probability $P(Z < 1.24)$ is shown in Figure 13.12. To find the area of the required region using the Standard Normal Distribution Table, we first locate the number 1.2 in the column and the number .04 in the row, both headed by z, and read off the number .8925 appearing in the body of the table. Thus,

$$P(Z < 1.24) = .8925$$

b. The region under the standard normal curve associated with the probability $P(Z > 0.5)$ is shown in Figure 13.13a. Observe, however, that the required area is, by virtue of the symmetry of the standard normal curve, equal to the shaded area shown in Figure 13.13b. Thus,

$$\begin{aligned} P(Z > 0.5) &= P(Z < -0.5) \\ &= .3085 \end{aligned}$$

c. The probability $P(0.24 < Z < 1.48)$ is equal to the shaded area shown in Figure 13.14. This area is obtained by subtracting the area under the curve to the left of $z = 0.24$ from the area under the curve to the left of $z = 1.48$; that is,

$$\begin{aligned} P(0.24 < Z < 1.48) &= P(Z < 1.48) - P(Z < 0.24) \\ &= .9306 - .5948 \\ &= .3358 \end{aligned}$$

d. The probability $P(-1.65 < Z < 2.02)$ is given by the shaded area in Figure 13.15. We have

$$\begin{aligned} P(-1.65 < Z < 2.02) &= P(Z < 2.02) - P(Z < -1.65) \\ &= .9783 - .0495 \\ &= .9288 \end{aligned}$$

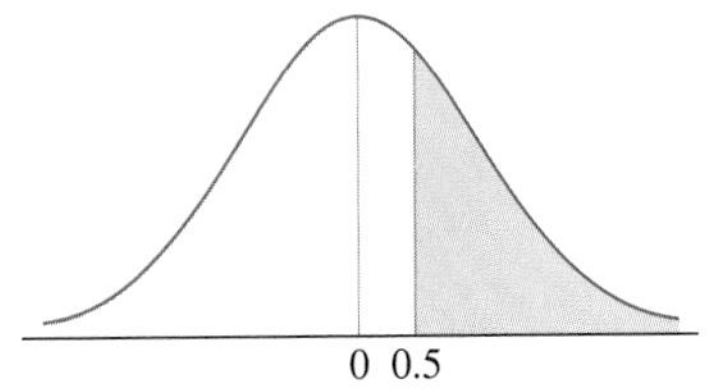

(a) $P(Z > 0.5)$

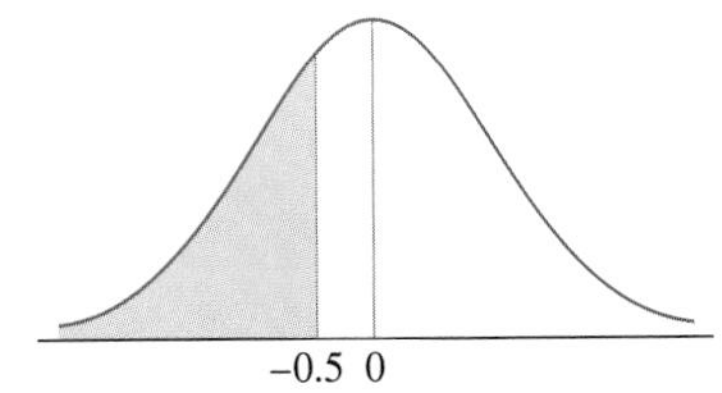

(b) $P(Z < -0.5)$

Figure 13.13

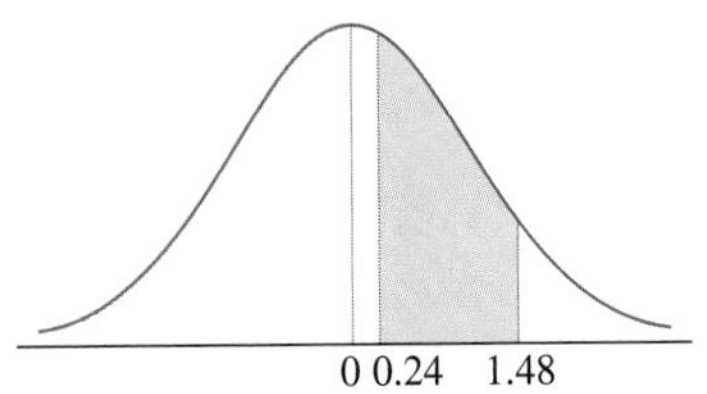

Figure 13.14
$P(0.24 < Z < 1.48)$

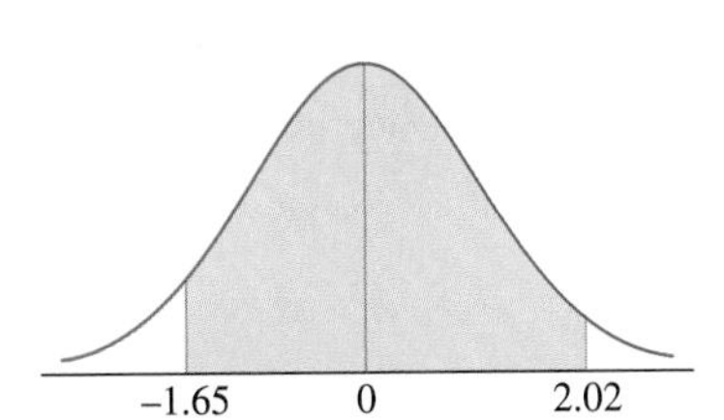

Figure 13.15
$P(-1.65 < Z < 2.02)$

Example 2

Let Z be the standard normal variable. Find the value of z if z satisfies the following:

a. $P(Z < z) = .9474$ **b.** $P(Z > z) = .9115$
c. $P(-z < Z < z) = .7888$

Solution

a. Refer to Figure 13.16. We want the value of z such that the area of the region under the standard normal curve and to the left of $Z = z$ is .9474. Locating the number .9474 in the Standard Normal Distribution Table and reading back, we find that $z = 1.62$.

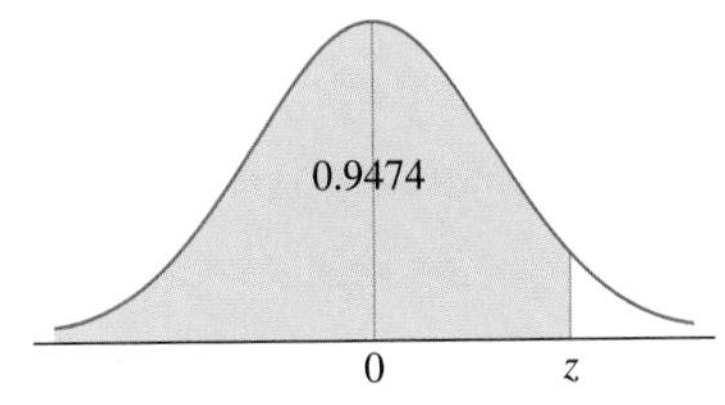

Figure 13.16
$P(Z < z) = .9474$

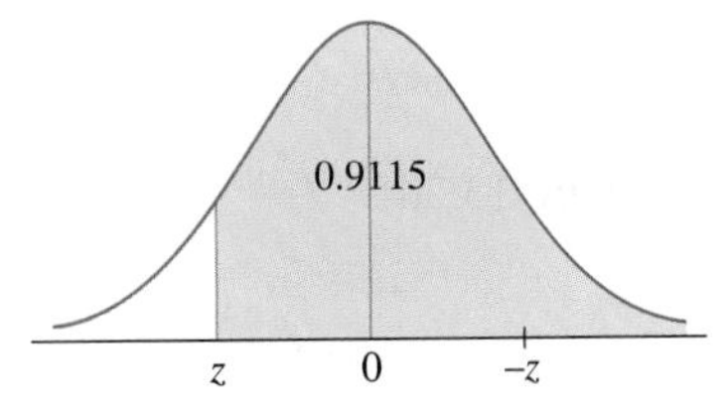

Figure 13.17
$P(Z > z) = .9115$

b. Since $P(Z > z)$ or, equivalently, the area of the region to the right of z is greater than 0.5, z must be negative (Figure 13.17). Therefore, $-z$ is positive. Furthermore, the area of the region to the right of z is the same as the area of the region to the left of $-z$:

$$\begin{aligned} P(Z > z) &= P(Z < -z) \\ &= .9115 \end{aligned}$$

Looking up the table, we find $-z = 1.35$, so $z = -1.35$.

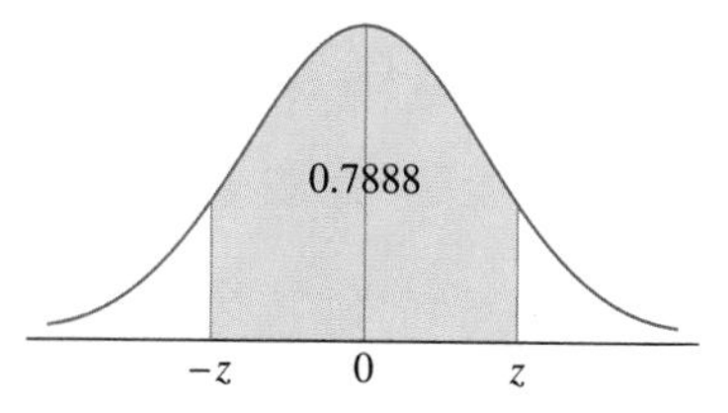

Figure 13.18
$P(-z < Z < z) = .7888$

c. The region associated with $P(-z < Z < z)$ is shown in Figure 13.18. Observe that by symmetry the area of this region is just double that of the area of the region between $Z = 0$ and $Z = z$; that is,

$$P(-z < Z < z) = 2P(0 < Z < z)$$

or

$$P(0 < Z < z) = P(Z < z) - \frac{1}{2}$$

(see Figure 13.19). Therefore,

$$\frac{1}{2}P(-z < Z < z) = P(Z < z) - \frac{1}{2}$$

or, solving for $P(Z < z)$,

$$\begin{aligned} P(Z < z) &= \frac{1}{2} + \frac{1}{2}P(-z < Z < z) \\ &= \frac{1}{2}(1 + .7888) \\ &= .8944 \end{aligned}$$

Consulting the table, we find $z = 1.25$.

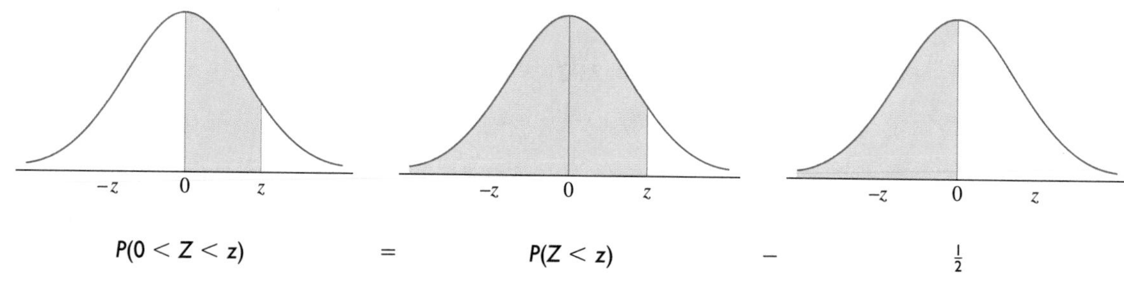

Figure 13.19

We now turn our attention to the computation of probabilities associated with normal distributions whose means and standard deviations are not necessarily equal to 0 and 1, respectively. As mentioned earlier, any normal curve may be transformed into the standard normal curve. In particular, it may be shown that if X is a normal random variable with mean μ and standard deviation σ, then it can be transformed into the standard normal random variable Z by means of the substitution

$$Z = \frac{X - \mu}{\sigma}$$

(see Figure 13.20).

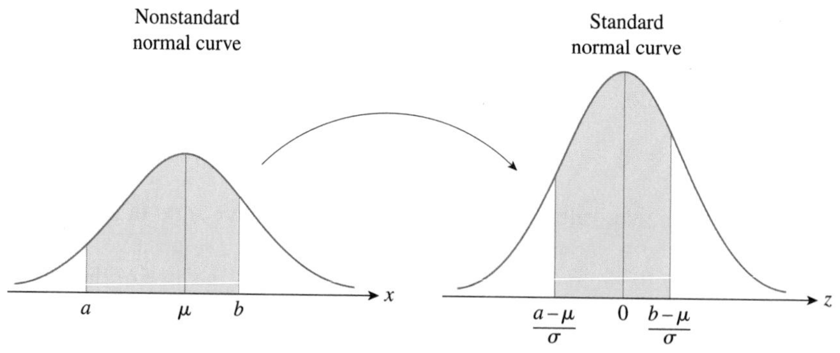

Figure 13.20

The area of the region under the normal curve (with random variable X) between $x = a$ and $x = b$ is *equal* to the area of the region under the standard normal curve between $z = (a - \mu)/\sigma$ and $z = (b - \mu)/\sigma$. In terms of probabilities associated with these distributions, we have

$$P(a < X < b) = P\left(\frac{a - \mu}{\sigma} < Z < \frac{b - \mu}{\sigma}\right) \tag{11}$$

Similarly, we have

$$P(X < b) = P\left(Z < \frac{b - \mu}{\sigma}\right) \tag{12}$$

$$P(X > a) = P\left(Z > \frac{a - \mu}{\sigma}\right) \tag{13}$$

Thus, with the help of (11)–(13), computations of probabilities associated with any normal distribution may be reduced to computations of areas of regions under the standard normal curve.

Example 3

Suppose X is a normal random variable with $\mu = 100$ and $\sigma = 20$. Find these values:

a. $P(X < 120)$ **b.** $P(X > 70)$ **c.** $P(75 < X < 110)$

Solution

a. Using (12) with $\mu = 100$, $\sigma = 20$, and $b = 120$, we have

$$P(X < 120) = P\left(Z < \frac{120 - 100}{20}\right)$$

$$= P(Z < 1) = .8413 \qquad \text{Using the table of values of } Z$$

b. Using (13) with $\mu = 100$, $\sigma = 20$, and $a = 70$, we have

$$P(X > 70) = P\left(Z > \frac{70 - 100}{20}\right)$$

$$= P(Z > -1.5)$$

$$= P(Z < 1.5) = .9332 \qquad \text{Using the table of values of } Z$$

c. Using (11) with $\mu = 100$, $\sigma = 20$, $a = 75$, and $b = 110$, we have

$$P(75 < X < 110)$$

$$= P\left(\frac{75 - 100}{20} < Z < \frac{110 - 100}{20}\right)$$

$$= P(-1.25 < Z < 0.5)$$

$$= P(Z < 0.5) - P(Z < -1.25) \qquad \text{See Figure 13.21}$$

$$= .6915 - .1056 = .5859 \qquad \text{Using the table of values of } Z$$

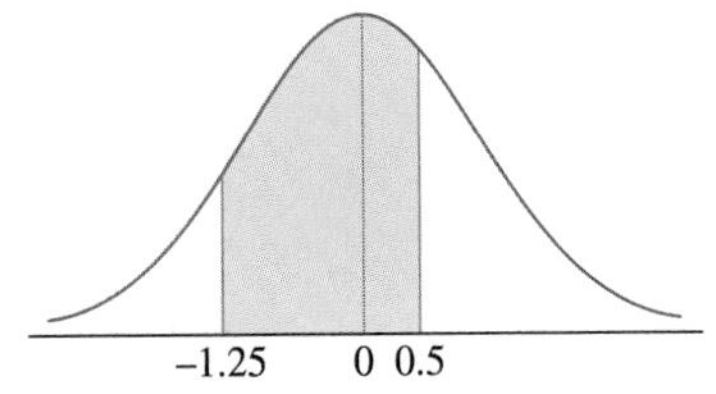

Figure 13.21

Applications Involving Normal Random Variables

Example 4

Infants' Birth Weights The medical records of infants delivered at Kaiser Memorial Hospital show that the infants' birth weights in kilograms are normally distributed with a mean of 3.36 and a standard deviation of 0.55. Find the probability that an infant selected at random from among those delivered at the hospital weighed more than 4.18 kilograms at birth.

Solution

Let X be the normal random variable denoting the birth weights of infants delivered at the hospital. Then the probability that an infant selected at random has a birth weight of more than 4.18 kilograms is given by $P(X > 4.18)$. To compute $P(X > 4.18)$, we use Formula (13) with $\mu = 3.36$, $\sigma = 0.55$, and $a = 4.18$. We find

$$\begin{aligned} P(X > 4.18) &= P\left(Z > \frac{4.18 - 3.36}{0.55}\right) \qquad P(X > a) = P\left(Z > \frac{a - \mu}{\sigma}\right) \\ &= P(Z > 1.5) \\ &= P(Z < -1.5) \\ &= 0.668 \end{aligned}$$

Thus, the probability that an infant delivered at the hospital weighed more than 4.18 kilograms is .0668.

Example 5

Packaging Manitoba Natural Produce ships potatoes to its distributors in bags with a mean weight of 50 kilograms and a standard deviation of 0.5 kilogram. If a bag of potatoes is selected at random from a shipment, what are the probabilities of these events?

a. The bag weighs more than 51 kilograms.
b. It weighs less than 48 kilograms.
c. It weighs between 49 and 51 kilograms.

Solution

Let X denote the weight of a bag of potatoes packed by the company. Then, the mean and standard deviation of X are $\mu = 50$ and $\sigma = 0.5$, respectively.

a. The probability that a bag selected at random weighs more than 51 kilograms is given by

$$\begin{aligned} P(X > 51) &= P\left(Z > \frac{51 - 50}{0.5}\right) \qquad P(X > a) = P\left(Z > \frac{a - \mu}{\sigma}\right) \\ &= P(Z > 2) \\ &= P(Z < -2) \\ &= 0.228 \end{aligned}$$

b. The probability that a bag selected at random weighs less than 48 kilograms is given by

$$\begin{aligned} P(X < 48) &= P\left(Z < \frac{48 - 50}{0.5}\right) \qquad P(X < b) = P\left(Z < \frac{b - \mu}{\sigma}\right) \\ &= P(Z < -4) \\ &= 0 \end{aligned}$$

c. The probability that a bag selected at random weighs between 49 and 51 kilograms is given by

$$\begin{aligned} &P(49 < X < 51) \qquad P(a < X < b) = P\left(\frac{a - \mu}{\sigma} < Z < \frac{b - \mu}{\sigma}\right) \\ &= P\left(\frac{49 - 50}{0.5} < Z < \frac{51 - 50}{0.5}\right) \\ &= P(-2 < Z < 2) \\ &= P(Z < 2) - P(Z < -2) \\ &= .9772 - .0228 \\ &= .9544 \end{aligned}$$

Example 6

University Admissions Eligibility The marks of the senior class of Wilfred Laurier High School are normally distributed with a mean of 80.2% and a standard deviation of 2.1 points. If a senior in the top 10% of his or her class is eligible for admission to any of the nine universities within driving distance, what is the minimum mark that a senior should have to ensure eligibility for admission?

Solution

Let X denote the marks of a randomly selected senior at Wilfred Laurier High School, and let x denote the minimum mark to ensure his or her eligibility for admission to the university. Since only the top 10% are eligible for admission, x must satisfy the equation

$$P(X \geq x) = .1$$

Using Formula (13) with $\mu = 80.1$ and $\sigma = 2.1$, we find

$$P(X \geq x) = P\left(Z \geq \frac{x - 80.1}{2.1}\right) = .1 \qquad P(X > a) = P\left(Z > \frac{a - \mu}{\sigma}\right)$$

But, this is equivalent to the equation

$$P\left(Z \leq \frac{x - 80.1}{2.1}\right) = .9 \qquad \text{Why?}$$

Consulting the Standard Normal Distribution Table, we find

$$\frac{x - 80.1}{2.1} = 1.28$$

Upon solving for x, we obtain

$$x = (1.28)(2.1) + 80.1$$
$$\approx 82.8$$

Thus, to ensure eligibility for admission to one of the nine universities, a senior at Wilfred Laurier High School should have a minimum mark of 82.8%.

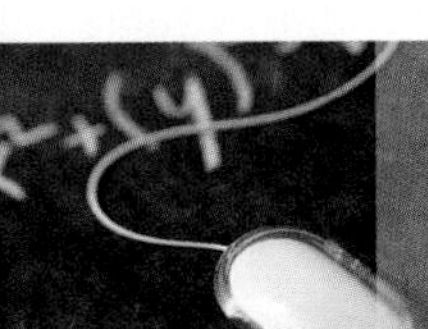

Self-Check Exercises 13.3

1. Let Z be a standard normal variable.
   a. Find the value of $P(-1.2 < Z < 2.1)$ by first sketching the appropriate region under the standard normal curve.
   b. Find the value of z if z satisfies

   $$P(-z < Z < z) = .8764$$

2. Let X be a normal random variable with $\mu = 80$ and $\sigma = 10$. Find these values:
   a. $P(X < 100)$
   b. $P(X > 60)$
   c. $P(70 < X < 90)$

3. The serum cholesterol levels (in mg/dL) in a current Mediterranean population are found to be normally distributed with a mean of 160 and a standard deviation of 50. Scientists at the National Heart, Lung, and Blood Institute consider this pattern ideal for a minimal risk of heart attacks. Find the percentage of the population who have blood cholesterol levels between 160 and 180 mg/dL.

Solutions to Self-Check Exercises 13.3 can be found on page 781.

13.3 Exercises

In Exercises 1–6, find the value of the probability of the standard normal variable Z corresponding to the shaded area under the standard normal curve.

1. $P(Z < 1.45)$

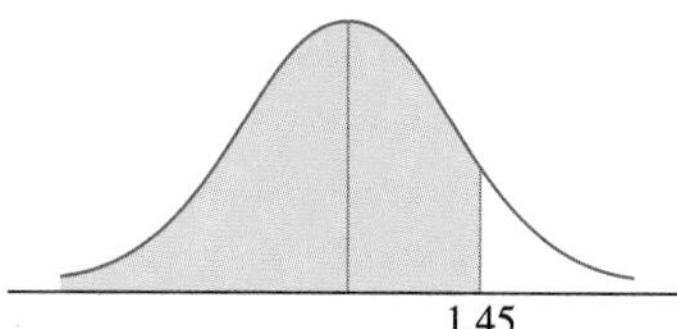

2. $P(Z > 1.11)$

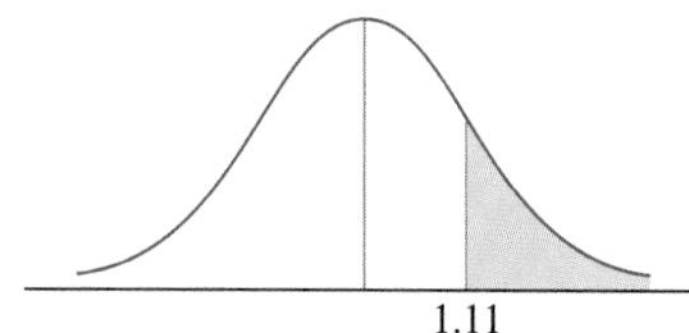

3. $P(Z < -1.75)$

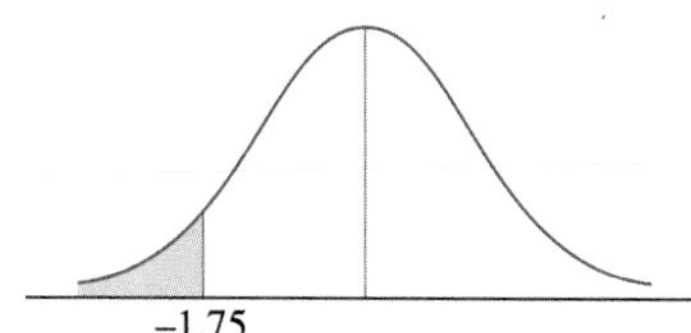

4. $P(0.3 < Z < 1.83)$

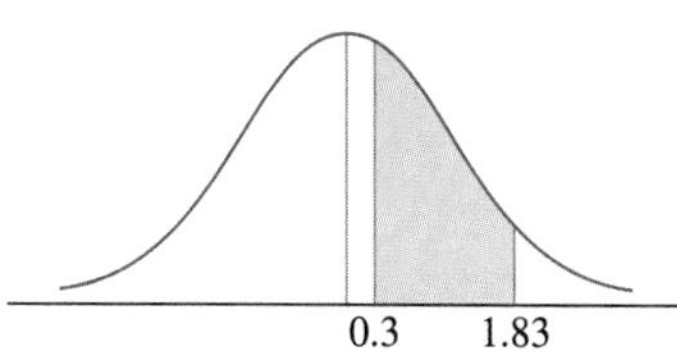

5. $P(-1.32 < Z < 1.74)$

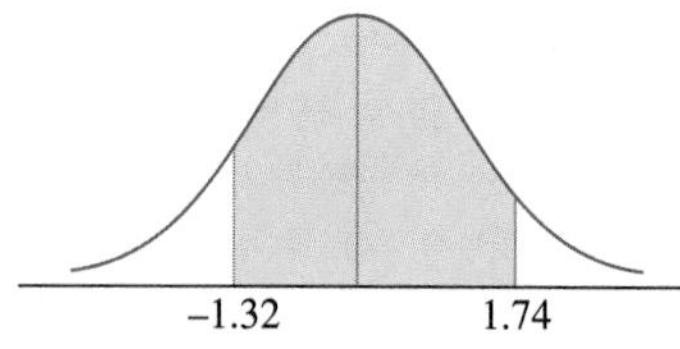

6. $P(-2.35 < Z < -0.51)$

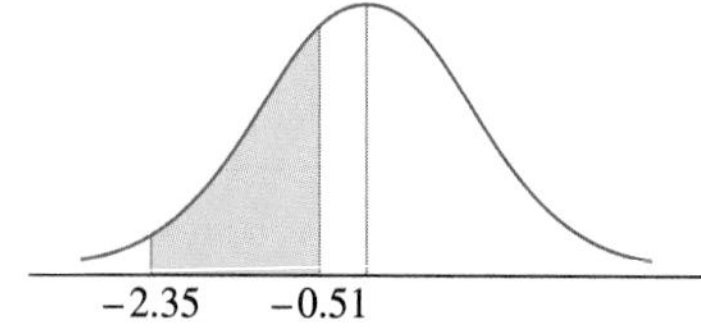

In Exercises 7–14, (a) sketch the area under the standard normal curve corresponding to the given probability and (b) find the value of the probability of the standard normal variable Z corresponding to this area.

7. $P(Z < 1.37)$
8. $P(Z > 2.24)$
9. $P(Z < -0.65)$
10. $P(0.45 < Z < 1.75)$
11. $P(Z > -1.25)$
12. $P(-1.48 < Z < 1.54)$
13. $P(0.68 < Z < 2.02)$
14. $P(-1.41 < Z < -0.24)$

15. Let Z be the standard normal variable. Find the values of z if z satisfies
 a. $P(Z < z) = .8907$
 b. $P(Z < z) = .2090$

16. Let Z be the standard normal variable. Find the values of z if z satisfies
 a. $P(Z > z) = .9678$
 b. $P(-z < Z < z) = .8354$

17. Let Z be the standard normal variable. Find the values of z if z satisfies
 a. $P(Z > -z) = .9713$
 b. $P(Z < -z) = .9713$

18. Suppose X is a normal random variable with $\mu = 380$ and $\sigma = 20$. Find these values:
 a. $P(X < 405)$
 b. $P(400 < X < 430)$
 c. $P(X > 400)$

19. Suppose X is a normal random variable with $\mu = 50$ and $\sigma = 5$. Find these values:
 a. $P(X < 60)$
 b. $P(X > 43)$
 c. $P(46 < X < 58)$

20. Suppose X is a normal random variable with $\mu = 500$ and $\sigma = 75$. Find these values:
 a. $P(X < 750)$
 b. $P(X > 350)$
 c. $P(400 < X < 600)$

21. **Product Reliability** TKK Products manufactures electric light bulbs in the 50-, 60-, 75-, and 100-W range. Laboratory tests show that the lives of these light bulbs are normally distributed with a mean of 750 hr and a standard deviation of 75 hr. Find the probability that a TKK light bulb selected at random will burn for the following hours:
 a. For more than 900 hr
 b. For less than 600 hr
 c. Between 750 and 900 hr
 d. Between 600 and 800 hr

22. **Education** On the average, a student takes 100 words/minute midway through an Advanced Shorthand course at the Institute of Stenography. Assuming that the students' dictation speeds are normally distributed and that the standard deviation

is 20 words/minute, find the probability that a student randomly selected from the course could take dictation at the following speeds:

a. More than 120 words/minute
b. Between 80 and 120 words/minute
c. Less than 80 words/minute

23. IQs The IQs of students at Wilson Elementary School were measured recently and found to be normally distributed with a mean of 100 and a standard deviation of 15. Find the probability that a student selected at random will have an IQ of the following:

a. 140 or higher
b. 120 or higher
c. Between 100 and 120
d. 90 or less

24. Product Reliability The tread lives of Super Titan radial tires under normal driving conditions are normally distributed with a mean of 40,000 km and a standard deviation of 2000 km. What is the probability that a tire selected at random will have a tread life of more than 35,000 km? If four new tires are installed on a car and they experience even wear, determine the probability that all four tires still have useful tread lives after 35,000 km.

25. Civil Service Exams To be eligible for further consideration, applicants for certain Civil Service positions must first pass a written qualifying examination on which a score of 70 or more must be obtained. In a recent examination, it was found that the scores were normally distributed with a mean of 60 points and a standard deviation of 10 points. Determine the percentage of applicants who passed the written qualifying examination.

26. Warranties The general manager of the Service Department of MCA Television Company has estimated that the time that elapses between the dates of purchase and the dates on which the 50-cm sets manufactured by the company first require service is normally distributed with a mean of 22 mo and a standard deviation of 4 mo. If MCA gives a 1-yr warranty on parts and labour for these sets, determine what percentage of sets manufactured and sold may require service before the warranty period runs out.

27. Grade Distributions The marks on an economics examination are normally distributed with a mean of 72 and a standard deviation of 16. If the instructor assigns a grade of A to 10% of the class, what is the lowest score a student may have and still get an A?

28. Grade Distributions The marks on a sociology examination are normally distributed with a mean of 70 and a standard deviation of 10. If the instructor assigns A's to 15%, B's to 25%, C's to 40%, D's to 15%, and F's to 5% of the class, find the cutoff points for these grades.

Business and Economics Applications

29. Factory Workers' Wages According to the data released by a city's Chamber of Commerce, the weekly wages of factory workers are normally distributed with a mean of \$400 and a standard deviation of \$50. Find the probability that a worker selected at random from the city has a weekly wage of the following:

a. Less than \$300
b. More than \$460
c. Between \$350 and \$450

30. Female Factory Workers' Wages According to data released by a city's Chamber of Commerce, the weekly wages (in dollars) of female factory workers are normally distributed with a mean of \$475 and a standard deviation of \$50. Find the probability that a female factory worker selected at random from the city has a weekly wage of \$450 to \$550.

Biological and Life Sciences Applications

31. Medical Records The medical records of infants delivered at Kaiser Memorial Hospital show that the infants' lengths at birth (in centimetres) are normally distributed with a mean of 50 and a standard deviation of 6.6. Find the probability that an infant selected at random from among those delivered at the hospital measures the following:

a. More than 56 cm
b. Less than 46 cm
c. Between 48 and 53 cm

Solutions to Self-Check Exercises 13.3

1. a. The probability $P(-1.2 < Z < 2.1)$ is given by the shaded area in the accompanying figure. We have

$$\begin{aligned} P(-1.2 < Z < 2.1) &= P(Z < 2.1) - P(Z < -1.2) \\ &= .9821 - .1151 \\ &= .867 \end{aligned}$$

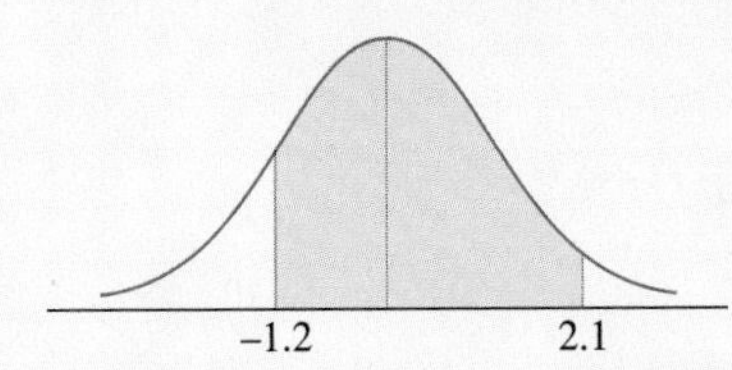

b. The region associated with $P(-z < Z < z)$ is shown in the accompanying figure. Observe that we have the following relationship:

$$P(Z < z) = \frac{1}{2}[1 + P(-z < Z < z)]$$

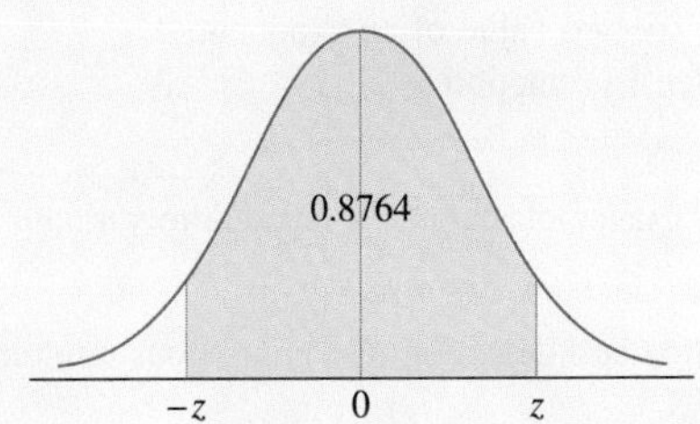

Solutions to Self-Check Exercises 13.3, *continued*

(see Example 2c). With $P(-z < Z < z) = .8764$, we find

$$P(Z < z) = \frac{1}{2}(1 + .8764)$$
$$= .9382$$

Consulting the table, we find $z = 1.54$.

2. Using the transformations (11), (12), and (13) and the table of values of Z, we have

a. $$P(X < 100) = P\left(Z < \frac{100 - 80}{10}\right)$$
$$= P(Z < 2)$$
$$= .9772$$

b. $$P(X > 60) = P\left(Z > \frac{60 - 80}{10}\right)$$
$$= P(Z > -2)$$
$$= P(Z < 2)$$
$$= .9772$$

c. $$P(70 < X < 90) = P\left(\frac{70 - 80}{10} < Z < \frac{90 - 80}{10}\right)$$
$$= P(-1 < Z < 1)$$
$$= P(Z < 1) - P(Z < -1)$$
$$= .8413 - .1587$$
$$= .6826$$

3. Let X be the normal random variable denoting the serum cholesterol levels (in mg/dL) in the current Mediterranean population under consideration. Thus, the percentage of the population having blood cholesterol levels between 160 and 180 mg/dL is given by $P(160 < X < 180)$. To compute $P(160 < X < 180)$, we use (11), with $\mu = 160$, $\sigma = 50$, $a = 160$, and $b = 180$. We find

$$P(160 < X < 180) = P\left(\frac{160 - 160}{50} < Z < \frac{180 - 160}{50}\right)$$
$$= P(0 < Z < 0.4)$$
$$= P(Z < 0.4) - P(Z < 0)$$
$$= .6554 - .5000$$
$$= .1554$$

so approximately 15.5% of the population has blood cholesterol levels between 160 and 180 mg/dL.

CHAPTER 13 Summary of Principal Formulas and Terms

Formulas

1. Expected value	$E(x) = x_1p_1 + x_2p_2 + \cdots + x_np_n$
2. Expected value of a continuous random variable	$E(x) = \int_a^b xf(x)\,dx$
3. Exponential density function	$f(x) = ke^{-kx}$
4. Expected value of an exponential density function	$E(x) = \frac{1}{k}$
5. Variance of a continuous random variable	$\text{Var}(x) = \int_a^b (x - \mu)^2 f(x)\,dx$
6. Standard deviation of a continuous random variable	$\sigma = \sqrt{\text{Var}(x)}$

Terms

experiment (750)
outcomes (sample points) (750)
sample space (750)
event (750)
probability of an event (750)
random variable (750)
finite discrete (750)
discrete probability function (750)
histogram (751)
continuous random variable (752)
probability density function (752)
exponential density function (755)
exponentially distributed (755)
average (mean) (761)
expected value (761, 763)
variance of a random variable (766, 769)
deviation (766)
standard deviation of a random variable (767, 769)
normal distribution (772)
normal probability density function (773)
normal curve (773)
standard normal curve (774)
standard normal distribution (774)
standard normal variable (774)

Review Exercises 

In Exercises 1–4, show that the function is a probability density function on the given interval.

1. $f(x) = \frac{1}{28}(2x + 3); [0, 4]$
2. $f(x) = \frac{3}{16}\sqrt{x}; [0, 4]$
3. $f(x) = \frac{1}{4}; [7, 11]$
4. $f(x) = \frac{4}{x^5}; [1, \infty)$

In Exercises 5–8, find the value of the constant k so that the function is a probability density function in the given interval.

5. $f(x) = kx^2; [0, 9]$
6. $f(x) = \frac{k}{\sqrt{x}}; [1, 16]$
7. $f(x) = \frac{k}{x^2}; [1, 3]$
8. $f(x) = \frac{k}{x^{2.5}}; [1, \infty)$

In Exercises 9–12, f is a probability density function defined on the given interval. Find the indicated probabilities.

9. $f(x) = \frac{2}{21}x; [2, 5]$
   a. $P(x \le 4)$ b. $P(x = 4)$ c. $P(3 \le x \le 4)$

10. $f(x) = \frac{1}{4}; [1, 5]$
   a. $P(2 \le x \le 4)$ b. $P(x \le 3)$ c. $P(x \ge 2)$

11. $f(x) = \frac{3}{16}\sqrt{x}; [0, 4]$
   a. $P(1 \le x \le 3)$ b. $P(x \le 3)$ c. $P(x = 2)$

12. $f(x) = \frac{1}{x^2}; [1, \infty)$
   a. $P(x \le 10)$ b. $P(2 \le x \le 4)$ c. $P(x \ge 2)$

In Exercises 13–16, find the mean, variance, and standard deviation of the random variable x associated with the probability density function f over the given interval.

13. $f(x) = \frac{1}{5}; [2, 7]$
14. $f(x) = \frac{1}{28}(2x + 3); [0, 4]$
15. $f(x) = \frac{1}{4}(3x^2 + 1); [-1, 1]$
16. $f(x) = \frac{4}{x^5}; [1, \infty)$

In Exercises 17–20, Z is the standard normal variable. Find the given probability.

17. $P(Z < 2.24)$
18. $P(Z > -1.24)$
19. $P(0.24 \le Z \le 1.28)$
20. $P(-1.37 \le Z \le 1.37)$

21. Show that the function

$$f(x, y) = \frac{1}{64}x^{1/2}y^{1/3}; D = \{0 \le x \le 4; 0 \le y \le 8\}$$

is a joint probability density function on D.

22. Given that

$$f(x, y) = x(2 - y); D = \{(x, y) | 0 \le x \le 1; 0 \le y \le 2\}$$

is a joint probability density function on D, find the indicated probabilities.
   a. $P(0 \le x \le \frac{1}{2}; 0 \le y \le 1)$
   b. $P(\{(x, y) | y \le 2x\})$

23. Suppose X is a normal random variable with $\mu = 80$ and $\sigma = 8$. Find these values:
   a. $P(X \le 84)$
   b. $P(X \ge 70)$
   c. $P(75 \le X \le 85)$

24. Suppose X is a normal random variable with $\mu = 45$ and $\sigma = 3$. Find these values:
   a. $P(X \le 50)$
   b. $P(X \ge 40)$
   c. $P(40 \le X \le 50)$

25. **Length of a Hospital Stay** Records at Centreville Hospital indicate that the length of time in days that a maternity patient stays in the hospital has a probability density function given by

$$P(t) = \frac{1}{4}e^{-(1/4)t}$$

a. What is the probability that a woman entering the maternity wing will be there longer than 6 days?
b. What is the probability that a woman entering the maternity wing will be there less than 2 days?
c. What is the average length of time that a woman entering the maternity wing stays in the hospital?

26. **Life Span of a Car Battery** The life span (in years) of a certain make of car battery is an exponentially distributed random variable with an expected value of 5. Find the probability that the life span of a battery is (a) less than 4 yr, (b) more than 6 yr, and (c) between 2 and 4 yr.

Appendix

The Standard Normal Distribution Table

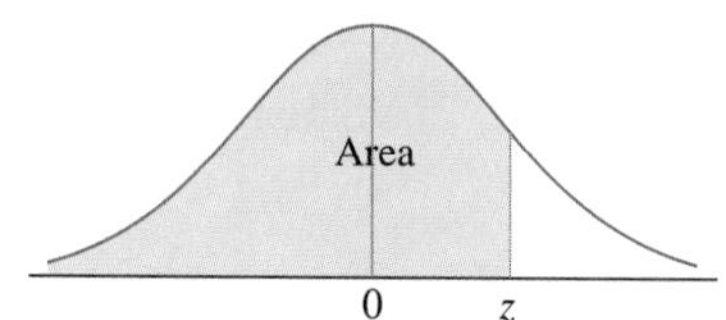

$F_z(z) = P[Z \leq z]$

z	0.00	0.01	0.02	0.03	0.04	0.05	0.06	0.07	0.08	0.09
−3.4	0.0003	0.0003	0.0003	0.0003	0.0003	0.0003	0.0003	0.0003	0.0003	0.0002
−3.3	0.0005	0.0005	0.0005	0.0004	0.0004	0.0004	0.0004	0.0004	0.0004	0.0003
−3.2	0.0007	0.0007	0.0006	0.0006	0.0006	0.0006	0.0006	0.0005	0.0005	0.0005
−3.1	0.0010	0.0009	0.0009	0.0009	0.0008	0.0008	0.0008	0.0008	0.0007	0.0007
−3.0	0.0013	0.0013	0.0013	0.0012	0.0012	0.0011	0.0011	0.0011	0.0010	0.0010
−2.9	0.0019	0.0018	0.0017	0.0017	0.0016	0.0016	0.0015	0.0015	0.0014	0.0014
−2.8	0.0026	0.0025	0.0024	0.0023	0.0023	0.0022	0.0021	0.0021	0.0020	0.0019
−2.7	0.0035	0.0034	0.0033	0.0032	0.0031	0.0030	0.0029	0.0028	0.0027	0.0026
−2.6	0.0047	0.0045	0.0044	0.0043	0.0041	0.0040	0.0039	0.0038	0.0037	0.0036
−2.5	0.0062	0.0060	0.0059	0.0057	0.0055	0.0054	0.0052	0.0051	0.0049	0.0048
−2.4	0.0082	0.0080	0.0078	0.0075	0.0073	0.0071	0.0069	0.0068	0.0066	0.0064
−2.3	0.0107	0.0104	0.0102	0.0099	0.0096	0.0094	0.0091	0.0089	0.0087	0.0084
−2.2	0.0139	0.0136	0.0132	0.0129	0.0125	0.0122	0.0119	0.0116	0.0113	0.0110
−2.1	0.0179	0.0174	0.0170	0.0166	0.0162	0.0158	0.0154	0.0150	0.0146	0.0143
−2.0	0.0228	0.0222	0.0217	0.0212	0.0207	0.0202	0.0197	0.0192	0.0188	0.0183
−1.9	0.0287	0.0281	0.0274	0.0268	0.0262	0.0256	0.0250	0.0244	0.0239	0.0233
−1.8	0.0359	0.0352	0.0344	0.0336	0.0329	0.0322	0.0314	0.0307	0.0301	0.0294
−1.7	0.0446	0.0436	0.0427	0.0418	0.0409	0.0401	0.0392	0.0384	0.0375	0.0367
−1.6	0.0548	0.0537	0.0526	0.0516	0.0505	0.0495	0.0485	0.0475	0.0465	0.0455
−1.5	0.0668	0.0655	0.0643	0.0630	0.0618	0.0606	0.0594	0.0582	0.0571	0.0559
−1.4	0.0808	0.0793	0.0778	0.0764	0.0749	0.0735	0.0722	0.0708	0.0694	0.0681
−1.3	0.0968	0.0951	0.0934	0.0918	0.0901	0.0885	0.0869	0.0853	0.0838	0.0823
−1.2	0.1151	0.1131	0.1112	0.1093	0.1075	0.1056	0.1038	0.1020	0.1003	0.0985
−1.1	0.1357	0.1335	0.1314	0.1292	0.1271	0.1251	0.1230	0.1210	0.1190	0.1170
−1.0	0.1587	0.1562	0.1539	0.1515	0.1492	0.1469	0.1446	0.1423	0.1401	0.1379
−0.9	0.1841	0.1814	0.1788	0.1762	0.1736	0.1711	0.1685	0.1660	0.1635	0.1611
−0.8	0.2119	0.2090	0.2061	0.2033	0.2005	0.1977	0.1949	0.1922	0.1894	0.1867
−0.7	0.2420	0.2389	0.2358	0.2327	0.2296	0.2266	0.2236	0.2206	0.2177	0.2148
−0.6	0.2743	0.2709	0.2676	0.2643	0.2611	0.2578	0.2546	0.2514	0.2483	0.2451
−0.5	0.3085	0.3050	0.3015	0.2981	0.2946	0.2912	0.2877	0.2843	0.2810	0.2776
−0.4	0.3446	0.3409	0.3372	0.3336	0.3300	0.3264	0.3228	0.3192	0.3156	0.3121
−0.3	0.3821	0.3783	0.3745	0.3707	0.3669	0.3632	0.3594	0.3557	0.3520	0.3483
−0.2	0.4207	0.4168	0.4129	0.4090	0.4052	0.4013	0.3974	0.3936	0.3897	0.3859
−0.1	0.4602	0.4562	0.4522	0.4483	0.4443	0.4404	0.4364	0.4325	0.4286	0.4247
−0.0	0.5000	0.4960	0.4920	0.4880	0.4840	0.4801	0.4761	0.4721	0.4681	0.4641

The Standard Normal Distribution Table *(continued)*

z	**0.00**	**0.01**	**0.02**	**0.03**	**0.04**	**0.05**	**0.06**	**0.07**	**0.08**	**0.09**
0.0	0.5000	0.5040	0.5080	0.5120	0.5160	0.5199	0.5239	0.5279	0.5319	0.5359
0.1	0.5398	0.5438	0.5478	0.5517	0.5557	0.5596	0.5636	0.5675	0.5714	0.5753
0.2	0.5793	0.5832	0.5871	0.5910	0.5948	0.5987	0.6026	0.6064	0.6103	0.6141
0.3	0.6179	0.6217	0.6255	0.6293	0.6331	0.6368	0.6406	0.6443	0.6480	0.6517
0.4	0.6554	0.6591	0.6628	0.6664	0.6700	0.6736	0.6772	0.6808	0.6844	0.6879
0.5	0.6915	0.6950	0.6985	0.7019	0.7054	0.7088	0.7123	0.7157	0.7190	0.7224
0.6	0.7257	0.7291	0.7324	0.7357	0.7389	0.7422	0.7454	0.7486	0.7517	0.7549
0.7	0.7580	0.7611	0.7642	0.7673	0.7704	0.7734	0.7764	0.7794	0.7823	0.7852
0.8	0.7881	0.7910	0.7939	0.7967	0.7995	0.8023	0.8051	0.8078	0.8106	0.8133
0.9	0.8159	0.8186	0.8212	0.8238	0.8264	0.8289	0.8315	0.8340	0.8365	0.8389
1.0	0.8413	0.8438	0.8461	0.8485	0.8508	0.8531	0.8554	0.8577	0.8599	0.8621
1.1	0.8643	0.8665	0.8686	0.8708	0.8729	0.8749	0.8770	0.8790	0.8810	0.8830
1.2	0.8849	0.8869	0.8888	0.8907	0.8925	0.8944	0.8962	0.8980	0.8997	0.9015
1.3	0.9032	0.9049	0.9066	0.9082	0.9099	0.9115	0.9131	0.9147	0.9162	0.9177
1.4	0.9192	0.9207	0.9222	0.9236	0.9251	0.9265	0.9278	0.9292	0.9306	0.9319
1.5	0.9332	0.9345	0.9357	0.9370	0.9382	0.9394	0.9406	0.9418	0.9429	0.9441
1.6	0.9452	0.9463	0.9474	0.9484	0.9495	0.9505	0.9515	0.9525	0.9535	0.9545
1.7	0.9554	0.9564	0.9573	0.9582	0.9591	0.9599	0.9608	0.9616	0.9625	0.9633
1.8	0.9641	0.9649	0.9656	0.9664	0.9671	0.9678	0.9686	0.9693	0.9699	0.9706
1.9	0.9713	0.9719	0.9726	0.9732	0.9738	0.9744	0.9750	0.9756	0.9761	0.9767
2.0	0.9772	0.9778	0.9783	0.9788	0.9793	0.9798	0.9803	0.9808	0.9812	0.9817
2.1	0.9821	0.9826	0.9830	0.9834	0.9838	0.9842	0.9846	0.9850	0.9854	0.9857
2.2	0.9861	0.9864	0.9868	0.9871	0.9875	0.9878	0.9881	0.9884	0.9887	0.9890
2.3	0.9893	0.9896	0.9898	0.9901	0.9904	0.9906	0.9909	0.9911	0.9913	0.9916
2.4	0.9918	0.9920	0.9922	0.9925	0.9927	0.9929	0.9931	0.9932	0.9934	0.9936
2.5	0.9938	0.9940	0.9951	0.9943	0.9945	0.9946	0.9948	0.9949	0.9951	0.9952
2.6	0.9953	0.9955	0.9956	0.9957	0.9959	0.9960	0.9961	0.9962	0.9963	0.9964
2.7	0.9965	0.9966	0.9967	0.9968	0.9969	0.9970	0.9971	0.9972	0.9973	0.9974
2.8	0.9974	0.9975	0.9976	0.9977	0.9977	0.9978	0.9979	0.9979	0.9980	0.9981
2.9	0.9981	0.9982	0.9982	0.9983	0.9984	0.9984	0.9985	0.9985	0.9986	0.9986
3.0	0.9987	0.9987	0.9987	0.9988	0.9988	0.9989	0.9989	0.9989	0.9990	0.9990
3.1	0.9990	0.9991	0.9991	0.9991	0.9992	0.9992	0.9992	0.9992	0.9993	0.9993
3.2	0.9993	0.9993	0.9994	0.9994	0.9994	0.9994	0.9994	0.9995	0.9995	0.9995
3.3	0.9995	0.9995	0.9995	0.9996	0.9996	0.9996	0.9996	0.9996	0.9996	0.9997
3.4	0.9997	0.9997	0.9997	0.9997	0.9997	0.9997	0.9997	0.9997	0.9997	0.9998

TAYLOR POLYNOMIALS AND INFINITE SERIES

Photo: Thinkstock/Jupiter Images

In this chapter we show how certain functions can be represented by a *power series*. A power series involves infinitely many terms, but when truncated it is just a polynomial. By approximating a function with a *Taylor polynomial,* we are often able to obtain approximate solutions to problems that we cannot otherwise solve.

Photo: Thinkstock/Jupiter Images

What percentage of the nonfarm workforce will be in the service industries one decade from now? In Example 4, page 792, you will see how a Taylor polynomial can be used to help answer this question.

Photo: Indu Ghuman

14.1 Taylor Polynomials

As we saw earlier, obtaining an exact solution to a problem is not always possible; in such cases we have to settle for an approximate solution. In this section we show how a function may be approximated near a given point by a polynomial. Polynomials, as we have seen time and again, are easy to work with; for example, they are easy to evaluate, differentiate, and integrate. Thus, by using polynomials rather than working with the original function itself, we can often obtain approximate solutions to a problem that we might otherwise not be able to solve.

Recall from Section 2.3 that a **polynomial of degree n** is a polynomial where the highest degree term is of the form ax^n. Thus $3x + 2$ is a polynomial of degree 1, $x^2 + 5x - 2$ is a polynomial of degree 2, and $x^7 + 4x^2 + 2x + 1$ is a polynomial of degree 7.

Taylor Polynomials

Suppose we are given a differentiable function f and a point $x = a$ in the domain of f. Then the polynomial of degree 0 that best approximates f *near* $x = a$ is the constant polynomial

$$P_0(x) = f(a)$$

which coincides with f at $x = a$ (Figure 14.1a).

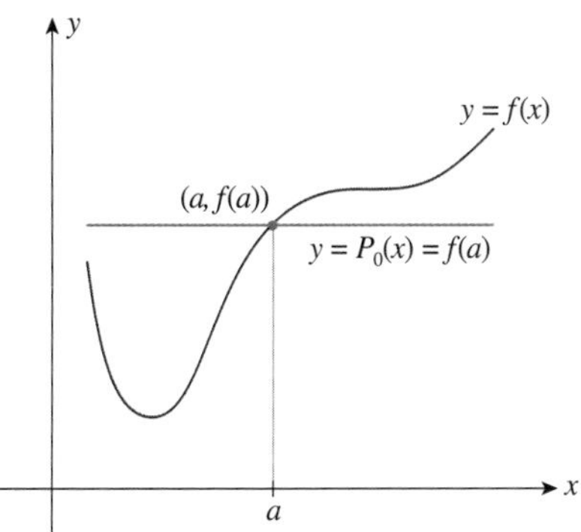

(a) $P_0(x) = f(a)$ is a zero-degree polynomial that approximates f near $x = a$.

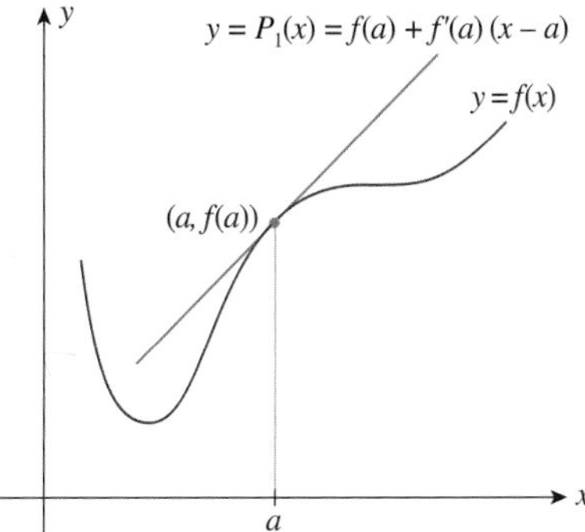

(b) $P_1(x) = f(a) + f'(a)(x - a)$ is a first-degree polynomial that approximates f near $x = a$.

Figure 14.1

Now, unless f itself is a constant function, it is possible in many cases to obtain a better approximation of f *near* $x = a$ by using a polynomial function of degree 1. Recall that the linear function

$$L(x) = f(a) + f'(a)(x - a)$$

is just an equation of the tangent line to the graph of the function f at the point $(a, f(a))$ (Figure 14.1b). As such, the value of the function L coincides with the value of f at $x = a$, and its slope coincides with the slope of f at $x = a$; that is,

$$L(a) = f(a) \qquad \text{and} \qquad L'(a) = f'(a)$$

Let's write

$$L(x) = P_1(x)$$

Then
$$P_1(x) = f(a) + f'(a)(x - a)$$

is the required polynomial approximation (of degree 1) of f near $x = a$.

This discussion suggests that yet a better approximation to f at $x = a$ may be found by using a polynomial of degree 2, $P_2(x)$, and requiring that its value, slope, and concavity coincide with those of f at $x = a$. In other words, $P_2(x)$ should satisfy the three conditions

$$P_2(a) = f(a) \quad P'_2(a) = f'(a) \quad P''_2(a) = f''(a)$$

The third condition ensures that the graph of the polynomial bends in the right way, at least near $x = a$. Pursuing this line of reasoning, we are led to the search for a polynomial of degree n in $x - a$,

$$\begin{aligned} P_n(x) &= a_0 + a_1(x - a) + a_2(x - a)^2 \\ &\quad + a_3(x - a)^3 + \cdots + a_n(x - a)^n \end{aligned}$$

(where $a_0, a_1, \ldots, a_n$ are constants), that satisfies the conditions

$$P_n(a) = f(a), P'_n(a) = f'(a), P''_n(a) = f''(a), \ldots, P_n^{(n)}(a) = f^{(n)}(a) \tag{1}$$

To determine the required polynomial, we compute

$$\begin{aligned} P'_n(x) &= a_1 + 2a_2(x - a) + 3a_3(x - a)^2 + \cdots + na_n(x - a)^{n-1} \\ P''_n(x) &= 2a_2 + 3 \cdot 2a_3(x - a) + \cdots + n(n - 1)a_n(x - a)^{n-2} \\ P'''_n(x) &= 3 \cdot 2a_3 + 4 \cdot 3 \cdot 2a_4(x - a) + \cdots + n(n - 1)(n - 2)a_n(x - a)^{n-3} \\ &\vdots \\ P_n^{(n)}(x) &= n(n - 1)(n - 2) \cdots (1)a_n \end{aligned}$$

Setting $x = a$ in each of the expressions for $P_n(x)$, $P'_n(x)$, $P''_n(x)$, $\ldots$, $P_n^{(n)}(x)$ in succession and using the conditions in (1), we find

$$\begin{aligned} P_n(a) &= a_0 = f(a) \\ P'_n(a) &= a_1 = f'(a) \\ P''_n(a) &= 2a_2 = f''(a) \\ P'''_n(a) &= 3 \cdot 2a_3 = f'''(a) \\ &\vdots \\ P_n^{(n)}(a) &= n(n - 1)(n - 2) \cdots (1)a_n = f^{(n)}(a) \end{aligned}$$

from which we deduce that

$$a_0 = f(a), a_1 = f'(a), a_2 = \frac{1}{2}f''(a), a_3 = \frac{1}{3 \cdot 2}f'''(a), \ldots,$$

$$a_n = \frac{1}{n(n-1)(n-2)\cdots(1)} f^{(n)}(a)$$

Let's introduce the expression $n!$ (read "n factorial"), defined by

$$n! = n(n-1)(n-2)(n-3)\cdots 3\cdot 2\cdot 1 \qquad (\text{for } n \geq 1)$$
$$0! = 1$$

Thus,

$$1! = 1 \qquad 4! = 4\cdot 3\cdot 2\cdot 1 = 24$$
$$2! = 2\cdot 1 = 2 \qquad 5! = 5\cdot 4\cdot 3\cdot 2\cdot 1 = 120$$
$$3! = 3\cdot 2\cdot 1 = 6$$

and so on. Using this notation, we may write the coefficients of $P_n(x)$ as

$$a_0 = f(a),\ a_1 = f'(a),\ a_2 = \frac{1}{2!} f''(a),\ a_3 = \frac{1}{3!} f'''(a), \ldots,\ a_n = \frac{1}{n!} f^{(n)}(a)$$

so that the required polynomial is

$$P_n(x) = f(a) + f'(a)(x-a) + \frac{f''(a)}{2!}(x-a)^2 + \cdots + \frac{f^{(n)}(a)}{n!}(x-a)^n$$

The *n*th Taylor Polynomial

Suppose that the function f and its first n derivatives are defined at $x = a$. Then the nth Taylor polynomial of f at $x = a$ is the polynomial

$$P_n(x) = f(a) + f'(a)(x-a) + \frac{f''(a)}{2!}(x-a)^2 + \cdots + \frac{f^{(n)}(a)}{n!}(x-a)^n \qquad (2)$$

which coincides with $f(x), f'(x), \ldots, f^{(n)}(x)$ at $x = a$; that is,

$$P_n(a) = f(a),\ P_n'(a) = f'(a), \ldots,\ P_n^{(n)}(a) = f^{(n)}(a)$$

Using a Taylor Polynomial to Approximate a Function

In many instances the Taylor polynomial $P_n(x)$ provides us with a good approximation of $f(x)$ near $x = a$.

Example 1

Find the first four Taylor polynomials of $f(x) = e^x$ at $x = 0$ and sketch the graph of each polynomial superimposed upon the graph of $f(x) = e^x$.

Solution

Here $a = 0$ and, since

$$f(x) = f'(x) = f''(x) = f'''(x) = f^{(4)}(x) = e^x$$

we find

$$f(0) = f'(0) = f''(0) = f'''(0) = f^{(4)}(0) = 1$$

Using Formula (2) with $n = 1, 2, 3,$ and 4 in succession, we find that the first four Taylor polynomials are

$$P_1(x) = f(0) + f'(0)(x-0) = 1 + x$$

$$P_2(x) = f(0) + f'(0)(x-0) + \frac{f''(0)}{2!}(x-0)^2 = 1 + x + \frac{1}{2}x^2$$

$$P_3(x) = f(0) + f'(0)(x - 0) + \frac{f''(0)}{2!}(x - 0)^2 + \frac{f'''(0)}{3!}(x - 0)^3$$
$$= 1 + x + \frac{1}{2}x^2 + \frac{1}{6}x^3$$
$$P_4(x) = f(0) + f'(0)(x - 0) + \frac{f''(0)}{2!}(x - 0)^2 + \frac{f'''(0)}{3!}(x - 0)^3$$
$$+ \frac{f^{(4)}(0)}{4!}(x - 0)^4 = 1 + x + \frac{1}{2}x^2 + \frac{1}{6}x^3 + \frac{1}{24}x^4$$

The graphs of these polynomials are shown in Figure 14.2. Observe that the approximation of $f(x)$ near $x = 0$ improves as the degree of the approximating Taylor polynomial increases.

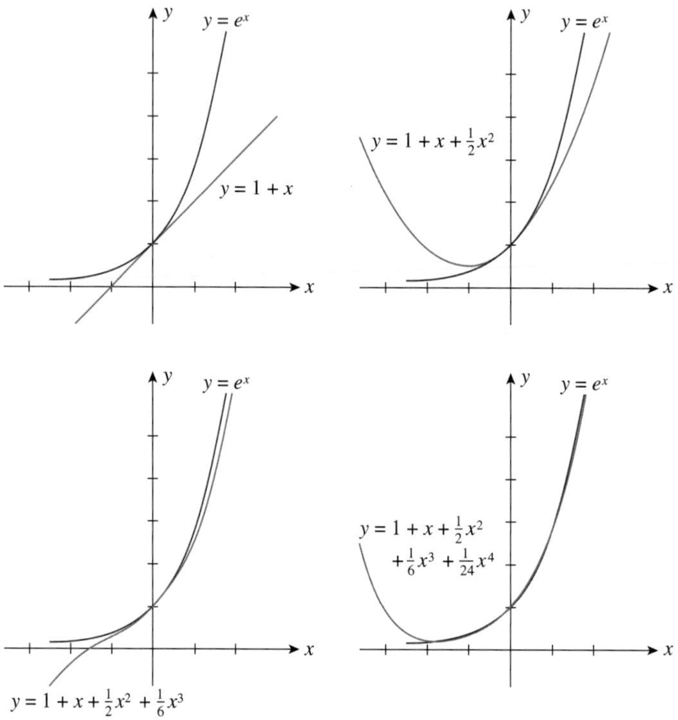

Figure 14.2
The graphs of the first four Taylor polynomials of $f(x) = e^x$ at $x = 0$ superimposed upon the graph of $f(x) = e^x$

EXPLORING WITH TECHNOLOGY

Let $f(x) = xe^{-x}$.

1. Find the first four Taylor polynomials of f at $x = 0$.
2. Use a graphing utility to plot the graphs of f, P_1, P_2, P_3, and P_4 on the same set of axes in the viewing window $[-0.5, 1] \cdot [-0.5, 0.5]$.
3. Comment on the approximation of f by the polynomial P_n near $x = 0$ for $n = 1, 2, 3$, and 4. What happens to the approximation if x is "far" from the origin?

Example 2

Find the nth Taylor polynomial of $f(x) = 1/(1 - x)$ at $x = 0$. Compute $P_5(0.1)$ and compare this result with $f(0.1)$.

Solution

Here $a = 0$. Next, we find

$$\begin{aligned}
f(x) &= \frac{1}{1-x} = (1-x)^{-1} \\
f'(x) &= -(1-x)^{-2}(-1) = (1-x)^{-2} \\
f''(x) &= -2(1-x)^{-3}(-1) = 2(1-x)^{-3} \\
f'''(x) &= 3\cdot 2(1-x)^{-4} \\
f^{(4)}(x) &= 4\cdot 3\cdot 2(1-x)^{-5} \\
&\vdots \\
f^{(n)}(x) &= n(n-1)(n-2)\cdots 1(1-x)^{-n-1}
\end{aligned}$$

so that

$$f(0) = 1, f'(0) = 1, f''(0) = 2, f'''(0) = 3!, f^{(4)}(0) = 4!, \ldots, f^{(n)}(0) = n!$$

Therefore, the required nth Taylor polynomial is

$$\begin{aligned}
P_n(x) &= 1 + \frac{1}{1}x + \frac{2}{2!}x^2 + \frac{3!}{3!}x^3 + \frac{4!}{4!}x^4 + \cdots + \frac{n!}{n!}x^n \\
&= 1 + x + x^2 + x^3 + x^4 + \cdots + x^n
\end{aligned}$$

In particular,

$$P_5(x) = 1 + x + x^2 + x^3 + x^4 + x^5$$

so that

$$\begin{aligned}
P_5(0.1) &= 1 + 0.1 + (0.1)^2 + (0.1)^3 + (0.1)^4 + (0.1)^5 \\
&= 1.11111
\end{aligned}$$

On the other hand, the actual value of $f(x)$ at $x = 0.1$ is

$$f(0.1) = \frac{1}{1-0.1} = \frac{1}{0.9} = 1.111\ldots$$

Thus, the error incurred in approximating $f(0.1)$ by $P_5(0.1)$ is 0.00000111. . . .

Applications

Example 3

Obtain the second Taylor polynomial of $f(x) = \sqrt{x}$ at $x = 25$ and use it to approximate the value of $\sqrt{26.5}$.

Solution

Here $a = 25$ and $n = 2$. Since

$$f(x) = x^{1/2} \quad f'(x) = \frac{1}{2}x^{-1/2} = \frac{1}{2\sqrt{x}} \quad f''(x) = -\frac{1}{4}x^{-3/2} = -\frac{1}{4x^{3/2}}$$

we find that

$$f(25) = 5 \qquad f'(25) = \frac{1}{10} \qquad f''(25) = -\frac{1}{500}$$

Using Formula (2), we obtain the required polynomial

$$P_2(x) = f(25) + f'(25)(x - 25) + \frac{f''(25)}{2!}(x - 25)^2$$
$$= 5 + \frac{1}{10}(x - 25) - \frac{1}{1000}(x - 25)^2$$

Next, using $P_2(x)$ as an approximation of $f(x)$ near $x = 25$, we find

$$\sqrt{26.5} = f(26.5)$$
$$\approx P_2(26.5)$$
$$= 5 + \frac{1}{10}(26.5 - 25) - \frac{1}{1000}(26.5 - 25)^2$$
$$\approx 5.14775$$

The exact value of $\sqrt{26.5}$, rounded off to five decimal places, is 5.14782. Thus, the error incurred in the approximation is 0.00007.

The next example shows how a Taylor polynomial can be used to approximate an integral that involves an integrand whose antiderivative cannot be expressed as an elementary function.

GROUP DISCUSSION

Suppose you want to find an approximation to the value of $\sqrt[3]{26.5}$ using a second Taylor polynomial of some function f expanded about $x = a$.

1. How would you choose f and a?
2. Obtain an approximation of $\sqrt[3]{26.5}$ and compare your result with that obtained using a calculator.

Example 4

Growth of the Service Industries It has been estimated that service industries, which currently make up 30% of the nonfarm workforce in a certain country, will continue to grow at the rate of

$$R(t) = 5e^{1/(t+1)}$$

percent per decade, t decades from now. Estimate the percentage of the nonfarm workforce in the service industries one decade from now.

Solution

The percentage of the nonfarm workforce in the service industries t decades from now will be given by

$$P(t) = \int 5e^{1/(t+1)}\,dt \qquad [P(0) = 30]$$

This integral cannot be expressed in terms of an elementary function. To obtain an approximate solution to the problem at hand, let's first make the substitution

$$u = \frac{1}{t+1}$$

so that

$$t + 1 = \frac{1}{u} \quad \text{and} \quad t = \frac{1}{u} - 1$$

giving

$$dt = -\frac{1}{u^2}\,du$$

The integral becomes

$$\tilde{P}(u) = 5\int e^u\left(-\frac{du}{u^2}\right) = -5\int \frac{e^u}{u^2}\,du$$

Next, let's approximate e^u at $u = 0$ by a fourth-degree Taylor polynomial. Proceeding as in Example 1, we find that

$$e^u \approx 1 + u + \frac{u^2}{2!} + \frac{u^3}{3!} + \frac{u^4}{4!}$$

Thus,

$$\begin{aligned}
\tilde{P}(u) &\approx -5\int \frac{1}{u^2}\left(1 + u + \frac{u^2}{2} + \frac{u^3}{6} + \frac{u^4}{24}\right) du \\
&= -5\int \left(\frac{1}{u^2} + \frac{1}{u} + \frac{1}{2} + \frac{u}{6} + \frac{u^2}{24}\right) du \\
&= -5\left(-\frac{1}{u} + \ln u + \frac{1}{2}u + \frac{u^2}{12} + \frac{u^3}{72}\right) + C
\end{aligned}$$

Therefore,

$$P(t) \approx -5\left[-(t+1) + \ln\left(\frac{1}{t+1}\right) + \frac{1}{2(t+1)} + \frac{1}{12(t+1)^2} + \frac{1}{72(t+1)^3}\right] + C$$

Using the condition $P(0) = 30$, we find

$$P(0) \approx -5\left(-1 + \ln 1 + \frac{1}{2} + \frac{1}{12} + \frac{1}{72}\right) + C = 30$$

or $C \approx 27.99$. So

$$P(t) \approx -5\left[-(t+1) + \ln\left(\frac{1}{t+1}\right) + \frac{1}{2(t+1)} + \frac{1}{12(t+1)^2} + \frac{1}{72(t+1)^3}\right] + 27.99$$

In particular, the percentage of the nonfarm workforce in the service industries one decade from now will be given by

$$\begin{aligned}
P(1) &\approx -5\left[-2 + \ln\left(\frac{1}{2}\right) + \frac{1}{4} + \frac{1}{48} + \frac{1}{576}\right] + 27.99 \\
&\approx 40.09
\end{aligned}$$

or approximately 40.1%.

Errors in Taylor Polynomial Approximations

The error incurred in approximating $f(x)$ by its Taylor polynomial $P_n(x)$ at $x = a$ is

$$R_n(x) = f(x) - P_n(x)$$

A formula for $R_n(x)$, called the *remainder*, associated with $P_n(x)$ is stated without proof in what follows.

THEOREM 1

The Remainder Theorem

Suppose a function f has derivatives up to order $(n + 1)$ on an interval I containing the number a. Then for each x in I, there exists a number c lying between a and x such that

$$R_n(x) = \frac{f^{(n+1)}(c)}{(n+1)!}(x-a)^{n+1}$$

Usually the exact value of c is unknown. But from a practical point of view, this is not a serious drawback because what we need is just a bound on the error incurred when a Taylor polynomial of f is used to approximate $f(x)$. For this purpose, we have the following result, which is a consequence of the remainder theorem.

Error Bound for Taylor Polynomial Approximations

Suppose a function f has derivatives up to order $(n + 1)$ on an interval I containing the number a. Then for each x in I,

$$|R_n(x)| \leq \frac{M}{(n+1)!}|x - a|^{n+1} \tag{3}$$

where M is any number such that $|f^{(n+1)}(t)| \leq M$ for all t lying between a and x.

Example 5

Refer to Example 2. Use Formula (3) to obtain an upper bound for the error incurred in approximating $f(0.1)$ by $P_5(0.1)$, where

$$f(x) = \frac{1}{1-x}$$

and $P_5(x)$ is the fifth Taylor polynomial of f at $x = 0$. Compare this result with the actual error incurred in the approximation.

Solution

First, we need to find a number M such that

$$|f^{(6)}(t)| \leq M$$

for all t lying in the interval [0, 0.1]. From the solution of Example 2, we find that

$$f^{(6)}(t) = (6)(5)(4)(3)(2)(1)(1-t)^{-6-1}$$
$$= \frac{720}{(1-t)^7}$$

Observe that

$$f^{(6)}(t) = \frac{720}{(1-t)^7}$$

is increasing on the interval [0, 0.1]. [Just look at $f^{(7)}(t)$.] Consequently, its maximum is attained at the right end point $t = 0.1$. Thus, we may take

$$M = \frac{720}{(1-0.1)^7} \approx 1505.3411$$

Next, with $a = 0$, $x = 0.1$, and $M = 1505.3411$, Formula (3) gives

$$|R_5(0.1)| \leq \frac{1505.3411}{(5+1)!}(0.1)^{5+1}$$
$$\approx 0.0000021$$

Finally, we refer once again to Example 2 to see that the actual error incurred in approximating $f(0.1)$ by $P_5(0.1)$ was 0.00000111. . . , and this is less than the error bound 0.0000021 just obtained.

GROUP DISCUSSION

Refer to the Group Discussion question on page 792. Obtain an upper bound for the error incurred in approximating $\sqrt[3]{26.5}$, using the Taylor polynomial you have chosen.

Example 6

Use the fourth Taylor polynomial to obtain an approximation of $e^{0.5}$. Find a bound for the error in the approximation.

Solution

The point $x = 0.5$ is near $x = 0$, so we use the fourth Taylor polynomial $P_4(x)$ at $x = 0$ to approximate $e^{0.5}$. From Example 1, we see that

$$P_4(x) = 1 + x + \frac{1}{2}x^2 + \frac{1}{6}x^3 + \frac{1}{24}x^4$$

Therefore,

$$\begin{aligned} e^{0.5} \approx P_4(0.5) &= 1 + 0.5 + \frac{1}{2}(0.5)^2 + \frac{1}{6}(0.5)^3 + \frac{1}{24}(0.5)^4 \\ &\approx 1.6484 \end{aligned}$$

To obtain an error bound for this approximation, we need to obtain an upper bound M for $|f^{(5)}(t)|$ for all t lying in the interval $[0, 0.5]$ (remember, $n = 4$). Now

$$f^{(5)}(t) = e^t$$

and since $e^t > 0$ for all values of t, $f^{(5)}(t) = |e^t| = e^t$. Next, we observe that e^t is an increasing function on $(-\infty, \infty)$ and in particular on $[0, 0.5]$. Therefore, e^t attains an absolute maximum at $x = 0.5$, the right end point of the interval in question; that is,

$$|f^{(5)}(t)| = |e^t| = e^t \le e^{0.5}$$

on $[0, 0.5]$. Now, $e < 3$, so certainly

$$|f^{(5)}(t)| \le e^{0.5} < 3^{0.5} = \sqrt{3} < 2$$

Thus, we may take $M = 2$. Finally, with $a = 0$, $x = 0.5$, and this value of M, Formula (3) gives

$$\begin{aligned} |R_4(0.5)| &\le \frac{2}{(4+1)!}|0.5 - 0|^{4+1} \\ &= \frac{2}{5!}(0.5)^5 \\ &\approx 0.00052 \end{aligned}$$

So our result, $e^{0.5} \approx 1.6484$, is guaranteed to be accurate to within 0.0005.

Remarks

1. Using a calculator, we see that $e^{0.5} \approx 1.6487$, rounded to four decimal places. Therefore, the error incurred in approximating this number by $P_4(0.5)$ is (1.6487 − 1.6484), or 0.0003, which is less than the upper error bound of 0.0005 obtained.
2. The error bound obtained in Example 6 is not the best possible. This stems partially from our choice of M. In practical applications, it is not essential that the error bound be the best possible as long as it is sharp enough. ◀

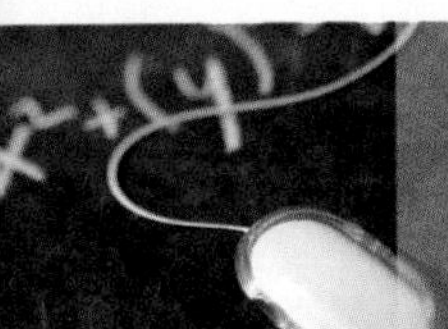

Self-Check Exercises 14.1

1. Let $f(x) = \dfrac{1}{1+x}$.
   a. Find the nth Taylor polynomial of f at $x = 0$.
   b. Use the third Taylor polynomial of f at $x = 0$ to obtain an approximation of $f(0.1)$. Compare this result with $f(0.1)$.
2. Refer to Exercise 1. Find a bound on the error in the approximation of $f(0.1)$.

Solutions to Self-Check Exercises 14.1 can be found on page 798.

14.1 Exercises

In Exercises 1–10, find the first three Taylor polynomials of the function at the indicated point.

1. $f(x) = e^{-x}$; $x = 0$

2. $f(x) = e^{2x}$; $x = 0$

3. $f(x) = \dfrac{1}{x+1}$; $x = 0$

4. $f(x) = \dfrac{1}{1-x}$; $x = 2$

5. $f(x) = \dfrac{1}{x}$; $x = 1$

6. $f(x) = \dfrac{1}{x+2}$; $x = 1$

7. $f(x) = \sqrt{1-x}$; $x = 0$

8. $f(x) = \sqrt{x}$; $x = 4$

9. $f(x) = \ln(1 - x)$; $x = 0$

10. $f(x) = xe^{x}$; $x = 0$

In Exercises 11–20, find the *n*th Taylor polynomial of the function at the indicated point.

11. $f(x) = x^4$ at $x = 2$, $n = 2$

12. $f(x) = x^5$ at $x = -1$, $n = 3$

13. $f(x) = \ln x$ at $x = 1$, $n = 4$

14. $f(x) = \dfrac{1}{x}$ at $x = 2$, $n = 4$

15. $f(x) = e^{x}$ at $x = 1$, $n = 4$

16. $f(x) = e^{2x}$ at $x = 1$, $n = 4$

17. $f(x) = \sqrt{1-x}$ at $x = 0$, $n = 3$

18. $f(x) = \sqrt{x+1}$ at $x = 0$, $n = 3$

19. $f(x) = \dfrac{1}{2x+3}$ at $x = 0$, $n = 3$

20. $f(x) = x^2e^{x}$ at $x = 0$, $n = 2$

21. Find the *n*th Taylor polynomial of $f(x) = \dfrac{1}{1+x}$ at $x = 0$. Compute $P_4(0.1)$ and compare this result with $f(0.1)$.

22. Find the third Taylor polynomial of $f(x) = \sqrt{3x+1}$ at $x = 0$. Compute $P_3(0.2)$ and compare this result with $f(0.2)$.

23. Find the fourth Taylor polynomial of $f(x) = e^{-x/2}$ at $x = 0$ and use it to estimate $e^{-0.1}$.

24. Find the third Taylor polynomial of $f(x) = \ln x$ at $x = 1$ and use it to estimate $\ln 1.1$.

25. Find the second Taylor polynomial of $f(x) = \sqrt{x}$ at $x = 16$ and use it to estimate the value of $\sqrt{15.6}$.

26. Find the second Taylor polynomial of $f(x) = x^{1/3}$ at $x = 8$ and use it to estimate the value of $\sqrt[3]{8.1}$.

27. Find the third Taylor polynomial of $f(x) = \ln(x + 1)$ at $x = 0$ and use it to estimate the value of

$$\int_0^{1/2} \ln(x+1)\,dx$$

Compare your result with the exact value found using integration by parts.

28. Use the fourth Taylor polynomial of $f(x) = \ln(1 + x)$ to approximate

$$\int_{0.1}^{0.2} \frac{\ln(x+1)}{x}\,dx$$

29. Use the second Taylor polynomial of $f(x) = \sqrt{x}$ at $x = 4$ to approximate $\sqrt{4.06}$ and find a bound for the error in the approximation.

30. Use the third Taylor polynomial of $f(x) = e^{-x}$ at $x = 0$ to approximate $e^{-0.2}$ and find a bound for the error in the approximation.

31. Use the third Taylor polynomial of $f(x) = \dfrac{1}{1-x}$ at $x = 0$ to approximate $f(0.2)$. Find a bound for the error in the approximation and compare your results to the exact value of $f(0.2)$.

32. Use the second Taylor polynomial of $f(x) = \sqrt{x+1}$ at $x = 0$ to approximate $f(0.1)$ and find a bound for the error in the approximation.

33. Use the second Taylor polynomial of $f(x) = \ln x$ at $x = 1$ to approximate $\ln 1.1$ and find a bound for the error in the approximation.

34. Use the second Taylor polynomial of $f(x) = \ln x$ at $x = 1$ to approximate $\ln 0.9$ and find a bound for the error in the approximation.

35. Let $f(x) = e^{-x}$.

a. Find a bound in the error incurred in approximating $f(x)$ by the third Taylor polynomial, $P_3(x)$, of f at $x = 0$ in the interval $[0, 1]$.

b. Estimate $\int_0^1 e^{-x}\,dx$ by computing $\int_0^1 P_3(x)\,dx$.

c. What is the bound on the error in the approximation in (b)?

Hint: Compute $\int_0^1 B\,dx$, where B is a bound on $|R_3(x)|$ for x in $[0, 1]$.

d. What is the actual error?

36. a. Find a bound on the error incurred in approximating $f(x) = \ln x$ by the fourth Taylor polynomial, $P_4(x)$, of f at $x = 1$ in the interval $[1, 1.5]$.

b. Approximate the area under the graph of f from $x = 1$ to $x = 1.5$ by computing

$$\int_1^{1.5} P_4(x)\,dx$$

c. What is the bound on the error in the approximation in part (b)?

Hint: Compute $\int_1^{1.5} B\,dx$, where B is a bound on $|R_4(x)|$ for x in $[1, 1.5]$.

d. What is the actual error?

Hint: Use integration by parts to evaluate $\int_1^{1.5} \ln x\,dx$.

37. Use the fourth Taylor polynomial at $x = 0$ to obtain an approximation of the area under the graph of $f(x) = e^{-x^2/2}$ from $x = 0$ to $x = 0.5$.

Hint: Use the substitution $u = -x^2/2$. See Example 4.

38. Use the fourth Taylor polynomial at $x = 0$ to obtain an approximation of the area under the graph of $f(x) = \ln(1 + x^2)$ from $x = -\frac{1}{2}$ to $x = \frac{1}{2}$.

Hint: First find the Taylor polynomial of $f(u) = \ln(1 + u)$; then use the substitution $u = x^2$.

Business and Economics Applications

39. Growth of Service Industries It has been estimated that service industries, which currently make up 30% of the nonfarm workforce in a certain country, will continue to grow at the rate of

$$R(t) = 6e^{1/(2t+1)}$$

percent per decade t decades from now. Estimate the percentage of the nonfarm workforce in service industries two decades from now.

40. Vacation Trends The Travel Data Centre of a European country has estimated that the percentage of vacations traditionally taken during the months of July, August, and September will continue to decline according to the rule

$$R(t) = -2e^{1/(t+1)}$$

percent per decade ($t = 0$ corresponds to the year 1994) as more and more vacationers switch to fall and winter vacations. There were 37% summer vacationers in that country in 1994. Estimate the percentage of vacations that were taken in the summer months in 2004.

41. Continuing Education Enrollment The registrar of Kellogg University estimates that the total student enrollment in the Continuing Education division will be given by

$$N(t) = -\frac{20{,}000}{\sqrt{1 + 0.2t}} + 21{,}000$$

where $N(t)$ denotes the number of students enrolled in the division t yr from now. Using the second Taylor polynomial of N at $t = 0$, find an approximation of the average enrollment at Kellogg University between $t = 0$ and $t = 2$.

Biological and Life Sciences Applications

42. Concentration of Carbon Monoxide in the Air According to a joint study conducted by an Environmental Management Department and a government agency, the concentration of carbon monoxide (CO) in the air due to automobile exhaust t yr from now is given by

$$C(t) = 0.01(0.2t^2 + 4t + 64)^{2/3}$$

parts per million. Use the second Taylor polynomial of C at $t = 0$ to obtain an approximation of the average level of concentration of CO in the air between $t = 0$ and $t = 2$.

43. Average Concentration A biochemical reaction exhibits rapidly increasing oscillation in the concentration of one of the reactants. For the first 30 minutes the reaction is run, the level of this reactant is

$$13.5 + 6.4\sin(t^2)$$

micrograms per minute. Recall that the average level of the reactant in the first 10 minutes of the reaction is given by the integral

$$\frac{1}{10 - 0}\int_0^{10} 13.5 + 6.4\sin(t^2)\,dt$$

micrograms. The function

$$\sin(t^2)$$

is not one we know how to integrate. Approximate the integrand with the tenth Taylor polynomial and use that to estimate the average concentration.

44. Estimating Mass A type of fast-growing moss is found to grow so that the new mass at time t is given by the formula

$$0.77t^2\,e^{-t^2}$$

grams per day. This means that the total mass of the moss at a time T is

$$\int_0^T 0.77t^2\,e^{-t^2}\,dt$$

but this is not a function that can be integrated symbolically. Find the eighth Taylor polynomial at $t = 0$ and integrate that polynomial to estimate the mass of the moss plant at the end of the first day.

In Exercises 43–46, determine whether the statement is true or false. If it is true, explain why it is true. If it is false, give an example to show why it is false.

45. If f is a polynomial of degree four and $P_4(x)$ is the fourth Taylor polynomial of f at $x = a$, then $f(x) \neq P_4(x)$ for at least one value of x.

46. The function $f(x) = \dfrac{1}{x - 2}$ has a Taylor polynomial at $x = a$ for any value of a except $a = 2$.

47. Suppose f is a polynomial of degree n and R_n is the remainder associated with P_n, the nth Taylor polynomial of f at $x = a$. Then $R_n(x) = 0$ for every value of x.

48. Let P_n denote the nth Taylor polynomial of the function $f(x) = e^{-x}$ at $x = a$. Then $R_n(x) = f(x) - P_n(x)$ is never equal to zero for all values of x.

Solutions to Self-Check Exercises 14.1

1. a. We rewrite $f(x)$ as

$$f(x) = (1 + x)^{-1}$$

and compute

$$f'(x) = -(1 + x)^{-2}$$

$$f''(x) = (-1)(-2)(1 + x)^{-3} = 2(1 + x)^{-3}$$

$$f'''(x) = 2(-3)(1 + x)^{-4} = -3 \cdot 2(1 + x)^{-4}$$

$$f^{(4)}(x) = -3 \cdot 2(-4)(1 + x)^{-5} = 4 \cdot 3 \cdot 2(1 + x)^{-5}$$

$$\vdots$$

$$f^{(n)}(x) = (-1)^n n!(1 + x)^{-(n+1)}$$

Evaluating $f(x)$ and its derivatives at $x = 0$ gives

$$f(0) = 1, f'(0) = -1, f''(0) = 2!, f'''(0) = -3!$$

$$f^{(4)}(0) = 4!, \ldots, f^{(n)}(0) = (-1)^n n!$$

Therefore, the required nth Taylor polynomial is

$$P_n(x) = 1 + \frac{(-1)}{1!}x + \frac{2}{2!}x^2 + \frac{-3!}{3!}x^3 + \frac{4!}{4!}x^4 + \cdots + \frac{(-1)^n n!}{n!}x^n$$

$$= 1 - x + x^2 - x^3 + x^4 - \cdots + (-1)^n x^n$$

b. The third Taylor polynomial of f at $x = 0$ is

$$P_3(x) = 1 - x + x^2 - x^3$$

So

$$P_3(0.1) = 1 - 0.1 + (0.1)^2 - (0.1)^3$$

$$= 0.909$$

The actual value of $f(x)$ at $x = 0.1$ is

$$f(0.1) = \frac{1}{1 + 0.1} = \frac{1}{1.1} = 0.909090\ldots$$

Thus, the error incurred in approximating $f(0.1)$ by $P_3(0.1)$ is $0.00009090\ldots$.

2. First, we need to find a number M such that

$$|f^{(4)}(t)| \leq M$$

for all t lying in the interval $[0, 0.1]$. From the solution of Self-Check Exercise 1, we see that

$$f^{(5)}(x) = (-1)^5 5!(1 + x)^{-6} = -5!(1 + x)^{-6}$$

Since $f^{(5)}(x)$ is negative for x in $[0, 0.1]$, $f^{(4)}(x)$ is decreasing on $[0, 0.1]$. Therefore, the maximum of the function

$$f^{(4)}(x) = 4!(1 + x)^{-5}$$

is attained at $x = 0$, and its value is

$$f^{(4)}(0) = 4! = 24$$

So we can take $M = 24$. Next, with $a = 0$, $x = 0.1$, and $M = 24$, Formula (3) gives

$$|R_3(0.1)| \leq \frac{24}{(3 + 1)!}|0.1 - 0|^{3+1}$$

$$= \frac{24}{24}(0.1)^4 = 0.0001$$

the required error bound.

14.2 Infinite Sequences

Infinite Sequences of Real Numbers

In Section 14.1 we saw that the approximation of a function f near a point $x = a$ by the nth Taylor polynomial of f at $x = a$ generally improves as n increases. This leads us to wonder what happens when the degree of a Taylor polynomial of a function at a point $x = a$ is allowed to increase without bound. Letting n go to infinity, however, leads to an expression involving the sum of "infinitely" many terms:

$$f(a) + f'(a)(x - a) + \frac{f''(a)}{2!}(x - a)^2 + \cdots$$

Clearly, this "sum" cannot be found by simply adding all the terms. Before we can define such a sum, we need to discuss sequences. (We will return to the problem posed here in Section 14.5.)

An infinite sequence of real numbers is a function whose domain is the set of positive integers. For example, the infinite sequence defined by

$$f(n) = \frac{1}{n}$$

is the ordered list of numbers

$$1, \frac{1}{2}, \frac{1}{3}, \ldots, \frac{1}{n}, \ldots$$

It is customary to represent the functional values called the terms of the sequence

$$f(1), f(2), f(3), \ldots, f(n), \ldots$$

by the notation

$$a_1, a_2, a_3, \ldots, a_n, \ldots$$

The term $a_n = f(n)$ is called the *n*th term, or general term, of the sequence. The sequence itself is abbreviated $\{a_n\}$. Observe that we can describe a sequence by specifying its *n*th term. For the example under consideration, the sequence is denoted by $\{1/n\}$, and its terms are

$$a_1 = 1, a_2 = \frac{1}{2}, a_3 = \frac{1}{3}, \ldots, a_n = \frac{1}{n}, \ldots$$

Infinite Sequence

An infinite sequence is a function whose domain is the set of positive integers. If f is a function defining the sequence $\{a_n\}$, then the functional values $a_1, a_2, a_3, \ldots, a_n, \ldots,$ where $a_n = f(n)$ $(n = 1, 2, 3, \ldots)$, are called the terms of the sequence, and a_n is called the *n*th term of the sequence.

Example 1

Write down the terms of the infinite sequence.

a. $\left\{\frac{1}{2^{n-1}}\right\}$ **b.** $\left\{\frac{1+n}{1+n^2}\right\}$ **c.** $\left\{\frac{\pi^n}{n!}\right\}$

Solution

a. Taking $n = 1, 2, 3, 4, \ldots$ in succession, we see that

$$a_1 = \frac{1}{2^0} = 1, a_2 = \frac{1}{2^1} = \frac{1}{2}, a_3 = \frac{1}{2^2} = \frac{1}{4}, a_4 = \frac{1}{2^3} = \frac{1}{8}, \ldots$$

so the required sequence is

$$1, \frac{1}{2}, \frac{1}{4}, \frac{1}{8}, \ldots, \frac{1}{2^{n-1}}, \ldots$$

b. The required sequence is

$$a_1 = \frac{1+1}{1+1^2} = 1, a_2 = \frac{1+2}{1+2^2} = \frac{3}{5}, a_3 = \frac{1+3}{1+3^2} = \frac{4}{10},$$

$$a_4 = \frac{1+4}{1+4^2} = \frac{5}{17}, \ldots, a_n = \frac{1+n}{1+n^2}, \ldots$$

or

$$1, \frac{3}{5}, \frac{4}{10}, \frac{5}{17}, \ldots, \frac{1+n}{1+n^2}, \ldots$$

c. Here

$$a_1 = \frac{\pi}{1!} = \pi, a_2 = \frac{\pi^2}{2!} = \frac{\pi^2}{2}, a_3 = \frac{\pi^3}{3!} = \frac{\pi^3}{6}, a_4 = \frac{\pi^4}{4!} = \frac{\pi^4}{24}, \ldots$$

so the required sequence is

$$\pi, \frac{\pi^2}{2!}, \frac{\pi^3}{3!}, \frac{\pi^4}{4!}, \ldots, \frac{\pi^n}{n!}, \ldots$$

Example 2

Find the general term of the sequence.

a. $1, \frac{1}{4}, \frac{1}{9}, \frac{1}{16}, \frac{1}{25}, \ldots$ **b.** $-1, 1, -1, 1, -1, 1, \ldots$

Solution

a. Observe that the terms may be written as

$$a_1 = \frac{1}{1^2}, a_2 = \frac{1}{2^2}, a_3 = \frac{1}{3^2}, a_4 = \frac{1}{4^2}, a_5 = \frac{1}{5^2}, \ldots$$

and we conclude that the nth term is $a_n = \frac{1}{n^2}$.

b. Here

$$a_1 = -1, a_2 = 1, a_3 = -1, a_4 = 1, a_5 = -1, \ldots$$

which may also be written as

$$a_1 = (-1)^1, a_2 = (-1)^2, a_3 = (-1)^3, a_4 = (-1)^4, a_5 = (-1)^5, \ldots$$

and so $a_n = (-1)^n$.

Graphs of Infinite Sequences

Since an infinite sequence is a function, we can sketch its graph. Because the domain of the function is the set of positive integers, the graph of a sequence consists of an infinite collection of points in the xy-plane.

Example 3

Sketch the graph of the sequence.

a. $\{n + 1\}$ **b.** $\left\{\frac{n+1}{n}\right\}$ **c.** $\{(-1)^n\}$

Solution

a. The terms of the sequence $\{n + 1\}$ are 2, 3, 4, 5, ... Recalling that these numbers are precisely the functional values of $f(n) = n + 1$ for $n = 1, 2, 3, \ldots$, we obtain the following table of values of f:

n	1	2	3	4	5	. . .
$f(n)$	2	3	4	5	6	. . .

from which we construct the graph of $\{n + 1\}$ shown in Figure 14.3a.

b. From the following table of values of f,

n	1	2	3	4	5	. . .
$f(n)$	2	$\frac{3}{2}$	$\frac{4}{3}$	$\frac{5}{4}$	$\frac{6}{5}$	. . .

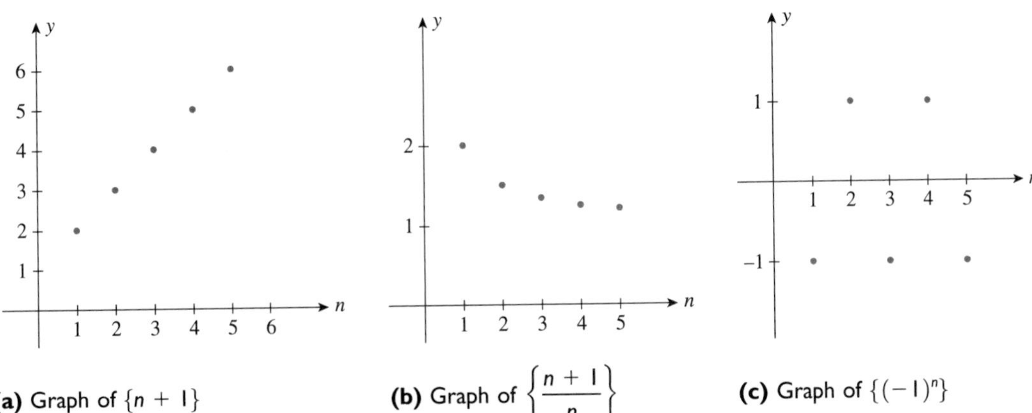

(a) Graph of $\{n + 1\}$

(b) Graph of $\left\{\frac{n+1}{n}\right\}$

(c) Graph of $\{(-1)^n\}$

Figure 14.3

We sketch the graph shown in Figure 14.3b.

c. We use the following table of values for the sequence

n	1	2	3	4	5	. . .
$f(n)$	−1	1	−1	1	−1	. . .

to sketch the graph of $\{(-1)^n\}$ shown in Figure 14.3c.

Remark

Notice that the function $f(n) = n + 1$ defining the sequence $\{n + 1\}$ may be viewed as the function $f(x) = x + 1$ $(-\infty < x < \infty)$ with x *restricted* to the set of positive integers. Similarly, the function defining the sequence $\{(n + 1)/n\}$ is just the function $f(x) = (x + 1)/x$ with a similar restriction on its domain. Finally, the function defining the sequence $\{(-1)^n\}$ is the function $f(x) = \cos \pi x$ with x a positive integer. These observations suggest that it is possible to study the properties of a sequence by analyzing the properties of a corresponding function defined for *all* values of x in some suitable interval. ◀

The Limit of a Sequence

Given a sequence $\{a_n\}$, we may ask whether the terms a_n of the sequence approach some specific number L as n gets larger and larger. If they do, we say that the sequence a_n *converges* to L. More specifically, we have the following informal definition.

Limit of a Sequence

Let $\{a_n\}$ be a given sequence. We say that the sequence $\{a_n\}$ converges and has the limit L, written

$$\lim_{n\to\infty} a_n = L$$

if the terms of the sequence, a_n, can be made as close to L as we please by taking n sufficiently large. If a sequence is not convergent, it is said to be divergent.

Example 4

Determine whether the infinite sequence converges or diverges.

a. $\left\{\frac{1}{n}\right\}$ **b.** $\{\sqrt{n}\}$ **c.** $\{(-1)^n\}$

Solution

a. The terms of the sequence,

$$1, \frac{1}{2}, \frac{1}{3}, \frac{1}{4}, \frac{1}{5}, \ldots, \frac{1}{n}, \ldots$$

approach the number $L = 0$ as n gets larger and larger. We conclude that

$$\lim_{n\to\infty} a_n = \lim_{n\to\infty} \frac{1}{n} = 0$$

Compare this with

$$\lim_{x\to\infty} \frac{1}{x} = 0$$

b. The terms of the sequence,

$$1, \sqrt{2}, \sqrt{3}, \sqrt{4}, \sqrt{5}, \ldots, \sqrt{n}, \ldots$$

get larger and larger as n gets larger and larger. Consequently, they do not approach any finite number L as n tends to infinity. Therefore, the sequence is divergent. Compare this with

$$\lim_{x\to\infty} \sqrt{x} = \infty$$

c. The terms of the sequence are

$$a_1 = -1, a_2 = 1, a_3 = -1, a_4 = 1, a_5 = -1, \cdots, a_n = (-1)^n, \ldots$$

Thus, no matter how large n is, there are terms that are equal to -1 (those with odd-numbered subscripts) and also terms that are equal to 1 (those with even-numbered subscripts). This implies that there cannot be a *unique* real number L such that a_n is arbitrarily close to L no matter how large n is. Therefore, the sequence is divergent.

EXPLORING WITH TECHNOLOGY

Refer to Example 4 and the REMARK on page 801.

1. Plot the graph of $f(x) = 1/x$, using the viewing window $[0, 10] \cdot [0, 3]$, and thus verify graphically that $\lim_{n\to\infty} (1/n) = 0$.
2. Plot the graph of $f(x) = \sqrt{x}$, using an appropriate viewing window, and thus verify graphically that $\lim_{n\to\infty} \sqrt{n}$ does not exist.
3. Plot the graph of $f(x) = \cos \pi x$, using the viewing window $[0, 10] \cdot [-2, 2]$, and thus verify graphically that $\lim_{n\to\infty} (-1)^n$ does not exist.

The following properties of sequences, which parallel those of the limit of $f(x)$ at infinity, are helpful in computing limits of sequences.

Limit Properties of Sequences

Suppose

$$\lim_{n\to\infty} a_n = A \quad \text{and} \quad \lim_{n\to\infty} b_n = B$$

Then

1. $\lim_{n\to\infty} ca_n = c \lim_{n\to\infty} a_n = cA$ c, a constant
2. $\lim_{n\to\infty} (a_n \pm b_n) = \lim_{n\to\infty} a_n \pm \lim_{n\to\infty} b_n = A \pm B$ Sum rule
3. $\lim_{n\to\infty} a_n b_n = (\lim_{n\to\infty} a_n)(\lim_{n\to\infty} b_n) = AB$ Product rule
4. $\lim_{n\to\infty} \dfrac{a_n}{b_n} = \dfrac{\lim_{n\to\infty} a_n}{\lim_{n\to\infty} b_n} = \dfrac{A}{B}$ provided $B \neq 0$ Quotient rule

Example 5

Evaluate

a. $\lim_{n\to\infty}\left(\frac{1}{2}\right)^n$ **b.** $\lim_{n\to\infty}\frac{2n^2+1}{3n^2+n+2}$

Solution

a.
$$\lim_{n\to\infty}\left(\frac{1}{2}\right)^2 = \lim_{n\to\infty}\frac{1}{2^n}$$
$$= \frac{1}{\lim\limits_{n\to\infty} 2n} = 0$$

b. Dividing the numerator and denominator by n^2, we find

$$\lim_{n\to\infty}\frac{2n^2+1}{3n^2+n+2} = \lim_{n\to\infty}\frac{2+\frac{1}{n^2}}{3+\frac{1}{n}=\frac{2}{n^2}}$$
$$= \frac{\lim\limits_{n\to\infty}\left(2+\frac{1}{n^2}\right)}{\lim\limits_{n\to\infty}\left(3+\frac{1}{n}+\frac{2}{n^2}\right)}$$
$$= \frac{2}{3}$$

GROUP DISCUSSION

Consider the sequences $\{a_n\}$ and $\{b_n\}$ defined by $a_n = (-1)^{n+1}$ and $b_n = (-1)^n$.

1. Show that $\lim\limits_{n\to\infty} a_n$ and $\lim\limits_{n\to\infty} b_n$ do not exist.
2. Show that $\lim\limits_{n\to\infty}(a_n + b_n) = 0$.
3. Do the results of parts 1 and 2 contradict the limit properties of sequences listed on page 802? Explain your answer.

Application

Example 6

Quality Control Of the spark plugs manufactured by the Parts Division of United Motors Corporation, 2% are defective. It can be shown that the probability of getting at least one defective plug in a random sample of n spark plugs is $f(n) = 1 - (0.98)^n$. Consider the sequence $\{a_n\}$ defined by $a_n = f(n)$.

a. Write down the terms a_5, a_{10}, a_{25}, a_{100}, and a_{200} of the sequence $\{a_n\}$.

b. Evaluate $\lim\limits_{n\to\infty} a_n$ and interpret your results.

Solution

a. The required terms of the sequence are

$$0.10, 0.18, 0.40, 0.87, 0.98$$

For example, the probability of getting at least one defective plug in a random sample of 25 is .4—that is, a 40% chance.

b. $\lim_{n\to\infty} a_n = \lim_{n\to\infty} [1 - (0.98)^n]$

$= \lim_{n\to\infty} 1 - \lim_{n\to\infty} (0.98)^n$

$= 1 - 0$

$= 1$

The result tells us that if the sample is large enough, we will almost certainly pick at least one defective plug!

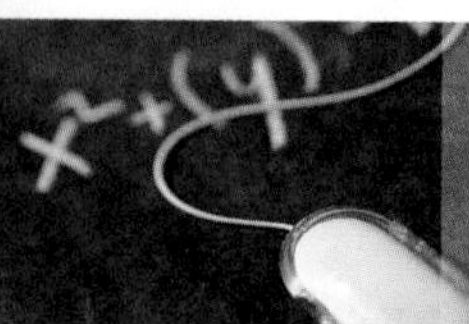

Self-Check Exercises 14.2

1. Determine whether the sequence $\left\{\frac{3n^2 + 2n + 1}{n^2 + 4}\right\}$ converges or diverges. If it converges, find its limit.

2. Consider the sequence $\left\{\frac{n}{n^2 + 1}\right\}$.
   a. Sketch the graph of the sequence.
   b. Show that the sequence is decreasing—that is, $a_1 > a_2 > a_3 > \cdots$.
   Hint: Consider $f(x) = x/(x^2 + 1)$ and show that f is decreasing by computing f'.

Solutions to Self-Check Exercises 14.2 can be found on page 805.

14.2 Exercises

In Exercises 1–9, write down the first five terms of the sequence.

1. $\{a_n\} = \{2^{n-1}\}$
2. $\{a_n\} = \left\{\frac{2n}{1 + n^2}\right\}$
3. $\{a_n\} = \left\{\frac{n - 1}{n + 1}\right\}$
4. $\{a_n\} = \left\{\left(-\frac{1}{3}\right)^n\right\}$
5. $\{a_n\} = \left\{\frac{2^{n-1}}{n!}\right\}$
6. $\{a_n\} = \left\{\frac{(-1)^n}{(2n)!}\right\}$
7. $\{a_n\} = \left\{\frac{e^n}{n^3}\right\}$
8. $\{a_n\} = \left\{\frac{\sqrt{n}}{\sqrt{n} + 1}\right\}$
9. $\{a_n\} = \left\{\frac{3n^2 - n + 1}{2n^2 + 1}\right\}$

In Exercises 10–21, find the general term of the sequence.

10. $\frac{1}{2}, \frac{1}{4}, \frac{1}{6}, \frac{1}{8}, \ldots$
11. $1, 4, 7, 10, \ldots$
12. $\frac{1}{3}, \frac{1}{7}, \frac{1}{11}, \frac{1}{15}, \ldots$
13. $1, \frac{1}{8}, \frac{1}{27}, \frac{1}{64}, \ldots$
14. $1, \frac{2}{3}, \frac{4}{9}, \frac{8}{27}, \ldots$
15. $2, \frac{8}{5}, \frac{32}{25}, \frac{128}{125}, \ldots$
16. $1, \frac{5}{4}, \frac{7}{5}, \frac{9}{6}, \ldots$
17. $1, -\frac{1}{2}, \frac{1}{4}, -\frac{1}{8}, \ldots$
18. $1 + \frac{1}{2}, 1 + \frac{1}{3}, 1 + \frac{1}{4}, 1 + \frac{1}{5}, \ldots$
19. $\frac{1}{2\cdot 3}, \frac{2}{3\cdot 4}, \frac{3}{4\cdot 5}, \frac{4}{5\cdot 6}, \ldots$
20. $1, \frac{2}{1\cdot 3}, \frac{4}{1\cdot 3\cdot 5}, \frac{8}{1\cdot 3\cdot 5\cdot 7}, \ldots$
21. $1, e, \frac{e^2}{2}, \frac{e^3}{6}, \frac{e^4}{24}, \frac{e^5}{120}, \ldots$

In Exercises 22–29, sketch the graph of the sequence.

22. $\{n^2\}$
23. $\left\{\frac{2n}{n + 1}\right\}$
24. $\left\{\frac{-1^n}{n}\right\}$
25. $\{\sqrt{n}\}$
26. $\{\ln n\}$
27. $\{e^n\}$
28. $\{ne^{-n}\}$
29. $\{n - \sqrt{n}\}$

In Exercises 30–44, determine the convergence or divergence of the sequence $\{a_n\}$. If the sequence converges, find its limit.

30. $a_n = \frac{n}{n^2 + 1}$
31. $a_n = \frac{n + 1}{2n}$
32. $a_n = \sqrt[3]{n}$
33. $a_n = \frac{(-1)^n}{\sqrt{n}}$
34. $a_n = \frac{1}{n + 1} - \frac{1}{n + 2}$
35. $a_n = \frac{\sqrt{n} - 1}{\sqrt{n} + 1}$
36. $a_n = \frac{3n^2 + n - 1}{6n^2 + n + 1}$
37. $a_n = \frac{2n^3 - 1}{n^3 + 2n + 1}$

38. $a_n = \dfrac{1 + (-1)^n}{3^n}$

39. $a_n = 2 - \dfrac{1}{2^n}$

40. $a_n = \dfrac{2n}{n!}$

41. $a_n = \dfrac{2^n}{3^n}$

42. $a_n = \dfrac{2^n - 1}{2^n}$

43. $a_n = \dfrac{n}{\sqrt{2n^2 + 3}}$

44. $a_n = \dfrac{2^n}{n!}$

45. **Automobile Microprocessors** Of the microprocessors manufactured by a microelectronics firm for use in regulating fuel consumption in automobiles, $1\frac{1}{2}\%$ are defective. It can be shown that the probability of getting at least one defective microprocessor in a random sample of n microprocessors is $f(n) = 1 - (0.985)^n$. Consider the sequence $\{a_n\}$ defined by $a_n = f(n)$.
   a. Write down the terms a_1, a_{10}, a_{100}, and a_{1000} of the sequence $\{a_n\}$.
   b. Evaluate $\lim_{n\to\infty} a_n$ and interpret your results.

Business and Economics Applications

46. **Savings Accounts** An amount of \$1000 is deposited in a bank that pays 8% interest/year compounded daily (take the number of days in a year to be 365). Let a_n denote the total amount on deposit after n days, assuming no deposits or withdrawals are made during the period in question.
   a. Find the formula for a_n.
   Hint: See Section 12.2.
   b. Compute a_1, a_{10}, a_{50}, and a_{100}.
   c. What amount is on deposit after 1 yr?

47. **Accumulated Amount** Suppose \$100 is deposited into an account earning interest at 12%/year compounded monthly. Let a_n denote the amount on deposit (called the accumulated amount or the future value) at the end of the nth mo.
   a. Show that $a_1 = 100(1.01)$, $a_2 = 100(1.01)^2$, and $a_3 = 100(1.01)^3$.
   b. Find the accumulated amount a_n.
   c. Find the 24th term of the sequence $\{a_n\}$ and interpret your result.

Biological and Life Sciences Applications

48. **Transmission of Disease** In the early stages of an epidemic, the number of persons who have contracted the disease on the $(n + 1)$st day, a_{n+1}, is related to the number of persons who have the disease on the nth day, a_n, by the equation

$$a_{n+1} = (1 + aN - b)a_n$$

where N denotes the total population and a and b are positive constants that depend on the nature of the disease. Put $r = 1 + aN - b$.
   a. Write down the first n terms of the sequence $\{a_n\}$.
   b. Evaluate $\lim_{n\to\infty} a_n$ for each of the three cases $r < 1$, $r = 1$, and $r > 1$.
   c. Interpret the results of part (b).

In Exercises 49–52, determine whether the statement is true or false. If it is true, explain why it is true. If it is false, give an example to show why it is false.

49. If $\lim_{n\to\infty} a_n = L$ and $\lim_{n\to\infty} b_n = 0$ then $\lim_{n\to\infty} a_nb_n = 0$.

50. If $\{a_n\}$ and $\{b_n\}$ are sequences such that $\lim_{n\to\infty}(a_n + b_n)$ exists, then both $\lim_{n\to\infty} a_n$ and $\lim_{n\to\infty} b_n$ must exist.

51. If $\{a_n\}$ is bounded (that is, $|a_n| \le M$ for some positive real numbers M and $n = 1, 2, 3, \ldots$) and $\{b_n\}$ converges, then $\lim_{n\to\infty} a_nb_n$ exists.

52. If $\lim_{n\to\infty} a_nb_n$ exists, then both $\lim_{n\to\infty} a_n$ and $\lim_{n\to\infty} b_n$ must exist.

Solutions to Self-Check Exercises 14.2

1. We compute

$$\lim_{n\to\infty} \frac{3n^2 + 2n + 1}{n^2 + 4} = \lim_{n\to\infty} \frac{3 + \frac{2}{n} + \frac{1}{n^2}}{1 + \frac{4}{n^2}} \quad \text{Divide numerator and denominator by } n^2.$$

$$= \frac{3}{1} = 3$$

2. **a.** From the table

n	1	2	3	4	5	...
$f(n)$	$\frac{1}{2}$	$\frac{2}{5}$	$\frac{3}{10}$	$\frac{4}{17}$	$\frac{5}{26}$	...

we obtain the graph shown in the figure.

b. We compute

$$f'(x) = \frac{(x^2 + 1)(1) - x(2x)}{(x^2 + 1)^2} = \frac{1 - x^2}{(x^2 + 1)^2} < 0$$

for $x > 1$, and so f is decreasing on $(1, \infty)$. We conclude that $\{a_n\}$ is decreasing.

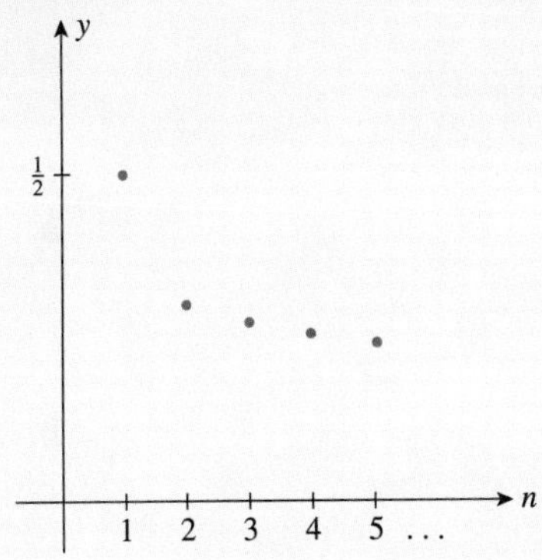

14.3 Infinite Series

The Sum of an Infinite Series

In Section 14.2 we asked the question, How do we find the "sum" of a series involving infinitely many terms? In this section we show how we define such a sum. Consider the **infinite series**

$$a_1 + a_2 + \cdots + a_n + \cdots$$

which may be abbreviated through the use of sigma notation as

$$\sum_{n=1}^{\infty} a_n = a_1 + a_2 + a_3 + \cdots + a_n + \cdots \tag{4}$$

and is read "the sum of the numbers a_n for n running from 1 to infinity."

Now, given an infinite series (4) whose terms are drawn from the sequence $\{a_n\}$, let's define the sums

$$S_1 = \sum_{n=1}^{1} a_n = a_1$$

$$S_2 = \sum_{n=1}^{2} a_n = a_1 + a_2$$

$$S_3 = \sum_{n=1}^{3} a_n = a_1 + a_2 + a_3$$

$$S_n = \sum_{n=1}^{N} a_n = a_1 + a_2 + a_3 + \cdots + a_N$$

Observe that each of these sums exists since each is obtained by adding together finitely many numbers. For each N, S_N is called the **Nth partial sum** of the series (4), and the sequence $\{S_n\}$ is called the **sequence of partial sums** of the series (4). We are now in a position to define the sum of an infinite series.

Sum of an Infinite Series

Let

$$\sum_{n=1}^{\infty} a_n = a_1 + a_2 + a_3 + \cdots$$

be an infinite series and let $\{S_n\}$ be the sequence of **partial sums of the infinite series.** If

$$\lim_{n\to\infty} S_n = S$$

we say that the infinite series $\sum_{n=1}^{\infty} a_n$ converges to S and write

$$\sum_{n=1}^{\infty} a_n = \lim_{n\to\infty} S_n = S$$

In this case, S is called the **sum of the series.** If $\{S_n\}$ does not converge, we say that the infinite series **diverges** and has no sum.

Example 1

a. Show that the following infinite series diverges:

$$\sum_{n=1}^{\infty} (-1)^{n+1} = 1 - 1 + 1 - 1 + 1 - 1 + \cdots$$

b. Show that the following infinite series converges:

$$\sum_{n=1}^{\infty} \frac{1}{n(n+1)} = \frac{1}{1\cdot 2} + \frac{1}{2\cdot 3} + \frac{1}{3\cdot 4} + \cdots$$

Solution

a. The partial sums of the given infinite series are

$$S_1 = 1, S_2 = 1 - 1 = 0, S_3 = 1 - 1 + 1 = 1, S_4 = 1 - 1 + 1 - 1 = 0, \ldots$$

The sequence of partial sums $\{S_n\}$ never settles toward a single value and therefore evidently diverges, and so the given infinite series diverges.

b. Let's write

$$\frac{1}{n(n+1)} = \frac{1}{n} - \frac{1}{n+1}$$

an equality that is easily verified. The Nth partial sum of the given series is

$$\begin{aligned} S_N &= \sum_{n=1}^{N} \frac{1}{n(n+1)} = \sum_{n=1}^{N} \left(\frac{1}{n} - \frac{1}{n+1}\right) \\ &= \left(1 - \frac{1}{2}\right) + \left(\frac{1}{2} - \frac{1}{3}\right) + \left(\frac{1}{3} - \frac{1}{4}\right) + \cdots + \left(\frac{1}{N} - \frac{1}{N+1}\right) \\ &= 1 + \left(-\frac{1}{2} + \frac{1}{2}\right) + \left(-\frac{1}{3} + \frac{1}{3}\right) + \cdots + \left(-\frac{1}{N} + \frac{1}{N}\right) - \frac{1}{N+1} \\ &= 1 - \frac{1}{N+1} \end{aligned}$$

Since

$$\lim_{N\to\infty} S_N = \lim_{N\to\infty} \left(1 - \frac{1}{N+1}\right) = 1$$

we conclude that the given series converges and has a sum equal to 1; that is,

$$\sum_{n=1}^{\infty} \frac{1}{n(n+1)} = 1$$

The series in Example 1(b) is called a *telescoping series* because all the terms between the first and the last in the expression for S_N "collapse."

EXPLORING WITH TECHNOLOGY

Refer to Example 1b.

1. Verify graphically the results of Example 1b by plotting the graphs of the partial sums

$$S_1 = \frac{1}{1\cdot 2}, S_2 = \frac{1}{1\cdot 2} + \frac{1}{2\cdot 3}, \ldots, S_6 = \frac{1}{1\cdot 2} + \frac{1}{2\cdot 3} + \cdots + \frac{1}{6\cdot 7}$$

and $S = 1$ in the viewing window $[0, 3] \cdot [0, 1]$.

2. Refer to the abbreviated form for S_N—namely, $S_N = \left(1 - \frac{1}{N+1}\right)$. By plotting the graphs of $y_1 = \left(1 - \frac{1}{x+1}\right)$ and $y_2 = 1$ in the viewing window $[0, 50] \cdot [0, 1.1]$, verify graphically that $\lim_{N\to\infty} S_N = 1$ and thus the result $\sum_{n=1}^{\infty} \frac{1}{n(n+1)} = 1$, as obtained in Example 1b.

GROUP DISCUSSION

1. An example of a divergent infinite series is

$$\sum_{n=0}^{\infty} 1 = 1 + 1 + 1 + \cdots$$

Show that the series is divergent by establishing the following:

a. The nth partial sum of the infinite series is $S_n = n$.

b. $\lim_{n \to \infty} S_n = \infty$ so that $\{S_n\}$ is divergent and the desired result follows.

2. Since the *terms* of the infinite series in part 1 do not decrease, it is evident that the partial sums of the series must grow without bound. Therefore, it is not difficult to see that the infinite series cannot converge. Now consider the infinite series

$$\sum_{n=1}^{\infty} \frac{1}{n} = 1 + \frac{1}{2} + \frac{1}{3} + \cdots$$

Even though the terms of this series (called the *harmonic series*) approach zero as n goes to infinity, it can be shown that the harmonic series is divergent. Show that this result is intuitively true by establishing the following:

a. Observe that $S_2 = 1 + \frac{1}{2} > \frac{1}{2} + \frac{1}{2} = 1$

and

$$S_4 = 1 + \frac{1}{2} + \frac{1}{3} + \frac{1}{4} = S_2 + \frac{1}{3} + \frac{1}{4} > 1 + \left(\frac{1}{4} + \frac{1}{4}\right) = \frac{3}{2}$$

Using a similar argument, show that $S_8 > 4/2$ and $S_{16} > 5/2$. Conclude that in general

$$S_{2^n} > \frac{n+1}{2}$$

b. Explain why $S_1 < S_2 < S_3 < \cdots < S_n < \cdots$.

c. Use the results of parts (a) and (b) to explain why the harmonic series is divergent.

Geometric Series

In general, it is no easy task to determine whether a given infinite series is convergent or divergent. It is an even more difficult problem to determine the sum of an infinite series that is known to be convergent. But there is an important and useful series whose sum, when it exists, is easy to find.

Geometric Series

A geometric series with ratio r is a series of the form

$$\sum_{n=0}^{\infty} ar^n = a + ar + ar^2 + ar^3 + \cdots + ar^n + \cdots \tag{5}$$

The ratio here refers to the ratio of two consecutive terms.

Note that, for this series, we begin the summation with $n = 0$ instead of $n = 1$. To determine the conditions under which the geometric series (5) converges and find its sum, let's consider the nth partial sum of the infinite series

$$S_n = a + ar + ar^2 + \cdots + ar^n$$

Multiplying both sides of the equation by r gives

$$rS_n = ar + ar^2 + ar^3 + \cdots + ar^{n+1}$$

Subtracting the second equation from the first yields

$$S_n - rS_n = a - ar^{n+1}$$

$$(1 - r)S_n = a(1 - r^{n+1})$$

$$S_n = \frac{a(1 - r^{n+1})}{1 - r}$$

provided $r \neq 1$. You are asked to show that if $|r| \geq 1$, then the series (5) diverges (Exercise 38). On the other hand, observe that if $|r| < 1$ (that is, if $-1 < r < 1$), then

$$\lim_{n\to\infty} r^{n+1} = 0$$

For example, if $r = \frac{1}{2}$, then

$$\lim_{n\to\infty} r^{n+1} = \lim_{n\to\infty} \left(\frac{1}{2}\right)^{n+1} = \lim_{n\to\infty} \frac{1}{2^{n+1}} = 0$$

Using this fact, together with the properties of limits stated earlier, we see that

$$\begin{aligned}\lim_{n\to\infty} S_n &= \lim_{n\to\infty} \frac{a(1 - r^{n+1})}{1 - r} \\ &= \frac{a}{1 - r} \lim_{n\to\infty} (1 - r^{n+1}) \\ &= \frac{a}{1 - r}\end{aligned}$$

These results are summarized in Theorem 2.

THEOREM 2

The geometric series

$$\sum_{n=0}^{\infty} ar^n = a + ar + ar^2 + \cdots$$

converges and its sum is $\frac{a}{1-r}$; that is,

$$\sum_{n=0}^{\infty} ar^n = a + ar + ar^2 + \cdots = \frac{a}{1 - r} \qquad \textbf{(6)}$$

if $|r| < 1$. The series diverges if $|r| \geq 1$.

Example 2

Show that the infinite series is a geometric series and find its sum if it is convergent.

a. $\sum_{n=0}^{\infty} \frac{1}{2^n}$ **b.** $\sum_{n=0}^{\infty} 3\left(\frac{5}{2}\right)^n$ **c.** $\sum_{n=1}^{\infty} 5\left(-\frac{3}{4}\right)^n$

Solution

a. Observe that

$$\frac{1}{2^n} = \left(\frac{1}{2}\right)^n$$

so that

$$\sum_{n=0}^{\infty} \frac{1}{2^n} = \sum_{n=0}^{\infty} \left(\frac{1}{2}\right)^n$$

is a geometric series with $a = 1$ and $r = \frac{1}{2}$. Since $|r| = \frac{1}{2} < 1$, we conclude that the series is convergent. Finally, using (6), we have

$$\sum_{n=0}^{\infty} \left(\frac{1}{2}\right)^n = 1 + \frac{1}{2} + \frac{1}{4} + \cdots = \frac{1}{1 - \frac{1}{2}} = 2$$

b. This is a geometric series with $a = 3$ and $r = \frac{5}{2}$. Since

$$|r| = \left|\frac{5}{2}\right| = \frac{5}{2} > 1$$

we deduce that the series is divergent.

c. The summation here begins with $n = 1$. However, we may rewrite the series as

$$\begin{aligned}\sum_{n=1}^{\infty} 5\left(-\frac{3}{4}\right)^n &= 5\left(-\frac{3}{4}\right) + 5\left(-\frac{3}{4}\right)^2 + 5\left(-\frac{3}{4}\right)^3 + \cdots \\ &= 5\left(-\frac{3}{4}\right)\left[1 + \left(-\frac{3}{4}\right) + \left(-\frac{3}{4}\right)^2 + \cdots\right] \\ &= \sum_{n=0}^{\infty}\left(-\frac{15}{4}\right)\left(-\frac{3}{4}\right)^n\end{aligned}$$

which is just a geometric series with $a = -\frac{15}{4}$ and $r = -\frac{3}{4}$. Since $|r| = |-\frac{3}{4}| = \frac{3}{4} < 1$, the series is convergent. Using (6), we find

$$\sum_{n=1}^{\infty} 5\left(-\frac{3}{4}\right)^n = \sum_{n=0}^{\infty}\left(-\frac{15}{4}\right)\left(-\frac{3}{4}\right)^n = \frac{-\frac{15}{4}}{1 - (-\frac{3}{4})} = -\frac{\frac{15}{4}}{\frac{7}{4}} = -\frac{15}{7}$$

Properties of Infinite Series

The following properties of infinite series enable us to perform algebraic operations on convergent series.

Properties of Infinite Series

If $\sum_{n=1}^{\infty} a_n$ and $\sum_{n=1}^{\infty} b_n$ are convergent infinite series and c is a constant, then

1. $\sum_{n=1}^{\infty} ca_n = c\sum_{n=1}^{\infty} a_n$
2. $\sum_{n=1}^{\infty} (a_n \pm b_n) = \sum_{n=1}^{\infty} a_n \pm \sum_{n=1}^{\infty} b_n$

Thus, we may multiply each term of a convergent series by a constant c, which results in a convergent series whose sum is c times the sum of the original series. We may also add (subtract) the corresponding terms of two convergent series, which gives a convergent series whose sum is the sum (difference) of the sums of the original series.

Example 3

Find the sum of the following series if it exists:

$$\sum_{n=0}^{\infty} \frac{2 \cdot 3^n - 2^n}{5^n} = 1 + \frac{4}{5} + \frac{14}{25} + \cdots$$

Solution

Note that this series starts with $n = 0$. We can write

$$\sum_{n=0}^{\infty} \frac{2 \cdot 3^n - 2^n}{5^n} = \sum_{n=0}^{\infty}\left(\frac{2 \cdot 3^n}{5^n} - \frac{2^n}{5^n}\right)$$

Now observe that

$$\sum_{n=0}^{\infty} \frac{2 \cdot 3^n}{5^n} \quad \text{and} \quad \sum_{n=0}^{\infty} \frac{2^n}{5^n}$$

are convergent geometric series with ratios $r = \frac{3}{5} < 1$ and $r = \frac{2}{5} < 1$, respectively. Therefore, using Property 2 of infinite series, we have

$$\begin{aligned}
\sum_{n=0}^{\infty} \frac{2 \cdot 3^n - 2^n}{5^n} &= \sum_{n=0}^{\infty} \frac{2 \cdot 3^n}{5^n} - \sum_{n=0}^{\infty} \frac{2^n}{5^n} \\
&= 2\sum_{n=0}^{\infty} \frac{3^n}{5^n} - \sum_{n=0}^{\infty} \frac{2^n}{5^n} \qquad \text{Using Property 1 on the first sum} \\
&= 2\sum_{n=0}^{\infty} \left(\frac{3}{5}\right)^n - \sum_{n=0}^{\infty} \left(\frac{2}{5}\right)^n \\
&= 2\left(\frac{1}{1 - \frac{3}{5}}\right) - \left(\frac{1}{1 - \frac{2}{5}}\right) \\
&= 2\left(\frac{5}{2}\right) - \frac{5}{3} \\
&= \frac{10}{3}
\end{aligned}$$

Applications

We now consider some applications of geometric series.

Example 4

Find the rational number that has the repeated decimal representation 0.222. . . .

Solution

By definition, the decimal representation

$$\begin{aligned}
0.222\ldots &= \frac{2}{10} + \frac{2}{100} + \frac{2}{1000} + \cdots \\
&= \frac{2}{10}\left(1 + \frac{1}{10} + \frac{1}{100} + \cdots\right) \\
&= \sum_{n=0}^{\infty} \left(\frac{2}{10}\right)\left(\frac{1}{10}\right)^n
\end{aligned}$$

which is a geometric series with $a = \frac{2}{10}$ and $r = \frac{1}{10}$. Since $|r| = r = \frac{1}{10} < 1$, the series converges. In fact, using Formula (6), we have

$$0.222\ldots = \frac{\left(\frac{2}{10}\right)}{1 - \frac{1}{10}} = \frac{\frac{2}{10}}{\frac{9}{10}} = \frac{2}{9}$$

The following example illustrates a phenomenon in economics known as the **multiplier effect.**

Example 5

The Multiplier Effect Suppose the average wage earner saves 10% of her take-home pay and spends the other 90%. Estimate the impact that a proposed \$20 billion tax cut will have on the economy over the long run in terms of the additional spending generated.

Solution

Of the \$20 billion received by the original beneficiaries of the proposed tax cut, (0.9)(20) billion dollars will be spent. Of the (0.9)(20) billion dollars reinjected into the economy, 90% of it, or (0.9)(0.9)(20) billion dollars, will find its way into the economy again. This process will go on ad infinitum, so this one-time proposed tax cut will result in additional spending over the years in the amount of

$$\begin{aligned}(0.9)(20) + (0.9)^2(20) + (0.9)^3(20) + \cdots &= (0.9)(20)[1 + 0.9 + 0.9^2 + 0.9^3 + \cdots] \\ &= 18\left[\frac{1}{1 - 0.9}\right] \\ &= 180\end{aligned}$$

or \$180 billion.

A **perpetuity** is a sequence of payments made at regular time intervals and continuing on forever. The **capital value of a perpetuity** is the sum of the present values of all future payments. The following example illustrates these concepts.

Example 6

Establishing a Scholarship Fund The Robinson family wishes to establish a scholarship fund at a college. If a scholarship in the amount of \$5000 is to be awarded on an annual basis beginning next year, find the amount of the endowment they are required to make now. Assume that this fund will earn interest at a rate of 10% per year compounded continuously.

Solution

The amount of the endowment, A, is given by the sum of the present values of the amounts awarded annually in perpetuity. Now, the present value of the amount of the first award is equal to

$$5000e^{-0.1(1)}$$

(see Section 8.7). The present value of the amount of the second award is

$$5000e^{-0.1(2)}$$

and so on. Continuing, we see that the present value of the amount of the nth award is

$$5000e^{-0.1(n)}$$

Therefore, the amount of the endowment is

$$A = 5000e^{-0.1(1)} + 5000e^{-0.1(2)} + \cdots + 5000e^{-0.1(n)} + \cdots$$

To find the sum of the infinite series on the right-hand side, let

$$r = e^{-0.1}$$

Then

$$\begin{aligned}A &= 5000r^1 + 5000r^2 + \cdots + 5000r^n + \cdots \\ &= 5000r(1 + r + r^2 + \cdots + r^n + \cdots)\end{aligned}$$

The series inside the parentheses is a geometric series with $r = e^{-0.1} \approx 0.905 < 1$, so that, using (6), we find

$$A = 5000r\left(\frac{1}{1 - r}\right) = \frac{5000e^{-0.1}}{1 - e^{-0.1}} = 47{,}541.66$$

Thus, the amount of the endowment is \$47,541.66.

Our final example is an application of a geometric series in the field of medicine.

Example 7

Residual Drug in the Bloodstream A patient is to be given 5 units of a certain drug daily for an indefinite period of time. For this particular drug, it is known that the fraction of a dose that remains in the patient's body after t days is given by $e^{-0.3t}$. Determine the residual amount of the drug that may be expected to be in the patient's body after an extended treatment.

Solution

The amount of the drug in the patient's body 1 day after the first dose is administered, and prior to administration of the second dose, is $5e^{-0.3}$ units. The amount of the drug in the patient's body 2 days later, and prior to administration of the third dose, consists of the residuals from the first two doses. Of the first dose, $5e^{-(0.3)2}$ units of the drug are left in the patient's body, and of the second dose, $5e^{-0.3}$ units of the drug are left. Thus, the amount of the drug 2 days later is given by

$$5e^{-0.3} + 5e^{-0.3(2)}$$

units. Continuing, we see that the amount of the drug left in the patient's body in the long run, and prior to administration of a fresh dose, is given by

$$R = 5e^{-0.3} + 5e^{-0.3(2)} + 5e^{-0.3(3)} + \cdots$$

To find the sum of the infinite series, we let

$$r = e^{-0.3}$$

Then

$$A = 5r + 5r^2 + 5r^3 + \cdots = 5r(1 + r + r^2 + \cdots)$$

$$= \frac{5r}{1 - r} = \frac{5e^{-0.3}}{1 - e^{-0.3}} \approx 14.29$$

Therefore, after an extended treatment, the residual amount of drug in the patient's body is approximately 14.29 units.

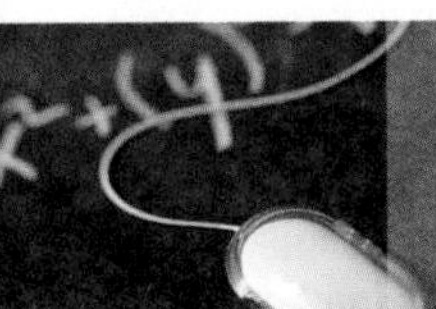

Self-Check Exercises 14.3

1. Determine whether the geometric series

$$\sum_{n=0}^{\infty} 5\left(-\frac{1}{3}\right)^n = 5 - 5\left(\frac{1}{3}\right) + 5\left(\frac{1}{9}\right) - \cdots$$

is convergent or divergent. If it is convergent, find its sum.

2. Suppose the average wage earner in a certain country saves 12% of his take-home pay and spends the other 88%. Estimate the impact that a proposed $10 billion tax cut will have on the economy over the long run due to the additional spending generated.

Solutions to Self-Check Exercises 14.3 can be found on page 815.

14.3 Exercises

In Exercises 1–4, find the *N*th partial sum of the infinite series and evaluate its limit to determine whether the series converges or diverges. If the series is convergent, find its sum.

1. $\sum_{n=1}^{\infty} (-2)^n$

2. $\sum_{n=1}^{\infty} \left(\frac{1}{n+1} - \frac{1}{n+2}\right)$

3. $\sum_{n=1}^{\infty} \frac{1}{n^2 + 3n + 2}$

   Hint: $\frac{1}{n^2 + 3n + 2} = \frac{1}{n+1} - \frac{1}{n+2}$

4. $\sum_{n=2}^{\infty} \left(\frac{1}{\ln n} - \frac{1}{\ln(n+1)}\right)$

In Exercises 5–16, determine whether the geometric series converges or diverges. If it converges, find its sum.

5. $\sum_{n=0}^{\infty} \left(\frac{1}{3}\right)^n$

6. $\sum_{n=0}^{\infty} 4\left(-\frac{2}{3}\right)^n$

7. $\sum_{n=0}^{\infty} 2(1.01)^n$

8. $\sum_{n=0}^{\infty} 3(0.9)^n$

9. $\sum_{n=0}^{\infty} \frac{(-2)^n}{3^n}$

10. $\sum_{n=0}^{\infty} \frac{3}{2^n}$

11. $\sum_{n=0}^{\infty} \frac{2^n}{3^{n+2}}$

12. $\sum_{n=0}^{\infty} \frac{3^{n+1}}{4^{n-1}}$

13. $\sum_{n=0}^{\infty} e^{-0.2n}$

14. $\sum_{n=0}^{\infty} 2e^{-0.1n}$

15. $\sum_{n=0}^{\infty} \left(-\frac{3}{\pi}\right)^n$

16. $\sum_{n=1}^{\infty} \frac{e^n}{3^{n+1}}$

In Exercises 17–26, determine whether the series converges or diverges. If it converges, find its sum.

17. $8 + 4 + \frac{1}{2} + \frac{1}{4} + \frac{1}{8} + \cdots$

18. $1 + 0.2 + 0.04 + 0.0016 + \cdots$

19. $3 - \frac{1}{3} + \frac{1}{9} - \frac{1}{27} + \cdots$

20. $5 - 1.01 + (1.01)^2 - (1.01)^3 + \cdots$

21. $\sum_{n=0}^{\infty} \frac{3 + 2^n}{3^n}$

22. $\sum_{n=0}^{\infty} \frac{2^n - 3^n}{4^n}$

23. $\sum_{n=0}^{\infty} \frac{3 \cdot 2^n + 4^n}{3^n}$

24. $\sum_{n=0}^{\infty} \frac{2 \cdot 3^n - 3 \cdot 5^n}{7^n}$

25. $\sum_{n=1}^{\infty} \left[\left(\frac{e}{\pi}\right)^n + \left(\frac{\pi}{e^2}\right)^n\right]$

26. $\sum_{n=1}^{\infty} \left[\frac{1}{2^n} - \frac{1}{n(n+1)}\right]$

Hint: $\frac{1}{n(n+1)} = \frac{1}{n} - \frac{1}{n+1}$

In Exercises 27–30, express the decimal as a rational number.

27. 0.3333...

28. 0.121212...

29. 1.213213213...

30. 6.2314314314...

In Exercises 31–34, find the values of x for which the series converges and find the sum of the series.

Hint: First show that the series is a geometric series.

31. $\sum_{n=0}^{\infty} (-x)^n$

32. $\sum_{n=0}^{\infty} (x-2)^n$

33. $\sum_{n=1}^{\infty} 2^n(x-1)^n$

34. $\sum_{n=0}^{\infty} \frac{x^{2n}}{3^n}$

35. **A Bouncing Ball** A ball is dropped from a height of 10 m. After hitting the ground, it rebounds to a height of 5 m and then continues to rebound at one-half of its former height thereafter. Find the total distance travelled by the ball before it comes to rest.

36. **Winning a Toss** Peter and Paul take turns tossing a pair of dice. The first to throw a 7 wins. If Peter starts the game, then it can be shown that his chances of winning are given by

$$p = \frac{1}{6} + \left(\frac{1}{6}\right)\left(\frac{5}{6}\right)^2 + \left(\frac{1}{6}\right)\left(\frac{5}{6}\right)^4 + \cdots$$

Find p.

37. **Residual Drug in the Bloodstream** Ten units of a certain drug are administered to a patient on a daily basis. The fraction of this drug that remains in the patient's bloodstream after t days is given by

$$f(t) = e^{(-1/4)t}$$

Determine the residual amount of the drug in the patient's bloodstream after a period of extended treatment with the drug.

38. Let

$$\sum_{n=0}^{\infty} ar^n = a + ar + ar^2 + ar^3 + \cdots + ar^n + \cdots$$

be a geometric series with common ratio r. Show that if $|r| \geq 1$, then the series diverges.

Business and Economics Applications

39. **Effect of a Tax Cut on Spending** Suppose the average wage earner saves 9% of her take-home pay and spends the other 91%. Estimate the impact that a proposed $30 billion tax cut will have on the economy over the long run due to the additional spending generated.

40. **Endowments** Hal Corporation wants to establish a fund to provide the art centre of a large metropolitan area with an annual grant of $250,000 beginning next year. If the fund will earn interest at a rate of 10%/year compounded continuously, find the amount of endowment the corporation must make at this time.

41. **Capital Value of a Perpetuity** Find a formula for the capital value of a perpetuity involving payments of P dollars each, paid at the end of each of m periods/year into a fund that earns interest at the nominal rate of r%/year compounded m times/year by verifying the following:

a. The present value of the nth payment is

$$P\left(1 + \frac{r}{m}\right)^{-n}$$

b. The capital value is

$$A = P\left(1 + \frac{r}{m}\right)^{-1} + P\left(1 + \frac{r}{m}\right)^{-2} + P\left(1 + \frac{r}{m}\right)^{-3} + \cdots$$

c. Using the fact that the series in part (b) is a geometric series, its sum is

$$A = \frac{mP}{r}$$

42. **Capital Value of a Perpetuity** Find a formula for the capital value of a perpetuity involving payments of P dollars paid at the end of each investment period into a fund that earns interest at the rate of r%/year compounded continuously.

Hint: Study Exercise 41.

Biological and Life Sciences Applications

43. **Transmission of Disease** Refer to Exercise 48, page 805. It can be shown that the total number of individuals, S_n, who have contracted the disease some time between the first and nth day is approximated by

$$s_n = a_1 + aNa_1 + aNa_2 + \cdots + aNa_{n-1}$$

Show that if $r < 1$, then the total number of persons who will have contracted the disease at some stage of the epidemic is no larger than $a_1b/(b - aN)$.

44. **Residual Drug in the Bloodstream** Suppose a dose of C units of a certain drug is administered to a patient and the fraction of the dose remaining in the patient's bloodstream t hr after the dose is administered is given by Ce^{-kt}, where k is a positive constant.
    a. Show that the residual concentration of the drug in the bloodstream after extended treatment when a dose of C units is administered at intervals of t hr is given by

    $$R = \frac{Ce^{-kt}}{1 - e^{-kt}}$$

    b. If the highest concentration of this particular drug that is considered safe is S units, find the minimal time that must exist between doses.

    Hint: $C + R \leq S$

45. **Estimating Total Number** Because of its ability to incorporate proteins from its host into its own protein coat, a virus manages to escape from the immune system and begins a new cycle of infection multiple times. Since the protein coat also affects the virus's ability to enter a cell, the efficiency of the virus to infect host cells declines steadily. If the first viral infection yields a litre of 1.1 million viral particles per cubic centimetre of blood and the amount of virus is 78% as large in each infection cycle as in the last, then sum a series to estimate the total number of viral particles in each cubic centimetre over the course of the infection.

46. **Comparing Estimates** The breeding bird survey finds that the number of nests present in an area near a new shopping mall declines geometrically because of habitat loss. The birds imprint on the location where they are raised and tend to return to it. It is decided to collect one chick from each nest and settle them in a new area. If the number of nests the first year this collection happens is 140 and the number of nests each year declines by 20% (there are 80% as many nests the next year), estimate the number of chicks collected in two ways. First compute the terms of the series until there is < 1 nest and add the figures. Second, sum the geometric series as if it were infinite. How much do the two estimates of the number of chicks collected disagree?

In Exercises 47–50, determine whether the statement is true or false. If it is true, explain why it is true. If it is false, give an example to show why it is false.

47. If $\sum_{n=0}^{\infty} (a_n + b_n)$ converges, then both $\sum_{n=0}^{\infty} a_n$ and $\sum_{n=0}^{\infty} b_n$ must converge.

48. If $\sum_{n=0}^{\infty} a_n$ converges and $\sum_{n=0}^{\infty} b_n$ converges, then $\sum_{n=0}^{\infty} (ca_n + db_n)$ also converges, where c and d are constants.

49. If $|r| < 1$, then $\sum_{n=0}^{\infty} |r^n| = \frac{1}{1 - |r|}$.

50. If $|r| > 1$, then $\sum_{n=1}^{\infty} \frac{1}{r^n} = \frac{1}{r - 1}$.

Solutions to Self-Check Exercises 14.3

1. This is a geometric series with $r = -\frac{1}{3}$ and $a = 5$. Since $0 < \left|-\frac{1}{3}\right| < 1$, we see that the series is convergent. Its sum is

$$\frac{5}{1 - \left(-\frac{1}{3}\right)} = \frac{5}{\frac{4}{3}} = \frac{15}{4} = 3\tfrac{3}{4}$$

2. Of the \$10 billion received by the original beneficiaries of the proposed cut, (0.88)(10) billion dollars will be spent. Of the (0.88)(10) billion dollars reinjected into the economy, 88% of it, or (0.88)(0.88)(10) billion dollars, will find its way into the economy again. This process will go on ad infinitum, so this one-time proposed tax cut will result in additional spending over the years in the amount of

$$\begin{aligned}
&(0.88)(10) + (0.88)^2(10) + (0.88)^3(10) + \cdots \\
&= (0.88)(10)[1 + 0.88 + (0.88)^2 + \cdots] \\
&= 8.8\left[\frac{1}{1 - 0.88}\right] \\
&= 73.3
\end{aligned}$$

or \$73.3 billion.

14.4 Series with Positive Terms

The convergence or divergence of a telescoping series or a geometric series is relatively easy to determine because we can find a simple formula for the nth partial sum S_n of these series. In fact, as we have seen earlier in this chapter, the knowledge of this formula helps us find the actual sum of a convergent series by simply evaluating $\lim_{n\to\infty} S_n$. More often than not, obtaining a simple formula for the nth partial sum of an infinite series is very difficult or impossible, and we are forced to look for alternative ways to investigate the convergence or divergence of the series.

In what follows, we look at several tests for determining the convergence or divergence of an infinite series by examining the nth term a_n of the series. These tests will confirm the convergence of a series without yielding a value for its sum. From the practical point of view, however, this is all that is required. Once it has been ascertained that a series is convergent, we can approximate its sum to any degree of accuracy desired by adding up the terms of its nth partial sum S_n, provided that n is chosen large enough.

Our first test tells us how to identify a divergent series.

The Test for Divergence

The following theorem tells us that the terms of a convergent series must ultimately approach zero.

THEOREM 3

If $\sum_{n=1}^{\infty} a_n$ converges, then $\lim_{n\to\infty} a_n = 0$.

To prove this result, let

$$S_n = a_1 + a_2 + \cdots + a_{n-1} + a_n = s_{n-1} + a_n$$

and so

$$a_n = S_n - S_{n-1}$$

Since $\sum_{n=1}^{\infty} a_n$ is convergent, the sequence $\{S_n\}$ is convergent. Let $\lim_{n\to\infty} S_n = S$. Then

$$\lim_{n\to\infty} a_n = \lim_{n\to\infty} (S_n - S_{n-1}) = \lim_{n\to\infty} S_n - \lim_{n\to\infty} S_{n-1} = S - S = 0$$

An important consequence of Theorem 3 is the following useful test for *divergence.*

THEOREM 4

The Test for Divergence

If $\lim_{n\to\infty} a_n$ does not exist or $\lim_{n\to\infty} a_n \neq 0$, then $\sum_{n=1}^{\infty} a_n$ diverges.

Observe that the test for divergence does *not* say

$$\text{If } \lim_{n\to\infty} a_n = 0, \text{ then } \sum_{n=1}^{\infty} a_n \text{ must converge.}$$

In other words, the converse of Theorem 4 is not true in general. For example, $\lim_{n\to\infty} \frac{1}{n} = 0$, and yet, as we will see later, the **harmonic series** $\sum_{n=1}^{\infty} \frac{1}{n}$ is divergent. (See Example 3.) In short, the test for divergence rules out convergence for a series whose nth term does not approach zero but yields no information if the nth term of a series does approach zero—that is, the series may or may not converge.

Example 1

Show that the following series are divergent:

a. $\sum_{n=1}^{\infty} (-1)^{n-1}$ **b.** $\sum_{n=1}^{\infty} \frac{2n^2 + 1}{3n^2 - 1}$

Solution

a. Here $a_n = (-1)^{n-1}$, and since

$$\lim_{n\to\infty} a_n = \lim_{n\to\infty} (-1)^{n-1}$$

does not exist, we conclude by the test for divergence that the series diverges.

b. Here

$$\lim_{n\to\infty} a_n = \lim_{n\to\infty} \frac{2n^2 + 1}{3n^2 - 1} = \lim_{n\to\infty} \frac{2 + \frac{1}{n^2}}{3 - \frac{1}{n^2}} = \frac{2}{3} \neq 0$$

and so, by the test for divergence, the series diverges.

We now look at several tests that tell us if a series is convergent. These tests apply only to series with positive terms.

The Integral Test

The integral test ties the convergence or divergence of an infinite series $\sum_{n=1}^{\infty} a_n$ to the convergence or divergence of the improper integral $\int_1^{\infty} f(x)\,dx$ where $f(n) = a_n$.

THEOREM 5

The Integral Test

Suppose f is a continuous, positive, and decreasing function on $[1, \infty)$, If $f(n) = a_n$ for $n \geq 1$, then

$$\sum_{n=1}^{\infty} a_n \quad \text{and} \quad \int_1^{\infty} f(x)\,dx$$

either both converge or both diverge.

Let's give an intuitive justification for this theorem. If you examine Figure 14.4a, you will see that the height of the first rectangle is $a_2 = f(2)$.

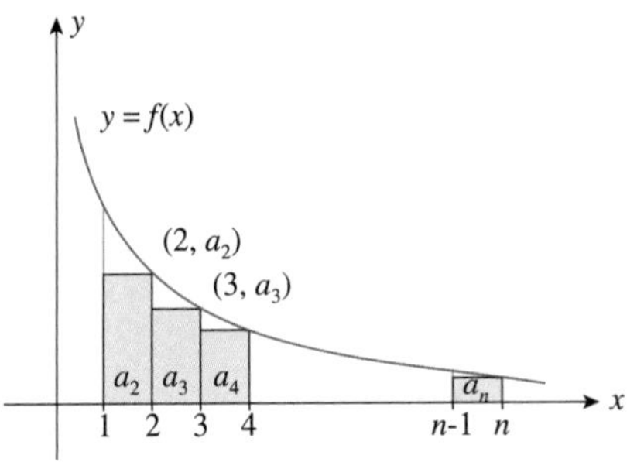

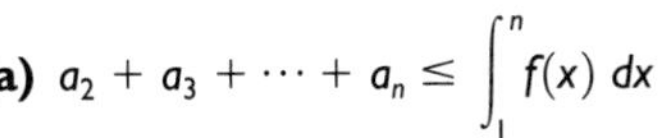

(a) $a_2 + a_3 + \cdots + a_n \leq \int_1^n f(x)\,dx$

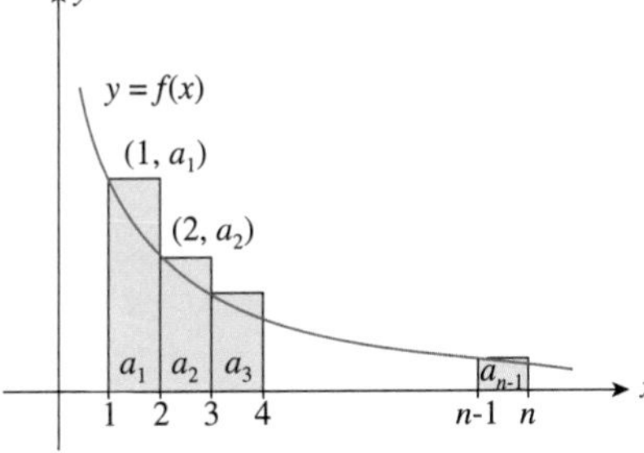

(b) $\int_1^n f(x)\,dx \leq a_1 + a_2 + \cdots + a_{n-1}$

Figure 14.4

Since this rectangle has width one, the area of the rectangle is also $a_2 = f(2)$. Similarly, the area of the second is a_3, and so on. Comparing the sum of the areas of the first $(n - 1)$ inscribed rectangles with the area under the graph of f over the interval $[1, n]$, we see that

$$a_2 + a_3 + \cdots + a_n \le \int_1^n f(x)\,dx$$

which implies that

$$S_n = a_1 + a_2 + a_3 + \cdots + a_n \le a_1 + \int_1^n f(x)\,dx$$

If $\int_1^\infty f(x)\,dx$ is convergent and has value L, then

$$S_n \le a_1 + \int_1^n f(x)\,dx \le a_1 + L$$

This shows that $\{S_n\}$ is bounded above. Also,

$$S_{n+1} = S_n + a_{n+1} \ge S_n \qquad \text{Because } a_{n+1} = f(n+1) \ge 0$$

shows that $\{S_n\}$ is increasing as well.

Now, intuitively, such a sequence must converge to a number no greater than its upper bound. (Although this result can be demonstrated with mathematical rigour, we will not do so here.) In other words $\sum_{n=1}^{\infty} a_n$ is convergent.

Next, by examining Figure 14.4b, you will see that

$$\int_1^n f(x)\,dx \le a_1 + a_2 + \cdots + a_{n-1} = S_{n-1}$$

So, if

$$\int_1^\infty f(x)\,dx$$

diverges to infinity (because $f(x) \ge 0$), then $\lim_{n\to\infty} S_n = \infty$ and so $\sum_{n=1}^{\infty} a_n$ is divergent.

Remarks

1. The integral test simply tells us whether a series converges or diverges. If it indicates that a series converges, we may not conclude that the (finite) value of the improper integral used in conjunction with the test is the *sum* of the convergent series.
2. Since the convergence of an infinite series is not affected by the omission or addition of a finite number of terms to the series, we sometimes study the series

$$\sum_{n=N}^{\infty} a_n = a_N + a_{N+1} + \cdots$$

rather than the series $\sum_{n=1}^{\infty} a_n$. In this case, the series is compared to the improper integral

$$\int_N^\infty f(x)\,dx$$

as we will see in Example 4. ◀

Example 2

Use the integral test to determine whether

$$\sum_{n=1}^{\infty} \frac{1}{n^2}$$

converges or diverges.

Solution

Here

$$a_n = f(n) = \frac{1}{n^2}$$

and so we consider the function $f(x) = 1/x^2$. Since f is continuous, positive, and decreasing on $[1, \infty)$, we may use the integral test. Now

$$\int_1^\infty \frac{1}{x^2}\,dx = \lim_{b\to\infty}\int_1^b x^{-2}\,dx = \lim_{b\to\infty}\left[-\frac{1}{x}\Big|_1^b\right]$$

$$= \lim_{b\to\infty}\left(-\frac{1}{b} + 1\right) = 1$$

Since

$$\int_1^\infty \frac{1}{x^2}\,dx$$

converges, we conclude that

$$\sum_{n=1}^{\infty} \frac{1}{n^2}$$

converges as well.

Example 3

Use the integral test to determine whether the harmonic series $\sum_{n=1}^{\infty} \frac{1}{n}$ converges or diverges.

Solution

Here $a_n = f(n) = 1/n$ and so we consider the function $f(x) = 1/x$. Since f is continuous, positive, and decreasing on $[1, \infty)$, we may use the integral test. But, as you may verify,

$$\int_1^\infty \frac{1}{x}\,dx = \infty$$

We conclude that $\sum_{n=1}^{\infty} \frac{1}{n}$ diverges.

Example 4

Use the integral test to determine whether $\sum_{n=2}^{\infty} \frac{\ln n}{n}$ converges or diverges.

Solution

Here $a_n = (\ln n)/n$ and so we consider the function $f(x) = (\ln x)/x$. Observe that f is continuous and positive on $[2, \infty)$. Next, we compute

$$f'(x) = \frac{x\left(\frac{1}{x}\right) - \ln x}{x^2} = \frac{1 - \ln x}{x^2}$$

Note that $f'(x) < 0$ if $\ln x > 1$; that is, if $x > e$. This shows that f is decreasing on $[3, \infty)$. (See **REMARK** on page 818.) Therefore, we may use the integral test. Now,

$$\int_3^\infty \frac{\ln x}{x}\,dx = \lim_{b\to\infty}\int_3^b \frac{\ln x}{x}\,dx = \lim_{b\to\infty}\left[\frac{1}{2}(\ln x)^2\Big|_3^b\right]$$

$$= \lim_{b\to\infty}\frac{1}{2}\left[(\ln b)^2 - (\ln 3)^2\right] = \infty$$

and we conclude that $\sum_{n=2}^{\infty} \frac{\ln n}{n}$ diverges.

The *p*-Series

The following series will play an important role in our work later on.

p-Series

A *p*-series is a series of the form

$$\sum_{n=1}^{\infty} \frac{1}{n^p} = 1 + \frac{1}{2^p} + \frac{1}{3^p} + \cdots + \frac{1}{n^p} + \cdots$$

where p is a constant.

Observe that for $p = 1$, the p-series is just the harmonic series $\sum_{n=1}^{\infty} \frac{1}{n}$. The conditions for the convergence or divergence of the p-series can be found by applying the integral test to the series. We have the following result.

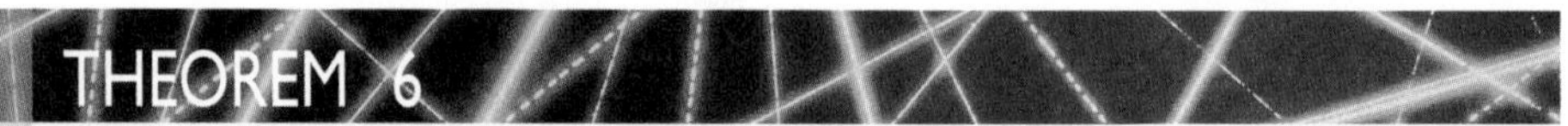

THEOREM 6

Convergence of *p*-Series

The p-series $\sum_{n=1}^{\infty} \frac{1}{n^p}$ converges if $p > 1$ and diverges if $p \le 1$.

Proof If $p < 0$, then

$$\lim_{n\to\infty} \frac{1}{n^p} = \infty$$

If $p = 0$, then

$$\lim_{n\to\infty} \frac{1}{n^p} = 1$$

In either case,

$$\lim_{n\to\infty} \frac{1}{n^p} \neq 0$$

and so the p-series diverges by the test for divergence. If $p > 0$, then the function $f(x) = 1/x^p$ is continuous, positive, and decreasing on $[1, \infty)$. It can be shown that

$$\int_1^{\infty} \frac{1}{x^p}\, dx \text{ converges if } p > 1 \text{ and diverges if } p \le 1$$

(see Exercise 53). Using this result and the integral test, we conclude that $\sum_{n=1}^{\infty} \frac{1}{n^p}$ converges if $p > 1$ and diverges if $0 < p \le 1$. Therefore, $\sum_{n=1}^{\infty} \frac{1}{n^p}$ converges if $p > 1$ and diverges if $p \le 1$.

Example 5

Determine whether each series converges or diverges.

a. $\sum_{n=1}^{\infty} \frac{1}{n^2}$ **b.** $\sum_{n=1}^{\infty} \frac{1}{\sqrt{n}}$ **c.** $\sum_{n=1}^{\infty} n^{-1.001}$

Solution

a. This is a p-series with $p = 2 > 1$, and so by Theorem 6, the series converges.

b. Rewriting the series in the form $\sum_{n=1}^{\infty} \frac{1}{n^{1/2}}$, we see that the series is a p-series with $p = \frac{1}{2} < 1$, and so by Theorem 6, it diverges.

c. We rewrite the series in the form $\sum_{n=1}^{\infty} \frac{1}{n^{1.001}}$, which we recognize to be a p-series with $p = 1.001 > 1$ and conclude accordingly that the series converges.

The Comparison Test

The convergence or divergence of a given series $\sum a_n$ can be determined by comparing its terms with the terms of a *test series* that is known to be convergent or divergent. This is the basis for the comparison test for series that follows. In the rest of this section, we assume that all series under consideration have positive terms.

Suppose the terms of a series $\sum a_n$ are smaller than the corresponding terms of a series $\sum b_n$. This situation is illustrated in Figure 14.5, where the respective terms are represented by rectangles, each of width one and appropriate height.

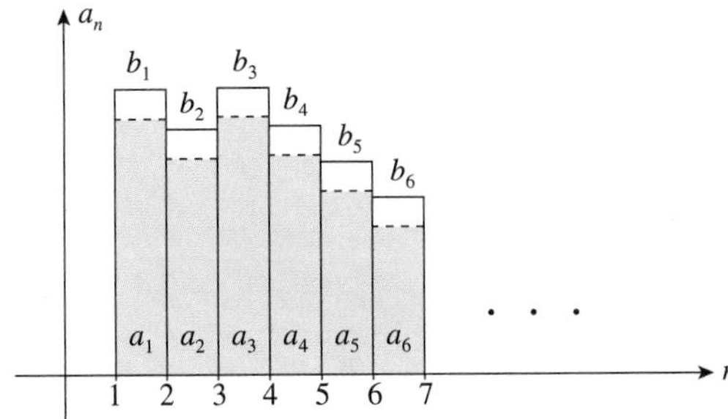

Figure 14.5
Each rectangle representing a_n is contained in the rectangle representing b_n.

If $\sum b_n$ is convergent, the total area of the rectangles representing this series is finite. Since each rectangle representing the series $\sum a_n$ is contained in a corresponding rectangle representing the terms of $\sum b_n$, the total area of the rectangles representing $\sum a_n$ must also be finite: that is, the series $\sum a_n$ must be convergent. A similar argument would seem to suggest that if all the terms of a series $\sum a_n$ are larger than the corresponding terms of a series $\sum b_n$ that is known to be divergent, then $\sum a_n$ must itself be divergent. These observations lead to the following theorem.

THEOREM 7

The Comparison Test

Suppose $\sum a_n$ and $\sum b_n$ are series with positive terms.

a. If $\sum b_n$ is convergent and $a_n \le b_n$ for all n, then $\sum a_n$ is also convergent.

b. If $\sum b_n$ is divergent and $a_n \ge b_n$ for all n, then $\sum a_n$ is also divergent.

Here is an intuitive justification for Theorem 7. Let

$$S_n = \sum_{k=1}^{n} a_k \quad \text{and} \quad T_n = \sum_{k=1}^{n} b_k$$

be the nth terms of the sequence of partial sums of $\sum a_n$ and $\sum b_n$, respectively. Since both series have positive terms, $\{S_n\}$ and $\{T_n\}$ are increasing.

1. If $\sum_{n=1}^{\infty} b_n$ is convergent, then there exists a number L such that $\lim_{n\to\infty} T_n = L$ and $T_n \le L$ for all n. Since $a_n \le b_n$ for all n, we have $S_n \le T_n$, and this implies that $S_n \le L$ for all n. We have shown that $\{S_n\}$ is increasing and bounded above, and so, as before, we can argue intuitively that S_n and therefore $\sum a_n$ converges.

2. If $\sum b_n$ is divergent, then $\lim_{n\to\infty} T_n = \infty$ since $\{T_n\}$ is increasing. But $a_n \geq b_n$ for all n, and this implies that $S_n \geq T_n$, which in turn implies that $\lim_{n\to\infty} S_n = \infty$. Therefore, $\sum a_n$ diverges.

Remark

Since the convergence or divergence of a series is not affected by the omission of a finite number of terms of the series, the condition $a_n \leq b_n$ $(a_n \geq b_n)$ for all n can be replaced by the condition that these inequalities hold for all $n \geq N$ for some integer N. ◀

To use Theorem 7, we need a catalogue of test series whose convergence and divergence are known. In what follows, we will use the geometric series and the p-series as test series.

Example 6

Determine whether the series

$$\sum_{n=1}^{\infty} \frac{1}{n^2 + 2}$$

converges or diverges.

Solution

Let

$$a_n = \frac{1}{n^2 + 2}$$

Observe that when n is large, $n^2 + 2$ behaves like n^2 and so a_n behaves like

$$b_n = \frac{1}{n^2}$$

This observation suggests that we compare $\sum a_n$ with the test series $\sum b_n$, which is a convergent p-series with $p = 2$. Now

$$0 < \frac{1}{n^2 + 2} < \frac{1}{n^2} \qquad (n \geq 1)$$

and the given series is indeed "smaller" than the test series

$$\sum_{n=1}^{\infty} \frac{1}{n^2}$$

Since the test series converges, we conclude by the comparison test that

$$\sum_{n=1}^{\infty} \frac{1}{n^2 + 2}$$

also converges.

Example 7

Determine whether the series

$$\sum_{n=1}^{\infty} \frac{1}{3 + 2^n}$$

converges or diverges.

Solution

Let

$$a_n = \frac{1}{3 + 2^n}$$

Observe that if n is large, $3 + 2^n$ behaves like 2^n, and so a_n behaves like $b_n = 1/2^n$. This observation suggests that we compare $\sum a_n$ with $\sum b_n$. Now the series

$\sum 1/2^n = \sum (1/2)^n$ is a geometric series with $r = 1/2 < 1$ and so it is convergent. Since

$$a_n = \frac{1}{3 + 2^n} < \frac{1}{2^n} = b_n \qquad (n \geq 1)$$

the comparison test tells us that the given series is convergent.

Example 8

Determine whether the series

$$\sum_{n=2}^{\infty} \frac{1}{\sqrt{n} - 1}$$

is convergent or divergent.

Solution

Let

$$a_n = \frac{1}{\sqrt{n} - 1}$$

If n is large, $\sqrt{n} - 1$ behaves like $\sqrt{n}$, and so a_n behaves like

$$b_n = \frac{1}{\sqrt{n}}$$

Now the series

$$\sum_{n=2}^{\infty} b_n = \sum_{n=2}^{\infty} \frac{1}{\sqrt{n}} = \sum_{n=2}^{\infty} \frac{1}{n^{1/2}}$$

is a p-series with $p = 1/2 < 1$ and so it is divergent. Since

$$a_n = \frac{1}{\sqrt{n} - 1} > \frac{1}{\sqrt{n}} = b_n \qquad (\text{for } n \geq 2)$$

See the remark following Theorem 7.

the comparison test implies that the given series is divergent.

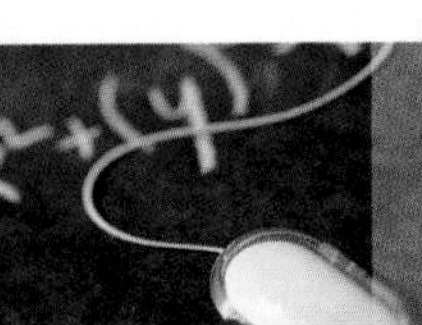

Self-Check Exercises 14.4

1. Show that the series $\sum_{n=1}^{\infty} \frac{2n^2 + 1}{3n^2 - 1}$ is divergent.

2. Use the integral test to determine whether the series $\sum_{n=1}^{\infty} \frac{\ln n}{n}$ converges or diverges.

Solutions to Self-Check Exercises 14.4 can be found on page 825.

14.4 Exercises

In Exercises 1–10, show that the series is divergent.

1. $\frac{1}{2} + \frac{2}{3} + \frac{3}{4} + \cdots$

2. $1 - \frac{3}{2} + \frac{9}{4} - \frac{27}{8} + \cdots$

3. $\sum_{n=1}^{\infty} \frac{2n}{3n + 1}$

4. $\sum_{n=1}^{\infty} \frac{n^2}{2n^2 + 1}$

5. $\sum_{n=1}^{\infty} 2(1.5)^n$

6. $\sum_{n=0}^{\infty} \frac{(-1)^n 3^n}{2^{n-1}}$

7. $\sum_{n=1}^{\infty} \frac{1}{2 + 3^{-n}}$

8. $\sum_{n=1}^{\infty} \frac{n}{\sqrt{2n^2 + 1}}$

9. $\sum_{n=0}^{\infty} \left(-\frac{\pi}{3}\right)^n$

10. $\sum_{n=1}^{\infty} \frac{3^{n+1}}{e^n}$

In Exercises 11–20, use the integral test to determine whether the series is convergent or divergent.

11. $\sum_{n=1}^{\infty} \frac{1}{n+1}$

12. $\sum_{n=1}^{\infty} \frac{3}{2n-1}$

13. $\sum_{n=1}^{\infty} \frac{n}{2n^2+1}$

14. $\sum_{n=1}^{\infty} ne^{-n^2}$

15. $\sum_{n=1}^{\infty} ne^{-n}$

16. $\sum_{n=1}^{\infty} \frac{1}{n(2n-1)}$

17. $\sum_{n=1}^{\infty} \frac{n}{(n^2+1)^{3/2}}$

18. $\sum_{n=2}^{\infty} \frac{1}{n\sqrt{\ln n}}$

19. $\sum_{n=9}^{\infty} \frac{1}{n \ln^3 n}$

20. $\sum_{n=0}^{\infty} \frac{1}{e^n+1}$

In Exercises 21–26, determine whether the *p*-series is convergent or divergent.

21. $\sum_{n=1}^{\infty} \frac{1}{n^3}$

22. $\sum_{n=1}^{\infty} \frac{1}{n^{2/3}}$

23. $\sum_{n=1}^{\infty} \frac{1}{n^{1.01}}$

24. $\sum_{n=1}^{\infty} \frac{1}{n^e}$

25. $\sum_{n=1}^{\infty} \frac{1}{n^{-\pi}}$

26. $\sum_{n=1}^{\infty} n^{-0.98}$

In Exercises 27–36, use the comparison test to determine whether the series is convergent or divergent.

27. $\sum_{n=1}^{\infty} \frac{1}{2n^2+1}$

28. $\sum_{n=1}^{\infty} \frac{1}{n^2+2n}$

29. $\sum_{n=3}^{\infty} \frac{1}{n-2}$

30. $\sum_{n=2}^{\infty} \frac{1}{n^{2/3}-1}$

31. $\sum_{n=2}^{\infty} \frac{1}{\sqrt{n^2-1}}$

32. $\sum_{n=0}^{\infty} \frac{1}{\sqrt{n^3+1}}$

33. $\sum_{n=0}^{\infty} \frac{2^n}{3^n+1}$

34. $\sum_{n=3}^{\infty} \frac{3^n}{2^n-4}$

35. $\sum_{n=2}^{\infty} \frac{\ln n}{n}$

36. $\sum_{n=1}^{\infty} \frac{1}{n^n}$

In Exercises 37–48, determine whether the series is convergent or divergent.

37. $\sum_{n=0}^{\infty} \frac{1}{\sqrt{n+1}}$

38. $\sum_{n=1}^{\infty} \frac{n}{\sqrt{2n^2+1}}$

39. $\sum_{n=2}^{\infty} \frac{1}{n\sqrt{n^2+1}}$

40. $\sum_{n=2}^{\infty} \frac{\sqrt{n^2+1}}{n^2}$

41. $\sum_{n=1}^{\infty} \left(\frac{1}{n\sqrt{n}} + \frac{2}{n^2}\right)$

42. $\sum_{n=1}^{\infty} \left[\left(\frac{2}{3}\right)^n + \frac{1}{n^{3/2}}\right]$

43. $\sum_{n=2}^{\infty} \frac{\ln n}{\sqrt{n}}$

44. $\sum_{n=2}^{\infty} \frac{\ln n}{n^{2.1}}$

45. $\sum_{n=2}^{\infty} \frac{1}{n(\ln n)^2}$

46. $\sum_{n=1}^{\infty} \frac{e^{1/n}}{n^2}$

47. $\sum_{n=1}^{\infty} \frac{1}{\sqrt{n}+4}$

Hint: $\frac{1}{\sqrt{n}+4} > \frac{1}{3\sqrt{n}}$ if $n > 4$

48. $\sum_{n=1}^{\infty} \frac{1}{4n^2-1}$

Hint: $\frac{1}{4n^2-1} < \frac{1}{2n^2}$ if $n \geq 1$

In Exercises 49 and 50, find the value of *p* for which the series is convergent.

49. $\sum_{n=2}^{\infty} \frac{1}{n(\ln n)^p}$

50. $\sum_{n=1}^{\infty} \frac{\ln n}{n^p}$

51. Find the value(s) of a for which the series

$$\sum_{n=1}^{\infty} \left(\frac{a}{n+1} - \frac{1}{n+2}\right)$$

converges. Justify your answer.

52. Consider the series

$$\sum_{n=0}^{\infty} e^{-n}$$

a. Evaluate

$$\int_0^{\infty} e^{-x}\,dx$$

and deduce from the integral test that the given series is convergent.

b. Show that the given series is a geometric series and find its sum.

c. Conclude that although the convergence of

$$\int_0^{\infty} e^{-x}\,dx$$

implies convergence of the infinite series, its value does not give the sum of the infinite series.

53. Show that $\int_1^{\infty} \frac{1}{x^p}\,dx$ converges if $p > 1$ and diverges if $p \leq 1$.

54. Suppose

$$\sum_{n=1}^{\infty} a_n$$

is a convergent series with positive terms. Let $f(n) = a_n$, where f is a continuous and decreasing function for $x \geq N$, where N is some positive integer. Show that the error incurred in approximating the sum of the given series by the Nth partial sum of the series

$$S_N = \sum_{n=1}^{N} a_n$$

is less than $\int_N^{\infty} f(x)\,dx$.

In Exercises 55–60, determine whether the statement is true or false. If it is true, explain why it is true. If it is false, give an example to show why it is false.

55. The series

$$\sum_{n=1}^{\infty} \frac{x}{n}$$

converges only for $x = 0$.

56. If $\lim_{n\to\infty} a_n = 0$, then $\sum_{n=0}^{\infty} a_n$ converges.

57. If $\sum_{n=0}^{\infty} a_n$ diverges, then $\lim_{n\to\infty} a_n \neq 0$.

58. $\int_1^{\infty} \frac{2}{(x^2+1)^{1.1}}\,dx$ converges.

59. Suppose $\sum a_n$ and $\sum b_n$ are series with positive terms. If $\sum a_n$ is convergent and $b_n \geq a_n$ for all n, then $\sum b_n$ is divergent.

60. Suppose $\sum a_n$ and $\sum b_n$ are series with positive terms. If $\sum b_n$ is divergent and $a_n \leq b_n$, for all n, then $\sum a_n$ may or may not converge.

Solutions to Self-Check Exercises 14.4

1. Here

$$\lim_{n\to\infty} a_n = \lim_{n\to\infty} \frac{2n^2+1}{3n^2-1} = \lim_{n\to\infty} \frac{2+\frac{1}{n^2}}{3-\frac{1}{n^2}} = \frac{2}{3} \neq 0$$

and so by the divergence test, the series diverges.

2. Here $a_n = (\ln n)/n$ and so we consider the function $f(x) = (\ln x)/x$. Observe that f is continuous and positive on $[1, \infty)$. Next, we compute

$$f'(x) = \frac{x\left(\frac{1}{x}\right) - \ln x}{x^2} = \frac{1 - \ln x}{x^2}$$

Note that $f'(x) < 0$ if $\ln x > 1$; that is, if $x > e$. This shows that f is decreasing on $[3, \infty)$. Therefore, we may use the integral test.

$$\begin{aligned}\int_3^{\infty} \frac{\ln x}{x}\,dx &= \lim_{b\to\infty} \int_3^b \frac{\ln x}{x}\,dx \\ &= \lim_{b\to\infty} \left[\frac{1}{2}(\ln x)^2\Big|_3^b\right] \\ &= \lim_{b\to\infty} \frac{1}{2}[(\ln b)^2 - (\ln 3)^2] \\ &= \infty\end{aligned}$$

and we conclude that $\sum_{n=1}^{\infty} (\ln n)/n$ diverges.

14.5 Power Series and Taylor Series

Power Series and Intervals of Convergence

Recall that one of our goals in this chapter is to see what happens when the number of terms of a Taylor polynomial is allowed to increase without bound. In other words, we wish to study the expression

$$f(a) + f'(a)(x-a) + \frac{f''(a)}{2!}(x-a)^2 + \cdots + \frac{f^{(n)}(a)}{n!}(x-a)^n + \cdots \quad (7)$$

called the **Taylor series of $f(x)$ at $x = a$.** If the series (7) is truncated after $(n + 1)$ terms, the result is a Taylor polynomial of degree n of $f(x)$ at $x = a$ (see Section 14.1). Note that when $a = 0$, (7) reduces to

$$f(0) + f'(0)x + \frac{f''(0)}{2!}x^2 + \cdots + \frac{f^{(n)}(0)}{n!}x^n + \cdots \quad (8)$$

which is referred to as the **Maclaurin series of $f(x)$.** Thus, the Maclaurin series is a special case of the Taylor series when $a = 0$.

Observe that the Taylor series (7) has the form

$$\sum_{n=0}^{\infty} a_n(x-a)^n = a_0 + a_1(x-a) + a_2(x-a)^2 + \cdots + a_n(x-a)^n + \cdots \quad \textbf{(9)}$$

In fact, comparing the Taylor series (7) with the infinite series (9), we see that

$$a_0 = f(a),\ a_1 = f'(a),\ a_2 = \frac{f''(a)}{2!}, \ldots, a_n = \frac{f^{(n)}(a)}{n!}, \ldots$$

The infinite series (9) is called a **power series centred at $x = a$.** Thus, the Taylor series is just a power series with coefficients that involve the values of some function f and its derivatives at the point $x = a$.

Let's examine some important properties of the power series given in (9). Observe that when x is assigned a value, then this power series becomes an infinite series with constant terms. Accordingly, we can determine, at least theoretically, whether this infinite series converges or diverges. Now the totality of all values of x for which this power series *converges* comprises a set of points in the real line called the **interval of convergence** of the power series. This observation suggests that we may view the power series (9) as a function f whose domain coincides with the interval of convergence of the series and whose functional values are the sums of the infinite series obtained by allowing x to take on all values in the interval of convergence. In this case, we also say the function f is represented by the power series (9).

Given such a power series, how does one determine its interval of convergence? The following theorem, whose proof we will omit, provides the answer to this question.

THEOREM 8

Suppose that we are given the power series.

$$\sum_{n=0}^{\infty} a_n(x-a)^n$$

Let

$$R = \lim_{n\to\infty}\left|\frac{a_n}{a_{n+1}}\right|$$

a. If $R = 0$, the series converges only for $x = a$.

b. If $0 < R < \infty$, the series converges for x in the interval $(a - R, a + R)$ and diverges for x outside this interval (Figure 14.6).

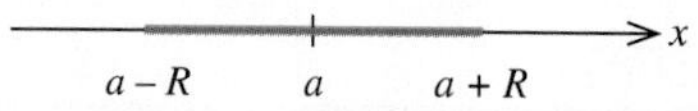

Figure 14.6

c. If $R = \infty$, the series converges for all x.

Remarks

1. The domain of convergence of a power series is an interval called the interval of convergence. This interval of convergence is determined by R, the **radius of convergence.** Depending on whether $R = 0$, $0 < R < \infty$, or $R = \infty$, the "interval" of convergence may just be the degenerate interval consisting of the point a, a bona fide interval $(a - R, a + R)$, or the entire real line.
2. As mentioned earlier, a power series, inside its interval of convergence, represents a function f whose domain of definition coincides with the interval of convergence; that is,

$$f(x) = \sum_{n=0}^{\infty} a_n(x-a)^n \qquad x \in (a - R, a + R)$$

3. The power series may or may not converge at an endpoint. In general, it is difficult to determine whether a power series is convergent at an end point. For this reason, we restrict our attention to points inside the interval of convergence of a power series. ◀

Example 1

Find the radius of convergence and the interval of convergence of the power series

$$\sum_{n=0}^{\infty} \frac{(x-1)^n}{2^n}$$

Show that $f(2)$ exists and find its value, where

$$f(x) = \sum_{n=0}^{\infty} \frac{(x-1)^n}{2^n}$$

Solution

Since $a_n = 1/2^n$, we have

$$R = \lim_{n\to\infty} \left| \frac{a_n}{a_{n+1}} \right| = \lim_{n\to\infty} \left| \frac{\frac{1}{2^n}}{\frac{1}{2^{n+1}}} \right|$$

$$= \lim_{n\to\infty} \frac{2^{n+1}}{2^n} = \lim_{n\to\infty} 2 = 2$$

Therefore, bearing in mind that $a = 1$, we see that the interval of convergence of the given series is $(-1, 3)$. Thus, in the interval $(-1, 3)$, the given power series defines a function

$$f(x) = \sum_{n=0}^{\infty} \frac{(x-1)^n}{2^n} \qquad x \in (-1, 3)$$

Since $2 \in (-1, 3)$, $f(2)$ exists. In fact,

$$f(2) = \sum_{n=0}^{\infty} \frac{(2-1)^n}{2^n} = \sum_{n=0}^{\infty} \frac{1}{2^n}$$

which is a geometric series with $a = 1$ and $r = \frac{1}{2}$, so that

$$f(2) = \frac{1}{1-\frac{1}{2}} = 2$$

Example 2

Find the interval of convergence of each of the following power series:

a. $\sum_{n=0}^{\infty} n^3(x+2)^n$ **b.** $\sum_{n=0}^{\infty} n!(x-1)^n$ **c.** $\sum_{n=0}^{\infty} \frac{x^n}{n!}$

Solution

a. Here $a_n = n^3$, so

$$R = \lim_{n\to\infty} \left| \frac{a_n}{a_{n+1}} \right| = \lim_{n\to\infty} \frac{n^3}{(n+1)^3}$$

$$= \lim_{n\to\infty} \frac{1}{\left(1+\frac{1}{n}\right)^3} = 1$$

Dividing numerator and denominator by n^3

Since $a = -2$, we find that the interval of convergence of the series is $(-3, -1)$.

b. Here $a_n = n!$, so

$$R = \lim_{n\to\infty}\left|\frac{a_n}{a_{n+1}}\right| = \lim_{n\to\infty}\left|\frac{n!}{(n+1)!}\right|$$

$$= \lim_{n\to\infty}\frac{1}{n+1} = 0$$

Therefore, the series converges only at the point $a = 1$.

c. Here $a_n = 1/n!$ and

$$R = \lim_{n\to\infty}\left|\frac{a_n}{a_{n+1}}\right| = \lim_{n\to\infty}\left|\frac{\frac{1}{n!}}{\frac{1}{(n+1)!}}\right| = \lim_{n\to\infty}(n+1) = \infty$$

Therefore, the interval of convergence of the series is $(-\infty, \infty)$; that is, it converges for any value of x.

GROUP DISCUSSION

Suppose the power series $\sum_{n=0}^{\infty} a_n(x-a)^n$ has a radius of convergence R. What can you deduce about the radius of convergence of the series $\sum_{n=0}^{\infty} a_n(x-a)^{2n}$?

Finding a Taylor Series

Theorem 8 guarantees that a power series represents a function whose domain is precisely the interval of convergence of the series. We now show that if a function is defined in this manner, then the power series representing this function must be a Taylor series. To see this, we need the following theorem, which we state without proof.

THEOREM 9

Suppose the function f is defined by

$$f(x) = \sum_{n=0}^{\infty} a_n(x-a)^n = a_0 + a_1(x-a) + a_2(x-a)^2 + \cdots$$

with radius of convergence $R > 0$. Then

$$f'(x) = \sum_{n=1}^{\infty} na_n(x-a)^{n-1} = a_1 + 2a_2(x-a) + 3a_3(x-a)^2 + \cdots$$

converges on the interval $(a - R, a + R)$.

Thus, the derivative of f may be found by differentiating the power series term by term.

We now prove the statement asserted earlier. Suppose that f is represented by a power series centred about $x = a$; that is,

$$f(x) = a_0 + a_1(x-a) + a_2(x-a)^2 + a_3(x-a)^3 + \cdots + a_n(x-a)^n + \cdots$$

Then, applying Theorem 9 repeatedly, we find

$$f'(x) = a_1 + 2a_2(x-a) + 3a_3(x-a)^2 + \cdots + na_n(x-a)^{n-1} + \cdots$$

$$f''(x) = 2a_2 + 3\cdot 2a_3(x-a) + 4\cdot 3a_4(x-a)^2 + \cdots$$
$$+ n(n-1)a_n(x-a)^{n-2} + \cdots$$

$$f'''(x) = 3\cdot 2a_3 + 4\cdot 3\cdot 2a_4(x-a) + \cdots$$
$$+ n(n-1)(n-2)a_n(x-a)^{n-3} + \cdots$$
$$= 3!a_3 + 4!a_4(x-a) + \cdots + n(n-1)(n-2)a_n(x-a)^{n-3} + \cdots$$
$$\vdots$$

$$f^{(n)}(x) = n!a_n + (n+1)!a_{n+1}(x-a) + \cdots$$

Evaluating $f(x)$ and each of these derivatives at $x = a$ yields

$$\begin{aligned} f(a) &= a_0 \\ f'(a) &= a_1 \\ f''(a) &= 2a_2 = 2!a_2 \\ f'''(a) &= 3 \cdot 2a_3 = 3!a_3 \\ &\vdots \\ f^{(n)}(a) &= n!a_n \end{aligned}$$

Thus,

$$a_0 = f(a), a_1 = f'(a), a_2 = \frac{f''(a)}{2!}, a_3 = \frac{f'''(a)}{3!}, \ldots, a_n = \frac{f^{(n)}(a)}{n!}, \ldots$$

as we set out to show.

Next we turn our attention to the converse problem. More precisely, suppose we are given a function f that has derivatives of *all* orders in an open interval I. Can we find a power series representation of f in that interval? Theorem 10 answers this question.

THEOREM 10

Taylor Series Representation of a Function

If a function f has derivatives of all orders in an open interval $I = (a - R, a + R)$ $(R > 0)$ centred at $x = a$, then

$$f(x) = \sum_{n=0}^{\infty} \frac{f^{(n)}(a)}{n!}(x - a)^n$$

if and only if

$$\lim_{n \to \infty} R_n(x) = 0$$

for all x in I, where $R_n(x) = f(x) - P_n(x)$ is as defined in Theorem 1.

Thus, f has a power series representation in the form of a Taylor series provided the error term associated with the Taylor polynomial tends to zero as the number of terms of the Taylor polynomial increases without bound. (A proof of this theorem is sketched in Exercise 33). In what follows, we assume that the functions under consideration have Taylor series representations.

Example 3

Find the Taylor series of the function $f(x) = \dfrac{1}{x - 1}$ at $x = 2$.

Solution

Here $a = 2$ and

$$\begin{aligned} f(x) &= \frac{1}{x - 1} \\ f'(x) &= -\frac{1}{(x - 1)^2} \\ f''(x) &= \frac{2}{(x - 1)^3} \\ f'''(x) &= -\frac{3 \cdot 2}{(x - 1)^4} \\ &\vdots \end{aligned}$$

$$f^{(n)}(x) = (-1)^n \frac{n!}{(x-1)^{n+1}}$$
$$\vdots$$

so that

$$f(2) = 1, f'(2) = -1, f''(2) = 2,$$
$$f'''(2) = -6 = -3!, \ldots, f^{(n)}(2) = (-1)^n n!, \ldots$$

Therefore, the required Taylor series is

$$\begin{aligned}
\frac{1}{x-1} &= f(2) + f'(2)(x-2) + \frac{f''(2)}{2!}(x-2)^2 + \frac{f'''(2)}{3!}(x-2)^3 + \cdots \\
&\quad + \frac{f^{(n)}(2)}{n!}(x-2)^n + \cdots \\
&= 1 - (x-2) + \frac{2!}{2!}(x-2)^2 - \frac{3!}{3!}(x-2)^3 + \cdots \\
&\quad + (-1)^n \frac{n!}{n!}(x-2)^n + \cdots \\
&= 1 - (x-2) + (x-2)^2 - (x-2)^3 + \cdots + (-1)^n(x-2)^n + \cdots \\
&= \sum_{n=0}^{\infty} (-1)^n (x-2)^n
\end{aligned}$$

To find the radius of convergence of the series, we compute

$$\begin{aligned}
R &= \lim_{n\to\infty} \left|\frac{a_n}{a_{n+1}}\right| = \lim_{n\to\infty} \left|\frac{(-1)^n}{(-1)^{n+1}}\right| \\
&= \lim_{n\to\infty} 1 = 1
\end{aligned}$$

Thus, the series converges in the interval (1, 3). By Theorem 10, we see that the function $f(x) = 1/(x-1)$ is represented by the Taylor series in the interval (1, 3).

The representation of the function $f(x) = 1/(x-1)$ by its Taylor series at $x = 2$ is valid only in the interval (1, 3), despite the fact that f itself has a domain that is the set of all real numbers except $x = 1$. This serves to remind us of the local nature of this representation.

EXPLORING WITH TECHNOLOGY

Refer to Example 3, where it was shown that the Taylor series at $x = 2$ representing $f(x) = 1/(x-1)$ is

$$P(x) = \sum_{n=0}^{\infty} (-1)^n (x-2)^n = 1 - (x-2) + (x-2)^2 - (x-2)^3 + \cdots$$

for $1 < x < 3$. This means that if c is any number satisfying $1 < c < 3$, then $f(c) = 1/(c-1) = P(c)$. In particular, this means that $\lim_{n\to\infty} |P_n(c) - f(c)| = 0$, where $\{P_n(x)\}$ is the sequence of partial sums of $P(x)$.

1. Plot the graphs of f, P_0, P_1, P_2, P_3, P_4, P_5, and P_6 on the same set of axes, using the viewing window $[2.5, 3.1] \cdot [-0.1, 1.1]$.
2. Do the results of part 1 give a visual confirmation of the statement $\lim_{n\to\infty} P_n(c) = f(c)$ for the special case where $c = 2.8$? Explain what happens when $c = 3$.

Example 4

Find the Taylor series of the function $f(x) = \ln x$ at $x = 1$.

Solution

Here $a = 1$ and

$$f(x) = \ln x$$

$$
\begin{aligned}
f'(x) &= \frac{1}{x} \\
f''(x) &= -\frac{1}{x^2} \\
f'''(x) &= \frac{2}{x^3} = \frac{2!}{x^3} \\
f^{(4)}(x) &= -\frac{3 \cdot 2}{x^4} = -\frac{3!}{x^4} \\
&\vdots \\
f^{(n)}(x) &= (-1)^{n+1}\frac{(n-1!)}{x^n}
\end{aligned}
$$

so that

$$
f(1) = 0, f'(1) = 1, f''(1) = -1, f'''(1) = 2!,
$$
$$
f^{(4)}(1) = -3!, \ldots, f^{(n)}(1) = (-1)^{n+1}(n-1)!, \ldots
$$

Therefore, the required Taylor series is

$$
\begin{aligned}
f(x) = \ln x &= f(1) + f'(1)(x-1) + \frac{f''(1)}{2!}(x-1)^2 + \frac{f''(1)}{3!}(x-1)^3 \\
&\quad + \cdots + \frac{f^{(n)}(1)}{n!}(x-1)^n + \cdots \\
&= (x-1) - \frac{1}{2!}(x-1)^2 + \frac{2!}{3!}(x-1)^3 - \frac{3!}{4!}(x-1)^4 + \cdots \\
&\quad + (-1)^{n+1}\frac{(n-1)!}{n!}(x-1)^n + \cdots \\
&= (x-1) - \frac{1}{2}(x-1)^2 + \frac{1}{3}(x-1)^3 - \frac{1}{4}(x-1)^4 + \cdots \\
&\quad + \frac{(-1)^{n+1}}{n}(x-1)^n + \cdots \\
&= \sum_{n=1}^{\infty}\frac{(-1)^{n+1}}{n}(x-1)^n
\end{aligned}
$$

Since

$$
\begin{aligned}
R = \lim_{n\to\infty}\left|\frac{a_n}{a_{n+1}}\right| &= \lim_{n\to\infty}\left|\frac{\dfrac{(-1)^{n+1}}{n}}{\dfrac{(-1)^{n+2}}{n+1}}\right| \\
&= \lim_{n\to\infty}\frac{n+1}{n} = \lim_{n\to\infty}\left(1 + \frac{1}{n}\right) = 1
\end{aligned}
$$

we see that the power series representation of $f(x) = \ln x$ is valid in the interval $(0, 2)$. It can be shown that the representation is valid at $x = 2$ as well.

Example 5

Find the power series representation about $x = 0$ for the function $f(x) = e^x$.

Solution

Here $a = 0$ and, since

$$
f(x) = f'(x) = f''(x) = \cdots = e^x
$$

we see that

$$
f(0) = f'(0) = f''(0) = \cdots = e^0 = 1
$$

Therefore, the required power series (Taylor series) is

$$f(x) = e^x = f(0) + f'(0)(x-0) + \frac{f''(!0)}{2!}(x-0)^2 + \cdots$$
$$+ \frac{f^{(n)}(0)}{n!}(x-0)^n + \cdots$$
$$= 1 + x + \frac{x^2}{2!} + \frac{x^3}{3!} + \cdots + \frac{x^n}{n!} + \cdots$$
$$= \sum_{n=0}^{\infty} \frac{x^n}{n!}$$

Since

$$R = \lim_{n\to\infty} \left| \frac{a_n}{a_{n+1}} \right| = \lim_{n\to\infty} \left| \frac{\frac{1}{n!}}{\frac{1}{(n+1)!}} \right|$$
$$= \lim_{n\to\infty} (n+1) = \infty$$

we see that the series representation is valid for all x.

Example 6

Find the power series representation around the point $x = 0$ for the function $f(x) = \sin(x)$.

Solution

In this case $a = 0$. The derivatives of $\sin(x)$ show a simple pattern.

$$\begin{aligned} f(x) &= \sin(x) & f(0) &= 0 \\ f'(x) &= \cos(x) & f'(0) &= 1 \\ f''(x) &= -\sin(x) & f''(0) &= 0 \\ f'''(x) &= -\cos(x) & f'''(0) &= -1 \\ f''''(x) &= \sin(x) & f''''(0) &= 0 \\ &\;\;\cdots & &\;\;\cdots \end{aligned}$$

From this we find that

$$f(x) = \sin(x)$$
$$= f(0) + f'(x)(x-0) + \frac{f''(x)(x-0)^2}{2!}$$
$$+ \frac{f'''(x)(x-0)^3}{3!} + \cdots + \frac{+f^{(n)}(x-0)^n}{n!} + \cdots$$
$$= 0 + x - 0 - \frac{x^3}{3!} + 0 + \frac{x^5}{5!} + \cdots$$

Notice that every other term of this series is zero. The remaining terms alternate in sign. Since the nonzero terms are the odd terms, we get the formula

$$f(x) = \sin(x) = \sum_{n=0}^{\infty} \frac{(-1)^n x^{2n+1}}{(2n+1)!}$$

Since

$$R = \lim_{n\to\infty} \left| \frac{a_n}{a_{n+1}} \right| = \lim_{n\to\infty} \left| \frac{\frac{1}{(2n+1)!}}{\frac{1}{(2n+3)!}} \right|$$

$$= \lim_{n\to\infty} |(2n+3)(2n+2)| = \infty$$

we see that the representation is valid for all x.

The power series representation for $g(x) = \cos(x)$ can be calculated in a similar way, but we leave computing this power series as an example of the techniques used in Section 14.6.

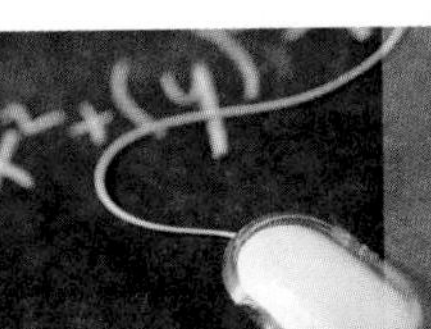

Self-Check Exercises 14.5

1. Find the interval of convergence of the power series $\sum_{n=0}^{\infty} \frac{x^n}{n^2+1}$ (disregard the end points).

2. Find the Taylor series of the function $f(x) = e^{-x}$ at $x = 1$ and determine its interval of convergence.

Solutions to Self-Check Exercises 14.5 can be found on page 834.

14.5 Exercises

In Exercises 1–20, find the radius of convergence and the interval of convergence of the power series.

1. $\sum_{n=0}^{\infty} (x-1)^n$

2. $\sum_{n=0}^{\infty} \left(\frac{x}{2}\right)^n$

3. $\sum_{n=1}^{\infty} n^2x^n$

4. $\sum_{n=0}^{\infty} \frac{(n+1)(x+2)^n}{2^n}$

5. $\sum_{n=0}^{\infty} \frac{(-1)^n x^n}{4^n}$

6. $\sum_{n=0}^{\infty} \frac{(2x)^n}{3^n}$

7. $\sum_{n=0}^{\infty} \frac{(x-1)^n}{n!2^n}$

8. $\sum_{n=0}^{\infty} (2n)!x^n$

9. $\sum_{n=0}^{\infty} \frac{(-1)^n n!(x+2)^n}{2^n}$

10. $\sum_{n=2}^{\infty} \frac{x^n}{n(n+1)}$

11. $\sum_{n=2}^{\infty} \frac{(x+3)^n}{(n+1)^2}$

12. $\sum_{n=0}^{\infty} \frac{n!(x+1)^n}{(3n)!}$

13. $\sum_{n=1}^{\infty} \frac{2n(x-3)^n}{(n+1)!}$

14. $\sum_{n=0}^{\infty} \frac{(-2x)^{2n}}{4^n(n+1)}$

15. $\sum_{n=1}^{\infty} \frac{n(-2x)^n}{n+1}$

16. $\sum_{n=0}^{\infty} \frac{x^{2n+1}}{(2n+1)!}$

17. $\sum_{n=0}^{\infty} \frac{n!(x+1)^n}{3^n}$

18. $\sum_{n=0}^{\infty} (n+1)(x+2)^n$

19. $\sum_{n=0}^{\infty} \frac{n^3(x-3)^n}{3^n}$

20. $\sum_{n=0}^{\infty} \frac{(-1)^{n+1}(x-2)^n}{n2^n}$

In Exercises 21–32, find the Taylor series of the function at the indicated point and give its radius and interval of convergence (disregard the end points).

21. $f(x) = \frac{1}{x}$; $x = 1$

22. $f(x) = \frac{1}{x+1}$; $x = 0$

23. $f(x) = \frac{1}{x+1}$; $x = 2$

24. $f(x) = \frac{1}{1-x}$; $x = 0$

25. $f(x) = \frac{1}{1-x}$; $x = 2$

26. $f(x) = \ln(x+1)$; $x = 0$

27. $f(x) = \sqrt{x}$; $x = 1$

28. $f(x) = \sqrt{1-x}$; $x = 0$

29. $f(x) = e^{2x}$; $x = 0$

30. $f(x) = e^{2x}$; $x = 1$

31. $f(x) = \frac{1}{\sqrt{x+1}}$; $x = 0$

32. $f(x) = \sqrt{x+1}$; $x = 1$

33. Prove Theorem 10.
Hint: The nth partial sum of the Taylor series is $S_n(x) = P_n(x)$ where $P_n(x)$ is the nth Taylor polynomial. Use this fact to show that $\lim_{n\to\infty} S_n(x) = f(x)$ if and only if $\lim_{n\to\infty} R_n(x) = 0$.

In Exercises 34–36, determine whether the statement is true or false. If it is true, explain why it is true. If it is false, give an example to show why it is false.

34. If $\sum_{n=0}^{\infty} a_n(x-2)^n$ converges for $x = 4$, then it converges for $x = 1$.

35. If $\sum_{n=0}^{\infty} a_n(x-a)^n$ has radius of convergence R, then $\sum_{n=0}^{\infty} na_n(x-a)^n$ has radius of convergence R.

36. If $\sum_{n=0}^{\infty} a_n(x-a)^n$ has radius of convergence R, then $\sum_{n=0}^{\infty} a_n^2(x-a)^n$ has radius of convergence $\sqrt{R}$.

Solutions to Self-Check Exercises 14.5

1. We first find the radius of convergence of the power series. Since $a_n = \dfrac{1}{n^2+1}$, we have

$$R = \lim_{n\to\infty}\left|\frac{a_n}{a_{n+1}}\right| = \lim_{n\to\infty}\frac{\dfrac{1}{n^2+1}}{\dfrac{1}{(n+1)^2+1}}$$

$$= \lim_{n\to\infty}\frac{(n+1)^2+1}{n^2+1}$$

$$= \lim_{n\to\infty}\frac{\left(1+\dfrac{1}{n}\right)^2+\dfrac{1}{n^2}}{1+\dfrac{1}{n^2}} \quad \text{Dividing numerator and denominator by } n^2$$

$$= 1$$

Therefore, the interval of convergence of the series is $(-1, 1)$.

2. Here $a = 1$ and

$$f(x) = e^{-x}$$

$$f'(x) = -e^{-x}$$

$$f''(x) = e^{-x}$$

$$f'''(x) = -e^{-x}$$

$$\vdots$$

$$f^{(n)}(x) = (-1)^n e^{-x}$$

so that

$$f(1) = e^{-1}, f'(1) = -e^{-1}, f''(1) = e^{-1},$$

$$f'''(1) = -e^{-1}, \ldots, f^{(n)}(1) = (-1)^n e^{-1}$$

Therefore, the required Taylor series is

$$e^{-x} = f(1) + f'(1)(x-1) + \frac{f''(1)}{2!}(x-1)^2 + \frac{f'''(1)}{3!}(x-1)^3 + \cdots + \frac{f^{(n)}(1)}{n!}(x-1)^n + \cdots$$

$$= e^{-1} - \frac{e^{-1}}{1!}(x-1) + \frac{e^{-1}}{2!}(x-1)^2 - \frac{e^{-1}}{3!}(x-1)^3 + \cdots + \frac{(-1)^n e^{-1}}{n!}(x-1)^n + \cdots$$

$$= \frac{1}{e} - \frac{1}{e}(x-1) + \frac{1}{2!e}(x-1)^2 - \frac{1}{3!e}(x-1)^3 + \cdots + \frac{(-1)^n}{n!e}(x-1)^n + \cdots$$

$$= \sum_{n=0}^{\infty}\frac{(-1)^n}{n!e}(x-1)^n$$

14.6 More on Taylor Series

A Useful Technique for Finding Taylor Series

In the last section, we showed how to find the power series representation of certain functions. This representation turned out to be the Taylor series of f at $x = a$. The method we used to compute the series relies solely on our ability to find the higher-order derivatives of the function f. This method, however, is rather tedious.

In this section, we show how the Taylor series of a function can often be found by manipulating some well-known power series. For this purpose, we first catalogue the power series of some of the most commonly used functions (Table 14.1). Recall that each of these representations was derived in Sections 14.1 and 14.5.

Taylor series and Taylor polynomials have an application that should be familiar to any student. In this section we take power series representations derived in previous sections and show how to quickly derive representations for other functions. Why, you may ask, do we need these when a calculator can compute these functions for us? The calculator is exactly a machine that embodies formulas created by people. Power series give us polynomial approximations to functions like $\ln(x)$, e^x, and $\sin(x)$. By making the polynomial higher degree, we can get polynomials that are as close as we wish to these transcendental functions. This in turn gives us the formulas needed to *build* calculators. The construction of machines that perform computations is itself a field of mathematics concerned with optimizing the computation of fundamental functions such as the log, exponential, and trig functions. Taylor series are a foundation of this branch of mathematics.

Table 14.1 Power Series Representations for Some Common Functions

1. $\dfrac{1}{1-x} = 1 + x + x^2 + \cdots + x^n + \cdots \qquad (-1 < x < 1)$

2. $e^x = 1 + x + \dfrac{1}{2!}x^2 + \dfrac{1}{3!}x^3 + \cdots + \dfrac{1}{n!}x^n + \cdots \qquad (-\infty < x < \infty)$

3. $\ln x = (x-1) - \dfrac{1}{2}(x-1)^2 + \dfrac{1}{3}(x-1)^3 - \cdots + \dfrac{(-1)^{n+1}}{n}(x-1)^n + \cdots \qquad (0 < x \le 2)$

4. $\sin(x) = x - \dfrac{x^3}{3!} + \dfrac{x^5}{5!} - \dfrac{x^7}{7!} + \cdots + \dfrac{(-1)^n x^{2n+1}}{(2n+1)!} + \cdots \qquad (-\infty < x < \infty)$

Example 1

Find the Taylor series of each function at the indicated point.

a. $f(x) = \dfrac{1}{1+x}; x = 0$

b. $f(x) = \dfrac{1}{1+x}; x = 2$

c. $f(x) = \dfrac{1}{1+3x}; x = 0$

d. $f(x) = \dfrac{x}{1+x^2}; x = 0$

Solution

a. We write

$$f(x) = \frac{1}{1-(-x)}$$

and use the power series representation of $1/(1-x)$ with x replaced by $-x$ to obtain

$$\begin{aligned} f(x) &= \frac{1}{1+x} \\ &= 1 + (-x) + (-x)^2 + \cdots + (-x)^n + \cdots \\ &= 1 - x + x^2 - x^3 + \cdots + (-1)^n x^n + \cdots \qquad (-1 < x < 1) \end{aligned}$$

b. We write

$$f(x) = \frac{1}{1+x} = \frac{1}{3+(x-2)} = \frac{1}{3\left[1 + \left(\frac{x-2}{3}\right)\right]} = \frac{1}{3}\left[\frac{1}{1+\left(\frac{x-2}{3}\right)}\right]$$

From the result of part (a), we have

$$\frac{1}{1+u} = 1 - u + u^2 - u^3 + \cdots + (-1)^n u^n + \cdots \qquad (-1 < u < 1)$$

Thus, with $u = \dfrac{x-2}{3}$, we find

$$f(x) = \frac{1}{3}\left[\frac{1}{1+\left(\frac{x-2}{3}\right)}\right]$$

$$= \frac{1}{3}\left[1 - \left(\frac{x-2}{3}\right) + \left(\frac{x-2}{3}\right)^2 - \left(\frac{x-2}{3}\right)^3 + \cdots + (-1)^n\left(\frac{x-2}{3}\right)^n + \cdots\right]$$

$$= \frac{1}{3} - \frac{1}{3^2}(x-2) + \frac{1}{3^3}(x-2)^2 - \frac{1}{3^4}(x-2)^3 + \cdots + \frac{(-1)^n}{3^{n+1}}(x-2)^n + \cdots$$

which converges when

$$-1 < \frac{x-2}{3} < 1 \qquad \text{or} \qquad -1 < x < 5$$

c. Here we simply replace x in the power series representation of $1/(1-x)$ by $3x$ to get

$$f(x) = \frac{1}{1-3x} = 1 + 3x + (3x)^2 + (3x)^3 + \cdots + (3x)^n + \cdots$$

$$= 1 + 3x + 3^2x^2 + 3^3x^3 + \cdots + 3^nx^n + \cdots$$

The series converges for $-1 < 3x < 1$, or $-\frac{1}{3} < x < \frac{1}{3}$.

d. We write

$$f(x) = \frac{x}{1+x^2} = x\left[\frac{1}{1-(-x^2)}\right]$$

$$= x[1 + (-x^2) + (-x^2)^2 + (-x^2)^3 + \cdots + (-x^2)^n + \cdots]$$

$$= x - x^3 + x^5 - x^7 + \cdots + (-1)^nx^{2n+1} + \cdots$$

The series converges for $-1 < x < 1$.

Example 2

Find the Taylor series of the function at the indicated point.

a. $f(x) = xe^{-x}$; $x = 0$

b. $f(x) = \ln(1+x)$; $x = 0$

Solution

a. First, we replace x with $-x$ in the expression

$$e^x = 1 + x + \frac{x^2}{2!} + \frac{x^3}{3!} + \cdots + \frac{x^n}{n!} + \cdots$$ See Table 11.1.

to obtain

$$e^{-x} = 1 + (-x) + \frac{(-x)^2}{2!} + \frac{(-x)^3}{3!} + \cdots + \frac{(-x)^n}{n!} + \cdots$$

$$= 1 - x + \frac{x^2}{2!} - \frac{x^3}{3!} + \cdots + \frac{(-1)^nx^n}{n!} + \cdots$$

Then, multiplying both sides of this expression by x gives the required expression

$$f(x) = xe^{-x} = x - x^2 + \frac{x^3}{2!} - \frac{x^4}{3!} + \cdots + \frac{(-1)^n x^{n+1}}{n!} + \cdots$$

b. From Table 14.1 we have

$$\ln x = (x-1) - \frac{1}{2}(x-1)^2 + \frac{1}{3}(x-1)^3 - \cdots$$

$$+ \frac{(-1)^{n+1}}{n}(x-1)^n + \cdots \qquad (0 < x \leq 2)$$

Replacing x with $1 + x$ in this expression, we obtain the required expression

$$f(x) = \ln(1+x) = x - \frac{1}{2}x^2 + \frac{1}{3}x^3 - \cdots + \frac{(-1)^{n+1}}{n}x^n + \cdots$$

valid for $-1 < x \leq 1$. (Why?)

GROUP DISCUSSION

The formula

$$\ln x = (x-1) - \frac{1}{2}(x-1)^2 + \frac{1}{3}(x-1)^3 - \cdots \qquad (0 < x \leq 2) \qquad \textbf{(A)}$$

from Table 14.1 can be used to compute the value of $\ln x$ for $0 < x \leq 2$. However, the restriction on x and the slow convergence of the series limit its effectiveness from the computational point of view. A more effective formula, first obtained by the Scottish mathematician James Gregory (1638–1675), follows:

1. Using Formula (A), derive the formulas

$$\ln(1+x) = x - \frac{x^2}{2} + \frac{x^3}{3} - \frac{x^4}{4} - \cdots \qquad (-1 < x \leq 1)$$

and

$$\ln(1-x) = -x - \frac{x^2}{2} - \frac{x^3}{3} - \frac{x^4}{4} - \cdots \qquad (-1 \leq x < 1)$$

2. Use part 1 to show that

$$\ln\left(\frac{1+x}{1-x}\right) = 2\left(x + \frac{x^3}{3} + \frac{x^5}{5} + \frac{x^7}{7} + \cdots\right) \qquad (-1 < x < 1)$$

3. To compute the natural logarithm of a positive number p, let

$$p = \frac{1+x}{1-x}$$

and show that

$$x = \frac{p-1}{p+1} \qquad (-1 < x < 1)$$

4. Use parts 2 and 3 to show that

$$\ln 2 = 2\left[\left(\frac{1}{3}\right) + \frac{(\frac{1}{3})^3}{3} + \frac{(\frac{1}{3})^5}{5} + \frac{(\frac{1}{3})^7}{7} + \cdots\right]$$

and this yields $\ln 2 \approx 0.6931$ when we add the first four terms of the series. This approximation of $\ln 2$ is accurate to four decimal places.

5. Compare this method of computing $\ln 2$ with that of using Formula (A) directly.

In Theorem 10 in the last section, we saw that a power series may be differentiated term by term to yield another power series whose interval of convergence coincides with that of the original series. The latter is of course the power series representation of the derivative f' of the function f represented by the first series. The following theorem tells us that we may integrate a power series term by term.

THEOREM 11

Suppose

$$f(x) = \sum_{n=0}^{\infty} a_n(x-a)^n$$
$$= a_0 + a_1(x-a) + a_2(x-a)^2 + a_3(x-a)^3 + \cdots$$
$$+ a_n(x-a)^n + \cdots \qquad x \in (a-R, a+R)$$

Then

$$\int f(x)\,dx = \sum_{n=0}^{\infty} \frac{a_n}{n+1}(x-a)^{n+1}$$
$$= a_0(x-a) + \frac{a_1}{2}(x-a)^2 + \frac{a_2}{3}(x-a)^3 + \cdots$$
$$+ \frac{a_n}{n+1}(x-a)^{n+1} + \cdots \qquad x \in (a-R, a+R)$$

Theorems 10 and 11 can also be used to help us find the power series representation of a function starting from the series representation of some appropriate function, as the next two examples show.

Example 3

Differentiate the power series for the function $1/(1-x)$ at $x = 0$ (see Table 14.1) to obtain a Taylor series representation of the function $f(x) = 1/(1-x)^2$ at $x = 0$.

Solution

From Table 14.1, we have

$$\frac{1}{1-x} = 1 + x + x^2 + x^3 + \cdots + x^n + \cdots \qquad (-1 < x < 1)$$

Differentiating both sides of the equation with respect to x, and using Theorem 10, we obtain

$$\frac{d}{dx}\left(\frac{1}{1-x}\right) = \frac{d}{dx}(1 + x + x^2 + x^3 + \cdots + x^n + \cdots)$$

or

$$f(x) = \frac{1}{(1-x)^2}$$
$$= 1 + 2x + 3x^2 + \cdots + nx^{n-1} + \cdots \qquad (-1 < x < 1)$$

Example 4

Find the power series representation around the point $x = 0$ for the function $f(x) = \cos(x)$.

Solution

This example is another application of Theorem 10. Since the derivative of $\sin(x)$ is $\cos(x)$, we can take the power series for $\sin(x)$ and differentiate it.

$$\sin(x) = x - \frac{x^3}{3!} + \frac{x^5}{5!} - \frac{x^7}{7!} + \cdots + \frac{(-1)^n x^{2n+1}}{(2n+1)!} + \cdots$$

$$\frac{d}{dx}\sin(x) = \frac{d}{dx}\left(x - \frac{x^3}{3!} + \frac{x^5}{5!} - \frac{x^7}{7!} + \cdots + \frac{(-1)^n x^{2n+1}}{(2n+1)!} + \cdots\right)$$

$$\cos(x) = 1 - \frac{3x^2}{3!} + \frac{5x^4}{5!} - \frac{7x^6}{7!} + \cdots + \frac{(-1)^n(2n+1)x^{2n}}{(2n+1)!} + \cdots$$

$$\cos(x) = 1 - \frac{x^2}{2!} + \frac{x^4}{4!} - \frac{x^6}{6!} + \cdots + \frac{(-1)^n x^{2n}}{(2n)!} + \cdots$$

$$\cos(x) = \sum_{n=0}^{\infty} \frac{(-1)^n x^{2n}}{(2n)!}$$

Since the power series for $\sin(x)$ had $R = \infty$, the power series for $\cos(x)$ is also valid for all x.

Example 5

Integrate the Taylor series for the function $1/(1 + x)$ at $x = 0$ (see Table 14.1) to obtain a power series representation for the function $f(x) = \ln(1 + x)$ centred at $x = 0$. Compare your result with that of Example 2b.

Solution

From Table 14.1 we see that

$$\frac{1}{1-x} = 1 + x + x^2 + x^3 + \cdots + x^n + \cdots \qquad (-1 < x < 1)$$

Replacing x by $-x$ gives

$$\frac{1}{1-(-x)} = 1 + (-x) + (-x)^2 + (-x)^3 + \cdots + (-x)^n + \cdots$$

or

$$\frac{1}{1+x} = 1 - x + x^2 - x^3 + \cdots + (-1)^n x^n + \cdots \qquad (-1 < x < 1)$$

Finally, integrating both sides of this equation with respect to x and using Theorem 11, we obtain

$$\int \frac{1}{1+x}\,dx = \int [1 - x + x^2 - x^3 + \cdots + (-1)^n x^n + \cdots]\,dx$$

or

$$f(x) = \ln(1 + x) = x - \frac{1}{2}x^2 + \frac{1}{3}x^3 - \cdots + \frac{(-1)^n}{n+1}x^{n+1} + \cdots$$

$$= x - \frac{1}{2}x^2 + \frac{1}{3}x^3 - \cdots + \frac{(-1)^{n+1}}{n}x^n + \cdots \qquad (-1 < x < 1)$$

This result is the same as that of Example 2b, as expected.

Example 6

Integrate the function

$$f(x) = \frac{1}{1+x^2}$$

to find a power series representation around the point $x = 0$ for the function $\arctan(x)$.

Solution

Substitute $-x^2$ for x in the Taylor series formula for $\frac{1}{1-x}$ (from Table 14.1) and then integrate.

$$\frac{1}{1-x} = 1 + x + x^2 + x^3 + \cdots + x^n + \cdots$$

$$\frac{1}{1-(-x^2)} = 1 + (-x)^2 + (-x^2)^2 + (-x^2)^3 + \cdots + (-x^2)^n + \cdots$$

$$\frac{1}{1+x^2} = 1 - x^2 + x^4 - x^6 + \cdots + (-1)^n x^{2n} + \cdots$$

$$\int \frac{dx}{1+x^2} = \int (1 - x^2 + x^4 - x^6 + \cdots + (-1)^n x^{2n} + \cdots)\, dx$$

$$\arctan(x) = x - \frac{x^3}{3} + \frac{x^5}{5} - \frac{x^7}{7} + \cdots + \frac{(-1)^n x^{2n+1}}{2n+1}$$

The power series for $\frac{1}{1-x}$ is valid for $-1 < x < 1$, and so the power series for $\arctan(x)$ has the same interval of convergence.

Application

In many practical applications, all that is required is an approximation of the actual solution to a problem. Thus, rather than working with the Taylor series of a function at a point that is *equal* to $f(x)$ inside its interval of convergence, one often works with the truncated Taylor series. The truncated Taylor series is a Taylor polynomial that gives an acceptable approximation to the values of $f(x)$ in the neighbourhood of the point about which the function is expanded, provided the degree of the polynomial or, equivalently, the number of terms of the Taylor series retained is large enough.

Before looking at an example, we want to point out the comparative ease with which one can obtain the Taylor polynomial approximation of a function using the method of this section, rather than obtaining it directly as was done in Section 14.1.

Example 7

Serum Cholesterol Levels The serum cholesterol levels (in mg/dL) in a current Mediterranean population are found to be normally distributed with a probability density function given by

$$f(x) = \frac{1}{50\sqrt{2\pi}} e^{-1/2[(x-160)/50]^2}$$

Scientists at the National Heart, Lung, and Blood Institute consider this pattern ideal for a minimal risk of heart attacks. Find the percentage of the population who have blood cholesterol levels between 160 and 180 mg/dL.

Solution

The required probability is given by

$$P(160 \le x \le 180) = \frac{1}{50\sqrt{2\pi}} \int_{160}^{180} e^{-1/2[(x-160)/50]^2}\, dx$$

(see Section 13.3). Let's approximate the integrand by a sixth-degree Taylor polynomial about $x = 160$. From Table 14.1, we have

$$e^x = 1 + x + \frac{1}{2!}x^2 + \frac{1}{3!}x^3 + \cdots$$

Replacing x with $-\frac{1}{2}[(x-160)/50]^2$, we obtain

$$e^{-1/2[(x-160)/50]^2} \approx 1 - \frac{1}{2}\left(\frac{x-160}{50}\right)^2 + \frac{1}{2!}\left[-\frac{1}{2}\left(\frac{x-160}{50}\right)^2\right]^2$$

$$+\frac{1}{3!}\left[-\frac{1}{2}\left(\frac{x-160}{50}\right)^2\right]^3$$

$$= 1 - \frac{(x-160)^2}{5000} + \frac{(x-160)^4}{5\cdot 10^7} - \frac{(x-160)^6}{7.5\cdot 10^{11}}$$

Therefore,

$$P(160 \le x \le 180)$$

$$\approx \frac{1}{50\sqrt{2\pi}}\int_{160}^{180}\left[1 - \frac{(x-160)^2}{500} + \frac{(x-160)^4}{5\cdot 10^7} - \frac{(x-160)^6}{7.5\cdot 10^{11}}\right]dx$$

$$\approx \frac{1}{50\sqrt{2\pi}}\left[x - \frac{(x-160)^3}{15{,}000} + \frac{(x-160)^5}{2.5\cdot 10^8} - \frac{(x-160)^7}{5.25\cdot 10^{12}}\right]\Bigg|_{160}^{180}$$

$$\approx \frac{1}{50\sqrt{2\pi}}[(180 - 0.53333 + 0.0128 - 0.00024) - 160]$$

$$\approx 0.1554$$

and so approximately 15.5% of the population has blood cholesterol levels between 160 and 180 mg/dL.

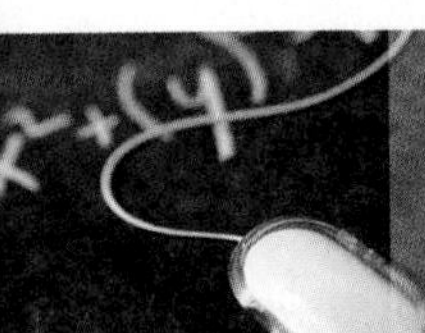

Self-Check Exercises 14.6

1. Find the Taylor series of the function $f(x) = xe^{-x^2}$ at $x = 0$.

2. Use the result from Self-Check Exercise 1 to write the seventh Taylor polynomial of f about $x = 0$, and use this polynomial to approximate

$$\int_0^{0.5} xe^{-x^2}\,dx$$

Compare this result with the exact value of the integral.

Solutions to Self-Check Exercises 14.6 can be found on page 842.

14.6 Exercises

In Exercises 1–24, find the Taylor series of each function at the indicated point. Give the interval of convergence for each series.

1. $f(x) = \dfrac{1}{1-x}$; $x = 2$
2. $f(x) = \dfrac{1}{1+x}$; $x = 1$
3. $f(x) = \dfrac{1}{1+3x}$; $x = 0$
4. $f(x) = \dfrac{x}{1-2x}$; $x = 0$
5. $f(x) = \dfrac{1}{4-3x}$; $x = 0$
6. $f(x) = \dfrac{1}{4-3x}$°; $x = 1$
7. $f(x) = \dfrac{1}{1-x^2}$°; $x = 0$
8. $f(x) = \dfrac{x^2}{1-x^3}$°; $x = 0$
9. $f(x) = e^{-x}$; $x = 0$
10. $f(x) = e^x$; $x = 1$
11. $f(x) = xe^{-x^2}$; $x = 0$
12. $f(x) = xe^{x/2}$; $x = 0$
13. $f(x) = \dfrac{1}{2}(e^x + e^{-x})$; $x = 0$
14. $f(x) = \dfrac{1}{2}(e^x - e^{-x})$; $x = 0$
15. $f(x) = \ln(1 + 2x)$; $x = 0$
16. $f(x) = \ln\left(1 + \dfrac{1}{2}\right)$; $x = 0$
17. $f(x) = \ln(1 + x^2)$; $x = 0$
18. $f(x) = \ln(1 + 2x)$; $x = 2$
19. $f(x) = (x - 2)\ln x$; $x = 2$
20. $f(x) = x^2\ln\left(1 + \dfrac{1}{2}\right)$; $x = 0$
21. $f(x) = \sin(2x)$; $x = 0$

22. $g(x) = \cos(-x); x = 0$ 23. $h(x) = x \sin(x); x = 0$

24. $f(x) = \arctan(\sqrt{x}); x = 0$

25. Differentiate the power series for $\ln(1 + x)$ at $x = 0$ to obtain a series representation for the function $f(x) = 1/(1 + x)$.

26. Differentiate the power series for $1/(1 + x)$ at $x = 0$ to obtain a series representation for the function $f(x) = 1/(1 + x)^2$.

27. Integrate the power series for $1/(1 + x)$ to obtain a power series representation for the function $f(x) = \ln(1 + x)$.

28. Integrate the power series for $2x/(1 + x^2)$ to obtain a power series representation for the function $f(x) = \ln(1 + x^2)$.

29. Use the sixth-degree Taylor polynomial to approximate

$$\int_0^{0.5} \frac{1}{\sqrt{1 + x^2}}\, dx$$

Hint: $P_6(x) = 1 - \frac{1}{2}x^2 + \frac{3}{8}x^4 - \frac{5}{16}x^6$

30. Use the sixth-degree Taylor polynomial to approximate

$$\int_0^{0.4} \ln(1 + x^2)\, dx$$

31. Use the eighth-degree Taylor polynomial to approximate

$$\int_0^{1} e^{-x^2}\, dx$$

32. Use the sixth-degree Taylor polynomial to approximate

$$\int_0^{0.5} \frac{\ln(1 + x)}{x}\, dx$$

33. Use the eighth-degree Taylor polynomial of $f(x) = 1/(1 + x^2)$ at $x = 0$ and the relationship

$$\pi = 4\int_0^{1} \frac{dx}{1 + x^2}$$

to obtain an approximation of π.

34. **TV Set Reliability** The General Manager of the Service Department of MCA Television Company has estimated that the time that elapses between the dates of purchase and the dates on which the 19-in. sets manufactured by the company first require service is normally distributed according to the probability density function

$$f(x) = \frac{1}{10\sqrt{2\pi}} e^{-1/2[(x-30)/10]^2}$$

where x is measured in months. Determine the percentage of sets manufactured and sold by MCA that may require service 28–32 mo after purchase.

Business and Economics Applications

35. **Factory Worker Wages** According to data released by a city's Chamber of Commerce, the weekly wages of factory workers are normally distributed according to the probability density function

$$f(x) = \frac{1}{50\sqrt{2\pi}} e^{-1/2[(x-500)/50]^2}$$

Find the probability that a worker selected at random from the city has a weekly wage of \$450–\$550.

36. **Demand for Wristwatches** The demand equation for the "Tempus" quartz wristwatch is given by

$$p = 50e^{-0.1(x+1)^2}$$

where x (measured in units of a thousand) is the quantity demanded per week and p is the unit wholesale price in dollars. National Importers, the supplier of the watches, will make x units (in thousands) available in the market if the unit wholesale price is

$$p = 10 + 5x^2$$

dollars. Use an appropriate Taylor polynomial approximation of the function

$$p = f(x) = 50e^{-0.1(x+1)^2}$$

to find the consumers' surplus (see Chapter 12).
Hint: The equilibrium quantity is 1689 units.

Solutions to Self-Check Exercises 14.6

1. First, we replace x in the expression

$$e^x = 1 + x + \frac{x^2}{2!} + \frac{x^3}{3!} + \cdots + \frac{x^n}{n!} + \cdots$$

(see Table 14.1) with $-x^2$ to obtain

$$e^{-x^2} = 1 + (-x^2) + \frac{(-x^2)^2}{2!} + \frac{(-x^2)^3}{3!} + \cdots + \frac{(-x^2)^n}{n!} + \cdots$$

$$= 1 - x^2 + \frac{x^4}{2!} - \frac{x^6}{3!} + \cdots + \frac{(-1)^n x^{2n}}{n!} + \cdots$$

Solutions to Self-Check Exercises 14.6, *continued*

Then multiplying both sides of this expression by x gives the required expression

$$f(x) = xe^{-x^2} = x - x^3 + \frac{x^5}{2!} - \frac{x^7}{3!} + \cdots + \frac{(-1)^n x^{2n+1}}{n!} + \cdots$$
$$= \sum_{n=0}^{\infty} \frac{(-1)^n x^{2n+1}}{n!}$$

2. Using the result from Exercise 1, we see that the seventh Taylor polynomial of f about $x = 0$ is

$$P_7(x) = x - x^3 + \frac{x^5}{2!} - \frac{x^7}{3!}$$
$$= x - x^3 + \frac{1}{2}x^5 - \frac{1}{6}x^7$$

Therefore,

$$\int_0^{0.5} xe^{-x^2}dx \approx \int_0^{0.5} \left(x - x^3 + \frac{1}{2}x^5 - \frac{1}{6}x^7\right)dx$$
$$= \frac{1}{2}x^2 - \frac{1}{4}x^4 + \frac{1}{12}x^6 - \frac{1}{48}x^8 \Big|_0^{0.5}$$
$$\approx 0.1105957$$

The exact value of the integral is

$$\int_0^{0.5} xe^{-x^2}dx = -\frac{1}{2}e^{-x^2}\Big|_0^{0.5} \qquad \text{Use the substitution } u = x^2.$$
$$= -\frac{1}{2}(e^{-0.25})$$

which is approximately 0.1105996. Thus, the error is 0.0000039.

CHAPTER 14 Summary of Principal Formulas and Terms

Formulas

1. nth Taylor polynomial of f at $x = a$

$$P_n(x) = f(a) + f'(a)(x - a) + \frac{f''a}{2!}(x - a)^2 + \cdots + \frac{f^{(n)}(a)}{n!}(x - a)^n$$

2. Infinite series

$$\sum_{x=1}^{\infty} a_n = a_1 + a_2 + a_3 + \cdots$$

3. Geometric series

$$\sum_{n=0}^{\infty} ar^n = a + ar + \cdots = \frac{a}{1 - r} \quad (|r| < 1)$$

4. Taylor series expansion of $f(x)$ at $x = a$

$$f(x) = f(a) + f'(a)(x - a) + \frac{f''(a)}{2!}(x - a)^2 + \frac{f'''(a)}{3!}(x - a)^3 + \cdots$$

Terms

polynomial of degree n (787)
error bound for Taylor polynomials (793)
term (799)
general (nth) term (799)
infinite sequence (799)
limit of a sequence (801)
convergent sequence (801)
divergent sequence (801)
infinite series (806)
Nth partial sum (806)
sequence of partial sums (806)
partial sum of an infinite series (806)
convergent series (806)
divergent series (806)
sum of the series (806)
geometric series (808)
multiplier effect (811)
perpetuity (812)
capital value of a perpetuity (812)
harmonic series (816)
p-series (820)
Taylor series of $f(x)$ (825)
Maclaurin series of $f(x)$ (825)
power series (826)
interval of convergence (826)
radius of convergence (826)

Review Exercises 

In Exercises 1–4, find the fourth Taylor polynomial of the function f at the indicated point.

1. $f(x) = \dfrac{1}{x+2}$ at $x = -1$

2. $f(x) = e^{-x}$ at $x = 1$

3. $f(x) = \ln(1 + x^2)$ at $x = 0$

4. $f(x) = \dfrac{1}{(1+x)^2}$ at $x = 0$

5. Find the second Taylor polynomial of $f(x) = \sqrt[3]{x}$ at $x = 8$ and use it to estimate the value of $\sqrt[3]{7.8}$.

6. Find the sixth Taylor polynomial of $f(x) = e^{-x^2}$ at $x = 0$ and use it to estimate the value of

$$\int_0^1 e^{-x^2}\,dx$$

7. Use the second Taylor polynomial of $f(x) = \sqrt[3]{x}$ at $x = 27$ to approximate $\sqrt[3]{26.98}$ and find a bound for the error in the approximation.

8. Use the third Taylor polynomial of $f(x) = 1/(1 + x)$ to approximate $f(0.1)$. Find a bound for the error in the approximation and compare your results to the exact value of $f(0.1)$.

9. Find an estimate of e^{-1} with an error less than 10^{-2}.

10. Estimate $\int_0^{0.2} e^{-(1/2)x^2}\,dx$ with an error less than 10^{-2}.

In Exercises 11–14, the nth term of a sequence is given. Determine whether the sequence converges or diverges. If the sequence converges, find its limit.

11. $a_n = \dfrac{2n^2+1}{3n^2-1}$

12. $a_n = \dfrac{(-1)^{n-1}n}{n+1}$

13. $a_n = 1 - \dfrac{1}{2^n}$

14. $a_n = \dfrac{1+\sqrt{n}}{1-\sqrt{n}}$

In Exercises 15–18, find the sum of the geometric series if it converges.

15. $\displaystyle\sum_{n=1}^{\infty} \frac{2^n}{3^n}$

16. $\displaystyle\sum_{n=1}^{\infty} 2^{-n}3^{-n+1}$

17. $\displaystyle\sum_{n=1}^{\infty} (-1)^{n-1}\left(\frac{1}{\sqrt{2}}\right)^n$

18. $\displaystyle\sum_{n=1}^{\infty} \left(\frac{1}{e}\right)^n$

19. Express the repeating decimal 1.424242. . . as a rational number.

20. Express the repeating decimal 3.142142142. . . as a rational number.

In Exercises 21–24, determine whether the series is convergent or divergent.

21. $\displaystyle\sum_{n=1}^{\infty} \frac{n^2+1}{2n^2-1}$

22. $\displaystyle\sum_{n=1}^{\infty} \frac{n+1}{2n^2+4n}$

23. $\displaystyle\sum_{n=1}^{\infty} \left(\frac{1}{n}\right)^{1.1}$

24. $\displaystyle\sum_{n=1}^{\infty} \frac{n^3}{n^5+2}$

In Exercises 25–28, find the radius of convergence and the interval of convergence of each power series.

25. $\displaystyle\sum_{n=0}^{\infty} \frac{x^n}{n^2+2}$

26. $\displaystyle\sum_{n=1}^{\infty} \frac{(-1)^{n-1}}{\sqrt{n}}x^n$

27. $\displaystyle\sum_{n=1}^{\infty} \frac{(x-1)^n}{n(n+1)}$

28. $\displaystyle\sum_{n=2}^{\infty} \frac{e^n}{n^2}(x-2)^n$

In Exercises 29–32, find the Taylor series of the function at the indicated point. Give the interval of convergence for each series.

29. $f(x) = \dfrac{1}{2x-1}$; $x = 0$

30. $f(x) = e^{-x}$; $x = 1$

31. $f(x) = \ln(1 + 2x)$; $x = 0$

32. $f(x) = x^2e^{-2x}$; $x = 0$

Business and Economics Applications

33. **Establishing a Trust Fund** Sam Simpson wishes to establish a trust fund that will provide each of his two children with an annual income of \$10,000 beginning next year and continuing throughout their lifetimes. Find the amount of money he is required to place in trust now if the fund will earn interest at the rate of 9%/year compounded continuously.

34. **Effect of a Tax Cut** Suppose the average wage earner in a certain country saves 8% of her take-home pay and spends the other 92%.
    - **a.** Estimate the impact that a proposed \$10 billion tax cut will have on the economy over the long run in terms of the additional spending generated.
    - **b.** Estimate the impact that a \$10 billion tax cut will have on the economy over the long run if, at the same time, legislation is enacted to boost the rate of savings of the average taxpayer from 8% to 10%.

Biological and Life Sciences Applications

35. **Heights of Women** The heights of 4000 women who participated in a recent survey were found to be normally distributed according to the probability density function

$$f(x) = \frac{1}{2.5\sqrt{2\pi}} e^{-1/2[(x-163.5)/6.4]^2}$$

What percentage of these women have heights between 160 and 167 cm?

Answers to Odd-Numbered Exercises

CHAPTER 1

1.1 Exercises, page 9

1. False **3.** False

5. 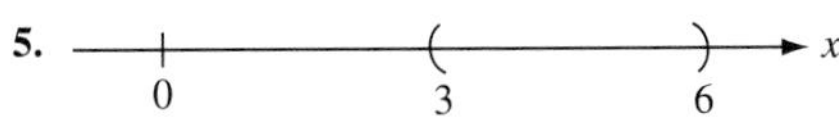

7. −1 4 x

9. 2 x

11. $(-\infty, 2)$ **13.** $(-\infty, -5]$ **15.** $(-4, 6)$

17. $(-\infty, -3) \cup (3, \infty)$ **19.** $(-2, 3)$ **21.** 4

23. 2 **25.** $5\sqrt{3}$ **27.** $\pi + 1$

29. 2 **31.** False **33.** False

35. True **37.** False **39.** True

41. False **43.** 9 **45.** 1

47. 4 **49.** 7 **51.** $\frac{1}{5}$

53. 2 **55.** 2 **57.** 1

59. True **61.** False **63.** False

65. False **67.** False **69.** $\frac{1}{(xy)^2}$

71. $\frac{1}{x^{5/6}}$ **73.** $\frac{1}{(s+t)^3}$ **75.** $x^{13/3}$

77. $\frac{1}{x^3}$ **79.** x **81.** $\frac{9}{x^2y^4}$

83. $\frac{y^8}{x^{10}}$ **85.** $2x^{11/6}$ **87.** $-2xy^2$

89. $2x^{4/3}y^{1/2}$ **91.** 2.828 **93.** 5.196

95. 31.62 **97.** 316.2 **99.** $\frac{3\sqrt{x}}{2x}$

101. $\frac{2\sqrt{3y}}{3}$ **103.** $\frac{\sqrt[3]{x^2}}{x}$ **105.** $\frac{2x}{3\sqrt{x}}$

107. $\frac{2y}{\sqrt{2xy}}$ **109.** $\frac{xz}{y\sqrt[3]{xz^2}}$ **111.** \$50.70

113. \$32,000 **115.** Between 1000 and 4000 units

117. [576, 792] **119.** $|x - 0.5| < 0.01$

121. False **123.** True

1.2 Exercises, page 21

1. $9x^2 + 3x + 1$ **3.** $4y^2 + y + 8$ **5.** $-x - 1$

7. $\frac{2}{3} + e - e^{-1}$ **9.** $6\sqrt{2} + 8 + \frac{1}{2}\sqrt{x} - \frac{11}{4}\sqrt{y}$

11. $x^2 + 6x - 16$ **13.** $a^2 + 10a + 25$ **15.** $x^2 + 4xy + 4y^2$

17. $4x^2 - y^2$ **19.** $-2x$ **21.** $2t(2\sqrt{t} + 1)$

23. $2x^3(2x^2 - 6x - 3)$ **25.** $7a^2(a^2 + 7ab - 6b^2)$

27. $e^{-x}(1 - x)$ **29.** $\frac{1}{2}x^{-5/2}(4 - 3x)$

31. $(2a + b)(3c - 2d)$ **33.** $(2a + b)(2a - b)$

35. $-2(3x + 5)(2x - 1)$ **37.** $3(x - 4)(x + 2)$

39. $2(3x - 5)(2x + 3)$ **41.** $(3x - 4y)(3x + 4y)$

43. $(x^2 + 5)(x^4 - 5x^2 + 25)$ **45.** $x^3 - xy^2$

47. $4(x - 1)(3x - 1)(2x + 2)^3$ **49.** $4(x - 1)(3x - 1)(2x + 2)^3$

51. $2x(x^2 + 2)^2(5x^4 + 20x^2 + 17)$

53. -4 and 3 **55.** -1 and $\frac{1}{2}$ **57.** 2 and 2

59. -2 and $\frac{3}{4}$ **61.** $\frac{1}{2} + \frac{1}{4}\sqrt{10}$ and $\frac{1}{2} - \frac{1}{4}\sqrt{10}$

63. $-1 + \frac{1}{2}\sqrt{10}$ and $-1 - \frac{1}{2}\sqrt{10}$

65. $\frac{x - 1}{x - 2}$ **67.** $\frac{3(2t + 1)}{2t - 1}$ **69.** $-\frac{7}{(4x - 1)^2}$

71. -8 **73.** $\frac{3x - 1}{2}$ **75.** $\frac{t + 20}{3t + 2}$

77. $-\frac{x(2x - 13)}{(2x - 1)(2x + 5)}$ **79.** $-\frac{x + 27}{(x - 3)^2(x + 3)}$

81. $\frac{x + 1}{x - 1}$ **83.** $\frac{4x^2 + 7}{\sqrt{2x^2 + 7}}$ **85.** $\frac{x - 1}{x^2\sqrt{x + 1}}$

87. $\frac{x - 1}{(2x + 1)^{3/2}}$ **89.** $\frac{\sqrt{3} + 1}{2}$ **91.** $\frac{\sqrt{x} + \sqrt{y}}{x - y}$

93. $\frac{(\sqrt{a} + \sqrt{b})^2}{a - b}$ **95.** $\frac{x}{3\sqrt{x}}$ **97.** $-\frac{2}{3(1 + \sqrt{3})}$

99. $-\frac{x + 1}{\sqrt{x + 2}(1 - \sqrt{x + 2})}$

101. True **103.** False

1.3 Exercises, page 27

1. (3, 3); Quadrant I **3.** (2, −2); Quadrant IV

5. (−4, −6); Quadrant III **7.** *A* **9.** *E*, *F*, and *G*

11. *F*

13–19. See the accompanying figure.

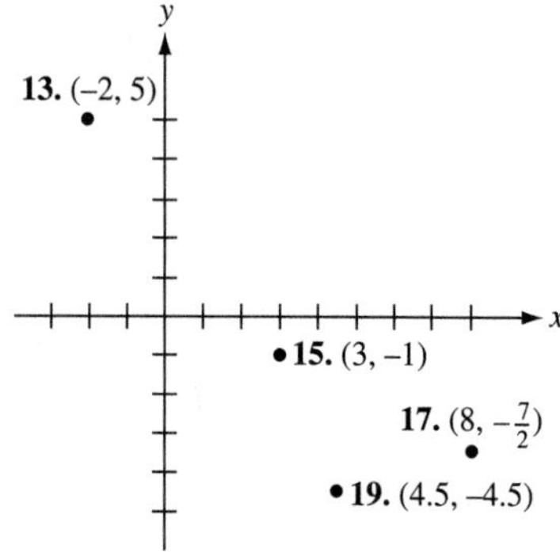

21. 5 **23.** $\sqrt{61}$ **25.** $(-8, -6)$ and $(8, -6)$

27.

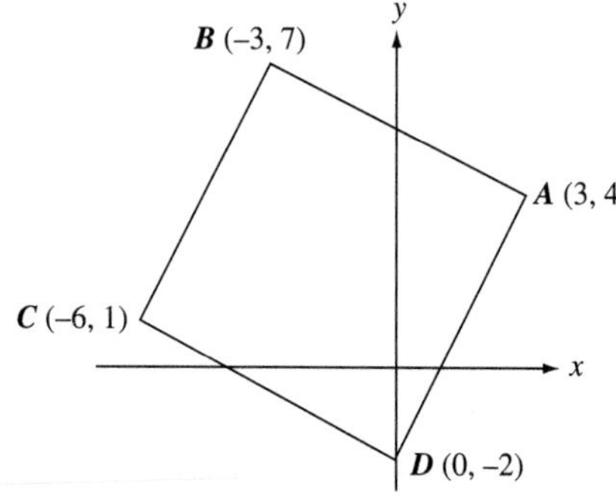

The four sides are equal: $d(A, B) = \sqrt{45}$; $d(B, C) = \sqrt{45}$; $d(C, D) = \sqrt{45}$; $d(A, D) = \sqrt{45}$ ΔABC and ΔBAD satisfy the Pythagorean Theorem: $d(A,C) = 3\sqrt{10}$; $d(B,D) = 3\sqrt{10}$

29. $(x - 2)^2 + (y + 3)^2 = 25$ **31.** $x^2 + y^2 = 25$

33. $(x - 2)^2 + (y + 3)^2 = 34$ **35.** $10\sqrt{13}t$; 72.1 km

37. $d = \sqrt{(x_2 - x_1)^2 + (y_2 - y_1)^2}$ **39.** 3400 km

41. Route 1 **43.** Model C **45.** True

47. False

1.4 Exercises, page 37

1. e **3.** a **5.** f

7. $\frac{1}{2}$ **9.** $\frac{3}{2}$ **11.** Not defined

13. 0 **15.** 5 **17.** $\frac{5}{6}$

19. $\frac{d - b}{c - a}$ **21. a.** 4 **b.** -8

23. Parallel **25.** Perpendicular

27. -5 **29.** $y = -3$ **31.** $y = 2x - 10$

33. $y = 2$ **35.** $y = 3x - 2$ **37.** $y = x + 1$

39. $y = 3x + 4$ **41.** $y = 5$

43. $y = \frac{1}{2}x$; $m = \frac{1}{2}$; $b = 0$

45. $y = \frac{2}{3}x - 3$; $m = \frac{2}{3}$; $b = -3$

47. $y = -\frac{1}{2}x + \frac{7}{2}$; $m = -\frac{1}{2}$; $b = \frac{7}{2}$

49. $y = \frac{1}{2}x + 3$ **51.** $y = -6$ **53.** $y = b$

55. $y = \frac{2}{3}x - \frac{2}{3}$ **57.** $k = 8$

59.

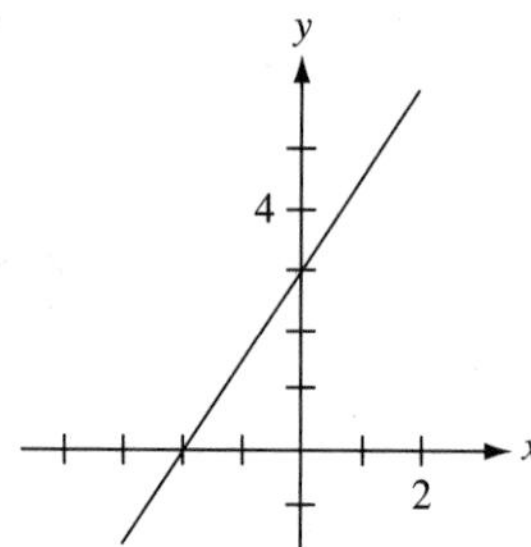

61.

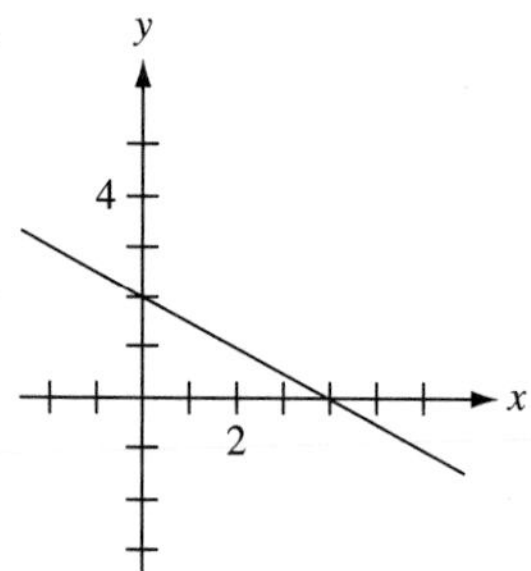

63.

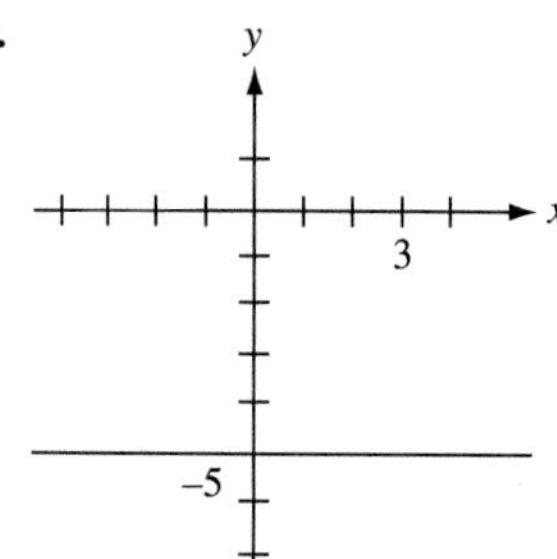

65. $\frac{b}{a}x + y = b$ **67.** $y = -2x - 4$ **69.** $y = \frac{1}{8}x - \frac{1}{2}$

71. Yes **73.** 33.6%

75. $y = -\frac{a_1}{b_1}x - \frac{c_1}{b_1}$ $(b_1 \neq 0)$; $y = -\frac{a_2}{b_2}x - \frac{c_2}{b_2}$ $(b_2 \neq 0)$

77. a. $y = 0.55x$ **b.** $x = 2000$

79. a.–b.

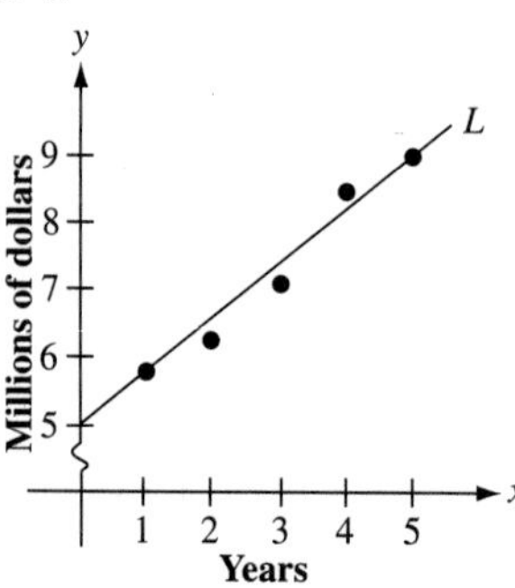

c. $y = 0.8x + 5$ **d.** $12.2 million

81. **a.**

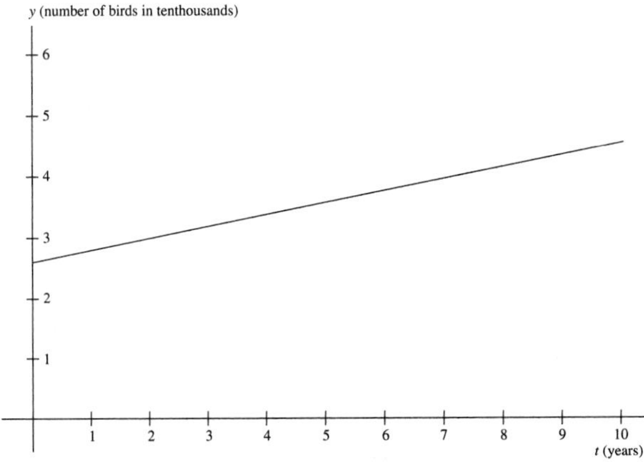

b. 1981.4; 25,928

c. The number of birds is increasing at the rate of 1981.4 birds/year; the number of birds at the beginning of 1990 was 25,928.

d. $t \approx 17$; 60,000 birds in 2007

83. **a.** 88.8 metric tons

b.

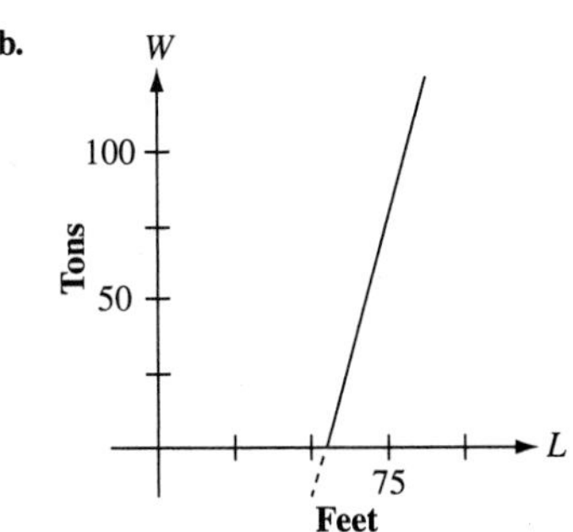

85. True **87.** True

Chapter 1 Review Exercises, page 42

1. $[-2, \infty]$ **3.** $(-\infty,-4) \cup (5,\infty)$ **5.** 4

7. $\pi - 6$ **9.** $\frac{27}{8}$ **11.** $\frac{1}{144}$

13. $\frac{1}{4}$ **15.** $4(x^2 + y)^2$ **17.** $\frac{2x}{3z}$

19. $6xy^7$ **21.** $-2\pi r^2(\pi r - 50)$ **23.** $(4 - x)(4 + x)$

25. $-\frac{3}{4}$ and $\frac{1}{2}$ **27.** 0, −3, 1 **29.** $1 + \sqrt{6}, 1 - \sqrt{6}$

31. $\frac{180}{(t + 6)^2}$ **33.** $\frac{78x^2 - 8x - 27}{3(2x^2 - 1)(3x - 1)}$

35. $\frac{1}{\sqrt{x} + 1}$ **37.** 5 **39.** −2

41. $y = -\frac{1}{10}x + \frac{19}{5}$ **43.** $y = \frac{5}{2}x + 9$ **45.** $y = -\frac{3}{4}x + \frac{9}{2}$

47. $y = 3x + 7$

49.

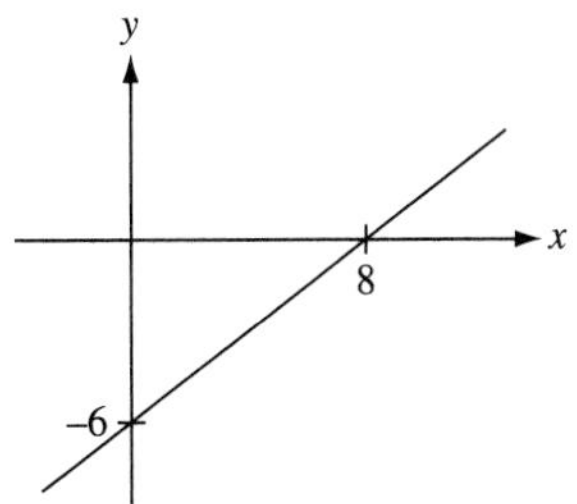

51. \$100

CHAPTER 2

2.1 Exercises, page 52

1. $21, -9, 5a + 6, -5a + 6, 5a + 21$

3. $-3, 6, 3a^2 - 6a - 3, 3a^2 + 6a - 3, 3x^2 - 6$

5. $2a + 2h + 5, -2a + 5, 2a^2 + 5, 2a - 4h + 5, 4a - 2h + 5$

7. $\frac{8}{15}, 0, \frac{2a}{a^2 - 1}, \frac{2(2 + a)}{a^2 + 4a + 3}, \frac{2(t + 1)}{t(t + 2)}$

9. $8, \frac{2a^2}{\sqrt{a - 1}}, \frac{2(x + 1)^2}{\sqrt{x}}, \frac{2(x - 1)^2}{\sqrt{x - 2}}$

11. 5, 1, 1 **13.** $\frac{5}{2}$, 3, 3, 9

15. **a.** −2 **b.** (i) $x = 2$; (ii) $x = 1$ **c.** $[0, 6]$ **d.** $[-2, 6]$

17. Yes **19.** Yes **21.** $(-\infty, \infty)$

23. $(-\infty, 0) \cup (0, \infty)$ **25.** $(-\infty, \infty)$ **27.** $(-\infty, 5]$

29. $(-\infty, -1) \cup (-1, 1) \cup (1, \infty)$

31. $[-3, \infty)$ **33.** $(-\infty, -2) \cup (-2, 1]$

35. **a.** $(-\infty, \infty)$ **b.** $6, 0, -4, -6, -\frac{25}{4}, -6, -4, 0$

c.

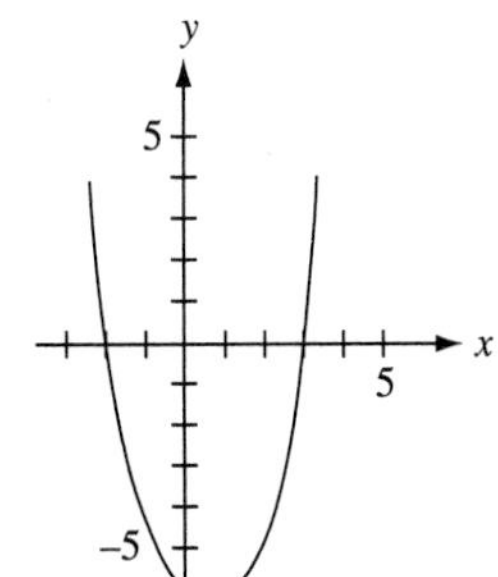

37.

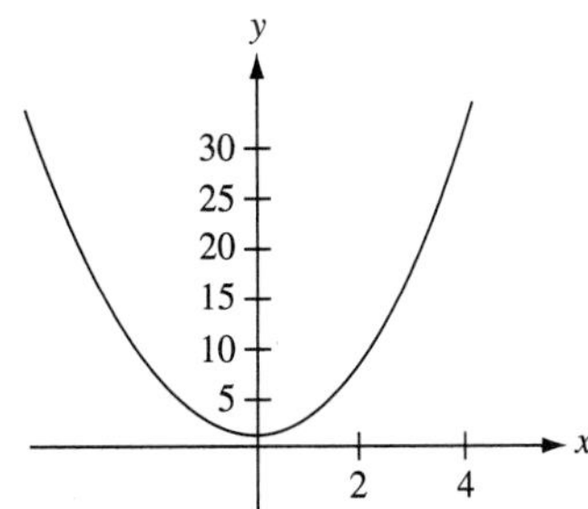

$(-\infty, \infty)$; $[1, \infty)$

39.

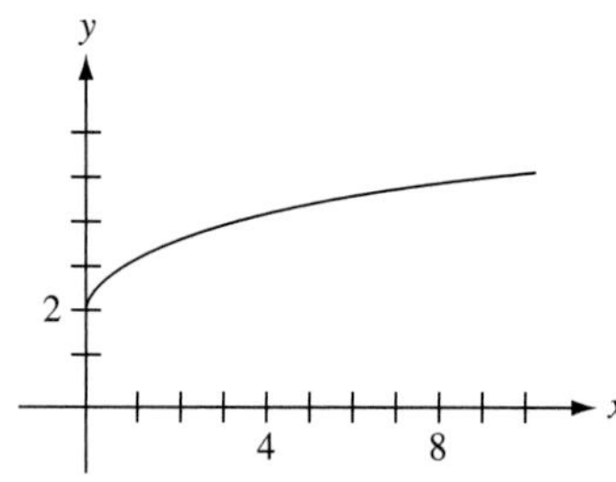

[0, ∞); [2, ∞)

41.

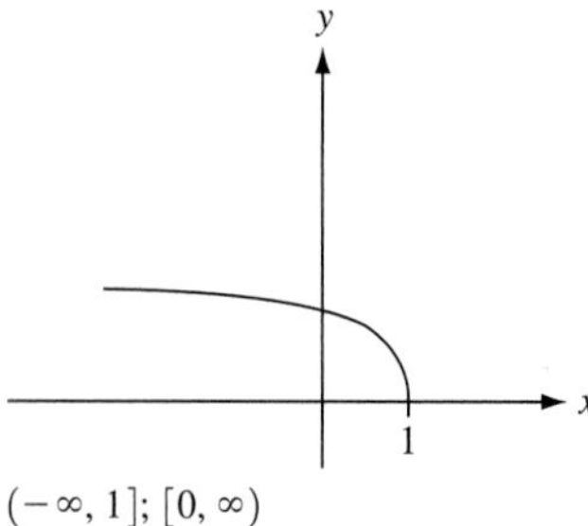

(−∞, 1]; [0, ∞)

43.

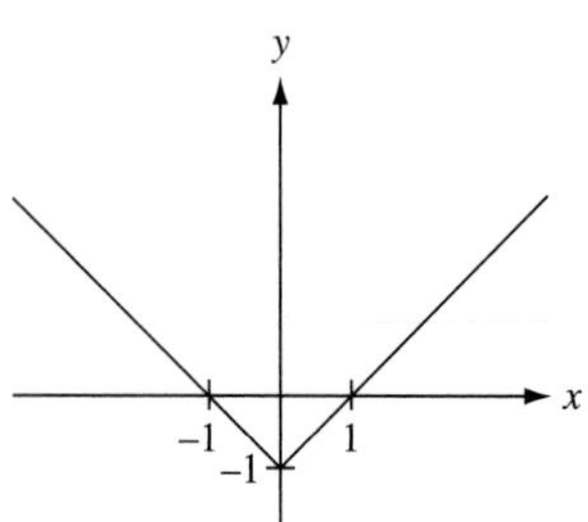

(−∞,∞); [−1,∞)

45.

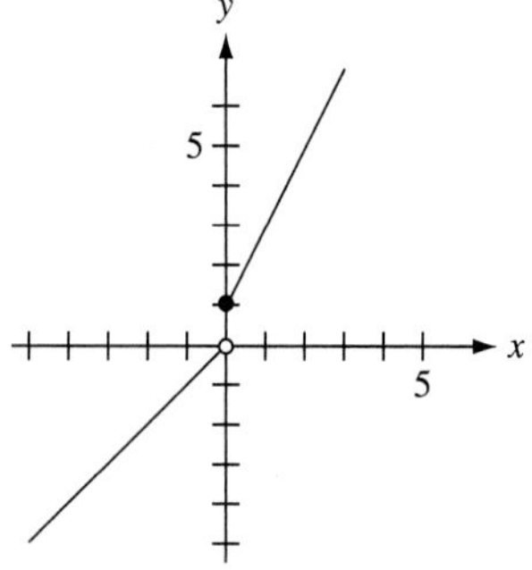

(−∞, ∞); (−∞, 0) ∪ [1, ∞)

47.

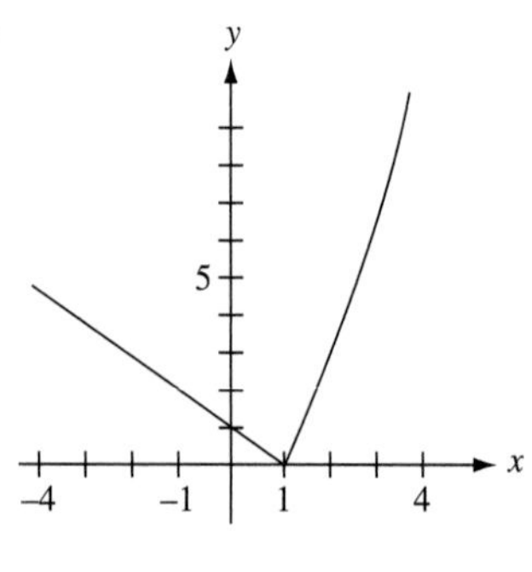

(−∞,∞); [0,∞)

49. Yes **51.** No **53.** Yes

55. Yes **57.** 10π cm

59. **a.** From 1985 to 1990

b. From 1990 on

c. 1990; $3.5 million

61. **a.** $f(t) = \begin{cases} 0.0185t + 0.58 & \text{if } 0 \le t \le 20 \\ 0.015t + 0.65 & \text{if } 20 < t \le 30 \end{cases}$

b. 0.0185/yr from 1960 through 1980; 0.015/yr from 1980 through 1990

c. 1983

63. **a.** $0.06x$ **b.** $12.00; $0.34

65. **a.** $30 + 0.45x$; $25 + 0.50x$

b.

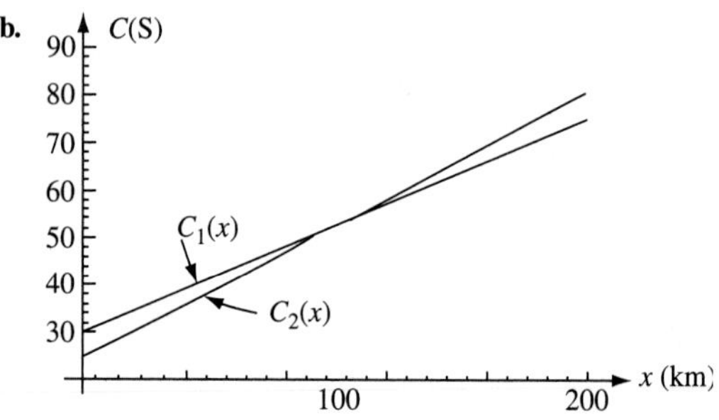

c. Acme

67. $700,000; $620,000; $540,000

69. 8 **71.** $S(r) = 4\pi r^2$

73.

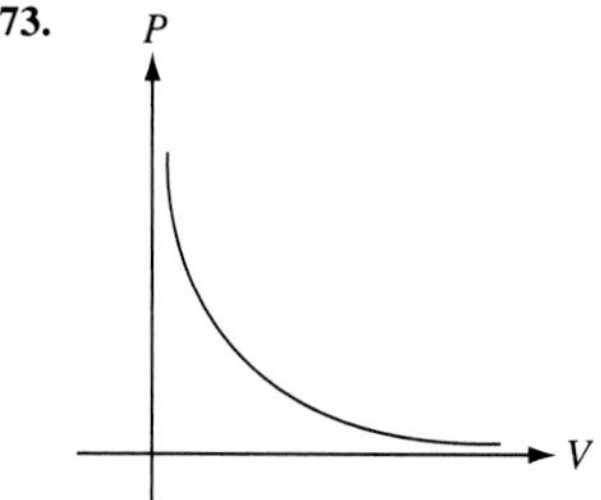

75. 50,000 or 990; 50,000 or 2240

77. 0.77 **79.** True **81.** False

2.2 Exercises, page 63

1.

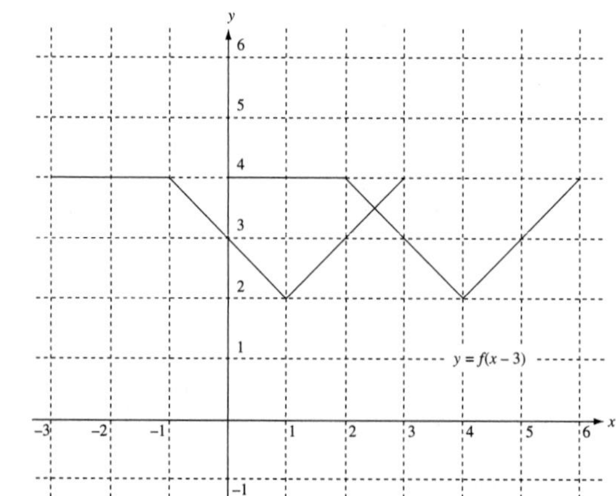

3.

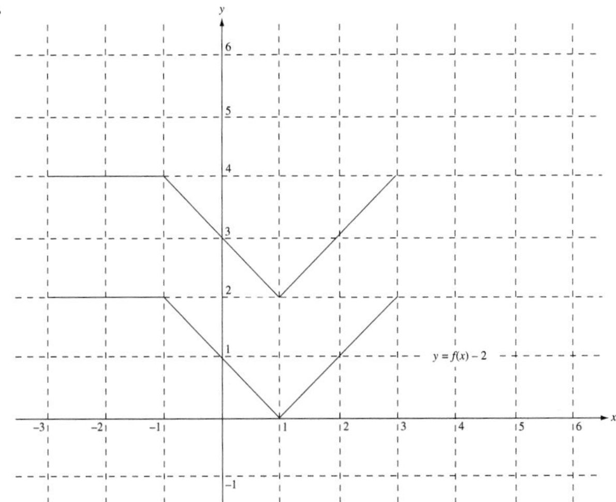

5.

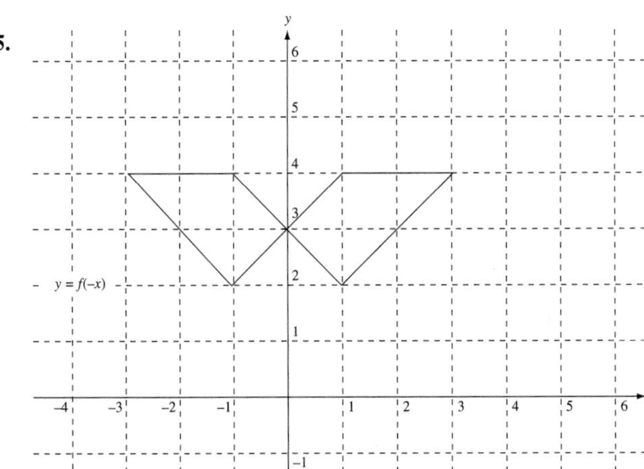

7.

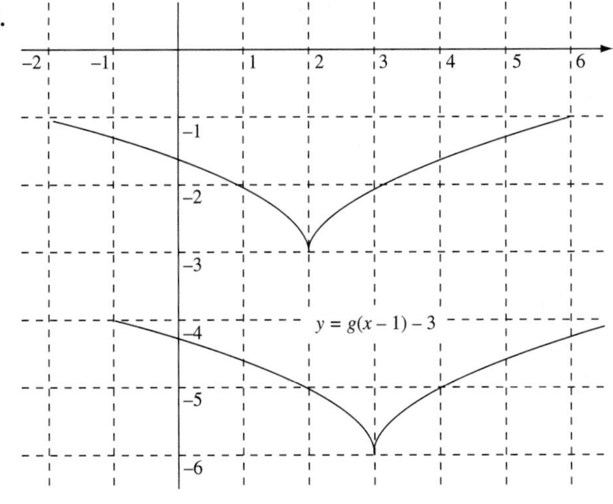

9.

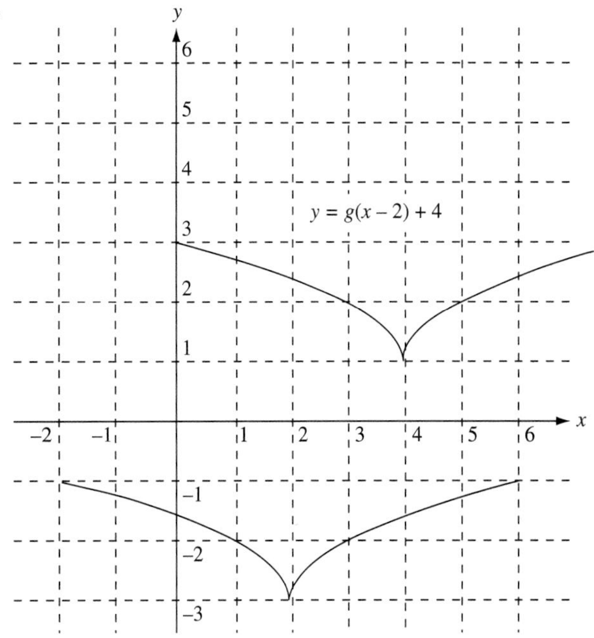

11.

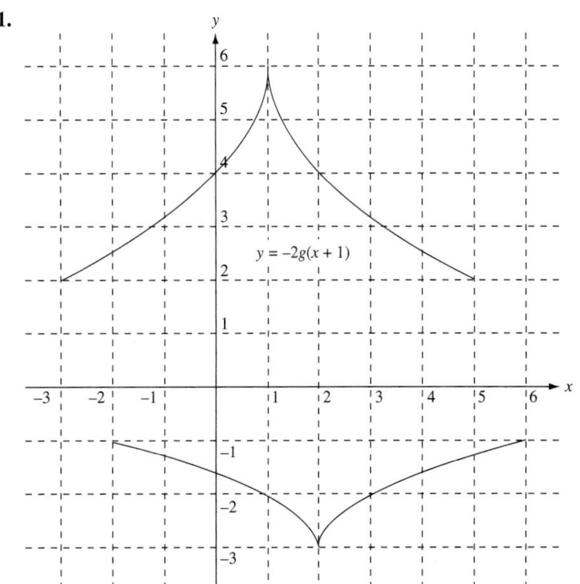

13.

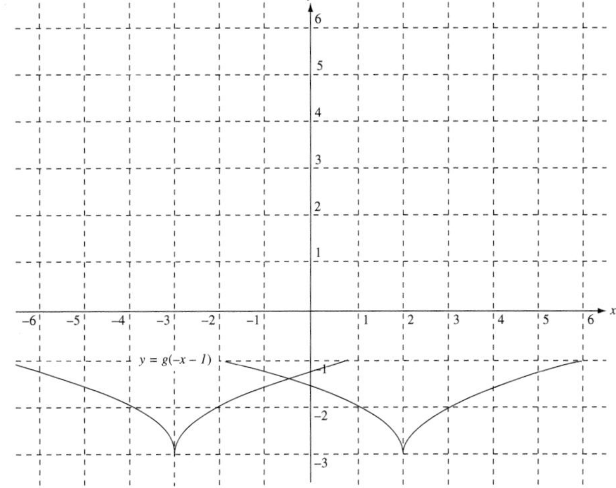

15. Horizontally translated left by 2 units, resulting in $y = f(x + 2)$; reflected in the x-axis, resulting in $y = -f(x + 2)$; vertically translated down by 5 units, resulting in $y = -f(x + 2) - 5$

17. Horizontally translated right by 5 units, resulting in $y = f(x - 5)$; vertically compressed by a factor of $\frac{1}{3}$, resulting in $y = \frac{1}{3}f(x - 5)$; vertically translated down by 1 unit, resulting in $y = \frac{1}{3}f(x - 5) - 1$

19. Horizontally compressed by a factor of $\frac{1}{3}$, resulting in $y = f(3x)$; reflected in the y-axis, resulting in $y = f(-3x)$; vertically translated down by 1 unit, resulting in $y = f(-3x) - 1$

21. $y = f(x + 5) - 2$

23. $y = 3f(x - 2) + 4$

25. $y = -\frac{3}{7}f(-x)$

2.3 Exercises, page 68

1. $a(x) = 2x^2$, E

3. $c(x) = 2(x - 3)^2$, D

5. $e(x) = -2(x - 3)^2 + 4$, C

7. $a(x) = -2(x + 4)^2 + 2$, F

9. $c(x) = \frac{1}{2}(x - 3)^2 - 6$, A

11. $e(x) = (x - 2)^2 + 2$, C

13. **a.** Upward **b.** $(-8, 0)$; lowest **c.** $x = -8$ **d.** $[0, \infty)$

15. **a.** Downward **b.** $(-1, 0)$; highest **c.** $x = -1$ **d.** $(-\infty, 0]$

17. **a.** Upward **b.** $(-7, 2)$; lowest **c.** $x = -7$ **d.** $[-\infty, 6)$

19. Horizontally translated right by 2 units, resulting in $f(x) = (x - 2)^2$; vertically translated down by 3 units, resulting in $f(x) = (x - 2)^2 - 3$

21. Horizontally translated right by 5 units, resulting in $f(x) = (x - 5)^2$; vertically expanded by a factor of 3, resulting in $f(x) = 3(x - 5)^2$

23. Horizontally translated right by 3 units, resulting in $f(x) = (x - 3)^2$; vertically expanded by a factor of 4, resulting in $f(x) = 4(x - 3)^2$; vertically translated down by 1 unit, resulting in $f(x) = 4(x - 3)^2 - 1$

25.

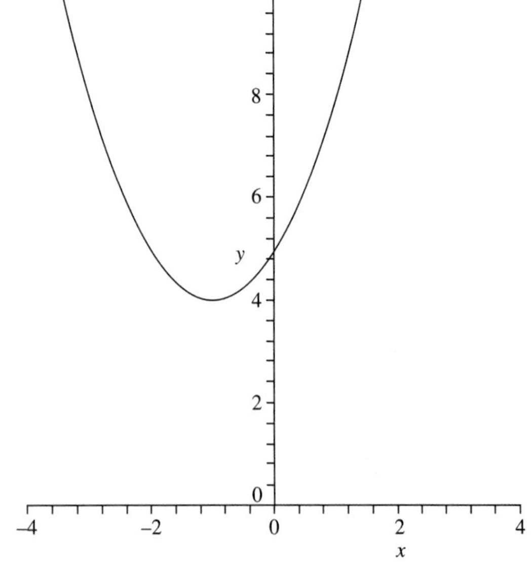

27.

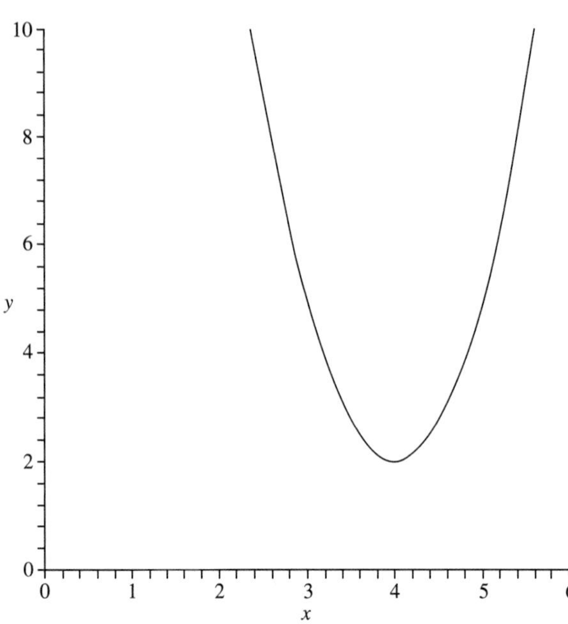

29.

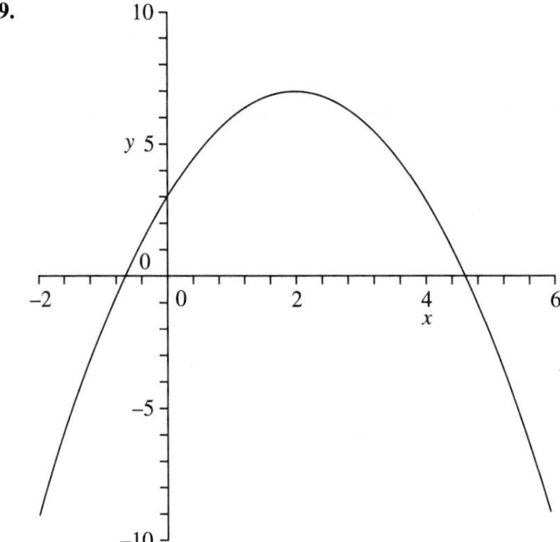

2.4 Exercises, page 75

1. **a.** $D_1 = (-\infty, -5) \cup (-5, 5) \cup (5, \infty)$ **b.** $x = -5$ and $x = 5$ **c.** $y = 0$ **d.** $f(0) = -\frac{3}{4}$

3. **a.** $D_1 = (-\infty, -3) \cup (-3, 3) \cup (3, \infty)$ **b.** $x = -3$ and $x = 3$ **c.** $y = 4$ **d.** $f(0) = \frac{1}{-2}$

5. **a.** $D_1 = (-\infty, -5) \cup (-5, 3) \cup (3, \infty)$ **b.** $x = -5$ and $x = 3$ **c.** $y = 3$ **d.** $f(0) = -2$

7. C

9. D

11. A

13. Horizontally translated left by 2 units, resulting in $g(x + 2)$; vertically translated down by 4 units, resulting in $f(x) = g(x + 2) - 4$

15. Horizontally translated left by 5 units, resulting in $g(x + 5)$; vertically expanded by a factor of 2, resulting in $f(x) = 2g(x + 5)$

17. Reflected in the x-axis, resulting in $-g(x)$; vertically translated up by 4 units, resulting in $f(x) = -g(x) + 4$

19.

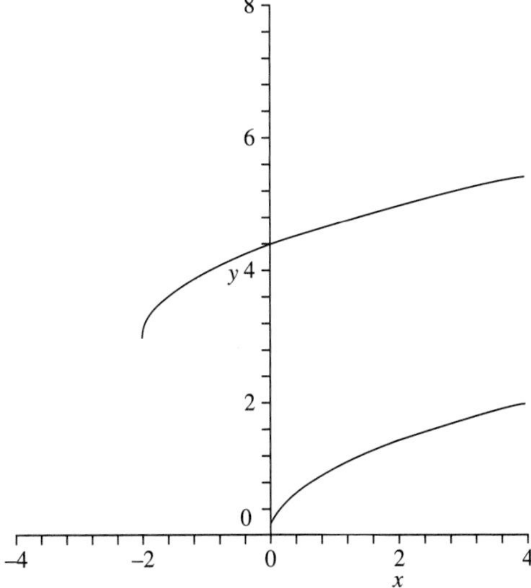

21.

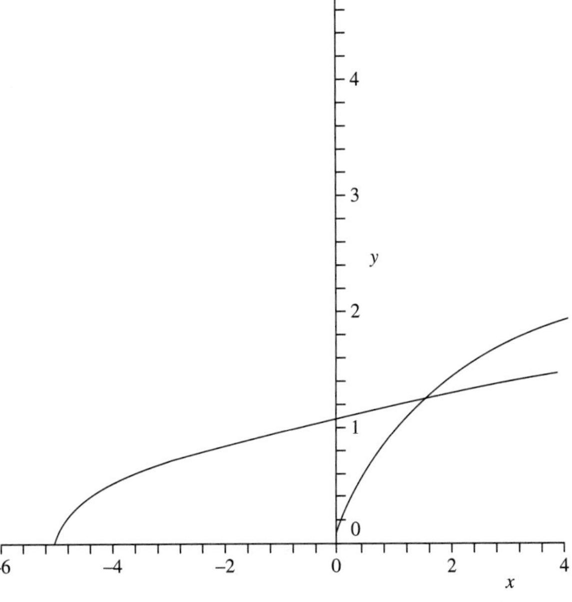

23.

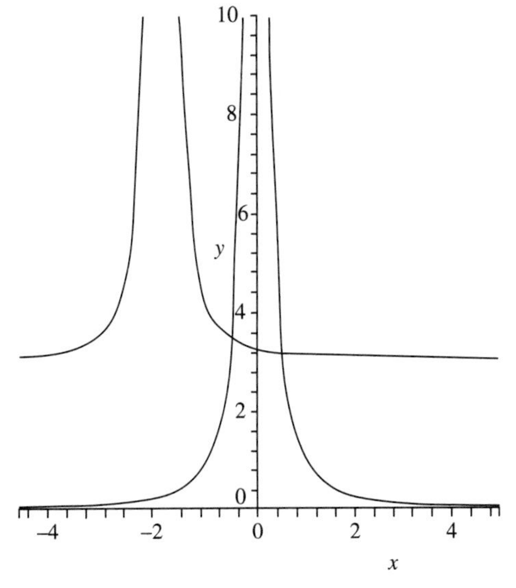

25.

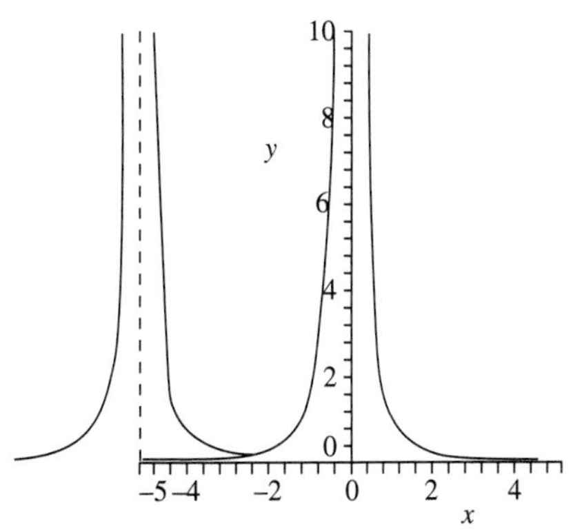

27. Polynomial of degree 6

29. Polynomial function of degree 6

31. Combination of two power functions

2.5 Exercises, page 82

1. $f(x) = |x + 3| = \begin{cases} -x - 3 & \text{if } x < -3 \\ x + 3 & \text{if } x \geq -3 \end{cases}$

3. $f(x) = |2x| + 4 = \begin{cases} -2x + 4 & \text{if } x < 0 \\ 2x + 4 & \text{if } x \geq 0 \end{cases}$

5. $f(x) = -|x - 4| = \begin{cases} x - 4 & \text{if } x < 4 \\ -x + 4 & \text{if } x \geq 4 \end{cases}$

7. $f(x) = |x - 1| + 3 = \begin{cases} -x + 4 & \text{if } x < 1 \\ x + 2 & \text{if } x \geq 1 \end{cases}$

9. $f(x) = \left|\frac{1}{2}x - 4\right| + 6 = \begin{cases} -\frac{1}{2}x + 10 & \text{if } x < 8 \\ \frac{1}{2}x + 2 & \text{if } x \geq 8 \end{cases}$

11.

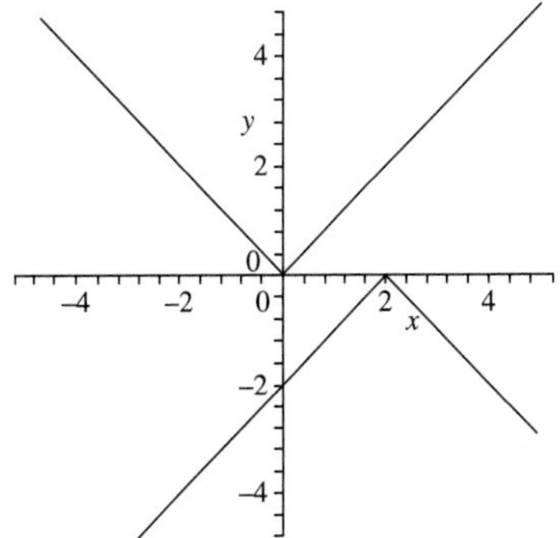

13.

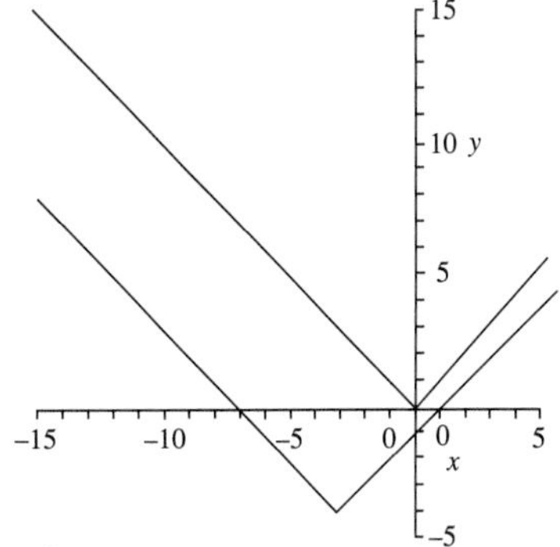

15.

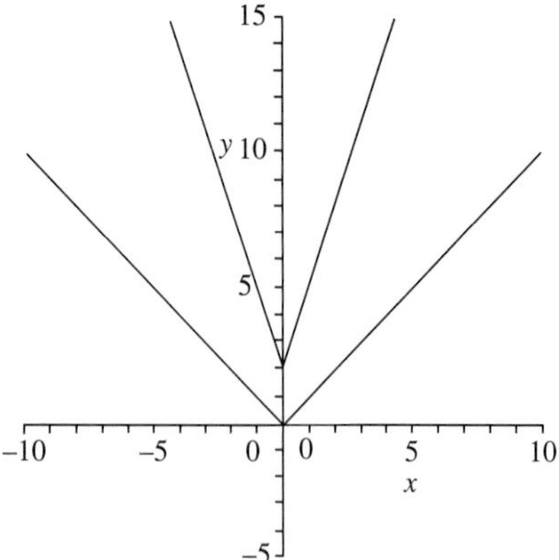

17. F **19.** D **21.** C

23. Horizontally translated right by 5 units, resulting in $|x - 5|$; vertically expanded by a factor of 2, resulting in $2|x - 5|$; vertically translated down by 1 unit, resulting in $f(x) = 2|x - 5| - 1$

25. Horizontally translated right by 1 unit, resulting in $|x - 1|$; vertically compressed by a factor of $\frac{2}{3}$, resulting in $\frac{2}{3}|x - 1|$; vertically translated up by 4 units, resulting in $f(x) = \frac{2}{3}|x - 1| + 4$

27. a. $D_1 = (-\infty, \infty)$ **b.** $x = -1, 3$

c.

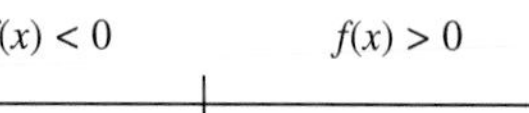

d.

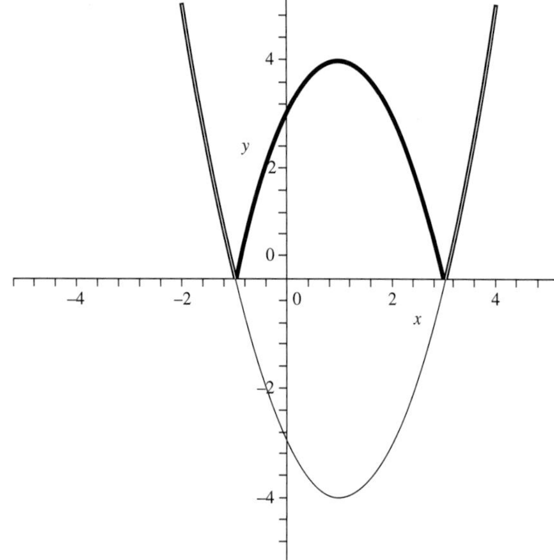

29. a. $D_1 = (-\infty, \infty)$ **b.** $x = -3, 1$

c.

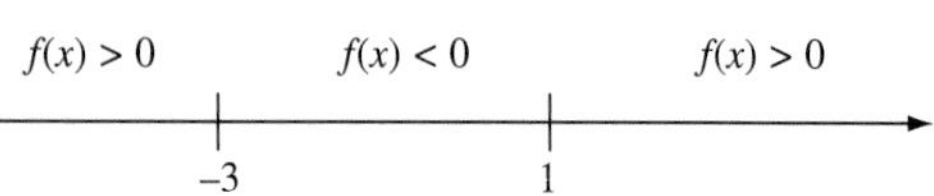

d.

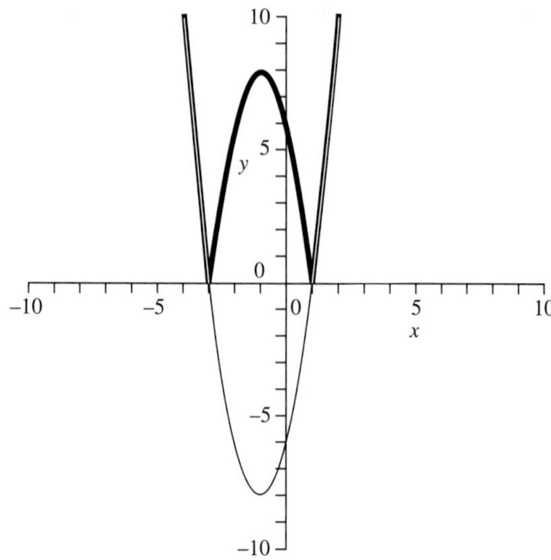

31. a. $D_1 = [-5, \infty)$

b. $x = -1$

c. f(x) < 0 f(x) > 0

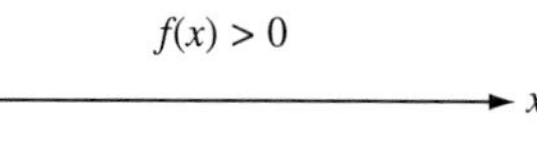

d.

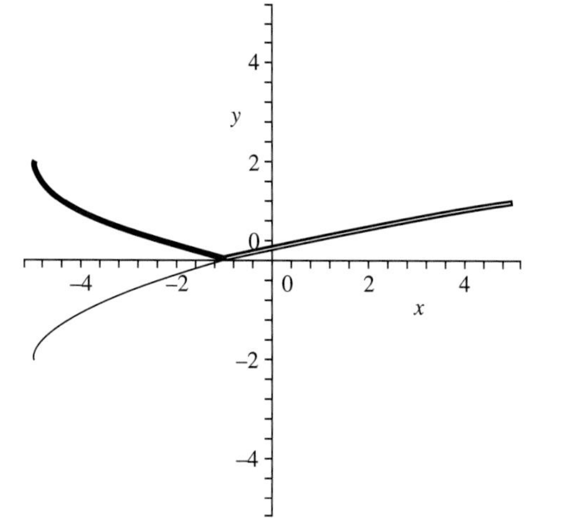

33. a. $D_1 = [3, \infty)$ **b.** $x = 7$

c. f(x) < 0 f(x) > 0

d.

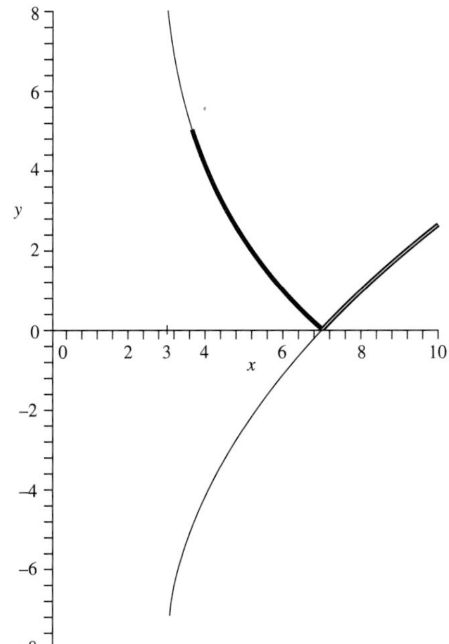

2.6 Exercises, page 90

1. $f(x) + g(x) = x^3 + x^2 + 3$ **3.** $f(x)g(x) = x^5 - 2x^3 + 5x^2 - 10$

5. $\dfrac{f(x)}{g(x)} = \dfrac{x^3 + 5}{x^2 - 2}$ **7.** $\dfrac{f(x)g(x)}{h(x)} = \dfrac{x^5 - 2x^3 + 5x^2 - 10}{2x + 4}$

9. $f(x) + g(x) = x - 1 + \sqrt{x + 1}$

11. $f(x)g(x) = (x - 1)\sqrt{x + 1}$ **13.** $\dfrac{g(x)}{h(x)} = \dfrac{\sqrt{x + 1}}{2x^3 - 1}$

15. $\dfrac{f(x)g(x)}{h(x)} = \dfrac{(x - 1)\sqrt{x + 1}}{2x^3 - 1}$ **17.** $\dfrac{f(x) - h(x)}{g(x)} = \dfrac{x - 2x^3}{\sqrt{x + 1}}$

19. $f(x) + g(x) = x^2 + \sqrt{x} + 3$; $f(x) - g(x) = x^2 = \sqrt{x} + 7$;

$f(x)g(x) = (x^2 + 5)(\sqrt{x} - 2)$;

$\dfrac{f(x)}{g(x)} = \dfrac{x^2 + 5}{\sqrt{x} - 2}$

21. $f(x) + g(x) = \dfrac{(x - 1)\sqrt{x + 3} + 1}{x - 1}$;

$f(x) - g(x) = \dfrac{(x - 1)\sqrt{x + 3} - 1}{x - 1}$;

$f(x)g(x) = \dfrac{\sqrt{x + 3}}{x - 1}$; $\dfrac{f(x)}{g(x)} = (x - 1)\sqrt{x + 3}$

23. $f(x) + g(x) = \dfrac{2(x^2 - 2)}{(x - 1)(x - 2)}$; $f(x) - g(x) = \dfrac{-2x}{(x - 1)(x - 2)}$;

$f(x)g(x) = \dfrac{(x + 1)(x + 2)}{(x - 1)(x - 2)}$;

$\dfrac{f(x)}{g(x)} = \dfrac{(x + 1)(x - 2)}{(x - 1)(x + 2)}$

25. $f(g(x)) = x^4 + x^2 + 1$; $g(f(x)) = (x^2 + x + 1)^2$

27. $f(g(x)) = \sqrt{x^2 - 1} + 1$; $g(f(x)) = x + 2\sqrt{x}$

29. $f(g(x)) = \dfrac{x}{x^2 + 1}$; $g(f(x)) = \dfrac{x^2 + 1}{x}$

31. 49 **33.** $\dfrac{\sqrt{5}}{5}$

35. $f(x) = 2x^3 + x^2 + 1$ and $g(x) = x^5$

37. $f(x) = x^2 - 1$ and $g(x) = \sqrt{x}$

39. $f(x) = x^2 - 1$ and $g(x) = \dfrac{1}{x}$

41. $f(x) = 3x^2 + 2$ and $g(x) = \dfrac{1}{x^{3/2}}$

43. $3h$ **45.** $-h(2a + h)$ **47.** $2a + h$

49. $3a^2 + 3ah + h^2 - 1$ **51.** $-\dfrac{1}{a(a + h)}$

53. The total revenue in dollars from both restaurants at time t

55. The unit cost of the commodity at time t

57. $C(x) = 0.6x + 12{,}100$

59. **a.** $P(x) = -0.000003x^3 - 0.07x^2 + 300x - 100{,}000$

b. \$182,375

61. **a.** $N(r(t)) = \dfrac{7}{1 + 0.02\left(\dfrac{10t + 150}{t + 10}\right)^2}$

b. $N(r(0)) \approx 1.27$; $N(r(12)) \approx 1.74$; $N(r(18)) \approx 1.85$

63. $N(t) = 1.42(x(t)) = \dfrac{9.94(t + 10)^2}{(t + 10)^2 + 2(t + 15)^2}$; 2.24 million jobs created 6 months from now; 2.48 million jobs created 12 months from now

65. The amount of carbon monoxide pollution at time t

67. $f(t) = g(t)$ **69.** $\dfrac{f(t)}{E(t)} = \dfrac{(0.74t + 26)(t + 3.14)}{t}$

71. False **73.** False

2.7 Exercises, page 96

1. Domain and range of both f and g are $(-\infty, \infty)$.

3. Domain and range of both f and g are $(-\infty, \infty)$.

5. $D_f = R_g = [-1, \infty)$ and $D_g = R_f = [0, \infty)$

7. Yes **9.** No **11.** No

13.

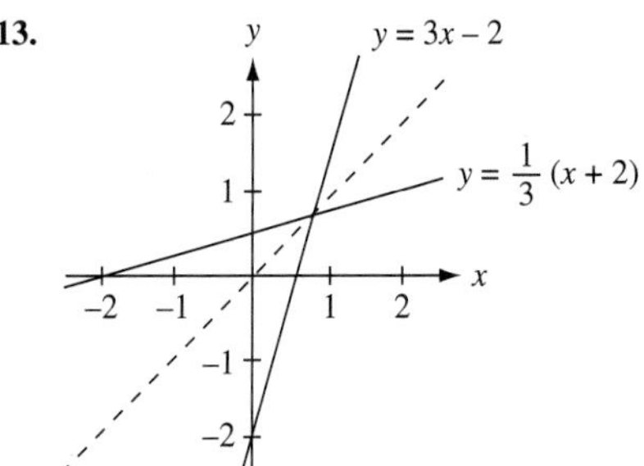

15.

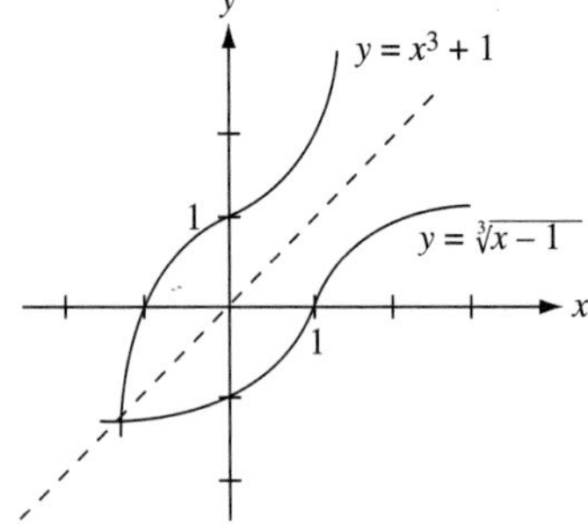

17.

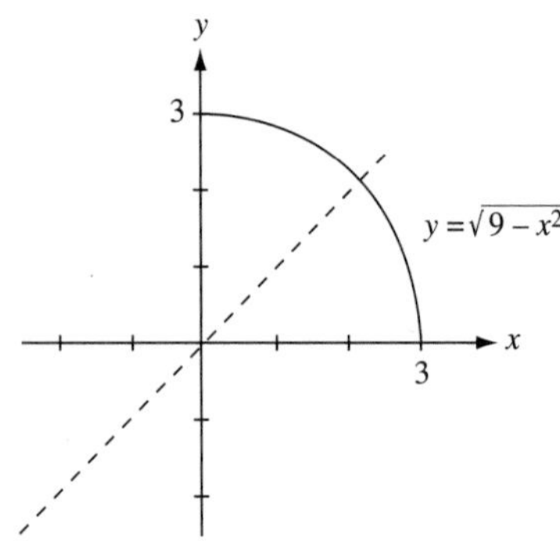

19. a. $-\frac{1}{2} + \frac{1}{2}\sqrt{1 + .8h}$; the time at which the balloon is at height h

b. Between 15 and 20 sec

21. $P = 3.1\sqrt[3]{k/17.4} \Rightarrow P/3.1 = \sqrt[3]{k/17.4} \Rightarrow (P/3.1)^3 = k/17.4 \Rightarrow k = 17.4(P/3.1)^3$.

So, $k = 17.4(10/3.1)^3 \approx 584$, or approximately 584 kilograms of fertilizer per acre.

23. $f = 216t^2 \Rightarrow \frac{f}{216} = t^2 \Rightarrow t = \sqrt{\frac{f}{216}}$. So, $t = \sqrt{\frac{572}{216}} \approx 1.6$, or approximately 1 hour and 36 minutes ago, the dandelions started sprouting.

2.8 Exercises, page 104

1. Yes; $y = -\frac{2}{3}x + 2$

3. Yes; $y = \frac{1}{2}x + 2$

5. Yes; $y = \frac{1}{2}x + \frac{9}{4}$

7. No

9. $m = -1$; $b = 2$

11. a.

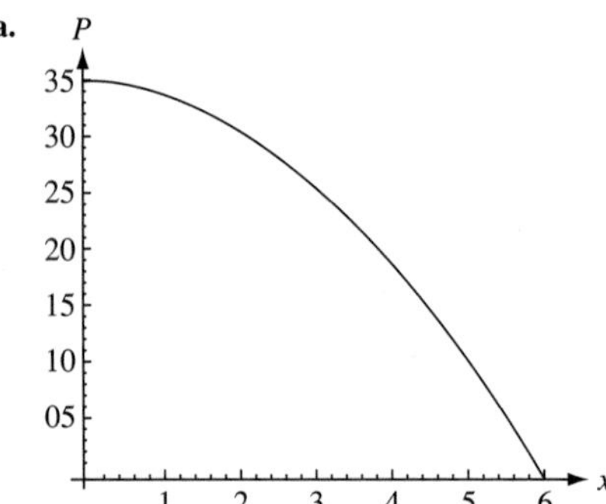

b. 5000 units

13. a.

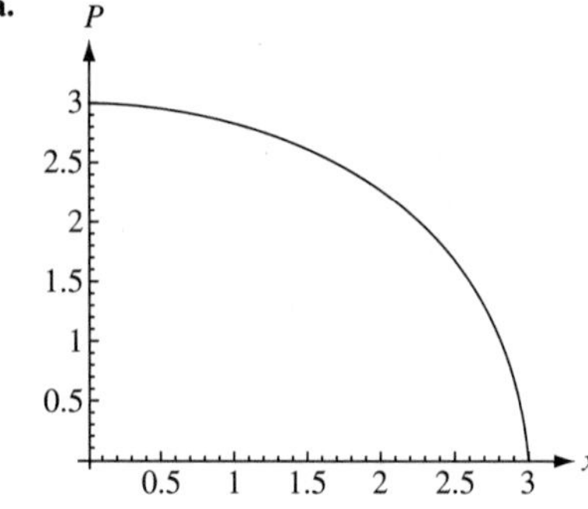

b. 2236 units

15. a.

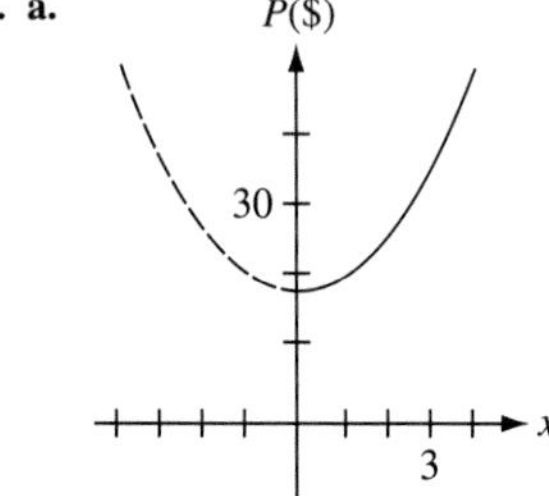

b. $26

17. a.

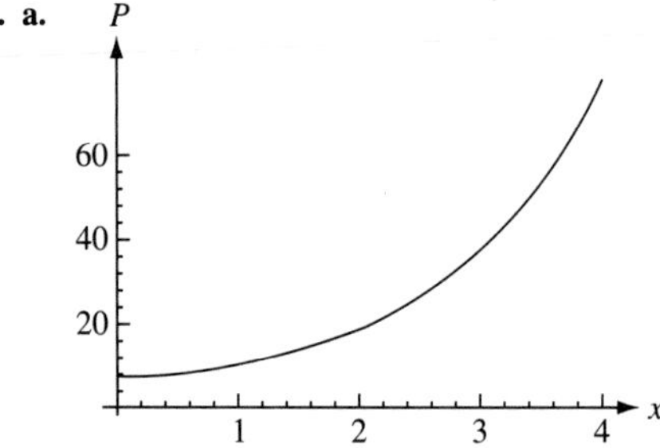

b. $20

19. 2500; $67.50

21. 11,000; $3

23. $f = 40x - x^2$; [0, 40]

25. $V = (15 - 2x)(8 - 2x)x$

27. $A = 28x - \frac{\pi}{2}x^2 - 2x^2$

29. $f(x) = -2x + 52 - \frac{50}{x}$; $\left[1, \frac{25}{2}\right]$

31. a. $R(x) = -4x^2 + 520x + 12{,}000$

b. 26,400; 28,800

33. a. $V = \frac{4}{3}\pi r^3$, $r^3 = \frac{3V}{4\pi}$, or $r = f(v) = \sqrt[3]{\frac{3V}{4\pi}}$

b. $g(t) = \frac{9}{2}\pi t$ **c.** $h(t) = \frac{3}{2}\sqrt[3]{t}$ **d.** 3 m

35. a. $C(x) = 14x + 100{,}000$ **b.** $R(x) = 12x$

c. $P(x) = 6x - 100{,}000$

d. Loss of $28,000; profit of $20,000

37. $400,000

39. **a.**

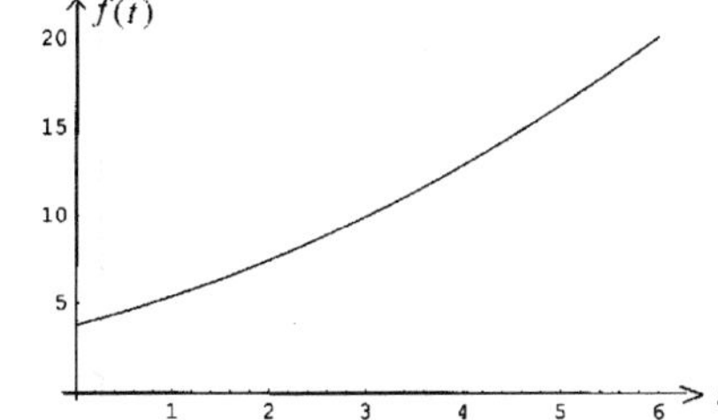

b. $20.09 million

41. $m = \dfrac{S - C}{n}$; $y - C = \dfrac{S - C}{n}(t - 0)$

43. **a.**

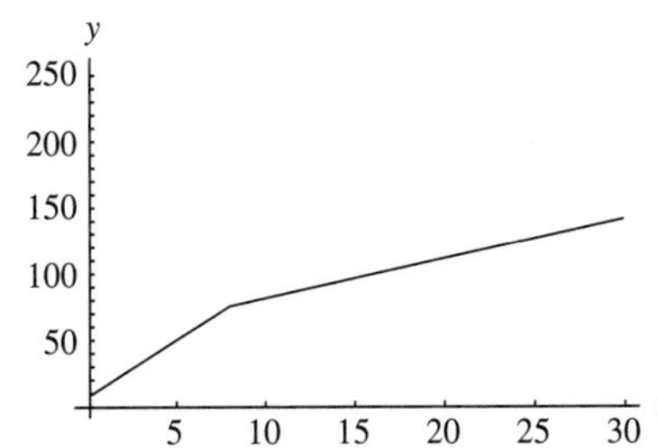

b. $7.44/kilo; $108.48/kilo

45. $4770 at the beginning of 1994; $6400 at the beginning of 1996; $7560 at the beginning of 1999

47. The slope of L_2 is greater than that of L_1; for each drop of a dollar in the price, the quantity demanded of model B is greater than that of model A.

49.

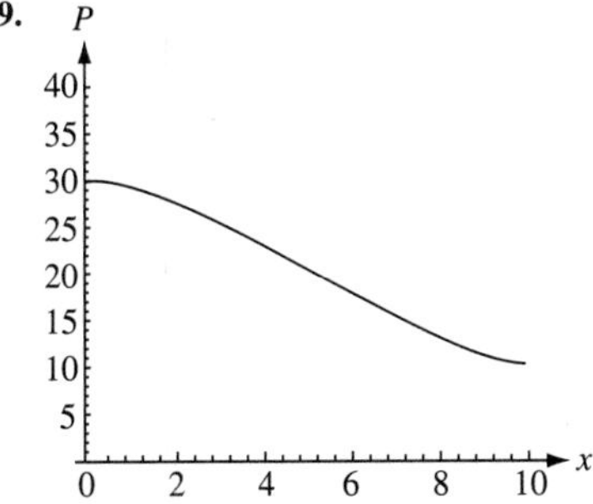

$p = \$10$

51.

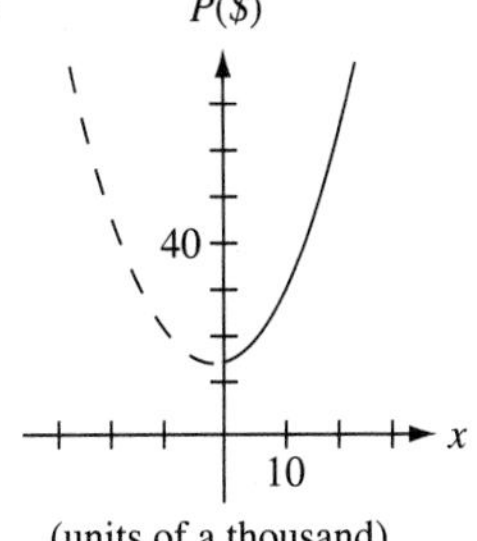

$20

53. **a.** $\dfrac{b - d}{c - a}$; $\dfrac{bc - ad}{c - a}$

b. If the unit price for producing the product is increased, equilibrium quantity decreases while equilibrium price increases.

c. If the upper bound for the unit price of a commodity is lowered, both equilibrium quantity and equilibrium price drop.

55. 8000; $80

57. 104 mg

59. **a.**

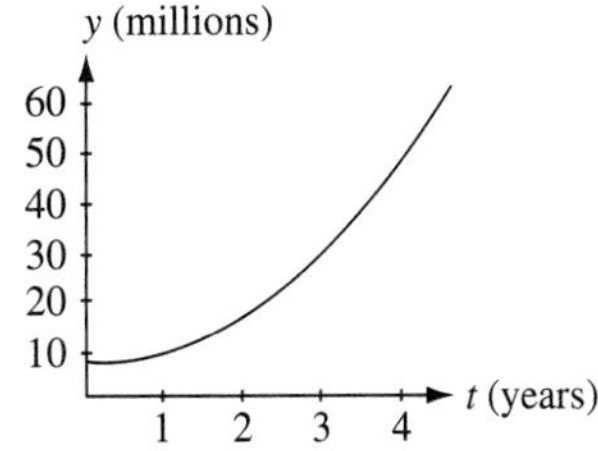

b. 59,910,000

61. **a.** $T = \dfrac{3}{20}N + 3$ **b.** $N = \dfrac{20}{3}(T - 3)$; 240 times/min

63. 2.9; 5.3; 1.2

65. 1,112,000; 1,342,000; 1,537,000

67. True

69. False

Chapter 2 Review Exercises, page 110

1. $D = (-\infty, \infty)$

3. $D = (-\infty, 9]$

5. $D = \left\{x \in \mathbb{R} \,/\, x \neq -1, \dfrac{3}{2}\right\}$

7. $D = (-\infty, \infty)$

9. **a.** $f(-2) = 3(-2)^2 + 5(-2) - 2 = 0$

b. $f(a + 2) = 3a^2 + 17a + 20$

c. $f(2a) = 12a^2 + 10a - 2$

d. $f(a + h) = 3a^2 + 6ah + 3h^2 + 5a + 5h - 2$

11.

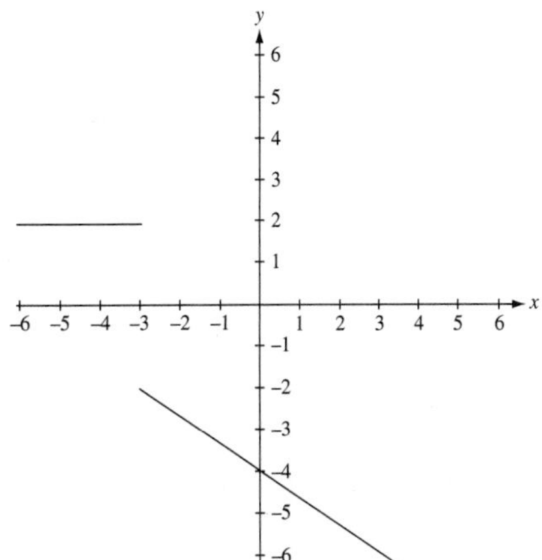

13.

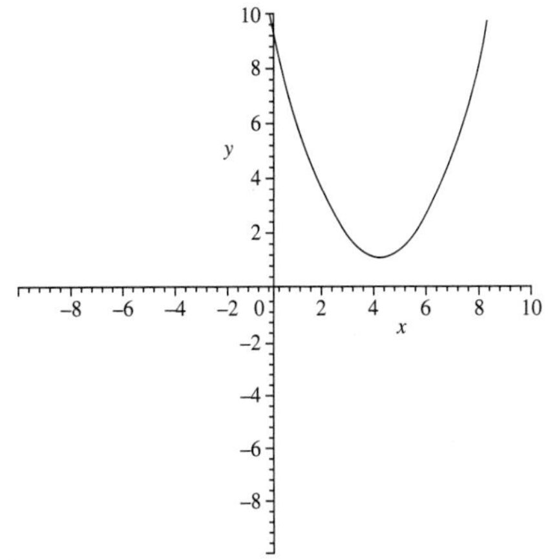

15.

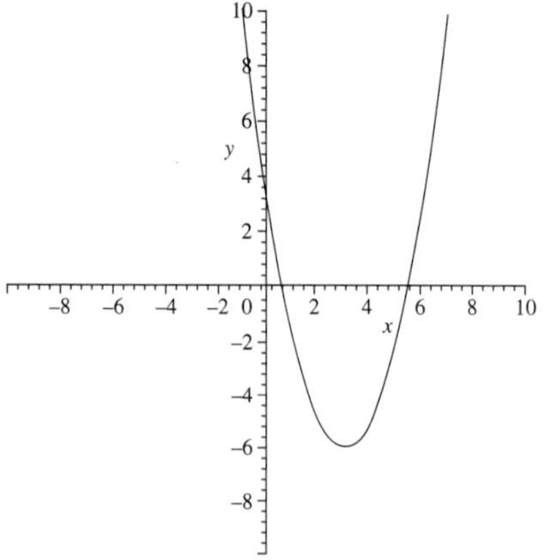

17.

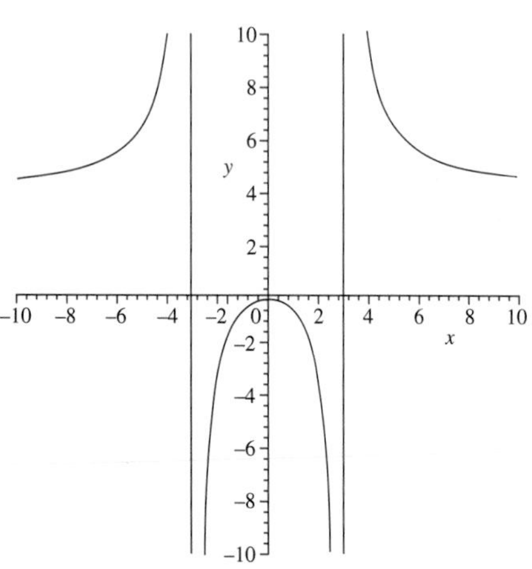

19.

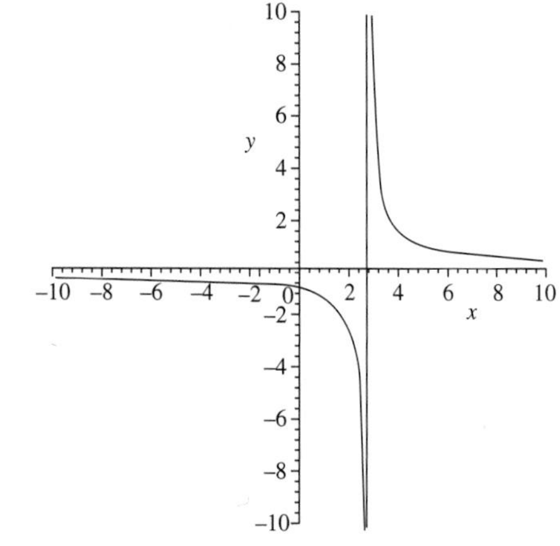

21.

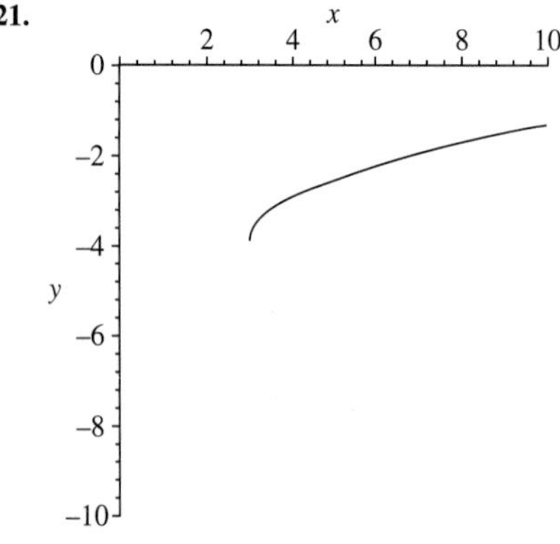

23.

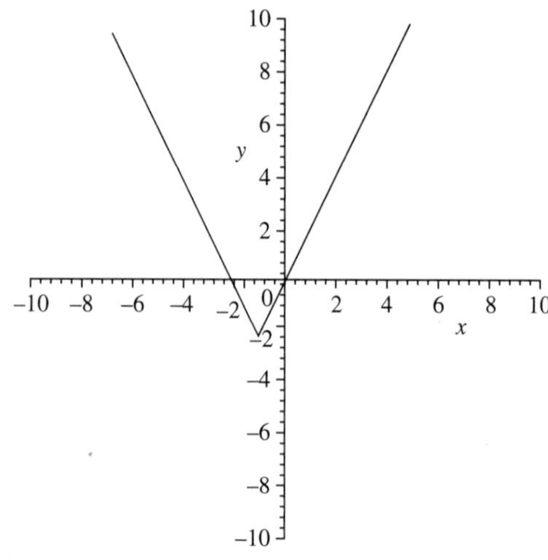

25. $f(x) = 2(x - 1)^2 - 4$

27. $f(x) = \frac{1}{x - 3}$

29. $\frac{2x + 3}{x}$

31. $\frac{1}{2x + 3}$

33. $f^{-1}(x) = -\frac{3}{2}x + \frac{15}{2}$

35. $f^{-1}(x) = (x - 3)^2 - 2$

37. (6, 2)

39. $x = 6$; $\left(6, \frac{21}{2}\right)$

41. a. (0, 2.4) and (5, 7.4); $m = 1$

43. $x = 2500$; (2500, 50,000)

45. $45,000

47. 117 mg

49. $f(4) = 0, f(5) = 20, \ldots, f(12) \approx 135.8$

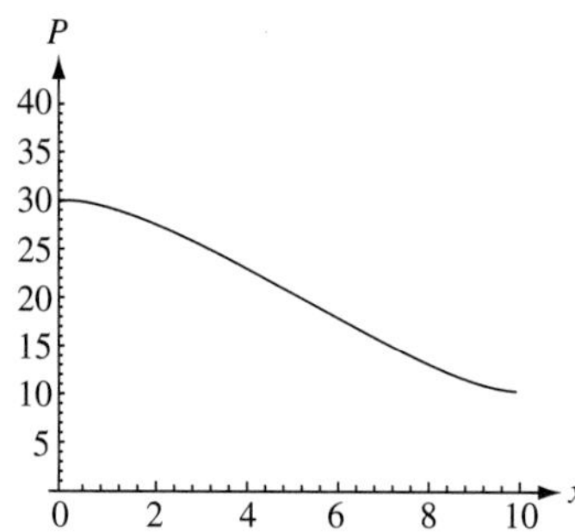

As the length of the list increases, the time taken to learn the list increases by a very large amount.

CHAPTER 3

3.1 Exercises, page 126

1. $\lim_{x \to -2} f(x) = 3$

3. $\lim_{x \to 3} f(x) = 3$

5. $\lim_{x \to -2} f(x) = 3$

7. The limit does not exist.

9. $\lim_{x \to 2} (x^2 + 1) = 5$

x	1.9	1.99	1.999	2.001	2.01	2.1
$f(x)$	4.61	4.9601	4.9960	5.004	5.0401	5.41

11.

x	−0.1	−0.01	−0.001	0.001	0.01	0.1
$f(x)$	−1	−1	−1	1	1	1

The limit does not exist.

13.

x	0.9	0.99	0.999	1.001	1.01	1.1
$f(x)$	100	10,000	1,000,000	1,000,000	10,000	100

The limit does not exist.

15.

x	0.9	0.99	0.999	1.001	1.01	1.1
$f(x)$	2.9	2.99	2.999	3.001	3.01	3.1

$\lim_{x\to 1} \frac{x^2 + x - 2}{x - 1} = 3$

17.

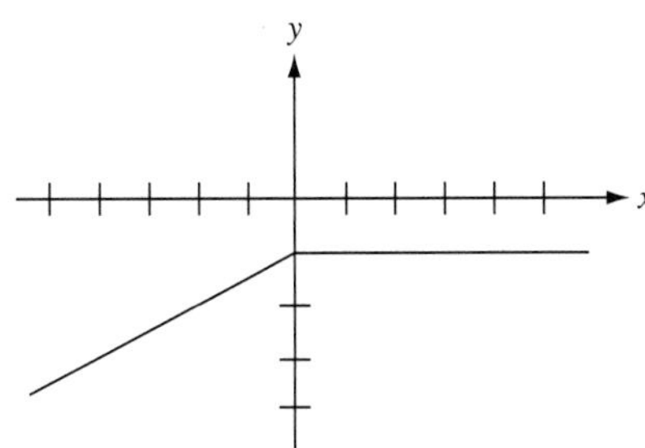

$\lim_{x\to 0} f(x) = -1$

19.

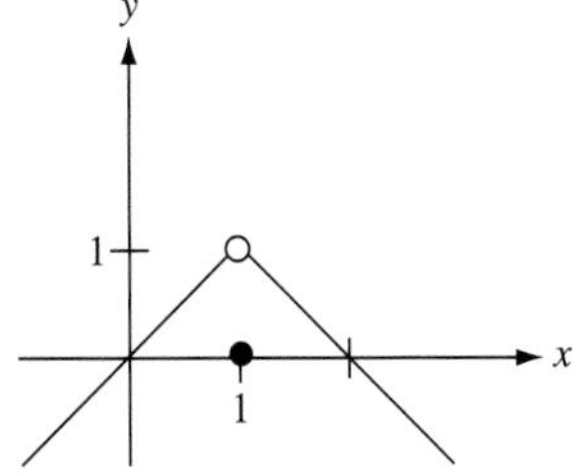

$\lim_{x\to 1} f(x) = 1$

21.

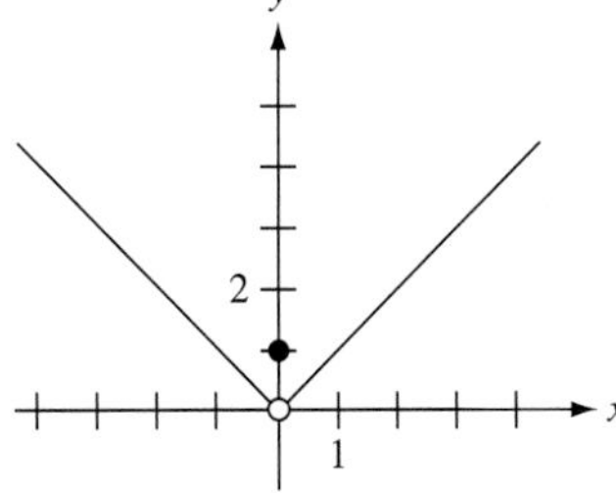

$\lim_{x\to 0} f(x) = 0$

23. 3 **25.** 3 **27.** -1

29. 2 **31.** -4 **33.** $\frac{5}{4}$

35. 2 **37.** $\sqrt{171} = 3\sqrt{19}$ **39.** $\frac{3}{2}$

41. -1 **43.** -6 **45.** 2

47. $\frac{1}{6}$ **49.** 2 **51.** -1

53. -10

55. The limit does not exist.

57. $\frac{5}{3}$ **59.** $\frac{1}{2}$ **61.** $\frac{1}{3}$

63. $\lim_{x\to\infty} f(x) = \infty$; $\lim_{x\to-\infty} f(x) = \infty$ **65.** 0; 0

67. $\lim_{x\to\infty} f(x) = -\infty$; $\lim_{x\to-\infty} f(x) = -\infty$

69.

x	1	10	100	1000
$f(x)$	0.5	0.009901	0.0001	0.000001

x	-1	-10	-100	-1000
$f(x)$	0.5	0.009901	0.0001	0.000001

$\lim_{x\to\infty} f(x) = 0$ and $\lim_{x\to-\infty} f(x) = 0$

71.

x	1	5	10	100	1000
$f(x)$	12	360	2910	2.99×10^6	2.999×10^9

x	-1	-5	-10	-100	-1000
$f(x)$	6	-390	-3090	-3.01×10^6	-3.0×10^9

$\lim_{x\to\infty} f(x) = \infty$ and $\lim_{x\to-\infty} f(x) = -\infty$

73. 3 **75.** 3 **77.** $\lim_{x\to-\infty} f(x) = -\infty$

79. 0 **81.** 5 **83.** 3

85. $\frac{2}{3}$ **87.** $\frac{1}{2}$ **89.** No

91. a. \$500,000; \$750,000; \$1,166,667; \$2 million; \$4.5 million; \$9.5 million

b. The limit does not exist; as the percent of pollutant to be removed approaches 100, the cost becomes astronomical.

93. $\lim_{t\to\infty} \frac{128t}{t + 3.6} = 128$ mm

95. a. \$24 million; \$60 million; \$83.1 million

b. \$120 million

97. a. 8 **b.** $\lim_{t\to 9} P(t) = \infty$

c.

P(t)
1200
1000
800
600
400
200
2 4 6 8
t

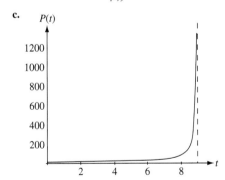

99. a. 5000 **b.** 25,000

101. a moles/litre/second **103.** True **105.** False

107. False

3.2 Exercises, page 140

1. 3; 2; the limit does not exist.

3. The limit does not exist; 2; the limit does not exist.

5. 0; 2; the limit does not exist.

7. -2; 2; the limit does not exist.

9. True   11. True   13. False

15. True   17. False   19. True

21. 6   23. $-\frac{1}{4}$

25. The limit does not exist.

27. -1 29. 0   31. -4

33. The limit does not exist.

35. 4   37. 0   39. 0; 0

41. 2; 3   43. $x = 0$; conditions 2 and 3

45. Continuous everywhere

47. $x = 0$; condition 3   49. $x = 0$; condition 3

51. $(-\infty, \infty)$ 53. $(-\infty, \infty)$ 55. $\left(-\infty, \frac{1}{2}\right) \cup \left(\frac{1}{2}, \infty\right)$

57. $(-\infty, -2) \cup (-2, 1) \cup (1, \infty)$

59. $(-\infty, \infty)$ 61. $(-\infty, \infty)$ 63. $(-\infty, \infty)$

65. $(-\infty, \infty)$ 67. -1 and 1   69. 1 and 2

71. **a.** f is a polynomial of degree 2.

b. $f(1) = 3$ and $f(3) = -1$

73. **a.** f is a polynomial of degree 3.

b. $f(-1) = -4$ and $f(1) = 4$

75. $x \approx 0.59$ 77. 3   79. $a = 1$

81. **a.** Yes   **b.** No

83. ≈ 1.34

85. **a.** $h(t) = 4$, and $h(2) = 68$

b. Joan must see the ball at least once during the time the ball is in the air.

c. $\frac{1}{2}$; $\frac{7}{2}$; Joan sees the ball on its way up $\frac{1}{2}$ sec after it was shot up and again $3\frac{1}{2}$ sec later.

87. No

89. **a.** f is a rational function whose denominator is never zero.

b. The numerator, x^2, is nonnegative and denominator is $x^2 + 1 \geq 1$ for all values of x.

c. $f(0) = 0$; No

91. Consider $f(-1) = -1$ and $f(1) = 1$ if $f(x) = \begin{cases} -1 & \text{if } x \leq x < 0 \\ 1 & \text{if } 0 \leq x < 1 \end{cases}$; but if we take the number $\frac{1}{2}$, which lies between $y = -1$ and $y = 1$, there is no values of x such that $f(x) = \frac{1}{2}$.

93. f is discontinuous at $t = 20$, 40, and 60. When $t = 0$, the inventory stands at 750 reams. The level drops to about 200 reams by the twentieth day at which time a new order of 500 reams arrives to replenish the supply.

95. The function P is discontinuous at $t = 12$, 16, and 28. At $t = 12$ and 16, the interest rate jumped from $10\frac{1}{2}$ to 11% and from 11% to $11\frac{1}{2}$, respectively. At $t = 18$, the interest rate jumped from $11\frac{1}{2}$ to 11%.

97.

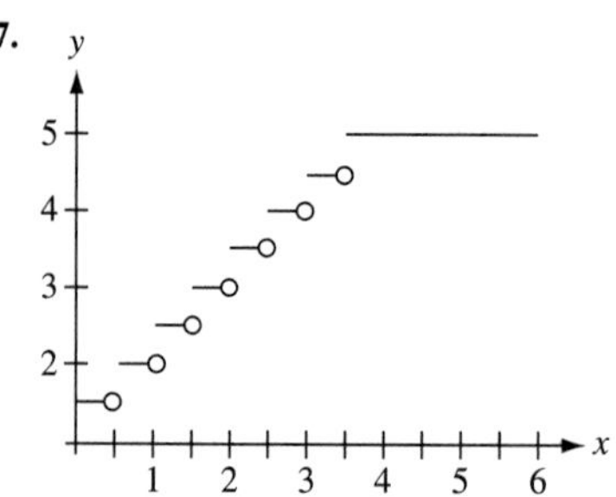

f is discontinuous at $x = \frac{1}{2}, 1, 1\frac{1}{2}, \ldots, 4$.

99. Michael makes progress toward solving the problem until $x = x_1$. Between $x = x_1$ and $x = x_2$ he makes no further progress; at $x = x_2$ he achieves a breakthrough, and at $x = x_3$ he proceeds to complete the problem.

101.

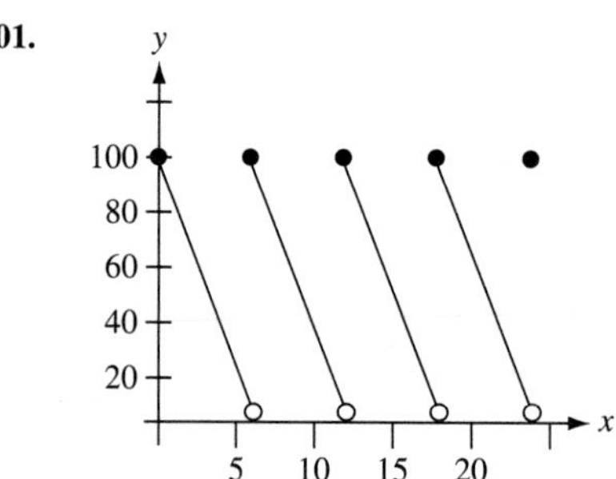

f is discontinuous at $x = 6, 12, 18, 24$.

103. **a.** $\lim_{v \to u^+} \frac{aLv^3}{v - u} = \infty$; when the speed of the fish is very close to that of the current, the energy expended by the fish will be enormous.

b. $\lim_{v \to \infty} \frac{aLv^3}{v - u} = \infty$; if the speed of the fish increases greatly, so does the amount of energy required to swim a distance of L m.

105. False   107. False

109. False   111. True

3.3 Exercises, page 157

1. **a.** Car A

b. They are travelling at the same speed.

c. Car B

d. Both cars covered the same distance.

3. 0   5. 2   7. $6x$

9. $-2x + 3$ 11. $\frac{1}{2\sqrt{x - 5}}$ 13. 2; $y = 2x + 7$

15. 6; $y = 6x - 3$

17. $\frac{1}{9}$; $y = \frac{1}{9}x - \frac{2}{3}$ 19. $f'(3) = \frac{1}{2}$; $y = \frac{1}{2}x + \frac{5}{2}$

21. **a.** $4x$

b. $y = 4x - 1$

c.

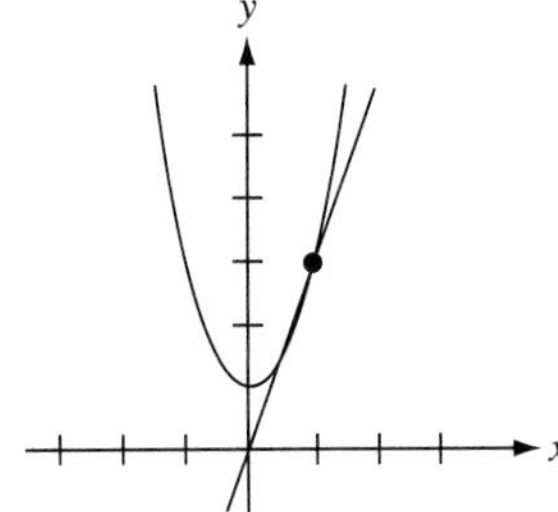

23. a. $2x - 2$

b. $(1, 0)$

c.

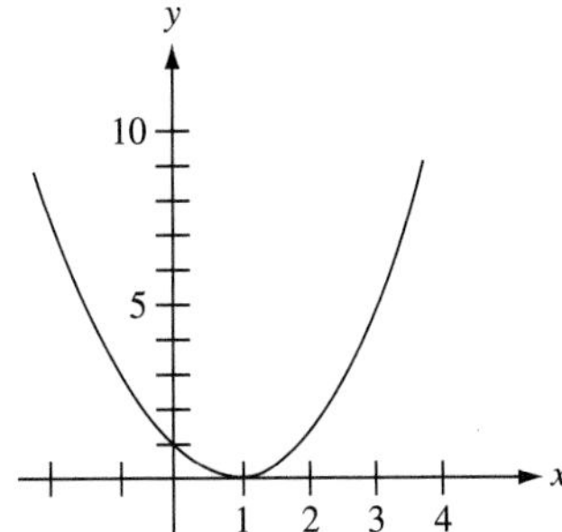

d. 0

25. a. 6; 5.5; 5.1

b. 5

c. The computations in part (a) show that as h approaches zero, the average velocity approaches the instantaneous velocity.

27. a. 130 km/h; 128.2 km/h; 128.02 km/h

b. 128 km/h

c. The computations in part (a) show that as the time intervals over which the average velocity are computed become smaller and smaller, the average velocity approaches the instantaneous velocity of the car at $t = 20$.

29. a. 5 sec

b. 80 ft/sec

c. 160 ft/sec

31. Average rate of change of the seal population over $[a, a + h]$; instantaneous rate of change of the seal population at $x = a$

33. Average rate of change of the country's industrial production over $[a, a + h]$; instantaneous rate of change of the country's industrial production at $x = a$

35. Average rate of change of atmospheric pressure over $[a, a + h]$; instantaneous rate of change of atmospheric pressure at $x = a$

37. a. Yes **b.** No **c.** No

39. a. Yes **b.** Yes **c.** No

41. a. No **b.** No **c.** No

43. 32.1, 30.939, 30.814, 30.8014, 30.8001, 30.8000; 30.8 m/sec

45. f does not have a derivative at $x = 1$ because it is not continuous there.

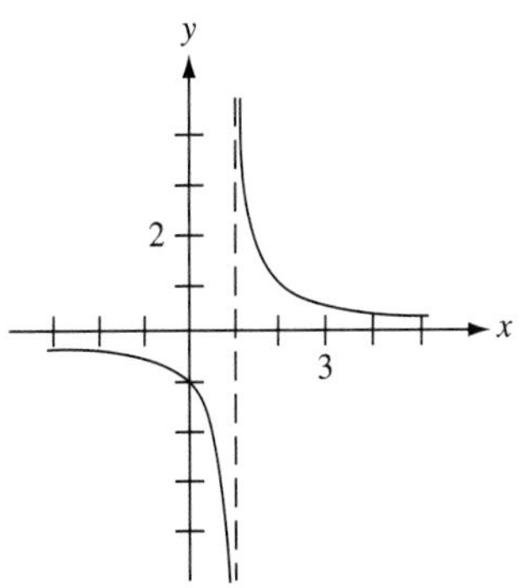

47. f is continuous at $x = 0$, but $f'(0)$ does not exist because the graph of f has a vertical tangent line at $x = 0$.

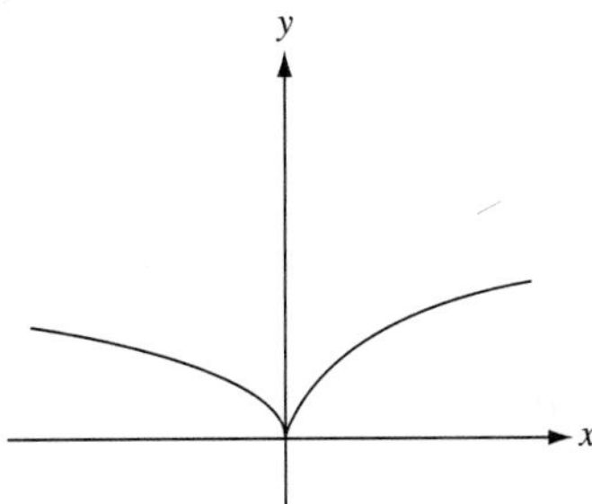

49. $\lim_{x \to a} f(x) = f(a)$

51. a. $C'(x) = -20x - 10h + 300$

b. \$100/surfboard

c. \$200/surfboard

53. a. $-\$2.00/1000$ tents

b. \$2/1000 tents

55. 5.06060, 5.06006, 5.060006, 5.0600006, 5.06000006; \$5.06

57. $\frac{1}{3}$ cubic metres/hectare/year; 1.25 cubic metres/hectare/year

59. -1.7; -0.5

61. a. $-\frac{1}{6}$ litre/atmosphere **b.** $-\frac{1}{4}$ litre/atmosphere

63. False

3.4 Exercises, page 167

1. 0 **3.** $5x^4$ **5.** $2.1x^{1.1}$

7. $6x$ **9.** $2\pi r$ **11.** $\frac{3}{x^{2/3}}$

13. $\frac{3}{2\sqrt{x}}$ **15.** $-84x^{-13}$ **17.** $10x - 3$

19. $-3x^2 + 4x$ **21.** $0.06x - 0.4$ **23.** $2x - 4 - \frac{3}{x^2}$

25. $16x^3 - 7.5x^{3/2}$ **27.** $-\frac{3}{x^2} - \frac{8}{x^3}$ **29.** $-\frac{16}{t^5} + \frac{9}{t^4} - \frac{2}{t^2}$

31. $2 - \frac{5}{2\sqrt{x}}$ **33.** $-\frac{4}{x^3} + \frac{1}{x^{4/3}}$

35. a. 20 **b.** -4 **c.** 20

37. 3 **39.** 11

41. $m = 5; y = 5x - 4$ **43.** $m = -2; y = -2x + 2$

45. a. (0, 0)

b.

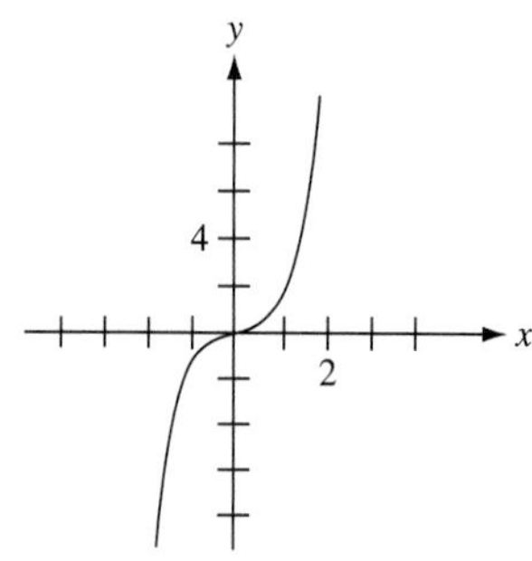

47. a. (−2, −7); (2, 9)

b. $y = 12x + 17$ and $y = 12x - 15$

c.

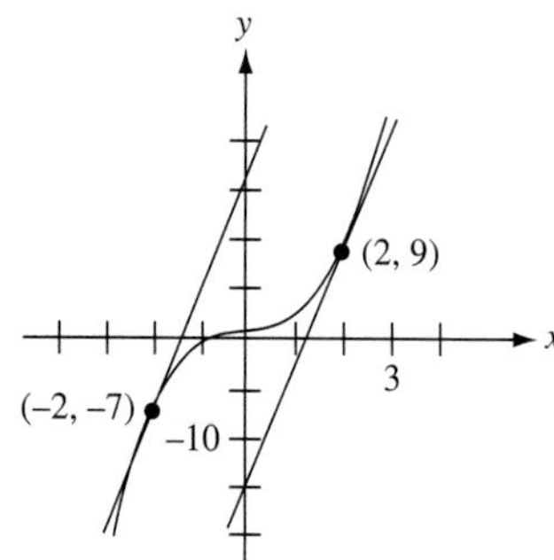

49. a. $(0, 0); \left(1, -\frac{13}{12}\right)$

b. $(0, 0); \left(2, -\frac{8}{3}\right); \left(-1, -\frac{5}{12}\right)$

c. $(0, 0); \left(4, \frac{80}{3}\right); -3, \frac{81}{4}$

51. −89.01; −4.36; if you make 0.25 stops/kilometre, your average speed will decrease at the rate of approximately 89.01 km/h per stop per kilometre. If you make 2 stops/kilometre, your average speed will decrease at the rate of approximately 4.36 km/h per stop per kilometre.

53. a. $120 - 30t$ **b.** 120 m/sec **c.** 240 m

55. a. 51.5% **b.** 1.95%/yr

57. a. 15 pts/yr; 12.6 pts/yr; 0 pts/yr **b.** 10 pts/yr

59. a. $f'(x) = -0.2x - 0.4$ **b.** $21

61. a. $(0.0001)\left(\frac{5}{4}\right)x^{1/4}$ **b.** $0.00125/radio

63. a. $S(t) = 0.08508t^2 - 0.10334t + 9.60881$

b. $27.20171 million/yr; $110.21491 million/yr

c. $270.1214 million; $1530.8476 million

65. $v(r) = 30$; $v'(r) = -200$; the velocity of blood 0.1 cm from the central axis is 30 cm/sec; at a point that is 0.1 cm from the central axis, the velocity of blood is decreasing at the rate of 200 cm/sec per cm along a line transverse to the central axis.

67. 155/mo; 200/mo

69. 32 turtles/yr; 428 turtles/yr; 3260 turtles

71. a. −0.9 thousand metric tons/yr; 20.3 thousand metric tons/yr

b. Yes

73. False

3.5 Exercises, page 177

1. $2x(2x) + (x^2 + 1)(2)$, or $6x^2 + 2$

3. $(t - 1)(2) + (2t + 1)(1)$, or $4t - 1$

5. $(3x + 1)(2x) + (x^2 - 2)(3)$, or $9x^2 + 2x - 6$

7. $(x^3 - 1)(1) + (x + 1)(3x^2)$, or $4x^3 + 3x^2 - 1$

9. $(w^3 - w^2 + w - 1)(2w) + (w^2 + 2)(3w^2 - 2w + 1)$, or $5w^4 - 4w^3 + 9w^2 - 6w + 2$

11. $(5x^2 + 1)(x^{-1/2}) + (2x^{1/2} - 1)(10x)$, or $\dfrac{25x^2 - 10x\sqrt{x} + 1}{\sqrt{x}}$

13. $\dfrac{(x^2 - 5x + 2)(x^2 + 2)}{x^2} + \dfrac{(x^2 - 2)(2x - 5)}{x}$, or $\dfrac{3x^4 - 10x^3 + 4}{x^2}$

15. $\dfrac{-1}{(x - 2)^2}$ **17.** $\dfrac{2x + 1 - (x - 1)(2)}{(2x + 1)^2}$, or $\dfrac{3}{(2x + 1)^2}$

19. $-\dfrac{2x}{(x^2 + 1)^2}$ **21.** $\dfrac{2x + 1 - (x - 1)(2)}{(2x + 1)^2}$, or $\dfrac{3}{(2x + 1)^2}$

23. $\dfrac{\left(\frac{1}{2}x^{-1/2}\right)[(x^2 + 1) - 4x^2]}{(x^2 + 1)^2}$, or $\dfrac{1 - 3x^2}{2\sqrt{x}(x^2 + 1)^2}$

25. $\dfrac{2x^3 + 2x^2 + 2x - 2x^3 - x^2 - 4x - 2}{(x^2 + x + 1)^2}$, or $\dfrac{x^2 - 2x - 2}{(x^2 + x + 1)^2}$

27. $\dfrac{(x - 2)(3x^2 + 2x + 1) - (x^3 + x^2 + x + 1)}{(x - 2)^2}$, or $\dfrac{2x^3 - 5x^2 - 4x - 3}{(x - 2)^2}$

29. $\dfrac{(x^2 - 4)(x^2 + 4)(2x + 8) - (x^2 + 8x - 4)(4x^3)}{(x^2 - 4)^2(x^2 + 4)}$, or $\dfrac{-2x^5 - 24x^4 + 16x^3 - 32x - 128}{(x^2 - 4)^2(x^2 + 4)^2}$

31. 8 **33.** −9 **35.** $2(3x^2 - x + 3)$; 10

37. $\dfrac{-3x^4 + 2x^2 - 1}{(x^4 - 2x^2 - 1)^2}; -\dfrac{1}{2}$ **39.** 60; $y = 60x - 102$

41. $-\dfrac{1}{2}; y = -\dfrac{1}{2}x + \dfrac{3}{2}$ **43.** $y = 7x - 5$

45. $\left(\frac{1}{3}, \frac{50}{27}\right)$; (1, 2) **47.** $\left(\frac{4}{3}, -\frac{770}{27}\right)$; (2, −30)

49. $y = -\dfrac{1}{2}x + 1; y = 2x - \dfrac{3}{2}$

51. a. $\dfrac{k(x + h) - k(x)}{h} = \dfrac{f(x + h)g(x) - f(x)g(x + h)}{hg(x + h)g(x)}$

b. $\dfrac{k(x + h) - k(x)}{h} = \dfrac{1}{g(x + h)g(x)}\left\{\left[\dfrac{f(x + h) - f(x)}{h}\right]g(x) - \left[\dfrac{g(x + h) - g(x)}{h}\right]f(x)\right\}$

c. $k'(x) = \dfrac{f'(x)g(x) - g'(x)f(x)}{[g(x)]^2}$

53. $-\$16$ per video game

55. 0.125, 0.5, 2 50; the cost of removing all of the pollutant is prohibitively high.

57. \$17.04 million/yr **59.** 18.5 mg/yr; 12.4 mg/yr

61. a. $\dfrac{180}{(t+6)^2}$ **b.** 3.7; 2.2; 1.8; 1.1 **c.** Yes

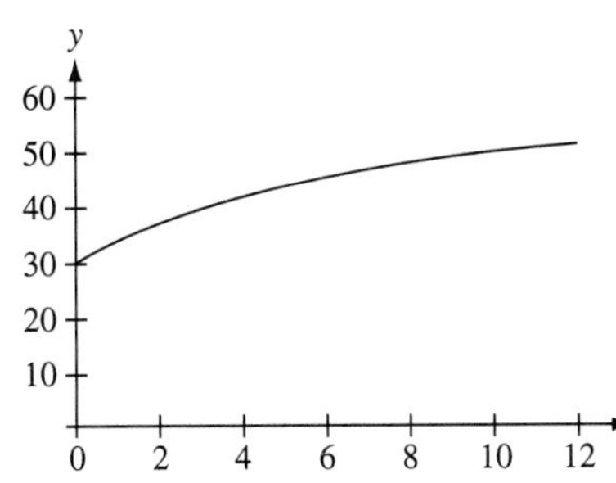

d. 50 words/min

63. a. $P'(t) = \dfrac{800(2t+5)}{(t^2+5t+40)^2}$ **b.** 20,790; 554 people/yr

65. True **67.** True

3.6 Exercises, page 190

1. $8(2x-1)^3$ **3.** $10x(x^2+2)^4$

5. $3(2x-x^2)^2(2-2x)$, or $6x^2(1-x)(2-x)^2$

7. $\dfrac{-4}{(2x+1)^3}$ **9.** $3x\sqrt{x^2-4}$

11. $\dfrac{3}{2\sqrt{3x-2}}$ **13.** $\dfrac{-2x}{3(1-x^2)^{2/3}}$

15. $-\dfrac{6}{(2x+3)^4}$ **17.** $\dfrac{-1}{(2t-3)^{3/2}}$

19. $-\dfrac{3(16x^3+1)}{2(4x^4+x)^{5/2}}$

21. $-2(3x^2+2x+1)^{-3}(6x+2) = -4(3x+1)(3x^2+2x+1)^{-3}$

23. $3(x^2+1)^2(2x) - 2(x^3+1)(3x^2)$, or $6x(2x^2-x+1)$

25. $3(t^{-1}-t^{-2})^2(-t^{-2}+2t^{-3})$

27. $\dfrac{1}{2\sqrt{x-1}} + \dfrac{1}{2\sqrt{x+1}}$

29. $2x^2(4)(3-4x)^3(-4) + (3-4x)^4(4x)$,

or $(-12x)(4x-1)(3-4x)^3$

31. $8(x-1)^2(2x+1)^3 + 2(x-1)(2x+1)^4$,

or $6(x-1)(2x-1)(2x+1)^3$

33. $3\left(\dfrac{x+3}{x-2}\right)^2\left[\dfrac{(x-2)(1)-(x+3)(1)}{(x-2)^2}\right]$, or $-\dfrac{15(x+3)^2}{(x-2)^4}$

35. $\dfrac{3}{2}\left(\dfrac{t}{2t+1}\right)^{1/2}\left[\dfrac{(2t+1)(1)-t(2)}{(2t+1)^2}\right]$, or $\dfrac{3t^{1/2}}{2(2t+1)^{5/2}}$

37. $\dfrac{1}{2}\left(\dfrac{u+1}{3u+2}\right)^{-1/2}\left[\dfrac{(3u+2)(1)-(u+1)(3)}{(3u+2)^2}\right]$, or

$-\dfrac{1}{2\sqrt{u+1}(3u+2)^{3/2}}$

39. $\dfrac{(x^2-1)^4(2x)-x^2(4)(x^2-1)^3(2x)}{(x^2-1)^8}$, or $\dfrac{(-2x)(3x^2+1)}{(x^2-1)^5}$

41. $\dfrac{2x(x^2-1)^3(3x^2+1)^2[9(x^2-1)-4(3x^2+1)]}{(x^2-1)^8}$, or

$-\dfrac{2x(3x^2+13)(3x^2+1)^2}{(x^2-1)^5}$

43. $\dfrac{(2x+1)^{-1/2}[(x^2-1)-(2x+1)(2x)]}{(x^2-1)^2}$, or $-\dfrac{3x^2+2x+1}{\sqrt{2x+1}(x^2-1)^2}$

45. $\dfrac{(t^2+1)^{1/2}\left(\frac{1}{2}\right)(t+1)^{-1/2}(1)-(t+1)^{1/2}\left(\frac{1}{2}\right)(t^2+1)^{-1/2}(2t)}{t^2+1}$, or

$-\dfrac{t^2+2t-1}{2\sqrt{t+1}(t^2+1)^{3/2}}$

47. $4(3x+1)^3(3)(x^2-x+1)^3 + (3x+1)^4(3)(x^2-x+1)^2(2x-1)$,
or
$3(3x+1)^3(x^2-x+1)^2(10x^2-5x+3)$

49. $-\dfrac{408\left(\dfrac{8}{x^3}+5\right)^{33}}{x^4\sqrt{\left(\dfrac{8}{x^3+5}\right)^{34}-1}}$

51. $-\dfrac{204\sqrt{4x^{-2}-1}\,[(\sqrt{4x^{-2}}-1)]^3+5]^{16}}{x^3}$

53. -0.5 **55.** -0.5

57. $g'[f(-2)]f'(-2)$, which does not exist, since $f'(-2)$ does not exist.

59. $\dfrac{4}{3}u^{1/3}$; $6x$; $8x(3x^2-1)^{1/3}$

61. $-\dfrac{2}{3u^{5/3}}$; $6x^2-1$; $-\dfrac{2(6x^2-1)}{3(2x^3-x+1)^{5/3}}$

63. $\dfrac{1}{2}u^{-1/2} - \dfrac{1}{2}u^{-3/2}$; $3x^2-1$; $\dfrac{(3x^2-1)(x^3-x-1)}{2(x^3-x)^{3/2}}$

65. -12 **67.** 6 **69.** No

71. $y = -33x + 57$ **73.** $y = \dfrac{43}{5}x - \dfrac{54}{5}$

75. $n[f(x)]^{n-1}f'(x) = 1$; $f'(x) = \dfrac{1}{n}x^{(1/n)-1}$

77. 0.333 thousand/wk; 0.305 thousand/wk; 16 thousand; 22.7 thousand

79. 2000 students/yr; 707 students/yr

81.

$(1.42)\left[\dfrac{(3t^2+80t+550)(14t+140)-(7t^2+140t+700)(6t+80)}{(3t^2+80t+550)^2}\right]$,

or

$$\frac{1.42(140t^2 + 3500t + 21{,}000)}{(3t^2 + 80t + 550)^2}; \text{ 87,322 jobs/yr}$$

83. 20.6%; 28.9%; 38.6%

85. 19 computers/month

87. −$5.86/passenger/yr

89. 0.6%/yr; 034%/yr; 26.1%

91.

$$300\left[\frac{(t+25)\frac{1}{2}\left(\frac{1}{2}t^2+2t+25\right)^{-1/2}(t+2) - \left(\frac{1}{2}t^2+2t+25\right)^{1/2}(1)}{(t+25)^2}\right],$$

or $\dfrac{3450t}{(t+25)^2\sqrt{\frac{1}{2}t^2+2t+25}}$; 2.9 beats/min², 0.7 beats/min², 0.2 beats/min², 179 beats/min

93. 160π m²/sec **95.** True **97.** True

Chapter 3 Review Exercises, page 195

1. −3 **3.** −21 **5.** −1

7. 7 **9.** 1 **11.** 1

13. $\frac{3}{2}$ **15.** $\lim_{x\to 2} f(x) = 1$

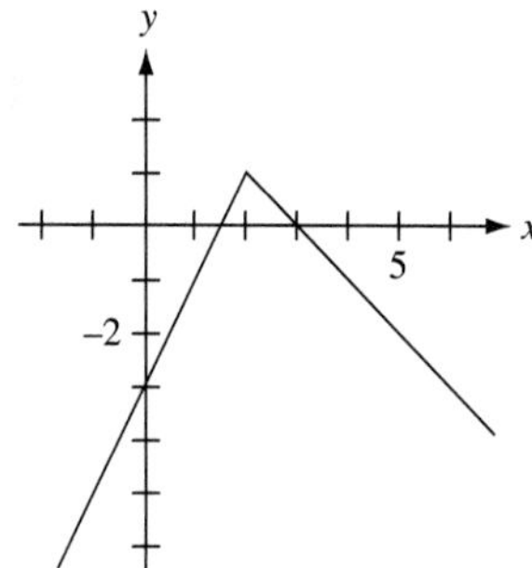

17. $x = 2$ **19.** $x = -1$

21. a. 3; 2.5; 2.1 **b.** 2

23. $\frac{1}{x^2}$ **25.** -4; $y = -4x + 4$

27. $15x^4 - 8x^3 + 6x - 2$ **29.** $\frac{6}{x^4} - \frac{3}{x^2}$

31. $-\frac{1}{t^{3/2}} - \frac{6}{t^{5/2}}$ **33.** $4s + \frac{4}{s^2} - \frac{1}{s^{3/2}}$

35. $\frac{1}{2\sqrt{t}(\sqrt{t}+1)^2}$ **37.** $1 - \frac{2}{t^2} - \frac{6}{t^3}$

39. $\frac{(2t^2+1)(2t) - t^2(4t)}{(2t^2+1)^2}$, or $\frac{2t}{(2t^2+1)^2}$

41. $\frac{(x^2-1)(4x^3+2x) - (x^4+x^2)(2x)}{(x^2-1)^2}$, or $\frac{2x(x^4-2x^2-1)}{(x^2-1)^2}$

43. $\frac{5(\sqrt{x}+2)^4}{2\sqrt{x}}$

45. $f(x) = (2x^3 - 3x^2 + 1)^{-3/2}$; $f'(x) = -9x(x-1)(2x^3-3x^2+1)^{-5/2}$

47. $h'(t) = 4t^2(5t+3)(t^2+t)^3$

49. $h(x) = \frac{(3x+2)^{1/2}}{4x-3}$; $h'(x) = -\frac{12x+25}{2\sqrt{3x+2}(4x-3)^2}$

51. $\frac{1}{2}(2t^2+1)^{-1/2}(4t)$, or $\frac{2t}{\sqrt{2t^2+1}}$

53. $h(x) = \frac{1+x}{(2x^2+1)^2}$; $h'(x) = -\frac{6x^2+8x-1}{(2x^2+1)^3}$

55. $f(x) = \frac{x}{(x^3+2)^{1/2}}$; $f'(x) = \frac{4-x^3}{2(x^3+2)^{3/2}}$

57. a. $(2, -25)$ and $(-1, 14)$

b. $y = -4x - 17$; $y = -4x + 10$

59. $y = -\frac{\sqrt{3}}{3}x + \frac{4}{3}\sqrt{3}$

61.

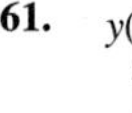

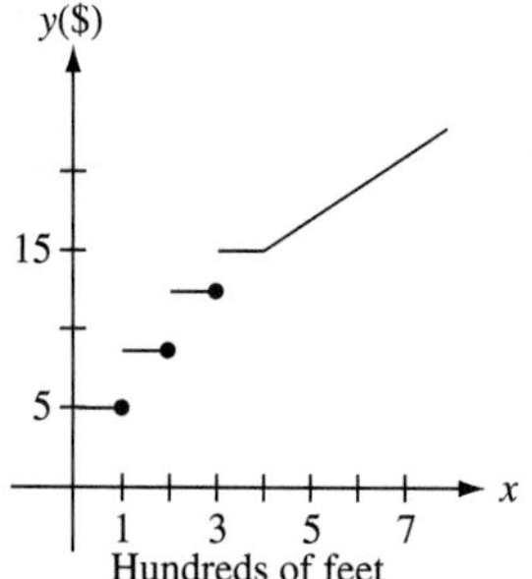

CHAPTER 4

4.1 Exercises, page 209

1. a. $C(x)$ is always increasing because as the number of units x produced increases, the amount of money that must be spent on production also increases.

b. 4000

3. a. $1.80; $1.60 **b.** $1.80; $1.60

5. a. $100 + \frac{200{,}000}{x}$ **b.** $-\frac{200{,}000}{x}$

c. $\bar{C}(x)$ approaches $100 if the production level is very high.

7. $\frac{2000}{x} + 2 - 0.0001x$; $-\frac{2000}{x^2} - 0.0001$

9. a. $8000 - 200x$ **b.** 200, 0, −200 **c.** $40

11. a. $-0.04x^2 + 600x - 300{,}000$

b. $-0.08x + 600$

c. 200; −40

d. The profit increases as production increase, peaking at 7500 units; beyond this level, profit falls.

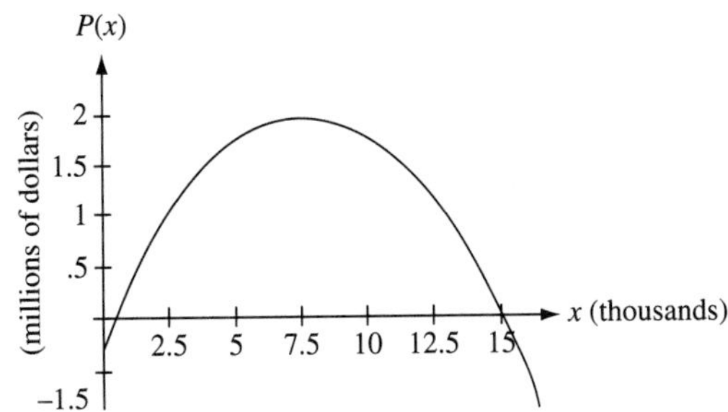

13. **a.** $600x - 0.05x^2$; $-0.000002x^3 - 0.02x^2 + 200x - 80{,}000$

b. $0.000006x^2 - 0.06x + 400$; $600 - 0.1x$; $-0.000006x^2 - 0.04x + 200$

c. 304; 400; 96

d.

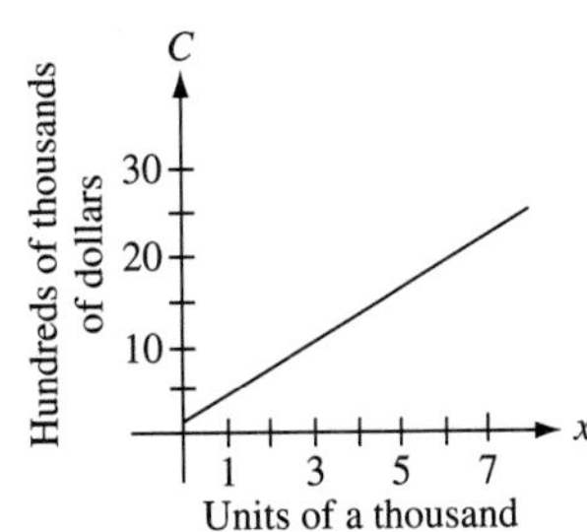

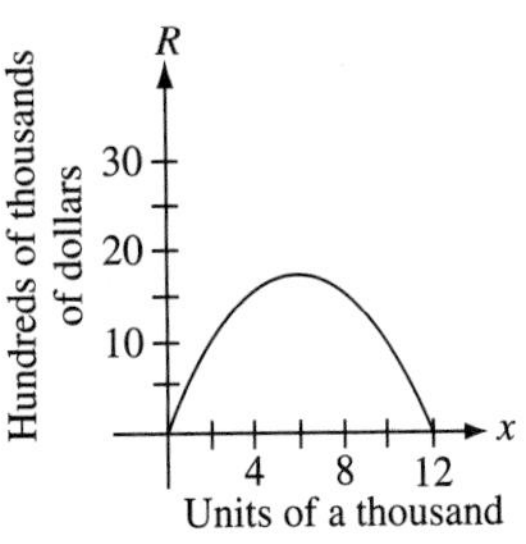

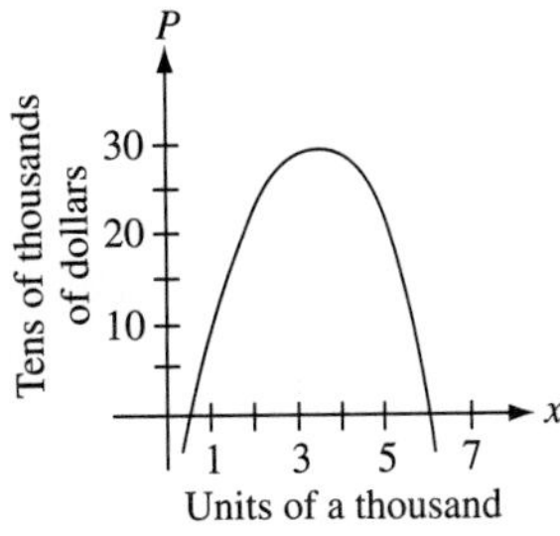

15. $0.000002x^2 - 0.03x + 400 + \dfrac{80{,}000}{x}$

a. $0.000004x - 0.03 - \dfrac{80{,}000}{x^2}$

b. -0.0132; 0.0092; the marginal average cost is negative (average cost is decreasing) when 5000 units are produced and positive (average cost is increasing) when 10,000 units are produced.

c.

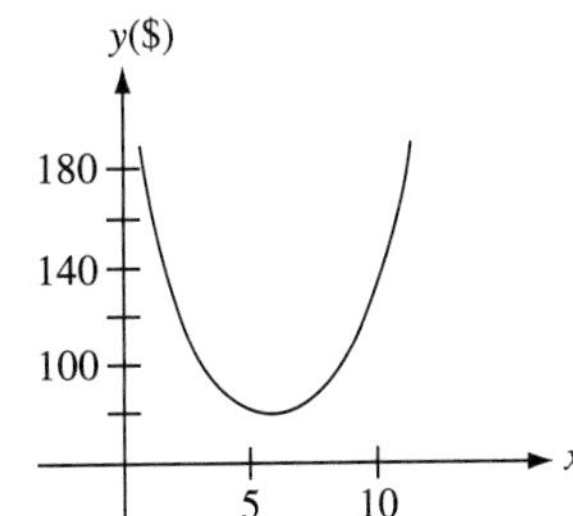

17. **a.** $\dfrac{50x}{0.01x^2 + 1}$ **b.** $\dfrac{50 - 0.5x^2}{(0.01x^2 + 1)^2}$

c. $44,380; when the level of production is 2000 units, the revenue increases at the rate of $44,380 per additional 1000 units produced.

19. $1.21 billion/billion dollars

21. $0.288 billion/billion dollars

23. $\frac{5}{3}$; elastic **25.** 1; unitary **27.** 0.104; inelastic

29. **a.** Inelastic; elastic **b.** When $p = 8.66$

c. Increase **d.** Increase

31. **a.** Inelastic **b.** Increase

33. $\dfrac{2p^2}{9 - p^2}$; for $p < \sqrt{3}$, demand is inelastic; for $p = \sqrt{3}$, demand is unitary; and for $p > \sqrt{3}$, demand is elastic.

35. True

4.2 Exercises, page 215

1.

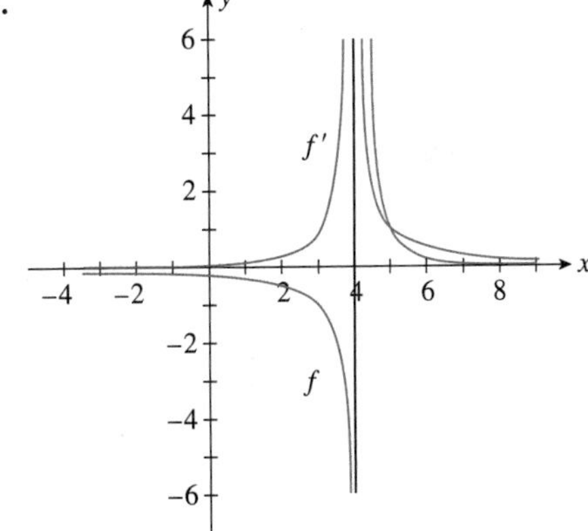

3.

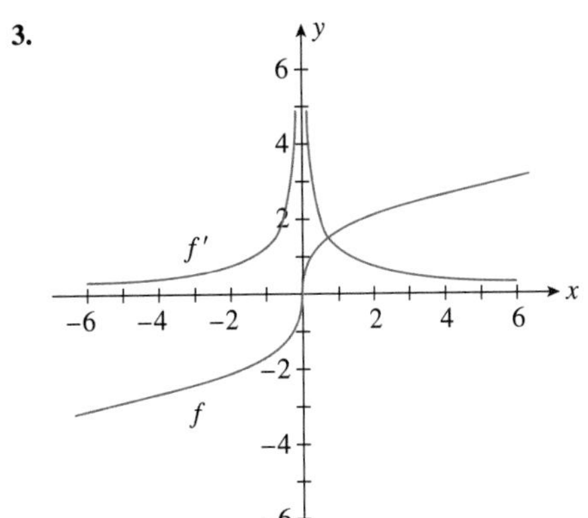

5.

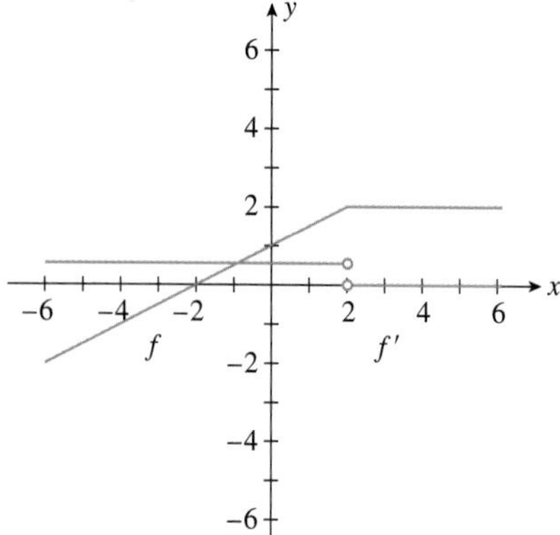

7.

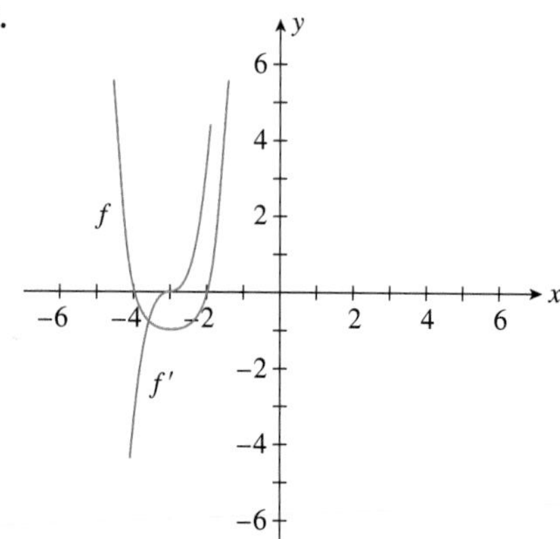

9.

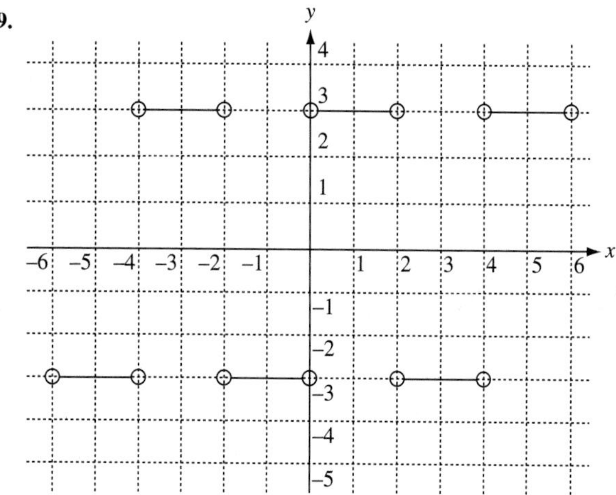

11.

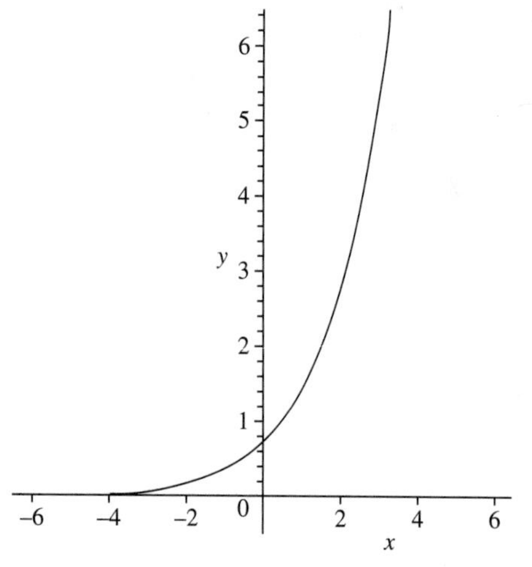

13.

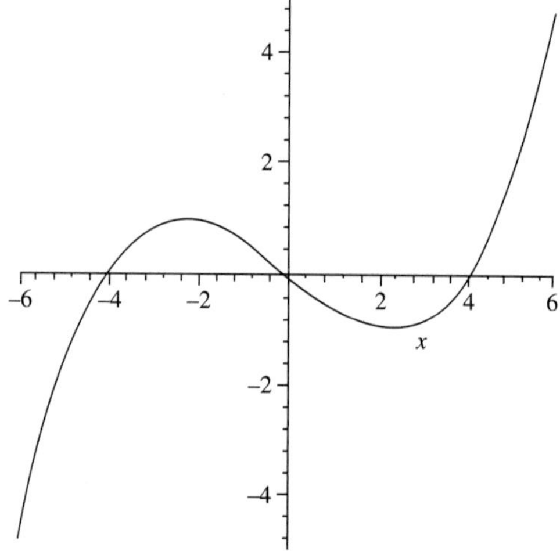

15.

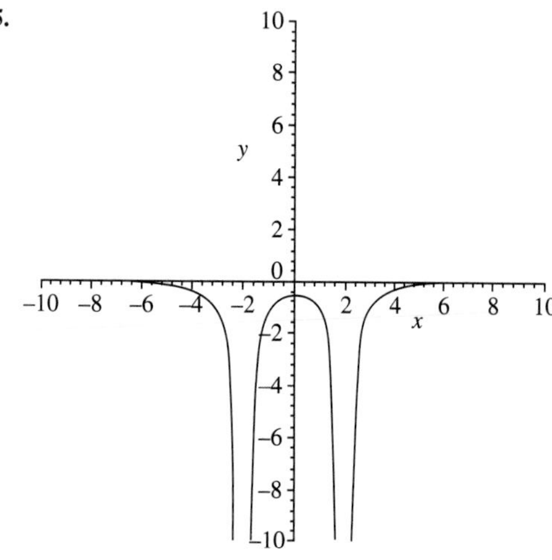

17. a. 4000; 2000; 0; −2000; −4000

b.

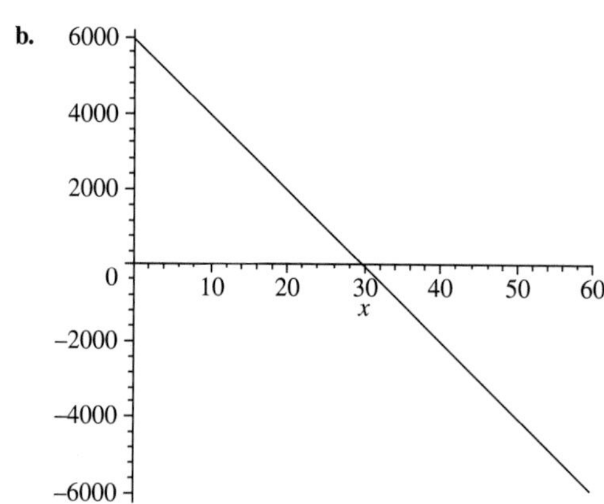

c. $30

19. a. 0.1; 0; −0.025

b.

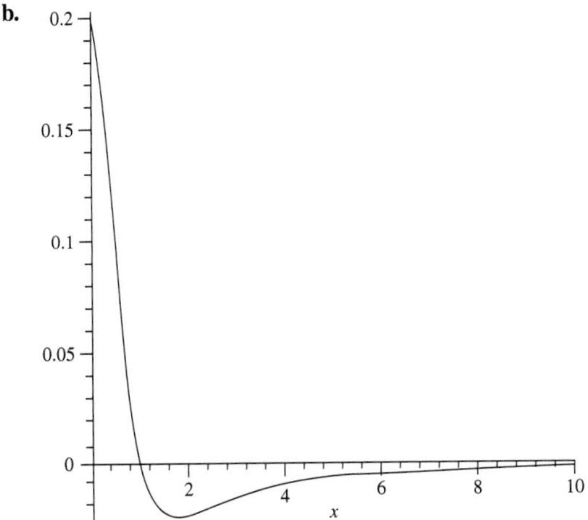

4.3 Exercises, page 222

1. $8x - 2; 8$

3. $6x^2 - 6x; 6(2x - 1)$

5. $4t^3 - 6t^2 + 12t - 3; 12(t^2 - t + 1)$

7. $10x(x^2 + 2)^4; 10(x^2 + 2)^3(9x^2 + 2)$

9. $6t(2t^2 - 1)(6t^2 - 1); 6(60t^4 - 24t^2 + 1)$

11. $14x(2x^2 + 2)^{5/2}; 28(2x^2 + 2)^{3/2}(6x^2 + 1)$

13. $(x^2 + 1)(5x^2 + 1); 4x(5x^2 + 3)$

15. $\dfrac{1}{(2x + 1)^2}; -\dfrac{4}{(2x + 1)^3}$

17. $\dfrac{2}{(s + 1)^2}; -\dfrac{4}{(s + 1)^3}$

19. $-\dfrac{3}{2(4 - 3u)^{1/2}}; -\dfrac{9}{4(4 - 3u)^{3/2}}$

21. $72x - 24$

23. $-\dfrac{6}{x^4}$

25. $\dfrac{81}{8}(3s - 2)^{-5/2}$

27. $192(2x - 3)$

29. f' and f'' exist everywhere; $f'''(x) = \dfrac{70}{27x^{1/3}}$ is not defined at $x = 0$.

31. $P(x) = a_0x^n + a_1x^{n-1} + a_2x^{n-2} + \cdots + a_n;$

$P'(x) = na_0x^{n-1} + (n - 1)a_1x^{n-2} + \cdots + a_{n-1}$

Eventually $P^{(n)}(x) = a_0, P^{(n+1)}(x) = P^{(n+2)}(x) = P^{(n+3)}(x) = \cdots = 0$

33. $S(t) = -0.3t^3 + 2.4t^2 + 6.1; v(t) = -0.9t^2 - 4.8t;$

$a(t) = -1.8t + 4.8 = 0; t = \dfrac{8}{3}$

The acceleration of the car at $t = \dfrac{8}{3}$ seconds is zero, and the car will start to decelerate at that point in time.

35. a. 0; 4.2; 7.2; 9; 9.6; 9; 7.2; 4.2; 0

b. 4.8; 3.6; 2.4; 1.2; 0; −1.2; −2.4; −3.6; −4.8

c. The GDP is increasing at an increasing rate in the first five years. Even though the GDP continues to rise from that point on, the negativity of $G'(t)$ shows that the rate of increase is slowing down.

37. a. and b.

t	0	1	2	3	4	5	6	7
$N'(t)$	0	2.7	4.8	6.3	7.2	7.5	7.2	6.3
$N''(t)$					0.6	0	− 0.6	−1.2

39. $A'(10) = -3.09; A''(10) = 0.35$; 10 minutes after the start of the test, the smoke remaining is decreasing at a rate of 3%/minute but the rate at which the rate of smoke is decreasing is increasing at the rate of 0.35%/minute/minute.

41. $P''(20) = -5.46781125(t + 5)^{-1.795}$; the rate of the rate of change of such mothers is decreasing at the rate of 0.02%/yr^2.

43. True

45. True

4.4 Exercises, page 231

1. a. $-\dfrac{1}{2}$ **b.** $-\dfrac{1}{2}$

3. a. $-\dfrac{1}{x^2}$ **b.** $-\dfrac{y}{x}$

5. a. $2x - 1 + \dfrac{4}{x^2}$ **b.** $3x - 2 - \dfrac{y}{x}$

7. a. $\dfrac{1 - x^2}{(1 + x^2)^2}$ **b.** $-2y^2 + \dfrac{y}{x}$

9. $-\dfrac{x}{y}$

11. $\dfrac{x}{2y}$

13. $1 - \dfrac{y}{x}$

15. $-\dfrac{y}{x}$

17. $-\dfrac{\sqrt{y}}{\sqrt{x}}$

19. $2\sqrt{x + y} - 1$

21. $-\dfrac{y^3}{x^3}$

23. $\dfrac{2\sqrt{xy} - y}{x - 2\sqrt{xy}}$

25. $\dfrac{6x - 3y - 1}{3x + 1}$

27. $\dfrac{2(2x - y)^{3/2}}{3x\sqrt{y} - 4y}$

29. $-\dfrac{2x^2 + 2xy + y^2}{x^2 + 2xy + 2y^2}$

31. $y = 2$

33. $y = -\dfrac{3}{2}x + \dfrac{5}{2}$

35. $\dfrac{2y}{x^2}$

37. $\dfrac{2y(y - x)}{(xy - x)^3}$

39. a. $\dfrac{dV}{dt} = \pi r\left(r\dfrac{dh}{dt} + 2h\dfrac{dr}{dt}\right)$

b. 3.6π cu cm/sec

41. 7.5 cu cm/sec

43. 17 m/sec

45. 2.3 m/sec

47. $\dfrac{dr}{dt} = -\dfrac{k}{\pi}$

49. 2.25 m/sec

51. 19.2 m/sec

53. Dropping at the rate of 111 tires/wk

55. Increasing at the rate of 44 ten packs/wk

57. Dropping at the rate of 3.7cents/carton/wk

59. 0.37; inelastic

61. 0.0064 cm/sec; 1.6 cm^2/sec

63. True

4.5 Exercises, page 241

1. $4x\,dx$ **3.** $(3x^2 - 1)dx$

5. $\dfrac{dx}{2\sqrt{x + 1}}$ **7.** $\dfrac{6x + 1}{2\sqrt{x}}\,dx$

9. $\dfrac{x^2 - 2}{x^2}\,dx$ **11.** $\dfrac{-x^2 + 2x + 1}{(x^2 + 1)^2}\,dx$

13. $\dfrac{6x - 1}{2\sqrt{3x^2 - x}}\,dx$ **15.** $L(x) = 3 + \dfrac{1}{6}(x - 9); f(10) = 3\dfrac{1}{6}$

17. $L(x) = \dfrac{1}{3} - \dfrac{1}{9}(x - 3); f(3.2) = \dfrac{14}{45}$

19. $L(x) = 28 + 10(x - 5); f(5.65) = 34.5$

21. **a.** $2x\,dx$ **b.** 0.04 **c.** 0.0404

23. **a.** $-\dfrac{dx}{x^2}$ **b.** -0.05 **c.** -0.05263

25. 3.167 **27.** 7.0358 **29.** 1.983

31. 0.298 **33.** 2.50146 **35.** ± 8.64 cm^2

37. 0.6 m^3 **39.** \$2000 dollars **41.** 75 cents

43. $\pm 64{,}800$

45. **a.** $P = \dfrac{10{,}000r}{1 - \left(1 + \dfrac{r}{12}\right)^{-360}}$;

$$dP = \frac{10{,}000\left[1 - \left(1 + \dfrac{r}{12}\right)^{-360} - 30r\left(1 + \dfrac{r}{12}\right)^{-361}\right]}{\left[1 - \left(1 + \dfrac{r}{12}\right)^{-360}\right]^2}\,dr$$

b. \$17.27; \$25.90; \$34.54

47. **a.** $dS = 24{,}000\left[\dfrac{(r)300\left(1 + \dfrac{r}{12}\right)^{299}\left(\dfrac{1}{12}\right) - \left(1 + \dfrac{r}{12}\right)^{300} + 1}{r^2}\right]dr$

b. \$37,342.02; \$74,684.05; \$112,026.00

49. Resistance will drop by 40%.

51. 0.133% **53.** True

4.6 Exercises, page 248

1. 1.732051 **3.** 2.645751 **5.** 2.410142

7. 2.30278 **9.** 1.11634 **11.** 1.61803

13. **a.** $f(1) = -1$ and $f(2) = 5$

b. 1.32472

15. **a.** $f(1) = 7$ and $f(2) = -6$

b. 1.61178

17. 0.7549 **19.** 0.426303

21. **a.** $f(x) = 0, f'(x) = nx^{n-1}$

b. 2.5813

23. 13.28%/yr

25. **a.** $C(1 + r)^N - R(1 + r)^{N-1} - R(1 + r)^{N-2} - \cdots - R = 0$

b. $f(r) = Cr + R[(1 + r)^{-N} - 1] = 0$

27. 9%/yr **29.** 12%/yr

31. 5871 units **33.** 0°C

Chapter 4 Review Exercises, page 251

1. b **3.** d **5.** a

7.

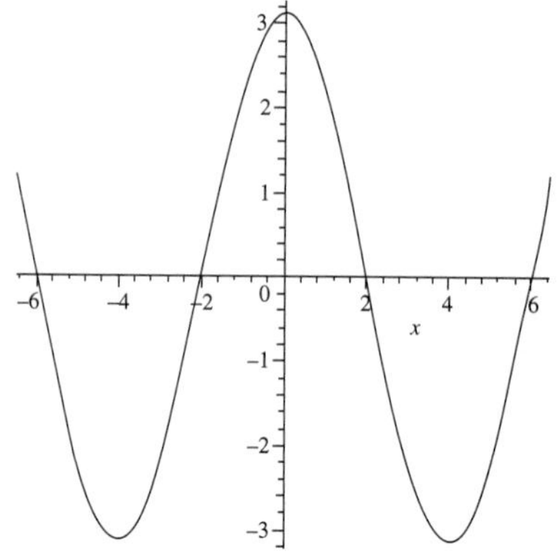

9.

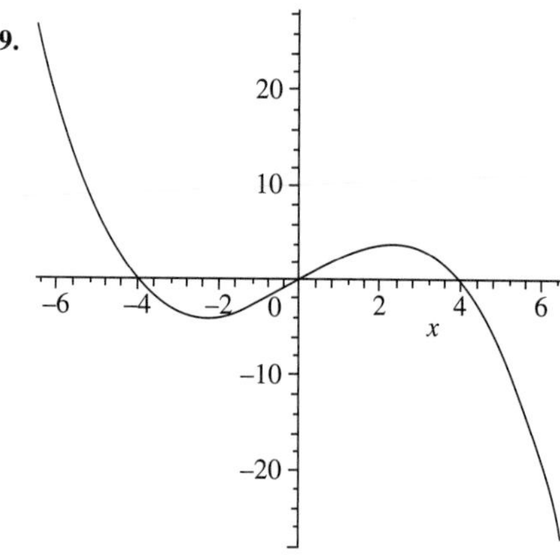

11. $2(12x^2 - 9x + 2)$

13. $\dfrac{(t^2 + 4)^2(-2t) - (4 - t^2)2(t^2 + 4)(2t)}{(t^2 + 4)^4}$, or $\dfrac{2t(t^2 - 12)}{(t^2 + 4)^3}$

15. $2(2x^2 + 1)^{-1/2} + 2x\left(-\dfrac{1}{2}\right)(2x^2 + 1)^{-3/2}(4x)$, or $\dfrac{2}{(2x^2 + 1)^{3/2}}$

17. $8\dfrac{7}{9}$ **19.** 2.1

21. $\dfrac{2x}{y}$ **23.** $-\dfrac{2x}{y^2 - 1}$

25. $\dfrac{x - 2y}{2x + y}$ **27.** $\dfrac{2(x^4 - 1)}{x^3}\,dx$

29. **a.** $\dfrac{2x}{\sqrt{2(x^2 + 2)}}\,dx$ **b.** 0.1333

c. 0.1335; differ by 0.0002

31. $-\dfrac{48}{(2x - 1)^4}$; $\left(-\infty, \dfrac{1}{2}\right) \cup \left(\dfrac{1}{2}, \infty\right)$

33. $\dfrac{25}{2(25 - \sqrt{p})}$; for $p > 156.25$, demand is elastic; for $p = 156.25$, demand is unitary; for $p < 156.25$, demand is inelastic.

35. a. Elastic **b.** Decrease

37. a. 2.2; 2.2 **b.** $2.2 + \frac{2500}{x}; -\frac{2500}{x^2}$ **c.** 2.2

39. a. $2000x - 0.04x^2$; $-0.000002x^3 - 0.02x^2 + 1000x - 120{,}000$; $0.000002x^2 - 0.02x + 1000 + \frac{120{,}000}{x}$

b. $0.000006x^2 - 0.04x + 1000$; $2000 - 0.08x$; $-0.000006x^2 - 0.04x + 1000$; $0.000004x - 0.02 - 120{,}000x^{-2}$

c. 934; 1760; 826

d. −0.0048; 0.0101; at a level of production of 5000 machines, the average cost of each additional unit is decreasing at a rate of 0.48 cents. At a level of production of 8000 machines, the average cost of each additional unit is increasing at a rate of 1 cent per unit.

41. 200 subscribers/wk

CHAPTER 5

5.1 Exercises, page 261

1. a. 16 **b.** 27

3. a. 3 **b.** $\sqrt{5}$

5. a. −3 **b.** 8

7. a. 25 **b.** $4^{1.8}$

9. a. $4x^3$ **b.** $5xy^2\sqrt{x}$

11. a. $\frac{2}{a^2}$ **b.** $\frac{1}{3}b^2$

13. a. $8x^9y^6$ **b.** $16x^4y^4z^6$

15. a. $\frac{64x^6}{y^4}$ **b.** $(x - y)(x + y)$

17. 2 **19.** 3 **21.** 3

23. $\frac{5}{4}$ **25.** 1 or 2

27.

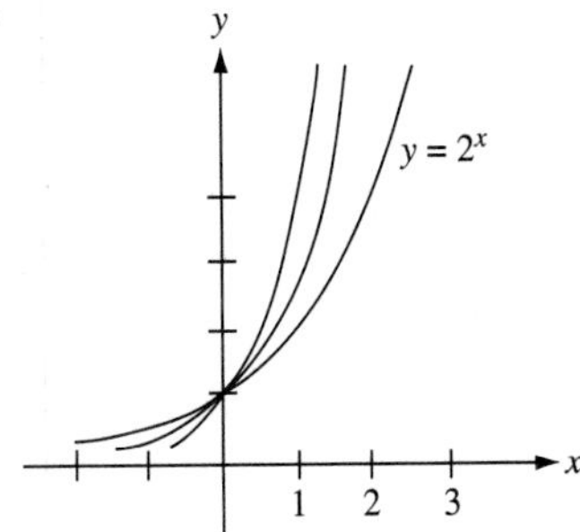

29.

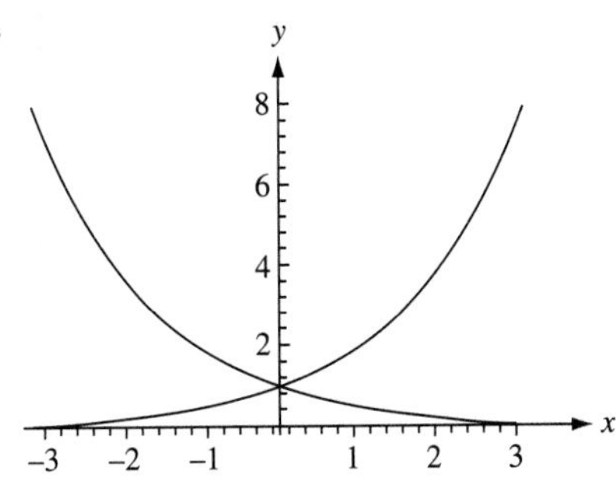

31.

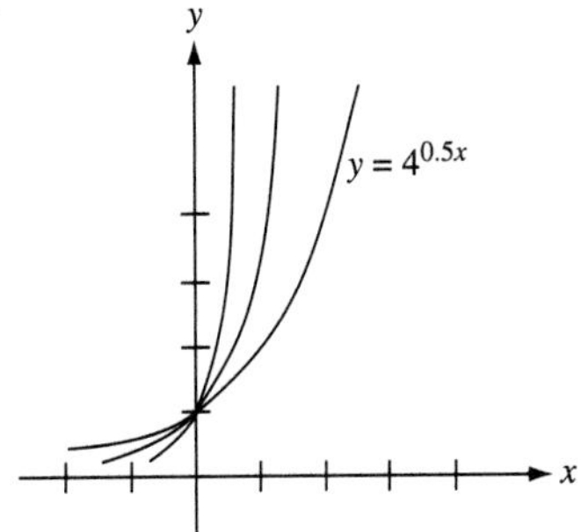

33.

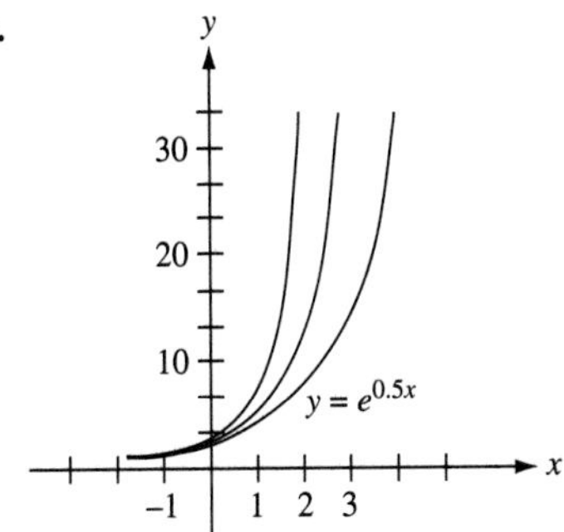

35.

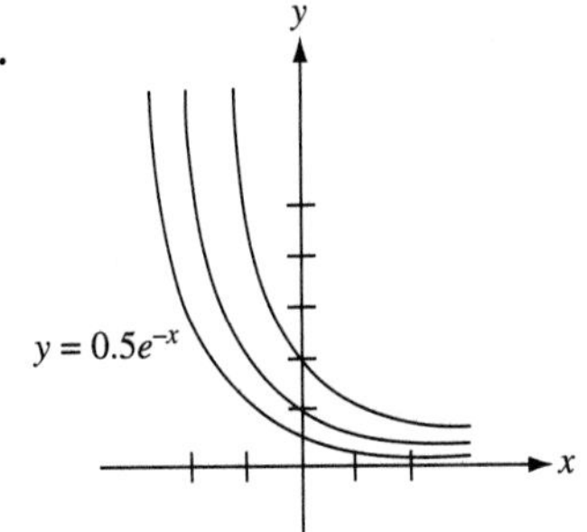

37. a.

Year	0	1	2	3	4	5
Number (billions)	0.45	0.80	1.41	2.49	4.39	7.76

b.

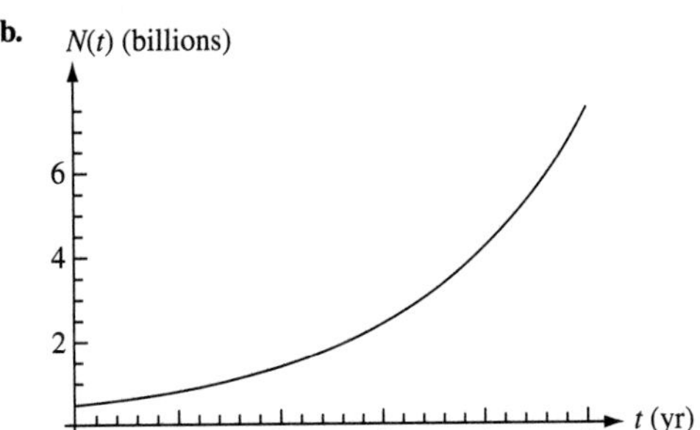

39. a. 0 **b.** 0.2367 g/cm^3

c. 0.7598 g/cm^3 **d.** 0

41. a. 0.08 g/cm³ **b.** 0.12 g/cm³ **c.** 0.2 g/cm³

d.

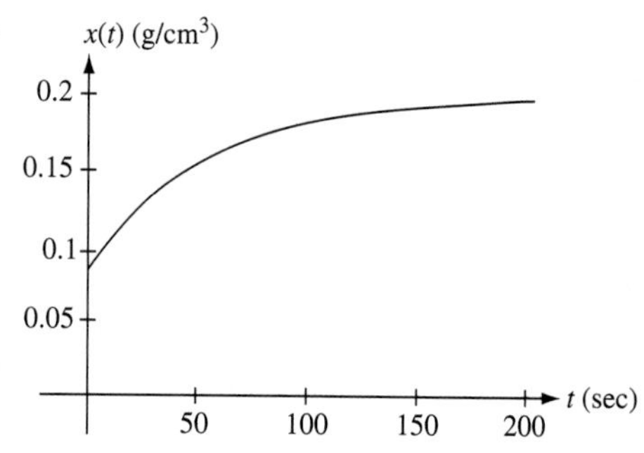

43. False **45.** True

5.2 Exercises, page 269

1. $\log_2 64 = 6$ **3.** $\log_3 \frac{1}{9} = -2$

5. $\log_{1/3} \frac{1}{3} = 1$ **7.** $\log_{32} 8 = \frac{3}{5}$

9. $\log_{10} 0.001 = -3$ **11.** 1.0792

13. 1.2042 **15.** 1.6813

17. $\log_3 64 \approx 3.7860$ **19.** $\log_4 36 \approx 2.5848$

21. $\ln a^2b^3$ **23.** $\ln \frac{3\sqrt{xy}}{\sqrt[3]{z}}$

25. $\log x + 4\log(x + 1)$ **27.** $\frac{1}{2}\log(x + 1) - \log(x^2 + 1)$

29. $\ln x - x^2$ **31.** $-\frac{3}{2}\ln x - \frac{1}{2}\ln(1 + x^2)$

33.

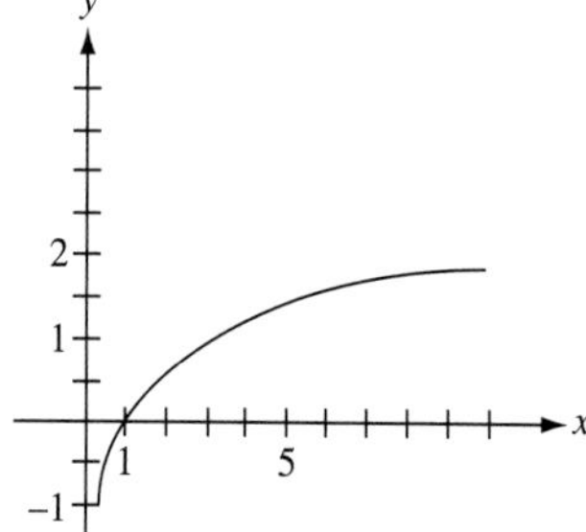

35.

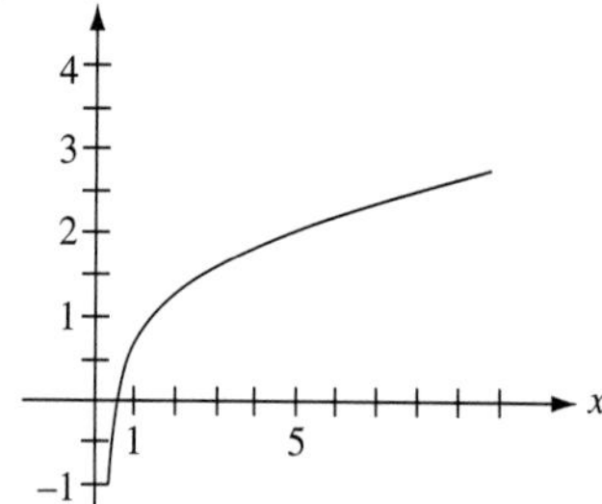

37.

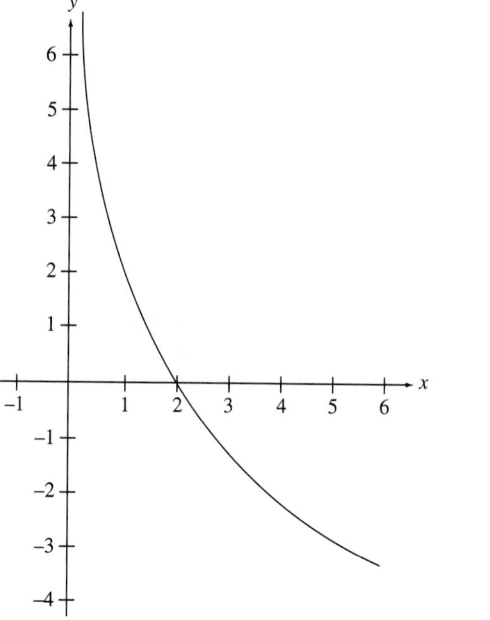

39.

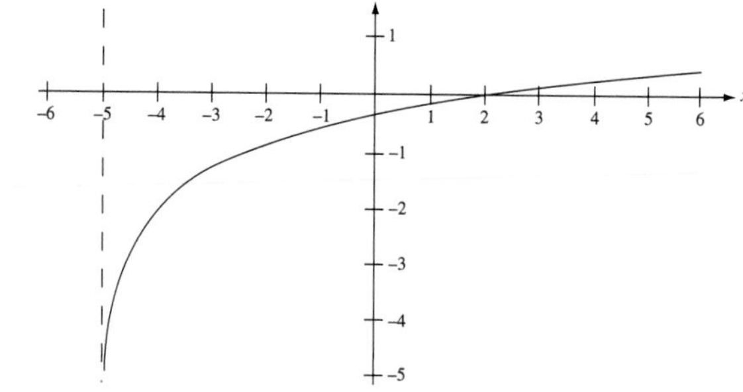

41.

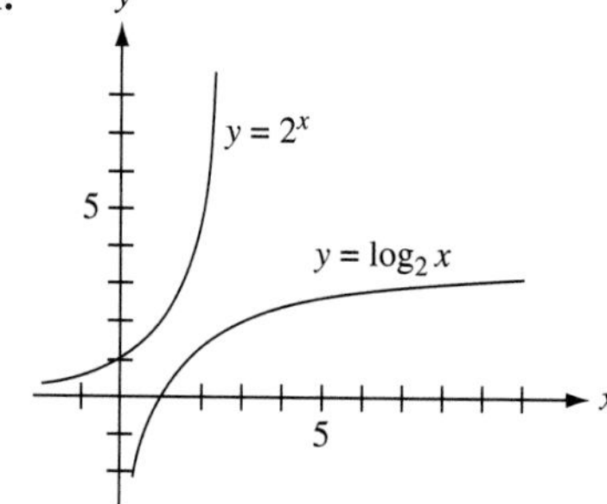

43. 5.19886 **45.** −0.0912

47. −8.0472 **49.** −4.9041

51. $-2\ln\left(\frac{A}{B}\right)$ **53.** 5

55. 1 **57.** −1, 4

59. a. $k = \ln 2$ **b.** $k = \ln b$

61. Let $\log_b m = p$, then $m = b^p$; therefore $m^n = b^{np}$.

63. $80 million

65. $82,864.24; $47,116.12

67. 105.7 mm

69. **a.** $10^3 I_0$

b. 100,000 times greater

c. 10,000,000 times greater

71. 27.40 yr　73. 6.44 yr

75. **a.** 9.12 sec　**b.** 20.27 sec

77. False　79. True

5.3 Exercises, page 277

1. $3e^{3x}$　3. $-e^{-1}$

5. $e^x + 1$　7. $x^2e^x(x + 3)$

9. $\dfrac{2e^x(x-1)}{x^2}$　11. $3(e^x - e^{-x})$

13. $-\dfrac{1}{e^w}$　15. $6e^{3x-1}$

17. $-2xe^{-x^2}$　19. $\dfrac{3e^{-1/x}}{x^2}$

21. $25e^x(e^x+1)^{24}$　23. $\dfrac{e^{\sqrt{x}}}{2\sqrt{x}}$

25. $e^{3x+2}(3x-2)$　27. $\dfrac{2e^x}{(e^x+1)^2}$

29. $\dfrac{3^x((\ln 3)x - 1)}{5x^2}$　31. $\dfrac{(\ln 2)2^{x-1}}{\sqrt{2^x - 4}}$

33. $1 - 4^x - (\ln 4)x4^x$

35. $10^t \ln 10\sqrt{t-5} + 10^t\dfrac{1}{2\sqrt{t-5}}$

37. $(\ln 3 + \ln 5)3^t5^t$

39. $\dfrac{(\ln 2)t2^t + (3\ln 2 - 1)2^t + 3}{(2^t + t)^2}$

41. $12e^{-2t} - 5e^{-t}$

43. $2(2t^2 - 4t + 1)e^{-2t}$

45. $(2\ln 3 - 5(\ln 3)^2)3^t + (\ln 3)^2t3^t$

47. $\dfrac{(\ln 6)^2\sqrt{x}\,6^{\sqrt{x}} - (\ln 6)6\sqrt{x}}{4(\sqrt{x})^3}$

49. $y - \dfrac{1}{e} = -\dfrac{2}{e}(x-1)$, or $y = -\dfrac{2}{e} + \dfrac{3}{e}$

51. 0.5671　53. 0.7531

55. $x_C = 2, x_1 = 1.5379, x_2 = 1.5571, x_3 = 1.5571$

57. **a.** $100xe^{-0.0001x}$

b. $100(1 - 0.0001x)e^{-0.0001x}$

c. \$99.80/thousand pair

59. **a.** $-\$1.68$ /case

b. \$40.36/case

61. **a.** \$12/unit

b. Decreasing at the rate of \$7/wk

c. \$8/unit

63. **a.** 15.6%　**b.** 12.1%/yr　**c.** 9.5%/yr/yr

65. **a.** 1986 kwh/yr　**b.** $-108.48e^{-0.073t}$　**c.** 899.35 kwh/yr

67. **a.** $N'(t) = 5.3e^{0.095t^2-0.85t}(0.19t - 0.85)$

b. 4505/yr; 273 cases/yr

69. **a.** 45.6 kg　**b.** 4.2 kg

71. False　73. False

5.4 Exercises, page 287

1. $\dfrac{5}{x}$　3. $\dfrac{1}{x+1}$

5. $\dfrac{8}{x}$　7. $\dfrac{1}{2x}$

9. $\dfrac{-2}{x}$　11. $\dfrac{2(4x-3)}{4x^2 - 6x + 3}$

13. $\dfrac{1}{x(x+1)}$　15. $x(1 + 2\ln x)$

17. $\dfrac{2(1-\ln x)}{x^2}$　19. $\dfrac{3}{u-2}$

21. $\dfrac{1}{2x\sqrt{\ln x}}$　23. $\dfrac{3(\ln x)^2}{x}$

25. $\dfrac{3x^2}{x^3+1}$　27. $\dfrac{(x\ln x + 1)e^x}{x}$

29. $\dfrac{e^{2t}[2(t+1)\ln(t+1) + 1]}{t+1}$

31. $\dfrac{1-\ln x}{x^2}$　33. $\dfrac{1-2x}{(\ln 3)x(1-x)}$

35. $-\dfrac{-1}{2(\ln 5)\sqrt{t}(6-\sqrt{t})}$　37. $\log_2 t + \dfrac{1}{\ln 2}$

39. $\dfrac{1 - t + (\ln 10)t\log 3t}{(\ln 10)t(1-t)^2}$　41. $\dfrac{1}{2(\ln 6)x\sqrt{\log_6 x}}$

43. $\dfrac{x^2 + 3 - 2(\ln 7)x^2\log_7 x}{(\ln 7)x(x^2+3)^2}$　45. $-\dfrac{1}{(x+5)^2}$

47. $\dfrac{2(1-\ln x)}{x^2}$　49. $-\dfrac{3}{(\ln 5)t^2}$

51. $\dfrac{-5 + 6(\ln 2)\log_2 x}{(\ln 2)x^4}$

53. $\dfrac{dy}{dx}\cdot\dfrac{1}{y} = \dfrac{2(45x+4)}{(3x+2)(5x-1)}; \dfrac{dy}{dx} = 2(3x+2)^3(5x-1)(45x+4)$

55. $\dfrac{1}{2}(2x-3)^3(54x+71)(3x+5)^{-1/2}$

57. $\dfrac{y'}{y} = \dfrac{9x(x^2+1) - 2x(4+3x^2)}{3(4+3x^2)(x^2+1)}$

59. $\dfrac{y'}{y} = \dfrac{x\ln x + x + 2}{x}; y' = \dfrac{(x\ln x + x + 2)x^{x+2}}{x}$

61. $\dfrac{y'}{y} = \dfrac{2\ln x}{x}; y' = 2(\ln x)x^{\ln x - 1}$

63. $y - \ln 4 = 1(x - 2)$, or $y = x + \ln 4 - 2$

65. a. $\frac{I'}{I} = \ln a; I' = (\ln a)I_0 a^x$

b. I' is proportional to I with $\ln a$ as the constant of proportion.

67. 0.0580%/kg; 0.0133%/kg

69. a. 0

b. $\frac{dR}{dS} = \frac{k}{S}$, and so $\frac{dr}{dS}$ is inversely proportional to S, with k as the constant of proportion; if the stimulus is small, then a small change in S is easily felt, but if the stimulus is larger, then a small change in S is not as discernable.

71. True

73. $f'(x) = \frac{1}{x}; f'(1) = \lim_{x \to 0} \frac{\ln(x + 1)}{x}$

5.5 Exercises, page 296

1. a. 0.05 **b.** 400

c.

t	0	10	20	100	1000
Q	400	660	1087	59365	2.07×10^{24}

3. a. 74% **b.** All will fail eventually.

5. 3%; 65.14%

7. $176,198

9.

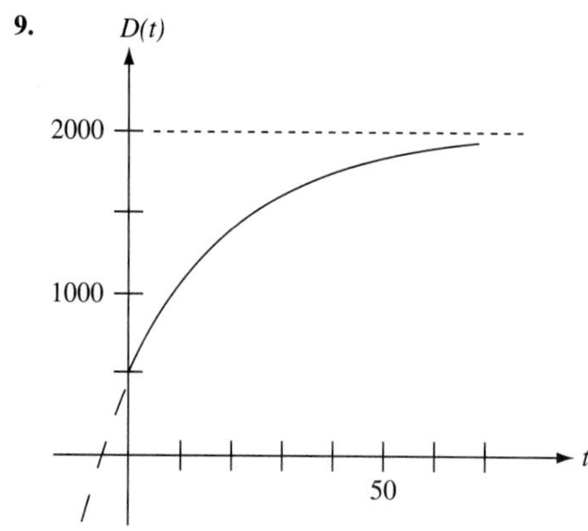

a. 573; 1177; 1548; 1925

b. 2000/month

c. 46/month

11. a. $Q(t) = 100e^{0.035t}$ **b.** 266 min **c.** $Q(t) = 1000e^{0.035t}$

13. a. 55.5 yr **b.** 14.23 billion

15. 80 mg **17.** 54 years **19.** 14,290 years

21. a. 122.3 cm **b.** 14 cm/yr **c.** 200 cm

23. a. 10 **b.** 400

c. 154 **d.** 15/day

25. 1080; 280 students/hour

27. a. bke^{-kt} **b.** a

Chapter 5 Review Exercises, page 300

1. a. and b.

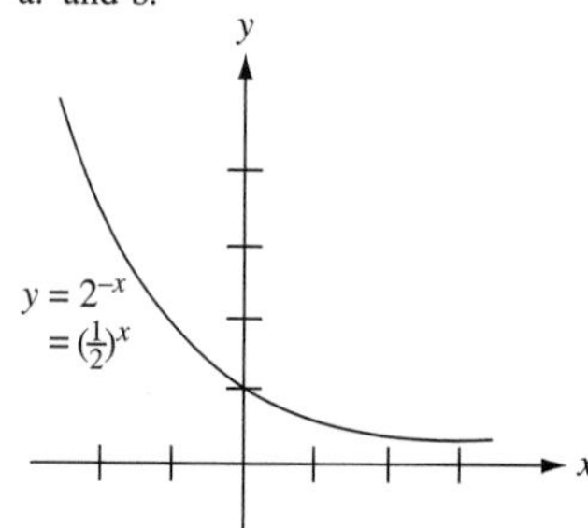

3. $\log_{16} 0.125 = -\frac{3}{4}$ **5.** $x = 2$

7. $x + 2y - z$

9.

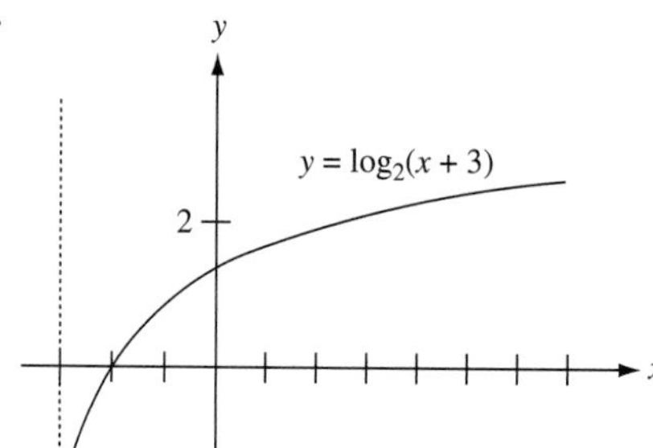

11. $(2x + 1)e^{2x}$ **13.** $\frac{1 - 4t}{2\sqrt{t}e^{2t}}$

15. $\frac{2(e^{2x} + 2)}{(1 + e^{-2x})^2}$ **17.** $(1 - 2x^2)e^{-x^2}$

19. $(x + 1)^2 e^x$ **21.** $\frac{2xe^{x^2}}{e^{x^2} + 1}$

23. $\frac{x - x\ln x + 1}{x(x + 1)^2}$ **25.** $\frac{4e^{4x}}{e^{4x} + 3}$

27. $\frac{1 + e^x(1 - x\ln x)}{x(1 + e^x)^2}$ **29.** $f'(x) = 2(\ln 10)x10^{x^2-1}$

31. $y' = \frac{4\ln 3 + 2x - \ln 3x^2}{3^x}$

33. $f'(x) = \frac{(\ln 2)x - 1}{2(\ln 2)x\sqrt{x - \log_2 x}}$

35. $y' = \frac{2x - 1}{(\ln 3)(x^2 - x + 5)}$

37. $\frac{d}{dx}(\ln y) = \frac{d}{dx(\sqrt{x}\ln x)}; \frac{dy}{dx}\frac{1}{y} = \frac{\ln x}{2\sqrt{x}} + \frac{\sqrt{x}}{x};$

$\frac{dy}{dx} = x^{\sqrt{x}}\left(\frac{\ln x}{2\sqrt{x}} + \frac{\sqrt{x}}{x}\right)$

39. $-\frac{9}{(3x + 1)^2}$ **41.** 0

43. $6x(x^2 + 2)^2(3x^3 + 2x + 1)$

45. $y = -(2x + 3)e^{-2}$

47. (0.35173, 0.70646)

49. a. 1175; 2540; 3289 **b.** 4000

51. $14,016/month **53.** 0.0004332

CHAPTER 6

6.1 Exercises, page 307

1. $\frac{5\pi}{2}$ radians

3. $-\frac{3\pi}{2}$ radians

5. **a.** III **b.** III **c.** II **d.** I

7. $\frac{5\pi}{12}$ radians

9. $\frac{8\pi}{9}$ radians

11. $\frac{7\pi}{2}$ radians

13. 120°

15. −270°

17. 220°

19.

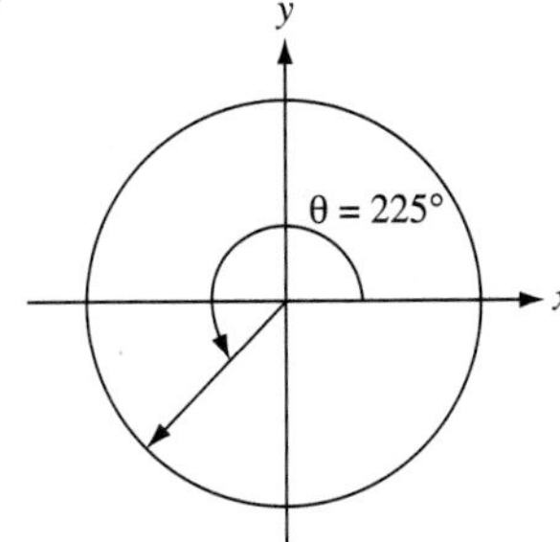

21.

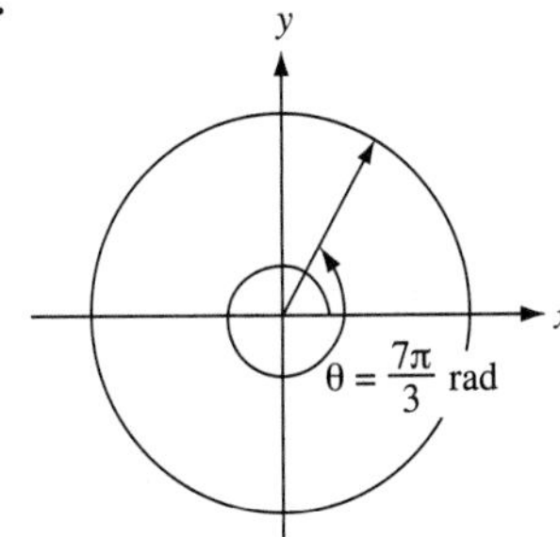

23. 150°, −210°

25. 3150°, −45°

27. True

29. True

6.2 Exercises, page 319

1. 0

3. 1

5. $\frac{\sqrt{3}}{2}$

7. $\frac{\sqrt{3}}{3}$

9. −2.6131

11. $\sin\frac{\pi}{2} = 1$, $\cos\frac{\pi}{2} = 0$, $\tan\frac{\pi}{2}$ is undefined, $\csc\frac{\pi}{2} = 1$, $\sec\frac{\pi}{2}$ is undefined, $\cot\frac{\pi}{2} = 0$

13. $\sin\frac{5\pi}{3} = -\frac{\sqrt{3}}{2}$, $\cos\frac{5\pi}{3} = \frac{1}{2}$, $\tan\frac{5\pi}{3} = -\sqrt{3}$,

$\csc\frac{5\pi}{3} = -\frac{2\sqrt{3}}{3}$, $\sec\frac{5\pi}{3} = 2$,

$\cot\frac{5\pi}{3} = -\frac{\sqrt{3}}{3}$

15. $\frac{7\pi}{6}, \frac{11\pi}{6}$

17. $\frac{5\pi}{6}, \frac{11\pi}{6}$

19. π

21. $\frac{\pi}{3}, \frac{2\pi}{3}$

23.

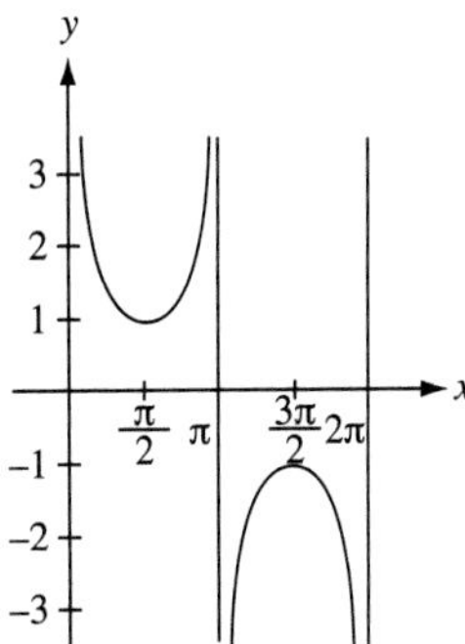

25.

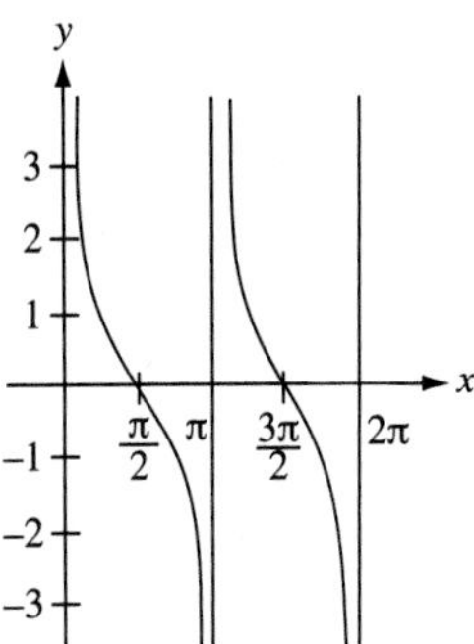

27.

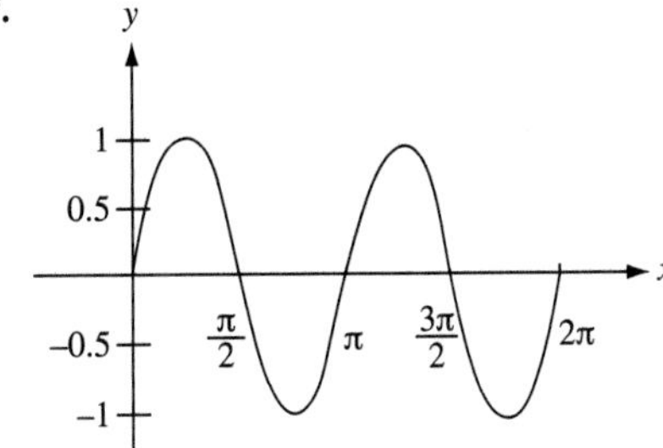

29.

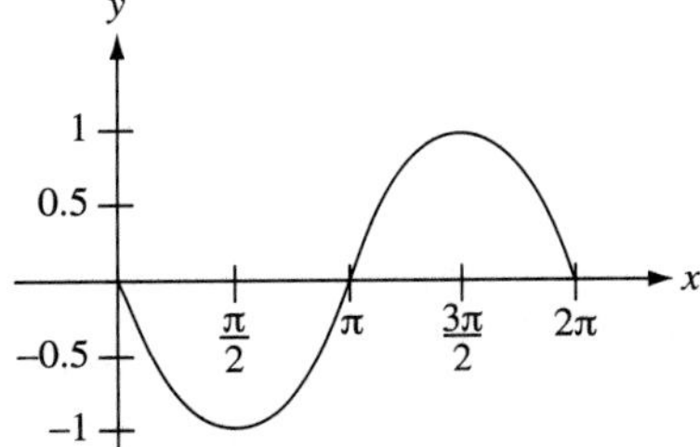

31.

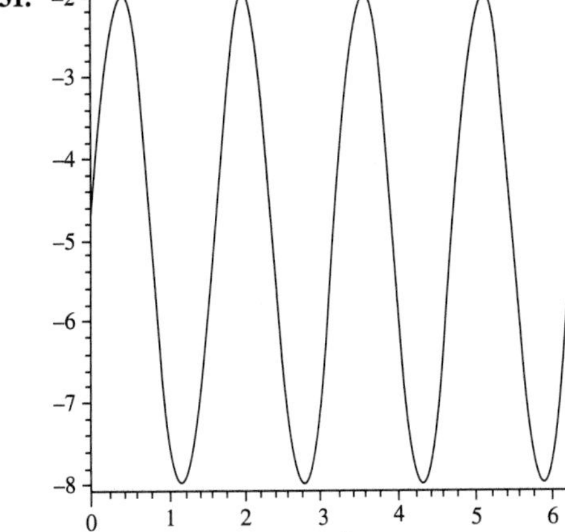

33.

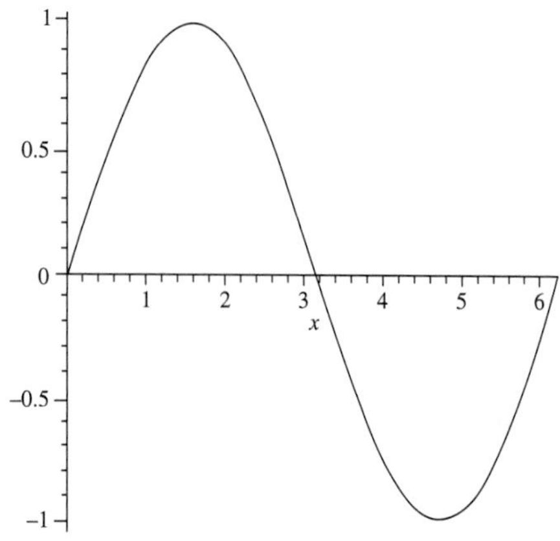

35.

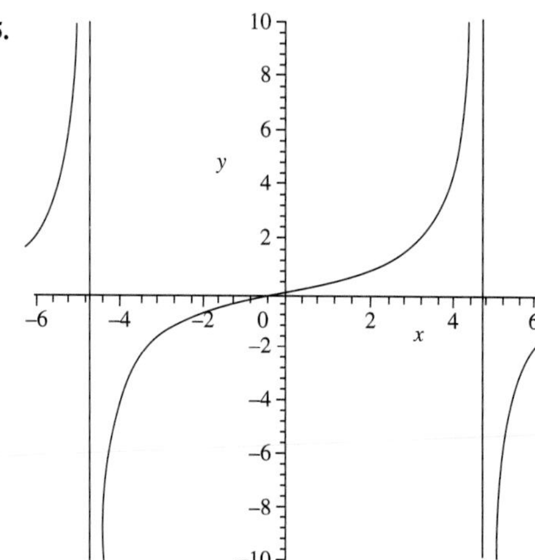

37.

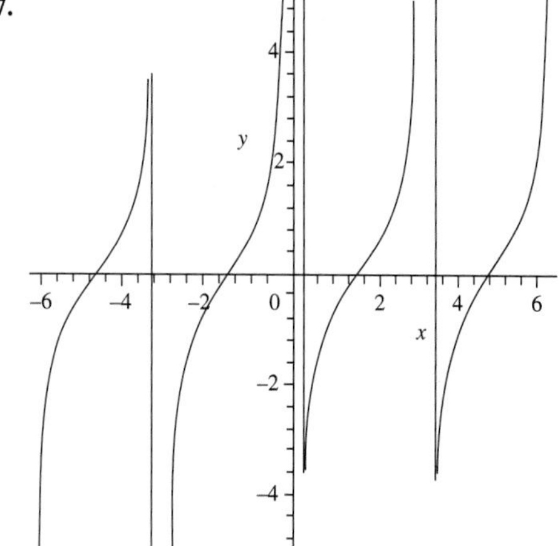

39. $2\cos^2\theta - 1$

41. $\cos\theta$

43. $\sec^2\theta$

45. $\cos\theta$

47. The results follow by using similar triangles.

49. $\sin\theta = \frac{5}{13}, \cos\theta = \frac{12}{13}, \tan\theta = \frac{5}{12}, \csc\theta = \frac{13}{5}, \sec\theta = \frac{13}{12},$
$\cot\theta = \frac{12}{5}$

51. a. $600,000; $400,000

b. $\frac{\pi}{2}(x+1) = \frac{\pi}{2}, \frac{5\pi}{2}, \ldots; x = 4n (n = 0, 1, 2, \ldots)$

$\frac{\pi}{2}(x+1) = \frac{3\pi}{2}, \frac{7\pi}{2}, \ldots; x = 4n + 2 (n = 0, 1, 2, \ldots)$

53. a. $\sin 6t = 1$ and $P(t) = 120$; $\sin 6t = -1$ and $P(t) = 80$

b. $\frac{\pi(4n+1)}{12} (n = 0, 1, 2, \ldots); \frac{\pi(4n+3)}{12} (n = 0, 1, 2, \ldots)$

55. False

57. True

6.3 Exercises, page 326

1. $\frac{\pi}{3}$

3. $\frac{2\pi}{3}$

5. $\frac{\pi}{6}$

7. $\frac{3\pi}{4}$

9. $\frac{\pi}{6}$

11. $\frac{\pi}{6}$

13. 0.644 radians

15. 1.168 radians

17. -1.384 radians

19. $\cos\left[\tan^{-1}\left(\frac{1}{\sqrt{3}}\right)\right] = \frac{\sqrt{3}}{2}$

21. $\sec\left[\cos^{-1}\left(-\frac{1}{\sqrt{2}}\right)\right] = -\sqrt{2}$

23. $\sin\theta = \frac{1}{\sqrt{5}}, \cos\theta = \frac{2}{\sqrt{5}}, \tan\theta = \frac{1}{2}, \csc\theta = \sqrt{5}, \cot\theta = 2$

25.
$\sin\theta = \frac{2\sqrt{10}}{7}, \tan\theta = \frac{2\sqrt{10}}{3}, \sec\theta = \frac{7}{3}, \csc\theta = \frac{7}{2\sqrt{10}}, \cot\theta = \frac{3}{2\sqrt{10}}$

27. False

29. False

31. True

6.4 Exercises, page 335

1. $-3\sin 3x$

3. $-2\pi\sin\pi x$

5. $2x\cos(x^2+1)$

7. $4x\sec^2 2x^2$

9. $x\cos x + \sin x$

11. $6(\cos 3x - \sin 2x)$

13. $2x(\cos 2x - x\sin 2x)$

15. $\frac{x\cos\sqrt{x^2-1}}{\sqrt{x^2-1}}$

17. $e^x \sec x(1 + \tan x)$

19. $\cos\frac{1}{x} + \frac{1}{x}\sin\frac{1}{x}$

21. $\frac{x\sin x}{(1+\cos x)^2}$

23. $\frac{\sec^2 x}{2\sqrt{\tan x}}$

25. $\frac{x\cos x - \sin x}{x^2}$

27. $2\tan x\sec^2 x$

29. $-\csc^2 x \cdot e^{\cot x}$

31. $3\sin x + (3x+1)\cos x$

33. $\frac{3\sec[\log_{10}(x^3)]\tan[\log_{10}(x^3)]}{(\ln 10)x}$

35. $-\frac{10}{(5x^2+1)\sqrt{(5x^2+1)^2-1}}$

37. $-\frac{\sqrt{e^{\cos^{-1}x}}}{2\sqrt{1-x^2}}$

39. $\frac{3}{2x\sqrt{(x^{3/2})^2-1}}$

41. $-\frac{1}{3(1+x^2)(\cot^{-1}x)^{2/3}}$

43. $\frac{dy}{dx} = -\frac{\cos x}{x \sin(xy)} - \frac{y}{x}$

45. $\frac{dy}{dx} = -\frac{\sqrt{1 - y^2}\cos(x + y)}{\sqrt{1 - y^2}\cos(x + y) - 1}$

47. $\frac{dy}{dx} = \frac{\left[1 - y - \frac{y}{\sqrt{1 - (xy)^2}}\right]}{\left[\frac{x}{\sqrt{1 - (xy)^2}} + x\right]}$

49. $y = -2x + \frac{\pi}{2}$

51. $-[\csc f(x) \cot f(x)]f'(x)$

53. $-\csc^2 f(x) f'(x)$

55. \$28 per share

57. −\$628.32 thousand/yr

59. Zero wolves/mo; −1571 caribou/mo

61. 1.26m/hr; 3.44 m

Chapter 6 Review Exercises, page 338

1. $\frac{2\pi}{3}$ radians

3. $-\frac{5\pi}{4}$ radians

5. −450°

7. $\frac{\pi}{3}$ or $\frac{5\pi}{3}$

9.

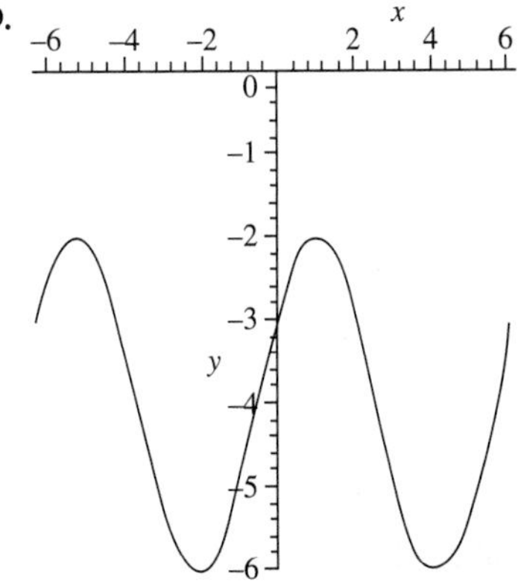

11. $\theta = \frac{\pi}{6}$

13. $\theta = \frac{5\pi}{6}$

15. $\cos\theta = \frac{4}{5}, \tan\theta = \frac{3}{4}, \sec\theta = \frac{5}{4}, \csc\theta = \frac{5}{3}, \cot\theta = \frac{4}{3}$

17. $3 \cos 3x$

19. $2 \cos x + 6 \sin 2x$

21. $e^{-x}(3 \sec^2 3x - \tan 3x)$

23. $4 \cos 2x$, or $4(\cos^2 x - \sin^2 x)$

25. $\frac{(\cot x - 1)\sec^2 x - (1 - \tan x)\csc^2 x}{(1 - \cot x)^2}$

27. $\cos(\sin x)\cos x$

29. $f'(x) = -\sin(3^x + \log_3 x)\left[3^x(\ln 3) + \frac{1}{(\ln 3)x}\right]$

31. $f'(x) = -\frac{3x^2}{\sqrt{1 - x^6}}$

33. $f'(x) = \frac{\tan^{-1}(e^x)}{x} + \frac{e^x \ln x}{1 + e^{2x}}$

35. $y = 4x + 1 - \pi$

37. −5 people/month

39. −6.5 foxes/month; −113.4 rabbits/month

CHAPTER 7

7.1 Exercises, page 350

1. Decreasing on $(-\infty, 0)$ and increasing on $(0, \infty)$

3. Increasing on $(-\infty, -1) \cup (1, \infty)$ and decreasing on $(-1, 1)$

5. Decreasing on $(-\infty, 0) \cup (2, \infty)$ and increasing on $(0, 2)$

7. Decreasing on $(-\infty, -1) \cup (1, \infty)$ and increasing on $(-1, 1)$

9. Increasing on $(20.2, 20.6) \cup (21.7, 21.8)$, constant on $(19.6, 20.2) \cup (20.6, 21.1)$, and decreasing on $(21.1, 21.7) \cup (21.8, 22.7)$

11. Increasing on $(-\infty, 0)$

13. Decreasing on $\left(-\infty, \frac{3}{2}\right)$ and increasing on $\left(\frac{3}{2}, \infty\right)$

15. Decreasing on $(-\infty, -\sqrt{3}/3) \cup (\sqrt{3}/3, \infty)$ and increasing on $(-\sqrt{3}/3, \sqrt{3}/3)$

17. Increasing on $(-\infty, -2) \cup (0, \infty)$ and decreasing on $(-2, 0)$

19. Increasing on $(-\infty, 3) \cup (3, \infty)$

21. Decreasing on $(-\infty, 0) \cup (0, 3)$ and increasing on $(3, \infty)$

23. Decreasing on $(-\infty, 2) \cup (2, \infty)$

25. Decreasing on $(-\infty, 1) \cup (1, \infty)$

27. Increasing on $(-\infty, 0) \cup (0, \infty)$

29. Increasing on $(-1, \infty)$

31. Increasing on $(-4, 0)$; decreasing on $(0, 4)$

33. Increasing on $(-\infty, 0) \cup (0, \infty)$

35. Increasing on $(-\infty, 1)$; decreasing on $(1, \infty)$

37. Increasing everywhere

39. Increasing on $(-\infty, 0)$ and decreasing on $(0, \infty)$

41. Increasing on $(-\infty, 3)$ and decreasing on $(3, \infty)$

43. Decreasing on $(-\infty, 0)$ and increasing on $(0, \infty)$

45. Increasing on $(\sqrt{10}, 0)$ and decreasing on $(0, \sqrt{10})$

47. Increasing on $\left(0, \frac{\pi}{2}\right) \cup \left(\frac{3\pi}{2}, 2\pi\right)$ and decreasing on $\left(\frac{\pi}{2}, \frac{3\pi}{2}\right)$

49. Increasing on $\left(0, \frac{\pi}{4}\right) \cup \left(\frac{3\pi}{4}, \frac{5\pi}{4}\right) \cup \left(\frac{7\pi}{4}, 2\pi\right)$ and decreasing on $\left(\frac{\pi}{4}, \frac{3\pi}{4}\right) \cup \left(\frac{5\pi}{4}, \frac{7\pi}{4}\right)$

51. Increasing on $(-\infty, \infty)$

53. Increasing on $(-\infty, 0)$ and decreasing on $(0, \infty)$

55. Relative maximum: $f(0) = 1$; relative minima: $f(-1) = 0$ and $f(1) = 0$

57. Relative maximum: $f(-1) = 2$; relative minimum: $f(1) = -2$

59. Relative maximum: $f(1) = 3$; relative minimum: $f(2) = 2$

61. Relative minimum: $f(0) = 2$

63. a

65. d

67. Relative minimum: $f(2) = -4$

69. Relative maximum: $f(3) = 15$

71. None

73. Relative maximum: $g(0) = 4$; relative minimum: $g(2) = 0$

75. Relative maximum: $f(0) = 0$; relative minima: $f(-1) = -\frac{1}{2}$ and $f(1) = -\frac{1}{2}$

77. Relative minimum: $F(3) = -5$; relative maximum: $F(-1) = \frac{17}{3}$

79. Relative minimum: $g(3) = -19$

81. None

83. Relative maximum: $f(-3) = -4$; relative minimum: $f(3) = 8$

85. Relative maximum: $f(1) = \frac{1}{2}$; relative minimum: $f(-1) = \frac{1}{2}$

87. Relative maximum: $f(0) = 0$

89. Relative maximum: $f(1) = 1$; relative minimum: $f(0) = 0$

91. Relative maximum: $f(2) = 2$; relative minimum: $f(0) = 0$

93. Relative minimum: $f(0) = 0$

95. Relative maximum: $f(0) = 0$; relative minimum: $f(2) = 2$

97. Critical points when $\cos(x) = \frac{1}{2}$ or $\cos(x) = -1$; no relative extrema

99. Relative maximum: $f(0) = 0$

101. Relative minimum: $f(1) = 0$

103.

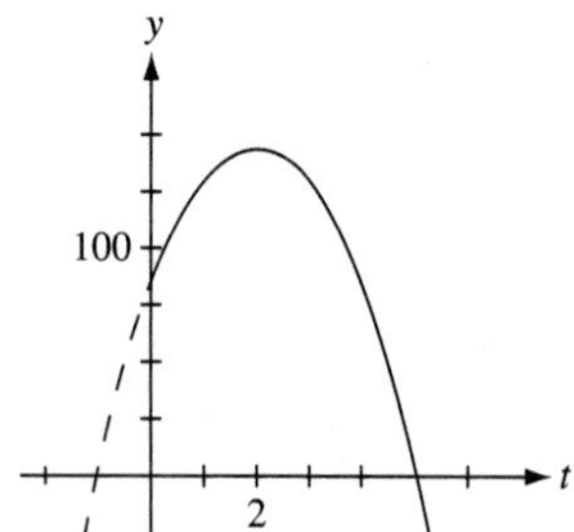

Rising in the time interval (0, 2); falling in the time interval (2, 5); when $t = 5$ sec

105. Decreases from 6 AM to 7 AM, and then picks up from 7 AM to 10 AM

107. If $m > 0, f'(x) > 0$ for all x and f is increasing; if $m < 0, f'(x) < 0$ for all x and f is decreasing; if $m = 0, f'(x) = 0$ for all x and f is a constant function.

109. $f'(x) = 3x^2 + 1$ is continuous on $(-\infty, \infty)$ and is always greater than or equal to 1, so f has no critical points in $(-\infty, \infty)$.

111. **a.** $-2x$ if $x \neq 0$

b. No

113. If $a < 0$, relative maximum: $x = -b/2a$

If $a > 0$, relative minimum: $x = -b/2a$

115. **a.** $f'(x) = 3x^2 + 1$ so $f'(x) > 1$ on $(0, 1)$

b. $f(0) = -1$ and $f(1) = 1$; the Intermediate Value Theorem guarantees that there is at least one root of $f(x) = 0$ in $(0, 1)$. Since f is increasing on $(0, 1)$, the graph of f can cross the x-axis at only one point in $(0, 1)$.

117. Increasing on (0, 4000) and decreasing on $(4000, \infty)$

119. Increasing from 1995 to 2035 and decreasing from 2035 to 2045

121. **a.** Increasing on (0, 6)

b. Increasing throughout the years from 1999 through 2005

123. **a.** Increasing on (0, 6)

b. Always increasing from 1997 through 2003

125. The percent of the population age 65 and over afflicted by the disease increases with age.

127. Increasing from 7 AM to 10 AM and decreasing from 10 AM to 2 PM

129. As the age of the driver increases from 16 to 27 years old, the predicted crash fatalities drop.

131. False

133. False

135. False

7.2 Exercises, page 365

1. Concave downward on $(-\infty, 0)$ and concave upward on $(0, \infty)$; inflection point: (0, 0)

3. Concave downward on $(-\infty, 0) \cup (0, \infty)$

5. Concave upward on $(-\infty, 0) \cup (1, \infty)$ and concave downward on (0, 1); inflection points: (0, 0) and (1, −1)

7. Concave downward on $(-\infty, -2) \cup (-2, 2) \cup (2, \infty)$

9. a

11. b

13. $f'(x) = 8x - 12; f''(x) = 8$ so $f''(x) > 0$ everywhere

15. $f'(x) = -\frac{4}{x^5}; f''(x) = \frac{20}{x^4} > 0$ for all values of x in $(-\infty, 0) \cup (0, \infty)$

17. $f'(x) = e^x; e^x > 0$ for all x

19. Concave upward on $(-\infty, \infty)$

21. Concave upward on $(-\infty, 0)$; concave upward on $(0, \infty)$

23. Concave upward on $(-\infty, 0) \cup (0, \infty)$; concave downward on (0, 3)

25. Concave downward on $(-\infty, 0) \cup (0, \infty)$

27. Concave downward on $(-\infty, 4)$

29. Concave downward on $(-\infty, 2)$; concave upward on $(2, \infty)$

31. Concave upward on $(-\infty, -\sqrt{6/3}) \cup (\sqrt{6/3}, \infty)$; concave downward on $(-\sqrt{6/3}, \sqrt{6/3})$

33. Concave downward on $(-\infty, 1)$; concave upward on $(1, \infty)$

35. Concave upward on $(-\infty, 0) \cup (0, \infty)$

37. Concave upward on $(-\infty, 2)$; concave downward on $(2, \infty)$

39. $f'(s) = -e^{-s}; f''(s) = e^{-s}$; any power of e is a positive number so the function is concave up everywhere.

41. $g'(x) = \frac{1}{x}; g''(x) = \frac{-1}{x^2}$; the second derivative is negative everywhere the function exists so the function is concave down on $(0, \infty)$.

43. $f'(x) = \cos(x); f''(x) = -\sin(x)$; $\sin(x)$ is concave up on all intervals of the form $((2n - 1)\pi, 2n\pi)$ and concave down on all intervals of the form $(2n\pi, (2n + 1)\pi)$, where n is any whole number.

45. $f'(x) = 2\cos(2x); f''(x) = -4\sin(2x)$; $\sin(2x)$ is concave up on all intervals of the form $\left(\frac{\pi}{2} + n\pi, (n + 1)\pi\right)$ and concave down on all intervals of the form $\left(n\pi, \frac{\pi}{2} + n\pi\right)$, where n is any whole number.

47. $(0, -2)$

49. $(1, -15)$

51. $(0, 1)$ and $\left(\frac{2}{3}, \frac{11}{27}\right)$

53. $(0, 0)$

55. $(1, 2)$

57. $(-\sqrt{3}/3, 3/2)$ and $(\sqrt{3}/3, 3/2)$

59. $\left(\frac{1}{2}, \frac{1}{2}e^{-2}\right)$

61. $\left(\frac{\pi}{4} + n\pi, 0\right)$

63. Relative maximum: $f(1) = 5$

65. None

67. Relative maximum: $f(-1) = -\frac{22}{3}$ f; relative minimum: $f(5) = -\frac{130}{3}$

69. Relative maximum: $f(-3) = -6$; relative minimum: $f(3) = 6$

71. None

73. Relative minimum: $f(-2) = 12$

75. Relative maximum: $g(1) = \frac{1}{2}$; relative minimum: $g(-1) = -\frac{1}{2}$

77. Relative maximum: $f(0) = 0$; relative minimum: $f\left(\frac{4}{3}\right) = \frac{256}{27}$

79. Relative maximum: $h(4) = 4$

81. Relative maximum: $g\left(\frac{\pi}{4} + \frac{n\pi}{2}\right) = \left(\frac{\pi}{4} + n\pi\right)$; relative minimum: $g\left(\frac{\pi}{4} + \frac{n\pi}{2}\right) = \frac{3\pi}{4} + n\pi$

83.

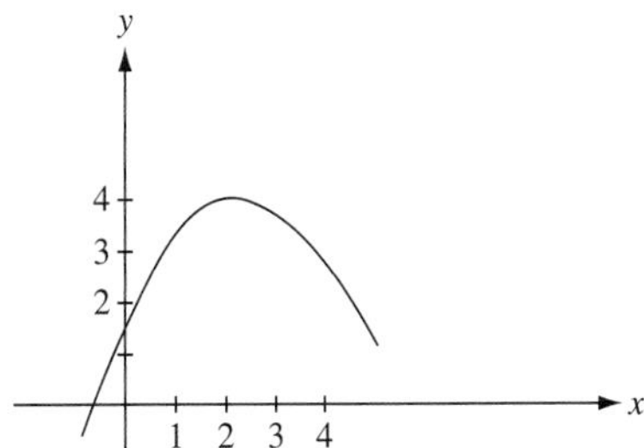

85.

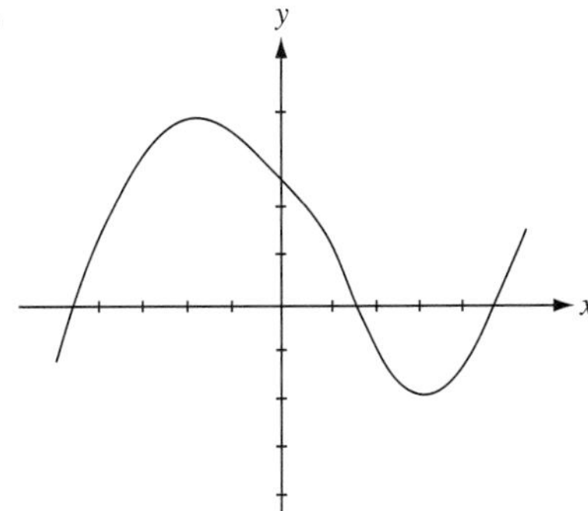

87.

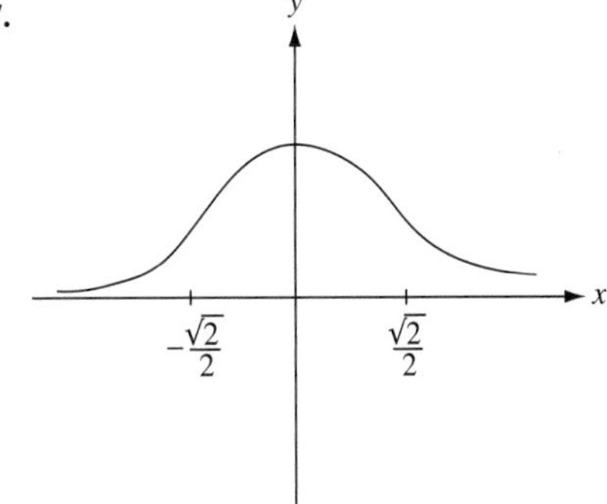

89. $f(t)$ increases at an increasing rate until the water level reaches the middle of the vase at which time (corresponding to the inflection point) $f(t)$ is increasing at the fastest rate. After that, $f(t)$ increases at a decreasing rate until the vase is filled.

91. $f'(x) = 2ax + b; f''(x) = 2a; f''(x) > 0$ if $a > 0$ and the parabola opens upward. If $a < 0, f''(x) < 0$ and the parabola opens downward.

93. The rumour spreads with increasing speed initially. The rate at which the rumour is spread reaches a maximum at the time corresponding to the t-coordinate of the point P on the curve. Therefore, the speed at which the rumour is spread decreases.

95. **a.** $D_1'(t) > 0, D_2'(t) > 0, D_1''(t) > 0, D_2''(t) < 0$ on $(0, 12)$

b. With or without the proposed promotional campaign, the deposits will increase, but with the promotion, the deposits will increase at an increasing rate whereas without the promotion, the deposits will increase at a decreasing rate.

97. **a.** $(150, 28{,}850)$

b. Total annual revenue is maximized at \$28,850,000 if the level of advertising expenditure is \$100,000.

99. $(3, 116)$; after declining the first 3 years, the growth rate of the company's profit is once again on the rise.

101. $(266.67, 11{,}874.08)$

103. (18, 6770); 484 m/sec

105. **a.** The population is always increasing.

b. $P''(t) = 0$; $t = 0.67$ is an inflection point of the graph of P

107. **a.** $f(t) = \dfrac{2t}{1 + t^4}$

b. 0.76 seconds

c. 1.14 radians/sec

109. 0.693 weeks

111. 303

113. False

115. True

7.3 Exercises, page 380

1. Horizontal asymptote: $y = 0$

3. Horizontal asymptote: $y = 0$; vertical asymptote: $x = 0$

5. Horizontal asymptote: $y = 0$; vertical asymptotes: $x = -1$ and $x = 1$

7. Horizontal asymptote: $y = 3$; vertical asymptote: $x = 0$

9. Horizontal asymptotes: $y = 1$ and $y = -1$

11. Horizontal asymptote: $y = 0$; vertical asymptote: $x = 0$

13. Horizontal asymptote: $y = 0$; vertical asymptote: $x = 0$

15. Horizontal asymptote: $y = 1$; vertical asymptote: $x = -1$

17. None

19. Horizontal asymptote: $y = 1$; vertical asymptotes: $t = -3$ and $t = 3$

21. Horizontal asymptote: $y = 0$; vertical asymptotes: $x = -2$ and $x = 3$

23. Horizontal asymptote: $y = 2$; vertical asymptote: $t = 2$

25. Horizontal asymptote: $y = 1$; vertical asymptotes: $x = -2$ and $x = 2$

27. None

29. f is the derivative function of the function g.

31.

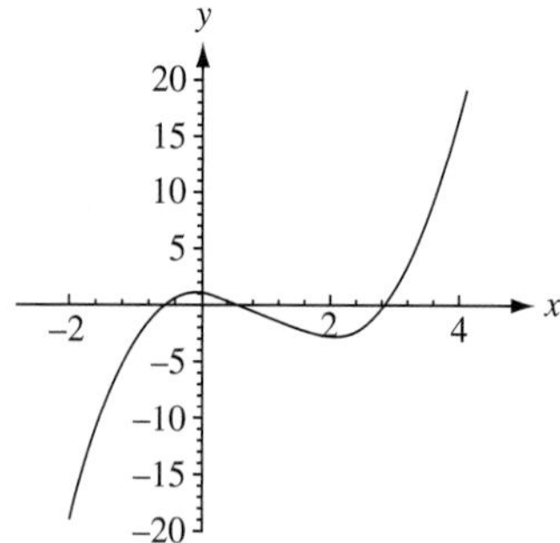

33.

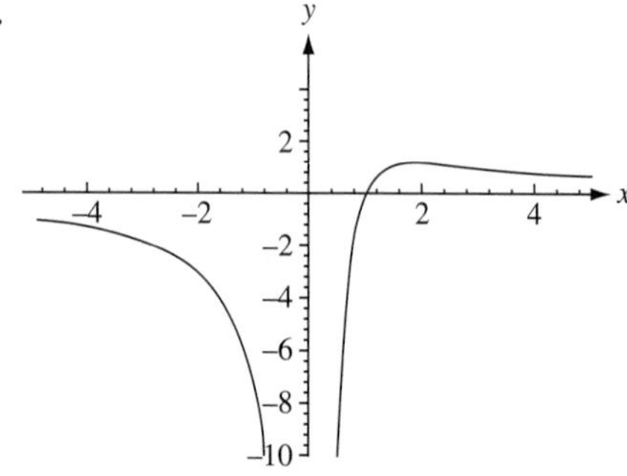

35.

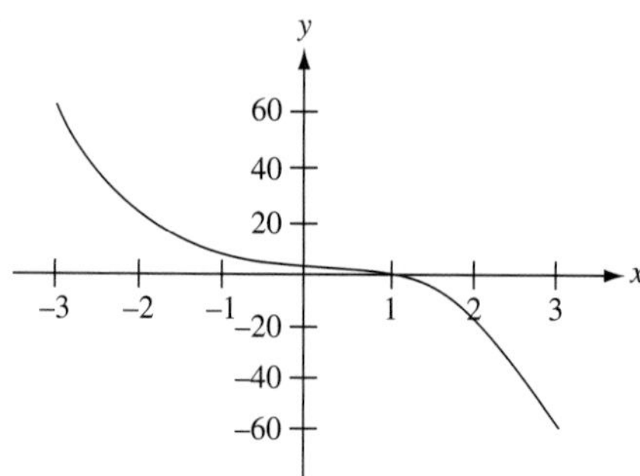

37.

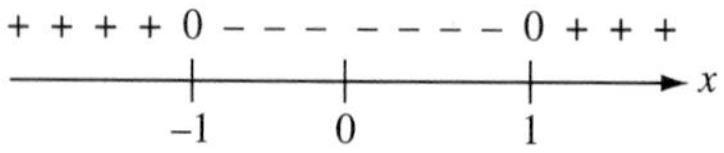

+ + + + 0 – – – – – – – 0 + + +

−1 0 1 x

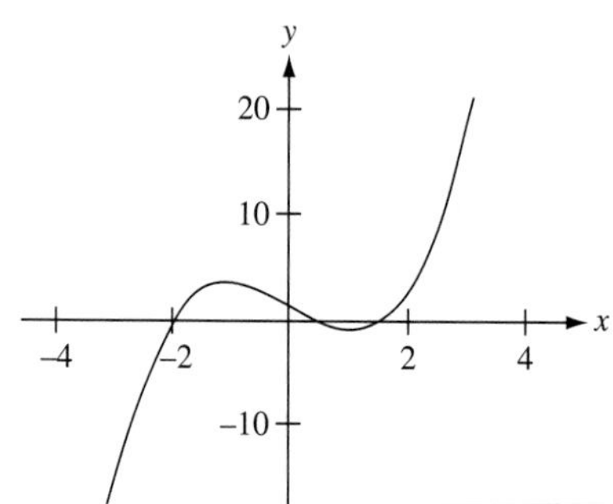

39.

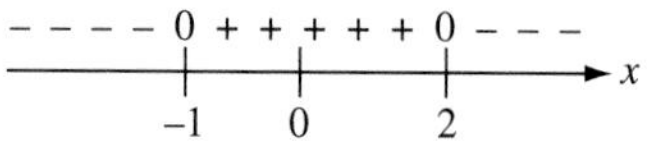

– – – – 0 + + + + + 0 – – –

−1 0 2 x

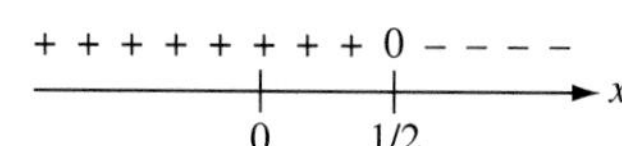

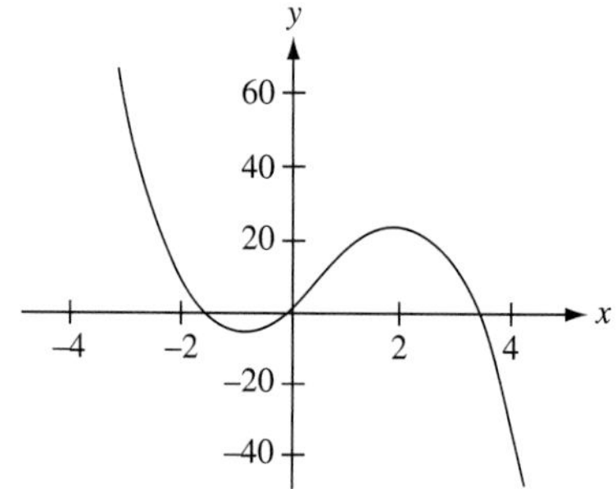

41.

– – – 0 + + 0 – – – 0 + + +

−1 0 2

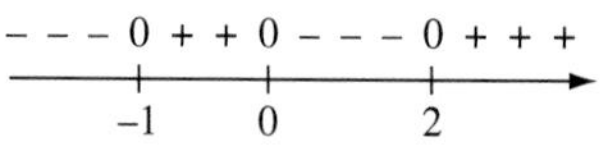

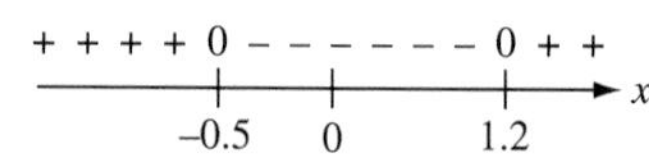

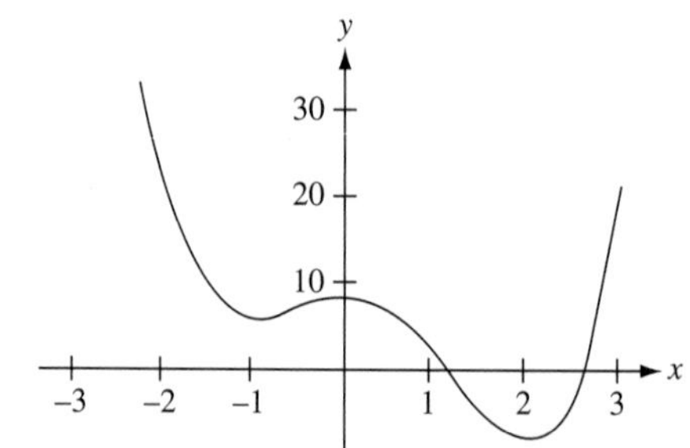

43.

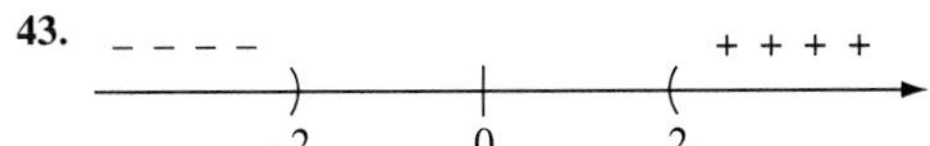

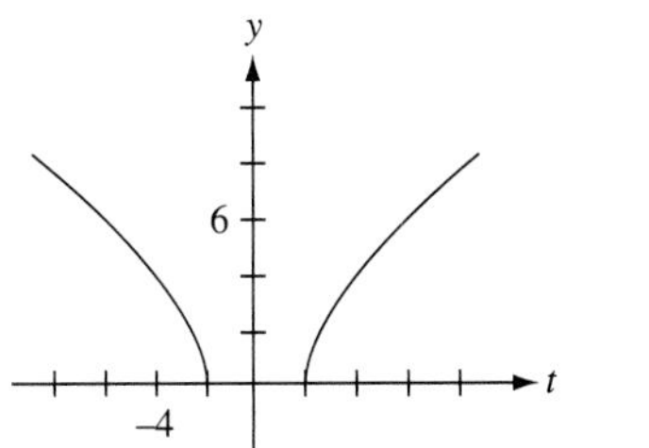

45.

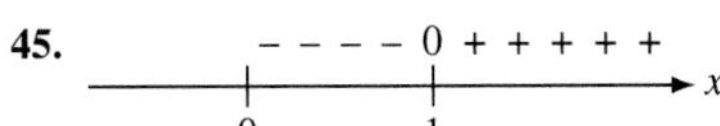

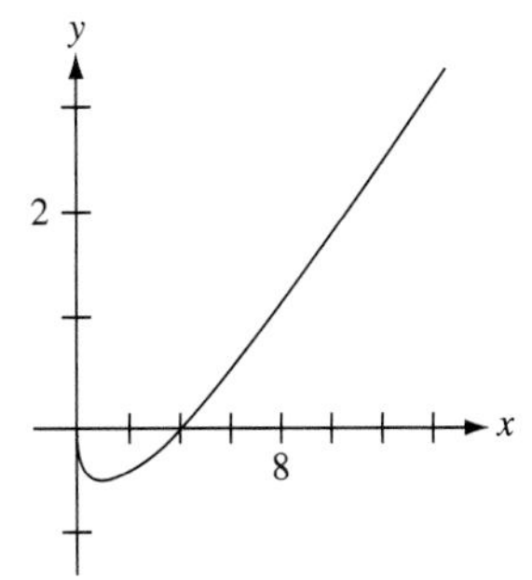

47.

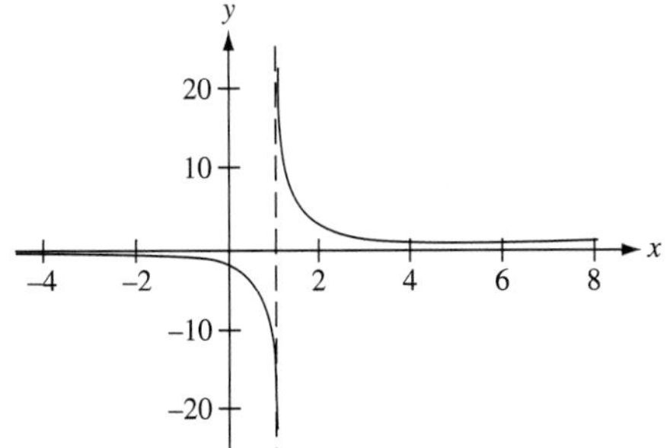

49.

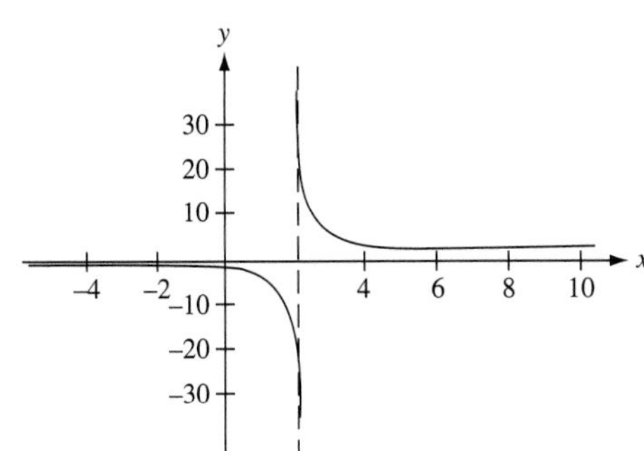

51.

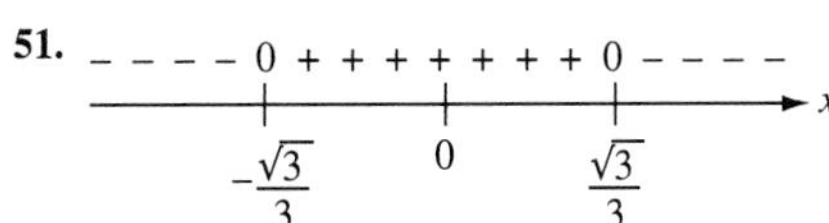

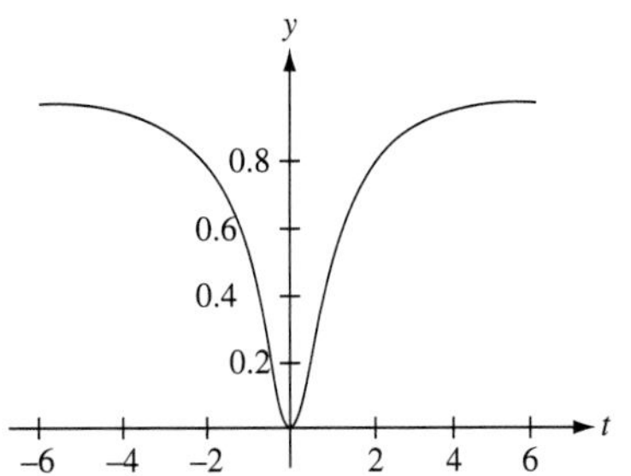

53.

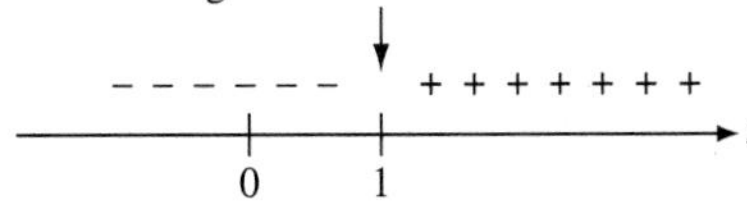

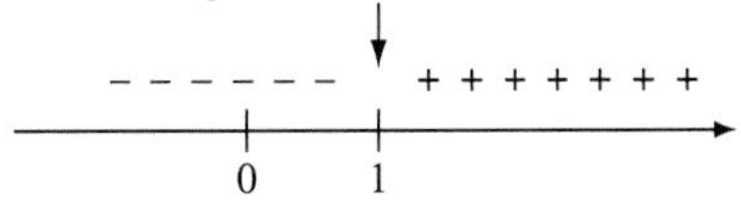

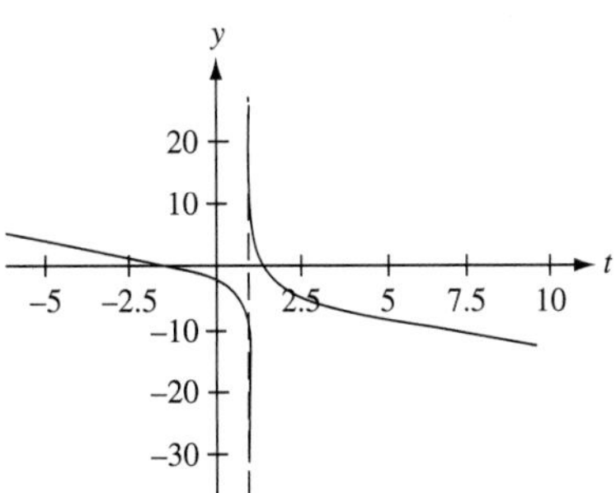

55.

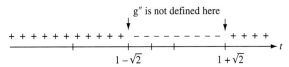

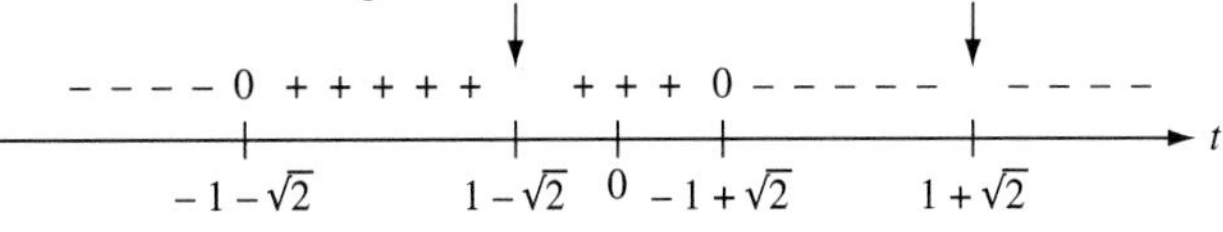

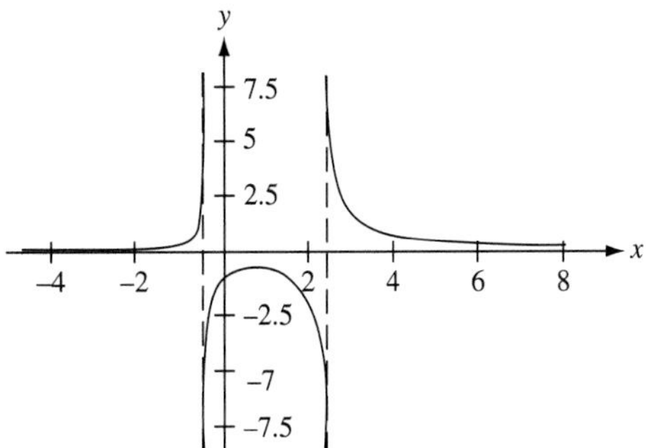

57.

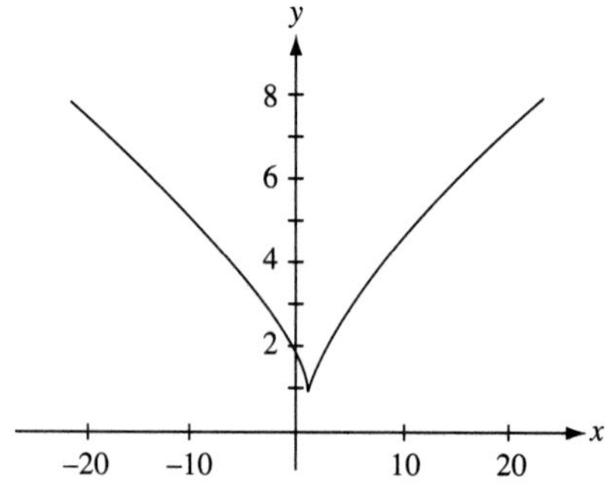

59.

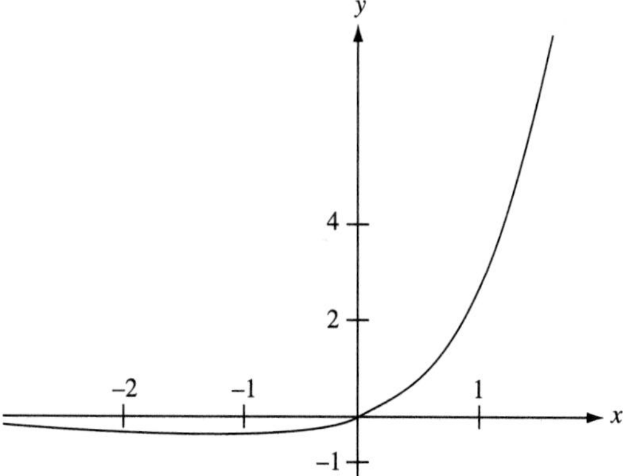

61.

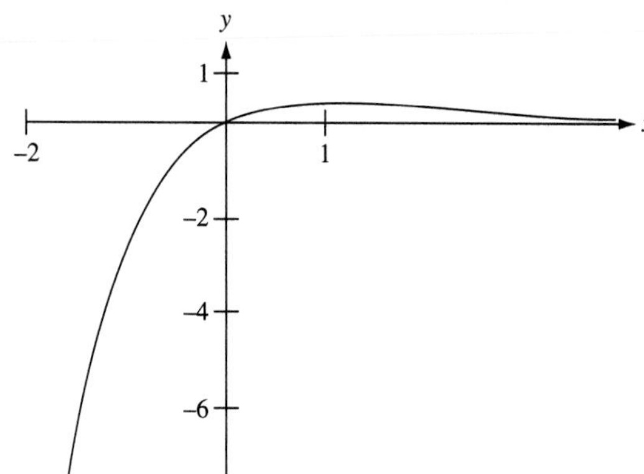

63.

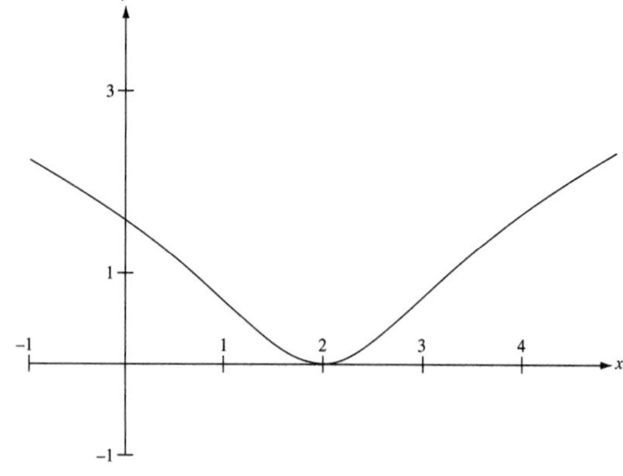

65.

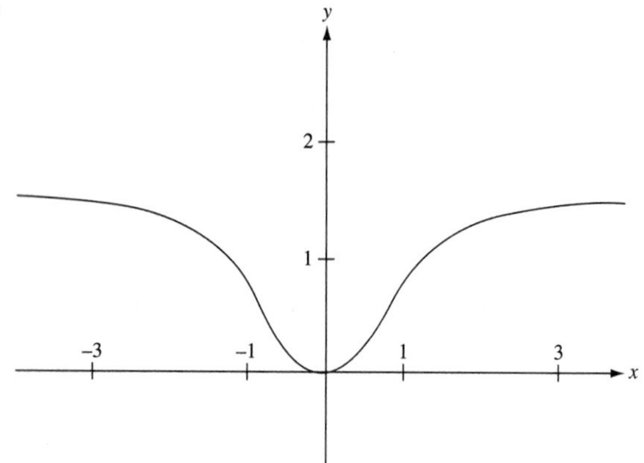

67.

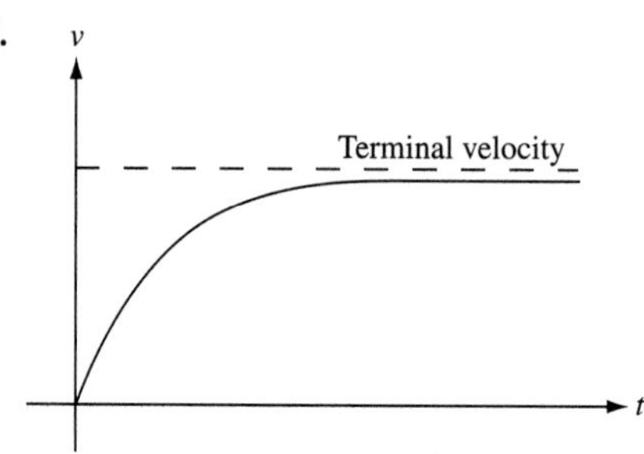

69.

+ + + + + +

0 8 t

+ + + + 0 − − −

0 4 t

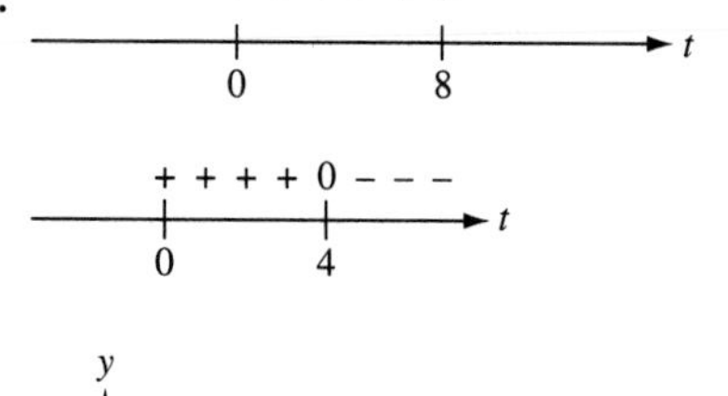

71.

+ + + 0 − − −

0 $\frac{2\sqrt{3}}{3}$ x

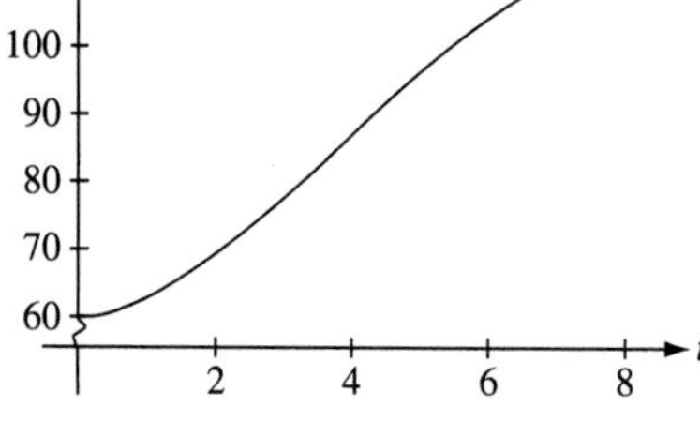

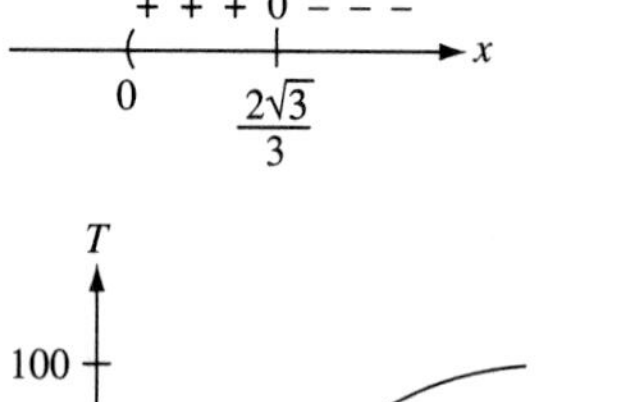

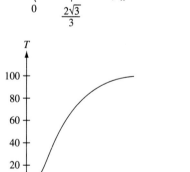

73. a. $x = 100$

b. No

75. a. $\lim_{x\to\infty} = a$

b. The initial speed of the reaction approaches a moles/litre/sec as the amount of substrate becomes arbitrarily large.

77.

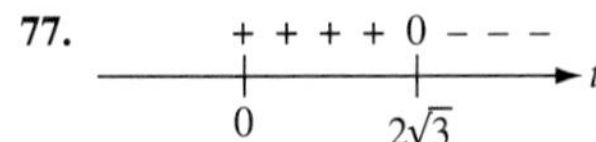

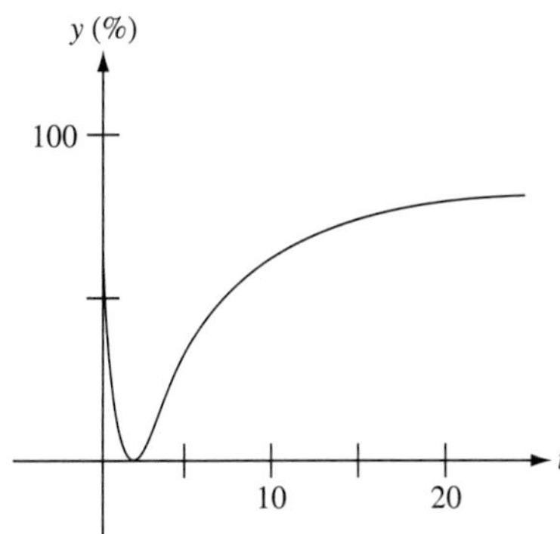

7.4 Exercises, page 394

1. None

3. Absolute minimum value: 0

5. Absolute maximum value: 3; absolute minimum value: -2

7. Absolute maximum value: 3; absolute minimum value: $-\frac{27}{16}$

9. Absolute minimum value: $-\frac{41}{8}$

11. No absolute extrema

13. Absolute maximum value: 1

15. Absolute maximum value: 5; absolute minimum value: -4

17. Absolute maximum value: 10; absolute minimum value: 1

19. Absolute maximum value: 19; absolute minimum value: -1

21. Absolute maximum value: 16; absolute minimum value: -1

23. Absolute maximum value: 3; absolute minimum value: $\frac{5}{3}$

25. Absolute maximum value: $\frac{37}{3}$; absolute minimum value: 5

27. Absolute maximum value: ≈ 1.04; absolute minimum value: -1.5

29. No absolute extrema

31. Absolute maximum value: 1; absolute minimum value: 0

33. Absolute maximum value: 0; absolute minimum value: -3

35. Absolute maximum value: $\sqrt{2}/4 \approx 0.35$; absolute minimum value: $-\frac{1}{3}$

37. Absolute maximum value: $\sqrt{2}/2$; absolute minimum value: $-\sqrt{2}/2$

39. Absolute maximum value: $f(0) = 1$; absolute minimum value: $f(20) \approx 2.06 \times 10^{-9}$

41. Absolute maximum value: $f(0) = 1$; absolute minimum value: $f(3)$

43. Absolute maximum value: $h(6) \approx 4.872$; absolute minimum value: $h(0) = 0$

45. Absolute maximum value: $f(0) = 2$; absolute minimum value: $f(3) \approx 0.357$

47. Absolute maximum value: $f(2) \approx 1.386$; absolute minimum value: $f\left(\frac{1}{e}\right) = -\frac{1}{e}$

49. Absolute maximum value: 1; absolute minimum value: -1

51. Absolute maximum value: $\cos\left(\frac{1}{26}\right)$ at $x = \pm 5$; absolute minimum value: $\cos(1)$ at $x = 0$

53. Absolute maximum value: $f(1) \approx 0.4636$; absolute minimum value: $f\left(\frac{1}{41}\right) = 0.0244$

55. Absolute maximum value: $e^{1/\sqrt{2}}$; absolute minimum value: $e^{-1/\sqrt{2}}$

57. Absolute maximum of g: $x = 0$; global maximum of f: $x = 0$ with a value of $f(1) = 0$

59. Absolute maximum value: $f(3) \cong 0.803$; absolute minimum value: $f(1) \cong -0.933$

61. (1) $\lim_{x\to-\infty} f(x) = -\infty$ and $\lim_{x\to\infty} f(x) = \infty$ and so f has no absolute maximum or absolute minimum.

(2) $\lim_{x\to-\infty} f(x) = \lim_{x\to\infty} f(x) = \infty$ so f cannot have an absolute minimum.

63. f has neither an absolute minimum value or an absolute maximum value.

65. The maximum altitude is 124.7 metres.

67. If 88 units are rented out, the maximum monthly profit realizable is \$27,440.

69. The largest possible profit is \$3600/day.

71. A level of production of 6000 rackets/day will yield a maximum profit.

73. A level of production of 3,333 pagers per week will yield a maximum profit of \$165,185 per week.

75. a. $\overline{C}(x) = \frac{C(x)}{x} = 0.0025x + 80 + \frac{10{,}000}{x}$.

b. $x = 2000$ yields a minimum.

c. $x = 2000$.

d. It appears that we can solve the problem in two ways.

77. The revenue is maximized by producing approximately 533 dresses.

79. By the Second Derivative Test, the critical point does give a maximum revenue.

81. $f(t)$ is maximized when $t = 23.6811$.

83. a. The oxygen content is the lowest 2 days after the organic waste has been dumped into the pond.

b. The rate of oxygen regeneration is greatest 3.5 days after the organic waste has been dumped into the pond.

85. v has an absolute maximum at $r = 0$. So the velocity is greatest along the central axis.

87. The fish must swim at $(3/2)u$ m/sec in order to minimize the total energy expended.

89. $kD = 2$ provides the relative (absolute) maximum.

91. The largest number of weeds are present at 6 weeks, and they number $P(6) = 296e^{-2} \cong 54$.

93. The minimum is 92 while the maximum is 148.

95. The maximum number of hectares covered is $A(\sqrt{2}) = 9.55$ acres at $t = 1.41$ weeks.

97. False.

99. True.

7.5 Exercises, page 407

1. 750 m × 1500 m; 1,125,000 m^2

3. $10\sqrt{2}$ m × $40\sqrt{2}$ m

5. $\frac{16}{3}$ cm × $\frac{16}{3}$ cm × $\frac{4}{3}$ cm

7. 5.04 cm × 5.04 cm × 5.04 cm

9. 42 cm × 42 cm × 84 cm; 148,176 cm^3

11. $r = \frac{84}{\pi}$ cm; $l = 84$ cm; $\frac{592,704}{\pi}$ cm^3

13. $\sqrt[3]{10/9}$ m × $\frac{3}{2}\sqrt[3]{10/9}$ m × $\sqrt[3]{4/15}$ m

15. $w \approx 13.86$ cm; $h \approx 19.60$ cm

17. $x = 867.5$

19. 97.97 km/h

21. 250; \$62,500; \$250

23. \$3.50/bottle

25. 44,445 bottles

27. 20 trees/acre

7.6 Exercises, page 421

1. Yes

3. Yes

5. Yes

7. Yes

9. Yes

11. $\frac{1}{2}$

13. $\frac{8}{0}$; the limit does not exist.

15. 0

17. 0

19. $\frac{1}{3}$

21. 0

23. The limit goes to zero without needed L'Hôpital's rule.

25. 1

27. 0

29. 1

31. 0

33. $-\frac{1}{6}$

35. 1

37. 0

39. 1; since this limit is a nonzero constant, the two functions grow at the same rate.

41. $\lim_{x\to\infty}\left(\frac{e}{2}\right)^x$; this limit diverges to infinity, therefore e^x grows faster.

43. 0; $g(x)$ grows faster.

45. **a.** Increasing A has no effect on the growth rate of the function; the two functions grow at the same rate.

b. The function with the higher exponent grows more quickly.

47. e^{ax} grows faster than x^n

49. 0; the purple loosestrife grows faster and will out-compete the native grass.

51. ∞; the purple loosestrife can be controlled so that the native grasses out-compete it by application of the flower-feeding weevil.

Chapter 7 Review Exercises, page 424

1. **a.** f is increasing on $(-\infty, 1) \cup (1, \infty)$.

b. No relative extrema

c. Concave down on $(-\infty, 1)$; concave up on $(1, \infty)$

d. $\left(1, -\frac{17}{3}\right)$

3. **a.** f is increasing on $(-1, 0) \cup (1, \infty)$and decreasing on $(-\infty, -1) \cup (0, 1)$.

b. Relative maximum value: 0; relative minimum value: -1

c. Concave up on $\left(-\infty, -\frac{\sqrt{3}}{3}\right) \cup \left(\frac{\sqrt{3}}{3}, \infty\right)$; concave down on $\left(-\frac{\sqrt{3}}{3}, \frac{\sqrt{3}}{3}\right)$

d. $\left(-\frac{\sqrt{3}}{3}, -\frac{5}{9}\right)$; $\left(\frac{\sqrt{3}}{3}, -\frac{5}{9}\right)$

5. **a.** f is increasing on $(-\infty, 0) \cup (2, \infty)$.

b. Relative maximum value: 0; relative minimum value: 4

c. Concave up on $(1, \infty)$; concave down on $(-\infty, 1)$

d. None

7. **a.** f is decreasing on $(-\infty, 1) \cup (1, \infty)$.

b. No relative extrema

c. Concave down on $(-\infty, 1)$; concave up on $(1, \infty)$

d. $(1, 0)$

9. **a.** f is increasing on $(-\infty, -1) \cup (-1, \infty)$.

b. No relative extrema

c. Concave down on $(-1, \infty)$; concave up on $(-\infty, -1)$

d. None

11. **a.** f is increasing on $(-\infty, 0) \cup (0, 6)$ and decreasing on $(6, \infty)$

b. Relative maximum value: 6

c. Concave up on $(0, 6 - 2\sqrt{3}) \cup (6 + 2\sqrt{3}, \infty)$; concave down on $(-\infty, 0) \cup (6 - 2\sqrt{3}, 6 + 2\sqrt{3})$

d. 0

13. **a.** f is increasing everywhere.

b. No relative extrema

c. Concave down on $(0, \infty)$; concave up on $(-\infty, 0)$

d. 0

15. **a.** $t = \dfrac{\ln\left(\dfrac{b}{a}\right)}{b - a}$

b. $\lim_{t\to\infty} C(t) = \dfrac{k}{b - a}$

17. **a.** $Q(t) = ACke^{-Ae^{-kt}-kt}$

b. $t = \dfrac{1}{k}\ln A$ is an inflection point; growth is most rapid at this time.

c. $\lim_{t\to\infty} Q(t) = C$

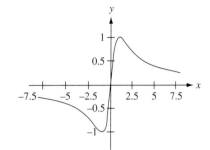

19.

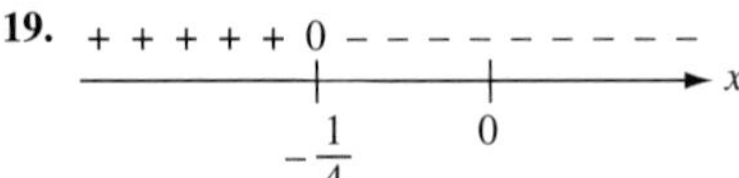

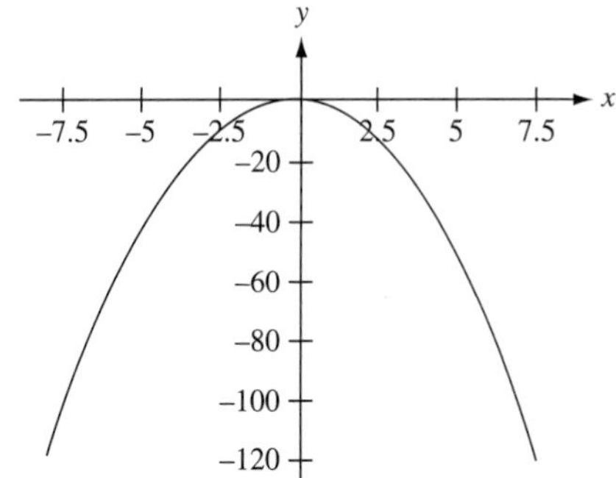

21.

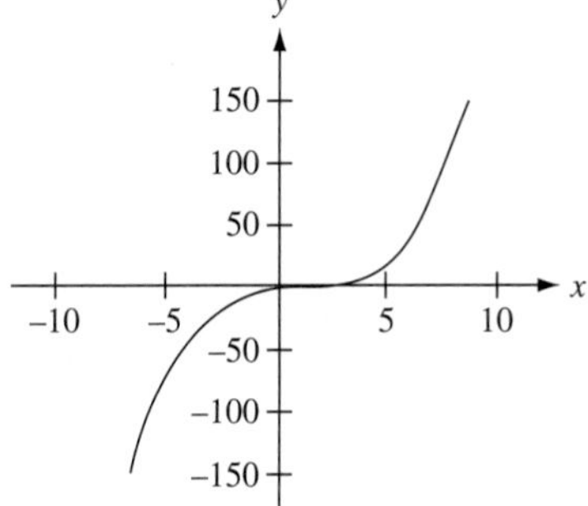

23.

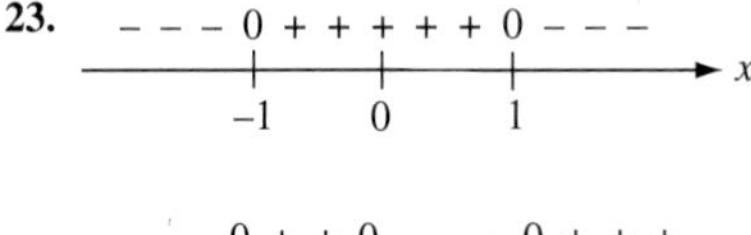

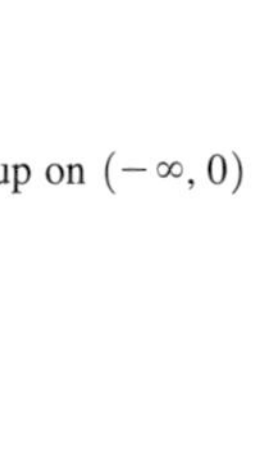

25.

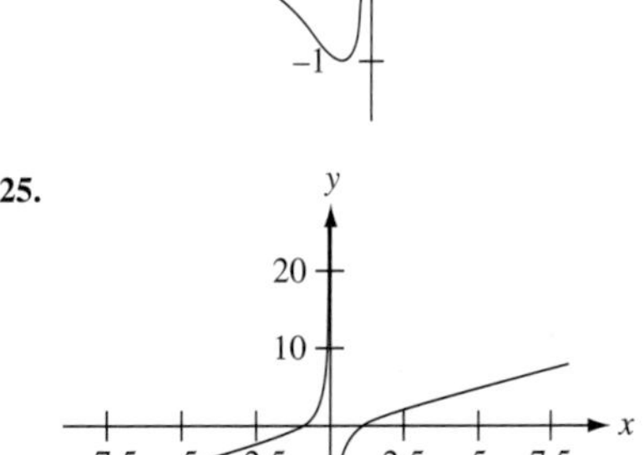

27.

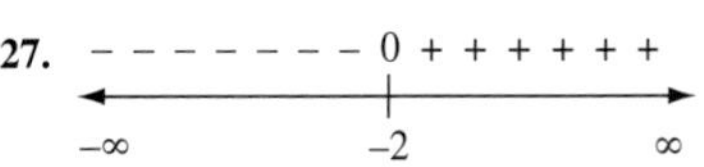

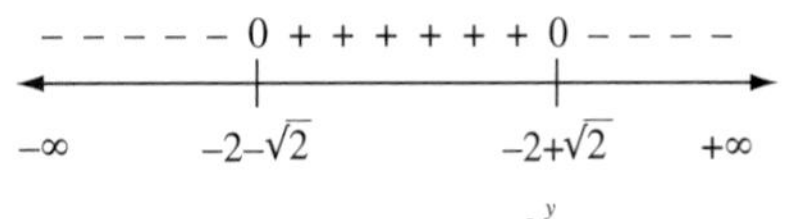

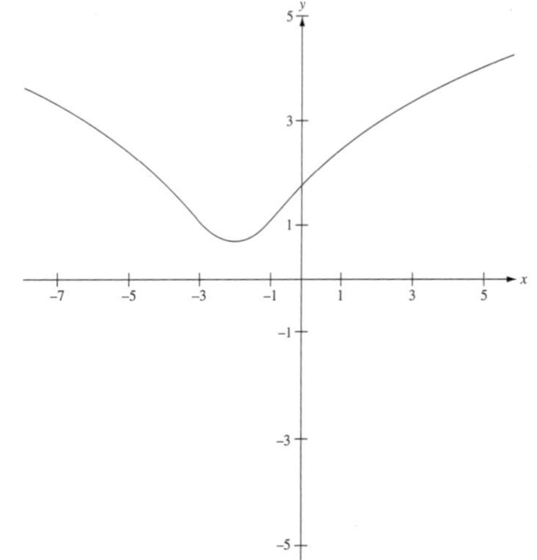

29.

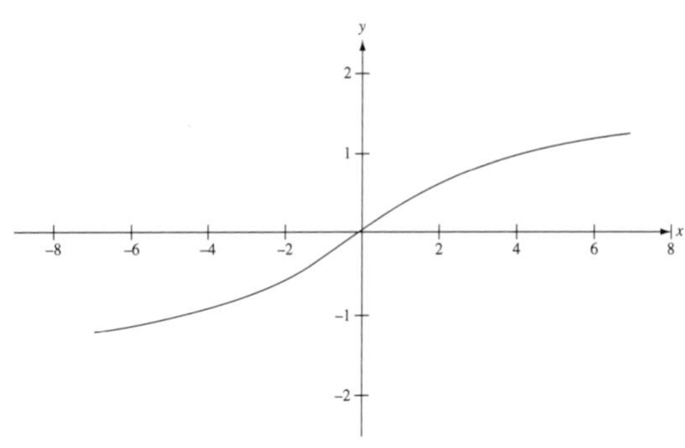

31. Horizontal asymptote: $y = 2$; vertical asymptote: $x = -1$

33. Horizontal asymptote: $y = 1$; vertical asymptote: $x = 1$

35. $\lim_{x\to\infty} \frac{x}{x-2} = 1$; $\lim_{x\to\infty} e^{x/x-2} = e$

37. Absolute minimum value: (0, 0)

39. Absolute maximum value: $f(2) = 5/3$; absolute minimum value: $f(0) = 1$

41. Absolute maximum value: $\left(1, \frac{1}{2}\right)$; absolute minimum value: (0, 0)

43. Absolute maximum value: $\left(3, \frac{215}{9}\right)$; absolute minimum value: (1, 7)

45. No absolute extrema

47. Absolute maximum value: $\ln\left(\frac{13}{14}\right)$; absolute minimum value: $\left(\ln\frac{1}{3}\right)$

49. Absolute maximum value: $2e^{-1/2} \cong 1.213$; absolute minimum value: $11e^{-5} \cong 0.0741$

51. Absolute maximum value: 1; absolute minimum value: -2

53. $\lim_{x\to 2} \frac{2x}{2x+1} = \frac{4}{5}$

55. 0

57. $\lim_{x\to+\infty} \frac{2e^{2x}}{-2e^{2x}} = -1$

59. $\lim_{x\to\infty} \frac{f(x)}{g(x)} = \lim_{x\to\infty} \frac{3.6}{1.7r}\left(\frac{e}{r}\right)^{2x}$; g grows faster if $r > e$ and f grows faster if $r < e$; if $r = e$, the functions grow at the same rate.

61. f is increasing on (0, 2) and decreasing on $(-\infty, 0) \cup (2, \infty)$

63. f is concave downward on $(-\infty, -2)$ and concave upward on $(-2, \infty)$

65. $\left(-\frac{\sqrt{2}}{2}, 2e^{-1/2}\right)$; $\left(\frac{\sqrt{2}}{2}, 2e^{-1/2}\right)$

67. Absolute maximum values: $h(-2) = 1$ and $h(2) = 1$; absolute minimum value: $h(0) = e^{-4}$

69. Absolute maximum value: $\left(\frac{\sqrt{2}}{2}, \frac{\sqrt{2}}{2}e^{-1/2}\right)$; absolute minimum value: (0, 0)

71. 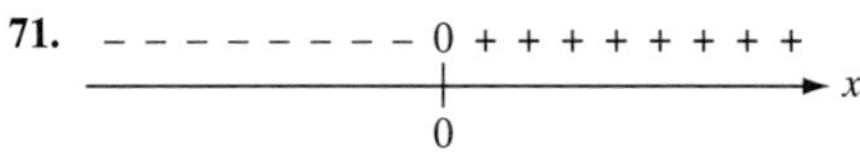

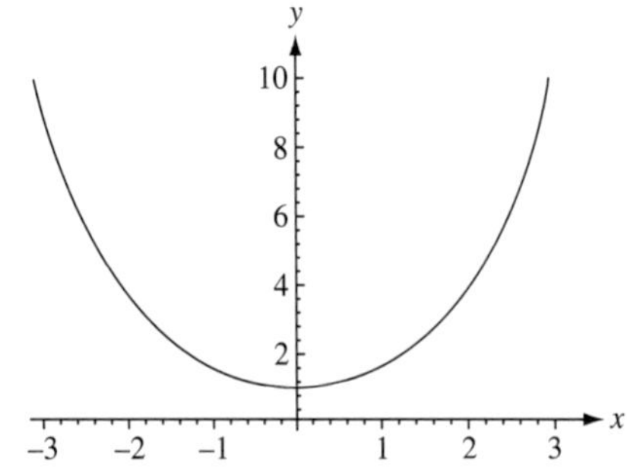

73. 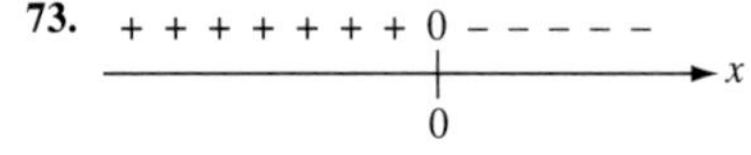

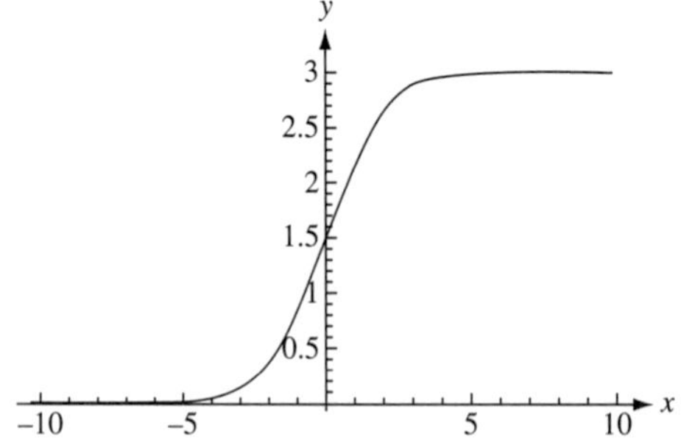

75. f is increasing on $(0, e)$ and decreasing on (e, ∞).

77. f is concave downward on $(0, e^{3/2})$ and concave upward on $(e^{3/2}, \infty)$.

79. $(e^{-3/2}, e^{-3})$

81. Absolute maximum value: (5, 3.1067); absolute minimum value: (e, e)

83. 74.07 cu cm

85. $\sqrt[3]{1/4}$ m $\times$ $2\sqrt[3]{1/4}$ m $\times$ $\sqrt[3]{2}$ m

87. \$4000

89. 168

91. 20,000 cases/order

93. $R''(x) = -2k < 0$ so $x = M/2$ affords a maximum, or R is greatest when half the population is infected.

95. $t = \frac{V}{a}\ln\left(\frac{c}{c-m}\right)$ minutes

97. **a.** 0.05 g/cm^3/sec

b. Decreasing at the rate of 0.01 g/cm^3/sec

c. $t = 20$

d. 0.90 g/cm^3

99. Absolute maximum value: $M(2.034) \cong 7.622$ m^2

CHAPTER 8

8.1 Exercises, page 442

1. $F(x) = \frac{1}{3}x^3 + 2x^2 - x + 2$; $F'(x) = x^2 + 4x - f(x)$

3. $F(x) = xe^x + \pi$; $F'(x) = xe^x + e^x = e^x(x+1) = f(x)$

5. $F(x) = \frac{1}{2}\cos(2x) + 4$; $F'(x) = \frac{1}{2}(2)(-\sin(2x)) = -\sin(2x) - f(x)$

7. **a.** $G'(x) = \frac{d}{dx}(2x) = 2 = f(x)$

b. $F(x) = G(x) + C = 2x + C$

c.

9. **a.** $G'(x) = \frac{d}{dx}\left(\frac{1}{3}x^3\right) = x^2 = f(x)$

b. $F(x) = G(x) + C = \frac{1}{3}x^3 + C$

c.

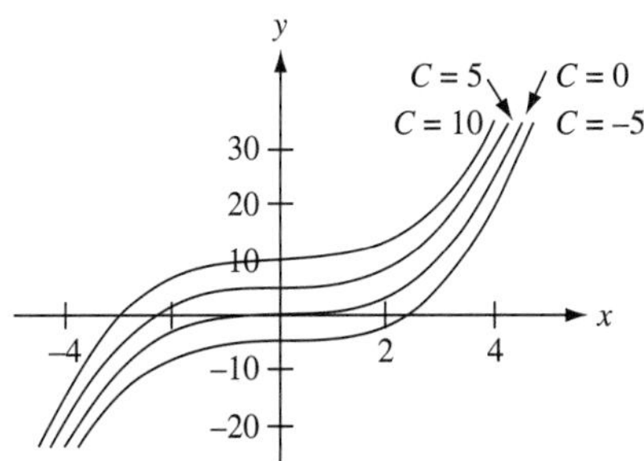

11. **a.** $G(x) = \cos(2x)$; $G'(x) = -2\sin(2x) = f(x)$

b. $F(x) = \cos(2x) + C$, C an arbitrary constant

c.

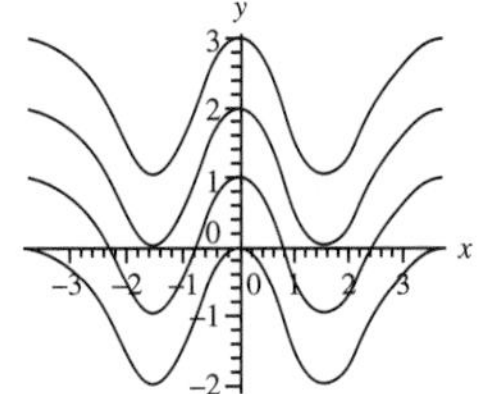

13. $6x + C$

15. $\frac{1}{4}x^4 + C$

17. $-\frac{1}{3}x^{-3} + C$

19. $\frac{3}{5}x^{5/3} + C$

21. $-4x^{-1/4} + C$

23. $-\frac{2}{x} + C$

25. $\frac{2\pi}{3}t^{3/2} + C$

27. $3x - x^2 + C$

29. $\frac{1}{3}x^3 + \frac{1}{3}x^2 - \frac{1}{3}x^{-2} + C$

31. $4e^x + C$

33. $x + \frac{1}{2}x^2 + e^x + C$

35. $x^4 + \frac{2}{x} - x + C$

37. $\frac{2}{7}x^{7/2} + \frac{4}{5}x^{5/2} - \frac{1}{2}x^2 + C$

39. $\frac{2}{3}x^{3/2} + 6x^{1/2} + C$

41. $\frac{1}{9}u^3 + \frac{1}{3}u^2 - \frac{1}{3}u + C$

43. $\frac{2}{3}t^3 - \frac{3}{2}t^2 - 2t + C$

45. $\frac{1}{3}x^3 - 2x - \frac{1}{x} + C$

47. $\frac{1}{3}s^3 + s^2 + s + C$

49. $e^1 + \frac{1}{e+1}t^{e+1} + C$

51. $\frac{1}{2}x^2 + x - \ln|x| - x^{-1} + C$

53. $\ln|x| + \frac{4}{\sqrt{x}} - \frac{1}{x} + C$

55. $2\sin(x) - 3\cos(x) + C$

57. $\frac{1}{2}s^2 + 3\arctan(s) + C$

59. $\frac{1}{2}s^2 - \tan(s) + C$

61. $x^2 + x + 1$

63. $x^3 + 2x^2 - x - 5$

65. $x - \frac{1}{x} + 2$

67. $x + \ln|x|$

69. $2\arctan(x)$

71. $\sqrt{x}$

73. $ex + \frac{1}{2}x^2 + 2$

75. $\frac{4}{9}t^{3/2}$

77. **a.** $\left(\frac{0.11}{3}\right)t^3 - 0.9t^2 + 2.2t + 4$

b. 0.74667°C

79. 21,000 m

81. 130.5 m

83. $\frac{5}{3}$ cm

85. Car A will be ahead of car B.

87.

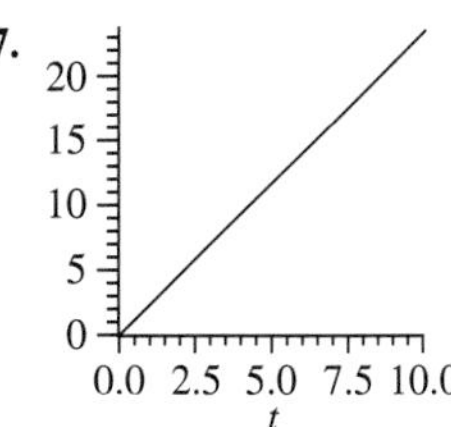

The functions are similar; sometimes A is a little more than B, sometimes it is a little less.

89. \$3370

91. 5000

93. **a.** $-t^3 + 6t^2 + 45t$ **b.** 212

95. 21,960

97. Branch A

99. 120,000

101. ≈ 1.9424

103. $v(R) = \frac{1}{2}k(R^2 - r^2)$

105. $\frac{1}{2}x + \frac{1}{4}\sin(2x) + C$

107. 3.61 g

109. False

111. False

8.2 Exercises, page 459

1. $\frac{1}{5}(4x + 3)^5 + C$

3. $\frac{1}{3}(x^3 - 2x)^3 + C$

5. $-\frac{1}{2(2x^2 + 3)^2} + C$

7. $\frac{2}{3}(t^3 + 2)^{3/2} + C$

9. $\frac{1}{20}(x^2 - 1)^{10} + C$

11. $-\frac{1}{5}\ln|1 - x^5| + C$

13. $\ln(x - 2)^2 + C$

15. $\frac{1}{2}\ln(0.3x^2 - 0.4x + 2) + C$

17. $\frac{1}{6}\ln|3x^2 - 1| + C$

19. $-\frac{1}{2}e^{-2x} + C$

21. $-e^{2-x} + C$

23. $-\frac{1}{2}e^{-x^2} + C$

25. $e^x + e^{-x} + C$

27. $\ln(1 + e^x) + C$

29. $2e^{\sqrt{x}} + C$

31. $-\frac{1}{6(e^{3x} + x^3)^2} + C$

33. $\frac{1}{8}(e^{2x} + 1)^4 + C$

35. $\frac{1}{2}(\ln 5x)^2 + C$

37. $\ln|\ln x| + C$

39. $\frac{2}{3}(\ln x)^{3/2} + C$

41. $\frac{1}{2}e^{x^2} - \frac{1}{2}\ln(x^2 + 2) + C$

43. $\frac{2}{3}(\sqrt{x} - 1)^3 + 3(\sqrt{x} - 1)^2 + 8(\sqrt{x} - 1) + 4\ln\left|\sqrt{x} - 1\right| + C$

45. $\frac{(6x + 1)(x - 1)^6}{42} + C$

47. $5 + 4\sqrt{x} - x - 4\ln(1 + \sqrt{x}) + C$

49. $-\frac{1}{252}(1 - v)^7(28v^2 + 7v + 1) + C$

51. $-\frac{1}{2}\cos(2x) + C$

53. $\frac{1}{3}\tan(3x + 1) + C$

55. $\sin(\ln(t)) + C$

57. $\ln|\sec(5x + 4)| + C$

59. $\tan(\sin(x)) + C$

61. $\frac{1}{2}\ln|\sin(x^2 + 4)| + C$

63. $\sin(e^x) + C$

65. $-\cos(x^2 + x + 1) + C$

67. $-\sin\left(\frac{1}{x}\right) + C$

69. $\ln|\sec(u)| + C$

71. $\frac{1}{2}[(2x - 1)^5 + 5]$

73. $e^{-x^2+1} - 1$

75. $\int \cot(x)dx = \ln|\sin(x)| + C$

77. $\int f\frac{\sqrt{\frac{4}{3}}}{u^2 + 1}\,du = \sqrt{\frac{4}{3}}\arctan\left(\sqrt{\frac{4}{3}}\left(x + \frac{1}{2}\right)\right) + C$

79. $\int \cos(x)dx - \int \sin^2(x)\cos(x)dx = \sin(x) - \frac{1}{3}\sin^3(x) + C$

81. 26.3 million viewers

83. $p(x) = \frac{250}{\sqrt{16 + x^2}}$

85. 24,555 pairs

87. 80.04

89. 59.6 cm

91. $C = \frac{ac}{b}$; $x(t) = \frac{ac}{b} + \left(x_0 - \frac{ac}{b}\right)e^{-bt/V}$

93. 72.4 mg

8.3 Exercises, page 470

1. 4.27 sq units

3. a. 6 sq units

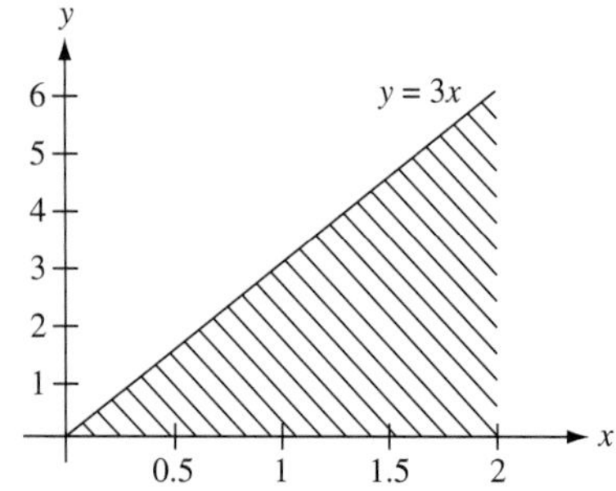

b. 4.5 sq units

c. 5.25 sq units

d. Yes

5. a. 4 sq units

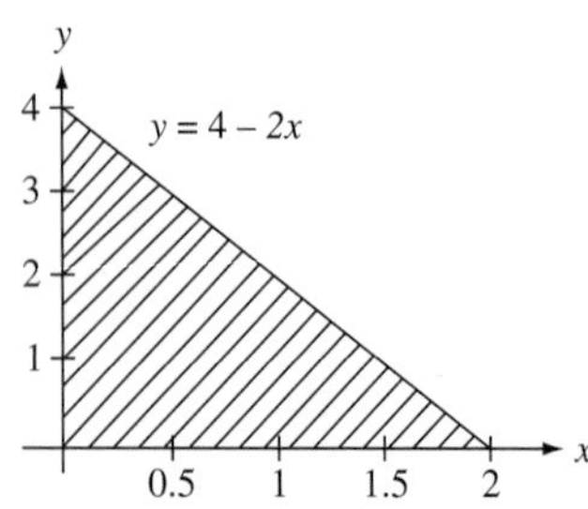

b. 4.8 sq units

c. 4.4 sq units

d. Yes

7. a. 18.5 sq units **b.** 18.64 sq units

c. 18.66 sq units **d.** $\approx$ 18.7 sq units

9. a. 25 sq units **b.** 21.12 sq units

c. 19.88 sq units **d.** $\approx$ 19.9 sq units

11. a. 0.0625 sq unit **b.** 0.16 sq unit

c. 0.2025 sq unit **d.** $\approx$ 0.2 sq unit

13. 4.64 sq units

15. 0.95 sq unit

17. 1.98 sq units

19. 9400 m^2

8.4 Exercises, page 480

1. 6 sq units

3. 8 sq units

5. 12 sq units

7. 9 sq units

9. ln 2 sq units

11. $17\frac{1}{3}$ sq units

13. $18\frac{1}{4}$ sq units

15. $(e^2 - 1)$ sq units

17. 2 sq units

19. $\approx$ 0.87 sq unit

21. 6

23. 14

25. $\frac{56}{3}$

27. $\frac{4}{3}$

29. 45

31. $\frac{7}{12}$

33. ln 2

35. 56

37. $\frac{256}{15}$

39. $\frac{2}{3}$

41. $\frac{8}{3}$

43. $\frac{39}{2}$

45. 0

47. $\frac{\pi}{3}$

49. 1

51. 4

53. 922.22 m

55. a. \$6000 **b.** \$1500

57. a. 96 **b.** $25\frac{1}{2}$

59. 46%

61. $\frac{kR^4}{4L}$

63. 508 mg produced from 2 AM to 4 AM; 5327 mg during the entire day

65. 5039

67. False

69. True

8.5 Exercises, page 491

1. 10

3. $\frac{19}{15}$

5. $32\frac{4}{15}$

7. $\sqrt{3} - 1$

9. $24\frac{1}{5}$

11. $\frac{32}{15}$

13. $18\frac{2}{15}$

15. $\frac{1}{2}(e^4 - 1)$

17. $\frac{1}{2}e^2 + \frac{5}{6}$

19. 0 **21.** 2 ln 4 **23.** $\frac{1}{3}(\ln 19 - \ln 3)$

25. $2e^4 - 2e^2 - \ln 2$ **27.** $\frac{1}{2}(e^{-4} - e^{-8} - 1)$ **29.** $\frac{\pi}{12}$

31. 0 **33.** $\frac{\pi}{2}$ **35.** ≈ 1.044476

37. $\left(\frac{1}{2}\right)\ln(2)$ **39.** 5 **41.** $\frac{17}{3}$

43. -1 **45.** $\frac{13}{6}$ **47.** $\frac{1}{4}(e^4 - 1)$

49. $\frac{2}{\pi}$ **51.** $du = -2x\,dx$

53. Cannot get average value of function

55. ≈ 0.245327 cm

57. $\int_a^a f(x)dx = 0$, where $F'(x) = f(x)$

59. $\int_1^3 x^2\,dx = \frac{26}{3}$ **61.** $\frac{104}{3}$ **63.** $\frac{93}{4}$

65. 0; Property 1

67. -4; Property 3

69. a. -1 **b.** 4

71. $21\frac{1}{3}$ m **73.** 4°C **75.** \$85/share

77. \$25.91

79. ≈ 2.24 million dollars

81. 343 thousand barrels

83. 16,863 **85.** 1367

87. $V = \frac{1.2}{\pi}\left(1 - \cos\frac{\pi t}{2}\right)$

89. 0.071 mg/cm^3

91. ≈ 23.604516; Yes

93. True **95.** False **97.** True

99. True **101.** False

8.6 Exercises, page 502

1. 108 sq units **3.** $\frac{2}{3}$ sq unit

5. $2\frac{2}{3}$ sq units **7.** $1\frac{1}{2}$ sq units

9. 3 **11.** $3\frac{1}{3}$

13. 27 **15.** $2(e^2 - e^{-1})$

17. 1 sq unit **19.** $\frac{38}{3}$

21. $\frac{10}{3}$ sq units **23.** $4\frac{3}{4}$ sq units

25. ≈ 10.6 sq units **27.** $(e^2 - e - \ln 2)$ sq units

29. $2\sqrt{2}$ sq units **31.** ≈ 0.23 sq unit

33. $2\frac{1}{2}$ sq units **35.** $\frac{22}{3}$ sq units

37. $\frac{3}{2}$ sq units **39.** ≈ 16.5 sq units

41. $\frac{125}{6}$ sq units **43.** $\frac{1}{12}$ sq units

45. $\frac{71}{6}$ **47.** 18 sq units

49. ≈ 2.83 sq units **51.** $\int_a^b [f(x) - g(x)]\,dx$

53. a. $A_2 - A$

b. The distance car 2 is ahead of car 1 after t seconds.

55. S gives the additional revenue that the company would realize if it used a different advertising agency; $S = \int_0^b [g(x) - f(x)]dx$.

57. 21,850 cars

59. S gives the difference between the total number of pulse beats between the present and that of six months ago; $S = \int_0^b [f(t) - g(t)]dt$.

61. 42.8 billion metric tons

63. False

8.7 Exercises, page 516

1. \$11,667 **3.** \$6667

5. \$11,667

7. Consumers' surplus: \$13,333; producer's surplus: \$11,667

9. \$824,200 **11.** \$148,239

13. \$52,203 **15.** \$76,645

17. \$111,869 **19.** \$20,964

21. a.

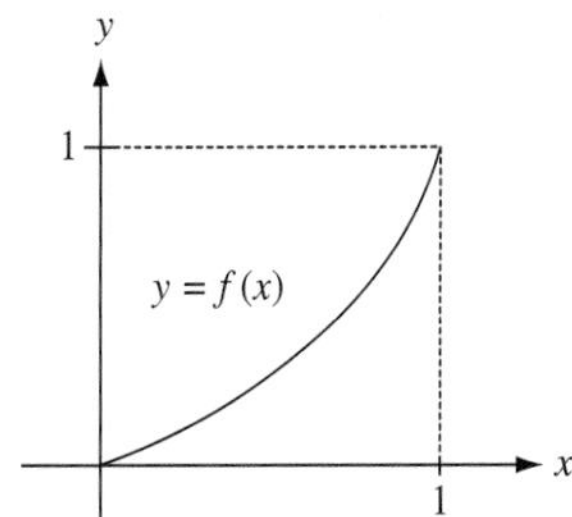

b. 0.175; 0.816

23. a.

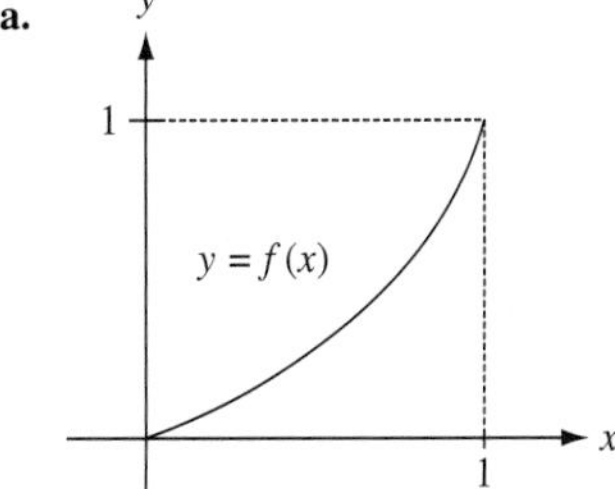

b. 0.104; 0.504

8.8 Exercises, page 523

1. 3π cu units **3.** $\frac{15\pi}{2}$ cu units

5. $\frac{4\pi}{3}$ cu units **7.** $\frac{16\pi}{15}$ cu units

9. $\frac{\pi}{2}(e^2 - 1)$ cu units

11. $\frac{\pi^2}{2}$ cu units

13. $\frac{\pi^2}{4}$ cu units

15. $\frac{2\pi}{15}$ cu units

17. $\frac{136\pi}{15}$ cu units

19. $\frac{64\sqrt{2}\pi}{3}$ cu units

21. $\frac{\pi}{2}(e^2 - 2^{+e} - 2)$ cu units

23. ≈ 6.7 cu units

25. $\frac{125\pi^2}{8}$ cu units

27. $\frac{\pi}{6}$ cu units

29. $\frac{6517\pi}{240}$ cu units

31. $\frac{64\sqrt{2}\pi}{3}$ cu units

33. $\frac{8(\sqrt{2} - 1)\pi}{3}$ cu units

35. $\frac{\pi^2}{2}$

37. $\frac{4}{3}\pi r^3$ cu units

39. $50{,}000\pi$ cu m

Chapter 8 Review Exercises, page 526

1. $\frac{1}{4}x^4 + \frac{2}{3}x^3 - \frac{1}{2}x^2 + C$

3. $\frac{1}{5}x^5 - \frac{1}{2}x^4 - \frac{1}{x} + C$

5. $\frac{1}{2}x^4 + \frac{2}{5}x^{5/2} + C$

7. $\frac{1}{3}x^3 - \frac{1}{2}x^2 + 2\ln|x| + 5x + C$

9. $\frac{3}{8}(3x^2 - 2x + 1)^{4/3} + C$

11. $\frac{1}{2}\ln(x^2 - 2x + 5) + C$

13. $\frac{1}{2}e^{x^2+x+1} + C$

15. $\frac{1}{6}(\ln x)^6 + C$

17. $\frac{(11x^2 - 1)(x^2 + 1)11}{264} + C$

19. $\frac{2}{3}(x + 4)\sqrt{x - 2} + C$

21. $-\frac{1}{3}\cos 3x + C$

23. $-3\cos x + 4\sin x + C$

25. $\frac{1}{2}\tan 2x + C$

27. $\frac{1}{2}\sin x^2 + C$

29. $-\frac{1}{\pi}\csc \pi x + C$

31. $\frac{1}{4}\sin^4 x + C$

33. $\frac{1}{\pi}\ln|\sec \pi x + \tan \pi x| + C$

35. $-\frac{2}{3}(\cos x)^{3/2} + C$

37. $-\frac{1}{9}(1 - 2\sin 3x)^{3/2} + C$

39. $\frac{1}{4}\tan^4 x + C$

41. $-\frac{1}{4}(\cot x - 1)^4 + C$

43. $\frac{1}{2}$

45. $\frac{17}{3}$

47. -80

49. $\frac{1}{2}\ln 5$

51. 4

53. $\frac{e - 1}{2(1 + e)}$

55. 2

57. $-\frac{1}{2}\ln\frac{1}{2}$

59. $\frac{1}{3}\ln(\sqrt{2} + 1)$

61. $I = \int \sin u\, du = -\cos(\ln x) + C;\ \int_1^{e^{\pi}} \frac{\sin(\ln x)}{x}\,dx = 2$

63. $f(x) = x^3 - 2x^2 + x + 1$

65. $f(x) = x + e^{-x} + 1$

67. -4.28

69. $2\sqrt{2}$ sq units

71. $-\ln\left(-\frac{\sqrt{2}}{2}\right) - \ln(\sqrt{2} - 1)$

73. $\frac{\sqrt{2}}{2} - \frac{\pi^2}{32}$

75. $\frac{1}{2}(e^4 - 1)$ sq units

77. $4\frac{1}{2}$

79. $\frac{3}{10}$ sq units

81. $\frac{1}{3}$ sq unit

83. $\frac{3\pi}{10}$ cu units

85. 19.4 billion metric tons

91. **a.** $R(x) = -0.015x^2 + 60x$

b. $-0.015x^2 + 60x = px$, or $p = -0.015x + 60$

93. 16,939

95. $270,000

97. $197,652

99. $505,696

101. $\approx 70{,}784$

CHAPTER 9

9.1 Exercises, page 536

1. $\frac{1}{4}e^{2x}(2x - 1) + C$

3. $4(x - 4)e^{x/4} + C$

5. $\frac{1}{2}e^{2x} - 2(x - 1)e^x + \frac{1}{3}x^3 + C$

7. $xe^x + C$

9. $\frac{2(x + 2)}{\sqrt{x + 1}} + C$

11. $\frac{2}{3}x(x - 5)^{3/2} - \frac{4}{15}(x - 5)^{5/2} + C$

13. $\frac{x^2}{4}(2\ln 2x - 1) + C$

15. $\frac{x^4}{16}(4\ln x - 1) + C$

17. $\frac{2}{9}x^{3/2}(3\ln\sqrt{x} - 1) + C$

19. $-\frac{1}{x}(\ln x + 1) + C$

21. $x(\ln x - 1) + C$

23. $-(x^2 + 2x + 2)e^{-x} + C$

25. $\frac{1}{4}x^2[2(\ln x)^2 - 2\ln x + 1] + C$

27. $-x\cos(x) + \sin(x) + C$

29. $\frac{x}{3}\sin(3x) + \frac{1}{9}\cos(3x) + C$

31. $\frac{-x}{2}\cos(2x) + \left(\frac{1}{4}\right)\sin c(2x) + C$

33. $\ln(|x^2 + 1|) + C$

35. $\frac{-x}{4}\cos(2x) + \frac{1}{8}\sin(2x) + C$

37. $\left(\frac{1}{4}\right)\sin(2x^2) + C$

39. $x^2\ln(5x) - \frac{1}{2}x^2 + C$

41. $\frac{1}{2}\arctan(2x + 1) + C$

43. $\left(\frac{1}{4}\right)e^u + C = \left(\frac{1}{4}\right)e^{2x^2} + C$

45. $-3e^{-2} + 1$

47. $\frac{1}{4}(8\ln 2 - 3)$

49. $-5e^{-1} + 2$

51. $2\arctan(2) - \frac{1}{2}\ln(5)$

53. $-\frac{1}{2}xe^{-2x} - \frac{1}{4}e^{-2x} + \frac{13}{4}$

55. $(5\ln 5 - 4)$ sq units

57. $\left(\frac{e^x}{2}\right)[\sin(x) - \cos(x)] + C$

59. 1485 m

61. 696

63. **a.** 0 kg **b.** 90 kg **c.** 45.56 kg

65. $2(-10te^{-0.1t} - 100e^{-0.1t}) + C$

67. $131,324

69. 0.17819 individuals

71. True

9.2 Exercises, page 546

1. $a = 3, b = 2, c = 7$

3. $a = 1, b = 7, c = 9$

5. $a = 2, b = 2$

7. $\frac{1}{2}\ln\left(\frac{|x-3|}{|x-1|}\right) + C$

9. $\left(\frac{1}{4}\right)\ln\left(\frac{|x+2|}{|x+6|}\right) + C$

11. $\left(-\frac{2}{3}\right)\ln(|x+1|) + \left(\frac{5}{3}\right)\ln(|x-2|) + C$

13. $-\ln(|x+1|) + 3\ln(|x+2|) + C$

15. $\left(\frac{-3}{4}\right)\ln(|x|) + \left(\frac{7}{4}\right)\ln(|x+4|) + C$

17. $\left(\frac{1}{3}\right)\ln\left(\frac{5}{4}\right)$

19. $\ln(3)$

21. $\left(\frac{1}{4}\right)\ln\left(\frac{|x^2-2|}{|x^2+2|}\right)$

23. $-\ln(|\ln(x) - 1|) + 2\ln(|\ln(x) - 3|) + C$

25. $\ln\left(\frac{|\sin(x) - 2|}{|\sin(x) - 1|}\right) + C$

27. $\left(\frac{1}{6}\right)\ln\left(\frac{|x^3|}{|x^3+2|}\right) + C$

29. $\ln\left(\frac{|\cos(x) + 3|}{|\cos(x) + 2|}\right) + C$

31. $\left(\frac{1}{2}\right)\arctan(x^2) + C$

33. $\left(\frac{1}{4}\right)\ln\left(\frac{|x^2|}{|x^2+2|}\right) + C$

35. $\left(\frac{1}{2}\right)\ln\left(\frac{|\sin(x) + 2|}{|\sin(x) + 4|}\right) + C$

37. 5.04614 g

39. 0.39879455 mg

41. $\frac{-1}{(x + -2)} + C$

43. $395.51

45. 0.150966 mg

47. 4.3654 g

9.3 Exercises, page 555

1. $\frac{2}{9}[2 + 3x - 2\ln|2 + 3x|] + C$

3. $\frac{3}{32}[(1 + 2x)^2 - 4(1 + 2x) + 2\ln|1 + 2x|] + C$

5. $2\left[\frac{x}{8}\left(\frac{9}{4} + 2x^2\right)\sqrt{\frac{9}{4} + x^2} - \frac{81}{128}\ln\left(x + \sqrt{\frac{9}{4} + x^2}\right)\right] + C$

7. $\ln\left(\frac{\sqrt{1+4x} - 1}{\sqrt{1+4x} + 1}\right) + C$

9. $\frac{1}{2}\ln 3$

11. $\frac{x}{9\sqrt{9 - x^2}} + C$

13. $\frac{x}{8}(2x^2 - 4)\sqrt{x^2 - 4} - 2\ln|x + \sqrt{x^2 - 4}| + C$

15. $\sqrt{4 - x^2} - 2\ln\left|\frac{2 + \sqrt{4 - x^2}}{x}\right| + C$

17. $\frac{1}{4}(2x - 1)e^{2x} + C$

19. $\ln|\ln(1 + x)| + C$

21. $\frac{1}{9}\left[\frac{1}{1 + 3e^x} + \ln(1 + 3e^x)\right] + C$

23. $6[e^{(1/2)x} - \ln(1 + e^{(1/2)x})] + C$

25. $\frac{1}{9}(2 + 3\ln x - 2\ln|2 + 3\ln x|) + C$

27. $e - 2$

29. $\frac{x^3}{9}(3\ln x - 1) + C$

31. $x[(\ln x)^3 - 3(\ln x)^2 + 6\ln x - 6] + C$

33. $\arctan(x + 3) + C$

35. $\left(\frac{1}{3}\right)\arctan\left(\frac{t + 2}{3}\right) + C$

37. $\left(\frac{2}{\sqrt{3}}\right)\arctan\left(\frac{(2x + 1)}{\sqrt{3}}\right) + C$

39. 27,136

41. 40%

43. $2,329

45. $1,901,510

47. ≈ 0.1722

49. Cannot solve for any specific $P(t)$ because unable to find the integration constant C

9.4 Exercises, page 567

1. 2.7037; 2.6667; $2\frac{2}{3}$

3. 0.2656; 0.2500; $\frac{1}{4}$

5. 0.6970; 0.6933; ≈ 0.6931

7. 0.5090; 0.5004; $\frac{1}{2}$

9. 5.2650; 5.3046; $\frac{16}{3}$

11. 0.6336; 0.6321; ≈ 0.6321

13. 0.3837; 0.3863; ≈ 0.3863

15. 1.95410; 2.00086

17. 1.32525; 1.32581

19. 1.1170; 1.1114

21. 1.3973; 1.4052

23. 0.8806; 0.8818

25. 3.7757; 3.7625

27. 0.744273; 0.806337

29. 0.747399; 0.749041

31. **a.** 3.6

b. 0.0324

33. **a.** 0.013

b. 0.00043

35. a. 0.0078125

b. 0.0002848

37. 1.983524; 2.000110

39. 574.4 km

41. 21.65

43. 7850 m^2

45. 50%

47. a. $142,373.56

b. $142,698.12

49. 40.1%

51. 6.42 litres/min

53. False

55. True

9.5 Exercises, page 579

1. $\frac{2}{3}$ sq unit

3. 1 sq unit

5. 2 sq units

7. $\frac{2}{3}$ sq unit

9. $\frac{1}{2}e^4$ sq units

11. 1 sq unit

13. a. $\frac{2}{3}b^{3/2}$

b. ∞

15. 1

17. 2

19. Divergent

21. $-\frac{1}{8}$

23. 1

25. 1

27. $\frac{1}{2}$

29. Divergent

31. -1

33. Divergent

35. 0

37. 0

39. Divergent

41. Convergent

43. $\frac{5}{e^2}$

45. $\frac{1}{9}$

47. $\frac{\left(\ln(2) + \frac{2}{3}\right)}{8}$

49. $\frac{1}{pe^{pa}}$ if $p > 0$ and is divergent if $p < 0$.

51. $83,333

53. $555,556

55. $300,000

57. 0.083

59. 4.52 mg

61. True

63. False

Chapter 9 Review Exercises, page 582

1. $-2(1 + x - e^{-x} + C)$

3. $x(\ln 5x - 1) + C$

5. $\frac{1}{4}(1 - 3e^{-2})$

7. $2\sqrt{x}(\ln x - 2) + 2$

9. $\left(\frac{1}{3}\right)\ln\left(\frac{|x-2|}{|x+1|}\right) + C$

11. $-3\ln(|x-1|) + 5\ln(|x-2|) + C$

13. $\ln\left(\frac{32}{27}\right)$

15. $\frac{1}{8}\left[3 + 2x - \frac{9}{3+2x} - 6\ln|3+2x|\right] + C$

17. $\frac{1}{32}e^{4x}(8x^2 - 4x + 1) + C$

19. $\frac{1}{4}\frac{\sqrt{x^2-4}}{x} + C$

21. $\frac{1}{2}$

23. Divergent

25. $\frac{1}{10}$

27. $\frac{1}{16}$

29. 0.8421; 0.8404

31. 2.2379; 2.1791

33. $1,157,641

35. 48,761

37. $111,111

39. The limit does not exist.

CHAPTER 10

10.1 Exercises, page 591

1. $f(0,0) = -4$; $f(1,0) = -2$; $f(0,1) = -1$; $f(1,2) = 4$; $f(2,-1) = -3$

3. $f(1,2) = 7$; $f(2,1) = 9$; $f(-1,2) = 1$; $f(2,-1) = 1$

5. $g(1,2) = 4 + 3\sqrt{2}$; $g(2,1) = 8 + \sqrt{2}$; $g(0,4) = 2$; $g(4,9) = 56$

7. $h(1,e) = 1$; $h(e,1) = -1$; $h(e,e) = 0$

9. $g(1,1,1) = e$; $g(1,0,1) = 1$; $g(-1,-1,-1) = -e$

11. $f\left(0, \frac{\pi}{2}\right) = 0$; $f\left(\frac{\pi}{3}, \frac{\pi}{4}\right) = \frac{\sqrt{3}}{2} + \frac{1}{\sqrt{2}}$; $f\left(0, -\frac{2\pi}{3}\right) = -\frac{1}{2}$

13. $f(1,0) = 1$; $f\left(\sqrt{3}, \frac{\pi}{3}\right) = \frac{\sqrt{3}}{2}$; $f\left(-1, -\frac{\pi}{4}\right) = \frac{-1}{\sqrt{2}}$

15. All real values of x and y

17. All real values of u and v except those satisfying the equation $u = v$

19. All real values of r and s satisfying $rs \geq 0$

21. All real values of x and y satisfying $x = y > 5$

23.

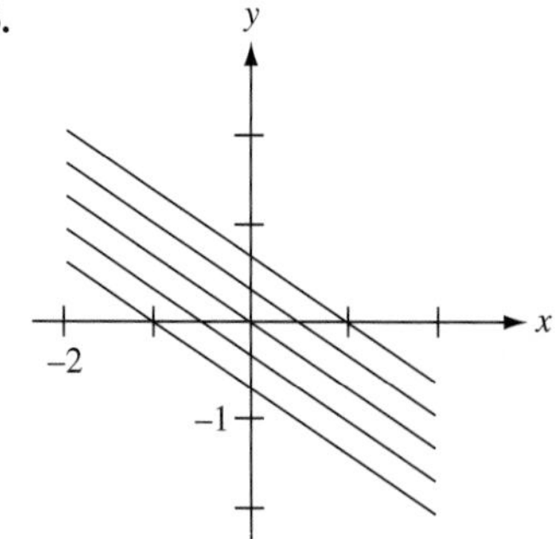

25.

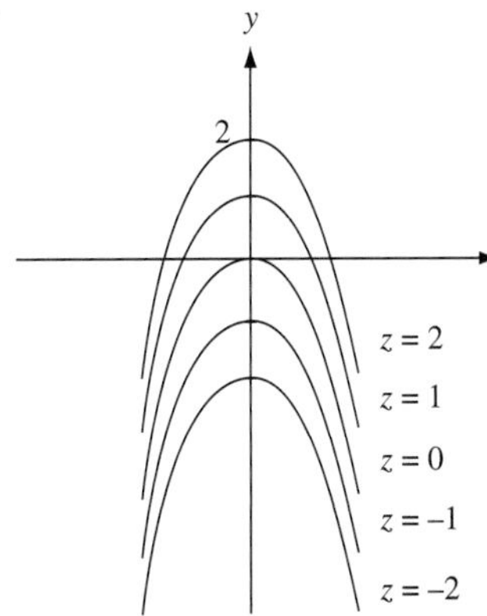

27.

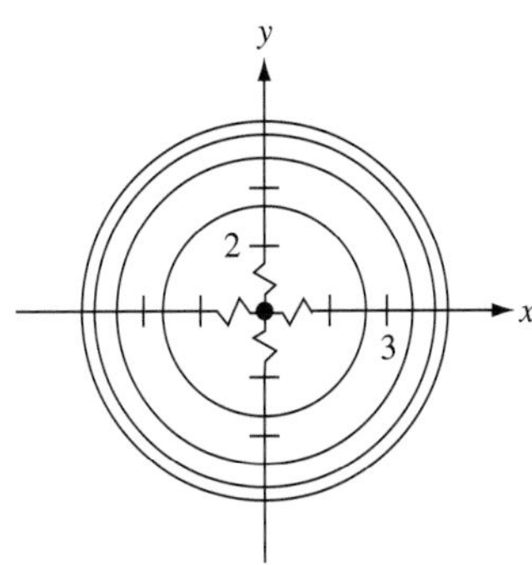

29. 9π cu m

31. a. The set of all ordered pairs (P, T) where P and T are positive real numbers

b. 11.10 litres

33. a. $-\frac{1}{5}x^2 - \frac{1}{4}y^2 - \frac{1}{5}xy + 200x + 160y$

b. The set of all points (x, y) satisfying $200 - \frac{1}{5}x - \frac{1}{10}y \geq 0,\ 160 - \frac{1}{10}x - \frac{1}{4}y \geq 0$

35. a. $-0.005x^2 - 0.003y^2 - 0.002xy + 20x + 15y$

b. The set of all ordered pairs (x, y) for which $20 - 0.005x - 0.001y \geq 0;\ 15 - 0.001x - 0.003y \geq 0$

37. 103

39. a. \$733.76; \$877.57

b. \$836.44

41. 18

43. 40,000k dynes

45. False

47. False

10.2 Exercises, page 604

1. 2; 3

3. $4x$; 4

5. $-\frac{4y}{x^3}; \frac{2}{x^2}$

7. $\frac{2v}{(u+v)^2}; -\frac{2u}{(u+v)^2}$

9. $3(2s - t)(s^2 - st + t^2)^2;\ 3(2t - s)(s^2 - st + t^2)^2$

11. $\frac{4x}{3(x^2 + y^2)^{1/3}}; \frac{4y}{3(x^2 + y^2)^{1/3}}$

13. $ye^{xy+1}; xe^{xy+1}$

15. $\ln y + \frac{y}{x}; \frac{x}{y} + \ln x$

17. $e^u \ln v; \frac{e^u}{v}$

19. $yz + y^2 + 2xz;\ xz + 2xy + z^2;\ xy + 2yz + x^2$

21. $ste^{rst};\ rte^{rst};\ rse^{rst}$

23. $\frac{\delta f}{\delta x} = \sin(y); \frac{\delta f}{\delta y} = x\cos(y)$

25. $\frac{\delta f}{\delta x} = e^{\sin(y)}; \frac{\delta f}{\delta y} = x\cos(y)e^{\sin(y)}$

27. $\frac{\delta f}{\delta x} = \frac{1}{y + x^2/y}; \frac{\delta f}{\delta y} = \frac{-x}{y^2 + x^2}$

29. $\frac{\delta h}{\delta s} = \left(\frac{\sin(t)}{t}\right)\left(\frac{\cos(s) + s\sin(s)}{\cos^2(s)}\right);$
$\frac{\delta h}{\delta t} = \left(\frac{s}{\cos(s)}\right)\left(\frac{t\cos(t) - \sin(t)}{t^2}\right)$

31. $f_x(1, 2) = 8; f_y(1, 2) = 5$

33. $f_x(2, 1) = 1; f_y(2, 1) = 3$

35. $f_x(1, 2) = \frac{1}{2}; f_y(1, 2) = -\frac{1}{4}$

37. $f_x(1, 1) = e; f_y = (1, 1) = e$

39. $f_x(1, 0, 2) = 0; f_y = (1, 0, 2) = 8; f_z(1, 0, 2) = 0$

41. $f_x = \left(\frac{\pi}{3}\right)\cos\left(\frac{\pi^2}{9}\right); f_y = \left(\frac{\pi}{3}\right)\cos\left(\frac{\pi^2}{9}\right)$

43. $h_a = \left(\frac{-5}{\sqrt{2}}\right)e^{(-1/\sqrt{2})}; h_b = \left(\frac{-\pi}{4\sqrt{2}}\right)e^{(-1/\sqrt{2})}$

45. $g_x = 0; g_y = \frac{1}{2\pi}$

47. $f_{xx} = 2y; f_{xy} = f_{yx}; f_{yy} = 6xy$

49. $f_{xx} = 2; f_{xy} = f_{yx} = -2; f_{yy} = 4$

51. $f_{xx} = \frac{y^2}{(x^2 + y^2)^{3/2}}; f_{xy} = f_{yx} = -\frac{xy}{(x^2 + y^2)^{3/2}}; fyy = \frac{x^2}{(x^2 + y^2)^{3/2}}$

53. $f_{xx} = \frac{1}{y^2}e^{-x/y}; f_{xy} = \frac{y - x}{y^3}e^{-x/y} = f_{yx}; f_{yy} = \frac{x}{y^3}\left(\frac{x}{y} - 2\right)e^{-x/y}$

55. $f_x = -y\sin(xy); f_{xy} = -\sin(xy) - yx\cos(xy); f_{xx} = -y^2\cos(xy);$
$f_y = -x\sin(xy); f_{yx} = -\sin(xy) - yx\cos(xy);$
$f_{yy} = -x^2\cos(xy); f_{xy} = f_{yx}$

57. $f_{xy} = \cos(y); f_{xx} = 0; f_y = x\cos(y); f_{yx} = \cos(y); f_{yy} = -x\sin(y);$
$f_{xy} = f_{yx}$

59. $f_x = \sec(y); f_{xy} = \sec(y)\tan(y); f_{xx} = 0; f_y = x\sec(y)\tan(y);$
$f_{yx} = \sec(y)\tan(y); f_{yy} = x(\sec(y)\tan^2(y) + \sec^3(y)); f_{xy} = f_{yx}$

61. a. -0.74°C

b. -0.038°C

63. −0.015 litres/mm

65. $\frac{\delta K}{\delta m} = \frac{1}{2}v^2; \frac{\delta K}{\delta v} = mv; \frac{\delta^2 K}{\delta v^2} = m$

67. a. 48 units/unit change in labour; $\frac{128}{81}$ units/unit change in capital

b. No

69. Substitute commodities

71. Substitute commodities

73. Monthly profit is increasing at the rate of $29 per thousand dollars increase in the inventory; monthly profit is increasing at the rate of $5 per thousand dollars increase in the inventory.

75. $b = 0, s = 2$

77. $t = 20, h = 0.78$

79. True

81. False

10.3 Exercises, page 615

1. (0, 0); relative maximum value: $f(0, 0) = 1$

3. (1, 2); saddle point: $f(1, 2) = 4$

5. (8, −6); relative minimum value: $f(8, -6) = -41$

7. (1, 2) and (2, 2); saddle point: $f(1, 2) = -1$; relative minimum value: $f(2, 2) = -2$

9. $\left(-\frac{1}{3}, \frac{11}{3}\right)$ and (1, 5); saddle point: $f\left(-\frac{1}{3}, \frac{11}{3}\right) = -\frac{319}{27}$; relative minimum value: $f(1, 5) = -13$

11. (0, 0) and (1, 1); saddle point: $f(0, 0) = -2$; relative minimum value: $f(1, 1) = -3$

13. (2, 1); relative minimum value: $f(2, 1) = 6$

15. (0, 0); saddle point: $f(0, 0) = -1$

17. (0, 0); relative minimum value: $f(0, 0) = 1$

19. (0, 0); relative minimum value: $f(0, 0) = 0$

21. $f(0, 0)$ is not defined; relative extrema cannot be obtained with the second derivative test.

23. $\left(0, \frac{\pi}{2}\right)$ and $\left(0, \frac{3\pi}{2}\right)$; relative minimum value: $\left(0, \frac{3\pi}{2}\right)$

25. (0, 0); no relative extrema

27. 6 cm × 6 cm × 3 cm

29. 18 cm × 36 cm × 18 cm

31. 200 finished units and 100 unfinished units; $10,500

33. Price of land ($200/m²) is highest at $\left(\frac{1}{2}, 1\right)$

35. $7,200

37. $f = 2.3; p = 3.05$

39. False

10.4 Exercises, page 625

1. a. $y = 2.3x + 1.5$

b.

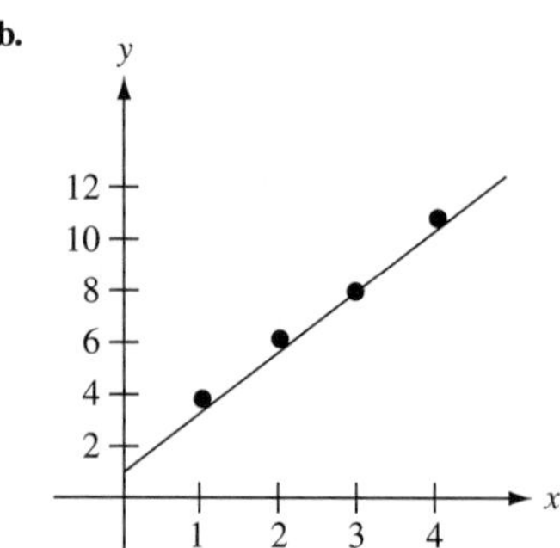

3. a. $y = -0.77x + 5.74$

b.

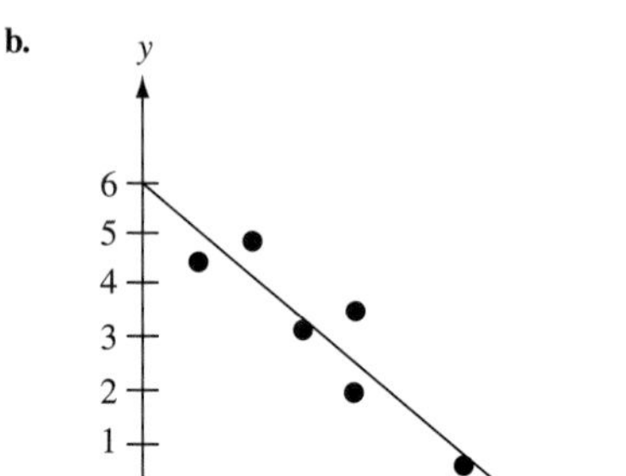

5. a. $y = 1.2x + 2$

b.

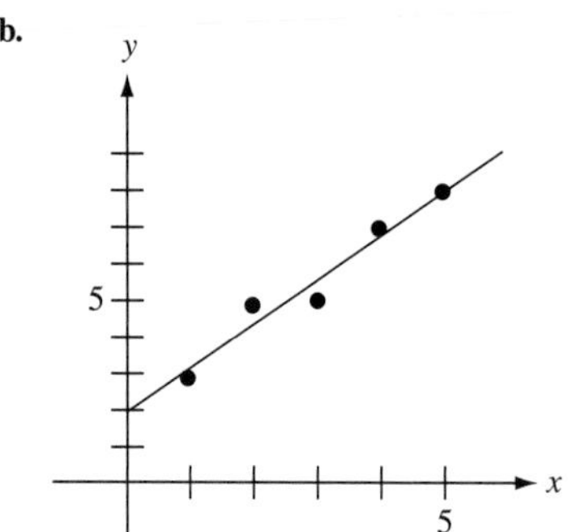

7. a. $y = 0.34x - 0.9$

b.

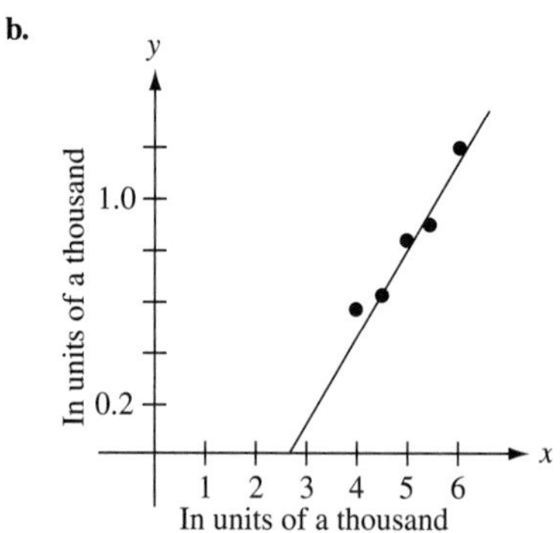

9. a. $y = 5.69x + 176$ **b.** 432 acres

11. a. $y = 12.2x + 20.9$ **b.** 81.9 million

13. a. $y = 10x + 90.34$ **b.** 150,000,000

15. a. $y = 380x + 901$ **b.** 380/yr

17. a. $y = 2.8x + 17.6$ **b.** $40 million

19. a. $y = 2.71x + 4.66$ **b.** 20,920,000

21. **a.** $y = 1.95x + 12.19$ **b.** \$23.89 billion

23. **a.** $y = 0.059x + 19.5$ **b.** 21.8 yrs **c.** 21.2 yrs

25. $y(x) = \left(\frac{132}{295}\right)x + \frac{-1200}{295}; \left(\frac{132}{295}\right)$

27. $G(t) = (5.003285937)e^{0.1823546461t}$; 3.8 hours

29. True

31. True

10.5 Exercises, page 637

1. Min. of $\frac{3}{4}$ at $\left(\frac{3}{4}, \frac{1}{4}\right)$

3. Max. of $-\frac{7}{4}$ at $\left(2, \frac{7}{2}\right)$

5. Min. of 4 at $(\sqrt{2}, \sqrt{2}/2)$ and $-(\sqrt{2}, -\sqrt{2}/2)$

7. Max. of $-\frac{3}{4}$ at $\left(\frac{3}{2}, 1\right)$

9. Max. of $2\sqrt{3}$ at $(\sqrt{3}/3, -\sqrt{6})$ and $(\sqrt{3}/3, \sqrt{6})$

11. Max. of 8 at $(2\sqrt{2}, 2\sqrt{2})$ and $(-2\sqrt{2}, -2\sqrt{2})$; min. of -8 at $(2\sqrt{2}, -2\sqrt{2})$ and $(-2\sqrt{2}, 2\sqrt{2})$

13. Max.: $\frac{2\sqrt{3}}{9}$; min.: $-\frac{2\sqrt{3}}{9}$

15. Min. of $\frac{18}{7}$ at $\left(\frac{9}{7}, \frac{6}{7}, \frac{3}{7}\right)$

17. \$7200

19. $\frac{512{,}000}{\pi}$ cm^3

21. $\sqrt{1.2}$ m $\times$ $\sqrt{1.2}$ m $\times$ $\frac{5}{3}\sqrt{1.2}$ m

23. 140 finished and 60 unfinished units

25. \$15,000; \$45,000

27. False

10.6 Exercises, page 643

1. $2x\,dx + 2\,dy$

3. $(4x - 3y + 4)dx - 3x\,dy$

5. $\frac{x}{\sqrt{x^2 + y^2}}dx + \frac{y}{\sqrt{x^2 + y^2}}dy$

7. $-\frac{5y}{(x - y)^2}dx + \frac{5x}{(x - y)^2}dy$

9. $(10x^4 + 3ye^{-3x})dx - e^{-3x}\,dy$

11. $\left(2xe^y + \frac{y}{x}\right)dx + (x^2e^y + \ln x)dy$

13. $y^2z^3\,dx + 2xyz^3\,dy + 3xy^2z^2\,dz$

15. $\frac{1}{y + z}dx - \frac{x}{(y + z)^2}dy - \frac{x}{(y + z)^2}dz$

17. $(yz + e^{yz})dx + xz(1 + e^{yz})dy + xy(1 + e^{yz})dz$

19. 0.04 **21.** -0.01

23. -0.10 **25.** -0.18

27. 0.06 **29.** -0.1401

31. $\pm 38.4\pi$ cc

33. An increase of \$19,250/mo

35. An increase of \$5000/mo

37. \$1.25 **39.** 1.875%

10.7 Exercises, page 650

1. $\frac{7}{2}$ **3.** 0

5. $4\frac{1}{2}$ **7.** $(e^2 - 1)(1 - e^{-2})$

9. 1 **11.** $\frac{2}{3}$

13. $\frac{188}{3}$ **15.** $\frac{84}{5}$

17. $2\frac{2}{3}$ **19.** 1

21. $\frac{1}{2}(3 - e)$ **23.** $\frac{1}{4}(e^4 - 1)$

25. $\frac{2}{3}(e - 1)$ **27.** False

10.8 Exercises, page 657

1. 42 cu units **3.** 20 cu units

5. $\frac{64}{3}$ cu units **7.** $\frac{1}{3}$ cu unit

9. 4 cu units **11.** $3\frac{1}{3}$ cu units

13. $2(e^2 - 1)$ cu units **15.** $\frac{2}{35}$ cu units

17. 54 **19.** $\frac{1}{3}$

21. 1 **23.** 43,329

25. 312,439 **27.** \$194/sq m

29. True

Chapter 10 Review Exercises, page 660

1. 0, 0, $\frac{1}{2}$; no

3. 2, $-(e + 1)$, $-(e + 1)$

5. The set of all ordered pairs (x, y) such that $y \neq -x$

7. The set of all triplets (x, y, z) such that $z \geq 0$ and $x \neq 1$, $y \neq 1$, and $z \neq 1$

9.

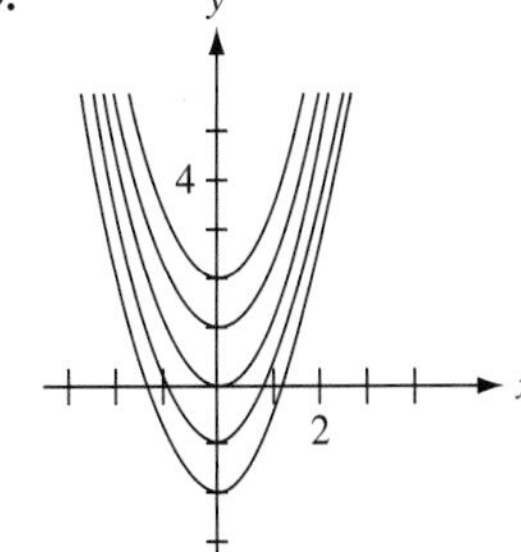

11.

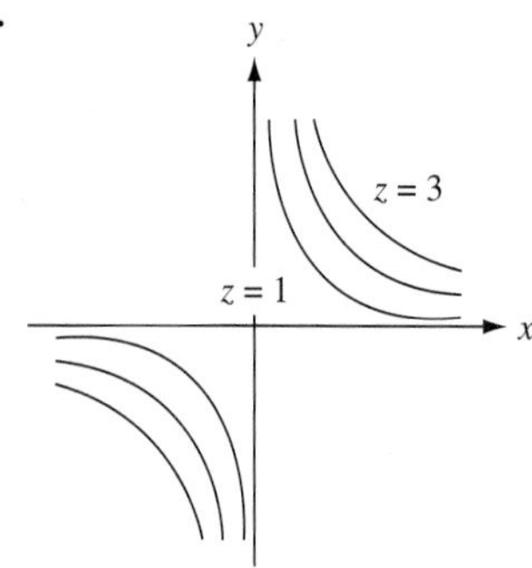

13. $f_x = \sqrt{y} + \dfrac{y}{2\sqrt{x}}; f_y = \dfrac{x}{2\sqrt{y}} + \sqrt{x}$

15. $f_x = \dfrac{3y}{(y + 2x)^2}; f_y = \dfrac{3x}{(y + 2x)^2}$

17. $h_x = 10y(2xy + 3y^2)^4$; $h_y = 10(x + 3y)(2xy + 3y^2)^4$

19. $f_x = 2x(1 + x^2 + y^2)e^{x^2+y^2}$; $f_y = 2y(1 + x^2 + y^2)e^{x^2+y^2}$

21. $f_x = \dfrac{2x}{x^2 + y^2}; f_y = -\dfrac{2x^2}{y(x^2 + y^2)}$

23. $f_{xx} = 12x^2 + 4y^2$; $f_{xy} = 8xy = f_{yx}$; $f_{yy} = 4x^2 - 12y^2$

25. $g_{xx} = \dfrac{-2y^2}{(x + y^2)^3}$; $g_{xy} = \dfrac{2y(x - y^2)}{(x + y^2)^3} = g_{yx}$; $g_{yy} = \dfrac{2x(3y^2 - x)}{(x + y^2)^3}$

27. $h_{xx} = -\dfrac{1}{s^2}; h_{st} = h_{ts} = 0; h_{tt} = \dfrac{1}{t^2}$

29. (2, 3); relative minimum value: $f(2, 3) = -13$

31. (0, 0) and $\left(\dfrac{3}{2}, \dfrac{9}{4}\right)$; saddle point at $f\left(-\dfrac{1}{3}, \dfrac{13}{3}\right) = \left(-\dfrac{4456}{27}\right)$; relative minimum value: $f(3, 11) = -35$

33. (0, 0); relative minimum value: $f(0, 0) = 1$

35. $f\left(\dfrac{12}{11}, \dfrac{20}{11}\right) = -\dfrac{32}{11}$

37. Relative maximum value: $f(5, -5) = 26$; relative minimum value: $f(-5, 5) = -24$

39. $45\,dx + 240\,dy$

41. 0.04

43. 48

45. $\dfrac{2}{63}$

47. $\dfrac{34}{3}$

49. 3

51. a. $-0.02x^2 - 0.2xy - 0.05y^2 + 80x + 60y$

b. The set of all points satisfying $0.02x + 0.1y \leq 80$; $0.1x + 0.05y \leq 60$; $x \geq 0, y \geq 0$

c. 100; 300

53. \$490

55. 337.5 m × 900 m

CHAPTER 11

11.1 Exercises, page 668

1. $x(2x) + x^2 = 3x^2$, $[xy' + y = 3x^2]$

3. $-2cxe^{-x^2} + 2x\left(\dfrac{1}{2} + ce^{-x}\right) = x$, $[y' + 2xy = x]$

5. $4e^{-2x} - 2e^{-2x} - 2e^{-2x} = 0$, $[y'' + y' - 2y = 0]$

7. $4C_1e^{-2x} - 4C_2e^{-2x} + 4C_2xe^{-2x} - 8C_1e^{-2x}$ $+ 4C_2e^{-2x} - 8C_2xe^{-2x+4C_1}e^{-2x} + 4C_2xe^{-2x} = 0$; solution: $y = C_1e^{-2x} + C_2xe^{-2x}$

9. $x^2[2C_1x^{-3} + C_2x^{-3}(2\ln x - 3)] + 3x[-C_1x^{-2} + C_2x^{-2}(1 - \ln x)]$ $+ C_1x^{-1} + C_2x^{-1}\ln x = 0$, $[x^2y'' + 3xy' + y = 0]$; solution: $y = \dfrac{C_1}{x} + \dfrac{\ln x}{x}$

11. $kAe^{-kt} = kAe^{-kt}$, $[y' = k(C - y)]$

13. $y = 12x^2 - 2x$

15. $y = \dfrac{1}{x}$

17. $y = -\dfrac{e^x}{x} + \dfrac{1}{2}xe^x$

19. $\dfrac{dQ}{dt} = -kQ; Q(0) = Q_0$

21. $\dfrac{dy}{dt} = -k(y - C)$

23. $\dfrac{dA}{dt} = k(C - A)$

25. $A' = C - kA$

27. $\dfrac{1}{x}\cdot\dfrac{dx}{dt} = k\dfrac{1}{y}\cdot\dfrac{dy}{dt}$

29. True

31. False

33. False

11.2 Exercises, page 674

1. $\dfrac{1}{3}y^3 = \dfrac{1}{2}x^2 + x + C$

3. $\dfrac{1}{3}y^3 = e^x + C$

5. $y = ce^{2x}$

7. $-\dfrac{1}{y} = \dfrac{1}{2}x^2 + C$

9. $y = -\dfrac{4}{3} + ce^{-6x}$

11. $y^3 = \dfrac{1}{3}x^3 + x + C$

13. $y^{1/2} - x^{1/2} = C$

15. $\ln|y| = \dfrac{1}{2}(\ln x)^2 + C$

17. $-\cos(x) + C$

19. $\sin(x) + C$

21. $\dfrac{1}{2}x^2 + C$

23. $y^2 = 2x^2 + 2$

25. $y = 2 - e^{-x}$

27. $y = \dfrac{2}{3} + \dfrac{1}{3}e^{(3/2)x^2}$

29. $y = \sqrt{x^2 + 1}$

31. $\ln|y| = (x - 1)e^x$

33. $y = \ln(x^3 + e)$

35. $2\sqrt{y} = \sin(x) + 4$

37. $\tan(y) = -\cos(x) + 1 + \dfrac{1}{\sqrt{2}}$

39. $\dfrac{dy}{dx} = 3xy; \dfrac{dy}{y} = -3x\,dx; \displaystyle\int\dfrac{dy}{y} = -\int 3x\,dx; \ln(y) = -\dfrac{3}{2}x^2 + C$

41. $\dfrac{dy}{dx} = \dfrac{y}{x^2 + 1}; \dfrac{dy}{y} = \dfrac{dx}{x^2 + 1}; \displaystyle\int\dfrac{dy}{y} = \int\dfrac{dx}{x^2 + 1}$; solution: $\ln(y) = \arctan(x) + C$

43. $\ln y = \dfrac{3}{2}x^2 + \ln 2, \ln\dfrac{y}{2} = \dfrac{3}{2}x^2$, or $y = 2e^{(3/2)s^2}$

45. $S(t) = D - (D - S_0)e^{-5t}$

47. $y = C - (C - y_0)e^{-(k/V)t}$; in the long run, the concentration of the solute inside the cell will approach that of the solution outside the cell.

49. $\ln y = \ln x^{1/k};\ y = \sqrt[k]{x}$

51. False

53. True

55. False

11.3 Exercises, page 685

1. $\frac{1}{2}$ in.

3. 3.6 minutes

5. 395

7. 20 kg

9. a. $16,487.21

b. 6.9 yrs

11. $\frac{dy}{dx} = -ky, y = y_0e^{-kt}$

13. 7.9 billion

15. 5.56 g

17. $P = -\frac{I}{k} + \left(P_0 + \frac{I}{k}\right)e^{kt}$

19. 1545

11.4 Exercises, page 697

1. $r + 4 = 0$

3. $r^2 + 3r - 7 = 0$

5. $r^2 + 4r + 2 = 0$

7. $r^2 + 4r + 4 = 0$

9. $y = Ae^{-2x}$

11. $y = Ae^x + Be^{3x}$

13. $y = Ae^{3x} + Be^{5x}$

15. $y = Ae^{3x} + Bxe^{3x}$

17. $y = A\sin(2t) + B\cos(2t)$

19. $y = Ae^{-2x} + \frac{1}{2}x - \frac{1}{4}$

21. $y = Ae^{-x} + Be^{-2x} + x - 2$

23. $12 = Ae^0$, or $y = 12e^{5x}$

25. $y = 4e^{-x} + 11xe^{-x}$

27. $y = \frac{9}{4}e^{2x} + \frac{1}{2}x + \frac{3}{4}$

29. 3050; 6050

31. $y(t) = 6.8e^{2t}$

33. $w(t) = -160e^{t/2} + 360e^{t/3}$

35. $y = \frac{A + Bx}{e^{t/2}}$; the mycoplasm will die out eventually.

11.5 Exercises, page 705

1. a. $y(1) = \frac{369}{128} \approx 2.8828$

b. $y(1) = \frac{70,993}{23,328} \approx 3.043$

3. a. $y(2) = \frac{51}{16} \approx 3.1875$

b. $y(2) = \frac{793}{243} \approx 3.2634$

5. a. $y(0.5) \approx 0.8324$

b. $y(0.5) \approx 0.8207$

7. a. $y(1.5) \approx 1.7831$

b. $y(1.5) \approx 1.7920$

9. a. $y(1) \approx 1.3390$

b. $y(1) \approx 1.3654$

11.

x	0	0.2	0.4	0.6	0.8	1
$\tilde{y}_n$	1	1	1.02	1.0608	1.2145	1.2144

13.

x	0	0.2	0.4	0.6	0.8	1
$\tilde{y}_n$	2	1.8	1.72	1.736	1.8288	1.9830

15.

x	0	0.1	0.2	0.3	0.4	5
$\tilde{y}_n$	1	1.1	1.211	1.3361	1.4787	1.64258

Chapter 11 Review Exercises, page 706

1. $4C_1e^{2x} + 9C_2e^{-3x} + 2C_2e^{2x} - 3C_2e^{-3x} - 6(C_1e^{2x} + C_2e^{-3x}) = 0$, $[y'' + y' - 6y = 0]$

3. $-\frac{4}{3}Cx^{-7/3} = -\frac{4}{3}Cx^{-7/3}$

5. $y = (9x + 8)^{-1/3}$

7. $y = 4 - Ce^{-2t}$

9. $y = -\frac{2}{2x^3 + 2x + 1}$

11. $y = 3e^{-x^3/2}$

13. $y = (2t + 4)e^t$

15. $y = \frac{4}{(x^2 + 1)^2}$

17. a. $y(1) \approx 0.2219$

b. $y(1) = 0.26280$

19. a. $y(1) \approx 1.9083$

b. $y(1) = 2.1722$

21.

x	0	0.2	0.4	0.6	0.8	1
$\tilde{y}_n$	1	1.2	1.496	1.9756	2. 8282	4.5560

23. 183

25. a. $S = 50,000(0.8)^t$

b. $16,384

27. $S = D - (D - S_0)e^{-kt}$

29. $R = CS^k$

31. $0.76e^{1.7t}$ million bacteria after t days

33. 1.14 g

CHAPTER 12

12.1 Exercises, page 712

1. $576

3. $287.50

5. $205.28

7. $12,576

9. $5,287.50

11. $3,655.28

13. $11,450.38

15. $4,728.13

17. $3,256.25

19. 3.71%

21. 10%

23. 19.0 yrs

25. 10.4 yrs

27. $4,912.00

29. 11.429%

31. 25%

33. 112 days

35. 7.7 yrs

37. $481.25

39. a. $160.00

b. $120; $130; $160; $4410

41. $98,806.20

43. $1767

12.2 Exercises, page 723

1. $4974.47
3. $223,403.11
5. **a.** 10.25%/yr **b.** 9.31%/yr
7. **a.** $29,227.61 **b.** $29,137.83
9. $6885.64
11. 13.59%/yr
13. 12.1%/yr
15. 14.87%/yr
17. 2.2 yr
19. 7.7 yr
21. 6.08%/yr
23. 2.06 yr
25. $112,926.52
27. $11,909.43; $11,905.58
29. $23,329.48
31. 9.58 yr
33. 12.75%
35. $40,000
37. **a.** $33,885.14 **b.** $33,565.38
39. **a.** $16,262.79 **b.** $12,047.77 **c.** $6611.96
41. $e^r - 1$
43. Bank B
45. 9.531%

12.3 Exercises, page 733

1. $315,005.42
3. $21,913.93
5. $259,081.24
7. $23.66
9. $335.67
11. $800.03
13. 40.2 yrs
15. 24.2 yrs
17. 8.8 yrs
19. $36,373.79
21. $51,965.43
23. 5.1 yrs
25. $17,863.34
27. $365,529.70
29. **a.** $13,041.21; $1,041.21; $14,690.83; $2,690.83
    b. 15 years and 5 months
31. $7,525.58
33. **a.** $29.17 **b.** $14.74 **c.** $966.08

12.4 Exercises, page 743

1. $153,659.83
3. $17,413.13
5. $244,146.33
7. $48.58
9. $421.92
11. $849.07
13. 41.6 yrs
15. 21.3 yrs
17. 10.7 yrs

19.

Payment Number	Payment	Interest Portion	Portion Applied to Principal	Principal (Unpaid Balance)
0				77,500.00
1	2388.42	823.44	1564.98	75,935.02
2	2388.42	806.81	1581.61	74,353.41
3	2388.42	790.00	1598.42	72,754.99
4	2388.42	773.02	1615.40	71.139.59
5	2388.42	755.86	1632.56	69,507.03
6	2388.42	738.51	1649.91	67,857.12

21.

Payment Number	Payment	Interest Portion	Portion Applied to Principal	Principal (Unpaid Balance)
0				520,000.00
1	1440.91	780.00	660.91	519,339.09
2	1440.91	779.01	661.90	518,677.19
3	1440.91	778.02	662.89	518,014.30
4	1440.91	777.02	663.89	517,350.41
5	1440.91	776.03	664.88	516,685.53
6	1440.91	775.03	665.88	516,019.65

23.

Payment Number	Payment	Interest Portion	Portion Applied to Principal	Principal (Unpaid Balance)
0				1,000,000.00
1	4219.10	1633.33	2585.77	997,414.23
2	4219.10	1629.11	2589.99	994,824.24
3	4219.10	1624.88	2594.22	992,230.02
4	4219.10	1620.64	2598.46	989,631.56
5	4219.10	1616.40	2602.70	987,028.86
6	4219.10	1612.15	2606.95	984,421.91

25. $332.38
27. $345.90
29. **a.** 276,151.48 **b.** $41,848.52
31. $49,049.20
33. **a.** $8681.43 **b.** $1,104,429 **c.** $7895.85
35. $152,333.66
37. $299.35
39. $315,347.30

Chapter 12 Review Exercises, page 747

1. $56.67
3. $16,260.00
5. $1453.49
7. 26 yrs
9. $17,110.02
11. $1224.70

13. 18 yrs

15. $27,887.93

17. $17.74

19. 11.2 yrs

21. $20,679.19

23. $52.60

25. 14.4 yrs

27. $87.50

29. Investment A

31. 4.9089%

33. $49,298.78

35. $82.93

37. $18,075.03; $2075.03

39. $160.94

41.

Payment Number	Payment	Interest Portion	Portion Applied to Principal	Principal (Unpaid Balance)
0				500,000.00
1	2072.13	1057.69	1014.44	498,985.56
2	2072.13	1055.55	1016.58	497,968.98
3	2072.13	1053.40	1018.73	496,950.25
4	2072.13	1051.24	1020.89	495,929.36
5	2072.13	1049.08	1023.05	494,906.31
6	2072.13	1046.92	1025.21	493,881.10

43. a. $57,750 **b.** $1857.00

c. $620,576.88 **d.** $1735.27

e. $92,618.02

45. $677,381.98

CHAPTER 13

13.1 Exercises, page 758

1. $\int_3^6 \frac{1}{3}\,dx = 1$

3. $\int_2^6 \frac{2}{32}x\,dx = 1$

5. $\int_0^3 \frac{2}{9}(3x - x^2)dx = 1$

7. $\int_3^6 \frac{1}{3}\,dx = 1$

9. $\int_0^{12} \frac{12 - x}{72}\,dx = 1$

11. $\int_0^{\infty} \frac{x}{(x^2 + 1)^{3/2}}\,dx = 1$

13. $k = \frac{1}{3}$

15. $k = \frac{1}{8}$

17. $k = \frac{3}{16}$

19. $k = 2$

21. a. $\frac{1}{2}$ **b.** $\frac{5}{8}$

c. $\frac{7}{8}$ **d.** 0

23. a. $\frac{11}{16}$ **b.** $\frac{1}{2}$

c. $\frac{27}{32}$ **d.** 0

25. a. $\frac{1}{2}$ **b.** $\frac{1}{2}(2\sqrt{2} - 1)$

c. 0 **d.** $\frac{1}{2}$

27. a. 1 **b.** .14

29. a. .375 **b.** .75

c. .5 **d.** .875

31. f is nonnegative on D; $\int_1^3\int_0^2 \frac{1}{4}xy\,dx\,dy = 1.$

33. f is nonnegative on D; $\frac{1}{3}\int_1^2\int_0^2 xy\,dx\,dy = 1.$

35. $k = 2$

37. $k = 12e^2$

39. a. $\frac{1}{4}$ **b.** $\frac{1}{96}$

41. a. $\frac{1}{56}(2\sqrt{2} - 1)$ **b.** $\approx .2575$

43. a. .63 **b.** .30 **c.** .30

45. a. .10 **b.** .30

47. .9355

49. .48148; .740741

51. $\frac{1}{2}$

53. ≈ 0.022

55. False

13.2 Exercises, page 771

1. $\mu = \frac{9}{2}$; $\text{Var}(x) = \frac{3}{4}$; $\sigma = 0.866$

3. $\mu = \frac{15}{4}$; $\text{Var}(x) = \frac{15}{16}$; $\sigma = 0.9682$

5. $\mu = 3$; $\text{Var}(x) = 0.8$; $\sigma = 0.8944$

7. $\mu \approx 2.3765$; $\text{Var}(x) \approx 2.3522$; $\sigma \approx 1.534$

9. $\mu = \frac{93}{35}$; $\text{Var}(x) \approx 0.7151$; $\sigma = 0.846$

11. $\mu = \frac{3}{2}$; $\text{Var}(x) = \frac{3}{4}$; $\sigma \approx \frac{1}{2}\sqrt{3}$

13. $\mu = 4$; $\text{Var}(x) = 16$; $\sigma = 4$

15. 100 days

17. $3\frac{1}{3}$ min

19. 0.5

21. 2500 kg/wk

23. $m = 5$

25. $m \approx 2.52$

27. $m \approx \frac{6}{5}$

29. False

13.3 Exercises, page 780

1. .9265

3. .0401

5. .8657

7. a.

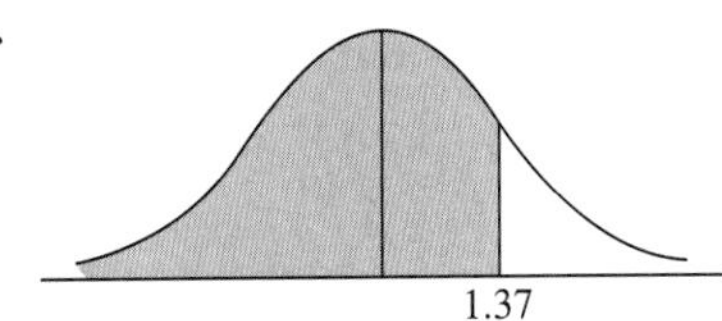

b. .9147

9. a.

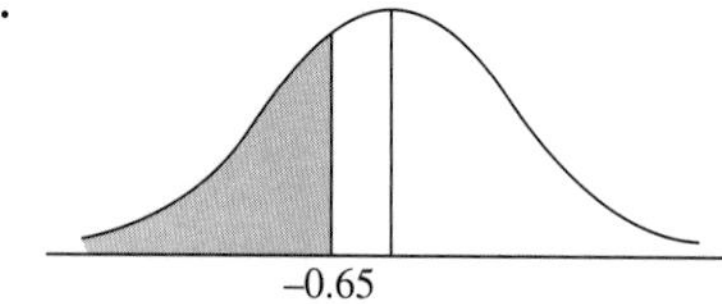

b. .2578

11. a.

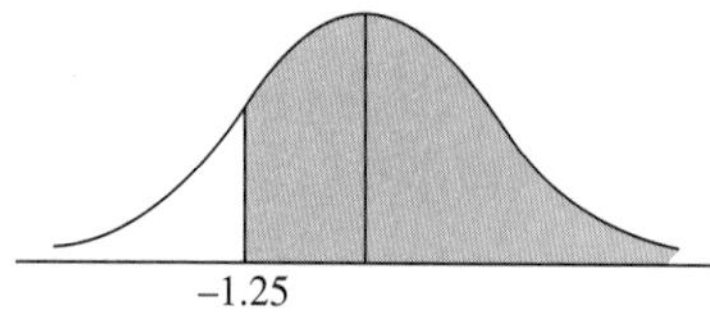

b. .8944

13. a.

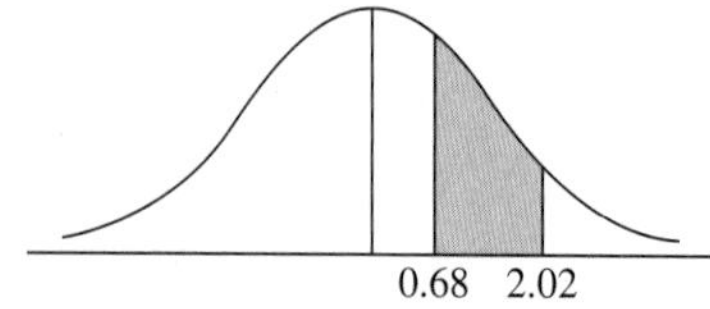

b. .2266

15. a. 1.23 **b.** −0.81

17. a. 1.9 **b.** −1.9

19. a. .9772 **b.** .9192 **c.** .7333

21. a. .0228 **b.** .0228 **c.** .4772

23. a. .0038 **b.** .0918

c. .4082 **d.** .2514

25. 15.87% **27.** 92.5

29. a. .0228 **b.** .1151 **c.** .6826

31. a. .2206 **b.** .2206 **c.** .3034

Chapter 13 Review Exercises, page 783

1. $f(x) = \frac{1}{28}(2x + 3)$ is nonnegative on [0, 4]; $\int_0^4 \frac{1}{28}(2x + 3)dx = 1.$

3. $f(x) = \frac{1}{4} > 0$ on [7, 11]; $\int_7^{11} \frac{1}{4}\,dx = 1$

5. $\frac{1}{243}$ **7.** $\frac{3}{2}$

9. a. $\frac{4}{7}$ **b.** 0 **c.** $\frac{1}{3}$

11. a. $\approx .52$ **b.** $\approx .65$ **c.** 0

13. $\mu = \frac{9}{2}$; $\text{Var}(x) \approx 2.083$; $\sigma \approx 1.44$

15. $\mu = 0$; $\text{Var}(x) \approx \frac{7}{15}$; $\sigma \approx 0.6831$

17. .9875 **19.** .3049

21. f is nonnegative in D; $\iint_D f(x, y)dA = 1.$

23. a. .6915 **b.** .8944 **c.** .4681

25. a. .22 **b.** .39 **c.** 4 days

CHAPTER 14

14.1 Exercises, page 796

1. $P_1(x) = 1 - x$; $P_2(x) = 1 - x + \frac{1}{2}x^2$; $P_3(x) = 1 - x + \frac{1}{2}x^2 - \frac{1}{6}x^3$

3. $P_1(x) = 1 - x$; $P_2(x) = 1 - x + x^2$; $P_3(x) = 1 - x + x^2 - x^3$

5. $P_1(x) = 1 - (x - 1)$; $P_2(x) = 1 - (x - 1) + (x - 1)^2$; $P_3(x) = 1 - (x - 1) + (x - 1)^2 - x^3$

7. $P_1(x) = 1 - \frac{1}{2}x$; $P_2(x) = 1 - \frac{1}{2}x - \frac{1}{8}x^2$; $P_3(x) = 1 - \frac{1}{2}x - \frac{1}{8}x^2 - \frac{1}{16}x^3$

9. $P_1(x) = -x$; $P_2(x) = -x - \frac{1}{2}x^2$; $P_3(x) = -x - \frac{1}{2}x^2 - \frac{1}{3}x^3$

11. $P_2(x) = 16 + 32(x - 2) + 24(x - 2)^2$

13. $P_4(x) = (x - 1) - \frac{1}{2}(x - 1)^2 + \frac{1}{3}(x - 1)^3 - \frac{1}{4}(x - 1)^4$

15. $P_4(x) = e + e(x - 1) + \frac{1}{2}e(x - 1)^2 + \frac{1}{6}e(x - 1)^3 + \frac{1}{24}e(x - 1)^4$

17. $P_3(x) = 1 - \frac{1}{2}x - \frac{1}{8}x^2 - \frac{1}{16}x^3$

19. $P_3(x) = \frac{1}{3} - \frac{2}{9}x + \frac{4}{27}x^2 - \frac{8}{81}x^3$

21. $P_n(x) = 1 - x + x^2 - x^3 + \cdots + (-1)^n x^n$; 0.9091; 0.909090...

23. $P_4(x) = 1 - \frac{1}{2}x + \frac{1}{8}x^2 - \frac{1}{48}x^3 + \frac{1}{384}x^4$; 0.90484

25. $P_2(x) = 4 + \frac{1}{8}(x - 16) - \frac{1}{512}(x - 16)^2$; ≈ 3.94969

27. $P_3(x) = x - \frac{1}{2}x^2 + \frac{1}{3}x^3$; 0.109; 0.108

29. 2.01494375; 0.00000042

31. 1.248; 0.00493; 1.25

33. 0.095; 0.00033

35. a. 0.04167 **b.** 0.625

c. 0.04167 **d.** 0.007121

37. 0.47995 **39.** 48%

41. 2600 **43.** $\approx$ 46,960,000 mg/min

45. False **47.** True

14.2 Exercises, page 804

1. 1, 2, 4, 8, 16 **3.** $0, \frac{1}{3}, \frac{2}{4}, \frac{3}{5}, \frac{4}{6}$

5. $1, 1, \frac{4}{6}, \frac{8}{24}, \frac{16}{120}$ **7.** $e, \frac{e^2}{8}, \frac{e^3}{27}, \frac{e^4}{64}, \frac{e^5}{125}$

9. $1, \frac{11}{9}, \frac{25}{19}, \frac{45}{33}, \frac{71}{51}$ **11.** $a_n = 3n - 2$

13. $a_n = \frac{1}{n^3}$ **15.** $a_n = \frac{2^{2n-1}}{5^{n-1}}$

17. $a_n = \frac{(-1)^{n+1}}{2^{n-1}}$ **19.** $a_n = \frac{n}{(n + 1)(n + 2)}$

21. $a_n = \frac{e^{n-1}}{(n - 1)!}$

23.

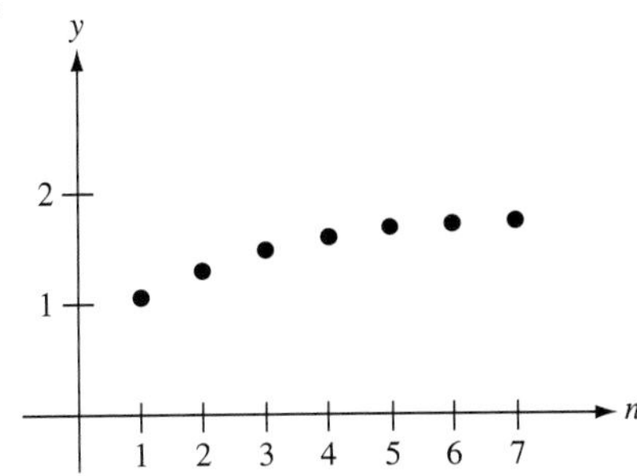

25.

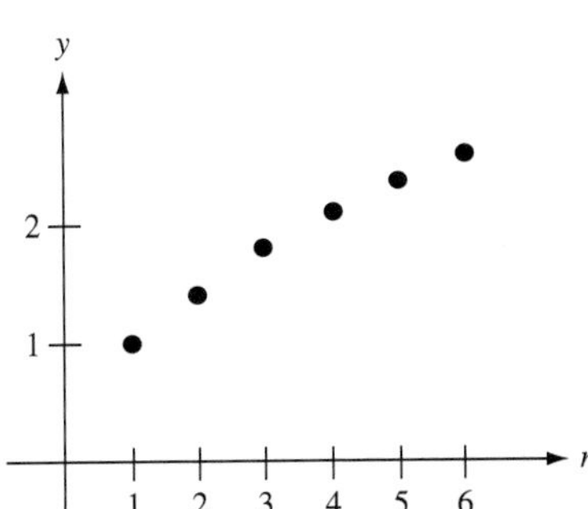

27.

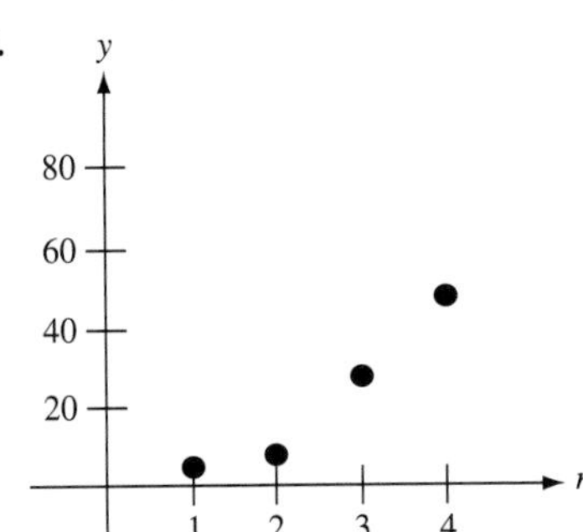

29.

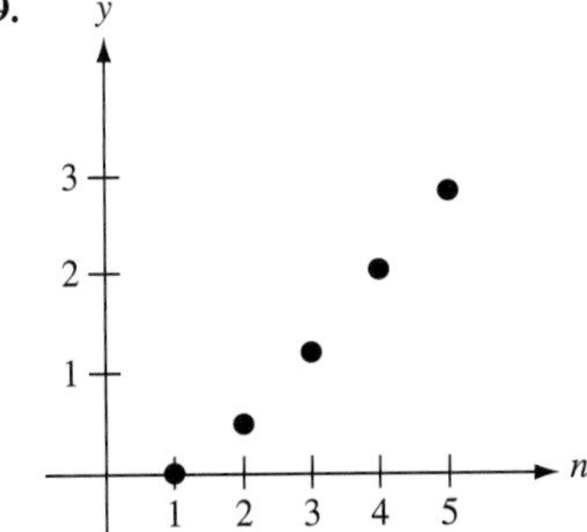

31. Converges; $\frac{1}{2}$

33. Converges; 0

35. Converges; 1

37. Converges; 2

39. Converges; 2

41. Converges; 0

43. Converges; $\frac{\sqrt{2}}{2}$

45. a. $a_1 = 0.015, a_{10} = 0.140, a_{100} = 0.77939, a_{1000} = 0.999999727$

b. 1

47. a. The result follows from the compound interest formula.

b. $a_n = 100(1.01)^n$

c. \$126.97

49. True

51. False

14.3 Exercises, page 813

1. $\lim_{N\to\infty} S_N$ does not exist; divergent

3. $S_N = \frac{1}{2} - \frac{1}{N+2}; \frac{1}{2}$

5. Converges; $\frac{3}{2}$

7. Diverges

9. Converges; $\frac{3}{5}$

11. Converges; $\frac{1}{3}$

13. Converges; 5.52

15. Converges; $\frac{\pi}{\pi+3}$

17. Converges; 13

19. Converges; $2\frac{3}{4}$

21. Converges; $7\frac{1}{2}$

23. Diverges

25. $\frac{e^3 - 2\pi e + \pi^2}{(\pi - e)(e^2 - \pi)}$

27. $\frac{1}{3}$

29. $\frac{404}{333}$

31. $-1 < x < 1; \frac{1}{1+x}$

33. $\frac{1}{2} < x < \frac{3}{2}; \frac{2(x-1)}{3-2x}$

35. 30 m

37. 35.21 units

39. \$303 billion

41. $\frac{mP}{r}$

43. $\frac{6}{11}$

45. 5 million

47. False

49. True

14.4 Exercises, page 823

1. $a_n = \frac{n}{n+1}$ and $\lim_{n\to\infty} \frac{n}{n+1} \neq 0$

3. $a_n = \frac{2n}{3n+1}$ and $\lim_{n\to\infty} \frac{n^2}{2n^2+1} \neq 0$

5. $a_n = 2(1.5)^n$ and $\lim_{n\to\infty} 2(1.5)^n \neq 0$

7. $a_n = \frac{1}{2+3^{-n}}$ and $\lim_{n\to\infty} \frac{1}{2+3^{-n}} \neq 0$

9. $a_n = \left(\frac{\pi}{3}\right)^n$; $\lim_{n\to\infty} a_n$ does not exist

11. Diverges

13. Diverges

15. Converges

17. Converges

19. Converges

21. Converges

23. Converges

25. Converges

27. Converges

29. Diverges

31. Diverges

33. Converges

35. Diverges

37. Diverges

39. Converges

41. Converges

43. Diverges

45. Converges

47. Diverges

49. $p > 1$

51. $a = 1$

53. $\int_1^\infty \frac{1}{x^p}\,dx = \frac{1}{p-1}$ if $p > 1$ and diverges to infinity if $p < 1$; if $p = 1$, $\int_1^\infty \frac{1}{x}\,dx = \infty$ and diverges in this case as well

55. True

57. False

59. False

14.5 Exercises, page 833

1. $R = 1$; $(0, 2)$

3. $R = 1$; $(-1, 1)$

5. $R = 4$; $(-4, 4)$

7. $R = \infty$; $(-\infty, \infty)$

9. $R = 0$; $x = -2$

11. $R = 1$;$(-4, -2)$

13. $R = \infty$; $(-\infty, \infty)$

15. $R = 1$; $\left(-\frac{1}{2}, \frac{1}{2}\right)$

17. $R = 0$; $x = -1$

19. $r = 3$; $(0, 6)$

21. $\sum_{n=0}^{\infty} (-1)^n (x-1)^n$; $R = 1$; $(0, 2)$

23. $\sum_{n=0}^{\infty} (-1)^n \frac{(x-2)^n}{3^{n+1}}$; $R = 3$; $(-1, 5)$

25. $\sum_{n=0}^{\infty} (-1)^{n+1} (x-2)^n$; $R = 1$; $(1, 3)$

27. $1 + \frac{1}{2}(x-1) + \sum_{n=2}^{\infty} (-1)^{n+1} \frac{1 \cdot 3 \cdot 5 \ldots (2n-3)}{n!2^n}(x-1)^n$; $R = 1$; $(0, 2)$

29. $\sum_{n=0}^{\infty} \frac{2^n}{n!} x^n$; $R = \infty$; $(-\infty, \infty)$

31. $\sum_{n=0}^{\infty} (-1)^n \frac{1 \cdot 3 \cdot 5 \ldots (2n-1)}{n!2^n} x^n$; $R = 1$; $(-1, 1)$

33. For fixed x, the sequence of partial sums of the Taylor series of f converges to f if and only if $\lim_{n \to \infty} R_n(x) = 0$.

35. True

14.6 Exercises, page 841

1. $\sum_{n=0}^{\infty} (-1)^{n+1} (x-2)^n$; $(1, 3)$

3. $\sum_{n=0}^{\infty} (-1)^n 3^n x^n$; $\left(-\frac{1}{3}, \frac{1}{3}\right)$

5. $\sum_{n=0}^{\infty} \frac{3^n}{4^{n+1}} x^n$; $\left(-\frac{4}{3}, \frac{4}{3}\right)$

7. $\sum_{n=0}^{\infty} x^{2n}$; $(-1, 1)$

9. $\sum_{n=0}^{\infty} (-1)^n \frac{x^n}{n!}$; $(-\infty, \infty)$

11. $\sum_{n=0}^{\infty} (-1)^n \frac{x^{2n+1}}{n!}$; $(-\infty, \infty)$

13. $f(x) = 1 + \frac{x^2}{2!} + \frac{x^4}{4!} + \frac{x^6}{6!} + \cdots + \frac{x^{2n}}{2n!} + \cdots$; $(-\infty, \infty)$

15. $\sum_{n=1}^{\infty} (-1)^{n-1} \frac{2^n x^n}{n}$; $\left(-\frac{1}{2}, \frac{1}{2}\right]$

17. $\sum_{n=1}^{\infty} (-1)^{n+1} \left(\frac{x^{2n}}{n}\right)$; $(-1, 1)$

19. $(\ln 2)(x-2) + \sum_{n=1}^{\infty} (-1)^{n-1} \left(\frac{1}{n2^n}\right)(x-2)^n$; $(0, 4]$

21. $\sin(2x) = 2x - \frac{2^3 x^3}{3!} + \frac{2^5 x^5}{5!} - \cdots$

23. $h(x) = x^2 - \frac{x^4}{3!} + \frac{x^6}{5!} - \cdots$

25. $f'(x) = 1 - x + x^2 + \cdots + (-1)^n x^n + \cdots$

29. 0.4812

31. 0.7475

33. 3.34

35. 0.6825

Chapter 14 Review Exercises, page 844

1. $f(x) = 1 - (x+1) + (x+1)^2 - (x+1)^3 + (x+1)^4$

3. $f(x) = x^2 - \frac{1}{2}x^4$

5. $f(x) = 2 + \frac{1}{12}(x-8) - \frac{1}{288}(x-8)^2$; 1.983

7. 2.9992591; 8×10^{-11}

9. 0.37

11. Converges; $\frac{2}{3}$

13. Converges; 1

15. 2

17. $\frac{1}{\sqrt{2}+1}$

19. $\frac{141}{99}$

21. Diverges

23. Converges

25. $R = 1$; $(-1, 1)$

27. $R = 1$; $(0, 2)$

29. $f(x) = -1 - 2x - 4x^2 - 8x^3 - \cdots - 2^n x^n - \cdots$; $\left(-\frac{1}{2}, \frac{1}{2}\right)$

31. $f(x) = 2x - 2x^2 + \frac{8}{3}x^3 - \cdots + \frac{(-1)^{n+1}2^n}{n!} + \cdots$; $(-\infty, \infty)$

33. \$106,186.10

35. 31.08%

INDEX

Basic Rules of Differentiation

1. $\frac{d}{dx}(c) = 0, \quad c$ a constant

2. $\frac{d}{dx}(u^n) = nu^{n-1}\frac{du}{dx}$

3. $\frac{d}{dx}(u \pm v) = \frac{du}{dx} \pm \frac{dv}{dx}$

4. $\frac{d}{dx}(cu) = c\frac{du}{dx}, \quad c$ a constant

5. $\frac{d}{dx}(uv) = u\frac{dv}{dx} + v\frac{du}{dx}$

6. $\frac{d}{dx}\left(\frac{u}{v}\right) = \frac{v\frac{du}{dx} - u\frac{dv}{dx}}{v^2}$

7. $\frac{d}{dx}(e^u) = e^u\frac{du}{dx}$

8. $\frac{d}{dx}(\ln u) = \frac{1}{u} \cdot \frac{du}{dx}$

9. $\frac{d}{dx}(\sin u) = \cos u\frac{du}{dx}$

10. $\frac{d}{dx}(\cos u) = -\sin u\frac{du}{dx}$

11. $\frac{d}{dx}(\tan u) = \sec^2 u\frac{du}{dx}$

12. $\frac{d}{dx}(\sec u) = \sec u \tan u\frac{du}{dx}$

13. $\frac{d}{dx}(\csc u) = -\csc u \cot u\frac{du}{dx}$

14. $\frac{d}{dx}(\cot u) = -\csc^2 u\frac{du}{dx}$

Basic Rules of Integration

1. $\int du = u + C$

2. $\int kf(u)\,du = k\int f(u)\,du, \quad k$ a constant

3. $\int [f(u) \pm g(u)]\,du = \int f(u)\,du \pm \int g(u)\,du$

4. $\int u^n\,du = \frac{u^{n+1}}{n+1} + C, \quad n \neq -1$

5. $\int e^u\,du = e^u + C$

6. $\int \frac{du}{u} = \ln|u| + C$